NIST RESEARCH LIBRARY
GAITHERSBURG, MD 20899-2500

AF699692

DATE DUE

APR 0 6 2001			
AUG 3 0 2001			

GAYLORD PRINTED IN U.S.A.

Handbook of Optics

OTHER McGRAW-HILL HANDBOOKS OF INTEREST

Baumeister and Marks · Standard Handook for Mechanical Engineers
Berry, Bollay, and Beers · Handbook of Meteorology
Brady and Clauser · Materials Handbook
Burington · Handbook of Mathematical Tables and Formulas
Burington and May · Handbook of Probability and Statistics with Tables
Chow · Handbook of Applied Hydrology
Condon and Odishaw · Handbook of Physics
Considine · Encyclopedia of Instrumentation and Control
Considine · Energy Technology Handbook
Considine · Process Instruments and Controls Handbook
Considine and Ross · Handbook of Applied Instrumentation
Etherington · Nuclear Engineering Handbook
Fink · Electronics Engineer's Handbook
Fink and Carroll · Standard Handbook for Electrical Engineers
Flügge · Handbook of Engineering Mechanics
Giacocetto · Electronics Designers' Handbook
Grant · Hackh's Chemical Dictionary
Hamsher · Communication System Engineering Handbook
Harris and Crede · Shock and Vibration Handbook
Hicks · Standard Handbook of Engineering Calculations
Ireson · Reliability Handbook
Juran · Quality Control Handbook
Kaelble · Handbook of X-rays
King and Brater · Handbook of Hydraulics
Klerer and Korn · Digital Computer User's Handbook
Korn and Korn · Mathematical Handbook for Scientists and Engineers
Lange · Handbook of Chemistry
Machol · System Engineering Handbook
Mantell · Engineering Materials Handbook
Markus · Electronics and Nucleonics Dictionary
Meites · Handbook of Analytical Chemistry
Merritt · Standard Handbook for Civil Engineers
Perry · Engineering Manual
Perry and Chilton · Chemical Engineers' Handbook
Raznjevic · Handbook of Thermodynamic Tables and Charts
Rohsenow and Hartnett · Handbook of Heat Transfer
Rothbart · Mechanical Design and Systems Handbook
Society of Mechanical Engineers · SME Handbooks:
Engineering Tables — Metals Engineering—Processes
Metals Engineering—Design — Metals Properties
Streeter · Handbook of Fluid Dynamics
Truxal · Control Engineers' Handbook
Tuma · Handbook of Physical Calculations
Urquhart · Civil Engineering Handbook

Handbook of Optics

Sponsored by the
OPTICAL SOCIETY OF AMERICA

Walter G. Driscoll
Editor

William Vaughan
Associate Editor

MCGRAW-HILL BOOK COMPANY

New York St. Louis San Francisco Auckland Bogotá Düsseldorf
Johannesburg London Madrid Mexico Montreal
New Delhi Panama Paris São Paulo
Singapore Sydney Tokyo Toronto

Library of Congress Cataloging in Publication Data
Main entry under title:

Handbook of optics.
Includes index.
1. Optics—Handbooks, manuals, etc. I. Driscoll, Walter G. II. Vaughan, William, date. III. Optical Society of America.
QC369.H35 535 77-22123
ISBN 0-07-047710-8

Copyright © 1978 by McGraw-Hill, Inc. All rights reserved. Printed in the United States of America. No part of this publication may be reproduced, stored in a retrieval system, or transmitted, in any form or by any means, electronic, mechanical, photocopying, recording, or otherwise, without the prior written permission of the publisher.

567890 KPKP 86543210

The editors for this book were Harold B. Crawford and Beatrice Carson, the designer was Naomi Auerbach, and the production supervisor was Teresa F. Leaden. It was set in Lino 21 by Bi-Comp, Incorporated.

Printed and bound by The Kingsport Press.

Contents

Contributors

MATHEW ALPERN *Department of Ophthalmology, University of Michigan Medical Center, Ann Arbor, Mich.* (SECTION 12)

HAROLD E. BENNETT *Michelson Laboratory, Naval Weapons Center, China Lake, Calif.* (COAUTHOR SECTION 10)

JEAN M. BENNETT *Michelson Laboratory, Naval Weapons Center, China Lake, Calif.* (COAUTHOR SECTION 10)

JAMES R. BENFORD *Former Director, Research, Development, and Engineering, Scientific Instruments Division, Bausch & Lomb, Incorporated, Rochester, N.Y.* (COAUTHOR SECTION 6)

J. A. DOBROWOLSKI *Division of Physics, National Research Council of Canada, Ottawa, Ontario* (SECTION 8)

ROBERT W. FENN *Atmospheric Optics Branch, Optical Physics Laboratory, Air Force Cambridge Research Laboratories, Bedford, Mass.* (COAUTHOR SECTION 14)

JOHN S. GARING *Director, Optical Physics Laboratory, Air Force Cambridge Research Laboratories, Bedford, Mass.* (COAUTHOR SECTION 14)

EDWARD HARTFIELD *Itek Corporation, Lexington, Mass.* (COAUTHOR SECTION 17)

STEPHEN F. JACOBS *Optical Sciences Center, University of Arizona, Tucson, Ariz.* (SECTION 4)

ANTHONY J. LAROCCA *Environmental Research Institute of Michigan, Ann Arbor, Mich.* (COAUTHOR SECTION 3)

D. J. LOVELL *Optical Consultant, Barton Road, Stow, Mass.* (SECTION 11)

ROBERT A. McCLATCHEY *Infrared Physics Branch, Optical Physics Laboratory, Air Force Cambridge Research Laboratories, Bedford, Mass.* (COAUTHOR SECTION 14)

KENNETH S. PETCHE *Coulter Information Systems, Inc., Bedford, Mass.* (SECTION 5)

HAROLD E. ROSENBERGER *Vice President, Scientific Instruments Development, Scientific Optical Products Division, Bausch & Lomb, Incorporated, Rochester, N.Y.* (COAUTHOR SECTION 6)

WILLIAM J. ROWAN *Electro-Optical Division, Kollmorgen Corporation, Northhampton, Mass.* (COAUTHOR SECTION 16)

JOHN E. A. SELBY *Infrared Physics Branch, Optical Physics Laboratory, Air Force Cambridge Research Laboratories, Bedford, Mass.* (COAUTHOR SECTION 14)

WALTER P. SIEGMUND *Technical Manager, Fiber-Optics Division, American Optical Corporation, Southbridge, Mass.* (SECTION 13)

WARREN J. SMITH *Director of Research and Development, Infrared Industries, Inc., Santa Barbara, Calif.* (SECTION 2)

JAY F. SNELL *Tektronix, Incorporated, Beaverton, Oreg.* (SECTION 1)

BRIAN J. THOMPSON *Institute of Optics, University of Rochester, Rochester, N.Y.* (COAUTHOR SECTION 17)

HERBERT E. TORBERG *Electro-Optical Division, Kollmorgen Corporation, Northampton, Mass.* (COAUTHOR SECTION 16)

JOHN E. TYLER *Visibility Laboratory, Scripps Institution of Oceanography, University of California, San Diego, Calif.* (SECTION 15)

FREDERIC E. VOLZ *Atmospheric Optics Branch, Optical Physics Laboratory, Air Force Geophysics Laboratory, Bedford, Mass.* (COAUTHOR SECTION 14)

J. RICHARD VYCE *Optical Systems Division, Itek Corporation, Lexington, Mass.* (COAUTHOR SECTION 16)

WILLIAM L. WOLFE *Optical Sciences Center, University of Arizona, Tucson, Ariz.* (SECTION 7)

GUNTER WYSZECKI *Division of Physics, National Research Council, Ottawa, Ontario* (SECTION 9)

GEORGE J. ZISSIS *Environmental Research Institute of Michigan, Ann Arbor, Mich.* (COAUTHOR SECTION 3)

Preface

The stimulus for a Handbook of Optics came quite directly from an effort by the Optical Society of America to learn what steps it might take to facilitate study and teaching and research in optics. In 1962 the Society acted on the recommendations of an ad hoc Committee on Needs in Optics and established "Optics: An Action Program," under the strong direction of Van Zandt Williams, to explore possible measures in a number of areas to strengthen the whole science of optics.

At a conference on undergraduate research programs held at Southwestern at Memphis in 1964, the participants, who were college teachers, placed a Handbook of Optics high on their list of needs.

In the same year, responses to a questionnaire that was sent to all members of the Society clearly indicated a need for a convenient compilation of optical information. Not only those engaged in optical physics, lens design, vision, color, and other specializations traditionally associated with optics but also chemists, engineering scientists, and medical scientists seemed to agree on the desirability of a handbook.

Dr. Williams discussed these matters with members of the New England Section of the Society, who agreed to make a thorough analysis of the suggestion. Their study confirmed the need for a handbook and produced guides for its content and style.

A main problem of making a handbook is obvious: by the time the book is completed, much of the information in it may have been improved upon. But there must be a beginning, and then further development and refinement in any future editions can follow, based on the comments and suggestions of the users. Tables and graphs contain data that are applicable to many optical problems involving the ultraviolet, visible, and infrared spectral regions. Authors of all chapters have supplied explanatory text and many references. While the book is not intended to be a textbook, it was our plan to speed the work of the user by providing more than tables of data. We trust that this book will prove valuable to users and that they will send us their comments and suggestions.

The guidance of Walter G. Driscoll, the editor, and William Vaughan, the associate editor, has been essential to the publication of this edition. The book is ample testimony to their efforts and those of the contributors.

The Optical Society of America

Acknowledgments

It is not without apprehension that one begins to cite acknowledgments. The efforts of our numerous contributors are recognized without exception. Their diligence, understanding, and patience have been beyond expectation.

To this group should be added the names of individuals who reviewed and carefully read sections of the manuscript or the manuscript in its entirety. In particular, we wish to recognize the participation of Stanley S. Ballard, John N. Howard, John A. Sanderson, and F. Dow Smith, who constituted the official review committee of the Society, and of Kenneth M. Baird, L. E. Barbrow, Bruce H. Billings, H. Richard Blackwell, Darrell E. Burch, Robert Burley, Richard Dixon, S. Q. Duntley, John W. Hicks, Robert E. Hopkins, W. Lewis Hyde, David Jeffries, Deane B. Judd, Karl G. Kessler, David L. MacAdam, Donald E. McCarthy, John C. Marchant, William W. Striker, Robert B. Tackaberry, A. Francis Turner, P. R. Wakeling, Adriaan Walther, Mary E. Warga, and the many others who have contributed to this compilation.

Also, we wish to acknowledge the assistance of Leois Conway, our secretary, without whose patience none of this would have been possible.

Walter G. Driscoll
William Vaughan

Handbook of Optics

Section 1

Radiometry and Photometry

JAY F. SNELL
Tektronix, Incorporated, Beaverton, Oreg.

(Numbers refer to paragraphs)

1. Introduction Radiometry is the measurement of quantities associated with radiant energy. Photometry is the measurement of quantities associated with light, i.e., visually evaluated radiant energy. Light is the aspect of radiant energy of which a human observer is aware through the stimulation of the retina of the eye. Thus, light is psychophysical, neither purely physical nor purely psychological. Unless specifically devised to evaluate radiant energy in the same way as the human retina, other detectors of radiation are properly considered only in the realm of radiometry.

This section treats the radiometry and photometry of unpolarized, incoherent radiation in the ultraviolet, visible, and infrared portions of the electromagnetic spectrum.

BASIC CONCEPTS

2. Table 1 presents eight quantities that are basic to radiometry and photometry. For comparison, they are arranged according as they are energy or flux, spectral or nonspectral, and radiometric or photometric quantities. The relationships between these quantities involve time, wavelength, and the response of the eye. For a given radiation field, they are similarly distributed over the field, having exactly similar angular and areal densities at each field point. Definitions, descriptions, and relationships for the quantities listed in Table 1 follow.

3. Energy and Flux *Radiant energy* Q_e is energy transported in the form of electromagnetic waves.

Radiant flux $\Phi_e \equiv \partial Q_e/\partial t$ is the time rate of change of radiant energy.

Luminous energy Q_v is that aspect of radiant energy which is light, i.e., visually evaluated radiant energy.

Luminous flux $\Phi_v = \partial Q_v/\partial t$ is the time rate of change of luminous energy. It is visually evaluated radiant flux.

4. Spectral and Nonspectral Quantities When applied to any of the quantities listed in Table 1 or its spatial densities, the term spectral denotes spectral density; e.g., spectral flux Φ_λ denotes the flux per unit wavelength at the wavelength λ and is defined as

$$\Phi_\lambda \equiv \frac{\partial \Phi}{\partial \lambda}$$

Spectral flux, or a close approximation to it, is most often determined by measurements made with a narrow-band spectroradiometer or spectrophotometer.

If Φ_λ is known for values of λ throughout the spectral band from λ_1 to λ_2, the flux

TABLE 1 Radiometric and Photometric, Spectral and Nonspectral Energy and Flux Terms, Symbols, and Units†

	Radiometric, subscript e		Photometric, subscript v	
	Nonspectral	Spectral, subscript λ	Nonspectral	Spectral, subscript λ
Energy Q	Radiant energy Q_e joules	Spectral radiant energy $Q_{e\lambda}$ joules per unit wavelength	Luminous energy Q_v lumen-seconds	Spectral luminous energy $Q_{v\lambda}$ lumen-seconds per unit wavelength
Flux Φ	Radiant flux Φ_e watts	Spectral radiant flux $\Phi_{e\lambda}$ watts per unit wavelength	Luminous flux Φ_v lumens	Spectral luminous flux $\Phi_{v\lambda}$ lumens per unit wavelength

† The terms, symbols, and units are in accordance with ANSI Z7.1-1967, American National Standard Nomenclature and Definitions for Illuminating Engineering, RP-16. The symbols for photometric quantities are the same as those for the corresponding radiometric quantities. When it is necessary to differentiate them, the subscripts v and e should be used. Quantities may be restricted to a narrow-wavelength band by adding the word spectral and indicating the wavelength. The corresponding symbols are changed by adding the subscript λ.

$\Phi(\lambda_1,\lambda_2)$ within that band can be computed by using the expression

$$\Phi(\lambda_1,\lambda_2) = \int_{\lambda_1}^{\lambda_2} \Phi_\lambda(\lambda)\, d\lambda$$

Such nonspectral quantities are sometimes called integrated or broadband quantities.

Total flux Φ is defined in terms of spectral flux by

$$\Phi = \int_0^\infty \Phi_\lambda(\lambda)\, d\lambda$$

the term total indicating that the integration is over all wavelengths.

Spectral-density functions, for example, $\Phi_\lambda(\lambda)$, can be expressed in the domains of wavelength λ, frequency ν, and wave number $\bar{\nu}$. The variables of these domains are related as follows:

$$\lambda = \frac{c}{\nu} = \frac{1}{\bar{\nu}}$$

where the speed of light $c = 2.997925 \pm 0.000003 \times 10^8$ m/s (see National Bureau of Standards, 1963). If $\bar{\nu}$ is expressed in waves per centimeter, as is usually the case, λ should be expressed in centimeters and c should be expressed in centimeters per second in the above relationship. The differentials of these variables are related as follows:

$$d\lambda = -\frac{c}{\nu^2}\, d\nu = -\frac{1}{\bar{\nu}^2}\, d\bar{\nu}$$

A relationship that is easy to remember is

$$\frac{d\lambda}{\lambda} = -\frac{d\nu}{\nu} = -\frac{d\bar{\nu}}{\bar{\nu}}$$

If the functions $\Phi_\lambda(\lambda)$, $\Phi_\nu(\nu)$, and $\Phi_{\bar{\nu}}(\bar{\nu})$, where different subscripts denote different functions, each represent the spectral density of the same flux but each in the spectral domain indicated, then these functions are related as follows:

$$\Phi_\lambda(\lambda) = \frac{c}{\lambda^2}\Phi_\nu\left(\frac{c}{\lambda}\right) = \frac{1}{\lambda^2}\Phi_{\bar{\nu}}\left(\frac{1}{\lambda}\right)$$

5. Radiometric and Photometric Quantities A photometric quantity expresses the magnitude of the visual stimulus to the human eye in the analogous radiometric quantity. The basis for evaluating this stimulus magnitude is the spectral luminous efficacy $K(\lambda)$ of radiation. Spectral radiant flux $\Phi_{e\lambda}$ contains the spectral luminous flux $\Phi_{v\lambda}$ given by

$$\Phi_{v\lambda} = K(\lambda)\Phi_{e\lambda}$$

$K(\lambda)$ is nonzero only in the visible region of the spectrum, 380 nm $\leq \lambda \leq$ 780 nm; within that region it is positive and has a peak value K_{max}. Spectral luminous efficacy normalized to a peak value of unity is called the spectral luminous efficiency $V(\lambda)$ of radiation. That is,

$$V(\lambda) = \frac{K(\lambda)}{K_{max}}$$

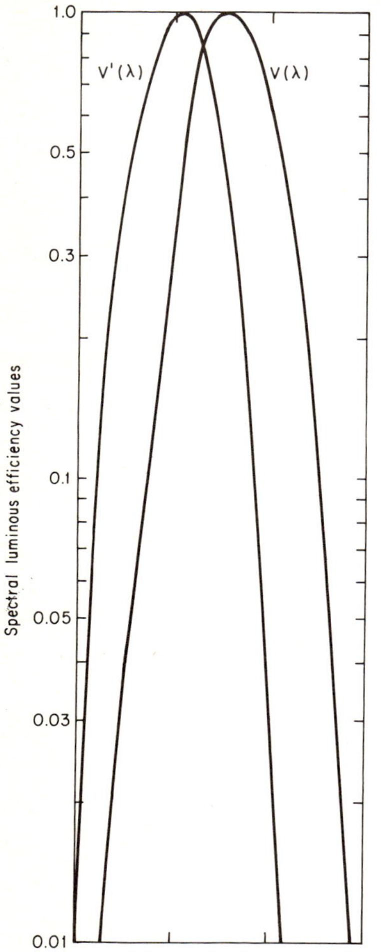

Fig. 1 Photopic spectral luminous efficiency $V(\lambda)$ and scotopic spectral luminous efficiency $V'(\lambda)$. (*Plotted from tables* 13.4 *and* 13.7 *of ANSI* Z7.1-1967.)

Spectral luminous efficiency for photopic (daylight-adapted) vision is denoted by $V(\lambda)$, as above, and spectral luminous efficiency for scotopic (night-adapted) vision is denoted by $V'(\lambda)$. Both these functions are presented graphically in Fig. 1. The peak value (unity) of $V(\lambda)$ occurs at λ = 555 nm, and that of $V'(\lambda)$ occurs at 510 nm. This shift of the peak response of the eye toward the shorter wavelengths as the eye becomes dark-adapted is called the *Purkinje phenomenon*.

The National Bureau of Standards (1970) gives the maximum value K_{max} of the photopic spectral luminous efficacy $K'(\lambda)$ of radiation as

$$K_{max} = 673 \text{ lm/W}$$

and the maximum value K'_{max} of the scotopic spectral luminous efficacy $K'(\lambda)$ of radiation as

$$K'_{max} = 1{,}725 \text{ lm/W}$$

The total magnitude of the visual stimulus contained in the radiant flux Φ_e having a spectral content $\Phi_{e\lambda}$ is given by the total luminous flux

$$\Phi_v = \int_0^\infty \Phi_{v\lambda}\, d\lambda = 673 \text{ lm/W}$$

$$\int_{380 \text{ nm}}^{780 \text{ nm}} V(\lambda)\Phi_{e\lambda}\, d\lambda$$

There is no general simple relationship for converting from radiant to luminous flux because detailed specification of the spectral content of the radiant flux is required for such a conversion, as the above relationship indicates. One radiant watt may contain 0 lm if it is distributed only in the infrared and ultraviolet regions, or it may contain as many as 673 lm if it is concentrated near 555 nm. However, radiant and luminous flux can be related by a constant under the restriction that all quantities of radiant flux to be considered have the same specified spectral distribution.

In general, it is not possible to convert from luminous to radiant flux. Exceptions to this rule are cases where the spectral distribution of the luminous flux is known and all wavelengths to be considered are in the visible region of the spectrum.

Radiant flux Φ_e containing luminous flux Φ_v is said to have a luminous efficacy K given by

$$K = \frac{\Phi_v}{\Phi_e} = \frac{\int_0^\infty K(\lambda)\Phi_{e\lambda}\, d\lambda}{\int_0^\infty \Phi_{e\lambda}\, d\lambda} \qquad \text{lm/W}$$

Spatial Densities

6. Concepts of areal and angular densities of radiant and luminous flux, i.e., radiance and luminance, radiant and luminous intensity, and irradiance and illuminance, are defined and described below. These geometric concepts apply as well to each of the quantities listed in Table 1, although the terms and symbols are specialized to densities of flux. Defining relations for the spatial densities as applied to each of the quantities listed in Table 1 are presented in Table 2. The terms and symbols are in accordance with ANSI Z7.1-1967.

7. Radiance (Luminance) The radiance (luminance) L at a given field point in a given direction is the flux per unit solid angle in the given direction and per unit projected area perpendicular to the given direction. Radiance (luminance) is defined by

$$L = \frac{\partial^2\Phi}{\partial\omega(\cos\theta\,\partial A)}$$

where dA = elemental area containing given point
$d\omega$ = elemental solid angle containing given direction
θ = angle between normal to elemental area and given direction

(see Fig. 2). If (x,y,z) is the given field point and (θ,ϕ) is the given direction, the dependence of L upon position and direction may be indicated by $L(x,y,z;\theta,\phi)$. The term $\cos\theta\, dA$ represents an element of projected area in the direction (θ,ϕ).

When a ray of flux extends through an isotropic medium in the absence of losses by absorption, scattering, or reflection, the radiance (luminance) in the direction of the ray at each point on the ray is the same. Hence, under these conditions, it is meaningful to speak of the radiance (luminance) of a ray. The quantity L/n^2, where n represents index of refraction, is invariant along a ray, even across a smooth boundary between two lossless media (Nicodemus, 1963).

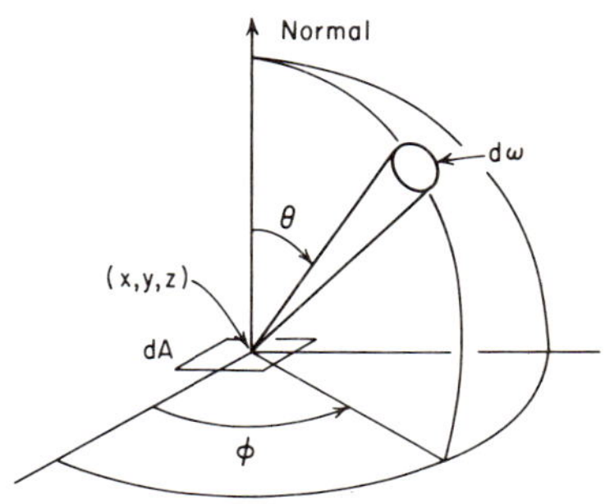

Fig. 2 Radiance (luminance).

If the elemental area dA containing the point (x,y,z) is contained in a given surface, the radiance at the point in the direction (θ,ϕ) may be regarded as being the radiance of the surface at that point and in that direction.

8. Radiant (Luminous) Intensity The radiant (luminous) intensity in a given direction is the flux per unit solid angle in the given direction. Intensity is defined by

$$I = \frac{\partial\Phi}{\partial\omega}$$

where $d\omega$ is an elemental solid angle containing the given direction.

The concept of intensity is used to describe the directional distribution of flux. The intensity $I(\theta,\phi)$ of a surface A in the direction (θ,ϕ) is the flux per unit solid angle emitted by the surface in that direction, and is given by†

$$I(\theta,\phi) = \int_A L(x,y,z;\theta,\phi) \cos \theta_s \, dA$$

where L is the radiance (luminance) of each point (x,y,z) in the direction (θ,ϕ), (x,y,z) being contained in the surface, and $\theta_s = \theta_s(x,y,z;\theta,\phi)$ is the angle between the direction (θ,ϕ) and the normal to the surface at each point (x,y,z).

The intensity of a surface can be measured directly by using an instrument which has an entrance aperture subtending small and nearly equal solid angles from every point on the surface and which responds only to nearly paraxial rays of flux. These conditions are fulfilled when the distance between the surface and the instrument is far greater than the diameters of surface and the entrance aperture of the instrument.

Intensity is particularly useful for describing the flux emitted by an effective point source. Such a source is so small and/or so distant from the observing instrument that the instrument, having only limited resolving power, necessarily integrates the source's radiance (luminance) over its entire projected area in the direction of the instrument.

Radiance (luminance) can be regarded as intensity per unit projected area:

$$L = \frac{\partial I}{\cos \theta \, \partial A}$$

This interpretation is used in photometry, where the candela is the unit of luminous intensity and the unit of luminance is the candela per square meter.

9. Irradiance (Illuminance) Irradiance (illuminance) E is flux per unit area:

$$E = \frac{\partial \Phi}{\partial A}$$

At each field point, and for each direction at each field point, there is a corresponding value of irradiance (illuminance). Consider an arbitrary field point (x,y,z). Through this point passes an arbitrarily oriented plane whose normal lies in the direction (θ_n,ϕ_n). Consider, also, an elemental area dA lying in that plane and containing the field point. The flux $d\Phi$ is that which passes through dA in all directions such that the cosine of the angle between the direction of passage and the direction (θ_n,ϕ_n) of the normal is positive. In general, the above expression for E therefore depends upon x, y, z, θ_n, and ϕ_n.

Irradiance (illuminance) at points on a particular smooth surface may be regarded as being dependent upon only position. At each point (x,y,z) on the surface, the only relevant surface element dA is that which is oriented so that its normal is coincident with the normal to the surface at (x,y,z). Thus both θ_n and ϕ_n are functions of x, y, and z.

The irradiance (illuminance) at a point (x,y,z) on a surface, if only flux incident from the directions (θ,ϕ) contained in the solid angle ω is considered, is given in terms of the radiance $L(x,y,z;\theta,\phi)$ of each incident ray by

$$E(x,y,z) = \int_\omega L(x,y,z;\theta,\phi) \cos \theta \, d\omega$$

where θ is the angle between the direction (θ_n,ϕ_n) of the normal to the surface at (x,y,z) and the direction (θ,ϕ) of each incident ray. The elemental solid angle $d\omega$ contains the direction (θ,ϕ).

The differential irradiance (illuminance)

$$dE = L \cos \theta \, d\omega$$

is sometimes called directional irradiance (illuminance) since it represents the elemental contribution of flux incident from only a single direction (θ,ϕ).

† Throughout Sec. 1 semicolons are used to group variables by type in a departure from conventional notation.

As is discussed in more detail in Par. 13, the standard terms and symbols of radiometry and photometry distinguish two cases of areal flux density (flux per unit area) at a surface. The terms irradiance and illuminance and the symbol E are used when flux is incident upon a surface. For flux emerging from a surface, the terms radiant and luminous exitance and the symbol M are used (see ANSI Z7.1-1967).

10. Radiant (Luminous) Density Radiant (luminous) density ω is radiant (luminous) energy per unit volume

$$\omega = \frac{\partial Q}{\partial V}$$

Radiant density ω_e is a concept useful to the theoretician in deriving blackbody and graybody formulas, but it is not useful to the experimental radiometrist. The elemental energy dQ is the sum of the energies of individual photons contained within an elemental volume dV at a given instant in time.

Summary of Basic Concepts

11. Table 2 presents a summary of definitions of the areal and angular densities of quantities basic to radiometry and photometry. As can be seen by comparison of the entries within each column, the spatial-density concepts defined and described above in terms of flux apply as well to all the quantities listed in Table 1.

The terms and symbols for the basic quantities are in accordance with ANSI Z7.1-1967. However, since there are no standard terms for the general spatial-density concepts, the column headings in Table 2 were chosen more or less arbitrarily.

TABLE 2 Spatial Densities of Basic Quantities

	Spatial density		
Basic quantity	Areal density at a surface (quantity per unit area at a given point on the surface)	Intensity (quantity per unit solid angle in a given direction)	Specific intensity (quantity per unit solid angle in a given direction per unit projected area at a given point)
Radiant energy Q_e	$\frac{\partial Q_e}{\partial A}$	$\frac{\partial Q_e}{\partial \omega}$	$\frac{\partial^2 Q_e}{\partial \omega(\cos\theta\,\partial A)}$
Spectral radiant energy $Q_{e\lambda}$	$\frac{\partial Q_{e\lambda}}{\partial A}$	$\frac{\partial Q_{e\lambda}}{\partial \omega}$	$\frac{\partial^2 Q_{e\lambda}}{\partial \omega(\cos\theta\,\partial A)}$
Radiant flux Φ_e	$\frac{\partial \Phi_e}{\partial A}$	$\frac{\partial \Phi_e}{\partial \omega}$	$\frac{\partial^2 \Phi_e}{\partial \omega(\cos\theta\,\partial A)}$
Spectral radiant flux $\Phi_{e\lambda}$	$\frac{\partial \Phi_{e\lambda}}{\partial A}$	$\frac{\partial \Phi_{e\lambda}}{\partial \omega}$	$\frac{\partial^2 \Phi_{e\lambda}}{\partial \omega(\cos\theta\,\partial A)}$
Luminous energy Q_v	$\frac{\partial Q_v}{\partial A}$	$\frac{\partial Q_v}{\partial \omega}$	$\frac{\partial^2 Q_v}{\partial \omega(\cos\theta\,\partial A)}$
Spectral luminous energy $Q_{v\lambda}$	$\frac{\partial Q_{v\lambda}}{\partial A}$	$\frac{\partial Q_{v\lambda}}{\partial \omega}$	$\frac{\partial^2 Q_{v\lambda}}{\partial \omega(\cos\theta\,\partial A)}$
Luminous flux Φ_v	$\frac{\partial \Phi_v}{\partial A}$	$\frac{\partial \Phi_v}{\partial \omega}$	$\frac{\partial^2 \Phi_v}{\partial \omega(\cos\theta\,\partial A)}$
Spectral luminous flux $\Phi_{v\lambda}$	$\frac{\partial \Phi_{v\lambda}}{\partial A}$	$\frac{\partial \Phi_{v\lambda}}{\partial \omega}$	$\frac{\partial^2 \Phi_{v\lambda}}{\partial \omega(\cos\theta\,\partial A)}$

[The term sterance, proposed by Jones (1963), could well have been used in place of the term specific intensity.] Standard terms and symbols exist for several of the entries in Table 2. The spatial densities of flux are all covered by standard terms. For example, what Table 2 indicates to be the specific intensity of radiant flux has the standard term radiance, and the intensity of spectral luminous flux has the standard term spectral luminous intensity. But there are currently no standard terms for the spatial densities of radiant and luminous energy. In the absence of standard terms or terms adopted by one's peers, use of concatenations of the terms in the row and column headings of Table 2 are suggested, but they should be accompanied with definitions.

Table 3 presents expressions in terms of spectral radiance $L_{e\lambda}$ for all the quantities listed in Table 2. For purposes of computation, spectral radiance is the most versatile of the quantities represented. If it is known with adequate precision as a function of its variables, it can be used to compute all the other quantities, as Table 3 indicates.

BASIC TERMS, SYMBOLS, AND UNITS

12. Although the basic concepts of radiometry and photometry constitute an elegant system, the conglomerate of terms, symbols, and units that have been applied to these concepts is inelegant. Ideally all current communications would be in accordance with the most recently reached international agreements on standards, and old terms, symbols, and units would be found only in the past literature. Unfortunately, much of the accumulation of present and past standard and nonstandard terms, symbols, and units continues to be used, including some invented before the electromagnetic spectrum was fully understood.

In this section, basic terms, symbols, and units are presented which are in accordance with the American National Standard Letter Symbols for Illuminating Engineering (ANSI Y10.18-1967) and the American National Standard Nomenclature and Definitions for Illuminating Engineering (ANSI Z7.1-1967). These documents are consistent with the international agreements on standards reached to date. Such standards are the result of a monumental effort on a national and international scale (Meyer-Arendt, 1968), an effort which is still small in comparison with the losses caused by a lack of standards.

For use in translating the previous standard terms, symbols, and units into the present ones, the previous system is also presented in this section. For translating other systems to the current standard one, it is recommended that translation tables be devised as needed for direct comparison of terms, symbols, and units on a concept-by-concept basis.

13. Terms and Symbols Table 4 presents current standard terms and symbols for radiometric and photometric quantities in accordance with ANSI Z7.1-1967.

Table 5 presents the previous standard terms and symbols. The current standard symbols give increased recognition to the analogy between radiometric and photometric quantities. The terms radiant and luminous emittance, standard in the past (Table 5), are now to be deprecated and have been replaced by the terms radiant and luminous exitance (Table 4). Different terms and symbols are used to distinguish between the areal density of flux leaving a surface and that incident upon a surface in both the current and previous standards. The preferred term for areal density of luminous flux incident upon a surface, in accordance with international standards, is illuminance rather than illumination.

Although some of the terms and all the symbols in Table 5 are now obsolete, they can be expected to fade from active use only gradually, and for some time the largest portion of existing literature will contain these terms and symbols.

14. Units Table 6 presents the SI units for basic radiometric and photometric quantities, along with some non-SI units in brackets. SI is the abbreviation for Système International d'Unités (International System of Units), an extension and refinement of the meter-kilogram-second (mks) system (Mechtly, 1969; see also Page and Vigoureux, 1971).

A particularly important unit in photometry is the candela, the unit of luminous intensity. It is defined as the luminous intensity of one-sixtieth of one square

TABLE 3 Spatial Densities of Basic Quantities Expressed in Terms of Spectral Radiance, $L_{e\lambda}$

Basic quantity	Spatial density: Areal density at a surface	Spatial density: Intensity	Spatial density: Specific intensity
Radiant energy $Q_e = \iiiint L_{e\lambda} \cos\theta \, d\omega \, dA \, d\lambda \, dt$	$\iiint L_{e\lambda} \cos\theta \, d\omega \, d\lambda \, dt$	$\iiint L_{e\lambda} \cos\theta \, dA \, d\lambda \, dt$	$\iint L_{e\lambda} \, d\lambda \, dt$
Spectral radiant energy $Q_{e\lambda} = \iiint L_{e\lambda} \cos\theta \, d\omega \, dA \, dt$	$\iint L_{e\lambda} \cos\theta \, d\omega \, dt$	$\iint L_{e\lambda} \cos\theta \, dA \, dt$	$\int L_{e\lambda} \, dt$
Radiant flux $\Phi_e = \iiint L_{e\lambda} \cos\theta \, d\omega \, dA \, d\lambda$	$\iint L_{e\lambda} \cos\theta \, d\omega \, d\lambda$	$\iint L_{e\lambda} \cos\theta \, dA \, d\lambda$	$\int L_{e\lambda} \, d\lambda$
Spectral radiant flux $\Phi_{e\lambda} = \iint L_{e\lambda} \cos\theta \, d\omega \, dA$	$\int L_{e\lambda} \cos\theta \, d\omega$	$\int L_{e\lambda} \cos\theta \, dA$	$L_{e\lambda}$
Luminous energy $Q_v = \iiiint K(\lambda) L_{e\lambda} \cos\theta \, d\omega \, dA \, d\lambda \, dt$	$\iiint K(\lambda) L_{e\lambda} \cos\theta \, d\omega \, d\lambda \, dt$	$\iiint K(\lambda) L_{e\lambda} \cos\theta \, dA \, d\lambda \, dt$	$\iint K(\lambda) L_{e\lambda} \, d\lambda \, dt$
Spectral luminous energy $Q_{v\lambda} = \iiint K(\lambda) L_{e\lambda} \cos\theta \, d\omega \, dA \, dt$	$\iint K(\lambda) L_{e\lambda} \cos\theta \, d\omega \, dt$	$\iint K(\lambda) L_{e\lambda} \cos\theta \, dA \, dt$	$\int K(\lambda) L_{e\lambda} \, dt$
Luminous flux $\Phi_v = \iiint K(\lambda) L_{e\lambda} \cos\theta \, d\omega \, dA \, dt$	$\iint K(\lambda) L_{e\lambda} \cos\theta \, d\omega \, d\lambda$	$\iint K(\lambda) L_{e\lambda} \cos\theta \, dA \, d\lambda$	$\int K(\lambda) L_{e\lambda} \, d\lambda$
Spectral luminous flux $\Phi_{v\lambda} = \iint K(\lambda) L_{e\lambda} \cos\theta \, d\omega \, dA$	$\int K(\lambda) L_{e\lambda} \cos\theta \, d\omega$	$\int K(\lambda) L_{e\lambda} \cos\theta \, dA$	$K(\lambda) L_{e\lambda}$

TABLE 4 Standard Terms and Symbols†

Basic quantity	Spatial density: Areal density at a surface	Intensity	Specific intensity	Volumetric energy density
Radiant energy Q (Q_e)				Radiant density w (w_e)
Radiant flux Φ (Φ_e)	Leaving surface: Radiant exitance M (M_e) (radiant emittance‡) Incident upon surface: Irradiance E (E_e)	Radiant intensity I (I_e)	Radiance L (L_e)	
Luminous energy Q (Q_v)				Luminous density w (w_v)
Luminous flux Φ (Φ_v)	Leaving surface: Luminous exitance M $(M)_v$ (luminance emittance‡) Incident upon surface: Illumination E (E_v) (illuminance)	Luminous intensity (candlepower) I (I_v)	Luminance (photometric brightness) L (L_v)	

† This table is in accordance with American National Standard Nomenclature and Definitions for Illuminating Engineering, ANSI Z7.1-1967. Alternative terms and symbols are in parentheses. The symbols for photometric quantities are the same as those for the corresponding radiometric quantities. When it is necessary to differentiate them, the subscripts v and e, respectively, should be used. Quantities may be restricted to a narrow-wavelength band by adding the word spectral and indicating the wavelength. The corresponding symbols are changed by adding the subscript λ.

‡ Term deprecated.

centimeter of projected area of a blackbody radiator operating at the temperature of solidification of platinum. (The radiometric output of this defined source has no special use.)

Table 6 refers to Table 7 for units of flux density at a surface and to Table 9 for units of luminance. Table 7 shows that the units of luminous flux density provide a distinction between luminous exitance and illuminance (illumination). Luminous exitance is expressed generally in lumens per unit area. The units of illuminance, on the other hand, have special names. The SI unit is the lux; the cgs unit is the phot; and the English unit of illuminance is the footcandle. The abbreviations for these units are also presented in Table 7, along with other equivalent units given in brackets.

In Fig. 3, units of illuminance produced by a small source with the intensity of 1 cd are represented for comparison. Table 8 gives conversion factors relating units of illuminance.

Table 9 presents units of luminance. The SI unit is the candela per square meter, which in the mks system was called the nit. The stilb is the cgs unit. No special name is given for an English unit of luminance, the candela per square foot, per square inch, etc., being used.

Also listed in Table 9 are the apostilb, lambert, and footlambert. The apostilb equals $1/\pi$ nit; the lambert equals $1/\pi$ stilb; and the footlambert equals $1/\pi$ candela per square foot. The apostilb, lambert, and footlambert are units of luminance that have been applied especially to lambertian (diffusely emitting) surfaces.

TABLE 5 Previous Standard Terms and Symbols†

Basic quantity	Spatial density: Areal density at a surface	Intensity	Specific intensity	Volumetric energy density
Radiant energy U				Radiant density u
Radiant flux P	Leaving surface: Radiant emittance W Incident upon surface: Irradiance H	Radiant intensity J	Radiance N	
Luminous energy Q				Luminous density q
Luminous flux F	Leaving surface: Luminous emittance L Incident upon surface: Illumination E	Luminous intensity I	Luminance (photometric brightness) B	

† See American Standard Nomenclature for Radiometry and Photometry, Z-58.1.1, American Standards Association, New York, 1953. Quantities may be restricted to a narrow-wavelength band by adding the word spectral and indicating the wavelength. The corresponding symbols are changed by adding the subscript λ.

TABLE 6 Basic Radiometric and Photometric Units

Basic concept	Spatial density: Areal density at a surface	Intensity	Specific intensity	Volumetric density
Radiant energy, joules J				Radiant density, joules per cubic meter J/m^3
Radiant flux, watts W	Radiant exitance, irradiance, watts per square meter W/m^2	Radiant intensity, watts per steradian W/sr	Radiance, watts per steradian–square meter $W/(sr)(m^2)$	
Luminous energy, lumen-second [=talbot] lm · s				Luminous density, lumen-seconds per cubic meter $lm \cdot s/m^3$
Luminous flux, lumen lm	Luminous exitance, illuminance (see Table 7)	Luminous intensity, candela cd [=lumen per steradian† (lm/sr)]	Luminance [photometric brightness] (see Table 9)	

† Non-SI form.

TABLE 7 Units of Luminous Flux Density at a Surface

Luminous exitance, lumens per square meter (lm/m^2)		
Illuminance (Illumination)		
SI and mks units	cgs units	English units
Lux (lx) [lumens per square meter (lm/m^2)]	Phot (ph) [lumens per square centimeter (lm/cm^2)]	Footcandle (fc) [lumens per square foot (lm/ft^2)]

TABLE 8 Conversion of Illuminance Units (ANSI Z7.1-1967)

	Footcandle	Lux	Phot	Milliphot
1 footcandle	1	10.76	0.00108	1.076
1 lux	0.0929	1	0.0001	0.1
1 phot	929	10,000	1	1,000
1 milliphot.	0.929	10	0.001	1

TABLE 9 Units of Luminance (Photometric Brightness)

General Units of Luminance		
SI and mks units	cgs units	English units
Candela per square meter (cd/m^2) [nit (nt)]	Stilb (sb) Candela per square centimeter (cd/cm^2)	Candela per square foot (cd/ft^2)
Units of Luminance That Apply Especially to Lambertian, Diffuse Surfaces with Luminous Exitance Equal to One Lumen per Unit Area		
mks units	cgs units	English units
[Apostilb (asb) = cd/(π) (m^2)]	Lambert (L) = cd/(π) (cm^2)	Footlambert (fL) = cd/(π) (ft^2)

Lambertian surfaces produce directionally uniform luminance (the same luminance in all outward directions). Such a surface emitting with luminous exitance M necessarily emits with luminance $L = M/\pi$. Thus a lambertian source of luminous exitance equal to one lumen per square meter (see Table 7) is a source of luminance equal to one apostilb. Similarly, a lambertian source of one lumen per square

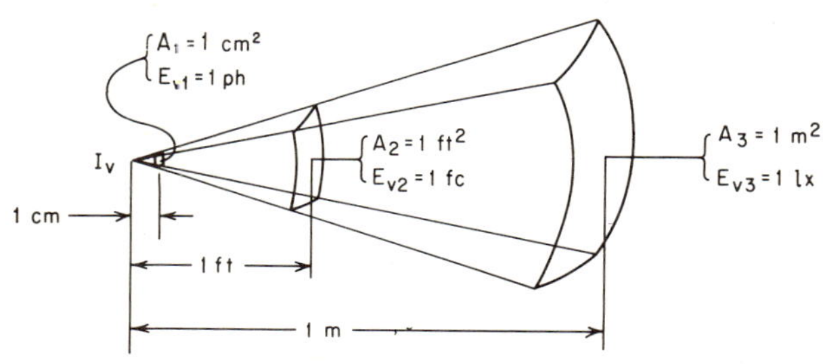

Fig. 3 Illumination E/v produced by a source of unit luminous intensity (1 cd). The solid angle indicated represents 1 sr, and the flux contained is 1 lm.

centimeter is a source of one lambert, and a lambertian source of one lumen per square foot is a source of one footlambert. Conversion factors are given in Table 10.

BLACKBODY RADIATION

15. A blackbody is an ideal body of uniform temperature that perfectly absorbs all incident radiation. A blackbody is a perfect temperature radiator in that all nonblackbody temperature radiators in equilibrium at the same temperature emit less radiant flux than the blackbody does. Whereas the radiant flux of a nonblackbody temperature radiator is determined by the material character of its surface as well as by its temperature, the blackbody's radiant flux is determined only by its temperature. The radiance of a blackbody is the same in all directions.

16. Total Blackbody Radiation: The Stefan-Boltzmann Law The Stefan-Boltzmann law expresses the total radiation emitted from a blackbody as a function of the body's absolute temperature. Stated in terms of radiant exitance, the law is

$$M^b(T) = \sigma T^4 \qquad \mathrm{W/m^2}$$

where the superscript b denotes blackbody radiation, T is the body's temperature in kelvins,† and σ is the Stefan-Boltzmann constant, given by

$$\sigma = \frac{2\pi^5 k^4}{15h^3c^2} = 5.6697 \times 10^{-8}\ \mathrm{W/(m^2)(K^4)}\dagger$$

where k = Boltzmann's constant
h = Planck's constant
c = speed of light

The value given for σ is that recommended by the National Bureau of Standards (1963).

The radiance of a blackbody is directionally uniform; i.e., it is the same in all outward directions. Thus the radiance is related to the radiant exitance by the expression $L = M/\pi$, and the Stefan-Boltzmann law stated in terms of radiance is given by

$$L^b = \frac{\sigma}{\pi} T^4 \qquad \mathrm{W/(m^2)(sr)}$$

17. Spectral Distribution of Blackbody Radiation: The Planck Radiation Law The spectral distribution of blackbody radiation is described by the Planck radiation law. Stated in terms of spectral radiant exitance, the law is

$$M_\lambda{}^b(\lambda,T) = \frac{c_1}{\lambda^5(e^{c_2/\lambda T} - 1)} \qquad \mathrm{W/m^3}$$

where the first radiation constant $c_1 = 2\pi hc^2 = 3.7415 \times 10^{-16}\ \mathrm{W\cdot m^2}$, and the second radiation constant $c_2 = hc/k = 1.43879 \times 10^{-2}\ \mathrm{m\cdot K}$. These values for c_1 and c_2 are recommended in National Bureau of Standards (1963) (corrected).

The spectral radiance of a blackbody is given by

$$L_\lambda{}^b = \frac{M_\lambda{}^b}{\pi} \qquad \mathrm{W/(m^3)(sr)}$$

Graphs of $L_\lambda{}^b$ are presented in Fig. 4.

For a generalized treatment of Planck's law see Smith et al. (1957).

18. Approximate Spectral Distribution of Blackbody Radiation: The Wien and the Rayleigh-Jeans Radiation-Law Approximations Since they have simpler algebraic forms than the Planck law, the Wien and Rayleigh-Jeans formulas for the spectral radiance of a blackbody provide useful approximations to the Planck function over limited ranges of temperature and wavelength.

† The unit kelvin (K) is the SI unit that replaces the degree Kelvin (°K).

TABLE 10 Luminance Conversion Factors (ANSI Z7.1-1967)

1 apostilb (international) = 0.1 mL = 1 blondel
1 apostilb (German Hefner) = 0.09 mL
1 lambert = 1,000 mL

	fL	cd/m^2	mL	cd/in^2	cd/ft^2	sb
1 footlambert (fL)	1	3.426	1.076	0.00221	0.3183	0.00034
1 candela per square meter† (nit) (cd/m^2)	0.2919	1	0.3142	0.000645	0.0929	0.0001
1 millilambert (mL)	0.929	3.183	1	0.00205	0.2957	0.00032
1 candela per square inch (cd/in^2)	452	1,550	487	1	144	0.155
1 candela per square foot (cd/ft^2)	3.142	10.76	3.382	0.00694	1	0.00108
1 stilb (sb)	2,919	10,000	3,142	6.45	929	1

† SI unit.

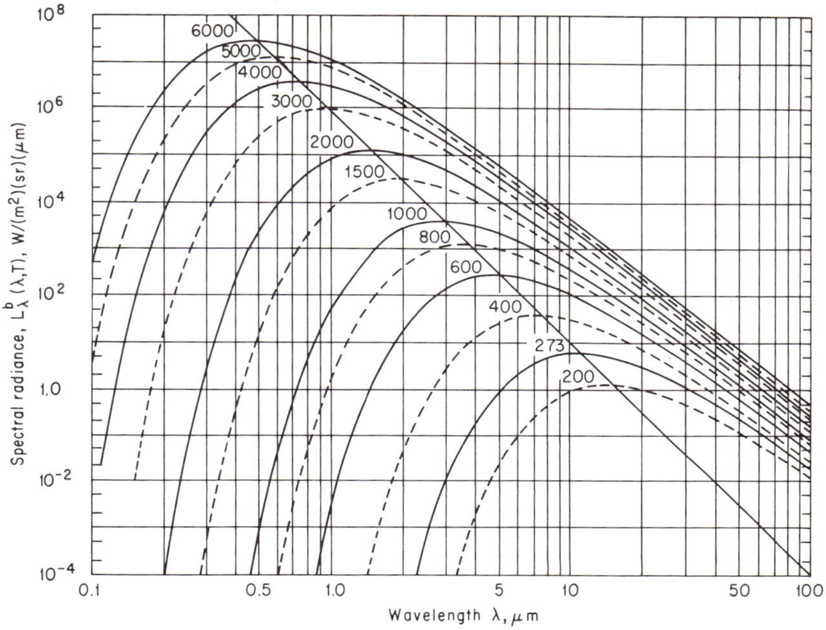

Fig. 4 Spectral radiance $L_\lambda{}^b$ of a blackbody at the temperature in kelvins shown on each curve. The diagonal line intersecting each curve at its maximum shows Wien's displacement law. Subdivisions of the ordinate scale are at 2 and 5. [*Adapted by permission from Valley* (1965.]

Wien's formula for the radiance of a blackbody is given by

$$L_\lambda{}^b \approx \frac{c_1}{\pi\lambda^5 e^{c_2/\lambda T}}$$

For values of λ and T such that $e^{c_2/\lambda T} \gg 1$ the Wien formula has nearly the same value as the Planck function. As indicated in Figs. 5 and 6, the Wien formula is accurate to within 1 percent for values of $\lambda T < 3000$ μm·K.

The Rayleigh-Jeans law for blackbody radiance is given by

$$L_\lambda{}^b \approx \frac{c_1}{c_2\pi\lambda^4 T}$$

For values of $\lambda T \gg c_2$, this formula can be obtained from the Planck function by expanding the term $e^{c_2/\lambda T} - 1$ in an infinite series and approximating the series by the single term $c_2/\lambda T$. The Rayleigh-Jeans formula is accurate to within 1 percent for values of $\lambda T > 7.7 \times 10^5$ μm·K. Since a blackbody emits more than 99.9 percent of its energy over values of λT less than 7.7×10^5 μm·K, the Rayleigh-Jeans formula has only limited application to ultraviolet, visible, and infrared wavelengths. The accuracy of the formula for various values of λT is indicated in Figs. 5 and 6.

19. Wien Displacement Law The Wien displacement law, in its general form, provides the following expressions for the maximum value $L_\lambda{}^b(\lambda_m,T)$ of the spectral-radiance function and for the wavelength λ_m at which this maximum occurs:

$$\lambda_m T = 2.8978 \times 10^{-3} \text{ m·K}$$
$$L_\lambda{}^b(\lambda_m,T) = 4.10T^5 \times 10^{-6} \text{ W/(m}^3\text{)(sr)(K}^5\text{)}$$

The first of these two expressions is often referred to as the Wien displacement law.

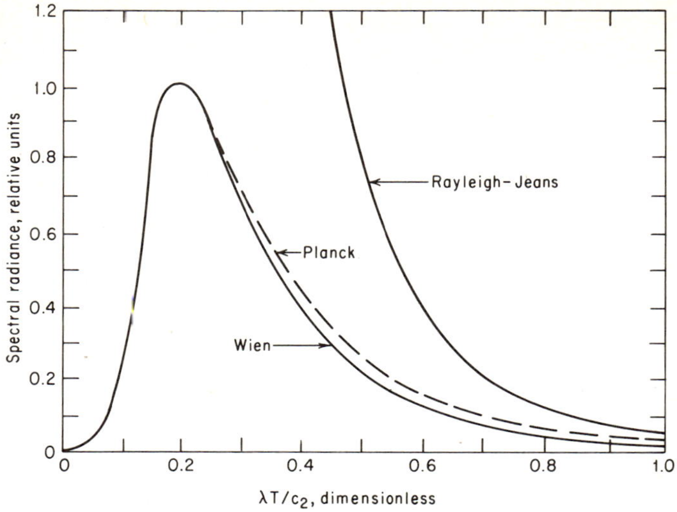

Fig. 5 Comparison between Planck, Wien, and Rayleigh-Jeans laws for the spectral radiance of a blackbody. [*Adapted from R. Kingslake (ed.), "Applied Optics and Optical Engineering," vol. IV, p. 276. Copyright* 1967 *by Academic Press, Inc.; used by permission.*]

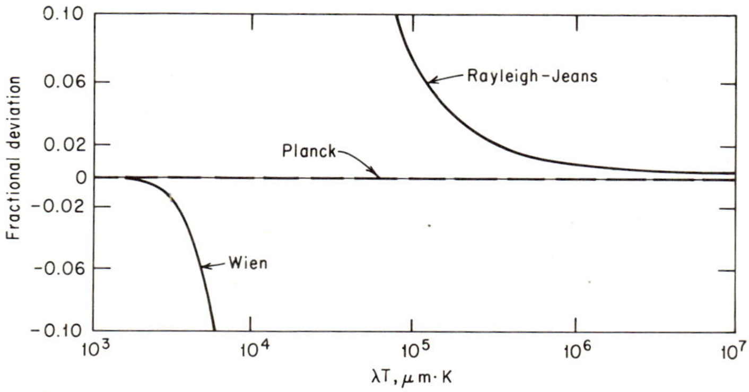

Fig. 6 Fractional deviations of Wien and Rayleigh-Jeans approximations to the Planck function for blackbody spectral radiation. [*Adapted from R. Kingslake (ed.), "Applied Optics and Optical Engineering," vol. IV, p. 276. Copyright* 1967 *by Academic Press, Inc.; used by permission.*]

20. Normalized Spectral-Distribution Function and Normalized Cumulative Spectral-Distribution Function for Blackbody Radiation The ratio of the Planck function and its maximum value is called the normalized spectral-distribution function for blackbody radiation. This ratio, $f(\lambda T)$, is a function only of the product λT, and is given by

$$f(\lambda T) = \frac{L_\lambda{}^b(\lambda, T)}{L_\lambda{}^b(\lambda_m, T)}$$

Thus, $f(\lambda T)$ is a universal function describing the radiation for any combination of temperature and wavelength.

The normalized cumulative spectral-distribution function $L(\lambda T)$ is also a universal function. It describes the fraction of radiation having wavelengths smaller than λ

and being emitted from a blackbody at a temperature T for any combination of λ and T. This function is given by

$$F(\lambda T) = \frac{\int_0^\lambda L_\lambda^b(\lambda,T)\, d\lambda}{\int_0^\infty L_\lambda^b(\lambda,T)\, d\lambda}$$

Thus $F(\lambda T)$ monotonically increases from zero toward unity as λT increases from zero.

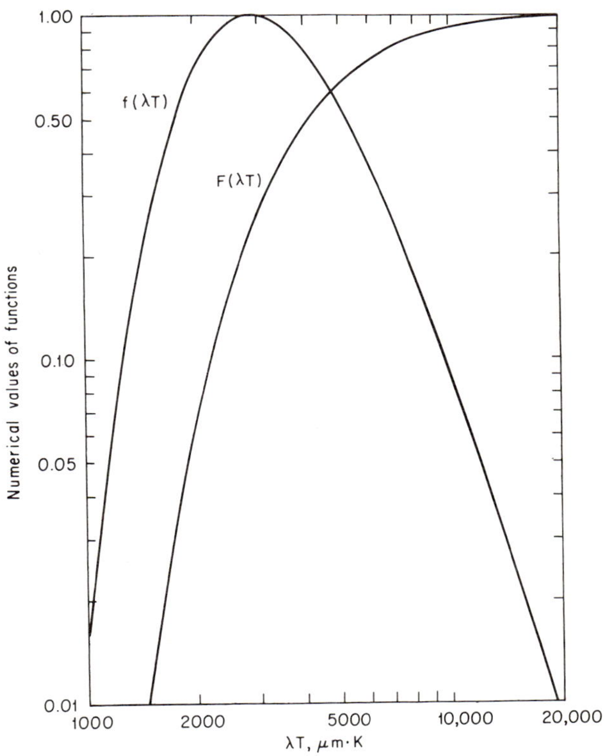

Fig. 7 Normalized spectral distribution $f(\lambda T)$ and normalized cumulative spectral distribution $F(\lambda T)$ of blackbody radiation. [*Data from Pivovonsky and Nagel* (1961).]

The functions $f(\lambda T)$ and $F(\lambda T)$ are graphically presented in Fig. 7. These normalized functions are often more convenient than the unnormalized Planck function. Some examples of the uses of $f(\lambda T)$ and $F(\lambda T)$ follow.

Radiance of a blackbody in a finite wavelength interval from λ_1 to λ_2:

$$\int_{\lambda_1}^{\lambda_2} L_\lambda^b(\lambda,T)\, d\lambda = \frac{\sigma T^4}{\pi} [F(\lambda_2 T) - F(\lambda_1 T)]$$

Radiance of the blackbody in a narrow wavelength interval $\Delta\lambda$:

$$\int_{\lambda_1}^{\lambda_1+\Delta\lambda} L_\lambda^b(\lambda,T)\, d\lambda \approx L_\lambda^b(\lambda_m,T) f(\lambda T)\, \Delta\lambda$$

Simple expressions can be formulated by use of $F(\lambda T)$ that indicate roughly the spectral distribution of blackbody radiation. These expressions have the same form as the Wien displacement law. Let $\lambda_{n\%}$ be such that $F(\lambda_{n\%}T) = n/100$, that is, such that n percent of the radiation is emitted at wavelengths shorter than $\lambda_{n\%}$. Then

$$T\lambda_{10\%} = 2199\ \mu\text{m}\cdot\text{K}$$
$$T\lambda_{50\%} = 4105\ \mu\text{m}\cdot\text{K}$$
$$T\lambda_{90\%} = 9371\ \mu\text{m}\cdot\text{K}$$

It can be shown in this way that $\lambda_{25.004\%} = \lambda_m$, which means that approximately 25 percent of the radiation from a blackbody is emitted on the short-wavelength side of the peak of the spectral-radiance function. A further observation is that nearly 95.2 percent of the radiation is emitted in the wavelength interval for which λT ranges over the decade from 1660 to 16,600 μm·K.

21. Blackbody-Radiation Data Several slide rules available for making quick radiation computations with an accuracy of one or two significant digits are described by Wolfe (1965) and Hudson (1969).

Extensive tables provide data for more precise computations.† The most recent in that list (Pivovonsky and Nagel and Czerny and Walther) together represent a thorough treatment of the Planck radiation function.

Figure 4 presents graphs of the spectral radiance $L_\lambda{}^b$ of a blackbody at selected temperatures ranging from 200 to 6000 K. As in this figure, the Planck function plotted on log-log scales has the same shape for each temperature. Indeed, one can construct an overlay by tracing the shape of the curve, and the overlay can then be slid back and forth along the line representing the Wien displacement law to represent the Planck function for any desired temperature. Valley (1965) contains numerous graphs of blackbody radiance and photon emission as functions of wavelength, wave number, and frequency.

Figure 7 presents the normalized spectral distribution $f(\lambda T)$ and the normalized cumulative spectral distribution $F(\lambda T)$ of blackbody radiation. These two functions were previously defined and discussed.

Luminous-Blackbody Radiation

22. Blackbody Luminance The luminance $L_v{}^b$ of a blackbody of temperature T is given by

$$L_v{}^b(T) = K_{\max} \int_0^\infty V(\lambda) L_{e\lambda}{}^b(\lambda,T)\, d\lambda$$

$$= 6 \times 10^5 \frac{\int_0^\infty V(\lambda) L_{e\lambda}{}^b(\lambda,T)\, d\lambda}{\int_0^\infty V(\lambda) L_{e\lambda}{}^b(\lambda, 2045\text{ K})\, d\lambda} \qquad \text{cd/m}^2$$

where 2045 K is the temperature of freezing platinum (National Bureau of Standards, 1970). A graph of this function is presented in Fig. 8.

23. Luminous Efficacy of Blackbody Radiation The luminous efficacy K_b of a blackbody radiator is equal to the ratio of its total luminance and its total radiance. Thus

$$K_b(T) = \frac{L_v{}^b(T)}{(\sigma/\pi)T^4} \qquad \text{lm/W}$$

† Howard (1962) has listed S. A. Golden, Spectral and Integrated Blackbody Radiation Functions, *North American Aviation, Inc., Rocketdyne Div., Canoga Park, Calif., Res. Rep.* 60-23, 1960; M. Pivovonsky and M. Nagel, "Tables of Blackbody Radiation Functions," Macmillan, New York, 1961; M. Czerny and A. Walther, "Tables of the Fractional Function for the Planck Radiation Law," Springer, Berlin, 1961; Parry Moon, *J. Opt. Soc. Am.*, vol. 38, p. 291, 1948; A. N. Lowen and G. Blanch, ibid., vol. 30, p. 70, 1940; E. Jahnke and F. Emde, "Tables of Functions," Dover, New York, 1945; A. G. DeBell, *North American Aviation, Inc., Rocketdyne Div., Canoga Park, Calif., Rocketdyne Res. Rep.* 59-32, 1959.

$K_b(T)$ reaches a maximum of approximately 95 lm/W at a temperature near 6600 K. A graph of $K_b(T)$ is presented in Fig. 9.

24. Temperature as a Parameter of Radiation Limited aspects of the radiation from nonblack sources can often be approximated by those of blackbody radiation. A blackbody temperature that results in a best approximation is said to be the equivalent, or apparent, temperature of the nonblack source. The convenience of such an approximation lies in specifying the nonblack radiation by a single parameter. However, an equivalent temperature is limited in its application to the conditions of the approximation and may differ substantially from the true temperature of the source. A number of commonly used concepts of equivalent temperature are described below.

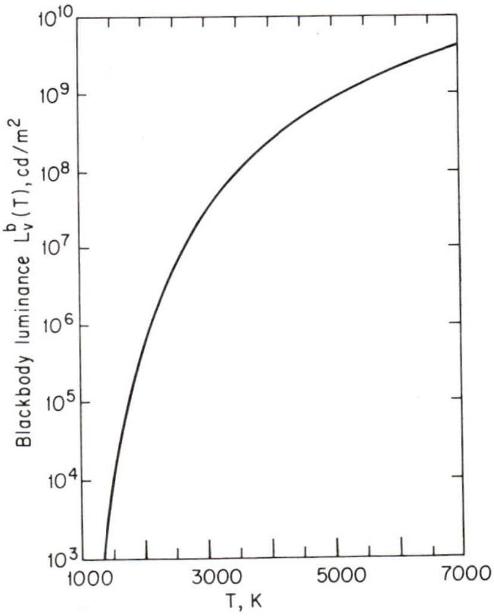

Fig. 8 Luminance $L_v{}^b$ of a blackbody as a function of its temperature. [*Data from Pivovonsky and Nagel* (1961).]

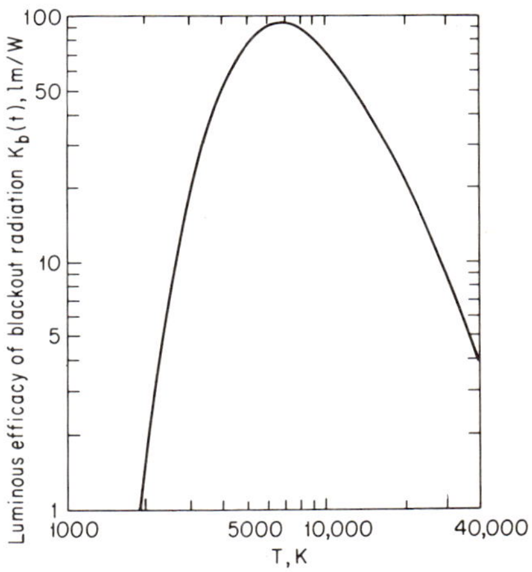

Fig. 9 Luminous efficacy K_b of blackbody radiation as a function of its temperature. [*Data from Pivovonsky and Nagel* (1961).]

True Temperature. The true temperature of a source is the temperature in kelvins measured by a suitable thermometer in thermal equilibrium with the source. The true temperature of a blackbody radiator is that temperature for which the Planck formula accurately represents spectral radiance.

Radiation Temperature. The radiation temperature of a source, not necessarily a blackbody, is the temperature of a blackbody having the same radiance.

Radiance Temperature. The radiance temperature of a source is the temperature of a blackbody that has the same radiance for some narrow spectral region (usually 0.665 μm). The wavelength should be specified when the radiance temperature is given.

Distribution Temperature of a Light Source. The distribution temperature of a light source is the temperature of a blackbody whose relative spectral distribution is the same or nearly the same in the visible region of the spectrum as that of the light source.

Color Temperature. The color temperature of a light source is the temperature at which a blackbody must be operated to have a chromaticity equal to that of the light source. Most thermal radiators, e.g., hot filaments of incandescent lamps, are fairly nonselective in the visible portion of the spectrum, and therefore light from such illuminants can be closely matched in chromaticity by the light from a blackbody (see Par. 25).

Correlated Color Temperature. The correlated color temperature of a light source is the temperature of a blackbody whose chromaticity most nearly resembles that of the light source. Correlated color temperatures may be specified for selective radiators whose chromaticities are only fairly close to the chromaticity of a blackbody (see Par. 25).

25. Chromaticity of Blackbody Radiation The chromaticity coordinates of a blackbody radiator as a function of its temperature are shown in Fig. 10. These

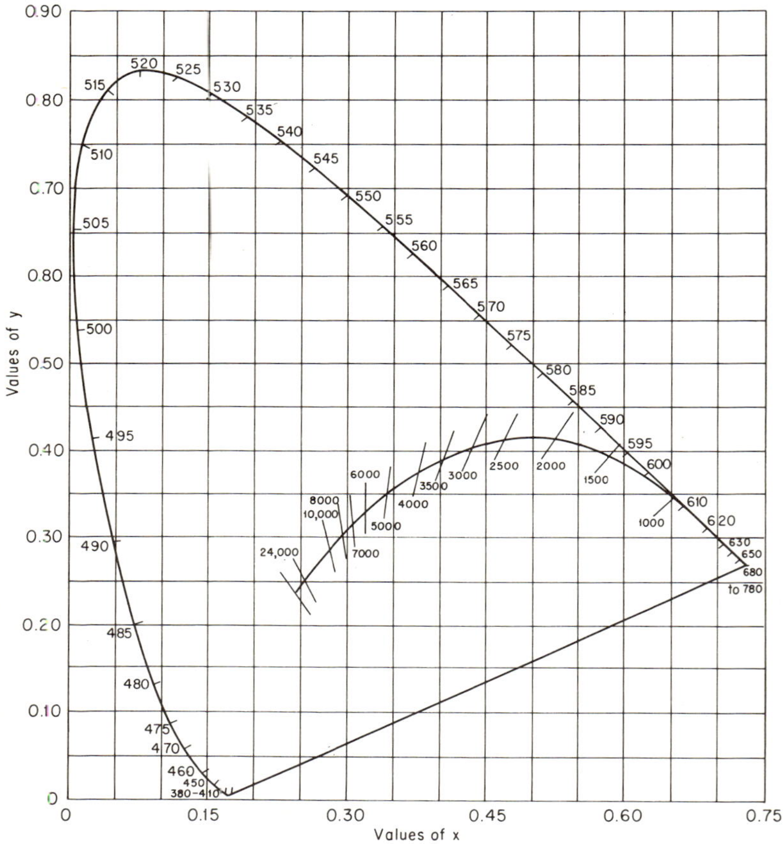

Fig. 10 Chromaticity diagram, showing locus of chromaticities of blackbody radiation as a function of temperature and lines of constant correlated color temperature. [*From Optical Society of America* (1963).]

coordinates can be computed in accordance with the CIE colorimetric system by use of Planck's formula. See Pivovonsky and Nagel (1961) for a comprehensive table of these coordinates. The correlation with color temperature is based on closest similarity and not exact equivalence with one of the blackbody chromaticities. The loci of points having equal correlated color temperatures have been defined and are indicated by the straight-line segments crossing the locus of blackbody-chromaticity coordinates in Fig. 10. Such lines are called isotemperature loci (see also Optical Society of America, 1963; Wyszecki and Stiles, 1967; Kelly, 1963; Walsh, 1958).

EMISSIVITY

26. A temperature radiator is one whose radiant flux density (radiant exitance) is determined by its temperature and the material character of its surface and is independent of its previous history. A blackbody is an ideal temperature radiator, its radiant exitance depending only upon its temperature.

Emissivity is a measure of how closely the flux radiated from a given temperature radiator approaches that from a blackbody at the same temperature. The emissivity ϵ of an element of surface is defined by

$$\epsilon = \frac{M}{M^b}$$

where M is the radiant exitance of the surface and M^b is that of a blackbody at the same temperature. (Notation indicating that these quantities are, in general, temperature-dependent has been omitted.) Similarly, the spectral emissivity $\epsilon(\lambda)$ of an element of surface is defined by

$$\epsilon(\lambda) = \frac{M_\lambda(\lambda)}{M_\lambda{}^b(\lambda)}$$

where the subscript λ denotes spectral radiant exitance.

Directional emissivity $\epsilon(\theta,\phi)$ is defined by

$$\epsilon(\theta,\phi) = \frac{L(\theta,\phi)}{L^b}$$

where $L(\theta,\phi)$ is the radiance of the surface element in the direction (θ,ϕ) and L^b is the radiance of a blackbody at the same temperature. Similarly, the directional spectral emissivity $\epsilon(\lambda;\theta,\phi)$ of an element of surface in the direction (θ,ϕ) and at the wavelength λ is defined by

$$\epsilon(\lambda;\theta,\phi) = \frac{L_\lambda(\lambda;\theta,\phi)}{L_\lambda{}^b(\lambda)}$$

In the spatial-spectral context, ϵ may be referred to as hemispherical total emissivity, $\epsilon(\lambda)$ as hemispherical spectral emissivity, and $\epsilon(\theta,\phi)$ as directional total emissivity; they are respectively related to directional spectral emissivity $\epsilon(\lambda;\theta,\phi)$ as follows:

$$\epsilon = \frac{1}{M^b}\int_0^\infty L_\lambda{}^b \int_0^{2\pi}\int_0^{\pi/2} \epsilon(\lambda;\theta,\phi)\cos\theta\sin\theta\,d\theta\,d\phi\,d\lambda$$

$$\epsilon(\lambda) = \frac{1}{\pi}\int_0^{2\pi}\int_0^{\pi/2} \epsilon(\lambda;\theta,\phi)\cos\theta\sin\theta\,d\theta\,d\phi$$

$$\epsilon(\theta,\phi) = \frac{\pi}{M^b}\int_0^\infty \epsilon(\lambda;\theta,\phi)L_\lambda{}^b\,d\lambda$$

For a uniformly diffuse spectrally selective surface,

$$\epsilon(\lambda;\theta,\phi) = \epsilon(\lambda)$$

For a directionally selective gray surface,

$$\epsilon(\lambda;\theta,\phi) = \epsilon(\theta,\phi)$$

For a uniformly diffuse gray surface,

$$\epsilon(\lambda;\theta,\phi) = \epsilon$$

ABSORPTANCE

27. When flux is incident upon a surface or medium, three processes can occur: absorption, reflection, or transmission. Absorptance is the fraction of incident flux that is absorbed.

The absorptance α of an element of surface is defined by

$$\alpha = \frac{E^a}{E^i}$$

where E^i is the irradiance of the element and E^a is the portion absorbed. Similarly spectral absorptance is defined by

$$\alpha(\lambda) = \frac{E_\lambda{}^a(\lambda)}{E_\lambda{}^i(\lambda)}$$

Directional absorptance $\alpha(\theta,\phi)$ is defined by

$$\alpha(\theta,\phi) = \frac{L^a(\theta,\phi)}{L^i(\theta,\phi)}$$

where $L^i(\theta,\phi)$ is the radiance of the rays incident upon the surface element from the direction (θ,ϕ) and $L^a(\theta,\phi)$ is the portion of that radiance which is absorbed. Similarly spectral directional absorptance is defined by

$$\alpha(\lambda;\theta,\phi) = \frac{L_\lambda{}^a(\lambda;\theta,\phi)}{L^i(\lambda;\theta,\phi)}$$

The absorptances α, $\alpha(\lambda)$, and $\alpha(\theta,\phi)$ depend upon the directional and spectral characters of both the incident radiation and the absorbing surface, whereas $\alpha(\lambda;\theta,\phi)$ is regarded as a property of the surface alone. For a given incident radiation, α, $\alpha(\lambda)$, and $\alpha(\theta,\phi)$ can be determined from $\alpha(\lambda;\theta,\phi)$, providing the radiation does not alter $\alpha(\lambda;\theta,\phi)$, for example, does not char the surface.

In the spatial spectral context, α may be referred to as hemispherical total absorptance if it applies to radiation of all wavelengths incident from all directions, i.e., if it is related to $\alpha(\lambda;\theta,\phi)$ as follows:

$$\alpha = \frac{\int_0^\infty \int_0^{2\pi} \int_0^{\pi/2} L_\lambda{}^i \alpha(\lambda;\theta,\phi) \cos\theta \sin\theta \, d\theta \, d\phi \, d\lambda}{E^i}$$

Similarly, $\alpha(\lambda)$ may be referred to as hemispherical spectral absorptance under the relationship

$$\alpha(\lambda) = \frac{\int_0^{2\pi} \int_0^{\pi/2} L_\lambda{}^i \alpha(\lambda;\theta,\phi) \cos\theta \sin\theta \, d\theta \, d\phi}{E_\lambda{}^i}$$

and $\alpha(\theta,\phi)$ may be referred to as directional total absorptance under the relationship

$$\alpha(\theta,\phi) = \frac{\int_0^\infty L_\lambda{}^i \alpha(\lambda;\theta,\phi) \, d\lambda}{L^i}$$

For a uniformly diffuse, spectrally selective surface

$$\alpha(\lambda;\theta,\phi) = \alpha(\lambda)$$

For a directionally selective, spectrally nonselective surface

$$\alpha(\lambda;\theta,\phi) = \alpha(\theta,\phi)$$

For a uniformly diffuse, spectrally nonselective surface

$$\alpha(\lambda;\theta,\phi) = \alpha$$

KIRCHHOFF'S LAW

28. Kirchhoff's law is a consequence of the necessary existence of an energy balance between emission and absorption for a body in an isothermal black enclosure and in temperature equilibrium with the enclosure. Since the radiation field in such an enclosure is isotropic, the directional spectral emissivity and the directional spectral absorptance of the body are equal, i.e.,

$$\epsilon(\lambda;\theta,\phi) = \alpha(\lambda;\theta,\phi)\dagger$$

For directionally uniform incident radiation, i.e., for $L_\lambda{}^i(\lambda;\theta,\phi) = L_\lambda{}^i(\lambda)$, and for a uniformly diffuse absorbing-emitting surface

$$\epsilon(\lambda) = \alpha(\lambda)$$

For directionally selective gray incident radiation, i.e., for $L_\lambda{}^i(\lambda;\theta,\phi) = K(\theta,\phi)L_\lambda{}^b(\lambda)$, where $K(\theta,\phi)$ is an arbitrary function except that $0 \le K(\theta,\phi) \le 1$, and for a directionally selective gray absorbing-emitting surface

$$\epsilon(\theta,\phi) = \alpha(\theta,\phi)$$

For (1) directionally uniform gray incident radiation, (2) directionally uniform radiation incident upon a gray surface, (3) gray radiation incident upon a diffuse surface, and (4) a diffuse-gray surface

$$\epsilon = \alpha$$

REFLECTANCE (Judd, 1967; Nicodemus, 1967; Siegel and Howell, 1968)

29. The reflectance of a surface or medium for a given incident flux is the fraction of that flux which is reflected. The reflectance of an element of opaque surface is defined by

$$\rho = \frac{M^r}{E^i}$$

where E^i is the (incident) irradiance and M^r is the radiant exitance due to reflection. Similarly, spectral reflectance $\rho(\lambda)$ is defined by

$$\rho(\lambda) = \frac{M_\lambda{}^r(\lambda)}{E_\lambda{}^i(\lambda)}$$

The reflectance of an element of opaque surface may be expressed in terms of the radiance of incident and reflected rays as follows:

$$\rho = \frac{\int_0^\infty \int_0^{2\pi} \int_0^{\pi/2} L_\lambda{}^r \cos\theta_r \sin\theta_r \, d\theta_r \, d\phi_r \, d\lambda}{\int_0^\infty \int_0^{2\pi} \int_0^{\pi/2} L_\lambda{}^i \cos\theta_i \sin\theta_i \, d\theta_i \, d\phi_i \, d\lambda}$$

where $L_\lambda{}^i$ is the spectral radiance of rays incident from the direction (θ_i,ϕ_i) and $L_\lambda{}^r$ is that of rays reflected in the direction (θ_r,ϕ_r). The above expression is written more briefly as

$$\rho = \frac{\int_0^\infty \int_{2\pi\ \mathrm{sr}} L_\lambda{}^r \, d\Omega_r \, d\lambda}{\int_0^\infty \int_{2\pi\ \mathrm{sr}} L_\lambda{}^i \, d\Omega_i \, d\lambda}$$

by using the notation

$$d\Omega_i = \cos\theta_i \, d\omega_i = \cos\theta_i \sin\theta_i \, d\theta_i \, d\phi_i$$
$$d\Omega_r = \cos\theta_r \, d\omega_r = \cos\theta_r \sin\theta_r \, d\theta_r \, d\phi_r$$
$$\int_{2\pi\ \mathrm{sr}} L(\theta,\phi) \, d\Omega = \int_0^{2\pi} \int_0^{\pi/2} L(\theta,\phi) \cos\theta \sin\theta \, d\theta \, d\phi$$

† This equation holds for each component of radiation polarization. To be valid for all incident flux, the polarization components of the incident flux must be equal (Siegel and Howell, 1968).

where $d\omega = \sin\theta\, d\theta\, d\phi$ is an element of solid angle having its apex at the reflecting surface element and containing the direction (θ,ϕ) and $d\Omega$ is an element of projected solid angle (Jones, 1962; Nicodemus, 1970). The subscript 2π sr on the integral sign denotes integration over all directions (θ,ϕ) contained in the hemisphere covering the surface element.

30. Bidirectional Spectral-Reflectance Distribution Function The reflected spectral radiance $L_\lambda{}^r$ is given in terms of the incident spectral radiance and the directional spectral-reflectance characteristics of the surface by

$$L_\lambda{}^r = \int_{2\pi\ \mathrm{sr}} L_\lambda{}^i f_r\, d\Omega_i$$

where f_r, the bidirectional spectral-reflectance distribution function for the surface element, is defined by

$$f_r(\lambda;\theta_i,\phi_i;\theta_r,\phi_r) = \frac{dL_\lambda{}^r}{L_\lambda{}^i\, d\Omega_i} \qquad \mathrm{sr}^{-1}$$

[f_r has also been symbolized by ρ', but this symbol is to be deprecated (Nicodemus, 1970).] The denominator in this defining relation is the spectral irradiance of a well-collimated beam incident from the direction (θ_i,ϕ_i), and the numerator $dL_\lambda{}^r(\lambda;\theta_i,\phi_i;\theta_r,\phi_r)$ is the differential contribution which the beam makes to the net reflected spectral radiance $L_\lambda{}^r(\lambda;\theta_r,\phi_r)$. The reflectance ρ can thus be expressed in terms of the incident radiation and the reflectance properties of the surface element as follows:

$$\rho = \frac{\int_0^\infty \int_{2\pi\ \mathrm{sr}} \int_{2\pi\ \mathrm{sr}} L_\lambda{}^i f_r\, d\Omega_i\, d\Omega_r\, d\lambda}{\int_0^\infty \int_{2\pi\ \mathrm{sr}} L_\lambda{}^i\, d\Omega_i\, d\lambda}$$

31. Partial Reflectance The term partial reflectance refers here to the intermediate case where only specified portions of the incident and reflected radiations are to be considered, although the term partial is seldom used elsewhere. The partial reflectance $\rho(\lambda;\omega_i;\omega_r)$ is the ratio of radiant exitance contained in the solid angle ω_r and in the spectral band $\Delta\lambda$ due to the reflection from an opaque surface element of irradiance contained in the solid angle ω_i to the corresponding partial irradiance; i.e.,

$$\rho(\Delta\lambda;\omega_i;\omega_r) = \frac{\int_{\lambda_1}^{\lambda_2} \int_{\omega_r} \int_{\omega_i} L_\lambda{}^i f_r\, d\Omega_i\, d\Omega_r\, d\lambda}{\int_{\lambda_1}^{\lambda_2} \int_{\omega_i} L_\lambda{}^i\, d\Omega_i\, d\lambda}$$

The subscript ω_i indicates that integration extends over all directions (θ_i,ϕ_i) contained in the solid angle ω_i, and likewise for ω_r, θ_r, and ϕ_r. The limits of the spectral band $\Delta\lambda$ of interest are denoted by λ_1 and λ_2.

Three-term concatenations are used to distinguish 27 categories of partial reflectance, including certain limiting cases. The first term describes the incident beam as directional, conical, or hemispherical. The second term likewise describes the reflected beam of interest, and the third term describes the spectral band by use of one of the three descriptors, spectral, total, or broadband. For example, conical hemispherical spectral reflectance $\rho(\lambda;\omega_i;2\pi)$, also denoted $\rho(\lambda;\omega_i)$, is given by

$$\rho(\lambda;\omega_i;2\pi) = \frac{\int_{2\pi\ \mathrm{sr}} \int_{\omega_i} L_\lambda{}^i f_r\, d\Omega_i\, d\Omega_r}{\int_{\omega_i} L_\lambda{}^i\, d\Omega_i}$$

Directional conical total reflectance $\rho(\theta_i,\phi_i;\omega_r)$ is given by

$$\rho(\theta_i,\phi_i;\omega_r) = \frac{\int_0^\infty \int_{\omega_r} L_\lambda{}^i f_r\, d\Omega_r\, d\lambda}{\int_0^\infty L_\lambda{}^i\, d\lambda}$$

Conical directional broadband (differential) reflectance, for example, is a differential quantity given by

$$d^2\rho(\Delta\lambda;\omega_i;\theta_r,\phi_r) = \frac{\int_{\lambda_1}^{\lambda_2}\int_{\omega_i} L_\lambda{}^i f_r \, d\Omega_i \, d\lambda \cos\theta_r \sin\theta_r \, d\theta_r \, d\phi_r}{\int_{\lambda_1}^{\lambda_2}\int_{\omega_i} L_\lambda{}^i \, d\Omega_i \, d\lambda}$$

Two-term concatenations are used when both the incident and reflected radiation fall into the same geometrical category, in which case the first two terms are replaced by whichever of the single terms (bidirectional, bihemispherical, or biconical) applies. The number of terms and symbols used is often further reduced when it is clear from context what reflectance category is under consideration. The system of terms and symbols outlined above is more detailed than that presented in ANSI Z7.1-1967. Standard terms and symbols for use in such detailed considerations remain to be set.

Most of the 27 types of reflectance distinguishable by the above system of terms and symbols are properties of the spectral and directional characteristics of both the incident radiation and the reflecting surface. An important exception to this rule is directional hemispherical spectral reflectance $\rho(\lambda;\theta_i,\phi_i;2\pi)$, which is also called directional spectral reflectance, or just directional reflectance. This quantity, given by

$$\rho(\lambda;\theta_i,\phi_i;2\pi) = \int_{2\pi\ \text{sr}} f_r \, d\Omega_r$$

is independent of the incident radiation. Since this reflectance is therefore a property solely of the surface, data describing it for the given surface have relatively wide application. Data regarding several other of the 27 types of reflectance have more limited utility in that they may be regarded as solely a surface property provided that the surface is uniformly diffuse and/or spectrally nonselective and in some cases also provided that the incident radiations to be considered are directionally uniform, that is, $L_\lambda{}^i(\lambda;\theta_i,\phi_i) = L_\lambda{}^i(\lambda)$ for all θ_i and ϕ_i of interest.

32. Reflectance Factor The reflectance factor R of an element of an opaque-sample surface is the ratio of the radiant exitance of the sample due to reflection of a given irradiance to the radiant exitance from an element of a perfect diffuse (standard) reflecting surface identically irradiated. A general expression for R is

$$R(\Delta\lambda;\omega_i;\omega_r) = \frac{\int_{\lambda_1}^{\lambda_2}\int_{\omega_r}\int_{\omega_i} L_\lambda{}^i f_r \, d\Omega_i \, d\Omega_r \, d\lambda}{\int_{\lambda_1}^{\lambda_2}\int_{\omega_r}\int_{\omega_i} L_\lambda{}^i \frac{1}{\pi} \, d\Omega_i \, d\Omega_r \, d\lambda}$$

where f_r is the bidirectional spectral-reflectance distribution function of the sample and $(1/\pi)$ sr^{-1} is that of the perfect diffuse surface.

33. Reflectance Reciprocity Roughly speaking, reflectance reciprocity refers to conditions under which the reflectance values measured for a given surface are the same before and after interchanging the positions of the source of incident radiation and the receiver of the reflected radiation.

The bidirectional spectral-reflectance distribution function is symmetrical with regard to the directions of incidence and reflection; i.e.,

$$f_r(\lambda;\theta_i,\phi_i;\theta_r,\phi_r) = f_r(\lambda;\theta_r,\phi_r;\theta_i,\phi_i)$$

where the function on the left represents an initial measurement and the function on the right represents a measurement made after interchange of the source and receiver. Thus on both sides of the equation, the first pair of angles denotes the direction of incidence, and the second pair, the direction of reflection.

The following reciprocity relation is important in the design of hemispherical reflectometers that measure directional hemispherical spectral reflectance (directional reflectance) $\rho(\lambda;\theta_i,\phi_i)$:

$$\rho(\lambda;\theta_i,\phi_i) = R(\lambda;\theta_r,\phi_r)$$

where the direction (θ_i,ϕ_i) and the direction (θ_r,ϕ_r) are defined to be the same. The reflectance $\rho(\lambda;\theta_i,\phi_i)$ applies when the reflecting surface is irradiated from the direction (θ_i,ϕ_i) and the reflected radiation is collected over an entire hemisphere. The reflectance factor $R(\lambda;\theta_r,\phi_r)$ applies when the element of surface is irradiated from all directions of the hemisphere and only that portion reflected in the direction (θ_r,ϕ_r) is collected. For the reciprocity relation to be valid, $R(\lambda;\theta_r,\phi_r)$ must be measured under directionally uniform incident radiation.

Numerous other reciprocity relations can be written under various assumptions of special spectral directional characteristics of the incident radiation and the reflecting surface (Siegel and Howell, 1968).

34. Diffuse Reflectance A uniformly diffuse reflecting surface irradiated from a single direction (θ_i,ϕ_i) reflects with uniform radiance in all directions (θ_r,ϕ_r). For such a surface, the bidirectional spectral-reflectance distribution function and the directional hemispherical spectral reflectance are related as follows:

$$f_r(\lambda;\theta_i,\phi_i;\theta_r,\phi_r) = \frac{1}{\pi}\,\rho(\lambda;\theta_i,\phi_i)$$

If, in addition, the surface is such that the reflected radiance is independent of the direction of incidence, then f_r and $\rho(\lambda)$, the spectral reflectance, i.e., the bihemispherical spectral reflectance, are related as follows:

$$f_r(\lambda;\theta_i,\phi_i;\theta_r,\phi_r) = \frac{1}{\pi}\,\rho(\lambda)$$

If, in addition, the surface is spectrally nonselective, then

$$f_r(\lambda;\theta_i,\phi_i;\theta_r,\phi_r) = \frac{\rho}{\pi}$$

where ρ is the reflectance, i.e., the bihemispherical total reflectance, of the surface. If the surface is perfect and uniformly diffuse, i.e., if $\rho = 1$, then

$$f_r(\lambda;\theta_i,\phi_i;\theta_r,\phi_r) = \frac{1}{\pi}$$

35. Regular (Specular) Reflectance For regular (specular), or mirrorlike, surfaces, there is a one-to-one correspondence between incident and reflected rays. An incident ray, the corresponding reflected ray, and the normal to the surface at the point of incidence all lie in the same plane, the direction (θ_i,ϕ_i) and the direction of reflection (θ_r,ϕ_r) being related as follows:

$$\theta_r = \theta_i$$

(the angle of reflection equals the angle of incidence), and

$$\phi_r = \phi_i + \pi$$

The particular θ_r such that $(\theta_r,\phi_r) = (\theta_i, \phi_i + \pi)$ is called the specular angle. The spectral radiance $L_\lambda{}^r(\lambda;\theta_r,\phi_r)$ of the reflected ray is related to the spectral radiance $L_\lambda{}^i(\lambda;\theta_i,\phi_i)$ of the incident ray and the directional spectral reflectance $\rho(\lambda;\theta_i,\phi_i)$ of the surface as follows:

$$L_\lambda{}^r(\lambda;\theta_r,\phi_r) = L_\lambda{}^i(\lambda;\, \theta_r,\, \phi_r - \pi)\rho(\lambda;\, \theta_r,\, \phi_r - \pi)$$

where $\rho(\lambda;\theta_i,\phi_i)$ is a measurable property of the surface. If the specular surface is directionally uniform, then

$$\rho(\lambda;\theta_i,\phi_i) = \rho(\lambda)$$

If, in addition, it is spectrally nonselective, then

$$\rho(\lambda;\theta_i,\phi_i) = \rho$$

For a perfect specular reflector, $\rho = 1$, and therefore

$$\rho(\lambda;\theta_i,\phi_i) = 1$$

The bidirectional spectral-reflectance distribution function for a specular surface is

$$f_r(\lambda;\theta_i,\phi_i;\theta_r,\phi_r) = 2\rho(\lambda;\theta_i,\phi_i)\ \delta(\sin^2\theta_r - \sin^2\theta_i)\ \delta(\phi_r - \phi_i + \pi)$$

where the two rightmost factors are Dirac delta functions for which the defining relations are $\delta(u) = 0$ for $u \neq 0$, $\int\delta(u)\,du = 1$, and $\int f(u)\delta(u)\,du = f(0)$ when the integrations are carried out over the full range of the variable, $0 \leq \theta \leq \pi/2$ and $0 \leq \phi \leq 2\pi$ in each case (Nicodemus, 1967).

TRANSMITTANCE

36. Transmission is a general term for the process by which incident flux leaves a surface or medium on a side other than the incident side. The transmittance of a medium is the ratio of the transmitted to the incident flux.

The transmittance of an element of surface is given by

$$\tau = \frac{M^t}{E^i}$$

where E^i is the (incident) irradiance and M^t is the radiant exitance from the opposite side of the surface. Similarly, spectral transmittance $\tau(\lambda)$ is given by

$$\tau(\lambda) = \frac{M_\lambda{}^t(\lambda)}{E_\lambda{}^i(\lambda)}$$

The transmittance τ may be expressed in terms of the radiance of incident and transmitted rays as follows:

$$\tau = \frac{\int_0^\infty \int_{2\pi\,\mathrm{sr}} L_\lambda{}^t\,d\Omega_t\,d\lambda}{\int_0^\infty \int_{2\pi\,\mathrm{sr}} L_\lambda{}^i\,d\Omega_i\,d\lambda}$$

where $L_\lambda{}^i$ represents the spectral radiance $L_\lambda{}^i(\lambda;\theta_i,\phi_i)$ of rays incident from the direction (θ_i,ϕ_i); $L_\lambda{}^t$ represents the spectral radiance $L_\lambda{}^t(\lambda;\theta_t,\phi_t)$ of rays transmitted in the direction (θ_t,ϕ_t); $d\Omega_i = \cos\theta_i\,d\omega_i = \cos\theta_i \sin\theta_i\,d\theta_i\,d\phi_i$, $d\omega_i$ being an elemental solid angle having its apex on the element of transmitting surface and containing the direction (θ_i,ϕ_i) and $d\Omega_i$ being the corresponding element of projected solid angle (Jones, 1962; Nicodemus, 1970); $d\Omega_t = \cos\theta_t \sin\theta_t\,d\theta_t\,d\phi_t$ is an elemental projected solid angle containing the direction (θ_t,ϕ_t); and 2π sr signifies integration over all directions (θ,ϕ) contained in the hemisphere covering the surface element on the side of incidence or the side of emergence.

For lossy transmission through a single surface, thin plate, or shell, where transmitted rays can be practically considered to emerge from the point of incidence, the concept of bidirectional spectral-transmittance distribution function f_t may be applied empirically. This function is defined by

$$f_t(\lambda;\theta_i,\phi_i;\theta_t,\phi_t) = \frac{dL_\lambda{}^t}{L_\lambda{}^i\,d\Omega_i} \qquad \mathrm{sr}^{-1}$$

and is analogous to the bidirectional spectral-reflectance distribution function f_r. Furthermore an expression for partial transmittance analogous to the one for partial reflectance is

$$\tau(\Delta\lambda;\omega_i,\omega_t) = \frac{\int_{\lambda_1}^{\lambda_2} \int_{\omega_t} \int_{\omega_i} L_\lambda{}^i f_t\,d\Omega_i\,d\Omega_t\,d\lambda}{\int_{\lambda_1}^{\lambda_2} \int_{\omega_i} L_\lambda{}^i\,d\Omega_i\,d\lambda}$$

where the subscript ω_t signifies integration over all directions (θ_t,ϕ_t) contained in the solid angle ω_t. In general, τ depends upon both $L_\lambda{}^i$ and f_t.

A uniformly diffuse transmitting surface when irradiated from a single direction (θ_i,ϕ_i) transmits with uniform radiance in all directions (θ_r,ϕ_r). For such a surface

$$f_T(\lambda;\theta_i,\phi_i;\theta_t,\phi_t) = \frac{1}{\pi}\tau(\lambda;\theta_i,\phi_i)$$

where $\tau(\lambda;\theta_i,\phi_i)$ is the directional spectral transmittance of the surface. If, in addition, the surface is such that the emergent radiance is independent of the direction of incidence, then

$$f_t(\lambda;\theta_i,\phi_i;\theta_t,\phi_t) = \frac{1}{\pi}\tau(\lambda)$$

If the surface is also spectrally nonselective,

$$f_t(\lambda;\theta_i,\phi_i;\theta_t,\phi_t) = \frac{\tau}{\pi}$$

For a perfect and uniformly diffuse transmitting surface, $\tau = 1$.

Regular transmission is that process by which incident flux passes through a surface or medium without scattering. For an element of transmitting surface, there is a one-to-one correspondence between incident and transmitted rays, the directions of each incident ray and the corresponding transmitted ray being governed by Snell's law of refraction. The spectral radiance of the transmitted ray is related to the spectral radiance of the incident ray and the directional spectral transmittance of the surface as follows:

$$L_\lambda{}^t(\lambda;\theta_t,\phi_t) = L_\lambda{}^i(\lambda;\theta_i,\phi_i)\tau(\lambda;\theta_i,\phi_i)$$

LUMINOUS REFLECTANCE AND TRANSMITTANCE

37. In the relationships below, the subscripts v denote photometric quantities; $L_{v\lambda}{}^i(\lambda;\theta_i,\phi_i) = K(\lambda)L_{e\lambda}{}^i(\lambda;\theta_i,\phi_i)$, where $K(\lambda)$ is the spectral luminous efficacy of radiant flux; f_r and f_t are the bidirectional spectral-reflectance distribution and transmittance distribution functions, respectively, of given elements of surface.

$$\rho_v = \frac{M_v{}^r}{E_v{}^i}$$

$$\rho_v(\omega_i,\omega_r) = \frac{\int_{380\text{ nm}}^{780\text{ nm}}\int_{\omega_r}\int_{\omega_i} L_{v\lambda}{}^i f_r\, d\Omega_i\, d\Omega_r\, d\lambda}{\int_{380\text{ nm}}^{780\text{ nm}}\int_{\omega_i} L_v{}^i\, d\Omega_i\, d\lambda}$$

$$\tau_v = \frac{M_v{}^t}{E_v{}^i}$$

$$\tau_v(\omega_i,\omega_r) = \frac{\int_{380\text{ nm}}^{780\text{ nm}}\int_{\omega_t}\int_{\omega_i} L_{v\lambda}{}^i f_t\, d\Omega_i\, d\Omega_t\, d\lambda}{\int_{380\text{ nm}}^{780\text{ nm}}\int_{\omega_i} L_{v\lambda}{}^i\, d\Omega_i\, d\lambda}$$

38. Luminance Factor Luminance factor β is the ratio of the luminance of a surface or medium under specified conditions of incidence, observation, and light source to the luminance of a perfectly reflecting (or transmitting) perfectly diffusing surface or medium under the same conditions. For an element of reflecting surface,

$$\beta = \frac{\int_{380\text{ nm}}^{780\text{ nm}}\int_{\omega_i} L_{v\lambda}{}^i f_r\, d\Omega_i\, d\lambda}{\int_{380\text{ nm}}^{780\text{ nm}}\int_{\omega_i} L_{v\lambda}{}^i \frac{1}{\pi}\, d\Omega_i\, d\lambda}$$

RELATIONSHIP BETWEEN ABSORPTANCE, TRANSMITTANCE, AND REFLECTANCE

39. When flux is incident upon a surface or medium, three processes can occur: absorption, reflection, or transmission. For irradiance E^i incident upon an element of semitransparent surface

$$E^i = E^a + M^r + M^t$$

where E^a = portion of E^i reflected
M^r = portion reflected
M^t = portion transmitted

Division of the above expression by E^i produces

$$\alpha + \rho + \tau = 1$$

Similarly,

$$\alpha(\lambda) + \rho(\lambda) + \tau(\lambda) = 1$$

and

$$\alpha(\lambda;\theta_i,\phi_i) + \rho(\lambda;\theta_i,\phi_i) + \tau(\lambda;\theta_i,\phi_i) = 1$$

When the conditions requisite to the application of Kirchhoff's law are satisfied, the absorptance functions in the above expression may be replaced by emissivity functions to obtain relationships of the form

$$\epsilon(\lambda;\theta_i,\phi_i) + \rho(\lambda;\theta_i,\phi_i) + \tau(\lambda;\theta_i,\phi_i) = 1$$

REFERENCES

American Society of Mechanical Engineers (1967): USA Standard Letter Symbols for Illuminating Engineering, USAS [ANSI] Y10.18-1967.

American Standards Association (1953): American Standard Nomenclature for Radiometry and Photometry, Z-58.1.1.

Czerny, M., and A. Walther (1961): "Tables of the Fractional Function for the Planck Radiation Law," Springer-Verlag, Berlin.

Howard, J. N. (1962): Books [a review of Czerny and Walther (1961)], *Appl. Opt.*, vol. 1, no. 5, p. 342, May.

Hudson, R. D., Jr. (1969): "Infrared System Engineering," Wiley, New York.

Illuminating Engineering Society (1968): American National Standard Nomenclature and Definitions for Illuminating Engineering, RP-16, ANSI Z7.1-1967.

Jones, R. C. (1962): Immersed Radiation Detectors, *Appl. Opt.*, vol. 1, no. 5, pp. 607–613, September.

——— (1963): Terminology in Photometry and Radiometry, *J. Opt. Soc. Am.*, vol. 53, no. 11, p. 1314, November.

Judd, D. B. (1967): Terms, Definitions, and Symbols in Reflectometry, *J. Opt. Soc. Am.*, vol. 57, no. 4, pp. 445–452, April.

Kelly, K. L. (1963): Lines of Constant Correlated Color Temperature Based on MacAdam's (u,v) Uniform Chromaticity Transformation of the CIE Diagram, *J. Opt. Soc. Am.*, vol. 53, p. 999.

Mechtly, E. A. (1969): The International System of Units: Physical Constants and Conversion Factors, *NASA Publ.* SP-7012.

Meyer-Arendt, J. R. (1968): Radiometry and Photometry: Units and Conversion Factors, *Appl. Opt.*, vol. 7, no. 10, pp. 2081–2084, October.

National Bureau of Standards (1963): New Values for the Physical Constants, *Natl. Bur. Stand. Tech. News Bull.*, vol. 47, no. 10, p. 175, October.

——— (1970): Color Temperature Luminous Efficacy and the International Practical Temperature Scale of 1968, *Natl. Bur. Stand. Tech. News Bull.*, September.

Nicodemus, Fred E. (1963): Radiance, *J. Am. Phys. Soc.*, vol. 31, pp. 368–377.

——— (1967): Radiometry, chap. 8, in R. Kingslake (ed.), "Applied Optics and Optical Engineering," vol. IV, pp. 263–307, Academic, New York.

——— (1970): Reflectance Nomenclature and Directional Reflectance and Emissivity, *Appl. Opt.*, vol. 9, no. 6, pp. 1474–1475, June.

Optical Society of America, Committee on Colorimetry (1963): "The Science of Color," Washington.

Page, C. H., and P. Vigoureux (1971): The International System of Units (SI), *Natl. Bur. Stand. (U.S.) Spec. Publ.* 330.

Pivovonsky, M., and M. R. Nagel (1961): "Tables of Blackbody Radiation Functions," Macmillan, New York.
Siegel, R., and J. R. Howell (1968): Thermal Radiation Heat Transfer, vol. 1, NASA SP-164.
Smith, R. A., F. E. Jones, and R. P. Chasmar (1957): "The Detection and Measurement of Infra-red Radiation," Oxford.
Valley, S. L. (1965): "Handbook of Geophysics and Space Environments," Office of Aerospace Research, Air Force Cambridge Research Laboratories, Bedford, Mass.
Walsh, W. T. (1958): "Photometry," Dover, New York.
Wolfe, W. L. (1965): "Handbook of Military Infrared Technology," Office of Naval Research, Washington, D.C.
Wyszecki, G., and W. S. Stiles (1967): "Color Science," Wiley, New York.

Section **2**

Image Formation: Geometrical and Physical Optics

WARREN J. SMITH

Director of Research and Development, Infrared Industries, Inc., Santa Barbara, Calif.

(Numbers refer to paragraphs)

SYMBOLS, CONVENTIONS, AND DEFINITIONS†

1. Standard Symbol and Abbreviation List

A_n = coefficient of the nth-order aspheric-deformation term
AA = angular aberration
AC = longitudinal astigmatism contribution (TOA)
B = linear diameter of image blur spot
β = angular diameter of image blur spot
BFL, bfl = back focal length
C, c = surface curvature, reciprocal of surface radius
CA = clear aperture
CC = coma contribution (TOA)
CC_s = sagittal coma contribution (TOA)
CC_t, CC3 = tangential coma contribution (TOA)
CDM = fractional chromatic difference of magnification
d = axial distance between (thin) elements or between principal points
$D_{j,k}$ = distance along ray from surface j to surface k
DC = distortion contribution (TOA)
EFL, efl = effective focal length
f = focal length (effective)
FFL, ffl = front focal length
f number = relative aperture, speed
G_n = function of index
H, h = object height
H', h' = image height
I = optical invariant
I, i = angle of incidence
I', i' = angle of refraction or reflection
K = conic constant
L, l = distance from surface to intersection of ray with axis, before refraction
L', l' = distance from surface to intersection of ray with axis, after refraction
LA = longitudinal (spherical) aberration
LchA = longitudinal chromatic aberration
LchC = longitudinal chromatic contribution (TOA)
λ = wavelength
m = magnification, lateral
$\bar{m}$ = magnification, longitudinal
M_j = distance (vector) from vertex of surface j (perpendicular) to ray
M_x = x component of M_j
MTF = modulation transfer function
N, n = index of refraction
NA = numerical aperture
OPD = optical-path difference
OSC = offense against sine condition
OTF = optical transfer function
PC = longitudinal Petzval contribution (TOA)
ϕ = power, reciprocal focal length
Q = stop-shift ratio
R, r = radius of curvature
s = semiaperture = $(y^2 + z^2)^{1/2}$
s = distance, first principal point to object
s' = distance, second principal point to image
SC = longitudinal spherical-aberration contribution (TOA)
t = thickness; axial spacing between surfaces
TA = transverse aberration

† References for specific topics are given in the appropriate paragraphs. For a more complete textbook treatment of this subject, the following references are suggested: Born and Wolf (1964), Conrady (1957), Cox (1964), Hardy and Perrin (1932), Hertzberger (1958), Kingslake (1965–1969), Martin (1960), Smith (1966), and Southall (1964).

TAC = transverse-aberration contribution of AC
TchC = lateral chromatic-aberration contribution
TchA = lateral chromatic aberration, chromatic difference of image size
TLchC = transverse-aberration contribution of LchC
TOA = third-order aberration
TPC = transverse-aberration contribution of PC
TSC = transverse-aberration contribution of SC
U, u = ray-slope angle before refraction
U', u' = ray-slope angle after refraction
V = Abbe ν number, reciprocal relative dispersion
x = distance from first focal point to object
x' = distance from second focal point to image
x_p = longitudinal curvature (sag) of Petzval field
x_s = longitudinal curvature (sag) of sagittal image field
x_t = longitudinal curvature (sag) of tangential image field
x,y,z = coordinate system: x is the optical axis; x and y define the meridional plane; the origin is at the vertex of the surface
X,Y,Z = direction cosines of a ray
y = height of intersection of ray with surface

In general, where both upper- and lowercase symbols are indicated for the same quantity, the uppercase symbols represent trigonometric (or "exact") quantities; lowercase symbols represent the corresponding paraxial, or first-order, values. Primed symbols refer to quantities after refraction or reflection by a surface or by a lens or to quantities associated with the image. Subscripts are used to indicate the surface or element with which a symbol is associated or to indicate a particular ray. The subscript p is used to denote a principal or chief ray, and the subscript k is frequently used to indicate the data of the last surface of a system.

2. Sign Conventions Light rays are assumed to progress from left to right. Radii and curvatures are positive when the center of curvature is to the right of the surface. Surfaces or elements have positive power if they converge light. Distances upward (or to the right) are positive; i.e., points which lie above the axis (or to the right of an element, surface, or another point) are considered to be a positive distance away. Slope angles are positive if the ray is rotated counterclockwise to reach the axis. (Note that this is the reverse of the usual geometric convention.) Angles of incidence, refraction, and reflection are positive if the ray is rotated clockwise to reach the normal to the surface.

The index of refraction is considered positive when the light ray travels in the normal left-to-right direction. When the ray travels from right to left, as after a reflection, the index is taken as negative (as is the thickness, since the distance to the "next" surface is to the left).

Unless otherwise indicated, optical systems are assumed to be axially symmetrical, i.e., composed of surfaces which are figures of rotation with axes of symmetry coincident with the x, or optical, axis.

3. Basic Definitions Many definitions are given in the balance of this section and not repeated here. The index should be used to find them.

Anamorphic System. An optical system with a different power (or magnification) in one meridian than in the other (Cooke, 1956).

Angle of Incidence. The angle between an incident ray and the normal to a surface at the point of incidence.

Angle of Refraction (Reflection). The angle between an emergent refracted (reflected) ray and the normal to the surface at the point of emergence.

Back Focal Length, or Back Focus. The distance from the last vertex of the optical system to the back focal point (see Figs. 6 and 7).

Back Focal Point, or Second Focal Point. The focal point to which paraxial rays parallel to the axis and incident on the optical system from the left are converged (see Figs. 6 and 7).

Brewster's Angle. The light reflected from a surface is completely polarized if the angle of incidence is Brewster's angle, $I = \tan^{-1}(n'/n)$, where $\tan^{-1}$ stands for arctangent.

Cardinal Points. See *Effective Focal Length, Back Focal Length, Back Focal Point, Front Focal Length, Front Focal Point, Principal Points,* and *Nodal Points* (see also Figs. 6 and 7).

Chief Ray. A ray directed toward the center of the entrance pupil of the optical system.

Component. One or more elements of an optical system which are treated as a unit, e.g., a cemented doublet.

Effective Focal Length. The distance from the second principal point to the back (or second) focal point. The distance from the front (or first) focal point to the first principal point (see Figs. 6 and 7).

Element. A single indivisible entity of an optical system, usually a lens, composed of refractive material bounded by two surfaces of revolution.

Focal Length. See *Effective Focal Length.*

Focal Point. The point to which (paraxial) rays parallel to the axis converge or appear to converge after passing through the optical system (see Figs. 6 and 7).

Front Focal Length. The distance from the front vertex of the optical system to the front focal point (see Figs. 6 and 7).

Front Focal Point, or First Focal Point. The focal point to which (paraxial) rays parallel to the axis and incident on the optical system from the right are converged (see Fig. 6).

Member. All the components either ahead of the aperture stop (front member) or behind it (the rear member).

Meridional. The *meridional plane* is any plane which includes the optical axis. A *meridional ray* is one which lies in the meridional plane and is thus coplanar with the axis.

Nodal Points. Two axial points so located that an oblique ray directed toward one appears to emerge from the other, parallel to its original direction. For optical systems in air, the nodal points coincide with the principal points.

Numerical Aperture. NA $= n' \sin U'$, where n' is the final index in an optical system and U' is the slope angle of the axial ray at the image. If the object is at infinity, NA $= 0.5/(f$ number).

Optical Axis. The common axis of symmetry of an optical system; in an element, the line between the centers of curvature of the two surfaces.

Paraxial. Pertaining to an infinitesimal threadlike region about the optical axis.

Plane of Incidence. The incident ray, the normal to the surface at the point of incidence, and the refracted (or reflected) ray all lie in the same plane, called the plane of incidence.

Power. The reciprocal of the effective focal length of an optical element or system.

Principal Planes. If each ray of a bundle incident on an optical system parallel to the axis is extended to meet the backward extension of the ray after passing through the system, the locus of the intersections of all the rays is called a principal plane. The second principal plane is formed by rays incident on the system from the left. The first principal plane is formed by rays from the right. The principal planes are true planes only in the paraxial region; at any finite distance from the axis they are figures of rotation, frequently approximating spherical surfaces (see Figs. 6 and 7).

Principal Points. The intersection of the principal planes with the optical axis (see Figs. 6 and 7).

Principal Ray. Strictly, a ray directed toward the first principal point; however, the term is ordinarily used to refer to the chief ray.

Relative Aperture, or f Number. The ratio of focal length to entrance-pupil diameter. If the object is at infinity, f number $= 0.5/$NA.

Snell's Law. The angles of incidence and refraction and the refractive indexes on either side of an optical surface are related by $n_I \sin I = n_R \sin R$.

Total Internal Reflection (TIR). When light is incident on a surface separating two media of indexes n and n', the light is totally reflected if the angle of incidence exceeds the critical angle, $I_c = \sin^{-1} (n'/n)$. Obviously, TIR can occur only when $n > n'$, that is, when the light is passing from a medium of higher index into that of a lower.

Varifocal or Zoom Lens. A lens whose effective focal length can be varied by moving one or more elements along the axis; the lens is designed so that the image position is relatively constant as the power is varied (Bergstein and Motz, 1962; Kingslake, 1960).

FIRST-ORDER OPTICS

4. System Imagery The following relationships apply rigorously to the paraxial characteristics of any optical system, simple or complex. Although the paraxial relationships are valid for only a thin threadlike infinitesimal region near the optical axis, most well-corrected systems perform in a manner closely approximated by these relationships. See Par. 65 for sample calculations.

Image position (see Fig. 1):

$$\frac{1}{s'} = \frac{1}{s} + \frac{1}{f} \tag{1}$$

$$x' = -\frac{f^2}{x} \tag{2}$$

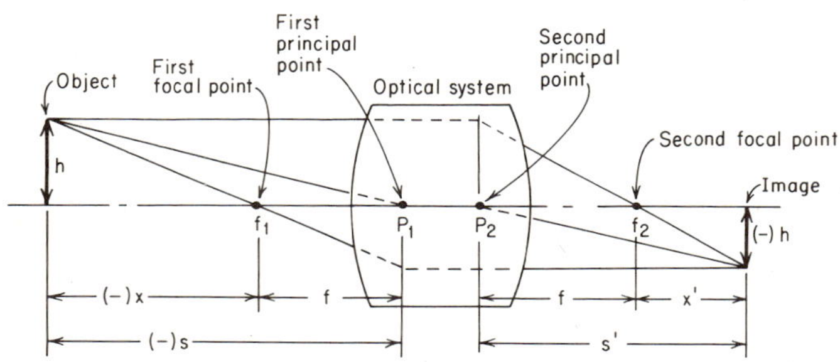

Fig. 1 The relationships between the cardinal points and the size and position of object and image.

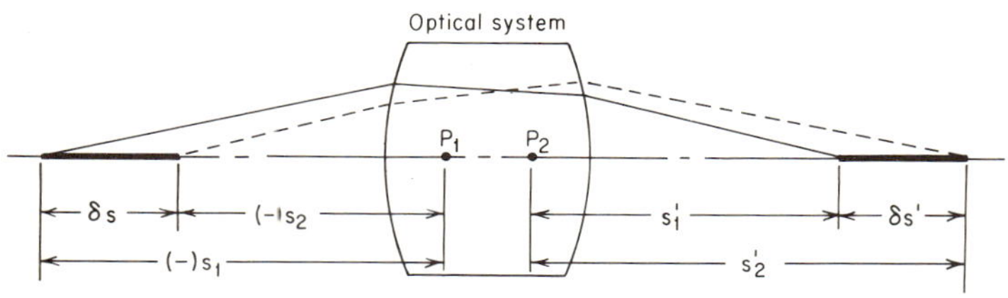

Fig. 2 Longitudinal magnification.

For image size and magnification see Figs. 1 and 2.

Lateral magnification:

$$m = \frac{h'}{h} = \frac{s'}{s} = \frac{f}{x} = -\frac{x'}{f} \tag{3}$$

Longitudinal magnification:

$$\bar{m} = \frac{\delta s'}{\delta s} = \frac{s_2' - s_1'}{s_2 - s_1} = \frac{s_1'}{s_1}\frac{s_2'}{s_2} = m_1 m_2 \approx m^2 \tag{4}$$

Equations (1) to (3) can be combined to give

$$s = f\left(\frac{1}{m} - 1\right) \qquad s' = f(1 - m)$$

Thin-lens overall track length $= -s + s'$

$$= -\frac{f}{m}(m-1)^2$$

5. Paraxial Equations The paraxial equations can be derived by reducing the exact ray-tracing equations to the limit (about the axis) where the angles are infinitesimals and are equal to their sines and tangents.

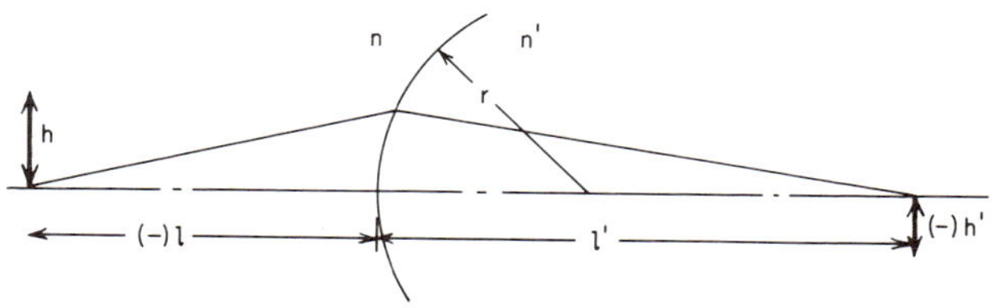

Fig. 3 Image formation by a single surface.

Paraxial image formation (see Fig. 3):

$$\frac{n'}{l'} = \frac{n}{l} + \frac{n'-n}{r} = \frac{n}{l} + (n'-n)c \tag{5}$$

$$m = \frac{h'}{h} = \frac{nl'}{n'l} \tag{6}$$

For paraxial ray tracing, see Fig. 4; see Pars. 67 and 68 for sample calculations. Note that the angles u and heights y in the following equations are fictitious

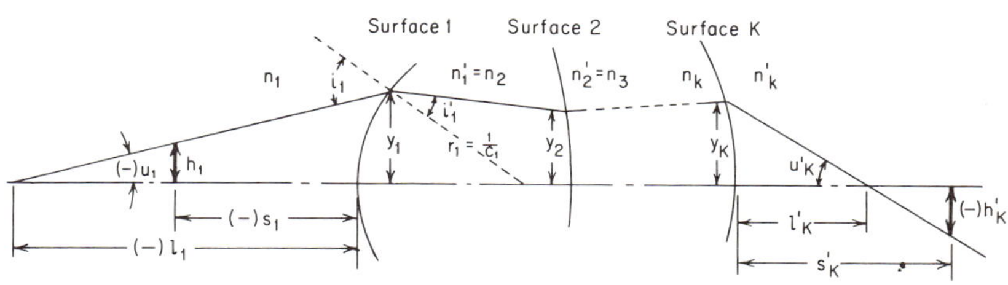

Fig. 4 The quantities used in paraxial ray tracing.

and may be algebraically canceled out to yield Eqs. (5) and (6). The following equations are more convenient for tracing ray paths and are also used in the third-order aberration calculations of Par. 17.

Opening equations (object to first surface):

$$n_1 u_1 = \frac{n_1 y_1}{l_1} \tag{7}$$

or

$$n_1 u_1 = \frac{n_1 h_1}{l_1 - s_1} \tag{8}$$

Iterative equations (applied to each surface in turn; $j = 1, 2, \ldots, k$):

$$n_j' u_j' = n_j u_j + (n_j' - n_j) y_j c_j \tag{9}$$

$$y_{j+1} = y_j - \frac{t_j' n_j' u_j'}{n_j'} \tag{10}$$

Alternate iterative equations:

$$i_j = y_j c_j - u_j \tag{11}$$

$$i_j' = \frac{n_j i_j}{n_j'} \tag{12}$$

$$u_j' = u_j + i_j - i_j' = u_j + i_j \left(1 - \frac{n_j}{n_j'}\right) \tag{13}$$

$$y_{j-1} = y_j - t_j' u_j' \tag{14}$$

Closing equations (last surface to image):

$$l_k' = \frac{n_k' y_k}{n_k' u_k'} = \frac{y_k}{u_k'} \tag{15}$$

$$h_k' = y_k - u_k' s_k' = u_k'(l_k' - s_k') \tag{16}$$

The Optical Invariant. If any two paraxial rays that are unrelated, i.e., have different axial intercepts, are traced through an optical system, their data are sufficient to define the system completely and can be used to determine the data of a third ray without additional ray tracing. If the data of the two rays are indicated by y, u and y_p, u_p, the expression

$$I = n(y_p u - y u_p) \tag{17}$$

is invariant across any surface or space of the optical system. Thus $I = I_j = I_j' = I_k$, etc. At an image plane, $y = 0$, $y_p = h$, and Eq. (17) reduces to

$$I = hnu = h'n'u' \tag{18}$$

which is the Lagrange invariant (see Fig. 5). The data $\bar{y}$, $\bar{u}$ of any third ray can be

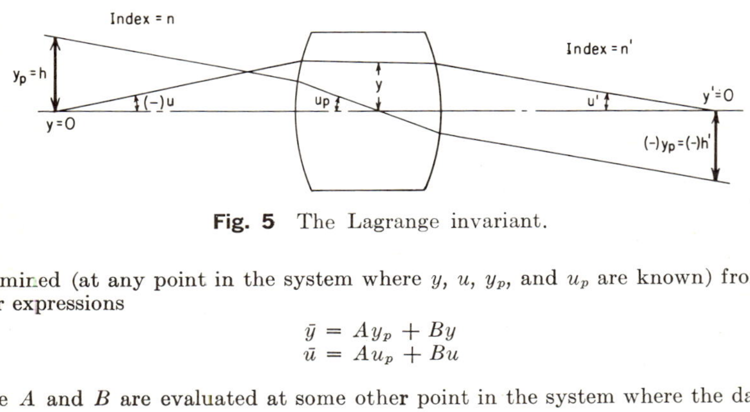

Fig. 5 The Lagrange invariant.

determined (at any point in the system where y, u, y_p, and u_p are known) from the linear expressions

$$\bar{y} = Ay_p + By \tag{19}$$

$$\bar{u} = Au_p + Bu \tag{20}$$

where A and B are evaluated at some other point in the system where the data for all three rays are known:

$$A = \frac{\bar{y}u - \bar{u}y}{y_p u - u_p y} = \frac{n(\bar{y}u - \bar{u}y)}{I} \tag{21}$$

$$B = \frac{y_p \bar{u} - u_p \bar{y}}{y_p u - u_p y} = \frac{n(y_p \bar{u} - u_p \bar{y})}{I} \tag{22}$$

With this technique, the focal lengths of an optical system can be determined from the initial data (y, u, y_p, and u_p at the first surface) and the final data (y', u', y_p', and u_p' at the last surface) of any two unrelated rays:

$$\text{EFL} = \frac{uy_p - yu_p}{uu_p' - u_p u'} \tag{23}$$

$$\mathrm{BFL} = \frac{uy_p' - u_py'}{uu_p' - u_pu'} \tag{24}$$

$$\mathrm{FFL} = \frac{u_p'y - u'y_p}{uu_p' - u_pu'} \tag{25}$$

6. The Cardinal Points See Fig. 6; sample calculations are given in Par. 63.

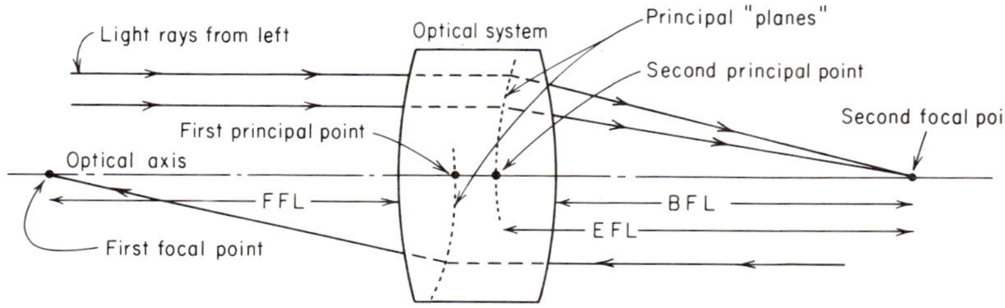

Fig. 6 The cardinal points of an optical system. [*After Smith* (1966).]

Complex Systems. To determine the cardinal points, two paraxial rays may be traced, using Eqs. (7) to (16) and the focal lengths determined from Eqs. (23) to (25). Or, following the definition of the focal points, a ray parallel to the axis ($u_1 = 0$) can be traced through the system. Then $\mathrm{EFL} = y_1/u_k'$, and $\mathrm{BFL} = y_k/u_k'$. The other focal point is determined by repeating the process with the system reversed.

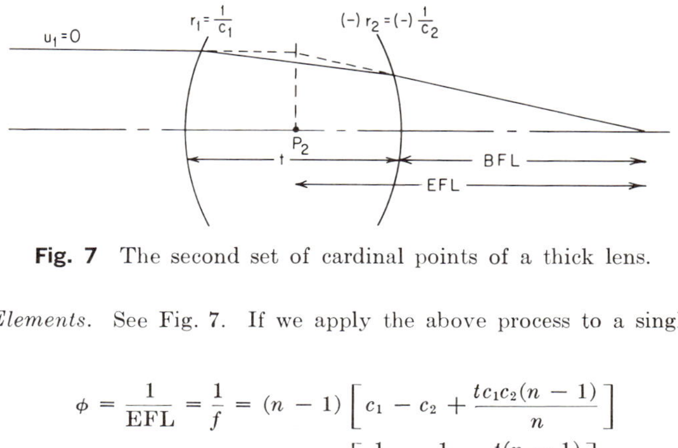

Fig. 7 The second set of cardinal points of a thick lens.

Thick Elements. See Fig. 7. If we apply the above process to a single element, we get

$$\phi = \frac{1}{\mathrm{EFL}} = \frac{1}{f} = (n-1)\left[c_1 - c_2 + \frac{tc_1c_2(n-1)}{n}\right]$$
$$= (n-1)\left[\frac{1}{R_1} - \frac{1}{R_2} + \frac{t(n-1)}{nR_1R_2}\right] \tag{26}$$

$$\mathrm{BFL} = f\left[1 - \frac{tc_1(n-1)}{n}\right]$$
$$= f\left[1 - \frac{t(n-1)}{nR_1}\right] \tag{27}$$

Thin Lenses. When the thickness of the element is negligible, Eq. (26) reduces to

$$\phi = (n-1)(c_1 - c_2) = (n-1)\left(\frac{1}{R_1} - \frac{1}{R_2}\right) \tag{28}$$

7. Multielement Systems See Fig. 8; sample calculations are given in Par 66. Paraxial rays can be traced through complete systems, one surface at a time, using Eqs. (7) to (16). However, it is frequently convenient to treat a system as a set of components separated by air; the object and image for each component (in turn) can be determined by using Eqs. (1) or (2) and (3). It is often more convenient to trace rays through the system component by component, using the

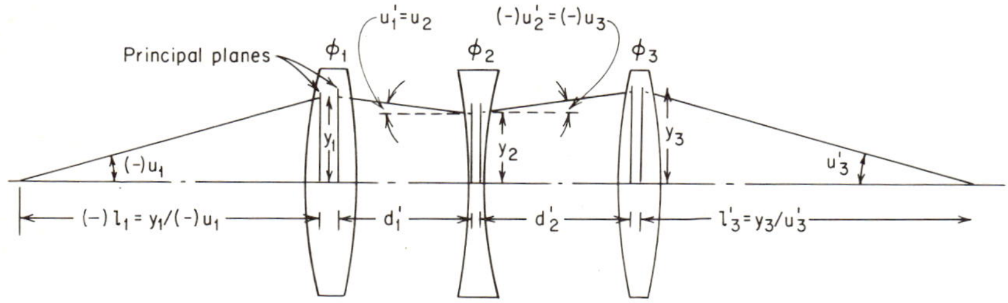

Fig. 8 Paraxial ray trace, component by component.

power ϕ_j, of the jth component (or element), the height y_j, at which the ray strikes the principal planes of the jth component, and the distance d_j', from the second principal plane of the jth component to the first principal plane of the $(j+1)$st component. The ray slope u_j' after refraction by the jth component is determined from

$$u_j' = u_j + y_j\phi_j \tag{29}$$

and the ray height at the next component is given by

$$y_{j+1} = y_j - d_j'u_j' \tag{30}$$

Note that if the elements are "thin," d is simply the space between the elements. Also if the ray from the axial intercept of the object has been traced, the magnification can be determined from Eq. (18).

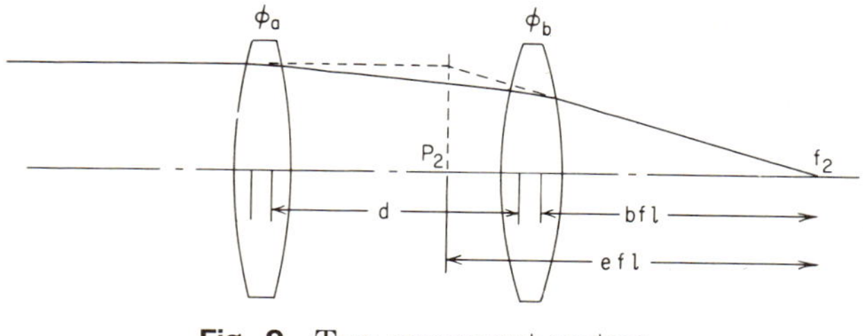

Fig. 9 Two-component system.

Two-Component Systems. See Fig. 9; sample calculations are given in Par. 64. When a system consists of two components, a and b, the following explicit expressions may be applied:

$$\phi_{ab} = \frac{1}{f_{ab}} = \phi_a + \phi_b - d\phi_a\phi_b = \frac{1}{f_a} + \frac{1}{f_b} - \frac{d}{f_af_b} \tag{31}$$

$$f_{ab} = \mathrm{EFL}_{ab} = \frac{f_af_b}{f_a + f_b - d} \tag{32}$$

$$\mathrm{bfl} = \mathrm{BFL}_{ab} = \frac{f_b(f_a - d)}{f_a + f_b - d} \tag{33a}$$

$$= \frac{f_{ab}(f_a - d)}{f_a} \tag{33b}$$

$$(-)\text{ffl} = \frac{f_{ab}(f_b - d)}{f_b} \tag{34}$$

$$f_a = d\,\frac{f_{ab}}{f_{ab} - \text{bfl}} \tag{35}$$

$$f_b = -d\,\frac{\text{bfl}}{f_{ab} - \text{bfl} - d} \tag{36}$$

$$d = f_b\,\frac{\text{bfl}}{f_b - \text{bfl}} = f_a + f_b - \frac{f_a f_b}{f_{ab}} \tag{37}$$

RAY TRACING

8. The General, or Skew, Ray† See Fig. 10; sample calculations are given in Par. 69. A general ray is defined by its direction cosines X, Y, and Z and

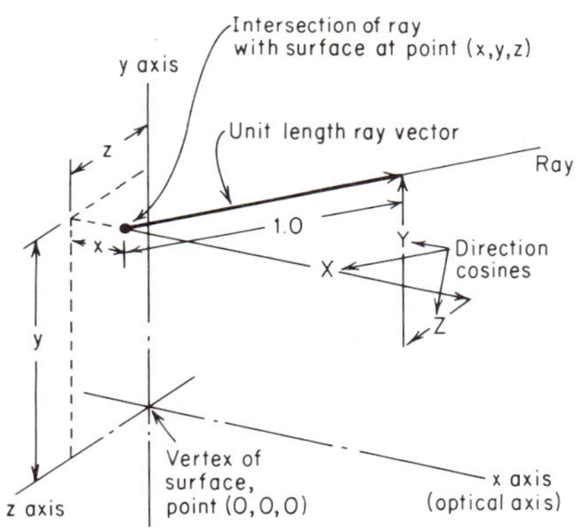

Fig. 10 The general ray is defined by its surface intersection coordinates (x,y,z) and its direction cosines (X,Y,Z). [*After Smith* (1966).]

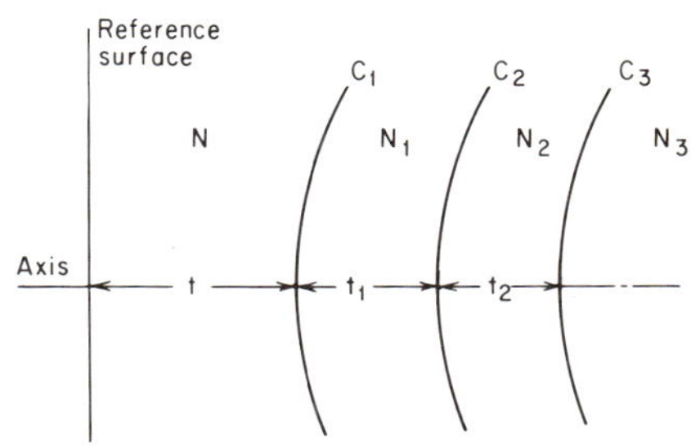

Fig. 11 Subscript notation for Par. 8. [*After Smith* (1966).]

by the coordinates x, y, and z of its intersection with a surface of the optical system. The subscript notation is illustrated in Fig. 11.

Spherical Surfaces. Opening (at the initial reference surface):

$$c(x^2 + y^2 + z^2) - 2x = 0 \tag{38}$$

$$X^2 + Y^2 + Z^2 = 1.0 \tag{39}$$

Intersection of ray with next surface:

$$e = tX - (xX + yY + zZ) \tag{40}$$

$$M_{1x} = x + eX - t \tag{41}$$

$$M_1^2 = x^2 + y^2 + z^2 - e^2 + t^2 - 2tx \tag{42}$$

$$\cos I_1 = E_1 = [X^2 - c_1(c_1M_1^2 - 2M_{1x})]^{1/2} \tag{43}$$

$$D_{0,1} = e + \frac{c_1M_1^2 - 2M_{1x}}{X + E_1} \tag{44}$$

$$x_1 = x + D_{0,1}X - t \tag{45}$$

$$y_1 = y + D_{0,1}Y \tag{46}$$

$$z_1 = z + D_{0,1}Z \tag{47}$$

† References for this paragraph are Feder (1951), MIL-HDBK-141 (1962), Montagnino (1968), and Spencer and Murty (1962).

Ray-direction cosines after refraction:

$$\cos I_1' = E_1' = \left[1 - (1 - E_1^2)\left(\frac{n}{n_1}\right)^2\right]^{1/2} \tag{48}$$

$$g_1 = E_1' - \frac{n}{n_1} E_1 \tag{49}$$

$$X_1 = \frac{n}{n_1} X - g_1 c_1 x_1 + g_1 \tag{50}$$

$$Y_1 = \frac{n}{n_1} Y - g_1 c_1 y_1 \tag{51}$$

$$Z_1 = \frac{n}{n_1} Z - g_1 c_1 z_1 \tag{52}$$

Equations (40) to (52) are repeated with the subscripts advanced by 1 for the next surface; the process is continued until the final (image) surface is reached.

Aspheric Surfaces. An aspheric surface of revolution may be represented as a sphere of curvature c deformed by a series of terms in even powers of the semidiameter s:

$$x = f(y,z) = \frac{cs^2}{1 + (1 - c^2s^2)^{1/2}} + A_2s^2 + A_4s^4 + \cdots + A_js^j \tag{53}$$

where $$s^2 = y^2 + z^2$$

Intersection of Ray with Aspheric. The sphere c is presumed to be a fair approximation to the aspheric; the intersection (x_0,y_0,z_0) of the ray with the sphere is found using Eqs. (40) to (47). The x coordinate of the aspheric surface corresponding to this distance from the axis is found by substituting the y and z coordinates of the ray intersection with the sphere into Eq. (53) to get $s_0^2 = y_0^2 + z_0^2$ and

$$\bar{x}_0 = f(y_0,z_0) \tag{54}$$

Then compute

$$l_0 = (1 - c^2s_0^2)^{1/2} \tag{55}$$

$$m_0 = -y_0[c + l_0(2A_2 + 4A_4s_0^2 + \cdots + jA_js_0^{(j-2)})] \tag{56}$$

$$n_0 = -z_0[c + l_0(2A_2 + 4A_4s_0^2 + \cdots + jA_js_0^{(j-2)})] \tag{57}$$

An improved approximation to the intersection of the ray with the aspheric (see

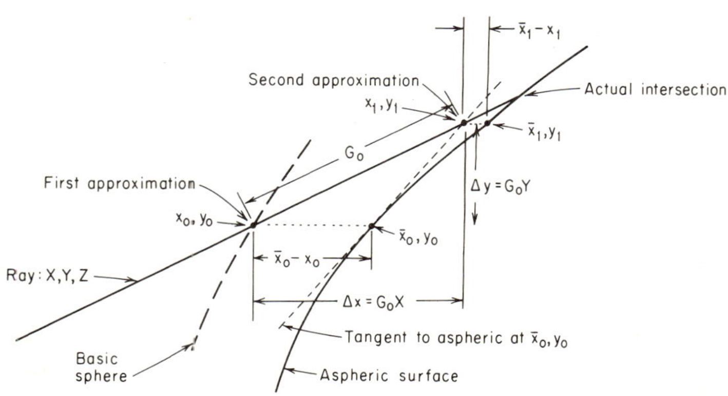

Fig. 12 The intersection point of a general ray with an aspheric surface is determined by successive approximations.

Fig. 12) is given by

$$G_0 = \frac{l_0(\bar{x}_0 - x_0)}{Xl_0 + Ym_0 + Zn_0} \tag{58}$$

$$x_1 = G_0X + x_0 \tag{59}$$

$$y_1 = G_0Y + y_0 \tag{60}$$

$$z_1 = G_0Z + z_0 \tag{61}$$

This approximation process is repeated, from Eq. (54) to Eq. (61), with the subscripts advanced by 1 at each iteration, until the error $\bar{x}_k - x_k$ after the kth iteration is negligible.

Refraction at the aspheric surface:

$$P^2 = l_k^2 + m_k^2 + n_k^2 \tag{62}$$

$$P \cos I = F = Xl_k + Ym_k + Zn_k \tag{63}$$

$$P \cos I' = F' = \left\{P^2\left[1 - \left(\frac{n}{n_1}\right)^2\right] + \left(\frac{n}{n_1}\right)^2 F^2\right\}^{1/2} \tag{64}$$

$$g = \frac{F' - (n/n_1)F}{P^2} \tag{65}$$

$$X_1 = X\frac{n}{n_1} + gl_k \tag{66}$$

$$Y_1 = Y\frac{n}{n_1} + gm_k \tag{67}$$

$$Z_1 = Z\frac{n}{n_1} + gn_k \tag{68}$$

9. Differential Ray Tracing (Feder, 1968; MIL-HDBK-141, 1962) Differential rays differ from an already traced ray by infinitesimal first-order quantities. Assuming that a ray has already been traced, using Eqs. (40) to (47) and (54) to (68), the problem is: given dX, dY, dZ, dx, dy, and dz (all but one of which may be zero), to find dX_1, dY_1, dZ_1, dx_1, dy_1, and dz_1. The computation can be carried through a complete optical system, applying the equations surface by surface.

$$D = \frac{x_1 - x + t}{X} \tag{69}$$

$$dD = [l(dx + D\,dX) + m(dy + D\,dY) + n(dz + D\,dZ)]/(-F) \tag{70}$$

$$dx_1 = dx + D\,dX + X\,dD \tag{71}$$

$$dy_1 = dy + D\,dY + Y\,dD \tag{72}$$

$$dz_1 = dz + D\,dZ + Z\,dD \tag{73}$$

$$dl = -\frac{c_1^2(y_1\,dy_1 + z_1\,dz_1)}{l} \tag{74}$$

$$B = c_1 + l(2A_2 + 4A_4s^2 + \cdots + jA_js^{(j-2)}) \tag{75}$$

$$dB = (y_1\,dy_1 + z_1\,dz_1)\left\{(c_1 - B)\left(\frac{c_1}{l}\right)^2 + l[8A_4 + 24A_6s^2 + \cdots + j(j-2)A_js^{(j-4)}]\right\} \tag{76}$$

$$dm = -y_1\,dB - B\,dy_1 \tag{77}$$

$$dn = -z_1\,dB - B\,dz_1 \tag{78}$$

$$dg = \left[X_1\left(dX\frac{n}{n_1} + g\,dl\right) + Y_1\left(dY\frac{n}{n_1} + g\,dm\right) + Z_1\left(dZ\frac{n}{n_1} + g\,dn\right)\right]\Big/(-F') \tag{79}$$

$$dX_1 = dX\frac{n}{n_1} + g\,dl + l\,dg \tag{80}$$

$$dY_1 = dY\frac{n}{n_1} + g\,dm + m\,dg \tag{81}$$

$$dZ_1 = dZ\frac{n}{n_1} + g\,dn + n\,dg \tag{82}$$

10. Meridional Ray Tracing; Coddington's Equations Meridional rays are those which lie in the same plane as the optical axis. Thus, by a suitable rotation of coordinates one can always set $z = 0$ and $Z = 0$ in the equations of Par. 8 when these equations are to be applied to meridional rays; a considerable simplification results. There are many specialized forms of equations for this purpose in the literature (Conrady, 1957; Cox, 1964; Smith, 1966).

The differential equations of Par. 9 can be applied to a meridional ray to determine the image-forming properties of the infinitesimal region about the ray. This is analogous to the paraxial region about the optical axis. A significant simplification of the equations is possible; the classical forms are known as *Coddington's equations:*

Sagittal:
$$\frac{n'}{s'} = \frac{n}{s} + \phi$$

Tangential:
$$\frac{n' \cos^2 I'}{t'} = \frac{n \cos^2 I}{t} + \phi$$

where s,t = object distances from ray-surface intersection along ray
s',t' = corresponding image distances
I = angle of incidence of ray
I' = angle of refraction of ray
n,n' = indexes of refraction
ϕ = oblique surface power, given by

$$\phi = c(n' \cos I' - n \cos I)$$

Notice that when the ray coincides with the optical axis, these equations reduce to Eq. (5).

11. Graphical Ray Tracing See Fig. 13. It is possible to trace meridional rays using only a scale, straightedge, and compass. The ray is drawn to the surface,

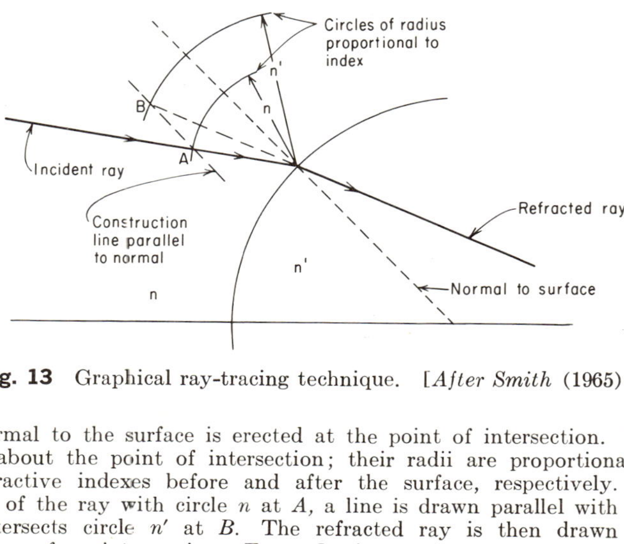

Fig. 13 Graphical ray-tracing technique. [*After Smith* (1965).]

and the normal to the surface is erected at the point of intersection. Two circles are drawn about the point of intersection; their radii are proportional to n and n', the refractive indexes before and after the surface, respectively. From the intersection of the ray with circle n at A, a line is drawn parallel with the normal until it intersects circle n' at B. The refracted ray is then drawn through B and the ray-surface intersection. For reflection, $n' = -n$, and a single circle is drawn; point B is located at the intersection of the parallel and the index circle on the opposite side of the surface.

12. Optical-Path Difference The optical path (OP) of a ray is the distance along the ray times the index of the medium in which the ray travels. The optical-path difference (OPD) is the difference between the optical paths of the ray in question and a reference ray, which is usually the principal, or chief, ray of a bundle. With the subscript p denoting the data of the reference ray, we have

$$\mathrm{OP}_{j\ \mathrm{to}\ j+1} = \frac{n_j'(t_j - x_j + x_{j+1})}{X_j'} \tag{83}$$

$$\begin{aligned}\mathrm{OPD} &= \mathrm{OP} - \mathrm{OP}_p \\ &= n_j'[t_j(X'_{pj} - X_j') + X'_{pj}(x_{j-1} - x_j) + X_j'(x_{pj} - x_{pj-1})]/X_j'X'_{pj} \\ &= n_j'(D - D_p)\end{aligned} \tag{84}$$

where D is from Eq. (44).

To avoid taking a small difference between two large numbers, the following are useful at the entrance and exit pupil. At the entrance pupil, where y and z are the ray coordinates in the entrance pupil, H is the object height, and L is the object-to-pupil distance

$$\frac{\mathrm{OPD}}{n} = \frac{z^2 + y^2 - 2yH}{[(y - H)^2 + z^2 + L^2]^{1/2} + (H^2 - L^2)^{1/2}} \tag{85}$$

At the exit pupil, where $x = 0$, y and z are the ray coordinates in the exit pupil, X, Y, and Z are the direction cosines of the ray, L is the axial distance from the pupil to the image surface, and x_p, y_p, and $z_p = 0$ are the coordinates of the center of the reference sphere (with respect to the image surface), i.e., the "ideal" image point,

$$\frac{\mathrm{OPD}}{n} = e + \frac{R}{B + (B^2 - R)^{1/2}} \tag{86}$$

where

$$\begin{aligned} e &= -(yY + zZ) \\ B &= X(L + x_p) + Yy_p \\ R &= y^2 + z^2 - e^2 - 2(yy_p + eB)\end{aligned}$$

13. Rotation and Translation of Axes Rays can be traced through a noncentered optical system using the standard ray-tracing equations of Par. 8 if a separate coordinate system (to which the surface is centered) is used for each surface.

If the origin, i.e., the vertex of the surface, is translated from its original locus by Δx, Δy, and Δz, the new coordinates of point x, y, z are $x' = x - \Delta x$, $y' = y - \Delta y$, $z' = z - \Delta z$.

When the axes are rotated (see Fig. 14), we have no choice for the direction of the x axis, but the y and z axes may be arbitrarily located. If we elect to

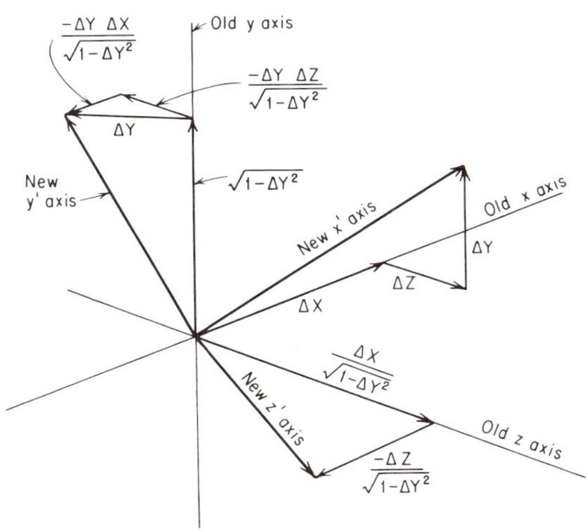

Fig. 14 Rotation of axes.

keep the new z axis in the old xz plane, the following can be used for the direction cosines of the new axes:

x axis:	ΔX	ΔY	ΔZ
y axis:	$\dfrac{-\Delta Y\,\Delta X}{(1-\Delta Y^2)^{1/2}}$	$(1-\Delta Y^2)^{1/2}$	$\dfrac{-\Delta Y\,\Delta Z}{(1-\Delta Y^2)^{1/2}}$
z axis:	$\dfrac{-\Delta Z}{(1-\Delta Y^2)^{1/2}}$	0	$\dfrac{\Delta X}{(1-\Delta Y^2)^{1/2}}$

where ΔX, ΔY, and ΔZ are the given rotation of the axes. The coordinates of a point (x,y,z) with respect to the new axes are

$$x' = \Delta X\,x + \Delta Y\,y + \Delta Z\,z$$
$$y' = -\frac{\Delta Y\,\Delta X\,x}{(1-\Delta Y^2)^{1/2}} + y(1-\Delta Y^2)^{1/2} - \frac{\Delta Y\,\Delta Z\,z}{(1-\Delta Y^2)^{1/2}}$$
$$z' = \frac{-\Delta Z\,x}{(1-\Delta Y^2)^{1/2}} + 0 + \frac{\Delta X\,z}{(1-\Delta Y^2)^{1/2}}$$

The direction cosines of a ray (X,Y,Z) with respect to the new axes are

$$X' = \Delta X\,X + \Delta Y\,Y + \Delta Z\,Z$$
$$Y' = \frac{-\Delta Y\,\Delta X\,X}{(1-\Delta Y^2)^{1/2}} + Y(1-\Delta Y^2)^{1/2} + \frac{\Delta Y\,\Delta Z\,Z}{(1-\Delta Y^2)^{1/2}}$$
$$Z' = \frac{-\Delta Z\,X}{(1-\Delta Y^2)^{1/2}} + 0 + \frac{\Delta X\,Z}{(1-\Delta Y^2)^{1/2}}$$

GEOMETRIC ABERRATION THEORY

14. Aberration Descriptions Optical aberrations are faults or defects of the image. In general they are described in terms of the amount by which a geometrically traced ray misses (or departs from) a desired location in the image formed by an optical system. Usually the desired location for a ray in the image is that indicated by the paraxial, or first-order, laws of image formation, as set forth in Pars. 4 to 7.

As indicated in Fig. 15, for the case of spherical aberration, an aberration may be measured in a lateral, longitudinal, or angular sense. Paragraph 15 describes

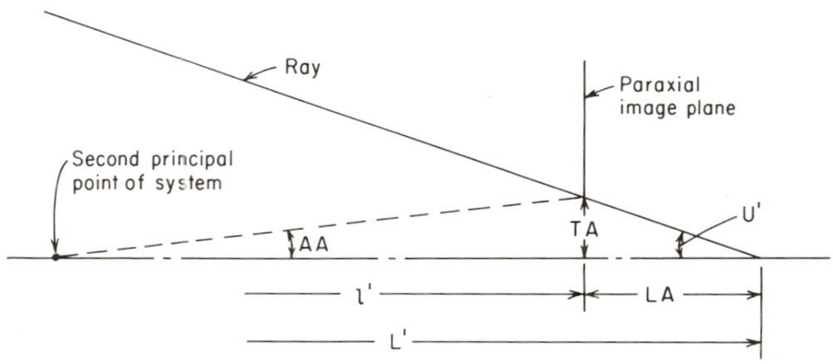

Fig. 15 The relationship between the longitudinal (LA), the transverse (TA), and the angular (AA) measures of spherical aberration, where TA = LA tan U' = AA·l'.

the aberrations as departures from an ideal spherical wavefront and indicates the relationship between the wavefront aberrations and the lateral-ray aberrations.

Spherical aberration can be defined as the (longitudinal) variation of focus with aperture. Figure 16 shows the undercorrected spherical aberration typical of a simple positive element in which the rays nearer the edge of the lens are more strongly refracted and come to a focus nearer the lens than the rays closer to the axis. Longitudinal spherical aberration is the distance from the paraxial focus to the axial intersection of the ray. Lateral or transverse spherical aberration is the vertical distance from the axis to the intersection of the ray with the paraxial image plane.

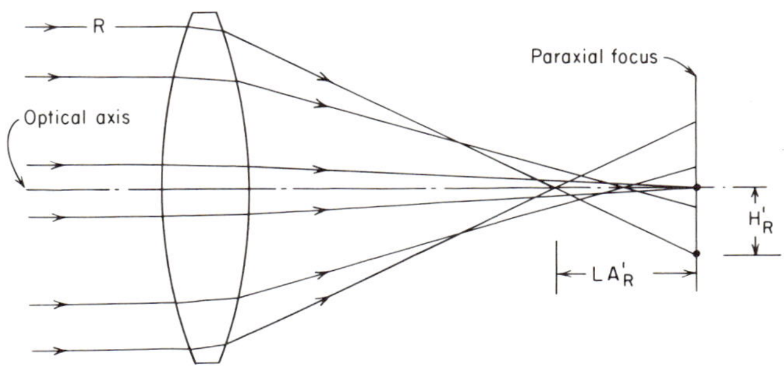

Fig. 16 A simple converging lens with undercorrected spherical aberration. [*After Smith* (1966).]

Coma is the variation of magnification, i.e., image size, with aperture. Figure 17 shows the rays through the outer portions of the lens aperture intersecting to form an "image" larger than that formed by rays through the center. Tangential coma is the vertical distance from the principal, or chief, ray to the intersection of the upper and lower rim rays. The appearance of a comatic point image is simulated in Fig. 25.

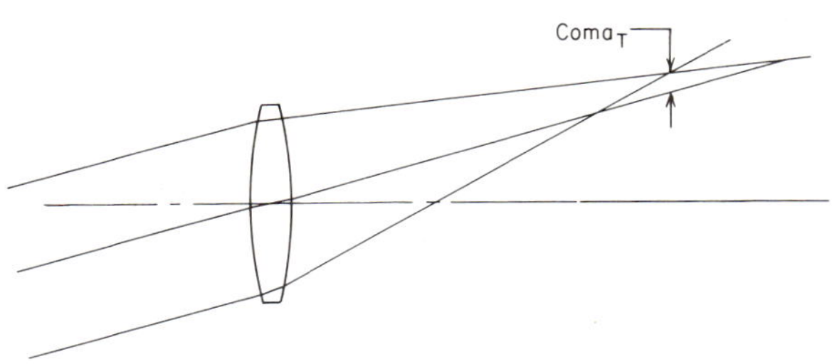

Fig. 17 Coma.

Field curvature describes the amount by which the off-axis image departs longitudinally from the surface in which it should (ideally) be located. It is usually indicated as a departure of the image from a flat image plane. The curvature of field may differ for fans of rays in different meridians. The focus of a fan of rays in the meridional plane is called the *tangential focus,* and the focus of a fan of rays lying in a plane normal to the meridional plane is called the *sagittal focus.* The distance from the sagittal to the tangential focus is the astigmatism, and the longitudinal distance from the paraxial image plane to the foci are the tangential and sagittal field curvatures x_t and x_s. The *Petzval curvature* is a sort of basic field curvature of an optical system; the Petzval surface lies 3 times as far from the tangential focal surface as from the sagittal focal surface, as indicated in Fig. 18. Field curvatures may be described as inward, undercorrected, and negative (as shown in Fig. 18) or as backward, overcorrected, and positive.

Distortion is the amount by which an image point is closer to or further from the axis than its ideal position as given by first-order optics. The linear amount of (simple) distortion varies with h^3; thus nonradial straight lines are imaged as curved lines, as shown in Fig. 19.

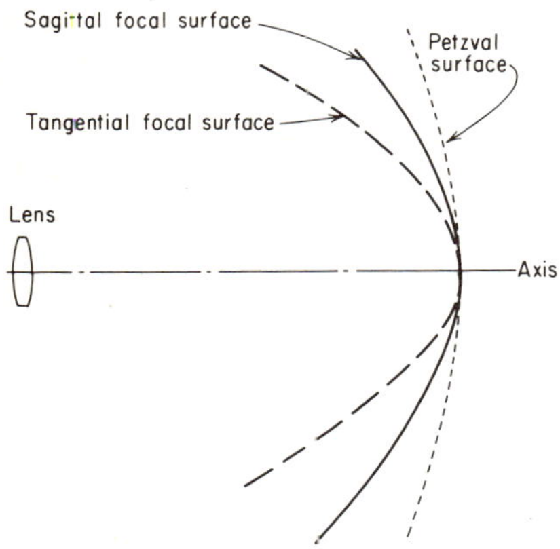

Fig. 18 The undercorrected astigmatism and field curvature of a simple lens. [*After Smith* (1966).]

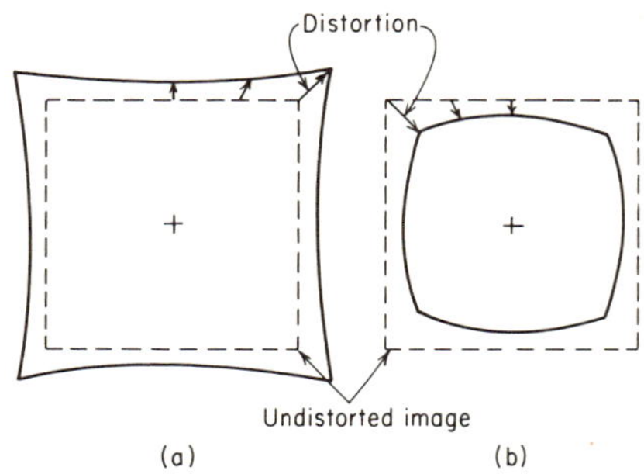

Fig. 19 (*a*) Pincushion and (*b*) barrel distortion. [*After Smith* (1966).]

The preceding are called monochromatic aberrations; they exist in monochromatic light, and although they vary with wavelength, the rate of change is usually relatively small.

Axial chromatic aberration is the longitudinal variation of focal position with wavelength. The longitudinal chromatic aberration is the distance from the long-wavelength focus to the short-wavelength focus (see Fig. 20).

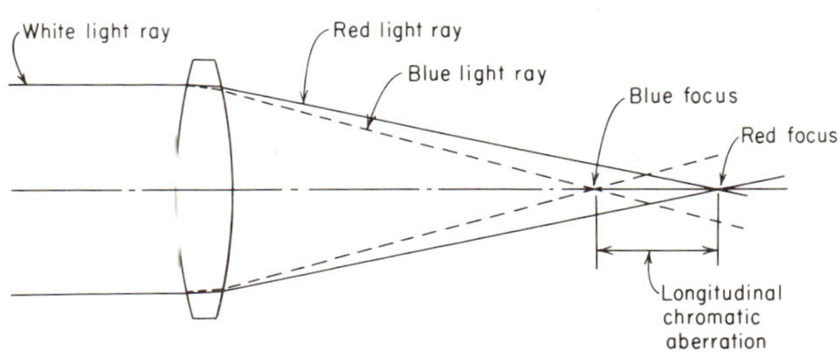

Fig. 20 A simple positive lens has undercorrected longitudinal chromatic aberration because the blue (shorter-wavelength) rays undergo a greater refraction. [*After Smith* (1966).]

Lateral chromatic aberration is the variation of image size with wavelength. It is the vertical distance from the off-axis image of a point in long-wavelength light to that in short-wavelength light, as shown in Fig. 21.

The monochromatic aberrations vary with wavelength. The chromatic variation of spherical aberration, or spherochromatism, is the most common. In the usual form the spherical aberration for the shorter wavelength is more overcorrected (or less undercorrected) than the longer wavelength. Thus a typical system might have the spherical aberration overcorrected in blue light, corrected in yellow light, and undercorrected in red light, as shown in Fig. 23. Chromatic variations of the other aberrations are less frequently found and are usually less serious, although

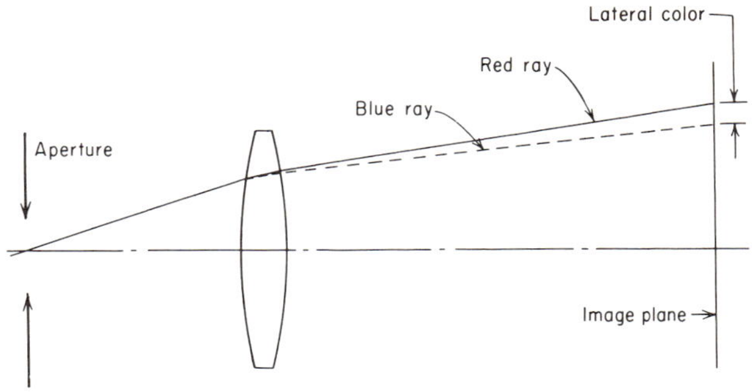

Fig. 21 Lateral color, or chromatic difference of magnification. [*After Smith* (1966).]

chromatic variation of distortion, which may also be regarded as a variation of CDM (lateral chromatic) with obliquity, can be troublesome in retrofocus or telephoto lenses.

15. Wave-Aberration Polynomial (Hopkins, 1950) If the aberration of an optical system is expressed as an optical-path difference (Par. 12), it can be shown (by reasons of symmetry) that the OPD can be expressed as a series of the form

$$\mathrm{OPD}(h,\rho,\phi) = \Sigma C_{ABC} h^A \rho^B \cos^C \phi \tag{87}$$

where C_{ABC} is a numerical coefficient, A,B,C,l, and n are positive integers related by $A = 2l + C$, $B = 2n + C$, h is the image height, and ρ and ϕ are the polar coordinates of the ray intersection with the system pupil. The term $C_{A00}h^A$ is zero by definition; the term $C_{020}\rho^2$ is equivalent to a simple longitudinal shift of the image (or reference) plane; the term $C_{111}h\rho \cos \phi$ is equivalent to a vertical shift of the reference point, i.e., a change in image height. These latter two terms are called the first-order terms. The "order" is given by $A + B - 1$.

The next five terms in the series are the third-order, or *Seidel,* aberrations, as follows:

Spherical aberration	$C_{040}\rho^4$
Coma	$C_{131}h\rho^3 \cos \phi$
Astigmatism and Petzval	$C_{220}h^2\rho^2 + C_{222}h^2\rho^2 \cos^2 \phi$
Distortion	$C_{311}h^3\rho \cos \phi$

The fifth-order aberrations are represented by:

Spherical aberration	$C_{060}\rho^6$
Linear coma	$C_{151}h\rho^5 \cos \phi$
Elliptical coma	$C_{331}h^3\rho^3 \cos \phi + C_{333}h^3\rho^3 \cos^3 \phi$
Oblique spherical	$C_{240}h^2\rho^4 + C_{242}h^2\rho^4 \cos^2 \phi$
Astigmatism and Petzval	$C_{420}h^4\rho^2 + C_{422}h^4\rho^2 \cos^2 \phi$
Distortion	$C_{511}h^5\rho \cos \phi$

There are 2 first-order terms, 5 third-order terms, 9 fifth-order terms, 14 seventh-order terms, 20 ninth-order terms, and $2 + 3 + \cdots + \frac{1}{2}(N + 3)$ Nth-order terms.

The wave-aberration polynomial, Eq. (87), represents the departure of the actual wavefront from a perfect spherical reference surface which passes through the axial intercept of the exit pupil and is centered on the ideal image point. Thus the y and z components of the transverse-ray aberration (see Fig. 22) can be expressed in terms of the wave aberration as

$$TA_y = -\frac{l}{n}\frac{\partial \mathrm{OPD}}{\partial y} \qquad TA_z = -\frac{l}{n}\frac{\partial \mathrm{OPD}}{\partial z} \tag{88}$$

where l = distance from exit pupil to image point
n = index of final medium
OPD = optical-path difference, Eq. (87)

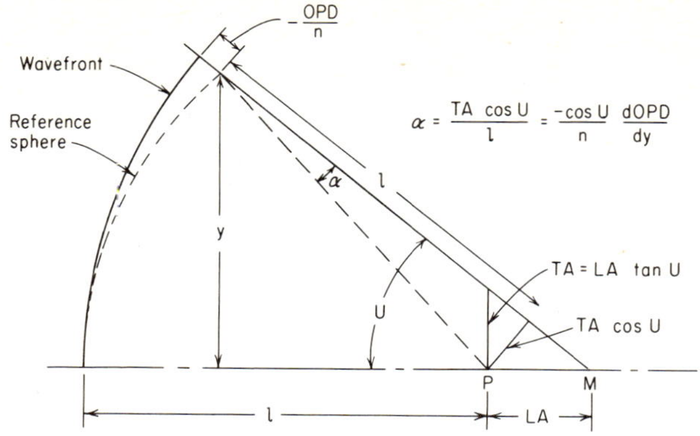

Fig. 22 The relationship between wave aberration (OPD) and transverse-ray aberration.

16. Ray-Aberration Polynomial When the operations indicated in Eq. (88) are carried out, the result is a pair of polynomial expressions for the ray aberration:

$$
\begin{aligned}
TA_y = -\frac{l}{n}[&C_{020}2\rho\cos\phi + C_{111}h \\
&+ C_{040}4\rho^3\cos\phi + C_{131}h\rho^2(2+\cos 2\phi) \\
&+ (C_{220}+C_{222})2h^2\rho\cos\phi + C_{311}h^3 \\
&+ C_{060}6\rho^5\cos\phi + C_{151}h\rho^4(3+2\cos 2\phi) \\
&+ C_{331}h^3\rho^2(2+\cos 2\phi) + C_{333}h^3\rho^2\tfrac{3}{2}(1+\cos 2\phi) \\
&+ C_{240}4h^2\rho^3\cos\phi + C_{242}h^2\rho^3\cos\phi(3+\cos 2\phi) \\
&+ (C_{420}+C_{422})2h^4\rho\cos\phi + C_{511}h^5 \\
&+ \text{7th and higher-order terms}]
\end{aligned}
\tag{89}
$$

$$
\begin{aligned}
TA_z = -\frac{l}{n}[&C_{020}2\rho\sin\phi \\
&+ C_{040}4\rho^3\sin\phi + C_{131}h\rho^2\sin 2\phi \\
&+ C_{220}2h^2\rho\sin\phi \\
&+ C_{060}6\rho^5\sin\phi + C_{151}2h\rho^4\sin 2\phi \\
&+ C_{331}h^3\rho^2\sin 2\phi \\
&+ C_{240}4h^2\rho^3\sin\phi + C_{242}h^2\rho^3\sin\phi(1+\cos 2\phi) \\
&+ C_{420}2h^4\rho\sin\phi \\
&+ \text{7th and higher-order terms}]
\end{aligned}
\tag{90}
$$

Note that in these expressions the sum of the exponents of the aperture and field terms (ρ and h) indicates the order of the aberration represented by that term; thus the first-order terms contain either ρ or h, the third-order aberration terms have ρ^3, ρ^2h, ρh^2, or h^3, the fifth-order terms are in ρ^5, ρ^4h, ρ^3h^2, ρ^2h^3, ρh^4, or h^5, and so forth.

17. Third-Order Aberrations The third-order portion of the aberration polynomials [Eqs. (87), (89), and (90)] can be evaluated from the data of two paraxial rays traced through the optical system.† There are many forms of the equations for this purpose; those given here directly indicate the amount of the transverse aberration at the image. See Par. 68 for sample calculations.

Two paraxial rays are traced through the system. The axial, or aperture, ray is traced from the axial intercept of the object and passes through the rim of the pupil of the system. The principal, chief, or field ray is traced from an appropriate off-axis point in the object, and passes through the center of the pupil. The axial-ray data are indicated by plain symbols; the principal-ray data are indicated by the subscript p. The rays are traced using the equations of Par. 5; the data i, y, u' and i_p, y_p, u_p' are thus available for each surface of

† Fifth- and higher-order aberrations can be calculated from the same data (Buchdahl, 1968).

the system. The optical invariant I is evaluated from the data of the two rays at the first (or any convenient) surface

$$I = n(y_p u - y u_p)$$

The final image height can be determined from the intersection height of the principal ray with the image plane or from $h = I/n_k' u_k'$, where n_k' and u_k' are the index and the slope of the axial ray after passing through the last surface of the system.

The third-order aberration contribution of each surface can be evaluated from the following:

Spherical: $$\text{TSC} = Bi^2h + Wy^4 \tag{91}$$
Sagittal coma: $$\text{CC} = Bii_ph + Wy^3y_p \tag{92}$$
Astigmatism: $$\text{TAC} = Bi_p^2h + Wy^2y_p^2 \tag{93}$$
Petzval: $$\text{TPC} = \frac{(n - n')IhC_e}{2nn'} \tag{94}$$
Distortion: $$\text{DC} = h[B_pii_p + \tfrac{1}{2}(u_p'^2 - u_p^2)] + Wyy_p^3 \tag{95}$$
Axial chromatic: $$\text{TLchC} = \frac{yi(\Delta n - \Delta n'\, n/n')}{u_k'} \tag{96}$$
Lateral chromatic: $$\text{TchC} = \frac{yi_p(\Delta n - \Delta n'\, n/n')}{u_k'} \tag{97}$$

where $$B = \frac{n(n' - n)y(u' - i)}{2n'I} \tag{98}$$

$$B_p = \frac{n(n' - n)y_p(u_p' - i_p)}{2n'I} \tag{99}$$

Δn = dispersion of medium = $n_F - n_C$ for visual work

$$W = \frac{4Kh(n - n')}{I}$$

$$K = A_4 - \tfrac{1}{4}A_2(4A_2^2 + 6CA_2 + 3C^2)$$
$$C_e = C + 2A_2$$

The surface equation is in the form of Eq. (53), and the equivalent curvature C_e is used in the paraxial ray tracing of any aspheric surface. Note that the chromatic aberrations TLchC and TchC are first-order aberrations; they are frequently calculated with the third-order aberrations.

The longitudinal values of the spherical aberration, astigmatism, Petzval curvature, and axial chromatic can be obtained by dividing the transverse contributions (TSC, TAC, TPC, or TLchC) by u_k', the final slope of the axial ray. The aberration at the image of the complete optical system is the sum of the contributions from each of the surfaces.

18. Stop-Shift Equations See Par. 70 for sample calculations. When the stop or pupil of an optical system is shifted, the changes in the third-order aberrations can be computed from the following equations, in which the starred terms are the aberration contributions after the stop is shifted and the unstarred terms are the contributions before the shift.

$$\text{TSC}^* = \text{TSC} \tag{100}$$
$$\text{CC}^* = \text{CC} + \text{TSC}\cdot Q \tag{101}$$
$$\text{TAC}^* = \text{TAC} + \text{CC}\cdot 2Q + \text{TSC}\cdot Q^2 \tag{102}$$
$$\text{TPC}^* = \text{TPC} \tag{103}$$
$$\text{DC}^* = \text{DC} + (\text{TPC} + 3\text{TAC})Q + \text{CC}\cdot 3Q^2 + \text{TSC}\cdot Q^3 \tag{104}$$
$$\text{TLchC}^* = \text{TLchC} \tag{105}$$
$$\text{TchC}^* = \text{TchC} + \text{TLchC}\cdot Q \tag{106}$$

Q represents the amount of the pupil shift and is evaluated from

$$Q = \frac{y_p^* - y_p}{y} \tag{107}$$

where y_p^* = ray height of new principal ray after stop is shifted
y_p = old principal-ray height
y = height of axial ray

Since Q is an invariant, it can be evaluated at any convenient surface of the system; a further consequence of the invariance of Q is that the starred and unstarred terms of Eqs. (100) to (106) may represent the contributions of an entire optical system or any part of it, e.g., a single surface or element.

19. Thin-Lens Aberrations See Par. 70 for sample calculation. If the elements of an optical system can be regarded as thin lenses, i.e., of zero thickness, surrounded by air, the following equations give the third-order aberration contribution of each element, assuming that the stop (pupil) is in contact with the element:

$$\text{TSC} = -y^4(G_1C^3 - G_2C^2C_1 + G_3C^2v + G_4CC_1^2 - G_5CC_1v + G_6Cv^2)/u_k' \quad (108)$$

$$= -y^4(G_1C^3 + G_2C^2C_2 - G_3C^2v' + G_4CC_2^2 - G_5CC_2v' + G_6Cv'^2)/u_k' \quad (108a)$$

$$\text{CC} = -hy^2(\tfrac{1}{4}G_5CC_1 - G_7Cv - G_8C^2) \quad (109)$$

$$= -hy^2(\tfrac{1}{4}G_5CC_2 - G_7Cv' + G_8C^2) \quad (109a)$$

$$\text{TAC} = -\tfrac{1}{2}h^2\phi u_k' \quad (110)$$

$$\text{TPC} = \frac{-\tfrac{1}{2}h^2\phi u_k'}{n} = \frac{\text{TAC}}{n} \quad (111)$$

$$\text{DC} = 0 \quad (112)$$

$$\text{TLchC} = \frac{-y^2\phi}{Vu_k'} \quad (113)$$

$$\text{TchC} = 0 \quad (114)$$

Paraxial rays are traced through the system using Eqs. (29) and (30); a principal and an axial ray (as defined in Par. 17) are traced. Equations (100) to (106) are used with $Q = y_p/y$ to determine the contributions of each element for the actual stop position. The contributions from the elements are summed to determine the aberrations of the whole system.

In Eqs. (108) to (114) C represents the total curvature of the element, and C_1 and C_2 are the curvatures of the left and right surfaces, so that $C = C_1 - C_2 = \phi/(n-1)$. The symbol v is the reciprocal object distance (for the element) and $v = u/y$; similarly $v' = u'/y$. V is the Abbe ν number of the lens material, equal to $(n_D - 1)/(n_F - n_C)$ in the visible region. G_1 to G_8 are functions of the index, as follows:

$$G_1 = \frac{n^2(n-1)}{2} \qquad G_2 = \frac{(2n+1)(n-1)}{2}$$

$$G_3 = \frac{(3n+1)(n-1)}{2} \qquad G_4 = \frac{(n+2)(n-1)}{2n}$$

$$G_5 = \frac{2(n+1)(n-1)}{n} \qquad G_6 = \frac{(3n+2)(n-1)}{2n} \quad (115)$$

$$G_7 = \frac{(2n+1)(n-1)}{2n} \qquad G_8 = \frac{n(n-1)}{2}$$

The aspheric terms of Eqs. (91) to (95) can be added to the thin-lens aberrations obtained from Eqs. (100) to (104) when the thin lenses have aspheric surfaces.

20. Interpretation of Third-Order Aberration Contributions To the extent that the third-order aberrations approximate the complete aberration polynomial, the values of the aberrations computed by means of the equations of Pars. 17 to 19 will correspond to the values of the aberrations as determined by actual ray tracing. Thus ΣTSC will approximate the intersection height (in the paraxial image plane) of a trigonometrically traced axial ray, i.e., the transverse spherical aberration. Similarly 3ΣCC will approximate the tangential coma and ΣDC the linear distortion, with 100ΣDC/h the percentage distortion. The longitudinal curvature (sag) of field is approximated by $x_s \approx (\Sigma\text{TAC} + \Sigma\text{TPC})/u_k'$ and $x_t \approx (3\Sigma\text{TAC} + \Sigma\text{TPC})/u_k'$ for the sagittal and tangential fields, respectively. Since the chromatic-aberration equations are first-order relationships, we have the relations

$$\frac{\Sigma \mathrm{TLchC}}{u_k'} = \mathrm{LchC} = l_F - l_C$$

$$\Sigma \mathrm{TchC} = h_F - h_C$$

where $l_F - l_C$ is the paraxial longitudinal axial chromatic aberration and $h_F - h_C$ is the paraxial lateral chromatic aberration. Note that the paraxial chromatic difference of magnification is given by $\mathrm{CDM} = (h_F - h_C)/h = \Sigma \mathrm{TchC}/h$.

21. Aberration Representation Because aberrations vary with aperture and image size, a simple single numerical value is seldom an adequate description of the state of correction of a system. Several more or less standard methods of representing the aberrations are in common use.

Longitudinal. Spherical aberration is frequently shown by a plot of the longitudinal position of the ray intercept with the axis as a function of the height at which the ray strikes the entrance pupil. It is convenient to plot the spherical aberration for several wavelengths to get a more complete indication of the correction (see Fig. 23). The variation of effective focal length with aperture is sometimes plotted with the spherical aberration (which is of course the variation of focal position with aperture); the distance between the two curves is a measure of the amount of low-obliquity coma (offense against the sine condition) in the system.

Field curvature is usually represented as a longitudinal aberration plotted as a function of image height or obliquity. The positions of the tangential and sagittal focal surfaces are plotted; the separation of the curves indicates the astigmatism.

Figure 23 shows examples of longitudinal representation of aberrations.

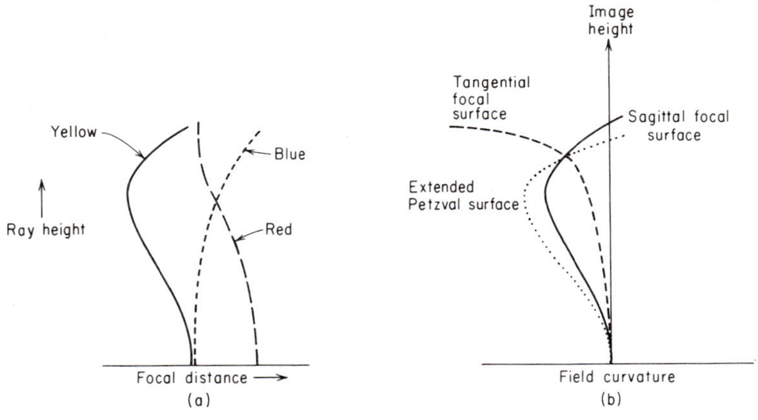

Fig. 23 Longitudinal representations of (*a*) zonal spherical aberration and spherochromatism and (*b*) residual astigmatism and curvature of field. [*After Smith* (1966).]

Lateral. The lateral representation of aberrations permits a more universal and complete description of the correction of the optical system. In this representation, a fan of rays from an object point is traced through the system, and the coordinates of the ray intersection with the reference (usually paraxial) image surface are plotted as a function of the ray position in the aperture, usually represented by the slope of the ray (tan U) at the image. The lateral presentation has two major advantages: (1) the effect of the aberrations can be estimated directly in terms of the size of the blur they cause; (2) the effects of refocusing, i.e., longitudinally shifting the reference surface, can be determined by simply rotating the tan U axis. Figure 24 illustrates the appearance of several common aberrations. These curves are called ray-intercept plots or (incorrectly) rim-ray curves.

Spot Diagrams and Related Representations. If the aperture of the optical system is divided into a large number of equal small areas and a ray from a selected object point is traced through the center of each small area, a plot of the intersection points of the rays (spots) with the image surface is an approximate representation of the (geometrical) illuminance distribution at the image. The more rays traced, the better the approximation becomes. Such a plot is called a *spot diagram.*

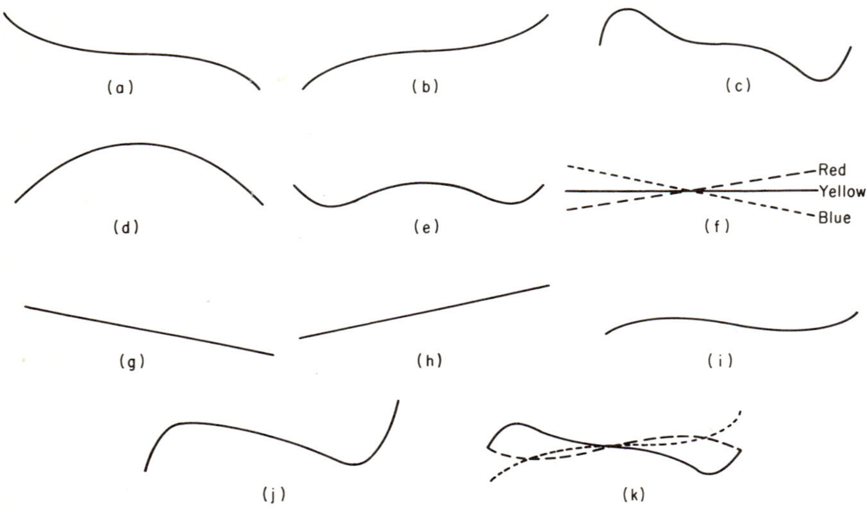

Fig. 24 Transverse representation of the aberrations as ray-intercept curves. The ordinate for each curve is H', the height at which the ray intersects the (paraxial) image plane; the abscissa is tan U', the final slope of the ray with respect to the optical axis. (*a*) Undercorrected spherical aberration; (*b*) overcorrected spherical aberration; (*c*) zonal spherical; (*d*) undercorrected coma; (*e*) coma (higher order); (*f*) longitudinal chromatic aberration (undercorrected); (*g*) inward tangential field curvature; (*h*) backward tangential field curvature; (*i*) inward field curvature plus overcorrected spherical; (*j*) a typical off-axis curve; (*k*) axial curve for an achromatic doublet, showing zonal spherical, secondary spectrum, and spherochromatism. [*After Smith* (1966).]

If the geometrical spot diagram is assumed to be a reasonable representation of the actual image illuminance distribution, several other representations can be derived from it. The *radial energy distribution* is obtained by selecting (arbitrarily) a center point and plotting the percentage of the energy, i.e., the number of spots, encircled within a radius R as a function of R. If the illuminance, i.e., the spot density, is represented as a function of the y and z coordinates of the image plane, this is the *point-spread function* of the system.

The *line-spread function* is the cross section of the illuminance distribution in the image of a line; it is derived from the point-spread function by summing it in one direction. Thus $L(y) = \int P(y,z)\,dz$ and $L(z) = \int P(y,z)\,dy$, where $L(\cdot)$ is the line-spread function and $P(y,z)$ is the point-spread function (see Fig. 25).

22. Aplanatic Surfaces There are three object positions for which the spherical aberration and coma of a spherical surface are identically zero:

1. Object (and image) at the surface:

$$L = L' = 0$$

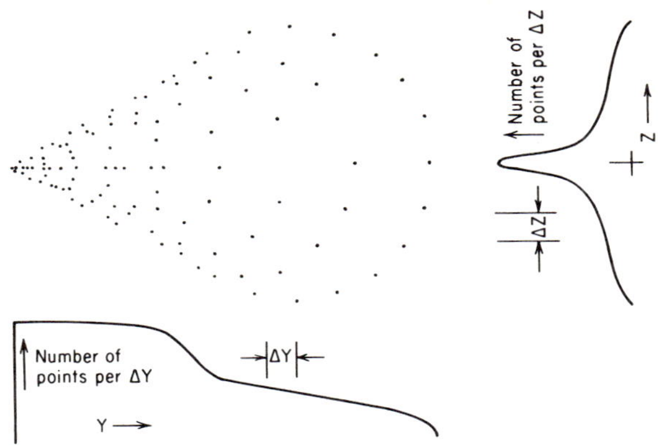

Fig. 25 A spot diagram (for a lens with third-order coma) and the line-spread functions derived from the spot diagram. [*After Smith* (1966).]

2. Object (and image) at the center of curvature of the surface:

$$L = L' = R$$

3. The "aplanatic case," where

$$L = \frac{R(n' + n)}{n} = \frac{n'L'}{n}$$

$$n \sin U' = n' \sin U$$

$$U = I' \qquad U' = I$$

The third-order aberrations of a single air-glass ($n' = 1.5$) surface are plotted as a function of object position in Fig. 26.

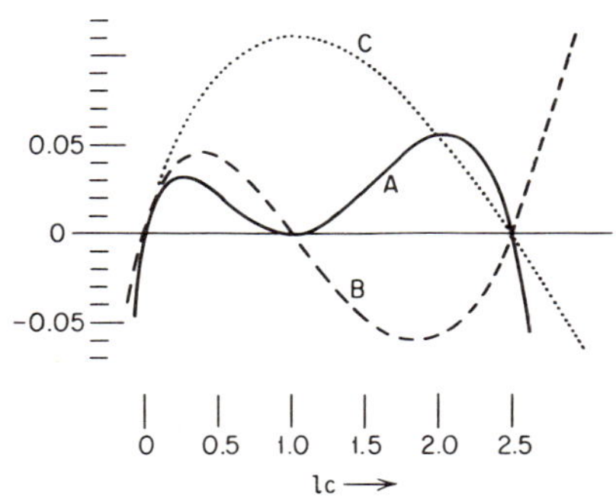

Fig. 26 The aberrations of a single refracting surface ($n = 1.0$, $n' = 1.5$) as a function of object position. Curve A, spherical aberration [TSC(c/u^3)]; curve B, sagittal coma [CC(c/u^2u_p)]; curve C, astigmatism [TAC(c/uu_p^2)], where l is the object position, c is the surface curvature, u is the axial ray-slope angle, and u_p is the slope of the principal ray.

23. Symmetrical Systems In an optical system which is completely symmetrical about the aperture stop, coma, distortion, and lateral color are identically zero. For complete symmetry the system must work at unit magnification and the optical system on the image side of the stop must be a mirror image of the system on the object side of the stop. Third-order coma and distortion are zero with flat object and image surfaces, but curved object and image surfaces may be necessary to satisfy completely the requirement for symmetry about the stop (in the presence of field curvature) and reduce all orders of coma and distortion to zero.

DIFFRACTION IMAGERY (Born and Wolf, 1964)

24. The Point-Spread Function; The Diffraction Integral The exact point-spread function $P(y,z)$ of an image (as contrasted to the approximate geometrical function derived from ray tracing) is given by

$$P(y,z) = |A(y,z)|^2 \tag{116}$$

where $A(y,z)$ is the complex phase amplitude of the wavefront emerging from the optical system, given by

$$A(y,z) = \iint_{-\infty}^{\infty} B(u,v) \exp\left[-i\frac{2\pi}{\lambda}\phi(u,v)\right] \exp\left(i\frac{2\pi}{\lambda}\frac{uy+vz}{F}\right) du\, dv \tag{117}$$

where y, z = coordinates in image plane normal to principal ray
u, v = coordinates in exit pupil normal to principal ray
λ = wavelength
$B(u,v)$ = amplitude factor proportional to square root of flux density, i.e., transmission, at point (u,v) in pupil; $B(u,v) = 0$ outside pupil
$\phi(u,v)$ = wave-aberration function (see Par. 15); $\phi(u,v)$ = OPD of the ray through point (u,v)
F = radius of reference sphere
i = imaginary $(-1)^{1/2}$

The terms $B(u,v)$ and $\exp[-(i2\pi/\lambda)\phi(u,v)]$ are often combined and referred to as the pupil function of the system. Equation (117) is sometimes written to include a focusing term, $\exp[i\pi x(u^2+v^2)/\lambda F^2]$.

25. The Diffraction Image If the pupil function is a constant, i.e., if the transmission of the system is uniform over the (circular) aperture and the system is aberration-free, the illuminance distribution in the image becomes

$$\begin{aligned} P(y,z) &= \pi\left(\frac{\text{NA}}{\lambda}\right)^2 P_t\left[\frac{2J_1(m)}{m}\right]^2 \\ &= I_0\left[\frac{2J_1(m)}{m}\right]^2 \end{aligned} \tag{118}$$

where $\text{NA} = n' \sin U'$ is the numerical aperture of the system, $J_1(\cdot)$ is the first-order Bessel function

$$J_1(x) = \frac{x}{2} - \frac{(x/2)^3}{1^2 2} + \frac{(x/2)^5}{1^2 2^2 3} - \cdots$$

P_t is the total power in the point image, and m is the normalized radial coordinate

$$m = \frac{2\pi}{\lambda}\text{NA}(y^2+z^2)^{1/2} = \frac{2\pi}{\lambda}\text{NA}\cdot s$$

The fraction of the total power falling within a radial distance s_0 of the center of the pattern is given by $1 - J_0^2(m_0) - J_1^2(m_0)$, where $J_0(\cdot)$ is the zero-order Bessel function

$$J_0(x) = 1 - \left(\frac{x}{2}\right)^2 + \frac{(x/2)^4}{1^2 2^2} - \frac{(x/2)^6}{1^2 2^2 3^2} + \cdots$$

Equation (118) is plotted in Fig. 27, which also shows the appearance of the diffraction pattern. The pattern consists of a circular patch of light (the Airy disk) surrounded by rings of rapidly decreasing intensity. Table 1 indicates the size and distribution of energy in the pattern for a circular aperture and a slit aperture.

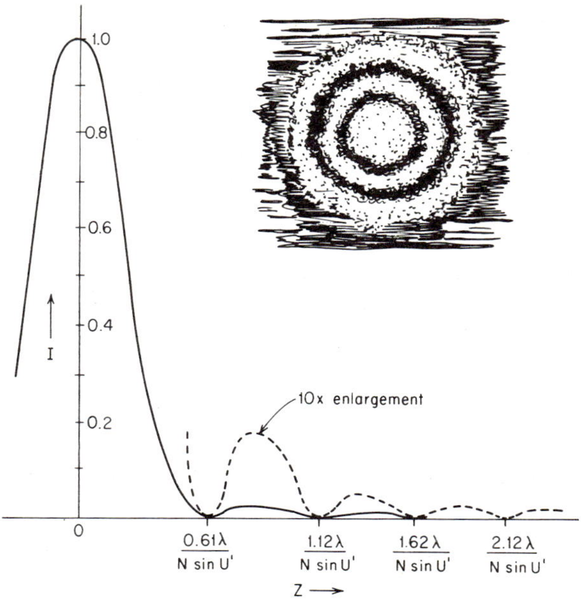

Fig. 27 The distribution of illumination in the diffraction pattern formed in the image plane of a perfect lens with a circular aperture. [*After Smith* (1966).]

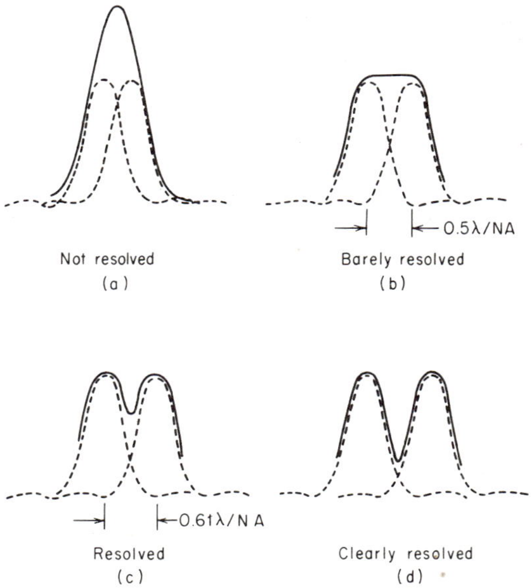

Fig. 28 The dashed lines represent the diffraction patterns of two point images at various separations. The solid line indicates the combined diffraction pattern; (*b*) is the Sparrow criterion for resolution, and (*c*) is the Rayleigh criterion. [*After Smith* (1966).]

TABLE 1 Distribution of Energy in the Diffraction Pattern at the Focus of a Perfect Lens as a Function of the Distance Z from the Pattern Center (Smith, 1965)

	Circular aperture			Slit aperture	
Ring (or band)	Z	Peak illumination	Energy in ring, %	Z	Peak illumination
Central maximum	0	1.0	83.9	0	1.0
First dark ring	$\frac{0.61\lambda}{N' \sin U'}$	0.0	....	$\frac{0.5\lambda}{N' \sin U'}$	0.0
First bright ring	$\frac{0.82\lambda}{N' \sin U'}$	0.017	7.1	$\frac{0.72\lambda}{N' \sin U'}$	0.047
Second dark ring	$\frac{1.12\lambda}{N' \sin U'}$	0.0	....	$\frac{1.0\lambda}{N' \sin U'}$	0.0
Second bright ring	$\frac{1.33\lambda}{N' \sin U'}$	0.0041	2.8	$\frac{1.23\lambda}{N' \sin U'}$	0.017
Third dark ring	$\frac{1.62\lambda}{N' \sin U'}$	0.0	....	$\frac{1.5\lambda}{N' \sin U'}$	0.0
Third bright ring	$\frac{1.85\lambda}{N' \sin U'}$	0.0016	1.5	$\frac{1.74\lambda}{N' \sin U'}$	0.0083
Fourth dark ring	$\frac{2.12\lambda}{N' \sin U'}$	0.0	....	$\frac{2.0\lambda}{N' \sin U'}$	0.0
Fourth bright ring	$\frac{2.36\lambda}{N' \sin U'}$	0.00078	1.0	$\frac{2.24\lambda}{N' \sin U'}$	0.0050
Fifth dark ring	$\frac{2.62\lambda}{N' \sin U'}$	0.0	....	$\frac{2.5\lambda}{N' \sin U'}$	0.0

When the aperture is rectangular,

$$P(y,z) = I_0 \left(\frac{\sin M_y}{M_y}\right)^2 \left(\frac{\sin M_z}{M_z}\right)^2 \tag{119}$$

where $M_y = 2\pi \mathrm{NA} \cdot y/\lambda$ and $M_z = 2\pi \mathrm{NA} \cdot z/\lambda$.

26. Point Resolution: The Rayleigh and Sparrow Criteria The images of two point sources are said to be resolved when the combined diffraction pattern can be identified as being due to two objects rather than one. The effect of various source spacings on the combined diffraction pattern is indicated in Fig. 28. The Rayleigh criterion assumes that the two point sources can be resolved if the first dark ring of one pattern coincides with the center of the other pattern. Thus the separation of the two points is equal to the radius of the first dark ring,

$$\text{Separation} = \frac{0.61\lambda}{n' \sin U'} = \frac{0.61\lambda}{\mathrm{NA}} \tag{120}$$

The Sparrow criterion assumes resolution at the point spacing for which the combined pattern has no minimum between the two pattern centers. This occurs when

$$\text{Separation} = \frac{0.5\lambda}{n' \sin U'} = \frac{0.5\lambda}{\mathrm{NA}} \tag{121}$$

where λ = wavelength

n' = index of refraction in image space

U' = angle to axis made by extreme marginal rays of image-forming cone

NA = $n' \sin U'$ = numerical aperture

The separations can be converted to object separations either (1) by dividing the image separation (above) by the system magnification or (2) by the use of the object-space NA in Eqs. (120) and (121). In object space the angular separation of the object points is frequently of interest and is given by

$$\text{Rayleigh angular separation} = \frac{1.22\lambda}{D} \quad \text{rad}$$

$$\text{Sparrow angular separation} = \frac{\lambda}{D} \quad \text{rad}$$

where D is the effective aperture (entrance-pupil diameter) of the optical system. For visual use, the respective resolution in seconds of arc is approximately $5.5/D$ and $4.5/D$ when D is expressed in inches.

Since these resolutions assume an aberration- and defect-free optical system, they are frequently used as a standard for excellence of design and construction. A system which meets the Rayleigh criterion is sometimes called diffraction-limited.

27. Apodization The diffraction pattern described in Pars. 25 and 26 assumed an aberration-free system and a uniformly transmitting aperture. If the transmission [$B(u,v)$ in Eq. (117)] is not constant across the aperture, the diffraction pattern will be modified. This is called apodization. In general, if the transmission at the center of the aperture is reduced, the diameter of the central (Airy) disk is reduced and the amount of light in the rings is increased (and vice versa); these effects tend to be small.

The term apodization is usually reserved for transmission effects; occasionally the term is applied to phase effects. This is, of course, a modification of $\phi(u,v)$ in Eq. (117) and is equivalent to the introduction of an aberration into the system.

IMAGE EVALUATION

28. In determining which of two optical systems is the better or whether a given system meets certain performance requirements, it is often necessary to evaluate the quality of the images formed. In many cases where one is testing an existing system, a simple functional test is suitable; in other instances a functional test may be impractical, or one may wish to evaluate a design without constructing a prototype. There are many schemes of evaluation; usually one is more suitable than the others. In some circumstances a simple numerical comparison of the aberrations is entirely satisfactory; this becomes less than ideal, however, when the systems under comparison have different aberration balances. A comparison between X units of coma in system A, Y units of astigmatism in system B, and Z units of chromatic in system C can be frustratingly inconclusive. Several commonly accepted criteria for image evaluation are described in the following paragraphs.

29. Blur-Spot Size The size of the blur spot into which the system images all the radiation from a point object is a common evaluation criterion. However, since the distribution of the energy within the spot is as important as the spot size in determining image quality, this simple criterion is completely satisfactory for only a limited number of applications. A variation is the use of the size of the spot into which the lens system images a certain portion of the total energy. The radial-energy-distribution plot described in Par. 21 is a logical extension of this technique for a radially symmetrical image. In general, the point-spread function is a complete description of the image. However, in all but the simpler cases the interpretation of blur-spot size information into terms of system performance is difficult.

Blur-spot size information can be determined by ray tracing, by evaluation of the diffraction integral, Eq. (116), or by measurement of the image. Although the geometrical ray-trace computation neglects the effects of diffraction, if the aberrations are large when compared with the diffraction effects, the results are of considerable use, as the widespread acceptance of the ray-intercept plot and spot diagram attest.

In the presence of simple third-order spherical aberration, the geometrical blur-spot size is minimized if the system is focused three-quarters of the distance from the paraxial focus to the marginal focus, i.e., the axial intercept of the marginal ray. At this point the diameter of the blur spot is

$$B = \tfrac{1}{2}\text{LA}_m \tan U_m = \tfrac{1}{2}\text{TLA}_m \tag{122}$$

where LA_m = marginal longitudinal spherical aberration
TLA_m = its transverse value
U_m = angle between axis and marginal ray

When there are both third- and fifth-order spherical aberration in the image, the smallest geometrical blur occurs when the correction is adjusted so that the zonal spherical $\text{LA}_z = 1.5\text{LA}_m$ (or $\text{TLA}_z \approx \text{TLA}_m$) and the system is focused 83 percent of the distance from the paraxial focus to the zonal focus (or 125 percent of the distance to the marginal focus). The blur-spot size is

$$\begin{aligned} B &= \tfrac{1}{2}\text{LA}_m \tan U_m = \tfrac{1}{2}\text{TLA}_m \\ &= \tfrac{1}{3}\text{LA}_z \tan U_m \approx 0.47\text{TLA}_z \end{aligned} \tag{123}$$

If the correction is such that the marginal spherical aberration is zero, the system can be focused 42 percent of the way from the paraxial to the zonal focus and the blur spot is

$$B = 0.84\text{LA}_z \tan U_m \approx 1.2\text{TLA}_z \tag{124}$$

Note that the preceding indicate the focal positions which yield the smallest-sized blur which contains all the energy in the image. While useful in many instances, e.g., in systems using detectors, this is not necessarily the "best" focus for all applications. For example, in most optical systems the marginal spherical aberration is fully corrected, despite the fact that a geometrically smaller blur is available. Further, the focus usually chosen is twice as far from the paraxial focus as indicated above; that is, 0.84LA_z vs. 0.42LA_z. Although the total blur size is larger, this focus yields an image with a bright core which is only about one-third the size indicated in Eq. (124) but which contains about 80 percent of the total energy (see Fig. 29).

For systems with simple third-order aberrations, evaluated in accordance with Pars. 17 to 19, the minimum blur diameters caused by each aberration are as follows:

Spherical aberration:	$B = \tfrac{1}{2}\text{TSC}$
Coma:	$B_z = 2\text{CC}$
	$B_y = 3\text{CC}$
Astigmatism and field curvature:	$B_z = 2(\text{TPC} + \text{TAC})$
	$B_y = 2(\text{TPC} + 3\text{TAC})$
Astigmatism (at best focus):	$B = 2\text{TAC}$
Axial chromatic aberration:	$B = \text{TLchC}$

30. The Rayleigh Quarter-Wave Limit The Rayleigh limit for image quality states that if the OPD (Pars. 12 and 15) does not vary more than one-quarter wavelength of light over the aperture of an optical system, the image will be sensibly perfect. In other words, the wavefront at the exit pupil must be contained between two spheres, centered on the image point, whose radii differ by less than one-quarter wavelength. When the wavefront is relatively smooth and free of high-order ripples, this is a very reliable limit and systems which meet this criterion are sometimes described as diffraction-limited.

The amounts of certain first-, third-, and fifth-order aberrations which correspond to a maximum OPD of one-quarter wave are as follows (Conrady, 1957):

Out of focus: $$\Delta l = \pm \frac{\lambda}{2n \sin^2 U_m}$$

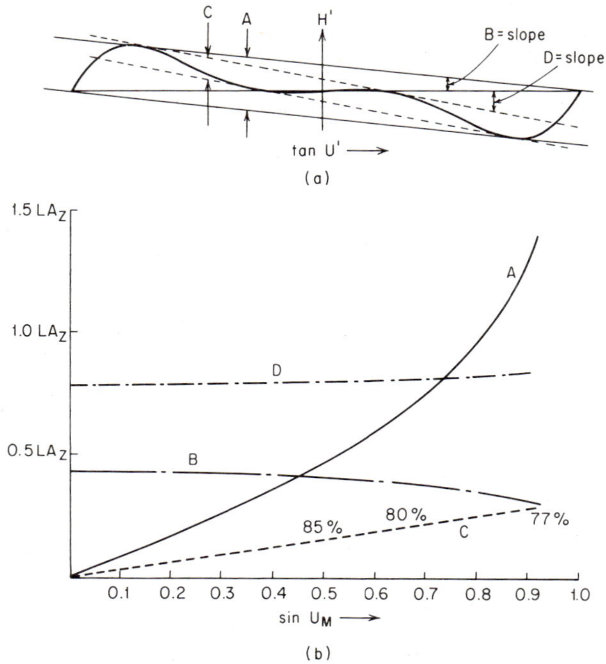

Fig. 29 The image blur-spot size in the presence of both third- and fifth-order spherical aberration ($LA_m = TA_m = 0$). (*a*) H' vs. tan U' ray-intercept plot. (*b*) Curve *A*, diameter of the smallest blur spot containing 100 percent of the energy in the image; curve *B*, distance from the paraxial focus to the plane of the smallest 100 percent energy spot; curve *C*, diameter of the bright core of the image; figures along the *C* curve indicate the percentage of the total energy in the bright core; curve *D*, distance from the paraxial focus to the plane for curve *C*. [*After Smith* (1966).]

Spherical aberration (third-order):

$$LA_m = \pm \frac{4\lambda}{n \sin^2 U_m}$$

Spherical aberration (zonal; $LA_m = 0$):

$$LA_z = \pm \frac{6\lambda}{n \sin^2 U_m}$$

Axial chromatic aberration:

$$\text{LchA} = \pm \frac{\lambda}{n \sin^2 U_m}$$

Sagittal coma:

$$\text{Coma}_s = \pm \frac{\lambda}{2n \sin U_m}$$

These values assume that the reference point is chosen to minimize the OPD. The third-order spherical aberration equation assumes that the focus, i.e., the center of the reference sphere, is located midway between the marginal and paraxial foci, since this is the position at which the OPD is minimized. Similarly the zonal spherical-aberration tolerance assumes that the marginal spherical is corrected to zero and that the focus is three-fourths of the distance from the paraxial focus to the zonal focus. For axial chromatic aberration, the focus is assumed midway between the extreme foci.

When the wavefront aberration is of low order, as in a simple defocusing or third-order spherical aberration, the effect of a quarter wave of aberration on the diffraction pattern (Par. 25) is to shift some light from the central patch to the rings. In a perfect system, 84 percent of the energy is in the central patch and 16 percent is in the rings. A quarter wave of defocusing produces a pattern with 68 percent in the central blur and 32 percent in the rings. This is of course a detectable change; however, the change is in practice difficult to detect, and for ordinary applications a system with less than a quarter wave of aberration is an excellent one.

Because the quarter-wave limit tacitly assumes a smooth, low-order wavefront, it is of limited value when the wavefront has high-order ripples. The root-mean-squared (rms) OPD is a somewhat more widely applicable measure of the quality of a system. An rms OPD of one-fourteenth of a wave ($\lambda/14$) is approximately the equivalent of the classical quarter-wave limit.

31. Strehl Definition (Born and Wolf; 1964) The Stehl definition is the ratio of the illuminance at the peak of the diffraction pattern of an aberrated point image to that at the peak of an aberration-free image. A Strehl ratio of 0.8 is the equivalent of the Rayleigh quarter-wave limit and has a much broader applicability.

The Strehl definition of a system can be evaluated by calculating the normalized illuminance at the center of the diffraction pattern

$$\mathrm{SD} = \left(\frac{1}{A}\iint_A e^{i2\pi\phi/\lambda}\,dA\right)^2 \tag{125}$$

where the integration is over the pupil area A and ϕ is the OPD of the wavefront, as in Par. 24. When the Strehl ratio is large enough, say 0.8 or greater, the following approximation is useful:

$$\mathrm{SD} \approx 1 - \left(\frac{2\pi}{\lambda}\right)^2 (\Delta\phi)^2 \tag{126}$$

where $(\Delta\phi)^2$ is the variance

$$(\Delta\phi)^2 = \overline{\phi^2} - (\overline{\phi^1})^2$$

$$\overline{\phi^n} = \frac{\iint \phi^n\,dA}{\iint dA}$$

The Strehl definition is also equal to the volume under the (three-dimensional) modulation transfer function (MTF, Par. 32) of the system divided, i.e., normalized, by the volume under the MTF for an aberration-free system.

32. The Optical Transfer Function (Born and Wolf, 1964; Perrin 1960) The optical transfer function [OTF(ν)] is a complex function describing the performance of an optical system in terms of its imagery of an object linear pattern whose luminance (brightness) varies sinusoidally according to the spatial frequency ν (in cycles per unit length). The modulus of the OTF is the modulation transfer function† [MTF(ν)], which describes the transfer of contrast (modulation) from object to image; the argument of the OTF is the spatial phase shift $\phi(\nu)$ of the sinusoidal image pattern from its nominal position.

We consider first the one-dimensional case. The luminance of the sinusoidal object can be described by

$$G(y) = b_0 + b_1 \cos 2\pi\nu y \tag{127}$$

where ν is the spatial frequency and y (or z) is the dimension or coordinate in which the luminance is varying. According to Eq. (127), the luminance varies from $b_0 - b_1$ to $b_0 + b_1$, and we define the modulation as the peak-to-trough variation in luminance divided by the sum of these values. Thus the object modulation is

$$M_0 = \frac{(b_0 + b_1) - (b_0 - b_1)}{(b_0 + b_1) + (b_0 - b_1)} = \frac{b_1}{b_0} \tag{128}$$

† Also known as the frequency response, sine-wave response, and contrast transfer.

Each line element in the object is imaged by the optical system as the line-spread function $L(y)$ (Par. 21), and if we assume unit magnification and 100 percent transmission to simplify the discussion, the illuminance at position y in the image is the summation of the product of $G(y)$ and $L(y)$; it can be expressed as

$$\begin{aligned} F(y) &= \int L(\delta)G(y-\delta)\,d\delta \\ &= b_0\int L(\delta)\,d\delta + b_1\int L(\delta)\cos[2\pi\nu(y-\delta)]\,d\delta \end{aligned} \tag{129}$$

After normalizing [by dividing by $\int L(\delta)\,d\delta$], Eq. (129) can be transformed into

$$\begin{aligned} F(y) &= b_0 + b_1[A_c(\nu)\cos 2\pi\nu y + A_s(\nu)\sin 2\pi\nu y] \\ &= b_0 + b_1|A(\nu)|\cos(2\pi\nu y - \phi) \end{aligned} \tag{130}$$

where

$$|A(\nu)| = [A_c^2(\nu) - A_s^2(\nu)]^{1/2} \tag{131}$$

$$A_c(\nu) = \frac{\int L(\delta)\cos(2\pi\nu\delta)\,d\delta}{\int L(\delta)\,d\delta} \tag{132}$$

$$A_s(\nu) = \frac{\int L(\delta)\sin(2\pi\nu\delta)\,d\delta}{\int L(\delta)\,d\delta} \tag{133}$$

$$\phi = \tan^{-1}\frac{A_s(\nu)}{A_c(\nu)} = \cos^{-1}\frac{A_c(\nu)}{|A(\nu)|} \tag{134}$$

Equation (130) is the illuminance in the image. Note that it is sinusoidally modulated at the same spatial frequency as the object in Eq. (127), although the pattern is shifted by a distance represented by the phase angle ϕ. The modulation of the image is

$$M_i = \frac{b_1}{b_0}|A(\nu)| = M_o|A(\nu)| \tag{135}$$

The modulation transfer function is the ratio of the modulation in the image to that in the object,

$$\mathrm{MTF}(\nu) = \frac{M_i}{M_o} = |A(\nu)| \tag{136}$$

33. Computation of the Optical Transfer Function The OTF can be computed in several ways. Paragraph 32 described the calculation based on the line-spread function $L(y)$. The line-spread function is calculated from the point-spread function $P(y,z)$ by

$$L(y) = \int P(y,z)\,dz \qquad \text{and} \qquad L(z) = \int P(y,z)\,dy \tag{137}$$

The point-spread function can be calculated from a ray-traced spot diagram, as indicated in Par. 21, or from Eq. (116), which includes diffraction effects and is based on the wavefront (which can be computed by ray tracing) (Cooley and Tukey, 1965).

Although many of these equations are shown as integrals, it should be noted that the integrations are usually carried out numerically. Thus, for example, Eq. (132) could be evaluated as

$$A_c(\nu) = \frac{\Sigma L(y)\cos(2\pi\nu y)\,\Delta y}{\Sigma L(y)\,\Delta y}$$

and Eq. (137) as

$$L(y) = \Sigma P(y,z)\,\Delta z$$

The two-dimensional expression for the optical transfer function is

$$\mathrm{OTF}(\nu_y,\nu_z) = \frac{\iint P(y,z)e^{-i2\pi(\nu_y y+\nu_z z)}\,dy\,dz}{\iint P(y,z)\,dy\,dz} \tag{138}$$

where ν_y and ν_z are the spatial frequencies in the y and z directions and OTF(·) is a complex function; thus

$$\mathrm{OTF}(\nu_y,\nu_z) = \mathrm{Re} - i\,\mathrm{Im}$$
$$\mathrm{MTF}(\nu_y,\nu_z) = |\mathrm{OTF}(\nu_y,\nu_z)| = (\mathrm{Re}^2 - \mathrm{Im}^2)^{1/2}$$
$$\phi = \tan^{-1}\frac{\mathrm{Im}}{\mathrm{Re}} = \cos^{-1}\frac{\mathrm{Re}}{\mathrm{MTF}}$$

Similarly in one dimension

$$\mathrm{OTF}(\nu_y) = \frac{\int L(y)e^{-i2\pi\nu_y y}\,dy}{\int L(y)\,dy} \tag{139}$$

It is also possible to calculate the OTF directly from the wavefront:

$$\mathrm{OTF}(\nu_y,\nu_z) = \frac{\iint C(u,v)C^*(u - F\lambda\nu_y,\ v - F\lambda\nu_z)\,du\,dv}{\iint B^2(u,v)\,du\,dv} \tag{140}$$

where C^* = complex conjugate of C
C = pupil function defined in Par. 24:

$$C(u,v) = B(u,v)e^{-i(2\pi/\lambda)\phi(u,v)}$$

u, v = coordinates in pupil
F = radius of reference sphere

The numerical evaluation of Eq. (140) can be carried out as follows. The aperture is divided into an array of $s \times s = s^2$ square (or rectangular) areas and a ray is traced through the center of each; typically $s = 20$. Then ϕ_{ab} is the OPD of the ray through area a, b and $\bar{Y}_{ab}$ and $\bar{Z}_{ab}$ are the orthogonal components of the ray aberration of the ray through a, b.

Then the optical transfer function is computed from

$$\mathrm{OTF}(\alpha,\beta) = \frac{1}{N}\sum_{a=1}^{s-\alpha}\sum_{b=1}^{s-\beta}[\mathrm{soa}\,P_u\,\mathrm{soa}\,P_v\,(\cos\Psi - i\sin\Psi)] \tag{141}$$

where u, v = pupil coordinates
α, β = normalized spatial frequencies; $\alpha = 1, 2, 3, \ldots, s$; $\beta = 1, 2, 3, \ldots, s$; actual frequencies are

$$\frac{\alpha}{s}\frac{2\mathrm{NA}}{\lambda} \quad \text{and} \quad \frac{\beta}{s}\frac{2\mathrm{NA}}{\lambda}$$

NA = numerical aperture = $n \sin U$
λ = wavelength
N = number of area elements in vignetted pupil ("vignetted" by itself displaced by α, β area units)

$$\mathrm{soa}\,P = \frac{\sin P}{P}$$

$$\Psi = \frac{2\pi}{\lambda}(\phi_{ab} - \phi_{cd})$$

$$c = a + \alpha \qquad d = b + \beta$$

$$P_u = -\frac{\pi}{\lambda}\frac{2\mathrm{NA}}{s}(\bar{Y}_{ab} - \bar{Y}_{cd})$$

$$P_v = -\frac{\pi}{\lambda}\frac{2\mathrm{NA}}{s}(\bar{Z}_{ab} - \bar{Z}_{cd})$$

2NA/s is the angular size of a pupil element. Note that

$$P_u = \tfrac{1}{2}\Psi_u\,\Delta u = \frac{1}{2}\frac{2\pi}{\lambda}\left(\frac{\partial\phi_{ab}}{\partial u} - \frac{\partial\phi_{cd}}{\partial u}\right)\Delta u$$
$$= \frac{\pi}{\lambda}\left(\frac{-\bar{Y}_{ab}}{F} + \frac{\bar{Y}_{cd}}{F}\right)\frac{\text{pupil diam}}{s}$$

34. Specific Modulation Transfer Functions.† The MTF graphs in this section are plotted as functions of ν_0, the limiting cutoff frequency. An optical system will not transmit spatial frequencies higher than ν_0, which can be evaluated from

$$\nu_0 = \frac{2\text{NA}}{\lambda} = \frac{1}{\lambda(f\text{ no.})} \tag{142}$$

where NA $= n \sin U$ = numerical aperture of optical system
λ = wavelength of light
f no. = effective speed or relative aperture of system

The limiting frequency can also be expressed in terms of the angular frequency of the object by $\nu_0 = D/\lambda$ c/rad, where D is the effective clear aperture of the system.

Perfect System. The MTF of an optical system with no oberrations and with a uniformly transmitting circular aperture is given by

$$\text{MTF}(\nu) = \frac{2}{\pi}(\phi - \cos\phi \sin\phi)$$

$$\phi = \cos^{-1}\frac{\lambda\nu}{2\text{NA}} \tag{143}$$

Equation (143) is plotted in Fig. 30.

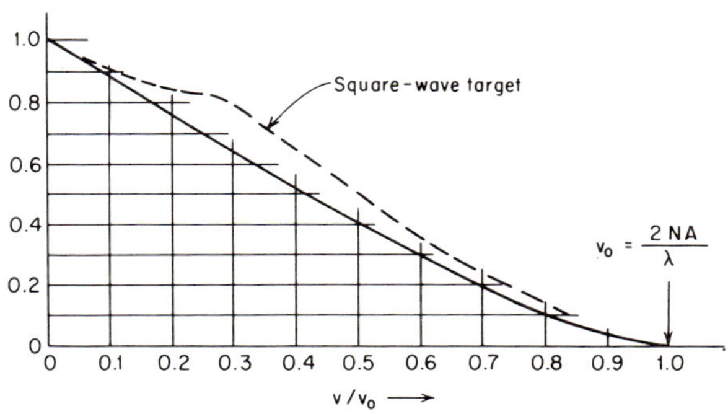

Fig. 30 The modulation transfer function of an aberration-free optical system with a circular aperture. [*After Smith* (1966).]

The MTF of a slit or rectangular aperture is given by

$$\text{MTF}(\nu) = 1 - \frac{\nu}{\nu_0} = 1 - \frac{\nu\lambda}{2\text{NA}}$$

Annular Aperture. In the presence of a central obscuration of the pupil, the MTF of an otherwise aberration-free system is reduced at low frequencies and increased at high frequencies. This is shown in Fig. 31. Note that this is a form of apodization.

Defocused System. The effects of various amounts of defocus of the image are shown in Fig. 32. The focus shift is given as a function of sin U so that the graph can be applied to any system. If the defocusing is large, say to the order of $4\lambda/(n \sin^2 U)$ or greater, the diffraction effects often can be neglected and

$$\text{MTF}(\nu) = \frac{2J_1(\pi B\nu)}{\pi B\nu} = \frac{J_1(2\pi\delta\text{NA}\nu)}{\pi\delta\text{NA}\nu} \tag{144}$$

† References for this paragraph are Barakat (1964), Barakat and Houston (1964, 1966), Linfoot (1964), O'Neill (1956, 1963), and Stephens and Sutton (1968).

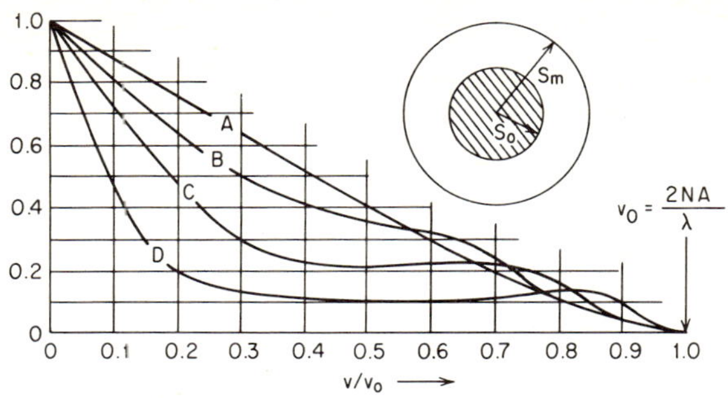

Fig. 31 The modulation transfer function of an aberration-free optical system with an annular aperture. Curve *A*, $S_o/S_m = 0.0$; curve *B*, $S_o/S_m = 0.25$; curve *C*, $S_o/S_m = 0.5$; curve *D*, $S_o/S_m = 0.75$. [*After Smith* (1966).]

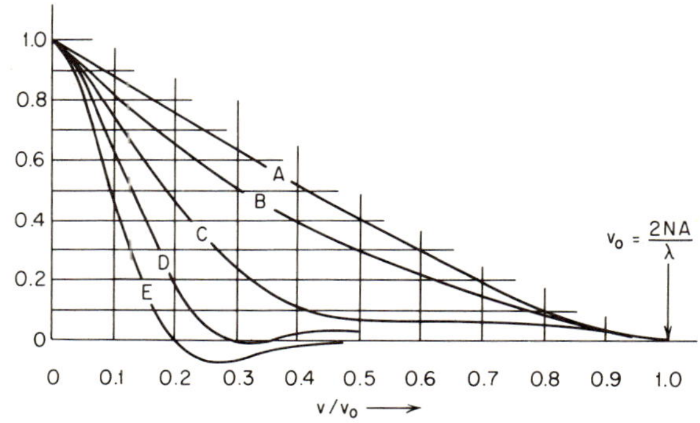

Fig. 32 The effect of defocusing on the MTF of an aberration-free optical system. Curve *A*, in focus, OPD = 0; curve *B*, defocus = $\lambda/(2N \sin^2 U)$, OPD = $\lambda/4$; curve *C*, defocus = $\lambda/(N \sin^2 U)$, OPD = $\lambda/2$; curve *D*, defocus = $3\lambda/(2N \sin^2 U)$, OPD = $3\lambda/4$; curve *E*, defocus = $2\lambda/(N \sin^2 U)$, OPD = λ. [*After Smith* (1966).]

where $J_1(\cdot)$ = first-order Bessel function (see Par. 25)
B = diameter of blur spot produced by defocusing
δ = longitudinal defocusing
NA = $n \sin U$ = numerical aperture
ν = spatial frequency, cycles per unit length

Equation (144) is the MTF for a system producing an image of a point which is a uniformly illuminated circular disk. A system whose image is a uniformly illuminated slit or band of light has an MTF given by

$$\text{MTF}(\nu) = \frac{\sin \pi W \nu}{\pi W \nu} \tag{145}$$

where W is the width of the slit or band and the "cutoff" frequency at which MTF drops to zero is obviously equal to 1/W. This is useful in the study of blurs caused by image motion.

Spherical Aberration. The MTF of a system with third-order spherical aberration is shown in Fig. 33; note that the focus for this graph is optimized by locating the reference plane midway between the marginal and paraxial foci. The MTF for zonal spherical aberration is shown in Fig. 34; the marginal spherical is assumed to be corrected and the focal plane is located 0.8LA_z from the paraxial focus.

Coma. The MTF for coma is shown in Fig. 35. Because coma is an asymmetrical aberration, the response is different in each meridian. The reference point for this graph was located above the principle-ray intersection.

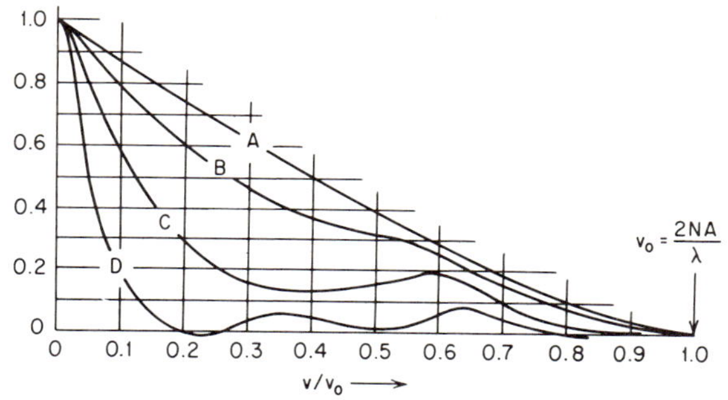

Fig. 33 The effect of third-order spherical aberration on the MTF when the reference plane is midway between the marginal and paraxial foci. Curve *A*, $LA_m = 0$, $OPD = 0$; curve *B*, $LA_m = 4\lambda/(N \sin^2 U)$, $OPD = \lambda/4$; curve *C*, $LA_m = 8\lambda/(N \sin^2 U)$, $OPD = \lambda/2$; curve *D*, $LA_m = 16\lambda/(N \sin^2 U)$, $OPD = \lambda$ [*After Smith* (1966).]

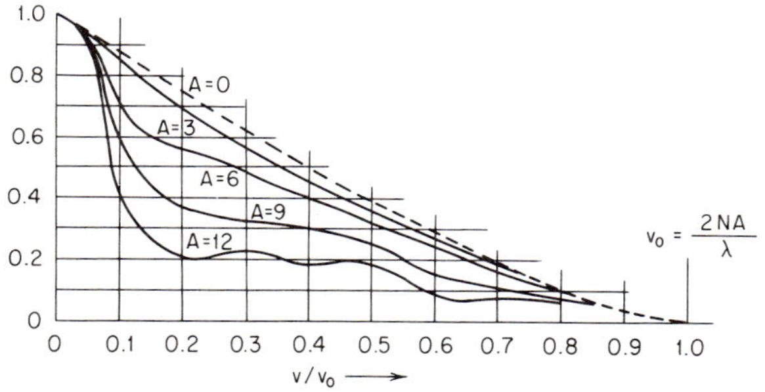

Fig. 34 The effect of residual zonal spherical aberration on the MTF. $LA_z = A\lambda/(\sin^2 U_m)$, $TA_{max} = 0.7436A\lambda/(\sin U_m)$, defocus $\delta l = 0.8LA_z = 0.8A\lambda/(\sin^2 U_m)$.

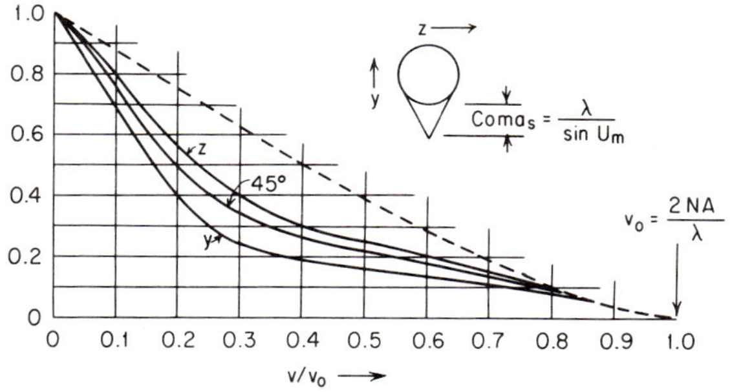

Fig. 35 The effect of third-order coma on the MTF.

Rippled Wavefront. There is a strong similarity between the MTF curves for the amounts of the various low-order aberrations corresponding to the Rayleigh quarter-wave limit, as can be seen by inspecting Figs. 32 to 35. The effect of similar amounts of higher-order aberrations is indicated in Fig. 36, which shows the response of a system whose emerging wavefront is concentrically rippled.

35. Response to Nonsinusoidal Targets (Coltman, 1954) The OTF and the MTF apply (by definition) to the imagery of an object or target whose luminance can be described by a sine function or by a combination of sine functions. They cannot be applied directly to other forms of targets.

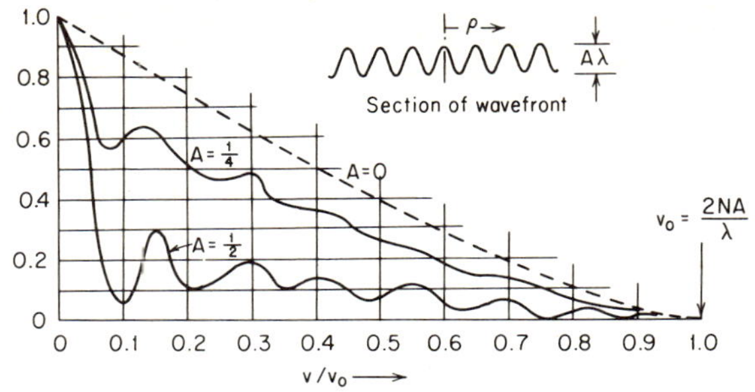

Fig. 36 The effect of a rippled wavefront on the MTF. The wavefront equation is $\mathrm{OPD}(\rho) = \frac{1}{2}A\lambda \cos 7\pi\rho$.

A common form of target used in testing optical systems is the square-wave, or castellated, target, in which the pattern consists of alternating bright and dark bars of equal width. When the MTF, that is, the sine-wave response, is known, the modulation transfer of a square wave can be calculated from

$$S(\nu) = \frac{4}{\pi}\left[M(\nu) - \frac{M(3\nu)}{3} + \frac{M(5\nu)}{5} - \frac{M(7\nu)}{7} + \cdots\right] \tag{146}$$

where $S(\nu)$ is the square-wave transfer factor for frequency ν and $M(\nu)$ is the MTF(ν). Conversely

$$M(\nu) = \frac{\pi}{4}\left[S(\nu) + \frac{S(3\nu)}{3} - \frac{S(5\nu)}{5} + \frac{S(7\nu)}{7} \cdots\right] \tag{147}$$

In general the response, or modulation transfer factor, is greater for a square-wave than for a sine-wave target. For example, the factor for a perfect optical system for frequencies between $0.25v_0$ and $0.5v_0$ is about 0.1 higher for a square-wave target than for a sinusoidal target.

The above applies to targets of considerable extent. The most common form of bar, i.e., square-wave, target consists of only three bright bars on an extended dark background (or the reverse). If the centerline spacing of the bars is taken as the reciprocal of the "frequency" of the target, the modulation transfer factor is again higher than that indicated by the sine-wave MTF. This is because the frequency content of a three-bar pattern is heavily concentrated in frequencies lower than the basic "frequency" of the target. Thus for a perfect lens, the three-bar response is about 0.07 at v_0, the cutoff frequency at which the sine-wave MTF is zero (Kelly, 1965).

PRISMS, PLATES, AND MIRRORS

36. Dispersing Prisms The angular deviation of a refracting prism (Fig. 37) is given by

$$D = I_1 + I_2' - A$$

where D = deviation of ray passing through prism
A = vertex angle of prism
I_1 = angle of incidence at first surface
I_2' = angle of refraction at second surface

With I_1 and A given and n as the index, I_2' can be evaluated from

$$\sin I_1' = \frac{1}{n}\sin I_1$$
$$I_2 = A - I_1'$$
$$\sin I_2' = n \sin I_2$$

The preceding can be combined to yield

$$D = I_1 - A + \sin^{-1} [(n^2 - \sin^2 I_1)^{1/2} \sin A - \cos A \sin I_1] \tag{148}$$

The dispersion of the prism is

$$dD = \frac{\cos I_2 \tan I_1' + \sin I_2}{\cos I_2'} dn \tag{149}$$

If both sides of Eq. (149) are divided by $d\lambda$, the angular dispersion with respect to wavelength $dD/d\lambda$ is given as a function of the (index) dispersion of the prism material $dn/d\lambda$.

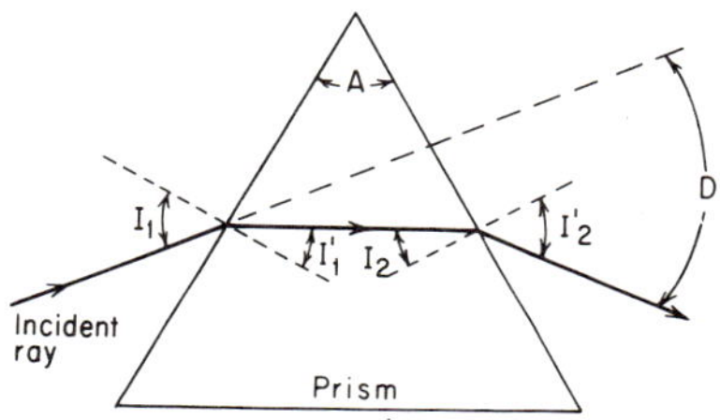

Fig. 37 The deviation of light by a refracting prism. [*After Smith* (1966).]

37. Thin Dispersing Prisms If the angle A is small, Eq. (148) can be expressed as

$$D \approx A(n-1)\left[1 + \frac{I^2(n+1)}{2n}\right] \tag{150}$$

where $I = I_1$ and the powers of I above the third order are neglected. If both A and I are small, the deviation is given by

$$D = A(n-1) \tag{151}$$

and the dispersion becomes

$$dD = D\frac{dn}{n-1} = \frac{D}{V} \tag{152}$$

where V is the reciprocal relative dispersion of the material; in the visible region V is the Abbe ν value, $(n_D - 1)/(n_F - n_C)$.

An achromatic prism (Fig. 38*a*), which deviates light rays through an angle D without dispersing the different wavelengths, will result when two prisms with

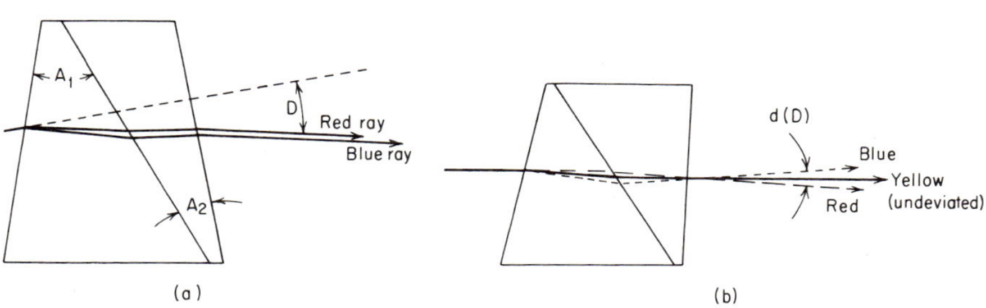

Fig. 38 Compound prisms: (a) achromatic prism; (b) direct-vision prism (dispersing). [*After Smith* (1966).]

the following vertex angles are combined:

$$A_1 = \frac{DV_1}{(n_1 - 1)(V_1 - V_2)} \qquad A_2 = \frac{DV_2}{(n_2 - 1)(V_2 - V_1)} \tag{153}$$

where n_1 and V_1 are the index and ν value of the material with vertex angle A_1; similarly n_2 and A_2 apply to the material associated with A_2.

A direct-vision prism (Fig. 38*b*) will disperse the incident light into an angular spread dD without deviating the ray of the central wavelength. The vertex angles of the component prisms are

$$A_1 = \frac{dD\, V_1V_2}{(n_1 - 1)(V_2 - V_1)} \qquad A_2 = \frac{dD\, V_1V_2}{(n_2 - 1)(V_1 - V_2)} \tag{154}$$

Note that Eqs. (153) and (154) are based on the approximate thin-prism equations (151) and (152) and that ray tracing by Snell's law is required for exact results when A and I are not infinitesimal. Also, each component may be composed of several prisms; for example, A_1 may be the sum of the vertex angles of several prisms made of material n_1, V_1.

38. Plane-Parallel Plates *Displacements.* When a paraxial image is formed through a plane-parallel plate (Fig. 39) of thickness t and index n, the image is

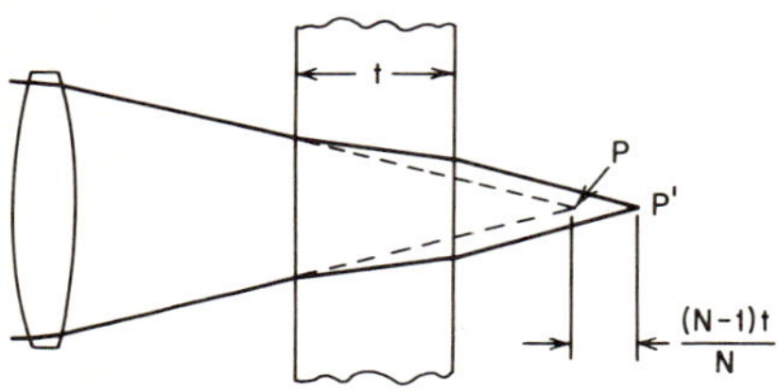

Fig. 39 The longitudinal displacement of an image by a plane-parallel plate. [*After Smith* (1966).]

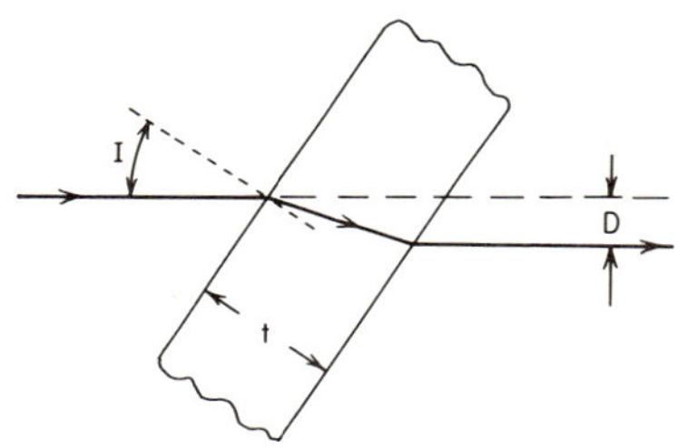

Fig. 40 The lateral displacement of a ray by a tilted plane-parallel plate. [*After Smith* (1966).]

longitudinally displaced by

$$\delta = \frac{(n-1)t}{n} \tag{155}$$

The equivalent air thickness of the plate is the actual thickness less this image shift, or

$$t_{\text{equiv}} = t - \delta = \frac{t}{n} \tag{156}$$

If the plate is rotated or tilted, as in Fig. 40, by an angle I with respect to the axis (or a ray), the axis (or ray) is displaced laterally by

$$\begin{aligned} D &= t \cos I(\tan I - \tan I') \\ &= t \sin I \left[1 - \left(\frac{1 - \sin^2 I}{n^2 - \sin^2 I} \right)^{1/2} \right] \end{aligned} \tag{157}$$

where $\sin I' = (n \sin I)/n'$. Equation (157) can be expanded into a power series in I to give

$$\begin{aligned} D = \frac{tI(n-1)}{n} \Bigg[1 &+ \frac{I^2}{6n^2}(-n^2 + 3n + 3) \\ &+ \frac{I^4}{120n^4}(n^4 - 15n^3 - 15n^2 + 45n + 45) + \cdots \Bigg] \end{aligned} \tag{158}$$

When angle I is very small, this reduces to

$$D \approx \frac{tI(n-1)}{n} \tag{159}$$

Aberrations. Longitudinal spherical aberration:

$$\mathrm{LA} = L' - l' = \frac{t}{n}\left[1 - \frac{n\cos U}{(n^2 - \sin^2 U)^{1/2}}\right] \approx \frac{tu^2(n^2-1)}{2n^3} \quad \text{third order} \tag{160}$$

Longitudinal astigmatism:

$$l_s' - l_t' = \frac{t}{(n^2 - \sin^2 U_p)^{1/2}}\left(\frac{n^2\cos^2 U_p}{n^2 - \sin^2 U_p} - 1\right)$$
$$\approx \frac{tu_p^2(n^2-1)}{n^3} \quad \text{third order} \tag{161}$$

Sagittal coma:

$$CC_s = \frac{tu^2u_p(n^2-1)}{2n^3} \quad \text{third order} \tag{162}$$

Longitudinal axial chromatic aberration:

$$\mathrm{LchA} = l_F' - l_C' = \frac{t(n-1)}{n^2V} \tag{163}$$

Lateral chromatic aberration:

$$\mathrm{TchA} = h_F - h_C = \frac{tu_p(n-1)}{n^2V} \tag{164}$$

where U, u = slope angle of axial ray to axis
U_p, u_p = tilt of plate
t = thickness of plate
n = index of plate material
V = Abbe ν value = $(n_D - 1)/(n_F - n_C)$

See Fig. 41 and note that the lowercase u's are the paraxial angles used in the third-order expressions.

39. Reflecting Prisms and Mirror Systems Reflecting systems are most frequently employed to modify the orientation of an image, e.g., to erect the inverted image formed by a simple telescope. They are also used to displace an image laterally or to redirect the light beam and occasionally to split or combine images.

Orientation. A simple way to determine the orientation of an image after it has passed through a prism or mirror system is to bounce a pencil through it as indicated in Fig. 42. The orientation in both meridians can be checked simultaneously by bouncing a card marked with the arrow and bar symbol as

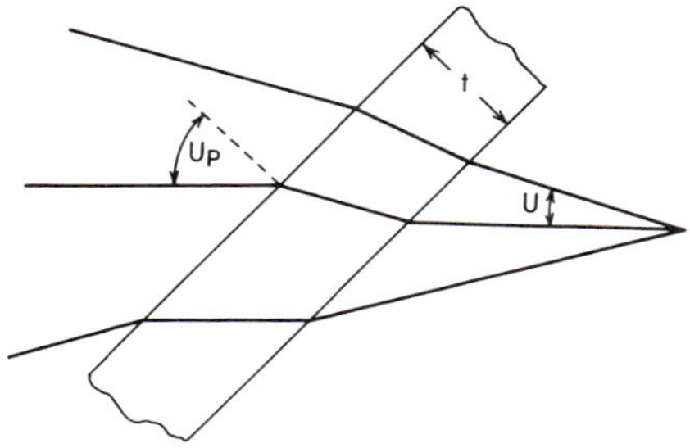

Fig. 41 The quantities used in Eqs. (160) to (164). [*After Smith* (1966).]

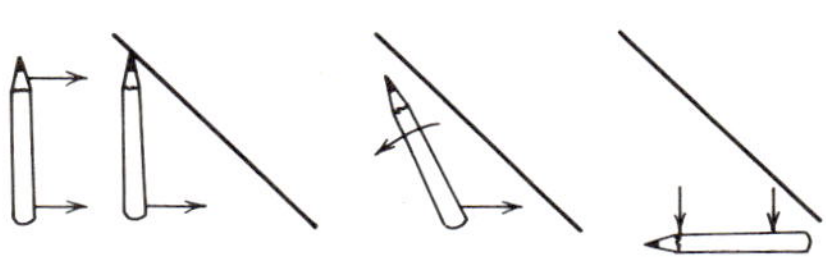

Fig. 42 The bouncing-pencil technique of determining the orientation of a reflected image. [*After Smith* (1966.]

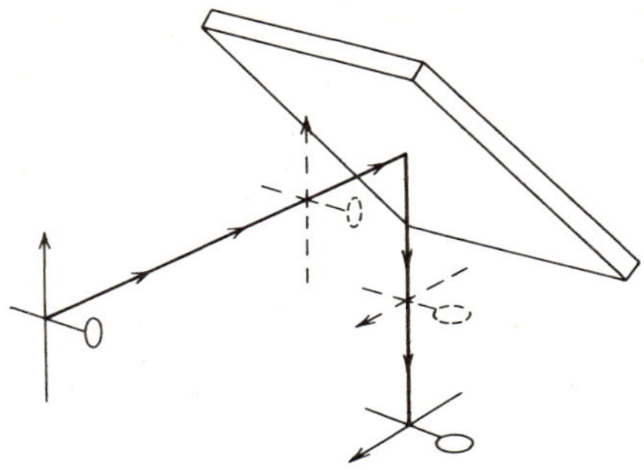

Fig. 43 Image orientation after reflection. [*After Smith* (1966).]

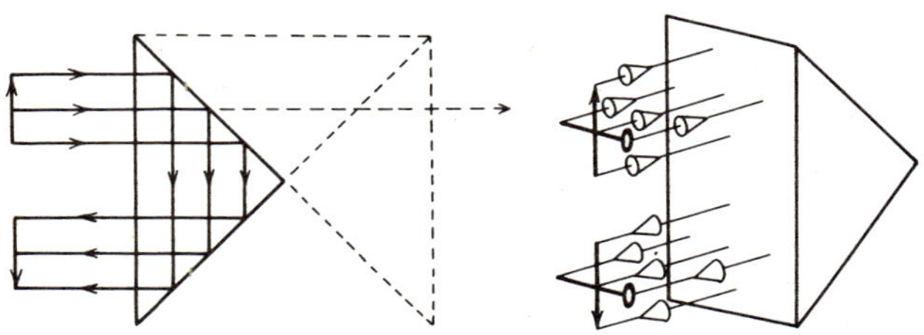

Fig. 44 A prism system can be unfolded (*left*) so that rays can be drawn through the system as continuous straight lines. [*After Smith* (1966).]

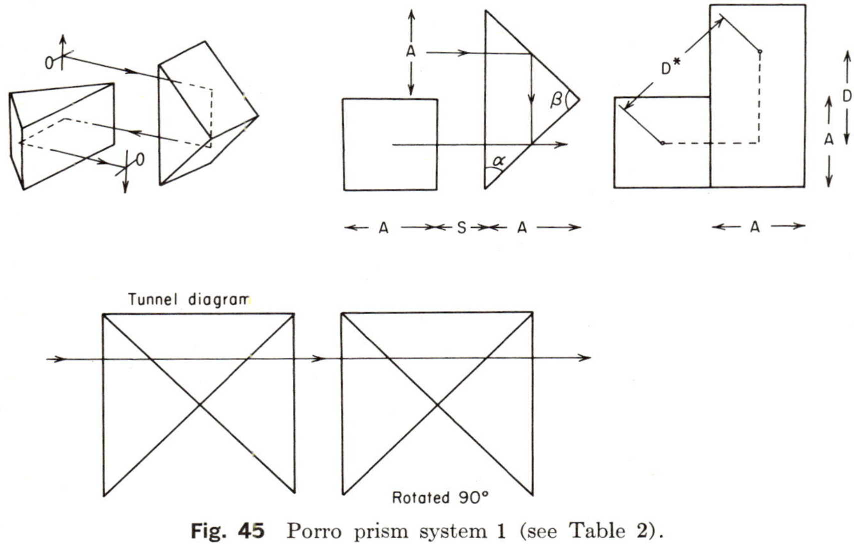

Fig. 45 Porro prism system 1 (see Table 2).

indicated in Fig. 43 and others. Note that both the pencil and the arrow-bar figure are initially oriented so that they lie in the plane of incidence or normal to it. When, as with a rotated Dove prism, subsequent reflections have the plane of incidence in a new, nonorthogonal orientation, the card can be marked with a second arrow-bar figure at such an angle to the original that it lies in the new plane of incidence. The new figure is then bounced through; the final image orientation is of course indicated by the original arrow-bar figure (Walles and Hopkins, 1964; MIL-HDBK-141, 1962).

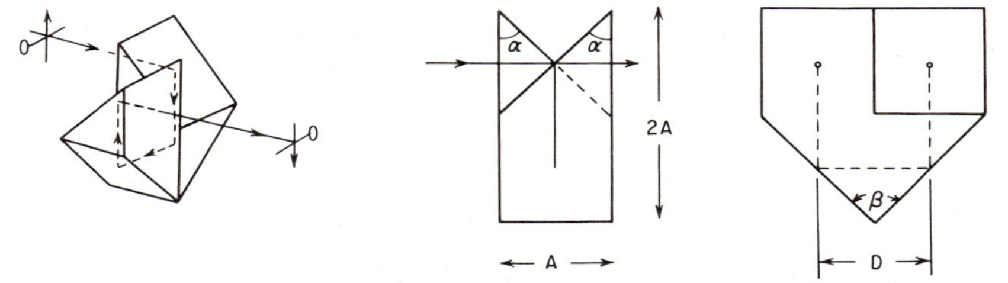

Fig. 46 Porro prism system 2, Abbe modification (see Table 2). Note that the tunnel diagram is basically the same as Fig. 45.

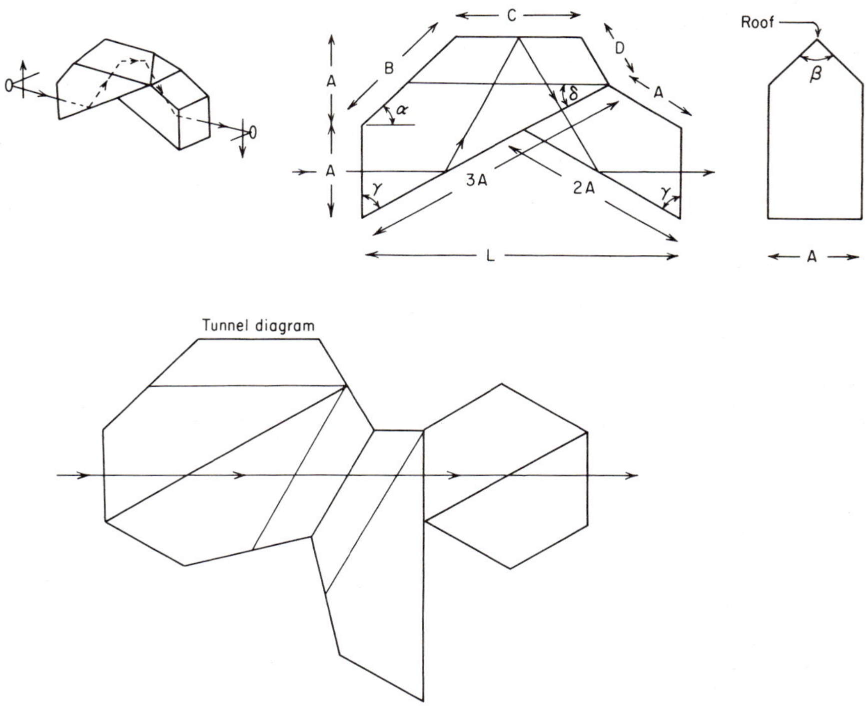

Fig. 47 Abbe prism (see Table 2).

Unfolding. Reflecting systems can be unfolded for the purpose of determining mirror sizes and to aid in visualizing or sketching ray paths through the system. Figure 44 illustrates the process for a Porro prism; the system "axis" is simply straightened at each reflecting surface by rotating the reflected axis and the balance of the system about a folding axis formed by the intersection of the plane of incidence with the surface (in most representations this is simply the line representing the section through the reflecting surface). A drawing of an unfolded prism system is called a *tunnel diagram;* its great usefulness derives from the fact that ray paths can be accurately constructed through the system by drawing straight lines on the tunnel diagram. Note that the rays must be refracted at the entrance and exit faces unless the tunnel length is reduced to t/n, the equivalent air thickness.

TABLE 2 Basic Prism Systems†

Dimensions	Function	Comment
Porro Prism System 1 (Fig. 45)		
$D = A \quad \alpha = 45°$ $D^* = A\sqrt{2} \quad \beta = 90°$ $T = 4A + s$ $\frac{T}{n} = \frac{4A}{n} + s = 2.637A + s$ $F_1 \approx \pm \tan^{-1} \frac{nA}{2T} = \pm 10.84° \quad n = 1.517,\ s = 0$ $F_2 \approx \pm n\left(45° - \sin^{-1}\frac{1}{n}\right) = \pm 5.71° \quad n = 1.517$	Erects and displaces image	Widely used as an erecting system in binoculars and telescopes
Porro Prism System 2, Abbe Modification (Fig. 46)		
$D = A \quad \alpha = 45°$ $T = 4A \quad \beta = 90°$ $\frac{T}{n} = \frac{4A}{n} = 2.637A$ $F_1 \approx \pm \tan^{-1} \frac{nA}{2T} = \pm 10.84° \quad n = 1.517$ $F_2 \approx \pm n\left(45° - \sin^{-1}\frac{1}{n}\right) = \pm 5.71° \quad n = 1.517$	Erects and displaces image	Used as an erecting system in binoculars, telescopes, and microscopes, especially for compactness
Abbe Prism (Fig. 47)		
$B = A\sqrt{2} \quad D = 0.5774A \quad \alpha = 45°$ $C = 1.3094A \quad L = 3.4644A \quad \beta = 90°$ $\gamma = 60°$ $\delta = 30°$ $T = 5.1962A$ $\frac{T}{n} = \frac{5.1962A}{n} = 3.425A$	Erects image without deviation or displacement	May be inserted into an in-line system with almost no change in the system (note that L is almost exactly equal to T/n)

$F_1 \approx \pm \tan^{-1} \frac{n}{10.3924} = \pm 8.35°$ $n = 1.517$		
Reversion Prism (Fig. 48, see also Fig. 47)		
$B = A\sqrt{2}$ $D = 0.5774A$ $\alpha = 45°$ $C = 1.3094A$ $L = 3.4644A$ $\beta = 180°$ $\gamma = 60°$ $\delta = 30°$ $T = 5.1962A$ $\frac{T}{n} = \frac{5.1962A}{n} = 3.425A$ $F_1 \approx \pm \tan^{-1} \frac{n}{10.3924} = \pm 8.35°$ $n = 1.517$	Inverts (or reverts) image without deviation or displacement	The same as the Abbe prism except that it has no roof; the top surface (dimension C in Fig. 47) is flat (and silvered), angle β is 180°; this surface must be silvered since, without the roof, the angle of incidence is too small to cause total internal reflection; if this prism is rotated, the image will rotate at twice the angular rate at which the prism is turned
Right-Angle Prism (45-90-45° Prism) (Fig. 49)		
$B = A\sqrt{2}$ $\alpha = 45°$ $T = A$ $\beta = 90°$ $\frac{T}{n} = \frac{A}{n} = 0.659A$ $F_1 \approx \pm \tan^{-1} \frac{n}{2} = \pm 42.7°$ $n = 1.517$ $F_2 \approx \pm n\left(45° - \sin^{-1} \frac{1}{n}\right) = \pm 5.71°$ $n = 1.517$	Deviates the beam 90°	Basic building block for many prism systems
Right-Angle Prism (Single Porro Prism) (Fig. 50)		
$T = 2A$ $\alpha = 45°$ $\frac{T}{n} = \frac{2A}{n} = 1.319A$ $\beta = 90°$ $F_1 \approx \pm \tan^{-1} \frac{n}{4} = \pm 21.6°$ $n = 1.517$ $F_2 \approx \pm n\left(45° - \sin^{-1} \frac{1}{n}\right) = \pm 5.71°$ $n = 1.517$	Deviates beam 180°	This is a constant-deviation prism and a retrodirector; the deviation angle of 180° is not a function of the ray entrance angle; the ray is directed back on a path parallel (in a plan view) to its original path

TABLE 2 Basic Prism Systems (Continued)

Dimensions	Function	Comment
Amici Prism (Roof Prism) (Fig. 51)		
$B = A\sqrt{2} \quad \alpha = 45°$ $C = \frac{A}{\sqrt{8}} \quad \beta = 90°$ $T = 1.707A$ $\frac{T}{n} = \frac{1.707A}{n} = 1.125A$ $F_1 \approx \pm \tan^{-1} \frac{n}{3.414} = \pm 23.96° \quad n = 1.517$	Deviates beam 90°	Roof adds an extra inversion to the effect of the right-angle prism
Pentaprism (Fig. 52)		
$B = 1.0824A \quad \alpha = 45°$ $T = 3.4142A \quad \beta = 90°$ $\frac{T}{n} = \frac{3.4142A}{n} = 2.251A$ $F_1 \approx \pm \tan^{-1} \frac{n}{6.828} = \pm 12.52° \quad n = 1.517$	Deviates beam 90°; does not change image orientation	A constant-deviation prism; the reflecting faces must be silvered or aluminized; sometimes made as a penta-roof with one of the reflecting faces as a roof in order to invert the image
Dove Prism (Fig. 53)		
$B = A\left[1 + \frac{(n^2 - \sin^2 \alpha)^{1/2} + \sin \alpha}{(n^2 - \sin^2 \alpha)^{1/2} - \sin \alpha}\right] = 4.2271A$ $T = \frac{nA}{\sin \alpha (n^2 - \sin^2 \alpha)^{1/2} - \sin \alpha} = 3.3787A \quad \alpha = 45°$ $\frac{T}{n} = \frac{3.3787A}{n} = 2.227A$ $F_1 \approx \pm \tan^{-1} \frac{n}{6.7574} = \pm 12.65° \quad n = 1.517$	Inverts image without deviation or displacement	The tunnel diagram indicates that this prism is equivalent to a tilted plane-parallel plate, which will introduce astigmatism into the image if used in a convergent or divergent beam; if the prism is rotated about the axis, the image rotates at twice the angular rate of the prism; the reflecting face is occasionally made as a roof reflector; a double Dove prism is two matched Dove prisms whose silvered reflecting faces are cemented together to obtain a more favorable aperture-to-length ratio

Dimensions	Function	Remarks
Pechan Prism (Fig. 54)		
$B = 1.8284A$ $\quad L = 1.2071A$ $\quad \alpha = 45°$ $C = 1.0824A$ $\quad T = 4.6213A$ $\quad \beta = 22\frac{1}{2}°$ $x = 0.2071A$ $\quad \gamma = 67\frac{1}{2}°$ $\frac{T}{n} = \frac{4.6213A}{n} = 3.046A$ $F_1 \approx \pm \tan^{-1}\frac{n}{9.2426} = \pm 9.45° \quad n = 1.517$ $F_2 \approx \pm n\left(45° - \sin^{-1}\frac{1}{n}\right) = \pm 5.17° \quad n = 1.517$	Inverts image without deviating or displacing beam	When this prism is rotated, the image rotates at twice the angular rate; surfaces C are silvered or aluminized; can be made with an unsilvered roof for one of the C faces to be an erecting prism
Schmidt Prism (Fig. 55)		
$B = 1.4142A$ $\quad \alpha = 45°$ $C = 1.0824A$ $\quad \beta = 90°$ $x = 0.1A$ (arbitrary) $y = 1.8478x = 0.1848A$ $T = 3.4142A$ $\frac{T}{n} = \frac{3.4142A}{n} = 2.251A$ $F_1 \approx \pm \tan^{-1}\frac{n}{6.8284} = \pm 12.52° \quad n = 1.517$ $F_2 \approx \pm n\left(45° - \sin^{-1}\frac{1}{n}\right) = \pm 5.17° \quad n = 1.517$	Erects image and deviates beam 45°	

† Prism dimensions are formulated in terms of a reference dimension A, which has been selected as a dimension of the prism closely related to the limiting aperture of the system. The actual clear aperture of the prism system will be equal to A minus any bevels or chamfers which intrude in the aperture. The length T of the mechanical path, i.e., the path of the axial ray, through the system and the equivalent air path T/n are given. Two first-order approximations to the zero-aperture limiting field of view are given. F_1 is the maximum possible total angular field which an aperture of size A at an equivalent air distance of T/n will permit to pass. F_2 is the size of the field as limited by total internal reflection. Note that although F_2 is given as a total field ($\pm$), the field is usually only limited in one direction by total internal reflection. Since the expressions for F_1 and F_2 are approximations, Snell's law should be applied to determine the exact values where required. The table also lists numerical values for T/n, F_1, and F_2 assuming a prism index of 1.517; these values of F_1 and F_2 are exact and will not agree with the approximate expressions.

Tabulation of Prism Systems. Table 2 presents the salient features of the basic prism systems. There are too many systems to include them all here; those shown illustrate basic features, and the reader may readily combine features or systems to meet his own particular requirements. MIL-HDBK-141 (1962) contains many excellent prism-system illustrations. Each prism in the table is illustrated in a figure; the table gives a brief description of its function and the important dimensional relationships. Additional prisms are shown in Figs. 56 to 58. The reader's attention is called to the explanatory footnote at the end of Table 2.

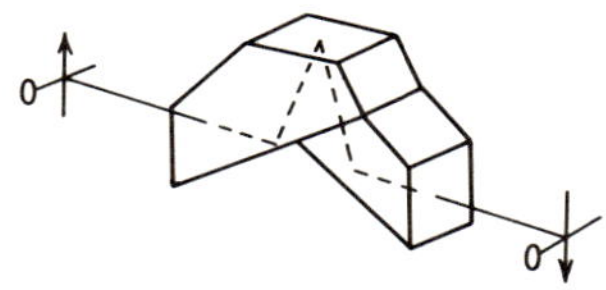

Fig. 48 Reversion prism (see Table 2).

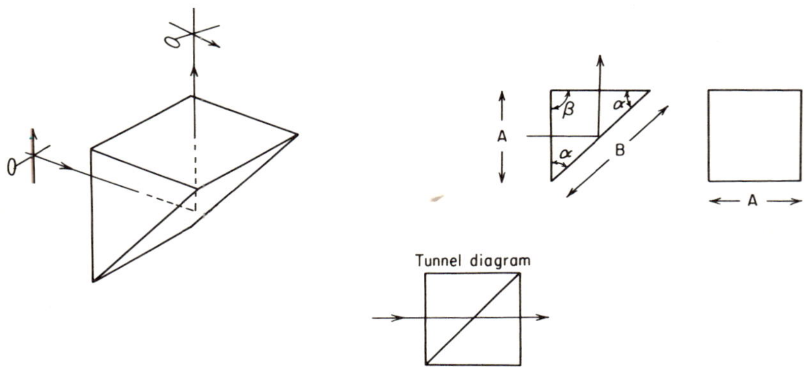

Fig. 49 Right-angle prism (see Table 2).

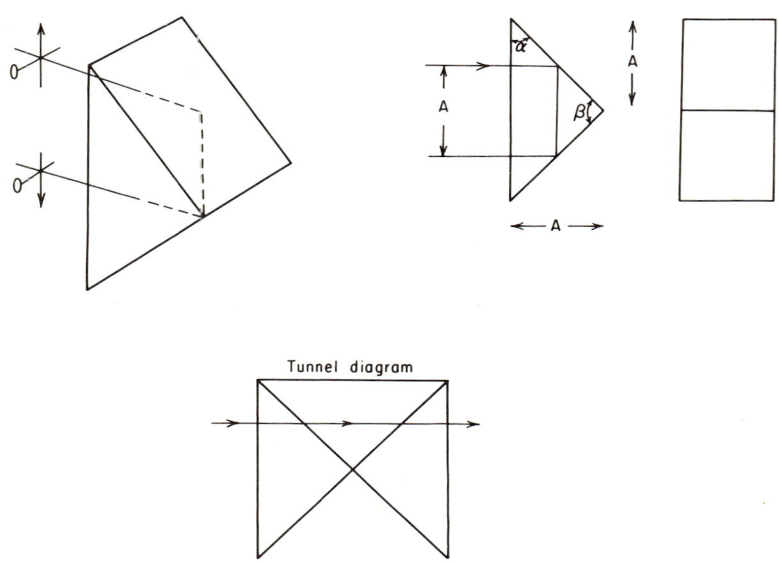

Fig. 50 Right-angle prism, or single Porro prism (see Table 2).

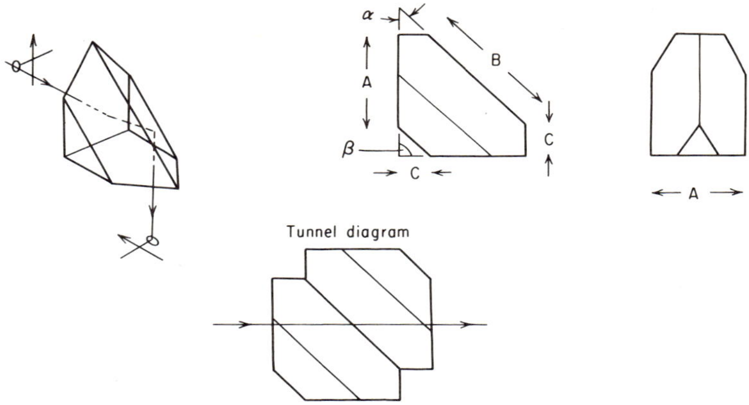

Fig. 51 Amici, or roof, prism (see Table 2).

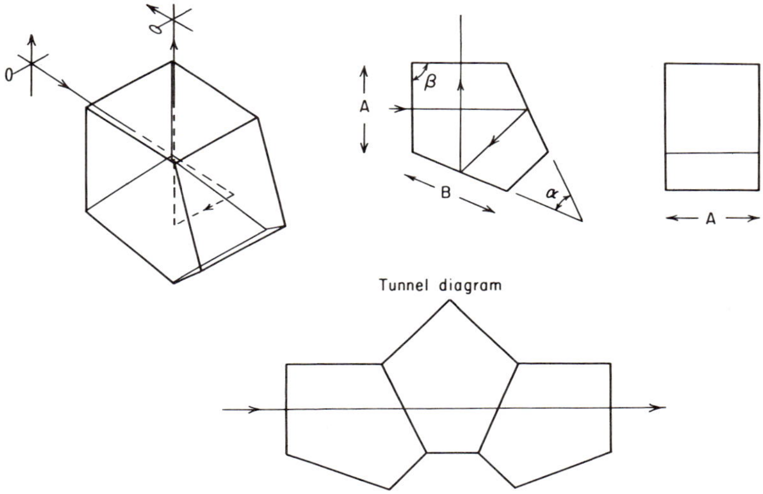

Fig. 52 Pentaprism (see Table 2).

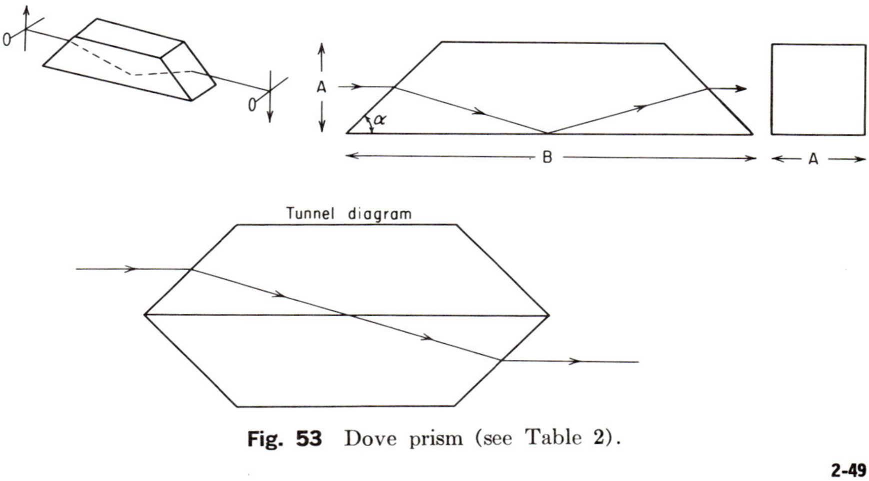

Fig. 53 Dove prism (see Table 2).

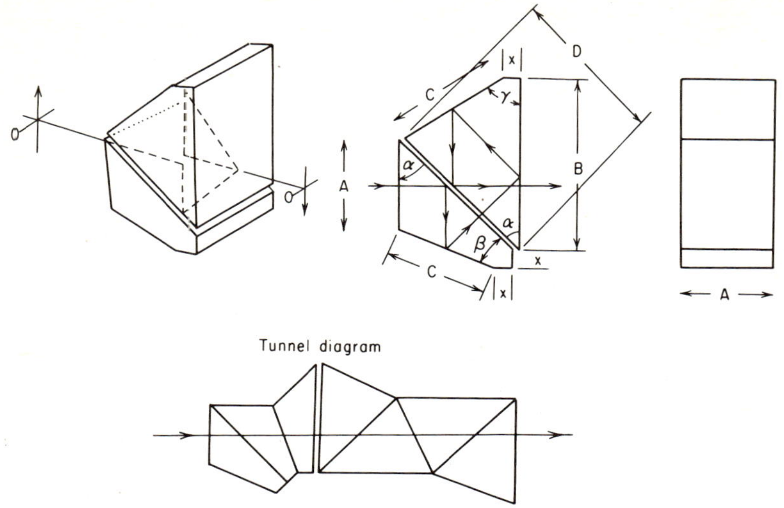

Fig. 54 Pechan prism (see Table 2).

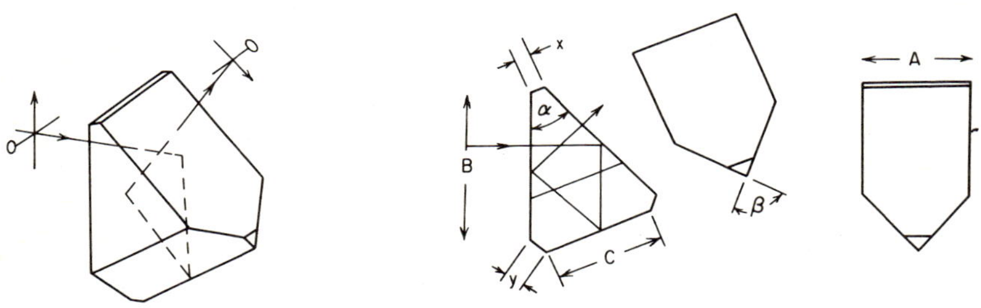

Fig. 55 Schmidt prism (see Table 2).

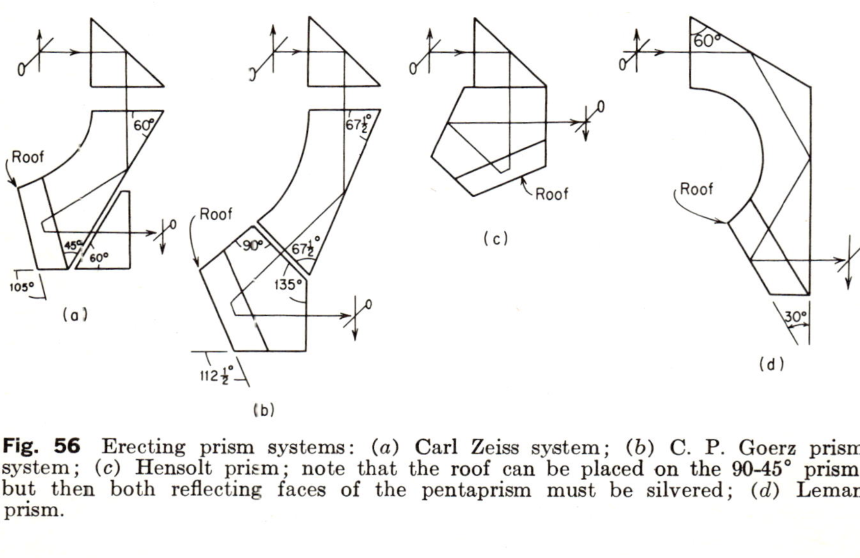

Fig. 56 Erecting prism systems: (*a*) Carl Zeiss system; (*b*) C. P. Goerz prism system; (*c*) Hensolt prism; note that the roof can be placed on the 90-45° prism, but then both reflecting faces of the pentaprism must be silvered; (*d*) Leman prism.

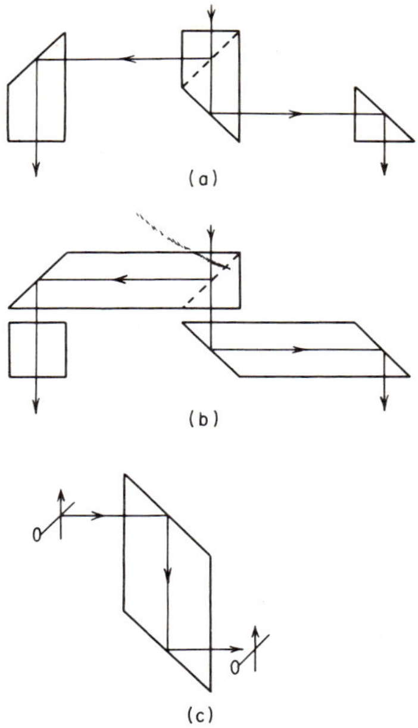

Fig. 57 (*a*) and (*b*) Prism systems used for binocular eyepieces. Interpupillary distances are adjustable in (*a*) by sliding the prisms in and out and in (*b*) by rotating the prisms about the common central axis. (*c*) The rhomboid prism displaces the image but does not affect the orientation of the image.

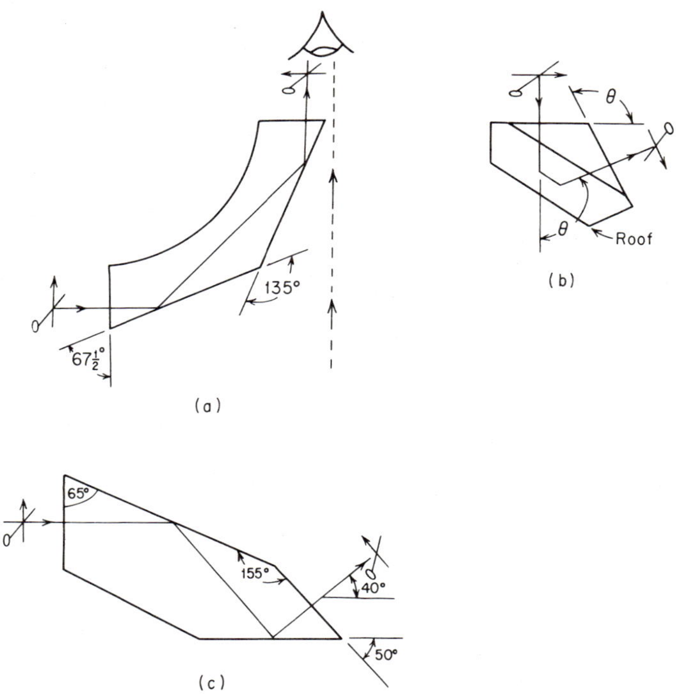

Fig. 58 (*a*) Wollaston prism (camera lucida), (*b*) Frankford Arsenal prism, (*c*) 40° deviation prism.

BASIC OPTICAL SYSTEMS

40. Stops and Apertures (Smith, 1966) *Aperture Stop.* The physical diameter which limits the size of the cone of radiation which an optical system will accept from an axial point on the object.

Entrance Pupil. The image of the aperture stop formed by the optical elements (if any) between the aperture stop and the object; the image of the aperture stop as seen from the object.

Entrance Pupil. The image of the aperture stop formed by the optical elements (if any) which are behind the aperture stop.

Eye Relief. In a visual instrument, e.g., telescope or microscope, the distance from the last optical surface to the (usually external) exit pupil; thus the clearance or relief between the instrument and the eye.

Field Stop. The physical diameter which limits the angular field of view of an optical system.

Entrance Window. The image of the field stop formed by the optical elements between the field stop and the object. If the field stop is located in an image plane, the entrance window will lie in the object plane.

Exit Window. The image of the field stop formed by the optical elements which are behind the field stop.

Although the apertures and field stops in a well-designed system are usually well defined, this is not a universal condition. For the field stop especially, there may be more than one physical diameter which limits the field of view; if these are not in an image plane, they produce vignetting, or gradual reduction or obscuration of the size of the exit pupil as the field angle is increased.

In some systems, especially anamorphic systems, the stops in one meridian may be completely different from those in the other.

41. Telescopes *Astronomical Telescope* (*Fig.* 59*a*). A telescope with a positive objective and a positive eyepiece which has a real internal image and produces a final image which is inverted (and reversed).

Galilean Telescope (*Fig.* 59*b*). A telescope with a positive objective and a (compact) negative eyepiece; the final image is erect. There is no real internal image.

Terrestrial Telescope (*Fig.* 59*c*). An astronomical telescope with an erecting system which produces an erect final image. The erecting system may be a lens system, as shown in Fig. 59*c*, or a prism system, as discussed in Pars. 36 to 39.

A telescope is usually considered to be an afocal system, i.e., one without a focal length. Both object and image are located at infinity. The angular magnification of a telescope is the ratio of the apparent angular size of the image to the angular size of the object. When the object is centered on the axis, this is the ratio of the tangents of the half-field angles.

In an ordinary telescope the angular magnification is given by

$$M = -\frac{f_o}{f_e} \tag{165}$$

where f_o is the focal length of the objective system, f_e is the focal length of the eyepiece system, and a positive magnification indicates an erect image. In an erecting telescope the erector may be considered as part of the objective ($f_o{}^* = f_o s_2/s_1$) or as part of the eyepiece ($f_e{}^* = f_e s_1/s_2$). Thus the magnification of the telescope is

$$M = -\frac{f_o}{f_e}\frac{s_2}{s_1}$$

where s_2/s_1 is the magnification of the erecting system.

Afocal System. For any afocal system, the magnification, the pupil diameters, and the fields of view are related as follows:

$$M = \frac{\mathrm{CA}_1}{\mathrm{CA}_2} = \frac{\tan \alpha_2}{\tan \alpha_1} \tag{166}$$

where M = magnification
CA_1 = entrance-pupil diameter
CA_2 = exit-pupil diameter
α_1 = half-field angle in object space
α_2 = half-field angle in image space
(see Fig. 59*d*).

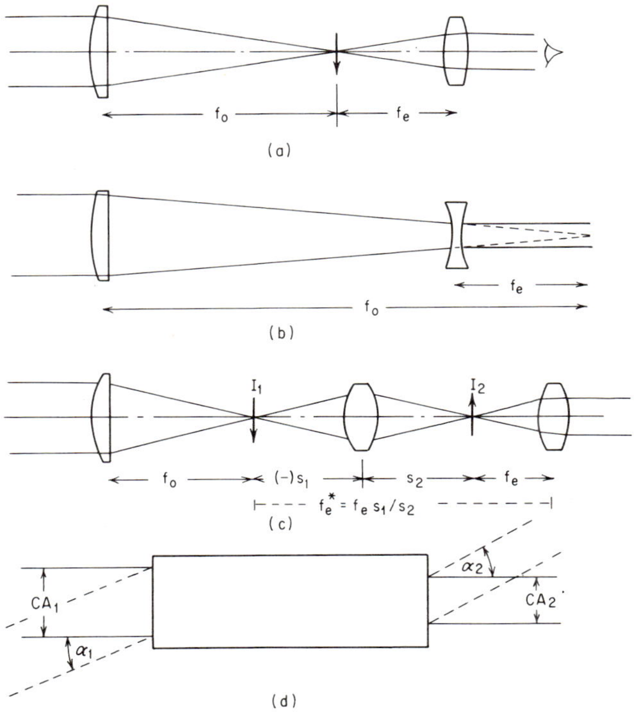

Fig. 59 Schematic diagrams of the basic telescope systems: (*a*) astronomical, (*b*) galilean, (*c*) erecting, (*d*) generalized afocal system.

Relay Lens. A relay lens or system is used to transfer an image longitudinally from one location to another. The center element in Fig. 59*c* is used to relay the image from I_1 to I_2; in the process it erects (inverts and reverses) the image. A chain of several relay lenses may be used to carry an image through a long hole of limited diameter.

Field Lens. A field lens is a (usually positive) lens placed at or near an image plane for the purpose of changing the location of the pupil without greatly changing the power of the system in which it is used. The field lens frequently images the pupil into subsequent elements of the system to prevent the escape of the diverging beams which form the images in the outer portions of the field.

Figure 60 shows a system of alternating field and relay lenses, illustrating both the transference of the image along the system by the relay lenses and the action of the field lenses in imaging (and thus relaying) the pupil from one relay lens to the next. This also illustrates the basic system used in periscopes.

42. Microscopes *Simple Microscope, or Magnifier.* A simple microscope is a single positive component used to magnify an object which is located at or within the focal point. The magnification of a microscope is somewhat arbitrarily defined as the ratio of the apparent angular size of the image to the angular size that

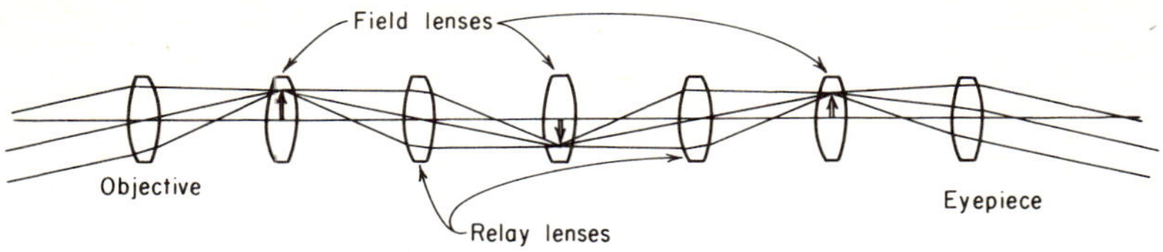

Fig. 60 A series of relay and field lenses can be used to carry an image through a long narrow space, as in a periscope.

the object would subtend if it were viewed from a distance of 10 in (which is assumed to be the distance of most distinct vision).

Under the assumption that the lens-to-eye distance is negligible, the simple microscope of Fig. 61 has a magnification given by

$$M = \frac{10\text{ in}}{f} - \frac{10\text{ in}}{s'} \tag{167}$$

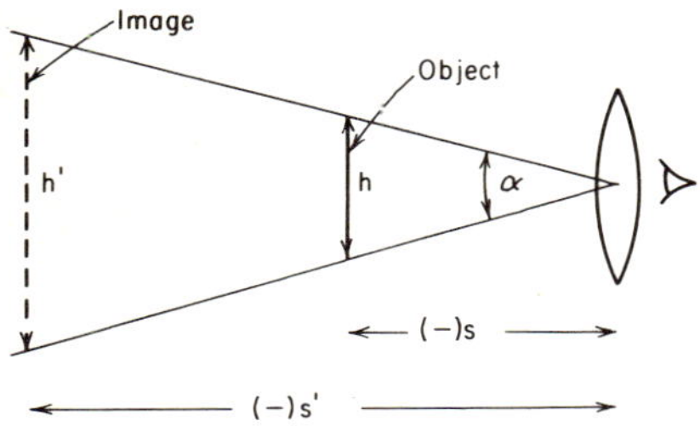

Fig. 61 A simple microscope, or magnifier.

If the object is located at the focal point, so that the image is located at $s' = \infty$, then $M = (10\text{ in})/f$. If the system is focused so that the image appears to be at $s' = -10$ in, then $M = 1 + (10\text{ in})/f$. These are the conventional relationships used to determine the power of magnifiers and eyepieces.

Compound Microscope. A compound microscope consists of an objective which forms an intermediate image and an eyepiece to view the image. The magnification is defined as in the simple microscope and (with reference to Fig. 62) is given by

$$M = \frac{10\text{ in}}{f_{eo}} = \frac{(f_e + f_o - d) \times 10\text{ in}}{f_e f_o} \tag{168}$$

or

$$M = M_o M_e = \frac{s_2}{s_1} \frac{10\text{ in}}{f_e} \tag{169}$$

Note that there is a gray area between the microscope and the telescope where the magnification is not clearly defined. Quite different magnification values are

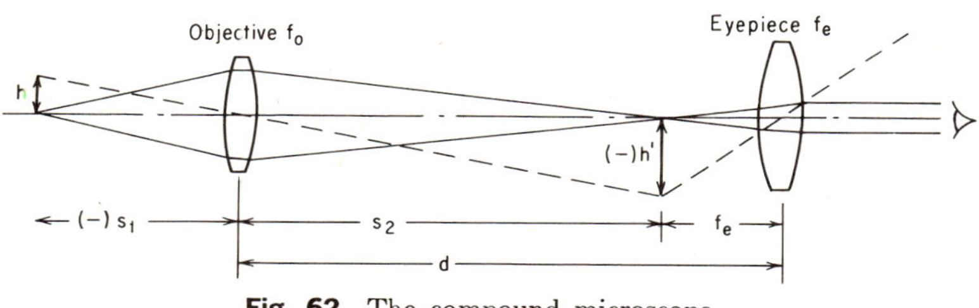

Fig. 62 The compound microscope.

obtained for systems in which the object is close (but not microscopically so), dpending on whether one defines the magnification with respect to the angle subtended by the object in its actual location (as in telescopic magnification) or with respect to the angle which the object would subtend if it were viewed from 10 in away (as in microscopic magnification).

43. Projectors and Condensers In projecting an illuminated image or area, a condensing system is frequently used to collect the usable energy from the light source and to provide smooth uniform illumination from a source which may be extremely nonuniform. Ordinarily the fraction of the radiant power from a source which can be utilized is severely limited. If the source has an average radiance (or luminance) L and the projection system has an exit pupil of area A, and if the illuminated area subtends a solid angle Ω from the pupil, it is obvious that the maximum possible output is $tLA\Omega$ W (or lumens), where t is the transmission efficiency of the system. There is an additional limit on the maximum size pupil which can be filled with light. Using the optical invariant, $I = n(y_p u - y u_p)$, and the sine condition, $(\sin U_0)/(\sin U_k) = \text{const}$, it can be shown that the absolute value of $D\alpha/s$ cannot exceed unity, where D is the usable exit-pupil linear dimension (diameter), α is the angular half field, and s is the source linear dimension. Thus the maximum useful pupil size D is limited to s/α, and in practice a pupil half this size is often difficult to utilize efficiently because of aberrations.

If the radiance (or luminance) of a source is sufficiently uniform, the source can be imaged directly on the area to be illuminated. This practice is followed in arc lamps for motion picture projection. The arc is imaged by an ellipsoidal mirror (to eliminate spherical aberration) almost directly on the film gate, which is imaged on the screen by the projection lens.

When the source is not uniform, even illumination can be obtained by imaging the source in the aperture stop of the projection lens, so that the source is a pupil of the system, as illustrated in Fig. 63.

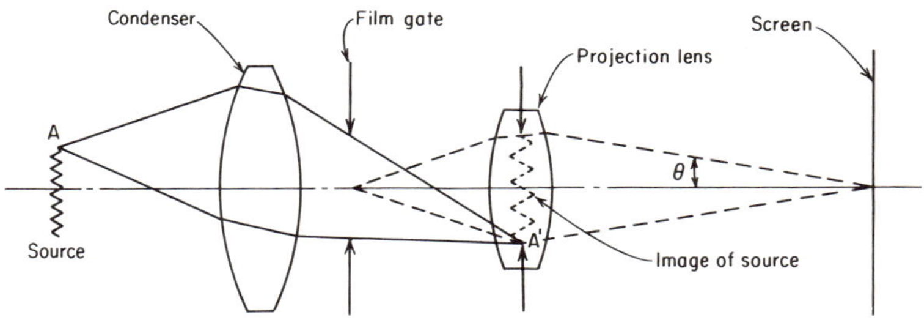

Fig. 63 Schematic of a projection condenser system. [*After Smith* (1966).]

Since the ordinary condenser collects only the radiant energy emitted into one hemisphere, it is often useful to place an auxiliary reflector behind the source to collect the energy emitted into the other hemisphere and redirect it into the primary system. A spherical reflector with its center of curvature at the source will image the source back on itself. Depending on the construction of the source, this may almost double its average radiance by filling in empty areas, or it may simply increase the temperature of an effectively solid source.

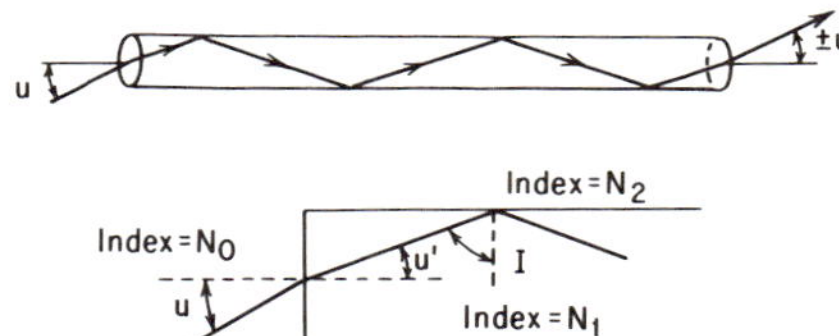

Fig. 64 Light can be very efficiently transmitted through a long polished cylinder by total internal reflection. [*After Smith* (1966).]

44. Light Pipes Light pipes conduct radiation from entrance to exit face by multiple reflections from polished walls. In many respects they function the same as a large-scale optical fiber [see Sec. 13 and Potter (1961)]. Figure 64 indicates the path of a light ray through a cylindrical light pipe, and Fig. 65 shows a convenient method of tracing ray paths through a pipe, using an unfolding technique which allows the reflected rays to be drawn as straight lines (Williamson, 1952). The light pipe of Fig. 65 is tapered and can be used to collect light at the large end and condense it to the small end, or vice versa. Note that the angle which ray A makes with the pipe axis is an inverse function of the diameter of a tapered pipe; at the large end the angle to the axis is small, and the angle increases with each reflection. This relationship is analogous to Eq. (166) for afocal systems. (Ray *B*, which enters at too large an angle, is actually turned around and comes back out of the pipe.) Thus a pipe can be used to expand or contract the size of a light beam, or it can be used to increase or reduce the angle of divergence of the beam. A light pipe can also be used as a scrambler to mix or decorrelate a light beam.

A light pipe may be conical or pyramidal, a hollow tube or a solid rod. In solid rods it is often possible to make use of total internal reflection so that the efficiency can be high. The refraction at the entrance face of a solid pipe reduces the internal angle and increases the acceptance angle of a pipe; however, a ray may be trapped in a solid pipe by total internal reflection at the exit face.

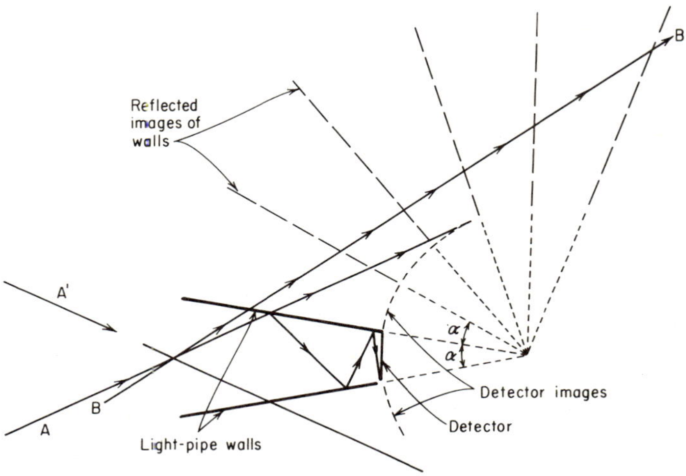

Fig. 65 Ray tracing through a tapered light pipe by means of an unfolded (tunnel) diagram. [*After Smith* (1966).]

45. Detector Optics Optics are frequently closely associated with radiation detectors or sensors (1) to spread the incident energy as uniformly as possible over the detector surface, (2) to collect as much energy as possible on as small a detector as possible, or (3) to limit the field of view.

There is a limit to how small the detector can be; for any optical system, $A\alpha/nD$ cannot exceed unity, where A is the effective diameter of the system entrance pupil, α is the half field of view of the system, D is the diameter of the detector, and n is the index of the medium in which the detector is immersed. A system

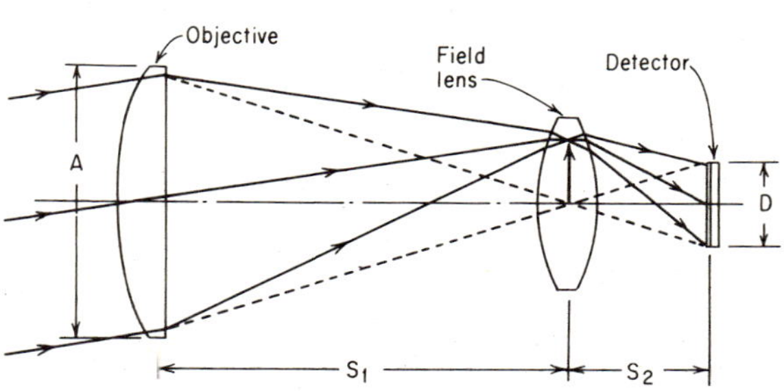

Fig. 66 Schematic of a radiometer with a field lens to increase the field of view with a small detector. [*After Smith* (1966).]

with $A\alpha/nD$ larger than unity is impossible; it is often difficult to design a satisfactory system wherein $A\alpha/nD$ exceeds 0.5.

Field Lenses. A field lens in a detector system is exactly analogous to a condenser in a projection system. Usually the field lens is located at or near the internal image and converges the radiant energy onto the detector. As shown in Fig. 66, a field lens which images the exit pupil of the objective system on the surface of the detector is ordinarily the most satisfactory arrangement.

Light Pipes. See Par. 44.

Immersion Lenses. If the detector is immersed, i.e., in optical contact, on the plano surface of a hemispherical or hyperhemispherical lens, a small detector may be used in place of a larger one. With the detector at the center of curvature (as in the hemisphere), the lens does not introduce any spherical aberration or coma (see Par. 22) and it increases the apparent size of the detector by a factor of n, as shown in Fig. 67. If the aplanatic case, $t = R(n + 1)/n$, is used (as in the hyperhemisphere), the apparent size may be increased by a factor as large as n^2, but this gain may not all be usable due to the limitation $1.0 \geq A\alpha/nD$, cited above.

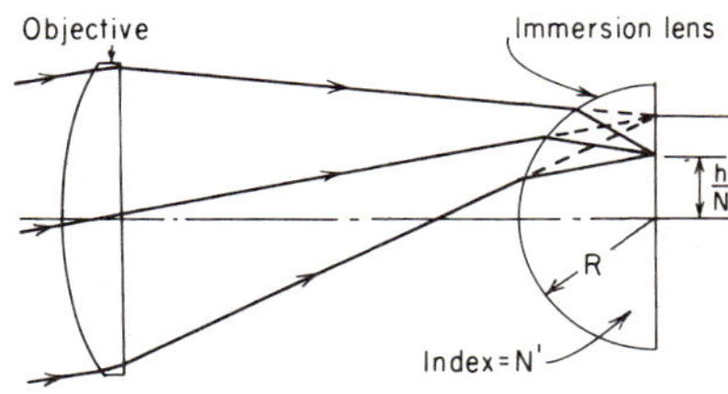

Fig. 67 A hemispherical immersion lens concentric with the focus of an optical system reduces the linear size of the image by a factor equal to the index of the lens. [*After Smith* (1966).]

46. Underwater Optics (McNeil, 1968; Schenck, 1957) An ordinary optical system, e.g., a camera, is frequently adapted for use underwater (index 1.33) by placing it behind a plano-plano window in a watertight enclosure. The index-of-refraction difference across the window introduces several effects. The angular field of view is reduced by approximately 25 percent. Distortion is introduced, as is lateral color. These are obvious from a consideration of Snell's law (Par. 3). For objects at finite distances, modest amounts of the other aberrations are also introduced.

An alternative window is a dome with spherical surfaces which are concentric to the entrance pupil of the optical system. This avoids the distortion, lateral color, and field narrowing of the plano window, and the optical system can be refocused to compensate for the negative power of the dome. If the dome is strongly curved, the aberrations introduced can be substantial.

Adapter lenses can be used. These often take the form of a pair of spherical elements, one of which may be the window. Of course the optimum solution is a system specifically designed to work with an immersed object.

IMAGE CHARACTERISTICS OF SELECTED OPTICAL SYSTEMS

47. In the following paragraphs the performance of a number of relatively simple optical systems is indicated, usually in terms of β, the angular diameter (in radians) of the best-focus blur spot caused by a specified aberration. In this context, best focus is taken to mean the smallest overall image diameter containing all the energy in the (geometrical) image. Note that this may imply a curved image surface for the off-axis case. With the exception of the diffraction limit, all the material in Pars. 47 to 56 is based on geometrical optics, and almost all of it is based on third-order aberration theory. Caution should be exercised in applying these data to systems of large aperture or wide field. In all cases the object is at infinity. Unless otherwise stated, the index is assumed to be 1.5 to 1.6. Many of these relationships are plotted as nomographs in Smith (1966). The following symbols are used:

B = linear diameter of blur spot
$\beta = B/f$ = angular diameter of blur spot, rad
f = effective focal length of optical system
n = index of refraction of lens material

V = Abbe ν value of the material or its equivalent

$$V = \frac{n_D - 1}{n_F - n_C} \qquad \text{or} \qquad V_{\text{eq}} = \frac{n - 1}{\Delta n}$$

R = radius of curvature
$c = 1/R$ = surface curvature
f no. $= f/D$ = relative aperture
D = diameter of entrance pupil
λ = wavelength of radiation
θ = half-field angle, rad
t = thickness of meniscus corrector plate
L_p = distance from mirror to aperture stop

When the blur sizes for several aberrations are applicable, the sum of all the blurs will usually provide a conservative, i.e., large, estimate of the image blur. When the values of the aberrations are given, the angular blur can be computed as follows:

For spherical aberration: $\beta = \dfrac{\text{SC}u_k'}{2f} = \dfrac{\text{TSC}}{2f}$

For coma: $\beta = \dfrac{\text{CC}}{f}$ sagittal

$\beta = 3\dfrac{\text{CC}}{f} = \dfrac{\text{CC3}}{f}$ tangential

For astigmatism: $\beta = 2\text{AC}\dfrac{u_k'}{f} = 2\dfrac{\text{TAC}}{f}$

For field curvature: $\beta = \dfrac{2(\text{PC} + \text{AC})u_k'}{f} = \dfrac{2(\text{TPC} + \text{TAC})}{f}$ sagittal

$\beta = \dfrac{2(\text{PC} + 3\text{AC})u_k'}{f} = \dfrac{2(\text{TPC} + 3\text{TAC})}{f}$ tangential

48. The Diffraction Limit For an aberration-free system, the diameter of the first dark ring of the diffraction pattern of a point image is

$$\beta = \frac{2.44\lambda}{D} \quad \text{rad}$$

This diameter contains 84 percent of the energy in the image. Note that when a value of β smaller than this is indicated as the blur-spot size in the following discussion, it simply means that the aberrations in question have an effect which is small in comparison with the diffraction effect; it does not mean that the actual image blur produced by the system will be correspondingly small.

Refracting Systems

49. The Simple Thin Lens Shaped for Minimum Spherical Abberation

Spherical aberration: $\beta = \dfrac{K}{(f\text{ no.})^3}$ rad

where
$$K = \begin{cases} 0.067 & n = 1.5 \\ 0.027 & n = 2.0 \\ 0.0129 & n = 3.0 \\ 0.0087 & n = 4.0 \end{cases}$$

Chromatic aberration: $\beta = \dfrac{1}{2V(f\text{ no.})}$ rad

Sagittal coma: $\beta = \dfrac{\theta}{16(n + 2)(f\text{ no.})^2}$ rad

Astigmatism: $\beta = \dfrac{\theta^2}{2(f\text{ no.})}$ rad

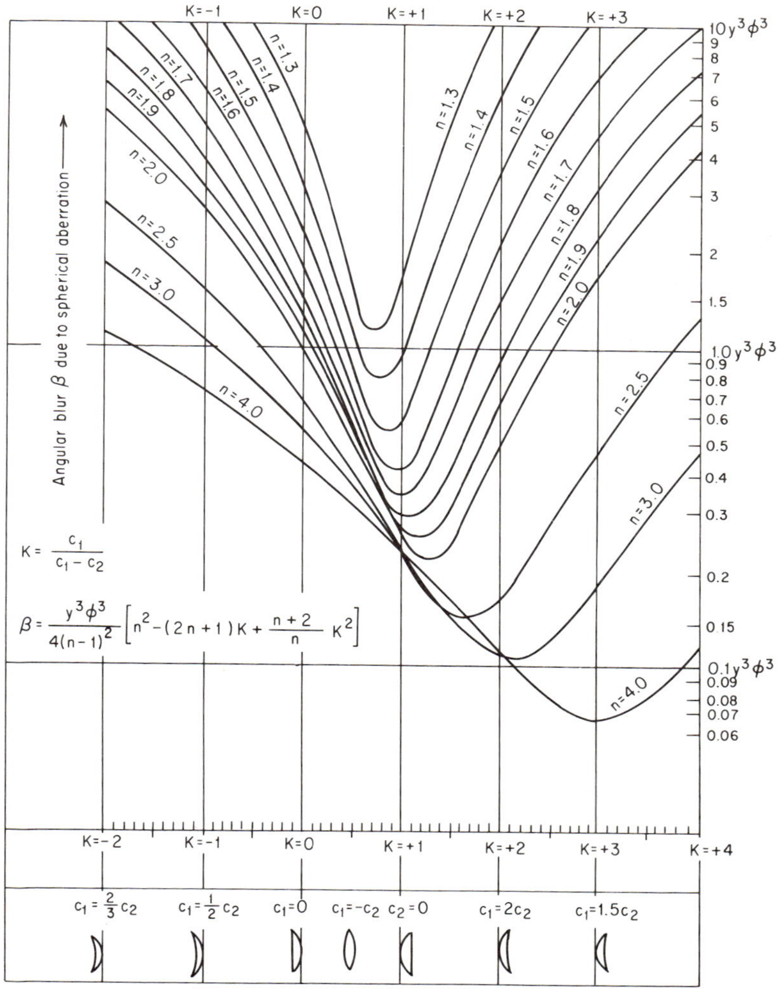

Fig. 68 The angular blur (due to spherical aberration) of a simple lens as a function of shape for various indexes of refraction. Object at infinity. [*After Smith* (1965).]

Figure 68 indicates the relationship between the shape of a thin lens and its spherical aberration.

50. The Simple Thin Lens Split into Several Elements Spherical aberration can be reduced by splitting a single element into two or more elements. Figure 69 shows the spherical aberration for systems of one, two, three, or four elements (of equal power) as a function of index when each element is shaped for minimum spherical aberration. This is a plot of

$$SC = \sum_{j=1}^{j=i} SC_j$$

$$SC_j = \frac{-y^2\phi_i n[4n - 1 - 4j(j-1)(n-1)^2]}{8i^3(n-1)^2(n+2)}$$

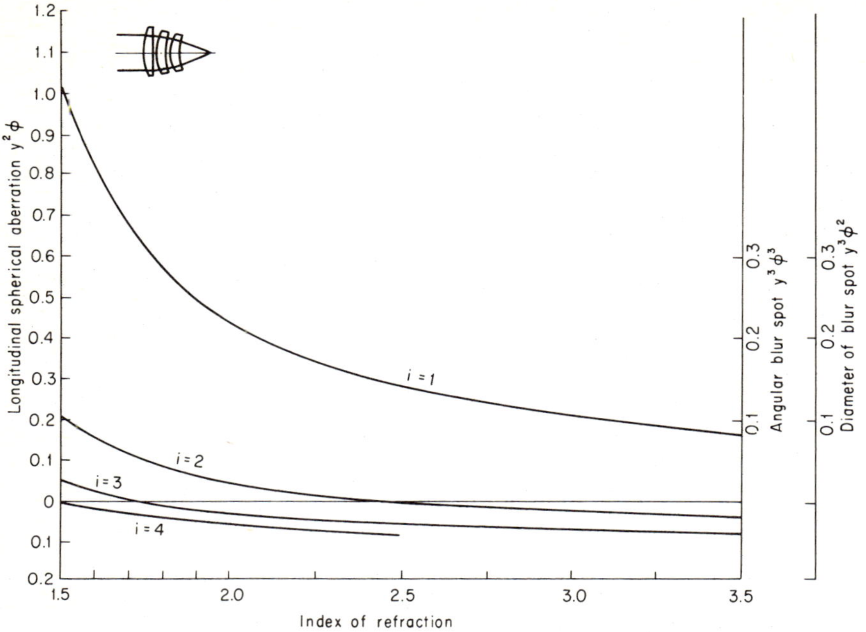

Fig. 69 The spherical aberration of a series of thin simple lenses, each bent for minimum spherical aberration. [*After Smith* (1965).]

where i = number of elements
y = semiaperture
$\phi_i = \Phi/i$ = power of each element
n = index of refraction of elements
j = element number

Reflective Systems

51. The Spherical Mirror This is illustrated in Fig. 70a.

Spherical aberration: $$\beta = \frac{1}{128(f\text{ no.})^3} \quad \text{rad}$$

Sagittal coma: $$\beta = \frac{(L_p - R)\theta}{16R(f\text{ no.})^2} \quad \text{rad}$$

Astigmatism: $$\beta = \frac{(L_p - R)^2\theta^2}{2R^2(f\text{ no.})} \quad \text{rad}$$

Note that if the stop is placed at the center of curvature, $L_p = R$ and both

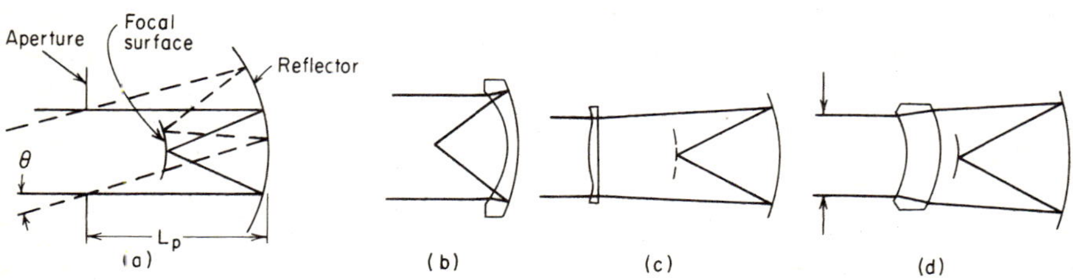

Fig. 70 Schematic sections of several reflecting optical systems: (a) spherical reflector, (b) Mangin mirror, (c) Schmidt system, (d) rear concentric. [*After Smith* (1965).]

coma and astigmatism are zero; the image is then located on a spherical surface of radius $R/2 = f$, and the only aberration is spherical aberration.

52. The Paraboloidal Mirror

Spherical aberration: $\beta = 0.0$

Sagittal coma: $\beta = \dfrac{\theta}{16(f\text{ no.})^2}$ rad

Astigmatism: $\beta = \dfrac{(L_p + f)^2\theta^2}{2f^2(f\text{ no.})}$ rad

Note that there is no astigmatism when the stop is located in the focal plane ($L_p = -f$).

53. Two-Reflector Systems See Fig. 71. The third-order aberrations of a generalized two-reflector system can be found using the equations of Par. 17. The

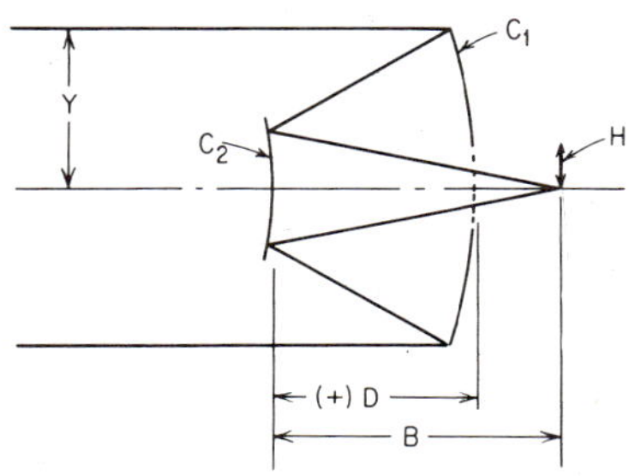

Fig. 71 Two-reflector optical system.

following equations are derived specifically for two-mirror systems with the stop at the primary mirror.

$$c_1 = \frac{B - F}{2DF} \qquad c_2 = \frac{B + D - F}{2DB}$$

$$\Sigma\text{TSC} = Y^3[F(B-F)^3 + 64D^3F^4K_1 + B(F-D-B)(F+D-B)^2 - 64B^4D^3K_2]/8D^3F^3$$

$$\Sigma\text{CC} = HY^2[2F(B-F)^2 + (F-D-B)(F+D-B)(D-F-B) - 64B^3D^3K_2]/8D^2F^3$$

$$\Sigma\text{TAC} = H^2Y[4BF(B-F) + (F-D-B)(D-F-B)^2 - 64B^3D^3K_2]/8BDF^3$$

$$\Sigma\text{TPC} = \frac{H^2Y[DF - (B-F)^2]}{2BDF^2}$$

where c_1 = primary mirror curvature
c_2 = secondary mirror curvature
F = system focal length
B = system back focus (mirror 2 to focus)
D = spacing between mirrors (use positive sign)
Y = system semiaperture (ray height at mirror 1)
K_1, K_2 = fourth-order aspheric deformation coefficients for mirrors 1 and 2
H = height of image

The equations for several common two-mirror systems are worked out below.

The classical Cassegrain has a paraboloidal primary and a hyperboloidal secondary, so that the spherical aberration is corrected.

$$K_1 = \frac{(F-B)^3}{64D^3F^3} \qquad K_2 = \frac{(F-D-B)(F+D-B)^2}{64B^3D^3}$$

$$\Sigma\text{TSC} = 0.0$$

$$\Sigma\text{CC} = \frac{HY^2}{4F^2}$$

$$\Sigma\text{TAC} = \frac{H^2Y(D-F)}{2BF^2}$$

The Ritchey-Chrétien Cassegrain has the aspheric surfaces chosen so that both spherical aberration and coma are corrected.

$$K_1 = \frac{2BD^2 - (B - F)^3}{64D^3F^3}$$

$$K_2 = [2F(B - F)^2 + (F - D - B)(F + D - B)(D - F - B)]/64B^3D^3$$

$$\Sigma\text{TSC} = 0.0 \qquad \Sigma\text{CC} = 0.0 \qquad \Sigma\text{TAC} = \frac{H^2Y(D - 2F)}{4BF^2}$$

Spherical Primary, Aspheric Secondary

$$\begin{aligned} K_1 &= 0.0 \\ K_2 &= [F(B - F)^3 - B(F - D - B)(F + D - B)^2]/64B^4D^3 \\ \Sigma\text{TSC} &= 0.0 \\ \Sigma\text{CC} &= HY^2[2BD^2 - (B - F)^3]/8BD^2F^2 \\ \Sigma\text{TAC} &= H^2Y[(F - B)^3 + 4BD(D - F)]/8B^2DF^2 \end{aligned}$$

Aspheric Primary, Spherical Secondary

$$\begin{aligned} K_1 &= [F(F - B)^3 - B(F - D - B)(F + D - B)^2]/64D^3F^4 \\ K_2 &= 0.0 \qquad \Sigma\text{TSC} = 0.0 \\ \Sigma\text{CC} &= HY^2[2F(B - F)^2 + (F - D - B)(F + D - B)(D - F - B)]/8D^2F^3 \\ \Sigma\text{TAC} &= H^2Y[4BF(B - F) + (F - D - B)(D - F - B)^2]/8DBF^3 \end{aligned}$$

Catadioptric Systems

54. Mangin Mirror Shaped for marginal spherical-aberration correction (see Fig. 70*b*) (Amon and Rosin, 1968).

Zonal spherical aberration: $\beta \approx \dfrac{0.25 \times 10^{-3}}{(f\text{ no.})^4}$ rad

Chromatic aberration: $\beta \approx \dfrac{1}{6V(f\text{ no.})}$ rad

Sagittal coma: $\beta \approx \dfrac{\theta}{32(f\text{ no.})^3}$ rad

Astigmatism: $\beta \approx \dfrac{\theta^2}{2(f\text{ no.})}$

55. Schmidt System The corrector neutral zone is at 0.866 (see Fig. 70*c*).

Spherical aberration: $\beta = 0.0$

Chromatic variation of spherical aberration: $\beta = \dfrac{1}{256V(f\text{ no.})^3}$ rad

High-order aberrations: $\beta_t = \dfrac{\theta^2}{24(f\text{ no.})^3}$ rad

Note that the high-order aberrations can often be reduced by a factor of 2 by refocusing. The value of β_t given is for the tangential aberration; the sagittal blur is about one-sixteenth as large. If the neutral zone of the corrector is at 0.707, the chromatic blur will be twice as large as indicated.

A very close approximation to the aspheric which will correct the spherical aberration and minimize the chromatic effects is given by

$$\pm x = \tfrac{1}{2}cy^2 - ky^4$$

where

$$k_0 = \frac{1}{32(n - 1)f^3}$$

$$c = \frac{-3}{128(n - 1)f(f\text{ no.})^2} = -3y_m^2k_0$$

$$k = k_0\left[1 - \frac{3}{64(f\text{ no.})^2}\right]^2$$

where x = sag of the aspheric
y = distance from axis to point at which x is measured
n = index of aspheric plate
f = focal length (negative) of mirror
y_m = half of clear aperture of aspheric
f no. = $-f/2y_m$ = (positive) relative aperture

56. Meniscus Corrector Systems† *Maksutov Color-corrected Meniscus*

$$c_2 = \begin{cases} \dfrac{n^2 c_1}{n^2 + tc_1(1 - n^2)} & \text{for } \dfrac{d\phi}{dn} = 0 \\ \dfrac{c_1[n^2 - c_1 t(n-1)^2]}{[c_1 t(n-1) - n]^2} & \text{for } \dfrac{dl'}{dn} = 0 \end{cases}$$

Bouwers Concentric System. Corrected for marginal spherical, see Fig. 70*d*.

Zonal spherical aberration: $\beta \approx \dfrac{10^{-4}}{(t/f + 0.06)(f\text{ no.})^5}$ rad

Chromatic aberration: $\beta = \dfrac{tf\,\Delta n}{2n^2 V R_1 R_2 (f\text{ no.})}$ rad

Coma and astigmatism are zero with the stop at the common center of curvature of all the surfaces of the Bouwers concentric.

Bouwers Corrected Concentric System. Aspheric corrector plate at the stop:

Higher-order aberrations: $\beta \approx \dfrac{9.75(\theta + 7.2\theta^3) \times 10^{-5}}{(f\text{ no.})^{6.5}}$ rad

OPTICAL-DESIGN TECHNIQUE

57. Optical design is the selection and arrangement of the materials, curvatures, and spacings of an optical system so that the characteristics of the image produced by the system are suitable for its intended application. In general, the process is essentially the reduction of the aberrations (Par. 14) to a negligible or at least an acceptable level (Pars. 28 to 35). A generalized outline of the process might include the following steps:

1. Selection of design type
2. Arrangement of element powers and spacings to control image size, image position, chromatic aberrations, and Petzval curvature
3. Correction of the primary aberrations
4. Reduction of residual (higher-order) aberrations

Each step may include several loops, and it may be necessary to loop through several steps repeatedly to complete the process (Conrady, 1957; Cox, 1964; Kingslake, 1965; MIL-HDBK-141, 1962; Smith, 1966).

58. Determination of Element Powers and Spacings The third-order aberration-contribution equations (Pars. 17 to 19) are very useful in determining the initial power and space layout. The thin-lens equations (Par. 19) are especially useful in this stage. Usually the problem can be expressed as the need to satisfy one or more of a set of equations, typified by the following, which are based on the thin-lens equations of Par. 19:

Desired power: $\Phi = \dfrac{\Sigma y_i \phi_i}{y_1}$

Desired back focus: $l' = \dfrac{y_k}{y_1 \Phi}$

Desired longitudinal chromatic: $l_F' - l_C' = -\dfrac{1}{u_k'^2} \sum \dfrac{y_i^2 \phi_i}{V_i}$

Desired lateral chromatic: $h_F' - h_C' = -\dfrac{1}{u_k'} \sum \dfrac{y_i y_{pi} \phi_i}{V_i}$

Desired Petzval curvature: $x_{\text{Petz}} = -\dfrac{h^2}{2u_k'} \sum \dfrac{\phi_i}{n_i}$

† References for this paragraph are Bouwers (1946), Maksutov (1944), and Lauroesch and Wing (1959).

Usually the requirements of this design stage can be formulated and solved on a reasonably straightforward basis, e.g., by the simultaneous solution of the set of typical equations outlined above. The equations are solved for the necessary values of the element powers ϕ_i and ray heights y_i and y_{pi} or the spacings implied by the ray heights.

Note that the chromatic aberrations, the Petzval curvature, and the system power can be controlled only by the powers and spacings of the elements; element shape is not effective for controlling these characteristics.

Achromatization. For a thin doublet component which is to be achromatic, the simultaneous solution of the first and third equations above will yield the element powers:

$$\phi_a = \frac{V_a \Phi}{V_a - V_b} \qquad \phi_b = \frac{V_b \Phi}{V_b - V_a}$$

59. Element Shapes The remaining four primary aberrations (spherical aberration, coma, astigmatism, and distortion) are controlled by the shape, or bending, of the elements (note, however, that element powers and ray heights also affect these aberrations). Shape means the distribution of the total surface curvature ($c = c_1 - c_2$) between the two surfaces of the element. As indicated most clearly by the equations of Par. 19, these aberrations tend to be quadratic functions of the element shapes, and for a system of any complexity, a straightforward solution is rare.

A useful approach to the problem consists of making a set of graphs of the aberration contribution of each element as a function of its shape. This aids in visualizing the problem, especially when there is no solution. It is also useful in locating the element shapes for which the individual element contributions are small; usually the best designs come from forms near these shapes.

One or more rough graphical solutions are found; usually the graph technique is not sufficiently accurate to yield a final design, and a technique of differential correction is applied. The partial derivatives of the aberrations A_j with respect to the variable parameters p_i are determined, either from the graphs or from an independent calculation. Then the present aberrations A_j are computed, and the desired values for the aberrations A_j' are determined. Since the graph solution will probably be close to the final solution, it is assumed that the relationship between the aberrations and variables is linear and that

$$A' = A + \frac{\partial A}{\partial p_1}\Delta p_1 + \frac{\partial A}{\partial p_2}\Delta p_2 + \cdots + \frac{\partial A}{\partial p_k}\Delta p_k$$

and thus the set of n linear equations in k unknowns

$$A_j' = A_j + \sum_{i=1}^{i=k} \frac{\partial A_j}{\partial p_i}\Delta p_i \qquad j = 1, 2, \ldots, n$$

can be solved for the parameter changes Δp_i to produce the desired set of aberration values A_j'.

This approximate solution is repeated until the system converges to a solution. Note that the use of the graphs of aberrations vs. bendings ensures that a solution does exist and that a sufficient number of partial derivatives $\partial A/\partial p$ are nonzero.

60. Residual Aberrations When the primary aberrations have been corrected, the image quality of the system is determined by the residual aberrations and it is often necessary to reduce the magnitude of these residuals in order to bring the system performance up to the desired level. Usually one or more of the following techniques are used:

1. The systematic variation of any parameters which are not determined by the necessity to correct the primary aberrations

2. Splitting an element (usually one which contributes large amounts of aberration) into two elements

3. Compounding an element by converting it to a doublet, which is usually at least partially achromatic
4. Reversing the order of the elements of a component
5. Drastically revising the system to respace or reshape or reduce the power of the elements or surfaces which contribute heavily to the troublesome aberrations

61. Automatic Design by Computer† The design processes outlined above can be carried out more or less automatically by a properly programmed electronic computer. The problem can be stated as follows. A set of aberrations is to be controlled, either to certain desired values, or else minimized, by the adjustment of a set of variables. The aberrations may be the conventional optical aberrations, the departures of the ray intercepts from a desired point, the focal length, back focal length, or edge thickness, or any characteristic of the optical system whose desired value can be specified. The variables are the construction parameters of the optical system; ordinarily these are the surface curvatures and aspheric deformations, the surface-to-surface spacings, the refractive indexes, and the chromatic variations thereof, and combinations of these variables.

In general, the aberrations are nonlinear functions of the variables. The usual method of approaching the problem is to assume that the relationships are linear and to solve for the changes to the variables which would produce the desired changes in the aberrations under the assumption of linearity. Since the relationships are actually nonlinear, the "solution" is usually incorrect. However, if the starting system was reasonably chosen, the "solution" is likely to be closer to the desired state of affairs than the starting point. The process is repeated until a satisfactory solution is obtained.

The mathematics of the solution are best described in terms of matrices. The following versatile formulation (Spencer, 1963) includes most methods as special cases.

F is the $n \times 1$ matrix of desired changes to the n aberrations $f_1, f_2, \ldots, f_n$. The system is to be modified so that the desired changes are realized exactly.

G^* is the $m \times 1$ matrix of the desired changes to the m aberrations $g_1, g_2, \ldots, g_m$. The g aberrations are to be minimized in a least-squares sense.

W is the $m \times m$ diagonal matrix of weighting factors $w_1, w_2, \ldots, w_m$ to be applied to the minimized aberrations.

C is the $j \times j$ diagonal matrix of weighting factors $c_1, c_2, \ldots, c_j$ to be applied to the changes to the parameters.

X is the $j \times 1$ matrix of the changes to the variables $x_1, x_2, \ldots, x_j$ necessary to produce a solution.

The desired solution is (1) that the values of the F aberrations be changed by $f_1, f_2, \ldots, f_n$ exactly, and (2) that the error function

$$E = \sum_{i=1}^{i=j} (c_i x_i)^2 + \sum_{k=1}^{k=m} w_k^2 (g_k - g_k')^2$$

be minimized, where $g - g'$ represents the difference between the desired change to the G^* aberrations and that expected.

A is the $n \times j$ matrix of the partial derivatives $a_{11}, \ldots, a_{1j}, \ldots, a_{n1}, \ldots, a_{nj}$ of the n "exact" aberrations with respect to the j variables.

B^* is the $m \times j$ matrix of the partial derivatives $b_{11}, \ldots, b_{1j}, \ldots, b_{m1}, \ldots, b_{mj}$ of the m minimized aberrations with respect to the j variables.

The solution is given by the matrix equation

$$X = (B^T B + C)^{-1}(A^T L + B^T G)$$

where

$$L = [A(B^T B + C)^{-1} A^T]^{-1}[F - A(B^T B + C)^{-1} B^T G]$$

L is an $n \times 1$ matrix of scalar multipliers. The superscript T indicates the transpose of the matrix, the superscript -1 indicates the inverse of the matrix, and

$$B = WB^* \qquad G = WG^*$$

† References for Par. 61 are Feder (1963), Glatzel and Wilson (1968), Meiron (1968), and Spencer (1963).

Exact Solution. If there are no aberrations to be minimized, B^*, B, G^*, G, and W do not exist and the solution reduces to

$$X = C^{-1}A^T(AC^{-1}A^T)^{-1}F$$

Note that the error function, $E = \Sigma(c_i x_i)^2$, is minimized so that the solution is as close as possible to the starting point.

Damped Least-Squares Solution. On the other hand, if all the aberrations are to be minimized, A and F do not exist and the solution reduces to

$$X = (B^TB + C)^{-1}B^TG$$

This is the damped least-squares solution, usually written

$$X = (B^TB + KI)^{-1}B^TG$$

where I is the identity matrix and K is a positive damping factor, which can be set large in regions where the aberration-variable relationships are highly nonlinear and set small where the problem is a nearly linear one. The damping feature can be used with weighted parameters by using KC in place of KI.

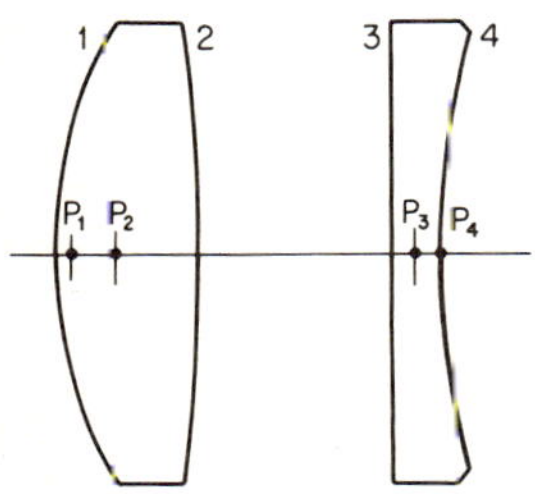

Fig. 72 Two-element optical system.

SAMPLE CALCULATIONS

62. Introduction The following paragraphs provide a practical aid to the application of the ray-tracing and image-locating formulas. Frequently, following through a worked-out sample computation will help to eliminate mistaken applications of the equations. The appropriate equations for each example are cited by number but are not restated with the example; thus it is necessary to refer back to the original equation.

All the first-order (paraxial) computations will be based on the same optical system, consisting of a biconvex lens followed by a plano-concave lens; these sample calculations will be primarily concerned with locating the image formed by this system of an object which is located 100 units to the left of the first surface. The system is sketched in Fig. 72, and its data are given in Table 3.

TABLE 3 Data for Two-Element Optical System of the Worked Examples

	Air	Surface 1	Medium 1	Surface 2	Medium 2	Surface 3	Medium 3	Surface 4	Air
Radius R	...	+10.0	...	−50.0	...	∞	...	+20.0	
Curvature C	...	+0.1	...	−0.02	...	0.00	...	+0.05	
Thickness t	...	.....	3.0	......	3.0	....	2.0	......	
Index n	1.0	.....	1.5	......	1.0	....	2.0	......	1.0

63. Focal Lengths and Cardinal Points of the Elements See Table 3 and Par. 62 for data. For the first element, the focal length is determined by Eq. (26):

$$\phi = (1.5 - 1)\left[+0.1 - (-0.02) + 3 \times 0.1(-0.02)\frac{1.5 - 1}{1.5}\right]$$

$$= 0.5(0.1 + 0.02 - 0.002) = +0.059$$

$$f = \frac{1}{\phi} = \frac{1}{0.059} = +16.949153$$

The back focal length is obtained from Eq. (27):

$$\text{bfl} = 16.949153\left(1 - 3 \times 0.1\frac{1.5 - 1}{1.5}\right) = +15.254237$$

The distance from the second principal point p_2 to the second surface is

$$f - \text{bfl} = 16.949153 - 15.254237 = +1.694915$$

The front focal length can be obtained by calculating the back focal length of the element in the reverse direction (note the changed sign of the curvature):

$$-\text{ffl} = 16.949153 \left[1 - 3(+0.02) \frac{1.5 - 1}{1.5} \right]$$

$$\text{ffl} = -16.610169$$

The distance from the first surface to the first principal point p_1 is

$$f + \text{ffl} = 16.949153 + (-16.610169) = +0.338983$$

Similarly for the second element; from Eq. (26):

$$\phi = (2.0 - 1) \left(0 - 0.05 + 2 \times 0 \times 0.05 \frac{2.0 - 1}{2.0} \right) = -0.05$$

$$f = \frac{1}{\phi} = -20.0$$

From Eq. (27),

$$\text{bfl} = -20 \left(1 - 2 \times 0 \frac{2 - 1}{2} \right) = -20.0$$

Thus the second principal point p_4 lies on surface 4.

$$-\text{ffl} = -20 \left[1 - 2(-0.05) \frac{2 - 1}{2} \right]$$

$$\text{ffl} = +21.0$$

Thus the first principal point p_3 lies 1.0 to the right of surface 3.

The focal lengths of the elements can also be determined by application of Eqs. (7) to (16), as described in Par. 5; see Par. 67 for a sample calculation using these equations.

64. The Cardinal Points of a Two-Component System See Table 3 and Par. 62 for data. The system can be considered to consist of two components separated by a distance $d = 1.694915 + 3.0 + 1.0 = 5.694915$ (the distance from p_2 to p_3). By Eq. (31)

$$\phi_{ab} = +0.059 + (-0.05) - 5.694915 \times 0.059(-0.05) = +0.0258$$

$$f_{ab} = \frac{1}{\phi_{ab}} = +38.759690$$

By Eq. (33)

$$\text{bfl} = 38.759690 \frac{16.949153 - 5.694915}{16.949153} = +25.736435$$

By Eq. (34)

$$-\text{ffl} = 38.759690 \frac{-20 - 5.694915}{-20} = +49.796347$$

These front and back focal lengths are with respect to the outer principal points of the elements, p_1 and p_4, respectively. The back focal length from the lens surface is the same as calculated above by Eq. (33) since p_4 lies on surface 4. The distance from the first surface to the first focal point of the system is $-49.796347 + 0.338983 = -49.457364$.

The first principal point of the system is located $49.457364 - 38.759690 = 10.697674$ to the left of the first surface, and the second principal point is located $38.759690 - 25.736434 = 13.023256$ to the left of the fourth surface.

65. System Imagery See Table 3 and Par. 62 for data. The object is located 100 units to the left of the first surface and is thus 100 — 10.697674 = 89.302326 to the left of the first principal point of the system. We can locate the image using Eq. (1):

$$\frac{1}{s'} = \frac{1}{-89.302326} + \frac{1}{38.759690} = +0.0146021$$

$$s' = +68.483378$$

Since this distance is measured from the second principal point of the system, the image is located 68.483378 — 13.023256 = 55.460122 to the right of surface 4.

The object is located 100 — 49.457364 = 50.542636 to the left of the first focal point. The image location can be determined by using Eq. (2):

$$x' = \frac{(-38.759690)^2}{-50.542636} = +29.723688$$

The distance from surface 4 to the image is 25.736435 + 29.723688 = 55.460123, in good agreement with the position found above from Eq. (1).

The system lateral magnification, from Eq. (3), is

$$m = \frac{h'}{h} = \frac{s'}{s} = \frac{68.483378}{-89.302326} = -0.7668712$$

$$= \frac{f}{x} = \frac{38.759690}{-50.542636} = 0.7668712$$

$$= \frac{-x'}{f} = \frac{-29.723689}{38.759690} = 0.7668712$$

66. Component-by-Component Ray Trace See Table 3 and Par. 62 for data.

Object distance:	$100 + 0.338983 = 100.338983$	(object to p_1)
Power of element A:	$\phi_A = +0.059$	
Spacing:	$d = 1.694915 + 3 + 1$	
	$= 5.694915$	(p_2 to p_3)
Power of element B:	$\phi_B = -0.05$	

A ray starting at the object with a slope $u_1 = -0.01$ will strike the first element (at its first principal plane) at

$$y_A = 0 - 100.338983(-0.01) = +1.0033898 \qquad \text{by Eq. (30)}$$
$$u_A' = -0.01 + 1.0033898 \times 0.059 = +0.0492 \qquad \text{by Eq. (29)}$$
$$y_B = 1.0033898 - 5.694915 \times 0.0492 = +0.7232 \qquad \text{by Eq. (30)}$$
$$u_B' = +0.0492 + 0.7232(-0.05) = +0.01304 \qquad \text{by Eq. (29)}$$

The ray will intersect the axis at $l_B' = 0.7232/0.01304 = 55.460123$, which (since the second principal plane of element B coincides with surface 4) is the distance from both surface 4 and p_4 to the location of the image.

The lateral magnification can be determined by rearranging Eq. (18) to get

$$m = \frac{h'}{h} = \frac{nu}{n'u'} = \frac{1(-0.01)}{1 \times 0.01304} = -0.7668712$$

Note that the image location and the magnification are in agreement with those determined in Par. 65.

67. Surface-by-Surface Paraxial Ray Trace See Table 3 and Par. 62 for data. Still another way to locate the image is to trace a paraxial ray through the system, using the equations of Par. 5. If we trace the ray from the axial intercept of the object (at 100 units from the first surface) and assume that the ray strikes the first surface at a distance 1.0 from the axis, Eq. (7) gives

$$n_1 u_1 = \frac{1.0 \times 1.0}{-100} = -0.01$$

Then

$$n_1'u_1' = -0.01 + (1.5 - 1.0) \times 1 \times 0.1 = +0.04$$
$$y_2 = 1.0 - 3\,\frac{0.04}{1.5} = 0.92$$
$$n_2'u_2' = +0.04 + (1 - 1.5) \times 0.92(-0.02) = +0.0492$$
$$y_3 = 0.92 - 3\,\frac{0.0492}{1.0} = 0.7724$$
$$n_3'u_3' = +0.0492 + (2 - 1) \times 0.7724 \times 0 = +0.0492$$
$$y_4 = 0.7724 - 2\,\frac{0.0492}{2} = 0.7232$$
$$n_4'u_4' = +0.0492 + (1 - 2) \times 0.7232 \times 0.05 = +0.01304$$

where the $n'u'$'s are given by Eq. (9) and the y's by Eq. (10). The final ray intercept (with the axis) is found from Eq. (15):

$$l_k' = 1.0\,\frac{0.7232}{0.01304} = +55.460123$$

Not only is the image location in agreement with the preceding calculations, but the ray slopes before, between, and after the elements are in agreement with Par. 66.

The focal lengths (effective, back, and front) and the cardinal points can readily be calculated using Eqs. (9) and (10). Simply trace a ray with $u_1 = 0.0$ at an arbitrary ray height y_1. Then if y_k and u_k' are the ray height and slope at and after the last surface of the system, the focal length is y_1/u_k' and the back focal length is y_k/u_k'. Or alternatively two unrelated rays may be traced and Eqs. (23) to (25) used to find the focal lengths.

68. Third-Order Aberrations; Surface Contributions For Pars. 68 and 69 the sample calculations will be based on a Schmidt system objective consisting of

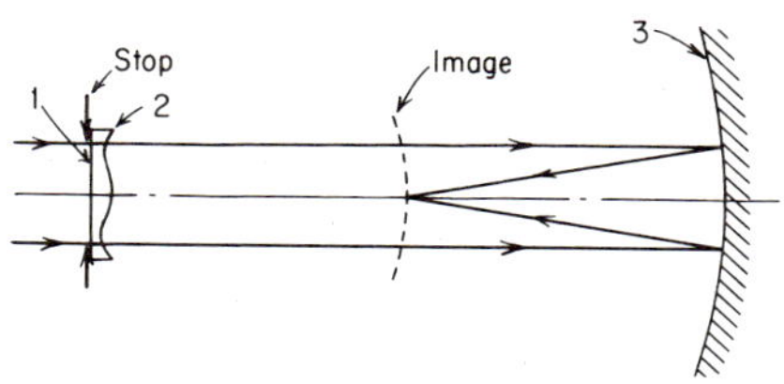

Fig. 73 Schmidt system.

a refracting aspheric corrector plate at the center of curvature of a spherical mirror, as shown in Fig. 73. When the equations of Par. 55 are used to lay out the system, the constructional data are as follows:

$$c_1 = 0.0$$
$$t_1 = 0.25 \qquad n_1 = 1.5 \qquad n_F = 1.5056 \qquad n_C = 1.4976$$
$$c_2 = -0.00075 \qquad A_4 = 0.61566 \times 10^{-4}$$
$$t_2 = 20.0 \qquad n_2 = 1.0$$
$$c_3 = -0.05$$
$$t_3 = -9.962359 \qquad n_3 = -1.0$$
$$c_{\text{image}} = -0.10$$

The stop (entrance pupil) is located at the first surface, and has a diameter of 4.0. The object is at infinity and subtends an angle of ±0.17633 (±10°). Thus the invariant

$$I = n(y_pu - yu_p) = 1.0[0 \times 0 - 2(-0.17633)] = +0.35266$$

The paraxial ray trace of the axial and chief rays and the calculation of the surface contributions are given in Table 4. Note that the paraxial trace through

TABLE 4 Sample Calculations for Schmidt Objective System

	Eq.	Air	Surface 1	Medium 1	Surface 2	Medium 2	Surface 3	Air
Curvature c			0.00		−0.00075		−0.05	
Thickness t				0.25		20.0		
Index n		1.0		1.5		1.0		−1.00
Δn		0.0		0.008		0.0		0.00
y	(14)		2.0		2.0		1.985	
i	(11)		0.0		−0.0015		−0.1	
u	(13)	0.0		0.0		+0.00075		−0.19925
$u' - i$			0.0		+0.00225		−0.09925	

EFL = −10.037641 BFL = −9.962359

	Eq.	Air	Surface 1	Medium 1	Surface 2	Medium 2	Surface 3	Air
y_p	(14)		0.0		+0.0293883		+3.555768	
i_p	(11)		+0.17633		+0.1175313		−0.0014694	
u_p	(13)	−0.17633		−0.1175533		−0.1763190		−0.1792578
$u_p' - i_p$			−0.2938833		−0.2938503		−0.1777884	

From Eqs. (15) and (16): $l_p' = -19.836057$ $h' = 1.769937$

	Eq.	Surface 1	Surface 2	Surface 3
$\frac{n(n'-n)}{2n'}$		+0.1666667	−0.375	1.00
B	(98)	0.00	−0.0047851	−0.5586436
B_p	(99)	0.00	+0.0091828	−1.7925886
$TSC_{sph} = Bi^2h$	(91)	0.00	-0.019×10^{-6}	−0.009888
$CC_{sph} = Bii_ph$	(92)	0.00	$+1.49 \times 10^{-6}$	−0.0001453
$TAC_{sph} = Bi_p^2h$	(93)	0.00	−0.000117	−0.00000213
TPC	(94)	0.00	−0.000078	+0.031209

$\Delta n - n \frac{\Delta n'}{n'}$	....	−0.005333	+0.008	0.0
TLchC	(96)	0.0	−0.000120	0.0
TchC	(97)	+0.009440	−0.009438	0.0
$u_p'^2 - u_p^2$	....	−0.0172735	+0.0172696	+0.0010450
DC_{sph}	(95)	−0.015286	+0.015280	+0.000459
Aspheric terms $A_4 = K$	....	0.0	$+0.61566 \times 10^{-4}$	0.0
$W = \frac{4Kh(n - n')}{I}$	....	0.0	$+6.1797731 \times 10^{-4}$	0.0
$TSC_\alpha = y^4W$	(91)	0.0	+0.009888	0.0
$CC_\alpha = y^3y_pW$	(92)	0.0	+0.000145	0.0
$TAC_\alpha = y^2y_p^2W$	(93)	0.0	+0.000002	0.0
$DC_\alpha = yy_p^3W$	(95)	0.0	+0.000000031	0.0

Total TOA Summary

	Eq.	Surface 1	Surface 2	Surface 3	Σ = 1 + 2 + 3
TSC	(91)	0.0	+0.009888	−0.009888	0.0
CC	(92)	0.0	+0.000146	−0.000145	+0.000001
TAC	(93)	0.0	−0.000115	−0.000002	−0.000117
TPC	(94)	0.0	−0.000078	+0.031209	+0.031131
DC	(95)	−0.015286	+0.015280	+0.000459	+0.000453
TLchC	(96)	0.0	−0.000120	0.0	−0.000120
TchC	(97)	+0.009440	−0.009438	0.0	+0.000002

the aspheric surface is treated as a spherical surface of curvature c when the surface is described by Eq. (53) and Par. 17. To the third-order approximation, the image plane ($c_{\text{image}} = -0.1$) has a sagittal height $= ch^2/2 = -0.156634$. Multiplying by $u_k' = -0.19925$ gives $+0.031209$ for the value of ΣTPC necessary to have the Petzval surface coincide with this image surface. Thus, with respect to this curved focal surface,

$$\Sigma\text{TPC} = +0.031131 - 0.031209 = -0.000078$$

69. Ray Tracing The path of a skew ray will be traced through the Schmidt system described in Par. 68. The ray will strike the pupil (at the first surface) at coordinates $y = 0$, $z = +2.0$, with a slope (tangent) of 0.17633 upward from the left. Thus, at the opening reference surface (the pupil) the ray data are

$$\begin{aligned} x &= 0 & X &= +0.9848072 \\ y &= 0 & Y &= +0.1736511\dagger \\ z &= +2.0 & Z &= 0 \end{aligned}$$

The transfer from the pupil to the next (first) surface is a formality in this case since both the reference and first surfaces are plane ($c_0 = c_1 = 0$) and are in contact ($t = 0$). By Eq. (40)

$$\begin{aligned} e &= 0 \times 0.9848072 - (0 \times 0.9848072 + 0 \times 0.1736511 + 2 \times 0) \\ &= 0.0 \end{aligned}$$

By Eq. (41)

$$M_{1x} = 0 + 0 \times 0.9848072 - 0 = 0.0$$

By Eq. (42)

$$M_1^2 = 0^2 + 0^2 + 2^2 - 0^2 + 0^2 - 2 \times 0 \times 0 = +4.0$$

By Eq. (43)

$$E_1 = [0.9848072^2 - 0(0 \times 4 - 2 \times 0)]^{1/2} = +0.9848072$$

By Eq. (44)

$$D_{0,1} = 0 + \frac{0 \times 4 - 2 \times 0}{0.9848072 + 0.9848072} = 0.0$$

By Eq. (45)

$$x_1 = 0 + 0 \times 0.9848072 - 0.0 = 0.0$$

By Eq. (46)

$$y_1 = 0 + 0 \times 0.1736511 = 0.0$$

By Eq. (47)

$$z_1 = 2.0 + 0 \times 0 = +2.0$$

Refraction at Surface 1. By Eq. (48)

$$E_1' = \left[1 - \left(\frac{1}{1.5}\right)^2 (1 - 0.9848072^2)\right]^{1/2} = +0.9932763$$

By Eq. (49)

$$g_1 = 0.9932763 - \frac{1}{1.5} \times 0.9848072 = +0.3367381$$

By Eq. (50)

$$\begin{aligned} X_1 &= \frac{1}{1.5} \times 0.9848072 - 0.3367381 \times 0 \times 0 + 0.3367381 \\ &= 0.9932763 \end{aligned}$$

By Eq. (51)

$$\begin{aligned} Y_1 &= \frac{1}{1.5} \times 0.1736511 - 0.3367381 \times 0 \times 0 \\ &= +0.1157674 \end{aligned}$$

By Eq. (52)

$$Z_1 = \frac{1}{1.5} \times 0 - 0.3367381 \times 0 \times 2 = 0.0$$

† This is the sine corresponding to a tangent of 0.17633.

Transfer to Surface 2. Transfer to the spherical surface $c = -0.00075$, by Eqs. (40) to (47) is

$$e = 0.25 \times 0.9932763 - (0 \times 0.9932763 + 0 \times 0.1157674 + 2 \times 0) = +0.2483191$$
$$M_{2x} = 0 + 0.2483191 \times 0.9932763 - 0.25 = -0.0033506$$
$$M_2{}^2 = 0^2 + 0^2 + 2^2 - 0.2483191^2 + 0.25^2 - 2 \times 0.25 \times 0 = +4.0008376$$
$$E_2 = \{0.9932763^2 - (-0.00075)[-0.00075 \times 4.0008376 - 2(-0.0033506)]\}^{1/2} = +0.9932777$$

$$D_{1,2} = +0.2483191 + \frac{+0.0037006}{0.9932763 + 0.9932777} = +0.2501819$$

$$\text{Prelim. } (x_2)_0 = 0 + 0.2501819 \times 0.9932763 - 0.25 = -0.0015002$$
$$\text{Prelim. } (y_2)_0 = 0 + 0.2501819 \times 0.1157674 = +0.0289629$$
$$\text{Prelim. } (z_2)_0 = 2 + 0.2501819 \times 0 = +2.0$$

Determine intersection with aspheric, $A_4 = 0.61566 \times 10^{-4}$:

$$x_0 = -0.0015002 \qquad y_0 = +0.0289629 \qquad z_0 = +2.0$$
$$s_0{}^2 = y_0{}^2 + z_0{}^2 = +4.0008388$$

By Eqs. (54) and (53)

$$\bar{x}_0 = \frac{-0.00075 \times 4.0008388}{1 + [1 - (-0.00075)^2(4.0008388)]^{1/2}} + 0 \times 4.0008388 + (0.61566 \times 10^{-4})(4.0008388^2)$$
$$= \frac{-0.0030006}{1.9999989} + 0 + 0.0009855 = -0.0005148$$
$$\text{Error} = \bar{x}_0 - x_0 = -0.0005148 - (-0.0015002) = +0.0009854$$

By Eq. (55)

$$l_0 = [1 - (-0.00075)^2(4.0008388)]^{1/2} = +0.9999989$$

By Eq. (56)

$$m_0 = -0.0289629\{-0.00075 + 0.9999989[2 \times 0 + 4(0.61566 \times 10^{-4})(4.0008388) + 0 \cdots]\} = -0.0289629(+0.0002353) = -0.0000068$$

By Eq. (57)

$$n_0 = -2(+0.0002353) = -0.0004705$$

By Eq. (58)

$$G_0 = \frac{0.9999989 \times 0.0009854}{0.9932763 \times 0.9999989 + 0.1157674(-0.0000068) + 0(-0.0004705)}$$
$$= \frac{+0.0009854}{0.9932744} = +0.0009920$$

Then the improved estimate of ray intersection coordinates is

$$x_1 = +0.0009920 \times 0.9932763 - 0.0015002 = -0.0005149 \qquad \text{by Eq. (59)}$$
$$y_1 = +0.0009920 \times 0.1157674 + 0.0289629 = +0.0290777 \qquad \text{by Eq. (60)}$$
$$z_1 = +0.0009920 \times 0 + 2.0 = +2.0 \qquad \text{by Eq. (61)}$$

$$s_1{}^2 = 4.0008455$$

By Eq. (54)

$$\bar{x}_1 = -0.0005148$$
$$\text{Error} = \bar{x}_1 - x_1 = +0.0000001$$
$$l_1 = +0.9999989 \qquad \text{by Eq. (55)}$$
$$m_1 = -0.0000068 \qquad \text{by Eq. (56)}$$
$$n_1 = -0.0004705 \qquad \text{by Eq. (57)}$$
$$G_1 = +0.0000001 \qquad \text{by Eq. (58)}$$

A second improved estimate of the ray intersection coordinates is

$$
\begin{aligned}
x_2 &= -0.0005148 && \text{by Eq. (59)}\\
y_2 &= +0.0290777 && \text{by Eq. (60)}\\
z_2 &= +2.0 && \text{by Eq. (61)}\\
\bar{x}_2 &= -0.0005148 && \text{by Eq. (54)}
\end{aligned}
$$

$$\text{Error} = \bar{x}_2 - x_2 = 0.0000000$$

Therefore the intersection coordinates are determined to a "sufficient" accuracy.

Refraction at Surface 2 (Aspheric). By Eq. (62)

$$P^2 = 0.9999989^2 + (-0.0000068)^2 + (-0.0004705)^2 = +0.9999980$$

By Eq. (63)

$$
\begin{aligned}
F &= 0.9932763 \times 0.9999989 + 0.1157674(-0.0000068) + 0(-0.0004705)\\
&= +0.9932744
\end{aligned}
$$

By Eq. (64)

$$F' = \left[0.9999980\left(1 - \frac{1.5^2}{1.0^2}\right) + \left(\frac{1.5}{1.0}\right)^2 \times 0.9932744^2\right]^{1/2} = +0.9848040$$

By Eq. (65)

$$g = \frac{0.9848040 - 1.5 \times 0.9932744/1.0}{0.9999980} = -0.5051086$$

By Eq. (66)

$$X_2 = 0.9932763\,\frac{1.5}{1.0} + (-0.5051086)(0.9999989) = +0.9848064$$

By Eq. (67)

$$Y_2 = 0.1157674\,\frac{1.5}{1.0} + (-0.5051086)(-0.0000068) = +0.1736545$$

By Eq. (68)

$$Z_2 = 0\,\frac{1.5}{1.0} + (-0.5051086)(-0.0004705) = +0.0002377$$

Transfer to Surface 3 (Mirror). By Eqs. (40) to (47)

$$
\begin{aligned}
e &= 20 \times 0.9848064 - [(-0.0005148)(0.9848064) + 0.0290777 \times 0.1736545\\
&\qquad + 2 \times 0.0002377]\\
&= +19.6911102
\end{aligned}
$$

$$
\begin{aligned}
M_{3x} &= -0.0005148 + 19.6911102 \times 0.9848064 - 20\\
&= -0.6085835
\end{aligned}
$$

$$
\begin{aligned}
M_3^2 &= (-0.0005148)^2 + 0.0290777^2 + 2^2 - 19.6911102^2 + 20^2\\
&\qquad - 2 \times 20(-0.0005148)\\
&= +16.2816169
\end{aligned}
$$

$$
\begin{aligned}
E_3 &= \{0.9848064^2 - (-0.05)[-0.05 \times 16.2816169 - 2(-0.6085835)]\}^{1/2}\\
&= +0.9949864
\end{aligned}
$$

$$D_{2,3} = 19.6911102 + \frac{-0.05 \times 16.2816169 - 2(-0.6085835)}{0.9848064 + 0.9949864} = 19.8947103$$

$$
\begin{aligned}
x_3 &= -0.0005148 + 19.8947103 \times 0.9848064 - 20 = -0.4080768\\
y_3 &= +0.0290777 + 19.8947103 \times 0.1736545 = +3.4838837\\
z_3 &= 2.0 + 19.8947103 \times 0.0002377 = +2.0047290
\end{aligned}
$$

Reflection at Surface 3 (Mirror). By Eqs. (48) to (52)

$$E_3' = \left[1 - \left(\frac{1}{-1}\right)^2 (1 - 0.9949864^2)\right]^{1/2} = +0.9949864$$

$$g_3 = +0.9949864 - \frac{1}{-1} \times 0.9949864 = +1.9899728$$

$$X_3 = \frac{1}{-1} \times 0.9848064 - 1.9899728(-0.05)(-0.4080768) + 1.9899728$$
$$= +0.9645633$$
$$Y_3 = \frac{1}{-1} \times 0.1736545 - 1.9899728(-0.05)(3.4838837) = +0.1729872$$
$$Z_3 = \frac{1}{-1} \times 0.0002377 - 1.9899728(-0.05)(2.0047290) = +0.1992301$$

The intersection with the image surface is, by Eqs. (40) to (47)

$$e = -9.962359 \times 0.9645633 - [(-0.4080768)(0.9645633) + 3.4838837 \times 0.1729872 + 2.0047290 \times 0.1992301] = -10.2177796$$
$$M_{4x} = -0.4080768 + (-10.2177796)(0.9645633) - (-9.962359) = -0.3014130$$
$$M_4^2 = (-0.4080768)^2 + 3.4838837^2 + 2.0047290^2 - (-10.2177796)^2 + (-9.962359)^2 - 2(-9.962359)(-0.4080768) = +3.0376724$$
$$E_4 = \{0.9645633^2 - (-0.1)[-0.1 \times 3.0376724 - 2(-0.3014130)]\}^{1/2}$$
$$= +0.9799430$$
$$D_{3,4} = -10.2177796 + \frac{-0.1 \times 3.0376724 - 2(-0.3014130)}{0.9645633 + 0.9799430} = -10.0639830$$
$$x_4 = -0.4080768 + (-10.0639830) \times 0.9645633 - (-9.962359) = -0.1530665$$
$$y_4 = +3.4838837 + (-10.0639830)(0.1729872) = +1.7429435$$
$$z_4 = 2.0047290 + (-10.0639830)(0.1992301) = -0.0003193$$

Thus the final intersection of the ray with the image surface ($c = -0.1$) is calculated to have the coordinates x_4, y_4, and z_4 with respect to the vertex of the image surface.

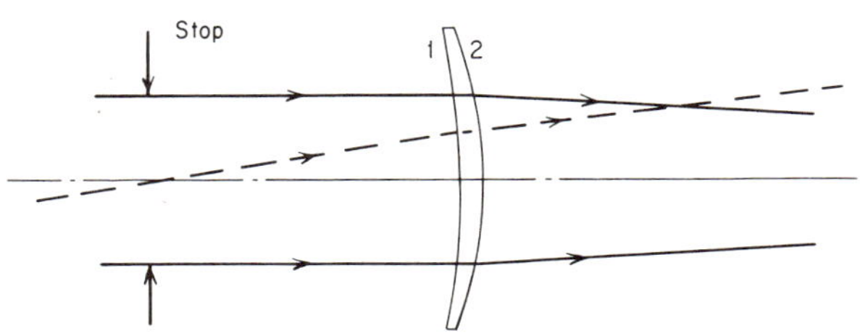

Fig. 74 Thin lens with displaced stop.

70. Thin-Lens Third-Order Aberrations The third-order aberrations of a thin ($t = 0$) meniscus lens (Fig. 74) will be computed using the equations of Par. 19. The lens data are:

$$t = 0 \qquad n = 1.5 \qquad c_1 = -0.2,\quad \Delta n = 0.008,\quad c_2 = -0.4 \qquad V = 62.5$$

Stop located 1.84465 to the left of c_1 ($l_p = -1.84465$).
Stop diameter is 1.0 ($f/10$).
Object is at infinity.
Field angle is ± 0.17633 ($\pm 10°$).
$\phi = (n - 1)(c_1 - c_2) = 0.5[-0.2 - (-0.4)] = +0.1$.
EFL = 10.0.

See Table 5 for data of the "axial" and "principal" rays.

The G numbers for $n = 1.5$ are evaluated from Eq. (115):

$$G_1 = 0.5625 \qquad G_2 = 1.0$$
$$G_3 = 1.375 \qquad G_4 = 0.58333\cdots$$
$$G_5 = 1.666\cdots \qquad G_6 = 1.08333\cdots$$
$$G_7 = 0.666\cdots \qquad G_8 = 0.375$$

TABLE 5 Ray Trace, Eqs. (29) and (30) plus (15) and (16)

	Airspace, object to stop	Stop	Airspace, stop to lens	Lens	Airspace, lens to image
ϕ	. . . .	. . .	. . . .	+0.1	
d	. . . .	. . .	1.84465		
y	. . . .	0.5	. . . .	0.5	
u	0.0	. . .	0.0	. . . .	+0.05
				$f = 10$	$l' = 10$
y_p	. . . .	0.0	. . . .	+0.3252671	
u_p	−0.17633	. . .	−0.17633	. . . .	−0.1438033
				$l_p' = -2.261888$	$h' = +1.7633$

The data required for the G-sum equations can be obtained from the system data and the two ray traces above; they are collected below for convenience.

$$c = -0.2 + 0.4 = +0.2 \qquad V = 62.5$$
$$c_1 = -0.2 \qquad y = 0.5$$
$$\phi = +0.1 \qquad Q = \frac{0.3252671}{0.5} = +0.6505342$$
$$v = \frac{0}{0.5} = 0.0 \qquad u_k' = +0.05$$
$$h = +1.7633$$

First the aberrations of the lens with the stop in contact with the lens are computed, using Eqs. (108) to (114).

$$\begin{aligned}
\text{TSC} &= -0.5^4[0.5625(0.2^3) - 1.0(0.2^2)(-0.2) + 1.375(0.2^2)(0) + 0.5833 \\
&\qquad \times 0.2(-0.2)^2 - 1.666 \times 0.2(-0.2)(0) + 1.0833 \times 0.2 \times 0^2]/0.05 \\
&= -1.25(+0.0045 + 0.008 + 0 + 0.00466 \cdots + 0 + 0) \\
&\qquad = -1.25(0.017166 \cdots) \\
&= -0.021458333 \\
\text{CC} &= -1.7633(0.5^2)[0.25 \times 1.666 \times 0.2(-0.2) - 0.666 \times 0.2 \times 0 - 0.375(0.2^2)] \\
&= -0.440825(-0.01666 \cdots - 0 - 0.015) = -0.440825(-0.031666 \cdots) \\
&= +0.013959458
\end{aligned}$$

$$\text{TAC} = -0.5(1.7633^2)(0.1)(0.05) = -0.00777305$$
$$\text{TPC} = -0.5(1.7633^2)(0.1)\frac{0.05}{1.5} = -0.00518205$$
$$\text{DC} = 0.0$$
$$\text{TLchC} = -(0.5^2)\frac{0.1}{62.5 \times 0.05} = -0.008$$
$$\text{TchC} = 0.0$$

To determine the aberrations with the stop located at a distance of 1.84465 to the left of the lens, Eqs. (100) to (106) are used with $Q = y_p/y = +0.6505342$.

$$\begin{aligned}
\text{TSC}^* &= \text{TSC} = -0.021458 \\
\text{CC}^* &= +0.013959 + (-0.021458)(0.6505342) \\
&= +0.013959 - 0.013959 = 0.0 \\
\text{TAC}^* &= -0.007773 + 0.013959 \times 2 \times 0.6505342 + (-0.021458)(0.6505342^2) \\
&= -0.007773 + 0.0181616 - 0.009081 = +0.001308 \\
\text{TPC}^* &= \text{TPC} = -0.005182 \\
\text{DC}^* &= 0 + [-0.005182 + 3(-0.007773)] \times 0.6505342 + 0.013959 \\
&\qquad \times 3(0.6505342^2) + (-0.021458)(0.6505342^3) \\
&= 0 - 0.018541 + 0.017722 - 0.005908 = -0.006727 \\
\text{TLchC}^* &= \text{TLchC} = -0.008 \\
\text{TchC} &= 0 + (-0.008)(0.6505342) = -0.005204
\end{aligned}$$

MATHEMATICAL TABLES

TABLE 6 Constants

π	=	3.14159 26535 89793 23846
$\frac{180}{\pi}$	=	57.29577 95130 82320 87540
$\frac{\pi}{180}$	=	0.01745 32925 19943 29577
e	=	2.71828 18284 59045 23536
$\sqrt{0.5}$	=	0.70710 67811 86547 52440
$\sqrt{2}$	=	1.41421 35623 73095 04880
$\sqrt{3}$	=	1.73205 08075 68877 29641
$\sqrt{\pi}$	=	1.77245 38509 05516 02729

TABLE 7 Series Expansions

$$\sin x = x - \frac{x^3}{3!} + \frac{x^5}{5!} - \frac{x^7}{7!} + \cdots$$

$$\sin^2 x = x^2 - x^4 \frac{2}{3!} + x^6 \left(\frac{2}{5!} + \frac{1}{3!3!}\right) - x^8 \left(\frac{2}{7!} + \frac{2}{3!5!}\right) + x^{10} \left(\frac{2}{9!} + \frac{2}{3!7!} + \frac{1}{5!5!}\right) - x^{12} \left(\frac{2}{11!} + \frac{2}{3!9!} + \frac{2}{5!7!}\right) + \cdots$$

$$\sin^3 x = x^3 - x^5 \frac{3}{3!} + x^7 \left(\frac{3}{5!} + \frac{3}{3!3!}\right) - x^9 \left(\frac{3}{7!} + \frac{6}{3!5!} + \frac{1}{3!3!}\right) + \cdots$$

$$\sin^4 x = x^4 - x^6 \frac{2}{3} + x^8 \frac{1}{5} - x^{10} \frac{34}{945} + \cdots$$

$$\sin^{-1} x = x + \frac{1}{2}\frac{x^3}{3} + \frac{1}{2}\frac{3}{4}\frac{x^5}{5} + \frac{1}{2}\frac{3}{4}\frac{5}{6}\frac{x^7}{7} + \cdots$$

$$\cos x = 1 - \frac{x^2}{2!} + \frac{x^4}{4!} - \frac{x^6}{6!} + \cdots$$

$$\cos^2 x = 1 - \sin^2 x = 1 - x^2 + \frac{x^4}{3} - \frac{2x^6}{45} + \frac{16x^8}{7!} - \cdots$$

$$\cos^3 x = 1 - \frac{3x^2}{2} + \frac{21x^4}{4!} - \frac{183x^6}{6!} + \frac{1{,}641x^8}{8!} - \cdots$$

$$\cos^4 x = 1 - 2x^2 + \frac{5x^4}{3} - \frac{34x^6}{45} + \frac{13x^8}{63} - \cdots$$

$$\cos^{-1} x = \frac{\pi}{2} - x - \frac{1}{2}\frac{x^3}{3} - \frac{1}{2}\frac{3}{4}\frac{x^5}{5} - \cdots$$

$$\tan x = x + \frac{x^3}{3} + \frac{2x^5}{15} + \frac{17x^7}{315} + \frac{62x^9}{2{,}835} - \cdots \qquad x^2 < \frac{\pi^2}{4}$$

$$\tan x = \sin x + \frac{1}{2}\sin^3 x + \frac{1 \cdot 3}{2^2 \cdot 2!}\sin^5 x + \frac{1 \cdot 3 \cdot 5}{2^3 \cdot 3!}\sin^7 x + \cdots$$

$$\tan^2 x = x^2 + \frac{3x^4}{2} + \frac{17x^6}{45} + \frac{62x^8}{315} + \frac{1{,}382x^{10}}{9^2 \cdot 7 \cdot 25} + \cdots$$

TABLE 7 Series Expansions (Continued)

$$\tan^{-1} x = x - \frac{x^3}{3} + \frac{x^5}{5} - \frac{x^7}{7} + \cdots \qquad x^2 < 1$$

$$\tan^{-1} x = \frac{\pi}{2} - \frac{1}{x} + \frac{1}{3x^3} - \frac{1}{5x^5} + \cdots \qquad x^2 > 1$$

$$e^x = 1 + x + \frac{x^2}{2!} + \frac{x^3}{3!} + \frac{x^4}{4!} - \cdots$$

$$e^z = e^x(\cos y + i \sin y) \qquad z = x + iy$$

$$a^x = 1 + x \ln a + \frac{(x \ln a)^2}{2!} + \frac{(x \ln a)^3}{3!} + \cdots$$

$$(x + y)^n = x^n + nx^{(n-1)}y + \frac{n(n-1)}{2!}x^{(n-2)}y^2 + \frac{n(n-1)}{3!}(n-2)x^{(n-3)}y^3 + \cdots \qquad y^2 < x^2$$

$$f(x + h) = f(x) + hf'(x) + \frac{h^2 f''(x)}{2!} + \frac{h^3 f'''(x)}{3!} + \cdots$$
$$= f(h) + xf'(h) + \frac{x^2 f''(h)}{2!} + \frac{x^3 f'''(h)}{3!} + \cdots$$

$$f(x) = f(0) + xf'(0) + \frac{x^2 f''(0)}{2!} + \frac{x^3 f'''(0)}{3!} + \cdots$$

If

$$y = a_1 x + a_2 x^2 + a_3 x^3 + \cdots$$

then

$$x = A_1 y + A_2 y^2 + A_3 y^3 + \cdots$$

where

$$A_1 = \frac{1}{a_1} \qquad A_2 = \frac{a_2}{-a_1^3} \qquad A_3 = \frac{2a_2^2 - a_1 a_3}{a_1^5}$$

$$A_4 = (5a_1 a_2 a_3 - a_1^2 a_4 - 5a_2^3)/a_1^7$$

$$A_5 = (6a_1^2 a_2 a_4 + 3a_1^2 a_3^2 + 14a_2^4 - a_1^3 a_5 - 21 a_1 a_2^2 a_3)/a_1^9$$

$$A_6 = (7a_1^3 a_2 a_5 + 7a_1^3 a_3 a_4 + 84 a_1 a_2^3 a_3 - a_1^4 a_6 - 28 a_1^2 a_2^2 a_4 - 28 a_1^2 a_2 a_3^2 - 42 a_2^5)/a_1^{11}$$

$$A_7 = (8a_1^4 a_2 a_6 + 8a_1^4 a_3 a_5 + 4a_1^4 a_4^2 + 120 a_1^2 a_2^3 a_4 + 180 a_1^2 a_2^2 a_3^2 + 132 a_2^6 - a_1^5 a_7 - 36 a_1^3 a_2^2 a_5 - 72 a_1^3 a_2 a_3 a_4 - 12 a_1^3 a_3^3 - 330 a_1 a_2^4 a_3)/a_1^{13}$$

TABLE 8 Equations of a Conic Section

Conic section through the origin:

$$y^2 - 2rx + px^2 = 0$$

$$x = \frac{r \pm (r^2 - py^2)^{1/2}}{p}$$

$$x = \frac{cy^2}{1 + (1 - pc^2 y^2)^{1/2}}$$

$$x = \frac{y^2}{2r} + \frac{1py^4}{2^2 \cdot 2! \cdot r^3} + \frac{1 \cdot 3 \cdot p^2 y^6}{2^3 \cdot 3! \cdot r^5} + \frac{1 \cdot 3 \cdot 5 \cdot p^3 y^8}{2^4 \cdot 4! \cdot r^7} + \cdots$$

where r = radius of osculation
$c = 1/r$
$p > 1$ = ellipse
$p = 1$ = circle
$1 > p > 0$ = ellipse
$p = 0$ = parabola
$p < 0$ = hyperbola

Conic section in polar coordinates, with focus at the origin:

$$\rho = \frac{r}{1 + e \cos \theta} \qquad \text{where } p = 1 - e^2$$

TABLE 9 Sine-Angle Conversion (MIL-HDBK-141, 1962)

Sine	Angle	Interpolation constant	Sine	Angle	Interpolation constant	Sine	Angle	Interpolation constant
.001	0.001000	1.000	.061	0.061036	1.002	.121	0.121296	1.009
.002	0.002000	1.000	.062	0.062038	1.002	.122	0.122305	1.007
.003	0.003000	1.000	.063	0.063040	1.002	.123	0.123312	1.008
.004	0.004000	1.000	.064	0.064042	1.003	.124	0.124320	1.007
.005	0.005000	1.000	.065	0.065045	1.002	.125	0.125327	1.008
.006	0.006000	1.000	.066	0.066047	1.002	.126	0.126335	1.008
.007	0.007000	1.000	.067	0.067049	1.003	.127	0.127343	1.008
.008	0.008000	1.000	.068	0.068052	1.002	.128	0.128351	1.008
.009	0.009000	1.000	.069	0.069054	1.003	.129	0.129359	1.009
.010	0.010000	1.000	.070	0.070057	1.002	.130	0.130368	1.009
.011	0.011000	1.000	.071	0.071059	1.003	.131	0.131377	1.009
.012	0.012000	1.000	.072	0.072062	1.002	.132	0.132386	1.009
.013	0.013000	1.000	.073	0.073064	1.003	.133	0.133395	1.009
.014	0.014000	1.000	.074	0.074067	1.003	.134	0.134404	1.009
.015	0.015000	1.000	.075	0.075070	1.003	.135	0.135413	1.009
.016	0.016000	1.000	.076	0.076073	1.003	.136	0.136422	1.010
.017	0.017000	1.000	.077	0.077076	1.003	.137	0.137432	1.010
.018	0.018000	1.000	.078	0.078079	1.003	.138	0.138442	1.009
.019	0.019000	1.000	.079	0.079082	1.003	.139	0.139451	1.010
.020	0.020000	1.001	.080	0.080085	1.003	.140	0.140461	1.010
.021	0.021001	1.000	.081	0.081088	1.003	.141	0.141471	1.010
.022	0.022001	1.000	.082	0.082091	1.004	.142	0.142481	1.010
.023	0.023001	1.001	.083	0.083095	1.003	.143	0.143491	1.011
.024	0.024002	1.000	.084	0.084098	1.004	.144	0.144502	1.010
.025	0.025002	1.000	.085	0.085102	1.004	.145	0.145512	1.011
.026	0.026002	1.000	.086	0.086106	1.003	.146	0.146523	1.011
.027	0.027002	1.001	.087	0.087109	1.004	.147	0.147534	1.011
.028	0.028003	1.000	.088	0.088113	1.004	.148	0.148545	1.011
.029	0.029003	1.001	.089	0.089117	1.004	.149	0.149556	1.012
.030	0.030004	1.000	.090	0.090121	1.004	.150	0.150568	1.011
.031	0.031004	1.001	.091	0.091125	1.004	.151	0.151579	1.012
.032	0.032005	1.001	.092	0.092129	1.005	.152	0.152591	1.011
.033	0.033006	1.000	.093	0.093134	1.004	.153	0.153602	1.012
.034	0.034006	1.001	.094	0.094138	1.004	.154	0.154614	1.013
.035	0.035007	1.000	.095	0.095142	1.005	.155	0.155627	1.012
.036	0.036007	1.001	.096	0.096147	1.005	.156	0.156639	1.013
.037	0.037008	1.001	.097	0.097152	1.005	.157	0.157652	1.012
.038	0.038009	1.000	.098	0.098157	1.004	.158	0.158664	1.013
.039	0.039009	1.001	.099	0.099161	1.006	.159	0.159677	1.013
.040	0.040010	1.001	.100	0.100167	1.005	.160	0.160690	1.013
.041	0.041011	1.001	.101	0.101172	1.005	.161	0.161703	1.013
.042	0.042012	1.001	.102	0.102177	1.006	.162	0.162716	1.013
.043	0.043013	1.001	.103	0.103183	1.005	.163	0.163729	1.014
.044	0.044014	1.001	.104	0.104188	1.006	.164	0.164743	1.014
.045	0.045015	1.001	.105	0.105194	1.005	.165	0.165757	1.014
.046	0.046016	1.001	.106	0.106199	1.005	.166	0.166771	1.014
.047	0.047017	1.001	.107	0.107204	1.006	.167	0.167785	1.014
.048	0.048018	1.001	.108	0.108210	1.006	.168	0.168799	1.015
.049	0.049019	1.001	.109	0.109216	1.005	.169	0.169814	1.015
.050	0.050020	1.001	.110	0.110221	1.007	.170	0.170829	1.014
.051	0.051021	1.001	.111	0.111228	1.006	.171	0.171843	1.016
.052	0.052022	1.001	.112	0.112234	1.007	.172	0.172859	1.015
.053	0.053023	1.002	.113	0.113241	1.006	.173	0.173874	1.015
.054	0.054025	1.001	.114	0.114247	1.007	.174	0.174889	1.016
.055	0.055026	1.002	.115	0.115254	1.006	.175	0.175905	1.015
.056	0.056028	1.001	.116	0.116260	1.007	.176	0.176920	1.016
.057	0.057029	1.002	.117	0.117267	1.007	.177	0.177936	1.016
.058	0.058031	1.002	.118	0.118274	1.007	.178	0.178952	1.016
.059	0.059033	1.002	.119	0.119281	1.008	.179	0.179968	1.017
.060	0.060035	1.001	.120	0.120280	1.007	.180	0.180985	1.016

TABLE 9 Sine-Angle Conversion (Continued)

Sine	Angle	Inter-polation constant	Sine	Angle	Inter-polation constant	Sine	Angle	Inter-polation constant
.181	0.182001	1.017	.241	0.243395	1.031	.301	0.305741	1.049
.182	0.183018	1.018	.242	0.244426	1.031	.302	0.306790	1.049
.183	0.184036	1.017	.243	0.245457	1.031	.303	0.307839	1.049
.184	0.185053	1.017	.244	0.246488	1.031	.304	0.308888	1.050
.185	0.186070	1.018	.245	0.247519	1.032	.305	0.309938	1.050
.186	0.187088	1.018	.246	0.248551	1.032	.306	0.310988	1.051
.187	0.188106	1.018	.247	0.249583	1.032	.307	0.312039	1.051
.188	0.189124	1.018	.248	0.250615	1.032	.308	0.313090	1.051
.189	0.190142	1.019	.249	0.251647	1.033	.309	0.314141	1.052
.190	0.191161	1.019	.250	0.252680	1.033	.310	0.315193	1.052
.191	0.192180	1.019	.251	0.253717	1.033	.311	0.316245	1.052
.192	0.193199	1.019	.252	0.254746	1.033	.312	0.317297	1.053
.193	0.194218	1.019	.253	0.255779	1.035	.313	0.318350	1.053
.194	0.195237	1.020	.254	0.256814	1.034	.314	0.319403	1.053
.195	0.196257	1.019	.255	0.257848	1.034	.315	0.320456	1.054
.196	0.197276	1.020	.256	0.258882	1.035	.316	0.321510	1.054
.197	0.198296	1.020	.257	0.259917	1.034	.317	0.322564	1.054
.198	0.199316	1.021	.258	0.260951	1.035	.318	0.323618	1.055
.199	0.200337	1.020	.259	0.261986	1.035	.319	0.324673	1.056
.200	0.201357	1.021	.260	0.263021	1.036	.320	0.325729	1.056
.201	0.202378	1.021	.261	0.264057	1.036	.321	0.326785	1.056
.202	0.203399	1.021	.262	0.265093	1.037	.322	0.327841	1.056
.203	0.204420	1.021	.263	0.266130	1.036	.323	0.328897	1.057
.204	0.205441	1.022	.264	0.267166	1.037	.324	0.329954	1.058
.205	0.206463	1.022	.265	0.268203	1.038	.325	0.331012	1.057
.206	0.207485	1.022	.266	0.269241	1.037	.326	0.332069	1.058
.207	0.208507	1.022	.267	0.270278	1.038	.327	0.333127	1.058
.208	0.209529	1.022	.268	0.271316	1.038	.328	0.334185	1.059
.209	0.210551	1.023	.269	0.272354	1.039	.329	0.335244	1.059
.210	0.211574	1.023	.270	0.273393	1.038	.330	0.336303	1.060
.211	0.212597	1.023	.271	0.274431	1.039	.331	0.337363	1.060
.212	0.213620	1.023	.272	0.275470	1.039	.332	0.338423	1.060
.213	0.214643	1.024	.273	0.276509	1.040	.333	0.339483	1.061
.214	0.215667	1.024	.274	0.277549	1.040	.334	0.340544	1.061
.215	0.216691	1.024	.275	0.278589	1.040	.335	0.341605	1.061
.216	0.217715	1.024	.276	0.279629	1.041	.336	0.342666	1.062
.217	0.218739	1.025	.277	0.280670	1.041	.337	0.343728	1.063
.218	0.219764	1.024	.278	0.281711	1.041	.338	0.344791	1.062
.219	0.220788	1.025	.279	0.282752	1.041	.339	0.345853	1.063
.220	0.221813	1.026	.280	0.283793	1.042	.340	0.346916	1.064
.221	0.222839	1.025	.281	0.284835	1.042	.341	0.347980	1.064
.222	0.223864	1.026	.282	0.285877	1.043	.342	0.349044	1.064
.223	0.224890	1.026	.283	0.286920	1.043	.343	0.350108	1.065
.224	0.225916	1.026	.284	0.287963	1.043	.344	0.351173	1.065
.225	0.226942	1.027	.285	0.289006	1.043	.345	0.352238	1.065
.226	0.227969	1.026	.286	0.290049	1.044	.346	0.353303	1.067
.227	0.228995	1.027	.287	0.291093	1.044	.347	0.354370	1.067
.228	0.230022	1.027	.288	0.292137	1.044	.348	0.355437	1.066
.229	0.231049	1.028	.289	0.293181	1.045	.349	0.356503	1.067
.230	0.232077	1.027	.290	0.294226	1.045	.350	0.357570	1.068
.231	0.233104	1.028	.291	0.295271	1.045	.351	0.358638	1.068
.232	0.234132	1.029	.292	0.296316	1.047	.352	0.359706	1.068
.233	0.235161	1.028	.293	0.297363	1.046	.353	0.360774	1.069
.234	0.236189	1.029	.294	0.298409	1.046	.354	0.361843	1.070
.235	0.237218	1.028	.295	0.299455	1.047	.355	0.362913	1.070
.236	0.238246	1.030	.296	0.300502	1.047	.356	0.363983	1.070
.237	0.239276	1.029	.297	0.301549	1.048	.357	0.365053	1.071
.238	0.240305	1.030	.298	0.302597	1.048	.358	0.366124	1.072
.239	0.241335	1.030	.299	0.303645	1.047	.359	0.367196	1.072
.240	0.242365	1.030	.300	0.304692	1.049	.360	0.368268	1.071

TABLE 9 Sine-Angle Conversion (Continued)

Sine	Angle	Interpolation constant	Sine	Angle	Interpolation constant	Sine	Angle	Interpolation constant
.361	0.369339	1.073	.421	0.434548	1.102	.481	0.501795	1.141
.362	0.370412	1.073	.422	0.435650	1.103	.482	0.502936	1.142
.363	0.371485	1.074	.423	0.436753	1.104	.483	0.504078	1.142
.364	0.372559	1.074	.424	0.437857	1.104	.484	0.505220	1.143
.365	0.373633	1.074	.425	0.438961	1.105	.485	0.506363	1.144
.366	0.374707	1.075	.426	0.440066	1.106	.486	0.507507	1.145
.367	0.375782	1.075	.427	0.441172	1.106	.487	0.508652	1.145
.368	0.376857	1.076	.428	0.442278	1.107	.488	0.509797	1.146
.369	0.377933	1.076	.429	0.443385	1.107	.489	0.510943	1.147
.370	0.379009	1.076	.430	0.444492	1.108	.490	0.512090	1.147
.371	0.380085	1.077	.431	0.445600	1.109	.491	0.513237	1.148
.372	0.381162	1.078	.432	0.446709	1.108	.492	0.514385	1.149
.373	0.382240	1.078	.433	0.447817	1.110	.493	0.515534	1.150
.374	0.383318	1.078	.434	0.448927	1.111	.494	0.516684	1.151
.375	0.384396	1.079	.435	0.450038	1.110	.495	0.517835	1.151
.376	0.385475	1.079	.436	0.451148	1.112	.496	0.518986	1.152
.377	0.386554	1.081	.437	0.452260	1.112	.497	0.520138	1.153
.378	0.387635	1.080	.438	0.453372	1.113	.498	0.521291	1.153
.379	0.388715	1.081	.439	0.454485	1.113	.499	0.522444	1.155
.380	0.389796	1.082	.440	0.455598	1.113	.500	0.523599	1.155
.381	0.390878	1.081	.441	0.456711	1.115	.501	0.524754	1.156
.382	0.391959	1.082	.442	0.457826	1.115	.502	0.525910	1.156
.383	0.393041	1.083	.443	0.458941	1.116	.503	0.527066	1.158
.384	0.394124	1.083	.444	0.460057	1.116	.504	0.528224	1.158
.385	0.395207	1.084	.445	0.461173	1.117	.505	0.529382	1.159
.386	0.396291	1.084	.446	0.462290	1.118	.506	0.530541	1.160
.387	0.397375	1.085	.447	0.463408	1.119	.507	0.531701	1.160
.388	0.398460	1.085	.448	0.464527	1.118	.508	0.532861	1.162
.389	0.399545	1.086	.449	0.465645	1.120	.509	0.534023	1.162
.390	0.400631	1.086	.450	0.466765	1.120	.510	0.535185	1.163
.391	0.401717	1.087	.451	0.467885	1.121	.511	0.536348	1.163
.392	0.402804	1.087	.452	0.469006	1.121	.512	0.537511	1.165
.393	0.403891	1.088	.453	0.470127	1.122	.513	0.538676	1.165
.394	0.404979	1.088	.454	0.471249	1.123	.514	0.539841	1.167
.395	0.406067	1.088	.455	0.472372	1.123	.515	0.541008	1.167
.396	0.407155	1.090	.456	0.473495	1.124	.516	0.542175	1.167
.397	0.408245	1.090	.457	0.474619	1.125	.517	0.543342	1.169
.398	0.409335	1.090	.458	0.475744	1.125	.518	0.544511	1.170
.399	0.410425	1.091	.459	0.476869	1.126	.519	0.545681	1.170
.400	0.411516	1.091	.460	0.477995	1.127	.520	0.546851	1.171
.401	0.412607	1.092	.461	0.479122	1.127	.521	0.548022	1.172
.402	0.413699	1.092	.462	0.480249	1.128	.522	0.549194	1.173
.403	0.414791	1.093	.463	0.481377	1.128	.523	0.550367	1.173
.404	0.415884	1.094	.464	0.482505	1.130	.524	0.551540	1.175
.405	0.416978	1.094	.465	0.483635	1.130	.525	0.552715	1.175
.406	0.418072	1.095	.466	0.484765	1.130	.526	0.553890	1.177
.407	0.419167	1.095	.467	0.485895	1.131	.527	0.555067	1.177
.408	0.420262	1.096	.468	0.487026	1.132	.528	0.556244	1.178
.409	0.421358	1.096	.469	0.488158	1.133	.529	0.557422	1.179
.410	0.422454	1.096	.470	0.489291	1.133	.530	0.558601	1.179
.411	0.423550	1.097	.471	0.490424	1.134	.531	0.559780	1.181
.412	0.424647	1.098	.472	0.491558	1.135	.532	0.560961	1.181
.413	0.425745	1.099	.473	0.492693	1.135	.533	0.562142	1.182
.414	0.426844	1.099	.474	0.493828	1.136	.534	0.563324	1.184
.415	0.427943	1.100	.475	0.494964	1.137	.535	0.564508	1.184
.416	0.429043	1.099	.476	0.496101	1.137	.536	0.565692	1.185
.417	0.430142	1.101	.477	0.497238	1.138	.537	0.566877	1.186
.418	0.431243	1.101	.478	0.498376	1.139	.538	0.568063	1.186
.419	0.432344	1.101	.479	0.499515	1.140	.539	0.569249	1.188
.420	0.433445	1.103	.480	0.500655	1.140	.540	0.570437	1.189

TABLE 9 Sine-Angle Conversion (Continued)

Sine	Angle	Inter-polation constant	Sine	Angle	Inter-polation constant	Sine	Angle	Inter-polation constant
.541	0.571626	1.189	.601	0.644752	1.251	.661	0.722151	1.333
.542	0.572815	1.191	.602	0.646003	1.253	.662	0.723484	1.335
.543	0.574006	1.191	.603	0.647256	1.255	.663	0.724819	1.337
.544	0.575197	1.192	.604	0.648511	1.255	.664	0.726156	1.338
.545	0.576389	1.193	.605	0.649766	1.256	.665	0.727494	1.340
.546	0.577582	1.194	.606	0.651022	1.258	.666	0.728834	1.341
.547	0.578776	1.195	.607	0.652280	1.259	.667	0.730175	1.343
.548	0.579971	1.196	.608	0.653539	1.260	.668	0.731518	1.345
.549	0.581167	1.197	.609	0.654799	1.262	.669	0.732863	1.346
.550	0.582364	1.198	.610	0.656061	1.262	.670	0.734209	1.348
.551	0.583562	1.199	.611	0.657323	1.264	.671	0.735557	1.349
.552	0.584761	1.200	.612	0.658587	1.265	.672	0.736906	1.351
.553	0.585961	1.200	.613	0.659852	1.266	.673	0.738257	1.353
.554	0.587161	1.202	.614	0.661118	1.268	.674	0.739610	1.355
.555	0.588363	1.203	.615	0.662386	1.269	.675	0.740965	1.356
.556	0.589566	1.203	.616	0.663655	1.270	.676	0.742321	1.358
.557	0.590769	1.205	.617	0.664925	1.271	.677	0.743679	1.359
.558	0.591974	1.205	.618	0.666196	1.273	.678	0.745038	1.362
.559	0.593179	1.207	.619	0.667469	1.274	.679	0.746400	1.363
.560	0.594386	1.207	.620	0.668743	1.275	.680	0.747763	1.364
.561	0.595593	1.209	.621	0.670018	1.276	.681	0.749127	1.367
.562	0.596802	1.209	.622	0.671294	1.278	.682	0.750494	1.368
.563	0.598011	1.211	.623	0.672572	1.279	.683	0.751862	1.370
.564	0.599222	1.211	.624	0.673851	1.281	.684	0.753232	1.372
.565	0.600433	1.213	.625	0.675132	1.281	.685	0.754604	1.373
.566	0.601646	1.213	.626	0.676413	1.283	.686	0.755977	1.376
.567	0.602859	1.215	.627	0.677696	1.284	.687	0.757353	1.377
.568	0.604074	1.215	.628	0.678980	1.286	.688	0.758730	1.378
.569	0.605289	1.217	.629	0.680266	1.287	.689	0.760108	1.381
.570	0.606506	1.217	.630	0.681553	1.289	.690	0.761489	1.383
.571	0.607723	1.219	.631	0.682842	1.289	.691	0.762872	1.384
.572	0.608942	1.220	.632	0.684131	1.291	.692	0.764256	1.386
.573	0.610162	1.220	.633	0.685422	1.293	.693	0.765642	1.388
.574	0.611382	1.222	.634	0.686715	1.293	.694	0.767030	1.390
.575	0.612604	1.223	.635	0.688008	1.296	.695	0.768420	1.392
.576	0.613827	1.224	.636	0.689304	1.296	.696	0.769812	1.393
.577	0.615051	1.225	.637	0.690600	1.298	.697	0.771205	1.396
.578	0.616276	1.226	.638	0.691898	1.299	.698	0.772601	1.397
.579	0.617502	1.227	.639	0.693197	1.301	.699	0.773998	1.400
.580	0.618729	1.228	.640	0.694498	1.302	.700	0.775398	1.401
.581	0.619957	1.229	.641	0.695800	1.304	.701	0.776799	1.403
.582	0.621186	1.230	.642	0.697104	1.305	.702	0.778202	1.405
.583	0.622416	1.232	.643	0.698409	1.306	.703	0.779607	1.407
.584	0.623648	1.232	.644	0.699715	1.308	.704	0.781014	1.409
.585	0.624880	1.234	.645	0.701023	1.310	.705	0.782423	1.411
.586	0.626114	1.234	.646	0.702333	1.310	.706	0.783334	1.413
.587	0.627348	1.236	.647	0.703643	1.313	.707	0.785247	1.415
.588	0.628584	1.237	.648	0.704956	1.313	.708	0.786662	1.417
.589	0.629821	1.238	.649	0.706269	1.315	.709	0.788079	1.419
.590	0.631059	1.239	.650	0.707584	1.317	.710	0.789498	1.421
.591	0.632298	1.240	.651	0.708901	1.318	.711	0.790919	1.423
.592	0.633538	1.242	.652	0.710219	1.320	.712	0.792342	1.426
.593	0.634780	1.242	.653	0.711539	1.321	.713	0.793768	1.427
.594	0.636022	1.244	.654	0.712860	1.323	.714	0.795195	1.429
.595	0.637266	1.245	.655	0.714183	1.324	.715	0.796624	1.432
.596	0.638511	1.245	.656	0.715507	1.325	.716	0.798056	1.433
.597	0.639756	1.247	.657	0.716832	1.328	.717	0.799489	1.436
.598	0.641003	1.249	.658	0.718160	1.329	.718	0.800925	1.438
.599	0.642252	1.249	.659	0.719489	1.330	.719	0.802363	1.439
.600	0.643501	1.251	.660	0.720819	1.332	.720	0.803802	1.442

TABLE 9 Sine-Angle Conversion (Continued)

Sine	Angle	Inter-polation constant	Sine	Angle	Inter-polation constant	Sine	Angle	Inter-polation constant
.721	0.805244	1.445	.781	0.896265	1.603	.841	0.999129	1.851
.722	0.806689	1.446	.782	0.897868	1.606	.842	1.000980	1.856
.723	0.808135	1.449	.783	0.899474	1.610	.843	1.002862	1.862
.724	0.809584	1.450	.784	0.901084	1.612	.844	1.004698	1.867
.725	0.811034	1.453	.785	0.902696	1.616	.845	1.006565	1.873
.726	0.812487	1.456	.786	0.904312	1.619	.846	1.008438	1.878
.727	0.813943	1.457	.787	0.905931	1.623	.847	1.010316	1.884
.728	0.815400	1.460	.788	0.907554	1.626	.848	1.012200	1.890
.729	0.816860	1.462	.789	0.909180	1.629	.849	1.014090	1.895
.730	0.818322	1.464	.790	0.910809	1.633	.850	1.015985	1.902
.731	0.819786	1.467	.791	0.912442	1.636	.851	1.017887	1.907
.732	0.821253	1.469	.792	0.914078	1.640	.852	1.019794	1.913
.733	0.822722	1.471	.793	0.915718	1.643	.853	1.021707	1.919
.734	0.824693	1.474	.794	0.917361	1.647	.854	1.023626	1.925
.735	0.825667	1.476	.795	0.919008	1.650	.855	1.025551	1.931
.736	0.827143	1.478	.796	0.920658	1.654	.856	1.027482	1.938
.737	0.828621	1.481	.797	0.922312	1.657	.857	1.029420	1.943
.738	0.830102	1.483	.798	0.923969	1.661	.858	1.031363	1.950
.739	0.831585	1.485	.799	0.925630	1.665	.859	1.033313	1.957
.740	0.833070	1.488	.800	0.927295	1.669	.860	1.035270	1.963
.741	0.834558	1.491	.801	0.928964	1.672	.861	1.037233	1.969
.742	0.836049	1.493	.802	0.930636	1.676	.862	1.039202	1.976
.743	0.837542	1.495	.803	0.932312	1.680	.863	1.041178	1.983
.744	0.839037	1.498	.804	0.933992	1.683	.864	1.043161	1.989
.745	0.840535	1.500	.805	0.935675	1.688	.865	1.045150	1.997
.746	0.842035	1.503	.806	0.937363	1.691	.866	1.047147	2.003
.747	0.843538	1.505	.807	0.939054	1.696	.867	1.049150	2.010
.748	0.845043	1.509	.808	0.940750	1.699	.868	1.051160	2.018
.749	0.846552	1.510	.809	0.942449	1.703	.869	1.053178	2.024
.750	0.848062	1.513	.810	0.944152	1.707	.870	1.055202	2.032
.751	0.849575	1.516	.811	0.945859	1.712	.871	1.057234	2.039
.752	0.851091	1.518	.812	0.947571	1.715	.872	1.059273	2.047
.753	0.852609	1.521	.813	0.949286	1.720	.873	1.061320	2.054
.754	0.854130	1.524	.814	0.951006	1.723	.874	1.063374	2.062
.755	0.855654	1.526	.815	0.952729	1.728	.875	1.065436	2.069
.756	0.857180	1.529	.816	0.954457	1.732	.876	1.067505	2.078
.757	0.858709	1.532	.817	0.956189	1.736	.877	1.069583	2.085
.758	0.860241	1.535	.818	0.957925	1.741	.878	1.071668	2.093
.759	0.861776	1.537	.819	0.959666	1.745	.879	1.073761	2.101
.760	0.863313	1.540	.820	0.961411	1.749	.880	1.075862	2.110
.761	0.864853	1.543	.821	0.963160	1.754	.881	1.077972	2.117
.762	0.866396	1.546	.822	0.964914	1.758	.882	1.080089	2.127
.763	0.867942	1.548	.823	0.966672	1.763	.883	1.082216	2.135
.764	0.869490	1.551	.824	0.968435	1.767	.884	1.084351	2.143
.765	0.871041	1.555	.825	0.970202	1.772	.885	1.086494	2.152
.766	0.872596	1.557	.826	0.971974	1.776	.886	1.088646	2.161
.767	0.874153	1.559	.827	0.973750	1.782	.887	1.090807	2.170
.768	0.875712	1.563	.828	0.975532	1.785	.888	1.092977	2.180
.769	0.877275	1.566	.829	0.977317	1.791	.889	1.095157	2.188
.770	0.878841	1.569	.830	0.979108	1.795	.890	1.097345	2.198
.771	0.880410	1.572	.831	0.980903	1.800	.891	1.099543	2.207
.772	0.881982	1.574	.832	0.982703	1.805	.892	1.101750	2.218
.773	0.883556	1.578	.833	0.984508	1.810	.893	1.103968	2.226
.774	0.885134	1.581	.834	0.986318	1.815	.894	1.106194	2.238
.775	0.886715	1.584	.835	0.988133	1.820	.895	1.108432	2.246
.776	0.888299	1.587	.836	0.989953	1.825	.896	1.110678	2.257
.777	0.889886	1.590	.837	0.991778	1.830	.897	1.112935	2.268
.778	0.891476	1.593	.838	0.993608	1.835	.898	1.115203	2.278
.779	0.893069	1.597	.839	0.995443	1.840	.899	1.117481	2.289
.780	0.894666	1.599	.840	0.997283	1.846	.900	1.119770	2.299

TABLE 9 Sine-Angle Conversion (Continued)

Sine	Angle	Inter-polation constant	Sine	Angle	Inter-polation constant	Sine	Angle	Inter-polation constant
.901	1.122069	2.311	.935	1.208259	2.831	.969	1.321151	4.080
.902	1.124380	2.322	.936	1.211090	2.852	.970	1.325231	4.148
.903	1.126702	2.333	.937	1.213942	2.874	.971	1.329379	4.219
.904	1.129035	2.345	.938	1.216816	2.896	.972	1.333598	4.294
.905	1.131380	2.356	.939	1.219712	2.919	.973	1.337892	4.373
.906	1.133736	2.368	.940	1.222631	2.943	.974	1.342265	4.457
.907	1.136104	2.381	.941	1.225574	2.967	.975	1.346722	4.545
.908	1.138485	2.393	.942	1.228541	2.992	.976	1.351267	4.640
.909	1.140878	2.406	.943	1.231533	3.018	.977	1.355907	4.742
.910	1.143282	2.419	.944	1.234551	3.044	.978	1.360649	4.849
.911	1.145703	2.431	.945	1.237595	3.071	.979	1.365498	4.964
.912	1.148134	2.444	.946	1.240666	3.099	.980	1.370462	5.089
.913	1.150578	2.458	.947	1.243765	3.127	.981	1.375551	5.223
.914	1.153036	2.472	.948	1.246892	3.157	.982	1.380774	5.370
.915	1.155508	2.486	.949	1.250049	3.187	.983	1.386144	5.529
.916	1.157994	2.500	.950	1.253236	3.218	.984	1.391673	5.702
.917	1.160494	2.514	.951	1.256454	3.251	.985	1.397375	5.894
.918	1.163008	2.529	.952	1.259705	3.284	.986	1.403269	6.108
.919	1.165537	2.544	.953	1.262989	3.317	.987	1.409377	6.345
.920	1.168081	2.559	.954	1.266306	3.354	.988	1.415722	6.614
.921	1.170640	2.575	.955	1.269660	3.390	.989	1.422336	6.921
.922	1.173215	2.591	.956	1.273050	3.428	.990	1.429257	7.275
.923	1.175806	2.607	.957	1.276478	3.467	.991	1.436532	7.689
.924	1.178413	2.623	.958	1.279945	3.508	.992	1.444221	8.186
.925	1.181036	2.640	.959	1.283453	3.550	.993	1.452407	8.791
.926	1.183676	2.658	.960	1.287003	3.593	.994	1.461198	9.557
.927	1.186334	2.675	.961	1.290596	3.639	.995	1.470755	10.569
.928	1.189009	2.693	.962	1.294235	3.686	.996	1.481324	11.993
.929	1.191702	2.711	.963	1.297921	3.736	.997	1.493317	14.225
.930	1.194413	2.730	.964	1.301657	3.787	.998	1.507542	18.529
.931	1.197143	2.749	.965	1.305444	3.840	.999	1.526071	44.725
.932	1.199892	2.769	.966	1.309284	3.896	1.000	1.570796	
.933	1.202661	2.789	.967	1.313180	3.955			
.934	1.205450	2.809	.968	1.317135	4.016			

TABLE 10 Angle-Sine Conversion (MIL-HDBK-141, 1942)

Angle, mrad	Sine	Interpolation constant	Angle, mrad	Sine	Interpolation constant	Angle, mrad	Sine	Interpolation constant
001	0.001000	1.000	061	0.060962	0.998	121	0.120705	0.993
002	0.002000	1.000	062	0.061960	0.998	122	0.121698	0.992
003	0.003000	1.000	063	0.062958	0.998	123	0.122690	0.992
004	0.004000	1.000	064	0.063956	0.998	124	0.123682	0.993
005	0.005000	1.000	065	0.064954	0.998	125	0.124675	0.992
006	0.006000	1.000	066	0.065952	0.998	126	0.125667	0.992
007	0.007000	1.000	067	0.066950	0.998	127	0.126659	0.992
008	0.008000	1.000	068	0.067948	0.997	128	0.127651	0.992
009	0.009000	1.000	069	0.068945	0.998	129	0.128643	0.991
010	0.010000	1.000	070	0.069943	0.997	130	0.129634	0.992
011	0.011000	1.000	071	0.070940	0.998	131	0.130626	0.991
012	0.012000	1.000	072	0.071938	0.997	132	0.131617	0.991
013	0.013000	1.000	073	0.072935	0.997	133	0.132608	0.991
014	0.014000	0.999	074	0.073932	0.998	134	0.133599	0.991
015	0.014999	1.000	075	0.074930	0.997	135	0.134590	0.991
016	0.015999	1.000	076	0.075927	0.997	136	0.135581	0.991
017	0.016999	1.000	077	0.076924	0.997	137	0.136572	0.990
018	0.017999	1.000	078	0.077921	0.997	138	0.137562	0.991
019	0.018999	1.000	079	0.078918	0.997	139	0.138553	0.990
020	0.019999	0.999	080	0.079915	0.996	140	0.139543	0.990
021	0.020998	1.000	081	0.080911	0.997	141	0.140533	0.990
022	0.021998	1.000	082	0.081908	0.997	142	0.141523	0.990
023	0.022998	1.000	083	0.082905	0.996	143	0.142513	0.990
024	0.023998	0.999	084	0.083901	0.997	144	0.143503	0.989
025	0.024997	1.000	085	0.084898	0.996	145	0.144492	0.990
026	0.025997	1.000	086	0.085894	0.996	146	0.145482	0.989
027	0.026997	0.999	087	0.086890	0.996	147	0.146471	0.989
028	0.027996	1.000	088	0.087886	0.997	148	0.147460	0.989
029	0.028996	1.000	089	0.088883	0.996	149	0.148449	0.989
030	0.029996	0.999	090	0.089879	0.995	150	0.149438	0.989
031	0.030995	1.000	091	0.090874	0.996	151	0.150427	0.988
032	0.031995	0.999	092	0.091870	0.996	152	0.151415	0.989
033	0.032994	0.999	093	0.092866	0.996	153	0.152404	0.988
034	0.033993	1.000	094	0.093862	0.995	154	0.153392	0.988
035	0.034993	0.999	095	0.094857	0.996	155	0.154380	0.988
036	0.035992	1.000	096	0.095853	0.995	156	0.155368	0.988
037	0.036992	0.999	097	0.096848	0.995	157	0.156356	0.987
038	0.037991	0.999	098	0.097843	0.995	158	0.157343	0.988
039	0.038990	0.999	099	0.098838	0.995	159	0.158331	0.987
040	0.039989	1.000	100	0.099833	0.995	160	0.159318	0.987
041	0.040989	0.999	101	0.100828	0.995	161	0.160305	0.987
042	0.041988	0.999	102	0.101823	0.995	162	0.161292	0.987
043	0.042987	0.999	103	0.102818	0.995	163	0.162279	0.987
044	0.043986	0.999	104	0.103813	0.994	164	0.163266	0.986
045	0.044985	0.999	105	0.104807	0.995	165	0.164252	0.987
046	0.045984	0.999	106	0.105802	0.994	166	0.165239	0.986
047	0.046983	0.999	107	0.106796	0.994	167	0.166225	0.986
048	0.047982	0.998	108	0.107790	0.994	168	0.167211	0.986
049	0.048980	0.999	109	0.108784	0.994	169	0.168197	0.985
050	0.049979	0.999	110	0.109778	0.994	170	0.169182	0.985
051	0.050978	0.999	111	0.110772	0.994	171	0.170167	0.986
052	0.051977	0.998	112	0.111766	0.994	172	0.171153	0.985
053	0.052975	0.999	113	0.112760	0.993	173	0.172138	0.985
054	0.053974	0.998	114	0.113753	0.994	174	0.173123	0.985
055	0.054972	0.999	115	0.114747	0.993	175	0.174108	0.985
056	0.055971	0.998	116	0.115740	0.993	176	0.175093	0.984
057	0.056969	0.998	117	0.116733	0.993	177	0.176077	0.985
058	0.057967	0.999	118	0.117726	0.993	178	0.177062	0.984
059	0.058966	0.998	119	0.118719	0.993	179	0.178046	0.984
060	0.059964	0.998	120	0.119712	0.993	180	0.179030	0.983

TABLE 10 Angle-Sine Conversion (Continued)

Angle, mrad	Sine	Interpolation constant	Angle, mrad	Sine	Interpolation constant	Angle, mrad	Sine	Interpolation constant
181	0.180013	0.984	241	0.238674	0.971	301	0.296475	0.955
182	0.180997	0.983	242	0.239645	0.971	302	0.297430	0.955
183	0.181980	0.984	243	0.240616	0.970	303	0.298385	0.954
184	0.182964	0.982	244	0.241586	0.970	304	0.299339	0.954
185	0.183946	0.983	245	0.242556	0.970	305	0.300293	0.954
186	0.184929	0.983	246	0.243526	0.970	306	0.301247	0.953
187	0.185912	0.983	247	0.244496	0.970	307	0.302200	0.953
188	0.186895	0.982	248	0.245466	0.969	308	0.303153	0.953
189	0.187877	0.982	249	0.246435	0.969	309	0.304106	0.953
190	0.188859	0.982	250	0.247404	0.969	310	0.305059	0.952
191	0.189841	0.982	251	0.248373	0.968	311	0.306011	0.952
192	0.190823	0.981	252	0.249341	0.969	312	0.306963	0.951
193	0.191804	0.981	253	0.250310	0.968	313	0.307914	0.952
194	0.192785	0.982	254	0.251278	0.967	314	0.308866	0.950
195	0.193767	0.980	255	0.252245	0.968	315	0.309816	0.951
196	0.194747	0.981	256	0.253213	0.967	316	0.310767	0.950
197	0.195728	0.981	257	0.254180	0.967	317	0.311717	0.950
198	0.196709	0.980	258	0.255147	0.967	318	0.312667	0.950
199	0.197689	0.980	259	0.256114	0.967	319	0.313617	0.950
200	0.198669	0.980	260	0.257081	0.966	320	0.314567	0.949
201	0.199649	0.980	261	0.258047	0.966	321	0.315516	0.948
202	0.200629	0.979	262	0.259013	0.966	322	0.316464	0.949
203	0.201608	0.980	263	0.259979	0.965	323	0.317413	0.948
204	0.202588	0.979	264	0.260944	0.965	324	0.318361	0.948
205	0.203567	0.979	265	0.261909	0.965	325	0.319309	0.947
206	0.204546	0.979	266	0.262874	0.965	326	0.320256	0.947
207	0.205525	0.978	267	0.263839	0.964	327	0.321203	0.947
208	0.206503	0.979	268	0.264803	0.965	328	0.322150	0.947
209	0.207482	0.978	269	0.265768	0.963	329	0.323097	0.946
210	0.208460	0.978	270	0.266731	0.964	330	0.324043	0.946
211	0.209438	0.978	271	0.267695	0.963	331	0.324989	0.945
212	0.210416	0.977	272	0.268658	0.964	332	0.325934	0.946
213	0.211393	0.977	273	0.269622	0.962	333	0.326880	0.945
214	0.212370	0.977	274	0.270584	0.963	334	0.327825	0.944
215	0.213347	0.977	275	0.271547	0.962	335	0.328769	0.944
216	0.214324	0.977	276	0.272509	0.962	336	0.329713	0.944
217	0.215301	0.976	277	0.273471	0.962	337	0.330657	0.944
218	0.216277	0.977	278	0.274433	0.961	338	0.331601	0.943
219	0.217254	0.976	279	0.275394	0.962	339	0.332544	0.943
220	0.218230	0.975	280	0.276356	0.961	340	0.333487	0.943
221	0.219205	0.976	281	0.277317	0.960	341	0.334430	0.942
222	0.220181	0.975	282	0.278277	0.961	342	0.335372	0.942
223	0.221156	0.975	283	0.279238	0.960	343	0.336314	0.941
224	0.222131	0.975	284	0.280198	0.959	344	0.337255	0.942
225	0.223106	0.975	285	0.281157	0.960	345	0.338197	0.941
226	0.224081	0.975	286	0.282117	0.959	346	0.339138	0.940
227	0.225056	0.974	287	0.283076	0.959	347	0.340078	0.940
228	0.226030	0.974	288	0.284035	0.959	348	0.341018	0.940
229	0.227004	0.974	289	0.284994	0.958	349	0.341958	0.940
230	0.227978	0.973	290	0.285952	0.958	350	0.342898	0.939
231	0.228951	0.973	291	0.286910	0.958	351	0.343837	0.939
232	0.229924	0.973	292	0.287868	0.958	352	0.344776	0.938
233	0.230897	0.973	293	0.288826	0.957	353	0.345714	0.939
234	0.231870	0.973	294	0.289783	0.957	354	0.346653	0.937
235	0.232843	0.972	295	0.290740	0.957	355	0.347590	0.938
236	0.233815	0.973	296	0.291697	0.956	356	0.348528	0.937
237	0.234788	0.971	297	0.292653	0.956	357	0.349465	0.937
238	0.235759	0.972	298	0.293609	0.956	358	0.350402	0.936
239	0.236731	0.972	299	0.294565	0.955	359	0.351338	0.936
240	0.237703	0.971	300	0.295520	0.955	360	0.352274	0.936

TABLE 10 Angle-Sine Conversion (Continued)

Angle, mrad	Sine	Inter-polation constant	Angle, mrad	Sine	Inter-polation constant	Angle, mrad	Sine	Inter-polation constant
361	0.353210	0.935	421	0.408673	0.913	481	0.462666	0.886
362	0.354145	0.935	422	0.409586	0.912	482	0.463552	0.886
363	0.355080	0.935	423	0.410498	0.912	483	0.464438	0.885
364	0.356015	0.934	424	0.411410	0.911	484	0.465323	0.885
365	0.356949	0.934	425	0.412321	0.911	485	0.466208	0.885
366	0.357883	0.934	426	0.413232	0.910	486	0.467093	0.884
367	0.358817	0.933	427	0.414142	0.910	487	0.467977	0.883
368	0.359750	0.933	428	0.415052	0.910	488	0.468860	0.883
369	0.360683	0.932	429	0.415962	0.909	489	0.469743	0.883
370	0.361615	0.933	430	0.416871	0.909	490	0.470626	0.882
371	0.362548	0.931	431	0.417780	0.908	491	0.471508	0.882
372	0.363479	0.932	432	0.418688	0.908	492	0.472390	0.881
373	0.364411	0.931	433	0.419596	0.907	493	0.473271	0.880
374	0.365342	0.931	434	0.420503	0.907	494	0.474151	0.881
375	0.366273	0.930	435	0.421410	0.907	495	0.475032	0.879
376	0.367203	0.930	436	0.422317	0.906	496	0.475911	0.880
377	0.368133	0.929	437	0.423223	0.906	497	0.476791	0.878
378	0.369062	0.930	438	0.424129	0.906	498	0.477669	0.879
379	0.369992	0.928	439	0.425035	0.904	499	0.478548	0.878
380	0.370920	0.929	440	0.425939	0.905	500	0.479426	0.877
381	0.371849	0.928	441	0.426844	0.904	501	0.480303	0.877
382	0.372777	0.928	442	0.427748	0.904	502	0.481180	0.876
383	0.373705	0.927	443	0.428652	0.903	503	0.482056	0.876
384	0.374632	0.927	444	0.429555	0.903	504	0.482932	0.875
385	0.375559	0.927	445	0.430458	0.902	505	0.483807	0.875
386	0.376486	0.926	446	0.431360	0.902	506	0.484682	0.875
387	0.377412	0.926	447	0.432262	0.902	507	0.485557	0.874
388	0.378338	0.925	448	0.433164	0.901	508	0.486431	0.873
389	0.379263	0.925	449	0.434065	0.901	509	0.487304	0.873
390	0.380188	0.925	450	0.434966	0.900	510	0.488177	0.873
391	0.381113	0.924	451	0.435866	0.900	511	0.489050	0.872
392	0.382037	0.924	452	0.436766	0.899	512	0.489922	0.871
393	0.382961	0.924	453	0.437665	0.899	513	0.490793	0.871
394	0.383885	0.923	454	0.438564	0.898	514	0.491664	0.871
395	0.384808	0.923	455	0.439462	0.898	515	0.492535	0.870
396	0.385731	0.922	456	0.440360	0.898	516	0.493405	0.869
397	0.386653	0.922	457	0.441258	0.897	517	0.494274	0.870
398	0.387575	0.922	458	0.442155	0.897	518	0.495144	0.868
399	0.388497	0.921	459	0.443052	0.896	519	0.496012	0.868
400	0.389418	0.921	460	0.443948	0.896	520	0.496880	0.868
401	0.390339	0.921	461	0.444844	0.895	521	0.497748	0.867
402	0.391260	0.920	462	0.445739	0.895	522	0.498615	0.866
403	0.392180	0.919	463	0.446634	0.895	523	0.499481	0.866
404	0.393099	0.920	464	0.447529	0.894	524	0.500347	0.866
405	0.394019	0.919	465	0.448423	0.893	525	0.501213	0.865
406	0.394938	0.918	466	0.449316	0.894	526	0.502078	0.865
407	0.395856	0.918	467	0.450210	0.892	527	0.502943	0.864
408	0.396774	0.918	468	0.451102	0.892	528	0.503807	0.863
409	0.397692	0.917	469	0.451994	0.892	529	0.504670	0.863
410	0.398609	0.917	470	0.452886	0.892	530	0.505533	0.863
411	0.399526	0.917	471	0.453778	0.891	531	0.506396	0.862
412	0.400443	0.916	472	0.454669	0.890	532	0.507258	0.861
413	0.401359	0.916	473	0.455559	0.890	533	0.508119	0.862
414	0.402275	0.915	474	0.456449	0.889	534	0.508981	0.860
415	0.403190	0.915	475	0.457338	0.890	535	0.509841	0.860
416	0.404105	0.914	476	0.458228	0.888	536	0.510701	0.860
417	0.405019	0.914	477	0.459116	0.888	537	0.511561	0.859
418	0.405933	0.914	478	0.460004	0.888	538	0.512420	0.858
419	0.406847	0.913	479	0.460892	0.887	539	0.513278	0.858
420	0.407760	0.913	480	0.461779	0.887	540	0.514136	0.857

TABLE 10 Angle-Sine Conversion (Continued)

Angle, mrad	Sine	Inter-polation constant	Angle, mrad	Sine	Inter-polation constant	Angle, mrad	Sine	Inter-polation constant
541	0.514993	0.857	601	0.565468	0.824	661	0.613907	0.789
542	0.515850	0.857	602	0.566292	0.824	662	0.614696	0.788
543	0.516707	0.856	603	0.567116	0.823	663	0.615484	0.788
544	0.517563	0.855	604	0.567939	0.823	664	0.616272	0.787
545	0.518418	0.855	605	0.568762	0.822	665	0.617059	0.787
546	0.519273	0.854	606	0.569584	0.822	666	0.617846	0.786
547	0.520127	0.854	607	0.570406	0.821	667	0.618632	0.785
548	0.520981	0.853	608	0.571227	0.821	668	0.619417	0.785
549	0.521834	0.853	609	0.572048	0.819	669	0.620202	0.784
550	0.522687	0.852	610	0.572867	0.820	670	0.620986	0.784
551	0.523539	0.852	611	0.573687	0.819	671	0.621770	0.782
552	0.524391	0.851	612	0.574506	0.818	672	0.622552	0.783
553	0.525242	0.851	613	0.575324	0.817	673	0.623335	0.781
554	0.526093	0.850	614	0.576141	0.818	674	0.624116	0.781
555	0.526943	0.850	615	0.576959	0.816	675	0.624897	0.781
556	0.527793	0.849	616	0.577775	0.816	676	0.625678	0.779
557	0.528642	0.849	617	0.578591	0.815	677	0.626457	0.780
558	0.529491	0.848	618	0.579406	0.815	678	0.627237	0.778
559	0.530339	0.847	619	0.580221	0.814	679	0.628015	0.778
560	0.531186	0.847	620	0.581035	0.814	680	0.628793	0.777
561	0.532033	0.847	621	0.581849	0.813	681	0.629570	0.777
562	0.532880	0.846	622	0.582662	0.812	682	0.630347	0.776
563	0.533726	0.845	623	0.583474	0.812	683	0.631123	0.775
564	0.534571	0.845	624	0.584286	0.811	684	0.631898	0.775
565	0.535416	0.844	625	0.585097	0.811	685	0.632673	0.774
566	0.536260	0.844	626	0.585908	0.810	686	0.633447	0.774
567	0.537104	0.843	627	0.586718	0.810	687	0.634221	0.773
568	0.537947	0.843	628	0.587528	0.808	688	0.634994	0.772
569	0.538790	0.842	629	0.588336	0.809	689	0.635766	0.771
570	0.539632	0.842	630	0.589145	0.807	690	0.636537	0.771
571	0.540474	0.841	631	0.589952	0.808	691	0.637308	0.770
572	0.541315	0.840	632	0.590760	0.806	692	0.638078	0.770
573	0.542155	0.840	633	0.591566	0.806	693	0.638848	0.769
574	0.542995	0.840	634	0.592372	0.806	694	0.639617	0.768
575	0.543835	0.839	635	0.593178	0.804	695	0.640385	0.768
576	0.544674	0.838	636	0.593982	0.804	696	0.641153	0.767
577	0.545512	0.838	637	0.594786	0.804	697	0.641920	0.767
578	0.546350	0.837	638	0.595590	0.803	698	0.642687	0.766
579	0.547187	0.837	639	0.596393	0.802	699	0.643453	0.765
580	0.548024	0.836	640	0.597195	0.802	700	0.644218	0.764
581	0.548860	0.836	641	0.597997	0.801	701	0.644982	0.764
582	0.549696	0.835	642	0.598798	0.801	702	0.645746	0.763
583	0.550531	0.834	643	0.599599	0.800	703	0.646509	0.763
584	0.551365	0.834	644	0.600399	0.799	704	0.647272	0.762
585	0.552199	0.834	645	0.601198	0.799	705	0.648034	0.761
586	0.553033	0.833	646	0.601997	0.798	706	0.648795	0.761
587	0.553866	0.832	647	0.602795	0.798	707	0.649556	0.760
588	0.554698	0.832	648	0.603593	0.797	708	0.650316	0.759
589	0.555530	0.831	649	0.604390	0.796	709	0.651075	0.759
590	0.556361	0.831	650	0.605186	0.796	710	0.651834	0.758
591	0.557192	0.830	651	0.605982	0.795	711	0.652592	0.757
592	0.558022	0.829	652	0.606777	0.795	712	0.653349	0.757
593	0.558851	0.829	653	0.607572	0.794	713	0.654106	0.756
594	0.559680	0.829	654	0.608366	0.793	714	0.654862	0.755
595	0.560509	0.828	655	0.609159	0.793	715	0.655617	0.755
596	0.561337	0.827	656	0.609952	0.792	716	0.656372	0.754
597	0.562164	0.827	657	0.610744	0.792	717	0.657126	0.754
598	0.562991	0.826	658	0.611536	0.791	718	0.657880	0.753
599	0.563817	0.825	659	0.612327	0.790	719	0.658633	0.752
600	0.564642	0.826	660	0.613117	0.790	720	0.659385	0.751

TABLE 10 Angle-Sine Conversion (Continued)

Angle, mrad	Sine	Interpolation constant	Angle, mrad	Sine	Interpolation constant	Angle, mrad	Sine	Interpolation constant
721	0.660136	0.751	781	0.703990	0.710	841	0.745310	0.667
722	0.660887	0.750	782	0.704700	0.709	842	0.745977	0.665
723	0.661637	0.750	783	0.705409	0.708	843	0.746642	0.665
724	0.662387	0.748	784	0.706117	0.708	844	0.747307	0.664
725	0.663135	0.749	785	0.706825	0.707	845	0.747971	0.663
726	0.663884	0.747	786	0.707532	0.707	846	0.748634	0.663
727	0.664631	0.747	787	0.708239	0.705	847	0.749297	0.662
728	0.665378	0.746	788	0.708944	0.705	848	0.749959	0.661
729	0.666124	0.746	789	0.709649	0.704	849	0.750620	0.660
730	0.666870	0.744	790	0.710353	0.704	850	0.751280	0.660
731	0.667614	0.745	791	0.711057	0.703	851	0.751940	0.659
732	0.668359	0.743	792	0.711760	0.702	852	0.752599	0.658
733	0.669102	0.743	793	0.712462	0.701	853	0.753257	0.657
734	0.669845	0.742	794	0.713163	0.701	854	0.753914	0.657
735	0.670587	0.742	795	0.713864	0.700	855	0.754571	0.656
736	0.671329	0.740	796	0.714564	0.699	856	0.755227	0.655
737	0.672069	0.741	797	0.715263	0.698	857	0.755882	0.654
738	0.672810	0.739	798	0.715961	0.698	858	0.756536	0.654
739	0.673549	0.739	799	0.716659	0.697	859	0.757190	0.653
740	0.674288	0.738	800	0.717356	0.696	860	0.757843	0.652
741	0.675026	0.738	801	0.718052	0.696	861	0.758495	0.651
742	0.675764	0.736	802	0.718748	0.695	862	0.759146	0.650
743	0.676500	0.736	803	0.719443	0.694	863	0.759796	0.650
744	0.677236	0.736	804	0.720137	0.694	864	0.760446	0.649
745	0.677972	0.735	805	0.720831	0.692	865	0.761095	0.649
746	0.678707	0.733	806	0.721523	0.692	866	0.761744	0.647
747	0.679440	0.734	807	0.722215	0.692	867	0.762391	0.647
748	0.680174	0.733	808	0.722907	0.690	868	0.763038	0.646
749	0.680907	0.732	809	0.723597	0.690	869	0.763684	0.645
750	0.681639	0.731	810	0.724287	0.689	870	0.764329	0.644
751	0.682370	0.731	811	0.724976	0.689	871	0.764973	0.644
752	0.683101	0.730	812	0.725665	0.687	872	0.765617	0.643
753	0.683831	0.729	813	0.726352	0.687	873	0.766260	0.642
754	0.684560	0.729	814	0.727039	0.687	874	0.766902	0.642
755	0.685289	0.728	815	0.727726	0.685	875	0.767544	0.640
756	0.686017	0.727	816	0.728411	0.685	876	0.768184	0.640
757	0.686744	0.726	817	0.729096	0.684	877	0.768824	0.639
758	0.687470	0.726	818	0.729780	0.683	878	0.769463	0.638
759	0.688196	0.725	819	0.730463	0.683	879	0.770101	0.638
760	0.688921	0.725	820	0.731146	0.682	880	0.770739	0.637
761	0.689646	0.724	821	0.731828	0.681	881	0.771376	0.636
762	0.690370	0.723	822	0.732509	0.680	882	0.772012	0.635
763	0.691093	0.722	823	0.733189	0.680	883	0.772647	0.634
764	0.691815	0.722	824	0.733869	0.679	884	0.773281	0.634
765	0.692537	0.721	825	0.734548	0.678	885	0.773915	0.633
766	0.693258	0.720	826	0.735226	0.677	886	0.774548	0.632
767	0.693978	0.720	827	0.735903	0.677	887	0.775180	0.631
768	0.694698	0.719	828	0.736580	0.676	888	0.775811	0.631
769	0.695417	0.718	829	0.737256	0.675	889	0.776442	0.630
770	0.696135	0.718	830	0.737931	0.675	890	0.777072	0.629
771	0.696853	0.717	831	0.738606	0.674	891	0.777701	0.628
772	0.697570	0.716	832	0.739280	0.673	892	0.778329	0.627
773	0.698286	0.715	833	0.739953	0.672	893	0.778956	0.627
774	0.699001	0.715	834	0.740625	0.672	894	0.779583	0.626
775	0.699716	0.714	835	0.741297	0.670	895	0.780209	0.625
776	0.700430	0.714	836	0.741967	0.670	896	0.780834	0.625
777	0.701144	0.712	837	0.742637	0.670	897	0.781459	0.623
778	0.701856	0.712	838	0.743307	0.668	898	0.782082	0.623
779	0.702568	0.711	839	0.743975	0.668	899	0.782705	0.622
780	0.703279	0.711	840	0.744643	0.667	900	0.783327	0.621

TABLE 10 Angle-Sine Conversion (Continued)

Angle, mrad	Sine	Interpolation constant	Angle, mrad	Sine	Interpolation constant	Angle, mrad	Sine	Interpolation constant
901	0.783948	0.621	935	0.804599	0.593	969	0.824320	0.566
902	0.784569	0.619	936	0.805192	0.593	970	0.824886	0.565
903	0.785188	0.619	937	0.805785	0.592	971	0.825451	0.564
904	0.785807	0.618	938	0.806377	0.591	972	0.826015	0.563
905	0.786425	0.617	939	0.806968	0.590	973	0.826578	0.562
906	0.787042	0.617	940	0.807558	0.589	974	0.827140	0.562
907	0.787659	0.616	941	0.808147	0.589	975	0.827702	0.561
908	0.788275	0.615	942	0.808736	0.588	976	0.828263	0.560
909	0.788890	0.614	943	0.809324	0.587	977	0.828823	0.559
910	0.789504	0.613	944	0.809911	0.586	978	0.829382	0.558
911	0.790117	0.613	945	0.810497	0.585	979	0.829940	0.557
912	0.790730	0.611	946	0.811082	0.585	980	0.830497	0.557
913	0.791341	0.611	947	0.811667	0.584	981	0.831054	0.556
914	0.791952	0.611	948	0.812251	0.582	982	0.831610	0.555
915	0.792563	0.609	949	0.812833	0.583	983	0.832165	0.554
916	0.793172	0.609	950	0.813416	0.581	984	0.832719	0.553
917	0.793781	0.607	951	0.813997	0.580	985	0.833272	0.553
918	0.794388	0.607	952	0.814577	0.580	986	0.833825	0.551
919	0.794995	0.607	953	0.815157	0.579	987	0.834376	0.551
920	0.795602	0.605	954	0.815736	0.578	988	0.834927	0.550
921	0.796207	0.605	955	0.816314	0.577	989	0.835477	0.549
922	0.796812	0.604	956	0.816891	0.576	990	0.836026	0.548
923	0.797416	0.603	957	0.817467	0.576	991	0.836574	0.548
924	0.798019	0.602	958	0.818043	0.575	992	0.837122	0.546
925	0.798621	0.601	959	0.818618	0.574	993	0.837668	0.546
926	0.799222	0.601	960	0.819192	0.573	994	0.838214	0.545
927	0.799823	0.600	961	0.819765	0.572	995	0.838759	0.544
928	0.800423	0.599	962	0.820337	0.571	996	0.839303	0.543
929	0.801022	0.598	963	0.820908	0.571	997	0.839846	0.543
930	0.801620	0.597	964	0.821479	0.570	998	0.840389	0.541
931	0.802217	0.597	965	0.822049	0.569	999	0.840930	0.541
932	0.802814	0.596	966	0.822618	0.568	1000	0.841471	0.540
933	0.803410	0.595	967	0.823186	0.567			
934	0.804005	0.594	968	0.823753	0.567			

REFERENCES

Amon, M., and S. Rosin (1967): Color Corrected Mangin Mirror, *Appl. Opt.,* vol 6, pp. 963–968.

——— and ——— (1968): Mangin Mirror Systems, *Appl. Opt.,* vol. 7, pp. 214–215.

Barakat, R. G. (1964): Numerical Results concerning the Transfer Functions and Total Illuminance for Optimum Balanced Fifth-Order Spherical Aberration, *J. Opt. Soc. Am.,* vol. 54, pp. 38–44.

——— and A. Houston (1964): Diffraction Effects of Coma, *J. Opt. Soc. Am.,* vol. 54, pp. 1084–1088.

——— and ——— (1966): The Effect of a Sinusoidal Wavefront on the Transfer Function of a Circular Aperture, *Appl. Opt.,* vol. 5, pp. 1850–1852.

Bergstein, L., and L. Motz (1962): Optically Compensated Varifocal Systems, *J. Opt. Soc. Am.,* vol. 52, pp. 353–388.

Born, M., and E. Wolf (1964): "Principles of Optics," Macmillan, New York.

Bouwers, A. (1946): "Achievements in Optics," Elsevier, New York.

Buchdahl, H. A. (1968): "Optical Aberration Coefficients," Dover, New York.

Coltman, J. W. (1954): The Specification of Imaging Properties by Response to a Sine Wave Input, *J. Opt. Soc. Am.,* vol. 44, pp. 468–471.

Conrady, A. E. (1957): "Applied Optics and Optical Design," vol. 1, Dover, New York.

——— (1960): "Applied Optics and Optical Design," vol. 2, Dover, New York.

Cooke, G. H. (1956): Recent Developments in Anamorphic Systems, *J. Soc. Motion Pict. Telev. Eng.,* vol. 65, pp. 151–154.

Cooley, J. W., and J. W. Tukey (1965): An Algorithm for the Machine Calculation of Complex Fourier Series, *Math. Comput.,* vol. 19, pp. 297–301.

Cox, A. (1964): "A System of Optical Design," Focal Press, New York.

Feder, D. (1951): Optical Calculations with Automatic Computing Machinery, *J. Opt. Soc. Am.,* vol. 41, pp. 630–635.

——— (1963): Automatic Optical Design, *Appl. Opt.,* vol. 2, pp. 1209–1226.

——— (1968): Differentiation of Ray-tracing Equations with Respect to Construction Parameters of Rotationally Symmetrical Optics, *J. Opt. Soc. Am.,* vol. 58, pp. 1494–1505.

Glatzel, E., and R. Wilson (1968): Adaptive Automatic Correction in Optical Design, *Appl. Opt.,* vol. 7, pp. 265–276.

Hardy, A., and F. Perrin (1932): "The Principles of Optics," McGraw-Hill, New York.

Herzberger, M. (1958): "Modern Geometrical Optics," Interscience, New York.

Hopkins, H. (1950): "Wave Theory of Aberrations," Oxford, London.

Kelly, D. (1965): Spatial Frequency, Bandwidth and Resolution, *Appl. Opt.,* vol. 4, pp. 435–437.

Kingslake, R. (1960): The Development of the Zoom Lens, *J. Soc. Motion Pict. Telev. Eng.,* vol. 69, pp. 534–544.

——— (1965–1969): "Applied Optics and Optical Engineering," 5 vols., Academic, New York.

Lauroesch, T., and C. Wing (1959): Bouwers Concentric Systems for Materials of High Refractive Index, *J. Opt. Soc. Am.,* vol. 49, pp. 410–411.

Linfoot, E. H. (1964): "Fourier Methods in Optical Design," Focal Press, New York.

Maksutov, D. D. (1944): New Catadioptric Meniscus Systems, *J. Opt. Soc. Am.,* vol. 34, pp. 270–284.

Martin, L. C. (1960): "Technical Optics," 2 vols., Pitman, London.

McNeil, G. T. (1968): "Optical Fundamentals of Underwater Photography," Photogrammetry, Rockville, Md.

Meiron, J. (1968): The Use of Merit Functions Based on Wavefront Aberrations in Automatic Lens Design, *Appl. Opt.,* vol. 7, pp. 667–672.

MIL-HDBK-141 (1962): "Handbook of Optical Design," U.S. Government Printing Office.

Montagnino, L. (1968): Ray Tracing in Inhomogenous Media, *J. Opt. Soc. Am.,* vol. 58, pp. 1667–1668.

O'Neill, E. L. (1956): Transfer Function for an Annular Aperture, *J. Opt. Soc. Am.,* vol. 46, pp. 285–288.

——— (1963): "Introduction to Statistical Optics," Addison-Wesley, Reading, Mass.

Perrin, F. (1960): Methods of Appraising Photographic Systems, *J. Soc. Motion Pict. Telev. Eng.,* vol. 69, pp. 151–156, 239–248.

Potter, R. (1961): Transmission Properties of Optical Fibers, *J. Opt. Soc. Am.*, vol. 51, pp. 1079–1089.

Schenck, H., and H. Kendall (1957): "Underwater Photography," Cornell Maritime Press, Cambridge, Md.

Smith, W. J. (1965): Optics, chap. 9, pp. 369–419, and Optical Systems, chap. 10, pp. 421–456, in W. L. Wolfe (ed.), "Handbook of Military Infrared Technology," Office of Naval Research, Washington.

——— (1966): "Modern Optical Engineering: The Design of Optical Systems," McGraw-Hill, New York.

Southall, J. P. C. (1964): "Mirrors, Prisms and Lenses," Dover, New York.

Spencer, G. (1963): A Flexible Automatic Lens Correction Procedure, *Appl. Opt.*, vol. 2, pp. 1257–1264.

——— and M. Murty (1962): Generalized Ray-tracing Procedure, *J. Opt. Soc. Am.*, vol. 52, pp. 672–678.

Stephens, R., and L. Sutton (1968): Diffraction Images of a Point in the Focal Plane and Several Out-of-Focus Planes, *J. Opt. Soc. Am.*, vol. 58, pp. 1001–1002.

Walles, S., and R. Hopkins (1964): The Orientation of the Image Formed by a Series of Plane Mirrors, *Appl. Opt.*, vol. 3, pp. 1447–1452.

Williamson, D. (1952): Cone Channel Condenser Optics, *J. Opt. Soc. Am.*, vol. 42, pp. 712–715.

Section 3

Optical Radiators and Sources

GEORGE J. ZISSIS

Environmental Research Institute of Michigan,
Ann Arbor, Mich.

ANTHONY J. LAROCCA

Environmental Research Institute of Michigan,
Ann Arbor, Mich.

1. Introduction† The literature is laden with useful information on sources. The purpose in this section is to consolidate much of this information to assist the optical-systems designer in making reasonable choices. Some of the literature already contains much of what is needed to fulfill this objective; to avoid duplication, such material is included here in abbreviated form, with reference to the original work for the researcher who requires the details.

It should be emphasized that the inclusiveness implied by the title of this section cannot possibly be attained. As indicated in the foreword to Cann (1967), variation in replies from manufacturers can result in unavoidable emphasis on some companies' products. The content of this section on commercial sources is the incorporation, extensively, of the work of other authors who have diligently collected the necessary information from the pertinent companies.

There should be no great fear of obsolescence because if we may assume that progress is as important in the manufacture of radiant sources as it is in other technologies, the absence of a particular source at some later time should signify either its uselessness or its replacement by a better, more modern one. Thus we

† For literature appearing after this section was written, the reader is referred to such standard reference sources as *Engineering Index, Science Abstracts,* and *Chemical Abstracts.* Similarly, progress in radiometric standardization may be found in the *Technical Notes* published by the National Bureau of Standards and available through the Superintendent of Documents.

include sources which may change in time but which will be representative of those obtainable at any time in the future, exclusive of completely new design and evolution.

Regarding the selection of a source, Worthing (1937) suggests that one ask the following questions:

1. Does it supply energy at such a rate or in such an amount as to make measurements possible?
2. Does it yield an irradiation that is generally constant or that may be varied with time as desired?
3. Is it reproducible?
4. Does it yield irradiations of the desired magnitudes over areas of the desired extent?
5. Has it the desired spectral distribution?
6. Has it the necessary operating life?
7. Has it sufficient ruggedness for the proposed problem?
8. Is it sufficiently easy to obtain and replace, or is its purchase price or its construction cost reasonable?

STANDARD SOURCES

2. There has been an unfortunate division historically between the standards of photometry and those used to establish thermal radiation and the thermodynamic temperature scale. Thus in photometry the standard has changed from the use of candles, the Carcel lamp, the Harcourt pentane lamp, and the Hefner lamp to the present primary standard of light adopted in January 1948 (Walsh, 1965).

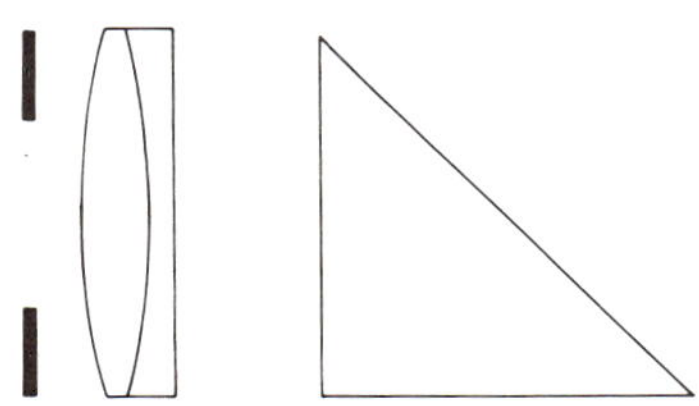

3. Primary Standard of Light The primary standard now established is the candela, corresponding to 1.0000 international lumens per steradian, 0.104 Carcel unit (approximately), 1 pentane candle (approximately), 1 English sperm candle (approximately), and 1.11 Hefner unit (approximately). The construction of the primary standard of light is shown in Fig. 1. The radiator itself consists of a small cylinder of pure fused thoria, about 45 mm long, with an internal diameter of about 2.5 mm and a wall thickness of 0.2 to 0.3 mm. This cylinder, the bottom of which is packed with powdered fused thoria to a depth of 10 to 15 mm, is supported vertically in a fused-thoria crucible of about 20 mm internal diameter nearly filled with pure platinum, as shown in Fig. 1. The crucible has a lid with a small hole in the center, about 1.5 mm in diameter; this hole, which is the source of light, is surrounded above by a funnel-shaped sheath forming part of the crucible lid. The crucible is embedded in powdered fused thoria in a larger refractory container and is heated by enclosing it in a high-frequency induction furnace. The power required to melt the platinum is about 7 kW at a frequency of the order of 1 MHz. With this arrangement it is possible to regulate the temperature so closely that the period required for the solidification of the platinum may exceed

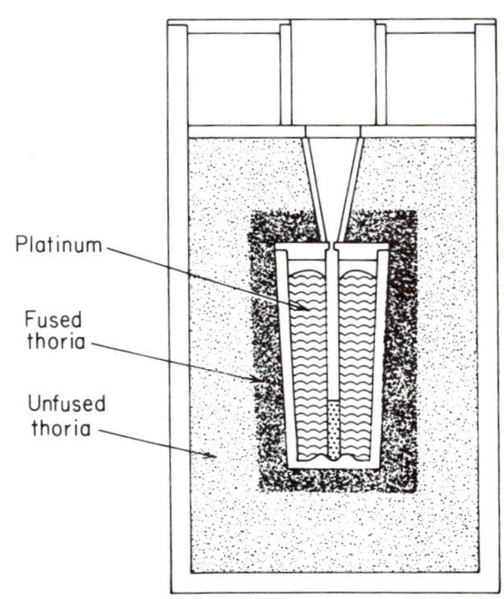

Fig. 1 The primary standard of light. [*From John W. T. Walsh, "Photometry," Dover Publications, Inc., New York,* 1965; *reprinted by permission of the publisher.*]

20 min. The purity of the platinum is controlled by taking samples before and after use and determining the ratio of the electric resistance at 100 and 0°C. A minimum value of 1.390 for this ratio is required to ensure that the temperature of solidification is sensibly the same as that of pure platinum. This corresponds to an impurity of less than 3 parts in 100,000.

The luminance (radiance) of the hole is given the value of 60 cd/cm². This unit, the candela, was introduced Jan. 1, 1948 by the action of the Comité International des Poids et Mesures to replace the international candle.

The procedure in using the standard is first to raise the temperature of the crucible and its container well above the melting point of platinum. Then it is allowed to cool slowly until the period of solidification of platinum is reached. This is used in preference to the period of fusion, as it has been found to give more consistent results. It should be noted that at this temperature of 2042 K a change of 10 K would vary the luminous intensity by about 5 to 6 percent.

The rest of the history of photometric standards is too complex and lengthy for inclusion here. An excellent treatment, especially with regard to standard lamps, is found in Jones and Preston (1969).

4. Thermal Radiation and Radiometry Two approaches have been taken to realize standards for thermal radiation and radiometry. One, taken by scientists at the National Physical Laboratory of England, has been to achieve a "black" detector with extremely stable response characteristics and to use it in an arrangement by which electrical heating is introduced to balance the heat gained by incoming radiation. The other, followed by workers in the U.S. National Bureau of Standards (NBS) and the National Research Council of Canada (NRC), has been to produce a "black" source or blackbody radiator.

The attainment of a black detector or source involves the parameters of emittance ϵ, absolute temperature T, and the Stefan-Boltzmann constant σ, as seen in the Stefan-Boltzmann equation for excitance M:

$$M = \epsilon\sigma T^4 \qquad \mathrm{W/cm^2}$$

The National Bureau of Standards standard of radiant energy or total thermal radiation was first established by Coblentz in 1914 with a ceramic wire-wound furnace used between about 1000 and 1150°C. This source was used to calibrate the carbon-filament lamp which was then issued by the Bureau. The NBS instructions on use of this now obsolete standard carefully define the required geometry of the setup (*Trans. Int. Astron.*, 1962). The instructions as revised in 1960 included statements to the effect that the values of radiant flux from this standard were based upon direct measurements and upon comparison to a blackbody using a value of 5.7 pW/cm² for the Stefan-Boltzmann constant as described in *Natl. Bur. Stand. Bull.* (1914) and *Natl. Bur. Stand. J. Res.* (1933). The results were believed to be accurate to about 1 percent for the absolute values of the primary standard and, for compared secondary standards, within about ½ percent.

The source used by Coblentz was dismantled. Subsequently another design was assembled with an oxidized nickel-chromium alloy (80% Ni, 20% Cr) cavity used up to 1400 K. The cavity had an aperture of 0.95 cm diameter and a depth of about 12.5 or 14 mm. A calculation by the DeVos method is said to have given a value of $\epsilon = 0.999$ for the cavity. At the operating temperature of 1300 K the uncertainty was said to be ±0.2 percent or ±½ K.

Using this blackbody for the 0.7- to 2.6-μm region, with a higher-temperature cylindrical graphite source (estimated emissivity by the Gouffe method of 0.996) for the 0.25- to 0.75-μm region, Stair et al. (1960) calibrated tungsten strip lamps (GE 30A/T24/17) to be used as working standards. The uncertainties ranged from 8 percent at the short wavelengths to 3 percent at the long wavelengths.

For spectral irradiance from 0.25 to 2.6 μm Stair et al. (1963) prepared, as working standards, tungsten-quartz-iodine 200-W lamps (GE 6.6A/T4Q 1 CL) and similar 1,000-W lamps. Again the uncertainties were estimated to range from 8 to 3 percent.

As a result of this and other work at NBS, standards of spectral radiance and spectral irradiance were established. The instructions issued read (quoted from Wolfe, 1965, p. 41):

Instructions for Using Standards of Spectral Radiance (Revised February 21, 1961)

GENERAL DISCUSSION

These instructions cover the use of ribbon-filament lamps issued as standards of spectral radiance for the wavelength ranges of 0.25 to 0.75, 0.5 to 2.5, and 0.25 to 2.5 microns. The lamps employed are commercial G.E. Type 30A/T24/7 lamps having a tungsten ribbon filament (SR-8A type) centered about 8 to 10 cm behind a fused silica window 3 cm in diameter.

The portion of the filament whose spectral radiance has been determined is the central portion visible through the fused-silica window. This determination was made by direct substitution of the lamps for working standard lamps which had in turn been calibrated by this substitution method relative to blackbodies (operated from 1400 to 2300°K) through the use of a double quartz prism spectroradiometer and associated electronic equipment.

In operation the lamp is mounted vertically and the beam of radiant flux with a horizontal axis passing through the center of the filament is measured. In the original determinations no portion of the beam measured departed from this axis by more than 2.5 degrees. Hence, if an aperture subtending a larger angle is required in any application of these standards of spectral radiance, it should be ascertained that the irradiance is constant over the whole aperture.

If there is excessive water vapor in the laboratory atmosphere errors may result at the wavelengths of water-vapor absorption bands. In the original calibrations the comparisons of the lamps with the blackbodies were made at the same distance and in such manner that the effect of water-vapor absorption canceled out.

In the calibrations wherein the blackbody was heated within a wirewound furnace and temperatures around 1400°K were reached, the temperatures were measured with Pt-Pt 10% Rh thermocouples. Tests using couples placed at various positions and observations with an optical pyrometer indicated closely uniform temperatures within the blackbody enclosure. A ratio of blackbody opening to total internal surface area equal to approximately 0.003 and an internal surface reflectance less than 0.10 indicate an emissivity of 0.999 or higher.

For the blackbody temperatures above 1400°K a graphite enclosure heated by a radio-frequency generator was employed and the temperatures were measured by an optical pyrometer. The physical characteristics of this blackbody indicated an emissivity approximating 0.996.

The spectral radiance of the blackbody is based upon the Planck radiation law in which the constants, based upon the most recent atomic and other information, are set down as follows:

$$c_1 = 1.19088 \times 10^{-12} \text{ watt cm}^2 \text{ per steradian}$$
$$c_2 = 1.4380 \text{ cm degree K}$$

Values of spectral radiance for these lamps are tabulated as a function of wavelength in microwatts per (steradian-millimicron-square millimeter of filament). Values of spectral radiance for slit-widths other than one millimicron, say x millimicrons, where x is less than 100, may be found by multiplying the tabulated values by x.

USE OF THE STANDARDS OF SPECTRAL RADIANCE

It is suggested that the auxiliary optics employed with these standards be composed of two units: namely, a plane mirror and a spherical mirror (each aluminized on the front surface). If the spherical mirror is placed at a distance from the lamp filament equal to its radius of curvature, and the plane mirror set about $\frac{1}{3}$ to $\frac{2}{5}$ this distance from the spherical mirror, facing it and so placed that the angle between incident and reflected beams is 10° or less, a good image of the filament itself may be focused upon the spectrometer slit. Little distortion of the filament image occurs provided precise optical surfaces are employed and angles between incident and reflected beams are kept to less than 10°. Various optical arrangements may be employed.

The solid-angular aperture of the auxiliary optics should be smaller than the solid-angular aperture of the spectrometer employed so that no loss of radiant energy will result through over-filling the spectrometer optics.

The spectral radiant flux, P_λ, in microwatts per millimicron, which enters the spectrometer slit is computed from the formula:

$$P_\lambda = R_\lambda N_\lambda sA/D^2$$

where R_λ is the spectral reflectance of the combination of mirrors used, N_λ is the reported spectral radiance of the standard, s is the area of the spectrometer slit in mm², A is the area of the limiting auxiliary optic, and D is the distance of this optic from the slit.

No diaphragm or other shielding is required in the use of these standards, except for a shield to prevent radiant energy from the lamp from entering the spectrometer directly without first falling on the concave mirror. An image of the filament should be focused upon the spectrometer slit, and only the energy by which this image is formed should enter the slit.

In order to calibrate a spectroradiometer with one of these standards of spectral radiance, a knowledge of the spectral reflectance of the mirror surfaces is required. A good aluminized surface should have a spectral reflectance considerably above 0.87 throughout the spectral region of 0.5 to 2.6 microns, increasing slightly with wavelength except possibly for a slight dip near 0.80 micron. In practice the proper reflectance losses can best be determined through the use of a third (similar) mirror (a second plane mirror) which may be temporarily incorporated into the optical set-up from time to time.

In measurements wherein two sources (a standard source and a test source) are being compared by the direct substitution method (use of the same auxiliary optics, slit-widths, areas and detector at any one wavelength), no knowledge of the spectral reflectance of the auxiliary mirrors, nor of the spectral transmittance of the spectrometer, nor of the spectral sensitivity of the detector is required. It is necessary, however, to make sure that the entrance slit of the spectrometer is fully and uniformly filled with radiant flux both from the standard and from the test source; and, if at any one wavelength the detector response for the standard is significantly different from that for the test source, the deviation from linearity of response of the detector must be evaluated and taken into account.

Operation of these standards should be on alternating current to obviate filament-crystallizing effects that occur when the operation is on direct current. The filaments are massive and "iron out" all effects of the normal fluctuations present in a commercial ac supply. All calibrations were made by means of alternating current. To reduce line voltage a step-down transformer (1 kva) having a ratio of 10 to 1 or a 50-ampere variable transformer may be employed. Then to give fine control a second variable transformer (10-ampere capacity) is wired into the circuit to control the input of the heavy duty transformer. For still finer control a third variable transformer may be employed with a radio-filament transformer to add (or subtract) voltage fed into the step-down transformer. It was found that this method is very effective for accurate control of large lamp currents. The heavy duty (1 kva) step-down transformer is preferred to the 50-ampere variable transformer since the latter is subject to contact damage when operated for long intervals of time at high currents.

The lamp standards of spectral radiance are expensive laboratory equipment and it is suggested that they be operated sparingly and with care in order to prolong their useful life. This precaution applies especially to the standards calibrated in the short-wave region and operated at 35 amperes. They should be turned on and off slowly and only for short intervals should they ever be operated at or above 30 amperes, and then only to calibrate a similar lamp as a working standard. In general even at lower currents a working standard should be prepared and used except for purposes of checking the operation of such a working standard.

Instructions for Using the NBS Standards of Spectral Irradiance (May 31, 1963)†

These instructions cover the use of tungsten-filament quartz-iodine lamps issued as standards of spectral irradiance for the wavelength range of 0.25 to 2.6 microns. The lamps employed are commercial G.E. type 6.6A/T4Q/1CL-200-watt lamps having a tungsten coiled-coil filament enclosed in a small ($\frac{1}{2}$ inch × 2 inches) quartz envelope containing a small amount of iodine.

The spectral radiant intensity of the entire lamp as mounted in the manner prescribed below is measured and reported. The spectral irradiance from these lamps is based upon the spectral radiance of a blackbody as defined by Planck's equation and has been determined through comparison of a group of quartz iodine lamps with (1) the NBS standards of spectral radiance, (2) the NBS standards of luminous intensity, and (3) the NBS standards of total irradiance.

The lamp is mounted vertically with the NBS-numbered end of the lamp down and the tip away from the detector. Measurements of distance (from lamp fila-

† Quoted from Wolfe (1965, p. 40).

ment) are made along a horizontal axis passing through the center of the lamp filament. The correct vertical position is determined by setting the centers of the upper and lower seals along a plumb line as viewed from one side of the lamp. The plane of the front surface of the lower press seal is set to contain the horizontal perpendicular to the line connecting the lamp filament center and detector.

The lamp is mounted in the supplied holder which is constructed in such a manner as to reflect a negligible amount of radiant flux in the direction of the radiometer or spectrometer slit. A black shield should be placed at a distance of about 3 feet to the rear of the lamp to intercept stray radiant flux along the radiometric axis and adequate shielding should be provided to intercept stray flux from other directions.

If there is excessive water vapor in the laboratory atmosphere, errors may result at the wavelengths of water-vapor absorption bands. In the original calibrations the comparisons of the lamps with the other NBS standards were made in such a manner that the effect of water-vapor absorption was eliminated.

Values of spectral irradiance for these lamps are tabulated as a function of wavelength in microwatts per (square centimeter-nanometer) at a distance of 43 centimeters from center of lamp to receiver. Values of spectral irradiance for wavelength intervals other than one nanometer, say x nanometers, may be found by multiplying the tabulated values by x.

USE OF THE STANDARDS OF SPECTRAL IRRADIANCE

These standards require no auxiliary optics. If any are employed proper correction must be made for their optical characteristics. The lamp is simply placed at a measured distance from the detector or spectrometer slit. If a distance other than 43 centimeters is used, the inverse-square law may be used to calculate the spectral irradiance (the inverse-square law should not, however, be used for distances shorter than about 40 centimeters).

In measurements wherein two sources (a standard source and a test source) are being compared by the direct substitution method (slit widths kept unchanged, use of the same detector) no knowledge of the spectral transmittance of the spectrometer, nor of the spectral sensitivity of the detector is required. It is necessary, however, to make sure that the entrance slit of the spectrometer is fully and uniformly filled with radiant flux both from the standard and from the test source; and if at any one wavelength the detector response for the standard is significantly different from that for the test source, the deviation from linearity of response of the detector must be evaluated and taken into account. Furthermore, if the standard and test source differ in geometrical shape, it must be ascertained that the instrument transmittance and detector response are not adversely affected thereby. Many detectors are highly variable in sensitivity over their surface area and may require diffusion of radiant flux over their surface to insure accurate radiant energy evaluations.

All calibrations were made by the use of alternating current and it is recommended that they be so used in service. To reduce the line voltage a 10-ampere variable autotransformer may be employed for coarse control. For fine control a second (5-ampere) variable autotransformer may be used to power a radio-filament transformer whose secondary (2.5–5 volt) winding is wired in series with the primary of the 10-ampere transformer. It was found that this method is very effective for accurate control of the 6.50-ampere current.

These standards of spectral irradiance are expensive laboratory equipment and it is suggested that they be operated sparingly and with care in order to prolong their useful life. They should be turned on and off at reduced current and great care should be taken so that at no time will the current appreciably exceed 6.50 amperes. It is recommended that for general use, working standards be prepared by calibrating them relative to the laboratory standard supplied by NBS.

These lamps operate at high temperature such that the quartz envelope is above the flammable point of organic materials. They may thus cause fires, and also the burning of lint, etc. on the enveloper which may result in optical damage to its surface. In no case should the fingers come into contact with the quartz envelope, either hot or cold, as the resulting fingerprints will burn into its surface during lamp operation.

A series of blackbodies using the melting and freezing points of several metals have been made by the NBS for use in a continuing research program by Kostkowski, Richmond, and their coworkers to increase the precision and accuracy of radiation standards. In particular, the gold-point blackbody (see Fig. 2) represents

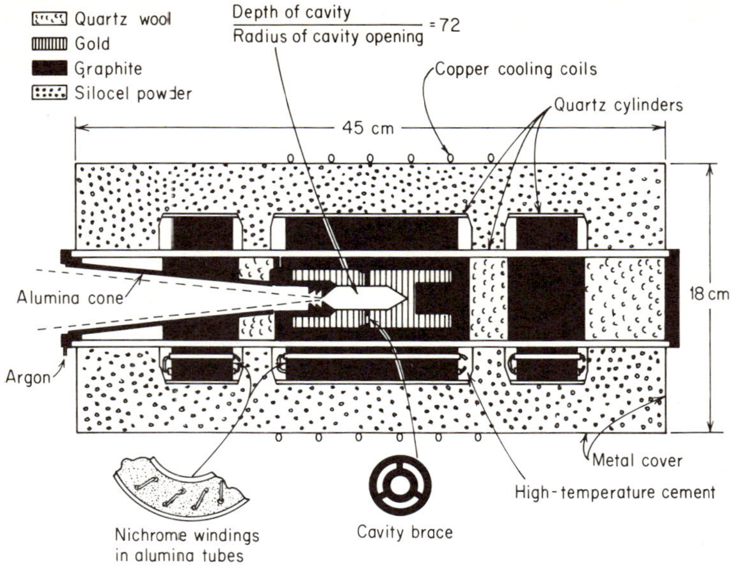

Fig. 2 Cross section of the National Bureau of Standards horizontal gold-point blackbody and furnace. [*Wolfe* (1965).]

the closest approach to a blackbody in any standard source existent today (Kostkowski and Lee, 1962).†

Work by Kostkowski, Erminy, and Hattenberg at NBS required a blackbody usable to 3000 K, stable to within 0.1 percent in temperature for 30 to 45 min, and with a uniformity of 5°C or better. A graphite cylinder used for this source is believed to have achieved these values with an $\epsilon = 0.999$. The temperature measurement required direct photoelectric pyrometry. The pyrometer was calibrated with respect to the International Practical Temperature Scale (IPTS) using the gold-point blackbody. In the vicinity of 6545 Å, the radiation is believed to be within 0.01 percent of that from a blackbody at the temperature of freezing gold. The resulting total uncertainty of the spectral-radiance determination was believed to be 0.35 percent relative to IPTS at 6500 Å and 1.5 percent at 2100 Å. These uncertainties are changed to 1 and 3.4 percent, respectively, for the thermodynamic temperature scale (TTS) due to present uncertainties of the temperature for freezing gold on the TTS.

The National Research Council of Canada has a standard based on work by Bedford (Bedford, 1960) in the Division of Applied Physics in calibrating thermal radiation detectors by use of a low-temperature (40 to 150°C) conical source, temperature-regulated by a stirred oil bath. The cone is made of 0.10-cm-thick copper sheet, is 56 cm long, and has an 18-cm cylindrical extension at the mouth of the cone. The interior is painted with a commercial flat black paint. Calculations by Sparrow's method and by the Gouffe method lead to an $\epsilon = 0.998 \pm 0.001$.

The detectors used were a set of Eppley circular-type 8- and 16-junction bismuth-silver thermopiles (Bedford and Ma, 1968). The cavity was operated vertically to minimize temperature gradients. The overall accuracy is believed to be better than 0.5 percent.

In 1962 Gillham (1959, 1961, 1962), working at the National Physical Laboratory in Great Britain (NPL), introduced three new radiometers each using an electric heating element attached to a radiation thermopile. One was similar to the Guild drift instrument. Another had a small, fast-response detector coated with a glossy

† Tin and zinc blackbodies are described in Nivert et al. (1967), and a copper-point blackbody in Lee (1969).

black whose reflectance is easily and accurately measurable. The third was also small but had a cavity receiver with an emissivity nearly equal to 1. Although the three systems agreed to within 2 or 3 parts in 1,000, they could differ from the original Guild instrument by as much as 0.6 percent.

The British working standards are Moll thermopiles, in which the thermopile itself receives the radiation directly. They are said to be stable to within $\pm \frac{1}{4}$ percent over long periods of time.

Another basis for standards of thermal radiation exists in the International Pyroheliometric Scale (IPS) of 1956, established by The International Radiation Commission of The World Meteorological Organization. It is based on the angstrom and Smithsonian scales. In 1964 international comparisons were made between primary working standard pyroheliometers in Switzerland. Recent investigations at Eppley indicate agreement between detector references to the IPS and the NBS references to within ± 1 percent. For our purposes the IPS is of less interest than the standards of NPS, NPL, and NRC.

The practice of transfer from the standards of NBS to practitioners of infrared technology has received special attention from the Department of Defense. In 1967 the Air Force, and in particular the Infrared and Laser Laboratory, Aerospace Guidance and Metrology Center, Newark Air Force Station, was designated the focal point for the Department of Defense work with NBS to achieve a family of traceable national standards of radiance and to provide assurance of the extension of the calibration chain from NBS to the working standards laboratories of defense installations and contractors. This program has incorporated the molten-point blackbodies (tin, zinc, and gold) of Richmond et al. at NBS with procedures allowing comparison calibrations of other radiation sources. These sources are certified as:

Tin: 505.11 ± 0.02 K

$$\epsilon = 0.9999 \begin{array}{l} + 0.0001 \\ - 0.0005 \end{array}$$

Zinc: 692.73 ± 0.04 K

$$\epsilon = 0.9999 \begin{array}{l} + 0.0001 \\ - 0.0005 \end{array}$$

The gold-point source is limited by the temperature value (1064.43°C) assigned to the melting point of gold.

5. Wavelength Standard One of the most important standard sources became important in 1960, when the international standard of length became the wavelength in vacuum of a spectral line, the orange line of ^{86}Kr. Thus the meter was defined as 1,650,763.73 wavelengths of the radiation from the transitions between the $2p_{10}$ and $5d_5$ levels in that isotope of krypton. The vacuum wavelength was stated as

$$\lambda_{vac} = 6057.80210_5 \text{ Å}$$

The attainment of this wavelength is discussed in *Trans. Int. Astron.* (1962) as follows:

The definition of the primary standard refers to the radiation from atoms unperturbed by external influences. Detailed investigations at the BIPM, the PTB (Germany), at NPL (United Kingdom), NRC (Canada) and NSL (Australia) of the various factors which influence the precise wave-length and the width of the line have led to the following recommendation of the International Committee for Weights and Measures for the most favourable conditions for the reproduction of the krypton line:†

"In accord with paragraph I of Resolution II adopted by the eleventh General Conference of Weights and Measures, the International Committee for Weights and Measures recommends that the krypton 86 radiation adopted as the fundamental length standard be produced in a hot cathode discharge tube conforming to the following specifications. The tube should be filled with enough krypton 86, having a purity of at least 99%, to ensure the presence of solid krypton at a temperature of 64 K. The tube should have a capillary with an internal diameter of 2 to 4 mm and a wall thickness of about 1 mm.

† Editor's note: The paragraphs that follow have been translated from the French.

"It is assumed that the wavelength of the radiation emitted by the positive column is equal to within 1 part in 10^8 to the wavelength corresponding to the transition between unperturbed levels when the following conditions are satisfied:

1. The capillary is observed so that the light rays go from the cathode to the anode.
2. The lower part of the discharge tube, including the capillary, is immersed in a cooling bath maintained to within 1 K at the triple point of nitrogen.
3. The current density in the capillary is 0.3 ± 0.1 A/cm^2."

Shifts in the wavelength of the ^{86}Kr line as a function of pressure, current density, and temperature are also discussed in *Trans. Int. Astron.* (1962). Humphreys and Paul (1970) used the 6047-Å line as generated in a microwave-excited lamp and quote the informally reported findings of Baird of the National Research Council of Canada that the 6057-Å line suffers a red shift in the microwave-excited lamp of 0.00006 Å, compared with the standard Engelhard lamp.

The situation with regard to standards continues to evolve. Recently an adjustment was made to the United States unit of luminous flux. The announcement issued in the *National Bureau of Standards Technical News Bulletin* of March 1970 is given here.

The bureau announces a change of 1.9 percent in the U.S. unit of total luminous flux for gas-filled tungsten-filament lamps. This change is necessary to bring the U.S. unit into agreement with the world mean. Owners of gas-filled tungsten-filament lamp standards of luminous flux supplied on the old basis should multiply the presently assigned values by the factor 1.019 to obtain values on the new basis.

Several international intercomparisons† of luminous flux measurements have been conducted since the establishment of the platinum-point black body‡ as the primary standard of luminous intensity. These intercomparisons indicated that the U.S. unit of total luminous flux at the color temperature of gas-filled tungsten-filament lamps (approx. 2800 K), maintained at NBS, is discrepant with respect to the world mean. Recent work at NBS has confirmed that this discrepancy represents a misassignment of the U.S. unit.

In the past 30 years NBS has set up standards of total luminous flux for several categories of gas-discharge lamps. These standards have not been referenced to the gas-filled tungsten lamp standards of luminous flux, but rather have been calibrated by the more fundamental method of distribution photometry based on the NBS standards of horizontal luminous intensity. Since no change is indicated for the other U.S. photometric units, flux standards derived directly from standards of luminous intensity will not change. Lamps falling specifically in this category for which no change in rated values is necessary include mercury vapor§ and fluorescent lamps.

Even more recent is the announced reorganization of the NBS programs in radiometry, pyrometry, photometry, and spectrophotometry (*Natl. Bur. Stand. Tech. News Bull.*, 1971). This announcement assesses present capability as follows:

Although there are many notable differences between radiometry and photometry, the present status of both can be characterized by two comprehensive statements:

1. Five percent accuracy is difficult to achieve in both fields and is realized only by experts.
2. One percent accuracy is very rare indeed and is realized only in a few laboratories and in a few types of measurements within these laboratories.

COMMERCIAL SOURCES

6. Blackbodies These sources of radiation constitute a class of radiators used to produce accurately known amounts of radiation in the optical spectrum but particularly in the infrared, the region in which they emit most efficiently from the temperatures to which they can be raised or lowered. Generally they will be considered here to consist of cavities with a hole, the emissivity of the hole being so nearly constant and so nearly equal to unity (for the purposes for which they

† *Com. Consult. Photom., Sess. 1952, P-V, Annexe* P3, pp. 46–75 and *Annexe* P4, pp. 76–100; *4e Sess., 1957, P-V, Annexe* P8, pp. 74–103; *5e Sess., 1962, Annexe* 9, pp. 63–87.

‡ Announcement of Changes in Electrical and Photometric Units, *Natl. Bur. Stand. (U.S.) Circ.* C459, May 1947.

§ *Natl. Bur. Stand. (U.S.), Tech. News Bull.*, vol. 35, no. 1, pp. 3–5, 1951.

are used) that any discrepancies can be neglected. The temperatures for which these blackbodies can be obtained range from that of liquid nitrogen to approximately 3000°C; those in the range between room temperature and about 1000°C are readily available commercially at a reasonable price.

Table 1 describes the typical sources (from companies which supplied pertinent data). The temperature range of these sources is (with one exception) from above room temperature to 3000°C.

Most of the manufacturers listed in Table 1 supply auxiliary equipment which can be purchased for use with high-temperature blackbodies particularly, e.g., an aperture plate to install in front of the unit. A selectable number of calibrated apertures can be rotated in front of a hole in the aperture plate to provide a variety of radiant intensities. For applications requiring a low-temperature background, the aperture can be water-cooled by using a special type of aperture plate also supplied by some manufacturers. Another, often essential, item of auxiliary equipment is a multispeed chopper to provide chopped radiation at levels between the value emitted by the blackbody cavity and that radiated by the chopper blade.

Blackbodies which provide radiation at temperatures below room temperature must be cooled; these are not so extensively available on a commercial basis. One published report (Karole et al., 1967) of a thermoelectrically cooled blackbody demonstrates the difficulty with which reliable cooled blackbodies can be made. In addition to the trouble attendant on keeping them stable, care must be taken to prevent moisture from forming.

Incandescent Nongaseous Sources†

7. Nernst Glower The Nernst glower is usually constructed in the form of a cylindrical rod or tube from refractory materials, usually zirconia, yttria, and thoria, in various sizes. Platinum leads at the ends of the tube conduct power to the glower from the source. Since the resistivity of the material at room temperature is quite high, the working voltage is insufficient to get the glower started. Once started, its negative temperature coefficient of resistance tends to increase current, which would cause its destruction, so that a ballast is required in the circuit. Starting is effected by applying external heat, either with a flame or adjacent electrically heated wire, until the glower begins to radiate.

Data from Ramsey and Alishouse (1968) for specific sized glower and operating conditions are:

Power consumption = 45 W, 0.6 A
Color temperature = 1980 K
Dimension = 3.1 by 12.7 mm

The spectral characteristics of a Nernst glower in terms of the ratio of its output to that of a 900°C blackbody are shown in Fig. 3.

The life of the Nernst glower diminishes as the operating temperature is increased. Beyond a certain point, depending on the particular glower, no great advantage is gained by increasing the current through the element. The glower is fragile, with low tensile strength, but can be maintained intact with rigid support. The life of the glower depends on the operating temperature, care in handling, and the like.

Since the Nernst glower is made in the form of a long thin cylinder, it is particularly useful for illuminating spectrometer slits. Its useful spectral range is from the visible region to almost 30 μm, although its usefulness compared with other sources diminishes beyond about 15 μm. As a rough estimate, the radiance of a glower is nearly that of a graybody at the operating temperature with an emissivity in excess of 75 percent, especially below about 15 μm.

The relatively low cost of the glower makes it a highly desirable source of moderate radiant power for optical uses in the laboratory.

Although glowers can be obtained from the makers of the material, the usual source of supply (or information about suppliers) seems to be the makers of spectroscopic equipment.

† Exclusive of high-temperature blackbodies.

TABLE 1 Blackbody Sources (Cann, 1967)

Manufacturer and model	Temperature range	Cavity length, in	Aperture† diameter, in	Emissivity	Comment
Astro-Industries:					
2570	300–3000°C	Cylinder, 4.0	0.5	0.99	
		Cone (30°), 3.36	0.75	0.99	Inert gas 2700°C
Barnes Engineering:					
11-101T	0–250°C	Cone (14°)	0.625	0.99 ± 0.01	Down to 20°C above ambient
11-110T	200–600°C	Cone (28°)	0.5	0.99 +	
11-119T	200–600°C	Cone (14°)	0.0086	0.99 +	
11-120T	200–600°C	Cone (14°)	0.015	0.99 +	
11-121T	200–600°C	Cone (14°)	0.040	0.99 +	
11-200T	200–1000°C	Cone (28°)	0.5	0.99 ± 0.01	
11-201T	200–1000°C	Cone (28°)	1.0	0.99 ± 0.01	
Electro-Optical Industries:					
111	50–600°C	Recessed cone	0.040	0.99 ± 0.01	
121	50–600°C	Recessed cone	0.080	0.99 ± 0.01	
131	50–600°C	Recessed cone	0.25	0.99 ± 0.01	
141	50–600°C	Recessed cone	0.50	0.99 ± 0.01	
181	50–600°C	Square	4	0.90 – 0.98	
191–194	50–600°C	Recessed cone	0.080	0.99 ± 0.01	
113	50–1000°C	Recessed cone	0.040	0.99 ± 0.1	
123	50–1000°C	Recessed cone	0.080	0.99 ± 0.1	
133	50–1000°C	Recessed cone	0.25	0.99 ± 0.1	
143	50–1000°C	Recessed cone	0.50	0.99 ± 0.1	
153	50–1000°C	Recessed cone	1.0	0.99 ± 0.1	
173	50–1000°C	Recessed cone	3.0	0.99 ± 0.1	
183	50–1000°C	Square	4	0.90 – 0.98	
115	1000–2000 K	Recessed cone	0.040	0.99 ± 0.1	
125	1000–2000 K	Recessed cone	0.080	0.99 ± 0.1	
135	1000–2000 K	Recessed cone	0.25	0.99 ± 0.1	
145	1000–2000 K	Recessed cone	0.50	0.99 ± 0.1	
155	1000–2000 K	Recessed cone	1.0	0.99 ± 0.1	
146	1000–3000°C	Recessed cone	0.50	0.99 ± 0.1	
Infrared Industries:					
403	325–1000 K	. . . .	0.5	0.99 ± 0.01	
404	50–1000°C	. . . .	0.5	0.99 ± 0.01	
405	325–1000 K	. . . .	1.0	0.99 ± 0.01	
406	50–1000°C	Recessed cone	1.0	0.99 ± 0.01	
407A	200–600°C	(20°)	0.08	0.99 ± 0.01	
408	200–600°C	. . . .	0.250	0.99 ± 0.01	
417	50–1000°C	. . . .	3.0	0.99 ± 0.01	
420	200–1200°C	. . . .	0.5	0.99 ± 0.01	
427	200–600°C	3.5	0.08	0.99 ± 0.01	
Rocketdyne:	Up to 3100 K	Cylinder	0.25	0.99	
	Up to 2200 K	. . . .	2.00	0.975	
	Up to 2500 K	Slit along tube length	0.25 × 0.03	0.975	Slit source

† Maximum is given. Some models have variable apertures.

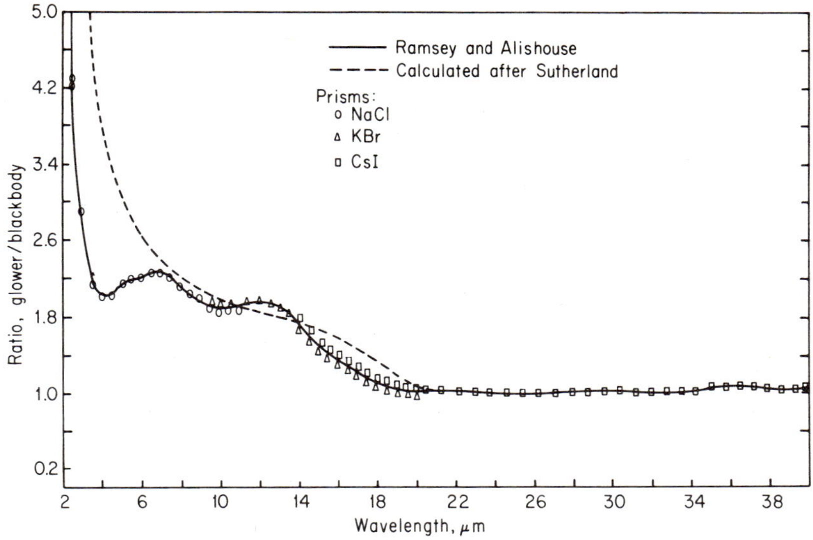

Fig. 3 The ratio of a Nernst glower to a 900°C blackbody vs. wavelength. (*Reprinted by permission from W. Y. Ramsey and J. C. Alishouse, A Comparison of Infrared Sources, Infrared Physics, vol.* 8, *p.* 148, *copyright* 1968 *by Pergamon Press.*)

8. Globar The globar is a rod of bonded silicon carbide capped with metallic caps which serve as electrodes for the conduction of current through the globar from the power source. The passage of current causes the globar to heat, yielding radiation at a temperature above 1000°C. A flow of water through the housing that contains the rod is needed to cool the electrodes (usually silver). This complexity makes the globar less convenient to use than the Nernst glower and necessarily more expensive. This source can be obtained already mounted from a number of manufacturers of spectroscopic equipment. In one case in particular, feedback in the controlled power source makes it possible to obtain high radiation output.

Ramsey and Alishouse (1968) provide information on a propane-heated sample globar:

Power consumption = 200 W, 6 A
Color temperature = 1470 K
Dimensions = 5.1 by 20.3 mm

From the same reference we find the spectral characteristics of the globar in terms of the ratio of its output to that of a 900°C blackbody. This ratio is plotted as a function of wavelength in Fig. 4.

In Fig. 5 is a representation of the spectral emittance (emissivity) of a globar as a function of wavelength. The emittance values are only representative and can be expected to change considerably with use.

9. Gas Mantle The Welsbach mantle is typified by the kind found in high-intensity gasoline lamps used where electricity is not available. The mantle is composed of thorium oxide with some additive to increase its efficiency in the visible region. Its near-infrared emittance is quite small, except for regions exemplified by gaseous emission, but increases considerably beyond 10 μm.

Ramsey and Alishouse (1968) provide information on a propane-heated sample from an experiment in which a comparison of several sources is made:

Color temperature = 1670 K
Dimensions = 25.4 by 38.1 mm

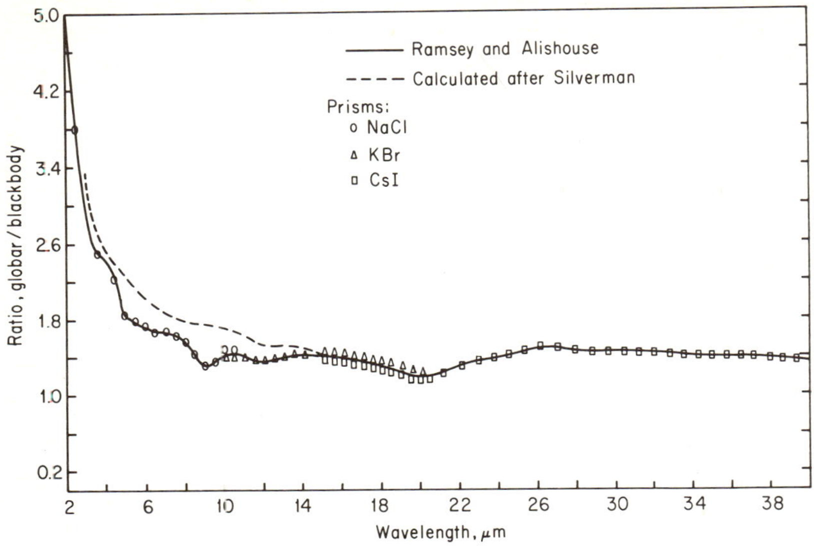

Fig. 4 The ratios of a globar to 900°C blackbody vs. wavelength. (*Reprinted by permission from W. Y. Ramsey and J. C. Alishouse, A Comparison of Infrared Sources, Infrared Physics, vol. 8, p. 148, copyright 1968 by Pergamon Press.*)

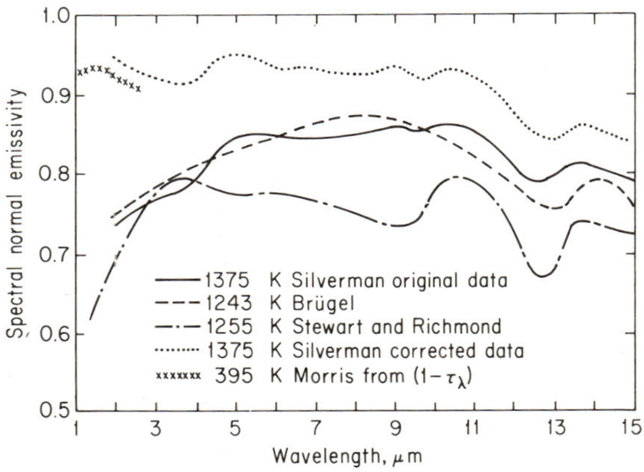

Fig. 5 The spectral emissivity of a globar. [*Morris* (1961).]

The spectral characteristics of the mantle in terms of the ratio of its output to that of a 900°C blackbody are shown in Fig. 6.

Pfund (1936) modified the gas mantle so that it became more a laboratory experimental source than an ordinary radiator. By playing a gas flame on an electrically heated mantle he was able to increase its radiation over that from the gas mantle itself. Figure 7 shows a comparison of the gas mantles and the gas electrically heated mantle with a Nernst glower. Strong (1938) points out that playing a flame against the mantle at an angle produces an elongated area of intense radiation useful for illuminating the slits of a spectrometer.

Comparison of Nernst, Globar, and Mantle. In the literature one generally finds the Nernst glower, globar, and gas mantle compared for the quality and quantity

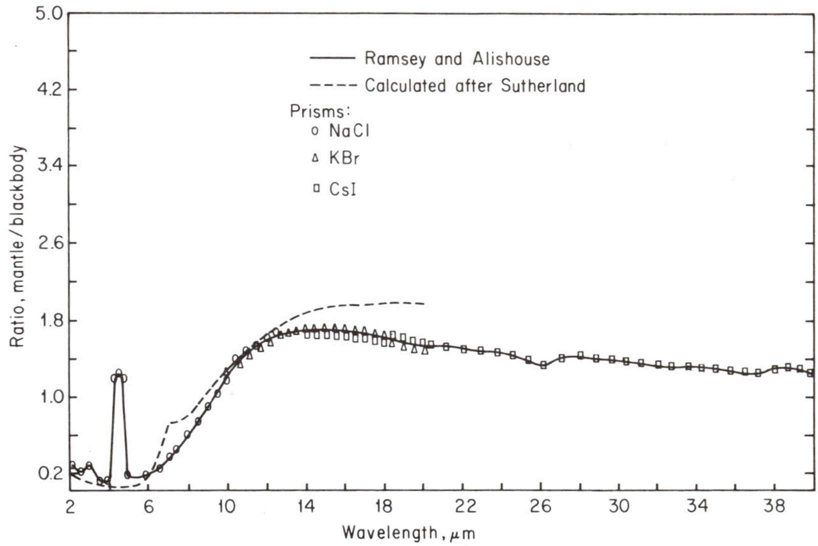

Fig. 6 The ratios of the gas mantle to a 900°C blackbody vs. wavelength. (*Reprinted by permission from W. Y. Ramsey and J. C. Alishouse, A Comparison of Infrared Sources, Infrared Physics, vol.* 8, *p.* 149, *copyright* 1968 *by Pergamon Press.*)

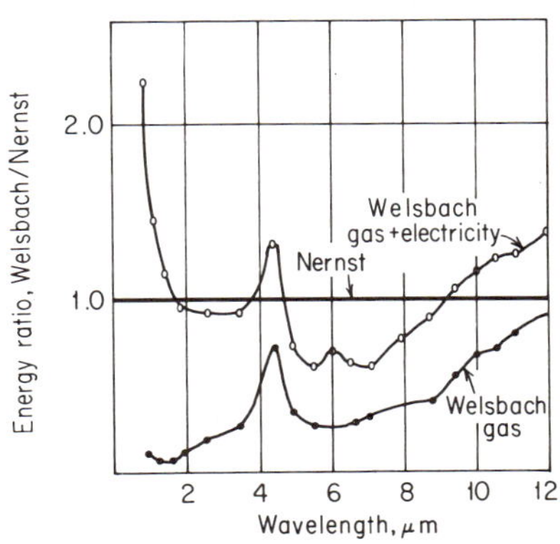

Fig. 7 Emission relative to that of a Nernst glower (2240 K) of the gas-heated mantle (*lower curve*) and that of the mantle heated by gas plus electricity (*upper curve*). [*Pfund* (1936).]

of their radiations. One of the latest of these comparisons is that of Ramsey and Alishouse (1968). Figure 8 compares the three types of source, omitting consideration of differences in instrumentation used in making measurements of the radiation from the sources.

One's choice of these sources is usually influenced by such factors as availability, convenience, and cost. At the very long wavelength regions in the infrared, the gas mantle and the globar have a slight edge over the Nernst glower, making this very convenient, small, inexpensive source vulnerable to the compromise that might have to be made for that little extra power.

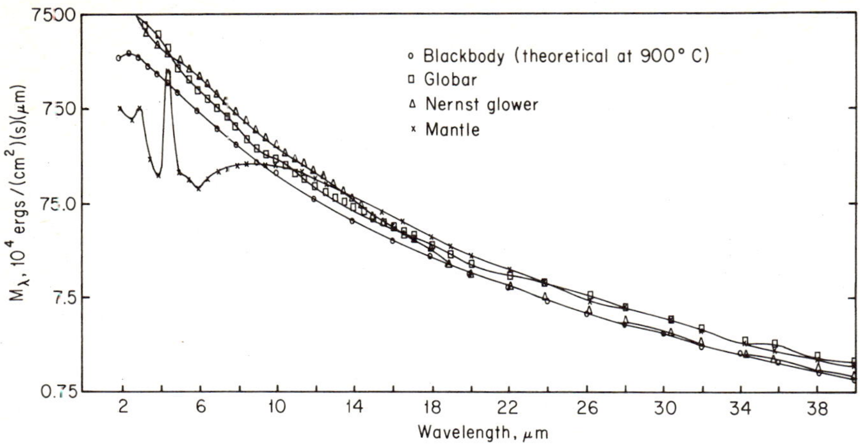

Fig. 8 The spectral radiant emittances of a globar, Nernst glower, 900°C blackbody, and gas mantle vs. wavelength. (*Reprinted by permission from W. Y. Ramsey and J. C. Alishouse, A Comparison of Infrared Sources, Infrared Physics, vol.* 8, *p.* 151, *copyright* 1968 *by Pergamon Press.*)

10. Tungsten-Filament Lamps A comprehensive discussion of tungsten-filament lamps is given by Carlson and Clark (1965), to which the reader is referred for more detailed information. Figures 9 to 11 show the configurations of lamp housings and filaments. Tables 2 to 4 give pertinent characteristics of tungsten and photoflood lamps.

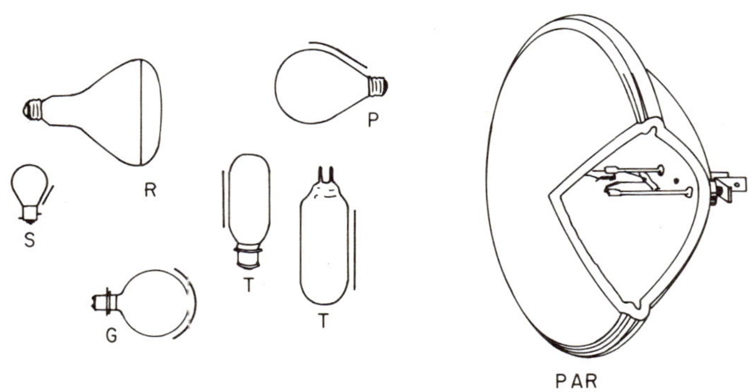

Fig. 9 Bulk shapes most frequently used for lamps in optical devices. Letter designations are for particular shapes. [*From F. E. Carlson and C. N. Clark, Light Sources for Optical Devices, chap.* 2 *in R. Kingslake (ed.), "Applied Optics and Optical Engineering," vol. I, Academic Press, New York,* 1965; *copyright by Academic Press, Inc.*]

Tungsten lamps have been designed for a variety of applications; few are directed toward scientific research, but some bear directly or indirectly on scientific pursuits, insofar as they can provide steady sources of numerous types of radiation. One set of sources cited here, particularly for what the manufacturer calls their scientific usefulness, is described in the Osram 1966 pamphlet Lamps for Scientific Purposes. Their filament structures are not dissimilar to those already described, but their designs provide envelope shapes, materials, etc., to reduce extraneous radiation and

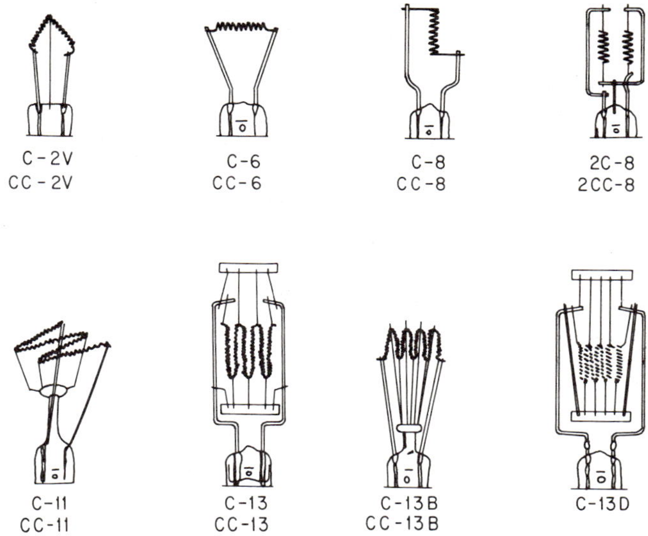

Fig. 10 Most commonly used filament forms. Letters designate the type of filament. [*From F. E. Carlson and C. N. Clark, Light Sources for Optical Devices, chap.* 2 *in R. Kingslake (ed.), "Applied Optics and Optical Engineering," vol. I, Academic Press, New York,* 1965; *copyright by Academic Press, Inc.*]

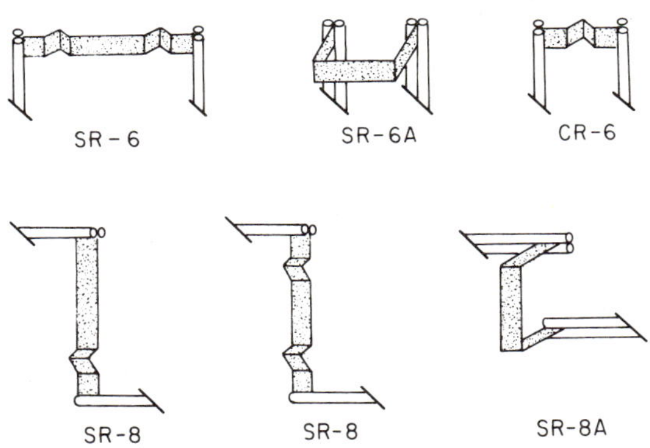

Fig. 11 Ribbon-type filaments. Type designations are by number. [*From F. E. Carlson and C. N. Clark, Light Sources for Optical Devices, chap.* 2 *in R. Kingslake (ed.), "Applied Optics and Optical Engineering," vol. I, Academic Press, New York,* 1965; *copyright by Academic Press, Inc.*]

ensure the quality and stability of the desired radiation. The lamps can be obtained with a certification of their calibration values.

The physical descriptions of some of these sources are given in Fig. 12. Applications according to the manufacturer are photometry, pyrometry, optical radiometry, sensitometry, spectroscopy, spectrometry, polarimetry, saccharimetry, spectrophotometry, colorimetry, microscopy, microphotography, microprojection, and stroboscopy.

Quartz envelope lamps (from General Electric and others) are particularly useful as standards (Par. 4) because they are longer-lasting (due to action of iodine in the quartz-iodine series), can be heated to higher temperatures than those of the glass-envelope series, are sturdier, and can transmit radiation to longer wavelengths

TABLE 2 Typical Tungsten-Filament Lamps with Approximately Square Sources[a]

Watts nominal	Source size,[b,c] mm		Filament form	Estimated luminance,[d] cd/mm²	Approximate average color temp. through life, K	Design values, V	Average life at design volts, h	Bulb	Base[e]	Ordering abbreviation[f]	Approx. initial lumens at design volts
	Width	Height									
4.8	1.2	1.2	CC-6	4.3	2650	8	50	G-4$\frac{1}{2}$	Min. Sc.	419	62
10	1.0	1.0	C-6	14	2850	3.7	100	T-5	S.C. Bay.	1874	140
17.9	2.0	1.1	C-6	13	2850	6.5	100	S-8	D.C. Pf.	1630	290
19	2.0	1.3	C-6	10	2850	6.5	100	RP-11	S.C. Pf.	1953	270
20	2.0	1.7	CC-6	8.9	2900	20	200	S-8	D.C. Pf.	1634	300
24.2	2.5	3.0	C-2V	5.3	2950	6.2	200	RP-11	S.C. Bay.	1621	400
28	2.5	1.2	CC-8	19	3050	14	50	T-8	S.C. Pf.	1933	570
30	2.9	3.2	C-2	4.1	2800	6	3,000	T-8	S.C. Pf.	5A/T8SCP	380
30	2.5	1.6	C-6	11	2900	6	250	S-8	D.C. Bay.	1594	450
50	7.0	7.0	CC-2V	1.6	2850	120	50	S-11	D.C. Bay.	BLX	780
60	3.8	2.2	CC-8	11	2850	11	1,000	T-4	Special	1960[g]	880
100	4.9	2.6	CC-6	20	3200	20	50	G-16$\frac{1}{2}$	S.C. Pf.	5A/G16$\frac{1}{2}$/3	2,600
100	5.5	5.1	CC-13	7.1	2975	118	50	T-8	S.C. Bay.	CDS	2,000
100	5.5	5.3	CC-13	6.5	3000	120	50	T-8$\frac{1}{2}$	Med. Pf.	100T8$\frac{1}{2}$/8	1,880
100	5.8	5.8	CC-13	4.8	2850	120	200	G-16$\frac{1}{2}$	D.C. Bay.	100G16$\frac{1}{2}$/29DC	1,630
150	5.5	6.1	2CC-8	10	3075	118	25	T-8	S.C. Bay.	CEW	3,500
150	6.6	3.7	CC-6	15	3100	20	50	T-8	S.C. Pf.	150T8SCP	3,600
300	8.5	9.3	C-13	9.5	3200	118	25	T-10	S.C. Bay.	CYC	7,500
400	8.2	7.7	C-13D	14	3050	65	200	T-12	Med. Pf.	400T12P	8,800
480	11.7	7.2	CC-6	12	3000	12	500	G-30	Med. Bip.	480G30	9,900
500	7.0	7.5	C-13D[h]	....	3300	120	25	T-10	4-Pin	CZA	
500	8.0	7.8	C-13D	20	3200	118	25	T-10	Med. Pf.	CZX	12,500
500	8.0	7.8	C-13D[h]	....	3200	120	25	T-12	4-Pin	DFW	
500	9.8	10.2	C-13D[i]	10	2950	120	200	T-14	Med. Bip.	500T14/8	10,200
750	9.4	9.3	C-13D	22	3250	118	25	T-12	Med. Pf.	DDB	19,500
750	14.2	13.7	C-13	8.7	3000	120	200	T-20	Med. Pf.	750T20P/SP	17,000
1,000	10.9	13.4	C-13D	16	3050	120	200	T-20	Med. Pf.	1M/T20P/SP	23,500
1,200	11.2	11.1	C-13D	29	3400	118	10	T-12	Med. Pf.	DHT	36,000
1,500	16.5	16.0	C-13D[i]	13	3000	120	200	T-24	Mog. Pf.	1500T24/6	33,500
1,500	20.	21.	C-13	10	3200	120	60	T-24	Med. Bip.	1500T24/15	44,000
2,000	18.2	18.0	C-13D[i]	15	3050	120	200	T-30	Mog. Bip.	2M/T30/1	48,000

2,000	24.	20.	C-13	12	3200	120	100	T-48	Mog. Bip.	2M/T48/1	57,500
2,100	13.6	14.5	C-13D	27	3100	60	50	T-24	Mog. Bip.	2100T24/8	54,000
3,000	20.	17.0	C-13D[i]	23	3150	120	100	T-32	Mog. Bip.	3M/T32/2	81,000
3,000	26.	15.6	C-13B	22	3250	32	100	T-32	Mog. Bip.	3M/T32	88,000
5,000	36.	27.	C-13	14	3200	120	150	T-64	Mog. Bip.	5M/T64/3	141,000
10,000	52.	34.	C-13	19	3350	120	75	G-96	Mog. Bip.	10K/G96	330,000

[a] From F. E. Carlson and C. N. Clark, Light Sources for Optical Devices, chap. 2 in R. Kingslake (ed.), "Applied Optics and Optical Engineering," vol. I, Academic Press, New York, 1965; copyright by Academic Press, Inc.

[b] Viewed in the direction perpendicular to the long axis of straight or coiled filaments and to the plane of ribbon or multisegmented coiled filaments.

[c] Source-aspect ratios (width:height) between 2:1 and 1:2.

[d] Equal to lamp lumens divided by 10 times the approximate filament width and height given in preceding columns.

[e] Abbreviations: Min. Sc., miniature screw; S.C. Bay., single-contact bayonet; D.C. Pf., double-contact prefocus; S.C. Pf., single-contact prefocus; D.C. Bay., double-contact bayonet; Med. Pf., medium prefocus; Med. Bip., medium bipost; Mog. Pf., mogul prefocus; Mog. Bip., mogul bipost.

[f] All-letter and all-numeral groups are ANSI standard identification codes used throughout the industry within some lamp categories. Letter-numeral combinations are the ordering abbreviations in common use for other categories; some manufacturers use different identifications for similar lamps.

[g] This lamp employs the iodine regenerative cycle.

[h] Proximity reflector.

[i] Designed for base-up operation.

TABLE 3 Typical Tungsten-Filament Lamps with Elongated Sources[a]

Watts nominal	Source size,[b] mm: Width	Source size,[b] mm: Height	Filament form	Estimated luminance,[c] cd/mm²	Approximate average color temperature through life, K	Design values, V or A	Average life at design volts or amperes, h	Bulb	Base[d]	Ordering abbreviation[e,f]	Approximate initial lumens at design volts or amperes
Source-Aspect Ratios (Width:Height) Greater than 10:1											
2	17	0.045	S-6	....	2550	4 V	200	G-10	Cand. Sc.	851K	215
6	5.3	9.3	C-6	4.4	2800	1 A	100	T-5	D.C. Pf.	BRX	70
10	16	0.10	S-6	....	2750	4 V	100	G-10	Cand. Sc.	855K	100
11	8.2	0.25	C-6	8.0	2900	11 V	200	T-8	D.C. Pf.	1926	165
16	6.3	0.6	C-6	....	2750	2 A	3,000+	T-8	S.C. Pf.	2A/T8SCP (8 V)	
22	16	0.25	C-6	8.8	2950	22 V	200	T-8	D.C. Pf.	1936	350
34	8.3	0.6	C-6	14	3050	4 A	100	T-8	S.C. Bay.	4A/T8/25 (8.5 V)	670
54	12.5	1.0	SR-6	....	3000	9 A	200	T-8½	Med. Pf.	9A/T8½P (6 V)	
500	60	1.4	C-8[g]	12	3000	120 V	2,000	T-3	R.S.C.	Q500T3/CL	10,200
1,000	65	2.0	C-8[g]	21	3200	120 V	500	T-3	R.S.C.	Q1000T3/4CL	27,000
1,500	170	1.6	C-8[g]	12	3000	240 V	2,000	T-3	R.S.C.	Q1500T3/CL	32,000
Source-Aspect Ratios (Width:Height) between 10:1 and 5:1											
3	2.5	0.3	C-6	3.7	2700	0.75 A	50	S-8	D.C. Pf.	BGB	28
6.4	2.0	0.4	C-6	12	2900	5.8 V	50	G-6	Min. Sc.	153	100
13.1	4.5	0.5	C-6	8.4	2900	6.7 V	500	S-8	S.C. Bay.	1619	190
25	4.5	0.8	C-6	13	3000	6 V	50	S-11	L.S. Med.	25S11/7	460
100	10	1.5	C-8	12	2900	6.6 A	1,000	T-3	R.S.C.	Q6.6A/T3/CL	1,900
2,500	37	5.3	C-6	41	3400	27 V	15	T-24	Mog. Bip.	2500T30	80,000
Source-Aspect Ratios (Width:Height) between 5:1 and 2:1											
7	1.9	0.5	C-6	11	2900	3.5 V	50	S-8	S.C. Pf.	2A/S8	105
7.2	1.8	0.6	C-6	10	2850	8 V	50	T-4½	Min. Sc.	870	115
18	3.4	0.7	C-6	13	2850	6.5 V	125	T-5	S.C. Bay.	1462	315
25	2.7	1.2	C-6	12	2950	6.1 V	125	S-11	S.C. Pf.	1763	400
28	3.6	1.1	C-6	10	2900	6.1 V	125	RP-11	S.C. Pf.	1954	400
36	4.8	1.2	C-6	10	2900	4 A	500	T-8	S.C. Pf.	4A/T8SCP (9 V)	575
45	4.0	1.5	C-8	10	2800	6.6 A	1,000	T-2½	R.S.C.	Q6.6A/T2½/CL	630
50	5.1	1.4	C-6	14	3050	5 A	100	T-8	S.C. Pf.	5A/T8SCP (10 V)	1,000
108	8	2.0	SR-6A	....	3000	18 A	370	T-10	Med. Pf.	18A/T10/2P (6 V)	2,200
125	15.3	6.2	C-13B	1.8	2800	120 V	500	T-10	Med. Pf.	125T10P	1,750

150	11.2	2.2	CC-8	13	3050	28 V	300	T-4	Special	1958	3,300
150	8.0	2.0	CC-6*h*	....	3150	120 V	15	T-12	4-Pin	DFA	
150	5.3	2.2	CC-6*i*	....	3300	21 V	15	T-12	4-Pin	DLS	
200	10	3.0	CC-8	15	3050	6.6 A	500	T-4	R.S.C.	Q6.6A/T4/CL	4,400
400	15	4.0	CC-8	17	3200	120 V	75	T-4	R.S.C.	FAL	10,000
650	16	4.2	CC-8	24	3200	120 V	100	T-4	R.S.C.	FAD	16,500
1,000	22	6.0	CC-8	20	3200	120 V	150	T-5	R.S.C.	DXW	27,000
Source-Aspect Ratios (Width:Height) between 1:2 and 1:5											
25	1.2	2.8	C-8	12	2950	6.1 V	125	S-11	S.C. Pf.	1759	400
36	1.2	4.4	C-8	11	2900	4 A	500	T-8	S.C. Pf.	4A/T8/87 (9 V)	575
75	1.8	5.2	C-8	17	3100	7.5 A	100	T-8	S.C. Pf.	7.5A/T8/94 (10 V)	1,575
108	1.9	5.5	C-8	22	3075	18 A	50	T-10	Med. Pf.	CPR	2,250
Source-Aspect Ratios (Width:Height) between 1:5 and 1:10											
25	0.8	4.5	C-8	13	3000	6 V	50	T-8	S.C. Pf.	25T8/SCP	460
108	2.0	14	SR-8A	....	3000	18 A	370	T-10	Med. Pf.	18A/T10/1P (6 V)	
144	1.7	11	C-8	17	3100	12 V	200	T-10	Med. Pf.	12A/T10P	3,170
500	2.5	25	C-8	....	3300	20 A	100	T-24	Med. Bip.	20A/T24/3	14,000
Source-Aspect Ratios (Width:Height) less than 1:10											
6.5	0.08	15	S-8	....	2850	3.6 V	100	T-5	S.C. Bay.	1874	140
11	0.25	8.3	C-8	8	2900	11 V	200	T-8	D.C. Pf.	1928	165
34	0.6	8.2	C-8	14	3100	8.5 V	150	T-8	S.C. Pf.	4.2A/T8/SCP	715
35	0.13	42	S-8	....	2900	10 V	500	T-8½	D.C. Pf.	1915	480
54	1.0	12	SR-8	....	3000	9 A	200	T-8½	Med. Pf.	9A/T8½/1P (6 V)	
180	3.0	33	SR-8	....	2300	30 A	...	T-24	Mog. Bip.	30A/T24/6 (6 V)	

a From F. E. Carlson and C. N. Clark, Light Sources for Optical Devices, chap. 2 in R. Kingslake (ed.), "Applied Optics and Optical Engineering," vol. I, Academic Press, New York, 1965; copyright by Academic Press, Inc.

b Viewed in the direction perpendicular to the long axis of straight or coiled filaments and to the plane of ribbon or multisegmented coiled filaments.

c Equal to lamp lumens divided by 10 times the approximate filament width and height given in preceding columns.

d Abbreviations: Cand. Sc., candelabra screw; D.C. Pf., double-contact prefocus; S.C. Pf., single-contact prefocus; S.C. Bay., single-contact bayonet; Med. Pf., medium prefocus; L.S. Med., long-screw medium; Mog. Bip., mogul bipost; Med. Bip., medium bipost; R.S.C., recessed single contacts on double-ended lamp.

e All-letter and all-numeral groups are ANSI standard identification codes used throughout the industry within some lamp categories. (Occasionally a suffix letter is added to the ANSI numeral codes; lamps 851K and 885K are examples.) Letter-numeral combinations are the ordering abbreviations in common use for other categories; some manufacturers use different identifications for similar lamps.

f Lamps 1958 FAL, FAD, and DXW and all lamps that include Q in the ordering abbreviation employ the iodine regenerative cycle.

g Coil axis must be horizontal. Essential filament supports may affect applicability.

h Internal reflector, designed for $f/1.5$ lens.

i Internal dichroic reflector, designed for $f/1.2$ lens.

TABLE 4 Photoflood Lamps[a,b]

Lamp identification	Approximate watts	Average life, h	Lumens or candelas[c]	Approx. beam spread,[d] deg	Bulb shape and size
			Nonreflectorized Bulbs		
BBA	250	3	8500 lm	....	A-21
EBV	500	6	17,000 lm	....	PS-25
DWY[e]	650	25	20,000 lm	....	T-4
DXR	1,000	10	31,000 lm	....	PS-35
			Reflectorized Bulbs		
DAN	200	4	10,500 cd	30–40	R-20
BEP	300	4	11,000 cd	30–40	R-30
EBR	375	4	14,000 cd	30–40	R-30
BFA	375	4	16,000 cd	30–40	R-40
DXC	500	6	6000 cd	90	R-40
DXB	500	6	50,000 cd	20	R-40
DWA	650	8	25,500 cd	30 × 40	PAR-36
DXK[e]	650	16	30,000 cd	30 × 40	PAR-36
FBJ[e]	650	16	60,000 cd	15 × 22	PAR-36
FBF[f]	650	8	20,000 cd or 40,000 cd	36 × 45 or 18 × 30	PAR-36

[a] From F. E. Carlson and C. N. Clark, Light Sources for Optical Devices, chap. 2 in R. Kingslake (ed.), "Applied Optics and Optical Engineering," vol. I, Academic Press, New York, 1965; copyright by Academic Press, Inc.
[b] 3400 K at nominal 120 V.
[c] Approximate mean candlepower in 10° axial cone.
[d] To one-half maximum candlepower.
[e] Iodine regenerative getter.
[f] Two-filament, dual-beam design.

in the infrared than glass-envelope lamps. Studer and Van Beers (1964) have shown the spectral deviation to be expected of those containing no iodine. The deviation, when known, is readily acceptable in lieu of the degradation in the lamp caused by the absence of iodine.

The particular tungsten-quartz-iodine lamps used in accordance with the National Bureau of Standards are described in Par. 4.

ARC AND GAS-DISCHARGE SOURCES

11. Carbon Arc The carbon arc has been passed down from early lighting applications in three forms: (1) low-intensity arc, (2) flame, and (3) high-intensity arc. The low- and high-intensity arcs are usually operated on direct current; the flame type adapts to either direct or alternating current. In all cases a ballast must be used. "In the alternating-current arc, the combined radiation from the two terminals is less than that from the positive crater of the direct-current arc of the same wattage" (Worthing, 1937).

Spatial variation in the amount of light energy across the crater of dc arcs for different currents is shown in Fig. 13.

The carbon arc is a good example of an open arc, widely used because of its very high radiation and color temperatures (from approximately 3800 to 6500 K or higher). The rate at which the material is consumed and expended during burning (5 to 30 cm/h) depends on the intensity of the arc. It is discharged between two electrodes that are moved to compensate for the rate of consumption of the

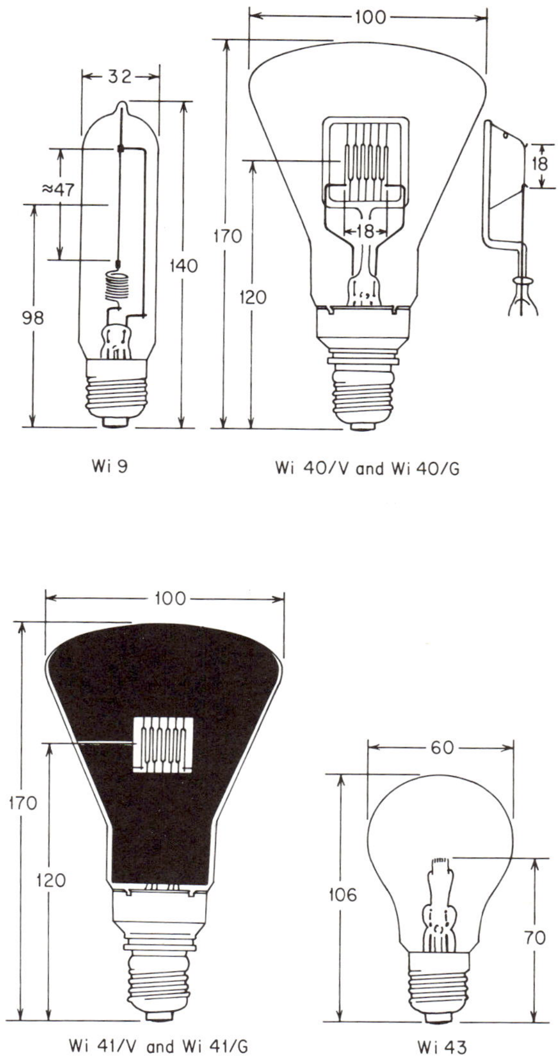

Fig. 12 See legend on page 3-24.

material. The anode forms a crater of decomposing material which provides a center of very high luminosity. Some electrodes are hollowed out and filled with a softer carbon material to help keep the arc fixed in the anode and prevent it from wandering around on the anode surface.

In some cored electrodes the center is filled with whatever material is needed to produce desired spectral characteristics in the arc. In such devices the flame between the electrodes becomes the important center of luminosity, and color temperatures reach values as high as 8000 K (Koller, 1965). An example of this so-called *flaming arc* is shown in Fig. 14*a*. Figure 14*b* and *c* shows the low-intensity dc carbon arc and the high-intensity dc carbon arc with rotating positive electrode.

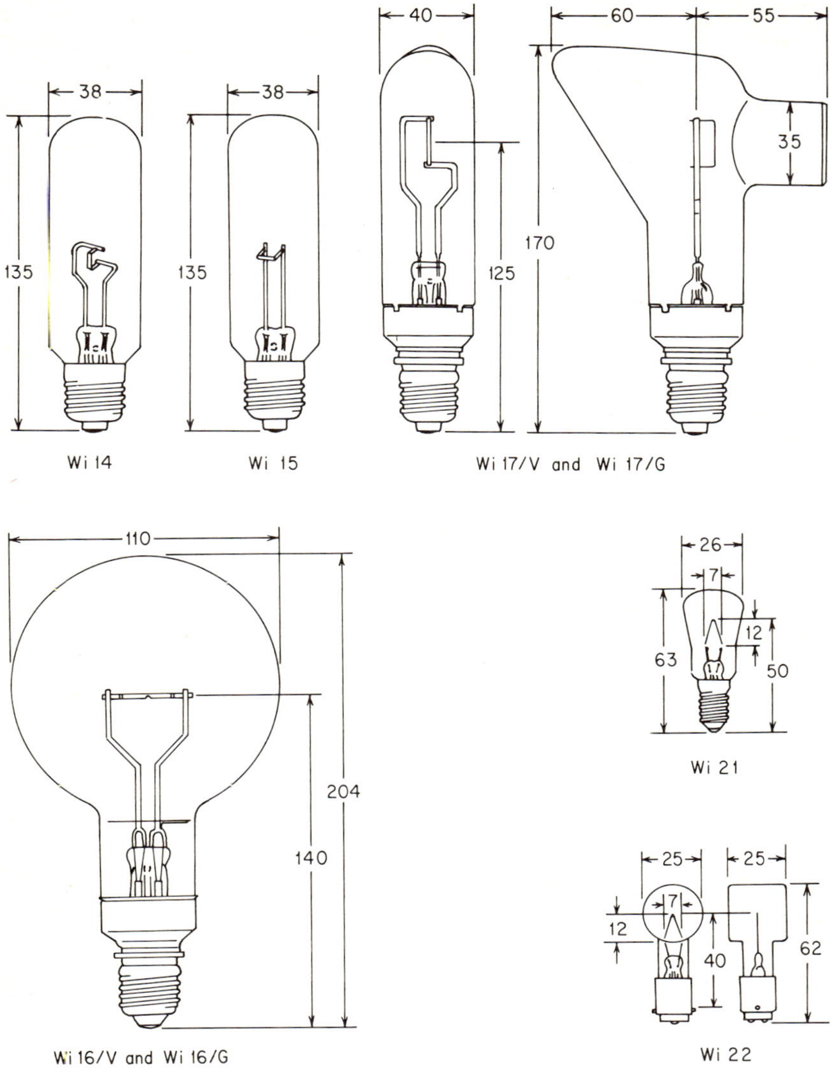

Fig. 12 Osram lamps for scientific purposes; dimensions in millimeters. (*Osram.*)

Tables 5 and 6 give characteristics of flame and dc high-intensity carbon arcs.

A spectrum of low-intensity arc (Fig. 15) shows the similarity between the radiation from it and a 3800-K blackbody except for the band structure at 0.25 and 0.39 μm. In Koller (1965) an assortment of spectra are given for cored carbons containing different materials. Those for a core of soft carbon and for a polymetallic core are shown in Figs. 16 and 17.

Because the radiation emitted from the carbon arc is very intense, it supplants, for many applications, sources which radiate at lower temperatures. Among the disadvantages in using the carbon arc are its inconvenience relative to the use of other sources, e.g., lamps, and its relative instability. However, Null and Lozier

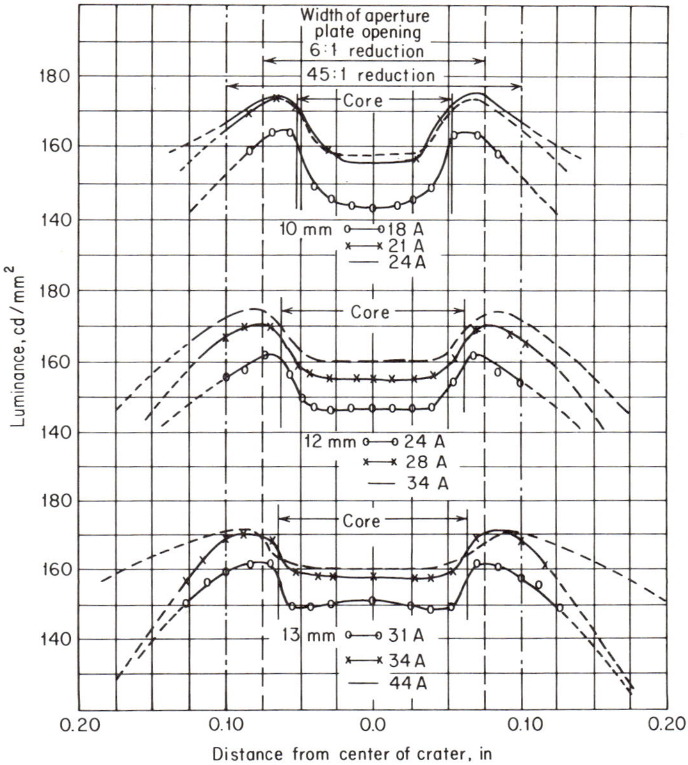

Fig. 13 Variations in brightness across the craters of 10-, 12-, and 13-mm positive carbons of dc plain arcs operated at different currents in the regions of recommended operation. [*Worthing* (1937), *by permission of the National Academy of Sciences.*]

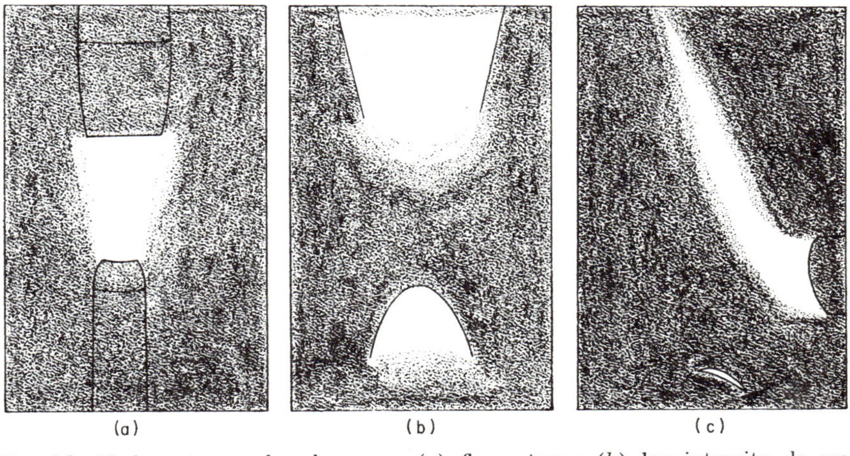

Fig. 14 Various types of carbon arc: (*a*) flame type; (*b*) low-intensity dc arc; (*c*) high-intensity dc arc with rotating positive carbon. (*National Carbon Co.*)

TABLE 5 Flame-Type Carbon Arcs (Kaufman, 1972)

	Application number[a]									
	1	2	3	4	5	6	7[b]	8[c,d]	9[d,e]	10[f]
Type of carbon	"C"	"E"	"Sunshine"	"Sunshine"	"W"	Enclosed arc	Photo[b]	"Sunshine"[f]	Photo[e]	Studio[f]
Flame materials	Polymetallic	Strontium	Rare earth	Rare earth	Polymetallic	None	Rare earth	Rare earth	Rare earth	Rare earth
Burning position[g]	Vertical	Vertical	Vertical	Vertical	Vertical	Vertical	Vertical	Horizontal	Horizontal	Vertical
Upper carbon[d]										
Diameter	22 mm	22 mm	22 mm	22 mm	22 mm	½ in	½ in	6 mm	9 mm	8 mm
Length, in	12	12	12	12	12	3–16	12	6⅝	8	12
Lower carbon[d]										
Diameter	13 mm	13 mm	13 mm	13 mm	13 mm	½ in	½ in	6 mm	9 mm	7 mm
Length	12 in	12 in	12 in	12 in	12 in	3–16 in	12 in	6⅝ in	8 in	9 in
Arc current, A	60	60	60	80	80	16	38	40	95	40
Arc voltage, ac[h]	50	50	50	50	50	138	50	24	30	37 dc[h]
Arc power, kW	3.0	3.0	3.0	4.0	4.0	2.2	1.9	1.0	2.85	1.5
Candlepower[i]	2,100	6,300	9,100	10,000	8,400	1,170	6,700	4,830	14,200	11,000
Lumens	23,000	69,000	100,000	110,000	92,000	13,000	74,000	53,000	156,000	110,000
Lumens per arc watt	7.6	23.0	33.3	27.5	23.0	5.9	39.8	53.0	54.8	73.5
Color temperature, K	[j]	[j]	12,800[j]	24,000[j]	[j]	[j]	7,420[j]	6,590	8,150	4,700
Spectral intensity, μW/cm² 1 m from arc axis:										
Below 270 nm	540	180	102	140	1,020	....	95	11	....	12
270–320 nm	540	150	186	244	1,860	....	76	49	100	48
320–400 nm	1,800	1,200	2,046	2,816	3,120	1,700	684	415	1,590	464
400–450 nm	390	1,100	1,704	2,306	1,480	177	722	405	844	726
450–700 nm	600	4,050	3,210	3,520	2,600	442	2,223	1,602	3,671	3,965
700–1125 nm	1,580	2,480	3,032	3,500	3,220	1,681	1,264	1,368	5,632	2,123
Above 1125 nm	9,480	10,290	9,820	11,420	14,500	6,600	5,189	3,290	8,763	4,593
Total	14,930	19,460	20,100	24,000	27,800	10,600	10,253	7,140	20,600	11,930
Spectral radiation (percent of input power):										
Below 270 nm	1.8	0.6	0.34	0.35	2.55	....	0.5	0.11	....	0.08
270–320 nm	1.8	0.5	0.62	0.61	4.65	....	0.4	0.49	0.35	0.32
320–400 nm	6.0	4.0	6.82	7.04	7.80	7.7	3.6	4.15	5.59	3.09
400–450 nm	1.3	3.7	5.68	5.90	3.70	0.8	3.8	4.05	2.96	4.84
450–700 nm	2.0	13.5	10.70	8.80	6.50	2.0	11.7	16.02	12.86	26.43
700–1125 nm	5.27	8.27	10.10	8.75	8.05	7.6	6.7	13.68	19.75	14.15
Above 1125 nm	31.6	34.3	32.70	28.55	36.25	29.9	27.3	32.90	30.69	30.62
Total	49.77	64.87	67.00	60.00	69.50	48.0	54.0	71.40	72.20	79.53

[a] Typical applications: 1 to 5 and 8, photochemical, therapeutic, accelerated exposure testing, or accelerated plant growth; 6, 7, and 9 blueprinting diazo printing, photocopying, and graphic arts; 10, motion-picture and television studio lighting.
[b] Photographic white-flame carbons.
[c] High intensity copper-coated sunshine carbons.
[d] Both carbons are same in horizontal, coaxial ac arcs.
[e] High-intensity photo 98 carbons.
[f] Motion-picture-studio carbons.
[g] All combinations shown are operated coaxially.
[h] All operated on alternating current except item 10.
[i] Horizontal candlepower, transverse to arc axis.
[j] Deviate enough from blackbody colors to make color temperature of doubtful meaning.

TABLE 6 DC Carbon Arcs (Kaufman, 1972)

	Low intensity	Nonrotating high intensity		Rotating high intensity						
	Application number†									
	1	2	3	4	5	6	7	8	9	10
Type of carbon	Microscope	Projector	Projector	Projector	Projector	Projector	Projector	Searchlight	Studio "yellow flame"	
Positive carbon:										
Diameter, mm	5	7	8	10	11	13.6	13.6	16	16	16
Length, in	8	12–14	12–14	20	20	22	22	22	22	22–30
Negative carbon										
Diameter	6 mm	6 mm	7 mm	$\frac{11}{32}$ in	$\frac{3}{8}$ in	$\frac{1}{2}$ in	$\frac{1}{2}$ in	11 mm	$\frac{17}{32}$ in	$\frac{7}{16}$ in
Length, in	$4\frac{1}{2}$	9	9	9	9	9	9	12	9	12–48
Arc current, A	5	50	70	105	120	160	180	150	225	400
Arc volts, dc	59	40	42	59	57	66	74	78	70	80
Arc power, W	295	2,000	2,940	6,200	6,840	10,600	13,300	11,700	15,800	32,000
Burning rate, in/hr										
Positive carbon	4.5	11.6	13.6	21.5	16.5	17.0	21.5	8.9	20.2	55.0
Negative carbon	2.1	4.3	4.3	2.9	2.4	2.2	2.5	3.9	2.2	3.5
Approximate crater diameter, in	0.12	0.23	0.28	0.36	0.39	0.50	0.50	0.55	0.59	0.59
Maximum luminance of crater, cd/cm²	15,000	55,000	83,000	90,000	85,000	96,000	95,000	65,000	68,000	145,000
Forward crater candlepower	975	10,500	22,000	36,000	44,000	63,000	78,000	68,000	99,000	185,000
Crater lumens‡	3,100	36,800	77,000	126,000	154,000	221,000	273,000	250,000	347,000	660,000
Total lumens§	3,100	55,000	115,000	189,000	231,000	368,000	410,000	374,000	521,000	990,000
Total lumens per arc watt	10.4	29.7	39.1	30.5	33.8	34.7	30.8	32.0	33.0	30.9
Color temperature, K¶	3,600	5,950	5,500–6,500	5,500–6,500	5,500–6,500	5,500–6,500	5,500–6,500	5,400	4,100	5,800–6,100

† Typical applications: 1, microscope illumination and projection; 2 to 7, motion-picture projection; 8, searchlight projection; 9, motion-picture-set lighting and motion-picture and television background projection.
‡ Includes light radiated in forward hemisphere.
§ Includes light from crater and arc flame in forward hemisphere.
¶ Crater radiation only.

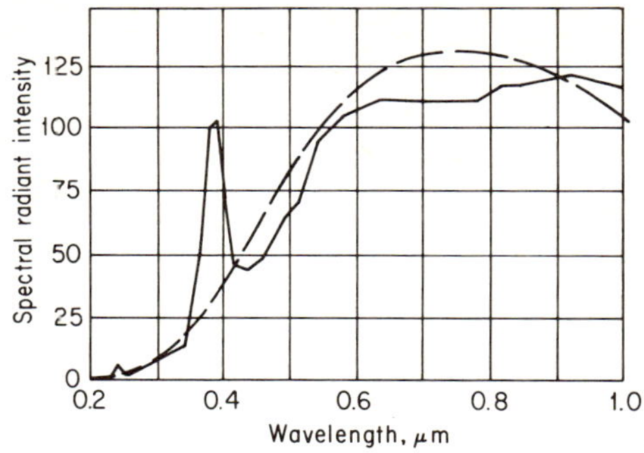

Fig. 15 Spectral distribution of radiant flux from 30-A 55-V dc low-intensity arc with 12-mm positive carbon (*solid line*) and of a 3800-K blackbody radiator (*broken line*). [*From F. E. Carlson and C. N. Clark, Light Sources for Optical Devices, chap. 2 in R. Kingslake (ed.), "Applied Optics and Optical Engineering," vol. I, Academic Press, New York,* 1965; *copyright by Academic Press, Inc.*]

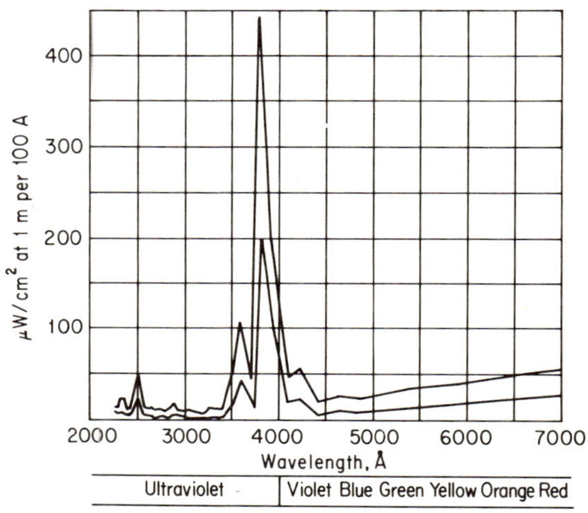

Fig. 16 Spectral energy distribution of carbon arc with core of soft carbon: *upper curve:* 60 A ac, 50 V across the arc; *lower curve:* 30 A ac, 50 V across the arc. (*National Carbon Co.*)

(1962) have studied the properties of the low-intensity carbon arc extensively and have found that under the proper operating conditions the carbon arc can be made quite stable; in fact, they recommended its use as a standard of radiation at high temperatures.

12. Enclosed Arc and Discharge Sources Koller (1965) states that the carbon arc is generally desired if a high intensity is required from a single unit but that it is less efficient than the mercury arc. Other disadvantages mentioned are the short life of the carbon with respect to mercury and combustion products which may be undesirable.

Worthing (1937) describes a number of the older enclosed metallic arc sources, many of which can be built in the laboratory for laboratory use. Today, however, it is rarely necessary to build one's own source unless it is highly specialized.

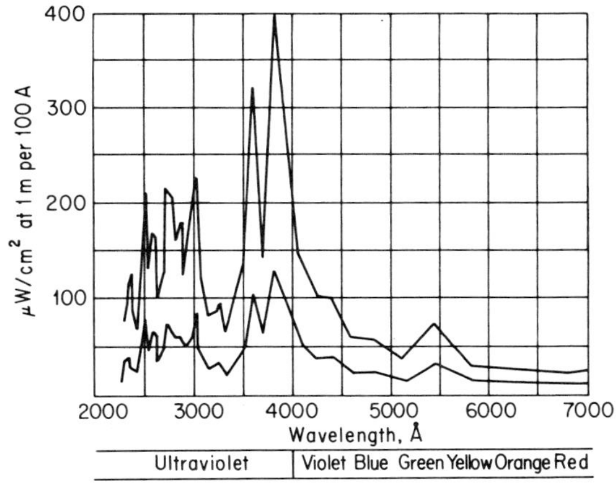

Fig. 17 Spectral energy distribution of carbon arc with polymetallic-cored carbons: *upper curve:* 60 A ac, 50 V across arc; *lower curve:* 30 A ac, 50 V across arc. (*National Carbon Co.*)

High-Pressure Sources

13. Uviarc This is a General Electric trade name for this type of lamp, which is an efficient radiator of ultraviolet radiation. The energy distribution of one type is given in Fig. 18. Since the pressure of this mercury-vapor lamp is intermediate between the usual high- and the low-pressure lamps, little background, or continuum, radiation is present.

In the truly high-pressure lamp, considerable continuum radiation results from greater molecular interaction. Figure 19 shows the dependence on pressure of the amount of continuum in mercury lamps of differing pressure. Bulb shapes and sizes are shown in Fig. 20.

14. Mercury Arcs A widely used type of high-pressure mercury-arc lamp and the components necessary for its successful operation are shown in Fig. 21. The coiled tungsten cathode is coated with a rare-earth material, e.g., thorium. The auxiliary electrode is used to help in starting. A high resistance limits the starting current. Once the arc is started, the operating current is limited by ballast supplied by the high reactance of the power transformer. Spectral data for a clear, 400-W mercury lamp of this type are given in Fig. 22.

15. Multivapor Arcs These lamps contain argon and mercury to provide the starting action, after which three other materials (sodium iodide, thallium iodide, and indium iodide) vaporize and dissociate to yield the bulk of the lamp radiation. The physical appearance is not unlike that of mercury lamps of the same general

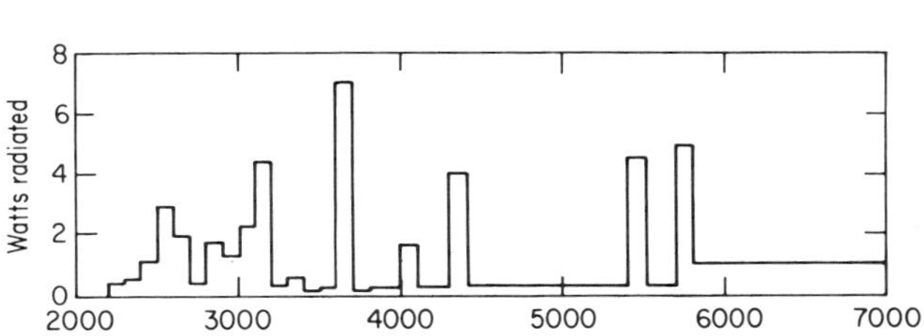

Fig. 18 Intensity distribution of UA-2 intermediate-pressure lamp. (*General Electric Bull.* LD-1.)

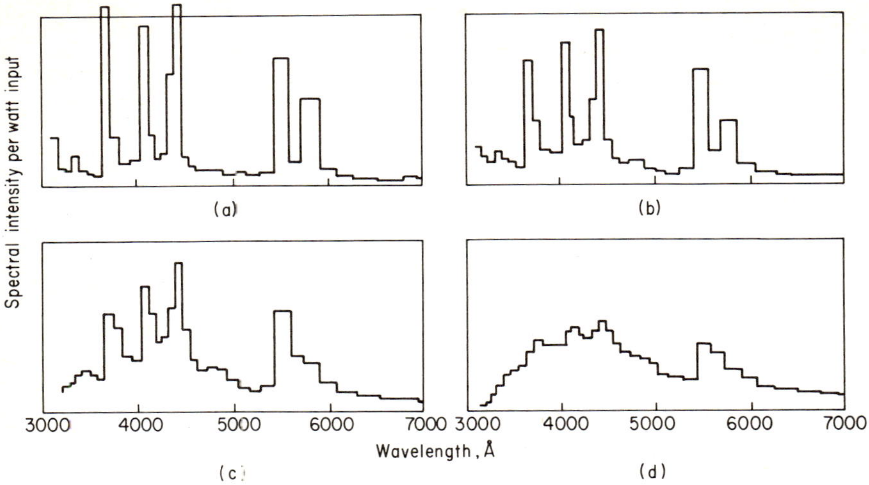

Fig. 19 Emission spectrum of high-pressure mercury-arc lamps showing continuous background: (*a*) 31 atm; (*b*) 75 atm; (*c*) 165 atm; (*d*) 285 atm. [*Noel* (1941).]

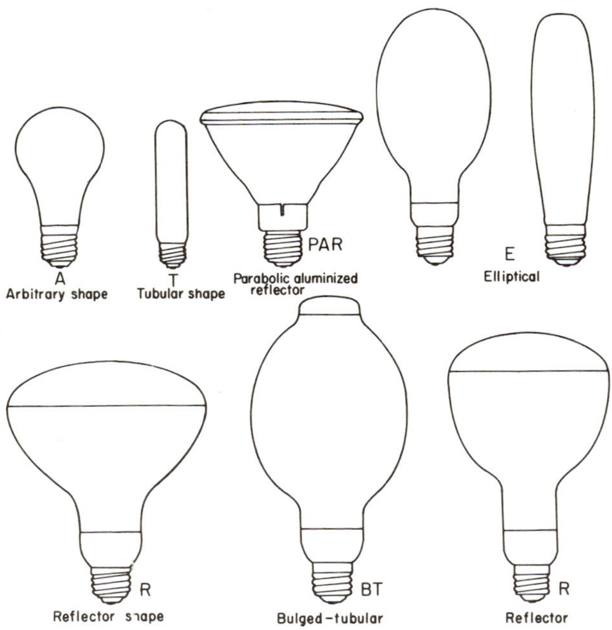

Fig. 20 Bulb shapes and sizes, not to scale. (*General Electric Bull.* TP-109.)

nature. Ballasts, which are needed with multivapor lamps as with mercury vapor, are similar to their counterparts for the mercury lamp. Lamp performance data are given in Table 7. Spectral characteristics of these lamps are given in Fig. 23.

16. Lucalox† Lamps The chief characteristics of this lamp are high-pressure sodium discharge and a high-temperature-withstanding ceramic, Lucalox (translucent

† Trade name of General Electric.

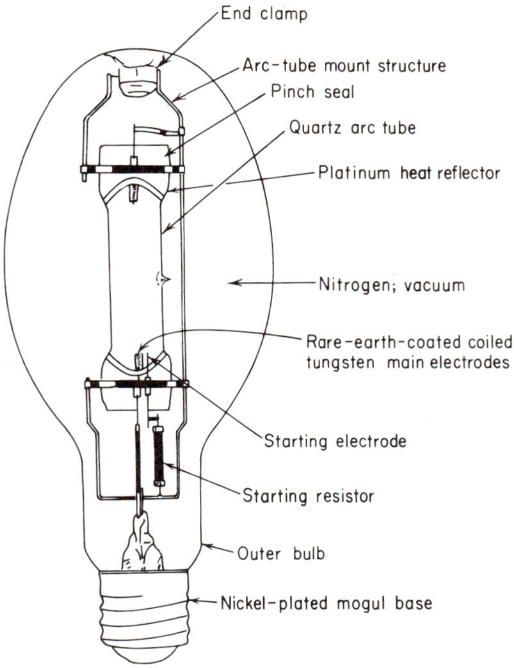

Fig. 21 High-pressure mercury lamp, showing various components. (*General Electric Bull.* TP-109.)

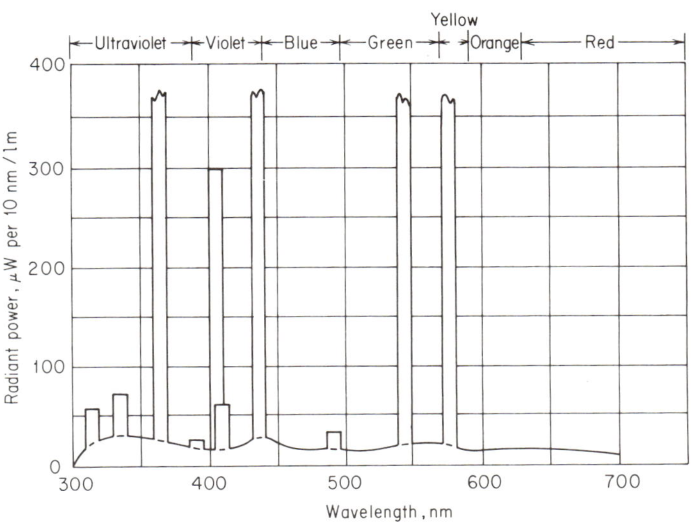

Fig. 22 Spectral energy distribution for clear mercury. (*General Electric Bull.* TP-109.)

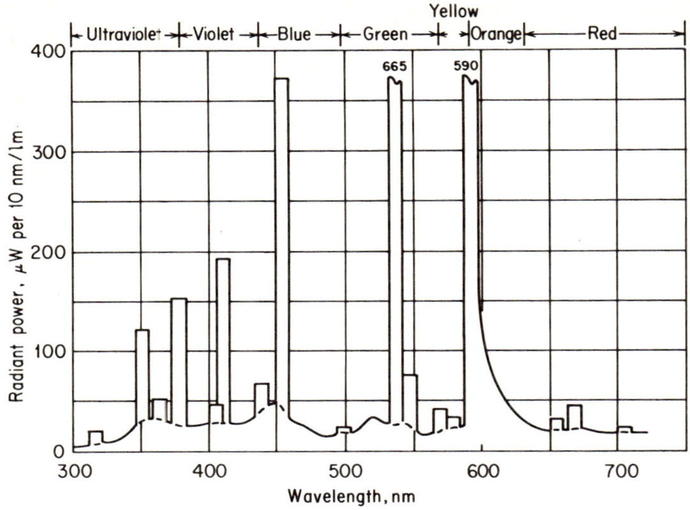

Fig. 23 Spectral energy distribution of multivapor lamp. (*General Electric Bull.* TP-109.)

TABLE 7 Lamp Performance Data†

	GE ordering code‡	
	MV-400/BU MV-400/BD	MV-1000/BU MV-1000/BD
Minimum open-circuit starting voltage with ballast that meets waveform requirements:		
To −20°F	280	375
To +50°F	230	300
Operating voltages	135	265
Finish	Clear	Clear
Bulb	E-37	BT-56
Base	Mogul	Mogul
Rated life, h	7,500	6,000
Approximate initial lumens:		
Vertical	31,500	90,000
Horizontal	27,500	80,000
Lumen maintenance, %:		
At 2,000 h	85	75
At 6,000 h	60	50
Light center length, in	7	$9\frac{3}{8}$
Maximum overall length, in	$11\frac{5}{16}$	$15\frac{1}{16}$

† *General Electric Bull.* TP-109.

‡ Burning position of BU lamps is base up to base 15° below horizontal; of BD lamps it is base down to 15° above horizontal.

aluminum oxide), to yield performance typified in Table 8. The spectral output of the 400-W Lucalox lamp is shown in Fig. 24. Ballasts for this lamp are described in *GE Bull.* TP-109.

17. Capillary Mercury-Arc Lamps The excellent description of these lamps given by Koller (1965) is summarized here. Some sort of cooling is required, as the pressure of the arc is increased, to avoid catastrophic effects to the tube. The A-H6 tube (Fig. 25) is constructed with a quartz bulb wall and a quartz outer jacket,

TABLE 8 Lamp Performance Data†

GE ordering code	Finish	Bulb	Average rated life, h‡	Initial lumens§	Lumen maintenance at 6,000 h, %	Maximum overall length, in	Light center length, in
LU-400/BD or BU	Clear	E-18	6,000	42,000	85	$9\frac{3}{4}$	$5\frac{3}{4}$
LU-275/BD or BU	Clear	E-18	6,000	27,500	80	$9\frac{3}{4}$	$5\frac{3}{4}$

† *General Electric Bull.* TP-109.
‡ At 5 or more burning hours per start.
§ Any burning position.

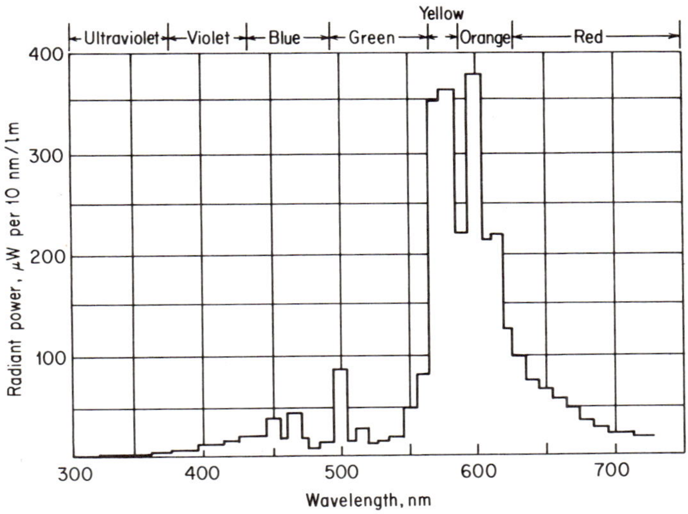

Fig. 24 Spectral output of a 400-W Lucalox lamp. (*General Electric Bull.* TP-109.)

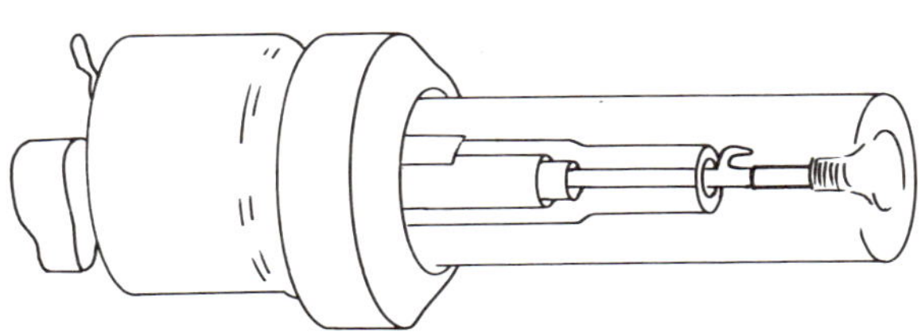

Fig. 25 Water-cooled high-pressure (110 atm) mercury arc showing lamp in water jacket. (*General Electric.*)

to allow 2800-Å radiation to pass, or a Pyrex outer jacket if the ultraviolet is not desired. Pure water is forced through at a rapid rate, while the tube is maintained at a potential of 840 V.

Table 9 shows the radiative characteristics of the tube with quartz or Pyrex jacket. The tube characteristics are similar when the tube is cooled by high-velocity air jets, but the input is slightly reduced and the tube is designated B-H6.

TABLE 9 Radiation from A-H6 Water-Cooled Capillary Lamp†

Lamp watts	1,000 W
Lamp operating volts	840 V
Lamp operating current	1.4 A

Wavelength, Å	Watts radiated: Pyrex water jacket	Quartz water jacket
Below 2800	......	33
2800–3165	4.2	56
3165–3800	53	80
Total ultraviolet below 3800	57.2	169
Total visible 3800–7000	236	244
	$\mu W/cm^2$ at 1 m	
Below 2800	.....	324
2800–3165	49	725
3165–3800	784	1,100

† *General Electric Bull.* GET-1248J.

18. Compact-Source Arcs Carlson and Clark (1965) and Cann (1967) devote considerable space to describing these sources and stress their importance as radiators. We include the salient features here. The original references should be consulted for details. Some of the common characteristics of currently available compact-source arc lamps are (Carlson and Clark, 1965):

A clear quartz bulb of roughly spherical shape with extensions at opposite ends constituting the electrode terminals. In some cases, the quartz bulb is then sealed within a larger glass bulb, which is filled with an inert gas.

A pair of electrodes with relatively close spacing (from less than 1 mm to about 1 cm)—hence the sometimes-used term "short-arc lamps."

A filling of gas or vapor through which the arc discharge takes place.

Extreme electrical loading of the arc gap, which results in very high luminance. It also results in internal pressures of many atmospheres and bulb temperatures as high as 900°C, which necessitate precautions to protect people and equipment in case the lamps should fail violently.

The need for a momentary high-voltage ignition pulse, and a ballast or other auxiliary equipment to limit current during operation.

Clean, attention-free operation for long periods of time.

Table 10 gives characteristics of commonly available compact-source arc lamps. These lamps are designated by the chief radiating gases enclosed as mercury, mercury-xenon, and xenon lamps.

Figure 26 shows the compact-source arc

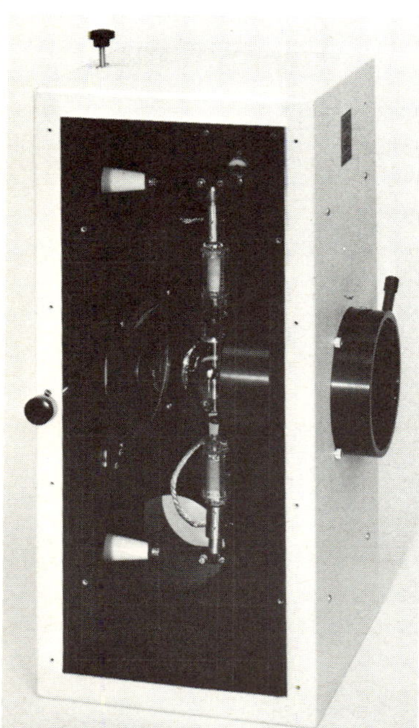

Fig. 26 Lamp housing for lamps up to 1,000 W, showing lamp, rear reflector, beam part, and reflective thermal shielding. (*Oriel Optics Corp.*)

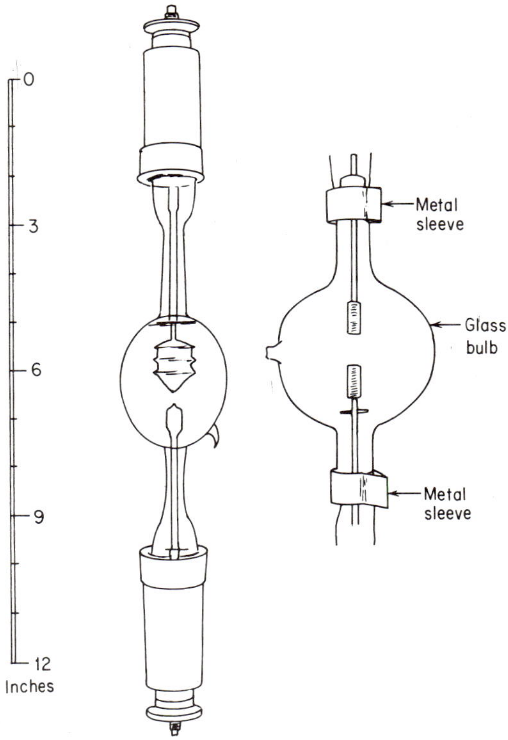

Fig. 27 Construction of different lamps showing differences in relative sizes of electrodes for dc (*left*) and ac (*right*) operation. [*Redrawn after F. E. Carlson and C. N. Clark, Light Sources for Optical Devices, chap. 2 in R. Kingslake (ed.), "Applied Optics and Optical Engineering," vol. I, Academic Press, New York,* 1965; *copyright by Academic Press, Inc.*]

lamp in a housing purchasable from one supplier. The one shown is for lamps up to 1,000 W. Typical of the construction of the lamps themselves are the ones shown in Fig. 27. Since starting may be a problem, some lamps (see Fig. 28) are constructed with a third, starting, electrode, to which a momentary high voltage is applied for starting and especially restarting while hot. The usual ballast is required for compact-source arcs; for stability these arcs, particularly mercury and mercury-xenon, should be operated near rated power on a well-regulated power supply. For more detailed information on stability, see Cann (1967).

The spatial distribution of luminance from these lamps is reported in the literature already cited, and typical contours are shown in Fig. 29. Polar distributions are similar to those shown in Fig. 30.

Spectral distributions are given in Figs. 31 to 33, for a 1,000-W ac mercury lamp, 5-kW dc xenon lamp, and 1,000-W dc mercury-xenon lamp.

Cann (1967) reports some interesting special lamps tested by Jet Propulsion Laboratories for the purpose of obtaining a good spectral match to the solar distribution. The types of lamps tested were Xe, Xe-Zn, Xe-Cd, Hg-Xe-Zn, Hg-Xe-Cd; Kr, Kr-Zn, Kr-Cd, Hg-Kr-Zn, Hg-Kr-Cd; Ar; Ne; and Hg-Xe with variable mercury-vapor pressure. For further details see Carlson and Clark (1965).

Low-Pressure Sources

19. Because a large amount of space is devoted in Koller (1965) to a description of lamps of this type, the data are only summarized here and the reader is referred

TABLE 10 Compact-Source Arc-Lamp Data[a]

Nominal wattage	Current	Approximate arc length,[b] mm	Approximate arc width,[c] mm	Approximate mean luminance,[d] cd/mm²	Approximate peak luminance, cd/mm²	Initial lumen output	Bulb diameter, in	Length overall, in	Ordering abbreviation[e]
Typical Mercury Lamps									
75	ac	2.8	0.7	65	....	2,500	1.2[f]	4.8[f]	(O) HB075W
100	dc	0.3	0.3	1,400	....	2,200	0.5	3.4	(P) PEK109
100	dc	0.25	0.25	1,700	....	2,000	0.4	3.6	(O) HB0100W/2
100	dc	0.4	....	1,000 (CL)	....	1,000	0.5	3.0	(H) 548A-1
100	dc	0.3	0.3	1,400	....	2,200	0.5[g]	3.4	(P) PEK107
200	Both	2.5	1.8	250	....	9,500	0.7[g]	4.1	(P) PEK200
200	dc	2.2	0.6	330	....	9,500	0.7	5.1	(O) HB0200W/2
250	ac	2.5	....	175 (CL)	250	10,000	1.5[f]	6.3[f]	(W) SAH250A
500	Both	4.3	2.5	220	....	24,500	1.3	6.8	(P) PEK500-2
800	ac	8.5	....	125 (CL)	180	40,000	1.6	9.5	(W) SAH800C
1,000	ac	6.5	....	240 (CL)	475	50,000	2.0	9.5	(W) SAH100A (G) H1000A19
2,500	ac	10	....	225 (CL)	325	125,000	2.8	13.0	(W) SAH2500A
Typical Mercury-Xenon Lamps									
225	ac	3.85	....	160 (CL)	....	9,000	0.8	4.0	(H) 510B-1
1,000	ac	6.5	....	240 (CL)	475	50,000	2.0	9.5	(W) SAHX1000A
1,000	dc	6.5	2.5	230	970	52,000	1.8	7.0	(H) 528B
2,500	ac	10	....	225 (CL)	325	125,000	2.8	13.0	(W) SAHX2500A
2,500	dc	4.5	....	....	2050	120,000	2.8	13.0	(W) SAHX2500B
2,500	dc	4.0	2.5	540	2050	120,000	2.6	12.5	(H) 929B
5,000	dc	5.0	3.0	780	2250	230,000	3.4	13.5	(H) 932B
Typical Xenon Lamps									
50	dc	0.16	....	800 (CL)	1800	200	0.5	4.3	(D)
50	dc	2.0	....	70 (CL)	560	500	0.5	4.3	(D)
75	dc	0.4	0.4	1,000	....	1,400	0.5	3.4	(P) X-75
100	dc	0.16	....	1,700 (CL)	4000	750	0.5	4.3	(D)
100	dc	2.0	....	170 (CL)	2000	1,500	0.5	4.3	(D)

150	ac	3.85	....	75 (CL)	....	3,000	0.8	4.0	(H) 510C-1
300	ac	3.85	....	125 (CL)	....	6,000	1.3	5.0	(H) 560C-1
450	dc	2.4	1.2	255	....	16,000	1.2	10.3	(O) XB0450W
800	ac	6.5	....	160 (CL)	....	24,000	1.6	7.0	(H) 507C
900	dc	6.5	....	180 (CL)	....	29,000	1.6	7.0	(H) 538C
900	dc	3.4	1.7	380	....	30,500	1.6	12.6	(O) XB0900W
1,600	dc	4.2	2.1	430	3300	56,000	2.1	14.6	(O) XB01600W
2,200	dc	4.0	2.5	440	3300	75,000	2.3	12.5	(H) 491C
2,500	dc	6.2	3.1	450	....	100,000	2.3	16.9	(O) XB02500W
5,000	dc	8.0	5.0	630	5900	275,000	3.5	19.5	(G) XE5000
5,000	dc	2.0	....	2,100 (CL)	6500	140,000	2.5	28	(D)
10,000	dc	8.3	....	1,450 (CL)	7600	460,000	4.0	28	(D)

a From F. E. Carlson and C. N. Clark, Light Sources for Optical Devices, chap. 2 in R. Kingslake (ed.), "Applied Optics and Optical Engineering," vol. I, Academic Press, New York, 1965; copyright by Academic Press, Inc.

b Values published by the manufacturers: generally, the distance between electrode tips.

c Values published by the manufacturers.

d Values followed by (CL) are averages along the centerline of the arc. Others are averages over rectangular areas defined by the arc width and length (see footnotes *b* and *c*).

e Letters in parentheses identify the following manufacturers, and are not a part of the ordering abbreviation: D, Durotest; G, General Electric; H, Hanovia; O, Osram; P, PEK Laboratories; W, Westinghouse.

f Tubular glass outer jacket, medium prefocus base.

g Three-terminal lamp; starting electrode extends radially outward from arc.

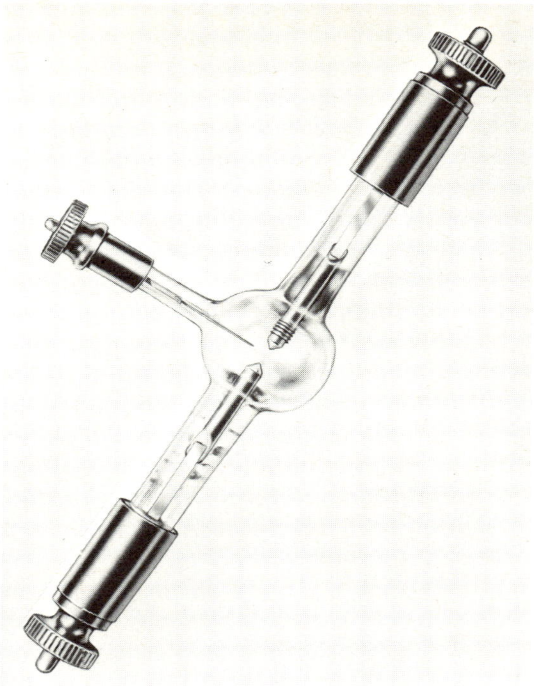

Fig. 28 Construction of a lamp with a third, starting, electrode. (*General Electic.*)

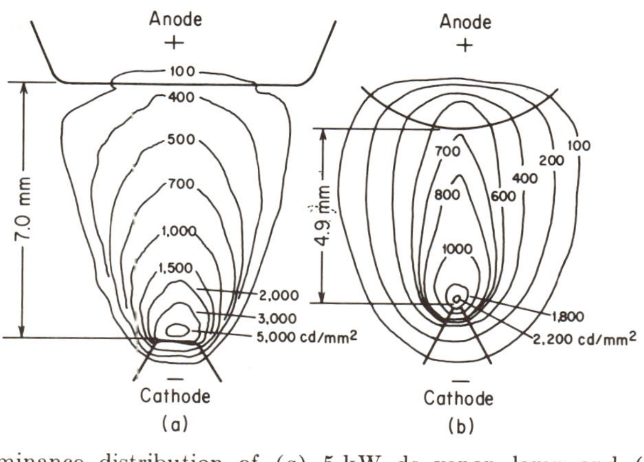

Fig. 29 Luminance distribution of (*a*) 5-kW dc xenon lamp and (*b*) 5-kW dc mercury-xenon lamp. [*From F. E. Carlson and C. N. Clark, Light Sources for Optical Devices, chap.* 2 *in R. Kingslake (ed.), "Applied Optics and Optical Engineering," vol. I, Academic Press, New York,* 1965; *copyright by Academic Press, Inc.*]

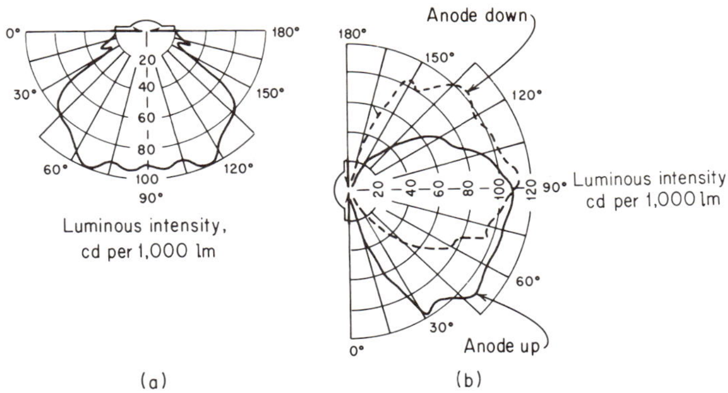

Fig. 30 Polar distribution of radiation in flames that include arc axis for (*a*) 7.5-kW ac mercury lamp and (*b*) 2.5-kW dc mercury-xenon lamp. Asymmetry in (*b*) is due to unequal size of electrodes. [*From F. E. Carlson and C. N. Clark, Light Sources for Optical Devices, chap. 2 in R. Kingslake (ed.), "Applied Optics and Optical Engineering," vol. I, Academic Press, New York,* 1965; *copyright by Academic Press, Inc.*]

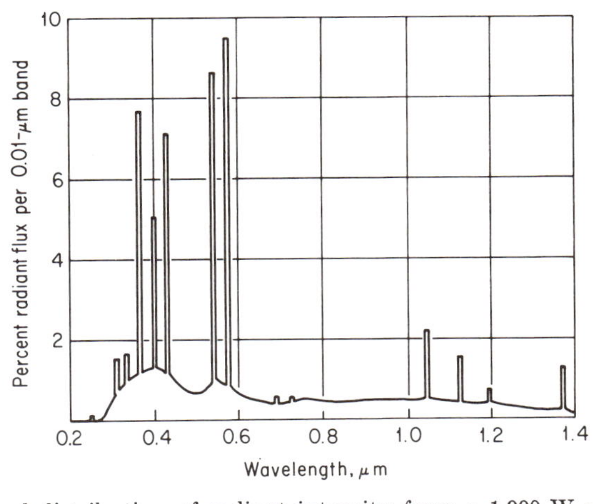

Fig. 31 Spectral distribution of radiant intensity from a 1,000-W ac mercury lamp perpendicular to the lamp axis. [*From F. E. Carlson and C. N. Clark, Light Sources for Optical Devices, chap. 2 in R. Kingslake (ed.), "Applied Optics and Optical Engineering," vol. I, Academic Press, New York,* 1965; *copyright by Academic Press, Inc.*]

to Koller (1965) or the equivalent for details. With the reduction of pressure in a tube filled with mercury vapor, the 2537-Å line becomes predominant, so that low-pressure mercury tubes are usually selected for their ability to emit ultraviolet radiation.

20. Germicidal Lamps These lamps are of the hot-cathode type and operate at relatively low voltages. Tables 11 and 12 give some characteristics of these lamps. The lamps described in this section differ from ordinary fluorescent lamps used in lighting in being designed to transmit ultraviolet, whereas the wall of the fluorescent lamp is coated with a material that absorbs ultraviolet and reemits visible light. The germicidal lamp is constructed of glass, 1 mm thick, which transmits

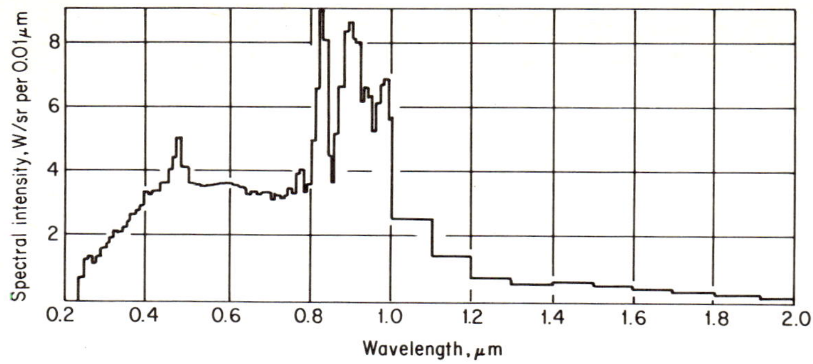

Fig. 32 Spectral distribution of radiant intensity from a 5-kW dc xenon lamp perpendicular to the lamp axis with electrode and bulb radiation excluded. [*From F. E. Carlson and C. N. Clark, Light Sources for Optical Devices, chap. 2 in R. Kingslake (ed.), "Applied Optics and Optical Engineering," vol. I, Academic Press, New York,* 1965; *copyright by Academic Press, Inc.*]

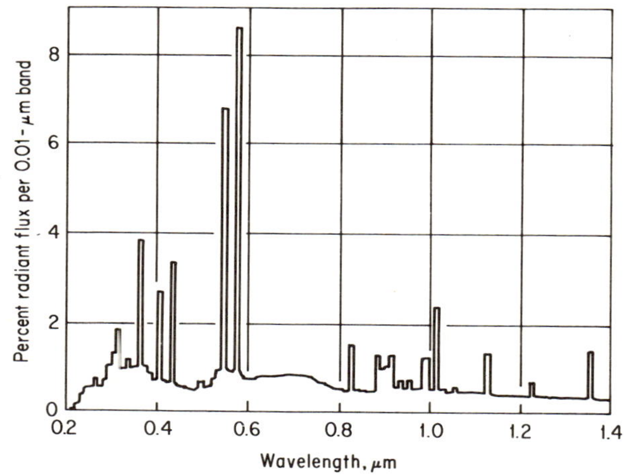

Fig. 33 Spectral distribution of radiant flux from a 1,000-W dc mercury-xenon *lamp.* [*From F. E. Carlson and C. N. Clark, Light Sources for Optical Devices, chap. 2 in R. Kingslake (ed.), "Applied Optics and Optical Engineering," vol. I, Academic Press, New York,* 1965; *copyright by Academic Press, Inc.*]

TABLE 11 Radiation from Low-Pressure Mercury Lamps†

Normal watts	Length, in	Diameter, in	Ultraviolet output at 2537 Å, W	Ultraviolet irradiance, μW/cm² at 1 m
30	36	1	8.3	85
15	18	1	3.6	38
8	12	$\frac{5}{8}$	1.3	14
4	$5\frac{3}{4}$	$\frac{1}{2}$	0.5	7

† *General Electric Lamp Bull.* LD-1.

TABLE 12 T-6 Slimline-Type Germicidal Lamp†

Length 36 in; diameter $\frac{3}{4}$ in

Lamp watts	Lamp operating volts	Lamp operating current	Ultraviolet watts	Ultraviolet irradiance, $\mu W/cm^2$ at 1 m
16	150	120	6	65
23	135	200	8	90
30	115	300	10	110
36	105	420	11	120

† *General Electric Lamp Bull.* LD-1.

about 65 percent of the 2537-Å radiation and virtually cuts off shorter-wavelength ultraviolet radiation.

21. Sterilamp† Types These cold-cathode lamps start and operate at higher voltages than the hot-cathode type and can be obtained in relatively small sizes, as shown in Fig. 34, which illustrates this category under a different name. Operating characteristics of the Sterilamp are given in Table 13.

22. Black-Light Fluorescent Lamps This fluorescent lamp is coated with a phosphor efficient in the absorption of 2537-Å radiation, reemitting ultraviolet radiation in a broad band around 3650 Å. The phosphor is a cerium-activated calcium phosphate, and the glass bulb is impervious to shorter-wavelength ultraviolet radiation. The characteristics of one type are given in Table 14.

23. Hollow-Cathode Lamps A device described early in this century (Paschen, 1916) and used for many years by spectroscopists is the hollow-cathode tube. The one used by Paschen (Worthing, 1937) consisted of a hollow metal cylinder and contained a small quantity of inert gas, yielding an intense cathode glow characteristic of the cathode constituents. Materials that vaporize easily can be incorporated into the tube so that their spectral characteristics predominate (Worthing, 1937).

A number of companies sell hollow-cathode lamps which do not differ significantly from those constructed in early laboratories. The external appearance of these modern tubes shows the marks of mass production and emphasis on convenience. They come with a large number of vaporizable elements, singly or multiply, and with Pyrex or quartz windows. A partial list of the characteristics of the lamps available from two manufacturers is given in Table 15, and their physical appearance is shown in Fig. 35. A schematic of the different elements obtainable in various lamps is shown in Fig. 36. Some sample spectra from those of another manufacturer are shown in Fig. 37 and Table 16.

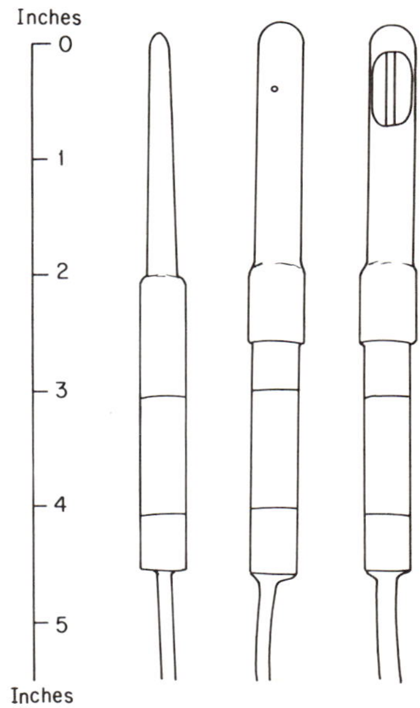

Fig. 34 Pen-Ray low-pressure lamp. (*Ultra-Violet Products.*)

24. Electrodeless Discharge Lamps The electrodeless lamp gained its greatest popularity when Meggers used it in his attempt to produce a highly precise standard of radiation. The simplicity of design makes it easy to construct this type of lamp in the laboratory. Some of the simplest lamps consist of a tube, containing the

† Trade name of Westinghouse Electric Corporation.

TABLE 13 Sterilamp Operating Characteristics†,‡

Lamp designation	782L-30	782-20	782H-10	793
Rated watts	17	14	12	3.5
Overall length, in	$34\frac{3}{4}$	$24\frac{3}{4}$	$14\frac{3}{4}$	$8\frac{1}{8}$
Starting ac voltage	750	575	400	200
Operating ac voltage	410	325	240	90
Operating current, A	0.050	0.005	0.060	0.040
Ultraviolet output (2537 Å), W	5.2	2.0	2.0	0.13
Ultraviolet, μW/cm² at 1 m (2537 Å)	46	20	20	1.3

† Westinghouse Data Sheet ACS-142.

‡ These lamps are made in a number of glasses having the same transmission for 2537 Å but having a transmission for 1849 Å of 10, 1.5, and 0.1 percent, respectively. Thus the quantity of ozone produced varies from a fairly large to a negligible amount.

TABLE 14 Spectral Energy Distribution for Black-Light (360 BL) Lamps†

Watts	Length, in	3200–3800 Å W	3200–3800 Å % W	Total ultraviolet below 3800 Å W	Total ultraviolet below 3800 Å % W	Total visible, W	3800–7600 Å, % W	Erythemal flux
6	9	0.55	9.1	0.56	9.4	0.1	1.7	250
15	18	2.1	14.0	2.2	14.6	0.4	2.7	950
30	36	4.6	15.3	4.7	15.8	0.9	3.0	2100
40	48	6.7	16.8	6.9	17.3	1.5	3.8	3000

† *General Electric Lamp Bull.* LD-1.

Fig. 35 Hollow-cathode spectral tubes. (*Westinghouse Electric Corporation, Electronic Tube Division.*)

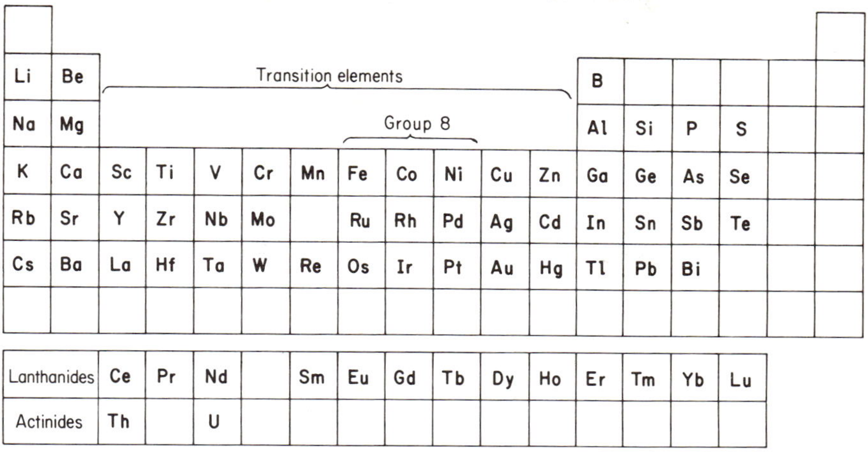

Fig. 36 Periodic table showing the prevalence of elements obtainable in hollow-cathode tubes.

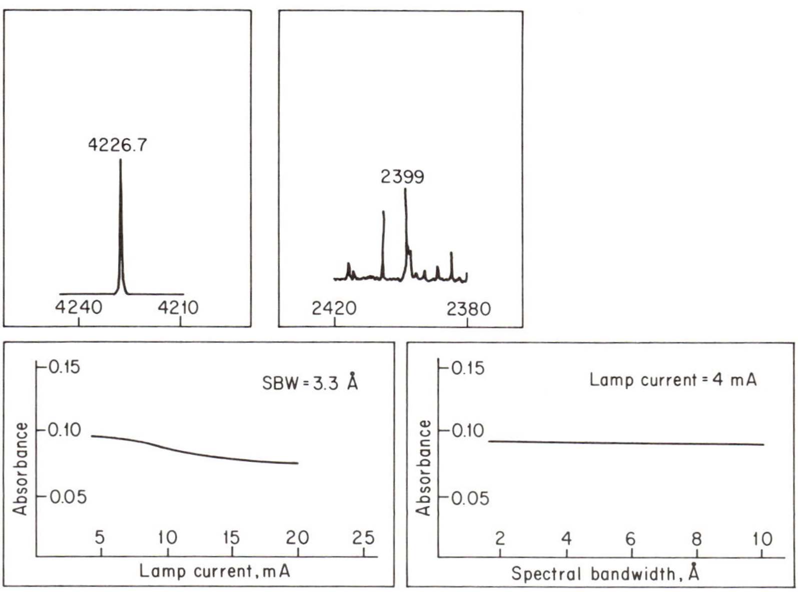

(a)

Fig. 37a See legend on page 3-47.

radiation-producing element, and a microwave generator for producing the electric field within the tube which excites the elemental spectra.

Lamps of this type can be purchased with specially designed microwave cavities for greater efficiency in coupling. Those made of fused quartz can transmit from ultraviolet to near infrared. The electrodeless lamp is better able than the arc to produce stable radiation of sharp spectral lines; this makes it useful in spectros-

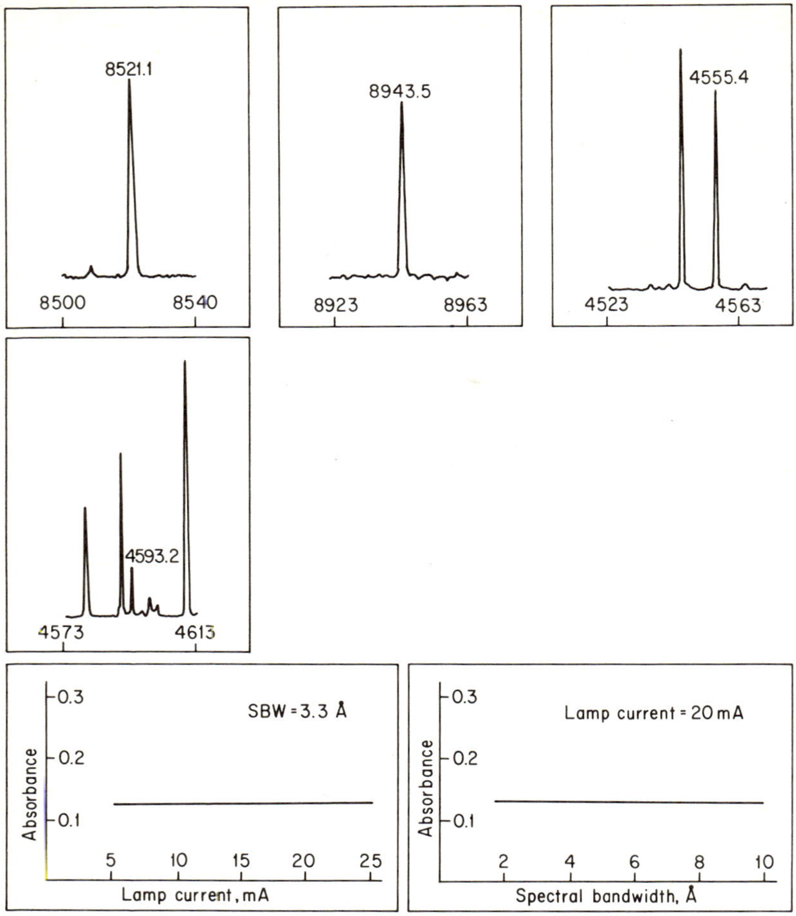

(b)

Fig. 37b

copy and interferometry. The Hg 198 lamp makes a suitable secondary standard of radiation particularly at the following selected (vacuum) wavelengths in angstroms:

2537.2687	3022.3796	4047.7145
2652.8324	3126.5761	4359.5622
2753.5968	3342.4422	5462.2704
2894.4464	3651.1966	5771.1982
2968.1498		

For further information the reader is referred to the literature, e.g., Worden (1963), Meggers and Westfall (1950), and Meggers (1960).

25. Spectral Lamps Some manufacturers produce groups of arc sources, similar in construction and filled with different elements and rare gases, to yield discontinuous or monochromatic radiation throughout most of the ultraviolet and visible

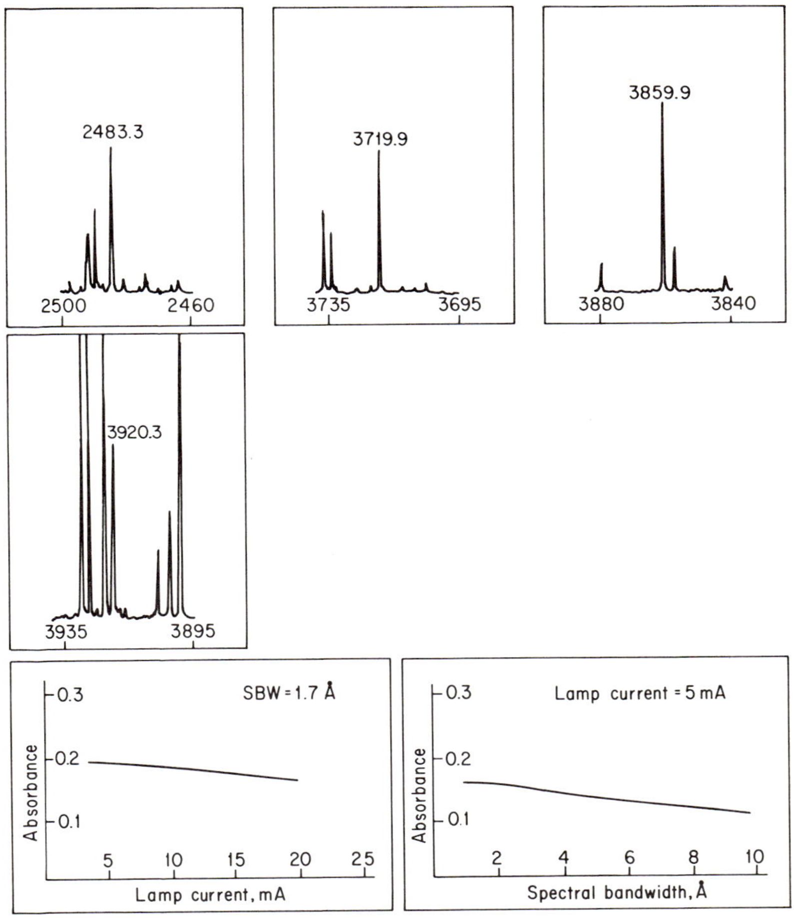

(c)

Fig. 37c

spectrum. They are called spectral lamps. The envelopes of these lamps are constructed of glass or quartz, depending on the part of the spectrum desired. Thus, discrete radiation can be obtained from around 2300 Å to the fringe of the visible region and, in fact, into the near infrared.

Figure 38 represents the various atomic lines observable from Osram spectral lamps. Figure 39 gives a physical description of various spectral lamps obtainable from Philips. Tables 17 and 18 list the characteristics of the various types of lamps obtainable respectively from Osram and Philips.

6. Pleucker Spectrum Tubes These are inexpensive tubes made of glass (see Fig. 40) with an overall length of 25 cm and capillary portion 8.5 to 10 cm long. They operate from an ordinary supply with a special transformer which supports the tubes in a vertical position and maintains the voltage and current values adequate to operate the discharge and regulate the spectral intensity. Table 19 lists the various gases in available tubes.

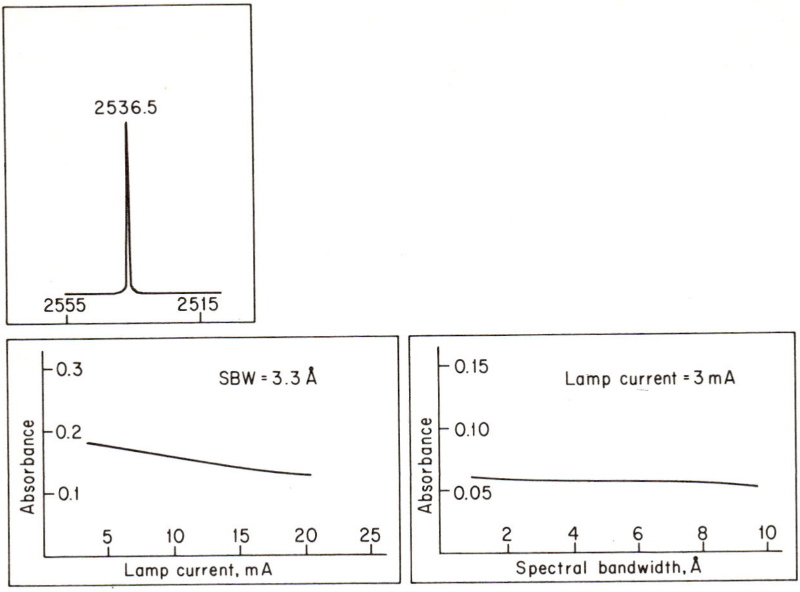

(d)

Fig. 37d

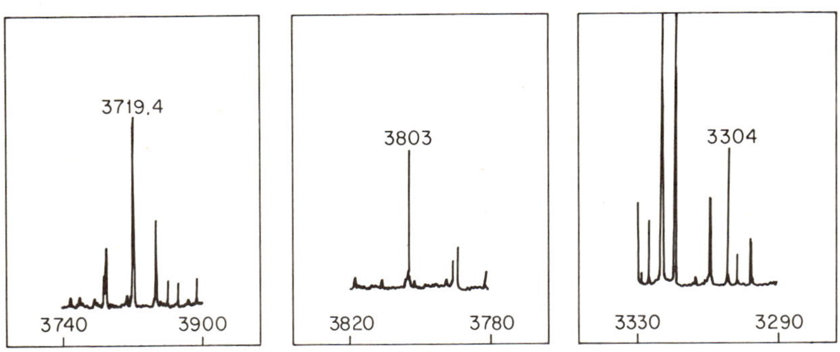

(e)

Fig. 37e

Concentrated-Arc Lamps

27. Zirconium Arc The cathodes of these lamps are made of a hollow refractory metal containing zirconium oxide. The anode, a disk of metal with an aperture, resides directly above the cathode with the normal to the aperture coincident with the longitudinal axis of the cathode. Argon gas fills the tube. The arc discharge causes the zirconium to heat (to about 3000 K) and produce an intense, very small source of light.

These lamps can be obtained from the Cenco Company in a number of wattages from 2 to 300 W. The end of the bulb through which the radiation passes comes with ordinary curvature or (for a slight increase in price) flat. Examples of a few types are shown in Fig. 41.

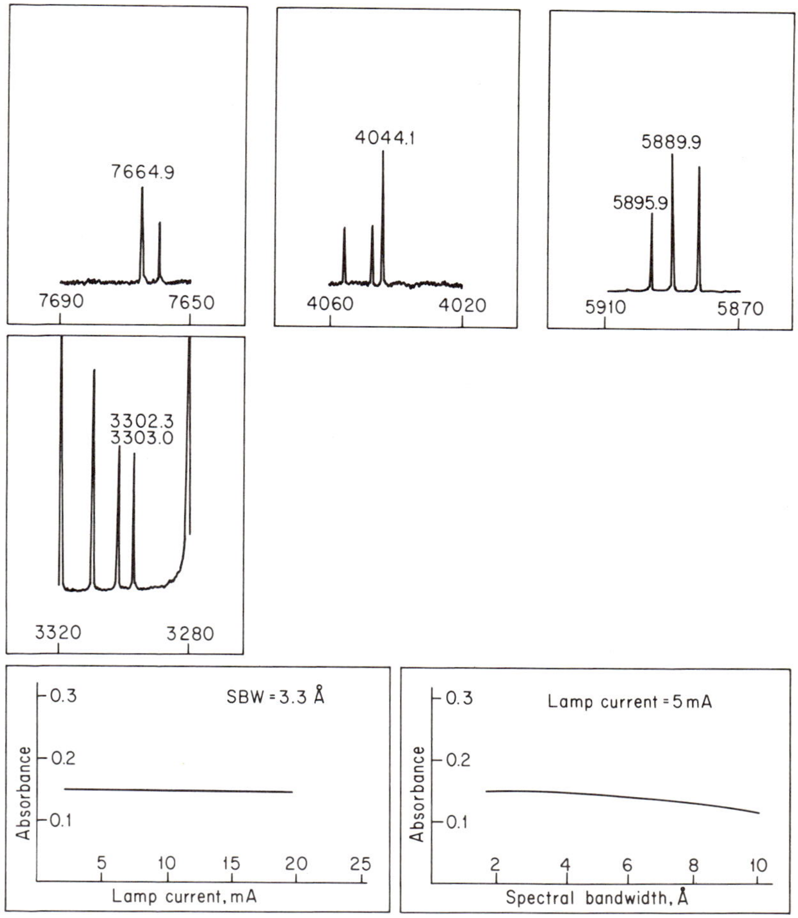

(f)

Fig. 37a–f Spectral characteristics of some hollow-cathode lamps: (*a*) calcium; (*b*) cesium; (*c*) iron; (*d*) mercury; (*e*) thorium; (*f*) sodium or potassium. (*Varian Techtron.*)

28. Tungsten-Arc (Photomicrographic) Lamp The essential elements of this discharge-type lamp (Fig. 42) are a ring electrode and a pellet electrode, both made of tungsten. The arc forms between these electrodes, causing the pellet to heat to incandescence. The ring also incandesces but to a lesser extent. Thus the hot pellet (approximately 3100 K) provides an intense source of small-area radiation. A plot of the spectral variation of this radiation is given in Fig. 43.

As with all tungsten sources, evaporation causes a steady erosion of the pellet surface with the introduction of gradients, which is not serious if the pellet is used as a point source.

General Electric, manufacturer of the 30A/PS22 photomicrographic lamp, which uses a 30-A operating current, states that it requires a special heavy-duty socket obtainable through certain manufacturers suggested in their brochure.†

† The General Electric brochure on the 30A/PS22 photomicrographic lamp is obtainable from General Electric, Large Lamp Division.

TABLE 15 Single-Element and Multiple-Element Hollow-Cathode Lamps

Element	Window†	Gas fill‡	Size §	Analytical line, Å	Catalog number¶
			Single-Element 22000 Series		
Aluminum	Q	A	B	3092	WL22804
	P	N	A	3092	JA-45-452
	Q	A	A	3092	WL22870
	Q	N	B	3092	WL22929
	Q	N	A	3092	WL22954
Antimony	Q	A	B	2311	WL22840
	Q	A	A	2311	WL22872
	Q	N	B	2311	JA-45-461
	Q	N	A	2311	WL22956
Arsenic	Q	N	B	1937	JA-45-317
	Q	N	A	1937	WL22873
	Q	A	B	1937	JA-45-315
Barium	P	N	A	5536	JA-45-480
Beryllium	Q	N	B	2349	WL23407
Bismuth	Q	A	B	3068	WL22841
	Q	A	A	3068	WL22874
	Q	N	B	3068	JA-45-469
	Q	N	A	3068	WL22957
Boron	Q	A	B	2497	JA-45-568
	Q	A	A	2497	WL22917
Cadmium	Q	A	B	3261	WL22816
	Q	A	A	3261	WL22875
	Q	N	B	3261	JA-45-462
	Q	N	A	3261	WL22958
Calcium	P	N	A	4227	JA-45-440
Cerium	Q	N	B	....	JA-45-569
	Q	N	A	....	WL22978
Cesium	P	A	A	4556	WL22817
	P	N	A	4556	JA-45-441
Chromium	P	A	A	3579	WL22812
	Q	A	B	3579	WL22821
	Q	A	A	3579	WL22877
	Q	N	B	3579	JA-45-454
	Q	N	A	3579	WL22959
Cobalt	P	A	A	3454	WL22813
	Q	A	B	3454	WL22814
	Q	A	A	3454	WL22878
	Q	N	B	3454	JA-45-456
	Q	N	A	3454	WL22953
Copper	P	N	A	3247	JA-45-458
	Q	A	B	3247	WL22606
	Q	A	A	3247	WL22879
	Q	N	B	3247	JA-45-490
	Q	N	A	3247	WL23042
Dysprosium	Q	N	B	4212	JA-45-595
	Q	N	A	4212	WL22880
Erbium	Q	N	B	4008	JA-45-571
	Q	N	A	4008	WL22881
Europium	Q	N	B	4594	JA-45-572
	Q	N	A	4594	WL22882
Gadolinium	P	N	A	4079	WL-22975
	Q	N	B	4079	JA-45-573
	Q	N	A	4079	WL22986
Gallium	Q	N	B	4172	JA-45-470
	Q	N	A	4172	WL22884
Germanium	Q	A	B	2651	JA-45-575
	Q	N	B	2651	JA-45-313
Gold	Q	A	B	2676	WL22839
	Q	A	A	2676	WL22883
	Q	N	B	2676	JA-45-467
	Q	N	A	2676	WL22960

TABLE 15 Single-Element and Multiple-Element Hollow-Cathode Lamps (Continued)

Element	Window†	Gas fill‡	Size §	Analytical line, Å	Catalog number¶
Hafnium	Q	N	B	3072	JA-45-303
Holmium	Q	N	A	4104	WL22885
	Q	N	B	4104	JA-45-576
Indium	Q	A	B	3040	WL22867
	Q	A	A	3040	WL22915
	Q	N	B	3040	JA-45-471
Iridium	Q	N	B	2850	JA-45-577
Iron	P	A	A	3720	WL22602
	Q	A	B	3720	WL22611
	Q	N	B	3720	JA-45-455
	P	N	A	3720	WL22820
	Q	A	A	3720	WL22886
	Q	N	A	3720	WL22887
Iron, high-purity	Q	N	B	3720	WL22837
	Q	N	A	3720	WL22888
Lanthanum	Q	A	B	5501	WL22846
	Q	A	A	5501	WL22889
	Q	N	B	5501	JA-45-495
Lead	Q	A	B	2833	WL22838
	Q	A	A	2833	WL22890
	Q	N	B	2833	JA-45-468
	Q	N	A	2833	WL22952
Lithium 6	P	N	A	6708	JA-45-579
	P	A	A	6708	WL22925
Lithium 7	P	N	A	6708	JA-45-580
	P	A	A	6708	WL22926
Lithium, natural	P	A	A	6708	WL22825
	P	N	A	6708	JA-45-444
	Q	A	B	6708	WL23115
Lutetium	Q	N	B	3282	JA-45-581
	Q	N	A	3282	WL23010
Magnesium	Q	A	B	2852	WL22609
	Q	A	A	2852	WL22891
	Q	N	A	2852	WL22951
	Q	N	B	2852	JA-45-451
Manganese	Q	A	B	2795	WL22608
	P	A	A	2795	WL22815
	Q	N	B	2795	JA-45-472
	Q	N	A	2795	WL22961
	Q	A	A	2795	WL22876
Mercury	Q	A	B	2537	JA-45-493
	Q	A	A	2537	WL22892
Molybdenum	Q	A	B	3133	WL22805
	Q	A	A	3133	WL22893
	Q	N	B	3133	JA-45-460
	Q	N	A	3133	WL22962
Neodymium	Q	N	B	4925	JA-45-582
	Q	N	A	4925	WL22980
Nickel	P	A	A	3415	WL22605
	Q	A	B	3415	WL22663
	Q	N	B	3415	JA-45-457
	Q	A	A	3415	WL22894
	Q	N	A	3415	WL22895
Niobium	Q	N	B	4059	JA-45-486
	Q	N	A	4059	WL22912
Osmium	Q	A	B	2909	JA-45-584
Palladium	Q	A	B	3404	WL22857
	Q	A	A	3404	WL22911
	Q	N	B	3404	JA-45-475
	Q	N	A	3404	WL22970
Phosphorus	Q	N	B	2136	JA-45-449
	Q	N	A	2136	WL22990

TABLE 15 Single-Element and Multiple-Element Hollow-Cathode Lamps (Continued)

Element	Window†	Gas fill‡	Size §	Analytical line, Å	Catalog number¶
Platinum	Q	A	B	2659	WL22851
	Q	A	A	2659	WL22896
	Q	N	B	2659	JA-45-466
Potassium	P	N	A	4044	JA-45-484
Praseodymium	Q	N	B	4951	JA-45-585
	Q	N	A	4951	WL22982
Rhenium	Q	N	B	3460	JA-45-489
	Q	N	A	3460	WL22967
Rhodium	Q	A	B	3435	WL22850
	Q	A	A	3435	WL22897
	Q	N	B	3435	JA-45-476
Rubidium	P	N	A	7800	JA-45-443
	Q	N	B	7800	WL23046
Ruthenium	Q	A	B	3499	JA-45-586
Samarium	Q	N	B	4760	JA-45-587
	Q	N	A	4760	WL22899
Scandium	Q	N	B	3912	JA-45-309
Selenium	Q	A	B	1960	WL22843
	Q	A	A	1960	WL22898
	Q	N	B	1960	JA-45-477
	Q	N	A	1960	WL22963
Silicon	Q	A	B	2516	WL22832
	Q	A	A	2516	WL22900
	Q	N	B	2516	JA-45-479
	Q	N	A	2516	WL22964
Silver	Q	A	B	3281	JA-45-483
	Q	A	A	3281	WL22901
Sodium	P	A	A	5890	WL22864
	P	N	A	5890	JA-45-485
Strontium	P	N	A	4607	JA-45-481
Sulphur	Q	N	B	....	JA-45-588
Tantalum	Q	A	B	2714	JA-45-488
	Q	A	A	2714	WL22913
	Q	N	B	2714	WL22971
	Q	N	A	2714	WL22972
Tellurium	Q	A	B	2143	WL22842
	Q	A	A	2143	WL22902
	Q	N	B	2143	JA-45-473
	Q	N	A	2143	WL22965
Terbium	Q	N	B	4326	JA-45-589
	Q	N	A	4326	WL22903
Thallium	Q	N	B	3776	WL23408
Thorium	Q	N	A	3245	WL23028
	Q	N	B	3245	JA-45-590
Thulium	Q	N	B	4105	JA-45-591
	Q	N	A	4105	WL23008
Tin	Q	A	B	2863	WL22822
	Q	A	A	2863	WL22904
	Q	N	B	2863	JA-45-463
	Q	N	A	2863	WL22966
Titanium	Q	N	B	3643	JA-45-592
	Q	N	A	3643	WL22992
Tungsten	Q	N	B	4009	JA-45-465
	Q	A	B	4009	WL22849
	Q	N	A	4009	WL22905
	Q	A	A	4009	WL22906
Uranium	Q	N	B	5027	JA-45-447
	Q	N	A	5027	WL22907
Vanadium	Q	A	B	3184	WL22856
	Q	A	A	3184	WL22910
	Q	N	B	3184	JA-45-453
	Q	N	A	3184	WL22974

TABLE 15 Single-Element and Multiple-Element Hollow-Cathode Lamps (Continued)

Element	Window†	Gas fill‡	Size §	Analytical line, Å	Catalog number¶
Ytterbium	Q	A	B	3988	JA-45-593
	Q	A	A	3988	WL22984
Yttrium	P	N	A	4102	WL22976
	Q	N	B	4102	JA-45-594
	Q	N	A	4102	WL22988
Zinc	Q	A	B	2139	WL22607
	Q	N	B	2139	JA-45-459
	Q	A	A	2139	WL22908
	Q	N	A	2139	WL22909
Zirconium	Q	A	B	3601	JA-45-482
	Q	A	A	3601	WL22914
	Q	N	B	3601	WL22998
Single-Element 36000 Series					
Aluminum	P	N	C	3092	JA-45-36009
Antimony	Q	N	C	2311	JA-45-36010
Arsenic	Q	A	C	1937	JA-45-36011
Barium	P	N	C	5536	JA-45-36012
Beryllium	Q	N	C	2349	JA-45-36013
Bismuth	Q	N	C	3068	JA-45-36014
Boron	Q	A	C	2497	JA-45-36015
Cadmium	Q	N	C	3261	JA-45-36016
Calcium	P	N	C	4227	JA-45-36017
Cerium	Q	N	C	....	JA-45-36019
Cesium	P	N	C	4556	JA-45-36020
Chromium	P	N	C	3579	JA-45-36021
Cobalt	Q	N	C	3454	JA-45-36022
Copper	P	N	C	3247	JA-45-36024
Dysprosium	P	N	C	4212	JA-45-36025
Erbium	P	N	C	4008	JA-45-36026
Europium	P	N	C	4594	JA-45-36027
Gadolinium	P	N	C	4079	JA-45-36028
Gallium	Q	N	C	4172	JA-45-36029
Germanium	Q	N	C	2651	JA-45-36030
Gold	Q	N	C	2676	JA-45-36031
Hafnium	Q	N	C	3072	JA-45-36032
Holmium	P	N	C	4104	JA-45-36033
Indium	Q	N	C	3040	JA-45-36034
Iridium	Q	N	C	2850	JA-45-36036
Iron	Q	N	C	3720	JA-45-36037
Lanthanum	P	N	C	5501	JA-45-36038
Lead	Q	N	C	2833	JA-45-36039
Lithium 6	P	N	C	6708	JA-45-36090
Lithium 7	P	N	C	6708	JA-45-36091
Lithium, natural	P	N	C	6708	JA-45-36040
Lutetium	P	N	C	3282	JA-45-36041
Magnesium	Q	N	C	2852	JA-45-36042
Manganese	Q	N	C	2795	JA-45-36043
Mercury	Q	A	C	2537	JA-45-36044
Molybdenum	Q	N	C	3133	JA-45-36045
Neodymium	P	N	C	4925	JA-45-36046
Nickel	Q	N	C	3415	JA-45-36047
Niobium	P	N	C	4059	JA-45-36023
Osmium	Q	A	C	2909	JA-45-36048
Palladium	Q	N	C	3404	JA-45-36049
Phosphorus	Q	N	C	2136	JA-45-36050
Platinum	Q	N	C	2659	JA-45-36051
Potassium	P	N	C	4044	JA-45-36052
Praseodymium	P	N	C	4951	JA-45-36053
Rhenium	P	N	C	3460	JA-45-36056
Rhodium	P	N	C	3435	JA-45-36057

TABLE 15 Single-Element and Multiple-Element Hollow-Cathode Lamps (Continued)

Element	Window†	Gas fill‡	Size §	Analytical line, Å	Catalog number¶
Rubidium	P	N	C	7800	JA-45-36058
Ruthenium	P	A	C	3499	JA-45-36059
Samarium	P	N	C	4760	JA-45-36060
Scandium	P	N	C	3912	JA-45-36061
Selenium	Q	N	C	1960	JA-45-36062
Silicon	Q	N	C	2516	JA-45-36063
Silver	P	A	C	3281	JA-45-36064
Sodium	P	N	C	5890	JA-45-36065
Strontium	P	N	C	4607	JA-45-36066
Sulphur	Q	N	C	....	JA-45-36067
Tantalum	Q	A	C	2714	JA-45-36068
Tellurium	Q	N	C	2143	JA-45-36069
Terbium	P	N	C	4326	JA-45-36070
Thallium	Q	N	C	3776	JA-45-36071
Thorium	Q	N	C	3245	JA-45-36072
Thulium	P	N	C	4105	JA-45-36073
Tin	Q	N	C	2863	JA-45-36074
Titanium	P	N	C	3643	JA-45-36075
Tungsten	Q	N	C	4009	JA-45-36076
Uranium	P	N	C	5027	JA-45-36077
Vanadium	Q	N	C	3184	JA-45-36078
Ytterbium	P	A	C	3988	JA-45-36079
Yttrium	P	N	C	4102	JA-45-36080
Zinc	Q	N	C	2139	JA-45-36081
Zirconium	P	A	C	3601	JA-45-36082
Multiple-Element 22000 Series					
Aluminum, calcium	Q	N	B	....	WL23246
Aluminum, calcium, magnesium	Q	A	B	....	WL22604
Aluminum, calcium, magnesium	Q	A	A	....	WL22871
Aluminum, calcium, magnesium	Q	N	B	....	JA-45-450
Aluminum, calcium, magnesium	Q	N	A	....	WL22955
Aluminum, calcium, magnesium, iron	Q	N	B	....	JA-45-310
Aluminum, calcium, magnesium, lithium	Q	N	B	....	JA-45-436
Aluminum, calcium, magnesium, lithium	Q	A	A	....	WL23036
Aluminum, calcium, strontium	P	N	A	....	WL23403
Antimony, arsenic, bismuth	Q	N	B	....	WL23147
Arsenic, nickel	Q	N	B	....	JA-45-434
Arsenic, selenium, tellurium	Q	N	B	....	JA-45-598
Barium, calcium, strontium	P	N	A	....	JA-45-437
Barium, calcium, silicon, magnesium	Q	N	B	....	JA-45-478
Cadmium, copper, zinc, lead	Q	N	B	....	JA-45-597
Cadmium, silver, zinc, lead	Q	N	B	....	JA-45-308
Calcium, magnesium, strontium	Q	N	B	....	WL23605
Calcium, magnesium, zinc	Q	N	B	....	JA-45-311
Calcium, magnesium, aluminum, lithium	Q	A	B	....	WL23158
Calcium, zinc	Q	N	B	....	JA-45-304
Chromium, iron, manganese, nickel	Q	N	B	....	JA-45-442
Chromium, cobalt, nickel	Q	N	B	....	WL23174
Chromium, copper	Q	N	B	....	JA-45-306
Chromium, manganese	Q	N	B	....	WL23499
Chromium, cobalt, copper, manganese, nickel	Q	N	B	....	WL23601
Chromium, cobalt, copper, iron, manganese, nickel	Q	N	B	....	JA-45-599

TABLE 15 Single-Element and Multiple-Element Hollow-Cathode Lamps (Continued)

Element	Window†	Gas fill‡	Size §	Analytical line, Å	Catalog number¶
Cobalt, copper	Q	N	B	....	JA-45-305
Cobalt, copper, gold, nickel	Q	N	B	....	WL23295
Cobalt, copper, zinc, molybdenum	Q	N	B	....	JA-45-596
Cobalt, iron	Q	N	B	....	WL23291
Cobalt, nickel	Q	N	B	....	WL23426
Copper, gallium	Q	N	B	....	JA-45-431
Copper, iron	Q	N	B	....	JA-45-312
Copper, iron, manganese	Q	N	B	....	JA-45-435
Copper, iron, molybdenum	Q	N	B	....	JA-45-301
Copper, iron, gold, nickel	Q	N	B	....	JA-45-307
Copper, iron, manganese, zinc	Q	N	B	....	JA-45-492
Copper, manganese	Q	N	B	....	JA-45-491
Copper, nickel	Q	N	B	....	WL23441A
Copper, nickel, zinc	Q	N	B	....	WL23405
Copper, zinc, molybdenum	Q	N	B	....	JA-45-496
Copper, zinc, lead, silver	Q	N	B	....	JA-45-448
Copper, zinc, lead, tin	Q	N	B	....	JA-45-438
Gold, nickel	Q	N	B	....	JA-45-433
Gold, silver	Q	N	B	....	WL23269
Indium, silver	Q	N	B	....	WL23294
Lead, silver, zinc	Q	N	B	....	WL23171
Magnesium, zinc	Q	N	B	....	WL23455
Sodium, potassium	P	N	A	....	JA-45-439
Sodium, potassium	P	A	A	....	WL23230
Zinc, lead, tin	Q	N	B	....	WL23404
Multiple-Element 36000 Series					
Aluminum, calcium, magnesium	Q	N	C	....	JA-45-36099
Aluminum, calcium, magnesium, lithium	Q	N	C	....	JA-45-36250
Antimony, arsenic, bismuth	Q	N	C	....	JA-45-36203
Barium, calcium, strontium, magnesium	Q	N	C	....	JA-45-36228
Cadmium, silver, zinc, lead	Q	N	C	....	JA-45-36205
Cadmium, copper, zinc, lead	Q	N	C	....	JA-45-36227
Calcium, magnesium	Q	N	C	....	JA-45-36092
Calcium, magnesium, zinc	Q	N	C	....	JA-45-36097
Calcium, zinc	Q	N	C	....	JA-45-36093
Chromium, iron, manganese, nickel	Q	N	C	....	JA-45-36201
Chromium, cobalt, copper, manganese, nickel	Q	N	C	....	JA-45-36094
Chromium, cobalt, copper, manganese, nickel	Q	N	C	....	JA-45-36103
Chromium, copper, nickel, silver	Q	N	C	....	JA-45-36096
Chromium, copper, iron, nickel, silver	Q	N	C	....	JA-45-36108
Cobalt, copper, iron, manganese, molybdenum	Q	N	C	....	JA-45-36102
Copper, zinc, lead, tin	Q	N	C	....	JA-45-36202
Copper, iron	Q	N	C	....	JA-45-36200
Copper, iron, nickel	Q	N	C	....	JA-45-36101
Copper, iron, lead, nickel, zinc	Q	N	C	....	JA-45-36204
Copper, iron, manganese, zinc	Q	N	C	....	JA-45-36105
Sodium, potassium	P	A	C	....	JA-45-36095

† P = Pyrex; Q = quartz.
‡ N = neon; A = argon.
§ A = 1½ in diameter; B = 1 in diameter; C = 2 in diameter.
¶ WL = Westinghouse; JA = Jarrell-Ash.

TABLE 16 Spectral Characteristics of Varian Hollow-Cathode Tubes

Lamp specifications	Calcium (see Fig. 37*a*)	Cesium (see Fig. 37*b*)	Iron (see Fig. 37*c*)	Mercury (see Fig. 37*d*)	Thorium (see Fig. 37*e*)	Sodium or potassium (see Fig. 37*f*)
Fill gas	Neon	Argon	Neon	Neon	Neon	Neon
Window	Quartz	Pyrex	Quartz	Quartz	Quartz	Pyrex
Operating current, mA	4	20	5	3	10	5
Strike voltage, V	260	. . . . .	260	310	270	280
Operating voltage, V	150	. . . . .	160	230	140	210
Most sensitive line, Å	4226.7	8521.1	2483.3	2536.5	[a]	5889.9
Spectral bandwidth, Å	3.3	9.9	1.7	3.3	0.8	3.3
Sensitivity, ppm	0.030[b]	0.16[c,d]	0.08[e]	2.0[e]	[a]	0.01[c]

Alternative Absorption Lines[f]

Element	Wavelength, Å	Sensitivity, ppm	Relative sensitivity	Relative signal intensity	Recommended spectral bandwidth, Å
Calcium	4226.7	0.03	1[g]	100[h]	3.3
	2398.6	6	200[g]	0.25[h]	0.8
Cesium	8521.1	0.16	1.0[i]	100[j,d]	9.9
	8943.5	0.22	1.2[i]	47[i,d]	9.9
	4555.4	10	5.3[i]	100[k]	3.3
	4593.2	40	21[i]	26[k,l]	0.8
Iron	2483.3	0.08	1.0[m]	1.0[n]	1.7
	3719.9	0.76	9[m]	100[n]	0.8
	3859.9	1.2	15[m]	53[n]	1.7
	3920.3	22	270[m]	4.0[n]	0.8
Thorium	3719.4	. . . .	. . . . .	100[o]	0.8
	3803.1	. . . .	. . . . .	57[o]	0.8
	3304.2	. . . .	. . . . .	32[o]	0.8
Sodium or potassium	5889.9	0.01	1.0[p]	100[q]	3.3
	5895.9	0.02	2.1[p]	56[q]	3.3
	3302.3, 3302.0 } D	2.9	360[p]	3.7[q]	3.3

[a] No absorption has been detected for this element with the Techtron burner system.
[b] With $N_2O-C_2H_2$ flame.
[c] With air-propane flame.
[d] With R196 photomultiplier tube.
[e] With air-C_2H_2 flame.
[f] No alternative lines for mercury have been examined.
[g] Measured relative to the 4226.7-Å line.
[h] Measured at 4 mA lamp current with spectral bandwidth of 3.3 Å relative to the 4226.7-Å line.
[i] Measured relative to the 8521.1-Å line.
[j] Measured at 20 mA lamp current with spectral bandwidth of 3.3 Å.
[k] With R213 photomultiplier tube.
[l] Measured relative to 4555.4 Å.
[m] Measured relative to 2483.3-Å line.
[n] Measured at 5 mA lamp current with spectral bandwidth of 0.8 Å relative to 3719.9-Å line.
[o] Measured at 20 mA lamp current with spectral bandwidth of 0.8 Å relative to 3719.4-Å line.
[p] Measured relative to 5889.9-Å line.
[q] Measured at 5 mA lamp current with spectral bandwidth of 0.8 Å relative to 5889.9-Å line.

TABLE 17 Specifications of Osram Lamps†

Catalog number	Type	Material		Operating potential, V	Operating current, A	Power dissipated in lamp, W	Light intensity, cd	Emitting height, mm	Surface width, mm	Surface brightness, stilb
		Burner	Envelope							
26-2204	Cadmium	Quartz	Glass‡	12	1.5	18	3.0	20	8	1.0
26-2220	Caesium	Glass	Glass	7	1.5	10	0.7	20	10	0.1
26-2246	Helium	Glass	Glass	60	1.3	65	3.5	15	10	2.3
26-2261	Mercury	Quartz	Glass‡	50	1.2	50	110	20	3	150
26-2287	Mercury and cadmium	Quartz	Glass‡	30	1.0	30	15	20	3	5.0
26-2295	Mercury (Hg3)	Quartz	Glass §	55	1.1	50	50	20	2	120
26-2303	Potassium	Glass	Glass	10	1.5	14	0.15	20	10	0.02
26-2329	Sodium	Glass	Glass	20	1.3	22	30	15	10	12
26-2360	Neon	Glass	Glass	30	1.5	40	3.5	20	10	0.8
26-2386	Rubidium	Glass	Glass	10	1.5	13	0.4	20	10	0.04
26-2402	Thallium	Quartz	Glass‡	14	1.0	13	0.8	9	3	3.0
26-2428	Zinc	Quartz	Glass‡	14	1.5	20	1.0	20	8	0.1

† Ealing Corp. catalog.

‡ Transmits to 2800 Å.

§ This lamp has an opening in the outer envelope to extend the range in the ultraviolet to 2300 Å and has a special burner of high constancy of light emission.

TABLE 18 Specifications of Philips Spectral Lamps†

Catalog number	Symbol	Type	Material: Burner	Material: Envelope	Operating current, A	Wattage	Arc length, mm
26-2709	Hg	Mercury (low-pressure)	Quartz	Glass	0.9	15	40
26-2717	Hg	Mercury (high-pressure)	Quartz	Glass	0.9	90	30
26-2725	Cd	Cadmium	Quartz	Glass	0.9	25	30
26-2733	Zn	Zinc	Quartz	Glass	0.9	25	30
26-2741	Hg, Cd, Zn	Mercury, cadmium, and zinc	Quartz	Glass	0.9	90	30
26-2758	He	Helium	Glass	Glass	0.9	45	32
26-2766	Ne	Neon	Glass	Glass	0.9	25	40
26-2774	A	Argon	Glass	Glass	0.9	15	40
26-2782	Kr	Krypton	Glass	Glass	0.9	15	40
26-2790	Xe	Xenon	Glass	Glass	0.9	10	40
26-2808	Na	Sodium	Glass	Glass	0.9	15	40
26-2816	Rb	Rubidium	Glass	Glass	0.9	15	40
26-2824	Cs	Caesium	Glass	Glass	0.9	10	40
26-2832	K	Potassium	Glass	Glass	0.9	10	40
26-2857	Hg	Mercury (low-pressure)	Quartz	Quartz	0.9	15	40
26-2865	Hg	Mercury (high-pressure)	Quartz	Quartz	0.9	90	30
26-2873	Cd	Cadmium	Quartz	Quartz	0.9	25	30
26-2881	Zn	Zinc	Quartz	Quartz	0.9	25	30
26-2899	Hg, Cd, Zn	Mercury, cadmium, and zinc	Quartz	Quartz	0.9	90	30
26-2907†	In	Indium‡	Quartz	Quartz	0.9	25	25
26-2915	Tl	Thallium	Quartz	Quartz	0.9	20	30
26-2923	Ga	Gallium	Quartz	Quartz	0.9	20	30

† Ealing Corp. catalog.
‡ Requires a Tesla coil to cause it to strike initially.

TABLE 19 Gas Fills in Pluecker Tubes†,‡

Cenco no.	Type
87210	Argon gas
87215	Helium gas
87220	Neon gas
87225	Carbonic acid gas
87230	Chlorine gas
87235	Hydrogen gas
87240	Nitrogen gas
87242	Air
87245	Oxygen gas
87255	Iodine vapor
87256	Krypton gas
87258	Xenon
87260	Mercury vapor
87265	Water vapor

† Cenco catalog.

‡ Consists of glass tube with overall length of 25 cm with capillary portion about 8.5 to 10 cm long. Glass-to-metal seal wires are welded in metal caps with loops for wire connection firmly sealed to the ends. Power supply no. 87208 is recommended as a source of excitation.

Glow Modulator Tubes

29. According to technical data supplied by Sylvania, these are cold-cathode light sources uniquely adaptable to high-frequency modulation. Pictures of two types are shown in Fig. 44. The cathode is a small hollow cylinder, and the high ionization density in the region of the cathode provides an intense source of radiation.

Figure 45 is a graph of the light output as a function of tube current. Figure 46 is a graph depicting the response of the tube to a modulating input. The spectral outputs of a variety of tubes are shown in Fig. 47. Table 20 gives some of the glow-modulator specifications.

Hydrogen and Deuterium Arcs

30. For applications requiring a strong continuum in the ultraviolet region, the hydrogen arc at a few millimeters pressure provides a useful source. It can be operated with cold or hot cathode. One hot-cathode type is shown in Fig. 48. Koller (1965) plots a distribution for this lamp down to about 200 Å.

Deuterium lamps (Fig. 49) provide a continuum in the ultraviolet with increased intensity (quoted by Ealing) over the hydrogen arc. Both lamps have quartz envelopes. The one on the left is designed for operation down to 2000 Å; the one on the right is provided with a Suprasil† window to increase the ultraviolet range down to 1650 Å.

OTHER COMMERCIAL SOURCES

31. Activated-Phosphor Sources Of particular importance and convenience in the use of photometers are sources composed of a phosphor activated by radioactive substances. Readily available, and not subject to Atomic Energy Commission licensing with small quantities of radioactive material, are the ^{14}C-activated phosphor light sources. These are relatively stable sources of low intensity, losing about 0.02 percent per year due to the half-life of ^{14}C and the destruction of phosphor centers.

† Trade name of Amersil, Inc.

TABLE 20 Glow-Modulator Specifications†

No.‡	Maximum operating voltage	Current, mA		Minimum starting voltage, V	Crater diameter, in	Approximate light center length, in	Light output, cd	Brightness, cd/in²	Rated life, h	Base type	Bulb type	Maximum overall length, in	Maximum diameter, in	Color of discharge
		Average	Peak											
GM-514	160	5–25	55	240	0.056	$1\frac{3}{4}$	.1 at 25 mA	41 at 25 mA	100 at 15 mA	3-pin miniature§	T-$4\frac{1}{2}$	$2\frac{5}{8}$	$\frac{41}{64}$	Blue-red
GM-514C	160	5–15	35	240	0.093	$1\frac{3}{4}$	.1 at 15 mA	15 at 15 mA	25 at 10 mA	3-pin miniature§	T-$4\frac{1}{2}$	$2\frac{5}{8}$	$\frac{41}{64}$	White
1B59/R-1130B	150	5–35	75	225	0.056	2	.13 at 30 mA	43 at 30 mA	250 at 20 mA	Intermediate shell oct.¶	T-9	$3\frac{1}{16}$	$1\frac{9}{32}$	Blue-red
R-1131C	150	3–25	55	225	0.093	2	.2 at 25 mA	29 at 25 mA	150 at 15 mA	Intermediate shell oct.¶	T-9	$3\frac{1}{16}$	$1\frac{9}{32}$	White
R-1166	150	3–25	55	225	0.093	2	.2 at 25 mA	29 at 25 mA	150 at 15 mA	Intermediate shell oct.¶	T-9	$3\frac{1}{16}$	$1\frac{9}{32}$	White
R-1168	150	5–15	30	225	0.015	2	.023 at 15 mA	132 at 15 mA	150 at 15 mA	Intermediate shell oct.¶	T-9	$3\frac{1}{16}$	$1\frac{9}{32}$	Blue-red
R-1169	150	5–25	45	225	0.025	2	.036 at 15 mA	72 at 15 mA	250 at 15 mA	Intermediate shell oct.¶	T-9	$3\frac{1}{16}$	$1\frac{9}{32}$	Blue-red

† Sylvania Technical Data Service.
‡ Type R-1166 is opaque-coated with the exception of a circle $\frac{3}{8}$ in in diameter at end of lamp. All other types have clear-finish bulb.
§ Pins 1 and 3 are anode; pin 2 cathode.
¶ Pin 7 anode; pin 3 cathode.

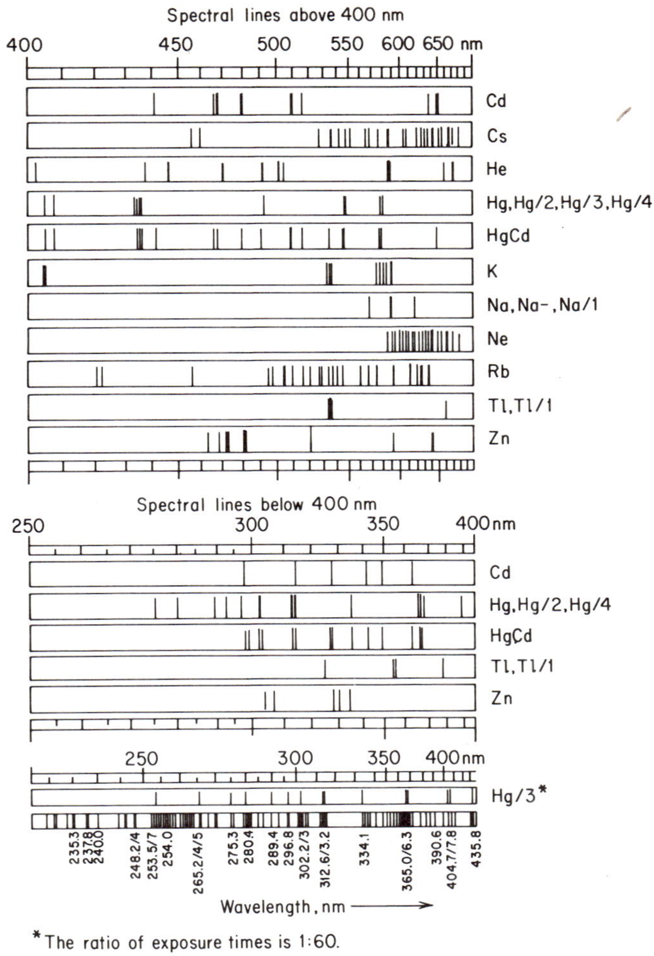

Fig. 38 Spectrograms of Osram spectral lamps. (*Osram.*)

Two types available† are described (New England Nuclear Corporation Catalog of Radioactive Sources, 1969–1970) as rod and disk standards. The rod standards are quartz cylinders 6 mm in diameter and 2 in high, with luminous centers at one end of approximately $\frac{1}{4}$ in diameter and $\frac{1}{4}$ in high. The disk standards are acrylic plastic with an aluminum housing 0.875 in in diameter and $\frac{1}{4}$ in high. Wavelength regions covered are 6300, 5200, and 4100 Å.

32. Other (High-Energy) Sources Radiation at very high powers can be produced in so many ways that it would not be feasible, even if it were possible, to cite them all, let alone describe them. Almost any source of high energy which can be efficiently and conveniently coupled radiationally can be considered. Just how useful these are for the optical scientist is another matter. The literature mentions such sources as synchrotrons, plasmatrons, arcs, sparks, exploding wires, shock tubes, and atomic and molecular beams, to name but a few.

† The reader should refer to the appropriate directories for sources to suit his purpose.

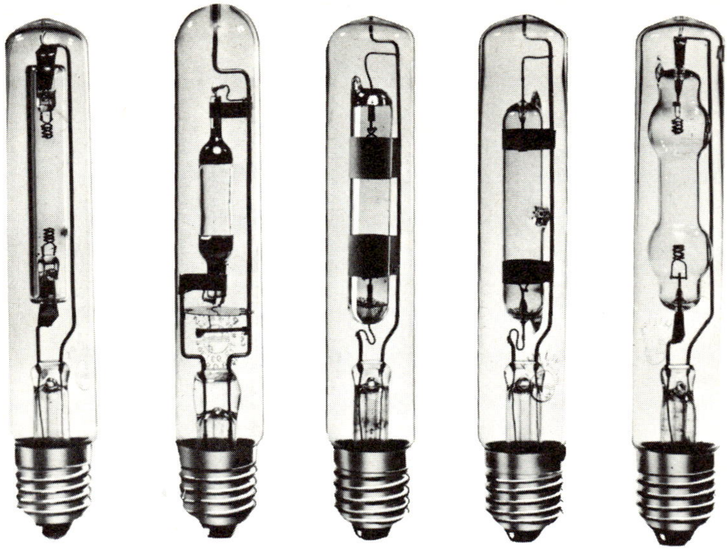

(a) (b) (c) (d) (e)

Fig. 39 Physical characteristics of spectral lamps: (*a*) mercury (low pressure); (*b*) zinc; (*c*) cadmium; (*d*) krypton; (*e*) helium. Not shown are thallium and gallium. The thallium lamp (26-2915), characterized by an extremely narrow (0.05-Å) line at 5350 Å suitable for interferometry; lines of the sodium doublet are also about 0.06 Å wide. Gallium (26-2923) exhibits intense sharp lines at 4033 and 4172 Å. (*Ealing Corporation.*)

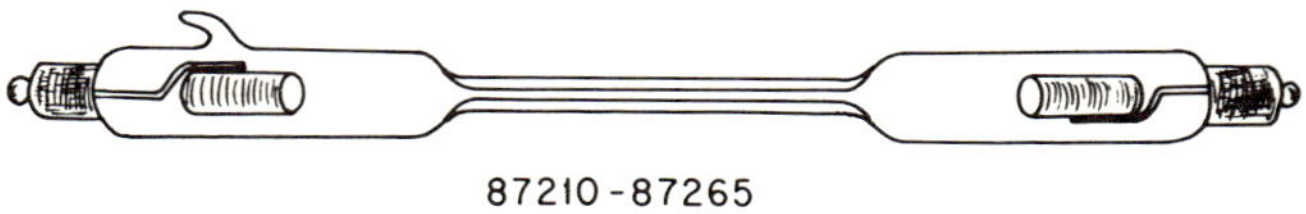

87210-87265

Fig. 40 Physical construction of Pluecker spectrum tubes. (*Cenco.*)

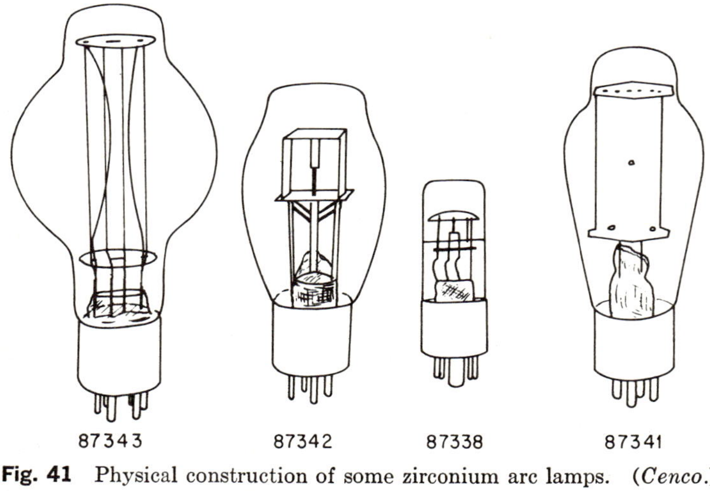

87343 87342 87338 87341

Fig. 41 Physical construction of some zirconium arc lamps. (*Cenco.*)

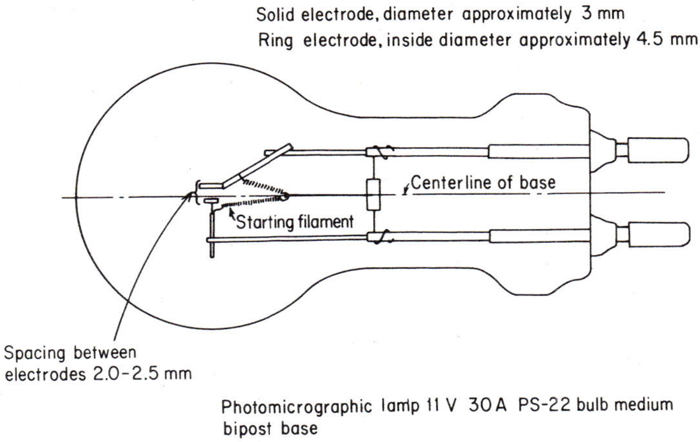

Fig. 42 Construction of tungsten arc lamp, drawn to scale. The lamp must be operated base up in a well-ventilated housing and using a special high-currrent socket which does not distort the position of the posts. (*General Electric.*)

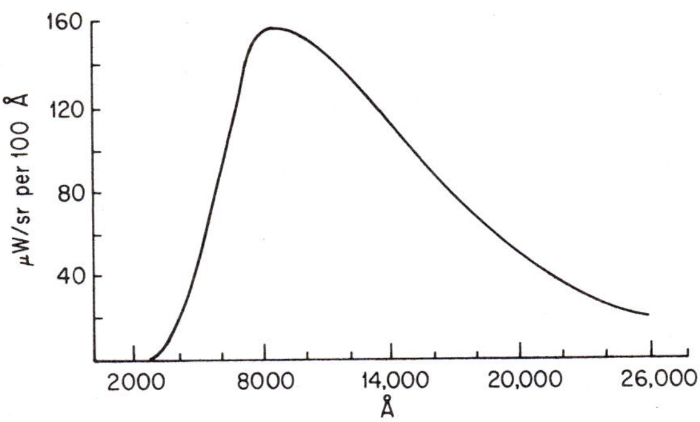

Fig. 43 Spectral distribution of a 30-A PS22 photomicrographic lamp. (*General Electric.*)

Among these, one can purchase at a reasonable price in convenient, usable form precisely controlled spark sources for yielding many joules of energy in a time interval of the order of microseconds. Also, consumable plasma tubes can be purchased as radiation sources emitting effectively at many thousands of degrees, yielding radiation rich in ultraviolet energy.

33. Other Special Sources Obtainable from different manufacturers and scientific instrument suppliers are an enormous number of special-purpose sources, many of which are among those described in this section. Because of the competitive nature of these components, they appear in different forms, under different trade names, and sometimes with slightly different characteristics.

One source that remains to be mentioned is the so-called miniature, sub- and microminiature, etc., lamps. These are small, even tiny, incandescent bulbs of glass or quartz, containing tungsten filaments. They serve excellently in certain applications where small, intense radiators of visible and near-infrared radiation are needed. One should consult the various manufacturers, given in, for example, the latest Optical Industry and Systems Directory.

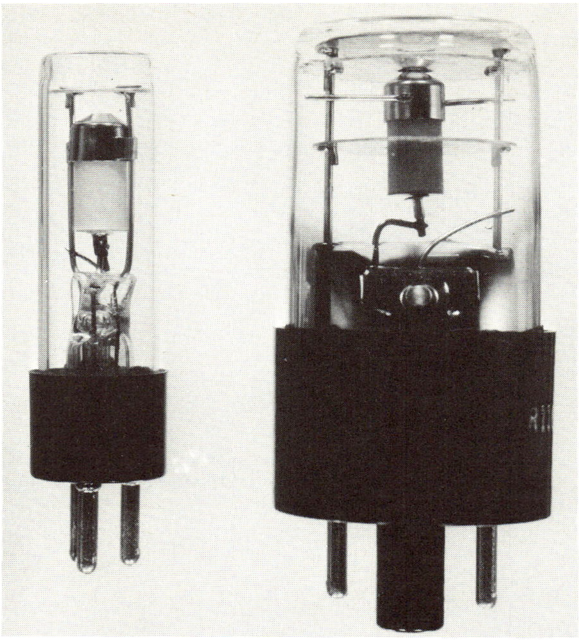

Fig. 44 Construction of two glow modulator tubes. (*Sylvania.*)

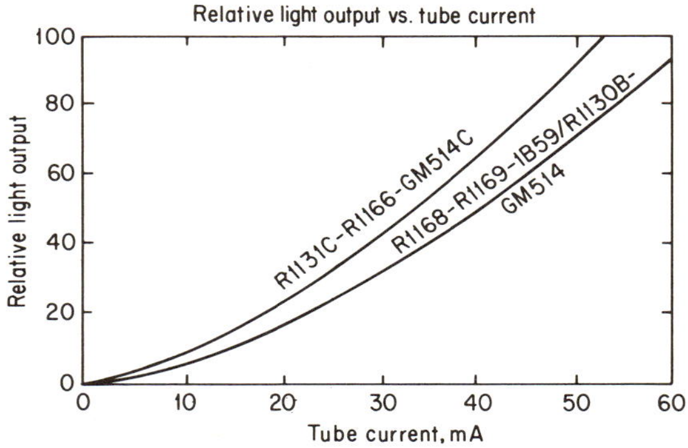

Fig. 45 Variations of the light output from a glow modulator tube as a function of tube current. (*Sylvania.*)

NATURAL SOURCES

34. There is hardly a facet of visible radiation from the environment that has not been known for many years. Some recent measurements have helped increase the confidence with which we can hope to improve upon some of the values obtained in the past. The one extraterrestrial source of greatest importance, the sun, has been measured accurately at high altitudes (Thekaekara et al., 1969), yielding results which are relatively uninfluenced by atmospheric effects. Most other environmental sources of radiation are intimately enough associated with the atmosphere to incorporate the effect of the atmosphere in their values. For these the older results are probably as reliable as any of recent vintage.

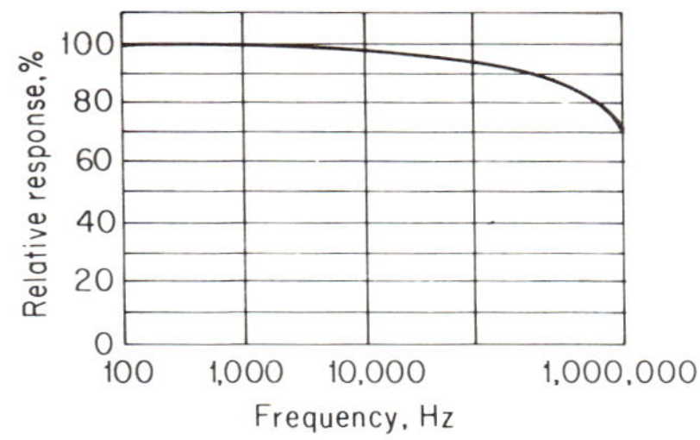

Fig. 46 Response of the glow modulator tube to a modulating input. (*Sylvania.*)

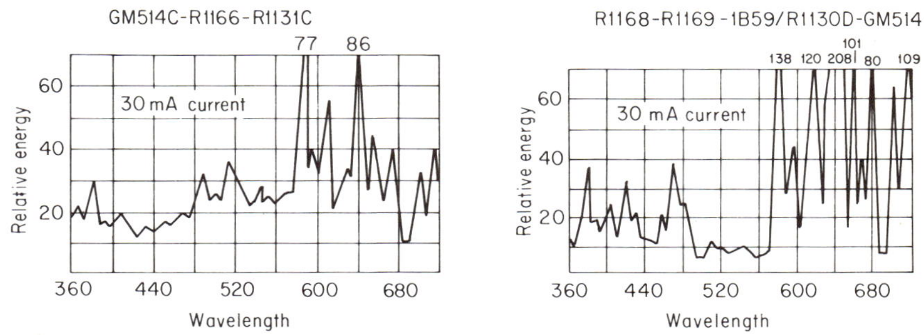

Fig. 47 Spectral variation of the output of glow modulator tubes. (*Sylvania.*)

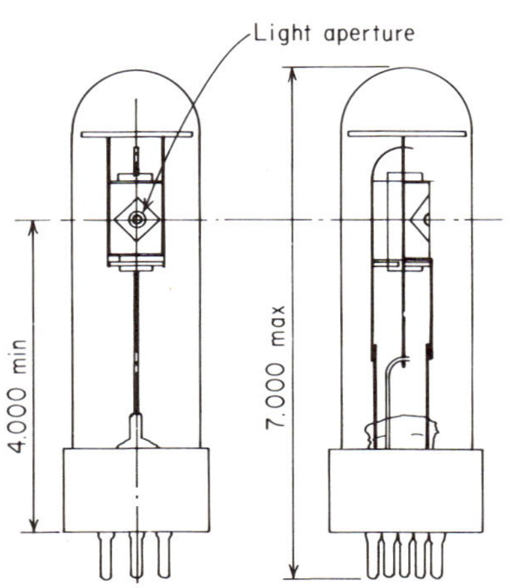

Fig. 48 Hydrogen arc lamp. (*Sylvania.*)

The part of the spectrum for which we depend on the more recent results is the infrared, because only few results appeared before the past two decades. One does not have to search far to find literature consolidating many of these results as well as those of older measurements. Since many of these works cover too broad a field to make specific data easy to find, the following paragraphs present a cross section of the information of interest to the optics designer.

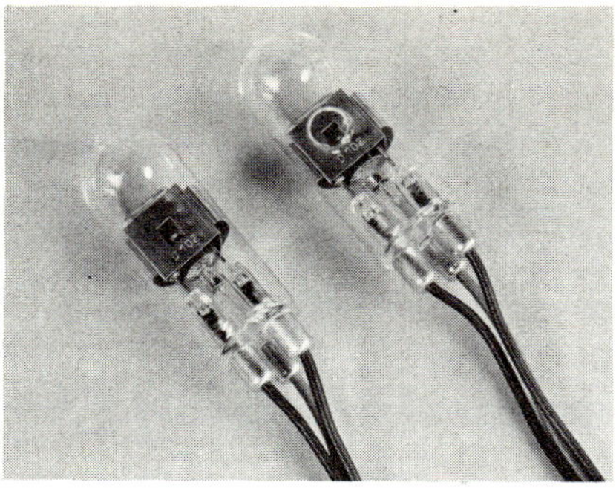

Fig. 49 Two types of deuterium arc lamp: *left:* 26-261; *right:* 26-262. (*Ealing Corporation.*)

Several references come close to satisfying these requirements, namely, Wolfe (1965, chap. 5), and, of more general importance, Air Research and Development Command (1960) and Valley (1965), from which certain data have been gathered.

The data are organized in two categories, downward-directed radiation originating from the sky, space, atmosphere, etc., and upward-directed radiation originating from the earth's surface.

Sky Radiation

35. The Sun The sun is the most influential source in the sky. The solar constant has been determined from measurements of sunlight radiation by different observers and represents the value of the total irradiance (in the whole spectrum) from the sun at a point outside the influence of the atmosphere. Allowing for differences in the earth-sun distance (3.4 percent larger at perihelion in January and 3.4 percent smaller for aphelion in July), the latest value reported (Thekaekara et al., 1969) is given as 135.1 mW/cm². Comparison of this value is made with others in Table 21.

TABLE 21 Significant Revisions of the Solar Constant (Thekaekara et al., 1969)

Author	Year	Solar constant, mW/cm²
P. Moore	1940	132.3
L. B. Aldrich and C. G. Abbot	1948	132.6
W. Schuepp	1949	136.7–141.6
C. W. Allen	1950	137.4
L. B. Aldrich and W. H. Hoover	1952	134.9
F. S. Johnson	1954	139.5
R. Stair and R. G. Johnston	1956	143.0
C. W. Allen	1958	138.0
E. G. Laue and A. J. Drummond	1968	136.1
M. P. Thekaekara, R. Kruger, and C. H. Duncan	1969	135.1

The differences are usually attributed to the inability to make precise corrections for the effect of the atmosphere. At least one reason for confidence in the later value is that measurements were made with a variety of instruments from an airplane flown at an altitude of 11.58 km (38,000 ft), where much of the atmospheric absorption is nonexistent.

The sun is usually considered to be represented nearly by a blackbody at a temperature of 6000 K. For ground-stationed observers this hardly matters, because after passage through the atmosphere, the spectrum barely resembles that of a blackbody. Two spectra of the expected extraterrestrial spectral irradiance are shown in Fig. 50.

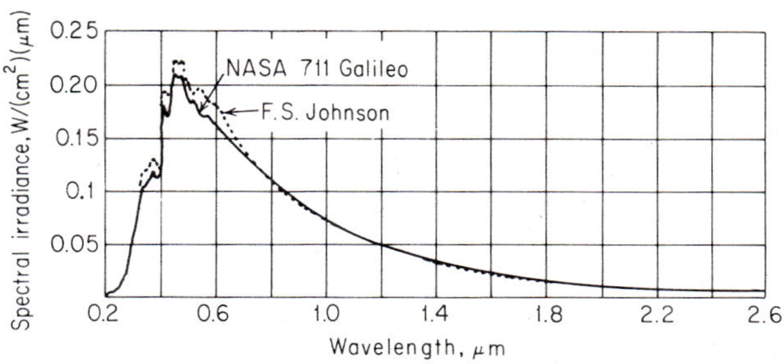

Fig. 50 Solar spectrum for zero air mass from NASA 711 flight. 0.2 to 2.6 μm, compared with curve of Johnson (1954). [*Thekaekara et al.* (1969).]

Figure 51 shows three spectra, a 5900-K blackbody, the extraterrestrial solar spectrum of Johnson, shown in Fig. 50, and the sea-level spectrum obtained through one air mass. Implicit in the last of these is the effect of the amount of absorbing constituents from the surface of the earth through the zenith. As the air mass is increased, i.e., as the sun is observed in higher latitudes at greater angles from

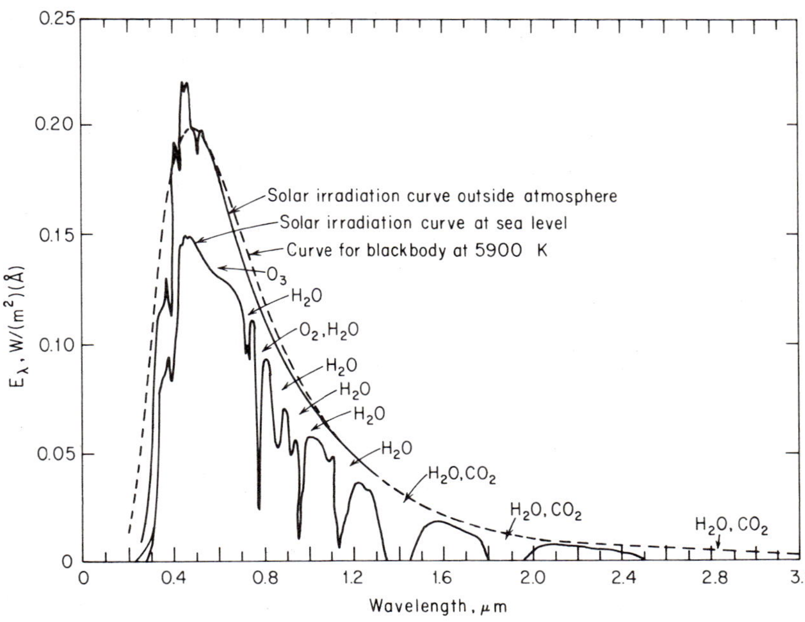

Fig. 51 Spectral distribution curves related to the sun; shaded areas indicate absorption at sea level due to the atmospheric constituents shown. [*Valley* (1965).]

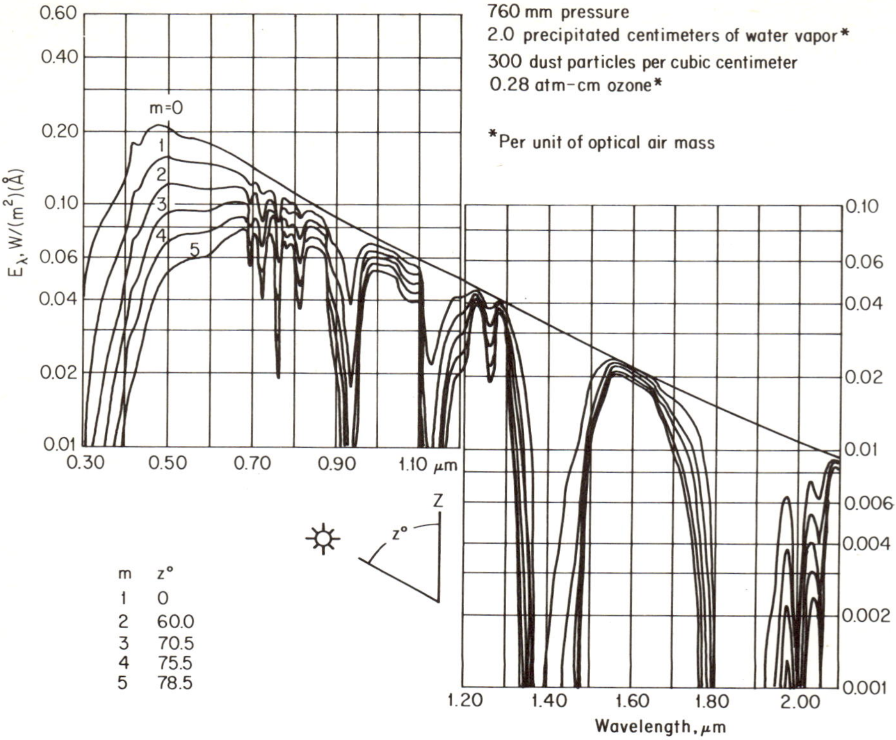

Fig. 52 Solar spectral irradiance curves at sea level for various optical air masses; the value of the solar constant is 1,322 W/m². [*Valley* (1965).]

the zenith, the effect of atmospheric absorption becomes more pronounced, and the spectra take on the appearances shown in Fig. 52. The influence of the sun at wavelengths in the infrared is shown in Fig. 53. One notes immediately from Fig. 51 that below 0.29 μm absorption by O_2 and O_3 presents sunlight from penetrating to the earth's surface. Table 22 gives a brief account of the effect of the atmosphere on solar radiation showing the constituents responsible and their regimes of effectiveness.

36. Moon and Planets Celestial bodies of less importance as sources of radiation are the moon, planets, and stars. The moon and planets, producing illumination as a result of the reflection of sunlight, are nearly (but not quite) replicas of sunlight, yielding radiation of an effective temperature of 5900 K at a very much reduced level from that of sunlight (Ramsay, 1962). Figure 54 shows the calculated distribution of radiation from the moon and planets at their brightest. Table 23 delineates the visual magnitudes and effective temperatures of these same bodies, along with similar quantities for the brightest visual and so-called red stars.

37. Stars Details on stellar radiation can be found in Wolfe (1965). Some data from that text are reproduced here. A few stars are bright enough to be considered as independent sources of radiation to most optical systems; their characteristics are shown in Table 23 and Fig. 55. The great majority of stars, however, yield irradiances individually which are well below the threshold of sensitivity of many wide field-of-view devices. The very large numbers and concentrations of these radiators cause a considerable background of radiation outside the influence of the earth's atmosphere. Importance of that radiation is indicated in Table 24, showing

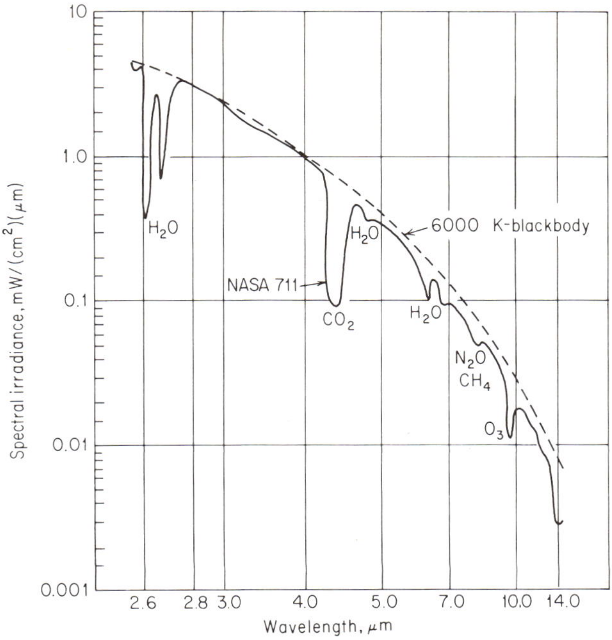

Fig. 53 Solar spectrum at 1.13 air mass from the I-4 interferometer. [*Thekaekara et al.* (1969).]

the distribution of stars as a function of magnitude. The definition of magnitude is

$$\log \frac{I_1}{I_2} = 0.4(m_2 - m_1)$$

where I is the intensity of the radiation within the spectral band of acceptance (the visual-response curve, for example, for the eye) and m is the magnitude.

Because the data of Table 24 have been collected over the years, the tumbers are reported in terms of photographic magnitude. The relationship between the photographic and visual magnitudes involves the center and width of the spectral response of the eye and of a prescribed filtered photographic image. To obtain more information on this relationship and similar quantities, see Allen (1955, 1963).

Table 25 gives a list of stars suitable for photometric standards. A broader list of stars (Table 26) includes the spectral irradiance (estimated) received at selected wavelengths, especially in the infrared.

38. Airglow and Aurorae These radiations, which are a result of interaction of energy and matter from the sun with the earth's atmosphere, have been the subject of extensive study. The results shown in works like Valley (1965) summarize part of this study. The bibliography in that reference should be consulted for sources of further information.

Tables 27 and 28 are examples of information available on radiation from airglow. Similar tables for the aurorae (Valley, 1965) are not given here since data on radiation from the aurorae are needed only for highly sophisticated optical system designs.

39. Other Atmospheric Emission (Bell et al., 1960) Most of what we have covered to this point is limited to the visible and ultraviolet part of the spectrum,

TABLE 22 Attenuation of Solar Radiation by the Atmosphere† (Johnson, 1959)

Pressure, mbar	Altitude, mi	Altitude, 1,000 ft	Altitude, km	Wavelength regions, μm: 0.12–0.20	0.20–0.29	0.29–0.32	0.32–0.35	0.35–0.55	0.55–0.9	0.9–2.5	2.5–7	7–20	Altitude, km	Nomenclature recommended by IUGG, km
0.2	37	200	60	O_2 absorbs almost completely	Solar irradiation intensity approximates extra atmospheric attenuation by scattering increases markedly toward shorter wavelengths								Above 60	110 ↑ Chemosphere
7.5	20	108	33		(0.20–0.21 μm; absorption by O_2); absorption by O_3 appreciable	O_3 absorption not important					Energy small	Energy very small	60–33	
227	6.8	36	11		No radiation penetrates below about 11 km	O_3 absorption attenuates more than loss by scattering	O_3 absorption significantly attenutates radiation	Irradiation diminished mostly by scattering by permanent gases in atmosphere	H_2O responsible for major absorption; CO_2 absorbs slightly at 2 μm; water vapor (or ice crystals) is found up to about 70,000 ft			Strong O_3 absorption at 9.6 μm; strong CO_2 absorption 12–17 μm	33–11	↓ 20 ↑ Stratosphere ↓
795	1.2	6.6	2				Highly variable dust, haze (H_2O) (and smoke) responsible for attenuation in regions 0.32 to 0.7 μm		Energy transmitted with small loss down to 2 km	Energy penetrates to sea level only through windows at approximately 1.2, 1.6, and 2.2 μm	No significant penetration below 2 km, except in windows at approximately 3.8 and 4.9 μm	Energy transmitted with moderate loss; many absorption bands due to atmospheric gases	11–2	↑ 11 Troposphere
1,013 (Sea level)						Appreciable penetration through "clear" atmosphere to sea level: About 7% / About 30%	Penetration through "clear" atmosphere to sea level about 40%	Dust may rise to more than 4 km					2–sea level	0 ↓

† Cf. Fig. 51.

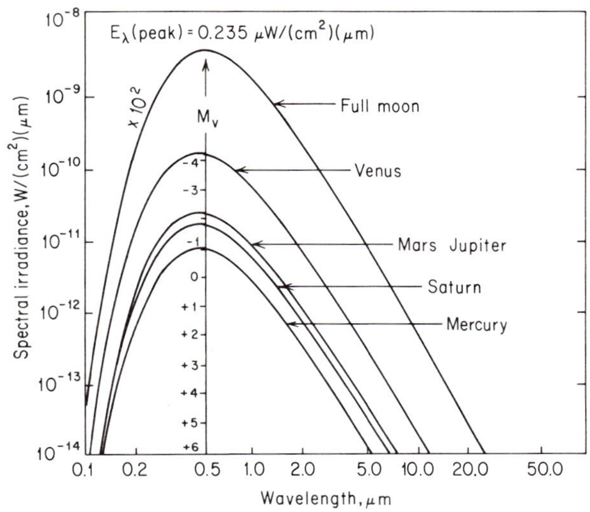

Fig. 54 Calculated irradiance from planets, at brightest, due to sun reflectance only. [*Ramsay* (1962).]

TABLE 23 Visual Magnitudes and Effective Temperatures of Planets and the Brightest Visual and Red Stars (Ramsay, 1962)

Name	Visual magnitude m_V	Effective temperature T, K
Planets:		
Moon (full)	−12.2	5,900
Venus (at brightest)	−4.28	5,900
Mars (at brightest)	−2.25	5,900
Jupiter (at brightest)	−2.25	5,900
Mercury (at brightest)	−1.8	5,900
Saturn (at brightest)	−0.93	5,900
Stars:		
Sirius	−1.60	11,200
Canopus	−0.82	6,200
Rigel Kent (double)	0.01	4,700
Vega	0.14	11,200
Capella	0.21	4,700
Arcturus	0.24	3,750
Rigel	0.34	13,000
Procyon	0.48	5,450
Achernar	0.60	15,000
β Centauri	0.86	23,000
Altair	0.89	7,500
Betelguex (variable)	0.92	2,810
Aldebaran	1.06	3,130
Pollux	1.21	3,750
Antares	1.22	2,900
α Crucis	1.61	2,810
Mira (variable)	1.70	2,390
β Gruis	2.24	2,810
R. Hydrae (variable)	3.60	2,250

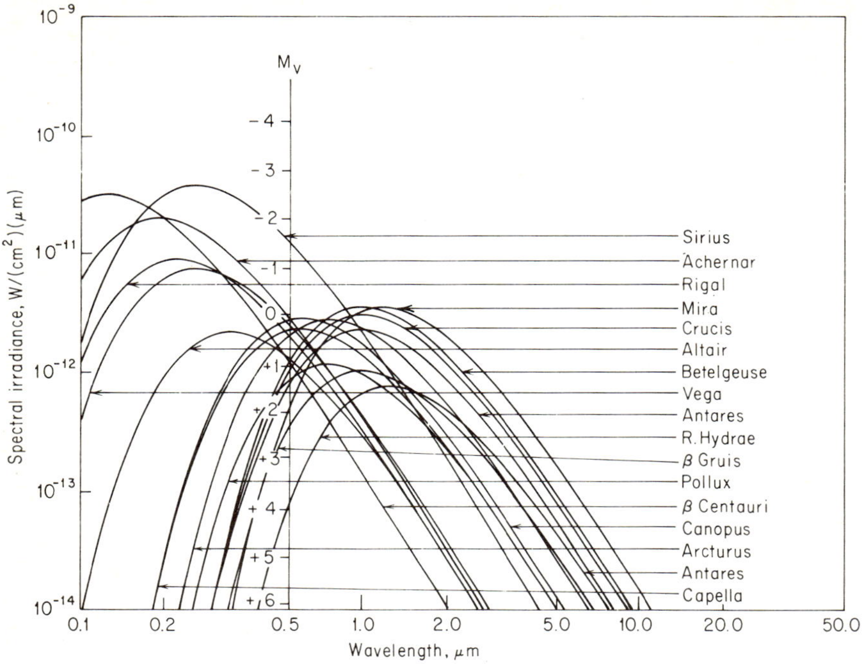

Fig. 55 Calculated irradiance from brightest stars at top of the atmosphere. [*Ramsay* (1962).]

TABLE 24 Estimated Total Number of Stars Brighter than a Given Magnitude (Wolfe, 1965)

Photographic magnitude	Number of stars	
	Photographic	Visual
0	. . . .	2
1	. . . .	2
1	. . . .	12
2	. . . .	40
3	. . . .	140
4	360	530
5	1,030	1,620
6	2,940	4,850
7	8,200	14,300
8	22,800	41,000
9	62,000	117,000
10	166,000	324,000
11	431,000	870,000
12	1,100,000	2,270,000
13	2,720,000	5,700,000
14	6,500,000	13,800,000
15	15,000,000	32,000,000
16	33,000,000	71,000,000
17	70,000,000	150,000,000
18	143,000,000	296,000,000
19	275,000,000	560,000,000
20	505,000,000	1,000,000,000
21	890,000,000	

TABLE 25 Stars Suitable for Photometric Standards (Valley, 1965)

Star	Spectral type	Visual magnitude	Wavelength, μm: 0.556	1.00	2.20	3.73	10.20
			Spectral irradiance, 10^{-14} W/(cm²)(μm)				
α Lyrae	A0V	0.03	371.0	62.9	3.63	0.501	0.19
α Aurigae	G8III	0.09	350.0	163.0	19.5	3.46	0.108
ε Cygni	K0III	2.45	40.1	21.9	2.99	0.459	
α Arietis	K2III	1.99	60.9	39.5	6.19	0.630	
α Tauri	K5III	0.78	186.0	206.0	49.4	8.85	0.289

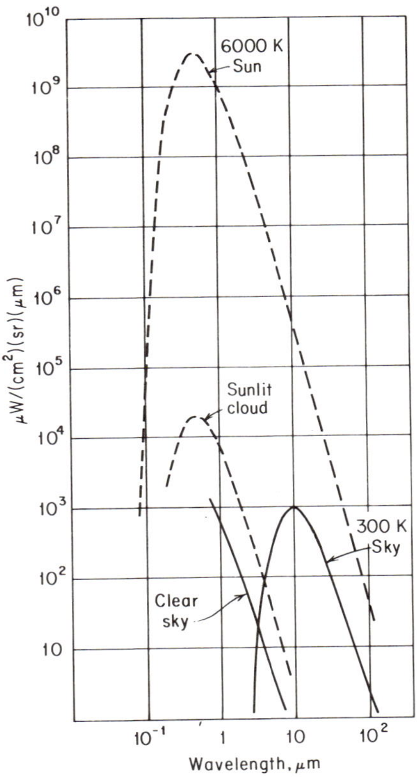

Fig. 56 An idealized spectral radiance of the sun, emitting atmosphere, sunlit cloud, and sunlight-scattering clear sky. [*Bell et al.* (1960).]

extending slightly beyond the visible region. Equally important in infrared system design is the thermal radiation from the environment, which is especially evident beyond the spectral region of 1 μm. This radiation originates in the molecular transitions in gases and in the scattering of radiation from aerosols and particulates. If one observes environmental radiation looking upward, the resulting daylight spectrum depends on the sun in the region out to roughly 3 μm and on the atmosphere for the thermally emitted component beyond 3 μm. Figure 56 gives an idealized

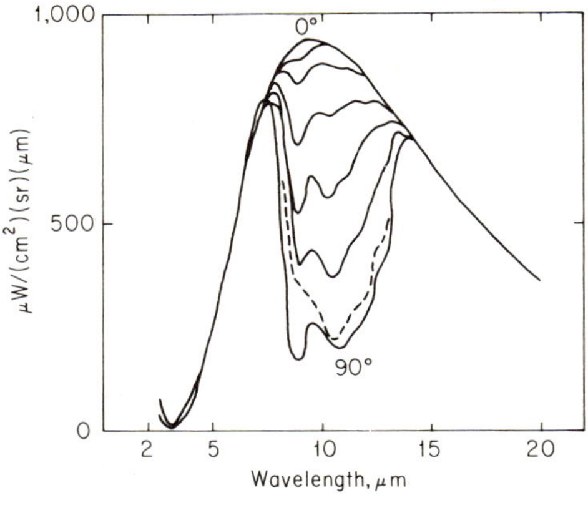

Fig. 57 Spectral radiance of a clear sky for several angles of elevation above the horizon, sea level. [*Bell et al.* (1960).]

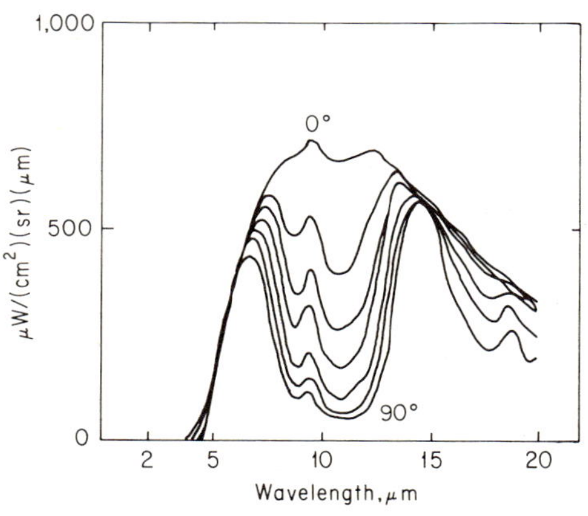

Fig. 58 Spectral radiance of a clear sky for several angles of elevation above the horizon, 3.5 km above sea level. [*Bell et al.* (1960).]

spectral radiance of the sun, emitting atmosphere, sunlit cloud, and a sunlight-scattering clear sky.

In reality, since the thermal radiation is from gaseous emitters, the spectral quality of the radiation depends greatly on the amount of atmosphere observed, i.e., on the direction of observation and the location of the measuring instrument. Figure 57 shows the spectral radiance of a clear sky taken at different elevation angles from a sea-level location. Note the similarities and slight differences between this figure and the curves of Fig. 58, in which similar measurements were taken 11,000 ft above sea level. In both figures the presence of an emitting ozone region in the atmosphere is evident at approximately 9.6 μm, more in Fig. 58 than in Fig. 57, because height above much of the absorbing H_2O and CO_2 helps to reveal its effect.

Murcray et al. (1969) reported the identification of absorption in high-altitude spectral data due to HNO_3. Considerable absorption of solar energy was observed with balloon-borne equipment by these investigators in the 7.5 μm spectral region in the period during sunset. In addition, absorption by HNO_3 occurs in the 10-

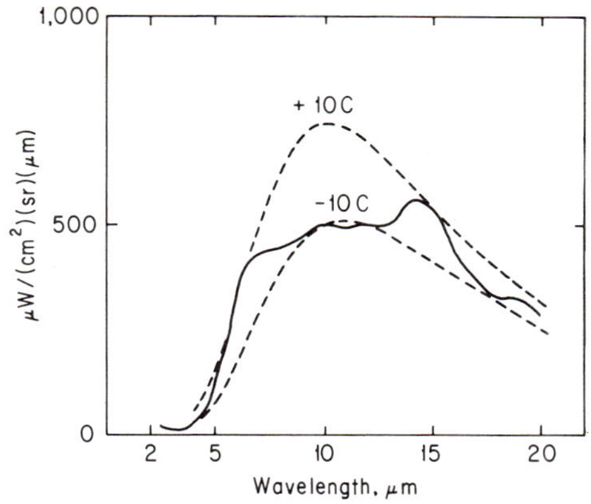

Fig. 59 Spectral radiance of the underside of a dark cumulus cloud. [*Bell et al.* (1960).]

to 12-μm region. In float-altitude spectra the prominence of HNO_3 seems to occur at 11.3 and 11.4 μm in observations at sunset, while this region is nearly transparent at small zenith angles. Murcray et al. (1973) give the results of measuring atmospheric radiance in the 10- to 12.5-μm region as function of altitude. These corroborate the data reported in showing a steep falloff in observed radiance with altitude from the ground to 12 km, an expected result of the decrease of H_2O vapor with altitude and a drop in temperature. Further increase in altitude is then accompanied

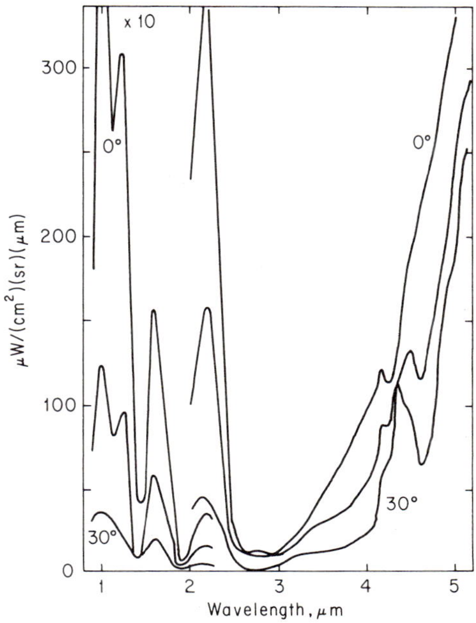

Fig. 60 Spectral radiance of a clear sky, showing the elevation-angle dependence of the scattered radiation. [*Bell et al.* (1960).]

TABLE 26 Spectral Irradiance, W/(cm²)(μm), at Selected Wavelengths
Values in parentheses are less reliable than others

Star	Spectral type†	Visual magnitude	10	5	3.33	2.5
α Tau	K5III	+0.78	0.28×10^{-14}	0.33×10^{-13}	0.13×10^{-12}	0.33×10^{-12}
α Ser	K3III	2.66	(0.14×10^{-15})	(0.19×10^{-14})	0.78×10^{-14}	0.20×10^{-13}
α Sco	Me+A	0.92	(0.70×10^{-14})	(0.85×10^{-13})	0.32×10^{-12}	0.79×10^{-12}
α Peg	F5Ib	1.80	0.17×10^{-15}	0.94×10^{-15}	0.45×10^{-14}	0.14×10^{-13}
χ Peg	M2II	4.80	. . . .	(0.85×10^{-15})	0.44×10^{-14}	0.12×10^{-13}
ε Peg	K2Ib	2.40	(0.43×10^{-15})	(0.57×10^{-14})	0.22×10^{-13}	0.54×10^{-13}
γ Ori	B2III	1.64	0.39×10^{-16}	0.31×10^{-15}	0.14×10^{-14}	0.38×10^{-14}
α Ori	M2II-III	0.10	0.75×10^{-14}	0.15×10^{-12}	0.58×10^{-12}	0.13×10^{-11}
β Oph	K2III	2.78	(0.13×10^{-15})	(0.17×10^{-14})	0.73×10^{-14}	0.19×10^{-13}
α Lep	F0Ib	2.56	. . . .	(0.36×10^{-15})	0.14×10^{-14}	0.40×10^{-14}
γ Leo	K0III	1.95	(0.29×10^{-15})	(0.40×10^{-14})	0.16×10^{-13}	0.44×10^{-13}
Lal 21185	M2V	7.47	(0.48×10^{-17})	(0.77×10^{-16})	0.31×10^{-15}	0.75×10^{-15}
α Hya	K4III	1.97	. . . .	. . . .	0.34×10^{-13}	0.86×10^{-13}
30 g Her	M6III	5.00	. . . .	(0.26×10^{-13})	0.89×10^{-13}	0.17×10^{-12}
ζ Her	G0IV	2.82	. . . .	(0.59×10^{-15})	0.26×10^{-13}	0.71×10^{-14}
γ Gem	A0IV	1.95	(0.19×10^{-16})	(0.28×10^{-15})	0.14×10^{-14}	0.51×10^{-14}
β Gem	K0III	1.15	0.63×10^{-15}	0.63×10^{-14}	0.26×10^{-13}	0.66×10^{-13}
α Gem	A1V	1.59	. . . .	. . . .	. . . .	0.59×10^{-14}
ε Eri	K2V	3.73	. . . .	(0.45×10^{-15})	0.19×10^{-14}	0.50×10^{-14}
σ Dra	K0V	4.68	. . . .	(0.15×10^{-15})	0.64×10^{-15}	0.17×10^{-14}
γ Dra	K5III	2.22	0.57×10^{-15}	0.78×10^{-14}	0.31×10^{-13}	0.75×10^{-13}
61 Cyg B	K7V	6.02	(0.14×10^{-16})	(0.17×10^{-15})	0.72×10^{-15}	0.19×10^{-14}
ε Cyg	K0III	2.45	(0.12×10^{-15})	(0.16×10^{-14})	0.69×10^{-14}	0.19×10^{-13}
β Cyg	gK1p	3.07	. . . .	(0.17×10^{-14})	0.70×10^{-14}	0.19×10^{-13}
α Cyg	A2Ia	1.26	(0.48×10^{-16})	0.72×10^{-15}	0.35×10^{-14}	0.10×10^{-13}
ε Cr B	K3III	4.15	(0.49×10^{-16})	(0.64×10^{-15})	0.27×10^{-14}	0.67×10^{-14}
β Com	G0V	4.30	. . . .	(0.16×10^{-15})	0.73×10^{-15}	0.20×10^{-14}
δ Cep	cF4	3.70	. . . .	. . . .	0.63×10^{-15}	0.22×10^{-14}
θ Cen	K0III-IV	1.98	. . . .	(0.27×10^{-14})	0.11×10^{-13}	0.30×10^{-13}
β Cas	F2IV	2.25	(0.25×10^{-16})	(0.44×10^{-15})	0.22×10^{-14}	0.65×10^{-14}
α CMa	A1V	−1.47	0.59×10^{-15}	0.65×10^{-14}	0.30×10^{-13}	0.84×10^{-13}
β Aur	A2IV	+1.90	. . . .	. . . .	(0.17×10^{-14})	0.44×10^{-14}
α Eri	B9V	0.06	. . . .	(0.44×10^{-15})	(0.26×10^{-14})	0.82×10^{-14}
α Leo	B7V	1.33	0.41×10^{-16}	0.35×10^{-15}	0.16×10^{-14}	0.50×10^{-14}
α Aql	A7IV	0.75	. . . .	. . . .	0.72×10^{-14}	0.19×10^{-13}
α And	A0p	2.13	. . . .	. . . .	. . . .	0.30×10^{-14}
β Ori	B8Ia	0.08	(0.98×10^{-16})	0.14×10^{-14}	0.67×10^{-14}	0.20×10^{-14}
α Lyr	A0V	0.03	0.11×10^{-15}	0.17×10^{-14}	0.79×10^{-14}	0.23×10^{-13}
ζ Pup	O5	2.25	. . . .	. . . .	0.37×10^{-15}	0.13×10^{-14}
Sun	G2V	−26.73	. . . .	0.41×10^{-03}	0.18×10^{-02}	0.49×10^{-02}
δ And	K3III	+3.25	. . . .	(0.12×10^{-14})	0.53×10^{-14}	0.14×10^{-13}
ζ Cep	K1Ib	3.36	. . . .	(0.17×10^{-14})	0.69×10^{-14}	0.22×10^{-13}
δ Sge	M2	3.80	. . . .	(0.18×10^{-14})	0.12×10^{-13}	0.43×10^{-13}
μ Cep	M2Ia	3.99	0.38×10^{-15}	0.11×10^{-13}	0.56×10^{-13}	0.13×10^{-12}
18 Cep	gM5	5.5	. . . .	(0.12×10^{-14})	0.80×10^{-14}	0.22×10^{-13}
α Her	M5II	3.04	0.39×10^{-14}	0.50×10^{-13}	0.21×10^{-12}	0.64×10^{-12}
η Gem	M3III	3.10	. . . .	0.46×10^{-14}	0.24×10^{-13}	0.66×10^{-13}
μ Gem	M3III	2.91	. . . .	(0.81×10^{-14})	0.47×10^{-13}	0.14×10^{-12}
2 Cen	cM6	4.20	. . . .	0.14×10^{-13}	0.51×10^{-13}	0.11×10^{-12}
β Peg	M2II	2.60	. . . .	0.14×10^{-13}	0.76×10^{-13}	0.19×10^{-12}
δ Lyr	M4	4.50	0.66×10^{-15}	0.80×10^{-14}	0.33×10^{-13}	0.77×10^{-13}
α Cet	M2II	2.52	0.77×10^{-15}	0.99×10^{-14}	0.39×10^{-13}	0.97×10^{-13}
Barnard's	M5V	9.54	. . . .	. . . .	(0.23×10^{-15})	0.45×10^{-15}
HR 5299	gM4	5.13	. . . .	(0.31×10^{-14})	0.12×10^{-13}	(0.32×10^{-13})
HR 483	G2V	4.94	. . . .	(0.89×10^{-16})	(0.39×10^{-15})	0.11×10^{-14}
γ Vir	F0V	2.74	. . . .	(0.30×10^{-15})	0.13×10^{-14}	0.38×10^{-14}
α Vir	B2V	0.98	. . . .	0.34×10^{-15}	0.15×10^{-14}	0.49×10^{-14}
μ UMa	M0III	3.00	0.47×10^{-15}	0.61×10^{-14}	0.23×10^{-13}	0.57×10^{-13}
α UMa	K0IV	1.79	0.36×10^{-15}	0.40×10^{-14}	0.17×10^{-13}	0.44×10^{-13}
δ Tau	K0III	3.73	(0.76×10^{-16})	(0.61×10^{-15})	0.25×10^{-14}	0.61×10^{-14}
α Cas	K0II	2.47	. . . .	0.29×10^{-14}	(0.12×10^{-13})	0.30×10^{-13}
α CMi	F5IV	0.34	0.41×10^{-15}	0.36×10^{-14}	0.16×10^{-13}	0.46×10^{-13}
β Cnc	K4III	3.53	(0.15×10^{-15})	(0.19×10^{-14})	0.76×10^{-14}	0.23×10^{-13}
η Boo	G0IV	2.70	. . . .	0.49×10^{-15}	0.22×10^{-14}	0.63×10^{-14}
α Boo	K2IIIp	−0.06	. . . .	0.42×10^{-13}	0.17×10^{-12}	0.44×10^{-12}
i Aur	K3II	+2.70	0.68×10^{-15}	0.50×10^{-14}	0.20×10^{-13}	0.48×10^{-13}
α Aur	G8III+F	0.09	0.11×10^{-14}	0.11×10^{-13}	0.49×10^{-13}	0.13×10^{-12}
α Ari	K2III	1.99	(0.26×10^{-15})	(0.35×10^{-14})	0.15×10^{-13}	0.38×10^{-13}
γ Aql	K3III	2.64	(0.34×10^{-15})	(0.45×10^{-14})	0.18×10^{-13}	0.46×10^{-13}
γ And	K2III	2.30	(0.32×10^{-15})	(0.44×10^{-14})	0.18×10^{-13}	0.45×10^{-13}
β And	MoIII	2.06	(0.11×10^{-14})	(0.13×10^{-13})	0.54×10^{-13}	0.12×10^{-12}

† For spectral type classification symbols see Valley (1965), secs. 21.2.8 and 21.4.1.
‡ Weighing factor: 1 is poor, 2 is fair, 3 is good, and 4 is suitable as a tentative standard.

and the Total Irradiance E in the Infrared Region (Valley, 1965)

Wavelength, μm					E	
2.00	1.67	1.25	1.0	0.566	W/cm^2	w‡
0.62×10^{-12}	0.98×10^{-12}	0.17×10^{-11}	0.22×10^{-11}	0.19×10^{-11}	0.32×10^{-11}	4
0.41×10^{-13}	0.69×10^{-13}	0.15×10^{-12}	0.21×10^{-12}	0.33×10^{-12}	0.31×10^{-12}	3
0.14×10^{-11}	0.24×10^{-11}	0.39×10^{-11}	0.50×10^{-11}	0.16×10^{-11}	0.76×10^{-11}	2
0.32×10^{-13}	0.61×10^{-13}	0.15×10^{-12}	0.26×10^{-12}	0.73×10^{-12}	0.44×10^{-12}	3
0.28×10^{-13}	0.41×10^{-13}	0.67×10^{-13}	0.76×10^{-13}	0.46×10^{-13}	0.11×10^{-12}	1
0.11×10^{-12}	0.17×10^{-12}	0.31×10^{-12}	0.40×10^{-12}	0.42×10^{-12}	0.55×10^{-12}	3
0.84×10^{-14}	0.16×10^{-13}	0.46×10^{-13}	0.10×10^{-12}	0.84×10^{-12}	0.36×10^{-11}	2
0.23×10^{-11}	0.35×10^{-11}	0.55×10^{-11}	0.66×10^{-11}	0.35×10^{-11}	0.10×10^{-10}	2
0.37×10^{-13}	0.63×10^{-13}	0.13×10^{-12}	0.19×10^{-12}	0.29×10^{-12}	0.27×10^{-12}	3
0.89×10^{-14}	0.17×10^{-13}	0.43×10^{-13}	0.89×10^{-13}	0.36×10^{-12}	0.25×10^{-12}	2
0.85×10^{-13}	0.15×10^{-12}	(0.30×10^{-12})	(0.45×10^{-12})	0.63×10^{-12}	0.60×10^{-12}	2
0.15×10^{-14}	0.24×10^{-14}	0.43×10^{-14}	0.63×10^{-14}	0.39×10^{-14}	0.80×10^{-14}	2
0.17×10^{-12}	0.26×10^{-12}	0.48×10^{-12}	0.66×10^{-12}	0.62×10^{-12}	0.87×10^{-12}	3
0.26×10^{-12}	0.36×10^{-12}	0.45×10^{-12}	0.41×10^{-12}	0.38×10^{-13}	0.85×10^{-12}	2
0.15×10^{-13}	0.27×10^{-13}	0.62×10^{-13}	0.11×10^{-12}	0.28×10^{-12}	0.20×10^{-12}	3
0.93×10^{-14}	0.18×10^{-13}	0.51×10^{-12}	0.11×10^{-12}	0.63×10^{-12}	0.53×10^{-12}	3
0.14×10^{-12}	0.23×10^{-12}	0.46×10^{-12}	0.71×10^{-12}	0.13×10^{-11}	0.11×10^{-11}	3
0.13×10^{-13}	0.25×10^{-13}	0.72×10^{-13}	0.17×10^{-12}	0.88×10^{-12}	0.73×10^{-12}	3
0.10×10^{-13}	0.18×10^{-13}	0.35×10^{-13}	0.57×10^{-13}	0.12×10^{-12}	0.92×10^{-13}	2
0.35×10^{-14}	0.61×10^{-14}	0.14×10^{-13}	0.23×10^{-13}	0.51×10^{-13}	0.37×10^{-13}	3
0.14×10^{-12}	0.23×10^{-12}	0.41×10^{-12}	0.54×10^{-12}	0.49×10^{-12}	0.78×10^{-12}	3
0.37×10^{-14}	0.61×10^{-14}	0.10×10^{-13}	0.15×10^{-13}	0.15×10^{-13}	0.23×10^{-13}	3
0.39×10^{-13}	0.66×10^{-13}	0.14×10^{-12}	0.22×10^{-12}	0.40×10^{-12}	0.35×10^{-12}	4
0.42×10^{-13}	0.66×10^{-13}	0.13×10^{-12}	0.19×10^{-12}	0.23×10^{-12}	0.22×10^{-12}	3
0.24×10^{-13}	0.46×10^{-13}	0.12×10^{-12}	0.26×10^{-12}	0.12×10^{-11}	0.94×10^{-12}	3
0.13×10^{-13}	0.21×10^{-13}	(0.44×10^{-13})	(0.67×10^{-13})	0.83×10^{-13}	0.91×10^{-13}	1
0.41×10^{-14}	0.73×10^{-14}	0.17×10^{-13}	0.29×10^{-13}	0.73×10^{-13}	0.53×10^{-13}	3
0.47×10^{-14}	0.87×10^{-14}	0.21×10^{-13}	0.38×10^{-13}	0.13×10^{-12}	0.10×10^{-12}	1
0.60×10^{-13}	0.10×10^{-12}	0.22×10^{-12}	0.35×10^{-12}	0.61×10^{-12}	0.53×10^{-12}	1
0.15×10^{-13}	0.28×10^{-13}	0.67×10^{-13}	0.12×10^{-12}	0.47×10^{-12}	0.34×10^{-12}	3
0.19×10^{-12}	0.39×10^{-12}	0.10×10^{-11}	0.22×10^{-11}	0.15×10^{-10}	0.14×10^{-10}	3
0.11×10^{-13}	0.20×10^{-13}	0.50×10^{-13}	0.11×10^{-12}	0.66×10^{-12}	0.56×10^{-12}	3
0.22×10^{-13}	0.56×10^{-13}	0.16×10^{-12}	0.94×10^{-12}	0.36×10^{-11}	0.81×10^{-11}	1
0.12×10^{-13}	0.24×10^{-13}	0.74×10^{-13}	0.16×10^{-12}	0.11×10^{-11}	0.14×10^{-11}	3
0.46×10^{-13}	0.84×10^{-13}	0.20×10^{-12}	0.41×10^{-12}	0.19×10^{-11}	0.14×10^{-11}	3
0.60×10^{-14}	0.13×10^{-13}	0.35×10^{-13}	0.77×10^{-13}	0.54×10^{-12}	0.68×10^{-12}	3
0.48×10^{-13}	0.92×10^{-13}	0.25×10^{-12}	0.62×10^{-12}	0.35×10^{-11}	0.45×10^{-11}	3
0.53×10^{-13}	0.11×10^{-12}	0.29×10^{-12}	0.64×10^{-12}	0.37×10^{-11}	0.31×10^{-11}	4
0.32×10^{-14}	0.65×10^{-14}	0.22×10^{-13}	0.52×10^{-13}	0.48×10^{-12}	0.13×10^{-10}	3
0.10×10^{-01}	0.19×10^{-01}	0.44×10^{-01}	0.71×10^{-01}	0.19×10^{-00}	0.13×10^{-00}	4
0.28×10^{-13}	0.51×10^{-13}	0.12×10^{-12}	0.18×10^{-12}	0.19×10^{-12}	0.22×10^{-12}	1
0.45×10^{-13}	0.75×10^{-13}	0.14×10^{-12}	0.21×10^{-12}	0.17×10^{-12}	0.25×10^{-12}	1
0.75×10^{-13}	0.11×10^{-12}	0.18×10^{-12}	0.22×10^{-12}	0.11×10^{-12}	0.37×10^{-12}	1
0.21×10^{-12}	0.31×10^{-12}	0.50×10^{-12}	0.61×10^{-12}	0.96×10^{-13}	0.64×10^{-12}	2
0.44×10^{-13}	0.67×10^{-13}	0.11×10^{-12}	0.14×10^{-12}	0.24×10^{-13}	. . . .	1
0.12×10^{-11}	(0.19×10^{-12})	(0.23×10^{-11})	0.23×10^{-11}	0.23×10^{-12}	0.35×10^{-11}	2
0.13×10^{-12}	0.23×10^{-12}	(0.39×10^{-12})	(0.57×10^{-12})	0.22×10^{-12}	0.70×10^{-12}	2
0.26×10^{-12}	(0.42×10^{-12})	(0.80×10^{-12})	0.10×10^{-11}	0.26×10^{-12}	0.87×10^{-12}	2
(0.17×10^{-12})	(0.24×10^{-12})	(0.30×10^{-12})	(0.29×10^{-12})	0.80×10^{-13}	0.63×10^{-12}	1
0.35×10^{-12}	0.50×10^{-12}	0.82×10^{-12}	0.97×10^{-12}	0.35×10^{-12}	0.13×10^{-11}	3
0.16×10^{-12}	0.23×10^{-12}	0.37×10^{-12}	0.39×10^{-12}	0.60×10^{-13}	0.52×10^{-12}	2
0.18×10^{-12}	0.27×10^{-12}	0.47×10^{-12}	0.62×10^{-12}	0.37×10^{-12}	0.90×10^{-12}	3
0.83×10^{-15}	0.11×10^{-14}	0.17×10^{-14}	0.20×10^{-14}	0.58×10^{-15}	0.34×10^{-14}	1
0.60×10^{-13}	(0.77×10^{-13})	(0.11×10^{-12})	(0.10×10^{-12})	0.34×10^{-13}	0.17×10^{-12}	1
0.22×10^{-14}	0.40×10^{-14}	0.92×10^{-14}	0.16×10^{-13}	0.40×10^{-13}	0.30×10^{-13}	3
0.89×10^{-14}	0.17×10^{-13}	0.40×10^{-13}	0.76×10^{-13}	0.31×10^{-12}	0.21×10^{-12}	2
0.12×10^{-13}	0.23×10^{-13}	0.73×10^{-13}	0.16×10^{-12}	0.15×10^{-11}	0.14×10^{-10}	3
0.10×10^{-12}	0.16×10^{-12}	0.27×10^{-12}	0.34×10^{-12}	0.24×10^{-12}	0.45×10^{-12}	3
0.88×10^{-13}	0.15×10^{-12}	0.30×10^{-12}	0.45×10^{-12}	0.73×10^{-12}	0.71×10^{-12}	3
0.12×10^{-13}	0.21×10^{-13}	0.45×10^{-13}	0.70×10^{-13}	0.12×10^{-12}	0.11×10^{-12}	2
0.60×10^{-13}	0.10×10^{-12}	0.20×10^{-12}	0.30×10^{-12}	0.39×10^{-12}	0.39×10^{-12}	1
0.10×10^{-12}	0.19×10^{-12}	0.45×10^{-12}	0.77×10^{-12}	0.28×10^{-11}	0.19×10^{-11}	3
0.43×10^{-13}	0.61×10^{-13}	0.99×10^{-13}	0.13×10^{-12}	0.15×10^{-12}	0.20×10^{-12}	2
0.13×10^{-13}	0.24×10^{-13}	0.58×10^{-13}	0.10×10^{-12}	0.32×10^{-12}	0.22×10^{-12}	2
0.85×10^{-12}	0.14×10^{-11}	0.26×10^{-11}	0.32×10^{-11}	0.40×10^{-11}	0.45×10^{-12}	3
0.93×10^{-13}	0.15×10^{-12}	0.26×10^{-12}	0.36×10^{-12}	0.32×10^{-12}	0.51×10^{-12}	3
0.26×10^{-12}	0.47×10^{-12}	0.10×10^{-11}	0.16×10^{-11}	0.35×10^{-11}	0.26×10^{-11}	4
0.78×10^{-13}	0.13×10^{-12}	0.26×10^{-12}	0.40×10^{-12}	0.61×10^{-12}	0.59×10^{-12}	4
0.88×10^{-13}	0.14×10^{-12}	0.25×10^{-12}	0.33×10^{-12}	0.33×10^{-12}	0.45×10^{-12}	3
0.87×10^{-13}	0.14×10^{-12}	0.27×10^{-12}	0.40×10^{-12}	0.46×10^{-12}	0.56×10^{-12}	3
0.23×10^{-12}	0.35×10^{-12}	0.60×10^{-12}	0.81×10^{-12}	0.57×10^{-12}	0.11×10^{-11}	3

TABLE 27 Atomic Emission Lines in Airglow (Valley, 1965)

Wavelength, Å	Element	Transition	Found in
1215	H Ly-α	$2p\ ^2P^0_{3/2}-1s\ ^2S_{1/2}$	Night glow
3933.7 3968.7	CaII	$4p\ ^2P^0_{3/2,1/2}-4s\ ^2S_{1/2}$	Twilight glow
5199.92 5202.29	[NI]	$2p^3\ ^2D_{3/2,5/2}-2p^3\ ^4S_{3/2}$	Twilight and night glow
5577.345	[OI]	$2p^4\ ^1S_0-2p^4\ ^1D_2$	Night and twilight glow
5889.953 5895.923	NaI	$3p\ ^2P^0_{3/2,1/2}-3s\ ^2S_{1/2}$	Night, twilight, and day glow
6300.308 6363.790	[OI]	$2p^4\ ^1D_2-2p^4\ ^3P_{2,1}$	Night, twilight, and day glow
6562.817	H-α	$3p\ ^2P^0_{3/2}-2s\ ^2S_{1/2}$ $3s\ ^2S_{1/2}-2p\ ^2P^0_{3/2}$ $3d\ ^2D_{5/2}-2p\ ^2P^0_{3/2}$	Night glow
6707.912 6707.761	LiI	$2p\ ^2P^0_{1/2,3/2}-2s\ ^2S_{1/2}$	Twilight glow
7664.907 7698.979	KI	$4p\ ^2P^0_{3/2,1/2}-4s\ ^2S_{1/2}$	Twilight glow
8446.26 8446.38 8446.76	OI	$3p\ ^3P_{0,2,1}-3s\ ^3S_1$	Twilight glow
10829.081 10830.250 10830.341	HeI	$2p\ ^3P^0_{0,1,2}-2s\ ^3S_1$	Twilight glow

TABLE 28 Molecular Bands in Airglow (Valley, 1965)

Molecule	System	Transition	Found in
O_2	Herzberg	$A^3\ \Sigma_u^+ \rightarrow X^3\ \Sigma_g^-$	Night glow
O_2	Atmospheric Kaplan-Meinel	$b^1\ \Sigma_g^+ \rightarrow X^3\ \Sigma_g^-$	Night glow
O_2	Infrared atmospheric	$a^1\ \Delta_g \rightarrow X^3\ \Sigma_g^-$	Twilight glow
N_2^+	First negative	$B^2\ \Sigma_u^+ \rightarrow X^2\ \Sigma_g^+$	Night and twilight glow
OH	Meinel	$X^2\ \pi$	Night glow

by a lower decrease in radiance, not inconsistent (as reported by Murray et al., in a progress report) with the fact that the major source of radiance in this region is HNO_3.

When the emission from the underside of a cloud is measured, two distinctly different sources of radiation become apparent. Figure 59 shows, with the help of an analysis of Bell et al., that the +10°C radiation from the atmosphere surrounding the instrument is evident in regions where the cloud radiation cannot penetrate the absorbing gaseous components of the environment. In the clear, transmitting regions of the atmosphere the −10°C radiation of the cloud bottom is evident.

Figure 60, the spectral radiance from a clear sky, shows the elevation-angle dependence of the scattered radiation in the visible region and the effect of path length on emitted radiation in the infrared.

Terrain Radiation

40. Whereas many studies have resulted in a large quantity of statistical data on terrain radiation, only one has been used for a spectral description of background radiation originating from the earth's surface. Most of this work is reported in an organizational report (Frederickson et al., 1957), cited frequently by Wolfe (1965) which drew on the report for data. What is covered in this section is representative of certain specific backgrounds and therefore limited in scope, but because of the nature of terrain, these data are not unlike what can be expected of a large variety of terrain radiation.

Figure 61 shows three effects evident in the results of spectral measurements of the materials observed, these being emissivity and temperature of the substances and the transmittance of the atmosphere. Because these are daylight measurements, the portion below 3 μm is due to reflected sunlight and that above 3 μm to thermally emitted radiation. Since these measurements were taken at relatively close range, the effect of the atmosphere (in both absorption and emission) is much less than in longer atmospheric paths (See data in Pars. 35 to 39.)

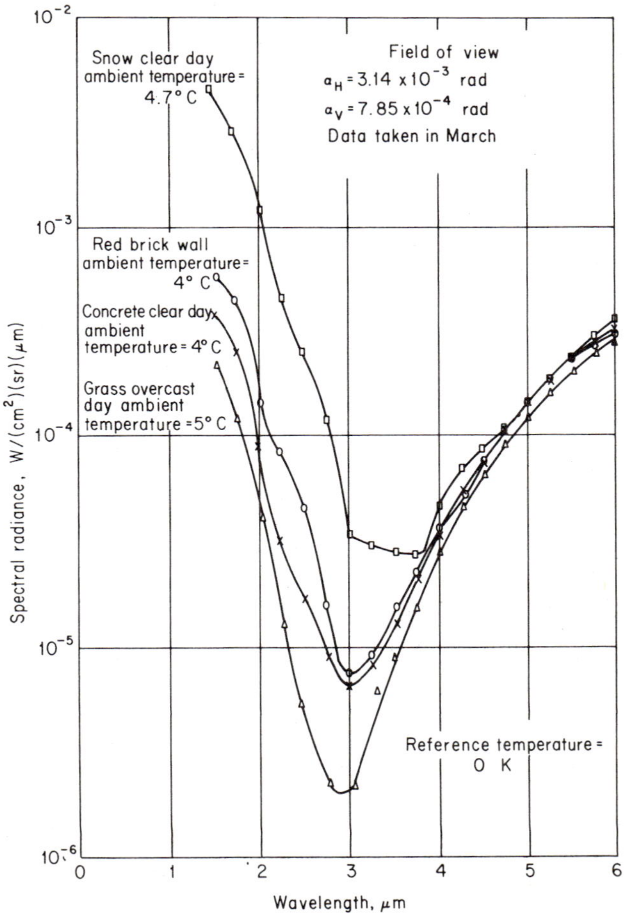

Fig. 61 Daytime spectral radiance of miscellaneous targets. [*Wolfe* (1965).]

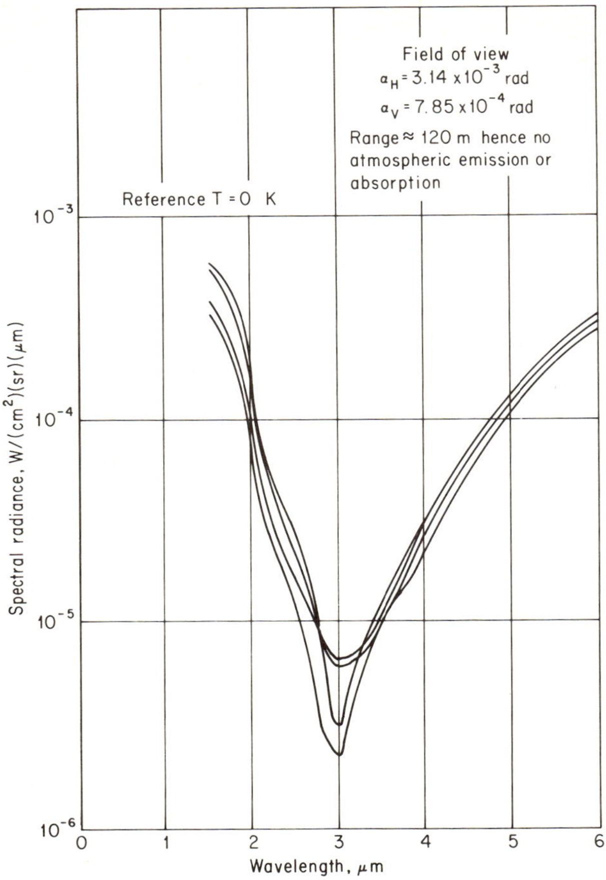

Fig. 62 Spectral radiance of concrete, winter day, clear. [*Wolfe* (1965).]

Other curves are presented here without comment. Essential points peculiar to a given curve are explained in the legends of Figs. 62 to 68 (Figs. 63–68 appear on the following pages).

REFERENCES

Air Research and Development Command (1960): "Handbook of Geophysics," rev. ed., Macmillan, New York.

Allen, C. W. (1955): "Astrophysical Quantities," Athlone Press, London.

——— (1963): "Astrophysical Quantities," 2d ed., Athlone Press, London.

Bedford, R. E. (1960): A Low Temperature Standard of Total Radiation, *Can. J. Phys.*, vol. 38, p. 1256.

——— and C. K. Ma (1968): International Comparison of Measurements of Irradiance, *Metrologia*, vol. 4, no. 3.

Bell, E. E., et al. (1960): Spectral Radiance of Sky and Terrain at Wavelengths between 1 and 20 Microns, II, *J. Opt. Soc. Am.*, vol. 50, no. 12, p. 1313, December.

Cann, M. W. P. (1967): Light Sources for Remote Sensing Systems, NASA CR-854, August.

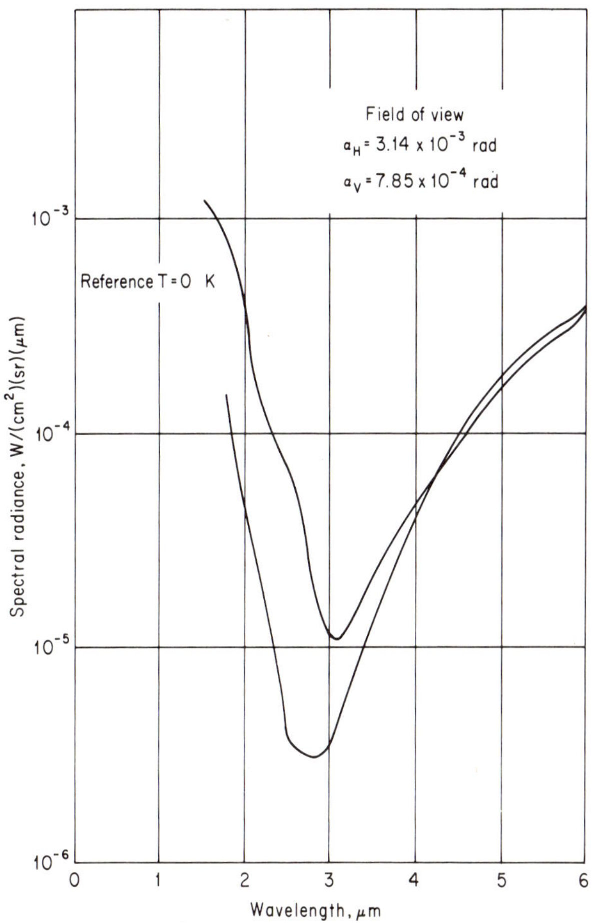

Fig. 63 Spectral radiance of concrete, summer day, overcast. [*Wolfe* (1965).]

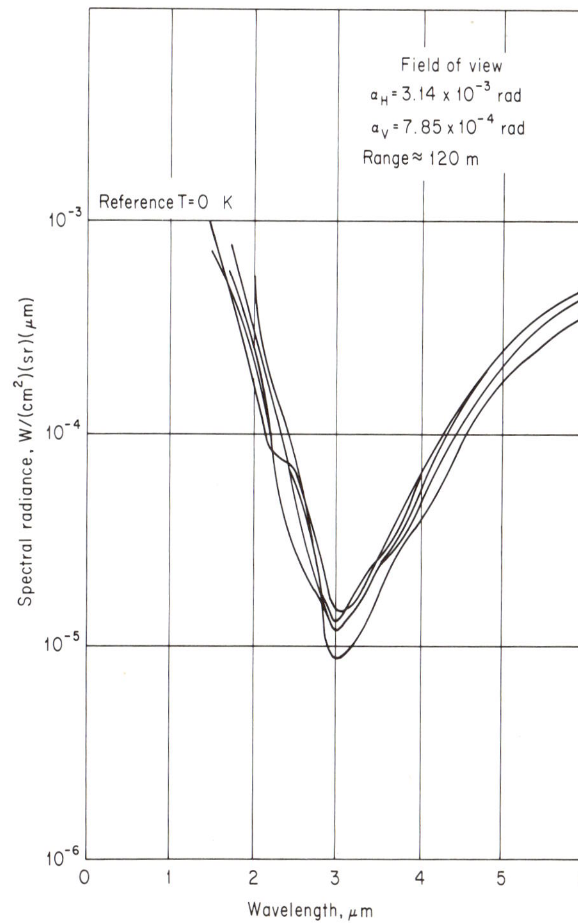

Fig. 64 Spectral radiance of damp concrete. [*Wolfe* (1965).]

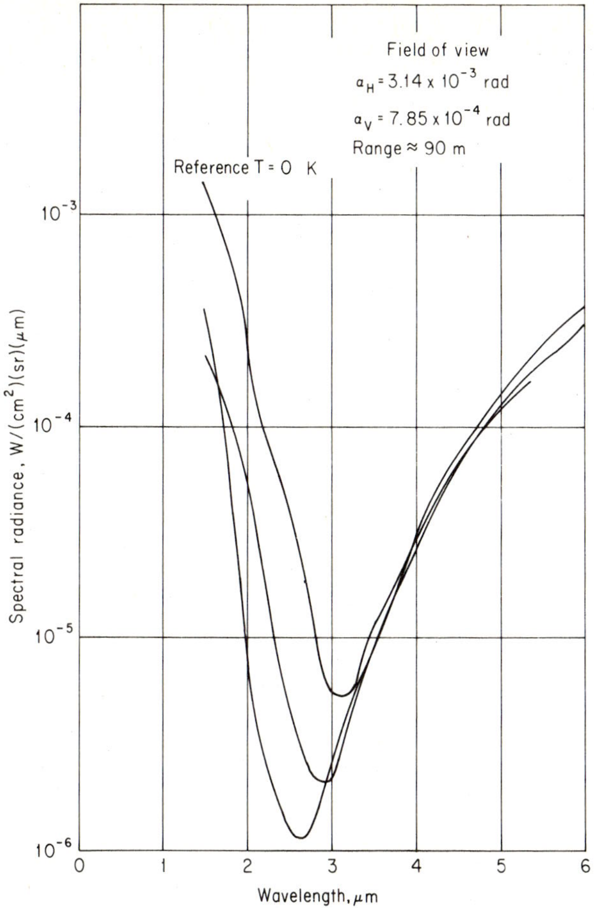

Fig. 65 Spectral radiance of grass, winter day, overcast. [*Wolfe* (1965).]

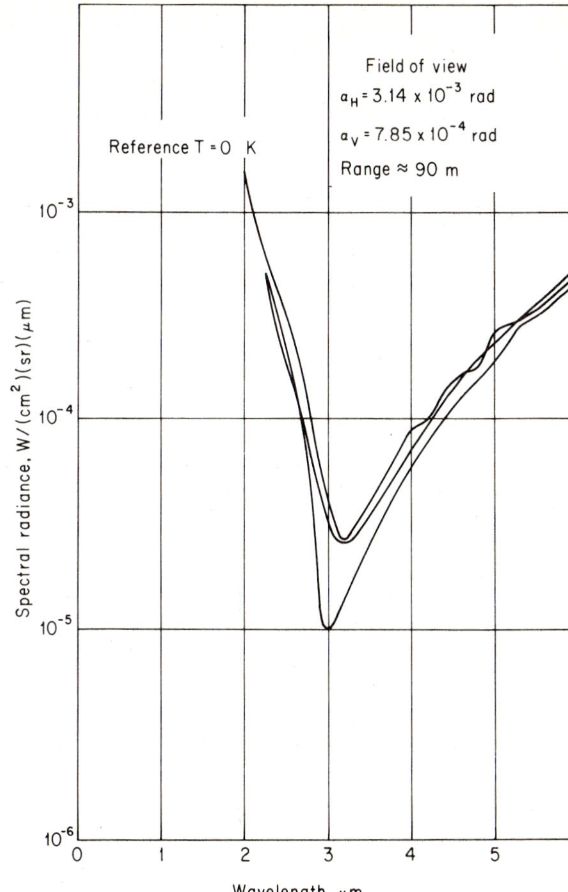

Fig. 66 Spectral radiance of grass, summer day, clear. [*Wolfe* (1965).]

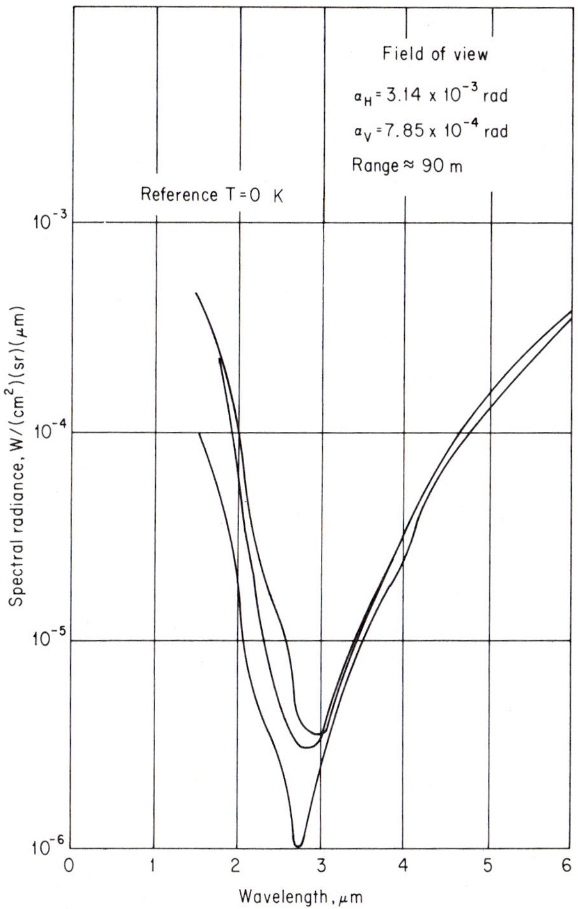

Fig. 67 Spectral radiance of grass, summer day, overcast. [*Wolfe* (1965.]

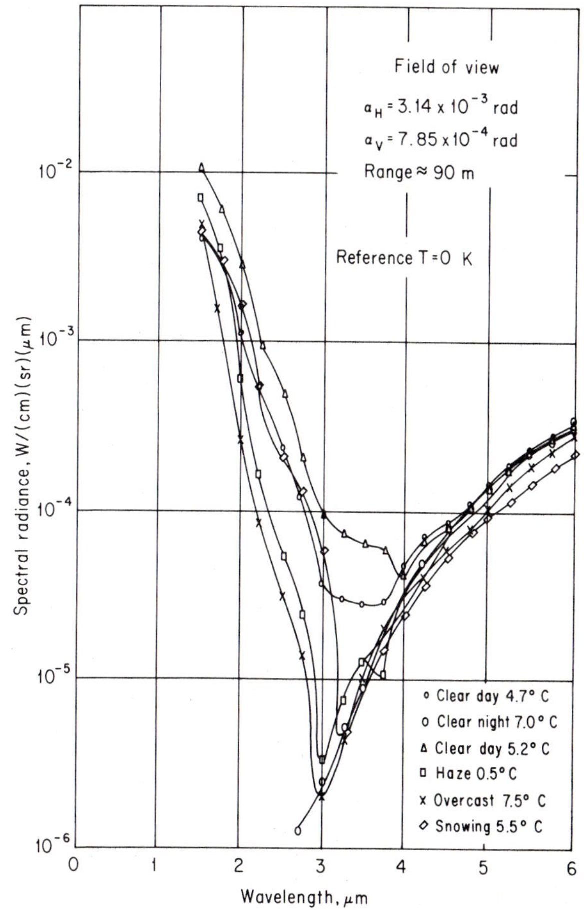

Fig. 68 Spectral radiance of snow cover under various weather conditions at various ambient temperatures. [*Wolfe* (1965).]

Carlson, F. E., and C. N. Clark (1965): Light Sources for Optical Devices, chap. 2 in R. Kingslake (ed.), "Applied Optics and Optical Engineering," vol. 1, Academic, New York.

Frederickson, W. R., et al. (1958): Infrared Spectral Emissivity of Terrain, *Syracuse Univ. Res. Inst., Syracuse, N.Y., Final Rep.* AD155552.

——— et al. (1957): Infrared Spectral Emissivity of Terrain, *Syracuse Univ. Res. Inst., Syracuse, N.Y., Int. Dev. Rep.* 2, AF33(616)-5034, Aug. 1.

Gillham, E. J. (1959): The Measurement of Optical Radiation, *Research,* vol. 12, pp. 404–411.

——— (1961): Radiometric Standards and Measurements, *Nat. Phys. Lab., Lond., Notes Appl. Sci.* 23.

——— (1962): Recent Investigations in Absolute Radiometry, *Proc. R. Soc.,* ser. A., vol. 269, pp. 240–276.

Humphreys, C. J., and E. Paul, Jr. (1970): Interferometric Observations in the Spectra of ^{86}Kr, *J. Opt. Soc. Am.,* vol. 60, p. 200.

Johnson, F. S. (1954): *J. Meteorol.,* vol. 11, p. 431.

Jones, O. C., and J. S. Preston (1969): "Photometric Standards and the Unit of Light," 2d ed., *Natl. Phys. Lab., Lond., Notes Appl. Sci.* 24.

Karoli, A. R., J. R. Hickey, and R. E. Nelson (1967): An Absolute Calibration Source for Laboratory and Satellite Infrared Spectrometers, *Appl. Opt.,* vol. 6, p. 1183.

Kaufman, J. E. (ed.) (1972): "IES Lighting Handbook," 5th ed., Illuminating Engineering Society, New York.

Koller, L. R. (1965): "Ultraviolet Radiation," 2d ed., Wiley, New York.

Kostkowski, H. J., and R. D. Lee (1962): Theory and Methods of Optical Pyrometry, *Natl. Bur. Stand. Monogr.* 41, March.

Lee, R. D. (1969): Construction and Operation of a Simple High Precision Copper Point Blackbody and Furnace, *Natl. Bur. Stand. Tech. Note* 483, May.

Meggers, W. F. (1960): *J. Opt. Soc. Am.,* vol. 50, p. 405.

——— and F. O. Westfall (1950): Lamps and Wave-lengths of Mercury 198, *J. Res. Natl. Bur. Stand.,* vol. 44, pp. 447–455.

Morris, J. C. (1961): Comments on the Measurement of Emittance of the Globar Radiation Source, *J. Opt. Soc. Am.,* vol. 51, p. 798, July.

Murcray, D. G., A. Goldman, A. Csoeke-Poeckh, et al. (1973): Nitric Acid Distribution in the Stratosphere, *J. Geophys. Res.,* vol. 78, no. 30, p. 7033.

———, F. H. Murcray, W. T. Williams, T. G. Kyle, and A. Goldman (1969): Variation of the Infrared Solar Spectrum between 700 cm^{-1} and 2240 cm^{-1} with Altitude, *Appl. Opt.,* vol. 8, no. 12, p. 2519, December.

Natl. Bur. Stand. Bull., vol. 11, no. 227, p. 87, 1914.

Natl. Bur. Stand. J. Res., vol. 11, no. 578, p. 79, 1933.

Natl. Bur. Stand. J. Res., vol. 53, p. 211, 1954.

Natl. Bur. Stand. Tech. News Bull., vol. 54, no. 3, p. 55, March 1970.

Nivert, L. J., M. Mulbrandon, and J. C. Richmond (1967): Primary Reference Standards of Spectral Radiance in the Infrared, abstr., *J. Opt. Soc. Am.,* vol. 57, p. 1417.

Noel, E. B. (1941): *Illum. Eng.,* vol. 36, p. 243.

Null, M. R., and W. W. Lozier (1962): Carbon Arc as a Radiation Standard, *J. Opt. Soc. Am.,* vol. 52, no. 10, pp. 1156–1162, October.

Paschen, F. (1916): *Ann. Phys.,* vol. 50, p. 901.

Pfund, A. H. (1936): The Electric Welsbach Lamp, *J. Opt. Soc. Am.,* vol. 26, p. 439, December.

Ramsay, R. C. (1962): Spectral Irradiance from Stars and Planets, above the Atmosphere, from 0.1 to 100.0 Microns, *Appl. Opt.,* vol. 1, no. 4, p. 465, July.

Ramsey, W. Y., and J. C. Alishouse (1968): A Comparison of Infrared Sources, *Infrared Phys.,* vol. 8, p. 143.

Silverman, S. (1968): The Emissivity of Globar, *J. Opt. Soc. Am.,* vol. 38, no. 11, p. 989, November.

Stair, Ralph, Russell G. Johnston, and E. W. Halbach (1960): Standard of Spectral Radiance for the Region of 0.25 to 2.6 Microns, *Natl. Bur. Stand. J. Res.,* vol. 64A, no. 4, July–August.

———, William E. Schneider, and John K. Jackson (1963): A New Standard of Spectral Irradiance, *Appl. Opt.,* vol. 2, no. 11, p. 1151, November.

Strong, J. (1938): "Procedures in Experimental Physics," Prentice-Hall, New York.

Studer, F. J., and R. F. Van Beers (1964): Modification of Spectrum of Tungsten Filament Quartz-Iodine Lamps Due to Iodine Vapor, *J. Opt. Soc. Am.,* vol. 54, no. 7, p. 945, July.

Thekaekara, M. P., R. Kruger, and C. H. Duncan (1969): Solar Irradiance Measurements from a Research Aircraft, *Appl. Opt.*, vol. 8, no. 8, p. 1713, August.
Trans. Int. Astron., vol. 11A, p. 97, 1962.
Valley, Shea L. (ed.) (1965): "Handbook of Geophysics and Space Environments," Air Force Cambridge Research Laboratories, Bedford, Mass.
Walsh, J. W. T. (1965): "Photometry," 3d ed., Dover, New York.
Wolfe, William L. (ed.) (1965): "Handbook of Military Infrared Technology," GPO, Washington.
Worden, E. F., R. G. Gutmacher, and J. F. Conway (1963): Use of Electrodeless Discharge Lamps in the Analysis of Atomic Spectra, *Appl. Opt.*, vol. 2, no. 7, pp. 707–713, July.
Worthing, A. G. (1937): Sources of Radiant Energy, chap. 2 in W. E. Forsythe (ed.), "Measurement of Radiant Energy," McGraw-Hill, New York.

Section **4**

Nonimaging Detectors

STEPHEN F. JACOBS

Optical Sciences Center, University of Arizona, Tucson, Arizona

(Numbers refer to paragraphs)

SCOPE

1. The discussion includes only commercially available detectors. It excludes detectors not yet in general use which are based on numerous photointeractions, e.g., thermal variation of length, capacitance, inductance, susceptibility, and polarizability. Many of these can be developed into extremely sensitive radiation detectors, but a discussion of their performance is beyond the scope of this handbook. The spectral range of detectors covered here is similar to that covered by journals of the Optical Society. At present this range extends from x-rays of about 5 nm wavelength to far infrared of submillimeter wavelength.

A brief description of the photosensitive mechanism is presented for each type of detector, but a detailed discussion of physical mechanisms is beyond the scope of this handbook. The usefulness of each detector is briefly summarized and followed by typical detector parameters. Techniques for evaluating detector performance are not covered in this treatment but can be found in references (Wolfe, 1965; IRE, 1962; and RCA, 1963).

The concept of sensitivity is presented, and a measure of ideal performance is derived. Manufacturers' specifications are given, often indicating a range of values for selected or unselected detectors. The closeness of approach to the state of the art depends on technological progress and the economics of supply and demand. A list of manufacturers is included for each detector.

The sensitivity of many detectors has reached the limit set by room-temperature background photon fluctuations (radiation shot noise). For these detectors, limiting sensitivity is enhanced by restricting their spatial and spectral field of view. Further progress will be called for in the future as lower-temperature backgrounds become more common.

Recent advances in amplifier technology have reduced preamplifier noise levels to a few $nV/Hz^{1/2}$ at high impedance. With these improved preamplifiers, many detectors here reported as amplifier-limited may no longer be so limited.

Thermal Detectors

2. Thermal detectors sense the change in temperature produced by absorption of incident radiation. Their spectral response can therefore be as flat as the absorption spectrum of their blackened coating and window† will allow. This makes them useful for spectroscopy and radiometry. These detectors are generally operated at room temperature, where their sensitivity is limited by thermodynamic considerations (Jones, 1949; Bauer, 1943) to 3 pW for 1 s measurement time and 1 mm^2 sensitive area. This limit has been very nearly reached in practice, whereas cooled bolometers have been made to reach the background photon noise limit.

† No windows exist without absorption bands somewhere between the visible and millimeter region. Some useful window materials for the far infrared are diamond, silicon, polyethylene, quartz, and CsI; see Sec. 7.

The response time τ of thermal detectors is generally slower than 1 ms, depending on thermal capacity K and heat loss per second per degree V, through the relation

$$\tau = \frac{K}{V}$$

A short time constant requires a small K. However, for room-temperature operation ultimate sensitivity is limited by the mean spontaneous temperature fluctuation

$$\Delta T = T\sqrt{\frac{k}{K}}$$

where k is Boltzmann's constant. There is thus a trade-off between time constant and ultimate sensitivity.

3. Pneumatic The receiver is a thin blackened membrane placed in a small airtight, gas-filled chamber. Heat absorbed by the foil is transmitted to the gas, causing a pressure rise, which is measurable in various ways (optically, by mirror motion; electrically, by change in capacitance). A commercially available form is the *Golay cell,* which has found wide use in far-infrared spectroscopy. The excellent impedance match to space of the blackened membrane makes high absorption possible over a wider spectrum than is attainable with other detectors. Golay cell sensitivity slightly exceeds that of thermocouples, but its construction is far more fragile.

4. Thermocouple The receiver is a thin blackened flake connected thermally to the junction of two dissimilar metals or semiconductors. Heat absorbed by the flake causes a temperature rise of the junction, and hence a thermoelectric emf is developed which can be measured, for example, with a potentiometer.

Thermocouples are limited in sensitivity by thermal (Johnson-Nyquist) noise but are nevertheless respectably sensitive. Their usefulness lies in the convenience of room-temperature operation and their rugged construction. Thermocouples are widely used in spectroscopy.

5. Bolometer The receiver is a thin blackened slab whose impedance is temperature-dependent. The dc impedance change is generally sensed using a bridge circuit with similar (compensating) material in the series or parallel arm. The ac operation requires only a single bolometer element in series with a load and voltage source. The signal is the change in voltage across the load.

Most bolometers in use today are of the thermistor type. Their sensitivity closely approaches that of the thermocouple for frequencies higher than 25 Hz. Below that there is excess noise. Construction can be very rugged for systems applications. Some extremely sensitive low-temperature bolometers are available commercially.

6. Pyroelectric Ferroelectric material exhibits a residual polarization in the absence of any electric field. Dipole moments, initially aligned by applying an external field, result in a surface charge which is normally slowly neutralized by leakage. This polarization is temperature-dependent (pyroelectric effect), and when a pulse of incident radiation heats a portion of an electroded sample, there is a change in surface charge (open-circuit voltage) which is proportional to the incident radiation power. Speed of response is inherently not limited by thermal inertia but by charge-reorientation time. Electrically, the device behaves like a capacitor, requiring no bias and therefore exhibiting no current noise. Sensitivity is limited either by amplifier noise or by loss-tangent noise.

Quantum Detectors

7. These detectors are photon detectors and respond in proportion to incident photon rates rather than to photon energies (heat). Thus, the spectral response of an ideal photon detector is flat on an incident-photon-rate basis but linearly rising with wavelength on an incident-power D^* basis. For this reason the sensitivity of efficient quantum detectors compares more favorably with that of efficient thermal detectors at long wavelengths. Quantum detectors generally have submicrosecond time constants. Their main disadvantage is the associated cooling required for optimum sensitivity. (These remarks do not apply to photographic detection, which measures cumulative photon numbers.)

8. Photoemissive The radiation is absorbed by a photosensitive surface which usually contains alkali metals (cesium, sodium, or potassium). Incident quanta release photoelectrons, via the photoelectric effect, which are collected by a positively biased anode. This is called a *diode phototube;* it can be made the basis for the multiplier phototube (photomultiplier phototube, or photomultiplier) by addition of a series of biased dynodes which serve as secondary emission multipliers.

In spectral regions where quantum efficiency is high ($\lambda < 550$ nm), the photoemissive detector is very nearly ideal. Sensitivity is high enough to count individual photons. Amplification does not degrade the signal-to-noise ratio. The sensitive area is conveniently large. Photomultiplier signal response time (transit spread time) can be made as short as 0.1 ns. Since the sensitivity in red-sensitive tubes is limited by thermally generated electrons, sensitivity can be improved by cooling.

9. Photoconductive The radiation is absorbed by a photoconductive material either in thin-film or bulk form. Each incident quantum may release an electron-hole pair, thereby increasing the electrical conductivity. The usual readout circuit is the same as that for the ac bolometer.

Photoconductors can have a high quantum efficiency from the visible region out to the far infrared but lack the nearly ideal high amplification of photomultipliers. They are therefore chiefly useful in the spectral region beyond 1 μm, where efficient photoemitters are unavailable. Photoconductors do, however, provide current gain which is equal to the electron-hole recombination time divided by the majority-carrier transit time. This current gain leads to higher responsivity than is possible with (nonavalanching) photovoltaic-detection detectors. For applications where photovoltaic detection would be amplifier-noise-limited, the larger photoconductive responsivity makes it possible to realize greater sensitivity with the photoconductor. In general, lower-temperature operation is associated with longer-wavelength sensitivity in order to suppress noise due to thermally induced transitions between close-lying energy levels. Ideally, photoconductors are limited by generation-recombination noise in the carriers, produced by incident radiation. Response time can be shorter than 1 μs. For broadband detection applications, reference is made to the microwave biasing technique (Sommers and Teutsch, 1964; Sommers and Gatchell, 1966; and Sun and Walsh, 1970).

10. Photovoltaic† The most widely used photovoltaic dectector is the *p-n* junction type, where a strong internal electric field exists across the junction even in the absence of radiation. Radiation quanta incident on the junction of this film or bulk material produce free hole-electron pairs which are separated by the internal electric field, causing a change in voltage across the open-circuited cell or a current to flow in the short-circuited case.

As with the photoconductor, quantum efficiency can be high from the visible to the far infrared. The limiting noise level can ideally be $\sqrt{2}$ times lower than that of the photoconductor, thanks to the absence of recombination noise. Lower temperatures are associated with longer-wavelength operation. Response times can be as short as several nanoseconds, which is much faster than photoconductors can operate without sacrificing sensitivity. The advantages of nearly ideal internal amplification have now become available in avalanche photodiodes sensitive out to 1.55 μm. This internal gain is most important for high-frequency operation, where external load resistance must be kept small and would otherwise introduce limiting thermal noise.

11. Photoelectromagnetic A thin slab of photoconductive material is oriented with radiation incident on a large face and a magnetic field perpendicular to it.

† The use of a photovoltaic detector at other than zero bias is often referred to as its *photoconductive mode* of operation because the circuit then is similar to the standard photoconductor circuit. This terminology is confusing with regard to detection mechanism and will not be used here.

Hole-electron pairs generated by the incident photons diffuse through the material and are separated by the magnetic field, causing a potential difference at opposite ends of the detector.

These detectors require no cooling or biasing electric field but do require a (permanent) magnetic field. Photoelectromagnetic InSb has response out to 7.5 μm, where it is as sensitive as a thermocouple of equal size, and has a response time less than 1 μs. Another competing uncooled detector is InAs, which is far more sensitive out to 3.5 μm. Cooled infrared detectors are one to two orders of magnitude more sensitive.

12. Photographic The receiver is an emulsion containing silver halide crystals. Incident photons are absorbed by the halide ion, which subsequently loses its electron. This electron eventually recombines with a silver ion and reduces it to a neutral silver atom. As more photons are absorbed, this process is repeated until a small but stable cluster of reduced silver atoms is formed within the crystal (latent image). Internal amplification is provided by introduction of an electron donor (photographic developer), which, using the latent image as a catalytic center, reduces to silver all the remaining silver ions within the exposed crystal. The density of reduced crystals is a measure of the total radiation exposure.

The spectral region of sensitivity for photographic detection coincides rather closely with that of the photoemissive detector. For $\lambda > 1.2$ μm there is too little energy in each photon to form a stable latent image. The basic detection process for both detectors operates well for higher and higher frequency radiation. The problem in ultraviolet and x-ray operation is one of eliminating nonessential materials, e.g., the emulsion which absorbs these radiations.

The photographic process is an integrating one in that the output (emulsion density) measures the cumulative effect of all the radiation incident during the exposure time. The efficiency of the photographic process can be very high, but it depends upon photon energy; e.g., in the visible region it takes only 10 to 100 photons to form a stable latent image (developable grain). The photographic process enjoys a large and efficient internal amplification ability (development) wherein the very small energy of the photons' interaction is converted into readily observed macroscopic changes. An extensive discussion of photographic detection is found in Sec. 5.

13. Photoionization The radiation is absorbed by a gas. If the photon energy exceeds the gas-ionization threshold, ion pairs can be produced with very high efficiency. They are collected by means of an applied voltage. Operating in a dc mode, these detectors are known as *ionization* and *gas-gain chambers*. When a pulse mode is used, the detectors are known as *proportional* and *photon* (*Geiger*) *counters*.

Photoionization detectors have a high sensitivity and a low noise level. They may also be quite selective spectrally since the choice of window and gas independently set upper and lower limits on detectable photon energies. Manufacturers' specifications are not discussed for these detectors as applications are still few enough to be treated as individual cases (Samson, 1967).

DEFINITIONS

14. The following definitions will be used:

Background temperature The effective temperature of all radiation sources viewed by the detector exclusive of the signal source.

Blackbody D star, D^*_{BB}, cm·Hz$^{1/2}$/W Similar to $D^*(\lambda)$ or $D^*(T_B,f)$ except that the source is a blackbody whose temperature must be specified.

Blackbody detectivity D_{BB}, W^{-1} A measure of detector sensitivity, defined as $D_{BB} = (\text{NEP}_{BB})^{-1}$.

Blackbody noise-equivalent power NEP$_{BB}$ Same as spectral NEP, except that the source has blackbody spectral character whose temperature must be specified, for example, NEP(500 K, 1, 800) means 500-K blackbody radiation incident, 1 Hz electrical bandwidth, and 800-Hz chopping rate.

Blackbody responsivity R_{BB} Same as spectral responsivity except that the incident signal radiation has a blackbody spectrum (whose temperature must be specified).

Blip detector or blip condition Originally meaning background-limited impurity photoconductor, this term has come to mean performance of any detector where the limiting noise is due to fluctuations in the arrival rate of background photons.

Cutoff wavelength λ_c The wavelength at which the detectivity has degraded to one-half its peak value.

Dark current The output current which flows even when input radiation is absent or negligible. (Note that although this current can be subtracted out for the dc measurements, the shot noise on the dark current can become the limiting noise.)

Detective quantum efficiency The square of the ratio of measured detectivity to the theoretical limit of detectivity.

Detective time constant τ_d $\tau_d = 1/2\pi f_d$ where f_d is the frequency at which D^* drops to 0.707 times its maximum value.

Dewar A container (cryostat) for holding detector coolant.

Equivalent noise input (ENI *or* NEI) A term used by photomultiplier manufacturers meaning nearly the same thing as $NEP_{BB}(287\ K, 1, f)$. The difference is that the peak-to-peak value of square-wave chopped input flux is used, rather than the rms value of the sinusoidally chopped input flux. (See recommendations IRE, 1962.)

Excess noise A term usually referring to noises other than generation-regeneration, shot, or thermal.

Extrinsic semiconductor transition Incident photons produce a free-electron–bound-hole pair or a bound-electron–free-hole pair by excitation of an impurity level (to valence or conduction band).

Field of view (fov) The solid angle from which the detector receives radiation.

Flicker noise (*see* Modulation noise).

Generation noise Noise produced by the statistical fluctuation in the rate of production of photoelectrons.

Generation-recombination (GR) noise Charge carriers are generated both by (optical) photons and (thermal) phonons. Fluctuations in these generation rates cause noise; fluctuations in carrier-recombination times cause recombination noise. The phonon contribution can be removed by cooling. The remaining photon contribution is indistinguishable from radiation shot noise (photon noise). With photovoltaic *p-n* junctions, carriers are swept away before recombination, so that recombination noise is absent.

Guard ring An electrically biased structure used in some photodiodes to reorient electric potential gradients and thus reduce the leakage current in the detection circuit (*see* Si photodiode).

Intrinsic semiconductor transition Incident photons produce a free-electron–free-hole pair by direct excitation across the forbidden energy gap (valence to conduction band).

Johnson noise Same as thermal noise.

Maximized D star, $D^*(\lambda_{pk}, f_0)$, $cm \cdot Hz^{1/2}W$ The value of $D^*(\lambda_{pk}, f)$ corresponding to wavelength λ_{pk} and chopping frequency of maximum D^*.

Modulation (or $1/f$) noise The mechanism is not well understood. As its name implies, it is characterized by a $1/f^n$ noise power spectrum, where $0.8 < n < 2$. This type of noise is prominent in thermal detectors.

Multiplier phototube or multiplier photodiode Phototube with built-in amplification via secondary emission from electrically biased dynodes.

Noise spectrum The electrical power spectral density of the noise.

Nyquist noise Same as thermal noise.

Photo cell *See* photodiode.

Photoconductive gain The ratio between carrier lifetime and carrier transit time in a biased photoconductor.

Photodiode The term photodiode has been applied both to vacuum- or gas-filled photoemissive detectors (diode phototubes, or photo cells) and to photovoltaic detectors (semiconductor *p-n* junction devices).

Photomultiplier Same as multiplier phototube.

Photon counting Digital counting of individual photoelectrons in contrast to averaging of the photocurrent. This technique leads to very great sensitivity but can be used only for extremely low light levels.

Quantum efficiency *See* Responsive quantum efficiency.

Rms noise $V_{n,\,\mathrm{rms}}$ That component of the electrical output which is not coherent with the radiation signal. (Generally measured with the signal radiation removed.)

Rms signal $V_{s,\,\mathrm{rms}}$ That component of the electrical output which is coherent with the input signal radiation.

Response time τ Same as time constant.

Responsive quantum efficiency The ratio of the number of countable output events to the number of incident photons, e.g., photoelectrons per photon.

Sensitivity Degree to which detector can sense small amounts of radiation.

Shot noise This current fluctuation results from the random arrival of charge carriers, as in a photodiode. Its magnitude is set by the size of the unit charge.

$$i_{n,\mathrm{rms}} = (2ei_{\mathrm{dc}}\,\Delta f)^{1/2}$$

Spectral D double star $D^{**}(\lambda,f)$ A normalization of D^* to account for detector field of view. It is used only when the detector is background-noise-limited. If the fov is 2π sr, $D^{**} = D^*$.

Spectral D star, $D^*(\lambda,f)$, cm·Hz$^{1/2}$/W A normalization of spectral detectivity to take into account the area and electrical bandwidth dependence, e.g., $D^*(1\ \mu\mathrm{m}, 800)$ means D^* at $\lambda = 1\ \mu$m and chopping frequency 800 Hz; unit area and electrical bandwidth are implied. For background-noise-limited detectors the fov and the background characteristics must be specified. For many types of detectors this normalization is not valid, so that care should be exercised in using D^*.

Spectral detectivity $D(\lambda)$, W^{-1} A measure of detector sensitivity, defined as $D(\lambda) = (\mathrm{NEP}_\lambda)^{-1}$. As with NEP, the chopping frequency, electrical bandwidth, sensitive area, and sometimes background characteristics should be specified.

Spectral noise equivalent power NEP_λ The rms value of sinusoidally modulated monochromatic radiant power incident upon a detector which gives rise to an rms signal voltage equal to the rms noise voltage from the detector in a 1-Hz bandwidth. The chopping frequency, electrical bandwidth, detector area, and sometimes the background fov characteristics should be specified. NEP (1 μm, 800) means noise equivalent power at 1 μm wavelength, 1 Hz electrical bandwidth, and 800 Hz chopping rate. Specification of electrical bandwidth is often simplified by expressing NEP in units of W/Hz$^{1/2}$.

Spectral responsivity $\mathcal{R}(\lambda)$ The ratio between rms signal output (voltage or current) and the rms value of the monochromatic incident signal power.

Temperature noise Fluctuations in the temperature of the sensitive element, due either to radiative exchange with the background or conductive exchange with a heat sink, produce a fluctuation in signal voltage. For thermal detectors, if the temperature noise is due to the former, the detector is said to be at its theoretical limit. For thermal detectors

$$\overline{(\Delta T)^2} = \frac{4kT^2K\,\Delta f}{K^2 + 4\pi^2f^2C^2}$$

where $\overline{(\Delta T)^2}$ = mean square temperature fluctuations
K = thermal conductance
C = heat capacity

Thermal noise Noise due to the random motion of charge carriers in a resistive element

$$V_{n,\mathrm{rms}} = (4kTR\,\Delta f)^{1/2} \qquad k = \text{Boltzmann's constant}$$

Thermopile A number of thermocouples mounted in series in such a way that their thermojunctions lie adjacent to one another in the plane of irradiation.

Time constant τ A measure of the detector's speed of response. $\tau = 1/2\pi f_c$, where f_c is that chopping frequency at which the responsivity has fallen to 0.707 of its maximum value:

$$\mathcal{R}(f) = \frac{\mathcal{R}_0}{(1 + 4\pi^2 f^2 \tau^2)^{1/2}}$$

IDEAL PERFORMANCE AND SENSITIVITY

15. Photon-Noise-limited Detectivity The ultimate limit in signal-to-noise ratio for any radiation power detector is set by the fluctuation in photon arrival times. We shall discuss limiting detectivity for three cases:

1. Photon detector where arrival rate of signal photons far exceeds that of background photons (all other noise being negligible)
2. Photon detector where background photon arrival rate exceeds signal photon rate (all other noise being negligible)
3. Thermal detector, background limited

The rate of signal-carrier generation is

$$n = \eta A N_s \tag{1}$$

where η = detector quantum efficiency
AN_s = average rate of arrival of signal photons

It can be shown (Fink, 1940) that in a bandwidth Δf, the rms fluctuation in carrier-generation rate is

$$\delta n_{\text{rms}} = (2P_N\,\Delta f)^{1/2} \tag{2}$$

where P_N is the frequency dependence of the mean square fluctuations in the rate of carrier generation, i.e.,

$$P_N = A \int_{\nu_0}^{\infty} \eta(\nu)(\Delta N)^2\,d\nu \tag{3}$$

where $(\Delta N)^2$ is the mean square deviation in the total rate of photon arrivals per unit area and frequency interval including signal and background photons. For thermally produced photons of frequency ν (see Smith et al., 1957)

$$(\Delta N)^2 = \bar{N}\,\frac{e^{h\nu/kT}}{e^{h\nu/kT} - 1} = \frac{2\pi\nu^2}{c^2}\,\frac{e^{h\nu/kT}}{(e^{h\nu/kT} - 1)^2} \tag{4}$$

where $\bar{N}$ is the average rate of photon arrivals per unit area and frequency interval. Then, for the special case $h\nu \gg kT$

$$\delta n_{\text{rms}} = (2A\eta\bar{N}\,\Delta f)^{1/2} \tag{5}$$

This is also the case for a laser well above threshold. Here the photon statistics become Poisson, and $(\Delta N)^2 = \bar{N}$ even when $h\nu$ is not greater than kT.

1. *Photon detector, strong-signal case* When signal photons arrive at a much faster rate than background photons

$$\delta n_{\text{rms}} \approx (2A\eta\bar{N}_s\,\Delta f)^{1/2} \tag{6}$$

then

$$\text{NEP} = \frac{N_s A h\nu}{(n/\delta n_{\text{rms}})(\Delta f)^{1/2}} = \left(\frac{2N_s A}{\eta}\right)^{1/2} h\nu \tag{7}$$

or the noise-equivalent quantum rate is

$$\text{NEQ} = \left(\frac{2 \times \text{incident photon rate}}{\text{quantum efficiency}}\right)^{1/2} \tag{8}$$

2. *Photon detector, background-limited case* When the background photon rate N_s exceeds the signal photon rate $(N_B \gg N_s)$

$$\delta n_{\text{rms}} \approx (2A\eta\bar{N}_B\,\Delta f)^{1/2} \tag{9}$$

the noise-equivalent power is

$$\mathrm{NEP} = \frac{N_s A h\nu}{(n/\delta n_{\mathrm{rms}})(\Delta f)^{1/2}} = \left(\frac{2\bar{N}_B A}{\eta}\right)^{1/2} h\nu \tag{10}$$

The noise-equivalent quantum rate is

$$\mathrm{NEQ} = \left(\frac{2 \times \text{incident background photon rate}}{\text{quantum efficiency}}\right)^{1/2} \tag{11}$$

or

$$D^* = \frac{A^{1/2}}{\mathrm{NEP}} = \left(\frac{\eta}{2N_B}\right)^{1/2} \frac{1}{h\nu} \tag{12}$$

or

$$\text{Area-normalized quantum detectivity} = \frac{A^{1/2}}{\mathrm{NEQ}} = \left(\frac{\eta}{2\bar{N}_B}\right)^{1/2} \tag{13}$$

For the general case of a detector with area A seeing 2π sr of blackbody background at temperature T, $(\Delta N)^2$ is that in Eq. (4)

$$\overline{(\delta n)^2} = 2A\,\Delta f \int_0^\infty \eta(\nu)(\Delta N)^2\,d\nu = \frac{4\pi A\,\Delta f}{c^2} \int_0^\infty \eta(\nu)\nu^2 \frac{e^{h\nu/kT}}{(e^{h\nu/kT}-1)^2}\,d\nu \tag{14}$$

Then for $\Delta f = 1$ Hz,

$$\mathrm{NEP} = \frac{h\nu}{c\eta}\left[4\pi A \int_0^\infty \eta(\nu) \frac{\nu^2 e^{h\nu/kT}}{(e^{h\nu/kT}-1)^2}\,d\nu\right]^{1/2} \tag{15}$$

or

$$D^*(T,\lambda) = \frac{c\eta}{2\pi^{1/2}h\nu}\left[\int_0^\infty \eta(\nu) \frac{\nu^2 e^{h\nu/kT}}{(e^{h\nu/kT}-1)^2}\,d\nu\right]^{-1/2} \tag{16}$$

Assuming $\eta(\nu)$ is independent of frequency but falls back to zero for $\nu < \nu_c$

$$D^*(T,\lambda_c) = \frac{c\eta^{1/2}}{2\pi^{1/2}h\nu}\left[\int_{\nu_c}^\infty \frac{\nu^2 e^{h\nu/kT}}{(e^{h\nu/kT}-1)^2}\,d\nu\right]^{-1/2} \tag{17}$$

Figure 1 shows photon-noise-limited D^* vs. cutoff wavelength λ_c for various thermal-background temperatures (Jacobs and Sargent, 1970). Note that these curves are

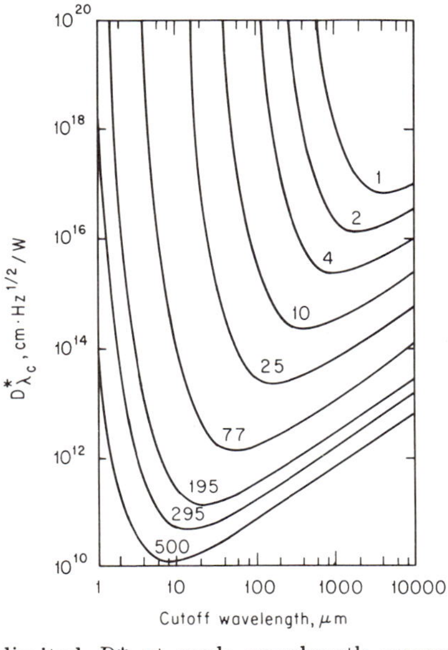

Fig. 1 Photon-noise-limited D^* at peak wavelength assumed to be cutoff wavelength, for background temperatures 1, 2, 4, 10, 25, 77, 195, 295, and 500 K (assumes 2π fov and $\eta = 1$). [*From Jacobs and Sargent* (1970).]

not independent. $D^*(T,\lambda_c)$ is related to $D^*(T',\lambda_c')$ by the formula

$$D^*(T,\lambda_c) = \left(\frac{T'}{T}\right)^{5/2} D^*(T',\lambda_c') \qquad \text{where } \lambda_c' = \frac{T}{T'}\lambda_c \tag{18}$$

This relation is useful for determining values of $D^*(T,\lambda_c)$, which do not appear in Fig. 1, in terms of a value $D^*(T',\lambda_c')$, which does appear. For example, to find $D^*(1{,}000\ K, 4\ \mu m)$ from the 500-K curve

$$D^*(1{,}000, 4) = (\tfrac{500}{1000})^{5/2} D^*(500, 4 \times \tfrac{1000}{500}) = 2.3 \times 10^9 \text{ cm} \cdot \text{Hz}^{1/2}/\text{W} \tag{19}$$

If higher accuracy is desired than can be determined from Fig. 1, one can use the formula above in combination with Table 1, which gives explicit values of $D^*(\lambda_c)$ vs. λ_c for $T = 295$ K.

TABLE 1 D^* **vs.** λ_c **for** $T = 295$ **K**

λ_c, μm	$D^*(\lambda_c)$	λ_c, μm	$D^*(\lambda_c)$	λ_c, μm	$D^*(\lambda_c)$	λ_c, μm	$D^*(\lambda_c)$
1	2.19×10^{18}	10	5.35×10^{10}	100	1.67×10^{11}	1,000	1.55×10^{12}
2	4.34×10^{13}	20	5.12×10^{10}	200	3.20×10^{11}	2,000	3.10×10^{12}
3	1.64×10^{12}	30	6.29×10^{10}	300	4.74×10^{11}	3,000	4.64×10^{12}
4	3.75×10^{11}	40	7.68×10^{10}	400	6.28×10^{11}	4,000	6.19×10^{12}
5	1.70×10^{11}	50	9.13×10^{10}	500	7.82×10^{11}	5,000	7.73×10^{12}
6	1.06×10^{11}	60	1.06×10^{11}	600	9.36×10^{11}	6,000	9.28×10^{12}
7	7.93×10^{10}	70	1.21×10^{11}	700	1.09×10^{12}	7,000	1.08×10^{13}
8	6.57×10^{10}	80	1.36×10^{11}	800	1.24×10^{12}	8,000	1.24×10^{13}
9	5.80×10^{10}	90	1.52×10^{11}	900	1.40×10^{12}	9,000	1.39×10^{13}

The photon-noise-limited sensitivity shown in Figs. 1 and 2 applies to photovoltaic and photoemissive detectors. For photoconductors, recombination noise results in a $\sqrt{2}$ reduction in D^* at all wavelengths. Figure 3 shows the relative

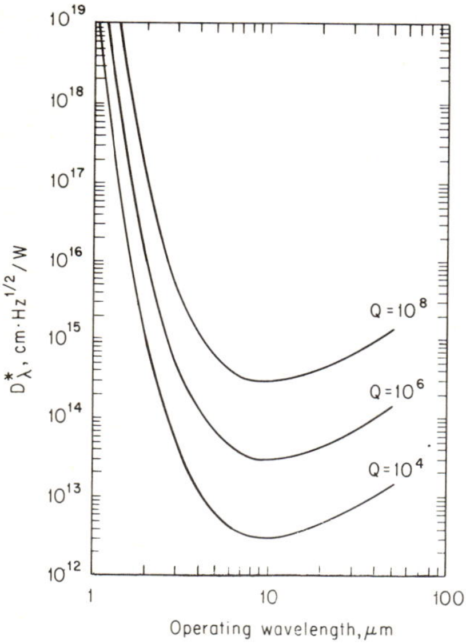

Fig. 2 Photon noise limit of a narrow-band quantum counter as a function of operating wavelength for a 290-K background, 2π fov, and $\eta = 1$. [*From Kruse et al.* (1963).]

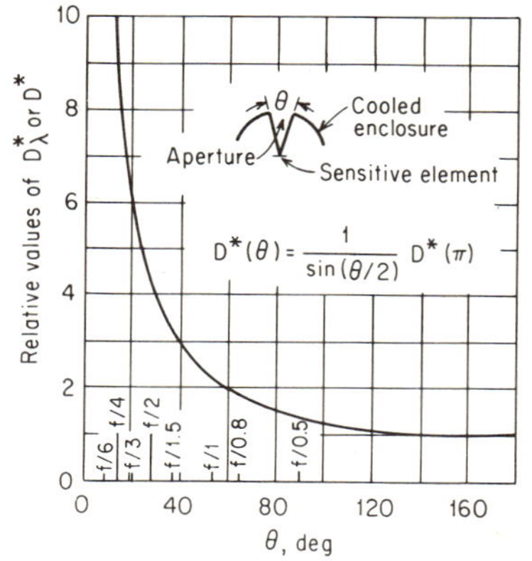

Fig. 3 Relative increase in photon-noise-limited $D^*(\lambda_{pk})$ or D^* achieved by using a cooled aperture in front of lambertian detector. [*From Kruse et al.* (1963).]

increase in photon-noise-limited D^* achievable by limiting the fov through use of a cooled aperture.

3. *Thermal detectors* Limiting sensitivity of an ideal thermal detector has been discussed previously (Smith et al., 1957; Felgett, 1949; Kruse et al., 1963). Assuming no shortwave or long-wavelength cutoffs exist,

$$D^* = \frac{\epsilon}{[8\epsilon\sigma k(T_1^5 + T_2^5)]^{1/2}} = \frac{4.0 \times 10^{16}\epsilon^{1/2}}{(T_1^5 + T_2^5)^{1/2}} \qquad \text{cm}\cdot\text{Hz}^{1/2}/\text{W}$$

where T_1 = detector temperature
T_2 = background temperature
ϵ = detector emissivity
σ = Stefan-Boltzmann constant
k = Boltzmann constant

D^* vs. T_2 is plotted for various T_1 in Fig. 4. Figure 5 shows the effect of both short- and long-wavelength cutoffs on bolometer sensitivity (Low and Hoffman, 1963), with the ideal photoconductor curve for reference.

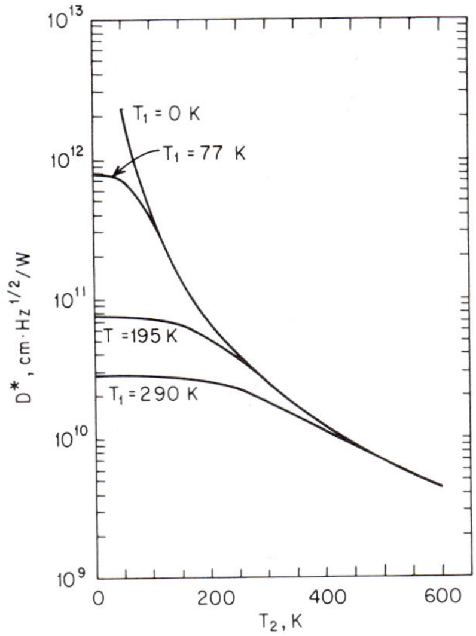

Fig. 4 Photon-noise-limited D^* for thermal detectors as a function of detector temperature T_1 and background temperature T_2 (2π fov; $\eta = 1$). [*From Kruse et al.* (1963).]

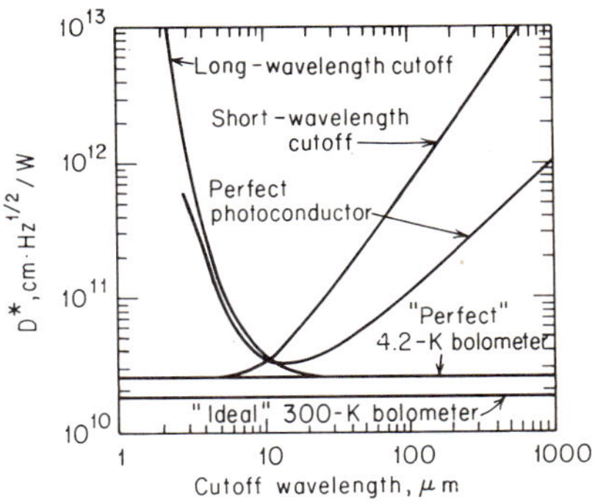

Fig. 5 The detectivity of a "perfect" bolometer plotted as a function of both short- and long-wavelength cutoffs. Plots for a perfect photoconductor and two other cases are included for comparison. The background temperature is 300 K. [*From Low and Hoffman* (1963).]

D VS. λ FOR SELECTED DETECTORS

16. See Fig. 6.

DETECTOR PERFORMANCE

Manufacturers' Specifications

17. During the past few years, infrared detectors have been improved greatly, and users have become more sophisticated in applying them. As a result, detectors frequently are limited in detectivity by background irradiance. Thus it

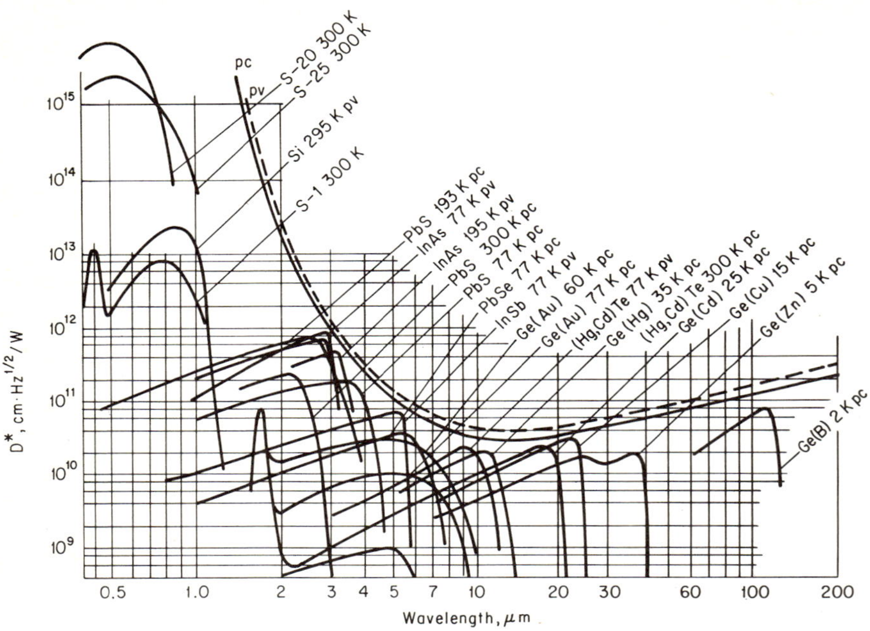

Fig. 6 D vs. λ for selected detectors.

has become important to include in the specification of cooled infrared detectors data on performance under reduced background conditions, e.g., specify background irradiance or temperature and geometry. At present, much of this information is unknown or classified; however, attention is called to this important frontier.

The following data are based on manufacturers' specification sheets and do not always represent the current state of the art. In many cases superior performance is obtainable. Reference is made to Fig. 6.

18. Golay Cell This thermal detector offers spectral coverage in the 250-nm to the millimeter region and high sensitivity, more or less independent of detector area. Its operating principle is thermal expansion of a gas upon heating. The device tends to be quite fragile and is fairly slow in response.

Sensitivity: (See Fig. 7)
Quantum efficiency: Depends on coating
Noise: Limited by amplifier (see Fig. 8)
Operating temperature: Ambient
Responsivity: Not applicable, since displacement measurement is a separate system from the detection system
Linearity: 1 percent in range 5×10^{-8} to 10^{-6} W
Time constant: 2 to 30 ms (shorter time constant implies poorer sensitivity)
Sensitive area: 2.4 by 3.2 mm standard; also 4 by 12 mm and 8×10^{-2} cm^2
Sensitivity profile: "Uniform" over 80 percent of diameter
Recommended circuit: See manufacturer
Stability: After 10-min warm-up, output varies less than 1 percent for 1-K ambient-temperature change
Manufacturers: Eppley Laboratory, Pye Unicam, Jobin

19. Thermocouple The thermocouple (or thermopile when used multiply) offers broad, uniform spectral response, a high degree of stability, and moderate sensitivity. Its slow response and relative fragility have limited its use to laboratory instruments, such as spectrometers.

Compared with thermistors, thermocouples are slower, require no bias, and have

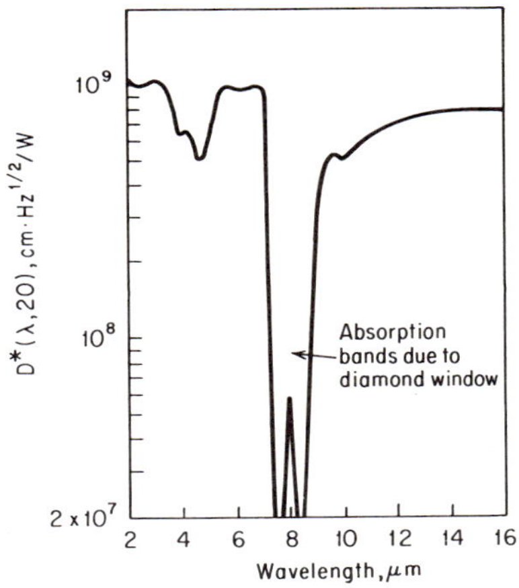

Fig. 7 Golay cell spectral sensitivity ($T = 300$ K, $T = 10$ ms, $A = 7.9 \times 10^{-2}$ cm^2, diamond window). [*From Wolfe* (1965).]

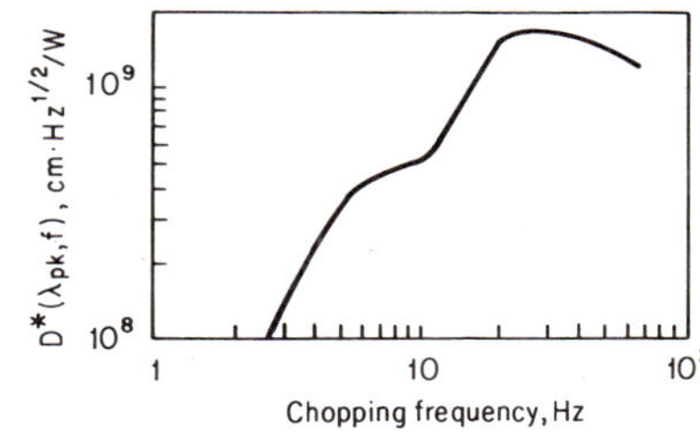

Fig. 8 Golay cell D^* vs. chopping frequency. [*From Wolfe* (1965).]

higher stability but much lower impedance and responsivity. This increases the amplification required for the thermocouple; however, the only voltage appearing is the signal voltage, so that the serious thermistor problem of bridge-circuit bias-fluctuations is avoided. With proper design, performance should not be amplifier-limited but limited instead by the Johnson noise of the thermocouple. Thermocouples perform stably in dc operation, although instability of dc amplifiers usually indicates ac operation.

The inherent dc stability of thermocouples is attractive for applications requiring no moving parts, and recently a relatively rugged solid-backed evaporated thermocouple has been developed whose sensitivity approaches that of the thermistor bolometer.

Sensitivity: $D^* \approx 1 \times 10^9$ cm·Hz$^{1/2}$/W for 20 ms response time; spectral response depends on black coating (usually gold black) (see Fig. 9)

Noise: White Johnson noise, falling off with responsivity (see Fig. 10)

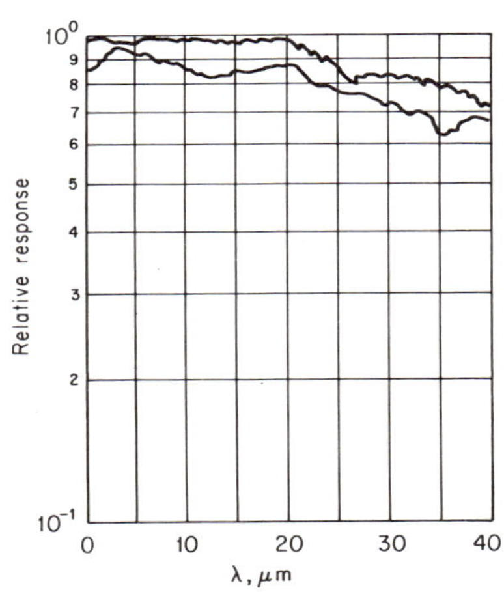

Fig. 9 Typical thermocouple spectral-response curves (CsI window) (two different manufacturers).

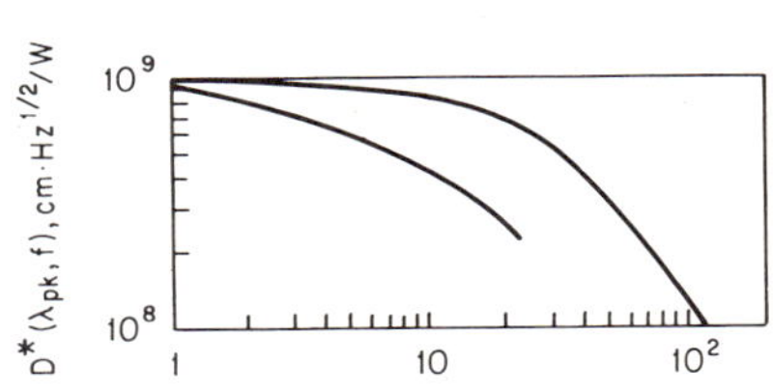

Fig. 10 Typical thermocouple D^* (noise) frequency spectra (two manufacturers). [*From Wolfe* (1965).]

Resistance: 5 to 15 Ω typical
Responsivity: 5 V/W (typical), 20 to 25 V/W (selected)
Time constant: 10 to 20 ms (typical)
Operating temperature: Normally ambient
Sensitive area: 0.1 by 1 to 0.3 by 3 or 0.6 by 2 mm (typical)
Linearity: <0.1 percent in region investigated (6×10^{-10} to 6×10^{-8} W incident)
Recommended circuit: Transformer coupled into low-noise (FET) amplifier with good low-frequency noise characteristics
Manufacturers: Perkin-Elmer, Charles Reeder, Beckman Instruments, Farrand, Eppley Laboratory, Santa Barbara Research Center, Barnes Engineering

20. Thermistor Bolometer Thermistors offer reliability, moderate sensitivity, and broad spectral coverage without cooling. Construction is rugged and highly resistant to vibration, shock, and other extreme environmental perturbations. Response is slower than 1 ms, and trade-off exists between speed and sensitivity.

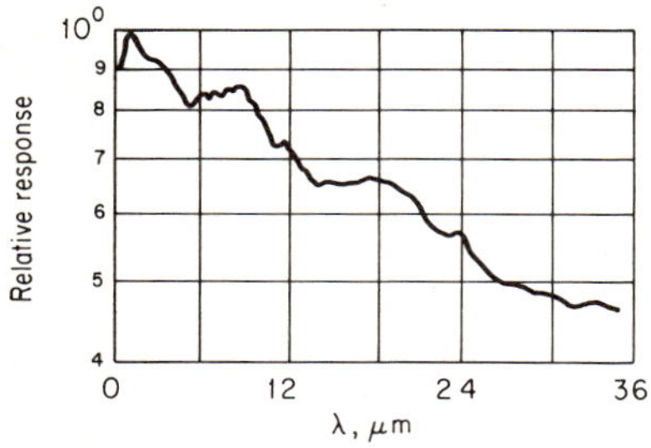

Fig. 11 Typical thermistor spectral response (no window).

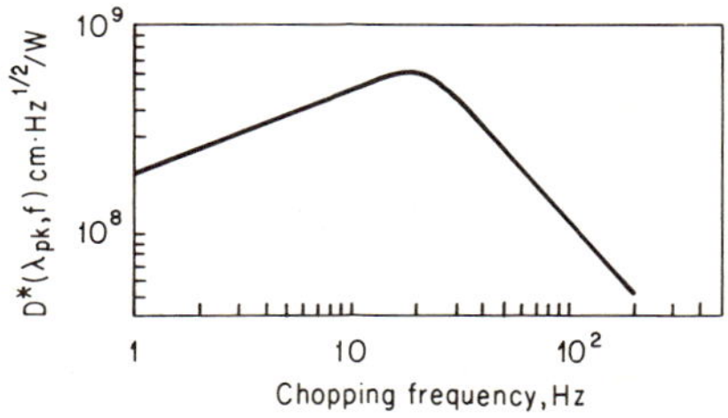

Fig. 12 Typical thermistor D^* (noise) frequency spectrum. [*From Wolfe* (1965).]

Thermistor elements are made of polycrystalline Mn, Ni, and Co oxides. In their final form they are semiconductor flakes 10 μm thick, which undergo a temperature resistance change of ~4 percent/K. Since thermistor resistance changes with ambient temperature enough to alter the biasing significantly, it is usually operated in a bridge circuit, with a nearly identical thermistor shielded from signal radiation and used for a balance resistor.

Sensitivity:

$$\text{NEP} = 8.9 \times 10^{-10} \sqrt{\frac{A\ (\text{mm}^2)}{\tau_{\text{rms}}}} \quad \text{W}$$

$$D^* = 1.1 \times 10^9 \sqrt{\tau_{\text{rms}}} \quad \text{cm} \cdot \text{Hz}^{1/2}/\text{W}$$

Spectral response: Depends on coating (usually Zapon lacquer); see Fig. 11
Quantum efficiency: Depends on blackening coating, typically 80 percent
Noise: Thermal-noise-limited above ~20 Hz ($V_{\text{noise}} = \sqrt{4kTR\ \Delta f}$); below that, $1/f$ type noise (see Fig. 12). Used in balanced-bridge circuit (two flakes in parallel); limiting noise due to thermal noise in *both* flakes
Resistance†: For standard 10-μm-thick flakes two different resistivities are available: 2.5 MΩ/sq or 250 kΩ/sq
Time constant: τ is 1 to 20 ms standard for nonimmersed detectors and 2 to 10 ms standard for immersed detectors
Sensitive area: 0.1 by 0.1 mm; 5 by 5 mm standard
Operating temperature: Normally ambient, 285 to 370 K

† In a bolometer bridge, resistance between output connection and ground is half that of single flake.

Responsivity†: Depends on bias, resistance, area, and time constant $\mathcal{R} \propto \sqrt{R\tau/A} \approx 10^3$ V/W for 0.1 by 0.1 mm area, 250 kΩ resistance, and $\tau = 4$ ms (see Fig. 13 for frequency–time-constant dependence with given area)
Sensitivity profile: Approximately 10 percent for 10-μm scan diameter (1 by 1 mm cell)
Linearity: ±5 percent 10^{-6} to 10^{-1} W/cm²
Recommended circuit: See Figs. 14 and 15
Manufacturers: Barnes Engineering, Servo Corporation of America, Polan Industries

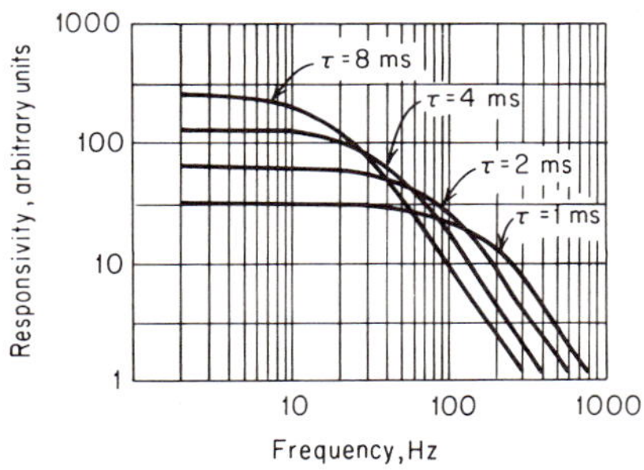

Fig. 13 Relative responsivity vs. frequency for various time-constant thermistor detectors (A = 1 by 1 mm). (*From Barnes Engineering, Bull.* 2-100.)

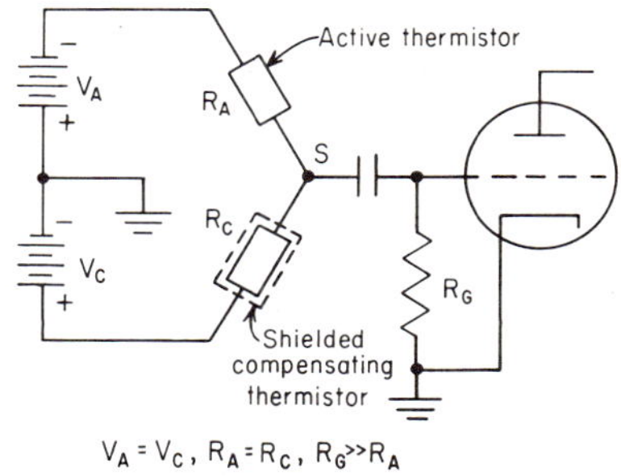

Fig. 14 Bolometer electrical circuit. (*From Barnes Engineering, Bull.* 2-100.)

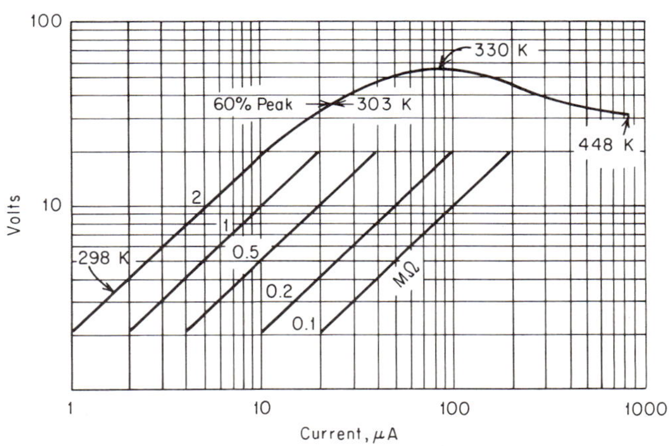

Fig. 15 Thermistor voltage-current characteristics, showing typical flake temperatures under different conditions. (*From Barnes Engineering, Bull.* 2-100.)

21. Pyroelectric Triglycine sulfate (TGS) is an unusually fast, uncooled thermal detector with high sensitivity. Its principle of operation is the pyroelectric effect, which is the change of electric polarization with temperature. It offers rugged construction and absence of $1/f$ noise because no bias is involved.

† Output voltage (responsivity) can be increased to a limited degree by raising bias voltage. Figure 15 shows the deviation from Ohm's law due to heating. Bias should be held below 60 percent of peak voltage. Listed responsivity is that of active flake. In the bridge circuit, responsivity is half this value.

Lack of $1/f$ noise, combined with moderate speed and sensitivity, makes TGS useful for scanning applications. In addition, the NEP is independent of area at low frequencies (<10 Hz), so that TGS is useful for large-area applications. Since the pyroelectric effect is an integrated volume effect, the output signal is unaffected by spatial or temporal distribution of the radiation,† which is useful for calorimetry. In some applications this detector is still developmental.

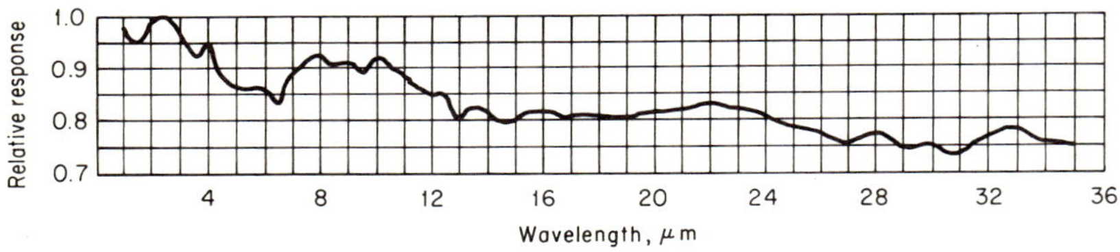

Fig. 16 Relative spectral response of TGS detectors with modified 3M black. (*From Barnes Engineering, Bull.* 2-220*A*.)

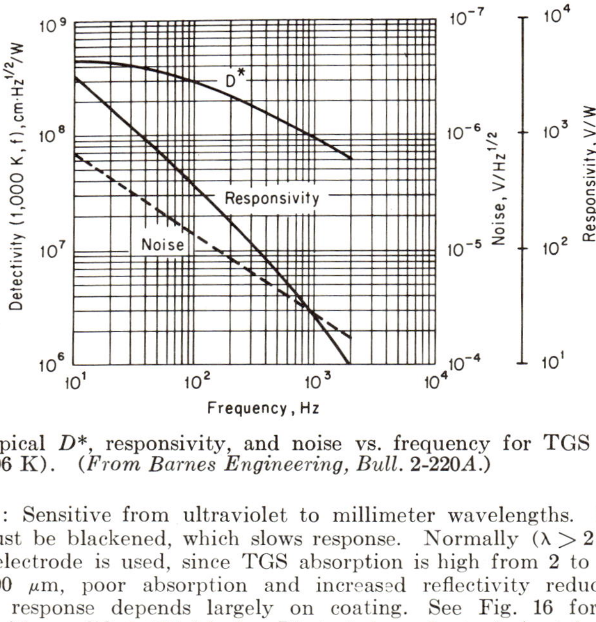

Fig. 17 Typical D^*, responsivity, and noise vs. frequency for TGS (A = 1 by 1 mm; T = 296 K). (*From Barnes Engineering, Bull.* 2-220*A*.)

Sensitivity: Sensitive from ultraviolet to millimeter wavelengths. For $\lambda < 2\mu$m, TGS must be blackened, which slows response. Normally ($\lambda > 2\ \mu$m) a transparent electrode is used, since TGS absorption is high from 2 to 300 μm. Beyond 300 μm, poor absorption and increased reflectivity reduce sensitivity. Spectral response depends largely on coating. See Fig. 16 for spectral response with modified 3M black. D^* is independent of A at low frequencies (<10 Hz) (see Figs. 17 and 18). Figure 19 shows NEP vs. A for various frequencies.

Quantum efficiency: Depends on coating absorptivity (for 3M black typically $\eta > 75$ percent).

Noise: (See Figs. 17 and 18.) Limited by loss-tangent noise up to frequencies where limited by amplifier short-circuit noise (see Fig. 20).

Operating temperature: Ambient, up to 315 K. Can be repolarized if $T > T_{\text{Curie}} = 322$ K. Irreversible damage at T = 394 K (see Fig. 21).

Output impedance: 10 K, set by built-in amplifier (see Fig. 22).

† Up to damage threshold or depolarizing temperature. Lead zirconate titanate ceramic (Clevite PZT-5) exhibits a much smaller pyroelectric effect than TGS, but its high Curie temperature of 638 K makes it more useful than TGS for high-energy applications.

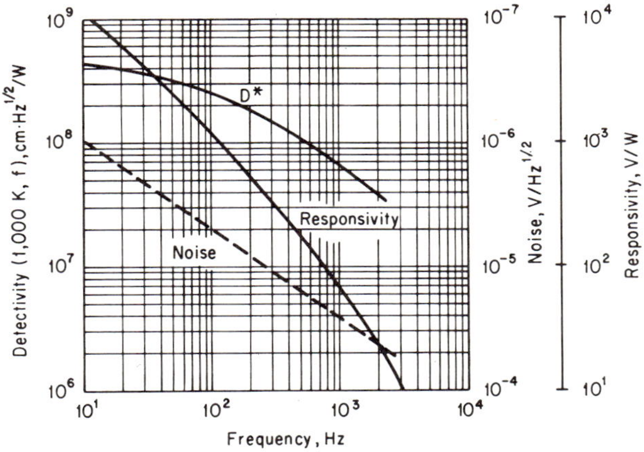

Fig. 18 Typical D^*, responsivity, and noise vs. frequency (A = 0.4 by 0.4 mm; T = 296 K). (*From Barnes Engineering, Bull.* 2-220*A*.)

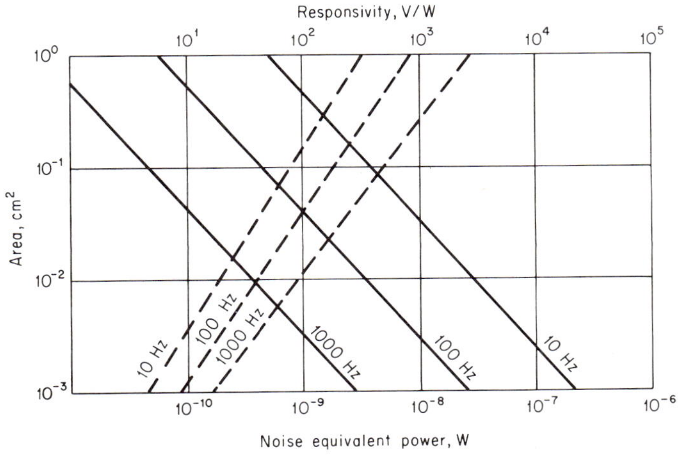

Fig. 19 Noise equivalent power in watts (broken lines) and responsivity (heavy lines) vs. TGS detector area for various frequencies. (*From Barnes Engineering, Bull.* 2-220*B*.)

Responsivity: See Figs. 14 to 19 and 21.
Capacitance: 5 pF for 0.5 by 0.5 mm; 20 pF for 1 by 1 mm; 100 pF for 5 by 5 mm.
Sensitive area: 0.5 by 0.5 mm, 10 by 10 mm standard.
Time constant: Not pertinent (see Fig. 19).
Linearity: 5 percent between 10^{-6} and 10^{-1} W/cm^2.
Sensitivity profile: Depends on coating or transparent electrode; 5 to 7 percent across 12 by 12 mm; spot size <250 μm.
Recommended circuit: See Fig. 22. FET amplification stage usually built-in. Since $\mathcal{R} \propto 1/f$, use of an amplifier with $1/f$ noise and gain $\propto f$ is recommended. Then output signal and signal-to-noise ratio are independent of frequency.
Manufacturers: Barnes Engineering, Mullard, Laser Precision Corp.

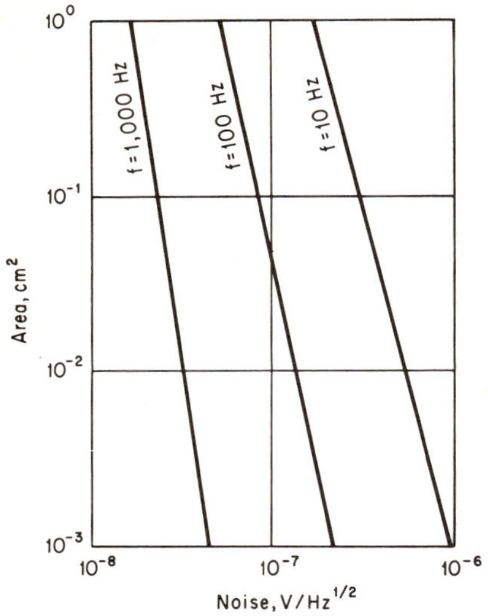

Fig. 20 TGS noise vs. detector area for various operating frequencies. (*From Barnes Engineering, Bull.* 2-220*B*.)

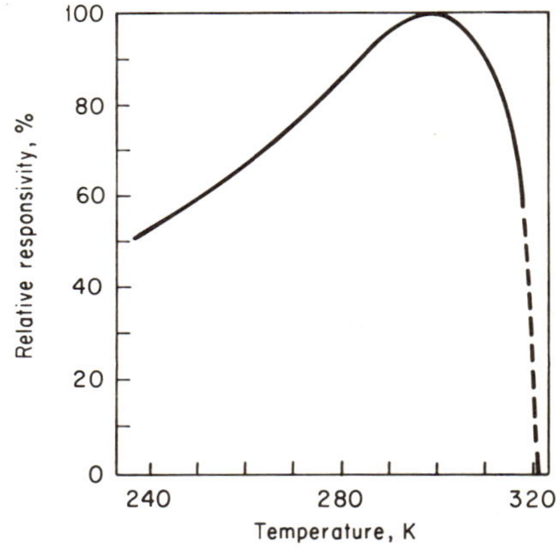

Fig. 21 Relative responsivity vs. temperature for TGS. (*From Barnes Engineering, Bull.* 2-220*B*.)

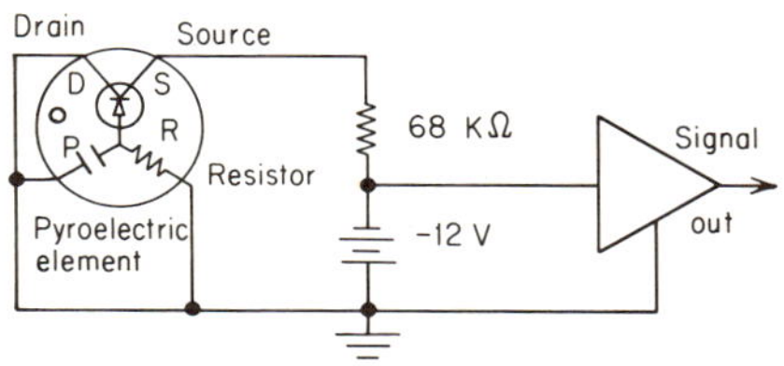

Fig. 22 Pyroelectric detector-amplifier circuit. (*From Barnes Engineering, Bull.* 2-220*A*.)

22. InSb Bolometer Indium antimonide has been made the basis for two closely related types of bolometer, one with and one without a magnetic field. Very few of these detectors have yet been made, so that information is still scanty; however, both are commercially available. They offer submicrosecond response and broad far-infrared coverage out to millimeter wavelengths but require liquid-helium cooling and very sophisticated receiver design.

Technically, these devices may be classed as bolometers, since incident radiation power produces a heating effect which causes a change in free-charge mobility. In the normal bolometer the crystal lattice absorbs energy and transfers it to the free carriers via collisions. However, in both InSb bolometers incident radiation power is absorbed directly by free carriers, the crystal lattice temperature remaining essentially constant. Hence the name *electron bolometer*. Note that this mechanism differs from photoconductivity in that free-electron mobility rather than electron number is altered by incident light (hence there is no photoconductive gain). In the form with $H \neq 0$ some photoconductivity takes place as well as mobility changes.

$H \neq 0$ (*Putley Detector*) This type requires cooling to 1.6 K (pumped liquid helium) and requires a magnet but has higher impedance and hence responsivity than the Rollin type ($H = 0$).

Sensitivity: $D^*(1\text{ mm}, 800) \geqq 1 \times 10^{12}$ cm·Hz$^{1/2}$/W; $D^*(373\text{ K}, 800)$† $\geqq 5 \times 10^{10}$ cm·Hz$^{1/2}$/W (see Fig. 23)
Noise: Depends on receiver
Time constant: <1 μs
Responsivity: 2,000 V/W at $\lambda = 1$ mm; 100 V/W (373 K, 800) (see Fig. 24)

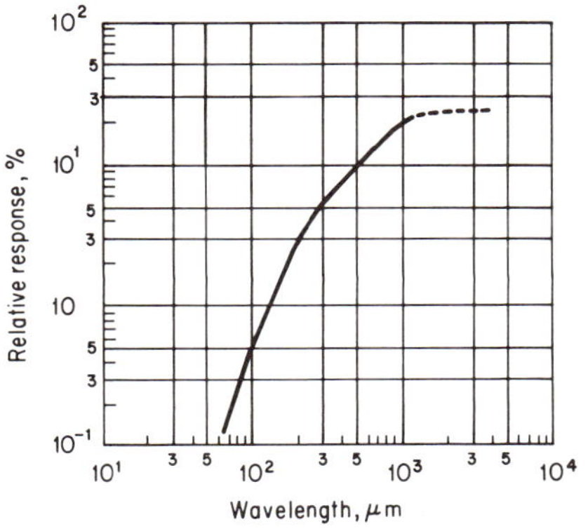

Fig. 23 Spectral response for Putley InSb bolometer (short-wavelength cutoff due to filter). (*From Mullard Ltd., Bull. TP*1080.)

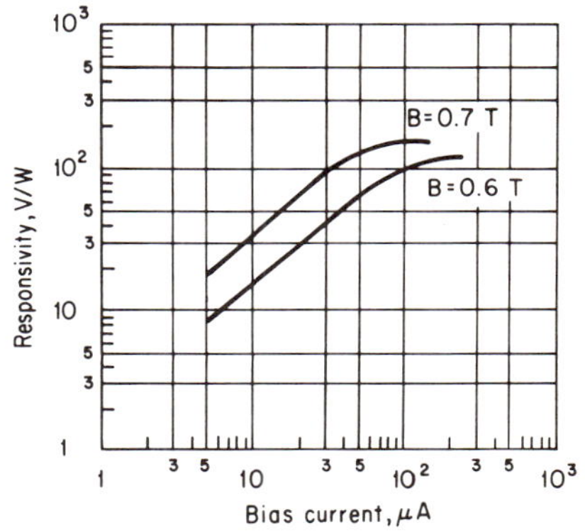

Fig. 24 Variation of responsivity with bias current for Putley InSb bolometer. (*From Mullard Ltd., Bull. TP*1080.)

Impedance: 500 to 5,000 Ω
Sensitive area: 5 by 5 mm
Recommended circuit: Superconducting solenoid used for magnet
Operating temperature: 1.6 K
Manufacturers: Mullard (Optoelectronics), Advanced Kinetics

$H = 0$ (*Rollin & Kinch Detector*) This type requires cooling to 5 K, needs no magnet, but has lower impedance than the Putley type ($H \neq 0$).

Sensitivity: $D^*(2\text{ mm}, 900) = 4 \times 10^{11}$ cm·Hz$^{1/2}$/W (see Fig. 25)
Noise: See Figs. 26 and 27
Responsivity: ~1,000 V/W
Time constant: $\tau = \frac{1}{4}$ μs
Sensitive area: 0.13 cm^2 typical
Operating temperature: ~5 K
Impedance: Without bias; 100 Ω; optimum bias; 150 Ω, depends on bias (see Fig. 26)
Recommended circuit: Optimum bias 0.5 mA (see Fig. 28)
Manufacturer: Santa Barbara Research Center

23. Ge(Ga) Low-Temperature Bolometer The Ge(Ga) bolometer offers very high sensitivity and broad spectral coverage in the region 1.7 to 2,000 μm. Liquid-helium cooling is required. A trade-off exists between response time (seconds) and sensitivity (10^{-14} W NEP). Operation at 1,000 Hz can be achieved still maintaining 2×10^{-13} W NEP.

Sensitivity: Depends on thermal conductance, G (see Figs. 29 and 30). NEP(λ, 10 Hz) $= V_n/S = 3 \times 10^{-14}$ W for $G = 1$ μW/K; $A = 1$ mm^2; $D^*(\lambda, 10\text{ Hz}) = 3 \times 10^{13}$ cm·Hz$^{1/2}$/W ($Q < 0.2$ μW). For this detector, NEP $\approx 4T(kG)^{1/2}$ and does not vary with $A^{1/2}$ (T is heat-sink temperature, and k is Boltzmann's constant). Thus D^* cannot be used as a valid means of comparison with other detectors; 300-K background-limited performance is achievable for 2π fov when the bolometer is operated at 4.25 K with $G = 10^{-3}$ W/K. (For $A = 0.1$ cm^2, the time constant is 50 s.)

† Blackbody was at 373 K, emissivity 0.8, with $\lambda < 0.150$ mm cut out by filter.

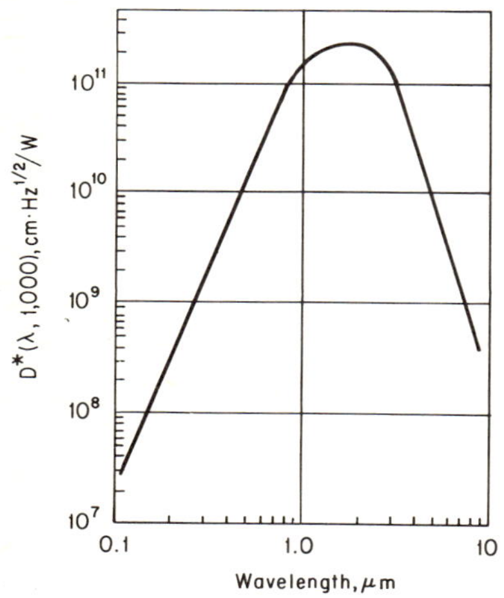

Fig. 25 D^* vs. λ for InSb electron bolometer ($H = 0$). (*From Raytheon, IR Millimeter Wave Detector,* 1967.)

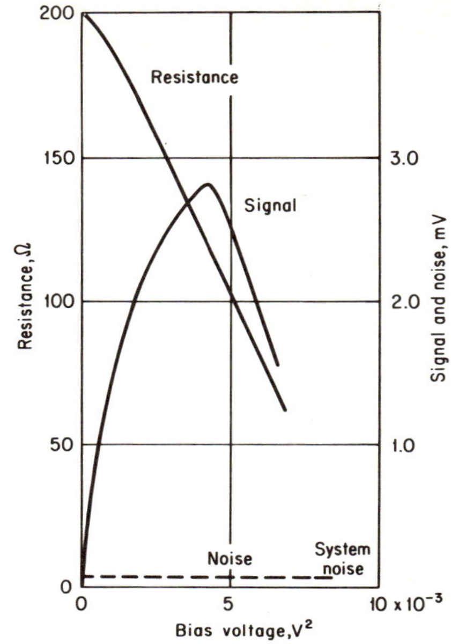

Fig. 26 InSb electron bolometer, typical resistance, signal and noise vs. bias voltage squared (T = 5 K; R_L = 200 Ω; gain = 2.4×10^4; F = 1,100 Hz). (*From Santa Barbara Research Center, Prelim. Res. Rep.,* 1967.)

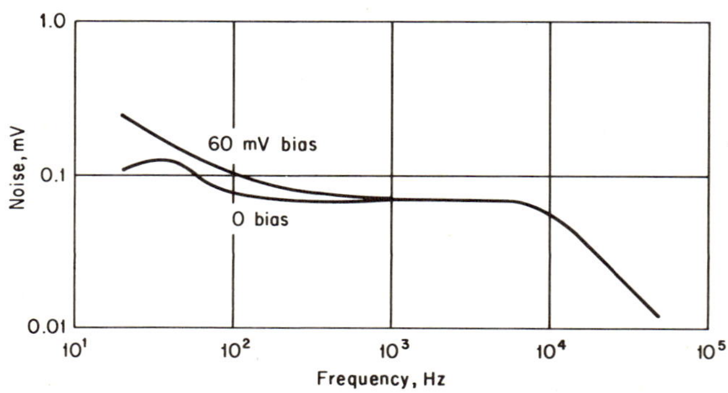

Fig. 27 InSb electron bolometer, typical noise spectrum (T = 5 K, R_L = 200 Ω; gain = 2.4×10^4, Δf = 5.6 Hz). (*From Santa Barbara Research Center, Prelim. Res. Rep.,* 1967.)

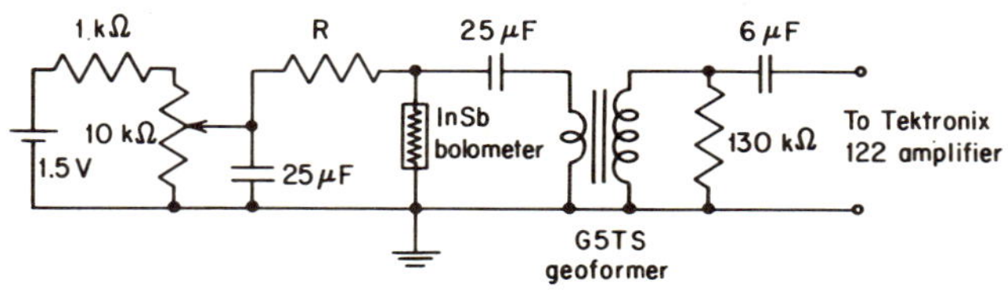

Fig. 28 Electron-bolometer biasing circuit. (*From Santa Barbara Research Center, Prelim. Res. Rep.,* 1967.)

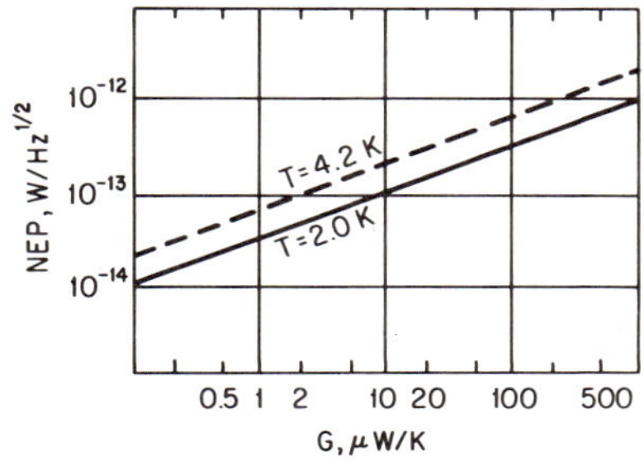

Fig. 29 Germanium bolometer NEP vs. conductance. (*IR Labs, Inc.*)

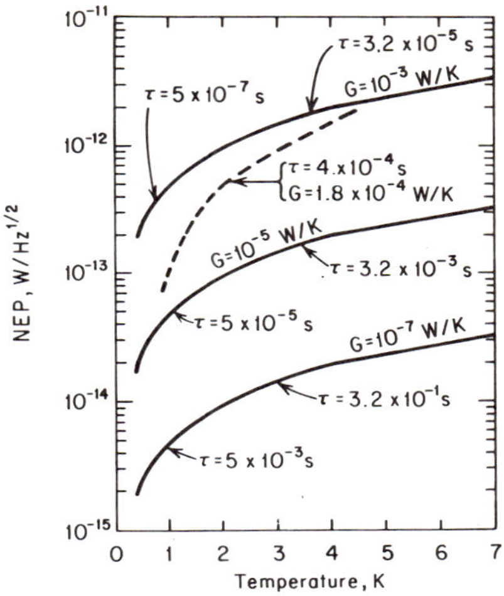

Fig. 30 Germanium bolometer NEP vs. temperature. Solid curves are theoretical: NEP $\approx 4T(kG)^{1/2}$; dashed curve is measured value for a typical bolometer. [*From Low* (1961).]

Responsivity: Typically, responsivity $= 2.5 \times 10^5$ V/W $= 0.7(R/TG)^{1/2}$, where R = resistance. Responsivity, and hence NEP, depends on thermal conductance G, which in turn is set by background power. G ranges from 0.4 to 1,000 μW/K,

Thermal conductance: Typically $G = 1$ μW/K for background $Q < 0.2$ μW; note that $Q < \frac{1}{2}$ (optimum bias power P).

Sensitive area: 0.25 by 0.25 to 10 by 10 mm.

Resistance: 0.5 MΩ.

Operating temperature: 2 K (see Fig. 29). In applications where radiation noise can be eliminated there is much to be gained by operating at the lowest possible temperature. Figure 30 shows the theoretical NEP and time constant at 0.5 K (the practical lower limit for existing cryostats), assuming that current noise remains unimportant.

Quantum efficiency: Depends on blackened coating and window. For $\lambda < 100$ μm absorptivity exceeds 95 percent. For $\lambda > 100$ μm efficiency varies with geometry.

Time constant: Response time constant is proportional to G^{-1}. Therefore, if G must be increased to accommodate larger background, the time constant is decreased proportionally. Responsivity and NEP, however, are degraded as $G^{-1/2}$.

Noise: $v_n = 1 \times 10^{-8}$ V/Hz$^{1/2}$; thermal noise is due to R and R_L.

Recommended circuit: Standard photoconductive circuit, with load resistor, grid resistor, and blocking capacitor at low temperature (see Fig. 31). See Fig. 32 for typical electrical characteristics. Bias power $P = 0.1TG$.

Manufacturer: Infrared Laboratories, Inc.

24. Photoemissive Detectors Photoemissive detectors are generally the detector of choice in all spectral regions where high quantum efficiency is available. In the spectral region $\lambda < 600$ nm, the photomultiplier, or multiplier phototube, has close to ideal sensitivity; i.e., selected photomultiplier tubes (PMT) are capable of detecting single photon arrivals (but at best only one of three) and amplifying the photocurrent (pulse) enormously without seriously degrading the signal-to-noise ratio. Time resolution can be as short as 0.1 ns. Only very specialized limitations have precluded their use for $\lambda < 800$ nm, for example, cost, size, ruggedness, uniformity of manufacture, or need for still faster response. Recently these limitations have all been met individually but generally not collectively. Where adequate light is available, the simple phototube has advantages over the multiplier phototube in that high voltages are not required, the output level is not sensitive to applied voltage, and dynode fatigue is eliminated.

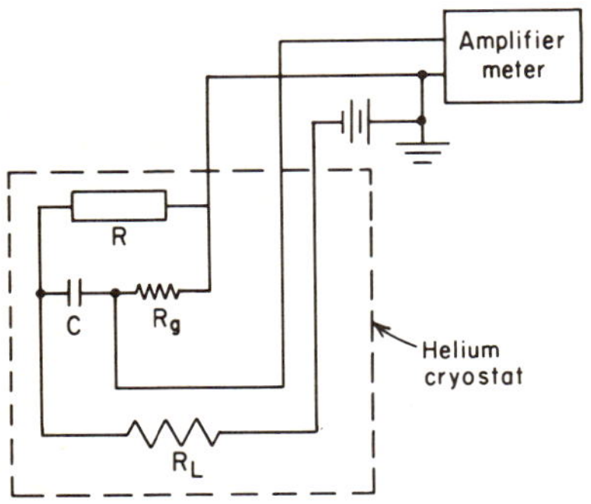

Fig. 31 Germanium bolometer circuit and cryostat.

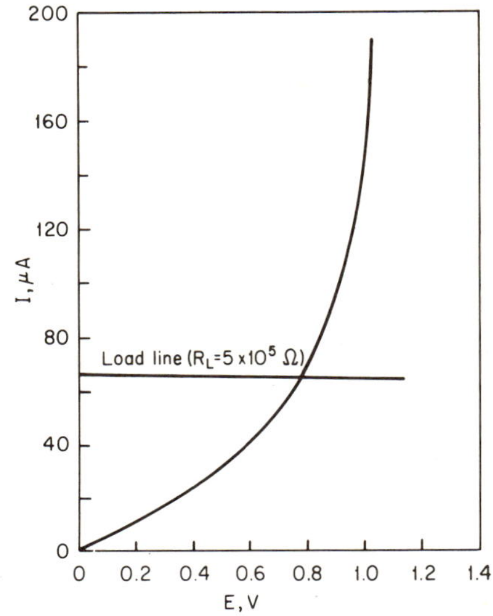

Fig. 32 Load curve for a typical germanium bolometer (T = 2.15 K) with load line, showing optimum operating point. [*From Low* (1961).]

Sensitivity. In modern phototubes shot noise due to the cathode dark current is by far the most important noise source. The most common descriptions of phototube sensitivity list both current responsivity (amperes per watt) and dark noise. Several useful measures of sensitivity are noise equivalent input (NEI) (see Par. 14), noise equivalent power (NEP), or its reciprocal $D \equiv 1/\text{NEP}$. NEP and NEI in the range 10^{-14} to 10^{-16} W/Hz$^{1/2}$ are not uncommon.

Detectivity is generally limited by dark-current shot noise. Dark current depends on photocathode material, area, and temperature. Thus the best detectivity is obtained with small effective sensitive area. Cooling is especially useful for red-sensitive tubes and is generally not worthwhile for others (see operating temperature, below).

The spectral response curves shown in Fig. 33 are for the combination of photocathode and window. Historically, this method of description gave rise to the S-response designations, most of which are now obsolete. It is often desirable to separate photocathode response from window transmission. Thus, Fig. 34 shows the quantum efficiency (electrons per incident photon) of a number of photocathodes without window losses. For $\lambda < 400$ nm each photocathode should maintain its peak quantum efficiency, up to photon energies where multiple photoemissions take place.

D^* is a meaningful figure of merit for phototubes whose sensitivity is limited by dark noise (shot noise on the dark current) and whose emitting photocathode area is clearly defined, but D^* must be used with caution because although modern phototubes are generally dark-noise-limited devices, they are often limited by noise in signal, i.e., the noise content of the signal itself (Eberhardt, 1967). Serious errors in predicting the detection capability of phototubes will arise if noise in signal is ignored and D^* is presumed to be the important limiting parameter. Very little reliable data are presently available on D^* for photoemitters. How-

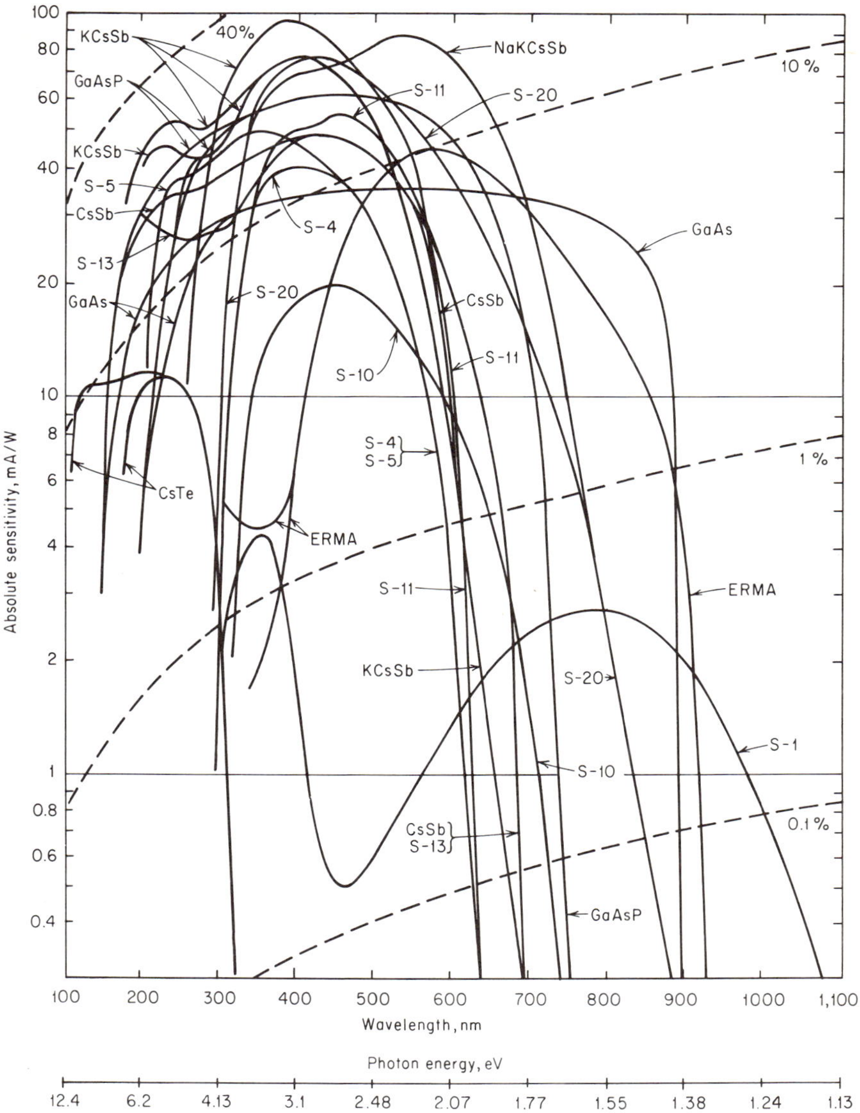

Fig. 33 Spectral sensitivity of various photoemitters. Dotted lines indicate photocathode quantum efficiency. Chemical formulas are abbreviated to conserve space. S-1 = AgOCs with lime or borosilicate crown-glass window; S-4 = Cs_3Sb with lime or borosilicate crown-glass window (opaque photocathode); S-5 = Cs_3Sb with ultraviolet-transmitting glass window; S-8 = Cs_3Bi with lime or borosilicate crown-glass window; S-10 = AgBiOCs with lime or borosilicate crown-glass window; S-11 = Cs_3Sb with lime or borosilicate crown-glass window (semitransparent photocathode); S-13 = Cs_3Sb with fused-silica window (semitransparent photocathode); S-19 = Cs_3Sb with fused-silica window (opaque semicathode); S-20 = Na_2KCsSb with lime or borosilicate glass window. ERMA = extended red multialkali (RCA; ITT uses MA for multialkali). This curve is representative of several manufacturers' products. Many variations of this response are available, e.g., trade-offs between short- and long-wavelength response. (*From RCA Electronic Components, chart PIT-701B.*)

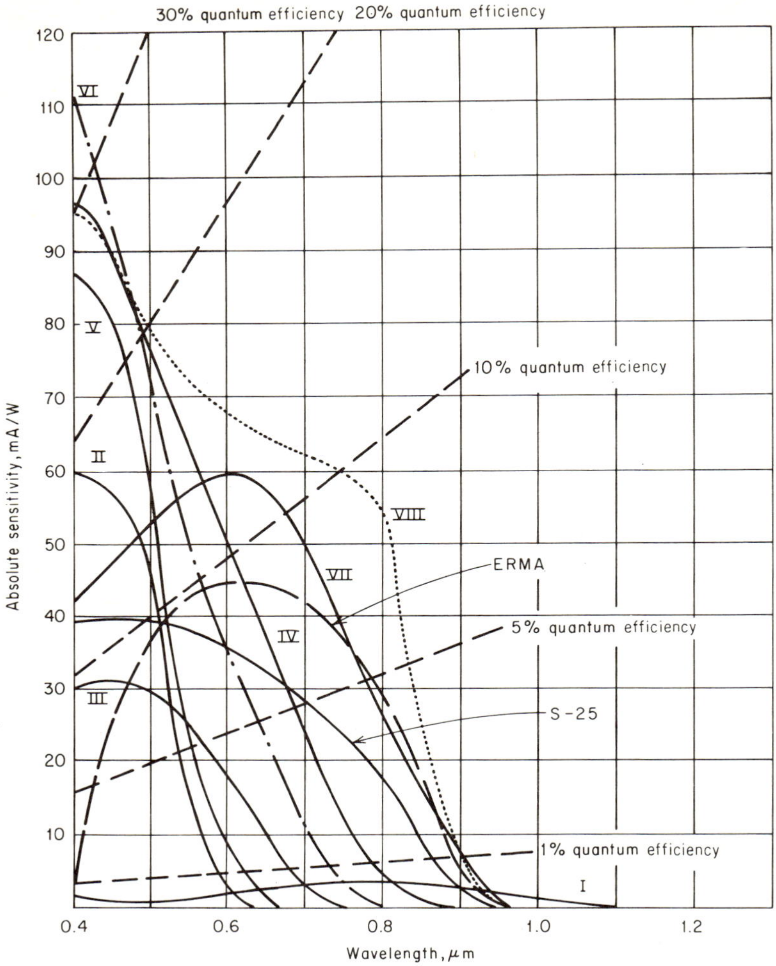

Fig. 34 Photocathode responsivity and quantum efficiency vs. wavelength (no windows). For abbreviations, see Fig. 33. I = S-1, II = S-11, III = S-10, IV = S-20, V = K_2CsSb, VI = $K_2CsSb(O)$, VII = $NaKCsSb_3$, VIII = GaAs(Cs). (*Based on material from RCA.*)

ever, curves for S-1, S-20, and S-25 are shown in Figs. 35 (300 K) and 36 (PMT cooled to 200 K).

Short-Wavelength Considerations. Window considerations are as follows:

For $\lambda > 200$ nm: windows are essential, as all useful photocathode materials are oxidized and performance would otherwise be destroyed.

200 nm $> \lambda >$ 105 nm: Photocathode materials are not oxidized by dry air (moisture degrades performance). Windows are optional. LiF windows have shortest known cutoff, 105 nm. For $\lambda < 180$ nm it is generally advisable to flush with dry nitrogen.

$\lambda < 105$ nm: No windows are available.

Since air absorbs radiation in the region 0.2 to 200 nm (ozone absorbs 200 to 300 nm), it is necessary to include the (windowless) detector in the vacuum with the source.

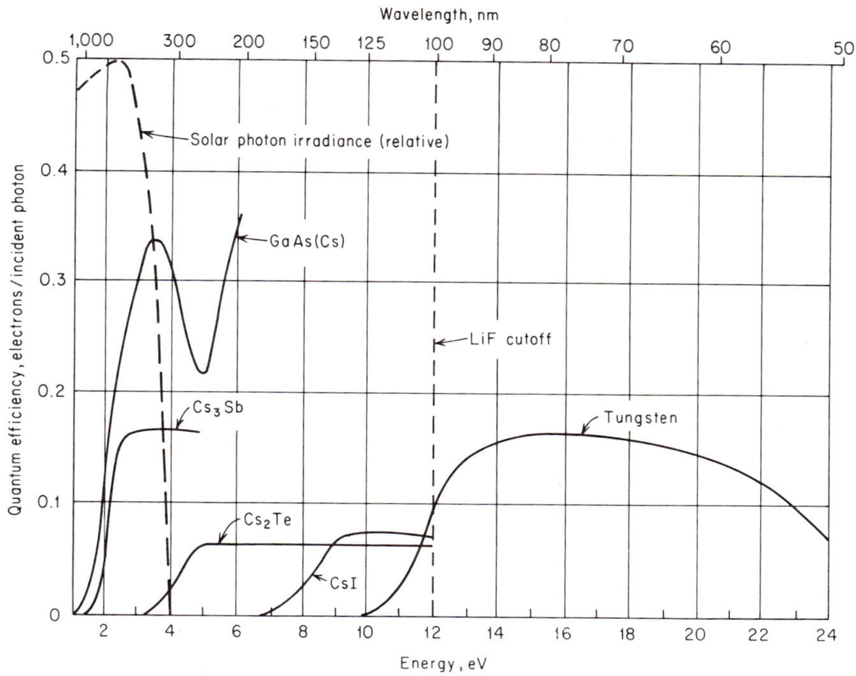

Fig. 37 Quantum efficiency vs. wavelength (photon energy) for several photoemitters.

Noise. The limiting noise in a PMT depends on the level of illumination. For low-level detection, limiting noise is the shot noise on the dark current,

$$i_n = (2ei_{\text{dark}}\,\Delta f)^{1/2}$$

For high illumination levels the shot noise on the signal photocurrent

$$i_n = (2ei_{\text{signal}}\,\Delta f)^{1/2}$$

far exceeds that on the dark current. Manufacturers usually express noise as photocathode dark current or anode dark current for given gain, which is therefore traceable to photocathode dark current.

Photocathode dark current is approximately proportional to photocathode area, so that small photocathode effective areas can be expected to have reduced noise.

Figure 38 shows how anode dark current and gain increase with applied voltage for a typical PMT.

Minimum detectable power is related to limiting noise through responsivity via

$$\text{NEP} = \frac{i_n}{\mathcal{R}}$$

where $\mathcal{R}$ is in amperes per watt.

Operating Temperature. Dark current due to thermionic emission, usually greater in red-sensitive tubes, can be reduced by cooling. The trialkali (S-20) performance does not benefit from cooling below 255 K. Maximum beneficial cooling (three to four dark counts per second) for AgOCs (S-1), (Cs)Na_2KSb (S-20), and Cs_3Sb (S-11) is 195, 255, and 239 K, respectively. Most photocathodes become noisier as temperature rises above ambient because of increased thermionic emission. Because its thermionic emission starts at a very low value, (Cs)Na_2KSb is a useful photocathode up to temperatures of approximately 373 K.

Response Time. The rise time of photomultiplier tubes depends chiefly on the spread in transit time during the multiplication process. For photomultiplier tubes this spread is about 10 ns. Some tubes with specially designed electron optics can give spread as low as 1 ns. The crossed-field PMT makes possible a spread as small as 0.1 ns.

For high-speed work (<1 ns rise time) good transmission-line technique must be used to obtain impedance match and to avoid reflection. The bandwidth of the output circuit will depend upon the total capacitance (PMT circuit plus stray capacitances) and the value of the load resistance.

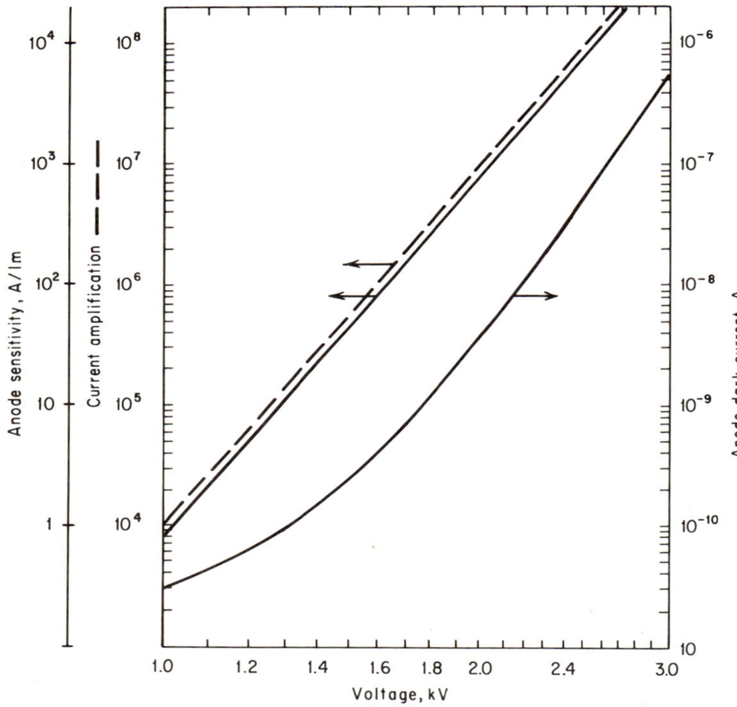

Fig. 38 Typical current amplification and anode dark current as a function of applied voltage. (*Based on E. H. Eberhardt, "D* of Photomultiplier Tubes and Image Detectors," ITT Industrial Labs,* 1969.)

Linearity. Photomultiplier tubes are nearly all linear to about 1 percent for cathode currents of 0.1 μA or less. Some tubes may be linear to better than 0.1 percent but must be individually selected (Bennett, 1966). Probably most of the nonlinearity results from the dynode structure.

Sensitive Area. No fundamental limitation. Only recently available with very small effective areas for extremely low dark current. Magnetic focusing has been used so that only a small fraction of the photocathode is used electron-optically.

Sensitivity Profile. Usually uniform within 20 to 50 percent.

Recommended Circuit. See Fig. 39.

1. Since a PMT is a current generator, increasing output resistance R_1 increases output voltage. An upper limit to R_1 may be imposed either by the time-constant limitation or by nonlinearity, which results from a space charge produced near the anode when the anode is left nearly floating electrically.

2. The rated photocathode current (referred to anode current through gain) should not be exceeded.

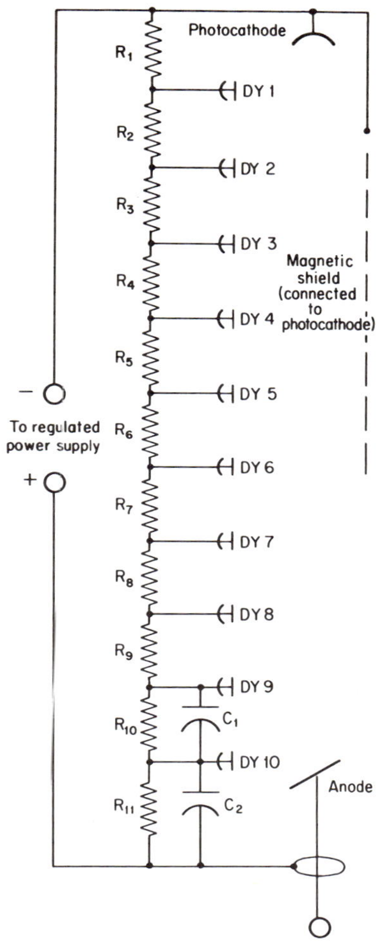

Fig. 39 $C_1 = 68$ pF $\pm$ 10 percent, 500 V (dc working); $C_2 = 270$ pF $\pm$ 10 percent, 500 V (dc working); $R_1 = 220$ kΩ $\pm$ 5 percent, $\frac{1}{4}$ W; $R_2 = 240$ kΩ $\pm$ 5 percent, $\frac{1}{4}$ W; $R_3 = 330$ kΩ $\pm$ 5 percent, $\frac{1}{4}$ W; R_4 to $R_{11} = 200$ kΩ $\pm$ 5 percent, $\frac{1}{4}$ W. (*Based on E. H. Eberhardt, "D* of Photomultiplier Tubes and Image Detectors," ITT Industrial Labs, Inc.,* 1969.)

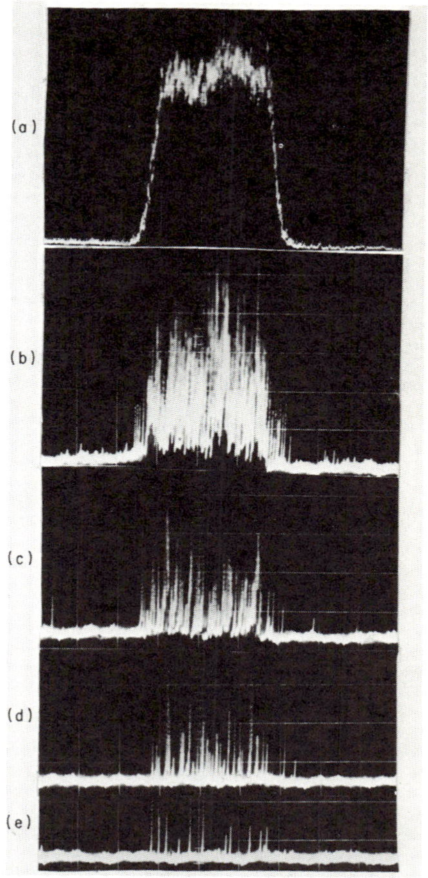

Fig. 40 Oscilloscope presentation of PMT output when viewing square-wave chopped light pulse. In (*a* to *e*) intensity is reduced and gain increased commensurately. (*Courtesy of E. H. Eberhardt.*)

3. Care should be taken not to destroy the photocathode with light (heating).
4. When large currents are drawn, it may affect later dynode interstage voltage and hence gain, causing nonlinearity; e.g., in Fig. 39, if the photocurrent from DY 10 to anode becomes comparable to the biasing current, through R_{11}, the gain of the final stage is reduced. This can be avoided by biasing the dynodes with constant-voltage sources.
5. To avoid dynode damage, final dynode current must not exceed the value suggested by the manufacturer.

Photon Counting. At the photocathode, the shot-noise-limited signal-to-noise ratio (with negligible dark current) is

$$\frac{i_s}{i_n} = \frac{i_s}{(2ei_s\,\Delta f)^{1/2}}\left(\frac{i_s}{2e\,\Delta f}\right)^{1/2} = \left(\frac{N_s}{2\Delta f}\right)^{1/2}$$

where N_s is the photoelectron rate at the photocathode. Thus, for extremely low levels of illumination the ideal signal-to-noise ratio becomes very poor. At this point there is much to be gained by abandoning attempts to measure the height of the fluctuating signal (Fig. 40*a*) in favor of digitally recording the presence or absence of individual pulses (Fig. 40*e*).

Single photoelectron counting can be achieved by using a pulse amplifier (see Fig. 41), which suppresses spurious dark-noise pulses not identical in amplitude and shape to those produced by photoelectrons.

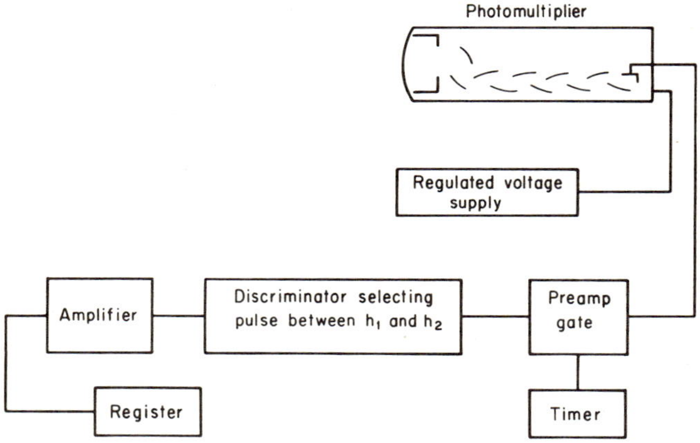

Fig. 41 Photomultiplier and associated circuits for photon counting. (*ITT Rep. E5.*)

An upper practical limit for (random) photon counting is set by convenient amplifier bandwidths at about $10^5\ s^{-1}$. For reasonable (1 percent) statistical accuracy this implies a 10-MHz bandwidth.

Gallium Phosphide Dynodes. The development of GaP dynodes for increased secondary-electron production (Simon et al., 1968; and Morton et al., 1968) makes possible unambiguous discrimination of small numbers of individual photoelectron counts which was not previously possible with lower dynode gains. This is shown in Fig. 42, where the spread in number of secondary electrons ($N \times$ gain) is just $(N \times \text{gain})^{1/2}$.

In addition to the above-mentioned fundamental advantage of high dynode gain, the large gain per stage in the first dynodes also helps discriminate against noise introduced by later stages of amplification. Also, fewer stages of amplification are required.

Manufacturers. Radio Corporation of America, Electro-Mechanical Research, International Telephone and Telegraph, EMI, Bendix, Amperex, Hamamatsu TV Co. Ltd.

25. Si Diffused-silicon *p-n* junction detectors are available with a variety of characteristics for various purposes. The main parameters of interest are spectral response, time constant, and low leakage current for large-area detectors. The effective time constant of silicon detectors is generally limited by *RC* considerations rather than by the inherent speed of the detection mechanism (diffusion). High reverse bias may or may not shorten charge-collection time, but it generally reduces cell capacitance. Both result in faster response. On the other hand, increased reverse bias causes increased noise, so that a trade-off exists between speed and sensitivity. For high-frequency applications, load resistance should be made small, and this, too, introduces (thermal) noise, which limits sensitivity. This problem has been met by the use of operational (current-mode) amplifiers, which can be built into the detector package, and by the avalanche photodiode, which introduces built-in gain before the load resistor is encountered. For the latter, very careful regulation of the avalanche bias is required for stable operation.

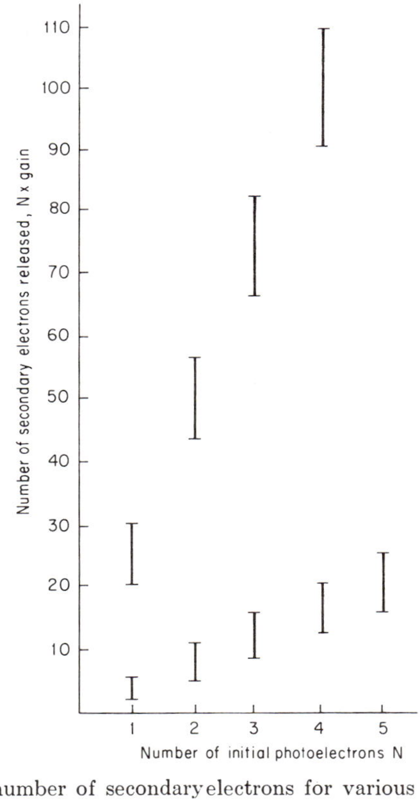

Fig. 42 Spread in number of secondary electrons for various phototube gains.

Five types of silicon detectors are described: type I, high-sensitivity; type II, faster, but somewhat less sensitive; type III, large-area (guard-ring construction); type IV, high-speed, with response shifted to red; type V, avalanche.†

Type I. This form was historically the first to be developed. It is generally used when high sensitivity is required and time constants slower than 1 μs are permissible (see types III and IV for modified spectral distributions).

Sensitivity: $D^*(\lambda_{pk}) \approx 10^{12}$, $D^*(2{,}800\text{ K}) \approx 2 \times 10^4$ cm · Hz$^{1/2}$/W, becoming amplifier-limited for small-area detectors (see Fig. 44).

† For relatively short-wavelength detectors, like Si and Ge, noise is not limited by 300-K background but is often limited instead by visible background from the sun or room lights or by amplifier noise. Under such circumstances D^* is not a very useful description.

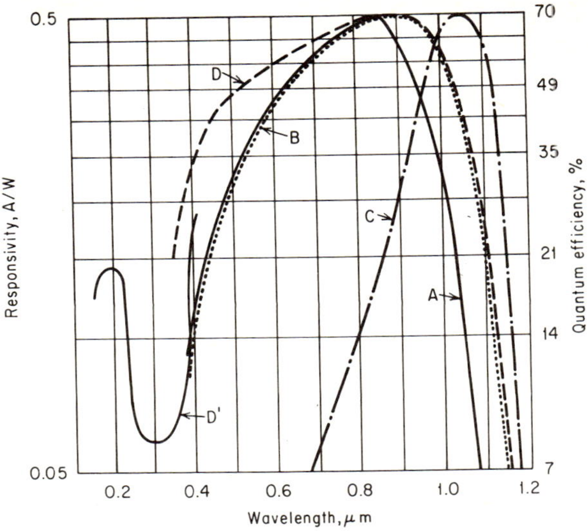

Fig. 43 Spectral response curves for silicon photodetectors. A = type I, B = types III, II, C = type IV, D = type III. For type IV see Fig. 64.

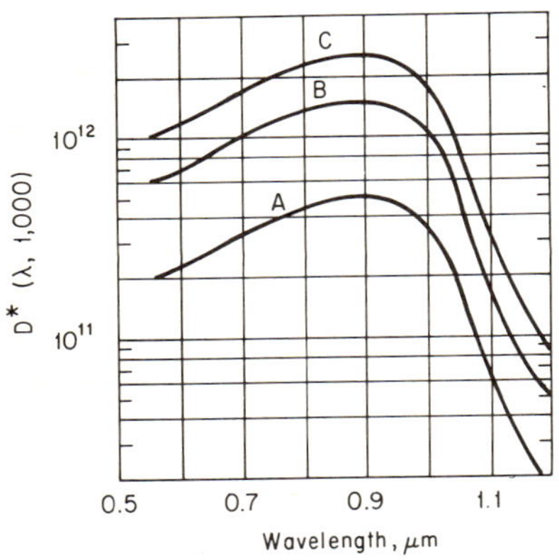

Fig. 44 D^* vs. λ for small-area type I silicon photodiodes: curves A, B, and C correspond to areas of 0.02, 0.2, and 1 cm². The lower D^* for smaller-area detector performance is due to amplifier limitations rather than intrinsically poorer D^* for small-area detectors. (*Texas Instruments, Infrared Devices, SC*-8385-366.)

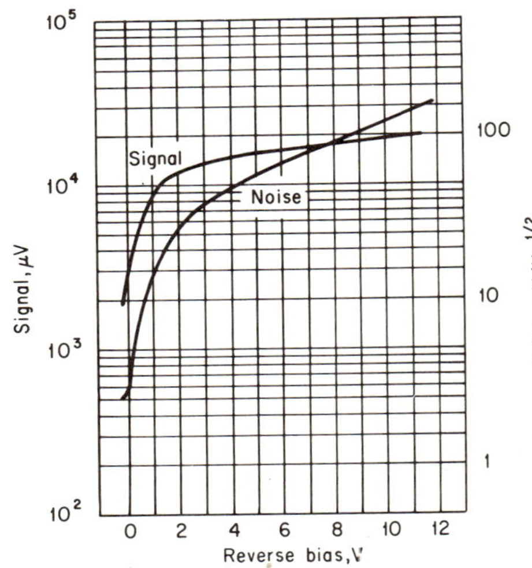

Fig. 45 Typical type I signal and noise vs. reverse bias (R_L = 10 MΩ). (*Electronuclear Laboratories, Bull.* 1053, 1966.)

Noise: See Fig. 45 (noise vs. bias) and Fig. 46 (noise vs. temperature); as T drops, impedance rises, so that decreasing noise current produces increasing noise voltage. However, the signal increases even faster, yielding an improved signal-to-noise ratio with cooling. Figure 48 (noise vs. frequency) shows dependence on bias (see Fig. 47).

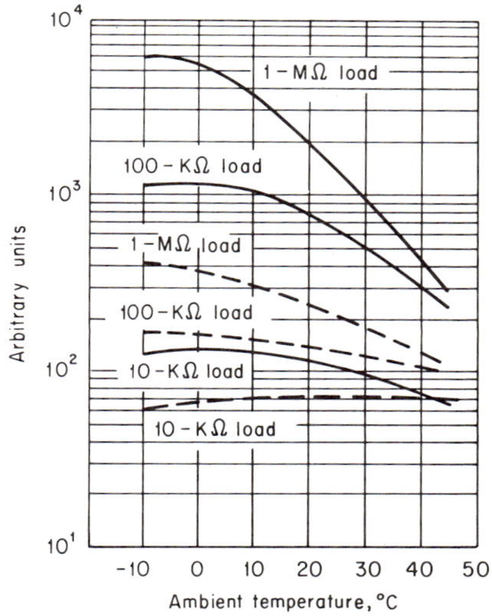

Fig. 46 Relative signal and noise vs. temperature for type I silicon photodiode (zero bias); —— = signal; ----- = noise. (*Electronuclear Laboratories, Bull.* 1052, 1966.)

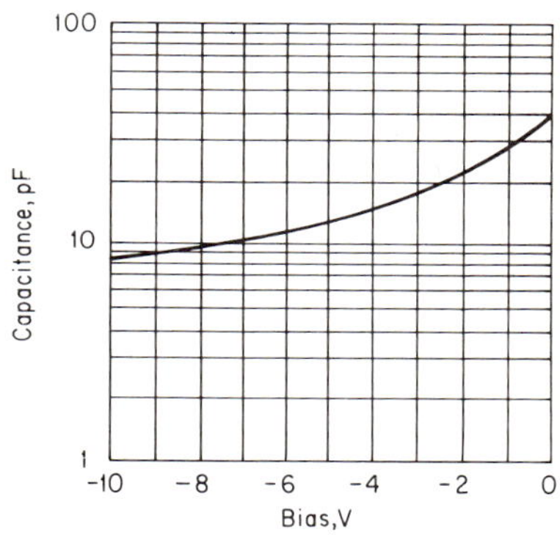

Fig. 47 Typical capacitance vs. bias for type I silicon photodiodes (A = 2 by 2 mm).

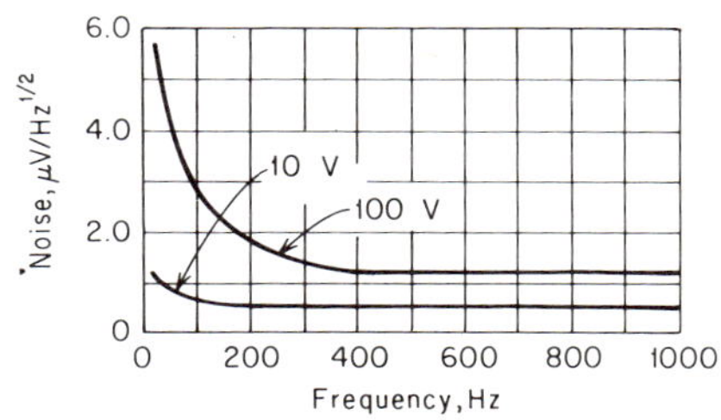

Fig. 48 Typical type I and type II noise-frequency spectrum for different reverse bias (A = 1 by 1 mm; R_L = 1 MΩ). (*Electronuclear Laboratories, Bull.* 1078, 1968.)

Capacitance: Capacitance is proportional to area and increases slightly with temperature.

Responsivity: See Fig. 45.

Quantum efficiency: See Fig. 43 (>90 percent quantum efficiency achievable with antireflection coating).

Sensitive area: 0.05 to 25 mm (linear dimensions).

Time constant: Inherently slower than 1 ns for high-sensitivity applications, limited by RC.

Operating temperature: Ambient (see Fig. 46 for typical signal and noise vs. temperature).

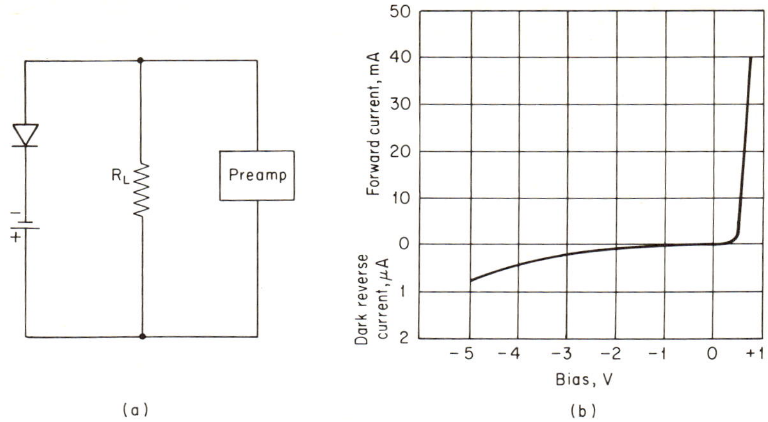

Fig. 49 Type I silicon photodiode: (*a*) recommended circuit; (*b*) typical electrical characteristics. (*Texas Instruments, Bull. SC*-8385-366.)

Recommended circuit: See Fig. 49. High-impedance FET current-mode amplifier to supply fixed bias voltage, regardless of current.

Manufacturers: Texas Instruments, Electro-Nuclear Labs, RCA, Solar Systems, Inc., Honeywell, Mullard (Optoelectronics).

Type II. This form (pin junction) is faster and somewhat less sensitive than type I, with slightly extended red response. In the normal *p-n* junction, charge-collection time $t = d^2/V\mu$, where d is the depletion layer or sensitivity thickness V is reverse bias, and μ is mobility. Increasing V to shorten the time constant is defeated by the fact that $D \propto \sqrt{V}$. It does, however, serve to decrease capacitance ($C \propto d^{-1}$), down to the limit where d equals total silicon-wafer thickness. For type II detectors d is effectively extended to that limit upon application of a relatively small bias, beyond which the relation to $d \propto \sqrt{V}$ no longer holds and collection time

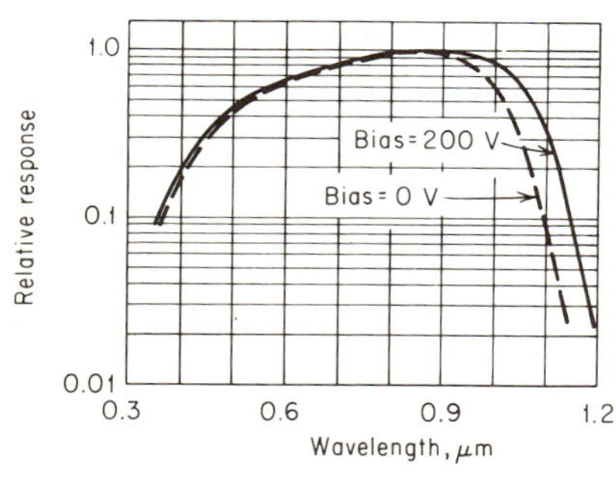

Fig. 50 Dependence of spectral response on bias for silicon photodiodes. (*Electronuclear Laboratories, Bull.* 1076, 1968.)

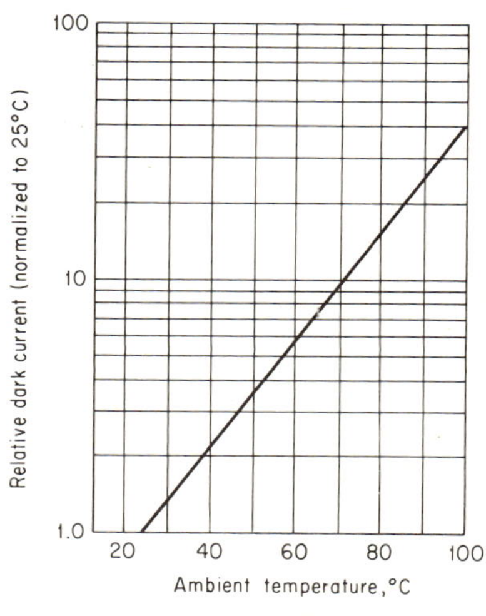

Fig. 51 Relative dark current vs. temperature for type II silicon photodiode (bias —100 V). (*Electronuclear Laboratories, Bull.* 1076, 1968.)

then decreases with increasing V, permitting very short rise time. Extended red response is a consequence of the thick depletion layer.

Sensitivity: $D^*(\lambda_{pk}) = 6 \times 10^{11}$ cm · Hz$^{1/2}$/W for 1 mm diameter (depends slightly on bias; see Fig. 50). For high-speed operation detectivity is approximately four times lower; $V_R = 10$ and $R_L = 1\ M\Omega$ (generally RC-limited).
Noise: See Fig. 48.
Quantum efficiency: See Fig. 43.
Responsivity: Similar to type I.
Capacitance: See Fig. 47.
Operating temperature: See Fig. 51.
Time constant: 3 ns rise time for $V_R = 100$; $R_L = 50\ \Omega$.
Sensitive area: 1 mm^2 standard.
Recommended circuit: Same as type I; (see Fig. 52).
Manufacturers: Electro-Nuclear Labs, Hewlett-Packard, United Detector Technology, Philco, Mullard (Optoelectronics), EG&G.

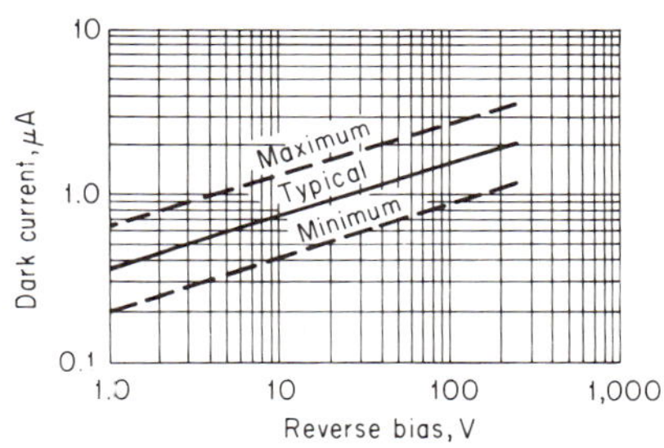

Fig. 52 Typical dark current vs. bias for type II silicon photodiode (A = 1 by 1 mm). (*Electronuclear Laboratories, Bull.* 1078, 1968.)

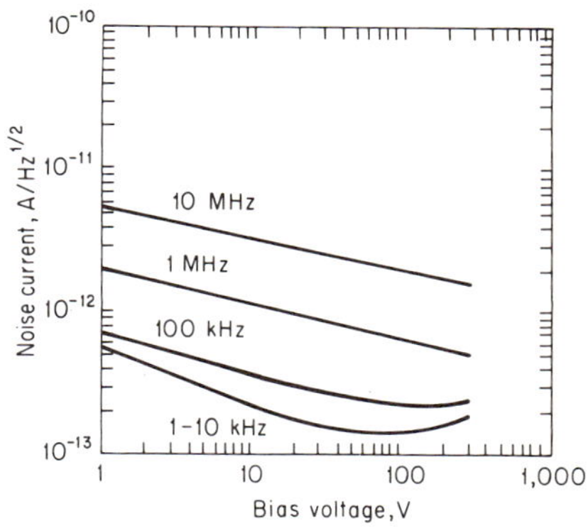

Fig. 53 Typical noise current vs. bias voltage and frequency for type III silicon photodiode (A = 0.05 cm^2). (*Electronuclear Laboratories.*)

Type III. This form has guard-ring construction, which greatly reduces surface leakage current, the main contributor to dark current. Two applications where this surface-leakage control is important are (1) for large-area detectors, especially where high-resistivity material is needed to withstand high bias for low capacitance, and (2) for extended ultraviolet response, where the depletion layer is made very thin, allowing large surface leakage.

The guard ring consists of an unilluminated separate part of the silicon wafer, which surrounds the sensitive region and which is maintained at the same potential (see Fig. 49*a*). The surface-leakage current is shunted around the load resistor and flows chiefly through the guard ring. Thus the much lower bulk leakage becomes the limiting dark current.

In addition to decreasing shot noise due to dark current, the guard ring also reduces thermal noise, since curtailment of leakage current effectively increases detector impedance. For high-speed (10 ns), large-sensitive-area applications, type III silicon offers low capacitance when high bias is applied, with the consequence of increased detector noise.

Sensitivity: Depends on bias (see Fig. 48). $D^*(0.9\ \mu m, 270) = 4 \times 10^{12}$ cm·Hz$^{1/2}$/W for 1 cm^2 ($V_R = 10$; $R_L = 1$ MΩ). Extended ultraviolet response can be had at the expense of additional leakage current.
Noise: See Figs. 53 and 54.
Quantum efficiency: See Fig. 43.
Responsivity: Similar to type I.
Capacitance: Depends on bias (see Fig. 55).

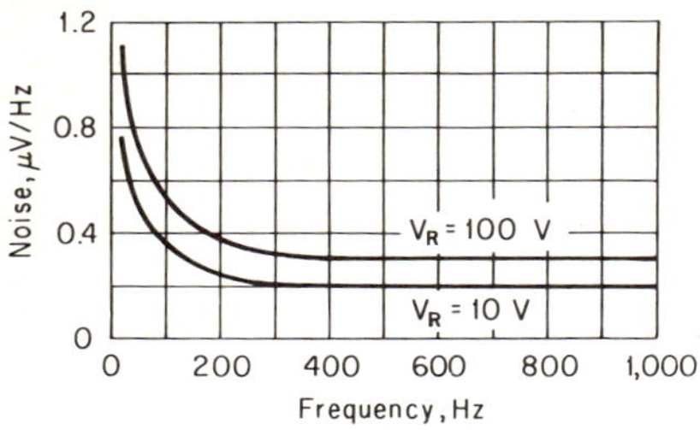

Fig. 54 Noise frequency spectrum for typical type III silicon photodiodes ($R_L = 1$ MΩ). (*Electronuclear Laboratories, Bull.* 1076, 1968.)

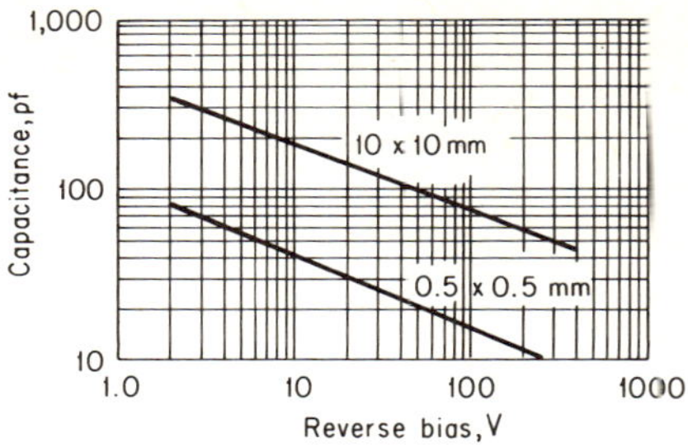

Fig. 55 Capacitance vs. bias for typical type III silicon photodiodes. (*Electronuclear Laboratories, Bull.* 1076, 1968.)

Time constant: Rise time is 10 ns with $V_R = 100$; $R_L = 50$ Ω.
Sensitive area: 2 by 2, 5 by 5, and 10 by 10 mm standard.
Linearity: ±5 percent, seven decades (up to 20 mW for 1 cm², 2 mW for 2 by 2 mm).
Sensitivity profile: ±2 percent for 0.6-mm light probe and ±10 percent for 1-μm light probe.
Recommended circuit: See Fig. 56.
Operating temperature: See Fig. 57.

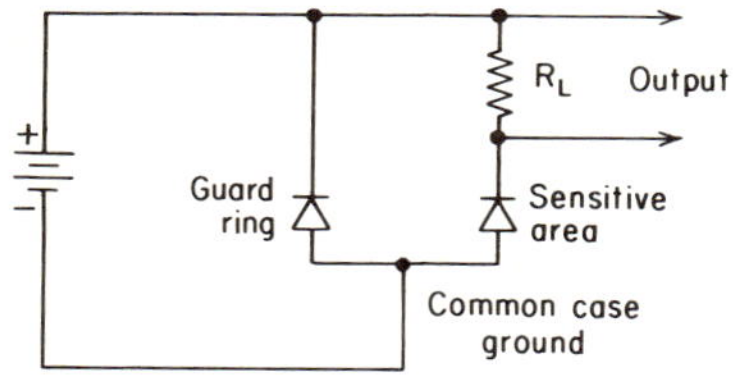

Fig. 56 Circuit for type III silicon photodiode. (*Electronuclear Laboratories, Bull.* 1076, 1968.)

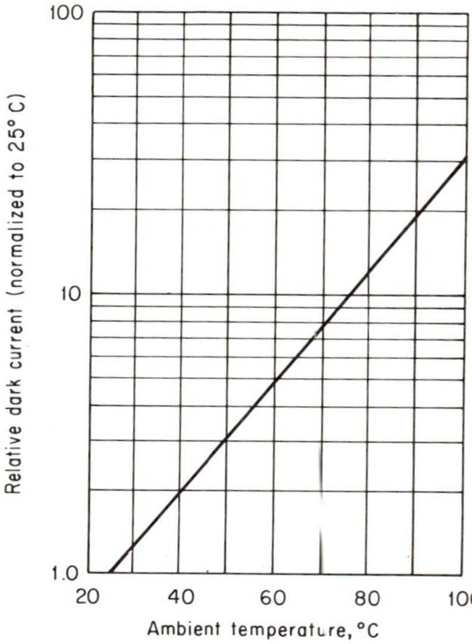

Fig. 57 Relative dark current vs. temperature for typical type III silicon photodiode (V_R = 100 V). (*Electronuclear Laboratories, Bull.* 1076, 1968.)

Manufacturers: Electro-Nuclear Labs, EG&G, Honeywell.

Type IV. This form is designed for fast response and fairly high sensitivity, peaking at 1.06 μm. Extended red response is obtained by forming a thick depletion layer underneath a dead layer (high surface recombination). Photons of $\lambda < 0.7$ μm are mostly absorbed by the dead layer. Low capacitance is achieved by making the depletion layer very thick. High-resistivity material is used to withstand high bias voltage for high-speed applications.

Sensitivity: See Fig. 43; $D^*(\lambda_{pk}) = 5 \times 10^{12}$ cm·Hz$^{1/2}$/W for $V_R = 0$, $R_L = 40$ MΩ; $D^* = 10^{12}$ cm · Hz$^{1/2}$/W for $V_R = 60$, $R_L = 50$ Ω.

Noise: See Figs. 58 to 60.

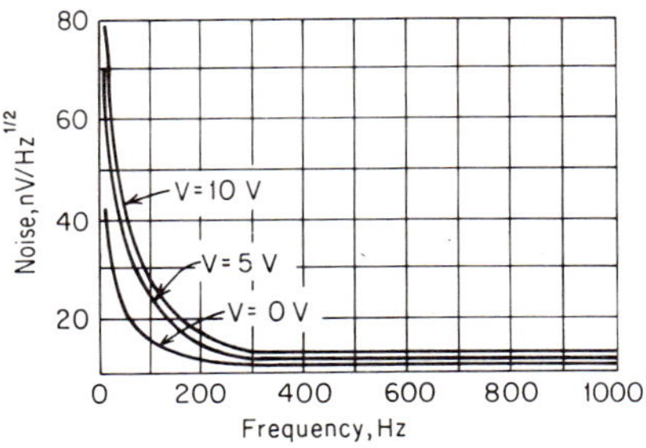

Fig. 58 Noise frequency spectrum for typical type IV silicon photodiode ($V_R = 0, 5, 10$ V; $R_L = 1$ kΩ). (*Electronuclear Laboratories, Bull.* 1078, 1968.)

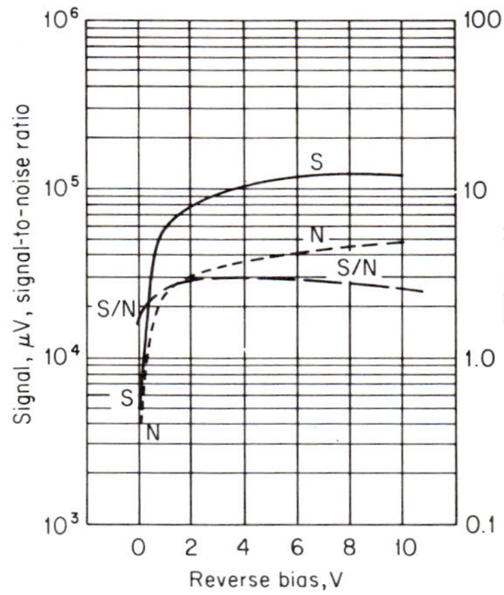

Fig. 59 Typical noise and signal vs. bias for type IV silicon photodiode ($R_L = 2.5$ MΩ; $\Delta f = 15$ Hz). (*Electronuclear Laboratories, Bull.* 1041, 1966.)

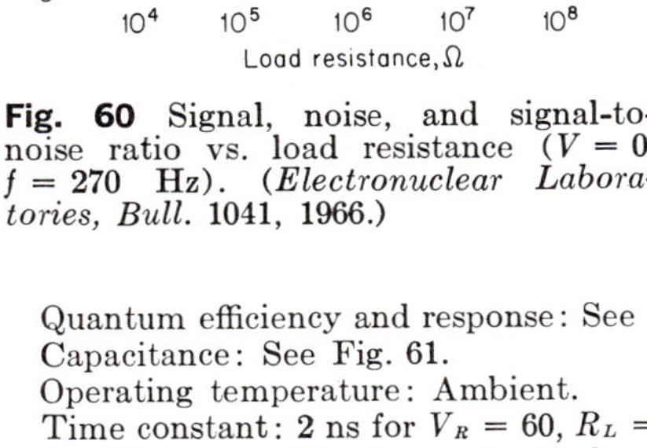

Fig. 60 Signal, noise, and signal-to-noise ratio vs. load resistance ($V = 0$, $f = 270$ Hz). (*Electronuclear Laboratories, Bull.* 1041, 1966.)

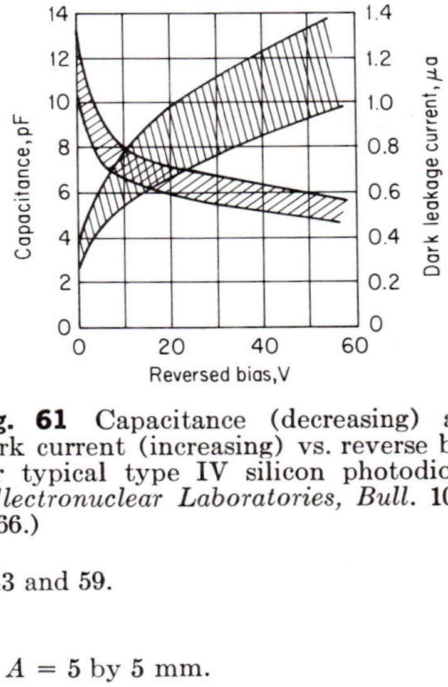

Fig. 61 Capacitance (decreasing) and dark current (increasing) vs. reverse bias for typical type IV silicon photodiode. (*Electronuclear Laboratories, Bull.* 1041, 1966.)

Quantum efficiency and response: See Figs. 43 and 59.

Capacitance: See Fig. 61.

Operating temperature: Ambient.

Time constant: 2 ns for $V_R = 60$, $R_L = 50$ Ω, $A = 5$ by 5 mm.

Sensitive area: 5 by 5 mm standard.

Recommended circuit: Same as type I.

Manufacturers: Electro-Nuclear Labs, Harshaw Chemical Company.

Type V. The avalanche photodiode is especially useful where both fast response and high sensitivity are required. Whereas normal photodiodes become thermal-noise-limited when heavily loaded for fast response, avalanche photodiodes make use of internal multiplication, associated with reverse breakdown in the *p-n* junction. Stable multiplication is made possible by a guard-ring construction, which prevents surface breakdown. However, very careful bias control is essential for stable performance. An optimum gain exists below which the system is limited by receiver noise and above which shot noise dominates receiver noise and the overall noise increases faster than the signal.

In addition to fast-response applications, avalanche photodiodes are useful whenever amplifier noise is limiting, e.g., small-area arrays. Signal-to-noise-ratio improvements of one to two orders of magnitude over nonavalanche case can be achieved.

Sensitivity: NEP $\approx$ 0.1 pW at 10 MHz; $A = 5 \times 10^{-4}$ cm^2; gain = 800.

Noise: Typically 0.2 to 10 nA. As gain increases, noise increases. Optimum gain is where avalanche noise equals system noise. Thus optimum gain is a function of system noise (see Fig. 62).

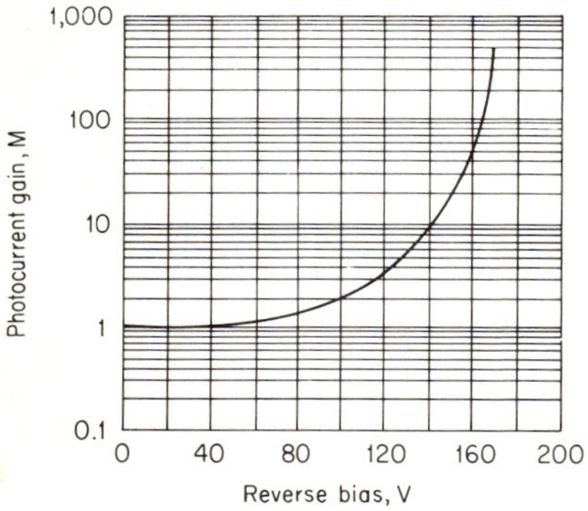

Fig. 62 Photocurrent gain vs. reverse voltage for silicon photodiode; T = 25°C. (*Texas Instruments, Bull. DL-S,* 1969.)

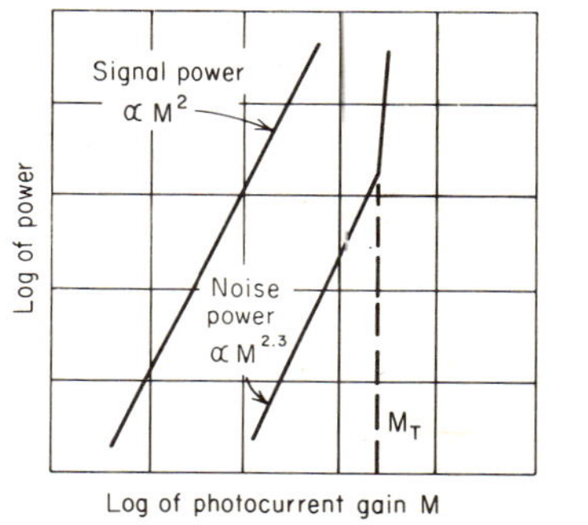

Fig. 63 Signal power and noise power vs. photocurrent gain for silicon avalanche photodiode. (*Texas Instruments, Bull. DL-S,* 1969.)

Responsivity: Depends on photocurrent gain (see Fig. 63). Gain-bandwidth product 80 GHz at 633 nm. Spectral response depends on operating frequency (see Fig. 64).

Quantum efficiency: Typically 35 percent peak (see Fig. 64).

Capacitance: Depends on bias (see Fig. 65) and area.

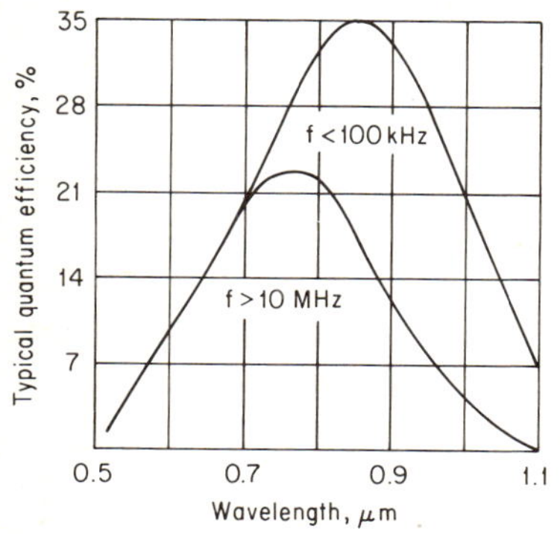

Fig. 64 Typical quantum efficiency vs. wavelength for silicon photodiode (M = 100, T = 298 K). (*Texas Instruments, Bull. DL-S,* 1969.)

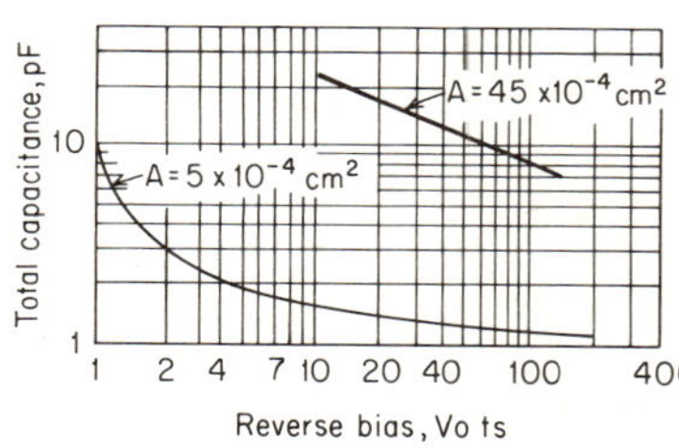

Fig. 65 Total capacitance vs. reverse voltage for silicon avalanche photodiodes of different areas (f = 1 MHz; T = 298 K). (*Texas Instruments, Bull. DL-S,* 1969.)

Sensitive area: 2 to 500 $\times 10^{-5}$ cm^2.
Series resistance: Depends on area.

A, 10^{-4} cm^2	f, GHz	R_S, Ω
5	90	50
45	0.9	5

Time constant: Depends on gain (gain bandwidth 80 GHz).
Recommended circuit: Constant-current operation can be achieved by using a reference diode to sense temperature and using its output to regulate a bias voltage, which properly varies with temperature.
Operating temperature: 208 to 398 K.
Manufacturers: Texas Instruments, General Electric Co., Space Technology Products, Honeywell.

26. Ge Germanium p-n junction photodiodes offer submicrosecond response or high sensitivity from the visible region to 1.8 μm. Zero bias is generally used for high sensitivity and large reverse bias for high speed.

As with silicon, spectral response and rise time can be modified. Three forms are described:

Type I: Peak response at 1.55 μm
Type II: Peak response at 1.75 μm
Type III: Avalanche

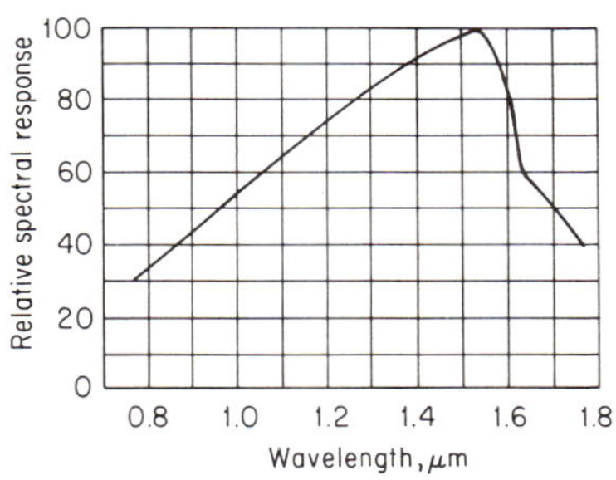

Fig. 66 Spectral response for type I germanium photodiode (T = 298 K). (*Electronuclear Laboratories, Bull.* 1052, 1966.)

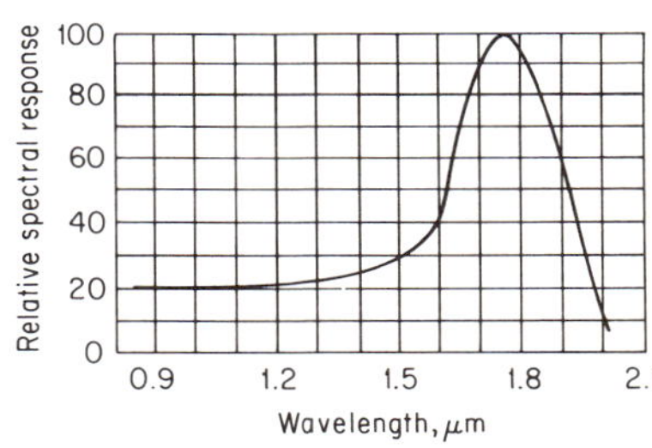

Fig. 67 Spectral response for type II germanium photodiode (T = 298 K). (*Electronuclear Laboratories, Bull.* 1042, 1966.)

Types I and II
Sensitivity: (See Figs. 66 and 67)

$$D^*(\lambda_{pk}, 270) = \begin{cases} 8 \times 10^{10} \text{ cm} \cdot \text{Hz}^{1/2}/\text{W} & \text{type I} \\ 5 \times 10^{10} \text{ cm} \cdot \text{Hz}^{1/2}/\text{W} & \text{type II} \end{cases}$$

Quantum efficiency: >50 percent with antireflection coating
Noise: $1/f$
Responsivity: 1.0 A/W
Capacitance: Similar to silicon
Time constant: Type I: 50 ns rise time, $V_R = 10$; 50 Ω; type II: 40 ns rise time, $V_R = 50$; 50 Ω
Sensitive area: Type I: 1 mm^2 standard, type II: 5 mm^2 standard
Operating temperature: Ambient
Profile: Similar to silicon
Linearity: Perfect over at least three decades
Recommended circuit: Standard pV, reverse-biased to 0 V across terminals
Manufacturers: Electro-Nuclear Labs, Texas Instruments

Type III. The germanium avalanche photodiode is similar to the silicon one but has lower optimum gain, longer cutoff wavelength (1.6 μm), and higher leakage current.

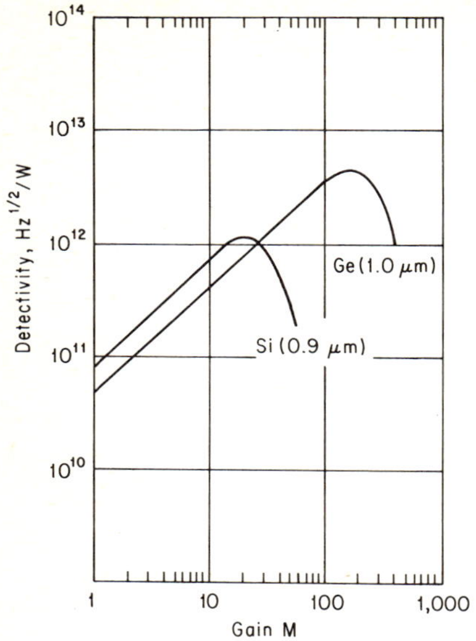

Fig. 68 Detectivity vs. gain for silicon and germanium avalanche photodiodes ($A = 5 \times 10^{-4}$ cm^2).

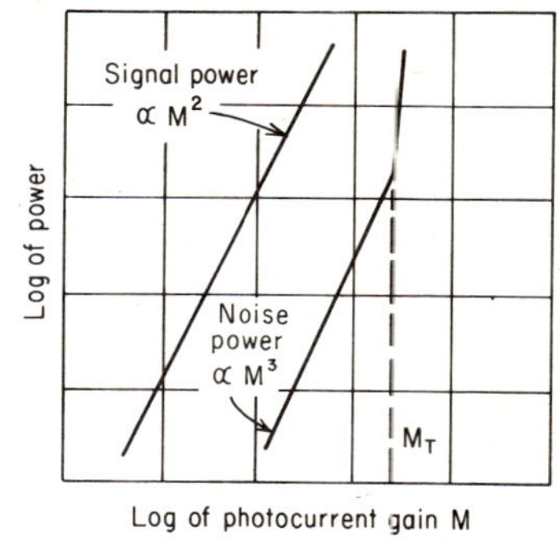

Fig. 69 Signal and noise power vs. photocurrent gain. For germanium avalanche photodiode $i_N{}^2 = 2ei_\beta M\ \Delta f$, where i_β = dark current, M = gain, M_T = threshold gain for excess noise, and Δf = bandwidth. (*Texas Instruments, Bull. DL-S,* 1968.)

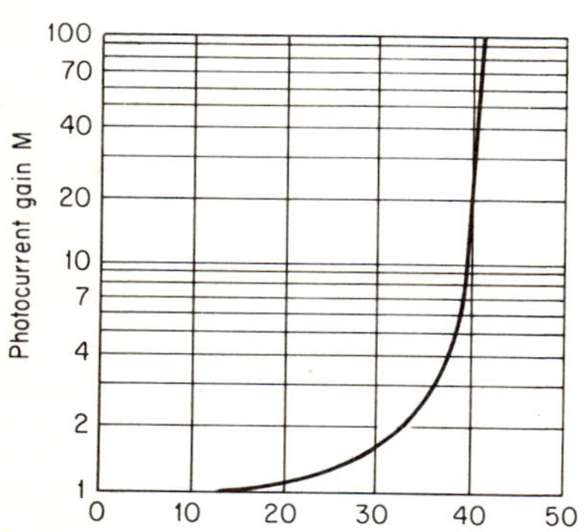

Fig. 70 Photocurrent gain vs. reverse voltage for germanium avalanche (T = 298 K). (*Texas Instruments, Bull. DL-S,* 1968.)

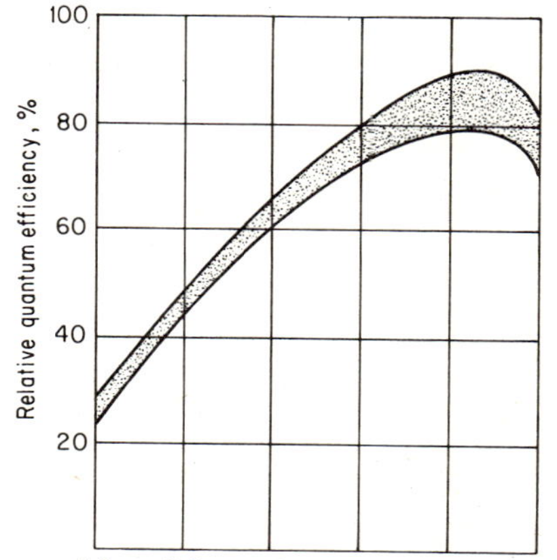

Fig. 71 Germanium avalanche photodiode, −25°C quantum efficiency vs. wavelength for antireflection coated detector. (*Texas Instruments, Bull. DL-S,* 1968.)

Sensitivity: NEP ≈ 0.1 pW/Hz$^{1/2}$ at 30 MHz, $A = 5 \times 10^{-4}$ cm^2; depends on gain (see Fig. 68)
Noise: See Fig. 69

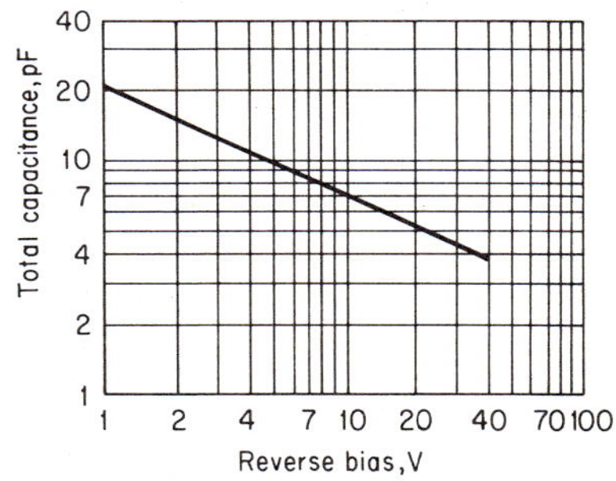

Fig. 72 Capacitance vs. reverse bias for germanium avalanche photodiode ($f = 25$ MHz; $T = 298$ K). (*Texas Instruments, Bull. DL-S,* 1968.)

Capacitance: Depends upon bias (see Fig. 72)
Quantum efficiency: ~85 percent at 1.5 μm with antireflection coating
Responsivity: $\mathcal{R} = \eta M/h\nu$ A/W
where η is quantum efficiency and M is gain (see Figs. 69 to 71)
Operating temperature: Ambient
Series resistance: 25 Ω
Gain-bandwidth product: 50 GHz
Sensitive area: 5×10^{-4} cm^2 standard
Profile: ±5 percent with $M = 150$ and 12-μm spot
Recommended circuit: V_B (breakdown voltage) ≈ 40 V; $\Delta V_B/T = 60$ mV/°C for reverse current 100 μA (see type V silicon recommended circuit)
Manufacturer: Texas Instruments

27. PbS Photoconductive lead sulfide was one of the earliest and most successful infrared detectors made. It remains the most sensitive uncooled detector in the region 1.3 to 3 μm. With cooling, PbS sensitivity is competitive with other detectors out to about 4.2 μm; however, its response time is slow.

Many PbS characteristics can be varied by special fabrication methods. These characteristics include spectral detectivity, time-constant resistance, and upper limit of operating temperature. PbS is generally made by chemical deposition, except for high-temperature (373 K) applications, where evaporation is used. By the former method unusually small sensitive areas can be made, and hence extremely small NEP is attainable.

Three distinct forms of PbS are described. ATO, ITO, and LTO, for ambient-, intermediate-, and low-temperature operation. They are manufactured differently for particular temperature ranges, as shown in Table 2.

TABLE 2

	Typical operating temperature, K			
	350	273 (ATO)	193 (ITO)	77 (LTO)
Sensitivity	†	Figs. 73, 76	Figs. 74, 76, 77	Figs. 75, 76
$\dfrac{D^*(\lambda_{max})}{D^*(500\ \text{K})}$		105	55	17
Noise, cm·Hz$^{1/2}$/W		Fig. 80	Fig. 80	Fig. 80
Dark resistance, MΩ/sq	<0.3	<2	<10	<20
Time constant, μs‡	50	100–500	5,000	3,000

† At 350 K, cutoff wavelength moves into ~2.4 μm, with $D^*(\lambda_{max}) \approx 10^{10}$ cm·Hz$^{1/2}$/W.
‡ These are typical values; the time constant can be adjusted over two orders of magnitude in fabrication, but D^* is affected.

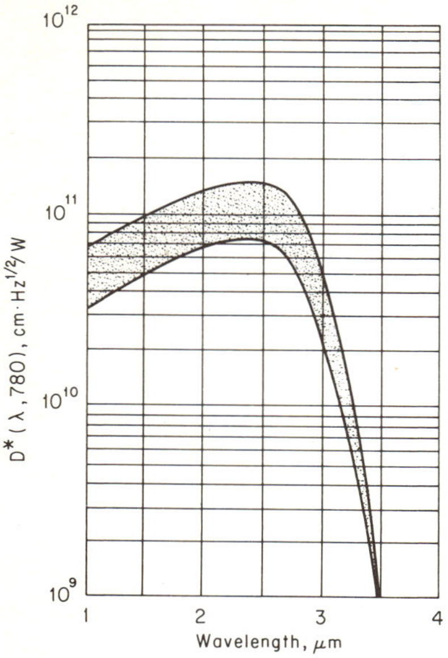

Fig. 73 Range of spectral detectivities for PbS (ATO) at 295 K; 2π fov, 295-K background. (*Santa Barbara Research Center, Brochure* 67.)

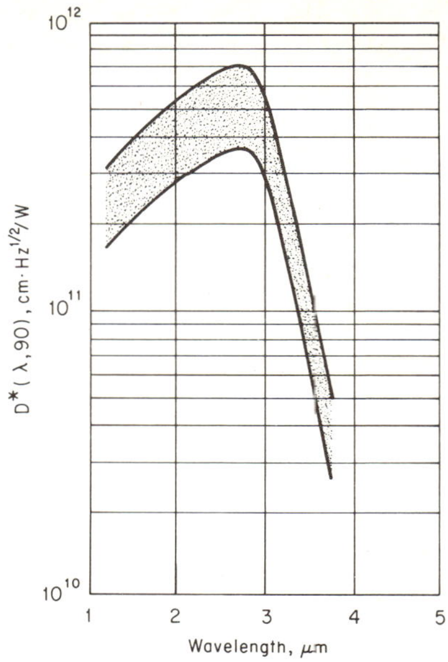

Fig. 74 Range of spectral detectivities for PbS (ITO) at 193 K; 2π fov, 295-K background. (*Santa Barbara Research Center, Brochure* 67.)

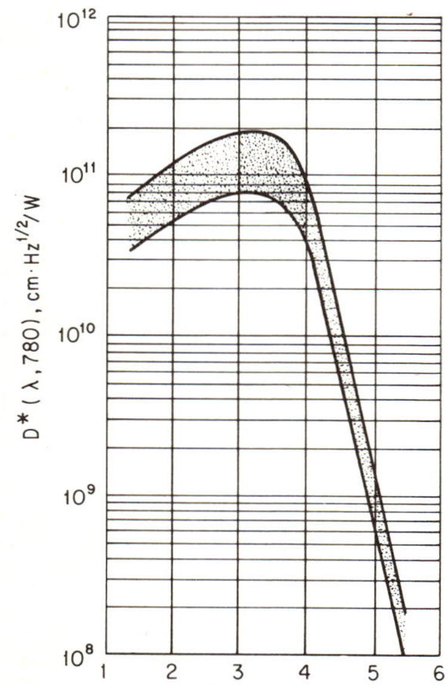

Fig. 75 Range of spectral detectivities for PbS (LTO) at 77 K; 2π fov, 295-K background. (*Santa Barbara Research Center, Brochure* 67.)

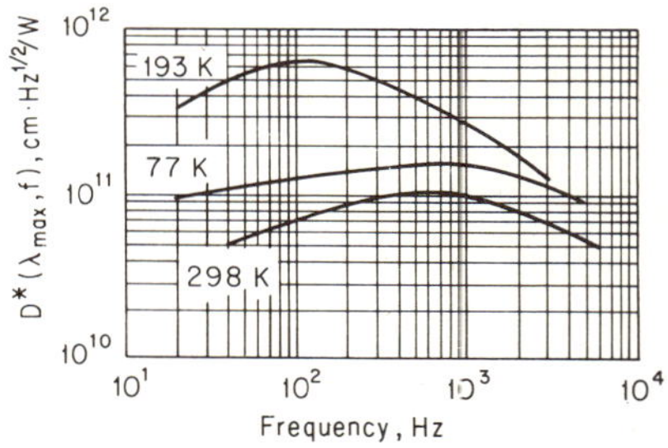

Fig. 76 Example of detectivity vs. frequency for PbS detectors at various operating temperatures; 2π fov, 295-K background. (*Santa Barbara Research Center, Brochure* 67.)

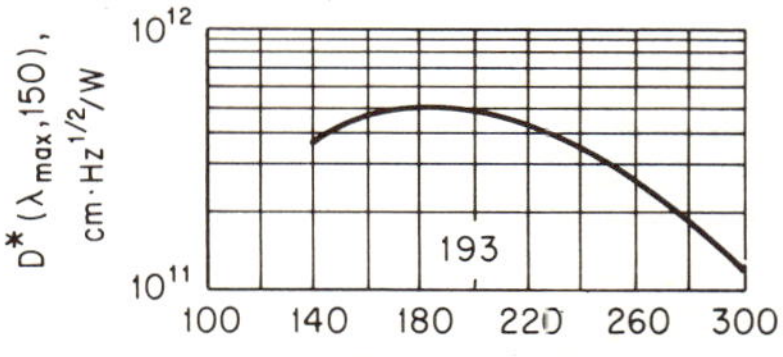

Fig. 77 Example of detectivity vs. temperature for PbS (ITO) detectors; 2π fov, 295-K background. (*Santa Barbara Research Center, Brochure* 67.)

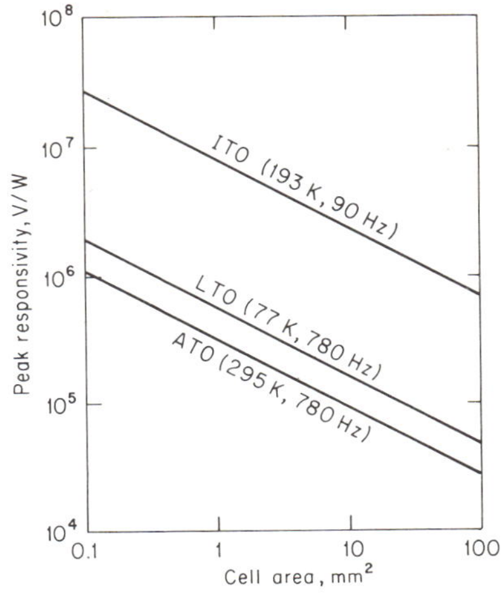

Fig. 78 PbS typical peak responsivity vs. cell area (actual values range within a factor of two of these shown).

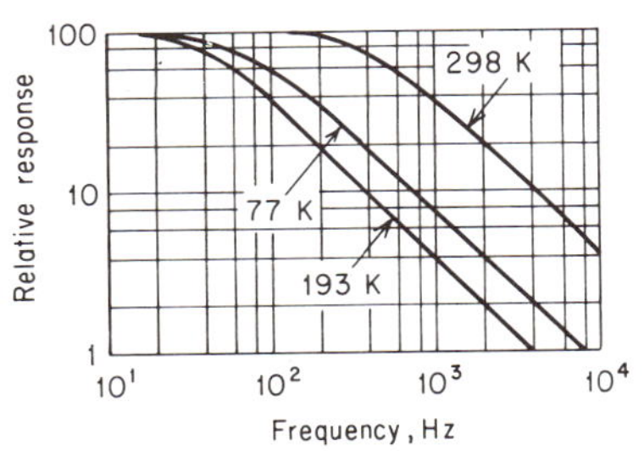

Fig. 79 Example of signal vs. frequency for PbS detectors. (*Santa Barbara Research Center, Brochure* 67.)

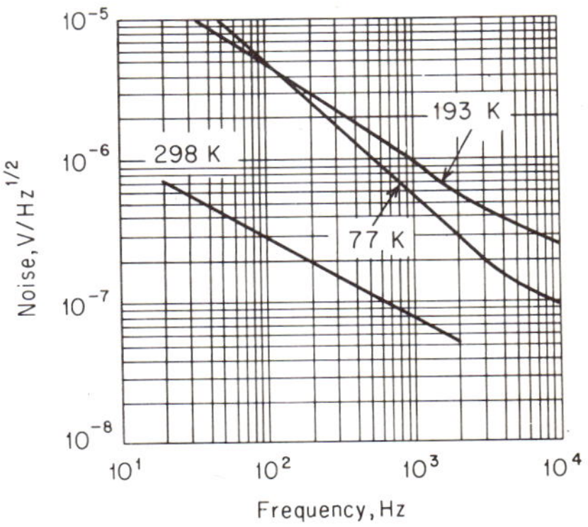

Fig. 80 Example of noise vs. frequency for PbS detectors (1 by 1 mm area). (*Santa Barbara Research Center, Brochure* 67.)

Responsivity: 8×10^4 V/W (see Figs. 78 and 79)
Sensitive area: Linear dimensions 10 to 7.5 μm
Capacitance: <1 pF (limited by mounting configuration)
Recommended circuit: Standard photoconductor
Sensitivity profile: Uniform within 10 percent
Linearity: Perfect 10^{-8} to 10^{-3} W
Manufacturers: Santa Barbara Research Center, Mullard, Sensor Precision Industries, Catron Electronics, Infrared Industries, Electronics Corp. of America

28. InAs (Photovoltaic) InAs detectors are single-crystal intrinsic photovoltaic devices for use in the 1- to 4-μm region. At room temperature InAs provides good sensitivity and submicrosecond response times. At 195 K InAs performance equals or betters the sensitivity of any other detector in the 1- to $3\frac{1}{2}$-μm region.

Sensitivity: At 298 K $D^*(\lambda_{pk})/D^*(500\ K) = 28$ (see Figs. 81 to 83)
Quantum efficiency: See Fig. 84
Noise: See Figs. 85, 86, 88, and 89
Time constant: Less than $\frac{1}{2}$ μs at all temperatures when loaded to lower circuit time constant

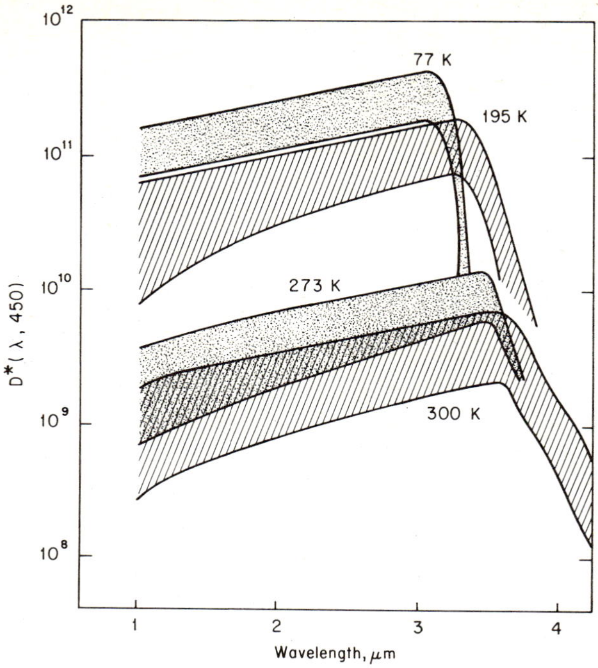

Fig. 81 InAs spectral sensitivity curves for several cell temperatures. (*Santa Barbara Research Center, Brochure* 67.)

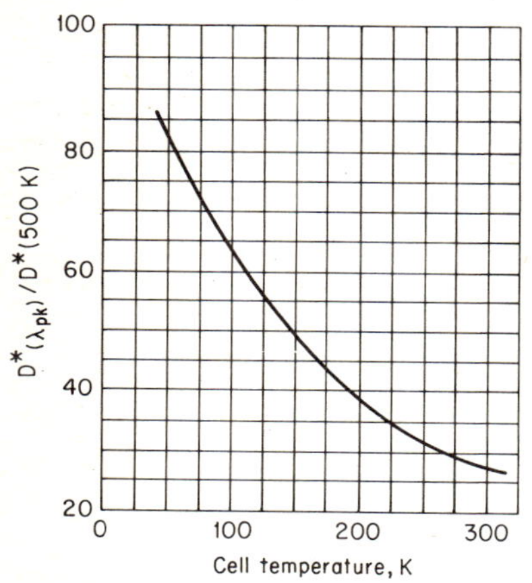

Fig. 82 Factor converting $D^*(500\ K)$ to $D^*(\lambda_{pk})$ vs. cell temperature. (*Texas Instruments, Bull. SC*-8385-366.)

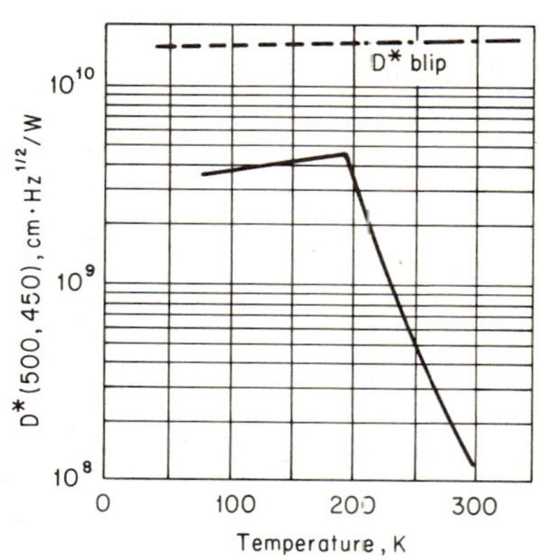

Fig. 83 InAs blackbody D^* vs. cell temperature; 2π fov. (*Texas Instruments, Bull. SC*-8385-366.)

Responsivity: Can exceed 10^6 V/W (see Figs. 87 to 89)

Dynamic resistance: ≤ 10 MΩ/sq

Diode capacitance: $\sim 0.01\ \mu$F/cm^2

Diode impedance: Inversely proportional to area and temperature (see Figs. 88 to 90)

Sensitive area: Standard sizes ~ 1 mm^2; sizes down to 0.008 mm^2 have been made, but it is difficult to obtain the high circuit dynamic impedance required to prevent the detector from being thermally limited

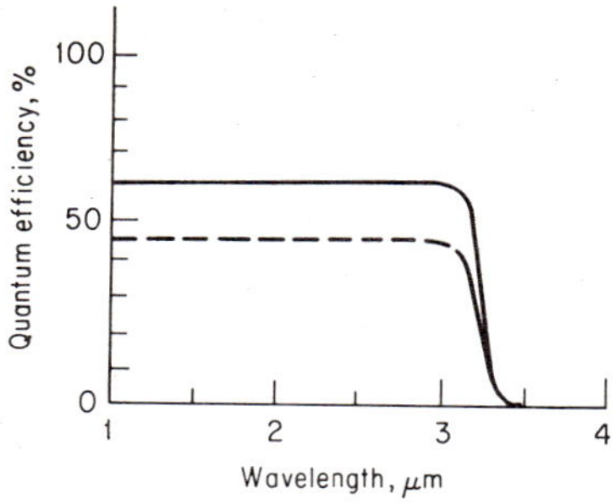

Fig. 84 InAs quantum efficiency vs. wavelength (77 K). (*Texas Instruments, Bull. SC*-8385-366.)

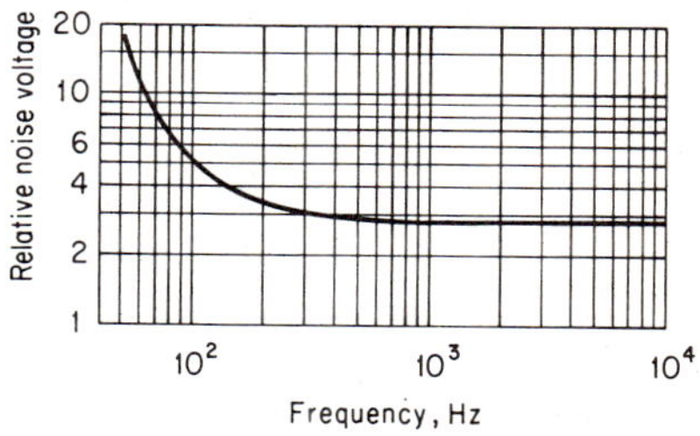

Fig. 85 Noise-frequency spectrum. (Variation with temperature is less than that between individuals at the same temperature.) (*Texas Instruments, Bull. SC*-8385-366.)

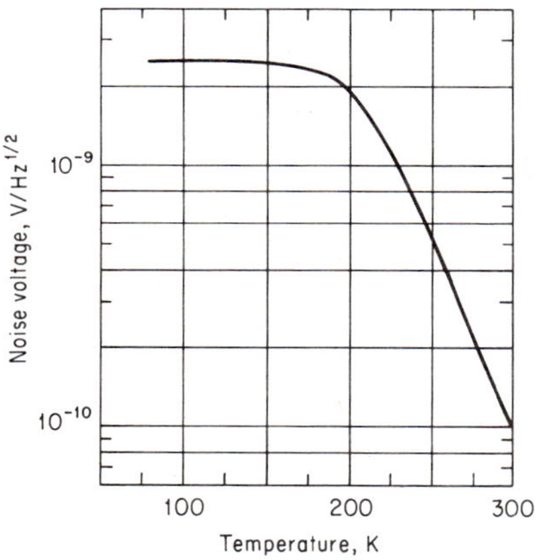

Fig. 86 Ac open-circuit noise voltage vs. temperature for typical InAs at dc short circuit ($A = 1$ cm²). (*Texas Instruments, Bull. SC*-8385-366.)

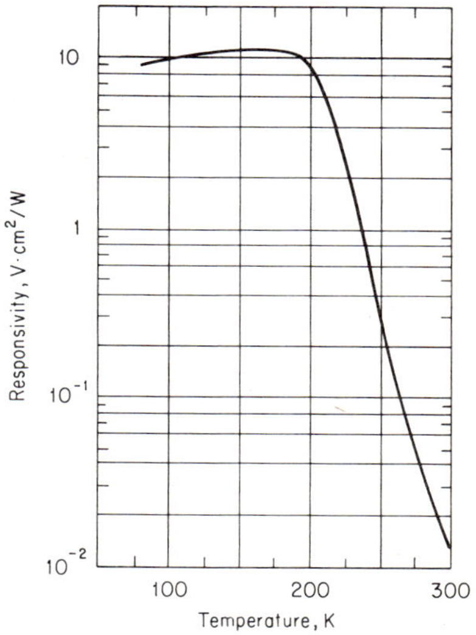

Fig. 87 InAs 500-K blackbody response vs. cell temperature at dc short circuit and ac open circuit ($A = 1$ cm²). (*Texas Instruments, Bull. SC*-8385-366.)

Operating temperatures: 77 to 300 K
Linearity: Perfect over at least three decades
Sensitivity profile: ±15 percent
Stability: Stable in all storage ambients so far tested

Recommended Circuits

Open circuit	*PV* InAs detectors with areas less than 2×10^{-2} cm² require no bias when operated and can be connected directly into the input stage of amplifier (capacitor ensures elimination of dc bias from amplifier); Fig. 91*a*
Transformer	A convenient means of using InAs at zero bias (Fig. 91*b*)
Reversed bias	At temperatures greater than 225 K considerable gain in impedance and responsivity is achieved by reverse-biasing (Fig. 91*c*)

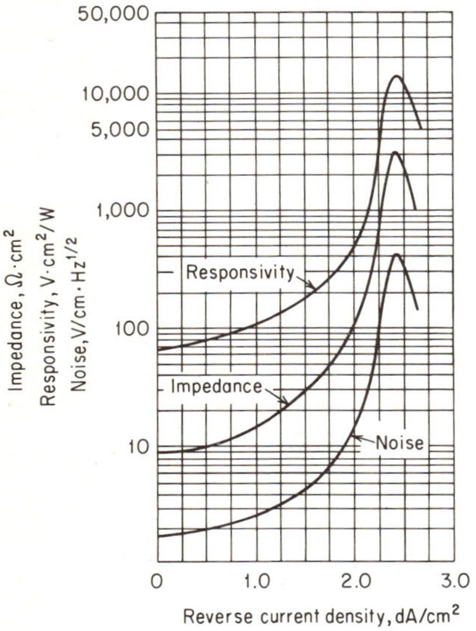

Fig. 88 Impedance, responsivity, and noise vs. reverse current density for InAs at 300 K. (*Texas Instruments, Bull. SC-8385-366.*)

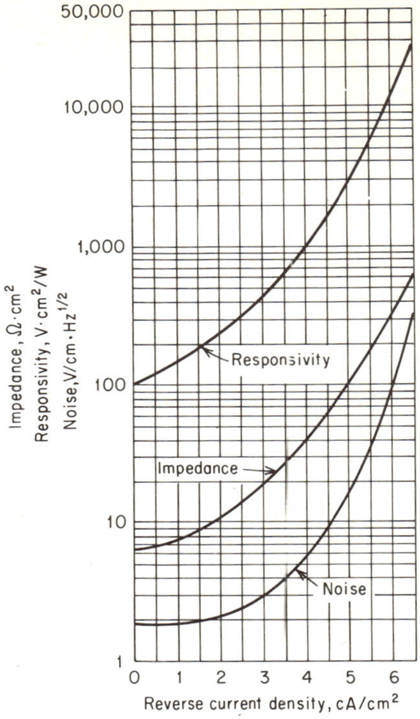

Fig. 89 Impedance, responsivity, and noise vs. reverse current density for InAs at 273 K. (*Texas Instruments, Bull. SC-8385-366.*)

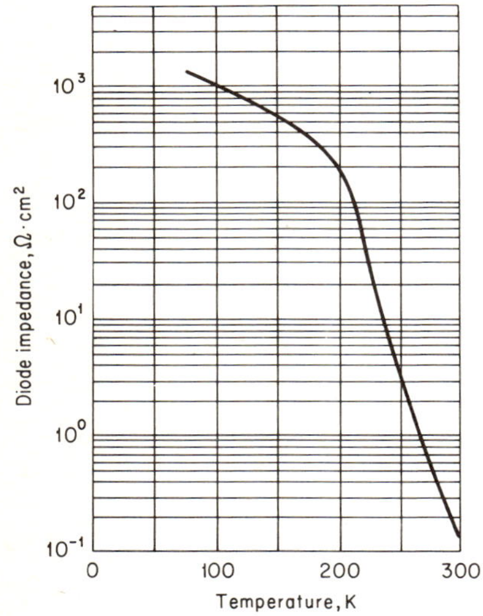

Fig. 90 Dc short-circuit impedance vs. cell temperature for InAs. (*Texas Instruments, Bull. SC-8385-366.*)

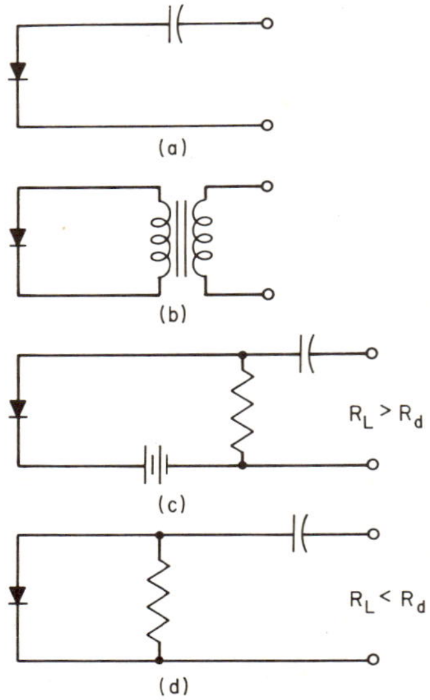

Fig. 91 Recommended circuits for InAs detectors: (*a*) open circuit, (*b*) transformer, (*c*) reversed bias, (*d*) fast response.

Fast response To utilize the short intrinsic time constant it is sometimes necessary to load the detector to lower the RC of the overall circuit (reverse bias will also lower detector capacitance) (Fig. 91*d*)

Manufacturers: Block Engineering, Texas Instruments, Santa Barbara Research Center, Barnes Engineering, Electro-Nuclear Labs, Spectronic

29. PbSe Lead selenide is an intrinsic, film-type photoconductor, whose speed of response exceeds that of PbS. At room temperature PbSe has D^*, exceeding that of room temperature PbS in the region 3.3 to 5 μm. At liquid-nitrogen temperature, its D^* is exceeded by a factor of 2 by InSb. However, for high-speed applications where operating temperature is maintained by thermoelectric cooling, InSb cannot presently be adequately cooled, and PbSe (~200 K) is then very useful. The long-wavelength response of 77 K PbSe extends approximately 1 μm past InSb and is competitive with 77 K Ge(Au) to 7 μm.

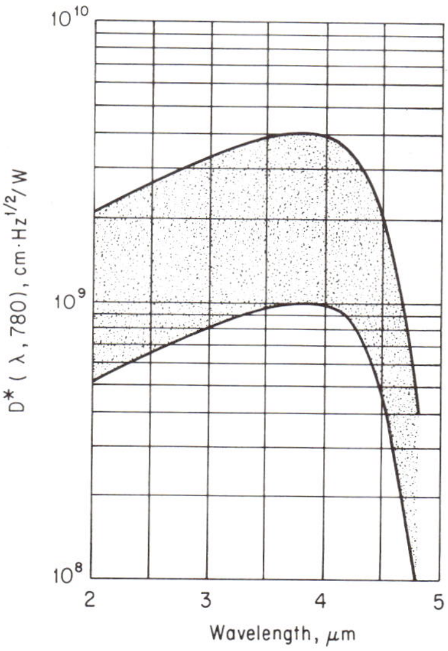

Fig. 92 Range of spectral detectivities for PbSe (ATO) detectors at 295 K. (*Santa Barbara Research Center, Brochure* 67.)

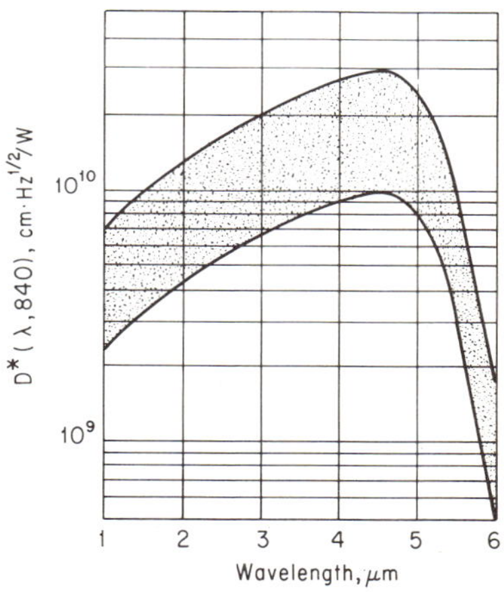

Fig. 93 Range of spectral detectivities for PbSe (ITO) infrared detectors at 193 K. (*Santa Barbara Research Center, Brochure* 67.)

TABLE 3

	Typical operating temperature, K		
	273 (ATO)	193 (ITO)	77 (LTO)
Sensitivity	Figs. 92, 97, 102	Figs. 93, 97, 102	Figs. 94–97, 102
$\frac{D^*(\lambda_{max})}{D^*(500\ K)}$	10	5.5	3.5†, 4.0‡
Time constant, μs	~2	30	40§
Responsivity	Figs. 98, 100	Figs. 99, 100	Figs. 99, 100
Dark resistance, MΩ/sq	<10	<10	<10
Sensitive area, mm diameter¶	≥0.05	≥0.05	≥0.05

† Silicon window. ‡ Sapphire window.

§ Improvement in D^* achieved with cold shields and cold filters also produces an increase in time constant.

¶ D^* for ATO depends somewhat on area; 5 by 5 mm and smaller have best D^*.

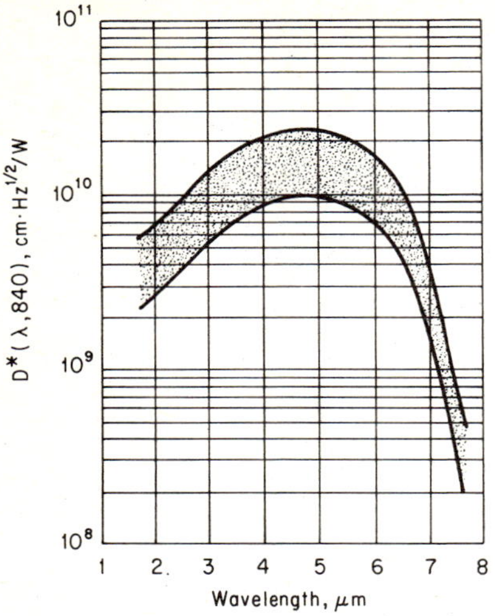

Fig. 94 Range of spectral detectivities, PbSe (LTO) infrared detectors with silicon window (no cold filter or shields). (*Santa Barbara Research Center, Brochure* 67.)

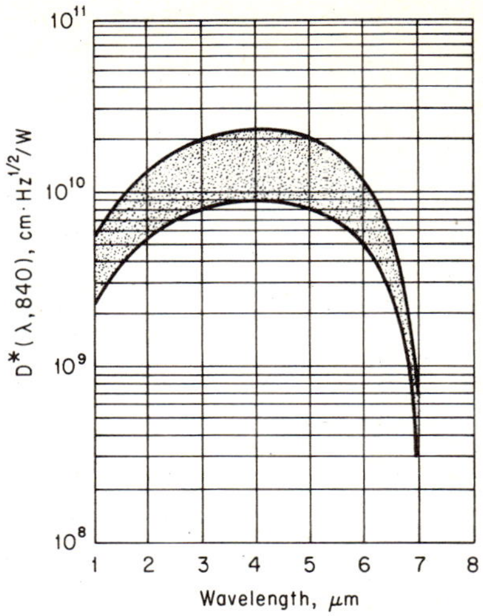

Fig. 95 Range of spectral detectivities, PbSe (LTO) infrared detectors with sapphire window (no cold filter or shields). (*Santa Barbara Research Center, Brochure* 67.)

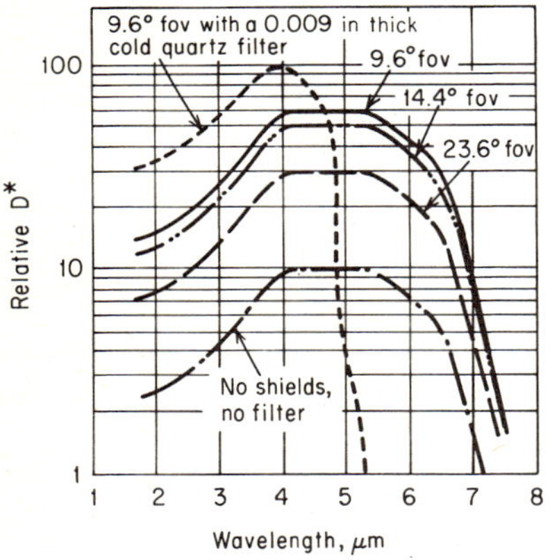

Fig. 96 Examples of spectral-detectivity improvement, PbSe (LTO) infrared detectors using radiation shielding and/or cold filters (with silicon windows). (*Santa Barbara Research Center, Brochure* 67.)

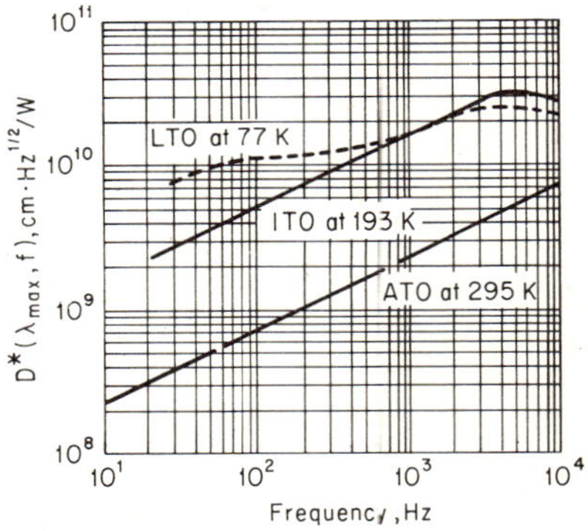

Fig. 97 Example of peak detectivity vs. frequency for PbSe (ATO, ITO, and LTO); 2π fov, 295-K background. (*Santa Barbara Research Center, Brochure* 67.)

As with PbS, three distinct forms are described, as shown in Table 3.

Noise: See Fig. 101

Capacitance: 1 pF (limited by mounting configuration)

Recommended circuit: Standard photoconductor

Operating temperature: See Fig. 102

Manufacturers: Santa Barbara Research Center, Infrared Industries, Optoelectronics

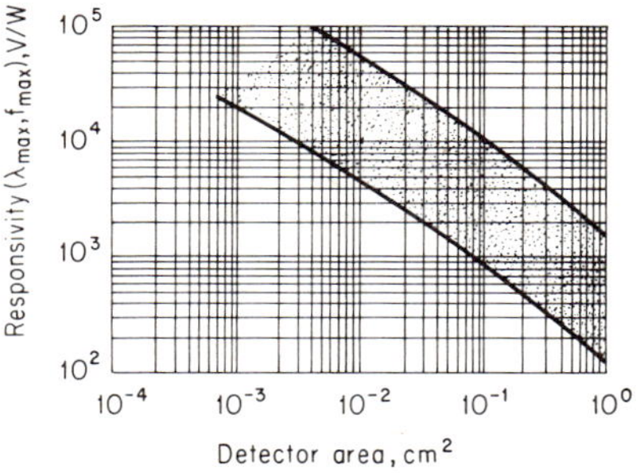

Fig. 98 Expected range of peak responsivities vs. detector size, typical PbSe (ATO) infrared detectors. (*Santa Barbara Research Center, Brochure* 67.)

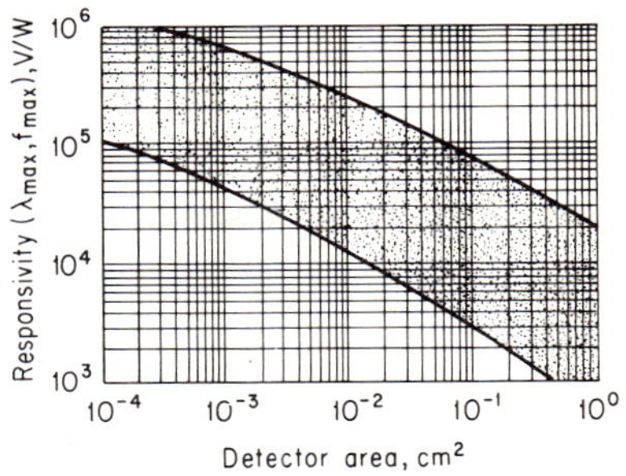

Fig. 99 Expected range of peak responsivities vs. detector size, PbSe (ITO and LTO) infrared detectors. (*Santa Barbara Research Center, Brochure* 67.)

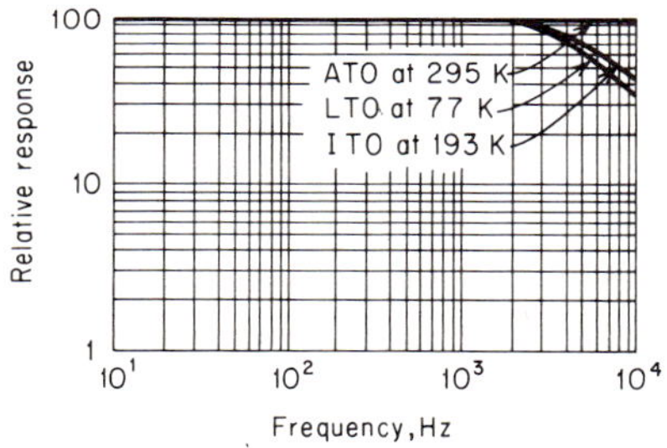

Fig. 100 Example of signal vs. frequency for PbSe (ATO, ITO, and LTO). (*Santa Barbara Research Center, Brochure* 67.)

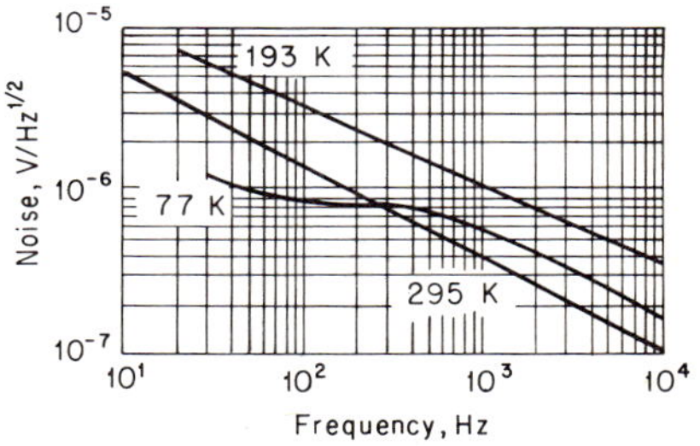

Fig. 101 Example of noise vs. frequency for PbSe cooled detectors (ATO, ITO, and LTO types) ($A = 1$ by 1 mm). (*Santa Barbara Research Center, Brochure* 67.)

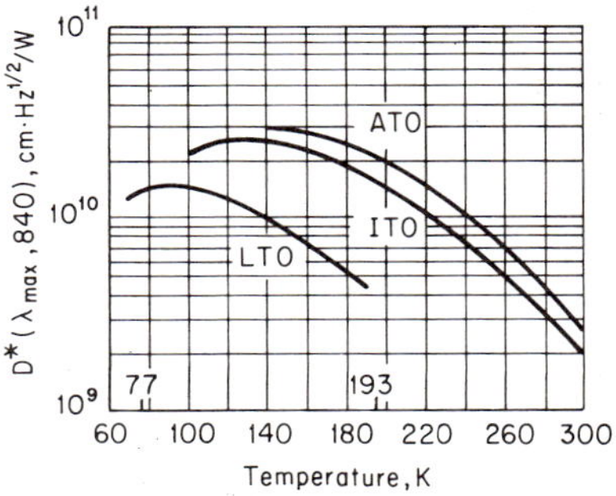

Fig. 102 Example of detectivity vs. temperature for PbSe (ATO, ITO, and LTO); 2π fov, 295-K background. (*Santa Barbara Research Center, Brochure* 67.)

30. InSb Indium antimonide has been made the basis for at least four different radiation detectors. We discuss here the intrinsic photoconductive and photovoltaic modes of operation. [The very-far-infrared bolometer (InSb bolometer) was discussed in Par. 22, and the intrinsic PEM detector is discussed separately below.]

Cooled photoconductive InSb has very nearly the same spectral detectivity as cooled photovoltaic InSb. The former offers simplicity of operation over a wide range of background radiation conditions and greater responsivity for large-area detectors ($A \geq 2$ by 2 mm). The latter is about 10 times faster under normal conditions and, being photovoltaic, has $\sqrt{2}$ times less limiting (generation) noise than photoconductive InSb (generation-recombination noise). Uncooled photoconductive InSb offers faster response and broader spectral coverage than either cooled InSb but at the expense of much reduced peak sensitivity.

Photovoltaic InSb. In order to secure the $\sqrt{2}$ additional detectivity available in photovoltaic InSb one must bias with care. Noise is smallest when the detector

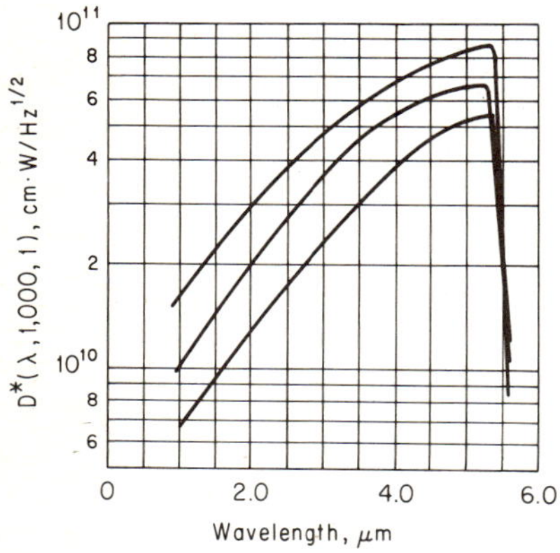

Fig. 103 Typical spectral detectivity for photovoltaic InSb, various sensitive areas ($T = 77$ K; 300-K background; 2π fov). *Upper curve,* 10^{-4} to 10^{-2} cm^2; *middle curve,* 10^{-2} to 10^{-1} cm^2; *lower curve,* 10^{-1} to 10 cm^2. (*Texas Instruments, Bull. SC*-8385-366.)

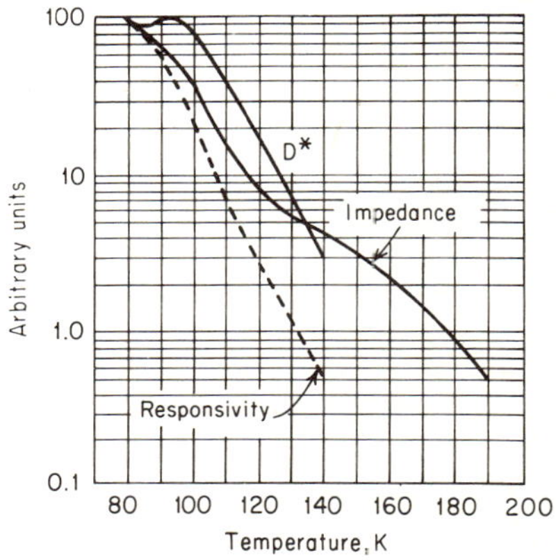

Fig. 104 Relative D^*, responsivity, and impedance vs. temperature for photovoltaic InSb; 2π fov, 300-K background. (*Texas Instruments, Bull. SC*-8385-366.)

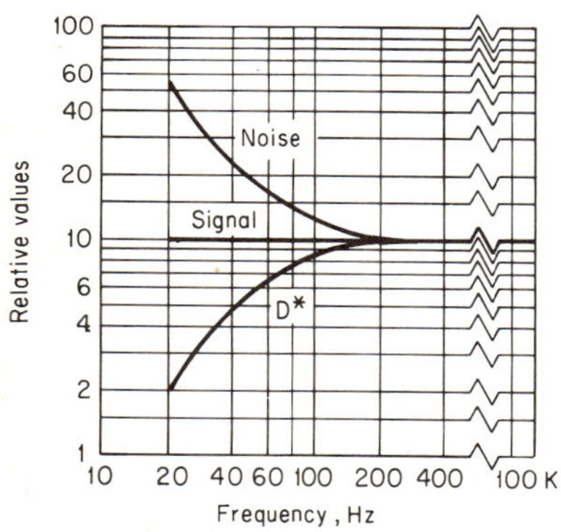

Fig. 105 Relative signal, noise, and D^* vs. frequency for photovoltaic InSb; 300-K background; 77 K. (*Texas Instruments, Bull. SC*-8385-366.)

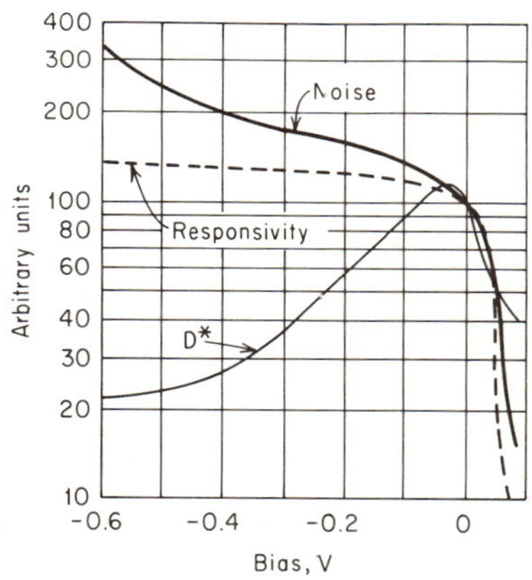

Fig. 106 Detectivity, responsivity, and noise vs. bias potential for photovoltaic InSb; 300-K background, 77 K; 2π fov, $A = 4 \times 10^{-2}$ cm^2. (*Texas Instruments, Bull. SC*-8385-366.)

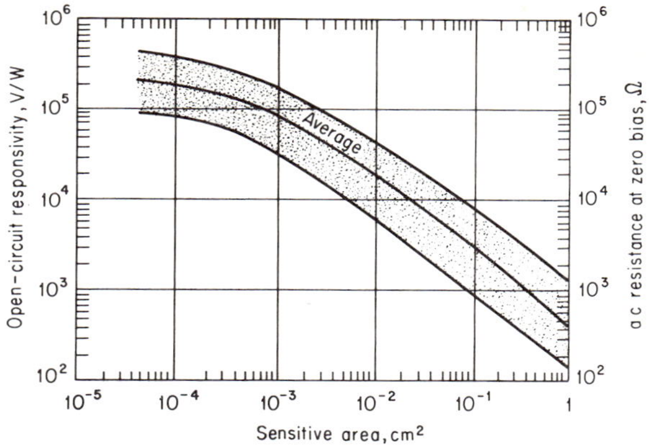

Fig. 107 Responsivity and ac resistance at zero bias vs. sensitive area for photovoltaic InSb; 500 K, 2π fov, 300-K background, 77 K. (*Texas Instruments, Bull. SC*-8385-366.)

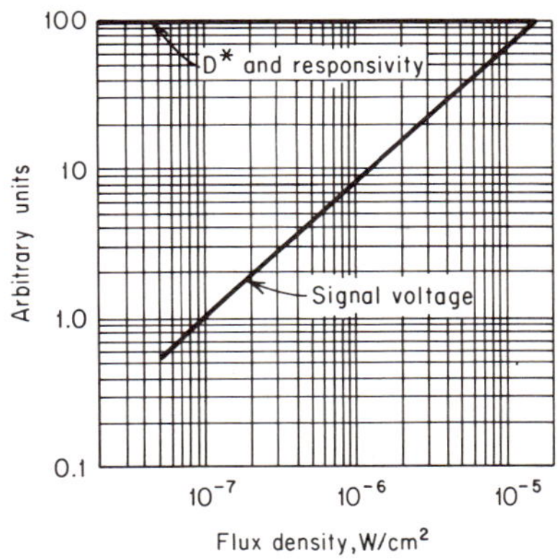

Fig. 108 Linearity of photovoltaic InSb; 2π fov, 300-K background, 77 K. (*Texas Instruments, Bull. SC*-8385-366.)

is maintained at small reverse bias. Thus the biasing depends on background flux density. The problem of varying ambient conditions can be taken care of by using current-mode amplification, which allows maximum possible D^*.

Sensitivity: In general, highest detectivity and responsivity are obtained by operating detectors near dc short-circuit condition (see Figs. 103 to 106 and 108); at 77 K $D^*(\lambda_{pk})/D^*(500\text{ K}) \approx 5$.

Quantum efficiency: ~0.6.

Noise: See Figs. 105 and 106.

Time constant: $<1\ \mu$s.

Responsivity: See Figs. 104 and 106 to 108.

Capacitance: Typically 0.01 μF/cm^2.

Impedance: See Figs. 107 and 109.

Sensitive area: 25 μm to 13 mm (range of linear dimensions).

Operating temperature: Normally 77 K (see Fig. 104); InSb can be fabricated especially for reduced temperature dependence of D^*, impedance, and responsivity. This permits greater flexibility in cooling requirements but involves some degradation of D^*.

Linearity: Linear to ~1 mW/cm^2.

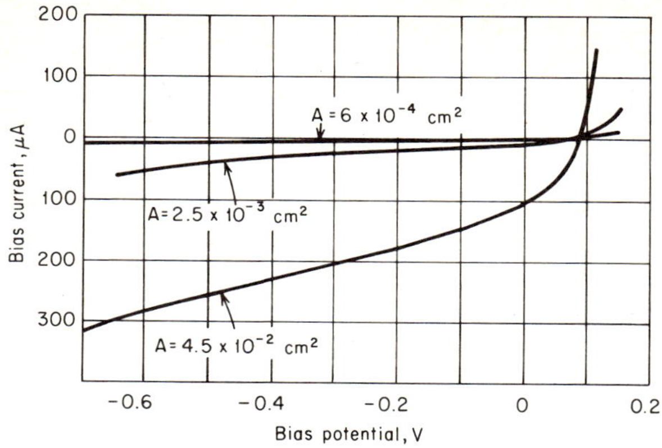

Fig. 109 Typical voltage-current curve for photovoltaic InSb for various areas; 2π fov, 77 K, 300-K background. (*Texas Instruments, Bull. SC*-8385-366.)

Sensitivity profile: ±15 percent.
Stability: Stable in vacuum.
Modes of operation:
1. Self-biased (no power supply needed)
2. Short-circuit (highest responsivity and detectivity)
a. Transformer coupling to preamplifier (dc short in primary and ac signal to secondary)
b. Low-drain dc bias supply adjusted to buck the detector photocurrent open-circuit voltage to the zero bias condition on its diode curve
3. Reverse bias from a low-drain supply

Recommended circuits: Recommended detector-preamplifier coupling techniques depend on the detector's properties and its ultimate use. One or more of the following general techniques are applicable in most cases: (1) The detector can be operated directly into the preamplifier through a dc blocking condenser. This method is satisfactory for small-area detectors, $<2 \times 10^{-3}$ cm², and in cases where spectral bandpass filtering has been used, reducing the short-circuit current to 1 μA or less. (2) The detector can be transformer-coupled to the amplifier. This method is satisfactory for all sizes of detectors provided that the primary dc resistance is sufficiently low to allow near-short-circuit operation and the ac impedance matches the amplifier optimum noise load. (3) The detector can be operated in series with a variable-bias supply and in parallel with a fixed-load resistor of value $R_L > R_d$ into a transistor preamplifier. The noise figure of the preamplifier should be minimized for the parallel load. (4) The detector can be coupled directly to a small current-mode preamplifier which will automatically maintain the zero-bias condition of the detector when changes in background radiation flux occur.

Manufacturers: Texas Instruments, Mullard (Optoelectronics), Barnes Engineering, Santa Barbara Research Center, Spectronics, Avco, Judson.

Photoconductive InSb. Liquid-nitrogen-cooled photoconductive InSb detectors are fast intrinsic photoconductors, highly sensitive in the spectral region 2 to 5.35 μm. Their detectivity depends somewhat upon sensitive area. For very high photon fluxes, photoconductive is faster than photovoltaic InSb because of the high recombination rate. However, this involves noise, which degrades D^*.

Room-temperature photoconductive InSb is sensitive from the visible region out to 7.5 μm, peaking at about 6 μm. Its time constant is <0.01 s, but peak D^* is typically as low as 2×10^8 cm·Hz$^{1/2}$/W.

Sensitivity: (See Figs. 110 to 113); at 77 K $D^*(\lambda_{pk})/D^*(500\text{ K}) \approx 5.5$
Quantum efficiency: ~0.6
Noise: 20 nV/Hz$^{1/2}$ (typical) (see Fig. 111)

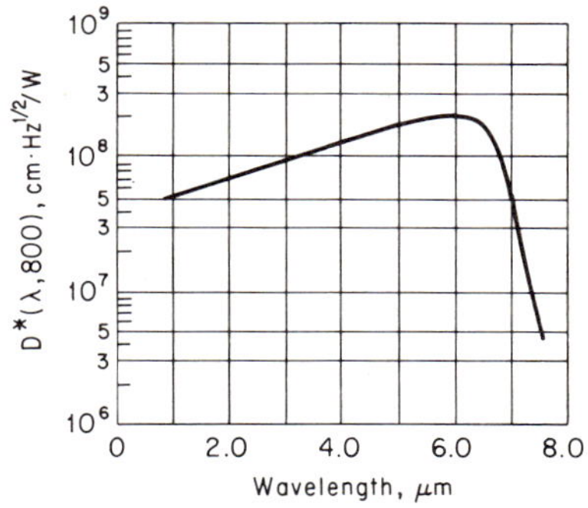

Fig. 110 Spectral sensitivity of uncooled photoconductive InSb. (*Mullard, Ltd., Bull. TP* 1080, 1969.)

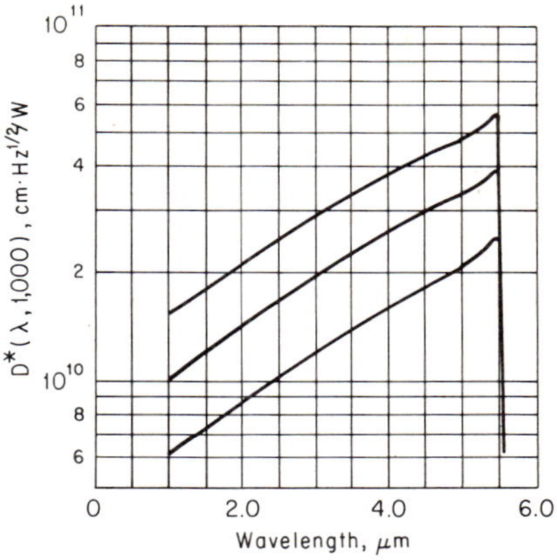

Fig. 111 Typical spectral detectivity range for cooled photoconductive InSb of various areas; 2π fov; 77 K, 300-K background. *Upper,* 0.1 by 0.1 to 0.5 by 0.5 mm; *middle,* 0.5 by 0.5 to 1.0 by 1.0 mm; *lower,* 1.0 by 1.0 to 5.0 by 1.0 mm. (*Texas Instruments, Bull. SC-*8385-366.)

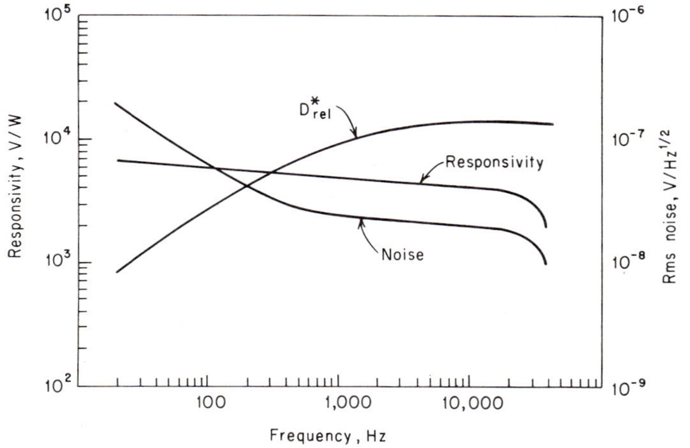

Fig. 112 Responsivity, noise, and relative D^* vs. frequency for photoconductive InSb. (*Honeywell Radiation Center, Brochure HRC* 67-65.)

Time constant: ~10 μs for 180° fov†; ~15 μs for 50° fov
Responsivity: ~10^4 V/W (see Figs. 111 to 113)
Dark resistance: ~5 kΩ/sq (typical)
Sensitive area: 0.1 by 0.1 to 10 by 10 mm (see Fig. 110)
Operating temperatures: Normally 77 K (see Fig. 112)
Linearity: Perfect between 10^{-3} and 10^{-8} W/cm^2
Sensitivity profile: ±15 percent
Stability: Stable in vacuum
Recommended circuit: See Fig. 114
Manufacturer: Santa Barbara Research Center

† With special construction <1 μs, but D^* is down by a factor of 2.

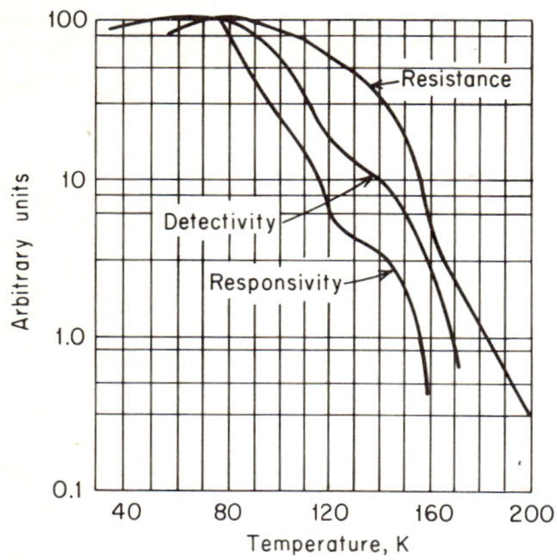

Fig. 113 Relative D^*, responsivity, and resistance vs. temperature for photoconductive InSb; 2π fov; 300-K background. (*Texas Instruments, Bull. SC*-8385-366.)

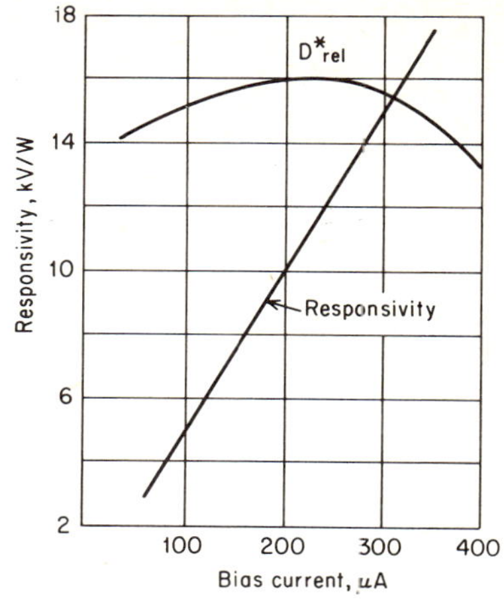

Fig. 114 Typical responsivity and relative D^* vs. bias current for photoconductive InSb; 0.5-mm² area, 300-K background, 180° fov, $f = 1$ kHz. (*Texas Instruments, Bull. SC*-8385-366.)

Uncooled Photoconductive InSb. Uncooled InSb offers submicrosecond response time and spectral response out to 7.5 μm. Compared with PEM InSb, the photoconductor is slower but more sensitive and with spectral response peaking at 6 μm instead of 5 μm.

Sensitivity: $D^*(6\ \mu m, 800\ K) = 1$ to 8.5×10^8 cm·Hz$^{1/2}$/W; $D^*(\lambda_{pk})/D^*(500\ K) \approx 3.3$ (see Figs. 115 to 117)

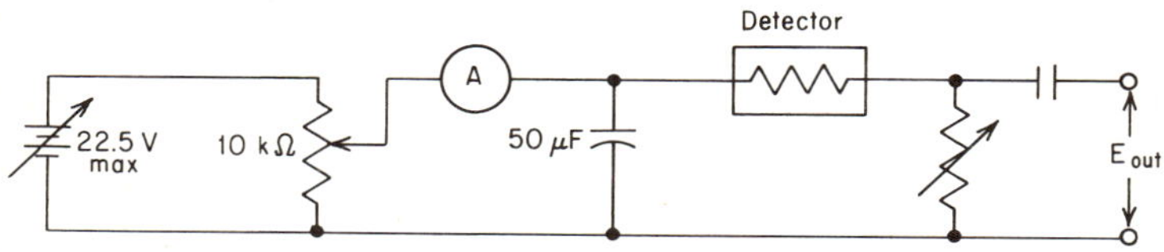

Fig. 115 Typical bias circuit for photoconductive detectors. E_{out} to be amplified by low-noise preamplifier. (*Texas Instruments, Bull. SC*-8385-366.)

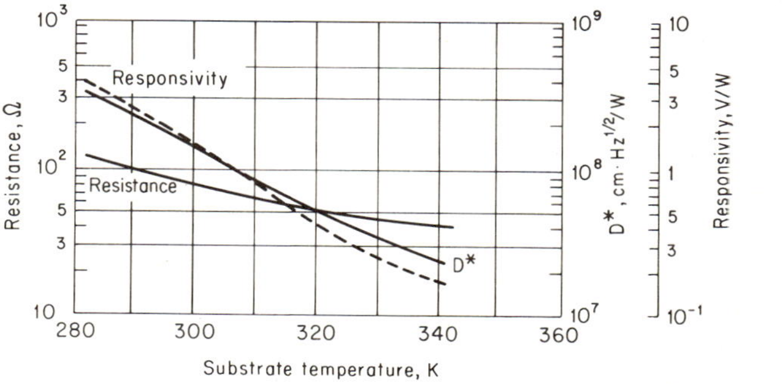

Fig. 116 Variation of $D^*(6\ \mu m, 800\ K)$, responsivity, and resistance with substrate temperature for room-temperature photoconductive InSb. (*Mullard Bull. TP* 1080.)

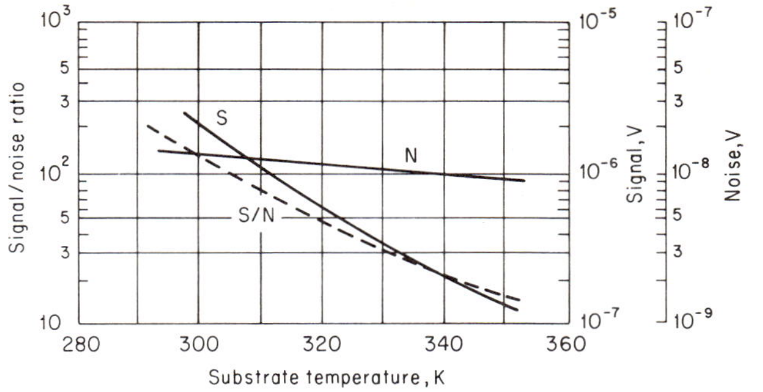

Fig. 117 Variation of signal, noise, and signal-to-noise ratio with substrate temperature for room-temperature photoconductive InSb. Test conditions: current, 10 mA; incident energy, 68 μW/cm^2 at 4.4 μm; modulation frequency, 800 Hz; amplifier bandwidth, 50 Hz. (*Mullard Bull. TP* 1080.)

Noise: ~0.3 nV/Hz$^{1/2}$ at 295 K (see Fig. 117)
Quantum efficiency: Same as cooled InSb
Responsivity: ~1 V/W
Time constant: ~40 ns
Sensitive area: 6 by ½ mm standard
Operating temperature: Normally 295 K (see Figs. 116 and 117)
Resistance: 30 to 130 Ω†
Recommended circuit: Standard photoconductive with transformer coupling (see Fig. 118)
Manufacturers: Electro-Nuclear Labs, Optoelectronics

InSb Photoelectromagnetic (PEM) Detector. PEM InSb offers high speed and modest sensitivity out to 7 μm without cooling. It is faster though less sensitive

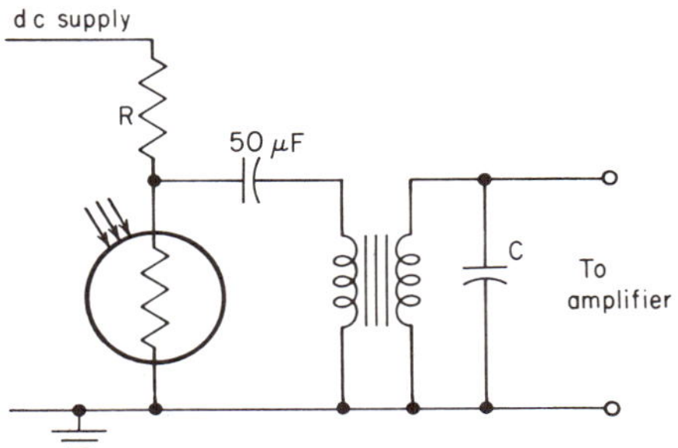

Fig. 118 Recommended circuit for uncooled photoconductive InSb. (*Mullard Bull. TP* 1080.)

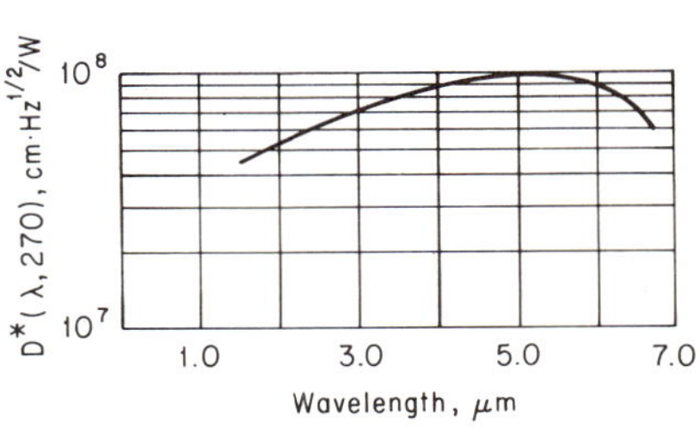

Fig. 119 PEM spectral response (A = 5 by 5 mm; T = 298 K). (*Electronuclear Laboratories, Bull.* 1049, 1966.)

than uncooled thermal detectors, e.g., the thermistor bolometer or thermocouple. In contrast to room-temperature photoconductive InSb, PEM needs no bias supply but does need a magnet; its responsivity is about 100 times greater, but it is about 10 times slower and somewhat less sensitive.

Sensitivity: See Fig. 119); $D^*(\lambda_{pk})/D^*(500\ K) \approx 3$
Noise: Thermal
Time constant: <1 μs rise time (R_L = 50 Ω)
Impedance: 5 to 10 Ω

† Higher-responsivity cells are available (up to 10 V/W) with somewhat reduced D^*.

Responsivity: 200 V/W (750 Hz)
Sensitive area: 5 by 5 mm standard
Operating temperature: 298 K
Recommended circuit: Transformer-coupled
Manufacturers: Electro-Nuclear Labs, Honeywell

31. Ge(Au) Gold-doped germanium detectors are relatively fast single-crystal *p*-type impurity-doped photoconductors for the 2- to 10-μm region. Although not

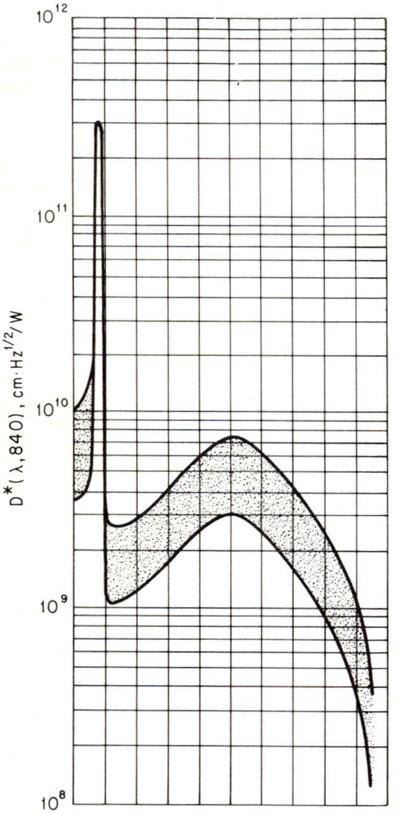

Fig. 120 D^* vs. λ for Ge(Au); T = 77 K, 2π fov; 295-K background. (*Santa Barbara Research Center, Brochure* 66.)

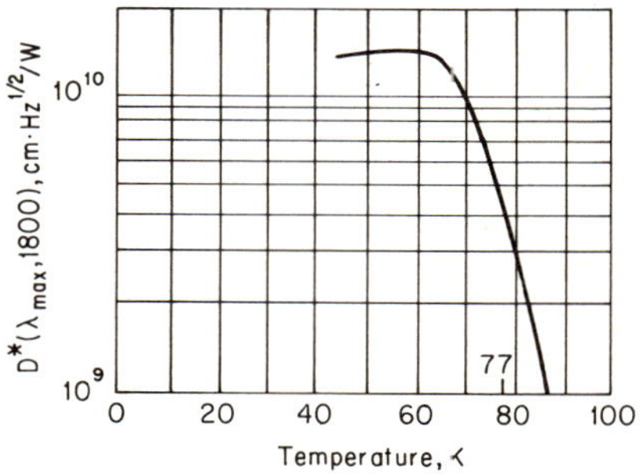

Fig. 121 Peak D^* vs. temperature for Ge(Au). (*Santa Barbara Research Center, Brochure* 66.)

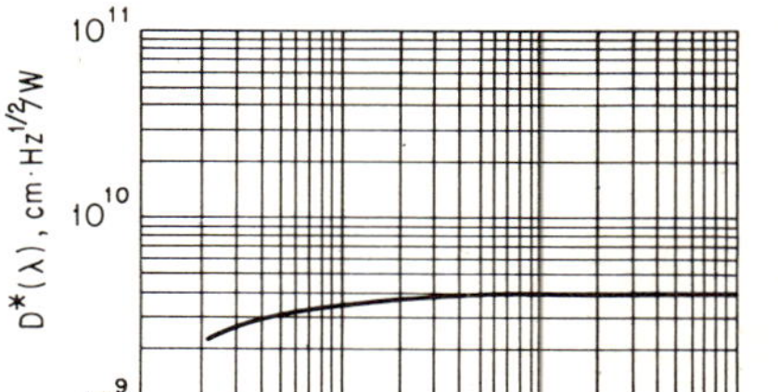

Fig. 122 Typical D^* vs. frequency (T = 77 K). (*Santa Barbara Research Center, Brochure* 66.)

the most sensitive detector anywhere in its broad region of spectral sensitivity, Ge(Au) offers respectable sensitivity over a broad spectral region without the need for coolants more troublesome than liquid nitrogen. Sensitivity can be improved by a factor of 2.5 by operating at $T < 65$ K (pumped liquid nitrogen). At these temperatures Ge(Au) becomes background-limited.

Sensitivity: See Figs. 120 to 122.
Quantum efficiency: $D^*(\lambda_{pk})/D^*(500\ K) = 2.7$ (see Fig. 123).
Noise: See Fig. 124.
Time constant: <50 ns with full D^* [shorter response times (<2 ns) can be

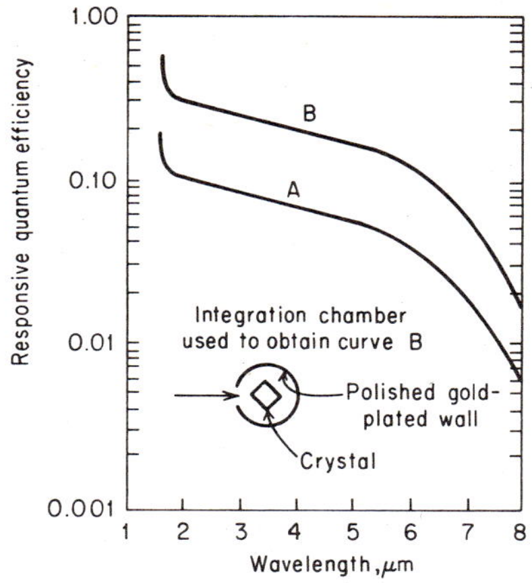

Fig. 123 Quantum efficiency vs. λ for Ge(Au) (T = 78 K). (*Santa Barbara Research Center, internal report.*)

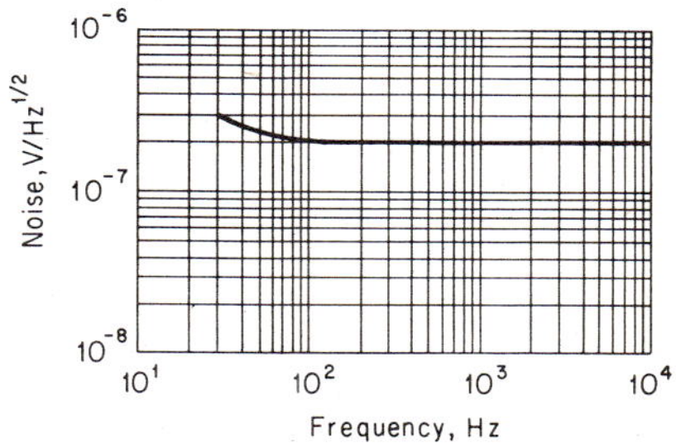

Fig. 124 Typical noise spectrum for Ge(Au) (T = 77 K; A = 1 by 1 mm). (*Santa Barbara Research Center, Brochure* 66.)

tailored by heavy concentration of dopant. This increases resistance, and hence the incoherent signal-to-noise ratio becomes limited by thermal noise of load (typically a factor of 2 degradation in the signal-to-noise ratio). Quantum efficiency, however, is not altered, so that concentration does not hurt coherent-detection signal-to-noise ratio.]

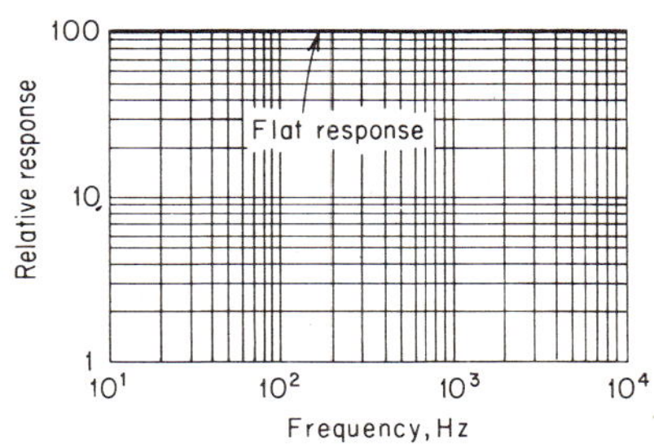

Fig. 125 Relative response vs. frequency for Ge(Au). (*Santa Barbara Research Center, Brochure* 67.)

Responsivity: $\mathcal{R}_{\lambda,\max} \sim 10^4\ V_{rms}/W_{rms}$ at 77 K (see Fig. 125).
Dark resistance: ≤500 kΩ/sq (see Time-constant discussion).
Capacitance: <1 pF.
Sensitive area: 0.25 to 5 mm linear dimensions standard.
Operating temperature: <85 K (normally 77 K) (see Fig. 121).
Recommended circuit: Standard photoconductive.
Manufacturers: Santa Barbara Research Center, Barnes.

32. (Hg,Cd)Te We discuss both photoconductive and photovoltaic (Hg,Cd)Te detectors designed for background-limited, high-speed, intrinsic photon detection in the 8- to 14-μm atmospheric window, requiring only liquid-nitrogen cooling. Characteristics of 2- to 5-μm (uncooled) photoconductive (Hg,Cd)Te and 1- to 3-μm photovoltaic (Hg,Cd)Te are also presented.

Mercury cadmium telluride is an alloy semiconductor whose peak sensitivity can be adjusted from 1 to 50 μm by varying the ratio of Hg to Cd (see Fig. 126). Both photovoltaic and photoconductive detectors are available. The former are low-impedance devices with subnanosecond response times whose sensitivity is frequently limited by amplifier noise. The latter make use of photoconductive gain to achieve greater sensitivities. However, this involves increased carrier lifetimes, resulting in reduced response time.

The high quantum efficiency of these intrinsic detectors (>70 percent) and their high speed of response makes them useful for infrared heterodyne detection.

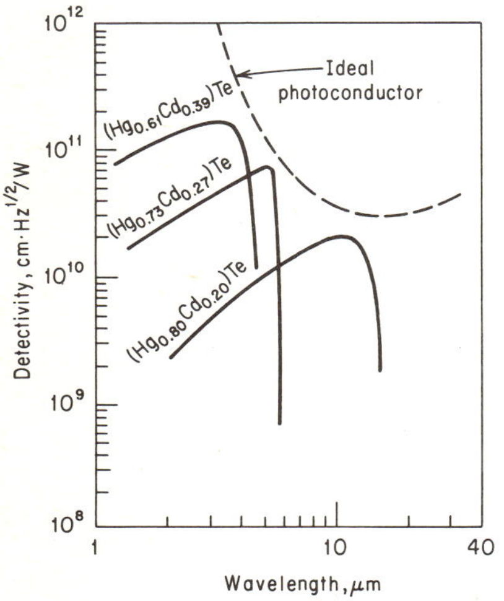

Fig. 126 Photoconductive (Hg,Cd)Te range of spectral sensitivity for various alloy mixtures; 77 K, 2π fov. (*Honeywell Radiation Center.*)

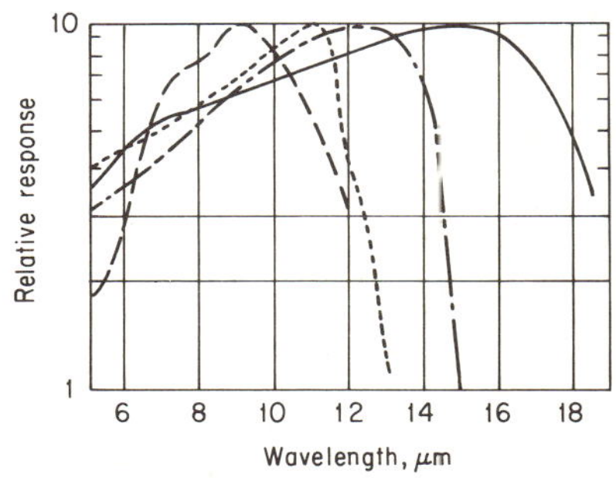

Fig. 127 Relative spectral response for different alloy photoconductive (Hg,Cd)Te; 77 K, 60° fov. (*Santa Barbara Research Center, Brochure 69CM.*)

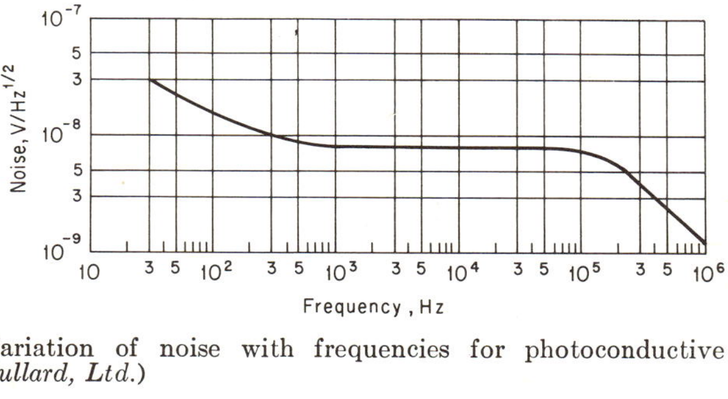

Fig. 128 Variation of noise with frequencies for photoconductive (Hg,Cd)Te (77 K). (*Mullard, Ltd.*)

When sufficient local oscillator power is available, detector cooling becomes unnecessary. Other things being equal, the photovoltaic detector has $\sqrt{2}$ sensitivity (signal-to-noise voltage) advantage over the photoconductor.

Photoconductive 8- to 14-μm

Sensitivity: Adjustable by varying alloy concentration (see Figs. 126 and 127); $D^*(\lambda_{pk})/D^*(500\ K) \approx 2$ when $\lambda_{pk} = 10\ \mu m$

Time constant†: ~1 μs, depending on area

Noise: See Fig. 128

Dark resistance: 20 to 200 Ω/sq

Responsivity: $\approx 10^4$ V/W

† Detectivity depends on time constant.

Operating temperature: 77 K
Sensitive area: 0.05 to 2 mm linear dimensions
Quantum efficiency: >70 percent
Capacitance: Limited by mounting configuration
Circuit: Standard photoconductive
Manufacturers: Mullard (Optoelectronics), Santa Barbara Research Center, Honeywell

Photovoltaic 8- to 14-μm

Sensitivity: $D^*(9\ \mu m, 900) = 3 \times 10^{10}$ cm·Hz$^{1/2}$/W, fov 30°. Can be adjusted by varying alloy concentration (see Fig. 129).

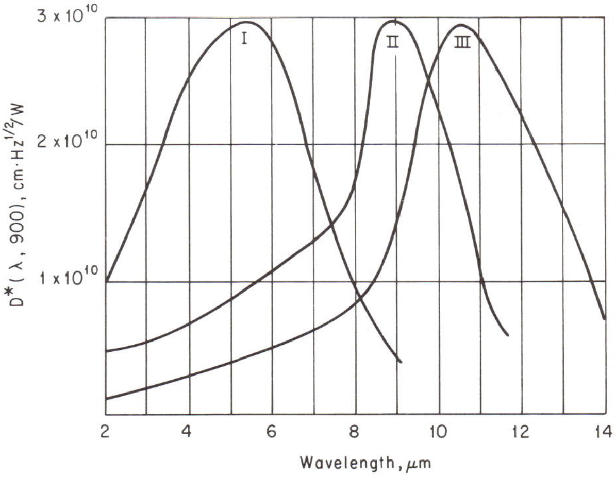

Fig. 129 D^* vs. λ for photovoltaic (Hg,Cd)Te; $T = 77$ K, 30° fov. Curves I, II, and III refer to different Cd/Hg ratios. *(Eltek Corp.)*

Time constant: 10 to 20 ns without bias; ~3 ns with reverse bias (some trade-off of sensitivity).
Resistance: 100 to 500 Ω for 0.1 by 0.1 mm (proportionately lower for larger areas).
Operating temperature: 77 K.
Noise: See Fig. 130.
Responsivity: 200 to 700 V/W (see Fig. 131).
Sensitive area: 0.01 to 0.25 mm linear dimensions.
Capacitance: <10 pF ($A \approx 0.04$ mm^2).

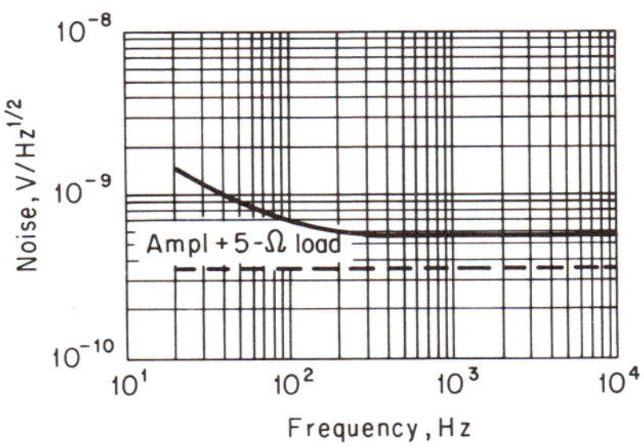

Fig. 130 Noise spectrum for photovoltaic (Hg,Cd)Te. *(Eltek Corp.)*

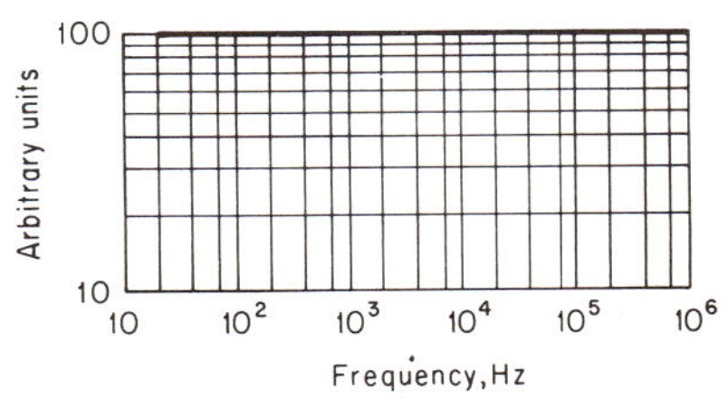

Fig. 131 Relative response vs. frequency in hertz for (Hg,Cd)Te. *(Societé Anonyme de Télécommunications.)*

Circuit: Because of low resistance, transformer coupling is recommended, provided response time permits.
Manufacturers: Societé Anonyme de Télécommunications, Honeywell.
2- to 5-μm Room-Temperature Photoconductive
Sensitivity: $D^*(\lambda_{max}) > 1 \times 10^9$ cm·Hz$^{1/2}$/W for 60° fov (see Fig. 132)
Time constant: ~400 ns
Noise: Similar to 8- to 14-μm (Hg,Cd)Te photoconductor
Dark resistance: >10 Ω
Responsivity: 30 V/W

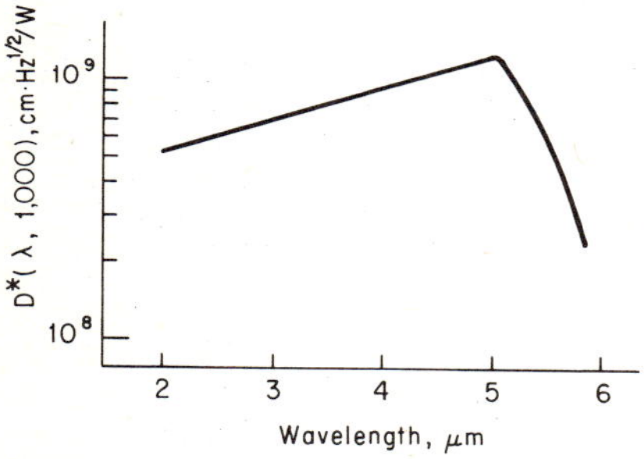

Fig. 132 Typical spectral D^* for 2- to 5-μm photoconductive (Hg,Cd)Te, 60° fov, 300 K. (*Honeywell Radiation Center.*)

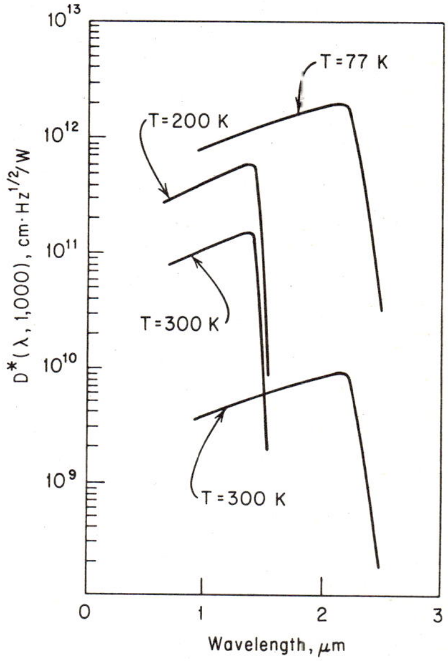

Fig. 133 D^* vs. λ for 1- to 3-μm photovoltaic (Hg,Cd)Te measured at various temperatures. (*Honeywell Radiation Center.*)

Operating temperature: 295 K (approximately 10 percent improvement in D^* per kelvin from 295 to 195 K)
Sensitive area: ½ by ½ mm typical
Quantum efficiency: >0.4
Profile, linearity, capacitance, stability, and circuit: Similar to other bulk single-crystal photoconductors
Manufacturer: Honeywell
1- to 3-μm Photovoltaic
Sensitivity: See Fig. 133
Time constant: ~10 ns
Noise: Similar to 8- to 14-μm photovoltaic (Hg,Cd)Te
Dark resistance: Junction resistance (zero bias at 1 kHz) ~ 1 MΩ series resistance <1 kΩ
Operating temperature: See Fig. 133
Noise: Similar to 8- to 14-μm photovoltaic (Hg,Cd)Te
Sensitive area: 2 by 2 mm
Quantum efficiency: >0.4
Capacitance: ~15 pF (junction capacitance at zero bias)
Profile, linearity, circuit: Similar to 8- to 14-μm photovoltaic (Hg,Cd)Te
Manufacturer: Honeywell

33. (Pb,Sn)Te (photovoltaic) At the time of writing, only photovoltaic (Pb,Sn)Te is available, and no clear-cut advantage over (Hg,Cd)Te is recognized except its greater ease of manufacture.

Sensitivity: $D^*(\lambda_{pk}, 900) > 10^{10}$ (cm · Hz$^{1/2}$/W); $D^*(\lambda_{pk})/D^*(500\ K) \approx 2$ (see Fig. 134)

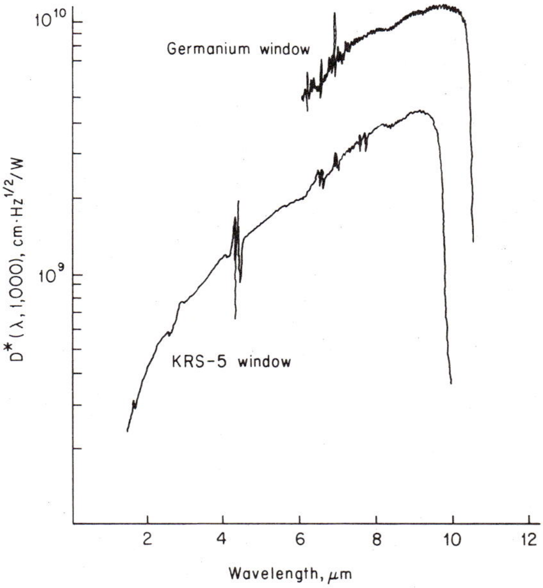

Fig. 134 Photovoltaic (Pb,Sn)Te; 77 K, 60° fov. (*Barnes Engineering Co.*)

Time constant: >50 ns
Sensitive area: 1 by 1 mm
Operating temperature: 77 K
Circuit: Standard photovoltaic
Manufacturers: Raytheon, Barnes, Plessey

34. Ge(Hg) Mercury-doped germanium detectors are fast single-crystal impurity-doped photoconductors, sensitive out to 14 μm. Ge(Hg) is especially well suited for detection through the 8- to 13-μm atmospheric window and for detection of near-ambient targets. Unfortunately its operating temperature must be kept less than 40 K, where it becomes 300-K background-limited.

Sensitivity: (See Figs. 135 to 138); $D^*(\lambda_{pk})/D^*(500\ K, 1{,}000) \approx 2$ at all temperatures.
Quantum efficiency: 25 to 30 percent.
Noise: See Figs. 139 and 140.
Time constant <50 ns with 50-Ω load for T <28 K and electric fields <30 V/cm. (Compensated material $t < 1$ ns with 50-Ω load. Responsivity then is reduced by 10^4 and detectivity is reduced by 2.)
Responsivity: See Figs. 136 and 140 to 142; ~10^5 V/W.
Dark resistance: Depends on fov; ~100 K for 180° fov (see Fig. 143).
Capacitance: <1 pF.
Sensitive area: 0.25 to 5 mm linear dimensions.
Operating temperature: See Fig. 136.
Linearity: 10^{-3} to 10^{-8} W (size-dependent).
Sensitivity profile: ±15 percent.
Recommended circuit: See Fig. 144 for electrical characteristics; standard photoconductive.

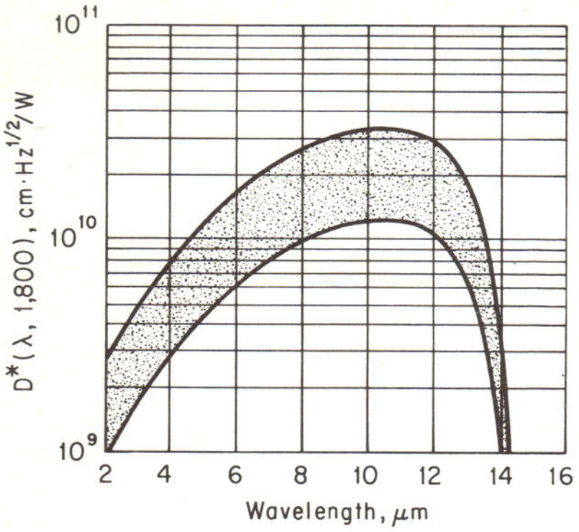

Fig. 135 Range of spectral sensitivities for Ge(Hg); $T = 5$ K, 60° fov, 300-K background. (*Santa Barbara Research Center.*)

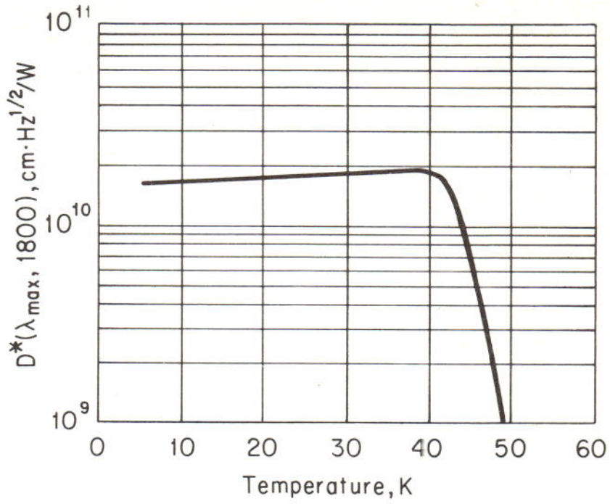

Fig. 136 Typical D^* vs. temperature for Ge(Hg). (*Santa Barbara Research Center.*)

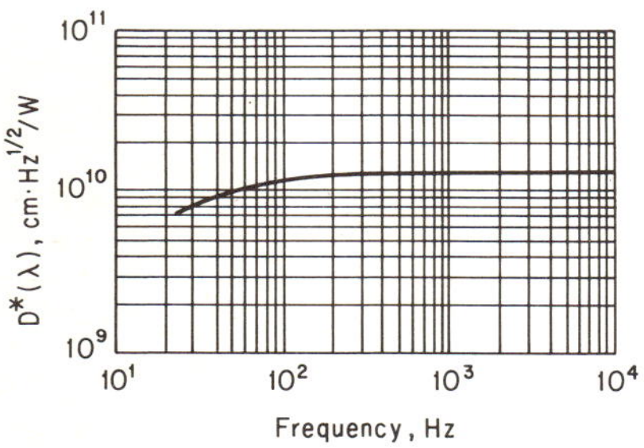

Fig. 137 Typical D^* vs. frequency at 30 K for Ge(Hg).

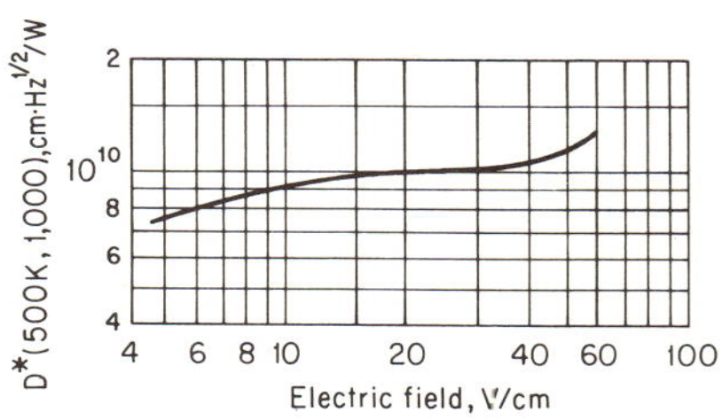

Fig. 138 D^* vs. electric field for Ge(Hg); $T = 5$ K, 90° fov; 300-K background; 6×10^{-4} cm^2 area; Irtran II window. (*Texas Instruments.*)

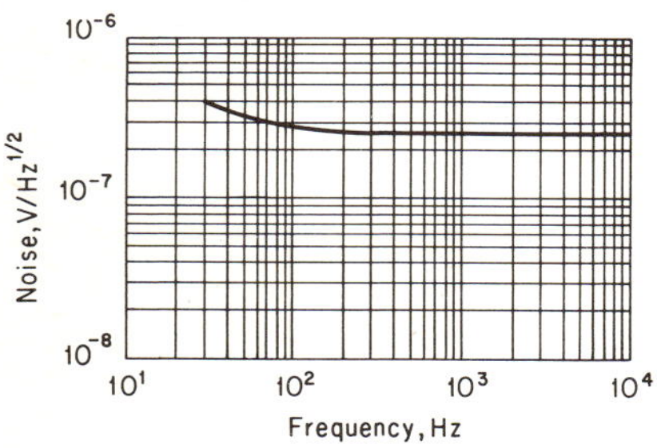

Fig. 139 Typical noise-frequency spectrum for Ge(Hg); 1 by 1 mm area; essentially constant with temperature. (*Santa Barbara Research Center.*)

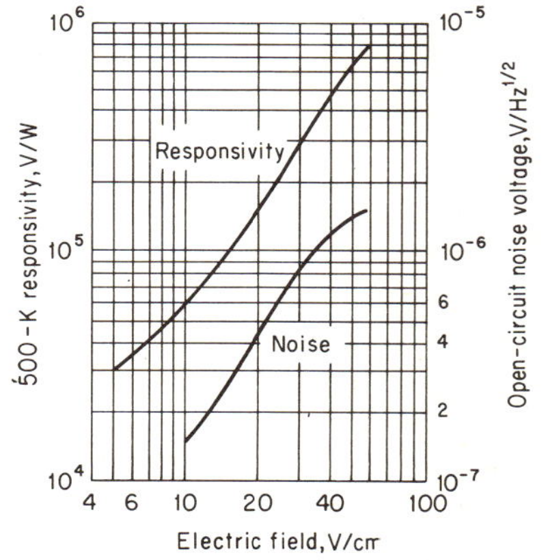

Fig. 140 Open-circuit responsivity and noise voltage vs. electric field for Ge(Hg); $T = 5$ K; 90° fov; 300-K background; $A = 6 \times 10^{-4}$ cm^2, Irtran II window. (*Texas Instruments.*)

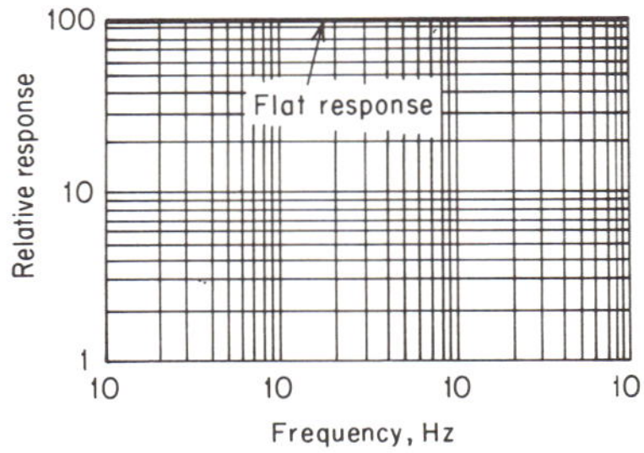

Fig. 141 Ge(Hg) typical relative response; $T = 30$ K. (*Santa Barbara Research Center.*)

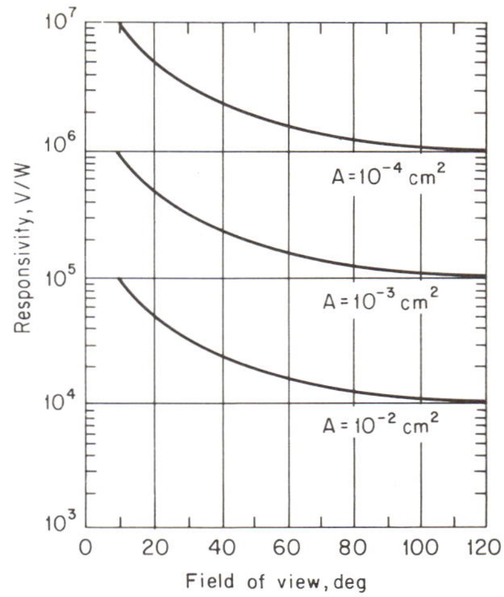

Fig. 142 Open-circuit responsivity vs. fov for Ge(Hg) at 5 K for various detector areas; 300-K background; 500-K blackbody source. (*Texas Instruments.*)

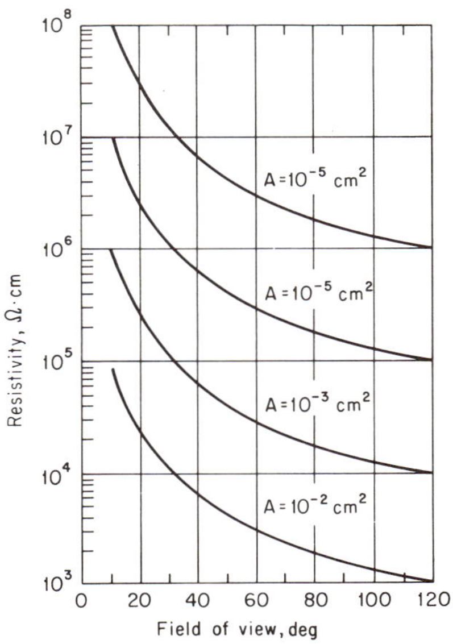

Fig. 143 Ge(Hg) resistivity vs. fov for several detector areas; $T = 5$ K; 300-K background, optimum bias. (*Texas Instruments.*)

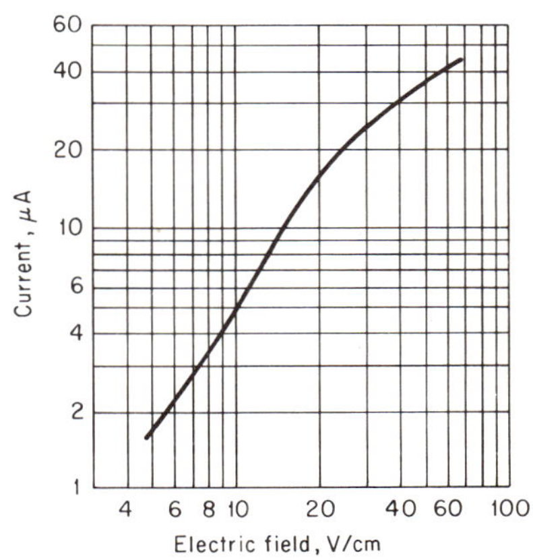

Fig. 144 Ge(Hg) bias current vs. electric field; 90° fov; 300-K background; $T = 5$ K; $A = 6 \times 10^{-4}$ cm^2; Irtran II window. (*Texas Instruments.*)

Manufacturers: Texas Instruments, Santa Barbara Research Center, Raytheon, Mullard (Optoelectronics), Aeroject, Autonetics.

35. Ge(Cd) Spectral response is very similar to Ge(Cu), except that cutoff wavelength moves in to 22 μm. The main advantage is that operating temperatures may be as high as 25 K, so that liquid-hydrogen cooling can replace liquid helium. Alternatively, if a liquefier is used for cooling, instead of a reservoir, a much less complex cryostat can be used.

Sensitivity: See Fig. 145; $D^*(\lambda_{pk})/D^*(500\text{ K}) \approx 2\frac{1}{2}$
Time constant: 50 ns (see discussion of time constant in Par. 33)
Responsivity: 10^4 V/W
Dark resistance: 10 to 50 kΩ/sq (2π for ambient background)
Capacitance: <1 pF (limited by mounting configuration)
Sensitive area: 0.25 to 5 mm linear dimensions
Operating temperature: 25 K
Recommended circuit: Standard photoconductive
Manufacturers: Santa Barbara Research Center, Raytheon, Texas Instruments, Autonetics

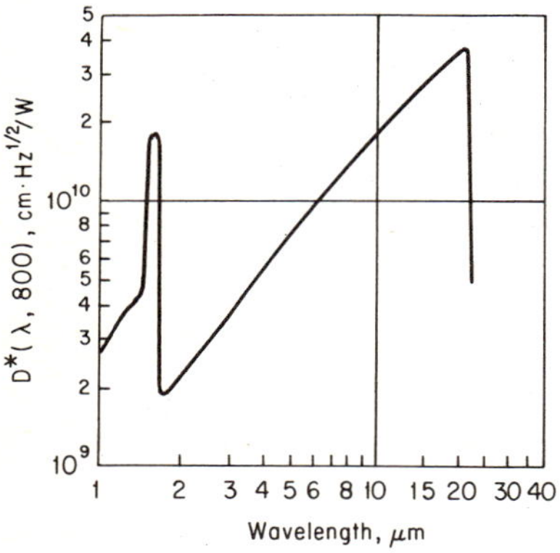

Fig. 145 Spectral detectivity for Ge(Cd) detector with 60° fov for an element containing 5×10^{15} Cd atoms per cubic centimeter. (*Santa Barbara Research Center.*)

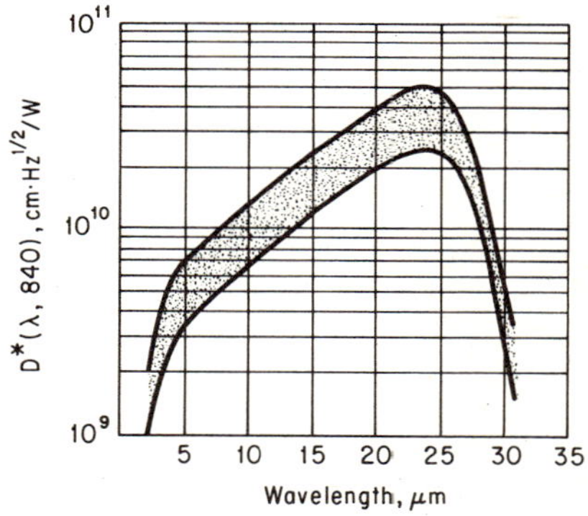

Fig. 146 Range of spectral detectivities for Ge(Cu); 60° fov. (*Santa Barbara Research Center.*)

36. Ge(Cu) Copper-doped germanium detectors are fast, single-crystal, impurity-doped photoconductors, with high sensitivity in the broad region 2 to 30 μm. Operating temperature must be maintained below 20 K (ideally <14 K). Ge(Cu) is then 300-K background-limited, and response time <50 ns.

Sensitivity: (See Figs. 146 to 148); $D^*(\lambda_{pk})/D^*(500\text{ K}) \approx 2\frac{1}{2}$
Noise: See Figs. 149 and 150
Time constant: <50 ns [can be doped to be faster; see Ge(Hg)]
Responsivity: 10^5 V/W (see Figs. 150 to 152)
Dark resistance: Depends on fov (~100 K for 180° fov)
Capacitance: <1 pF
Sensitive area: 0.25 to 5 mm linear dimensions
Operating temperature: See Figs. 147 and 152
Linearity: 10^{-3} to 10^{-8} W/cm^2 (depends on size)
Sensitivity profile: ±15 percent
Stability: Stable in all ambient storage environments tested
Recommended circuit: See Fig. 153
Manufacturers: Santa Barbara Research Center, Texas Instruments, Mullard (Optoelectronics), Raytheon, Aerojet, Autonetics

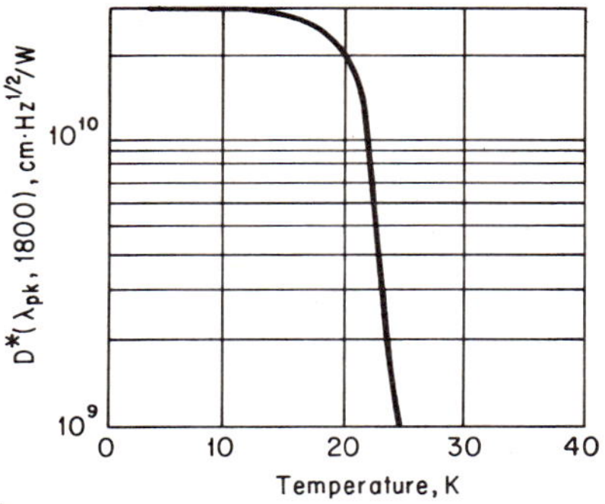

Fig. 147 Typical D^* vs. operating temperature for Ge(Cu). (*Santa Barbara Research Center.*)

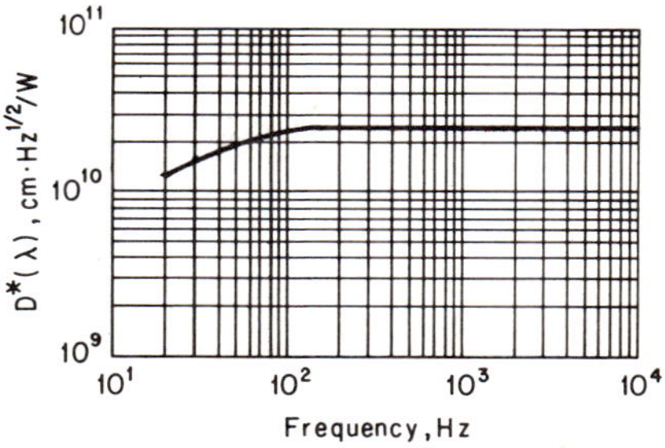

Fig. 148 Typical D^* vs. frequency for Ge(Cu) at 4.2 K. (*Santa Barbara Research Center.*)

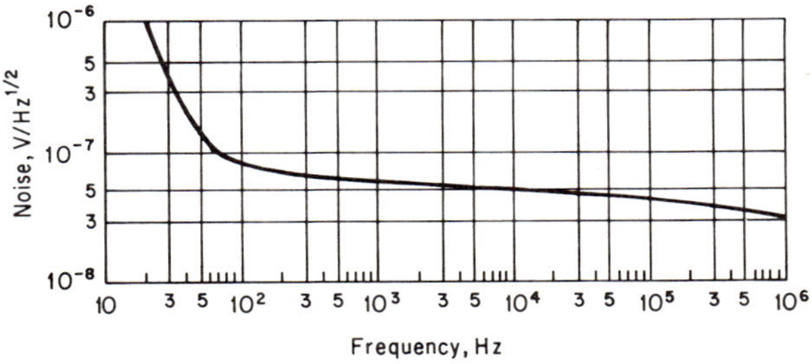

Fig. 149 Typical noise frequency spectrum for Ge(Cu). (*Mullard, Ltd.*)

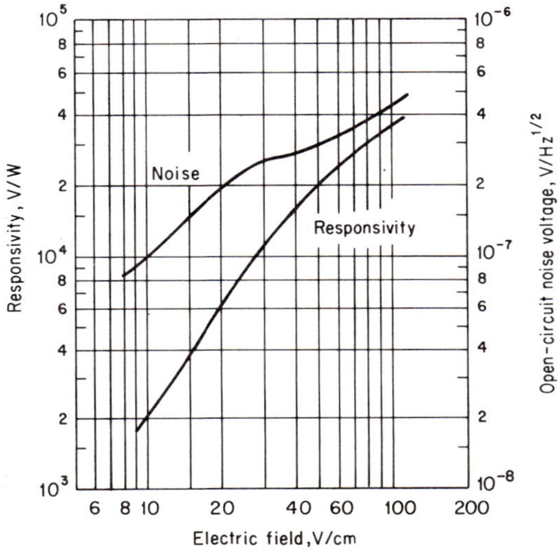

Fig. 150 Typical noise and responsivity vs. biasing field for Ge(Cu); 5 K, 60° fov; 300-K background, $A = 10^{-2}$ cm^2; 500-K blackbody. (*Texas Instruments.*)

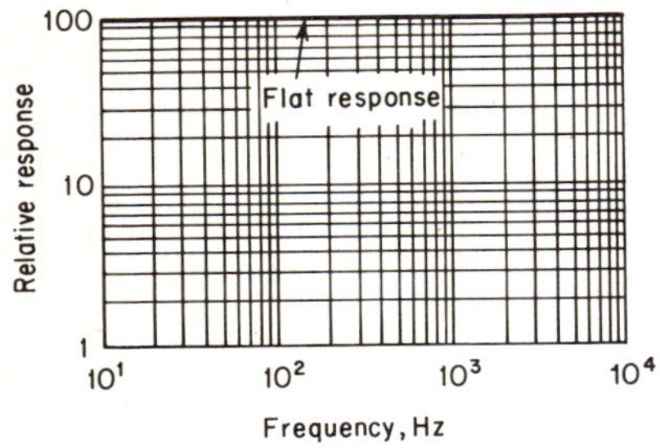

Fig. 151 Relative response vs. frequency for Ge(Cu) at 4.2 K. (*Santa Barbara Research Center.*)

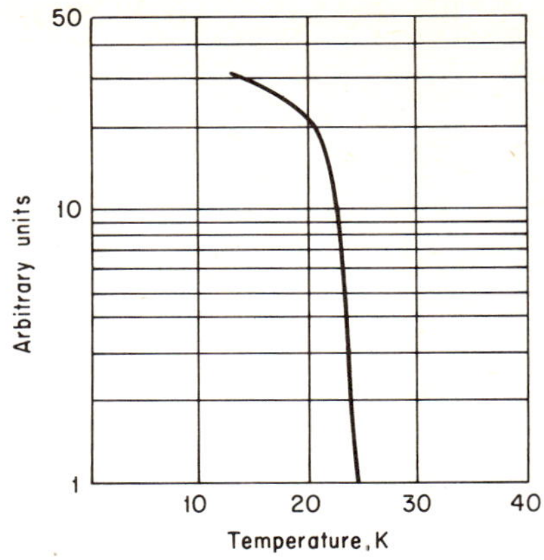

Fig. 152 Relative responsivity vs. temperature for Ge(Cu).

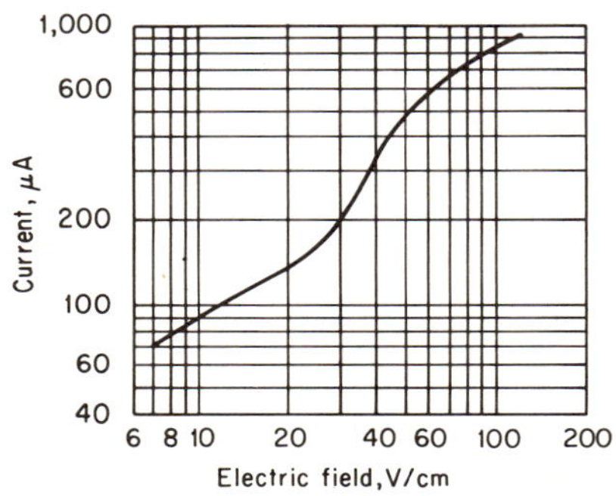

Fig. 153 Typical current-voltage curve for Ge(Cu); 60° fov; 300-K background: $A = 10^{-2}$ cm^2. (*Texas Instruments.*)

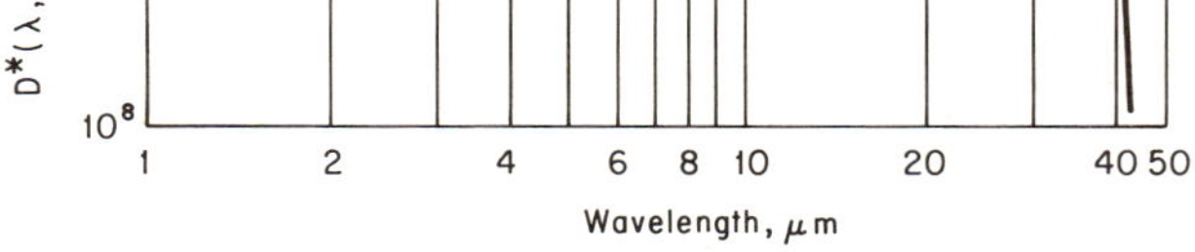

Fig. 154 D^* vs. wavelength for Ge(Zn).

37. Ge(Zn) Very similar to Ge(Cu), except that cutoff wavelength moves out to 40 μm and operating temperature should be <10 K. A relatively low field breakdown limits the responsivity.

Sensitivity: (See Fig. 154); $D^*(\lambda_{pk})/D^*(500\text{ K}) = 2$

Noise: See Fig. 155

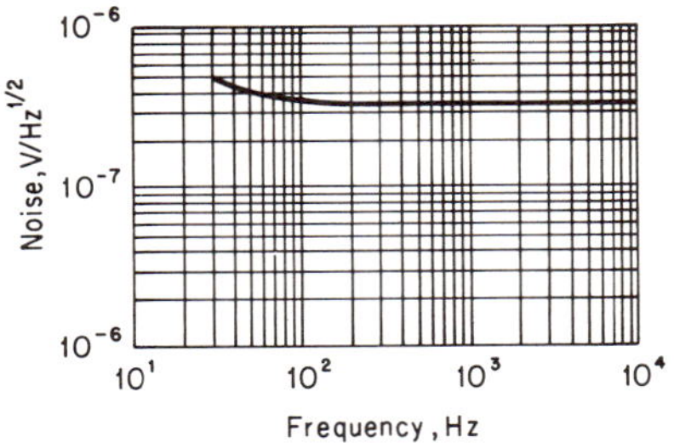

Fig. 155 Typical noise-frequency spectrum at $T = 4.2$ K for Ge(Zn); $A = 1$ by 1 mm. (*Santa Barbara Research Center.*)

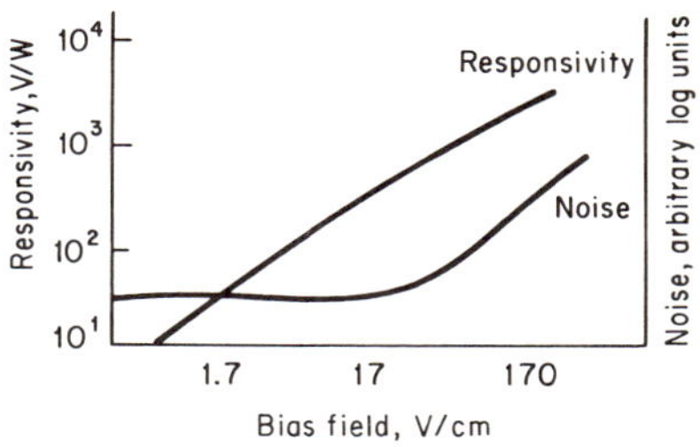

Fig. 156 Signal and noise for Ge(Zn).

Time constant: <50 ns (see discussion of time constant in Par. 33)
Responsivity: 10^3 V/W (see Fig. 156)
Dark resistance: 10–300 kΩ/sq (2π fov ambient background)
Capacitance: <1 pF (limited by mounting configuration)
Sensitive area: 0.25 to 5 mm linear dimensions
Operating temperature: <10 K
Recommended circuit: Standard photoconductive
Manufacturer: Santa Barbara Research Center

38. Ge(In) Indium-doped germanium is typical of a large group of *p*-type germanium extrinsic photoconductors which are useful for far-infrared detection, for example, Ge(Be), Ge(B), Ge(Ga). There is little to be said in favor of any one over the others. All require cooling to liquid-helium temperatures. All have submicrosecond response times and similar electrical characteristics.

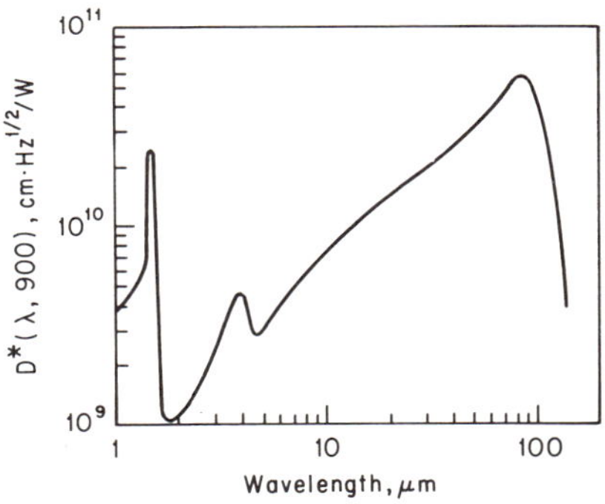

Fig. 157 D^* vs. λ for Ge(In) (polyethylene window); $T = 5$ K; 90° fov. (*Raytheon.*)

Sensitivity: (See Fig. 157); $D^*(\lambda_{pk}, 900) = 6 \times 10^{10}$ cm·Hz$^{1/2}$/W, fov 90°. $D^*(\lambda_{pk})/D^*(500\text{ K}) = 60$
Noise: Same as Ge(Cu)
Responsivity: ~10^2 V/W for 500-K blackbody source
Time constant: <1 μs
Dark resistance: ~100 K
Operating temperature: ~5 K
Sensitive area: 0.25 to 5 mm linear dimensions
Capacitance: <1 pF (limited by mounting configuration)
Recommended circuit: Standard photoconductive
Manufacturers: Raytheon, Santa Barbara Research Center

39. Photographic In this paragraph we present only the spectral sensitivity of some typical photographic emulsions. See Sec. 5 for a more extensive coverage.

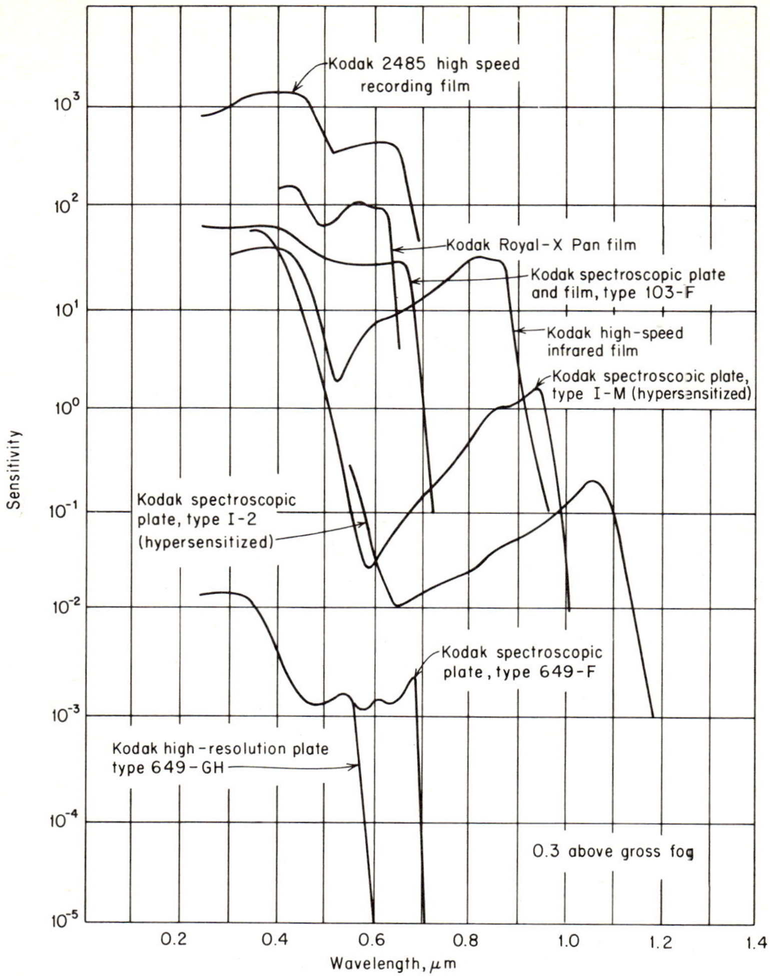

Fig. 158 Sensitivity vs. λ for typical photographic emulsions. (*Eastman Kodak.*)

The term spectral sensitivity generally has a different meaning when applied to photographic detectors than it does when applied to the other detectors described in this chapter. It comes closer to responsivity than to minimum detectable power or energy.† In Fig. 158 sensitivity is the reciprocal of exposure, expressed in ergs per square centimeter, required to produce

$$\text{Density} \equiv \log \frac{1}{\text{transmittance}} = 0.3$$

above gross fog in the emulsion when processed as recommended.

Manufacturers: Eastman Kodak Co., Ilford, Agfa, Ansco, Pruntz, 3M.

† Work is in progress to evaluate photographic materials in terms of minimum detectable energy, a concept involving the average number of photons necessary to produce a change in density (signal) equal to that of the fog-density fluctuations (noise); see Marchant (1964), Jones (1958), Bird et al. (1969).

REFERENCES

Bauer, G. (1943): Ein halbleiter Hochohmbolometer mit Tafel, IV, *Phys. Z.*, vol. 44, p. 53.

Bennett, H. E. (1966): Accurate Method for Determining Photometric Linearity, *Appl. Opt.*, vol. 5, p. 1265.

Bird, G. R., R. C. Jones, and A. E. Ames (1969): The Efficiency of Radiation Detection by Photographic Films: State-of-the-Art and Methods of Improvement, *Appl. Opt.*, vol. 8, p. 2389.

Eberhardt, E. H. (1967): Threshold Sensitivity and Noise Ratings of Multiplier Phototubes, *Appl. Opt.*, vol. 6, p. 251.

Felgett, P. B. (1949): Threshold Sensitivity and Noise Ratings of Multiplier Phototubes, *Appl. Opt.*, vol. 6, p. 251.

Fink, D. J. (1940): "Principles of Television Engineering," McGraw-Hill, New York.

Gunter, W. D., G. R. Grant, and S. A. Shaw (1970): Optical Devices to Increase Photocathode Quantum Efficiency, *Appl. Opt.*, vol. 9, p. 251.

Institute of Radio Engineers (1962): IRE Standards on Electron Tubes: Methods of Testing, publ. 62, pp. 7–51.

Jacobs, S. F., and M. Sargent III (1970): Photon Noise Limited D^* for Low Temperature Backgrounds and Long Wavelengths, *Infrared Phys.*, vol. 10, no. 4, pp. 233–235.

Jones, R. C. (1949): Factors of Merit for Radiation Detectors, *J. Opt. Soc. Am.*, vol. 39, p. 344.

——— (1958): On the Quantum Efficiency of Photographic Negatives: On the Minimum Energy Detectable by Photographic Materials, *Photogr. Sci. Eng.*, vol. 2, pp. 57–65j, 191–204.

Kruse, P. W., L. D. McGlaughlin, and R. B. McQuistan (1963): "Elements of Infrared Technology," Wiley, New York.

Livingston, W. C. (1966): Enhancement of a Photocathode Sensitivity by Total Internal Reflection as Applied to an Image Tube, *Appl. Opt.*, vol. 5, p. 1335.

Low, F. J. (1961): Low-Temperature Germanium Bolometer, *J. Opt. Soc. Am.*, vol. 51, p. 1300.

——— and A. R. Hoffman (1963): The Detectivity of Cryogenic Bolometers, *Appl. Opt.*, vol. 2, p. 649.

Marchant, J. C. (1964): Exposure Criteria for the Photographic Detection of Threshold Signals, *J. Opt. Soc. Am.*, vol. 54, p. 79.

Morton, G. A., H. M. Smith, and H. R. Krall (1968): Pulse Height Resolution of High Gain First Dynode Photomultipliers, *Appl. Phys. Lett.*, vol. 13, p. 356.

Radio Corporation of America (1963): Phototubes and Photocells, *Tech. Man.* PT-60.

Samson, J. A. R. (1967): "Techniques of Ultraviolet Spectroscopy," Wiley, New York.

Simon, R. E., A. H. Sommer, J. J. Tietjen, and B. F. Williams (1968): New High-Gain Dynode for Photomultipliers, *Appl. Phys. Lett.*, vol. 13, p. 355.

Smith, R. A., F. E. Jones, and R. P. Chasmar (1957): "Detection and Measurement of Infrared Radiation," Oxford University, London.

Sommer, A. H. (1968): "Photoemissive Materials," Wiley, New York.

Sommers, H. S., Jr., and E. K. Gatchell (1966): Demodulation of Low-Level Broad-Band Optical Signals with Semiconductors, *Proc. IEEE*, vol. 54, p. 1553.

——— and W. B. Teusch (1964): Demodulation of Low-Level Broadband Optical Signals with Semiconductors, II: Analysis of the Photoconductive Detector, *Proc. IEEE*, vol. 52, p. 144.

Sun, C., and T. E. Walsh (1970): Performance of Broadband Microwave-biased Extrinsic Photoconductive Detectors at 10.6 μm, *IEEE J. Quantum Electron.*, vol. 6, p. 450.

Wolfe, W. L. (ed.) (1965): "Handbook of Military and Infrared Technology," Office of Naval Research, Washington.

Section 5

Photosensitive Materials

KENNETH S. PETCHE

Coulter Information Systems, Inc., Bedford, Mass.†

(Numbers refer to paragraphs)

1. Introduction Photosensitive materials are thin layers of organic and/or inorganic preparations which are usually dispersed in a binder and coated on flexible film or paper substrates or rigid glass or metal substrates and which form a direct image upon exposure to radiation to which they are sensitive. Usually a means

† This material was written when the author was affiliated with Itek Corporation, Lexington, Mass.

of converting the exposed material to produce a visible image is required. In some cases, this step may also yield major improvements in bulk amplification or speed. This is particularly true of silver halide materials, which form the largest and most diversified class of photosensitive materials. Other important classes of photosensitive materials include such non-silver halide systems as diazo, vesicular, electrographic, thermographic, free-radical, and metal oxides. At the present state of the art, silver halide photosensitive materials are the only practical camera-speed materials, exceeding all others in sensitivity by several orders of magnitude. However, with the limited supply of silver that will become more costly in the future, new impetus has been given to the research and development of materials to replace silver halide. Photosensitive materials in the free-radical or metal oxide classes show promise for the future if their sensitivity can be significantly increased.

PROPERTIES OF SILVER HALIDE SYSTEMS

2. The outstanding characteristic of the silver halide photographic process is its flexibility. By judicious control of the manufacturing process, the speed (light sensitivity), contrast, and spectral sensitivity can be altered over a tremendous range, as Table 1 shows. No other photographic system has been developed to this extent, and no contending system is yet in sight.

TABLE 1 Flexibility of Silver Halide Process

Characteristic	Range
Speed	Printout proof paper to Polaroid ASA 10,000
Spectral sensitivity	Ultraviolet (300 nm) to near infrared (1200 nm); can be selected within these extremes by optical sensitization
Contrast	Low contrast with wide exposure latitude (pictorial photography) to extremely high contrast with little exposure latitude (lithography)
Resolution	20 line pairs/mm (paper prints) to more than 2,000 line pairs/mm (high-resolution plates for resolution targets, integrated circuits, etc.)
Processing	3 to 5 s (special thin, hardened emulsions processed with specialized chemistry and equipment) to several hours (thick nuclear-track emulsions)

Specialized emulsions that incorporate part of the processing chemicals to permit rapid and simplified processing include Fotorite and Itek Reader-Printer paper. In the last 30 years, tremendous improvements have been made in (1) the trade-off of resolving power (image quality) for speed (sensitivity), (2) emulsion hardening to permit high-temperature, rapid machine processing, (3) color emulsions, (4) high-resolution emulsions required for targets, reticles, and integrated-circuit master negatives, and (5) rapid-access lithographic materials (Project-a-Lith). Diffusion transfer, observed for many years as a darkroom curiosity, was transformed into the Polaroid process.

3. Physical Characteristics of Silver Halide Materials Black-and-white silver halide emulsions are typically colloidal suspensions of silver salts (bromide, iodide, chloride) in carefully selected and prepared gelatin or synthetic binders coated on paper, glass, or flexible supports such as cellulose acetates or polyesters. Depending on its intended application, the emulsion may range in thickness from several micrometers to several millimeters. It is usually overcoated with a very thin layer of clear gelatin for protection against scratches and pressure. The emulsion may vary widely in its silver salt content.

The gelatin binders contribute to the photographic properties (speed sensitizer) and provide an ideal medium for rapid penetration of the processing solutions. A gelatin binder has good optical transparency; with the addition of preservatives and flexing agents it has long life under adverse handling and provides the extremely rapid transport rates necessary in high-speed photography, search and retrieval microfilm systems, etc.

In addition to serving as a binder and sensitizer, gelatin plays an important role as a halogen acceptor. Exposure to ordinary light decomposes silver halide into metallic silver and halogen. In the absence of any compound capable of reacting with liberated halogen, the latter can recombine with silver and the effect of the radiation is lost.

Emulsion making is based largely on trade secrets and techniques developed by a photosensitive-material manufacturer. Since most of the attributes of a silver halide emulsion are determined by the manner in which it was manufactured, the various stages of emulsion preparation are worth noting.

Precipitation. The silver halide is precipitated in a dilute gelatin solution. Negative emulsions are usually composed of silver bromide with a small quantity of silver iodide. Developing papers and print film usually employ silver chloride, silver bromide, or both. A silver chloride paper is the lowest in sensitivity, while the presence of silver bromide increases its speed. A silver bromide paper or print film is the fastest of the group (Neblette, 1952).

Ripening. The emulsion is heated at a temperature of about 50 to 80°C as one of the methods to increase the grain size and set the size-frequency distribution of the silver halide crystals. Larger grain size helps increase emulsion sensitivity; a narrower size-frequency distribution increases the inherent contrast of the emulsion.

Gelatin. When ripening has reached the desired stage, additional gelatin is added to the solution to form a firm gel when cooled.

Washing. The gelled emulsion is shredded and washed with cold water to remove the excess salts. As an alternative, the washing stage can be eliminated by the addition of a coagulating agent to the liquefied emulsion. Decanting the supernatant liquid leaves the coagulum, which is then redispersed in pure water or in a gelatin solution. In general, printing papers tend to be of the washed-emulsion type where an excess of salts is desirable while modern negative materials are of the coagulated, low-salt type.

Afterripening (Chemical Sensitizing). The emulsion is melted and heated to about 50°C. Since most of the excess salts have been removed, little or no change occurs in grain size and shape. Chemical sensitizers such as sulfur, silver, and gold salts as well as other noble metals are added in very small quantities to enhance the sensitivity and modify the contrast of the material.

Other Additions. Optical sensitizers are added to the emulsion to obtain the desired spectral sensitivity. Normally, a silver halide grain without an optical sensitizer is not sensitive beyond 490 nm, but the addition of optical sensitizers can extend sensitivity to 1300 nm. Preservatives and antifoggants are included before coating to stabilize the product and extend its shelf life (James and Higgins, 1960).

Coating. The finished light-sensitive emulsion is coated upon a suitable support. Ordinary photographic-film supports now used are various forms of cellulose acetate or polyester films (polyethylene terephthalate), known by such names as Mylar, Cronar, or Estar. In applications where improved dimensional stability, minimum absorption of processing liquids, and high resistance to physical wear and tear are desired, these polyester films are rapidly replacing the acetate varieties. When the maximum in dimensional stability is required in such critical applications as spectroscopic photography or integrated-circuit masters, the emulsion is coated on special blanks of optical-grade glass. Cellulose nitrate is no longer used for the manufacture of photographic film because of its flammability and long-term chemical instability. Table 2 details the approximate dimensional change characteristics of typical Kodak aerial films on polyester base.

Usually a subbing layer is required to ensure good adhesion between the emulsion and substrate. Other layers such as a clear, thin protective gelatin overcoat are applied to resist scratches and abrasions. A thin layer which may be applied between the emulsion and support or on the reverse side of the base material contains light-absorbing dyes to minimize the undesirable light scattering (halation).

Most photographic papers employ high-quality, high-rag-content paper and frequently have a layer between the emulsion and the paper to improve whiteness. This layer may consist of barium sulfate, also known as baryta. Other materials such as wood, cloth, plastic, or metal can be coated with an emulsion for special purposes.

TABLE 2 Approximate Dimensional-Change Characteristics of Some Kodak Aerial Films (Eastman Kodak, 1976)

Base and film	Nominal base thickness, mils	Dimensional change					
		Temporary			Permanent		
		Humidity coefficient of linear expansion, % per 1% RH, unprocessed	Thermal coefficient of linear expansion, %		Processing dimensional change, %	Aging shrinkage of processed film, %	
			Per °F	Per °C		1 wk at 120°F, 20% RH	1 yr at 78°F, 60% RH
Estar polyester:							
Double-X Aerographic 2405†	4.0	0.0020	0.001	0.002	−0.03 to +0.01	0.03	0.03
Plus-X Aerocon 3411‡	2.5	0.0035	0.001	0.002	−0.004 to +0.03	0.08	0.04
Aerographic Duplicating 4421§	7.0	0.0015	0.001	0.002	−0.01 to +0.02	0.02	0.02
Aerocolor Negative 2445†	4.0	0.0030	0.001	0.002	−0.03 to +0.03	0.04	0.03
Aerochrome Infrared 2443 .	4.0	0.0025	0.001	0.002	−0.03 to +0.02	0.06	0.03

†Estar base. ‡Estar thin base. § Estar thick base.

4. Spectral Response A main advantage of the silver halide emulsion is that it can be sensitized to a broad range of the spectrum from the inherent sensitivity in the near ultraviolet and blue portions up to the near-infrared region of the spectrum (1300 nm). Many non-silver halide systems in addition to low overall sensitivity are limited in their application because it is difficult or impossible to extend the sensitivity beyond the near-ultraviolet and blue portions of the spectrum.

The spectral sensitivity is the wavelengths of the spectrum to which the photographic emulsion will respond. There are five common classifications of film sensitivity.

Blue. Sensitive to the near ultraviolet and blue portions of the spectrum. This is the inherent sensitivity of silver halide emulsions without spectral sensitizers.

Ortho. Sensitive to the near ultraviolet, blue, and green portions of the spectrum.

Panchromatic. Sensitive to the near-ultraviolet, blue, green, and red (to 660 nm) portions of the spectrum. Films with this sensitivity are used for a wide range of applications.

Extended red. Sensitive to the near-ultraviolet, blue, green, and red (to 720 nm) portions of the spectrum. Aerial reconnaissance films such as Kodak High-Definition Aerial Film (3404) are sensitized to the extended-red regions to permit more efficient use of red or minus-blue filters for penetration of atmospheric haze.

Infrared. Sensitive to the near-ultraviolet, blue, red, and near-infrared (not more than 1300 nm) portions of the spectrum.

Two common methods are used for determining the spectral sensitivity of an emulsion. In the first a wedge spectrograph uses a diffraction grating or prism to display a continuous spectrum onto the film under test. Because of the linear dispersion of the diffraction grating, it is preferred by most workers. By placing a stepped or continuous light modulator in font o fthe film, the relative sensitivity at each wavelength can be determined from the resultant processed density contour. Only relative comparisons between different films can be made since the method is qualitative. These comparisons are valid only for the particular light source used to make the exposure.

The second method employs a spectral sensitometer which provides known amounts of energy at various wavelength bands. The absolute sensitivity, in ergs per square centimeter, can be determined to produce a specified density, for example, $D = 1.0$. By making a series of sensitometric exposures throughout the spectrum, a plot of the energy required to produce a given density against wavelength results in a curve of the spectral sensitivity (Hall and Howell, 1966).

The results of the wedge-spectrograph determinations of spectral sensitivity can be observed in the wedge spectrograms shown in Figs. 1 to 4. Figure 1 shows

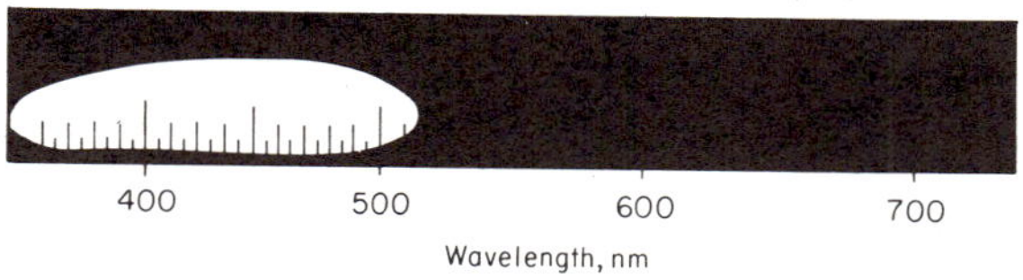

Fig. 1 Wedge spectrogram of blue-sensitive material, Du Pont Fine-Grain Positive Film (type 225). (*E. I. du Pont de Nemours & Co.*)

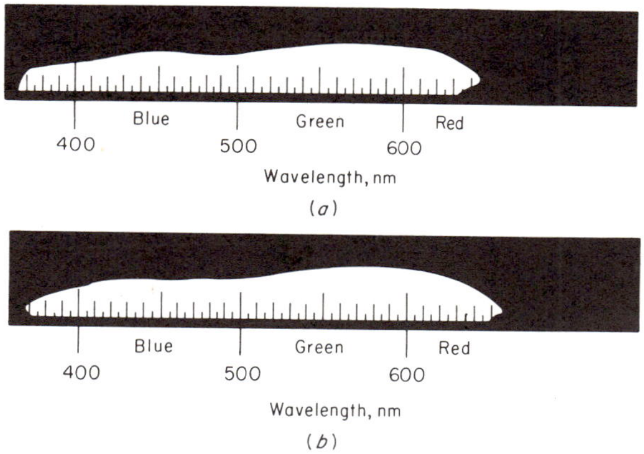

Fig. 2 Wedge spectrogram of panchromatic-sensitized material; Kodak Panatomic-X (35-mm): (*a*) to tungsten light; (*b*) to daylight. (*Eastman Kodak Co.*)

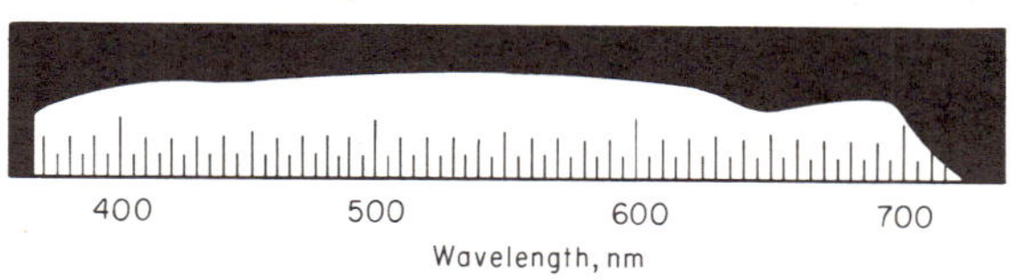

Fig. 3 Wedge spectrogram to daylight of extended-red-sensitivity material; Kodak Special Fine-Grain Aerial (SO-190). (*Itek Corporation.*)

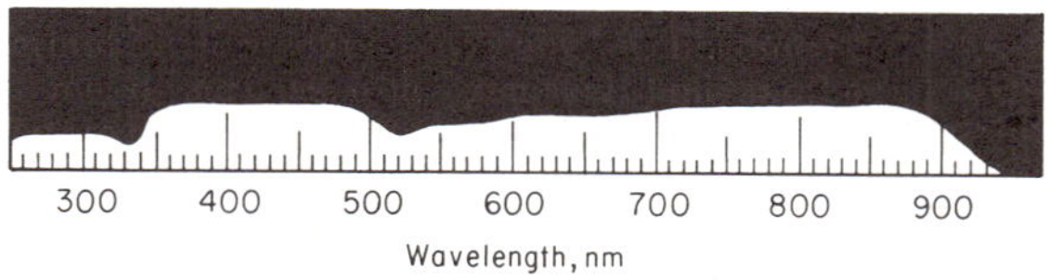

Fig. 4 Wedge spectrogram of material with sensitivity extended into infrared region; Kodak Infrared Aerographic Type 2424. (*Eastman Kodak Co.*)

a near-ultraviolet- and blue-sensitive film; Fig. 2 is a panchromatic-sensitized film; Fig. 3 depicts the spectral response of a film with extended red sensitivity; and Fig. 4 is a film with special response into the infrared region. These examples cannot be quantitatively compared since the wedge spectrograms were not produced with the same wedge spectrograph or light source. Examples of absolute determinations of spectral sensitivity with a spectral sensitometer are presented in Figs. 5 and 6, showing the spectral sensitivity of a blue-sensitive film and an extended-red-sensitized emulsion.

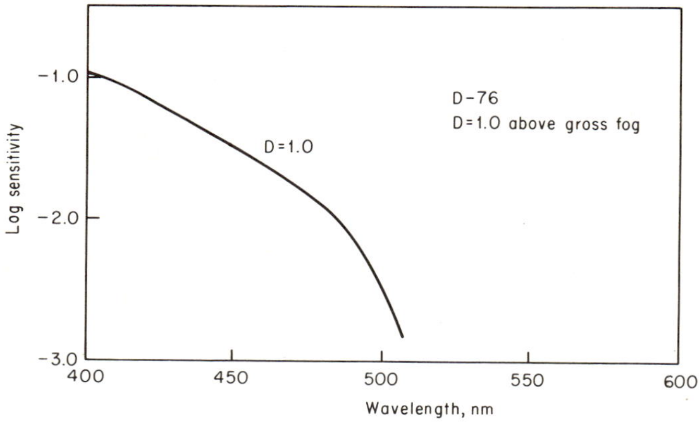

Fig. 5 Absolute spectral sensitivity of Kodak Fine-Grain Aerial Duplication Film 2430. (*Eastman Kodak Co.*)

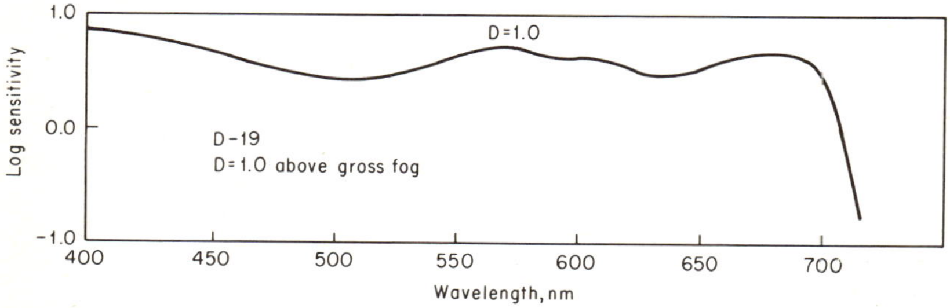

Fig. 6 Absolute spectral sensitivity of Kodak High-Definition Aerial Film 3404. (*Eastman Kodak Co.*)

A summary of spectral sensitizings of the sensitizing classes of Kodak spectroscopic materials and a sampling of the broad range of their emulsions are given in Fig. 7. All the materials exhibit the fundamental blue sensitivity, but the various spectral dye sensitizers used during emulsion manufacture extend the films' response to the desired longer wavelengths.

When an optical filter is used to modify or limit the spectral nature of the image, the exposure must be increased to compensate for the amount of light absorbed by the filter. The filter factor represents the increase in exposure required to produce the same density level in the film that would have been achieved without the filter. The filter factor depends on the spectral sensitivity of the film, the spectral distribution of the illumination, and the filter characteristics (Hall and Howell, 1966).

5. Image Quality and Measurement Most photographic images look homogeneous until they are viewed with magnification, which shows that the image is composed of a random distribution of silver grains within the emulsion. The size and distribution of these grains is a major factor in a material's capability to record minute detail (James and Higgins, 1960). Since terms like definition, sharpness, and graininess are subjective evaluations of photographic imagery, a number of objective methods are used to measure the image qualities of the processed film.

Resolving Power. Resolving power is a measure of the ability of a photographic material to record an image of recognizable configuration at a minimum size. The resolving power assigned for a film type is the reciprocal of twice the smallest target bar width that can be recorded with optimum focus, exposure, and processing

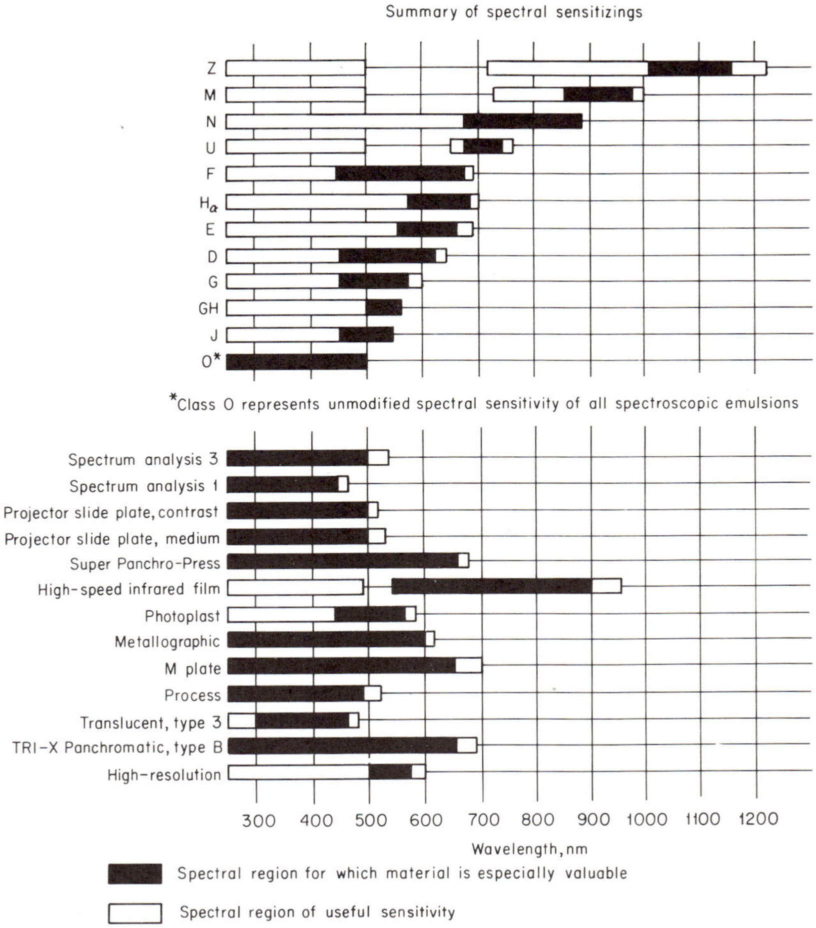

Fig. 7 Summary of spectral sensitizing classes of Kodak spectroscopic materials. (*Eastman Kodak Co.*)

conditions. The resolving power of a given film decreases as the contrast of the target is decreased. Some manufacturers' films are conveniently ranked in the following resolving-power classifications:

L = low (below 56 lines/mm)
ML = moderately low (56–68 lines/mm)
M = medium (64–95 lines/mm)
H = high (96–135 lines/mm)
VH = very high (136–225 lines/mm)
EH = extremely high (above 225 lines/mm)

The data are for a test-object contrast of 1,000:1. As a general rule, resolving power increases as the sensitivity of the photographic emulsion decreases.

Resolving power is a relatively simple technique to measure image quality and is widely used because it relates directly to the capabilities of a film to record fine detail. Furthermore, resolving power can give an account of the total system performance. However, it has a drawback in that it may not be able to pinpoint the cause of image degradation in a photographic system.

Granularity. Granularity refers to the random, microscale spatial variations in the transmitting properties of the film due to the granular structure of the developed photographic image (James and Higgins, 1960). Photographically, *granularity* is an objective measurement that correlates with the subjective appearance of graininess. In general, increasing the speed (light sensitivity) of an emulsion results in an increase of the granularity of the film. Granularity can be referred to as the "noise level" of a photographic system. A comparison of microdensitometer traces of a fine-grain emulsion with a coarse-grain emulsion is shown in Fig. 8.

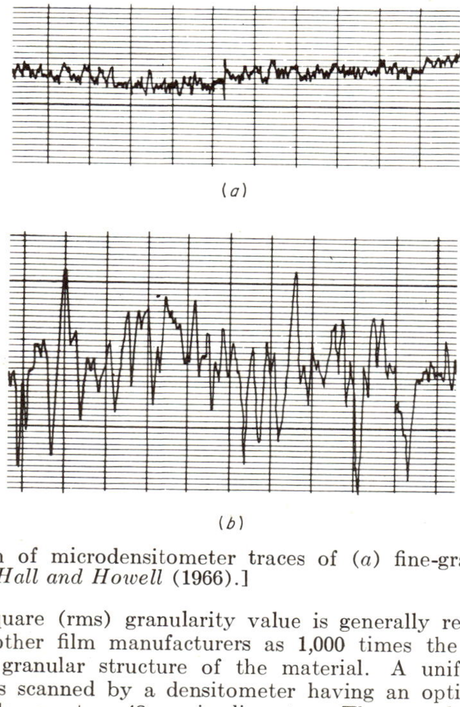

(*a*)

(*b*)

Fig. 8 Comparison of microdensitometer traces of (*a*) fine-grain and (*b*) coarse-grain emulsions. [*Hall and Howell* (1966).]

The root-mean-square (rms) granularity value is generally reported by Eastman Kodak and some other film manufacturers as 1,000 times the standard deviation in density by the granular structure of the material. A uniformly exposed and developed sample is scanned by a densitometer having an optical system aperture of *f*/2.0 and a circular aperture 48 μm in diameter. The granularity value indicates the magnitude of the impression of graininess that would be produced if the sample were examined at a 12$\times$ magnification. The rms granularity values contained in manufacturers' data sheets are for negative films having a net density of 1.0 excluding the density of the support. For reversal materials, the rms granularity values are obtained at a gross density of 1.0 (Eastman Kodak, 1969).

Acutance. The measure of the shape of the density gradient across the edge of an image is associated with the psychophysical concept of sharpness. Acutance can be enhanced by special processing techniques.

Modulation Transfer Function (MTF). If a point or narrow slit is imaged on an emulsion layer, it is not confined to its nominal boundaries but diffuses and spreads out. This effect is caused by the turbidity of the emulsion, and the changed intensity profile of the image is called the *spread function* (Fig. 9).

Figures 10 and 11 compare the modulation transfer functions of two films. Figure

10 is Royal-X Pan Recording (5475), a very fast, coarse-grained emulsion. Primarily because of the large, coarse silver halide grain structure, this film's response to detail of increasing spatial frequency (cycles per millimeter) falls off rapidly. Panatomic-X (Fig. 11) is an extremely fine-grain coating and therefore has an inherent response to detail at higher spatial frequencies. Panatomic-X offers moderate speed with high definition.

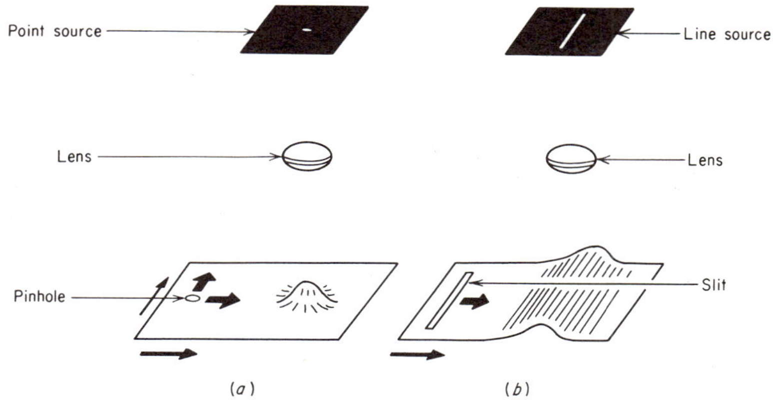

Fig. 9 (*a*) Point-spread function formed and scanned with a pinhole and (*b*) line-spread function formed and scanned with a slit. [*Hall and Howell* (1966).]

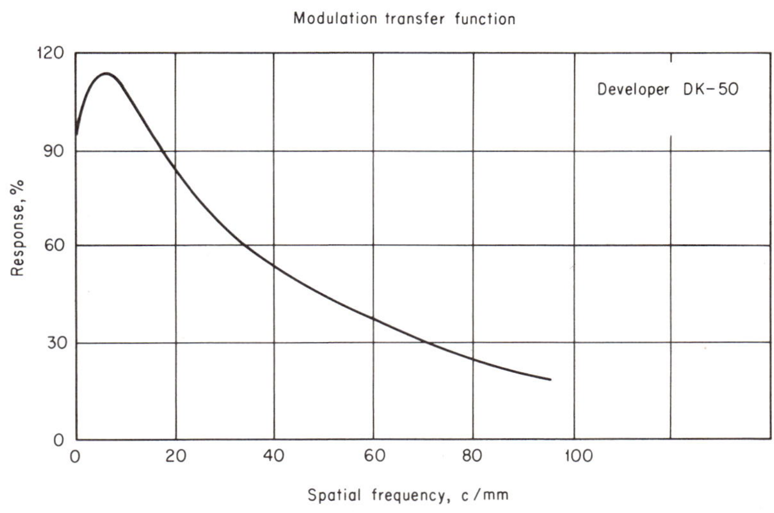

Fig. 10 Modulation transfer function of Kodak Royal-X Pan Recording (5475). (*Eastman Kodak Co.*)

Modulation-transfer characteristics indicate the effects on the microstructure of the image caused by the diffusion of light within the image. The modulation transfer is the ratio of the modulation of the developed image to the modulation of the exposing target, or M_i/M_o. MTF can effectively evaluate system performance or that of any of the individual components.

6. Latent-Image Phenomena The term *latent image* describes the invisible pattern produced by exposure. The pattern is then made visible by the chemical development (Mees and James, 1966). In typical emulsions each crystal of silver halide contains 10^6 to 10^9 silver ions. Exposing the silver halide crystal to reduce

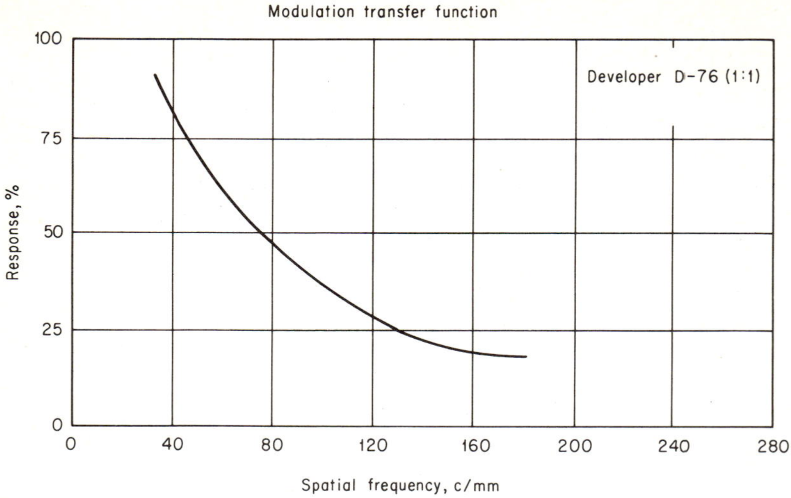

Fig. 11 Modulation transfer function of Kodak Panatomic-X. (*Eastman Kodak Co.*)

only a few of the silver ions to metallic silver is all that is necessary to make the whole crystal developable:

$$AgX \xrightarrow{h\nu} Ag^0 + X^-$$

Model of Latent-Image Formation. Three or four atoms are sufficient to form a latent-image site. A tremendous advantage of the silver halide process is that the few atoms of latent-image silver on a crystal can be amplified by action of developer to form 10^6 to 10^9 atoms of metallic silver. No other photosensitive system to date has this magnitude of amplification. Furthermore, the latent image is usually quite persistent, and the visible image can be retrieved by development at a later time. However, critical users of color and black-and-white materials should develop the exposed goods as soon as practical. With most non-silver halide systems, the latent image is transitory and must be processed to a visible image shortly after exposure.

Reciprocity-Law Failure Characteristics. The reciprocity law of Bunsen and Roscoe states that the amount of photochemical reaction depends simply upon the actual light energy absorbed by the photosensitive material and not upon the rate of absorption (James and Higgins, 1960). Exposure E is defined as the product of intensity I and time t:

$$E = It$$

Reciprocity-law failure is deviation from the equation and occurs because the effect of exposure (density) depends on the rate at which energy is absorbed by the photosensitive surface (Neblette, 1952). Some effects of reciprocity are noted with all photographic emulsion, but usually they are serious only at very short or very long exposure times. Reciprocity failure must be compensated for in such applications as astronomical photography, where exposures can be of several hours' duration, or high-speed photography, with exposures as short as 1 ns. The photographer may be troubled with this phenomenon when (1) exposing multilayer color film, (2) when using deep "cutting" filters, since filter factors depend on the reciprocity failure of the material, and (3) under extremes of temperature.

Reciprocity-law failure is described graphically by plotting the log exposure required to produce a fixed density against log intensity. Lines of constant exposure time are drawn at an angle to the coordinates to help in interpreting the results.

For most photographic materials, there is an optimum combination of intensity and time, a combination which produces the greatest photographic efficiency. The reciprocity curve exhibits a minimum exposure in this region. Most photographic emulsions are manufactured to yield the highest efficiency at commonly used exposure times. Some specialized films are manufactured for the extremes of exposure time. The reciprocity curves in Figs. 12 and 13 are for a number of representative

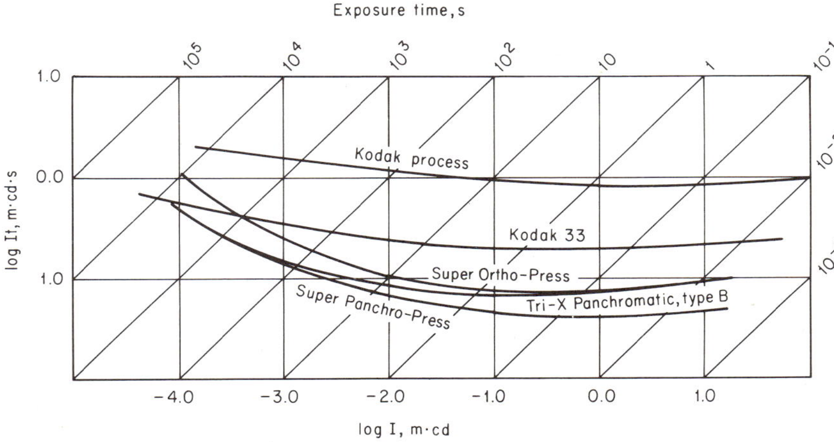

Fig. 12 Reciprocity-effect curves for several Kodak films. (*Eastman Kodak Co.*)

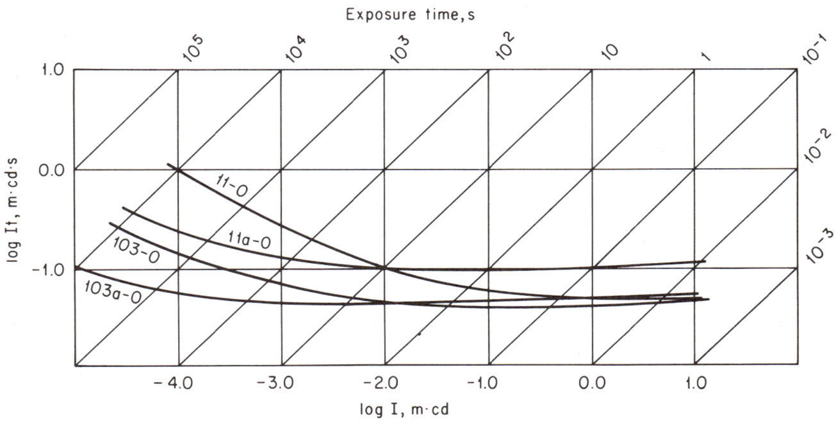

Fig. 13 Reciprocity-effect curves for Kodak spectroscopic plates. (*Eastman Kodak Co.*)

Kodak emulsions. It should be noted that reciprocity failure is most severe at exposures of long duration.

Intermittency Effect. The intermittency effect is another form of reciprocity failure in which a continuous exposure does not produce the same density as another exposure in which the same total energy is applied to the material in a number of separate installments. The intermittency effect can be of considerable importance where intermittent exposures are used as a matter of practice, as in absorption spectrophotometry. When the frequency of interruption is above a certain critical level, the effects of the intermittent exposure become equal to that of a continuous exposure of the same average intensity. This critical level of frequency varies with the intensity. A good practical rule to follow is to divide the total exposure into at least 100 installments, regardless of the overall exposure time, so that the intermittent exposure will produce the same effect as a continuous exposure of the same average intensity (Eastman Kodak, 1967).

Specialized latent-image effects such as solarization, the Herschel effect, the Clayden effect, and others noted in the literature involve reversal or densitization phenomena and find limited application (Mees and James, 1966).

7. Processing Processing silver halide materials converts the invisible latent image into a visible silver image. The processing steps can be summarized as follows:

1. Develop the exposed silver halide grains to form a visible image
2. Fix the image to dissolve the undeveloped silver halide.
3. Wash to remove the fixing by-products
4. Dry

A photographic developer forms a visible image related to the exposing radiation incident on the film. The latent-image site acts as a catalyst, increasing the rate at which the developer reduces the silver halide crystal to metallic silver. Unexposed crystals are also reduced by the developer, but much slower; this is photographic fog, the inherent noise level of the system. Allowed sufficient time, the developer will reduce all the silver halide in an emulsion, producing an overall high density (Hall and Howell, 1966).

The rate of development is affected by a number of factors:

1. Emulsion composition, thickness, and hardness
2. Developer formulation (selection and combination of developing agents, solution pH)
3. Developer exhaustion (oxidation of developing agents, by-products, e.g., bromide buildup)
4. Time, temperature, and agitation of developer

Uniformity. The uniformity of the photographic image depends upon proper agitation to remove the developer by-products evenly. Poor agitation technique can produce mottling and uneven densities (Society of Motion Picture and Television Engineers, 1960).

Immersion Processing. A popular method of processing is the immersion configuration, e.g., roll- and sheet-film hangers in a tank, multiple racks in a deep tank, or continuous feedthrough in a deep tank. Hand agitation is normally used for small batch quantities of roll- or sheet-film processing. The more sophisticated sheet, rack, and continuous-immersion processors are usually equipped with nitrogen-burst agitation. Some processes, e.g., lithographic film, require stagnant development to enhance the acutance of the image.

Spray Agitation. Many high-volume continuous processors are equipped with spray agitation. The force of the spray striking the film's surface quickly and thoroughly removes the developer by-products, thereby increasing the rate of development. Spray is the most efficient method of agitation. It may be necessary to use developers expressly formulated for spray application to minimize the effects of aerial oxidation on the solution.

Viscous Processing. Viscous processing finds application when conventional wet-processing techniques are not feasible, e.g., in-flight reconnaissance processing or remote locations without water. Specialized developers, fixers, and monobaths combining both developer and fixer are prepared with various polymers to form thick, jellylike solutions with viscosities ranging from 2,000 to 15,000 cP. The viscous solution is extruded onto the film in a layer up to 0.25 mm thick. By this procedure acutance and resolving power are usually enhanced.

8. Diffusion Transfer The conventional process is often cumbersome for many applications. It is often important to have quick access to the photographic image without complex processing techniques. Polaroid photosensitive products and the Kodak Bimat process are based on the diffusion-transfer mechanism. The photosensitive and image-receiving layers are brought in contact with a silver halide solvent-developer simultaneously and immediately pressed together. The negative and positive images are formed concurrently (Neblette, 1952). The Kodak Bimat process is modified so that the image-receiving layer is presoaked with the silver halide solvent-developer.

9. Color Processes Many practical applications have proved that color photography can provide more information than normal black-and-white photography. Color film has a distinct advantage in that it distinguishes between different-color objects that would be displayed in the same shade of gray in black-and-white photograhy.

Since color films of the integral tripack construction, whether reversal or negative (Figs. 14 and 15) are the most widely used, only color materials of this class

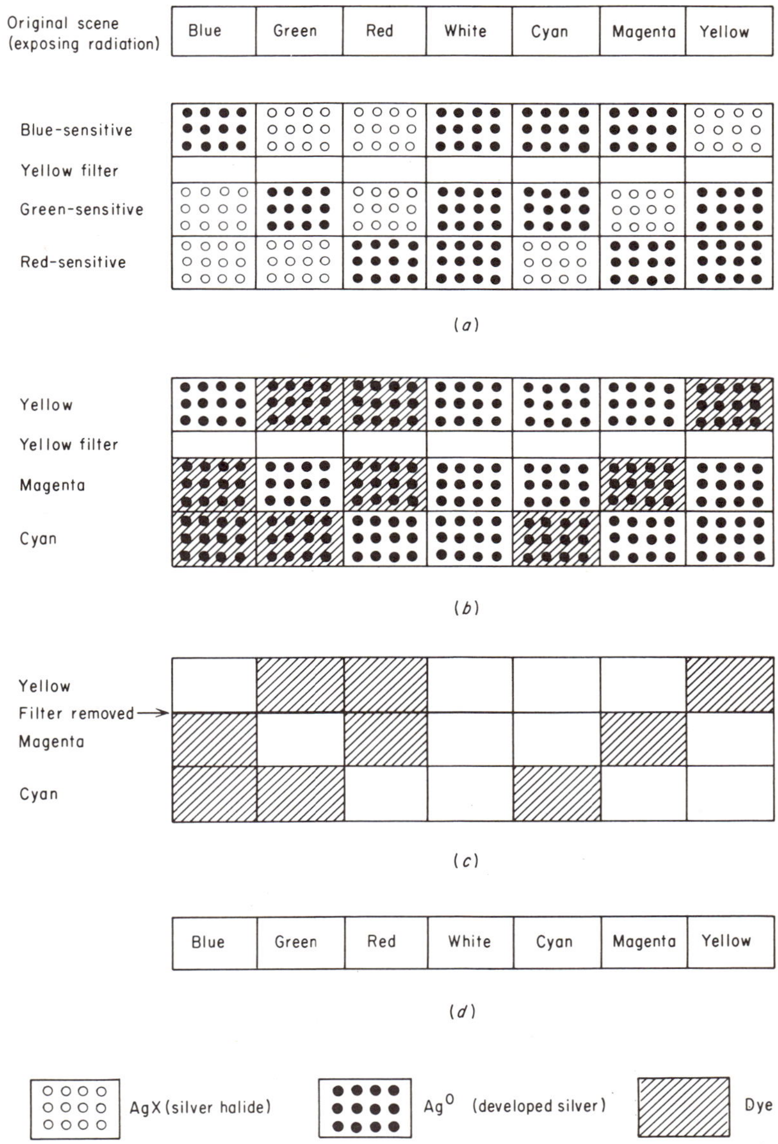

Fig. 14 Typical color-reversal process: (*a*) after color exposure and first developer; (*b*) after fogging exposure and color developer; (*c*) after bleaching and fixing; (*d*) reproduction of original scene. (*Itek Corporation.*)

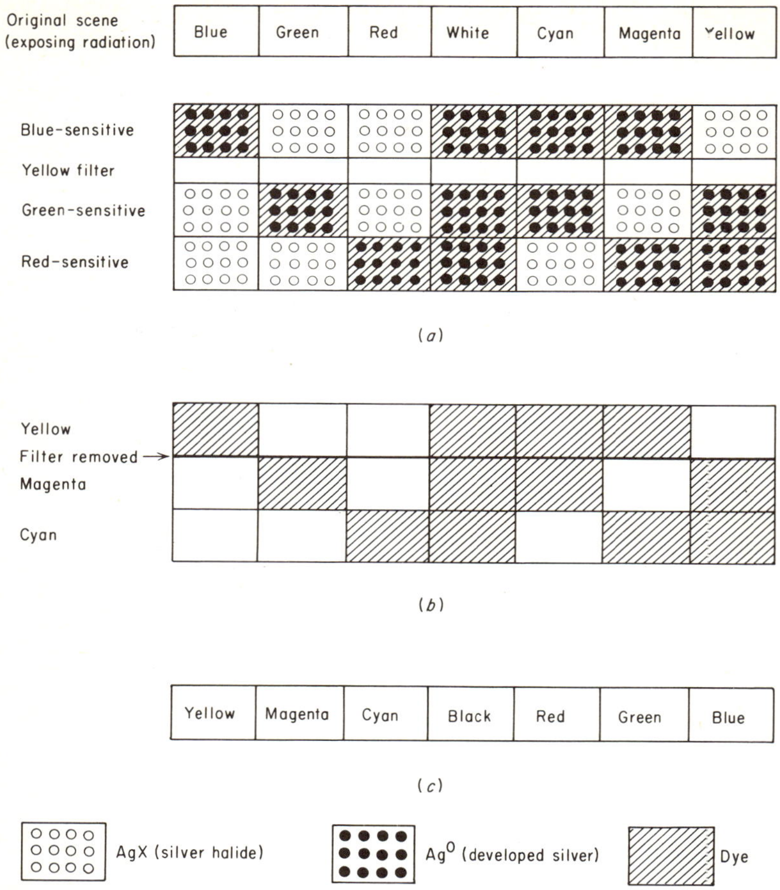

Fig. 15 Typical color-negative process: (*a*) after color exposure and color developer; (*b*) after bleaching and fixing; (*c*) color negative of original scene. (*Itek Corporation.*)

will be discussed. Three light-sensitive emulsion layers are placed on one side; each layer responds to a separate region of the spectrum. The three emulsions are not separable and must be handled and processed together.

Anscochrome 100, Ektachrome-X, Kodacolor-X, Agfachrome CT-18, and Fujicolor are examples of color film containing couplers put into the emulsion layers during manufacture. These incorporated couplers form dyes during processing. The three dye images in the emulsion layers are simultaneously produced by one color developer. These films may be user-processed. Reversal or negative emulsions are available for amateur professional, professional, and aerial applications. Speeds range from ASA 25 to 500.

Anscochrome II 8 mm, Kodachrome 25, Agfachrome 8 mm, and Dynachrome Daylight are a class of reversal color films that do not contain dye-forming particles in the emulsion layers. The dyes are formed during processing from three separate color-developer solutions. Processing is much more complex than the incorporated coupler films and is generally not done by the user. These films are generally intended for slides or movies. Speeds range from ASA 25 to 64.

10. Sources of Sensitometric Data From manufacturers' data sheets the user can determine important characteristics of the photographic material such as speed, spectral response, and processing conditions. Other data dealing with resolution, acutance, light sources, and special handling procedures may be found in most manufacturers' literature.

NON-SILVER HALIDE SYSTEMS

11. Introduction Research and development and introduction of new non-silver halide systems has been increasing steadily. The reasons are varied—to eliminate or reduce the use of costly silver, to improve proprietary marketing positions, to achieve the technical objective of dry "instantaneous" image formation, or to provide inexpensive photo systems (document copying). In general, non-silver halide systems serve specialized applications and lack the versatility of the silver halide system.

The absolute speeds in ergs per square centimeter of several photosensitive systems are compared in Table 3. Table 4 compares minimum energy requirements, resolving power, access time, and processing conditions for several photosensitive systems.

TABLE 3 Speed of Several Recording Processes (Nitka, 1968)

Process	Absolute speed, ergs/cm^2
Fast silver halide film	2×10^2 †
Medium-speed silver halide film	1×10†
Electrophotography	1†
Photopolymerization	2×10^{-6} ‡
Fast diazo	2×10^{-7} ‡
Standard diazo	2×10^{-8} ‡

† For wavelength of 430 nm.
‡ For wavelength of 350 to 450 nm (for 3200 K tungsten light).

TABLE 4 Minimum Energy Requirements, Resolving Power, Access Time, and Processing Conditions for Several Recording Processes (Nitka, 1968)

Process	Minimum energy required for recording image, ergs/cm^2	Resolving power,† lines/mm	Access time	Processing
Vidicon tube	1×10^{-4} ‡	5	Micro-seconds	Dry
Fast silver halide film	5×10^{-3} §	50	Seconds to minutes	Wet
Medium-speed silver halide film		150	Seconds to minutes	Wet
Electrophotography	1§	20–200	Seconds	Dry
Photopolymerization	6×10^5	500	Seconds to minutes	Wet
Fast diazo	4×10^4 ¶	1,000	Seconds to minutes	Dry

† Order-of-magnitude values only.
‡ For visible spectrum.
§ For wavelength of 430 nm.
¶ For wavelength of 350 to 450 nm (for 3200 K tungsten light).

12. Electrophotography Electrophotography is foremost among the nonchemical processes which have revolutionized the photocopy field. Two important commercial document-copying processes depend on the selective formation of an electrostatic charge during the exposure sequence. The Xerox process uses a metal surface (such as selenium) to form a charge image and pick up the toner particles. The toner is then transferred by offset to ordinary paper and fused. The other approach, based on RCA patents (Electrofax), eliminates the selenium surface and uses a special paper coated with a conductive metal layer and a layer of photoconductor (ZnO). The toned image is formed directly on the paper. The SCM, APECO, and Dennison photocopiers are based on this system. Figure 16 details the relative spectral response of selenium and zinc oxide photoreceptors. The spectral output of common light sources and the spectral sensitivity of the eye are included.

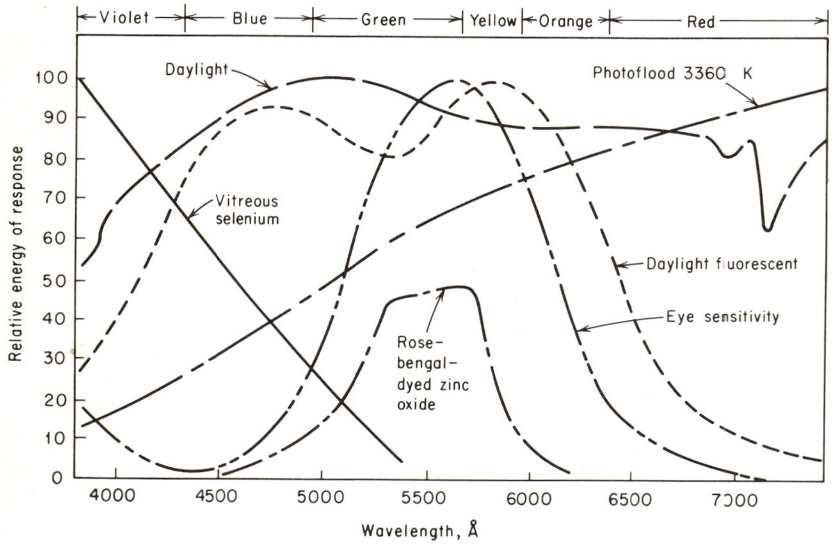

Fig. 16 Relative spectral energy of typical emitters and response of typical xerographic photoreceptors. [*Dessaurer and Clark* (1965).]

13. Diazo The diazotype process (positive) depends on the destruction by light of a diazonium salt to prevent coupler dye formation when the material is placed in an alkaline environment (usually ammonia gas). The diazo-reversal process is negative-working. The diazo processes depend upon ultraviolet energy (peak sensitivity 380 to 410 nm) for their quantum yield; there is no analogous amplification process as with silver halide materials. Since diazo materials are very slow and require large amounts of exposure to ultraviolet energy, they are unsuited for camera applications.

Negative-working diazo films are available for microfilm-distribution applications. Since the diazo image is essentially molecular, it has no granularity. Resolution capabilities are reported to be in excess of 1,000 lines/mm. Figures 17 and 18 give the sensitometric characteristics, image spectral-absorption characteristics, and spectral sensitivity for two Technifax products, coated on 0.125-mm polyester base and 0.125-mm acetate base, respectively. Image color is black for the first and sepia for the second.

14. Vesicular Diazo (Kalvar) Kalvar Corporation manufacturers heat-developable films. Kalvar materials consist of a thermoplastic resin with an ultraviolet-sensitive compound coated upon transparent film or an opaque black base. The reaction of ultraviolet energy (peak sensitivity at 385 to 405 nm) with the diazonium compound produces a molecular gas (under pressure). When the material is heated for a few seconds to 93 to 140°C, the thermoplastic resin softens to permit the molecular gas to expand, forming light-scattering image elements 0.5 to 2 μm in diameter. Film resolution generally exceeds 350 c/mm to a high-contrast target.

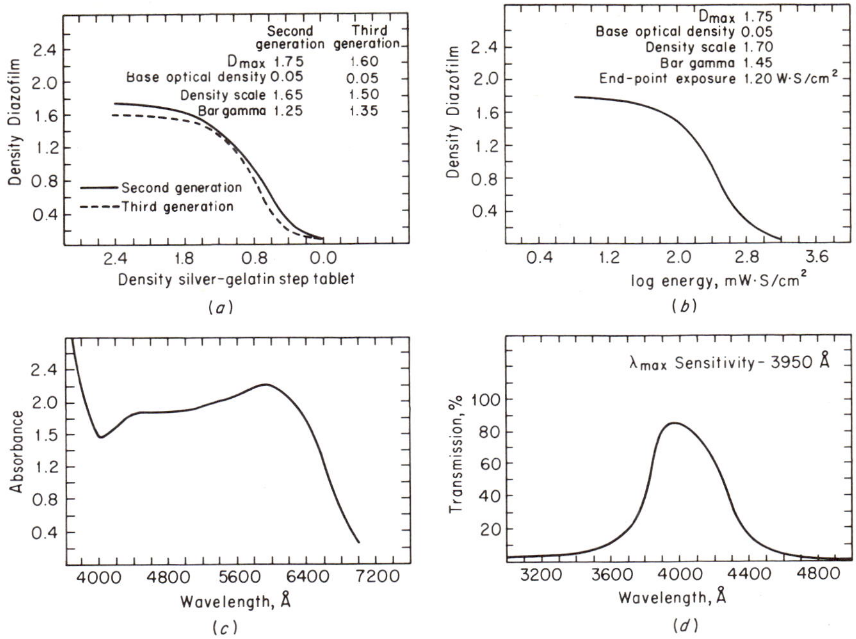

Fig. 17 Characteristics of Technifax Diazofilm K-Tone, series 100, P5-100: (*a*) sensitometric; (*b*) NABDC-controlled mercury-arc sensitometer; (*c*) image spectral absorption; (*d*) spectral sensitivity. (*Technifax Corporation.*)

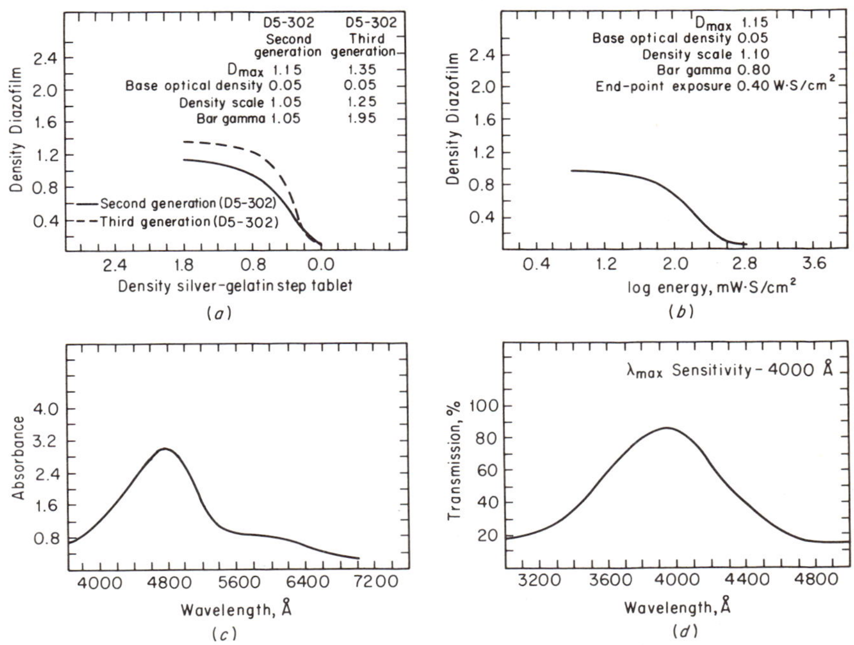

Fig. 18 Characteristics of Technifax Diazofilm S-Tone, series 300: (*a*) sensitometric; (*b*) NABDC-controlled mercury-arc sensitometer; (*c*) image spectral absorption; (*d*) spectral sensitivity. (*Technifax Corporation.*)

Current applications of Kalvar products include rapid, high-quality duplication of both line and continuous-tone materials. With Kalvar films the image contrast tends to be high unless the film is viewed over a diffuse light source. Data of representative Kalvar duplicating film, MiKrolith 200 and MiKrolith 500, are presented in Figs. 19 and 20 and Table 5. Technical data of Kalvar type 91, a

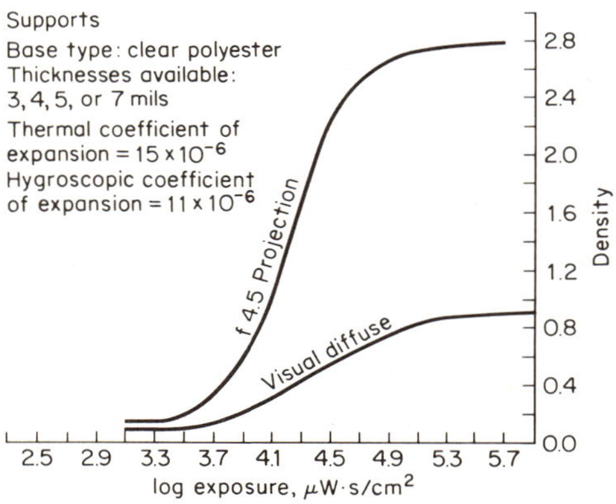

Fig. 19 Sensitometric data for Kalvar MiKrolith 200. (*Kalvar Corporation.*)

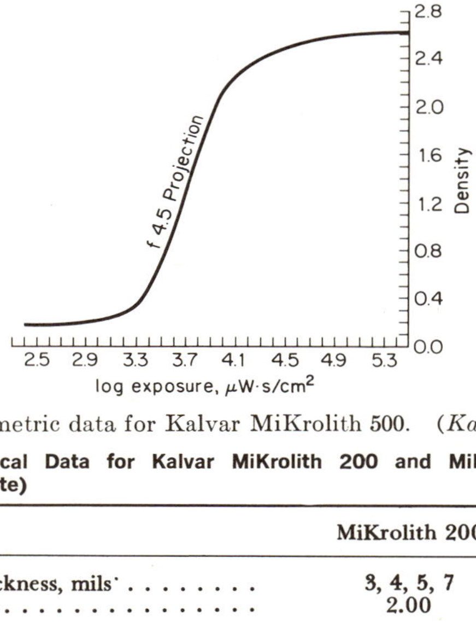

Fig. 20 Sensitometric data for Kalvar MiKrolith 500. (*Kalvar Corporation.*)

TABLE 5 Technical Data for Kalvar MiKrolith 200 and MiKrolith 500 (Kalvar Corporation, no date)

	MiKrolith 200	MiKrolith 500
Polyester base thickness, mils	3, 4, 5, 7	3, 4, 5, 7
Relative speed†	2.00	5.00
Gamma:		
Visual diffuse	0.55	
Projection	3.30	3.10
Log exposure scale	1.60	1.60
Resolution, c/mm:		
High-contrast target (100:1 or greater)	350	350
Medium-contrast target (6.3:1)	250	250
Low-contrast target (1.6:1)	150	150

†Kalvar type 10 is used as a base for all speed comparisons.

high-contrast, wide-latitude autopositive film designed for reflection viewing, are presented in Fig. 21.

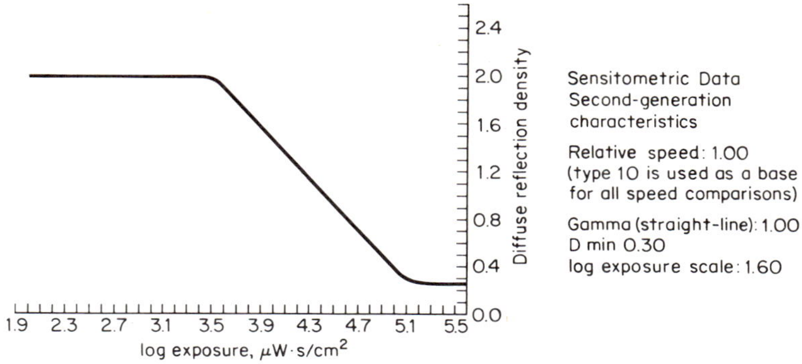

Fig. 21 Sensitometric and technical data for Kalvar Film Type 91. (*Kalvar Corporation.*)

15. Itek Physical-Development Photography Itek photographic products using physical-development, photoconductor photography include the Itek RS Reader-Printer paper intended for microfilm blow-back applications and the Itek type 462 Aerial Duplication Film.

Itek type 462 film is a high-resolution (over 900 lines/mm with contact-printing techniques) duplicating material coated on a 0.10-mm polyester base. Spectral sensitivity is in the ultraviolet and short-blue regions with peak sensitivity at about 380 nm (Fig. 22). Additional photoconductors, for example, ZnO, have been used in experimental emulsions.

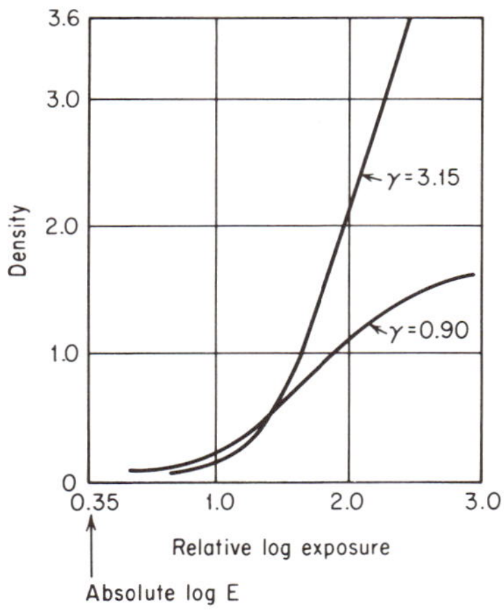

Fig. 22 Typical sensitometric curves on Itek film. (*Itek Corporation.*)

Because the photosensitive coating forms a latent image that gradually decays, processing must follow exposure within a few minutes to avoid a decline in density and contrast. Raw-stock type 462 film contains no silver or silver salts, but the processed image is formed of silver supplied by the processing solutions.

Processing is accomplished rapidly in three basic steps: an activator treatment, image development, and fixing. The total processing cycle including drying can be completed in less than 1 min. Gamma can be controlled from about 0.8 to 3.6 by simple changes in processing. Because the latent image of Itek type 462 film is not permanent until the activation step has been completed, this material is not permanently affected by accidental exposure to light, heat, or high-energy radiation. Dark storage for about 24 h will restore its original sensitivity. This property is unique with type 462 film among high-resolution photosensitive materials. The modulation transfer function is shown in Fig. 23.

16. Dry Silver Process Dry silver films and papers for copying and duplicating applications, such as those manufactured by the Microfilm Products Division of

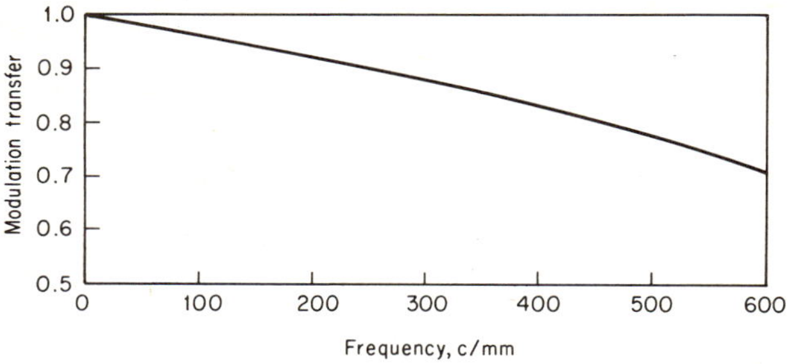

Fig. 23 Modulation transfer response of Itek emulsion. (*Itek Corporation.*)

3M, need only a treatment of heat in the range of 95 to 150°C to produce a fully developed image. All ingredients are coated on a single substrate (U.S. Pat. 3,152,903). Spectral sensitivity is presently in the ultraviolet and blue range. Sensitivity is comparable to slow contact-printing silver halide materials. Figure 24 gives sensitometric curves of a dry silver paper. Contrast variation is achieved by control of the time of the heat-development step.

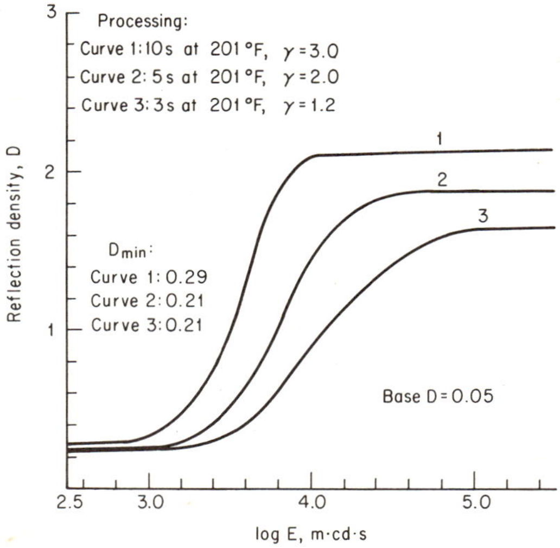

Fig. 24 Sensitometric response of 3M Dry Silver Paper, type 773. (*3M Co.*)

17. Thermography Photothermographic processes are light-sensitive only in a limited sense. Such systems depend upon a reaction to long-wavelength infrared energy which changes the visual appearance of a surface layer. The best-known example of such a paper is 3M Thermofax, for contact copying of documents, which uses a salt or soap of a heavy metal, such as iron or nickel, and another component which react to form a colored image (Kosar, 1965). Since thermographic document-copying papers exhibit poor image sharpness and a tendency for the background to darken, they have been largely superseded by the various systems of electrostatic copying.

18. Free-Radical Photosystem The free-radical photosystem comprises (1) an oxidizable leuco (colorless) compound or an indicator-type dye base and (2) an

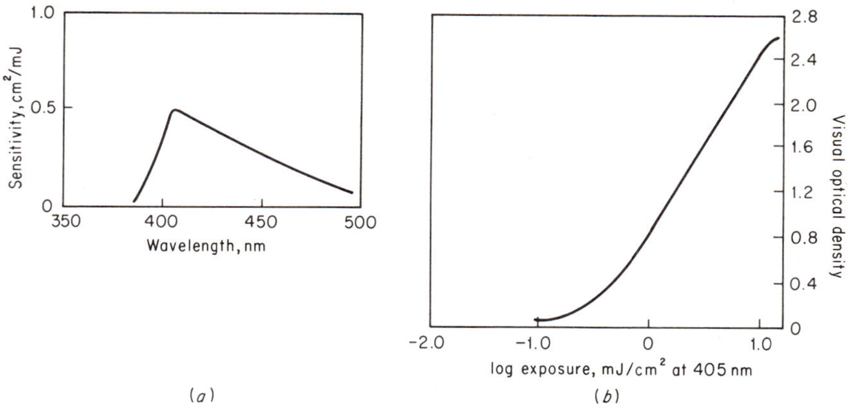

Fig. 25 (*a*) Spectral sensitivity and (*b*) sensitometric characteristics of Photohorizons type 2000 duplicating film. (*Photohorizons.*)

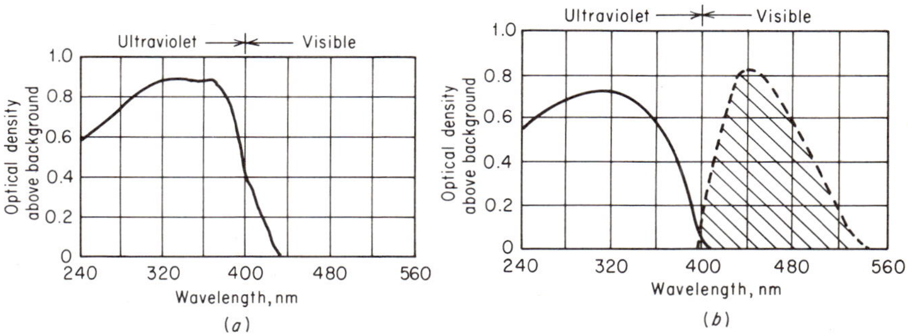

Fig. 26 Du Pont Dylux: (*a*) color formation and (*b*) color formation and deactivation. (*E. I. du Pont de Nemours & Co.*)

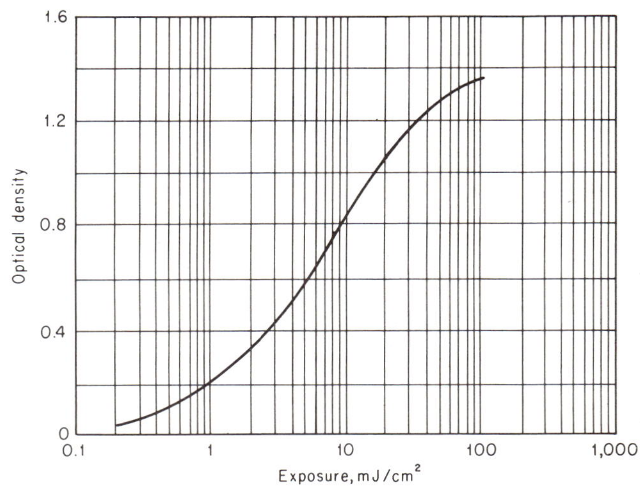

Fig. 27 Du Pont Dylux sensitometric curve for imaging composition. (*E. I. du Pont de Nemours & Co.*)

activator or sensitizer consisting of mercaptans or polyhalogen compounds. Upon irradiation, chain-propagating free radicals are formed to produce a direct printout image.

Photohorizons Type 2000 Duplicating Film. Photohorizons, a division of Horizons Research, Inc., has introduced type 2000 free-radical duplicating film, a medium-contrast, high-resolution, dry-processed, nonsilver material coated on a 75- to 125-μm polyester base. The material is negative-working and produces a continuous-tone blue image by direct printout. It is designed for making copies of original silver halide microfilm or for printing original aerial negatives. This material can be optically developed to reduce initial exposure levels by one-fortieth or more by flashing the initial latent image to energy in the range of 600 nm. Fixing is required by about a 90-s exposure to heat of about 170°C. Figure 25 shows a spectral sensitivity plot and a typical sensitometric curve for this film. Gamma can be varied from 0.9 to over 2 by control of film temperature during optical development.

Du Pont Dylux. Du Pont Dylux instant-access photosensitive coatings consist of free-radical coatings applied to film or paper substrates (see Figs. 26 to 28 and Table 6). No external processing, heating, or other physical treatments are

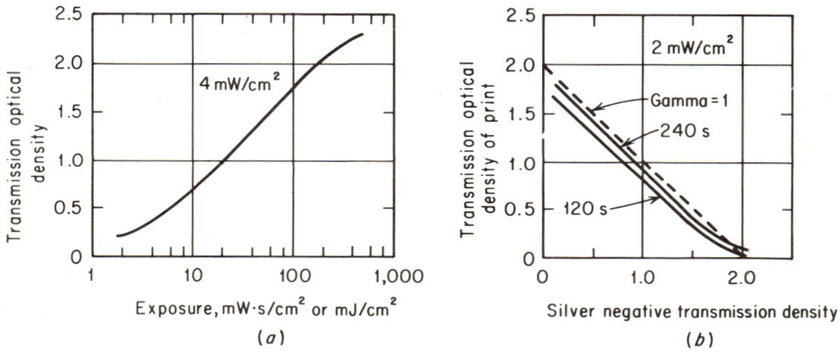

Fig. 28 (*a*) Sensitometric curve and (*b*) tonal reproduction of Du Pont Dylux experimental film XF-203. (*E. I. du Pont de Nemours & Co.*)

TABLE 6 Typical Properties of Du Pont Dylux Experimental Film XF-203 (E. I. du Pont, 1969)

Sensitometry, color formation:	
Exposure time to maximum optical density, s	<90
Speed:	
$S_{0.4}$	250
$S_{1.8}$	7
Gamma	0.9–1.1
Optical density:	
Maximum	2.0–2.3
Minimum	0.04–0.06
Spectral sensitivity, nm:	
Range	200–400
Peak	360
Tonal reproduction	20 of 21 steps
Contact resolution capture, line pairs/mm:	
Uncollimated source	80
Collimated source	400
Imaged film:	
Image color	Deep blue-purple (absorption maxima at 510 and 650 nm)
Background color	Clear
Resolution retention, line pairs/mm:	
After 2 days	>50
After 4 days	>40
After 18 days	>25

required to form the image. Energy in the ultraviolet end of the spectrum forms permanent dye images. Exposure of Dylux to energy in the 400- to 500-nm region deactivates the coating to further color formation. Since the image is molecular, resolution can be high, but it degrades with storage time.

REFERENCES

Dessaurer, John, and Harold B. Clark (1965): "Xerography and Related Processes," Focal Press, New York.

Eastman Kodak Company (1960): "Kodak Photographic Films and Plates for Scientific and Technical Use," 8th ed., Rochester, N.Y.

——— (1965): "Kodak Films in Rolls for Black-and-White Photography," 2d ed., Rochester, N.Y.

——— (1967): "Kodak Plates and Films for Science and Industry," Rochester, N.Y.

——— (1969): "Kodak Data for Aerial Photography," 2d ed., Rochester, N.Y.

E. I. du Pont de Nemours & Co. (1968): "Du Pont Films for Instrumentation/Motion Pictures," Wilmington, Del.

——— (1969): Dylux Experimental Film XF-203, Preliminary Data.

Hall, H. J., and H. K. Howell (1966): "Photographic Considerations for Aerospace," 2d ed., Itek Corporation, Lexington, Mass.

Itek Corporation (1968): Color Tone Reproduction, pt. 1, Theory Manual, *Tech. Rep.* AFAL-TR-67-164.

James, T. H., and George C. Higgins (1960): "Fundamentals of Photographic Theory," 2d ed., Morgan & Morgan, New York.

Kalvar Corporation (no date): The Kalvar Handbook, *Tech. Bull.* 104.

Kosar, Jaromir (1965): "Light Sensitive Systems: Chemistry and Application of Nonsilver Halide Photographic Processes," Wiley, New York.

Mees, C. E. K., and T. H. James (1966): "The Theory of the Photographic Process," 3d ed., Macmillan, New York.

Neblette, C. B. (1952): "Photography: Its Materials and Processes," 5th ed., Van Nostrand, Princeton, N.J.

Nitka, Heinz F. (1968): Introduction to Photographic Systems, *Image Technol.,* vol. 11, no. 5, October-November.

Society of Motion Picture and Television Engineers (1960): "Control Techniques in Film Processing," New York.

Technifax Corporation (1967): Technifax Diazofilm for Microfilm Distribution Systems.

Section **6**

Microscope Objectives and Eyepieces

JAMES R. BENFORD

Former Director, Research, Development, and Engineering, Scientific Instruments Division, Bausch & Lomb, Incorporated, Rochester, N.Y.

HAROLD E. ROSENBERGER

Vice President, Scientific Instruments Development, Scientific Optical Products Division, Bausch & Lomb, Incorporated, Rochester, N.Y.

(*Numbers refer to paragraphs*)

1. Introduction Although most microscope objectives use the standard Royal Microscopical Society thread and most eyepieces employ an almost identical tube diameter so that they are physically interchangeable, for the most part they are not truly interchangeable optically, for a variety of reasons. In the first place, the objective shoulder-to-specimen distance is not the same for various objective types, so that these different types are not parfocal. Of greater importance, the optical tube length for which various objective types are corrected varies markedly from manufacturer to manufacturer and even for different objective series of the same manufacturer. Another point of difference is in the correction for off-axis imagery, notably lateral color. Modern practice is to correct all powers of a series of objectives to have the same lateral color, so that a single eyepiece type compensates lateral color uniformly throughout the series. This is not true for older types of objectives.

This leads us to a related factor in modern microscope design, the almost universal presence of a relay-lens system within the body tube of the microscope. The relay lenses serve to locate the field and pupil planes in positions that allow comfortable inclined binocular viewing and locate viewing screens or photographic film planes in optimum positions. The optical design of these relay lenses, whether zooming or fixed-power, is tied in with that of objectives and eyepieces to give optimum image correction and proper eye relief.

Thus today's microscope, in contrast to that of the past, is an optical *system* design and not a collection of interchangeable parts held together by a body tube. Indiscriminate assembly of objectives, eyepieces, and bodies will generally result in poor image quality. The tables of objectives and eyepieces that follow offer some guidance to intelligent selection of compatible components, but space limitations mean that much is left unsaid.

2. Objectives All data on objectives are collected in Table 1 (pp. 6-10–6-36), the numbered columns of which are explained in the paragraphs that follow.

Column 1. Objectives in these tables are listed in order of increasing *power,* or magnification. This is the ratio of image height to object height and is a nominal value usually good to within 1 or 2 percent.

Column 2. The numerical aperture (NA) is a measure of the resolving power of the objective, defined as

$$\mathrm{NA} = N \sin u$$

where N is the refractive index of the object space and u is the half angle of the extreme ray from the object which the objective can utilize (see Fig. 1). From

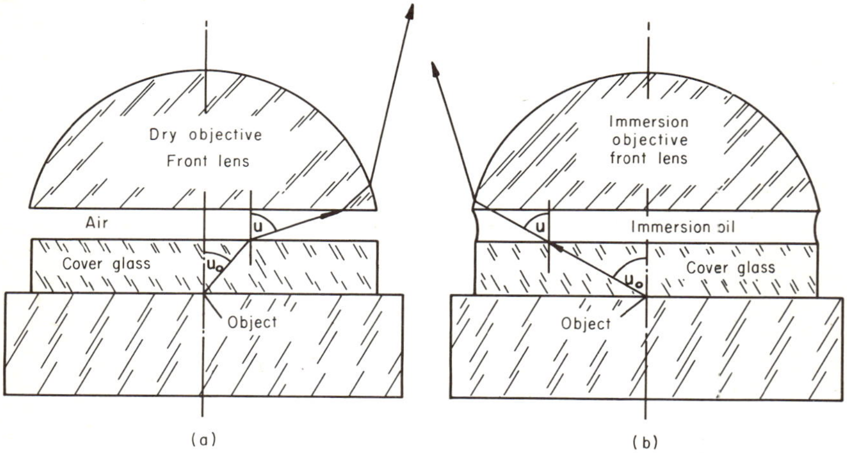

Fig. 1 The path of light entering (*a*) a dry objective and (*b*) an immersion objective. [*From R. Kingslake (ed.), "Applied Optics and Optical Engineering," vol.* 3, *p.* 147. *Copyright* 1967 *by Academic Press, Inc.; used by permission.*]

the law of refraction, the product $N \sin u$ is an invariant in the object space, whether the object is covered or uncovered, dry (Fig. 1*a*) or oil immersion (Fig. 1*b*). Thus

$$N_0 \sin u_0 = N \sin u = \mathrm{NA}$$

The NA of an objective is an extremely important design characteristic, since image brightness, resolving power, and depth of focus are intimately related to the NA. Image brightness is directly proportional to $(\mathrm{NA})^2$. Resolving power is directly proportional to NA. Thus

$$\text{Resolving power} = 3{,}600(\mathrm{NA}) \qquad \text{line pairs/mm}$$

Depth of focus is inversely proportional to $(NA)^2$. Thus

$$\text{Depth of focus} = \pm \frac{N\lambda}{2(NA)^2}$$

where λ is the wavelength of light and N is the lowest index in the object space.

Column 3. A *cover glass* (Fig. 1) is used to cover biological specimens. Since no cover glass is required for metallurgical specimens, this forms one major distinction between biological and metallurgical objectives. Although a cover glass is only 0.17 or 0.18 mm thick, it has a profound influence on the image quality, particularly of high "dry" objectives, due to the extreme amount of spherical aberration it introduces at high NA. For objectives below 0.30 NA it has little effect, but since the aberration increases exponentially with NA, its effect upon image quality becomes rapidly worse, and at 0.50 NA it becomes intolerable if not compensated in the objective design. Oil-immersion objectives are less sensitive to cover glass since the oil index (1.516) is close to the cover-glass index (1.522).

Column 4. The *tube-length* rating of an objective is the manufacturer's specification for the length of the body tube for which the objective is designed. It defines the image distance. We use the term to specify the distance from objective shoulder to objective image although this is neither the "optical tube length" nor the "mechanical tube length" of classical literature. In our terminology *tube length* plus *shoulder height* (see below) equals the object-to-image distance of the objective.

Infinity-corrected objectives are corrected for an infinite tube length; hence they require a telescope objective in addition to an eyepiece to bring the image into focus for the eye. Such systems permit a convenient modular, or building-block, concept of design since fairly sizable accessories can be inserted into the infinity space without upsetting tube length, magnification, parfocality, working distance, or axial image quality.

Column 5. Shoulder height is the design distance from specimen plane to objective shoulder when the objective is in focus. The definition is valid for both covered and uncovered specimens. For covered specimens it is measured from the specimen plane, not from the cover glass.

Column 6. The mounting *thread* is traditionally the Royal Microscopical Society thread established many years ago. This is listed as 0.797 in by 36 TPI (threads per inch) in the older texts but is now set up as a standard 0.800 in by 36 AMO, 55°, and is labeled "RMS" thread in Table 1. Some objectives, notably dark-field incident-light objectives, require a larger thread; these are labeled "Spec" for special. The abbreviation AMO preceded by the basic major diameter in inches and the number of threads per inch designates the American National Standard Institute standard for microscope objectives; see ANSI B1.11-1958(R1972).

Column 7. For metallurgical objectives, *working distance* is specified as the distance between the specimen plane on which the objective is focused and the nearest metal or glass surface of the objective. For biological objectives the distance is specified from the top of the cover glass rather than from the specimen itself.

Column 8. We abbreviate the manufacturers' names as follows:

AO	American Optical Co.	R	Reichert
BL	Bausch & Lomb	U	Unitron
L	Leitz	V	Vickers
N	Nikon	W	Wild
O	Olympus	Z	Zeiss

Column 9. The manufacturer's catalog number needs no explanation.

Column 10. As comments we list the basic optical design type followed by data to indicate the field of application of the objective. The terminology of Table 1 is explained in the following paragraphs.

Achromat, an objective having all-glass lens components and designed to favor the yellow-green region for optimum correction of lens aberrations. The residual spherical aberration generally goes slightly undercorrected in the red region (spherochromatism). Under critical examination a variation in focus between the blue-red and the green-yellow wavelength regions can be detected (secondary spectrum). Figure 2 shows typical construction of several achromats.

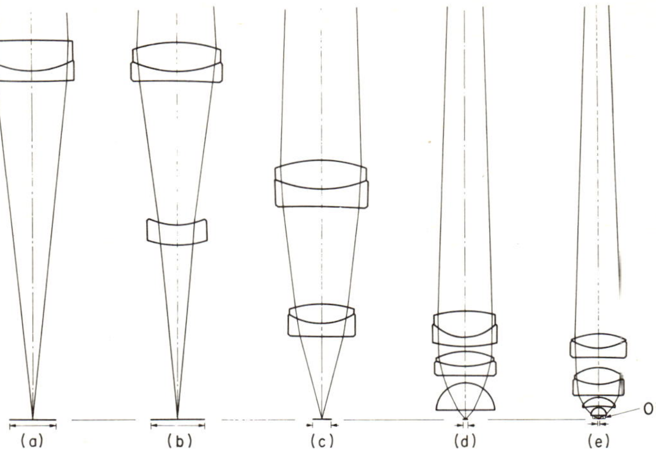

Fig. 2 Achromatic objectives: (*a*) 4.50-mm field, 4.0×, 0.10 NA; (*b*) 5.20-mm field, 3.5×, 0.09 NA; (*c*) 1.80-mm field, 10×, 0.25 NA; (*d*) 0.42-mm field, 43×, 0.65 NA; (*e*) 0.18-mm field, 97×, 1.30 NA. [*From R. Kingslake (ed.), "Applied Optics and Optical Engineering," vol.* 3, *p.* 168. *Copyright* 1967 *by Academic Press, Inc.; used by permission.*]

Apo., abbreviation for apochromat. An objective having some of its components made of fluorite and well corrected for the entire visual spectrum. The optical characteristics of fluorite are such that when this material is used in combination with certain select optical glasses, both the secondary spectrum and spherochromatism can be reduced to a fraction of the residuals of an achromat. Since there are no rigorous standards which describe the precise state of correction which an objective must have in order to be classified as an apochromat, some variation from objective to objective may be noted. However, the manifestation of a variation in focus with wavelength will rarely be detected except under the most severe and critical examination. Apochromats generally have higher NAs than achromats of equal power. Figure 3 shows several typical apochromats.

Semiapo., an objective having chromatic corrections which lie somewhere between those of an achromat and an apochromat. These objectives usually are made with one or two fluorite elements and are sometimes called *fluorites, neofluars,* or *fluotars.*

FF, flat-field, a modern class of objective which is designed to show little or no field curvature over the useful field. This improvement in performance has been achieved through the application of new optical materials combined with improved design and manufacturing techniques. Flat-field objectives are substantially more complex in their construction than the older objectives, but their greatly enlarged useful field often justifies the added cost.

Depending upon their state of color correction, flat-field objectives are classified as achromats, semiapochromats, and apochromats, in the same manner as non-flat-field objectives.

In the flat-field objectives made by Bausch & Lomb a common negative lens doublet built within the nosepiece acts as the rear lens component for all objectives on the nosepiece. This special construction means that these objectives are not interchangeable with other objectives. Figure 4 shows several objectives of this type.

Met., an abbreviation for metallurgical. If the objective is for biological or transmitted light, no notation to this effect is made, but if for use in vertical

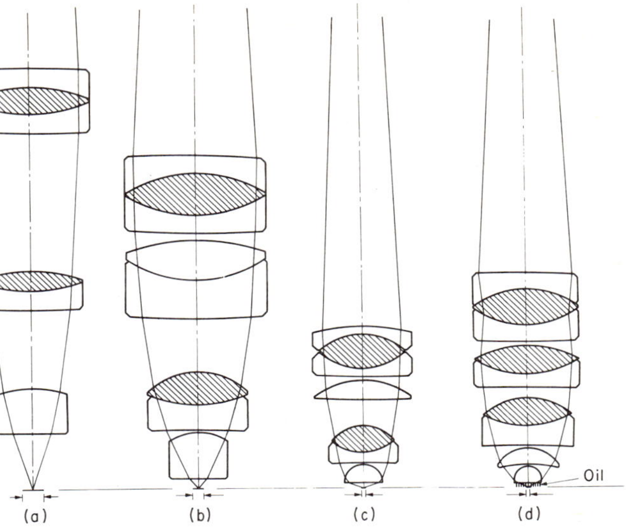

Fig. 3 Apochromatic objectives with (*shaded*) elements of fluorite: (*a*) 1.8-mm field, 10×, 0.30 NA; (*b*) 0.9-mm field, 20×, 0.65 NA; (*c*) 0.38-mm field, 47×, 0.95 NA; (*d*) 0.30-mm field, 61×, 1.40 NA. [*From R. Kingslake (ed.), "Applied Optics and Optical Engineering," vol.* 3, *p.* 169. *Copyright* 1967 *by Academic Press, Inc.; used by permission.*]

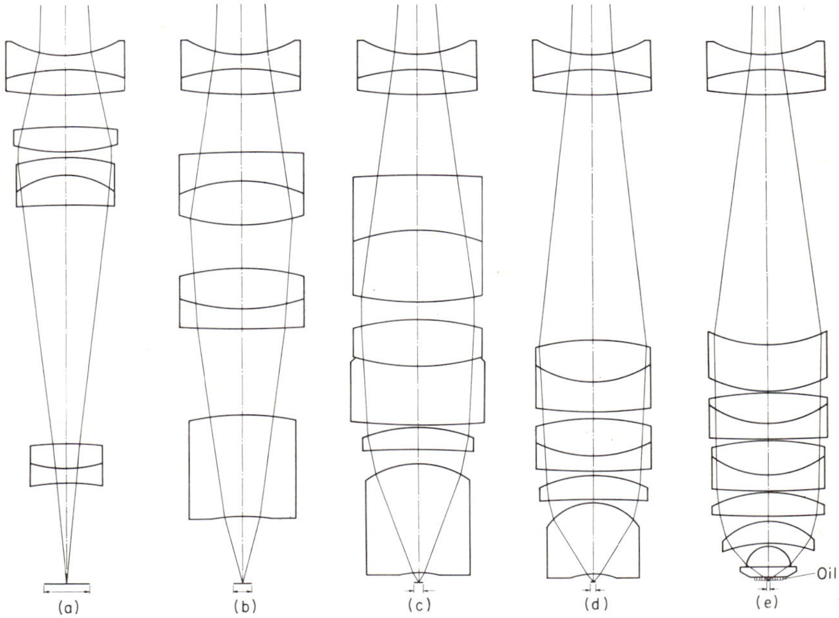

Fig. 4 Flat-field achromatic objectives with rear negative doublet mounted separately in the nosepiece: (*a*) 4.50-mm field, 4.0×, 0.09 NA; (*b*) 1.80-mm field, 10×, 0.25 NA; (*c*) 0.90-mm field, 20×, 0.50 NA; (*d*) 0.45-mm field, 40×, 0.65 NA; (*e*) 0.18-mm field, 100×, 1.25 NA. Other manufacturers do not separate the negative power but use different lenses for each objective. (*Bausch & Lomb, Inc.*)

illumination and corrected for no cover glass, it is labeled "Met.," although this does not mean it cannot be used for other opaque materials.

Polar., objectives for *polarized-light* application. Although glass and fluorite are nominally isotropic, they are very susceptible to strain introduced during manufacture. When strained, these materials become birefringent. This induced birefringence has no perceptible effect on performance except when the objective is used in polarized light. For this application special processing and mounting are required to remove all vestiges of strain from the lenses.

DF, in vertical dark-field illumination a hollow illuminating cone must be provided by means of a special dark-field annulus which is an integral part of the objective (sometimes called a *catoptric illuminator*). Such objectives are called dark-field objectives although they are of course also useful in normal bright-field conditions. Figure 5 shows the construction of such an objective. (Since in transmitted light

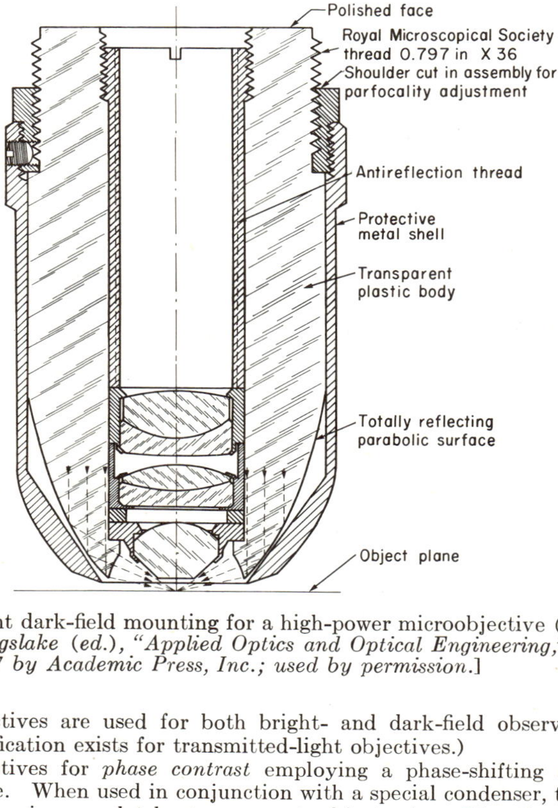

Fig. 5 Incident dark-field mounting for a high-power microobjective (40×, 0.65 NA). [*From R. Kingslake (ed.), "Applied Optics and Optical Engineering," vol.* 3, *p.* 164. *Copyright* 1967 *by Academic Press, Inc.; used by permission.*]

identical objectives are used for both bright- and dark-field observation, no such separate classification exists for transmitted-light objectives.)

Phase, objectives for *phase contrast* employing a phase-shifting annulus in the real focal plane. When used in conjunction with a special condenser, these objectives reveal structure in completely transparent objects (see Fig. 6). Phase-contrast objectives are manufactured with two types of phase-shifting elements. In the more popular type, referred to as *positive* or *dark contrast,* the specimen detail is imaged as darker areas against a bright background. In the *negative* or *bright-contrast* type, the specimen is imaged as bright against a dark background. The Leitz company differentiates between the normal Zernike phase and Heine phase, the latter using a pancratic condenser instead of a series of condenser annuli. They also employ some objective annuli with greater than average absorption in the phase annulus, called high-contrast, which we abbreviate *HC*. The Reichert company employs an absorbing phase annulus of a scattering nature and calls this *anoptral phase*.

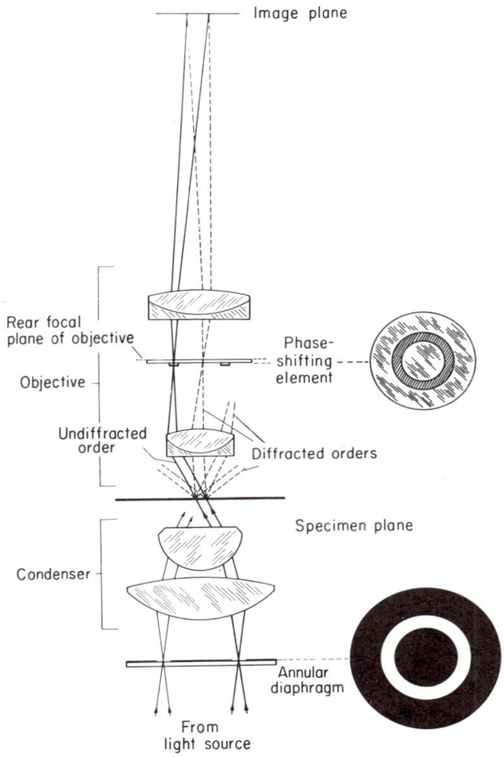

Fig. 6 A phase-contrast objective showing the arrangement by which the clear annulus in the substage condenser is imaged into the phase-shifting annulus in the rear focal plane of the objective. [*From R. Kingslake (ed.), "Applied Optics and Optical Engineering," vol.* 4, *p.* 68. *Copyright* 1967 *by Academic Press, Inc.; used by permission.*]

Fluoresc., objectives of low-fluorescing glasses for use in studying fluorescent materials. The light-gathering ability of the higher-NA apochromats makes them peculiarly suitable for this purpose.

Oil, objectives designed for oil immersion, for high NA and high resolving power.

Water, objectives designed for water immersion, for use on liver preparations.

Glycerin, objectives designed for glycerin immersion, for use in ultraviolet work.

Corr. collar, abbreviation for correction collar. When the NA of a dry objective ranges from 0.85 to 0.95, small variations in cover-glass thickness significantly change the amount of spherical aberration introduced by the cover glass. Objectives in this NA range often include a *correction collar* by means of which a lens spacing in the objective can be altered to compensate for variations in the amount of spherical aberration introduced by the cover glass.

Iris. Immersion objectives may be used in dark field by stopping them down to a lower NA. This is most conveniently done by means of a built-in *iris diaphragm.*

Nonbrowning objectives employ only radiation-resistant glasses, for use in the study of radioactive materials. These glasses are more resistant to browning or darkening under the exposure to atomic radiation.

U stage. The Federoff *universal stage* for the angular scanning of petrographic specimens in polarized light requires special long-working-distance low-NA objectives designed to work through the large hemispheres of glass supplied with the universal stage.

UV. The ultraviolet region of the spectrum requires special optical materials for the lens elements, notably fluorite and very pure fused quartz, inasmuch as glass fails to transmit very far into the ultraviolet.

High temp. Hot stages require objectives with especially long working distance, to prevent heat damage to the lenses and to lessen thermal effects on the mounts. These are sometimes corrected for use with special furnace windows.

Catadioptric, objectives employing mirror elements as well as lenses. They are sometimes used in ultraviolet work or for work requiring a long working distance. Figure 7 shows the optical elements of a catadioptric objective designed for ultraviolet work.

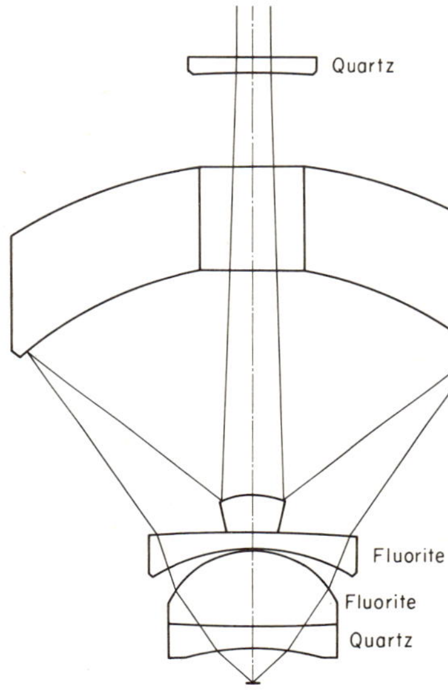

Fig. 7 Catadioptric objective for ultraviolet (53×, 0.72 NA). [*From R. Kingslake (ed.), "Applied Optics and Optical Engineering," vol.* 3, *p.* 173. *Copyright* 1967 *by Academic Press, Inc.; used by permission.*]

LWD, long-working-distance objectives designed for work requiring extraordinary clearance between the object and the objective.

Various mechanical designs of objectives are identified as follows:

1. *Short mount.* Extraordinarily short specimen-to-shoulder distance, usually for metallography and usually nonparfocal
2. *Quick change.* Individual objectives each carrying the equivalent of a dovetail mount, usually centerable for use in polarized light
3. *Magn. cent.* A magnetic-centering objective mount employed by Vickers
4. *Slotted mount.* A mount employing a slot into which accessories can be inserted
5. *Spring.* A mount employing a light resilient spring to avoid accidental damage to objective or specimen
6. *Immersion cap.* A plano window cap to permit insertion of the objective into a liquid specimen, usually water

TO. Infinity-corrected objectives work in conjunction with a telescope objective and to obtain the correct magnification must be used with the correct focal length (EFL) telescope objective. This focal length is stated in the tables.

Interfer. or *Interfer. contrast* are objectives for use in interference microscopy.†

† For further description of these terms and design categories, the reader is referred to R. Kingslake, "Applied Optics and Optical Engineering," vol. 1, chap. 6; vol. 3, chaps. 1 and 4; and vol. 4, chap. 2, Academic, New York, 1963.

3. Eyepieces Table 2 (pp. 6-38–6-41) for eyepieces is much less extensive than Table 1 since the eyepiece is less important and less critical than the objective. Many eyepiece subgroups, e.g., crossline eyepieces and micrometer eyepieces have been omitted, the emphasis being largely on the various lens design types. Manufacturers' trade names have been used since there seemed to be no way of avoiding them in identifying the eyepiece types. Table 2 covers the following data.

Column 1. Power. Computed as a magnifier, i.e., power = $250/f$, where f is the focal length of the eyepiece in millimeters.

Column 2. Type. Three basic types are shown in Fig. 8. The Huygens is the least complex and most common, consisting of two separated singlets. Its

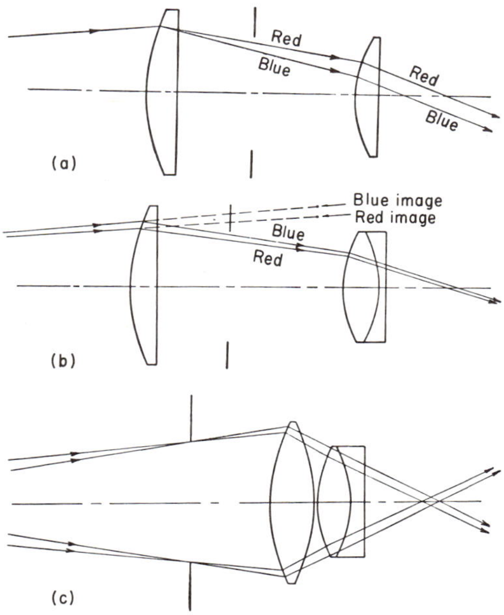

Fig. 8 Three eyepiece types: (*a*) Huygens, corrected for essentially zero lateral color; (*b*) general construction of compensating type of eyepiece corrected to work with objectives having some lateral color; (*c*) typical wide-field eyepiece.

chromatic difference of magnification (CDM) is essentially zero and combines well with low- and medium-power achromats. It has a comparatively small field and short eye relief. In the next category the in-between eyepieces, corrected to compensate the CDM of the high-power achromats and the semiapochromats, have trade names such as Periplanatic, Complanat, and Hyperplane. The wide-field eyepieces also sometimes fall into this category of lateral color correction. Finally come the *compensating* eyepieces, whose CDM correction compensates the apochromatic series of objectives. Zeiss terminology for this type of eyepiece is KPL (from the German *Kompensplan,* compensating plane).

Although CDM cannot be eliminated from the higher-power objectives, it can be introduced deliberately into the lower powers in the same degree and with the same sign as that in the higher powers. Hence, the modern trend is to maintain a nearly constant amount of CDM (about 1 percent) in all objectives and to provide compensating CDM in the body-tube relay systems and the eyepieces, giving an image substantially free from lateral color.

Older microscopes may have lateral color varying from zero in the lower powers

(*text continued on page* 6-37)

TABLE 1 Objectives *(see text for explanation of abbreviations)*

Power (1)	NA (2)	Cover glass, mm (3)	Tube length, mm (4)	Shoulder height, mm (5)	Thread (6)	Working distance, mm (7)	Mfr. (8)	Catalog No. (9)	Comments (10)
0.9	....	....	160	33.7	RMS	173.0	AO	123	Achromat
1.0	0.04	0.17	150	......	RMS	4.4	Z	46 20 10	FF achromat
	0.04	0.17	150	......	RMS	4.4	Z	46 20 11	FF achromat, polar.
	0.04	0.17	152	45.0	RMS	29.7	L	519 050	FF achromat, iris
1.2	0.03	....	160	60.6	RMS	29.7	N	200-30-000	FF achromat
1.3	0.03	0.17	160	36.77	RMS	38.4	O	1-B501	FF achromat
	0.03	0	200	......	RMS	.....	O	....	FF achromat, met., DF
1.5	....	....	160	......	RMS	9.0	AO	124	Achromat
	....	0	167	130	Spec	110.0	L	810 343	Achromat, met, DF
2.0	0.08	0.18	160	33.7	RMS	71.2	AO	C 101	Achromat
	0.08	0.18	145	72.2	RMS	59.0	BL	31-10-05-01	Achromat
	0.06	....	146	......	....	49.0	W	5000	Achromat
	0.05	....	160	60.6	RMS	35.6	N	200-30-001	FF achromat
	0.05	0.17	160	36.77	RMS	42.2	O	1-B510	FF achromat
	....	....	167	111	Spec	90.0	L	810 344	Achromat, met., DF
2.5	0.08	0.17	151	37.0	RMS	5.0	R	9223101	FF achromat
	0.08	0.18	140	44.0	RMS	17.0	V	M025611	FF achromat
	0.08	0.17	150	45.0	RMS	9.0	Z	46 01 10	FF achromat
	0.08	0.17	150	45.0	RMS	9.0	Z	46 01 18	FF achromat, polar.
	0.08	0.17	152	45.0	RMS	13.6	L	519 049	FF achromat
	0.07	0.17	152	45.0	RMS	14.0	L	519 044	Achromat
	0.05	....	200	......	RMS	.....	O	....	FF achromat
2.6	0.08	0.18	145	53.7	RMS	43.5	BL	31-10-07-01	Achromat
2.8	0.08	....	181	38.0	RMS	5.0	R	9228201	FF achromat
3.0	0.10	0	156	69.0	RMS	50.0	U	M 3X	Achromat
	0.10	0.18	140	44.0	RMS	40.0	V	M 022 011	Achromat
	0.10	0.18	140	44.0	RMS	40.0	V	M 022 013	Achromat, polar.
	0.10	0.18	140	50.2	....	40.0	V	M 022 031	Achromat, magn. cent.
	0.10	0.18	140	50.2	....	40.0	V	M 022 033	Achromat, polar., magn. cent.
	0.10	....	146	......	RMS	8.8	W	5100	FF semiapo.
	0.08	....	160	60.6	RMS	28.6	N	200-30-001	FF achromat
	0.08	0.18	160	33.7	RMS	52.1	AO	C102	Achromat

	0.07	0.17	145	45.0	RMS	38.0	V	M058141	Achromat
	....	0	167	83.9	Spec	60.0	L	810 345	Achromat, met., DF
3.2	0.12	0.17	152	37.0	RMS	35.0	L	559046	Achromat, polar.
	0.07	0.17	150	45.0	RMS	30.0	Z	46 01 00	Achromat
	0.06	0	∞	45.0	RMS	12.1	L	569 007	FF achromat, met.
3.5	0.10	0.17	152	37.0	RMS	23.0	L	519 036	Achromat
	0.10	0.17	152	37.0	RMS	23.0	L	559 037	Achromat, polar.
	0.09	0.18	145	35.7	RMS	17.8	BL	31-10-06-01	Achromat
	0.09	0.18	160	33.7	RMS	25.0	AO	C 105	Achromat
	0.05	....	∞	73.5	RMS	66.0	V	M023052	Achromat, magn. cent., TO 250 EFL
	0.05	....	∞	73.5	RMS	66.0	V	M023054	Achromat, polar., magn. cent., TO 250 EFL
4.0	0.16	0.17	150	45.0	RMS	2.5	Z	46 02 40	FF apo.
	0.14	0.17	152	45.0	RMS	15.0	L	519 176	FF achromat
	0.12	0.18	∞	33.7	RMS	8.5	AO	1075	Achromat, TO 182.7 EFL
	0.10	0.17	160	36.77	RMS	5.8	O	1B 520	FF achromat
	0.10	0.17	160	36.77	RMS	35.0	O	1B 220	Achromat, polar.
	0.10	0.17	160	36.77	RMS	35.0	O	1B 221	Achromat, polar.
	0.10	0	150	33.0	Spec	1.0	Z	46 02 69	Achromat, met., DF
	0.10	0	150	33.0	Spec	9.05	Z	46 20 01	Achromat, met., polar.
	0.10	0	150	33.0	Spec	0.3	Z	46 20 06	Achromat, met., polar., oil
	0.10	0	150	33.0	RMS	9.0	Z	46 20 31	Achromat, met.
	0.10	0	167	45.0	Spec	13.2	L	513 182	Achromat, met., DF
	0.10	0	146	36.5	RMS	16.0	W	5002	Achromat
	0.10	0	146	36.5	Spec	16.0	W	8524	Achromat, met., DF
	0.10	0	146	36.5	RMS	16.0	W	3150	Achromat, polar.
	0.10	0.18	145	35.7	RMS	38.0	BL	31-10-09-01	Achromat
	0.10	....	160	33.6	RMS	9.5	N	200-10-000	Achromat
	0.10	....	160	33.6	RMS	18.2	N	200-31-000	FF achromat
	0.10	0.17	160	36.77	RMS	35.0	O	1B 120	Achromat
	0.10	0.18	∞	33.7	RMS	33.2	AO	103	Achromat
	0.10	0.18	160	33.7	RMS	40.05	AO	C 104	Achromat
	0.10	0.17	151	37.0	RMS	11.0	R	9223201	FF achromat
	0.10	0.17	151	37.0	RMS	11.0	R	9223251	FF achromat, polar.
	0.10	....	160	33.6	RMS	9.5	N	200-11-600	Achromat, polar.
	0.10	....	160	33.6	RMS	18.2	N	200-31-600	FF achromat, polar.
	0.10	....	160	33.6	RMS	9.5	N	200-11-601	Achromat, polar.
	0.10	....	160	33.6	RMS	9.5	N	200-10-000	Achromat
	0.09	0.18	145	43.0	RMS	10.0	BL	31-10-62	FF achromat
	0.08	0	241	38.0	RMS	5.0	R	9228203	FF achromat

TABLE 1 Objectives (Continued)

Power (1)	NA (2)	Cover glass, mm (3)	Tube length, mm (4)	Shoulder height, mm (5)	Thread (6)	Working distance, mm (7)	Mfr. (8)	Catalog No. (9)	Comments (10)
5.0	0.20	0.18	160	33.7	RMS	20.0	AO	C107	Achromat
	0.15	0.18	140	34.0	RMS	17.0	V	M 022101	Achromat
	0.15	0.18	140	34.0	RMS	17.0	V	M 022103	Achromat, polar.
	0.15	0.18	140	44.0	RMS	17.0	V	M 022111	Achromat
	0.15	0.18	140	44.0	RMS	17.0	V	M 022113	Achromat, polar.
	0.15	0.18	140	36.5	Spec	17.0	V	M 022121	Achromat, short mount
	0.15	0.18	140	40.2	....	17.0	V	M 022131	Achromat, magn. cent.
	0.15	0.18	140	40.2	....	17.0	V	M 022133	Achromat, polar., magn. cent.
	0.15	0.18	140	62.0	RMS	17.0	V	M 022143	Achromat, slotted mount
	0.15	....	140	39.0	RMS	17.0	V	M 022144	Achromat, U stage
	0.10	0	200	42.5	RMS	27.0	O	1M121	FF achromat, met.
	0.10	0	200	42.5	RMS	4.7	O	1M520	FF achromat, met.
	0.10	0	200	42.5	Spec	22.5	O	1M420	Achromat, met., DF
	0.10	....	152	45.4	RMS	....	L	559 002	Achromat, U stage
	0.10	0	200	36.7	RMS	6.4	BL	42-35-01	FF achromat, met., polar.
	0.10	0	200	36.7	RMS	2.4	BL	42-33-51	Achromat, met., polar.
	0.10	0	200	36.7	Spec	3.2	BL	42-35-15	FF achromat, met., polar.
	0.10	0	200	36.7	RMS	3.2	BL	42-35-50	FF achromat, met., polar.
	0.10	0	200	36.7	Spec	3.2	BL	42-35-29	FF achromat, met., polar., DF
	0.10	0	160	33.6	RMS	15.0	N	201-11-601	Achromat, met., polar.
	0.10	0	160	33.6	RMS	15.0	N	201-11-002	Achromat, met.
	0.10	0	160	45.0	RMS	5.3	N	202-71-000	Achromat, U stage
	0.10	0	156	36.0	RMS	12.0	U	M5X	Achromat, met.
	0.10	0.17	156	36.0	RMS	12.0	U	5X	Achromat
	0.09	0	∞	45.0	RMS	12.7	L	569 049	Achromat, met.
	0.09	0	∞	45.0	RMS	12.7	L	569 019	Achromat, met., phase
	0.08	0.18	145	43.0	RMS	2.2	BL	31-10-47	FF achromat
5.5	0.15	0	181	38.0	RMS	17.2	R	9225401	Achromat, met.
	0.15	0	181	38.0	RMS	17.2	R	9225451	Achromat, met., polar.
	0.15	0	181	38.0	RMS	17.2	R	9225401	Achromat, met., DF
	0.15	0	181	38.0	RMS	17.2	R	9225452	Achromat, met., polar., DF
5.6	0.15	0	197	53.3	RMS	26.0	L	559 007	Achromat, met., polar.

6.0	0.20	0	146	36.5	RMS	5.5	W	5105	FF semiapo.
	0.18	0.17	152	37.0	RMS	17.0	L	519 001	Achromat
	0.18	0.17	152	37.0	RMS	17.5	L	559 044	Achromat, polar.
	0.17	0.18	145	35.7	RMS	15.5	BL	31-10-18-01	Achromat
	0.12	0	200	29.6	RMS	29.1	O	1-M120	Achromat, met.
	0.12	0	200	29.6	RMS	29.1	O	1-M220	Achromat, met., polar.
	0.10	....	230	48.3	....	35.0	V	M022034	Achromat, polar., magn. cent.
	0.10	....	∞	46.0	RMS	35.0	V	M022052	Achromat, TO 250 EFL
	0.10	....	∞	46.0	RMS	35.0	V	M022054	Achromat, polar., TO 250 EFL
	0.10	....	230	48.3	....	35.0	V	M022032	Achromat, magn. cent.
6.3	0.20	0.17	150	45.0	RMS	10.8	Z	46 03 20	Semiapo.
	0.16	0.17	150	45.0	RMS	4.9	Z	46 03 10	Achromat
	0.12	0	150	33.0	RMS	19.0	Z	46 20 42	Achromat, U stage
6.5	0.20	0.18	∞	33.7	RMS	17.2	AO	3006	Achromat
7.5	0.20	0.18	145	43.0	RMS	1.5	BL	31-11-50	FF apo.
8.0	0.20	0.18	140	34.0	RMS	0.05	V	M024501	Achromat, oil
	0.20	0	140	34.0	RMS	0.23	V	M024502	Achromat, met., oil
	0.20	0.18	140	44.0	RMS	0.05	V	M024511	Achromat, oil
	0.20	0	140	44.0	RMS	0.23	V	M024512	Achromat, met., oil
	0.20	0.18	140	44.0	RMS	0.05	V	M024513	Achromat, polar., oil
	0.20	0	200	36.7	RMS	14.2	BL	42-33-63	Achromat, met., polar., DF
	0.20	0	200	36.7	RMS	14.2	BL	42-33-52	Achromat, met., polar.
	0.20	0	150	33.0	Spec	1.0	Z	46 03 69	Achromat, met., DF
	0.20	0	150	33.0	Spec	7.1	Z	46 20 02	Achromat, met., polar.
	0.20	0	150	33.0	Spec	0.3	Z	46 20 07	Achromat, met., polar., oil
	0.20	0	150	33.0	RMS	7.2	Z	46 20 32	Achromat, met
	0.18	0	197	35.8	RMS	16.5	L	559 015	Achromat, met., polar.
	0.18	0	∞	45.0	RMS	13.9	L	569 026	FF achromat, met.
	0.18	0	∞	45.0	RMS	0.22	L	569 040	FF achromat, met., oil
	0.18	0	∞	45.0	RMS	13.9	L	569 032	FF achromat, met., phase
	0.15	0	241	38.0	RMS	17.2	R	9225403	Achromat
	0.15	0	241	38.0	RMS	17.2	R	9225453	Achromat, polar.
	0.15	0	241	38.0	RMS	17.2	R	9225404	Achromat, DF
	0.15	0	241	38.0	RMS	17.2	R	9225454	Achromat, DF, polar.
9.0	0.15	....	230	40.2	....	16.0	V	M022132	Achromat, magn. cent.
	0.15	....	230	40.2	....	16.0	V	M022134	Achromat, magn. cent., polar.
10.0	0.40	....	146	36.5	RMS	4.9	W	5050	Semiapo.
	0.40	....	146	36.5	RMS	4.9	W	5250	Semiapo., phase
	0.32	0.17	150	45.0	RMS	0.35	Z	46 04 40	FF apo.

TABLE 1 Objectives (Continued)

Power (1)	NA (2)	Cover glass, mm (3)	Tube length, mm (4)	Shoulder height, mm (5)	Thread (6)	Working distance, mm (7)	Mfr. (8)	Catalog No. (9)	Comments (10)
10.0	0.30	0.17	152	45.0	RMS	7.5	L	519 175	FF achromat
	0.30	0.17	150	45.0	RMS	4.0	Z	46 04 20	Semiapo.
	0.30	0.18	145	35.7	RMS	4.85	BL	31-11-71-01	Apo.
	0.30	....	160	33.6	RMS	6.4	N	200-11-200	Achromat, phase (pos.)
	0.30	....	160	33.6	RMS	6.4	N	200-11-300	Achromat, phase (neg.)
	0.30	....	160	33.6	RMS	6.4	N	202-61-200	Achromat, phase (pos.), fluoresc.
	0.30	0	156	36.0	RMS	8.0	U	MPL 10X	FF achromat, met.
	0.30	0	156	36.0	RMS	8.0	U	M 10X	Achromat, met.
	0.30	0.17	156	36.0	RMS	7.9	U	BPL 10X	FF achromat
	0.30	0.17	156	36.0	RMS	7.9	U	10X	Achromat
	0.30	0.18	140	34.0	RMS	3.6	V	M024001	Apo.
	0.30	0.18	140	34.0	RMS	3.6	V	M024005	Apo., phase
	0.30	0.18	140	44.0	RMS	3.6	V	M024011	Apo.
	0.30	0.18	140	44.0	RMS	3.6	V	M024015	Apo., phase
	0.30	0.18	140	40.2	....	3.6	V	M024031	Apo., magn. cent.
	0.25	0.17	160	36.77	RMS	5.6	O	1 B 130	Achromat
	0.25	0.17	160	36.77	RMS	0.95	O	1 B 350	FF achromat, spring
	0.25	....	160	36.65	RMS	6.5	O	1 W 130	Achromat, water
	0.25	0.17	160	36.77	RMS	5.6	O	1 P 130	Achromat, phase
	0.25	0.17	160	36.77	RMS	5.6	O	1 B 231	Achromat, polar.
	0.25	0	200	42.5	RMS	7.5	O	1 M 120	Achromat, met.
	0.25	0	200	42.5	RMS	1.2	O	1 M 530	FF achromat, met.
	0.25	0	200	42.5	RMS	7.5	O	1 M 130	Achromat, phase
	0.25	0	200	42.5	RMS	7.5	O	1 M 230	Achromat, met., polar.
	0.25	0.17	151	37.0	RMS	1.4	R	9223451	FF achromat, interfer. contrast
	0.25	0.18	160	33.7	RMS	7.3	AO	C109	Achromat
	0.25	0.18	∞	33.7	RMS	7.8	AO	113	Achromat, TO 165.2 EFL
	0.25	0.18	∞	33.7	RMS	10.1	AO	1076	Achromat, TO 182.7 EFL
	0.25	0.18	∞	33.7	RMS	7.9	AO	1109	Achromat, TO 182.7 EFL
	0.25	0.18	∞	33.7	RMS	9.1	AO	1224	Achromat, phase, TO 182.7 EFL
	0.25	0.18	∞	33.7	RMS	10.1	AO	1234	Achromat, phase, TO 182.7 EFL
	0.25	0.18	140	34.0	RMS	5.0	V	M022301	Achromat

10.0	0.25	0.18	140	34.0	RMS	5.0	V	M022303	Achromat, polar.
	0.25	0.18	140	34.0	RMS	5.0	V	M022305	Achromat, phase
	0.25	0.18	140	34.0	RMS	5.0	V	M022308	Achromat, phase
	0.25	0.18	140	44.0	RMS	5.0	V	M022311	Achromat
	0.25	0.18	140	44.0	RMS	5.0	V	M022313	Achromat, polar.
	0.25	0.18	140	44.0	RMS	5.0	V	M022315	Achromat, phase
	0.25	0.18	140	44.0	RMS	5.0	V	M022317	Achromat, phase (neg.)
	0.25	0.18	140	44.0	RMS	5.0	V	M022318	Achromat, phase
	0.25	0.18	140	44.0	RMS	5.0	V	M022319	Achromat, phase
	0.25	0.18	140	36.5	Spec	5.0	V	M022321	Achromat, short mount
	0.25	0.18	140	40.2	....	5.0	V	M022331	Achromat, magn. cent.
	0.25	0.18	140	42.8	....	5.0	V	M022332	Achromat, magn. cent.
	0.25	0.18	140	50.3	....	5.0	V	M022333	Achromat, magn. cent., polar.
	0.25	0.18	140	27.4	....	5.0	V	M022334	Achromat, magn. cent., polar.
	0.25	0.18	140	31.7	....	5.0	V	M022336	Achromat, 16-mm quick-change adapter
	0.25	0.18	140	21.0	RMS	5.0	V	M022341	Achromat
	0.25	0.18	140	34.0	RMS	5.0	V	M022342	Achromat, for Dyson interfer.
	0.25	0.18	140	21.0	RMS	5.0	V	M022343	Achromat, polar.
	0.25	0.18	140	34.0	RMS	3.0	V	M025101	FF achromat
	0.25	0	200	42.5	Spec	7.5	O	1M430	Achromat, met., DF
	0.25	0	200	....	RMS	....	O	1M131	FF achromat, met.
	0.25	0.17	152	37.0	RMS	5.7	L	519 004	Achromat
	0.25	0.17	152	45.0	RMS	7.7	L	519 165	Achromat, phase (Zernike)
	0.25	0.17	152	37.0	RMS	5.8	L	519 149	Achromat, phase (Heine)
	0.25	0.17	152	37.0	RMS	5.8	L	051 920	Achromat, phase (Heine, HC)
	0.25	0.17	152	37.0	RMS	5.7	L	559 045	Achromat, polar.
	0.25	....	146	36.5	RMS	4.1	W	5005	Achromat
	0.25	....	146	36.5	RMS	3.0	W	5106	FF semiapo.
	0.25	0	146	36.5	Spec	5.1	W	8525	Achromat, met., DF
	0.25	....	146	36.5	RMS	4.1	W	3151	Achromat, polar.
	0.25	....	146	36.5	RMS	4.1	W	5200	Achromat, phase
	0.25	0.18	145	35.7	RMS	6.2	BL	31-10-19	Achromat
	0.25	0.18	145	35.7	RMS	7.7	BL	31-10-20-01	Achromat
	0.25	0.18	145	35.7	RMS	4.5	BL	31-10-22-01	Achromat, divisible to 4X
	0.25	0.18	145	43.0	RMS	6.8	BL	31-10-63	FF achromat
	0.25	0	200	36.7	RMS	10.6	BL	42-33-53	Achromat, met., polar.
	0.25	0	200	36.7	RMS	10.6	BL	42-33-64	Achromat, met., polar., DF
	0.25	0.18	145	43.0	RMS	2.3	BL	31-10-66	FF achromat
	0.25	0.18	145	43.0	RMS	2.4	BL	31-10-48	FF achromat

TABLE 1 Objectives (Continued)

Power (1)	NA (2)	Cover glass, mm (3)	Tube length, mm (4)	Shoulder height, mm (5)	Thread (6)	Working distance, mm (7)	Mfr. (8)	Catalog No. (9)	Comments (10)
10.0	0.25	2.0	201	99.0	Spec	20.9	U	LWD10X	Achromat, quartz cover glass, high temp.
	0.25	0.18	140	44.0	RMS	5.0	V	M020311	Achromat
	0.25	0.17	151	37.0	RMS	6.8	R	9220401	Achromat
	0.25	0.17	151	37.0	RMS	6.8	R	9220451	Achromat, polar.
	0.25	0.17	151	37.0	RMS	6.8	R	9220406	Achromat, phase
	0.25	0.17	151	37.0	RMS	6.8	R	9220405	Achromat, phase (anoptral)
	0.25	0.17	151	37.0	RMS	1.4	R	9223401	FF achromat
	0.25	0.18	140	34.0	RMS	3.0	V	M025101	FF achromat, polar.
	0.25	0.18	140	34.0	RMS	3.0	V	M025108	FF achromat, phase
	0.25	0.18	140	44.0	RMS	3.0	V	M025111	FF achromat
	0.25	0.18	140	44.0	RMS	3.0	V	M025113	FF achromat, polar.
	0.25	0.18	140	44.0	RMS	3.0	V	M025115	FF achromat, phase
	0.25	0.18	140	44.0	RMS	3.0	V	M025117	FF achromat, phase
	0.25	0.18	140	44.0	RMS	3.0	V	M025118	FF achromat, phase
	0.25	0.18	140	44.0	RMS	3.0	V	M025119	FF achromat, phase
	0.25	0.18	140	34.0	RMS	5.0	V	M022201	Achromat
	0.25	0.18	140	34.0	RMS	5.0	V	M022205	Achromat, phase
	0.25	0.18	140	44.0	RMS	5.0	V	M022211	Achromat
	0.25	0.18	140	44.0	RMS	5.0	V	M022215	Achromat, phase
	0.25	0.18	140	40.2	....	5.0	V	M022235	Achromat, phase, magn. cent.
	0.25	0.18	140	34.0	RMS	12.0	V	M808413	UV monochromat, 2536A, quartz
	0.25	0.18	140	34.0	RMS	12.0	V	M808414	UV monochromat, 2750A, quartz
	0.25	0	160	33.6	RMS	7.1	N	201-11-003	Achromat, met.
	0.25	0	160	33.6	RMS	7.0	N	201-31-000	Achromat, met.
	0.25	....	160	33.6	RMS	7.1	N	200-11-602	Achromat, polar.
	0.25	....	160	33.6	RMS	7.1	N	200-11-000	Achromat
	0.25	....	160	33.6	RMS	7.0	N	200-31-001	FF achromat
	0.25	....	160	33.6	RMS	7.0	N	200-31-601	FF achromat, polar.
	0.25	....	160	33.6	RMS	7.1	N	200-11-603	Achromat, polar.
	0.25	....	160	33.6	RMS	7.1	N	201-11-602	Achromat, polar., met.
	0.25	....	160	33.6	RMS	6.5	N	200-11-002	Achromat
	0.22	0.17	150	45.0	RMS	5.0	Z	46 04 00	Achromat

10.0	0.22	0.17	150	45.0	RMS	4.8	Z	46 04 10	FF achromat
	0.22	0.17	150	45.0	RMS	5.0	Z	46 04 01	Achromat, phase
	0.22	0.17	150	45.0	RMS	5.0	Z	46 04 08	Achromat, polar.
	0.22	0.17	150	45.0	RMS	2.3	Z	47 44 03	Achromat, interfer.
	0.22	....	152	45.4	RMS	....	L	559 003	Achromat, U stage
	0.22	....	160	45.0	RMS	2.9	N	202-71-001	Achromat, polar., U stage
	0.22	....	...	....	....	....	N	200-18-000	Achromat, water, var. work. dist.
	0.20	0	200	36.7	Spec	3.0	BL	42-35-16	FF achromat, met., polar.
	0.20	0	200	36.7	RMS	3.0	BL	42-35-51	FF achromat, met., polar.
	0.20	0	200	36.7	Spec	3.0	BL	42-35-30	FF achromat, met., polar, DF
	0.20	0.35	150	45.0	RMS	7.4	Z	46 20 58	UV
	0.20	0.35	150	45.0	RMS	7.4	Z	46 20 59	UV, polar.
	0.20	0.18	140	14.0	Spec	7.0	V	M024421	Achromat, short mount
	0.18	0	∞	45.0	RMS	13.0	L	569 050	Achromat, met.
	0.18	0	∞	45.0	RMS	13.0	L	569 020	Achromat, met., phase
	0.18	0	∞	45.0	RMS	13.7	L	569 086	Achromat, met., nonbrowning
	0.17	0.17	145	45.0	RMS	12.0	V	M058241	Achromat
	0.17	0.17	145	45.0	RMS	12.0	V	M058248	Achromat, phase
	0.16	0	241	38.0	RMS	11.6	R	9226403	Achromat
	0.15	....	∞	24.0	RMS	14.0	V	M022154	Achromat, polar., TO 250 EFL
	0.15	....	∞	24.0	RMS	14.0	V	M022152	Achromat, TO 250 EFL
11	0.25	....	167	35.0	Spec	5.7	L	513 005	Achromat, met., DF
	0.25	0	181	38.0	RMS	4.6	R	9225501	Achromat
	0.25	0	181	38.0	RMS	4.6	R	9225551	Achromat, polar.
	0.25	0	181	38.0	RMS	4.6	R	9225502	Achromat, DF
	0.25	0	181	38.0	RMS	4.6	R	9225552	Achromat, polar. DF
	0.25	0	181	38.0	RMS	4.8	R	9228501	FF achromat
	0.25	....	140	29.0	....	1.7	V	M023202	Achromat, DF w/catopt. illum., magn. cent.
	0.25	....	140	44.0	RMS	1.7	V	M412060	Achromat, DF w/catopt. illum., magn. cent.
12.5	0.30	0.17	152	37.0	RMS	2.5	L	519 009	Apochromat
	0.30	0.18	145	43.0	RMS	1.8	BL	31-11-51	FF apochromat
	0.25	0.17	197	33.0	RMS	0.35	L	559 010	Achromat, met., polar., oil or water
13	0.40	0.17	152	37.0	RMS	3.4	L	519 237	Achromat
15	0.25	....	∞	24.0	RMS	5.0	V	M022352	Achromat, magn. cent., TO 250 EFL
	0.25	....	∞	24.0	RMS	5.0	V	M022354	Achromat, magn. cent., TO 250 EFL, polar.
	0.25	....	∞	24.0	Spec	2.5	V	M023252	Achromat, DF w/catopt. illum., TO 250 EFL
16	0.40	0.17	150	45.0	RMS	0.9	Z	46 05 20	Semiapo.
	0.40	0.17	150	45.0	RMS	0.9	Z	46 05 21	Semiapo., phase
	0.40	0	150	33.0	Spec	0.85	Z	46 20 08	Achromat, met., polar., DF, oil

TABLE 1 Objectives (Continued)

Power (1)	NA (2)	Cover glass, mm (3)	Tube length, mm (4)	Shoulder height, mm (5)	Thread (6)	Working distance, mm (7)	Mfr. (8)	Catalog No. (9)	Comments (10)
16	0.40	0	197	26.0	RMS	3.4	L	559 008	Achromat, met., polar.
	0.35	0.17	150	45.0	RMS	2.7	Z	46 05 10	FF achromat
	0.35	0	150	33.0	Spec	0.23	Z	46 07 69	Achromat, met., DF
	0.35	0	150	33.0	Spec	2.7	Z	46 20 03	Achromat, met., polar.
	0.35	0	150	33.0	RMS	2.7	Z	46 20 27	Achromat, met., phase
	0.35	0	150	33.0	RMS	2.8	Z	46 20 33	Achromat, met.
	0.35	0	150	33.0	RMS	2.8	Z	46 20 37	Achromat, met., phase
	0.32	0.17	151	37.0	RMS	1.2	R	9220501	Achromat
	0.32	0.17	151	37.0	RMS	1.2	R	9220551	Achromat, polar.
	0.32	0.17	151	37.0	RMS	1.2	R	9220506	Achromat, phase
	0.32	0.17	151	37.0	RMS	1.2	R	9220505	Achromat, phase (anoptral)
	0.30	0	∞	45.0	RMS	7.2	L	569 069	FF achromat, met.
	0.30	0	∞	45.0	RMS	0.29	L	569 041	FF achromat, met., oil
	0.30	0	∞	42.0	Spec	7.3	L	569 054	FF achromat, met., DF
	0.25	0	241	38.0	RMS	4.6	R	9225503	Achromat
	0.25	0	241	38.0	RMS	4.6	R	9225553	Achromat, polar.
	0.25	0	241	38.0	RMS	4.6	R	9225504	Achromat, DF
	0.25	0	241	38.0	RMS	4.6	R	9225554	Achromat, DF, polar.
	0.25	0	241	38.0	RMS	6.6	R	9220464	Achromat, DF, polar., for Vacutherm
	0.25	0	241	38.0	RMS	4.8	R	9228503	FF achromat
	0.17	....	150	33.0	RMS	13.5	Z	46 20 44	Achromat, U stage
17	0.30	....	230	40.2	....	3.8	V	M024032	Achromat, magn. cent.
	0.30	....	∞	24.0	RMS	3.8	V	M024052	Achromat, magn. cent., TO 250 EFL
	0.25	....	∞	34.0	RMS	3.1	V	M025152	FF achromat, TO 250 EFL
	0.25	....	∞	34.0	RMS	3.1	V	M025154	FF achromat, TO 250 EFL, polar.
20	0.65	0.18	145	35.7	RMS	0.50	BL	31-11-78-01	Apo.
	0.65	0.18	140	34.0	RMS	0.30	V	M023901	Apo.
	0.65	0.18	140	34.0	RMS	0.30	V	M023905	Apo., phase
	0.65	0.18	140	44.0	RMS	0.30	V	M023911	Apo.
	0.65	0.18	140	44.0	RMS	0.30	V	M023915	Apo., phase
	0.65	0.18	140	44.0	RMS	0.30	V	M023917	Apo., phase
	0.65	0.18	140	44.0	RMS	0.30	V	M023918	Apo., phase

20	0.60	0.17	146	36.5	RMS	1.26	W	5055	Semiapo.
	0.60	0.17	146	36.5	RMS	1.26	W	5251	Semiapo., phase
	0.60	0.17	151	37.0	RMS	0.67	R	9211101	Apo.
	0.50	....	150	33.0	RMS	....	Z	46 20 45	Achromat, U stage
	0.50	0.18	145	43.0	RMS	0.54	BL	31-10-61	FF achromat
	0.50	0.18	160	33.7	RMS	1.50	AO	C 112	Achromat
	0.50	0.18	∞	33.7	RMS	1.37	AO	1077	Achromat, TO 182.7 EFL
	0.50	0.18	∞	33.7	RMS	1.35	AO	1225	Achromat, phase, TO 182.7 EFL
	0.50	0.18	∞	33.7	RMS	1.35	AO	1235	Achromat, phase, TO 182.7 EFL
	0.50	0.18	140	34.0	RMS	1.2	V	M022401	Achromat
	0.50	0	140	34.0	RMS	1.3	V	M022402	Achromat, met.
	0.50	0.18	140	34.0	RMS	1.2	V	M022403	Achromat, polar.
	0.50	0	140	34.0	RMS	1.3	V	M022404	Achromat, polar., met.
	0.50	0.18	140	34.0	RMS	1.2	V	M022405	Achromat, phase
	0.50	0.18	140	44.0	RMS	1.2	V	M022411	Achromat
	0.50	0	140	44.0	RMS	1.3	V	M022412	Achromat, met.
	0.50	0.18	140	44.0	RMS	1.2	V	M022413	Achromat, polar.
	0.50	0	140	44.0	RMS	1.3	V	M022414	Achromat, polar., met.
	0.50	0.18	140	44.0	RMS	1.2	V	M022415	Achromat, phase
	0.50	0.18	140	44.0	RMS	1.2	V	M022417	Achromat, phase
	0.50	0.18	140	44.0	RMS	1.2	V	M022418	Achromat, phase
	0.50	0.18	140	40.2	....	1.2	V	M022431	Achromat, magn. cent.
	0.50	0.18	140	40.2	....	1.2	V	M022433	Achromat, magn. cent., polar.
	0.50	0.18	140	40.2	....	1.2	V	M022435	Achromat, magn. cent., phase
	0.50	0.18	140	34.0	RMS	1.2	V	M022443	Achromat, slotted mount
	0.50	0	140	34.0	RMS	1.3	V	M022444	Achromat, slotted mount, met.
	0.50	0.18	140	68.0	RMS	14.0	V	M022445	Achromat, catadioptric, LWD phase
	0.50	0	140	22.8	....	1.3	V	M022446	Achromat, met., quick change
	0.50	0.18	140	68.0	RMS	14.0	V	M022441	Achromat, catadioptric, LWD
	0.50	0.18	140	34.0	RMS	0.31	V	M025401	FF achromat
	0.50	0	140	34.0	RMS	0.43	V	M025402	FF achromat, met.
	0.50	0.18	140	34.0	RMS	0.31	V	M025403	FF achromat, polar.
	0.50	0	140	34.0	RMS	0.43	V	M025404	FF achromat, polar., met.
	0.50	0.18	140	44.0	RMS	0.31	V	M025411	FF achromat
	0.50	0	140	44.0	RMS	0.43	V	M025412	FF achromat, met.
	0.50	0.18	140	44.0	RMS	0.31	V	M025413	FF achromat, polar.
	0.50	0	140	44.0	RMS	0.43	V	M025414	FF achromat, polar., met.
	0.45	0.17	146	36.5	RMS	1.55	W	5010	Achromat
	0.45	0.17	146	36.5	RMS	1.05	W	5107	FF semiapo.

TABLE 1 Objectives (Continued)

Power (1)	NA (2)	Cover glass, mm (3)	Tube length, mm (4)	Shoulder height, mm (5)	Thread (6)	Working distance, mm (7)	Mfr. (8)	Catalog No. (9)	Comments (10)
20	0.45	0	146	36.5	RMS	1.55	W	5160	Achromat, met.
	0.45	0	146	36.5	RMS	1.05	W	5120	FF semiapo., met.
	0.45	0	146	36.5	Spec	1.55	W	8526	Achromat, met., DF
	0.45	0.17	146	36.5	RMS	4.10	W	3151	Achromat, polar.
	0.45	0.17	146	36.5	RMS	1.55	W	5201	Achromat, phase
	0.45	....	140	34.0	RMS	0.38	V	M024302	Achromat, immersion cap
	0.45	....	140	44.0	RMS	0.38	V	M024314	Achromat, immersion cap, polar.
	0.45	0.18	140	14.0	Spec	2.16	V	M024321	Achromat, short mount
	0.45	0.18	140	14.0	Spec	2.16	V	M024323	Achromat, short mount, polar.
	0.45	0.18	140	14.0	Spec	2.16	V	M024325	Achromat, short mount, phase
	0.45	0	156	36.0	RMS	1.7	U	M20X	Achromat, met.
	0.45	0.17	156	36.0	RMS	1.6	U	20X	Achromat
	0.40	....	160	36.65	RMS	2.00	O	1W140	Achromat, water
	0.40	0.17	160	36.77	RMS	1.55	O	1P140	Achromat, phase
	0.40	0	200	42.5	RMS	2.1	O	1M141	FF achromat, met.
	0.40	0	200	42.5	RMS	0.76	O	1M540	FF achromat, met.
	0.40	0	200	42.5	RMS	2.1	O	1M140	Achromat, phase
	0.40	0.17	160	42.5	RMS	1.55	O	1B240	Achromat, polar.
	0.40	1.80	∞	45.57	....	8.3	L	569 001	Achromat, high temp.
	0.40	1.80	∞	45.57	....	8.3	L	569 005	Achromat, high temp., phase
	0.40	0	∞	45.0	....	6.6	L	569 003	Achromat, high temp.
	0.40	0	200	36.7	RMS	2.8	BL	42-33-54	Achromat, met., polar.
	0.40	0	200	36.7	RMS	2.8	BL	42-33-07	Achromat, met., polar., DF
	0.40	0	200	36.7	Spec	0.60	BL	42-35-17	FF achromat, met., polar.
	0.40	0	200	36.7	RMS	0.60	BL	42-35-52	FF achromat, met., polar.
	0.40	0	200	36.7	Spec	0.60	BL	42-35-31	FF achromat, met., polar., DF
	0.40	2.0	201	99.0	Spec	18.3	U	LWD20X	Achromat, high temp., quartz cover glass, LWD
	0.40	0	156	36.0	RMS	2.2	U	MP1 20X	FF achromat, met.
	0.40	0	156	36.0	RMS	2.1	U	BP1 20X	FF achromat
	0.40	0.17	160	36.77	RMS	1.55	O	1B140	Achromat
	0.40	0.17	160	36.77	RMS	0.73	O	1B540	FF achromat, spring
	0.40	....	160	33.6	RMS	4.5	N	200-12-200	Achromat, phase (pos.)

20	0.40	....	160	33.6	RMS	4.5	N	200-12-300	Achromat, phase (neg.)
	0.40	....	160	33.6	RMS	5.7	N	201-12-000	Achromat
	0.40	....	160	33.6	RMS	5.7	N	200-12-601	Achromat, polar.
	0.40	0	160	33.6	RMS	5.7	N	201-12-600	Achromat, polar., met.
	0.40	....	160	33.6	RMS	5.7	N	200-12-000	Achromat
	0.40	....	160	33.6	RMS	5.7	N	200-12-600	Achromat, polar.
	0.35	0	∞	45.0	RMS	1.2	L	569 051	Achromat, met.
	0.35	0	∞	45.0	RMS	1.2	L	569 020	Achromat, met., phase
	0.35	0	∞	45.0	RMS	1.56	L	569 087	Achromat, met., nonbrowning
	0.33	....	160	45.0	RMS	1.1	N	202-72-000	Achromat, U stage
	0.33	....	...	....	....	0–2.0	N	200-18-001	Achromat, water, var. work. dist.
	0.30	0.17	145	45.0	RMS	2.0	V	M058341	Achromat
	0.25	....	140	36.4	RMS	12.0	V	M007817	Achromat, U stage
21	0.50	0.18	145	35.7	RMS	1.6	BL	31-10-27-01	Achromat
22	0.65	0	152	37.0	RMS	0.32	L	519 059	Achromat, oil or water, spring
	0.65	0	152	37.0	RMS	2.3	L	519 041	Achromat, oil, spring
	0.50	....	140	29.0	....	1.4	V	M023402	Achromat, DF, catoptric illum., magn. cent.
	0.50	....	140	44.0	RMS	1.4	V	M412050	Achromat, DF, catoptric illum., magn. cent.
	0.45	0	167	35.0	Spec	2.1	L	513 006	Achromat, met., DF
25	0.65	0.18	145	43.0	RMS	0.35	BL	31-11-52	FF apo.
	0.65	0.17	197	28.7	RMS	0.43	L	559 011	Achromat, met., polar., oil or water
	0.65	0	152	37.0	RMS	0.86	L	519 040	Apo., fluoresc., spring
	0.65	0.17	152	37.0	RMS	0.85	L	519 007	Apo., spring
	0.65	0.17	150	45.0	RMS	0.14	Z	46 06 41	FF apo., phase
	0.65	0.17	150	45.0	RMS	0.14	Z	46 06 40	FF apo.
	0.65	0	140	34.0	RMS	1.0	V	M024201	Achromat, immersion cap
	0.60	0.17	150	45.0	RMS	0.54	Z	46 06 20	Semiapo.
	0.60	0.17	150	45.0	RMS	0.54	Z	46 06 21	Semiapo., phase
	0.60	0.17	150	45.0	RMS	0.54	Z	46 06 28	Semiapo., polar.
	0.50	0	200	36.7	RMS	1.70	BL	42-33-55	Achromat, met., polar.
	0.50	0.17	152	37.0	RMS	0.92	L	559 036	Achromat, polar.
	0.50	0.17	152	37.0	RMS	0.92	L	051 924	Achromat, phase (Heine, HC, neg.), spring
	0.50	0.17	152	37.0	RMS	0.92	L	051 923	Achromat, phase (Heine, HC), spring
	0.50	0.17	152	38.5	RMS	0.92	L	519 150	Achromat, phase (Heine), spring
	0.50	0.17	152	45.0	RMS	0.90	L	519 163	FF achromat, spring
	0.50	0.17	152	37.0	RMS	0.92	L	519 006	Achromat, spring
	0.45	0.17	150	45.0	RMS	0.4	Z	46 06 10	FF achromat
	0.45	0.17	150	45.0	RMS	0.4	Z	46 06 11	FF achromat, phase
	0.45	0.17	151	37.0	RMS	0.85	R	9220601	Achromat

TABLE 1 Objectives (Continued)

Power (1)	NA (2)	Cover glass, mm (3)	Tube length, mm (4)	Shoulder height, mm (5)	Thread (6)	Working distance, mm (7)	Mfr. (8)	Catalog No. (9)	Comments (10)
25	0.45	0.17	151	37.0	RMS	0.85	R	9220651	Achromat, polar.
	0.45	0.17	151	37.0	RMS	0.85	R	9220606	Achromat, phase
	0.45	0.17	151	37.0	RMS	0.85	R	9220605	Achromat, phase (anoptral)
	0.45	0.17	151	37.0	RMS	0.27	R	9223601	FF achromat
27	0.70	0.18	140	34.0	RMS	1.0	V	M808421	UV monochromat, 2536A, quartz
	0.70	0.18	140	34.0	RMS	1.0	V	M808422	UV monochromat, 2750A, quartz
	0.70	0	140	34.0	RMS	1.1	V	M808423	UV monochromat, 2750A, quartz
28	0.55	0	181	38.0	RMS	0.5	R	9228701	FF achromat
	0.55	0	181	38.0	RMS	0.7	R	9225701	Achromat
	0.55	0	181	38.0	RMS	0.7	R	9225751	Achromat, polar.
	0.55	0	181	38.0	RMS	0.7	R	9225702	Achromat, DF
	0.55	0	181	38.0	RMS	0.7	R	9225752	Achromat, DF, polar.
	0.50	0	230	20.3	....	1.3	V	M022432	Achromat, met., magn. cent.
	0.50	0	230	20.3	....	1.3	V	M022434	Achromat, met., magn. cent., polar.
	0.50	0	230	17.8	....	1.3	V	M022436	Achromat, met., quick change
	0.50	0	230	29.0	....	1.4	V	M023432	Achromat, DF, catoptric illum., magn. cent.
30	0.50	0	∞	24.0	Spec	1.4	V	M023452	Achromat, DF, catoptric illum., TO 250 EFL
	0.50	0	∞	24.0	RMS	0.46	V	M025452	FF achromat, TO 250 EFL
	0.50	0	∞	24.0	RMS	0.46	V	M025454	FF achromat, TO 250 EFL, polar.
	0.50	0	∞	24.0	RMS	1.3	V	M022454	Achromat, polar., TO 250 EFL
32	0.65	0.17	150	45.0	RMS	0.3	Z	46 20 16	FF achromat
	0.60	1.80	∞	45.57	....	5.7	L	569 002	Achromat, high temp.
	0.60	1.80	∞	45.57	....	5.7	L	569 006	Achromat, high temp., phase
	0.60	0	∞	45.0	....	4.0	L	569 004	Achromat, high temp.
	0.60	....	152	37.0	RMS	....	L	559 016	Achromat, U stage, iris
	0.55	0	167	35.0	Spec	1.0	L	513 119	Achromat, met., DF
	0.50	0	∞	45.0	RMS	2.3	L	569 028	FF achromat, met.
	0.50	0	∞	42.0	Spec	2.3	L	569 038	FF achromat, met., DF
	0.50	0	∞	45.0	RMS	2.3	L	569 033	FF achromat, met., phase
	0.40	0.35	150	45.0	RMS	0.45	Z	46 20 60	UV glycerin
	0.40	0.35	150	45.0	RMS	0.45	Z	46 20 70	UV glycerin, phase
	0.40	0.35	150	45.0	RMS	0.46	Z	46 20 61	UV glycerin, polar.

32	0.30	....	152	45.4	RMS	....	L	559 005	Achromat, U stage
40	1.00	0.17	150	45.0	RMS	0.22	Z	46 17 36	Apo., iris, oil
	1.00	0.17	150	45.0	RMS	0.22	Z	46 17 46	FF apo., iris, oil
	1.00	0.17	150	45.0	RMS	0.22	Z	46 17 47	FF apo., iris, oil, phase
	1.00	0.18	145	35.7	RMS	0.27	BL	31-10-59-02	Semiapo., oil
	0.95	0.17	152	37.0	RMS	0.12	L	519 038	Apo., corr. collar, spring
	0.95	0.18	140	34.0	RMS	0.20	V	M023801	Apo., corr. collar
	0.95	0.18	140	44.0	RMS	0.20	V	M023811	Apo., corr. collar
	0.95	0.18	140	40.2	....	0.20	V	M023831	Apo., magn. cent.
	0.85	0.17	157	37.0	RMS	0.38	L	519 025	Semiapo., spring
	0.85	0	150	33.0	RMS	0.23	Z	46 20 38	Achromat, met., phase
	0.85	0.17	150	45.0	RMS	0.35	Z	46 17 06	Achromat, oil
	0.85	1.50	150	45.0	RMS	0.35	Z	46 16 08	Achromat, oil
	0.85	0.17	150	45.0	RMS	0.36	Z	46 07 08	Achromat, polar.
	0.85	0	150	45.0	Spec	0.23	Z	46 07 69	Achromat
	0.85	0	150	33.0	Spec	0.23	Z	46 20 04	Achromat, met., polar.
	0.85	0	150	33.0	Spec	0.50	Z	46 20 09	Achromat, met., polar., DF, oil
	0.85	0	150	33.0	RMS	0.23	Z	46 20 28	Achromat, met., phase
	0.85	0	150	33.0	RMS	0.23	Z	46 20 34	Achromat, met.
	0.85	0	152	37.0	RMS	0.36	L	519 011	Semiapo., fluoresc., spring
	0.85	0.18	140	34.0	RMS	0.30	V	M022901	Achromat
	0.85	0	140	34.0	RMS	0.35	V	M022902	Achromat, met.
	0.85	0.18	140	34.0	RMS	0.30	V	M022903	Achromat, polar.
	0.85	0	140	34.0	RMS	0.35	V	M022904	Achromat, polar., met.
	0.85	0.18	140	44.0	RMS	0.30	V	M022911	Achromat
	0.85	0	140	44.0	RMS	0.35	V	M022912	Achromat, met.
	0.85	0.18	140	44.0	RMS	0.30	V	M022913	Achromat, polar.
	0.85	0	140	44.0	RMS	0.35	V	M022914	Achromat, polar., met.
	0.85	0.18	140	14.0	Spec	0.30	V	M022921	Achromat, short mount
	0.85	0.18	140	14.0	Spec	0.30	V	M022923	Achromat, short mount, polar.
	0.85	0.18	140	40.2	....	0.30	V	M022931	Achromat, magn. cent.
	0.85	0.18	140	40.2	....	0.30	V	M022933	Achromat, magn. cent., polar.
	0.85	0	140	53.0	....	0.40	V	M022941	Achromat, Dyson interfer., micros.
	0.85	0.18	140	34.0	RMS	0.30	V	M022943	Achromat, slotted mount
	0.80	0.17	160	33.6	RMS	0.19	N	200-62-000	Apo., spring
	0.75	0.17	160	33.6	RMS	0.45	N	200-52-000	Semiapo., spring
	0.75	0.17	146	36.5	RMS	0.45	W	5060	Semiapo.
	0.75	0.17	146	36.5	RMS	0.45	W	5252	Semiapo., phase
	0.75	0.17	150	45.0	RMS	0.33	Z	46 07 20	Semiapo.

TABLE 1 Objectives (Continued)

Power (1)	NA (2)	Cover glass, mm (3)	Tube length, mm (4)	Shoulder height, mm (5)	Thread (6)	Working distance, mm (7)	Mfr. (8)	Catalog No. (9)	Comments (10)
40	0.75	0.17	150	45.0	RMS	0.33	Z	46 07 21	Semiapo., phase
	0.75	0.17	150	45.0	RMS	1.60	Z	46 17 03	Achromat, phase, water
	0.75	0.17	150	45.0	RMS	1.60	Z	46 17 02	Achromat, water
	0.75	0.17	160	36.77	RMS	0.49	O	1B351	Semiapo.
	0.75	0.17	160	36.77	RMS	0.49	O	1F150	Semiapo., phase
	0.75	0	200	42.5	RMS	0.40	O	1M153	Semiapo., met.
	0.75	0	160	36.77	RMS	0.58	O	1B350	Semiapo., met.
	0.70	0.17	152	37.0	RMS	0.38	L	519 151	Apo., phase (Heine), corr. collar
	0.70	0.17	152	37.0	RMS	0.38	L	051 927	Apo., phase (Heine, HC), corr. collar
	0.70	0.17	152	37.0	RMS	0.38	L	051 928	Apo., phase (Heine, HC, neg.), corr. collar
	0.70	0.18	140	34.0	RMS	0.18	V	M025201	FF achromat
	0.70	0.18	140	34.0	RMS	0.18	V	M025203	FF achromat, polar.
	0.70	0.18	140	34.0	RMS	0.18	V	M025205	FF achromat, phase
	0.70	0.18	140	34.0	RMS	0.18	V	M025208	FF achromat, phase
	0.70	0.18	140	44.0	RMS	0.18	V	M025211	FF achromat
	0.70	0.18	140	44.0	RMS	0.18	V	M025213	FF achromat, polar.
	0.70	0.18	140	44.0	RMS	0.18	V	M025215	FF achromat, phase
	0.70	0.18	140	44.0	RMS	0.18	V	M025217	FF achromat, phase
	0.70	0.18	140	44.0	RMS	0.18	V	M025218	FF achromat, phase
	0.70	0.18	140	44.0	RMS	0.18	V	M025219	FF achromat, phase
	0.70	0.18	140	14.0	Spec	0.18	V	M025221	FF achromat, short mount
	0.66	0.18	∞	33.7	RMS	0.50	AO	1023	FF achromat
	0.65	0.17	160	36.77	RMS	0.56	O	1B 151	Achromat
	0.65	0.17	160	36.77	RMS	0.56	O	1B 550	Achromat, spring
	0.65	0	160	36.77	RMS	0.65	O	1B 152	Achromat, met.
	0.65	0.17	160	36.77	RMS	0.24	O	1B 352	Achromat, spring
	0.65	0	160	36.65	RMS	0.44	O	1W150	Achromat, water
	0.65	0.17	160	36.77	RMS	0.56	O	1P150	Achromat, phase, spring
	0.65	0.17	160	36.77	RMS	0.56	O	1B250	Achromat, polar.
	0.65	0	200	42.5	RMS	0.6	O	1M150	Achromat, met.
	0.65	0	200	42.5	RMS	0.44	O	1M550	FF achromat, met.
	0.65	0	200	42.5	RMS	0.6	O	1M251	Achromat, phase, met.

40	0.65	0	200	42.5	RMS	0.6	O	1M250	Achromat, met., polar.
	0.65	0	200	42.5	Spec	0.53	O	1M450	Achromat, met., DF
	0.65	0	200	42.5	RMS	....	O	1M151	FF achromat, met.
	0.65	0	200	36.7	RMS	0.91	BL	42-33-56	Achromat, met., polar.
	0.65	0	200	36.7	RMS	0.91	BL	42-33-66	Achromat, met., polar., DF
	0.65	0.17	152	45.0	RMS	0.71	L	519 166	Achromat, phase (Zernike), spring
	0.65	0.17	146	36.5	RMS	0.58	W	5015	Achromat
	0.65	0.17	146	36.5	RMS	0.35	W	5110	FF semiapo.
	0.65	0	146	36.5	RMS	0.58	W	5165	Achromat, met.
	0.65	0	146	36.5	RMS	0.35	W	5125	FF semiapo., met.
	0.65	0	146	36.5	Spec	0.58	W	8527	Achromat, met., DF
	0.65	0.17	146	36.5	RMS	0.58	W	3153	Achromat, polar.
	0.65	0.17	146	36.5	RMS	0.58	W	5202	Achromat, phase
	0.65	0.18	145	43.0	RMS	0.20	BL	31-10-64	FF achromat
	0.65	0	200	36.7	Spec	0.60	BL	42-35-18	FF achromat, met., polar.
	0.65	0	200	36.7	RMS	0.60	BL	42-35-53	FF achromat, met., polar.
	0.65	0	200	36.7	Spec	0.60	BL	42-35-32	FF achromat, met., polar., DF
	0.65	0.17	150	45.0	RMS	0.47	Z	46 07 00	Achromat
	0.65	0.17	150	45.0	RMS	0.18	Z	46 07 10	FF achromat
	0.65	0.17	150	45.0	RMS	0.47	Z	46 07 01	Achromat, phase
	0.65	0.17	150	45.0	RMS	0.18	Z	46 07 11	FF achromat, phase
	0.65	....	150	33.0	RMS	6.8	Z	46 20 46	Achromat, U stage
	0.65	0.17	150	45.0	RMS	0.2	Z	47 44 06	Achromat, interfer.
	0.65	0.17	152	37.0	RMS	0.67	L	519 003	Achromat, spring
	0.65	0.17	152	45.0	RMS	0.58	L	519 161	FF semiapo., spring
	0.65	0.18	140	34.0	RMS	0.68	V	M022501	Achromat
	0.65	0	140	34.0	RMS	0.81	V	M022502	Achromat, met.
	0.65	0.18	140	34.0	RMS	0.68	V	M022503	Achromat, polar.
	0.65	0	140	34.0	RMS	0.81	V	M022504	Achromat, polar., met.
	0.65	0.18	140	34.0	RMS	0.68	V	M022505	Achromat, phase
	0.65	0.18	140	34.0	RMS	0.68	V	M022508	Achromat, phase
	0.65	0.18	140	44.0	RMS	0.68	V	M022511	Achromat
	0.65	0	140	44.0	RMS	0.81	V	M022512	Achromat, met.
	0.65	0.18	140	44.0	RMS	0.68	V	M022513	Achromat, polar.
	0.65	0.17	160	33.6	RMS	0.54	N	200-12-001	Achromat
	0.65	0.17	160	33.6	RMS	0.54	N	200-22-000	Achromat, spring
	0.65	0.17	160	33.6	RMS	0.24	N	200-32-000	FF achromat
	0.65	0	160	33.6	RMS	1.3	N	200-32-001	FF achromat, met.
	0.65	0.17	160	33.6	RMS	0.54	N	200-22-200	Achromat, phase (pos.)

TABLE 1 Objectives (Continued)

Power (1)	NA (2)	Cover glass, mm (3)	Tube length, mm (4)	Shoulder height, mm (5)	Thread (6)	Working distance, mm (7)	Mfr. (8)	Catalog No. (9)	Comments (10)
40	0.65	0.17	160	33.6	RMS	0.54	N	200-22-201	Achromat, phase (pos.)
	0.65	0.17	160	33.6	RMS	0.54	N	200-22-300	Achromat, phase (neg.)
	0.65	0	160	33.6	RMS	0.52	N	201-12-003	Achromat, met.
	0.65	0	160	33.6	RMS	1.3	N	201-32-000	FF achromat, met.
	0.65	0.17	160	33.6	RMS	0.54	N	200-22-600	Achromat, polar.
	0.65	0.17	160	33.6	RMS	0.24	N	200-32-600	FF achromat, polar.
	0.65	0.17	160	33.6	RMS	0.54	N	200-12-602	Achromat, polar.
	0.65	0.17	160	33.6	RMS	0.52	N	201-12-601	Achromat, met., spring
	0.65	0.17	160	33.6	RMS	0.55	N	200-21-003	Achromat for type G
	0.65	0.17	160	33.6	RMS	0.55	N	200-12-003	Achromat for type E
	0.65	....	...	....	....	0.74	N	200-18-002	Achromat, water
	0.65	0.17	151	37.0	RMS	0.41	R	9220701	Achromat
	0.65	0.17	151	37.0	RMS	0.41	R	9220751	Achromat, polar.
	0.65	0.17	151	37.0	RMS	0.41	R	9220706	Achromat, phase, iris
	0.65	0.17	151	37.0	RMS	0.41	R	9220705	Achromat, phase (anoptral), iris
	0.65	0.17	151	37.0	RMS	0.20	R	9223701	FF achromat
	0.65	0.17	151	37.0	RMS	0.20	R	9223751	FF achromat, interfer. contrast
	0.65	0	156	36.0	RMS	0.35	U	MP1 40X	FF achromat, met.
	0.65	0.17	156	36.0	RMS	0.35	U	BP1 40X	FF achromat
	0.65	0	156	36.0	RMS	0.33	U	M40X	Achromat, met.
	0.65	0.17	156	36.0	RMS	0.33	U	40X	Achromat
	0.65	0	140	44.0	RMS	0.81	V	M022514	Achromat, polar., met.
	0.65	0.18	140	44.0	RMS	0.68	V	M022515	Achromat, phase
	0.65	0.18	140	44.0	RMS	0.68	V	M022517	Achromat, phase
	0.65	0.18	140	44.0	RMS	0.68	V	M022518	Achromat, phase
	0.65	0.18	140	44.0	RMS	0.68	V	M022519	Achromat, phase
	0.65	0.18	140	14.0	Spec	0.68	V	M022521	Achromat, short mount
	0.65	0	140	35.5	Spec	0.81	V	M022522	Achromat, short mount, met.
	0.65	0.18	140	14.0	Spec	0.68	V	M022523	Achromat, short mount, polar.
	0.65	0	140	35.5	Spec	0.81	V	M022524	Achromat, short mount, polar., extension tube
	0.65	0.18	140	14.0	Spec	0.68	V	M022525	Achromat, short mount, phase
	0.65	0.18	140	40.2	....	0.68	V	M022531	Achromat, magn. cent.

40	0.65	0.18	140	40.2	. . . .	0.68	V	M022533	Achromat, magn. cent., polar.
	0.65	0.18	140	44.0	RMS	0.94	V	M020511	Achromat
	0.65	0	140	44.0	RMS	1.06	V	M020512	Achromat, met.
	0.65	0.18	140	40.2	. . . .	0.68	V	M022535	Achromat, magn. cent., phase
	0.65	0	140	15.9	RMS	0.81	V	M022542	Achromat, GKN adapter
	0.65	0.18	140	34.0	RMS	0.68	V	M022543	Achromat, slotted mount
	0.65	0.18	140	17.8	RMS	0.68	V	M022546	Achromat, quick change
	0.65	0	140	34.0	RMS	0.81	V	M022544	Achromat, slotted mount, polar., met.
	0.65	0	140	34.0	RMS	0.25	V	M025202	FF achromat, met.
	0.65	0	140	34.0	RMS	0.25	V	M025204	FF achromat, met., polar.
	0.65	0	140	44.0	RMS	0.25	V	M025212	FF achromat, met.
	0.65	0	140	44.0	RMS	0.25	V	M025214	FF achromat, met., polar.
	0.65	0	230	34.0	RMS	0.35	V	M025301	FF achromat
	0.65	0	230	40.2	. . . .	0.35	V	M025332	FF achromat, magn. cent.
	0.65	0	∞	24.0	RMS	0.35	V	M025352	FF achromat, TO 250 EFL
	0.65	0	∞	24.0	RMS	0.35	V	M025354	FF achromat, TO 250 EFL, polar.
	0.60	1.5	160	33.6	RMS	2.0	N	. . . .	Achromat, corr. collar, LWD
	0.60	1.5	160	33.6	RMS	2.0	N	200-12-000	Achromat, corr. collar, phase (pos.), spring
	0.57	0.18	140	68.0	RMS	14.0	V	M022545	Achromat, catadioptric, LWD phase
	0.57	0.18	140	68.0	RMS	14.0	V	M022541	Achromat, catadioptric, LWD
	0.57	0.18	140	68.0	RMS	14.0	V	M022548	Achromat, catadioptric, LWD phase
	0.55	0	241	38.0	RMS	0.7	R	9225703	Achromat, met.
	0.55	0	241	38.0	RMS	0.7	R	9225753	Achromat, met., polar.
	0.55	0	241	38.0	RMS	0.7	R	9225704	Achromat, met., DF
	0.55	0	241	38.0	RMS	0.7	R	9225754	Achromat, met., DF, polar.
	0.55	0	241	38.0	RMS	0.5	R	9228703	FF achromat
	0.50	0.17	145	45.0	RMS	0.5	V	M058448	Achromat, phase
	0.50	0.17	145	45.0	RMS	0.5	V	M058441	Achromat
	0.45	1.0	156	81.0	RMS	4.6	U	FF40X	Achromat, LWD, high temp.
	0.40	1.5	200	60.0	RMS	7.3	O	1M152	Achromat, LWD, phase
43	0.66	0.18	160	33.7	RMS	0.73	AO	C115	Achromat
	0.65	0.18	145	35.7	RMS	0.6	BL	31-10-29-01	Achromat
	0.55	0.18	145	35.7	RMS	0.4	BL	31-10-24	Achromat
	0.55	0.18	∞	33.7	RMS	0.6	AO	114	Achromat, TO 165.2 EFL
44	0.65	0	197	14.0	RMS	0.70	L	559 033	Achromat, met., polar.
45	0.95	0.18	140	34.0	RMS	0.21	V	M023601	Semiapo., oil
	0.95	0.18	140	34.0	RMS	0.21	V	M023605	Semiapo., oil, phase
	0.95	0.18	140	34.0	RMS	0.21	V	M023608	Semiapo., oil, phase
	0.95	0.18	140	14.0	Spec	0.21	V	M023621	Semiapo., oil, short mount

TABLE 1 Objectives (Continued)

Power (1)	NA (2)	Cover glass, mm (3)	Tube length, mm (4)	Shoulder height, mm (5)	Thread (6)	Working distance, mm (7)	Mfr. (8)	Catalog No. (9)	Comments (10)
45	0.95	0.18	140	40.2	....	0.21	V	M023635	Semiapo., oil, magn. cent.
	0.95	0.18	140	34.0	RMS	1.07	V	M023647	Semiapo., oil, LWD
	0.95	0.18	140	34.0	RMS	1.33	V	M023648	Semiapo., oil, LWD
	0.95	0.18	140	34.0	RMS	0.21	V	M023649	Semiapo., glycerin
	0.90	0.17	151	37.0	RMS	0.15	R	9211201	Apo., corr. collar
	0.85	0	197	14.0	RMS	0.43	L	559 009	Semiapo., met., polar.
	0.85	0.18	145	35.7	RMS	0.30	BL	31-10-31-01	Achromat
	0.75	0	181	38.0	RMS	0.29	R	9225801	Semiapo.
	0.75	0	181	38.0	RMS	0.29	R	9225851	Semiapo., polar.
	0.75	0	181	38.0	RMS	0.21	R	9228801	FF achromat
	0.66	0.18	∞	33.7	RMS	0.75	AO	1116	Achromat, TO 182.7 EFL
	0.66	0.18	∞	33.7	RMS	0.75	AO	1238	Achromat, TO 182.7 EFL, phase
	0.66	0.18	∞	33.7	RMS	0.75	AO	1239	Achromat, TO 182.7 EFL, phase
	0.65	0	152	37.0	RMS	0.68	L	519 042	Achromat, oil, spring
	0.65	0	140	29.0	....	0.7	V	M024102	Achromat, DF, catoptric illum., magn. cent.
	0.65	0	140	44.0	RMS	0.7	V	M412040	Achromat, DF, catoptric illum., magn. cent.
	0.55	0.18	145	43.0	RMS	0.46	BL	31-10-49	FF achromat
47.5	0.95	0.18	145	35.7	RMS	0.18	BL	31-11-81-01	Apo., corr. collar
50	1.00	....	146	36.5	RMS	0.20	W	5070	Semiapo., oil
	1.00	....	146	36.5	RMS	0.20	W	5253	Semiapo., oil, phase
	1.00	0	200	36.7	RMS	0.45	BL	42-33-61	Semiapo., oil, met., polar.
	1.00	0.18	140	34.0	RMS	0.50	V	M022701	Achromat, water
	1.00	0.18	140	34.0	RMS	0.50	V	M022705	Achromat, water, phase
	1.00	0.18	140	44.0	RMS	0.50	V	M022711	Achromat, water
	1.00	0.18	140	44.0	RMS	0.50	V	M022715	Achromat, water, phase
	1.00	0.18	140	14.0	Spec	0.50	V	M022721	Achromat, water, short mount
	0.95	0	200	36.7	RMS	0.18	BL	42-33-71	Apo., met., polar.
	0.95	0.18	140	44.0	RMS	0.21	V	M023611	Semiapo., oil
	0.95	0.18	140	44.0	RMS	0.21	V	M023615	Semiapo., oil, phase
	0.95	0.18	140	44.0	RMS	0.21	V	M023617	Semiapo., oil, phase
	0.95	0.18	140	44.0	RMS	0.21	V	M023618	Semiapo., oil, phase
	0.95	0.18	140	44.0	RMS	0.21	V	M023619	Semiapo., oil, phase

50	0.90	0	∞	24.0	RMS	0.20	V	M023852	Apo., TO 250 EFL
	0.85	0	200	36.7	RMS	0.38	BL	42-33-57	Achromat, met., polar.
	0.85	0	200	36.7	RMS	0.39	BL	42-33-70	Semiapo., met., polar.
	0.85	0.17	152	37.0	RMS	0.42	L	559 038	Achromat, polar.
	0.85	0	∞	45.0	RMS	0.30	L	569 052	Semiapo., met.
	0.85	0	∞	45.0	RMS	0.30	L	569 022	Semiapo., met., phase
	0.85	0.17	146	36.5	RMS	0.22	W	5017	Achromat
	0.85	0.17	146	36.5	RMS	0.22	W	3155	Achromat, polar.
	0.85	0	∞	34.0	RMS	0.48	V	M027152	FF apo.
	0.85	0.18	∞	33.7	RMS	0.30	AO	1026	Achromat, iris, TO 182.7 EFL
	0.85	0.18	∞	33.7	RMS	0.30	AO	1029	Achromat, TO 182.7 EFL
	0.85	0.18	∞	33.7	RMS	0.30	AO	1228	Achromat, TO 182.7 EFL, phase
	0.80	0	200	36.7	Spec	0.30	BL	42-35-22	FF semiapo., met., polar.
	0.80	0	∞	24.0	RMS	0.35	V	M022952	Achromat, TO 250 EFL
	0.80	0	∞	24.0	RMS	0.35	V	M022954	Achromat, TO 250 EFL, polar.
	0.75	0	∞	45.0	RMS	0.62	L	569 088	Achromat, met., nonbrowning
	0.70	0	∞	24.0	RMS	0.37	V	M025[illegible]2	FF achromat, TO 250 EFL
	0.70	0	∞	24.0	RMS	0.37	V	M02[illegible]54	FF achromat, TO 250 EFL
	0.65	0	167	35.0	Spec	0.66	L	513 007	Achromat, met., DF
	0.65	0	∞	24.0	Spec	0.7	V	M024152	Achromat, met., DF, catoptric illum., TO 250 EFL
	0.60	....	152	37.0	RMS	....	L	559 040	Achromat, U stage, iris
53	0.95	0	152	37.0	RMS	1.07	L	519 043	Semiapo., oil, spring
54	0.95	0.17	152	37.0	RMS	0.22	L	519 027	Semiapo., oil, spring
55	0.95	0	230	20.3	....	0.32	V	M023832	Apo., magn. centr.
	....	0	167	35.0	Spec	0.59	L	513 008	Achromat, met., DF, water
56	0.90	0	181	38.0	RMS	0.23	R	9225901	Semiapo.
	0.90	0	181	38.0	RMS	0.23	R	9225951	Semiapo., polar.
	0.90	0	181	38.0	RMS	0.15	R	9228901	FF achromat
	0.65	0	230	29.0	....	0.7	V	M024132	Achromat, met., DF, catoptric illum., magn. cent.
60	0.95	0	197	14.0	RMS	0.38	L	559 012	Semiapo., met., polar., oil
	0.95	0.17	151	37.0	RMS	0.36	R	9220908	Achromat, glycerin, iris
	0.85	0.18	145	35.7	RMS	0.20	BL	31-10-35-01	Achromat
	0.85	0	167	35.0	Spec	0.57	L	513 010	Achromat, met., DF, oil
	0.85	0	230	20.3	....	0.30	V	M022932	Achromat, short mount, magn. cent.
	0.85	0	230	20.3	....	0.30	V	M022934	Achromat, short mount, magn. cent., polar.
	0.80	0.17	160	36.77	RMS	0.26	O	1B170	Achromat
	0.80	0.17	160	33.6	RMS	0.37	N	200-13-000	Achromat for type E
	0.80	0	156	36.0	RMS	0.35	U	M60X	Achromat, met.
	0.80	0.17	156	36.0	RMS	0.28	U	60X	Achromat

TABLE 1 Objectives (Continued)

Power (1)	NA (2)	Cover glass, mm (3)	Tube length, mm (4)	Shoulder height, mm (5)	Thread (6)	Working distance, mm (7)	Mfr. (8)	Catalog No. (9)	Comments (10)
60	0.70	0.17	164	36.5	RMS	0.29	W	5020	Achromat
	0.65	0	230	17.8	....	0.68	V	M022536	Achromat, met., quick change
	0.65	0	230	20.3	....	0.68	V	M022532	Achromat, met., magn. cent.
	0.65	0	230	20.3	....	0.68	V	M022534	Achromat, met., magn. cent., polar.
61	1.40	0.18	145	35.7	RMS	0.12	BL	31-11-89-02	Apo., oil
63	1.25	0.17	150	45.0	RMS	0.65	Z	46 18 20	Semiapo., oil
	1.25	0.17	150	45.0	RMS	0.65	Z	46 18 21	Semiapo., oil, phase
	0.95	0.17	152	37.0	RMS	0.12	L	519 039	Apo., corr. collar, spring
	0.90	0.17	150	45.0	RMS	0.12	Z	46 08 22	Semiapo., corr. collar
	0.90	0.17	150	45.0	RMS	0.09	Z	46 08 13	FF achromat, phase
	0.90	0.17	150	45.0	RMS	0.12	Z	46 08 23	Semiapo., phase
	0.90	0.17	150	45.0	RMS	0.12	Z	46 08 28	Semiapo., polar.
	0.90	0	150	45.0	RMS	0.09	Z	46 08 60	FF achromat
	0.85	0.17	152	37.0	RMS	0.29	L	519 002	Achromat, spring
	0.85	0.17	152	37.0	RMS	0.29	L	519 047	Achromat, spring, iris
	0.85	0.17	152	37.0	RMS	0.29	L	559 043	Achromat, polar.
	0.85	0.17	151	37.0	RMS	0.18	R	9210603	Semiapo., corr. collar
	0.85	0.17	151	37.0	RMS	0.18	R	9210653	Semiapo., corr. collar, polar.
	0.80	0.17	151	37.0	RMS	0.15	R	9220806	Achromat, phase, iris
	0.80	0.17	151	37.0	RMS	0.15	R	9220805	Achromat, phase (anoptral), iris
	0.80	0.17	151	37.0	RMS	0.15	R	9220801	Achromat
	0.80	0.17	151	37.0	RMS	0.15	R	9220851	Achromat, polar.
	0.80	0.17	151	37.0	RMS	0.11	R	9223801	FF achromat
	0.80	0.17	151	37.0	RMS	0.11	R	9223851	FF achromat, interfer. contrast
	0.75	0	241	38.0	RMS	0.29	R	9225803	Semiapo.
	0.75	0	241	38.0	RMS	0.29	R	9225853	Semiapo., polar.
	0.75	0	241	38.0	RMS	0.21	R	9228803	FF achromat
	0.70	0.17	152	37.0	RMS	0.30	L	519 155	Apo., phase (Heine), spring
	0.70	0.17	152	37.0	RMS	0.30	L	051 931	Apo., phase (Heine, HC), spring
	0.70	0.17	152	37.0	RMS	0.30	L	051 932	Apo., phase (Heine, HC, neg.)
70	1.15	0.17	152	38.5	RMS	0.20	L	519 154	Semiapo., phase, oil, spring
	1.15	0.17	152	38.5	RMS	0.20	L	051 943	Semiapo., phase, oil, HC

70	1.15	0.17	152	38.5	RMS	0.20	L	051 944	Semiapo., phase, oil, neg., HC
75	1.40	0	200	36.7	RMS	0.29	BL	42-33-79	Apo., met., polar., oil
	1.25	0	200	36.7	RMS	0.30	BL	42-33-58	Achromat, met., polar., oil
	1.25	0	200	36.7	RMS	0.51	BL	42-33-68	Semiapo., met., polar., oil
	1.00	0	167	35.0	Spec	0.51	L	513 011	Semiapo., met., DF, oil
	0.90	0	167	35.0	Spec	0.48	L	513 009	Achromat, met., DF, water
76	1.25	0.18	140	34.0	RMS	....	V	M 808429	UV monochromat, 2750A, glycerin
	1.25	0.18	140	34.0	RMS	....	V	M 808 480	UV monochromat, 2750A, glycerin, phase
79	0.95	0	230	20.3	....	0.05	V	M023332	Apo., short mount, magn. cent.
80	1.32	0.18	140	34.0	RMS	0.10	V	M023701	Apo., oil
	1.32	0.18	140	44.0	RMS	0.10	V	M023711	Apo., oil
	1.32	0	140	44.0	RMS	0.28	V	M023712	Apo., oil, met.
	1.32	0.18	140	40.2	....	0.10	V	M023731	Apo., oil, magn. cent.
	1.30	0	197	14.0	RMS	0.36	L	559 013	Semiapo., met., polar., oil, short mount
	0.95	0	∞	45.0	RMS	0.13	L	569 034	FF achromat, met., phase
	0.95	0	∞	45.0	RMS	0.13	L	569 029	FF achromat, met.
	0.95	0	150	33.0	RMS	0.09	Z	46 20 35	Achromat, met.
	0.95	0	150	33.0	Spec	0.09	Z	46 08 69	Achromat, met., DF
	0.90	0	200	36.7	Spec	0.30	BL	42-35-23	FF semiapo., met., polar.
	0.90	0	241	38.0	RMS	0.23	R	9225903	Semiapo.
	0.90	0	241	38.0	RMS	0.23	R	9225953	Semiapo., polar.
	0.90	0	241	38.0	RMS	0.15	R	9228903	FF achromat
	0.90	0	156	36.0	RMS	0.26	U	M P1 80X	FF achromat, met.
	0.85	0	200	36.7	Spec	0.60	BL	42-35-19	FF achromat, met., polar.
	0.85	0	200	36.7	RMS	0.60	BL	42-35-54	FF achromat, met., polar.
	0.85	0	200	36.7	Spec	0.60	BL	42-35-33	FF achromat, met., polar., DF
	0.75	0	∞	42.0	Spec	0.18	L	569 070	FF achromat, met., DF
81	1.25	0.18	140	34.0	RMS	0.1	V	M808428	UV monochromat, 2536A, glycerin
	1.25	0.18	140	34.0	RMS	0.1	V	M808475	UV monochromat, 2536A, glycerin, phase
85	1.25	....	164	36.5	RMS	0.17	W	5030	Achromat, oil
	0.95	0	∞	24.0	RMS	0.16	V	M023352	Apo., TO 250 EFL
	0.85	0	∞	24.0	RMS	0.15	V	M023152	Achromat, TO 250 EFL
	0.85	0	∞	24.0	RMS	0.15	V	M023154	Achromat, TO 250 EFL, polar.
90	1.40	0.18	145	35.7	RMS	0.07	BL	31-11-93-02	Apo., oil
	1.40	0.17	152	37.0	RMS	0.06	L	519 010	Apo., oil, spring
	1.32	0.17	152	37.0	RMS	0.12	L	519 008	Apo., oil, spring
	1.32	0.17	152	38.5	RMS	0.12	L	519 158	Apo., oil, spring, phase
	1.32	0.17	152	38.5	RMS	0.12	L	051 949	Apo., oil, spring, phase (HC)
	1.30	0.18	145	35.7	RMS	0.12	BL	31-11-91-02	Apo., oil

TABLE 1 Objectives (Continued)

Power (1)	NA (2)	Cover glass, mm (3)	Tube length, mm (4)	Shoulder height, mm (5)	Thread (6)	Working distance, mm (7)	Mfr. (8)	Catalog No. (9)	Comments (10)
90	1.25	0	181	38.0	RMS	0.26	R	9229001	FF achromat, oil
	1.20	0.17	152	37.0	RMS	0.32	L	519 171	Achromat, water, spring
	1.15	0.17	152	38.5	RMS	0.12	L	519 152	Apo., oil, phase, spring
	1.15	0.17	152	38.5	RMS	0.12	L	051 947	Apo., oil, phase, spring, HC
95	1.32	0.17	152	37.0	RMS	0.15	L	519 026	Semiapo., oil, spring
	1.32	0.17	152	37.0	RMS	0.15	L	519 046	Semiapo., oil, spring, iris
	1.32	0	152	37.0	RMS	0.11	L	519 012	Semiapo., oil, spring, fluoresc.
	1.32	0.17	152	37.0	RMS	0.14	L	559 039	Semiapo., oil, polar.
	1.30	0.18	140	34.0	RMS	0.15	V	M022601	Achromat, oil
	1.30	0	140	34.0	RMS	0.33	V	M022602	Achromat, oil, met.
	1.30	0.18	140	34.0	RMS	0.15	V	M022603	Achromat, oil, polar.
	1.30	0	140	34.0	RMS	0.33	V	M022604	Achromat, oil, polar., met.
	1.30	0.18	140	34.0	RMS	0.15	V	M022605	Achromat, oil, phase
	1.30	0.18	140	34.0	RMS	0.15	V	M022608	Achromat, oil, phase
	1.30	0.18	140	34.0	RMS	0.15	V	M022609	Achromat, oil, phase
	1.30	0.18	140	14.0	Spec	0.15	V	M022621	Achromat, oil, short mount
	1.30	0	140	35.5	Spec	0.33	V	M022622	Achromat, oil, short mount, extension tube
	1.30	0.18	140	14.0	Spec	0.15	V	M022623	Achromat, oil, short mount, polar.
	1.30	0.18	140	14.0	Spec	0.15	V	M022625	Achromat, oil, short mount, phase
	1.30	0.18	140	34.0	RMS	0.15	V	M022640	Achromat, glycerin
	1.30	0	140	53.0	RMS	0.40	V	M022641	Achromat, oil, Dyson interfer., micros.
	1.30	0.18	140	34.0	RMS	0.15	V	M022645	Achromat, glycerin, phase
	1.30	0.18	140	17.8	....	0.15	V	M022646	Achromat, oil, quick change
	1.30	0.18	140	34.0	RMS	0.10	V	M023501	Semiapo., oil
	1.30	0.18	140	34.0	RMS	0.10	V	M023505	Semiapo., oil, phase
	1.30	0.18	140	34.0	RMS	0.10	V	M023508	Semiapo., oil, phase
	1.18	0.17	151	37.0	RMS	0.11	R	9221008	Achromat, glycerin, iris
	1.18	0.17	151	37.0	RMS	0.11	R	9221011	Achromat, glycerin, phase (anoptral)
97	1.25	0.18	145	35.7	RMS	0.13	BL	31-10-71-02	Achromat, oil
	1.25	0.18	145	35.7	RMS	0.13	BL	31-10-75-02	Achromat, oil, iris
	1.25	0.18	160	33.7	RMS	0.13	AO	C 127	Achromat, oil
98	1.30	0.18	145	35.7	RMS	0.13	BL	31-10-73-02	Semiapo., oil

100	1.40	0.17	160	33.6	RMS	0.1	N	200-65-000	Apo., oil, spring
	1.36	0	∞	45.0	RMS	0.36	L	569 046	Semiapo., met., oil
	1.32	0.17	152	45.0	RMS	0.27	L	519 160	FF apo., oil, spring
	1.32	0	152	37.0	RMS	0.38	L	519 037	Semiapo., oil, spring
	1.30	0.17	160	36.57	RMS	0.14	O	1B 190	Achromat, oil
	1.30	0.17	160	36.57	RMS	0.14	O	1B 192	Achromat, oil, iris
	1.30	0.17	160	36.57	RMS	0.14	O	1B 191	Achromat. oil, spring
	1.30	0.17	160	36.57	RMS	0.11	O	1B 390	Semiapo., oil, iris
	1.30	0.17	160	36.57	RMS	0.14	O	1P 190	Achromat, phase, oil
	1.30	0.17	160	36.57	RMS	0.11	O	1F 190	Semiapo., phase, oil, spring
	1.30	0.17	160	36.57	RMS	0.14	O	1B 290	Achromat, polar., oil
	1.30	0	200	42.3	RMS	0.33	O	1M 190	Achromat, met., oil
	1.30	0	200	42.3	RMS	0.17	O	1M 193	Semiapo., met., oil
	1.30	0	200	42.3	RMS	0.33	O	1M 192	Achromat, met., oil, phase
	1.30	0	200	42.3	RMS	0.33	O	1M 290	Achromat, met., oil, polar.
	1.30	0	200	42.3	RMS	....	O	1M 191	FF achromat, met., oil
	1.30	0.17	152	37.0	RMS	0.13	L	559 042	Achromat, polar., oil
	1.30	....	146	36.5	RMS	0.14	W	5075	Semiapo., oil
	1.30	....	146	36.5	RMS	0.14	W	5076	Semiapo., oil, iris
	1.30	....	146	36.5	RMS	0.35	W	5115	FF semiapo., oil
	1.30	....	146	36.5	RMS	0.14	W	5254	Semiapo., phase, oil
	1.30	0.17	150	45.0	RMS	0.12	Z	46 19 20	Semiapo., oil
	1.30	0.17	150	45.0	RMS	0.09	Z	46 19 40	FF apo., oil
	1.30	0.17	150	45.0	RMS	0.09	Z	46 19 46	FF apo., oil, iris
	1.30	0.17	150	45.0	RMS	0.12	Z	46 19 21	Semiapo., phase, oil
	1.30	0.17	150	45.0	RMS	0.09	Z	46 19 41	FF apo., phase, oil
	1.30	0.17	152	37.0	RMS	0.13	L	519 060	Achromat, oil, spring
	1.30	0.17	152	37.0	RMS	0.13	L	519 045	Achromat, oil, spring, iris
	1.30	0.17	152	45.0	RMS	0.17	L	519 167	Achromat, oil, spring, phase (Zernike)
	1.30	0.17	151	37.0	RMS	0.10	R	9223002	Semiapo., oil, iris
	1.30	0.17	151	37.0	RMS	0.10	R	9223052	Semiapo., oil, iris, polar.
	1.30	0.17	151	37.0	RMS	0.10	R	9223006	Semiapo., oil, iris, phase
	1.30	0.17	151	37.0	RMS	0.10	R	9211401	Apo., oil
	1.30	0.17	160	33.6	RMS	0.12	N	200-35-000	FF achromat, oil
	1.30	0.17	160	33.6	RMS	0.12	N	200-75-000	FF achromat, oil, spring
	1.30	0.17	160	33.6	RMS	0.12	N	200-35-600	FF achromat, oil, spring, polar.
	1.30	0	160	33.6	RMS	0.12	N	201-35-000	FF achromat, oil, met.
	1.30	0.18	140	44.0	RMS	0.15	V	M022611	Achromat, oil
	1.30	0	140	44.0	RMS	0.33	V	M022612	Achromat, oil, met.

TABLE 1 Objectives (Continued)

Power (1)	NA (2)	Cover glass, mm (3)	Tube length, mm (4)	Shoulder height, mm (5)	Thread (6)	Working distance, mm (7)	Mfr. (8)	Catalog No. (9)	Comments (10)
100	1.30	0.18	140	44.0	RMS	0.15	V	M022613	Achromat, oil, polar.
	1.30	0	140	44.0	RMS	0.33	V	M022614	Achromat, oil, polar., met.
	1.30	0.18	140	44.0	RMS	0.15	V	M022615	Achromat, oil, phase
	1.30	0.18	140	44.0	RMS	0.15	V	M022617	Achromat, oil, phase (neg.)
	1.30	0.18	140	44.0	RMS	0.15	V	M022618	Achromat, oil, phase
	1.30	0.18	140	44.0	RMS	0.15	V	M022619	Achromat, oil, phase
	1.30	0.18	140	40.2	....	0.15	V	M022631	Achromat, oil, magn. cent.
	1.30	0.18	140	40.2	....	0.15	V	M022633	Achromat, oil, magn. cent., polar.
	1.30	0.18	140	40.2	....	0.15	V	M022635	Achromat, oil, magn. cent., phase
	1.30	0.18	140	44.0	RMS	0.15	V	M022643	Achromat, oil, iris
	1.30	0.18	140	44.0	RMS	0.10	V	M023511	Semiapo., oil
	1.30	0.18	140	44.0	RMS	0.10	V	M023515	Semiapo., oil, phase
	1.30	0.18	140	44.0	RMS	0.10	V	M023517	Semiapo., oil, phase
	1.30	0.18	140	44.0	RMS	0.10	V	M023518	Semiapo., oil, phase
	1.30	0.18	140	44.0	RMS	0.10	V	M023519	Semiapo., oil, phase
	1.30	0.18	140	14.0	Spec	0.10	V	M023521	Semiapo., oil, short mount
	1.30	0.18	140	40.2	....	0.10	V	M023531	Semiapo., oil, magn. cent.
	1.30	0.18	140	44.0	RMS	0.10	V	M023541	Semiapo., oil, iris
	1.25	0.18	∞	33.7	RMS	0.14	AO	1027	Achromat, iris, TO 182.7 EFL, oil
	1.25	0.18	∞	33.7	RMS	0.14	AO	1079	Achromat, TO 182.7 EFL, oil
	1.25	0.18	∞	33.7	RMS	0.14	AO	1227	Achromat, TO 182.7 EFL, oil, phase
	1.25	0.18	∞	33.7	RMS	0.14	AO	1229	Achromat, TO 182.7 EFL, oil, phase
	1.25	0.17	151	37.0	RMS	0.12	R	9221001	Achromat, oil
	1.25	0.17	151	37.0	RMS	0.12	R	9221002	Achromat, oil, iris
	1.25	0.17	151	37.0	RMS	0.12	R	9221051	Achromat, oil, polar.
	1.25	0.17	151	37.0	RMS	0.12	R	9221005	Achromat, oil, phase (anoptral)
	1.25	0.17	151	37.0	RMS	0.12	R	9221009	Achromat, oil, phase (anoptral), iris
	1.25	0.17	151	37.0	RMS	0.09	R	9224002	FF achromat, oil, iris
	1.25	0.17	151	37.0	RMS	0.09	R	9224052	FF achromat, oil, iris, interfer. contrast
	1.25	0.17	160	33.6	RMS	0.16	N	200-15-000	Achromat, oil
	1.25	0.17	160	33.6	RMS	0.16	N	200-15-50	Achromat, oil, iris
	1.25	0.17	160	33.6	RMS	0.16	N	200-25-000	Achromat, oil, spring

100	1.25	0.17	160	33.6	RMS	0.16	N	200-25-200	Achromat, oil, phase (pos.)
	1.25	0.17	160	33.6	RMS	0.16	N	200-25-201	Achromat, oil, phase (pos.)
	1.25	0.17	160	33.6	RMS	0.16	N	200-25-30	Achromat, oil, phase (neg.)
	1.25	0.17	160	33.6	RMS	0.16	N	202-65-20	Achromat, oil, phase (pos.), iris
	1.25	0	160	33.6	RMS	0.16	N	201-25-000	Achromat, oil, met.
	1.25	0.17	160	33.6	RMS	0.16	N	200-25-600	Achromat, oil, polar.
	1.25	0.17	160	33.6	RMS	0.16	N	200-15-600	Achromat, oil, polar.
	1.25	0.17	160	33.6	RMS	0.16	N	201-25-601	Achromat, oil, met., spring
	1.25	0.17	160	33.6	RMS	0.16	N	200-15-000	Achromat, oil, for type G
	1.25	0	200	36.7	Spec	0.30	BL	42-35-20	FF achromat, met., polar., oil
	1.25	0	200	36.7	RMS	0.30	BL	42-35-55	FF achromat, met., polar., oil
	1.25	0.18	145	43.0	RMS	0.12	BL	31-10-65	FF achromat, oil
	1.25	0.17	150	45.0	RMS	0.09	Z	46 19 00	Achromat, oil
	1.25	0.17	150	45.0	RMS	0.09	Z	46 19 06	Achromat, oil, iris
	1.25	0.17	150	45.0	RMS	0.09	Z	46 19 10	FF achromat, oil
	1.25	0.17	150	45.0	RMS	0.09	Z	46 19 16	FF achromat, oil, iris
	1.25	0.17	150	45.0	RMS	0.09	Z	46 19 01	Achromat, phase, oil
	1.25	0.17	150	45.0	RMS	0.09	Z	46 19 11	FF achromat, phase, oil
	1.25	0.17	150	45.0	RMS	0.09	Z	46 19 08	Achromat, polar., oil
	1.25	0.35	150	45.0	RMS	0.07	Z	46 20 64	UV, glycerin
	1.25	0	150	33.0	Spec	0.25	Z	46 19 69	Achromat, met., DF, oil
	1.25	0	150	33.0	Spec	0.28	Z	46 20 05	Achromat, met., DF, oil, polar.
	1.25	0	150	33.0	Spec	0.25	Z	46 20 29	Achromat, met., phase, oil
	1.25	0	150	33.0	RMS	0.25	Z	46 20 36	Achromat, met., oil
	1.25	0	150	33.0	RMS	0.25	Z	46 20 39	Achromat, met., oil, phase
	1.25	0.17	160	36.57	RMS	0.15	O	1B590	FF achromat, oil, spring
	1.25	0	200	42.3	RMS	0.16	O	1M591	FF achromat, oil, met.
	1.25	0.17	146	36.5	RMS	0.14	W	5025	Achromat, oil
	1.25	0.17	146	36.5	RMS	0.14	W	5026	Achromat, oil, iris
	1.25	0	146	36.5	Spec	0.14	W	8528	Achromat, oil, met., DF
	1.25	0.17	146	36.5	RMS	0.14	W	3154	Achromat, oil, polar.
	1.25	0.17	146	36.5	RMS	0.14	W	5203	Achromat, oil, phase
	1.25	0	156	36.0	RMS	0.25	U	M100X	Achromat, met., oil
	1.25	0	156	36.0	RMS	0.08	U	100X	Achromat, oil
	1.25	0.18	140	44.0	RMS	0.15	V	M025511	FF achromat, oil
	1.25	0	140	44.0	RMS	0.33	V	M025512	FF achromat, oil, met.
	1.25	0.18	140	44.0	RMS	0.15	V	M025517	FF achromat, oil, phase
	1.25	0.18	140	44.0	RMS	0.15	V	M025518	FF achromat, oil, phase
	1.00	0.17	150	45.0	RMS	0.08	Z	47 44 08	Achromat, oil, interfer.

TABLE 1 Objectives (Continued)

Power (1)	NA (2)	Cover glass, mm (3)	Tube length, mm (4)	Shoulder height, mm (5)	Thread (6)	Working distance, mm (7)	Mfr. (8)	Catalog No. (9)	Comments (10)
100	1.00	0.18	145	43.0	RMS	0.12	BL	31-10-50	FF achromat, oil
	0.95	0	∞	45.0	RMS	0.12	L	569 047	Semiapo., met., phase
	0.95	0	∞	45.0	RMS	0.12	L	569 053	Semiapo., met.
	0.85	0.35	150	45.0	RMS	0.12	Z	46 20 63	UV, glycerin
	0.85	0.35	150	45.0	RMS	0.12	Z	46 20 73	UV, glycerin, phase
105	1.32	0	197	14.0	RMS	0.30	L	559 014	Semiapo., met., polar., oil
112	1.32	0	230	20.3	. . . . .	0.28	V	M023732	Apo., oil, magn. cent.
115	1.32	0	∞	24.0	RMS	0.28	V	M023752	Apo., oil, TO 250 EFL
125	1.30	0	241	38.0	RMS	0.23	R	9226003	Semiapo., oil, met.
	1.30	0	241	38.0	RMS	0.23	R	9226053	Semiapo., oil, met., polar.
	1.25	0	241	38.0	RMS	0.26	R	9229003	FF achromat, oil, met.
130	1.30	0	230	20.3	. . . .	0.28	V	M023532	Semiapo., oil, magn. cent.
135	1.30	0	230	20.3	. . . .	0.33	V	M022632	Achromat, oil, magn. cent.
	1.30	0	230	20.3	. . . .	0.33	V	M022634	Achromat, oil, magn. cent., polar.
	1.30	0	230	17.8	. . . .	0.33	V	M022636	Achromat, oil, quick change
140	1.30	0	∞	24.0	RMS	0.39	V	M022652	Achromat, oil, quick change, TO 250 EFL
	1.30	0	∞	24.0	RMS	0.39	V	M022654	Achromat, oil, polar., TO 250 EFL
	1.30	0	∞	24.0	RMS	0.28	V	M023552	Semiapo., oil, TO 250 EFL
	1.25	0	∞	34.0	RMS	0.30	V	M025552	FF achromat, oil, TO 250 EFL
160	1.40	0	∞	45.0	RMS	0.30	L	569 031	FF achromat, oil, met.
	1.40	0	∞	45.0	RMS	0.30	L	569 036	FF achromat, oil, met., phase
	0.95	0	∞	45.0	RMS	0.16	L	569 030	FF achromat, met.
	0.95	0	∞	45.0	RMS	0.16	L	569 035	FF achromat, met., phase
170	0.50	0.18	382	45.0	RMS	2.1	L	520 108	Catadioptric, UV
300	0.85	0.18	382	45.0	RMS	0.4	L	520 109	Catadioptric, UV, glycerin

up to as much as 2 percent in the higher-power apochromats, and suitable selection of eyepiece types is required to obtain optimum correction for CDM in each case.

Column 3. Field-of-view number gives the size of the primary image which the eyepiece can cover. When this is divided by the objective power, it gives the real field on the specimen.

Column 4. Image-to-shoulder distance gives the distance from the primary image to the eyepiece mounting shoulder.

Column 5. Eye relief is the distance from the upper surface of the eyepiece to the exit pupil, where the eye pupil is normally located.

Column 6. Diameter is the diameter of the eyepiece sleeve.

Column 7. Manufacturer is expressed in the same code as in Table 1.

Column 8. The manufacturer's catalog number needs no explanation.

Column 9. Recommended use is highly abbreviated. Reference should be made to the manufacturer's literature where optimum results are important.

Column 10. In this column the chromatic difference in magnification (CDM) is given for most eyepieces. According to the sign convention adopted, CDM is almost always either zero or plus for microscope objectives; hence the CDM of a compensating eyepiece is minus. The back focal lengths (BFL) for most eyepieces manufactured by American Optical Co. are listed in this column instead of listing the eye relief in column 5. In general, the eye relief is slightly longer than the BFL.

TABLE 2 Eyepiece Data *(see text for explanation of abbreviations)*

Power (1)	Type (2)	Field of view number, mm (3)	Image to shoulder distance, mm (4)	Eye relief, mm (5)	Diameter, mm (6)	Mfr. (7)	Catalog No. (8)	Recommended use (9)	Comments (10)
4	Comp. plano	25.0	9.0	9	23.2	R	9250102	Photo	CDM −1.0%
	Huygens	20.0	20.3	22	23.2	V	M001501	....	CDM 0%
	Compensating	19.0	20.3	16	23.2	V	M001519	....	CDM −1.0%
5	Huygens	30.0	20.3	11	29.1	V	M606501	Stereo micros.	CDM 0%
	Huygens	25.0	20.3	10	29.1	V	M045005	Polar. micros.	CDM 0%
	Compensating	21.4	11.0	16	23.2	BL	31-05-70	Apochromats	CDM −0.7%
	Huygens	21.0	13.0	13	23.2	O	2C 101	Achromats	CDM 0%
	Comp. plano	21.0	9.0	18	23.2	R	9250202	Photo	CDM −1.0%
	Huygens	20.6	11.0	25	23.2	BL	31-15-08	Achromats	CDM +0.1%
	Hyperplane	20.1	11.0	16	23.2	BL	31-05-40	Achromats, semiapo	CDM −0.8%
	Compensating	20.0	10.0	9	23.2	Z	46 37 10	Achromats	CDM −0.9%
	Huygens	20.0	20.3	10	23.2	V	M045000	Polar. micros.	CDM 0%
	Comp. plano	19.0	9.0	18	23.2	R	9250201	....	CDM −1.0%
	Periplan	19.0	14.0	8	23.2	U	P 5X	....	CDM −0.9%
	Huygens	18.0	20.3	14	23.2	V	M808503	UV micros.	Quartz
	Huygens	11.1	11.3	....	23.2	AO	C 133	....	BFL 28.6
6	Huygens	19.0	18	....	23.2	L	519 034	Achromats	
	Huygens	18.0	14	11.6	23.2	W	5502	Achromats	
	Compensating	18.0	14	11.9	23.2	W	5550	Achromats, semiapo	
	Complanat	18.0	14	11.8	23.2	W	5660	Achromats, semiapo	Photographic
	Huygens	18.0	20.3	11	23.2	V	M0401000	....	CDM 0%
	Compensating	17.0	20.3	11	23.2	V	M041100	....	CDM −1.0%
6.3	Periplanatic	28.0	10	....	30	L	519 131	Achromats, semiapo	Extra large field
	Huygens	21.5	18	....	30	L	559 001	Achromats	Adjustable eye lens
	Comp. plano	19.0	9	16.6	23.2	R	9250301	....	CDM −1.0%
	Compensating	18.0	9	12.0	23.2	R	9254301	....	CDM −1.0%
	Huygens	18.0	18	....	23.2	L	519 180	Achromats	
	Huygens	18.0	18	....	23.2	L	519 182	Achromats	Adjustable eye lens
	Periplanatic	18.0	18	....	23.2	L	519 188	Achromats	Adjustable eye lens
	Compensating	18.0	10	9.0	23.2	Z	46 38 10	Achromats	CDM −0.9%
6.4	Huygens	19.4	11.0	16	23.2	BL	31-15-06	Achromats	CDM −0.2%

7	....	25.0	20.3	10	29.1	V	M044205	Stereo micros.	CDM 0%
	....	22.0	20.3	10	29.1	V	M007506	Polar. micros.	CDM 0%
	Photo	18.5	13	12.5	23.2	O	2C201	Achromats, semiapo	CDM –0.7%, photomicro.
	Huygens	18.0	13	10	23.2	O	2C110	Achromats	CDM 0%
	Huygens	18.0	20.3	10	23.2	V	M007508	Polar. micros.	CDM 0%
	Huygens	18.0	20.3	10	23.2	V	M808508	UV micros.	Quartz
7.5	Hyperplane	18.7	11.0	9	23.2	BL	31-05-42	Achromats, semiapo	CDM –0.7%
	Compensating	17.5	11.0	10	23.2	BL	31-05-72	Apochromats	CDM –1.8%
	Huygens	17.1	11.0	11	23.2	BL	31-15-07	Achromats	CDM –0.3%
	Huygens	15.4	11.0	11	23.2	BL	31-15-17	Achromats	Adjustable eye lens, CDM –0.1%
8	KPL	18.0	10.0	9	23.2	Z	46 39 20	FF, apochromats	CDM –1.4%
	KPL Br	18.0	10.0	17	23.2	Z	46 39 22	FF, apochromats	CDM –1.4%
	Huygens	18.0	20.3	9	23.2	V	M040700		
	Comp. plano	17.5	9.0	14	23.2	R	9250401	....	CDM –1.0%
	Compensating	16.0	10.0	8	23.2	Z	46 39 10	Achromats	
	Compensating	15.0	20.3	8	23.2	V	M041700	....	CDM –1.0%
10	Periplanatic	24.0	10	....	30	L	519 133	Achromats, semiapo	Extra large field
	Periplanatic	24.0	10	....	30	L	519 234	Achromats, semiapo	Extra large field, adj. eye lens
	KPL	20.0	10	7	23.2	Z	46 40 20	FF, apochromats	CDM –1.4%
	Kellner	20.0	20.3	13	29.1	V	M044208	Stereo micros.	CDM 0%
	Kellner	20.0	20.3	13	29.1	V	M045305	Polar. micros.	CDM 0%
	Wide-field	20.0	11.3	....	23.2	AO	134	....	BFL 33.6 mm
	Wide-field	20.0	11.3	21	23.2	AO	145		
	Wide-field	18.5	11.3	....	23.2	AO	176	....	BFL 35.5
	Wide-field	18.5	10.8	....	23.2	AO	437	....	BFL 35.5
	Wide-field	18.44	11.3	....	23.2	AO	146	....	
	Wide-field	18.4	11.0	25	23.2	BL	31-05-23	FF achromats	CDM –0.1%, for nonzoom bodies
	Wide-field	18.4	11.0	24	23.2	BL	31-05-60-01	Achromats	CDM +0.3%
	Periplanatic	18.0	18	....	23.2	L	519 128	Achromats, semiapo	
	Periplanatic	18.0	18	....	23.2	L	519 126	Achromats, semiapo	Adjustable eye lens
	Wide-field	18.0	13	26	23.2	O	2C320	Achromats, semiapo	CDM –0.7%
	Compensating	18.0	13	19	23.2	O	2C321	Apochromats	CDM –1.0%
	Wide-field	18.0	11.0	25	23.2	BL	31-05-22	FF achromats	CDM –1.0%, for zoom bodies
	KPL Br	18.0	10.0	20	23.2	Z	46 40 42	FF, apochromats	CDM –1.4%
	Periplanatic	18.0	18	....	23.2	L	519 137	Achromats, semiapo	
	Wide-field	18.0	11.3	....	23.2	AO	161A	....	BFL 39.3
	Comp. plano	18.0	9.0	21	23.2	R	9253501	Spectacle wearers	
	Complanat	18.0	20.3	20	23.2	V	M041301		
	Wide-field	18.0	14.0	19	23.2	U	WFH10XR	....	CDM –0.8%, focusable eye lens

TABLE 2 Eyepiece Data (Continued)

Power (1)	Type (2)	Field of view number, mm (3)	Image to shoulder distance, mm (4)	Eye relief, mm (5)	Diameter, mm (6)	Mfr. (7)	Catalog No. (8)	Recommended use (9)	Comments (10)
10	Hyperplane	16.7	11.0	8	23.2	BL	31-05-44	Achromats, semiapo	CDM −0.5%
	Compensating	16.0	10.0	....	23.2	Z	46 40 10	Achromats	CDM −0.9%
	Compensating	16.0	9.0	8.5	23.2	R	9254501	....	CDM −1.0%
	Huygens	16.0	20.3	8	23.2	V	M045300	Polar. micros.	CDM 0%
	Kellner	16.0	20.3	12	23.2	V	M042308	Photovisual	CDM 0%
	Huygens	16.0	20.3	8	23.2	V	M040300	....	CDM 0%
	Compensating	16.0	20.3	14	23.2	V	M001525	Spectacle wearers	CDM −1.0%, long eye relief
	Huygens	16.0	20.3	8	....	V	M808516	UV micros.	Quartz
	Comp. plano	15.5	9.0	11.9	23.2	R	9250501	....	CDM −1.0%
	Huygens	15.1	11	11	23.2	BL	31-05-24	FF achromats	CDM +0.3%
	Periplanatic	15.0	18	....	23.2	L	519 135	Achromats, semiapo	
	Periplanatic	15.0	18	....	23.2	L	519 134	Achromats, semiapo	Adjustable eye lens
	Huygens	14.8	11	8	23.2	BL	31-15-09	Achromats	CDM +0.1%
	Compensating	14.8	11	9	23.2	BL	31-05-75	Apochromats	CDM −1.0%
	Periplanatic	14.6	18	....	23.2	L	519 170	Achromats, semiapo	Adjustable eye lens
	Periplanatic	14.6	18	....	23.2	L	519 122	Achromats, semiapo	Adjustable eye lens
	Huygens	14.4	11	8	23.2	BL	31-15-18	Achromats	CDM +0.1%, adjustable eye lens
	....	14.0	18	....	23.2	L	519 221	Achromats	
	High-point	13.4	11	26	23.2	BL	31-15-14	Achromats	CDM −0.3%, spectacle wearers
	Huygens	13.0	13	7.3	23.2	O	2C120	Achromats	
	Photo	13.0	13	6	23.2	O	2C220	Achromats, semiapo	CDM −0.7%
	Compensating	13.0	20.3	6	23.2	V	M041300	....	CDM −1.0%
	Periplan	12	14.0	8	23.2	U	P10X	....	CDM −0.5%
	Kellner	12	14.0	13	23.2	U	Ke 10XR	....	CDM 0%, focusable eye lens
	Huygens	11.1	11.3	....	23.2	AO	C 142	....	BFL 16.7
	Huygens	10.4	11.3	....	23.2	AO	160A	....	BFL 16.0
	Huygens	10.1	11.3	....	23.2	AO	177	....	BFL 16.0
12.5	KPL Br	18.0	10.0	15	23.2	Z	46 41 42	FF, apochromats	CDM −1.4%
	Kellner	16.0	20.3	11	29.1	V	M606516	Stereo micros.	CDM 0%
	Wide-field	15.0	14.0	20	23.2	U	WF 12.5X	Spectacle wearers	CDM −0.4%, long eye relief
	Comp. plano	15.0	9.0	17.8	23.2	R	9250601	....	

12.5	Hyperplane	14.3	11.0	5	23.2	BL	31-05-46	Achromats, semiapo	CDM −0.2%
	Huygens	13	20.3	8	23.2	V	M001521	....	CDM 0%
	Compensating	12.9	11.0	20	23.2	BL	31-05-86	Apochromats	CDM −1.5%
	Huygens	12.8	11.0	6	23.2	BL	31-15-12	Achromats	CDM +0.2%
	Compensating	12.5	10	6	23.2	Z	46-41-10	Achromats	
	KPL Br	12.5	10	16	23.2	Z	46-41-20	FF, apochromats	CDM −1.4%
	Compensating	12.0	20.3	15	23.2	V	M001529	Spectacle wearers	CDM −1.5%, long eye relief
	Compensating	11.0	20.3	6	23.2	V	M001538	....	CDM −1.0%
14	....	15.0	20.3	14	29.1	V	M044405	Stereo micros.	CDM 0%
15	Wide-field	17.2	11.0	16.5	23.2	AO	184		
	Wide-field	16.8	11.3	....	23.2	AO	147	....	BFL 44.2
	Wide-field	16.8	11.0	12	23.2	BL	31-05-62	Achromats	CDM +0.5%
	Wide-field	14.0	14.0	20	23.2	U	WF15X	Spectacle wearers	CDM −1.1%, long eye relief
	Kellner	13.0	20.3	8	23.2	V	M001546	....	CDM 0%
	Kellner	13.0	20.3	8	29.1	V	M606521	Stereo micros.	CDM 0%
	Huygens	11.4	11.0	5	23.2	BL	31-15-15	Achromats	CDM +0.2%
	Hyperplane	11.3	11.0	14	23.2	BL	31-05-48	Achromats, semiapo	CDM −0.8%
	Compensating	11.2	11.0	15	23.2	BL	31-05-88	Apochromats	CDM −1.4%
	Compensating	11.0	20.3	11	23.2	V	M001533	....	CDM −1.5%
	Huygens	11.0	20.3	6	23.2	V	M001523	....	CDM 0%
	Kellner	9.0	14.0	11	23.2	U	Ke 15X	....	CDM +0.3%
	Kellner	9.0	14.0	11	23.2	U	Ke 15XR	....	CDM +0.3%, focusable eye lens
16	Comp. plano	12.0	9.0	13.7	23.2	R	9250701	....	CDM −1.0%
	KPL Br	10.0	10.0	11	23.2	Z	46-42-20	FF, apochromats	CDM −1.4%
20	Wide-field	12.2	11.3	....	23.2	AO	157	....	BFL 16.9
	Wide-field	12.0	11.0	11	23.2	BL	31-05-63	Achromats	CDM 0%
	Wide-field	9.0	14.0	9	23.2	U	WF20X	....	CDM −0.4%
	Wide-field	8.5	20.3	11	29.1	V	M044604	Stereo micros.	CDM 0%
	Compensating	8.5	20.3	11	23.2	V	M041602	....	CDM −1.5%
	Hyperplane	8.4	11.0	10	23.2	BL	31-05-50	Achromats, semiapo	CDM −0.7%
	KPL Br	8.0	10.0	8.5	23.2	Z	46 43 20	FF, apochromats	CDM −1.4%
	Compensating	6.2	14.0	10	23.2	U	K20X	....	CDM −1.0%
	Compensating	6.2	14.0	10	23.2	U	K20XR	....	CDM −1.0%, focusable eye lens
25	Wide-field	12.2	10.8	14.4	23.2	AO	186		
	Comp. plano	7.7	9.0	8	23.2	R	9250901		
	Compensating	6.4	7.9	9	23.2	BL	31-05-90	Apochromats	CDM −1.3%
	KPL Br	6.3	10.0	6.5	23.2	Z	46 44 20	FF, apochromats	CDM −1.4%

Section 7

Properties of Optical Materials

WILLIAM L. WOLFE
University of Arizona

(Numbers refer to paragraphs)

1. Introduction The field of optical materials is immense. Because optical materials can be defined as any and all substances that interact with light, this chapter could be the complete description of all the properties of all materials. However, no such universal catalog is found here. We could greatly restrict the field by defining optical materials as those substances which are used in optical systems for "handling" photons, but the subject would still be quite large. For instance, photographic film fits into this more restrictive definition. For these reasons, this section does not include materials used as intentional transducers of light (or optical radiation). Nor does it include polarizing materials, optically active devices, modulators, or the properties of air and other gases.

The principal subject of this section is the description of the properties of materials useful for lenses, windows, mirrors, and radiometric or photometric standards.

SPECIAL CRYSTALS AND GLASSES

2. Paragraphs 2 to 7 describe special materials used in the infrared or ultraviolet spectral regions or for some unusual purposes in the visible region.

3. Absorption, Reflection, and Transmission When light is incident on a plane-parallel plate of material, some is reflected, some is absorbed, and some is transmitted, as shown in Fig. 1.†

In general, the light is reflected many times from side to side. If the sum of all light is considered,

$$I_0 = I_R + I_A + I_T$$

$$1 = \frac{I_R}{I_0} + \frac{I_A}{I_0} + \frac{I_T}{I_0} = \rho + \alpha + \tau$$

where I_0 = incident flux density‡
I_R = reflected flux density
I_A = absorbed flux density
I_T = transmitted flux density
ρ = reflectance
α = absorptance
τ = transmittance

Specifically, τ is the external transmittance, a measurable quantity. The ending *-ance* is used to indicate that these are properties of a given sample. If the sample does not scatter but only absorbs and reflects, the relationship between I_1 and I_2 is exponential:

$$\frac{I_2}{I_1} = \tau_{\text{int}} = e^{-ax}$$

where a = absorption coefficient
x = sample thickness
τ_{int} = internal transmittance

Then it follows that

$$\tau = \frac{(1-\rho)^2 e^{-ax}}{1-\rho^2 e^{-2ax}}$$

A plane wave incident normally on a flat, specular, absorbing surface will have the reflectivity given by Fresnel

$$\rho = \left(\frac{n - ik - 1}{n - ik + 1}\right)^2$$

where k = extinction coefficient $= 4\pi a/\lambda = 4\pi a/n\lambda_0$
λ = wavelength of light in medium
n = refractive index of medium
λ_0 = wavelength of light in vacuum

The substitution of these equations into each other leads to complicated expressions omitted here. Transmission data given in this paragraph are usually

† In this section all the figures and tables follow the text.

‡ The recommended symbol for incident flux density E is not used here to avoid confusion with the electromagnetic field strength.

external transmittance; the few that are internal transmittance are so indicated. Absorption-coefficient data are for a, as in the equations above. Some extinction coefficients are given as defined above and not the quantity determined by

$$\tilde{n} = n(1 - ik)$$

Data are representative, but they are usually for a single sample.

External-transmittance data are usually those found experimentally with a double-beam spectrophotometer, or they are corrected data from a single-beam system. Absorption-coefficient data are usually obtained by measuring the transmittance of two samples:

$$\tau_1 = \frac{(1 - \rho_1)^2 e^{-a_1 x_1}}{1 - \rho_1{}^2 e^{-2a_1 x}} \qquad \tau_2 = \frac{(1 - \rho_2)^2 e^{-a_2 x_2}}{1 - \rho_2{}^2 e^{-2a_2 x_2}}$$

If the reflectances and absorption coefficients are the same ($\rho_1 = \rho_2$ and $a_1 = a_2$), then

$$a = \frac{\ln \tau_1 - \ln \tau_2}{x_1 - x_2}$$

but only if

$$\frac{1 - \rho^2 e^{-2ax_2}}{1 - \rho^2 e^{-2ax_1}} = 1$$

This correction factor can be accounted for by checking with the original equations.

A more elegant way to obtain an absorption coefficient is through the measurement of emittance. From the expressions given by McMahon (1950), sample emittance is given by

$$\epsilon^* = \frac{(1 - \rho)(1 - \tau_{\text{int}})}{1 - \rho\tau_{\text{int}}}$$

This equation for small a reduces to the simple relationship

$$a = \frac{\epsilon^*}{x}$$

Thus emittance data can be converted into absorption-coefficient data by dividing by the thickness.

Another method should be mentioned although the data reported here were not obtained in this way. The sample is illuminated with a single beam from a laser. The rise in temperature of the sample is related to the amount of energy absorbed and the thermal conductance, convective losses, and reradiation by the sample. The method is most useful for large input in a short time.

4. Refractive Index For a nonabsorbing medium the refractive index n is given simply as the ratio of the velocity of light in vacuum c to that in the sample v

$$n = \frac{c}{v}$$

The relative refractive index is the ratio of the index of one medium to that of another; for instance,

$$n_{\text{rel}} = \frac{n_{\text{mat}}}{n_{\text{air}}}$$

This is the refractive index of a material relative to that of air. All the data in this paragraph are refractive indexes of the material relative to the refractive index of air rather than to that of vacuum. Most values of n have been obtained by the *minimum-deviation method* and are accurate to the fifth decimal place. Of course materials vary from sample to sample, and these data are only representative.

For anisotropic materials, data are given for the direction of principal interest within the crystal.

Data for the change in refractive index with temperature are sometimes obtained by measuring the index at different temperatures and sometimes by interferometric measurements.

The dispersion curves are usually obtained by fitting the data to a Cauchy, Sellmeier, or Herzberger equation. Least-squares fits are generally used.

5. Elastic Coefficients Hooke's law states that for small deformations the strain on a solid is proportional to the stress. The constants of proportionality in the different directions are the elastic stiffness constants c_{hk}, which have units of pressure. Table 58 gives them in bars. Temperatures are given when they are known; if no temperature is given one can *assume* the measurement was made at room temperature.

6. Elastic Moduli Table 59 includes data for Young's modulus, the modulus of rigidity, the bulk modulus, the modulus of rupture, and the apparent elastic limit. The units are all pounds per square inch. *Young's modulus* is usually defined as the ratio of stress to strain. Here the stress is the force per unit normal area and the strain is the fractional change in length. *Modulus of rigidity* is the tangential force per unit area divided by the angle of shear in radians. *Bulk modulus* is the ratio of pressure applied to the body and the resulting change in volume. The *apparent elastic limit* is the force per unit area for which stress is no longer proportional to strain. For most optical materials this is only an approximate value.

7. Thermal Properties It generally requires different amounts of energy to increase equal masses of different materials by the same temperature increment. The energy required to raise a unit mass of material 1 degree is called the *specific heat* or *specific-heat capacity*. In the cgs system, the units are cal/(g)(°C). In the SI system they are J/(kg)(°C). In the English system they are Btu/(lb)(°F). The heat capacity can be defined thermodynamically as

$$C_v = \left(\frac{dQ}{dT}\right)_v \quad \text{or} \quad C_p = \left(\frac{dQ}{dT}\right)_p$$

$$c_v = \frac{1}{m} C_v \quad \text{and} \quad c_p = \frac{1}{m} C_p$$

The heat capacity at constant volume is the change in energy of a substance for a given change in temperature when the substance is held at constant volume. The heat capacity at constant pressure is the change in heat divided by the change in temperature at constant pressure. The corresponding specific heats normalize these to unit mass, as indicated. In practice, all specific-heat values for optical materials are at constant pressure, not constant volume. Thus, symbolically, all measurements given herein are for c_p. Since specific heats are by definition equal in the cgs, mks [in which one uses kcal/(kg)(°C)], and English systems, no units are specified in Tables 60 and 61. To obtain the units for the SI system one uses 1 J = 0.23833 cal.

When heat is applied to one end of a bar of material, the heat is conducted by the bar along its length (ignoring end affects). The thermal conductivity of the material is a measure of its tendency to allow this heat to flow. The defining equation is

$$\frac{dQ}{dt} = kA \frac{dT}{dx}$$

where Q = heat energy
t = time
A = cross-sectional area
T = temperature
x = length coordinate of bar
k = thermal conductivity

The most common units for thermal conductivity are cal/(s)(cm)(°C) and

Btu/(in)(h)(R); another useful unit is W/(m)(K). The relationships between them are

$$1 \text{ Btu/(in)(h)(R)} = 20.75 \text{ W/(cm)(K)} = 178.6 \text{ cal/(s)(cm)(K)}$$

When heat is applied uniformly to a bar of material, the bar changes its length. The change in length may be divided by the original length to obtain a relative change. This relative change in length per degree temperature change is the *coefficient of linear expansion,* usually designated α:

$$\alpha = \frac{1}{L}\frac{\partial L}{\partial T}$$

It can be shown that the bulk expansion, usually designated by β, is just 3 times α. The linear expansion can be expressed as a series:

$$\alpha = A \times 10^{-6} + B \times 10^{-8}T + C \times 10^{-11}T^2$$

The coefficient A corresponds to the linear change of the coefficient of linear expansion and has the dimensions of reciprocal temperatures; B and C have units of reciprocal square and reciprocal cube of temperature, respectively. These coefficients are given in Table 61.

8. Hardness Values of hardness for several materials are given in Table 62. Knoop values with the indenter aligned in either the [100] or the [110] direction are tabulated, and the indenter load is given when known. Where Knoop values were not available, Mohs or Vickers values are given.

9. Permittivity, or Dielectric Constant The dielectric constant is a variable depending upon frequency and temperature, among other things; permittivity may be a better name. The displacement D is related to the electric field strength E by the permittivity; $D = \epsilon E$. For free space $D = \epsilon_0 E$, and for most materials $\epsilon = \epsilon_r \epsilon_0$. The data presented in Table 63 are relative data, ϵ_r.

10. Solubility, Molecular Weight, and Specific Gravity The solubility of an optical material is a measure of its tendency to be attacked by water or other chemicals. When a sample is immersed in water long enough to establish a saturated solution, the sample's difference in weight (before and after) is a measure of the amount of solute, which is specified as the number of grams of the material in 100 g of water (100 ml). Many substances with values of the parameter less than 10^{-3} have been listed as insoluble in Table 64.

Specific gravity is the ratio of the density of a substance to that of water. Since water has a density of about 1 g/ml, or 62.5 lb/ft^3, one can easily calculate the density in either system of units, but specific gravity is dimensionless.

BLACK AND WHITE STANDARDS

11. The design and construction of precise sources for radiometric calibration and similar applications require the use of highly emitting, relatively diffuse materials, generally called *blacks*. Data should include not only the spectral specular emittance but also the integrated emittance and some information on goniometric properties. Unfortunately, these are not all available. Data on the quality of various blacks are further complicated by how thick the black really is and in some cases how it was deposited. In Figs. 190 through 201 and Tables 65 through 68 we try to give representative data.

OPTICAL PROPERTIES OF CHALCOGENIDE GLASSES†

12. In general, the normal oxide-based optical glasses do not transmit beyond 3 to 5 μm because of the strong absorption of chemical bonds formed with the

† Par. 12 was written by A. R. Hilton, Texas Instruments, Dallas, Tex.

light chalcogen element, oxygen. The first attempt to extend the wavelength range of glasses using the heavier chalcogen element sulfur as a composition base was carried out by Frerichs (1950, 1953) in his studies of As-S glasses. Development of useful optical materials based on the As-S system was accomplished by Fraser and Jerger (1953) of the Servo Corporation in the early 1950s. The studies of chalcogenide glasses based on the heavier chalcogen elements Se and Te began in Russia ("The Structure of Glass," 1960) and the United States (Black et al., 1957) in the late 1950s as an outgrowth of exploratory programs searching for new semiconducting materials useful for thermoelectric applications. Optical-materials investigations began in earnest in the early 1960s in the United States (Billian and Jerger, 1963; Hilton and Brau, 1963; Hilton et al., 1964, 1966; Jerger and Sherwood, 1964), Great Britain (Nielson, 1962; Edmond et al., 1963; Savage and Nielson, 1965), and Russia (Kolomiets, 1964; Myroller et al., 1962). Most of the results have been qualitative, and only a few compositions have been prepared in the amount required for quantitative evaluation.

In Table 69 are listed the qualitative results obtained for glasses made from systems based on the elements listed in the first column. Glasses are evaluated for softening point, refractive index, and suitability for application in the wavelength regions 3 to 5 μm and 8 to 14 μm. The results obtained in this country (Jerger and Sherwood, 1964; Jones and Hafner, 1969; Hilton and Jones, 1967; Rosberry, 1968) in Great Britain (Gornsworth, 1967), and Russia (Kokorina, 1967) indicate that at present, Ge-As-Se is the glass system that provides the best infrared optical materials.

Table 70 contains a list of the properties measured for specific glass compositions. These glasses represent the best composition for their particular system. Also included are other infrared optical materials for comparison: Si and Ge for crystalline semiconductors, NaCl for the alkali halides, Irtran 4 for the hot-pressed materials, and oxide-based optical glasses.

Table 70 also lists chalcogenide glasses that have good infrared transmission out to 20 μm. The glass with apparently the best overall physical and optical properties is the Ge-As-Se glass, the only addition to this table since it was first prepared and published (Hilton, 1966).

There are only four potential suppliers (excluding As_2S_3) of infrared transmitting glasses: the Servo Corporation of Hicksville, New York (Jerger, 1969), Texas Instruments of Dallas, Texas (Jones, 1969), British Drug House, Ltd., of Poole, Dorset, England (Gornsworth, 1967), and a production factory in Leningrad (Kokorina, 1967).

The two most promising glasses, according to literature from Texas Instruments (Jones and Hafner, 1969), are described in Pars. 3 to 10 for the standard material properties (alphabetical by material in the appropriate paragraph).

VISIBLE-SPECTRUM GLASSES†

13. No lens designer ever works without some constraints. In addition to a specific level of optical performance, a lens must be transparent and of reasonable size and weight. Generally it should cost no more than necessary, and it must be manufacturable. While modern computer programs enable the designer to exploit every degree of freedom, his capability is limited by the sizes and types of glass at his disposal. Some of the glasses he would like to have simply do not exist. More commonly, no glass is precisely reproducible, and uncertainties in its characteristics put a practical limit on the final product.

Many glasses with diverse optical and physical properties have been developed and can be found listed in manufacturers' catalogs. Only the 28 types most commonly used in the design of moderate-aperture lenses, selected by a special NATO committee, are described here. They are available from most suppliers of optical glass.

† Par. 13 was written by R. A. Buchroeder, Optical Sciences Center, University of Arizona.

Earlier, when glass types were far less numerous, the terms *crown* and *flint* were sufficient. The creation of a variety of glasses resulted in the inclusion of the chemical that gave the glass its special character. Eventually the number became so large that numerical suffixes were added, generally in chronological order. Consequently, a move to switch to a six-digit code has been made in order to restore some meaning to glass names.

The six-digit code is composed of two parts. The first three digits describe the nominal index of refraction, the last three the Abbe number or V number (or ν number), which is the reciprocal of the dispersive power

$$V = \frac{N_M - 1}{N_S - N_L}$$

where N_M = refractive index at mean wavelength
N_S = refractive index at short wavelength
N_L = refractive index at long wavelength

Thus, one may actually design a lens with nothing more than the six-digit code. Except for certain glasses that differ only in partial dispersion, the process of computer retrieval can be simplified to a numerical code:

$$N_d = 1.5168 \qquad W = 64.17 \qquad 517642 \text{ (BK-7)}$$

Traditionally glasses were intended for the visible spectrum; thus glass catalogs list their glasses in terms of a nominal index of refraction and a relative dispersion in the visible region. The common nominal index is N_d or N_D at wavelengths of 587.6 or 589.3 nm. The reciprocal dispersive power, or Abbe number, is given in terms of the indexes N_F and N_C at 486.1 and 656.3 nm, respectively. Many designers prefer to work at other wavelengths; nonetheless, glasses are cataloged in this popular scheme

$$V_d = \frac{N_d - 1}{N_F - N_C} \qquad \text{Abbe number}$$

For the visible region, the Corning and Schott $n\nu^2$ plots are shown in Figs. 210 and 211, respectively.

The partial dispersion of an optical material is a measure of how much its refractive index changes in a small spectral region compared with the total region of design. For the popular Abbe value V_d that makes use of the F, d, and C lines, one partial dispersion is

$$P_{Fd} = \frac{N_F - N_d}{N_F - N_C}$$

The secondary color that exists in a well-corrected achromat is due to the imperfect matching of the nonlinear dispersion curves of the two elements. For a thin achromatic doublet and an object at infinity, its value is given by

$$\text{Secondary spectrum} = \text{focal length}\ \frac{P_a - P_b}{V_a - V_b}$$

where P_a and P_b are the partial dispersions of lenses a and b and V_a and V_b are their Abbe numbers. Typically, the secondary spectrum for normal glasses will be the focal length divided by 2,500. Table 72 permits accurate determination of this.

The manufacturer tests his glasses and rates them according to resistance to alkaline and acidic solutions. Among common types, flints are most affected by acidic solution and may acquire an interference film or even be etched. Crowns are more susceptible to alkaline solution, and in time, usually over several years, may acquire a chalky coat that lowers transmission and scatters light. Ordinarily, flints are more troublesome to process since they are more stainable as well as being soft and subject to scratching. While coatings usually alleviate the problem, it is always desirable to use a durable glass since coatings may develop pinholes, subjecting the glass to attack. Many designers incorporate a durable glass in the exposed part of the lens to act partly as a protective window.

A class 0 stain (Table 71) indicates practically no stain even after 100 h exposure to a standard acid solution. The stain scale ascends to class 5, an indication that stain will occur in less than 12 min.

Phosphate crowns and short flints are particularly troublesome, often resisting efforts to give them a durable coating. Some of them stain in a manner that is not amenable to specification by the standard testing procedures.

Because of their complexity, their instability in the glassy matrix, or simply because of their viscosity in the molten state, some glasses are difficult to produce without some inclusions like bubbles and stria. A class 0 bubble designation means that the total cross section of all apparent inclusions will not exceed 0.029 mm^2 in a volume of 100 cm^3; this value is 2 mm^2 for class 5.

Density is specified in terms of the dimensionless quantity specific gravity. For most of the optical glasses, transmission at wavelengths longer than 400 nm is relatively invariant with wavelength. Table 71 gives internal transmittance for a wavelength of 350 nm and a thickness of 5 mm.

The cost of optical glass is due only in part to the cost of the ingredients; the difficulty of manufacture (hence quality and availability) is perhaps the major component of the price. In practice it might be said that the quality of a glass is inversely proportional to its price. The prices in Table 71 are only representative. The glass is Grade A, for lots under 199 lb. Thus, the prices are for best quality, nonselect grades of glass.

METALS SOMETIMES USED FOR MIRROR SURFACES

14. The subject of specially reflecting mirror surfaces is a very detailed one. Many mirrors are now coated with multilayer dielectrics or metals and dielectrics. Some still use evaporated coatings of aluminum, silver, gold, etc. In Figs. 215 through 222 and Tables 72 and 73 are given the optical properties of a variety of materials that might be considered for this application. The properties of all these materials are functions of the evaporation conditions.

OPTICAL CEMENTS

15. Lenses and other components are often bonded together by glue, epoxy, or cement. Ishmuratova and Sergeyev (1967) present the situation on infrared glues.

We have investigated the transmission of polymeric films, prepared by polymerization or polycondensation, in the 1–13 μm range. In addition to experimental studies, we collated literature data. In all, 70 polymers were investigated, including ethylene-, vinyl, styrene, acrylic, and methacrylic plastics, as well as ester-, amino-, urethane-, and silicone-based resins, as well as natural resins and solvents. Just as expected, at film thicknesses of 10 to 15 μm many of these compounds exhibited strong absorption in the 2.8 to 3.5 μm, 5 to 6 μm, and 6.5 to 8 μm regions.

The hydroxyl, methyl, ethyl, carboxyl, and ester groups of many high molecular weight compounds absorbed in this region. In other words, none of the polymers we investigated, nor any of the existing optical cements (balsam, balsamine, acrylic and OK-50) can be used as ir-transmitting cements.

The cementing layers were deposited on polished sodium chloride or potassium bromide crystals. The measurements were made on an IKS-12 spectrophotometer after the cement had completely hardened. The cement transmissions [in Figs. 220 through 223 show] that the cements are transparent only below 2.8 μm.

Better results were obtained with an adhesive solution of an organosilicone resin. It wets the crystal surface well, spreading uniformly over it. The initial viscosity of this cement is 0.2 to 0.28 N-sec/m^2 at 20°C; the refractive index of the cold cement is 1.514.

[Figure 224 shows] the transmission spectrum of a 10 μm thick, dry, organosilicone resin film in the 1 to 11 μm range. The figure and appropriate calculations show that the total transmission amounts to 82.1% in the 1 to 8 μm region. At 3.4 μm, there is a narrow absorption band where the transmission drops to 63%.

There is a second band at 8 μm (transmission of 36%). The hardened adhesive layer has the physicomechanical properties listed [in Table 73].

The organosilicone resin solution can be stored indefinitely at room temperature in a closed container and does not require any preparation prior to use. For joining components a drop of the solution is placed on the surface of the flint lens, which is covered with crown. The excess cement, as well as air bubbles, are removed via the normal procedures.

Preliminary and sparse material, published by Packard (1969), is summarized in Figs. 225 and 226.

PROPERTIES OF MIRROR-BLANK MATERIALS†

16. The requirements of mirror materials are all related to the necessity of maintaining a reflecting surface unchanged in shape in the presence of a gravitational field of uncertain orientation and serious thermal gradients and transients. It would be pleasant if mirror-blank materials had high reflectance. However, the reflecting surface is usually deposited as a thin layer (metallic) or a series of layers (dielectric, metal dielectric, or metal-dielectric stack). The mirror blank must furnish a smooth backing for this layer or layers. The ideal mirror blank is perfectly smooth, perfectly stable, infinitely rigid, and has zero coefficient of thermal expansion.

17. Smoothness The requirement that a mirror blank accept a very smooth polish has led most people to consider only noncrystalline solids. Until recently vitreous (or glassy) materials were the only candidates. The development of an amorphous coating (Kanigen; General American Transportation Company) that can be added to metal surfaces has now made it possible to use almost any material.

Little ordering is possible in choosing a material for smoothness. Any solid that does not contain voids can be made sufficiently smooth if enough time is spent on polishing. Applications and systems for shorter waves, e.g., ultraviolet wavelengths, need longer polishing periods to produce smoother, low-scattering surfaces.

18. Stability Once an optical surface has been fabricated, it is most important that the mirror blank not change shape for any reason. The most common source of instability is the release of internal stresses left by production or forming processes.

Vitreous materials, which are usually transparent, are free of internal stress when properly annealed. These are made by manufacturers who have learned that high purity and careful annealing are indispensable in producing glass for refractive elements. In addition, there is a good test for detecting internal stress in transparent materials. Birefringence is produced, which is proportional to the amount of stress and the thickness of the material. Stress birefringence is displayed as fringes when a stressed material is placed between crossed polarizers. Amounts less than one fringe can be described quantitatively by comparison of the fringe colors with tint plates made from a known birefringent material such as mica. Blanks containing less than 5 nm color shift per centimeter of thickness are quite stable.

The greatest obstacle to wider use of opaque materials, such as metals, for precision mirror blanks is the lack of a good test for internal stress.

19. Rigidity Deformation in a mirror depends on its dimensions, method of support, and orientation in a gravitational field as well as on the material. A simple case shows how each of these contributes. Assume that a circular plate of radius r and thickness t is supported within a centered ring of radius b. The deformation δ_r at the edge is a function of the density ρ, Young's modulus E, and a term K, which is a complicated function of Poisson's ratio and the ratio of b to r:

$$\delta_r = K \frac{r^4}{t^2} \frac{\rho}{E}$$

† Pars. 16 to 20 were written by Robert H. Noble, Optical Sciences Center, University of Arizona.

Poisson's ratio ν enters as $1 - \nu^2$. Fused silica has a ratio equal to 0.14:

$$1 - 0.14^2 = 0.98$$

Most metals have a ratio near 0.3:

$$1 - 0.3^2 = 0.91$$

One can see that a change of material does not have a large effect on the term K. However, the ratio ρ/E may change quite markedly with a change in material, as seen in Table 74.

It is clear that for similar dimensions and support, a beryllium mirror blank deforms the least of those listed. Differences between the rest are scarcely significant.

20. Thermal Deformation Changes in optical figure due to a shift from one temperature to another are generally proportional to the thermal coefficient of expansion. A deformation due to a temperature gradient involves the ratio of this coefficient and the thermal conductivity. Effects of these characteristics in several mirror materials may be noted in Table 74.

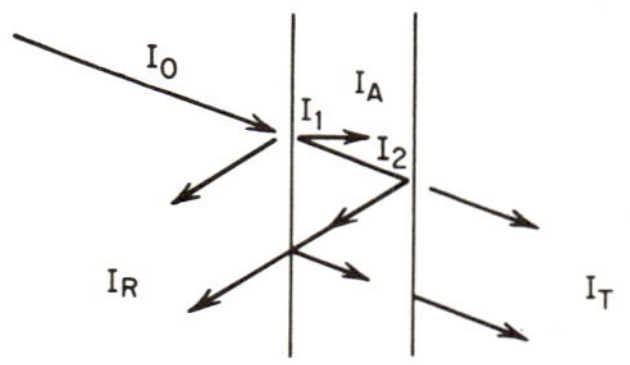

Fig. 1 Reflectance, absorptance, and transmittance of light incident on a plane-parallel plate of material.

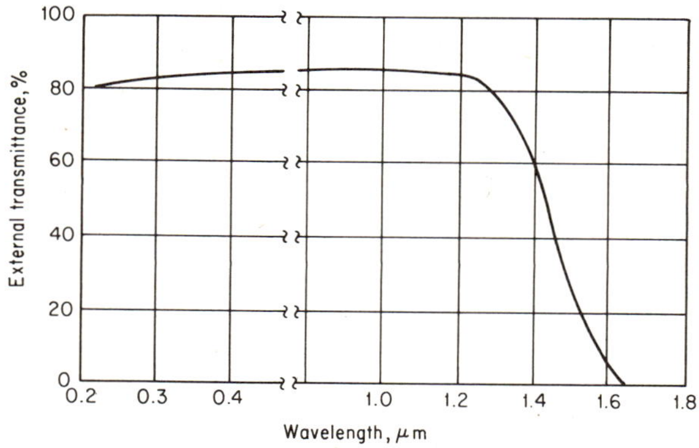

Fig. 2 Transmission of ammonium dihydrogen phosphate (ADP); thickness 1.6 mm (short-wavelength portion) and 7.8 mm (long-wavelength portion). [*From Deshotels* (1960), *McDonough* (*no date*).]

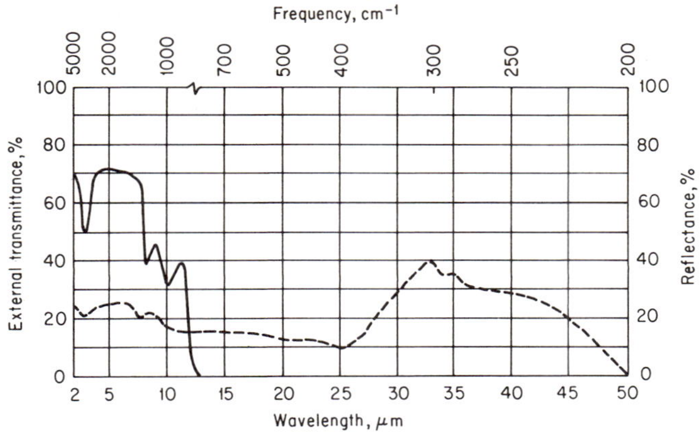

Fig. 3 Transmission (*solid curve*) and reflection (*dashed curve*) of arsenic trisulfide glass; thickness 5 mm. [*From McCarthy* (1963*a*).]

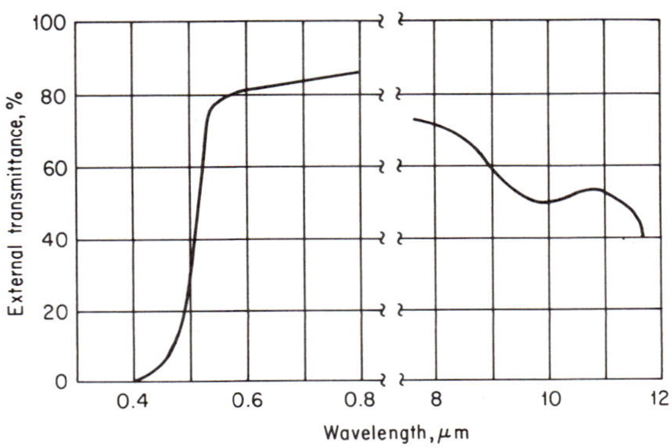

Fig. 4 Transmission of arsenic trisulfide glass; thickness 5 mm long-wavelength portion). [*Visible curve from Schlosser* (1957).]

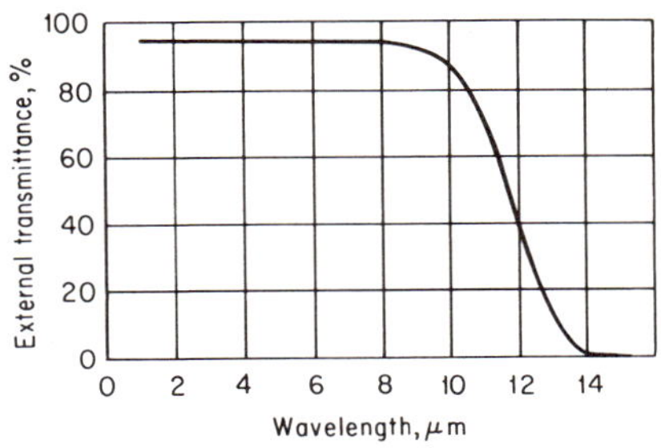

Fig. 5 Transmission of barium fluoride; thickness 9.1 mm. [*From Phillippi and Beardsley* (1955).]

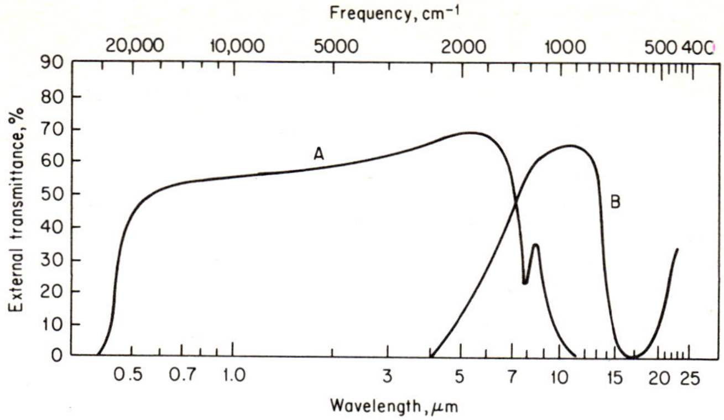

Fig. 6 Transmission of barium titanate; curve *A*, thickness 0.25 mm (crystal); curve *B*, thickness 10 mm (powder). [*From Iatsenko* (1957).]

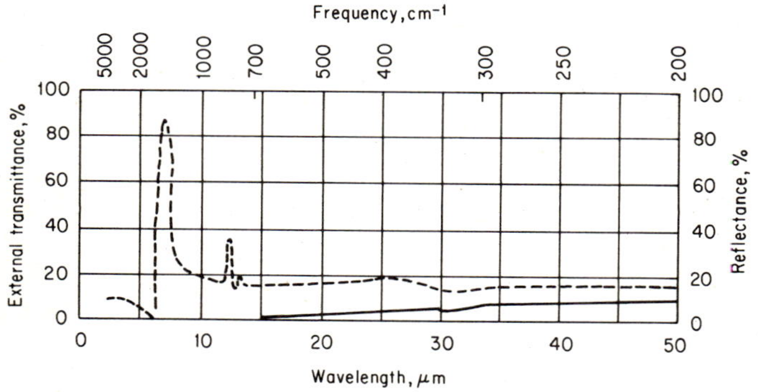

Fig. 7 Transmission (*solid curve*) and reflection (*dashed curve*) of boron nitride; thickness 6.0 mm. [*From McCarthy* (*no date*).]

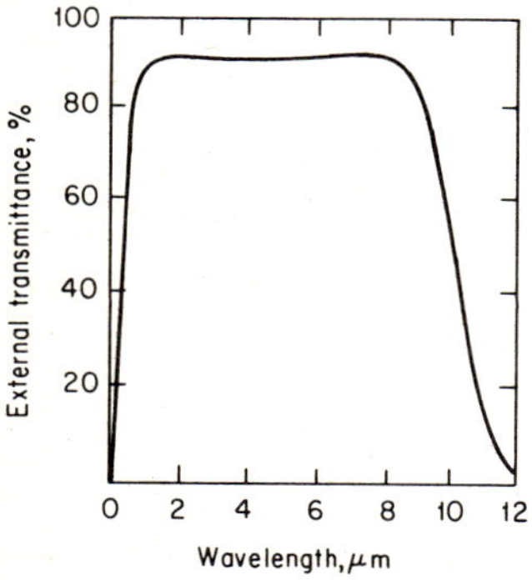

Fig. 8 Transmission of cadmium fluoride; thickness 5.0 mm. [*From Jones et al.* (1952); *copyright by the Institute of Physics.*]

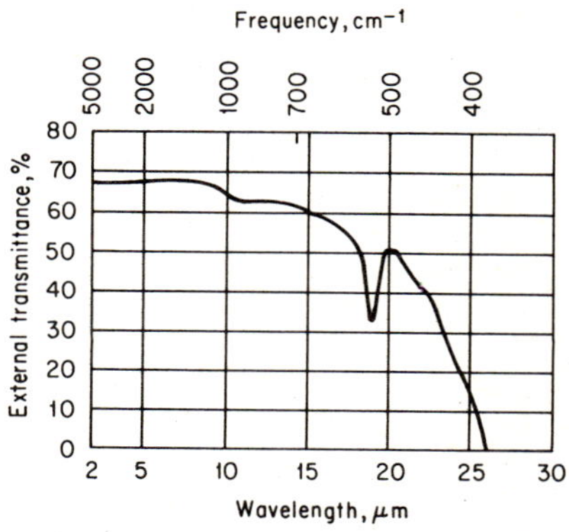

Fig. 9 Transmission of cadmium selenide; thickness 1.67 mm. [*From McCarthy* (1965*b*).]

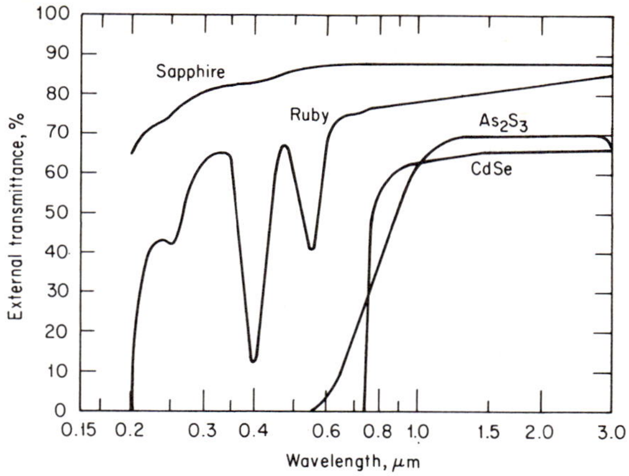

Fig. 10 Transmission of cadmium selenide, thickness 1.67 mm; ruby, thickness 6.10 mm; sapphire, thickness 3.0 mm; and arsenic trisulfide glass, thickness 5.0 mm. [*From McCarthy* (1967).]

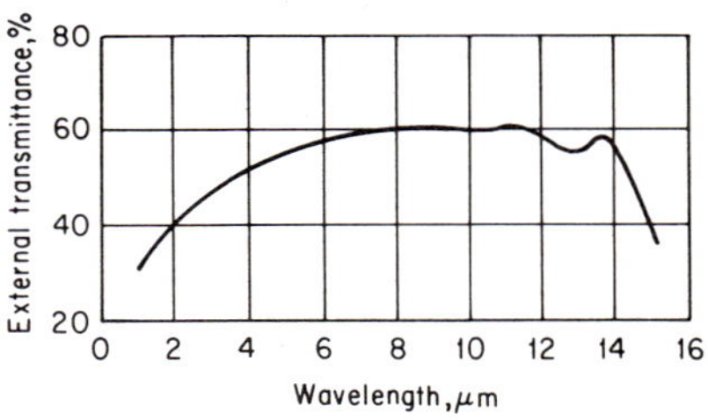

Fig. 11 Transmission of cadmium sulfide; thickness 3.9 mm. The dip at about 13 μm is probably due to impurities; the gradual decrease from 6 to 2 μm is due to scattering from a poorly polished sample. [*From Eagle Picher Co.* (*no date*).]

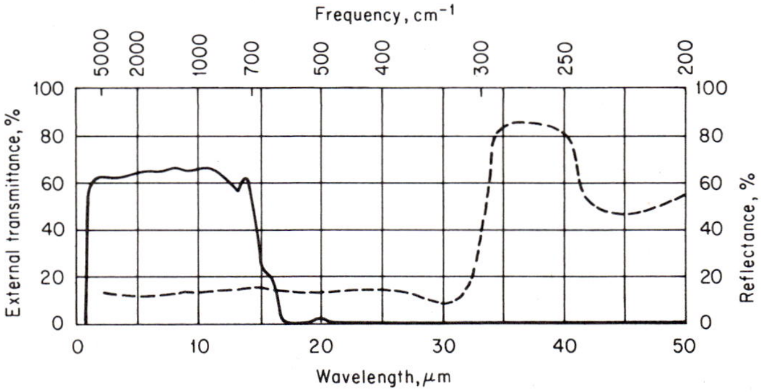

Fig. 12 Transmission (*solid curve*) and reflection (*dashed curve*) of cadmium sulfide. [*From McCarthy* (*no date*).]

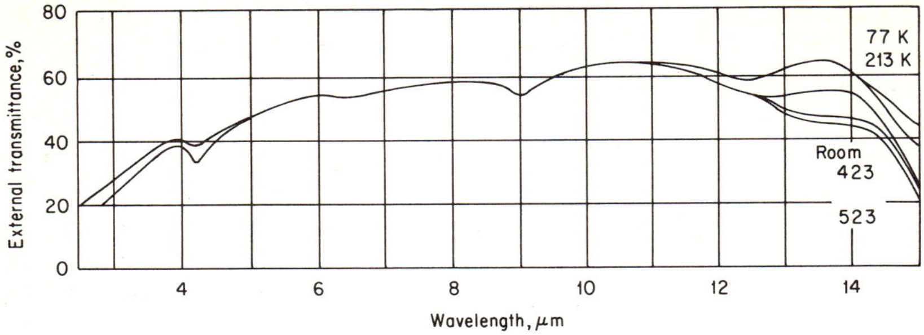

Fig. 13 Transmission of hot-pressed cadmium sulfide at several temperatures. [*From Platt* (1976).]

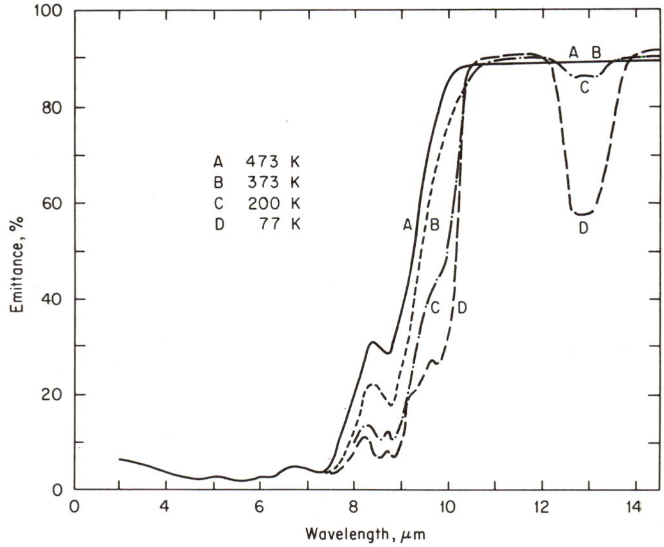

Fig. 14 Spectral emittance of cadmium sulfide at 77, 200, 373, and 473 K; thickness 5.1 mm. [*From Stierwalt* (1966*a*).]

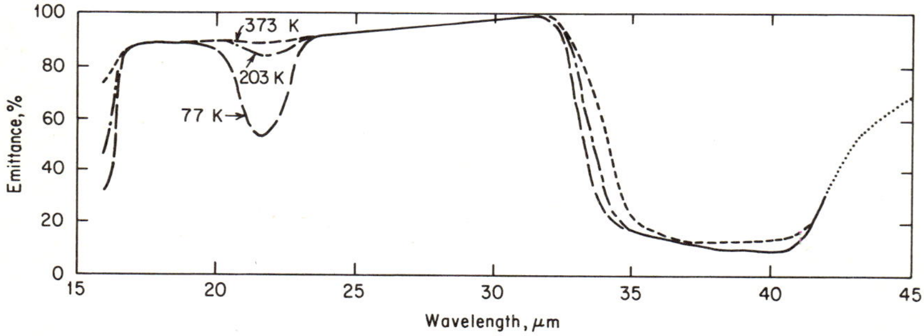

Fig. 15 Spectral emittance of cadmium sulfide at 77, 203, and 373 K; thickness 5.1 mm. [*From Stierwalt* (1966*a*).]

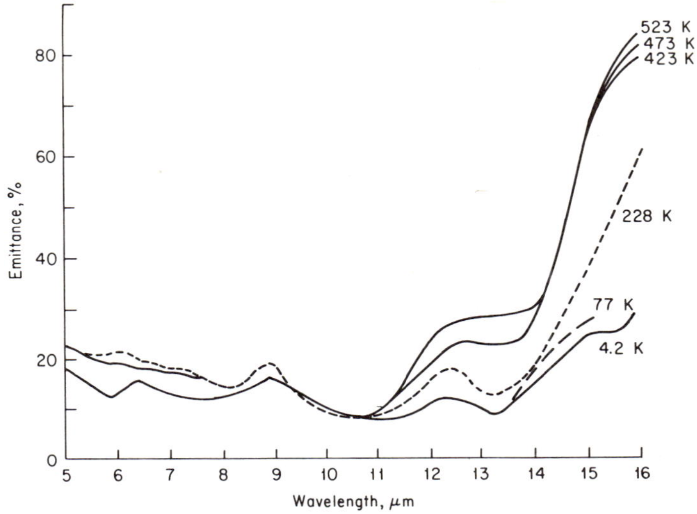

Fig. 16 Emittance of hot-pressed cadmium sulfide for several temperatures; thickness 4.01 mm. [*From Stierwalt (no date).*]

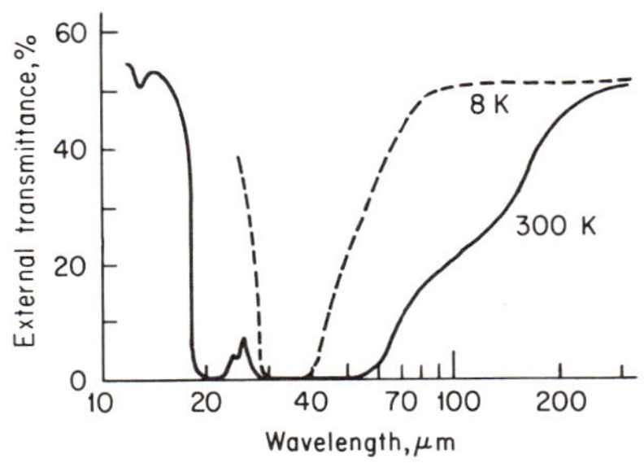

Fig. 17 Transmission of high-resistivity cadmium telluride at 300 and 8 K; thickness 2.26 mm. [111] single crystal plate optically polished. Resistivity 0.35 MΩ/cm. [*From Johnson et al.* (1969)]

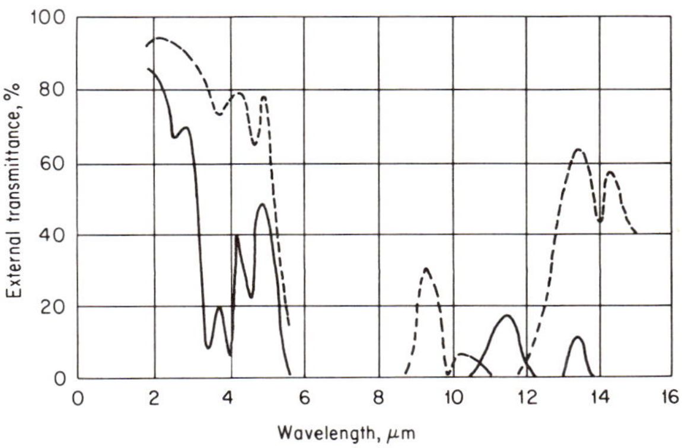

Fig. 18 Transmission of calcite for the ordinary ray (*solid curve*) and the extraordinary ray (*dashed curve*); thickness 1 mm. [*From Nysander* (1909).]

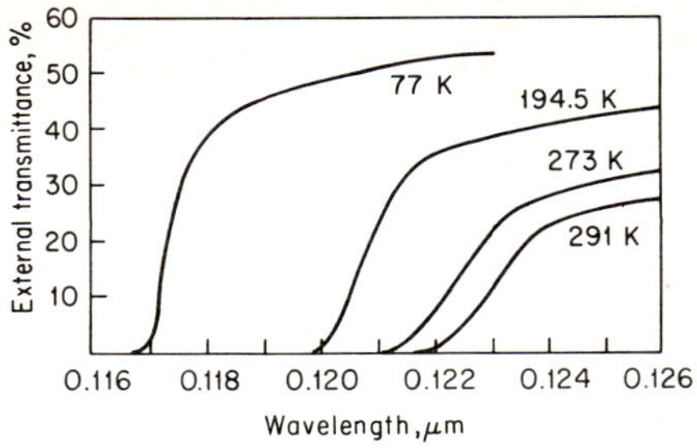

Fig. 19 Ultraviolet transmission of cleaved calcium fluoride crystals at various temperatures; thickness 0.71 mm. [*From Knudson and Kupperian* (1957).]

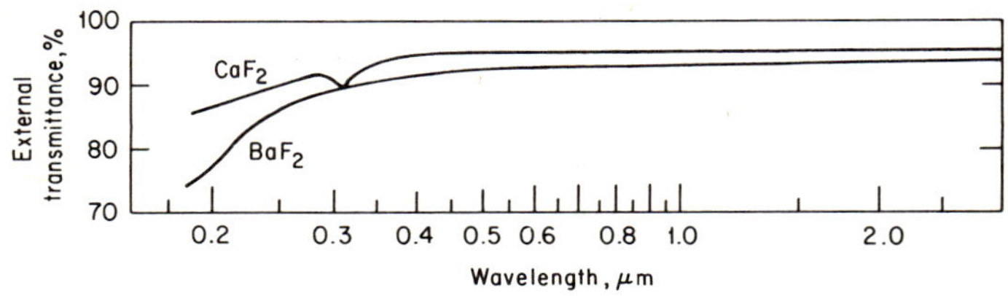

Fig. 20 Transmission of calcium fluoride and barium fluoride (short-wavelength portion). [*From McCarthy* (*no date*).]

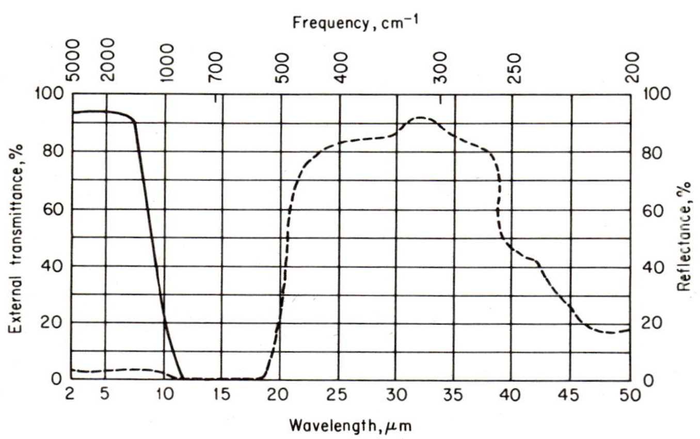

Fig. 21 Transmission (*solid curve*) and reflection (*dashed curve*) of calcium fluoride; thickness 5 mm. [*From McCarthy* (1963*a*).]

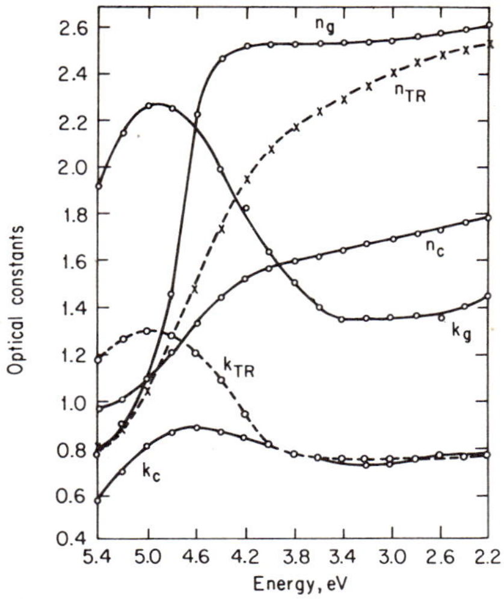

Fig. 22 The optical constants of glassy carbon, graphite, and evaporated carbon foil. $n_g k_g$: graphite (Taft and Phillipp, 1965), $n_c k_c$: glassy carbon (Phillipp, no date), n_{TR}, k_{TR}: this experiment; solid curve: carbon foil. [*Nardo and Goland* (1971).]

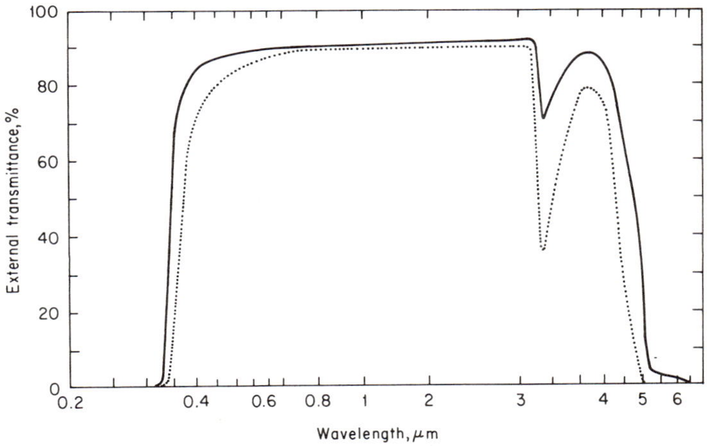

Fig. 23 Transmission of Cer-Vit† material type C-101 for 0.29 mm (*solid curve*) and 1.13 mm (*dotted curve*). [*From Fuller et al.* (1968).]

† Trade name of a material made by Owens-Illinois, Inc., Toledo, Ohio, which belongs to the family known as glass ceramics. In the process of manufacturing Cer-Vit material products, a special glass is first melted, then formed into monolithic structures of the desired shape. Finally, a controlled heat treatment converts the glass into a nonporous polycrystalline material.

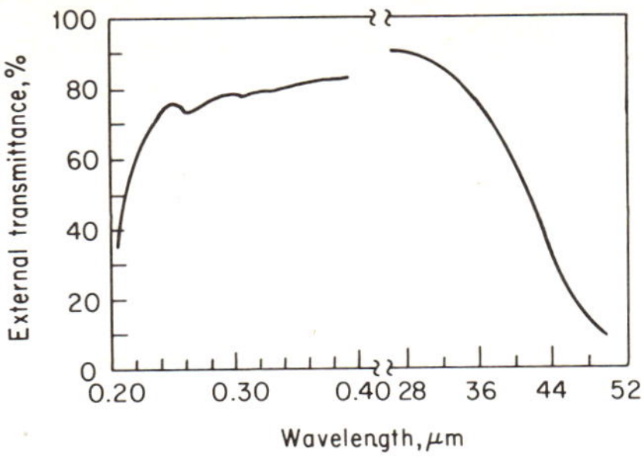

Fig. 24 Transmission of cesium bromide; thickness 5 mm (short-wavelength portion) and 5.10 mm (long-wavelength portion). Ultraviolet curve measured by A. Sutton, Polaroid Corporation. Infrared curve from data of Plyler (no date); similar data for 7-mm-thick sample in Plyler and Phelps (1951).

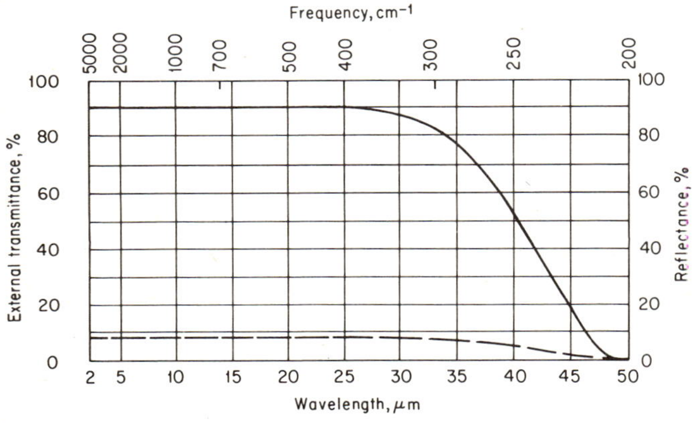

Fig. 25 Transmission (*solid curve*) and reflection (*dashed curve*) of cesium bromide; thickness 10 mm. [*From McCarthy* (1963*b*).]

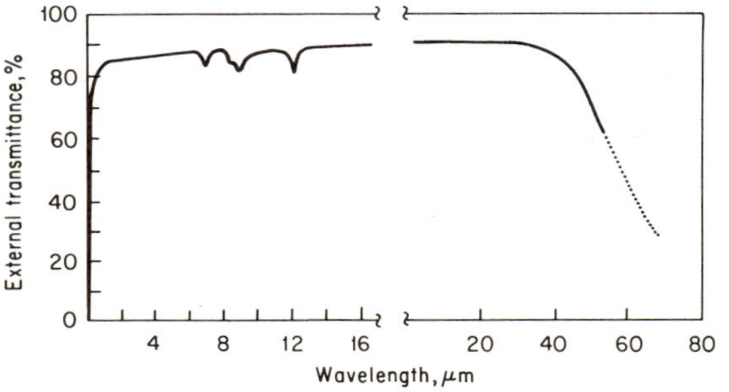

Fig. 26 Transmission of cesium iodide; thickness 3 mm (short-wavelength portion) and 5 mm (long-wavelength portion). [*From Plyler and Phelps* (1952); *Acquista and Plyler* (1953, 1958).]

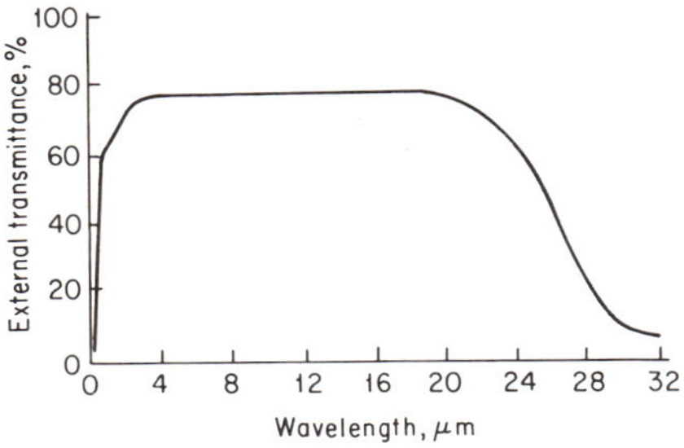

Fig. 27 Transmission of copper bromide; thickness 2.92 mm. [*From Dobrshanskii et al.* (1964).]

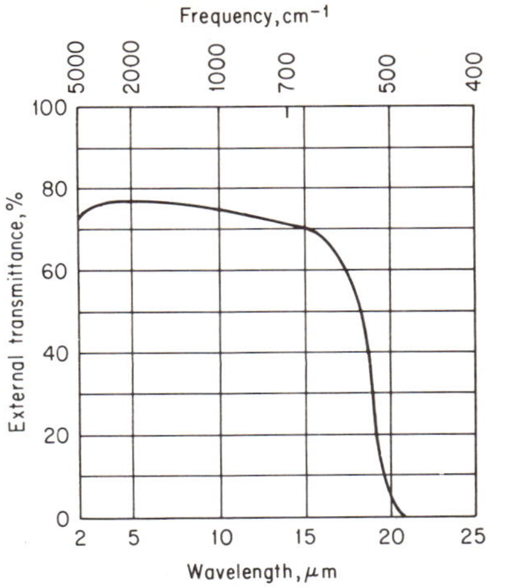

Fig. 28 Transmission of cuprous chloride; thickness 9.1 mm. [*From McCarthy* (1965*a*).]

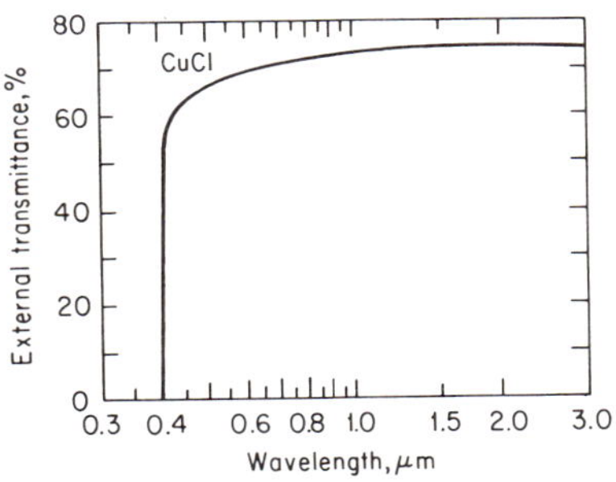

Fig. 29 Transmission of cuprous chloride; thickness 9.1 mm. [*From McCarthy* (1967).]

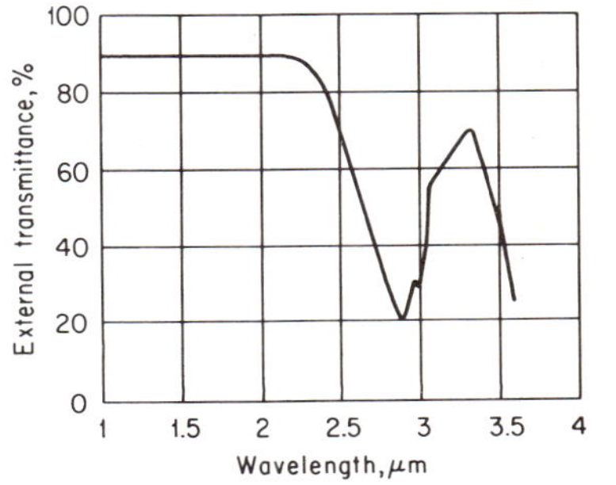

Fig. 30 Transmission of crystal quartz for the ordinary ray; thickness 10 mm. [*From Smith et al.* (1957).]

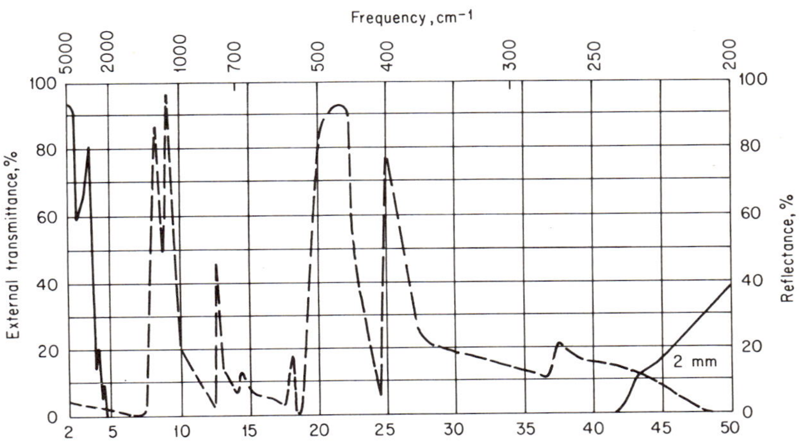

Fig. 31 Transmission (*solid curve*) and reflection (*dashed curve*) of quartz (ordinary or extraordinary ray not specified). [*From McCarthy* (1963*a*).]

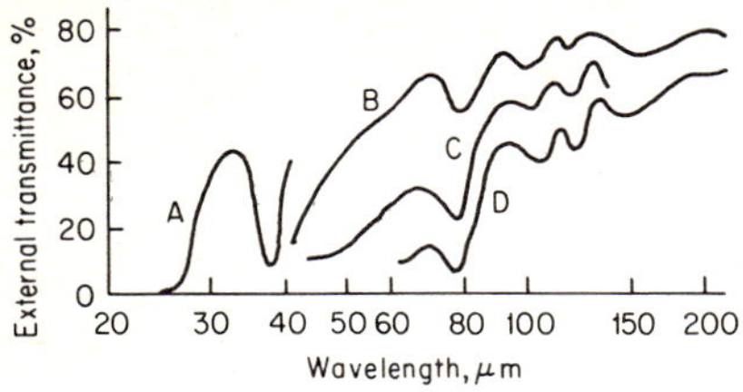

Fig. 32 Transmission of quartz. Curve *A*: 0.1 mm thick, perpendicular to optic axis; curve *B*: 1.03 mm thick, parallel; curve *C*: 3.49 mm thick, perpendicular; curve *D*: 6.05 mm thick, perpendicular. [*From Czerny and Röder* (1938).]

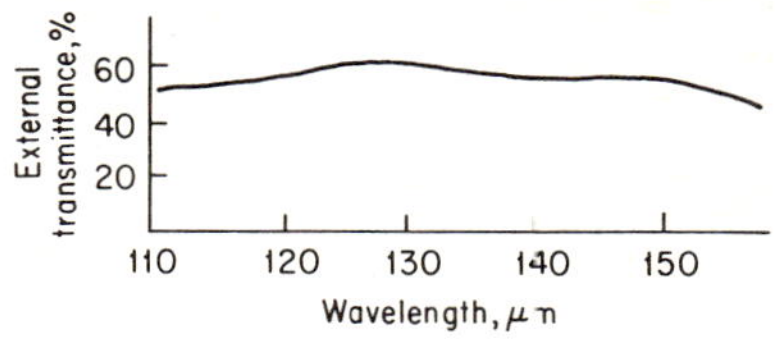

Fig. 33 Transmission of quartz; thickness 0.6 mm. [*From Hadni et al.* (1959).]

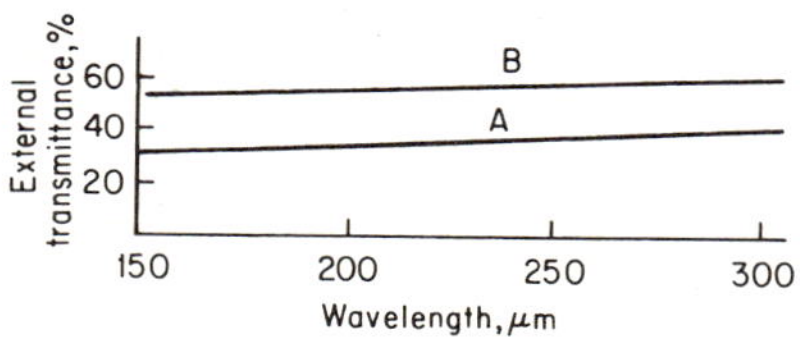

Fig. 34 Transmission of quartz; curve *A*: thickness 25 mm (Brazilian); curve *B*: thickness 0.6 mm. [*From Hadni et al.* (1959).]

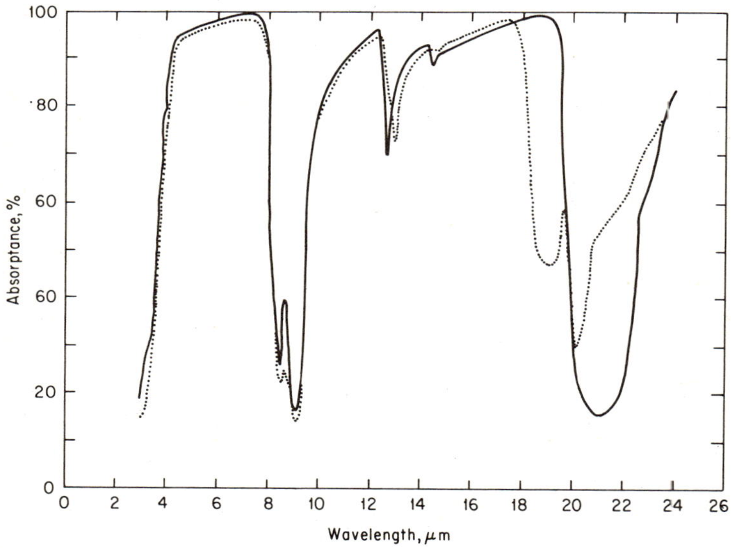

Fig. 35 Spectral absorption of crystalline quartz at 100°C; solid curve: ordinary ray; dotted curve: extraordinary ray. [*From Stierwalt et al.* (1963).]

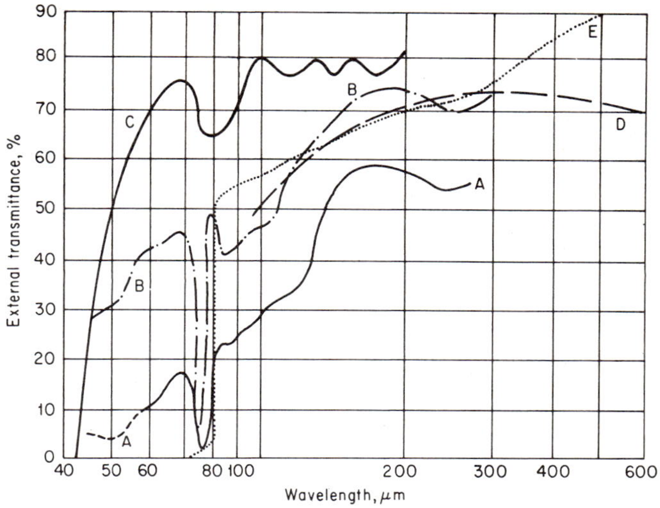

Fig. 36 Transmission of crystal quartz. Curve *A*: 5 mm at 20°C. [*Data from Stavenich and Yaroslavskii* (1961).] Curve *B*: 5 mm at −175°C. [*Data from Stavenich and Yaroslavskii* (1971).] Curve *C*: 1 mm. [*Data from Lord and McCubbin* (1957).] Curve *D*: 4.5 mm. [*Data from McCubbin and Sinton* (1950).] Curve *E*: 5 mm. [*Data from Decamps and Hadni* (1960).]

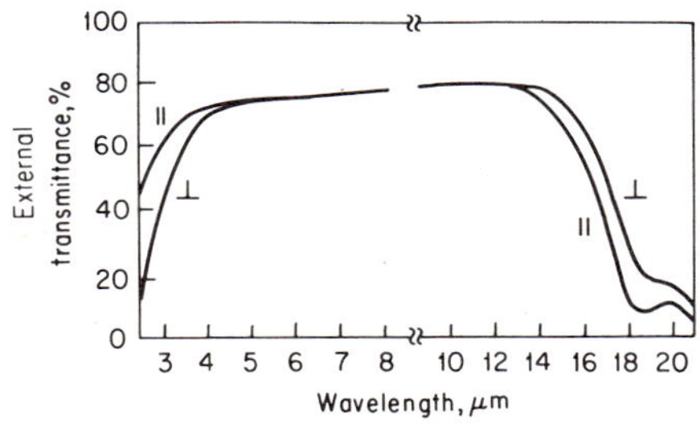

Fig. 37 Transmission of a 1.85-mm-thick sample of cesium cupric chloride; ⊥ ordinary ray, || extraordinary ray. [*From Day and Gruzensky* (1970).]

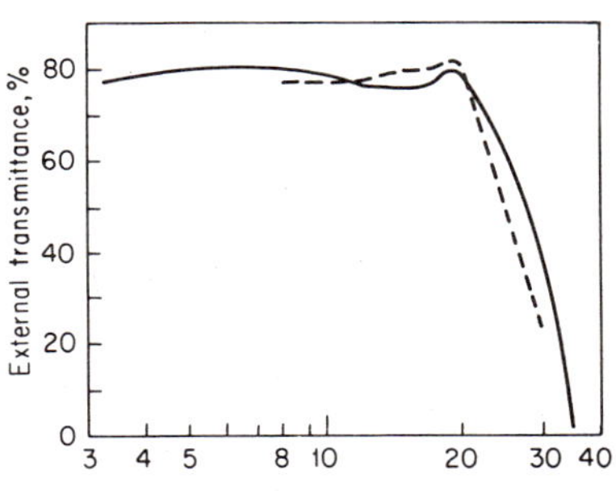

Fig. 38 Transmission of copper(I) chloride at 300 K for 90-μm sample (*solid curve*) and 200-μm sample (*dashed curve*). [*From Alonas et al.* (1969).]

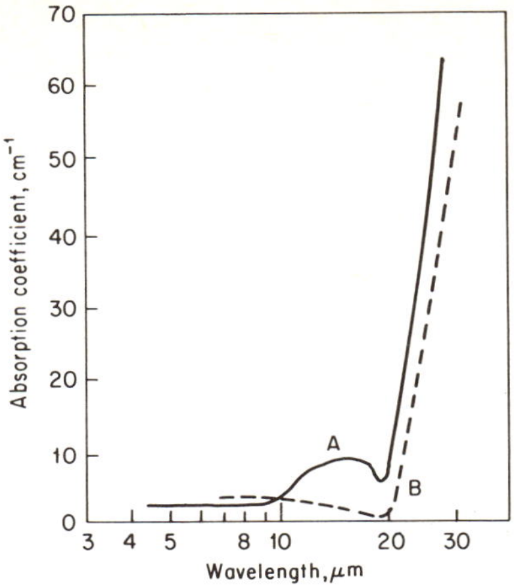

Fig. 39 Absorption coefficient of copper(I) chloride at 300 K for 90-μm sample (*solid curve*) and 200-μm sample (*dashed curve*). [*From Alonas et al.* (1969).]

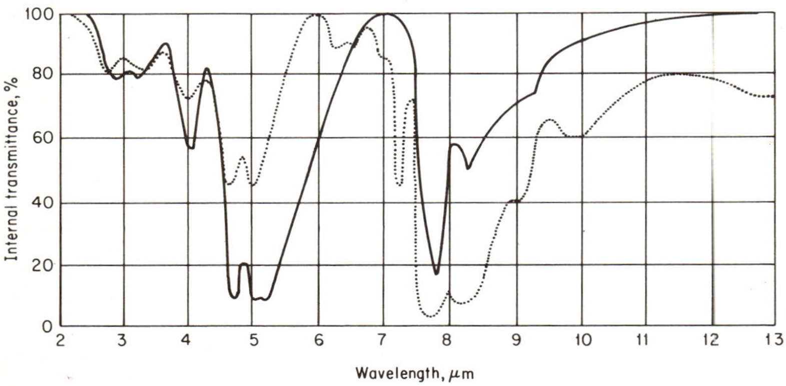

(*a*)

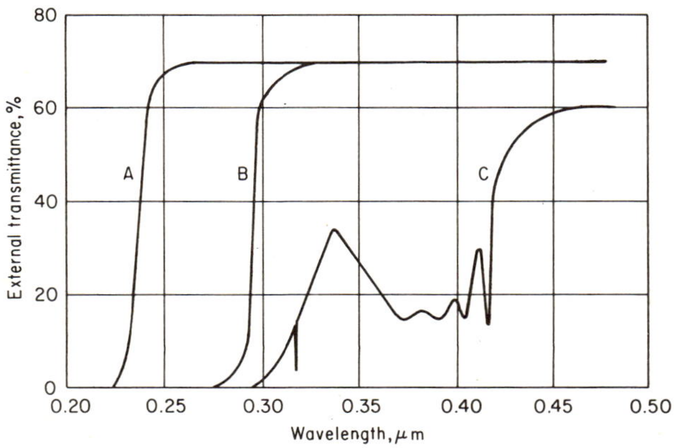

(*b*)

Fig. 40 (*a*) Transmission of two types of diamond. *Dotted curve:* GM 71 medium infrared; *solid curve:* SLO 44P2, strong infrared I. (*b*) Ultraviolet and visible transmission of three types of diamond: *A*: BP-2 (ultraviolet II), *B*: M-4 (weak ultraviolet), *C*: SLF 127 (medium ultraviolet I). [*From Simeral* (1953).]

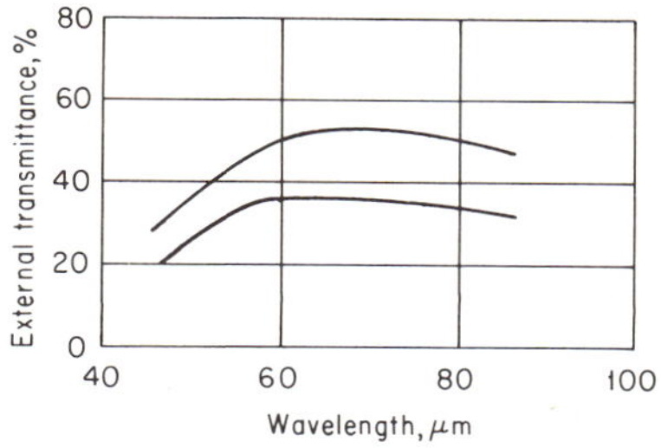

Fig. 41 Transmission of diamond for two thicknesses, 0.096 in (*lower*) and 0.066 in (*upper*). Figures 40 and 41 show the transmission of diamond in three different parts of the spectrum. [*From Hausler* (1952).]

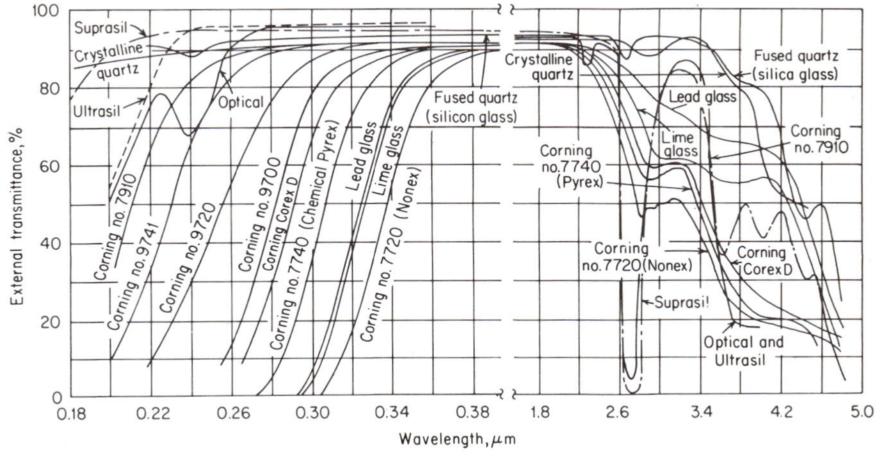

Fig. 42 Transmission of several ultraviolet and infrared transmitting glasses. [*From General Electric Company* (1957); *Engelhard Industries, Inc.* (*no date*).]

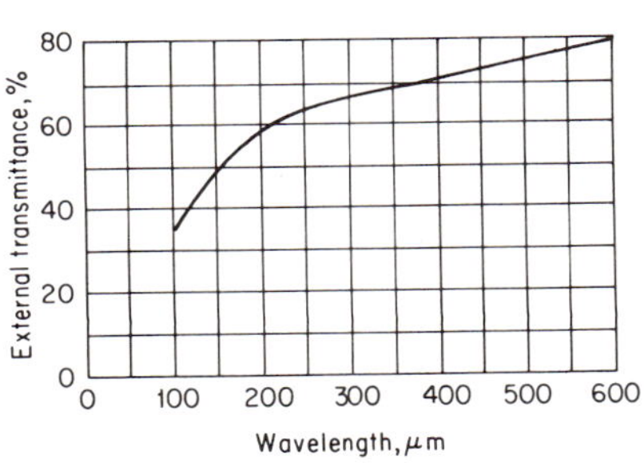

Fig. 43 Transmission of fused silica; thickness 0.56 mm. [*From McCubbin and Sinton* (1950).]

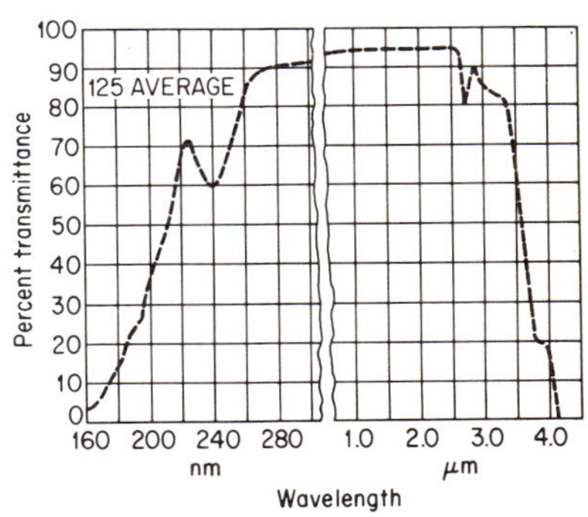

Fig. 44 Transmission of ultraviolet and visible radiation by General Electric type 125 fused quartz, 1 cm thick. Curve allows for surface reflection losses. [*From General Electric Company catalog* 5100 (*no date*).]

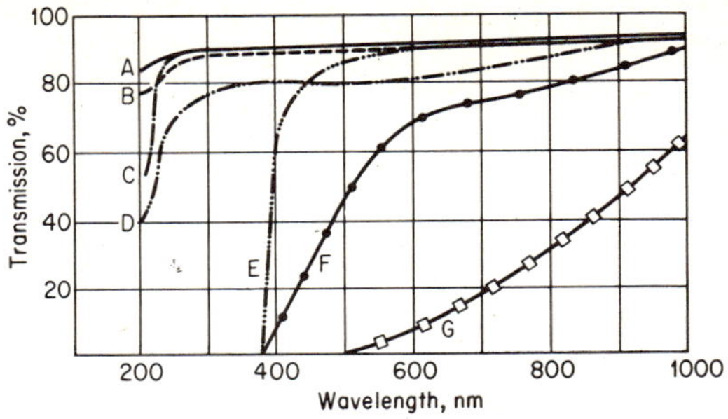

Fig. 45 Transmission and resistance to discoloration caused by radiation for Suprasil (fused silica) and Ultrasil (fused quartz) compared to lead silicate glass, using 5-mm-thick plates. Curve *A*: Suprasil before radiation; curve *B*: Suprasil after Röntgen radiation with 3×10^8 R; curve *C*: Ultrasil before radiation; curve *D*: Ultrasil after Röntgen radiation with 3×10^8 R; curve *E*: lead silicate glass before radiation; curve *F*: lead silicate glass after γ-Co-radiation of 10^8 R; curve *G*: lead silicate glass with cerium stabilization after γ-Co-radiation of 10^8 R. [*From Amersil Inc., brochure* EM-9227-2 (*no date*).]

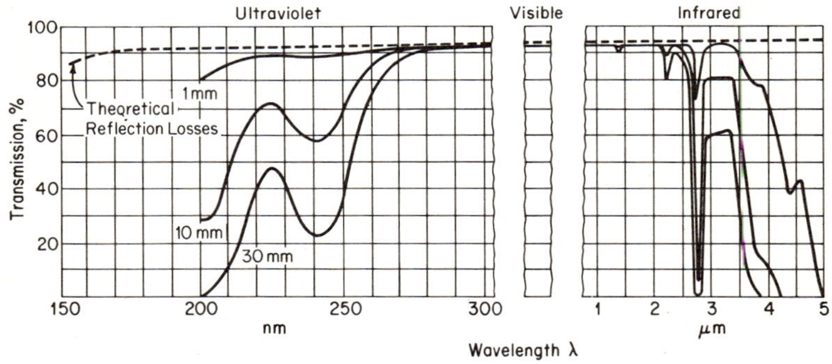

Fig. 46 Transmission of ultraviolet, visible, and infrared radiation by Homosil fused quartz, 1 cm thick. [*From Amersil Inc., brochure* EM-9227-2 (*no date*).]

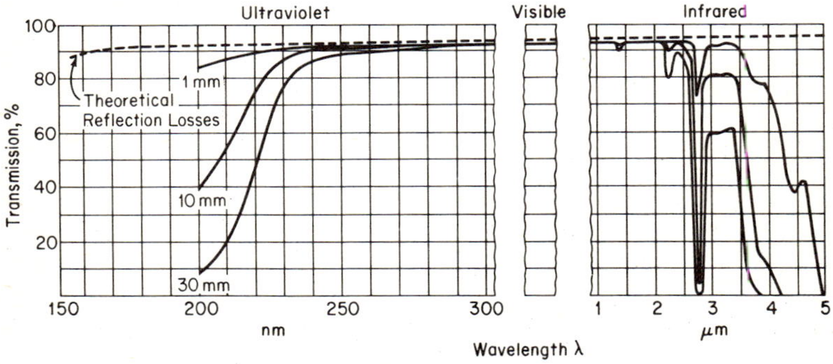

Fig. 47 Transmission of ultraviolet, visible, and infrared radiation by Ultrasil fused quartz, 1 cm thick. [*From Amersil Inc., brochure* EM-9227-2 (*no date*).]

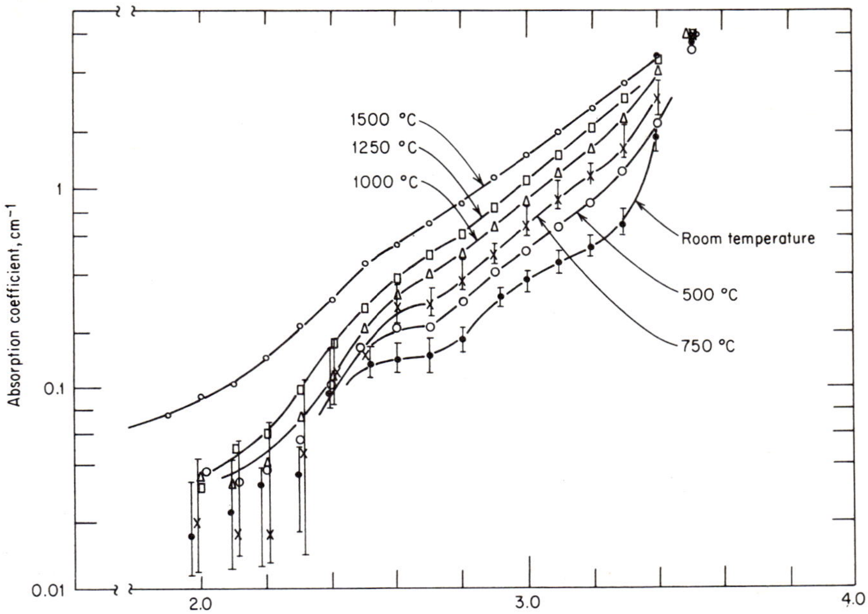

Fig. 48 Infrared absorption for 0.953-cm thickness and absorption coefficient of Corning Vycor infrared transmitting glass 7905 (96 percent silica).

No. of samples tested	9	2	8	2	2	2
Temp., °C	room	500	750	1000	1250	1500

Spread bars include at least 80 percent of data. Solid symbols (♦, 25°C; ●, 1000°C) show 0.953-cm transmissivities derived from 0.20-cm data of Corning *Bull.* CJR-2. [*Beder et al.* (1971).]

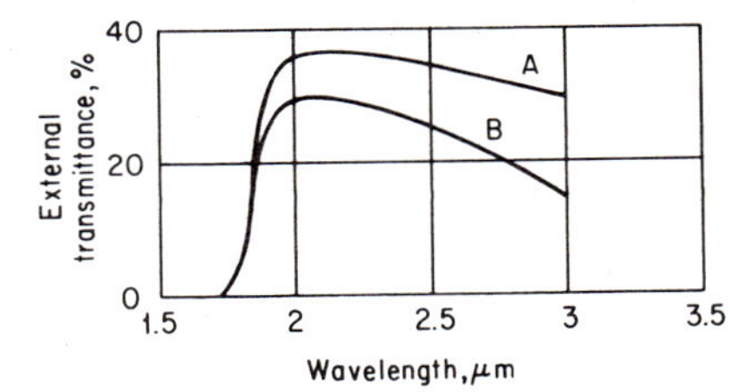

Fig. 49 Transmission of gallium antimonide; curve A: 0.33 mm; curve B: 0.66 mm. [*From Edwards (no date).*]

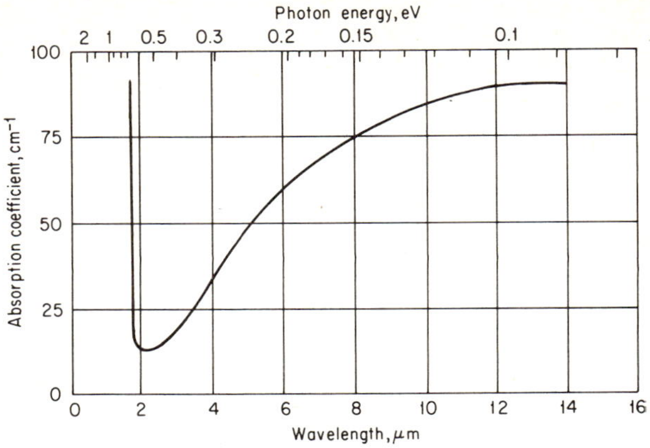

Fig. 50 Absorption coefficient of gallium antimonide; resistivity, 0.08 Ω·cm. Apparently there was a considerable impurity contribution in the sample measured. [*From Oswald and Schade* (1954).]

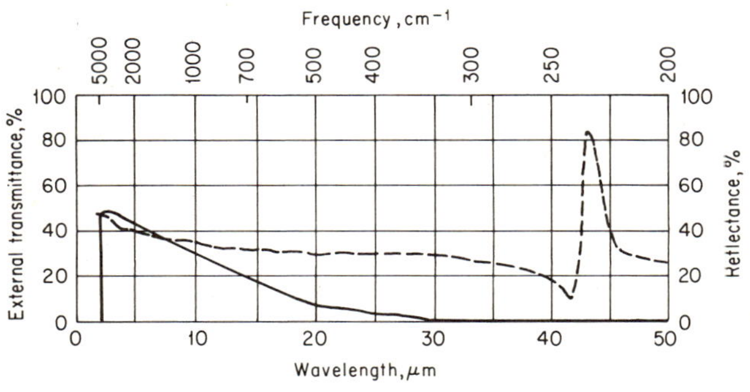

Fig. 51 Transmission (*solid curve*) and reflection (*dashed curve*) of gallium antimonide; thickness 0.5 mm. [*From McCarthy (no date)*.]

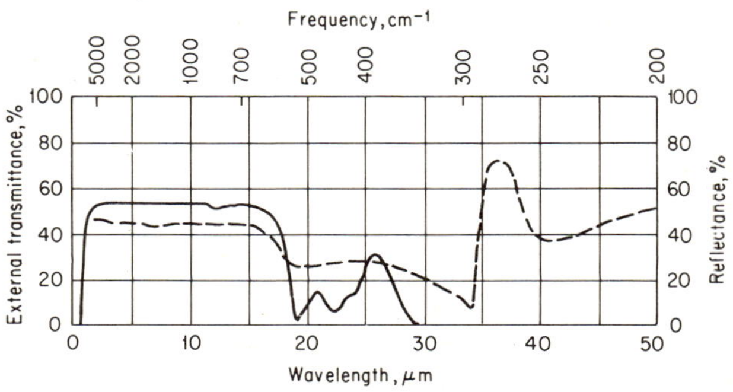

Fig. 52 Transmission (*solid curve*) and reflection (*dashed curve*) of gallium arsenide; thickness 0.5 mm. [*From McCarthy (no date)*.]

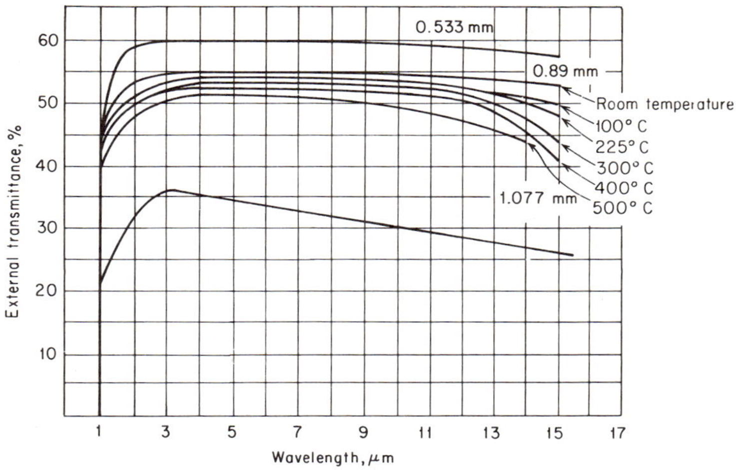

Fig. 53 Transmission of gallium arsenide at several temperatures and for several thicknesses. [*From Philco Corporation* (1958).]

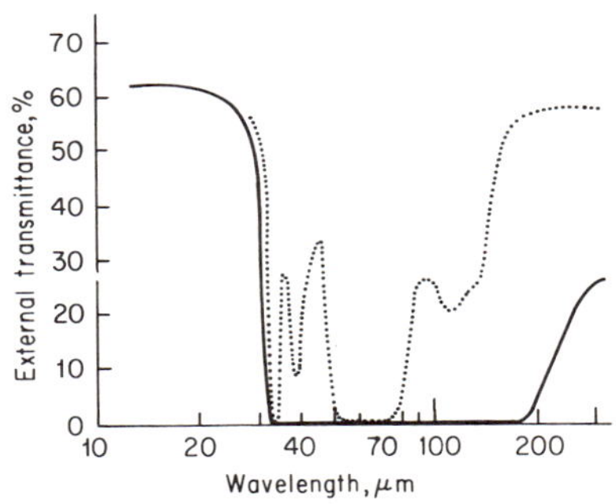

Fig. 54 Far-infrared transmission of high-resistivity gallium arsenide at 300 K (*solid curve*) and 8 K (*dotted curve*) (measured). [110] single crystal plate, optically polished, 2.56 mm thick, resistivity 0.85 MΩ/cm. [*From Johnson et al.* (1969).]

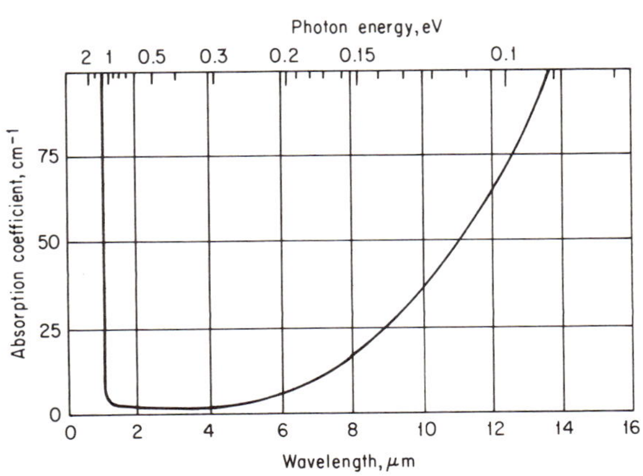

Fig. 55 Absorption coefficient of gallium arsenide; resistivity 1 mΩ·cm. Apparently there was a considerable impurity contribution in the sample measured. [*From Oswald and Schade* (1954).]

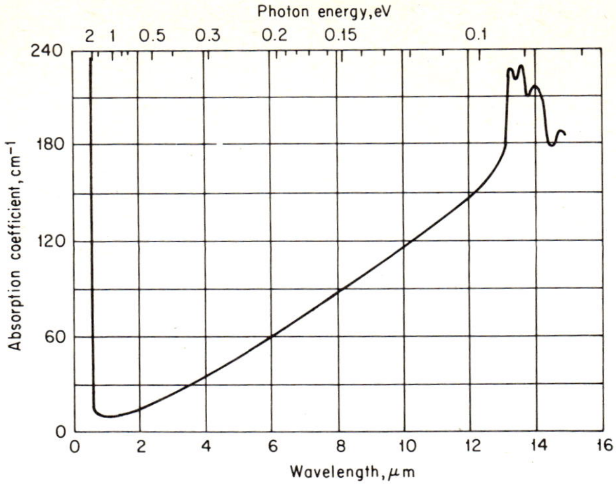

Fig. 56 Absorption coefficient of gallium phosphide. Sample reported as slightly impure; workers at Philco Corporation report a sample with about half the absorption at 4 μm. [*From Oswald and Schade* (1954).]

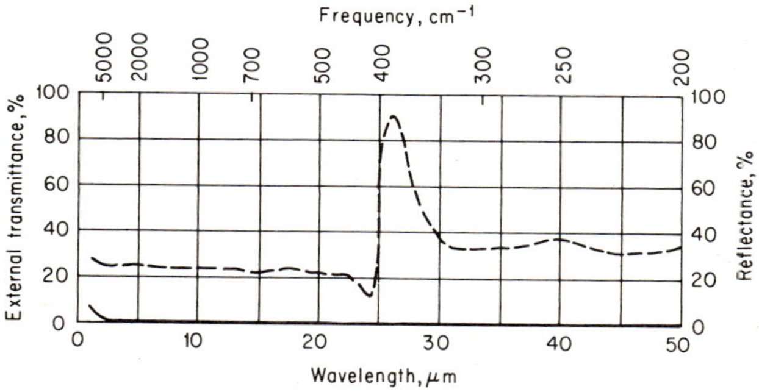

Fig. 57 Transmission (*solid curve*) and reflection (*dashed curve*) of gallium phosphide; thickness 1.0 mm. [*From McCarthy* (*no date*).] See also Par. 4, which has values for absorption coefficient at longer wavelengths.

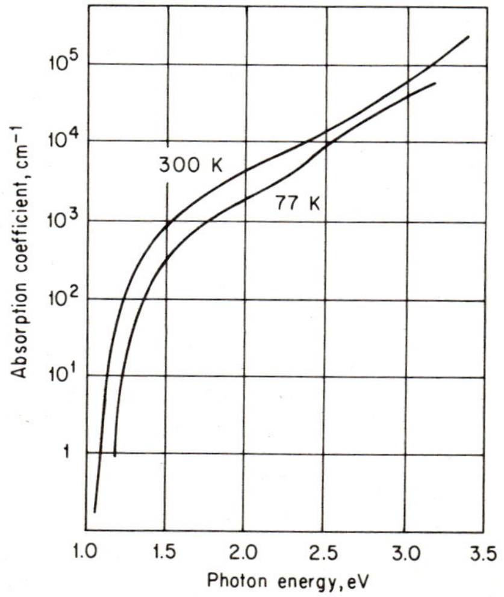

Fig. 58 Intrinsic absorption in germanium. [*From Dash and Newman* (1955).]

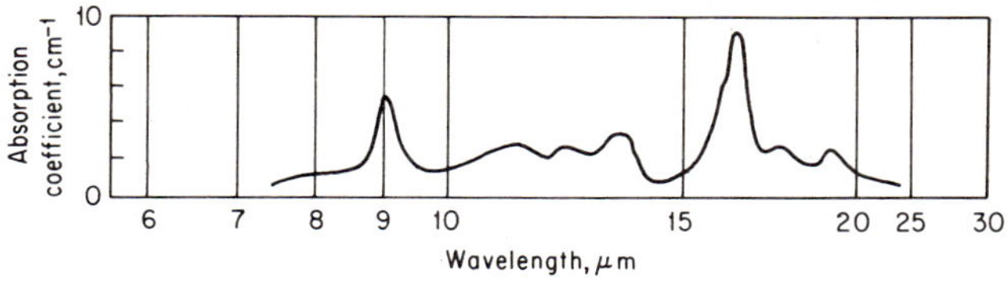

Fig. 59 Absorption in germanium due to lattice vibrations. [*From Lax and Burstein* (1955).] A summary of the short-wavelength absorption in germanium is given by Moss (1959*b*).

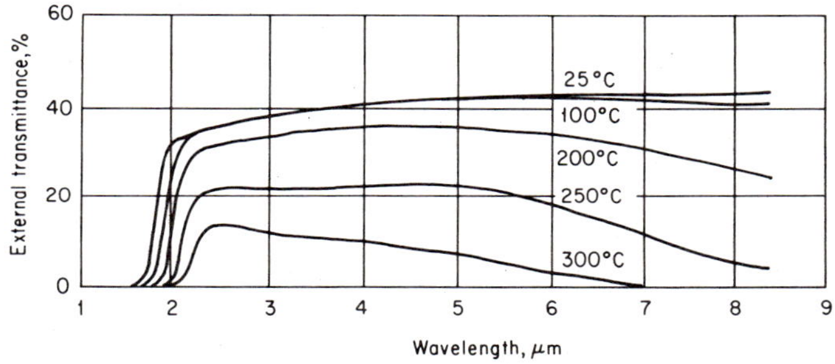

Fig. 60 Transmission of germanium for several temperatures; thickness 1.15 mm; resistivity 30 Ω·cm. [*From Phillippi and Beardsley* (1955).]

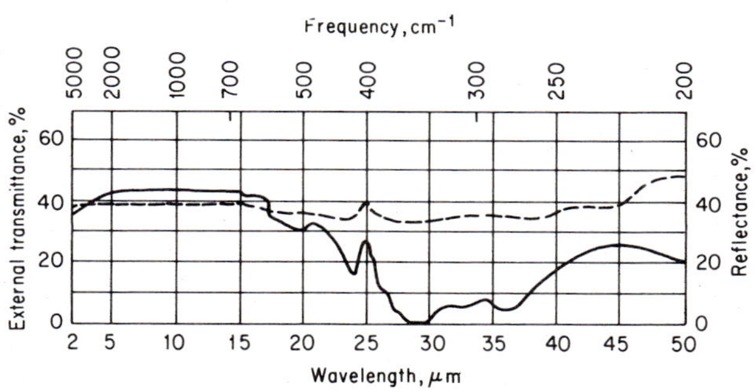

Fig. 61 Transmission (*solid curve*) and reflection (*dashed curve*) of germanium; thickness 1.6 mm. [*From McCarthy* (1963*a*).]

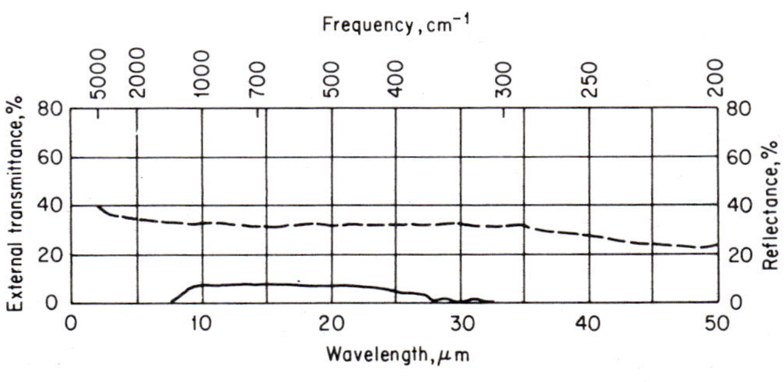

Fig. 62 Transmission (*solid curve*) and reflection (*dashed curve*) of indium antimonide; thickness 1.0 mm. [*From McCarthy* (*no date*).]

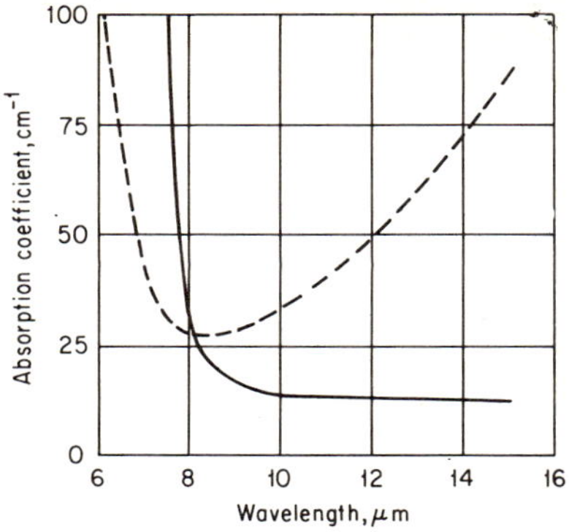

Fig. 63 Absorption coefficient of indium antimonide; "clean" (*solid curve*), resistivity 2 cΩ·cm; "impure" (*dashed curve*), resistivity 1 cΩ·cm. [*From Oswald and Schade* (1954).]

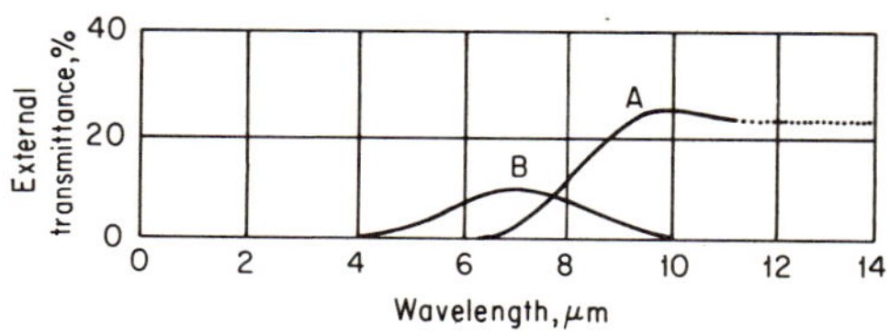

Fig. 64 Transmission of indium antimonide; thickness 0.007 in; resistivity of *A*, 7 mΩ·cm; resistivity of *B*, 0.25 mΩ·cm. [*From Tanenbaum and Briggs* (1953).]

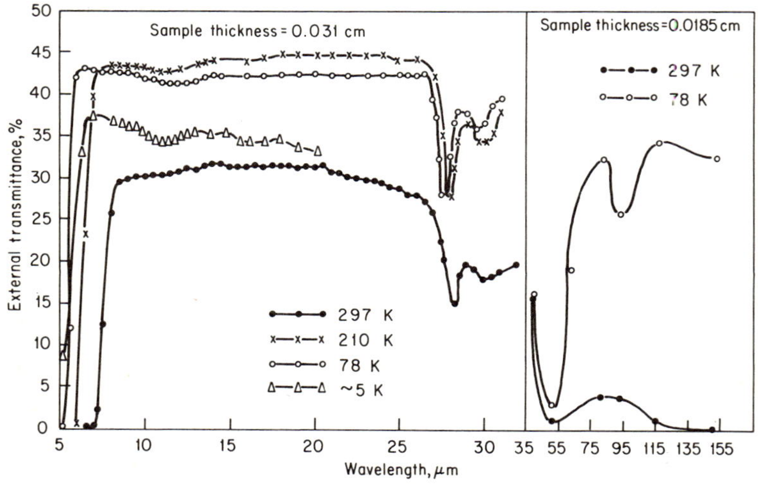

Fig. 65 Transmission of indium antimonide at 5, 78, 210, and 297 K. [*From Spitzer and Fan* (1955).]

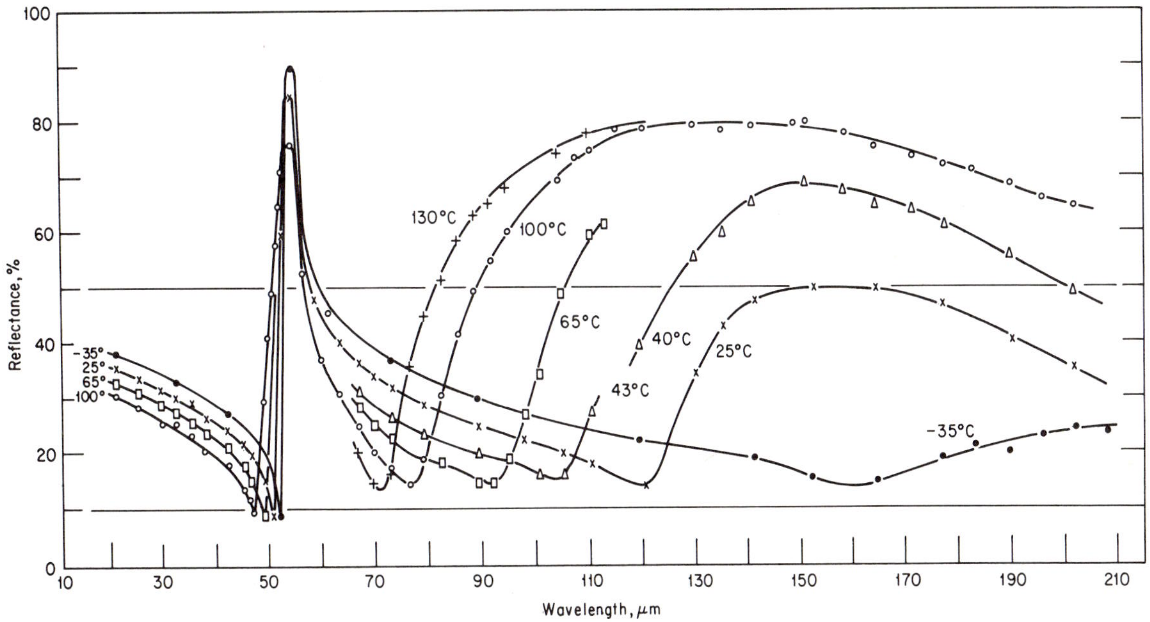

Fig. 66 Far-infrared reflection of indium antimonide at the temperatures indicated. [*From Yoshinago and Oetjen* (1956).]

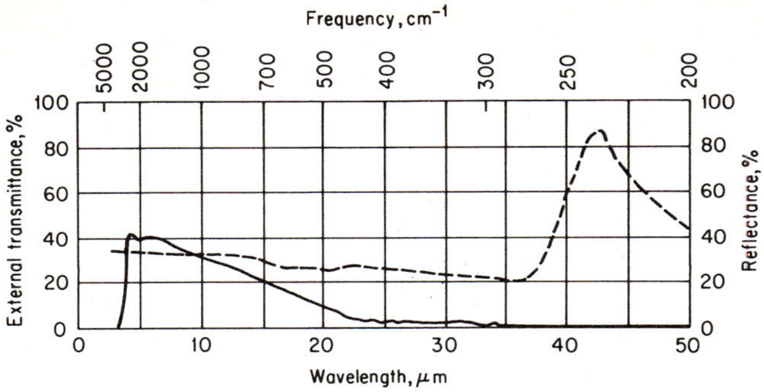

Fig. 67 Transmission (*solid curve*) and reflection (*dashed curve*) of indium arsenide; thickness 0.17 mm. [*From McCarthy* (*no date*).]

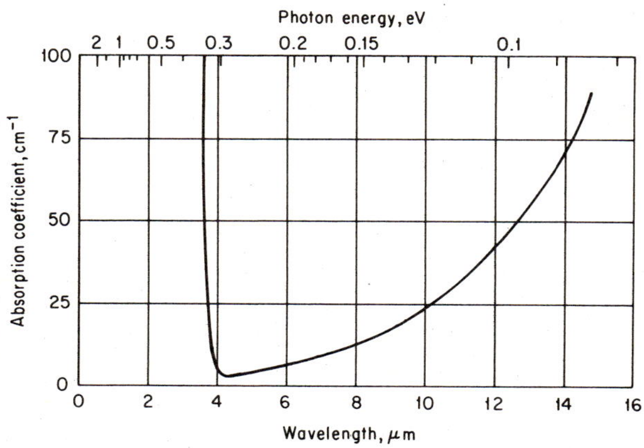

Fig. 68 Absorption coefficient of indium arsenide; resistivity 1 mΩ·cm. Apparently there was a considerable impurity contribution in the sample measured. [*From Oswald and Schade* (1954).]

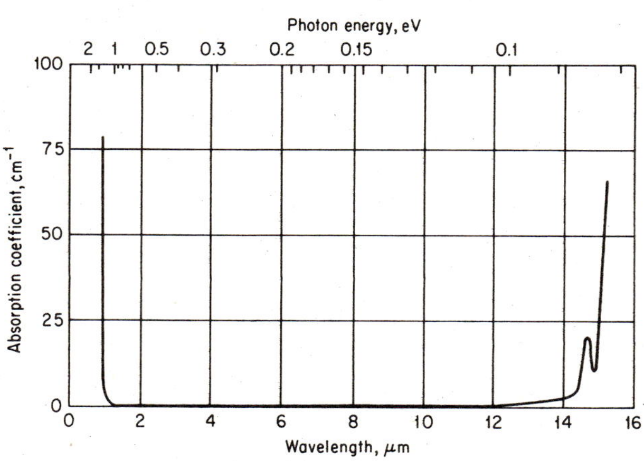

Fig. 69 Absorption coefficient of indium phosphide; no purity data given. [*From Oswald and Schade* (1954).]

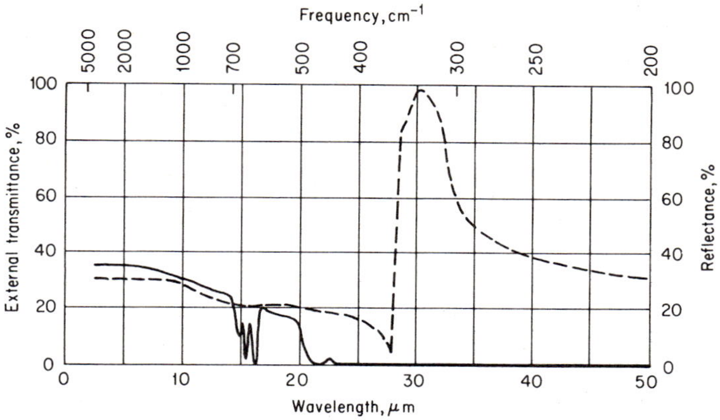

Fig. 70 Transmission (*solid curve*) and reflection (*dashed curve*) of indium phosphide; thickness 1.0 mm. [*From McCarthy* (*no date*).]

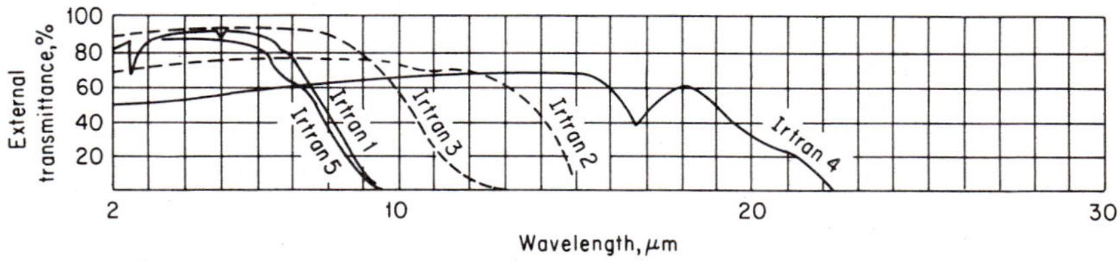

Fig. 71 Transmission of Irtran materials; thickness 2.0 mm. [*From Eastman Kodak* (*no date*).]

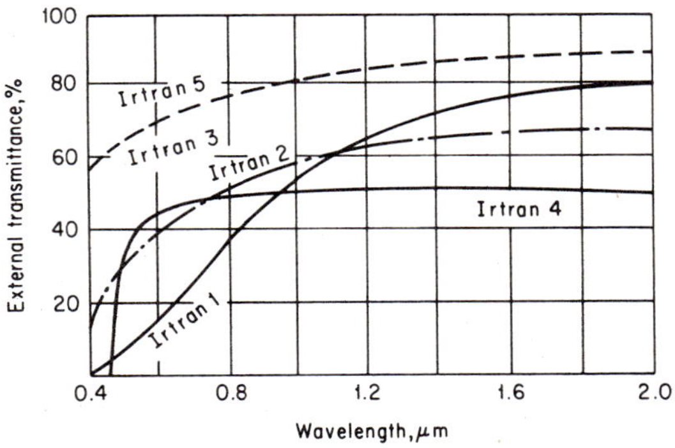

Fig. 72 Transmission of Irtrans; thickness 2.0 mm (short-wavelength portion). [*From Eastman Kodak* (*no date*).]

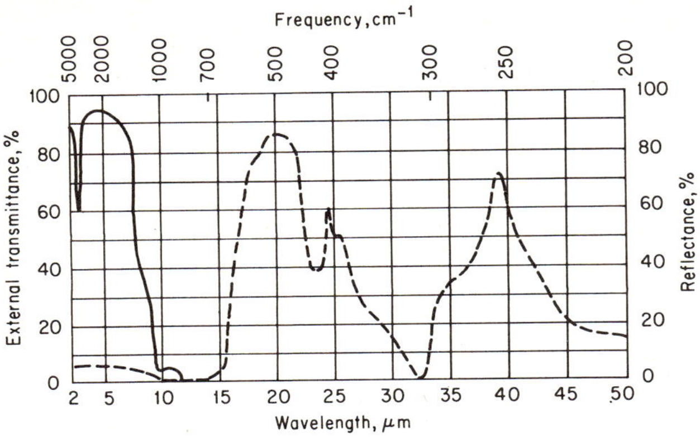

Fig. 73 Transmission (*solid curve*) and reflection (*dashed curve*) of Irtran 1; thickness 2 mm. [*From McCarthy* (1963*a*).]

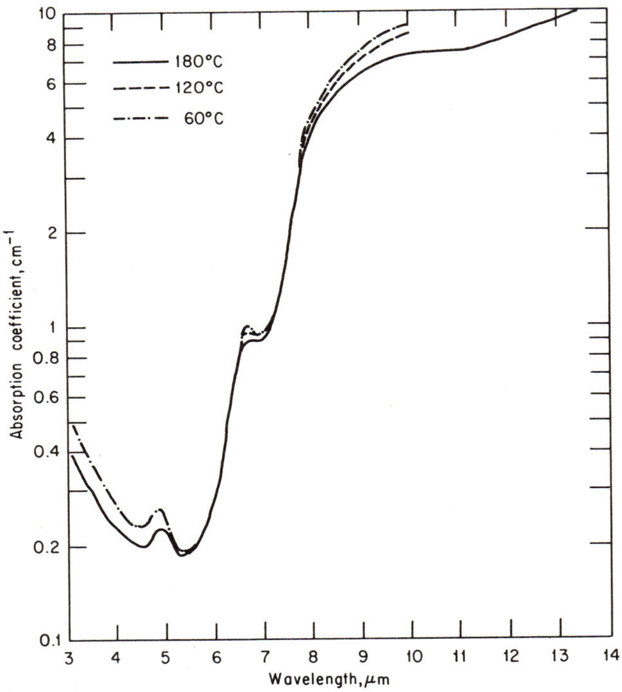

Fig. 74 Absorption coefficient of Irtran 1 at three temperatures. [*From Stierwalt et al.* (1963).]

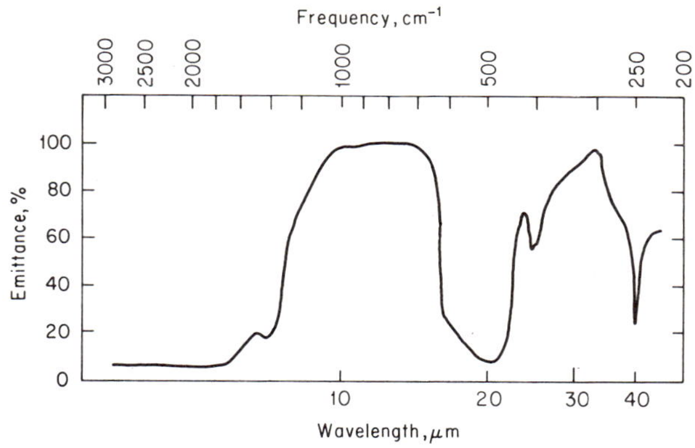

Fig. 75 Spectral emission of Irtran 1 at 77 K; thickness 2 mm. [*From Stierwalt* (1966*b*).]

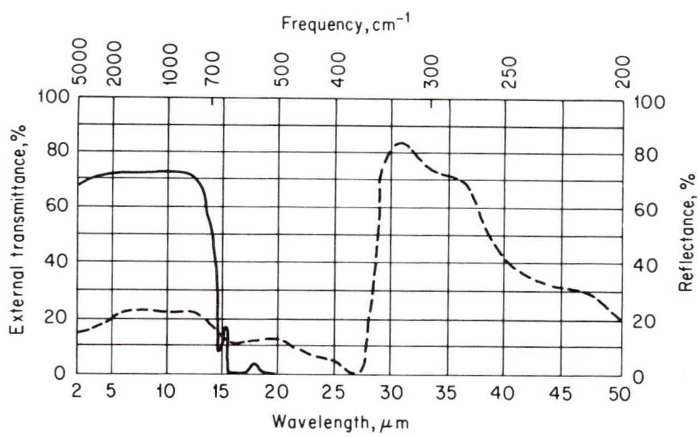

Fig. 76 Transmission (*solid curve*) and reflection (*dashed curve*) of Irtran 2; thickness 1 mm. [*From McCarthy* (1963*a*).]

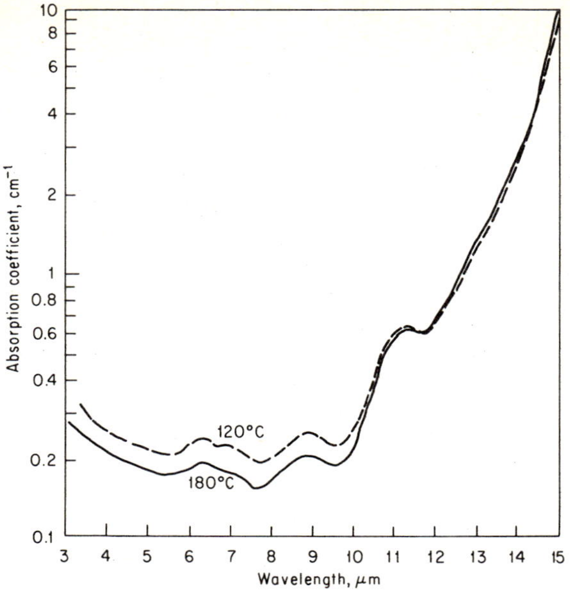

Fig. 77 Absorption coefficient of Irtran 2 at 120 and 180°C. [*From Stierwalt et al.* (1963).]

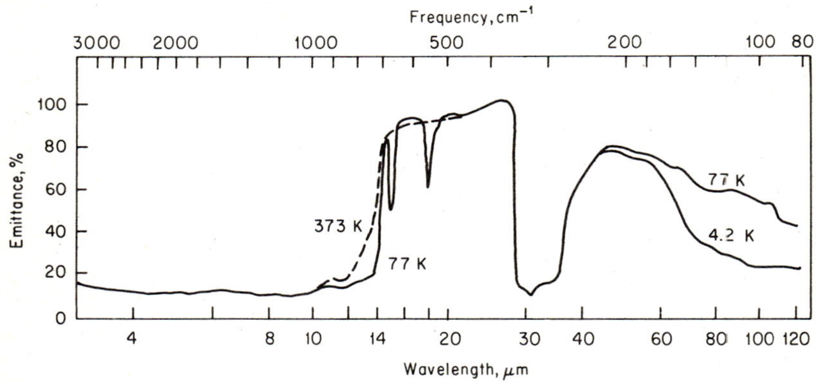

Fig. 78 Spectral emission of Irtran 2 at 4.2, 77, and 373 K; thickness 2 mm. [*From Stierwalt* (1966*b*).]

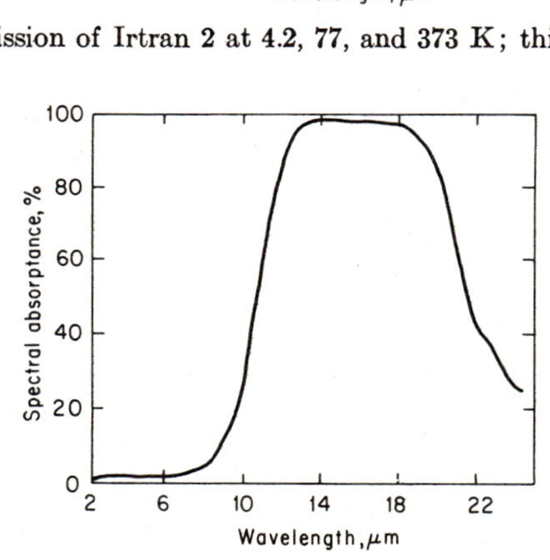

Fig. 79 Spectral absorption of Irtran 3 at 100°C. [*From Stierwalt et al.* (1963).]

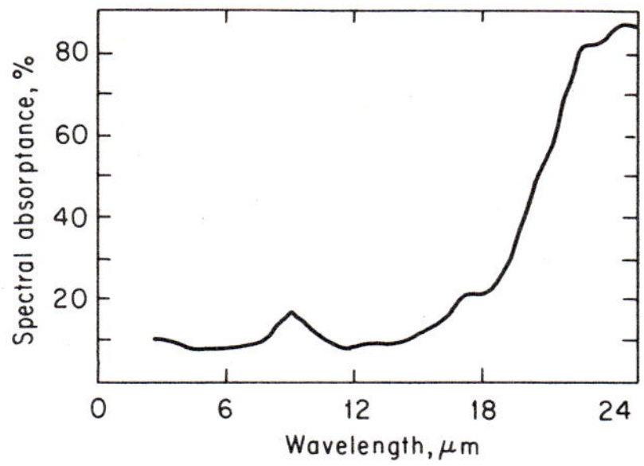

Fig. 80 Spectral absorption of Irtran 4 at 100°C. [*From Stierwalt et al.* (1963).]

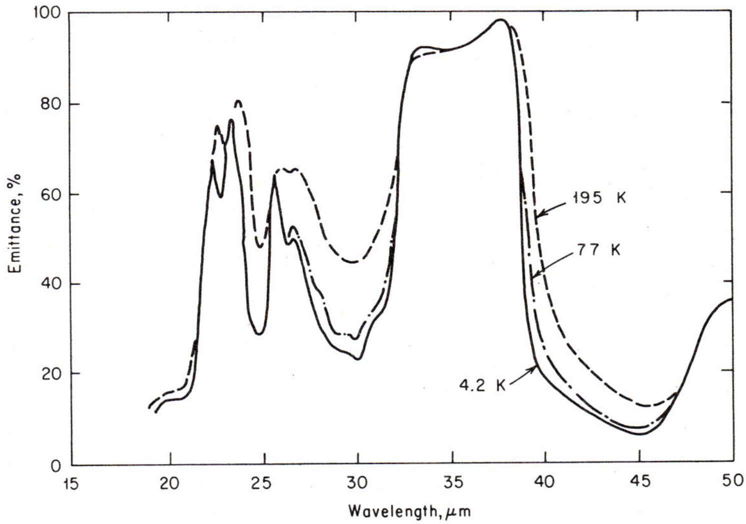

Fig. 81 Spectral emission of Irtran 4 at 4.2, 77, and 195 K; thickness 0.37 mm; d = 37 mm. [*From Stierwalt* (1966*b*).]

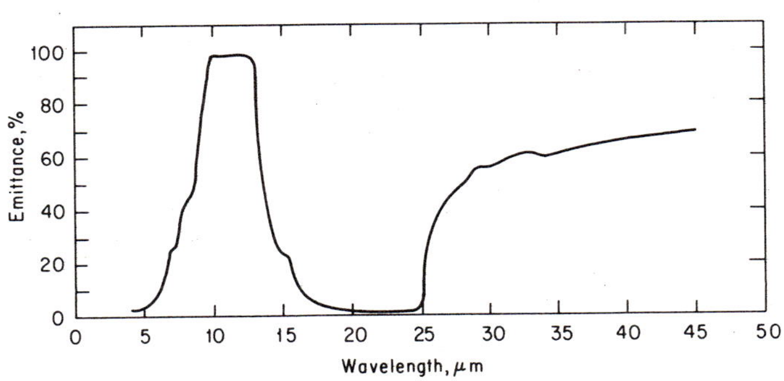

Fig. 82 Spectral emission of Irtran 5 at 77 K; thickness 2.11 mm. [*From Stierwalt* (1966*b*).]

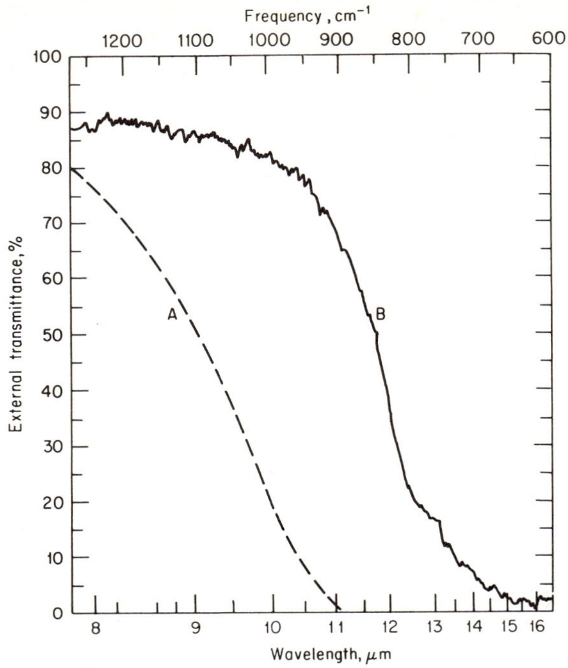

Fig. 83 Transmission of lanthanum fluoride; thickness 0.43 mm (curve *A*) and 11.7 mm (curve *B*). [*From Mooney* (1966).]

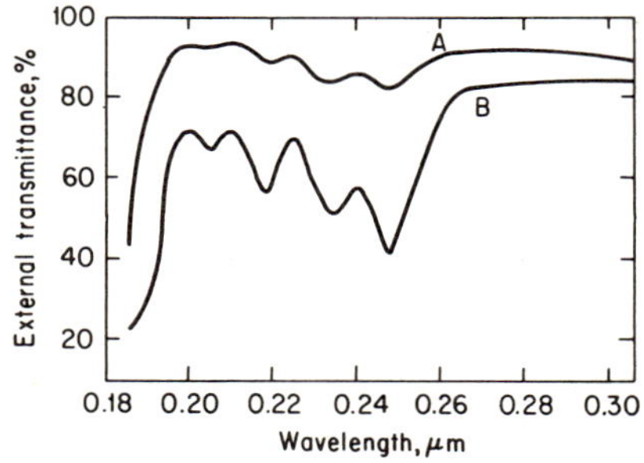

Fig. 84 Ultraviolet transmission of lanthanum fluoride; curve *A*: 1.3 mm thick, optical axis perpendicular to light; curve *B*: 8 mm thick, optical axis parallel to light. [*From Wirick* (1966).]

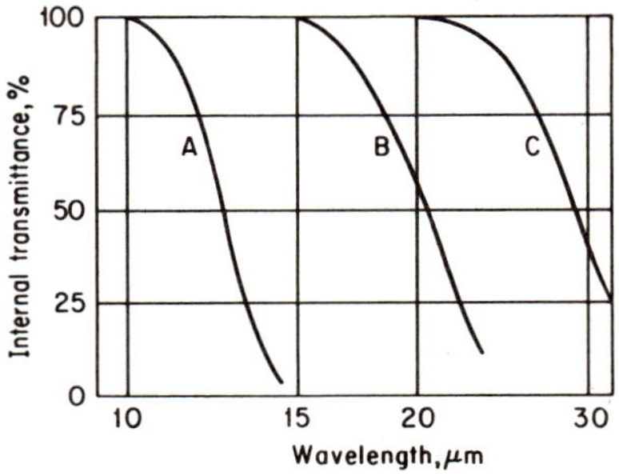

Fig. 85 Internal transmission of lead fluoride (curve *A*), 6 to 7 mm thick; lead chloride (curve *B*), 2 to 4 mm thick; and lead bromide (curve *C*), 3 to 9 mm thick. [*From Moss and Peacock* (1961).]

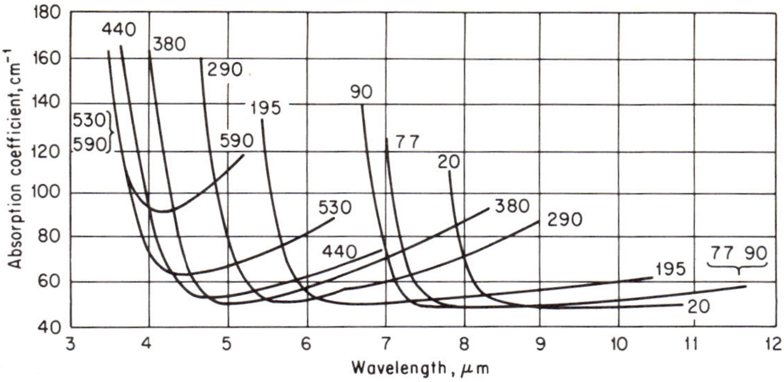

Fig. 86 Absorption coefficient of lead selenide for several Kelvin temperatures; thickness 0.68 mm; purity unknown. [*From Gibson* (1952); *copyright by the Institute of Physics.*]

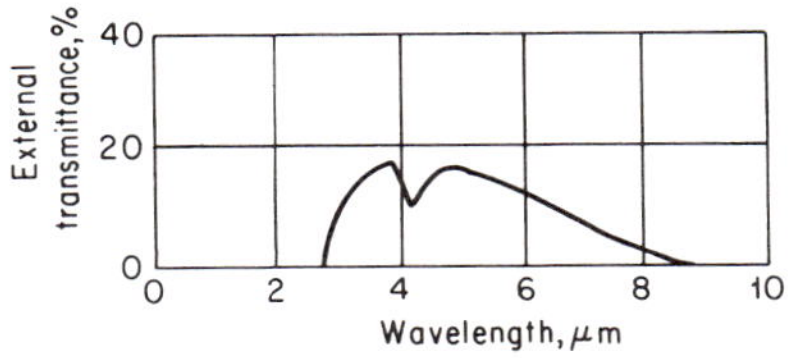

Fig. 87 Transmission of lead sulfide; thickness unknown; no detectable impurity by x-ray analysis. [*From Jones* (*no date*).]

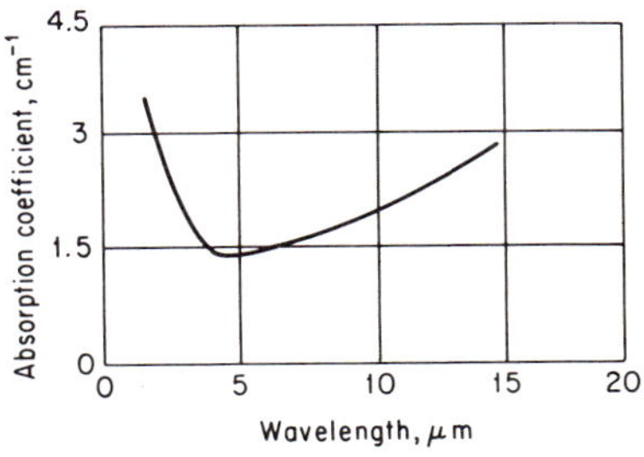

Fig. 88 Absorption of lead sulfide. [*From Jones* (*no date*).]

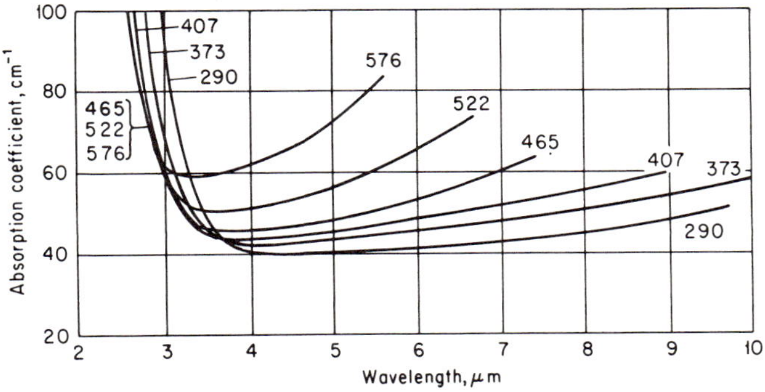

Fig. 89 Absorption of *n*-type lead sulfide for several Kelvin temperatures; purity unknown. [*From Gibson* (1952); *copyright by the Institute of Physics.*]

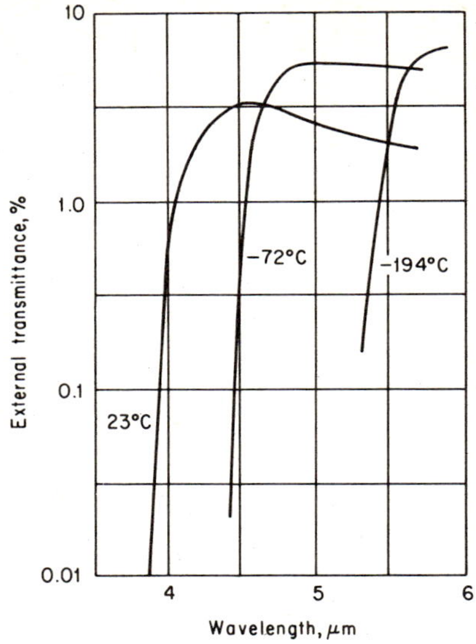

Fig. 90 Transmission of a lead telluride layer for 23, −72, and −194°C; thickness 0.11 mm. [*From Clark and Cashman* (1952).]

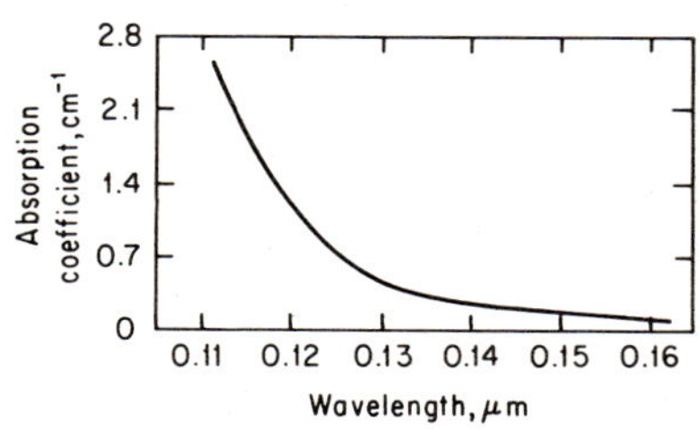

Fig. 91 Ultraviolet absorption of lithium fluoride. [*From Schneider* (1936).]

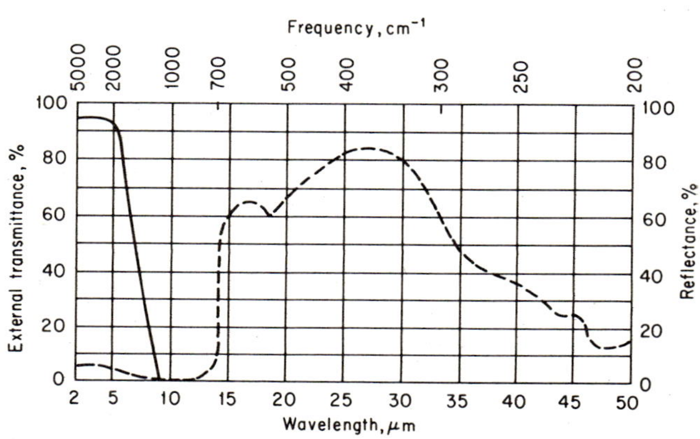

Fig. 92 Transmission (*solid curve*) and reflection (*dashed curve*) of lithium fluoride; thickness 5 mm. [*From McCarthy* (1963*a*).]

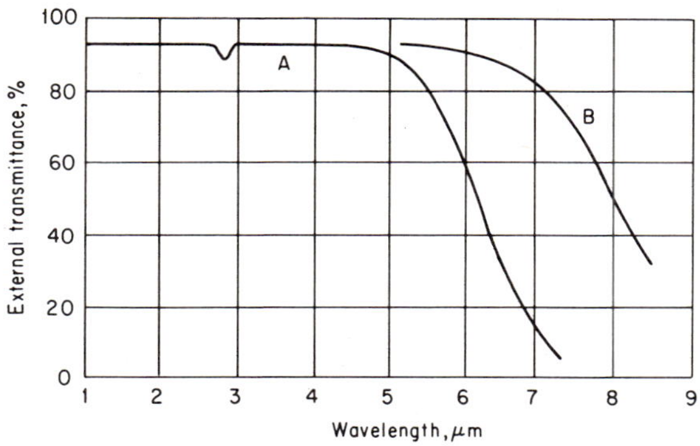

Fig. 93 Transmission of lithium fluoride; curve A: 12 mm; curve B: 1 mm. [*From Smith et al.* (1957).]

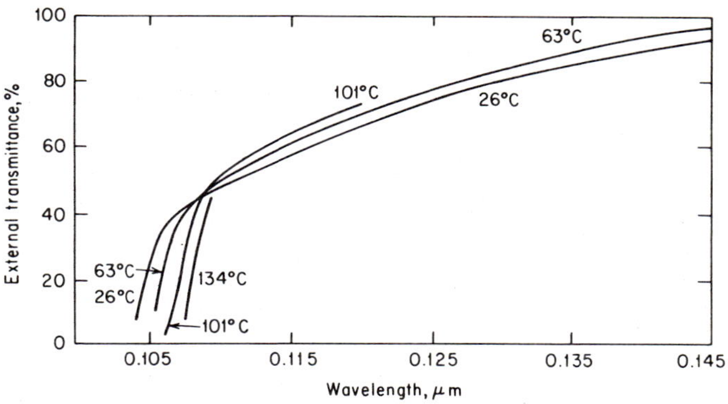

Fig. 94 Transmission of lithium fluoride at 26, 63, 101, and 134°C; thickness 1.55 mm (short-wavelength portion). [*From Laufer et al.* (1965).]

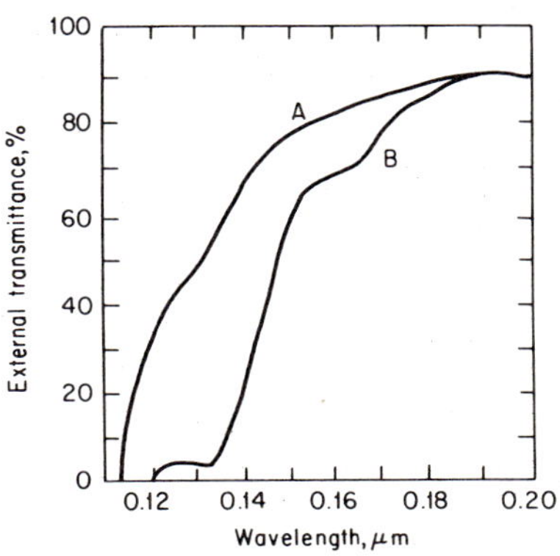

Fig. 95 Ultraviolet transmission of single-crystal magnesium fluoride; thickness 14.82 mm (curve A) and 1.40 mm (curve B). [*From Steinmetz et al.* (1967).]

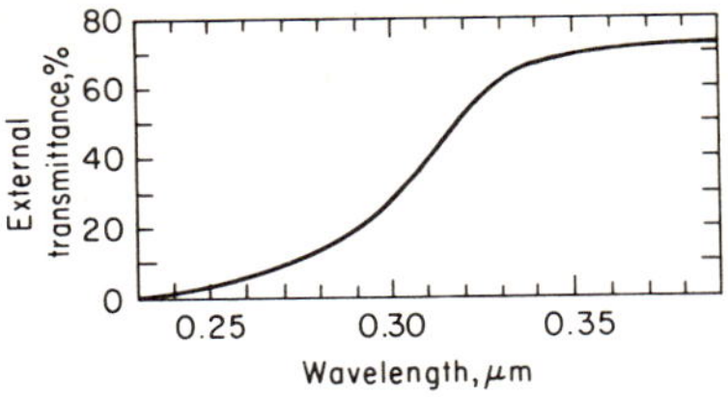

Fig. 96 Transmission of magnesium oxide in the ultraviolet; thickness 0.49 mm. [*From Sutton (no date).*]

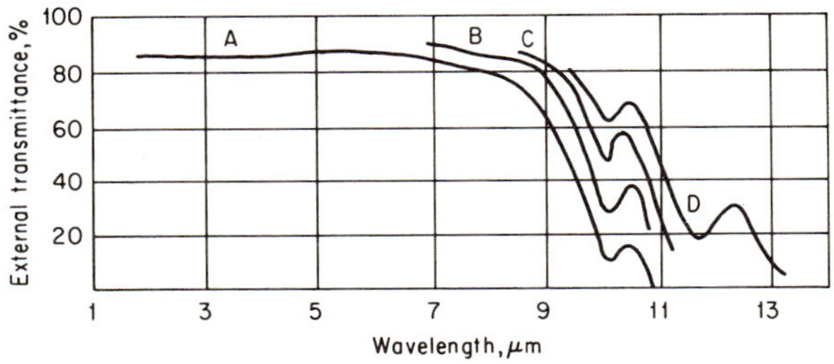

Fig. 97 Transmission of magnesium oxide; curve *A*: 0.468 mm; *B*: 0.235 mm; *C*: 0.124 mm; *D*: 0.076 mm thick. [*From Willmott* (1950); *copyright by the Institute of Physics.*]

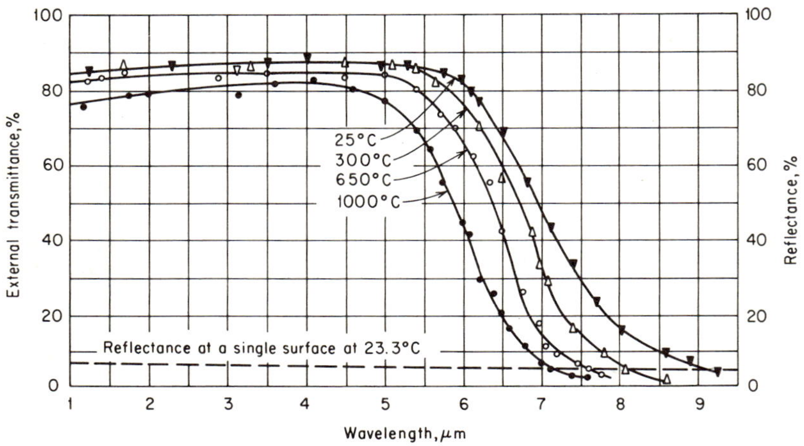

Fig. 98 Transmission of magnesium oxide at several temperatures; thickness 5.5 mm. [*From Oppenheim and Goldman* (1964).]

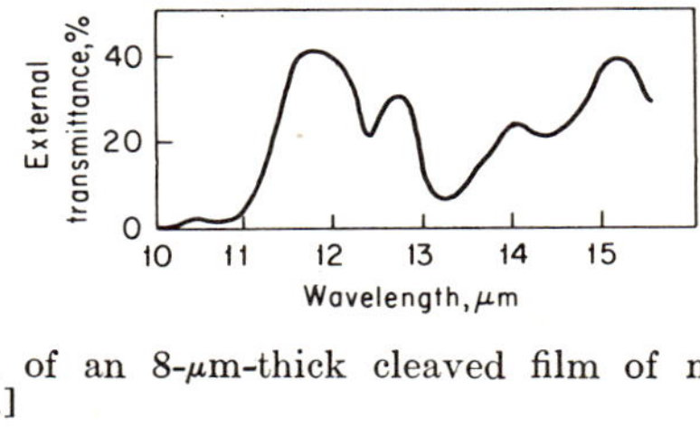

Fig. 99 Transmission of an 8-μm-thick cleaved film of muscovite mica. [*From Ruthberg et al.* (1963).]

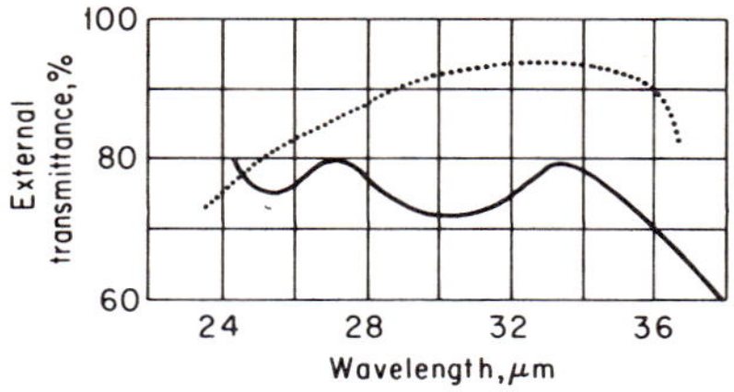

Fig. 100 Transmission of polyethylene (*dashed curve*), thickness 0.24 mm, and polystyrene (*solid curve*), thickness 0.025 mm. [*From Plyler* (1948*b*).]

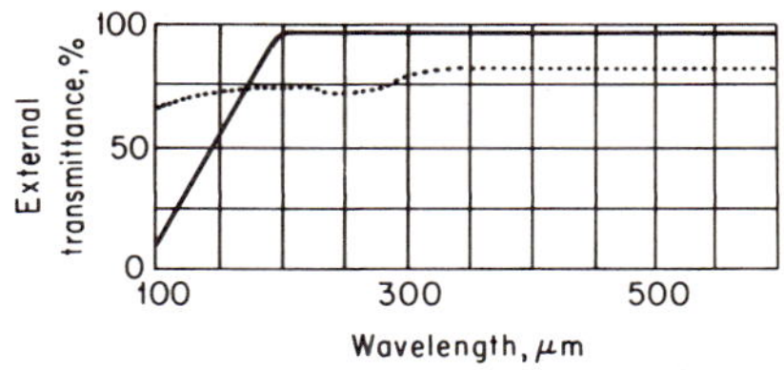

Fig. 101 Transmission of polystyrene (*solid curve*), thickness 5 mm. [*Data from Decamps and Hadni* (1960).] Transmission of polyethylene (*dashed curve*), thickness 3.21 mm. [*Data from McCubbin and Sinton* (1950).]

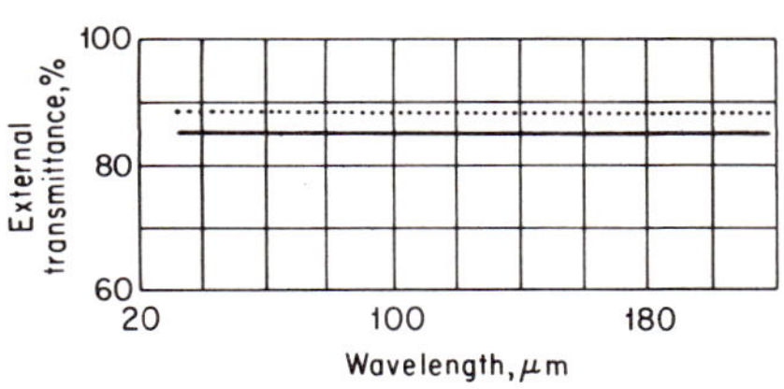

Fig. 102 Transmission of polyethylene 0.64 mm thick (*dotted curve*) and 1.0 mm thick (*solid curve*). [*Data from Ballard et al.* (*no date*).]

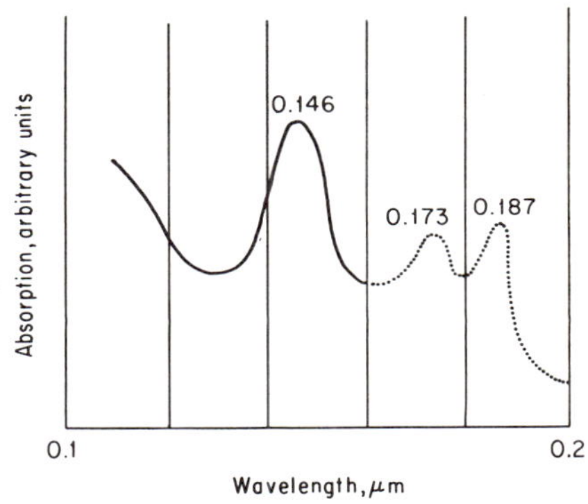

Fig. 103 Ultraviolet absorption of potassium bromide. [*From Schneider and O'Bryan* (1937); *Hilsch and Pohl* (1930).]

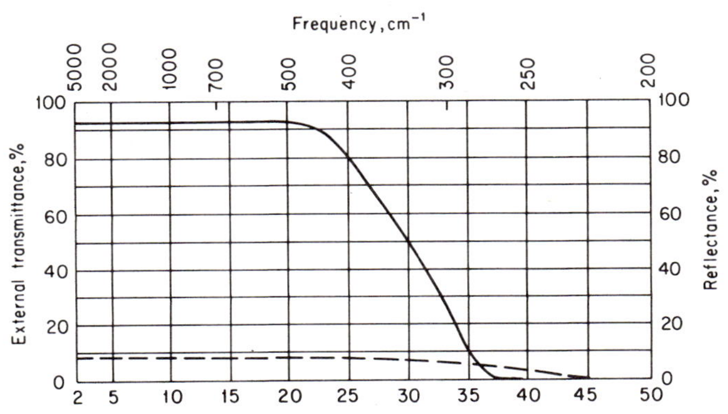

Fig. 104 Transmission (*solid curve*) and reflection (*dashed curve*) of potassium bromide, thickness 5 mm. [*From McCarthy* (1963*a*).]

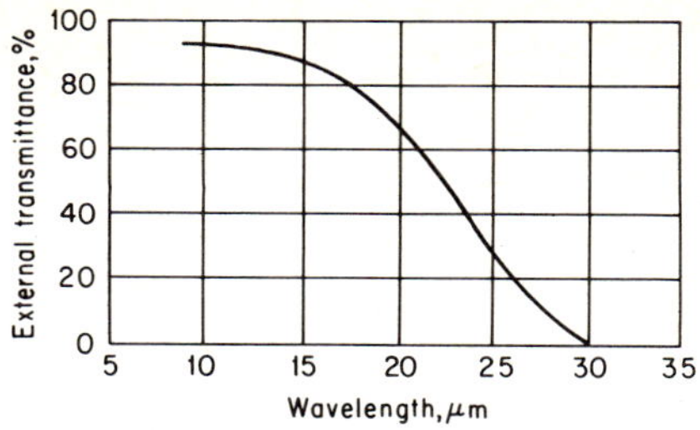

Fig. 105 Transmission of potassium chloride, thickness 5.30 mm. [*From Plyler* (*no date*).]

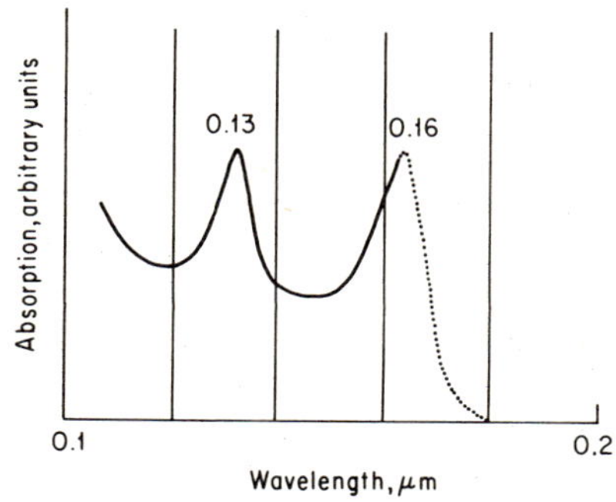

Fig. 106 Ultraviolet absorption of potassium chloride. [*From Schneider and O'Bryan* (1937); *Hilsch and Pohl* (1930).]

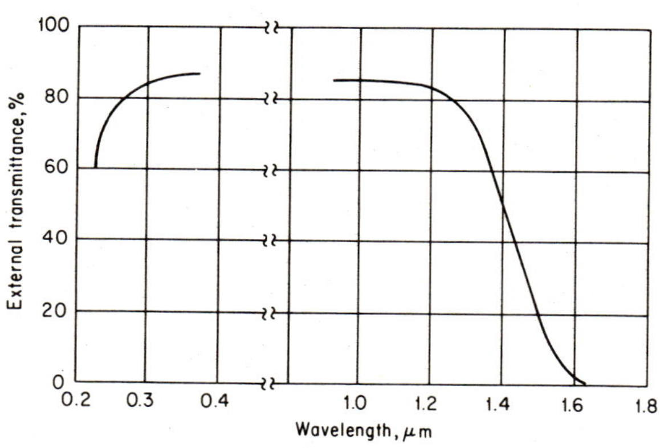

Fig. 107 Transmission of potassium dihydrogen phosphate (KDP), thickness 1.6 mm (short-wavelength portion) and 12.5 mm (long-wavelength portion). [*From Deshotels* (1960); *McDonough* (*no date*).]

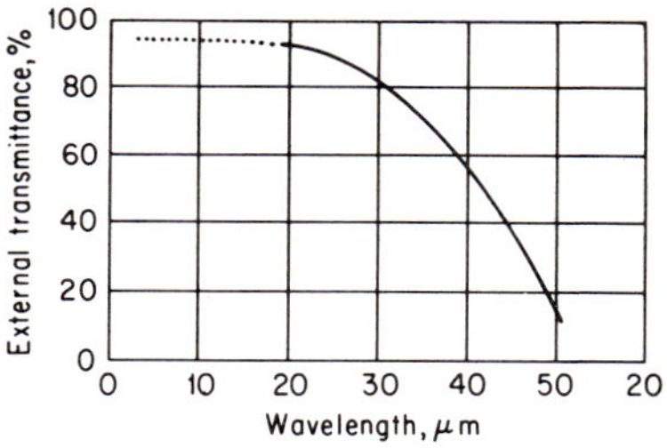

Fig. 108 Infrared transmission of potassium iodide, thickness 0.83 mm. [*From Strong* (1931).]

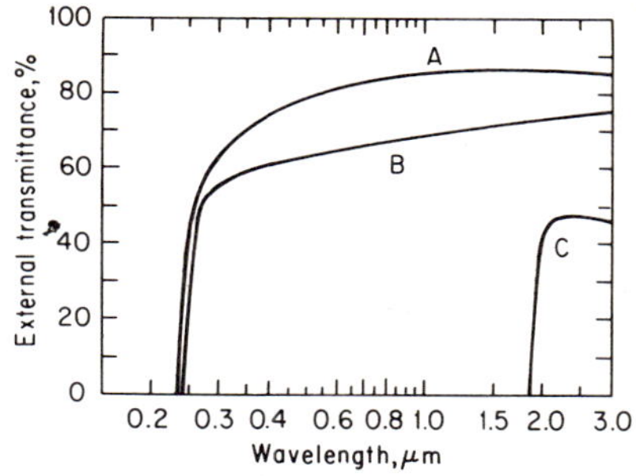

Fig. 109 Transmission of potassium iodide (curve *A*), thickness 4.3 mm; gallium antimonide (curve *B*), thickness 0.5 mm; and rubidium iodide (curve *C*), thickness 1.36 mm. [*From McCarthy* (1968).]

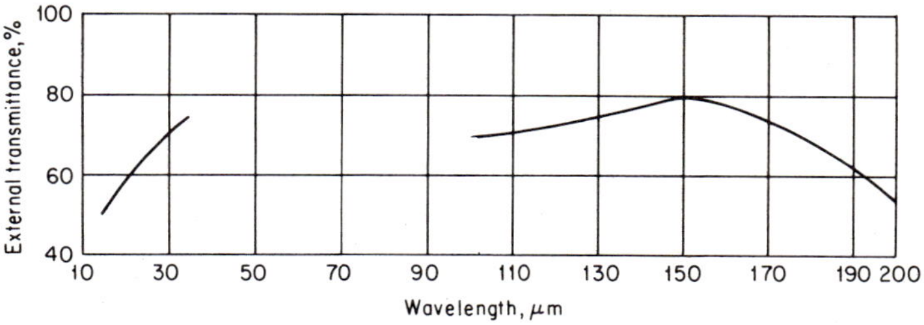

Fig. 110 Transmission of pyrographite; thickness 0.0076 mm. [*From Landers et al. (no date), based on unpublished data of Raytheon Company.*]

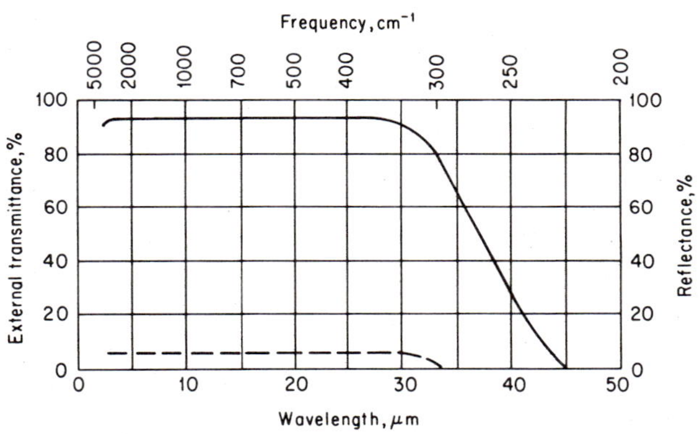

Fig. 111 Transmission (*solid curve*) and reflection (*dashed curve*) of rubidium bromide, thickness 5.31 mm. [*From McCarthy* (*no date*).]

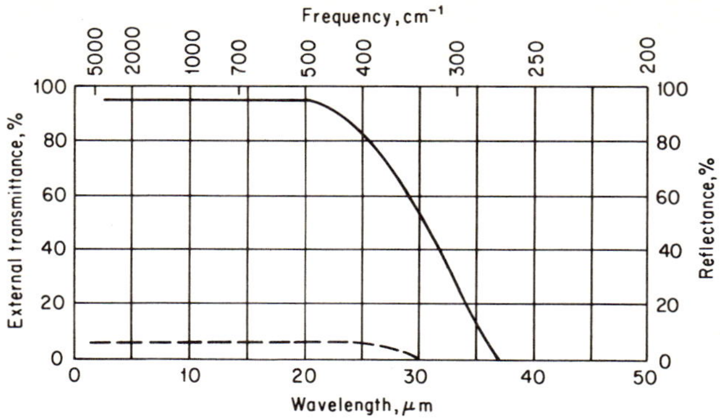

Fig. 112 Transmission (*solid curve*) and reflection (*dashed curve*) of rubidium chloride, thickness 3.0 mm. [*From McCarthy* (*no date*).]

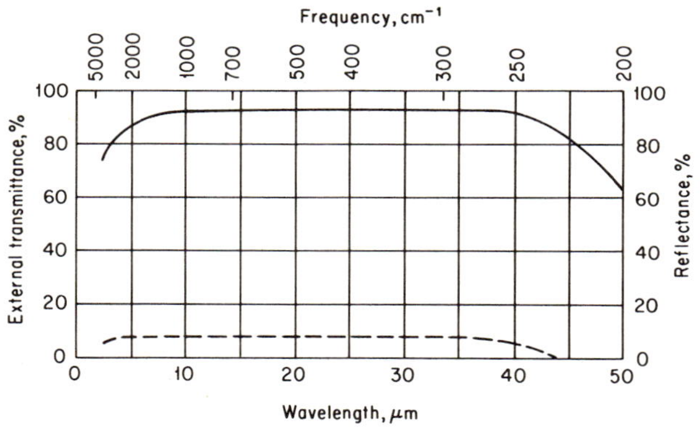

Fig. 113 Transmission (*solid curve*) and reflection (*dashed curve*) of rubidium iodide, thickness 3.91 mm. [*From McCarthy* (*no date*).]

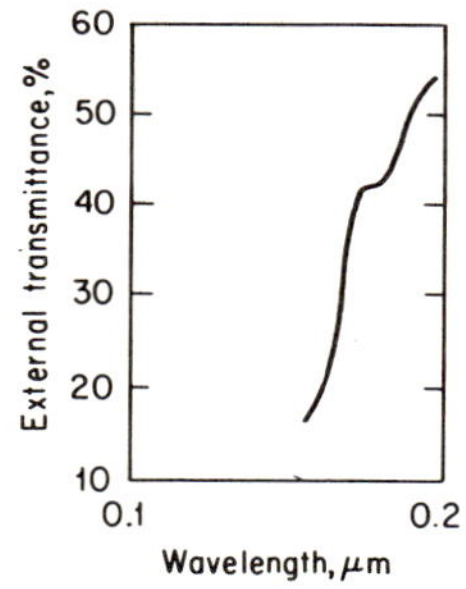

Fig. 114 Transmission of sapphire in the ultraviolet, thickness 1.5 mm. [*From Bauple et al.* (1950).]

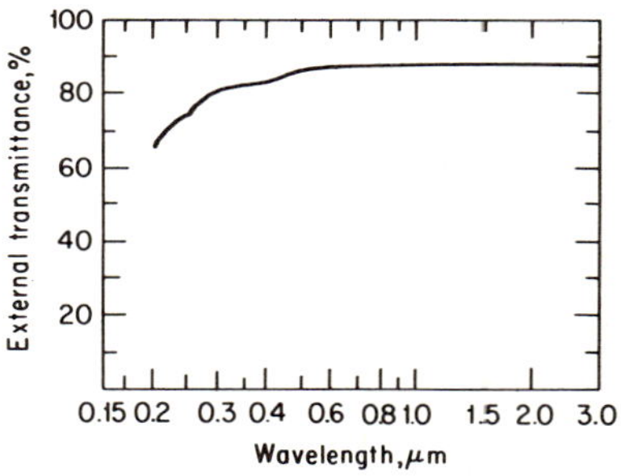

Fig. 115 Short-wavelength transmission of sapphire, thickness 3 mm. [*From McCarthy* (1967).]

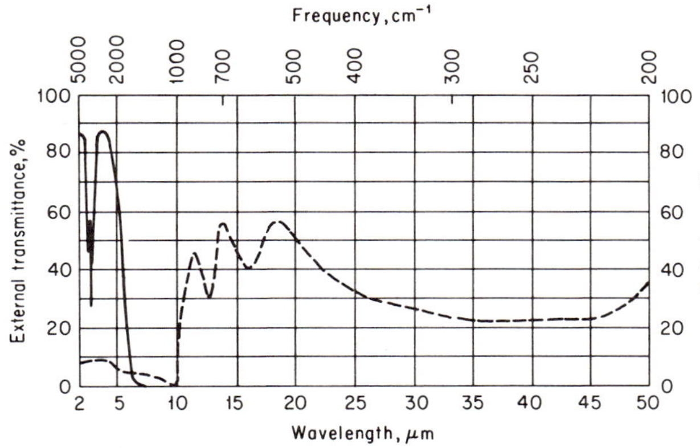

Fig. 116 Transmission (*solid curve*) and reflection (*dashed curve*) of sapphire, thickness 2 mm. [*From McCarthy* (1963*b*).]

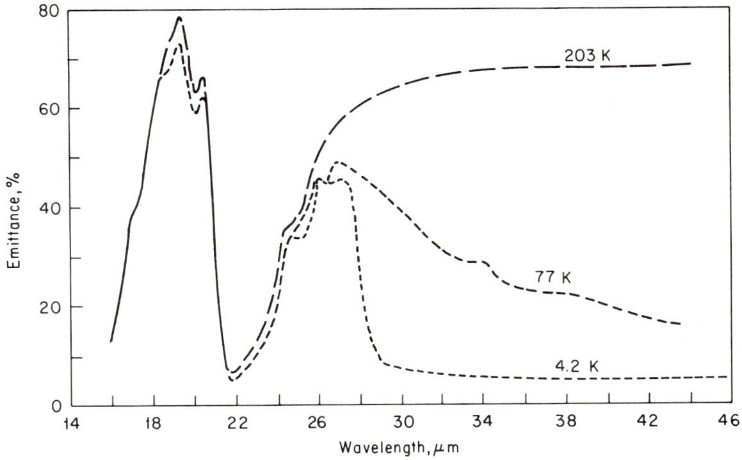

Fig. 117 Spectral emission of sapphire at three temperatures; thickness 0.79 mm. [*From Stierwalt* (1966*a*).]

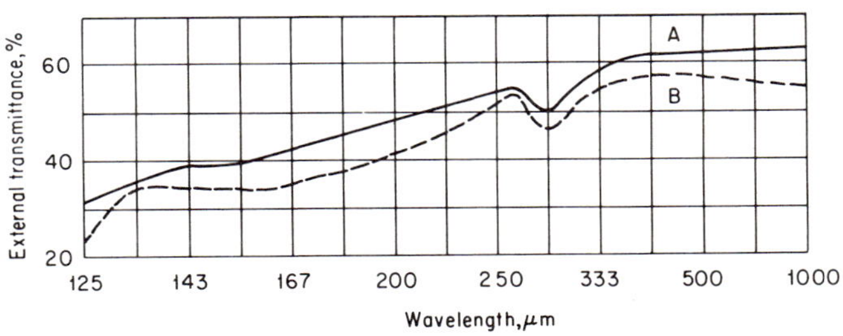

Fig. 118 Long-wavelength transmission of sapphire, thickness 1.0 mm; curve *A*: sample with unoriented axis; curve *B*: sample with the optical axis parallel to the face. [*Data from Loewenstein* (1961).]

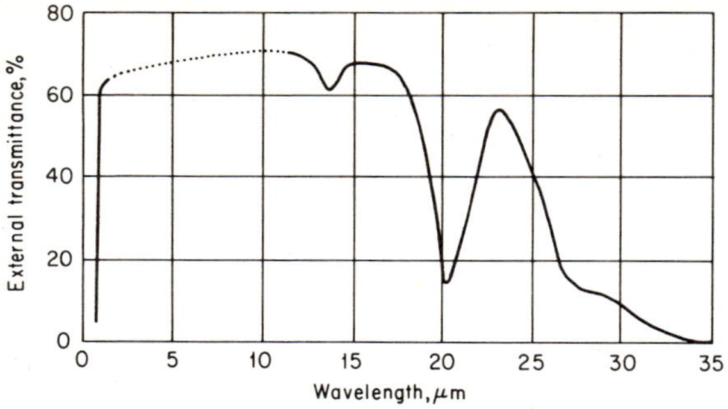

Fig. 119 Transmission of amorphous selenium, thickness 1.69 mm. [*From Caldwell* (1958).]

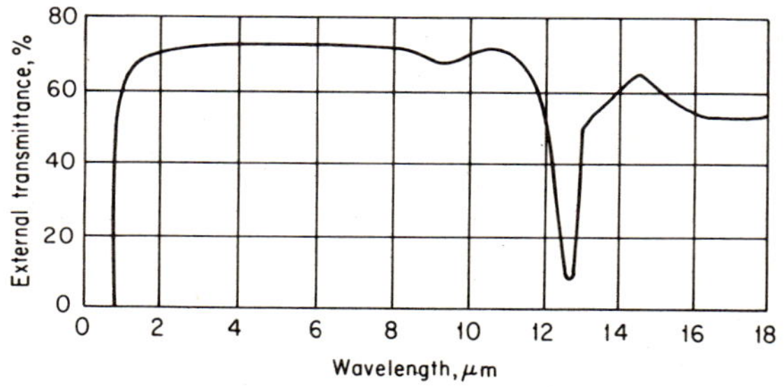

Fig. 120 Transmission of arsenic-modified selenium glass, thickness 2 mm. The properties of glass vary from batch to batch; this curve can be regarded as typical. [*From Salzberg and Villa* (1957).]

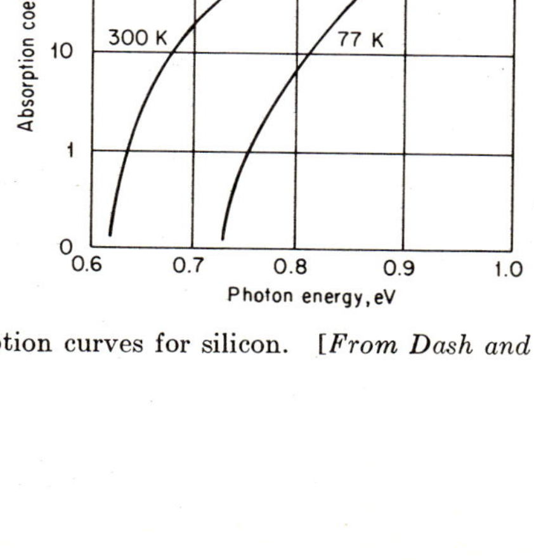

Fig. 121 Absorption curves for silicon. [*From Dash and Newman* (1955).]

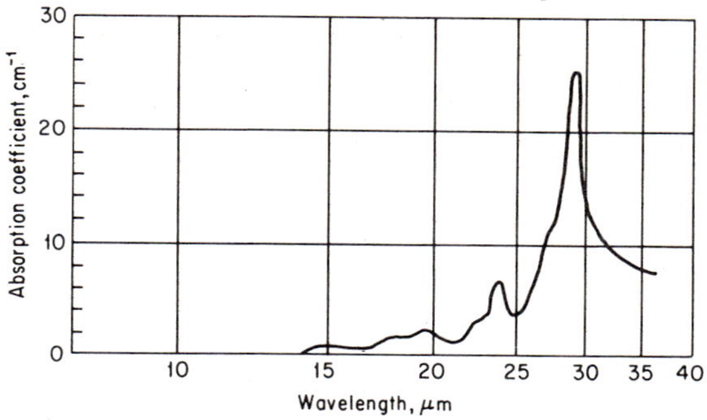

Fig. 122 Lattice absorption curve for silicon. [*From Lax and Burstein* (1955).] A summary of the short-wavelength absorption in silicon is given by Moss (1959*a*).

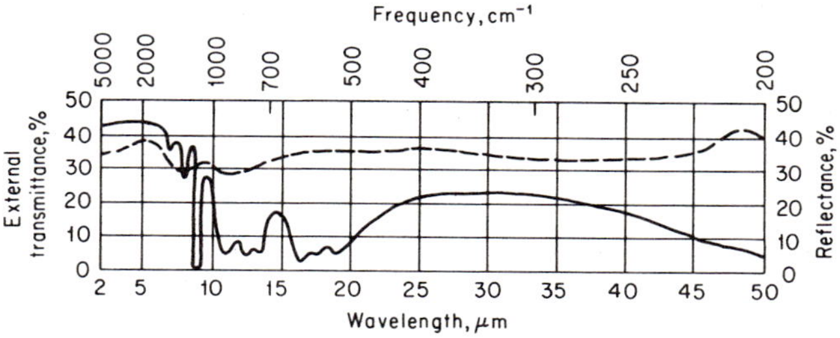

Fig. 123 Transmission (*solid curve*) and reflection (*dashed curve*) of silicon, thickness 10 mm. [*From McCarthy* (1963*a*).]

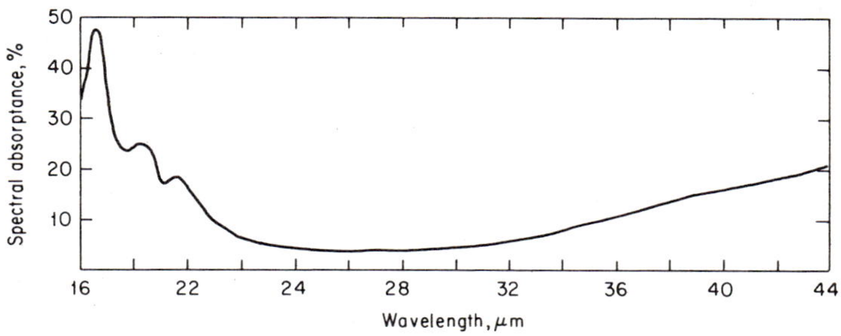

Fig. 124 Spectral absorption of n-type silicon at 77 K. [*From Stierwalt et al.* (1963).]

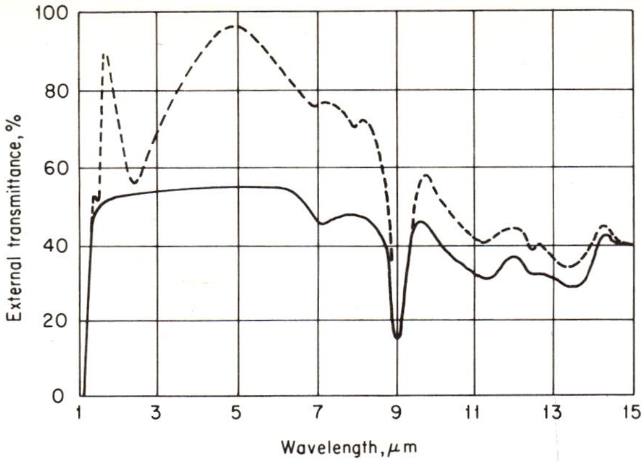

Fig. 125 Transmission of silicon, thickness 2.5 mm. Dashed curve is for a sample coated to reduce reflection loss. [*From Texas Instruments* (*no date*).]

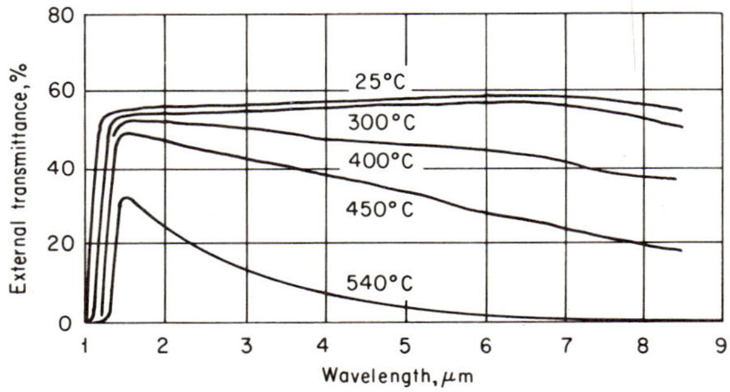

Fig. 126 Transmission of silicon for several temperatures, thickness 5.89 mm; resistivity at least 5 Ω·cm. [*From Phillippi and Beardsley* (1955).]

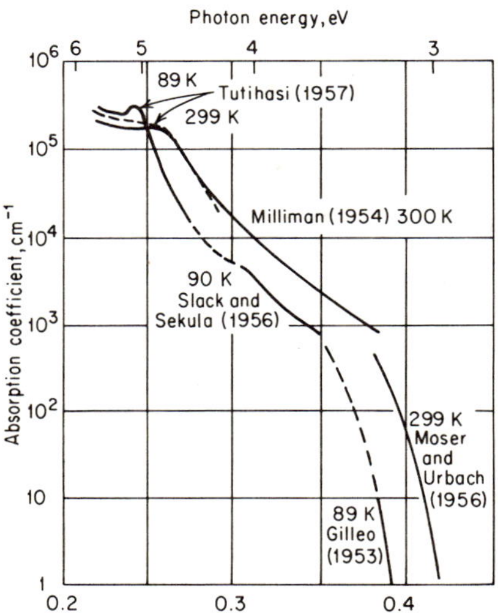

Fig. 127 Absorption of silver chloride in the ultraviolet and the blue portion of the visible spectrum. Data of Milliman and of Slack and Sekula were taken on fused film; other data were taken on single crystals. [*From Tutihasi* (1957).]

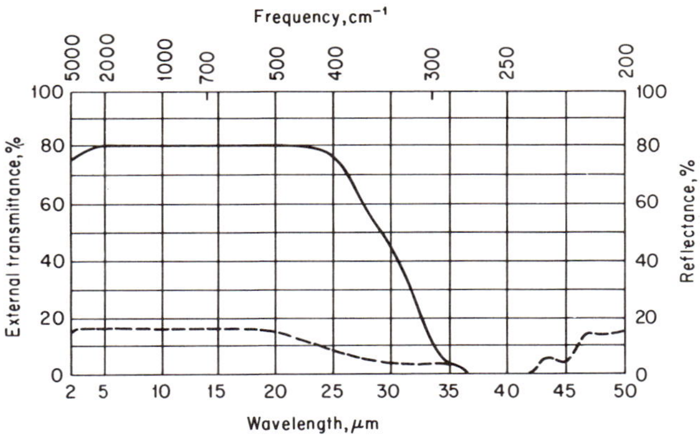

Fig. 128 Transmission (*solid curve*) and reflection (*dashed curve*) of silver chloride; thickness 0.5 mm. [*From McCarthy* (1963*a*).]

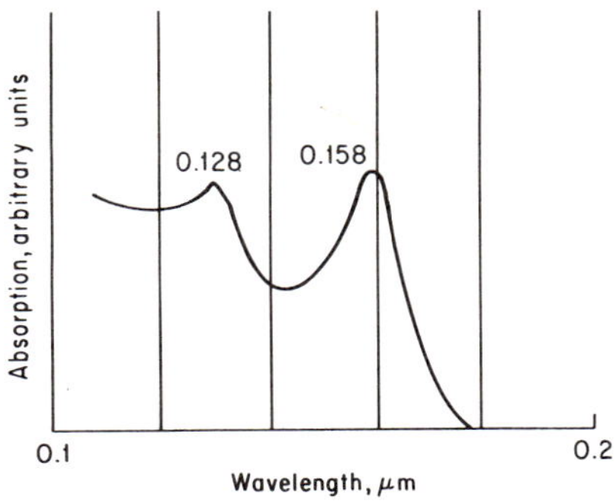

Fig. 129 Ultraviolet absorption of sodium chloride. [*From Schneider and O'Bryan* (1937).]

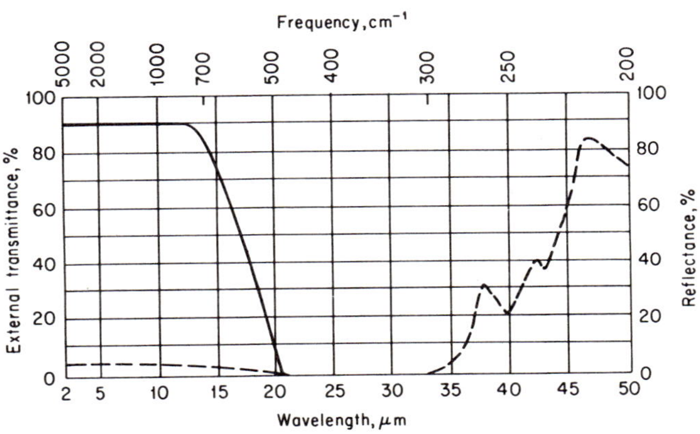

Fig. 130 Transmission (*solid curve*) and reflection (*dashed curve*) of sodium chloride; thickness 5 mm. [*From McCarthy* (1963*a*).]

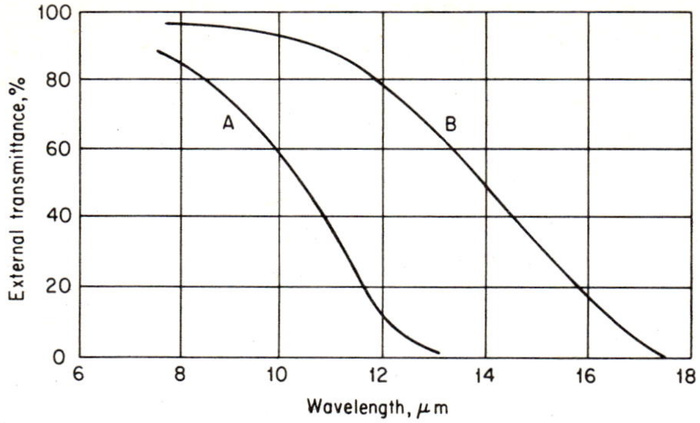

Fig. 131 Infrared transmission of sodium fluoride for (*A*) 10 mm and (*B*) 1 mm. [*From Ballard (no date).*]

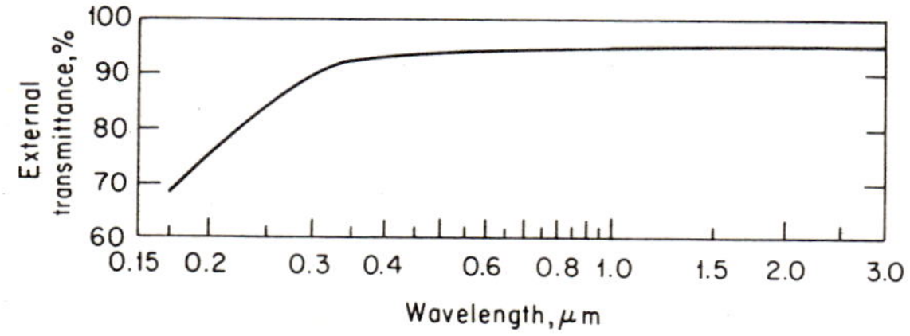

Fig. 132 Transmission of sodium fluoride, thickness 2.16 mm (short-wavelength portion). [*Adapted from McCarthy* (1967).]

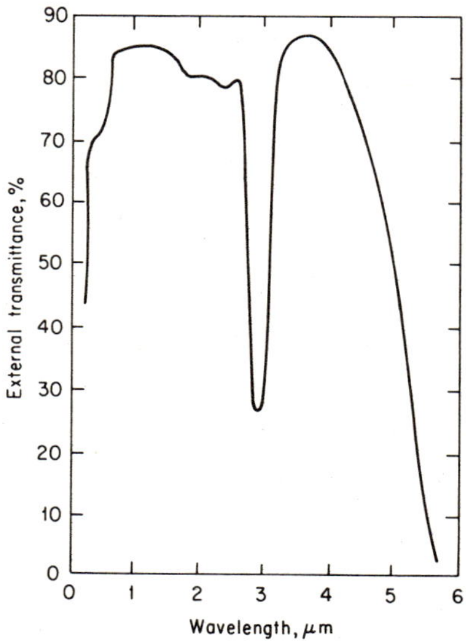

Fig. 133 Transmission of spinel, thickness 5.45 mm. [*From Calingaert et al.* (1936).] The absorption band at 2.8 μm may be due to entrapped water. The mixture ratio 1:3.5 magnesium oxide to aluminum dioxide is not unique; other ratios that exist seem to give essentially the same physical properties. The 1:3.5 mixture has been an easy one from which to grow gemstones and is usually homogeneous.

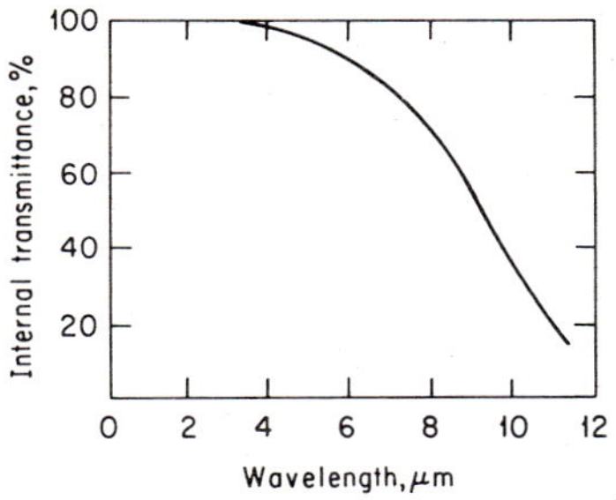

Fig. 134 Internal transmission of strontium fluoride, thickness 10 mm. [*From Jones et al.* 1952); *copyright by the National Institute of Physics.*]

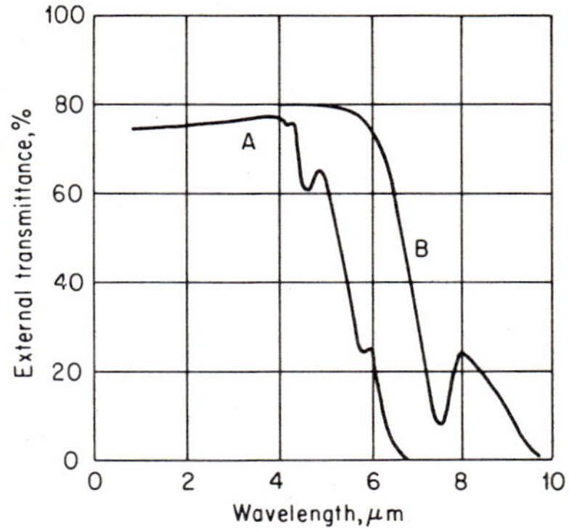

Fig. 135 Transmission of strontium titanate for (*A*) 3.7 mm and (*B*) 0.26 mm. [*From Parsons (no date).*]

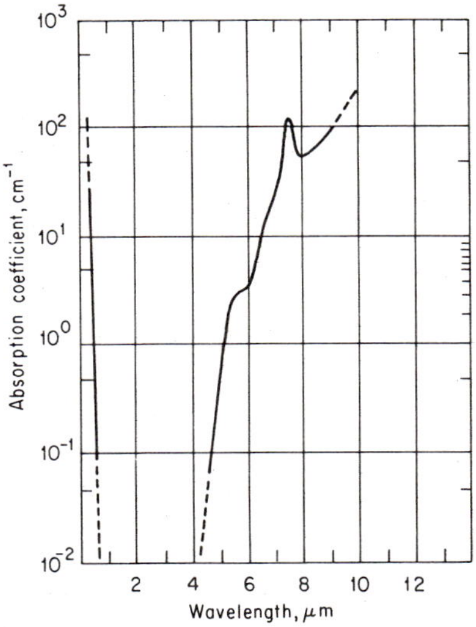

Fig. 136 Absorption coefficient of synthetic strontium titanate in the ultraviolet and infrared regions. [*From Levin et al.* (1955).] The same authors also report the impurity content and spectrographic analysis of the boule feed material.

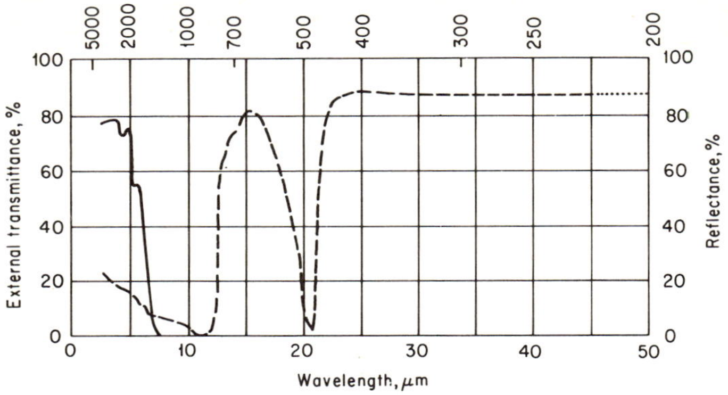

Fig. 137 Transmission (*solid curve*) and reflection (*dashed curve*) of strontium titanate, thickness 1.0 mm. [*From McCarthy (no date).*]

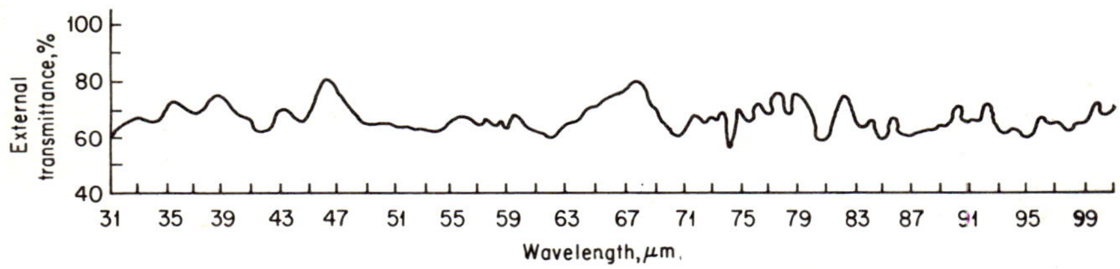

Fig. 138 Transmission of rhombic sulfur, thickness 0.7 mm. [*From MacNeill (1969).*]

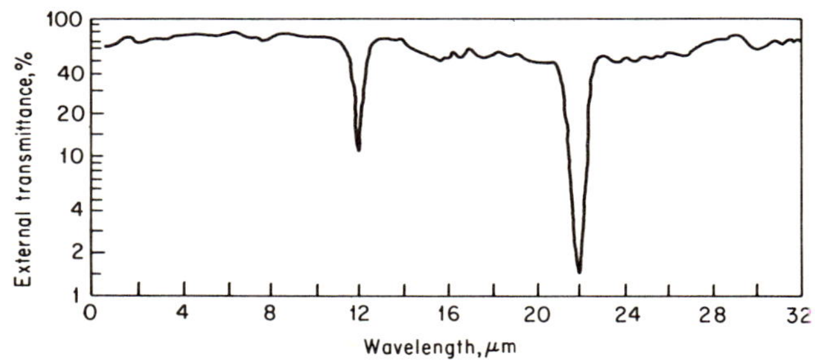

Fig. 139 Transmission of rhombic sulfur, thickness 0.4 mm. [*From MacNeill (1963).*]

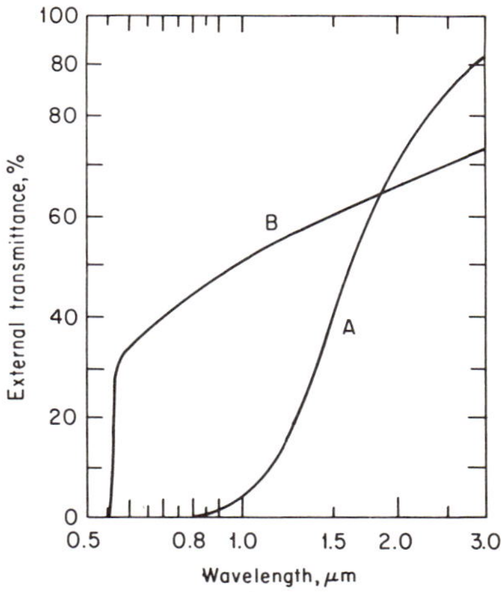

Fig. 140 Transmission of T-12,† thickness 4.0 mm (curve *A*) and KRS-5, thickness 2.0 mm (curve *B*). [*From McCarthy* (1967).]

† T-12 is trade name of the Harshaw Chemical Company, Cleveland, Ohio, for an optically integral two-phase polycrystalline body consisting of barium fluoride and calcium fluoride in nearly equal molar proportions.

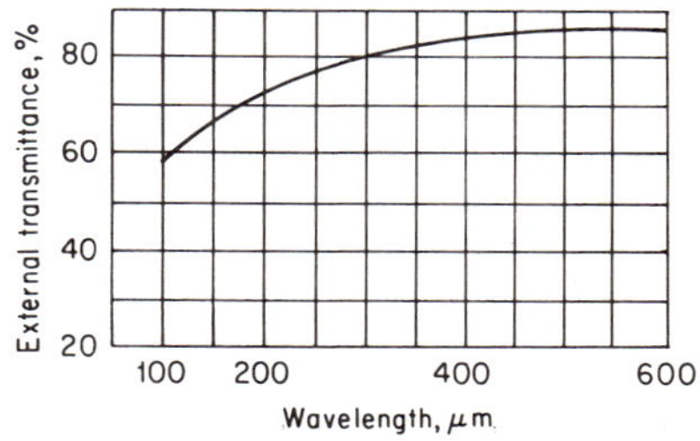

Fig. 141 Transmission of Teflon‡ (polytetrafluorethylene, thickness 1.81 mm. [*Data from McCubbin and Sinton* (1950).]

‡ Trademark of E. I. du Pont de Nemours & Co.

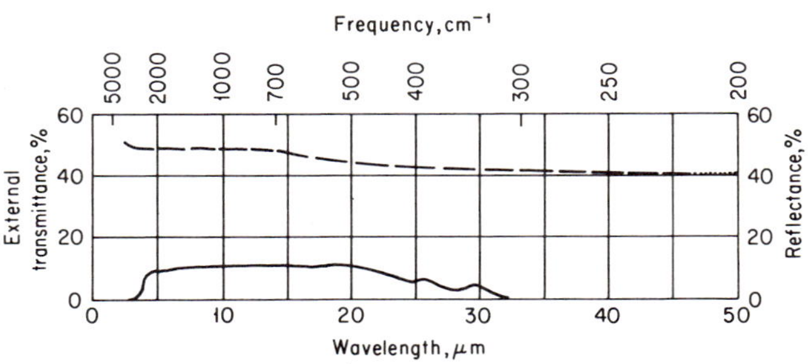

Fig. 142 Transmission (*solid curve*) and reflection (*dashed curve*) of tellurium, thickness 3.5 mm. [*From McCarthy* (*no date*).]

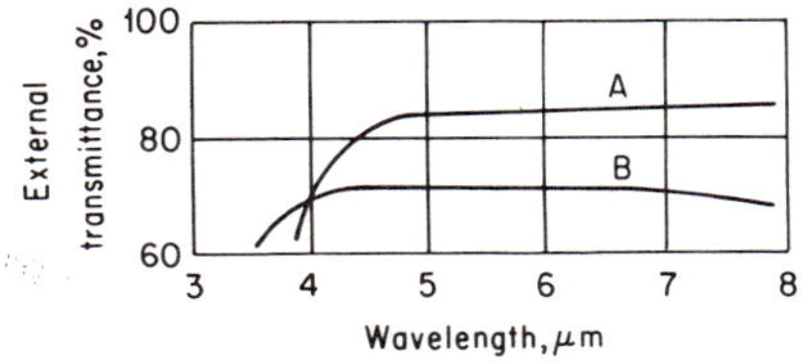

Fig. 143 Transmission of tellurium (*A*) **E** perpendicular to *c* and (*B*) **E** parallel to *c*; thickness 0.85 mm. [*From Loferski* (1954).] Tellurium is an interesting anisotropic crystal. Loferski gives more data on the polarization effects, and Caldwell (1958) has also given data. Optical activity has been observed by Nomura (1960).

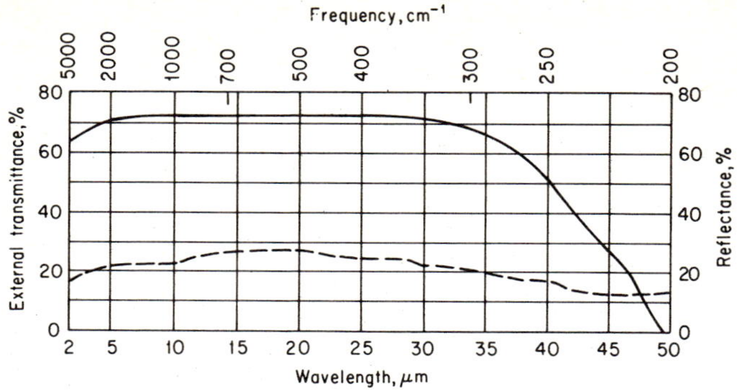

Fig. 144 Transmission (*solid curve*) and reflection (*dashed curve*) of thallium bromide, thickness 1.65 mm. [*Adapted from McCarthy* (1965*b*).]

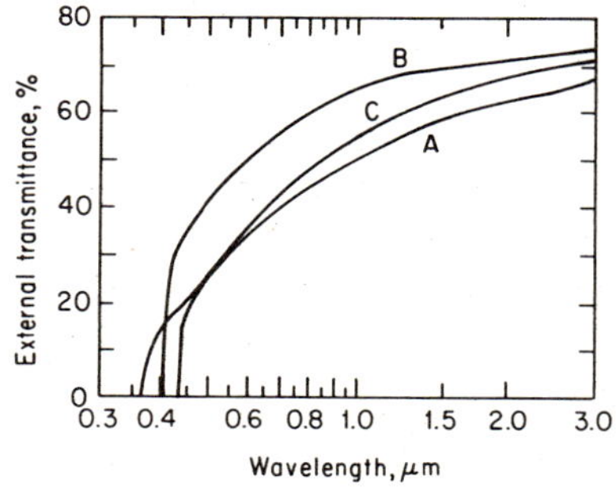

Fig. 145 Transmission of (*A*) thallium bromide, (*B*) thallium bromide-chloride (KRS-6), and (*C*) thallium chloride, all 1.65 mm thick. [*From McCarthy* (1967).] Thallium bromide can be ground a very small amount at a time without cracking or chipping. It bends like lead and is slightly soluble in water.

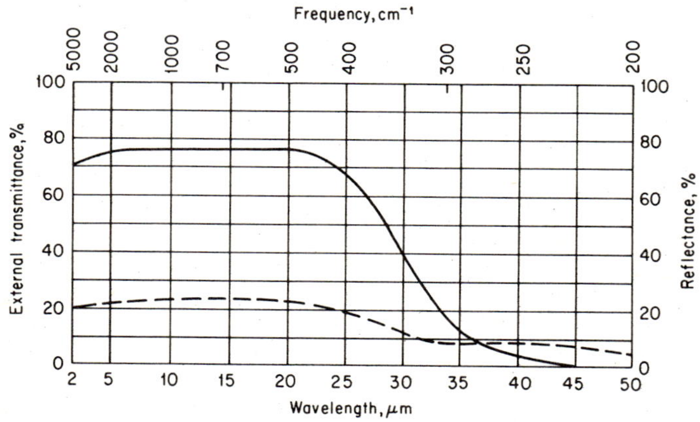

Fig. 146 Transmission (*solid curve*) and reflection (*dashed curve*) of thallium bromide-chloride (KRS-6), thickness 1.65 mm. [*From McCarthy* (1965*b*).]

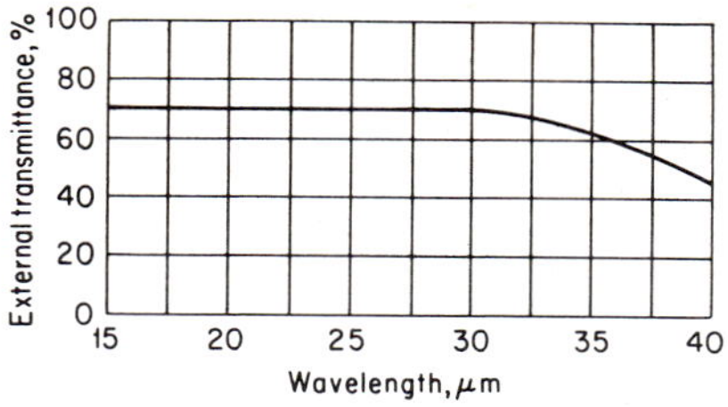

Fig. 147 Transmission of KRS-5, thickness 5.15 mm. [*From Plyler* (*no date*).]

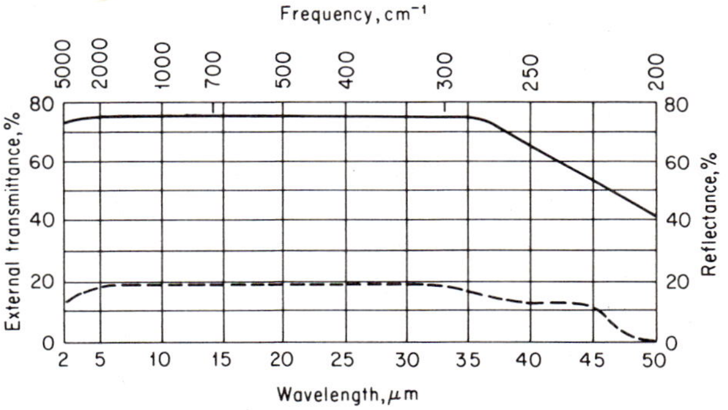

Fig. 148 Transmission (*solid curve*) and reflection (*dashed curve*) of KRS-5, thickness 2 mm. [*From McCarthy* (1963*a*).]

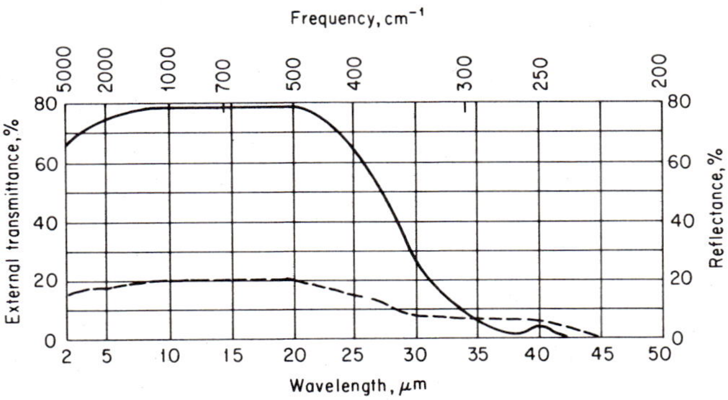

Fig. 149 Transmission (*solid curve*) and reflection (*dashed curve*) of thallium chloride, thickness 1.65 mm. [*From McCarthy* (1965*b*).] Thallium chloride can be cut easily on a diamond saw (melted beeswax is used as a lubricant while it is being sawed). It is difficult to grind to dimension since the material flows to the edge and fills the grinding wheel. Thin strips bend like lead. It is soluble in water.

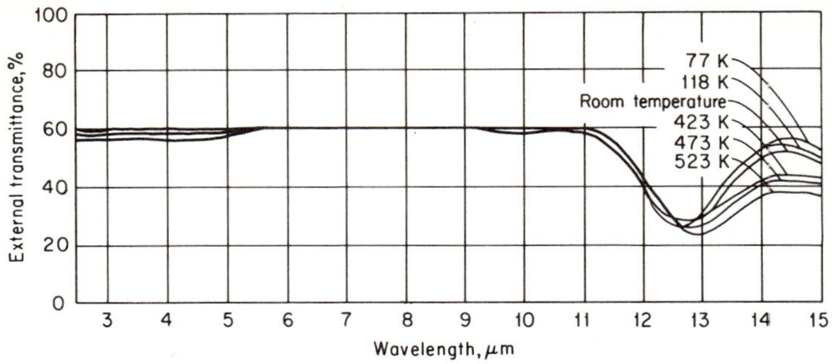

Fig. 150 Transmission of TI-20, $Ge_{33}Se_{55}As_{12}$ glass, at a variety of temperatures, thickness 6 mm. [*From Wolfe and Platt (no date)*.]

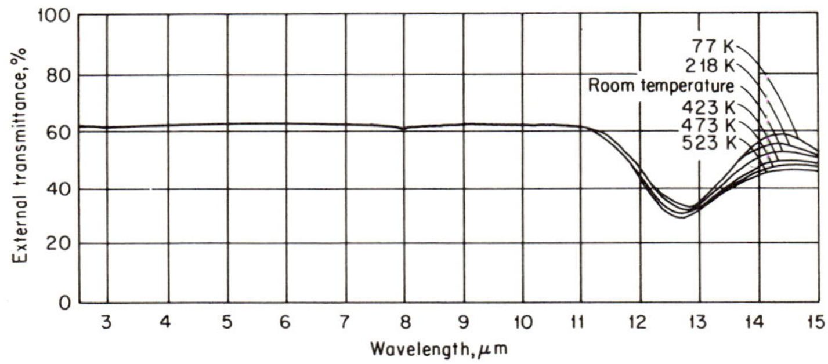

Fig. 151 Transmission of TI-20, tempered $Ge_{33}Se_{55}As_{12}$ glass, at a variety of temperatures, thickness 6 mm. [*From Wolfe and Platt (no date)*.]

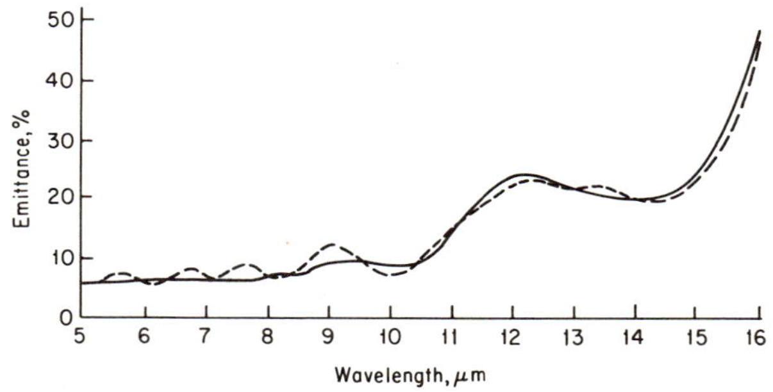

Fig. 152 Emission of TI-20, $Ge_{33}Se_{55}As_{12}$, at 228 K (*dashed curve*) and at 4.2, 2.8, and 77 K (*solid curve*). [*From Stierwalt (no date)*.]

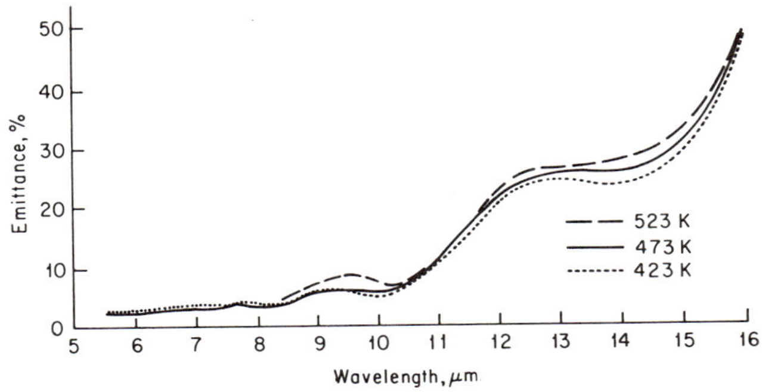

Fig. 153 Emission of TI-20, $Ge_{33}Se_{55}As_{12}$, at 523, 473, and 423 K. [*From Stierwalt (no date).*]

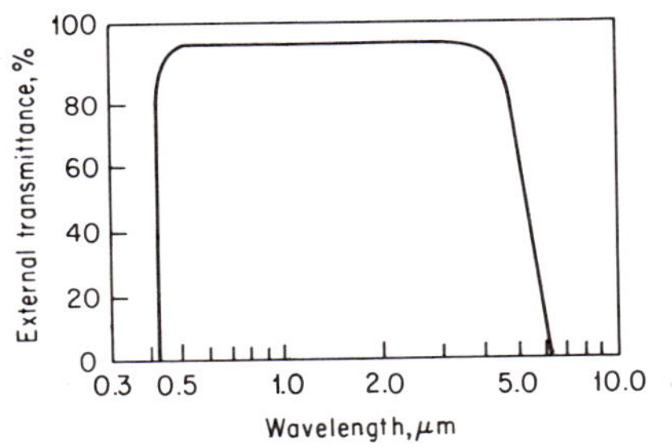

Fig. 154 Transmission of titanium dioxide, thickness 2 mm. [*From Beals and Merker* (1960).]

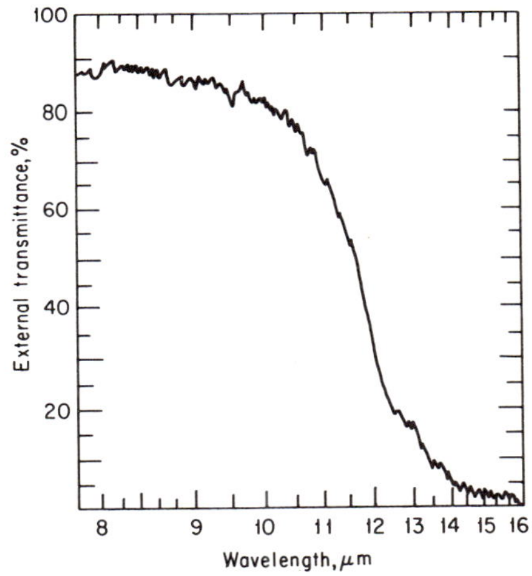

Fig. 155 Reflectance of titanium dioxide. [*From Spitzer et al.* (1962).]

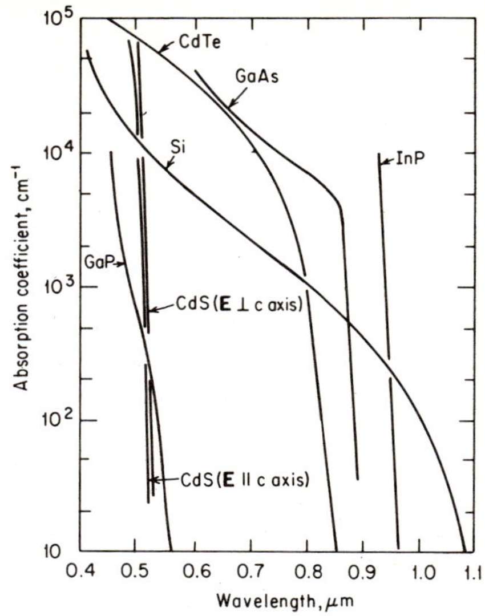

Fig. 156 Absorption of various semiconductors at 300 K. [*From Acta Electron.* (1961).]

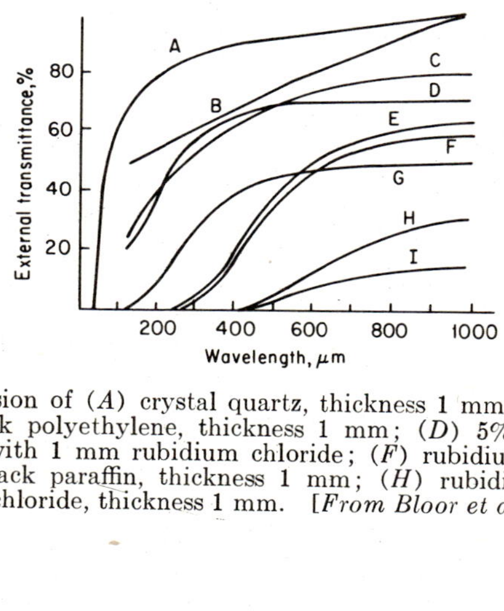

Fig. 157 Transmission of (*A*) crystal quartz, thickness 1 mm; (*B*) mica, thickness 0.015 mm; (*C*) black polyethylene, thickness 1 mm; (*D*) 5% black paraffin; (*E*) potassium bromide with 1 mm rubidium chloride; (*F*) rubidium bromide, thickness 1 mm; (*G*) 15% black paraffin, thickness 1 mm; (*H*) rubidium iodide, thickness 1 mm; (*I*) thallium chloride, thickness 1 mm. [*From Bloor et al.* (1961).]

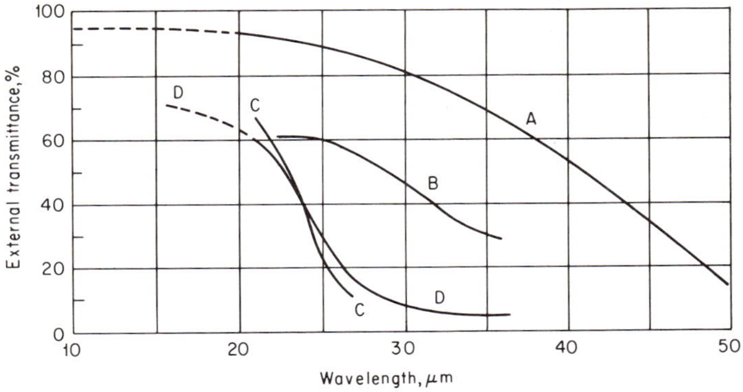

Fig. 158 Transmission of (*A*) potassium iodide, thickness 0.83 mm [*Strong* (1931)] and of (*B*) thallium bromide, (*C*) thallium chloride, and (*D*) thallium bromide-chloride (KRS-6), thickness 6 mm [*Plyler* (1948).]

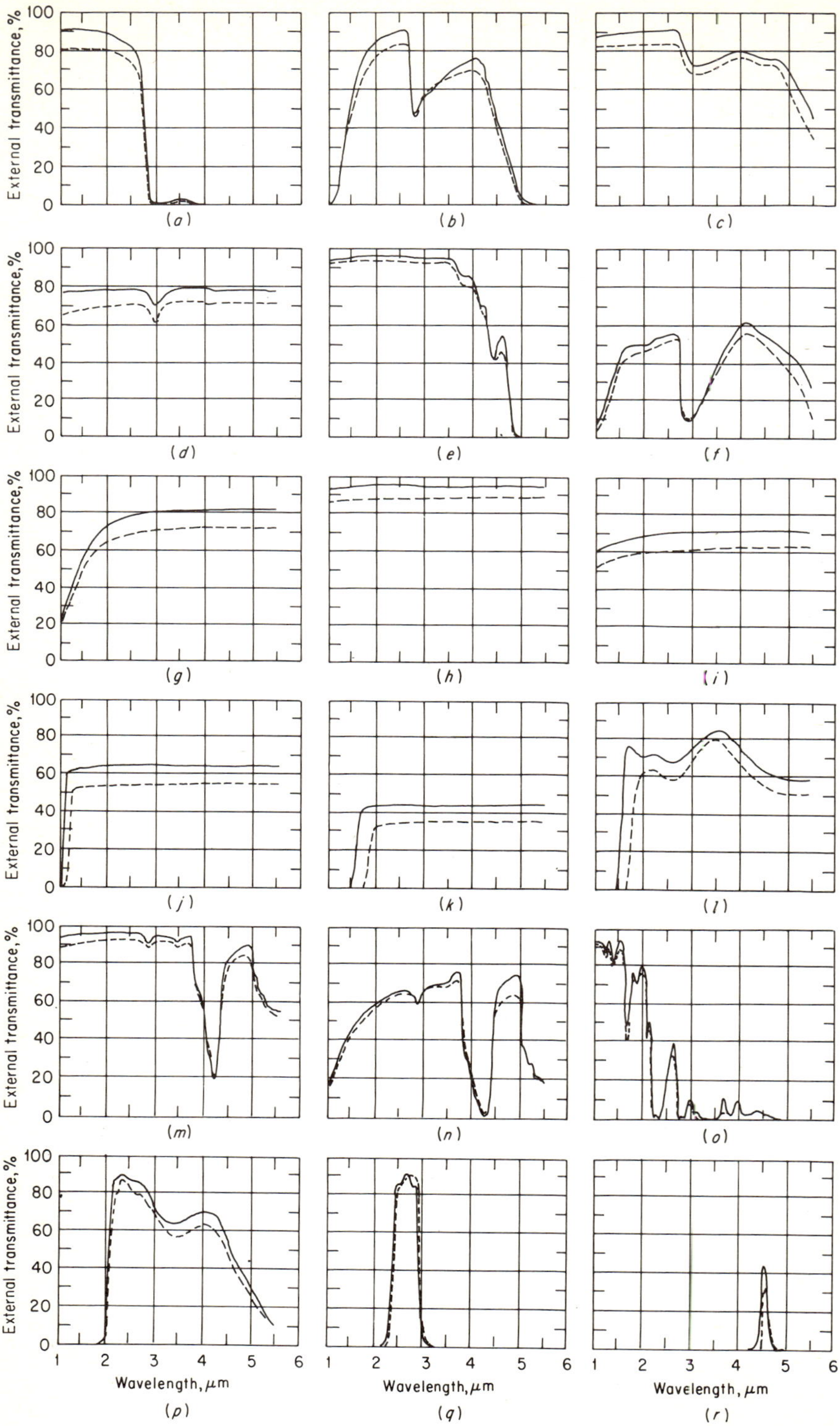
External transmittance, %
100
80
60
40
20
0
(a)
(b)
(c)
(d)
(e)
(f)
(g)
(h)
(i)
(j)
(k)
(l)
(m)
(n)
(o)
(p)
(q)
(r)
1 2 3 4 5 6
Wavelength, μm

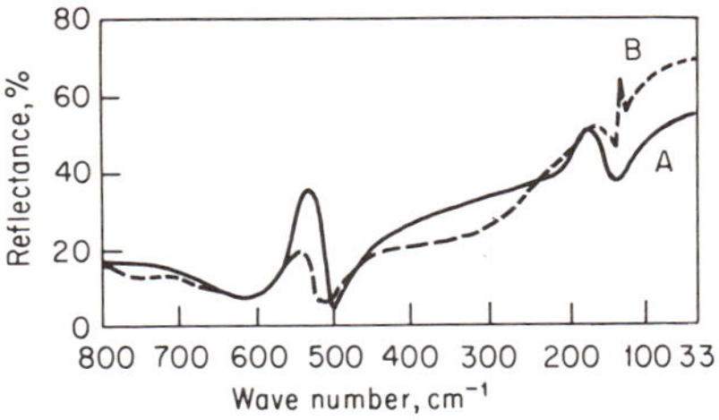

Fig. 160 Infrared reflectance of (*A*) ammonium dihydrogen phosphate (ADP) and (*B*) potassium dihydrogen phosphate (KDP). [*McCarthy* (1971).]

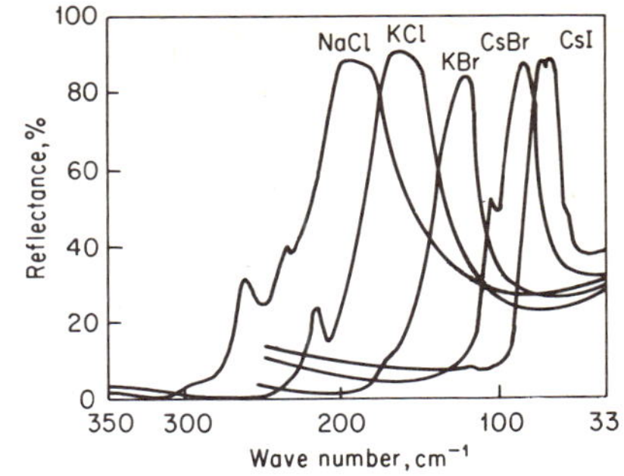

Fig. 161 Infrared reflectance of NaCl, KCl, KBr, CsBr, and CsI. [*McCarthy* (1971).]

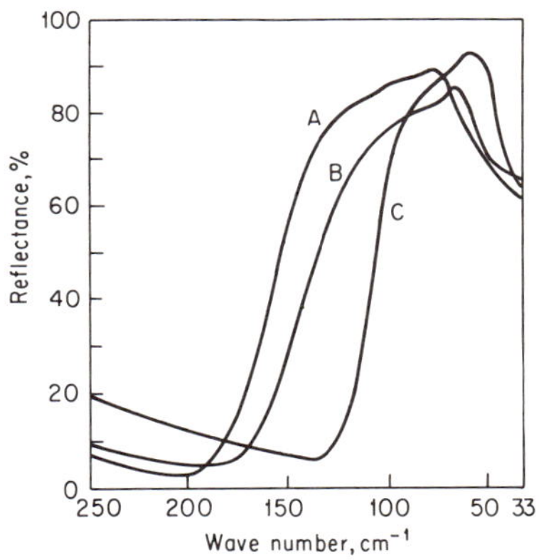

Fig. 162 Infrared reflectance of (*A*) thallium chloride, (*B*) thallium bromide-chloride (KRS-6), and (*C*) thallium bromide. [*McCarthy* (1971).]

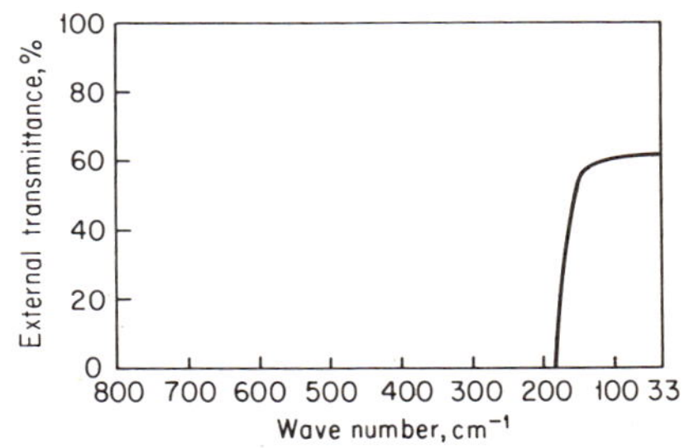

Fig. 163 Infrared transmittance of Kel-F, thickness 1.0 mm. [*McCarthy* (1971).]

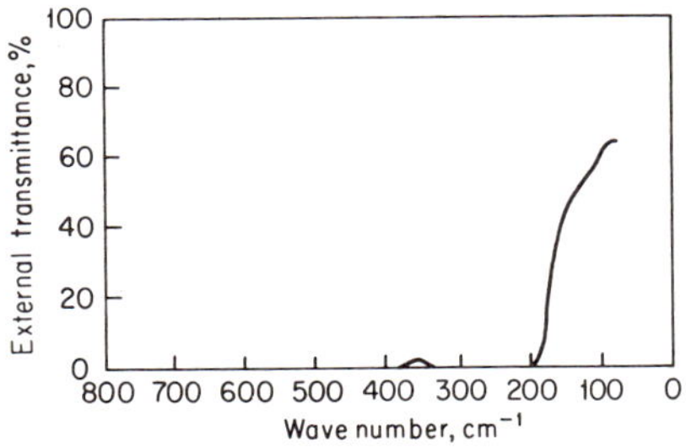

Fig. 164 Transmittance of Teflon, thickness 1.6 mm. [*McCarthy* (1971).]

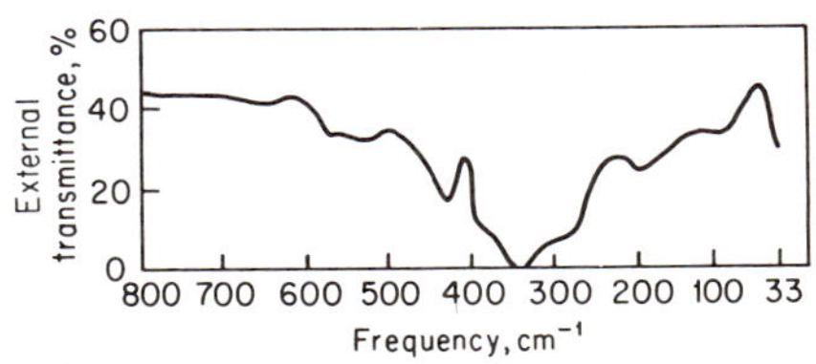

Fig. 165 Transmittance of germanium, thickness 1.6 mm. [*McCarthy* (1971).]

Fig. 159 Low-temperature transmission data for selected materials. Solid curve = cooled = −186°C except in parts (*b*), (*j*), (*k*), (*p*), and (*r*), where it is −195°C. Dashed curve = uncooled = 26°C. (*a*) Corning 8363 (1.8 mm thick); (*b*) RIR 1 (1.0 mm); (*c*) RIR 20 (2.4 mm); (*d*) arsenic trisulfide glass (4.4 mm); (*e*) fused quartz (1.0 mm); (*f*) $CaAlO_3$ with Fe_2O_3 (4.0 mm); (*g*) Irtran ABC II (2.0 mm); (*h*) magnesium oxide (3.2 mm); (*i*) selenium (2.0 mm); (*j*) silicon (6.2 mm); (*k*) germanium (3.4 mm); (*l*) germanium with a silicon monoxide antireflection coat (0.8 mm); (*m*) Teflon (0.45 mm); (*n*) Kel-F (0.5 mm); (*o*) Lucite (1.45 mm); (*p*) bandpass interference filter; (*q*) interference filter; (*r*) narrowband filter. [*Linsteadt and Leet* (1961).]

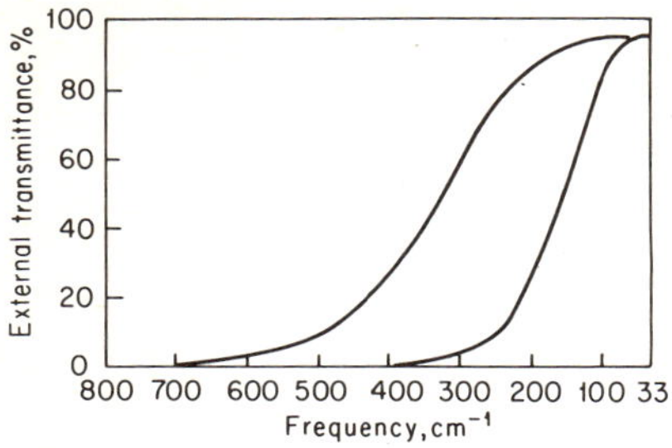

Fig. 166 Transmittance of Millipore† filters. [*McCarthy* (1971).]

† Trade name of the Millipore Corporation.

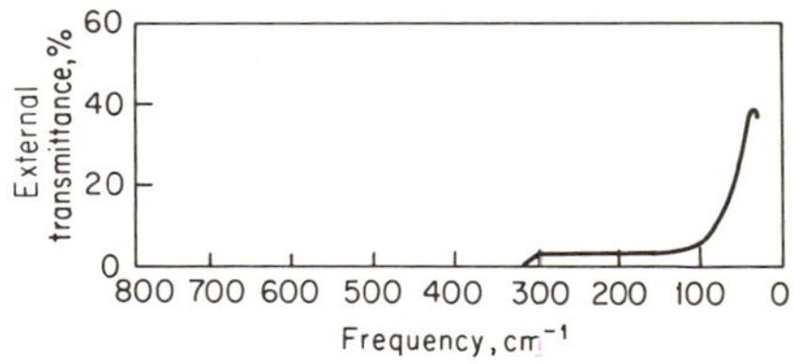

Fig. 167 Transmittance of Kodak Wratten filter ND10. [*McCarthy* (1971).]

TABLE 1 Absorption Coefficient of Calcite and Crystal Quartz

Wavelength, μm	Absorption coefficient, cm^{-1}: Ordinary ray	Absorption coefficient, cm^{-1}: Extraordinary ray	Wavelength, μm	Absorption coefficient, cm^{-1}: Ordinary ray	Absorption coefficient, cm^{-1}: Extraordinary ray
Calcite (Nysander, 1909)					
1.6	0.05		3.6	18	1.8
1.7	0.09		3.7	14	2.1
1.8	0.16		3.8	14	2.1
1.9	0.23		3.9	24.0	1.8
2.0	0.37		4.0	27.0	1.4
2.1	0.62	0.02	4.1	22	1.2
2.2	1.1	0.05	4.2	8.1	1.3
2.3	1.7	0.07	4.3	9.7	1.4
2.4	2.3	0.09	4.4	12	2.1
2.5	2.7	0.14	4.5	14	2.5
2.6	2.5	0.07	4.6	12	2.1
2.7	2.3	0.07	4.7	6.5	1.4
2.8	2.3	0.09	4.8	5.5	1.4
2.9	2.8	0.18	4.9	6.0	1.8
3.0	4.0	0.28	5.0	7.1	3.0
3.1	6.7	0.46	5.1	9.0	4.8
3.2	10.0	0.69	5.2	12	7.1
3.3	15	0.92	5.3	16	11
3.4	19	1.2	5.4	20	16
3.5	20	1.4	5.5	28	28
Crystal Quartz (Drummond, 1935)					
2.7	0.02	0.05	3.9	2.8	2.8
2.8	0.05	0.09	4.0	3.9	3.9
2.9	2.2	0.30	4.1	4.8	4.8
3.0	1.1	0.37	4.2	6.4	6.4
3.1	0.60	0.34	4.3	8.1	8.1
3.2	0.30	0.21	4.4	10	10
3.3	0.18	0.14	4.5	15	15
3.4	0.18	0.18	4.6	20	20
3.5	0.44	0.44	4.7	30	30
3.6	0.83	0.83	4.8	51	51
3.7	1.3	1.3	4.9	91	91
3.8	2.0	2.0	5.0	150	150

TABLE 2 Absorption Coefficients in the Infrared

Wavelength, μm	Absorption coefficient, cm^{-1}	Wavelength, μm	Absorption coefficient, cm^{-1}	Wavelength, μm	Absorption coefficient, cm^{-1}
Lithium Fluoride†					
4.5	0.05	7.0	2.0	9.0	13
5.0	0.07	7.5	4.1	9.5	20
5.5	0.16	8.0	6.4	10.0	32
6.0	0.44	8.5	9.4	11.0	67
6.5	0.99				
Magnesium Oxide (Willmott, 1948; Burstein et al., 1948)					
5.5	0.05	7.5	2.1	9.5	10.0
6.0	0.16	8.0	3.2	10.0	51
6.5	0.48	8.5	4.5	11	69
7.0	1.1	9.0	6.4	12	
Potassium Bromide (Mentzel, 1934)					
18	0.02	26	0.44	34	3.2
19	0.05	28	0.76	36	4.6
20	0.09	30	1.2	38	7.2
22	0.11	32	2.0	40	12
24	0.25				
Potassium Chloride (Mentzel, 1934)					
14	0.02	20	0.39	32	20
15	0.05	22	1.3	34	39
16	0.07	24	2.5	36	69
17	0.09	26	4.4	38	115
18	0.16	28	6.7	40	200
19	0.28	30	12		
Sapphire (Calingaert et al., 1936)					
2.5	0.02	4.0	0.02	5.0	0.92
3.0	2.1	4.5	0.25	5.5	3.9
3.5	0.0				
Silver Chloride (Plyler, 1947)					
19	0.02	22	0.41	26	1.8
20	0.05	24	0.92		
Sodium Chloride (Mentzel, 1934)					
13	0.02	17	0.67	22	7.1
14	0.07	18	1.3	24	13
15	0.16	19	2.3	26	23
16	0.41	20	3.9	28	44

† Calculated from the data of Hohls (1937).

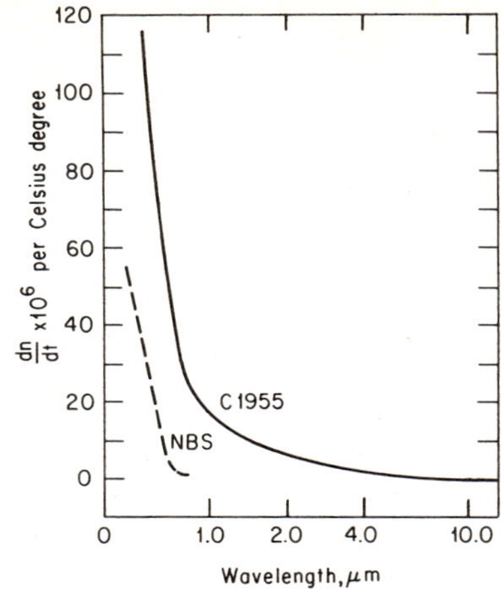

Fig. 168 Change of refractive index with temperature, *dn/dt*, for two types of arsenic sulfur glass. [*From Rodney et al.* (1958).]

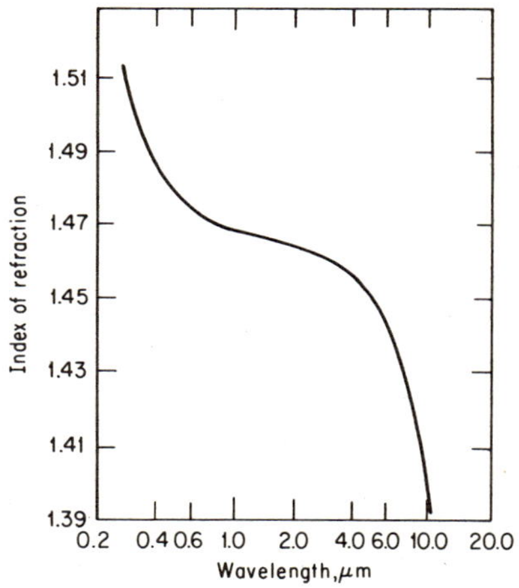

Fig. 169 Refractive index of barium fluoride. [*From Malitson* (1964).]

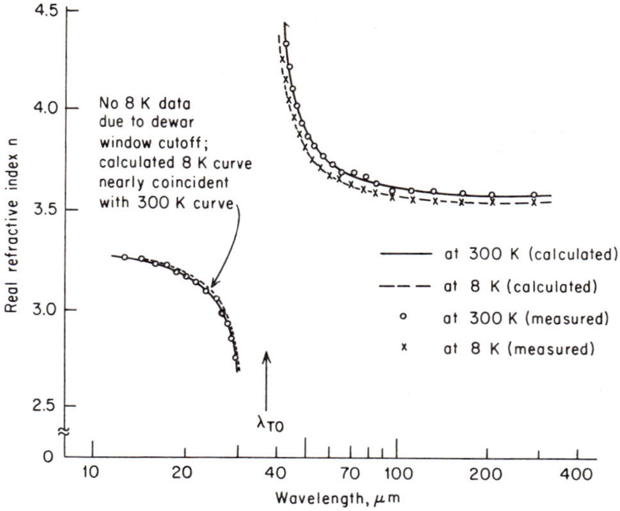

Fig. 170 Far-infrared real refractive index of high-resistivity cadmium telluride at 8 and 300 K. [*From Johnson et al.* (1969).]

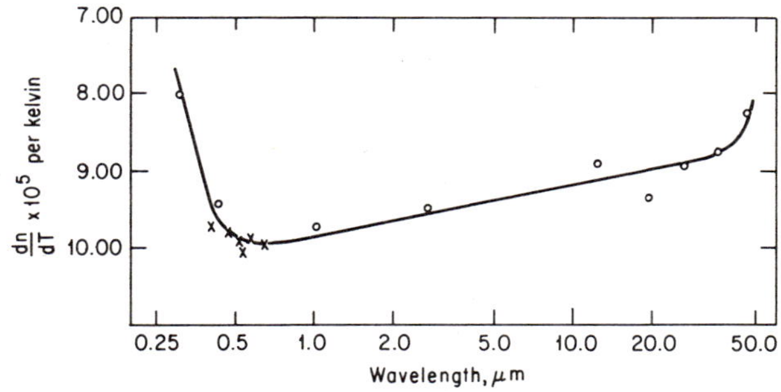

Fig. 171 Temperature coefficient of refractive index of cesium iodide. Circles = average for five wavelengths; crosses = visible data, average of five values at each wavelength. [*From Rodney* (1955).]

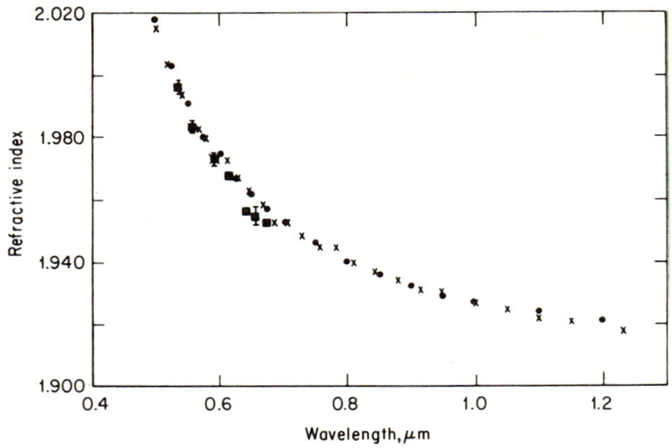

Fig. 172 Refractive-index measurements of copper(I) chloride obtained from fringe measurements on two samples. The estimated uncertainty of n over most of the range is ± 0.005. The thickness of sample 1 (dotted) is 54.13 μm and of sample 2 (X) is 14.99 μm. The dispersion of n for the wavelength range 0.43 to 2.5 μm is given by

$$n^2 = 3.580 + \frac{0.03162\lambda^2}{\lambda^2 - 0.4052^2} + \frac{0.09288}{\lambda^2}$$

where λ is the wavelength expressed in micrometers. [*From Feldman and Horowitz* (1969).]

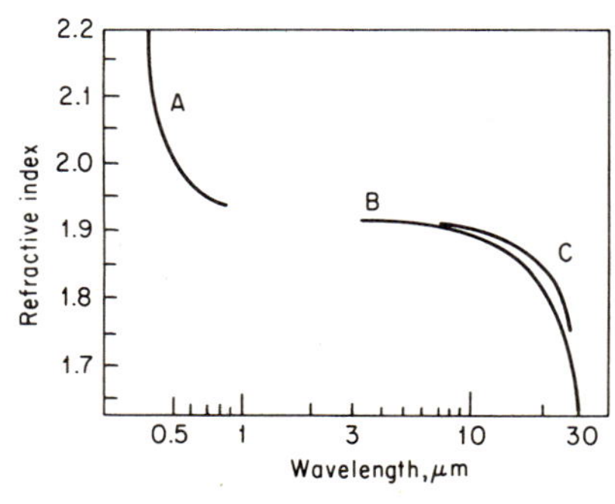

Fig. 173 Refractive index of copper(I) chloride at 300 K. Curve A: Kaifu and Komatsu (1968); curve B: 90-μm sample; curve C: 200-μm sample. [*From Alonas et al.* (1969).]

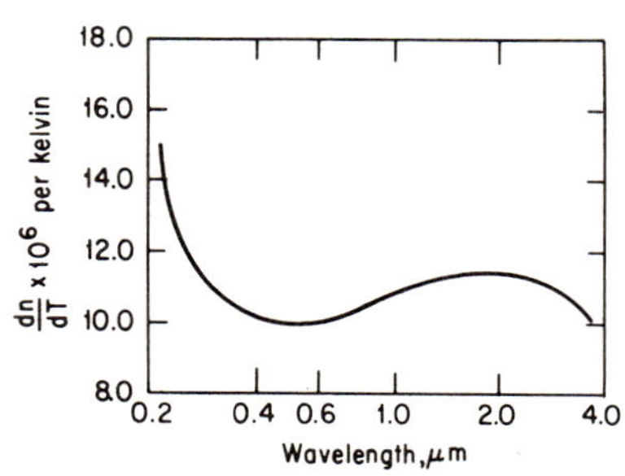

Fig. 174 Thermal coefficient of index of fused silica for a mean temperature of 25°C. [*From Malitson* (1965).]

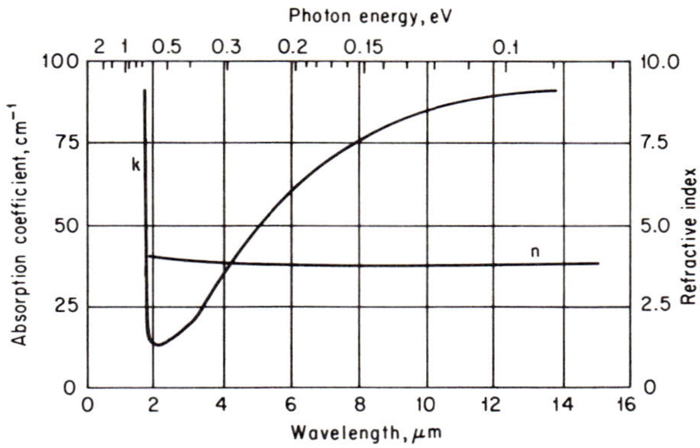

Fig. 175 Refractive index and absorption coefficient of gallium antimonide; resistivity 0.08 Ω·cm. [*From Oswald and Schade* (1954).]

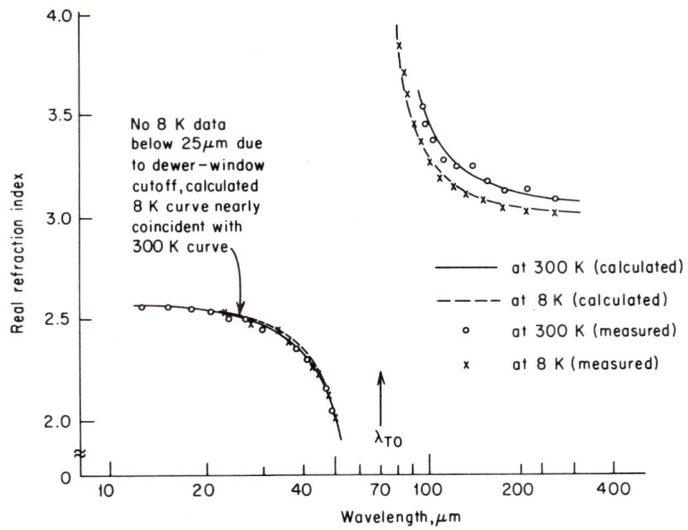

Fig. 176 Far-infrared real refractive index of high-resistivity gallium arsenide at 8 and 300 K. [*From Johnson et al.* (1969).]

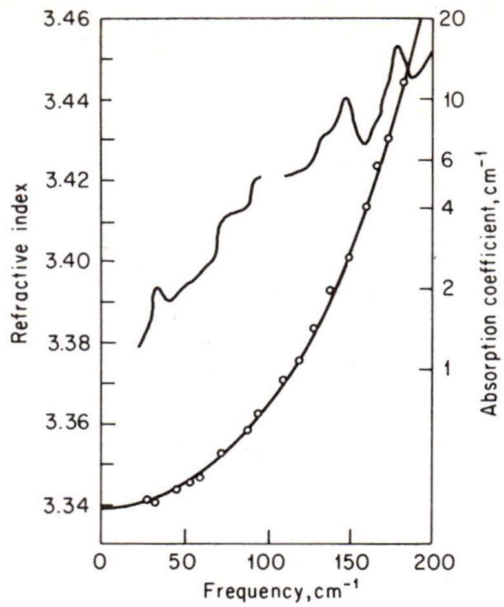

Fig. 177 Experimental absorption coefficient α and experimental refractive index curve n compared to theoretical classical oscillator equation modified to include a frequency-dependent damping term (solid line through circles) for gallium phosphide. One of the absorption peaks occurs at 181 cm^{-1} and the other broad flat peak at 148 to 151 cm^{-1}. [*From Parsons and Coleman* (1971).]

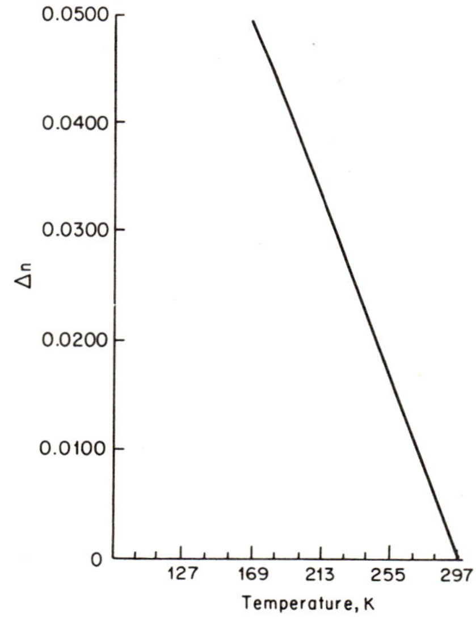

Fig. 178 Temperature variation of the refractive index of germanium. [*From Wolfe and Platt* (1976).]

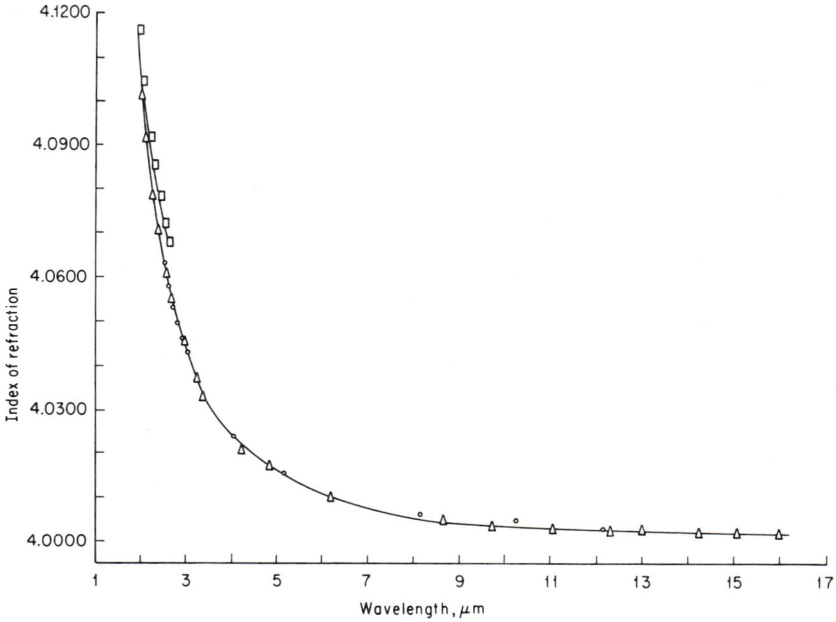

Fig. 179 The data of Wolfe and Platt (1976) for germanium compared with previously reported data, all corrected to 27°C. △ Salzberg and Villa (1953, 1957), □ Briggs, (1950), ○ Wolfe and Platt (1976).

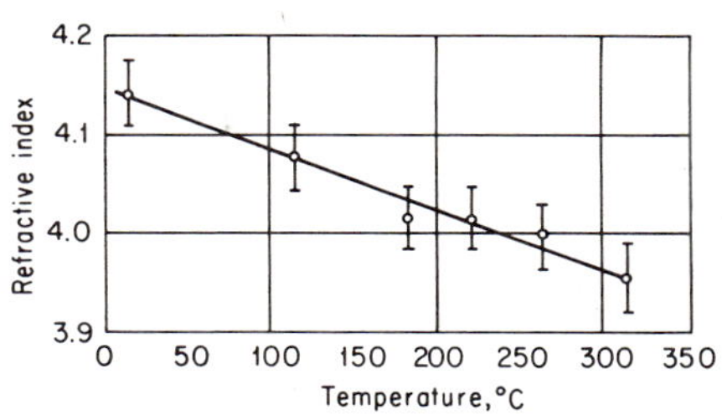

Fig. 180 Variation of refractive index of lead sulfide with temperature. Values of refractive index for the wavelength region of 0.5 to 3.0 μm of lead sulfide have been reported by Avery (1953). His values are for measurements made on several cleaved samples and one polished sample, and for two different types of polishing. Avery (1954) quotes 4.10 ± 0.06 for all samples at 3.0 μm, 4.0 ± 0.1 at 1.0 μm, and 4.19 ± 0.06 at 6 μm.

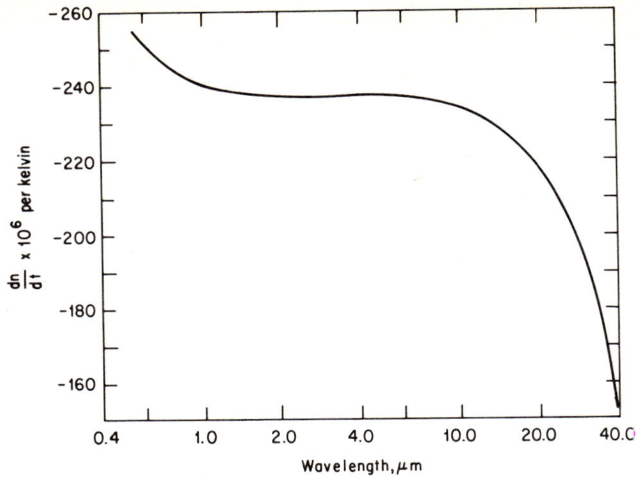

Fig. 181 Average thermal coefficient of refractive index of potassium chloride near room temperature. [*From Paschen* (1908).]

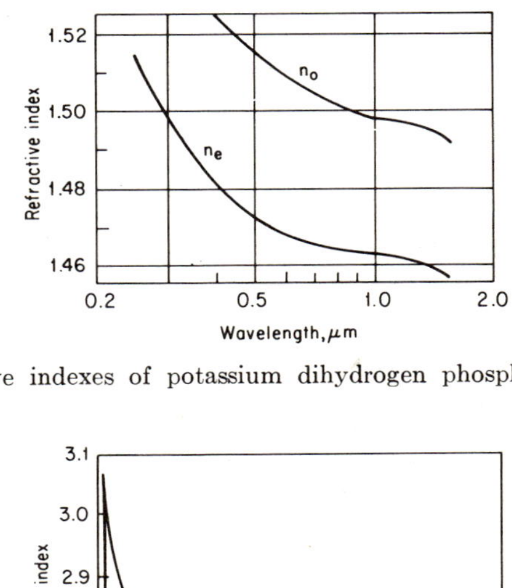

Fig. 182 Refractive indexes of potassium dihydrogen phosphate. [*From Zernike* (1964).]

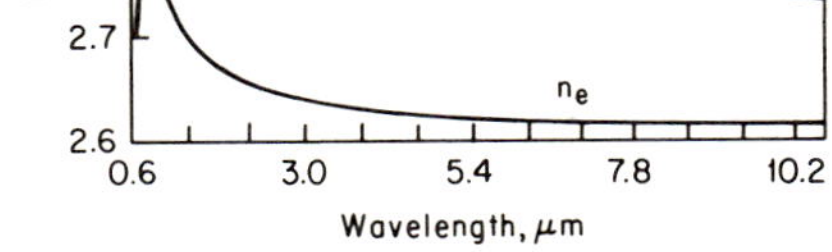

Fig. 183 Refractive indexes of single-crystal pyrargyrite, Ag_3SbS_3. The dispersion equations

$$n_o^2 = 1 + \frac{6.585\lambda^2}{\lambda^2 - 0.4^2} + \frac{0.1133\lambda^2}{\lambda^2 - 15^2}$$

$$n_e^2 = 1 + \frac{5.845\lambda^2}{\lambda^2 - 0.4^2} + \frac{0.0202\lambda^2}{\lambda^2 - 15^2}$$

are accurate from 1.5 to 10.6 μm within ± 0.005. [*From Feichtner et al.* (1970).]

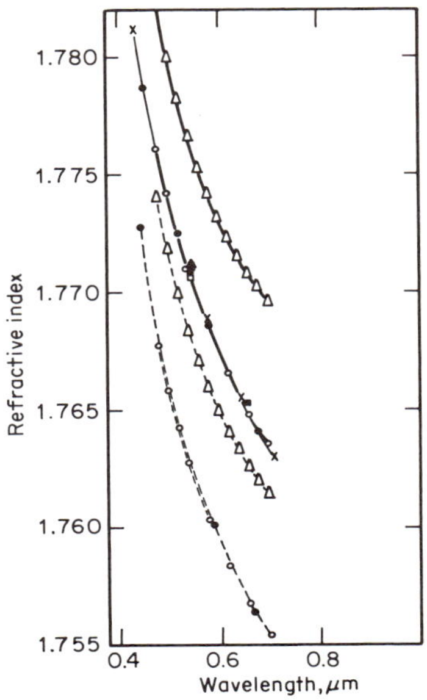

Fig. 184 Comparison of the refractive index of sapphire and ruby as a function of wavelength. The solid and dashed lines indicate the ordinary and extraordinary rays, respectively. X, sapphire, Malitson (1962); □, sapphire, Dowell (1964); ▲, 0.01% Cr_2O_3, Houston et al. (1963); ●, 0.062% Cr_2O_3, NBS; ■, 0.077% Cr_2O_3, Dowell (1964); ○, 0.11% Cr_2O_3, Mandarino (1959); △, 1.4% Cr_2O_3, Mandarino (1959). [*Dodge et al.* (1969).]

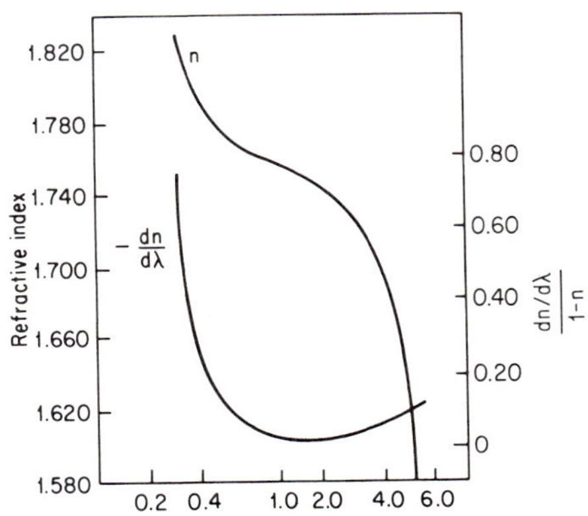

Fig. 185 Refractive index and relative dispersion of synthetic sapphire, Al_2O_3, for the ordinary ray n_o at 24°C from Malitson (1962); Loewenstein (1961) gives 3.14 ± 0.13 for n_o and 3.61 ± 0.14 for n_e from 167 to 500 μm.

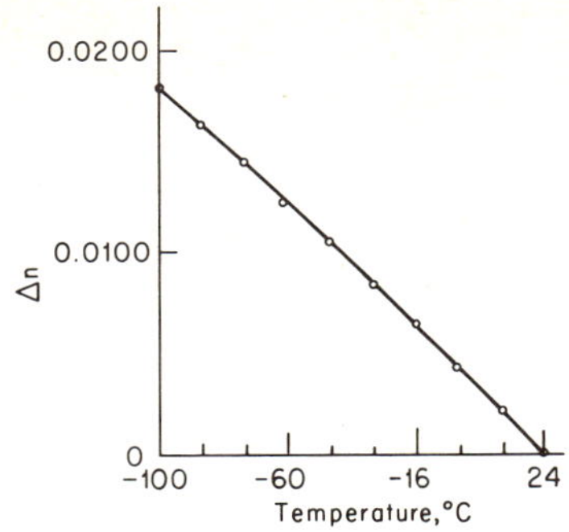

Fig. 186 Temperature variation of the refractive index of silicon. [*From Wolfe and Platt* (1976).]

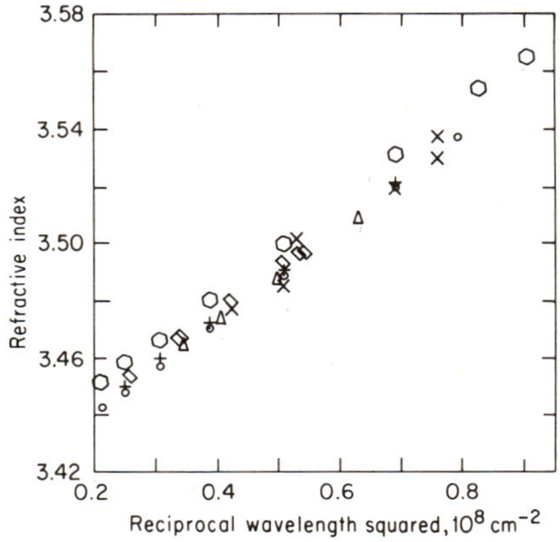

Fig. 187 Refractive index of silicon at 24°C as a function of reciprocal wavelength squared; small cross, set II; large cross, set III; large X, set IV; small circle, set V. [*From Primak* (1971).]

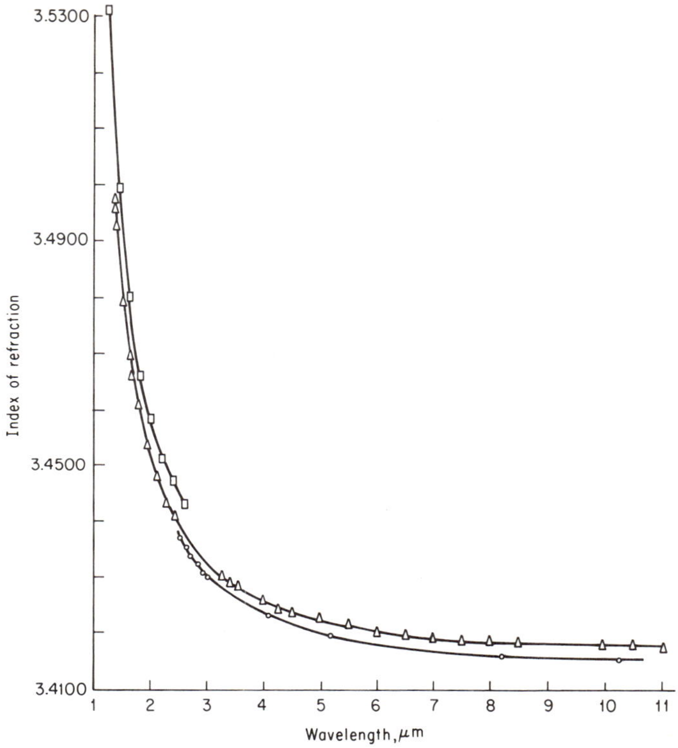

Fig. 188 The data of Wolfe and Platt (1976) for silicon compared with the previously reported data, all corrected to 26.1°C. △ Salzberg and Villa (1953, 1957), □ Briggs (1950), ○ Wolfe and Platt (1976).

TABLE 3 Ammonium Dihydrogen Phosphate: Refractive Index at 24.8°C for the Ordinary and Extraordinary Rays (Zernike, 1964)

λ, μm	n_o	n_e
0.4047	1.53969	1.49159
0.4077	1.53925	1.49123
0.4358	1.53578	1.48831
0.4916	....	1.48390
0.5461	1.52662	1.48079
0.5769	1.52478	1.47939
0.5791	1.52466	1.47930
0.6328	1.52166	1.47685
1.0139	1.50835	1.46895
1.1287	1.50446	1.46704
1.1523	1.50364	1.46666

Dispersion Equation (Bishnevskii and Domanok, 1960)

$$n^2 = 1 + \frac{1.298\lambda^2}{\lambda^2 - \alpha^2} + \frac{0.179\lambda^2}{\lambda^2 - b^2}$$

where $\alpha = 9.9 \times 10^{-6}$ cm $\quad b = 2.5 \times 10^{-4}$ cm

TABLE 4 Arsenic Sulfur Glass: Refractive Index at 25°C (Rodney et al., 1958)†

λ, μm	n	λ, μm	n	λ, μm	n
0.560	2.68689	1.800	2.43000	7.000	2.39899
0.580	2.65934	2.000	2.42615	7.200	2.39806
0.600	2.63640	2.200	2.42318	7.400	2.39709
0.620	2.61708	2.400	2.42086	7.600	2.39610
0.640	2.60043	2.600	2.41898	7.800	2.39508
0.660	2.58594	2.800	2.41742	8.000	2.39403
0.680	2.57323	3.000	2.41008	8.200	2.39294
0.700	2.56198	3.200	2.41491	8.400	2.39183
0.720	2.55195	3.400	2.41386	8.600	2.39068
0.740	2.54297	3.600	2.41290	8.800	2.38949
0.760	2.53488	3.800	2.41200	9.000	2.38827
0.780	2.52756	4.000	2.41116	9.200	2.38700
0.800	2.52090	4.200	2.41035	9.400	2.38570
0.820	2.51483	4.400	2.40956	9.600	2.38436
0.840	2.50928	4.600	2.40878	9.800	2.38298
0.860	2.50418	4.800	2.40802	10.000	2.38155
0.880	2.49949	5.000	2.40725	10.200	2.38007
0.900	2.49515	5.200	2.40649	10.400	2.37855
0.920	2.49114	5.400	2.40571	10.600	2.37698
0.940	2.48742	5.600	2.40493	10.800	2.37536
0.960	2.48396	5.800	2.40414	11.000	2.37369
0.980	2.48074	6.000	2.40333	11.200	2.37196
1.000	2.47773	6.200	2.40250	11.400	2.37018
1.200	2.45612	6.400	2.40166	11.600	2.36833
1.400	2.44357	6.600	2.40079	11.800	2.36643
1.600	2.43556	6.800	2.39991	12.000	2.36446

Dispersion Equation

$$n^2 - 1 = \sum_{i=1}^{i=5} \frac{K_i \lambda^2}{\lambda^2 - \lambda_i^2}$$

Constants

i	λ_i^2	K_i
1	0.0225	1.8983678
2	0.0625	1.9222979
3	0.1225	0.8765134
4	0.2025	0.1188704
5	750	0.9569903

† The properties of arsenic trisulfide glass vary with different batches (Fraser and Fraser, 1959). This difference is probably caused by various quantities of arsenic disulfide and pentasulfide as well as free sulfur.

TABLE 5 Barium Fluoride: Refractive Index at 25°C† and Temperature Coefficient (Malitson, 1964)

λ, μm	n_{obs}	n_{cal}	Δn	λ, μm	n_{obs}	n_{cal}	Δn
0.2652	1.51217	1.51216	+1	1.12866	1.46779	1.46778	+1
0.28035	1.50668	1.50669	−1	1.36728	1.46673	1.46671	+2
0.28936	1.50390	1.50390	0	1.52952	1.46613	1.46611	+2
0.296728	1.50186	1.50184	+2	1.681	1.46561	1.46561	0
0.30215	1.50044	1.50043	+1	1.7012	1.46554	1.46555	−1
0.3130	1.49782	1.49786	−4	1.97009	1.46472	1.46470	+2
0.32546	1.49521	1.49526	−5	2.1526	1.46440	1.46442	−2
0.334448	1.49363	1.49363	0	2.32542	1.46356	1.46356	0
0.340365	1.49257	1.49256	+1	2.5766	1.46262	1.46271	−9
0.34662	1.49158	1.49154	+4	2.6738	1.46234	1.46237	−3
0.361051	1.48939	1.48940	−1	3.2434	1.46018	1.46017	+1
0.366328	1.48869	1.48869	0	3.422	1.45940	1.45941	−1
0.404656	1.48438	1.48439	−1	5.138	1.45012	1.45014	−2
0.465835	1.48173	1.48174	−1	5.3034	1.44904	1.44905	−1
0.486433	1.47855	1.47856	−1	5.343	1.44878	1.44878	0
0.546074	1.47586	1.47586	0	5.549	1.44732	1.44736	−4
0.589262	1.47443	1.47443	0	6.238	1.44216	1.44217	−1
0.643847	1.47302	1.47301	+1	6.6331	1.43899	1.43890	+9
0.656279	1.47274	1.47273	+1	6.8559	1.43694	1.43696	−2
0.706519	1.47177	1.47176	+1	7.0442	1.43529	1.43526	+3
0.85211	1.46984	1.46981	+3	7.268	1.43314	1.43317	−3
0.89435	1.46942	1.46940	+2	9.724	1.40514	1.40511	+3
1.01398	1.46847	1.46846	+1	10.346	1.39636	1.39639	−3

Dispersion Equation

$$n^2 - 1 = \sum \frac{A_i \lambda^2}{\lambda^2 - \lambda_i^2}$$

Constants at 25°C

$A_1 = 0.63356$ $\lambda_1^2 = 0.0033396$
$A_2 = 0.506762$ $\lambda_2^2 = 0.012030$
$A_3 = 3.8261$ $\lambda_3^2 = 2{,}151.70$

Temperature Coefficient

λ, μm	$\frac{-dn}{dT}$, 10^{-6}/°C	λ, μm	$\frac{-dn}{dT}$, 10^{-6}/°C
0.4046563	15.05	0.6562793	15.23
0.4358342	15.00	0.6678149	15.25
0.4861327	15.15	0.7065188	15.28
0.5460740	15.20	0.767858	15.45
0.589262	15.22		

† For low-temperature data see Houston et al. (1963).

TABLE 6 Cadmium Sulfide (Hexagonal): Refractive Index for Ordinary and Extraordinary Rays (Zyzak et al., 1957)

	n_o		n_e			n_o		n_e	
λ, μm	Calc.	Obs.	Calc.	Obs.	λ, μm	Calc.	Obs.	Calc.	Obs.
0.5500	2.561	2.565	2.598	2.597	0.9000	2.349	2.353	2.358	2.358
0.5750	2.517	2.518	2.548	2.545	0.9500	2.341	2.340	2.350	2.352
0.6000	2.484	2.483	2.511	2.511	1.0000	2.363	2.338	2.343	2.341
0.6250	2.458	2.458	2.482	2.478	1.0500	2.330	2.332	2.338	2.338
0.6500	2.438	2.438	2.459	2.459	1.1000	2.326	2.325	2.334	2.333
0.6750	2.421	2.421	2.438	2.437	1.1500	2.321	2.322	2.328	2.330
0.7000	2.407	2.407	2.425	2.425	1.2000	2.319	2.317	2.324	2.322
0.7500	2.386	2.386	2.400	2.403	1.2500	2.317	2.316	2.321	2.320
0.8000	2.371	2.371	2.383	2.383	1.3000	2.314	2.315	2.319	2.319
0.8500	2.359	2.359	2.369	2.372	1.4000	2.310	2.311	2.314	2.314

Dispersion Equation

$$n_o^2 = 5.235 + \frac{1.819 \times 10^7}{\lambda^2 - 1.651 \times 10^7} \qquad n_e^2 = 5.239 + \frac{2.076 \times 10^7}{\lambda^2 - 1.651 \times 10^7}$$

TABLE 7 Calcite: Refractive Index for the Ordinary and Extraordinary Rays and Temperature Coefficient

Refractive Index†

λ, μm	n_o	n_e	λ, μm	n_o	n_e	λ, μm	n_o	n_e
0.198	....	1.57796	0.410	1.68014	1.49640	1.220	1.63926	1.47870
0.200	1.90284	1.57649	0.434	1.67552	1.49430	1.273	1.63849	
0.204	1.88242	1.57081	0.441	1.67423	1.49373	1.307	1.63789	1.47831
0.208	1.80733	1.56640	0.508	1.66527	1.48956	1.320	1.63767	
0.211	1.85692	1.56327	0.533	1.66277	1.48841	1.369	1.63681	
0.214	1.84558	1.55976	0.560	1.66046	1.48736	1.396	1.63637	1.47780
0.219	1.83075	1.55496	0.589	1.65835	1.48640	1.422	1.63590	
0.226	1.81309	1.54921	0.643	1.65504	1.48490	1.479	1.63490	
0.231	1.80233	1.54541	0.656	1.65437	1.48459	1.497	1.63457	1.47744
0.242	1.78111	1.53782	0.670	1.65367	1.48426	1.541	1.63381	
0.257	1.76038	1.53005	0.706	1.65207	1.48353	1.609	1.63261	
0.263	1.75343	1.52736	0.768	1.61974	1.48259	1.615	....	1.47695
0.267	1.74864	1.52547	0.795	1.64880	1.48216	1.682	1.63127	
0.274	1.74139	1.52261	0.801	1.64869	1.48216	1.749	....	1.47638
0.291	1.72774	1.51705	0.833	1.64772	1.48176	1.761	1.62974	
0.303	1.71959	1.51365	0.867	1.64676	1.48137	1.849	1.62800	
0.312	1.71425	1.51140	0.905	1.64578	1.48098	1.900	....	1.47573
0.330	1.70515	1.50746	0.946	1.64480	1.48060	1.946	1.62602	
0.340	1.70078	1.50562	0.991	1.64380	1.48022	2.053	1.62372	
0.346	1.69833	1.50450	1.042	1.64276	1.47985	2.100	....	1.47492
0.361	1.69317	1.50228	1.097	1.64167	1.47948	2.172	1.62099	
0.394	1.68374	1.49810	1.159	1.64051	1.47910	3.324	....	1.47392

Temperature Coefficient (Micheli, 1902*b*)

λ, μm	$\frac{dn_o}{dT}$, 10^{-5}/°C	$\frac{dn_e}{dT}$, 10^{-5}/°C	λ, μm	$\frac{dn_o}{dT}$, 10^{-5}/°C	$\frac{dn_e}{dT}$, 10^{-5}/°C
0.211	2.150		0.467	0.319	
0.231	1.397	2.198	0.480	0.305	1.287
0.298	0.604	1.641	0.508	0.287	1.234
0.361	0.360	1.449	0.589	0.240	1.213
0.441	0.325	1.318	0.643	0.208	1.185

† From Martens (1901*b*) for 0.198 to 0.340 μm; Gifford (1902) and Carvallo (1896*b*) for 0.226, 0.303, 0.330, and 0.361 μm; *J. Phys. Radium* (1900) for 0.346 to 1.541 μm. The data fit together to within a few parts in the fifth decimal place. Data for the ordinary ray at about 0.21 μm appear a little out of line.

TABLE 8 Calcium Fluoride: Refractive Index and Temperature Coefficient†

Refractive Index‡

λ, μm	n	λ, μm	n	λ, μm	n	λ, μm	n
0.19	1.50500	1.0140	1.42884	2.1608	1.42306	4.000	1.40963
0.20	1.49531	1.08304	1.42843	2.250	1.42258	4.1252	1.40847
0.22	1.48119	1.1000	1.42834	2.3573	1.42198	4.2500	1.40722
0.24	1.47133	1.1786	1.42789	2.450	1.42143	4.4000	1.40568
0.26	1.46397	1.250	1.42752	2.5537	1.42080	4.6000	1.40357
0.28	1.45841	1.3756	1.42689	2.6519	1.42018	4.7146	1.40233
0.30	1.45400	1.4733	1.42642	2.700	1.41988	4.8000	1.40130
0.35	1.44658	1.5715	1.42596	2.750	1.41956	5.000	1.39908
0.40	1.441857	1.650	1.42558	2.800	1.41923	5.3036	1.39522
0.48615	1.43704	1.7680	1.42502	2.880	1.41890	5.8932	1.38712
0.58758	1.43388	1.8400	1.42468	2.9466	1.41823	6.4825	1.37824
0.58932	1.43384	1.8688	1.42454	3.0500	1.41750	7.0718	1.36805
0.65630	1.43249	1.900	1.42439	3.0980	1.41714	7.6612	1.35675
0.68671	1.43200	1.9153	1.42431	3.2413	1.41610	8.2505	1.34440
0.72818	1.43143	1.9644	1.42407	3.4000	1.41487	8.8398	1.33075
0.76653	1.43093	2.0582	1.42360	3.5359	1.41367	9.4291	1.31605
0.88400	1.42980	2.0626	1.42357	3.8306	1.41119		

Temperature Coefficient (Liebreich, 1911)

λ, μm	$\frac{dn}{dT}$, 10^{-5}/°C	T, °C	λ, μm	$\frac{dn}{dT}$, 10^{-5}/°C	T, °C
0.185	−0.296	61.25	0.325	−0.948	61.25
0.186	−0.313	61.25	0.340	−0.964	61.25
0.193	−0.402	61.25	0.361	−0.979	61.25
0.197	−0.451	61.25	0.441	−1.028	61.25
0.198	−0.464	61.25	0.480	−1.035	61.25
0.200	−0.493	61.25	0.508	−1.056	61.25
0.204	−0.538	61.25	0.589	−1.111	60.5
0.208	−0.582	61.25	0.640	−1.113	59.2
0.211	−0.601	61.25	0.900	−1.031	59.9
0.214	−0.637	61.25	1.2	−1.040	59.9
0.219	−0.655	61.25	1.25	−1.029	60.0
0.224	−0.696	61.25	1.30	−1.018	60.2
0.231	−0.732	61.25	2.0	−0.932	60.2
0.257	−0.811	61.25	3.16	−0.881	59.4
0.274	−0.855	61.25	4.2	−0.831	59.6
0.288	−0.884	61.25	5.3	−0.821	59.0
0.298	−0.904	61.25	6.5	−0.787	56.8

† For fluorite see Table 15.
‡ From Kohlrausch (1968) for the values from 0.19 to 0.40 μm, and from W. W. Coblentz (1914*a*) for the other values. Coblentz quotes the earlier values of Langley, Paschen, and Rubens; both sets of data were measured at 20°C. The four sets of data agree to within several parts in the fifth decimal place through the region of overlap (0.50 to 9.00 μm).

TABLE 9 Calcium Molybdate: Refractive Index at 19.5°C (Malitson, no date)

λ, μm	n_o	n_e	λ, μm	n_o	n_e
0.40466	2.04452	2.06156	0.57906	1.98049	1.99066
0.43584	2.02553	2.04029	0.58756	1.97893	1.98897
0.46782	2.01091	2.02409	0.64385	1.97033	1.97961
0.47999	2.00629	2.01898	0.66781	1.96737	1.97639
0.50858	1.99697	2.00873	0.70652	1.96321	1.97189
0.54607	1.98730	1.99810	1.0830	1.94317	1.95030
0.57696	1.98089	1.99109			

TABLE 10 Calcium Titanate: Refractive Index (Smakula, 1962*a*)

λ, μm	n_o	n_e	λ, μm	n_o	n_e
0.434	1.6126	1.3404	0.578	1.5860	1.3363
0.436	1.6121	1.3403	0.589	1.5848	1.3360
0.486	1.5998	1.3384	0.656	1.5791	1.3347
0.501	1.5968	1.3379	0.668	1.5783	1.3345
0.546	1.5899	1.3365			

TABLE 11 Cesium Bromide: Refractive Index at 27°C (Rodney and Spindler, 1953)

λ, μm	0.0	0.1	0.2	0.3	0.4	0.5	0.6	0.7	0.8	0.9
0	......	......	......	......	73,519	70,896	69,588	68,825	68,345	68,022
1	67,793	67,624	67,496	67,397	67,318	67,254	67,201	67,157	67,120	67,088
2	67,061	67,036	67,015	66,996	66,979	66,963	66,948	66,935	66,923	66,911
3	66,904	66,890	66,881	66,871	66,862	66,853	66,845	66,837	66,829	66,821
4	66,813	66,805	66,798	66,790	66,782	66,775	66,767	66,760	66,752	66,745
5	66,737	66,730	66,722	66,715	66,707	66,699	66,691	66,683	66,675	66,667
6	66,659	66,651	66,643	66,634	66,626	66,617	66,609	66,600	66,591	66,582
7	66,573	66,564	66,555	66,545	66,536	66,526	66,517	66,507	66,497	66,487
8	66,477	66,467	66,457	66,446	66,436	66,425	66,414	66,403	66,392	66,381
9	66,370	66,359	66,347	66,335	66,324	66,312	66,300	66,288	66,276	66,263
10	66,251	66,238	66,226	66,213	66,200	66,187	66,174	66,160	66,147	66,134
11	66,120	66,106	66,092	66,078	66,064	66,050	66,035	66,021	66,006	65,991
12	65,976	65,961	65,946	65,931	65,915	65,900	65,884	65,868	65,852	65,836
13	65,820	65,804	65,787	65,770	65,754	65,737	65,720	65,703	65,685	65,668
14	65,651	65,633	65,615	65,597	65,579	65,561	65,543	65,524	65,505	65,487
15	65,468	65,449	65,430	65,411	65,391	65,372	65,352	65,332	65,312	65,292
16	65,272	65,251	65,231	65,210	65,190	65,169	65,148	65,126	65,105	65,084
17	65,062	65,040	65,018	64,996	64,974	64,952	64,929	64,907	64,884	64,861
18	64,838	64,815	64,792	64,768	64,745	64,721	64,697	64,673	64,649	64,625
19	64,600	64,576	64,551	64,526	64,501	64,476	64,450	64,425	64,390	64,374
20	64,348	64,322	64,295	64,269	64,243	64,216	64,189	64,162	64,135	64,108
21	64,080	64,053	64,025	63,997	63,969	63,941	63,913	63,884	63,856	63,827
22	63,798	63,769	63,739	63,710	63,681	63,651	63,621	63,591	63,561	63,530
23	63,500	63,469	63,438	63,407	63,376	63,345	63,313	63,282	63,250	63,218
24	63,186	63,154	63,121	63,089	63,056	63,023	62,990	62,957	62,923	63,890
25	62,856	62,822	62,788	62,754	62,719	62,685	62,650	62,615	62,580	62,545
26	62,509	62,474	62,438	62,402	62,366	62,330	62,293	62,256	62,220	62,183
27	62,146	62,108	62,071	62,033	61,995	61,957	61,919	61,881	61,842	61,803
28	61,764	61,725	61,686	61,646	61,607	61,567	61,527	61,487	61,446	61,406
29	61,365	61,324	61,283	61,242	61,200	61,158	61,116	61,074	61,032	60,990
30	60,947	60,904	60,861	60,818	60,775	60,731	60,687	60,643	60,599	60,555
31	60,510	60,465	60,420	60,375	60,330	60,284	60,238	60,192	60,146	60,100
32	60,053	60,007	59,960	59,912	59,865	59,817	59,770	59,722	59,673	59,625
33	59,576	59,527	59,478	59,429	59,380	59,330	59,280	59,230	59,179	59,129
34	59,078	59,027	58,976	58,924	58,873	58,821	58,769	58,717	58,664	58,611
35	58,558	58,505	58,452	58,398	58,344	58,290	58,236	58,181	58,126	58,071
36	58,016	57,960	57,905	57,849	57,792	57,736	57,679	57,622	57,565	57,503
37	57,450	57,392	57,334	57,276	57,217	57,158	57,099	57,040	56,980	56,920
38	56,860	56,809	56,739	56,678	56,617	56,556	56,494	56,432	56,370	56,308
39	56,245	56,182	56,110							

Dispersion Equation

$$n^2 = 5.640752 - 0.000003338\lambda^2 = \frac{0.0018612}{\lambda^2} + \left(\frac{41{,}110.49}{\lambda^2} - 14{,}390.4\right) + \left(\frac{0.0290764}{\lambda^2} - 0.024964\right)$$

Temperature Coefficient

The average temperature coefficient of refractive index for two samples of different origin is given by Rodney and Spindler as $7.9 \times 10^{-5}/°\text{C}$.

TABLE 12 Cesium Iodide: Refractive Index at 24°C (Rodney, 1955)

λ, μm	n	λ, μm	n	λ, μm	n	λ, μm	n	λ, μm	n
0.280	103,939	1.84	74,683	16.6	73,242	27.8	71,289	39.0	68,159
0.300	97,872	1.88	74,664	16.8	73,210	28.0	71,244	39.2	68,091
0.320	93,700	1.92	74,647	17.0	73,190	28.2	71,199	39.4	68,023
0.340	90,649	1.96	74,631	17.2	73,164	28.4	71,153	39.6	67,954
0.360	88,321	2.00	74,616	17.4	73,137	28.6	71,107	39.8	67,881
0.380	86,497	2.20	74,551	17.6	73,111	28.8	71,061	40.0	67,814
0.400	85,927	2.40	74,500	17.8	73,088	29.0	71,014	40.2	67,744
0.420	83,823	2.60	74,460	18.0	73,056	29.2	70,967	40.4	67,673
0.440	82,820	2.80	74,427	18.2	73,028	29.4	70,919	40.6	67,601
0.460	81,975	3.00	74,400	18.4	72,999	29.6	70,871	40.8	67,530
0.480	81,255	3.50	74,346	18.6	72,971	29.8	70,823	41.0	67,457
0.500	80,635	4.00	74,305	18.8	72,942	30.0	70,774	41.2	67,381
0.520	80,097	4.50	74,270	19.0	72,913	30.2	70,725	41.4	67,311
0.540	79,626	5.00	74,239	19.2	72,833	30.4	70,676	41.6	67,237
0.560	79,213	5.50	74,210	19.4	72,853	30.6	70,626	41.8	67,163
0.580	78,846	6.00	74,181	19.6	72,823	30.8	70,576	42.0	67,088
0.600	78,520	6.50	74,152	19.8	72,793	31.0	70,525	42.2	67,013
0.620	78,229	7.00	74,122	20.0	72,762	31.2	70,474	42.4	66,937
0.640	77,967	7.50	74,091	20.2	72,731	31.4	70,422	42.6	66,861
0.660	77,731	8.00	74,059	20.4	72,699	31.6	70,371	42.8	66,781
0.680	77,517	8.50	74,026	20.6	72,667	31.8	70,318	43.0	66,707
0.700	77,323	9.00	73,991	20.8	72,635	32.0	70,266	43.2	66,620
0.720	77,146	9.50	73,954	21.0	72,602	32.2	70,213	43.4	66,551
0.740	76,985	10.0	73,916	21.2	72,570	32.4	70,150	43.6	66,472
0.760	76,836	10.2	73,901	21.4	72,536	32.6	70,105	43.8	66,392
0.780	76,700	10.4	73,885	21.6	72,503	32.8	70,051	44.0	66,312
0.800	76,575	10.6	73,868	21.8	72,469	33.0	69,996	44.2	66,232
0.820	76,459	10.8	73,852	22.0	72,435	33.2	69,941	44.4	66,151
0.840	76,352	11.0	73,835	22.2	72,400	33.4	69,886	44.6	66,070
0.860	76,252	11.2	73,818	22.4	72,365	33.6	69,830	44.8	65,988
0.880	76,159	11.4	73,800	22.6	72,330	33.8	69,774	45.0	65,905
0.900	76,074	11.6	73,783	22.8	72,291	34.0	69,717	45.2	65,822
0.920	75,993	11.8	73,765	23.0	72,258	34.2	69,660	45.4	65,739
0.940	75,918	12.0	73,746	23.2	72,222	34.4	69,602	45.6	65,655
0.960	75,848	12.2	73,728	23.4	72,185	34.6	69,544	45.8	65,570
0.980	75,782	12.4	73,700	23.6	72,148	34.8	69,486	46.0	65,485
1.00	75,721	12.6	73,690	23.8	72,111	35.0	69,427	46.2	65,399
1.04	75,608	12.8	73,670	24.0	72,073	35.2	69,368	46.4	65,313
1.08	75,508	13.0	73,650	24.2	72,035	35.4	69,398	46.6	65,226
1.12	75,419	13.2	73,630	24.4	71,997	35.6	69,248	46.8	65,138
1.16	75,339	13.4	73,610	24.6	71,958	35.8	69,188	47.0	65,051
1.20	75,268	13.6	73,589	24.8	71,919	36.0	69,127	47.2	64,962
1.24	75,203	13.8	73,568	25.0	71,880	36.2	69,065	47.4	64,873
1.28	75,144	14.0	73,547	25.2	71,840	36.4	69,004	47.6	64,783
1.32	75,091	14.2	73,525	25.4	71,800	36.6	68,911	47.8	64,693
1.36	75,042	14.4	73,504	25.6	71,759	36.8	68,879	48.0	64,602
1.40	74,997	14.6	73,481	25.8	71,718	37.0	68,815	48.2	64,511
1.44	74,956	14.8	73,459	26.0	71,677	37.2	68,752	48.4	64,419
1.48	74,919	15.0	73,436	26.2	71,635	37.4	68,688	48.6	64,326
1.52	74,884	15.2	73,413	26.4	71,593	37.6	68,623	48.8	64,233
1.56	74,852	15.4	73,389	26.6	71,551	37.8	68,558	49.0	64,139
1.60	74,822	15.6	73,366	26.8	71,508	38.0	68,493	49.2	64,045
1.64	74,795	15.8	73,342	27.0	71,465	38.2	68,427	49.4	63,950
1.68	74,769	16.0	73,317	27.2	71,422	38.4	68,361	49.6	63,855
1.72	74,745	16.2	73,292	27.4	71,378	38.6	68,294	49.8	63,759
1.76	74,723	16.4	73,267	27.6	71,334	38.8	68,227	50.0	63,662
1.80	74,702								

Dispersion Equation

$$n^2 - 1 = \sum_i \frac{K_i \lambda^2}{\lambda^2 - \lambda_i^2}$$

Constants

i	λ_i^2	K_i	i	λ_i^2	K_i
1	0.00052701	0.34617251	4	0.044944	0.39743178
2	0.02140156	1.0080886	5	25,921.0	3.3605359
3	0.032761	0.28551800			

TABLE 13 Crystal Quartz: Refractive Index and Temperature Coefficient

Refractive Index†

λ, μm	n_o	n_e	λ, μm	n_o	n_e
0.185	1.05751	1.68988	1.5414	1.52781	1.53630
0.198	1.65087	1.66394	1.6815	1.52583	1.53422
0.231	1.61395	1.62555	1.7614	1.52468	1.53301
0.340	1.56747	1.57737	1.9457	1.52184	1.53004
0.394	1.55846	1.56805	2.0531	1.52005	1.52823
0.434	1.55396	1.56330	2.30	1.51561	
0.508	1.54822	1.55746	2.00	1.50986	
0.5893	1.54424	1.55335	3.00	1.49953	
0.768	1.53903	1.54794	3.50	1.48451	
0.8325	1.53773	1.54661	4.00	1.46617	
0.9914	1.53514	1.54392	4.20	1.4569	
1.1592	1.53233	1.54152	5.00	1.417	
1.3070	1.53090	1.53951	6.45	1.274	
1.3958	1.52977	1.53832	7.0	1.167	
1.4792	1.52865	1.53716			

Temperature Coefficient (Micheli, 1902*a*)

λ, μm	$\frac{dn_o}{dT}$, $10^{-5}/°C$	$\frac{dn_e}{dT}$, $10^{-5}/°C$	λ, μm	$\frac{dn_o}{dT}$, $10^{-5}/°C$	$\frac{dn_e}{dT}$, $10^{-5}/°C$
0.202	+0.321	+0.267	0.293	−0.311	−0.415
0.206	0.253	0.193	0.313	−0.348	−0.450
0.210	0.193	0.143	0.325	−0.352	−0.469
0.214	0.124	0.083	0.340	−0.393	−0.591
0.219	0.074	0.027	0.361	−0.418	−0.521
0.224	0.017	−0.048	0.441	−0.475	−0.593
0.226	−0.008	−0.075	0.467	−0.485	−0.601
0.228	−0.027	−0.093	0.480	−0.499	−0.610
0.231	−0.052	−0.112	0.508	−0.514	−0.616
0.257	−0.186	−0.265	0.589	−0.530	−0.642
0.274	−0.235	−0.323	0.643	−0.549	−0.653
0.283	−0.279	−0.385			

† The data for the region 0.185 to 0.76 μm were taken at 23°C by Martens (1901*a*). Similar values are given by Gifford (1902) and Trommsdorff (1901). Sosman (1927) gives a collation of the above data. The data for the wavelength region from 0.8325 to 2.30 μm (at 20°C) are taken from Carvallo (1896*a*). For the region from 2.60 to 7.0 μm (at 18°C) the data are taken from Rubens (1895). The values fit together well (to a few parts in the fifth decimal place). They are, however, questionable at the extreme ends of the range (0.185, 5.00, 6.45, and 7.0 μm).

TABLE 14 Diamond: Refractive Index (Rosch, 1965)

λ, μm	n	λ, μm	n
0.480	2.4368	0.589	2.4175
0.486	2.4354	0.644	2.4114
0.546	2.4235	0.656	2.4104

TABLE 15 Fluorite (Calcium Fluoride): Refractive Index and Temperature Coefficient (Malitson, 1963)†

λ, μm	n_{cal}	Δn Synthetic	Δn Natural	$\frac{-dn}{dT}$, 10^{-6}/°C	λ, μm	n_{cal}	Δn Synthetic	Δn Natural	$\frac{-dn}{dT}$, 10^{-6}/°C
0.228803	1.47635	−2	+1	6.2	1.39506	1.42675	+1	+6	9.9
0.24827	1.46793	+3	+5	7.0	1.52952	1.42642	+4	+4	9.6
0.2537	1.46602	+9	+12	7.5	1.7012	1.42531	+2	+4	9.4
0.26520	1.46233	−1	0	8.1	1.81307	1.42478	0	+9	9.1
0.28035	1.45828	−1	+1	8.4	1.97009	1.42401	+3	+3	8.9
0.296728	1.45467	−2	0	8.8	2.1526	1.42306	−1	+1	8.7
0.334448	1.44852	−1	+2	9.2	2.32542	1.42212	+3	+4	8.5
0.34662	1.44691	−3	0	9.4	2.4374	1.42147	0	+2	8.5
0.365015	1.44490	−4	0	9.6	3.3026	1.41561	0	+3	8.2
0.4046563	1.44451	−3	+1	9.8	3.422	1.41467	+2	+2	8.1
0.4358312	1.43949	−3	0	10.0	3.5070	1.41398	−1	+2	8.0
0.4864327	1.43703	−4	0	10.2	3.7067	1.41229	+2	+2	7.8
0.546074	1.43494	−3	+1	10.4	4.258	1.40713	+4	+4	7.5
0.580262	1.43381	−2	+2	10.4	5.01882	1.39873	+1	+5	7.3
0.643847	1.43268	−2	+3	10.4	5.3034	1.39520	+3	+3	7.2
0.6562793	1.43246	−2	+1	10.4	6.0140	1.38539	+5	+5	7.0
0.6678449	1.43226	−1	+1	10.5	6.238	1.38200	−6	0	7.0
0.7065188	1.43167	−2	+2	10.5	6.63306	1.37565	0	+1	6.9
0.767858	1.43088	−2	+2	10.6	6.8559	1.37186	−8	+2	6.7
0.85212	1.43002	−1	+4	10.6	7.268	1.36443	+2	+7	6.5
0.8944	1.42966	0	+1	10.6	7.4644	1.36070	+5	+6	6.4
1.01398	1.42879	−2	0	10.5	8.662	1.33500	−4	+3	6.0
1.3622	1.42691	+1	+8	10.0	9.724	1.30756	+1	+5	5.6

Dispersion Equation

$$n^2 - 1 = \frac{A_i\lambda^2}{\lambda^2 - \lambda_i^2}$$

Constants at 25°C

$\lambda_1 =$ 0.050263605	$\lambda_1^2 = 0.002526430$	$A_1 = 0.5675888$
$\lambda_2 =$ 0.1003909	$\lambda_2^2 = 0.01007833$	$A_2 = 0.4710914$
$\lambda_3 =$ 34.649040	$\lambda_3^2 = 1{,}200.5560$	$A_3 = 3.8484723$

† It is useful to compare these with the data of earlier investigators.

TABLE 16 Fused Silica: Refractive Index Data (Malitson, 1965)

Refractive Index and Residuals at 20°C for Three Specimens

Wave-length, μm	Spectral source	Com-puted index	Residual × 10^6 Mean†	Corning	Dynasil	General Electric
0.213856	Zn	1.534307	−27	−29	−43	−31
0.214438	Cd	1.533722	− 2	−11	−21	−22
0.226747	Cd	1.522750	+70	+71	+68	+73
0.230209	Hg	1.520081	−21	−28	−31	−23
0.237833	Hg	1.514729	+ 1	+13	+23	+19
0.239938	Hg	1.513367	+ 3	+ 6	+ 2	+ 9
0.248272	Hg	1.508398	+ 2	+ 6	− 1	+ 7
0.265204	Hg	1.500029	−29	−32	−25	−13
0.269885	Hg	1.498047	+ 3	+ 7	− 4	+11
0.275278	Hg	1.495913	− 3	+ 2	+ 8	+12
0.280347	Hg	1.494039	+ 1	− 4	− 9	−11
0.289360	Hg	1.490990	+20	+18	+22	+20
0.296728	Hg	1.488734	−14	− 7	−12	− 4
0.302150	Hg	1.487194	− 4	− 9	− 2	+ 4
0.330259	Zn	1.480539	− 9	+ 1	+10	+ 3
0.334148	Hg	1.479763	− 3	− 8	− 1	+ 9
0.340365	Cd	1.478584	+ 6	+ 9	+ 2	− 8
0.346620	Cd	1.477468	+ 2	−17	−12	−14
0.361051	Cd	1.475129	+ 1	+ 3	− 9	− 8
0.365015	Hg	1.474539	−19	−11	−15	−21
0.404656	Hg	1.469618	+ 2	+ 1	− 1	+ 2
0.435835	Hg	1.466693	− 3	+ 5	+ 1	+ 3
0.467816	Cd	1.464292	+ 8	+ 5	+ 3	+ 6
0.486133	H	1.463126	+ 4	+ 6	+ 5	+ 7
0.508582	Cd	1.461863	+ 7	+ 4	+ 1	+ 5
0.546074	Hg	1.460078	+ 2	+ 4	+ 1	− 5
0.576959	Hg	1.458846	+ 4	+ 5	+ 3	+ 4
0.579065	Hg	1.458769	+ 1	+ 6	+ 6	+ 6
0.587561	He	1.458464	+ 6	+ 3	− 2	+ 1
0.589262	Na	1.458404	− 4	+ 6	+ 3	+ 7
0.643847	Cd	1.456704	+ 6	+ 9	+ 4	+ 7

Wave-length, μm	Spectral source	Com-puted source	Residual × 10^6 Mean†	Corning	Dynasil	General Electric
0.656272	H	1.456367	+ 3	+ 7	+ 5	+ 7
0.667815	He	1.456067	+ 3	+ 8	+ 6	+ 3
0.706519	He	1.455145	+ 5	+10	+12	+ 7
0.852111	Cs	1.452465	+ 5	+ 8	+ 3	+ 5
0.894350	Cs	1.451835	+ 5	+11	+ 5	+10
1.01398	Hg	1.450242	+ 8	+ 6	+ 3	+ 6
1.08297	He	1.449405	− 5	+ 8	+ 1	+ 9
1.12866	Hg	1.448869	+ 1	+ 7	+ 8	+ 9
1.3622	Hg	1.446212	−12	− 6	−14	−12
1.39506	Hg	1.445836	+ 4	− 1	+ 4	− 3
1.4695	Cs	1.444975	− 5	+ 3	+ 9	+10
1.52952	Hg	1.444268	+ 2	+ 8	+ 6	0
1.6606	TCB‡	1.442670	−20	−14	−19	−11
1.681	Poly §	1.442414	+ 6	− 2	−10	+ 8
1.6932	Hg	1.442260	0	+ 7	− 6	+ 1
1.70913	Hg	1.442057	+ 3	0	+ 3	− 1
1.81307	Hg	1.440699	+21	− 7	− 7	+ 6
1.97009	Hg	1.438519	+ 1	+ 6	+12	+12
2.0581	He	1.437224	− 4	− 3	− 9	−11
2.1526	TCB‡	1.435769	−29	−22	−25	−24
2.32542	Hg	1.432928	−18	−10	− 3	− 6
2.4374	TCB‡	1.430954	−24	−23	−21	−14
3.2439	Poly §	1.413118	+32	+21	+29	+25
3.2668	Poly §	1.412505	+25	+20	+30	+25
3.3026	Poly §	1.411535	+25	+32	+30	+28
3.422	Poly §	1.408180	+20	+40	+42	+37
3.5070	Poly §	1.405676	−16	−26	−20	−10
3.5564	TCB‡	1.401174	−24	−27	−29	−18
3.7067	TCB‡	1.399389	−19	−22	−14	− 9
Average of absolute values of residuals			10.5	11.9	12.2	11.7

Dispersion Equation

$$n^2 - 1 = \frac{0.6961663\lambda^2}{\lambda^2 - 0.0684043^2} + \frac{0.4079426\lambda^2}{\lambda^2 - 0.1162414^2} + \frac{0.8974794\lambda^2}{\lambda^2 - 9.896161^2}$$

where λ is expressed in micrometers

Variation of the Refractive Index at 20°C between Specimens from Different Sources

	Residual × 10⁶											
	Corning¶				Dynasil¶				General Electric¶			
λ, μm	1	2	3	4	1	2	3	4	1	2	3	4
0.4047	11	26	24	20	4	27	21	21	11	12	−12	14
0.4861	16	27	16	21	5	29	19	21	7	15	− 7	12
0.5461	14	24	22	17	5	33	25	16	11	12	−13	12
0.5893	12	27	23	27	0	18	20	22	12	13	− 8	12
0.6563	12	23	19	19	1	27	22	23	13	10	− 7	14
0.7065	17	28	20	18	2	31	23	23	13	15	−11	17
0.8944	11	26	22	19	3	31	19	25	11	12	− 8	13
1.014	16	21	20	20	6	30	26	22	14	13	−11	14
1.083	15	24	25	25	7	35	25	25	13	16	− 9	12
Av. Δn	13.5	24.9	20.0	21.6	3.5	27.4	20.4	20.5	11.3	12.9	− 9.6	13.2
ν value	67.78	67.78	67.86	67.80	67.78	67.80	67.85	67.85	67.88	67.77	67.82	67.84

† Residuals for arithmetic-mean table of values from experimental data of Corning, Dynasil, and General Electric.
‡ 1,4,4-Trichlorobenzene.
§ Polystyrene.
¶ Numbered columns under each brand indicate individual specimens. The residuals are differences between measured values and those computed by the dispersion equation. Each Δn is an average for the 18 wavelengths used.

TABLE 17 Gallium Antimonide: Refractive Index and Extinction Coefficient (Seraphin and Bennett, 1967*b*)†

λ, μm	*n*	*k*	λ, μm	*n*	*k*
1.49	.....	0.0970	1.97	.....	0.000110
1.51	.....	0.0945	2.00	3.789	0.000108
1.53	.....	0.0905			0.0000987
1.55	.....	0.0867	2.03	.....	0.000109
1.56	.....	0.0852	2.07	.....	0.000113
1.57	.....	0.0816	2.1	3.780	
1.58	.....	0.0790	2.2	3.764	
1.59	.....	0.0768	2.3	3.758	
1.60	.....	0.0739	2.4	3.755	0.000143
1.61	.....	0.0710	2.5	3.749	0.000165
1.62	.....	0.0680	2.8	.....	0.000265
1.63	.....	0.0649	3.0	3.898	0.000365
1.64	.....	0.0614	3.4	.....	0.000666
1.65	.....	0.0582	3.5	3.861	0.00746
1.66	.....	0.0547	3.7	.....	0.000925
1.68	.....	0.0510	4.0	3.833	0.00126
1.69	.....	0.0470	4.5	.....	0.00188
1.70	.....	0.0406	5.0	3.824	0.00253
1.71	.....	0.0251	5.4	.....	0.00313
1.72	.....	0.00748	5.8	.....	0.00366
1.73	.....	0.00668	6.0	3.824	0.00394
		0.00214	6.2	.....	0.00422
1.74	.....	0.00305	6.7	.....	0.00490
1.75	.....	0.00183	7.0	3.843	0.00533
1.76	.....	0.00124	7.4	.....	0.00590
1.77	.....	0.000918	8.0	3.843	0.00668
		0.00123	8.4	.....	0.00721
1.80	3.820	0.000540	9.0	3.843	0.00799
		0.000551	9.5	.....	0.00863
1.82	.....	0.000356	10.0	3.843	0.00926
		0.000355	10.6	.....	0.00995
1.84	.....	0.000196	11.1	.....	0.0106
1.85	.....	0.000252	12.0	3.843	0.0116
1.88	.....	0.000200	12.4	.....	0.0121
		0.000141	12.8	.....	0.0126
1.90	3.802		13.4	.....	0.0141
1.91	.....	0.000141	14.0	3.861	0.0140
1.94	.....	0.000118	14.9	3.880	

† Two or more values for one wavelength are from different sources; for identification of all sources the original book should be consulted.

TABLE 18 Gallium Arsenide: Refractive Index† (Barcus, 1958, no date)

λ, μm	*n*	λ, μm	*n*
0.78 ± 0.01	3.34 ± 0.04	14.5 ± 0.05	2.82 ± 0.04
8.0 ± 0.05	3.34 ± 0.04	15.0 ± 0.05	2.73 ± 0.04
10.0 ± 0.05	3.135 ± 0.04	17.0 ± 0.05	2.59 ± 0.04
11.0 ± 0.05	3.045 ± 0.04	19.0 ± 0.05	2.41 ± 0.04
13.0 ± 0.05	2.97 ± 0.04	21.9 ± 0.1	2.12 ± 0.04
13.7 ± 0.05	2.895 ± 0.04		

† The experimental data seem to be somewhat more scattered than the reported experimental errors indicate.

TABLE 19 Gallium Phosphide: Refractive-Index Data

Refractive Index and Extinction Coefficient (Seraphin and Bennett, 1967*a*)

λ, μm	*n*	*k*	λ, μm	*n*	*k*
0.049	1.028	0.168	0.504	....	1.85×10^{-3}
0.050	1.000	0.201	0.510	....	1.42×10^{-3}
0.052	0.991	0.231	0.514	....	1.19×10^{-3}
0.054	0.999	0.280	0.519	....	1.04×10^{-3}
0.056	1.023	0.290	0.520	3.65	
0.059	1.026	0.278	0.521	....	8.57×10^{-4}
0.063	0.972	0.243	0.530	....	5.34×10^{-4}
0.065	0.901	0.286	0.534	....	3.34×10^{-4}
0.069	0.839	0.362	0.538	....	2.41×10^{-4}
0.073	0.790	0.474	0.544	....	1.18×10^{-4}
0.077	0.801	0.588	0.547	....	6.90×10^{-5}
0.083	0.825	0.691	0.553	....	3.44×10^{-5}
0.089	0.854	0.796	0.555	....	2.19×10^{-5}
0.095	0.896	0.905	0.561	....	1.39×10^{-5}
0.103	0.951	1.023	0.565	....	9.38×10^{-6}
0.113	1.025	1.138	0.571	....	6.88×10^{-6}
0.124	1.143	1.303	0.574	....	5.70×10^{-6}
0.135	1.250	1.230	0.579	....	4.96×10^{-6}
0.139	1.195	1.234	0.582	....	4.71×10^{-6}
0.146	1.116	1.331	0.600	3.46	
0.157	1.075	1.639	0.620	3.42	
0.165	1.187	1.841	0.640	3.36	
0.177	1.390	2.119	0.660	3.34	
0.180	1.479	2.188	0.680	3.34	
0.188	1.727	2.125	0.700	3.33	
0.190	1.732	2.073	0.750	3.28	
0.194	1.696	2.085	0.800	3.27	
0.207	1.709	2.353	0.850	3.24	
0.229	2.246	3.295	0.900	3.20	
0.234	2.631	3.472	0.950	3.17	
0.239	3.191	3.462	1.00	3.17	
0.264	4.130	2.102	1.10	3.13	
0.288	3.862	1.481	1.20	3.10	
0.295	3.785	1.494	1.30	3.07	
0.302	3.743	1.541	1.40	3.07	
0.310	3.770	1.649	1.60	3.04	4.99×10^{-4}
0.318	3.906	1.801	1.80	3.03	7.56×10^{-4}
0.330	4.653	1.959	2.00	3.02	1.03×10^{-3}
0.335	5.078	1.710	2.50	2.99	2.11×10^{-3}
0.344	5.192		3.00	2.97	2.53×10^{-3}
0.354	4.848		3.50	2.95	2.23×10^{-3}
0.375	4.252		4.00	2.95	2.57×10^{-3}
0.452	.....	3.12×10^{-2}	4.50	2.95	3.21×10^{-3}
0.454	.....	2.36×10^{-2}	5.00	2.94	4.22×10^{-3}
0.456	.....	1.87×10^{-2}	6.00	2.92	7.26×10^{-3}
0.457	.....	1.59×10^{-2}	7.00	2.95	1.11×10^{-2}
0.459	.....	1.43×10^{-2}	8.00	2.94	1.64×10^{-2}
0.463	.....	1.16×10^{-2}	9.00	2.91	2.69×10^{-2}
0.464	.....	9.49×10^{-3}	10.00	2.90	
0.470	.....	7.68×10^{-3}	16.00	2.750	
0.473	.....	6.37×10^{-3}	16.65	2.714	
0.477	.....	5.47×10^{-3}	17.23	2.679	
0.480	.....	4.66×10^{-3}	18.00	2.643	
0.482	.....	4.29×10^{-3}	18.87	2.600	
0.488	.....	3.41×10^{-3}	20.00	2.529	
0.492	.....	3.06×10^{-3}	20.87	2.500	
0.496	.....	2.54×10^{-3}	21.72	2.386	
0.500	.....	2.47×10^{-3}	22.51	2.207	
0.501	.....	2.18×10^{-3}	23.29	1.979	

TABLE 19 Gallium Phosphide: Refractive-Index Data (Continued)

Refractive Index and Extinction Coefficient (Seraphin and Bennett, 1967*a*)

λ, μm	*n*	*k*	λ, μm	*n*	*k*
24.01	1.650		27.34	.....	10.84
24.57	1.129		27.36	15.19	
24.82	.....	0.064	27.37	.....	8.20
25.00	0.100		27.44	.....	5.13
25.26	.....	1.379	27.45	13.47	
25.42	0.043		27.46	.....	2.57
26.23	.....	3.400	27.53	.....	1.04
26.51	0.129		27.54	10.87	
26.66	.....	5.000	27.57	....	0.68
26.89	.....	6.614	27.65	....	0.45
26.98	0.457		27.69	8.50	
27.04	....	8.893	27.82	....	0.200
27.14	....	11.50	28.09	....	0.114
27.18	1.83		28.21	6.19	
27.19	4.07		28.64	....	0.107
27.21	7.62		29.12	....	0.050
27.26	13.12		29.34	4.67	
27.26	.....	14.68	29.69	....	0.050
27.29	.....	17.33	29.94	4.39	
27.32	16.58	13.89	30.00	4.36	0.021

Comparison of Experimental and Calculated Index Data (Parsons and Coleman, 1971)

ν, cm⁻¹	Exp.	Cal.	ν, cm⁻¹	Exp.	Cal.	ν, cm⁻¹	Exp.	Cal.
12,500	3.1830	3.18286	183.93	3.4440	3.4435	94.21	3.3621	3.3609
9,091	3.0981	3.09821	175.29	3.4302	3.4311	87.92	3.3584	3.3579
5,000	3.0379	3.03754	167.78	3.4238	3.4214	72.07	3.3522	3.3513
4,167	3.0296	3.02925	162.00	3.4134	3.4145	59.36	3.3463	3.3472
3,125	3.0197	3.01982	149.96	3.4010	3.4015	52.97	3.3450	3.3454
2,778	3.0166	3.01658	139.27	3.3922	3.3915	46.17	3.3438	3.3438
2,500	3.0137	3.01375	128.54	3.3830	3.3826	33.76	3.3399	3.3414
			117.70	3.3754	3.3747	27.32	3.3407	3.3405
			109.90	3.3706	3.3697			

TABLE 20 Germanium: Refractive Index and Temperature Coefficient

Refractive Index at 27°C† (Salzberg and Villa, 1953)

λ, μm	n	λ, μm	n	λ, μm	n
2.0581	4.1016	2.998	4.0452	8.66	4.0043
2.1526	4.0919	3.3033	4.0369	9.72	4.0034
2.3126	4.0786	3.4188	4.0334	11.04	4.0026
2.4374	4.0708	4.258	4.0216	12.20	4.0023
2.577	4.0609	4.866	4.0170	13.02	4.0021
2.7144	4.0552	6.238	4.0094		

Dispersion Equation

$$n = A + BL + CL^2 + D\lambda^2 + E\lambda^4$$

where $A = 3.99931$, $B = 0.391707$, $C = 0.163492$, $D = -0.0000060$, $E = 0.000000053$, $L = \dfrac{1}{\lambda^2 - 0.028}$

Temperature Coefficient of Refractive Index at 24.5°C (Rank and Bennett, 1954)

λ, μm	$\frac{dn}{dT}$, 10^{-4}/°C	λ, μm	$\frac{dn}{dT}$, 10^{-4}/°C
1.934	5.919	2.246	5.251
2.174	5.285	2.401	5.037

† The resistivity of the germanium sample is about 50 Ω/cm. These data are about five parts in the third decimal place lower than those reported by Briggs (1950). Salzberg and Villa (1957) report corrections to their original measurements. They also report that refractive indexes of cast polycrystalline germanium differ from those of the single crystal by only a few parts in the fourth decimal place. The relative temperature coefficient of refractive index $(1/n)(dn/dT)$ has been given as $(6.7 \pm 0.4) \times 10^{-5}$ per Celsius degree by Cardona et al. (no date).

TABLE 21 Indium Antimonide: Refractive Index and Extinction Coefficient (Seraphin and Bennett, 1967*e*)

λ, μm	*n*	*k*	λ, μm	*n*	*k*	λ, μm	*n*	*k*
0.049	1.15	0.15	0.477	3.42	2.06	7.00	.....	0.025
0.052	1.15	0.18	0.517	3.82	2.25	7.50	.....	0.0052
0.054	1.15	0.19	0.539	4.05	2.09	7.87	4.001	
0.056	1.16	0.20	0.564	4.18	1.94	8.00	3.995	0.00231
0.059	1.17	0.21	0.590	4.22	1.82	8.50	.....	0.0020
0.062	1.17	0.21	0.620	4.29	1.83	9.00	.....	0.0019
0.065	1.18	0.20	0.656	4.71	1.88	9.01	3.967	
0.069	1.15	0.18	0.677	5.08	1.65	9.50	.....	0.0019
0.073	1.11	0.16	0.689	5.13	1.37	10.06	3.953	
0.080	1.02	0.17	0.708	5.06	1.07	11.01	3.937	
0.083	0.97	0.19	0.775	4.72	0.60	12.06	3.920	
0.089	0.88	0.26	0.886	4.40	0.40	12.98	3.912	
0.095	0.80	0.37	1.03	4.24	0.32	13.90	3.902	
0.103	0.75	0.51	1.24	4.15	0.26	15.13	3.881	
0.113	0.72	0.69	1.55	4.08	0.20	15.79	3.873	
0.124	0.74	0.88	1.60	....	0.18	16.96	3.866	
0.138	0.80	1.08	1.80	....	0.17	17.85	3.850	
0.155	0.88	1.32	2.00	....	0.17	18.85	3.843	
0.163	0.94	1.41	2.07	4.03		19.98	3.826	
0.182	1.06	1.61	2.50	....	0.15	20.0	.....	0.0020
0.207	1.23	1.91	3.00	....	0.13	21.15	3.814	
0.218	1.36	1.97	3.50	....	0.12	22.20	3.805	
0.221	1.38	2.00	4.00	....	0.11	25.0	3.78	0.0051
0.234	1.48	2.11	4.50	....	0.10	26.0	3.74	0.0075
0.238	1.53	2.14	5.00	....	0.091	27.0	3.66	0.0097
0.243	1.57	2.13	6.00	....	0.074	28.0	3.56	0.011
0.248	1.56	2.15	6.10	....	0.072	29.0	3.50	0.010
0.282	1.66	2.70	6.20	....	0.070	30.0	3.47	0.010
0.302	2.19	3.26	6.30	....	0.068	31.0	3.44	0.010
0.310	2.62	3.36	6.40	....	0.066	32.0	3.39	0.010
0.344	3.51	2.44	6.50	....	0.063	33.0	3.34	0.011
0.365	3.51	2.15	6.60	....	0.059	34.0	3.30	0.012
0.413	3.37	1.81	6.70	....	0.055	35.0	3.25	0.014
0.428	3.32	1.86	6.80	....	0.049	40.0	2.98	0.026
0.443	3.32	1.91	6.90	....	0.037	45.0	2.57	0.054

Relative Temperature Coefficient of Refractive Index (Cardona et al., 1958)

$$\frac{1}{n}\frac{dn}{dT} = (3.9 \pm 0.4) \times 10^{-5}/°\mathrm{C} \qquad \text{for 77 to 400 K}$$

TABLE 22 Indium Arsenide: Refractive Index and Extinction Coefficient (Seraphin and Bennett, 1967*d*)

λ, μm	*n*	*k*	λ, μm	*n*	*k*	λ, μm	*n*	*k*
0.049	1.139	0.168	0.248	1.987	2.647	1.38	3.516	0.047
0.051	1.135	0.195	0.259	2.288	3.086	1.68	....	0.200
0.054	1.135	0.207	0.264	2.617	3.264	1.80	....	0.185
0.056	1.133	0.215	0.269	3.060	3.274	2.00	.....	0.168
0.059	1.131	0.222	0.222	3.800	2.735	2.07	....	0.162
0.062	1.125	0.225	0.310	3.678	1.508	2.25	....	0.149
0.064	1.120	0.224	0.335	3.359	1.340	2.50	....	0.133
0.067	1.110	0.215	0.344	3.271	1.363	2.76	....	0.119
0.070	1.047	0.189	0.354	3.227	1.411	3.00	....	0.102
0.077	0.948	0.272	0.387	3.484	1.547	3.35	....	0.064
0.082	0.894	0.336	0.413	3.331	1.787	3.40	....	0.052
0.089	0.829	0.426	0.443	3.817	1.954	3.44	....	0.037
0.095	0.766	0.563	0.451	3.980	1.865	3.50	....	0.018
0.103	0.745	0.727	0.459	4.087	1.748	3.65	....	0.002
0.108	0.751	0.830	0.468	4.119	1.644	3.74	3.52	
0.112	0.775	0.905	0.477	4.192	1.618	4.00	3.51	
0.123	0.835	1.071	0.496	4.489	1.452	5.00	3.46	
0.136	0.890	1.260	0.517	4.558	1.047	6.67	3.45	
0.153	0.967	1.552	0.563	4.320	0.554	10.0	3.42	
0.172	1.184	1.889	0.620	4.101	0.348	14.3	3.39	
0.180	1.332	1.998	0.689	3.934	0.231	16.7	3.38	
0.188	1.483	2.020	0.775	3.800	0.157	20.0	3.35	
0.195	1.583	2.120	0.885	3.696	0.100	25.0	3.26	
0.211	1.782	2.011	1.03	3.613	0.076	33.3	2.95	
0.225	1.765	2.202	1.24	3.548	0.051			

Dispersion Equation

$$n^2 = 11.1 + \frac{0.71}{1 - (1/3922\lambda)^2} + \frac{2.75}{1 - (1/218.9\lambda)^2} - 6 \times 10^4\lambda^2$$

TABLE 23 Indium Phosphide: Refractive Index and Extinction Coefficient (Seraphin and Bennett, 1967c)

λ, μm	n	k	λ, μm	n	k	λ, μm	n	k
0.062	0.793	0.494	0.197	1.526	1.991	0.945	3.374	0.000967
0.064	0.815	0.499	0.200	1.525	1.982	0.950	3.369	0.000527
0.065	0.834	0.493	0.203	1.508	2.005	0.953	.....	0.000281
0.067	0.843	0.487	0.207	1.500	2.063	0.955	3.364	
0.069	0.846	0.477	0.210	1.516	2.130	0.957	.....	0.000145
0.071	0.840	0.469	0.214	1.544	2.191	0.960	3.359	0.0000739
0.073	0.824	0.454	0.217	1.573	2.267	0.965	3.355	0.0000392
0.075	0.785	0.457	0.221	1.616	2.349	0.968	.....	0.0000246
0.077	0.742	0.491	0.225	1.668	2.442	0.970	3.351	
0.079	0.719	0.529	0.229	1.737	2.553	0.972	.....	0.0000166
0.083	0.695	0.574	0.234	1.834	2.675	0.975	3.346	0.0000113
0.085	0.675	0.645	0.238	1.960	2.801	1.00	3.327	
0.089	0.688	0.706	0.243	2.132	2.982	1.10	3.268	
0.092	0.701	0.765	0.248	2.451	3.166	1.20	3.231	
0.095	0.726	0.820	0.253	2.885	3.144	1.30	3.205	
0.099	0.754	0.861	0.258	3.335	3.039	1.40	3.186	
0.103	0.771	0.899	0.264	3.729	2.635	1.50	3.172	
0.108	0.781	0.946	0.269	3.849	2.117	1.60	3.161	
0.113	0.793	0.996	0.275	3.655	1.691	1.70	3.152	
0.118	0.797	1.056	0.282	3.473	1.549	1.80	3.145	
0.124	0.806	1.154	0.288	3.347	1.468	1.90	3.139	
0.125	0.820	1.172	0.295	3.248	1.415	2.00	3.134	
0.126	0.832	1.185	0.302	3.162	1.389	5.00	3.08	
0.128	0.840	1.198	0.310	3.105	1.392	6.00	3.07	
0.129	0.847	1.210	0.318	3.054	1.401	7.00	3.07	
0.130	0.852	1.225	0.326	3.027	1.440	8.00	3.06	
0.132	0.859	1.237	0.335	3.024	1.489	9.00	3.06	
0.133	0.861	1.253	0.344	3.047	1.550	10.00	3.05	
0.135	0.865	1.269	0.354	3.082	1.622	12.00	3.05	0.000527
0.136	0.868	1.287	0.364	3.192	1.747	13.08	....	0.000667
0.138	0.872	1.304	0.375	3.441	1.857	14.00	3.04	0.000886
0.139	0.874	1.324	0.387	3.835	1.804	14.40	....	0.00128
0.141	0.875	1.346	0.399	4.100	1.439	14.85	3.03	0.00300
0.142	0.877	1.375	0.413	4.083	1.056	15.00	....	0.00371
0.144	0.885	1.403	0.427	3.982	0.816	15.24	....	0.00525
0.146	0.894	1.433	0.443	3.833	0.670	15.32	....	0.00617
0.148	0.909	1.458	0.459	3.754	0.599	15.45	....	0.00626
0.149	0.919	1.486	0.477	3.675	0.531	15.52	....	0.00563
0.151	0.934	1.512	0.496	3.621	0.480	15.74	....	0.00522
0.153	0.947	1.539	0.517	3.567	0.430	15.85	....	0.00613
0.155	0.960	1.566	0.539	3.521	0.389	16.00	....	0.00712
0.157	0.973	1.594	0.563	3.472	0.358	16.14	....	0.00746
0.159	0.984	1.627	0.590	3.450	0.334	16.21	....	0.00667
0.161	1.000	1.664	0.620	3.430	0.298	16.28	....	0.00516
0.163	1.022	1.700	0.652	3.410	0.253	16.39	....	0.00333
0.165	1.046	1.736	0.689	.....	0.206	16.55	....	0.00231
0.167	1.072	1.771	0.729	.....	0.176	17.00	....	0.00177
0.170	1.100	1.812	0.775	.....	0.163	18.00	....	0.00181
0.172	1.136	1.847	0.826	.....	0.140	18.93	....	0.00232
0.175	1.174	1.882	0.885	.....	0.0906	19.51	....	0.00384
0.177	1.215	1.915	0.921	.....	0.0571	19.62	....	0.00473
0.180	1.261	1.941	0.925	3.396	0.0355	19.78	....	0.00602
0.182	1.307	1.966	0.928	.....	0.0204	20.00	....	0.00794
0.185	1.354	1.986	0.930	3.390	0.0109	20.19	....	0.00949
0.188	1.402	2.004	0.935	3.385	0.00590	20.34	....	0.0108
0.191	1.453	2.010	0.940	3.379	0.00318	20.42	....	0.0115
0.194	1.496	2.008	0.942	.....	0.00171	20.57	....	0.0130

TABLE 24 Irtran† Materials: Refractive Index (Eastman Kodak, 1968)

λ, μm	Irtran 1	Irtran 2	Irtran 3	Irtran 4	Irtran 5
1.0000	1.3778	2.2907	1.4289	2.491	1.7227
1.2500	1.3763	2.2777	1.4275	2.469	1.7188
1.5000	1.3749	2.2706	1.4263	2.458	1.7156
1.7500	1.3735	2.2662	1.4251	2.452	1.7123
2.0000	1.3720	2.2631	1.4239	2.447	1.7089
2.2500	1.3702	2.2608	1.4226	2.444	1.7052
2.5000	1.3683	2.2589	1.4211	2.442	1.7012
2.7500	1.3663	2.2573	1.4196	2.440	1.6968
3.0000	1.3640	2.2558	1.4179	2.438	1.6920
3.2500	1.3614	2.2544	1.4161	2.437	1.6868
3.5000	1.3587	2.2531	1.4141	2.436	1.6811
3.7500	1.3558	2.2518	1.4120	2.435	1.6750
4.0000	1.3526	2.2504	1.4097	2.434	1.6684
4.2500	1.3492	2.2491	1.4072	2.433	1.6612
4.5000	1.3455	2.2477	1.4047	2.432	1.6536
4.7500	1.3416	2.2462	1.4019	2.431	1.6455
5.0000	1.3374	2.2447	1.3990	2.430	1.6368
5.2500	1.3329	2.2432	1.3959	2.429	1.6275
5.5000	1.3282	2.2416	1.3926	2.429	1.6177
5.7500	1.3232	2.2399	1.3892	2.428	1.6072
6.0000	1.3179	2.2381	1.3856	2.427	1.5962
6.2500	1.3122	2.2363	1.3818	2.426	1.5845
6.5000	1.3063	2.2344	1.3778	2.425	1.5721
6.7500	1.3000	2.2324	1.3737	2.424	1.5590
7.0000	1.2934	2.2304	1.3693	2.423	1.5452
7.2500	1.2865	2.2282	1.3648	2.422	1.5307
7.5000	1.2792	2.2260	1.3600	2.420	1.5154
7.7500	1.2715	2.2237	1.3550	2.419	1.4993
8.0000	1.2634	2.2213	1.3498	2.418	1.4824
8.2500	1.2549	2.2188	1.3445	2.417	1.4646
8.5000	1.2460	2.2162	1.3388	2.416	1.4460
8.7500	1.2367	2.2135	1.3330	2.414	1.4265
9.0000	1.2269	2.2107	1.3269	2.413	1.4060
9.2500	......	2.2078	1.3206	2.412	
9.5000	......	2.2048	1.3141	2.410	
9.7500	......	2.2018	1.3073	2.409	
10.0000	......	2.1986	1.3002	2.407	
11.0000	......	2.1846	1.2694	2.401	
12.0000	......	2.1688	......	2.394	
13.0000	......	2.1508	......	2.387	
14.0000	......	......	......	2.378	
15.0000	......	......	......	2.369	
16.0000	......	......	......	2.359	
17.0000	......	......	......	2.348	
18.0000	......	......	......	2.337	
19.0000	......	......	......	2.324	
20.0000	......	......	......	2.311	

† Irtran 1 to 5 are hot-pressed polycrystalline compacts of MgF_2, ZnS, CdF_2, ZnSe, MgO, and CdTe, respectively. Index-of-refraction values were experimentally determined at selected wavelengths between 1 and 10 μm. Coefficients of an interpolation formula were established and reduced by least-squares methods, and the values computed. All values beyond 10 μm were extrapolated.

TABLE 25 Lanthanum Fluoride: Refractive Index (Wirick, 1966)

	n_e		n_o	
λ, μm	Obs.	Calc.	Obs.	Calc.
0.25365	1.64866	1.64866	1.65587	1.65587
0.31315	1.61803	1.61694	....	1.63639
0.36633	1.61803	1.61694	....	1.62520
0.40465	1.61184	1.61216	1.61797	1.61733
0.43583	1.60950	1.60916	1.61664	1.61546
0.54607	1.60223	1.60223	1.60597	1.60597

Dispersion Equation

$$n_e = 1.58330 + \frac{77.850}{\lambda - 1346.5} \qquad n_o = 1.57376 + \frac{153.137}{\lambda - 686.2}$$

TABLE 26 Lead Fluoride: Refractive Index at 20°C (Malitson and Dodge, no date)

λ, μm	n	λ, μm	n	λ, μm	n	λ, μm	n
0.30	1.93665	0.80	1.74879	3.50	1.72030	8.00	1.67125
0.35	1.85422	0.85	1.74648	4.00	1.71663	8.50	1.66340
0.40	1.81804	0.90	1.74455	4.50	1.71255	9.00	1.65504
0.45	1.79644	0.95	1.74291	5.00	1.70805	9.50	1.64615
0.50	1.78220	1.00	1.74150	5.50	1.70310	10.00	1.63674
0.55	1.77221	1.50	1.73371	6.00	1.69769	10.50	1.62679
0.60	1.76489	2.00	1.72983	6.50	1.69181	11.00	1.61629
0.65	1.75934	2.50	1.72672	7.00	1.68544	11.50	1.60523
0.70	1.75502	3.00	1.72363	7.50	1.67859	11.90	1.59597
0.75	1.75158						

TABLE 27 Lead Molybdate: Refractive Index at 19.5°C (Malitson, no date)

λ, μm	n_o	n_e	$n_o - n_e$	λ, μm	n_o	n_e	$n_o - n_e$
0.40466	....	2.44317		0.57906	2.41515	2.28073	0.13442
0.43584	2.60487	2.39011	0.21476	0.58756	2.40977	2.27729	0.13248
0.46782	2.53415	2.35258	0.18157	0.64385	2.38119	2.25835	0.12284
0.47999	2.51398	2.34119	0.17279	0.66781	2.37170	2.25197	0.11973
0.50858	2.47589	2.31877	0.15712	0.70652	2.35878	2.24311	0.11567
0.54607	2.43929	2.29618	0.14311	1.0830	2.30177	2.20267	0.09910
0.57696	2.41651	2.28162	0.13489				

TABLE 28 Lithium Fluoride: Refractive Index†

λ, μm	n	λ, μm	n	λ, μm	n	λ, μm	n
0.1935	1.4450	0.254	1.41792	1.50	1.38320	5.50	1.31287
0.1990	1.4413	0.280	1.41188	2.00	1.37875	6.00	1.29745
0.2026	1.4390	0.302	1.40818	2.50	1.37327	6.91	1.260
0.2063	1.4367	0.366	1.40121	3.00	1.36660	7.53	1.239
0.2100	1.4346	0.391	1.39937	3.50	1.35868	8.05	1.215
0.2144	1.4319	0.4861	1.39480	4.00	1.34942	8.60	1.190
0.2194	1.4300	0.50	1.39430	4.50	1.33875	9.18	1.155
0.2265	1.4268	0.80	1.38896	5.00	1.32661	9.79	1.109
0.231	1.4244	1.00	1.38711				

Dispersion Equation (Herzberger and Salzberg, 1962)

$$n = A + BL + CL^2 + D\lambda^2 + E\lambda^4$$

where

$A = 1.38761$ $D = -0.0023045$
$B = 0.001796$ $E = -0.00000557$
$C = -0.000041$ $L = (\lambda^2 - 0.028)^{-1}$

† Data at a temperature of 20°C for wavelengths 0.193 to 0.231 μm are taken from Gyulai (1927); at 20°C for 0.254 to 0.486 μm from Harting (1948); at 23.6°C for 0.50 to 6.0 μm from Tilton and Plyler (1951); at 18°C for 6.91 to 9.79 μm from Hohls (1937). The data for the four spectral regions reported here fit together to within a few parts in the fourth decimal place.

TABLE 29 Magnesium Fluoride: Refractive Index and Temperature Coefficient

Refractive Index at 19.5°C (Malitson and Dodge, no date)

λ, μm	n_e	n_o	$n_e - n_o$	λ, μm	n_e	n_o	$n_e - n_o$
0.21386	1.42897	1.41566	0.01331	0.43584	1.39408	1.38208	0.01200
0.22675	1.42251	1.40942	0.01309	0.46782	1.39276	1.38082	0.01194
0.24827	1.41615	1.40329	0.01286	0.47999	1.39232	1.38040	0.01192
0.25763	1.41382	1.40106	0.01276	0.50858	1.39142	1.37954	0.01188
0.27528	1.40967	1.39707	0.01260	0.54607	1.39043	1.37859	0.01184
0.29673	1.40592	1.39345	0.01247	0.58756	1.38955	1.37774	0.01181
0.31315	1.40364	1.39124	0.01240	0.64385	1.38858	1.37682	0.01176
0.33415	1.40116	1.38889	0.01227	0.66781	1.38823	1.37649	0.01174
0.36501	1.39834	1.38614	0.01220	0.79476	1.38679	1.37512	0.01167
0.40466	1.39567	1.38359	0.01208	1.0830	1.38465	1.37307	0.01158

Dispersion Equation (Duncanson and Stevenson, 1958)

$$n_o = 1.36957 + \frac{35.821}{(\lambda - 1492.5)^{-1}} \qquad n_e = 1.38100 + \frac{37.415}{(\lambda - 1494.7)^{-1}}$$

where the wavelength must be given in angstroms

Temperature Coefficients of Refractive Index (Duncanson and Stevenson, 1958)†

λ, μm	$\frac{dn_o}{dT}$, 10^{-5}/°C	$\frac{dn_e}{dT}$, 10^{-5}/°C
0.4047	0.23	0.17
0.7065	0.19	0.10

†Copyright by the Institute of Physics.

TABLE 30 Magnesium Oxide: Refractive Index and Temperature Coefficient

Refractive Index at 23.3°C (Stephens and Malitson, 1952)†

λ, μm	*n*	λ, μm	*n*	λ, μm	*n*	λ, μm	*n*
0.36117	1.77318	1.52952	1.71496	1.97009	1.70885	3.5078	1.68055
0.365015	1.77186	1.6932	1.71281	2.24920	1.70470	4.258	1.66039
1.01398	1.72259	1.7092	1.71258	2.32542	1.70350	5.138	1.63138
1.12866	1.72059	1.81307	1.71108	3.3033	1.68526	5.35	1.62404
1.36728	1.71715						

Dispersion Equation

$$n^2 = 2.956362 - 0.01062387\lambda^2 - 0.0000204968\lambda^4 - \frac{0.02195770}{\lambda^2 - 0.01428322}$$

Temperature Coefficients of Refractive Index (Stephens and Malitson, 1952)†

	$\frac{dn}{dT}$, 10^{-6}/°C				
λ, μm	20°C	25°C	30°C	35°C	40°C
7.679	13.6	13.7	13.8	13.9	14.0
7.665	14.1	14.2	14.3	14.4	14.5
6.678	14.4	14.5	14.6	14.7	14.8
6.563	14.5	14.6	14.7	14.8	14.9
5.893	15.3	15.4	15.5	15.6	15.7
5.461	15.9	16.0	16.1	16.2	16.3
4.861	16.9	17.0	17.1	17.2	17.3
4.358	18.0	18.1	18.2	18.3	18.4
4.047	18.9	19.0	19.1	19.2	19.3

† The earlier measurements of Strong and Brice (1935) are discussed in Stephens and Malitson (1952). A systematic difference in values of *n* appears to exist, the data of Strong and Brice being about 37 × 10^{-5} higher.

TABLE 31 Muscovite Mica: Refractive Index (Jeppesen and Taylor, 1966)

λ_o, μm	λ_e, μm	n_o	n_e	λ_o, μm	λ_e, μm	n_o	n_e
t = 5.24 μm				t = 20.82 μm			
0.6665	6675	1.590	1.592	0.6960	6985	1.598	1.594
0.6188	6210	1.594	1.600	0.6316	6336	1.593	1.598
0.5573	5590	1.595	1.600	0.5539	5555	1.596	1.601
0.5082	5090	1.600	1.603	0.4935	4950	1.600	1.605
0.4538	4555	1.602	1.608	0.4600	4615	1.062	1.607
0.4320	4330	1.608	1.611	0.4310	4326	1.604	1.610
t = 13.91 μm				t = 48.68 μm			
0.6910	6930	1.590	1.594	0.6914	6935	1.591	1.596
0.5995	6010	1.595	1.599	0.6110	6125	1.594	1.598
0.5293	5308	1.598	1.603	0.5470	5487	1.596	1.601
0.4740	4754	1.602	1.606	0.4958	4971	1.599	1.603
0.4300	4308	1.607	1.611	0.4667	4680	1.601	1.606
				0.4408	4425	1.603	1.609

TABLE 32 Potassium Bromide: Refractive Index at 22°C (Stephens et al., 1953)†

λ, μm	n	λ, μm	n	λ, μm	n	λ, μm	n
0.404656	1.589752	1.01398	1.54408	6.238	1.53288	17.40	1.50390
0.435835	1.581479	1.12866	1.54258	6.692	1.53225	18.10	1.50076
0.486133	1.571791	1.36728	1.54061	8.662	1.52903	19.01	1.49703
0.508582	1.568475	1.7012	1.53901	9.724	1.52695	19.91	1.49288
0.546074	1.563928	2.44	1.53733	11.035	1.52404	21.18	1.48655
0.587562	1.559965	2.73	1.53693	11.862	1.52200	21.83	1.48311
0.643847	1.555858	3.419	1.53612	14.29	1.51505	23.86	1.47140
0.706520	1.552447	4.258	1.53523	14.98	1.51280	25.14	1.46324

Dispersion Equation

$$n^2 = 2.361323 - 0.00311497\lambda^2 - 0.000000058613\lambda^4 + \frac{0.007676}{\lambda^2} + \frac{0.0156569}{\lambda^2 - 0.0324}$$

† The average value of the temperature coefficient of refractive index is given in Stephens et al. (1953) as 4.0×10^{-5} per Celsius degree.

TABLE 33 Potassium Chloride: Refractive-Index Data†

λ, μm	n	λ, μm	n	λ, μm	n
0.185409	1.82710	0.410185	1.50907	5.3039	1.470013
0.186220	1.81853	0.434066	1.50503	5.8932	1.468804
0.197760	1.73120	0.444587	1.50390	8.2505	1.462726
0.198900	1.72438	0.467832	1.50044	8.8398	1.460858
0.200000	1.71870	0.486149	1.49841	10.0184	1.45672
0.204470	1.69817	0.508606	1.49620	11.786	1.44919
0.208216	1.68308	0.53383	1.49410	12.965	1.44346
0.211078	1.67281	0.54610	1.49319	14.141	1.43722
0.21445	1.66188	0.56070	1.49218	15.912	1.42617
0.21946	1.64745	0.58931	1.49044	17.680	1.41403
0.22400	1.63612	0.58932	1.490443	18.2	1.409
0.23129	1.62043	0.62784	1.48847	18.8	1.401
0.242810	1.60047	0.64388	1.48777	19.7	1.398
0.250833	1.58970	0.656304	1.48727	20.4	1.389
0.257317	1.58125	0.67082	1.48669	21.1	1.379
0.263200	1.57483	0.76824	1.48377	22.2	1.374
0.267610	1.57044	0.78576	1.483282	23.1	1.363
0.274871	1.56386	0.88308	1.481422	24.1	1.352
0.281640	1.55836	0.98220	1.480081	24.9	1.336
0.291368	1.55140	1.1786	1.478311	25.7	1.317
0.308227	1.54136	1.7680	1.475890	26.7	1.300
0.312280	1.53926	2.3573	1.474751	27.2	1.275
0.340358	1.52726	2.9466	1.473834	28.2	1.254
0.358702	1.52115	3.5359	1.473049	28.8	1.226
0.394415	1.51219	4.7146	1.471122		

Dispersion Equation

$$n^2 = \begin{cases} a^2 + \left(\dfrac{M_1}{\lambda^2} - \lambda_1{}^2\right) + \left(\dfrac{M_2}{\lambda^2} - \lambda_2{}^2\right) - k\lambda^2 - h\lambda^4 & \text{ultraviolet} \\ b^2 + \left(\dfrac{M_1}{\lambda^2} - \lambda_1{}^2\right) + \left(\dfrac{M_2}{\lambda^2} - \lambda_2{}^2\right) - \left(\dfrac{M_3}{\lambda_3{}^2} - \lambda^2\right) & \text{visible} \end{cases}$$

where

$a^2 = 2.174967$ $\quad k = 0.000513495$
$M_1 = 0.008344206$ $\quad h = 0.06167587$
$\lambda_1{}^2 = 0.0119082$ $\quad b^2 = 3.866619$
$M_2 = 0.00698382$ $\quad M_3 = 5{,}567.715$
$\lambda^2 = 0.0255550$ $\quad \lambda_3{}^2 = 3{,}292.47$

Temperature Coefficient of Refractive Index (Paschen, 1908)

$$n_T = 1.490443 - (T - 15)(0.000034)$$

† Refractive-index data for 0.185409 to 0.76824 μm at 18°C, Martens (1901*c*); 18.2 to 28.8 μm, Hohls (1937); 0.58932 to 17.680 μm at 15°C, Paschen (1908). The data of Paschen and Martens overlap in a small region, and both sets are presented. There is less spread between Hohl's data and Paschen's data than there is between Paschen's data in the region where they join. The fit in this region is good (~0.0005). Paschen also presents two dispersion curves that fit the data of Martens to about five parts in the fifth decimal place.

TABLE 34 Potassium Iodide: Refractive Index†,‡

λ, μm	n	λ, μm	n	λ, μm	n	λ, μm	n
0.248	2.0548	0.436	1.70350	1.083	1.6381	15.91	1.6085
0.254	2.0105	0.486	1.68664	1.18	1.6366	18.10	1.6030
0.265	1.9424	0.546	1.67310	1.77	1.6313	19	1.5997
0.270	1.9221	0.588	1.66654	2.36	1.6295	20	1.5964
0.280	1.8837	0.589	1.66643	3.54	1.6275	21	1.5930
0.289	1.85746	0.656	1.65809	4.13	1.6268	22	1.5895
0.297	1.83967	0.707	1.6537	5.89	1.6252	23	1.5858
0.302	1.82769	0.728	1.6520	7.66	1.6235	24	1.5819
0.313	1.80707	0.768	1.6494	8.84	1.6218	25	1.5775
0.334	1.77664	0.811	1.6471	10.02	1.6201	26	1.5720
0.366	1.74416	0.842	1.6456	11.79	1.6172	27	1.5681
0.391	1.72671	0.912	1.6427	12.97	1.6150	28	1.5629
0.405	1.71843	1.014	1.6396	14.14	1.6127	29	1.5571

† Data for 0.248 through 1.083 μm from Harting (1948); from 1.18 through 29 μm from Korth (1933).

‡ The temperature coefficient of refractive index for 0.546 μm in the region 38 to 90°C is -5.0×10^{-5} per Celsius degree.

TABLE 35 Potassium Dihydrogen Phosphate (KDP): Refractive Index (Zernike, 1964)

λ, μm	n_o	n_e	λ, μm	n_o	n_e
0.4047	1.52341	1.47927	0.6328	1.50737	1.46685
0.4077	1.52301	1.47898	1.0139	1.49535	1.46041
0.4358	1.51990	1.47640	1.1287	1.49205	1.45917
0.4916	....	1.47254	1.1523	1.49135	1.45893
0.5461	1.51152	1.46982	1.3570	1.48455	
0.5769	1.50987		1.5231	....	1.45521
0.5791	1.50977	1.46856	1.5295	....	1.45512

TABLE 36 Ruby: Refractive Index at 22°C (Dodge et al., 1969)

λ, μm	n_o	n_e	λ, μm	n_o	n_e
0.4358	1.78115	1.77276	0.6678	1.76445	1.75611
0.5461	1.77071	1.76258	0.7065	1.76302	1.75501
0.5876	1.76822	1.76010			

TABLE 37 Rutile (TiO_2): Refractive Index (DeVore, 1951)

λ, μm	n_o	n_e	λ, μm	n_o
0.4358	2.853	3.216	2.0000	2.399
0.4916	2.725	3.051	2.5000	2.387
0.4960	2.718	3.042	3.0000	2.380
0.5461	2.652	2.958	3.5000	2.367
0.5770	2.623	2.921	4.0000	2.350
0.5791	2.621	2.919	4.5000	2.322
0.6907	2.555	2.836	5.0000	2.290
0.7082	2.548	2.826	5.5000	2.200
1.0140	2.484	2.747		
1.5296	2.454	2.710		

Dispersion Equation

$$n^2 = \begin{cases} 5.913 + \dfrac{2.441 \times 10^7}{\lambda^2 - 0.803 \times 10^7} & \text{ordinary ray} \\ 7.197 + \dfrac{3.322 \times 10^7}{\lambda^2 - 0.843 \times 10^7} & \text{extraordinary ray} \end{cases}$$

where the wavelength must be in angstroms

Temperature Changes of Indexes

$$\frac{dn_o}{dT} \approx -4 \times 10^{-5}/°\text{C} \qquad \frac{dn_e}{dT} \approx -9 \times 10^{-5}/°\text{C}$$

TABLE 38 Sapphire (Synthetic): Refractive Index for the Ordinary Ray (Malitson, 1962)†

λ, μm	n_{obs}	n_{cal}	$n_{obs} - n_{cal} \times 10^5$	λ, μm	n_{obs}	n_{cal}	$n_{obs} - n_{cal} \times 10^5$
0.26520	1.83360	1.83365	−5	1.39506	1.74888	1.74888	0
0.28035	1.82427	1.82426	+1	1.52952	1.74660	1.74659	+1
0.28936	1.81949	1.81947	+2	1.6932	1.74368	1.74368	0
0.29673	1.81595	1.81593	+2	1.70913	1.74340	1.74339	+1
0.30215	1.81351	1.81351	0	1.81307	1.74144	1.74144	0
0.3130	1.80906	1.80909	−3	1.9701	1.73833	1.73835	−2
0.33415	1.80184	1.80181	+3	2.1526	1.73444	1.73449	−5
0.34662	1.79815	1.79820	−5	2.24929	1.73231	1.73232	−1
0.361051	1.79450	1.79451	−1	2.32542	1.73057	1.73055	+2
0.365015	1.79358	1.79358	0	2.4374	1.72783	1.72784	−1
0.39064	1.78826	1.78828	−2	3.2439	1.70437	1.70433	+3
0.404656	1.78582	1.78582	0	3.2668	1.70356	1.70354	+2
0.435834	1.78120	1.78120	0	3.3026	1.70231	1.70231	0
0.546071	1.77078	1.77077	+1	3.3303	1.70140	1.70136	+4
0.576960	1.76884	1.76884	0	3.422	1.69818	1.69814	+4
0.579066	1.76871	1.76871	0	3.5070	1.69504	1.69501	+3
0.64385	1.76547	1.76547	0	3.7067	1.68746	1.68743	+3
0.706519	1.76303	1.76303	0	4.2553	1.66371	1.66363	+8
0.85212	1.75885	1.75887	−2	4.954	1.62665	1.62669	−4
0.89440	1.75796	1.75790	+6	5.1456	1.61514	1.61510	+4
1.01398	1.75547	1.75546	−1	5.349	1.60202	1.60204	−2
1.12866	1.75339	1.75339	0	5.419	1.59735	1.59736	−1
1.36728	1.74936	1.74935	+1	5.577	1.58638	1.58642	−4

Dispersion Equation

$$n^2 - 1 = \sum_i \frac{A_i \lambda^2}{\lambda^2 - \lambda_i^2}$$

where at 24°C

$\lambda_1^2 = 0.00377588$ $\quad A_1 = 1.023798$
$\lambda_2^2 = 0.0122544$ $\quad A_2 = 1.058264$
$\lambda_3^2 = 321.3616$ $\quad A_3 = 5.280792$

Temperature Coefficients of Index

Temperature coefficients of refractive index dn/dT were determined from the differences between the indexes at 19 and 24°C. The results indicate that the coefficient is positive and decreases from about 20×10^{-6} per Celsius degree at the short wavelengths to about 10×10^{-6} per Celsius degree near 4 μm. The average value of 13×10^{-6} per Celsius degree for the visible region was determined from additional measurements made at 17, 24, and 31°C on a Wild precision spectrometer.

† Loewenstein (1961) gives 3.14 ± 0.13 for n_o and 3.61 ± 0.14 for n_e from 167 to 500 μm.

TABLE 39 Selenium (Trigonal): Refractive Index (Gampel and Johnson, 1969)†

λ, μm	n_o	n_e	λ, μm	n_o	n_e
1.06	2.790 ± 0.008	3.608 ± 0.008	3.39	2.65 ± 0.01	3.46 ± 0.01
1.15	2.737 ± 0.008	3.573 ± 0.008	10.6	2.64 ± 0.01	3.41 ± 0.01

† Trigonal selenium: for **E** perpendicular to *c*, *n* is 2.78 from 9 to 23 μm at 27°C; for **E** parallel to *c*, *n* is 3.58 ± 0.0 from 9 to 23 μm (Caldwell, 1958). Amorphous selenium: *n* is 2.44 to 2.46 at 2.5 μm (Saker, 1952). *n* varies from 2.42 to 2.38 between 5 and 15 μm (Henry, 1953). Liquid selenium: refractive index of 2.44 from 30 to 152 μm (Saker, 1952).

TABLE 40 Selenium Glass (Arsenic-modified): Refractive Index at 27°C (Salzberg and Villa, 1957)

Data for two different prisms illustrate the variability to be expected

λ, μm	n Prism *a*	n Prism *b*	λ, μm	n Prism *a*	n Prism *b*
1.0140	2.5774	2.5783	7.00	2.4778	2.4787
1.1286	2.5554	2.5565	7.50	......	2.4784
1.3622	2.5285	2.5294	8.10	2.4772	2.4778
1.5295	2.5173	2.5183	8.50	......	2.4775
1.7012	2.5089	2.5100	9.10	2.4765	2.4771
2.1526	2.4950	2.4973	9.50	......	2.4768
3.00	2.4861	2.4882	10.00	2.4756	2.4767
3.4188	2.4841	2.4858	10.50	......	2.4759
4.00	2.4825	2.4835	11.00	2.4752	2.4758
4.50	......	2.4822	11.50	......	2.4753
5.00	2.4803	2.4811	12.00	......	2.4749
5.50	......	2.4804	13.00	......	2.4760 [*sic*]
6.00	2.4789	2.4798	13.50	......	2.4748
6.50	......	2.4792	14.00	......	2.4743

TABLE 41 Silicon: Refractive Index at 26°C (Salzberg and Villa, 1957)†

λ, μm	n	λ, μm	n	λ, μm	n	λ, μm	n
1.3570	3.4975	2.1526	3.4476	4.00	3.4255	7.00	3.4189
1.3673	3.4962	2.3254	3.4430	4.258	3.4242	7.50	3.4186
1.3951	3.4929	2.4373	3.4408	4.50	3.4236	8.00	3.4184
1.3295	3.4795	2.7144	3.4358	5.00	3.4223	8.50	3.4182
1.6606	3.4696	3.00	3.4320	5.50	3.4213	10.00	3.4179
1.7092	3.4664	3.3033	3.4297	6.00	3.4202	10.50	3.4178
1.8134	3.4608	3.4188	3.4286	6.50	3.4195	11.04	3.4176
1.9704	3.4537	3.50	3.4284				

Dispersion Equation

$$n = A + BL + CL^2 + D\lambda^2 + E\lambda^4$$

where

$A = 3.41696$ $\quad D = -0.0000209$

$B = 0.138497$ $\quad E = 0.000000148$

$C = 0.013924$ $\quad L = \dfrac{1}{\lambda^2 - 0.028}$

† The purity of the silicon sample is not specified. These data are about five parts in the third decimal place lower than those reported by Briggs (1950). The refractive-index values for adequately pure (30 Ω/cm) cast polycrystal silicon should be very near those for single crystals.

TABLE 42 Silicon Carbide (Beta): Refractive Index (Shaffer and Naum, 1969)

λ, μm	n	λ, μm	n
0.467	2.7104	0.589	2.6525
0.498	2.6916	0.616	2.6446
0.515	2.6823	0.691	2.6264
0.568	2.6600		

TABLE 43 Silver Chloride: Refractive Index at 23.9°C (Tilton et al., 1950)

λ, μm	n	λ, μm	n	λ, μm	n	λ, μm	n
0.5	2.09648	2.3	2.00465	4.0	1.99983	12.5	1.96742
0.6	2.06385	2.4	2.00424	4.5	1.99866	13.0	1.96444
0.7	2.04590	2.5	2.00386	5.0	1.99745	13.5	1.96133
0.8	2.03485	2.6	2.00351	5.5	1.99618	14.0	1.95807
0.9	2.02752	2.7	2.00318	6.0	1.99483	14.5	1.95467
1.0	2.02239	2.8	2.00287	6.5	1.99339	15.0	1.95113
1.1	2.01865	2.9	2.00258	7.0	1.99185	15.5	1.94743
1.2	2.01582	3.0	2.00230	7.5	1.99021	16.0	1.94358
1.3	2.01363	3.1	2.00203	8.0	1.99847	16.5	1.93958
1.4	2.01189	3.2	2.00177	8.5	1.98661	17.0	1.93542
1.5	2.01047	3.3	2.00151	9.0	1.98464	17.5	1.93109
1.6	2.00931	3.4	2.00126	9.5	1.98255	18.0	1.92660
1.7	2.00833	3.5	2.00102	10.0	1.98034	18.5	1.92194
1.8	2.00750	3.6	2.00078	10.5	1.97801	19.0	1.91710
1.9	2.00678	3.7	2.00054	11.0	1.97556	19.5	1.91208
2.0	2.00615	3.8	2.00030	11.5	1.97297	20.0	1.90688
2.1	2.00559	3.9	2.00007	12.0	1.97026	20.5	1.90149
2.2	2.00510						

Dispersion Equation

$$n^2 = 4.00804 - 0.00085111\lambda^2 - 0.0000001976\lambda^4 + \frac{0.079086}{\lambda^2 - 0.04585}$$

Temperature Coefficient of Index

$$\frac{dn}{dT} \approx 6.1 \times 10^{-5}/^\circ\text{C} \quad \text{at } 0.61\ \mu\text{m}$$

TABLE 44 Sodium Chloride: Refractive Index and Temperature Coefficient

Refractive Index†

λ, μm	n	λ, μm	n	λ, μm	n	λ, μm	n
0.19	1.85343	1.1786	1.53031	4.0	1.52190	12.50	1.47568
0.20	1.79073	1.2016	1.53014	4.1230	1.52156	12.9650	1.47160
0.22	1.71591	1.2604	1.52971	4.7120	1.51979	13.0	1.47141
0.24	1.67197	1.3126	1.52937	5.0	1.51890	14.0	1.46189
0.26	1.64294	1.4	1.52888	5.0092	1.51883	14.1436	1.46044
0.28	1.62239	1.4874	1.52845	5.3009	1.51790	14.7330	1.45427
0.30	1.60714	1.5552	1.52815	5.8932	1.51593	15.0	1.45145
0.35	1.58232	1.6	1.52798	6.0	1.51548	15.3223	1.44743
0.40	1.56769	1.6368	1.52781	6.4825	1.51347	15.9116	1.44090
0.50	1.55175	1.6848	1.52764	6.80	1.51200	16.0	1.44001
0.589	1.54427	1.7670	1.52736	7.0	1.51136	17.0	1.42753
0.6400	1.54141	1.8	1.52728	7.0718	1.51093	17.93	1.4149
0.6874	1.53930	2.0	1.52670	7.22	1.51020	18.0	1.41393
0.70	1.53881	2.0736	1.52649	7.59	1.50850	19.0	1.39914
0.7604	1.53682	2.1824	1.52621	7.6611	1.50822	20.0	1.38307
0.7858	1.53607	2.2464	1.52606	7.9558	1.50665	20.57	1.3735
0.80	1.53575	2.3	1.52594	8.0	1.50655	21.0	1.36563
0.8835	1.53395	2.3560	1.52579	8.04	1.5064	21.3	1.352
0.90	1.53366	2.6	1.52525	8.8398	1.50192	22.3	1.3403
0.9033	1.53361	2.6505	1.52512	9.0	1.50105	22.8	1.318
0.9724	1.53253	2.9466	1.52466	9.00	1.50100	23.6	1.299
1.0	1.53216	3.0	1.52434	9.50	1.49980	24.2	1.278
1.0084	1.53206	3.2736	1.52371	10.0	1.49482	25.0	1.254
1.0540	1.53153	3.5	1.52317	10.0184	1.49462	25.8	1.229
1.0810	1.53123	3.5359	1.52312	11.0	1.48783	26.6	1.203
1.1058	1.53098	3.6288	1.52286	11.7864	1.48171	27.3	1.175
1.1420	1.53063	3.8192	1.52238	12.0	1.48004		

Temperature Coefficient of Refractive Index at about 60°C‡

λ, μm	$\frac{dn}{dT}$, 10^{-5}/°C	λ, μm	$\frac{dn}{dT}$, 10^{-5}/°C	λ, μm	$\frac{dn}{dT}$, 10^{-5}/°C
0.202	3.134	0.325	−2.987	4.96	−3.172
0.206	2.229	0.340	−3.068	6.4	−3.149
0.210	1.570	0.361	−3.194	8.85	−2.405
0.214	0.861	0.441	−3.425	10.02	−2.2
0.219	0.235	0.467	−3.454	11.79	−1.6
0.224	−0.187	0.480	−3.468	12.97	−1.4
0.226	−0.382	0.508	−3.517	14.14	−1.2
0.229	−0.598	0.589	−3.622	14.73	−1.0
0.231	−0.757	0.643	−3.636	15.32	−0.8
0.257	−1.979	0.656	−3.652	15.91	−0.7
0.274	−2.396	1.1	−3.642	17.93	−0.5
0.288	−2.602	1.6	−3.557	20.57	0
0.298	−2.727	2.7	−3.427	22.3	0
0.313	−2.862	3.96	−3.286		

† Data for wavelengths given with two-figure accuracy (0.19, 0.50, . . .) or three-figure accuracy (10.0, 11.0, . . .) are reported for 20°C by Kohlrausch (1968); the data reported to four figures (1.299) in index at 18°C (even though they are three figures in wavelength) are from Hohls (1937); other data at 20°C are from Coblentz (1914*b*). Still more data have been published by Langley, Martens, Paschen, Rubens, Trowbridge, Nichols, and others, but all have apparently measured natural crystals of undetermined purity; the data all agree to the fifth decimal place.

‡ From Micheli (1902*a*) for 0.202 through 0.643 μm; Liebreich (1911) for 0.656 through 15.91 μm; and Rubens and Nichols (1897) for 17.93 to 22.3 μm.

TABLE 45 Sodium Fluoride: Refractive Index (Smakula, 1952*b*)†

λ, μm	n	λ, μm	n	λ, μm	n	λ, μm	n
0.186	1.3930	0.546	1.32640	4.1	1.308	9.8	1.241
0.193	1.3854	0.588	1.32552	4.5	1.305	10.3	1.233
0.199	1.3805	0.589	1.32549	4.7	1.303	10.8	1.222
0.203	1.3772	0.656	1.32436	4.9	1.302	11.3	1.200
0.206	1.3745	0.707	1.32372	5.1	1.301	11.7	1.193
0.210	1.3718	0.720	1.32349	5.3	1.299	12.5	1.180
0.214	1.3691	0.768	1.32307	5.5	1.297	13.2	1.163
0.219	1.3665	0.811	1.32272	5.7	1.295	13.8	1.142
0.227	1.3630	0.842	1.32247	5.9	1.294	14.3	1.118
0.231	1.3606	0.912	1.32198	6.1	1.292	15.1	1.093
0.237	1.3586	1.014	1.32150	6.3	1.290	15.9	1.065
0.240	1.35793	1.083	1.32125	6.5	1.288	16.7	1.034
0.248	1.35500	1.27	1.320	6.7	1.286	17.3	1.000
0.254	1.35325	1.48	1.319	6.9	1.284	18.1	0.963
0.265	1.34999	1.67	1.318	7.1	1.281	18.6	0.924
0.270	1.34881	1.83	1.318	7.3	1.279	19.3	0.881
0.280	1.34645	2.0	1.317	7.5	1.277	19.7	0.838
0.289	1.34462	2.2	1.317	7.7	1.274	20.0	0.82
0.297	1.34328	2.4	1.316	7.9	1.272	20.5	0.75
0.302	1.34232	2.6	1.315	8.1	1.269	21.0	0.70
0.313	1.34062	2.8	1.314	8.3	1.266	21.5	0.65
0.334	1.33795	3.1	1.313	8.5	1.263	22.0	0.55
0.366	1.33482	3.3	1.312	8.7	1.261	22.5	0.45
0.391	1.33290	3.5	1.311	8.9	1.258	23.0	0.33
0.405	1.33194	3.7	1.309	9.1	1.252	23.5	0.25
0.436	1.33025	3.9	1.309	9.4	1.251	24.0	0.24
0.486	1.32818						

† 0.186 to 0.237 μm, Kublitsky (1934); 0.240 to 1.083 μm, Harting (1948); and 1.27 to 24.0 μm, Hohls (1937). The data fit together well (within a few parts in the fifth decimal place).

TABLE 46 Sodium Nitrate: Refractive Index at Room Temperature (International Critical Tables, 1929)†

λ, μm	n_o	n_e	λ, μm	n_o	n_e
0.434	1.6126	1.340	0.578	1.5860	1.336
0.436	1.6121	1.340	0.589	1.5848	1.336
0.486	1.5998	1.338	0.656	1.5791	1.334
0.501	1.5968	1.337	0.668	1.5783	1.334
0.546	1.5899	1.336			

†Reproduced with the permission of the National Academy of Sciences.

TABLE 47 Sphalerite: Refractive Index (for Ordinary Ray) (DeVore, 1951)

λ, μm	n_o	λ, μm	n_o	λ, μm	n_o	λ, μm	n_o
0.3650	2.679	0.3906	2.585	0.4358	2.490	0.5780	2.372
0.3654	2.677	0.4047	2.549	0.4916	2.425	0.5296	2.284
0.3663	2.673	0.4077	2.542	0.5461	2.388		

Dispersion Equation

$$n_o^{\,2} = 5.164 + \frac{1.208 \times 10^7}{\lambda^2 - 0.732 \times 10^7}$$

where the wavelength must be given in angstroms

TABLE 48 Spinel: Refractive Index (Linde Air Products Co., no date)

λ, μm	n	λ, μm	n
0.486	1.7261	0.656	1.7143
0.589	1.7182	1.00	1.73

TABLE 49 Strontium Titanate: Refractive Index at 20°C (Malitson, no date)

λ, μm	n	λ, μm	n	λ, μm	n	λ, μm	n
0.40	2.66386	0.68	2.37135	0.96	2.32035	3.40	2.21370
0.44	2.56007	0.72	2.35998	1.00	2.31633	3.80	2.19428
0.48	2.49751	0.76	2.35055	1.40	2.29073	4.20	2.17265
0.52	2.45553	0.80	2.34260	1.80	2.27498	4.60	2.14865
0.56	2.42544	0.84	2.33583	2.20	2.26109	5.00	2.12212
0.60	2.40285	0.88	2.32997	2.60	2.24676	5.40	2.09290
0.64	2.38532	0.92	2.32486	3.00	2.23111		

Dispersion Equation

$$n = A + BL + CL^2 + D\lambda^2 + E\lambda^4$$

where

$A = 2.28355$ $\quad D = -0.0061335$

$B = 0.035906$ $\quad E = -0.00001502$

$C = 0.001666$ $\quad L = \dfrac{1}{\lambda^2 - 0.028}$

TABLE 50 Tellurium: Refractive Index (Caldwell, 1958)

	n			n			n	
λ, μm	**E** ⊥ c	**E** ∥ c	λ, μm	**E** ⊥ c	**E** ∥ c	λ, μm	**E** ⊥ c	**E** ∥ c
4.0	4.929	6.372	7.0	4.821	6.270	10.0	4.796	6.246
5.0	4.864	6.316	8.0	4.809†	6.257	12.0	4.789	6.237
6.0	4.838	6.286	9.0	4.802	6.253	14.0	4.785	6.230

† This value is probably about 0.002 too low.

TABLE 51 Thallium Bromide: Refractive Index†

λ, μm	n	λ, μm	n	λ, μm	n
0.438	2.652	0.650	2.384	13.95	2.321
0.546	2.452	0.750	2.350	19.76	2.321
0.578	2.424	9.98	2.338	24.39	2.321
0.589	2.418				

Dispersion Equation

$$\frac{n^2-1}{n^2+1} = 0.48484 + \frac{0.10279\lambda^2}{\lambda^2 - 0.0900} - 0.0047896\lambda^2$$

† From Barth (1929) for the visible and from McCarthy (1965*c*) for the infrared.

TABLE 52 Thallium Bromide–Thallium Chloride (KRS-6): Refractive Index (Hettner and Leisegang, 1948)†

λ, μm	n	λ, μm	n	λ, μm	n	λ, μm	n
Na,D	2.3367	1.6	2.2124	4.0	2.1956	14.0	2.1563
0.6	2.3294	1.7	2.2103	4.5	2.1942	15.0	2.1504
0.7	2.2982	1.8	2.2086	5.0	2.1928	16.0	2.1442
0.8	2.2660	1.9	2.2071	6.0	2.1900	17.0	2.1377
0.9	2.2510	2.0	2.2059	7.0	2.1870	18.0	2.1309
1.0	2.2404	2.2	2.2039	8.0	2.1839	19.0	2.1236
1.1	2.2321	2.4	2.2024	9.0	2.1805	20.0	2.1154
1.2	2.2255	2.6	2.2011	10.0	2.1767	21.0	2.1067
1.3	2.2212	2.8	2.2001	11.0	2.1723	22.0	2.0976
1.4	2.2176	3.0	2.1990	12.0	2.1674	23.0	2.0869
1.5	2.2148	3.5	2.1972	13.0	2.1620	24.0	2.0752

† 44% TlBr, 56% TlCl.

TABLE 53 Thallium Bromide–Thallium Iodide (KRS-5): Refractive Index at 25°C (Rodney and Malitson, 1956)†

λ	*n*	λ	*n*	λ	*n*	λ	*n*
0.540	2.68059	1.76	2.39974	11.8	2.36669	22.0	2.33206
0.560	2.64959	1.78	2.39926	12.0	2.36622	22.1	2.33161
0.580	2.62390	1.80	2.39881	12.2	2.36573	22.2	2.33116
0.600	2.60221	1.82	2.39837	12.4	2.36523	22.3	2.33070
0.620	2.58261	1.84	2.39794	12.6	2.36473	22.4	2.33025
0.640	2.56748	1.86	2.39753	12.8	2.36422	22.5	2.32979
0.660	2.55337	1.88	2.39713	13.0	2.36371	22.6	2.32933
0.680	2.54092	1.90	2.39674	13.2	2.36318	22.7	2.32887
0.700	2.52986	1.92	2.39637	13.4	2.36265	22.8	2.32840
0.720	2.51998	1.94	2.39600	13.6	2.36211	22.9	2.32793
0.740	2.51110	1.96	2.39565	13.8	2.36157	23.0	2.32746
0.760	2.50309	1.98	2.39531	14.0	2.36101	23.1	2.32699
0.780	2.49583	2.00	2.39498	14.2	2.36045	23.2	2.32652
0.800	2.48922	2.20	2.39214	14.4	2.35988	23.3	2.32604
0.820	2.48318	2.40	2.38997	14.6	2.35930	23.4	2.32556
0.840	2.47766	2.60	2.38826	14.8	2.35871	23.5	2.32508
0.860	2.47258	2.80	2.38688	15.0	2.35812	23.6	2.32460
0.880	2.46790	3.00	2.38574	15.2	2.35751	23.7	2.32411
0.900	2.46358	3.20	2.38478	15.4	2.35690	23.8	2.32362
0.920	2.45958	3.40	2.38396	15.6	2.35629	23.9	2.32313
0.940	2.45587	3.60	2.38325	15.8	2.35566	24.0	2.32264
0.960	2.45242	3.80	2.38261	16.0	2.35502	24.1	2.32215
0.980	2.44920	4.00	2.38204	16.2	2.35438	24.2	2.32165
1.00	2.44620	4.20	2.38153	16.4	2.35373	24.3	2.32115
1.02	2.44339	4.40	2.38105	16.6	2.35307	24.4	2.32065
1.04	2.44076	4.60	2.38061	16.8	2.35240	24.5	2.32014
1.06	2.43830	4.80	2.38019	17.0	2.35173	24.6	2.31964
1.08	2.43598	5.00	2.37979	17.2	2.35104	24.7	2.31913
1.10	2.43380	5.20	2.37940	17.4	2.35035	24.8	2.31861
1.12	2.43175	5.40	2.37903	17.6	2.34965	24.9	2.31810
1.14	2.42981	5.60	2.37867	17.8	2.34894	25.0	2.31758
1.16	2.42798	5.80	2.37832	18.0	2.34822	25.1	2.31707
1.18	2.42625	6.00	2.37797	18.2	2.34750	25.2	2.31655
1.20	2.42462	6.20	2.37763	18.4	2.34676	25.3	2.31602
1.22	2.42307	6.40	2.37729	18.6	2.34602	25.4	2.31550
1.24	2.42159	6.60	2.37695	18.8	2.34527	25.5	2.31497
1.26	2.42020	6.80	2.37661	19.0	2.34451	25.6	2.31444
1.28	2.41887	7.00	2.37627	19.2	2.34374	25.7	2.31391
1.30	2.41760	7.20	2.37592	19.4	2.34296	25.8	2.31337
1.32	2.41640	7.40	2.37558	19.6	2.34217	25.9	2.31283
1.34	2.41525	7.60	2.37523	19.8	2.34138	26.0	2.31229
1.36	2.41416	7.80	2.37488	20.0	2.34058	26.1	2.31175
1.38	2.41312	8.00	2.37452	20.1	2.34017	26.2	2.31121
1.40	2.41212	8.20	2.37416	20.2	2.33976	26.3	2.31066
1.42	2.41117	8.40	2.37380	20.3	2.33935	26.4	2.31011
1.44	2.41025	8.60	2.37343	20.4	2.33894	26.5	2.30956
1.46	2.40938	8.80	2.37305	20.5	2.33853	26.6	2.30900
1.48	2.40854	9.00	2.37267	20.6	2.33811	26.7	2.30844
1.50	2.40774	9.20	2.37229	20.7	2.33770	26.8	2.30789
1.52	2.40697	9.40	2.37190	20.8	2.33727	26.9	2.30732
1.54	2.40623	9.60	2.37150	20.9	2.33685	27.0	2.30676
1.56	2.40552	9.80	2.37110	21.0	2.33643	27.1	2.30619
1.58	2.40484	10.0	2.37069	21.1	2.33600	27.2	2.30562
1.60	2.40419	10.2	2.37027	21.2	2.33557	27.3	2.30505
1.62	2.40355	10.4	2.36985	21.3	2.33514	27.4	2.30448
1.64	2.40295	10.6	2.36942	21.4	2.33471	27.5	2.30390
1.66	2.40236	10.8	2.36898	21.5	2.33427	27.6	2.30332
1.68	2.40180	11.0	2.36854	21.6	2.33383	27.7	2.30274
1.70	2.40125	11.2	2.36809	21.7	2.33339	27.8	2.30216
1.72	2.40073	11.4	2.36763	21.8	2.33295	27.9	2.30157
1.74	2.40022	11.6	2.36717	21.9	2.33251	28.0	2.30098

TABLE 53 Thallium Bromide–Thallium Iodide (KRS-5): Refractive Index at 25°C (Rodney and Malitson, 1956)† (Continued)

λ	n	λ	n	λ	n	λ	n
28.1	2.30039	31.1	2.28145	34.1	2.26011	37.1	2.23621
28.2	2.29979	31.2	2.28078	34.2	2.25936	37.2	2.23536
28.3	2.29920	31.3	2.28011	34.3	2.25860	37.3	2.23452
28.4	2.29860	31.4	2.27943	34.4	2.25784	37.4	2.23367
28.5	2.29800	31.5	2.27875	34.5	2.25708	37.5	2.23281
28.6	2.29739	31.6	2.27807	34.6	2.25631	37.6	2.23196
28.7	2.29679	31.7	2.27738	34.7	2.25554	37.7	2.23110
28.8	2.29618	31.8	2.27669	34.8	2.25477	37.8	2.23024
28.9	2.29556	31.9	2.27600	34.9	2.25400	37.9	2.22937
29.0	2.29495	32.0	2.27531	35.0	2.25322	38.0	2.22850
29.1	2.29433	32.1	2.27461	35.1	2.25244	38.1	2.22763
29.2	2.29371	32.2	2.27391	35.2	2.25166	38.2	2.22676
29.3	2.29309	32.3	2.27321	35.3	2.25087	38.3	2.22588
29.4	2.29247	32.4	2.27251	35.4	2.25008	38.4	2.22500
29.5	2.29184	32.5	2.27180	35.5	2.24929	38.5	2.22412
29.6	2.29121	32.6	2.27109	35.6	2.24849	38.6	2.22323
29.7	2.29058	32.7	2.27038	35.7	2.24769	38.7	2.22234
29.8	2.28994	32.8	2.26966	35.8	2.24689	38.8	2.22145
29.9	2.28931	32.9	2.26895	35.9	2.24609	38.9	2.22055
30.0	2.28867	33.0	2.26823	36.0	2.24528	39.0	2.21965
30.1	2.28802	33.1	2.26750	36.1	2.24447	39.1	2.21875
30.2	2.28738	33.2	2.26678	36.2	2.24366	39.2	2.21784
30.3	2.28673	33.3	2.26605	36.3	2.24284	39.3	2.21693
30.4	2.28608	33.4	2.26532	36.4	2.24202	39.4	2.21602
30.5	2.28543	33.5	2.26458	36.5	2.24120	39.5	2.21510
30.6	2.28477	33.6	2.26384	36.6	2.24038	39.6	2.21418
30.7	2.28411	33.7	2.26310	36.7	2.23955	39.7	2.21326
30.8	2.28345	33.8	2.26236	36.8	2.23872	39.8	2.21233
30.9	2.28279	33.9	2.26161	36.9	2.23788	39.9	2.21140
31.0	2.28212	34.0	2.26087	37.0	2.23705	40.0	2.21047

Dispersion Equation

$$n^2 - 1 = \sum_i \frac{K_i \lambda^2}{\lambda^2 - \lambda_i^2}$$

Constants of Dispersion Equation at 25°C

i	λ_i^2	K_i	i	λ_i^2	K_i
1	0.0225	1.8293958	4	0.2025	0.04513366
2	0.0625	1.6675593	5	27,089.737	12.380234
3	0.1225	1.1210424			

† Data are also given in Rodney and Malitson (1956) for temperatures of 19 and 31°C. A comparison of these data, taken with the 45.7 to 54.3 mol percent mixed crystal, is made with the older data taken with 42 to 58 crystal material by Tilton et al. (1949). The 45.7 to 54.3 composition, which has the lowest freezing temperature of the binary system, should give the best optical homogeneity.

TABLE 54 Thallium Chloride: Refractive Index†

λ, μm	n	λ, μm	n
0.436	2.400	0.750	2.198
0.546	2.270	10.0	2.193
0.578	2.253	12.47	2.191
0.589	2.247	18.35	2.182
0.650	2.223		

† The data from 0.436 to 0.750 are from Barth (1929). Data from 10.0 to 18.35 are from McCarthy (1965*c*).

TABLE 55 Zinc Sulfide: Refractive Index

Hexagonal (Bieniewski and Czyzak, 1963)

λ, μm	n_e	n_o	λ, μm	n_e	n_o	λ, μm	n_e	n_o
0.3600	2.709	2.705	0.4700	2.453	2.448	0.6250	2.358	2.354
0.3750	2.640	2.637	0.4750	2.449	2.445	0.6500	2.350	2.346
0.4000	2.564	2.560	0.4800	2.443	2.438	0.6750	2.343	2.339
0.4100	2.544	2.539	0.4900	2.433	2.428	0.7000	2.337	2.332
0.4200	2.525	2.522	0.5000	2.425	2.421	0.8000	2.328	2.324
0.4250	2.514	2.511	0.5250	2.407	2.402	0.9000	2.315	2.310
0.4300	2.505	2.502	0.5500	2.392	2.386	1.0000	2.303	2.301
0.4400	2.488	2.486	0.5750	2.378	2.375	1.2000	2.294	2.290
0.4500	2.477	2.473	0.6000	2.368	2.363	1.4000	2.288	2.285
0.4600	2.463	2.459						

Cubic (Czyzak et al., 1957)

λ, μm	n	λ, μm	n	λ, μm	n
0.4400	2.488	0.5500	2.384	0.9000	2.306
0.4600	2.458	0.5750	2.375	1.0500	2.293
0.4800	2.435	0.6000	2.359	1.2000	2.282
0.5000	2.414	0.6500	2.346	1.4000	2.280
0.5250	2.395	0.7000	2.334		

Dispersion Equation

$$n^2 = 5.131 + \frac{1.275 \times 10^7}{\lambda^2 - 0.732 \times 10^7}$$

TABLE 56 Refractive Index vs. Temperature for Corning Glass 7913 (Vycor), Optical Grade (Wray and Neu, 1969)

λ, μm	n 28°C	n 526°C	$\frac{dn}{dT}$, 10^{-6}/°C	n 826°C	$\frac{dn}{dT}$, 10^{-6}/°C
0.26520	1.49988	1.50799	16.3	1.51438	18.2
0.28936	1.49074	1.49831	15.2	1.50418	16.8
0.29673	1.48851	1.49587	14.8	1.50164	16.5
0.30215	1.48694	1.49423	14.6	1.49990	16.2
0.3130	1.48416	1.49121	14.2	1.49679	15.8
0.33415	1.47949	1.48622	13.5	1.49158	15.2
0.36502	1.47415	1.48065	13.1	1.48570	14.5
0.40466	1.46925	1.47547	12.5	1.48027	13.8
0.43584	1.46628	1.47234	12.2	1.47708	13.5
0.54607	1.45960	1.46544	11.7	1.46992	12.9
0.5780	1.45831	1.46407	11.6	1.46849	12.8
1.01398	1.44968	1.45526	11.2	1.45924	12.0
1.12866	1.44831	1.45373	10.9	1.45779	11.9
1.254†	1.44677	1.45222	10.9	1.45627	11.9
1.36728	1.44554	1.45095	10.9	1.45504	11.9
1.570†	1.44422	1.44965	10.9	1.45370	11.9
1.52952	1.44356	1.44896	10.8	1.45306	11.9
1.660†	1.44206	1.44750	11.0	1.45157	11.9
1.701	1.44137	1.44677	10.8	1.45088	11.9
1.981†	1.43750	1.44291	10.9	1.44702	11.9
2.262†	1.43298	1.43839	10.9	1.44258	12.0
2.553†	1.42825	1.43373	11.0	1.43824	12.5

† Wavelength determination by narrow-bandwidth interference filters.

TABLE 57 Miscellaneous Refractive-Index Data

Material	λ, μm	n	Ref.†
Arsenic triselenide glass	4	2.796	14
	2–16	2.812–2.768	14
Barium titanate	Visible and infrared	2.40	
Cadmium fluoride	2	1.63	1
	5.5	1.53	1
	11	1.45	1
Cadmium iodide	0.6	2.7	2
Cadmium selenide	1–8	2.45	3
	>8	2.42	3
Cer-Vit C-101	0.589 at 25°C	1.540	17
Cesium copper (III) chloride	3.39	$n_e = 1.798 \pm 0.011$	18
		$n_o = 1.706 \pm 0.011$	18
	10.6	$n_e = 1.769 \pm 0.007$	18
		$n_o = 1.682 \pm 0.007$	18
Copper (III) chloride	0.4–20.5	1.93	4
Gallium phosphide	<1	3.5	12
	1–8	3.2–2.9	12
	>8	2.8	12
Indium arsenide	4–15	3–3.5	13
Indium phosphide	2–15	3–3.5	13
Lead bromide	White light	2.53	5
Lead chloride	Yellow light	2.2	6
Lead selenide	1–3.5	3.5–4.6	7
	5	4.6	8
Lead telluride	1–3.5	4.1–5.3	7
	3.9–20	5.10	9
Rubidium bromide	1–8	1.53	10
Rubidium chloride	1–8	1.48	10
Rubidium iodide	1–8	1.62	10
T-12	Near infrared	1.41	11
Telluride glass (Ge-As-Te)	5	3.5	15
Texas Instruments glass 1173‡	3.3	2.63	15
	5	2.62	16

† 1 = adapted from Welber (1967); 2 = adapted from Nilsson (1968); 3 = Vitrikhovsky et al. (1967); 4 = Heilmeir (1964); 5 = Moss and Peacock (1961); 6 = "Handbook of Chemistry and Physics" (1960); 7 = Avery (1953); 8 = Smakula (1962*b*); 9 = Smakula et al. (1964); 10 = Mott and Gurney (1950) from the values given for dielectric constants; 11 = Ballard (1965); 12 = Willardson and Beer (1967); 13 = Oswald and Schade (1954); 14 = Savage and Nielson (1965); 15 = Ballard and Browder (1966); 16 = Hilton and Jones (1967); 17 = Monnier (1967); 18 = Day and Gruzensky (1970).

‡ $dn/dT = 79 \times 10^{-6}$ per Celsius degree in the region 3 to 13 μm.

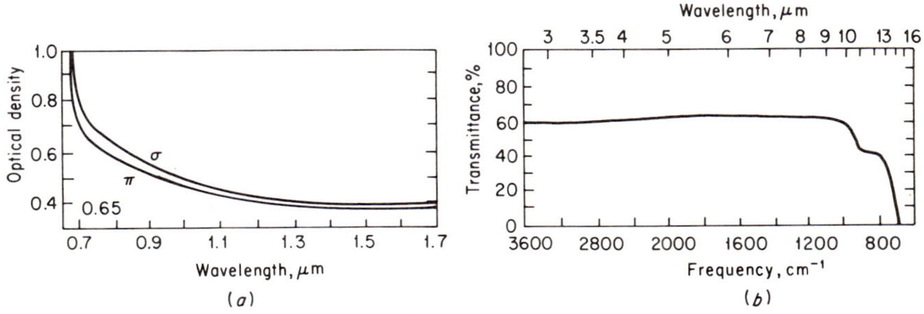

Fig. 189 (*a*) Optical density of pyrargyrite in the region 0.65 to 1.7 μm. Sample 0.66 cm thick; curve uncorrected for reflection losses. (*b*) Transmission of Ag_3SbS_3 in the region 2.5 to 15 μm. [*From Feichtner et al.* (1970).]

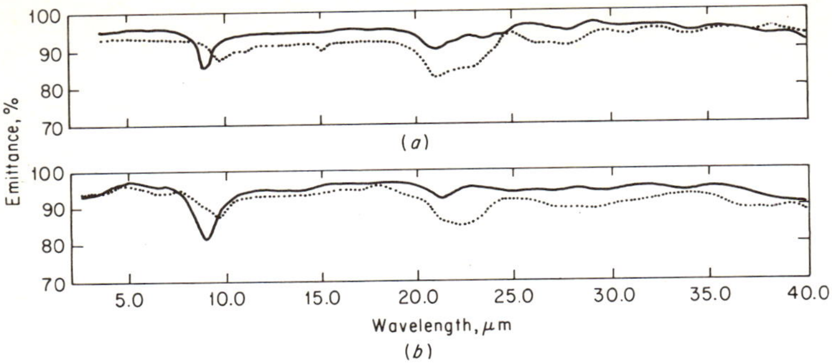

Fig. 190 Spectral emission of Cat-a-lac† (*dashed curve*) and 3M‡ black paint at (*a*) 77 K and (*b*) 373 K. [*From Stierwalt* (1966*b*).]

† Trade name of Finch Paint and Chemical, Torrance, Calif.
‡ Trade name of Minnesota Mining and Manufacturing Company.

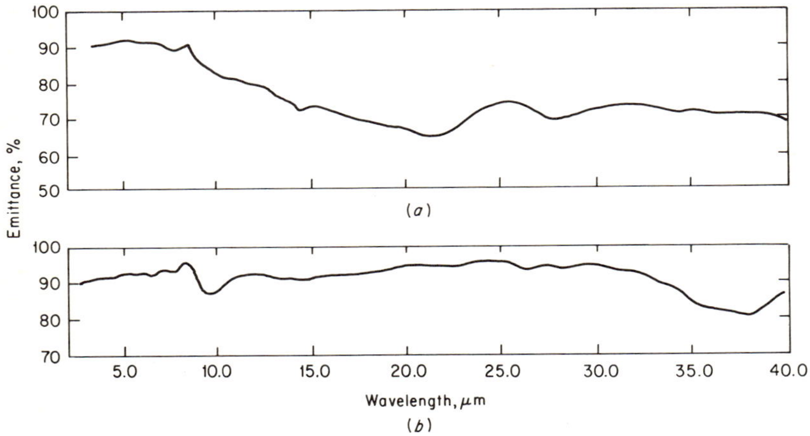

Fig. 191 Spectral emission of Sicon† black paint at (*a*) 77 K and (*b*) 373 K. [*From Stierwalt* (1966*b*).]

† Trade name of Midland Industrial Finishes, Waukegan, Ill.

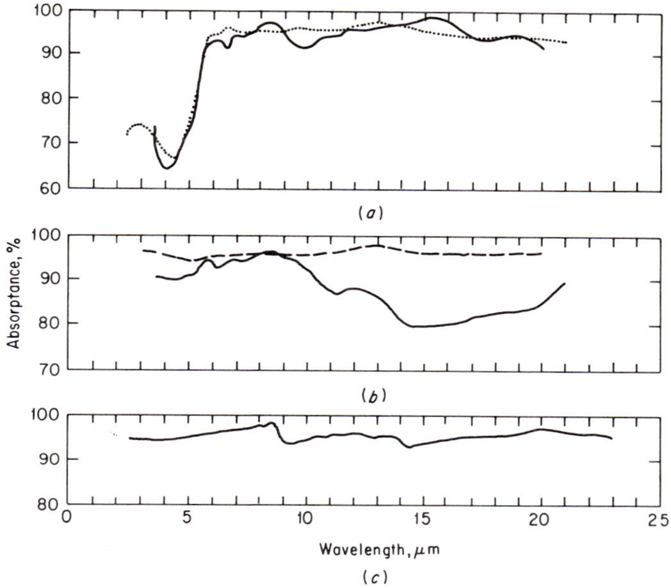

Fig. 192 Spectral absorption of (*a*) BEC-1 (*solid curve*) and BEC-4 (*dotted curve*); (*b*) iron oxide (*solid curve*) and BEC-3 (*dotted curve*); and (*c*) BEC-2. [*From Stierwalt et al.* (1963).]

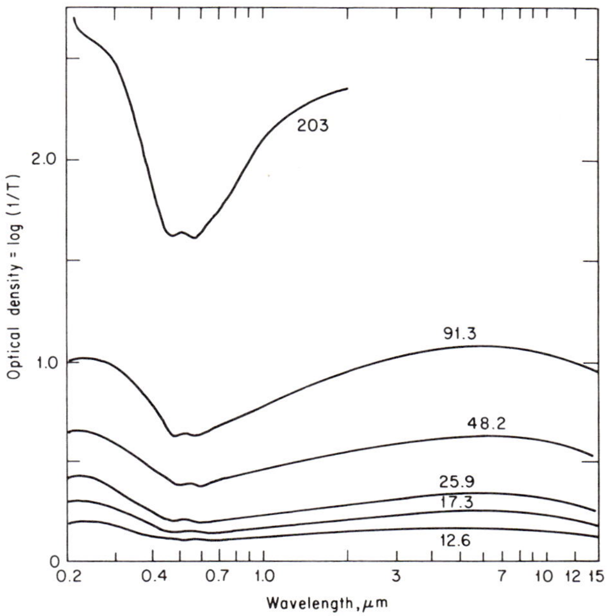

Fig. 193 Optical density of gold blacks for different depositions (micrograms per square centimeter) based on direct transmission. [*From Harris* (1967).]

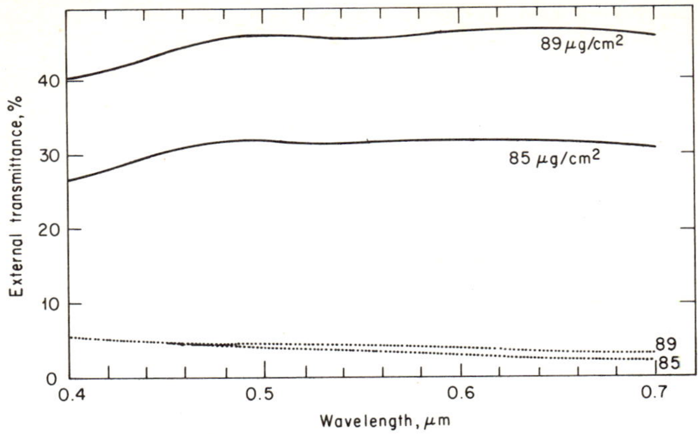

Fig. 194 Total (*solid curve*) and diffuse (*dotted curve*) transmission of gold black on cellulose nitrate; depositions as indicated. [*From Harris* (1967).]

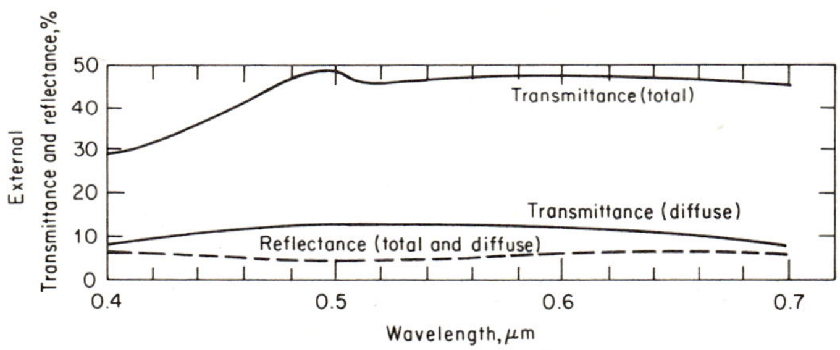

Fig. 195 Transmission and reflection of gold black on glass, 205 μg/cm². [*From Harris* (1967).]

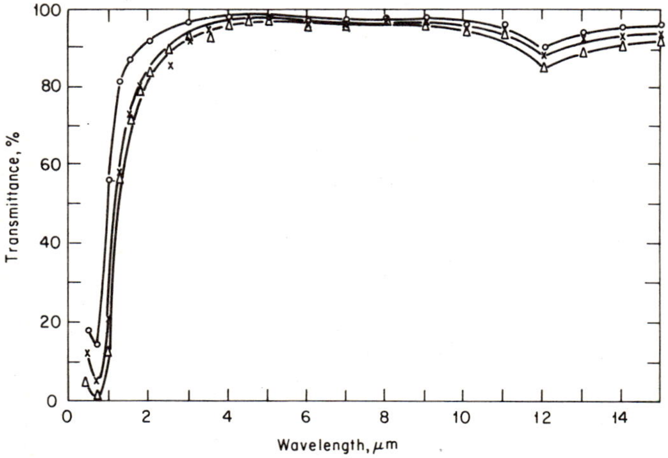

Fig. 196 Transmission of gold smokes with nitrogen containing 3 percent oxygen; weight of gold filter: ○ = 65 μg/cm²; X = 80 μg/cm²; △ = 129 μg/cm². [*From Harris* (1967).]

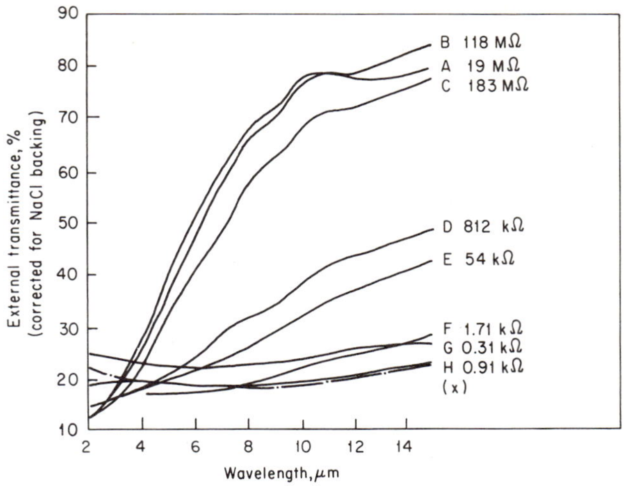

Fig. 197 Transmission of a gold smoke deposit containing ~1 percent of tungsten oxides, upon heat treatment: (*A*) immediately after preparation, (*B*) after 72 h at room temperature, (*C*) after 24 h a 69°C, (*D*) after 3 h at 170°C, (*E*) 3 h at 180°C after (*D*), (*F*) 15 h at 180°C after (*E*), (*H*) 18 h at 180°C after (*F*), (*G*) 18 h at 180° C after (*H*), (*X*) a high infrared-absorbing gold black (62 μg/cm²), after stabilization at 69°C. [*From Harris* (1967).]

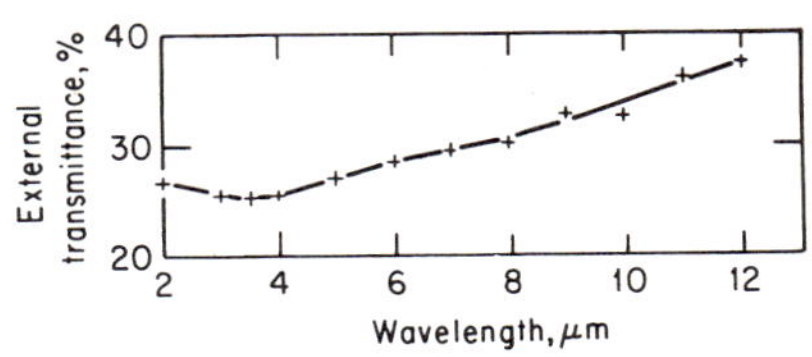

Fig. 198 Infrared transmission of a silver black deposit, estimated thickness 36 μg/cm². [*From Harris* (1967).]

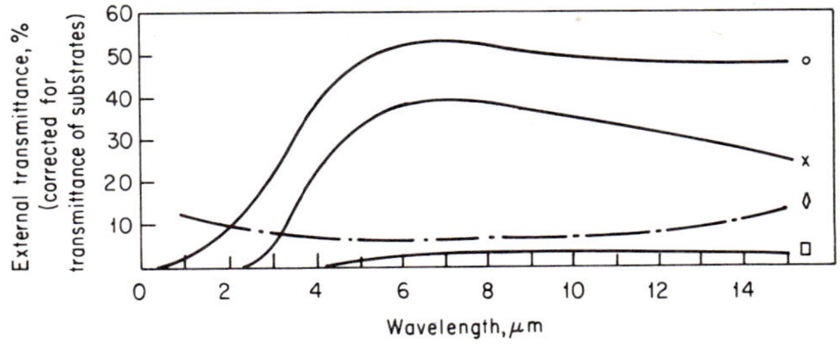

Fig. 199 Transmission of antimony black deposits: ○ antimony black on cellulose nitrate (122 μg/cm²), X antimony black on KRS-5, □ antimony black on KRS-5, ◇ gold black on cellulose nitrate (100 μg/cm²). [*From Harris* (1967).]

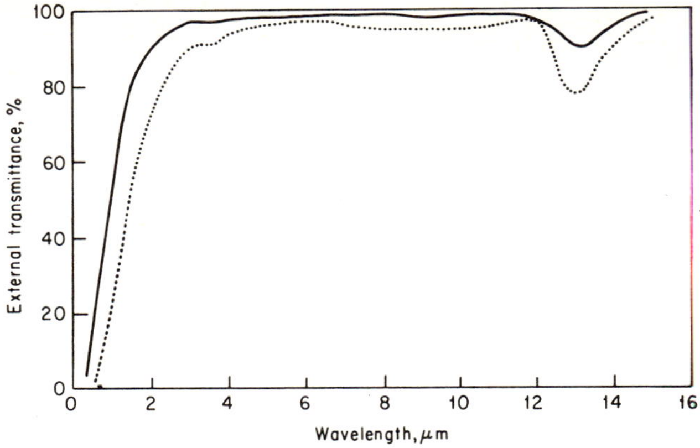

Fig. 200 Transmission on antimony smoke deposits on cellulose nitrate substrates for 6-min deposition (*solid curve*) and 11-min deposition (*dotted curve*). [*From Harris* (1967).]

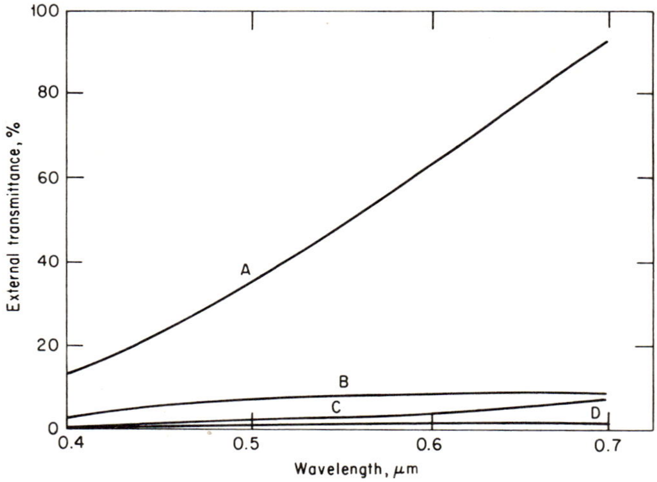

Fig. 201 Visible transmission of thin acetylene black deposits: curve A: total transmission (sample 7); curve B: diffuse transmission (sample 7); curve C: total transmission (sample 3); curve D: diffuse transmission (sample 3). [*From Harris* (1967).]

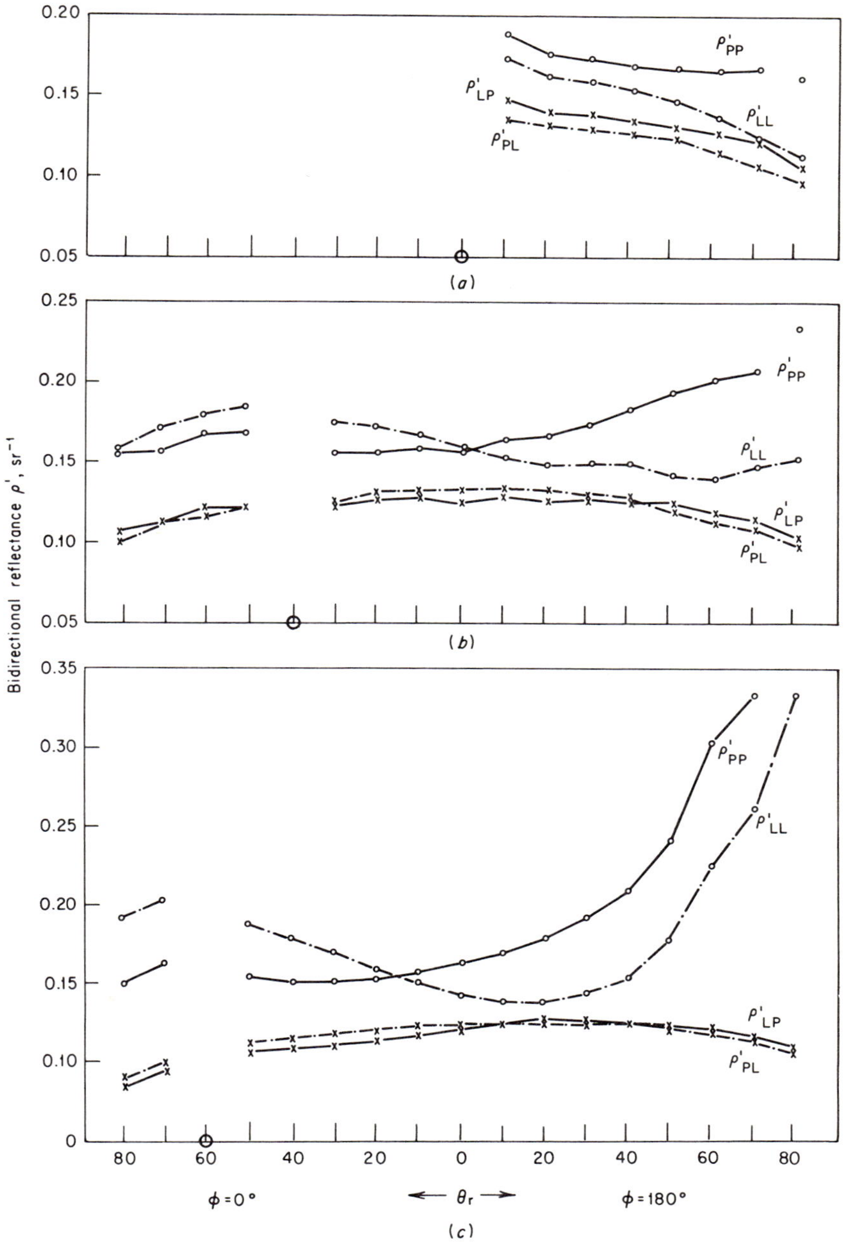

Fig. 202 p' vs. θ_γ as a function of polarization for smoke-deposited magnesium oxide. Incidence angle: (*a*) 0°, (*b*) 40°, (*c*) 60°. [*From Carmer and Baer* (1969*a*).]

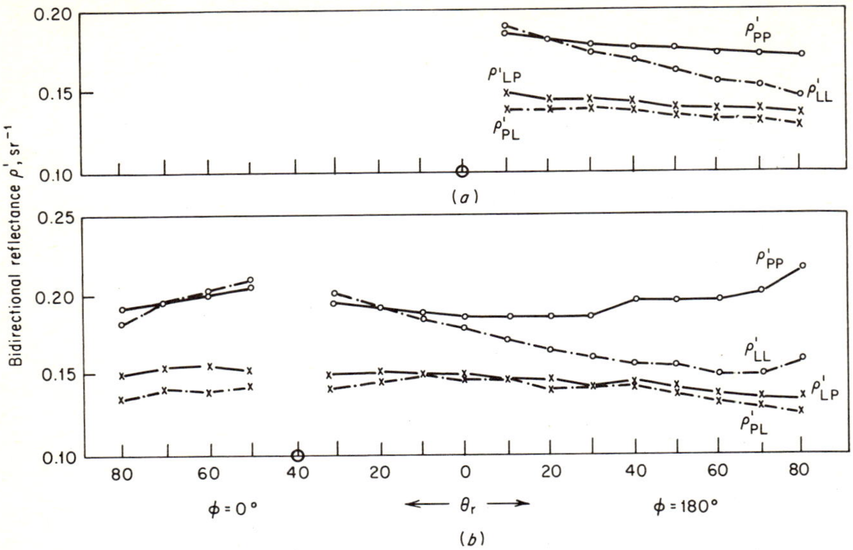

Fig. 203 p' vs. θ_γ as a function of polarization for pressed powder magnesium oxide. Incidence angle: (*a*) 0°, (*b*) 40°. [*From Carmer and Baer* (1969*b*).]

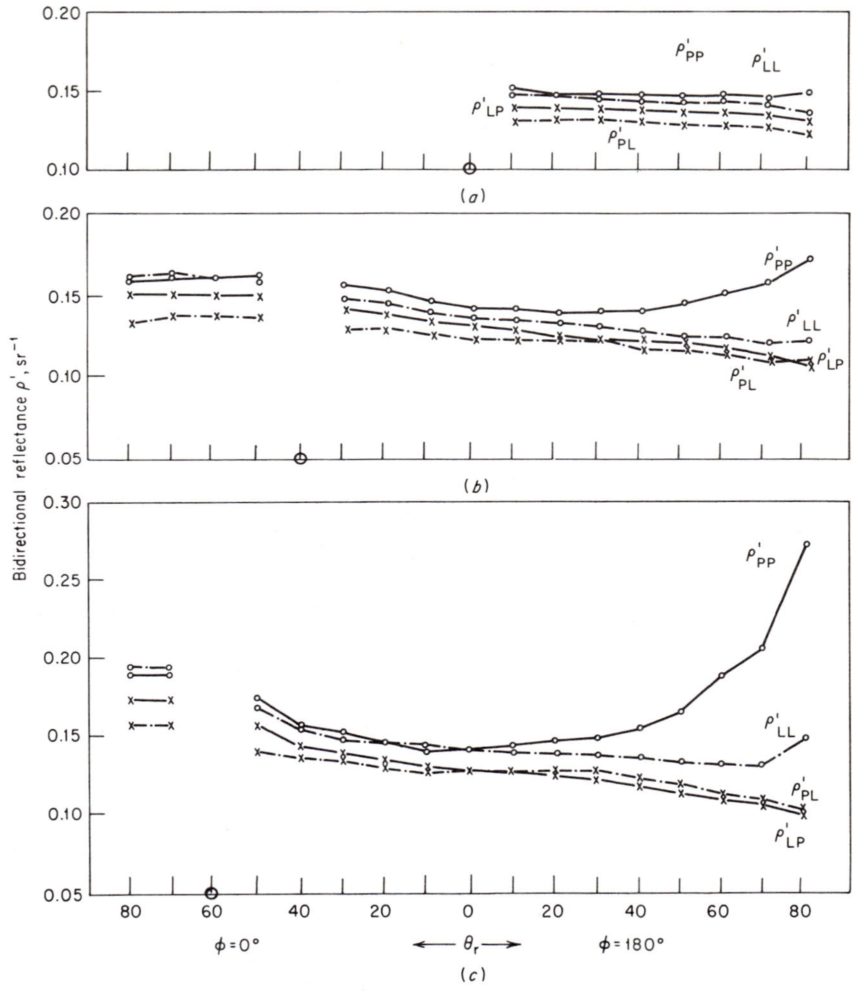

Fig. 204 p' vs. θ_γ as a function of polarization for 3M white velvet paint. Incidence angle: (*a*) 0°, (*b*) 40°, (*c*) 60°. [*From Carmer and Baer* (1969*b*).]

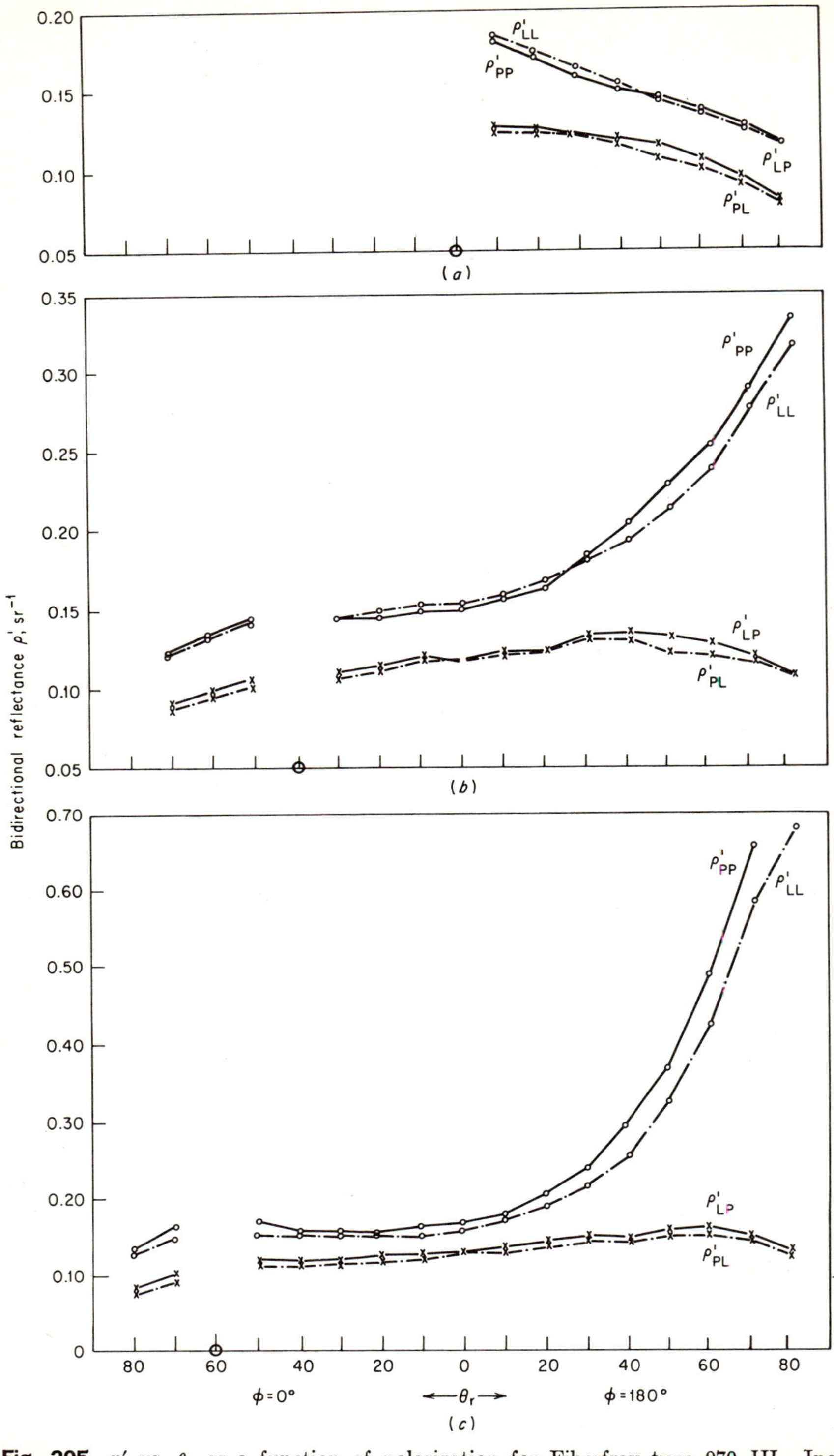

Fig. 205 p' vs. θ_γ as a function of polarization for Fiberfrax type 970 JH. Incidence angle: (*a*) 0°, (*b*) 40°, (*c*) 60°. [*From Carmer and Baer* (1969*c*).]

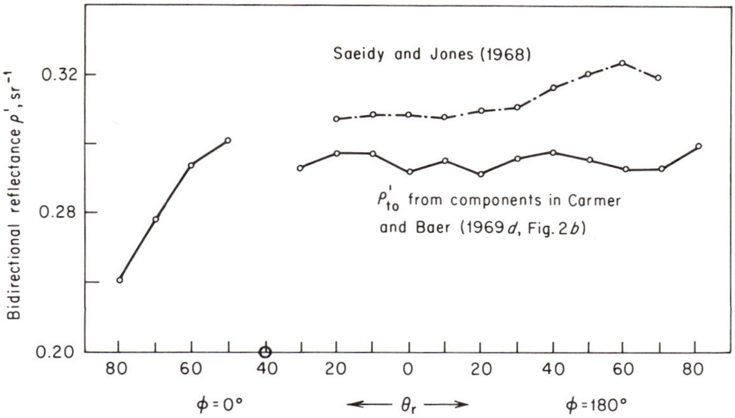

Fig. 206 p' vs. θ_γ for two magnesium oxide samples. For identification of the reference on the figure, see the original source. [*From Carmer and Baer* (1969*d*).]

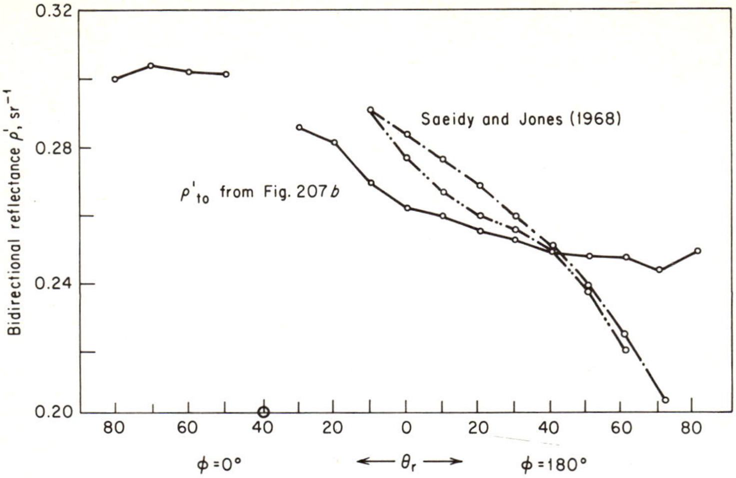

Fig. 207 p' vs. θ_γ for two 3M white velvet paint samples. [*From Carmer and Baer* (1969*c*).]

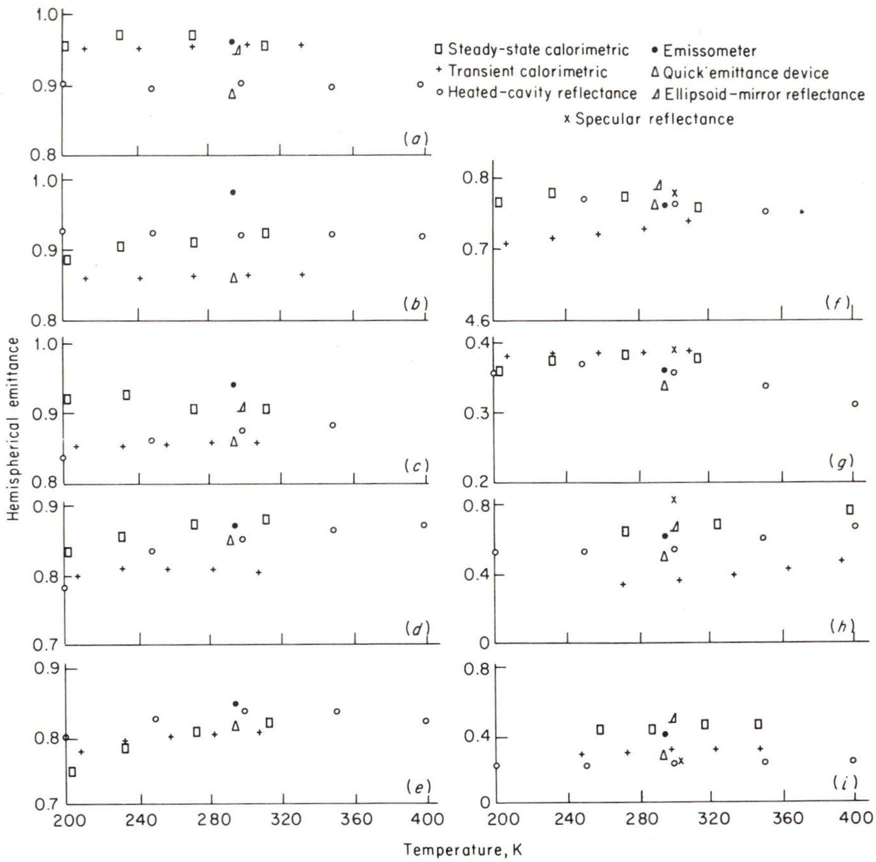

Fig. 208 The hemispherical emittance as a function of temperature for the coatings. (*a*) Parsons black, (*b*) V groove, (*c*) Z-93, (*d*) S-13G, (*e*) S-13, (*f*) optical solar reflector, (*g*) barrier layer anodic, (*h*) cobalt oxide, (*i*) 2000-Å aluminum on stainless steel. [*From Millard and Streed* (1969*a*).] See the reference for definitions of the coatings.

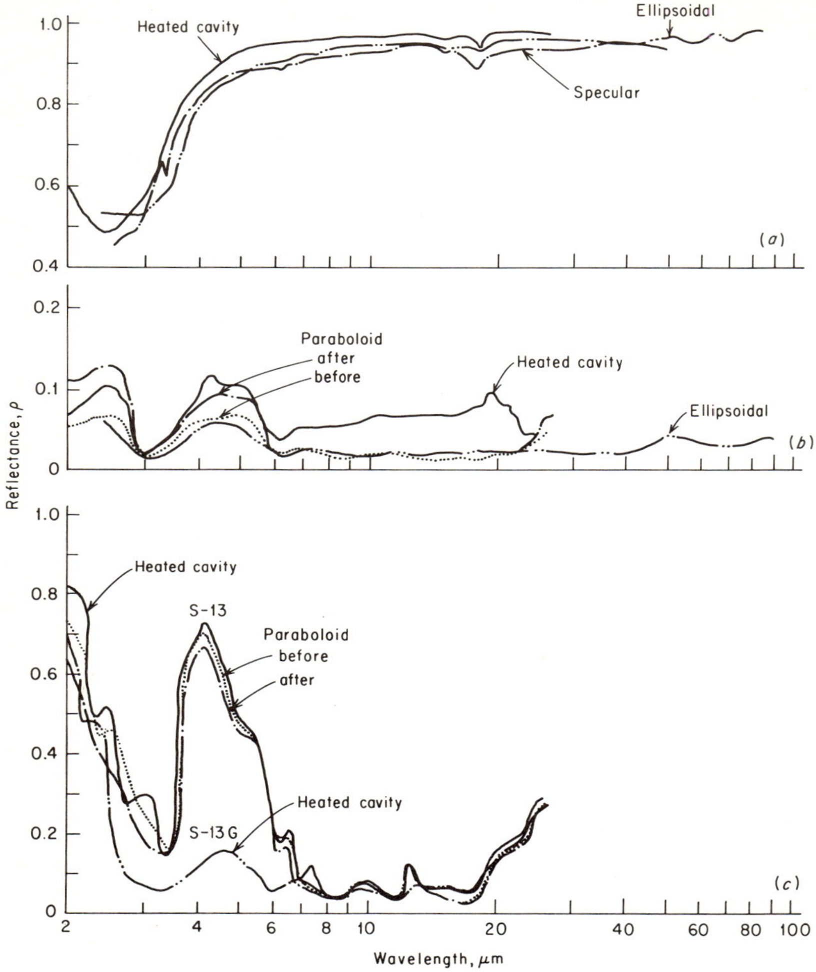

Fig. 209 Spectral reflectance as a function of wavelength for coatings: (*a*) cobalt oxide, (*b*) Parsons black, (*c*) S-13 and S-13G, (*d*) barrier layer anodic and Z-93, (*e*) optical solar reflector. [*From Millard and Streed* (1969*b*).]

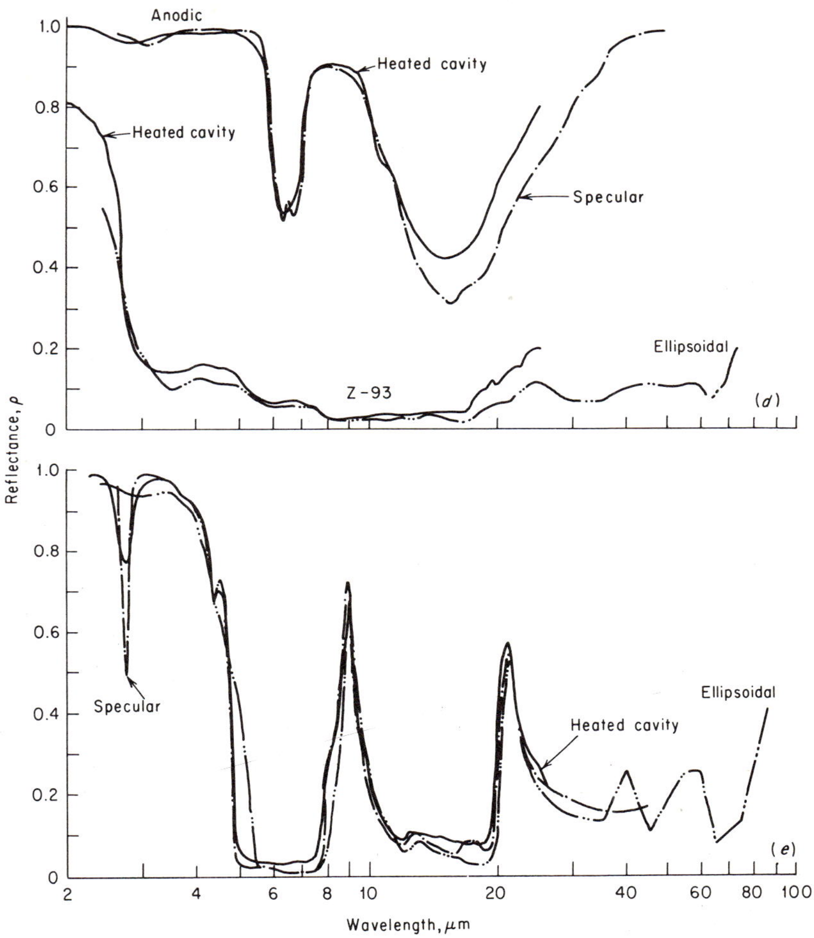
Anodic
Heated cavity
Heated cavity
Specular
Ellipsoidal
Z-93
(d)
Specular
Ellipsoidal
Heated cavity
(e)
Reflectance, ρ
Wavelength, μm
1.0
0.8
0.6
0.4
0.2
0
2
4
6
8
10
20
40
60
80
100

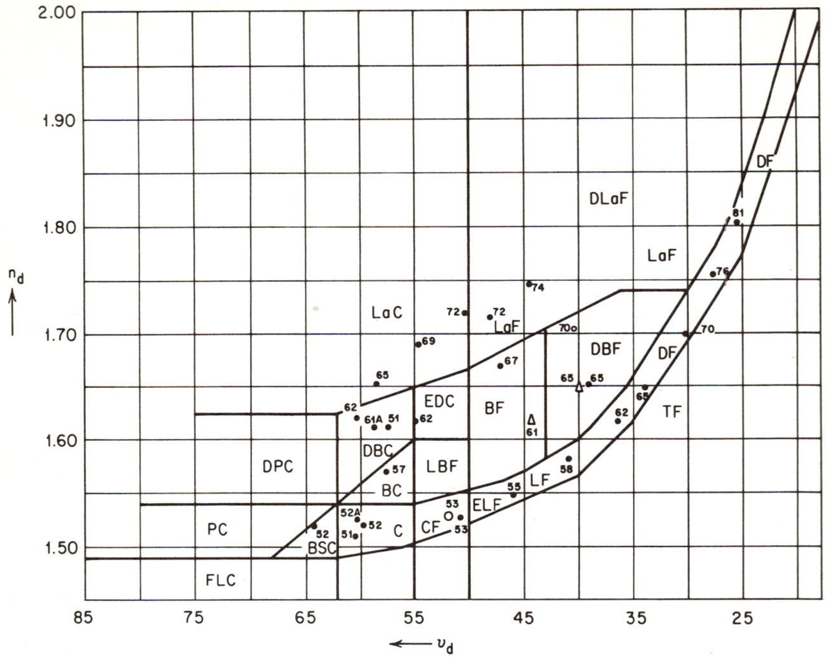

Fig. 210 Refractive index and dispersion value of precision optical glasses. Open circles = flint special (FS), triangle = borate flint (BoF), solid circles = all other glasses. E = extra, D = dense, L = light. Abbreviations of glass types used by this manufacturer vary slightly from one bulletin to another. [*From Corning Glass Works* (1967).]

TABLE 58 Elastic Coefficients

Material	T, K	Coefficient, bars				
		c_{11}	c_{12}	c_{13}	c_{33}	c_{44}
ADP	...	617	0.72	1.94	3.28	0.85
AgCl	...	6.01	3.62	....	....	0.625
Al_2O_3	298	49.68	16.36	....	49.81	14.74
BaF	...	9.01	4.03	....	....	2.49
$BaTiO_3$	298	8.18	2.98	1.95	6.76	18.3
$CaCO_3$	...	13.71	4.56	4.51	7.97	3.42
CaF_2	...	16.4 ± 0.1	5.3 ± 0.2	....	....	3.370 ± 0.01
CdS	...	8.432	5.212	4.638	9.379	1.489
CdTe	...	5.351	3.681	....	....	1.994
CsBr	...	3.097	0.403	....	....	0.7500
CsI	...	2.46	0.67	....	....	0.624
GaAs	...	1.192	0.5986	....	....	0.538
GaP	...	14.7				
GaSb	...	8.849	4.037	....	....	4.325
Ge	...	1.29	4.83	....	....	6.71
InAs	...	8.329	4.526	....	....	3.959
InP	...	10.7				
InSb	300	6.472	3.625	....	....	3.071
KBr	...	3.45	0.54	....	....	0.508
KCl	...	3.98	0.62			
KDP	...	7.14	−0.49	1.29	5.62	1.27
KI	...	2.69	0.43	....	....	0.362
KRS-5	...	3.31	1.32	....	....	0.597
KRS-6	...	3.85	1.49	....	....	0.737
LiF	...	9.74	4.04	....	....	5.54
$MgAl_2O_3$	...	30.05	15.37	....	....	15.86
MgO	...	2.90	0.876	....	....	1.55
NaCl	...	4.85	1.23	....	....	1.26
NaF	...	9.09	2.64	....	....	1.27
$NaNO_3$	...	8.67	1.63	1.60	3.74	2.13
PbS	...	12.7	2.98	....	....	2.48
Quartz, crystal	...	8.675	0.687	1.13	10.68	5.786
Si	...	1.67	0.65	....	....	0.80
$SrTiO_3$	...	31.56	10.27	....	....	12.15
Te	300	3.265	0.195	2.493	7.22	3.121
TiO_2	...	35.8	26.7	17.0	49.9	12.5
TlBr	...	3.78	1.48	....	...	0.756
TlCl	...	4.01	1.53	....	...	0.760

TABLE 59 Elastic Moduli

Material	Young's, 10^6 lb/in^2	Rigidity, 10^6 lb/in^2	Bulk, 10^6 lb/in^2	Rupture, lb/in^2	Apparent elastic limit, lb/in^2
AgCl	0.02	1.03	6.39	....	3.8×10^3
					7.4×10^2
Al_2O_3	50.0	21.5	0.30		
As_2S_3	2.30	0.94	.....	2.4×10^3	
BaF_2	7.7	......	.....	3.9×10^7	3.9×10^7
$BaTiO_3$	4.9	18.3	23.5		
$CaCO_3$	10.5∥	......	18.8		
	12.8⊥				
CaF_2	11.0	4.9	12.0	5.3×10^3	5.3×10^3
CdTe	......	......	.....	850.0	
CsBr	2.3	......	.....	23.9	12.2×10^2
CsI	0.769	......	.....	....	8.1×10^2
GaSb	9.19	6.28	8.19		
Ge	14.9	9.73	11.3		
InSb	6.21	4.45	6.28		

Irtran	Temp., °C	Young's, 10^6 lb/in^2	Rupture, lb/in^2	Irtran	Temp., °C	Young's, 10^6 lb/in^2	Rupture, lb/in^2
1	25	16.6	21,800	3	25	14.3	5,300
	500	16.6	10,000		500	14.0	9,000
2	25	14.0	14,100	4	25	10.3	7,500
	250	10.6	13,500	5	25	48.2	19,200

Material	Young's, 10^6 lb/in^2	Rigidity, 10^6 lb/in^2	Bulk, 10^6 lb/in^2	Rupture, lb/in^2	Apparent elastic limit, lb/in^2
KBr	3.9	0.737	2.18	4.8×10^2	1.6×10^2
KCl	4.3	0.906	2.52	6.4×10^2	3.3×10^2
KI	4.57	0.90	124.0		
KRS-5	2.3	0.840	2.87	1.81×10^4	3.8×10^4
KRS-6	3.0	1.23	3.31	....	3.05×10^3
LiF	9.4	8.0	9.0	2.0×10^6	16.2×10^6
MgO	36.1	22.4	22.4		
NaCl	5.8	1.83	3.53	5.7×10^2	3.5×10^2
$NaNO_3$	....	......	3.8		
Quartz, crystal	11.1⊥ to *c* axis	5.28			
	14.1∥ to *c* axis				
Fused	10.6	4.52			
Si	1.90	11.6	14.8		
TlBr	4.28	1.10	3.26		
TlCl	4.60	1.10	3.42		

TABLE 60 Thermal Properties

Material	m.p., K	Specific heat: Value	Specific heat: T, K	Thermal conductivity: cal/(cm)(s)(K)	Thermal conductivity: T, K
ADP	.....	......	...	0.0017	273
AgCl	730.7	0.848	273	0.0026	273
		0.0906	323	0.00271	
Al_2O_3	2303	0.18	298		
		0.174	273		
As_2S_3	483	.....	...	0.0004	
BaF_2	1553	.....	...	0.028	286
				0.017	311
$BaTiO_3$	1600	0.01799	55	0.0016	401
		0.03004	75	0.0032	Room
		0.04421	100		
		0.05709	125		
		0.06868	150		
		0.07813	175		
$CaCO_3$	1612	0.203	273	0.0132	273
		0.214	373	0.0111	273
CaF_2	1633	0.204	273	0.0932	83
		0.212	373	0.0360	200
				0.0247	273
				0.0191	373
				0.0232	298
CdF_2	1047				
CdS	1500	0.0882	273	0.038	287
		0.922	323		
Pressed	....	.....	...	0.10	10
				0.10	100
				0.40	50
CdTe	1041–1050	0.01875	323	0.015	
CsBr	909	0.63	293	0.0023	298
				0.0022	318
				0.0026	338
CsI	621	0.048	293	0.0027	298
CuBr	777				
CuCl	695				
GaAs	1238	.....	...	0.125	300
GaP	1350	.....	...	0.13	300
GaSb	720	0.01828	...	0.105	300
Ge	936–942	0.074	273	0.14	293
InAs	942	3.4	78–290		
		5.2	290–573		
		7.01	573–673		
InP	1050–1070				
InSb	523	0.0231	180	0.085	293
		0.248	300		
Irtran 1	1528	.....	...	0.035	329
				0.026	452
Irtran 2	2103†	.....	...	0.037	327
				0.026	447
Irtran 3	1692	.....	...	0.019	353
				0.015	449
Irtran 4	1788	.....	...	0.031	327
				0.016	695
Irtran 5	3220	.....	...	0.104	298
				0.070	441
				0.0085	417
KBr	1003	0.104	273	0.0115	319
		0.108	373	0.00698	299
KCl	776	0.162	273	0.0156	315
		0.168	373		

TABLE 60 Thermal Properties (Continued)

Material	m.p., K	Specific heat		Thermal conductivity		
		Value	T, K	cal/(cm)(s)(K)		T, K
KDP	252.6	....	...	0.0029		312
				0.0032		319
KI	723	0.75	270	0.005 ± 3%		299
		0.73	200			
		0.75	250			
KRS-5	687.5	....	...	0.0013		
KRS-6	423.5	0.0482	293	0.00171		329
LiF	1143	0.373	283			
		0.03	308			
MgO	3073	0.209	273	2.62		10
				6.38		100
				0.077		500
				7.56		30
				0.140		300
$MgO \cdot Al_2O_3$	2030	0.026	441	0.033		
		0.028	443			
Muscovite	1473–1573	0.208	293–373	0.0006–0.0014		
NaCl	1074	0.204	273	0.0155		289
		0.217	373			
NaF	980	0.26	273	0.124		83
				0.0252		273
				0.0220		298
$NaNO_3$	306.8	0.247	273			
		0.270	373			
$PbCl_2$	774	0.0649	273			
		0.0681	373			
PbF_2	1128					
	822 ± 2					
$PbMoO_4$	1333–1343	0.100	288			
PbS	1114	0.0502	273	0.0016		
		0.0511	373			
PbSe	1065	.....	...	0.0100		
PbTe	917	.....	...	0.0120		
Quartz, fused	<1743	0.188	285–373			
				∥	⊥	
				0.117	0.586	83
				0.0467	0.0249	195
				0.0273	0.0163	273
				0.0224	0.0135	323
				0.0190	0.0118	373
				0.0168	0.0160	423
				0.0151	0.00967	473
				0.0136	0.00895	523
				0.0123	0.0084	573
				0.0113	0.0079	623
Crystal	2000	0.1657	273	0.00282		314
		0.201	373	0.00264		
				0.00450		
Se, amorphous	308	0.068	85–291	0.0031		
Crystal	490	0.072	276	0.0026		
		0.077	293.5	0.0004		
		0.085	302.5			
		0.127	305			
		0.131	311			
		0.95	291–311			
Se(As)	~343	....	...	0.00033		
Si	1693	0.168	298	0.39		313
SrF_2	1190					

TABLE 60 Thermal Properties (Continued)

Material	m.p., K	Specific heat		Thermal conductivity	
		Value	*T*, K	cal/(cm)(s)(K)	*T*, K
$SrTiO_3$	2353				
Te	722.8	0.0483	288–373	0.015	
TiO_2	2093	0.17	298	0.030	309∥
				0.033	340∥
				0.021	317⊥
				0.017	340⊥
TlBr	460	0.045	20	0.0014	
TlCl	430	0.0520	273	0.0019	

† At 150 lb/in^2.

TABLE 61 Thermal Expansion

$\alpha = A \times 10^{-6} + B \times 10^{-8}T + C \times 10^{-11}T^2$

Material	T, K	A	B	C	Remarks
ADP	297–407	39.3			
AgCl	298	30.01			
	473	34.59			
	623	52.09			
	653	58.37			
	673	63.19			
	698	69.99			
	293–333	30			
Al_2O_3	323	6.7	......	......	‖ to c axis
	323	5.0	......	......	⊥
As_2S_3	306–438	24.62			
BaF_2	272–573	1.9			
$BaTiO_3$	193–253	16			
	283–343	19			
	393–453	13			
$CaCO_3$	123–273	24.39	0.533	−30.7	‖
	123–273	−5.68	0.0333	−4.58	⊥
	323	26.6	......	......	‖
	323	5.2	......	......	‖
	638	−3.8	......	......	⊥
	348–673	24.71	3.775	−3.653	‖
CaF_2	181–280	18.38	2.511	−21.10	
CdF_2	293–393	27			
CdS	300–343	4.2			
	323–773	3.5	......	......	‖
Pressed	10	0			
	50	−2.4			
	110	0			
	200	2.5			
	300	4.2			
CdTe	323	4.5			
	873	5.9			
	316–900	1.851	1.481	21.52	
CsBr	293–323	47.9			
	134–573	46.6	4.67	−1.78	
CsI	298–323	50			
CuBr	293–423	19			
CuCl	313–413	10			
GaAs	40	−0.5			
	491	0.00			
	78–290	3.64			
	291–560	5.74			
	560–680	7.44			
GaP	....	5.3			
GaSb	....	6.9			
Ge	40	0.07			
	50	0.20			
	60	0.39			
	70	0.67			
	80	1.05			
	90	1.54			
	100	2.20			
	110	2.79			
	120	3.25			
	130	3.62			
	140	3.91			
	150	4.12			
	160	4.29			
	170	4.45			
	180	4.58			
	190	4.70			

TABLE 61 Thermal Expansion (Continued)

Material	T, K	A	B	C	Remarks
Ge	200	4.82			
	210	4.93			
	220	5.03			
	230	5.13			
	240	5.23			
	250	5.32			
	260	5.42			
	270	5.50			
	280	5.59			
	290	5.67			
	300	5.75			
InP	. . .	4.5			
InSb	10	−0.06			
	30	−1.72			
	50	−0.33			
	70	0.89			
	100	2.76			
	160	4.08			
	190	4.35			
	220	4.58			
	253	4.78			
	270	4.89			
	280	4.95			
	300	5.04			
Irtran 1	298–573	11.0			
Irtran 2	298–573	6.9			
Irtran 3	298–573	20.0			
Irtran 4	298–573	7.7			
Irtran 5	298–573	12.0			
KBr	113–573	27.6	4.1		
	318–953	37.99	1.263	5.256	
	293–333	43			
KCl	293–333	36			
KDP	123–293	21.6			
KI	313	42.6			
KRS-5	223–293	61			
	293–373	58			
KRS-6	223	55			
	233	56			
	253	56			
	273	55			
	293	51			
	313	48			
	333	49			
	353	51			
	373	53			
	393	56			
	413	57			
	433	58			
	453	59			
	473	59			
LiF	273–373	37			
	123–273	31.95	5.049	−4.070	
	320–1067	33.17	3.075	2.399	
MgO	323–988	10.98			
	300	11.2			
	481	12.3			
	659	13.5			
	825	14.6			
	964	15.4			
	1061	16.0			
	293–1000	13.8			

TABLE 61 Thermal Expansion (Continued)

Material	*T*, K	*A*	*B*	*C*	Remarks
$MgO \cdot Al_2O_3$	313	5.9			
Muscovite	324	8.1	7.5		
NaCl	223–473	44			
NaF	Room	36			
$NaNO_3$	323	12			
	323	11			
$PbCl_2$	293–393	31			
PbSe	303	7.65			
	313	10.55			
	323	12.92			
	333	14.55			
	343	15.63			
	353	16.41			
	363	16.97			
	373	17.37			
	383	17.66			
	393	17.89			
	403	18.09			
	413	18.97			
	423	18.43			
	433	18.57			
	433	18.57			
	453	18.79			
	473	18.94			
	493	19.06			
	513	19.16			
	533	19.26			
	553	19.34			
	573	19.40			
	593	19.46			
	613	19.50			
PbTe	303	9.02			
	313	12.08			
	323	14.30			
	333	15.57			
	343	15.38			
	353	16.42			
	363	17.31			
	373	17.70			
	383	18.04			
	393	18.33			
	403	18.57			
	413	18.78			
	423	18.97			
	433	19.42			
	453	19.62			
	473	19.74			
	493	19.79			
	513	19.80			
	533	19.80			
	553	19.80			
	573	19.80			
	593	19.80			
	613	19.80			
Quartz, crystal	173–310	7.067	2.11		$\parallel$
	283–607				$\parallel$
	273–633	7.067	1.6742		$\parallel$
	633–723	25.80		20.163	$\parallel$
	273–353	7.97			$\parallel$
	273–353	13.37			$\perp$
Fused	293–1173	0.5			
Se	195–292	20.3			

TABLE 61 Thermal Expansion (Continued)

Material	T, K	A	B	C	Remarks
Se	195–273	42.7			
	273–294	48.7			
	293–373	22.9			
	478	45.2			
Se(As)	....	34			
Si	50–100	2.5			[111]
		2.7			[110]
	100–200	3.1			[111]
		3.5			[110]
	200–300	3.9			[111]
		3.8			[110]
	400–500	4.3			[111]
		4.1			[110]
	500–600	4.7			[111]
		4.4			[110]
	600–700	5.0			[111]
		4.5			[110]
	25–900	3.0024	0.1544	0.20576	[111]
$SrTiO_3$	....	9.4			
Te	313	16.75			
	293	−1.6			∥
	293	27.2			⊥
	293–333	−1.7			∥
	293–333	27.0			⊥
TiO_2	313	9.19	2.25		∥
	313	7.14	1.10		⊥
TlCl	293–333	53			

TABLE 62 Hardness

Material	Hardness, Knoop no.	Direction	Indenter load, g	Remarks
AgCl	9.5	.....	200	
Arsenic trisulfide glass, As_2S_3	109	.....	100	
BaF_2	82	.....	500	
$BaTiO_3$	.....	.....	...	200–580 Vickers, single crystal
CaF_2	158.3	[110]	500	
	158.3	[100]	500	
Calcite, $CaCO_3$	.....	.....	...	3 Mohs' no.
CdTe	.....	.....	...	43.5 Vickers
CsBr	19.5	.....	200	
KBr	5.9	[110]	200	
	7.0	[100]	200	
KCl	7.2	[110]	200	
	9.3	[100]	200	
KRS-5 (thallium bromide-iodide)	40.2	.....	200	Machined surface
	39.8	[100]	500	
	33.2	[110]	500	
KRS-6 (thallium bromide-chloride)	29.9	[110]	500	
	38.5	[100]	500	
LiF	102–113	.....	600	Vacuum-grown
MgO	692	⊥ to cleavage planes	600	
NaCl	15.2	[110]	200	
	18.2	[100]	200	
$NaNO_3$	19.2	⊥ to cleavage planes	200	
Sapphire, Al_2O_3	1370	Random	1,000	
Si	1150	....	.....	7 Mohs' no.
SiO_2:				
Crystal quartz	741	⊥ to *x*- and *y*-cut faces	500	
Fused silica	461	....	200	
Spinel, $MgO \cdot 3.5Al_2O_3$	1140	Random	1,000	
TiO_2	879	Random	500	
TlBr	11.9	[110]	500	
	11.9	[100]	500	
TlCl	12.8	[110]	500	
	12.8	[100]	500	

TABLE 63 Permittivity (Dielectric Constant)

Material	ϵ_r	f, Hz	T, K
AgCl	12.3	10^6	293
Al_2O_3	10.55	10^2–10^8	298
	8.6	10^2–10^{10}	298
As_2S_3	8.1	10^3–10^6	
BaF_2	7.33	2×10^6	
$CaTiO_3$	140	1.5×10^6	294
CuBr	8.0	3×10^6	293
CuCl	10.0	5×10^5	293
KBr	4.90	10^2–10^{10}	298
	4.97	10^2–10^{10}	360
LiF	9.00	10^2–10^{10}	298
	9.11	10^2–10^{10}	353
MgO	9.65	10^2–10^8	298
Muscovite	5.4	10^2–3×10^9	299
NaCl	5.90	10^2–10^{10}	298
	6.35–5.97	10^2–10^{10}	358
PbCl	33.5	5×10^5	293
PbF	3.6	10^6	
$PbMoO_4$	26.8	4×10^8	
SiO_2:			
Crystal	4.34	3×10^7	290–295
	4.27	3×10^7	290–295
Fused	378	10^{12}–10^{10}	300
SrF_2	7.69	2×10^6	
$SrTiO_3$	306	1×10^5	25
TiO_2	200–160	10^4–10^7	298
	87.3–85.8	10^2–10^7	298

TABLE 64 Solubility, Molecular Weight, and Specific Gravity

Material	Solubility g/100 g H_2O	Solubility T, K	Mol. wt.	Specific gravity Value	Specific gravity T, K
ADP	22.7	273	115.04	1.803	293
AgCl	8.9×10^{-5}	283	143.34	5.589	273
				5.56	293
Al_2O_3	Insol.	...	101.94	3.98 (3.95–4.10 for natural)	
As_2S_3	Insol.	...	364.02	3.198	
BaF_2	0.17	...	175.36	4.83	293
$BaTiO_3$	....	...	232.96	5.90 single crystal	
$CaCO_3$	1.4×10^{-3}	298	100.09	2.7102	293
	1.8×10^{-3}	398			
CaF_2	0.0017[a]	299	78.08	3.179	298
$CaTiO_3$	......	...	135.98	4.10	293
CdF_2	......	...	150.41	6.382 ± 0.006	293
CdS	Insol.	...	144.48	4.82	293
CdTe	Probably insol.[b]	...	240.02	5.854	
CsBr	124.3[c]	298	212.83	4.44	293
CsI	.....	...	259.83	4.526	
CuBr	Insol.	...	143.46	4.718	293
CuCl	0.0062	293	99.00	3.53	293
GaAs	Insol.	...	144.63	5.3161 ± 0.0002	298
GaP	.....	...	100.70		
GaSb	Insol.	...	191.48		
Ge	Insol.[d]	...	72.60	5.327	298
InAs	Insol.	...	189.73	5.66	
InP	.....	...	145.80	4.8	
InSb	Insol.	...	237	5.78	
Irtran 1	Insol.	...	62.32	3.18	
Irtran 2	Insol.	...	97.45	4.09	
Irtran 3	Insol.	...	78.08	3.18	
Irtran 4	Insol.	...	144.34	5.27	
Irtran 5	0.00062	...	40.32	3.58	
KBr	53.48[e]	273	119.01	2.75	298
	102	373			
KCl	34.7	293	74.55	1.984	293
KDP	.....	...	136.09	2.338	
KI	127.5	273	166.02	3.13	
KRS-5	0.05	Room	......	7.371	289
KRS-6	0.32[f]	293	......	7.192	289
LiF	0.27	291	25.94	2.639	298
MgO	Insol.[g]	...	40.32	3.567	298
$MgO \cdot Al_2O_3$	Insol.[h]	...	356.74	3.61	
Muscovite	Insol.	...	......	2.8–2.9	
NaCl	35.7	273	58.45	2.164	293
	39.12[i]	373			
NaF	4.22	291	42.00	2.79	293
				2.558	314
$NaNO_3$	73	273	85.01	2.261	
	180	373			
$PbCl_2$	0.673	273	278.12	5.85	293
	0.99	283			
	3.34	373			
PbF_2	0.064	293	245.21	8.24	293
				7.763 ± 0.001	291
$PbMoO_4$	Insol.	...	367.16	6.03–7.01	293
PbS	Insol.	...	239.28	7.5	
PbSe	Insol.	...	286.17	8.10	288
PbTe	Insol.	...	334.82	8.16	
Quartz, crystal	Insol.	...	60.06	2.648	298
Fused	Insol.[j]	...	60.06	2.202	293

TABLE 64 Solubility, Molecular Weight, and Specific Gravity (Continued)

Material	Solubility g/100 g H_2O	Solubility T, K	Mol. wt.	Specific gravity Value	Specific gravity T, K
Se	Insol.	...	......	4.82	
				4.26	
SeAs	Insol.	...	......	Not applicable	
SrF_2	0.011	273	125.63	4.24	293
	0.012	27			
$SrTiO_3$	Insol.	...	183.53	5.122	293
Te	Insol.	...	......	6.24	293
TiO_2	Insol.[k]	...	79.90	4.25	
				4.18–5.13 for rutile	
TlBr	0.05	298	284.31	7.453	298
	0.25	341			
TlCl	0.32	293	238.85	7.018	298

[a] Soluble in ammonia salt solutions. [b] No deterioration in transmittance and no observable corrosion of the sample after it had been exposed to 95% relative humidity at 322 K for 360 h. [c] Soluble in acid. [d] Soluble in hot sulfuric acid and aqua regia; etched in CP-4. [e] Slightly hygroscopic. [f] The solubility of a microcrystal is that of the more soluble component, in this case, TlCl. [g] Soluble in acids and ammonia salts. [h] Not attacked by common acid or NaOH; slightly etched by hydrofluoric acid after 65 days at temperature. [i] Soluble in glycerin, slightly soluble in alcohol and liquid ammonia, insoluble in hydrochloric acid. [j] Very slightly soluble in alkalis; soluble in hydrofluoric acid. [k] Soluble in concentrated sulfuric acid.

TABLE 65 Direct Transmittance of Carbon-Black Deposits, Corrected for Substrate Transmittance (Harris, 1967)

	Wavelength, μm 2	6	8.8	12	15	18	24
Acetylene black, 0.0075 cm thick	....	....	0.25				
Sperm-candle soot	0.07	0.38	0.46	0.58			
Camphor soot	.....	0.14	0.26	0.36	0.46	0.56	0.65
Soot, 1.8×10^{-4} g/cm^2	0.005	0.16	0.28	0.38	0.48	0.59	0.70
Lampblack paint, g/cm^2:							
$2.6\ (\pm 0.5) \times 10^{-4}$	.....	~0.002	0.024	0.056	0.089	0.14	0.197
$5.2\ (\pm 0.3) \times 10^{-4}$	.....	.....	.....	.....	0.01	0.022	0.097
-10×10^{-4}	.....	.....	.....	.....	~0.002		
Parsons optical-black lacquer, g/cm^3:							
Topcoat alone							
3.7×10^{-3}	.....	.....	.....	.....	0.0013	0.002	0.007
7.3×10^{-3}	.....	.....	.....	.....	......	.....	~0.002
Undercoat and topcoat	.....	.....	.....	.....	......	.....	<0.002

TABLE 66 Diffuse Hemispherical Reflectance of Carbon Blacks for the Ultraviolet and Visible Regions (Harris, 1967)

	Wavelength, μm					
	0.254	0.366	0.436	0.546	0.600	0.700
Acetylene black:						
Goniophotometer	0.009	0.012	0.010	0.010		
Integrating sphere	.....	.....	0.012	0.012	0.011	0.010
Hemisphere	.....	.....	.....	.....	0.006	
Lampblack paint:						
Goniophotometer	0.045	0.045	0.045	0.042		
Integrating sphere	.....	.....	0.045	0.042	0.041	0.039
Hemisphere	.....	0.045	0.044	0.049	0.033	
Lampblack paint plus soot (sperm candle):						
Goniophotometer	0.014	0.015	0.015	0.016		
Integrating sphere	.....	.....	0.017	0.016	0.016	0.016
Parsons optical-black lacquer:						
Integrating sphere	.....	.....	0.014	0.013	0.012	0.012

TABLE 67 Diffuse Hemispherical Reflectance of Carbon Blacks for the Infrared (Harris, 1967)

	Wavelength, μm						
	0.95	2.4	4.4	8.8	12	18	24
Acetylene black:							
Hemisphere	0.004	.....	0.007	0.012	.....		0.030
Goniophotometer	.....	.....	.....	<0.005			
Camphor soot, hemisphere	0.014	.....	0.010	0.017	.....		0.057
Sperm-candle soot, hemisphere	0.010	.....	0.009	0.013	.....		0.041
Lampblack paint, hemisphere	0.047	0.047	0.032	0.042	0.056		0.044
Parsons optical-black lacquer:							
goniophotometer	.....	.....	.....	0.022			

TABLE 68 Infrared Absorptance of Lampblack Paint, Parsons Optical-Black Lacquer, and Gold Black on Metal Receivers (Harris, 1967)

	Wavelength, μm						
	1	2	6	8.8	12	15	18
Lampblack paint:							
Calculated from reflectance and transmittance measurements	0.96	0.96	0.96	0.96	0.94		
Response† compared with cavity detector	1.00	....	0.975	0.925	0.895	0.89	0.88
Parsons optical-black lacquer:							
Calculated from reflectance and transmittance measurements	0.99	0.99	.....	0.98	.....	....	0.98
Response† compared with cavity detector	1.00	....	.....	....	.....	....	0.99
Gold black:							
Calculated from reflectance and transmittance measurements	0.97	0.978	0.982	0.98	0.978	0.95	0.935
Response† compared with cavity detector	1.00	.....	0.99	0.98	0.972	0.970	0.96

† The responses were normalized to unity at 1 μm.

TABLE 69 Qualitative Evaluation of Glasses from Nonoxide Chalcogenide Systems

System	Softening point, °C	Refractive index at ~5 μm	Absorption† 3–5 μm	Absorption† 8–14 μm	Ref.‡
As-S	200	2.4	W	M	1–3
As-Se	200	2.8	—	W	4, 5
Ge-S	420	2.3	W	S	6, 7
Ge-Se	360	2.6	—	S	6, 8
Si-P-Te	180	3.4	—	M	6
Si-Sb-Se	270	3.3	S	S	6
Si-Sb-S	280	...	S	S	6
Ge-P-Se	420	2.4–2.6	M	S	6
Ge-P-S	520	2.0–2.3	W	S	6
Si-As-Te	475	2.9–3.1	—	M	9
Ge-As-Te	270	3.5	—	—	10, 8
Ge-P-Te	380	3.5	—	—	10, 8
Ge-As-Se	450	2.9	—	—	11, 5
Ge-Sb-Se	325	2.6	—	—	12
As-Se-Te	200	2.6–3.1	—	M	13
As-S-Se-Te	195	2.1–2.9	—	M	13
Si-Ge-As-Te	325	3.1	—	M	10

† — = no appreciable absorption; W = weak absorption; M = medium absorption; S = strong absorption.

‡ 1 = Frerichs (1950); 2 = Frerichs (1953); 3 = Fraser and Jerger (1953); 4 = Black et al. (1957); 5 = Billian and Jerger (1963); 6 = Hilton et al. (1964); 7 = Nielson (1962); 8 = Savage and Nielson (1965); 9 = Hilton and Brau (1963); 10 = Hilton et al. (1966); 11 = "The Structure of Glass" (1960); 12 = Patterson and Brau (1966); 13 = Jerger and Sherwood (1964).

TABLE 70 Quantitative Evaluation of Glasses from Nonoxide Chalcogenide Systems

Chemical composition	Transmission range, μm	Refractive index at 5 μm	Thermal change in refractive index at 5 μm $\frac{\Delta N}{\Delta T}$, 10⁶/°C	Softening point, °C	Thermal coefficient of expansion $\frac{\Delta l}{l}$, 10⁶/°C	Hardness, Knoop	Young's modulus, 10⁶ lb/in²	Ref.†
$Si_{25}As_{25}Te_{50}$	2-9	2.93	+110	317	13	167	....	1
$Ge_{10}As_{20}Te_{70}$	2-20	3.55	....	178	18	111	....	1
$Si_{15}Ge_{10}As_{25}Te_{50}$	2-12.5	3.06	+169	320	10	179	....	1
$Ge_{30}P_{10}S_{60}$	2-8	2.15	....	520	15	185	....	1
$Ge_{40}S_{60}$	0.9-12	2.30	....	420	14	179	....	1
$Ge_{28}Sb_{12}Se_{60}$	1-15	2.62	+ 80	326	15	150	4.2†	2, 3
$Ge_{33}As_{12}Se_{55}$	1-16	2.49	....	474	...	171	4.5†	2
$As_{50}S_{20}Se_{60}$	1-13	2.53	....	218	20	121	2.1	4
$As_{50}S_{20}Se_{20}Te_{10}$	1-13	2.51	....	195	27	94	1.5	4
$As_{35}S_{10}Se_{35}Te_{20}$	1-12	2.70	....	176	25	106	2.5	4
$As_{38.7}Se_{61.3}$	1-15	2.79	....	202	19	114	2.5	5

† 1 = Hilton et al. (1966); 2 = Jerger (1969); 3 = Patterson and Brau (1966); 4 = Jerger and Sherwood (1964); 5 = Billian and Jerger (1963).

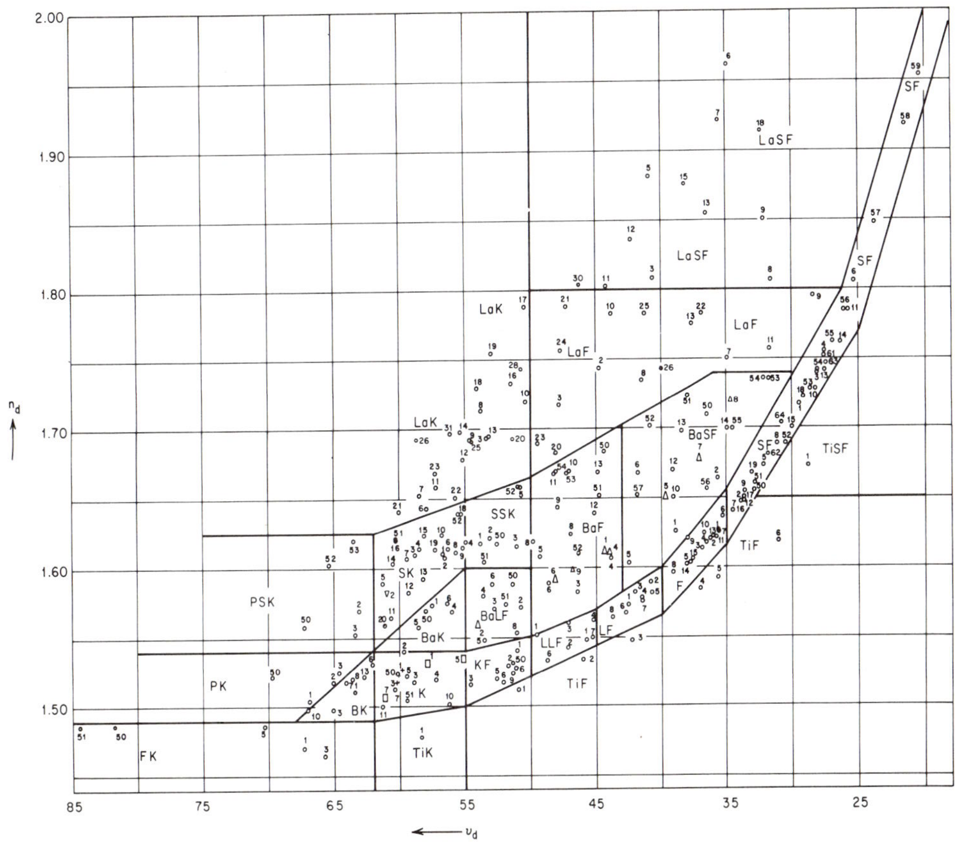

Fig. 211 Refractive index and dispersive value of Schott optical glasses. Squares = zinc crown (ZK), crosses = barium light crown (BalK), open circles = short flint (KzF), triangles = special short flint (KzFS), inverted triangles = special long crowns, solid circles = all other glasses. Each area represents the same glass type as the comparable area in Fig. 210 although the abbreviations differ. [*Fish-Schurman Corporation.*]

TABLE 71 NATO Optical Glasses

Six-digit code	Schott name	N_d	V_d	P_{Fd}	Stain	Bubble	Density	Internal transmittance	1969 cost per lb, $
511604	K-7	1.51112	60.41	0.6950	0	0–1	2.53	0.986	4.25
517642	BK-7	1.51680	64.17	0.6928	0	0	2.51	0.972	4.70
522595	K-5	1.52249	59.48	0.6956	0	2	2.63	0.98	4.10
523602	K-50	1.52257	60.18	0.6956	0	0	2.62	0.986	4.50
527511	KzF-6	1.52682	51.13	0.6969	2	1	2.54	0.88	8.30
529517	KzF-2	1.52944	51.68	0.6970	2	0–1	2.55	0.91	7.65
548458	LLF-1	1.54814	45.75	0.7020	0	0–1	2.94	0.985	5.75
573575	BaK-1	1.57250	57.55	0.6966	1	1	3.19	0.97	5.75
581409	LF-5	1.58144	40.85	0.7040	0	0–1	3.22	0.988	4.75
613443	KzFS-N4	1.61340	44.30	0.6997	2	0–1	3.20	0.65	22.40
613574	SK-19	1.61342	57.37	0.6967	1	0–1	3.56	0.91	8.55
613586	SK-4	1.61722	58.63	0.6966	2	0–1	3.57	0.92	8.35
618551	SSK-4	1.61765	55.14	0.6982	1	0–1	3.63	0.94	7.50
620364	F-2	1.62004	36.37	0.7062	0	0	3.61	0.90	3.85
620603	SK-16	1.62041	60.33	0.6953	4	0–1	3.58	0.88	9.00
648339	SF-2	1.64769	33.85	0.7076	0	0–1	3.86	0.84	5.35
650392	BaSF-10	1.65016	39.15	0.7057	2–3	2	3.91	0.80	7.90
652585	LaK-N7	1.65160	58.52	0.6961	2	1	3.84	0.845	20.40
653397	KzFS-5	1.65332	39.71	0.7013	4–5	0–1	3.44	0.14	23.75
670471	BaF-N10	1.67003	47.11	0.7017	3	0–1	3.76	0.29	8.75
691547	LaK-N9	1.69100	54.71	0.6959	1	1	3.55	0.745	30.55
699301	SF-15	1.69895	30.07	0.7102	0	1	4.06	0.01	8.55
702410	BaSF-52	1.70181	41.01	0.7030	5	1	3.96	0.43	17.85
717480	LaF-N3	1.71700	47.98	0.7007	5	1	4.34	0.62	20.40
720504	LaK-10	1.72000	50.41	0.6974	2	1	3.81	0.755	36.25
744448	LaF-N2	1.74400	44.77	0.7023	4–5	1	4.46	0.47	24.50
755276	SF-4	1.75520	27.58	0.7114	2	1	4.78	0.31	10.25
805254	SF-6	1.80518	25.43	0.7129	3–4	0–1	5.18	0.03	11.95

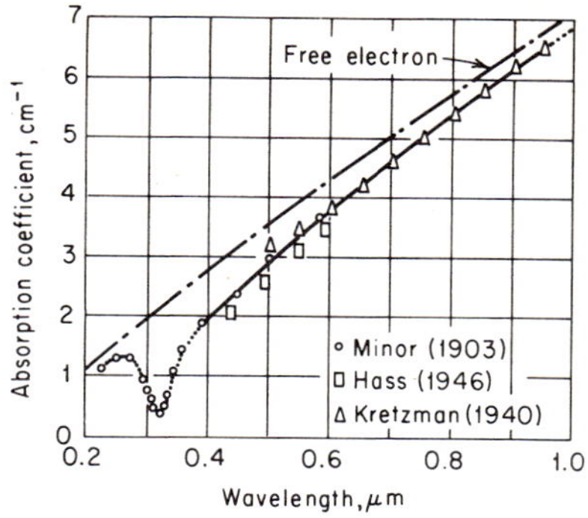

Fig. 212 Absorption coefficient of silver. For identification of references cited on the figure, see the original source. [*From Schulz* (1954).]

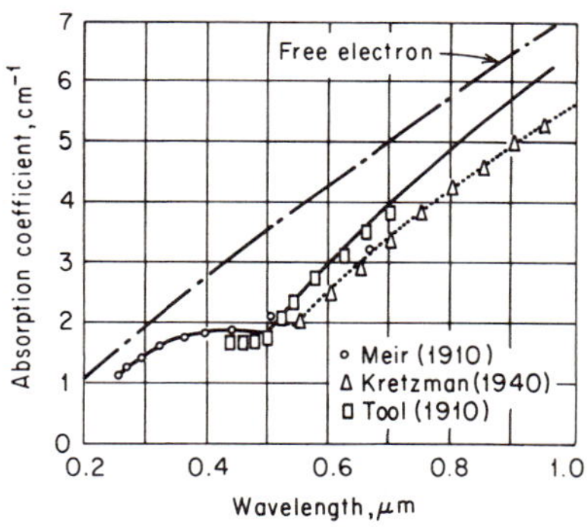

Fig. 213 Absorption coefficient of gold. For identification of references cited on the figure, see the original source. [*From Schulz* (1954).]

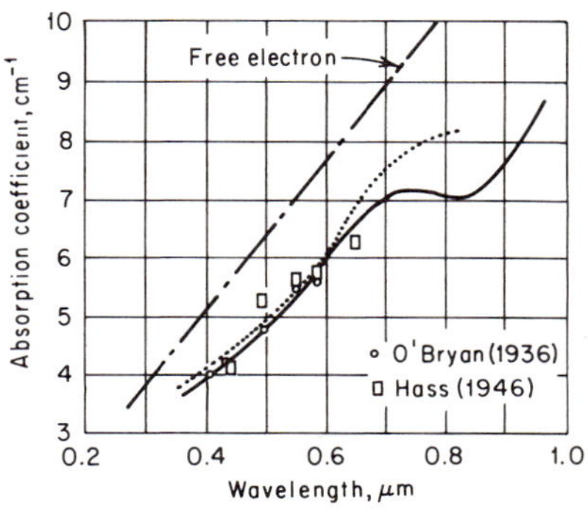

Fig. 214 Absorption coefficient of aluminum. For identification of references cited on the figure, see the original source. [*From Schulz* (1954).]

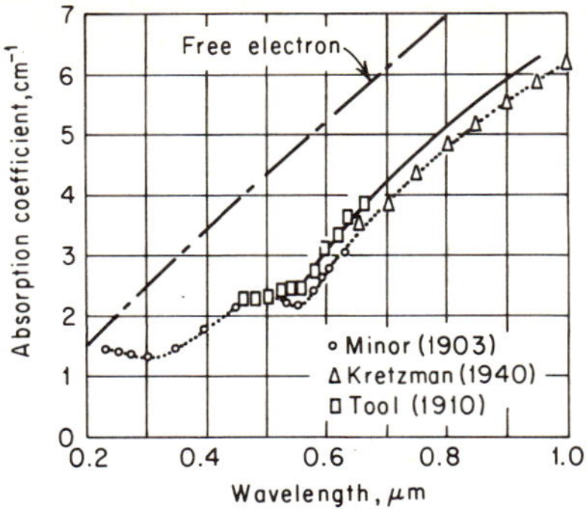

Fig. 215 Absorption coefficient of copper. For identification of references cited on the figure, see the original source. [*From Schulz* (1954).]

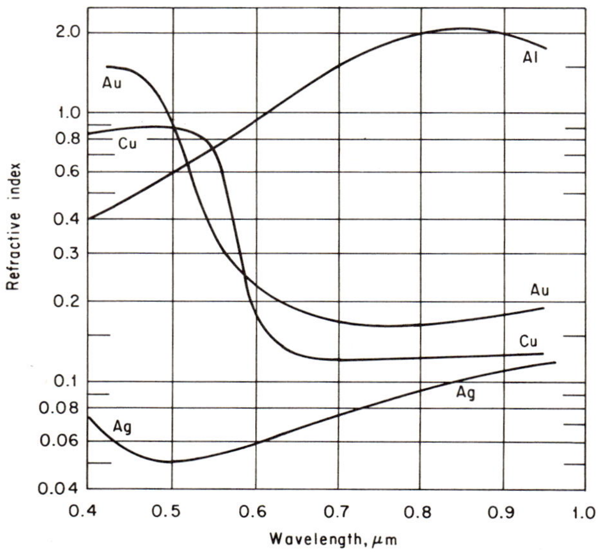

Fig. 216 Refractive index of silver, gold, copper, and aluminum. It is interesting to compare the dispersion shown here in the index values with that of the absorption coefficient values in Figs. 212 to 215. [*From Schulz and Tangherlini* (1954).]

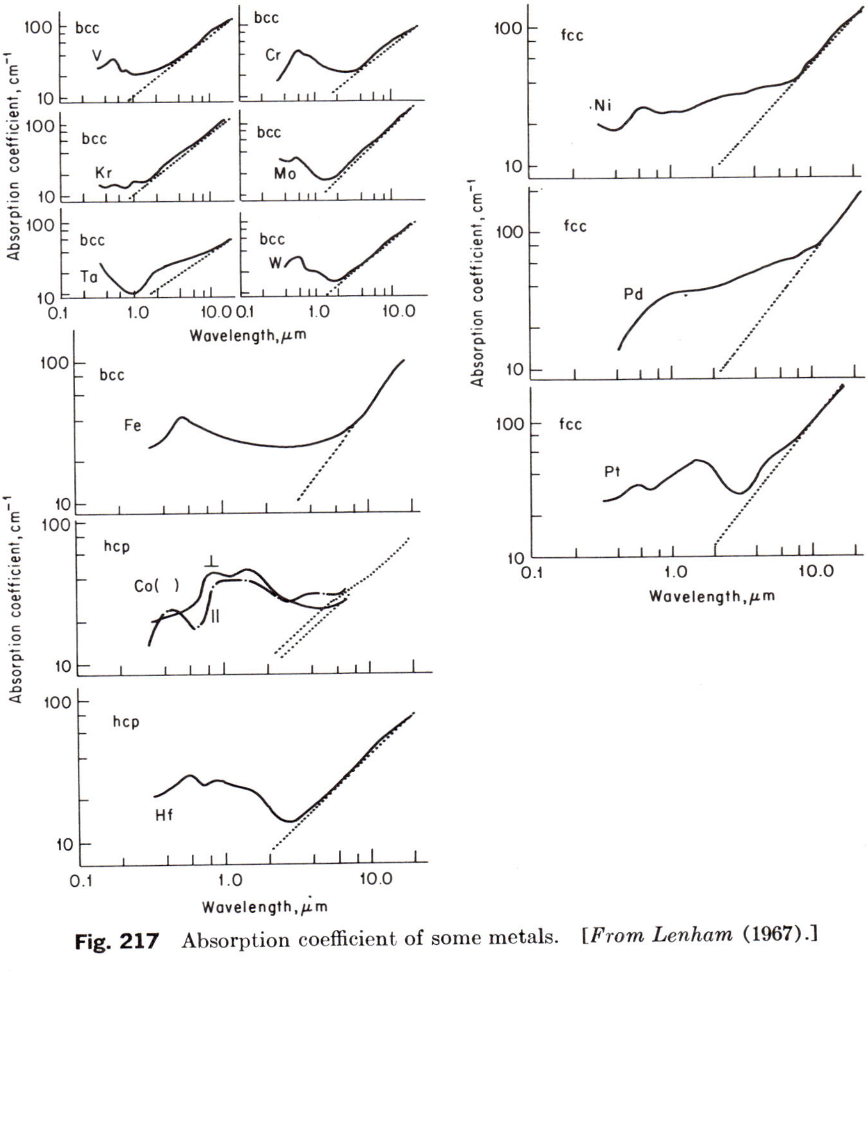

Fig. 217 Absorption coefficient of some metals. [*From Lenham* (1967).]

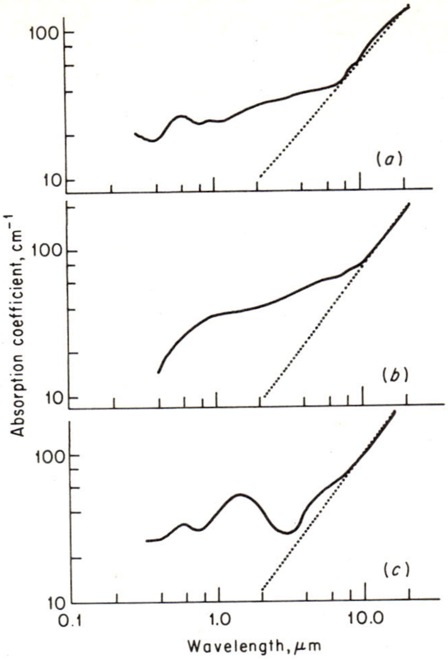

Fig. 218 Absorption coefficient vs. wavelength in micrometers for face-centered cubic crystals of (*a*) nickel, (*b*), palladium, (*c*), platinum. The dotted lines indicate the intraband contribution adopted. [*From Lenham* (1967).]

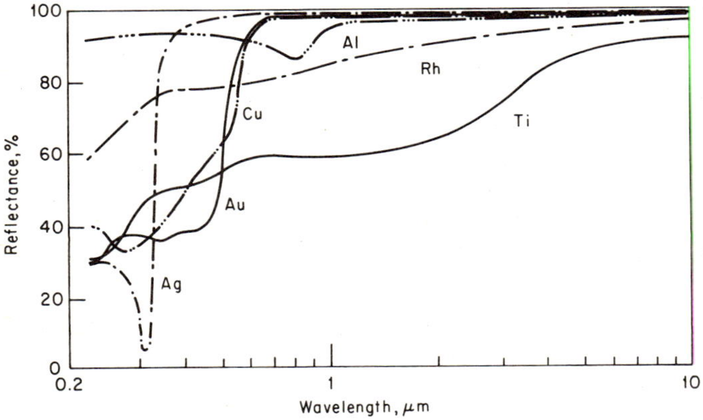

Fig. 219 Reflectance of various films of silver, gold, aluminum, copper, rhodium, and titanium. [*From Hass and Turner* (1957).]

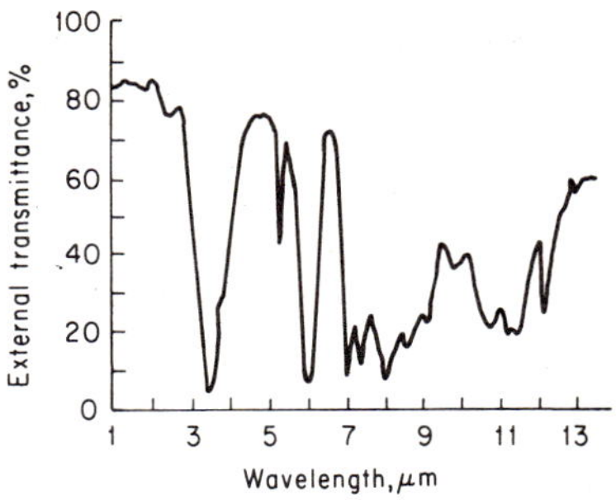

Fig. 220 Transmission of 12-μm-thick balsam cement. [*From Ishmuratova and Sergeyev* (1967).]

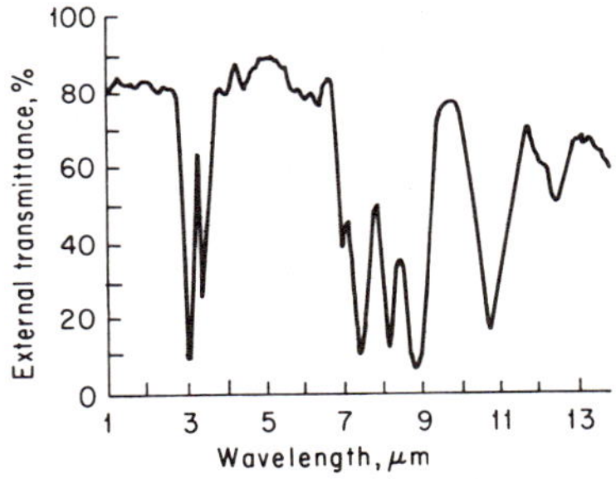

Fig. 221 Transmission of 15-μm-thick balsamine cement. [*From Ishmuratova and Sergeyev* (1967).]

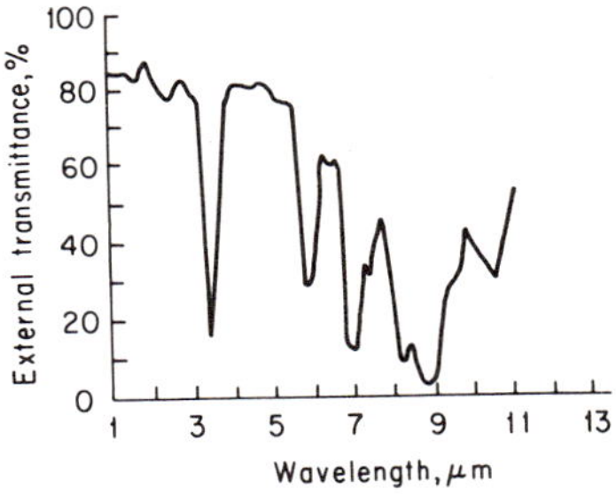

Fig. 222 Transmission of 14-μm-thick acrylic cement. [*From Ishmuratova and Sergeyev* (1967).]

TABLE 72 Absorption Coefficient and Index of Refraction for Silver, Gold, Copper, and Aluminum

λ, μm	Ag	Au	Cu	Al
Absorption coefficient, cm^{-1} (Schulz, 1954)				
0.40	1.93	....	....	3.92
0.45	2.42	1.88	2.20	4.32
0.50	2.87	1.84	2.42	4.80
0.55	3.32	2.37	2.42	5.32
0.60	3.75	2.97	3.07	6.00
0.65	4.20	3.50	3.65	6.60
0.70	4.62	3.97	4.17	7.00
0.75	5.05	4.42	4.62	7.12
0.80	5.45	4.84	5.07	7.05
0.85	5.85	5.30	5.47	7.15
0.90	6.22	5.72	5.86	7.70
0.95	6.56	6.10	6.22	8.50
Index of Refraction† (Schulz and Tangherlini, 1954)				
0.40	0.075	1.45	0.85	0.40
0.45	0.055	1.40	0.87	0.49
0.50	0.050	0.84	0.88	0.62
0.55	0.055	0.34	0.72	0.76
0.60	0.060	0.23	0.17	0.97
0.65	0.070	0.19	0.13	1.24
0.70	0.075	0.17	0.12	1.55
0.75	0.080	0.16	0.12	1.80
0.80	0.090	0.16	0.12	1.99
0.85	0.100	0.17	0.12	2.08
0.90	0.105	0.18	0.13	1.96
0.95	0.110	0.19	0.13	1.75

† These values are based on glass-metal reflectivities measured with a Beckman spectrophotometer on samples that had been annealed or aged and that satisfied the condition $R_s^2 = R_p$. The number of samples (a sample consists of four metallized prisms) for each metal was four for silver and aluminum and two for gold and copper.

TABLE 73 Physicomechanical Properties of Organosilicone Resin Film (Ishmuratova and Sergeyev, 1967)

Type of test	Results obtained
Mechanical strength	35 kg/cm^2
Resistance to freezing	−60°C
Heat stability	+150°C
Cyclical thermal shock at +50 and −60°C (in 2 h)	Up to 5 cycles without damaging the layer

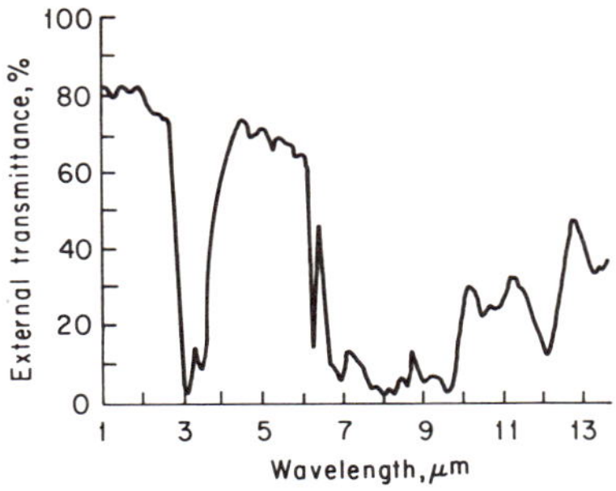

Fig. 223 Transmission of 15-μm-thick OK-50 cement. [*From Ishmuratova and Sergeyev* (1967).]

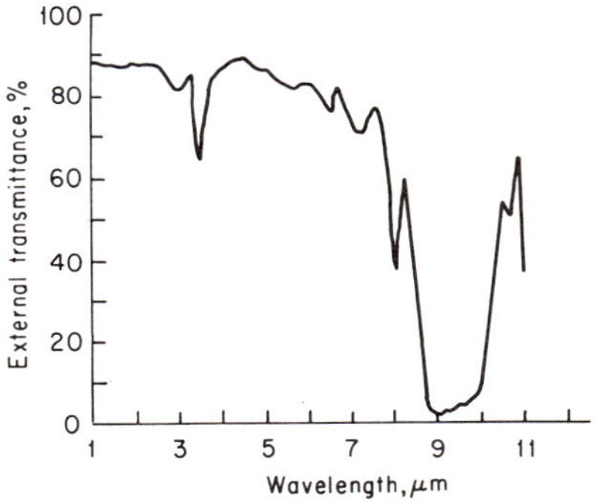

Fig. 224 Transmission of 13-μm-thick organosilicone resin. [*From Ishmuratova and Sergeyev* (1967).]

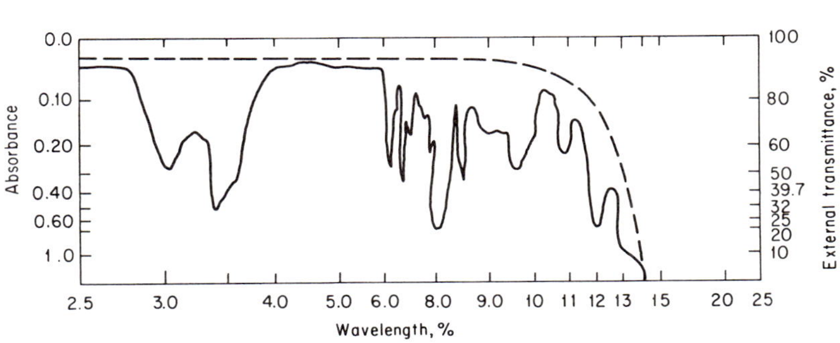

Fig. 225 Absorption of barium fluoride with (*solid curve*) and without (*dashed curve*) Araldite 6010 epoxy resin. Two pieces of 3-mm-thick barium fluoride were cemented together with Araldite 6010 epoxy resin and cured with 13 parts by weight of hardener HN 951/100 parts of resin. Care was taken to remove all air bubbles and to make the epoxy layer thin. [*From Packard* (1969).]

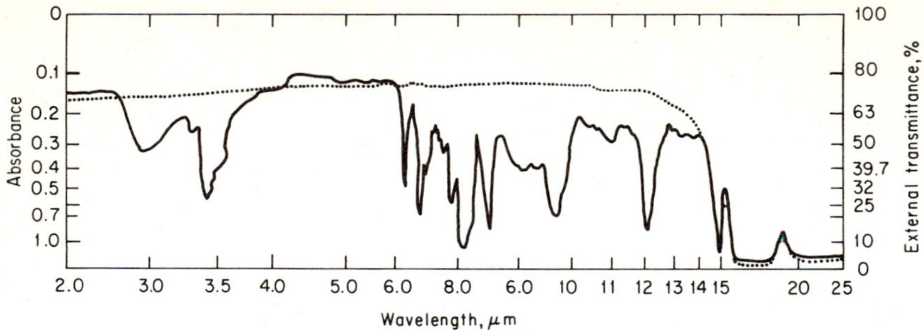

Fig. 226 Absorption of Irtran 2 with (*solid curve*) and without (*dotted curve*) Araldite 6010 epoxy resin. An optically flat polished window of Kodak Irtran 2 (hot-pressed polycrystalline zinc sulfide, 1 mm thick) was measured for transmission properties. It was then coated with a 1-mil-thick layer of Araldite 6010 epoxy cured at room temperature with 13 parts of hardener HN 951/100 parts resin by weight. The transmission properties of the window plus epoxy are shown in the figure. A calculation of the absorption coefficient of the epoxy cement at 10.6 μm gives $\alpha_c = 101\ \text{cm}^{-1}$, indicating a 1-mil-thick film transmittance of about 77 percent at this wavelength. [*From Packard* (1969).]

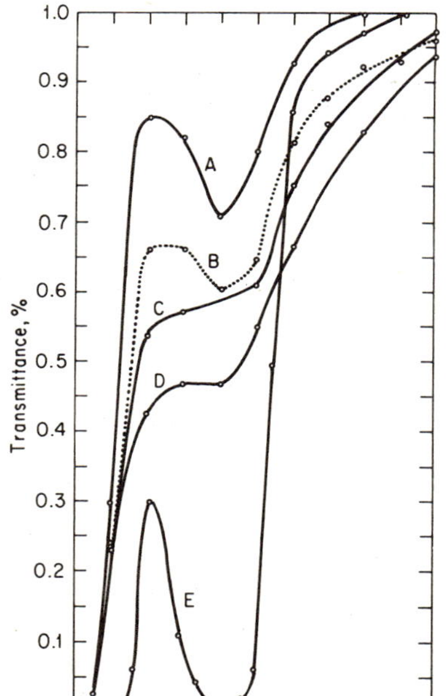

Fig. 227 Ultraviolet transmittances for silicone elastomers 7 months after irradiation with 2 Mrd of 10-MeV electrons, except for Dow-Corning 182, which had no irradiation. Curve *A*: Dow-Corning 93-500, thickness 0.066 mm; curve *B*: Dow-Corning 182, thickness 0.061 mm; curve *C*: Dow-Corning 184, thickness 0.064 mm; curve *D*: Dow-Corning 488, thickness 0.056 mm; curve *E*: General Electric 655, thickness 0.064 mm. [*From Pellicori* (1970).] Only one of the samples, Dow-Corning 93-500, did not outgas appreciably after 2 days at 90°C and 1 μTorr. The others lost up to 1 percent of their weight. The disadvantages of silicone elastomers as optical bonding agents are their low refractive index (1.43 green light) and low adhesive strength. The low adhesive strength is not a significant problem where weight support is not required. In thin layers, the adhesion is great enough, for example, to prevent easy separation of two lenses by hand.

TABLE 74 Properties of Selected Mirror-Blank Materials

Material	Density ρ, g/cm^3	Young's modulus E, g/(cm)(s^2) × 10^{-12}	$\frac{\rho}{E}$, s^2/cm^2 × 10^{12}	Expansion α, K^{-1} × 10^6	Conductivity k, cal/(cm)(s)(k)	$\frac{\alpha}{k}$, cm · s/cal × 10^6
Alumina	3.85	3.5	1.1	6.0	0.041	150
Aluminum	2.70	0.69	3.9	24	0.53	45
Beryllia.	3.03	. . . .	. . . .	9.5	0.42	23
Beryllium	1.82	2.8	0.65	12	0.38	32
Cer-Vit	2.50	0.92	2.7	0.05	0.0040	13
Fused silica:						
7940	2.20	0.73	3.0	0.55	0.0033	170
ULE 7971	2.21	0.68	3.2	0.035	0.0031	11
Magnesium	1.74	0.45	3.9	26	0.38	68
Pyrex 7740	2.35	0.68	3.5	3.2	0.0027	1,200

REFERENCES

Acquista, N., and E. K. Plyler (1953): *J. Opt. Soc. Am.*, vol. 42, p. 432.

——— and ——— (1958): *J. Opt. Soc. Am.*, vol. 48, p. 668.

Acta Electron., vol. 5, p. 364, 1961.

Alonas, P., G. Sherman, C. Whittig, and P. D. Coleman (1969): *Appl. Opt.*, vol. 8, p. 2557.

Appl. Opt., vol. 8, no. 8, pp. 1598–1599, August 1969.

Appl. Opt., vol. 10, no. 11, pp. 2540–2541, November 1971.

Avery, D. G. (1953): *Proc. Phys. Soc.*, vol. B66, p. 134.

——— (1954): *Proc. Phys. Soc.*, vol. B67, p. 2.

Ballard, S. S. (no date): personal communication.

——— (1965): *Jap. J. Appl. Phys.*, vol. 4, suppl. 1, p. 23.

——— and J. S. Browder (1966): *Appl. Opt.*, vol. 5, p. 1873.

———, K. A. McCarthy, and W. C. Wolfe (no date); Optical Materials for Infrared Instrumentation, Univ. Mich. Willow Run Lab. Rep. 2389-11-5.

Barcus, L. C., Lowell Institute of Technology (no date): personal communication.

——— (1958): *Phys. Rev.*, vol. 111, p. 167.

Barth, T. F. W. (1929): *Am. Mineral.*, vol. 14, p. 358.

Bauple, R., A. Gilles, J. Romand, and B. Vodar (1950): *J. Opt. Soc. Am.*, vol. 40, p. 788.

Beals, M. D., and L. Merker (1960): "Materials in Design Engineering," Reinhold, New York.

Beder, E. C., C. D. Bass, and W. L. Shackleford (1971): *Appl. Opt.*, vol. 10, p. 2263.

Bieniewski, T. M., and S. J. Czyzak (1963): *J. Opt. Soc. Am.*, vol. 53, p. 496.

Billian, C. J., and J. Jerger, Jr., Servo Corporation of America (1963): Final Tech. Rep. contract NONR3647(00), Jan. 2.

Bishnevskii, B. H., and H. A. Domanok (1960): *Opt. Spectrosc.*, vol. 5, p. 736.

Black, J., E. M. Conwell, L. Sleigle, and L. W. Spencer (1957): *J. Phys. Chem. Solids*, vol. 2, p. 240.

Bloor, D., T. J. Dean, D. H. Martin, P. A. Mawer, and C. H. Perry (1961): *Proc. R. Soc.*, vol. A260, p. 510.

Briggs, H. B. (1950): *Phys. Rev.*, vol. 77, p. 287.

Buchroeder, R. A., Optical Sciences Center, Univ. of Arizona (no date): personal communication.

Burstein, E., J. J. Oberly, and E. K. Plyler (1948): *Proc. Ind. Acad. Sci.*, vol. 38, p. 388.

Caldwell, R. S. (1958): *Purdue Univ. Phys. Dept. Spec. Rep.*, contract DA 36-039-SC-71131, January.

Calingaert, G., S. D. Heron, and R. Stair (1936): *Trans. Soc. Automot. Eng.*, vol. 39, p. 448.

Cardona, M., W. Paul, and H. Brooks (no date): *Rochester Conf. Solid State Phys.*

———, ———, and ——— (1958): *Adv. Semicond. Sci.: Proc. 3d Int. Conf. Semicond., Univ. Rochester, Aug. 13–22, 1958*, Pergamon, New York, p. 205.

Carmer, D. C., and M. E. Baer (1969*a*): *Appl. Opt.*, vol. 8, no. 8, p. 1598.

——— and ——— (1969*b*): ibid., p. 1599.

——— and ——— (1969*c*): ibid., p. 1600.
——— and ——— (1969*d*): ibid., p. 1601.
Carvallo, A. (1896*a*): *C. R.*, vol. 126, p. 728.
——— (1896*b*): ibid., p. 950.
Clark, M. A., and R. D. Cashman (1952): *Phys. Rev.*, vol. 85, p. 1043.
Coblentz, W. W. (1914*a*): *J. Opt. Soc. Am.*, vol. 4, p. 441.
——— (1914*b*): ibid., p. 443.
Corning Glass Works *Bull.* CJR-2, Nov. 20, 1959.
Corning Glass Works *Bull.* MO-1.1, Aug. 21, 1967.
Czerny, M., and H. Röder (1938): *Ergeb. Exakten Naturwiss.*, vol. 17, p. 70.
Czyzak, S. J., W. M. Baker, R. C. Crane, and J. B. Howe (1957): *J. Opt. Soc. Am.*, vol. 47, p. 240.
Dash, W. C., and R. Newman (1955): *Phys. Rev.*, vol. 99, p. 1151.
Day, G. W., and P. M. Gruzensky (1970): *Appl. Opt.*, vol. 9, p. 2794.
Decamps, E., and A. Hadni (1960): *C. R. Acad. Sci. Paris*, vol. 250, no. 10.
Deshotels, W. J. (1960): *J. Opt. Soc. Am.*, vol. 50, p. 865.
DeVore, J. R. (1951): *J. Opt. Soc. Am.*, vol. 41, p. 417.
DiNardo, R. P., and L. N. Goland (1971): *J. Opt. Soc. Am.*, vol. 61, p. 1321.
Dobrshanskii, G. F., L. M. Belyev, and E. P. Petrov (1964): *Crystallographia*, vol. 9, p. 928.
Dodge, M. J., I. H. Malitson, and A. I. Mahan (1969): *Appl. Opt.*, vol. 8, p. 1705.
Dowell, H. G. (1964): refractive-index data, private communication of F. R. Charvat, Union Carbide Corporation, Linde Division.
Drummond, D. G. (1935): vol. A153, p. 328.
Duncanson, A., and R. W. H. Stevenson (1958): *Proc. Phys. Soc.*, vol. 72, p. 1001.
Eagle Picher Co. (no date): personal communication.
Eastman Kodak (no date): data sheets.
Eastman Kodak Pam. V-71, November 1968.
Edmond, T., A. Anderson, and H. A. Gebbie (1963): *Proc. Phys. Soc.*, vol. 81, p. 378.
Edwards, D. F. (no date): personal communication.
Engelhard Industries, Inc., Amersil Quartz Division (no date): "Optical Fused Quartz."
Feichtner, J. D., R. Johannes, and G. W. Roland (1970): *Appl. Opt.*, vol. 9, p. 1716.
Feldman, A., and D. Horowitz (1969): *J. Opt. Soc. Am.*, vol. 59, p. 1406.
Fraser, W. A., and D. M. Fraser (1959): *J. Opt. Soc. Am.*, vol. 49, p. 497.
——— and J. Jerger (1953): *J. Opt. Soc. Am.*, vol. 43, p. 332.
Frerichs, R. (1950): *Phys. Rev.*, vol. 78, p. 643.
——— (1953): *J. Opt. Soc. Am.*, vol. 43, p. 1153.
Fuller, R. M., D. G. Rathburn, and R. J. Bell (1968): *Appl. Opt.*, vol. 7, p. 1243.
Gampel, L., and F. M. Johnson (1969): *J. Opt. Soc. Am.*, vol. 59, p. 72.
General Electric Company (1957): "Fused Quartz Catalog," Q-6.
General Electric Lamp Glass Department (no date): data sheets.
Gibson, A. F. (1952): *Proc. Phys. Soc.*, vol. 65B, p. 378.
Gifford, J. W. (1902): *Proc. R. Soc. (Lond.)*, vol. 70, p. 329.
Gilleo, M. A. (1953): *Phys. Rev.*, vol. 91, p. 534.
Gillespie, D. T., A. L. Olsen, and L. W. Nichols (1965): *Appl. Opt.*, vol. 4, p. 1488.
Gornsworth, C., British Drug House Ltd., Poole, Dorset, England (1967), personal communication.
Gyulai, Z. (1927): *Z. Phys.*, vol. 46, p. 84.
Hadni, A., C. Janot, and E. Decamps (1959): *Rev. Opt.*, vol. 38, p. 463.
"Handbook of Chemistry and Physics," Chemical Rubber Publishing Co., Cleveland, 1960.
Harris, L. (1967): "The Optical Properties of Metal Blacks and Carbon Blacks," MIT and Eppley Foundation for Research, Cambridge, Mass.
Harting, H. (1948): *Sitzungsber. Dtsch. Akad. Wiss. Ber.*, vol. 4, pp. 1–25.
Hass, G. (1946): *Optik*, vol. 1, p. 2.
——— and A. F. Turner (1957): Coatings for Infrared Optics, p. 143 in M. Auwarter (ed.), "Ergebnisse der Hochvacuumtechnik und der Physik dünner Schichten," Wissenschaftliche Verlagsgesellschaft, Stuttgart.
Hausler, R. L. (1952): thesis, Ohio State University, Columbus.
Heilmeir, G. H. (1964): *Appl. Opt.*, vol. 3, p. 1281.
Henry, L. (1953): *C. R.*, vol. 273, p. 148.
Herzberger. M., and C. D. Salzberg (1962): *J. Opt. Soc. Am.*, vol. 52, p. 420.
Hettner, G., and G. Leisegang (1948): *Optik*, vol. 3, p. 305.
Hilsch, R., and R. W. Pohl (1930): *Z. Phys.*, vol. 59, p. 812.
Hilton, A. R., Texas Instruments, Dallas, Texas (no date): personal communication.
——— (1966): *Appl. Opt.*, vol. 5, p. 1877.

——— and M. J. Brau (1963): *Infrared Phys.*, vol. 3, p. 69.
——— and C. E. Jones (1967): *Appl. Opt.*, vol. 6, p. 1513.
———, ———, and M. J. Brau (1964): *Infrared Phys.*, vol. 4, p. 213.
———, ———, and ——— (1966): *Infrared Phys.*, vol. 6, p. 183.
Hohls, H. W. (1937): *Ann. Phys.*, vol. 29, p. 433.
Houston, T., L. F. Johnson, P. Kisliuk, and D. J. Walsh (1963): *J. Opt. Soc. Am.*, vol. 53, p. 1286.
Iatsenko, A. F. (1957): *Sov. Phys.-Tech. Phys.*, vol. 2, p. 2257.
"International Critical Tables" (1929): McGraw-Hill, New York, vol. 3, p. 26.
Ishmuratova, M. S., and L. V. Sergeyev (1967): *Sov. J. Opt. Tech.*, vol. 34, p. 801.
Jena Glassworks Schott and Gen., Mainz, August 1965.
Jeppesen, M. A., and A. M. Taylor (1966): *J. Opt. Soc. Am.*, vol. 56, p. 451.
Jerger, J., Servo Corporation, Hicksville, N.Y. (1969): personal communication.
——— and R. Sherwood (1964): *Servo Corp. Am. Final Tech. Rep.*, contract NONR4212(00), August.
Johnson, C. J., G. H. Sherman, and R. Weil (1969): *Appl. Opt.*, vol. 8, p. 1667.
Jones, C. E., Texas Instruments, Inc., Dallas, Texas (1969): personal communication.
——— and H. Hafner (1969): *Texas Instrum., Inc., Tech. Rep.* AFAL-TR-68-348, contract AF33(615)-3963, January.
Jones, D. A., R. V. Jones, and R. W. H. Stevenson (1952): *Proc. Phys. Soc.*, vol. B65, p. 906.
Jones, R. V. (no date): personal communication.
J. Phys. Radium, ser. 3, vol. 9, pp. 465, 1900.
Kaifu, Y., and T. Komatsu (1968): *J. Phys. Soc. Jap.*, vol. 25, p. 664.
Knudson, A. R., and J. E. Kupperian, Jr. (1957): *J. Opt. Soc. Am.*, vol. 47, p. 440.
Kohlrausch, F. (1968): "Praktische Physik," vol. 3, 22d ed., p. 74, Teubner, Leipzig.
Kokorina, V. F. (1967): *Symp. Vitreous Chalcogenide Semicond. Leningrad.*
Kolomiets, T. (1964): *Phys. Status Solidi*, vol. 7, pp. 359, 713.
Korth, K. (1933): *Z. Phys.*, vol. 84, p. 677.
Kretzman, R. (1940): *Ann. Phys.*, vol. 37, p. 303.
Kublitsky, A. (1934): *Ann. Phys.*, vol. 20, p. 793.
Landers, D., et al., Raytheon Corporation, Burlington, Mass. (no date): unpublished data.
Laufer, A. H., J. A. Pirog, and J. R. McNesby (1965): *J. Opt. Soc. Am.*, vol. 55, p. 64.
Lax, M., and E. Burstein (1955): *Phys. Rev.*, vol. 97, p. 39.
Lenham, A. P. (1967): *J. Opt. Soc. Am.*, vol. 57, p. 473.
Levin, S. B., N. J. Field, F. M. Plock, and L. Merker (1955): *J. Opt. Soc. Am.*, vol. 45, p. 737.
Liebreich, E. (1911): *Verh. Dtsch. Phys. Ges.*, vol. 13, p. 709.
Linde Air Products Company (no date): technical data sheets.
Linsteadt, G. F., and H. P. Leet (1961): *Proc. IRIS*, vol. 6, p. 159.
Loewenstein, E. V. (1961): *J. Opt. Soc. Am.*, vol. 51, p. 108.
Loferski, J. J. (1954): *Phys. Rev.*, vol. 93, p. 707.
Lord, R. C., and T. K. McCubbin, Jr. (1957): *J. Opt. Soc. Am.*, vol. 47, no. 8, pp. 689–697.
MacNeill, C. E. (1963): *J. Opt. Soc. Am.*, vol. 53, p. 398.
——— (1969): *J. Opt. Soc. Am.*, vol. 59, p. 1491.
Malitson, I. H., National Bureau of Standards (no date): personal communication.
——— (1962): *J. Opt. Soc. Am.*, vol. 52, p. 1377.
——— (1963): *Appl. Opt.*, vol. 2, p. 1103.
——— (1964): *J. Opt. Soc. Am.*, vol. 54, p. 628.
——— (1965): *J. Opt. Soc. Am.*, vol. 55, p. 1205.
——— and M. J. Dodge, National Bureau of Standards (no date): personal communication.
Mandarino, J. A. (1959): *Am. Mineral.*, vol. 44, p. 961.
Martens, F. F. (1901*a*): *Ann. Phys.*, vol. 6, p. 602.
——— (1901*b*): ibid., p. 603.
——— (1901*c*): ibid., pp. 619–626.
McCarthy, D. E. (no date): personal communication.
——— (1963*a*): *Appl. Opt.*, vol. 2, p. 591.
——— (1963*b*): ibid., p. 293.
——— (1965*a*): *Appl. Opt.*, vol. 4, p. 316.
——— (1965*b*): ibid., p. 317.
——— (1965*c*): ibid., p. 878.
——— (1967): *Appl. Opt.*, vol. 6, p. 1896.
——— (1968): *Appl. Opt.*, vol. 7, p. 1243.
——— (1971): *Appl. Opt.*, vol. 10, pp. 2539 ff.

McCubbin, T. K., Jr., and W. M. Sinton (1950): *J. Opt. Soc. Am.,* vol. 40, no. 8, pp. 537–539.
McDonough, R. C. (no date): personal communication.
McMahon, H. O. (1950): *J. Opt. Soc. Am.,* vol. 40, p. 376.
Meier, W. (1910): *Ann. Phys.,* vol. 31, p. 1017.
Mentzel, A. (1934): *Z. Phys.,* vol. 88, p. 178.
Micheli, F. J. (1902*a*): *Ann. Phys.,* vol. 4, p. 7.
——— (1902*b*): ibid., pp. 7–772.
Millard, J. P., and E. R. Streed (1969*a*): *Appl. Opt.,* vol. 8, no. 7, p. 1488, July.
——— and ——— (1969*b*): ibid., p. 1489.
Milliman, P. D. (1954): master's thesis, Cornell University, Ithaca, N.Y. (unpublished).
Minor, R. (1903): *Ann. Phys.,* vol. 10, p. 581.
Monnier, R. C. (1967): *Appl. Opt.,* vol. 6, p. 1437.
Mooney, J. B. (1966): *Infrared Phys.,* vol. 6, p. 153.
Moser, F., and F. Urbach (1956): *Phys. Rev.,* vol. 102, p. 1519.
Moss, T. S. (1959*a*): "Optical Properties of Semi-conductors," pp. 116–126, Academic, New York.
——— (1959*b*): ibid., pp. 133–151.
——— and A. G. Peacock (1961): *Infrared Phys.,* vol. 1, p. 104.
Mott, N. F., and R. W. Gurney (1950): "Electronic Processes in Ionic Crystals," p. 12, The Clarendon Press, Oxford.
Myuller, R. L., L. A. Bordakov, and Z. U. Borisova (1962): *J. Univ. Leningr.,* vol. 10, p. 94.
Nardo, D., and A. N. Goland (1971): *J. Opt. Soc. Am.,* vol. 61, p. 1321.
Nielson, S. (1962): *Infrared Phys.,* vol. 2, p. 117.
Nilsson, P. O. (1968): *Appl. Opt.,* vol. 7, p. 435.
Noble, R. H., Optical Sciences Center, Univ. of Arizona (no date): personal communication.
Nomura, K. C. (1960): *Phys. Rev. Lett.,* vol. 5, pp. 500–501.
Nysander, R. E. (1909): *Phys. Rev.,* vol. 28, p. 291.
O'Bryan, H. (1936): *J. Opt. Soc. Am.,* vol. 26, p. 122.
Oppenheim, U. P., and A. Goldman (1964): *J. Opt. Soc. Am.,* vol. 54, p. 127.
Oswald, F., and R. Schade (1954): *Z. Naturforsch.,* vol. 9a, p. 611.
Packard, R. D. (1969): *Appl. Opt.,* vol. 8, p. 1901.
Parsons, D. F., and D. D. Coleman (1971): *Appl. Opt.,* vol. 10, p. 1683.
Parsons, W. F. (no date): personal communication.
Paschen, F. (1908): *Ann. Phys.,* vol. 26, p. 120.
Patterson, R. J., and M. J. Brau (1966): *129th Electrochem. Soc. Meet., Cleveland, May.*
Pellicori, S. F. (1970): *Appl. Opt.,* vol. 9, no. 11, p. 2581.
Philco Corporation (1958): *Final Rept.* 2120-F, November.
Phillipp, H. R. (no date): personal communication cited in Nardo and Goland (1971).
Phillippi, C. M., and N. F. Beardsley (1955): Infrared Window Studies, pt. I, p. 16, *Wright Air Dev. Cent. Tech. Note* 55-194, June.
Platt, B. F., and W. L. Wolfe (1976): *Appl. Opt.,* vol. 15, no. 10.
Plyler, E. K., National Bureau of Standards (no date): personal communication.
——— (1947): cited in Smakula (1952*a*).
——— (1948*a*): *Natl. Bur. Stand. J. Res.,* vol. 41, p. 128.
——— (1948*b*): ibid., p. 125.
——— and F. R. Phelps (1951): *J. Opt. Soc. Am.,* vol. 41, p. 209.
——— and ——— (1952): *J. Opt. Soc. Am.,* vol. 42, p. 432.
Primak, W. (1971): *Appl. Opt.,* vol. 10, p. 759.
Rank, D. H., and H. E. Bennett (1954): *J. Opt. Soc. Am.,* vol. 44, p. 13.
Rodney, W. S. (1955): *J. Opt. Soc. Am.,* vol. 45, p. 987.
——— and I. H. Malitson (1956): *J. Opt. Soc. Am.,* vol. 46, p. 956.
———, ———, and T. A. King (1958): *J. Opt. Soc. Am.,* vol. 48, p. 633.
——— and R. J. Spindler (1953): *Natl. Bur. Stand. J. Res.,* vol. 51, p. 123.
Rosberry, F. W., National Bureau of Standards (1968): personal communication.
Rosch, S. (1965): *Opt. Acta,* vol. 12, p. 253.
Rubens, H. (1895): *Wied. Ann.,* vol. 54, p. 488.
——— and E. F. Nichols (1897): *Wied. Ann.,* vol. 60, p. 454.
Ruthberg, S., et al. (1963): *Appl. Opt.,* vol. 2, p. 177.
Saiedy, F., and G. D. Jones (1968): *Appl. Opt.,* vol. 7, p. 429.
Saker, E. W. (1952): *Proc. Phys. Soc.,* vol. B65, p. 785.
Salzberg, C. D., and J. J. Villa (1953): *J. Opt. Soc. Am.,* vol. 43, p. 579.
——— and ——— (1957): *J. Opt. Soc. Am.,* vol. 47, p. 244.

Savage, J. A., and S. Nielson (1965): *Infrared Phys.*, vol. 5, p. 195.
Schlosser, H. (1957): *J. Appl. Phys.*, vol. 28, p. 512.
Schneider, E. G. (1936): *Phys. Rev.*, vol. 49, p. 341.
——— and H. M. O'Bryan (1937): *Phys. Rev.*, vol. 51, p. 293.
Schulz, L. G. (1954): *J. Opt. Soc. Am.*, vol. 44, p. 357.
——— and F. R. Tangherlini (1954): *J. Opt. Soc. Am.*, vol. 44, p. 362.
Seraphin, B. O., and H. E. Bennett (1967*a*): Optical Constants, chap. 12 in R. K. Willardson and A. C. Beer (eds.), "Semiconductors and Semimetals," vol. 3, pp. 510–512, Academic, New York.
——— and ——— (1967*b*): ibid., pp. 525–526.
——— and ——— (1967*c*): ibid., pp. 529–531.
——— and ——— (1967*d*): ibid., p. 535.
——— and ——— (1967*e*): ibid., pp. 541–542.
Shaffer, P. T. B., and R. G. Naum (1969): *J. Opt. Soc. Am.*, vol. 59, p. 1498.
Simeral, W. G. (1953): dissertation, Univ. of Mich., Ann Arbor, June.
Slack and Sekula (1956): see J. A. Krunhansl, *Proc. Conf. Photocond., Atlantic City, Nov. 4–6, 1954*, p. 450, Wiley, New York, 1956.
Smakula, A. (1952*a*): U.S. Dept. Comm. Off. Tech. Serv. Doc. 111,052, p. 119, October.
——— (1952*b*): ibid., pp. 88–89.
——— (1962*a*): "Einkristalle," Springer-Verlag, Berlin.
——— (1962*b*): *Opt. Acta*, vol. 9, p. 205.
———, J. Kalnajs, and M. J. Redman (1964): *Appl. Opt.*, vol. 3, p. 323.
Smith, R. A., F. E. Jones, and R. P. Chasmar (1957): "The Detection and Measurement of Infra-Red Radiation," The Clarendon Press, Oxford.
Sosman, R. B. (1927): "The Properties of Silica," Chemical Catalog, New York.
Spitzer, W. G., and H. Y. Fan (1955): *Phys. Rev.*, vol. 99, p. 1893.
———, R. C. Miller, D. A. Kleinman, and L. E. Howarth (1962): *Phys. Rev.*, vol. 126, p. 1710.
Stavenich, A. E., and N. G. Yaroslavskii (1961): *Opt. Spectrosc. USSR*, vol. 11, no. 1, July.
Steinmetz, D. L., W. G. Phillips, M. Wirick, and F. F. Forbes (1967): *Appl. Opt.*, vol. 6, p. 1001.
Stephens, R. E., and I. H. Malitson (1952): *Natl. Bur. Stand. J. Res.*, vol. 49, pp. 249–252.
———, E. K. Plyler, W. S. Rodney, and R. J. Spindler (1953): *J. Opt. Soc. Am.*, vol. 43, p. 100.
Stierwalt, D. (no date): unpublished data.
——— (1966*a*): Navy Ordnance Lab. Rep. 667, Corona, Calif.
——— (1966*b*): *Appl. Opt.*, vol. 5, p. 1911.
———, J. B. Bernstein, and D. D. Kirk (1963): *Appl. Opt.*, vol. 2, p. 1169.
Strong, J. (1931): *Phys. Rev.*, vol. 38, p. 1818.
——— and R. T. Brice (1935): *J. Opt. Soc. Am.*, vol. 25, p. 207.
"The Structure of Glass," vol. 1. *Proc. 3d All Union Conf. Glassy State, Leningr., 1959*. Consultants Bureau Enterprises, New York, 1960.
Sutton, A., Polaroid Corporation (no date): personal communication.
Taft, E. A., and H. R. Phillipp (1965): *Phys. Rev.*, vol. 138, p. A197.
Tanenbaum, M., and H. B. Briggs (1953): *Phys. Rev.*, vol. 91, p. 1561.
Texas Instruments (no date): technical data sheets.
Tilton, L. W., and E. K. Plyler (1951): *Natl. Bur. Stand. J. Res.*, vol. 47, p. 25.
———, ———, and R. E. Stephens (1949): *Natl. Bur. Stand. J. Res.*, vol. 43, pp. 81–86.
———, ———, and ——— (1950): *J. Opt. Soc. Am.*, vol. 40, p. 540.
Tool, A. (1910): *Phys. Rev.*, vol. 31, p. 1.
Trommsdorf, H. (1901): *Z. Phys.*, vol. 2, p. 576.
Tutihasi, S. (1957): *Phys. Rev.*, vol. 105, p. 882.
Vitrikhovsky, N. I., L. F. Gudymenko, A. F. Maznichenko, V. N. Malinko, E. V. Pidlinsu, and S. F. Terekhova (1967): *Ukr. Fiz. Zh.*, vol. 12, p. 796.
Welber, B. (1967): *Appl. Opt.*, vol. 6, p. 925.
Willardson, R. K., and A. C. Beer (1967): "Semiconductors and Semimetals," vol. 3, Academic, New York.
Willmott, J. C. (1948): *Nature*, vol. 162, p. 996.
——— (1950): *Proc. Phys. Soc.*, vol. 63, p. 389.
Wirick, M. (1966): *Appl. Opt.*, vol. 5, p. 1966.
Wolfe, W., and B. Platt (1976): unpublished data.
Wray, J. H., and J. T. Neu (1969): *J. Opt. Soc. Am.*, vol. 59, p. 774.
Yoshinaga, H., and R. A. Oetjen (1956): *Phys. Rev.*, vol. 101, p. 526.
Zernike, F. (1964): *J. Opt. Soc. Am.*, vol. 54, p. 1215.

Section 8

Coatings and Filters

J. A. DOBROWOLSKI

Division of Physics, National Research Council of Canada, Ottawa, Ontario

(Numbers refer to paragraphs)

INTRODUCTION†

1. Definitions and Scope of Section In the broadest sense of the word, an optical filter is any device or material which deliberately changes the spectral-intensity distribution or the state of polarization of the electromagnetic radiation incident upon it. The change in the spectral-intensity distribution may be either selective or nonselective with respect to the wavelength. The filter may act in transmission or in reflection.

It is neither possible nor desirable to discuss in this chapter all the devices that are embraced by the above general definition of an optical filter. Some that are described in other parts of the book are prisms (Par. 35 of Sec. 2), reflectors (Par. 14 of Sec. 7), paints (Par. 11 of Sec. 7), photochromic materials and media whose spectral transmission or polarizing properties can be changed by external electric or magnetic fields (Pars. 2ff. and 20ff. of Sec. 17). In this section the discussion is largely confined to generic filters, such as cutoff filters, narrow-band transmission or rejection filters, reflectors, beam splitters, etc. Filters for very specific applications, e.g., filters for colorimeters and other scientific instruments, eye-protective filters, and signal glasses, are, as a rule, not treated. The technologically more important filters are discussed in greater detail than those of mainly historical interest. Of the filters described many are available commercially while others have to be made in the laboratory.

The spectral region covered extends from about 0.01 to 1000 μm (10 to 10^6 nm), although the main emphasis in this chapter is on filters for the visible and adjacent spectral regions.

The subject matter of this chapter has, of course, been discussed before in several reviews and books. The interested reader is particularly encouraged to consult the review articles by Geffcken (1957), Mohler (1952), and Scharf (1965). Filters for the vacuum ultraviolet are discussed at length by Samson (1967) and Gerasimova (1970), and for the infrared spectral region by Lecompte (1958), Smith et al. (1968*a*), and Möller and Rothschild (1971).

2. A Guide to Filters Based on Their Spectral Properties For a discussion of their properties, filters are grouped in this section according to their mode of operation. The main mechanisms responsible for filter action are absorption, reflection, refraction, interference, polarization, scattering, and diffraction. Filters based on the same material but on different processes are discussed in different sections (see, for example, Fig. 1). If more than one process is involved in any particular filter, the filter is classified under the predominant one. Exceptions are made whenever they facilitate the discussion.

The above arrangement is not necessarily the most convenient one from the point of view of someone wishing to locate a filter for a particular application. To simplify this task filters are classified in this paragraph according to their spectral transmission and reflection characteristics. It will be seen that often there exist several alternative solutions.

Short- and Long-Wavelength Cutoff Filters, Filters for Isolating and Rejecting Spectral Regions, High-Reflectance Coatings. Filters useful as short- and long-wavelength cutoff filters, filters for the isolation and rejection of spectral regions, and those giving a high reflectance are listed in Tables 1 and 2. They are listed separately for each of the following spectral ranges: extreme ultraviolet, middle and near ultraviolet, visible region, near infrared, intermediate infrared, and far infrared. The filters are further subdivided into transmission and reflection filters. The latter should be given serious consideration since they are often capable of a performance superior to that of transmission filters.

A few comments concerning Tables 1 and 2 should be made. The listing of a filter does not imply that it is available throughout the spectral region in question. Nor should it be assumed that the spectral characteristics attainable with one particular type of filter are uniformly good in all parts of the spectrum. In

† The author would like to express his thanks to Dr. A. F. Turner and Dr. G. R. Hanes for many valuable discussions and suggestions and to Mr. G. E. Marsh and Mrs. J. M. Paris for their untiring help in the preparation of the figures.

the visible region, sophisticated filters with closely controlled spectral characteristics are available, whereas filters for the extreme ultraviolet are crude by comparison. Even in the same spectral region the performance of different types of filters varies greatly. Thus, for example, transmission bandpass filters with half widths of 0.0025 and 60 percent of center wavelength are listed under the same heading. Similar comments can be made concerning the cutoff filters.

Beam Splitters. Achromatic and color-selective beam splitters are discussed in Pars. 89 and 90.

Neutral-Density Filters. Neutral-density filters and devices are discussed in Pars. 29 to 32. Spectral-transmittance curves of neutral-density glasses and gelatin filters are given in Figs. 6*g* and *h* and 10*i*.

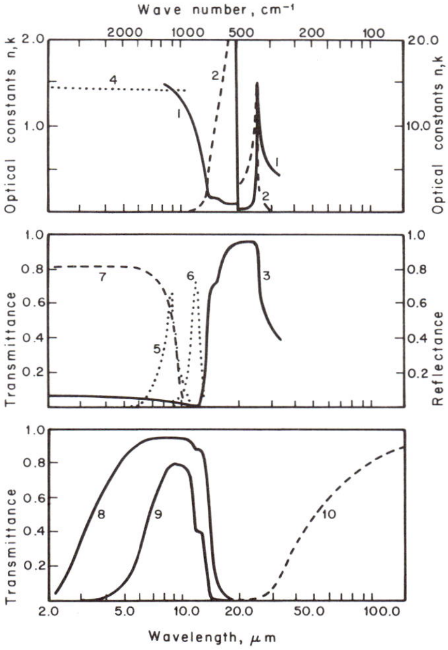

Fig. 1 Filters made from magnesium oxide. Curves 1 to 3: dispersion of the optical constants *n, k* and the Reststrahlen reflectance (Par. 99) of MgO [*after Piriou and Cabannes* (1968)]; curve 4: refractive index of CCl_4 [*after Pfund* (1935)]; curves 5 and 6: MgO-CCl_4 Christiansen filter (Par. 102) and MgO Christiansen powder filter (Par. 103) [*after Barnes and Bonner* (1936*b*)]; curves 7 to 9: transmission of a 0.67-mm-thick MgO plate (Par. 3 of Sec. 7) [*Smith et al.* (1968)] and of two MgO scatter filters (Par. 101) [*after Greenler* (1958)]; curve 10: MgO Yoshinaga powder filter (Par. 20) [*after Yamada et al.* (1962)].

Polarizers. Thin-film interference polarizers are treated in Pars. 69 to 71. For all other types of polarizers, see Sec. 10.

Filters for Color-Temperature Correction. Several series of gelatin filters suitable for color-temperature correction and light balancing are shown in Fig. 10*b*, *g*, and *h*. The corresponding glass filters are shown in Fig. 6*d* to *q*. More accurate correction filters can be made by combining thin-film interference filters with colored-glass filters, e.g., Dobrowolski (1970).

General Theory of Filters

3. Specification of Filters There are many different ways of describing the performance of optical filters. Thus, for example, for some purposes transmission and reflection filters for visual use are adequately described by a color name

TABLE 1 A Guide to Filters Based on Their Spectral Characteristics

T and R stand for transmission and reflection filters, respectively; paragraph numbers indicate where information on a particular type of filter can be found

Wavelength region, μm	Short-wavelength cutoff filters	Long-wavelength cutoff filters	Isolation of spectral regions	Energy range or wave number
0.01-0.2	T Metal films, 26	T Metal films, 26 R Metal-dielectric reflectors, 95	T Metal films, 26 Thin-film interference filters, 72-88	124-6.2 eV
0.2-0.4	T Transmitting optical materials, Par. 3 of Sec. 7 Colored glasses, 10-17 Solid organic materials, 21-23 Liquid and gaseous filters, 24 Absorbing thin films, 25-27 Periodic multilayers, 58-68 R Multiple-reflection filters, 96-98	T Solid inorganic materials, 18-20 Solid organic materials, 21-23 Liquid and gaseous filters, 24 Metal films, 26 Periodic multilayers, 58-68 R Metal-dielectric reflectors, 95 Multiple-reflection filters, 96-98	T Colored glasses, 10-17 Solid organic materials, 21-23 Metal films, 26 Absorbing filter combinations, 28 Thin-film interference filters, 72-88 Christiansen filters, 102-103 Polarization interference filters, 108-117 R Multiple-reflection filters, 98	6.2-3.1 eV
0.4-0.75	T Transmitting optical materials, Par. 3 of Sec. 7 Colored glasses, 10-17 Solid inorganic materials, 18-19 Solid organic materials, 21-23 Liquid and gaseous filters, 24 Absorbing thin films, 25-27 Periodic multilayers, 58-68 R Multiple-reflection filters, 96-99	T Colored glasses, 10-17 Solid organic materials, 21-23 Liquid and gaseous filters, 24 Periodic multilayers, 58-68 R Multiple-reflection filters, 96-99	T Colored glasses, 10-17 Solid organic materials, 21-23 Absorbing filter combinations, 28 Thin-film interference filters, 72-88 Christiansen filters, 102-103 Polarization interference filters, 108-117 R Thin-film interference filters, 95 Multiple-reflection filters, 98	
0.75-2.5	T Transmitting optical materials, Par. 3 of Sec. 7 Colored glasses, 10-17 Solid inorganic materials, 18-19 Solid organic materials, 21-22 Absorbing thin films, 27	T Transmitting optical materials, Par. 3 of Sec. 7 Colored glasses, 10-17 Periodic multilayers, 58-68 R Multiple-reflection filters, 96-98	T Thin-film interference filters, 72-88 Christiansen filters, 102-103 Polarization interference filters, 108-117 R Thin-film interference filters, 95 Multiple-reflection filters, 98	13,333-4000 cm^{-1}

		Periodic multilayers, 58-68 Scatter filters, 101					
	R	Multiple-reflection filters, 96-98 Scatter filters, 101					
2.5-50	T	Transmitting optical materials, Par. 3 of Sec. 7 Yoshinaga powder filters, 20 Periodic multilayers, 58-68 Scatter filters, 101 Echelette transmission gratings, 104	T	Transmitting optical materials, Par. 3 of Sec. 7 Periodic multilayers, 58-68	T	Thin-film interference filters, 72-88 Christiansen filters, 102-103	4000-200 cm^{-1}
	R	Multiple-reflection filters, 96-99 Scatter filters, 101 Echelette reflection gratings, 104	R	Multiple-reflection filters, 96-99	R	Multiple-reflection filters, 98 Reststrahlen filters, 99	
50-1000	T	Yoshinaga powder filters, 20 Echelette transmission gratings, 104 Metallic meshes, 106	T	Transmitting optical materials, Par. 3 of Sec. 7 Metallic meshes, 106	T	Thin-film interference filters, 72-88 Multimesh interference filters, 107	200-10 cm^{-1}
	R	Echelette reflection gratings, 104 Metallic grids and meshes, 105-106	R	Metallic meshes, 106			

TABLE 2 Guide to Fliters Based on Their Spectral Characteristics

T and R stand for transmission and reflection filters, respectively; paragraph numbers indicate where information on a particular type of filter can be found

Wavelength region, μm	Rejection of spectral regions	High reflectance	Energy range or wave number
0.01–0.2		Reflecting materials, Par. 3 of Sec. 7 Metal and metal-dielectric reflectors, 91–95	124–6.2 eV
0.2–0.4	T Solid inorganic materials, 19 Periodic multilayers, 52–53 R Multiple-reflection filters, 98	Reflecting materials, Par. 3 of Sec. 7 Periodic reflectors, 47–51 Metal and metal-dielectric reflectors, 91–95	6.2–3.1 eV
0.4–0.75	T Periodic multilayers, 52-53 R Multiple-reflection filters, 98	Reflecting materials, Par. 3 of Sec. 7 Periodic reflectors, 47–51 Resonant reflectors, 54 Broadband reflectors, 55–57 Metal and metal-dielectric reflectors, 91–95	
0.75–2.5	T Periodic multilayers, 52–53 R Multiple-reflection filters, 98	Reflecting materials, Par. 3 of Sec. 7 Periodic reflectors, 47–51 Broadband reflectors, 55–57 Metal and metal-dielectric reflectors, 91–95	13,333–4000 cm^{-1}
2.5–50	R Multiple-reflection filters, 98	Reflecting materials, Par. 3 of Sec. 7 Periodic reflectors, 47–51 Broadband reflectors, 55–57 Metal and metal-dielectric reflectors, 91–95 Reststrahlen filters, 99	4000–200 cm^{-1}
50–1000	T, R Metallic meshes, 106	Reststrahlen filters, 99 Metallic grids and meshes, 105–107	200–10 cm^{-1}

alone or by reference to one of the several existing color systems (see Par. 20 of Sec. 9). A description of color filters based on the CIE chromaticity diagram is described in detail in Par. 14 of Sec. 9. There exist also a number of specialized specifications of filter performance for particular applications. However, the most complete information on the performance of a filter is provided by spectral-transmittance, reflectance, and density curves. This is the method that will be adopted in this section.

Referring to Fig. 2 the specular spectral transmittance T_λ of a filter at a wavelength λ is equal to the ratio of the radiant energy of that wavelength transmitted $I_{2,\lambda}$ by the filter to that incident $I_{1,\lambda}$ upon it,

$$T_\lambda = \frac{I_{2,\lambda}}{I_{1,\lambda}} \tag{1}$$

The specular spectral reflectance R_λ of a filter is defined in a similar way,

$$R_\lambda = \frac{I_{R,\lambda}}{I_{1,\lambda}} \tag{2}$$

The relationship between the transmittance T_λ and the density of a filter is given by

$$D_\lambda = \log \frac{1}{T_\lambda} \tag{3}$$

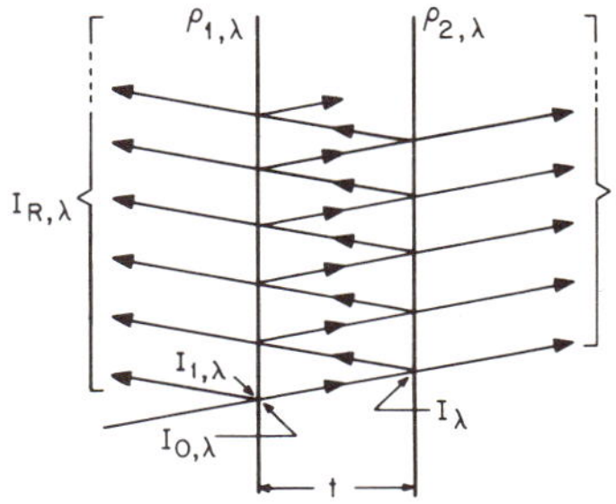

Fig. 2 Specular transmission and reflection of radiation by a plane-parallel plate (see Par. 3).

In this section transmittances will be plotted using linear and logarithmic scales. Another representation that will be used extensively is the logarithm of the density, log D_λ. For illustration, these three scales are used in Fig. 3 to plot the calculated transmittance of a bandpass interference filter. The linear scale is the scale most commonly used. The logarithmic scale, like the density representation, is particularly well suited whenever accurate information about the low-transmission region

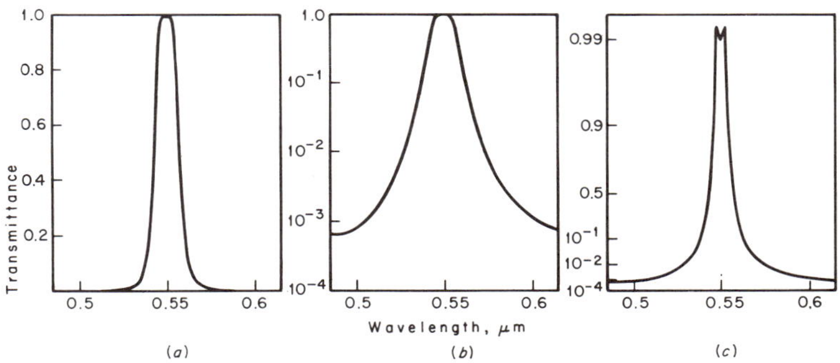

Fig. 3 Calculated transmittance of a square-top interference filter (Par. 74) plotted on linear (*a*), logarithmic (*b*), and log (density) (*c*) scales.

is to be conveyed. However, this is done at the expense of detail in the high-transmission region. The log density scale is almost linear in the 0.05 to 0.80 transmittance region, but it expands both the low- and the high-transmission regions and has certain other advantages. It will be used extensively in this section and will be explained in detail in Par. 9. Wavelengths are specified in micrometers (μm) because this is the most convenient unit for the whole spectral range covered in this section.

4. Surface Reflection Losses; Internal Transmittance In most filters in addition to the attenuation due to the main physical mechanisms responsible for the filter action, losses are incurred due to secondary mechanisms, e.g., reflections at the boundaries between different media. Since the latter are constant and can often be reduced by antireflection coatings or by cementing several components together, it is convenient at times to distinguish between the transmittance of the primary filter mechanism and the overall transmittance of the filter.

In absorption and dispersion filters reflections occur on both sides of the filter medium, as shown in Fig. 2, and it is customary to define an internal transmittance τ_λ which is the ratio of the intensity of the radiation just before leaving the filter medium to the intensity just after entering it,

$$\tau_\lambda = \frac{I_\lambda}{I_{0,\lambda}} \tag{4}$$

The corresponding quantity for interference filters is the transmittance of the multilayer when the surrounding media are semi-infinite.

The overall transmittance T_λ of a filter can be expressed in terms of the internal, or intrinsic, transmittance τ_λ and the reflectances $\rho_{1,\lambda}$, $\rho_{2,\lambda}$ of the two surfaces. Assuming that the surfaces are plane-parallel and that the incident radiation is not highly monochromatic, so that interference effects can be ignored, a summation of all the interreflections leads to the following expression for the spectral transmittance:

$$T_\lambda = \frac{(1 - \rho_{1,\lambda})(1 - \rho_{2,\lambda})\tau_\lambda}{1 - \rho_{1,\lambda}\rho_{2,\lambda}\tau_\lambda^2} \tag{5}$$

The reflection coefficients can be calculated from Eq. (76) providing that the complex refractive indexes of all three media are known. If the surrounding media are the same, $\rho_\lambda = \rho_{1,\lambda} = \rho_{2,\lambda}$ and Eq. (5) reduces to

$$T_\lambda = \frac{(1 - \rho_\lambda)^2\tau_\lambda}{1 - \rho_\lambda^2\tau_\lambda^2} \tag{6}$$

For most absorption filters in the visible part of the spectrum ρ_λ is of the order of 0.04, and one may write with sufficient accuracy for most purposes

$$T_\lambda \approx (1 - \rho_\lambda)^2\tau_\lambda \tag{7}$$

This last approximation is not valid in general; some filter materials for the infrared have high reflection coefficients, and in such cases Eq. (6) must be used.

If the plate of Fig. 2 represents an interference filter on a substrate, and if the reflection coefficient from the substrate side is $\rho_{2,\lambda}$, the quantity $1 - \rho_{2,\lambda}$ must be replaced by t_λ, which now represents the intrinsic transmittance of the filter. Usually the substrate material is transparent, so that $\tau_\lambda = 1$. We then get

$$T_\lambda = \frac{(1 - \rho_{1,\lambda})t_\lambda}{1 - \rho_{1,\lambda}\rho_{2,\lambda}} \tag{8}$$

In interference filters $\rho_{2,\lambda}$ is usually a rapidly varying function of λ and may assume any values between 0 and 1.0; hence if $\rho_{1,\lambda}$ is small, an appropriate approximation for T_λ is

$$T_\lambda \approx [1 - \rho_{1,\lambda}(1 - \rho_{2,\lambda})]\tau_\lambda \tag{9}$$

5. Transmission Filters in Series and Parallel To obtain a desired spectral transmittance curve it is frequently necessary to combine several filters of different absorption characteristics. One common way of combining filters is to place them in series (Fig. 4*a*). Unfortunately, because of the many different interreflections that may take place between the various surfaces, precise formulas for the resulting transmittance are very complicated (Tuckerman, 1947). Accurate calculations are best carried out using matrix methods (Dioffo, 1968).

When the surface reflections are small, e.g., in air-spaced absorption filters, to a first approximation the resultant transmittance is given by

$$T_\lambda = T_{1,\lambda}T_{2,\lambda}T_{3,\lambda} \cdots T_{k,\lambda} \tag{10}$$

If transmittance-density conversion tables are readily available, the density version of the above equation,

$$D_\lambda = D_{1,\lambda} + D_{2,\lambda} + D_{3,\lambda} + \cdots + D_{k,\lambda} \tag{11}$$

may be more convenient for this type of calculation. If all the filters are cemented (Fig. 4*b*), and if the interfaces do not give rise to any significant reflections, the appropriate formula is

$$T_\lambda = (1 - \rho_{1,\lambda})(1 - \rho_{k,\lambda})\tau_{1,\lambda}\tau_{2,\lambda}\tau_{3,\lambda} \cdots \tau_{k,\lambda} \tag{12}$$

If ρ_λ represents the mean reflectance of the filters, the gain in transmittance by cementing amounts to $(1 - \rho_\lambda)^{2(k-1)}$, a quantity that may be quite significant.

The application of Eqs. (10) to (12) to interference filters can lead to serious errors. Consider the case of two dielectric interference filters with $\tau_{1,\lambda} = \tau_{2,\lambda} = 0.5$ deposited onto opposite sides of the same plane-parallel substrate. According to Eq. (10), the resulting transmittance will be $T_\lambda = 0.25$. If, however, we use Eq. (5) and put $\rho_{1,\lambda} = \rho_{2,\lambda} = 0.5$ and $t_\lambda = 1.0$, we get $T_\lambda = 0.33$. If in the last calculation the internal transmittance of the substrate were assumed to be 0.9, the resultant transmittance would be $T_\lambda \approx 0.27$. Equation (10) is valid for interference filters only if the latter are slightly inclined to one another and the optics are arranged in such a way that the detector sees only the direct beam.

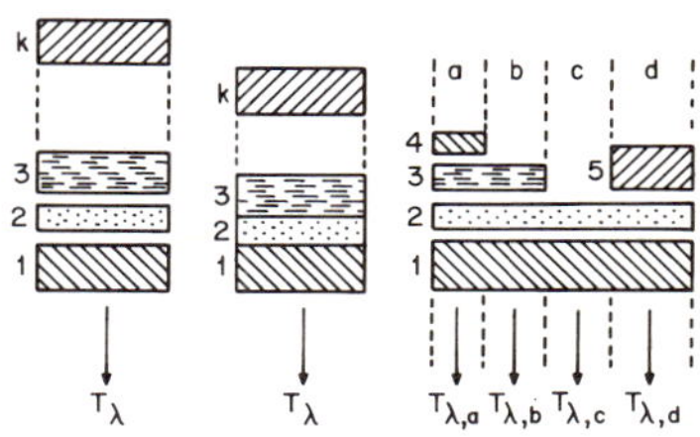

(*a*) (*b*) (*c*)

Fig. 4 Transmission filters in series (*a*), (*b*) and in parallel-series (*c*) arrangements. The filters can be air-spaced (*a*), (*c*) or cemented (*b*).

Some spectral-transmittance curves cannot be synthesized by placing filters in series alone. For certain applications it is quite acceptable to place filters not only in series but also in parallel (Dresler, 1933). This introduces areas as additional design parameters. Thus, for example, the effective spectral transmittance T_λ of the filter shown in Fig. 4*c* would be given by

$$T_\lambda = \left(\frac{a}{A} T_{\lambda,a} + \frac{b}{A} T_{\lambda,b} + \frac{c}{A} T_{\lambda,c} + \frac{d}{A} T_{\lambda,d}\right) \tag{13}$$

where A = overall area of filter

a, b, c, d = areas of the four zones

$T_{\lambda,a}, T_{\lambda,b}, T_{\lambda,c}, T_{\lambda,d}$ = transmittances of four zones

The latter are given by

$$\begin{aligned} T_{\lambda,a} &= T_{1,\lambda}T_{2,\lambda}T_{3,\lambda}T_{4,\lambda} \\ T_{\lambda,b} &= T_{1,\lambda}T_{2,\lambda}T_{3,\lambda} \\ T_{\lambda,c} &= T_{1,\lambda}T_{2,\lambda} \\ T_{\lambda,d} &= T_{1,\lambda}T_{2,\lambda}T_{5,\lambda} \end{aligned} \tag{14}$$

Great care must be exercised when using such filters. Because the spectral trasmittance of each zone of the filter is different, errors will result unless the incident radiation illuminates the filter uniformly. Similar care must be used when employing the filtered radiation. One way proposed to alleviate these problems (Nagel, 1954) is to break the filter down into a large number of small, regular elements and to reassemble it in the form of a mosaic.

6. Reflection Filters in Series If radiation is reflected from k different reflecting surfaces, the resultant reflectance R_λ will be given by

$$R_\lambda = R_{1,\lambda}R_{2,\lambda}R_{3,\lambda} \cdots R_{k,\lambda} \tag{15}$$

which is completely analogous to Eq. (10) for the resultant transmittance of a series of filters (Par. 5). Many of the other considerations of that paragraph apply here also. Thus, for instance, R_λ will be significant only at those wavelengths at which every one of the reflectors has a significant reflectance. Metal layers (Pars. 91 to 95), thin-film interference coatings (Pars. 41 to 88), Reststrahlen materials (Par. 99), and metallic grids and meshes (Pars. 105 to 107) can all be used exclusively or in combination.

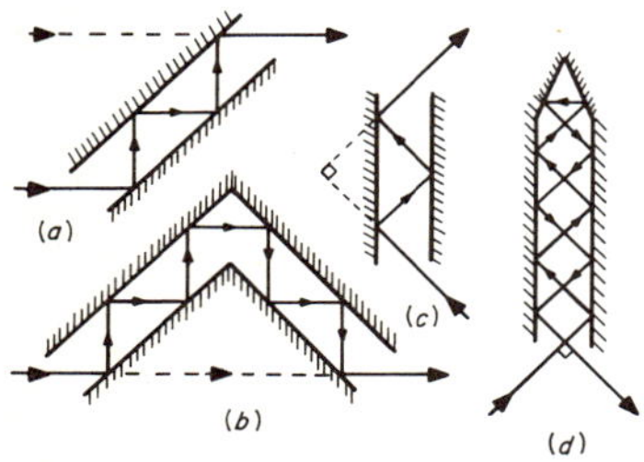

Fig. 5 Various arrangements for multiple-reflection filters.

For the sake of convenience the number of different reflecting surfaces used is normally restricted. The outlines of some possible reflector arrangements given in Fig. 5 are self-explanatory. The arrangement shown in Fig. 5*b* does not deviate or displace the incident beam. The number of reflections depends in each case on the lengths of the plates and the angle of incidence of the beam. Other arrangements are possible.

Clearly multiple-reflection filters are larger and more complicated to use than transmission filters. But if in a given application these shortcomings can be accepted, multiple-reflection filters offer great advantages, which stem mainly from the nature of the reflectors available for their construction (see Pars. 96 to 99).

FILTERS BASED ON ABSORPTION

Basic Theory

7. General Comments on Absorption Filters Absorption filters are still the most widely used type of filter. They may take the form of colored glass, crystals, sintered materials, gelatin films, inorganic and organic liquids, gases, and thin films. Their main advantages are that they are simple to use, insensitive to angle of incidence, and cheap or moderate in cost, depending on the actual material used.

There exists an enormous number of solids, liquids, and gases with absorption features in the spectral region considered in this section. A very useful compendium of the numerous atlases with absorption spectra of materials in existence has been prepared by CODATA (1969).

Of course, not every material which exhibits spectrally selective absorption is a potential filter material; another important requirement is that its optical behavior be stable with time and insensitive to reasonable fluctuations in the operating conditions.

The radiation absorbed by a material is usually dissipated in the form of heat or fluorescence or by photochemical reactions within the material. Under some circumstances this can lead to the rejection of a material: photochemical reactions cause changes in the absorption properties, and fluorescence introduces an unwanted radiation. Such materials must be avoided unless they can be shielded from the radiation which gives rise to these effects. Heat, unless removed in some other way, can cause the material to emit infrared radiation. This will be troublesome if the detector used is sensitive to this radiation.

8. The Laws of Bouguer, Beer, and Lambert When electromagnetic radiation passes through a homogeneous, isotropic, but absorbing material, its intensity is reduced. Bouguer's law (Bouguer, 1729) relates the internal transmittance (Par. 4) of an absorbing material to its thickness t

$$\tau_\lambda = 10^{-\alpha_\lambda t} \tag{16}$$

The spectral absorption coefficient α_λ depends on the nature of the absorber and is a function of the wavelength.

The internal transmittance of a liquid filter is given by Beer's law (Beer, 1852)

$$\tau_\lambda = 10^{-\epsilon_\lambda ct} \tag{17}$$

where c is the concentration of the solution. The spectral absorption coefficient per unit concentration ϵ_λ depends not only on the solute but also on the solvent used. Departures from Beer's law have been observed for some solutes in very high or very low concentrations, but they all seem to be due to changes in the absorbing material that occur under such conditions (see, for example, Hardy and Young, 1948).

The internal transmittance of a certain thickness of an absorbing material can be readily calculated from that for a different thickness using Lambert's law (Lambert, 1760):

$$\tau_{\lambda,t_2} = (\tau_{\lambda,t_1})^{t_2/t_1} \tag{18}$$

Bouguer's and Lambert's laws, expressed in terms of internal densities,

$$d_\lambda = \alpha_\lambda t \tag{19}$$

$$d_{\lambda,t_2} = \frac{t_2}{t_1} d_{\lambda,t_1} \tag{20}$$

are particularly convenient if density-transmittance tables are available (see Appendix Table B18).

Exact and approximate relationships between the spectral transmittance T_λ, the internal transmittance τ_λ, and the reflection coefficient ρ_λ of the surface were given in Par. 4. For these cases in which the use of the approximate expression (7) is justified, Lambert's law may also be written in terms of the spectral transmittances

$$T_{\lambda,t_1} = (T_{\lambda,t_2})^{t_1/t_2} \tag{21}$$

9. Log (Density) Method of Plotting Transmittance and Reflectance Data (Hardy, 1936; Mellon, 1950; Geffcken, 1957) The above formulas enable one to calculate the transmittance of different thicknesses of a filter one wavelength at a time. But if, for a given material, the quantity $1 - \log\log(1/\tau_\lambda)$ rather than τ_λ or T_λ is plotted against wavelength, then since

$$1 - \log\log\frac{1}{\tau_\lambda} = 1 - \log\alpha_\lambda - \log t \tag{22}$$

the shape of the curve is independent of the thickness and a change in the latter is equivalent to a relative vertical displacement of the curve and the ordinate scale. This method of presentation is used by one glass-filter manufacturer who supplies a transparent scale that can be moved across any diagram in the catalog to provide the spectral-transmittance curve for any desired thickness of the filter (Schott & Gen., 1959).

In this section a simpler, though somewhat less convenient version of this plotting procedure is used extensively. A series of displaced vertical scales, each corresponding to a different thickness of the filter, is printed at the side of the graph. A transparent ruler should be used to read off the internal transmittance from the scale corresponding to the desired thickness. Should the amount of use justify it, it might be worthwhile to draw a set of horizontal internal-transmittance grid lines onto a piece of tracing paper and to slide this across the diagram.

To obtain the spectral transmittance of a filter the internal transmittances must be multiplied by T_0, the correction factor for reflection losses. Since

$$1 - \log\log\frac{1}{T_\lambda} = 1 - \log\log\frac{1}{T_0} - \log\alpha_\lambda - \log t \tag{23}$$

it is also possible to draw grids corresponding to the spectral transmittance T_λ. The additional term in T_0 is constant with wavelength if dispersion is neglected. It follows from Eq. (23) that a reliable knowledge of the refractive index of the filter material is necessary if serious errors in the high-transmission regions are not to result when the log (internal-density) scale is used to present filter data.

The above representation will also be used in cases where it may be of interest to read off directly the resultant spectral transmittance of several identical filters or coatings placed in series, providing that the interreflections do not contribute to the resultant transmittance.

Colored Filter Glasses

10. Colored filter glasses consist of a base glass to which colorants have been added. Typical base glasses are potassium, sodium, phosphate, borate, borosilicate, and silicate glasses. The colorants may take the form of metal ions in solution in the base glass, e.g., di- and trivalent iron, manganese, chromium, vanadium, cupric or cuprous ions, cobalt, nickel, uranium, titanium, tungsten, molybdenum, neodymium, praseodymium, or cerium. Alternatively, they may consist of non-metallic elements or their compounds, e.g., sulfur, sulfides, selenium, selenides, tellurium, or phosphorus, or metal atoms, e.g., gold, silver, or copper, suspended in the glass in the form of submicroscopic crystals that reach their active size only after a special heat treatment (called *striking*). The spectral properties of the filter glasses depend not only on the combination of the colorant and base glass used and on the furnace atmosphere but also, in the case of struck glasses, on the temperature and duration of the subsequent heat treatment (Weyl, 1959; Kreidl, 1960).

11. Repeatability of the Optical Properties Because of the complicated manufacturing process a certain variation from one melt to another is unavoidable, and published spectral-transmittance curves should not be regarded as being more than the performance of a typical sample. In all but the least demanding applications the transmittance of the actual color glass to be used should be measured before the final thickness of the filter is decided upon. Because of the variation in the colorant concentration, the thickness necessary to achieve a specified transmittance curve will often differ significantly from that expected on the basis of the information in the catalog.

The measurements should preferably be made under conditions which match those of the intended use as closely as possible. This is particularly advisable for struck glasses, where, in addition, slight variations in transmittance across the area of the filter may occur. Struck cutoff filters, when used in thicknesses less than 3 mm, may show a measurable transmittance in the cutoff region.

12. Spectral Transmittances of Colored Filter Glasses Colored filter glasses in sheet and block form are made by many manufacturers and in hundreds of different glass types. For the purpose of this section specifications of about 400 different glasses currently marketed by nine manufacturers were examined and compared. Of these about 310 were included in Fig. 6*a* to *s* and the accompanying tables. Internal-spectral-transmittance curves of similar glasses are grouped together in the same diagram. They are plotted on a log (internal-density) scale so that spectral-transmittance curves for thicknesses other than 2 mm can be read off in the manner described in Par. 9. The glasses are identified in the tables, which also list the correction factor by which the internal transmittances must be multiplied to convert them to external transmittances (Par. 4).

The data used for this work were in tabular or in graphical form,† the latter, if necessary, being enlarged photographically before digitization. When plotted on a log (internal-density) scale, many of the curves were almost identical except for a vertical displacement, indicating that the glasses were equivalent except for the colorant concentration. In such cases only one curve is given, and the glass selected was usually that which could best be presented in the same diagram with the other filters to be shown. The other equivalent glasses, together with the thicknesses for which their internal-transmittance curves coincide with that in the figure, are listed in the tables. In the majority of cases the mean departure from one or another of the shapes of the internal-transmittance curves of equivalent glasses, plotted on a log (internal-density) scale, amounted to less than 5 percent of the length of the 10^{-5} to 0.99 transmittance interval. Only in a few cases did it amount to more than 10 percent. By selecting equivalent glasses with the

† The author is indebted to Mr. I. Molyneux, of Chance-Pilkington, Dr. R. C. Saxton, of Corning Glass Works, and Dr. H. Schroeder, of Jenaer Glaswerk, Schott & Gen., for unpublished data on filter glasses manufactured by their respective companies, and to Dr. G. Wyszecki, Division of Physics, National Research Council of Canada, for the use of the results of his measurements on filters of different makes.

Fig. 6 Spectral-transmittance characteristics of some commercial filter glasses (Par. 12). The internal transmittances of 2-mm-thick filter glasses are plotted on a log (internal-density) scale. Transmittances for other thicknesses between 0.25 and 8 mm can be read off with the aid of the auxiliary scales (Par. 9). The filters are identified in the accompanying tables, which also list the thicknesses of other equivalent glasses and the approximate multipliers necessary to convert from internal to external transmittances. The tabulated abbreviations precede the manufacturers' glass type names.

AO	American Optical Co., Southbridge, Mass.
BL	Bausch & Lomb, Inc., Rochester, N.Y.
C	Corning Glass Works, Corning, N.Y.
CP	Chance-Pilkington Optical Works, St. Asaph, U.K.
DS	Deutsche Spiegelglas AG, Grünenplan, West Germany
J	Jena, East Germany
PM	Parra Mantois, Paris, France
PPG	Pittsburgh Plate Glass Co., Pittsburgh, Pa.
S	Jenaer Glaswerk, Schott & Gen., Mainz, West Germany

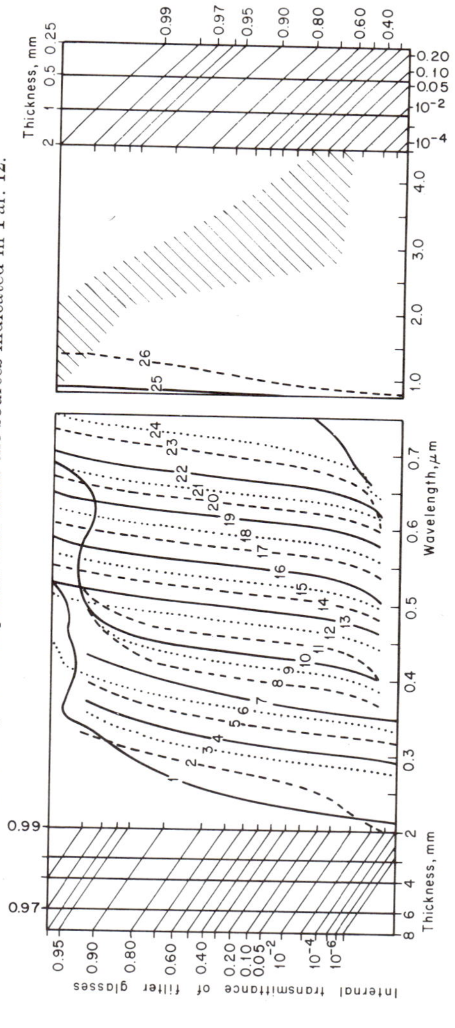

(a)

The graphs and tables are based on information obtained from catalogs and data sheets issued by the companies and from the sources indicated in Par. 12.

Fig. 6(*a*)

Curve	Correction factor	Glass type	Thickness, mm
1	0.925	C-7910	2.00
2	0.925	S-WG-280	2.00
		S-WG-295	1.31
3	0.920	S-WG-305	2.00
		C-7740	1.24
4	0.910	S-WG-320	2.00
		C-0160	1.65
5	0.885	S-WG-345	2.00
		S-WG-335	2.79
6	0.920	S-WG-9	2.00
		PPG-2876	0.43
		PM-AUV-2	1.38
		C-7380	2.84
		S-WG-360	3.22
7	0.920	S-GG-375	2.00
		S-GG-3[illegible]5	1.15
		S-GG-385	1.51
		CP-OY-10	3.36
8	0.910	S-GG-400	2.00
		C-3391	1.10
9	0.910	S-GG-420	2.00
		C-3389	0.96
10	0.910	S-GG-435	2.00
		PM-VJ-44	1.28
11	0.915	S-GG-455	2.00
12	0.915	S-GG-475	2.00
13	0.915	S-GG-495	2.00
		C-3385	0.96
		PPG-1397	3.00
		PM-VJ-49	3.74
14	0.915	S-OG-515	2.00
15	0.915	S-OG-530	2.00
		C-3484	0.67
		C-3486	1.03
		PM-VJ-53	1.05
		BL-Y-8	4.04
16	0.915	S-OG-550	2.00
		C-3482	0.97
		PM-VJ-55	1.00
		BL-Y-4	5.35
17	0.915	S-OG-570	2.00
		C-2434	0.33
		BL-R-6	0.77
		C-3480	1.04
		PM-VR-57	1.10
18	0.915	S-OG-590	2.00
		C-2418	0.44
		C-2424	0.86
		BL-R-2	0.89
		PM-VR-59	1.06
19	0.915	S-RG-610	2.00
		PM-VR-62	0.49
		C-2412	0.81
		BL-R-5	0.92
		BL-R-1	5.34
20	0.915	S-RG-630	2.00
		C-2403	0.53
		C-2404	0.79
		BL-R-3	1.02
		C-2408	3.36
21	0.915	S-RG-645	2.00
		C-2030	0.41
22	0.915	S-RG-665	2.00
		BL-R-7	0.59
23	0.915	S-RG-695	2.00
24	0.915	S-RG-715	2.00
		BL-R-4	4.82
25	0.915	S-RG-780	2.00
26	0.910	PM-VR-7	2.00
		C-2540	1.23
		S-RG-1000	1.45
		CP-OX-5	3.18

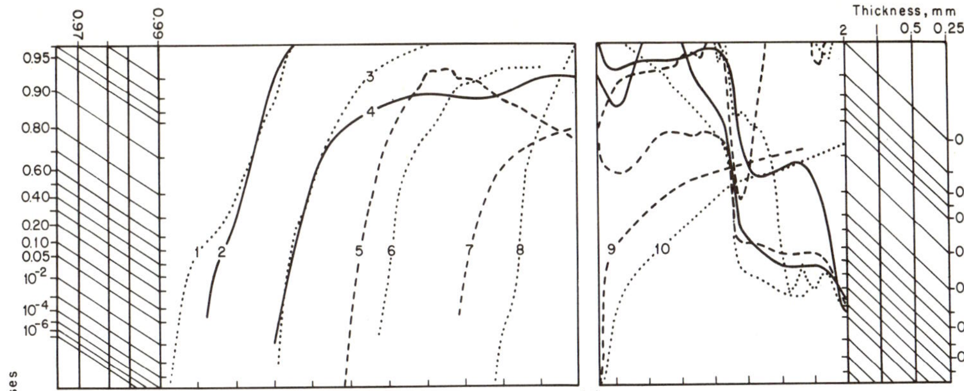

(b)

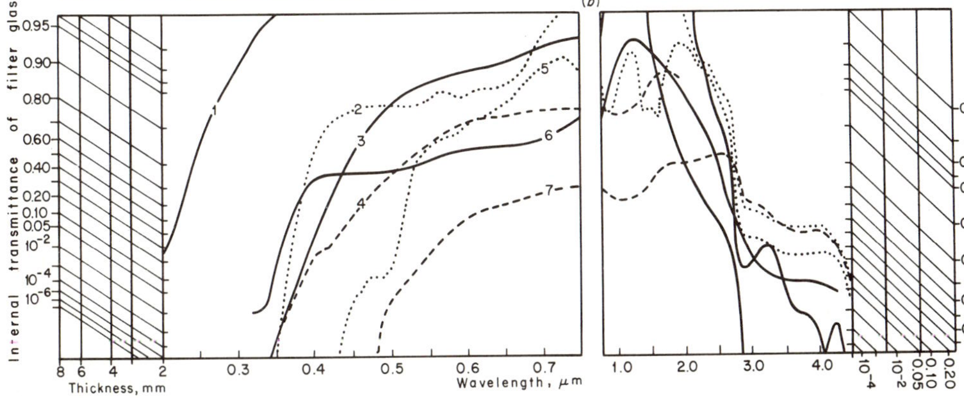

(c)

Fig. 6(*b*)

Curve	Correction factor	Glass type	Thickness, mm
1	0.931	C-7905	2.00
2	0.930	S-UR-2	2.00
3	0.836	PM-VIR-3	2.00
4	0.918	C-3850	2.00
5	0.917	C-3387	2.00
6	0.918	BL-Y-6	2.00
7	0.710	S-UR-1	2.00
8	0.913	PM-VR-67	2.00
9	0.662	PM-As_2Se_1	2.00
10	0.662	PM-As_2Se_1 (P)	2.00

Fig. 6(*c*)

Curve	Correction factor	Glass type	Thickness, mm
1	0.911	CP-OW-1	2.00
		S-WG-230	1.55
2	0.917	CP-OW-12	2.00
3	0.915	S-FG-10	2.00
4	0.918	DS-GELB-376	2.00
5	0.917	C-3780	2.00
6	0.918	AO-CRUXITE-AX	2.00
		AO-CRUXITE-A	3.20
	0.918	C-3304	2.00

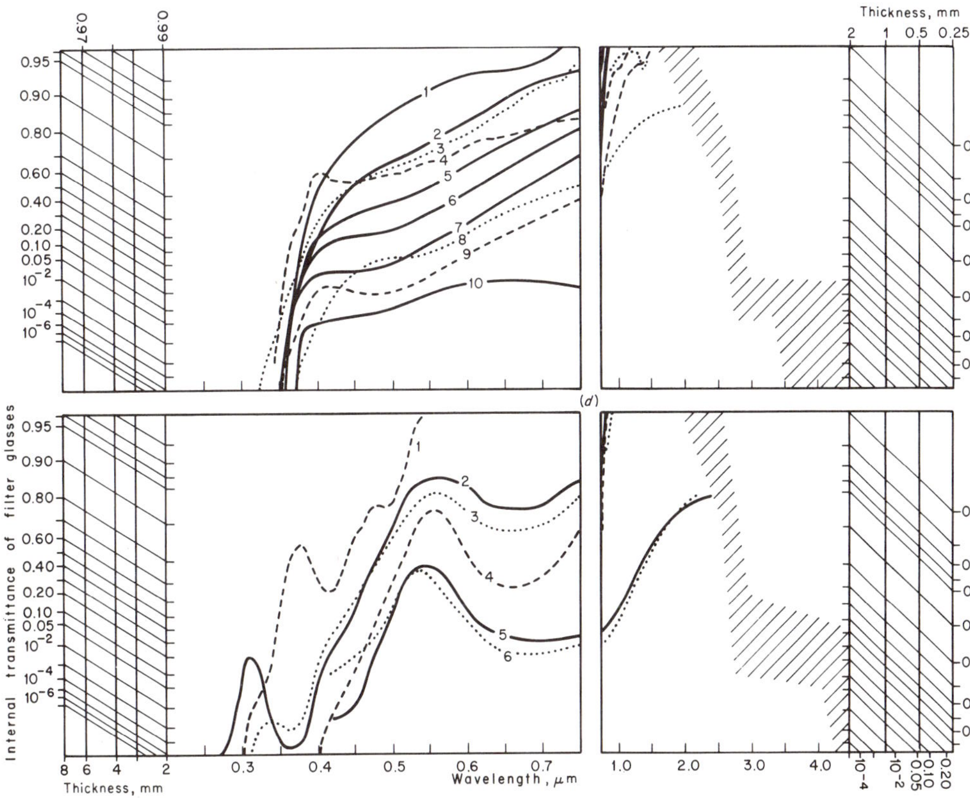

(d)

(e)

Fig. 6(*d*)

Curve	Correction factor	Glass type	Thickness, mm
1	0.918	DS-UVR	2.00
2	0.906	PM-AUV-1	2.00
3	0.920	S-FG-11	2.00
4	0.918	DS-PINK-B	2.00
		DS-ROSALIN-355	1.57
5	0.906	PM-VCR-15	2.00
		DS-KR-12	0.42
		PM-VCR-120	0.44
		DS-KR-9	0.55
6	0.906	PM-VCR-30	2.00
		DS-KR-15	6.62
7	0.906	PM-VCR-60	2.00
8	0.920	S-FG-9	2.00
		S-FG-8	5.30
		S-FG-7	7.45
9	0.918	CP-OY-21	2.00
		DS-PINK-C	7.88
		DS-KR-3	8.27
10	0.918	AO-COSMETAN	2.00
		DS-BRAUN-328F	1.58
		DS-BRAUN-424	1.83
		DS-BRAUN-4010F	2.39
		DS-BRAUN-345F	2.41

Fig. 6(*e*)

Curve	Correction factor	Glass type	Thickness, mm
1	0.925	S-GG-21	2.00
		CP-OY-14A	0.40
		S-GG-17	0.70
		C-3750	0.71
		C-3718	1.43
2	0.915	S-GG-19	2.00
		S-GG-9	0.56
3	0.915	S-GG-4	2.00
4	0.891	CP-OY-13	2.00
		PPG-5353	0.73
		CP-OY-12	0.80
		DS-GELB-GRÜN-C	0.92
		PM-VJV-4	1.07
		S-GG-10	1.48
		DS-GELB-GRÜN-B2	2.55
		PM-VJV-3	3.61
5	0.918	C-4015	2.00
		S-VG-4	1.28
		PM-VJV-6	1.30
		BL-G-9	1.49
		DS-GRÜN	1.50
		S-VG-5	1.61
		S-VG-10	4.76
6	0.902	PM-VJV-2	2.00

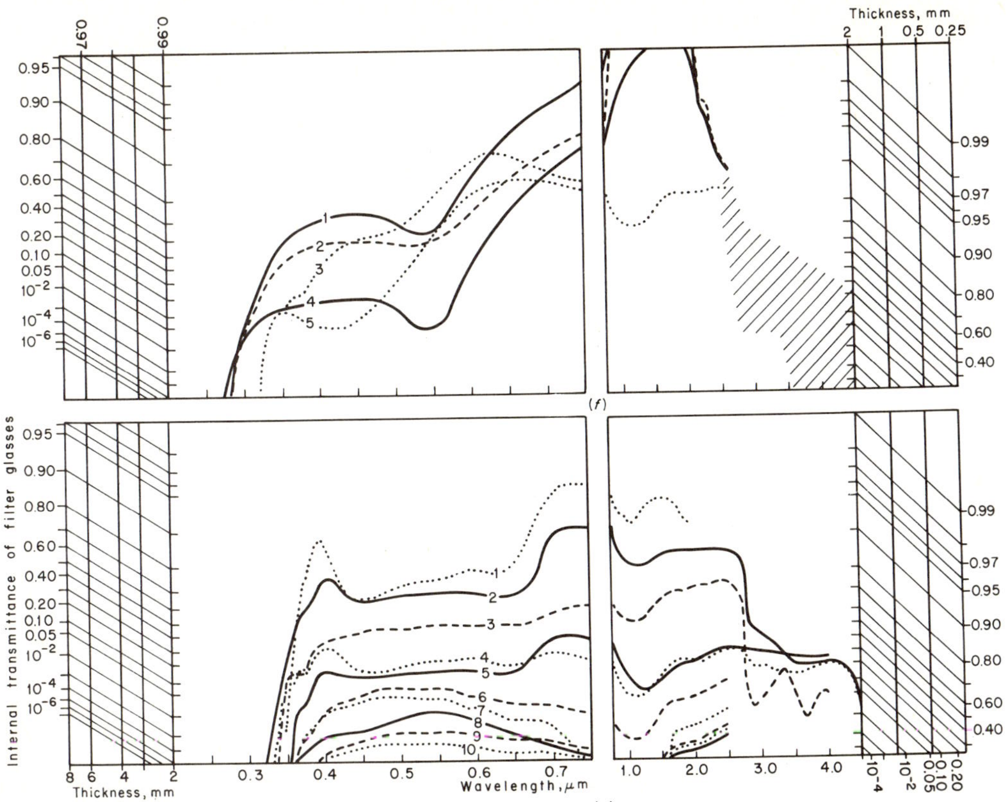

(f)

(g)

Fig. 6(*f*)

Curve	Correction factor	Glass type	Thickness, mm
1	0.915	S-FG-2	2.00
2	0.915	S-FG-1	2.00
3	0.915	DS-KR-6	2.00
4	0.920	S-RG-6	2.00
5	0.918	C-3307	2.00

Fig. 6(*g*)

Curve	Correction factor	Glass type	Thickness, mm
1	0.918	DS-GRAUBRAUN-425	2.00
		DS-ROSAL-041-Z1	0.26
		DS-SOMBRA	0.31
		DS-ROSAL-027-Z1	0.40
		DS-GRAUBRAUN-426	0.94
2	0.918	CP-ON-33	2.00
		CP-ON-10	0.28
		CP-ON-11A	0.46
		CP-ON-31A	0.57
		DS-OPHTASAN-BII	0.64
		CP-ON-32A	1.02
		CP-ON-33	1.53
3	0.923	PM-VN-3	2.00
		BL-N-5	0.42
		BL-N-4	0.86
		BL-N-3	1.25
		BL-N-2	3.38
		BL-N-1	9.58
4	0.918	C-8364	2.00
		AO-TRUE COLOR	1.72
		AO-TRUE COLOR-C	2.50
5	0.917	PM-OMBRANE	2.00
		AO-TRUE COLOR-D	2.28
		DS-GRAU-093	8.21
		DS-GRAU-254	8.78
6	0.918	CP-ON-8	2.00
7	0.918	CP-ON-7	2.00
8	0.918	CP-ON-6A	2.00
		DS-GRÜN-041 (Z2)	4.84
		DS-GRÜN-018 (Z1)	9.78
9	0.918	CP-ON-5	2.00
10	0.918	CP-ON-4	2.00

Note: DS makes many glasses with properties intermediate to those shown in this figure.

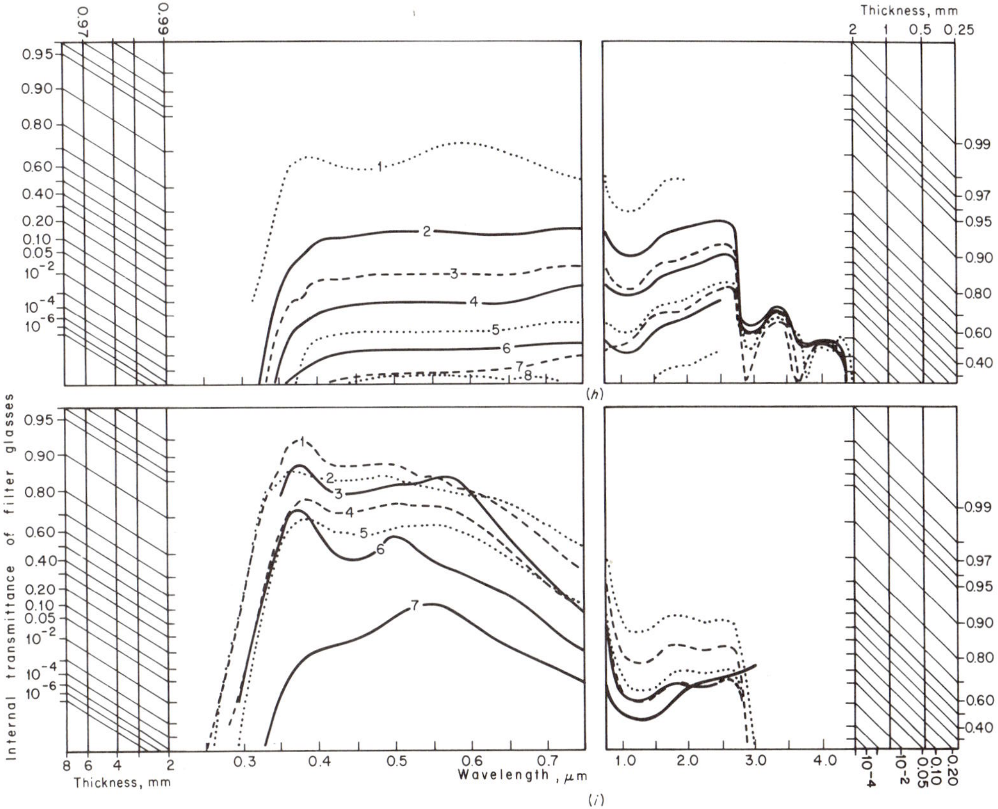

Fig. 6(*h*)

Curve	Correction factor	Glass type	Thickness, mm
1		DS-UROPA-F	2.00
2	0.920	S-NG-11	2.00
3	0.920	S-NG-5	2.00
4	0.920	S-NG-4	2.00
5	0.918	CP-ON-30	2.00
6	0.918	CP-ON-2	2.00
		S-NG-9	1.76
		S-NG-3	2.42
7	0.920	S-NG-10	2.00
8	0.918	CP-ON-3	2.00

Fig. 6(*i*)

Curve	Correction factor	Glass type	Thickness, mm
1	0.920	S-KG-2	2.00
2	0.920	S-KG-4	2.00
3	0.918	CP-HA-3	2.00
		PPG-2043	7.08
		PPG-9844	14.8
4	0.920	S-KG-1	2.00
		C-4600	1.11
		S-KG-3	1.40
5	0.920	S-BG-22	2.00
6	0.921	J-BG-30	2.00
7	0.918	J-SG-1	2.00

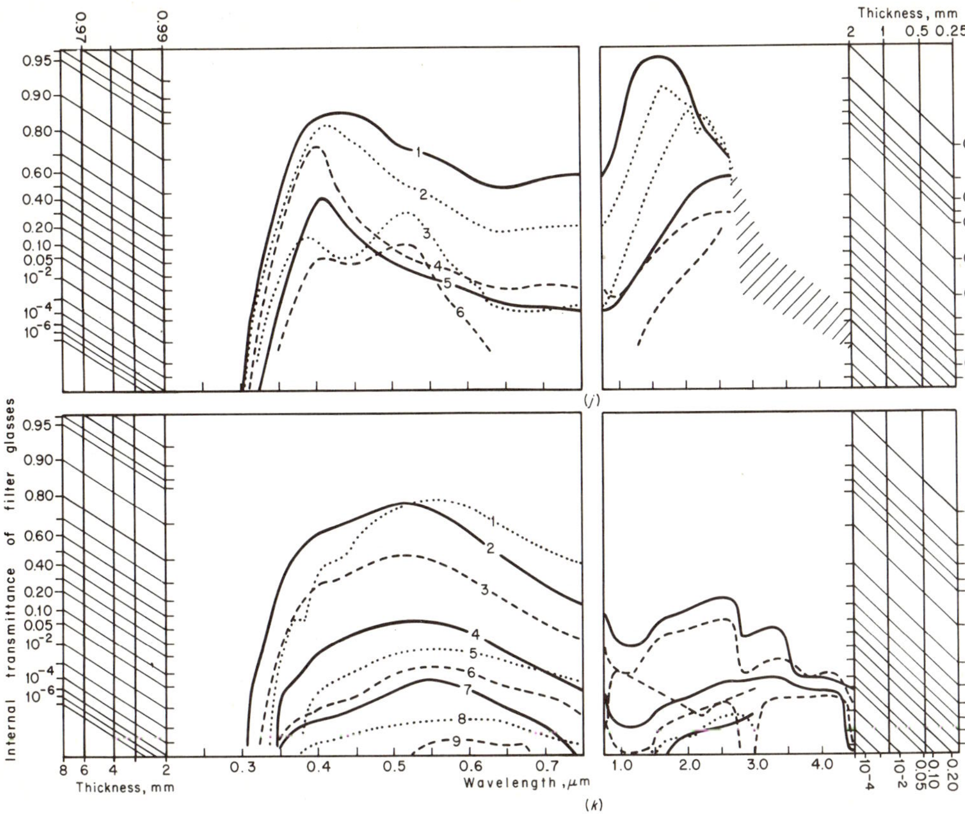

Fig. 6(*j*)

Curve	Correction factor	Glass type	Thickness, mm
1	0.906	PM-VCB-15	2.00
		S-FG-12	2.94
2	0.906	PM-VCB-30	2.00
		PM-VCB-120	0.72
		PM-VCB-60	1.17
3	0.910	C-4084	2.00
4	0.918	CP-OB-8	2.00
5	0.900	S-BG-34	2.00
		C-5900	3.86
		S-FG-5	4.00
		S-FG-4	5.67
		S-FG-6	9.68
6	0.910	J-VG-12	2.00

Fig. 6(*k*)

Curve	Correction factor	Glass type	Thickness, mm
1	0.918	AO-CALOBAR-A	2.00
2	0.922	C-3966	2.00
		C-3962	0.54
		C-3965	0.96
3	0.920	S-BG-10	2.00
		DS-ATHERMAL-OB	1.57
		DS-ATHERMAL-OD	5.08
4	0.918	C-3961	2.00
		DS-ATHERMAL-264F	5.13
5	0.918	AO-CALOBAR-EX-D	2.00
		DS-ATHERMAL-365	3.07
		DS-ATHERMAL-272	4.31
6	0.917	J-SG-2	2.00
7	0.916	J-SG-3	2.00
8	0.917	J-SG-5	2.00
9	0.918	C-3390	2.00

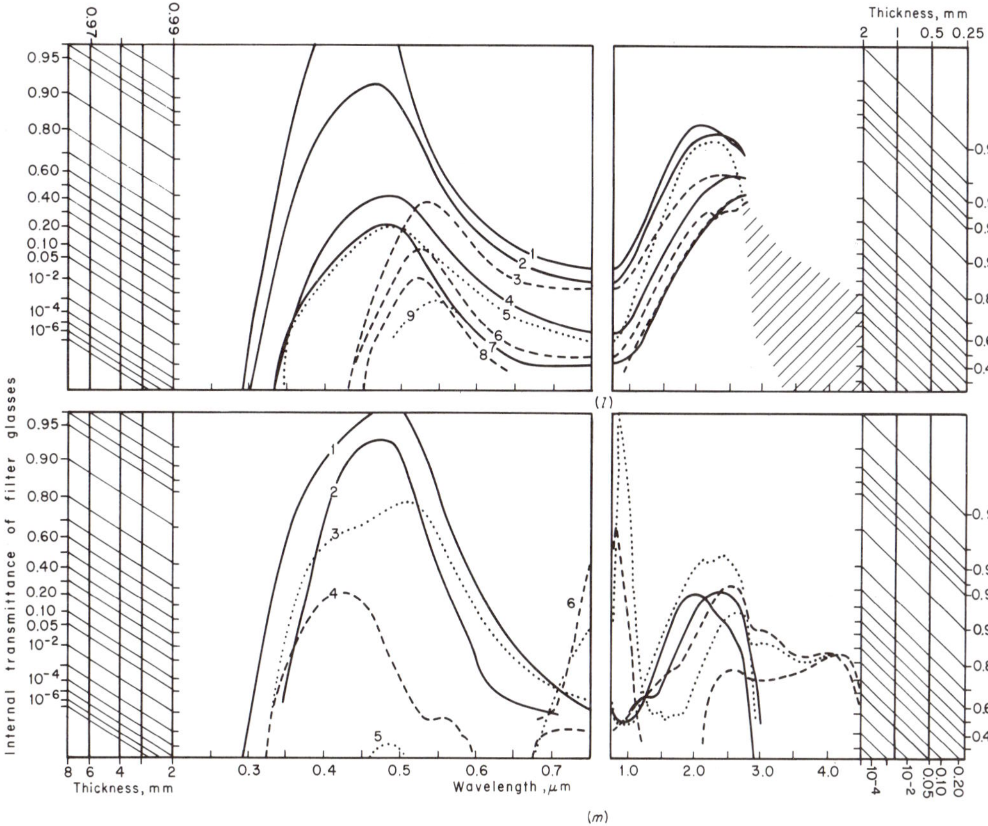

Fig. 6(*l*)

Curve	Correction factor	Glass type	Thickness, mm
1	0.920	S-BG-26	2.00
		S-BG-14	0.66
		C-4309	1.03
2	0.917	CP-OB-4	2.00
		C-4303	0.33
		S-BG-23	0.53
		C-4305	0.67
		C-4308	1.20
		DS-ASURBLAU-05	1.51
		DS-ASURBLAU-00	3.21
		PM-VB-1	4.08
3	0.918	CP-OGR-2	2.00
		CP-OGR-1	0.70
4	0.915	S-BG-13	2.00
		BL-G-1	0.32
		C-4060	0.68
5	0.915	PM-VB-3	2.00
6	0.910	S-VG-9	2.00
		S-VG-6	3.63
7	0.920	S-BG-7	2.00
		C-4445	3.43
8	0.907	PM-VV-2	2.00
		S-VG-14	1.69
		C-4010	3.45
		PM-VV-1	5.82
		CP-OGR-3	24.8
9	0.918	BL-G-19	2.00

Fig. 6(*m*)

Curve	Correction factor	Glass type	Thickness, mm
1	0.915	S-BG-38	2.00
2	0.905	C-9782	2.00
		C-9780	4.05
		C-9788	6.67
3	0.917	C-4784	2.00
		S-BG-18	0.56
4	0.918	CP-OB-10	2.00
		S-BG-28	3.18
		BL-B-10	20.4
5	0.910	PM-VR-6	2.00
6	0.918	C-2600	2.00

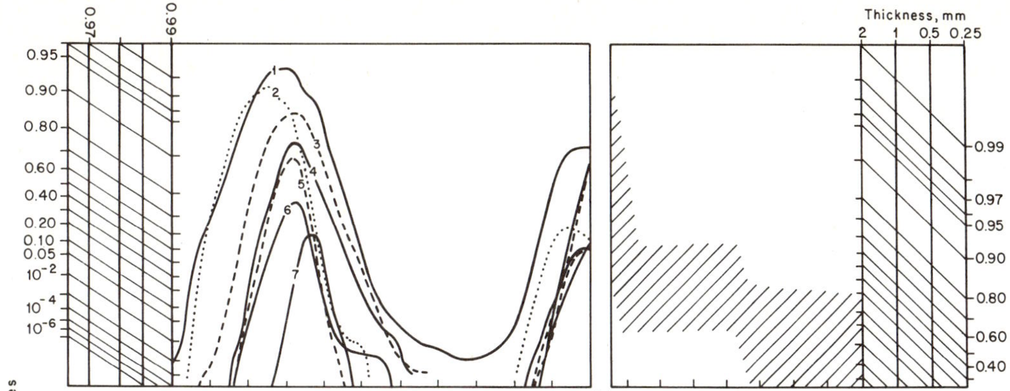

(n)

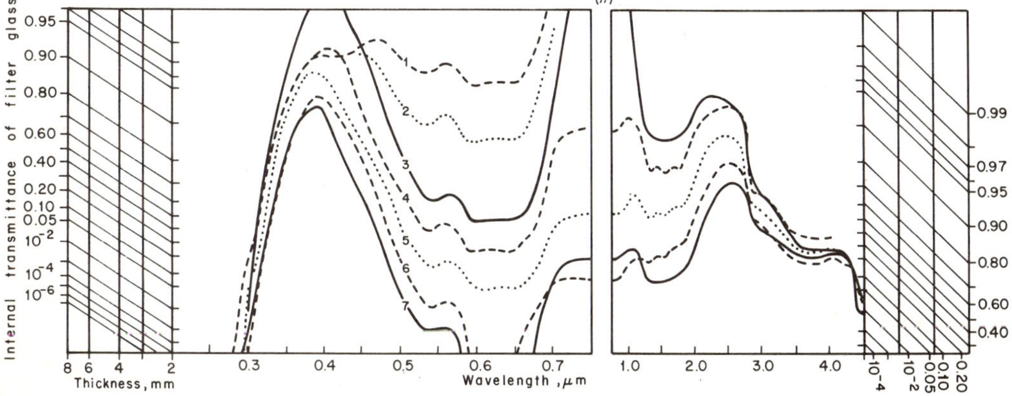

(o)

Fig. 6(*n*)

Curve	Correction factor	Glass type	Thickness, mm
1	0.915	S-BG-24	2.00
2	0.918	C-9863	2.00
3	0.920	S-BG-3	2.00
4	0.918	PM-VB-2	2.00
5	0.915	S-UG-12	2.00
6	0.918	CP-OX-1A	2.00
7	0.918	CP-OV-1	2.00

Note: AO makes many glasses with properties intermediate to those shown in this figure.

Fig. 6(*o*)

Curve	Correction factor	Glass type	Thickness, mm
1	0.918	BL-B-6	2.00
2	0.918	BL-B-7	2.00
		DS-BLAU-7	0.21
3	0.899	C-5572	2.00
		DS-KB-12	0.91
		DS-KB-9	1.17
		DS-KB-6	1.27
		DS-KB-3	3.81
4	0.919	PM-VB-7	2.00
		BL-B-3	0.48
		BL-B-2	0.66
		BL-B-1	2.56
		BL-B-8	4.97
5	0.919	PM-VB-6	2.00
		S-BG-12	11.8
6	0.920	S-BG-37	2.00
7	0.920	S-BG-25	2.00

Note: AO makes many glasses with properties intermediate to those shown in this figure.

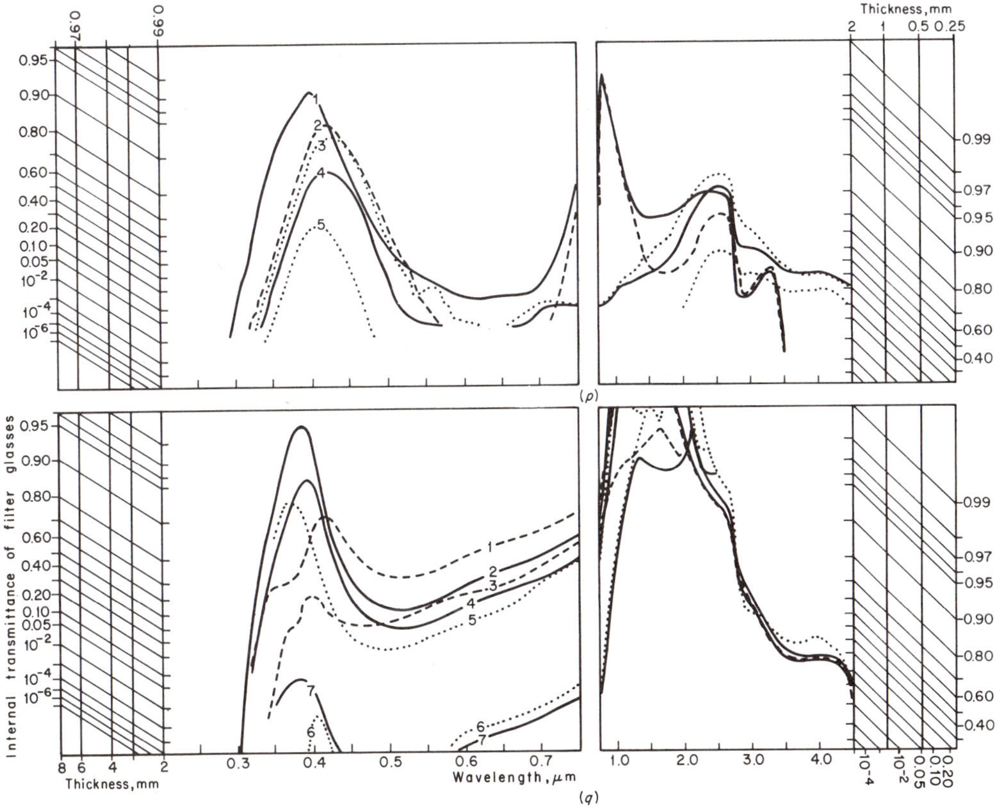

Fig. 6(*p*)

Curve	Correction factor	Glass type	Thickness, mm
1	0.925	C-5330	2.00
2	0.924	C-5030	2.00
3	0.908	C-5562	2.00
		C-5433	1.30
4	0.916	C-5543	2.00
5	0.918	C-5113	2.00

Fig. 6(*q*)

Curve	Correction factor	Glass type	Thickness, mm
1	0.895	C-5071	2.00
2	0.900	S-UG-10	2.00
		C-5070	0.36
		S-UG-3	0.38
3	0.910	DS-PINK-D	2.00
4	0.895	C-5073	2.00
5	0.918	CP-OV-2	2.00
6	0.910	C-2550	2.00
7	0.918	CP-OX-2	2.00

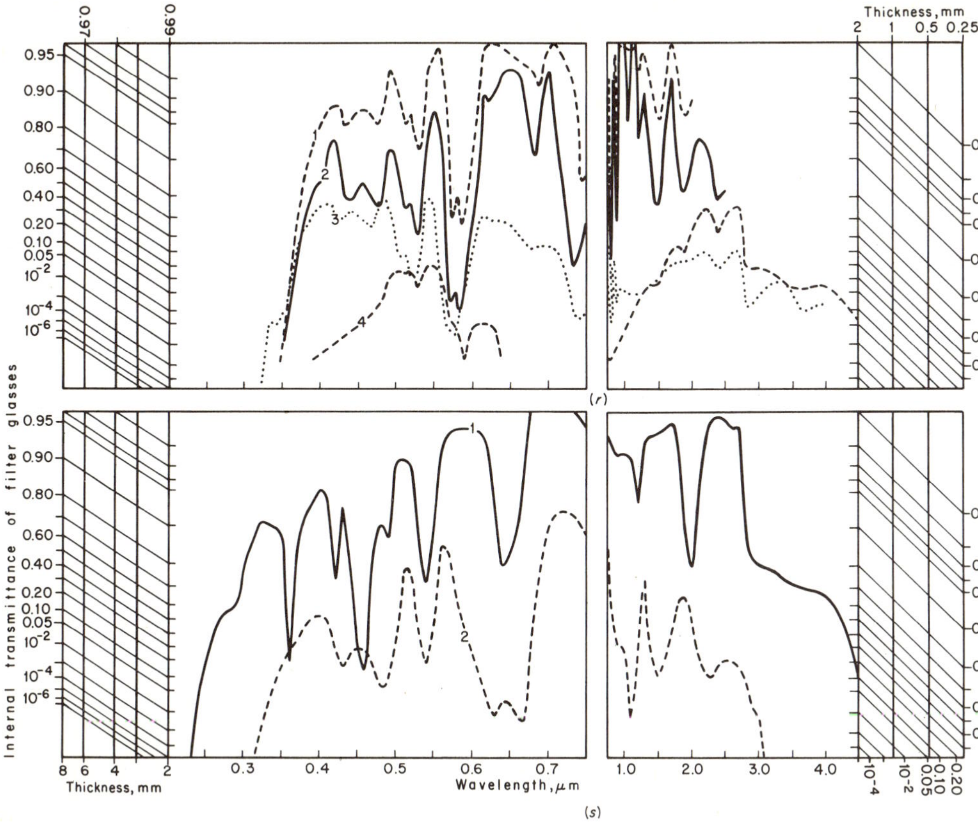

Fig. 6(*r*)

Curve	Correction factor	Glass type	Thickness, mm
1	0.918	DS-OPHTASAN-324	2.00
2	0.915	CP-ON-12	2.00
3	0.911	PM-VB-9	2.00
4	0.918	C-5300	2.00

Note: DS makes many glasses with properties intermediate to those shown in this figure.

Fig. 6(*s*)

Curve	Correction factor	Glass type	Thickness, mm
1	0.917	C-3130	2.00
2	0.910	S-VG-13	2.00

Note: Various manufacturers make glasses with properties intermediate to those shown in this figure.

most appropriate colorant concentration excessively thick or thin filters can often be avoided.

Special criteria were used for the selection of neutral-glass filters and short-wavelength cutoff filters. For the former, glasses with the most uniform spectral transmittance were chosen (Fig. 6*h*) and for the latter, glasses with the steepest cutoff (Fig. 6*a*).

13. Optical Quality of Glass Filters Although on the whole the optical quality of colored glasses is good, striae and bubbles may occur in some glasses, and filters should be checked for them.

14. Temperature Stability Transmittance curves of glass filters shift when the temperature of the filter is changed (Fig. 7), and this should be allowed for in precise work (Gibson and Belknap, 1950). The effect is largest in glasses that contain selenium or sulfur. Thus for example, a yellow CdS glass becomes transparent when immersed in liquid air and red on heating. The temperature coefficients for a typical line of short-wavelength cutoff filters varies between 0.2 and 4.0 Å/°C (Schott & Gen., 1967*a*). Information on the temperature variation of the spectral transmittance of certain selected filter glasses is given, for example, by Gibbs (1910), Lütge (1932), Holland and Turner (1941), Gibson (1949), Ångström and Drummond (1959), Dodd and West (1961), and Hickey et al. (1962).

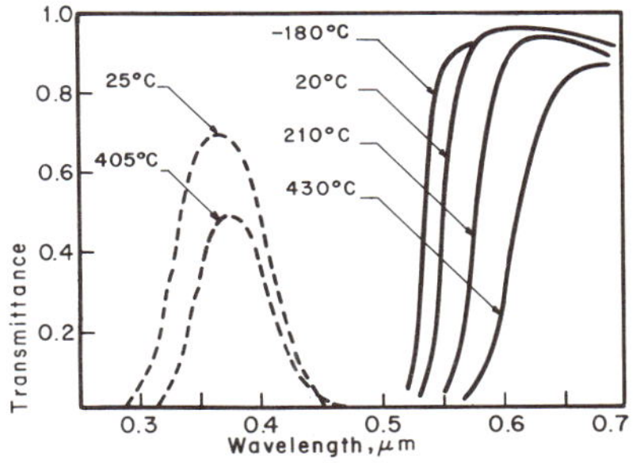

Fig. 7 Effect of temperature on the spectral-transmittance curves of two colored glasses. *Broken curves:* violet glass [*after Meyer* (1936)]; *full curves:* selenium orange glass [*after Gibson* (1916)].

15. Environmental Stability Some filter glasses, e.g., those based on a phosphate glass, tarnish when exposed for prolonged periods to high humidity or temperature. They can be specially hardened by the manufacturer or, alternatively, protected by cementing between two stable glasses or by the deposition of protective quartz or other metal oxide layers. Coatings can now be prepared (Heinrich et al., 1967) which are sufficiently impervious to protect even highly soluble optical materials from the atmosphere.

16. Fluorescence (Gibson, 1949) Some glass filters exhibit fluorescence in the ultraviolet and visible parts of the spectrum. If this is troublesome and a similar glass without fluorescence cannot be found, the effect can be reduced by using an auxiliary filter to remove the exciting radiation or the fluorescence itself.

Some manufacturers provide information on fluorescence in their catalogs. A list of colored glasses made by Chance-Pilkington, Corning, and Schott that are fluorescent in the visible part of the spectrum is given by Wyszecki and Stiles (1967).

17. Radiation Damage Like most other glasses, glass filters are discolored by nuclear irradiation. The amount of discoloration depends on the base glass. There is as yet little quantitative information on this subject.

Solid Inorganic Materials

18. Crystals, Semiconductors, and Sintered Materials There exist a large number of glasses, crystals, semiconductors, and sintered materials which must be regarded as being primarily optical materials. As such they are dealt with in Sec. 7. However, many of them have transmission properties that make them interesting from the point of view of filter construction. Section 7 should be consulted, especially for cutoff filters for the ultraviolet and infrared spectral regions.

Nickel sulfate hexahydrate is a crystal especially developed for filter construction (McBride and Olsen, 1962). It is grown from a supersaturated aqueous solution of nickel sulfate (curve 4, Fig. 13), and, like it, has a broad transmission band in the middle ultraviolet (Fig. 14). This makes it useful for suppressing near-ultraviolet and visible radiation (Childs, 1961; McBride, 1963*a*; Bennett and McBride,

1964). Crystals large enough to yield disks exceeding 5 cm in diameter can be grown. Since nickel sulfate hexahydrate is hygroscopic, it must be protected by hermetically sealing it between other filter elements.

19. Crystals with Color Centers and Doped Materials Changes in the spectral-transmission (McBride and Olsen, 1962) characteristics produced by the introduction of impurities have been observed in a legion of materials. Thus, for example, alkali halide crystals with high enough F- and X-center concentrations (see, for example, Friedman and Glover, 1949; Udod, 1969) become opaque in the visible region yet continue to transmit in the ultraviolet and infrared spectral regions (Fig. 8). Doped crystals exhibit sharp absorption bands and the cutoff wavelengths of semiconductors can be modified by doping (Fig. 8). So far, however, few filters utilizing these and similar effects have been described.

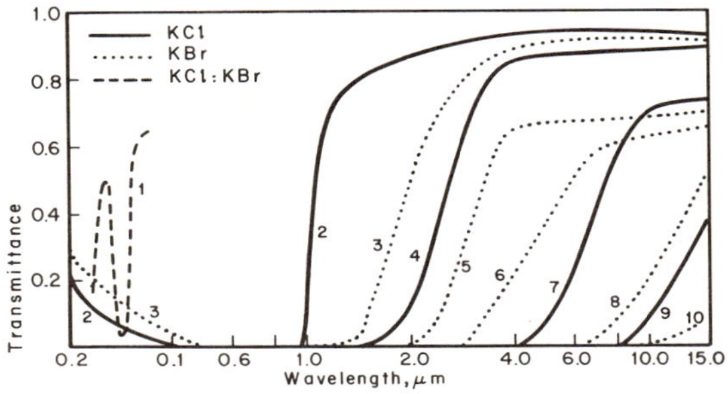

Fig. 8 Effect of doping and color-center concentration on the transmission of alkali halide crystals. Curve 1: 2-mm-thick 0.05 percent-lead-doped KCl:KBr crystal of 1:1 mol ratio [*after Childs* (1961)]; curves 2 and 3: 10-mm-thick KCl and KBr crystals with 5×10^{17} color centers per cubic centimeter [*after Friedman and Glover* (1949)]; curves 4 to 10: KCl and KBr crystals with increasing color-center concentration [*after Gaunt and Burns* (1954)].

20. Yoshinaga Powder Filters The filters to be described in this paragraph are based on the powder method of spectral analysis (Lecomte, 1958), in which a particulate sample of the absorbing material under test is suspended in a nonreactive medium. The medium must be transparent and not contribute to the resultant absorption spectrum. In the past the specimen was commonly suspended in a sheet of paraffin wax or mixed with some powdered transparent alkali halide and either used between two transparent plates or pressed into a pellet. More recently Yoshinaga and Oetjen (1955) proposed polyethylene as a host material. Polyethylene has no absorption bands at wavelengths higher than 14 μm and can be readily pressed into uniform thin sheets. By embedding in it powders of suitable Reststrahlen materials one can construct short-wavelength cutoff filters for the infrared spectral region (Yamada et al., 1962). Because such filters work in transmission, they are more convenient to use than Reststrahlen reflection filters (Par. 99). Furthermore, the number of potential materials (Table 4) for filter construction is much larger because crystal size is unimportant and only small quantities of the material are needed.

Although scattering (Pars. 100 to 101) also plays a part, the primary filtering mechanism of Yoshinaga filters is absorption. For some materials the cutoff occurs at a wavelength appreciably shorter than that corresponding to the lattice vibration. This cutoff wavelength is not very sensitive to powder size. Reststrahlen materials with the steepest absorption bands yield the sharpest cutoff filters. For the sharpest transition between the pass and rejection regions the radiation should pass through the minimum thickness of the Reststrahlen material necessary to achieve the desired rejection, and the powder size should be small compared to the cutoff wavelength. This is in contrast with scatter filters (Par. 101) and Christiansen powder filters (Par. 103) for which the particle sizes should be respectively of the order of the filter wavelength and considerably greater.

Several methods for preparing homogeneous suspensions of powders in thin polyethylene films have been described (see, for example, Yamada et al., 1962; Manley and Williams, 1965; Russell and Strauss, 1965; Möller et al., 1966; Chernyavskaya et al., 1971). The Reststrahlen powder content of such films varies between 10 and 40 percent, the particle size is of the order of 5 or 10 μm, and the thickness ranges between 0.25 and 1.0 mm. Thinner films cannot be made because they are too fragile. Quintard and Delorme (1967) developed a technique in which the powder is localized only at the surfaces of the polyethylene films. This results in a much thinner layer of the Reststrahlen material and an improvement in the cutoff characteristics of the filter.

The measured spectral-transmittance curves of various Reststrahlen powder suspensions in polyethylene are shown in Fig. 9*a* to *c*. In the long-wavelength transmission region interference fringes will appear if the film surfaces are accurately parallel. To the short-wavelength side of the Reststrahlen peaks the transmittance usually rises at first, only to be reduced again by the absorption of the polyethylene and the scattering of the suspension. The rejection region can be extended by combining several powders with contiguous absorption bands in the same film. A thin quartz plate, camphor soot on celluloid, or various thicknesses of black paper or black polyethylene, i.e., polyethylene containing a suspension of carbon (see, for example, Blea et al., 1970), are used at times to effect the final blocking. The spectral-transmittance curves of several blocked Yoshinaga powder filters are shown in Fig. 9*d*. When the temperature is lowered, the transmittance of the filters increases in the passband region and the cutoff edge becomes steeper, moving at the same time somewhat toward shorter wavelengths (Sakai et al., 1968; Zwerdling and Theriault, 1968). This is in accordance with the behavior of bulk Reststrahlen crystals.

Narrow-band rejection filters based on the same principle are described by Carlon (1962).

Solid Organic-Dye Filters

21. There exist a large number of organic dyes with spectral-absorption characteristics that can be utilized for filter construction (see, for example, Geffcken, 1957). Organic dyes can be used in both solid and liquid (Par. 24) filters. Although such filters are generally less stable than glass filters, their spectral features are often much sharper.

22. Gelatin Filters This is the most common type of solid organic filter. The filters are usually prepared by mixing the dyes in a gelatin solution, which is then poured over a glass surface and allowed to dry to form a uniform film approximately 0.1 mm thick. After stripping from the glass the gelatin film is either lacquered or cemented between glass for protection. Even so, the filters are quite fragile; they will be destroyed if they are heated beyond 50 to 55°C or immersed in water. Since their transmission characteristics may change on exposure to light, they should be stored in a dark, dry place.

Gelatin filters are relatively cheap. In the lacquered version they can be cut to any size and are thin, flexible, and convenient to use. They find various applications in photography, lighting, and related fields and are used in many scientific disciplines for cutoff purposes and for the isolation of spectral lines.

Gelatin filters are made by several companies, e.g., Eastman Kodak Company (1968*a*), Ilford, Ltd. (1960), and Göttinger Farbfilter, Göttingen, Germany. The internal transmittances of some typical Wratten gelatin filters made by Eastman Kodak are shown in Fig. 10*a* to *i*. The filter designations are those of the Wratten catalog. A log (density) scale is used, and the internal transmittance of one, two, three, or four identical filters in series can be read off directly (Par. 9). To obtain the external transmittance, the internal transmittance must be multiplied by a factor of 0.91 for each filter in the filter train.

The individual filters in several series of filters used for light balancing, color compensation, and photometric purposes differ from one another mainly in the concentration of the dye. Their internal-transmittance curves are therefore almost the same except for a vertical displacement. Only one curve is given in such a case, but the remaining filters are listed in the figure together with the thickness of the film (in units of standard gelatin-film thickness) necessary for their transmission curve to coincide with that illustrated.

Caution should be exercised when using the transmittance curves shown in Fig.

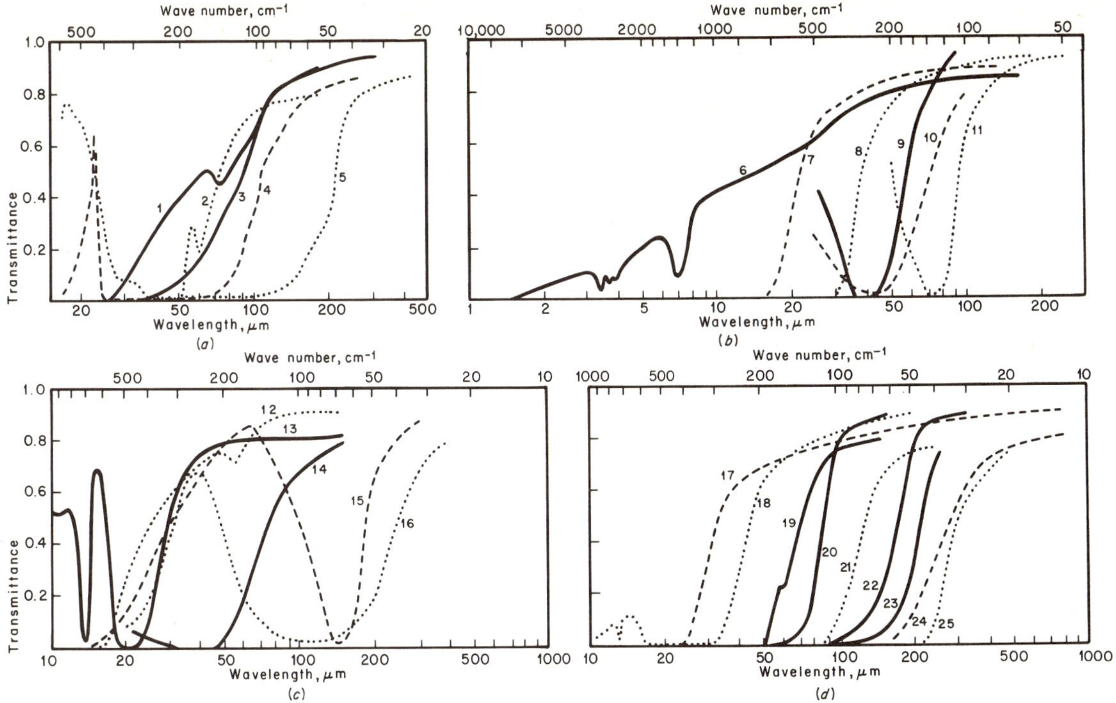

Fig. 9 Yoshinaga powder filters. (*a*) to (*c*) Suspensions of individual Reststrahlen materials in polyethylene. Curve 1: Cu_2O; curve 2: U_3O_8; curve 3: WO_3; curve 4: $Cr_2O_3 + Al_2O_3$; curve 5: PbF_2 [*after Manley and Williams* (1965)]; curve 6: black polyethylene [*after Möller et al.* (1966)]; curve 7: LiF; curve 8: BaF_2; curve 9: RbCl; curve 10: KBr; curve 11: TlI [*after Quintard and Delorme* (1967)]; curve 12: BeO; curve 13: ZnO; curve 14: NaF; curve 15: TlI; curve 16: TlCl [*after Yamada et al.* (1962)]. (*d*) Blocked short-wavelength cutoff filters consisting of several of the above powders in one film. Curve 17: BeO + ZnO; curve 24: TlCl + TlI + NaF [*after Yamada et al.* (1962)]; curve 19: $U_3O_8 + Cu_2O$; curve 20: $CuO_2 + WO_3$; curve 22: $Cr_2O_3 + Al_2O_3 + PbF_2$; curve 23: $TlI + TlCl + WO_3 + Cu_2O$ [*after Manley and Williams* (1965)]; curve 18: LiF; curve 21: $LiF + BaF_2 + KBr$; curve 25: $LiF + BaF_2 + KBr + TlI$ [*after Quintard and Delorme* (1967)].

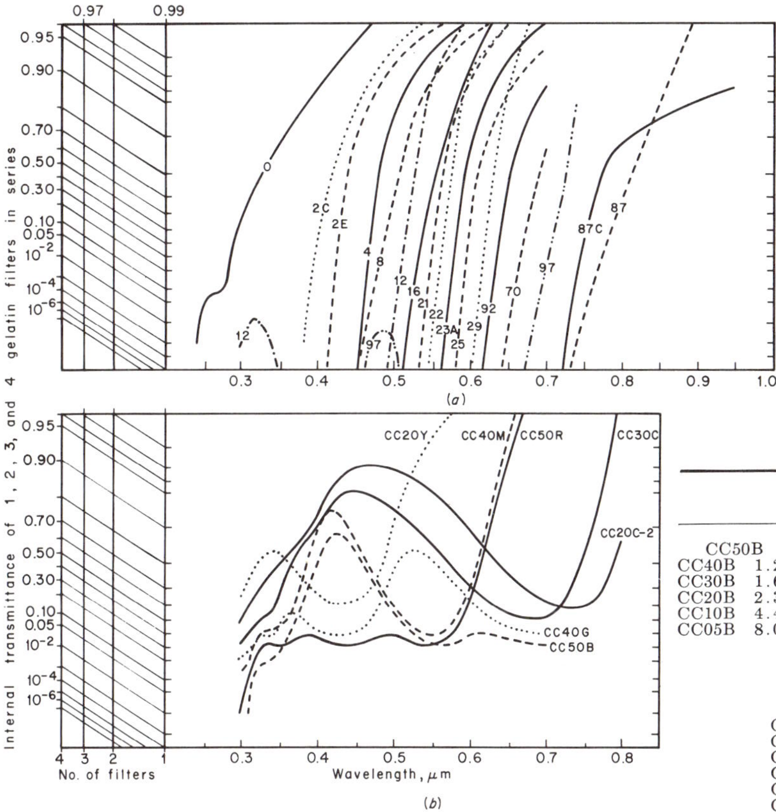

Fig. 10 Spectral-transmission characteristics of some Wratten gelatin filters. The internal transmittances are plotted on a log (internal-density) scale. The transmittances of two, three, or four identical filters in series can be read off with the aid of the auxiliary scales (Par. 9). To convert from internal to external transmittances the former must be multiplied by a factor of 0.91 for each filter. Equivalent filters and the thicknesses necessary to match the transmittance in the figure are listed in the tables.

(*a*) Short-wavelength cutoff filters

	Equivalent	
Filter	Filter	Unit gelatin-film thickness
2C	2B	0.65
2E	2A	2.09
4	3	2.25
16	15	3.63
25	26	0.72
	24	2.02
87	88A	0.06

(*b*) Color-compensating filters

Equivalent filter thicknesses, unit gelatin-film thickness

CC50B		CC30C		CC20C-2		CC40G	
CC40B	1.25	38A	0.09	CC50C-2	0.46	CC50G	0.81
CC30B	1.66	38	0.23	CC40C-2	0.57	CC30G	1.30
CC20B	2.35	CC50C	0.65	CC30C-2	0.72	CC20G	1.90
CC10B	4.46	CC40C	0.78	CC10C-2	1.70	CC10G	3.64
CC05B	8.01	CC20C	1.35	CC05C-2	2.36	CC05G	6.30
		CC10C	2.51	CC025C-2	3.00		
		CC05C	4.28				
		CC025C	6.55				

CC40M		CC50R		CC20Y	
CC50M	0.81	CC40R	1.25	CC50Y	0.47
CC30M	1.32	CC30R	1.68	CC40Y	0.57
CC20M	1.87	CC20R	2.70	CC30Y	0.71
CC10M	3.22	CC10R	4.63	CC10Y	1.78
CC05M	5.55	CC05R	7.85	CC05Y	2.50
CC025M	7.80	CC025R	9.50	CC025Y	3.04

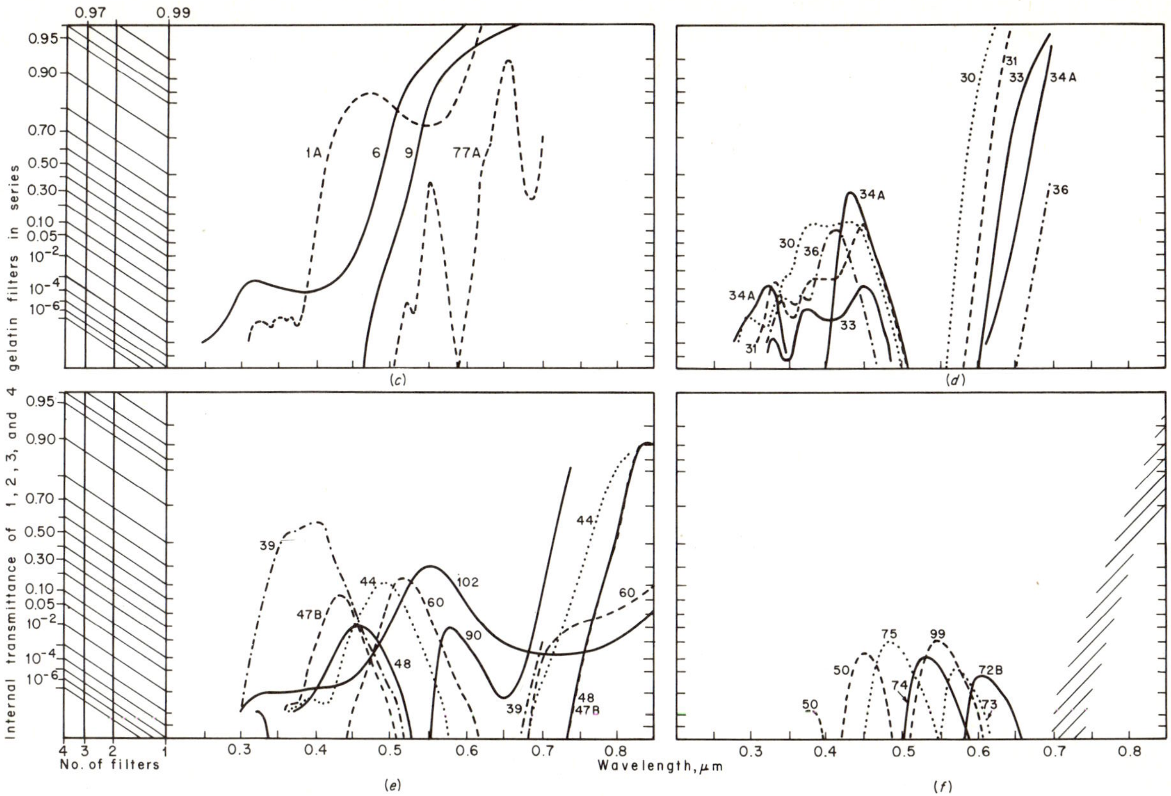

Fig. 10 *(Continued)*.

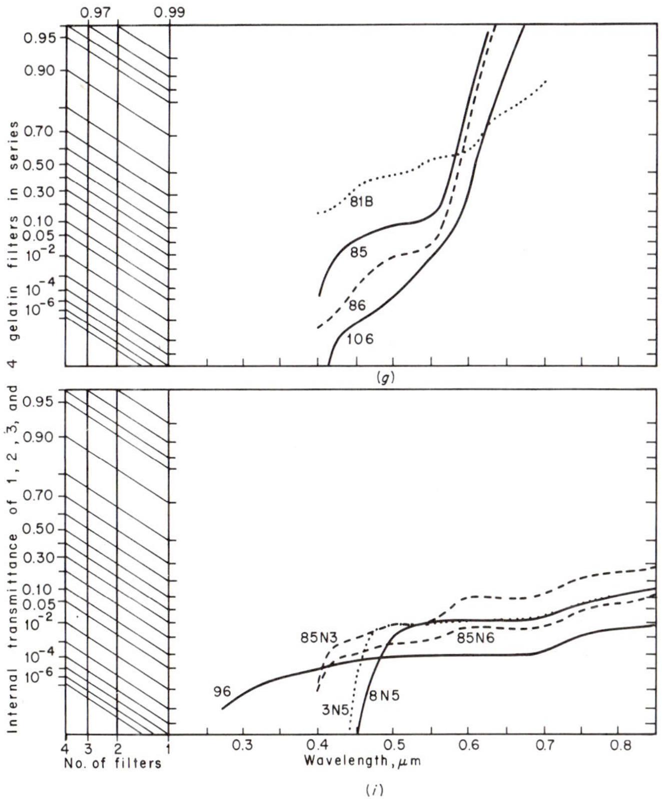

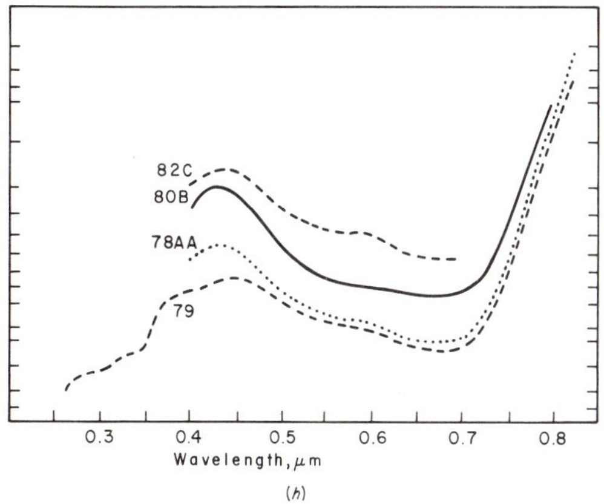

(*g*) to (*i*) Photometric and light-balancing filters

Filter	Equivalent	Unit gelatin-film thickness	Filter	Equivalent	Unit gelatin-film thickness
81B	81EF	0.46	78AA	78	0.84
	81D	0.57		78A	1.82
	81C	0.73		78B	2.95
	81A	1.61		78C	7.02
	81	3.50			
			80B	80A	0.89
85	85B	0.68		80C	1.42
	85C	1.58		80D	2.38
86	86A	2.20	82C	82B	1.32
	86B	3.61		82A	1.94
	86C	10.20		82	3.50

10*a* to *i*. Many of the filters shown transmit appreciably outside the spectral region depicted. Furthermore, if lacquered gelatin filters are used, the transmittance curve may be modulated in high-transmission regions by fringes due to the interference of light reflected at the two filter-air boundaries.

The reader is referred to the manufacturers' catalogs for a description of the full line of filters and for information on the dominant wavelengths, excitation purity, x and y colorimetric coordinates, and stability of the individual filters.

The spectral-transmittance curves of a number of narrow-band gelatin filters made by Ilford, Ltd. are shown in Fig. 23.

23. Other Organic Filters Some other host materials suggested in the past for solid organic-dye filters are beetle plastic, cellophane, cellulose acetate, ethylene terephthalate, methyl methacrylate, nylon, polystyrene, polytetrafluoroethylene, and polyvinyl alcohol. Thin films of some of these materials can serve as short-wavelength cutoff filters for the ultraviolet region. Additional cutoff filters are obtained by adding organic dyes (Fig. 11). Spectral-transmittance curves of many

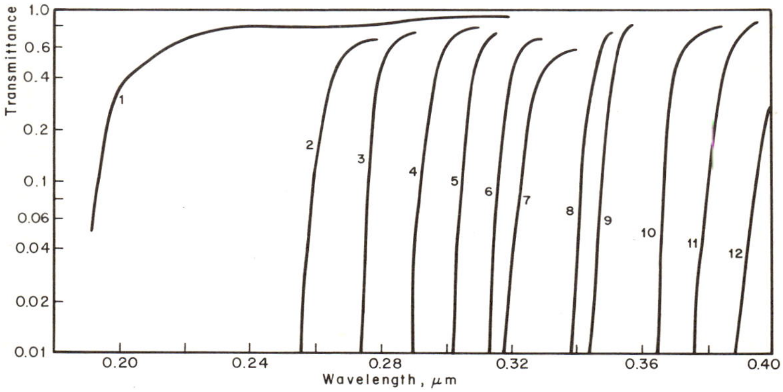

Fig. 11 Some organic short-wavelength cutoff filters. Curve 1: Owens-Illinois type 650 glass resin (1 mm); curve 2: thiourea in polyvinyl alcohol (0.0015 in); curve 3: polystyrene (0.002 in); curve 4: cellulose triacetate (0.003 in); curve 5: potassium biphthalate in polyvinyl alcohol (0.002 in); curve 6: ethylene terephthalate (0.003 in, Mylar, Du Pont); curve 7: 2-amino-4-methylpyridine in polyvinyl alcohol (0.002 in); curve 8: methyl methacrylate (0.025 in, Plexiglas, Rohm and Haas); curve 9: potassium biphthalate and 2,7-dimethyl-3,6-diazocyclohepta-1,6-diene perchlorate in polyvinyl alcohol (0.004 in); curve 10: 1,4-diphenylbutadiene in polystyrene (0.002 in); curve 11: 2(2′-hydroxy-5′-methylphenyl)benzotriazole in polystyrene (0.002 in); curve 12: 2,2′-dihydroxy-4-methoxybenzophenane in polystyrene (0.003 in). [*Curve 1 after Pellicori et al.* (1966); *others after McBride* (1963*b*).]

dyed plastics have been published by the American National Standards Institute (1955). Thin dyed films of polyvinyl chloride with spectral transmittances resembling those of some gelatin filters can be made in widths of up to 54 in and lengths of several hundred feet (Sherr et al., 1969). Since the films can be made to have high luminosities but very high optical densities at some laser wavelengths, they can be used to make protective shades and curtains.

A new silicone, the Owens-Illinois glass resin type 650 (curve 1, Fig. 11), is a host material for organic dyes for the 0.2- to 0.3-μm region; its thickness can be adjusted in much the same way as that of glass, and it is unaffected by exposure to vacuum or ultraviolet and gamma radiations (Pellicori et al., 1966). This development would be particularly useful if it were applicable to a larger number of dyes (Fig. 26).

Liquid and Gaseous Filters

24. Only a few of the large number of liquids and gases investigated in the past are suitable for the construction of filters. Many lack adequate stability with time or are decomposed by the radiation to be filtered. In others the absorption characteristics are very sensitive to temperature and the amount of dissolved oxygen. Some solutions show a marked fluorescence. Still other solutions become turbid, and the material precipitates out. These and other factors that may disqualify a material for filter construction are discussed by Stearns (1950). The marked departures of many solutions from Beer's law [Eq. (17)] are treated in Mellon (1950).

Geffcken (1957) and Mohler and Loofbourow (1952) give information on a number of materials that have been successively used in the past for the construction of liquid and gaseous filters. However, for instrumental and technological applications such filters have been largely replaced by other, newer types that are often more convenient to use and give a better performance. Liquid and gaseous filters are still used when large, cheap, angle-insensitive filters are needed. Such filters continue to be very useful for the isolation of the ultraviolet spectral region and for cutoff purposes within it. Materials for these latter applications are given in Figs. 12 to 14. Further gas-window material combinations for the use in the

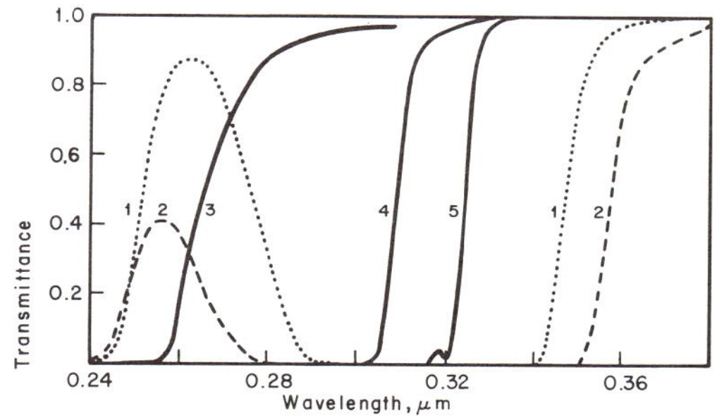

Fig. 12 Spectral transmittance of liquid filters. Curve 1: 1-cm thickness of 20 mg of iodine in 100 ml of water; curve 2: 1 cm of 4.24 mg of 1,4-diphenylbutadiene in 100 ml of ethyl ether; curve 3: 0.5 cm of CCl_4; curve 4: 1 cm of 5.0 g of potassium acid phthalate in 1 l of water. [*After Kasha* (1948).]

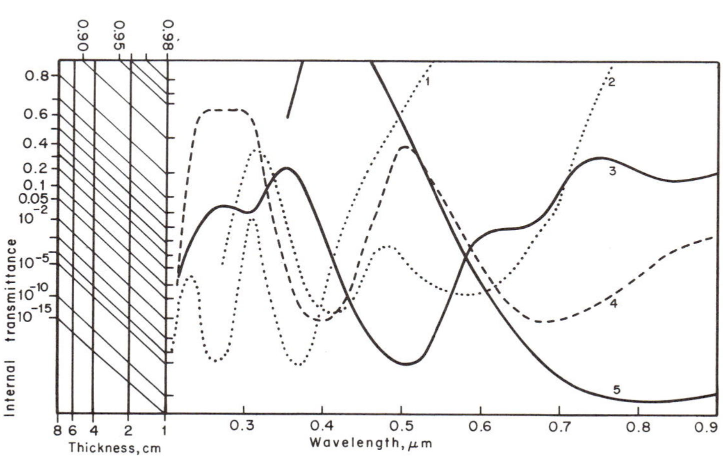

Fig. 13 Spectral transmittance of aqueous solutions of inorganic materials, plotted on a log (internal-density) scale. The auxiliary scales can be used to read off directly the transmittances of 1-, 2-, 4-, 6-, and 8-cm thicknesses of these filters (Par. 9). Curve 1: $KCrO_4$, 1 mmol/l; curve 2: $KCr(SO_4)_2$, 0.1 mol/l; curve 3: $CoSO_4$, 1 mol/l; curve 4: $NiSO_4$, 1 mol/l; curve 5: $CuSO_4$, 1 mol/l. [*After Geffcken* (1957).]

vacuum ultraviolet are described in Gerasimova (1970). Narrow-bandpass filters employing liquid filters are described in Par. 28.

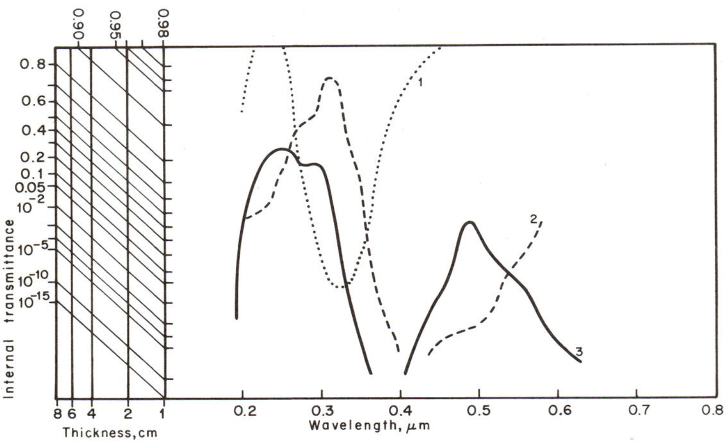

Fig. 14 Internal spectral transmittance of chlorine (*curve* 1) and bromine (*curve* 2) at 1 atm pressure [*after Alberti* (1970)] and of a $NiSO_4(H_2O)_6$ crystal (*curve* 3) [*after Childs* (1961)].

Absorbing Thin Films

25. Paragraphs 25 to 27 deal with thin films in which the primary filtering process is absorption, although at times interference within the film may have to be considered to explain the transmission characteristics fully. Such films can be produced by different processes and may or may not be self-supporting.

26. Metal Films The fact that some metals have narrow regions of low absorption (curves 1 and 2, Fig. 15; Fig. 70*b*) and that others transmit freely below 0.2 μm is well known. Wood and Larkens (1938) state, for instance, that the

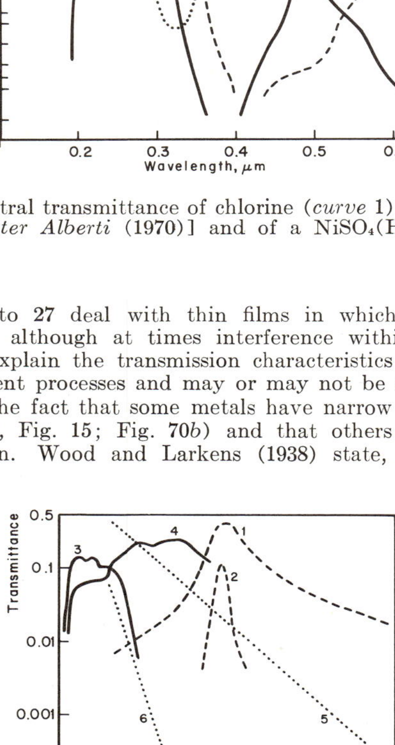

Fig. 15 Transmission of various metal films. Curves 1 and 2: 0.022- and 0.055-μm-thick silver films [*after Gibson* (1949)]; curves 3 and 4: 0.10- and 0.45-μm-thick sodium and potassium films [*after Robin and Robin* (1958)]; curves 5 and 6: 0.3- and 0.5-μm-thick films of a 25% Na, 75% K alloy between quartz plates [*after Shaw and Foreman* (1959)].

alkali metals Cs, Rb, K, Na, and Li transmit at wavelengths shorter than 0.44, 0.36, 0.315, 0.210, and 0.205 μm, respectively (see also Fig. 15). Unfortunately the alkali metals are chemically very active and must be sealed between quartz (see, for example, O'Bryan, 1935; Novikov, 1968) if they are not to corrode on exposure to air. This prevents their application in the extreme-ultraviolet region.

The measured and expected transmission properties in the extreme ultraviolet of certain promising metals that can be prepared in thin-film form are summarized in Fig. 16. The measured spectral transmittances of some of these materials are shown in Figs. 17 and 18. By increasing the thicknesses of the films higher rejection ratios could be obtained at the expense of peak transmissions, and vice versa. The transmittance of the most promising material, aluminum, would be higher if it were not for the formation of absorbing oxide layers.

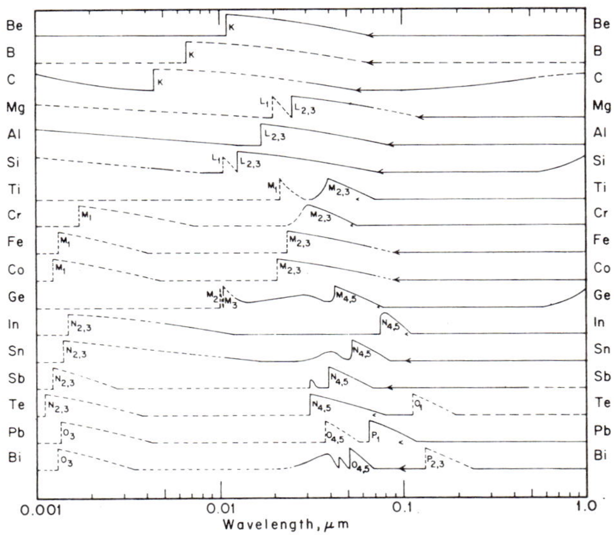

Fig. 16 Summary of the transmission properties of several metals, most of which can be obtained in thin-film form. X-ray edges are shown as steps; the arrowheads show the critical wavelengths as calculated by Pines (1956); dotted sections indicate ranges for which no experimental data are available. [*Hunter et al.* (1965), *reproduced with the permission of the authors.*]

Many of the layers are self-supporting (Fig. 17). Others (Fig. 18) must be deposited onto a suitable transparent substrate. Thin aluminum films are sometimes used for this purpose. Other materials used in the past are Zapon (cellulose acetate); collodion, Parlodion, and Celluloid (cellulose nitrates); Mylar (polyethylene tetraphthalate); and Formvar (polyvinyl formal) (Fig. 19). Any residual absorption in the substrate contributes, of course, to the overall-transmission curve. The preparation of self-supporting thin films is described by Novikov (1968) and Sorokin and Blank (1970). Because of their fragility such films are usually mounted on a very fine mesh screen.

27. Dielectric and Semiconductor Films All materials used in multilayer interference coatings possess short- and long-wavelength absorption edges. These can often be used to assist in the blocking of such filters. The admixture of small amounts of absorbing materials to an evaporant or organic coating solution is used at times to tune this absorption edge (Fig. 20*a* and *c*). For example, antireflection coatings containing such ultraviolet-absorbing materials can be used to protect works of art.

A series of commercial short-wavelength cutoff filters for the infrared spectral region consisting of chemically deposited silver sulfide coatings on silver chloride substrates are also shown in Fig. 20*b*. These filters are quite delicate. When

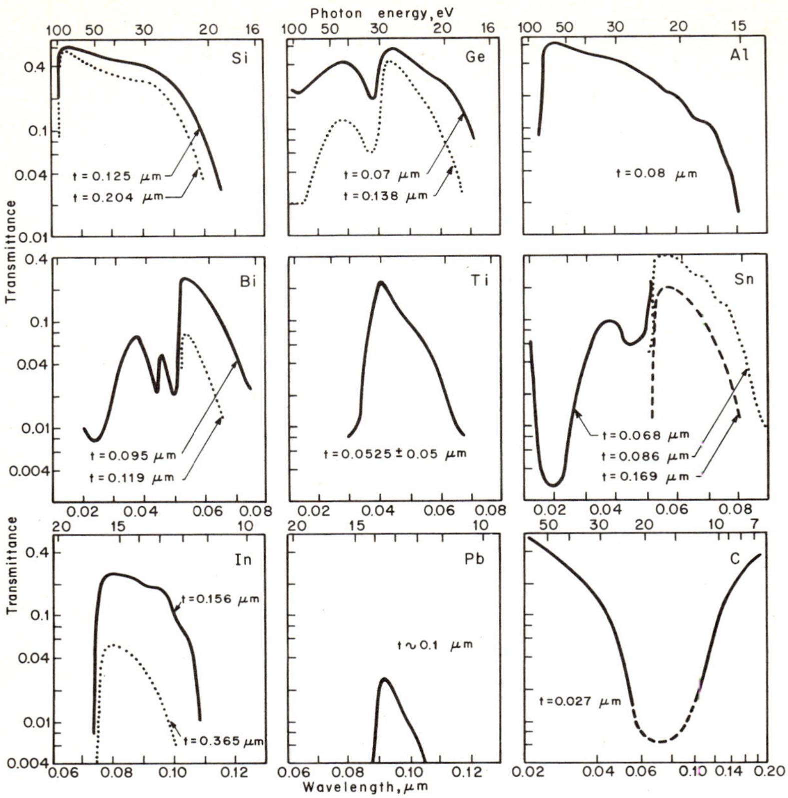

Fig. 17 Measured extreme-ultraviolet transmittance of several self-supporting metal films of indicated thicknesses. [Si, Ge, AP, Bi, Sn. In, *after Hunter et al.* (1965); Ti, *after Rustgi* (1965); Pb, *after Walker* (1963); C, *after Samson and Cairns* (1965).] Transmittance curves for Mg, Se, and Te films are given by Sorokin and Blank (1970).

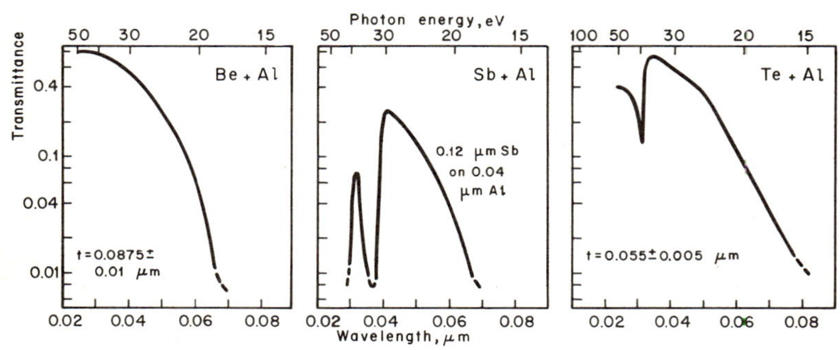

Fig. 18 Measured extreme-ultraviolet transmittance of metals deposited onto thin aluminum films. [*After Rustgi* (1965).]

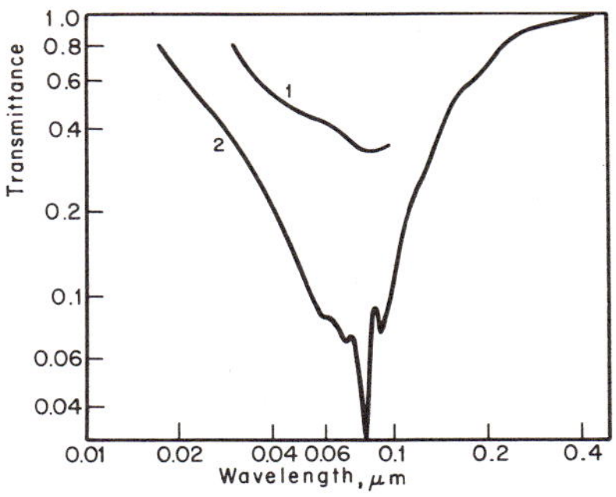

Fig. 19 Measured transmittance curves of two organic films used for supporting fragile thin-metal-film filters for the extreme ultraviolet. Curve 1: 0.01-μm-thick Celluloid film [*after O'Bryan* (1935)]; curve 2: 0.027-μm-thick Parlodion film [*after Hunter et al.* (1965)].

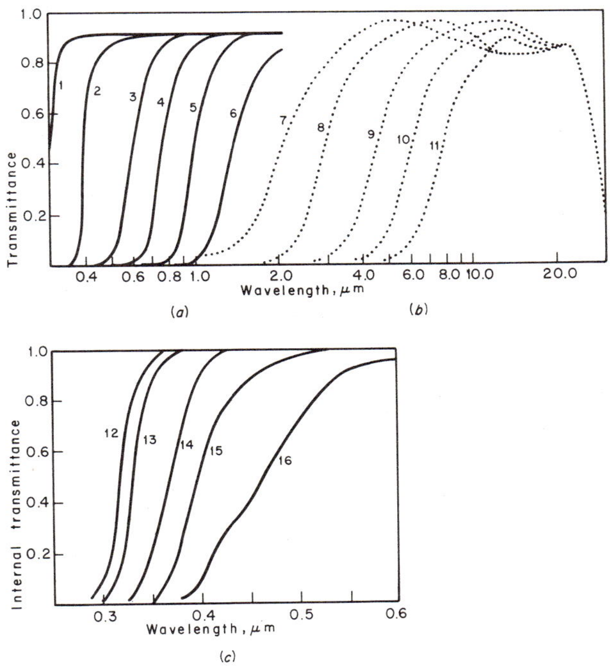

Fig. 20 Spectral transmittance of absorbing films produced in various ways. (*a*) Envelopes of the transmission maxima of thick evaporated films of ZnS (*curve* 2), Ge (*curve* 6), and various mixtures of ZnS and Ge (*curves* 3 to 5) on a glass substrate (*curve* 1). [*After Chang* (1964).] (*b*) Spectral transmittances of chemically deposited silver sulfide coatings on silver chloride substrates. [*Curves* 7 *to* 11, *Eastman Kodak* (1968*a*).] (*c*) Intrinsic transmittance of thin films of titanium dioxide with admixtures of heavy metal oxides, deposited from organic solutions. Curve 12: $TiO_2 + 1.5SiO_2$; curve 13: TiO_2; curve 14: $TiO_2 + 0.5PbO$; curve 15: $TiO_2 + 0.15Fe_2O_3$; curve 16: $TiO_2 + 5.7UO_3$. [*After Schröder* (1969).]

protected with a polystyrene layer, their transmittance is reduced and sharp absorption bands appear. The filters should not be used outdoors unless additionally protected, nor should they be exposed to ultraviolet radiation or temperatures in excess of 110°C.

In architecture absorbing films on glass are used for visual appearance and thermal control, in tinted window panes for illumination control and in "one-way mirrors" (Fig. 21) which depend for their effectiveness on unequal illumination.

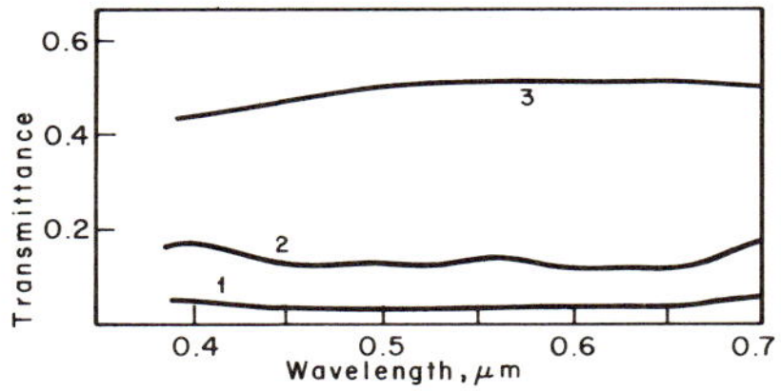

Fig. 21 Spectral characteristics of a thin chromium alloy film on glass. Curve 1: transmittance; curves 2 and 3: reflectance from the glass and air sides, respectively. [*Liberty Mirror* (1966).]

Narrow-Transmission-Band Filters Composed of Absorbing Materials

28. Interference filters (Pars. 72 to 88) combined with suitable absorption filters have several advantages over narrow-band filters that are composed entirely of absorption filters. Their peak transmittance is usually higher, and they can be constructed for any wavelength within a broad spectral region. Much narrower filters can be produced, and the shapes and half widths of their transmission bands can be varied almost at will. Nevertheless, narrow-bandpass absorption filters continue to be used because they are normally cheaper than interference filters and because for many purposes their performance is entirely satisfactory. For some applications the fact that their spectral properties are much less dependent on the angle of incidence is a decided advantage.

A set of commercially produced colored-glass filters with transmission bands in the 0.3- to 0.8-μm spectral region is shown in Fig. 22. Each filter consists of two or more components. Many similar glass-filter combinations have been described in the past (for example, Wyszecki, 1960). By the use of more components and a careful selection of the available colored glasses (Fig. 6*a* to *s*) filters with appreciably higher peak transmittances and narrower half widths can often be designed. One such filter is included in Fig. 22 for comparison.

A set of gelatin filters with narrow transmission bands produced by Ilford (1960) is shown in Fig. 23. The properties of such filters are described in greater detail in Par. 22. The transmittance curves of the corresponding Wratten filters are given in Fig. 10*d* to *f*.

Many publications describe bandpass filters in which several different types of absorption filters, i.e., liquids, glasses, crystals, gelatin films, etc., are combined. A particularly useful set of filters for the ultraviolet spectral region is shown in Fig. 24. A corresponding set of filters for the visible region is described by Winther (1937).

Readers contemplating the construction of absorbing filters with narrow transmission bands should also consult the review article by Geffcken (1957) for the description of additional filters of this type. McBride and Olsen (1962) discuss several new types of absorbing materials that have been suggested more recently for the construction of the narrow-band filters in the ultraviolet spectral region. Thus, for example, Childs (1961) used $NiSO_4(H_2O)_6$ crystals, cation-X, and lead-doped KCl:KBr crystals to construct filters for 0.26 and 0.27 μm (Fig. 25). Pellicori (1964) used organic solutions and glasses to construct filters for 0.27 and 0.29 μm. It was found later that the same dyes could be used in a glass resin to form solid organic filters with similar special characteristics (Fig. 26).

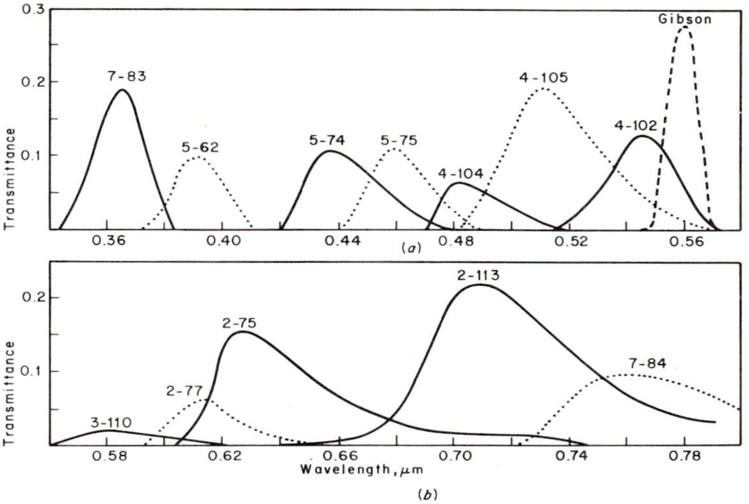

Fig. 22 (*a*) and (*b*) Glass filters with narrow transmission bands made by Corning 1965. These filters transmit again in the 1.0- to 4.0-μm region. A special four-component filter is shown for comparison. [*Gibson* (1935).]

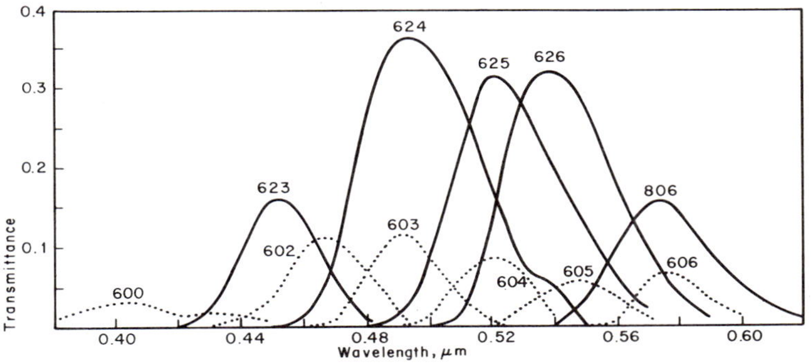

Fig. 23 Gelatin filters with narrow transmission bands. [*Ilford* (1960).]

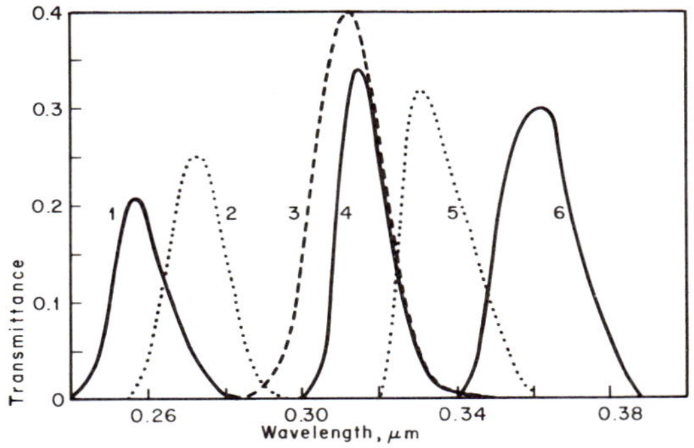

Fig. 24 Narrow-bandpass filters for the ultraviolet spectral region [*Kasha* (1948)]. The filters are composed of glass filters and of several organic and inorganic liquid solutions shown in Figs. 12 and 13. The original reference should be consulted for details.

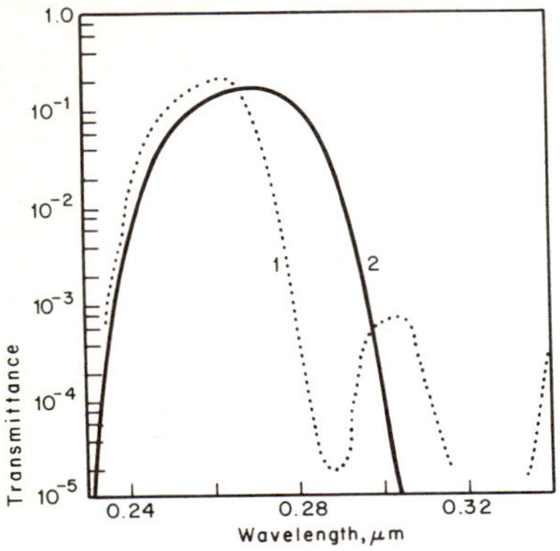

Fig. 25 Ultraviolet filters with narrow transmission bands composed of $NiSO_4(H_2O)_6$ crystals, glass filters, cation-X in polyvinyl alcohol films, and a lead-doped KCl:KBr crystal. [*Childs* (1961).]

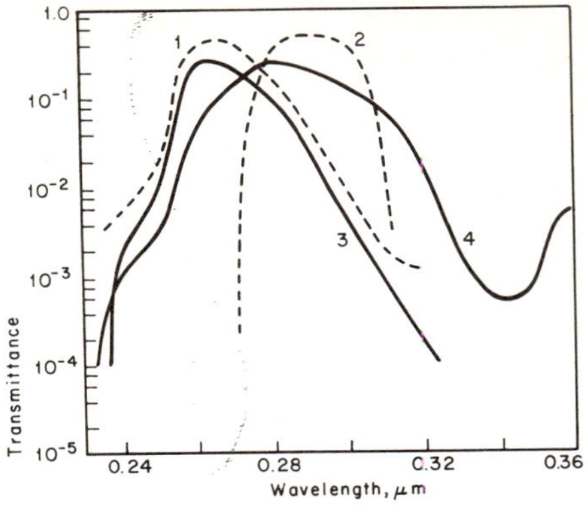

Fig. 26 Ultraviolet filters with narrow transmission bands composed of glass filters and organic compounds in liquid solutions (*full curves*) [*Pellicori* (1964)] and in solid glass resin (*broken curves*) [*Pellicori et al.* (1966)]. The filters transmit above 0.36 μm.

Neutral-Density Filters and Devices

29. These devices are used whenever the intensity of the incident radiation is to be reduced uniformly throughout an extended part of the spectrum.

30. Mechanical Methods Self-supporting opaque grids and screens offer one of the most uniform means of attenuating radiation (Harrison, 1929) providing that the ratio of the filter aperture to the wavelength is large. Densities of up to 2.0 can be achieved in this way. Higher densities can be obtained with evaporated or etched metallic-film grids on transparent substrates, but then the neutrality and spectral range depend on the substrate material. Unless the diameter of the beam of radiation incident onto the screen is wide enough to contain a large number of screen elements, an uncertainty in the transmission will result. This can be reduced by vibrating the screen.

Rotating choppers can be used to obtain densities of up to about 2 for those applications in which the intermittent nature of the resulting signal can be tolerated (see, for example, Jones and Clarke, 1961; Stewart, 1962). Various electromechanical shutters have also been proposed; e.g., one such device, with a transmission range of 5 to 95 percent, has been described by Berry and Jones (1964).

Photographic emulsions and colloidal carbon suspensions also remove radiation by obstruction. Wedge and step filters with densities up to 4 or 6 can be readily made. The range and neutrality again depend on the materials used (Fig. 27). Photographic emulsions scatter much of the transmitted radiation, and for this reason the actual density depends somewhat on the angles and apertures of the associated optical system (Fig. 27, curves 2 and 3).

31. Absorption Filters A large number of glasses are suitable for making neutral-density filters with densities of up to 5.0 (Fig. 6*g* and *h*). Gelatin filters (Fig. 10*i*) offer an alternative to glass filters.

Evaporated films of metals such as aluminum, chromium, palladium, platinum, rhodium, and tungsten and alloys such as Chromel and Inconel have been used for a long time to produce filters with densities of up to 6.0. A disadvantage of such filters is their high specular reflection. At present Inconel is commonly used for high-precision neutral-density filters. Chromium is favored when tough, unprotected coatings are required (Fig. 27). The operating range and neutrality depend again on the substrate materials. The spectral-transmittance curves of two sets of commercial filters on glass and quartz substrates are shown in Fig. 28*a* and *b*. Linear and circular metal-film neutral-density wedges and step filters are available commercially.

For the highest-precision neutral-density filters one or more neutral glasses and/or metal films can be combined. A series of such filters with excellent characteristics has been described by Theissing and Goebert (1940).

32. Methods Based on Polarized Light Polarized beams of radiation can be attenuated with polarizers (Par. 3 of Sec. 10). Densities of up to about 3.0 can be obtained in this way. For best neutrality a nicol prism, a pile of plates (Pars. 56ff. of Sec. 10), or a wide-band interference polarizer (Pars. 69 to 71) should be used.

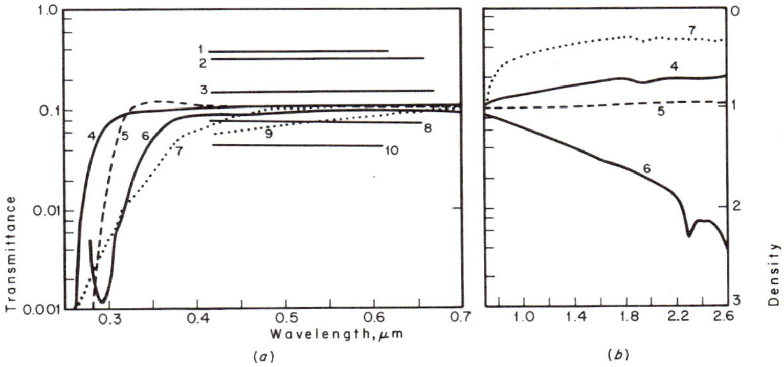

Fig. 27 (*a*), (*b*) Spectral transmittance of various neutral-density materials. Curve 1: tungsten film on glass; curves 2 and 3: diffuse and specular transmittance of a photographic emulsion; curve 4: M-type carbon suspension in gelatin; curve 5: Inconel film on glass; curve 6: photographic silver density; curve 7: Wratten 96 density filter; curve 8: chromium film; curves 9 and 10: Chromel A film on glass evaporated at pressures of 10^{-3} and 10^{-4} Torr, respectively. [*Curves* 1 *to* 3, 9 *and* 10 *after Banning* (1947*a*); *curves* 4 *to* 7, *Eastman Kodak* (1969); *curve* 8, *Optical Coating Laboratory* (1964).]

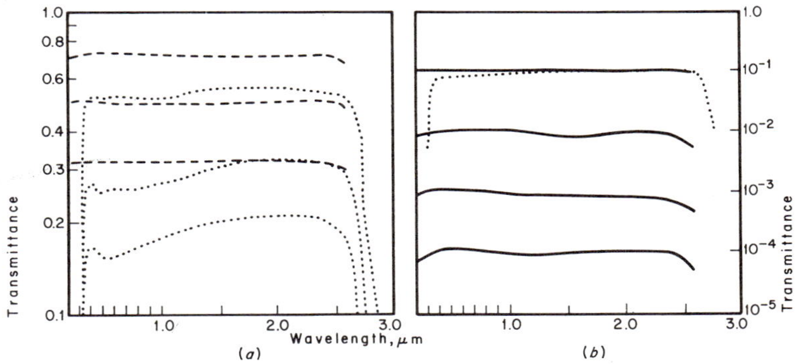

Fig. 28 (*a*), (*b*) Neutral-density attenuators. Inconel films on glass substrates (*dotted curves*) [*Bausch & Lomb* (1968)] and on quartz substrates (*full and broken curves*) [*Corion* (1969)].

Very small and accurately controlled attenuations (0.05 percent and more) of a beam of radiation can be achieved by inserting a plane-parallel plate into the beam and tilting it away from the Brewster angle. The neutrality of this device depends only on the dispersion of the optical constants of the plate material. Since the inherent loss of such a plate can be very small, it can be used within a laser cavity (Schleusner and Read, 1966; Sporton, 1967). In a modification of this device the plate is maintained at the Brewster angle with respect to the incident radiation, but its azimuth is rotated away from the plane of polarization of the latter. This arrangement is discussed by Allen et al. (1968), Evans and Morgan (1968), Dändliker (1968), and Dobrowolski (1969).

FILTERS AND COATINGS BASED ON INTERFERENCE IN THIN FILMS

Introduction to the Theory of Optical Thin Films (Berning, 1963)

33. If a beam of electromagnetic radiation falls onto a structure consisting of films of several different materials, multiple reflections will take place within the structure. Should the distances between the various boundaries be small enough, the reflected beams will be coherent with one another and will interfere. This effect can be utilized to construct optical filters with widely varying spectral characteristics. In this paragraph the basic equations for thin-film calculations are presented and used to describe some general properties of interference filters.

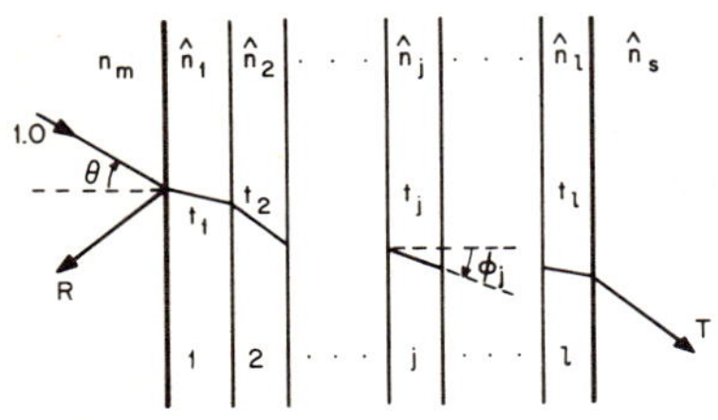

Fig. 29 Construction parameters of a multilayer.

Consider the thin-film system consisting of l layers shown in Fig. 29. The construction parameters comprise not only the refractive indexes n_j and the thicknesses t_j of the layers but also the refractive indexes n_s and n_m of the substrate and the incident medium. The angle of incidence θ and the wavelength λ and plane of polarization ($\parallel$ or $\perp$) of the incident radiations are the external variables of the system.

The most general method of calculating the transmittance T and the reflectance R of a multilayer from the above quantities is based on a matrix formulation (Herpin, 1947; Weinstein, 1954) of the boundary conditions of the film surfaces derived from Maxwell's equations (Born and Wolf, 1959).

It can be shown that the jth film can be represented by the 2×2 matrix

$$M_j = \begin{bmatrix} \cos \delta_j & \dfrac{i}{u_j} \sin \delta_j \\ iu_j \sin \delta_j & \cos \delta_j \end{bmatrix} \tag{24}$$

where

$$\delta_j = \frac{2\pi}{\lambda} (n_j t_j \cos \phi_j) \tag{25}$$

the quantity $n_j t_j \cos \phi_j$ often being called the *effective optical thickness* of the layer for an angle of refraction ϕ_j, and where u_j, the *effective refractive index*, is given by

$$u_j = \begin{cases} \dfrac{n_j}{\cos \phi_j} & \text{parallel} \\ n_j \cos \phi_j & \text{perpendicular} \end{cases} \tag{26}$$

depending on whether the incident radiation is polarized parallel or perpendicular to the plane of incidence.

The angle ϕ_j is related to the angle of incidence θ by Snell's law

$$n_m \sin \theta = n_j \sin \phi_j \tag{27}$$

The complete multilayer of Fig. 29 is represented by the product matrix M,

$$M = M_1 M_2 \cdots M_j \cdots M_{l-1} M_l \tag{28}$$

$$M = \begin{bmatrix} m_{11} & im_{12} \\ im_{21} & m_{22} \end{bmatrix} \tag{29}$$

In the above equations the refractive index of any absorbing material in the system must be replaced by its complex refractive index $\hat{n}$, defined by

$$\hat{n} - n - ik \tag{30}$$

where k is the extinction coefficient of the material. Even though all the elements of the layer matrix for such a material are now complex, its determinant will still be unity.

In terms of the elements of the product matrix, the amplitude transmittance and reflectance coefficients t and r are given by

$$t = \frac{2n_m}{(X + W) + i(Y + V)} \tag{31}$$

$$r = \frac{(X - W) + i(Y - V)}{(X + W) + i(Y + V)} \tag{32}$$

where

$$X = n_m m_{11} + n_m k_s m_{12} \qquad Y = n_m n_s m_{12}$$
$$W = n_s m_{22} \qquad V = m_{21} - k_s m_{22} \tag{33}$$

The intensity transmittance and reflectance coefficients are

$$T = \frac{n_s}{n_m} |t|^2 \tag{34}$$

$$R = |r|^2 \tag{35}$$

and the phase changes on transmission ϵ_T and reflection ϵ_R are given by

$$\epsilon_T = \arg t \tag{36}$$
$$\epsilon_R = \arg r \tag{37}$$

The absorption of a multilayer is calculated from

$$A = 1 - T - R \tag{38}$$

Certain important general conclusions about the properties of multilayer filters can be drawn from the above equations.

1. The properties of thin-film systems vary with angle of incidence [Eq. (25)]. For some applications this is the major disadvantage of interference over absorption filters.

2. This variation depends on the plane of polarization of the incident radiation [Eq. (26)]. The following equations define T and R for obliquely incident nonpolarized radiation:

$$T = \tfrac{1}{2}(T^{\parallel} + T^{\perp}) \tag{39}$$
$$R = \tfrac{1}{2}(R^{\parallel} + R^{\perp}) \tag{40}$$

3. Transmittance curves of nonabsorbing multilayers composed of layers whose optical thicknesses are all multiples of $\lambda_0/4$ show symmetry about λ_0 when plotted on a wave-number scale [Eqs. (24) and (25)].

4. A proportional change of all the thicknesses of a nonabsorbing multilayer results merely in a displacement of the transmittance curve plotted on a wave-number scale [Eq. (25)]. Thus a thin-film design can be utilized in any part of the spectrum subject only to the limitations imposed by the dispersion and transmittances of the materials used.

5. The reflectance and absorption of a layer system containing absorbing layers will depend, in general, on which side of the coating the radiation is incident.

The analysis of optical thin-film systems [with computer programs based in the most part on Eqs. (24) to (40)] has become relatively simple. The design of filters with any but the simplest spectral characteristics remains a complicated problem. A detailed description of the various theoretical and numerical tools available for this task is outside the scope of this section.†

† The interested reader is referred to the books on this topic (Anders, 1967; Heavens, 1955; Macleod, 1969; Mayer, 1950; Rozenberg, 1958; Vacisek, 1960) and comprehensive review articles (Abelès, 1950 and 1967; Baumeister, 1962, Berning, 1963; Heavens, 1960; Thelen, 1969; Weinstein, 1954; Wolter, 1956).

Antireflection Coatings

34. Effect of Surface Reflections on Performance of Optical Systems The intensity of the radiation reflected at the boundary between two optical media of refractive indexes n_1 and n_2 is given by

$$R = \left[\frac{n_1 - n_2}{n_1 + n_2}\right]^2 \tag{41}$$

When multiple internal reflections are taken into account, the total transmittance T_0 of a nonabsorbing plane-parallel plate is given in terms of R by (Par. 4)

$$T_0 = \frac{1 - R}{1 + R} \tag{42}$$

Of this radiation only a fraction $(1 - R)^2$ is transmitted without undergoing any reflections. In the corresponding expressions for a number of identical plates placed in series it is necessary to allow for the effect of interreflections that occur between the various plates. It can be shown (Dioffo, 1968) that the total transmittance T_{total} of m plates placed in series is given by

$$T_{\text{total}} = \frac{T_0}{m - T_0(m - 1)} \tag{43}$$

The amount of radiation transmitted directly T_{dir} is

$$T_{\text{dir}} = (1 - R)^{2m} \tag{44}$$

and so the fraction T_{refl} of the radiation that suffers reflections before transmission and which therefore contributes to spurious images and stray radiation is

$$T_{\text{refl}} = T_{\text{total}} - T_{\text{dir}} \tag{45}$$

The variation of T_{dir} and T_{refl} with R, for several values of m, is shown in Fig. 30. The refractive indexes of the plate material plotted against the upper x axis assume that the plates are in air.

It will be seen that even for a relatively small number of low-refractive-index plates the ratio $T_{\text{refl}}/T_{\text{dir}}$ becomes significant. This means that in an image-forming system under unfavorable conditions the stray radiation can completely obscure the true signal (Mussett, 1970). Second, even in nonimaging optical systems the loss in radiation $(1 - T_{\text{dir}} - T_{\text{refl}})$ can become quite prohibitive.

Both these problems can be overcome by reducing the surface reflection through the application of suitable antireflection coatings to the plate boundaries. Since antireflection coatings with zero reflectance across the whole spectrum cannot be constructed, the spectral reflectance R_λ of antireflection coatings is chosen to minimize the integral

$$\int_\lambda R_\lambda I_\lambda S_\lambda \, d\lambda \tag{46}$$

where I_λ and S_λ are the spectral-intensity distribution of the incident radiation and the spectral sensitivity of the detector, respectively. A low reflectance is thus needed only in the spectral region in which $I_\lambda S_\lambda$ is significant. It should be emphasized here that in addition to this requirement, for most applications antireflection coatings must be mechanically very tough, be able to withstand drastic climatic and thermal variations, and stand up to the usual lens-cleaning procedures. Some examples of improvements in the performance of image-forming and non-image-forming optical systems obtained through the use of antireflection coatings are given by Mussett (1970) and Faber et al. (1961).

Because of their industrial importance antireflection coatings for the visible and the infrared (Turner et al., 1953*b*) spectral regions have been the subject of much research and development. Two books have been written on this topic (Grebenshchekov, 1946; Sawaki, 1960), and there exists a very extensive literature in the

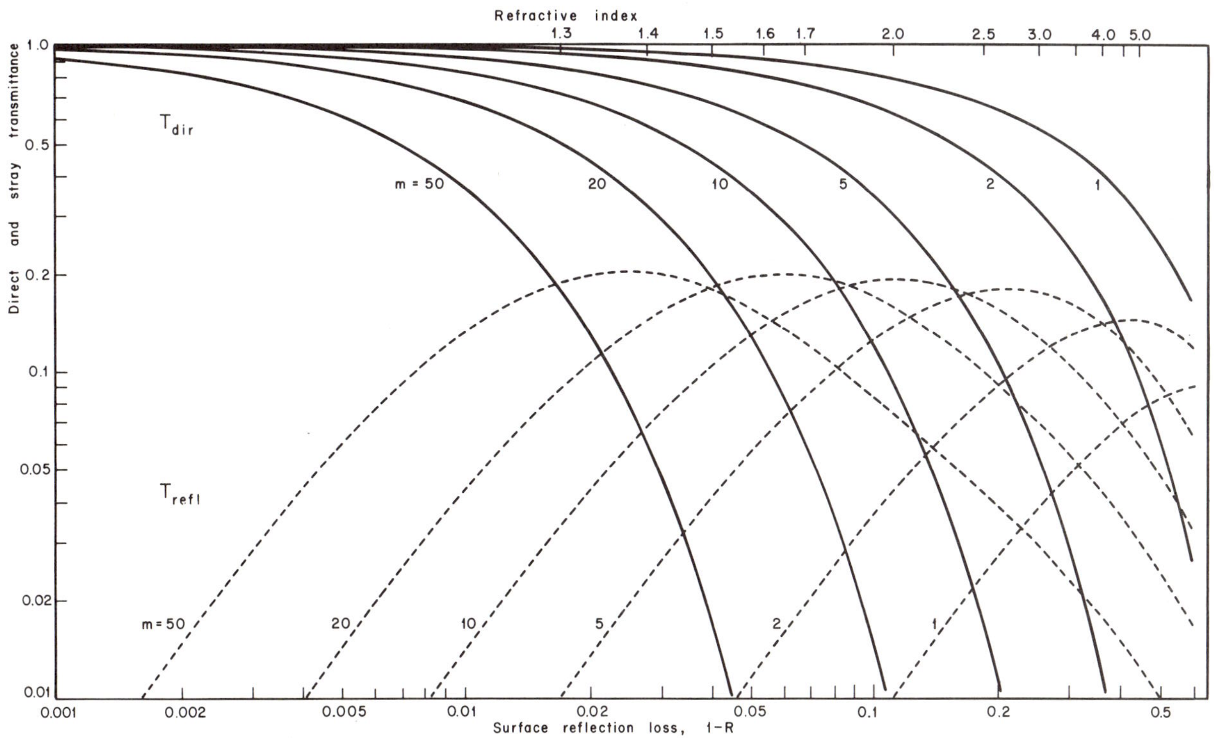

Fig. 30 Transmittance of a system of m plane-parallel plates of surface reflectance R for directly transmitted radiation T_{dir} and for radiation that suffers multiple reflections before transmission T_{refl}. The upper x axis is calibrated in terms of the refractive index of the plates for the special case when the plates are in air.

scientific and technical journals. For a review of this literature and a systematic discussion of antireflection coatings the reader is referred to the excellent review articles by Cox and Hass (1964) and Mussett and Thelen (1970). In this section only a brief summary will be given of the results obtained thus far, intended to aid in the selection of antireflection coatings for particular applications. The calculated data are presented on a log (density) scale (Par. 9) to enable the transmittance of a number of identical antireflection-coated surfaces placed in series to be read off directly. The relative wave-number scale facilitates the calculations of the width of the effective region of a coating in different parts of the spectrum.

35. Antireflection Coatings Made of Homogeneous Layers The single-homogeneous-layer antireflection coating is the first antireflection coating and probably still the most widely used. Theoretically it should be possible to obtain with it a zero reflectance at one wavelength, but because of a lack of suitable low-index coating materials, this cannot be realized in practice for materials with indexes less than about 1.9. Nevertheless, even with the available materials a very useful reduction in reflection in a broad spectral region is obtained for all common glass types, the reflectance never rising above that of the uncoated surface.

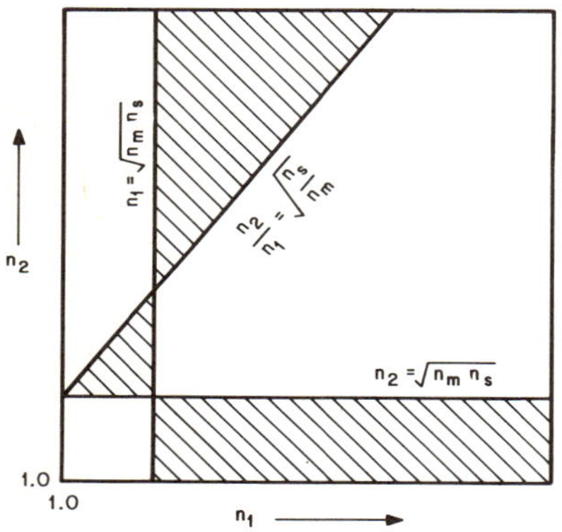

Fig. 31 Refractive-index combinations (*shaded areas*) of two-layer antireflection coatings with which a zero reflectance at one wavelength can be achieved.

If more than one layer is used, all the degrees of freedom could be used (1) to obtain a more complete antireflection in one particular spectral region, (2) to increase the width of the spectral region over which the reflectance is generally low, or (3) to obtain a coating in which the low reflectance is very uniform across the spectrum. But, for example, Vlasov has shown (Grebenshchekov, 1946), with the aid of a diagram of the type shown in Fig. 31, that even for a two-layer antireflection coating there exists a large number of refractive-index combinations which will yield zero reflectance at one wavelength. It therefore becomes possible, as the number of layers increases, to find solutions for a particular problem which not only fully meet the most important above desiderata and almost satisfy the others but are also based on the use of the mechanically most satisfactory coating materials.

The conditions that must be satisfied by the refractive indexes and thicknesses of various types of antireflection coatings are given in Table 3, unless they are very complicated, in which case reference is made to a paper in which they are set out in full. The calculated transmittance curves of antireflection-coated surfaces of glass (Fig. 32*a* to *c*), quartz (Fig. 32*d*), germanium (Fig. 32*e* and *f*), silicon (Fig. 32*g*), and other infrared materials (Fig. 32*h*) utilize refractive indexes that for the most part correspond to real coating materials, and hence the curves represent realistic solutions rather than the theoretically best possible ones. The actual refractive indexes used in the calculations (and optical thicknesses where they do not correspond to a multiple of $\lambda/4$) are given in Table 3.

Figure 32*a* shows the performance of several antireflection coatings on glass for applications which require the highest possible efficiency for a limited wavelength region. Of these the coating 2.1 has probably found the greatest acceptance. A typical measured-performance curve for such a coating is shown in Fig. 34.

Antireflection coatings with a low reflectance in a broad spectral region are shown in Figs. 32*b* and 35. Coatings of the type 3.4 probably find the widest application in practice (Fig. 33). It is possible to trade off the width of the low-reflectance region for the maximum reflectance within it and to design similar antireflection coatings for other substrate indexes. For example, Furman (1966) gives a series

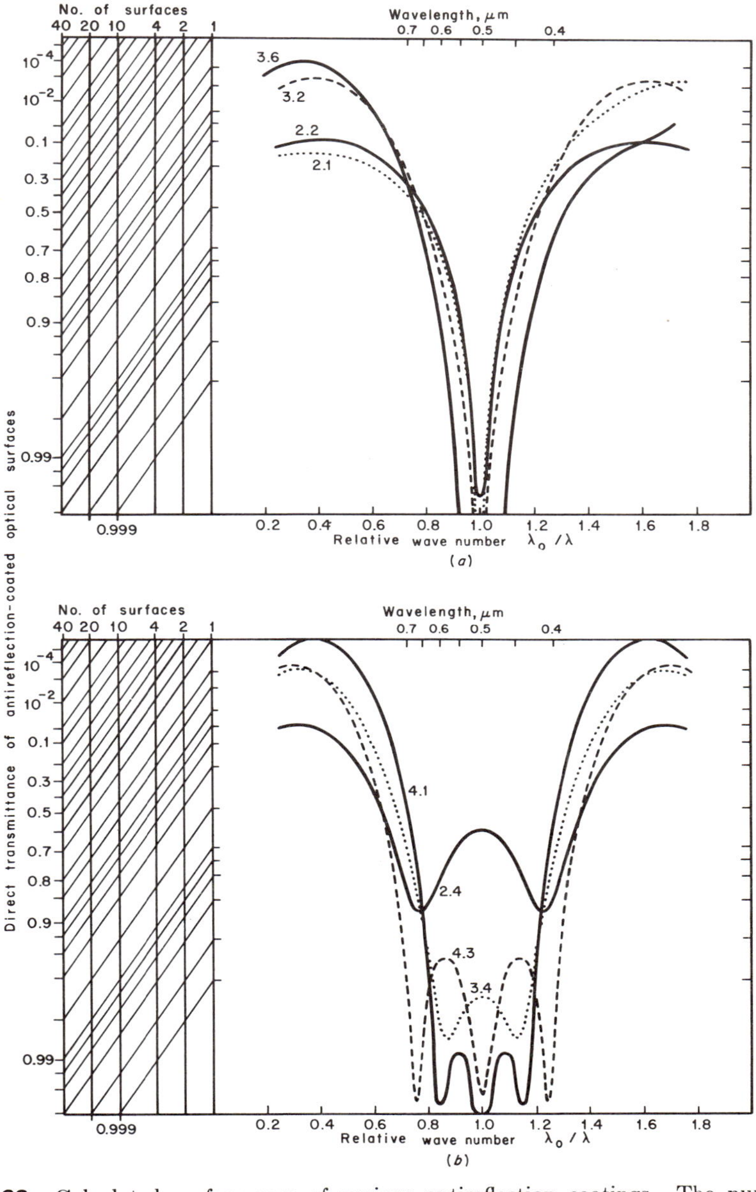

Fig. 32 Calculated performance of various antireflection coatings. The numbers identifying the individual curves refer to Table 3. The direct transmittances of 1, 2, 4, 10, 20, and 40 identical antireflection-coated surfaces can be read off directly with the aid of the auxiliary scales (Par. 9). (*a*) High-efficiency antireflection coatings for glass; (*b*) broadband antireflection coatings for glass. (*Figure 32 continues on pp. 8-50 through 8-52.*)

TABLE 3 Some Antireflection Coatings

The medium in all cases is air with $n_m - 1.00$

Type	Conditions or reference	Substrate material	n_s	n_1 (n_1t_1/λ)	n_2 (n_2t_2/λ)	n_3 (n_3t_3/λ)	n_4 (n_4t_4/λ)
1.1	$n_1 = \sqrt{n_s n_m};\ n_1t_1 = \frac{\lambda}{4}$	Glass	1.51	1.38			
		Irtran II	2.20	1.59			
		InSb	4.0	2.20			
		InAs	3.4	1.85			
		Ge	4.1	2.20			
		Si	3.5	1.85			
2.1	$\tan^2 \delta_1 = \frac{n_1^2(n_m - n_s)(n_s n_m - n_2^2)}{(n_s n_1^2 - n_m n_2^2)(n_s n_m - n_1^2)}$ $\tan^2 \delta_2 = \frac{n_2^2(n_m - n_s)(n_s n_m - n_1^2)}{(n_s n_1^2 - n_m n_2^2)(n_s n_m - n_2^2)}$	Glass	1.51	1.38 (0.3250)	2.30 (0.0524)		
		Quartz	1.48	1.48 (0.3255)	2.09 (0.0947)		
		Ge	4.1	4.10 (0.0586)	1.35 (0.0951)		
2.2	$n_1^2 n_s = n_2^2 n_m \qquad n_1t_1 = n_2t_2 = \frac{\lambda}{4}$	Glass	1.51	1.38	1.7		
		Ge	4.1	1.57	3.3		
2.3	$n_1 n_2 = n_m n_s \qquad n_1t_1 = n_2t_2 = \frac{\lambda}{4}$	Si	3.5	1.35	2.20		
2.4	$n_2^2 - \frac{n_2 n_s}{2n_1 n_m}(n_m^2 + n_1^2)(n_1 + n_2) + n_1 n_s^2 = 0$ $n_1t_1 = \frac{1}{2}n_2t_2 = \frac{\lambda}{4}$	Glass	1.51	1.38	1.70		
2.5	$n_1t_1 = n_2t_2 = \frac{\lambda}{4}$ [see Kard (1959*a*)]	Glass	1.55	1.32	1.484		

3.1	$n_1n_3 = n_2^2 = n_mn_s \quad n_1t_1 = n_2t_2 = n_3t_3 = \frac{\lambda}{4}$	Ge	4.1	1.35	2.20	3.30	
3.2	$n_1n_3 = n_2\sqrt{n_mn_s} \quad n_1t_1 = n_2t_2 = n_3t_3 = \frac{\lambda}{4}$	Glass	1.53	1.47	2.14	1.80	
3.3	$n_2^2 = n_mn_s \quad n_mn_3^2 = n_1^2n_s$ $n_1t_1 = n_2t_2 = n_3t_3 = \frac{\lambda}{4}$	Si	3.45	1.38	1.86	2.56	
3.4	$n_1^2n_s = n_mn_3^2 \quad n_1t_1 = \frac{1}{2}n_2t_2 = n_3t_3 = \frac{\lambda}{4}$	Glass	1.51	1.38	2.10	1.65	
		Quartz	1.48	1.38	2.10	1.65	
3.5	$n_1^2n_s = n_mn_3^2 \quad n_1t_1 = \frac{1}{2}n_2t_2 = \frac{1}{3}n_3t_3 = \frac{\lambda}{4}$						
3.6	Thetford (1969)	Glass	1.52	1.38 (0.2402)	2.20 (0.4105)	1.80 (0.1799)	
3.7	$n_1t_1 = n_2t_2 = n_3t_3 = \frac{\lambda}{4}$ [Kard (1959*a*)]	Glass	1.55	1.32	1.454	1.53	
4.1	$n_1n_4 = n_3\sqrt{n_mn_s} \quad n_1t_1 = \frac{1}{2}n_2t_2 = n_3t_3 = n_4t_4 = \frac{\lambda}{4}$	Glass	1.51	1.38	2.35	1.548	1.38
4.2	$n_1n_4 = n_2n_3 = n_mn_s \quad n_1t_1 = n_2t_2 = n_3t_3 = n_4t_4 = \frac{\lambda}{4}$	Ge	4.0	1.38	1.82	2.20	2.96
4.3	$n_1t_1 = n_2t_2 = n_3t_3 = n_4t_4 = \frac{\lambda}{4}$ [Kard (1959*a*)]	Glass	1.55	1.32	2.014	2.289	1.846
4.4	Design derived from Kard (1959*a*)	Glass	1.55	1.38 (0.2424)	1.832 (0.2390)	1.888 (0.2463)	1.656 (0.2417)
10.1	$n_l = n_s - l\frac{n_s - n_2}{10} \quad l = 1, 2 \ldots, 10$ $n_1t_1 = n_2t_2 = \cdots = n_{10}t_{10} = \frac{\lambda}{4}$	Ge	4.0	$n_1 = 1.35, n_2 = 1.614, \ldots, n_{10} = 3.735$			

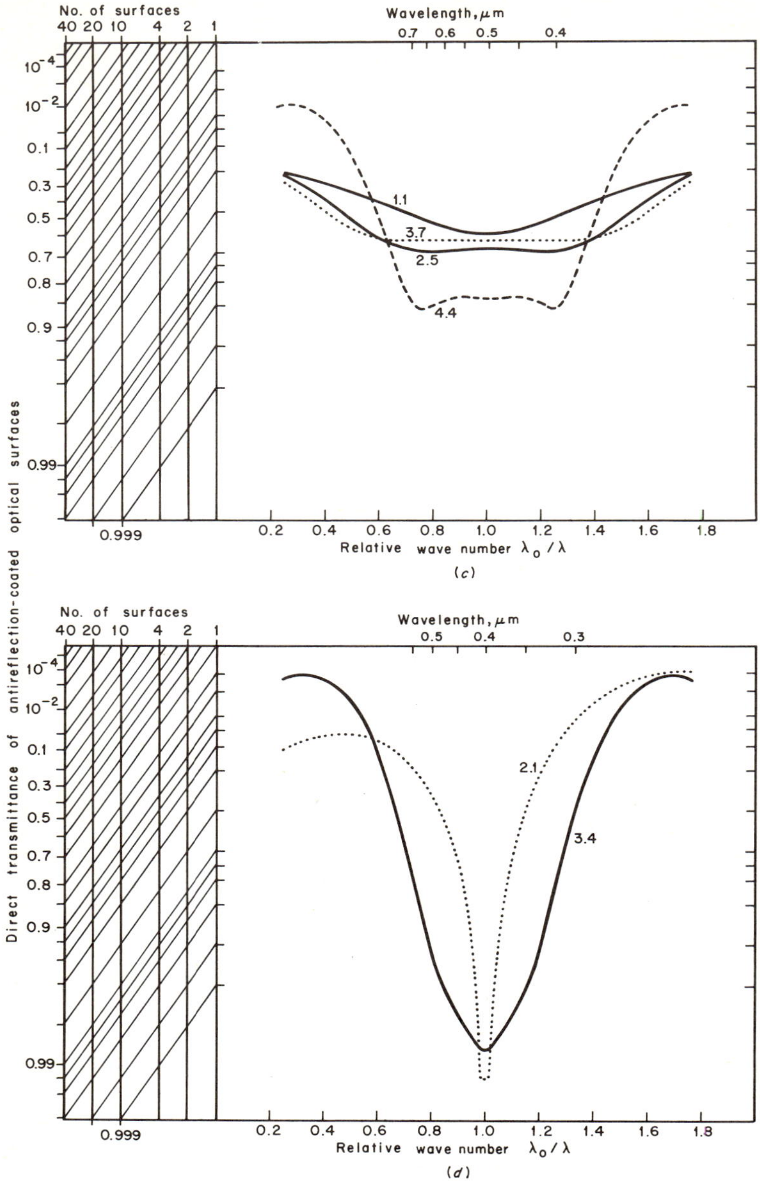

Fig. 32 (*Continued*) (*c*) Highly achromatic antireflection coatings for glass; (*d*) antireflection coatings for quartz.

of two-, three-, and four-layer solutions for 10 different values of the substrate index ranging between 1.5 and 4.0.

Since for some applications the color introduced into an optical system by antireflection coatings is of paramount importance, it has been the subject of many studies (Scharf, 1952; Murray, 1956; Kubota, 1961; Obodovsky et al., 1967). One way to avoid the problem is to utilize coatings that are particularly achromatic (Fig.

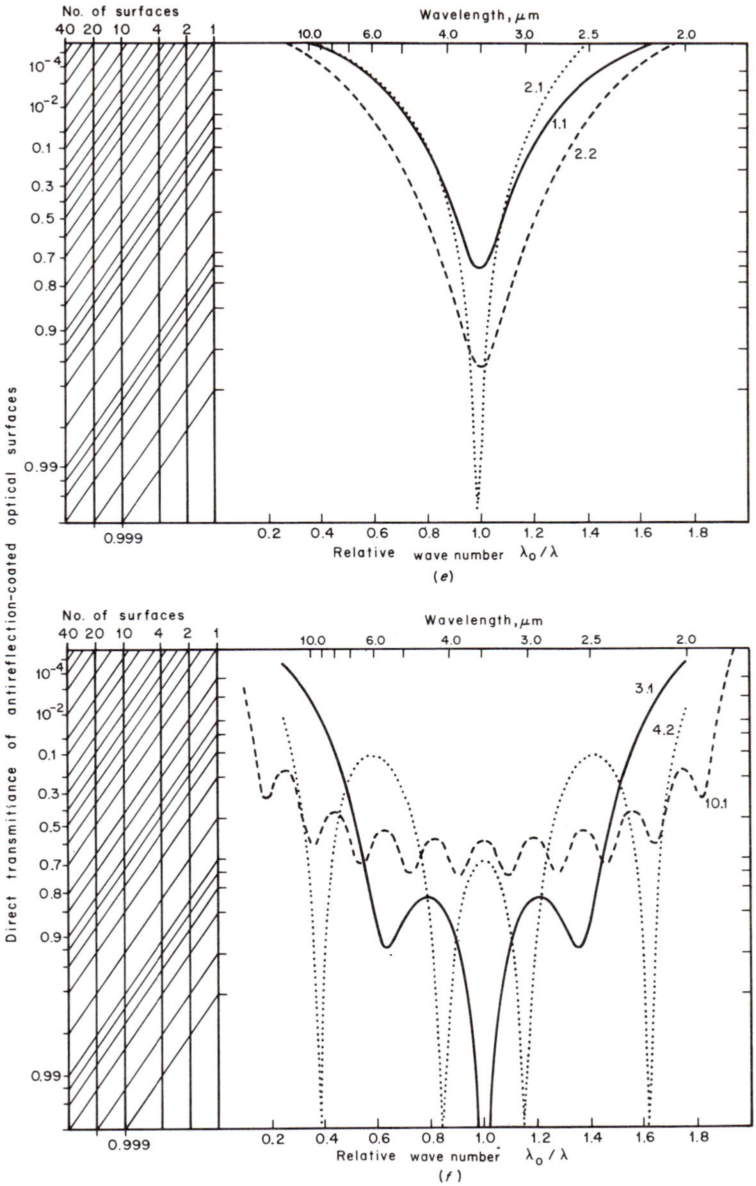

Fig. 32 (*Continued*) (*e*) Antireflection coatings for germanium; (*f*) broadband antireflection coatings for germanium.

32*c*). Thus, for example, 50 surfaces coated with antireflection coating 4.4 would have a transmittance of 78 ± 3 percent across the whole visible spectrum. Nevertheless, of the coatings shown in Fig. 32*c* only the single-layer antireflection coating is being used extensively (Fig. 33).

Often transmittance rather than reflectance measurements are performed to evaluate antireflection coatings. In general it is incorrect to assume that $T = 1 - R$. Thus, for instance, the transmittance of antireflection-coated infrared materials

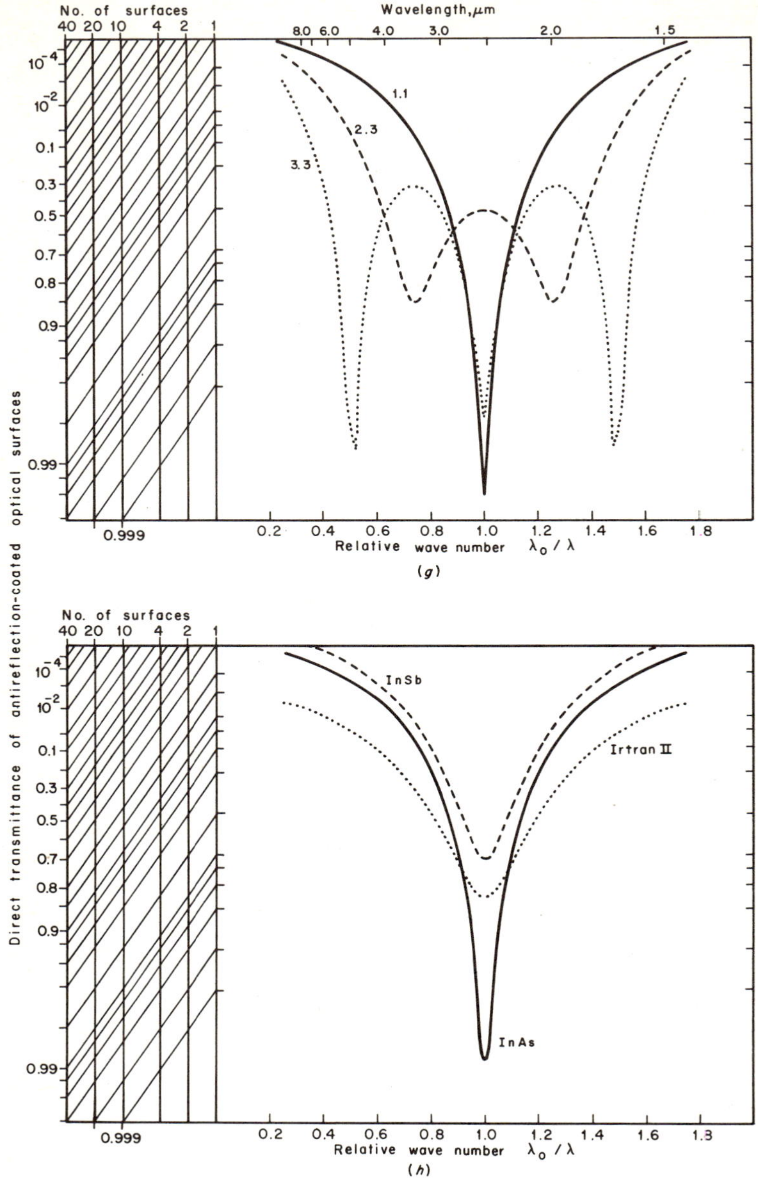

Fig. 32 *(Continued)* (*g*) Antireflection coatings for silicon; (*h*) single-layer antireflection coatings for Irtran II, InAs, and InSb.

depends not only on the efficiency of the antireflection coating but also on the thickness and temperature of the material. This is because of the finite scatter and absorption in such materials and because of the dependence, in some cases, of the latter on temperature. The measured spectral transmittances of three common antireflection-coated infrared materials at room temperature are shown in Fig. 37. Curves for other materials are given by Cox and Hass (1964).

36. Inhomogeneous and Many-Layer Antireflection Coatings The interface between two media with refractive indexes n_1 and n_2 can be completely antireflection-coated by the application of a transition layer with an index that changes continuously from n_1 to n_2. Linear, cosine, and hyperbolic refractive-index profiles have been investigated theoretically, and it has been found that the coatings remain effective over that spectral region for which the thickness is of the order of one wavelength or more (Fig. 38*a*).

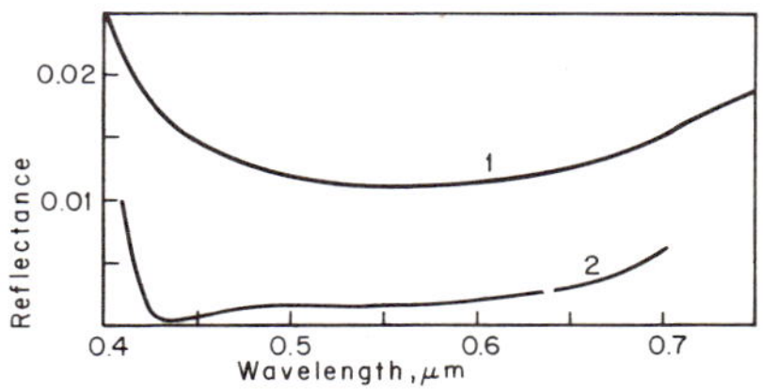

Fig. 33 Reflectance of a single layer and a broadband antireflection coating on glass. [*Curve* 1, *Balzers* (1962); *curve* 2, *after Costich* (1968).]

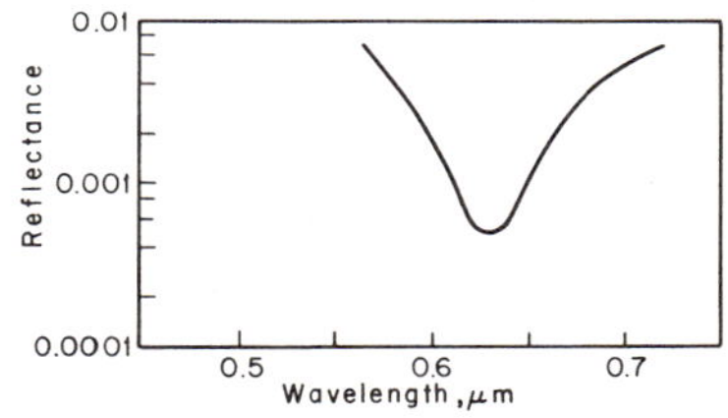

Fig. 34 Reflectance of a high-efficiency antireflection coating on glass. [*After Costich* (1968).]

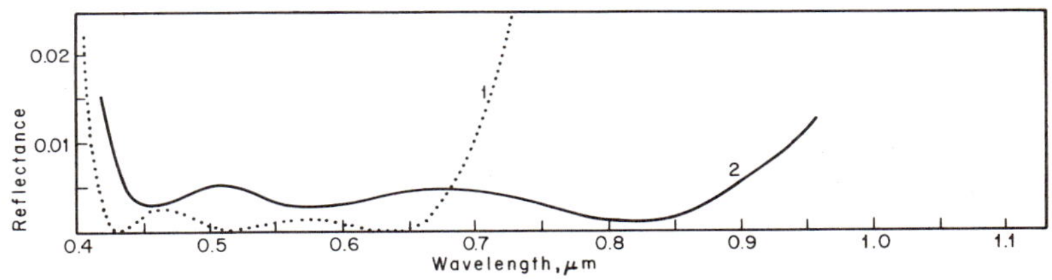

Fig. 35 Reflectances of two broadband antireflection coatings. [*Curve* 1, *after Turner* (1970); *curve* 2, *Optical Coating Laboratory* (1965).]

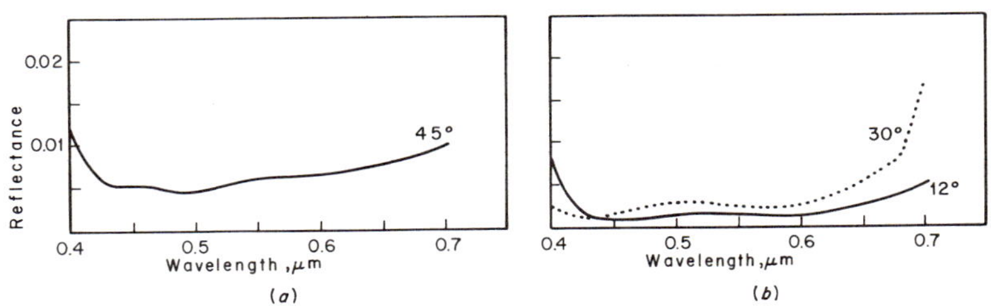

Fig. 36 Average reflectances for unpolarized light of antireflection coatings designed (*a*) for 45° incidence and (*b*) for use with a convergent light cone of semiangle of 30°. [*Optical Coating Laboratory* (1965).]

Because of a lack of coating materials with refractive indexes less than about 1.35, inhomogeneous layers are not very suitable for the antireflection coating of air-glass interfaces. The several different inhomogeneous antireflection coatings suggested for this purpose in the past (see, for example, Sawaki, 1960; Lessman, 1966) do not offer any special advantages over those composed of homogeneous layers.

The situation is different for high-index materials (Fig. 39). An even lower reflectance can be achieved by ending the inhomogeneous layer when its index is equal to the square of that of the lowest-index coating material available. It is then possible to complete the coating by depositing an additional homogeneous quarter-wave-thick layer of that material (Fig. 39*a*).

Inhomogeneous layers are prepared by the simultaneous evaporation from two separate sources whose rates of evaporation can be controlled individually. Alternatively, the inhomogeneous layer can be approximated by a series of homogeneous layers of gradually decreasing refractive indexes (Berning, 1962; curve 10.1 of Fig. 32*f*). Such layers can be prepared by evaporating a series of appropriate mixtures

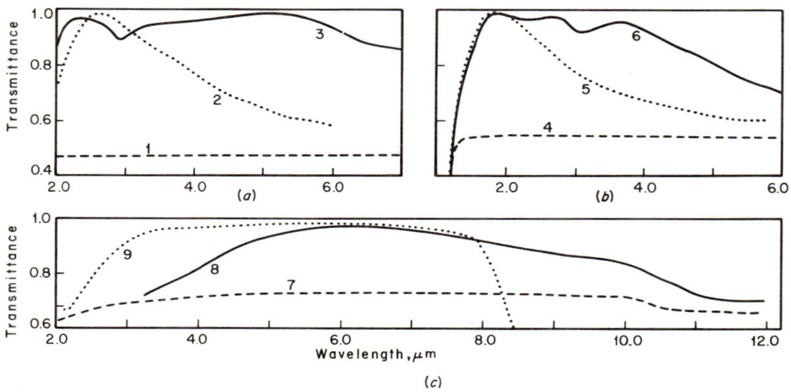

Fig. 37 Transmittance of plates of infrared materials antireflection-coated on both sides. (*a*) Germanium. Curve 1: uncoated plate; curve 2: single $\lambda_0/4$ layer of SiO ($\lambda_0 = 2.7$ μm); curve 3: $\lambda_0/4$ coatings of MgF_2, CeO_2, and Si ($\lambda_0 = 3.5$ μm). (*b*) Silicon. Curve 4: uncoated plate of 1.5-mm thickness; curve 5: single $\lambda_0/4$ layer of SiO_2 ($\lambda_0 = 1.8$ μm); curve 6: $\lambda_0/4$ coatings of MgF_2 and CeO_2 ($\lambda_0 = 2.2$ μm). (*c*) Irtran II. Curve 7: uncoated plate of 2-mm thickness; curve 8: single $\lambda_0/4$ coating of CeF_3; curve 9: $\lambda_0/4$ layers of MgF_2 and SiO ($\lambda_0 = 4.2$ μm). [*After Cox and Hass* (1964).]

of two coating materials (Fig. 39*b*) or, without mixing, by using the Herpin equivalent-index concept (Epstein, 1952) to simulate intermediate refractive indexes.

37. Antireflection Coatings Consisting of a Thin Metal Layer (Pohlack, 1956*b*) A suitable thin metal layer deposited onto a glass surface can act as a very efficient achromatic antireflection coating for light incident from the glass

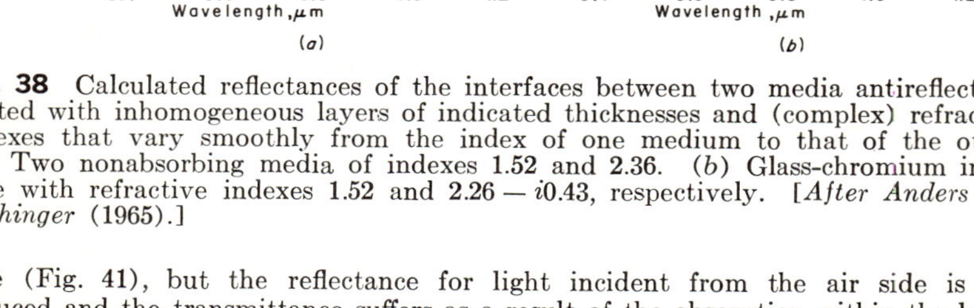

Fig. 38 Calculated reflectances of the interfaces between two media antireflection-coated with inhomogeneous layers of indicated thicknesses and (complex) refractive indexes that vary smoothly from the index of one medium to that of the other. (*a*) Two nonabsorbing media of indexes 1.52 and 2.36. (*b*) Glass-chromium interface with refractive indexes 1.52 and $2.26 - i0.43$, respectively. [*After Anders and Eichinger* (1965).]

side (Fig. 41), but the reflectance for light incident from the air side is not reduced and the transmittance suffers as a result of the absorption within the layer.

38. Antireflection Coating of Absorbing and Amplifying Media Antireflection coatings for glasses and semiconductors in regions of weak absorption and for present-day laser materials (Smiley, 1968), in which k_1, the imaginary part of the complex refractive index [Eq. (30)], is small and negative, differ little from those described in Pars. 35 and 36.

The reduction of the reflectance of opaque materials for architectural, decorative, and technical purposes leads to a corresponding increase in the absorption (Lupashko and Sklyarevskii, 1964; Park, 1964). This can be utilized to improve the efficiency of radiation detectors and to control the solar-absorptance and thermal-emittance

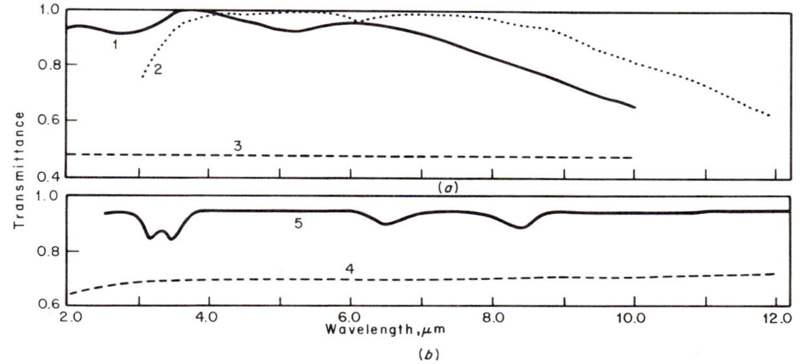

Fig. 39 (*a*) Measured transmittance of a germanium plate coated on both sides with inhomogeneous antireflection coatings. Curve 1: 1.2-μm-thick film with an index that changes gradually from that of Ge to that of MgF_2 [*after Jacobsson* (1966)]; curve 2: 1.76-μm-thick film with an index that varies from 4.0 to 1.5, overcoated with a 0.74-μm-thick homogeneous MgF_2 layer [*after Scheuerman* (1968)]; curve 3: transmittance of an uncoated plate. (*b*) Measured transmittance of a TBI-5 plate. Curve 4: uncoated; curve 5: coated on both sides with ten 2.5-μm-thick homogeneous layers with refractive indexes that vary between 1.3 and 2.29 and that are obtained through the evaporation of suitable NaF-CdTe mixtures. [*After Kuznetsov and Perveyev* (1967).]

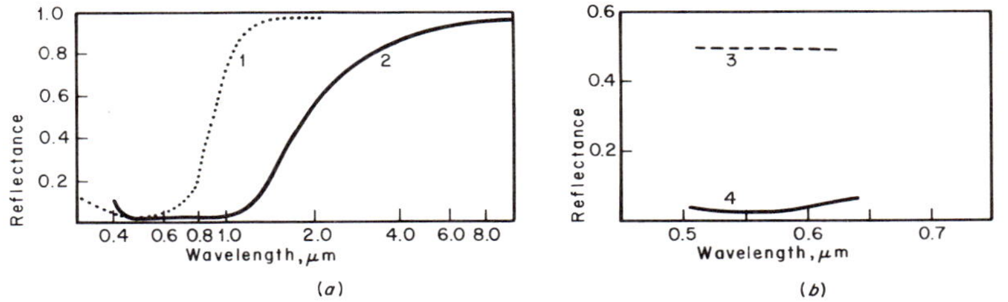

Fig. 40 (*a*), (*b*) Antireflection coating of opaque metals. Curves 1 and 2: three-layer coatings on aluminum [*after Drummeter and Hass* (1964)]; curves 3 and 4: reflectance of chromium and of chromium with a ZnS layer [*after Lupashko and Sklyarevskii* (1964)].

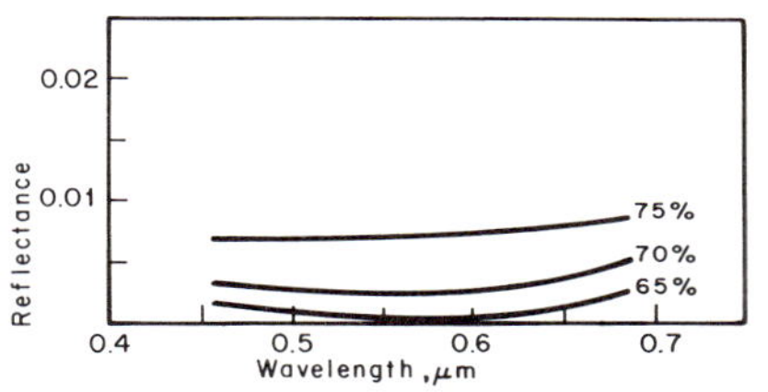

Fig. 41 Spectral reflectance of thin chromium layers on glass for light incident from the substrate side. The transmittance of the layers at $\lambda = 0.565$ μm is indicated. [*After Pohlack* (1956*b*).]

characteristics of surfaces (see, for example, Drummeter and Hass, 1964; Schmidt and Park, 1965). Three measured examples of the reduction in the reflectance of metal surfaces attainable with homogeneous absorbing and nonabsorbing layers are shown in Fig. 40.

Inhomogeneous transition layers whose refractive index and extinction coefficient change gradually from the values of one of the media to those of the other are also very effective (Anders and Eichinger, 1965). The calculated reflectance of a glass-chromium interface coated in this way, useful for blackening prism faces, lens edges, scales, etc., is shown in Fig. 38*b*. Metal-air interfaces can be treated a little less effectively because of the lack of suitable low-index materials.

39. Antireflection Coating of Surfaces Carrying a Thin Film For some applications it is necessary to deposit a certain film onto a glass surface. The objectionable reflectance that such a film would normally introduce can be avoided by incorporating it into an antireflection coating. Thus, for example, Fig. 42 shows the calculated

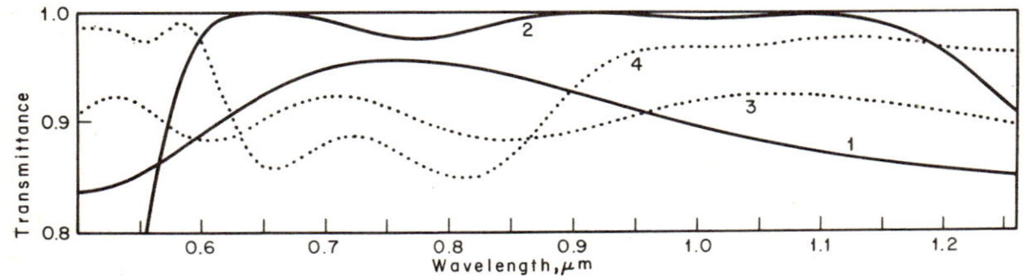

Fig. 42 Calculated single-surface transmittance of antireflection coatings incorporating a conducting layer. Curves 1 and 2: glass of refractive index 1.755 coated with a 0.566-μm-thick SnO_2 layer before and after three dielectric layers were added to complete the antireflection coating; curves 3 and 4: as above, except that the refractive index of the glass and the metric thickness of the SnO_2 film are 1.516 and 0.200 μm, respectively. [*After Veremey and Gorbunova* (1968).]

transmittances of two conducting layers before and after inclusion in antireflection coatings. The use of homogeneous and inhomogeneous antireflection coatings with absorbing layers for ophthalmic purposes is described by Katsube et al. (1968) and Anders (1966).

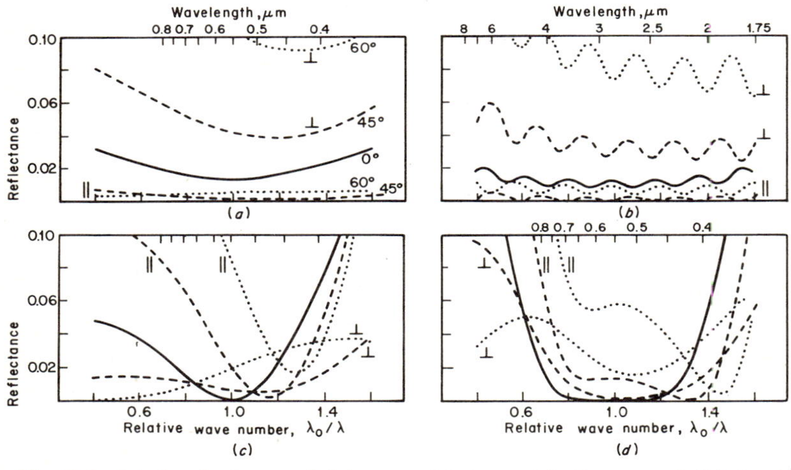

Fig. 43 Calculated reflectance of (*a*) one-, (*c*) two-, (*d*) three-, and (*b*) ten-layer antireflection coatings (systems 1.1, 2.1, 3.4, and 10.1 in Table 3) at angles of incidence of 0° (*solid curve*), 45° (*dashed curve*), and 60° (*dotted curve*) for light polarized parallel and perpendicular to the plane of incidence.

40. Antireflection Coatings at Nonnormal Angle of Incidence (Pohlack, 1952; Baumeister, 1961) The calculated performances at 0, 45, and 60° incidence of the three commercially most important antireflection coatings and of a 10-layer coating for germanium are shown in Fig. 43. The deterioration with angle of incidence is particularly severe for the narrow-band high-efficiency antireflection coating. Figure 43*d* suggests that the closer the design of an antireflection coating approximates an inhomogeneous transition layer (Par. 36) the less angle-dependent its performance.

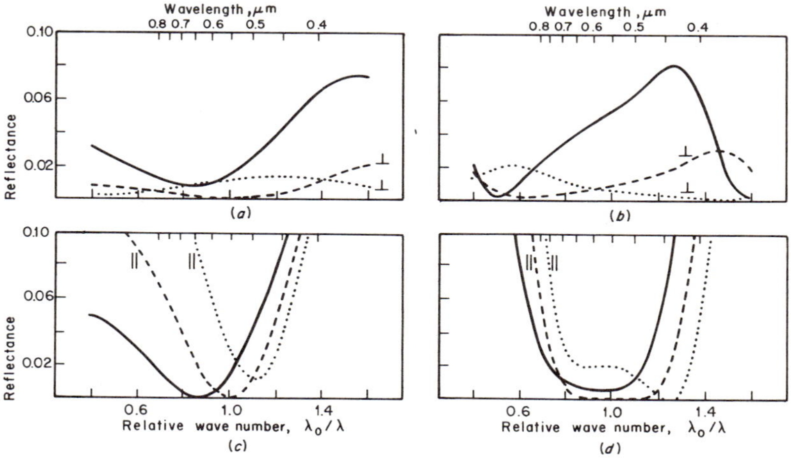

Fig. 44 Calculated reflectance at 0° (*solid curve*), 45° (*dashed curve*), and 60° (*dotted curve*) of (*a*) and (*c*) two-layer and (*b*) and (*d*) three-layer antireflection coatings designed for light incident at 45° and polarized parallel or perpendicular to the plane of incidence. [*After Turbadar* (1964*a*,*b*).]

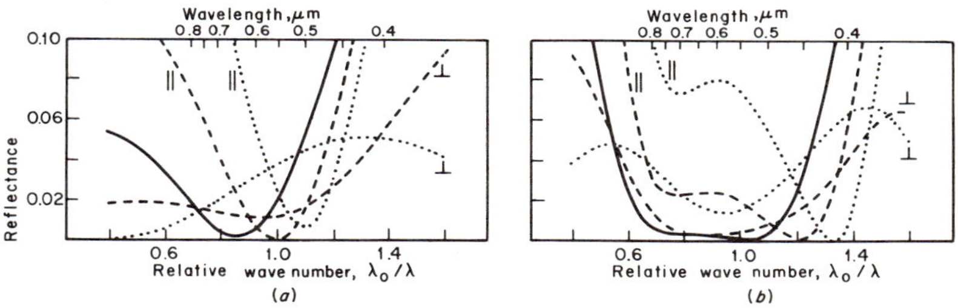

Fig. 45 Calculated reflectance at 0° (*solid curve*), 45° (*dashed curve*), and 60° (*dotted curve*) of (*a*) two-layer and (*b*) three-layer antireflection coatings (systems 2.1 and 3.4 in Table 3) for use with unpolarized light, with the thicknesses of the layers optimized for an angle of incidence of 45°.

To design an antireflection coating for one angle of incidence and one plane of polarization the effective refractive indexes and thicknesses [Eqs. (25) and (26)] of its layers should satisfy the relations set out in Table 3. In practice small departures from those conditions are made to optimize the performance using good coating materials. Calculated curves for two sets of two- and three-layer coatings designed for use at 45° are shown in Fig. 44.

If the obliquely incident radiation is not polarized, a compromise is necessary. The effective thicknesses are matched for the design angle, but the refractive-index conditions set out in Table 3 are satisfied for normal incidence since they cannot be satisfied for both polarizations at the same time (Fig. 45).

Achromatic antireflection coatings can be designed that are optimized for both polarizations at the same time or that are suitable for use over a range of angles of incidence (Fig. 36*b*). But the problem becomes more difficult the greater the spectral and angular ranges required, especially if angles of incidence greater than 60° are involved.

Two-Material Periodic Multilayers; Theory (Giacomo, 1956; Baumeister, 1962)

41. Reflectance Envelopes Let a periodic multilayer be composed of N periods AB, where A, B represent layers of refractive indexes n_A, n_B and optical thicknesses n_At_A, n_Bt_B. The most general representation of the complete multilayer is

$$\underset{1}{AB} \cdot \underset{2}{AB} \cdot \underset{3}{AB} \cdot \cdot \cdot \underset{N}{AB} = [AB]^N \tag{47}$$

In practice it is customary to write $[HL]^N$ or $[LH]^N$, depending on whether n_A is greater or less than n_B. The spectral-reflectance curve of the multilayer $[AB]^N$ will lie within a pair of envelopes which, at normal incidence, depend only on n_At_A, n_Bt_B, n_A, n_B, n_s, and n_m. For $n_s = n_m$ the lower of the envelopes becomes $R = 0$. The envelopes contain high-reflectance zones within which the reflectance at each wavelength increases monotonically with the number of periods, approaching 1.0 as $N \rightarrow \infty$ (Fig. 46). Outside

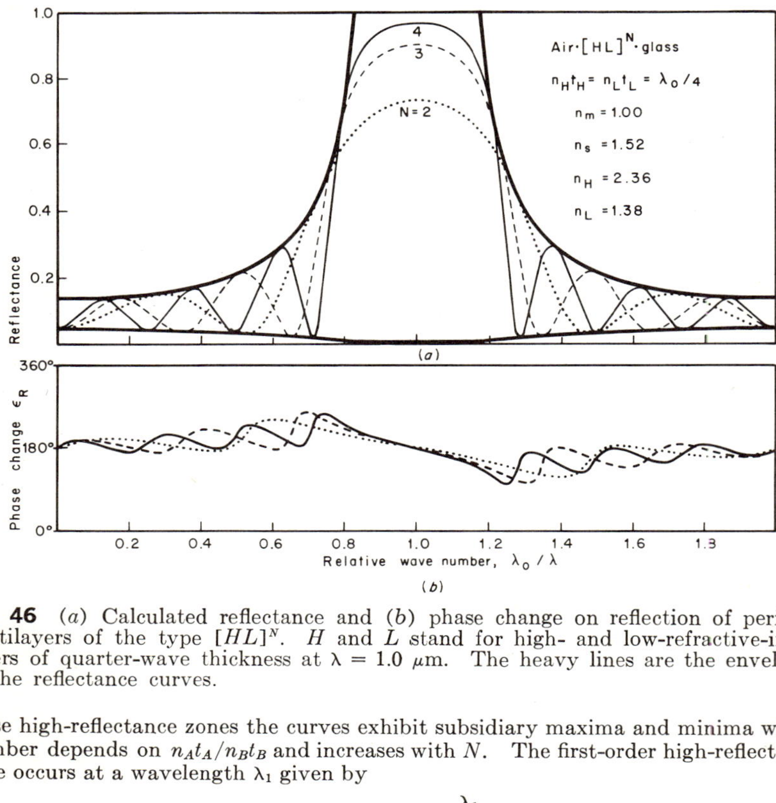

Fig. 46 (*a*) Calculated reflectance and (*b*) phase change on reflection of periodic multilayers of the type $[HL]^N$. H and L stand for high- and low-refractive-index layers of quarter-wave thickness at $\lambda = 1.0$ μm. The heavy lines are the envelopes of the reflectance curves.

these high-reflectance zones the curves exhibit subsidiary maxima and minima whose number depends on n_At_A/n_Bt_B and increases with N. The first-order high-reflectance zone occurs at a wavelength λ_1 given by

$$n_At_A + n_Bt_B = \frac{\lambda_1}{2} \tag{48}$$

and subsequent zones occur at wavelengths λ_q ($\lambda_1 > \lambda_2 > \lambda_3 \cdot \cdot \cdot$) given by

$$N(n_At_A + n_Bt_B) = q\frac{\lambda_q}{2} \qquad q = 2, 3, 4 \cdot \cdot \cdot \tag{49}$$

providing that at these wavelengths

$$n_A t_A,\ n_B t_B \neq p\,\frac{\lambda_q}{2} \qquad p = 1, 2, 3 \cdots \tag{50}$$

It is thus possible, by choosing suitable thickness ratios, to arrange for or suppress high-reflectance zones in several spectral regions at the same time. This is useful in the design of broadband reflectors (Pars. 55 to 57), cutoff filters (Pars. 58 to 68), hot and cold mirrors (Par. 65), laser reflectors, etc. Typical curves for thickness ratios 1:1 and 2:1 are given in Fig. 47. Plotted on a λ_0/λ scale, they show symmetry about all wavelengths λ for which $n_A t_A$ and $n_B t_B$ are both equal to some integral multiple of $\lambda/4$.

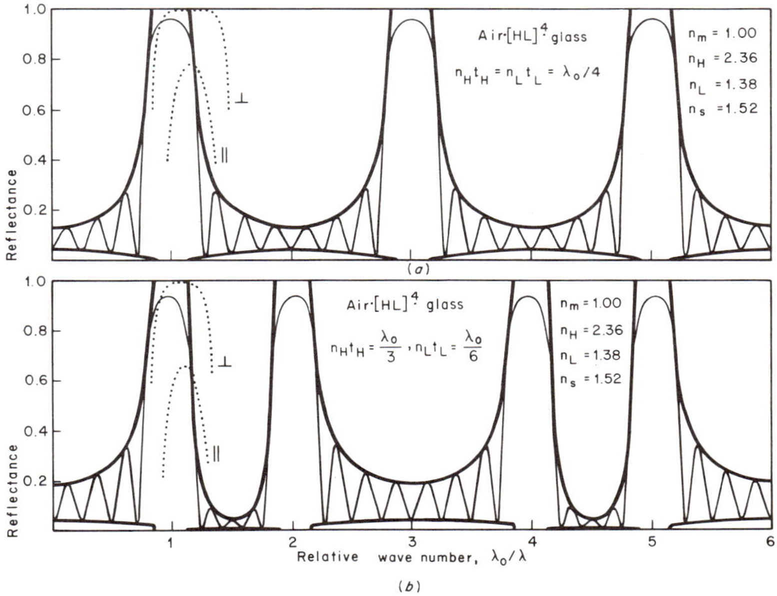

Fig. 47 Calculated reflectance curves of periodic multilayers with (*a*) 1:1 and (*b*) 2:1 thickness ratios. The dotted curves represent the reflectance of polarized radiation incident at an angle of 60°.

42. Maximum Reflectance For a given refractive index ratio n_A/n_B and number of periods N the highest reflectance occurs whenever $n_A t_A$, $n_B t_B$ are each equal to an odd multiple of $\lambda/4$. It is given by

$$R_{\max} = \left[\frac{n_m/n_s - (n_A/n_B)^{2N}}{n_m/n_s + (n_A/n_B)^{2N}}\right]^2 \tag{51}$$

Intermediate reflectances are obtained for the related symmetrical multilayers $[AB]^N A$:

$$R_{\max} = \left[\frac{n_m n_s/n_A{}^2 - (n_A/n_B)^{2N}}{n_m n_s/n_A{}^2 + (n_A/n_B)^{2N}}\right]^2 \tag{52}$$

Results for a number of cases are given in Fig. 48. Intermediate reflectances can be obtained by changing the refractive index of any of the layers in the stack.

Explicit expressions for R for other thicknesses are complicated.

43. Width of the High-Reflectance Zones For a given value of n_A/n_B the width $\Delta\lambda_R/\lambda$ of the high-reflectance zone is greatest when $n_A t_A = n_B t_B = \lambda_1/4$, and it is

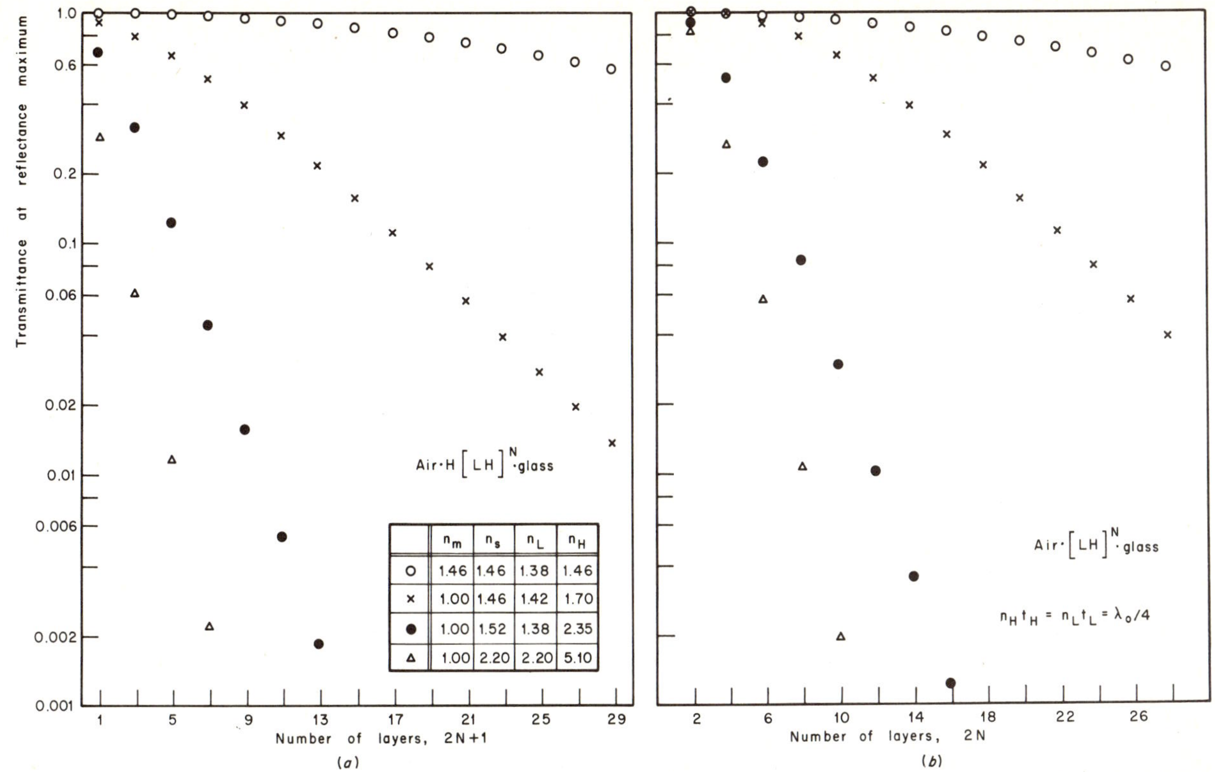

Fig. 48 Transmittance at the reflection maximum of quarter-wave stacks composed of MgF_2 and SiO_2 in the visible region (○), MgF_2 and MgO in the ultraviolet (×), MgF_2 and ZnS in the visible region (●), and ZnS and PbTe in the infrared (Δ).

then given by

$$\frac{\Delta\lambda_R}{\lambda} = \frac{4}{\pi}\sin^{-1}\frac{1 - n_A/n_B}{1 + n_A/n_B} \tag{53}$$

This width is reduced by a factor of $2N - 1$ if Nth-order quarter-wavelength layers are used. Graphs of the widths of the high-reflectance zones vs. refractive-index ratio for $\lambda/4$, $3\lambda/4$, and $5\lambda/4$ layer stacks are given in Fig. 49. Periodic multilayers with equal refractive-index ratios have high-reflectance zones of equal widths, but their reflectance curves will not be the same unless the refractive indexes of the surrounding media are also increased by the same ratio (Fig. 50).

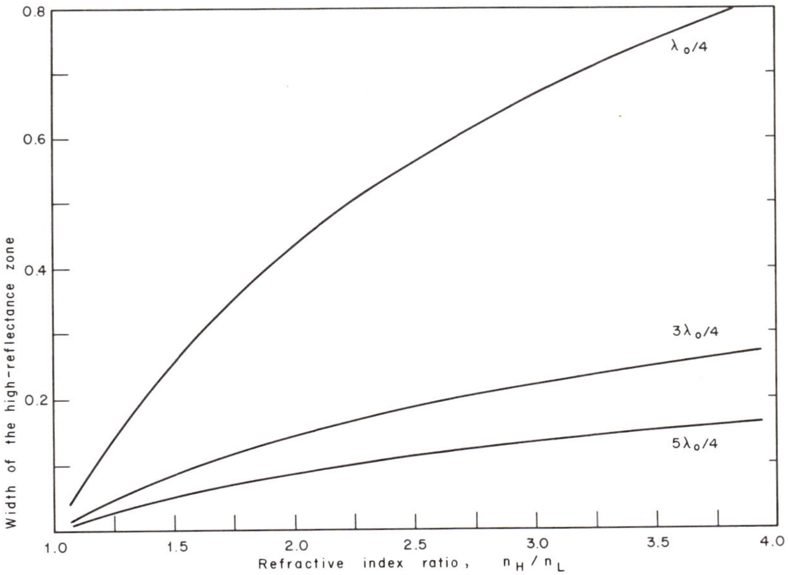

Fig. 49 Calculated widths of high-reflectance zones of two-material periodic stacks of layers of optical thicknesses $\lambda_0/4$, $3\lambda_0/4$, and $5\lambda_0/4$ for different refractive-index ratios.

44. Phase Change on Reflection The dispersion of the phase change on reflection from periodic all-dielectric multilayers is much greater than that of metal reflectors. Unless corrected for, it will lead to errors in some metrological and interferometric applications. Like reflectance, it varies rapidly outside the high-reflection zone (Fig. 46*b*). Within the high-reflectance zone it changes almost linearly with wave number and is 180.0° at $\lambda_0/\lambda = 1$. The slope of this portion of the graph increases and approaches a limiting value as N tends to infinity. The limiting value (in degrees per unit wavelength) is given by

$$\left[\frac{d\epsilon}{d\lambda}\right]_{\lambda_0} = \begin{cases} \dfrac{180 n_m}{\lambda_0(n_H - n_L)} & \text{for } n_H \\ \dfrac{180 n_L n_H}{\lambda_0 n_m(n_H - n_L)} & \text{for } n_L \end{cases} \tag{54}$$

depending on whether the radiation is first incident on a high- (n_H) or low- (n_L) refractive-index layer (Seeley, 1964). The above values should be multiplied by 3, 5, . . . in stacks composed of $3\lambda/4$, $5\lambda/4$, . . . layers.

45. Angular Sensitivity For some applications it is important to reduce the angular variation of the reflectance curve. This can be done by using materials with high refractive indexes (Fig. 50) (see also Par. 67). Another way is to use periods in which the high-index film is thicker than the low-index film (Figs. 47*b* and 71*a*).

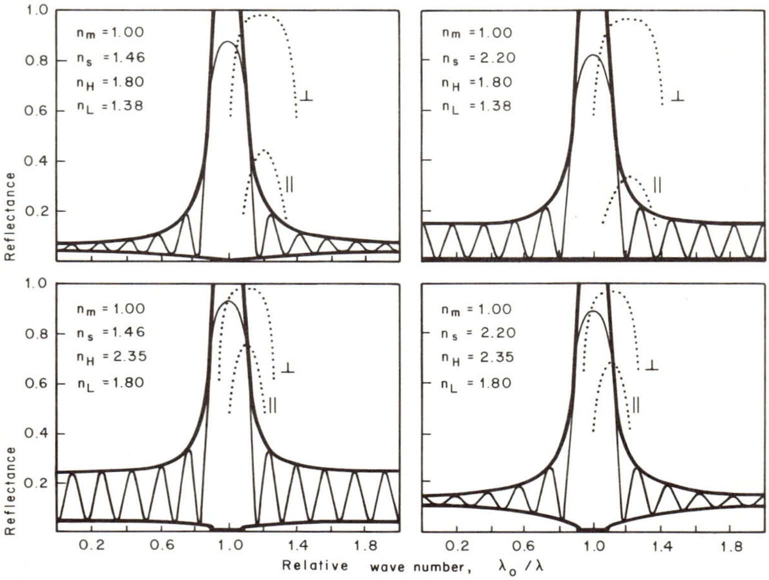

Fig. 50 Calculated reflectance curves of two quarter-wave stacks of the type $H[LH]^5$ with different values of n_H and n_L but with the same ratio n_H/n_L, deposited onto two different substrate materials. The dotted curves represent the reflectances for light incident at 60°.

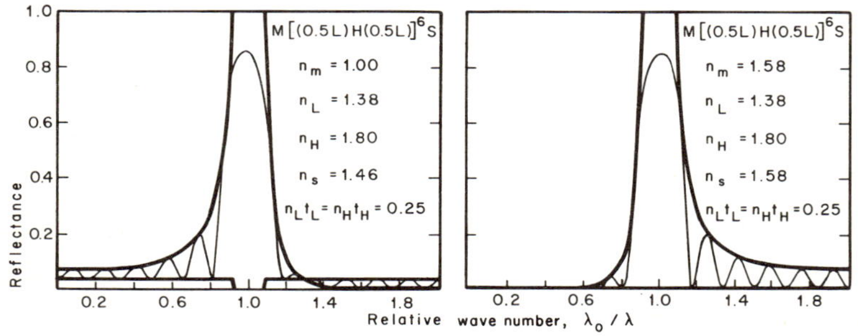

Fig. 51 Calculated reflectance curves of two periodic multilayers with symmetrical periods of the type $[(0.5L)H(0.5L)]^6$ in which the subsidiary reflectance maxima on the short- and the long-wavelength side of the high-reflectance zone are reduced.

46. Periodic Multilayers of the Type $[(0.5A)B(0.5A)]^N$ (Epstein, 1952) The construction of such multilayers differs from that of the type $[AB]^NA$ discussed in Pars. 41 and 42 by having outermost layers of only half the thickness of the remaining layers. The position and width of the high-reflection zones are the same in both cases, but in coatings of the type $[(0.5A)B(0.5A)]^N$ it is possible to reduce the height of the subsidiary maxima on either one or the other side of the main reflectance maximum if $n_At_A = n_Bt_B$. Kard et al. (1968) describe the optimum choice of all the construction parameters (N, n_A, n_B, n_s, n_m) of such a coating. If the refractive indexes n_s and n_m of the surrounding media must be chosen on some other basis, then for maximum improvement of the short- or long-wavelength transmission the refractive indexes n_A and n_B of the coating materials must satisfy

$$n_s n_m = \frac{n_A{}^3}{n_B} \qquad (55)$$

or

$$n_s n_m = n_A n_B \qquad (56)$$

respectively. First estimates of the maximum reflectance for larger values of N can be obtained from Fig. 48. Multilayers of this type are of importance in the design of cutoff filters (Pars. 58 to 68), and are illustrated in Fig. 51, which should be compared with the curves of Fig. 50.

All-Dielectric Reflectors for Interferometers and Lasers

47. Introduction In the calculations of the data for Figs. 46 to 51 the dispersion of the optical constants of the materials was ignored, and it was assumed that the films were absorption- and scatter-free and that their thicknesses had precisely the required values. In practice none of these assumptions is strictly valid, and hence there are departures from the calculated values. In general the agreement is better within the high-reflectance zone than outside.

48. Experimental Results The measured reflectances and transmittances of a number of quarter-wave reflectors suitable for use in Fabry-Perot interferometers are shown in Fig. 52. The transmission of a typical commercial laser reflector for $\lambda = 0.6328$ is shown in Fig. 53. The measured reflectances of a number of highly reflecting coatings for the infrared spectral region are shown in Fig. 54.

Both "soft" laser coatings that can be dissolved in weak acids and "hard" coatings that can be removed only through polishing are available commercially.

49. Thickness Errors Small errors in the thickness of the layers of a quarter-wave stack have only a very small effect on the reflectance of the multilayer within the high-reflectance zone (Heavens, 1954) but do affect the reflectance curve outside the zone. The effect of the phase change on reflection within the high-reflectance zone is also small (Rank and Bennett, 1955). Outside the high-reflectance zone the effect may be quite serious: thickness variations can give rise to an apparent lack of flatness of the substrate surface (Giacomo, 1958; Bouwhuis, 1962; Ramsay and Ciddor, 1967).

50. Dispersion The most noticeable effects of dispersion in quarter-wave stacks of a given multilayer type are the increase in the maximum reflectance with a decrease in wavelength (see Fig. 52) and the asymmetry of the maxima on either side of the main reflectance maximum.

51. Absorption and Scattering The separation between the transmission and reflectance curves in Fig. 52 is due to absorption and scattering. These losses can limit the usefulness of the reflectors for some applications. Thus, for example, in interference filters and Fabry-Perot interferometers they lead to a reduction in the peak transmissions and limit the attainable half widths. In optical information-storage devices they set a limit to the highest reflectance attainable. In lasers the losses directly counteract the gain in the laser medium. In addition, Becker et al. (1966) found that the threshold-energy densities of 3 to 45 J/cm^2 required to damage common dielectric coating materials are correlated with their ultraviolet cutoff wavelengths. This suggests that absorption within the layers may also be responsible for damage to laser reflectors (Bonch-Bruevich et al., 1964; Meneely, 1967; Persun et al., 1968). Failure of reflector coatings is at present setting a limit to the powers attainable with TEA CO_2 lasers. Consequently there has been a constant search for new and better coating materials (Penselin and Steudel, 1955; Honcia and Krebs, 1961; Hefft et al., 1963; Heitmann, 1966; Apfel, 1966*a*).

If the two materials used for the construction of periodic quarter-wave stacks have small but finite absorption coefficients, the resulting absorption at first reduces both the transmission and reflection coefficients (Giacomo, 1956, 1958; Koppelman, 1960). But with an increase in the reflectance of the multilayer, the absorption occurs more and more at the expense of the reflection coefficient, approaching

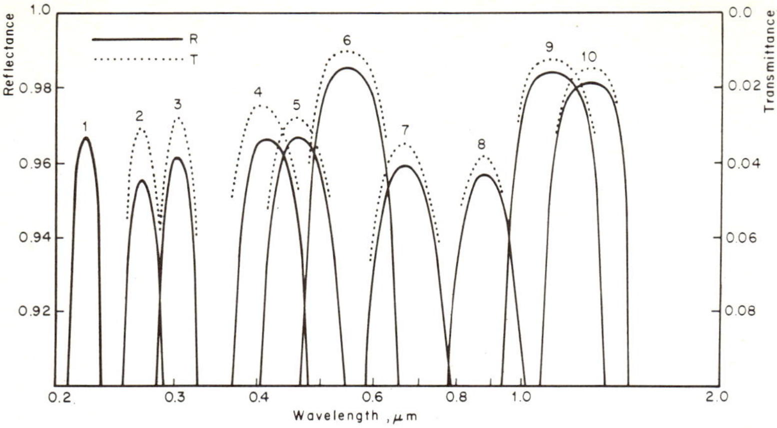

Fig. 52 Measured spectral reflectance and transmittance curves of periodic all-dielectric reflectors of the type $[HL]^N H$ for the ultraviolet, visible, and near-infrared parts of the spectrum. Curve 1: 27 layers of MgO and MgF_2; curves 2 and 3: 11 and 13 layers of PbF_2 and cryolite, respectively; curves 4, 5, 7, and 8: 7 layers of ZnS and cryolite; curves 6, 9, and 10: 9 layers of ZnS and cryolite. [*Curve* 1 *after Apfel* (1966*a*), *curves* 2 *to* 10 *after Hefft et al.* (1963).]

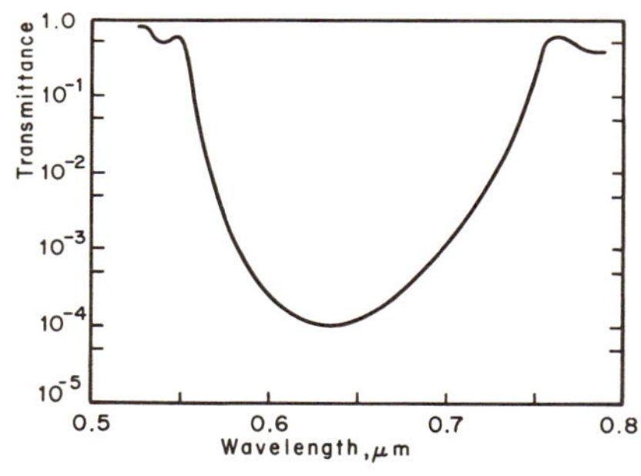

Fig. 53 Measured spectral transmittance of a laser reflector. [*After Costich* (1968).]

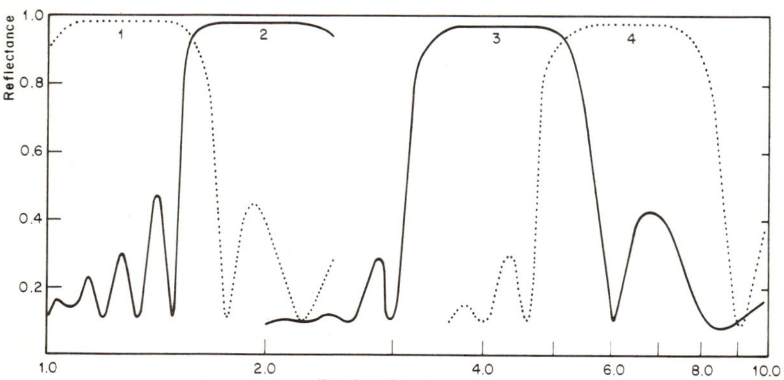

Fig. 54 Measured spectral-reflectance curves of all-dielectric reflectors for the infrared made of stibnite and chiolite. Curves 1 and 2: mutilayers of the type $[HL]^4 H$ on glass; curves 3 and 4: type $[(0.5L)H(0.5L)]^4$ on barium fluoride. [*After Turner and Baumeister* (1966).]

a limiting value of

$$A = -\delta R = \frac{2\pi n_m}{n_A{}^2 - n_B{}^2}(k_A + k_B) \tag{57}$$

that is independent of the number of layers. Figure 55 shows the spectral variation of experimentally determined maximum reflectances and limiting losses of various quarter-wave multilayer reflectors.

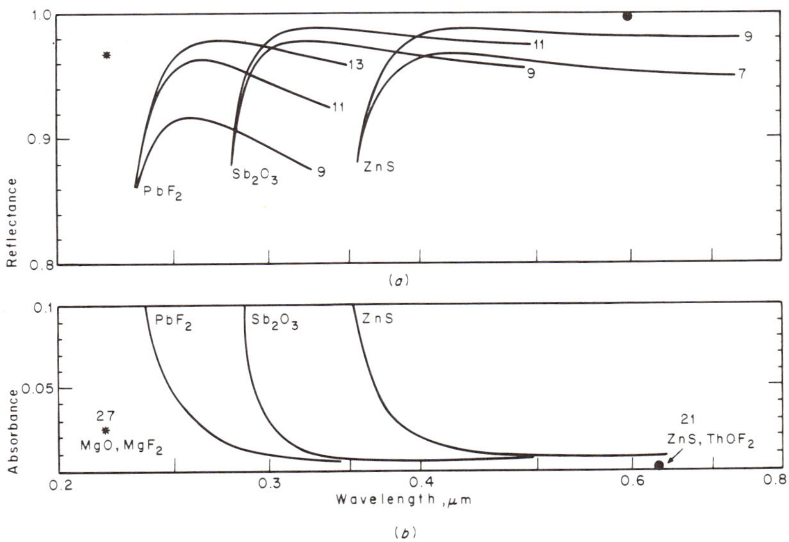

Fig. 55 Variation with wavelength of the observed peak reflectance of 7-, 9-, 11-, and 13-layer quarter-wave stacks made of PbF_2, Sb_2O_3, or ZnS with cryolite as the low-index material. The lower curves indicate the limiting value of the absorbance with these materials. [*After Honcia and Krebs* (1961).] Results obtained by Apfel (1966*a*) (*) and Behrndt and Doughty (1967) (●) with other materials are included for comparison.

Losses due to scatter can also be significant, but with special manufacturing techniques practically loss-free reflectors can be made. Mirrors with a reflectance of 99.89 percent and scattering losses of 0.09 percent have been reported (Behrndt and Doughty, 1967). Investigations have shown (Giacomo, 1956; Gloge et al., 1969) that when scatter does occur, most of the light is scattered in directions close to that of the specularly reflected light. The experimental results for a typical mirror are shown in Fig. 56.

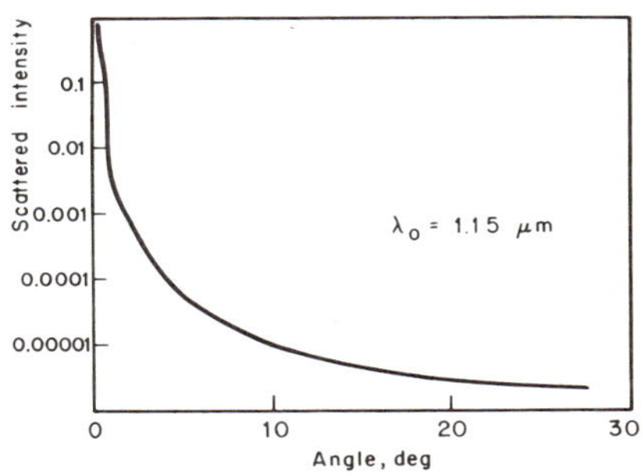

Fig. 56 Measured intensity of the scattered radiation as a function of the angle away from the direction of specular reflection. [*After Blazey* (1967).]

Narrow-Band Rejection Filters (Young, 1967)

52. Narrow-band rejection filters transmit freely all the radiation incident upon them except in one narrow spectral region in which the radiation is either wholly or partially reflected. Lord Rayleigh (1888) observed a corresponding natural phenomenon in potassium chlorate crystals. Subsequent experimenters (Wood, 1940; Schröder, 1951; Strong, 1961) reported crystals with rejection bands varying between 0.001 and 0.038 μm in width and reflectances between 33 and 99.9 percent. But unfortunately at present the size of the crystals that can be grown is insufficient, and the position and width of the rejection region cannot be controlled. With thin films both these limitations can be overcome, but it is difficult to achieve the extremely narrow widths and high rejections observed in the crystals.

A quarter-wave stack of the type $[AB]^N A$ has been suggested as a model for the construction of such filters (see, for example, Schröder, 1951; Heavens et al., 1957). It follows from Figs. 48 and 49 that the closer the refractive index ratio n_A/n_B to unity, the narrower the width of the reflectance zone and the more layers required to achieve a certain rejection. To reduce the number of layers, films with higher n_A/n_B ratios could be utilized and the width deduced by using layers with thicknesses that are odd multiples of a quarter wavelength. But this can be done only at the expense of bringing the adjacent higher- and lower-order reflection peaks closer. Resonant reflectors (Par. 54) are an extreme example of this.

Figure 57 shows the measured transmittances of a number of rejection filters of various widths and attenuations. Filters with relatively low rejections find application as correction filters (Dobrowolski, 1970).

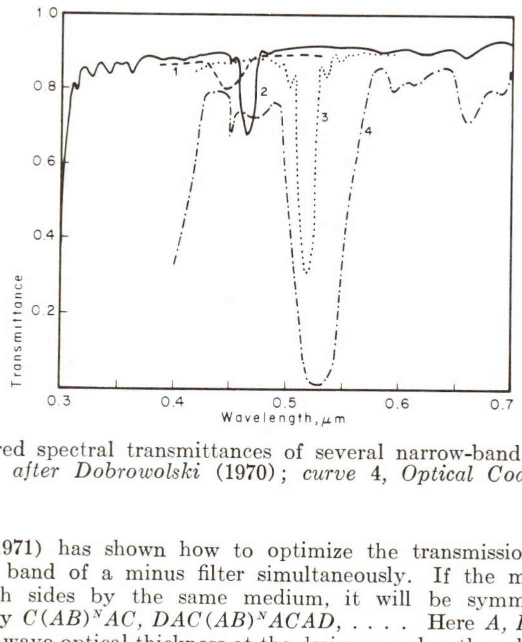

Fig. 57 Measured spectral transmittances of several narrow-band rejection filters. [*Curves* 1 *to* 3 *after Dobrowolski* (1970); *curve* 4, *Optical Coating Laboratory* (1964).]

53. Thelen (1971) has shown how to optimize the transmission on both sides of the rejection band of a minus filter simultaneously. If the multilayer is surrounded on both sides by the same medium, it will be symmetrical and can be represented by $C(AB)^N AC$, $DAC(AB)^N ACAD$, ... Here A, B, C, D, . . . are layers of quarter-wave optical thickness at the design wavelength and

$$n_m = n_s = n_A \tag{58}$$

The refractive indexes n_C, n_D, . . . depend on n_A and n_B in a more complicated way. The larger the number of different materials used in the construction of these multilayers the better the performance in the transmission region.

Young (1967) described two other design methods for narrow-band rejection filters with improved transmission characteristics. The methods are based on analogies with antenna theory, and they yield nonperiodic equiripple designs in which all the layers either have equal thicknesses but different indexes or are made of two materials only but have many different thicknesses.

Resonant Reflectors

54. Even "hard" evaporated coatings cannot survive the power densities present in some high-power lasers. In such cases resonant reflectors may be used (Herscher, 1965; Magyar, 1967; Watts, 1968; Mahlein and Schollmeier, 1969). They consist of one or more air-spaced plane-parallel plates of thicknesses of the order of millimeters made of a tough, high-optical-quality material. Because of the long coherence length of the laser radiation incident upon them, interference takes place in the same way as in thin films. Resonant reflectors may be regarded as quarter-wave reflectors of enormously high order of interference, and all the equations given in Pars. 41 to 46 apply.

Resonant reflectors made of quartz and sapphire are commercially available. Since the refractive index of quartz is lower than that of sapphire, a larger number of plates is required to attain the same reflectance. On the other hand, quartz is much cheaper and is less temperature-sensitive. The calculated reflectances of one-, two-, three-, and four-plate sapphire resonant reflectors are shown in Fig. 58. In a new development (Anon., 1969) diffusion-doped quartz plates are used

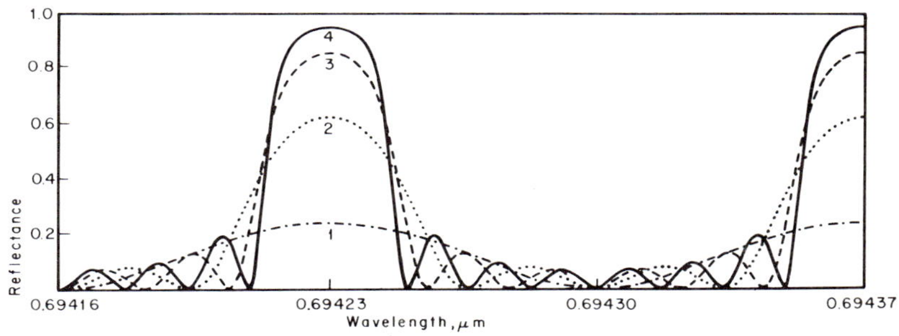

Fig. 58 Calculated spectral reflectance of resonant reflectors consisting of one, two, three, and four sapphire plates ($n = 1.7$). The optical thicknesses of the plates and of the air spaces between them were assumed to be 1.7 mm.

in the construction of resonant reflectors. The doping process causes the refractive index of the plate to increase smoothly in a $\frac{1}{2}$-μm region from that of quartz to about 2.0 at the surface. As a consequence fewer elements are needed to achieve a given reflectance.

All-Dielectric Broadband Reflectors and Reflectors for Two Widely Separated Wavelengths

55. There are several ways of obtaining a coating with a broad high-reflection region should the width attainable with quarter-wave stacks be inadequate. For a broad region with a very high reflectance one can superimpose two quarter-wave stacks tuned to two different wavelengths. The widest continuous high-reflectance region is attained when materials with the highest available refractive-index ratios are used and the thicknesses of the layers are so chosen that the two high-reflectance zones are contiguous (see Fig. 59). For an even broader region further quarter-wave stacks can be superimposed. If high-reflection regions overlap, special precautions must be taken to prevent the appearance of sharp reflectance minima in the high-reflectance region (Turner and Baumeister, 1966). It is not easy to obtain a very uniform, moderately high reflectance in this way.

Another approach is to deposit onto the substrate a series of alternating high- and low-refractive-index films of gradually increasing or decreasing thicknesses (Fig. 60).

A broad high-reflection region can also be obtained with a multilayer in which the layers are different multiples of $\lambda/4$ of a selected wavelength (see Fig. 61).

If only a relatively small increase in the high-reflectance region is required, or if a uniform but only moderately high reflectance is required, the desired result can be achieved by modifying the thicknesses and refractive indexes of a quarter-wave stack using a computer refinement program or by the addition of achromatizing $\lambda/2$ layers. The measured performances of several such reflectors are shown in Figs. 62 and 63.

Broadband reflectors with moderate and high reflectances for the ultraviolet spectral region have been reported by Korolev et al. (1970) and Sokolova (1971).

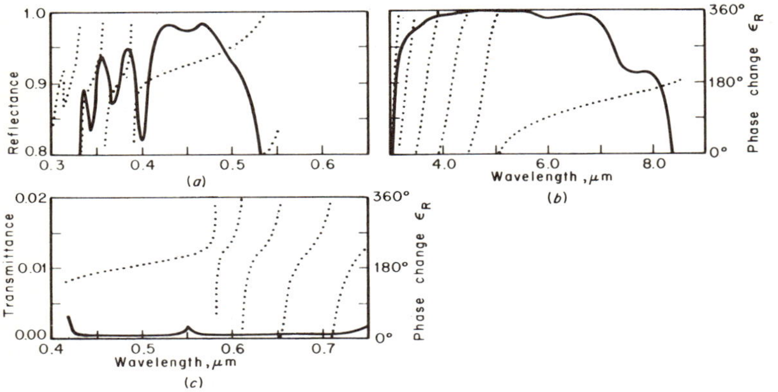

Fig. 59 Measured reflectance of all-dielectric broadband reflectors consisting of two superimposed quarter-wave stacks with contiguous high-reflectance zones. The dotted curves in Figs. 59 to 63 represent the calculated phase changes on reflection [(*a*) and (*b*) *after Turner and Baumeister* (1966); (*c*) *after Perry* (1965)].

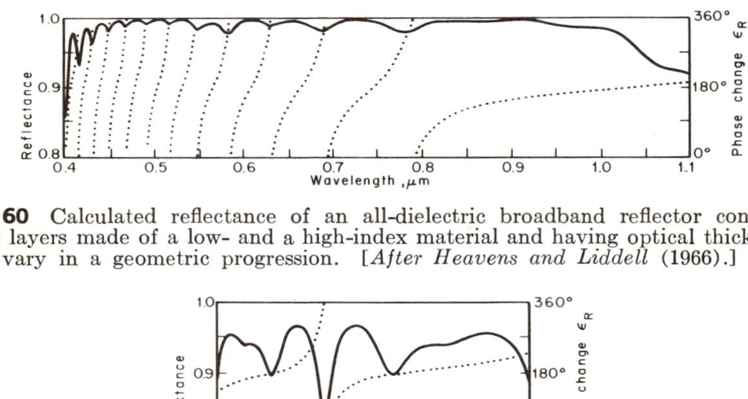

Fig. 60 Calculated reflectance of an all-dielectric broadband reflector consisting of 35 layers made of a low- and a high-index material and having optical thicknesses that vary in a geometric progression. [*After Heavens and Liddell* (1966).]

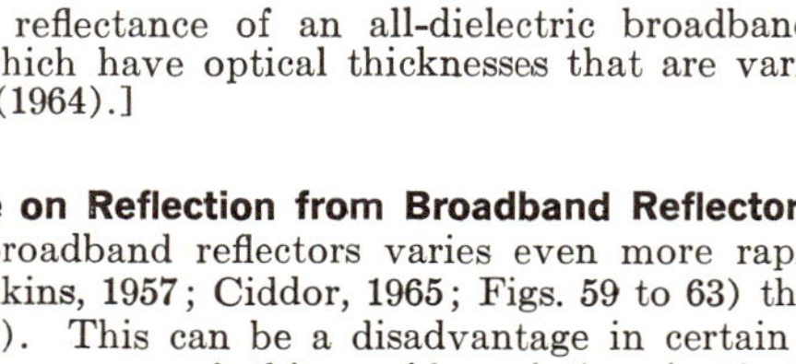

Fig. 61 Calculated reflectance of an all-dielectric broadband reflector consisting of 11 layers all of which have optical thicknesses that are various multiples of 0.13 μm. [*After Elsner* (1964).]

56. Phase Change on Reflection from Broadband Reflectors The phase change on reflection from broadband reflectors varies even more rapidly with wavelength (Baumeister and Jenkins, 1957; Ciddor, 1965; Figs. 59 to 63) than that from quarter-wave stacks (Fig. 46). This can be a disadvantage in certain metrological applications. Another consequence of this rapid variation is that in the presence of systematic nonuniformities of the layers it can give an impression of lack of flatness of the substrate (Ramsay and Ciddor, 1967). Figures 62*c* and 63*b* represent designs of broadband reflectors in which an effort was made to reduce this effect.

57. Reflectors for Two Widely Separated Wavelengths For some applications all that is needed are high reflectances at two isolated and widely separated wavelengths. Solutions other than the broadband reflectors described above are often easier to implement in such cases. By way of example the calculated reflectances of three different types of reflectors for simultaneous laser action at

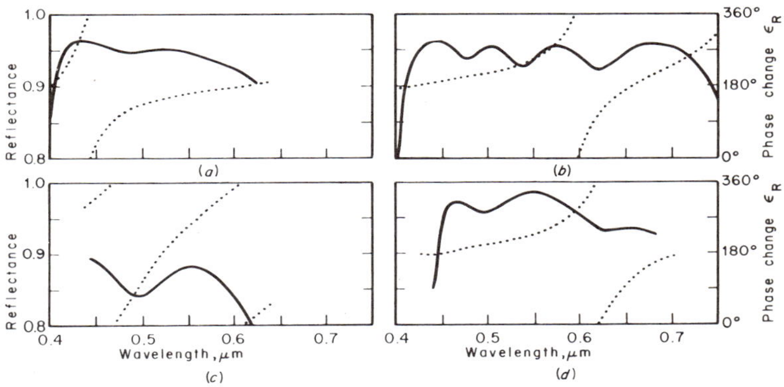

Fig. 62 Measured reflectances of all-dielectric broadband reflectors designed with refinement programs [(*a*) *after Penselin and Steudel* (1955), (*b*) *after Baumeister and Stone* (1956), (*c*) *after Ciddor* (1968), (*d*) *after Ramsay and Ciddor* (1967)].

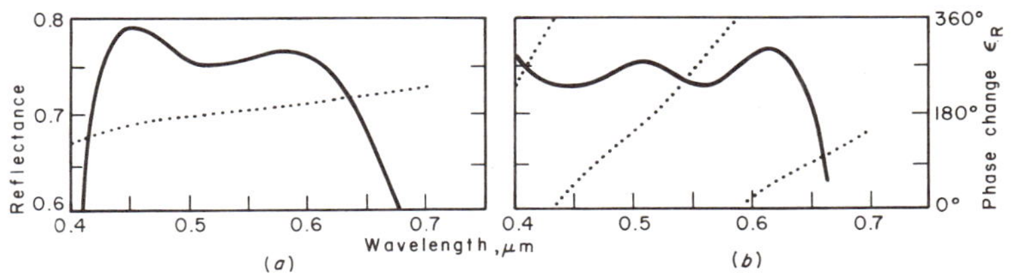

Fig. 63 Measured reflectances of achromatic all-dielectric reflectors with an intermediate reflectance derived from a quarter-wave stack through (*a*) the addition of half-wavelength achromatizing layers [*after Turner* (1954)] and (*b*) designed with a refinement program [*after Ciddor* (1968)].

0.6328 and 1.15 μm are shown in Fig. 64. A fourth solution, consisting of a third-order λ/4 stack at 1.15 μm is not shown because the total optical thickness of such a multilayer is about 2.5 times that of the other three systems.

All-Dielectric Cutoff Filters, Heat-Control Filters, and Solar-Cell Covers (Epstein, 1955)

58. Ideal cutoff filters would reject all the radiation below and transmit all that above a certain wavelength, or vice versa. Real cutoff filters, of course, are not perfect, and so, in addition to the cutoff wavelength, the slope of the transition region and the extent and average transmission values of the transmission and rejection regions must be specified. The tolerable departures of these quantities from the ideal values depend greatly on the application. Most all-dielectric cutoff filters are based on periodic multilayers, whose basic properties were described in Pars. 41 to 46.

59. Transmission in the Passband Region The usual way of avoiding the secondary transmission minima in the transmission band of a quarter-wave stack is through the use of eighth-wave layers on both sides of the stack (Par. 46).

Other, less frequently used methods of smoothing the transmission in the passband are the adjustment of the thicknesses of all the layers of a quarter-wave stack (Baumeister and Jenkins, 1957); the use of homogeneous (Gisin, 1969) and inhomogeneous (Jacobsson, 1964) layers on either side of the stack; the choice of an optimum set of refractive-index values for the substrate and films (Nesmelov, 1971); and the use of an equiripple design (Young, 1967) in which thicknesses are kept constant but indexes are varied.

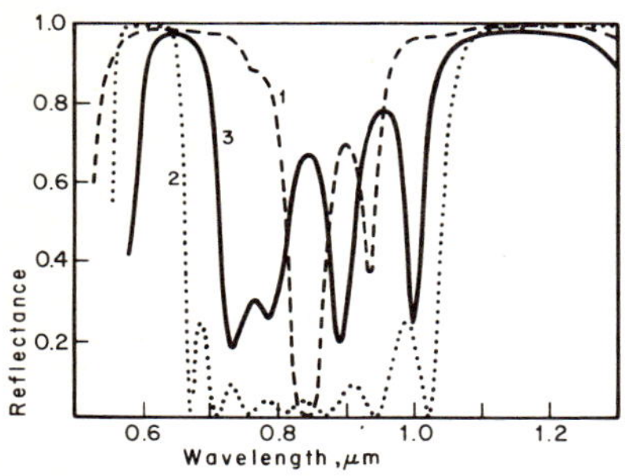

Fig. 64 Coatings with a high reflectance at 0.63 and 1.15 μm. Curve 1: calculated reflectance of two superimposed quarter-wave stacks centered at the two wavelengths [*Rigden and White* (1962)]; curve 2: calculated reflectance of a 2:1 stack (Par. 41); curve 3: measured reflectance of a two-material coating designed with an automatic synthesis program (Dobrowolski, 1965) in which the optical thicknesses of all the layers are multiples of $\lambda/4$ at 0.6328 μm.

60. The Width of the Transmission Region For short-wavelength cutoff filters of the type described above the transmission region is limited only by the transmission characteristics of the materials used for their construction. In long-wavelength cutoff filters the transmission regions can be limited by the appearance of higher-order reflectance maxima (Par. 41). Should this be a serious limitation, it is possible to suppress reflectance maxima at wave numbers equal to 2, 3, and 4 times that of the first-order maximum by using multilayers with periods composed of three different materials (Fig. 65). By using a period consisting of an inhomogeneous layer with a special refractive-index profile an even larger number of consecutive reflectance maxima can be suppressed (Jacobsson, 1966). Figure 66 shows the measured results for two such

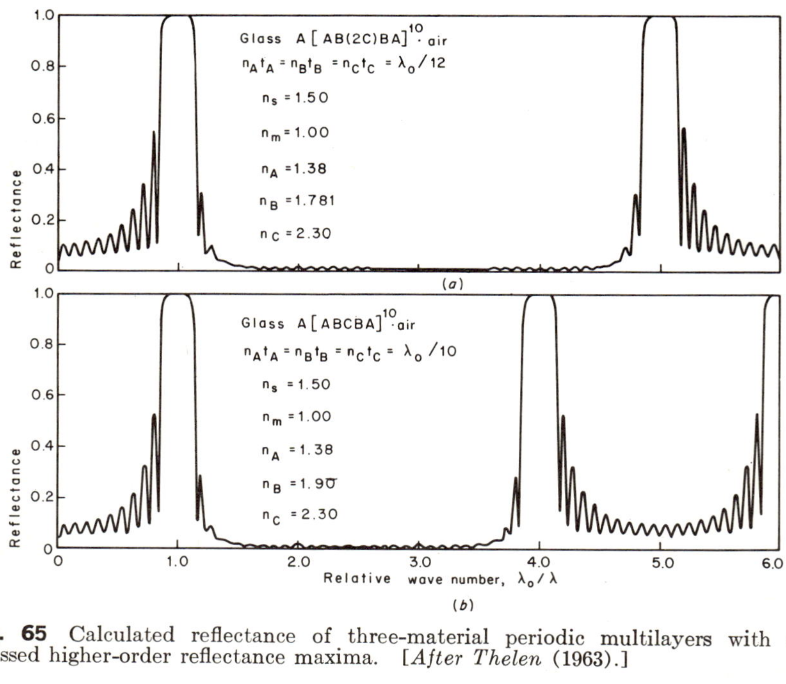

Fig. 65 Calculated reflectance of three-material periodic multilayers with suppressed higher-order reflectance maxima. [*After Thelen* (1963).]

experimental coatings in which reflectance maxima are suppressed at three and nine consecutive integer multiples of $1/\lambda_0$, a fact obscured in the case of the second filter by the absorption of the materials used.

61. Transmission in the Rejection Region Figure 48 can be used for a first estimate of the effect of the multilayer (Par. 42). Through the use of suitable substrate materials the transmittances throughout the rejection region are typically below 0.01 and 0.1 percent for short- and long-wavelength cutoff filters, respectively. Rejection filters with higher rejections can be provided, or, alternatively, two or more filters in series can be used if they are placed at a small angle to one another.

62. The Width of the Rejection Region The width of the high-reflection-region $[(0.5A)B(0.5A)]^N$ coatings can be estimated from Fig. 49. There is no shortage of materials (Par. 12; Par. 3 of Sec. 7) should it be necessary to extend the rejection region of shortwave cutoff filters. The number of suitable materials for long-wavelength cutoff filters is more restricted (Par. 3 of Sec. 7), and often it is necessary to use thin films to extend the cutoff region. In addition to the superposition of two or more cutoff filters tuned to different wavelengths (Par. 55) one can deposit such coatings onto different substrates or onto the opposite sides of the same substrate (Ploke, 1967). The resulting transmission will be governed by the considerations of Par. 5.

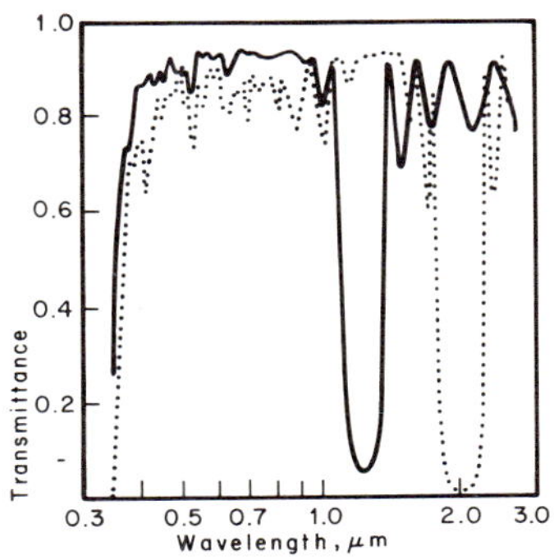

Fig. 66 Measured transmittance of periodic multilayers in which higher-order reflectance maxima are suppressed through the use of periods that consist of an inhomogeneous layer. [*After Scheuerman* (1968).]

63. Slope of the Cutoff This quantity is defined in a number of ways. One practical definition is

$$\left|\frac{\lambda_{0.8} - \lambda_{0.05}}{\lambda_{0.5}}\right| \times 100\% \tag{59}$$

where $\lambda_{0.8}$, $\lambda_{0.5}$, and $\lambda_{0.05}$ refer to the wavelengths at which the transmittances are 0.8, 0.5, and 0.05 of the maximum transmittance of the filter. Explicit formulas for the slope are complicated (Kard et al., 1968). The slope increases with the number of periods and with the refractive-index ratio. Slopes with values of the order of 5 percent are readily available in practice.

64. Experimental Results The spectral-transmittance curves of a number of commercially available short- and long-wavelength cutoff filters are shown in Figs. 67 and 68. Similar filters for intermediate wavelengths can, of course, be constructed.

65. Heat Reflectors and Cold Mirrors Only 39 percent of the total radiation from carbon arcs and 13 percent from tungsten lamps operated at 3250 K represent visible light. Most of the remaining energy is infrared radiation, which is converted into heat on absorption. The use of heat reflectors and cold mirrors in film projectors (Schroeder and Turner, 1960), in spot lamps for television and film studios (Carlson et al., 1956), and other optical instruments can lead to a very significant reduction of this unwanted heat.

Heat reflectors (also called hot mirrors) are special long-wavelength cutoff filters with a cutoff at 0.7 μm which transmit the visible radiation from 0.4 to 0.7 μm without disturbing the color balance. The width of the rejection region depends on the light source to be used and on whether a heat-absorbing glass is also to be used. The spectral-transmittance curves of three typical commercial heat reflectors are shown in Fig. 69*a*. The measured spectral-transmittance and reflectance curves of two heat-reflecting coatings not based on periodic multilayers are shown in Fig. 70.

Cold mirrors reflect as much as possible of the visible light incident upon them and transmit the remaining radiation. The reflectance curves of two commercial cold mirrors are shown in Fig. 69*b*.

66. Solar-Cell Covers Solar-cell covers remove the incident solar energy that does not contribute to the electrical output of the cell and protect it from possible deterioration of its performance through the action of ultraviolet radiation (Thelen, 1961; Reynard and Andrew, 1966). The spectral transmittance of a blue-red solar-cell cover is shown in Fig. 69*c*. The earlier blue solar-cell covers (curve 2, Fig. 67) protected the cell only from the adverse effects of ultraviolet radiation.

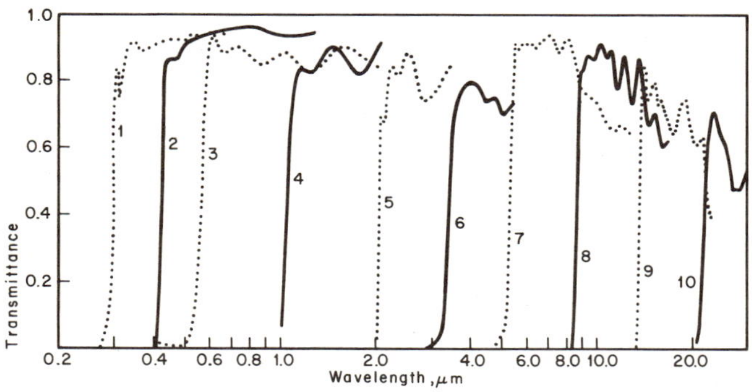

Fig. 67 A series of commercial short-wavelength cutoff filters. [*Curves* 1, 5, 8, 9, 10, *Optical Coating Laboratory* (1964, 1970), *curve* 2, *Bausch & Lomb* (1968), *curves* 3 *and* 6, *after Turner* (1970), *curve* 4, *Eastman Kodak* (1968*a*), *curve* 5, *Infrared Industries* (1966).]

67. Angle-of-Incidence Effects The edge of cutoff filters moves toward shorter wavelengths as the angle of incidence is increased. The use of higher-index materials reduces the effect (Par. 45). Measured results for a cutoff filter in which the shift is reduced by using high-refractive-index layers that have 3 times the thickness of the low-index layers are shown in Fig. 71*a*.

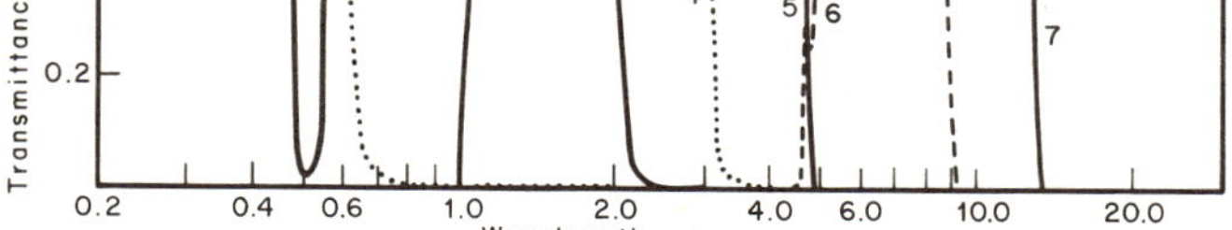

Fig. 68 A series of commercial long-wavelength cutoff filters. [*Curve* 1, *after Apfel* (1966*a*), *curve* 2, *Eastman Kodak* (1968*a*), *curves* 3, 4, 7, *Optical Coating Laboratory* (1970), *curves* 5 *and* 6, *Infrared Industries* (1966).]

68. Temperature Effects Refractive indexes of optical materials increase almost linearly with increase in temperature, thus causing cutoff edges to move toward longer wavelengths. In actual filters the fractional-wavelength shift varies between 3×10^{-5} and 10^{-4} per Celsius degree. Higher-index materials tend to be more temperature-sensitive, thus making it difficult to construct filters that are insensitive

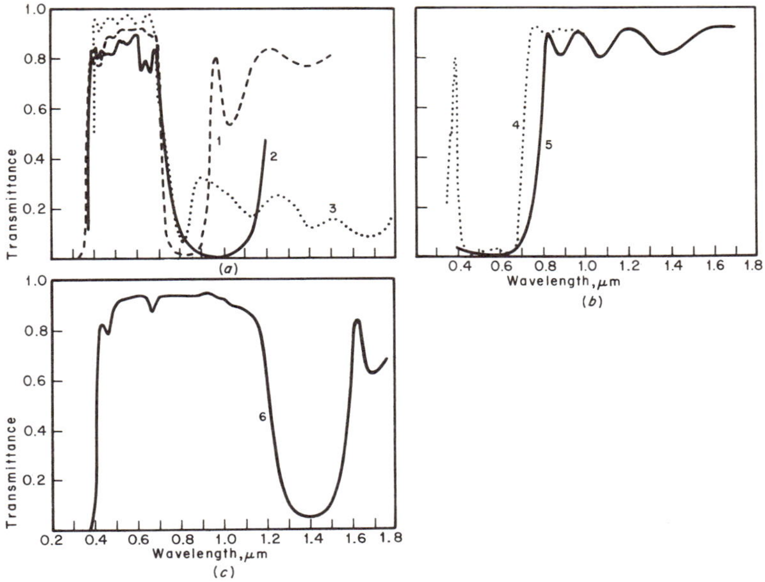

Fig. 69 Measured performance of commercial multilayer coatings for heat control: (*a*) heat-reflecting coatings; (*b*) cold mirrors; (*c*) blue-red solar-cell cover. [*Curves* 1 *and* 4, *Optical Coating Laboratory* (1964), *curve* 2, *Balzers* (1962), *curve* 3, *Bausch & Lomb* (1968), *curve* 5, *Heliotek* (1969), *curve* 6, *after Thelen* (1961).]

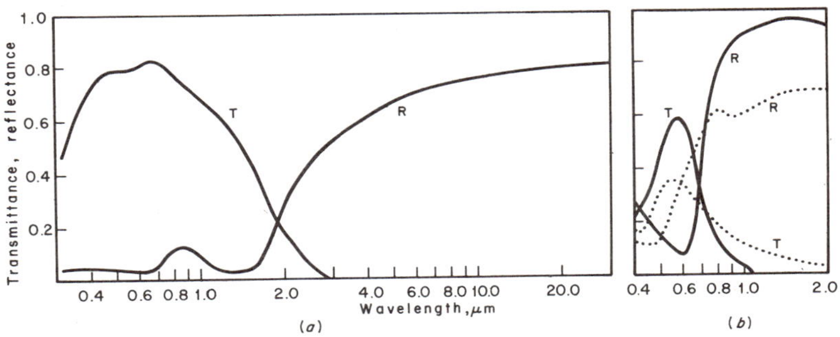

Fig. 70 Two nonperiodic coatings with heat-reflecting properties: (*a*) spray-deposited conducting coating of tin oxide; (*b*) transmittance and reflectance of a gold film (*dotted curves*) and of a broadband metal dielectric filter (*solid curves*, Par. 73) with semitransparent gold films of the same total thickness. [*After Turner* (1962).]

both to angle of incidence and temperature changes. The measured performance of a cutoff filter at two temperatures is shown in Fig. 71*b*.

Interference Polarizers

69. The dependence of the optical properties of thin-film systems on the plane of polarization of obliquely incident radiation (Par. 33; Baumeister, 1961) can be exploited to design interference polarizers with properties that augment those attainable by other means (Sec. 10). Presently they can be constructed for the ultraviolet, visible, and infrared spectral regions, and they are particularly of interest whenever large-area low-loss polarizers are required.

70. Single-Layer Polarizers It is always possible to find an angle of incidence at which a substrate coated with a film of quarter-wave effective thickness will reflect only radiation polarized perpendicular to the plane of incidence (Schröder, 1948). This property can be used to construct efficient transmitting polarizers using far fewer plates than necessary in the conventional pile-of-plates polarizer (Pars. 56ff. of Sec. 10) of equal performance (Fig. 72*a*). The calculated degree of polarization *P* attainable with different numbers of films is shown in Fig. 73

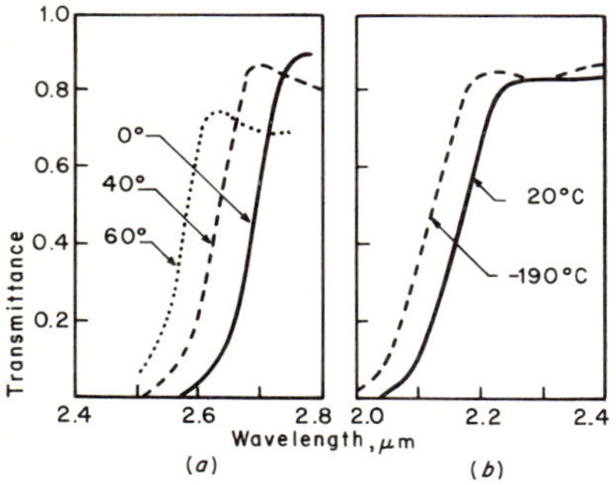

Fig. 71 Effect of (*a*) angle of incidence and (*b*) temperature on the performance of cutoff filters. [*Optical Coating Laboratory* (1967).]

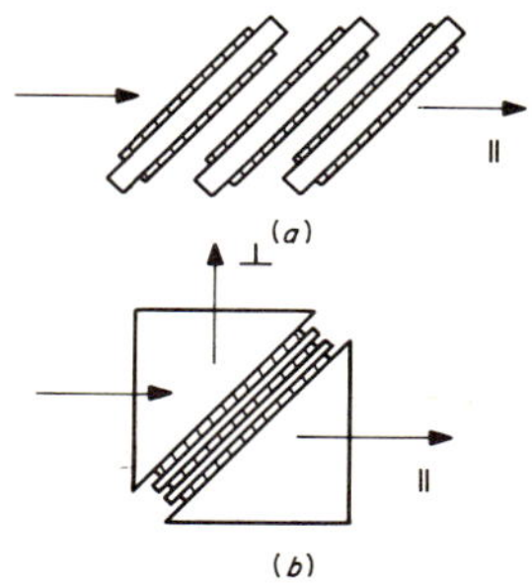

Fig. 72 Schematic outline of (*a*) single and (*b*) multilayer thin-film polarizers.

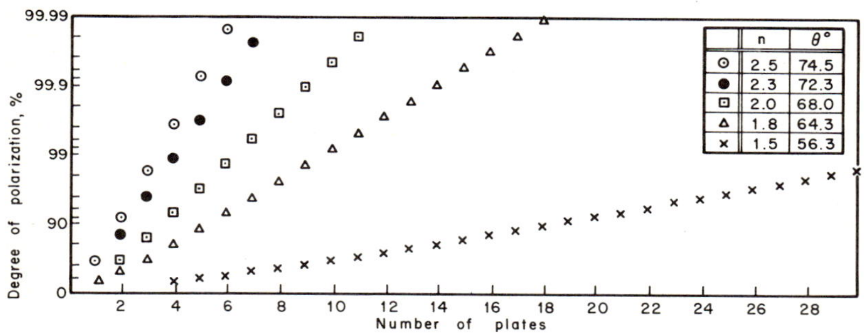

Fig. 73 Calculated degree of polarization of different numbers of plates of refractive index 1.5 coated on both sides with films of different refractive indexes *n*. The films have effective optical thicknesses of a quarter wavelength at the appropriate polarizing angle *θ*.

for a series of film indexes and polarizing angles. Experimental results agree closely with the calculations. The variation of the degree of polarization is small over a wavelength span of one octave and for angular apertures of up to 10° in polarizers of the above type having high degrees of polarization (Schröder, 1948).

71. Multilayer Polarizers (Clapham et al., 1969) The number of coated plates in the polarizer described above can be reduced without compromising the performance by depositing more than one high-refractive-index layer onto each surface and spacing them with quarter-wavelength-thick low-index films (Messner, 1953). In particular, it is possible to minimize the surface scatter, plate absorption, and lateral beam displacement by combining all the layers into one coating of the type $[HL]^N H$. The higher the refractive-index ratio of the two film materials used, the fewer the layers needed to achieve a certain degree of polarization and the wider the spectral region over which the polarizer will act. As before, a certain relationship between the refractive indexes of the materials and the angle of incidence must be satisfied (Kard, 1960). Also, the V number of the substrate material can be chosen to decrease the disturbing effect of the dispersion of the high-index material which will otherwise tend to decrease the useful spectral range of the polarizer (Banning, 1947*b*). A particularly convenient polarizer with no lateral beam displacement results when the multilayer is embedded between two right-angled prisms, as shown in Fig. 72*b* (Geffcken, 1953/1944; MacNeille, 1946). A 45° angle of incidence is normally chosen when both the transmitted and reflected beams are to be utilized.

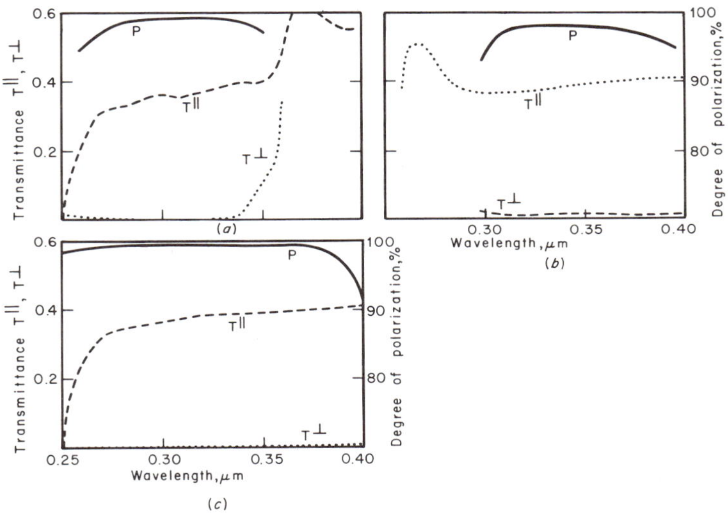

Fig. 74 The measured degree of polarization P and transmittance of parallel and crossed multilayer interference polarizers for the ultraviolet spectral region. (*c*) represents the results obtained when the polarizers shown in (*a*) and (*b*) are placed in series. [*After Sokolova and Krylova* (1963).]

The measured results for some interference polarizers are shown in Fig. 74. If polarizers for a wider spectral region are required, it is possible to use the technique of superposition of stacks with contiguous high-reflectance zones (Par. 55) or of two periodic multilayers with thickness ratios of 1:1 and 1:2 (Turner et al., 1965).

Bandpass Interference Filters

72. An ideal bandpass filter transmits all the incident radiation in one spectral region and rejects all the other radiation. Such a filter is completely described by the width of the transmission region and the wavelength at which it is centered. Practical filters are not perfect and require more parameters to describe their performance adequately. No uniform terminology has yet developed for this purpose. Care should be taken when reading and writing specifications since often different terms are used to describe different types of filters, and sometimes quantities bearing the same name are defined differently.

The position of the transmission band is variously specified by the wavelength λ_{max} at which the maximum transmission occurs, the wavelength λ_0 about which the filter passband is symmetrical, or the spectral center of gravity λ_c of the band. When specifying the tolerance on λ_0 it should be remembered that the peak can be moved only toward shorter wavelengths by tilting (see Par. 33).

The peak transmittance T_0 may or may not take into account the absorption within the substrate and/or blocking filters used to remove the unwanted transmission of the interference filter away from the principal passband (Fig. 75).

The half width (HW) $\Delta\lambda_{0.5}$ of the filter is the difference between the wavelengths at which the transmittance is half of T_0. It is often expressed as a percentage of

λ_0. The base width (BW) $\Delta\lambda_{0.01}$ is similarly defined. The ratio $\Delta\lambda_{0.01}/\Delta\lambda_{0.5}$, sometimes called the *shape factor,* indicates how "square" the transmission band is. Sometimes widths corresponding to other fractions of the transmittance are used to define it.

The minimum transmittance T_{min} of the filter does not take into account the effect of blocking filters. The quantity T_{min}/T_0 is called the *rejection ratio.* An accurate measure of how successful a blocked filter is in isolating the desired radiation is the ratio of the integrals of the transmittance outside and inside the transmission band.

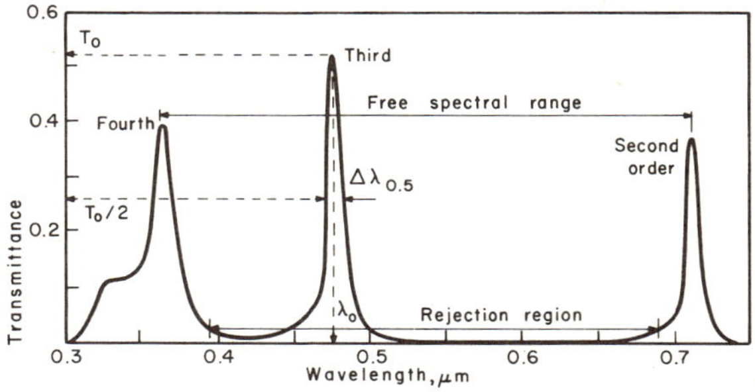

Fig. 75 Definition of some of the terms used to describe the properties of narrow-bandpass filters (Pars. 72 to 88). The curve represents the measured transmittance of an unblocked second-order metal-dielectric interference filter of the Fabry-Perot type (Par. 73). [*After Bausch & Lomb* (1968).]

In all-dielectric transmission-band filters the transmittance rises at some distance on either side of the transmission band. The distance over which the transmittance is low is called the *rejection region.* The distance between the two transmission maxima adjacent to the principal transmission band is called the *free spectral range.* Should either of the above two quantities be inadequate, auxiliary blocking filters must be provided.

Narrow- and Medium-Bandpass Filters (0.1 to 35 Percent HW)

73. Fabry-Perot Interference Filters (0.1 to 10 Percent HW) (Furman, 1968) Even though the essential components of Fabry-Perot (FP) interference filters, i.e., the spacer and the two reflectors that surround it, can take on many different forms (see Fig. 76 and Pars. 73 and 76 to 81), the filters are essentially low-order FP interferometers and hence the theory developed for the latter (see, for example, Born and Wolf, 1959) applies in full.

The transmittance of an FP filter, not allowing for absorption and multiple reflections within the substrate, is given by

$$T = \frac{T_R^2}{(1 - R)^2 + 4R \sin^2 \delta} \tag{60}$$

where

$$T_R = \sqrt{T_1 T_2} \qquad R = \sqrt{R_1 R_2} \tag{61}$$

T_1, R_1 and T_2, R_2 are the transmittances and reflectances of the first and second reflectors, respectively, as seen from within the spacer medium, and δ is given by

$$\delta = \frac{2\pi}{\lambda} nt \cos \phi + \epsilon \tag{62}$$

$$\epsilon = \frac{\epsilon_1 + \epsilon_2}{2} \tag{63}$$

n, t, and ϕ are the refractive index, thickness, and angle of refraction of the spacer, and ϵ_1, ϵ_2 are the phase changes on reflection from the spacer side of the first and second

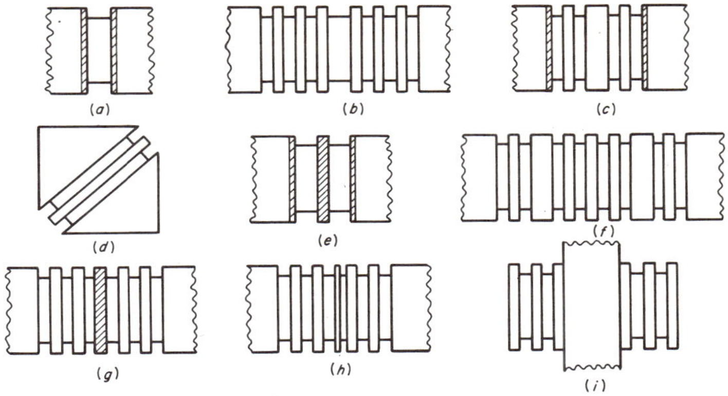

Fig. 76 Schematic representations of various types of bandpass interference filters: (*a*) to (*c*) Fabry-Perot interference filters with metal, dielectric, and metal-dielectric reflectors; (*d*) frustrated-total-internal-reflection filter; (*e*) and (*f*) square-top multicavity filters with metal and dielectric reflectors; (*g*) induced-transmission filter; (*h*) phase-dispersion (spacerless) interference filter; (*i*) Fabry-Perot filter with a mica or quartz spacer.

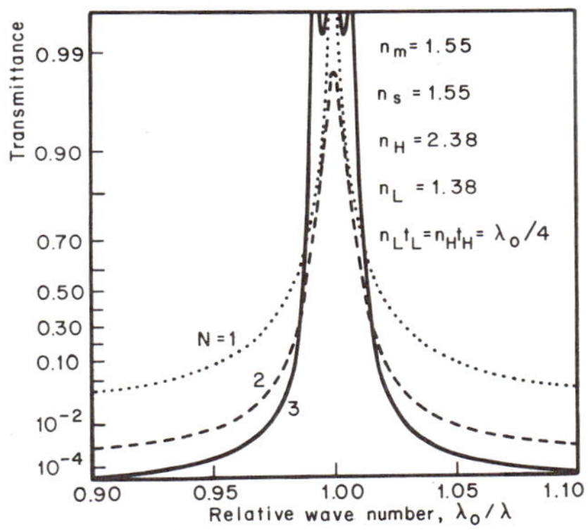

Fig. 77 Calculated transmittance on a log (density) scale of the bandpass filters: air$\cdot\{[(0.5H)L(0.5H)]^3H[(0.5H)L(0.5H)]^3\}^N\cdot$glass, $N = 1$, 2, and 3.

reflectors for radiation of wavelength λ. Maxima of T occur at wavelengths

$$\lambda_0 = \frac{2nt\cos\phi}{k - \epsilon/\pi} \qquad k = 0, 1, 2 \cdots \tag{64}$$

and are given by

$$T_0 = \left(\frac{T_R}{1 - R}\right)^2 = \left(\frac{1}{1 + A/T_R}\right)^2 \tag{65}$$

A $(= 1 - T_R - R)$ is the mean absorption of the reflectors. The minimum transmittance

$$T_{\min} = \left(\frac{T_R}{1 + R}\right)^2 \tag{66}$$

occurs at

$$\lambda_{\min} = \frac{2nt\cos\phi}{k - \epsilon/\pi} \qquad k = \tfrac{1}{2}, \tfrac{3}{2}, \tfrac{5}{2} \cdots \tag{67}$$

If T_R and R are essentially the same at λ_0 and $\lambda_{\min}$, the rejection ratio is given by

$$\frac{T_{\min}}{T_0} = \left(\frac{1-R}{1+R}\right)^2 \tag{68}$$

For $R > 0.7$ the half width of the transmission band (expressed as a percentage of λ_0) is given by

$$\frac{\Delta\lambda_{0.5}}{\lambda_0} \times 100 \approx \frac{1-R}{\sqrt{R}} \frac{100}{\dfrac{2\pi nt \cos\phi}{\lambda_0} - \lambda_0 \dfrac{\partial\epsilon}{\partial\lambda}} \tag{69}$$

For a given order of interference the half width and rejection ratio cannot be varied independently. The formula

$$T \approx \frac{T_0}{1 + 4[(\lambda - \lambda_0)/\Delta\lambda_{0.5}]^2} \tag{70}$$

valid for FP filters with small values of $\partial\epsilon/\partial\lambda$ in the neighborhood of λ_0, represents a lorentzian line shape with $\Delta\lambda_{0.1} = 3\Delta\lambda_{0.5}$ and $\Delta\lambda_{0.01} \approx 10\Delta\lambda_{0.5}$. The shape factor of all FP interference filters is therefore of the order of 10.

Filters with Metallic Reflecting Coatings. This was the first interference bandpass filter ever made (Geffcken, 1939) and the simplest. It consists of two partially transmitting, highly reflecting metal layers separated by a dielectric film and is symbolically represented by *MDM* (Fig. 76*a*). The best metallic reflectors currently available are aluminum and silver for the 0.125- to 0.34- and 0.34- to 3.0-μm spectral ranges, respectively. The measured spectral-transmittance curves of a number of typical filters are shown in Fig. 78.

The phase changes on reflection at the spacer–metal-reflector surface are finite and hence affect the position of the transmission maxima [Eq. (64)]. But the dispersion of the phase change on reflection can be neglected, and so the half width depends only on the reflectance of the metal layers and the order of the spacer. Filters with half widths of 1 to 8 percent are common. Because the ratio A/T is not very small for metals, the maximum transmittance is limited [Eq. (65)]. Maximum transmittances of 40 percent are relatively common. For filters with the narrower half widths or for shorter wavelengths, transmittances of the order of 20 percent have to be accepted. This is much less than the transmittances of all-dielectric filters of comparable half widths. Nevertheless, filters of this type are useful because their transmittances remain low except at wavelengths at which the first- and higher-order transmission maxima occur [Eq. (64)]. Blocking is thus easy. In particular, first-order filters usually do not require any blocking on the high-wavelength side—a difficult task at all times. The rejection of the filters is good, though not spectacular. If a better rejection is required and a lower transmittance can be tolerated, two identical filters may be cemented together (curve 6, Fig. 78). This is possible because of the finite absorption in the metal films. Alternatively, metal square-top filters (Par. 74) or filters with even more complicated structures (Par. 75) can be used.

Filters with All-Dielectric Reflectors (Geffcken, 1954/1944). Above 0.2 μm the metallic reflectors can be replaced by all-dielectric quarter-wave stacks (Pars. 47 to 51). The symbolic representation of such a filter (Fig. 76*b*) is, for example, $[HL]^N 2H [LH]^N$ or $H[LH]^N 2L [HL]^N H$, H and L being quarter-wavelength layers of high and low refractive indexes, respectively, and N the number of full periods in the reflecting stacks. The phase change on reflection at the boundary between the spacer and such a reflector does not affect the position of λ_0 [Eq. (64)], but the dispersion of the phase change on reflection is finite, depends on the materials used, and for lower-order spacers contributes very significantly to the reduction of the half width of the transmission band [Eq. (69)]. By choosing a suitable combination of the reflectance and order of the spacer almost any half width between 0.1 and 5 percent can be achieved in the visible part of the spectrum while maintaining a useful rejection ratio.

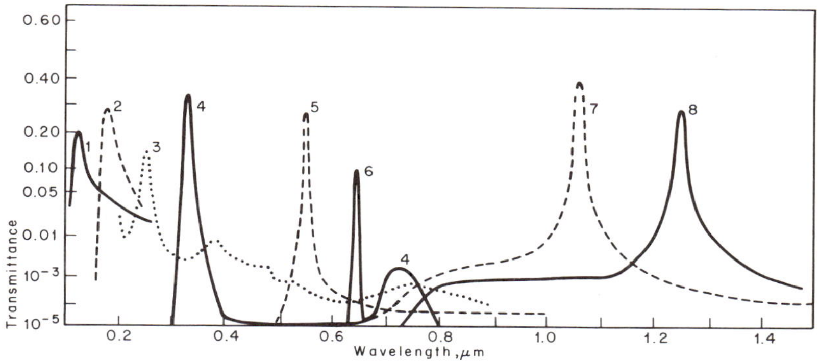

Fig. 78 Measured transmittance of FP filters with metallic reflecting coatings. Curve 6 represents the transmission of two identical filters cemented together. [*Curve* 1, *after Harrison* (1968); *curve* 2, *after Bradley et al.* (1967); *curves* 3 *and* 5 *to* 7, *Balzers* (1968); *curves* 4 *and* 8, *Schott & Gen.* (1965).]

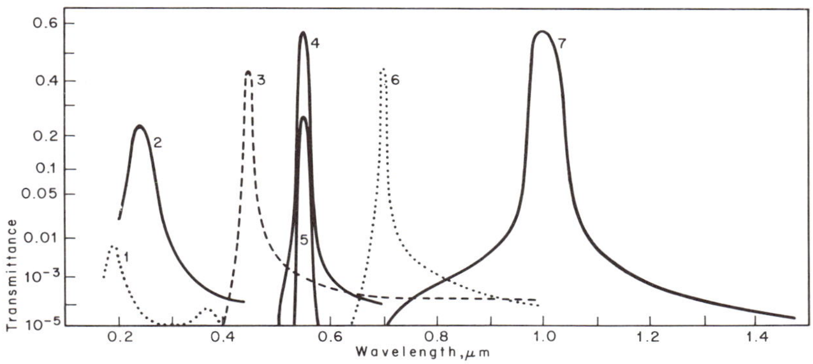

Fig. 79 Measured transmittance of square-top bandpass filters with metallic reflecting coatings. Curve 5 corresponds to the transmission of two identical filters cemented together. [*Curves* 1 *and* 2, *Schroeder* (1962); *curves* 3 *and* 6, *Balzers* (1968); *curves* 4, 5, *and* 7, *Schott & Gen.* (1965).]

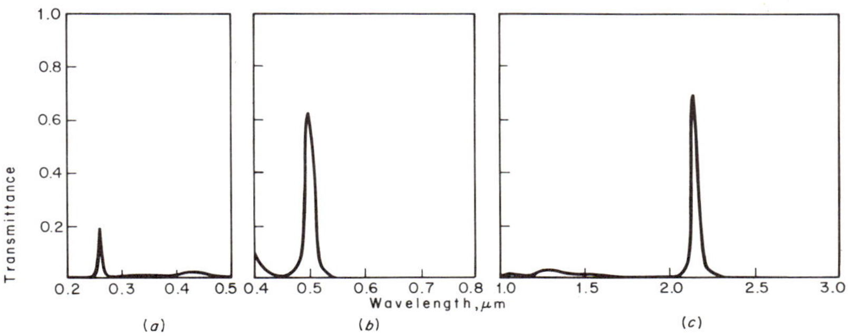

Fig. 80 Measured transmittance of induced transmission filters for the ultraviolet, visible, and infrared spectral regions. [(*a*) *after Baumeister et al.* (1965), (*b*) *after Berning and Turner* (1957), (*c*) *after Holloway and Lissberger* (1969).]

The maximum transmittance of all-dielectric FP filters depart from unity because of the finite absorption, scattering, and errors in the thicknesses and refractive indexes of the films. In the central part of the visible spectrum maximum transmittances of 0.8 are readily achieved for unblocked filters with a half width of 1 percent. This figure is gradually reduced as λ_0 approaches 0.2 or 20 μm, and filters with narrower half widths become impractical for lack of adequate transmittance.

The transmittance away from the transmission maximum is low only over the extent of the rejection region of the two materials used for the construction of the reflectors (Fig. 49), and additional blocking is often required on both the long- and short-wavelength sides. This can result in a considerable lowering of the maximum transmittance of the blocked filter, a 30 to 40 percent loss being not uncommon for filters peaked in the ultraviolet or infrared spectral regions.

For those parts of the visible and infrared for which nonabsorbing mechanically robust coating materials abound, square-top interference filters (Par. 74) are often preferred because of their better shape factor and higher rejection ratio. All-dielectric FP filters are still attractive in the ultraviolet, where there is a lack of such materials and where thickness monitoring is difficult, and also in the far infrared, where very thick layers are required.

The measured transmittance curves of a number of typical all-dielectric FP filters are shown in Fig. 82. All are plotted on the same logarithmic wavelength scale to facilitate comparison. An additional curve on a smaller scale is given, whenever necessary, to show the transmittance away from the passband.

Filters with Metal-Dielectric Reflectors (Turner, 1950). In these filters the reflectors consist of metal layers whose reflectance has been enhanced through the addition of several dielectric layers (Par. 93) (Fig. 76*c*). The properties of such filters are intermediate to those described in the two previous subparagraphs.

Frustrated-Total-Internal-Reflection Filters (Baumeister, 1967). These are essentially FP filters in which the spacer layer is surrounded by two frustrated-total-reflection surfaces (Turner, 1950; Fig. 76*d*). They have not found wide applications as bandpass filters because the finite absorption and scattering within the layers have prevented the theoretically expected high transmittance and small half widths from being realized and because the angular variation of the wavelength of the transmission peaks is very high.

74. Square-Top Multicavity Bandpass Filters (0.1 to 35 Percent HW) (Seeley, 1964; Warren, 1968; Thelen, 1969) A filter with a "squarer" shape that does not suffer from some of the disadvantages of the FP filters results when the basic FP structure is repeated two or more times. Such filters may be based on metal (Geffcken, 1954) or all-dielectric (Turner, 1952; Smith, 1958) reflectors (Fig. 76*e* and *f*). Thus, for example, $[HL]^N(2H)[LH]^N C[HL]^N(2H)[LH]^N$ represents an all-dielectric square-top filter in which the FP structure $[HL]^N(2H)[LH]^N$ is repeated twice. The quarter-wavelength layer C is called a *coupling* or *tie layer*, and the half-wavelength-thick spacer layers $2H$ are often called *cavities*. Small departures from the above model are made at times to improve the transmittance in the passband or the angular properties of the filter.

The half widths of the narrower multicavity filters of the above type do not differ very significantly from those of the basic FP structure [Eq. (69)]. The shape factors decrease with an increase in the number of the cavities and do not seem to depend on the materials (Warren, 1968). They are approximately 11, 3.5, 2.0, and 1.5 for one, two, three, and four cavities, respectively. The minimum transmittance in the rejection region is roughly that which could be obtained if the filter were composed entirely of $\lambda/4$ layers [Eq. (52)]. Unlike in the FP filter, the half width and rejection ratio can therefore be varied independently. These various points are illustrated in Fig. 77. The peak transmittance of multicavity square-top filters is less affected by the residual absorption in the layers than that of FP filters. As in the case of their FP counterparts, metal-dielectric square-top filters can be cemented together to enhance the rejection (curve 5, Fig. 79).

The improvement in the performance of square-top bandpass filters over that of the FP type is such that despite their more critical and expensive production,

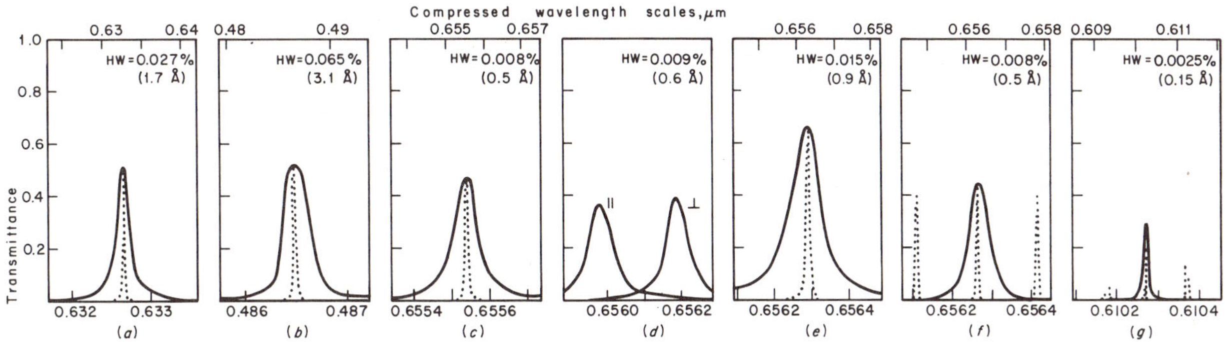

Fig. 81 Measured transmittance of very narrow bandpass interference filters with half widths less than 0.1 percent. Evaporated spacers: (*a*) [*after Meltzer* (1968)]; (*b*) [*after Eather and Reasoner* (1969)]; (*c*) mica interference filter for H_α; (*d*) mica interference filter with transmission bands polarized at right angles to one another [*Heliotek* (1969)]; (*e*) single and (*f*) and (*g*) double quartz-spacer interference filters [*after Austin* (1968*b*)]. The dotted curves in Figs. 81 and 82 represent the transmittances of the filters plotted over an extended spectral region (upper wavelength scales).

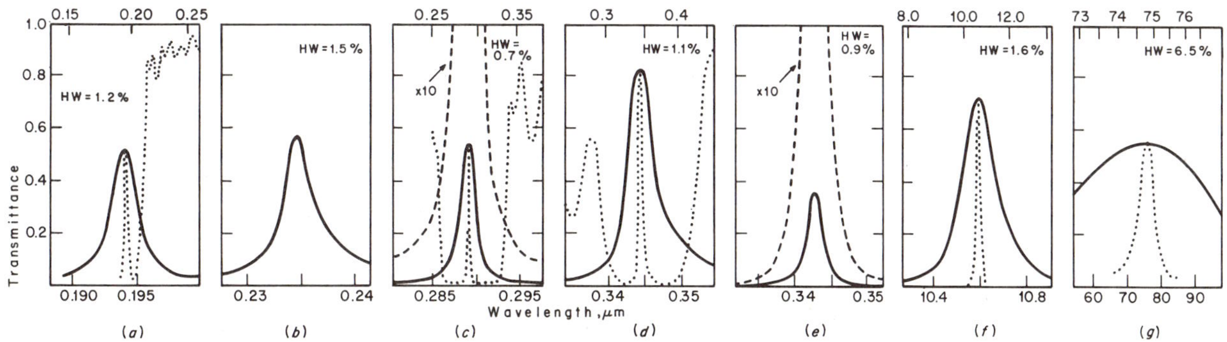

Fig. 82 Measured transmittance of FP all-dielectric interference filters with narrow half widths. [(*a*) *after Cohendet and Saudreau* (1967), (*b*) *and* (*d*) *after Motovilov* (1967), (*c*) *and* (*e*) *after Neilson and Ring* (1967), (*f*) *after Turner and Walsh* (1961), (*g*) *after Smith and Seeley* (1968).]

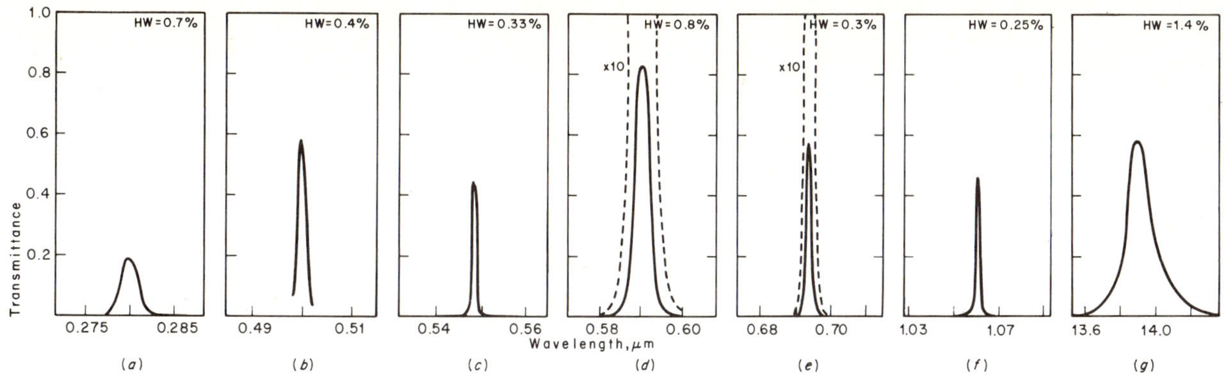

Fig. 83 Measured transmittances of fully blocked square-top interference filters with half widths between 0.25 and 1.4 percent. [(*a*), (*c*), *and* (*f*) *Corion* (1969), (*b*) *after Blifford* (1966), (*d*) *Heliotek* (1969), (*e*) *Baird Atomic* (1968).] (*g*) Only filter in the series that is not blocked [*after Smith and Seeley* (1968)].

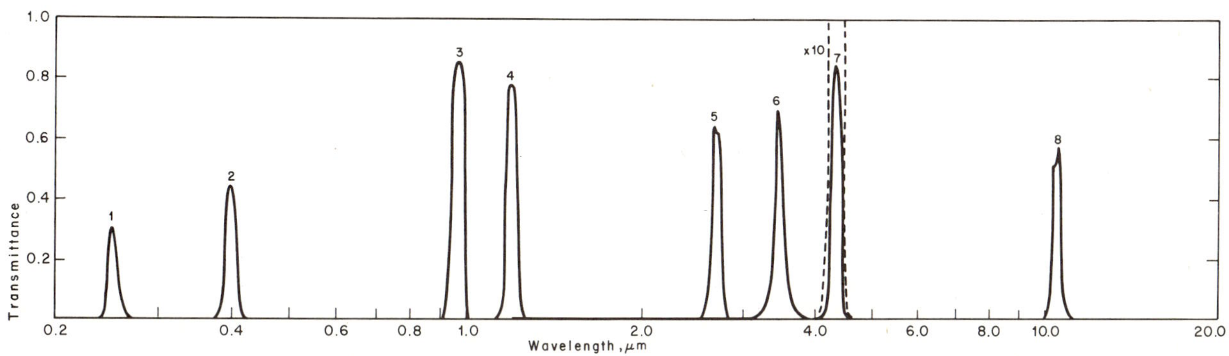

Fig. 84 Measured transmittances of fully blocked square-top interference filters with half widths of the order of 5 percent. [*Curves* 1 *and* 2, *Corion* (1969), *curve* 3, *Spectrum Systems* (1967), *curves* 4 *and* 6, *after Turner* (1970), *curve* 5, *Eastman Kodak* (1968*a*), *curves* 7 *and* 8, *Optical Coating Laboratory* (1964, 1970).]

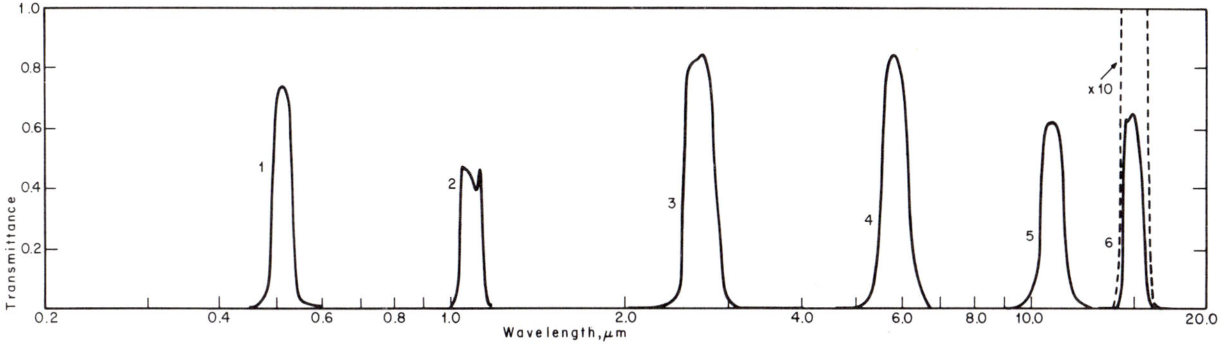

Fig. 85 Measured transmittances of fully blocked square-top interference filters with half widths of the order of 10 percent. [*Curve* 1, *Bausch & Lomb* (1968), *curve* 2, *Baird Atomic* (1968), *curve* 3, *Infrared Industries* (1967), *curve* 4, *after Turner* (1970), *curve* 5, *Eastman Kodak* (1968*a*), *curve* 6, *Optical Coating Laboratory* (1970).]

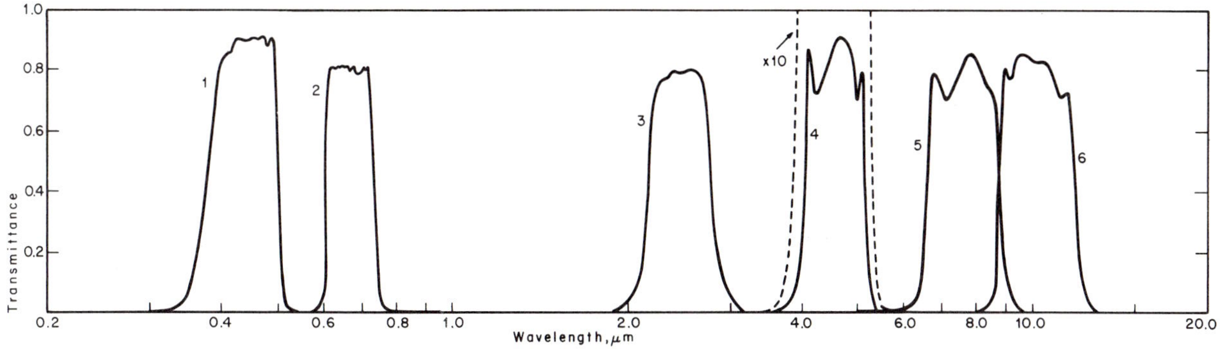

Fig. 86 Measured transmittances of blocked square-top interference filters with half widths of the order of 25 percent. [*Curve* 1, *Heliotek* (1969), *curves* 2 *and* 4, *Optical Coating Laboratory* (1964), *curves* 3 *and* 6, *after Turner and Walsh* (1961) *and Turner* (1970), *curve* 5, *Infrared Industries* (1967).]

many manufacturers regard them as their standard line of filters. The spectral-transmittance characteristics of commercially produced metal-dielectric and all-dielectric bandpass interference filters of different half widths are shown in Figs. 79 and 83 to 86, respectively. Most of these filters are commercially produced, and filters with intermediate half widths and peak wavelengths can readily be obtained.

75. Induced-Transmission Filters The transmittance of a metal layer can be considerably enhanced by surrounding it with suitable multilayer structures (Fig. 76*g*). Thus, for example, it is possible to induce a transmittance of 65 percent at $\lambda = 0.25\ \mu m$ in a 0.03-μm-thick aluminum film which, when deposited directly onto a quartz substrate, would transmit only 2.5 percent of the same radiation (Baumeister et al., 1965). The induced transmittance is highly wavelength-sensitive and can be used to construct bandpass filters containing one or more metal layers (Berning and Turner, 1957; Maier, 1967; Baumeister, 1969; Holloway and Lissberger, 1969). Induced-transmission filters combine the good long-wavelength attenuation properties of the more common types of metal-dielectric filters (Pars. 73 and 74) with peak transmittances that are closer to those of all-dielectric filters. The performance of some experimentally produced induced-transmission filters is shown in Fig. 80.

76. Very Narrow Bandpass Filters (HW < 0.1 Percent) It follows from Eq. (69) that the half widths of interference filters can be reduced by increasing the reflectance of the reflectors, the order of interference of the spacer, or the dispersion of the phase change on reflection. All these approaches have been tried in the past.

77. Filters with Evaporated Spacers (HW > 0.03 Percent) In narrow-band filters of conventional construction both high reflectance and higher-order spacers are used. The manufacturing process is quite critical, and attention must be paid to details. The films must be very uniform over the filter area, and they must not absorb or scatter. They must not age, or, alternatively, their aging must be capable of being accelerated or arrested. Monitoring must be precise, so that the peak occurs at or close to the desired wavelength.

Both FP and square-top filters of very narrow bandwidths can be made, the latter being preferable for most applications. The present limit on the half widths of this type of filter seems to be of the order of 0.03 percent (Korolev et al., 1960). The performance of two commercially produced filters of this type is shown in Fig. 81*a* and *b*.

78. Phase-Dispersion Filters (HW > 0.1 Percent) The dispersion of the phase change on reflection enters into Eq. (69) for the half width of FP filters. Typical values of this quantity at $\lambda = 0.5\ \mu m$ for a silver reflector, a nine-layer quarter-wave stack with $n_H/n_L = 1.75$, and for a broadband reflector are −0.5, −6.8, and −112.0, respectively (Baumeister and Jenkins, 1957). As a result, the half widths of FP filters constructed with such reflectors should be reduced by factors of about 1.05, 2, and 20. In the last case the contribution of the spacer to the half width is negligible, and a spacerless design is possible (Baumeister et al., 1959; Fig. 76*h*). Unfortunately, the expected reduction in half width has so far not been fully realized in practice, probably because of errors in the monitoring and lack of uniformity of the layers (Giacomo et al., 1959).

79. Fabry-Perot Filters with Mica Spacers (HW > 0.01 Percent) In practice the half width of interference filters cannot be reduced indefinitely by increasing the optical thickness of an evaporated spacer [Eq. (69)] because when the latter exceeds about two wavelengths, it becomes too rough to be useful (Eather and Reasoner, 1969). A high-order filter can, however, be constructed by evaporating reflecting coatings on either side of a thin sheet of mica (Fig. 76*i*). Transmission bands in silvered mica were probably first observed by Wood (1904), but the deliberate use of mica to construct filters came much later (Kartashev and Syromyatnikova, 1949; Ring et al., 1958; Dobrowolski, 1959). The construction of mica interference filters with transmittances of 30 to 80 percent per polarization for half widths of the order of 0.01 to 0.1 percent in the 0.45- to 2.0-μm wavelength region is relatively straightforward (Dobrowolski, 1959). The position of the transmittance peak can be located within a fraction of an angstrom, does not change with time, and can be sufficiently uniform over areas of 2 to 5 cm diameter. Because of the very high order of interference (70 to 700 orders) the free spectral range is quite small, and auxiliary filtering is necessary for most applications. Unless the thickness of the mica is specially selected, the birefringence of mica will result in two mutually perpendicularly polarized sets of transmission bands, a fact that can be used to advantage in some applications. The spectral-transmittance curve of a fully blocked mica interference filter for H_α is shown in Fig. 81*c*.

80. Filters with Fused-Silica Spacers (HW > 0.002 Percent) (Title, 1970) It is possible to construct very narrow bandpass filters having thin fused-quartz spacers (Herriott et al., 1965; Rehnberg, 1970; Fisher, 1971). A good fused-quartz flat is coated with an all-dielectric reflector, and this coated surface is optically contacted to another flat (Austin, 1968*a*). The flat is then ground down and polished to form a spacer layer of the required thickness, and the second reflector layers are applied to complete the filter. As in mica filters, the position of the transmission band is very stable, and auxiliary blocking filters are needed because of the small free spectral range. The transmittance of filters with silica spacers is higher than that of corresponding mica filters because fused quartz is highly transparent and not birefringent. A typical unblocked filter with a clear aperture of 3.5 cm and a half width of 0.007 percent had a transmittance of 45 percent for nonpolarized light. An important advantage of filters with fused-silica spacers is that it has been found possible, by repeating the process described above, to construct square-top filters with rejection ratios of the order of 5×10^4 (Fig. 81*f*) and filters with half widths as low as 0.002 percent (Fig. 81*g*).

81. Tunable Air-spaced Filters (HW > 0.001 Percent) (Smith and Heavens, 1957). These are air-spaced FP interferometers, often provided with elaborate automatic plate-parallelism and spacing control, which are more akin to spectrometers than to filters. The position of the passband can be tuned quite significantly by changing the separation between the reflector plates. This type of tuning, unlike the tuning of filters by tilting, does not affect the angular field or shape of the transmission band. Ramsey (Ramsey, 1968; Ramsey et al., 1970) reviews the various problems associated with the construction and use of such instruments.

82. Wide-Bandpass Filters Filters with half widths ranging from 10 to 40 percent can be constructed using techniques described in Par. 74 (Figs. 85 and 86). Filters with wider transmission bands are usually obtained by combining short- and long-pass filters with cutoffs at the desired wavelengths. The cutoff filters may be all-dielectric (Figs. 67 and 68), glass or gelatin cutoff filters (Figs. 6 and 10), or antireflection-coated infrared materials (Par. 3 of Sec. 7; Fig. 7). Some of the short- or long-pass filters may be regarded as wide-bandpass filters in their own right. The cutoff filters may be combined into a single filter, or they may be deposited onto separate substrates. In the latter arrangement the cutoff positions of all dielectric short- and long-pass filters can be tuned individually by tilting to coincide with the desired wavelength. Further, it is possible to assemble a number of individual filters into sets that make it possible to form wide-passband filters of different half widths and peak wavelengths.

Filters with very broad transmission bands are also obtained when a multilayer is formed from two suitably displaced long-wavelength cutoff filters separated by an appropriate matching layer (McKenny and Slater, 1970; Fig. 87*a*). Automatic computer programs can be used to design high-transmission broadband filters with a high rejection and a shape factor close to unity (Dobrowolski and Bastien, 1966; Fig. 87*b*).

83. Wedge Filters If the thicknesses of all the layers of a bandpass filter vary in proportion across the surface of a substrate, the position of the transmission peak will vary in the same way (Par. 33). Such wedge filters are as old as the interference filter itself and are available in versions in which the wavelength variation occurs along a straight line or a circle. The latter arrangement is particularly useful because it lends itself well to the construction of low-cost, small and lightweight rapid-scan monochromators of moderate resolution that are robust and environmentally stable (Hovis et al., 1967). Thelen (1965) and Apfel (1966*b*) describe in detail one method of production of circular variable filters with a linear dependence of wavelength on angle and give references to some of the previous work on wedge filters. Yen (1969) describes circular variable square-top filters for the 0.4- to 25-μm spectral region with several different half widths between 1 and 12 percent. The maximum transmittances for fully blocked filters vary between 15 and 75 percent, depending on the spectral region and the half width of the filter. Rejection levels of 0.1 or 0.01 percent are possible. Typical transmission curves for several angular positions on two circular variable square-top filters

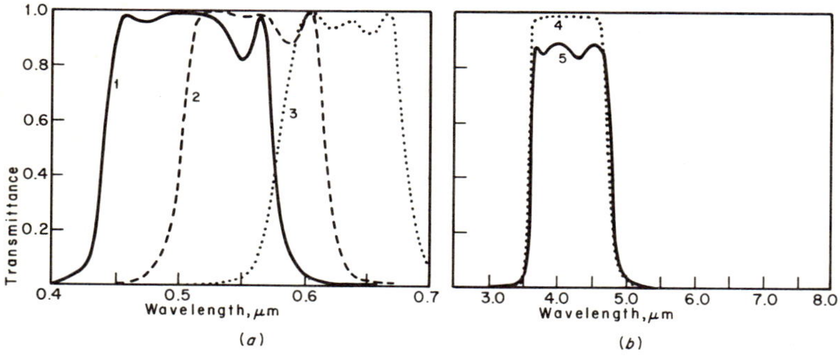

Fig. 87 Broadband transmission filters: (*a*) calculated transmittance of three filters consisting essentially of two superimposed suitably tuned long-wavelength cutoff filter structures [*after McKenney and Turner* (1967)]; (*b*) calculated (*curve* 4) and measured (*curve* 5) transmittances of a filter designed with an automatic synthesis program [*after Michael* (1968)].

are shown in Fig. 88. The ratio of maximum to minimum wavelength available in one wheel varies between 1.11 and 16 (Yen, 1969). The angular width of the slit used in conjunction with a circular variable filter, expressed as a percentage of the angular size of the filter wedge, should not exceed the nominal half width of the filter (expressed in percent) if it is not to cause a marked reduction in the resolution.

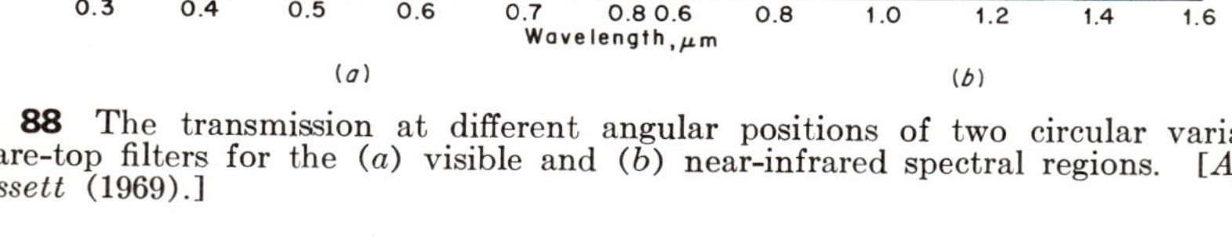

Fig. 88 The transmission at different angular positions of two circular variable square-top filters for the (*a*) visible and (*b*) near-infrared spectral regions. [*After Mussett* (1969).]

84. Angular Properties of Bandpass Interference Filters With the gradual increase of the angle of incidence the transmittance maximum of a typical bandpass filter moves toward shorter wavelengths. On further increase the maximum transmittance and the half width deteriorate; the transmission band becomes asymmetric and eventually splits up into two perpendicularly polarized components (Fig. 89*a*). The deterioration is more rapid with nonparallel radiation.

85. Properties of Bandpass Filters for Angles of Incidence Less than 20° The behavior of bandpass filters for such angles θ can be described quantitatively using the concept of an *effective index* μ^* of the filter. In terms of μ^* the transmittance T in the neighborhood of the transmission peak of any FP filter is given by

$$T \approx \frac{T_0}{1 + \left[\frac{2(\lambda - \lambda_0)}{\Delta\lambda_{0.5}} + \frac{\lambda_0}{\Delta\lambda_{0.5}} \frac{\theta^2}{\mu^{*2}}\right]^2} \tag{71}$$

(Lissberger, 1968). $\Delta\lambda_{0.5}$ and T_0 are the half widths and the maximum transmittance (at λ_0) for normal incidence of the radiation. Formulas for μ^* in terms of the construction parameters have been found for the all-dielectric FP filter and the double-spacer filter (Lissberger, 1959; Pidgeon and Smith, 1964) and for the metal-dielectric FP and induced-transmission filters (Hemmingway and Lissberger, 1967). The change in position of the transmission peak $(\delta\lambda)_0$ and the half width $(\Delta\lambda_{0.5})_\theta$ at angle θ are

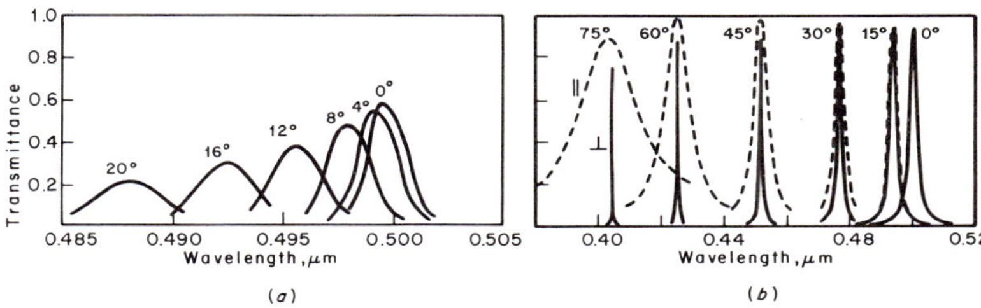

Fig. 89 Angular properties of all-dielectric interference filters. (*a*) Measured variation with angle of incidence of the spectral transmittance of a typical commercial interference filter [*after Blifford* (1966)]. (*b*) Calculated transmittance of a filter in which the peaks of the two polarized transmission bands at nonnormal incidence coincide. The filter is of the type air$\cdot[HL]^4(2A)[LH]^4\cdot$glass, where $n_H t_H = n_L t_L = n_A t_A = \lambda_0/4$ and $n_S = 1.52$, $n_M = 1.00$, $n_H = 2.30$, $n_L = 1.38$, and $n_A = 1.825$.

$$\left(\frac{\delta\lambda}{\lambda_0}\right)_\theta = -\frac{\theta^2}{2\mu^{*2}} \tag{72}$$

and

$$\frac{(\Delta\lambda_{0.5})_\theta}{\Delta\lambda_{0.5}} = \left[1 + \left(\frac{\theta^2\lambda_0}{\mu^{*2}\,\Delta\lambda_{0.5}}\right)^2\right]^{1/2} \tag{73}$$

For convergent radiation of semiangle α the corresponding expressions are

$$\left(\frac{\delta\lambda}{\lambda_0}\right)_\alpha = -\frac{\alpha^2}{4\mu^{*2}} \tag{74}$$

$$\frac{(\Delta\lambda_{0.5})_\alpha}{\Delta\lambda_{0.5}} = \left[1 + \left(\frac{\alpha^2\lambda_0}{2\mu^{*2}\,\Delta\lambda_{0.5}}\right)^2\right]^{1/2} \tag{75}$$

Linder (1967) and Lissberger (1968) discuss the requirements and design of filters for this case.

Small tilts are commonly used to tune the peak of a filter to the desired wavelength even though they have an adverse effect on the angular field of the filter.

86. Wide-Angle Filters Figure 90 shows the variation with effective index μ^* of the angular field of FP filters, defined as being twice the angle of tilt necessary to reduce to $0.8T_0$ the transmittance of the filter for radiation of wavelength λ_0. To increase the angular field, μ^* must be increased. Thus, for example, in an all-dielectric FP filter the expression for μ^* shows that $n_L < \mu^* < n_H$ and that with increasing spacer order μ^* approaches the refractive index of the spacer (see also Ring, 1956). The upper limits for μ^* for an all-dielectric FP filter in the ultraviolet, visible, and infrared parts of the spectrum are of the order of 2.0, 2.35, and 5.0, respectively. Little can be done about the angular field of mica (Par. 79) and quartz (Par. 80) spacer filters.

For metal-dielectric FP and induced-transmission filters effective indexes μ^* of up to 3.2 and 2.0 have been reported (Hemmingway and Lissberger, 1967).

Wilmot and Schineller (1966) and Schineller and Flam (1968) have announced a filter consisting of a thin, plane-parallel fiber-optic faceplate coated on both sides with all-dielectric mirrors. Since in such a filter the half width is determined only by the thickness of the plate and the reflectivity of the coatings, and since the field of view depends on the ratio of the wavelength to the fiber diameter, the two quantities are independent. The measured transmittance of 15-Å-half-width, 6-mm-diameter filter composed of 1.5-μm-diameter fibers was 30 percent, and the shift in wavelength with angle of incidence was one-eighth that of a conventional filter.

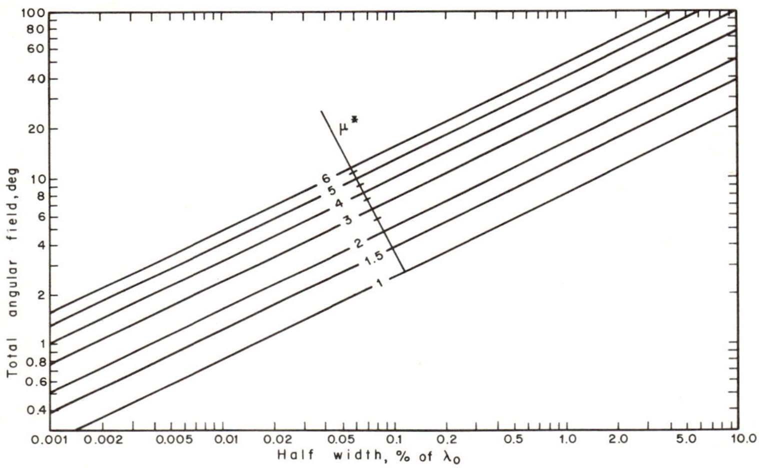

Fig. 90 Angular field of FP filters as a function of the half width for different effective indexes μ^*.

87. Filters with No Polarization Splitting at High Angles of Incidence It has been shown numerically for the phase-dispersion filter (Baumeister et al., 1959), for the frustrated-total-reflection filter (Kard, 1959*b*), and for the metal-dielectric (Costich, 1970) and all-dielectric (Konyukhov and Nesmelov, 1969) FP filters that it is possible to arrange for the two polarized transmission bands, which may have different widths, to coincide at high angles of incidence (Fig. 89*b*). The narrow, symmetrical high-transmittance bands that result may be useful for some applications even though the position of the maximum is still displaced with angle.

88. Stability and Temperature Dependence of Bandpass Filters The stability of the position of the transmission peak has been studied by many workers (e.g., Meaburn, 1966; Schild and Walther, 1967; Warren, 1968; Eather and Reasoner, 1969; Lazareva, 1970; Furman and Levina, 1971*a*,*b*). The observed changes (up to 1 percent of λ_0) seem to depend greatly on the materials and manufacturing conditions. In filters with evaporated spacers both irreversible changes, probably due to changes in the structure of the films, and reversible changes due to the adsorption of water vapor have been observed. Many manufacturers are now able to minimize these effects through the use of more stable materials, improved manufacturing techniques, and accelerated artificial aging processes. No changes were observed in filters with mica (Par. 79) or quartz (Par. 80) spacers.

Changes in the operating temperature normally do not significantly affect the half widths and peak transmittances of bandpass interference filters (see, for example, Furman, 1970; Furman and Levina, 1970; Chaikin and Pukhonin, 1970; Golubeva et al., 1971). An exception are filters containing semiconductors, e.g., germanium, that start to absorb significantly at elevated temperatures. The positions of the transmission peaks shift linearly toward longer wavelengths with an increase in temperature, the magnitude of the shift depending largely on the spacer material. The temperature coefficient, expressed as a percentage change in λ_0 per degree Celsius change in temperature, lies between 2×10^{-4} and 3×10^{-3} for filters with evaporated spacers for the 0.3- to 1.0-μm spectral region (Blifford, 1966) and between 2×10^{-3} and 2×10^{-2} for infrared filters (Warren, 1968; Optical Coating Laboratory, 1967). It is of the order of 1×10^{-3} for filters with mica (Dobrowolski, 1959) and quartz (Austin, 1968*b*) spacers. For very narrow band filters (Pars. 76 to 81) under adverse conditions all these temperature coefficients could lead to serious errors unless temperature control is provided. This latter can take the form of an external constant-temperature enclosure, or it may be built right into the filter. Eather and Reasoner (1969) and Mark et al. (1970) describe arrangements of the latter type in which two sensors embedded in the filters are used to control the current flowing through transparent conducting coatings that surround the filter.

Deliberate changes in the temperature can be used for a fine tuning of the transmission wavelength without having an adverse effect on the angular field of the filter.

Achromatic and Color-selective Beam Splitters

89. Achromatic Beam Splitters (Anders, 1967*a*) These devices are introduced into an incident beam of radiation when it is desired to divide it into two beams of approximately equal relative spectral composition but propagating in two different directions. Four different physical forms of beam splitters are illustrated in Fig. 91. The simplest (Fig. 91*a*) consists of a coating on a transparent plane-parallel substrate. If the two derived beams are to traverse identical paths, a cemented beam splitter is used (Fig. 91*b*). The lateral displacement of the transmitted beam introduced by the above forms can be avoided with a beam-splitting cube (Fig. 91*c*). To reduce the stray reflected light in the system, the free surfaces

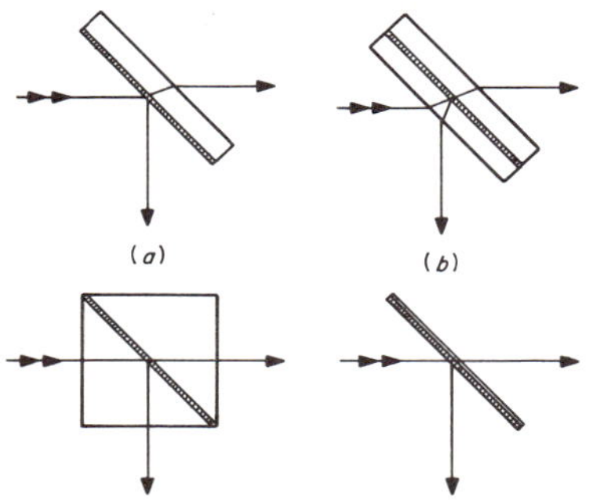

Fig. 91 Beam-splitter types: (*a*) single plate, (*b*) two cemented plates, (*c*) beam-splitter cube, (*d*) pellicle beam splitter.

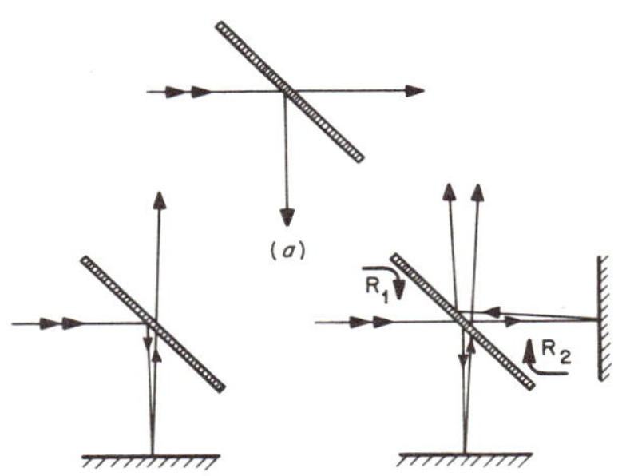

Fig. 92 Three different ways of using beam splitters.

of the above beam splitters can be antireflection-coated. Alternatively the coating can be deposited onto an approximately 1-μm-thick pellicle (Fig. 91*d*). The latter is an integral part of the coating and may introduce an interference pattern into the spectral reflectance and transmittance characteristics. Pellicle beam splitters are very light and yet quite sturdy (Scheuerman, 1968*a*; Vande Kleft et al., 1969; Hancock, 1970). The optomechanical design of rugged, environmentally stable mounts for the above types of beam splitter have been discussed by Heinrich et al. (1967) and Lipshutz (1968).

In general, the transmission and reflection coefficients T and R depend on the polarization of the incident radiation. However, in neutral beam splitters the quantity $\frac{1}{2}(R^{\parallel} + R^{\perp})_{\theta=45°}$ is always close to the reflectance at normal incidence. The reflectance of absorbing, uncemented beam splitters depends also on the direction of incidence (Par. 33). The optimum values of T and R depend on the application. For example, for a binocular eyepiece of a nonpolarizing microscope the most important requirement is $T^{\parallel} + T^{\perp} = R^{\parallel} + R^{\perp}$ (Fig. 92*a*). For a vertical illuminator $R^{\parallel}T^{\parallel} + R^{\perp}T^{\perp}$ should be a maximum (Fig. 92*b*). The condition for maximum fringe contrast in some interferometers requires that $R_1^{\parallel}T^{\parallel} = R_2^{\parallel}T^{\parallel}$ and $R_1^{\perp}T^{\perp} = R_2^{\perp}T^{\perp}$ (Fig. 92*c*). This is satisfied automatically by all nonabsorbing and by absorbing cemented beam splitters. The occasional requirement that the phase change on reflection be the same for radiation incident onto the beam splitter from opposite sides is automatically satisfied at the design wavelength by all-dielectric coatings composed of $\lambda/4$ layers but not by uncemented metal beam splitters (Rowley, 1966).

For maximum efficiency with unpolarized radiation $R^{\parallel}$, $R^{\perp}$, $T^{\parallel}$, and $T^{\perp}$ should all approach 0.5. However, such a coating will not necessarily exhibit the best ratio of the intensities of the directly transmitted or reflected radiation to that which first undergoes multiple reflections (Anders, 1967*a*). Often beam splitters are required that are uniform over a broad spectral region. Inconel films satisfy this requirement although about one-third of the incident radiation is lost through absorption (see

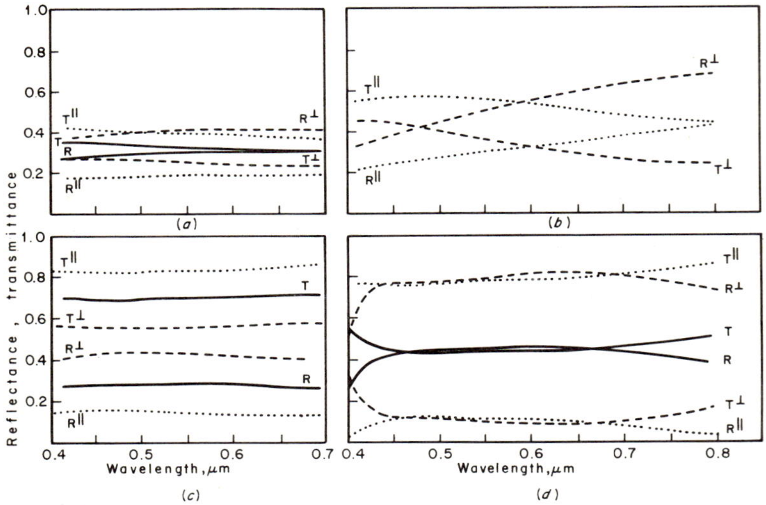

Fig. 93 Measured spectral transmittance for polarized and unpolarized light of (*a*) Inconel- and (*c*) dielectric-coated beam-splitting plates [*Oriel* (1968)] and of (*b*) silver- and (*d*) dielectric-coated beam-splitting cubes [*after Anders* (1967)].

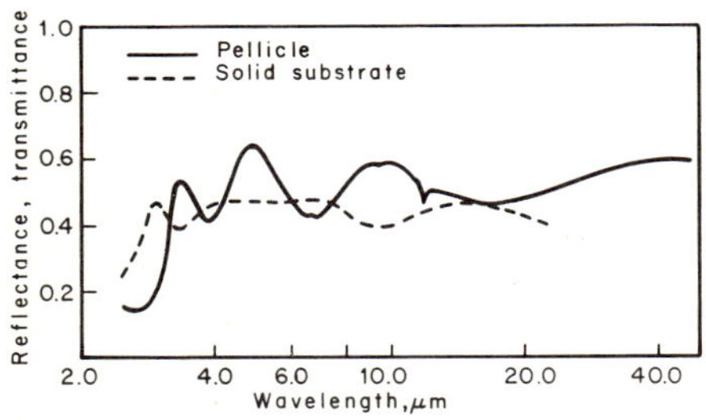

Fig. 94 Infrared beam splitters consisting of a suitably coated plate [*after Heinrich et al.* (1967)] and pellicle [*after Scheuerman and Beardsley* (1968)].

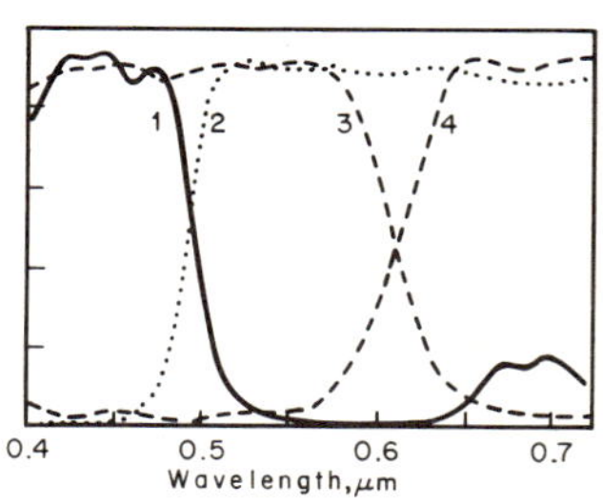

Fig. 95 Measured spectral transmittance of four color-selective beam splitters. [*Optical Coating Laboratory* (1964).]

also Par. 31). The design of achromatic all-dielectric beam splitters has been discussed, for example, by Pohlack (1956*a*), Catalan and Putner (1961), Miyake (1964), Sokolova (1970), Sergeeva (1971), and Clapham et al. (1971). Knittl (1964) considered the design of beam splitters in which both the reflectance and the phase change on reflection are achromatized.

The measured performance of several beam splitters is shown in Figs. 93 and 94.

90. Color-selective Beam Splitters For various technological applications a beam of light must be divided into several components of different color. All-dielectric color-selective beam splitters (*dichroics*) are used for this purpose because they are practically lossless and because their transition wavelengths can be selected at will. They are essentially cutoff filters (Pars. 58 to 68) designed for use at 45° incidence, and their spectral characteristics normally depend somewhat on the polarization of the incident radiation. The effect of this and of variations in the angle of incidence and thickness of the coatings on the chromaticity coordinates of dichroic beam splitters for television cameras was investigated by Pohlack (1957). If necessary, the polarization effect can be reduced by passing the separated components through auxiliary cutoff filters (Turner, 1954).

Typical transmittance curves of several color-selective beam splitters are shown in Fig. 95.

REFLECTION FILTERS AND COATINGS

Metal and Metal-Dielectric Reflectors

91. Metal Reflectors The Fresnel reflection coefficient of a material with a complex refractive index $n - ik$ is given by

$$R = \frac{(n-1)^2 + k^2}{(n+1)^2 + k^2} \tag{76}$$

If the material is opaque, the above represents the total energy reflected, the remaining energy being absorbed within the material. Metal reflectors are most commonly made by vacuum deposition of the material onto a suitable glass or quartz substrate. If aluminum or beryllium mirror surfaces are to be coated, they are sometimes first chemically plated with a nickel-phosphorus alloy (Kanigen process). Such deposits have excellent adhesion to the base metal and have a very hard surface that can be optically polished before coating (ASTM, 1959).

The visible, infrared, and ultraviolet spectral-reflectance curves of some of the more frequently used metals are shown in Figs. 96 to 98. Silver has the highest visible and infrared reflectance, and hence it is used for interferometer mirrors and interference filters. Exposed silver films tarnish readily. Aluminum has the broadest high-reflectance region of all metals and is commonly used in front-surface mirrors. It would reflect highly down to 0.1 μm were it not for the absorption below 0.18 μm of the thin oxide layer that starts to form seconds after deposition (Madden et al., 1963). In some parts of the extreme ultraviolet platinum and iridium have the highest known reflectances. Measured values of the reflection coefficient in the 0.0023- to 0.019-μm spectral region are given by Lukirskii et al. (1964, 1965) for a number of materials.

92. Protected-Front-Surface Mirrors For many applications the thin oxide layer on aluminum does not offer sufficient protection against abrasion and chemical attack, and therefore aluminum mirrors are often overcoated with thin films of SiO_2 (or MgF_2). Such protected coatings can be repeatedly cleaned with water and even withstand boiling in salt water (Hass, 1955). The protective layer reduces the reflectance (Fig. 101; Hass et al., 1969*b*). If necessary this can be avoided by using protective coatings consisting of two or more layers (Fig. 100).

93. Enhancement of Reflection (Turner, 1950) By depositing a quarter-wave stack (Pars. 41 to 51) onto the metal mirror its reflectance can be enhanced considerably. The thickness of the first layer should be adjusted to allow for the phase change on reflection at the metal surface (Young, 1963). The spectral-reflectance curve has dips on either side of the high-reflectance region whose width

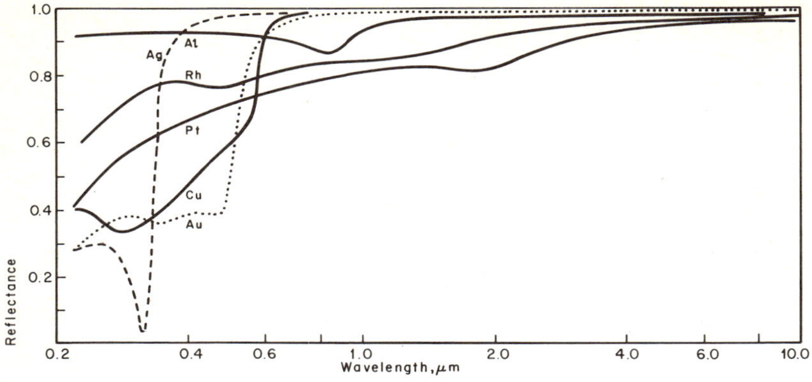

Fig. 96 Spectral reflectance of certain metals. [*After Drummeter and Hass* (1964).]

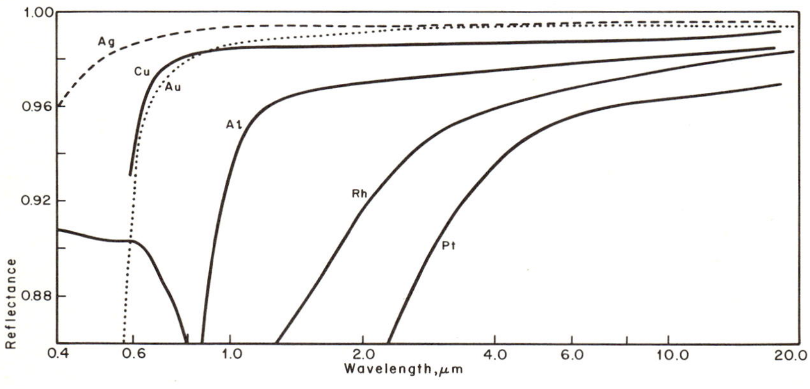

Fig. 97 Visible and infrared reflectance of certain metals. [Al, *after Bennett et al.* (1963*b*), Ag *and* Au, *after Bennett and Ashley* (1965), Cu, *after Hass and Hadley* (1972), Rh *and* Pt, *after Hass and Fowler* (*see Drummeter and Hass,* 1964).]

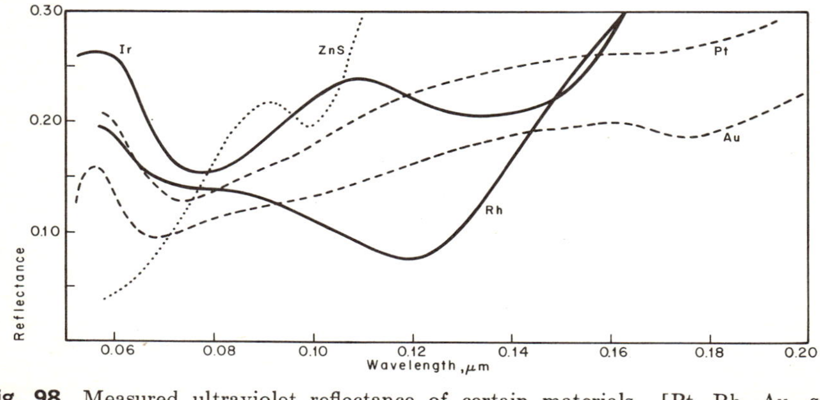

Fig. 98 Measured ultraviolet reflectance of certain materials. [Pt, Rh, Au, *and* ZnS, *after Hunter* (1962), Ir, *after Hass et al.* (1967).]

is governed by the considerations of Par. 43. The measured spectral characteristics of three metal-dielectric reflectors for the ultraviolet region are shown in Fig. 99.

In the extreme ultraviolet, reflection enhancement of a different kind has been reported. By depositing semitransparent platinum films onto different substrates the opaque-film reflectances of 19.3 and 12.8 percent at 0.0584 and 0.0736 μm were increased by up to 2.8 and 3.8 percent, respectively (Hass et al., 1969*b*). For space applications (where oxidation can be discounted) suitably thick aluminum films on iridium are expected to yield reflectances as high as 40 and 52 percent at the same wavelengths (Fig. 102). Other proposed reflectance-increasing combinations will be found in Madden et al. (1963*a*).

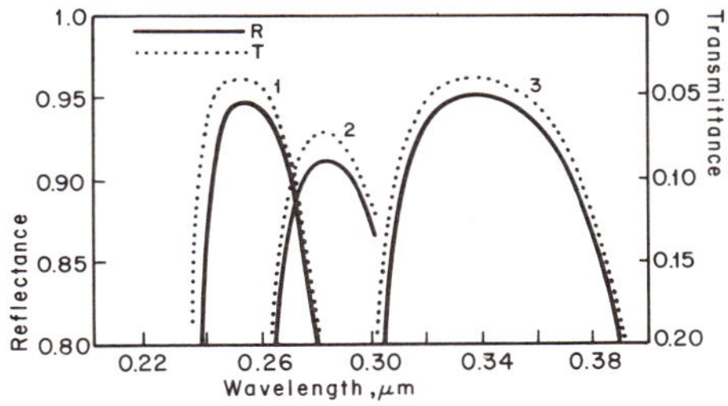

Fig. 99 Enhancement of the ultraviolet reflectances of semitransparent aluminum films by the deposition of quarter-wave stacks. Curves 1 and 2: eleven layers of PbF_2 and MgF_2 [*after Leś et al.* (1963)]; curve 3: nine layers of Sb_2O_3 and MgF_2 [*after Leś and Leś* (1962)].

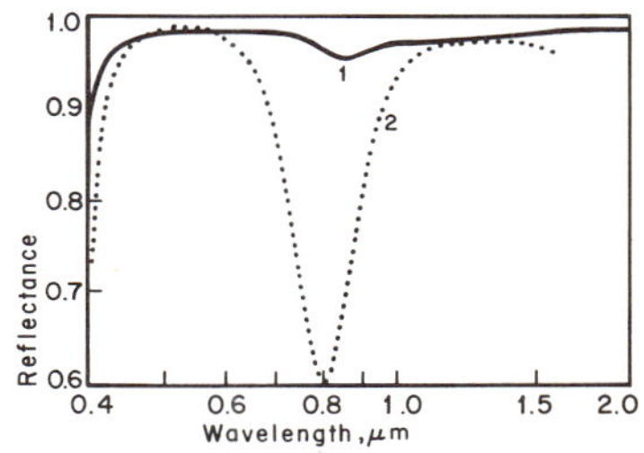

Fig. 100 Reflectance of very durable overcoated metal mirrors. Curve 1: protected front-surface silver mirror [*after Denton* (1970)]; curve 2: aluminum with four layers of MgF_2 and CeO_2 [*after Hass* (1955)].

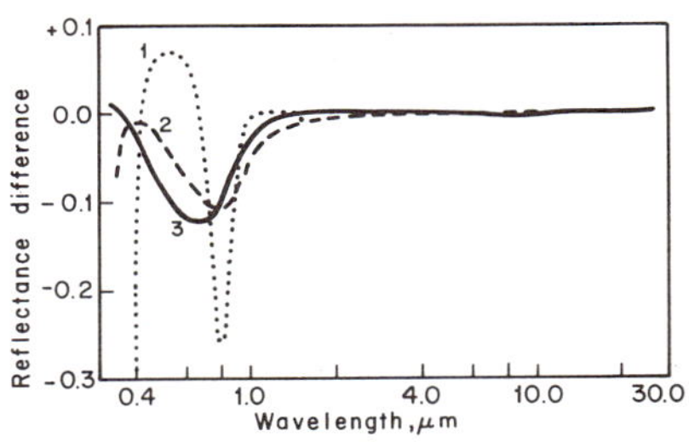

Fig. 101 Measured difference between the reflectances of protected and unprotected aluminum mirrors. Curve 1: coating 2 of Fig. 100; curves 2 and 3: 0.1122 ± 0.002 and 0.0752 ± 0.001 μm thick films of MgF_2 and SiO_x, respectively, on aluminum. [*After Bennett* (1963*a*).]

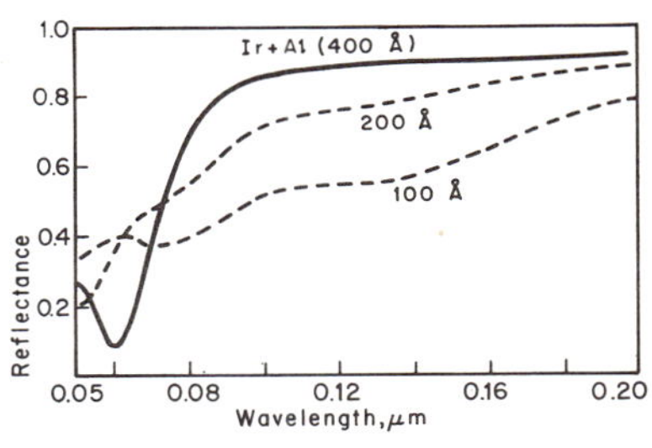

Fig. 102 Calculated spectral-reflectance curves of iridium overcoated with different thicknesses of unoxidized films of aluminum. [*After Hass and Hunter* (1967).]

94. Protection against Oxidation The deterioration of ultraviolet reflectance of aluminum mirrors due to oxidation can be partially avoided by covering the freshly deposited aluminum layer immediately with a suitable coating of MgF_2 (Hass and Tousey, 1959; Canfield et al., 1966; Adriaens and Feuerbacker, 1971) or LiF (Angel et al., 1961; Cox et al., 1968; Adriaens and Feuerbacker, 1971). The reflectances of such overcoated aluminum reflectors are shown in Fig. 103. Their variation with angle of incidence in the 0.03- to 0.16-μm spectral region is discussed by Hunter (1972).

95. Selective Metal-Dielectric Reflectors Several types of coatings that reflect highly in one spectral region but not in another have been developed in the past for different applications. Hadley and Dennison (1947, 1948) presented the theory and experimental results of reflection interference filters for the isolation of narrow spectral regions (Fig. 105). High-infrared and low-visible reflectance coatings (Fig. 40) are used to control the temperature of satellites (Drummeter and Hass, 1964). These coatings could also be used to remove visible light from infrared optical systems. Several reflectors designed to reduce stray visible light in ultraviolet systems are shown in Fig. 104.

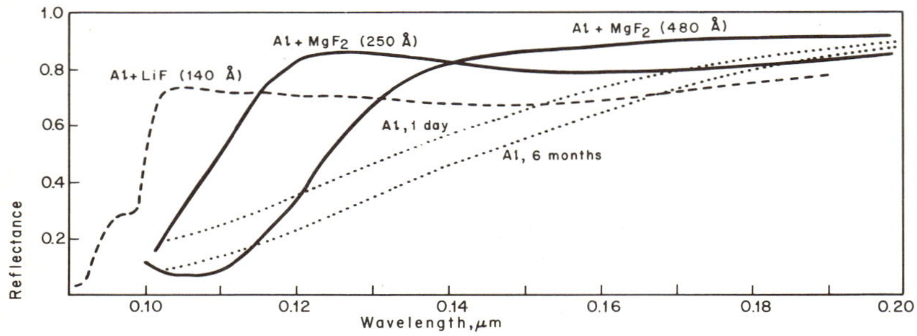

Fig. 103 Measured spectral-reflectance curves of unprotected aluminum and aluminum overcoated with MgF_2 and LiF films of indicated thicknesses. [Al + LiF *coating, after Cox et al.* (1968), *all other curves after Canfield et al.* (1966).]

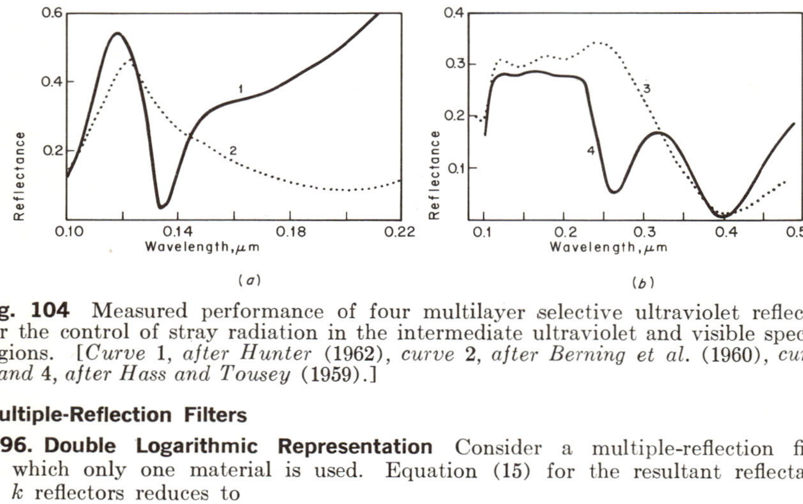

Fig. 104 Measured performance of four multilayer selective ultraviolet reflectors for the control of stray radiation in the intermediate ultraviolet and visible spectral regions. [*Curve* 1, *after Hunter* (1962), *curve* 2, *after Berning et al.* (1960), *curves* 3 *and* 4, *after Hass and Tousey* (1959).]

Multiple-Reflection Filters

96. Double Logarithmic Representation Consider a multiple-reflection filter in which only one material is used. Equation (15) for the resultant reflectance of k reflectors reduces to

$$R_\lambda = (R_{1\lambda})^k \tag{77}$$

This may also be written

$$\log \log R_\lambda = \log k + \log \log R_{1\lambda} \tag{78}$$

It is evident from this that a plot of log log $R_{1\lambda}$ is particularly useful for depicting the reflectance of materials to be used in multiple-reflection filters and that it is equivalent to the log (density) representation of transmission filters (Par. 9). Reflection data of Reststrahlen materials will be presented this way (Par. 99).

97. Metal and Metal-Dielectric Multiple-Reflection Filters Metals such as silver, copper, gold, and metal-dielectric coatings of the type shown in Fig. 40 used in a multiple-reflection arrangement should make low-wavelength cutoff filters with excellent rejection, sharp transition, and a long, unattenuated pass region far superior to those available with transmission filters.

98. Multiple-Reflection Filters Composed of Thin-Film Interference Coatings Interference coatings for use in a multiple-reflection filter need not be deposited onto substrates that transmit well in the spectral region of interest, but they should be used at small angles of incidence if disturbing effects due to polarization are not to occur (van Rooyen and Therou, 1969). The following are examples of some of the difficult filtering problems that can be readily solved with multiple-reflection filters composed of interference coatings.

It is difficult to provide adequate blocking with transmission filters for narrow-bandpass filters of the type shown in Figs. 81 to 86 without considerably reducing the peak transmittance. This is done readily with a multiple-reflection filter composed of quarter-wave stacks of the same materials used for the construction of the bandpass filter and centered at the same wavelength (Fig. 106).

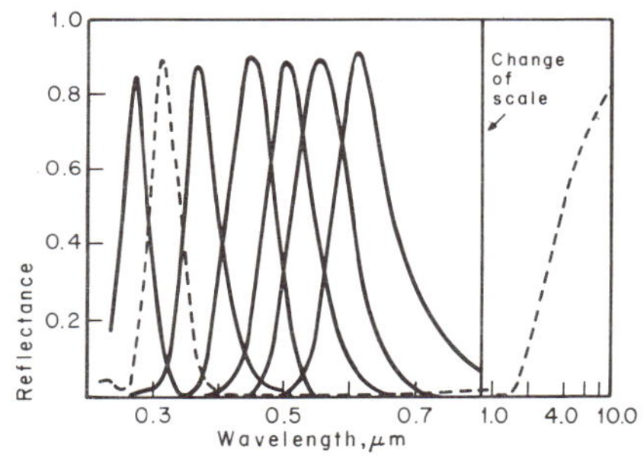

Fig. 105 Measured performance of selective reflection filters for the ultraviolet and visible spectral regions. The filters consist of three half-wave cryolite or MgF_2 spacer layers separated by Inconel films of suitable transmission deposited onto an opaque aluminum film. [*After Turner and Hopkinson* (1953).]

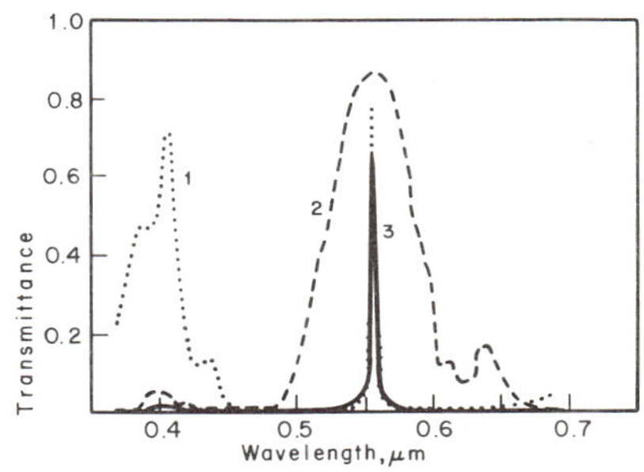

Fig. 106 Blocking an all-dielectric narrow-band interference filter with a multiple-reflection filter. Curve 1: transmittance of interference filter alone; curve 2: spectral reflectance of a quarter-wave stack after fourfold reflection; curve 3: transmission of blocked filter. [*After Cohendet and Saudreau* (1967).]

It should be possible to construct highly efficient short-pass filters with a very long and low rejection region by using broadband reflectors consisting of several contiguous stacks (Pars. 55 to 57).

The use of a narrow-band transmission filter in multiple reflection should result in a narrow-band rejection filter.

The transmittance curves of three multiple-reflection bandpass filters are shown in Figs. 107 to 108.

99. Reststrahlen Filters For many inorganic materials the refractive index remains constant and the absorption coefficient is small or zero over an extended spectral region except in the neighborhood of wavelengths which excite atomic or crystal-lattice vibrations in the compound. There the extinction coefficient k rises steeply, the refractive index n undergoes a rapid variation, and the Fresnel reflection coefficient R [Eq. (76)] may become quite high (Fig. 1). This effect, called the *Reststrahlen* (residual-rays) *effect,* occurs notably in the 6- to 300-μm region, for which there exist few satisfactory filters. It can therefore be used to extract a more or less narrow spectral region from a radiation continuum. Crystals, sintered powders, and evaporated films of Reststrahlen materials are introduced into infrared instruments in the form of plane or concave mirrors. Reststrahlen powder filters, which are used in transmission, are described in Par. 20.

The following properties of Reststrahlen materials should be borne in mind. The maximum reflectance increases markedly with a lowering of the temperature (Fig. 109*a*). Thin films of Reststrahlen materials (see, for example, Parodi, 1937; Gottlieb, 1960; Martin et al., 1962; Chang, 1964; Martin and Turner, 1966; Riccius, 1968) exhibit interference maxima and minima on the short-wavelength side of the reflectance peak, the location and number of the extrema depending on the thickness of the film (Fig. 109*b*). However, they can be applied to large mirror surfaces. Mixtures of polar crystals show reflectance curves that are intermediate to those of the components (Fig. 109*c*). While the Reststrahlen reflectance is unaffected by the presence of free charge carriers, high carrier concentrations in semiconductors can significantly affect the free-carrier reflection curve (Fig. 109*d*). In birefringent crystals the reflectance curves depend on the orientation of the plane of polarization of the incident radiation with respect to the crystallographic axes (Fig. 109*e* to *h*).

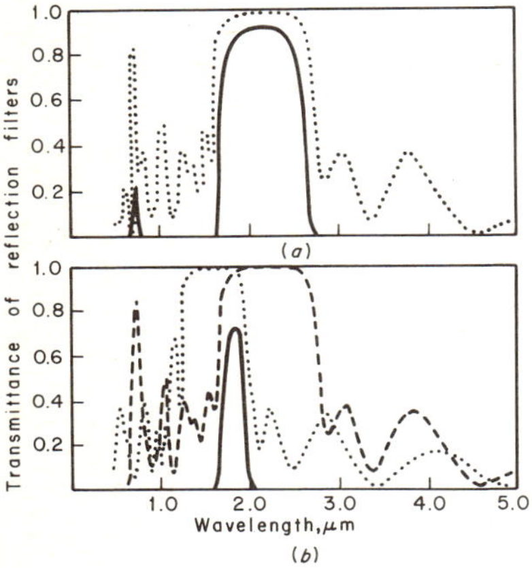

Fig. 107 Measured spectral characteristics of broad- and narrow-band multiple-reflection filters for the infrared spectral region. (*a*) Eight reflections (*solid curves*) from identical quarter-wave stacks (*dotted*); (*b*) six reflections (*solid*) from each of two quarter-wave stacks (*dotted, dashed*) tuned to different wavelengths. [*After Valeyev* (1967).]

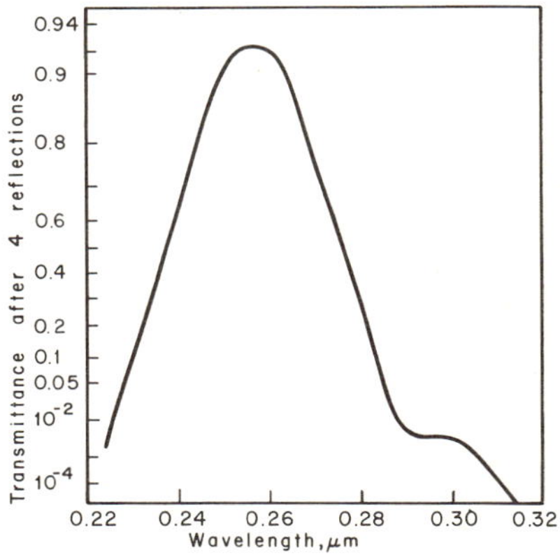

Fig. 108 Measured spectral transmittance of a commercial multiple-reflection interference filter for the ultraviolet spectral region. [*Schott & Gen.* (1967*b*).]

There exist several compilations of Reststrahlen materials, the most recent one being by Genzel (Kneubühl, 1969), but data on new materials are being published at such a rate that no list can be considered to be complete or up to date. Certainly no such claim is made for the list of materials given in Table 4. Generally only materials with a rather simple structure of the reflectance curve, a maximum reflectance in excess of 70 percent, and a high maximum-to-minimum reflectance ratio were considered for this list since such materials are the most useful ones for filter purposes. Although some of the materials listed in Table 4 have not yet been obtained in sizes large enough for conventional Reststrahlen methods, they might be of interest for the production of thin films or Yoshinaga powder filters (Par. 20).

Unless indicated otherwise, the materials are arranged in the table in ascending order of the wavelength at which the maximum reflectance occurs. Also given are the value of this reflectance and the wavelengths at which the reflectance is equal to half the peak reflectance on the short- and (where applicable) the long-wavelength side of the maximum. Other information includes the figure in which the reflectance curve can be found and the source of the data.

For most of the materials many different references could be quoted. However, no attempt was made to evaluate their relative merit. The references selected were generally rather recent and gave measured reflectance curves for wide spectral regions for materials at or near room temperature.

TABLE 4 Selected Reststrahlen Materials

λ_{max}, μm	$\lambda_{0.5\,max}$ Low	$\lambda_{0.5\,max}$ High	Maximum reflectance	Material	Fig.	Ref.†	σ_{max}, cm^{-1}
8.8	7.5	10.2	0.83	BN	110*a*	1	1141
12.2	10.4	13.7	0.97	SiC	110*d*	2	819
12.9	9.1	14.9	0.98	BeO	110*d*	3	776
13.3	9.2	16.2	0.95	BeO	110*d*	3	750
13.3	10.9	16.2	0.94	AlN	110*a*	4	753
15.6	12.3	20.8	0.85	TiO_2	109*h*	5	640
20.8	17.2	27.3	0.90	ZnO	110*b*	4	481
21.3	18.0	29.5	0.91	NiO	110*d*	1	469
22.2	15.0	37.6	0.82	MgO	110*e*	6	450
25.0	14.9	37.8	0.90	LiF	110*f*	6	400
26.6	25.1	28.9	0.94	GaP	110*a*	7	376
26.6	17.6	38.9	0.80	CoO	110*g*	1	376
29.9	17.8	56.4	0.97	ThO_2	110*e*	8	335
30.5	28.6	39.8	0.83	ZnS	110*b*	9	328
30.9	29.5	32.7	0.90	AlSb	110*a*	10	324
33.8	26.2	63.7	0.75	CdF_2	110*g*	11	296
34.1	18.7	92.2	0.94	UO_2	110*f*	8	293
35.0	21.2	42.6	0.92	CaF_2	110*e*	12	286
35.8	27.2	43.1	0.92	NaF	110*c*	13	279
36.1	34.5	40.5	0.91	GaAs	110*a*	7	277
39.9	33.7	46.9	0.91	CdS	110*b*	9	250
40.0	26.8	49.5	0.95	SrF_2	110*d*	12	250
41.3	23.9	51.8	0.93	$KTiO_3$	110*i*	14	242
43.8	24.5	65.3	0.84	LiCl	110*f*	15	228
45.7	40.2	50.9	0.88	ZnSe	110*b*	9	219
47.0	31.5	60.3	0.95	BaF_2	110*d*	12	213
48.5	36.5	57.5	0.86	KF	110*e*	15	206
50.1	43.8	66.2	0.90	NaCl	110*c*	16	199
54.7	49.0	61.1	0.88	ZnTe	110*b*	9	183
55.4	36.3	76.4	0.70	RbF	110*f*	15	181
63.1	53.0	75.1	0.88	KCl	110*c*	16	159
68.8	60.9	83.1	0.77	CdTe	110*b*	9	145
77.6	66.4	91.6	0.80	KBr	110*c*	16	129
91.5	80.3	106.2	0.76	KI	110*c*	16	109
121.2	91.3	147.2	0.88	CsBr	110*c*	16	83
145.8	122.8	177.8	0.86	CsI	110*c*	16	69
21.0	12.9	....	0.91	TiO_2	109*h*	5	476
28.5	21.3	....	0.95	$SrTiO_3$	110*i*	5	351
23.2	21.5	....	0.94	$BaTiO_3$	110*i*	5	432
20.5	26.8	....	0.78	ZrO_2	110*g*	17	488
48.0	29.0	....	0.74	LiBr	110*g*	15	208
72.0	31.1	....	0.86	PbF_2	110*h*	11	139
51.8	44.0	....	0.87	Mg_2Sn	110*d*	18	193
93.8	47.5	....	0.94	PbS	110*h*	19	107
77.0	52.0	....	0.66	PbSe	110*g*	20	130
131.1	61.6	....	0.90	TlCl	110*h*	16	76
54.4	62.7	....	0.84	InSb	110*a*	21	184
83.1	77.5	....	0.89	HgTe	110*b*	22	120
172.8	93.5	....	0.91	TlBr	110*h*	16	58
154.2	123.1	....	0.75	TlI	110*h*	23	65
8.9	8.0	9.4	0.92	Quartz	109*f*	24	1124
14.2	11.0	18.1	0.93	Al_2O_3	110*h*	25	703
20.0	18.1	20.6	0.92	Quartz	109*f*	24	501
21.7	19.8	29.6	0.93	Al_2O_3	110*h*	25	461

† 1 = Gielisse et al. (1967); 2 = Spitzer et al. (1959); 3 = Loh (1968); 4 = Collins and Kleinman (1959); 5 = Spitzer et al. (1962); 6 = Jasperse et al. (1966); 7 = Verleur and Barker (1966); 8 = Axe et al. (1968); 9 = Manabe et al. (1967); 10 = Turner and Reese (1962); 11 = Axe et al. (1965); 12 = Kaiser et al. (1962); 13 = McCarthy (1965); 14 = Miller and Spitzer (1963); 15 = Hass and Hadley (1963); 16 = Mitsuishi et al. (1962); 17 = Piriou and Tsakiris (1965); 18 = Geick et al. (1966); 19 = Dixon and Riedl (1965); 20 = Yoshinaga (1955); 21 = Sanderson (1965); 22 = Dickey and Mavroides (1964); 23 = Sinton and Davis (1954); 24 = Spitzer and Kleinman (1961); 25 = Barker (1963).

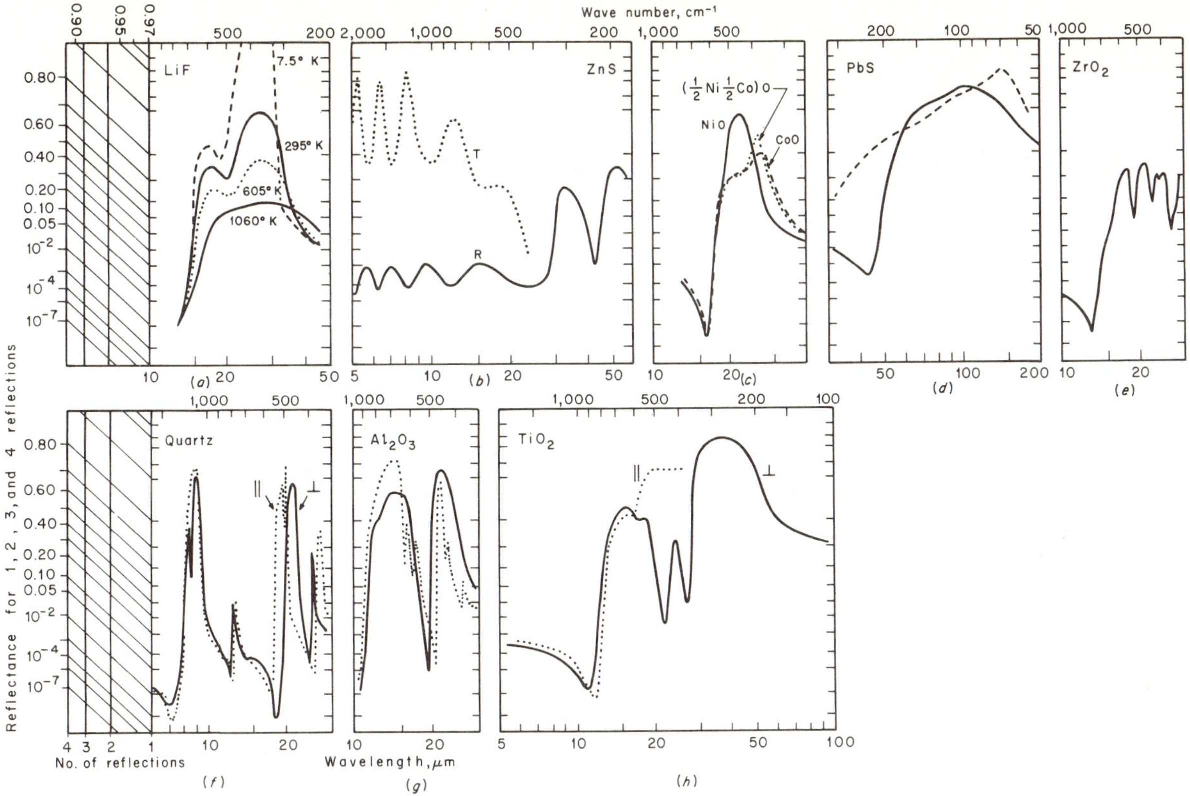

Fig. 109 Properties affecting the reflectance of Reststrahlen materials: (*a*) temperature [*after Jasperse et al.* (1966)]; (*b*) interference in thin films [*after Riccius* (1969)]; (*c*) composition of a mixture [*after Gielisse et al.* (1965)]; (*d*) free-carrier concentration in semiconductors [*after Dixon and Riedl* (1965)]; (*e*) to (*h*) birefringence [(*e*) *after Piriou and Tsakiris* (1965), (*f*) *after Spitzer and Kleinman* (1961), (*g*) *after Barker* (1963), (*h*) *after Spitzer et al.* (1962)]. The data in Figs. 109 to 111 are plotted on a log-log scale (Par. 96) to permit the resultant reflectance after one, two, three, and four reflections to be read off directly.

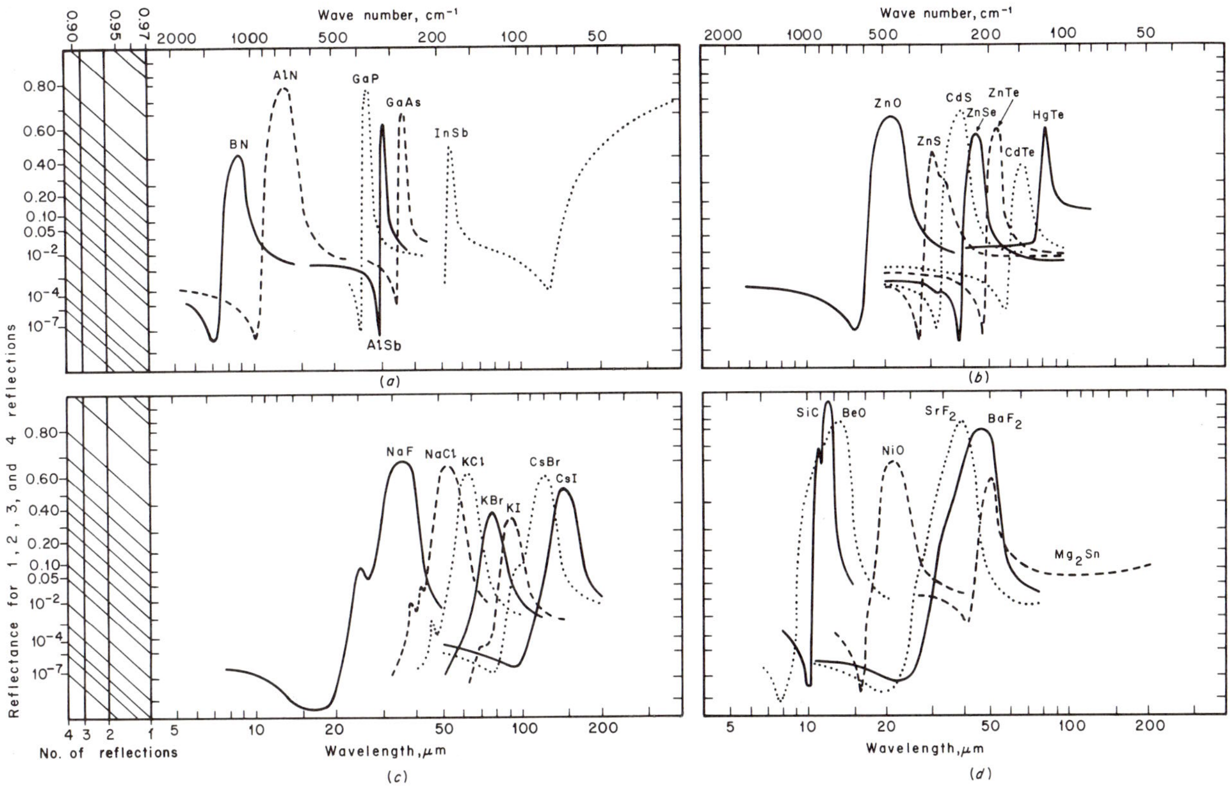

Fig. 110 Reflectance of some Reststrahlen materials. Further information on the materials shown is given in Table 4.

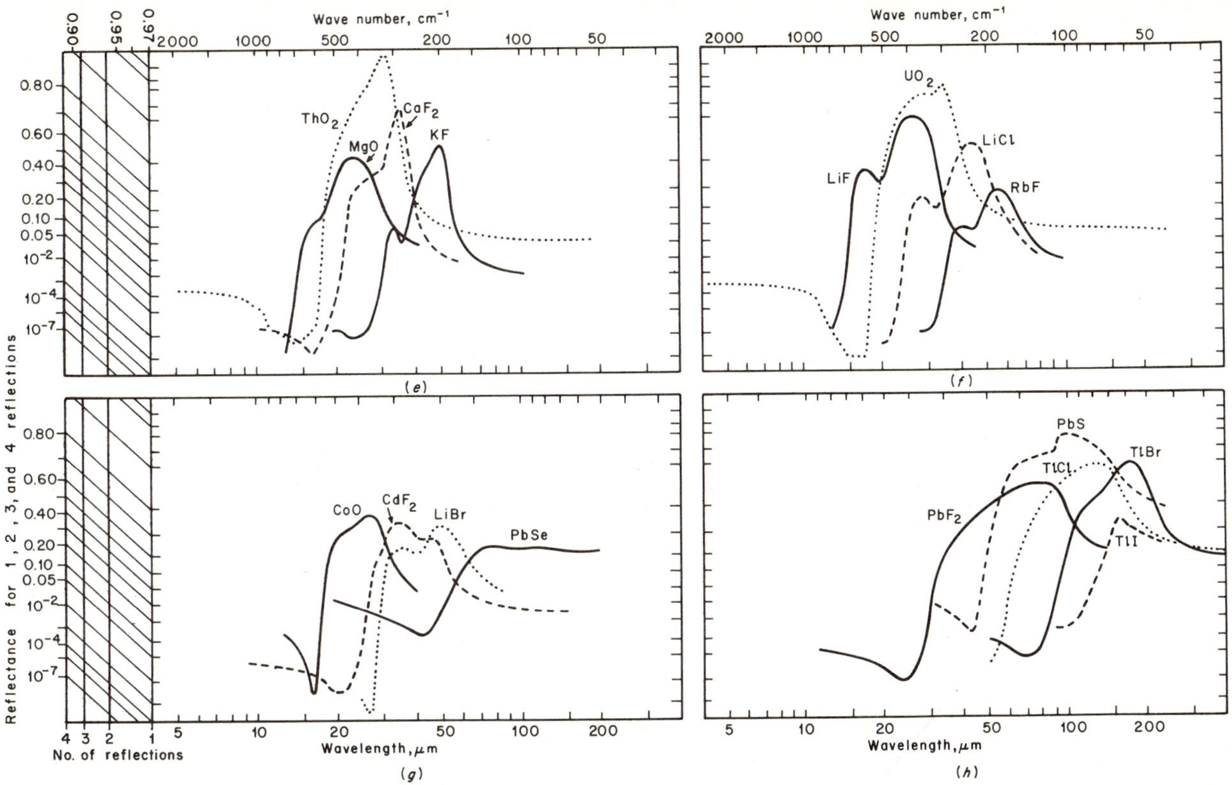

Fig. 110 *(Continued)*

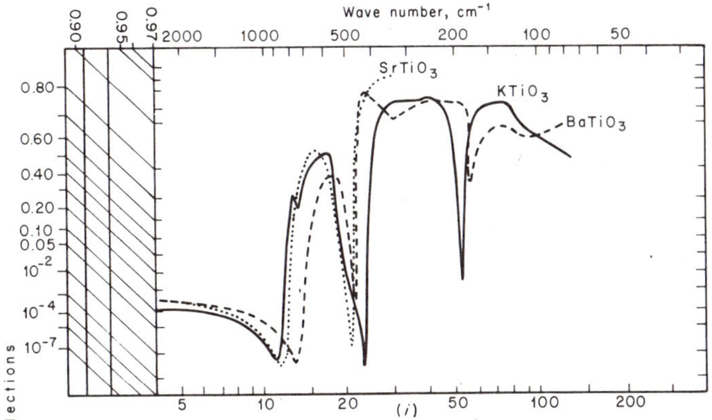

Fig. 110 *(Continued)*

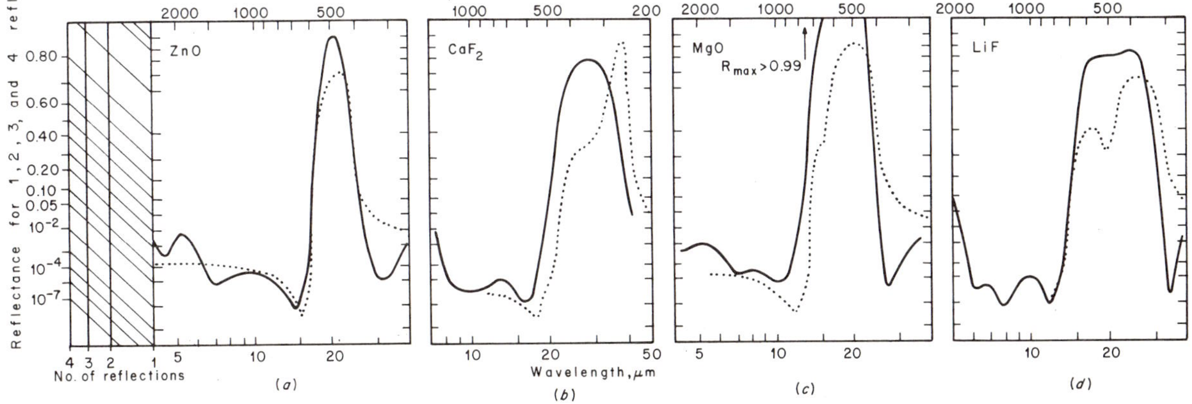

Fig. 111 Measured spectral reflectance curves of some Reststrahlen crystals before (*dotted curves*) and after (*full curves*) enhancement with thin-film interference coatings. [*After Turner et al.* (1965).]

The materials arranged in Fig. 110 should be used in conjunction with Table 4. For most applications a high rejection ratio and a sharp transition between the low- and high-reflectance regions is desired. This can be accomplished by causing the radiation to undergo multiple reflections from the Reststrahlen material. To facilitate the visualization of the effect of multiple reflections the reflectance curves are plotted on a double logarithmic scale (see Par. 96). They may not be too accurate for reflectances above 95 percent and below 5 percent since the original data were given on a linear scale, but this should not detract from their usefulness for selecting materials for filter purposes.

The number of reflections needed to achieve a certain spectral purity can be reduced by auxiliary filtering techniques or by the improvement of the performance of the Reststrahlen filters. One way to remove unwanted short-wavelength radiation is to introduce into the system a reflection from a roughened surface (see Par. 101). Another method is to arrange two surfaces of the Reststrahlen material so that the radiation is incident on both surfaces, at the Brewster angle for the short-wavelength region to be removed, but in such a way that the azimuth of one surface is perpendicular to that of the other (Czerny, 1923).

The shape and height of the reflectance maximum of a Reststrahlen material, as well as its long- and short-wavelength cutoff characteristics, can be enhanced considerably if it is peaked at a wavelength for which adequate thin-film coatings can be deposited. The reflectances of four Reststrahlen filters before and after such treatment are shown in Fig. 111.

FILTERS BASED ON SCATTERING AND DIFFRACTION

100. When radiation passes through a thickness L of a uniform suspension of small particles in a medium of different refractive index, some of the radiation is lost through scattering. Assuming that there is no absorption, the transmission is given by

$$T = \exp(-LS) \tag{79}$$

where S, the scattering coefficient, depends on the refractive index n_0 of the medium and n_p of the particles, on the size α of the latter, as well as on the wavelength λ of the incident radiation. This dependence on wavelength can be exploited to construct filters.

Clewell (1941) gave the following useful empirical formula for the scattering coefficient:

$$S = \frac{[(n^2 - 1)/(n^2 + 2)]^2}{\alpha\left[\left(\dfrac{0.61\lambda}{n_0\alpha}\right)^2 - 1\right]^2 + \dfrac{\lambda^2}{n_0^2\alpha[(n-1)^2a + 1]}} \tag{80}$$

in which $n = n_p/n_0$ and a is a suitable constant. For $\alpha/\lambda \gg 1$ the scattering coefficient S is wavelength-independent and is proportional to $1/\alpha$. A maximum of S occurs for particle diameters of the order of $\lambda/2$, as required by the theory of Mie (1908). For small particles, and for particles with $\alpha \approx \lambda$ but with $n \approx 1$, Eq. (80) is essentially equivalent to the equations given by Rayleigh (1914):

$$S = 9\left(\frac{n^2 - 1}{n^2 + 2}\right)^2 \frac{\alpha^3}{(\lambda/n_0)^4} \qquad \alpha/\lambda \ll 1 \tag{81}$$

and

$$S = 9\left(\frac{n^2 - 1}{n^2 + 2}\right)^2 \frac{\alpha}{(\lambda/n_0)^2} \qquad n \approx 1 \tag{82}$$

Scatter Filters

101. Scattering can be used to remove shorter wavelengths from a beam of radiation. Thus, for example, a transmission filter for the near-infrared spectral region based on this principle might consist of a NaCl plate whose surface has been fogged by humidity. Alternatively, MgO smoke can be deposited onto the

NaCl surface (Fig. 112). However, the Reststrahlen band of MgO occurs at a shorter wavelength than that of NaCl (Table 4), and this limits the transmission region (Fig. 1).

Similar filter action can be obtained by reflecting the beam from a roughened surface (Rayleigh, 1901). The reflection of radiation from an aluminized ground-glass surface has been investigated by Bennett (1963*c*) and Porteus (1963). An approximate expression for the variation with wavelength λ of the reflectance R_s of such a surface is

$$R_s = R_0 \exp\left[-\left(\frac{4\pi\alpha\cos\theta}{\lambda}\right)^2\right] \tag{83}$$

where R_0 = reflectance of similarly coated smooth surface
α = rms roughness
θ = angle of incidence of radiation

The cutoff wavelength of such a filter can be controlled by choosing the particle size of the grinding powder (Fig. 113). For optimum results the particle size of scatter filters must be as uniform as possible. Relationship (83) precludes any sharpening of the cutoff edge through multiple reflections.

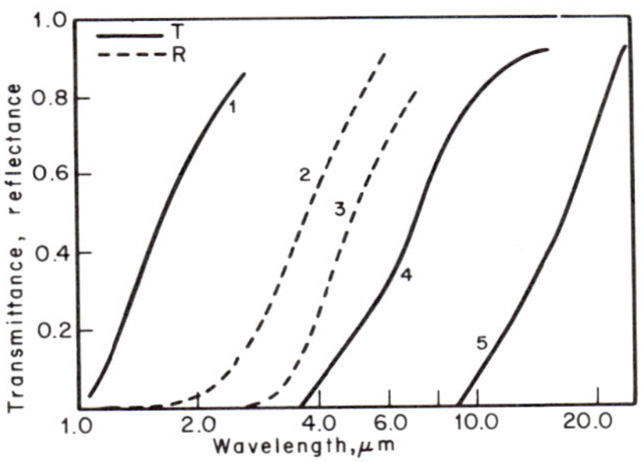

Fig. 112 Scatter filters composed of thin powder layers on different host surfaces. Curves 1, 4, and 5: transmittance of ZnO powder on glass and of NaCl and KBr powders on NaCl; curves 2 and 3: reflectance of MgO powder on speculum. [*Curves* 1 *to* 3 *after Pfund* (1930); *curves* 4 *and* 5 *after Plummer* (1936).]

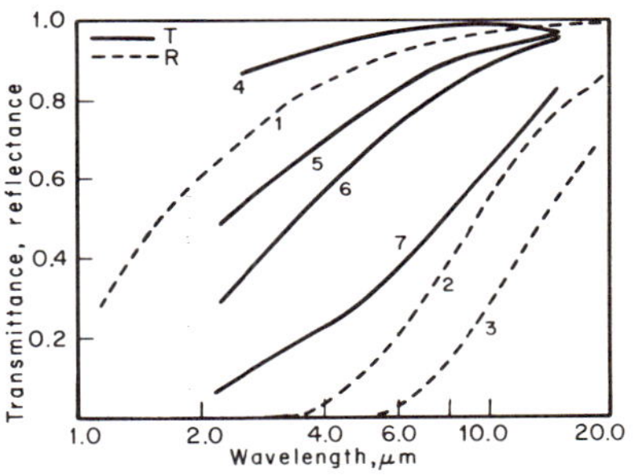

Fig. 113 Scatter filters composed of ground surfaces. Curves 1 to 3: reflectance of aluminized ground-glass surfaces [*after Bennett* (1963)]; curves 4 to 7: transmittance of rock-salt plates ground on both sides [*after McGovern and Fredel* (1947)].

The surfaces of Reststrahlen crystals are sometimes roughened to reduce the reflection on the short-wavelength side of the reflectance peak (see, for example, Verbitskaya et al., 1967).

Christiansen Filters (McAlister, 1935)

102. Liquid Christiansen Filters In the filter first described by Christiansen (1884) small particles of a transparent optical material are suspended in a liquid of similar refractive index but of widely different dispersion (Fig. 1). This suspension is homogeneous only to the radiation of wavelength at which the two dispersion curves cross. Other radiation is scattered according to Eq. (82) and may be removed with an aperture. Solid-liquid filters of this type and the solid-air filters described in Par. 103 are the most common forms of the Christiansen filter, although solid-solid and liquid-liquid filters have also been suggested (see, for example, Sogani, 1926; McAlister, 1935; Wood, 1940; Gobrecht and Klose, 1965).

The theory of the Christiansen filter has been discussed in several papers (e.g., Raman, 1949, 1955; Schrader, 1958). Clarke (1968) applied multiple-scattering theory to derive the following expressions for the transmittance T and half width

TABLE 5 Some Solid-Liquid Combinations for Christiansen Filters†

Wavelength, μm	Solid phase	Liquid phase	Reference
0.2–0.3	Chloride	Hexane	Kohn and Fragstein (1932)
	Quartz	Decalene	Kohn and Fragstein (1932)
		Benzene	Minkoff and Gordon (1946)
		Decahydronaphthalene in cyclohexane	Sinsheimer and Loofbourow (1947)
0.3–0.4	Quartz	Benzene-alcohol mixture	Kohn and Fragstein (1932), Fragstein (1933)
Visible	Crown glass	Methyl benzoate	Weigert et al. (1929)
	Glass	Ethyl salicilate	Denmark and Cody (1935)
	Borosilicate crown glass	Carbon disulfide in benzene	McAlister (1935)
	Calcium fluoride	Benzene–ethyl alcohol mixture	Barnes and Bonner (1936)
	Gypsum	Benzonitrile	Auerbach (1957)
	Orthoclase		
	Beryllonite	Nitrobenzene	
	Cesium nitrate		
	Quartz		
	Rubidium bromide		
	Glass	Polyvinyl chloride in xylol	Schrader (1958)
	Sodium chloride	Styrene	Auerbach (1960)
	Quartz	Cinnamyl acetate	
		Dibromochloromethane	
		Isobutyl cinnamate	
		Styrene	
	Fused silica	Pinene	Matovich (1965)
		Amyl acetate	
3.3	Quartz	Carbon disulfide-carbon tetrachloride mixture	Barnes and Bonner (1936*b*)
4.6	Quartz	Carbon tetrachloride	
5.5	Sodium chloride	Carbon disulfide	Mohler and Loofbourow (1952)
9.0	Magnesium oxide	Carbon tetrachloride	

† Many additional combinations for the ultraviolet and visible spectral regions will be found in Ahier (1937) and Auerbach (1957, 1960).

$\Delta\lambda_{0.5}/\lambda_c$ of a Christiansen filter:

$$T = \exp\left[-KcL\alpha(\Delta n')^2\left(\frac{\Delta\lambda}{\lambda_c}\right)^2\right] \tag{84}$$

and

$$\frac{\Delta\lambda_{0.5}}{\lambda_c} = \left(\frac{2.77}{K}\right)^{1/2}\frac{1}{(cL\alpha)^{1/2}\,\Delta n'} \tag{85}$$

where c = particle volume per unit volume
n_0 = refractive index of liquid
n_p = refractive indexes of particles
L = length of cell
α = average particle size
$\Delta n' = \dfrac{d}{d\lambda}(n_0 - n_p)$
λ_c = wavelength at which $n_0 = n_p$
$\Delta\lambda$ = shift in wavelength away from λ_c

The constant K may assume a value between 8 and 49, depending on the particular mechanism for scattering assumed.

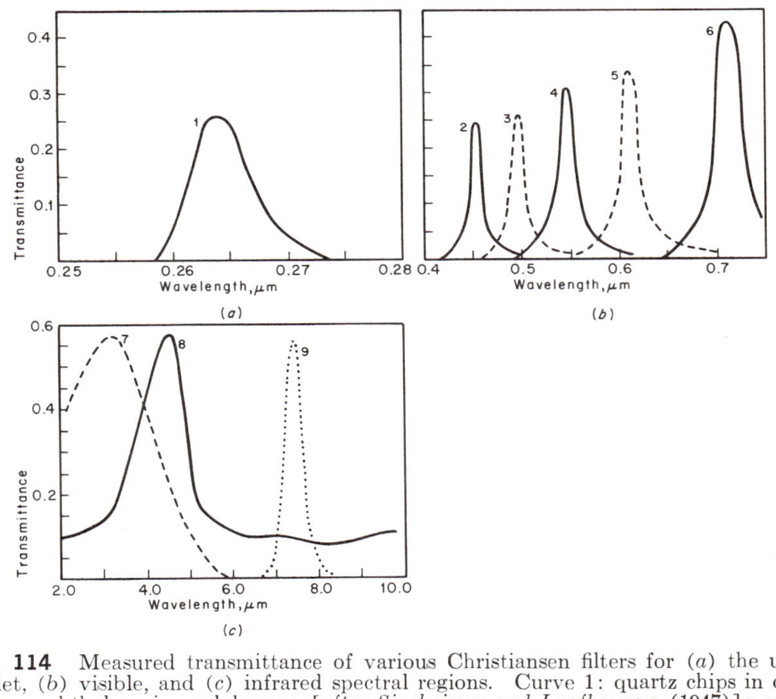

Fig. 114 Measured transmittance of various Christiansen filters for (*a*) the ultraviolet, (*b*) visible, and (*c*) infrared spectral regions. Curve 1: quartz chips in decahydronaphthalene in cyclohexane [*after Sinsheimer and Loofbourow* (1947)]; curves 2 to 6: borosilicate crown glass particles in different mixtures of carbon disulfide and benzene [*after McAlister* (1935)]; curves 7 to 9: quartz powder in 50 percent by volume mixture of CS_2 and CCl_4, in CCl_4, and in air [*after Barnes and Bonner* (1936*b*)].

Some of the solid-liquid combinations proposed in the past are listed in Table 5. Experimentally measured transmittance curves for some ultraviolet, visible, and infrared Christiansen filters are shown in Fig. 114. Although filter half widths of 0.0002 μm in the visible part of the spectrum have been reported (Ahier, 1937), typical values are one or two orders larger.

Because refractive indexes of liquids change more rapidly with temperature than those of solids, temperature control of the Christiansen filter is necessary to maintain a constant peak wavelength. Conversely, temperature changes can be used to tune the filter (McAlister, 1935). Other ways of tuning are based on the change in the composition of the liquid and on the pressure change of the refractive index of the liquid. The latter effect was used by Matovich (1965) to construct a 0.015-μm half-width filter of 70 percent transmittance that could be scanned across the visible spectrum very rapidly.

103. Christiansen Powder Filters (Barnes, 1936*a,b*; Henry, 1948) Crystals with absorption bands in the infrared part of the spectrum exhibit anomalous dispersion (Fig. 1), and consequently at some wavelength shorter than that of the absorption maximum they have a refractive index of 1.0. At this wavelength a powder of the crystal will form a Christiansen filter with air (Fig. 1). Such filters are much less temperature-sensitive than the solid-liquid filters described above. In powder filters absorption is involved in addition to the scattering process at wavelengths higher than that of the transmission peak. Most of the Reststrahlen materials listed in Table 4 could be used for the construction of powder filters.

Christiansen wavelengths for a number of crystals in air given by Barnes and Bonner (1936*b*), Gurinovich and Borisevich (1936), and Vereshchagin and Borisevich (1968).

Gratings and Metallic Grids and Meshes

104. Diffraction Gratings Used as Filters (Rohrbaugh and Hatcher, (1958) In general, radiation incident onto a diffraction grating will be diffracted into several orders. But radiation whose wavelength λ is such that

$$\lambda > g(1 + \sin \theta) \tag{86}$$

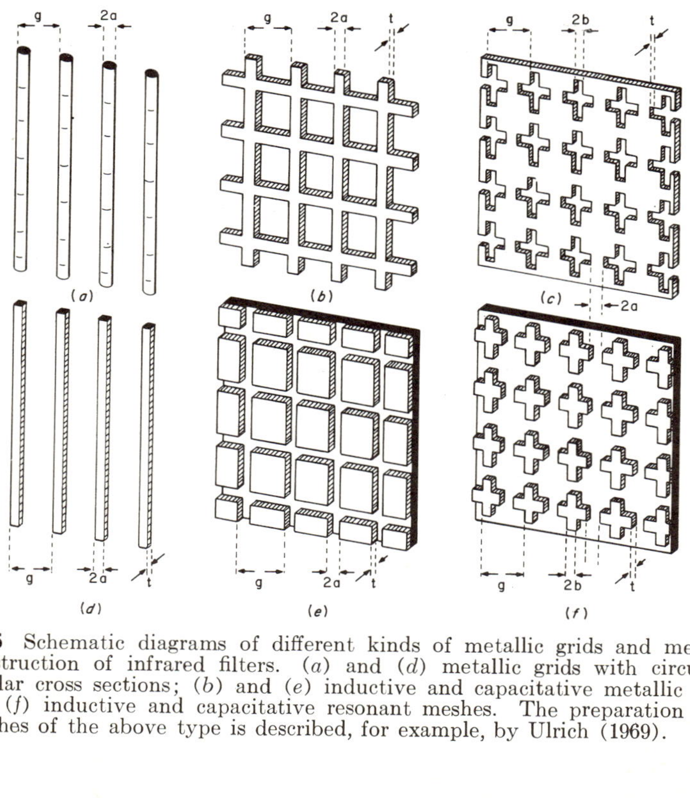

Fig. 115 Schematic diagrams of different kinds of metallic grids and meshes for the construction of infrared filters. (*a*) and (*d*) metallic grids with circular and rectangular cross sections; (*b*) and (*e*) inductive and capacitative metallic meshes; (*c*) and (*f*) inductive and capacitative resonant meshes. The preparation of grids and meshes of the above type is described, for example, by Ulrich (1969).

where g is the grating constant and θ the angle of incidence, can propagate only in the direction of the zero transmitted or reflected orders. This property can be used to construct short-wavelength cutoff filters. White (1947) utilized for this purpose echelette reflection gratings with blaze angles such that the zero-order content of radiation not satisfying Eq. (86) was minimized. The cutoff characteristics of such a filter after one, two, and four reflections are shown in Fig. 116.

Möller and McKnight (1963, 1965), Levine (1957), and Tishchenko (1971) describe echelette transmission gratings formed by pressing polyethylene onto heated metal forms with machined grooves. Filters formed from such gratings have a comparable performance to the reflection echelettes, but they are easier to utilize in an optical system and do not show any polarizing effects (Fig. 117).

Harris (Harris et al., 1969; Harris and Wallace, 1969, 1970) describes filters based on the diffraction of light by regular changes of refractive index induced in $LiNbO_3$ and $CaMoO_4$ crystals by acoustic waves. Typically the filters had passbands of 0.0002- to 0.0008-μm widths and 50 to 80 percent peak transmittances. They could be tuned from about 0.5 to 0.7 μm by changing the acoustic driving frequency.

Schmidtke (1968, 1969) describes a filter-collimator composed of a number of suitably spaced identical diffraction gratings acting in transmission. The filter acted as a long-wavelength cutoff filter transmitting in the ultraviolet but could be constructed for other spectral regions by a suitable scaling of all the dimensions.

105. Metallic Grids The theory of the electromagnetic properties of grids consisting of parallel metal wires or thin strips has been reviewed by Larsen (1962). As in the case of echelette gratings (Par. 104), incident radiation of wavelengths satisfying Eq. (86) can be propagated only in the directions of the zero-order transmitted and reflected waves. Grids of this type have long been used as polarizers in the far infrared because although they are practically transparent to radiation polarized parallel to the grid elements, they strongly reflect radiation of the other polarization at wavelengths λ greater than the grid constant g. Under these conditions the grids may be regarded as continuous, partially reflecting surfaces of low loss. Therefore for middle- and far-infrared interference filter construction they represent an alternative to homogeneous, semitransparent metal films whose optical properties in this region are no longer attractive (Schröder, 1957).

The phase change on reflection ϵ_R, the transmittance T, and reflectance R of a grid for radiation polarized perpendicular to the grid elements depend only on its dimensions (Casey and Lewis, 1952):

$$\epsilon_R = \pi - \tan^{-1} X \tag{87}$$
$$T = 1 - R \approx \tan^2 \epsilon_R = X^2 \tag{88}$$

where for wires with a circular cross section of radius a (Fig. 115a)

$$X = \frac{2g}{\lambda} \ln \frac{g}{2\pi a} \tag{89}$$

and for strips with rectangular cross section of dimensions $2a$ and t (Fig. 115b)

$$X = \frac{2g}{\lambda} \ln \left(\sin \frac{\pi a}{g} \right) \tag{90}$$

These approximate equations are valid for grids made from a material with a good electrical conductivity and for the case when $\lambda > 2g > 16a$, a being large compared to the skin depth. In addition, for metal strips t/g should be small.

The absorption loss A due to the finite conductivity σ is small compared to R and is given by

$$A = \frac{2gR}{u} \left(\frac{c}{\sigma\lambda} \right)^{1/2} \tag{91}$$

where u is the circumference of the cross section of a grid element.

106. Metallic Meshes In the construction of filters for the infrared part of the spectrum one-dimensional metallic grids (Par. 105) have been largely replaced by two-dimensional meshes. The chief advantages of the latter are that they

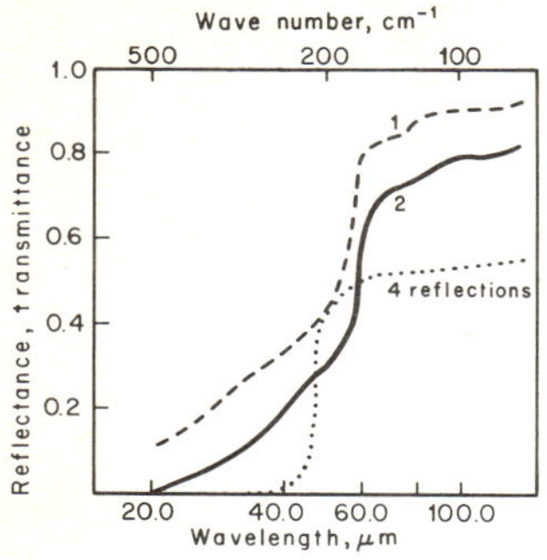

Fig. 116 Resultant reflectivity after one, two, and four reflections at 45° incidence from 7,500 lines/in echelette gratings. [*After White* (1947).]

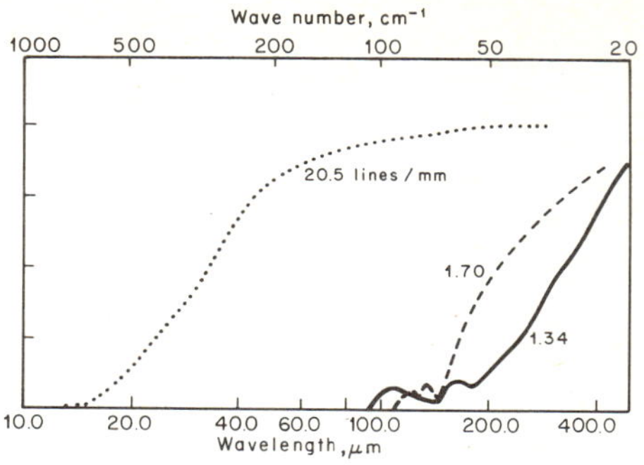

Fig. 117 Measured performance of echelette transmission gratings made of grooved polyethylene sheets with 20.5 lines/mm on one side [*after Möller and McKnight* (1963)] and with 1.7 and 1.34 lines/mm orthogonal grooves on both sides of the sheet [*after Russell and Strauss* (1965)].

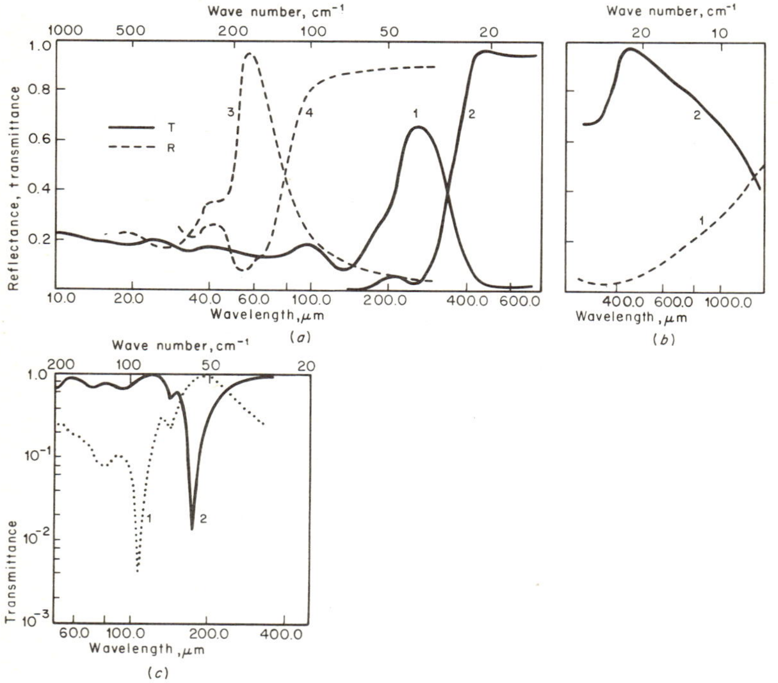

Fig. 118 Measured spectral characteristics of different types of metallic meshes. (a) Inductive brass-wire meshes (curves 1 and 2: $g = 212$ μm, $a/g = 0.2$) [*after Mitsuishi et al.* (1963)] and electroformed nickel meshes (curves 3 and 4: $g = 50$ μm, $a/g = 0.15$) [*after Sakai et al.* (1969)]. (b) Two nearly complementary meshes. Curve 1: capacitative photoetched copper layer on Mylar, $g = 368$ μm, $a/g = 0.051$, $t = 5$ μm. Curve 2: inductive nickel mesh, $g = 368$ μm, $a/g = 0.046$, $t = 14$ μm [*after Ulrich* (1967)]. (c) Resonant meshes. Curve 1: inductive resonant mesh, $g = 102$ μm, $a/g = 0.06$, $b/g = 0.14$. Curve 2: capacitative resonant mesh, $g = 102$ μm, $a/g = 0.13$, $b/g = 0.06$ [*after Ulrich* (1968)].

are more readily available and easier to handle and because of their rectangular symmetry their properties do not depend on the plane of polarization of the incident radiation.

Inductive Meshes. Inductive meshes are the two-dimensional equivalents of the metallic grids described in Par. 105. They commonly consist of a wire mesh or of an electroformed metallic plate (Fig. 115*c*). Equations for the phase change on reflection, reflectance, transmittance, and absorbance for such meshes have not yet been given. However, Eqs. (87) to (91) for a one-dimensional grid can also be used to roughly predict the properties of a two-dimensional mesh for an unpolarized beam of radiation.

The measured spectral transmittances and reflectances of a wire mesh and of an electroformed mesh are shown in Fig. 118*a*. The curves outside the range of validity of Eqs. (87) to (91) depend greatly on the geometrical configuration of the mesh used. Thus, for instance, transmission maxima between 0.6 and 1.0 located at wavelengths between $1.15g$ and $1.3g$ have been reported (Renk and Genzel, 1962; Mitsuishi et al., 1963; Ressler and Möller, 1967; Sakai et al., 1969; Demeshina et al., 1970). Meshes of this type can be used in transmission to isolate fairly broad spectral regions (Fig. 119) or in reflection as short-wavelength cutoff filters (Fig. 120). The filtering action can be improved by reflection or transmission through several nonparallel or widely spaced meshes.

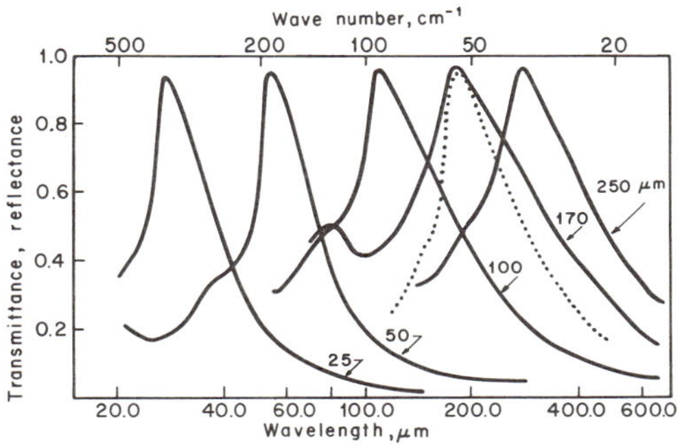

Fig. 119 Measured transmission curves of electroformed inductive nickel meshes with $a/g \approx 3$ for different values of g. The dotted curve represents the measured transmittance of two meshes slightly inclined to one another. [*After Sakai et al.* (1969).]

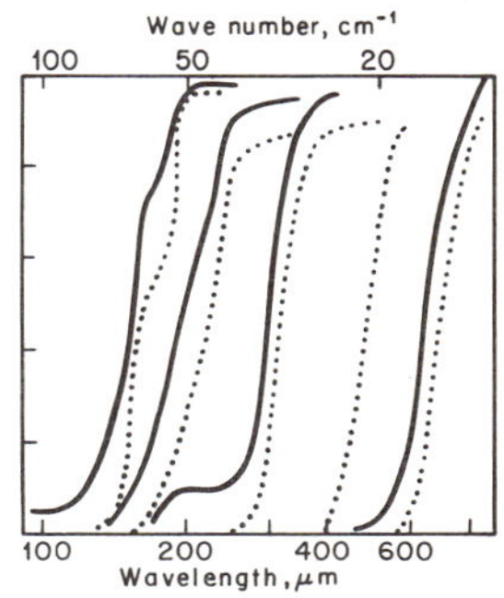

Fig. 120 The measured cutoff characteristics of different brass-wire meshes after one (*solid*) and two (*dotted*) reflections at 15° incidence. [*After Mitsuishi et al.* (1963) *and Sakai et al.* (1969).]

Capacitive Meshes. Capacitive meshes are the complementary or reciprocal structure to the inductive meshes described above. They consist of small metallic squares deposited in rectangular symmetry onto a suitable thin transparent substrate (Fig. 115*d*). Ulrich (1967) has shown that for $\lambda/g > (1 + \sin\theta)$ the transmittance of a capacitive mesh is equal to the reflectance of an inductive mesh, and vice versa. The absorbance of a capacitive mesh is given by

$$A = \frac{gR}{g - 2a}\left(\frac{c}{\sigma\lambda}\right)^{1/2} \tag{92}$$

This quantity does not include the dielectric loss in the substrate. Capacitive meshes can be used as long-wavelength cutoff filters (Fig. 118*b*; Pradhan, 1970).

Resonant Meshes. By using cross-shaped rather than square holes or platelets (Fig. 115*e* and *f*) it is possible to construct meshes which generally reflect or transmit in the nondiffraction region, except in the neighborhood of one particular frequency (Fig. 118*c*). Since rectangular symmetry has been maintained, such resonant inductive and capacitive meshes are nonpolarizing. Their properties can be varied widely by adjusting the parameters g, a, and b. However, expressions for the resonant wavelength and for ϵ, T, R, and A in terms of the above parameters have not yet been given.

107. Multimesh Interference Filters (Ulrich, 1968) Metallic meshes of the type described in Par. 106 can be used to construct a variety of interference filters for the infrared part of the spectrum. The more meshes used the greater the control over the spectral characteristics of the resultant filter. In practice the performance and the wavelength for which filters can be constructed are limited by the flatness and parallelism of the meshes, the ohmic losses in the meshes, and the dielectric losses in the substrates.

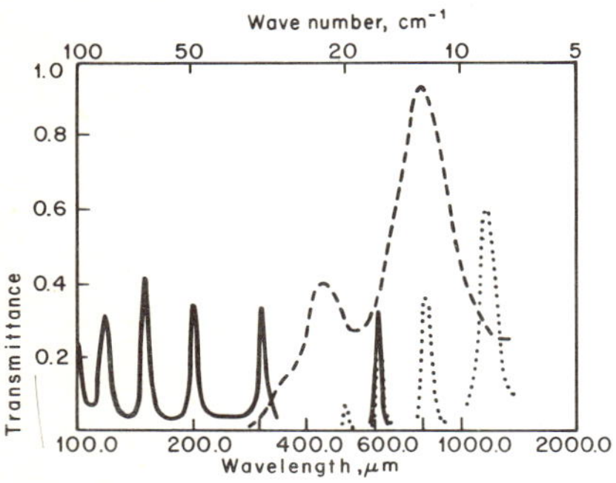

Fig. 121 The measured transmittance of narrow-band interference filters formed from two inductive meshes on quartz (*solid curve*) [*after Renk and Genzel* (1962)], two capacitive meshes on Mylar (*dotted curve*), and a combination of one capacitive and one inductive wire mesh (*broken curve*) [*after Ulrich* (1967)]. The curves are not corrected for the slit width of the measuring device.

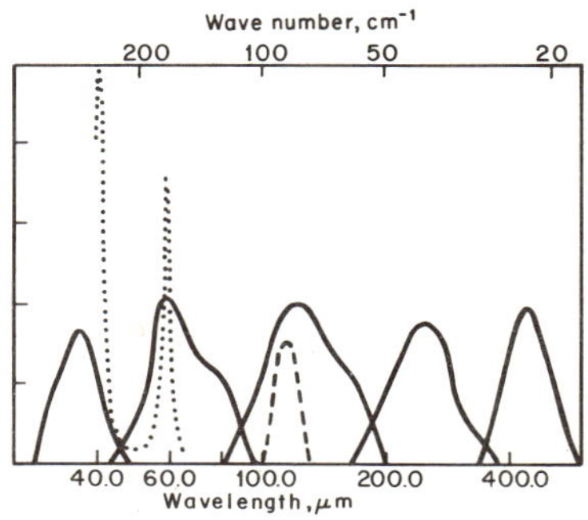

Fig. 122 Measured transmission curves of a set of fully blocked bandpass filters (*solid curves*) [*after Varma and Möller* (1969)] and of first-order Fabry-Perot and square-top interference filters made of inductive meshes [*after Smith and Seeley* (1968)].

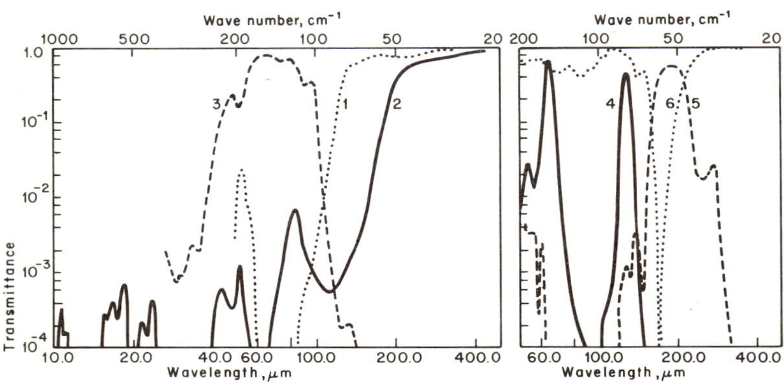

Fig. 123 Measured transmission curves of multimesh interference filters for the infrared part of the spectrum. [*After Ulrich* (1968)]. (*a*) Short- and long-wavelength cutoff filters. Curve 1: four equal capacitive meshes; curve 2: four different capacitive meshes; curve 3: four identical inductive meshes. (*b*) Bandpass and band-rejection filters. Curve 4: four inductive meshes of two different types; curve 5: four equal inductive resonant meshes; curve 6: two equal capacitive resonant grids.

Both thin-film and microwave (see, for example, Ulrich, 1967; Bell and Romero, 1970; Bell et al., 1970) analogies are useful in the design of the filters. For example, Eq. (60) is directly applicable and Fabry-Perot and square-top filters (Pars. 73 and 74) can be constructed with metallic meshes (Figs. 121 and 122). A set of fully blocked bandpass filters for the infrared is shown in Fig. 122. The performances of other multimesh filters are shown in Fig. 123.

POLARIZATION INTERFERENCE FILTERS

108. Polarization interference filters are used whenever very narrow transmission-band filters (HW < 0.1 percent) with wide angular fields or with tuning capability are required. They are composed of birefringent crystal plates, fractional wave plates, and polarizers (Sec. 10). Both biaxial crystals (with smallest, intermediate, and largest refractive indexes α, β, and γ) and uniaxial crystals (with ordinary and extraordinary refractive indexes ω and ϵ) can be used. Uniaxial crystals are cut with entrance and exit faces parallel to the optic axis; biaxial crystals must be cut parallel to the plane containing both principal axes. When radiation passes through such a plate of thickness t, a retardation ξ,

$$\xi = \frac{\mu t}{\lambda} \tag{93}$$

is introduced between the components of the beam polarized parallel to the fast and slow axes of the crystal, where λ is the wavelength of the incident radiation and μ is the birefringence of the crystal,

$$\mu = \begin{cases} \gamma - \alpha & \text{biaxial} \\ \epsilon - \omega & \text{uniaxial} \end{cases} \tag{94}$$

Desirable attributes of crystals for filter construction are a large birefringence, a high transmittance and chemical stability, a low temperature coefficient of μ, and availability of large crystals of good optical quality. Currently the most commonly used materials are quartz ($\mu = 0.009$), calcite ($\mu = 0.172$), and ADP ($\mu = 0.045$).

The Lyot-Öhman Filter (Evans, 1949a)

109. The Basic Filter The simplest version of the polarization interference filter first described by Lyot (1933, 1944) and Öhman (1938) consists of a series of N parallel crystal plates of thicknesses t_r,

$$t_r = 2^{r-1} t_1 \qquad r = 1, 2, \ldots, N \tag{95}$$

interspersed between $N + 1$ polarizers. The axes of all the crystals are parallel, and the planes of polarization of the polarizers are inclined at an angle of 45° to them (Fig. 124a). The transmittance of the rth element of the filter, consisting of one crystal and one polarizer, is

$$T_r = \cos^2 \pi\xi_r \tag{96}$$

for incident light plane-polarized parallel to the azimuth of the polarizer. The transmittance of the complete filter for unpolarized radiation is

$$T = \tfrac{1}{2}T_0 \prod_{r=1}^{N} \cos^2 \pi\xi_r = \tfrac{1}{2}T_0 \left(\frac{\sin 2^N \pi\xi_1}{2^N \sin \pi\xi_1}\right)^2 \tag{97}$$

T_0 depends on the light losses within the filter. The transmittance curve represented by Eq. (97) consists of a series of narrow transmission bands (Fig. 125). Successive bands are separated by a wavelength interval $\delta\lambda$,

$$\delta\lambda \approx \frac{\lambda}{\xi_1} \tag{98}$$

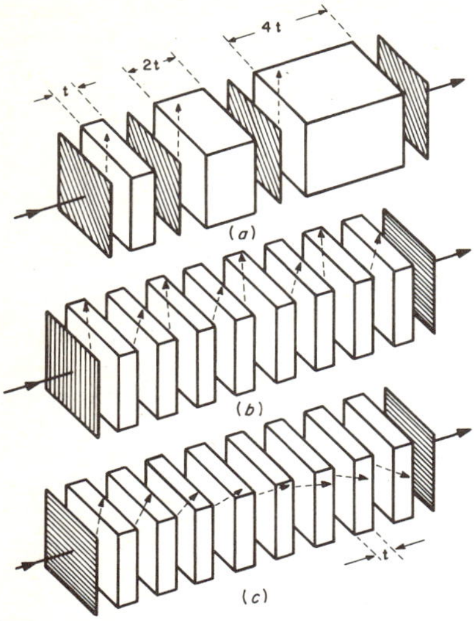

Fig. 124 Schematic drawing of (*a*) three-element Lyot-Öhman polarization interference filter and the equivalent Šolc (*b*) folded and (*c*) fan-type filters. The principal axes of the birefringent elements are indicated by arrows. The plane of polarization of the polarizers is parallel to the cross-hatching.

and their half widths $\Delta\lambda_{0.5}$ are

$$\Delta\lambda_{0.5} \approx \frac{\lambda}{2\xi_N} \tag{99}$$

There are $2^N - 2$ secondary maxima between any two adjacent bands giving rise to a parasitic transmission equal to 11 percent of the transmission of one band. Giovanelli and Jefferies (1954) and Öhman (1958) describe ways of reducing the parasitic radiation.

Vinogradova (1970) discusses the effect of the dispersion of the birefringent media on the performance of polarization interference filters.

110. Angular Field Marked improvements in the angular properties of Lyot filters can be obtained by replacing the thickest elements of the simple filter by more complicated structures. The three wide-angle elements proposed by Lyot (1944) are shown in Fig. 126. The original reference or the article by Evans (1949*a*) should be consulted for full details of the construction and theory of the devices. Here we shall only quote the expressions for $\epsilon - \epsilon_0$, the change in retardation brought about by nonnormal incidence of the radiation.

For the simple Lyot-Öhman filter (Fig. 126*a*) $\xi - \xi_0$ depends not only on the angle of incidence ϕ but also on the azimuth θ of the incident radiation:

$$(\xi - \xi_0)_{L\ddot{O}} = \xi_0 k \left(\frac{\cos^2 \theta}{\gamma} - \frac{\sin^2 \theta}{\alpha} \right) \phi^2 \tag{100}$$

k is given by

$$k = \frac{\alpha\gamma - \beta^2}{2\mu\beta^2} \tag{101}$$

its values for quartz, calcite, and ADP being 0.324, −0.302, and −0.328, respectively.

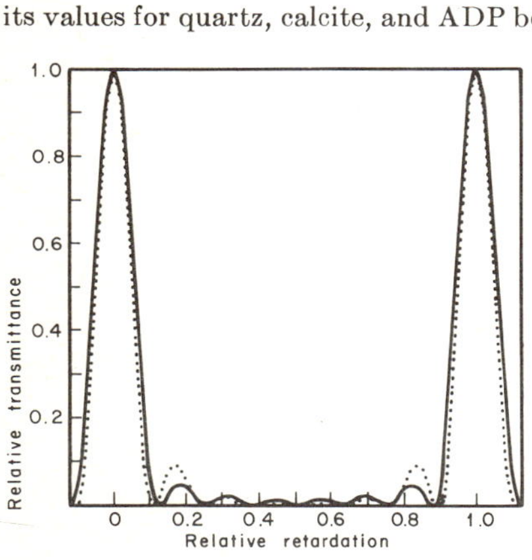

Fig. 125 Calculated relative transmittance of a three-element Lyot-Öhman filter (*full curve*) and the equivalent Šolc filter (*dotted curve*).

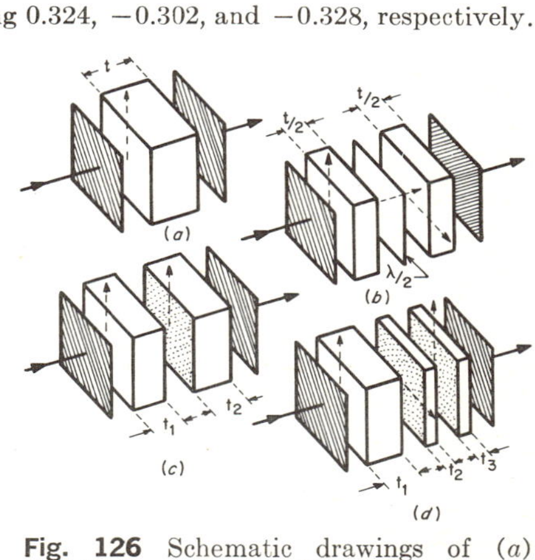

Fig. 126 Schematic drawings of (*a*) simple Lyot-Öhman filter element and Lyot's wide-angle elements of (*b*) the first, (*c*) second, and (*d*) third types.

In Lyot's first wide-angle element (Fig. 126*b*) $\xi - \xi_0$ is independent of the azimuth:

$$(\xi - \xi_0)_{L1} = \tfrac{1}{2}\xi_0 k \left(\frac{1}{\gamma} - \frac{1}{\alpha}\right) \phi^2 \tag{102}$$

The element can be used over an extended spectral region only if an achromatic half-wave plate is employed.

Lyot's second wide-angle element (Fig. 126*c*) is made of two materials with values of k of opposite sign. With a suitable choice of the thicknesses the change in retardation can also be made independent of the azimuth,

$$(\xi - \xi_0)_{L2} = \left(\frac{\xi_1 k_1}{\gamma_1} + \frac{\xi_2 k_2}{\gamma_2}\right) \phi^2 \tag{103}$$

and the device can be used over a wide spectral region.

By the use of two materials with values of k of opposite sign, and with a suitable choice of the thicknesses of the three component crystals, Lyot's third wide-angle element (Fig. 126*d*) can be made to have a retardation which, within the accuracy of Eq. (100), is independent of the angle of incidence,

$$(\xi - \xi_0)_{L3} = 0 \tag{104}$$

Equations (100) to (102) were used to calculate the angular field of the various types of elements as a function of the half width of the filter (Fig. 127). The definition of the angular field was the same as that used in Par. 86 to facilitate a comparison with the corresponding Fig. 90 for thin-film interference filters.

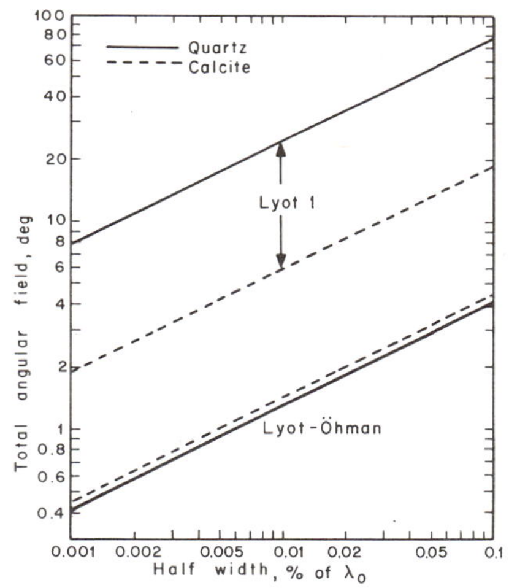

Fig. 127 Total angular field attainable for a given half width of the transmission band, with simple Lyot-Öhman filter elements (curves 1 and 2) and with Lyot type 1 wide-angle elements (curves 3 and 4) made of quartz (1 and 4) and calcite (2 and 3).

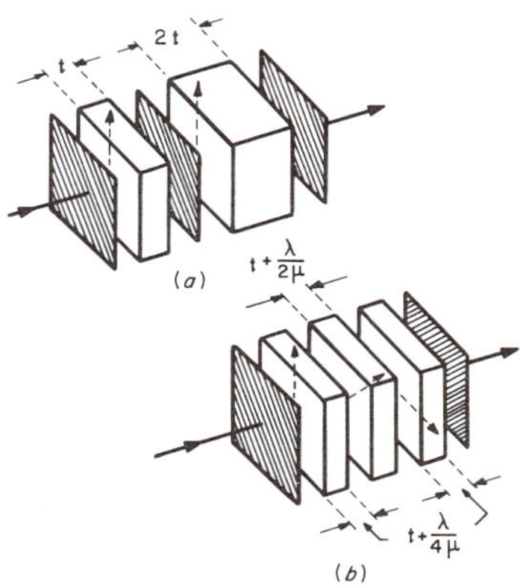

Fig. 128 Schematic drawing of two elements of (*a*) simple Lyot-Öhman filter and (*b*) the equivalent Evans split-element stage.

111. Maximum Transmittance The maximum transmittance of a simple Lyot filter for unpolarized light is $\frac{1}{2}T_0$ [Eq. (97)]. The factor T_0 takes into account all the losses in the filter due to absorption, scattering, and reflection. But since the losses due to scattering and reflection can be kept down to very low values by careful polishing and by cementing and/or immersion of the components in an index-matching liquid, and since little light is lost by absorption in the retardation plates, T_0 is essentially given by

$$T_0 = \tau^{N+1} \tag{105}$$

where τ is the internal transmittance of the polarizers used. The use of sheet polarizers, for which τ is of the order of 0.85, can result in serious transmission losses.

Evans (1949*a*) proposed a split-element stage (Fig. 128) which is equivalent to two stages of the simple Lyot filter. A filter composed entirely of such elements and equivalent to an N-element Lyot filter requires only $N/2 + 1$ polarizers. If, for example, $N = 14$, the value of T_0 for the Evans split-element and the simple Lyot filters would be 0.27 and 0.10, respectively. The angular fields of the two types of filters are essentially the same (Evans, 1949*b*). For a wider field the methods referred to in Par. 110 can be applied to each plate of the split element.

112. Temperature Stability The variation $d\lambda/dT$ of wavelength with temperature is given by

$$\frac{d\lambda}{dT} = \lambda \frac{\dfrac{1}{\mu}\dfrac{\partial u}{\partial T} + \dfrac{1}{t}\dfrac{\partial t}{\partial T}}{1 - \dfrac{\lambda}{\mu}\dfrac{\partial \mu}{\partial \lambda}} \tag{106}$$

For quartz, calcite, and ADP the values of $d\lambda/dT$ at $\lambda = 0.5$ μm are of the order of -0.5×10^{-4}, -0.3×10^{-4}, and -5×10^{-4} μm/°C, respectively. Constant-temperature enclosures for all but the widest-half-width polarization interference filters are therefore a necessity. The requirements on the temperature control can be considerably relaxed but only at the expense of an increase in the complexity of the filters themselves; it is possible to replace any crystal plate by a thermooptically compensated element consisting of two different materials (see, for example, Ioffe and Smirnova, 1967*a*; Drichko, 1969*a*; Kimura and Saruwatari, 1971).

113. Tuning the Passband The ability to tune a polarization interference filter is an advantage in many applications. The transmission wavelength can be adjusted somewhat by changing the filter temperature (Par. 112). However, the thermal inertia of the filter precludes rapid tuning, and other methods have to be used.

To tune a filter it is necessary to make proportional changes in the retardation of each component. One way to do this is to fashion each plate from two movable wedges, as in a Babinet compensator (Lyot, 1944). Alternatively, an adjustable phase shifter can be added to each retardation plate. This could consist of a sheet of photoelastic material under adjustable tension or an electrooptic material in which the required retardation is induced by the application of an electric field (Billings, 1947; Drichko, 1969*a*; Ahmed and Ley, 1970).

With the above techniques a continuous adjustment of the passband over an extended wavelength region requires some discontinuous adjustments of the phase shifters. By adding a quarter-wave plate to each element of a simple Lyot filter and by a suitable relative rotation of all the elements the passband can be tuned continuously (Lyot, 1944; Evans, 1949*a*). But the transmission band and the rejection remain satisfactory over an extended region only if the quarter-wave plates used are achromatic (Gilvarg and Dolgov-Savelev, 1965; Ioffe and Smirnova, 1967*b*). A more complicated phase shifter consisting of two quarter-wave plates and one rotating half-wave plate is needed for the split element (Evans, 1949*a*). This latter device can be simplified by replacing the rotating half-wave plate by an electrooptical half-wave plate in which the applied fields rotate (Willisen, 1966).

Drichko and Ioffe (1969) and Drichko (1970) described an element in which a rotating phase shifter is combined with an electrooptic phase shifter of the type described by Billings. With such a device it is possible to tune the transmission peak over a broad spectral region while at the same time superimposing a high-frequency, low-amplitude oscillation of the wavelength.

114. Measured Performance Öhman (1938), Lyot (1944), Evans (1949*a*), Billings and Draisin (1951), Dollfus (1956, 1957), Ioffe (1961), and Steel et al. (1961) are but a few of the many authors who describe Lyot-Öhman filters. Steel's filter appears to have the highest overall performance. It consists of 57 quartz or calcite retardation plates, fractional-wave plates, and sheet polarizers arranged in seven wide-angle split elements of the type described by Evans. The filter is centered on H_α but can be tuned through ±15 Å. It has a half width of

$\frac{1}{8}$ Å, the narrowest reported to date for a Lyot filter. The peak transmittance is 12 percent, and the full angular field is 2°15′. The filter operates at a temperature of 40 ± 0.01°C. The distortion of the transmitted wavefront is less than $\lambda/2$ across the 35-mm aperture of the filter and is due largely to the windows of the enclosure.

The Šolc Filter (Evans, 1958)

115. Šolc (1953, 1955) describes a polarization interference filter composed of a number of identical retardation plates placed between two polarizers. Two arrangements of the plates are possible. In the *folded filter* (Fig. 124*b*) the polarizers are crossed, and the rth plate makes an angle $(-1)^r\pi/4N$ with the first polarizer, N being the number of plates in the filter. In the *fan filter* (Fig. 124*c*) the polarizers are parallel and the rth plate is inclined at an angle $(r - \frac{1}{2})\pi/2N$ to them.

The transmittance of both types of filters is given by

$$T = \tfrac{1}{2}T_0 \left(\frac{\sin N\chi}{\sin \chi} \cos \chi \tan \frac{\pi}{2N}\right)^2 \tag{107}$$

where

$$\cos \chi = \begin{cases} \cos \pi\xi \cos \dfrac{\pi}{2N} & \text{fan filter} \\ \cos \left(\pi\xi + \dfrac{\pi}{2}\right) \cos \dfrac{\pi}{2N} & \text{folded filter} \end{cases} \tag{108}$$

The transmittance curve of the Šolc filter generally resembles that of a Lyot filter whose thinnest plate has the same thickness as the Šolc plates and whose overall thickness is approximately equal to that of the Šolc filter. An advantage of the Šolc filter is its much higher transmittance. Thus, for example, $T_0 = 0.72$ with sheet polarizers of internal transmittance 0.85. An important disadvantage of the Šolc filter is that its parasitic transmission (Fig. 125) is about 2.5 times that of a Lyot filter (Evans, 1958). This can be rectified by tuning the angles between the Šolc retardation plates (Harris et al., 1964; Šolc, 1965; Schiffman, 1967; Fredga, 1969).

Miscellaneous Hybrid Filters

116. Various combinations of different types of filters or filter elements have been made in the past in an effort to obtain a simpler filter or one with an improved performance. Drichko and Ioffe (1967), for example, tried to obtain a filter with comparatively high peak and low parasitic transmittances by combining an Evans type split element with a Šolc filter. Zirin (1966) describes filters in which the thickest stages of a Lyot filter are combined with narrow-band thin-film interference filters (Pars. 76 to 81). The resultant combinations often have a higher transmittance and better optical performance besides being cheaper. But care must be exercised when selecting the thin-film filter to ensure that its angular field does not limit that of the combination.

Katzenstein (1968; Katzenstein and Ward, 1971) proposed a filter consisting of a birefringent plate placed in the cavity of a Fabry-Perot interferometer. Such a device combines the relatively large angular field of polarization filters with the high resolution of Fabry-Perot interferometers, though at the expense of the transmittance (Williamson, 1971).

Synthesis of Polarization Interference Filters

117. Ammann and his coworkers in a series of papers (Harris et al., 1964; Ammann and Chang, 1965; Ammann, 1966; Ammann and Yarborough, 1966, 1967) developed powerful methods for the design of birefringent crystal filters with prescribed spectral characteristics. These should lead to improvements in the band shape and rejection characteristics of narrow-band polarization interference filters.

REFERENCES

Abelès, F. (1950): *Ann. Phys.*, vol. 5, pp. 596, 706.

——— (1967): p. 144 in A. C. S. Van Heel (ed.), "Advanced Optical Techniques," North-Holland, Amsterdam.

Adriaens, M. R., and B. Feuerbacker (1971): *Appl. Opt.*, vol. 10, p. 958.
Ahier, G. (1937): *C. R.*, vol. 205, p. 37.
Ahmed, M. S., and J. M. Ley (1970): *J. Phys.*, ser. D, *Appl. Phys.*, vol. 3, p. 1737.
Alberti, F., Division of Physics, National Research Council of Canada (1970): private communication.
Allen, L., D. G. D. Jones, and M. D. Sayers (1968): *J. Sci. Instrum.*, vol. 1, p. 133.
American National Standards Institute (1955): The Spectral-transmissive Properties of Plastics for Use in Eye Protection, New York.
Ammann, E. O. (1966): *J. Opt. Soc. Am.*, vol. 56, pp. 943, 952.
——— and I. C. Chang (1965): *J. Opt. Soc. Am.*, vol. 55, p. 835.
——— and J. M. Yarborough (1966): *J. Opt. Soc. Am.*, vol. 56, p. 1746.
——— and ——— (1967): *J. Opt. Soc. Am.*, vol. 57, p. 349.
Anders, H. (1966): *Vak. Tech.*, vol. 15, p. 123.
——— (1967): "Thin Films in Optics," Focal Press, New York.
——— and R. Eichinger (1965): *Appl. Opt.*, vol. 4, p. 899.
——— and ——— (1967): *Optik*, vol. 25, p. 17.
Angel, D. W., W. R. Hunter, R. Tousey, and G. Hass (1961): *J. Opt. Soc. Am.*, vol. 51, p. 913.
Ångström, A. K., and A. J. Drummond (1959): *J. Opt. Soc. Am.*, vol. 49, p. 1096.
Anon. (1969): *Microwaves*, September, p. 106.
Apfel, J. H. (1966*a*): *J. Opt. Soc. Am.*, vol. 56, p. 553.
——— (1966*b*): *Appl. Opt.*, vol. 4, p. 983.
ASTM (1959): Symposium on Electrodeless Nickel Plating, *Spec. Tech. Publ.* 265.
Auerbach, L. (1957): *Am. J. Phys.*, vol. 25, p. 440.
——— (1960): *Am. J. Phys.*, vol. 28, p. 743.
Austin, R. R., Perkin-Elmer Corporation (1968*a*): private communication.
——— (1968*b*): *Prod. News Opt. Coating Dept.*, Perkin-Elmer Corporation, November.
Axe, J. D., J. W. Gaglianello, and J. E. Scardefield (1965): *Phys. Rev.*, vol. 139, p. A1211.
——— and G. D. Pettit (1968): *Phys. Rev.*, vol. 151, p. 676.
Baird Atomic, Inc. (1968): Optical Components.
Balzers Aktiengesellschaft für Hochvakuumtechnik und dünne Schichten (1962): Kurvenblätter und ihre Bezeichnungen.
——— (1968): Interference Filters.
Banning, M. (1947*a*): *J. Opt. Soc. Am.*, vol. 37, p. 686.
——— (1947*b*): *J. Opt. Soc. Am.*, vol. 37, p. 792.
Barker, A. S. (1963): *Phys. Rev.*, vol. 132, p. 1474.
Barnes, R. B., and L. G. Bonner (1936*a*): *J. Opt. Soc. Am.*, vol. 26, p. 428.
——— and ——— (1936*b*): *Phys. Rev.*, vol. 49, p. 732.
Bass, A. M. (1948): *J. Opt. Soc. Am.*, vol. 38, p. 977.
Baumeister, P. W. (1958): *J. Opt. Soc. Am.*, vol. 48, p. 955.
——— (1961): *Opt. Acta*, vol. 8, p. 105.
——— (1962): in "Military Standardization Handbook: Optical Design," MIL-HDBK-141, Defense Supply Agency, Washington.
——— (1967): *Appl. Opt.*, vol. 6, p. 897.
——— (1969): *Appl. Opt.*, vol. 8, p. 423.
———, V. R. Costich, and S. C. Pieper (1965): *Appl. Opt.*, vol. 4, p. 911.
——— and F. A. Jenkins (1957): *J. Opt. Soc. Am.*, vol. 47, p. 57.
———, ———, and M. A. Jeppesen (1959): *J. Opt. Soc. Am.*, vol. 49, p. 1188.
——— and J. M. Stone (1956): *J. Opt. Soc. Am.*, vol. 46, p. 228.
Bausch & Lomb, Inc. (1958): *Infrared Prog. Rep.* 3.
——— (1968): Bausch and Lomb Multi-Films.
Becker, J., W. F. Coombs, and A. F. Turner (1966): *J. Opt. Soc. Am.*, vol. 56, p. 1435.
Beer, A. (1852): *Ann. Phys. Chem.*, vol. 86, no. 2, p. 78.
Behrndt, K. H., and D. W. Doughty (1967): *J. Vac. Sci. Technol.*, vol. 4, p. 199.
Bell, R. J., and H. V. Romero (1970): *Appl. Opt.*, vol. 9, p. 2341.
———, ———, and J. M. Blea (1970): *Appl. Opt.*, vol. 9, p. 2350.
Bennett, H. E. (1963): *J. Opt. Soc. Am.*, vol. 53, p. 1389.
———, J. M. Bennett, and E. J. Ashley (1963*a*): *Appl. Opt.*, vol. 2, p. 156.
——— and W. R. McBride (1964): *Appl. Opt.*, vol. 3, p. 919.
———, M. Silver, and E. J. Ashley (1963*b*): *J. Opt. Soc. Am.*, vol. 53, p. 1089.
Bennett, J. E., and E. J. Ashley (1965): *Appl. Opt.*, vol. 4, p. 221.
Berning, P. H. (1962): *J. Opt. Soc. Am.*, vol. 52, p. 431.

——— (1963): p. 69 in G. Hass (ed.), "Physics of Thin Films," vol. 1, Academic, New York.
———, G. Hass, and R. P. Madden (1960): *J. Opt. Soc. Am.*, vol. 50, p. 586.
——— and A. F. Turner (1957): *J. Opt. Soc. Am.*, vol. 47, p. 230.
Berry, R. G., and O. C. Jones (1964): *J. Sci. Instrum.*, vol. 41, p. 92.
Billings, B. H. (1947): *J. Opt. Soc. Am.*, vol. 37, p. 738.
———, S. Sage, and W. Draisin (1951): *Rev. Sci. Instrum.*, vol. 22, p. 1009.
Blazey, R. (1967): *Appl. Opt.*, vol. 6, p. 831.
Blea, J. M., W. F. Parks, P. A. R. Ade, and R. J. Bell (1970): *J. Opt. Soc. Am.*, vol. 60, p. 603.
Blifford, I. H., Jr. (1966): *Appl. Opt.*, vol. 5, p. 105.
Bonch-Bruevich, A. M., Ya. A. Imas, G. S. Malika, V. N. Rozhalestrenskii, and E. I. Fadeeva (1964): *Zh. Prikl. Spektrosk.*, vol. 1, p. 265.
Born, M., and E. Wolf (1959): "Principles of Optics," Pergamon, London.
Bouguer, P. (1729): "Essai d'optique sur la gradation de la luminère," Claude Tombert, Paris.
Bouwhuis, G. (1962): *Philips Res. Rep.*, vol. 17, p. 130.
Bradley, D. J., B. Bates, C. O. L. Juulman, and T. Kohno (1967): *J. Phys.*, vol. 28, p. C2-280.
Canfield, L. R., G. Hass, and J. E. Waylonis (1966): *Appl. Opt.*, vol. 5, p. 45.
Carlon, H. R. (1962): *Appl. Opt.*, vol. 1, p. 603.
Carlson, F. E., G. T. Howard, A. F. Turner, and H. H. Schroeder (1956): *J. Soc. Motion Pict. Telev. Eng.*, vol. 65, p. 136.
Casey, J. P., and E. A. Lewis (1952): *J. Opt. Soc. Am.*, vol. 42, p. 971.
Catalan, L. A., and T. Putner (1961): *Br. J. Appl. Phys.*, vol. 12, p. 499.
Chaikin, A. S., and V. V. Pukhonin (1970): *Zh. Prikl. Spektrosk.*, vol. 13, p. 894.
Chang, L. (1964): Infrared Dispersion of ZnS-Metal Films, Ph.D. dissertation, Colorado State University, July.
Chernyavskaya, N. A., I. G. Rudyavskaya, A. E. Stanevich, M. A. Okatov, and T. N. Fomina (1971): *Zh. Prikl. Spektrosk.*, vol. 15, p. 86.
Childs, C. B. (1961): *J. Opt. Soc. Am.*, vol. 51, p. 895.
Christiansen, C. (1884): *Ann. Phys. Chem.*, vol. 23, p. 298.
Ciddor, P. E. (1965): *Opt. Acta*, vol. 12, p. 177.
——— (1968): *Appl. Opt.*, vol. 7, p. 2328.
Clapham, P. B. (1971): *Opt. Acta*, vol. 18, p. 563.
———, M. J. Downs, and R. J. King (1969): *Appl. Opt.*, vol. 8, p. 1965.
Clarke, R. H. (1968): *Appl. Opt.*, vol. 7, p. 861.
Clewell, D. H. (1941): *J. Opt. Soc. Am.*, vol. 31, p. 521.
CODATA, Committee on Data for Science and Technology of the International Council of Scientific Unions (1969): "International Compendium of Numerical Data Projects," Springer, Berlin.
Cohendet, M. A., and B. Saudreau (1967): SEAVOM Contract Report DRME-66-34-443.
Collins, A. T., E. C. Lightowlers, and P. J. Dean (1967): *Phys. Rev.*, vol. 158, p. 833.
Collins, R. J., and D. A. Kleinman (1959): *J. Phys. Chem. Solids*, vol. 11, p. 190.
Corion Instrument Corp. (1969): Thin Film Optical Filters.
Corning Glass Works (1965): Glass Color Filters.
Costich, V. R., Spectra-Physics (1968): private communication.
——— (1970): *Appl. Opt.*, vol. 9, p. 866.
Cox, J. T., and G. Hass (1964): p. 239 in G. Hass and R. E. Thun (eds.), "Physics of Thin Films," vol. 2, Academic, New York.
———, ———, and J. E. Waylonis (1968): *Appl. Opt.*, vol. 7, p. 1535.
Czerny, M. (1923): *Z. Phys.*, vol. 16, p. 321.
Dändliker, R. (1968): *J. Opt. Soc. Am.*, vol. 58, p. 1062.
Demeshina, A. I., V. A. Zayats, V. I. Lapshin, and V. N. Murzin (1970): *Zh. Prikl. Spektrosk.*, vol. 13, p. 346.
Denmark, H. S., and W. M. Cady (1935): *J. Opt. Soc. Am.*, vol. 25, p. 330.
Denton, R., Denton Vacuum Inc. (1970): private communication.
Dickey, D. N., and J. G. Mavroides (1964): *Solid State Commun.*, vol. 2, p. 213.
Dioffo, A. M. (1968): *Rev. Opt.*, vol. 47, p. 117.
Dixon, J. R., and H. R. Riedl (1965): *Phys. Rev.*, vol. 140, p. A1283.
Dobrowolski, J. A. (1959): *J. Opt. Soc. Am.*, vol. 49, p. 794.
——— (1965): *Appl. Opt.*, vol. 4, p. 937.
——— (1969): *J. Sci. Instrum.*, vol. 2, p. 429.

Dobrowolski, J. A. (1970): *Appl. Opt.*, vol. 9, p. 1396.
——— and R. C. Bastien (1966): *J. Opt. Soc. Am.*, vol. 55, p. 554.
Dodd, C. X., and T. W. West (1961): *J. Opt. Soc. Am.*, vol. 51, p. 915.
Dollfus, A. (1956): *Rev. Opt.*, vol. 35, pp. 539, 625.
——— (1957): *C. R.*, vol. 245, p. 32.
Dresler, A. (1933): *Licht*, vol. 3, p. 41.
Drichko, N. M. (1969): *Sov. J. Opt. Technol.*, vol. 36, p. 386.
——— (1970): *Sov. J. Opt. Technol.*, vol. 36, p. 90.
——— and S. B. Ioffe (1967): *Sov. J. Opt. Technol.*, vol. 34, p. 360.
——— and ——— (1969): *Sov. J. Opt. Technol.*, vol. 36, p. 300.
Drummeter, L. F., Jr., and G. Hass (1964): p. 305 in G. Hass and R. E. Thun (eds.), "Physics of Thin Films," vol. 2, Academic, New York.
Eastman Kodak Company (1968*a*): Special Filters from Kodak for Technical Applications.
——— (1968*b*): Kodak Wratten Filters.
——— (1969): Kodak Neutral Density Attenuators.
Eather, R. H., and D. L. Reasoner (1969): *Appl. Opt.*, vol. 8, p. 227.
Elsner, Z. N. (1964): *Opt. Spectrosc.*, vol. 17, p. 238.
Epstein, L. I. (1952): *J. Opt. Soc. Am.*, vol. 42, p. 806.
——— (1955): *J. Opt. Soc. Am.*, vol. 45, p. 360.
Evans, J. W. (1949*a*): *J. Opt. Soc. Am.*, vol. 39, p. 229.
——— (1949*b*): *J. Opt. Soc. Am.*, vol. 39, p. 412.
——— (1958): *J. Opt. Soc. Am.*, vol. 48, p. 142.
Evans, L. R., and C. Grey Morgan (1968): *J. Sci. Instrum.*, vol. 1, p. 856.
Faber, W. A., P. W. Kruse, and W. D. Saur (1961): *J. Opt. Soc. Am.*, vol. 51, p. 115.
Fisher, R. (1971): *Solar Phys.*, vol. 18, p. 253.
Fragstein, K. von (1933): *Ann. Phys.*, vol. 17, p. 22.
Fredga, K. (1969): *Appl. Opt.*, vol. 8, p. 333.
Friedman, H., and C. P. Glover (1949): *J. Opt. Soc. Am.*, vol. 39, p. 795.
Furman, Sh. A. (1966): *Sov. J. Opt. Technol.*, vol. 33, p. 559.
——— (1968): *Opt.-Mekh. Prom.*, vol. 35, no. 9, p. 50.
——— (1970): *Opt. Spectrosc.*, vol. 28, p. 218.
——— and M. D. Levina (1970): *Opt. Spectrosc.*, vol. 28, p. 412.
——— and ——— (1971*a*): *Opt. Spectrosc.*, vol. 30, p. 404.
——— and ——— (1971*b*): *Opt.-Mekh. Prom.*, vol. 38, no. 5, p. 18.
Gaunt, J. T., and W. G. Burns (1954): *J. Sci. Instrum.*, vol. 31, p. 315.
Geffcken, W. (1939): DRPat. 716,153.
——— (1953/1944): DBPat. 899,120.
——— (1954/1944): DBPat. 913,005.
——— (1954): *Z. Angew. Phys.*, vol. 6, p. 249.
——— (1957): Lichtfilter, p. 925 in Landolt-Börnstein, "Zahlenwerte und Funktionen," 6th ed., vol. 4, pt. 3, Springer, Berlin.
Geick, R., W. J. Hakel, and C. H. Perry (1966): *Phys. Rev.*, vol. 148, p. 824.
Gerasimova, N. G. (1970): *Opt.-Mekh. Prom.*, vol. 37, p. 58.
Giacomo, P. (1956): *Rev. Opt.*, vol. 35, p. 317.
——— (1958): *J. Phys. Radium*, vol. 19, p. 307.
———, P. W. Baumeister, and F. A. Jenkins (1959): *Proc. Phys. Soc.*, vol. 73, p. 480.
Gibbs, R. C. (1910): *Phys. Rev.*, vol. 31, p. 463.
Gibson, K. S. (1916): *Phys. Rev.*, vol. 7, p. 194.
——— (1935): *J. Opt. Soc. Am.*, vol. 25, p. 131.
——— (1949): Spectrophotometry, *Natl. Bur. Stand. Circ.* 484.
——— and M. A. Belknap (1950): *J. Res. Natl. Bur. Stand.*, vol. 44, p. 463.
Gielisse, P. J., S. S. Mitra, J. N. Plendl, R. D. Griffis, L. C. Mansur, R. Marshall, and E. A. Pascoe (1967): *Phys. Rev.*, vol. 155, p. 1039.
———, J. N. Plendl, L. C. Mansur, R. Marshall, S. S. Mitra, R. Mykolajewycz, and A. Smakula (1965): *J. Appl. Phys.*, vol. 36, p. 2446.
Gilvarg, A. B., and G. G. Dolgov-Savelyev (1965): *Opt. Spectrosc.*, vol. 18, p. 497.
Giovanelli, R. G., and J. T. Jefferies (1954): *Austr. J. Phys.*, vol. 7, p. 254.
Gisin, M. A. (1969): *Sov. J. Opt. Technol.*, vol. 36, p. 191.
Glick, R., W. J. Hakel, and C. H. Perry (1966): *Phys. Rev.*, vol. 148, p. 824.
Gloge, D., E. L. Chinnock, and H. E. Earl (1969): *Bell Syst. Tech. J.*, vol. 48, p. 511.
Gobrecht, H., S. Seeck, and T. Klose (1965): *C. R.*, vol. 261, p. 373.
Golubeva, G. I., M. S. Abakumova, and A. M. Klochkov (1971): *Opt.-Mekh. Prom.*, vol. 38, no. 4, p. 48.

Gottlieb, M. (1960): *J. Opt. Soc. Am.*, vol. 50, p. 350.
Grebenshchikov, I. V. (ed.) (1946): "Prosvetlenie Optiki" (Antireflection coating of optical surfaces), State Publishers of Technical and Theoretical Literature, Moscow-Leningrad.
Greenler, R. G. (1958): in J. Strong, "Concepts of Classical Optics," Freeman, San Francisco.
Gurinovich, I. F., and N. A. Borisevich (1963): *Acad. Sci. USSR Bull. Phys. Ser.*, vol. 27, p. 31.
Hadley, L. N., and D. M. Dennison (1947): *J. Opt. Soc. Am.*, vol. 37, p. 451.
——— and ——— (1948): *J. Opt. Soc. Am.*, vol. 38, p. 483.
Hancock, K. (1970): p. 231 in K. A. Kopetzky (ed.), *Proc. Tech. Program, Electro-Opt. Syst. Des. Conf., New York, Sept. 16–18, 1969,* Industrial and Scientific Conference Management, Chicago.
Hardy, A. C. (1936): "Handbook of Colorimetry," p. 24, Technology Press of M.I.T., Cambridge, Mass.
——— and F. M. Young (1948): *J. Opt. Soc. Am.*, vol. 38, p. 854.
Harris, S. E., E. O. Ammann, and I. C. Chang (1964): *J. Opt. Soc. Am.*, vol. 54, p. 1267.
———, S. T. K. Nieh, and R. S. Feigelson (1970): *Appl. Phys. Lett.*, vol. 17, p. 223.
———, ———, and D. K. Winslow (1969): *Appl. Phys. Lett.*, vol. 15, p. 325.
——— and R. W. Wallace (1969): *J. Opt. Soc. Am.*, vol. 59, p. 744.
Harrison, D. H. (1968): *Appl. Opt.*, vol. 7, p. 210.
Harrison, G. R. (1929): *J. Opt. Soc. Am.*, vol. 18, p. 492.
Hass, G. (1955): *J. Opt. Soc. Am.*, vol. 45, p. 949.
——— and L. Hadley (1972): p. 6-118 in O. F. Gray (ed.), "American Institute of Physics Handbook," 3d ed., McGraw-Hill, New York.
——— and W. R. Hunter (1967): *Appl. Opt.*, vol. 6, p. 2097.
———, G. F. Jacobus, and W. R. Hunter (1967): *J. Opt. Soc. Am.*, vol. 57, p. 758.
———, J. B. Ramsey, J. B. Heany, and J. J. Triolo (1969*a*): *Appl. Opt.*, vol. 8, p. 275.
———, ———, and W. R. Hunter (1969*b*): *Appl. Opt.*, vol. 8, p. 2255.
——— and R. Tousey (1959): *J. Opt. Soc. Am.*, vol. 49, p. 593.
Hass, M. (1963): *J. Phys. Chem. Solids,* vol. 24, p. 1159.
Heavens, O. S. (1954): *J. Opt. Soc. Am.*, vol. 44, p. 371.
——— (1955): "Optical Properties of Thin Solid Films," Butterworths, London.
——— (1960): *Rep. Prog. Phys.*, vol. 23, p. 1.
——— and H. M. Liddell (1966): *Appl. Opt.*, vol. 5, p. 373.
———, J. Ring, and S. D. Smith (1957): *Spectrochim. Acta,* vol. 10, p. 179.
Hefft, K., R. Kern, G. Nöldeke, and A. Steudel (1963): *Z. Phys.*, vol. 175, p. 391.
Heinrich, P. L., R. C. Bastien, A. Dos Santos, and M. Ostrelich (1967): Development of an All-Dielectric Infrared Beamsplitter Operating in the 5 to 30 Micron Region, *NASA Rep.* NASA CR-703, February.
Heitmann, W. (1966): *Z. Angew. Phys.*, vol. 21, p. 503.
Heliotek, A Division of Textron, Inc. (1969): data sheets.
Hemmingway, D. J., and P. H. Lissberger (1967): *Appl. Opt.*, vol. 6, p. 471.
Henry, R. L. (1948): *J. Opt. Soc. Am.*, vol. 38, p. 775.
Herpin, A. (1947): *C. R.*, vol. 225, p. 182.
Herriott, D. R., J. R. Wimperis, and D. L. Perry (1965): *J. Opt. Soc. Am.*, vol. 55, p. 546.
Herscher, M. (1965): *Appl. Phys. Lett.*, vol. 7, p. 39.
Hickey, J. R., D. A. Brett, and A. J. Drummond (1962): *Tellus,* vol. 14, p. 451.
Holland, A. J., and W. E. S. Turner (1941): *J. Soc. Glass Technol.*, vol. 25, p. 164.
Holloway, R. J., and P. W. Lissberger (1969): *Appl. Opt.*, vol. 8, p. 653.
Honcia, G., and K. Krebs (1961): *Z. Phys.*, vol. 165, p. 202.
Hovis, W. A., Jr., W. A. Kley, and M. G. Strange (1967): *Appl. Opt.*, vol. 6, p. 1057.
Hunter, W. R. (1962): *Opt. Acta,* vol. 9, p. 255.
———, D. W. Angel, and R. Tousey (1965): *Appl. Opt.*, vol. 4, p. 891.
Ilford, Ltd. (1960): Ilford Colour Filters.
Infrared Industries, Inc. (1966): Infratron Interference Filters.
——— (1967): Thin Film Products at Infrared Industries, Inc.
Ioffe, S. B. (1961): *Sov. Astron. J.*, vol. 4, p. 1032.
——— and T. A. Smirnova (1967*a*): *Opt. Spectrosc.*, vol. 22, p. 157.
——— and ——— (1967*b*): *Opt. Spectrosc.*, vol. 23, p. 74.

Jacobsson, R. (1964): *J. Opt. Soc. Am.*, vol. 54, p. 422.
——— (1966): *J. Opt. Soc. Am.*, vol. 56, p. 1435.
——— and J. O. Mårtensson (1966): *Appl. Opt.*, vol. 5, p. 29.
Jasperse, J. R., A. Kahan, J. N. Plendl, and S. S. Mitra (1966): *Phys. Rev.*, vol. 146, p. 526.
Jones, O. C., and F. J. J. Clarke (1961): *Nature*, vol. 191, p. 1290.
Kaiser, W., W. C. Spitzer, R. H. Kaiser, and L. E. Howarth (1962): *Phys. Rev.*, vol. 127, p. 1950.
Kard, P. (1959*a*): *Loodus Mat.*, vol. 1, p. 67.
——— (1959*b*): *Opt. Spectrosc.*, vol. 6, p. 244.
——— (1960): *Eesti NSV Tead. Akad. Toim.*, vol. 9, p. 26.
———, E. Nesmelov, and G. Konyukhov (1968): *Eesti NSV Tead. Akad. Toim., Fuus.-Mut.*, vol. 17, p. 314.
Kartashev, A. I., and N. M. Syromyatnikova (1947): in Wavelength of Light as a Standard of Length, *Proc. Mendeleev State Sci. Res. Inst., Leningr.*, no. 7.
Kasha, M. (1948): *J. Opt. Soc. Am.*, vol. 38, p. 929.
Katsube, S., Y. Katsube, K. Mitone, and S. Furuta (1968): *Oyo Buturi*, vol. 37, p. 225.
Katzenstein, J. (1968): *J. Opt. Soc. Am.*, vol. 58, p. 1348.
——— and S. Ward (1971): *Appl. Opt.*, vol. 10, p. 1119.
Kimura, T., and M. Saruwatari (1971): *Proc. IEEE*, vol. 59, p. 1273.
Kneubühl, F. (1969): *Appl. Opt.*, vol. 8, p. 505.
Knittl, Z. (1964): *J. Phys.*, vol. 25, p. 245.
Kohn, H., and K. von Fragstein (1932): *Phys. Z.*, vol. 33, p. 929.
Konyukhov, G. P., and E. A. Nesmelov (1969): *Zh. Prikl. Spectrosk.*, vol. 11, p. 468.
Koppelmann, G. (1960): *Ann. Phys.*, vol. 5, p. 388.
Korolev, F. A., A. Yu. Klement'eva, and T. F. Meshcheryakova (1960): *Opt. Spectrosc.*, vol. 9, p. 341.
———, ———, ———, and I. A. Ramazina (1970): *Opt. Spectrosc.*, vol. 28, p. 416.
Kreidl, N. J. (1960): in F. V. Tooley (ed.), "Handbook of Glass Manufacture," vol. II, Ogden, New York.
Kubota, H. (1961): *Prog. Opt.*, vol. 1, p. 211.
Kuznetsov, A. Ya., and A. F. Perveyev (1967): *Sov. J. Opt. Technol.*, vol. 34, p. 593.
Lambert, J. H. (1760): "Photometria sine de mensura et gradibus luminis colorum et umbrae," Eberhard Klett, Augsburg.
Larsen, T. (1962): *IRE MTT*, vol. 10, p. 191.
Laser Systems, Inc. (1966): Resonant Reflector Assemblies, *Bull.* S601-1.
Lazareva, L. D. (1970): *Sov. J. Opt. Technol.*, vol. 36, p. 801.
Lecomte, J. (1958): p. 297 in S. Flügge (ed.), "Handbuch der Physik," vol. 26, Springer, Berlin.
Leś, F., and Z. Leś (1962): *Acta Phys. Polon.*, vol. 21, p. 523.
Leś, Z., F. Leś, and L. Gabła (1963): *Acta Phys. Polon.*, vol. 33, p. 211.
Lessman, G. (1966): *J. Opt. Soc. Am.*, vol. 56, p. 554.
Liberty Mirror, A Division of Libbey-Owens Ford Glass Co. (1966): Coatings.
Linder, S. L. (1967): *Appl. Opt.*, vol. 6, p. 1201.
Lipshutz, M. L. (1968): *Appl. Opt.*, vol. 7, p. 2326.
Lissberger, P. H. (1959): *J. Opt. Soc. Am.*, vol. 49, p. 121.
——— (1968): *J. Opt. Soc. Am.*, vol. 58, p. 1586.
Loh, E. (1968): *Phys. Rev.*, vol. 166, p. 673.
Lukirskii, A. P., E. P. Savinov, O. A. Ershov, and Yu. F. Shepelev (1964): *Opt. Spectrosc.*, vol. 16, p. 168.
———, ———, ———, I. I. Zhukova, and V. A. Fomichev (1965): *Opt. Spectrosc.*, vol. 19, p. 237.
Lupashko, E. A., and I. N. Sklyarevskii (1964): *Opt. Spectrosc.*, vol. 16, p. 279.
Lütge, H. (1932): *Glastechnol. Ber.*, vol. 10, p. 374.
Lyot, B. (1933): *C. R.*, vol. 197, p. 1593.
——— (1944): *Ann. Astrophys.*, vol. 7, p. 31.
Macleod, H. A. (1969): "Thin Film Optical Filters," American Elsevier, New York.
MacNeille, S. M. (1946): U.S. Pat. 2,403,731 (July 9).
Madden, R. P. (1963): p. 123 in G. Hass (ed.), "Physics of Thin Films," vol. 1, Academic, New York.
———, L. R. Canfield, and G. Hass (1963): *J. Opt. Soc. Am.*, vol. 53, p. 620.
Magyar, G. (1967): *Rev. Sci. Instrum.*, vol. 38, p. 517.
Mahlein, A. H. F., and G. Schollmeier (1969): *Appl. Opt.*, vol. 8, p. 119.
Maier, R. L. (1967): *Thin Solid Films*, vol. 1, p. 31.

Manabe, A., A. Mitsuishi, and H. Yoshinaga (1967): *Jap. J. Appl. Phys.,* vol. 6, p. 593.
Manley, T. R., and D. A. Williams (1965): *Spectrochim. Acta,* vol. 21, p. 737.
Mark, R., D. Morand, and S. Waldstein (1970): *Appl. Opt.,* vol. 9, p. 2305.
Martin, T. R., and A. F. Turner (1966): *J. Appl. Phys.,* vol. 37, p. 1749.
———, ———, J. Masso, and F. Sulzbach (1962): *J. Opt. Soc. Am.,* vol. 52, p. 1309.
Matovich, E. (1965): *ISA J.,* vol. 12, p. 53, December.
Mayer, H. (1950): "Physik dünner Schichten," vol. 1, Wissenschaftliche Verlagsgesellschaft, Stuttgart.
McAlister, E. D. (1935): *Smithson. Misc. Collect.,* vol. 93, no. 7.
McBride, W. R. (1963*a*): *J. Opt. Soc. Am.,* vol. 53, p. 519.
——— (1963*b*): *J. Opt. Soc. Am.,* vol. 53, p. 1352.
——— and A. L. Olsen (1962): Survey of Optical Materials, *Proc. Symp. Mil. Appl. Ultra-violet Radiat., Wash., D.C., July 14–15, 1960, Univ. Chicago Rep.* LAS-TR-199-37, November.
McCarthy, D. E. (1965): *Appl. Opt.,* vol. 4, p. 317.
McGovern, J. J., and R. A. Fredel (1947): *J. Opt. Soc. Am.,* vol. 37, p. 660.
McKenney, D., and A. F. Turner (1967): *Univ. Ariz. Opt. Sci. Cent. Newsl.,* vol. 1, no. 5.
McKenney, D. B., and P. N. Slater (1970): *Appl. Opt.,* vol. 9, p. 2435.
McLeod, J. H. (1945): *J. Opt. Soc. Am.,* vol. 35, p. 185.
Meaburn, J. (1966): *Appl. Opt.,* vol. 5, p. 1757.
Mellon, M. G. (ed.) (1950): "Analytical Absorption Spectroscopy," Wiley, New York.
Meltzer, R. S., Industrial Optics Division, Yardney Razdow Laboratories, Inc. (1968): private communication.
Meneely, C. T. (1967): *Appl. Opt.,* vol. 6, p. 1434.
Messner, R. (1953): *Feinwerktech.,* vol. 57, p. 142.
Meyer, S. (1936): *Glastech. Ber.,* vol. 14, p. 305.
Michael, J., Perkin-Elmer Corporation (1968): private communication.
Mie, G. (1908): *Ann. Phys.,* 4th ser., vol. 25, p. 377.
Miller, R. C., and W. G. Spitzer (1963): *Phys. Rev.,* vol. 129, p. 94.
Minkoff, G. J., and A. G. Gordon (1946): *Nature,* vol. 158, p. 788.
Mitsuishi, A., Y. Otsuka, S. Fujita, and H. Yoshinaga (1963): *Jap. J. Appl. Phys.,* vol. 2, p. 574.
———, Y. Yamada, and H. Yoshinaga (1962): *J. Opt. Soc. Am.,* vol. 52, p. 14.
Miyake, K. P. (1964): *J. Phys.,* vol. 25, p. 255.
Mohler, N. M., and J. R. Loofbourow (1952): *Am. J. Phys.,* vol. 20, pp. 499, 579.
Möller, K. D., and R. V. McKnight (1963): *J. Opt. Soc. Am.,* vol. 53, p. 760.
——— and ——— (1965): *J. Opt. Soc. Am.,* vol. 55, p. 1075.
———, D. J. McMahon, and D. R. Smith (1966): *Appl. Opt.,* vol. 5, p. 403.
———, and W. G. Rothschild (1971): "Far-Infrared Spectroscopy," Wiley-Interscience, New York.
Motovilov, O. A. (1967): *Opt. Spectrosc.,* vol. 22, p. 537.
Murray, A. E. (1956): *J. Opt. Soc. Am.,* vol. 46, p. 790.
Mussett, A., Optical Coating Laboratory (1969): private communication.
——— and A. Thelen (1970): *Prog. Opt.,* vol. 8, p. 203.
Nagel, M. R. (1954): *J. Opt. Soc. Am.,* vol. 44, p. 621.
Neilson, R. G. T., and J. Ring (1967): *J. Phys.,* vol. 28, p. C2-270.
Nesmelov, E. A., and G. P. Konyukhov (1968): *Opt. Spektrosk.,* vol. 31, p. 133.
Novikov, V. M. (1968): *Sov. J. Opt. Technol.,* vol. 35, p. 121.
——— and G. N. Vasni (1968): *Sov. J. Opt. Technol.,* vol. 35, p. 522.
Obodovskiy, V. V., V. N. Rozhdestvenskiy, and I. Chernyy (1967): *Sov. J. Opt. Technol.,* vol. 34, p. 324.
O'Bryan, H. M. (1935): *Rev. Sci. Instrum.,* vol. 6, p. 328.
Öhman, Y. (1938): *Nature,* vol. 41, pp. 157, 291.
——— (1958): *Ark. Astron.,* vol. 2, p. 165.
Optical Coating Laboratory, Santa Rosa, Calif. (1964): Catalog.
——— (1965): Multilayer Antireflection Coatings.
——— (1967): Effects of the Variation of Angle of Incidence and Temperature on Infrared Filter Characteristics.
——— (1970): Infrared Handbook.
Oriel Optics Corporation (1968): Catalog Sec. C: Optical Filters.
Park, K. C. (1964): *Appl. Opt.,* vol. 3, p. 877.
Parodi, M. (1937): *C. R.,* vol. 204, p. 1111.
Pellicori, S. F. (1964): *Appl. Opt.,* vol. 3, p. 361.

Pellicori, S. F., C. A. Johnson, and F. T. King (1966): *Appl. Opt.*, vol. 5, p. 1916.
Penselin, S., and A. Steudel (1955): *Z. Phys.*, vol. 142, p. 21.
Perry, D. L. (1965): *Proc. IEEE*, vol. 53, p. 76.
Peršun, A., B. Marković, and A. Kuntarić (1968): *Thin Solid Films*, vol. 2, p. 377.
Pfund, A. H. (1930): *Phys. Rev.*, vol. 36, p. 71.
——— (1935): *J. Opt. Soc. Am.*, vol. 25, p. 351.
Pidgeon, C. R., and S. D. Smith (1964): *J. Opt. Soc. Am.*, vol. 54, p. 1459.
Pines, D. (1956): *Rev. Mod. Phys.*, vol. 28, p. 184.
Piriou, D., and F. Cabannes (1968): *Opt. Acta*, vol. 15, p. 271.
——— and J. Tsakiris (1965): *C. R.*, vol. 261, p. 3079.
Ploke, M. (1967): *Zeiss-Mitt.*, vol. 4, p. 279.
Plummer, J. H. (1936): *J. Opt. Soc. Am.*, vol. 26, p. 434.
Pohlack, H. (1952): p. 103 in P. Görlich (ed.), "Jenaer Jahrbuch 1952," Fischer, Jena.
——— (1956*a*): p. 79 in P. Görlich (ed.), "Jenaer Jahrbuch 1956," Fischer, Jena.
——— (1956*b*): ibid., p. 87.
——— and H. Koch (1957): p. 74 in P. Görlich (ed.), "Jenaer Jahrbuch 1957," Fischer, Jena.
Porteus, J. O. (1963): *J. Opt. Soc. Am.*, vol. 53, p. 1394.
Pradhan, M. M. (1970): *Infrared Phys.*, vol. 10, p. 199.
Quintard, P., and P. Delorme (1967): *Rev. Phys. Appl.*, vol. 2, p. 255.
Raman, C. V. (1949): *Proc. Indian Acad. Sci.*, vol. A29, p. 381.
——— and K. S. Viswanathan (1955): *Proc. Indian Acad. Sci.*, vol. A41, p. 55.
Ramsay, J. V. (1968): *Austr. Phys.*, vol. 5, p. 87.
——— and P. E. Ciddor (1967): *Appl. Opt.*, vol. 6, p. 2003.
———, H. Kobler, and E. G. V. Mugridge (1970): *Solar Phys.*, vol. 12, p. 492.
Rank, D. H., and H. E. Bennett (1955): *J. Opt. Soc. Am.*, vol. 45, p. 69.
Rayleigh, Lord (1888): *Phil. Mag. J. Sci.*, vol. 26, p. 258.
——— (1901): *Nature*, vol. 64, p. 385.
——— (1914): *Proc. R. Soc.*, vol. A90, p. 219.
Rehnberg, J. D. (1970): p. 784 in K. A. Kopetzky (ed.), *Proc. Tech. Program, Electro-Opt., Syst. Des. Conf., New York, Sept. 16–18, 1969*, Industrial and Scientific Conference Management, Inc., Chicago.
Renk, K. F., and L. Genzel (1962): *Appl. Opt.*, vol. 1, p. 643.
Ressler, G. M., and K. D. Möller (1967): *Appl. Opt.*, vol. 6, p. 893.
Reynard, D. L., and A. Andrew (1966): *Appl. Opt.*, vol. 5, p. 23.
Riccius, H. D. (1968): *J. Appl. Phys.*, vol. 39, p. 4381.
———, Division of Physics, National Research Council of Canada (1969): private communication.
Rigden, J. D., and A. D. White (1962): *Proc. IRE*, vol. 50, p. 2366.
Ring, J. (1956): p. 381 in Z. Kopal (ed.), "Astronomical Optics," North-Holland, Amsterdam.
———, R. Beer, and V. Hewison (1958): *Prog. recent. spectroscop. interferentielle, Bellevue, 9–13 Sept. 1957*, Éditions du Centre National de la Recherche Scientifique, Paris.
Robin, S., and S. Robin (1958): *J. Phys. Radium*, vol. 19, p. 913.
Rohrbaugh, J. H., and R. D. Hatcher (1958): *J. Opt. Soc. Am.*, vol. 48, p. 704.
Rooyen, E., van, and E. Therou (1969): *Appl. Opt.*, vol. 8, p. 832.
Rowley, W. R. C. (1966): *IEEE Trans. Instrum. Meas.*, vol. IM-15, p. 146.
Rozenberg, G. V. (1958): "Optika Tonkosloinykh Pokrytii," State Publishers of Physical-Mathematical Literature, Moscow.
Russell, J. W., and H. L. Strauss (1965): *Appl. Opt.*, vol. 4, p. 1131.
Rustgi, O. P. (1965): *J. Opt. Soc. Am.*, vol. 55, p. 630.
Sakai, K., T. Fukui, Y. Tsunawaki, and H. Yoshinaga (1969): *Jap. J. Appl. Phys.*, vol. 8, p. 1046.
———, Y. Nakagawa, and H. Yoshinaga (1968): *Jap. J. Appl. Phys.*, vol. 7, p. 792.
Samson, J. A. R. (1967): "Techniques of Vacuum Ultraviolet Spectroscopy," Wiley, New York.
——— and R. B. Cairns (1965): *Appl. Opt.*, vol. 4, p. 915.
Sanderson, R. B. (1965): *J. Phys. Chem. Solids*, vol. 26, p. 803.
Sawaki, T. (1960): Studies on Antireflection Films, *Rep. Osaka Ind. Res. Inst.* 315, September.
Scharf, P. T. (1952): *J. Soc. Motion Pict. Telev. Eng.*, vol. 59, p. 191.
——— (1965): p. 111 in R. Kingslake (ed.), "Applied Optics and Optical Engineering," vol. 1, Academic, New York.
Scheuerman, R. J., Perkin-Elmer Corporation (1968): private communication.
——— and J. H. Beardsley (1968): Unsupported Thin Film Beamsplitter, *Perkin-Elmer Tech. Rep.* 9337, Contract NAS 5-10292, June.

Schiffman, B. M. (1967): *J. Opt. Soc. Am.*, vol. 57, p. 1390.
Schild, J., A. Steudel, and H. Walther (1967): *J. Phys.*, vol. 28, p. C2-276.
Schineller, E. R., and R. P. Flam (1968): *J. Opt. Soc. Am.*, vol. 58, p. 736.
Schleusner, S. A., and A. A. Read (1966): *Rev. Sci. Instrum.*, vol. 37, p. 287.
Schmidt, R. N., and K. C. Park (1965): *Appl. Opt.*, vol. 4, p. 917.
Schmidtke, G. (1968): *Z. Angew. Phys.*, vol. 25, p. 314.
——— (1968–1969): *Optik,* vol. 28, p. 578.
Schott & Gen. (1959): Mainz Farb- und Filterglas.
——— (1965): Monochromatic Interference Filters.
——— (1967*a*): New Color Filter Glasses, pt. I, Data Sheet 3504/Ie.
——— (1967*b*): Interference Reflection Filter UV-R-250, Data Sheet 3703e.
Schrader, W. (1958): *Optik,* vol. 15, p. 265.
Schröder, H. (1948): *Optik,* vol. 3, p. 499.
——— (1951): *Z. Angew. Phys.*, vol. 3, p. 53.
——— (1957): in M. Auwärter (ed.), "Ergebnisse der Hochvakuum Technik und der Physik dünner Schichten," Wissenschaftliche Verlagsgesellschaft, Stuttgart.
——— (1962): *Opt. Acta,* vol. 9, p. 249.
——— (1969): p. 87 in G. Hass and R. E. Thun (eds.), "Physics of Thin Films," vol. 5, Academic, New York.
Schroeder, D. J. (1962): *J. Opt. Soc. Am.*, vol. 52, p. 1380.
Schroeder, H. H., and A. F. Turner (1960): *J. Soc. Motion Pict. Telev. Eng.*, vol. 69, p. 351.
Seeley, J. S. (1964): *J. Opt. Soc. Am.*, vol. 54, p. 342.
Sergeeva, A. I. (1970): *Opt.-Mekh. Prom.*, vol. 38, no. 3, p. 65.
Shaw, C. N., and W. T. Foreman (1959): *J. Opt. Soc. Am.*, vol. 49, p. 724.
Sherr, A. E., R. J. Tucker, and R. A. Greenwood (1969): *Laser Focus,* p. 46, September.
Sinsheimer, R. L., and J. R. Loofbourow (1947): *Nature,* vol. 160, p. 674.
Sinton, W. M., and W. C. Davis (1954): *J. Opt. Soc. Am.*, vol. 44, p. 503.
Smiley, V. N. (1968): *J. Opt. Soc. Am.*, vol. 58, p. 1469.
Smith, R. A., F. E. Jones, and R. P. Chasmar (1968): "The Detection and Measurement of Infrared Radiation," p. 369, Clarendon, Oxford.
Smith, S. D. (1958): *J. Opt. Soc. Am.*, vol. 48, p. 43.
——— and O. S. Heavens (1957): *J. Sci. Instrum.*, vol. 34, p. 492.
——— and J. S. Seeley (1968): Multilayer Filters for the Region 0.8 to 100 Microns, *Univ. Reading, Final Sci. Rep. Contr.* AF61 (052)-833, May 7.
Sogani, C. M. (1926): *Phil. Mag.*, vol. 1, p. 321.
Sokolova, R. S. (1970): *Sov. J. Opt. Technol.*, vol. 37, p. 318.
——— (1971): *Opt.-Mekh. Prom.*, vol. 38, no. 5, p. 46.
——— and T. N. Krylova (1963): *Opt. Spectrosc.*, vol. 14, p. 213.
Šolc, I. (1953): *Čs. Čas. Fys.*, vol. 3, p. 366.
——— (1955): *Czech. J. Phys.*, vol. 5, pp. 86, 97, 272.
——— (1965): *J. Opt. Soc. Am.*, vol. 55, p. 621.
Sorokin, O. M., and V. A. Blank (1970): *Sov. J. Opt. Technol.*, vol. 37, p. 343.
Spectrum Systems, Inc. (1967): Visible and Near Infrared Filters, *Bull.* SS-14.
Spitzer, W. G., and D. A. Kleinman (1961): *Phys. Rev.*, vol. 121, p. 1324.
———, ———, and D. Walsh (1959): *Phys. Rev.*, vol. 113, p. 127.
———, R. C. Miller, D. A. Kleinman, and L. E. Howarth (1962): *Phys. Rev.*, vol. 126, p. 1710.
Sporton, T. M. (1967): *J. Sci. Instrum.*, vol. 44, p. 720.
Stafsudd, O. M. (1966): *Appl. Opt.*, vol. 5, p. 1957.
Stearns, E. I. (1950): chap. 7 in M. G. Mellon (ed.), "Analytical Absorption Spectroscopy," Wiley, New York.
Steel, W. H., R. N. Smartt, and R. G. Giovanelli (1961): *Austr. J. Phys.*, vol. 14, p. 201.
Stewart, J. E. (1962): *Appl. Opt.*, vol. 1, p. 75.
Strong, J. (1936): *J. Opt. Soc. Am.*, vol. 26, p. 73.
——— (1961): *J. Opt. Soc. Am.*, vol. 51, p. 853.
Theissing, H., and M. Goebert (1940): *Z. Tech. Phys.*, vol. 21, p. 148.
Thelen, A. (1961): Optimum Heat Reflectors for Silicon Solar Cell Power Converters, Optical Coating Laboratory, Inc., Santa Rosa, Calif.
——— (1963): *J. Opt. Soc. Am.*, vol. 53, p. 1266.
——— (1965): *Appl. Opt.*, vol. 4, p. 977.
——— (1969): p. 47 in G. Hass and R. E. Thun (eds.), "Physics of Thin Films," vol. 5, Academic, New York.
——— (1971): *J. Opt. Soc. Am.*, vol. 61, p. 365.
Thetford, A. (1969): *Opt. Acta,* vol. 16, p. 37.

Tishchenko, E. A. (1971): *Opt. Spectrosc.*, vol. 30, p. 82.
Title, A. (1970): Fabry-Perot Interferometers as Narrow-Band Optical Filters, *Harv. Coll. Obs. Solar Satell. Proj. Rep.* TR-18, June.
Tuckerman, L. B. (1947): *J. Opt. Soc. Am.*, vol. 37, p. 818.
Turbadar, T. (1964*a*): *Opt. Acta*, vol. 11, p. 159.
——— (1964*b*): *Opt. Acta*, vol. 11, p. 195.
Turner, A. F. (1950): *J. Phys. Radium*, vol. 11, p. 444.
——— (1952): *J. Opt. Soc. Am.*, vol. 42, p. 878.
——— (1954): *J. Opt. Soc. Am.*, vol. 44, p. 352.
——— (1962): in H. H. Blau and H. Fischer (eds.), "Radiative Transfer from Solid Materials," Macmillan, New York.
———, Bausch & Lomb (1970): private communication.
——— and P. W. Baumeister (1966): *Appl. Opt.*, vol. 5, p. 69.
———, L. Chang, and T. P. Martin (1965): *Appl. Opt.*, vol. 4, p. 927.
——— and H. R. Hopkinson (1953): *J. Opt. Soc. Am.*, vol. 43, p. 819.
———, C. F. Mooney, and L. Bartle, Scientific Bureau, Bausch & Lomb Optical Company (1953): Infrared Transmission Filters, *ERDL Fort Belvoir Q. Tech. Rep.* 6, July 23, 1953 to Oct. 22, 1953, Contract DA -44-009-eng-1113.
——— and R. Walsh (1961): *J. Opt. Soc. Am.*, vol. 51, p. 484.
Turner, W. J., and W. E. Reese (1962): *Phys. Rev.*, vol. 127, p. 126.
Udod, V. V. (1969): *Opt. Spectrosc.*, vol. 27, p. 379.
Ulrich, R. (1967): *Infrared Phys.*, vol. 7, p. 37.
——— (1968): *Appl. Opt.*, vol. 7, p. 1987.
——— (1969): *Appl. Opt.*, vol. 8, p. 319.
———, K. F. Renk, and L. Genzel (1963): *IEEE Trans. Microwave Theory Tech.*, vol. MTT-11, p. 363.
Vacisek, A. (1960): "Optics of Thin Films," North-Holland, Amsterdam.
Valeyev, A. S. (1967): *Sov. J. Opt. Technol.*, vol. 34, p. 317.
Vande Kleft, L. J., E. A. Murray, W. M. Frey, and P. W. Yunker (1969): Construction and Evaluation of Thin Film Beam Dividers, *U.S. Army Aberdeen Res. Dev. Cent. Ballist. Res. Lab. Tech. Note* 1723, AD698025, October.
Varma, S. P., and K. D. Möller (1969): *Appl. Opt.*, vol. 8, p. 2151.
Verbitskaya, T. N., L. D. Kuslovskii, E. K. Galanov, and L. S. Sokolova (1967): *Zh. Prikl. Spektrosk.*, vol. 7, p. 276.
Veremey, V. V., and T. A. Gorbunova (1968): *Sov. J. Opt. Technol.*, vol. 35, p. 519.
Vereshchagin, V. G., and N. A. Borisevich (1968): *Zh. Prikl. Spektrosk.*, vol. 8, p. 612.
Verleur, H. W., and A. S. Barker, Jr. (1966): *Phys. Rev.*, vol. 149, p. 715.
Vinogradova, T. A. (1970): *Opt. Spektrosk.*, vol. 29, p. 395.
Walker, W. C. (1963): *J. Phys. Chem. Solids*, vol. 24, p. 1667.
Warren, S. W. (1968): *Infrared Phys.*, vol. 8, p. 65.
Watts, J. K. (1968): *Appl. Opt.*, vol. 7, p. 1621.
Weigert, F., H. Stande, and E. Elvegard (1929): *Z. Phys. Chem.*, vol. B2, p. 149.
Weinstein, W. (1954): *Vacuum*, vol. 4, p. 3.
Weyl, W. A. (1959): "Coloured Glasses," Dawsons, London.
White, J. V. (1947): *J. Opt. Soc. Am.*, vol. 37, p. 713.
Williamson, J. H. (1971): *J. Opt. Soc. Am.*, vol. 61, p. 767.
Willisen, F. K., von (1966): *Appl. Opt.*, vol. 5, p. 97.
Wilmot, W., and E. R. Schineller (1966): *J. Opt. Soc. Am.*, vol. 56, p. 549.
Winther, C. (1937): *Z. Elektrochem.*, vol. 43, p. 691.
Wolter, H. (1956): p. 461 in S. Flügge (ed.), "Handbuch der Physik," vol. 24; Springer, Berlin.
Wood, R. W. (1904): *Phil. Mag.*, 6th ser., vol. 7, p. 376.
——— (1940): "Physical Optics," 3d ed., p. 198, Macmillan, New York.
——— and C. Larkens (1938): *Phys. Rev.*, vol. 54, p. 332.
Wyszecki, G. (1960): *J. Opt. Soc. Am.*, vol. 50, p. 992.
——— and W. S. Stiles (1967): "Color Science," Wiley, New York.
Yamada, Y., A. Mitsuishi, and H. Yoshinaga (1962): *J. Opt. Soc. Am.*, vol. 52, p. 17.
Yen, V. L. (1969): *Opt. Spect.*, vol. 3, no. 3, p. 78.
Yoshinaga, H. (1955): *Phys. Rev.*, vol. 100, p. 753.
——— and R. A. Oetjen (1955): *J. Opt. Soc. Am.*, vol. 45, p. 1085.
Young, L. (1963): *Appl. Opt.*, vol. 2, p. 445.
——— (1967): *Appl. Opt.*, vol. 6, p. 297.
Zirin, H. (1966): *Appl. Opt.*, vol. 5, p. 474.
Zwerdling, S., and J. P. Theriault (1968): *Appl. Opt.*, vol. 1, p. 209.

Section 9

Colorimetry

GUNTER WYSZECKI

Division of Physics, National Research Council, Ottawa, Ontario

(Numbers refer to paragraphs)

1. Introduction Contemporary methods and tools used in colorimetry are based essentially on well-established laws of visual color matching and internationally agreed-upon standards. The Commission Internationale de l'Eclairage (CIE), an international organization, makes recommendations regarding colorimetric methods and basic standards through its Colorimetry Committee, whose members are drawn from many different countries.

This section is largely devoted to colorimetric methods and standards presently recommended by the CIE. Other data of importance to colorimetry are also included.

BASIC COLORIMETRIC CONCEPTS

2. The colorimetric concepts listed below have been selected as basic and of most general use. A more comprehensive listing may be found in the "International Lighting Vocabulary" (CIE, 1957).

The concepts are divided in two categories, psychological concepts and psychophysical concepts. The *psychological concepts* of color refer to color perception. The terms which apply to these concepts enable the individual observer to describe his color perception. The *psychophysical concepts* of color refer to the color matching of one photometric field with another and to judgments of similarities and degree of difference between two such fields.

3. Psychological Concepts *Color.* That aspect of visual perception by which an observer can distinguish differences between two fields of view of the same size, shape, and structure, such as may be caused by differences in the spectral composition of the radiation concerned in the observation. In this sense the term color is often preceded by the adjective *perceived* to distinguish it from the same term used in the *psychophysical* sense (see *Color* in Par. 4).

Hue. That attribute of visual perception which has given rise to color names, such as blue, green, yellow, red, purple, and so on.

Saturation. That attribute of a visual perception which permits a judgment to be made of the proportion of pure chromatic color (perceived color possessing a hue) in the total perception.

Chromaticness. That attribute of visual perception which is composed of the attributes hue and saturation.

Chroma. That attribute of a visual perception which permits a judgment to be made of the amount of pure chromatic color (perceived color possessing a hue) present, irrespective of the amount of a chromatic color (perceived color not possessing a hue). For colors of the same brightness and hue, equal intervals of saturation and chroma are identical. In a series of perceived colors of constant saturation, the chroma increases with the brightness.

Brightness (Luminosity). That attribute of a visual perception which permits an area perceived as self-luminous to be classed as equivalent to some member of the series of achromatic colors ranging from very dim to very bright or dazzling. *Luminosity* is another term for the same concept.

Lightness. That attribute of a visual perception which permits an object perceived as non-self-luminous to be classed as equivalent to some member of the series of achromatic object colors ranging for light-diffusing objects from black to white and ranging for regularly transmitting objects from black to perfectly clear and colorless.

4. Psychophysical Concepts *Color.* That characteristic of a visual stimulus by which an observer can distinguish differences between two fields of view of the same size, shape, and structure, such as may be caused by differences in the spectral composition of the stimulus concerned in the observation. In this sense the term color is often referred to as *psychophysical color* to distinguish it from color used in the sense of *perceived color* (see Par. 3). Psychophysical color is specified by the tristimulus values of the radiation entering the eye.

Color Stimulus. Radiant flux of given spectral composition entering the eye and producing a sensation of color.

Primary Colors. The colors of three reference lights by whose additive mixture nearly all other colors can be produced. These colors are often chosen to be red, green, and blue. In accordance with the laws of additive color mixture, nonreal primaries can be defined which have the useful property that any real color can be represented by an additive mixture of positive amounts of the primaries (linear combination with positive coefficients).

Tristimulus Values. The amounts of the three primary colors (matching stimuli) required to give by additive mixture a match with the color considered.

Color-matching Functions. The tristimulus values, with respect to three given primary colors, of monochromatic stimuli of equal radiant flux, regarded as a function of wavelength.

Chromaticity Coordinates. The ratios of each tristimulus value of the color to their sum. The *chromaticity* of a color is the color quality of a stimulus definable by its chromaticity coordinates. A diagram in which any one of the three chromaticity coordinates is plotted against any other is called a *chromaticity diagram.* In this diagram the chromaticity of a color plots as a *chromaticity* point.

Dominant Wavelength. The wavelength of the monochromatic color which when additively mixed in suitable proportions with a specified achromatic color yields a match with the color considered.

Complementary Wavelength. The wavelength of the monochromatic color which when additively mixed in suitable proportions with the color considered yields a match with a specified achromatic color. Every color has either a complementary wavelength or a dominant wavelength. Some, but not all, colors have both.

Excitation Purity. The ratio of two lengths on a chromaticity diagram. The first length is the distance between the chromaticity point of a specified achromatic color and that of the color considered; the second length is the distance along the same direction and in the same sense from the first point to the boundary of real colors in the chromaticity diagram.

Metameric Colors. Color stimuli of identical tristimulus values but different spectral distributions of radiant flux.

THE CIE COLORIMETRIC SYSTEM

5. In colorimetry, the main objective is to obtain color specifications in terms of tristimulus values valid for observers with normal color vision. For this, the color-matching properties of an ideal observer are defined by specifying two sets of three independent functions of wavelength that are identified with the ideal observer's color-matching functions for two different ranges of the angular subtense of the color-matching field. By definition, the ideal observer makes color matches in strict accord with the laws of additive color mixture, and his matches are subject to the principle of persistence of color matches; i.e., a color match is unaffected by his previous exposure to white or colored light or by varying light adaptations of the eye making the match.

The CIE colorimetric system also includes specifications for standards of illuminants, sources, and reflectance, as well as methods of calculating the tristimulus values and chromaticity coordinates of given color-stimulus functions. These and related items are given in Pars. 6 to 15.

Detailed accounts of the CIE colorimetric system and its development may be found in the literature (Optical Society of America, 1953; Judd and Wyszecki, 1963; Wright, 1965; Wyszecki and Stiles, 1967; CIE, 1970).

6. CIE 1931 Standard Colorimetric Observer The color-matching properties of the CIE 1931 standard colorimetric observer are defined by spectral tristimulus values (color-matching functions) $\bar{x}(\lambda)$, $\bar{y}(\lambda)$, $\bar{z}(\lambda)$, whose values at 5-nm intervals are tabulated in Table 1 and illustrated in Fig. 1. These functions are to be used whenever a correlation is desired with visual color matching of fields of angular subtense between 1 and 4° at the eye of the observer.

The spectral tristimulus values $\bar{x}(\lambda)$, $\bar{y}(\lambda)$, $\bar{z}(\lambda)$ refer to a set of reference stimuli (primary colors) **X, Y, Z**, which are nonreal and are chosen for reasons of convenience in colorimetric computations.

The $\bar{y}(\lambda)$ function is identical with the $V(\lambda)$ function, the photopic luminous efficiency function defining the standard observer for photometry (see Sec. 2).

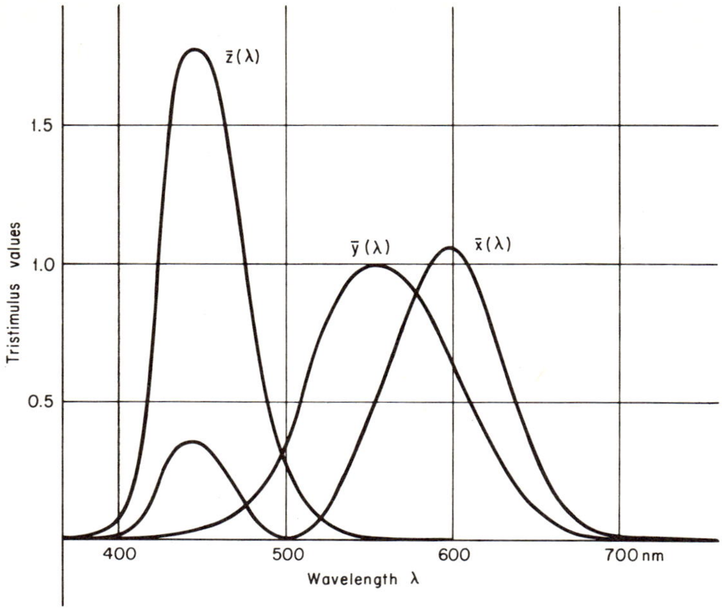

Fig. 1 Spectral tristimulus values (color-matching functions) $\bar{x}(\lambda)$, $\bar{y}(\lambda)$, $\bar{z}(\lambda)$ of CIE 1931 standard colorimetric observer.

The chromaticity coordinates $x(\lambda)$, $y(\lambda)$, $z(\lambda)$ of the monochromatic stimuli are derived from $\bar{x}(\lambda)$, $\bar{y}(\lambda)$, $\bar{z}(\lambda)$ by forming the ratios

$$x(\lambda) = \frac{\bar{x}(\lambda)}{\bar{x}(\lambda) + \bar{y}(\lambda) + \bar{z}(\lambda)}$$

$$y(\lambda) = \frac{\bar{y}(\lambda)}{\bar{x}(\lambda) + \bar{y}(\lambda) + \bar{z}(\lambda)}$$

$$z(\lambda) = \frac{\bar{z}(\lambda)}{\bar{x}(\lambda) + \bar{y}(\lambda) + \bar{z}(\lambda)}$$

The values of $x(\lambda)$, $y(\lambda)$, and $z(\lambda)$ are also given in Table 1; the resulting *spectrum locus* in the CIE 1931 chromaticity diagram is illustrated in Fig. 2.

Table 1 contains the recommended *abridged* spectral tristimulus values and chromaticity coordinates. An unabridged table for the wavelength range 360 to 830 nm at wavelength intervals of 1 nm has been published (CIE, 1970; Wyszecki and Stiles, 1967); however, in most colorimetric computations it should suffice to use the rounded-off values at 5-nm intervals from 380 to 780 nm given in Table 1. In some colorimetric computations, it might be sufficient to use the spectral tristimulus values at 10-nm intervals. If this is desired, all values of Table 1 at wavelengths 385, 395, . . . , 775 nm are to be ignored.

7. CIE 1964 Supplementary Standard Colorimetric Observer The color-matching properties of the CIE 1964 supplementary standard colorimetric observer are defined by spectral tristimulus values (color-matching functions) $\bar{x}_{10}(\lambda)$, $\bar{y}_{10}(\lambda)$, and $\bar{z}_{10}(\lambda)$ whose values at 5-nm intervals are tabulated in Table 2 and illustrated in Fig. 3. These functions are to be used whenever a correlation is desired with visual color matching of fields of angular subtense more than 4° at the eye of the observer.

The spectral tristimulus values $\bar{x}_{10}(\lambda)$, $\bar{y}_{10}(\lambda)$, and $\bar{z}_{10}(\lambda)$ refer to a set of reference stimuli (primary colors) $\mathbf{X}_{10}$, $\mathbf{Y}_{10}$, $\mathbf{Z}_{10}$ which are nonreal and, as for the CIE 1931 standard observer (see Par. 6), chosen for reasons of convenience in colorimetric computations.

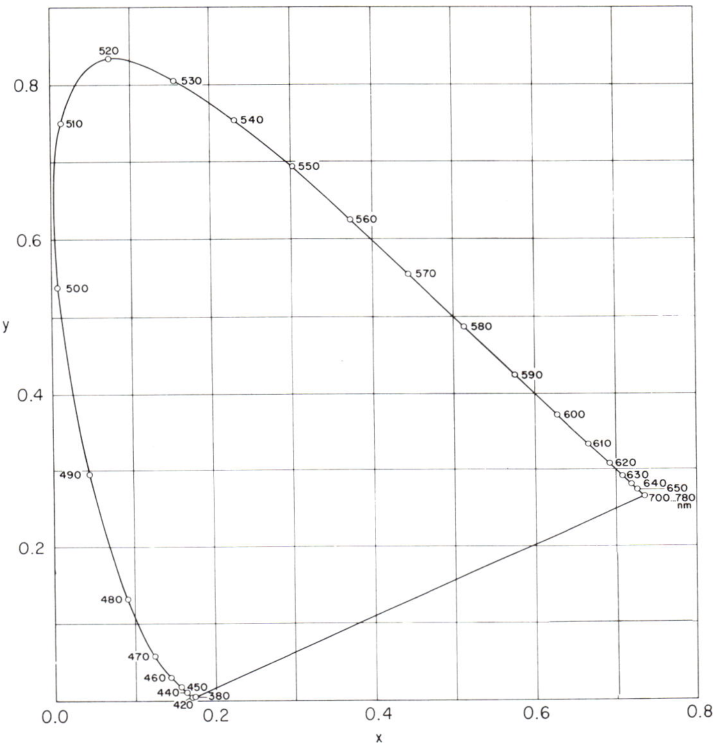

Fig. 2 CIE 1931 (x,y) chromaticity diagram showing spectrum locus (chromaticity points of monochromatic stimuli of $\lambda = 380$ to 780 nm) and purple line (line connecting the ends of the spectrum locus).

The $\bar{y}_{10}(\lambda)$ function is *not* to be identified with a photopic luminous efficiency function for large-field observations.

The chromaticity coordinates $x_{10}(\lambda)$, $y_{10}(\lambda)$, $z_{10}(\lambda)$ of the monochromatic stimuli are derived from $\bar{x}_{10}(\lambda)$, $\bar{y}_{10}(\lambda)$, $\bar{z}_{10}(\lambda)$ by forming the ratios

$$x_{10}(\lambda) = \frac{\bar{x}_{10}(\lambda)}{\bar{x}_{10}(\lambda) + \bar{y}_{10}(\lambda) + \bar{z}_{10}(\lambda)}$$

$$y_{10}(\lambda) = \frac{\bar{y}_{10}(\lambda)}{\bar{x}_{10}(\lambda) + \bar{y}_{10}(\lambda) + \bar{z}_{10}(\lambda)}$$

$$z_{10}(\lambda) = \frac{\bar{z}_{10}(\lambda)}{\bar{x}_{10}(\lambda) + \bar{y}_{10}(\lambda) + \bar{z}_{10}(\lambda)}$$

The values of $x_{10}(\lambda)$, $y_{10}(\lambda)$, $z_{10}(\lambda)$ are also given in Table 2, and the resulting spectrum locus in the CIE 1964 (x_{10},y_{10}) chromaticity diagram is illustrated in Fig. 4.

Table 2 contains the recommended *abridged* spectral tristimulus values and chromaticity coordinates. An unabridged table for the wavelength range 360 to 830 nm at wavelength intervals of 1 nm has been published (CIE, 1970; Wyszecki and Stiles, 1967); however, in most colorimetric computations it should suffice to use the rounded-off values at 5-nm intervals from 380 to 780 nm given in Table 2. In some colorimetric computations, it might be sufficient to use the spectral tristimulus values at 10-nm intervals. If so, all values of Table 2 at wavelengths 385, 395, . . . , 775 nm are to be ignored.

TABLE 1 Spectral Tristimulus Values Defining CIE 1931 Standard Colorimetric Observer and Chromaticity Coordinates Derived from Them

λ, nm	Spectral tristimulus values			Chromaticity coordinates		
	$\bar{x}(\lambda)$	$\bar{y}(\lambda)$	$\bar{z}(\lambda)$	$x(\lambda)$	$y(\lambda)$	$z(\lambda)$
380	0.0014	0.0000	0.0065	0.1741	0.0050	0.8209
385	0.0022	0.0001	0.0105	0.1740	0.0050	0.8210
390	0.0042	0.0001	0.0201	0.1738	0.0049	0.8213
395	0.0076	0.0002	0.0362	0.1736	0.0049	0.8215
400	0.0143	0.0004	0.0679	0.1733	0.0048	0.8219
405	0.0232	0.0006	0.1102	0.1730	0.0048	0.8222
410	0.0435	0.0012	0.2074	0.1726	0.0048	0.8226
415	0.0776	0.0022	0.3713	0.1721	0.0048	0.8231
420	0.1344	0.0040	0.6456	0.1714	0.0051	0.8235
425	0.2148	0.0073	1.0391	0.1703	0.0058	0.8239
430	0.2839	0.0116	1.3856	0.1689	0.0069	0.8242
435	0.3285	0.0168	1.6230	0.1669	0.0086	0.8245
440	0.3483	0.0230	1.7471	0.1644	0.0109	0.8247
445	0.3481	0.0298	1.7826	0.1611	0.0138	0.8251
450	0.3362	0.0380	1.7721	0.1566	0.0177	0.8257
455	0.3187	0.0480	1.7441	0.1510	0.0227	0.8263
460	0.2908	0.0600	1.6692	0.1440	0.0297	0.8263
465	0.2511	0.0739	1.5281	0.1355	0.0399	0.8246
470	0.1954	0.0910	1.2876	0.1241	0.0578	0.8181
475	0.1421	0.1126	1.0419	0.1096	0.0868	0.8036
480	0.0956	0.1390	0.8130	0.0913	0.1327	0.7760
485	0.0580	0.1693	0.6162	0.0687	0.2007	0.7306
490	0.0320	0.2080	0.4652	0.0454	0.2950	0.6596
495	0.0147	0.2586	0.3533	0.0235	0.4127	0.5638
500	0.0049	0.3230	0.2720	0.0082	0.5384	0.4534
505	0.0024	0.4073	0.2123	0.0039	0.6548	0.3413
510	0.0093	0.5030	0.1582	0.0139	0.7502	0.2359
515	0.0291	0.6082	0.1117	0.0389	0.8120	0.1491
520	0.0633	0.7100	0.0782	0.0743	0.8338	0.0919
525	0.1096	0.7932	0.0573	0.1142	0.8262	0.0596
530	0.1655	0.8620	0.0422	0.1547	0.8059	0.0394
535	0.2257	0.9149	0.0298	0.1929	0.7816	0.0255
540	0.2904	0.9540	0.0203	0.2296	0.7543	0.0161
545	0.3597	0.9803	0.0134	0.2658	0.7243	0.0099
550	0.4334	0.9950	0.0087	0.3016	0.6923	0.0061
555	0.5121	1.0000	0.0057	0.3373	0.6589	0.0038
560	0.5945	0.9950	0.0039	0.3731	0.6245	0.0024
565	0.6784	0.9786	0.0027	0.4087	0.5896	0.0017
570	0.7621	0.9520	0.0021	0.4441	0.5547	0.0012
575	0.8425	0.9154	0.0018	0.4788	0.5202	0.0010
580	0.9163	0.8700	0.0017	0.5125	0.4866	0.0009
585	0.9786	0.8163	0.0014	0.5448	0.4544	0.0008
590	1.0263	0.7570	0.0011	0.5752	0.4242	0.0006
595	1.0567	0.6949	0.0010	0.6029	0.3965	0.0006
600	1.0622	0.6310	0.0008	0.6270	0.3725	0.0005
605	1.0456	0.5668	0.0006	0.6482	0.3514	0.0004
610	1.0026	0.5030	0.0003	0.6658	0.3340	0.0002
615	0.9384	0.4412	0.0002	0.6801	0.3197	0.0002
620	0.8544	0.3810	0.0002	0.6915	0.3083	0.0002
625	0.7514	0.3210	0.0001	0.7006	0.2993	0.0001
630	0.6424	0.2650	0.0000	0.7079	0.2920	0.0001
635	0.5419	0.2170	0.0000	0.7140	0.2859	0.0001
640	0.4479	0.1750	0.0000	0.7190	0.2809	0.0001
645	0.3608	0.1382	0.0000	0.7230	0.2770	0.0000

TABLE 1 Spectral Tristimulus Values Defining CIE 1931 Standard Colorimetric Observer and Chromaticity Coordinates Derived from Them (Continued)

λ, nm	Spectral tristimulus values			Chromaticity coordinates		
	$\bar{x}(\lambda)$	$\bar{y}(\lambda)$	$\bar{z}(\lambda)$	$x(\lambda)$	$y(\lambda)$	$z(\lambda)$
650	0.2835	0.1070	0.0000	0.7260	0.2740	0.0000
655	0.2187	0.0816	0.0000	0.7283	0.2717	0.0000
660	0.1649	0.0610	0.0000	0.7300	0.2700	0.0000
665	0.1212	0.0446	0.0000	0.7311	0.2689	0.0000
670	0.0874	0.0320	0.0000	0.7320	0.2680	0.0000
675	0.0636	0.0232	0.0000	0.7327	0.2673	0.0000
680	0.0468	0.0170	0.0000	0.7334	0.2666	0.0000
685	0.0329	0.0119	0.0000	0.7340	0.2660	0.0000
690	0.0227	0.0082	0.0000	0.7344	0.2656	0.0000
695	0.0158	0.0057	0.0000	0.7346	0.2654	0.0000
700	0.0114	0.0041	0.0000	0.7347	0.2653	0.0000
705	0.0081	0.0029	0.0000	0.7347	0.2653	0.0000
710	0.0058	0.0021	0.0000	0.7347	0.2653	0.0000
715	0.0041	0.0015	0.0000	0.7347	0.2653	0.0000
720	0.0029	0.0010	0.0000	0.7347	0.2653	0.0000
725	0.0020	0.0007	0.0000	0.7347	0.2653	0.0000
730	0.0014	0.0005	0.0000	0.7347	0.2653	0.0000
735	0.0010	0.0004	0.0000	0.7347	0.2653	0.0000
740	0.0007	0.0002	0.0000	0.7347	0.2653	0.0000
745	0.0005	0.0002	0.0000	0.7347	0.2653	0.0000
750	0.0003	0.0001	0.0000	0.7347	0.2653	0.0000
755	0.0002	0.0001	0.0000	0.7347	0.2653	0.0000
760	0.0002	0.0001	0.0000	0.7347	0.2653	0.0000
765	0.0001	0.0000	0.0000	0.7347	0.2653	0.0000
770	0.0001	0.0000	0.0000	0.7347	0.2653	0.0000
775	0.0001	0.0000	0.0000	0.7347	0.2653	0.0000
780	0.0000	0.0000	0.0000	0.7347	0.2653	0.0000

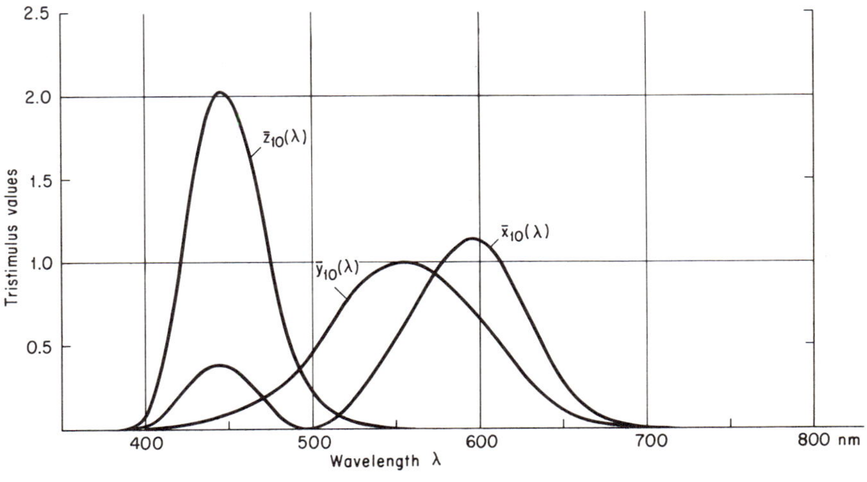

Fig. 3 Spectral tristimulus values (color-matching functions) $\bar{x}_{10}(\lambda)$, $\bar{y}_{10}(\lambda)$, and $\bar{z}_{10}(\lambda)$ of the CIE 1964 supplementary standard colorimetric observer.

TABLE 2 Spectral Tristimulus Values Defining CIE 1964 Supplementary Standard Colorimetric Observer and Chromaticity Coordinates Derived from Them

λ, nm	Spectral tristimulus values			Chromaticity coordinates		
	$\bar{x}_{10}(\lambda)$	$\bar{y}_{10}(\lambda)$	$\bar{z}_{10}(\lambda)$	$x_{10}(\lambda)$	$y_{10}(\lambda)$	$z_{10}(\lambda)$
380	0.0002	0.0000	0.0007	0.1813	0.0197	0.7990
385	0.0007	0.0001	0.0029	0.1809	0.0195	0.7996
390	0.0024	0.0003	0.0105	0.1803	0.0194	0.8003
395	0.0072	0.0008	0.0323	0.1795	0.0190	0.8015
400	0.0191	0.0020	0.0860	0.1784	0.0187	0.8029
405	0.0434	0.0045	0.1971	0.1771	0.0184	0.8045
410	0.0847	0.0088	0.3894	0.1755	0.0181	0.8064
415	0.1406	0.0145	0.6568	0.1732	0.0178	0.8090
420	0.2045	0.0214	0.9725	0.1706	0.0179	0.8115
425	0.2647	0.0295	1.2825	0.1679	0.0187	0.8134
430	0.3147	0.0387	1.5535	0.1650	0.0203	0.8147
435	0.3577	0.0496	1.7985	0.1622	0.0225	0.8153
440	0.3837	0.0621	1.9673	0.1590	0.0257	0.8153
445	0.3867	0.0747	2.0273	0.1554	0.0300	0.8146
450	0.3707	0.0895	1.9948	0.1510	0.0364	0.8126
455	0.3430	0.1063	1.9007	0.1459	0.0452	0.8088
460	0.3023	0.1282	1.7454	0.1389	0.0589	0.8022
465	0.2541	0.1528	1.5549	0.1295	0.0779	0.7926
470	0.1956	0.1852	1.3176	0.1152	0.1090	0.7758
475	0.1323	0.2199	1.0302	0.0957	0.1591	0.7452
480	0.0805	0.2536	0.7721	0.0728	0.2292	0.6980
485	0.0411	0.2977	0.5701	0.0452	0.3275	0.6273
490	0.0162	0.3391	0.4153	0.0210	0.4401	0.5389
495	0.0051	0.3954	0.3024	0.0073	0.5625	0.4302
500	0.0038	0.4608	0.2185	0.0056	0.6745	0.3199
505	0.0154	0.5314	0.1592	0.0219	0.7526	0.2256
510	0.0375	0.6067	0.1120	0.0495	0.8023	0.1482
515	0.0714	0.6857	0.0822	0.0850	0.8170	0.0980
520	0.1177	0.7618	0.0607	0.1252	0.8102	0.0646
525	0.1730	0.8233	0.0431	0.1664	0.7922	0.0414
530	0.2365	0.8752	0.0305	0.2071	0.7663	0.0267
535	0.3042	0.9238	0.0206	0.2436	0.7399	0.0165
540	0.3768	0.9620	0.0137	0.2786	0.7113	0.0101
545	0.4516	0.9822	0.0079	0.3132	0.6813	0.0055
550	0.5298	0.9918	0.0040	0.3473	0.6501	0.0026
555	0.6161	0.9991	0.0011	0.3812	0.6182	0.0007
560	0.7052	0.9973	0.0000	0.4142	0.5858	0.0000
565	0.7938	0.9824	0.0000	0.4469	0.5531	0.0000
570	0.8787	0.9556	0.0000	0.4790	0.5210	0.0000
575	0.9512	0.9152	0.0000	0.5096	0.4904	0.0000
580	1.0142	0.8689	0.0000	0.5386	0.4614	0.0000
585	1.0743	0.8256	0.0000	0.5654	0.4346	0.0000
590	1.1185	0.7774	0.0000	0.5900	0.4100	0.0000
595	1.1343	0.7204	0.0000	0.6116	0.3884	0.0000
600	1.1240	0.6583	0.0000	0.6306	0.3694	0.0000
605	1.0891	0.5939	0.0000	0.6471	0.3529	0.0000
610	1.0305	0.5280	0.0000	0.6612	0.3388	0.0000
615	0.9507	0.4618	0.0000	0.6731	0.3269	0.0000
620	0.8563	0.3981	0.0000	0.6827	0.3173	0.0000
625	0.7549	0.3396	0.0000	0.6898	0.3102	0.0000
630	0.6475	0.2835	0.0000	0.6955	0.3045	0.0000
635	0.5351	0.2283	0.0000	0.7010	0.2990	0.0000
640	0.4316	0.1798	0.0000	0.7059	0.2941	0.0000
645	0.3437	0.1402	0.0000	0.7103	0.2898	0.0000

TABLE 2 Spectral Tristimulus Values Defining CIE 1964 Supplementary Standard Colorimetric Observer and Chromaticity Coordinates Derived from Them (Continued)

λ, nm	Spectral tristimulus values			Chromaticity coordinates		
	$\bar{x}_{10}(\lambda)$	$\bar{y}_{10}(\lambda)$	$\bar{z}_{10}(\lambda)$	$x_{10}(\lambda)$	$y_{10}(\lambda)$	$z_{10}(\lambda)$
650	0.2683	0.1076	0.0000	0.7137	0.2863	0.0000
655	0.2043	0.0812	0.0000	0.7156	0.2844	0.0000
660	0.1526	0.0603	0.0000	0.7168	0.2832	0.0000
665	0.1122	0.0441	0.0000	0.7179	0.2821	0.0000
670	0.0813	0.0318	0.0000	0.7187	0.2813	0.0000
675	0.0579	0.0226	0.0000	0.7193	0.2807	0.0000
680	0.0409	0.0159	0.0000	0.7198	0.2802	0.0000
685	0.0286	0.0111	0.0000	0.7200	0.2800	0.0000
690	0.0199	0.0077	0.0000	0.7202	0.2798	0.0000
695	0.0138	0.0054	0.0000	0.7203	0.2797	0.0000
700	0.0096	0.0037	0.0000	0.7204	0.2796	0.0000
705	0.0066	0.0026	0.0000	0.7203	0.2797	0.0000
710	0.0046	0.0018	0.0000	0.7202	0.2798	0.0000
715	0.0031	0.0012	0.0000	0.7201	0.2799	0.0000
720	0.0022	0.0008	0.0000	0.7199	0.2801	0.0000
725	0.0015	0.0006	0.0000	0.7197	0.2803	0.0000
730	0.0010	0.0004	0.0000	0.7195	0.2806	0.0000
735	0.0007	0.0003	0.0000	0.7192	0.2808	0.0000
740	0.0005	0.0002	0.0000	0.7189	0.2811	0.0000
745	0.0004	0.0001	0.0000	0.7186	0.2814	0.0000
750	0.0003	0.0001	0.0000	0.7183	0.2817	0.0000
755	0.0002	0.0001	0.0000	0.7180	0.2820	0.0000
760	0.0001	0.0000	0.0000	0.7176	0.2824	0.0000
765	0.0001	0.0000	0.0000	0.7172	0.2828	0.0000
770	0.0001	0.0000	0.0000	0.7169	0.2831	0.0000
775	0.0000	0.0000	0.0000	0.7165	0.2835	0.0000
780	0.0000	0.0000	0.0000	0.7161	0.2839	0.0000

8. Standard Illuminants An *illuminant* is defined by its spectral irradiance distribution, which may or may not be provided directly by an actual *source* in the sense of a physical emitter of light, e.g., a lamp or the sun and sky. However, the illuminants recommended by the CIE as standard illuminants for colorimetry can be represented by actual sources to close approximations (see Par. 9). The CIE standard illuminants are the following.

Illuminant A represents light from the full radiator (blackbody) at absolute temperature 2855.6 K (approximately) according to the International Practical Temperature Scale, 1968.

Table 3 gives the relative spectral irradiance distribution of illuminant A from 300 to 830 nm at 5-nm intervals. This distribution, illustrated in Fig. 5, has been calculated in accordance with Planck's radiation law, which gives the spectral concentration of radiant emittance of a full radiator as a function of wavelength and temperature:

$$M_{e,\lambda}(\lambda,T) = c_1\lambda^{-5}(e^{c_2/\lambda T} - 1)^{-1} \qquad \mathrm{W/m^3}$$

where the radiation constants are taken as

$$c_1 = 3.74150 \times 10^{-16}\ \mathrm{W \cdot m^2}$$
$$c_2 = 1.4388 \times 10^{-2}\ \mathrm{m \cdot K}$$

and the temperature T is set equal to

$$T = \frac{1.4388}{1.4350}\,2848\ \mathrm{K} \approx 2856\ \mathrm{K}$$

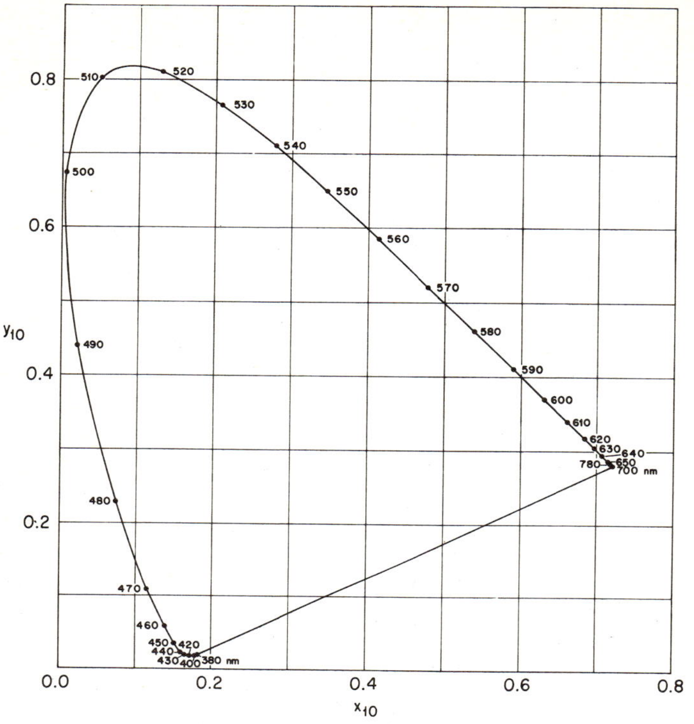

Fig. 4 CIE 1964 (x_{10},y_{10}) chromaticity diagram showing spectrum locus (chromaticity points of monochromatic stimuli of λ = 380 to 780 nm) and purple line (line connecting the ends of the spectrum locus).

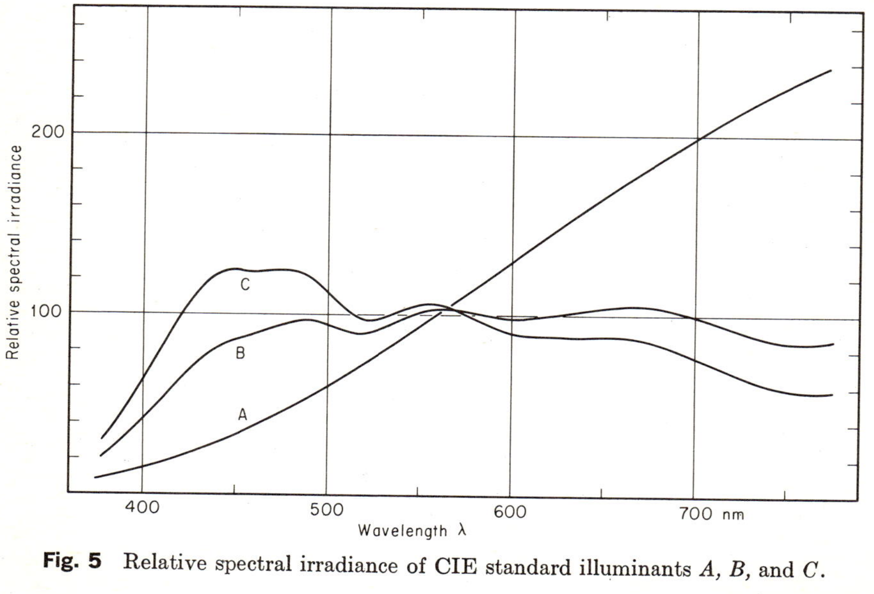

Fig. 5 Relative spectral irradiance of CIE standard illuminants *A*, *B*, and *C*.

TABLE 3 Relative Spectral Irradiance Distributions of CIE Standard Illuminants A, B, C, and D_{65}

λ, nm	(A) $S(\lambda)$	(B) $S(\lambda)$	(C) $S(\lambda)$	(D_{65}) $S(\lambda)$
300	0.93	......	......	0.03
305	1.13	......	......	1.7
310	1.36	......	......	3.3
315	1.62	......	......	11.8
320	1.93	0.02	0.01	20.2
325	2.27	0.26	0.20	28.6
330	2.66	0.50	0.40	37.1
335	3.10	1.45	1.55	38.5
340	3.59	2.40	2.70	39.9
345	4.14	4.00	4.85	42.4
350	4.74	5.60	7.00	44.9
355	5.41	7.60	9.95	45.8
360	6.14	9.60	12.90	46.6
365	6.95	12.40	17.20	49.4
370	7.82	15.20	21.40	52.1
375	8.77	18.80	27.50	51.0
380	9.80	22.40	33.00	50.0
385	10.90	26.85	39.92	52.3
390	12.09	31.30	47.40	54.6
395	13.35	36.18	55.17	68.7
400	14.71	41.30	63.30	82.8
405	16.15	46.62	71.81	87.1
410	17.68	52.10	80.60	91.5
415	19.29	57.70	89.53	92.5
420	20.99	63.20	98.10	93.4
425	22.79	68.37	105.80	90.1
430	24.67	73.10	112.40	86.7
435	26.64	77.31	117.75	95.8
440	28.70	80.80	121.50	104.9
445	30.85	83.44	123.45	110.9
450	33.09	85.40	124.00	117.0
455	35.41	86.88	123.60	117.4
460	37.81	88.30	123.10	117.8
465	40.30	90.08	123.30	116.3
470	42.87	92.00	123.80	114.9
475	45.52	93.75	124.09	115.4
480	48.24	95.20	123.90	115.9
485	51.04	96.23	122.92	112.4
490	53.91	96.50	120.70	108.8
495	56.85	95.71	116.90	109.1
500	59.86	94.20	112.10	109.4
505	62.93	92.37	106.98	108.6
510	66.06	90.70	102.30	107.8
515	69.25	89.65	98.81	106.3
520	72.50	89.50	96.90	104.8
525	75.79	90.43	96.78	106.2
530	79.13	92.20	98.00	107.7
535	82.52	94.46	99.94	106.0
540	85.95	96.90	102.10	104.4
545	89.41	99.16	103.95	104.2
550	92.91	101.00	105.20	104.0
555	96.44	102.20	105.67	102.0
560	100.00	102.80	105.30	100.0
565	103.58	102.92	104.11	98.2
570	107.18	102.60	102.30	96.3

TABLE 3 Relative Spectral Irradiance Distributions of CIE Standard Illuminants A, B, C, and D_{65} (Continued)

λ, nm	(A) $S(\lambda)$	(B) $S(\lambda)$	(C) $S(\lambda)$	(D_{65}) $S(\lambda)$
575	110.80	101.90	100.15	96.1
580	114.44	101.00	97.80	95.8
585	118.08	100.07	95.43	92.2
590	121.73	99.20	93.20	88.7
595	125.39	98.44	91.22	89.3
600	129.04	98.00	89.70	90.0
605	132.70	98.08	88.83	89.8
610	136.35	98.50	88.40	89.6
615	139.99	99.06	88.19	88.6
620	143.62	99.70	88.10	87.7
625	147.24	100.36	88.06	85.5
630	150.84	101.00	88.00	83.3
635	154.42	101.56	87.86	83.5
640	157.98	102.20	87.80	83.7
645	161.52	103.05	87.99	81.9
650	165.03	103.90	88.20	80.0
655	168.51	104.59	88.20	80.1
660	171.96	105.00	87.90	80.2
665	175.38	105.08	87.22	81.2
670	178.77	104.90	86.30	82.3
675	182.12	104.55	85.30	80.3
680	185.43	103.90	84.00	78.3
685	188.70	102.84	82.21	74.0
690	191.93	101.60	80.20	69.7
695	195.12	100.38	78.24	70.7
700	198.26	99.10	76.30	71.6
705	201.36	97.70	74.36	73.0
710	204.41	96.20	72.40	74.3
715	207.41	94.60	70.40	68.0
720	210.36	92.90	68.30	61.6
725	213.27	91.10	66.30	65.7
730	216.12	89.40	64.40	69.9
735	218.92	88.00	62.80	72.5
740	221.67	86.90	61.50	75.1
745	224.36	85.90	60.20	69.3
750	227.00	85.20	59.20	63.6
755	229.59	84.80	58.50	55.0
760	232.12	84.70	58.10	46.4
765	234.59	84.90	58.00	56.6
770	237.01	85.40	58.20	66.8
775	239.37	.....	.....	65.1
780	241.68	.....	.....	63.4
785	243.92	.....	.....	63.8
790	246.12	.....	.....	64.3
795	248.25	.....	.....	61.9
800	250.33	.....	.....	59.5
805	252.35	.....	.....	55.7
810	254.31	.....	.....	52.0
815	256.22	.....	.....	54.7
820	258.07	.....	.....	57.4
825	259.86	.....	.....	58.9
830	261.60	.....	.....	60.3

The choice of this particular temperature assures a spectral irradiance distribution identical to the one adopted for illuminant A in 1931, when c_2 was taken as equal to 1.4350×10^{-2} m·K and T was set equal to 2848 K, in accordance with the International Practical Temperature Scale, 1927. The new value of c_2, quoted above, is chosen in accordance with the International Practical Temperature Scale, 1968.

Table 3 gives values of $S(\lambda)$, the relative spectral irradiance distribution of illuminant A, computed as $100M_{e,\lambda}/M_{e,560}$ and rounded to two decimals. $S(\lambda) = 100.00$ for $\lambda = 560 \times 10^{-9}$ m = 560 nm.

Illuminant B represents direct sunlight with a correlated color temperature of approximately 4874 K. Its relative spectral irradiance distribution is given in Table 3 and illustrated in Fig. 5.

Illuminant C represents daylight with a correlated color temperature of approximately 6774 K. Its relative spectral irradiance distribution is given in Table 3 and illustrated in Fig. 5.

Illuminant D_{65} (originally called D_{6500}) represents a phase of daylight with a correlated color temperature of approximately 6504 K. Its relative spectral irradiance distribution is given in Table 3 and illustrated in Fig. 6. Illuminant D_{65}

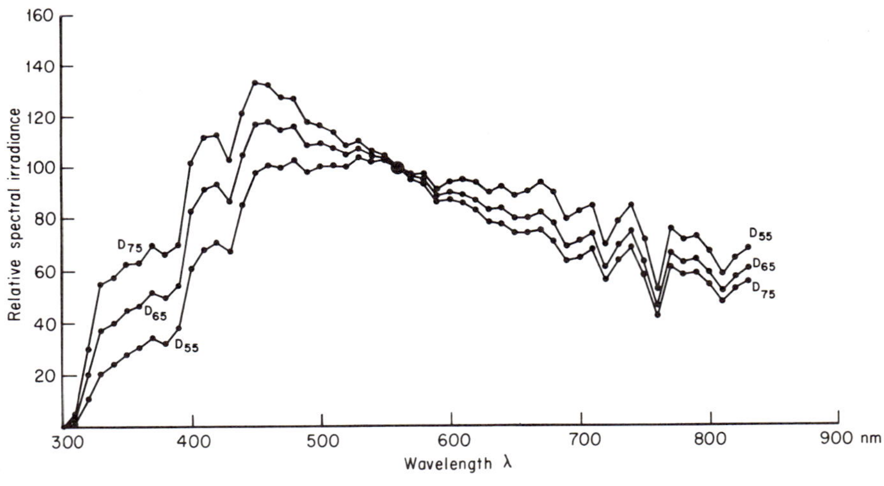

Fig. 6 Relative spectral irradiance of CIE daylight illuminants D_{55}, D_{65}, and D_{75}.

supplements the illuminants A, B, and C. For general use in colorimetry illuminants A and D_{65} should suffice. Illuminants B and C do not adequately represent common phases of daylight, as originally intended. It is anticipated that at some future date, illuminants B and C will be eliminated from the list of CIE recommended standard illuminants for colorimetry.

Other Illuminants D. The CIE recommends a method of calculating the relative spectral irradiance distributions of other phases of daylight whenever it becomes desirable to choose one different from D_{65} specified in Table 3. The requirement is the correlated color temperature T_c of the phase of daylight whose relative spectral irradiance distribution is to be calculated. The method is specified in two steps.

Chromaticity. The 1931 (x,y) chromaticity coordinates of the daylight D to be calculated must satisfy the relation

$$y_D = -3.000x_D^2 + 2.870x_D - 0.275$$

x_D being within the range 0.250 to 0.380. The correlated color temperature T_c ($c_2 = 1.4388 \times 10^{-2}$ m · K) of daylight D is related to x_D by two formulas. For correlated color temperatures from approximately 4000 to 7000 K,

$$x_D = -4.6070 \frac{10^9}{T_c^3} + 2.9678 \frac{10^6}{T_c^2} + 0.09911 \frac{10^3}{T_c} + 0.244063$$

For correlated color temperatures from 7000 to approximately 25,000 K,

$$x_D = -2.0064 \frac{10^9}{T_c^3} + 1.9018 \frac{10^3}{T_c^2} + 0.24748 \frac{10^3}{T_c} + 0.237040$$

Relative Spectral Irradiance Distribution. The relative spectral irradiance distribution $S(\lambda)$ of daylight D is to be calculated from

$$S(\lambda) = S_0(\lambda) + M_1 S_1(\lambda) + M_2 S_2(\lambda)$$

where $S_0(\lambda)$, $S_1(\lambda)$, and $S_2(\lambda)$ are functions of wavelength λ, given in Table 4 and illustrated in Fig. 7, and M_1 and M_2 are factors whose values are related to the

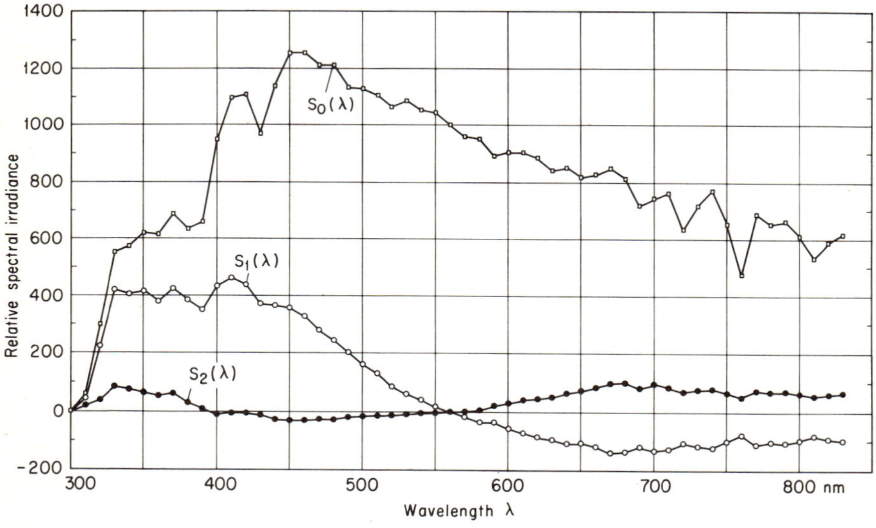

Fig. 7 Relative spectral irradiance of the daylight components $S_0(\lambda)$, $S_1(\lambda)$, and $S_2(\lambda)$.

chromaticity coordinates x_D, y_D, as follows:

$$M_1 = \frac{-1.3515 - 1.7703x_D + 5.9114y_D}{0.0241 + 0.2562x_D - 0.7341y_D}$$

$$M_2 = \frac{0.0300 - 31.4424x_D + 30.0717y_D}{0.0241 + 0.2562x_D - 0.7341y_D}$$

To facilitate the practical use of this method of calculating the relative spectral irradiance distribution of D illuminants for correlated color temperatures in the range 4000 to 25,000 K Table 5 has been prepared. However, it is recommended that in the interests of standardization, D_{65} ($T_c \approx 6504$ K) be used whenever possible. When D_{65} cannot be used, the CIE recommends that either standard illuminant D_{55} ($T_c \approx 5503$ K) or standard illuminant D_{75} ($T_c \approx 7504$ K) be used whenever possible.

TABLE 4 Spectral Irradiance Distributions of Standard Daylight Components

λ, nm	$S_0(\lambda)$	$S_1(\lambda)$	$S_2(\lambda)$	λ, nm	$S_0(\lambda)$	$S_1(\lambda)$	$S_2(\lambda)$
300	0.04	0.02	0.0	600	90.5	−5.8	3.2
310	6.0	4.5	2.0	610	90.3	−7.2	4.1
320	29.6	22.4	4.0	620	88.4	−8.6	4.7
330	55.3	42.0	8.5	630	84.0	−9.5	5.1
340	57.3	40.6	7.8	640	85.1	−10.9	6.7
350	61.8	41.6	6.7	650	81.9	−10.7	7.3
360	61.5	38.0	5.3	660	82.6	−12.0	8.6
370	68.8	42.4	6.1	670	84.9	−14.0	9.8
380	63.4	38.5	3.0	680	81.3	−13.6	10.2
390	65.8	35.0	1.2	690	71.9	−12.0	8.3
400	94.8	43.4	−1.1	700	74.3	−13.3	9.6
410	104.8	46.3	−0.5	710	76.4	−12.9	8.5
420	105.9	43.9	−0.7	720	63.3	−10.6	7.0
430	96.8	37.1	−1.2	730	71.7	−11.6	7.6
440	113.9	36.7	−2.6	740	77.0	−12.2	8.0
450	125.6	35.9	−2.9	750	65.2	−10.2	6.7
460	125.5	32.6	−2.8	760	47.7	−7.8	5.2
470	121.3	27.9	−2.6	770	68.6	−11.2	7.4
480	121.3	24.3	−2.6	780	65.0	−10.4	6.8
490	113.5	20.1	−1.8	790	66.0	−10.6	7.0
500	113.1	16.2	−1.5	800	61.0	−9.7	6.4
510	110.8	13.2	−1.3	810	53.3	−8.3	5.5
520	106.5	8.6	−1.2	820	58.9	−9.3	6.1
530	108.8	6.1	−1.0	830	61.9	−9.8	6.5
540	105.3	4.2	−0.5				
550	104.4	1.9	−0.3				
560	100.0	0.0	0.0				
570	96.0	−1.6	0.2				
580	95.1	−3.5	0.5				
590	89.1	−3.5	2.1				

These two preferred alternatives are specified in Table 6 in terms of their relative spectral irradiance distributions and are illustrated in Fig. 6.

The CIE recommendations for daylight illuminants D are based on numerous experimental observations over the range 330 to 700 nm and on extrapolation in the ranges 300 to 330 and 700 to 830 nm (Judd et al., 1964). The extrapolated values are believed to be accurate enough for normal colorimetric purposes but should not be used for other purposes if high accuracy in these regions is necessary.

The correlated color temperature of an illuminant is defined by the temperature corresponding to the point on the planckian locus which is nearest to the point representing the chromaticity of the illuminant in the CIE 1960 UCS diagram (see Par. 18) when based on the data of the CIE 1931 standard colorimetric observer.

The correlated color temperatures given for illuminants B, C, and D_{65} have been determined by commonly used graphical and numerical methods (Kelly, 1963; Wyszecki and Stiles, 1967; Robertson, 1968). Figure 8 shows a portion of the CIE 1931 (x,y) chromaticity diagram with lines of constant correlated color temperature (isotemperature lines) from 0 to 660 micro reciprocal degrees (mireds). Correlated color temperature T_c is related to mireds by $T_c = 10^6 \text{ mired}^{-1}$.

The correlated color temperatures are affected by the numerical value of the radiation constant c_2. In accordance with the International Practical Temperature Scale, 1948, amended 1960, the value of c_2 was equal to 1.4380×10^{-2} m·K. With this value the correlated color temperatures of illuminants B, C, and D_{65} are approximately equal to 4871, 6770, and 6500 K, respectively. The change of c_2

TABLE 5 CIE 1931 (x,y) Chromaticity Coordinates, Corresponding CIE 1960 (u,v) Chromaticity Coordinates, and Factors M_1, M_2 Used in the Calculation of the Relative Spectral Irradiance Distribution of Daylight Illuminants D for Correlated Color Temperatures T_c in the Range 4000 to 25,000 K

T_c†	x_D	y_D	u_D	v_D	M_1	M_2
4,000	0.3823	0.3838	0.2236	0.3366	−1.505	2.827
4,100	0.3779	0.3812	0.2217	0.3354	−1.464	2.460
4,200	0.3737	0.3786	0.2200	0.3343	−1.422	2.127
4,300	0.3697	0.3760	0.2183	0.3331	−1.378	1.825
4,400	0.3658	0.3734	0.2168	0.3320	−1.333	1.550
4,500	0.3621	0.3709	0.2153	0.3308	−1.286	1.302
4,600	0.3585	0.3684	0.2139	0.3297	−1.238	1.076
4,700	0.3551	0.3659	0.2126	0.3286	−1.190	0.871
4,800	0.3519	0.3634	0.2114	0.3275	−1.140	0.686
4,900	0.3487	0.3610	0.2102	0.3265	−1.090	0.518
5,000	0.3457	0.3587	0.2091	0.3254	−1.040	0.367
5,100	0.3429	0.3564	0.2081	0.3244	−0.989	0.230
5,200	0.3401	0.3541	0.2071	0.3234	−0.939	0.106
5,300	0.3375	0.3519	0.2062	0.3225	−0.888	−0.005
5,400	0.3349	0.3497	0.2053	0.3215	−0.837	−0.105
5,500	0.3325	0.3476	0.2044	0.3206	−0.786	−0.195
5,600	0.3302	0.3455	0.2036	0.3196	−0.736	−0.276
5,700	0.3279	0.3435	0.2028	0.3187	−0.685	−0.348
5,800	0.3258	0.3416	0.2021	0.3179	−0.635	−0.412
5,900	0.3237	0.3397	0.2014	0.3170	−0.586	−0.469
6,000	0.3217	0.3378	0.2007	0.3162	−0.536	−0.519
6,100	0.3198	0.3360	0.2001	0.3154	−0.487	−0.563
6,200	0.3179	0.3342	0.1995	0.3146	−0.439	−0.602
6,300	0.3161	0.3325	0.1989	0.3138	−0.391	−0.635
6,400	0.3144	0.3308	0.1983	0.3130	−0.343	−0.664
6,500	0.3128	0.3292	0.1978	0.3123	−0.296	−0.688
6,600	0.3112	0.3276	0.1973	0.3116	−0.250	−0.709
6,700	0.3097	0.3260	0.1968	0.3109	−0.204	−0.726
6,800	0.3082	0.3245	0.1963	0.3102	−0.159	−0.739
6,900	0.3067	0.3231	0.1959	0.3095	−0.114	−0.749
7,000	0.3054	0.3216	0.1955	0.3088	−0.070	−0.757
7,100	0.3040	0.3202	0.1950	0.3082	−0.026	−0.762
7,200	0.3027	0.3189	0.1946	0.3076	0.017	−0.765
7,300	0.3015	0.3176	0.1943	0.3069	0.060	−0.765
7,400	0.3003	0.3163	0.1939	0.3063	0.102	−0.763
7,500	0.2991	0.3150	0.1935	0.3057	0.144	−0.760
7,600	0.2980	0.3138	0.1932	0.3052	0.184	−0.755
7,700	0.2969	0.3126	0.1928	0.3046	0.225	−0.748
7,800	0.2958	0.3115	0.1925	0.3041	0.264	−0.740
7,900	0.2948	0.3103	0.1922	0.3035	0.303	−0.730
8,000	0.2938	0.3092	0.1919	0.3030	0.342	−0.720
8,100	0.2928	0.3081	0.1916	0.3025	0.380	−0.708
8,200	0.2919	0.3071	0.1913	0.3020	0.417	−0.695
8,300	0.2910	0.3061	0.1911	0.3015	0.454	−0.682
8,400	0.2901	0.3051	0.1908	0.3010	0.490	−0.667
8,500	0.2892	0.3041	0.1906	0.3006	0.526	−0.652
9,000	0.2853	0.2996	0.1894	0.2984	0.697	−0.566
9,500	0.2818	0.2956	0.1884	0.2964	0.856	−0.471
10,000	0.2788	0.2920	0.1876	0.2946	1.003	−0.369
10,500	0.2761	0.2887	0.1868	0.2930	1.139	−0.265
11,000	0.2737	0.2858	0.1861	0.2915	1.266	−0.160
12,000	0.2697	0.2808	0.1850	0.2890	1.495	0.045
13,000	0.2664	0.2767	0.1841	0.2868	1.693	0.239
14,000	0.2637	0.2732	0.1834	0.2850	1.868	0.419
15,000	0.2614	0.2702	0.1828	0.2835	2.021	0.586
17,000	0.2578	0.2655	0.1818	0.2809	2.278	0.878
20,000	0.2539	0.2603	0.1809	0.2781	2.571	1.231
25,000	0.2499	0.2548	0.1798	0.2751	2.907	1.655
5,503‡	0.3324	0.3475	0.2044	0.3205	−0.785	−0.198
6,504§	0.3127	0.3291	0.1978	0.3123	−0.295	−0.689
7,504¶	0.2990	0.3150	0.1935	0.3057	0.145	−0.760

$u_D = 4x_D/(-2x_D + 12y_D + 3) \quad v_D = 6y_D/(-2x_D + 12y_D + 3)$

† All correlated color temperatures T_c are based on $c_2 = 1.4388 \times 10^{-2}$ m·K.
‡ Standard illuminant D_{55}; $T_c = 5{,}500(1.4388/1.4380) \approx 5503$ K.
§ Standard illuminant D_{65}; $T_c = 6{,}500(1.4388/1.4380) \approx 6504$ K.
¶ Standard illuminant D_{75}; $T_c = 7{,}500(1.4388/1.4380) \approx 7504$ K.

TABLE 6 Relative Spectral Irradiance Distributions of CIE Standard Illuminants D_{55} and D_{75}

λ, nm	(D_{55}) S(λ)	(D_{75}) S(λ)	λ, nm	(D_{55}) S(λ)	(D_{75}) S(λ)
300	0.02	0.04	600	94.4	87.2
310	2.1	5.1	610	95.1	86.1
320	11.2	29.8	620	94.2	83.6
330	20.6	54.9	630	90.4	78.7
340	23.9	57.3	640	92.3	78.4
350	27.8	62.7	650	88.9	74.8
360	30.6	63.0	660	90.3	74.3
370	34.3	70.3	670	93.9	75.4
380	32.4	66.7	680	90.0	71.6
390	38.1	70.0	690	79.7	63.9
400	61.0	101.9	700	82.8	65.1
410	68.6	111.9	710	84.8	68.1
420	71.6	112.8	720	70.2	56.4
430	67.9	103.1	730	79.3	64.2
440	85.6	121.2	740	85.0	69.2
450	98.0	133.0	750	71.9	58.6
460	100.5	132.4	760	52.8	42.6
470	99.9	127.3	770	75.9	61.4
480	102.7	126.8	780	71.8	58.3
490	98.1	117.8	790	72.9	59.1
500	100.7	116.6	800	67.3	54.7
510	100.7	113.7	810	58.7	47.9
520	100.0	108.7	820	65.0	52.9
530	104.2	110.4	830	68.3	55.5
540	102.1	106.3			
550	103.0	104.9			
560	100.0	100.0			
570	97.2	95.6			
580	97.7	94.2			
590	91.4	87.0			

to the value of 1.4388×10^{-2} m·K (International Practical Temperature Scale, 1968) increases the correlated color temperatures of illuminants B, C, and D_{65} each by the factor 1.4383/1.4380. Thus the correlated color temperatures of these illuminants increase by approximately 3, 4, and 4 K, respectively.

9. Natural Sources of Importance in Colorimetry In recent years, numerous spectroradiometric measurements have been made of natural daylight (Henderson and Hodgkiss, 1963; Condit and Grum, 1964; Das and Sastri, 1965; Winch et al., 1966; Sastri and Das, 1966; Nayatani et al., 1967; Tarrant, 1968), all of which are aimed at providing basic data for colorimetry. There is a remarkable agreement between the data obtained by the various investigators, particularly in the visible region of the spectrum. In the ultraviolet region some significant variations have been noted; in the main these are interpreted as seasonal variations and differences in altitude of measurement location and sun altitude.

The most representative set of data on natural daylight is considered to be that on which the CIE recommendations on illuminants D are based (see Par. 8 and Judd et al., 1964).

The spectral irradiance distributions of D_{55} and D_{75} (Table 6) are representative of measurements taken at every 10 nm and averaged over the wavelength ranges from −5 to +5 nm from the nominal values of $\lambda = 300, 310, \ldots, 830$ nm. The spectral irradiance distribution of D_{65} (Table 3) is to be interpreted similarly, but the relative irradiances at $\lambda = 305, 315, \ldots, 825$ nm have been interpolated linearly.

A 10-nm spectral bandwidth as adopted in these measurements has a smoothing

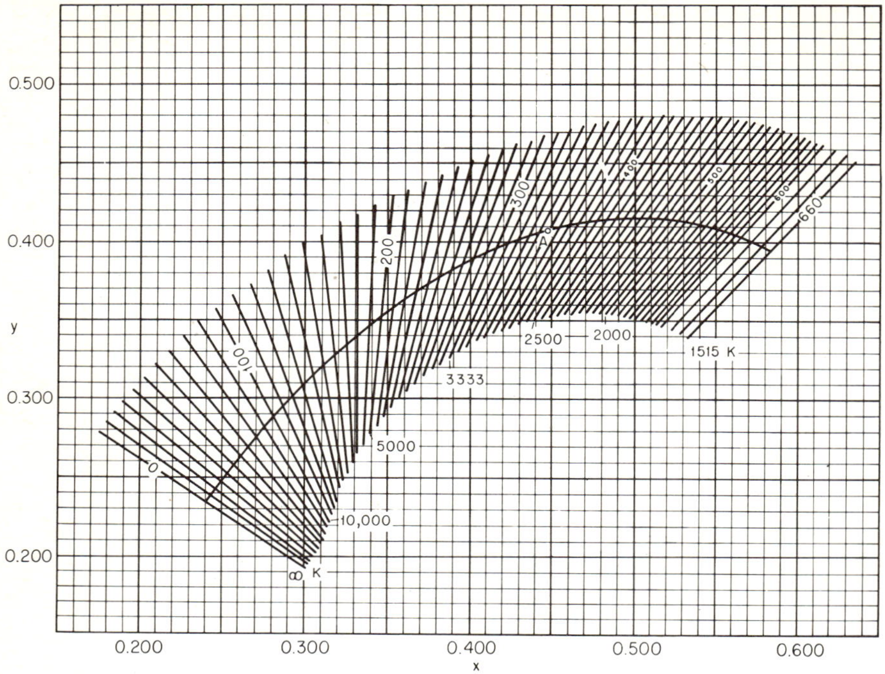

Fig. 8 CIE 1931 (x,y) chromaticity diagram showing isotemperature lines for the graphical determination of correlated color temperatures of illuminants. The curved line is the locus of chromaticities of planckian radiators of different temperatures. [*From Kelly* (1963).]

effect on the spectral distribution curves. When the monochromator slits are opened to give spectral bandwidths ranging between approximately 0.25 nm in the ultraviolet and 2.0 nm in the red, an extremely fine structure of the daylight spectrum is revealed, as shown in Fig. 9. However, for the colorimetrist the fine structure shown here is of academic interest rather than practical importance.

10. Artificial Sources of Importance in Colorimetry The CIE recommends three standard sources representative of CIE standard illuminants *A*, *B*, and *C*, respectively. These sources are defined as follows.

Source A. Illuminant *A* (see Table 3) is to be realized by a gas-filled coiled-tungsten-filament lamp operating at a correlated color temperature of 2856 K ($c_2 = 1.4388 \times 10^{-2}$ m·K). A lamp with a fused-quartz envelope or window is recommended if the spectral irradiance distribution of illuminant *A* is to be realized more accurately in the ultraviolet.

Source B. Illuminant *B* (see Table 3) is to be realized by source *A* combined with a filter consisting of a 1-cm-thick layer of each of two solutions B_1 and B_2, contained in a double cell made of colorless optical glass. The solutions are to be made up as follows:

Solution B_1		Solution B_2	
Copper sulfate, $CuSO_4 \cdot 5H_2O$	2.452 g	Cobalt ammonium sulfate, $CoSO_4 \cdot (NH_4)_2SO_4 \cdot 6H_2O$	21.71 g
Mannite, $C_6H_8(OH)_6$	2.452 g	Copper sulfate, $CuSO_4 \cdot 5H_2O$	16.11 g
Pyridine, C_5H_5N	30.0 ml	Sulfuric acid, density 1.835 g/ml	10.0 ml
Distilled water to make	1,000.0 ml	Distilled water to make	1,000.0 ml

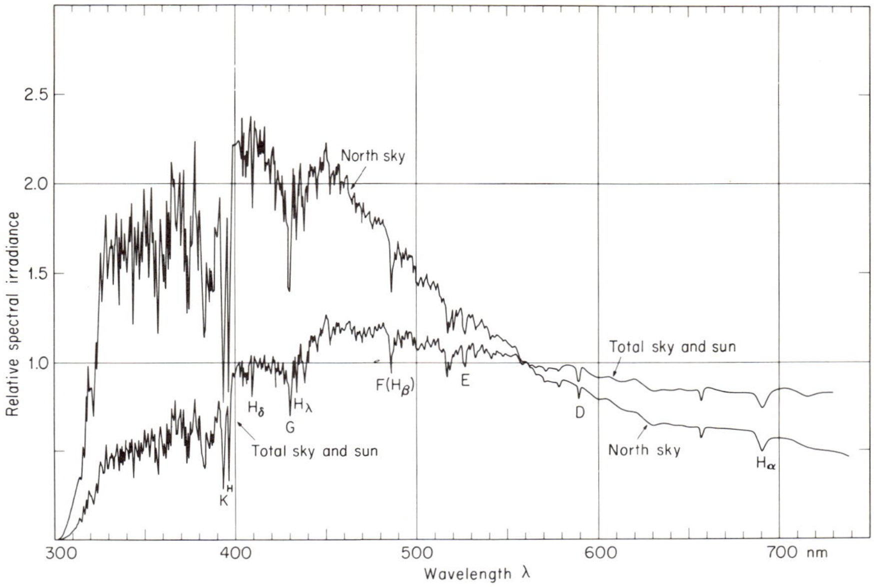

Fig. 9 Relative spectral irradiance of two phases of daylight measured with monochromator slits opened to give spectral bandwidth ranging from approximately 0.25 nm in the ultraviolet to 2.0 nm in the red. Some of the major Fraunhofer lines are indicated by their conventional symbols. [*Budde, private communication* (1963).]

Source C. Illuminant *C* (see Table 3) is to be realized by source *A* combined with a filter consisting of a 1-cm-thick layer of each of two solutions C_1 and C_2, contained in a double cell made of colorless optical glass. The solutions are to be made up as follows:

Solution C_1		Solution C_2	
Copper sulfate, $CuSO_4 \cdot 5H_2O$	3.412 g	Cobalt ammonium sulfate, $CoSO_4 \cdot (NH_4)_2SO_4 \cdot 6H_2O$	30.58 g
Mannite, $C_6H_8(OH)_6$	3.412 g	Copper sulfate, $CuSO_4 \cdot 5H_2O$	22.52 g
Pyridine, C_5H_5N	30.0 ml	Sulfuric acid, density 1.835 g/ml	10.0 ml
Distilled water to make	1,000.0 ml	Distilled water to make	1,000.0 ml

At present, no artificial source is recommended by the CIE to realize illuminant D_{65} (see Table 3) or any other illuminant *D* of different correlated color temperature (see Tables 5 and 6). It is hoped that new developments in light sources and filters will soon offer a sufficient basis for a CIE recommendation.

In colorimetric practice, several types of sources are frequently used to simulate daylight with a fair degree of approximation. Some examples are illustrated in Figs. 10 to 12, where the relative spectral distributions of irradiance are compared with that of the corresponding CIE standard illuminant D_{65}.

In Table 7 a number of common natural and artificial sources are listed together with their respective nominal luminances, CIE 1931 (x,y) chromaticity coordinates, and correlated color temperatures.

11. Color-rendering Index of Selected Sources In 1965 the CIE recommended a method for measuring and specifying the color-rendering properties of light sources (CIE, 1965). The CIE recommendation aims at rating sources in terms of a color-rendering index *R*. This index is limited to considerations of the difference

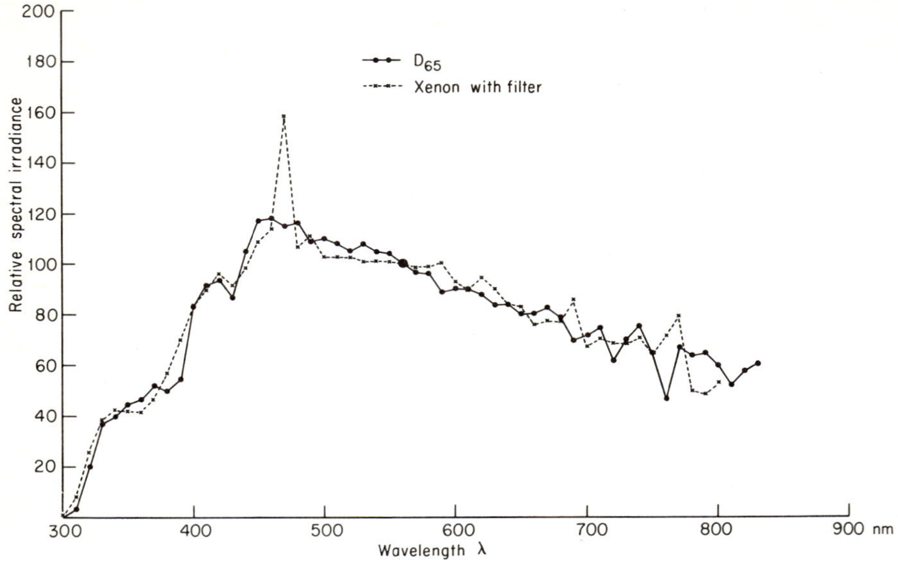

Fig. 10 Relative spectral irradiance of a high-pressure xenon arc filtered to approximate CIE standard illuminant D_{65}.

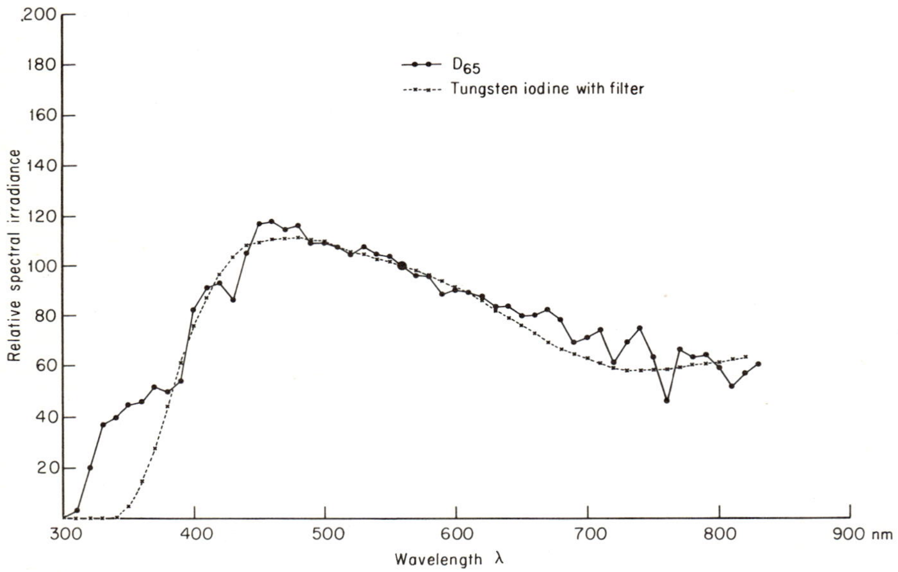

Fig. 11 Relative spectral irradiance of a tungsten-iodine lamp with filter to approximate CIE standard illuminant D_{65}.

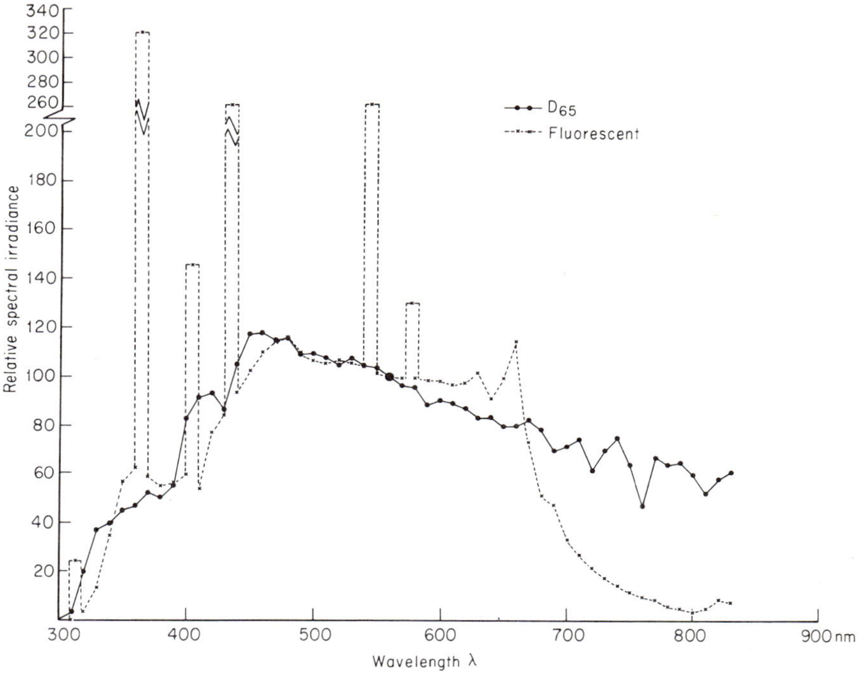

Fig. 12 Relative spectral irradiance of a fluorescent lamp to approximate CIE standard illuminant D_{65}.

between the color of a test object under a test source and the color of the same test object under a reference source whose chromaticity is identical to that of the test source. Computing a *general index* R_a requires a set of eight test objects (paint samples of known spectral reflectance characteristics) whose chromaticities cover the hue circuit and whose luminous reflectances are approximately equal. A special index R_i can be computed for each test object of the basic set of eight ($i = 1$ to 8) and for each of six additional test objects ($i = 9$ to 14) of more saturated colors and different luminous reflectances. The indexes are based on a comparison of the color differences computed in accordance with the CIE 1964 color-difference formula (see Par. 19). If small color differences are found, a high color-rendering index is obtained for the test source; "perfect" color-rendering properties are indicated when the index is 100. "Very poor" color-rendering properties are indicated when the index is 50 or below.

Fluorescent lamps, commonly used for general lighting purposes, have color-rendering indexes ranging from as low as 53 (usually "warm white" fluorescent lamps) to as high as 85 ("deluxe"). For critical visual color-matching purposes, special fluorescent lamps, filtered high-pressure xenon-arc lamps, and filtered tungsten-halogen lamps are commercially available from various manufacturers. These special lamps have color-rendering indexes ranging from 90 to 98.

12. Standards of Reflectance The CIE recommends the use of the perfect reflecting diffuser as the reference standard of reflectance. It is defined as the ideal uniform diffuser with a reflectance equal to unity (CIE, 1970). This recommendation has been in force since Jan. 1, 1969. Before that date, smoked magnesium oxide, MgO, was adopted as the reference standard. The spectral reflectance of MgO in terms of the ideal uniform diffuser varies with wavelength and is affected by many factors, including impurities in the sample, its thickness, and age.

In recent years, pressed disks of MgO powder or barium sulfate, $BaSO_4$, powder

TABLE 7 Luminance, CIE 1931 (x,y) Chromaticity Coordinates, and Correlated Color Temperature of Selected Natural and Artificial Sources (from Wyszecki and Stiles, 1967)

Source	Luminance, cd/cm²	Chromaticity coordinates x	Chromaticity coordinates y	Correlated color temperature, K
Sun's disk:[a]				
Measured above atmosphere (m = 0)	200,000	0.318	0.330	6200
Measured near sea level for air masses:[b]				
m = 1	150,000	0.331	0.344	5600
m = 2	125,000	0.343	0.357	5100
m = 3	100,000	0.356	0.369	4700
m = 4	80,000	0.368	0.379	4400
m = 5	65,000	0.380	0.388	4100
Sky:[c]				
Clear blue	0.06–0.4	0.262	0.270	15000
		0.247	0.251	30000
Partly cloudy	0.1–0.4	0.294	0.309	8000
		0.279	0.291	10000
Overcast	0.2–0.5	0.313	0.329	6500
Blackbody:[d]				
Gold point	0.1073	0.607	0.381	1336
Platinum point	60.0	0.522	0.414	2042
Rhodium point	170.1	0.502	0.415	2233
Iridium point	1,253.0	0.458	0.410	2716
Tungsten filament:[e]				
Vacuum lamp (10 lm/W)	200	0.477	0.414	2400–2600
Gas-filled lamp (20 lm/W)	1,200	0.444	0.406	2700–3100
750-W projection lamp (25 lm/W)	2,400	0.430	0.402	3000–3200
1,200-W projection lamp (30 lm/W)	3,300	0.417	0.396	3200–3400
60-W inside frosted bulb (14 lm/W)	12	0.452	0.409	2800
60-W "white" bulb (14 lm/W)	3	0.452	0.409	2800
CIE source A (500 W)	800	0.448	0.407	2854
CIE source B (500 W)	200	0.349	0.352	4870
CIE source C (500 W)	125	0.310	0.316	6770
Carbon arc	15,000–100,000	.....	.....	2900–6500
Mercury arc (H6) (1,000 W, 65 lm/W)	30,000			
Fluorescent lamps (T-12, 40 W, 0.430 A):				
Standard warm white	0.7	0.437	0.405	3000
White	0.7	0.409	0.398	3500
Standard cool white	0.7	0.372	0.374	4200
Daylight	0.6	0.311	0.338	6500
Xenon (high-pressure, XBO 1,600-W dc)	65,000	.....	.....	6000
Zirconium (300-W)	4,500	0.423	0.399	3200

[a] Chromaticity coordinates from Moon (1940).

[b] The air masses m = 1, 2, . . . , 5 are equivalent to the secants of the following angles between the zenith and the sun: 0° (m = 1), 60° (m = 2), 70.5° (m = 3), 75.5° (m = 4), 78.5° (m = 5).

[c] For clear blue and partly cloudy skies the range of minimum luminance is given (Paix 1963). Chromaticity coordinates and correlated color temperature are estimated from results reported by Judd et al. (1964).

[d] From Pivovonski and Nagel (1961).

[e] From Illuminating Engineering Society (1959).

have been used successfully as secondary standards of reflectance. Suitable powders are available commercially† and can be calibrated in terms of absolute reflectance by most major national standardizing laboratories. Special powder-pressing devices can also be purchased.‡

† For example, from Carl Zeiss Company, Oberkochen, Württemberg, Germany, and Eastman Kodak Co., Distillation Products Ind., Rochester, N.Y. 14650.

‡ For example, from Carl Zeiss Company, of Germany, or from Bausch & Lomb Company, Rochester, N.Y.

In some cases more rugged working standards are being used in the form of glazed ceramic tiles. These can also be calibrated in terms of absolute reflectance.

13. Illuminating and Viewing Conditions The CIE recommends that the colorimetric specification of opaque specimens be so given that it corresponds to one of the following illuminating and viewing conditions (CIE, 1970).

45°/Normal (Abbreviation, 45/0). The specimen is illuminated by one or more beams whose axes are at an angle of $45 \pm 5°$ from the normal to the specimen surface. The angle between the direction of viewing and the normal to the specimen should not exceed 10°. The angle between the axis and any ray of an illuminating beam should not exceed 5°. The same restriction should be observed in the viewing beam.

Normal/45° (Abbreviation, 0/45). The specimen is illuminated by a beam whose axis is at an angle not exceeding 10° from the normal to the specimen. The specimen is viewed at an angle of $45 \pm 5°$ from the normal. The angle between the axis and any ray of the illuminating beam should not exceed 5°. The same restriction should be observed in the viewing beam.

*Diffuse/Normal (Abbreviation, d/*0). The specimen is illuminated diffusely by an integrating sphere. The angle between the normal to the specimen and the axis of the viewing beam should not exceed 10°. The integrating sphere may be of any diameter, provided that the total area of the ports does not exceed 10 percent of the internal reflecting area of the sphere. The angle between the axis and any ray of the viewing beam should not exceed 5°.

Normal/Diffuse (Abbreviation, 0/*d*). The specimen is illuminated by a beam whose axis is at an angle not exceeding 10° from the normal to the specimen. The reflected flux is collected by means of an integrating sphere. The angle between the axis and any ray of the illuminating beam should not exceed 5°. The integrating sphere may be of any diameter, provided that the total area of the ports does not exceed 10 percent of the internal reflecting area of the sphere.

The following points should be noted:

1. For the conditions diffuse/normal and normal/diffuse the influence of the specular component of specimens with mixed reflection may be reduced by using a gloss trap. If a gloss trap is used, details of its size, shape, and position should be given.
2. In the normal/45° and the normal/diffuse conditions specimens with mixed reflection should not be measured with strictly normal illumination.
3. The 45°/normal condition gives the radiance factor $\beta_{45/0}$. The normal/45° condition gives the radiance factor $\beta_{0/45}$. The diffuse/normal condition gives the radiance factor $\beta_{d/0}$. The normal/diffuse condition gives the reflectance ρ.
4. It is important that the particular illuminating and viewing conditions used be specified even if they are within the range of one of the recommended standard conditions.

14. Calculation of Tristimulus Values and Chromaticity Coordinates The CIE tristimulus values of a color stimulus can be obtained by multiplying the color-stimulus function $\varphi(\lambda)$ by the CIE spectral tristimulus values and integrating these products over the whole spectrum:

$$X = k \int_\lambda \varphi(\lambda)\bar{x}(\lambda)\,d\lambda \qquad X_{10} = k_{10} \int_\lambda \varphi(\lambda)\bar{x}_{10}(\lambda)\,d\lambda$$

$$Y = k \int_\lambda \varphi(\lambda)\bar{y}(\lambda)\,d\lambda \qquad Y_{10} = k_{10} \int_\lambda \varphi(\lambda)\bar{y}_{10}(\lambda)\,d\lambda$$

$$Z = k \int_\lambda \varphi(\lambda)\bar{z}(\lambda)\,d\lambda \qquad Z_{10} = k_{10} \int_\lambda \varphi(\lambda)\bar{z}_{10}(\lambda)\,d\lambda$$

The symbols X, Y, Z refer to tristimulus values in the CIE 1931 standard colorimetric system, and the symbols X_{10}, Y_{10}, Z_{10} refer to tristimulus values in the CIE 1964 supplementary standard colorimetric system. Similarly, $\bar{x}(\lambda)$, $\bar{y}(\lambda)$, $(\bar{z}\lambda)$ are the spectral tristimulus values defining the CIE 1931 standard colorimetric observer, and $\bar{x}_{10}(\lambda)$, $\bar{y}_{10}(\lambda)$, $\bar{z}_{10}(\lambda)$ are the spectral tristimulus values defining the CIE 1964 supplementary standard colorimetric observer. The constants k and k_{10} are normalizing factors specified below.

For object colors, the color-stimulus function $\varphi(\lambda)$ is the product of the spectral reflectance $\rho(\lambda)$, or the spectral radiance factor $\beta(\lambda)$, or the spectral transmittance $\tau(\lambda)$ of the object and the relative spectral irradiance distribution $S(\lambda)$ of the illuminant irradiating the object; thus either

$$\varphi(\lambda) = \rho(\lambda)S(\lambda) \qquad \text{or} \qquad \varphi(\lambda) = \beta(\lambda)S(\lambda) \qquad \text{or} \qquad \varphi(\lambda) = \tau(\lambda)S(\lambda)$$

For illuminants the color-stimulus function $\varphi(\lambda)$ is simply the relative spectral irradiance distribution of the illuminant; thus $\varphi(\lambda) = S(\lambda)$.

It is recommended that for calculating the tristimulus values of object colors, the illuminant be one of the standard illuminants whenever possible (see Par. 8).

The constants k and k_{10} are defined as

$$k = \frac{100}{\int_\lambda S(\lambda)\bar{y}(\lambda)\,d\lambda} \qquad k_{10} = \frac{100}{\int_\lambda S(\lambda)\bar{y}_{10}(\lambda)\,d\lambda}$$

With k so defined, Y becomes the luminous reflectance of the object when $\varphi(\lambda) = \rho(\lambda)S(\lambda)$, the luminance factor when $\varphi(\lambda) = \beta(\lambda)S(\lambda)$, and the luminous transmittance when $\varphi(\lambda) = \tau(\lambda)S(\lambda)$, in each case expressed as a percentage. The Y value of the illuminant $[\varphi(\lambda) = S(\lambda)]$ becomes 100. The calculated value of Y_{10} has no significance with regard to standard photometric quantities. To calculate the luminous reflectance (or luminance factor or luminous transmittance) the function $\bar{y}(\lambda)$, which is identical to $V(\lambda)$, must be used.

In all practical calculations of tristimulus values, the integration is approximated by a summation. Thus, for example,

$$X = k\sum_\lambda \varphi(\lambda)\bar{x}(\lambda)\,\Delta\lambda \qquad Y = k\sum_\lambda \varphi(\lambda)\bar{y}(\lambda)\,\Delta\lambda \qquad Z = k\sum_\lambda \varphi(\lambda)\bar{z}(\lambda)\,\Delta\lambda$$

with $k = 100\Big/\Big[\sum_\lambda S(\lambda)\bar{y}(\lambda)\,\Delta\lambda\Big]$.

The wavelength interval $\Delta\lambda$ is either 1, 5, or 10 nm. The spectral tristimulus values $\bar{x}(\lambda)$, $\bar{y}(\lambda)$, $\bar{z}(\lambda)$ for 5- and 10-nm intervals are those defined in Table 1. Tables for 1-nm intervals can be found elsewhere (Wyszecki and Stiles, 1967; CIE, 1970). Similar formulas are used for the calculation of X_{10}, Y_{10}, Z_{10}, in which case the spectral tristimulus values $\bar{x}_{10}(\lambda)$, $\bar{y}_{10}(\lambda)$, $\bar{z}_{10}(\lambda)$ of Table 2 are used for 5- and 10-nm intervals. Tables for 1-nm intervals can be found elsewhere (Wyszecki and Stiles, 1967; CIE, 1970).

The chromaticity coordinates (x,y,z) are derived from the tristimulus values (X,Y,Z) as follows:

$$x = \frac{X}{X+Y+Z} \qquad y = \frac{Y}{X+Y+Z} \qquad z = \frac{Z}{X+Y+Z}$$

Because $x + y + z = 1$, it suffices to use x and y.

The chromaticity coordinates x_{10}, y_{10}, z_{10} are computed similarly from the tristimulus values X_{10}, Y_{10}, Z_{10}.

Figures 2 and 4 show the appropriate diagrams in which chromaticities are commonly plotted.

15. Excitation and Colorimetric Purity The CIE recommends that *excitation purity* of a given color stimulus be represented by the symbol p_e and be calculated by the formula (CIE, 1970)

$$p_e = \frac{y - y_w}{y_d - y_w} \qquad \text{or} \qquad p_e = \frac{x - x_w}{x_d - x_w}$$

whichever has the greater arithmetic value in the numerator.

The CIE further recommends that *colorimetric purity* for a given stimulus be represented by the symbol p_c and be calculated by

$$p_c = \frac{p_e y_d}{y}$$

In these expressions, x and y are the chromaticity coordinates of the given stimulus with respect to the CIE 1931 standard colorimetric system; x_w and y_w are the coordinates of the achromatic stimulus, which is conventionally taken to have zero purity. For stimuli referring to self-luminous objects, x_w and y_w are the coordinates of the equienergy stimulus.† For stimuli referring to object colors, x_w and y_w are the coordinates of the light used to illuminate the object. The chromaticity coordinates x_d and y_d specify in the chromaticity diagram the point λ_d on the spectrum locus or the point P on the boundary joining its extremes (purple line) representing the stimulus that must be mixed with the achromatic stimulus to match the stimulus considered (see Fig. 13).

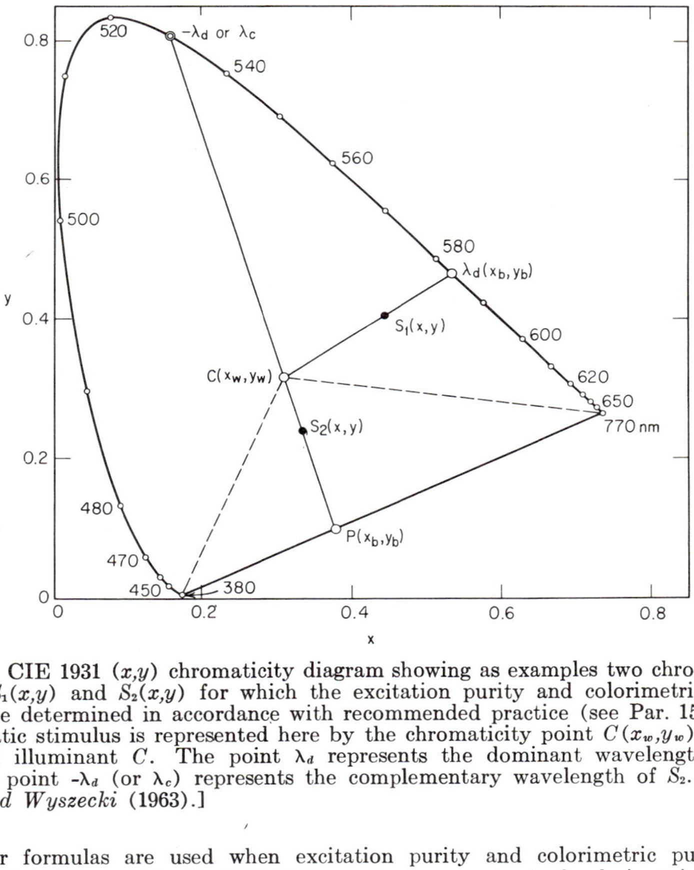

Fig. 13 CIE 1931 (x,y) chromaticity diagram showing as examples two chromaticity points $S_1(x,y)$ and $S_2(x,y)$ for which the excitation purity and colorimetric purity are to be determined in accordance with recommended practice (see Par. 15). The achromatic stimulus is represented here by the chromaticity point $C(x_w,y_w)$ of CIE standard illuminant C. The point λ_d represents the dominant wavelength of S_1, and the point $-\lambda_d$ (or λ_c) represents the complementary wavelength of S_2. [*From Judd and Wyszecki* (1963).]

Similar formulas are used when excitation purity and colorimetric purity are calculated with respect to the CIE 1964 supplementary standard colorimetric system.

UNIFORM COLOR SCALES

16. In the derivation and application of uniform color scales, the typical experimental situation is provided by a fixed illuminant, e.g., daylight, under which material color samples (painted chips, colored papers, and so forth) are exposed against a background surface. The latter is usually nonselectively reflecting but

† The equienergy stimulus may be defined as a stimulus whose total radiant flux at all wavelengths between any two limiting wavelengths within the visible spectrum is a constant multiple of the difference between these limiting wavelengths.

has an assigned reflectance that may range from high to low. All the color stimuli are then object colors whose tristimulus values define representative points contained in the object-color solid appropriate to the relative spectral irradiance distribution of the illuminant being used. Points in the object-color solid which mark out the graduations of any derived scale are realizable, at least in principle, by actual samples.

The usual judgments being made by the observer often go beyond mere color matching; he is assessing differences or equalities in certain attributes of his perception of the colors of the test samples, attributes he himself isolates subjectively. The principal attributes for object colors are hue, saturation, and lightness (see Pars. 3 and 4). Hue and saturation may be conceived as forming together an attribute of chromaticness which is a complex attribute but which can be assessed for magnitude differences as if it were simple. Attributes are initially undefined in any quantitative or numerical sense, and their scaling must be constructed from the subjective judgments of observers, obtained in extended studies.

17. Lightness Scales Most of the data available on lightness scales are subject to the restriction that they deal with achromatic colors only, and the uniform lightness scales then developed are commonly called *gray scales*. A gray scale is usually specified by stating the numerical relationship between the lightness step Λ (varying from black to white) of a given scale and the luminous reflectance Y of the corresponding gray sample. Table 8 contains a number of such relationships reported by different investigators. Figure 14 illustrates these relationships.

TABLE 8 Uniform Lightness Scales (Wyszecki and Stiles, 1967)†

Name and source	Equation	No.	Comment
Square-root formula, Priest et al. (1920)	$\Lambda = 10Y^{1/2}$	(1)	Used in connection with original Munsell system; applies to observations with white background (Munsell et al., 1933; Godlove, 1933)
Munsell value function, Newhall et al. (1943)	$Y = 1.2219\Lambda - 0.23111\Lambda^2 + 0.23951\Lambda^3 - 0.021009\Lambda^4 + 0.0008404\Lambda^5$	(2)	Used in connection with Munsell renotation system; applies to observations with middle-gray background ($Y_b \approx 20$); defines Munsell value ($V \equiv \Lambda$) in Munsell renotation system; luminous reflectance Y is relative to magnesium oxide taken as 97.5 %; this gives a value $Y = 102.568$ for $\Lambda = 10$; Munsell value function tabulated in literature, e.g., Judd and Wyszecki (1963), Wyszecki and Stiles (1967)
CIE lightness index, Wyszecki (1963)	$10\Lambda \equiv W^* = 25Y^{1/3} - 17$, $1 \leq Y \leq 100$	(3)	Approximates Munsell value function; similar expression with slightly different coefficients used by Ladd and Pinney (1955) and Glaser et al. (1958)
Scale of color-harmony manual, Foss et al. (1944)	$\Lambda = 10.25 + 5 \log Y$	(4)	Based on Weber's law; $\Lambda = 0$ corresponds to good pigment black ($Y = 0.009$); $\Lambda = 10$ corresponds to good pigment white ($Y = 0.891$); applies to observations with gray background whose reflectance is close to that of gray chips being compared for lightness difference
Scale of DIN system, Richter (1953)	$\Lambda = 6.1723 \log (40.7Y + 1)$	(5)	Based on modified Delboeuff formula; applies to observations with gray background of reflectance $Y_b \approx 50$ %

† Luminous reflectances (Y in percent) of the gray chips making up the 10-step lightness scales defined by Eqs. (1) to (5) are given in Table 9.

The adaptive state of the observer's eye, largely controlled by the reflectance of the background, has a pronounced effect on the lightness scale. The divergencies between the various curves in Fig. 14 arise principally from the fact that the different investigators used different backgrounds in their experimental studies (Judd and Wyszecki, 1963).

18. Chromaticness Scales Typical chromaticness scales are organized in terms of loci of constant hue and saturation (or chroma), lightness being kept constant. These loci form a polar coordinate system (Fig. 15). The central point represents a gray; points on a concentric circle represent colors of constant saturation (or

TABLE 9 Luminous Reflectance† Y, %

Number of steps from black Λ	Equation in Table 8 (1)	(2)	(3)	(4)	(5)
0	0.0	0.0	....	0.9	0.0
1	1.0	1.2	1.3	1.4	1.2
2	4.0	3.1	3.2	2.2	2.7
3	9.0	6.6	6.6	3.6	5.1
4	16.0	12.0	11.9	5.6	8.5
5	25.0	19.8	19.3	8.9	13.3
6	36.0	30.0	29.2	14.1	20.5
7	49.0	43.1	42.1	22.4	31.2
8	64.0	59.1	58.4	35.5	46.1
9	81.0	78.7	78.4	56.2	68.0
10	100.0	102.6	....	89.1	100.0

† The luminous reflectances Y refer to CIE illuminant C and the CIE 1931 standard observer. In calculating Λ by means of Eqs. (1) to (5) the luminous reflectance Y is to be inserted in percent (ideal white: $Y = 100$) in Eqs. (2) and (3) and is to be inserted in absolute ratio values (ideal white: $Y = 1.0$) in Eqs. (1), (4), and (5). For convenience, this table gives the Y values in percent throughout.

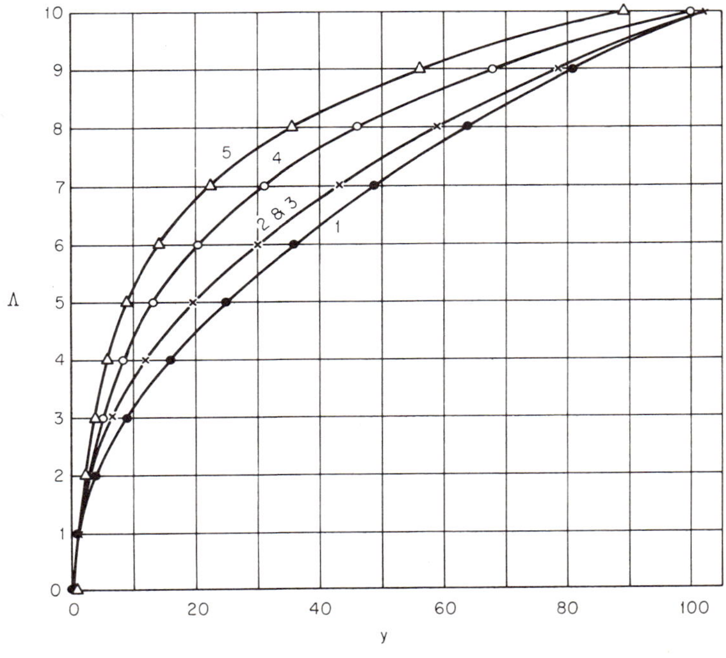

Fig. 14 Relationship between steps Λ of lightness scale and luminous reflectance Y based on different formulas listed in Table 8. Curve 1: square-root formula; curve 2: Munsell value function; curve 3: CIE lightness index; curve 4: scale of color harmony manual; curve 5: scale of DIN system. [*From Wyszecki and Stiles* (1967).]

chroma). The circles are evenly spaced, corresponding to uniform steps of saturation. Points on a radial line starting at the center represent colors of the same hue. Equal angles between these radial lines correspond ideally to equal differences of hue. Typical examples of chromaticness scales of this type are given in the Munsell color system (Newhall et al., 1943; Judd and Wyszecki, 1963; Wyszecki and Stiles, 1967).

Another type of chromaticness scale is a network of equilateral triangles, each lattice point of the network representing a sample showing the same apparent difference of chromaticness from each of its six nearest neighbors (Fig. 16). Such a set of lattice-point samples provides an ideal uniform scaling of equally light colors. Present evidence is that observers' judgments of chromaticness difference do not satisfy the requirements for constructing such an ideal uniform scaling system in a plane; the geometry of color-perception space is not euclidean. A fuller discussion of this may be found elsewhere (for example, Wyszecki and Stiles, 1967). The Committee on Uniform Color Scales of the Optical Society of America is actively engaged in the construction of chromaticness scales of this type with the closest possible approximation to uniformity.

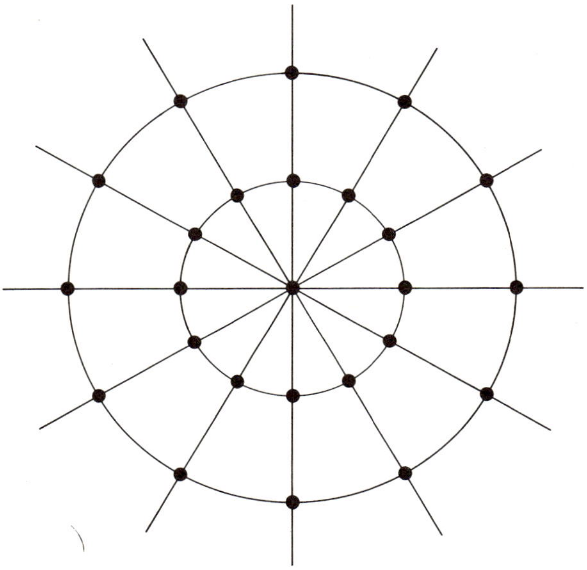

Fig. 15 Polar coordinate system representing a chromaticness-scale diagram. Points on a circle represent colors of constant saturation (or chroma); points on a radial line represent colors of the same hue. The center represents a gray having the same lightness as the chromatic colors.

With the Munsell system taken as basis, a number of attempts at acceptable but simple analytical transformations have been made which provide a quantitative description of chromaticness scales. A selection of such transformations may be found elsewhere (Judd and Wyszecki, 1963; Wyszecki and Stiles, 1967). Formulas presently favored by the CIE are the *cube-root formula* and the *Munsell renotation formula,* given in Par. 19.

In a different approach to uniform scaling, all those color stimuli (not necessarily object colors) are considered which lie in a plane of constant luminance of the CIE tristimulus space. Only linear transformations of the CIE space are allowed, corresponding to projective transformations of the CIE chromaticity diagram [either the 1931 (x,y) version or the 1964 (x_{10},y_{10}) version]. The objective is to make equal displacements in the transformed diagram correspond to perceptually equal color differences. It is known that such a uniform chromaticity scale (UCS) diagram can only be a crude approximation to perfectly uniform spacing, but it has been found to be of considerable practical value. The CIE recommends the following transformation (MacAdam, 1937):

$$u = \frac{4x}{-2x + 12y + 3} \qquad v = \frac{6y}{-2x + 12y + 3}$$

The resulting diagram is known as the CIE 1960 UCS diagram (Fig. 17). Other projective transformations proposed by different investigators have been compiled elsewhere (Judd and Wyszecki, 1963; Wyszecki and Stiles, 1967).

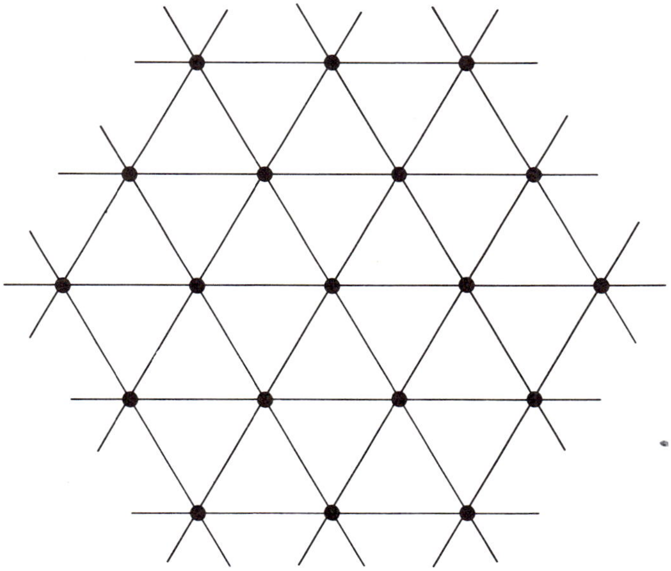

Fig. 16 Triangular lattice used for uniform scaling of chromaticness.

19. Combined Lightness and Chromaticness Scales and Color-Difference Formulas The combination of chromaticness scales and lightness scales leads to a complete sampling of the object-color solid, and analytical expressions can be formulated that permit the calculation of perceived color differences between any two colors of the solid. Many proposals have been made (see, for example, Wyszecki and Stiles, 1967), but so far none has been universally accepted as fully satisfactory. At present, the CIE recommends further study of four formulas (Wyszecki, 1968) including the one proposed tentatively by the CIE in 1964.

CIE 1964 Formula. The recommended coordinate system is formed by plotting the variables U^*, V^*, and W^* along orthogonal axes, where U^*, V^*, and W^* are defined in terms of the tristimulus values X, Y, Z as

$$W^* = 25Y^{1/3} - 17 \qquad 1 \leq Y \leq 100$$
$$U^* = 13W^*(u - u_0)$$
$$V^* = 13W^*(v - v_0)$$

where

$$u = \frac{4X}{X + 15Y + 3Z} \qquad v = \frac{6Y}{X + 15Y + 3Z}$$

and u_0 and v_0 are values of these variables for the nominally achromatic color placed at the origin of the (U^*,V^*) system. [This is an extension of the CIE 1960 UCS diagram (see Par. 18) to three dimensions.]

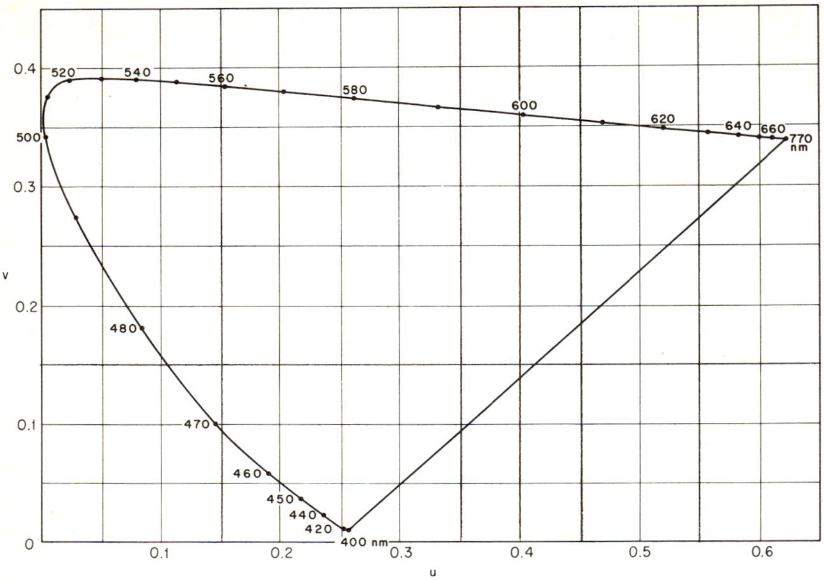

Fig. 17 CIE 1960 UCS diagram. Projective transformation of the CIE 1931 (x,y) chromaticity diagram.

For object colors the choice of u_0, v_0 to correspond to the illuminant is satisfactory.

If the angle subtended at the eye by the pairs of object colors being compared is more than 1° and less than or equal to 4°, the tristimulus values X, Y, Z calculated with respect to the CIE 1931 standard colorimetric observer (see Par. 6) should be used for the calculation of U^*, V^*, and W^*. If this angular subtense is greater than 4°, the tristimulus values X_{10}, Y_{10}, Z_{10} calculated with respect to the CIE 1964 supplementary standard colorimetric observer (see Par. 7) should be used for the calculation of U_{10}^*, V_{10}^*, and W_{10}^*.

It is recommended that the measure ΔE of the perceptual size of the difference between color (U_1^*, V_1^*, W_1^*) and color (U_2^*, V_2^*, W_2^*) be calculated by means of the formula

$$\Delta E_{\mathrm{CIE}} = [(U_1^* - U_2^*)^2 + (V_1^* - V_2^*)^2 + (W_1^* - W_2^*)^2]^{1/2}$$

which is the distance in the CIE 1964 uniform color space between colors (U_1^*, V_1^*, W_1^*) and (U_2^*, V_2^*, W_2^*).

This measure ΔE_{CIE} is intended to apply to comparisons of differences between object colors of the same size and shape, viewed in identical white to middle-gray surroundings by an observer photopically adapted to a field of chromaticity not too different from that of average daylight.

Cube-Root Formula

$$L = 25.29G^{1/3} - 18.38 \qquad a = 106.0(R^{1/3} - G^{1/3}) \qquad b = 42.34(G^{1/3} - B^{1/3})$$

where

$$\begin{aligned} R &= 1.1084X + 0.0852Y - 0.1454Z \\ G &= -0.0010X + 1.0005Y + 0.0004Z \\ B &= -0.0062X + 0.0394Y + 0.8192Z \end{aligned}$$

The measure ΔE of the perceptual size of the difference between color (a_1,b_1,L_1) and color (a_2,b_2,L_2) is calculated by means of

$$\Delta E_{CR} = [(a_1 - a_2)^2 + (b_1 - b_2)^2 + (L_1 - L_2)^2]^{1/2}$$

The transformation of X, Y, Z to R, G, B was proposed by Reilly at the Colorimetry Committee meeting in June 1963 in Vienna as an improvement of the original proposal of 1958 (Glasser et al., 1958).

Munsell Renotation Formula. The measure ΔE of the perceptual size of the difference between color (H_1,V_1,C_1) and color (H_2,V_2,C_2) is calculated by means of

$$\Delta E_{GM} = f_s\{2f_hC_1C_2[1 - \cos 3.6(H_1 - H_2)] + (C_1 - C_2)^2 + 16(V_1 - V_2)^2\}^{1/2}$$

where H, V, and C are Munsell renotation hue, value, and chroma, respectively. The difference $H_1 - H_2$ is the hue difference expressed in percent of the hue circuit; $3.6(H_1 - H_2)$ expresses the same angle in degrees; f_h is a hue-weighting factor, and f_s is a surround factor. When the surround color is approximately equal to the average of the two colors being compared, $f_h = 1$ and $f_s = 3$. For a white to middle-gray surround,

$$f_h = \left[\frac{4}{3 - \cos 3.6(H_1 - H_2)}\right]^2$$

and

$$f_s = \frac{15 + [\bar{C}^2 + 16(\bar{V} - V_s)^2]^{1/2}}{5 + [\bar{C}^2 + 16(\bar{V} - V_s)^2]^{1/2}}$$

where $\bar{V} = (V_1 + V_2)/2$, $\bar{C} = (C_1 + C_2)/2$, and V_s is the Munsell value of the gray or white surround.

When $f_h = 1$, the formula becomes the Godlove formula (Godlove, 1951). For a white to middle-gray surround, f_h weights hue differences by twice the weight specified by the Godlove formula. The factor f_s expresses the reduction in perceived size of the color difference caused by use of a surround that differs in color from the average of the two colors being compared. (Proposed by Judd at the Colorimetry Committee meeting in June 1967 in Washington.)

MacAdam-modified Friele Formula. The measure ΔE of the perceptual size of the difference between color (X_1,Y_1,Z_1) and color (X_2,Y_2,Z_2) is calculated by means of

$$\Delta E_{FC} = [(\Delta C)^2 + (\Delta L)^2]^{1/2}$$

where $\Delta C = K_1\,\Delta C_1 \qquad \Delta L = K_2\,\Delta L_2$

$$\Delta C_1 = \left[\left(\frac{\Delta C_{rg}}{a}\right)^2 + \left(\frac{\Delta C_{yb}}{b}\right)^2\right]^{1/2} \qquad \Delta L_1 = \frac{P\,\Delta P + Q\,\Delta Q}{(P^2 + Q^2)^{1/2}}$$

$$\Delta C_{rg} = \frac{Q\,\Delta P - P\,\Delta Q}{(P^2 + Q^2)^{1/2}} \qquad \Delta C_{yb} = \frac{S\,\Delta L_1}{(P^2 + Q^2)^{1/2}} - \Delta S$$

$$\Delta L_2 = 0.279\frac{\Delta L_1}{a} \qquad K_1 = 0.054 + 0.46Y^{1/3} \qquad K_2 = 0.465K_1 - 0.062$$

$$a^2 = 17.3 \times 10^{-6}\frac{P^2 + Q^2}{1 + 2.73P^2Q^2/(P^4 + Q^4)}$$

$$b^2 = 3.098 \times 10^{-4}(S^2 + 0.2015Y^2)$$

$$P = 0.724X + 0.382Y - 0.098Z$$

$$Q = -0.48X + 1.37Y + 0.1276Z$$

$$S = 0.686Z$$

X, Y, Z being the tristimulus values for either of the two colors constituting the difference to be evaluated and ΔP, ΔQ, and ΔS being the differences between the values of P, Q, and S for these two colors.

The introduction of the additional parameters K_1 and K_2 was proposed by Chickering at the Colorimetry Committee meeting in June 1967 in Washington as an improvement on the published formula (Chickering, 1967). In subsequent discussions MacAdam proposed simpler alternatives for K_1 and K_2, which are given here.

20. Color-Order Systems A color-order system is a rational method or a plan of ordering and specifying all object colors or all within a limited domain by means of a set of material standards selected and displayed so as to represent adequately the whole set of object colors under consideration (Wyszecki and Stiles, 1967).

Of the many conceivable color-order systems a number have been developed and are used. Fairly detailed accounts of the more important systems are given in the literature (Judd and Wyszecki, 1963; Wyszecki, 1960; Wyszecki and Stiles, 1967).

COLORIMETRIC INSTRUMENTATION

21. Visual Colorimeters A visual colorimeter is a device in which a test color can be matched visually with another, adjustable color. The field of view contains one or more areas occupied by the test color and similar, symmetrically disposed areas occupied by the comparison color. The observer adjusts the color of the

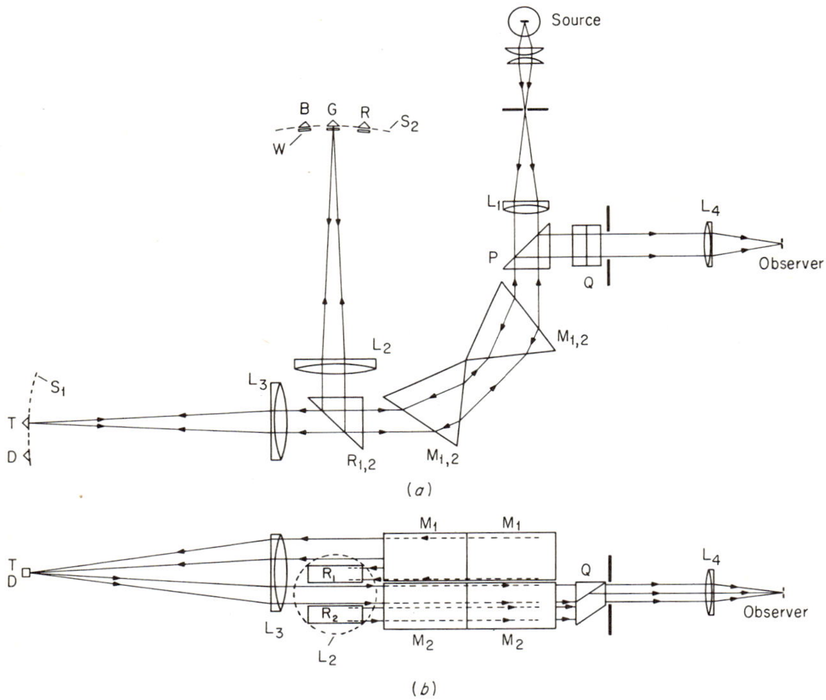

Fig. 18 (*a*) Top view and (*b*) side view of the Wright colorimeter. [*Wyszecki and Stiles* (1967).]

comparison field until test and comparison areas match. In general, the comparison color can be modified in three different ways so that the scales on the controls provide a three-number specification of color in a direct and simple way. The settings are, at least potentially, convertible into CIE tristimulus values (Optical Society of America, 1953; Wyszecki and Stiles, 1967).

Wright Colorimeter (Wright, 1927–1928). This instrument (Fig. 18) incorporates a double monochromator whose two pairs of dispersing prisms, $M_1 - M_1$ and $M_2 - M_2$, are mounted one above the other. Two small totally reflecting prisms T and D, placed in the spectrum S_1 formed by the upper dispersing system $(M_1 - M_1)$, serve as two slits for the lower dispersing system $(M_2 - M_2)$. After reflection in the prism P, the two monochromatic components are brought to a common focus in the pupil of the eye of the observer. The prism R_1 splits the beam from the upper monochromator to form a second spectrum at S_2. Three small totally reflecting prisms R, G, and B are positioned at S_2. The three monochromatic components are reflected by these prisms and redirected by the prism R_2 through the lower dispersing system ($M_2 - M_2$) and the prisms P and Q and brought to a common focus in the pupil of the eye of the observer. The mixtures of the two sets of monochromatic components isolated from the spectra S_1 and S_2 fill adjacent areas of the field prism Q, forming the field of view. The monochromatic lights reflected from the prisms R, G, B serve as primaries in a color-matching project and may be fixed at wavelengths 650, 530, and 460 nm, respectively. Calibrated neutral-density wedges in front of each of these prisms control the radiances of the primaries. The prism T at S_1 reflects the monochromatic test light, which can be varied along S_1. The prism D is placed at S_1 to reflect whatever primary light may be necessary to desaturate the test color before it can be matched by a mixture of the other two primaries.

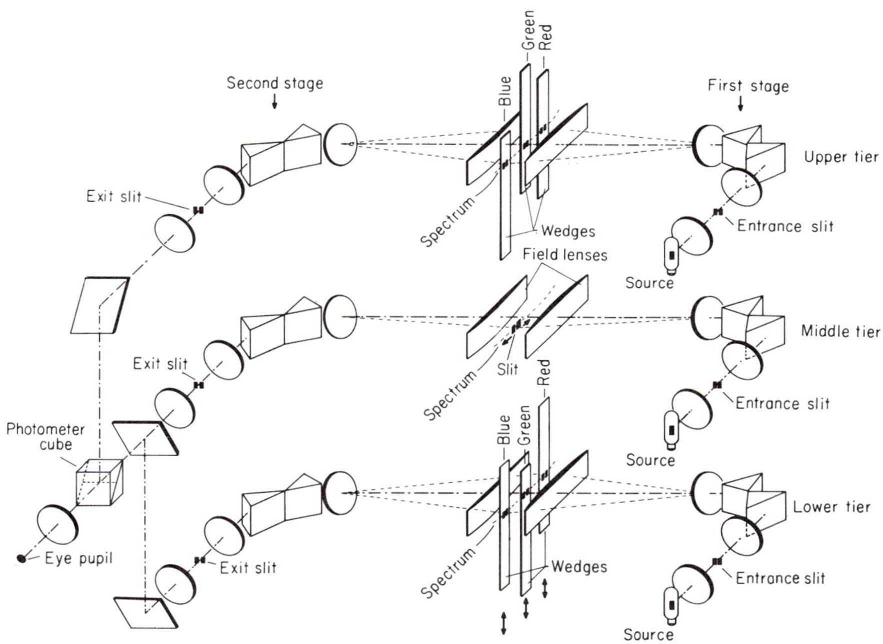

Fig. 19 Schematic diagram of the Stiles (NPL) trichromator. [*Wyszecki and Stiles* (1967).]

Stiles Trichromator (Stiles, 1955). Three double monochromators of van Cittert type with canceling (subtractive) dispersion are mounted vertically, one above the other (see Fig. 19). In the central spectrum of the middle tier a single movable narrow slit selects the monochromatic test color, which fills one-half of the test field provided in the photometer cube in front of the eye of the observer. Three fixed slits in the upper tier select the monochromatic primaries. These, after recombination in the second stage of the monochromator, provide a mixture which fills the comparison field in the photometer cube. The radiances of the primaries are independently variable by neutral-density wedges inserted immediately after the slits in the central spectrum. The lower tier provides a similar mixture, in variable amounts, of three primaries of the same wavelengths as those in the upper tier. These primaries can be added to the test color to desaturate it. The exit slits of all three tiers are imaged at a common point in space; the eye of the observer is placed so that the pupil center coincides with this point, and all the light in the slit images, which lie within a square area of about 2 mm, is collected by the pupil. The observer sees the bipartite visual field by the method of maxwellian view. Different photometer cubes can be provided to give different field sizes and shapes.

Donaldson Six-Primary Colorimeter (Donaldson, 1947). The six primary lights (red, orange, yellow-green, green, blue-green, and blue) are provided by incandescent-lamp light passed separately through each of six colored filters (see Fig. 20). The additive mixture of the primary lights taking place in the integrating sphere fills one half of the photometric field seen by the observer. The other half (test field) is filled by light coming from a second source and from the specimen, which is mounted outside the main instrument. The instrument can be used as an ordinary tristimulus colorimeter by employing only three of the

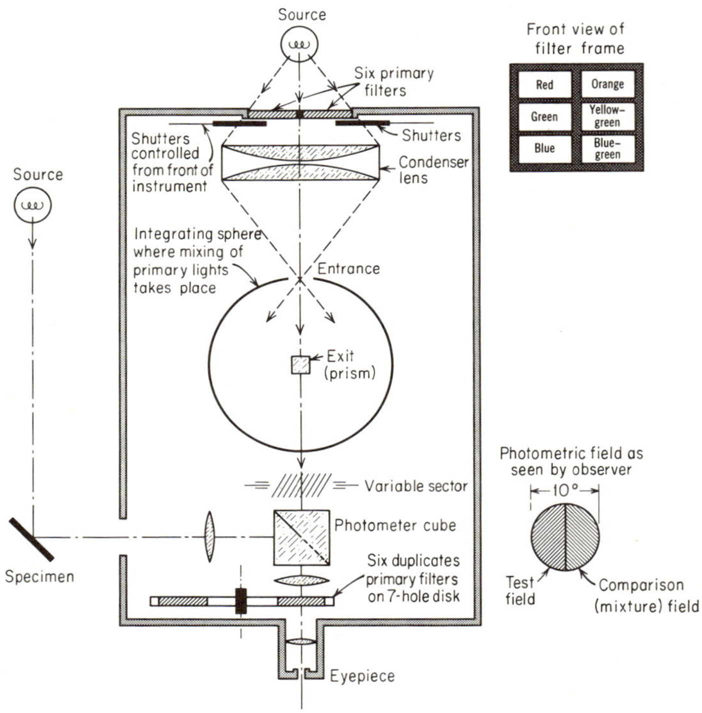

Fig. 20 Schematic diagram of Donaldson six-primary colorimeter. [*Wyszecki and Stiles* (1967).]

six primary filters, i.e., red, green, and blue or orange, yellow-green, and blue-green. In practical color-matching work, however, it is advantageous to use all six primaries, which reduces the metamerism that would occur between a three-color match and the test color. Color matching with six primaries is carried out with the aid of six duplicate primary filters mounted near the eyepiece in a seven-hole disk. The observer looks through one of six duplicate filters and adjusts the corresponding primary in the comparison field to an approximate brightness match with the test field. This is repeated with each of the remaining five duplicate filters. Then the observer looks at the photometric field without a duplicate filter in the beam. Since test field and comparison field have been adjusted to approximate matches in brightness for each of six parts of the spectrum separately, they will be found to be an approximate color match. Final adjustments to obtain a perfect match are made by using three primaries only. The Donaldson colorimeter can be calibrated in terms of the 1931 or 1964 CIE standard observer and coordinate system.

MacAdam Binocular Colorimeter (MacAdam, 1950). See Fig. 21. The condensing lenses and the field lens are plastic Fresnel lenses. The large bipartite visual field provided by the instrument can be viewed with both eyes simultaneously. The instrument consists of two symmetrical parts, each of which can be used to provide a match with the color provided by the other part.

Burnham Colorimeter (Burnham, 1952). Light from an incandescent lamp passes through a frame holding the primary filters (red, green, and blue), an aperture

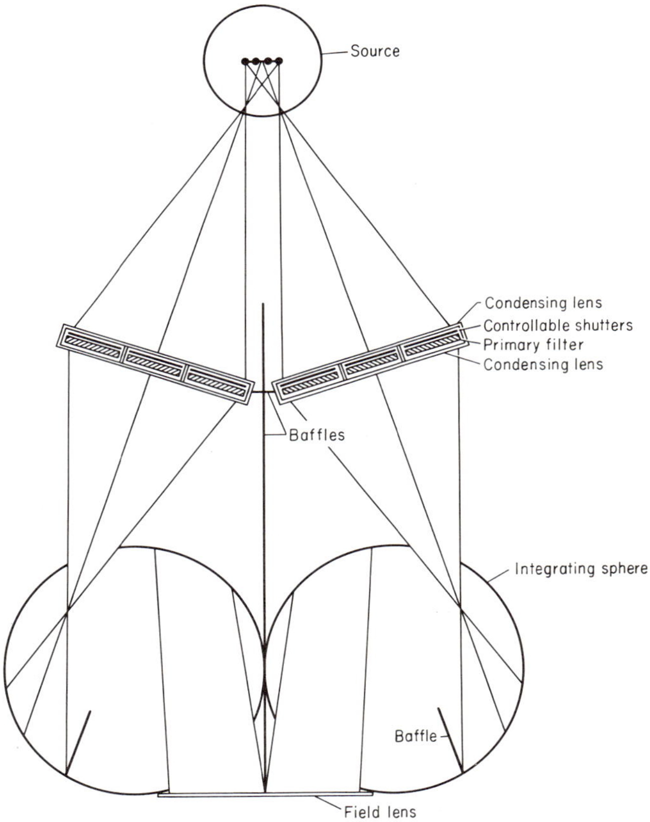

Fig. 21 Schematic diagram of the MacAdam binocular colorimeter. [*Wyszecki and Stiles* (1967).]

plate with an opening one-quarter of the filter area, and an integrating bar whose front and back faces are covered with opal-glass plates (Fig. 22). The back face of the bar appears uniformly filled with the mixture of the primary lights. The mixture color can be controlled by moving the filter frame in both horizontal and vertical directions in front of the fixed aperture plate. A light attenuator may be placed between the source and the filter frame to control the luminance of the visual field. Other filter arrangements than that illustrated incorporate four or more primary filters. A second similar colorimeter assembly is generally used to provide another visual field for color-matching purposes or for studying other colorimetric problems. The two fields are brought into juxtaposition to provide a bipartite field of view.

22. Photoelectric Colorimeters A photoelectric colorimeter is a device designed to measure the CIE tristimulus values (or some convenient transforms of them) of a test color directly and by physical means.

Two main methods are used to modify the spectral response of the physical detector, e.g., photocell or photomultiplier, so that it approximates to the desired spectral tristimulus values of one of the CIE standard colorimetric observers. One method employs a double monochromator with spectrum templates; the other simply colored filters. Template colorimeters may be considered to provide a more accurate duplication of the standard spectral tristimulus values, $\bar{x}(\lambda)$, $\bar{y}(\lambda)$, $\bar{z}(\lambda)$, than most filter colorimeters. However, the template colorimeters are usually much more complicated than filter colorimeters, and no template colorimeters

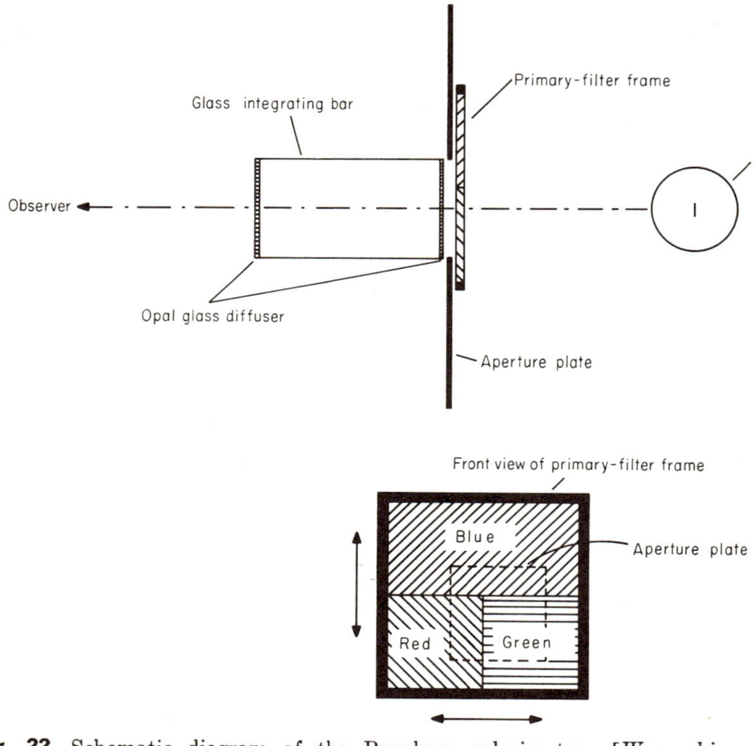

Fig. 22 Schematic diagram of the Burnham colorimeter. [*Wyszecki and Stiles* (1967).]

are commercially available as of 1970 (Winch, 1951; Mahr, 1958). A schematic diagram of a photoelectric template colorimeter is given in Fig. 23.

None of the photoelectric filter colorimeters presently available on the market incorporates spectral response functions in perfect agreement with the $\bar{x}(\lambda)$, $\bar{y}(\lambda)$, $\bar{z}(\lambda)$ functions. This deficiency limits the usefulness of these instruments for reading the CIE tristimulus values X, Y, Z directly. However, as color-difference meters, they can be quite useful (Judd and Wyszecki, 1963). If one of the two colors to be compared in such an instrument has known tristimulus values and the other differs only slightly from the first, the differences in tristimulus values will usually be measured with satisfactory accuracy. The essential components of a photoelectric filter colorimeter are illustrated in Fig. 24.

More detailed descriptions of photoelectric colorimeters can be obtained from the manufacturers of such instruments. A list of better-known commercial color-difference meters and their manufacturers is given in Table 10. More recent reviews

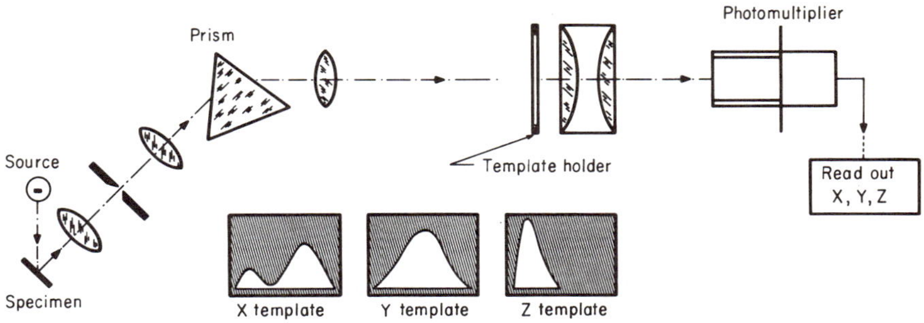

Fig. 23 Schematic diagram of the essential components of a photoelectric template colorimeter.

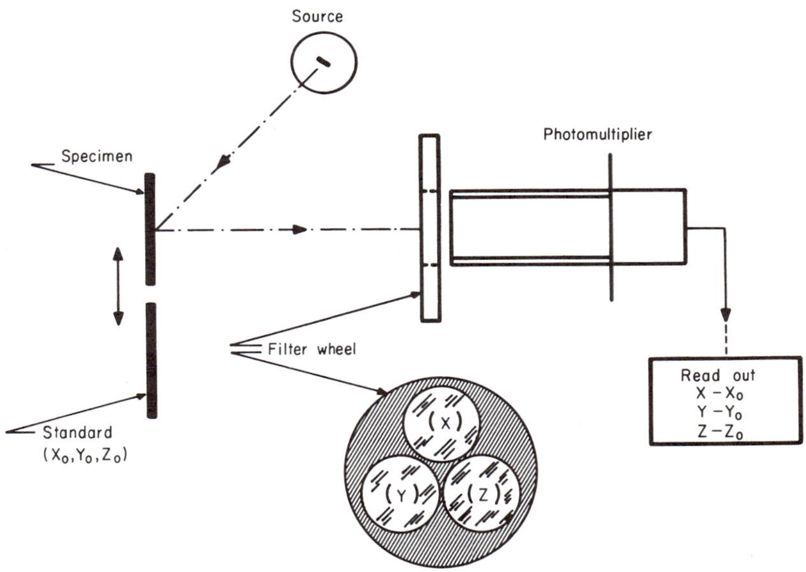

Fig. 24 Schematic diagram of the essential components of a photoelectric filter colorimeter.

TABLE 10 Selected List of Photoelectric Color-Difference Meters and Their Manufacturers

Color-Eye	Kollmorgen Corp., Box 950, Newburgh, N.Y. 12550
Colormaster	Manufacturers Engineering and Equipment Corporation, 250 Titus Avenue, Warrington, Pa. 18976
Colourmeter T40	Hilger & Watts, Ltd., 98 St. Pancras Way, Camden Road, London N.W.1, England
Elrepho	Carl Zeiss, Oberkochen, Württemberg, Germany
Gardner	Gardner Laboratory Inc., 5521 Landy Lane, Bethesda, Md. 20014
Hunter	Hunter Associates Laboratory, Inc., 9529 Lee Highway, Fairfax, Va. 22030

of such instruments may be found in the literature (Bassemir, 1959; Wurzburg, 1963; Thurner, 1965; Billmeyer and Saltzman, 1966).

23. Spectroradiometers and Spectrophotometers A spectroradiometer is an apparatus designed to measure the spectral distribution of irradiance (or any other radiometric quantity) of a source. Its principal components are a monochromator and a radiant power detector such as a thermopile or a photoemissive cell. Radiant power enters the entrance slit of the monochromator, which selects a narrow spectral band of radiant power and transmits it through the exit slit to the photosensitive surface of the detector.

A direct determination of the absolute spectral irradiance distribution of a test source is rarely made because it entails knowledge of the spectral transmittance of the optical system between source and detector and the absolute sensitivity of the detector itself. In general a spectroradiometer is used to compare the spectral irradiance distribution of the test source with the known distribution of a reference source. The reference source is usually an incandescent lamp which has been calibrated against a blackbody radiator operating at a fixed temperature.

Detailed descriptions of typical spectroradiometers may be found in the literature (Mitsuhashi, 1955; MacAdam, 1958; Hammond et al., 1960; Mori et al., 1963; Henderson and Hodgkiss, 1963; Grum, 1963; Thorington et al., 1965; Sanders and Gaw, 1967).

A spectrophotometer is an apparatus designed to measure the spectral transmittance and the spectral reflectance of objects. It is similar in design to the spectroradiometer but has a rather different purpose, namely, to compare at each wavelength the radiant power leaving the object with that incident on it or with that reflected by a white working standard equivalently irradiated. Usually a spectrophotometer has a built-in illuminating device incorporating a source which provides adequate irradiance of both sample and standard at all wavelengths of interest. The actual spectral irradiance distribution is not critical, since only power ratios at the various wavelengths are measured. On the other hand, it is important to note that the spectral transmittances and spectral reflectances measured depend on the illuminating and viewing conditions adopted. For measurements of spectral transmittance, the incident power is usually arranged to be incident normal to the surface of the object, and the angles of view are confined close to the same

TABLE 11 Selected List of Manufacturers of Spectrophotometers Useful for Colorimetry

Manufacturer	Instrument
Applied Physics Corp., 2724 South Peck Rd., Monrovia, Calif. 91016	Cary Md. 14, Cary Md. 15
Bausch & Lomb Optical Company, Rochester, N.Y. 14602	Spectronic 505
Beckman Instruments, Inc., 2500 Fullerton Rd., Fullerton, Calif. 92634	Beckman DU, Beckman DK1, DK2, Beckman DB
General Electric Company, Instrument Department 40 Federal St., West Lynn, Mass. 01905	GE recording spectrophotometer
Hilger & Watts, Ltd., 98 St. Pancras Way, Camden Rd., London N.W.1, England	Uvispek
Perkin-Elmer Corp., Norwalk, Conn. 06852	Perkin-Elmer Md. 350, Md. 13-U
Carl Zeiss, Oberkochen, Württemberg, Germany	PMQ II, RMQ 20

direction. For measurements of spectral reflectance of opaque objects, the illuminating and viewing conditions recommended by the CIE are to be preferred (see Par. 13).

Detailed descriptions of spectrophotometers can be obtained from the manufacturers of such instruments. Table 11 gives a selected list of manufacturers of spectrophotometers useful for colorimetry.

REFERENCES

Bassemir, R. W. (1959): Color Measuring Instruments for the Printing Ink Industry, *Am. Ink Maker,* vol. 37, p. 52.

Billmeyer, F. W., Jr., and M. Saltzman (1966): "Principles of Color Technology," Interscience-Wiley, New York.

Budde, W., National Research Council of Canada, Ottawa (1963): private communication.

Burnham, R. W. (1952): A Colorimeter for Research in Color Perception, *Am. J. Psychol.,* vol. 65, p. 603.

Chickering, K. D. (1967): Optimization of the MacAdam-modified 1965 Friele Color-Difference Formula, *J. Opt. Soc. Am.,* vol. 57, p. 537.

CIE (Commission Internationale de l'Eclairage) (1957): "International Lighting Vocabulary," 2d ed., vol. 1, *Publ. CIE* 1.1.

——— (1965): Method of Measuring and Specifying Colour Rendering Properties of Light Sources, *Publ.* 13.

——— (1970): CIE Document on Colorimetry (Official Recommendations), *Publ.* 15.

Condit, H. R., and F. Grum (1964): Spectral Energy Distribution of Daylight, *J. Opt. Soc. Am.,* vol. 54, p. 937.

Das, S. R., and V. D. P. Sastri (1965): Spectral Distribution and Color of Tropical Daylight, *J. Opt. Soc. Am.,* vol. 55, p. 319.

Donaldson, R. (1947): A Colorimeter with Six Matching Stimuli, *Proc. Phys. Soc. Lond.,* vol. 59, p. 554.

Foss, C. E., D. Nickerson, and W. C. Granville (1944): Analysis of the Ostwald Color System, *J. Opt. Soc. Am.,* vol. 34, p. 361.

Glasser, L. G., A. H. McKinney, C. D. Reilly, and P. D. Schnelle (1958): Cube-Root Color Coordinate System, *J. Opt. Soc. Am.,* vol. 48, p. 736.

Godlove, I. H. (1933): Neutral Value Scales, II: A Comparison of Results and Equations Describing Value Scales, *J. Opt. Soc. Am.,* vol. 23, p. 419.

——— (1951): Improved Color-Difference Formula, with Applications to the Perceptibility and Acceptability of Fadings, *J. Opt. Soc. Am.,* vol. 41, p. 760.

Grum, F. (1963): Modification of a Beckman Spectrophotometer for Direct Measurements of Spectral Energy Distribution of Light Sources, *Appl. Opt.,* vol. 2, p. 237.

Hammond, H. K., III, W. L. Holford, and M. L. Kuder (1960): Ratio-recording Spectroradiometer, *J. Res. Natl. Bur. Stand.,* vol. 64C, p. 151.

Henderson, S. T., and D. Hodgkiss (1963): The Spectral Energy Distribution of Daylight, *Br. J. Appl. Phys,* vol. 14, p. 125.

Illuminating Engineering Society (1959): "IES Lighting Handbook," 3d ed., New York.

Judd, D. B., D. L. MacAdam, and G. Wyszecki (1964): Spectral Distribution of Typical Daylight as a Function of Correlated Color Temperature, *J. Opt. Soc. Am.,* vol. 54, p. 1031.

——— and G. Wyszecki (1963): "Color in Business, Science, and Industry," 2d ed., Wiley, New York.

Kelly, K. L. (1963): Lines of Constant Correlated Color Temperature Based on MacAdam's *(u,v)* Uniform Chromaticity Transformation of the CIE Diagram, *J. Opt. Soc. Am.,* vol. 53, p. 999.

Ladd, J. H., and J. E. Pinney (1955): Empirical Relationships with the Munsell Value Scale, *Proc. Inst. Radio Eng.,* vol. 43, p. 1137.

MacAdam, D. L. (1937): Projective Transformations of ICI Color Specifications, *J. Opt. Soc. Am.,* vol. 27, p. 294.

——— (1950): Loci of Constant Hue and Brightness Determined with Various Surrounding Colors, *J. Opt. Soc. Am.,* vol. 40, p. 589.

——— (1958): Continuously Recording Spectroradiometer, *J. Opt. Soc. Am.,* vol. 48, p. 832.

Mahr, K. (1958): Über ein lichtelektrisches Spektralschablonen-Gerät für Licht- und Farbmessungen, *Farbe,* vol. 7, p. 283.

Mitsuhashi, H. (1955): Some Considerations of the Balanced Spectro-Radiometry, *Sci. Light,* vol. 4, p. 61.

Moon, P. (1940): Proposed Standard Solar-Radiation Curves for Engineering Use, *J. Franklin Inst.,* vol. 230, p. 583.

Mori, L., I. Niikura, and T. Azuma (1963): A New Spectrophotometer with a Digital Tristimulus Integrator, *CIE Proc. 15th Sess., Vienna, 1963,* vol. A, p. 69.

Munsell, A. E. O., L. L. Sloan, and I. H. Godlove (1933): Neutral Value Scales, I: Munsell Neutral Value Scale, *J. Opt. Soc. Am.,* vol. 23, p. 394.

Nayatani, Y., M. Hitani, and H. Minato (1967): Chromaticity and Spectral Energy Distribution of Daylight, from North Sky at Amagasaki, Japan, *Bull. Electrotech. Lab. Jap.,* vol. 31, p. 55.

Newhall, S. M., D. Nickerson, and D. B. Judd (1943): Final Report of the O.S.A. Subcommittee on Spacing of the Munsell Colors, *J. Opt. Soc. Am.,* vol. 33, p. 385.

Nickerson, D. (1960): Light Sources and Color Rendering, *J. Opt. Soc. Am.,* vol. 50, p. 57.

Optical Society of America, Committee on Colorimetry (1953): "The Science of Color," Optical Society of America, Washington.

Paix, D. (1963): Measurements of the Luminance Distribution of the Sky near Sydney, Australia, *IES Monogr.* 6.

Pivovansky, M., and M. R. Nagel (1961): "Tables of Blackbody Radiation Functions," Macmillan, New York.

Priest, I. G., K. S. Gibson, and H. J. McNicholas (1920): An Examination of the Munsell Color System, I: Spectral and Total Reflection and the Munsell Scale of Value, *U.S. Natl. Bur. Stand. Tech. Pap.* 167.

Richter, M. (1953): Das System der DIN-Farbenkarte, *Farbe,* vol. 1, p. 85.

Robertson, A. R. (1968): Computation of Correlated Color Temperature and Distribution Temperature, *J. Opt. Soc. Am.,* vol. 58, p. 1528.

Sanders, C. L., and W. Gaw (1967): A Versatile Spectroradiometer and Its Applications, *Appl. Opt.,* vol. 6, p. 1639.

Sastri, V. D. P., and S. R. Das (1966): Spectral Distribution and Color of North Sky at Delhi, *J. Opt. Soc. Am.,* vol. 56, p. 829.

Stiles, W. S. (1955): 18th Thomas Young Oration: The Basic Data of Color-Matching, *Phys. Soc. Yearb.,* p. 44.

Tarrant, A. W. S. (1968): The Spectral Power Distribution of Daylight, *Trans. Illum. Eng. Soc. Lond.,* vol. 33, p. 75.

Thorington, L., J. Parascandola, and G. Schiazzano (1965): Chromaticity and Color Rendition of Light Sources from Fundamental Spectroradiometry, *Illum. Eng.,* vol. 60, p. 227.

Thurner, K. (1965): Colorimetry in Textile Dyeing: Theory and Practice, Badische Anilin & Soda Fabrik AG, Ludwigshafen am Rhein.

Winch, G. T. (1951): The Physical Eye Photometer, Colorimeter, and Spectrophotometer, *Instrum. Pract.,* January; also Instruction Manual M1495, Cat. no. M908B, General Electric Company, Ltd., Wembley, England.

———, M. C. Boshoff, C. J. Kok, and A. G. du Toit (1966): Spectroradiometric and Colorimetric Characteristics of Daylight in the Southern Hemisphere: Pretoria, South Africa, *J. Opt. Soc. Am.,* vol. 56, p. 457.

Wright, W. D. (1927–1928): A Trichromatic Colorimeter with Spectral Stimuli, *Trans. Opt. Soc.,* vol. 29, p. 225.

——— (1964): "The Measurement of Colour," 3d ed., Hilger & Watts, London.

Wurzburg, F. L., Jr. (1963): Survey of Instruments for Color Specification, *Tappi,* vol. 46, no. 7, p. 155A, July.

Wyszecki, G. (1960): "Farbsysteme," Musterschmidt, Göttingen.

——— (1963): Proposal for a New Color-Difference Formula, *J. Opt. Soc. Am.,* vol. 53, p. 1318.

——— (1968): Recent Agreements Reached by the Colorimetry Committee of the Commission Internationale de l'Eclairage, *J. Opt. Soc. Am.,* vol. 58, p. 290.

——— and W. S. Stiles (1967): "Color Science: Concepts and Methods, Quantitative Data and Formulas," Wiley, New York.

Section 10

Polarization

JEAN M. BENNETT AND HAROLD E. BENNETT
Michelson Laboratory, Naval Weapons Center, China Lake, Calif.

(Numbers refer to paragraphs)

THEORY, TERMINOLOGY, AND APPLICATIONS OF POLARIZATION

1. Basic Concepts and Conventions Optical polarization was discovered by E. L. Malus in 1808. A major triumph of nineteenth- and early twentieth-century theoretical physics was the development of electromagnetic theory and the demonstration that optical polarization is completely described by it. This theory is phenomenological in that instead of trying to explain why materials have certain fundamental characteristics, it concentrates on the resulting properties which any material with those characteristics will display. In the optical case, the polarization and all other optical properties of a material are determined by two or more phenomenological parameters called *optical constants.* Electromagnetic theory has little or nothing to say about *why* a material should have these particular optical constants or *how* they are related to its atomic character. This problem has been extensively investigated in twentieth-century solid-state physics and is still only partially understood. It is clear, however, that the optical constants are a function not only of the atomic nature of the material, i.e., its position in the periodic table, but are also quite sensitive to how it is prepared. Perhaps optical parameters would be a better term than optical constants. Nevertheless, the concept of optical constants is an extremely useful one and makes it possible to predict quantitatively the optical behavior of a material and, under certain conditions, to relate this behavior to nonoptical parameters.

Since the optical constants are so fundamental, differences in their definition are particularly unfortunate. The most damaging of these differences arises from an ambiguity in the initial derivation. Maxwell's equations, which form the basis of electromagnetic theory, result in the wave equation, which in gaussian units is

$$\nabla^2\mathbf{E} = \frac{\epsilon}{c^2}\frac{\partial^2\mathbf{E}}{\partial t^2} + \frac{4\pi\sigma}{c^2}\frac{\partial\mathbf{E}}{\partial t} \tag{1}$$

where ∇^2 = laplacian operator
$\mathbf{E}$ = electric field vector of traveling wave
t = time
c = velocity of light
σ = conductivity of material at frequency of wave motion
ϵ = dielectric constant of material at frequency of wave motion

A solution to this equation is

$$\mathbf{E} = \mathbf{E}_0 \exp [i(\omega t + \delta)] \exp (-i\mathbf{k} \cdot \mathbf{r}) \exp \left(-\frac{\alpha z}{2}\right) \tag{2}$$

where $\mathbf{E}_0$ = amplitude of wave
ω = angular frequency of wave
δ = phase angle
$\mathbf{k}$ = wave vector
$\mathbf{r}$ = position vector
z = direction wave is traveling
α = absorption coefficient

The wave vector $\mathbf{k}$ is assumed to be real and equal to $(2\pi/\lambda_m)\mathbf{n}$, where λ_m is the wavelength in the medium in which the wave is traveling and $\mathbf{n}$ is a unit vector in the k direction.† Equation (2) can also be written in terms of $\bar{n}$, the complex index of refraction, defined as

$$\bar{n} = n - ik \tag{3}$$

where n is the index of refraction and k the extinction coefficient. In this form, Eq. (2) is

$$E = E_0 \exp \left[i\omega \left(t - \frac{\bar{n}z}{c} \right) \right] \tag{4}$$

when $\delta = 0$. By setting the imaginary part of the exponent equal to zero one obtains

$$z = \frac{c}{n} t \tag{5}$$

showing that Eq. (4) represents a wave traveling in the positive z direction with velocity c/n. The amplitude of the wave at z is, from Eq. (4),

$$|E| = E_0 e^{-2\pi kz/\lambda} \tag{6}$$

where λ is the wavelength in vacuum. The wave is thus exponentially damped, and the amplitude penetration depth, or distance below an interface at which the *amplitude* of the wave falls to $1/e$ times its initial value, is $z = \lambda/2\pi k$. The absorption coefficient α, or the reciprocal of the distance in which the *intensity* of the wave falls to $1/e$ times its initial value, is

$$\alpha = \frac{4\pi k}{\lambda} \tag{7}$$

This development follows that commonly given by those working at optical or radio frequencies. The confusion in the definition of the optical constants arises because an equally valid solution to Eq. (1) is

$$E' = E_0 \exp \left[-i\omega \left(t - \frac{\bar{n}'z}{c} \right) \right] \tag{8}$$

which also represents an exponentially damped wave traveling in the $+z$ direction *provided that the complex index of refraction is defined to be*

$$\bar{n}' = n + ik \tag{9}$$

† Frequently the wave vector is taken to be complex, that is, $\mathbf{k}' = (2\pi/\lambda_m - i\alpha/2)\mathbf{n}$, and Eq. (2) is written $\mathbf{E} = \mathbf{E}_0 \exp [i(\omega t + \delta)] \exp (-i\mathbf{k}'\cdot\mathbf{r})$.

Unfortunately, when the wave equation arises in quantum mechanics, the solution chosen is usually the negative exponential, i.e., Eq. (8) rather than Eq. (4). Solid-state physicists working in optics thus often define the complex index of refraction as the form given in Eq. (9) rather than that in Eq. (3). Equally valid, self-consistent theories can be built up using either definition, and as long as only intensities are considered, the resulting expressions are usually identical. However, when phase differences are calculated, the two conventions usually lead to contradictory results. Even worse, unless an author is extremely careful, he may not consistently follow either convention, and the result may be pure nonsense. Some well-known books might be cited in which the authors are not even consistent from chapter to chapter.

There are several other cases in optics in which alternative conventions are possible and both are found in the literature. Among these, the most distressing are the use of a left-handed rather than a right-handed coordinate system, which makes the p and s components of polarized light have the same phase change at normal incidence (see Par. 2), and defining the optical constants so that they

TABLE 1 Preferred Conventions for Optical Parameters

Electric field strength..........	$\mathbf{E} = \mathbf{E}_0 \exp i(\omega t + \delta) \exp(-i\mathbf{k} \cdot \mathbf{r}) \exp\left(-\frac{\alpha z}{2}\right)$
Complex index of refraction....	$\bar{n} = n - ik$
Complex relative amplitude attenuation on reflection.....	$\rho = \frac{r_p}{r_s} = \frac{\lvert E_p''\rvert/\lvert E_p\rvert}{\lvert E_s''\rvert/\lvert E_s\rvert} e^{i(\delta_p - \delta_s)} = \tan\psi\, e^{i\Delta}$
Relative amplitude attenuation on reflection.................	$\tan\psi = \frac{\lvert r_p\rvert}{\lvert r_s\rvert}$
Relative phase change on reflection....................	$\Delta = \delta_p - \delta_s \quad 0 \leq \Delta \leq \pi$ for a bare surface

depend on the angle of incidence, which makes the angle of refraction given by Snell's law real for an absorbing medium. The resulting confusion is particularly acute in the field of ellipsometry, which deals primarily with phase differences and amplitudes. It prompted R. H. Muller, of the University of California, Berkeley, to make a study of this problem (Muller, 1969). He found that, depending on the conventions used (all of which have appeared in the literature), it was possible to obtain 16 different combinations of optical constants, all of which are valid within their respective frameworks, from a single measurement of the two ellipsometric parameters ψ and Δ.† He also made a survey of those conventions most frequently used in the optical, radio, and electromagnetic-theory literature and, after consultation with other participants (including these authors) at the Second International Conference on Ellipsometry, suggested the preferred set of conventions given in Table 1. These conventions are widely used in all regions of the spectrum, making it possible, for example, to use expressions from transmission-line theory developed originally for microwave frequencies to solve multilayer-film problems at optical frequencies. The advantages to be gained by using a single set of conventions in electromagnetic theory are appealing, and it is hoped that most new work in this field will adhere to them, as this section does. In any event, an author should *clearly* state the conventions he is using and then *stay with them.*

One final word may be said in regard to terminology. The complex index of refraction is sometimes written

$$\bar{n} = n(1 - i\kappa) \tag{10}$$

† A brief discussion of ellipsometry and the ellipsometric parameters is given in Par. 6.

In this formulation the symbol κ is almost universally used instead of k, which is reserved for the imaginary part of the refractive index. Although k is more directly related to the absorption coefficient α than κ [see Eq. (7)] and usually makes the resulting expressions slightly simpler, in areas such as attenuated total reflection the use of κ results in a simplification. To avoid confusion between k and κ, if Eq. (10) is used, Muller suggests that κ be called the *absorption index,* not the extinction index or extinction coefficient, which is reserved for k, nor the absorption coefficient, which is reserved for α.

2. Fresnel Equations The Fresnel equations are expressions for the reflection and transmission coefficients of light at nonnormal incidence. In deriving these

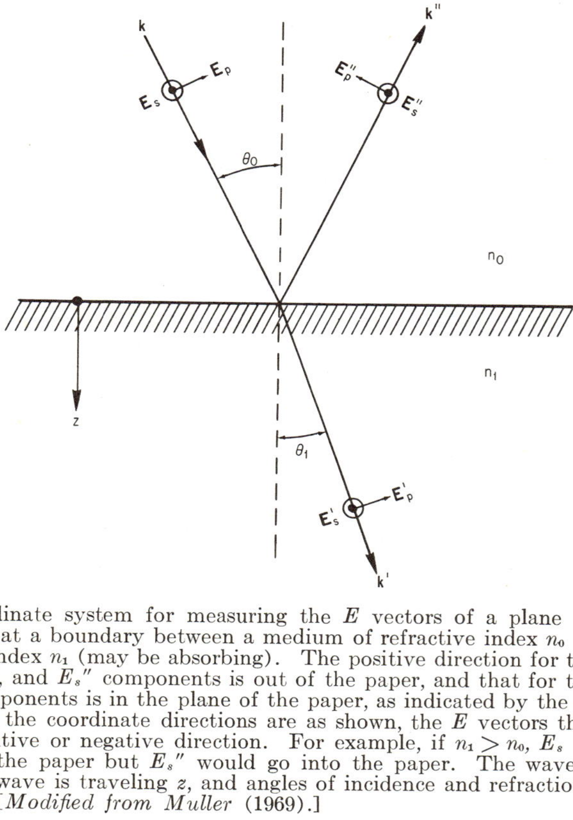

Fig. 1 Coordinate system for measuring the E vectors of a plane wave reflected and refracted at a boundary between a medium of refractive index n_0 and a medium of refractive index n_1 (may be absorbing). The positive direction for the coordinates of the E_s, E_s', and E_s'' components is out of the paper, and that for the coordinates of the E_p components is in the plane of the paper, as indicated by the arrows. Note that although the coordinate directions are as shown, the E vectors themselves may be in the positive or negative direction. For example, if $n_1 > n_0$, E_s and E_s' would come out of the paper but E_s'' would go into the paper. The wave vector **k**, the direction the wave is traveling z, and angles of incidence and refraction θ_0 and θ_1 are also shown. [*Modified from Muller* (1969).]

equations, the coordinate system assumed determines the signs in the equations and therefore the phase changes on reflection for the p and s components. In accordance with the Muller convention (Par. 1), we shall assume that the coordinate system is as shown in Fig. 1. In this system, the angle of incidence is θ_0, and the angle of refraction is θ_1. The *s component* of polarization is the plane of vibration of the E wave which is perpendicular to the plane of the paper, and the *p component* is the plane of vibration which is in the plane of the paper.† (The plane of incidence

† Unfortunately, when, early in the nineteenth century, Malus discovered that light reflected at a certain angle from glass is, as he said, "polarized," he defined the "plane of polarization" of the reflected light as the plane of incidence. Since the reflected light in this case has its E vector perpendicular to the plane of incidence, the "plane of polarization" is perpendicular to the plane in which the E vector vi-

is in the plane of the paper.) The positive directions for the vibrations are indicated by the dots for E_s, E_s', and E_s'' and by the arrows for the corresponding p components. Note that the positive direction for E_p'' is as shown in the figure because of the *mirror-image effect*. By convention, one always looks *against the direction of propagation of the wave* so that the positive direction of E_p is to the right and the positive direction of E_p'' is also to the right. The positive directions of the reflected E vectors are not the same as the actual directions of the reflected E vectors. These latter directions will depend on the refractive index of the material and may be either positive or negative. For example, if $n_1 > n_0$, at normal incidence E_s'' will be in the negative direction and E_p'' will be in the positive direction. Thus we say that there is a phase change on reflection of 180° for the s wave and a phase change of 0° for the p wave.

With this coordinate system, the Fresnel amplitude reflection coefficients for a single interface, obtained from Eq. (4) by setting up and solving the boundary-value problem, can be written

$$\frac{E_s''}{E_s} \equiv r_s = \frac{n_0 \cos \theta_0 - n_1 \cos \theta_1}{n_0 \cos \theta_0 + n_1 \cos \theta_1} \tag{11}$$

and

$$\frac{E_p''}{E_p} \equiv r_p = \frac{n_1 \cos \theta_0 - n_0 \cos \theta_1}{n_1 \cos \theta_0 + n_0 \cos \theta_1} \tag{12}$$

The amplitude transmission coefficients are

$$\frac{E_s'}{E_s} \equiv t_s = \frac{2n_0 \cos \theta_0}{n_0 \cos \theta_0 + n_1 \cos \theta_1} \tag{13}$$

and

$$\frac{E_p'}{E_p} \equiv t_p = \frac{2n_0 \cos \theta_0}{n_1 \cos \theta_0 + n_0 \cos \theta_1} \tag{14}$$

Other forms of the Fresnel amplitude reflection and transmission coefficients containing only the angles of incidence and refraction are somewhat more convenient. These relations can be derived using Snell's law

$$\frac{\sin \theta_0}{\sin \theta_1} = \frac{n_1}{n_0} \tag{15}$$

to eliminate n_0 and n_1 from Eqs. (11) to (14):

$$r_s = \frac{-\sin (\theta_0 - \theta_1)}{\sin (\theta_0 + \theta_1)} \tag{16}$$

$$r_p = \frac{\tan (\theta_0 - \theta_1)}{\tan (\theta_0 + \theta_1)} \tag{17}$$

$$t_s = \frac{2 \sin \theta_1 \cos \theta_0}{\sin (\theta_0 + \theta_1)} \tag{18}$$

$$t_p = \frac{2 \sin \theta_1 \cos \theta_0}{\sin (\theta_0 + \theta_1) \cos (\theta_0 - \theta_1)} \tag{19}$$

For nonabsorbing materials the intensity reflection coefficients R_s and R_p are simply the squares of Eqs. (16) and (17):

$$R_s = r_s^2 = \frac{\sin^2 (\theta_0 - \theta_1)}{\sin^2 (\theta_0 + \theta_1)} \tag{20}$$

$$R_p = r_p^2 = \frac{\tan^2 (\theta_0 - \theta_1)}{\tan^2 (\theta_0 + \theta_1)} \tag{21}$$

---

brates. This nomenclature causes considerable confusion and has been partially resolved in modern terminology by discussing the *plane of vibration* of the E vector and avoiding, insofar as possible, the term plane of polarization. We shall follow this practice. When specifying the direction in which light is polarized, we shall give the direction of vibration, *not* the direction of polarization.

and, at normal incidence,

$$R_s = R_p = \frac{(n_0 - n_1)^2}{(n_0 + n_1)^2} \tag{22}$$

from Eqs. (11) and (12). In the lower part of Fig. 2, R_s and R_p are given as a function of angle of incidence for various values of the refractive-index ratio n_1/n_0 with k_1 for the material equal to zero. The curves for $n_1/n_0 = 1.3$, 1.8, and 2.3 show that the normal-incidence reflectance increases as n_1 increases. The curves for $n_1/n_0 = 0.3$ and 0.8 and $k_1 = 0$ have no physical significance as long as the incident medium is air. However, they are representative of *internal reflections* in materials of refractive index $n_0 = 3.33$ and 1.25, respectively, when the *other* medium is air ($n_1 = 1$).

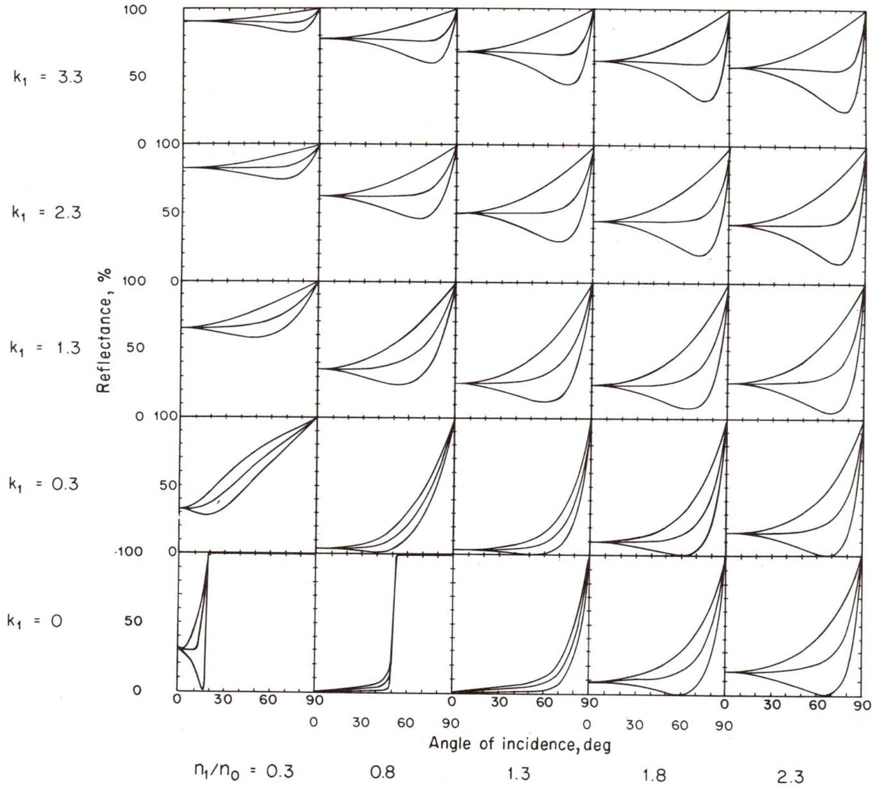

Fig. 2 R_s *(upper curves)*, R_p *(lower curves)*, and $R_{av} = (R_s + R_p)/2$ as a function of angle of incidence for various values of the refractive-index ratio n_1/n_0 and k_1. The incident medium, having refractive index n_0, is assumed to be nonabsorbing. [*Modified from Hunter* (1965*b*).]

The intensity transmission coefficients T_s and T_p are obtained from the Poynting vector and for nonabsorbing materials are

$$T_s = 1 - R_s = \frac{n_1 \cos\theta_1}{n_0 \cos\theta_0} t_s^2 = \frac{4n_0 n_1 \cos\theta_0 \cos\theta_1}{(n_0 \cos\theta_0 + n_1 \cos\theta_1)^2} = \frac{\sin 2\theta_0 \sin 2\theta_1}{\sin^2(\theta_0 + \theta_1)} \tag{23}$$

$$T_p = 1 - R_p = \frac{n_1 \cos\theta_1}{n_0 \cos\theta_0} t_p^2 = \frac{4n_0 n_1 \cos\theta_0 \cos\theta_1}{(n_1 \cos\theta_0 + n_0 \cos\theta_1)^2} = \frac{\sin 2\theta_0 \sin 2\theta_1}{\sin^2(\theta_0 + \theta_1)\cos^2(\theta_0 - \theta_1)} \tag{24}$$

These coefficients are for light passing through a single boundary and hence are of limited usefulness. In actual cases, the light is transmitted through a slab of material where there are two boundaries, generally multiple reflections within the material, and sometimes interference effects when the boundaries are smooth and plane-parallel.

The intensity transmission coefficient T_{sample} for a slab of transparent material in air is given by the well-known Airy equation (Williams, 1950) when the sample has smooth, plane-parallel sides and coherent multiple reflections occur within it:

$$T_{\text{sample}} = \frac{1}{1 + [4R_{s,p}/(1 - R_{s,p})^2] \sin^2 \gamma} \tag{25}$$

where

$$\gamma = \frac{2\pi n_1 d \cos \theta_1}{\lambda} \tag{26}$$

The values of R_s and R_p can be determined from Eqs. (20) to (22); d is the sample thickness, λ the wavelength, n_1 the refractive index of the material, and θ_1 the angle of refraction. Equation (25) holds for all angles of incidence including the Brewster angle (see Par. 53). The Airy equation predicts that at a given angle of incidence the transmission of the sample will vary from a maximum value of 1 to a minimum value of $(1 - R_{s,p})^2/(1 + R_{s,p})^2$ as the wavelength or the thickness is changed. If the sample is very thick, the oscillations in the transmittance will occur at wavelengths very close together and hence will be unresolved. A complete oscillation occurs every time γ changes by π, so that the wavelength interval $\Delta\lambda$ between oscillations is

$$\Delta\lambda \approx \frac{\lambda^2}{2n_1 d \cos \theta_1} \tag{27}$$

As an example, a sample 1 mm thick with an index of 1.5 at 5000 Å will have transmission maxima separated by 0.83 Å when measured at normal incidence ($\cos \theta_1 = 1$). These maxima would not be resolved by most commercial spectrophotometers. In such a case, one would be measuring the average transmission $T_{\text{sample, av}}$:

$$T_{\text{sample, av}} = \frac{1 - R_{s,p}}{1 + R_{s,p}} \tag{28}$$

For nonabsorbing materials, this is the same value as that which would be obtained if the multiply reflected beams did not coherently interfere within the sample. If the sample is wedge-shaped, so that no multiply reflected beams contribute to the measured transmittance, T_{sample} is simply T_s^2 or T_p^2 and can be calculated from Eq. (23) or (24).

When the material is absorbing, i.e., has a complex refractive index, it is not so easy to calculate the reflectance and transmittance since the angle of refraction is complex. However, Snell's law [Eq. (15)] and Fresnel's equations (11) and (12) are sometimes used with complex values of n_1 and θ_1. The resulting amplitude reflection coefficients are written

$$r_s = |r_s|e^{i\delta_s} \tag{29}$$

and

$$r_p = |r_p|e^{i\delta_p} \tag{30}$$

where $|r_s|$ and $|r_p|$ are the magnitudes of the reflectances and δ_s and δ_p are the phase changes on reflection. The intensity reflection coefficients are

$$R_{s,p} = r_{s,p}r_{s,p}^* \tag{31}$$

An alternate approach is to use the method of effective indexes to calculate R_s and R_p. In the medium of incidence, which is assumed to be nonabsorbing, the effective indexes η_{0s} and η_{0p} for the s and p components are

$$\eta_{0s} = n_0 \cos \theta_0 \tag{32}$$

$$\eta_{0p} = \frac{n_0}{\cos \theta_0} \tag{33}$$

where n_0 generally equals 1 for air. In the absorbing material both η's are complex and can be written (Berning and Berning, 1960; Berning, 1963)

$$\bar{\eta}_{1s} = \bar{n}_1 \cos \Theta_1 \tag{34}$$

$$\bar{\eta}_{1p} = \frac{\bar{n}_1}{\cos \Theta_1} \tag{35}$$

where $\bar{n}_1 = n_1 - ik_1$ is the complex refractive index of the material, and

$$\cos \Theta_1 = \left[\frac{(\alpha_1^2 + \beta_1^2)^{1/2} + \alpha_1}{2}\right]^{1/2} - i\left[\frac{(\alpha_1^2 + \beta_1^2)^{1/2} - \alpha_1}{2}\right]^{1/2} \tag{36}$$

$$\alpha_1 = 1 + \left(\frac{n_0 \sin \theta_0}{n_1^2 + k_1^2}\right)^2 (k_1^2 - n_1^2) \tag{37}$$

and

$$\beta_1 = -2n_1k_1 \left(\frac{n_0 \sin \theta_0}{n_1^2 + k_1^2}\right)^2 \tag{38}$$

Abelès' method (Abelès, 1963) also uses effective indexes for the absorbing material, but they are calculated differently:

$$\bar{\eta}_{1s} = a - ib \tag{39}$$

$$\bar{\eta}_{1p} = c - id \tag{40}$$

where

$$a^2 - b^2 = n_1^2 - k_1^2 - n_0^2 \sin^2 \theta_0 \tag{41}$$

$$ab = n_1k_1 \tag{42}$$

$$c = a\left(1 + \frac{n_0^2 \sin^2 \theta_0}{a^2 + b^2}\right) \tag{43}$$

$$d = b\left(1 - \frac{n_0^2 \sin^2 \theta_0}{a^2 + b^2}\right) \tag{44}$$

In both methods employing effective indexes, the amplitude reflection coefficients are

$$r_s = \frac{\eta_{0s} - \eta_{1s}}{\eta_{0s} + \eta_{1s}} \tag{45}$$

$$r_p = \frac{\eta_{1p} - \eta_{0p}}{\eta_{1p} + \eta_{0p}} \tag{46}$$

which are equivalent to Eqs. (29) and (30) and reduce to Eqs. (11) and (12) when $k_1 = 0$. The intensity reflection coefficients are given by Eq. (31), as before. At normal incidence,

$$R_s = R_p = \frac{(n_0 - n_1)^2 + k_1^2}{(n_0 + n_1)^2 + k_1^2} \tag{47}$$

Values of R_s and R_p are plotted as a function of angle of incidence in Fig. 2 for various values of n_1 and k_1. (The incident medium is assumed to be air with $n_0 = 1$ unless otherwise noted.) As n_1 increases with k_1 held constant, the magnitudes of R_s and R_p both decrease unless $k_1 \ll n_1$, that is, $k_1 = 0.3$ in Fig. 2. As k_1 increases with n_1 held constant, the magnitudes of R_s and R_p both increase. Tables of R_s and R_p for values of n_1 from 0.1 to 4.0 (in steps of 0.1) and k_1 from 0.1 to 6.0 (in steps of 0.1) are given for angles of incidence from 0 to 85° (in steps of 5°) by Holl (1963). He also gives graphs of R_s, R_p, and $R_{av} = (R_s + R_p)/2$ for selected values of the optical constants.

The absolute phase changes on reflection δ_s and δ_p are also of interest in problems involving polarization. When the material is nonabsorbing, the phase changes can be determined from the amplitude reflection coefficients, Eqs. (11) and (12); when $\theta_0 = 0$ and $n_1 > n_0$, $\delta_s = 180°$ and $\delta_p = 360°$.† This is an apparent contradiction since at normal incidence the s and p components should be indistinguishable. However, the problem is resolved by recalling that by convention we are always looking against the direction of propagation of the light (see Fig. 1 and the discussion of the mirror-image effect in Par. 2). To avoid complications, the phase change on reflection at normal incidence (often defined as β) is identified with δ_s.

For a dielectric, if $n_0 < n_1$, δ_s remains 180° for all angles of incidence from 0 to 90°, as can be seen from the numerator of Eq. (11). However, there is an abrupt discontinuity in δ_p, as can be seen from Eq. (12). If $n_0 < n_1$, $\delta_p = 360°$† at normal incidence and at larger angles for which the numerator of Eq. (12) is positive. Since $\cos\theta_0$ becomes increasingly less than $\cos\theta_1$ as θ_0 increases, and since $n_1 > n_0$, there will be an angle for which $n_1 \cos\theta_0 = n_0 \cos\theta_1$. At this angle δ_p undergoes an abrupt change from 360 to 180°, and it remains 180° for larger angles of incidence. At the transition value of θ_0, which is called the *Brewster angle* θ_B since $R_p = 0$,

$$\tan\theta_B = \frac{n_1}{n_0} \tag{48}$$

(This angle is also called the *polarizing angle* since $\theta_0 + \theta_1 = 90°$.) Equation (48) is the basis for a sensitive technique for determining the refractive index of a dielectric film or substrate (Par. 5).

The phase changes δ_s and δ_p are not simply 360 or 180° for an absorbing material At normal incidence it follows from Eq. (45) that

$$\tan\delta_s = \frac{2n_0k_1}{n_0^2 - n_1^2 - k_1^2} \tag{49}$$

so that $\delta_s = 180°$ only if $k_1 = 0$. As before, $\delta_p = \delta_s + 180°$, as seen by comparing Eqs. (45) and (46). At nonnormal incidence

$$\tan\delta_s = \frac{2\eta_{0s}b}{\eta_{0s}^2 - a^2 - b^2} \tag{50}$$

and

$$\tan\delta_p = \frac{-2\eta_{0p}d}{c^2 + d^2 - \eta_{0p}^2} \tag{51}$$

where the relations for a, b, c, and d have been given in Eqs. (41) to (44). The following relations between these quantities may also prove helpful:

$$a^2 + b^2 = [(n_1^2 - k_1^2 - n_0^2\sin^2\theta_0)^2 + 4n_1^2k_1^2]^{1/2} \tag{52}$$

$$c^2 + d^2 = \frac{(n_1^2 + k_1^2)^2}{a^2 + b^2} \tag{53}$$

$$b^2 = -\frac{n_1^2 - k_1^2 - n_0^2\sin^2\theta_0}{2} + \frac{a^2 + b^2}{2} \tag{54}$$

Figures 3 and 4 show how δ_s and δ_p change as a function of angle of incidence for an absorbing material. At normal incidence they are 180° apart because of the mirror-image effect, mentioned previously. As the angle of incidence increases, δ_p approaches δ_s, and at the *principal angle* $\bar{\theta}$ the two quantities differ by only 90°. At grazing incidence they coincide.

The reflectance R_p does not reach zero for an absorbing material as it does for a dielectric, but the angle for which it is a minimum is called the *pseudo Brewster angle* θ_B'. Two other angles closely associated with the pseudo Brewster angle are also of interest. The angle for which the ratio R_p/R_s is a minimum is sometimes called the

† Since 360° and 0° are indistinguishable, many optics books state that $\delta_p = 0°$ for dielectrics at normal incidence. This statement is misleading since it is incompatible with $\Delta = \delta_p - \delta_s > 0$ and with the change in Δ as the material becomes absorbing (see Figs. 3 and 4 and Par. 6).

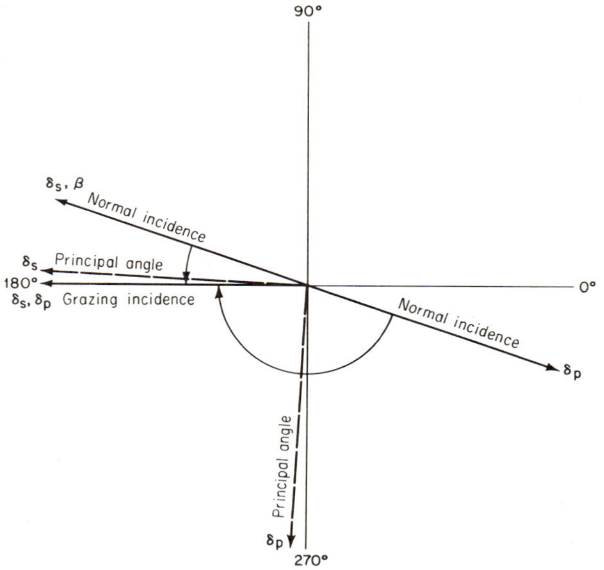

Fig. 3 Rotation of δ_s and δ_p with angle of incidence for an absorbing material. [*Bennett and Bennett* (1967*a*).]

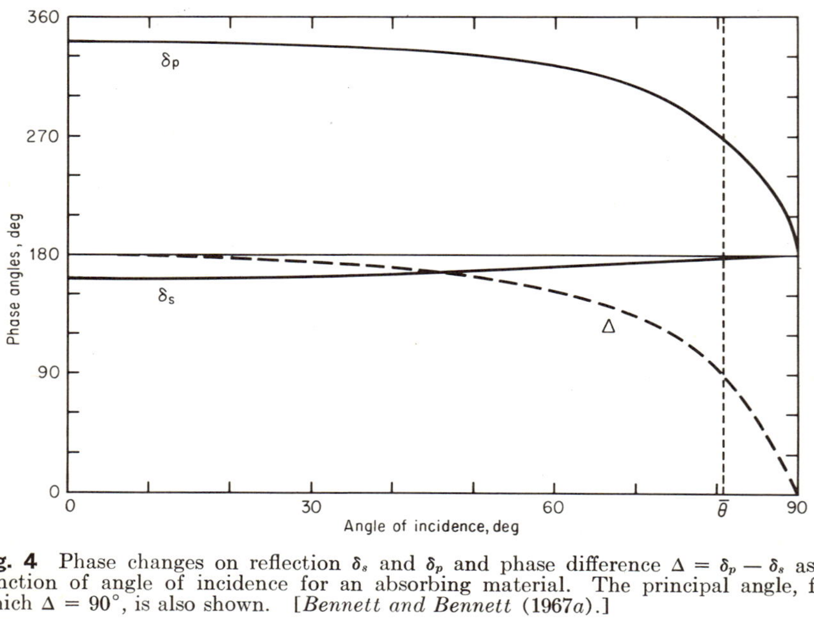

Fig. 4 Phase changes on reflection δ_s and δ_p and phase difference $\Delta = \delta_p - \delta_s$ as a function of angle of incidence for an absorbing material. The principal angle, for which $\Delta = 90°$, is also shown. [*Bennett and Bennett* (1967*a*).]

second Brewster angle. It is generally only slightly larger than θ_B'. The *principal angle* $\bar{\theta}$, at which $\delta_p - \delta_s = 90°$, is always larger than the second Brewster angle and θ_B'. For most metals θ_B' and $\bar{\theta}$ are only a fraction of a degree apart, but it is possible for them to differ by as much as 45° (Holl, 1963). There is no polarizing angle as such for an absorbing material because the angle of refraction is complex.

3. Basic Relations for Polarizers A linear† polarizer is anything which when placed in an incident unpolarized beam produces a beam of light whose electric vector is vibrating primarily in one plane, with only a small component vibrating in the plane perpendicular to it. If a polarizer is placed in a plane-polarized beam and is rotated about an axis parallel to the beam direction, the transmittance T will vary between a maximum value T_1 and a minimum value T_2 according to the law

$$T = (T_1 - T_2)\cos^2\theta + T_2 \tag{55}$$

Although the quantities T_1 and T_2 are called the *principal transmittances*, in general $T_1 \gg T_2$; θ is the angle between the plane of the principal transmittance T_1 and the plane of polarization of the incident beam. If the polarizer is placed in a beam of unpolarized light, its transmittance is

$$T = \tfrac{1}{2}(T_1 + T_2) \tag{56}$$

so that a perfect polarizer would transmit only 50 percent of an incident unpolarized beam‡.

When two identical polarizers are placed in an unpolarized beam, the resulting transmittance will be

$$T_{\parallel} = \tfrac{1}{2}(T_1^2 + T_2^2) \tag{57}$$

when their principal transmittance directions are parallel and will be

$$T_{\perp} = T_1T_2 \tag{58}$$

when they are perpendicular. In general, if the directions of principal transmittance are inclined at an angle θ to each other, the transmittance of the pair will be

$$T_\theta = \tfrac{1}{2}(T_1^2 + T_2^2)\cos^2\theta + T_1T_2\sin^2\theta \tag{59}$$

The polarizing properties of a polarizer are generally defined in terms of its *degree of polarization* P§,||

$$P = \frac{T_1 - T_2}{T_1 + T_2} \tag{60}$$

or its *extinction ratio* ρ_p

$$\rho_p = \frac{T_2}{T_1} \tag{61}$$

When one deals with nonnormal-incidence reflection polarizers, one generally writes P and ρ_p in terms of R_p and R_s, the reflectances of light polarized parallel and perpendicular to the plane of incidence, respectively. As will be shown in Par. 51, R_s can be

† Circular polarizers are discussed in Pars. 56, 74, and 75.

‡ Jones (1962) has pointed out that a perfect polarizer can transmit more than 50 percent of an incident unpolarized beam under certain conditions. In a doubly refracting material such as calcite, two beams polarized at mutually perpendicular angles to each other are transmitted. If a suitably oriented half-wave plate is placed in one of the beams, the plane of vibration can be rotated to coincide with that of the other beam. However, the two beams will either be parallel but separated or will be overlapping at one point but traveling in different directions.

§ Bird and Shurcliff (1959) distinguish between *degree of polarization*, which is a constant of the light beam, and *polarizance*, which is a constant of the polarizer. The polarizance is defined as being equal to the degree of polarization the polarizer produces in an incident monochromatic beam that is unpolarized. In practice, incident beams are often slightly polarized, so that the polarizance values differ slightly from the ideal degree of polarization. Other authors have not followed this distinction.

|| Authors dealing with topics such as scattering from aerosols sometimes define *degree of polarization* (of the scattered light) in terms of the Stokes vectors (Par. 4) as $P = (S_1^2 + S_2^2 + S_3^2)^{1/2}/S_0$.

equated to T_1 and R_p to T_2, so that Eqs. (60) and (61) become $P = (R_s - R_p)/(R_s + R_p)$ and $\rho_p = R_p/R_s$. If either ρ_p or P is known, the other can be deduced since

$$P = \frac{1 - \rho_p}{1 + \rho_p} \tag{62}$$

and

$$\rho_p = \frac{1 - P}{1 + P} \tag{63}$$

If one is determining the degree of polarization or the extinction ratio of a polarizer, the ratio of $T_\perp$ to $T_\parallel$ can be measured for two identical polarizers in unpolarized light. From Eqs. (57) and (58),

$$\frac{T_\perp}{T_\parallel} = \frac{T_1 T_2}{(T_1^2 + T_2^2)/2} \approx \frac{2T_2}{T_1} = 2\rho_p \tag{64}$$

if $T_2^2 \ll T_1^2$. If a perfect polarizer or a source of perfectly plane-polarized light is available, T_2/T_1 can be determined directly by measuring the ratio of the minimum to the maximum transmittance of the polarizer. Other relations for two identical partial polarizers are given by West and Jones (1951), as well as the transmittance $T_{\theta ab}$ of two dissimilar partial polarizers a and b whose principal axes are inclined at an angle θ with respect to each other. This latter expression is

$$T_{\theta ab} = \tfrac{1}{2}(T_{1a}T_{1b} + T_{2a}T_{2b}) \cos^2 \theta + \tfrac{1}{2}(T_{1a}T_{2b} + T_{1b}T_{2a}) \sin^2 \theta \tag{65}$$

where the subscripts 1 and 2 refer to the principal transmittances, as before.

In spectrophotometric measurements involving polarizers and dichroic† (optically anisotropic) samples, one should know the degree of polarization of the polarizer and how to correct for instrumental polarization. This latter quantity may arise from nonnormal-incidence reflections from a grating, dispersing prism, or mirrors. Light sources are also sometimes polarized. Šimon (1951) and Charney (1955) suggest methods for dealing with imperfect polarizers, dichroic samples, and instrumental polarization. In addition, when a dichroic sample is placed between a polarizer and a spectrophotometer which itself acts like an imperfect polarizer, one has effectively three polarizers in series. This situation has been treated by Jones (1956), who showed that anomalies can arise when the phase retardation of the polarizers takes on certain values.

4. Matrix Methods for Computing Polarization In dealing with problems involving polarized light, it is often necessary to determine the effect of various types of polarizers (linear, circular, elliptical, etc.), rotators, retardation plates, and other polarization-sensitive devices on the state of polarization of a light beam. The Poincaré sphere construction is helpful for giving a qualitative understanding of the problem; for quantitative calculations, one of several forms of matrix calculus can be used. The matrix methods are based on the fact that the effect of a polarizer or retarder is to perform a linear transformation (represented by a matrix) on the vector representation of a polarized light beam. The advantage of these methods over conventional techniques is that problems are reduced to simple matrix operations; thus since one does not have to think through the physics of every problem, the probability of making an error is greatly reduced. The most common forms of matrix calculus are the Mueller calculus and the Jones calculus, but the coherency-matrix formulation is also gaining popularity for dealing with problems involving partially polarized light. We give here a brief description of the Poincaré sphere and the different matrix methods, indicating how they are used, the different types of problems for which they are helpful, and where complete descriptions of each may be found.

The *Poincaré sphere* is a useful device for visualizing the effects of polarizers and retarders on a beam of polarized light. The various states of polarization are represented on the sphere as follows. The equator represents various forms of linear polarization, the poles represent right and left circular polarization,‡

†See Par. 44 for definition of the term dichroic.

‡ Right circularly polarized light is defined as a *clockwise* rotation of the electric vector when the observer is looking *against the direction the wave is traveling.*

and other points on the sphere represent elliptically polarized light. Every point on the sphere corresponds to a different polarization form. The radius of the sphere indicates the intensity of the light beam (which is usually assumed to be unity). The effects of polarizers and retarders are determined by appropriate displacements on the sphere. Partially polarized light or absorption may be dealt with approximately by ignoring the intensity factor, since one is generally interested only in the state of polarization; however, the construction is most useful when dealing with nonabsorbing materials. A good introductory description of the Poincaré sphere, including references, can be found in Shurcliff (1962*a*),† and illustrative examples and problems are given in Sutton and Panati (1969). More comprehensive treatments are given by Ramachandran and Ramaseshan (1961*a*)‡ and Jerrard (1954*b*)‡ and include numerous examples of applications to various types of problems. The main advantage of the Poincaré sphere, like other graphical methods, is to reveal by essentially a physical argument which terms in exceedingly complex equations are negligible or can be made negligible by modifying the experiment. It is characteristic of problems in polarized light that the trigonometric equations are opaque to inspection and yield useful results only after exact calculation with the aid of a computer or after complex manipulation and rather clever trigonometric identities. The Poincaré sphere thus serves as a guide to the physical interpretation of otherwise obscure polarization phenomena. It can be used for solving problems involving retarders or combinations of retarders (Shurcliff, 1962*a*; Koester, 1959*a*; Pancharatnam, 1955*a,b*), compensators, half-shade devices, and depolarizers (Ramachandran and Ramaseshan, 1961*a*), and it has also been applied to ellipsometric problems (McCrackin et al., 1963) and stress-optical measurements (Michael, 1968).

The Poincaré sphere is based on the Stokes vectors, which are sometimes designated S_0, S_1, S_2, and S_3. The physical interpretation of the vectors is as follows. S_0 is the intensity of the light beam, corresponding to the radius of the Poincaré sphere. S_1 is the difference in intensities between the horizontal and vertical polarization components of the beam; when S_1 is positive, the preference is for horizontal polarization, and when it is negative, the preference is for vertical polarization.§ S_2 indicates preference for $+45°$ or $-45°$ polarization, depending upon whether it is positive or negative, and S_3 gives the preference for right or left circular polarization. The Stokes vectors S_1, S_2, and S_3 are simply the three cartesian coordinates of a point on the Poincaré sphere: S_1 and S_2 are at right angles to each other in the equatorial plane, and S_3 points toward the north pole of the sphere.† Thus, any state of polarization of a light beam can be specified by these three Stokes vectors. The intensity vector S_0 is related to the other three by the relation $S_0^2 = S_1^2 + S_2^2 + S_3^2$ when the beam is completely polarized. If the beam is partially polarized, $S_0^2 > S_1^2 + S_2^2 + S_3^2$. Good introductory material on Stokes vectors is given by Shurcliff (1962*a*), Sutton and Panati (1969), and Walker (1954). Rigorous definitions of the simple vectors and those for partially coherent light can be found in Born and Wolf (1965*a*); other authors are cited by Shurcliff (1962*a*). Stokes vectors are generally used in conjunction with the Mueller calculus, and examples of applications will be given. We only note here that Budde (1962) has demonstrated a method for experimentally determining the Stokes vectors and other polarization parameters from a Fourier analysis of measured quantities. Ioshpa and Obridko (1963) have proposed a photoelectric method for simultaneously and independently measuring the four Stokes parameters.

The matrix methods for solving problems involving polarized light have certain properties in common. All use some type of representation for the original light beam (assumed to be a plane wave traveling in a given direction) that uniquely describes its state of polarization. Generally the beam is completely polarized,

† Shurcliff has the S_3 axis pointing down, so that the upper pole represents left circular polarization. The more logical convention, followed by most others, is for the upper pole to represent right circular polarization.

‡ The notation is similar to that used by Shurcliff (1962*a*), with the upper pole representing left circular polarization.

§ Some authors dealing with light scattering from aerosols define S_1 as positive when the preference is for vertical polarization.

but for some of the matrix methods it can also be unpolarized or partially polarized or its phase may be specified. The beam encounters one or more devices which change its state of polarization. These are called *instruments* and are represented by appropriate matrices. After the instruments operate on the light beam, it emerges as an outgoing plane wave in an altered state of polarization. The basic problem for all the methods is to find a suitable representation for the incident plane wave (usually a 2- or 4-component column vector), and the correct matrices (2×2 or 4×4) to represent the instruments. Once the problem is set up, one can perform the appropriate matrix operations to obtain a representation for the outgoing plane wave. Its properties are interpreted in the same way as the properties of the incident plane wave.

An introduction to the Jones and Mueller calculus is given by Shurcliff (1962*a*), and an excellent systematic and rigorous discussion of all the matrix methods has been given by O'Neill (1963). Both references contain tables of vectors for the various types of polarized beams and tables of instrument matrices. More complete tables are given by Sutton and Panati (1969). In the Mueller calculus the beam is represented by the 4-component Stokes vector, written as a column vector. This vector has all real elements and gives information about *intensity* properties of the beam. Thus it is not able to handle problems involving phase changes or combinations of two beams that are coherent. The instrument matrix is a 4×4 matrix with all real elements. In the Jones calculus, the Jones vector is a 2-component column vector that generally has complex elements. It contains information about the *amplitude* properties of the beam and hence is well suited for handling coherency problems. However, it cannot handle problems involving depolarization, as the Mueller calculus can. The Jones instrument matrix is a 2×2 matrix whose elements are generally complex.

Shurcliff (1962*a*) has noted some additional differences between Jones calculus and Mueller calculus. The Jones calculus is well suited to problems involving a large number of similar devices arranged in series in a regular manner and permits an investigator to arrive at an answer expressed explicitly in terms of the number of such devices. The Mueller calculus is not suited for this type of problem. The Jones instrument matrix of a train of transparent or absorbing nondepolarizing polarizers and retarders contains no redundant information. The matrix contains 4 elements each of which has 2 parts, so that there are a total of 8 constants, none of which is a function of any other. The Mueller instrument matrix of such a train contains much redundancy; there are 16 constants but only 7 of them are independent.

In order to handle problems involving partially coherent, partially polarized light, coherency-matrix formalism has been developed. In this system the beam is represented by a 4×4 matrix called a *coherency* or *density matrix,* which is the time average of the product of the Jones vector with its hermitian conjugate. The instrument matrices are the same as those used in the Jones calculus. O'Neill (1963) and Born and Wolf (1965*a*) have good basic descriptions of coherency-matrix formalism; later extensions of the theory are given by Marathay (1965, 1966).

There have been some modifications of the various matrix methods. Priebe (1969) has introduced an operational notation for the Mueller matrices that facilitates the analysis by simplifying the functional description of a train of optical components. Collins and Steele (1962) have suggested a modification of the Jones calculus in which the light vector is expressed as the sum of two circularly polarized (rather than linearly polarized) components. Schmieder (1969) has given a unified treatment of Jones calculus and Mueller calculus including the coherency matrix and has shown that if the Stokes parameters are ordered in a different way from that customarily used, familiar relationships are preserved and the rotation matrix looks like a rotation matrix rather than like a rearranged one. Tewarson (1966) presents a generalized reciprocity equation expressing an algebraic relationship between the parameters of an optical system and its reciprocal system and has verified the equation for both plane-polarized and circularly polarized light beams. Since his equation follows from the reciprocity law in the Mueller calculus, that law is verified also. Cernosek (1971) presents a simple geometric method based on the properties of quaternions to give a quick, quantitative analysis of the effect of any combination of linear retarders and rotators on the state of polarization of a system.

Among the applications of Mueller calculus and Jones calculus to problems involving polarized light, McCrackin (1970) has used both matrix methods to analyze instrumental errors in ellipsometry, and Hellerstein (1963) has used Mueller calculus to study the passage of linearly, circularly, and elliptically polarized light through a Sénarmont polariscope. Azzam and Bashara (1971) have used Jones calculus to give a unified analysis of errors in ellipsometry, including effects of birefringence in cell windows, imperfect components, and incorrect azimuth angles.

5. Nonnormal-Incidence Photometry Methods for determining optical constants from nonnormal-incidence reflectance measurements may be divided into two general categories: (1) those in which the reflected intensity or intensity ratio R_p/R_s is measured at the Brewster or other angle of incidence and (2) those in which angular measurements leading to the reflected amplitude ratio $|r_p|/|r_s|$ or to the phase difference between the reflected p and s components are made at a specified angle of incidence. The first category, which involves intensity, i.e., photometric, measurements and is sometimes called polarimetry,† will be covered in this paragraph; the second, ellipsometry, will be discussed in Par. 6. Various types of photometric measurements are possible (Bennett and Bennett, 1967*a*; Wendlandt and Hecht, 1966), but we shall mention only three: measurements of the refractive indexes of dielectric films and substrates at the Brewster angle, determination of n and k from reflectance ratio or related measurements, and attenuated-total-reflection measurements of optical constants.

When one has a film-free dielectric substrate, the reflectance R_p is zero at the Brewster angle and Eq. (48) holds. Thus, by measuring the angle at which the reflectance of the p component is a minimum, one can determine n for the substrate. Since the minimum is rather broad, one of the photoelectric methods described by Bennett and Bennett (1967*a*) can be used to give an accurate value. The Abelès Brewster angle method for measuring the refractive index of a thin dielectric film (Abelès, 1949, 1950*b*, 1963; Heavens, 1964) is both simple and accurate. If a film-covered dielectric substrate is illuminated with p-component light, at the Brewster angle for the film no light is reflected from the air-film interface, and so the reflectance of the film-covered surface is the same as that of the uncoated substrate. Equation (48) then gives the refractive index of the *film* ($n_0 = 1$ for air). Either visual or photoelectric techniques may be used to match reflectances of the unfilmed and film-covered substrate; the method has its greatest sensitivity when the refractive index of the film is within ± 0.3 of that of the substrate. Variations of the Abelès method, described in Bennett and Bennett (1967*a*) can be used to obtain comparable accuracy when the refractive index of the substrate differs by more than 0.3 from that of the film.

If one is measuring the optical constants of a metal or other absorbing material, various combinations of R_s, R_p, $R_{av} = (R_s + R_p)/2$, or R_p/R_s at various angles of incidence are possible. One method refined by Potter (1964, 1965) involves measurements of the pseudo Brewster angle θ_B' and R_p/R_s at this angle. Figure 5 gives values of θ_B' and R_p/R_s as a function of the reflectance ratio n_1/n_0 and k_1, the extinction coefficient for the material. (The incident medium is generally assumed to be air, so that $n_0 = 1$.) This figure shows that θ_B' increases as both n_1 and k_1 increase. To a first approximation, lines of constant θ_B' are circles centered at the origin, and lines of constant R_p/R_s are radii. Since n_1 and k_1 are functions of wavelength, it is possible to estimate the approximate pseudo Brewster angle and the value of R_p/R_s to use at any given wavelength if one knows roughly how n_1 and k_1 vary with wavelength.

Methods using reflectance-ratio measurements are most sensitive for values of $R_p/R_s < 0.9$. When the ratio approaches 1, as it does for many metals, the polarization-modulation technique developed by Jasperson and Schnatterly (1969) is much more sensitive (see also Jasperson et al., 1973). The quantity $(1 - R_p/R_s)/$

† Chemists use the term *polarimetry* to denote a quantitative investigation of the change in the direction of vibration of linearly polarized light during its passage through an optically anisotropic liquid or solution, i.e., *optical rotation*. This subject has been comprehensively covered by Heller (1960*a*).

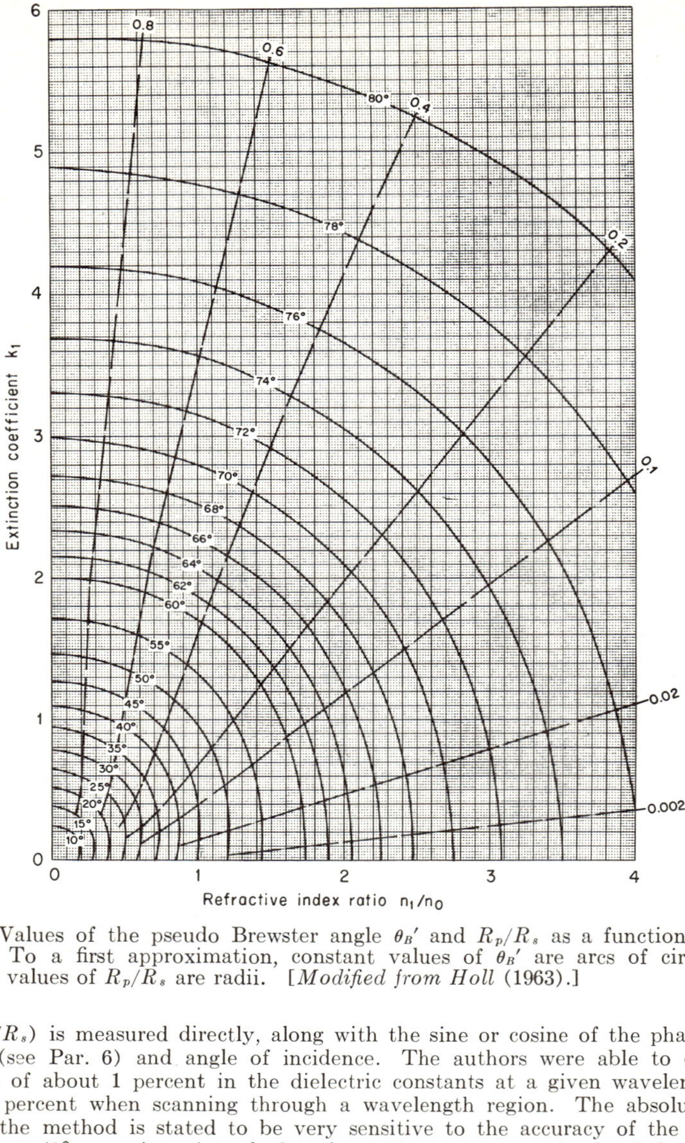

Fig. 5 Values of the pseudo Brewster angle θ_B' and R_p/R_s as a function of n_1/n_0 and k_1. To a first approximation, constant values of θ_B' are arcs of circles, and constant values of R_p/R_s are radii. [*Modified from Holl* (1963).]

$(1 + R_p/R_s)$ is measured directly, along with the sine or cosine of the phase difference Δ (see Par. 6) and angle of incidence. The authors were able to obtain a precision of about 1 percent in the dielectric constants at a given wavelength and about 4 percent when scanning through a wavelength region. The absolute accuracy of the method is stated to be very sensitive to the accuracy of the angle of incidence θ_0 (1° error in θ_0 introducing about 10 percent error in the dielectric constants). Thus the angle of incidence should be known to better that 0.1° to translate the excellent precision of the method into absolute accuracy.

Attenuated total reflection has become a popular technique for measuring the optical constants or absorption coefficient of materials which are too highly absorbing to be measured by transmission (Harrick, 1967). This technique has the advantages that the sample surface is protected, roughness of the sample does not affect the measurements, and highly accurate measurements of the absorption of very thin or weakly absorbing films can be made. Generally the greatest sensitivity in n and k is obtained if R_s is measured at two angles of incidence, one less and one greater than the critical angle. The relationships are complicated, and a computer program (Horton and Hansen, 1967) is essential to reduce the data.

Recent developments in internal-reflection spectroscopy are given in the published proceedings of a symposium on reflectance spectroscopy (Wendlandt, 1968).

6. Ellipsometry The phase change on reflection is much more sensitive to the presence of thin surface films than the reflectance is. It is difficult to measure δ_s or δ_p directly, although precise absolute-phase-change measurements at normal incidence have been made using interferometric techniques (Bennett, 1964). Fortunately the difference between δ_p and δ_s is also sensitive to surface films; both this relative phase change Δ, defined as

$$\Delta = \delta_p - \delta_s \tag{66}$$

and the relative amplitude attenuation, defined as

$$\tan \psi = \frac{|r_p|}{|r_s|} \tag{67}$$

can be determined by a technique known as *ellipsometry*. A graphical representation of the angles ψ and Δ is shown in Fig. 6 both for reflection at angles less

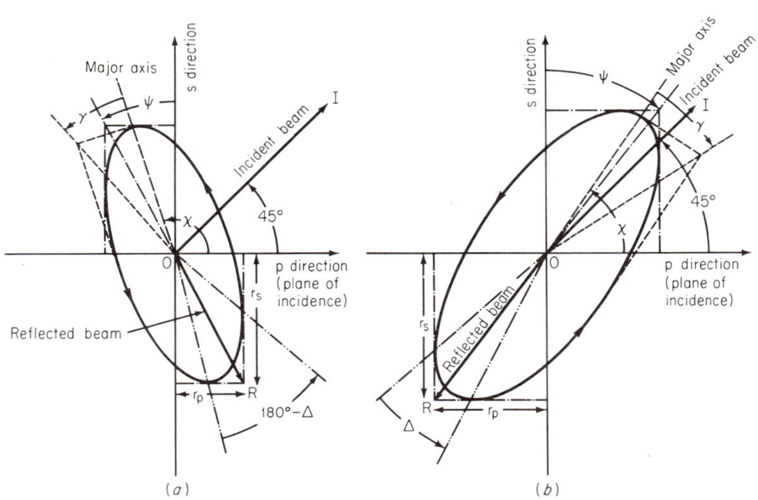

Fig. 6 Graphical representation of various ellipsometric parameters for reflection at an angle of incidence (a) less than and (b) geater than the principal angle. [*Bennett and Bennett* (1967a).]

than and greater than the principal angle. For convenience the incident light has been assumed to be linearly polarized at 45° to the plane of incidence. At the principal angle the s and p directions are along the semimajor and semiminor axes, respectively. If the sample were a single dielectric interface, r_s and r_p should be 180° out of phase for angles less than the principal angle, Δ would equal π, and the ellipse would collapse into line OR. If the surface were absorbing, however, or if there were a surface film, r_s and r_p would no longer be 180° out of phase and elliptical polarization would result.

A typical ellipsometric setup is shown in Fig. 7. Here a quarter-wave plate is fixed at 45° (angles are measured from the plane of incidence in a counterclockwise direction when looking against the beam) and is mounted preceding the sample in the optical path. By rotating the polarizer and analyzer until the light is extinguished and then noting their settings, the values of ψ and Δ for the sample can be determined from the relations

$$\tan \psi = \cot L \tan(-A) \tag{68}$$

$$\tan \Delta = \sin \Delta_c \tan(90° - 2P) \tag{69}$$

$$\cos 2L = -\cos \Delta_c \cos 2P \tag{70}$$

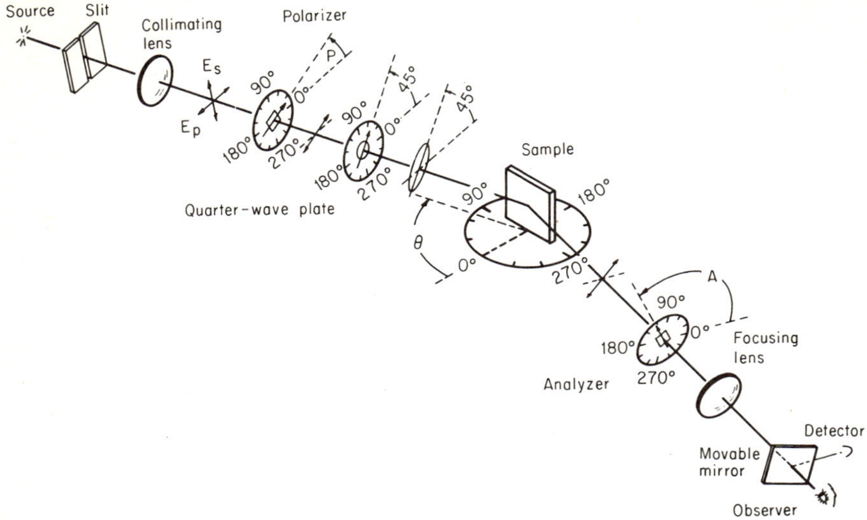

Fig. 7 Ellipsometric apparatus for measuring ψ and Δ with the quarter-wave plate set at 45° and the polarizer and analyzer rotated to produce extinction.

where P and A are the polarizer and analyzer angles, respectively, and Δ_c is the retardation of the quarter-wave-plate compensator.† Extinction positions can be found for A in both the first and fourth quadrants, but Eqs. (68) to (70) refer to the settings in the fourth quadrant. Measurements made with A in these two quadrants are not redundant and not only make it possible to determine Δ_c but also to cancel certain systematic errors which otherwise would affect the accuracy of the measurement (McCrackin et al., 1963; Yolken et al., 1967; Oldham, 1967; Holmes and Feucht, 1967; Archer and Shank, 1967). Values for both quadrants should thus always be obtained when making absolute ellipsometric measurements. Fortunately, for settings near extinction the positions of the polarizer and analyzer are nearly independent, and one does not obtain the local spurious minima which we and others have observed when the polarizer is fixed at 45° and the quarter-wave plate and analyzer are rotated to obtain extinction. McCrackin (1970) and Aspnes (1971) give comprehensive analyses of instrumental errors in ellipsometry and suggest methods for correcting them.

Although the arrangement shown in Fig. 7 is more commonly used, an alternate procedure is to place the quarter-wave plate between the sample and the analyzer. In this case the functions of the polarizer and analyzer are interchanged. Various other techniques for making ellipsometric measurements have also been reported (Bennett and Bennett, 1967*b*). Properly used, ellipsometry not only provides a means for accurately determining the optical constants of a clean surface but is one of the most sensitive techniques for measuring the growth of surface films. It can provide accurate average thickness values even when the surface "film" is highly discontinuous (Bennett et al., 1969). Changes in average film thickness of 0.07 Å, or less than 0.03 monolayer, have been measured using this technique (Archer, 1964; Archer and Gobeli, 1965).

† These relations are the same as those given by Bennett and Bennett (1967*b*). However, they should refer to the ellipsometric arrangement in which the quarter-wave plate precedes the sample (Fig. 7) rather than that shown in Fig. 43 of Bennett and Bennett (1967*b*), where the quarter-wave plate follows the sample. In the latter arrangement the functions of the polarizer and analyzer are interchanged.

PRISM POLARIZERS

7. Double Refraction in Calcite Although many minerals, specifically those which do not have a cubic crystal structure, are doubly refracting, nearly all polarizing prisms used in the visible, near-ultraviolet, and near-infrared regions of the spectrum are made from optical calcite, which exhibits strong birefringence over a wide wavelength range. Polarizing prisms made from other birefringent crystals are used primarily in the ultraviolet at wavelengths for which calcite is opaque. Prism materials used successfully in this region include crystalline quartz, magnesium fluoride, sodium nitrate, and ammonium dihydrogen phosphate. Rutile polarizing prisms have been used beyond the calcite cutoff in the infrared. A new prism material, yttrium orthovanadate, has been used to make high-transmission polarizers for the visible and near-infrared spectral regions. Many of the noncalcite polarizers are described in Par. 43.

Next to quartz, calcite is the most widely distributed of all minerals and usually occurs in an impure polycrystalline form as marble, limestone, or chalk. However, optical calcite, or Iceland spar, is quite rare. Before the twentieth century it came almost entirely from a cavity in basalt 11 by 4.6 m^2 in area and about 3 m deep, located by the farm Helgustadir near Reydar Fjördur on the east coast of Iceland. When first found, the cavity was filled with a pure crystallized mass containing enormous crystals, some a yard across. This source is now exhausted, but crystals from it described in detail by Erasmus Bartholinus in 1669, led to the discovery of the laws of double refraction by Christian Huygens in 1690 and to the discovery of the polarization of light by Malus in 1808. Optical calcite now comes principally from Mexico, Africa, and Siberia. It has been grown artificially by a hybrid gel-solution method (Nickl and Henisch, 1969) but maximum edge lengths are only 3 to 4 mm.

Although calcite is much softer than glass, with care it can be worked to an excellent polish. Surfaces flat to one-fifth fringe or even, with care, one-tenth fringe which are free from surface defects or perceptible turned edges can be produced using more or less conventional pitch-polishing techniques (Smartt, 1961). Such techniques fail only for surfaces normal to the optic axis, in which case pitch polishing tends to cleave out small tetrahedra. Such surfaces can be polished to a lower surface quality using cloth polishers.

Crystals of calcite are negative uniaxial and display a prominent double refraction. The material can easily be cleaved along three distinct planes, making it possible to produce rhombs of the form shown in Fig. 8. At points B and H, a given face makes an angle of 101°55′ with each of the other two. At all the other points, two of the angles are 78°5′ and one is 101°55′. The *optic axis HI,* the direction in the crystal along which the two sets of refracted waves travel at the same velocity, makes equal angles with all three faces at point H.† Any plane, such as $DBFH$, which contains the optic axis and is perpendicular to the two opposite faces of the rhomb $ABCD$ and $EFGH$ is called a *principal section.* An end view of the rhomb showing the edge of the principal section (DB) is seen in Fig. 9, and a side view of the principal section $DBFH$ is shown in Fig. 10. If light is incident on the rhomb so that the plane of incidence coincides with a principal section, the light is broken up into two components polarized at right angles to each other. One of these, the ordinary ray o, obeys Snell's law and has its plane of vibration perpendicular to the principal section.‡ The second, the extraordinary ray e, has its plane of vibration parallel to the principal section. The refraction of the extraordinary ray in some cases violates Snell's law, at least in its simple form. The anomalous deflection of the ray is caused by the wavefront's becoming ellipsoidal, so that the direction of propaga-

† The direction of the optic axis in a uniaxial crystal such as calcite or crystalline quartz can be determined by observing the crystal between crossed polarizers. If the alignment is correct, so that the optic axis is parallel to the line of sight, there will be concentric colored circles with a black cross superimposed (Wahlstrom, 1969).

‡ We are following the convention mentioned in Par. 2 of specifying the *plane of vibration* of the E vector, *not* the *plane of polarization.*

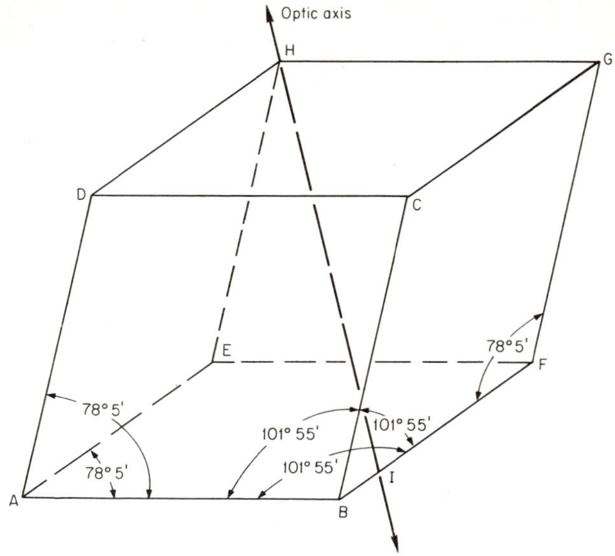

Fig. 8 Schematic representation of a rhombohedral calcite crystal showing the angles between faces. The optic axis passes through corner *H* and point *I* on side *BF*.

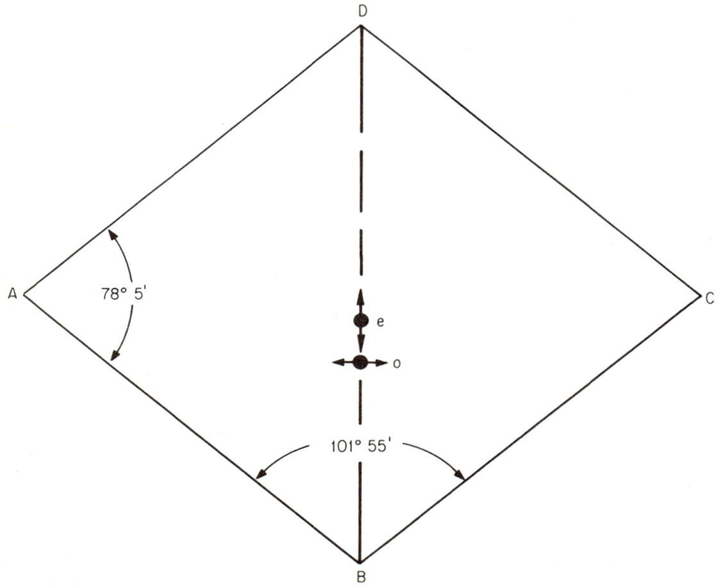

Fig. 9 End view of the calcite rhomb in Fig. 8, showing the edge of the principal section (*DB*) and the emergence positions of the *o* and *e* rays when the incident ray is centered on face *EFGH* (see Fig. 10).

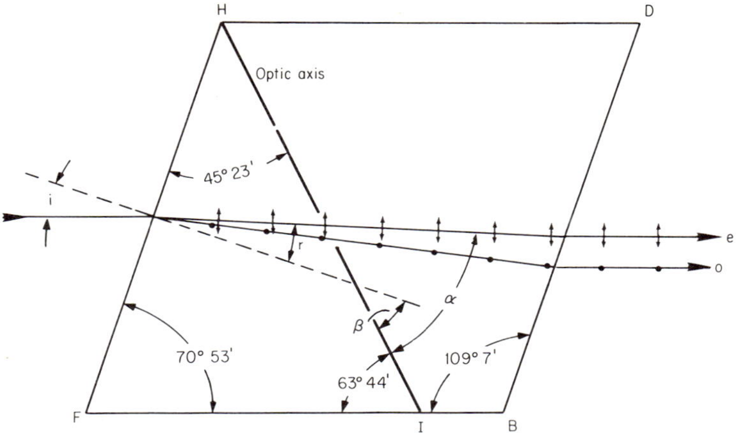

Fig. 10 Side view of a principal section for the calcite rhomb in Fig. 8. The direction of the optic axis and the angles of the principal section are indicated. The angle of incidence is i, angle of refraction is r, angle between the e ray and the optic axis is α, and angle between the normal to the surface and the optic axis is β. The directions of vibration of the e and o rays are in the plane of the paper and perpendicular to it, respectively.

tion of the light is not along the wave normal. This ellipticity causes the velocity of the light in the crystal, and hence its refractive index, to be a function of angle. If light is incident on rhomb face *EFGH* parallel to edge *BF* of the rhomb, the o and e rays, both of which lie in a principal section, are as shown in Figs. 9 and 10. As the angle of incidence is changed in Fig. 10 so that the direction taken by the o ray approaches that of the optic axis *HI*, the separation between the e and o rays decreases. If the rhomb is rotated about an axis parallel to *HD*, the e ray will precess about the o ray. However, unlike the o ray, it will not remain in the plane of incidence unless this plane coincides with the principal section.

The plane containing the o ray and the optic axis is defined as the *principal plane of the o ray* and that containing the e ray and the optic axis as the *principal plane of the e ray.* In the case discussed above, the two principal planes and the principal section coincide. In the general case they may all be different. However, in all cases the o ray is polarized with its plane of vibration perpendicular to its principal plane and the e ray with its plane of vibration in its principal plane (see Fig. 10). It may be helpful to think of the crystal as a block of wood having a grain, the optic axis. The o ray always vibrates perpendicular to the grain and hence has only one velocity and thus one index of refraction n_o. The e ray vibrates in a plane containing the grain, and depending on its inclination to the grain, the vibration direction can vary from parallel to perpendicular to it. The index of the e ray can thus vary from the o-ray value, when the vibration is perpendicular to the grain, to a maximum difference from it when the vibration direction is parallel to the grain. In all cases, the vibration direction of the e ray remains perpendicular to that of the o ray.

The value of the index of refraction of the e ray which differs most from that of the o ray, i.e., the index when the e ray vibrations are parallel to the optic axis, is called the *principal index for the extraordinary ray* n_e. Figure 11 shows a situation in which the e ray takes on its principal index. Snell's law can be used to calculate the path of the e ray through a prism for this case. In Fig. 11 the plane of the surface is perpendicular to the paper, and the optic axis is also perpendicular to the paper. The principal section and the two principal planes are all different; the e ray vibrates in its principal plane, but the o ray vibrates at right angles to its principal plane.

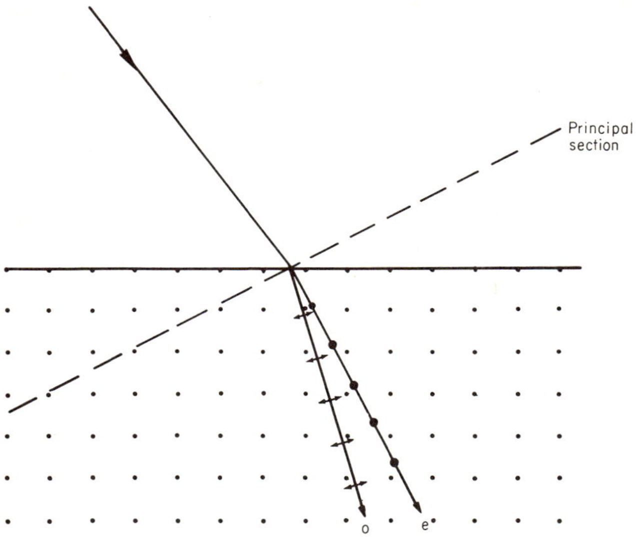

Fig. 11 Refraction of a ray which is not in the principal section of a calcite rhomb. The principal section is a plane perpendicular to the plane of the paper; the direction of the optic axis, indicated by dots, is also perpendicular to the paper. The directions of vibration of the *o* and *e* rays are in the plane of the paper and perpendicular to it, respectively. The principal planes of the *o* and *e* rays are both perpendicular to the paper but are at an angle to each other and to the principal section.

Table 2 lists values of n_o and n_e for calcite, along with the two absorption coefficients α_o and α_e, all as a function of wavelength. Since $n_e < n_o$ in the ultraviolet, visible, and infrared regions, calcite is a negative uniaxial crystal. However, at wavelengths shorter than 1520 Å in the vacuum ultraviolet, the birefringence $n_e - n_o$ becomes positive, in agreement with theoretical predictions (Uzan et al., 1969; Damany, 1970).† The range of transparency of calcite is approximately from 0.214 to 3.3 μm for the extraordinary ray but only from about 0.23 to 2.2 μm for the ordinary ray.

The direction of propagation and the refraction of the ordinary ray in calcite are determined simply by Snell's law. However, calculations for the extraordinary ray may be more complicated. For the special case in which the optic axis is at right angles to the plane of incidence (Fig. 11), the extraordinary ray follows Snell's law also, except with index n_e rather than n_o. For other cases the *normal to the wavefront* and the *ray direction,* i.e., the direction of propagation of the wave as given by the Poynting vector, no longer coincide.

If the principal plane of the *e* ray and the principal section coincide (Fig. 10), the wave normal (*but not the e ray*) obeys Snell's law, except that the index of refraction n_ϕ of this wave is given by (Born and Wolf, 1965*b*; Martin, 1948)

$$\frac{1}{n_\phi^2} = \frac{\sin^2 \phi}{n_e^2} + \frac{\cos^2 \phi}{n_o^2} \tag{71}$$

where ϕ is the angle between the direction of the *wave normal* and the optic axis ($\phi \leq 90°$). When $\phi = 0°$, $n_\phi = n_o$, and when $\phi = 90°$, $n_\phi = n_e$. The angle of refraction for the wave normal is $\phi - \beta$, where β is the angle the normal to the surface

† For more recent data in the 0.17- to 0.19-μm region, see Uzan et al. (1970).

makes with the optic axis. Snell's law for the extraordinary-ray *wave normal* then becomes

$$n \sin i = \frac{n_e n_o \sin(\phi - \beta)}{(n_o^2 \sin^2 \phi + n_e^2 \cos^2 \phi)^{1/2}} \tag{72}$$

where i is the angle of incidence of light in a medium of refractive index n. Since all other quantities in this equation are known, ϕ is uniquely determined but often must be solved for by iteration. Once ϕ is known, the angle of refraction r for the extraordinary *ray* can be determined as follows. If α is the angle the *ray* makes with the optic axis ($\alpha \leq 90°$), then $r = \alpha - \beta$ and (Martin, 1948)

$$\tan \alpha = \frac{n_o^2}{n_e^2} \tan \phi \tag{73}$$

Although the angle of refraction of the extraordinary ray determines the path of the light beam through the prism, one must use the angle of refraction of the *wave normal*, $\phi - \beta$, in Fresnel's equation (21) when calculating the reflection loss of the e ray at the surface of the prism.

For the special case in which the optic axis is parallel to the surface as well as in the plane of incidence, α and ϕ are the complements of the angles of refraction of the ray and wave normal, respectively. If the light is normally incident on the surface, ϕ and α are both 90° and the extraordinary ray is undeviated and has its minimum refractive index n_e. In other cases for which the optic axis is not parallel to the surface, the extraordinary ray is refracted even for normal incidence.

If the plane of incidence is neither in a principal section nor perpendicular to the optic axis, it is more difficult to determine the angle of refraction of the extraordinary ray. In such cases, Huygens' construction is helpful (Martin, 1948; Ditchburn, 1963*a*; Schuster, 1920).

8. Types of Polarizing Prisms and Definitions In order to make a polarizing prism out of calcite, some way must be found to separate the two polarized beams. In wavelength regions where calcite is absorbing (and hence only a minimum thickness of calcite can be used), this separation has been made simply by using a very thin calcite wedge cut so that the optic axis is parallel to the faces of the wedge to enable the e and o rays to be separated by a maximum amount. The incident light beam is restricted to a narrow pencil. Calcite polarizers of this type can be used at wavelengths as short as 1900 Å (Müller, 1963). In more favorable wavelength regions, where the amount of calcite through which the light passes is not so critical, more sophisticated designs are usually employed. Such prisms can be divided into two main categories, conventional polarizing prisms (Pars. 9 to 32) and polarizing beam-splitter prisms (Pars. 33 to 41), and a third category, Feussner prisms (Par. 42).

In conventional polarizing prisms, only light polarized in one direction is transmitted. This is accomplished by cutting and cementing the two halves of the prism together in such a way that the other beam suffers total internal reflection at the cut. It is usually deflected to the side, where it is absorbed by a coating containing lampblack, etc. Since the ordinary ray, which has the higher index, is the one usually deflected, the lampblack is often mixed in a matching high-index binder such as resin of aloes ($n_D = 1.634$) or balsam of Tolu ($n_D = 1.628$) to minimize reflections (Dévé, 1954). When high-powered lasers are used, the coating is omitted to avoid overheating the prism and the light is absorbed externally.

Conventional polarizing prisms fall into two general categories: Glan types (Pars. 9 to 23) and Nicol types (Pars. 24 to 31), which are illustrated in Fig. 12. Glan types have the optic axis in the plane of the entrance face. If the principal section is parallel to the plane of the cut, the prism is a Glan-Thompson design (sometimes called a Glazebrook design); if perpendicular, a Lippich design; and if 45°, a Frank-Ritter design. In Nicol-type prisms, which include the various Nicol designs and the Hartnack-Prazmowsky, the principal section is perpendicular to the entrance face, but the optic axis is neither parallel nor perpendicular to the face.

Air-spaced prisms can be used at shorter wavelengths than cemented prisms, and special names have been given to some of them. An air-spaced Glan-Thompson

TABLE 2 Refractive Indexes[a] and Absorption Coefficients[a] for Calcite

λ, μm	n_o	α_o	n_e	α_e
0.1318	1.56[b]	534,000[b]	1.80[b]	477,000[b]
0.1355	1.48	473,000	1.84	380,000
0.1411	1.40	561,000	1.82	196,000
0.1447	1.48	669,000	1.80	87,000
0.1467	1.51	711,000	1.75	20,500
0.1478$_5$	1.54	722,000	1.75	17,000
0.1487	1.58	735,000	1.75	14,400
0.1495$_5$	1.62	714,000	1.75	12,600
0.1513	1.68	756,000	1.75	8,300
0.1518$_5$	1.72	753,000	1.74	10,700
0.1536	1.80	761,000	1.74	9,000
0.1544$_5$	1.87	748,000	1.74	6,500
0.1558$_5$	1.92	766,000	1.74	8,100
0.1581$_5$	2.02	715,000	1.73	11,100
0.1596	2.14	669,000	1.72	12,600
0.1608	2.20	594,000	1.70	13,300
0.1620	2.10	566,000	1.65	14,000
0.1633	2.00	608,000	1.65	10,800
0.1662	2.00	559,000	1.64	7,500
0.1700	1.94	414,000	1.63	4,400
0.1800	1.70	391,000	1.61	≤ 1,400
0.1900	1.72	278,000	1.59	≤ 321[d]
0.198	....	....	1.57796[c]	
0.200	1.90284[c]	257,000	1.57649	≤ 133
0.204	1.88242	....	1.57081	
0.208	1.86733	149,000	1.56640	
0.211	1.85692	....	1.56327	
0.214	1.84558	....	1.55976	~0.1
0.219	1.83075	....	1.55496	
0.226	1.81309	....	1.54921	
0.231	1.80233	....	1.54541	
0.242	1.78111	....	1.53782	
0.2475	....	0.159[e]		
0.2520	....	0.125		
0.256	....	0.109		
0.257	1.76038	....	1.53005	

λ, μm	n_o	α_o	n_e	α_e
0.397	....	0.000		
0.410	1.68014[c]	.....	1.49640[c]	
0.434	1.67552	.....	1.49430	
0.441	1.67423	.....	1.49373	
0.508	1.66527	.....	1.48956	
0.533	1.66277	.....	1.48841	
0.560	1.66046	.....	1.48736	
0.589	1.65835	.....	1.48640	
0.643	1.65504	.....	1.48490	
0.656	1.65437	.....	1.48459	
0.670	1.65367	.....	1.48426	
0.706	1.65207	.....	1.48353	
0.768	1.64974	.....	1.48259	
0.795	1.64886	.....	1.48215	
0.801	1.64869	.....	1.48216	
0.833	1.64772	.....	1.48176	
0.867	1.64676	.....	1.48137	
0.905	1.64578	.....	1.48098	
0.946	1.64480	.....	1.48060	
0.991	1.64380	.....	1.48022	
1.042	1.64276	.....	1.47985	
1.097	1.64167	.....	1.47948	
1.159	1.64051	.....	1.47910	
1.229	1.63926	.....	1.47870	
1.273	1.63849			
1.307	1.63789	.....	1.47831	
1.320	1.63767			
1.369	1.63681			
1.396	1.63637	.....	1.47789	
1.422	1.63590			
1.479	1.63490			
1.497	1.63457	.....	1.47744	
1.541	1.63381			
1.6	....	0.05[f]		
1.609	1.63261			
1.615	....	.....	1.47695	

0.2605	....	0.102		
0.263	1.75343	....	1.52736	
0.265	....	0.096		
0.267	1.74864	....	1.52547	
0.270	....	0.096		
0.274	1.74139	....	1.52261	
0.275	....	0.102		
0.2805	....	0.096		
0.286	....	0.102		
0.291	1.72774	....	1.51705	
0.2918	....	0.109		
0.2980	....	0.118		
0.303	1.71959	....	1.51365	
0.305	....	0.118		
0.312	1.71425	0.096	1.51140	
0.3195	....	0.059		
0.327	....	0.028		
0.330	1.70515	....	1.50746	
0.3355	....	0.028		
0.340	1.70078	....	1.50562	
0.3450	....	0.0170		
0.346	1.69833	....	1.50450	
0.3565	....	0.0112		
0.361	1.69316	....	1.50224	
0.3685	....	0.0056		
0.3820	....	0.0056		
0.394	1.68374	....	1.49810	

1.682	1.63127			
1.7	....	0.09		
1.749	....	....	1.47638	
1.761	1.62974			
1.8	....	0.16		
1.849	1.62800			
1.9	....	0.23		
1.909	....	....	1.47573	
1.946	1.62602			
2.0	....	0.37		
2.053	1.62372			
2.100	....	0.62	1.47492	0.02[f]
2.172	1.62099			
2.2	....	1.1	....	0.05
2.3	....	1.7	....	0.07
2.4	....	2.3	....	0.09
2.5	....	2.7	....	0.14
2.6	....	2.5	....	0.07
2.7	....	2.3	....	0.07
2.8	....	2.3	....	0.09
2.9	....	2.8	....	0.18
3.0	....	4.0	....	0.28
3.1	....	6.7	....	0.46
3.2	....	10.6	....	0.69
3.3	....	15.0	....	0.92
3.324	....	....	1.47392	
3.4	....	19.0	....	1.2

[a] Refractive indexes n_o and n_e are for the ordinary and extraordinary rays, respectively, and the corresponding absorption coefficients are $\alpha_o = 4\pi k_o/\lambda$ cm^{-1} and $\alpha_e = 4\pi k_e/\lambda$ cm^{-1}, where the wavelength λ is in centimeters. In the table, the wavelength is in micrometers.

[b] Uzan et al. (1969); α_o and α_e were calculated from the reported values of k_o and k_e.

[c] Wolfe et al. (1963*a*).

[d] Schellman et al. (1965); α_e was calculated from the optical density for the extraordinary ray.

[e] Bouriau and Lenoble (1957); reported absorption coefficient in this paper was for both *o* and *e* rays. α_o was calculated by assuming $\alpha_e = 0$.

[f] McCarthy et al. (1963).

prism is called a Glan-Foucault, and an air-spaced Lippich prism, a Glan-Taylor. In common practice, either of these may be called a Glan prism. An air-spaced Nicol prism is called a Foucault prism. Double prisms can also be made, thus increasing the prism aperture without a corresponding increase in length. Most double prisms are referred to as double Frank-Ritter, etc., but a double Glan-Thompson is called an Ahrens prism.

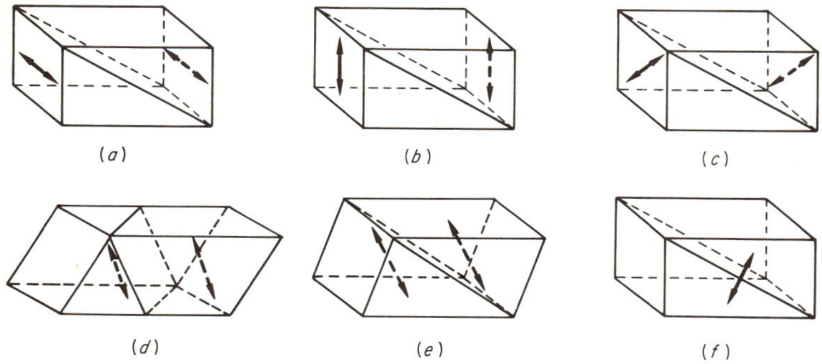

Fig. 12 Types of conventional polarizing prisms. Glan types: (*a*) Glan-Thompson, (*b*) Lippich, and (*c*) Frank-Ritter; Nicol types: (*d*) conventional Nicol, (*e*) Nicol, Halle form, and (*f*) Hartnack-Prazmowsky. The optic axes are indicated by the double-pointed arrows.

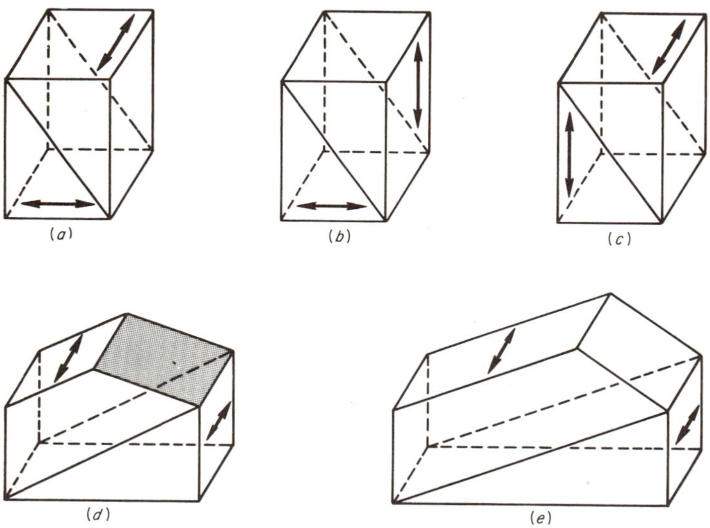

Fig. 13 Types of polarizing beam-splitter prisms: (*a*) Rochon, (*b*) Sénarmont, (*c*) Wollaston, (*d*) Foster (shaded face is silvered), and (*e*) beam-splitting Glan-Thompson. In each case the light is incident on the left face. The paths of the rays through the prisms are shown in Fig. 37.

In polarizing beam-splitter prisms two beams, which are polarized at right angles to each other, emerge but are separated spatially. The prisms have usually been used in applications for which both beams are needed, e.g., in interference experiments, but they can also be used when only one beam is desired. These prisms are also of two general types, illustrated in Fig. 13, those having the optic axis in the two sections of the prism perpendicular and those having them parallel.

Prisms of the first type include the Rochon, Sénarmont, Wollaston, double Rochon, and double Sénarmont. Prisms of the second type are similar to the conventional polarizing prisms but usually have their shape modified so that the two beams emerge in special directions. Examples are the Foster, the beam-splitting Glan-Thompson, and the beam-splitting Ahrens.

The Feussner-type prisms, shown in Fig. 14, are made of isotropic material, and the film separating them is birefringent. For negative uniaxial materials the ordinary ray rather than the extraordinary ray is transmitted. These prisms have the advantage that much less birefringent material is required than for the other types of polarizing prisms, but they have a more limited wavelength range when calcite or sodium nitrate is used because, for these materials, the extraordinary ray is transmitted over a wider wavelength range than the ordinary ray.

The amount of flux which can be transmitted through a prism or other optical element depends on both its angular aperture and its cross-sectional area. The greater the amount of flux which can be transmitted the better the *throughput* or *light-gathering power* (sometimes called *étendue* or *luminosity*) of the system (Jacquinot, 1954; Mertz, 1965). If a pupil or object is magnified, the convergence angle of the light beam is reduced in direct ratio to the increase in size of the

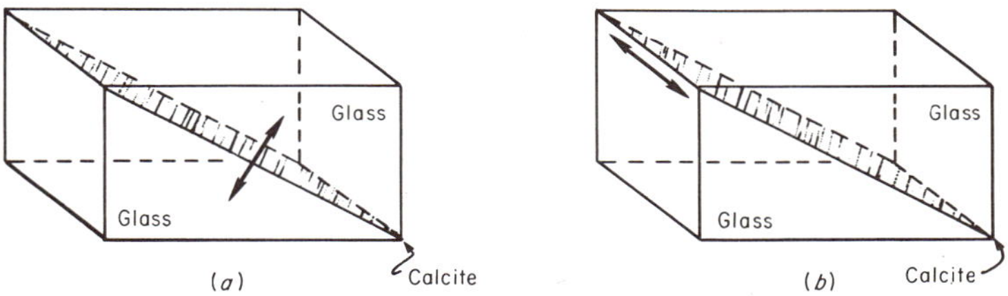

Fig. 14 Types of Feussner prisms: (*a*) original Feussner prism and (*b*) Bertrand type. The arrows indicate the orientation of the optic axis in the calcite (or other birefringent material).

image. The maximum throughput of a prism is thus proportional to the product of the prism's solid angle of acceptance and its cross-sectional area perpendicular to the prism axis. Hence, a large Glan-Taylor prism having an 8° field angle may, if suitable magnification is used, have a throughput comparable to a small Glan-Thompson prism with a 26° field angle. In general, to maximize prism throughput in an optical system, both the angular aperture and clear aperture (diameter of the largest circle perpendicular to the prism axis† which can be included by the prism) should be as large as possible.

The quantities normally specified for a prism are its clear aperture, field angle, and length-to-aperture (L/A) ratio. The *semi-field angle* is defined as the maximum angle to the prism axis† at which a ray can strike the prism and still be completely polarized *when the prism is rotated about its axis.* The field angle is properly twice the semi-field angle.‡ (Some manufacturers quote a "field angle" for their polarizing prisms which is not symmetric about the prism axis and is thus in most cases unusable.) The *length-to-aperture* (L/A) *ratio* is the ratio of the length of the prism base (parallel to the prism axis) to the minimum dimension of the prism measured perpendicular to the prism base. For a square-ended prism, the L/A ratio is thus the ratio of prism length to width.

† The prism axis, which is parallel to its base, is not to be confused with the optic axis of the calcite.

‡ In many prism designs, there is asymmetry about the prism axis, so that although light transmitted at a given angle to the prism axis may be completely polarized for one prism orientation, it will not be completely polarized when the prism is rotated about its axis. Thus, the semi-field angle is not necessarily the largest angle at which completely polarized light can be transmitted by the prism in any orientation.

In determining the maximum angular spread a light beam can have and still be passed by the prism, both the field angle and the L/A ratio must be considered, as illustrated in Fig. 15. If the image of a point source were focused at the center of the prism, as in Fig. 15*a*, the limiting angular divergence of the beam would be determined by the field angle $2i$ of the prism.† However, if an extended source were focused there (Fig. 15*b*), the limiting angular divergence would be determined by the L/A ratio, not the field angle.

The field angle of a polarizing prism is strongly wavelength-dependent. For example, a Glan prism having an 8° field angle at 0.4 μm has only a 2° field angle at 2 μm (see Fig. 25). In designing optical systems in which polarizing prisms are to be used, the designer must allow for this variation in field angle. If

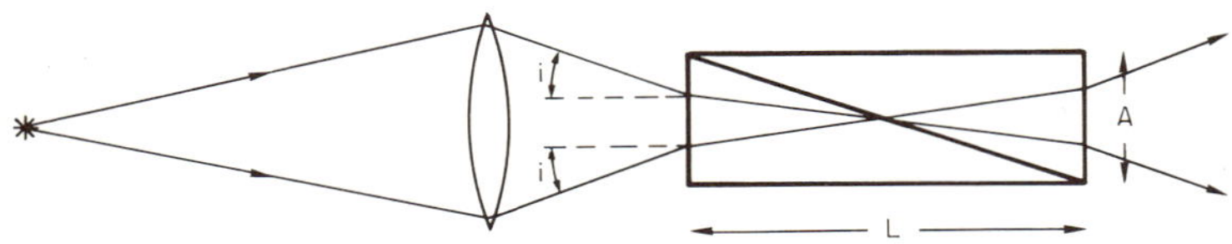

(*a*)

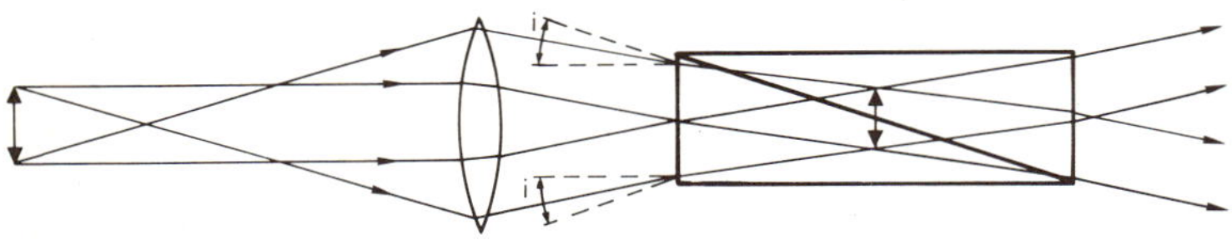

(*b*)

Fig. 15 The effect of field angle and length-to-aperture ratio of a prism polarizer on the maximum angular beam spread for (*a*) a point source and (*b*) an extended source. The field angle is $2i$, and $L/A = 3$. The field angle is exaggerated for clarity.

he does not, serious systematic errors may occur in measurements made with the system.

GLAN-TYPE PRISMS

9. Most prisms used at the present time are of the Glan type. Although they require considerably more calcite than Nicol types of comparable size, they are optically superior in several ways. (1) Since the optic axis is perpendicular to the prism axis, the index of the extraordinary ray differs by a maximum amount from that of the ordinary ray. Thus, a wider field angle or a smaller L/A ratio is possible than with Nicol types. (2) The light is nearly uniformly polarized over the field; it is not for Nicol types. (3) There is effectively no lateral displacement in the apparent position of an axial object viewed through a (perfectly constructed) Glan-type prism. Nicol types give a lateral displacement. (4) Since off-axis wander results in images which have astigmatism when the prism is placed in a converging beam, Glan types have slightly better imaging qualities than Nicol types.

Two other often stated advantages of Glan-type prisms over Nicol types appear to be fallacious. One is that the slanting end faces of Nicol-type prisms have higher reflection losses than the square-ended faces of Glan types. Since the extraordinary ray vibrates in the plane of incidence and hence is in the p direction, increasing

† We are assuming that the prism is wide enough to ensure that the sides of the prism do not limit the angular width of the beam.

the angle of incidence toward the polarizing angle should decrease the reflection loss. However, the index of refraction for the extraordinary ray is higher in Nicol-type prisms (Glan types have the minimum value of the extraordinary index) so the reflection losses are actually almost identical in the two types of prisms. The second "advantage" of Glan-type prisms is that the slanting end faces of the Nicol type supposedly induce elliptical polarization. This widely stated belief probably arises because in converging light the field in Nicol-type polarizers is not uniformly polarized, an effect which could be misinterpreted as ellipticity (see Par. 32). It is possible that strain birefringence could be introduced in the surface layer of a calcite prism by some optical polishing techniques resulting in ellipticity in the transmitted light, but there is no reason why Nicol-type prisms should be more affected than Glan types.

Glan-Thompson-Type Prisms

10. Glan-Thompson-type prisms may be either cemented or air-spaced. Since, as was mentioned previously, an air-spaced Glan-Thompson-type prism is called a Glan-Foucault or simply a Glan prism,† the name Glan-Thompson prism implies that the prism is cemented. Both cemented and air-spaced prisms, however, have the

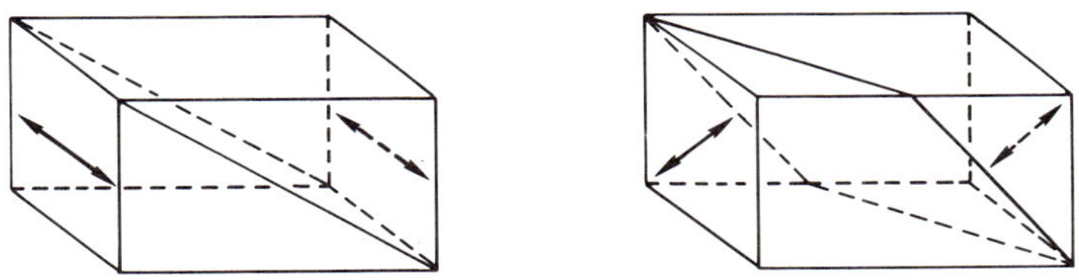

(a) (b)

Fig. 16 Methods of cutting a Glan-Thompson prism: (*a*) cut is parallel to two sides, or (*b*) cut is along a diagonal of the crystal. In both cases the optic axis (indicated by a double arrow) is in the plane of the end faces, and the cut is parallel to the optic axis.

same basic design. The cemented prisms are optically the better design for most applications and are the most common type of prisms in use today. The Glan-Thompson prism is named for P. Glan, who described an air-spaced Glan-Thompson-type prism in 1880 (Glan, 1880), and for S. P. Thompson, who constructed a cemented version in 1881 (Thompson, 1881) and modified it to its present square-ended design in 1882 (Thompson, 1905). These prisms are also sometimes called Glazebrook prisms because R. T. Glazebrook demonstrated analytically in 1883 (Glazebrook, 1883) that when rotated about its axis, this prism gives the most uniform rotation of the plane of polarization for a conical beam of incident light. The cut in a Glan-Thompson-type prism is made parallel to the optic axis, which may either be parallel to two sides, as in Figs. 12*a* and 16*a*, or along a diagonal, as in Fig. 16*b*. The end faces are always perpendicular to the axis of the prism and contain the optic axis.

The extinction ratio [Eq. (61)] obtainable with a good Glan-Thompson-type prism equals or exceeds that of any other polarizer. Ratios of 5 parts in 100,000 to 1 part in 1 million can be expected, although values as high as 1 part in 3×10^7 have been reported for small selected apertures of the prism (Moeller and Grieser, 1969). The small residuals result mainly from imperfections in the calcite or from depolarization by scattering from the prism faces (Moeller and Grieser, 1969), although if the optic axis is not strictly in the plane of the end face, or if the optic axes in the two halves of the prism are not accurately parallel, the extinction ratio will be reduced. Also, the extinction ratio may depend strongly

† An air-spaced Lippich prism, the Glan-Taylor (Par. 19), has similar optical properties to the Glan-Foucault prism but better transmission. It is also called a Glan prism.

upon which end of the prism the light is incident. When prisms are turned end for end, changes in the extinction ratio of as much as a factor of 6 have been reported (Moeller and Grieser, 1969).

When measuring the extinction ratio, it is essential that none of the unwanted ordinary ray, which is internally reflected at the interface and absorbed or scattered at the blackened side of the prism, reach the detector. King and Talim (1971) found that they had to use two 4-mm-diameter apertures and a distance of 80 mm between the photomultiplier detector and prism to eliminate the *o*-ray scattered light. With no limiting apertures and a 20-mm distance, their measured extinction ratio was in error by a factor of 80.

The field angle of the prism depends both on the cement used between the two halves and on the angle of the cut, which is determined by the L/A ratio

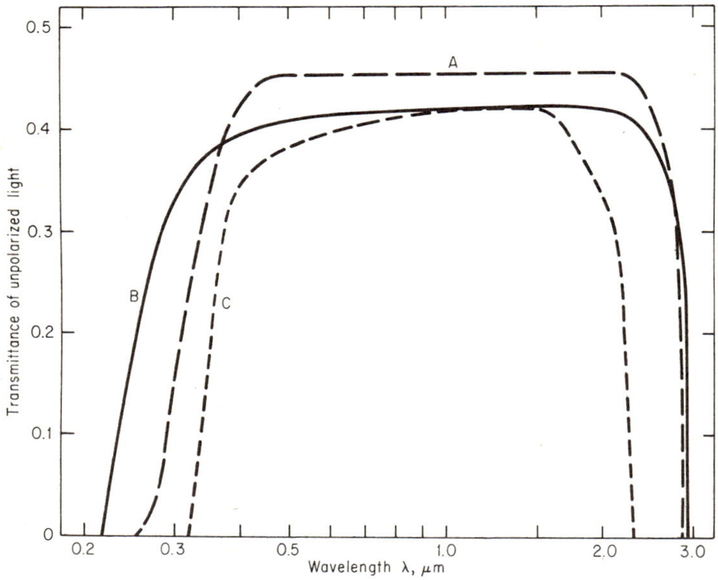

Fig. 17 Transmittance curves for typical polarizing prisms: *A*, Glan-Thompson, *B*, Glan-Taylor, and *C*, Nicol prism. (*Measured by D. L. Decker, Michelson Laboratory.*) In the visible and near-infrared regions the Glan-Thompson has the best energy throughput. In the near ultaviolet the Glan-Thompson may still be superior because the Glan-Taylor has such an extremely small field angle that it may cut out most of the incident beam.

(Par. 8). Calculation of the field angle is discussed in Par. 13. Very large field angles can be obtained with Glan-Thompson prisms. For example, if the L/A ratio is 4, the field angle can be nearly 42°. Normally, however, smaller L/A ratios are used. The most common types of cemented prisms are the long form, having an L/A ratio of 3 and a field angle of 26°, and the short form, having an L/A ratio of 2.5 and a field angle of 15°.

11. Transmission In Fig. 17 the transmission of a typical Glan-Thompson prism is compared with curves for a Glan-Taylor prism and a Nicol prism. The Glan-Thompson is superior over most of the range, but its transmission decreases in the near ultraviolet, primarily because the cement begins to absorb. Its usable transmission range can be extended to about 2500 Å by using an ultraviolet-transmitting cement. Highly purified glycerin, mineral oil, castor oil, and Dow Corning DC-200 silicone oil, which because of its high viscosity is not as subject to seepage as lighter oils, have been used as cements in the ultraviolet, as have dextrose, glucose, and *gédamine* (a urea formaldehyde resin in butyl alcohol). Transmission curves for 1-mm thicknesses of several of these materials are shown in Fig. 18, along with the curve for canada balsam, a cement formerly widely used for polarizing prisms in the visible region (Bouriau and Lenoble, 1957). *Gédamine,* one of the best of the ultraviolet-transmitting cements, has an index of refraction $n_D = 1.465_7$ and can be fitted to the dispersion relation (Bouriau and Lenoble, 1957)

$$n = 1.464 + \frac{0.0048}{\lambda^2} \tag{74}$$

where the wavelength λ is in micrometers.

Figure 19 shows ultraviolet transmission curves for Glan-Thompson prisms with L/A ratios of 2.5 and 3 which are probably cemented with *n*-butyl methacrylate, a low-index polymer that has largely replaced canada balsam (see Par. 13). Better

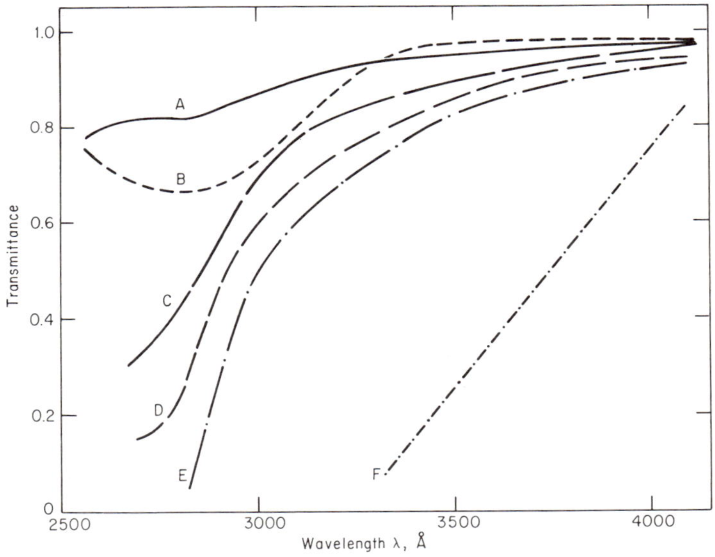

Fig. 18 Transmittance curves for 1-mm thicknesses of various cements: *A*, crystalline glucose, *B*, glycerin, *C*, *gédamine* (urea formaldehyde resin in butyl alcohol), *D*, Rhodopas N60A (polymerized vinyl acetate in alcohol), *E*, urea formaldehyde, and *F*, canada balsam. [*Modified from Bouriau and Lenoble* (1957).] The transmittance of these materials is adequate at longer wavelengths.

ultraviolet transmission is obtained with a Glan-Thompson prism cemented with DC-200 silicone oil. Air-spaced prisms can be used to nearly 2140 Å in the ultraviolet, where calcite begins to absorb strongly. Transmission curves for two such prisms are shown in Fig. 19. The Glan-Taylor, which is an air-spaced prism of the Lippich design, has a higher ultraviolet transmission than the Glan-Foucault, an air-spaced Glan-Thompson prism. The reason for this difference is that multiple reflections occur between the two halves of the Glan-Foucault prism, resulting in a lowered transmission, but are largely absent in the Glan-Taylor design (see Pars. 19 and 16).

The infrared transmission limit of typical Glan-Thompson prisms is about 2.7 μm although they have been used to 3 μm (Ellis and Bath, 1938). The same authors report using a 2.5-cm-long Glan-Thompson prism in the 4.4- to 4.9-μm region.

12. Common Defects Although in principle Glan-Thompson polarizers should not displace an axial beam as they are rotated, in practice it is difficult to manufacture them so that no displacement occurs. This defect, known as *squirm,* results

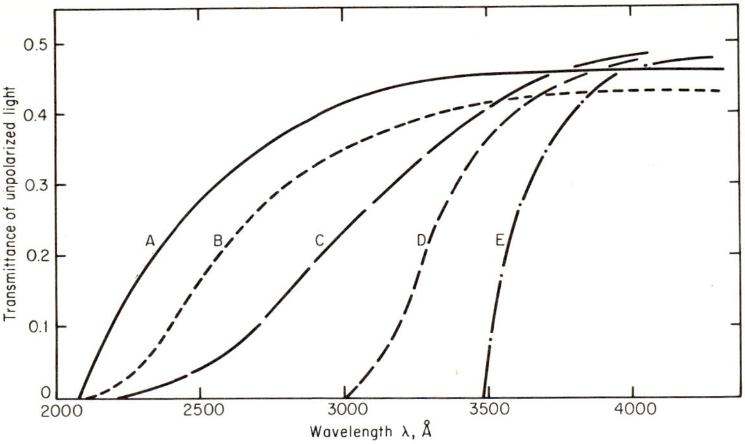

Fig. 19 Ultraviolet transmittance curves for various Glan-Thompson and air-spaced prisms: *A*, Glan-Taylor (air-spaced Lippich-type prism), *B*, Glan-Foucault (air-spaced Glan-Thompson prism), *C*, Glan-Thompson prism with L/A ratio of 2 cemented with DC-200 silicone oil, *D*, Glan-Thompson prism with L/A ratio of 2.5 probably cemented with *n*-butyl methacrylate, and *E*, Glan-Thompson prism similar to *D* except with $L/A = 3$. (*Modified from curves supplied by Karl Lambrecht Corporation, Chicago.*)

when the optic axes in the two halves of the prism are not quite parallel (Twyman, 1952). In addition, if the two prism faces are nonparallel, the axial ray will be deviated. A residual deviation of 3 minutes of arc is a normal tolerance for a good Glan-Thompson prism; deviations of 1 minute or less can be obtained on special order. (The testing of Glan-Thompson prisms is described in Par. 23.)

It is worth pointing out that even if very good Glan-Thompson prisms are used, the transmittance is not necessarily the same when the prism is rotated through 180° (Bennett and Bennett, 1967*a*). One reason for this anomaly is the presence of additional light outside the entrance or exit field angle, as will be discussed in Par. 22. Even when this requirement is met, the effect may still be present. Prisms should thus be checked by rotating them through 180° at various angular settings if they are to be used in photometric applications.

Two factors which limit polarizer performance in addition to the extinction ratio are *axis wander,* i.e., variation of the azimuth of the transmitted beam over the polarizer aperture, and the ellipticity of the emergent polarized beam (King and Talim, 1971). In polarimetric as opposed to photometric applications, these factors may be more significant than the extinction ratio. Also, although the ellipticity may be too small to affect extinction-ratio measurements, its effect can be amplified by residual birefringence in other components, leading to significant errors in precision polarimetry. The axis wander and the ellipticity are both caused by defects in the material constituting the second half of the prism. The ellipticity is enhanced by strain in the cement layer and the actual mounting of the prism. King and Talim (1971) found that the ellipticity across a prism aperture is similar in magnitude to the axis wander for the same prism. The best prisms had ellipticities of less than 0.005°; typical maximum values were 0.01 to 0.04°. (The ellipticity θ is defined by $\tan\theta$, which is equal to the ratio of the minor to major axis of the emergent elliptical vibration.) The best prisms had a measured peak-to-peak axis wander of only 0.003°; more typical values were in the range 0.01 to 0.02°. Close visual inspection of the prisms revealed that those with the poorer surface finish tended to exhibit the larger axis wander. The exit face of one such polarizer was given a good optical polish, resulting in an improvement in axis wander from an original value of over 0.02° to less than 0.005°, with a corresponding reduction in ellipticity of the polarized beam.

13. Field Angle The semi-field angle and field angle were defined in Par. 8. The field angle of a Glan-type prism is determined in principle by the minimum angle at which the unwanted ray (generally the o ray) is totally reflected at the cut or by the maximum angle at which the desired ray (e ray) is transmitted through the cut, as shown in Fig. 20. The prism is assumed to have plane-parallel end faces, a gap of zero thickness between the two prism halves, and a cut angle S that is the same in both halves. Furthermore, the cut is assumed to extend to the face of the prism. Even though these assumptions are not strictly valid for real prisms, the relations derived below hold surprisingly well for well-constructed prisms, as will be discussed below.

In order for the ordinary ray to be totally internally reflected at the cut, the angle of incidence i_2 at the cut for ray A in Fig. 20 should be greater than the value given by the relation

$$\sin i_2 = \frac{n_2}{n_o} \tag{75}$$

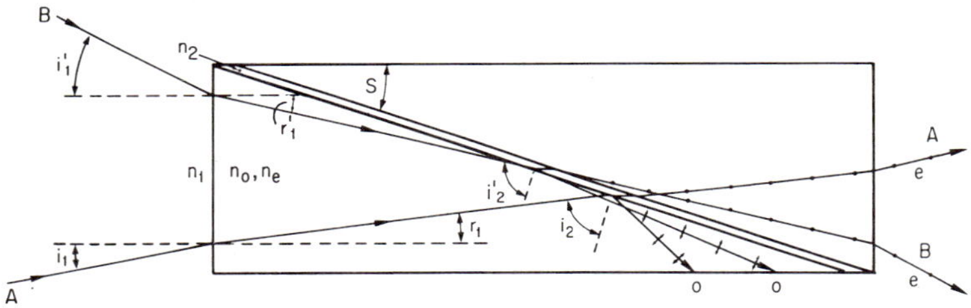

Fig. 20 Determination of the field angle for a Glan-Thompson prism with $L/A = 3$, optic axis perpendicular to the plane of incidence, i.e., perpendicular to the plane of the paper, in an incident medium of refractive index n_1. The index of the ordinary ray is n_o, the extraordinary ray has its minimum index n_e, and both follow Snell's law (see Par. 7); the index of the cement is n_2. Ray A is incident on the polarizer at an angle i_1 such that the angle i_2 on the cut is the smallest angle for which the o ray is totally internally reflected. Ray B is incident at an angle i_1' such that the angle of refraction r_1' is essentially equal to S, so that the e ray just passes through the cut. The field angle is twice the smaller of i_1 and i_1'.

This minimum value of i_2 corresponds to a maximum value of i_1, the angle of incidence of ray A at the air-calcite interface, given by the expression

$$n_1 \sin (i_1)_{\max} = (n_o^2 - n_2^2)^{1/2} \cos S - n_2 \sin S \tag{76}$$

where S is the acute angle the cut makes with the prism base. Equation (76), obtained from Snell's law and the equality

$$r_1 + S + i_2 = 90° \tag{77}$$

(see Fig. 20), can thus be used to calculate the semi-field angle beyond which the ordinary ray is no longer totally internally reflected.

In computing the angle beyond which the extraordinary ray is not transmitted, there are two cases: (1) when $n_2 \geq n_e$, total internal reflection never occurs, and (2) when $n_2 < n_e$, there is a critical angle for the e ray. For the first case, the limiting angle is assumed to be the one for which the angle of refraction r_1' for ray B is equal to S, so that the refracted e ray travels parallel to the cut.† Thus

$$\frac{\sin (i_1')_{\max}}{\sin S} = \frac{n_e}{n_1} \tag{78}$$

† This assumption is merely a mathematical convenience. In an actual prism, the limiting ray will be affected by the geometry of the prism, but the angle will differ from $(i_1')_{\max}$ by only a negligible amount.

For the second case, the critical angle i' for the e ray can be calculated from Eq. (76) by substituting $-i_1'$ for i_1 and n_e for n_o:

$$n_1 \sin (i_1')_{\max} = n_2 \sin S - (n_e^2 - n_2^2)^{1/2} \cos S \tag{79}$$

The semi-field angle is the smaller of the values of i_1 or i_1' in Eqs. (76) and (78) or (79).

We can determine the maximum field angle possible for a cemented calcite Glan-Thompson prism as follows. If $n_2 \geq n_e$, the values of $(i_1)_{\max}$ for ray A will *increase* as S decreases (L/A ratio increases) but the value of $(i_1')_{\max}$ for ray B will *decrease* as S decreases. These relationships are shown by the curve for $n_2 = 1.540$ in Fig. 21*a*. The maximum field angle will be obtained when $(i_1)_{\max} = (i_1')_{\max}$, and the corresponding value of S can be determined by eliminating the angles of incidence from Eqs. (76) and (78):

$$n_e \sin S = (n_o^2 - n_2^2)^{1/2} \cos S - n_2 \sin S \tag{80}$$

or

$$\tan S = \frac{(n_o^2 - n_2^2)^{1/2}}{n_e + n_2} \tag{81}$$

If $n_2 < n_e$, the value of $(i_1)_{\max}$ for ray A will increase as S decreases since the ordinary ray is totally internally reflected as before (see curve for $n_2 = 1.485$ in Fig. 21*a*). However, the limiting angle for ray B is now the angle for which the extraordinary ray is totally internally reflected. For a given value of n_2, $(i_1')_{\max}$ decreases as S decreases so that the maximum field angle will again be obtained when $(i_1)_{\max} = (i_1')_{\max}$. This value of S is determined by eliminating the angles of incidence from Eqs. (76) and (79):

$$(n_o^2 - n_2^2)^{1/2} \cos S - n_2 \sin S = -(n_e^2 - n_2^2)^{1/2} \cos S + n_2 \sin S \tag{82}$$

or

$$\tan S = \frac{(n_o^2 - n_2^2)^{1/2} + (n_e^2 - n_2^2)^{1/2}}{2n_2} \tag{83}$$

The largest field angle occurs when Eqs. (81) and (83) reduce to the same form, i.e., when $n_2 = n_e$. At a wavelength of 5893 Å, these equations yield a value of $S = 13.89°$, and L/A, the length-to-aperture ratio, is equal to $1/(\tan S) = 4.04$. The maximum value of the semi-field angle is equal to 20.91° from Eq. (76) or (78), making the field angle for a calcite Glan-Thompson prism 41.82° at this wavelength. Feussner (1884) concluded that this configuration gives the maximum field angle obtainable for any single Glan or Nicol type of prism design. The field angle can be exceeded in double prisms, however (Par. 15).

The curve of limiting angles vs. cut angle for the case when $n_2 = n_o$ is shown in Fig. 21*a* by the short-dashed lines. This curve closely follows the limiting o-ray curve for $n_2 = 1.485$ and the limiting e-ray curve for $n_2 = 1.540$. Surprisingly, these curves form two families of parallel straight lines to a very good approximation, even though the equations from which they were calculated are all transcendental.

Figure 21*b* shows the field angles corresponding to the allowed values of n_2 for two common Glan-Thompson prisms. This graph shows clearly that (1) for a constant value of L/A the field angle, determined by the ordinary ray, increases with decreasing n_2, so that the maximum field angle is obtained when the minimum-index cement is used. This minimum index is calculated from the appropriate relation for the cut angle when the limiting angles for the o and e rays are equal [see Eqs. (81) and (83), and Fig. 22]. (2) The field angle increases with increasing L/A ratio for a constant value of n_2. The maximum attainable field angle is for $L/A = 4.04$ and $n_2 = n_e$, as discussed above.

The relation between the allowed refractive index of the cement and S is shown in Fig. 22. In this figure, the lower solid curve to the left of $L/A = 4$ was calculated from Eq. (81) and gives the minimum values of n_2 when $n_2 \geq n_e$. The lower solid curve to the right of $L/A = 4$ gives minimum values of n_2 when $n_2 < n_e$, and was calculated from Eq. (83). There is also a maximum allowed value of n_2, represented by the upper dashed curve, which depends on S as follows. As can be seen from Fig. 21*a* and Eqs. (76), (78), and (79), the limiting angle is generally not symmetric about the longitudinal axis of the prism and is determined by the angle i_2 at which the ordinary ray is totally internally reflected (ray A in Fig. 20). As n_2 increases,

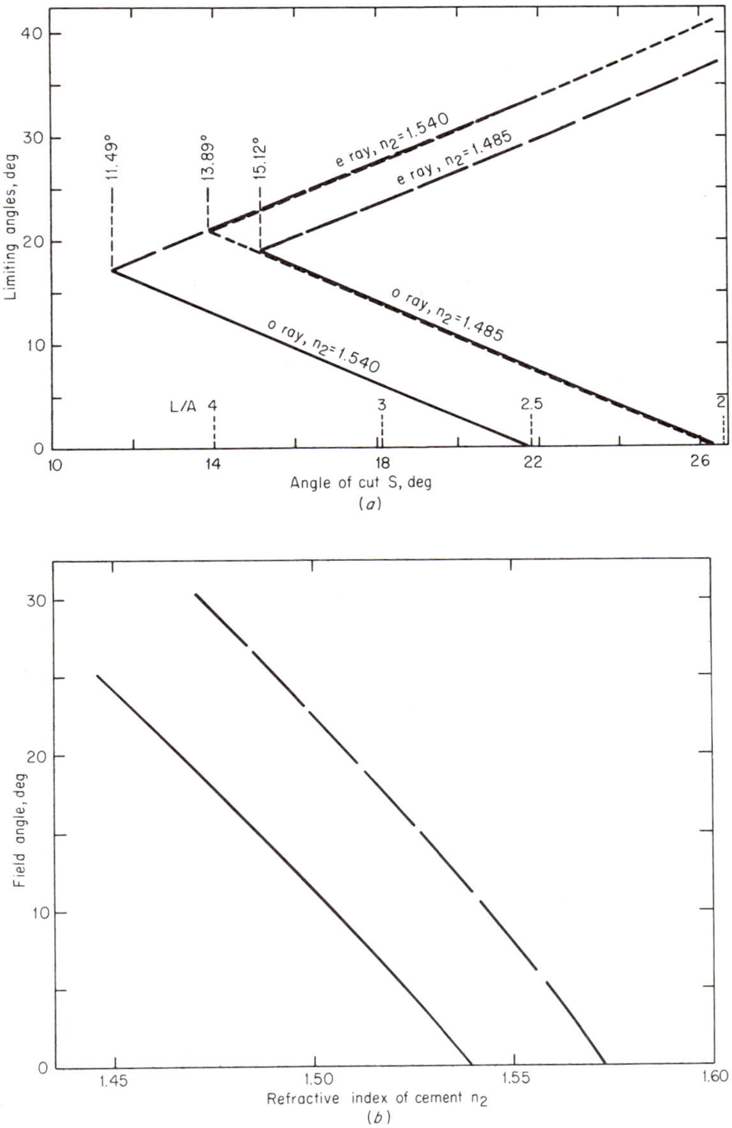

Fig. 21 (*a*) Limiting angles $(i_1)_{max}$ and $(i_1')_{max}$ calculated from Eqs. (76), (78), and (79) as a function of cut angle S for two common cements: canada balsam with $n_2 = 1.540$ and n-butyl methacrylate polymer with $n_2 = 1.485$ at 5893 Å (n_o = 1.65835 and n_e = 1.48640). The semi-field angles are the lower solid curves in both cases. The short-dashed curve is for $n_2 = n_e$ and gives the maximum attainable semi-field angle when S = 13.89°. The values of S corresponding to the largest semi-field angles for the two cements are indicated and are calculated from Eqs. (81) and (83), respectively. The limiting angles for the o ray go to zero when S reaches values given by Eq. (84). (*b*) Field angles for Glan-Thompson prisms with L/A = 3.0 (*dashed line*) and L/A = 2.5 (*solid line*) as a function of the refractive index of the cement. The allowed values for n_2 are indicated in Fig. 22.

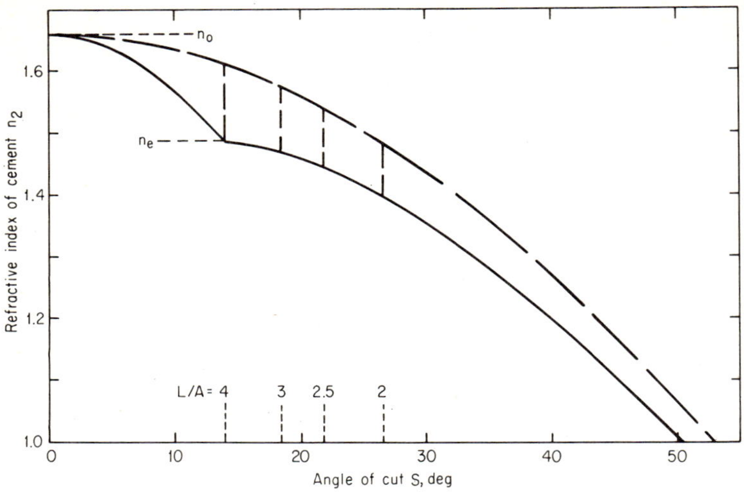

Fig. 22 Range of allowed values of the refractive index n_2 of the cement as a function of S for a Glan-Thompson prism with $n_o = 1.65835$ and $n_e = 1.48640$ at 5893 Å. The maximum values (*dashed curve*) were calculated from Eq. (84) and the minimum values (*solid curve*) from Eq. (83) for $n_2 < n_e$ and from Eq. (81) for $n_2 \geq n_e$. Several common L/A ratios for Glan-Thompson prisms are also indicated, along with the allowed values for the refractive index of the cement.

$\sin(i_1)_{max}$ decreases and at some value of n_2 becomes zero. Equation (76) then yields the relation for $(n_2)_{max}$

$$(n_2)_{max} = n_o \cos S \tag{84}$$

Since the field angle is zero for the maximum allowed value of n_2, one should not design a prism with S very close to the limiting value for a given index cement or, conversely, choose a value of n_2 close to the maximum value for a prism with a given cut angle unless the prism is to be used with a light source that is very parallel. In summary, n_2 can take on only a very limited range of values in a prism with a given L/A ratio, as indicated by the vertical dashed lines in Fig. 22 and by the solid and dashed curves in Fig. 21*b*.

Canada balsam has been the traditional cement to use with calcite polarizing prisms and has a refractive index which generally lies between 1.534 and 1.540 at 5893 Å although values between 1.528 and 1.550 have been reported (Johannsen, 1968*a*). It can be used for a prism with $L/A = 3.0$, but the field angle is then only about 12°; if $L/A = 2.5$, it cannot be used at all. Cements with lower refractive indexes make much larger field angles possible. Linseed oil, for which $n_D = 1.485$, increases the field angle to 26.7° if $L/A = 3.0$ and to 15.4° if $L/A = 2.5$. However, it is subject to seepage and has been largely replaced by a permanent cement, *n*-butyl methacrylate, which as a monomer has an index of 1.426 and when fully polymerized, 1.485, the same as linseed oil. A disadvantage of this cement is that permanent strains can be set up during polymerization (Twyman, 1952). Another solution is to cement a low-index glass plate between the two halves of the prism using canada balsam. Before the introduction of *n*-butyl methacrylate, this technique, first suggested by Bertrand (Johannsen, 1968*a*), was used extensively by Hilger & Watts, Ltd. (Twyman, 1952).

If a cement with a lower index than that of *n*-butyl methacrylate is used, the field angle becomes even larger. For example, if $L/A = 2.5$ and the cement has $n_D = 1.445$, the field angle would be about 25°, which is nearly as large as the angle obtained for a prism with $L/A = 3.0$ and a cement with $n_D = 1.485$. However, we are now near the value for which the extraordinary ray will be totally internally reflected (see Fig. 22), so that the field angle will be limited for this reason.

All the preceding comments about specific field angles apply only at the average wavelength of the sodium D lines, 5893 Å, at which the acceptance angle is usually specified. Since the refractive indexes of calcite and the cement are wavelength-dependent, the field angle will change with wavelength. In principle, it can be calculated from Eq. (76) or (79) at any wavelength so that it should be possible to optimize the prism for use in a specific wavelength region. In actual fact, however, the refractive indexes of the cements are usually given only for the sodium D line, so that calculations of the field angle cannot be made at other wavelengths. For air-spaced prisms such as the Glan-Foucault and Glan-Taylor the refractive index of the "cement" equals 1. Field-angle calculations for these prisms will be found in Pars. 16 and 19.

The theory presented thus far applies strictly to ideal prisms, i.e., ones which have plane-parallel end faces, zero-thickness gaps between the two prism halves, equal cut angles in both halves, and cuts which extend to the prism faces. Fortunately, some types of prism defects have a negligible effect on the semi-field angle. Decker et al. (1970) have shown analytically that for an air-spaced prism the nonzero thickness of the air gap and a wedge angle in the air gap have a negligible effect on the semi-field angle. However, a wedge angle in the air gap may indicate different cut angles in the two prism halves or nonparallel end faces, both of

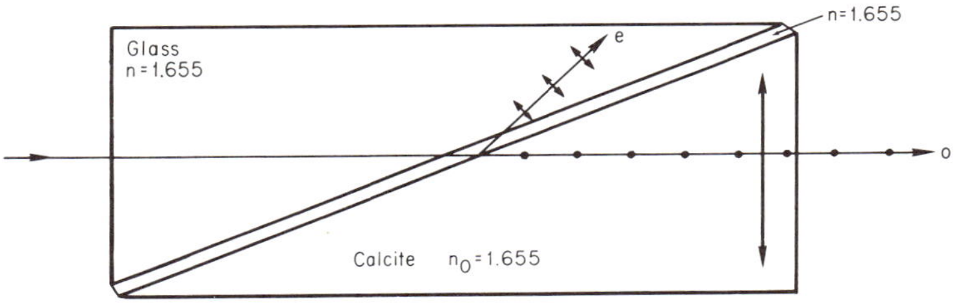

Fig. 23 Ammann-Massey prism, a modified Lippich design, composed of glass and calcite. The ordinary ray is transmitted and the extraordinary ray is reflected at the cement-calcite interface. The optic axis is in the plane of the paper.

which may cause a deviation of the emerging beams (see Par. 23). In some prisms the cut does not extend to the prism face, and so it would appear that relations (76) and (79) for the semi-field angles would have to be modified. Fortunately, Decker et al. have also shown that, to a good approximation, the position of the cut does not affect the calculated or measured semi-field angle.

14. Glass-Calcite Construction Large, optically perfect calcite crystals are difficult to obtain, and any technique which reduces the amount of calcite required in a polarizing prism is of interest. Several early designs were proposed in which glass of matching index was to be used in place of one of the two calcite halves, but the prisms that were actually built performed poorly because glass of the proper index was not available (Ammann and Massey, 1968). It is available now, and Ammann and Massey (1968) have discussed the design problems in regard to optimizing calcite-glass combinations. Although they used the name Glan-Thompson, the Ammann-Massey prism, shown in Fig. 23, is actually a modified Lippich design. Their discussion, however, may also be applied to true Glan-Thompson prisms. A modification of the usual Glan-Thompson prism design is required, since if the extraordinary ray is transmitted, rays traveling at an angle to the optic axis will be refracted differently in the two halves of the prism and severe image distortion will result. However, if glass of a high index approximately matching that of the ordinary ray in calcite is used and the light is incident from the glass end of the prism, the ordinary ray rather than the extraordinary ray will be transmitted. Since the index of the ordinary ray does not depend on the angular direction of the ray in the calcite, no image distortion will result.

In addition to saving calcite, glass-calcite prisms offer another advantage. Since one half is glass, the problem of aligning the optic axes in the two halves of the prism is eliminated, and there should be no squirm. Glass-calcite prisms are not reciprocal, however. If light is incident on the glass end, the prism acts like a Glan-Thompson or Lippich prism, but if it is incident on the calcite end, both ordinary and extraordinary rays are transmitted and the prism acts like a Rochon prism (Pars. 34 to 37). At least one company† is now manufacturing glass-calcite prisms.

Despite their advantages, glass-calcite prisms will probably not become popular until better cements are found. Since the thermal expansion of glass is isotropic and cannot match that of calcite, which differs in directions parallel and perpendicular to the optic axis by over a factor of 4, strains and strain birefringence will occur unless the cement flows readily. A thixotropic cement which flows more readily with increasing stress would be an ideal solution to this problem.

15. Ahrens Prism For a given field angle, the L/A ratio of a conventional Glan-Thompson prism may be halved if a double prism such as that shown in

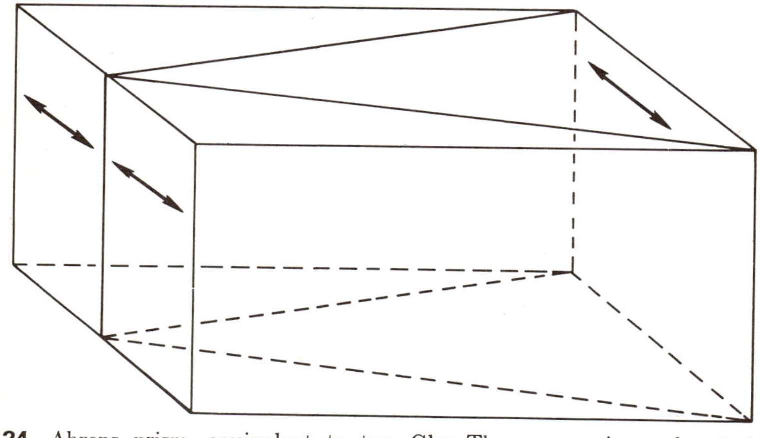

Fig. 24 Ahrens prism, equivalent to two Glan-Thompson prisms placed side by side. It has twice the field angle of a Glan-Thompson prism for a given L/A ratio.

Fig. 24 is used. This design is called an Ahrens prism after C. D. Ahrens, who described it in 1886 (Ahrens, 1886). His prism had an L/A ratio of 1.8 and a field angle of 26°, but ratios of 1.75 and 1.25 are more common today. The field angles of Ahrens prisms are equal to those of Glan-Thompson prisms having twice the L/A values. Good Ahrens prisms are difficult to make since the optic axes in the three sections must coincide and the leading face must be carefully polished down so that the cut is just visible. Often a thin cover glass is cemented to the entrance face to improve the optical continuity. Alternately, a thin black line is painted along the section line to ensure that no unpolarized light will be transmitted. To prevent disturbing multiple reflections, the end of the prism containing the section line should be always turned toward the source of light.

The principal advantage of the Ahrens prism is its large angular field for a given L/A ratio. As a complete polarizer, however, it is less desirable than a simple Glan-Thompson prism (K. Lambrecht, 1969). Extinction ratios are typically between 2 and 10 times worse than that of a good Glan-Thompson prism. The difference may be due in part to residual strains set up in the two cemented interfaces and in the cemented front glass cover plate of the Ahrens prism.

16. Glan-Foucault Prism Since there is no satisfactory cement which can be used at wavelengths as short as 2140 Å, where calcite begins to absorb strongly,

† Industrial Optics Corporation, Newark, N.J.

Glan-Thompson-type prisms used in this range must be air-spaced. Air-spaced prisms are also convenient for applications involving an extended wavelength range from the ultraviolet to the near infrared. The first air-spaced polarizing prism, a Nicol type, was reported by L. Foucault in 1857 (Foucault, 1857) (Par. 31); P. Glan described the first air-spaced Glan-Thompson-type prism in 1880 (Glan, 1880). Therefore, air-spaced Glan-Thompson prisms are called Glan-Foucault or simply Glan prisms. Since the refractive index n_2 of the "cement" now equals 1, Fig. 22 indicates that the prism must be cut at a large angle, which for use at a wavelength of 5893 Å will lie between 50.46 and 52.91°. The effective field angle is zero for the larger value of S, and even for the smaller value the maximum field angle, calculated from Eq. (76), is only 8.15°. The L/A ratio is 0.85, so that the prism has a relatively large entrance face for its length and requires less calcite than most other prism designs.

The calculations described above were made for a wavelength of 5893 Å and should be repeated for other wavelengths at which the prism is to be used in order to select an optimum angle for the cut. If the prism is intended for the visible and ultraviolet regions, Archard and Taylor (1948) recommend an angle of 51.25°. The limiting semiangle $(i_1)_{max}$ at 5893 Å determined by the total reflection of the ordinary ray (ray A in Fig. 20) is 2.76°, as calculated from Eq. (76). On the other side, the limiting semiangle $(i_1')_{max}$ is determined by the critical angle of the extraordinary ray (ray B in Fig. 20) and is 5.25°, as calculated from Eq. (79). Thus the angular field is not symmetric about the longitudinal axis of the prism, and although the total angle is $2.76° + 5.25° = 8.01°$, the field angle is only $2 \times 2.76° = 5.52°$.

Figure 25*a* shows a plot of the semi-field angle vs. wavelength for a Glan-Foucault prism and also for a Marple-Hess prism (Par. 20). For the Glan-Foucault prism, both the ordinary and extraordinary rays are totally reflected at larger angles than the semi-field angle in the ultraviolet wavelength region (0.214 to 0.355 μm) until half angles larger than 5° are reached (where the ordinary ray starts being transmitted). If the semi-field angle is only slightly exceeded, the prism acts merely as a stop, reducing the speed of the optical system, and any light in a slightly convergent beam which is transmitted by the prism *will be properly polarized.* In the longer-wavelength region, both ordinary and extraordinary rays are transmitted for angles larger than the semi-field angle, so that a convergent beam *will not be linearly polarized if the semi-field angle is exceeded.* Thus, serious systematic errors can arise, particularly in the near infrared, if air-spaced Glan-Foucault prisms are used in convergent light.

Since there is relatively little wastage of calcite involved in making air-spaced prisms, apertures as large as 2 or 3 cm are possible. In designing optical systems in which air-spaced Glan-Foucault prisms are to be used, the image of the source can therefore be magnified to reduce the convergence angle and focused near the center of the prism. In this way the convergence angle on the prism can be reduced to an acceptable value (see Par. 22 and Fig. 28). Excellent extinction ratios, between 1×10^{-5} and 5×10^{-6}, can then be obtained, and the prism is usable from about 0.214 μm in the ultraviolet to 2.3 μm in the infrared (Karl Lambrecht, 1973).

The major drawbacks of the Glan-Foucault prism are the presence of multiple reflections between the two sections of the prism and the lowered transmission which results from this reflection loss. Both these weaknesses are corrected in the Glan-Taylor design (Par. 19), which otherwise performs almost exactly like the Glan-Foucault prism and has largely superseded it. Comparative ultraviolet transmittance curves for the two prisms are shown in Fig. 19.

17. Grosse Prism Just as an Ahrens prism (Par. 15) decreases the L/A ratio of a Glan-Thompson prism by a factor of about 2, so the L/A ratio of a Glan-Foucault prism can be decreased by an analogous construction. An air-spaced Ahrens prism is called a Grosse prism (Grosse, 1890) and has an L/A ratio of only 0.5. As with the Glan-Foucault prism, it can be used only in essentially parallel light. Grosse prisms are particularly valuable when the available space is very limited in the direction of the instrument axis.

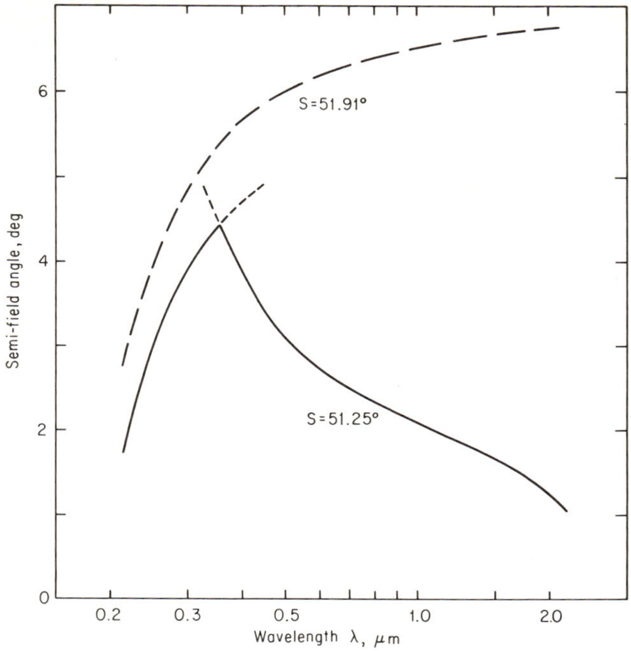

(a)

Fig. 25 See legend opposite page.

Lippich-Type Prisms

18. In 1885, Lippich (1885) suggested a polarizing-prism design similar to the Glan-Thompson but with the optic axis in the entrance face and at right angles to the intersection of the cut with the entrance face (Fig. 12*b*)†. For this case, the index of refraction of the extraordinary ray is a function of angle of incidence and can be calculated from Eq. (71) after ϕ, the complement of the angle of refraction of the wave normal, is determined from Eq. (72). In the latter equation, β, the angle the normal to the surface makes with the optic axis, is 90° since the optic axis is parallel to the entrance face. Since the directions of the ray and the wave normal no longer coincide, the ray direction must be calculated from Eq. (73).

The analysis given in Par. 13 for the field angle of a Glan-Thompson prism also applies to a Lippich prism except that since the optic axis is now in the plane of the paper in Fig. 20 (rather than perpendicular to it), values of n_ϕ must be used in place of n_e in Eqs. (78) to (83). The semi-field angle $(i_1)_{max}$ or $(i_1')_{max}$ can be determined for any angle of cut S for the prism from Eq. (76) or (79). Since $n_\phi > n_e$ for all oblique angles, it is clear from Eq. (79) that the semi-field angle for a Lippich prism will always be less than that of a Glan-Thompson prism with the same value of S. Hence, Lippich prisms are little used except for one, the air-spaced Lippich, often called a Glan-Taylor prism.

† The Lippich prism should not be confused with the Lippich half-shade prism, which is a device to determine a photometric match point. The half-shade prism consists of a Glan-Thompson or Nicol prism placed between the polarizer and analyzer such that it intercepts half the beam and is tipped slightly in the beam. The prism edge at the center of the field is highly polished to give a sharp dividing line. The eye is focused on this edge; the disappearance of the edge gives the photometric match point (Hardy and Perrin, 1932*a*).

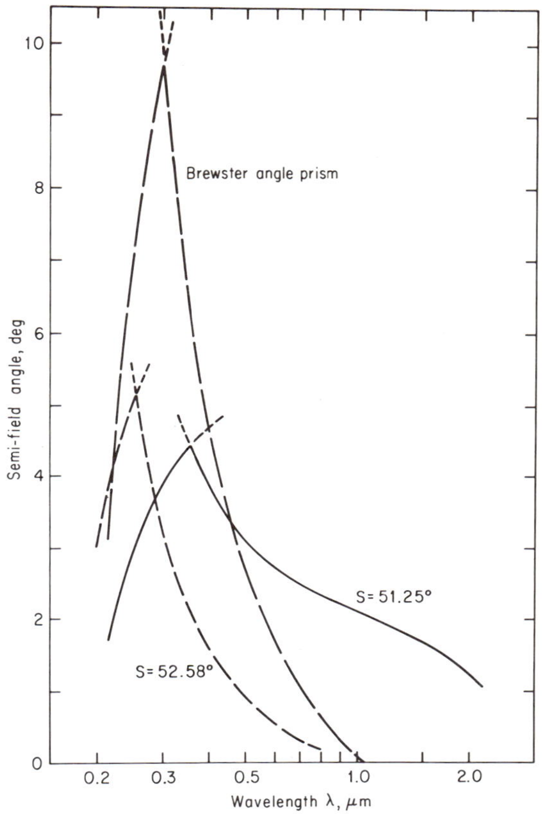

(*b*)

Fig. 25 (*a*) Semi-field angles as a function of wavelength for a Glan-Foucault prism with $S = 51.25°$ (*solid curve*) and for a Marple-Hess prism with $S = 51.91°$ (*long-dashed curve*) (Par. 20). For the Glan-Foucault prism, the semi-field angle is $(i_1')_{max}$ calculated from Eq. (79) in the wavelength range 0.214 to 0.355 μm and is $(i_1)_{max}$ calculated from Eq. (76) in the rest of the wavelength range. For the Marple-Hess prism, the semi-field angle is $(i_1')_{max}$ calculated from Eq. (79) with n_d substituted for n_e. The semi-field angle of the Glan-Taylor prism (Par. 19) is essentially equal to that of the Glan-Foucault prism. (*b*) Semi-field angle as a function of wavelength for the Glan-Foucault (or a Glan-Taylor) prism with $S = 51.25°$ as shown in Fig. 25*a* (*solid curve*), laser-optimized Glan-Taylor prism with $S = 52.58°$ (*medium-dashed curve*), and Brewster angle polarizing prism (*long-dashed curve*). The long-wavelength portion of the curves is determined by the angle at which the *o* ray is transmitted and the short-wavelength portion by the angle at which the *e* ray is totally internally reflected.

19. Glan-Taylor Prism The Glan-Taylor prism, first described in 1948 (Archard and Taylor, 1948), has substantial advantages over its Glan-Thompson design counterpart, the Glan-Foucault prism (Par. 16). Since air-spaced prisms have a very small field angle, the light must be nearly normally incident on the prism face, so that the difference in field angles between the Glan-Taylor and Glan-Foucault prisms (caused by the difference in the refractive index of the extraordinary ray) is negligible. For example, if the Glan-Foucault prism discussed in Par. 16 had been a Glan-Taylor prism, the field angle would have been decreased by only about two minutes of arc. Thus, the semi-field angle for the Glan-Taylor prism is also represented by the solid curve in Fig. 25*a*, and the comments on depolarization effects in convergent light also apply. The smallness of the field angle also makes possible the approximation $n_\phi \approx n_e$, so that we can use the equations derived for the Glan-Foucault prism in calculating prism angles for the Glan-Taylor prism.

The major advantages of the Glan-Taylor prism are that its calculated transmission is between 60 and 100 percent higher than that of the Glan-Foucault prism and the intensity of multiple reflections between the two sides of the cut, always a principal drawback with air-spaced prisms, is reduced to less than 10 percent of the value for the Glan-Foucault prism. To understand how this situation occurs,

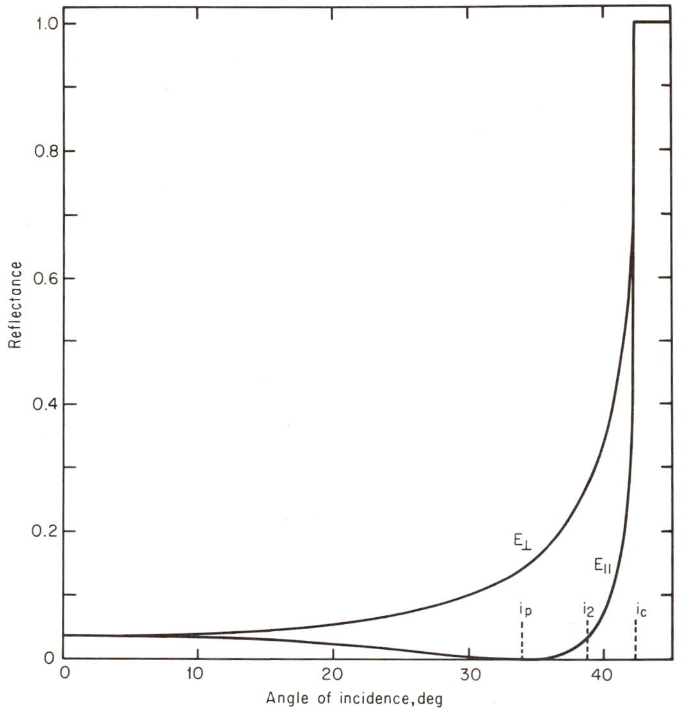

Fig. 26 Calculated reflectance loss for the extraordinary ray at the cut for an air-spaced calcite prism as a function of angle of incidence on the cut ($n_e = 1.48640$ at 5893 Å). The polarizing angle i_p in the calcite is 33.93° at this wavelength, and the angle of incidence i_2 on the cut is 38.75° for light normally incident on the prism face. The critical angle i_c in the calcite is 42.28°. In a Glan-Foucault prism, the plane of vibration of the extraordinary ray is perpendicular to the plane of incidence ($E_\perp$); in a Glan-Taylor prism, it is in the plane of incidence ($E_\parallel$).

consider Fig. 26, in which the reflection loss for the extraordinary ray is plotted as a function of angle of incidence on the cut. If the Glan-Taylor prism has a cut angle S of 51.25° (the same value used for the Glan-Foucault prism), a beam normally incident on the prism face will have an angle of incidence of $90° - 51.25° = 38.75°$ on the cut (see Fig. 20). In a Glan-Foucault prism the extraordinary ray vibrates perpendicular to the plane of incidence, so that its reflection loss at the cut is about 0.270. However, in a Glan-Taylor prism, the extraordinary ray is vibrating in the plane of incidence (because the optic axis is perpendicular to its direction in the Glan-Foucault prism), and the reflection loss is only about 0.031. Not only is the transmission of the Glan-Taylor prism enhanced by the low reflectance of the extraordinary ray, but multiple reflections in the air space are greatly reduced. Only about 0.1 percent of the transmitted light in the near infrared (0.9 μm) and 3 percent in the ultraviolet arise from multiple reflections, as contrasted with 7 and 30 percent, respectively, for the Glan-Foucault prism (Archard and Taylor, 1948).

The calculated and measured transmittances of a Glan-Taylor prism are in reasonable agreement, but the measured transmittance of a Glan-Foucault prism (Fig. 19) may be considerably higher than its theoretical value (Archard and Taylor, 1948). Even so, the transmission of the Glan-Taylor prism is definitely superior to that of the Glan-Foucault prism, as can be seen in Fig. 19. Extinction ratios of better than 1 part in 10^5 are obtainable for the Glan-Taylor prism (Karl Lambrecht, 1970*b*).

A final advantage of the Glan-Taylor prism is that it can be cut in such a way as to conserve calcite. Archard and Taylor (1948) used the Ahrens method of spar cutting described by Thompson (1905) and found that 35 percent of the original calcite rhomb could be used in the finished prism.

In a modified version of the Glan-Taylor prism becoming popular for laser applications, the cut angle is increased, the front and back faces are coated with antireflection coatings, and portions of the sides are either covered with absorbing black glass plates or highly polished to let the unwanted beams escape (Karl Lambrecht, 1973). The effect of increasing the cut angle is twofold: a beam normally incident on the prism face will have a smaller angle of incidence on the cut and hence a smaller reflection loss at the cut than a standard Glan-Taylor prism (see Figs. 20 and 26), but, at the same time, the semi-field angle will be reduced throughout most of the visible and near-infrared regions, as can be seen from Eq. (76). In Fig. 25*b* the semi-field angle for a laser-optimized Glan-Taylor prism with $S = 52.58°$ is shown as the medium-dashed curve. This curve has been verified by measurements at several wavelengths in the visible and ultraviolet region (Decker et al., 1970).

A new type of air-spaced prism (Swartz et al., 1971) has a very high transmittance for the extraordinary ray. It resembles the Glan-Taylor prism in that the optic axis is parallel to the entrance face and at right angles to the intersection of the cut with the entrance face. However, instead of striking the prism face at normal incidence, the light is incident at the Brewster angle for the extraordinary ray (54.02° for the 6328-Å helium-neon laser wavelength), so that there is no reflection loss for the *e* ray at this surface. Since the ordinary ray is deviated about 3° more than the extraordinary ray and its critical angle is over 4° less, it can be totally reflected at the cut with tolerance to spare while the extraordinary ray can be incident on the cut at only a few degrees beyond its Brewster angle. Thus this prism design has the possibility of an extremely low light loss caused by reflections at the various surfaces. A prototype had a measured transmission of 0.985 for the extraordinary ray at 6328 Å (Swartz et al., 1971).

If the prism is to be used with light sources other than lasers, its semi-field angle is also of interest. This quantity has been calculated for a prism having a 67° angle at the apex (corresponding to the angle $90° - S$ in Fig. 20) and is shown as the long-dashed curve in Fig. 25*b*. For wavelengths longer than 0.300 μm, the semi-field angle is determined by the angle at which the ordinary ray starts to be transmitted (calculated from the *o*-ray refractive index and Snell's law), while for wavelengths shorter than 0.300 μm the semi-field angle is determined from the critical angle at which the extraordinary ray is totally internally reflected [calculated from Eq. (72) with $\beta = 90°$ at the first surface and $90° - 67° = 23°$ at the cut]. Note that the Brewster angle prism has a much larger semi-field angle than either of the other two prisms for wavelengths between 0.22 and 0.45 μm. However, its semi-field angle is less than that of the standard Glan-Taylor prism at longer wavelengths.

A major drawback to the Brewster angle prism is that since the light beam passes through a plane-parallel slab of calcite at nonnormal incidence, it is displaced by an amount that is proportional to the total thickness of the calcite. Some of the prisms are made with glass in place of calcite for the second element. In this case, the beam will usually be deviated in addition to being displaced. Measurements on a calcite-glass prototype at 6328 Å showed that the output beam was laterally displaced by several millimeters with an angular deviation estimated to be less than 0.5° (Swartz et al., 1971).

20. Marple-Hess Prism If a larger field angle is required than can be obtained with a Glan-Taylor prism, a Marple-Hess prism may be used (Fig. 27). This prism, which was first proposed in 1960 as a double Glan-Foucault by D. T. F. Marple of the General Electric Research Laboratories and modified to the Taylor design by Howard Hess of the Karl Lambrecht Corporation (A. Lambrecht, 1969), is effectively two Glan-Taylor prisms back to back. The analysis for this prism is made in the same way as for the Glan-Taylor prism (Par. 19) and Lippich-type prisms in general (Pars. 18 to 20), keeping in mind that the refractive index of the "cement" is 1 since the components are air-spaced. At the second interface, the angle S in Eq. (76) is negative, which means physically that any ordinary ray not totally reflected at the first interface will be totally reflected at the second. It follows that S need only be small enough for the ordinary

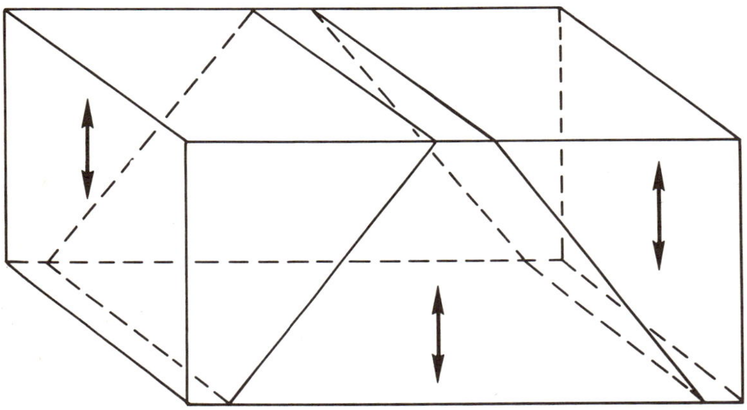

Fig. 27 Marple-Hess prism made of three pieces of calcite cut with their optic axes parallel. The light is incident on the left face, and the extraordinary ray passes through both cuts. The L/A ratio is 1.8.

ray normally incident on the entrance face of the prism to be totally reflected. For normally incident light, $r_1 = 0$, and total internal reflection occurs when $\sin i_2 = 1/n_o$ (Fig. 20). S is then given by Eq. (77) as

$$S = 90° - \sin^{-1}\frac{1}{n_o} \tag{85}$$

The value of S is determined at the wavelength for which n_o is a minimum. If the prism is to be used out as far as 2.17 μm in the infrared, the minimum value of n_o is 1.62099, and thus S will be 51.91°. If the prism is designed to be used only in the visible and ultraviolet regions, the value of S, and hence the field angle, may be increased.

Since the ordinary ray is totally reflected for all angles of incidence by one or the other of the two cuts, the field angle is symmetric about the longitudinal axis of the prism and is determined entirely by the angle at which the extraordinary ray is totally reflected at one of the two cuts. [This angle can be calculated from Eq. (79) with n_ϕ substituted for n_e.] The field angle is thus considerably larger than for the Glan-Foucault or Glan-Taylor prism and does not decrease as the wavelength increases. The semi-field angle for a Marple-Hess prism for which $S = 51.91°$ is compared with that of a Glan-Foucault prism in Fig. 25*a*.

Unlike the Glan-Foucault or Glan-Taylor prisms, which stop being efficient polarizers when the angle of incidence on the prism face becomes too large, the Marple-

Hess prism continues to be an efficient polarizer as long as the axial ordinary ray is not transmitted. If the prism is used at a longer wavelength than the longest one for which it was designed (smallest value of n_o used to determine the cut angle S), the value of n_o will be still smaller and the critical angle for the axial ordinary ray will not be exceeded. Thus the axial o ray will start to be transmitted before off-axis ordinary rays get through. When this situation occurs, it only makes matters worse to decrease the convergence angle. Thus, there is a limiting long wavelength, depending on the cut angle S, beyond which the Marple-Hess prism is not a good polarizer. At wavelengths shorter than the limiting wavelength, the Marple-Hess prism has significant advantages over other air-spaced prism designs.

It is not easy to make a Marple-Hess prism, and the extinction ratio in the commercial model is given as between 1×10^{-4} and 5×10^{-5}, somewhat lower than for a Glan-Taylor prism (Karl Lambrecht, 1973). On the other hand, even though the Marple-Hess prism has an increased L/A ratio, 1.8 as compared to 0.85 for a Glan-Taylor prism, its ultraviolet transmission is still superior to commercially available ultraviolet transmitting Glan-Thompson prisms of comparable aperture.

Frank-Ritter-Type Prisms

21. The third general category of Glan-type polarizing prisms is the Frank-Ritter design. Prisms of this type are characterized by having the optic axis in the plane of the entrance face, as in other Glan-type prisms, but having the cut made at 45° to the optic axis (Fig. 12*c*) rather than at 0°, as in Glan-Thompson prisms, or at 90°, as in Lippich prisms. Frank-Ritter prisms are particularly popular in the Soviet Union, and over 80 percent of the polarizing prisms made there have been of this design (Shustov, 1967). Usually double prisms comparable to the Ahrens modification of the Glan-Thompson (Par. 15) are used, primarily because from a rhombohedron of Iceland spar two Frank-Ritter double prisms can be obtained but only one Ahrens of the same cross section or one Glan-Thompson of smaller cross section (Shustov, 1967). However, this apparent advantage can be illusory since Iceland spar crystals often are not obtained as rhombs. For example, if the natural crystal is in the form of a plate, it may be less wasteful of material to make a Glan-Thompson or Ahrens prism than a Frank-Ritter prism (Shustov, 1967).

Optically, Frank-Ritter prisms should be similar to Glan-Thompson and Ahrens types, although the acceptance angle for a given L/A ratio is somewhat smaller since the refractive index of the extraordinary ray is larger than n_e in the prism section containing the longitudinal axis and perpendicular to the cut. In practice, the degree of polarization for a Frank-Ritter prism seems to be quite inferior to that of a good Glan-Thompson or even an Ahrens prism (Shustov, 1967).

Use of Glan-Type Prisms in Optical Systems

22. Several precautions should be taken when using Glan-type prisms in optical systems: (1) the field angle of the prism should not be exceeded; (2) there should be an adequate entrance aperture so that the prism does not become the limiting aperture of the optical system; and (3) baffles should be placed preceding and following the prism to avoid incorrect collection of polarized light or extraneous stray light.

Precaution 1 needs no explanation. To illustrate precaution 2, consider the two common situations shown in Fig. 28. In Fig. 28*a*, parallel beams of light from an extended source are transmitted by the prism. Rays from different parts of the source are not parallel, however, but diverge. Thus, the prism face must often be considerably larger than the collimating lens, or an aperture stop must be placed in front of the lens, as shown in the figure. In Fig. 28*b*, the extended source is focused near the center of the prism in order to achieve maximum light transmission. The chief ray† from each end of the extended source must

† The chief ray is the one passing through the center of the limiting aperture, which in this case is the lens.

pass through the prism to avoid vignetting, and for photometric applications the entire beam (including the extreme rays from each edge of the limiting aperture) should be transmitted. Since the beam spreads more following the image position, it is often desirable to image the extended source toward the back of the prism rather than at its midpoint.

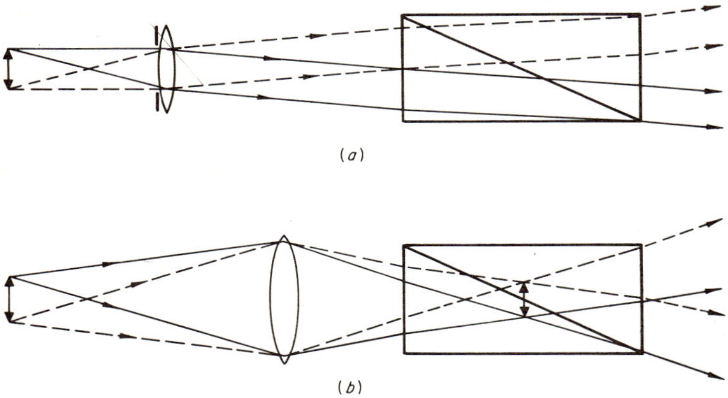

Fig. 28 Transmission of light from an extended source by a Glan prism. (*a*) Since the source is at the focus of the lens, parallel beams from each point on the source pass through the prism. An aperture stop must be placed in front of the lens to make it (rather than the prism) the limiting aperture. (*b*) The source is imaged at the center of the prism. The extreme rays from the edge of the limiting aperture should be transmitted to avoid vignetting the beam.

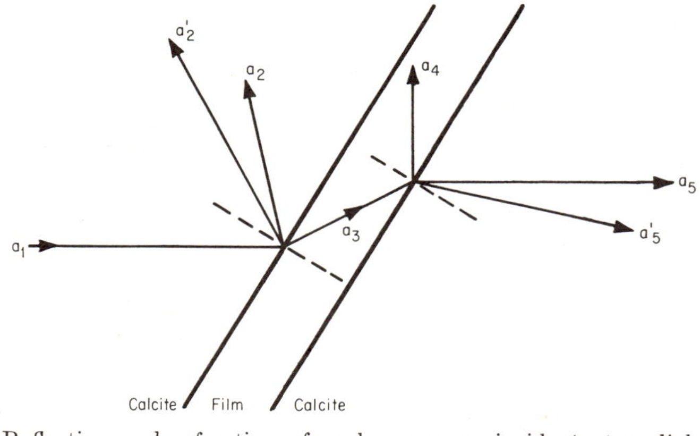

Fig. 29 Reflection and refraction of a skew ray a_1, incident at a slight angle to the plane of the paper. The optic axis is perpendicular to the paper (parallel to the cut). Reflected ray a_2 is an *e* ray and a_2' an *o* ray. Transmitted ray a_3 has its plane of vibration rotated with respect to the plane of incidence, and a_4 is the reflected component thereof. Transmitted ray a_5 is an *e* ray traveling parallel to a_1, and a_5' is an *o* ray propagated in a different direction from that of the *o* ray in the first section. [*After Archard* (1949).]

Precaution 3 was pointed out by Archard (1949) and may be understood by considering the skew rays incident on the cut of a Glan-type prism, one of which, ray a_1, is illustrated in Fig. 29. Assume that this ray is an extraordinary ray incident from a direction slightly out of the plane of the paper and that the optic axis is parallel to the cut, i.e., perpendicular to the paper. Ray a_1 will have a large vibration component parallel to the optic axis and a small component at right angles to it. The reflection coefficients for these two components will be different since their refractive indexes are different, and hence the plane of vibration of the transmitted ray a_3 will be rotated with respect to the plane of incidence. There are then two refracted rays at the second boundary, an *e* ray, a_5, vibrating parallel to a_1, and a more sharply refracted *o* ray, a_5'. The net result, as shown in Fig. 30*a*, is that a skewed *e* ray incident inside the semi-field angle gives rise to a spurious *o* ray emerging from the prism outside the semi-field

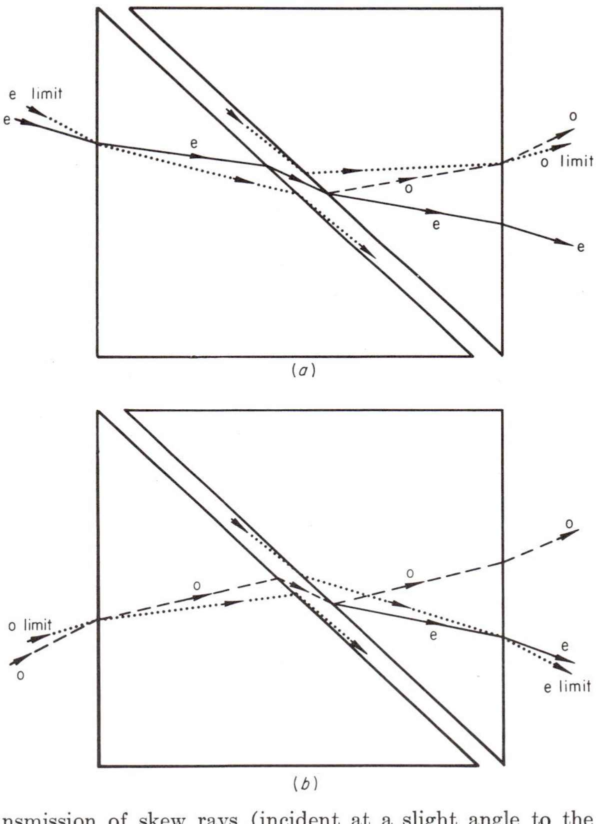

Fig. 30 Transmission of skew rays (incident at a slight angle to the plane of the paper) in a Glan-Foucault prism whose optic axis is perpendicular to the paper. (*a*) A skew *e* ray incident inside the semi-field angle gives rise to a weak *o* ray outside the semi-field angle in addition to the *e* ray traveling parallel to the incident *e* ray. (*b*) A skew *o* ray incident outside the semi-field angle gives rise to a weak *e* ray inside the semi-field angle in addition to an *o* ray parallel to the incident *o* ray. [*After Archard* (1949).]

angle. Thus a baffle must be used to limit the angular field of the emergent light in order to keep from decreasing the extinction ratio of the prism.

A similar effect, shown in Fig. 30*b*, is caused by a skewed *o* ray incident outside the semi-field angle. This ray gives rise to a spurious *e* ray that is inside the semi-field angle. In this case, unless the beam is limited by a baffle on the entrance side, light scattered into the prism at angles greater than the semi-field angle may cause spurious (but correctly polarized) beams emerging from the prism within the semi-field angle.

It should be noted that multiple reflections at the cut do not have a serious effect on the polarization of the transmitted light since, as discussed above, problems arise only when rays are incident outside the semi-field angle.

Testing Glan-Type Prisms

23. Three major errors are found in the construction of Glan-type prisms: (1) the optic axis does not lie in the end face, (2) the cut angle is incorrect or is different in the two halves of the prism, and (3) the optic axes in the two sections of the prism are not parallel. Error 1 is often the most serious, since if the optic axis is not in the end face and the prism is illuminated with convergent light, the planes of vibration of the transmitted light are no longer parallel across the face of the prism. This effect, which in Nicol-type prisms gives rise to the Landolt fringe (Par. 32), is illustrated in the following practical case (Archard, 1949). For a convergent beam of light of semi-cone angle i, the maximum variation of the plane of vibration of the emergent beam is $\pm\gamma$, where, approximately,

$$\tan \gamma = n_e \sin i \tan \phi \tag{86}$$

and ϕ is the angle of inclination of the optic axis to the end face, caused by a polishing error. For $i = 3°$ and $\phi = 5°$, the plane of vibration of the emergent beam varies across the prism face by ± 23 minutes of arc. Thus, good extinction cannot be achieved over the entire aperture of this prism even if nearly parallel light is incident on it. The field angle is also affected if the optic axis is not in the end face or is not properly oriented in the end face, but these effects are small.

If the cut angle is slightly incorrect, the field angle may be decreased. This error is particularly important in Glan-Foucault or Glan-Taylor prisms, for which the angular tolerances are quite severe, and a small change in cut angle for these prisms may greatly alter the field angle, as discussed in Pars. 16 and 19. If the cut angles are different in the two halves of the prism, the field angle will change when the prism is turned end for end. The field angle is determined by the cut angle in the half of the prism toward the incident beam. Differences in the two cut angles may also cause a beam deviation. If the angles in the two halves differ by a small angle α that makes the end faces nonparallel, the beam will be deviated by an angle $\delta = \alpha(n_e - 1)$ (Archard, 1949). If instead, the end faces are parallel and the difference in cut angle is taken up by the cement layer which has a refractive index of approximately n_e, there will be no deviation. However, if the prism is air-spaced, the deviation δ' caused by a nonparallel air film is approximately $\delta' = \alpha n_e$, illustrating one reason why air-spaced prisms are harder to make than conventional Glan-Thompson prisms (Archard, 1949).

If the optic axes in the two prism sections are not strictly parallel, the prism will suffer from *squirm,* i.e., a line object viewed through the completed prism will oscillate as the prism is turned around the line of sight (Twyman, 1952). A potentially more serious problem when one is making photometric measurements is that the transmission of the prism may not be the same in two orientations exactly 180° apart (Bennett and Bennett, 1967*a*). This effect may be caused by strain birefringence in the calcite.

In order to determine the cut angle, field angle, parallelism of the prism surfaces, thickness and parallelism of the air film or cement layer, and other prism parameters, one can use the testing procedures outlined by Decker et al. (1970), which require a spectrometer with a Gauss eyepiece, laser source, and moderately good polarizer. [Other testing procedures have been suggested by Archard (1949).] The prism is set on the table of a spectrometer and aligned so that its faces are approximately perpendicular to the light beam. The parallelism of the faces may be measured by using a Gauss eyepiece and observing the vertical difference in the reflected images from the two ends of the prism. The prism is then rotated on its axis through 90° and the measurement repeated to obtain the deviation of the prism faces from parallelism in the other direction. If the prism is mounted in a tube, the accuracy with which the axis of the prism coincides with the axis of the tube can be determined by placing the tube in a V block on the prism table, rotating it, and observing the oscillation in the image position.

The beam deviation caused by the prism can be determined by aligning the telescope and collimator without the prism in place, inserting the prism with its faces perpendicular to the beam, and measuring the angular shift in the image of the collimator slit when viewed through the prism. Prisms are generally supplied with a beam deviation of 6 minutes or less, but a beam deviation of less than 1 minute is possible on special order.

In testing air-spaced prisms, it is easy to distinguish between the Glan-Foucault and Glan-Taylor designs by determining the plane of vibration of the transmitted beam (*e* ray). If the top view of the prism appears as in Fig. 20 and the beam is vibrating in a horizontal plane, the prism is a Glan-Taylor; if the plane of vibration is vertical, the prism is a Glan-Foucault (see Pars. 16 and 19).

The limiting acceptance angles can be determined by rotating the prism from its normal incidence position until either the *o* ray starts to be transmitted or the *e* ray starts to be vignetted and then measuring the rotation angle with a Gauss eyepiece by autocollimating on the back face of the prism. If either of the limiting angles is measured, the cut angle can be calculated from Eq. (76) or (79) using the appropriate refractive index for the calcite and letting $n_2 = 1$ for an air-spaced prism.

In order to determine the thickness of the air film, the prism is rotated until multiple images of the *e* ray reflected from the cut appear. The air-film thickness can be calculated directly from the measured separation of the images and the angle of incidence of the light on the prism face. The wedge angle of the air film can be determined from the variation in the spacing of the multiple images. Decker et al. (1970) report measuring an air-film thickness of approximately 0.006 in to two significant figures for a Glan-Taylor prism of 1-in aperture and detecting a wedge angle of less than 1 minute of arc using the above-mentioned techniques. They demonstrated, however, that neither of these quantities significantly affects the acceptance angle of the prism, and the measured limiting angles for the *o* and *e* rays for this prism agree with the values calculated from Eq. (76) or (79) to within approximately 0.05° over the entire visible, near-infrared, and ultraviolet wavelength regions.

Nicol-Type Prisms

24. Nicol-type prisms are not generally used at the present time, as Glan types are optically preferable. However, they were the first kind made and were once so common that Nicol became a synonym for polarizer. There is much more calcite wastage in making Glan-type prisms than in making the simpler Nicol types so that, even though Glan polarizers were developed in the nineteenth century, it was only following the recent discoveries of new calcite deposits that they became popular. Many of the older instruments are still equipped with Nicol prisms, and a discussion of polarizing prisms which omitted Nicols would not be complete.

25. Conventional Nicol Prism The first polarizing prism was made in 1828 by William Nicol (Nicol, 1828–1829), a teacher of physics in Edinburgh. By cutting a calcite rhomb diagonally and symmetrically through its blunt corners and then cementing the pieces together with canada balsam, he could produce a better polarizer than any known up to that time. A three-dimensional view of Nicol's prism is shown in Figs. 12*d* and 31*a*. The cut is made perpendicular to the principal section (defined in Par. 7), and the angle is such that the ordinary ray is totally reflected and only the extraordinary ray emerges. The end of a Nicol prism thus looks like Fig. 9, except that only the *e* ray emerges, so that the plane of vibration of the polarized light is parallel to the short diagonal of the rhomb. When the rhomb is intact, the direction of polarization can be determined by inspection. However, the corners are sometimes cut off, making the rhomb difficult to recognize.

The principal section of Nicol's original prism is similar to that shown in Fig. 10 except that the ordinary ray is internally reflected at the cut along diagonal *BH*. The cut makes an angle of 19°8′ with edges *BF* and *DH* and an angle of about 90° with the end face (line FH) of the rhomb. Since the obtuse angle DHF is 109°7′, the angle between the cut and the optic axis is 44°36′. The field of the prism is limited on one side by the angle at which the ordinary ray is no longer totally reflected from the balsam film, about 18.8° from the axis of rotation of the prism, and on the other by the angle at which the extraordinary ray is totally reflected by the film, about 9.7° from the axis. Thus the total angle is about 28.5° but is not by any means symmetric about the axis of rotation; the field angle (Par. 8) is only 2 × 9.7° = 19.4°.

In order to produce a somewhat more symmetric field and increase the field angle, the end faces of Nicol prisms are usually trimmed to an angle of 68°. This

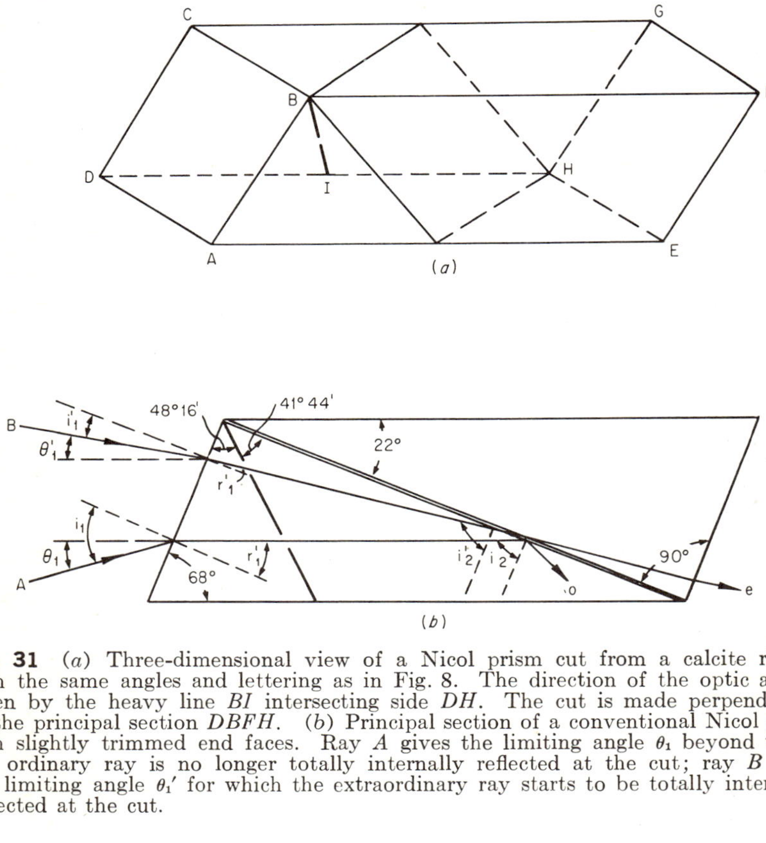

Fig. 31 (*a*) Three-dimensional view of a Nicol prism cut from a calcite rhomb with the same angles and lettering as in Fig. 8. The direction of the optic axis is given by the heavy line BI intersecting side DH. The cut is made perpendicular to the principal section $DBFH$. (*b*) Principal section of a conventional Nicol prism with slightly trimmed end faces. Ray A gives the limiting angle θ_1 beyond which the ordinary ray is no longer totally internally reflected at the cut; ray B gives the limiting angle θ_1' for which the extraordinary ray starts to be totally internally reflected at the cut.

practice was apparently started by Nicol himself (Thompson, 1905). If the cut is made at 90° to the new face, as shown in Fig. 31*b*, the new field angle is twice the smaller of θ_1 and θ_1'. The angles are computed as follows. The limiting angle for the ordinary ray (ray A) can be determined simply from Snell's law. The refractive index of the canada balsam is approximately 1.54, and n_o is given in Table 2 (Par. 7). At 5893 Å, the critical angle i_2 at the cut is

$$\sin i_2 = \frac{1.54}{1.65835} \qquad \text{or} \qquad i_2 = 68.22°$$

The cut is made normal to the entrance face of the prism, so that the angle of refraction r_1 at the entrance face is $90° - i_2 = 21.78°$; the angle of incidence $i_1 = 37.97°$ from Snell's law. Since the entrance face makes an angle of 68° with the longitudinal axis of the prism, the normal to the entrance face is 90° — 68° = 22° to the longitudinal axis. Thus, the limiting angle at which the ordinary ray is totally reflected by the balsam results in a half angle θ_1 of 37.97° — 22° = 15.97° relative to the longitudinal axis.

The limiting angle at which the extraordinary ray is not totally reflected at the cut is computed in a similar manner except that now the refractive index of the extraordinary ray is a funcion of the angle ϕ between the wave normal and the optic axis. Ray *B* in Fig. 31*b* indicates the path of the wave normal for the *e* ray in the prism, for which Snell's law applies (see Par. 7). As before, $i_2' = 90° - r_1'$, so that at the critical angle at the cut

$$\sin (90° - r_1') = \cos r_1' = \frac{1.54}{n_\phi} \tag{87}$$

But n_ϕ is given by Eq. (71), for which $\phi = r_1' + 41°44'$, so that one obtains the transcendental equation

$$\frac{\cos^2 r_1'}{1.54^2} = \frac{\sin^2 (r_1' + 41.73°)}{n_e{}^2} + \frac{\cos^2 (r_1' + 41.73°)}{n_o{}^2} \tag{88}$$

By using values for n_o and n_e at 5893 Å from Table 2, r_1' is determined to be 7.44° and $i_1' = 11.61°$. The half angle θ_1' is then 22° — 11.61° = 10.39°. Since θ_1' is smaller than θ_1, the semi-field angle is 10.39° and the total field angle is 2 × 10.39° = 20.78°. At first one might wonder why the extraordinary ray is totally internally reflected at the cut since n_e is less than 1.54, the refractive index of the cement. However, $n_\phi = 1.5531$, as calculated from Eq. (87) once r_1' is known, and thus the *e* ray can be totally internally reflected.

The field edges are easily observed by looking through a Nicol prism. Thompson (1905) has described what one will see:

On looking through any ordinary Nicol prism, towards an illuminated surface such as a white sheet or a white cloud, and turning it about to peer through it in different directions while it is held close to the eye, one at once observes that the useful field of view is a region varying (in ordinary Nicols) from about 15° to 20° in width; being delimited between two slightly curved boundaries, the one a blue band beyond which is a region of darkness, the other a rainbow-tinted set of fringes beyond which there is a region of greater brightness. The blue band marks the limit where the useful (extraordinary) ray vanishes by total reflexion for internal incidences, at the balsam film, greater than the critical angle for the extraordinary index. The rainbow-tinted fringes mark the limit beyond which the ordinary ray is no longer eliminated by total reflexion, and the light beyond this limit, containing both ordinary and extraordinary beams, is not polarized.

The rainbow-tinted fringes Thompson refers to are really interference fringes, which he explained (Thompson, 1877).

The ratio of the length of one side of the prism to the slant height of the entrance face is determined by the angle at which the cut is made. For a cut at 90° to the face, such as is shown in Fig. 31*b*, the ratio is 2.7. In commercial specifications, this so-called length-to-aperture ratio is often said to be between 2.5 and 3. In actually mounting the prism, however, one needs to know the total length of the prism and the height of the aperture perpendicular to the prism axis. The ratio of these two quantities is considerably larger than the numbers given above; for example, 3.3 for the prism shown in Fig. 31*b*.

Trimmed Nicol-Type Prisms

26. The angle at which the cut is made in a Nicol-type prism is not critical. The field angle is affected, but a useful prism will probably result even if the cut is made at an angle considerably different from 90°. The conventional trimmed Nicol, discussed in Par. 25, is shown again in Fig. 32*a*. In this and the other five parts of the figure, principal sections of various prisms are shown superimposed on the principal section of the basic calcite rhomb (Fig. 10). Thus, it is clear how much of the original rhomb is lost in making the different types of trimmed Nicols.

In the Steeg and Reuter Nicol shown in Fig. 32*b*, the rhomb faces are not trimmed, and the cut is made at 84° to the faces instead of 90°, giving a smaller L/A ratio. The asymmetry of the field which results is reduced by using a cement having a slightly higher index than canada balsam.

Alternately, in the Ahrens Nicol shown in Fig. 32*c*, the ends are trimmed in the opposite direction, increasing their angles with the long edges of the rhomb from 70°53′ to 74°30′ or more. By also trimming the long edges by 3°30′, the limiting angles are made more symmetric about the prism axis.

27. Thompson Reversed Nicol In the Thompson reversed Nicol shown in Fig. 32*d*, the ends are heavily trimmed so that the optic axis lies nearly in the end

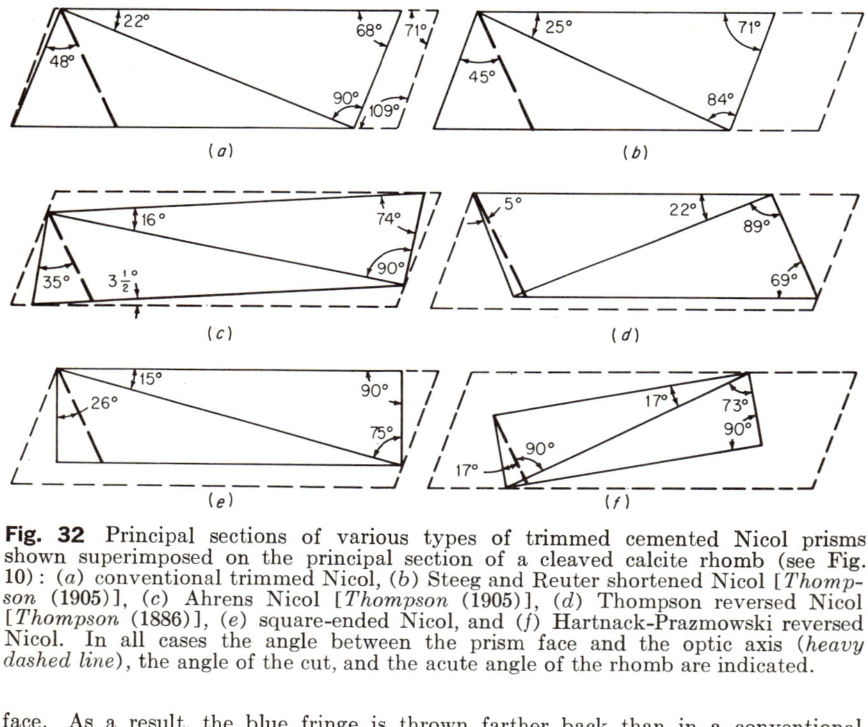

Fig. 32 Principal sections of various types of trimmed cemented Nicol prisms shown superimposed on the principal section of a cleaved calcite rhomb (see Fig. 10): (*a*) conventional trimmed Nicol, (*b*) Steeg and Reuter shortened Nicol [*Thompson* (1905)], (*c*) Ahrens Nicol [*Thompson* (1905)], (*d*) Thompson reversed Nicol [*Thompson* (1886)], (*e*) square-ended Nicol, and (*f*) Hartnack-Prazmowski reversed Nicol. In all cases the angle between the prism face and the optic axis (*heavy dashed line*), the angle of the cut, and the acute angle of the rhomb are indicated.

face. As a result, the blue fringe is thrown farther back than in a conventional Nicol, and although the resulting prism is shorter, its field angle is actually increased.

28. Nicol Curtate, or Halle, Prism The sides of the calcite rhomb may also be trimmed so that they are parallel or perpendicular to the principal section. Thus, instead of the end of the prism having the shape shown in Fig. 9, it is square (or sometimes octagonal) with the sides of the square parallel to line *DB* in Fig. 9. This prism is of the Halle type (Halle, 1908*a*,*b*) and was shown in Fig. 12*e*. Halle, in addition, used thickened linseed oil instead of canada balsam and altered the angle of the cut. In this way he reduced the length-to-aperture ratio from about 2.7 to 1.8 and the total acceptance angle from 25° to about 17°. Such shortened prisms cemented with low-index cements are often called Nicol curtate prisms (curtate means shortened).

29. Square-ended Nicol The slanting end faces on conventional Nicol prisms introduce some difficulties, primarily because the image is slightly displaced as the prism is rotated. To help correct this defect, the slanting ends of the calcite rhomb can be squared off, as in Fig. 32*e*, producing the so-called square-ended

Nicol prism. The angle at which the cut is made must then be altered since the limiting angle θ_1 for the ordinary ray depends on the angle of refraction at the end face in a conventional prism, in which the limiting ray travels nearly parallel to the prism axis inside the prism (ray *A* in Fig. 31*b*). If the cut remained the same, the limiting value of θ_1 would thus be zero. However, if the cut is modified to be 15° to the sides of the prism, the total acceptance angle is in the 24 to 27° range, depending on the type of cement used (Thompson, 1905).

Some image displacement will occur even in square-ended Nicol prisms since the optic axis is not in the plane of the entrance face. Therefore, the extraordinary ray will be bent even if light strikes the entrance face of the prism at normal incidence. There is considerable confusion on this point in the literature (Thompson, 1905; Dale, 1923*b*).

30. Hartnack-Prazmowski Prism A reversed Nicol which has the cut at 90° to the optic axis (Hartnack and Prazmowski, 1866) is shown in Figs. 12*f* and 32*f*. If it is cemented with linseed oil, the optimum cut angle calculated by Hartnack is 17° to the long axis of the prism, giving a total acceptance angle of 35° and an *L*/*A* ratio of 3.4 (Thompson, 1905). If canada balsam is used, the cut should be 11°, in which case the total acceptance angle is 33° and the *L*/*A* ratio is 5.2.

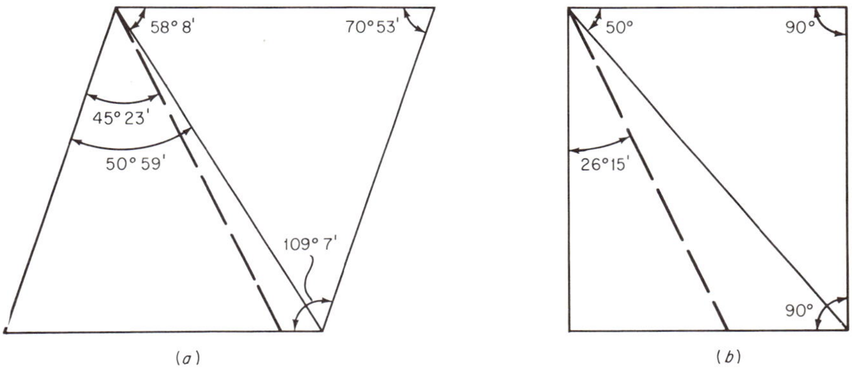

Fig. 33 Principal sections of air-spaced Nicol prisms: (*a*) Foucault prism and (*b*) Hofmann square-ended prism. The optic axis is indicated by the heavy dashed line and the various angles are shown.

31. Foucault Prism A modified Nicol prism in which an air space is used between the two prism halves instead of a cement layer (Foucault, 1857) consists of a natural-cleavage rhombohedron of calcite which has been cut at an angle of 51° to the face, as shown in Fig. 33*a*. The cut nearly parallels the optic axis. Square-ended Foucault-type prisms, such as the Hofmann prism shown in Fig. 33*b*, have also been reported (Thompson, 1905). The angle at which the cut is made can be varied slightly in both the normal Foucault prism and the Hofmann variation of it. In all designs the *L*/*A* ratio is 1.5 or less, and the total acceptance angle about 8° or less. The prisms suffer somewhat from multiple reflections, but the principal trouble, as with all Nicol prisms, is that the optic axis is not in the plane of the entrance face. This defect causes various difficulties, including nonuniform polarization across the field and the occurrence of a Landolt fringe (Par. 32) when two Nicol-type prisms are crossed.

32. Landolt Fringe If an intense extended light source is viewed through crossed polarizing prisms, careful observation will reveal that the field is not uniformly dark. In Nicol-type prisms the darkened field is crossed by a darker line whose position is an extremely sensitive function of the angle between the polarizer and analyzer. Other types of polarizing prisms also exhibit this anomaly but to a lesser extent. The effect was first reported by Landolt (1879) and was explained by Lippich (1882), who used it to construct a polarimeter in which the analyzer could be set to 2 or 3 seconds of arc by referring the fringe to a pair of fixed cross hairs (Wood, 1934*a*). The Landolt fringe is probably the reason for the frequently reported but questionable observation that Nicol prisms introduce ellipticity into the transmitted light.

A rigorous discussion of the origin of the Landolt fringe is given (in German) by Groosmuller (1926), and the phenomenon is explained qualitatively in Figs. 34 and 35. Lines of constant polarization† are shown in Fig. 34*a* for a Nicol prism and in Fig. 34*b* for a Glan-Thompson type. The planes of vibration of the E vector, which we now use to characterize polarized light, are at right angles to the solid lines in Fig. 34. The circles represent constant values of the angle of incidence measured from the normal to the entrance face; equal increments of tan i are plotted. The Nicol prism has a maximum limiting angle of 38°, corresponding to a value of 0.78 for tan i. Thus, we are concerned only with

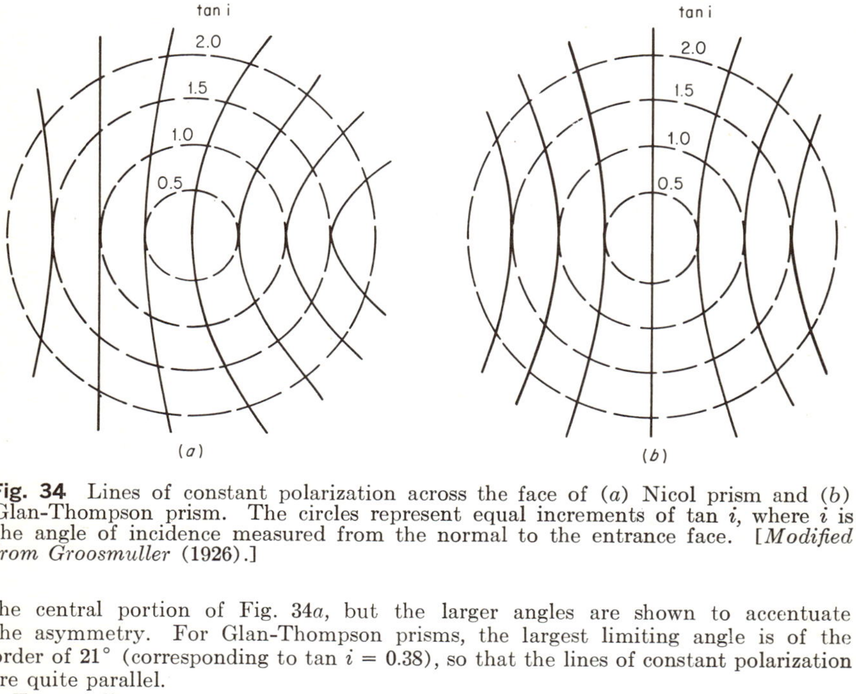

Fig. 34 Lines of constant polarization across the face of (*a*) Nicol prism and (*b*) Glan-Thompson prism. The circles represent equal increments of tan i, where i is the angle of incidence measured from the normal to the entrance face. [*Modified from Groosmuller* (1926).]

the central portion of Fig. 34*a*, but the larger angles are shown to accentuate the asymmetry. For Glan-Thompson prisms, the largest limiting angle is of the order of 21° (corresponding to tan i = 0.38), so that the lines of constant polarization are quite parallel.

For small field angles in a Nicol prism, one can approximate the planes of vibration of the E vector by straight lines which, if extended, would meet at a point to one side of the field of view. Such a situation is shown in Fig. 35*a*. If two such Nicol prisms are crossed, we would have the situation shown in Fig. 35*b*. Complete extinction can occur only at points where the lines are mutually perpendicular, so that there will be a very dark curved region surrounded by a lighter area where extinction is not quite complete. The dark curved region is the Landolt fringe.

Polarizing Beam-Splitter Prisms

33. The three classic polarizing beam-splitter prisms are the **Rochon, Sénarmont,** and **Wollaston,** shown in perspective in Fig. 36*a* to *c* and in side view in Fig. 37*a* to *c*. In addition, any polarizing prism can be used as a polarizing beam splitter by changing the shape of one side and removing the absorbing coating from its surface.

† See the first footnote in Par. 2.

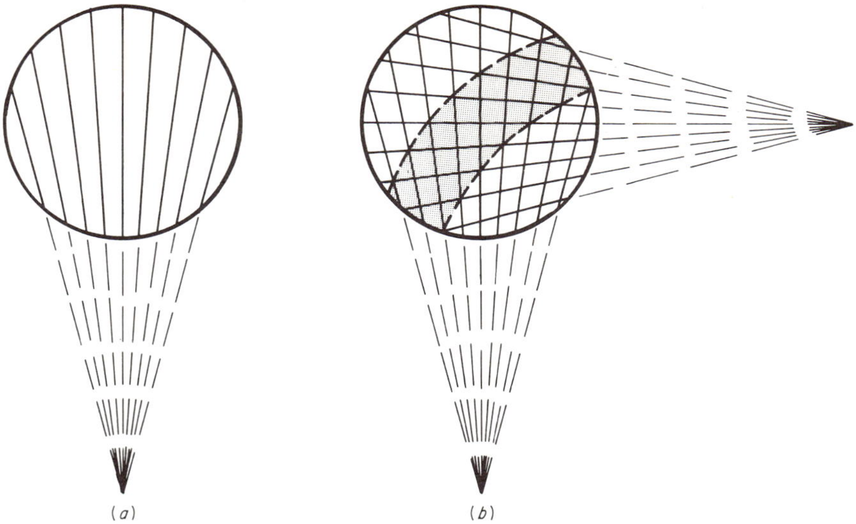

Fig. 35 Lines of constant vibration of the E vector in a small angular field of view across the entrance face of (*a*) one Nicol prism and (*b*) two crossed Nicols. The lines, if extended, would meet at a point to one side of the entrance face for each prism. In (*b*) the Landolt fringe occurs at the points where the lines intersect at strictly right angles; its limits are indicated by the curved dashed lines. Other portions of the field are less dark. [*Modified from Wood* (1934*a*).]

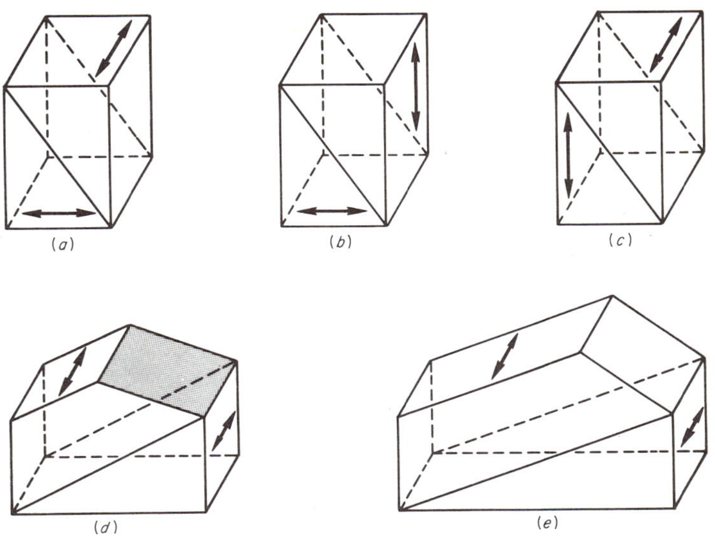

Fig. 36 Three-dimensional views of various types of polarizing beam-splitter prisms: (*a*) Rochon, (*b*) Sénarmont, (*c*) Wollaston, (*d*) Foster (shaded face is silvered), and (*e*) beam-splitting Glan-Thompson.

Two examples of such prisms are the Foster prism, in which the ordinary and extraordinary rays emerge at right angles to each other, and the beam-splitting Glan-Thompson prism, in which the ordinary ray emerges normal to one side (Figs. 36*d* and *e* and 37*d* and *e*). Another prism of this type, the beam-splitting Ahrens prism, is a double beam-splitting Glan-Thompson prism (see Par. 15).

In polarizing prisms, the optic axes are always parallel to each other in the two halves of the prism. By contrast, the optic axes in the two halves of the Rochon, Sénarmont, and Wollaston polarizing beam-splitter prisms are at right angles to each other. Crystal quartz is often used to make these beam splitters, and such prisms can be used down to the vacuum ultraviolet. In applications not requiring such short wavelengths, calcite is preferable because it gives a greater angular separation of the beams (typically 10° as compared to 0.5° for quartz) and does not produce optical rotation.

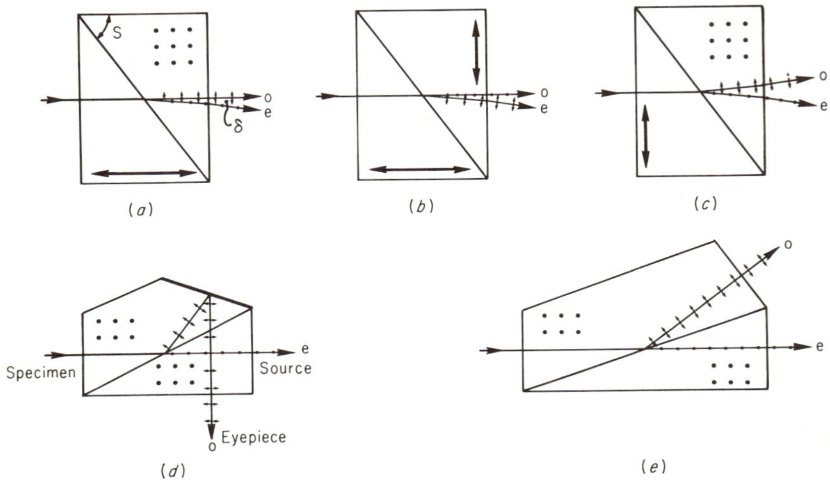

Fig. 37 Side views of the polarizing beam-splitter prisms in Fig. 36. The directions of the optic axes are indicated by the dots and the heavy double-pointed arrows. The angle of the cut for the Rochon prism is *S*. When the Foster prism is used as a microscope illuminator, the source, specimen, and eyepiece are in the positions indicated.

34. Rochon Prism The Rochon prism, invented in 1783 (West and Jones, 1951), is the most common type of polarizing beam splitter. It is often used in photometric applications in which both beams are utilized. It is also used as a polarizing prism in the ultraviolet, in which case one of the beams must be eliminated, e.g., by imaging the source beyond the prism and blocking off the deviated image.

The paths of the two beams through the prism are shown in Fig. 37*a*. A ray normally incident on the entrance face travels along the optic axis in the first half of the prism, so that both ordinary and extraordinary rays are undeviated and have the same refractive index n_o. The second half of the prism has its optic axis at right angles to that in the first half, but the ordinary ray is undeviated since its refractive index is the same in both halves. The extraordinary ray, however, has its minimum index in the second half, so that it is refracted at the cut according to Snell's law (see Par. 7 and Fig. 11). Since the deviation angle depends on the ratio n_e/n_o, it is a function of wavelength. If the angle of the cut is S, to a good approximation the beam deviation δ of the extraordinary ray depends on the cut angle in the following manner (Steinmetz et al., 1967):

$$\tan S = \frac{n_o - n_e}{\sin \delta} + \frac{\sin \delta}{2n_e} \tag{89}$$

This relation holds for light normally incident on the prism face. The semi-field angle i_{max} is given (Steinmetz et al., 1967) by

$$\tan i_{max} = \tfrac{1}{2}(n_e - n_o) \cot S \qquad (90)$$

If the prism is to be used as a polarizer, the light should be incident as shown. Rochon prisms also act as polarizing beam splitters when used backward, but the deviation of the two beams is then slightly less. The smaller beam deviation can be understood by calculating the angle of refraction for the *e* ray at the cut when light is normally incident on the prism face. The angle of incidence on the cut will be 90° — S so that, from Snell's law,

$$\sin r = \frac{n_o}{n_e} \cos S \qquad (91)$$

for a normal Rochon and

$$\sin r = \frac{n_e}{n_o} \cos S \qquad (92)$$

for a backward Rochon. The angle r is the angle of refraction in the half of the prism the light encounters second. Since $n_o > n_e$ for calcite, Eq. (91) gives the larger value of r. When a Rochon prism is used backward, both the dispersion and the optical activity (for quartz) will adversely affect the polarization. Thus one generally uses a Rochon in the normal manner. However, an exception occurs when a quartz Rochon is to be used as an analyzer. In this case it is best to reverse the prism and use a detector that is insensitive to polarization to monitor the relative intensities of the two transmitted beams.

A Rochon prism is achromatic for the ordinary ray but chromatic for the extraordinary ray. Since total internal reflection does not occur for either beam, the type of cement used between the two halves of the prism is less critical than that used for conventional polarizing prisms. Canada balsam is generally used, although the two halves are sometimes optically contacted for high-power laser applications or for use in the ultraviolet at wavelengths shorter than 3500 Å. Optically contacted crystalline-quartz Rochon prisms can be used to wavelengths as short as 1700 Å, and a double Rochon of MgF_2 has been used to 1300 Å in the vacuum ultraviolet (Steinmetz et al., 1967). Optically contacted single Rochon prisms of MgF_2 have also been constructed, and the transmission of one has been measured from 1400 Å to 7μm (Morris and Abramson, 1969). Ultraviolet-transmitting cements such as *gédamine* can be used to extend the short-wavelength limit of calcite prisms to about 2500 Å (see Par. 11).

35. Defects Quartz and calcite Rochon prisms suffer from several defects. Quartz exhibits optical activity when light is transmitted through it parallel to the optic axis, and although two mutually perpendicular, polarized beams will emerge from a quartz Rochon prism used in the conventional direction, their spectral composition will not faithfully reproduce the spectral compositions of the horizontal and vertical components of the input. If such a prism is used backward, different wavelengths emerge from the prism vibrating in different planes. Hence the output consists of many different polarizations instead of the desired two (Ammann and Massey, 1968).

Calcite Rochon prisms do not exhibit optical activity but are difficult to make, since, as mentioned in Par. 7, when calcite surfaces are cut normal to the optic axis, small tetrahedra tend to cleave out from the surface during pitch polishing. These tetrahedra may also cleave out during attempts to clean the prisms, and occasionally glass plates are cemented to such surfaces to prevent damage. Some image distortion will occur in calcite prisms; if nonnormally incident rays pass through the prism, both beams will be distorted along their directions of vibration; i.e., the undeviated beam (*o* ray), which vibrates in a vertical plane, will be distorted vertically, and the deviated beam (*e* ray), which vibrates in a horizontal plane, will be distorted horizontally (Ammann and Massey, 1968).

36. Glass-Calcite Rochons Some of the difficulties mentioned in Par. 35 can be minimized or eliminated by making the entrance half of the Rochon prism out of glass of matching index instead of quartz or calcite. Both *o* and *e* rays travel along the same path and have the same refractive index in this half of the prism, so that the birefringent qualities of the quartz or calcite are not being used and an isotropic medium would serve just as well. By properly choosing the index of the glass, either the ordinary or the extraordinary ray can be deviated, and glasses are available for matching either index of calcite reasonably well over much of the visible region (Ammann and Massey, 1968). The extraordinary ray always suffers some distortion in its direction of vibration, but the distortion of the ordinary ray can be eliminated in the glass-calcite construction. By properly choosing the refractive index of the glass, we can determine whether the *e* ray will be the deviated or the undeviated beam. (Some distortion also arises for deviated beams in the direction of the deviation because of Snell's law and cannot be corrected in this way.) Another method of obtaining an undeviated beam was used by Hardy (1935); unable to find a glass with refractive index and dispersion matching those of calcite, he selected a glass with the correct dispersive power and then compensated for the difference in refractive index by putting a slight wedge angle on the calcite surface. Now a wider selection of glasses is available, but glass-calcite prisms cannot be made strictly achromatic over an extended wavelength range, and thermally induced strains caused by the difference in expansion coefficients in the two parts of the prism may be expected unless the cement yields readily.

37. Total Internal Reflection in Rochons When normal Rochon prisms are used as polarizers, one of the beams must be screened off and eliminated. This restriction might be removed by making the cut between halves of the prism at a sufficiently small angle for the extraordinary ray to be totally reflected. Calculations indicate that this approach should be feasible (Bouhet and Lafont, 1949), but it has apparently not been followed.

38. Sénarmont Prism The Sénarmont polarizing beam splitter, shown in Figs. 36*b* and 37*b*, is similar to the Rochon prism except that the optic axis in the exit half of the prism is coplanar with the optic axis in the entrance half, i.e., at right angles to the Rochon configuration. As a result, light whose plane of vibration is initially vertical is deviated in the Sénarmont prism, while in the Rochon prism the deviated beam has its plane of vibration horizontal (assuming no optical activity in either case) (compare Fig. 37*a* and *b*). The amount of the deviation in the Sénarmont prism is slightly less than in the Rochon because the extraordinary ray does not have its minimum refractive index [Eq. (71)].

An alternate form of Sénarmont prism, the right-angle Sénarmont or Cotton polarizer (Cotton, 1931) consists of only the first half of the Sénarmont prism. Unpolarized light normally incident on the prism face is totally internally reflected at the hypotenuse and is then resolved into two planes of vibration, one parallel to the optic axis and the other at right angles to it. Double refraction will then occur just as in a normal Sénarmont prism. Such a prism has a transmission equivalent to that of an optically contacted Sénarmont or Rochon but is much less expensive.

39. Wollaston Prism The Wollaston prism (Figs. 36*c* and 37*c*) is a polarizing beam splitter, also used as a polarizing prism in the vacuum ultraviolet (Johnson, 1964), that deviates both transmitted beams. The deviations, indicated in Fig. 37*c*, are nearly symmetrical about the incident direction, so that the Wollaston has about twice the angular separation of a Rochon or Sénarmont prism. A normally incident beam is undeviated upon entering the prism, but the *o* ray, vibrating perpendicular to the optic axis, has a refractive index n_o while the *e* ray, vibrating parallel to the optic axis, has its minimum (or principal) index n_e. At the interface the *e* ray becomes the *o* ray and vice versa because the direction of the optic axis in the second half is at right angles to its direction in the first half. Thus the original *o* ray enters a medium of lower refractive index and is refracted away from the normal at the cut, while the original *e* ray passes into a medium of higher refractive index and is refracted toward the normal. On leaving the second half of the prism, both rays are refracted away from the normal, so that their divergence increases.

The deviation of each beam is chromatic in Wollaston prisms, which are most commonly used to determine the relative intensities of two plane-polarized components. Since the light never travels along the optic axis, optical activity does not occur and the relative intensities of the two beams are always proportional to the intensities of the horizontal and vertical polarization components in the incident beam. For an L/A ratio of 1.0, the angular separation between beams is about 1° for a crystalline-quartz Wollaston prism; it can be as high as 3°30′ for an L/A ratio of 4.0. With a calcite prism, the beams would have an angular separation of about 19° for an L/A ratio of 1.0, but severe image distortion and lateral chromatism results when such large angular separations are used. These effects can be minimized or the angular separation can be increased for a given L/A ratio by using a three-element Wollaston prism such as is shown in Fig. 38. With this modification, apparently suggested by Karl Lambrecht (Karl Lambrecht, 1973), divergences as large as 30° can be obtained.

The ellipticity in the emergent polarized beams has been measured by King and Talim (1971). For calcite Wollaston prisms, the ellipticities were in the 0.004 to 0.025° range, comparable to those of Glan-Thompson prisms (Par. 12). Larger values, between 0.12 and 0.16°, were measured for crystalline-quartz Wollaston prisms. The major contribution, which was from the combined optical activity and birefringence in the quartz rather than from defects within the crystal, cannot be avoided in quartz polarizers.

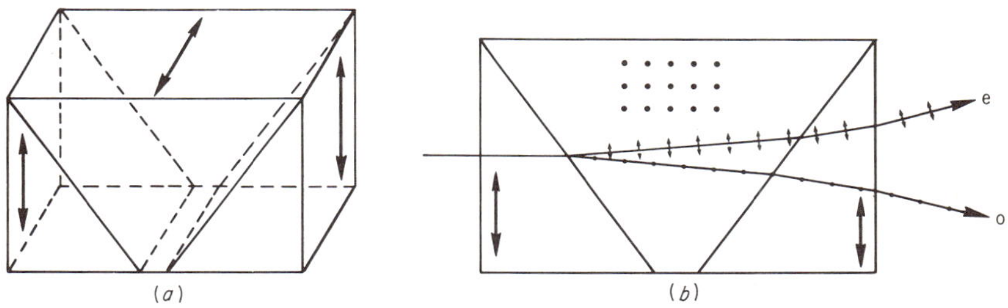

Fig. 38 Three-element Wollaston prism: (*a*) three-dimensional view and (*b*) side view. The designation *o* and *e* on the rays is for the final prism element.

40. Foster Prism This prism, shown in a three-dimensional view in Fig. 36*d* and in cross section in Fig. 37*d*, can be used to form two plane-polarized beams separated by 90° from each other (Foster, 1938). Its construction is similar to that of a Glan-Thompson prism except that one side is cut at an angle and silvered to reflect the ordinary ray out the other side.

The Foster prism is often used backward as a polarizing microscope illuminator for observing reflecting specimens. For this application, the light source is at *e* in Fig. 37*d*, and unpolarized light enters the right-hand face of the prism. The ordinary ray (not shown) is reflected at the cut and absorbed in the blackened side of the prism, while the extraordinary ray is transmitted undeviated out the left face of the prism. It then passes through the microscope objective and is reflected by the specimen, returning on its same path to the prism. Light that is unchanged in polarization will be transmitted undeviated by the prism along the path to the light source. If, however, the plane of vibration has been rotated so that it is at right angles to the optic axis (in the plane of the figure), the light will be reflected into the eyepiece. The prism thus acts like a crossed polarizer-analyzer combination.

If a correctly oriented quarter-wave plate is inserted in the beam between the prism and the microscope objective, the light striking the sample will be circularly polarized, and after being reflected back through the quarter-wave plate, it will be linearly polarized again but with the plane of vibration rotated by 90°. This light is vibrating perpendicular to the optic axis and will be reflected into the eyepiece, giving bright-field illumination. Foster prisms used in this manner introduce no astigmatism since the light forming the image enters and leaves the prism normal to the prism faces and is reflected only by plane surfaces.

41. Beam-splitting Glan-Thompson Prism If a prism design similar to the Foster is used but the side of the prism is cut at an angle so that the ordinary ray, which is deflected, passes out normal to the surface of the prism rather than being reflected, the prism is called a beam-splitting Glan-Thompson prism (Figs. 36*e* and 37*e*). Since no refraction occurs for either beam, the prism is achromatic and nearly free from distortion. The angle between the two emerging beams is determined by the angle of the cut between the two halves of the prism and hence depends on the L/A ratio of the prism. For an L/A ratio of 2.414, the angle is 45°. The field angle around each beam is calculated for different L/A ratios just as for a conventional Glan-Thompson prism (see Par. 13). By making the prism double, i.e., a beam-splitting Ahrens prism, the incident beam can be divided into three parts, one deflected to the left, one to the right, and one undeviated.

Feussner Prisms

42. The polarizing prisms discussed so far require large pieces of birefringent material, and the extraordinary ray is the one usually transmitted. Feussner (1884) suggested an alternate prism design in which only thin plates of birefringent material are required and the ordinary ray rather than the extraordinary ray is transmitted for negative uniaxial materials. A similar suggestion was apparently made by Sang in 1837 although he did not publish it until 1891 (Johannsen, 1968*a*). In essence, Feussner's idea was to make the prisms isotropic and the film separating them birefringent, as shown in Fig. 14. The isotropic prisms should have the same refractive index as the higher index of the birefringent material so that for negative uniaxial materials, e.g., calcite or sodium nitrate, the ordinary ray is transmitted and the extraordinary ray totally internally reflected. Advantages of this design are that (1) since the ordinary ray is transmitted, the refractive index does not vary with angle of incidence and hence the image is anastigmatic, (2) large field angles or prisms of compact size can be obtained, and (3) the birefringent material is used economically. Furthermore, because the path length of the ray through the birefringent material is short, a lower-quality material can be used.

Disadvantages are that (1) for both calcite and sodium nitrate, the extraordinary ray is transmitted over a larger wavelength range than the ordinary ray so that Feussner prisms do not transmit over as large a wavelength range as conventional prisms and (2) the thermal-expansion coefficients of the isotropic and birefringent materials are different, making thermally induced strains likely. Solutions to the second problem are to use a thixotropic cement, which flows more readily with increasing stress, or to enclose the system in a metal sleeve and use oil instead of cement. If the ordinary index is matched by the oil, the birefringent material does not even need to be polished very well. Even a cleavage section of calcite can be used, with only a tolerable loss in angular field (Hyde, 1970).

Feussner suggested orienting the optic axis of the birefringent slab perpendicular to the cut, as indicated in Fig. 14*a*. Since the thermal expansion of the slab is the same in all directions perpendicular to the optic axis, thermally induced strains are minimized in this way. Field angles for Feussner prisms employing calcite and sodium nitrate slabs are given in Table 3 for various cut angles S.

Shortly after Feussner's article was published, Bertrand (1884) pointed out that the optic axis of the birefringent slab should be parallel to the entrance face of the prism to give the maximum difference between the refractive indexes of the ordinary and extraordinary rays. A prism made in this way, sometimes called a Bertrand-type Feussner prism, is shown in Fig. 14*b*.

Since sodium nitrate is easily obtainable and has a birefringence even larger than that of calcite, repeated attempts have been made to produce polarizing prisms of this material (Wulff, 1896; Stöber, 1924, 1928, 1930; Tzekhnovitzer, 1934; West, 1945; Huot de Longchamp, 1947; Yamaguti, 1955*a*; Yamaguti et al., 1959). However, it is not only deliquescent but also very soft, so that although large single crystals can be obtained, they are difficult to work. They can be crystallized in the desired orientation from a melt using a technique discovered by West

TABLE 3 Field Angles for Feussner Prisms (Feussner, 1884)

Slab material	Field angle, deg	Cut angle S, deg	L/A ratio
Calcite	44 (max)	13.2	4.26
	30	17.4	3.19
	20	20.3	2.70
Sodium nitrate	54 (max)	16.7	3.53
	30	24	2.25
	20	27	1.96

(1945). When sodium nitrate crystallizes from a melt on a mica cleavage surface, one of its basal planes is oriented parallel to the mica cleavage and hence its optic axis is perpendicular to the mica surface. West reports growing single crystals as large as 38 by 19 by 2 cm using this technique. Yamaguti (1955*a*; Yamaguti et al., 1959) has produced polarizing prisms of sodium nitrate by placing thin, closely spaced glass plates on edge on a mica sheet and then immersing the assembly in a melt of sodium nitrate. The thin single crystal thus formed was annealed and cemented between glass prisms to form a Bertrand-type Feussner prism. Conceivably, the sodium nitrate could have been grown directly between the glass prisms themselves, but when such thick pieces of glass are used, it is difficult to avoid setting up strains in the crystal and consequently reducing the polarization ratio. Yamaguti used SK5 glass prisms ($n_D = 1.5889$) cut at an angle of 23° to form his polarizing prism and reports a field of view of 31°, symmetric about the normal to the entrance face.

Another possible birefringent material suitable for a Feussner prism is muscovite mica, and such prisms have actually been constructed and tested (Dumont, 1969, 1970). A 6° field angle can be obtained (Hyde, 1970), which is adequate for many optical systems illuminated by lasers.

Noncalcite Polarizing Prisms

43. As mentioned in Par. 7, polarizing prisms made of materials other than calcite have been used primarily in regions where calcite is opaque. An exception is a new material, yttrium orthovanadate, which shows promise for high-transmission polarizers for the visible region (DeShazer, 1971). This material can be grown as a positive uniaxial single crystal with good optical transmission from approximately 0.380 to 3.7 μm. The ordinary and extraordinary refractive indexes have been measured at selected laser wavelengths from 0.4880 to 3.39 μm and have been fitted to a Sellmeier equation of the form $n^2 = 1 + A\lambda^2/(\lambda^2 - B)$, with $A = 2.7665$, $B = 0.026884$ for the ordinary ray and $A = 3.5930$, $B = 0.032103$ for the extraordinary ray (λ is in micrometers). The values given by this equation are accurate to a few parts in the fourth decimal for n_o and about 1 part in the third decimal for n_e for wavelengths between 0.4880 and 1.0 μm. At 0.6328 μm, $n_o = 1.9915$ and $n_e = 2.2148$. The birefringence is only slightly less than that of rutile and larger than that of calcite (see Table 15 in Par. 56), and the optical quality is much better than that of rutile. Yttrium orthovanadate has a Knoop hardness of 480 (compared with 135 for calcite) and takes a good optical polish. Optical-quality single crystals as large as 2.5 by 2.5 by 1.3 cm are currently available,† and larger ones are expected in the near future. The primary advantage of yttrium orthovanadate is that in a Glan-Foucault-type prism, the ordinary ray (the one transmitted for a material having a positive birefringence) is incident on the cut at an angle very close to the Brewster angle, so that the reflection loss is well under 1 percent. If the front and rear faces of the prism are antireflected, the transmission should be much larger than that of a calcite Glan-Taylor prism. Several prisms of yttrium

† Crystal Products Division, Union Carbide Corporation, 8888 Balboa Ave., San Diego, Calif. 92123.

orthovanadate are currently operating in laser systems under conditions for which calcite prisms are unsatisfactory (DeShazer, 1971).

Other noncalcite polarizing prisms have been made for use in the ultraviolet. Rochon or Wollaston prisms (Pars. 34 and 39) are sometimes made of crystalline quartz for use in the far ultraviolet. A spectropolarimeter which uses crystalline-quartz prisms both as dispersing elements and as polarizers in the ultraviolet has also been reported (Gillham and King, 1961; Gates and King, 1962). The short-wavelength cutoff of the quartz is variable, depending on the impurities present, but can be as low as 1600 Å.

By utilizing magnesium fluoride instead of quartz for the polarizing prisms, the short-wavelength limit can be extended to 1300 Å. Magnesium fluoride transmits to about 1125 Å, but below 1300 Å its birefringence decreases rapidly and changes sign at 1194 Å (Johnson, 1964; Chandrasekharan and Damany, 1969). Although it is the most birefringent material available in this region, MgF_2 has a much smaller birefringence than that of calcite; hence a small cut angle and large L/A ratio for the prism are unavoidable. Since absorption does occur, it is desirable to minimize the length of the prism. Johnson (1964) solved this problem by constructing a MgF_2 Wollaston prism which requires only half the path length necessary for a Rochon prism. However, both beams are deviated, creating instrumental difficulties.

Steinmetz et al. (1967) constructed a double Rochon prism of MgF_2 which has the same L/A ratio as the Wollaston prism but does not deviate the desired beam. The prism was 16.25 mm long with an L/A ratio of about 2, and the apex was blackened to reduce scattered light. Fluorescence created difficulties below 1450 Å when a sodium salicylate window detector was used but not with a solar-blind detector. Scattered light from the polished prism surfaces was also a problem; it might be reduced by using a different polishing technique. The optic axes of the two side pieces in the prism were not strictly parallel, so that the double Rochon acted like two separate polarizers out of phase by about 2° and the polarization efficiency was reduced. In principle, however, a MgF_2 double Rochon polarizing prism should be an efficient, high-extinction-ratio, on-axis polarizer for the 1300- to 3000-Å wavelength range and should also be useful at longer wavelengths. Morris and Abramson (1969) report on the characteristics of optically contacted MgF_2 single Rochon prisms with L/A ratios of about 5. These had a deviation δ of the ordinary ray (Fig. 37*a*) of 3° at 6328 Å and 5.2° at 1800 Å.

A different type of polarizer suggested by Chandrasekharan and Damany (1971) to take the place of a Rochon or Wollaston prism in the vacuum ultraviolet consists of a combination of two MgF_2 lenses, one plano-concave and the other plano-convex of the same radius of curvature, combined so that their optic axes are crossed. The combination acts as a convergent lens for one polarization and as a divergent lens for the other. It has the advantage that the polarized beam remains on axis and is focused. A measured degree of polarization of 98.5 percent was obtained at 1608 Å, in good agreement with the calculated value.

Prism polarizers can also be constructed for use in the infrared at wavelengths longer than those transmitted by calcite. Rutile, TiO_2, a positive uniaxial mineral with a large birefringence and good transmittance to 5 μm in the infrared, has been used by Landais (1968) to make the Glan-Foucault-type crystal polarizer (Par. 16) shown in Fig. 39. Since rutile has a positive birefringence (in contrast to the negative birefringence of calcite), the ordinary ray is transmitted undeviated and the extraordinary ray is reflected out the side. The (symmetric) field angle (Par. 8) is rather small: 2.4° at 2.6 μm, 4.4° at 3.7 μm, and 2° at 5 μm; and the extinction ratio (Par. 3) is less than 1 part in 2,000 at 2.6 μm and less than 1 part in 1,000 at longer wavelengths. The transmission is between 20 and 35 percent in the 2.5- to 5-μm wavelength region except for a sharp dip at 3.1 μm, attributed to water vapor.

DICHROIC AND DIFFRACTION-TYPE POLARIZERS

44. Some of the most useful polarizers available employ either dichroism or diffraction effects. These polarizers come in sheet form, sometimes in large sizes,

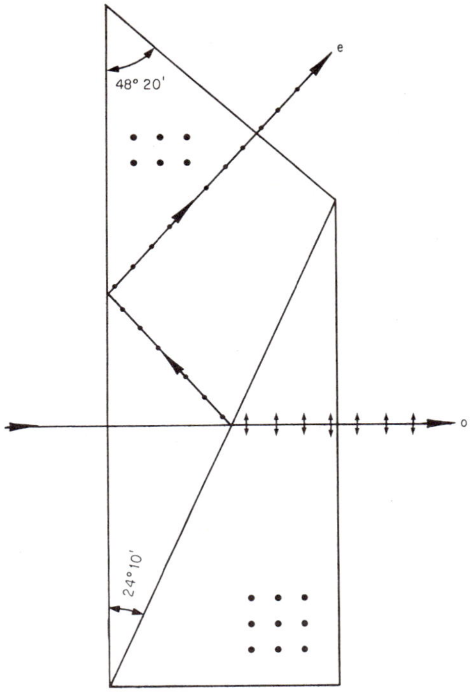

Fig. 39 Rutile Glan-Foucault-type polarizing prism for use in the infrared. The optic axis of the positive-birefringent rutile is perpendicular to the paper. [*Modified from Landais* (1968).]

are easily rotated, and produce negligible beam deviation. Also, they are usually thin, lightweight, and rugged, and most can be made in any desired shape. The cost is generally much less than that of a prism-type polarizer. Furthermore, both types are insensitive to the degree of collimation of the beam, so that dichroic or diffraction-type polarizers can be used in strongly convergent or divergent light.

A dichroic† material is one which absorbs light polarized in one direction more strongly than light polarized at right angles to that direction. Dichroic materials are to be distinguished from birefringent materials, which may have different refractive indexes for the two electric vectors vibrating at right angles to each other but similar (usually negligible) absorption coefficients. Various materials are dichroic, either in their natural state or in a stretched condition. The most common materials used as dichroic polarizers are stretched polyvinyl alcohol sheets treated with absorbing dyes or polymeric iodine, commonly marketed under the trade name Polaroid. These and similar materials are discussed in Par. 45. Another type of dichroic polarizer is prepared by rubbing a glass or plastic surface in a single direction and then treating it with an appropriate dye. Polarizers of this type are sold under the trade name Polacoat and will be described in Par. 46. In certain portions of the infrared spectral region, calcite is strongly dichroic and makes an excellent high-extinction polarizer (Bridges and Kluver, 1965). Pyro-

† The term *dichroic* is also used in three other ways: (1) to denote the change in color of a dye solution with change in concentration, (2) to denote a color filter that has two transmission bands in very different portions of the visible region and hence changes color when the spectral distribution of the illuminating source is changed, and (3) to denote an interference filter that appears to be of a different color when viewed in reflected or transmitted light.

lytic graphite is electrically and optically anisotropic and has been successfully used as an infrared polarizer; it is described in Par. 47. Other materials which exhibit dichroism in the infrared include single-crystal tellurium (Loferski, 1952), ammonium nitrate (Newman and Halford, 1948), mica, rubber films under tension, polyvinyl alcohol, and polyethylene (Hyde, 1948). In the visible region, gold, silver, and mercury in the form of microcrystals (Land and West, 1946), needles of tellurium (Land, 1951), graphite particles (Shurcliff, 1962*b*), and glasses containing small elongated silver particles (Stookey and Araujo, 1968) are all dichroic.

A sodium nitrate polarizer described by Yamaguti (1955*b*) is not dichroic in the strict sense of the word but acts like a dichroic polarizer. Roughened plates of SK5 glass are bonded together by a single crystal of sodium nitrate, which has a refractive index for the ordinary ray nearly equal to that of the glass. The extraordinary ray has a much lower index, so that it is scattered out of the beam by the rough surfaces, leaving the ordinary ray to be transmitted nearly undiminished. (Yamaguti has also made Feussner prisms out of single-crystal sodium nitrate, described in Par. 42.)

Diffraction-type polarizers include diffraction gratings, echelettes, and wire grids. These are all planar sturctures that have properties similar to those of dichroic polarizers except that they transmit one component of polarization and reflect the other when the wavelength of the radiation is much longer than the grating or grid spacing. Wire grid and grating polarizers are covered in Par. 48.

None of these polarizers has as high a degree of polarization as the prism polarizers of Pars. 7 to 43. Thus it is frequently necessary to measure the polarizing properties of the particular polarizer used. A source of plane-polarized light is desirable for such a measurement. Lacking that, one of the procedures described in Par. 49 can be followed if there are two identical imperfect polarizers. Alternate methods are also described which are applicable to two nonidentical imperfect polarizers.

45. Sheet Polarizers Various types of sheet polarizers have been developed by Edwin H. Land and coworkers at the Polaroid Corporation, Cambridge, Mass. Sheet polarizers are also available from several European companies. The J sheet polarizer, the first type available in America (around 1930), consisted of submicroscopic needles of herapathite oriented parallel to one another in a sheet of cellulose acetate. Since this type of polarizer, being microcrystalline, had some tendency to scatter light, it was superseded by H and K sheet molecular polarizers, which exhibit virtually no scattering. The most widely used sheet polarizer is the H type, which consists of a sheet of polyvinyl alcohol that has been unidirectionally stretched and stained with iodine in a polymeric form. The K type is made by heating a sheet of polyvinyl alcohol in the presence of a catalyst to remove some of the water molecules and produce the dichromophore polyvinylene. It was developed primarily for applications where resistance to high temperature and high humidity are necessary. Another type of polarizing sheet, made from a combination of the H and K types, has an absorption maximum at about 1.5 μm in the infrared and is designated as HR Polaroid.

The history of the development of the various kinds of sheet polarizers has been given by Land (1951), their chemical composition by Land and West (1946), and their optical performance by Shurcliff (1962*b*; Shurcliff and Billings, 1963), Land and West (1946), and Land (1951). In addition, Blake et al. (1949) mention the HR infrared polarizer, and Makas (1962) describes the modified H-film polarizer for use in the near ultraviolet. Baxter et al. (1956) describe a technique for measuring the optical density of high-extinction polarizers in the presence of instrumental polarization.

Figure 40 shows the principal transmittance† T_1 and extinction ratio† T_2/T_1 of various types of H and K sheet polarizers used in the visible and near ultraviolet (Shurcliff 1962*b*; Shurcliff and Billings, 1963; Jones, 1970). [In addition, curves for two sheet polarizers manufactured by Zeiss and two types of polarizing filters from Polacoat (Par. 46) are shown.] The letter N in the designation of the Polaroid sheets stands for neutral (to distinguish them from sheet polarizers prepared from colored dyes), and the number 22, 32, etc., indicates the approximate transmit-

† These quantities were defined in Par. 3.

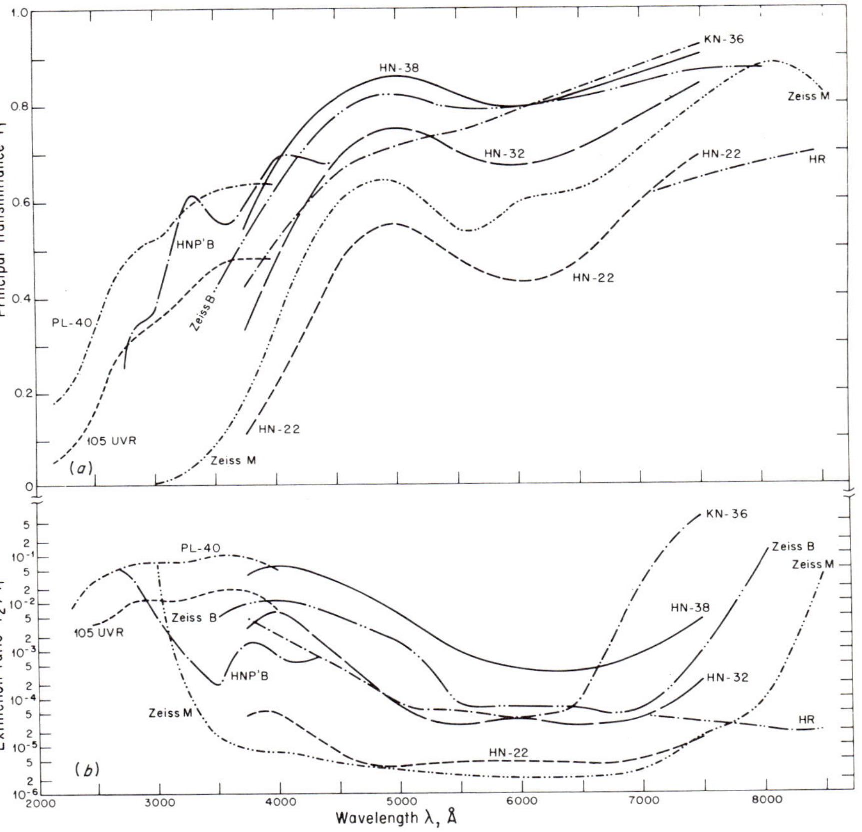

Fig. 40 (*a*) Principal transmittance and (*b*) extinction ratio for various types of dichroic polarizers: Polaroid sheet polarizers HN-22, HN-32, HN-38, and KN-36; Zeiss (Oberkochen) Bernotar and Micro Polarization filters; and Polacoat PL-40 and 105 UVR polarizing filters. The last is stated to have a transmittance (for unpolarized light) of 32 percent at 5460 Å. [*Modified from curves of Shurcliff* (1962*b*), *and Shurcliff and Billings* (1963), *Jones* (1970), *Haase* (1961), *and McDermott and Novick* (1961).]

tance of unpolarized visible light. Figure 41 gives the principal transmittance and extinction ratio of a typical plastic laminated HR infrared polarizer (Shurcliff, 1962*b*; Jones, 1970). In order to correlate the degree of polarization (Par. 3) with the extinction ratio, the relation between these quantities is shown graphically in Fig. 42. Sometimes the optical density D of a polarizer is plotted instead of its transmittance. The relation between these two quantities is

$$D = \log \frac{1}{T} \tag{93}$$

The extinction ratio of the HN-22 Polaroid compares favorably with that of Glan-Thompson prisms throughout the visible region, but the transmission of the Glan-Thompson is superior. In the ultraviolet, the new HNP′B material has a reasonably good extinction ratio (about 10^{-3} or better) for wavelengths longer than 3200 Å. It is a specially purified form of HN-32, and its properties match those

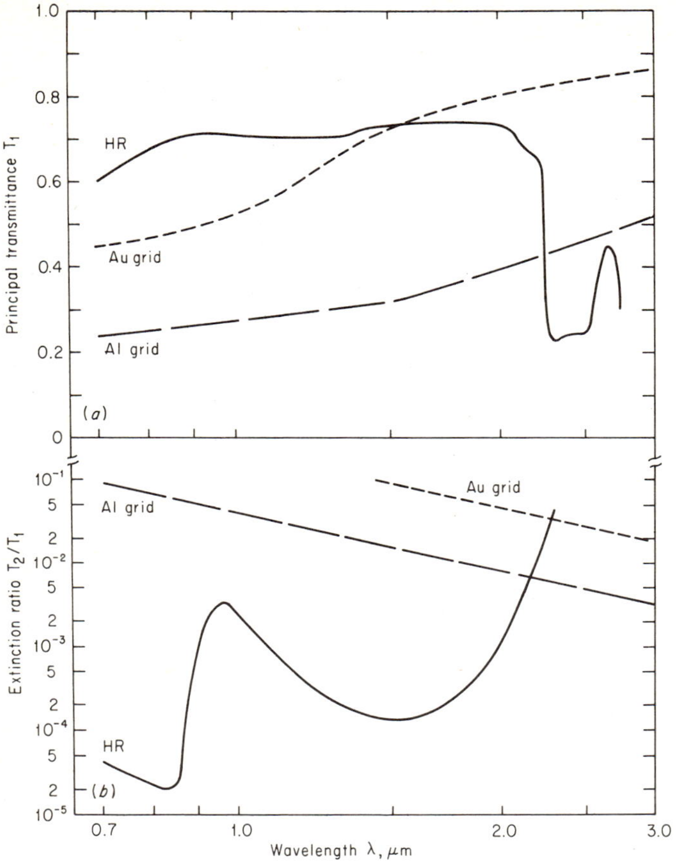

Fig. 41 (*a*) Principal transmittance and (*b*) extinction ratio for plastic laminated HR infrared polarizer [*modified from curves of Shurcliff* (1962*b*) *and Jones* (1970)] and two wire grid polarizers with 0.463-μm grating spacings [*Bird and Parrish* (1960)].

of the standard HN-32 Polaroid at wavelengths longer than 4500 Å. According to West and Jones (1951), the extinction ratio for a dichroic polarizer of the Polaroid type has a practical limit of about 10^{-5} because as the concentration of dichromophore is increased beyond a certain value, the optical density no longer increases proportionately.

If Polaroids are used in applications where beam deviation is important, they should be checked for possible deviation. Most Polaroids, which are laminated in plastic sheets, do produce a slight beam deviation that can be observed through a telescope as a shift in the image position when the Polaroid is rotated. The amount of the deviation varies from point to point on the Polaroid and can be much worse if the material is mounted between glass plates. It is possible to order specially selected sheet Polaroid laminated between polished glass plates that deviates the beam by only about 5 seconds of arc.

Sheet polarizers made of stretched polyvinyl alcohol that has been stained with iodine or various dyes are also made in countries outside the United States. In West Germany they are produced by Carl Zeiss, Oberkochen, and Erwin Käsemann GmbH, Oberaudorf; in East Germany by Carl Zeiss, Jena; and in the United Kingdom by Barr and Stroud, Ltd., Glasgow, although this company does not

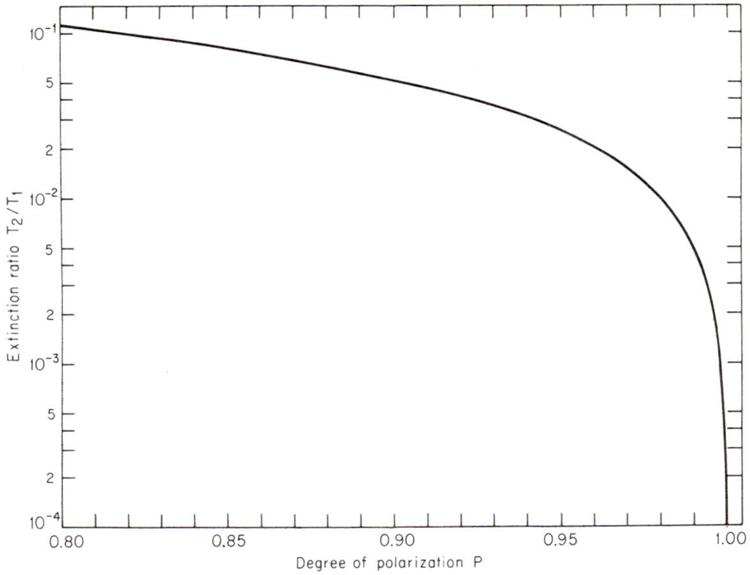

Fig. 42 Relation between extinction ratio and degree of polarization calculated from Eq. (63).

advertise its sheet polarizers. The development work on sheet polarizers in Germany was done by Käsemann, Bernauer, Drechsel, and others, and is summarized by Haase (1961). Käsemann (1962) published curves of the optical density of his Ks-MIK visible polarizer and Ks-W58 visible and ultraviolet polarizer. The former polarizer has an excellent extinction ratio and good transmission characteristics in the 4000- to 7000-Å region, and the latter has some transmission out to 2000 Å with a sharply deteriorating extinction ratio from 3500 to 2000 Å. Ludwig Drechsel at Carl Zeiss, Jena, has published articles on infrared and ultraviolet sheet polarizers (Drechsel, 1957, 1958, 1962). The infrared polarizers were initially stated to polarize in the region 0.6 to 2.8 μm and 3.5 to 6.5 μm (Drechsel, 1957), but later papers mention only the performance out to 3 μm. There are also articles by Distler et al. (1966, 1967) about Russian infrared sheet polarizers, which are stated to have a degree of polarization greater than 99.95 percent and a transmission between 25 and 40 percent in the wavelength region 1.5 to 6.5 μm.

The curves for the two Zeiss (Oberkochen) filters shown in Fig. 40 were obtained from a paper by Haase (1961) and are presumably representative of currently available products. The polarization filter designated Bernotar is intended for photographic work to eliminate glare and unwanted reflections. The primary requirement for this application is a high transmission that is nearly constant over the entire visible region (for color photography) with less emphasis on a good extinction ratio. For precise optical measurements for which a better extinction ratio is required, the Micro Polarization filter is superior. It is comparable to the HN-22 Polaroid with a somewhat higher transmission and better extinction ratio than the latter product. All Zeiss polarizing filters are laminated between glass plates that are selected for freedom from strain, and the cementing is performed so as to give good image quality and minimum beam deviation (Haase, 1961).

King and Talim (1971) have measured the axis wander and ellipticity of beams transmitted by various types of sheet polarizers in the same way as for Glan-Thompson prisms (Par. 12). The materials investigated were HN-32 Polaroid, Käsemann Ks-MIK and Ks-DEM sheet polarizers, and Barr and Stroud high-quality polarizers, no longer available. In contrast to the results for the prism polarizers, the axis wander and ellipticity scans from the same region of a sheet polarizer showed no similarities. For example, whereas some traces exhibited a smooth gradual change in ellipticity across the polarizer aperture, the corresponding axis-wander trace varied irregularly. Axis wander is presumably determined by the orientation of the dichroic molecules, while ellipticity is produced by residual birefringence in the polarizer plastic or glass support so that the lack of correspondence between axis-wander and ellipticity traces in sheet polarizers is understandable.

The axis wander varied considerably from one type of sheet polarizer to another and also over a single sheet. A 10-mm scan gave an overall variation of the transmission-plane azimuth as high as 0.4° on some areas of glass-mounted HN-32 Polaroid, whereas with selected regions of butyrate-laminated HN-32 material, the axis wander was as low as 0.015° for the same length scan. Typical values for a 10-mm scan for Käsemann Ks-MIK sheet polarizer were in the range 0.01 to 0.05°, and selected areas of Käsemann Ks-DEM material gave similar figures, although other regions were poorer, typically 0.1° for a 10-mm scan. Relatively low ellipticities, typically between 0.01 and 0.03°, were recorded for high-quality sheet polarizers, the values often being almost constant over a 10-mm scan. Ellipticities from HN-32 Polaroid varied considerably, however, and it was not uncommon to find values ranging between 0.01 and 0.5° from different areas of the same sheet.

46. Dichroic Polarizing Coatings Beilby-layer polarizers (Shurcliff, 1962*b*) are dichroic coatings that can be applied to the surface of glass or plastic. The process was developed by Dreyer (1947), who founded the company which manufactures Polacoat polarizing filters. There are three main steps in the production of these polarizers. First, the substrate (quartz, glass, plastic, etc.) is rubbed along parallel lines with filter paper, cotton, or rouge to produce a preferred surface orientation. (The affected region of minute scratches extends to a depth of less than 1 μm.) Then the sheet is rinsed and treated with a solution of dichroic molecules, e.g., a 0.5% solution of methylene blue in ethanol or one or more azo dyes, and then dried in a controlled fashion. Presumably the molecules line up preferentially along the rubbing direction, resulting in a greater absorption for light polarized in that direction. As a final step, the surface is treated with an acidic solution, often that of a metallic salt such as stannous chloride, which can increase the dichroism and produce a more neutral color. A protective coating over the polarized surface provides mechanical protection for the fragile layer with no loss in transmission. McDermott and Novick (1961) give a somewhat more complete description of the Polacoat process, and Anderson (1949) has investigated the absorption of methylene blue molecules on a unidirectionally polished surface. References to patents and related work are given by Shurcliff (1962*b*).

The principal transmittance and extinction ratio of two standard Polacoat coatings, PL-40 and 105 UVR (32 percent transmission of unpolarized light at 5460 Å), are shown in Fig. 40. These curves are taken from the data of McDermott and Novick (1961); to make them comparable with the curves of the other materials, the correction for the reflection losses at the surfaces of the quartz made by those authors was removed. Polacoat 105 UVR coating comes in various densities; the data shown are for the highest-density material with the best extinction ratio.† The transmittance of these materials is roughly constant throughout the visible region to about 6500 Å, where it starts increasing slightly. A major advantage of Polacoat over sheet Polaroid is that it does not bleach upon exposure to intense ultraviolet radiation. McDermott and Novick (1961) observed no significant change in its characteristics at 2280 Å after almost 50 h exposure to a radiant flux of about 0.003 W/cm^2. On the other hand, they state that a somewhat lesser exposure produced significant bleaching and consequent reduction in the polarizing properties of modified HN ultraviolet Polaroid.

† The company literature (Polacoat, 1967) is somewhat misleading in that the transmittance of this material is stated to be 35 percent, but the transmission curve (for unpolarized light) given in the bulletin does not rise above 30 percent until the wavelength becomes longer than 6500 Å.

The transmittance of a Polacoat PL-40 film without the protective varnish applied to a 2.5-cm-diameter 1.6-mm-thick Suprasil† optical flat was measured at Michelson Laboratory in the 1500- to 3000-Å range of the vacuum ultraviolet to determine whether it would have any transmission at the shorter wavelengths. Its transmittance (in unpolarized light) is given in Table 4. (Since these values are comparable to the transmittance of standard PL-40 filters, the varnish seems to be transparent in this wavelength region.) The transmittance of the Suprasil is also given in Table 4.

The average measured extinction ratios for two such PL-40 films on Suprasil were 0.028 at 5461 Å, 0.0087 at 4358 Å, and 0.104 at 3650 Å, corresponding to degrees of polarization of 0.945, 0.983, and 0.812, respectively (Stanford, 1970). The value at 3650 Å is essentially the same as that reported by McDermott and Novick (1961), but the other values cannot be compared because the published measurements did not extend into the visible region. However, it is clear that PL-40 Polacoat does not have an extinction ratio in the visible region comparable to that of any of the polyvinyl alcohol sheet polarizers discussed previously. The PL-40 polarizers also introduce ellipticity into the light transmitted and rotate its plane of polarization. Another difficulty was found by Kyser (1970), who

TABLE 4 Transmittance of Polacoat PL-40 and Suprasil

Wavelength, Å	Transmittance (Stanford, 1970)	
	Polacoat	Suprasil
3000	0.268	0.925
2400	0.146	0.910
2000	0.066	0.804
1700	0.028	0.788
1608	0.003	0.108

tested a stock PL-40 Polacoat polarizing filter on fused quartz and found that it produced a large quantity of scattered light of the unwanted component. This light was dispersed spectrally and was scattered at angles up to about 20° as though the scratches on the rubbed surface were acting like rulings on a diffraction grating. There was relatively little of the unwanted component on axis; most of it was scattered at larger angles. Despite these difficulties, Polacoat PL-40 polarizers appear to be the best large-aperture transmission-type polarizers available for work in the 2000- to 3000-Å wavelength range in the ultraviolet.

47. Pyrolytic-Graphite Polarizers Pyrolytic graphite has a large anisotropy in both the electric conductivity and in the optical properties. If the E vector of an electromagnetic wave is pointing in the direction of the c axis of the graphite, the absorption coefficient is a minimum, the reflectance is also a minimum, and hence the transmittance is a maximum. If the E vector lies in the a plane perpendicular to the c direction, the absorption is a maximum, reflectance is a maximum, and transmittance is a minimum. Thus, pyrolytic graphite should be a good material from which to make a dichroic polarizer if a thin foil is cut and polished to contain the c axis. Several such polarizers have been made by Rupprecht et al. (1962); two had thicknesses of 9.2 μm, and a third was 4.2 μm thick. The transmittances T_1 of the thinner one and T_1 and T_2 of the two thicker ones were determined using one of the methods described in Par. 49. The principal transmittance T_1 and extinction ratio T_2/T_1 for one of the 9.2-μm-thick ones are shown in Fig. 43 for infrared wavelengths from 2 to 16 μm, along with curves for various wire-grid polarizers (Par. 48). In the far infrared out to 600 μm, T_1 gradually increases to 0.50, and T_2/T_1 drops down to the 10^{-3} range (Rupprecht et al., 1962). The transmittance of the thinner pyrographite polarizer was larger

† Ultraviolet-grade fused silica manufactured by Amersil, Inc., Hillside, N.J.

than the curve shown, but its extinction ratio, although not given, was probably poorer. Pyrolytic-graphite polarizers have the advantages of being planar and thus easily rotatable, having large acceptance angles, and having reasonably high transmittances and good extinction ratios in the far infrared. However, in the shorter-wavelength region shown in Fig. 43, they are inferior to all the wire-grid polarizers. In addition, they are fragile, and the largest clear aperture obtained by Rupprecht et al. (1962) was about 12 mm diameter.

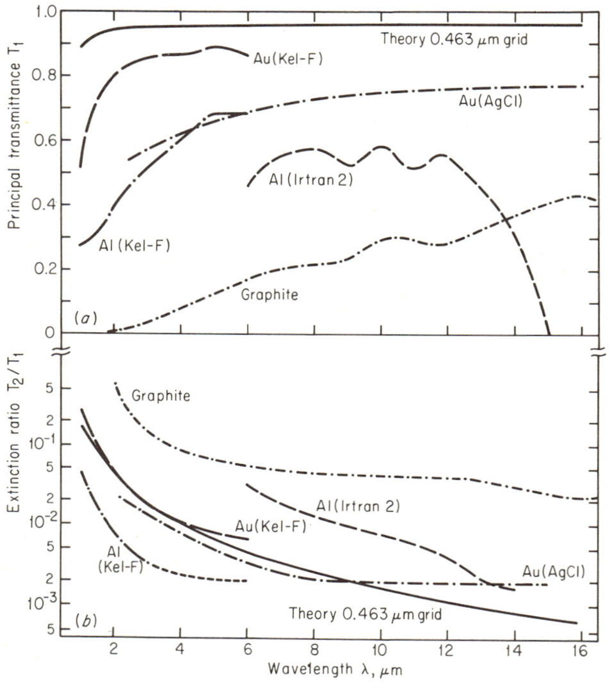

Fig. 43 (*a*) Principal transmittance and (*b*) extinction ratio for a pyrolytic-graphite polarizer [*Rupprecht et al.* (1962)] and various wire-grid polarizers [*Bird and Parrish* (1960), *Perkin-Elmer* (1966), *and Young et al.* (1965)]. The substrate materials and metals used for the grids are indicated. Theoretical curves (*solid lines*) calculated from Eqs. (94), (95), (99), and (100) with $n = 1.5$ and $d = 0.463$ are also shown for comparison.

48. Wire-Grid and Grating Polarizers Wire grids† have a long history of use as optical elements to disperse radiation and detect polarization in far-infrared radiation and radio waves (Bird and Parrish, 1960). They transmit radiation whose E vector is vibrating perpendicular to the grid wires and reflect radiation with the E vector vibrating parallel to the wires when the wavelength is much longer than the grid spacing. When the wavelength is comparable to the grid spacing, both components are transmitted. For grids made of good conductors, absorption is negligible.

† *Wire grid* is being used here, as is customary, to denote a planar structure composed of a series of parallel wires or strips. Renk and Genzel (1962) and a few others use the term to designate a two-dimensional array with two series of elements arranged at right angles to each other. They call a one-dimensional array a wire or strip grating.

Various aspects of the theory of reflection and transmission of radiation by wire grids have been presented; Larsen (1962) lists 83 references on the subject, and Bird and Parrish (1960) mention some papers particularly dealing with polarization by wire grids. Kerr and Palmer (1971) investigated the anomalous behavior of thin-wire gratings and found good agreement between theory and experiment. Since much of the work has been done in connection with microwaves, the equations are generally given in the representation for transmission lines. The appropriate formulas can be found in Marcuvitz (1951). Following the treatment given by Auton (1967) but simplifying the relations slightly and changing to optical notation, we see that the transmittances T_1 and T_2 for the grid are

$$(T_1)_\perp = \frac{4nA^2}{1 + (1 + n)^2A^2} \tag{94}$$

$$(T_2)_\parallel = \frac{4nB^2}{1 + (1 + n)^2B^2} \tag{95}$$

where n = refractive index of (transparent) substrate material
$(T_1)_\perp$ = transmittance for radiation polarized perpendicular to grid wires
$(T_2)_\parallel$ = transmittance for radiation polarized parallel to the grid wires

The general expressions for A and B are

$$\frac{1}{A} = \frac{4d}{\lambda}\left\{\ln\left[\csc\frac{\pi(d-a)}{2d}\right] + \frac{Q_2\cos^4[\pi(d-a)/2d]}{1 + Q_2\sin^4[\pi(d-a)/2d]} + \frac{1}{16}\left(\frac{d}{\lambda}\right)^2\left[1 - 3\sin^2\frac{\pi(d-a)}{2d}\right]^2\cos^4\frac{\pi(d-a)}{2d}\right\} \tag{96}$$

$$B = \frac{d}{\lambda}\left[\ln\left(\csc\frac{\pi a}{2d}\right) + \frac{Q_2\cos^4(\pi a/2d)}{1 + Q_2\sin^4(\pi a/2d)} + \frac{1}{16}\left(\frac{d}{\lambda}\right)^2\left(1 - 3\sin^2\frac{\pi a}{2d}\right)^2\cos^4\frac{\pi a}{2d}\right] \tag{97}$$

where
$$Q_2 = \frac{1}{[1 - (d/\lambda)^2]^{1/2}} - 1 \tag{98}$$

These relations hold for grids made of thin metal strips of width a and spacing d when $\lambda > 2d$. According to Marcuvitz (1951), Eqs. (96) and (97) are in error by less than 1 percent when $\lambda > 2d$; for the condition $\lambda > d$, the error is supposed to be less than 5 percent but increases for still shorter wavelengths.

When the width of the metal strips is equal to the width of the spaces between strips ($d = 2a$), Eqs. (96) and (97) can be considerably simplified since $d - a = a = d/2$:

$$B = \frac{d}{\lambda}\left[0.3466 + \frac{0.25Q_2}{1 + 0.25Q_2} + 0.003906\left(\frac{d}{\lambda}\right)^2\right] \tag{99}$$

$$A = \frac{1}{4B} \tag{100}$$

This case was shown by Auton (1967) to be optimum for providing as wide a useful wavelength range as possible. Equations (99) and (100) are adequate for most calculations of wire-grid performance. However, many of the wire grids in actual use are more like wires than like the flat metal strips assumed in the derivation. Expressions analogous to Eqs. (96) and (97) are available for wires (Hass and O'Hara, 1965; Marcuvitz, 1951), but they involve sums of infinite series and are much less convenient to use for computational purposes.

Figure 44 shows values of the principal transmittance and extinction ratio for various values of n as a function of λ/d calculated from Eqs. (99) and (100). It is clear that the shortest wavelength for which a given grid will act as a useful polarizer is $\lambda \approx 2d$. Also, the best performance is obtained with the lowest-index substrate.

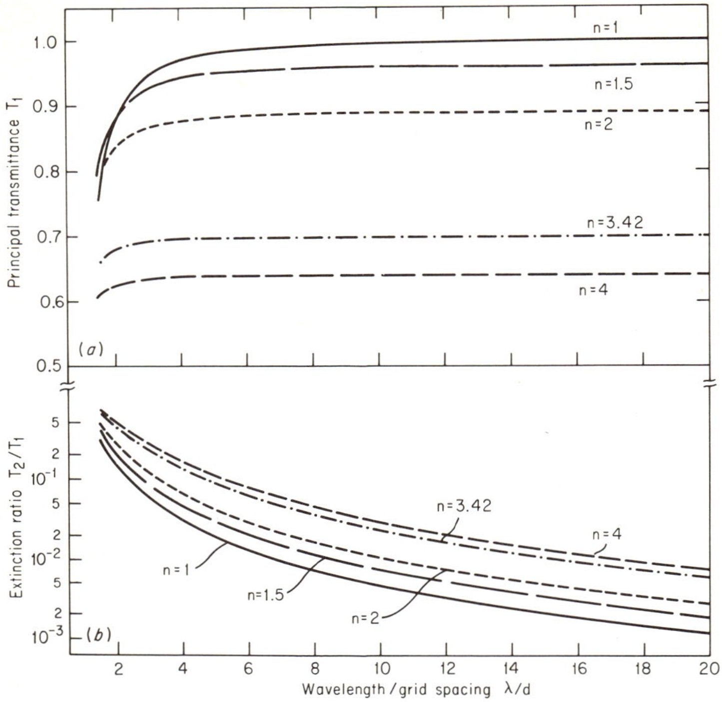

Fig. 44 (*a*) Principal transmittance and (*b*) extinction ratio as a function of λ/d calculated from Eqs. (94), (95), (99), and (100) for various values of n for the substrate. Substrate indexes correspond approximately to an antireflected substrate of air, organic plastic, silver chloride, silicon, and germanium.

Since absorption in the substrate material has been neglected, principal transmittances measured for real materials will be lower than the calculated values but the extinction ratios should be unaffected. If one must use a high-index substrate such as silicon or germanium, the performance of the grid can be considerably improved by applying an antireflection coating to the substrate *before* depositing the conducting strips, since a perfectly antireflected substrate acts like an unsupported grid (Auton, 1967). However, if the antireflecting layer is laid down *over* the grid strips, the expressions for $(T_1)_\perp$ and $(T_2)_\parallel$ become (Auton, 1967)

$$(T_1)_\perp = \frac{4n^2A^2}{1 + 4n^2A^2} \tag{101}$$

$$(T_2)_\parallel = \frac{4n^2B^2}{1 + 4n^2B^2} \tag{102}$$

If $\lambda \gg d$, $(T_1)_\perp$ approaches unity and $(T_2)_\parallel$ approaches n^2 times the value of $(T_2)_\parallel$ for an unsupported grid. The extinction ratio is also n^2 times its value for the grid on the antireflected substrate, so that a polarizer made of metal strips and then antireflected should be much inferior to one made with the antireflecting layer between the grid and the substrate.

Several people have built and tested wire-grid polarizers (Bird and Parrish, 1960; Young et al., 1965; Hass and O'Hara, 1965; Hilton and Jones, 1966; Auton, 1967). In addition, two types of wire grids are manufactured commercially by Buckbee Mears (see Hass and O'Hara, 1965) and Perkin-Elmer (1966), and a third type composed of 152-μm-diameter tungsten wires spaced 800 to the inch has been mentioned, but no performance characteristics have been given (Roberts and Coon, 1962). The different wire-grid polarizers are listed in Table 5, and the principal transmittances and extinction ratios of several are shown in Figs. 41 and 43.

The polarizers with grid spacings of 1.69 μm and less were all made by evaporating the grid material at a very oblique angle onto a grating surface which had been prepared either by replicating a diffraction grating with the appropriate substrate material (silver bromide, Kel-F, polymethyl methacrylate, etc.) or by ruling a series of lines directly onto the substrate (Irtran 2 and Irtran 4). The oblique evaporation (8 to 12° from the surface) produced metallic lines on the groove tips which acted like the conducting strips of the theory, while the rest of the surface was uncoated and became the transparent region between strips. The larger grid spaces (4 to 25.4 μm) were produced by a photoetching process, and one 25.4-μm grid was made by an electroforming process.

If a wire-grid polarizer is to be used in the near infrared, it is desirable to have the grid spacing as small as possible. Bird and Parrish (1960) succeeded in obtaining a very good extinction ratio in the 2- to 6-μm wavelength region with

TABLE 5 Types of Wire-Grid Polarizers

Grid spacing, μm	Grid material	Substrate	Wavelength range, μm	Reference
0.347	Evaporated Au	Silver chloride	2.5–30	Perkin-Elmer (1966)
0.463	Evaporated Au	Kel-F	1.5–10†	Bird and Parrish (1960)
0.463	Evaporated Al	Kel-F	0.7–15†	Bird and Parrish (1960)
0.463	Evaporated Al	Polymethyl methacrylate	1–4000‡	Hass and O'Hara (1965)
1.67	Evaporated Al	Irtran 2	6–14	Young et al. (1965)
1.67	Evaporated Al	Irtran 4	8–19	Young et al. (1965)
1.69	Evaporated Al	Polyethylene	2.9–200§	Hass and O'Hara (1965)
2	Evaporated Cr	Silicon	10.6	Cheo and Bass (1971)
4	Photoetched Al	Polyethylene	>16	Auton (1967)
5.1	Photoetched Al	Silicon	54.6	Hilton and Jones (1966)
10	Photoetched Al	Polyethylene	>16	Auton (1967)
25.4	Photoetched Al	Silicon	54.6	Hilton and Jones (1966)
25.4	Evaporated Au	Mylar	>60	Hass and O'Hara (1965)
317	W wires 152 μm diam	Air	40–300	Roberts and Coon (1962)

† Strong absorption bands near 8.3 and 10.5 μm.
‡ Strong absorption bands between 5.7 and 12.5 μm.
§ Absorption bands between 6 and 15.5 μm.

an aluminum-coated Kel-F substrate (Figs. 41 and 43). Unfortunately, Kel-F, $(CF_2CFCl)_n$, has absorption bands at 7.7 to 9.2 and 10.0 to 11.0 μm, making the polarizer useless in these regions, but it can be used at longer wavelengths out to 25 μm (Bird and Parrish, 1960). Polyethylene would be an excellent substrate material since it has fewer absorption bands than Kel-F, but its insolubility in common solvents makes it much more difficult to use for replicating gratings (Hass and O'Hara, 1965). It does, however, make an excellent substrate material for photoetched grids (Auton, 1967).

For infrared wavelengths longer than about 24 μm, a photoetched grid with 1-μm-wide lines (close to the present limit for the photoetching process) and a 2-μm spacing should have an extinction ratio of 5×10^{-3} or better if the refractive index of the substrate is about 1.5, for example, polyethylene. The extinction ratio would continue to decrease; i.e., the polarization properties would improve, as the wavelength increased. At very long wavelengths, grids with a larger spacing would have a high degree of polarization. The important factor is the ratio of wavelength to grid spacing, which should be kept as large as possible (Fig. 44*b*).

One definite advantage of the wire-grid polarizer is that it can be used in sharply converging beams, i.e., systems with high numerical apertures. Young et al. (1965) found no decrease in percent of polarization for an Irtran 2 polarizer at 12 μm used at angles of incidence from 0 to 45°. They did find, however, that the transmittance decreased from 0.55 at normal incidence to less than 0.40 at 45° incidence.

If a grid were to be used at a single wavelength, one might possibly make use of interference effects in the substrate to increase the transmission (Auton, 1967). If the substrate has perfectly plane-parallel surfaces, it will act like a Fabry-Perot interferometer and transmit a maximum amount of light when twice the product of the thickness and refractive index is equal to an integral number of wavelengths. The 0.25-mm-thick pressed polyethylene substrates used by Auton (1967) were not uniform enough to show interference effects, but the Mylar film backing on the Buckbee Mears electroformed grid did show interference effects (Hass and O'Hara, 1965).

Lamellar eutectics of two phases have recently been developed which consist of thin needles of a conducting material embedded in a transparent matrix. The material is made by a controlled cooling process in which there is a unidirectional temperature gradient. This method of cooling orients conducting needles parallel to the temperature gradient, and hence the material can act like a wire-grid polarizer. Weiss and coworkers (Weiss and Wilhelm, 1963; Paul et al., 1964; Mueller and Wilhelm, 1965) have grown eutectic alloys of InSb and NiSb in which the conducting needles of NiSb are approximately 1 μm in diameter and approximately 50 μm long. A degree of polarization of more than 99 percent has been reported. Other eutectic alloys of InAs, GaSb, and InSb containing conducting needlelike crystals of Ni, Fe, Mn, Cr, and Co (or their compounds) have also been investigated. Wieder and coworkers (see Davis et al., 1969) have taken evaporated InSb films and grown antimony lamellae by electron-beam microzone crystallization. The conducting lamellae inside the InSb matrix act like a wire grid and produce plane-polarized light beyond the InSb absorption edge. The transmitted light is polarized perpendicular to the lamellae for wavelengths longer than about 5 μm, and typical films have $a \approx 1.2$ μm and $d \approx 2.7$ μm. Davis et al. (1969) measured a maximum principal transmittance of 0.19 and extinction ratio of 0.10 at a wavelength of 11 μm for one of the films but feel that these values can be improved by obtaining a higher degree of ordering of the lamellae. An advantage of this type of polarizer is that its performance can be optimized at a specific wavelength, e.g., that of a CO_2 laser line, by choosing the thickness of the InSb film so that there will be an interference maximum at the desired wavelength (Davis et al., 1969).

In summary, wire grids are very useful infrared polarizers, particularly for wavelengths much greater than the grid spacing. They are compact and easily rotatable and can be used with sharply converging beams. A major advantage is the extreme breadth of the wavelength band over which they have good polarizing properties. The long-wavelength limit is set by the transmission of the substrate material rather than by the loss of polarization of the grid. The short-wavelength limit is determined by the grid spacing; if gratings with smaller spacings could be successfully replicated and coated, the short-wavelength limit could be pushed closer to the visible region.

Another possible method of producing plane-polarized light is by using diffraction gratings or echelette gratings. Light reflected from diffraction gratings has long been known to be polarized, but the effect is generally small and extremely wavelength-dependent (Hidalgo et al., 1962; Rawlins, 1964). There is experimental evidence that the degree of polarization, $(I_\perp - I_\parallel)/(I_\perp + I_\parallel)$, is usually positive and increases with increasing grating angle for a blazed grating operating at angles greater than the blaze angle ($I_\perp$ and $I_\parallel$ are the fluxes of diffracted energy polarized respectively perpendicular and parallel to the grating grooves). Lee and Neo (1971) have shown that this empirical rule is observed for a 100 lines/mm grating in a Beckman IR-12 spectrometer.

In contrast to diffraction gratings, echelette gratings have been found to produce an appreciable amount of plane-polarized light. In a study with 3-cm microwaves reflected in the first order from model echelette gratings, Peters et al. (1953) found that the waves are more highly reflected when the E vector is perpendicular to the grooves than when it is parallel. (This effect is counter to that obtained with wire grids, where the reflected component is polarized with the E vector vibrating parallel to the grid wires.) Hadni et al. (1958) used an echelette grating in the zero order as a filter in the 100- to 300-μm far-infrared region because it strongly reflects light polarized with the E vector perpendicular to the rulings but only weakly reflects light with the E vector parallel to the rulings. (The main dispersing grating also produced plane-polarized light in the first order but with the E vector vibrating parallel to the rulings.) Mitsuishi et al. (1960) and Hadni et al. (1960) have also observed polarization effects with echelettes used in the zero and first orders in the infrared, but their observations do not entirely agree.

The theory of the polarization of light reflected by echelette gratings in the far-infrared and microwave regions has been given by Janot and Hadni (1963) and Rohrbaugh et al. (1958). A general numerical technique published by Kalhor and Neureuther (1971) should be useful for calculating the polarization effects of echelette gratings of arbitrary groove shape used in the visible region. Azzam and Bashara (1972) have measured the polarization characteristics of radiation diffracted by gratings, and other articles dealing with the effects of polarization on the energy distribution in diffraction gratings have been listed by Stroke (1964). In general, one tries to avoid polarization in the diffracted light to obtain high efficiencies in a blazed grating (Stroke, 1963) since polarization effects are frequently associated with grating anomalies.

49. Measuring Polarization of Imperfect Polarizers In determining the principal transmittance, extinction ratio, and other properties of an imperfect polarizer, the effects of source polarization, instrumental polarization, and sensitivity of the detector to the plane of polarization must either be measured or eliminated from the calculations. This is easy if an auxiliary polarizer is available that has a much higher degree of polarization than the one to be measured. In such a case, the "perfect" polarizer can be placed in the beam, and the transmittances T_1 and T_2 for the unknown polarizer can be measured directly.† As will be shown below, source polarization, instrumental polarization, and variation of detector response with plane of polarization can all be lumped together as a product. If this product is different in the horizontal and vertical planes, the ratio of the signals obtained when the "perfect" polarizer is oriented horizontally and vertically will not equal unity. One should always take more than the minimum number of measurements, i.e., introduce redundancy, to make sure that no systematic errors are present.

If a high-quality polarizer is not available, two polarizers having unknown properties may be used instead. Several procedures have been given which yield the polarizing properties of the polarizers and the instrumental polarization. These include two general cases: (1) identical polarizers (Rupprecht et al., 1962; Hass and O'Hara, 1965; Young et al., 1965; Charney, 1955; Horton et al., 1969) and (2) nonidentical polarizers (Hamm et al., 1965; Kudo et al., 1970; Young et al., 1965; Schledermann and Skibowski, 1971). In this section two methods will be described for determining the properties of identical polarizers, the method of Rupprecht et al. (1962) for determining T_1 and T_2 and the method of Arakawa and coworkers (see Horton et al., 1969) for determining the extinction ratio T_2/T_1 and the instrumental polarization. The methods required to determine the properties of nonidentical polarizers are more involved. The method of Hamm et al. (1965), which yields the extinction ratio of each polarizer and the instrumental polarization, seems straightforward and will be described in detail; a brief summary of the method of Kudo et al. (1970) will also be given.

The methods are applicable to any type of imperfect polarizer, including the nonnormal-incidence reflection and transmission polarizers described in Pars. 50 to 55, but are particularly useful for the planar dichroic and diffraction-type polarizers described in Pars. 45 to 47, which are easily rotatable. The principal transmittance of a dichroic polarizer is the fraction of the light transmitted when the E vector is perpendicular to the direction of stretch; for a pyrolytic-graphite polarizer, T_1 is for the E vector parallel to the c direction, and for a wire-grid polarizer, T_1 is for the E vector perpendicular to the direction of the wires when $\lambda \gg d$ (the grid spacing).

† When using an air-spaced polarizing prism, extreme care should be taken not to exceed the acceptance angle of the prism (see Pars. 16 and 19).

In an ideal experimental setup, the light source is unpolarized, the spectrometer or monochromator does not produce any polarization, and the detector is not sensitive to the plane of polarization of light falling on it. Most light sources (except for lasers) are only slightly polarized, but gratings, prisms, and off-axis mirrors in optical instruments can produce appreciable amounts of polarization. The percent polarization in several commercial spectrophotometers has been measured by Grum and Costa (1974), and certain types of photomultipliers have been shown to be polarization-sensitive (Clancy, 1952). Assume that a nearly unpolarized source of intensity I_0 can be considered to emit plane-polarized light of intensity $\frac{1}{2}I_0\alpha$ (where $\alpha \approx 1$) in the horizontal plane and plane-polarized light of intensity $\frac{1}{2}I_0(2 - \alpha)$ in the vertical plane. Assume further that in the horizontal plane the instrument transmits a fraction β and the detector response function is γ, where γ is the response to horizontally polarized light divided by the response to vertically polarized light. The detector signal is then proportional to $\frac{1}{2}I_0\alpha\beta\gamma$, compared with $\frac{1}{2}I_0(2 - \alpha)\delta$ for light polarized in the vertical plane (δ is the instrument transmittance for the vertical plane). The factors α, β, γ, and δ can be lumped into the terms T_h and T_v, the effective transmittances of polarized light in the horizontal and vertical planes, respectively. Thus, one would measure a signal S_0:

$$S_0 = \tfrac{1}{2}I_0(T_h + T_v) \tag{103}$$

when there is no polarizer in the beam. The term I_0 denotes the total intensity of the source; $\frac{1}{2}I_0T_h$ and $\frac{1}{2}I_0T_v$ are the transmitted horizontal and vertical components, respectively.

In the following derivation, the two polarizers will be assumed to be identical, and we shall obtain expressions for T_1 and T_2, from which the extinction ratio or the degree of polarization can be determined. *One* of the polarizers is placed in the beam with its principal transmittance direction horizontal. The detector signal that would be measured is S_h:

$$S_h = \tfrac{1}{2}I_0(T_hT_1 + T_vT_2) \tag{104}$$

When the polarizer is oriented vertically, the detector signal S_v is

$$S_v = \tfrac{1}{2}I_0(T_vT_1 + T_hT_2) \tag{105}$$

Frequently the ratios S_h/S_0 and S_v/S_0 are measured to eliminate problems of source stability. Then, from Eqs. (103) to (105),

$$\frac{S_h}{S_0} + \frac{S_v}{S_0} = \frac{S_h + S_v}{S_0} = T_1 + T_2 \tag{106}$$

Two polarizers are next inserted in the beam with their maximum transmission axes crossed. The detector signal S_{hv} is

$$S_{hv} = \tfrac{1}{2}I_0(T_hT_1T_2 + T_vT_2T_1) \tag{107}$$

and the ratio S_{hv}/S_0 becomes

$$\frac{S_{hv}}{S_0} = T_1T_2 \tag{108}$$

T_1 and T_2 can be found from Eqs. (106) and (108) using the solution for a quadratic equation:

$$T_1 = \frac{S_h + S_v}{2S_0}\left\{1 + \left[1 - \frac{4S_0S_{hv}}{(S_h + S_v)^2}\right]^{1/2}\right\} \approx \frac{S_h + S_v}{S_0} - \frac{S_{hv}}{S_0}\frac{S_0}{S_h + S_v} \tag{109}$$

$$T_2 = \frac{S_h + S_v}{2S_0}\left\{1 - \left[1 - \frac{4S_0S_{hv}}{(S_h + S_v)^2}\right]^{1/2}\right\} \approx \frac{S_{hv}}{S_0}\frac{S_0}{S_h + S_v} \tag{110}$$

The approximations hold when $4S_0S_{hv}/(S_h + S_v)^2 \ll 1$.

Generally one is most interested in the quality of the polarizer as specified by the extinction ratio or, alternately, by the degree of polarization. The extinction ratio ρ_p is

$$\rho_p = \frac{T_2}{T_1} \approx \frac{1}{[(S_h + S_v)/S_0]^2(S_0/S_{hv}) - 1} \tag{111}$$

and the degree of polarization P is†

$$P = \frac{T_1 - T_2}{T_1 + T_2} \approx 1 - 2\frac{S_{hv}}{S_0}\left(\frac{S_0}{S_h + S_v}\right)^2 \tag{112}$$

Thus, to determine the principal transmittance and extinction ratio, only the quantities S_h/S_0, S_v/S_0, and S_{hv}/S_0 need be measured if the two polarizers are strictly identical. However, as a double check, the quantities S_{hh}/S_0 and S_{vv}/S_0 can also be measured. These are the ratios obtained when *two* polarizers are oriented horizontally or vertically in the beam with their axes parallel. The pertinent relations are

$$S_{hh} = \tfrac{1}{2}I_0(T_hT_1^2 + T_vT_2^2) \tag{113}$$

and

$$S_{vv} = \tfrac{1}{2}I_0(T_vT_1^2 + T_hT_2^2) \tag{114}$$

so that

$$\frac{S_{hh}}{S_0} + \frac{S_{vv}}{S_0} = \frac{S_{hh} + S_{vv}}{S_0} = T_1^2 + T_2^2 \tag{115}$$

From Eqs. (106), (108), and (115) it can be easily shown that

$$\left(\frac{S_h + S_v}{S_0}\right)^2 - 2\frac{S_{hv}}{S_0} = \frac{S_{hh} + S_{vv}}{S_0} \tag{116}$$

and if this relation is satisfied, added confidence can be placed in the measurements.

An alternate method for determining the extinction ratio and instrumental polarization when the two polarizers are identical has been given by Horton et al. (1969). *One* polarizer is mounted in the instrument, and the ratio of S_h to S_v is measured directly. Assuming that the intensity of the light source does not change,

$$\frac{S_h}{S_v} = \frac{T_hT_1 + T_vT_2}{T_vT_1 + T_hT_2} = \frac{\rho_{hv} + \rho_p}{1 + \rho_{hv}\rho_p} \tag{117}$$

where the instrumental polarization $\rho_{hv} = T_h/T_v$ and the extinction ratio $\rho_p = T_2/T_1$. Then *two* polarizers are placed in the instrument, and the ratio S_{hv}/S_{vv} is measured:

$$\frac{S_{hv}}{S_{vv}} = \frac{T_hT_1T_2 + T_vT_2T_1}{T_vT_1^2 + T_hT_2^2} = \frac{\rho_{hv}\rho_p + \rho_p}{1 + \rho_{hv}\rho_p^2} \tag{118}$$

There are two equations and two unknowns, ρ_p and ρ_{hv}, so that each quantity can, in principle, be determined. The equation for ρ_p is a cubic, but an approximate relation can be obtained since $\rho_p \ll 1$ and $\rho_{hv} \approx 1$. Equation (117) yields

$$\rho_{hv} \approx \frac{S_h}{S_v} \tag{119}$$

and from Eq. (118) we obtain

$$\rho_{hv}\rho_p + \rho_p \approx \frac{S_{hv}}{S_{vv}} \tag{120}$$

making

$$\rho_p \approx \frac{S_{hv}/S_{vv}}{1 + S_h/S_v} \tag{121}$$

† Rupprecht et al. (1962), Young et al. (1965), and some others define the degree of polarization as $T_1/(T_1 + T_2)$ so that their values of P range from 0.5 to 1.0 instead of from 0 to 1.0, the range of the values using the definition in Eqs. (60) and (112).

Instrumental polarization effects can also be approximately eliminated by taking measurements at 45° to the horizontal and vertical axes. Charney (1955) describes a method of obtaining the transmittances of a pile-of-plates polarizer using ratios of intensities S_{vh}/S_{vv} for two orientations: (1) when one polarizer is oriented parallel to the spectrometer slit and (2) when the polarizers are oriented at 45° to the slit.

If the two polarizers are not identical, or if they cannot be easily removed from the beam, a method suggested by Hamm et al. (1965) can be used to determine the extinction ratio of each polarizer and the instrumental polarization. Four measurements are taken:

$$\begin{aligned} S_{vv} &= \tfrac{1}{2}I_0(T_vT_1T_1' + T_hT_2T_2') \qquad & S_{hh} &= \tfrac{1}{2}I_0(T_vT_2T_2' + T_hT_1T_1') \\ S_{vh} &= \tfrac{1}{2}I_0(T_vT_1T_2' + T_hT_2T_1') \qquad & S_{hv} &= \tfrac{1}{2}I_0(T_vT_2T_1' + T_hT_1T_2') \end{aligned} \tag{122}$$

The transmittances T_1 and T_2 refer to polarizer A; T_1' and T_2' refer to polarizer B. The quantities S_{vh} and S_{hv} are measurements when polarizers A and B are crossed: the principal transmittance axis of polarizer A is vertical for the measurement of S_{vh} while it is horizontal for S_{hv}. (The light source is assumed to have constant light output while all four measurements are taken so that the term I_0 is identical in the equations.) Equations (122) can be expressed in terms of the extinction ratios $\rho_p = T_2/T_1$ and $\rho_p' = T_2'/T_1'$ for the two polarizers and the instrumental polarization $\rho_{hv} = T_h/T_v$ as follows:

$$\begin{aligned} S_{vv} &= \tfrac{1}{2}I_0T_vT_1T_1'(1 + \rho_{hv}\rho_p\rho_p') \qquad & S_{hh} &= \tfrac{1}{2}I_0T_vT_1T_1'(\rho_p\rho_p' + \rho_{hv}) \\ S_{vh} &= \tfrac{1}{2}I_0T_vT_1T_1'(\rho_p' + \rho_{hv}\rho_p) \qquad & S_{hv} &= \tfrac{1}{2}I_0T_vT_1T_1'(\rho_p + \rho_{hv}\rho_p') \end{aligned} \tag{123}$$

These equations contain four unknowns, ρ_p, ρ_p', ρ_{hv}, and the first term on the right-hand side, so that solutions can be obtained as follows:

$$\rho_p = a - (a^2 - 1)^{1/2} \qquad \rho_p' = b - (b^2 - 1)^{1/2} \qquad \rho_{hv} = c \pm (c^2 - 1)^{1/2} \tag{124}$$

where

$$\begin{aligned} a &= \frac{1}{2}\,\frac{S_{vv}^2 - S_{hh}^2 - S_{vh}^2 + S_{hv}^2}{S_{vv}S_{hv} - S_{hh}S_{vh}} \\ b &= \frac{1}{2}\,\frac{S_{vv}^2 - S_{hh}^2 + S_{vh}^2 - S_{hv}^2}{S_{vv}S_{vh} - S_{hh}S_{hv}} \\ c &= \frac{1}{2}\,\frac{S_{vv}^2 + S_{hh}^2 - S_{vh}^2 - S_{hv}^2}{S_{vv}S_{hh} - S_{vh}S_{hv}} \end{aligned} \tag{125}$$

The extinction ratios ρ_p and ρ_p' are both less than 1, but the instrumental polarization ρ_{hv} may be either less than or greater than 1. The positive (negative) sign in the expression for ρ_{hv} is used if S_{vv} is smaller (larger) than S_{hh}. If the two polarizers are identical,

$$\rho_p = \rho_p' \qquad S_{vh} = S_{hv} \qquad a = b = \frac{1}{2}\,\frac{S_{vv} + S_{hh}}{S_{vh}}$$

and

$$c = \frac{1}{2}\,\frac{S_{vv}^2 + S_{hh}^2 - 2S_{vh}^2}{S_{vv}S_{hh} - S_{vh}^2} \tag{126}$$

In making the measurements of S_{vv}, S_{hh}, S_{vh}, and S_{hv}, good signal-to-noise ratios are desirable because a, b, and c in Eqs. (124) and (125) all contain differences of measured quantities in both the numerator and denominator. Also, as already mentioned, the intensity of the source must remain constant during all four measurements to satisfy the conditions of the simultaneous equations.

Once the extinction ratios of two polarizers are determined, their transmittances are easily obtained. From Eq. (106), $T_1 + T_2$ and $T_1' + T_2'$ can be obtained by measuring S_h, S_v, S_h', S_v', and S_0 (primed quantities refer to polarizer B). Then we can solve for T_1 and T_2:

$$T_1 = \frac{(S_h + S_v)/S_0}{1 + \rho_p} \qquad T_2 = \frac{\rho_p(S_h + S_v)/S_0}{1 + \rho_p} \tag{127}$$

Similar expressions are obtained for T_1' and T_2':

$$T_1' = \frac{(S_h' + S_v')/S_0}{1 + \rho_p'} \qquad T_2' = \frac{\rho_p'(S_h' + S_v')/S_0}{1 + \rho_p'} \tag{128}$$

An alternate method for determining T_1, T_2, T_1', and T_2' from Eqs. (122), (103) to (105), and similar equations for S_h' and S_v' is given by Young et al. (1965). Combinations of the same nine quantities can be used to calculate ρ_p, ρ_p', ρ_{hv}, and the transmittances of the two polarizers using the method of Kudo et al. (1970). The extinction ratios and instrumental polarization are determined from the ratios S_h/S_v, S_h'/S_v', and either S_{hh}/S_{vh} or S_{hv}/S_{vv}, whichever has the better signal-to-noise ratio. In order to eliminate fluctuations in the source intensity, the actual measurements were of S_h/S_0, S_v/S_0, S_h'/S_0, and S_v'/S_0 in a double-beam instrument. The signal-to-noise ratio in the quantity S_{hh}/S_{vh} was improved by measuring S_{hh}/S_h and S_{vh}/S_h and then dividing one quantity by the other. Similarly, S_{hv}/S_{vv} was obtained from the ratios S_{hv}/S_v' and S_{vv}/S_v'. Kudo suggests using a relatively poor quality polarizer for polarizer B when measuring the extinction ratio of a better polarizer. In this way, more light will be transmitted when the two polarizers are crossed, so that the signal-to-noise ratio will be markedly improved. Kudo's method also gives accurate values for ρ_{hv} even when it is fluctuating drastically, e.g., in a grating monochromator.

NONNORMAL-INCIDENCE REFLECTION AND TRANSMISSION POLARIZERS

50. By far the largest class of polarizers used in the infrared and ultraviolet spectral regions (where dichroic sheet polarizers and calcite polarizing prisms cannot be used) are so-called *pile-of-plates polarizers* from which light is reflected (or transmitted) at nonnormal incidence. Since most of these polarizers operate at angles near the Brewster or polarizing angle [see Eq. (48)], they are frequently called Brewster angle polarizers. The plane-parallel plates which are used for Brewster angle transmission polarizers (Par. 52) are generally thick enough to ensure that although multiple reflections occur within each plate, the coherence of the light beam is lost and there are no interference effects. However, another class of nonnormal-incidence transmission polarizers makes use of interference effects to enhance their polarizing properties (Pars. 53 to 55). These include polarizing beam splitters (Par. 54) and frustrated-total-reflection filters (Par. 55). A relation which is frequently used in connection with nonnormal-incidence reflectance measurements is the Abelès condition, discussed in Par. 51.

Most pile-of-plates polarizers are used to illuminate instruments such as spectrographs, spectrophotometers, etc., which contain prisms or gratings and have several nonnormal-incidence reflections. These instruments may produce appreciable polarization of the light beam. The polarizing properties of the instrument should thus be considered when choosing the orientation of the polarizer so that the two augment each other. In instruments, such as a Littrow spectrograph, in which the light strikes the prism four times at nearly Brewster's angle, the polarization may be high enough for a polarizer to be unnecessary if qualitative results will suffice.

51. Brewster Angle Reflection Polarizers Most reflection-type polarizers are made of plates which are either nonabsorbing or only slightly absorbing. Examples are the germanium, selenium, silver chloride, and sheet-plastic polarizers used in the infrared and LiF, CaF_2, Al_2O_3, MgF_2, and similar materials used in the vacuum ultraviolet. Metallic films, particularly Au, Ag, and Al, are sometimes used as polarizers in the extreme ultraviolet, where all materials absorb strongly. The polarizing efficiency of reflection-type polarizers can be experimentally determined using any of the methods given in Par. 49; the methods of Hamm et al. (1965), Horton et al. (1969), and Schledermann and Skibowski (1971) were specifically developed for polarizers of this type. The extinction ratio and "transmittance" can also be calculated directly from the Fresnel equations. Fortunately some simplifications occur for nonabsorbing or slightly absorbing plates, as described below.

Equations (20) and (21) give the values of the intensity reflection coefficients R_s and R_p for light vibrating perpendicular to the plane of incidence (s component) and parallel to the plane of incidence (p component). The angle of refraction θ_1 in those equations is related to the refractive index n of the material by Snell's law [Eq. (15)†]. At the Brewster angle $R_p = 0$, so that the reflected light is, in principle, completely plane-polarized. This is the basis for all Brewster angle reflection polarizers.

Let us now see how the characteristics of a reflection polarizer depend on its refractive index. In Fig. 45 the reflectances R_s and R_p have been plotted for different values of the refractive index, roughly representing alkali halides in the

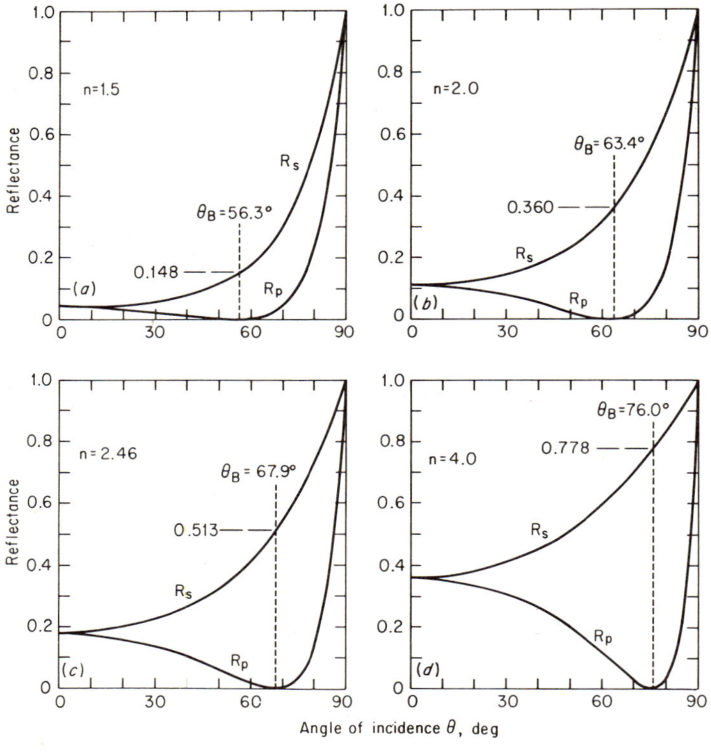

Fig. 45 Reflectance of light polarized parallel R_p and perpendicular R_s to the plane of incidence from materials of different refractive index n as a function of angle of incidence: (a) $n = 1.5$ (alkali halides in ultraviolet and sheet plastics in infrared), (b) $n = 2.0$ (AgCl in infrared), (c) $n = 2.46$ (Se in infrared), and (d) $n = 4.0$ (Ge in infrared). The Brewster angle θ_B (at which R_p goes to 0) and the magnitude of R_s at θ_B are also indicated.

ultraviolet and sheet-plastic materials, silver chloride, selenium, and germanium in the infrared. The Brewster angle, given by Eq. (48), is also indicated, as well as the magnitude of R_s at the Brewster angle. Figure 46 gives the complete range of values of R_s at the Brewster angle as a function of refractive index. We note from these graphs that if light is polarized by a single reflection from a nonabsorbing material, the polarizer with the highest refractive index will have the largest throughput. In reflection polarizers, the quantity R_s is essentially the principal "transmittance" of the polarizer [T_1 in Eqs. (55) to (65)] except that it must be

† Since we are assuming that the medium of incidence is air, $n_0 = 1$ and $n_1 = n$, the refractive index of the material.

multiplied by the reflectance of any other mirrors used to return the beam to its axial position.

The reflectance R_p can be equated to T_2, the minimum "transmittance" of the polarizer, so that the extinction ratio ρ_p of a reflection polarizer [Eq. (61)] is $\rho_p = R_p/R_s$. If R_p is really zero at the Brewster angle, the extinction ratio will be zero for all materials *independent of the value of n.* These results are shown in another way in Fig. 47, in which R_s and R_p/R_s are plotted vs. angle of incidence for the same four values of n. If a given extinction ratio is desired, for example, 10^{-3} [corresponding to 99.8 percent polarization; see Eq. (62) and Fig. 42], then the convergence angle of the light beam must be small so that all the angles of incidence lie within about $\pm 1°$ of the Brewster angle. The convergence angle depends only weakly on the refractive index for this case, varying from $\pm 1.2°$ for $n = 1.5$ to $\pm 0.8°$ for $n = 4.0$.

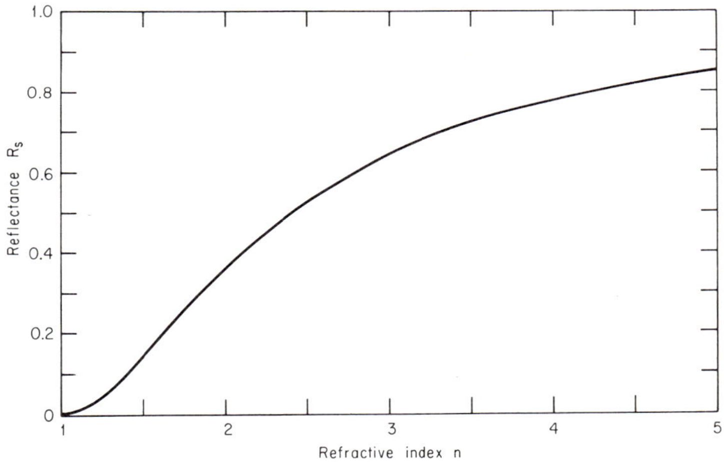

Fig. 46 Reflectance R_s at the Brewster angle as a function of the refractive index n of the material.

If a good extinction ratio is required for a beam of larger convergence angle, two polarizing reflections may be used. Then all the exponents in Fig. 47*b* are doubled, and the convergence angles for a given extinction ratio are greatly increased. To obtain a value of 10^{-3} with two reflections, the angle of incidence must be within about $\pm 6°$ of the Brewster angle for values of n less than 3.5; for $n = 4$ it is reduced slightly and becomes more asymmetric ($+4.0$ and $-5.2°$). A disadvantage of having two reflections from the polarizing material is that the throughput is reduced. All the values of R_s in Fig. 47*a* are squared, so that for $n = 4$, $R_s = 0.78$ but $R_s^2 = 0.61$; for smaller refractive indexes the reduction in throughput is much greater.

Although in many cases multiple reflections within a plate degrade its polarizing properties, this is not true for Brewster angle reflection polarizers. In Fig. 48 we have plotted R_s and R_p/R_s for $n = 1.5$ and 4.0 for a single reflection at the surface and for multiple reflections within a plane-parallel plate of material. The relation for the latter case is

$$(R_{s,p})_{\text{plate}} = \frac{2R_{s,p}}{1 + R_{s,p}} \tag{129}$$

assuming multiple internal reflections and no interference or absorption, where R_s and R_p are given by Eqs. (20) and (21). Note that multiple reflections have

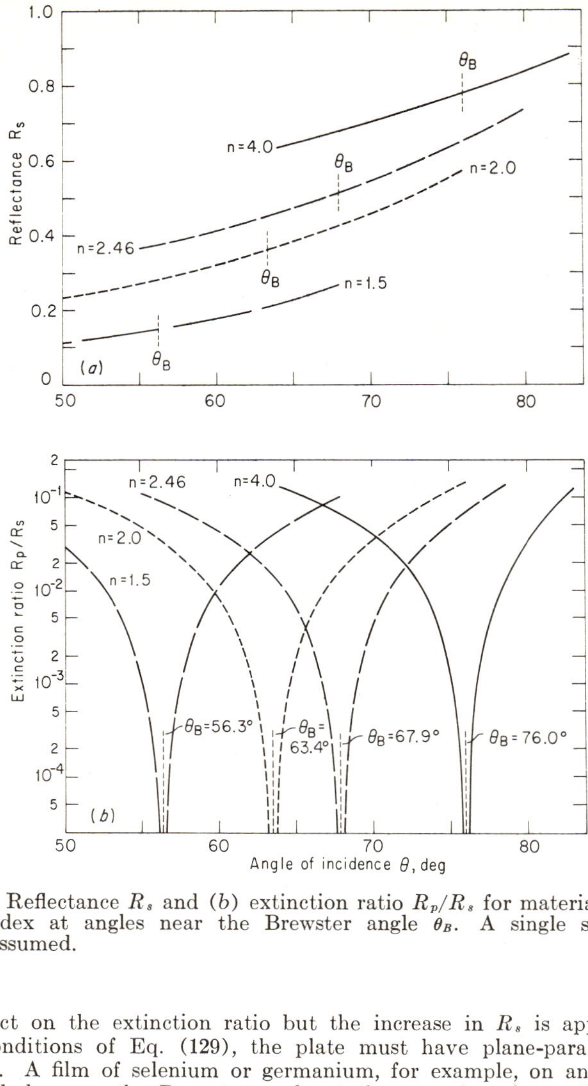

Fig. 47 (*a*) Reflectance R_s and (*b*) extinction ratio R_p/R_s for materials of different refractive index at angles near the Brewster angle θ_B. A single surface of the material is assumed.

a minor effect on the extinction ratio but the increase in R_s is appreciable. To fulfill the conditions of Eq. (129), the plate must have plane-parallel sides and be unbacked. A film of selenium or germanium, for example, on another backing will not work because the Brewster angle at the second surface will be different from that at the front surface. We are also assuming that the plate is thick or nonuniform enough for interference effects within it to be neglected.

All the preceding discussion applies strictly only to nonabsorbing materials. If a small amount of absorption is present, R_p will have a minimum that is very close to zero and the material will still make a good reflection polarizer. However, if the extinction coefficient k becomes appreciable, the minimum in R_p will increase and the polarizing efficiency will be degraded. By referring to Fig. 2 one can see roughly what the ratio of R_p to R_s will be for a given set of optical constants. Exact values of R_p and R_s can be calculated from n and k using Eqs. (45), (46), (31), and the other pertinent relations in Par. 2. When choosing materials for possible use as metallic reflection polarizers, one wants the largest difference between R_s and R_p and the smallest magnitude of R_p at the minimum. Thus, ideally n should be much larger than k.

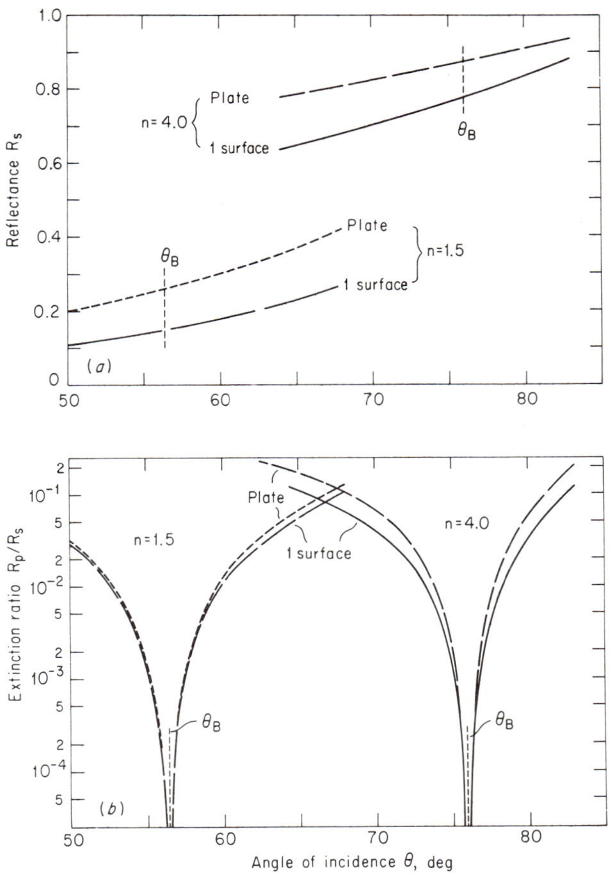

Fig. 48 (*a*) Reflectance R_s and (*b*) extinction ratio R_p/R_s for materials of refractive index 1.5 and 4.0 at angles near the Brewster angle θ_B. Lower curves are for reflection from a single surface, and upper curves are for reflection from a plane-parallel plate of the material, assuming multiple reflections but no interference within the material.

Brewster angle reflection polarizers for the infrared are mostly made from the semiconductors selenium and germanium, which are transparent beyond their absorption edges and have high refractive indexes. Table 6 lists various infrared polarizers which have been described in the literature. All involve external reflections except the Ge-Hg polarizer described by Harrick (1959), in which light undergoes two or four reflections within a bar of germanium. While this type of polarizer has attractive features, it depends on maintaining polarization in the medium, so that great care must be taken to obtain germanium with a minimum of strain birefringence.

In the ultraviolet, materials such as LiF, MgF_2, CaF_2, Al_2O_3, etc., can be used as polarizers in wavelength regions where they are transparent. Biotite, a form of mica, has also been found to perform very well in the 1000- to 6000-Å region. At shorter wavelengths, metals such as Au, Ag, and Al have been used as polarizers. Table 7 lists various nonnormal-incidence ultraviolet reflection polarizers and calculations and measurements on various materials used (or proposed) as ultraviolet polarizers.

TABLE 6 Infrared Brewster Angle Reflection Polarizers

Material	Description	Reference
Ge-Hg	Multiple internal reflections in Ge immersed in Hg	Harrick (1959)
Ge	Single external reflection from 1-cm-thick polished Ge single crystal	Edwards and Bruemmer (1959)
Ge	Proposed parallel and antiparallel arrangements of two Ge plates	Křížek (1963)
Ge	Double-beam system: beam 1, single reflection; beam 2, one transmission, one reflection	Craig et al. (1964)
Ge	Axial arrangement with reflections from two Ge wedges and two Al mirrors	Bor and Brooks (1966)
Se	Reflections from two cast-Se films on roughened glass plates	Pfund (1947)
Se	Axial arrangement with reflections from two Se films evaporated on NaCl and one Ag mirror	Barchewitz and Henry (1954)
Se	Large-aperture, axial, venetian-blind arrangement with one or two reflections from evaporated Se films on roughened glass plates (additional reflections from Al mirrors)	Takahashi (1961)
Si	Single reflection from polished single crystal Si	Walton and Moss (1961)
Si	Axial arrangement with reflection from two Al mirrors and polished Si plate with roughened back	Baumel and Schnatterly (1971)
PbS	Axial arrangement with reflections from two chemically deposited PbS films and one Al film	Grechushnikov and Petrov (1963)

The most versatile nonnormal-incidence reflection polarizer would be one which does not deviate or displace the beam from its axial position. One convenient arrangement would be a symmetric three-reflection system in which the light is incident on one side of a triangle, reflected to a plane mirror opposite the apex, and back to the other side of the triangle (as in Horton et al., 1969, or Barchewitz and Henry, 1954). If the polarizer must have a good extinction ratio and the light beam is highly convergent, two of the reflections could be from the polarizing material and the third from a silvered or aluminized mirror. If the beam is highly collimated or more throughput is required, only one reflection may be from the polarizing material. The throughput can also be increased by using a plane-parallel plate for the polarizing reflection (Fig. 48). The major drawback to a reflection polarizer is the extreme length of the device required to accommodate a beam of large cross-sectional area. For example, if a germanium polarizer were used at the Brewster angle (76°) and the beam width were about 25 mm, each Ge plate would have to be about 25 by 100 mm and the overall length of the polarizer would be greater than 200 mm if a three-reflection axial arrangement such as that described above were used.

No section on Brewster angle reflection polarizers would be complete without mentioning the so-called Abelès condition (Abelès, 1950*a*) and some of the applications to which it has been put. It applies to the amplitude reflectances r_s and r_p for either dielectrics or metals at 45° angle of incidence. At this angle

$$r_s^2 = r_p \tag{130}$$

and

$$2\delta_s = \delta_p \tag{131}$$

where the δ's are the absolute phase changes on reflection for the p and s components (see Par. 2). Relation (130) is frequently applied to the intensity reflectances R_s and R_p, which are directly related to the amplitude reflectances [Eqs. (20), (21), and (31)].

TABLE 7 Ultraviolet Reflection Polarizers and Polarization Measurements

Material	Description	Wavelength range, Å	Reference
Al_2O_3, Al, Au, ZnS, glass, and others	Calculated values of R_s and $(R_s/R_p)_{max}$ vs. wavelength for a single reflection	500–2000	Hunter (1965*a*)
Al_2O_3, Al, glass, and others	Calculated values of R_s and $(R_s - R_p)/(R_s + R_p)$ vs. angle of incidence for a single reflection; also principal angle and related angles	584	Damany (1965)
Al_2O_3, CaF_2, LiF, and Pyrex	Measured optical constants; calculated R_s and $(R_s - R_p)/(R_s + R_p)$ vs. angle of incidence and wavelength for a single reflection	200–2000	Stephan et al. (1966)
Al_2O_3 and CaF_2	Measured $(R_s - R_p)/(R_s + R_p)$ vs. angle of incidence at selected wavelengths for a single reflection; used both materials as single reflection polarizers	1026–1600	de Chelle and Merdy (1967)
LiF, Al_2O_3, MgF_2, SiO, ZnS	Used single reflection LiF polarizer at Brewster angle to measure R_s and R_s/R_p for various materials; best polarizers were Al_2O_3 and MgF_2	1216	McIlrath (1968)
Al, Ag, Au, MgF_2, SiO, ZnS	Measured polarization of uncoated aluminum grating and optical constants of all materials listed	304–1216	Cole and Oppenheimer (1962)
Al, Au	Determined polarization of Au- and Al-coated gratings by measuring reflectance of Au and fused-silica mirrors at 45°	600–2000	Rabinovitch et al. (1965)
Al, Au, glass	Measured average reflectance and degree of polarization of Al, Au, and glass as a function of angle of incidence; measured polarization of a glass grating and an Al-coated grating	584	Uzan et al. (1965)
MgF_2	Measured R_p/R_s at 60° for a single reflection and compared it with calculated values	916, 1085, 1216	Sasaki and Fukutani (1964)
Au, Ag	Measured R_p/R_s at 45° for a single reflection and compared it with calculated values; measured R_p/R_s for a platinized grating	500–1300	Hamm et al. (1965)
Au	Used single-reflection Au mirror at 60° as polarizer (Brewster angle about 55°)	600–1200	Ejiri (1967)
Au	Used axial arrangement of eight Au mirrors at 60° as polarizer and analyzer to measure polarization of synchrotron radiation; determined polarizing properties of each polarizer	500–1000	Rosenbaum et al. (1968)
Au	Constructed axial triple-reflection Au polarizer; measured extinction ratio and transmission for different angles of incidence on Au plates	500–5000	Horton et al. (1969)
Biotite	Constructed axial polarizer and analyzer each with 61° Brewster angle reflection from biotite and two reflections from MgF_2-coated Al mirrors; measured transmission and extinction ratio	1100–6000	Robin et al. (1966)
Biotite	Constructed two polarizers: (1) axial polarizer with two 60° reflections from biotite and 30° reflection from MgF_2-coated Al mirror; (2) displaced-beam polarizer with 60° reflections from two biotite plates; measured degree of polarization of various gratings	1000–2000	Matsui and Walker (1970)

Schulz and Tangherlini (1954) apparently rediscovered the Abelès condition and used the ratio $R_s^2/R_p = 1$ as a test to evaluate their reflecting surfaces. They found that surface roughness made the ratio too small but annealing the metal films at temperatures higher than 150°C made the ratio larger than unity. Rabinovitch et al. (1965) made use of the Abelès condition to determine the polarization of their Seya-Namioka vacuum-ultraviolet monochromator. They measured the reflectance at 45° of a sample whose plane of incidence was perpendicular to the exit slit and parallel to the exit slit. From these measurements they deduced the instrumental polarization by assuming the Abelès condition. Values of instrumental polarization obtained using carefully prepared gold and fused-silica samples were in excellent agreement, showing that neither of these materials had surface films which invalidated the Abelès condition. [Surface films usually have relatively little effect on the Abelès condition in the visible region (Burge and Bennett, 1964) but become important in the vacuum ultraviolet.] Hamm et al. (1965) made use of the Abelès condition to determine the reflectance of a sample in unpolarized light at 45° angle of incidence; i.e., they eliminated the effect of instrumental polarization from their measurements. Although McIlrath (1968) did not refer to the Abelès condition as such, he used it to calibrate his vacuum-ultraviolet apparatus to measure absolute reflectances by determining the instrumental polarization using a sample at 45° angle of incidence.

52. Brewster Angle Transmission Polarizers To help overcome the beam-deviation problem and the extreme length of reflection-type polarizers, Brewster angle polarizers are often used in transmission, particularly in the infrared, where transparent materials are available. At the Brewster angle, all of the *p* component and an appreciable fraction of the *s* component are transmitted. Thus, several plates must be used to achieve a reasonable degree of polarization. The higher the refractive index of the plates the fewer required.

The theory of Brewster angle transmission polarizers follows directly from that given for reflection polarizers in Par. 51. Table 8 lists the relations giving the *s* and *p* transmittances of the polarizers with various assumptions about multiple reflections, interference, absorption, etc.† All these relations contain R_s and R_p, the reflectances at a single interface, which are given at the bottom of the table.

At the Brewster angle, R_p at a single interface equals zero, and the transmittances of the plates can be expressed in terms of the refractive index of the material and the number of plates. The relations for the *s* and *p* transmittances at this angle are given in Table 9. Most references that contain an expression for the degree of polarization of a pile of plates give the formula of Provostaye and Desains (1850), which assumes an infinite series of multiple reflections between all surfaces, i.e., multiple reflections within and between plates. This assumption is not valid for most real transmission polarizers. Conn and Eaton (1954) have shown that the formulas which assume incoherent multiple reflections within each plate and *none* between plates give the correct degree of polarization for a series of Zapon lacquer films ($n = 1.54$) and also for a series of eight selenium films, whereas the formula of Provostaye and Desains (1850) predicted values which were much too low. These authors also point out that the number of multiply reflected beams between plates that enter the optical system depends on the spacing between plates and the diaphragm used to limit the number of beams. One can use a fanned arrangement, as suggested by Bird and Shurcliff (1959), to eliminate these multiply reflected beams. Internal reflections within each plate can be removed by wedging the plates (Bird and Shurcliff, 1959).

For most parallel-plate polarizers it is reasonable to assume incoherent multiple reflections within each plate and no reflections between plates. Figure 49 shows the principal transmittance (*p* component) and extinction ratio for several four-plate polarizers having the refractive indexes indicated.‡ The extinction ratio improves considerably with increasing refractive index. It is also improved by using the

† Transmission polarizers in which the multiply internally reflected beams are coherent and produce interference effects are discussed in Par. 53.

‡ The extinction ratio of a pile of *m* plates (no multiple reflections between plates) is simply the product of the extinction ratios of the individual plates.

TABLE 8 Transmittances and Degree of Polarization for a Single Plate and Multiple Plates at Any Angle of Incidence in Terms of R_s and R_p for a Single Surface†

	One plate (two surfaces) $(T_{s,p})_{\text{sample}}$	*m* plates (2*m* surfaces) $(T_{s,p})_{\text{sample}}$	*m* plates (2*m* surfaces) $P = \frac{T_p - T_s}{T_p + T_s} = \frac{1 - T_s/T_p}{1 + T_s/T_p}$
Single transmitted beam, no multiple reflections, no absorption	$(1 - R_{s,p})^2$	$(1 - R_{s,p})^{2m}$‡	$\frac{1 - \cos^{4m}(\theta_0 - \theta_1)}{1 + \cos^{4m}(\theta_0 + \theta_1)}$‡
Multiple reflections within plate, no interference effects, no absorption	$\frac{1 - R_{s,p}}{1 + R_{s,p}}$‖	$\left(\frac{1 - R_{s,p}}{1 + R_{s,p}}\right)^m$‡,‖	$\frac{1 - \left[\cos^2(\theta_0 - \theta_1)\frac{1 + R_p}{1 + R_s}\right]^m}{1 + \left[\cos^2(\theta_0 - \theta_1)\frac{1 + R_p}{1 + R_s}\right]^m}$‡,‖
		$\frac{1 - R_{s,p}}{1 + (2m - 1)R_{s,p}}$§	$\frac{1 - \cos^2(\theta_0 - \theta_1)\frac{1 + (2m - 1)R_p}{1 + (2m - 1)R_s}}{1 + \cos^2(\theta_0 - \theta_1)\frac{1 + (2m - 1)R_p}{1 + (2m - 1)R_s}}$§
Multiple reflections within plate, no interference effects, small absorption	$\frac{(1 - R_{s,p})^2 e^{-\alpha d}}{1 - R_{s,p}{}^2 e^{-2\alpha d}}$	$\frac{(1 - R_{s,p})^{2m} e^{-m\alpha d}}{(1 - R_{s,p}{}^2 e^{-2\alpha d})^m}$‡	$\frac{1 - \cos^{4m}(\theta_0 - \theta_1)\left(\frac{1 - R_p{}^2 e^{-2\alpha d}}{1 - R_s{}^2 e^{-2\alpha d}}\right)^m}{1 + \cos^{4m}(\theta_0 - \theta_1)\left(\frac{1 - R_p{}^2 e^{-2\alpha d}}{1 - R_s{}^2 e^{-2\alpha d}}\right)^m}$‡
Multiple reflections within plate, interference within plate, no absorption	$\frac{1}{1 + \frac{4R_{s,p}}{(1 - R_{s,p})^2}\sin^2\gamma}$	$\left[\frac{1}{1 + \frac{4R_{s,p}}{(1 - R_{s,p})^2}\sin^2\gamma}\right]^m$‡ $\begin{cases} \max & 1 \\ \min & \left(\frac{1 - R_{s,p}}{1 + R_{s,p}}\right)^{2m} \end{cases}$	
Single surface	$R_s = \frac{\sin^2(\theta_0 - \theta_1)}{\sin^2(\theta_0 + \theta_1)}$	$R_p = \frac{\tan^2(\theta_0 - \theta_1)}{\tan^2(\theta_0 + \theta_1)}$	
	$T_s = 1 - R_s = \frac{\sin 2\theta_0 \sin 2\theta_1}{\sin^2(\theta_0 + \theta_1)}$	$T_p = 1 - R_p = \frac{\sin 2\theta_0 \sin 2\theta_1}{\sin^2(\theta_0 + \theta_1)\cos^2(\theta_0 - \theta_1)}$	$P = \frac{1 - \cos^2(\theta_0 - \theta_1)}{1 + \cos^2(\theta_0 - \theta_1)}$

† $\alpha = 4\pi k/(\lambda \cos\theta_1)$, $\gamma = 2\pi nd \cos\theta_1/\lambda$, θ_0 = angle of incidence, θ_1 = angle of refraction, n = refractive index = $(\sin\theta_0)/(\sin\theta_1)$, k = extinction coefficient, d = plate thickness, λ = wavelength. ‡ No multiple reflections between plates. § Multiple reflections between plates. ‖ Also holds for coherent multiple reflections averaged over one period of $\sin^2\gamma$.

TABLE 9 Transmittances and Degree of Polarization for a Single Plate and Multiple Plates at the Brewster Angle θ_B, where $\tan\theta_B = n$ and $\theta_0 + \theta_1 = 90°$†

	One plate (two surfaces)		*m* plates (2*m* surfaces)		
	$(T_p)_{\text{sample}}$	$(T_s)_{\text{sample}}$	$(T_p)_{\text{sample}}$	$(T_s)_{\text{sample}}$	$P = \dfrac{T_p - T_s}{T_p + T_s} = \dfrac{1 - T_s/T_p}{1 + T_s/T_p}$
Single transmitted beams, no multiple reflections, no absorption	1	$\left(\dfrac{2n}{n^2+1}\right)^4$	1	$\left[\dfrac{2n}{n^2+1}\right]^{4m}$‡	$\dfrac{1 - [2n/(n^2+1)]^{4m}}{1 + [2n/(n^2+1)]^{4m}}$‡
Multiple reflections within plate, no interference effects, no absorption	1	$\dfrac{2n^2}{n^4+1}$¶	1	$\left[\dfrac{2n^2}{n^4+1}\right]^{m}$‡,¶	$\dfrac{1 - [2n^2/(n^4+1)]^{m}}{1 + [2n^2/(n^4+1)]^{m}}$‡,¶
			1	$\dfrac{1}{1 + m(n^2-1)^2/2n^2}$§	$\dfrac{m}{m + [2n/(n^2-1)]^2}$§,‖
Multiple reflections within plate, no interference effects, small absorption	$e^{-\alpha d}$	$\dfrac{\left(\dfrac{2n}{n^2+1}\right)^4 e^{-\alpha d}}{1 - \left(\dfrac{n^2-1}{n^2+1}\right)^4 e^{-2\alpha d}}$	$e^{-m\alpha d}$	$\dfrac{\left(\dfrac{2n}{n^2+1}\right)^{4m} e^{-m\alpha d}}{\left[1 - \left(\dfrac{n^2-1}{n^2+1}\right)^4 e^{-2\alpha d}\right]^m}$‡	$\dfrac{1 - \dfrac{[2n/(n^2+1)]^{4m}}{\{1 - [(n^2-1)/(n^2+1)]^4 e^{-2\alpha d}\}^m}}{1 + \dfrac{[2n/(n^2+1)]^{4m}}{\{1 - [(n^2-1)/(n^2+1)]^4 e^{-2\alpha d}\}^m}}$‡
Multiple reflections within plate, interference within plate, no absorption	1	$\dfrac{1}{1 + \dfrac{(n^4-1)^2}{4n^4}\sin^2\gamma}$	1	$\left[\dfrac{1}{1 + \dfrac{(n^4-1)^2}{4n^4}\sin^2\gamma}\right]^m$‡	$\begin{cases}\max & 1 \\ \min & \left(\dfrac{2n^2}{n^4+1}\right)^{2m}\end{cases}$
Single surface		$R_p = 0$		$R_s = \left(\dfrac{n^2-1}{n^2+1}\right)^2$	
		$T_p = 1 - R_p = 1$		$T_s = 1 - R_s = \left(\dfrac{2n}{n^2+1}\right)^2$	$P = \dfrac{1 - [2n/(n^2+1)]^2}{1 + [2n/(n^2+1)]^2}$

† Where $\alpha = 4\pi k(n^2+1)^{1/2}/\lambda n$, $\gamma = 2\pi n^2 d/\lambda(n^2+1)^{1/2}$, n = refractive index, k = extinction coefficient, d = plate thickness, λ = wavelength.

‡ No multiple reflections between plates.

§ Multiple reflections between plates.

‖ Formula of Provostaye and Desains (1850).

¶ Also holds for coherent multiple reflections averaged over one period of $\sin^2\gamma$.

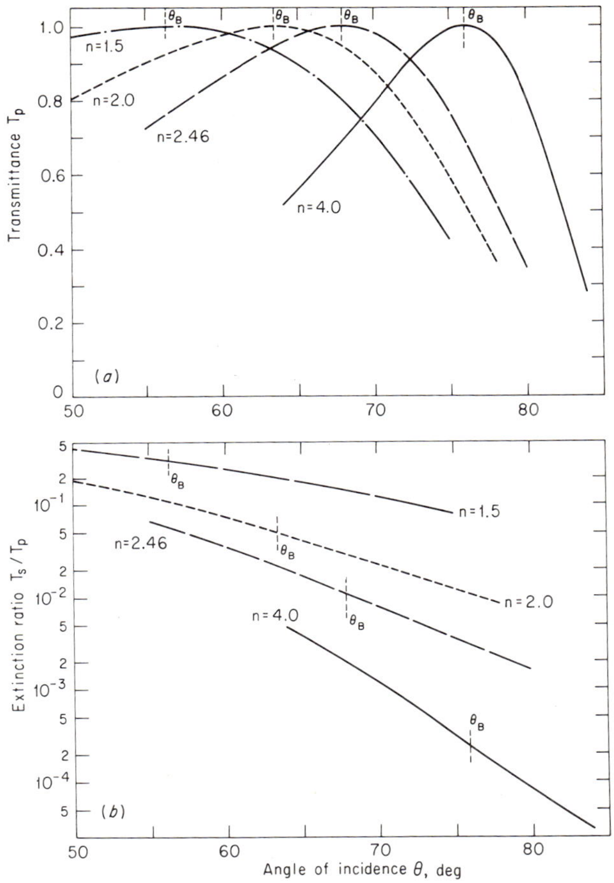

Fig. 49 (*a*) Transmittance and (*b*) extinction ratio of four plane-parallel plates of refractive index n as a function of angle of incidence, for angles near the Brewster angle. Assumptions are multiple reflections but no interference within each plate and no reflections between plates.

plates at an angle of incidence slightly above the Brewster angle. This procedure, which is most helpful for high-index plates, reduces the transmission per plate so that a trade-off is required between losses resulting from absorption or scattering when many plates are used and the reflectance loss per plate when only a few plates are used above the Brewster angle. In some cases significant improvements have been achieved by following the latter course (Bennett et al., 1961).

The effect of increasing the number of plates of a given refractive index is shown in Figs. 50 and 51 for low- and high-index materials. In the absence of absorption, comparable transmittances and extinction ratios are obtained with a large number of low-index plates or a small number of high-index plates. Figures 52 and 53 show the effect of small amounts of absorption αd per plate on an assembled polarizer. Figure 52 shows the transmittance and extinction ratio for a polarizer consisting of 16 slightly absorbing plates of refractive index 1.5. A polarizer having a comparable extinction ratio but consisting of only two slightly absorbing plates of index 4.0 is shown in Fig. 53. In these figures, a value of $\alpha d = 0.10$ corresponds to an extinction coefficient $k = 0.001$ for a plate 12.7 μm

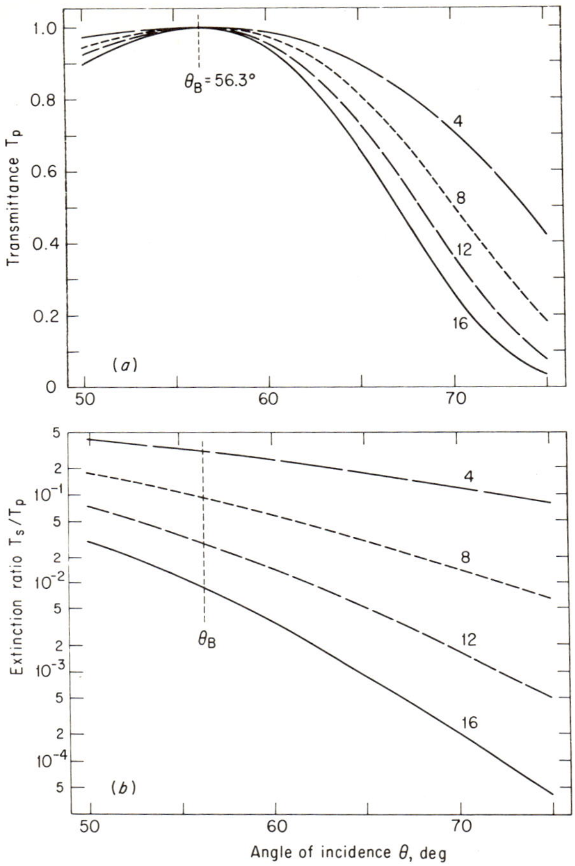

Fig. 50 (*a*) Transmittance and (*b*) extinction ratio of various numbers of plane-parallel plates of refractive index 1.5 as a function of angle of incidence, for angles near the Brewster angle. The number of plates and θ_B are indicated. Same assumptions as in Fig. 49.

(0.005 in) thick at a wavelength of 2 μm. A thicker plate of another material having the same value of αd would have a much smaller extinction coefficient at the same wavelength. This absorption may be real, or it may be caused by internal scattering or scattering at the plate surfaces. It is clear that for a large number of low-index plates, very small amounts of absorption per plate can drastically reduce the transmittance of a polarizer without improving its extinction ratio. With fewer numbers of higher-index plates, the extinction ratio is marginally improved, and the transmittance is not drastically reduced.

There is very little information in the literature about the acceptance angle of Brewster angle transmission polarizers beyond statements to the effect that they are very sensitive to angle of incidence. A figure of merit giving the variation of the extinction ratio with angle of incidence can be defined as in Fig. 54, where the ordinate is the extinction ratio at a given angle of incidence divided by the extinction ratio at the Brewster angle. The angles of incidence are referred to the Brewster angle, and curves for different values of the refractive index are

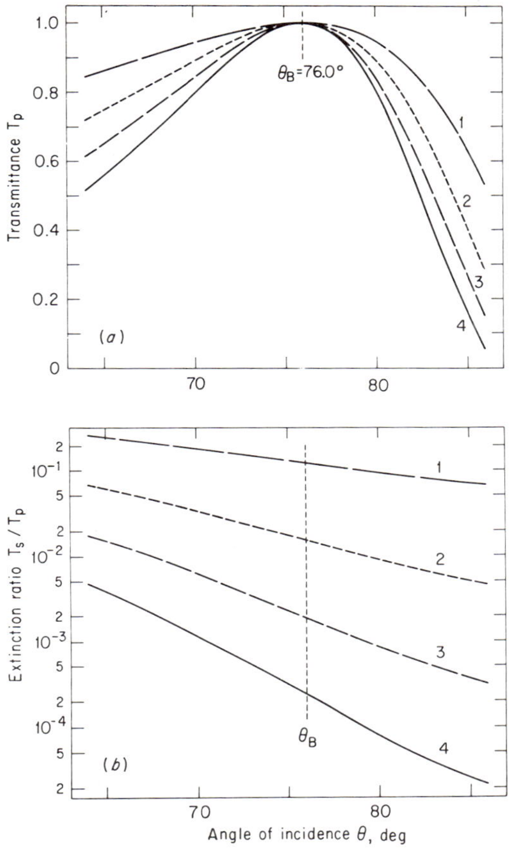

Fig. 51 Same as Fig. 50 except that $n = 4.0$.

shown. These curves are calculated from the ratio

$$\frac{\left(\frac{T_s}{T_p}\right)_\theta}{\left(\frac{T_s}{T_p}\right)_{\theta_B}} = \frac{\left(\frac{1 - R_s}{1 + R_s}\frac{1 + R_p}{1 - R_p}\right)_\theta}{\left(\frac{1 - R_s}{1 + R_s}\right)_{\theta_B}} \tag{132}$$

and are for a single transparent film or plate having multiple incoherent internal reflections within the material. As an example of how to use the graphs, consider an optical system having a two-plate germanium polarizer with a refractive index of 4.0. If the angles of incidence vary from −1.4 to +1.5° around the Brewster angle, the ratio of the extinction ratios will vary between $1.10^2 = 1.21$ and $0.90^2 = 0.81$, respectively. (For m plates it would be 1.10^m and 0.90^m.) Thus, in order to restrict the percent variation of the extinction ratio to a given value, one must use a smaller acceptance angle when using more plates.

It might appear from Fig. 54 that one would be better off using a larger number of lower-index plates to achieve a minimum variation of the extinction ratio. However, this is not true, as can be seen by comparing the 16-plate polarizer with

$n = 1.5$ (Fig. 52*b*) with the 2-plate polarizer with $n = 4$ (Fig. 53*b*). Both these polarizers have an extinction ratio in the neighborhood of 2×10^{-2} at the Brewster angle. An acceptance angle of about $\pm 1.6°$ around the Brewster angle for the germanium gives the same percent variation as an acceptance angle of $\pm 1.0°$ for the polarizer with $n = 1.5$ [as calculated from Eq. (132)].

We have assumed that there are multiple incoherent reflections within each plate and no multiple reflections between plates. The difference in extinction ratios

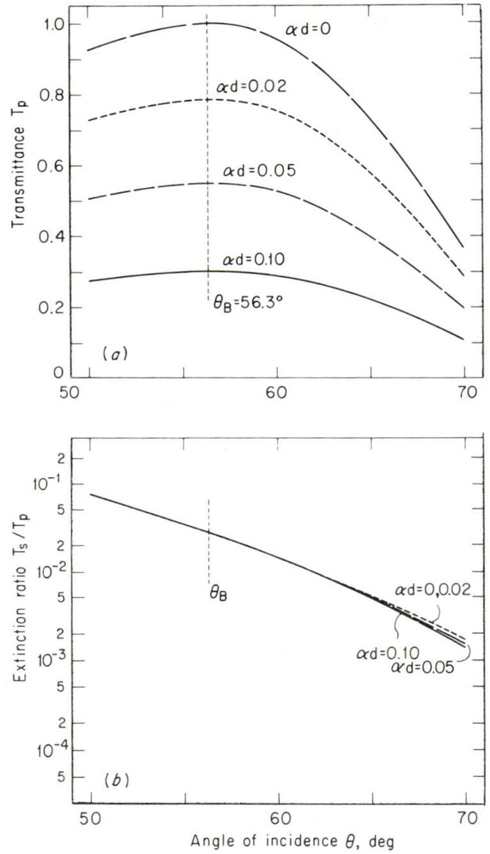

Fig. 52 (*a*) Transmittance and (*b*) extinction ratio of 16 plane-parallel plates of refractive index 1.5 having varying amounts of absorption αd per plate, where $\alpha = 4\pi k/(\lambda \cos \theta)$ and d is the plate thickness. Same assumptions as in Fig. 49.

for a series of four plates with and without internal reflections is shown in Fig. 55. The principal transmittance is essentially the same as in Fig. 49 for values of T_p above 0.70 (and only about 0.025 lower when T_p drops to 0.30). However, the extinction ratio of high-index materials is much better without multiple internal reflections; for low-index materials the difference in extinction ratios is small.

The effect of multiple reflections on the extinction ratio can readily be seen from the three relations for the transmittances of the p and s components:

No multiple reflections:

$$(T_{s,p})_{\text{sample}} = (1 - R_{s,p})^{2m} = 1 - 2mR_{s,p} + 2m^2R_{s,p}^2 - mR_{s,p}^2 + \cdots \quad (133)$$

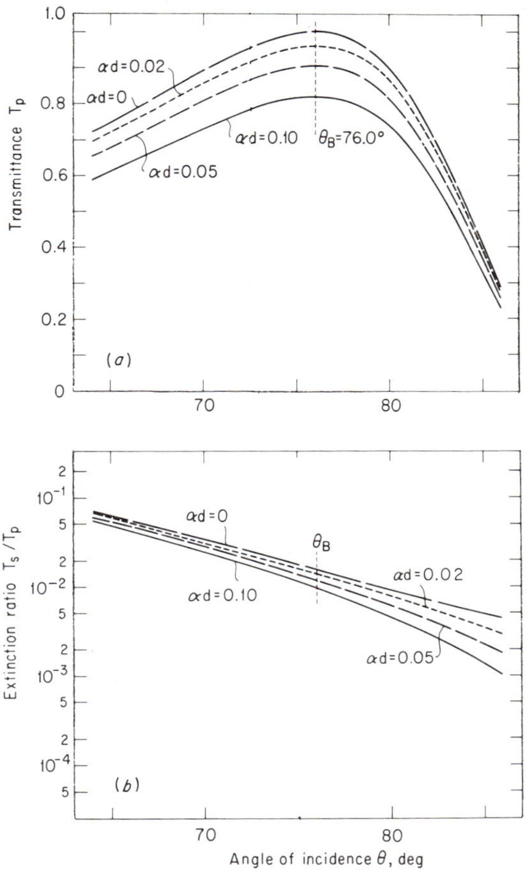

Fig. 53 Same as Fig. 52 except there are two plane-parallel plates of refractive index 4.0.

Multiple reflections within plates:

$$(T_{s,p})_{\text{sample}} = \left(\frac{1 - R_{s,p}}{1 + R_{s,p}}\right)^m = 1 - 2mR_{s,p} + 2m^2R_{s,p}{}^2 + \cdots \tag{134}$$

Multiple reflections within and between plates:

$$(T_{s,p})_{\text{sample}} = \frac{1 - R_{s,p}}{1 + (2m - 1)R_{s,p}} = 1 - 2mR_{s,p} + 4m^2R_{s,p}{}^2 - 2mR_{s,p}{}^2 + \cdots \tag{135}$$

At the Brewster angle, $R_p = 0$, $T_p = 1$, and the extinction ratio will be smallest, i.e., highest degree of polarization, for the smallest values of the s transmittance. The first three terms in Eqs. (133) and (134) are identical, but Eq. (133) has an additional negative term in $R_s{}^2$ and so it will give a slightly smaller value of the s transmittance. Equation (135), from which the formula of Provostaye and Desains was derived, has twice as large a third term as the other two equations, and the negative fourth term is only $1/2m$ of the third term, so that it does not reduce the overall value of the expression appreciably. Thus, Eq. (135) gives

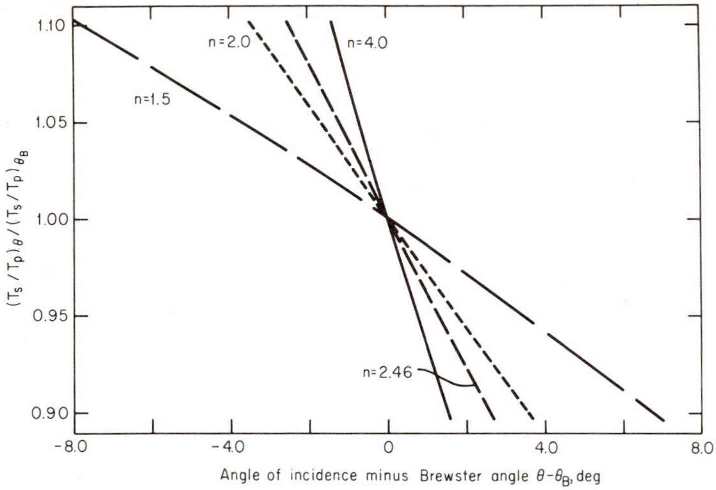

Fig. 54 Variation of extinction ratio (per film) as a function of angle near the Brewster angle $\theta - \theta_B$. The ordinate is the extinction ratio at θ divided by the extinction ratio at θ_B.

an appreciably larger value of the s transmittance, but fortunately it is a limiting case and is rarely encountered experimentally.

Most of the infrared Brewster angle transmission polarizers described in the literature have been made of selenium, silver chloride, or polyethylene sheet; they are briefly described in Table 10. In the wavelength range from beyond 3 μm, where calcite polarizing prisms become highly absorbing, to about 10 μm, beyond which wire-grid polarizers have good extinction ratios, the Brewster angle transmission polarizers described in the literature are the most useful, since the better-extinction, on-axis reflection-type polarizers (Par. 51) are impossibly long. Some of the interference polarizers described in Pars. 53 to 55 are superior if the beam-convergence angle is small. Ultraviolet Brewster angle transmission polarizers are not nearly as common; lithium fluoride and calcium fluoride have mainly been used from about 1500 to 2500 Å (see Table 11). In the region where calcite polarizing prisms are usable (>2140 Å) the Brewster angle polarizers have the advantage of a larger linear aperture and less absorption.

Low-absorption glass pile-of-plates polarizers have been used in the visible region in preference to more absorbing Glan-Thompson prism polarizers to increase the power output of giant-pulse ruby lasers (Weiser, 1964). Also in the visible region, Weinberg (1964) has calculated the degree of polarization of glass and silver chloride plates, but unfortunately he did not include the most physically reasonable case or calculate the transmittance of his polarizers.

Interference Polarizers

53. When the sheets or films constituting a nonnormal-incidence transmission polarizer are thin and have very smooth surfaces, the internally reflected beams can interfere constructively or destructively. In this case the transmittance of the p component remains unity at the Brewster angle (where $R_p = 0$) and only oscillates slightly (with respect to wavelength) for angles close to the Brewster angle. However, the s transmittance varies from a maximum of unity to a minimum of $(1 - R_s)^2/(1 + R_s)^2$ whenever λ changes by an amount that will make the quantity $(nd \cos \theta_1)/\lambda$ in Eq. (26) change by $\frac{1}{2}$.† These transmittance oscillations are only ± 0.225 for a single film of refractive index 1.5 but can

† The approximate expression for this wavelength interval $\Delta\lambda$ (assuming that the oscillations are sufficiently close together for $\lambda_1\lambda_2 \approx \lambda^2$) is given in Eq. (27).

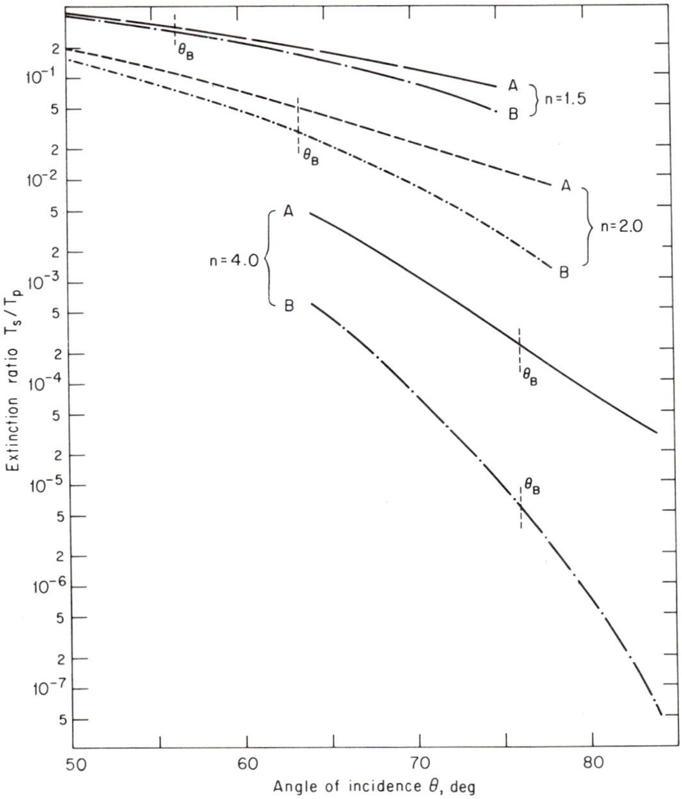

Fig. 55 Extinction ratio of four plane-parallel plates of refractive index n as a function of angle of incidence for angles near the Brewster angle. Assumptions are A, multiple reflections but no interference within each plate and no reflections between plates; B, no multiple reflections within each plate or between plates. The transmittances for conditions A and B are essentially identical (see Fig. 49a).

become as large as ± 0.492 when $n = 4.0$. Since the p transmittance remains essentially constant, the extinction ratio will vary cyclically with the s transmittance, as can be seen in the upper curve of Fig. 56 for a 2.016-μm-thick selenium film.

If a transmission polarizer with a good extinction ratio is needed for use over a limited wavelength range, it can be made of several uniform films of a thickness that yields a minimum extinction ratio in the given wavelength region. [The extinction ratio for a series of m films is $(T_s/T_p)^m$ when there are no multiple reflections between them.] In this way only *half* as many films would be needed to achieve a given extinction ratio as would be necessary if interference effects were not present. This rather surprising result can be seen from the expressions for $(T_s)_{\text{sample}}$ for m plates with and without interference effects in Table 9. Assuming no multiple reflections between plates, the expressions are $[2n^2/(n^4+1)]^{2m}$ and $[2n^2/(n^4+1)]^m$, respectively. Hertz (1959) achieved a degree of polarization of 99.5 percent in the 6- to 17-μm region using three unbacked selenium films 0.95 μm thick. Conn and Eaton (1954) obtained only a slightly better performance with eight thicker nonuniform selenium films.

As can be seen in Fig. 56, the calculated extinction ratio for the 2.016-μm-thick film goes to unity at 3.064, 4.596, and 9.191 μm, indicating that the s as well as the p transmittance at these wavelengths is unity. This ratio will remain unity

TABLE 10 Infrared Brewster Angle Transmission Polarizers

Material	Description	Wavelength range, μm	Reference
Se	5 or 6 unbacked films (4 μm thick) at 65° angle of incidence (Brewster angle 68.5°)	2–14	Elliott and Ambrose (1947); Elliott et al. (1948)
Se	5 unbacked films at 65° incidence (different method of preparation from above)		Ames and Sampson (1949)
Se	8 unbacked films at the Brewster angle		Conn and Eaton (1954)
Se	Se films (3–8 μm thick) evaporated on one side of collodion films; 68.5° angle of incidence	1–15	Barchewitz and Henry (1954)
Se	1 to 6 unbacked films (1.44–8 μm thick) at 68° angle of incidence	Visible–20	Duverney (1959)
Se	3 unbacked films (0.95 μm thick) at 71° angle of incidence	6–17	Hertz (1959)
Se	5 Formvar films coated on both sides with Se (various thicknesses) at 65° angle of incidence	1.8–3.2	Buijs (1960)
Se	Unbacked films [different method of preparation from Elliott et al. (1948)]		Bradbury and Elliott (1962)
Se	4 to 6 pleated unsupported films (4–8 μm thick) at the Brewster angle	2.5–25	Greenler et al. (1966)
AgCl	3 plates (1 mm thick) at 63.5° angle of incidence	Visible–15	Wright (1948)
AgCl	6 to 12 plates (0.05 mm thick) at 60–75° angle of incidence	2–20	Newman and Halford (1948)
AgCl	6 plates (0.5 mm thick) at 63.5° stacked in alternate directions		Makas and Shurcliff (1955)
AgCl	Suggests 6 wedge-shaped plates at 68° stacked in alternate directions in a fanned arrangement		Bird and Shurcliff (1959)
AgCl	2 V-shaped plates (3.2 mm thick) at 77° angle of incidence; large aperture		Bennett et al. (1961)
KRS-5	1 to 3 thallium bromide–iodide plates (1 and 4 mm thick) at polarizing angle	1–15	Lagemann and Miller (1951)
ZnS	4 glass plates (0.1 mm thick) coated on both sides with uniform ZnS films of same thickness (several sets of plates to cover extended wavelength range)	Visible–6	Huldt and Staflin (1959)
Ge	1 single-crystal Ge plate (0.8 mm thick) at 76° angle of incidence		Meier and Günthard (1959)
Ge	2 plates (1 mm thick) in an X-shaped arrangement at 76° angle of incidence		Harrick (1964)
Polyethylene	12 sheets (8 μm thick) at the Brewster angle	6–20 (absorption bands 6–14)	Smith et al. (1959)
Polyethylene	9 to 15 sheets (20–50 μm thick) at the Brewster angle (55°)	30–200	Mitsuishi et al. (1960)
Polyethylene	4 sheets at the Brewster angle	200–350	Hadni et al. (1960)
Polyethylene	12 sheets (5 μm thick) at the Brewster angle	1.5–13 (selected wavelengths)	Walton and Moss (1961)
Polyethylene	20 sheets (30 μm thick) at the Brewster angle	45–200	Munier et al. (1962)
Polyethylene	25 to 30 sheets at the Brewster angle	54.6	Hilton and Jones (1966)
Melinex	11 to 13 polyethylene terephthalate (Melinex) sheets (4.25–9 μm thick) at the Brewster angle	1–5	Walton et al. (1964)

TABLE 11 Ultraviolet Brewster Angle Transmission Polarizers

Material	Description	Wavelength range, Å	Reference
LiF	4 to 8 plates (0.3–0.8 mm thick) at 60° angle of incidence (Brewster angle 55.7–58.7°) stacked in alternate directions	1200–2000	Walker (1964)
LiF	8 plates	1100–3000	Hinson (1966)
LiF	8 plates (0.25–0.38 mm thick) at 60° angle of incidence stacked in groups of 4 in alternate directions	1200–2000	Heath (1968)
CaF_2	4 and 8 wedged plates stacked in alternate directions in fanned arrangement at 65° angle of incidence (Brewster angle 56.7°)	1500–2500	Schellman et al. (1964)
Al	Calculations of polarizing efficiency for 1000-Å-thick unbacked Al film, 1000- and 500-Å Al films each covered with 30-Å Al_2O_3 and 100-Å Au films	300–800	Hunter (1965*a*)

at the above wavelengths if there are several films of the same thickness. Even if the films have slightly different thicknesses, or if their surfaces are somewhat rough, interference effects may still persist, adversely affecting polarizer performance. Such effects have been observed by various authors (Elliott et al., 1948; Barchewitz and Henry, 1954; Duverney, 1959; Mitsuishi et al., 1960; and Walton et al., 1964).

By choosing films of appropriate thicknesses, interference effects can be used to advantage. The lower curve in Fig. 56 shows the extinction ratio obtained if four selenium films of thicknesses 1.080, 1.440, 1.800, and 2.016 μm are used at the Brewster angle as a transmission polarizer. (The wavelengths at which

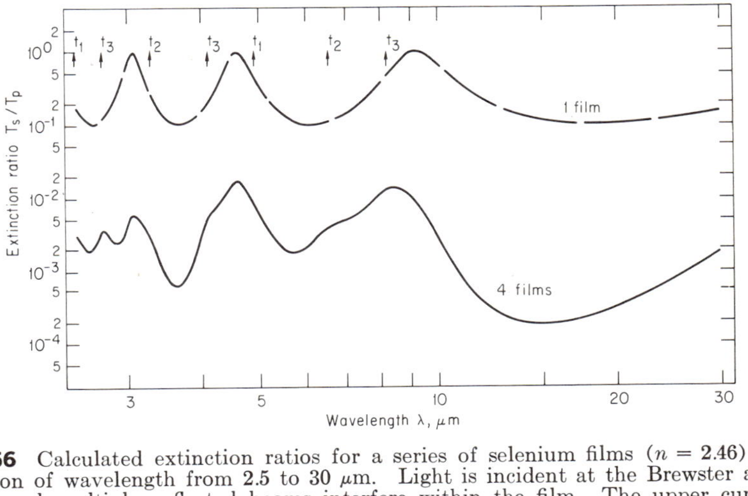

Fig. 56 Calculated extinction ratios for a series of selenium films (n = 2.46) as a function of wavelength from 2.5 to 30 μm. Light is incident at the Brewster angle, 67.9°, and multiply reflected beams interfere within the film. The upper curve is for a single film 2.016 μm thick; arrows indicate positions of maxima for three thinner films: t_1 = 1.080 μm, t_2 = 1.440 μm, and t_3 = 1.800 μm. The lower curve is the extinction ratio for the four films in series assuming no reflections between films. The calculated p transmittance for each film (and for four films in series) is unity at the Brewster angle.

maxima occur for the three thinner films are indicated by arrows in the upper portion of the figure.) In this example the extinction ratio for the four films in series is better than 2×10^{-2} from 2.5 to 30 μm and at most wavelengths is better than 10^{-2} (corresponding to a degree of polarization in excess of 98 percent). In the 11- to 27-μm wavelength region the extinction ratio is better than 10^{-3}. Four thick or nonuniform selenium films without interference effects have a calculated extinction ratio of about 10^{-2}, and six films are required to change this ratio to 10^{-3}. Thus, in the 11- to 27-μm wavelength region, *four* selenium films of appropriate thicknesses *with interference* have a superior extinction ratio to *six* selenium films *without interference*. If one wishes to optimize the extinction ratio over a more limited wavelength range, the film thicknesses can be adjusted accordingly and the extinction ratio improved. Unfortunately, the gain in extinction ratio is offset by a more sensitive angular function than that shown in Fig. 54, so that the incident beam must be very well collimated.

Interference effects can also be used to advantage in other types of nonnormal-incidence polarizers. Bennett et al. (1970) made a transmission polarizer from a series of four germanium films (ranging in thickness from 0.164 to 0.593 μm), evaporated onto strain-free plates of sodium chloride. The plates were inclined at the Brewster angle for germanium and arranged in the form of an X so that the polarizer would have a large square aperture and would not deviate the beam. An extinction ratio better than 3×10^{-3} was measured at 2.5 μm, and the plates transmitted from 2 to 13 μm. (Calculated extinction ratios in this wavelength range vary from 1×10^{-3} to 2×10^{-4} for radiation incident at the Brewster angle.)

Polarizers consisting of a high-index transparent film on a lower-index transparent substrate have been suggested for use in the visible (Schröder, 1948; Abelès, 1950*c*). These still have a Brewster angle where $R_p = 0$, and furthermore R_s at this angle is greatly increased over its value for an uncoated low-index substrate. Thus, one can obtain a large-aperture, high-efficiency polarizer with no absorption losses, which should find numerous applications in laser systems. One polarizer of this type, suggested independently by Schröder and Abelès, would consist of high-index titanium dioxide films ($n \approx 2.5$) evaporated onto both sides of a glass substrate ($n = 1.51$). At the Brewster angle, 74.4°, $R_s \approx 0.8$, making this polarizer equivalent to one made from a material of refractive index 4 ($\theta_B = 76.0°$, as shown in Fig. 45).† Two glass plates coated on both sides with TiO_2 films should have an extinction ratio of about 1.6×10^{-3} at 5500 Å and about twice that value at the extreme ends of the visible region (Abelès, 1950*c*). Schröder (1948) measured the degree of polarization as a function of angle of incidence for one such TiO_2-coated glass plate and found values comparable to the calculated ones. Kubo (1953) has calculated the degree of polarization, reflectance, and transmittance (as a function of angle of incidence and wavelength) of a glass plate ($n = 1.50$) covered with a thin transparent film of index 2.20. His results are similar to those of Abelès and Schröder.

Schopper (1953), Ruiz-Urbieta and Sparrow (1972), and Abelès (1965) have also investigated making nonnormal-incidence reflection polarizers from a thin transparent or absorbing film deposited onto an absorbing substrate. Abelès has designed some specialized reflection polarizers for use in the vacuum ultraviolet. Unfortunately the wavelength range covered is very narrow; for one polarizer it was 25 Å at a wavelength of 1500 Å. However, possibly the spectral range could be increased by using several thin films instead of one.

Multilayer film stacks have also been used to produce nonnormal-incidence reflection or transmission polarizers (Buchman et al., 1971). In several of these, both the reflected and the transmitted beams are used; they are discussed in Par. 54.

54. Polarizing Beam Splitters Polarizing beam splitters are a special form of nonnormal-incidence interference polarizer in which the beam is incident on a multilayer dielectric stack at 45°. The transmitted beam is almost entirely plane-polarized in the *p* direction, while the reflected beam is nearly all plane-polarized in the *s* direction. Generally the alternating high- and low-index dielectric layers are deposited onto the hypotenuses of two right-angle prisms, which are then

† We are assuming no multiply reflected beams within the substrate in either case.

cemented together to form a cube. The beam enters a cube face normally and strikes the multilayers on the hypotenuse (high-index layer is next to the glass), and the reflected and transmitted beams emerge normal to cube faces, being separated by 90°. Clapham et al. (1969) have a good discussion of polarizing beam splitters, which were invented by S. M. MacNeille (1946) and developed by Banning (1947). Banning's beam splitter was made with three zinc sulfide and two cryolite layers on each prism; the polarization for white light was greater than 98 percent over a 5° angle on each side of the normal to the cube face for both the reflected and transmitted beams.

Various high- and low-index materials have been successfully used in the multilayer stacks. In addition to zinc sulfide and cryolite on glass (Banning, 1947; Schröder and Schlafer, 1949) layers of a controlled mixture of silicon dioxide and titanium dioxide have been alternated with pure titanium dioxide on fused-silica prisms (Pridatko and Krylova, 1958), thorium dioxide and silicon dioxide on fused-silica prisms (Sokolova and Krylova, 1963), chiolite (a mixture of sodium and aluminum fluorides) and lead fluoride on fused-silica prisms (Turner and Baumeister, 1966), bismuth oxide and magnesium fluoride on EDF glass prisms (Clapham et al., 1969), and zirconium oxide and magnesium fluoride on dense flint-glass prisms (Clapham et al., 1969). The calculations involved in optimizing these beam splitters for good polarizing characteristics, achromaticity, and relative insensitivity to angle of incidence are quite involved. Clapham et al. (1969) and Turner and Baumeister (1966) discuss various calculational techniques frequently used. Clapham (1971) also gives the measured characteristics of a high-performance achromatic polarizing beam splitter made with zirconium oxide and magnesium fluoride multilayers.

Although polarizing beam splitters are generally designed so that the *s* and *p* polarized beams emerge at right angles to each other, Schröder and Schlafer (1949) have an ingenious arrangement in which a half-wave plate and mirror are introduced into the path of the reflected beam to make it parallel to the transmitted beam and of the same polarization. Other optical schemes to accomplish the same purpose have been described in later papers (Schröder, 1956).

For some purposes it is desirable to have a beam splitter that is insensitive to the polarization of the incident beam. Baumeister (1961) has discussed the design of such beam splitters made from multilayer dielectric stacks of alternating low- and high-index materials. One of his designs is composed of six dielectric layers for which the extinction ratio T_s/T_p varies from 0.93 to 0.99 in a bandwidth of about 800 Å, with a $\pm 1°$ variation in the angle of incidence. In principle, any multilayer filter which is nonreflecting at normal incidence will be nonpolarizing at all angles of incidence (Baumeister, 1971). Costich (1970) has described filter designs for use in the near infrared which are relatively independent of polarization at 45° angle of incidence.

55. Frustrated-Total-Reflection Filters Frustrated-total-reflection (FTR) filters are not usually considered as polarization devices because their bandpasses are extremely narrow. However, the light transmitted by these filters is almost completely polarized and for this reason they could be used as narrow-band polarizers. The special type of FTR filter made with a metal film, also discussed in this section, has not yet proved useful even as a narrow-band transmission filter; further work is in order to demonstrate the effectiveness of this type of filter design.

In contrast to the polarizing beam splitters (Par. 54), which have a high-index film next to a lower-index prism and transmit throughout the transmission range of the prism material, FTR filters, which were invented by A. F. Turner (see Leurgans and Turner, 1947), are made with high-index glass prisms and alternating low-, high-, and low-index dielectric films sandwiched between them. The light is incident on the prism–low-index-film interface at an angle greater than the critical angle, so that if the film is very thick, total reflection occurs. As the film becomes thinner, the total reflection is frustrated and some of the incident energy penetrates into the middle high-index layer. By adjusting the thickness of the low-index film, any reflectance-to-transmittance ratio can be obtained, and since the film is transparent, no energy is lost. However, the phase change on reflection is different for the *s* and *p* components in this film, and so two narrow transmission peaks occur; the one at the shorter wavelength is polarized in the p direction, and the other is polarized in the s direction for an FTR filter composed of magnesium fluoride and zinc sulfide.

The maximum transmission at the peak wavelength is 100 percent for the appropriate type of polarization, and the half width is determined primarily by the thickness of the low-index layer (the thicker the layer the narrower the passband). The thickness of the high-index layer controls the wavelengths of the transmission peaks and secondarily their half widths. It is very critical since a change in optical thickness of only 10 Å will shift the peak wavelength by 20 Å in FTR filters composed of MgF_2 and ZnS (Turner, 1950). Sensitive methods have been developed for testing the optical uniformity of the spacer layers (Gee and Polster, 1949). Typical half widths of the s and p transmission bands obtained for FTR filters are in the 30- to 100-Å range, with most falling around 60 Å (Turner, 1950). The theory of FTR filter design has been given by Billings and Pittman (1949) and Bergstein and Shulman (1966), although the latter underestimate the effect of thickness variations on filter performance.

Frustrated-total-reflection filters have also been used in the infrared. Billings and Pittman (1949) made such a filter out of low-index sodium fluoride and high-index silver chloride films sandwiched between sodium chloride prisms. This filter had alternately polarized passbands at about 4.5 and 5.0 μm, which could be shifted to shorter wavelengths by tilting the filter. Unfortunately the filter had a very small field of view so that the apparent peak transmittances and half widths were a function of the convergence angle of the beam in the spectrophotometer used for the measurements. An FTR filter made of germanium prisms and sodium chloride and germanium films has been proposed for use at 10.6 μm (Bergstein, 1970). Filters have also been designed to operate in the 250-μm far-infrared region with a bandwidth of approximately 0.25 μm (Billings, 1949*a*). These would be made of polystyrene prisms with air for the low-index spacer layers and polystyrene for the high-index layer (Turner, 1950).

A modified FTR filter design has been proposed (Bergstein, 1971; Bergstein and DiSapato, 1971) in which the s and p passbands have been widened so that the filter will function effectively as a polarizer over an extended spectral region with incident beams of large divergence angle. An infrared filter of this design consisting of two germanium prisms and three thin-film layers has a theoretical extinction ratio of better than 1 part in 10^3 (Bergstein, 1971; Bergstein and DiSapato, 1971).

By replacing the high-index dielectric layer in the FTR filter with a silver film, one can theoretically obtain a spectral variation of the filter transmission. This effect can be explained as an excitation of plasma oscillations in the electron gas of the silver (or other metal) (Otto, 1969; Salwén and Stensland, 1970) or as a special case of induced transmission because of the particular optical constants of silver (Turner, 1970). Otto (1969) made a series of calculations of various filter designs containing thin layers of silver or aluminum (approximately 100 to 150 Å thick), low-index dielectrics such as LiF, NaF, or NaCl, and Suprasil quartz or As_2S_3 prisms. Salwén and Stensland (1970) made further calculations on some of Otto's filter designs and showed that in addition to the sharp peak expected for the p component there was a broader peak of the same polarization occurring at a longer wavelength. However, the filters constructed by Stensland showed only the broader peak. Possible improvement in evaporation conditions and better control of thickness, uniformity, and roughness of the dielectric and metal layers may result in filters which also have the sharp peak, in agreement with the calculations.

RETARDATION PLATES

56. Introduction and Theory A retardation plate is a piece of birefringent, uniaxial (or uniaxial-appearing) material in which the ordinary and extraordinary rays travel at different velocities. Thus, one ray is retarded relative to the other, and the path difference $N\lambda$ between the two rays is given by

$$N\lambda = \pm d(n_e - n_o) \tag{136}$$

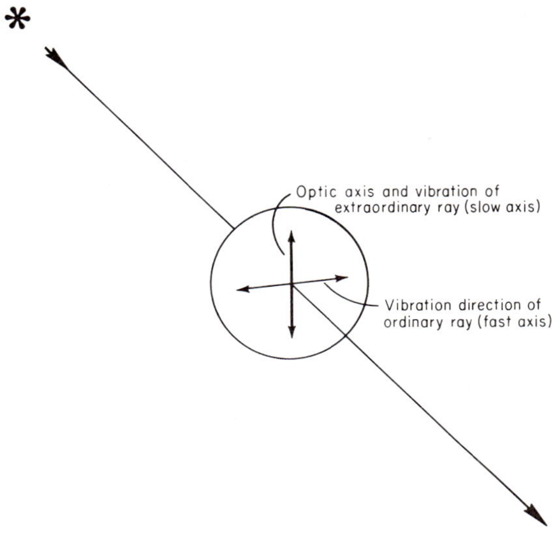

Fig. 57 Light incident normally on the front surface of a retardation plate showing the vibration directions of the ordinary and extraordinary rays. In a positive uniaxial crystal, the fast and slow axes are as indicated in parentheses; in a negative uniaxial crystal, the two axes are interchanged.

where n_o = refractive index of ordinary ray
n_e = refractive index of extraordinary ray
d = physical thickness of plate
λ = wavelength

The positive sign is used when $n_e > n_o$, that is, a positive uniaxial crystal, and the negative sign is used for a negative uniaxial crystal, for which $n_e < n_o$. Since $N\lambda$ is the path difference between the two rays, N can be considered the retardation expressed in fractions of a wavelength. For example, $N = \frac{1}{4}$ for a quarter-wave (or $\lambda/4$) plate, $\frac{1}{2}$ for a half-wave (or $\lambda/2$) plate, $\frac{3}{4}$ for a three-quarter-wave (or $3\lambda/4$) plate, etc.

The phase difference between two rays traveling through a birefringent material is $2\pi/\lambda$ times the path difference, so that the phase retardation δ is

$$\delta = 2\pi N = \pm \frac{2\pi d(n_e - n_o)}{\lambda} \tag{137}$$

Thus, phase differences of $\pi/2$, π, and $3\pi/2$ are introduced between the two beams in quarter-wave, half-wave, and three-quarter-wave plates, respectively.

A retardation plate can be made from a crystal which is cut so that the optic axis lies in a plane parallel to the face of the plate, as shown in Fig. 57. Consider a beam of unpolarized or plane-polarized light normally incident on the crystal. It can be resolved into two components traveling along the same path through the crystal but vibrating at right angles to each other. The ordinary ray vibrates in a direction at right angles to the optic axis, while the extraordinary ray vibrates in a direction parallel to the optic axis. In a positive uniaxial crystal $n_e > n_o$, so that the extraordinary ray travels more slowly than the ordinary ray. The fast axis is defined as the direction in which the faster-moving ray vibrates; thus in a positive uniaxial crystal, the fast axis (ordinary ray) is at right angles to the optic axis, while the slow axis (extraordinary ray) coincides with the optic axis. For a negative uniaxial crystal, the fast axis coincides with the optic axis.

Figure 58 shows how the state of polarization of a light wave changes after passing through retardation plates of various thicknesses when the incident light is plane-polarized at an azimuth of 45° to the fast axis of the plate. If the plate has a retardation of $\lambda/8$, which means that the ordinary and extraordinary

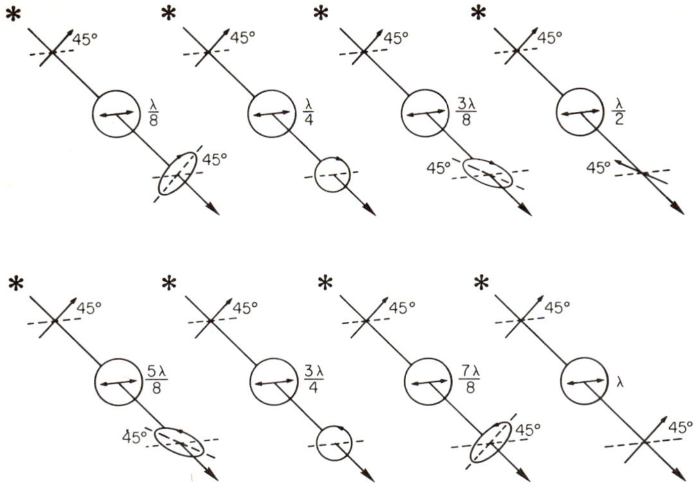

Fig. 58 State of polarization of a light wave after passing through a crystal plate whose retardation is indicated in fractions of a wavelength (phase retardation $2\pi/\lambda$ times these values) and whose fast axis is indicated by the double arrow. In all cases the incident light is plane-polarized at an azimuth of 45° to the direction of the fast axis.

waves are out of phase by $\pi/4$ with each other, the transmitted light will be elliptically polarized with the major axis of the ellipse coinciding with the axis of the original plane-polarized beam. As the retardation gradually increases (plate gets thicker for a given wavelength or wavelength gets shorter for a given plate thickness), the ellipse gradually turns into a circle, but its major axis remains at 45° to the fast axis of the retardation plate. For a retardation of $\lambda/4$, the emerging light is right circularly polarized. As the retardation continues to increase, the transmitted light becomes elliptically polarized with the major axis of the ellipse lying at right angles to the plane of the incident polarized beam, and then the minor axis of the ellipse shrinks to zero and plane-polarized light is produced when the retardation becomes $\lambda/2$. As the retardation increases further, the patterns change in opposite order and the polarized light is left circularly polarized when the retardation equals $3\lambda/4$. Finally, when the retardation is a full wave, the incident plane-polarized light is transmitted unchanged although the slow wave has now been retarded by a full wavelength relative to the fast wave.

The most common type of retardation plate is the quarter-wave plate. Figure 59 shows how this plate affects the state of polarization of light passing through it when the fast axis is positioned in the horizontal plane and the azimuth of the incident plane-polarized light is changed from $\theta = 0°$ to $\theta = 90°$. When $\theta = 0°$, only the ordinary ray (for a positive birefringent material) passes through the plate, so that the state of polarization of the beam is unchanged. When θ starts increasing, the transmitted beam is elliptically polarized with the major axis of the ellipse lying along the fast axis of the $\lambda/4$ plate; $\tan\theta = b/a$, the ratio of the minor to the major axis of the ellipse. In the next case, $\theta = 15°$ and $\tan\theta = 0.268$, and so the ellipse is long and narrow. When the plane of vibration has rotated to an azimuth of 45°, the emerging beam is right circularly polarized (the same situation as that shown in the second part of Fig. 58). For values of θ between 45 and 90°, the light is again elliptically polarized, this time with the major axis of the ellipse lying along the direction of the slow axis of the $\lambda/4$ plate. The angle shown in the figure is 60°, and $\tan 60° = 1.732$, so that b/a (referred to the fast axis) is greater than unity. When θ increases to 90°, the plane of vibration coincides with the slow axis and the transmitted light is again

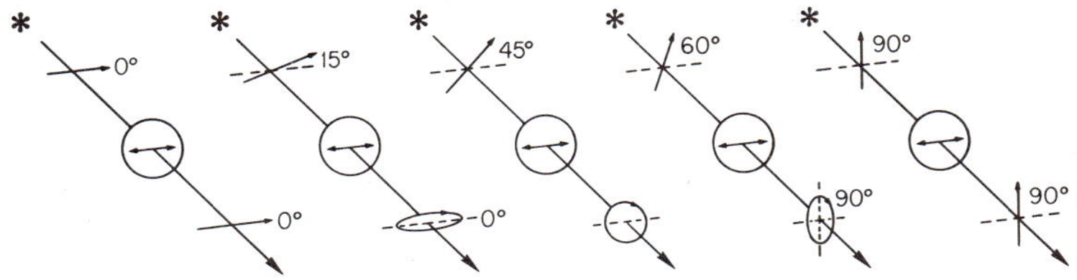

Fig. 59 State of polarization of a light wave after passing through a $\lambda/4$ plate (whose fast axis is indicated by the double arrow) for different azimuths of the incident plane-polarized beam.

plane-polarized. As θ continues to increase, the transmitted patterns repeat those already described and are symmetric about the slow axis, but the direction of rotation in the ellipse changes from right-handed to left-handed, so that left circularly polarized light is produced when $\theta = 135°$.

The definition of right and left circularly polarized light should be clear from Figs. 58 and 59. When the rotation is *clockwise* with the observer looking *opposite to the direction of propagation,* the light is called *right circularly polarized;* if the rotation is *counterclockwise,* the light is called *left circularly polarized* (Jenkins and White, 1957*a*). When circularly polarized light is reflected from a mirror, the direction of propagation is reversed, so that the sense of the circular polarization changes; i.e., left circularly polarized light changes on reflection into right circularly polarized light and vice versa. Therefore, in experiments involving magnetic fields in which the sense of the circularly polarized light is important (Palik, 1963; Richards and Smith, 1964) it is important to know which kind one started with and how many mirror reflections occurred in the rest of the light path. Cyclotron resonance experiments can sometimes be used to determine the sense of the circular polarization (Richards and Smith, 1964). Another method utilizing a polarizer and $\lambda/4$ plate has been described by Wood (1934*b*).

The behavior of a half-wave plate in a beam of plane-polarized light is completely different from that of a quarter-wave plate; the transmitted light is *always* plane-polarized. If the incident plane of vibration is at an azimuth θ with respect to the fast axis of the $\lambda/2$ plate, the transmitted beam will be rotated through an angle 2θ relative to the azimuth of the incident beam. The case showing $\theta = 45°$ where the phase of vibration is rotated through 90° is illustrated in the fourth part of Fig. 58. In this situation the extraordinary beam is retarded by half a wavelength relative to the ordinary beam (for a positive birefringent material), hence the name, half-wave plate. If the polarizer is fixed and the $\lambda/2$ plate is rotated (or vice versa), the plane of vibration of the transmitted beam will rotate at twice the frequency of rotation of the $\lambda/2$ plate.

Quarter-wave plates are useful for analyzing all kinds of polarized light. In addition, they are widely employed in experiments using polarized light, e.g., measurements of the thickness and refractive index of thin films by ellipsometry or measurements of optical rotary dispersion, circular dichroism, or strain birefringence. Polarizing microscopes, interference microscopes, and petrographic microscopes are usually equipped with $\lambda/4$ plates. In some applications the $\lambda/4$ plate is needed only to produce circularly polarized light, e.g., for optical pumping in some laser experiments, or to convert a partially polarized light source into one which appears unpolarized, i.e., has equal amplitudes of vibration in all azimuths. For these and similar applications, one can sometimes use a circular polarizer which does not have all the other properties of a $\lambda/4$ plate (Pars. 73 to 76).

The customary application for a $\lambda/2$ plate is to rotate the plane of polarization through an angle of 90°. Examples are $\lambda/2$ plates mounted in the condenser of one type of interference microscope and in a wide-angle compensator (Koester, 1959*a*). In other applications the angle of rotation is variable. One such device is the polarization rectifier designed for a polarizing microscope (Inoué and Hyde, 1957). In this case the light incident on the $\lambda/2$ plate has a plane of vibration whose azimuth depends on the position of the ray within the aperture, and the λ/2 plate gives each plane of vibration a rotation equal to twice the incident azimuth. Automatic setting ellipsometers or polarimeters sometimes employ rotating λ/2 plates in which the azimuth of the transmitted beam rotates at twice the frequency of the λ/2 plate.

TABLE 12 Refractive Indexes of Selected Materials at 5893 Å (Billings, 1963)

Material	n_o	n_e
Positive Uniaxial Crystals		
Ice, H_2O	1.309	1.313
Sellaite, MgF_2	1.378	1.390
Apophyllite, $2[KCa_4Si_8O_{20}(F,OH) \cdot 8H_2O]$	1.535±	1.537±
Crystalline quartz, SiO_2	1.544	1.553
Dioptase, $CuSiO_3 \cdot H_2O$	1.654	1.707
Zircon, $ZrSiO_4$	1.923±	1.968±
Rutile, TiO_2	2.616	2.903
Negative Uniaxial Crystals		
Beryl (emerald), $Be_3Al_2(SiO_3)_6$	1.581±	1.575±
Sodium nitrate, $NaNO_3$	1.584	1.336
Muscovite mica (complex silicate)	1.5977±	1.5936±
Apatite, $Ca_{10}(F,Cl)_2(PO_4)_6$	1.634	1.631
Calcite, $CaCO_3$	1.658	1.486
Tourmaline (complex silicate)	1.669±	1.638±
Sapphire, Al_2O_3	1.768	1.760

TABLE 13 Refractive Indexes of Two Mica Samples (Einsporn, 1936)

λ, μm	Sample 1			Sample 2		
	n_o	n_e	$n_e - n_o$	n_o	n_e	$n_e - n_o$
0.4358	1.61092	1.60637	−0.00455	1.60986	1.60487	−0.00499
0.4780	1.60592	1.60131	−0.00461	1.60495	1.59993	−0.00502
0.5461	1.60213	1.59734	−0.00479	1.60112	1.59613	−0.00499
0.5893	1.59978	1.59500	−0.00478	1.59885	1.59389	−0.00496
0.6600	1.59667	1.59184	−0.00483	1.59602	1.59092	−0.00510
0.7065	1.59505	1.59023	−0.00482	1.59451	1.58950	−0.00501
		Average	−0.00473		Average	−0.00501

TABLE 14 Birefringence of Apophyllite (Françon et al., 1965)

λ, μm	$n_e - n_o$	$\frac{n_e - n_o}{\lambda}$, μm^{-1}
0.4358	0.0018	0.0041
0.5086	0.0022	0.0043
0.5461	0.0024	0.0044
0.5893	0.0026	0.0044
0.6438	0.0028	0.0043

Retardation plates are generally made of mica, stretched polyvinyl alcohol, and quartz, although other stretched plastics such as cellophane, Mylar, cellulose acetate, cellulose nitrate, etc. (West and Makas, 1949), sapphire, magnesium fluoride, and other materials can also be used. Polyvinyl alcohol in sheet form transmits well

TABLE 15 Birefringence $n_e - n_o$ of Various Optical Materials[a]

Wave-length, μm	Rutile[b] TiO_2	CdSe[b]	Crystalline quartz[c-g]	MgF_2[c,d,h,i]	CdS[j-l]	Apophyllite[m]	ZnS[j] (Wurtzite)	Calcite[c]	$LiNbO_3$[n]	$BaTiO_3$[o]	ADP[p]	KDP[p]	Sapphire[d,h,q,r] (Al_2O_3)	Mica[s]
0.15	.....	.....	0.0214	0.0143										
0.20	.....	.....	0.0130	0.0134	.....	.....	.....	-0.326	....	.....	-0.0613	-0.0587	-0.0111	
0.25	.....	.....	0.0111	0.0128	.....	.....	.....	-0.234	....	.....	-0.0543	-0.0508	-0.0097	
0.30	.....	.....	0.0103	0.0125	.....	.....	.....	-0.206	....	.....	-0.0511	-0.0474	-0.0091	
0.35	.....	.....	0.0098	0.0122	.....	.....	.....	-0.193	....	.....	-0.0492	-0.0456	-0.0087	
0.40	.....	.....	0.0096	0.0121	.....	.....	0.004	-0.184	....	.....	-0.0482	-0.0442	-0.0085	
0.45	0.338	.....	0.00937	0.0120	.....	0.0019	0.004	-0.179	-0.1049	-0.097	-0.0473	-0.0432	-0.0083	-0.00457
0.50	0.313	.....	0.00925	0.0119	.....	0.0022	0.004	-0.176	-0.0998	-0.079	-0.0465	-0.0424	-0.0082	-0.00468
0.55	0.297	.....	0.00917	0.0118	0.014	0.0024	0.004	-0.173	-0.0947	-0.070	-0.0458	-0.0417	-0.0081	-0.00476
0.60	0.287	.....	0.00909	0.0118	0.018	0.0026	0.004	-0.172	-0.0919	-0.064	-0.0451	-0.0410	-0.0081	-0.00480
0.65	0.279	.....	0.00903	0.0117	0.018	0.0028	0.004	-0.170	-0.0898	.....	-0.0444	-0.0403	-0.0080	-0.00482
0.70	0.274	.....	0.00898	0.0117	0.018	.....	0.004	-0.169	-0.0882	.....	-0.0438	-0.0396	-0.0080	-0.00483
0.80	0.265	.....	0.0089	0.0116	0.018	.....	0.004	-0.167	-0.0857	.....	-0.0425	-0.0382	-0.0079	
0.90	0.262	0.0195	0.0088	0.0115	0.018	.....	0.004	-0.165	-0.0840	.....	-0.0411	-0.0367	-0.0079	
1.00	0.259	0.0195	0.0088	0.0114	0.018	.....	0.004	-0.164	-0.0827	.....	-0.0396	-0.0350		
1.10	0.256	0.0195	0.0087	0.0114	0.018	.....	0.004	-0.162	-0.0818	.....	-0.0379	-0.0332		
1.20	0.254	0.0195	0.0087	0.0114	0.017	.....	0.004	-0.161	-0.0810	.....	-0.0361	-0.0313		
1.30	0.252	0.0195	0.0086	0.0113	0.017	.....	0.004	-0.160	-0.0804	.....	-0.0342	-0.0292		
1.40	0.251	0.0195	0.0085	0.0113	0.017	.....	0.004	-0.158	-0.0798	.....	-0.0321	-0.0269		
1.50	0.250	0.0195	0.0085	0.0113	.....	.....	.....	-0.157	-0.0793	.....	-0.0298	-0.0245		
1.60	0.249	0.0195	0.0084	0.0112	.....	.....	.....	-0.156	-0.0788	.....	-0.0274	-0.0219		
1.70	0.248	0.0195	0.0084	0.0112	.....	.....	.....	-0.154	-0.0782	.....	-0.0248	-0.0191		
1.80	0.247	0.0195	0.0083	0.0112	.....	.....	.....	-0.153	-0.0777	.....	-0.0221	-0.0162		
1.90	0.246	0.0195	0.0082	0.0112	.....	.....	.....	-0.151	-0.0774	.....	-0.0192	-0.0130		
2.00	0.246	0.0195	0.0081	0.0111	.....	.....	.....	-0.150	-0.0771	.....	-0.0161	-0.0097		
2.10	0.245	0.0195	0.0081	0.0111	.....	.....	.....	-0.148	-0.0766					
2.20	0.244	0.0195	0.0080	0.0111	.....	.....	.....	.....	-0.0761					
2.30	0.243	0.0195	0.0079	0.0110	.....	.....	.....	.....	-0.0752					
2.40	0.241	0.0195	0.0078	0.0110	.....	.....	.....	.....	-0.0744					
2.50	.....	0.0195	0.0077	0.0110	.....	.....	.....	.....	-0.0739					
2.60	.....	0.0195	0.0076	0.0110	.....	.....	.....	.....	-0.0734					

[a] Calculated values at 24.8°C obtained from analytical expressions are given for crystalline quartz, MgF_2, calcite, ADP, and KDP by Beckers (1971).
[b] Bond (1965).
[c] Wolfe et al. (1963*b*).
[d] Chandrasekharan and Damany (1968).
[e] Shields and Ellis (1956).
[f] Ennos and Opperman (1966).
[g] Maillard (1971).
[h] Chandrasekharan and Damany (1969).
[i] Palik (1968).
[j] Bieniewski and Czyzak (1963).
[k] Palik and Henvis (1967).
[l] Gobrecht and Bartschat (1959).
[m] Françon et al. (1965).
[n] Boyd et al. (1967).
[o] Shumate (1966).
[p] Zernike (1964).
[q] Jeppeson (1958).
[r] Loewenstein (1961).
[s] Einsporn (1936).

into the ultraviolet beyond the cutoff for natural mica and is thus particularly useful for ultraviolet retardation plates (McDermott and Novick, 1961). Refractive indexes and birefringences of some materials are listed in Tables 12 to 15. The birefringences reported for mica and apophyllite should be considered as approximate, since they are measurements made on single samples. There is good reason to believe that the birefringence of apophyllite may be different for other samples (see Par. 66). Although calcite would seem at first to be a good material for retardation plates, its birefringence is so high that an extremely thin piece, less than 1 μm, would be required for a single $\lambda/4$ retardation plate. If a "first-order" or multiple-order plate were constructed (Pars. 61 and 63), or if calcite were used as one component of an achromatic retardation plate (Par. 66), the tolerance on the thickness would be very stringent.

Retardation plates are generally made of a single piece of material, although when the thickness required for a plate is too small, two thicker pieces may be used with the fast axis of one aligned parallel to the slow axis of the other to cancel out all but the desired retardation. Plates which are a little too thin or a little too thick may be rotated about an axis parallel or perpendicular to the optic axis to change the retardation to the desired amount (Gieszelmann et al., 1969; Daniels, 1967). There are also some novel circular polarizers and polarization rotators for use in the far ultraviolet (McIlrath, 1968), far infrared (Richards and Smith, 1964), and visible region (Lostis, 1957).

Achromatic retardation plates which have the same retardation over a range of wavelengths can be made from two or more different materials or from two or more plates of the same material whose axes are oriented at appropriate angles with respect to each other. These latter devices are known as composite plates (Pars. 73 to 76), and although they can change plane-polarized light into circularly polarized light, they do not have all the other properties of true retardation plates. By far the most achromatic $\lambda/4$ retarders are devices, such as the Fresnel rhomb, which obtain their retardation from internal reflections at angles greater than the critical angle.

Mica retardation plates are discussed in detail in Pars. 57 to 59, including the theory of multiple reflections; Pars. 60 to 65 are devoted to various types of crystalline-quartz retardation plates, and Par. 66 covers all achromatic retardation plates except those of the rhomb-type; the latter are described in Pars. 67 to 72. Various types of composite plates are described in Pars. 73 to 76, while unusual retardation plates are presented in Par. 77.

Methods for making and testing quarter-wave plates will be discussed in detail in Pars. 78 to 82. These include ways of splitting mica, how to distinguish between fast and slow axes, methods for measuring retardations close to $\lambda/4$, and the tolerance on plate thickness.

Mica Retardation Plates

57. Mica quarter-wave plates are perhaps the most common type in use today. They can be made by splitting thick sheets of mica down to the appropriate thickness (Par. 79). Since the difference between the velocities of the ordinary and extraordinary rays is very small, the mica sheets need not be split too thin; typical thicknesses lie in the range 0.032 to 0.036 mm for yellow light. The fast and slow axes of a mica quarter-wave plate can be distinguished using Tutton's test (Par. 80), and the retardation can be measured using one of several rather simple tests (Par. 81).

If the mica sheets are used without glass cover plates, multiply reflected beams in the mica can cause the retardation to oscillate around the value calculated from the simple theory (Par. 59). Fortunately this effect can be eliminated in one of the ways described in Par. 59. Mica does have one serious drawback. There are zones in the cleaved mica sheets which lie at angles to each other and which do not extinguish at the same angle (Smith, 1969). Thus extinction cannot be obtained over the whole sheet simultaneously. In very critical applications such as ellipsometry, much better extinction can be obtained using quarter-wave plates made of crystalline quartz (Pars. 60 to 65), which do not exhibit this effect.

58. Properties of Mica Mica is a complex sheet silicate and occurs in many forms (Deer et al., 1962*a*). It is actually a negative biaxial crystal; the angle between the optic axes may be almost anything between 0 and 42° (Jenkins and White, 1957*d*). Muscovite, the most common form of mica (pale brown), has an angle of 42° between the optic axes. When viewed externally, this angle looks much larger because of refraction. The refractive indexes vary from sample to sample, but representative values for sodium light (5893 Å) are $n_\alpha = 1.5601$, $n_\beta = 1.5936$, and $n_\gamma = 1.5977$ (Jenkins and White, 1957*d*). The cleavage planes along which mica splits are not parallel to either optic axis but along crystallographic axes such that $n_o = n_\gamma$ and $n_e = n_\beta$. Values of n_o and n_e in the visible region are given in Table 13 for two mica samples, and smoothed values of $n_e - n_o$ for sample 1 are listed in Table 15. Since these latter values are for a single sample, they are certainly not representative of the birefringence of all muscovite mica samples. Bradford et al. (1967) have measured the average of n_e and n_o for mica and give a dispersion curve for the wavelength region from 0.35 to 5.5 μm. Thin mica sheets (< 30 μm thick) have a very high transmittance from the ultraviolet to the near infrared, while the transmittance of thicker samples is influenced by impurities which vary from sample to sample. Bradford et al. (1967) found that the limit at which the transmittance decreased to less than 10 percent in the ultraviolet was approximately 2800 Å for a 10.9-μm thick sample and approximately 2100 Å for a 4.7-μm-thick sample.† There is a strong but narrow water absorption band in the near infrared at about 2.7 μm. Otherwise the material transmits out to about 8 μm. There are also regions farther in the infrared (out to 40 μm) in which thin sheets of mica exhibit rather high transmittance, making them suitable for various types of spectroscopic studies. Some mica sheets are pleochroic; i.e., they exhibit different absorption for light vibrating along the fast and slow axes (Randall, 1954; Archer and Shank, 1967). If present, pleochroism can affect the performance of mica retardation plates since all theories assume that absorption is either zero or the same along the two axes.

59. Multiple Reflections in Mica The thinness of mica retardation plates is a disadvantage in that the multiply reflected beams in the mica are coherent and can interfere with each other. The effect of this interference is to produce oscillations in the phase retardation and in the transmittance, as shown in Fig. 60. The mica sheet is assumed to have exactly one-quarter-wave retardation at 5893 Å, as calculated from the simple theory [Eq. (136)], which assumes that the light passes through the plate only once. Although the simple theory predicts that the retardation should change slowly (dashed line in Fig. 60*a*) and the transmittance should be nearly constant, the complete theory [see Eq. (141)] predicts oscillations in both the phase retardation and transmittance. The magnitude of the oscillations is determined by the average refractive index of the mica; the separation between adjacent maxima is a function of the total optical thickness.

Oscillations in the retardation have been observed by Smith (1969), and Weinberger and Harris (1964) detected the oscillations in the transmitted intensity. Yolken et al. (1967) measured the intensity variations in their mica retardation plate and found them to be reasonably close to those given in Fig. 60*b*. Their measured separation between adjacent maxima was approximately correct for a mica plate which was a quarter-wave retarder at 5461 Å.

If an uncoated mica λ/4 plate is used with a narrow-bandwidth source, the phase retardation can be expected to lie somewhere between 84 and 96° (Fig. 60*a*) but not necessarily be 90°. The ratio of the light intensities transmitted along the fast and slow axes will lie somewhere between 1.00/0.81 = 1.23 and 0.81/1.00 = 0.81 (Fig. 60*b*); both phase and intensity variations must be taken into account in the calculations or else be eliminated. One way of eliminating multiple reflections is to coat the mica surfaces with antireflection films (Cotton and Manigault, 1944). By using a λ/4 (optical thickness) coating of SiO_2 and a λ/4 coating of MgF_2 on the SiO_2, Yolken et al. (1967) were able to eliminate the multiple reflections completely. An alternate solution is to embed the mica in canada balsam on an annealed cover glass (Jerrard, 1952). Canada balsam has

† Chandrasekharan and Damany (1966) reported that a 4-μm-thick piece of synthetic mica has a limit of transparency around 1600 Å. However, this material cleaves with the optic axis perpendicular to the cleavage plane so that it is not birefringent and thus cannot be used for retardation plates.

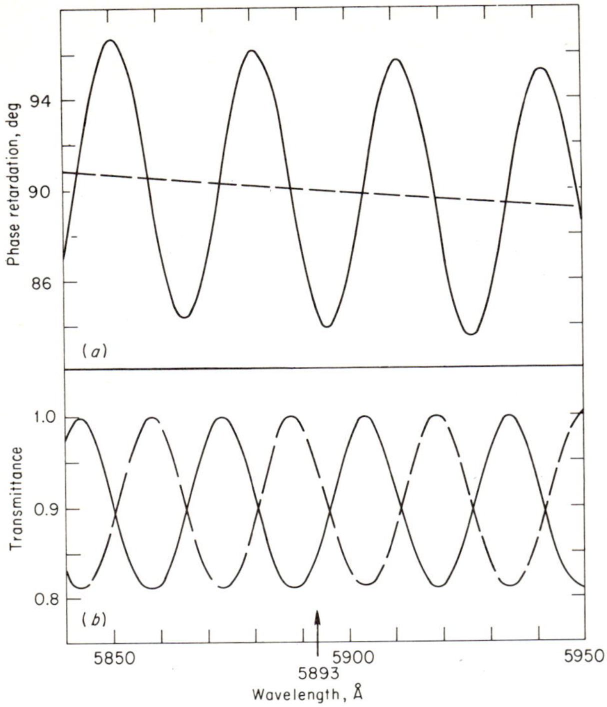

Fig. 60 (*a*) Phase retardation for a mica $\lambda/4$ plate calculated from the simple theory, Eq. (136) (*dashed curve*) and from the complete theory, Eq. (141), which assumes multiple reflections within the mica (*solid curve*). The plate has exactly 90° retardation at 5893 Å according to the simple theory. (*b*) Calculated transmittances for the ordinary (*dashed curve*) and extraordinary rays (*solid curve*) assuming multiple reflections in the mica [see Eq. (138) and text].

an index of about 1.54, intermediate between the index of mica and glass, so that the reflection at the mica-glass interface is effectively eliminated. The method used in making commercial mica quarter-wave plates is similar. The mica sheet is encased in pieces of glass using a cement having a refractive index intermediate between the indexes of mica and glass; reflections then occur only at the outer air-glass interfaces. Since the thickness of the sandwich is large (4.83 mm for one such plate), the coherence properties of the multiply reflected beams are generally lost and the transmission assumes its average value. Even if the plates were accurately flat and plane-parallel (wedge angles are frequently purposely introduced), the separation of adjacent intensity maxima would be only about 0.2 Å, too small to be resolved with an ordinary spectrometer.

The exact theory of multiple coherent reflections in a retardation plate has been given by Holmes (1964*a*) and will be briefly sketched here. The retardation plate acts like a plane-parallel-plate (Fabry-Perot) interferometer, and the light intensity I transmitted along either the fast or the slow axis is given by the well-known Airy formula [Eq. (25)], which can be written

$$I = \frac{1}{1 - (1 - 1/T^2)\sin^2(2\pi nd/\lambda)} \tag{138}$$

for a nonabsorbing material. The transmission T of the plate excluding interference effects is equal to $2n/(n^2 + 1)$. The refractive index n can be either n_e or n_o, but since they are nearly equal for mica, an average value can be used. The transmitted intensity varies from a maximum of 1 to a minimum of T^2. The transmission along the fast and slow axes will have the same range of values, but for a $\lambda/4$ plate the transmissions will be exactly out of phase with each other. This can be shown from the quantity $2\pi n_e d/\lambda - 2\pi n_o d/\lambda$, which, from Eq. (137), equals $\pi/2$ for a $\lambda/4$ plate. When $2\pi n_e d/\lambda$ equals $M\pi$ (M a positive integer), the $\sin^2$ term in Eq. (138) becomes zero and I takes on its maximum value of 1. For this case the quantity $2\pi n_o d/\lambda$ differs by $\pi/2$, so that the transmission is a minimum for the ordinary ray.

The oscillation of the phase retardation is not so easily explained but comes out of the interference equations. The phase change on transmission γ through the plate is obtained from the ratio of the transmitted electric vector to the incident electric vector and is given by the relation (Heavens, 1955)

$$\tan\gamma = -\frac{1}{T}\tan\frac{2\pi nd}{\lambda} \tag{139}$$

where the symbols have the same meanings as before. [Holmes (1964*a*) uses a positive sign in this equation, but the physical result is independent of sign.] We are interested in the phase retardation δ, which equals the difference between γ_e and γ_o (the phase changes for the extraordinary and ordinary rays). This difference is obtained from the trigonometric relation

$$\tan(a-b) = \frac{\tan a - \tan b}{1+\tan a \tan b} \tag{140}$$

Thus,

$$\tan\delta \equiv \tan(\gamma_e - \gamma_o) = \frac{-(1/T_e)\tan(2\pi n_e d/\lambda) + (1/T_o)\tan(2\pi n_o d/\lambda)}{1+(1/T_e T_o)\tan(2\pi n_e d/\lambda)\tan(2\pi n_o d/\lambda)} \tag{141}$$

It is clear that this function oscillates with a change in wavelength, but the exact nature of the oscillation is difficult to recognize.

For an exact $\lambda/4$ plate, $\tan(2\pi n_o d/\lambda) = -\cot(2\pi n_e d/\lambda)$. Also, $T_e \approx T_o \approx T$, and $n_e \approx n_o \approx n$, so that Eq. (141) can be simplified to

$$\tan\delta \equiv \tan(\gamma_e - \gamma_o) = \frac{T}{1-T^2}\left(\tan\frac{2\pi nd}{\lambda} + \cot\frac{2\pi nd}{\lambda}\right) \tag{142}$$

Whenever $2\pi nd/\lambda$ equals $M\pi/2$, δ will equal $\pi/2$. However, when $2\pi nd/\lambda = (\pi/2)(M+\frac{1}{2})$, $\tan(2\pi nd/\lambda) = \cot(2\pi nd/\lambda) = \pm 1$, so that $\tan\delta = \pm 2T(1-T^2)$. For these angles, δ will deviate from $\pi/2$ by an amount which depends on the value of T, which in turn depends on the refractive index of the material.

In Fig. 60*a*, the solid curve gives the phase retardation as a function of wavelength for a mica $\lambda/4$ plate calculated using Eq. (141). Note that the wavelength 5893 Å where $\delta = \pi/2$ from Eq. (136) does not correspond to a maximum, minimum, or inflection point. This can be explained as follows. The plate thickness d was calculated from Eq. (136) letting $n_e - n_o = -0.0041$, $\lambda = 5893$ Å, and $N = \frac{1}{4}$; thus $d = 0.0359329$ mm (keeping the digits for computational purposes only), making $2\pi n_e d/\lambda = 194.34\pi$ and $2\pi n_o d/\lambda = 194.84\pi$ since $n_e = 1.5936$ and $n_o = 1.5977$. Although the fractional parts of these numbers are not simple, they do differ from each other by 0.50π. Also, $n_e d/\lambda - n_o d/\lambda = 0.25$, as it should since d was determined from Eq. (136) by assuming this relationship.

The intensity maxima for the extraordinary ray (solid curve in Fig. 60*b*) occur for values of λ such that $2\pi n_e d/\lambda = 193.00\pi$, 194.00π, 195.00π, and 196.00π, and the intensity minima for 192.50π, 193.50π, 194.50π, and 195.50π. The phase-retardation maxima are displaced by $\pi/4$ from the intensity maxima, a relationship which holds as long as the retardation remains near $\lambda/4$.

The separation between adjacent maxima in either phase retardation or transmittance can be obtained from the relation

$$\frac{2\pi nd}{\lambda_1} - \frac{2\pi nd}{\lambda_2} = \pi \tag{143}$$

Rearranging terms we obtain

$$\lambda_2 - \lambda_1 = \frac{\lambda_1\lambda_2}{2nd} \simeq \frac{\lambda^2}{2nd} \tag{144}$$

For this example, $\lambda_2 - \lambda_1 = 30.3$ Å if $\lambda = 5893$ Å, $n = 1.5956$, and $d = 0.0359329$ mm.

Crystalline-Quartz Retardation Plates

60. Crystalline quartz is also frequently used for retardation plates, particularly those of the highest quality. It escapes the problem of zones with different orientations like those found in mica (Pars. 57 to 59). The thickness of quartz required for a single quarter-wave retardation at the 6328 Å helium-neon laser line is about 0.017 mm, much too thin for convenient polishing. If the plate is to be used in the infrared, single-order quarter-wave plates are feasible (Par. 62). Two types of quartz retardation plates are generally employed in the visible and ultraviolet regions: so-called "first-order" plates made of two pieces of material (Par. 61), which are the best for critical applications, and multiple-order plates made of one thick piece of crystalline quartz (Pars. 63 to 65). The multiple-order plates are generally not used for work of the highest accuracy since they are extremely sensitive to small temperature changes (Par. 64) and to angle of incidence (Par. 65). Also they have $\lambda/4$ retardation only at certain wavelengths; at other wavelengths the retardation may not even be close to $\lambda/4$.

When using any of the different types of retardation plates at a single wavelength, the methods for measuring the retardation given in Par. 81 and for distinguishing between fast and slow axes given in Par. 80 can be used.

61. "First-Order" Plates A so-called "first-order" plate is made by cementing together two nearly equal thicknesses of quartz such that the fast axis of one is aligned parallel to the slow axis of the other (both axes lie in planes parallel to the polished faces). The plate is then polished until the difference in thickness between the two pieces equals the thickness of a single $\lambda/4$ plate. The retardation of this plate can be calculated from Eq. (136) by setting d equal to the *difference in thickness* between the two pieces. The "first-order" plate acts strictly like a single-order quarter-wave plate with respect to the variation of retardation with wavelength, temperature coefficient of retardation, and angle of incidence.

The change in phase retardation with temperature at 6328 Å, as calculated from Eq. (147) or (148), is 0.0091°/°C, less than one-hundredth that of the 1.973-mm multiple-order plate discussed in Par. 64. The change in retardation with angle of incidence† at this wavelength is also small: $(\Delta N)_{10^\circ} = 0.0016$, as compared with 0.18 for the thick plate (see Par. 65).

A "first-order" quartz $\lambda/4$ plate has several advantages over a mica $\lambda/4$ plate. (1) Crystalline quartz has a uniform structure, so that extinction can be obtained over the entire area of the plate at a given angular setting. (2) Since the total plate thickness is generally large, of the order of 1 mm or so, the coherence of the multiple, internally reflected, beams is lost and there are no oscillations in the transmitted light or in the phase retardation (see Par. 59). (3) Crystalline quartz is not pleochroic, except in the infrared, so that the intensity transmitted along the two axes is the same. (4) Crystalline quartz transmits farther into the ultraviolet than mica, so that "first-order" plates can be used from about 0.185 to 2.0 μm (see Table 15).

Bruhat and Weil (1936) describe a somewhat different type of "first-order" plate made of a 0.1077-mm-thick piece of right-handed quartz and a 0.1013-mm-thick piece of left-handed quartz, each cut parallel to the optic axis. The pieces are placed so that their axes are opposed and are inclined to each other at an angle of about 1° to eliminate multiple reflections. The authors show that the right- and left-handed optical rotations cancel, making the resultant combination act like a bire-

† Grechushnikov (1962) has an incorrect relation for the change in phase retardation with angle of incidence [his eq. (2)]. He assumed that the retardations in the two halves of the plate add rather than subtract, yielding a retardation comparable to that of a thick quartz plate.

fringent plate devoid of rotary power. In the ultraviolet region between 2537 and 3130 Å the measured retardation varies between 98°47′ and 73°40′, in excellent agreement with the calculated values.

62. Single-Order Plates in the Infrared Although a crystalline-quartz retardation plate which is $\lambda/4$ in the visible is too thin to make from a single piece of material, the thickness required for such a plate is larger in the infrared. Jacobs and coworkers (see Gieszelmann et al., 1969) describe such a $\lambda/4$ plate for use at the 3.39-μm helium-neon laser line. They measured the birefringence of quartz at this wavelength and found it to be 0.0065 ± 0.0001, so that the thickness required for the plate was 0.1304 mm. The actual plate was slightly thinner (0.1278 mm), so that it was tipped at an angle of 10° (rotating it about an axis parallel to the optic axis) to give it exactly $\lambda/4$ retardation (see Par. 65). Maillard (1971) has also measured the birefringence of quartz at 3.391 and 3.508 μm and obtained values of 0.00659 and 0.00642, respectively (both ± 0.00002), in agreement with Jacobs' value. These data lie on a smooth curve extrapolated from the values of Shields and Ellis (1956).

A problem encountered when using crystalline quartz in the infrared is that, in general, the ordinary and extraordinary rays have different absorption coefficients; thus it may be impossible to construct a perfect wave plate regardless of the relative retardation between the rays. For an absorbing wave plate to have a retardation of exactly $\lambda/4$, the requirement

$$\left(\frac{n_o + 1}{n_e + 1}\right)^2 \exp\left[-\frac{(\alpha_e - \alpha_o)\lambda}{8(n_e - n_o)}\right] = 1 \tag{145}$$

must be met (Gieszelmann et al., 1969); α_e and α_o are the absorption coefficients for the extraordinary and ordinary rays, respectively. At wavelengths shorter than 3.39 μm, the birefringence is small enough for it to be possible to approximate this condition closely whenever $\alpha_e \approx \alpha_o$. [Values of these quantities are given by Drummond (1936).] Unfortunately, the birefringence for $\lambda > 3.51$ μm has not yet been measured, so that wave plates cannot be precisely designed for this region yet.

Another problem which occurs for crystalline quartz (and also for other materials) in the infrared is that the Fresnel reflection coefficients are slightly different for the ordinary and extraordinary rays since the refractive indexes and absorption coefficients are in general different. One possible solution is to deposit isotropic thin films on the crystal surfaces. The refractive index of these films is chosen to balance the anisotropic absorption effect by making the Fresnel reflection coefficients appropriately anisotropic. On the other hand, if anisotropic Fresnel reflection proves to be undesirable, it can be greatly diminished by using an antireflection coating (Gieszelmann et al., 1969).

If a single-order, crystalline-quartz wave plate is to be used for a continuous range of wavelengths, both the phase retardation and the transmittance of the ordinary and extraordinary rays will oscillate as a function of wavelength in the manner described in Par. 59 because of multiple coherent reflections in the quartz. The separation between adjacent maxima in the phase retardation can be calculated from Eq. (144). Using $\lambda = 3.3913$ μm, $n \approx 1.4881$, and $d = 127.8$ μm, $\Delta\lambda = 0.03024$ μm, an amount which should be well resolved with most infrared instruments. Thus, if a wave plate is to be used over a range of wavelengths, it would be well to antireflect the surfaces to eliminate the phase oscillations.

63. Multiple-Order Plates Thick plates made from crystalline quartz are sometimes used to produce circularly polarized light at a single wavelength or at a discrete series of wavelengths. The plate thickness is generally of the order of one or more millimeters so that the retardation is an integral number of wavelengths plus $\lambda/4$, hence the name multiple-order wave plate. This plate acts like a single $\lambda/4$ plate providing it is used only at certain specific wavelengths; at other wavelengths it may not even approximate the desired retardation. For example, a 1.973-mm-thick quartz plate was purchased which had an order of interference $N = 28.25$ at 6328 Å. From Eq. (136) and Table 15, this plate would have $N = 30.52$ at 5890 Å, and would thus be an almost perfect half-wave plate at this latter wavelength.

If a multiple-order plate is used to produce circularly polarized light at unspecified discrete wavelengths, e.g., to measure circular or linear dichroism, it can be placed following a polarizer and oriented at 45° to the plane of vibration of the polarized beam. When the wavelengths are such that N in Eq. (136) equals $\frac{1}{4}$, $\frac{3}{4}$, or in general $(2M - 1)/4$ (where M is a positive integer), the emerging beam will be alternately right and left circularly polarized. The frequency interval $\Delta\nu$ between wavelengths at which circular polarization occurs is

$$\Delta\nu = \frac{1}{2d(n_e - n_o)} \tag{146}$$

where $\nu = 1/\lambda$. If the birefringence is independent of wavelength, the retardation plate will thus produce circularly polarized light at *equal intervals* on a frequency scale and can conveniently be used to measure circular dichroism (Holzwarth, 1965).

In order to approximately calibrate a multiple-order retardation plate at a series of wavelengths, it can be inserted between crossed polarizers and oriented at 45° to the polarizer axis. Transmission maxima will occur when the plate retardation is $\lambda/2$ or an odd multiple thereof; minima will occur when the retardation is a full wave or multiple thereof. If the axes of the two polarizers are parallel, maxima in the transmitted beam will occur when the plate retardation is a multiple of a full wavelength. The birefringence of the retardation plate can be determined by measuring the wavelengths at which maxima or minima occur if the plate thickness is known. Otherwise d can be measured with a micrometer, and an approximate value of $n_e - n_o$ can be obtained.

Palik (1963) made and tested a 2.070-mm-thick CdS plate for the 2- to 15-μm infrared region and also made thick retardation plates of SnSe, sapphire, and crystalline quartz to be used in various parts of the infrared. Holzwarth (1965) used a cultured-quartz retardation plate 0.8 mm thick to measure circular dichroism in the 1850- to 2500-Å region of the ultraviolet; Jaffe et al. (1967) measured linear dichroism in the ultraviolet using a thick quartz plate and linear polarizer.

64. Sensitivity to Temperature Changes Small temperature changes can have a large effect on the retardation of a multiple-order plate. We can calculate this effect as follows. From Eq. (136) the change in path difference $(\Delta N)\lambda$ caused by a change in temperature is

$$(\Delta N)\lambda = \pm[\Delta d(n_e - n_o) + d(\Delta n_e - \Delta n_o)] \tag{147}$$

where Δd, Δn_e, and Δn_o are changes in the thickness and refractive indexes with temperature at a given wavelength. (The second-order terms in the equation have been neglected.) We can write the equation in terms of the coefficient of linear expansion α (perpendicular to the optic axis), where $\alpha = \Delta d/d$, solve for d from Eq. (136), and obtain an expression for ΔN in terms of N:

$$\Delta N = N\left(\alpha + \frac{\Delta n_e - \Delta n_o}{n_e - n_o}\right) \tag{148}$$

If the change in phase retardation is desired in radians or in degrees, Eq. (148) should be multiplied by 2π or 360°, respectively.

For crystalline quartz, the coefficient of linear expansion α perpendicular to the optic axis is 13.37×10^{-6} per Celsius degree ("Handbook of Chemistry and Physics," 1955), $\Delta n_e = -0.651 \times 10^{-5}$ per Celsius degree, $\Delta n_o = -0.547 \times 10^{-5}$ per Celsius degree, and $n_e - n_o = 0.00906$ at 6328 Å (Wolfe et al., 1963*b*), so that Eq. (148) becomes

$$\Delta N = -1.014 \times 10^{-4} N/°\text{C} \tag{149}$$

and

$$360°\,\Delta N = -0.0365 N°/°\text{C} \tag{150}$$

Thus, for the 1.973-mm-thick quartz plate ($N = 28.25$ at 6328 Å), the phase retardation will decrease 1.03° for each Celsius degree increase in temperature. If the temperature of the wave plate is not controlled extremely accurately, the large temperature coefficient of retardation can introduce sizable errors in precise ellipsometric measurements in which polarizer and analyzer settings can be made to ±0.01° (see Par. 6). When temperature drifts occur, analyzer readings taken at different times cannot be reduced to give a correct value of ψ.

65. Sensitivity to Angle of Incidence The effect of angle of incidence (and hence field angle) on the retardation can also be determined. If the plate is rotated about an axis parallel to the optic axis, both ordinary and extraordinary rays travel at the same speeds as at normal incidence but they must pass through a greater thickness of material. If the angle of incidence θ is small, the angle of refraction in the quartz is, from Snell's law, approximately θ/n, where n is the geometric mean refractive index, $(n_e n_o)^{1/2}$.† The path length of both beams in the quartz at nonnormal incidence is approximately $d/[\cos(\theta/n)]$, so that Eq. (136) becomes

$$(N\lambda)_\theta = \pm \frac{d}{\cos(\theta/n)}(n_e - n_o) \approx \pm d(n_e - n_o)\left(1 + \frac{\theta^2}{2n^2}\right) \tag{151}$$

where θ is measured in radians. At normal incidence, the expression for $N\lambda$ is given by Eq. (136), so that ΔN, the change in N with θ, is

$$(\Delta N)_\theta = \frac{N\theta^2}{2n^2} \tag{152}$$

Thus the change in phase retardation with angle of incidence, $2\pi(\Delta N)_\theta$, is proportional to the total thickness of the plate (which is incorporated in N) and the square of the angle of incidence when the rotation is about an axis parallel to the optic axis.

We now calculate the effect of a 10° angle of incidence (rotation parallel to the optic axis) at a wavelength of 6328 Å for the 1.973-mm-thick plate mentioned previously. At normal incidence $N = 28.25$; $\theta = 0.174$ rad, and $n_{av} = 1.5475$ (Wolfe et al., 1963*b*), making $(\Delta N)_{10°} = 0.18$. This means that the total retardation has changed from 28.25 to 28.43, so that the $\lambda/4$ plate is now nearly a $\lambda/2$ plate.

If the plate had been rotated about an axis *perpendicular* to the direction of the optic axis, in the limit when the angle of incidence is 90°, the beam would have been traveling along the optic axis; in this case the ordinary and extraordinary rays would be traveling with the same velocities, and there would have been *no retardation* of one ray relative to the other. For any intermediate angle of incidence the retardation would have been *less than* the value at normal incidence. The relation for the retardation as a function of angle of incidence is not simple, but the retardation will be approximately as angle-sensitive as it was in the other case. An advantage of rotation about either axis is that, with care, one can adjust the retardation of an inexact wave plate to a desired value. Rotation about an axis *parallel* to the optic axis will *increase* the retardation, while rotation about an axis *perpendicular* to the optic axis will *decrease* the retardation.

Achromatic Retardation Plates

66. Achromatic retardation plates are those for which the phase retardation is independent of wavelength. The name arose because when a plate of this type is placed between polarizers, it does not appear colored and hence is achromatic (Gaudefroy, 1929). In many applications a truly achromatic retardation plate is not required. Since the wavelength of light changes by less than a factor of 2 across the visible region, a quarter- or half-wave mica plate often introduces only tolerable errors even in white light. The errors that do occur cancel out in many kinds of experiments. For instance, photoelectric experiments carried out in white light often produce the same result that pure circular polarized light would have produced, because the ellipticity at the blue end of the spectrum is opposite in sign to that at the red end of the spectrum and thus the transmittance is independent of azimuth. This result is not automatic; one must adjust the quarter-

† Actually the angles of refraction are strictly θ/n_o and θ/n_e for the ordinary and extraordinary rays in this case, and one should also take into account the path length the extraordinary ray travels in air after leaving the plate to catch up with the ordinary ray. However, for small angles these corrections are negligible. See also Gieszelmann et al. (1969).

wave plate to some "center of gravity" of the spectral distribution, but when this is done, white light can be used (Hyde, 1970).

Achromatic retardation plates can be made in various ways. The most achromatic are based on the principle of the Fresnel rhomb, in which the phase retardation occurs when light undergoes two or more total internal reflections (Pars. 67 to 72). A material with the appropriate variation of birefringence with wavelength can also be used. Such materials are uncommon, but plates of two or more different birefringent materials can be combined to produce a reasonably achromatic combination. Composite plates, consisting of two or more plates of the same material whose axes are oriented at the appropriate angles, can be used as achromatic circular polarizers (Par. 75) or achromatic polarization rotators (Par. 76) although they do not have all the properties of true λ/4 or λ/2 plates. One unusual achromatic half-wave plate is described in Par. 77.

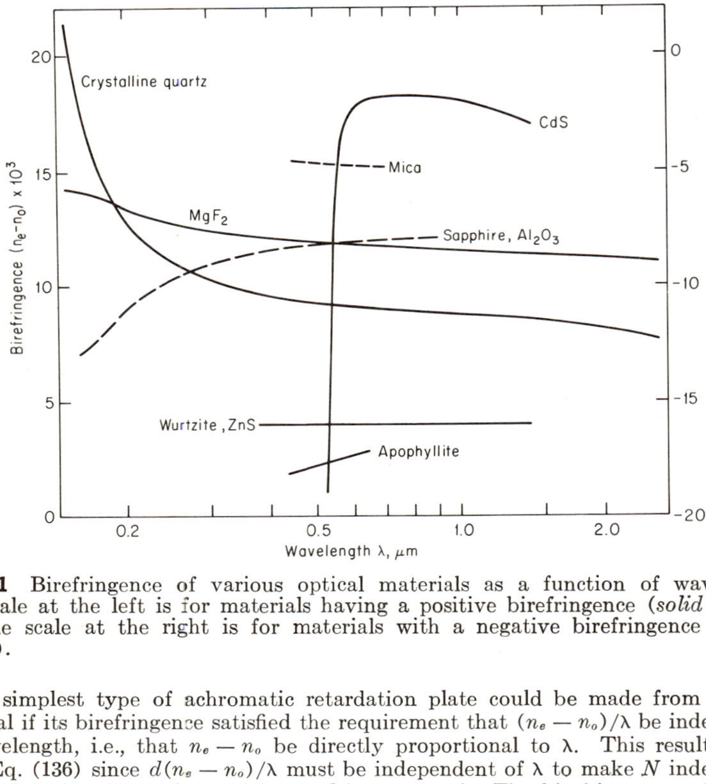

Fig. 61 Birefringence of various optical materials as a function of wavelength. The scale at the left is for materials having a positive birefringence (*solid curves*), and the scale at the right is for materials with a negative birefringence (*dashed curves*).

The simplest type of achromatic retardation plate could be made from a single material if its birefringence satisfied the requirement that $(n_e - n_o)/\lambda$ be independent of wavelength, i.e., that $n_e - n_o$ be directly proportional to λ. This result follows from Eq. (136) since $d(n_e - n_o)/\lambda$ must be independent of λ to make N independent of wavelength. (The plate thickness d is constant.) The birefringences of various materials are listed in Table 15 and plotted in Figs. 61 and 62. Only one material, the mineral apophyllite, has a birefringence which increases in the correct manner with increasing wavelength† (Fig. 61 and Table 14) (Françon et al., 1965). A curve of the phase retardation vs. wavelength for a quarter-wave apophyllite plate is shown as curve D in Fig. 63. Also included are curves for other so-called achromatic λ/4 plates as well as for simple λ/4 plates of quartz and mica. The phase retardation of apophyllite is not as constant with λ as that of the rhomb-type retarders, but it is considerably more constant than that of the other "achromatic" λ/4 plates. Since the birefringence of apophyllite is small, a λ/4 plate needs a thickness of about

† For materials having a negative birefringence the requirement is that $-(n_e - n_o)$ be proportional to λ.

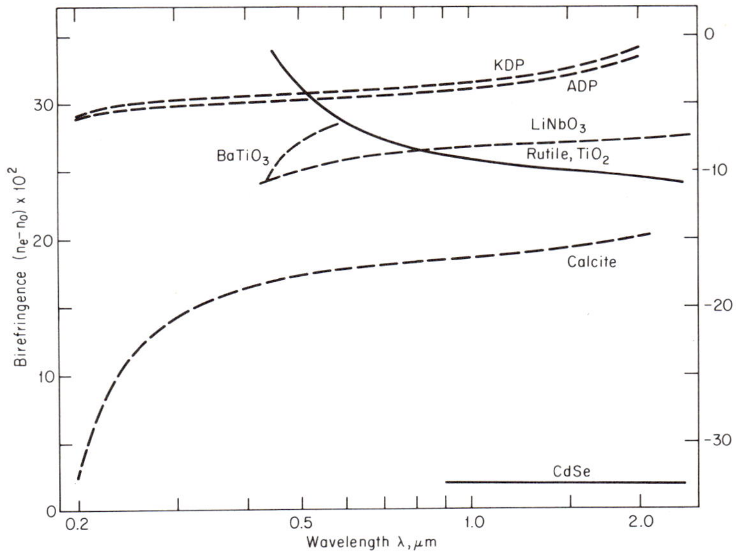

Fig. 62 Birefringence of various optical materials which have larger birefringences than those shown in Fig. 61. The scale at the left is for materials having a positive birefringence (*solid curves*), and the scale at the right is for materials with a negative birefringence (*dashed curves*).

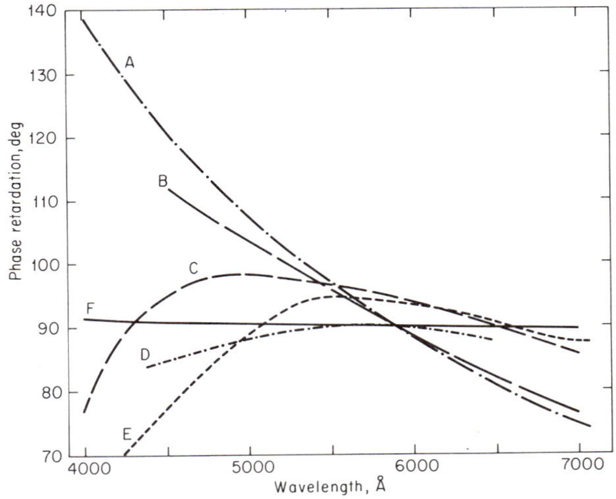

Fig. 63 Curves of the phase retardation vs. wavelength for λ/4 plates; *A*, quartz; *B*, mica; *C*, stretched plastic film; *D*, apophyllite; and *E*, quartz-calcite achromatic combination. Curve *F* is for a Fresnel rhomb but is representative of all the rhomb-type devices. [*Bennett* (1970).]

56.8 μm, which is enough for it to be made as a single piece rather than as a "first-order" plate. Unfortunately optical-grade apophyllite is rare, the sample for which data are reported here having come from Sweden. There is some indication that the optical properties of other apophyllite samples may be different. Isotropic, positive, and negative-birefringent specimens have been reported (Deer et al., 1962*b*). According to Deer et al. (1962*b*), the optical properties of apophyllite are often anomalous, some specimens being isotropic, uniaxial negative, or even biaxial with crossed dispersion of optic axial planes. Whether many samples have the favorable birefringence of the Swedish sample is uncertain.

Certain types of plastic film stretched during the manufacturing process have birefringences which are nearly proportional to wavelength and can serve as achromatic retardation plates if they have the proper thickness (West and Makas, 1949). Curve C in Fig. 63 is the retardation of a stretched cellulose nitrate film as measured by West and Makas (1949). A combination of stretched cellulose acetate and cellulose nitrate sheets with their axes parallel will also make a reasonably achromatic $\lambda/4$ plate over the visible region. The advantages of using stretched plastic films for retardation plates are that they are cheap, readily available, have a retardation which is uniform over large areas, and can be used in strongly convergent light. However, each sheet must be individually selected since the birefringence is a strong function of the treatment during the manufacturing process and the sheets come in various thicknesses, with the result that their retardations are not necessarily $\lambda/4$ or $\lambda/2$. Also, Ennos (1963*b*) found that while the magnitude of the retardation was uniform over large areas of the sheets, the direction of the effective crystal axis varied from point to point by as much as 1.5° on the samples he was testing. Thus, film retarders appear to be excellent for many applications but are probably not suitable for measurements of the highest precision.

A reasonably achromatic retardation plate can be constructed from pairs of readily available birefringent materials such as crystalline quartz, sapphire, magnesium fluoride, calcite, or others whose birefringences are listed in Table 15. Assume that the plate is to be made of materials a and b having thicknesses d_a and d_b, respectively (to be calculated), and that it is to be achromatized at wavelengths λ_1 and λ_2. From Eq. (136) we can obtain the relations

$$\begin{aligned} N\lambda_1 &= d_a\,\Delta n_{1a} + d_b\,\Delta n_{1b} \\ N\lambda_2 &= d_a\,\Delta n_{2a} + d_b\,\Delta n_{2b} \end{aligned} \qquad (153)$$

where $N = \frac{1}{4}$ for a $\lambda/4$ plate, $\frac{1}{2}$ for a $\lambda/2$ plate, etc., and the Δn's are values of $n_e - n_o$ for the particular materials at the wavelengths specified; Δn will be positive for a positive uniaxial crystal and negative for a negative uniaxial crystal. (A positive uniaxial material can be used with its fast axis crossed with that of another positive uniaxial material; in this case the first material will have a negative Δn.) Equations (153) can be solved for d_a and d_b:

$$d_a = \frac{N(\lambda_1\,\Delta n_{2b} - \lambda_2\,\Delta n_{1b})}{\Delta n_{1a}\,\Delta n_{2b} - \Delta n_{1b}\,\Delta n_{2a}} \qquad d_b = \frac{N(\lambda_2\,\Delta n_{1a} - \lambda_1\,\Delta n_{2a})}{\Delta n_{1a}\,\Delta n_{2b} - \Delta n_{1b}\,\Delta n_{2a}} \qquad (154)$$

As an example of a compound plate, let us design a $\lambda/4$ plate of crystalline quartz and calcite and achromatize it at wavelengths $\lambda_1 = 0.508$ μm and $\lambda_2 = 0.656$ μm. Quartz has a positive birefringence and calcite a negative birefringence (Table 15) so that Δn_{1a} and Δn_{2a} (for quartz) are positive and Δn_{1b} and Δn_{2b} (for calcite) are negative. Equations (154) are satisfied for $d_{qtz} = 426.2$ μm and $d_{calc} = 21.69$ μm; thus the phase retardation is exactly 90° at these two wavelengths. An equation of the form of those in Eq. (153) is now used to calculate N for all wavelengths in the visible region using birefringence values listed in Table 15, and the results are plotted as curve E in Fig. 63. Although the achromatization for this quartz-calcite combination is not as good as can be obtained with a rhomb-type device or apophyllite, the phase retardation is within ±5° of 90° in the wavelength region 4900 to 7000 Å and is thus much more constant than the retardation of a single mica or quartz $\lambda/4$ plate. Better two-plate combinations have been calculated by Beckers (1971), the best being MgF_2-ADP and MgF_2-KDP, which have maximum deviations of ±0.5 and ±0.4 percent, respectively, compared with ±7.2 percent for a quartz-calcite combination over the same 4000- to 7000-Å wavelength region. The thicknesses of the materials which are required to produce $\lambda/4$ retardation†

† Beckers' tables II to V give the thickness of materials required to produce one full-wave retardation. To obtain values of thicknesses for $\lambda/4$ retardation, for example, multiply all d values in the table by $\frac{1}{4}$. The percent deviations should remain unchanged.

are $d_{MgF_2} = 113.79\ \mu m$, $d_{ADP} = 26.38\ \mu m$, and $d_{MgF_2} = 94.47\ \mu m$, $d_{KDP} = 23.49\ \mu m$. Since the ADP and KDP must be so thin, these components could be made in two pieces as "first-order" plates (Par. 61).

Other two-component compound plates have been proposed (Chandrasekharan and Damany, 1968; Gaudefroy, 1929; Ioffe and Smirnova, 1964; and Mitchell, 1966). The paper by Ioffe and Smirnova describes a quartz-calcite combination similar to the one illustrated above, but it contains various numerical errors which partially invalidate the results.

If better achromatization is desired and one does not wish to use a rhomb-type $\lambda/4$ device, three materials can be used which satisfy the relations

$$\begin{aligned} N\lambda_1 &= d_a\,\Delta n_{1a} + d_b\,\Delta n_{1b} + d_c\,\Delta n_{1c} \\ N\lambda_2 &= d_a\,\Delta n_{2a} + d_b\,\Delta n_{2b} + d_c\,\Delta n_{2c} \\ N\lambda_3 &= d_a\,\Delta n_{3a} + d_b\,\Delta n_{3b} + d_c\,\Delta n_{3c} \end{aligned} \tag{155}$$

where the Δn's are birefringences of the various materials at wavelengths λ_1, λ_2, and λ_3.

Instead of using only three wavelengths, the thicknesses can be optimized such that the maximum deviations from achromatization are minimized over the entire wavelength interval desired (Beckers, 1971). In this way, Beckers obtained a three-component combination of quartz, calcite, and MgF_2 which has a retardation of a full wavelength and a maximum deviation of only ± 0.2 percent over the 4000- to 7000-Å wavelength region. The maximum deviation of slightly different thicknesses of these same three materials rises to ± 2.6 percent if the wavelength interval is extended to 3000 to 11,000 Å. Chandrasekharan and Damany (1968) have designed a three-component $\lambda/4$ plate from quartz, MgF_2, and sapphire for use in the vacuum ultraviolet.

Rhombs as Achromatic $\lambda/4$ Retarders†

67. The simplest stable, highly achromatic $\lambda/4$ retarder with a reasonable acceptance angle and convenient size appears to be a rhomb-type retarder. Several of these will be described here; the choice of which one to use for a specific application depends on (1) the geometry of the optical system (can a deviated or displaced beam be tolerated?), (2) wavelength range, (3) degree of collimation of the beam, (4) beam diameter (determining the aperture of the retarder), (5) space available, and (6) accuracy required. The devices are shown schematically in Fig. 64, where they are drawn approximately to scale to give an estimate of the length-to-aperture ratios. They will be described in detail in Pars. 68 to 72.

All the rhombs derive their phase retardation from the difference in the phase changes on reflection for the p and s components when an incident beam of ligh plane-polarized at an azimuth of 45° to the plane of incidence undergoes two or more internal reflections.‡ If δ_p and δ_s are the phase changes, the difference Δ is given by

$$\tan\frac{\Delta}{2} \equiv \tan\left(\frac{\delta_p}{2} - \frac{\delta_s}{2}\right) = \frac{\cos\theta\sqrt{n^2\sin^2\theta - 1}}{n\sin^2\theta} \tag{156}$$

where n is the refractive index of the material and θ the internal angle of incidence. The material from which the rhombs are made must be homogeneous and isotropic since the phase retardation must not be modified by strain birefringence. Equation (156) gives the phase retardation for a single internal reflection. For a rhomb with two equal reflection angles, Δ should be 45° for each reflection in order for the light to emerge circularly polarized. In general, the total phase retardation of the rhomb is the sum of the phase retardations produced at each reflection.

† The material in Par. 67 is condensed from Bennett (1970), by permission.

‡ If the refractive index of the material is high enough, a quarter-wave retarder can be made from a "rhomb" having a single internal reflection; examples are diamond, suggested by Harle (1950), and germanium and other high-index infrared-transmitting materials, suggested by Ukhanov (1969).

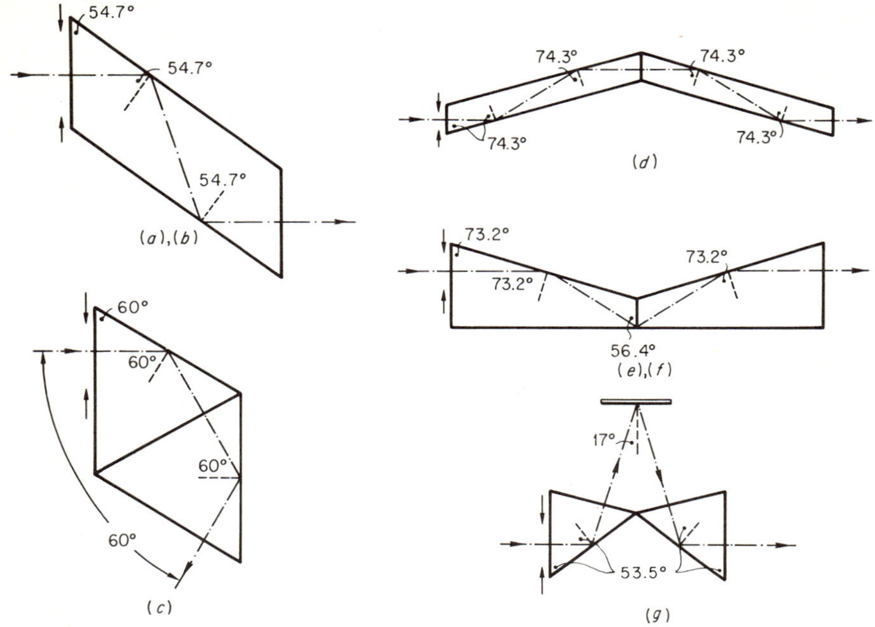

Fig. 64 Types of rhombs: (*a*) Fresnel rhomb, (*b*) King optimized rhomb with angle of 51.5° and one reflecting surface coated with 200 Å of MgF_2, (*c*) Mooney rhomb, (*d*) achromatic device 1 (AD-1), (*e*) achromatic device 2 (AD-2), (*f*) AD-2 with angle changed to 72.2° and base surface coated with 275 Å of MgF_2, and (*g*) achromatic device (AD) (two glass prisms and aluminum mirror). The useful apertures are indicated by arrows. [*Bennett* (1970).]

To determine how achromatic the rhomb will be, values of n can be inserted into Eq. (156) for the appropriate wavelengths. In this way, curves showing the variation of the phase retardation with wavelength have been calculated for all the rhombs in Fig. 64 and are shown in Figs. 65 and 66. The wavelength at which the retardation of each device is exactly 90° depends on the refractive index of the rhomb material and angle of incidence as well as the rhomb design. If the angle of incidence is changed slightly, the entire curve will be translated parallel to the y axis and will cross the 90° line at another wavelength. The increase in the retardation in the ultraviolet is caused by the increase in the refractive index of the glass in the ultraviolet region.

To find the acceptance angle of a rhomb for which the phase retardation is within a specified amount of 90°, different values of θ have been substituted into Eq. (156) to obtain the curves shown in Fig. 67. In this figure an angle of 0° means that the beam is incident normally on the rhomb face and has an internal angle of incidence equal to the value shown in Table 16. Negative angles of incidence correspond to smaller internal angles and positive angles to larger internal angles. The wavelength has been chosen so that the phase retardation will be exactly 90° when the external angle of incidence is 0°.

68. Fresnel Rhomb The classical version of the Fresnel rhomb (Fig. 64*a*) is made of crown glass with $n_D = 1.51$; thus there are two angles of incidence, 48°37′ and 54°37′, for which a 45° phase retardation is introduced at each reflection (Jenkins and White, 1957*c*). The larger angle is generally chosen because the wavelength achromatization is considerably better. In order to calculate the variation of the phase retardation with wavelength, we have chosen a borosilicate crown glass with $n_D = 1.51100$ (Wolfe et al., 1963*c*) so that the larger angle of incidence is

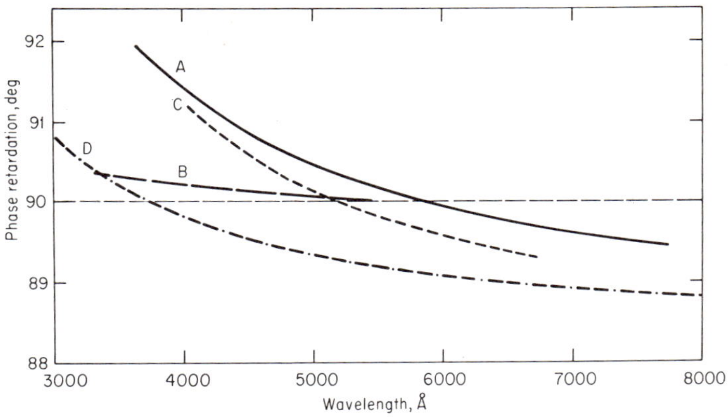

Fig. 65 Curves of the phase retardation vs. wavelength for some of the rhombs shown in Fig. 64: *A,* Fresnel rhomb; *B,* coated Fresnel rhomb; *C,* Mooney rhomb; and *D,* AD-1. [*Bennett* (1970).]

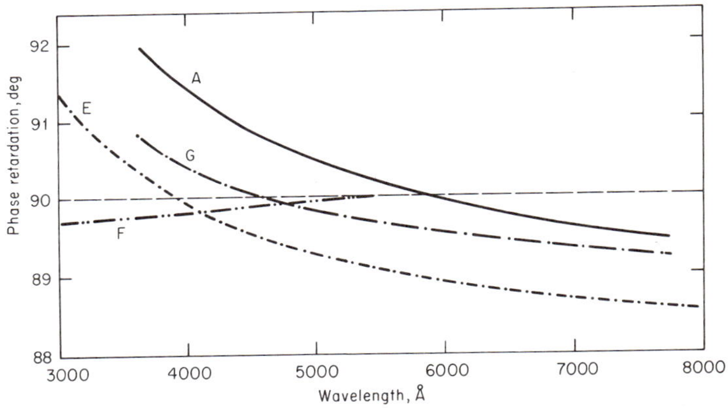

Fig. 66 Curves of the phase retardation vs. wavelength for some of the rhombs shown in Fig. 64: *A,* Fresnel rhomb; *E,* AD-2; *F,* coated AD-2; and *G,* AD. [*Bennett* (1970).]

actually 54.72°. (The acute angles of the rhomb should also be 54.72° so that light will be normally incident on the rhomb face.) Curves *A* in Figs. 65 and 66 show that the phase retardation changes by about 2° over the visible spectral region, and curve *A* in Fig. 67 shows that the phase retardation is not symmetric around 0° external angle of incidence.

Palik (1963) has suggested using rhombs made of NaCl and CsI in the infrared. In the wavelength region 1 to 13 μm the phase retardation of an NaCl rhomb having an acute angle of 53°20′ lies in the range 93.7 to 86.3°. If the rhomb is placed at the focus of a beam in an *f*/4.5 system, the phase retardation will have values between 85.2 and 90.5° at 9 μm. The retardation of a CsI rhomb having an acute angle of 60°30′ should vary from 92.0 to 89.6° in the 10- to 40-μm region. However, the change in retardation with acceptance angle is much larger than for the NaCl rhomb; in order to keep the retardation within 5° of the normal-incidence value, the system should be stopped down to about *f*/9. Polyethylene, paraffin, and crystalline quartz have also been suggested as possible materials for infrared rhombs (Palik, 1963).

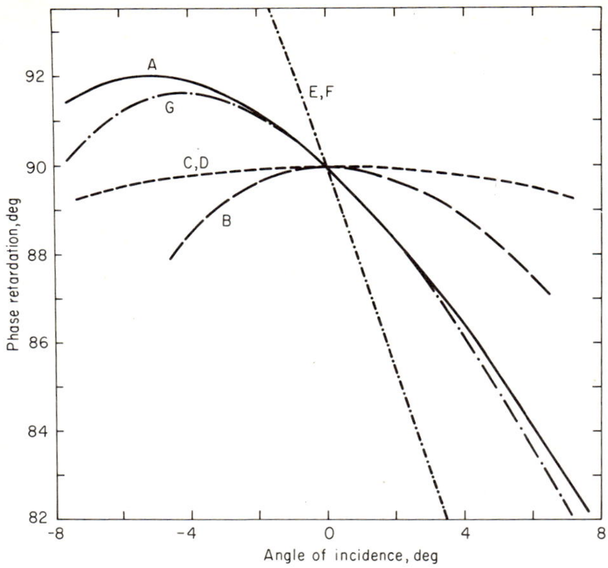

Fig. 67 Curves of the phase retardation vs. external angle of incidence on the rhomb face for the rhombs shown in Fig. 64. Negative angles correspond to smaller internal angles, positive angles to larger internal angles. [*Bennett* (1970).]

69. Coated Fresnel Rhomb King (1966) and Clapham et al. (1969) have optimized the Fresnel rhomb to make the retardation symmetric about 0° external angle of incidence; in this way the phase retardation for a given acceptance angle can be decreased. This has been accomplished by evaporating appropriate thicknesses of various materials on one or both of the internally reflecting surfaces. The evaporated films can adjust the maximum value of the retardation to be exactly 90° and can sometimes decrease its variation with wavelength. More importantly, they can compensate for strain birefringence in the glass and for modified surface layers remaining from the polishing process. This is an especially important advantage, particularly when working in the ultraviolet, since, to a first approximation, the phase retardation resulting from residual strain is inversely proportional to wavelength (Clapham et al., 1969).

The measured achromatization curve for a crown-glass rhomb with one reflecting surface coated with 200 Å of MgF_2 is shown as curve *B* in Fig. 65. The retardation varies by only 0.4° in the wavelength range 3341 to 5461 Å, making this rhomb and King's coated AD-2 the best achromatized of any of the rhombs. The measured variation of the retardation with acceptance angle (curve *B* in Fig. 67) is also much less than that of the uncoated rhomb and is nearly symmetric about the 0° point. If the acceptance angle were limited to $\pm 2°$, a reasonable amount, the retardation would vary by only 0.4° and would thus be comparable to the wavelength variation of the retardation.

Clapham et al. (1969) have also coated fused-silica rhombs with various thicknesses of NdF_3 and MgF_2 and have achieved excellent achromatization in the ultraviolet wavelength region.

All Fresnel-type rhombs have the disadvantage that they displace (but do not deviate) the light beam. One method of restoring the beam to its on-axis position is to add two additional non-phase-retarding reflections.

70. Mooney Rhomb The rhomb shown in Fig. 64*c*, proposed by Mooney (1952), consists of two 60° dense-flint-glass prisms in contact. Light is incident normally on the face of the first prism, is totally internally reflected twice at 60° angles, and leaves the rhomb deviated by 120° from the straight-through position. The

TABLE 16 Properties of Achromatic Rhombs (Bennett, 1970)

Name	Light path	Internal angle of incidence, deg	Material	Refractive index		Variation of phase retardation			
						With wavelength		With angle of incidence	
				n	Wavelength, Å	Variation, deg	Wavelength, Å	Variation, deg	Angle, deg
Fresnel rhomb	Translated	54.7	Crown glass	1.511	5893	2.5	3650–7682	9.1	−7 to +7
Coated Fresnel rhomb	Translated	51.5	Crown glass	1.5217	5461	0.4	3341–5461	2.5	−4 to +6
	Translated	54.0	Fused quartz	1.4880	3000	0.7	2148–3341	<0.5	−1.5 to +1.5
Mooney rhomb	Deviated	60.0	Flint glass	1.650	5893	1.9	4047–6708	0.7	−7 to +7
AD-1	Undeviated	74.3	Fused quartz	1.4702	4000	2.0	3000–8000	0.7	−7 to +7
AD-2	Undeviated	73.2, 56.4	Fused quartz	1.4702	4000	2.9	3000–8000	13.2	−3 to +3
Coated AD-2	Undeviated	72.2	Fused quartz	1.4601	5461	0.3	2140–5461	6.0	−1.5 to +1.5
AD	Undeviated	53.5	Crown glass	1.511	5893	1.6	3650–7682	9.4	−7 to +7

beam deviation makes it difficult to use the Mooney rhomb in a conventional on-axis optical system. However, this rhomb has the largest aperture in proportion to its length of any of the rhombs and also has a very large acceptance angle because of the compensating feature of the two internal reflections. In a Fresnel rhomb, if the internal angle of incidence is increased on the first reflection, it will be increased by the same amount on the second reflection because the two internally reflecting faces are parallel. With the Mooney rhomb, if the first internal angle of incidence is increased, the second will be decreased by the same amount, so that, to a first approximation, the total phase retardation will be unchanged (curve *C* in Fig. 67).

The variation of the phase retardation with wavelength (curve *C* in Fig. 65) is the largest of any of the rhombs discussed here because the refractive index of the dense flint glass is changing most rapidly with wavelength. It would be better to make the rhomb out of a different material, e.g., fused quartz, and change the angles accordingly.

71. Achromatic Devices 1 and 2 (AD-1 and AD-2) These devices, which consist of pairs of rhombs in series, were originally proposed by Oxley (1911) but more recently have been reinvented by Kizel et al. (1964). They are made of fused quartz and employ four and three internal reflections, respectively, at the angles shown in Fig. 64*d* and *e*. They have a major advantage over the rhombs previously discussed in that they do not deviate or displace the light beam. Also, since they are made of fused quartz, they have quite flat achromatization curves (curve *D* in Fig. 65 and curve *E* in Fig. 66). The AD-1 has the same advantage as the Mooney rhomb: its reflections are compensating; so it has the same small variation of retardation with angle of incidence (curve *D* in Fig. 67). However, its useful aperture is very small compared to its overall length, in the ratio of approximately 1:17.5. Furthermore, the long path length of the beam in quartz will inevitably lead to several degrees of phase retardation caused by residual birefringence of the quartz, and therefore its good theoretical performance may never be attained in practice (King, 1966).

The AD-2 has a more favorable length-to-aperture ratio, so that it can be considerably shorter and should therefore be less affected by strain birefringence. However, its retardation is extremely sensitive to angle of incidence because of the three uncompensating reflections (curve *E* in Fig. 67). Alignment is also critical; it is essential that the incident beam be exactly normal to the end face to prevent deviation of the emerging beam.

Clapham et al. (1969) have modified the AD-2 by changing the prism angle to 72.2° and coating the base surface with a 275-Å-thick film of MgF_2. They have obtained an extremely small variation of the phase retardation with wavelength (curve *F* in Fig. 66), but have been unable to decrease the sensitivity of the phase retardation to angle of incidence (curve *F* in Fig. 67).

72. Achromatic Device (AD) The final $\lambda/4$ retarder (Fig. 64*g*) was designed by Krasilov (1967) and consists of two total internal reflections in isosceles prisms and an external reflection from an aluminized surface. Unfortunately Krasilov made several calculational errors in his paper, so that the angles have all been recalculated (Bennett, 1970). The advantages of the AD are that it does not deviate or displace the beam and it has a relatively large aperture in comparison to its length. The wavelength achromatization (curve *G*, Fig. 66) is nearly as good as that of the AD-1 and better than for the AD-2 even though all these devices are made of fused quartz. Its acceptance-angle characteristics (curve *G*, Fig. 67) are very similar to those of the classical version of the Fresnel rhomb. The major drawback of the device is that it may be difficult to align because of the three independent parts.

Composite Retardation Plates

73. A composite retardation plate is made up of two or more elements of the same material combined so that their optic axes are at appropriate angles to each other. Some of the composite plates have nearly all the properties of a true retardation plate, whereas others do not. In Pars. 74 to 76 we describe composite plates which produce circularly polarized light at a given wavelength, those which act as achromatic circular polarizers, and those which act as achromatic polarization rotators or pseudo $\lambda/2$ plates. The effect of combining several birefringent plates with their axes at arbitrary angles to each other can be easily understood using the Poincaré sphere. A general treatment of this subject has been given by Ramachandran and Ramaseshan (1961*c*).

74. Nonachromatic Circular Polarizers Before we discuss two-element composite plates used as circular polarizers, we should mention the circular polarizer† used to minimize light reflected from surfaces such as a phosphor-illuminated oscilloscope screen and at the same time transmit the light from the screen. This device consists of a linear polarizer (usually a sheet of Polaroid) cemented to a sheet of stretched cellulose acetate butyrate (which acts as a $\lambda/4$ retardation plate), with the axis of the polarizer at 45° to the fast and slow axes of the retarder. This combination transmits light passing through it once but extinguishes light which passes through it twice with a reflection in between passes. The device works in a manner similar to that of the test setup of Emberson (1936) described in Par. 81 for checking retardation of $\lambda/4$ plates. This circular polarizer is not achromatic because the retardation of the stretched plastic film increases with increasing wavelength (West and Makas, 1949). However, it does reduce the reflected light throughout the visible region to under 10 percent of the value that would be obtained without the device (Polaroid, 1965).

A two-element composite plate can also be used as a retardation plate to produce circularly polarized light. To distinguish the composite plate from a true $\lambda/4$ retardation plate, we shall call the former a circular polarizer. (When using this definition, we are assuming that the incident beam is plane-polarized.) Although the theory for a nonachromatic two-element composite plate has been worked out by numerous authors (many of whom were unaware of the work of others) and is given in the appendix of Jessop and Harris (1950), the articles by Childs (1956), Adams and Waxler (1965), and Aben (1962) seem to be particularly helpful. These authors have shown that *any two plates having retardations whose sum lies between* 90 *and* 270° *can be used in combination to produce circularly polarized light*. Thus, a composite plate having a retardation of *exactly* $\lambda/4$ can be made from two plates of nearly any retardation. This is a distinct advantage since it is difficult, for example, to cleave mica to exactly the right thickness.

If both plates have the same retardation δ (between 45 and 135°), certain simplifications occur in the general expressions. In order to produce circularly polarized light from an incident plane-polarized beam, there is a unique angle ψ_c between the fast axes of the two plates which is given by the relation

$$\cos 2\psi_c = \pm \cot^2 \delta \tag{157}$$

The angle θ_1 between the vibration direction of the incident plane-polarized beam and the fast axis of the first plate is also specified. When circularly polarized light is produced, the angle is designated θ_{1c} and satisfies the relation

$$\sin 2\theta_{1c} = \pm \cot \delta \tag{158}$$

The same composite plate will also produce plane-polarized light when it is rotated so that $\theta_1 = \theta_{1p}$ satisfies the relation

$$\cot 2\theta_{1p} = \tan \psi_c \cos \delta \tag{159}$$

Thus, the angle θ_{1p} defines one of the axes of the composite plate, since an incident beam plane-polarized at an azimuth of θ_{1p} to the fast axis of the first element of the composite plate will emerge plane-polarized.‡ However, the plane of vibration will be rotated through an angle $\phi_2 - \theta_{2p}$, where ϕ_2 is the angle between the plane of vibration after the light has passed through both the plates and the fast axis of the second plate and θ_{2p} is the angle between the vibration direction of the *incident* plane-polarized beam and the fast axis of the second plate. Note that all these

† Sold by Polaroid Corporation, Cambridge, Mass.

‡ For a simple $\lambda/4$ plate, plane-polarized light whose plane of vibration is parallel to either the fast or slow axis will emerge plane-polarized and unrotated.

angles must be measured in the proper direction. Diagrams are given, for example, by Adams and Waxler (1965).

In a simple $\lambda/4$ plate, the azimuth of the incident plane-polarized light must be 45° with respect to the fast or slow axis in order for the light to emerge circularly polarized. For a composite plate, the corresponding angle $\theta_1 - \theta_{1p}$ also equals 45° (Adams and Waxler, 1965). The composite plate further resembles a simple $\lambda/4$ plate in that when a beam of plane-polarized light passes through the plate at any arbitrary azimuth, the azimuth of the ellipse of the emerging beam is *independent* of the azimuth of the incident beam.†

If one has a composite quarter-wave plate and wishes to measure the retardation of an unknown sample, it can be done in the following manner (Aben, 1962). The sample is placed between the polarizer and the composite plate, and the polarizer axis is set parallel to one of the "axes" of the composite plate. The fast and slow axes of the sample must be determined and oriented at an azimuth of 45° to the polarizer axis. In this configuration the composite plate converts the unknown elliptical vibration into a linear vibration whose azimuth can be determined from the extinction orientation of the analyzer. The phase difference of the sample is then calculated from the analyzer azimuth angle.

In constructing a composite plate to produce circularly polarized light, one can start with two plates having identical retardations, which can be measured using one of the methods given in Par. 81. The angle ψ_c between the two fast axes should be approximately determined, e.g., from Eq. (157) or from table 5 in Adams and Waxler (1965), and the plates mounted accordingly. (The directions of the fast axes of the plates are assumed to be known. If they are not, they can be determined by a method given in Par. 80.) By placing the combination between fixed and rotating polarizers [fixed polarizer placed so that θ_1 has the appropriate value given by Eq. (158) or obtained from Adams and Waxler's table 5; they call the angle i_{c1}] adjust ψ_c and θ_1 until the sine wave at the rotation frequency of the polarizer disappears (Daniels, 1967). This combination is *not* achromatic, so that the light source used for the assembling process should have the same wavelength as that to be used for the actual experiment. In order to achieve optimum performance from the composite plate, the individual plates should be coated with antireflection films to minimize spurious reflections between plates. Alternately, the plates can be cemented together to form one plate (Osterberg and Carlan, 1957*b*).

Aben (1962) describes an alternate method for aligning a composite plate which requires a polarizer and analyzer but not a rotating Polaroid or photoelectric detector. With the polarizer and analyzer crossed, the first birefringent plate is inserted and rotated until the light is completely extinguished. In this configuration, one of the axes of the plate will coincide with the polarizer axis. Next, the polarizer is rotated through the angle θ_{1p}, and the analyzer is rotated through the angle $\psi_c + \phi_2$. The second birefringent plate is then inserted and rotated until the light is completely extinguished. The composite plate is now set so that the plane-polarized beam is passing through it and emerging plane-polarized and rotated; the analyzer extinguishes the emerging plane-polarized beam. If the entire composite plate is now rotated through an angle 45°, the light emerging from the plate will be circularly polarized and cannot be extinguished by the analyzer. In this alignment procedure, care must be taken to measure all the angles in the proper direction.

Violino (1965) has described a variable circular polarizer made from two plates which were $\lambda/4$ at wavelength λ_0. He states that this system can be adjusted to produce completely circularly polarized light for any monochromatic wavelength in the range between $2\lambda_0/3$ and $2\lambda_0$. In an experimental test of the system, he used pairs of retardation plates which were $\lambda/4$ for the D line of sodium or for the D_1 line of rubidium. With the composite plate set at either wavelength, he measured a retardation of $90° \pm 10'$, compared with $90° \pm 3°$, the best he could do with a single plate.

† With a simple $\lambda/4$ plate, the azimuth of the emerging elliptically polarized beam always lies along either the fast or slow axis of the plate.

75. Achromatic Circular Polarizers The preceding discussion was limited to nonachromatic composite plates. We now describe different achromatic or quasi-achromatic combinations made from two or more elements of the same material. To distinguish these from true $\lambda/4$ retardation plates, we shall call them achromatic circular polarizers (assuming, as in Par. 74, that the incident beam is plane-polarized). Usually the degree of achromaticity is variable depending on the wavelength interval desired; the smaller the wavelength interval the better the achromatization.

Destriau and Prouteau (1949) were apparently the first to suggest a way of making a quasi-achromatic circular polarizer by superimposing two plates of the same material, one a $\lambda/2$ plate and the other a $\lambda/4$ plate for a wavelength in the middle of the range to be achromatized. The angle between the fast axes of the two plates is $60° - \beta$, where β is an angle less than 3.5° which depends on the nature of the plates and the spectral region to be achromatized. The incident plane-polarized beam makes an angle of 15° with the *slow* axis of the first $\lambda/2$ plate. In an experimental test of an achromatic retarder of this type made from mica plates $\lambda/2$ and $\lambda/4$ for the D line of sodium, the authors obtained retardations varying from 83 to 95° over the visible spectral region (0.44 to 0.64 μm) and stated that the achromatization could have been improved if the wavelength interval had been reduced.

Osterberg and Carlan (1957*b*) independently discovered the same $\lambda/2$, $\lambda/4$ achromatic combination proposed by Destriau and Prouteau and worked out the theory in detail. They found that if the retardations of the plates are accurately in the ratio of 2:1, there are two wavelengths λ_1 and λ_2 where the retardation of the combination is exactly $\lambda/4$. These can be made equal to each other at λ_0 (where the individual plates have exactly $\lambda/2$ and $\lambda/4$ retardation) or varied so that $\lambda_1 < \lambda_0$ and $\lambda_2 > \lambda_0$. Thus, one can place λ_1 and λ_2 at a variety of wavelengths in the visible or infrared region simply by varying the angle between the axes of the two plates and obtain an achromatic circular polarizer over the desired wavelength region. This combination is thus more flexible than the three-element combination of Pancharatnam described below.

Pancharatnam (1955*a*) made an achromatic circular polarizer out of three superimposed plates of the same material; the first two plates were half-wave plates, and the third was a quarter-wave plate, all for the wavelength λ_0 in the center of the spectral range to be covered. He reported that if the dispersion of the birefringence is neglected, the combination can produce exactly circularly polarized light at three wavelengths: $(1 - \epsilon/90°)\lambda_0$, λ_0, and $(1 + \epsilon/90°)\lambda_0$, where ϵ is the difference in retardation from 90° for a simple $\lambda/4$ plate at a wavelength near λ_0. The extreme wavelengths can supposedly be made as far apart as desired, but then the achromatization will not be as good as when the wavelengths are closer. Pancharatnam stated that this three-element combination does not act like an ordinary $\lambda/4$ plate, since when the polarizer is turned through an azimuth of 45° from the angle needed to produce circular polarization, all the emerging wavelengths are nearly plane-polarized but with different azimuths. (In a true $\lambda/4$ plate, plane-polarized light vibrating parallel to either the fast or slow axis emerges plane-polarized in the incident azimuth.)

A second design of achromatic circular polarizer which supposedly acts more like a true $\lambda/4$ plate has also been described by Pancharatnam (1955*b*). This combination consists of three superimposed birefringent plates; the outer two have identical retardations and are mounted with their fast axes parallel. The central plate is a $\lambda/2$ plate for a wavelength λ_0 in the middle of the spectral range to be covered. The author states that the design is strictly achromatic for two wavelengths $(1 - f)\lambda_0$ and $(1 + f)\lambda_0$, where f is a fraction multiplying the retardation of each individual plate. If the dispersion of the birefringence in the material is appreciable, the wavelength range over which the combination can be considered achromatic will change, but the basic theory supposedly remains the same. Pancharatnam gives no values for the magnitude of the retardation as a function of wavelength, only the statement that plane-polarized white light can be almost completely extinguished when passed through the composite plate twice. Osterberg and Carlan (1957*a*) worked out the theory for the combination and verified Pancharatnam's assertions. However, they conclude that their two-element combination has as good performance and is more flexible than the more specialized three-element combination.

Retardation plates with a very high degree of achromaticity or any desired retardation as a function of wavelength can be designed using the synthesis procedures described by Harris et al. (1964). The number of birefringent plates required depends on the form of the retardation vs. wavelength function. McIntyre and Harris (1968) designed, built, and tested an achromatic circular polarizer covering the wavelength region from 4000 to 8000 Å. It consisted of 10 sapphire plates each 1 mm thick and 8 mm square and having a retardation of $\pi/2$ (not $\pi/2 \pm 2N\pi$) at 5333 Å, the center of the interval as measured on a frequency scale. Each plate was ground and polished until it was within one-hundredth wave of the calculated retardation. The plates were then placed in special holders allowing rotation around three orthogonal axes, and the entire unit was placed in an index-matching oil bath to eliminate multiple reflections between plates. The calculated retardation of the circular polarizer was $90° \pm 15'$ over the entire visible region (4000 to 8000 Å), and measurements confirmed this value to within the limit of experimental error. This device is by far the most achromatic of any of the composite plates, but drawbacks are its extremely long length (457 mm in the experimental model, which was later reduced to 51 mm) and the difficulty of fabricating and aligning the plates. If an achromatic circular polarizer with this degree of achromaticity is required for an experiment, it would be well to consider one of the rhomb-type designs described in Pars. 67 to 72.

76. Achromatic Polarization Rotators Two superimposed birefringent plates can also be used, in restricted cases, as polarization rotators. Adams and Waxler (1965) and Aben (1962) have pointed out that for any combination of two birefringent plates there can be found two mutually perpendicular azimuths for incident plane-polarized light for which the light emerging from the composite plate is linearly polarized. However, the plane of vibration of the light is rotated through an angle $\phi_2 - \theta_{2p}$, where the angles have the same meanings as in Par. 74. A true half-wave plate can rotate the plane of vibration of an incident plane-polarized beam through a continuous range of angles, so that the composite plate is much less versatile.

It is possible to make a composite plate which produces a 180° phase difference and thus can rotate the plane of vibration of an incident plane-polarized beam through an angle of 90°. Furthermore, it is possible to make such a system achromatic. Koester (1959*a*) has pointed out that if two $\lambda/2$ plates are placed so that their slow axes make angles of 22.5 and 67.5° respectively with the incident plane of vibration, each plate will rotate the plane of vibration by 45° and the combination will thus produce a 90° rotation of the incident beam. However, this combination is not achromatic (Clarke, 1967). If the angles are changed to $22.5° + \delta$ and $67.5° - \delta$, where δ is a small angle of the order of 1°, there will be two wavelengths λ_1 and λ_2 at which incident plane-polarized light will have its plane of vibration rotated by 90° and will have only a very small component of elliptically polarized light. The system can be improved by using three or four half-wave plates. When four mica plates with retardations of $\lambda/2$ at 0.546 μm are assembled so that their slow axes make angles of 8, 30, 60, and 82° respectively with the plane of vibration of the incident beam, the wavelengths at which a 90° rotation is produced are 0.325, 0.468, 0.625, and 0.768 μm, respectively (Koester, 1959*a*). This composite plate will thus be an achromatic rotator over the entire visible and near-ultraviolet spectral region. Koester worked out the theory of two- and three-element achromatic combinations of $\lambda/2$ plates using Jones calculus and checked them by constructions on the Poincaré sphere. The four-element case was done entirely on the Poincaré sphere, and thus the achromatic wavelengths are only approximate.

Grechushnikov et al. (1964) considered a system composed of three half-wave plates for a wavelength λ_0 which were oriented at equal angles ψ with respect to each other. These formed a composite plate with a 180° phase retardation for the wavelengths λ_0, λ_1, and λ_2, where $\lambda_1 = \lambda_0/(1 + 2\epsilon/180°)$ and $\lambda_2 = \lambda_0/(1 - 2\epsilon/180°)$ (ignoring the dispersion of the birefringence in the interval). The phase retardation of a single plate is $180° + 2\epsilon$ at wavelength λ_1 and $180° - 2\epsilon$ at wavelength λ_2. For example, if the plates have a retardation of 180° at $\lambda_0 = 0.520$ μm, and $\epsilon = 20°$, then $\lambda_1 = 0.425$ μm and $\lambda_2 = 0.670$ μm. In this case ψ was said to be 66°10′, but the orientation of the plate relative to the incident plane-polarized beam was not given. This composite plate is said to rotate the plane of vibration of an incident plane-polarized beam through 90° over nearly the entire visible spectrum, and the transmission of the plate between crossed polarizers at intermediate wavelengths should not differ from full transmission by more than a few percent (Grechushnikov et al., 1964).

77. Unusual Retardation Plates Several retardation plates obtain their retardations in unusual ways. A circular polarizer for use at the 1216-Å Lyman α line in the vacuum ultraviolet has been described by McIlrath (1968). Unpolarized Lyman α radiation is linearly polarized by reflection from an Al_2O_3 or MgF_2 mirror at a 60° angle of incidence and is then circularly polarized by reflection from an aged aluminum mirror, i.e., one having a thin Al_2O_3 coating formed by exposure to air, at an angle of incidence of $52° \pm 2°$.† McIlrath found that at almost all angles of incidence $r_p = r_s$, the r's being amplitude reflection coefficients for the p and s components; at 52° incidence $r_p^2 = r_s^2 = 0.50 \pm 0.05$.

A reflection-type circular polarizer has also been proposed for the infrared by Palik (1963). He suggested using a semiconductor such as indium antimonide and calculated the angles for the orientation of the InSb plate in the plane-polarized beam. Since the optical constants change with wavelength, the plate would have to be reset as the wavelength was changed.

In the far infrared, Richards and Smith (1964) have shown that a circular polarizer can be constructed from a magnetorotary solid operated near its cyclotron resonance frequency. At the resonance frequency the material absorbs circularly polarized light of one sense, e.g., right-handed circularly polarized light, and transmits the other sense. InSb was used at a frequency of 51 cm^{-1} ($\lambda \approx 0.196$ mm) and a magnetic field strength of 7.12 kOe; the unwanted sense of circularly polarized light was found to be at most 2 percent of the transmitted power over a bandwidth of ± 5 cm^{-1}. Although the experiment was done at 4.2 K in order to use a superconducting magnet, the InSb circular polarizer could be used at higher temperatures since the carrier concentration remains constant and the mobility increases for temperatures up to about 100 K.

A novel type of achromatic half-wave retarder for use in the visible region has been described by Lostis (1957). It consists of a total internal reflection from a glass surface coated with a thin layer of a high-index material. For the particular case chosen, a layer of titanium dioxide of optical thickness $\lambda/11$ for a 45° angle of incidence was deposited on the hypotenuse of a fused-silica prism. The calculated variation of the retardation was $\pm\lambda/50$ (or 7.2° in phase retardation) in the wavelength region 4000 to 7000 Å. Since the reflection was a total internal reflection, both r_s and r_p were equal to unity and the only light losses were at the air-fused silica interfaces where the beam entered and left the prism. This method is not limited to producing a phase retardation of 180°. According to Lostis, it is possible to obtain any desired phase retardation by choosing the characteristics of the film and the refractive index of the prism. In particular, one could make a quarter-wave retarder or a system producing a zero phase retardation.

Methods for Making and Testing $\lambda/4$ Plates

78. Since mica is the most common material used for retardation plates, methods for splitting mica and for distinguishing between its fast and slow axes will be described. Most of the procedures for measuring retardation (Par. 81) are applicable to plates made of any material.

79. Splitting Mica Muscovite mica is the most common type used for optical components (Pars. 57 to 59). It can be split into pieces of the desired thickness by the method described by Strong (1938). The starting sheet is trimmed with sharp tin snips in order to have clean edges, one edge is frayed by rubbing it, and a clean needle inserted to divide the sheet approximately in half. A drop

† The azimuth relation between the two reflections is shown in McIlrath (1968, Fig. 9).

of water is introduced into the cavity so produced, and the needle is then worked along the edge, point first, at an angle of about 30° so that the first cleavage starts inside the sheet. This avoids a terraced cleavage. After the needle has gone around the circumference, a second water drop is introduced and the plates are drawn apart, separating almost as easily as the pages of a book. The pieces are split in half a second time using the same technique, and the process is continued until sheets close to the desired thickness are obtained, as measured by a micrometer. Since different pieces of mica have different refractive indexes and birefringences (Einsporn, 1936), final selection must be based on an optical test, several of which are described in Par. 81.

Dobrowolski (1959) used a modification of Strong's method in which the mica is immersed in water; after cleavage has been started with a needle, it is completed with a knife-shaped sheet of Teflon. This is supposed to avoid the scratches caused by a metal blade. Another method for splitting mica under water using a needle is described in detail by Benedicks (1931), who claims that sheets as thin as 10 μm can be produced. Wright (1939) has a more complicated method involving the use of spatulas of various thicknesses, which is supposed to yield uniform, scratch-free mica sheets.

The chance of finding a piece of mica close enough to the proper thickness for a retardation plate by random splitting of mica samples is quite small. However, Emberson (1936) describes a technique for making $\lambda/4$ plates from sheets of mica which have been split down a little too thin by rotating the sheet about an axis parallel to the optic axis (see Par. 65). A sheet is mounted in a tube held in a larger tube by a ring at one end and set screws at the other. The inner tube holding the mica can thus be tilted relative to the outer tube. The light paths of both rays are increased at nonnormal incidence, but, in addition, the light path of the extraordinary ray will be increased by a greater amount than the path of the ordinary ray so that some angle of tilt can be found for which the retardation of the mica is exactly $\lambda/4$.

80. Distinguishing Fast and Slow Axes To tell where the fast and slow axes of a $\lambda/4$ plate lie, the plate is placed between crossed polarizers and rotated until the field is as dark as without the $\lambda/4$ plate. Then the fast and slow axes are aligned parallel to the axes of the polarizers, but it is not possible to distinguish between them. If the $\lambda/4$ plate is made of mica, Tutton's test (Strong, 1938) can be used to distinguish the fast axis from the slow axis. The plate is placed between crossed polarizers in white light and rotated until the field has its maximum brightness (axes at 45° to the plane of vibration of the incident light). The $\lambda/4$ plate is now rotated first about one axis and then about the other so that, in each case, the light passes through a thicker layer of mica. When rotation is about the fast axis, the color passes from bluish gray through iron gray to black. If it is about the slow axis, the color passes from white to yellow and then to interference colors of a higher order.

Another easy method for distinguishing between the fast and slow axes if one has an ellipsometer has been described by Alfano and Woodruff (1966). The ellipsometer is arranged so that the light passes through the polarizer and quarter-wave plate, reflects off a metallic surface near its principal angle of incidence, and then passes through an analyzer. The $\lambda/4$ plate is placed so that its axes are at 45° to the plane of incidence (all angles are read counterclockwise from the right-hand side of this plane when facing *against* the direction of the light beam). By alternately rotating the polarizer and analyzer, one attempts to obtain extinction with the polarizer angle lying between 0 and 45°. If extinction can be obtained in this range, the fast axis is at $-45°$ to the plane of incidence; otherwise it is at $+45°$. Other methods for distinguishing between the fast and slow axes have been given by Wood (1934*c*) and Childs (1956). However, since these methods require a Michelson interferometer and are tedious, they will not be described here.

81. Measuring Retardations Close to $\lambda/4$ Many of the tests described in this section are relatively simple, and qualitative or quantitative determinations of the retardation can be made with relatively simple equipment. However, if one has an ellipsometer (see Par. 6), the method of Alfano and Woodruff (1966) can be used to measure retardations with an accuracy of 0.5° or so over approximately the range of retardations from $\pi/4$ to $3\pi/4$.

When making a $\lambda/4$ plate out of mica, a test setup is needed which will show when the retardation is close to $\lambda/4$ (after a micrometer reading shows the thickness to be near 0.036 mm—to prevent getting plates that are $3\lambda/4$, $5\lambda/4$, or $7\lambda/4$). Emberson (1936) describes such a setup which can also be used to test quarter-wave plates made out of other materials. It consists of a collimated source of monochromatic light (of the wavelength for which the retardation is to be $\lambda/4$), beam splitter, polarizer test retardation plate, and mirror arranged as shown in Fig. 68. Light striking the beam splitter and passing through the polarizer will be plane-polarized. If the test plate has nearly $\lambda/4$ retardation and has one of its axes at 45° to the plane of vibration of the incident beam, the transmitted light will be nearly circularly polarized. Reflection from the mirror will change the sense

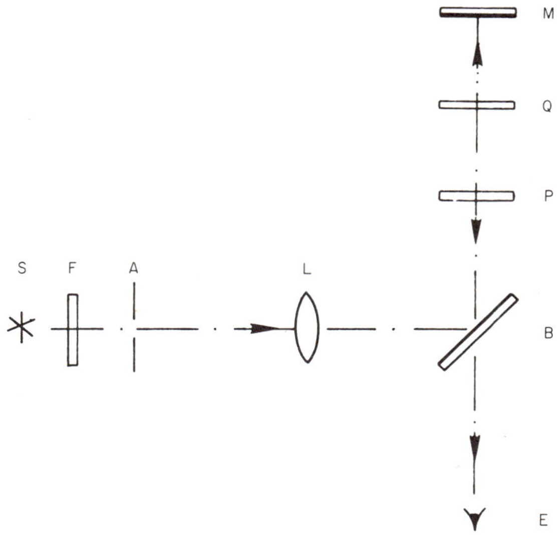

Fig. 68 Experimental arrangement for observing retardations near $\lambda/4$: *S*, source; *F*, filter; *A*, aperture; *L*, collimating lens; *B*, beam splitter; *P*, polarizer; *Q*, test plate (near $\lambda/4$); *M*, mirror; and *E*, eye of observer.

of the circularly polarized light, e.g., from right-handed to left-handed, so that when the beam again passes through the test plate, it will be almost linearly polarized but with the plane of vibration at right angles to the plane transmitted by the polarizer; thus an observer at *E* will see a dark field. If the test plate is rotated, there will be four positions per cycle where the field will be dark: where either axis of the plate makes an angle of 45° with the transmission axis of the polarizer. If the test plate is slightly thicker than $\lambda/4$, the twice-transmitted light will be elliptically polarized with the major axis of the ellipse lying at right angles to the transmission axis of the polarizer when the plate is oriented at 45° to the polarizer axis (see Fig. 58). Upon rotation of the test plate, the intensity will pass through a series of minima, and only light lying along the minor axis of the ellipse will be transmitted. If the retardation of the plate is $\lambda/4 + \epsilon$, it can be shown that the minor axis of the ellipse is proportional to $2\sqrt{2}\epsilon$. This method is not quantitative but it has the advantage of doubling the effect of any discrepancy of the test plate from a true quarter-wave plate since the light traverses the plate twice.

Emberson's method can also be used to determine the retardation of an achromatic $\lambda/4$ plate by substituting a white-light source for the monochromatic source (Pancharatnam, 1955*a*). If the $\lambda/4$ plate is really achromatic, the field will appear black at the four extinction positions, but if the achromaticity is only partial, the "extinction" image will be colored.

A different test with the same apparatus, proposed by Kubota et al. (1951), makes use of the sensitive color which arises when birefringent plates are placed between linearly polarizing devices. The light source is chosen to be a definite color temperature T such that a sensitive color appears when the test plate is exactly $\lambda/4$ for the desired wavelength. For 5461 Å, the value of T is 3940 K, and the sensitive color is reddish-purple. The method seems to suffer from the difficulty of finding the correct value of T and then of obtaining a suitable source; its sensitivity is also limited by the ability of the eye to distinguish the correct color (Jerrard, 1952).

Another test of $\lambda/4$ plates can be made with a modification of Emberson's apparatus in which a calcite rhomb is substituted for the polarizer (Hall, 1963). When the specimen has exactly a quarter-wavelength retardation and is oriented correctly, the central spot disappears from the group of three spots generally observed. The same experimental arrangement can also be used for testing $\lambda/2$ plates; if the specimen is an exact half-wave plate, only the central spot is observed and its brightness does not vary as the specimen is rotated.

The preceding tests are qualitative and give only a visual indication of when a plate has a retardation close to $\lambda/4$ or $\lambda/2$. A straightforward quantitative test of a $\lambda/4$ plate has been described by Ennos (1963*a*). The experimental setup consists of a monochromatic light source (filtered mercury arc), polarizer, and lens to focus the source onto a small aperture. The test plate (fast axis oriented at 45° to the polarizer axis), uniformly rotating analyzer, and detector follow. If the test plate has exactly $\lambda/4$ retardation (or an odd multiple thereof), the transmitted light will be circularly polarized and hence the signal transmitted by the rotating analyzer will be constant. If the retardation is slightly different from $\lambda/4$, an oscillating signal will be obtained, the amplitude of which is a measure of the ellipticity of the light and hence of the deviation of the plate retardation from 90°. When the plate retardation is less than 90°, the axis of the elliptically polarized light transmitted by the test plate is parallel to the polarizer axis; when more than 90°, the ellipse axis is perpendicular to the polarizer axis (Fig. 58). When the elliptically polarized light passes through a rotating analyzer, the transmitted intensity is (Ennos, 1963*a*)

$$I = 0.5I_0(1 - \sin 2\alpha \sin \delta) \tag{160}$$

where α is the angle between the major axis of the ellipse and the analyzer axis and δ is the deviation of the test plate from 90° retardation. Reflection losses at the surfaces of the test plate and absorption in the analyzer can be lumped into the term I_0, as will be shown below. If $\delta = 0$, that is, a perfect $\lambda/4$ plate, $I = 0.5I_0$, so that the light is perfectly circularly polarized and the transmitted intensity does not vary as the analyzer rotates. If $\delta \neq 0$, the observed signal will vary from $0.5I_0$ $(1 - \sin\delta)$ to $0.5I_0$ $(1 + \sin\delta)$ at twice the rotation frequency of the analyzer. The amplitude of the oscillation is thus $0.5I_0 \sin\delta$ and is superimposed on a steady signal of $0.5I_0$. The value of I_0 can be determined by aligning the fast axis of the retardation plate parallel to the polarizer axis and then rotating the analyzer. In this situation the light transmitted by the test plate is plane-polarized and I varies from 0 to I_0 twice each cycle, so that the amplitude of oscillation is $0.5I_0$. Since I_0 is the maximum light transmitted by the test plate and analyzer, reflection losses at the surfaces of the test plate and absorption in the analyzer are taken into account.

Several precautions must be taken to ensure a constant signal when the retardation of the test plate is exactly 90°. If the detector is a photomultiplier or other polarization-sensitive detector, the light striking it should be depolarized or approximately circularly polarized. A convenient way to achieve circular polarization is to attach an approximate $\lambda/4$ plate to the detector side of the rotating analyzer. Also the surface of the rotating analyzer should be focused on the detector to reduce the effect of nonparallelism of the analyzer faces; otherwise the transmitted light would describe a circle on the detector surface and produce a fluctuating signal. By taking these precautions, Ennos (1963*a*) was able to reduce the oscillating component of the signal from a perfect $\lambda/4$ plate to less than 1 part in 1,000 of the steady component.

Another method for accurately measuring retardation angles near 90° has been given by Grunstra and Perkins (1966). This method consists of measuring the *azimuth* of the elliptically polarized light produced when plane-polarized light passes through a nearly exact $\lambda/4$ plate. When an exact $\lambda/4$ plate is oriented at an azimuth β almost equal to 45° with respect to the polarizer axis, elliptically polarized light is produced. The major axis of the ellipse lies along the fast axis of the $\lambda/4$ plate if $\beta < 45°$ and along the slow axis of the plate if $\beta > 45°$ (Fig. 59). For $\beta = 45°$, the ellipse becomes a circle. If the plate is not a perfect $\lambda/4$ plate but has a retardation less than 90°, when $\beta = 45°$, elliptically polarized light is produced; however, the axis of the ellipse is *parallel to the polarizer axis* (not parallel to the fast or slow axes of the $\lambda/4$ plate; see Fig. 58). For retardations greater than 90°, the axis of the ellipse is *perpendicular* to the polarizer axis. Thus, when β takes on values close to but less than 45°, the azimuth of the elliptically polarized light is very sensitive to both the magnitude of the retardation and to β (see Grunstra and Perkins, 1966, Fig. 3). Measurements of the azimuth of the ellipse then afford an accurate method for determining the retardation of the test plate. In practice the azimuth of the *minor axis* of the ellipse is measured photoelectrically, since intensity minima are easier to determine than intensity maxima. This method was used by Grunstra and Perkins to measure the retardation of two plastic sheets and had a sensitivity of ±0.3° in retardation angle.

There are numerous other methods for measuring the retardation of quarter-wave plates and plates of other thicknesses (Goranson and Adams, 1933; Jerrard, 1952; Randall, 1954; Aben, 1963; and Adams and Waxler, 1965). Of these, the third Friedel method described by Goranson and Adams is supposedly a very accurate method for measuring any magnitude of phase retardation, but it requires a compensator which is almost exactly $\lambda/4$ for the particular wavelength used. Randall's photoelectric method is claimed to have an accuracy of better than 0.25 percent in retardation angle when the angle is near 90°. The method is supposedly least sensitive when used to calibrate $\lambda/2$ plates directly, but if an auxiliary plate about $\lambda/4$ in retardation is used in conjunction with the $\lambda/2$ plate (exact value need not be known), the retardation of the $\lambda/2$ plate can be determined with an accuracy of about 0.25°. Randall's method and also those of Ennos (1963*a*) and Grunstra and Perkins (1966) can be used to measure the dispersion of the phase retardation with wavelength and thus will yield the change of $n_e - n_o$ with λ [from Eq. (136)] if the thickness of the retardation plate can be measured accurately by an independent method. [We are of course assuming that there are no multiple reflections within the retardation plate, so that Eq. (136) holds exactly.]

Some of the calibration methods given by Jerrard (1952) seem tedious in comparison with methods already described. One of his methods, which is based on the "fact" that when a $\lambda/4$ plate is placed in any azimuth following a polarizer, the axes of the elliptical vibration always coincide with the main vibration directions of the plate, should be carefully scrutinized. As has already been noted, the azimuth of the ellipse is sensitive to the angle between the polarizer axis and the fast axis of the retardation plate and to the magnitude of the retardation. According to Randall (1954), Jerrard's channeled-spectrum method seems to be the best and supposedly fixes the wavelength at which the plate has a retardation of exactly $\lambda/4$ to within 10 Å. Osterberg (1970) has used the channeled-spectrum method and has been able to determine the $\lambda/4$ retardation wavelength with an accuracy of ±6 Å.

When $\lambda/4$ plates are used in ellipsometry to compensate for the ellipticity introduced by nonnormal-incidence reflection of light from a metal surface or from a dielectric film, the retardation of the plate need not be exactly 90° nor need it be known (Bennett and Bennett, 1967*b*). However, polarizer and analyzer settings must be taken in several quadrants so that the deviation of the retardation from 90° can be eliminated from the measurements.

82. Tolerance on Plate Thickness When making or purchasing a retardation plate, particularly one made from crystalline quartz, the tolerance on the plate thickness, degree of plane parallelism of the surfaces, and surface flatness should be specified. From Eq. (136), ΔN, the change in N caused by a change in thickness Δd, is

$$(\Delta N)\,\lambda = \pm\Delta d(n_e - n_o) \tag{161}$$

Written in terms of a phase angle, the expression becomes

$$\Delta\delta = 2\pi\,\Delta N = \pm 2\pi\frac{\Delta d(n_e - n_o)}{\lambda} \tag{162}$$

If one wants to make a crystalline-quartz plate which would have a retardation within 1° of a specified amount at 6328 Å, Eq. (162) predicts that the thickness must be within 1940 Å of the correct thickness. Thus the variation in total thickness, surface parallelism, and surface flatness must all be less than this amount. Since the standard tolerance on flatness of an optical surface is of the order of a quarter wavelength of mercury green light (approximately 1365 Å), there should be no difficulty in making the plate flat to the specified amount. However, it is more difficult to hold the total thickness and surface parallelism to this tolerance.

VARIABLE RETARDATION PLATES AND COMPENSATORS

83. Variable retardation plates can be used to modulate or vary the phase of a beam of plane-polarized light, to measure birefringence in mineral specimens, flow birefringence, or stress in transparent materials, or to analyze a beam of elliptically polarized light such as might be produced by transmission through a birefringent material or by reflection from a metal or film-covered surface. The term compensator is frequently applied to a variable retardation plate since it can be used to compensate for the phase retardation produced by a specimen. Common types of variable compensators include the Babinet and Soleil compensators (Pars. 84 and 85), in which the total thickness of birefringent material in the light path is changed, the Sénarmont compensator (Par. 86), which consists of a fixed quarter-wave plate and rotatable analyzer to compensate for varying amounts of ellipticity in a light beam, and tilting-plate compensators (Par. 87), with which the total thickness of birefringent material in the light beam is changed by changing the angle of incidence. Electrooptic and piezooptic modulators (Pars. 88 and 89) can also be used as variable retardation plates since their birefringence can be changed by varying the electric field or pressure. However, they are generally used for modulating the amplitude, phase, frequency, or direction of a light beam, in particular a laser beam, at frequencies too high for mechanical shutters or moving mirrors to follow. Since much information is already available elsewhere on electrooptic materials and devices, the treatment here will be abbreviated.

84. Babinet Compensator There are many devices which compensate for differences in phase retardation by having a variable thickness of a birefringent material (such as crystalline quartz) in the light beam (Johannsen, 1968*b*; Jerrard, 1948). The most common of these are the Babinet compensator and the Soleil compensator (Wood, 1934*d*; Ditchburn, 1963*b*). The Babinet compensator was proposed by Babinet in 1837 and later modified by Jamin; references to the voluminous early literature are given by Partington (1953). Ellerbroek and Groosmuller (1926) have a good description of the theory of operation (in German), and Jerrard (1949*a,b*; 1950*a*) and Archard (1950) describe various optical and mechanical defects of Babinet compensators.

The Babinet compensator, shown schematically in Fig. 69, consists of two crystalline-quartz wedges, each with its optic axis in the plane of the face but with the two optic axes exactly 90° apart. One wedge is stationary, and the other is movable by means of a micrometer screw in the direction indicated by the arrow, so that the total amount of quartz through which the light passes can be varied uniformly. In the first wedge, the extraordinary ray vibrates in a horizontal plane and is retarded relative to the ordinary ray (crystalline quartz has a positive birefringence; see Table 15). When the rays enter the second wedge, the ray vibrating in the horizontal plane becomes the ordinary ray and is advanced relative to the ray vibrating in the vertical plane. Thus, the total retardation is proportional to the difference in thickness between the two wedges:

$$N\lambda = (d_1 - d_2)(n_e - n_o) \tag{163}$$

where N = retardation in integral and fractional parts of a wavelength
d_1, d_2 = thickness of the first and second wedges where light passes through
n_o, n_e = ordinary and extraordinary refractive indexes for crystalline quartz

If light polarized at an angle of 45° to one of the axes of the compensator passes through it, the field will appear as shown in Fig. 69; the wedges have been set so there is zero retardation at the center of the field. (If the angle α of the incident plane-polarized beam were different from 45°, the beam retarded or advanced by 180° in phase angle would make an angle of 2α instead of 90° with

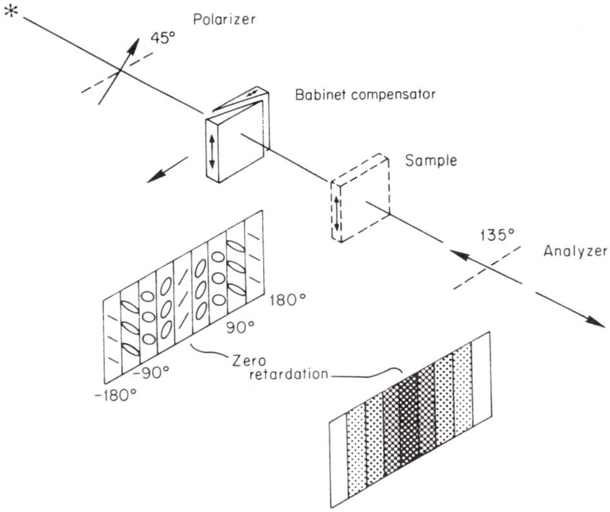

Fig. 69 Arrangement of a Babinet compensator, polarizer, and analyzer for measuring the retardation of a sample. The appearance of the field after the light has passed through the compensator is shown to the left of the sample position. Retardations are indicated for alternate regions. After the beam passes through the analyzer, the field is crossed by a series of dark bands, one of which is shown to the left of the analyzer.

the original beam.) When an analyzer whose axis is crossed with that of the polarizer is used to observe the beam passing through the compensator, a series of light and dark bands is observed in monochromatic light. In white light only one band, that for which the retardation is zero, remains black. All the other bands are colored. These are the bands for which the retardation is multiples of 2π (or, expressed in terms of path differences, integral numbers of wavelengths). On one side of the central black band one ray is advanced in phase relative to the other ray; on the other side it is retarded. If one wedge is moved, the whole fringe system translates across the field of view. The reference line is scribed on the stationary wedge so that it remains in the center of the field.

When using the compensator, the orientation of the axes should be determined and also the direction of wedge motion corresponding to a phase advance (or retardation). This is done as follows. With the axes of the polarizer and analyzer crossed, insert the compensator and rotate it until the entire field appears dark. In this orientation either the fast or slow axis of the compensator is parallel to the axis of the polarizer. If the compensator is constructed and oriented correctly, the entire field will appear black. However, certain optical and mechanical defects can produce fringes instead of a uniform field, as described by Jerrard (1949*a,b*; 1950*a*). Unfortunately, there are generally no adjustments for reorienting one wedge relative to the other in commercially available compensators.

The direction of the wedge motion which produces a phase advance or retardation can be found by using the arrangement shown in Fig. 69 with white light. The zero-retardation black band is set in the center of the field under the cross hair, and an auxiliary plate whose retardation is known is inserted in the place indicated by the dashed lines. One then notes in which direction the micrometer screw must be turned to make the zero-retardation band return to the center of the field. This experiment tells whether an increased or decreased thickness of the quartz wedge compensates for the retardation produced by the sample.

Next, the movable wedge must be calibrated to find out how much linear motion corresponds to a change in retardation of a full wavelength. Some compensators

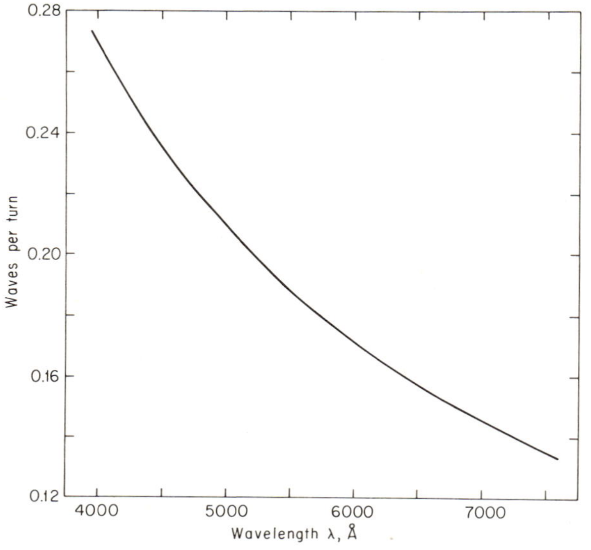

Fig. 70 Calibration curve for a Babinet compensator. The number of turns K of the micrometer screw required to change the retardation by 2π (or one wavelength) is the reciprocal of the ordinate.

come with a card giving the fraction of wavelength per turn of the micrometer screw for a series of wavelengths in the visible spectral region. A graph of values for one such compensator is shown in Fig. 70. This type of curve can easily be determined experimentally by orienting the polarizer, compensator, and analyzer as shown in Fig. 69 and first finding the zero-retardation band in white light. Next, using monochromatic light of a known wavelength, set on the black bands on either side of the central band and record the micrometer readings. The bands should be equally spaced, and by measuring positions on both sides of the central band one can determine whether the micrometer reading for the central band is correct. If the bands are not equally spaced within the precision of the settings, one of the wedges may not be uniform.

The Babinet compensator is now calibrated. It has generally been used in the following manner to measure the retardation of a sample whose fast or slow axis is known. The polarizer, compensator, and analyzer are oriented as indicated in Fig. 69, and the sample is placed in the position indicated by the dashed lines. Assuming the cross hair of the compensator was initially set on the zero-retardation band, the retardation of the sample can be determined directly by turning the micrometer screw until the zero-retardation band is once more in the center of the field under the cross hair. From a graph like Fig. 70, find the number of turns K per wavelength retardation (reciprocal of the ordinate) for the wavelength used. If the wedge (controlled by the micrometer screw) has to be moved a distance x to restore the central black band, the phase retardation introduced by the sample is $(x/K)2\pi$ rad, or x/K wavelengths. The sign of the birefringence of the sample can be determined by noting in which direction the micrometer screw must be turned to restore the central black band.

Another use for the Babinet compensator is to determine the phase retardation and ellipticity of a light beam which has, for example, been reflected from a

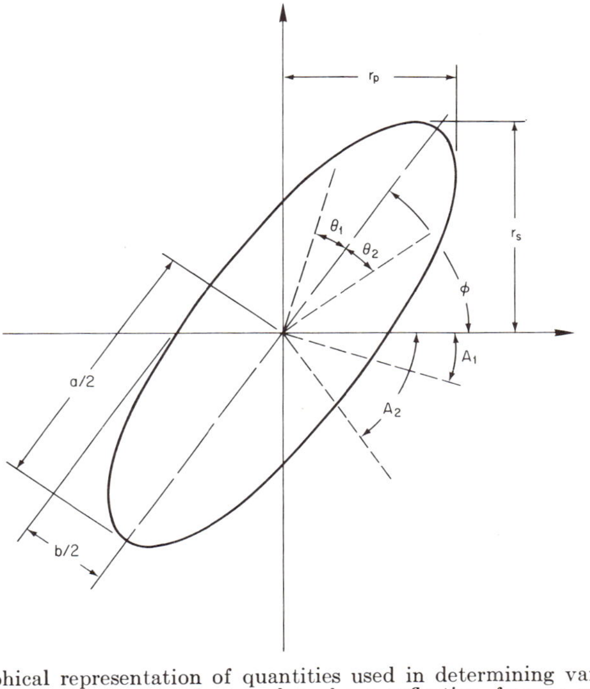

Fig. 71 Graphical representation of quantities used in determining various parameters of an elliptical vibration (assumed to be a reflection from a metal surface having amplitude components r_s and r_p). The angles θ_1 and θ_2 correspond to the orientation of the plane-polarized beam when the fast axis of the compensator is set on the major and minor axes of the ellipse. A_1 and A_2 are the analyzer settings necessary to extinguish the plane-polarized beam.

metal mirror. Such an elliptically polarized beam is shown in Fig. 71. The two rectangular coordinates correspond to the horizontal and vertical directions in Fig. 69. First, without the sample, set the polarizer, analyzer, and compensator as shown in Fig. 69 with the black zero-retardation band under the cross hair. Next, insert the sample in the beam. (If the sample happens to be a reflection sample, the analyzer will have to be swung around to the appropriate angle of incidence.) Now, reset the compensator so that the black band is again centered in the field and note the difference in micrometer readings. The phase retardation (or advance) is $(x/K)2\pi$ rad, where K is the linear distance between dark bands determined previously.

To determine the orientation of the ellipse and the ratio b/a of the minor to the major axis, set the compensator a distance $K/4$ away from the central dark band with the sample out. At this setting the center of the compensator field has a $\lambda/4$ retardation. Now, insert the sample and rotate the compensator *and analyzer* until the dark band is again under the cross hair. The compensator axis now coincides with the major or minor axis of the ellipse. Since the compensator is acting as a quarter-wave retarder, it is converting the elliptically polarized light into linearly polarized light, which can then be extinguished by the analyzer. From Fig. 59 (assume the light is passing backward through the plate) it is seen that the orientation of the plane-polarized beam depends on the ellipticity of the elliptically polarized beam: $\tan\theta = b/a$, where θ is the angle between the plane-polarized beam and the major axis of the ellipse.

There are actually four positions of the compensator (and analyzer) where the dark line will be centered on the cross hair. These correspond to alignment of the compensator fast axis with either the major or the minor axis of the ellipse. Two pairs of the compensator positions should be exactly 90° apart if the compensator is constructed correctly and is accurately set at $\lambda/4$ retardation. From Fig. 71 it is seen that the corresponding analyzer angles A_1 and A_2 can be used to determine the orientation and ellipticity of the ellipse since ϕ, the angle the major axis of the ellipse makes with the horizontal coordinate, equals $(A_1 + A_2)/2 + 90°$. Also, $A_2 - A_1 = 2\theta$, so that $\tan (A_2 - A_1)/2 = b/a$.

In ellipsometry (Par. 6) there is a method for determining the phase difference Δ from the value of the compensator retardation (Hartman, 1951). For this and other applications, it is helpful to know what kind of setting accuracy can be expected using a Babinet compensator. Jerrard (1948) suggests that a lower limit of about $1.4 \times 10^{-3}(2\pi)$ or about 0.5° is possible at a wavelength of 5890 Å. However, this number is only the uncertainty associated with setting on fringes spaced approximately 5 mm apart along a quartz wedge having an angle of about 30 minutes of arc. One should also calculate reasonable optical and mechanical tolerances which could be expected when making a set of Babinet wedges and see how they compare with the setting precision. The thickness uniformity required to produce an error of no more than 0.5° in phase angle can be calculated from Eq. (163) letting $n_e - n_o = 0.00911$ at 5893 Å and $N = 1.4 \times 10^{-3}$. Then $d_1 - d_2 = 0.15\lambda$. Surfaces flat to $\lambda/4$ are reasonably common, and it is possible to obtain optical flats good to better than 0.1λ. However, the optic axes must still be oriented properly in each wedge, and the wedges must be assembled correctly. Thus if the construction of the Babinet wedges were optimum, one would expect the compensator to perform with errors comparable to the visual-setting uncertainty.

85. Soleil Compensator The Soleil compensator, sometimes called a Babinet-Soleil compensator, is shown in Fig. 72. It is similar to the Babinet compensator in the way it is used, but instead of having a field crossed with alternating light and dark bands in monochromatic light, the field has a uniform tint if the compensator is constructed correctly. This is because the ratio of the thicknesses of the two quartz blocks (one composed of a fixed and a movable wedge) is the same over the entire field. The Soleil compensator will produce light of varying ellipticity depending on the position of the movable wedge. Calibration of the Soleil compensator is similar to that of the Babinet compensator. The zero-retardation position is found in the same manner as before except that now the entire field is dark. The compensator is used in the same way as a Babinet compensator with the uniformly dark field (in white light) of the Soleil corresponding to the black zero-retardation band in the Babinet.

The major advantage of the Soleil compensator is that a photoelectric detector can be used to make the settings. The compensator is offset a small amount on each side of the null position so that equal-intensity readings are obtained. The average of the two drum positions gives the null position. Photoelectric setting can be much more precise than visual setting, but this will not necessarily imply increased accuracy unless the compensator is properly constructed. Since Soleil compensators are composed of three pieces of crystalline quartz all of which must be very accurately made, they are subject to more optical and mechanical defects than Babinet compensators. Jerrard (1950*b*, 1951, 1953) has described many of these defects in detail. Ives and Briggs (1936) found random departures of about

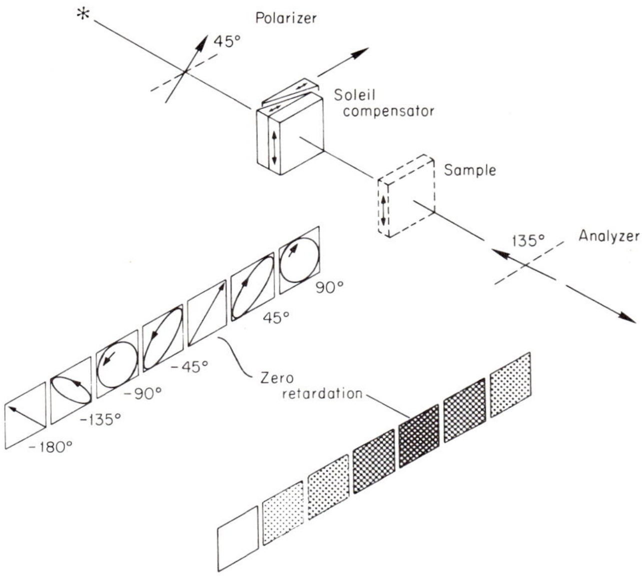

Fig. 72 Arrangement of a Soleil compensator, polarizer, and analyzer for measuring the retardation of a sample. The appearance of the field after the light has passed through the compensator is shown to the left of the sample position. After the beam passes through the analyzer, the field appears as one of the shades of gray shown to the left of the analyzer.

$\pm 1.5°$ from their straight-line calibration curve of micrometer reading for extinction vs. wedge position. This variation was considerably larger than the setting error with a half-shade plate and was attributed to variations in thickness of the order of $\pm\lambda/4$ along the quartz wedges.

Soleil compensators have been used for measurements of retardation in the infrared. They have been made of crystalline quartz, cadmium sulfide, and magnesium fluoride (Palik, 1965, 1968; Palik and Henvis, 1967). A by-product of this work was the measurement of the birefringence of these materials in the infrared.

Two other uniform-field compensators have been proposed. Jerrard (1951), following a suggestion by Soleil, has taken the Babinet wedges and reversed one of them so the light passes through the thicker portions of each wedge. This reversed Babinet compensator is less subject to mechanical imperfections than the Soleil compensator but does produce a small deviation of the main beam. Harihan and Sen (1960) suggest double-passing a Babinet compensator (with a reflection between the two passes) to obtain a uniform field.

86. Sénarmont Compensator A Sénarmont compensator is frequently used to determine the ellipticity of a beam of plane-polarized light in which the axis of the ellipse remains fixed. It consists of a quarter-wave plate whose fast axis is oriented parallel to the major axis of the ellipse followed by a rotatable analyzer, as shown in Fig. 73. When either axis of a quarter-wave plate is oriented parallel to the major axis of the ellipse, the elliptically polarized light is converted into plane-polarized light. The azimuth θ, measured relative to the fast axis of the $\lambda/4$ plate, depends on the ellipticity of the original beam according to the relation $\tan\theta = b/a$, where a and b are the lengths of the major and minor axes of the ellipse, respectively. The phase retardation between the two beams producing the elliptical vibration is simply 2θ, as can be seen from Fig. 59. An analyzer is used to extinguish the plane-polarized beam, so that its orientation (measured relative to the fast axis of the $\lambda/4$ plate) will be 90° from the desired angle θ.

Sénarmont compensators have been used to measure phase differences of light passing through birefringent materials such as electrooptic crystals (see, for example, Johnston and Weingart, 1965; Lenzo et al., 1966*b*). For this application the optic axis of the birefringent material is oriented at 45° to the azimuth of the plane-polarized beam and also at 45° to the fast axis of the $\lambda/4$ plate (Fig. 73). A detailed mathematical treatment of the Sénarmont compensator is given by Jerrard (1948), who also describes its operation using the Poincaré sphere (Jerrard, 1954*b*).

87. Tilting-Plate Compensators Another method for changing the thickness of birefringent material in the light path is to tilt a birefringent plate about an axis perpendicular to the direction of propagation, thus varying the angle of incidence of the light on the plate. Variable retardation plates operating on this principle include the Berek compensator and the Ehringhaus compensator, both discussed by Holmes (1964*b*).

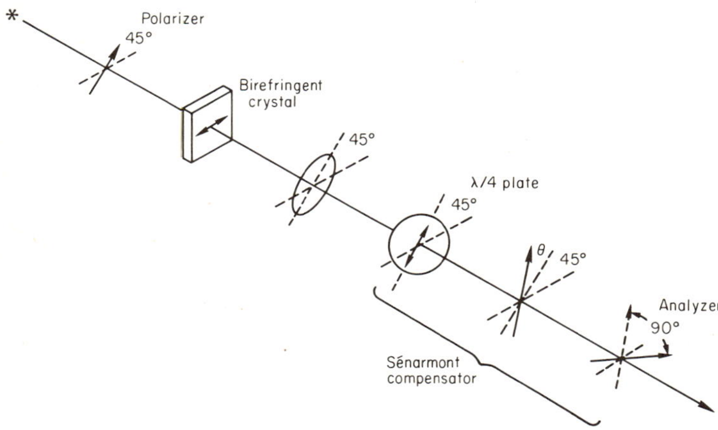

Fig. 73 Use of a Sénarmont compensator for determining the ellipticity or phase retardation produced by light passing through a birefringent material. The axis of the $\lambda/4$ plate is fixed, and the axis of the analyzer is rotatable.

The typical Berek compensator consists of a single calcite plate, usually about 0.1 mm thick, with the optic axis normal to the plate surfaces. When multiple internal reflections are neglected, the retardation is (Holmes, 1964*b*)

$$N\lambda = d\left[(n_o^2 - \sin^2\theta)^{1/2} - \frac{n_o}{n_e}(n_e^2 - \sin^2\theta)^{1/2}\right] \tag{164}$$

where N = retardation in wavelengths λ of light
d = plate thickness
θ = angle of incidence

A typical Ehringhaus compensator uses two plates of crystalline quartz of equal thickness d, usually 1.0 mm, with the optic axes parallel to the plate surfaces and perpendicular to each other. (This arrangement is similar to a Soleil compensator used in the zero-retardation position.) The axis of rotation is parallel to one of the optic axes, and the retardation is (Holmes, 1964*b*)

$$N\lambda = d\left[(n_e^2 - \sin^2\theta)^{1/2} - \frac{n_e}{n_o}(n_o^2 - \sin^2\theta)^{1/2}\right] \tag{165}$$

again neglecting multiple internal reflections. The retardation can be expressed in terms of phase angles by multiplying Eqs. (164) and (165) by $2\pi/\lambda$.

Holmes (1964*b*) has shown that multiple reflections within the compensator can affect the retardation, causing it to vary in an oscillatory manner somewhat similar to the way the phase retardation varies in thin sheets of mica (see Par. 59). Methods of eliminating this oscillation are similar to those proposed for mica: reduce the reflectance at the air-calcite (or quartz) surfaces. One method for accomplishing this is to immerse the compensator in an index-matching fluid (Holmes, 1965). Unfortunately, fluid immersion decreases the setting accuracy, and the refractive index of the immersion medium can be temperature-sensitive.

Other types of proposed tilting-plate compensators use a stressed plate of CR 39 resin (Fisher, 1948), lepidolite (a uniaxial variety of mica) (Record, 1948), or a pair of selenite plates (Smith, 1949). These compensators have the advantage of a large field of view and are suitable for observing stresses in transparent materials and other applications for which moderate accuracy is required.

88. Electrooptic Modulators An electrooptic material is one whose birefringence changes when an electric field is applied. When the field is applied in certain directions relative to the symmetry axes of a crystal that lacks a center of symmetry, the electrooptic effect is linear; i.e., the change in birefringence is directly proportional to the magnitude of the electric field. In some other cases the electrooptic effect is quadratic, the change in birefringence being proportional to the square of the applied field. The terms *Pockels effect* and *Kerr effect* are sometimes used for the linear and quadratic electrooptic effects, respectively, in crystals, although the latter term is preferably reserved for centrosymmetric or isotropic materials. (The term Kerr effect is more often used for the quadratic electrooptic effect in liquids.)

The electrooptic effect is generally described in terms of various coefficients related to the symmetry properties of the crystal. However, when using an electrooptic material as a variable retardation plate or light modulator, the half-wave field-distance product $V_{\lambda/2} = (El)_{\lambda/2}$ is a convenient quantity to use. This is the amount of voltage which must be applied across a thickness $d = 1$ cm to make the retardation $\Gamma = 2\pi l(n_e - n_o)/\lambda$ change by a half wavelength when the light passes through a length $l = 1$ cm of material. In other words, $V_{\lambda/2}$ represents the voltage required to produce a half-wavelength retardation in a geometry $l/d = 1$. For most materials $V_{\lambda/2}$ is proportional to wavelength, so that a larger voltage is required to produce a half-wave retardation at a longer wavelength. Conversely, for a given applied voltage, the retardation varies inversely with wavelength.

In a material exhibiting a longitudinal electrooptic effect, the applied field is in the same direction as the light path. Thus l/d must always be unity, and the applied field cannot be reduced by changing the aspect ratio. The longitudinal electrooptic effect is particularly useful for light beams of large cross-sectional area. However, since most materials having a longitudinal effect are uniaxial, the field angle is severely restricted. It can be shown that the field angle is inversely proportional to the square root of the crystal length for uniaxial materials (Billings, 1949c). Furthermore, since the light must pass through the electrodes, they must be transparent. Conducting Nesa-coated glass is frequently used for this purpose (Billings, 1949c). Other types of transparent electrodes have been described by Ploss (1969) and include metal grids, transparent conducting liquids, e.g., glycerol, and a new type of electrode consisting of a pile of glass plates viewed edge on onto which conducting films have been deposited.

The use of transparent electrodes is avoided when the electrooptic effect is transverse. In this case the directions of the applied field and the light path are at right angles. The applied voltage necessary to produce a given retardation can sometimes be significantly reduced by making l very large and d very small. Ratios of l/d as large as 250 have been obtained (Spencer et al., 1967). However, the reduction in voltage is obtained at the expense of the field angle. For applications which require large field angles, crystals with cubic symmetry are necessary with l/d ratios near unity.

Table 17 lists a number of common electrooptic materials, their half-wave voltages, transmission regions, and crystal structures. The most widely used are the first three in the table, ammonium dihydrogen phosphate (ADP), potassium dihydrogen phosphate (KDP), and potassium dideuterium phosphate (KD*P). The ADP-type materials have tetragonal crystal structure and exhibit a longitudinal electrooptic effect. Some of these materials have also been shown to exhibit a transverse electro-

TABLE 17 Properties of Electrooptic Materials at Room Temperature

Crystal structure and material	Optical symmetry	Refractive index			Half-wave voltage[a]		Direction[b]	λ_1 to λ_2,[c] μm	Transmission region, μm	Ref.[d]
		n_o	n_e	Wave-length λ, μm	$V_{\lambda/2}$, kV	Wave-length λ, μm				
Tetragonal:										
Ammonium dihydrogen phosphate (ADP), $NH_4H_2PO_4$	Uniaxial	1.526	1.481	0.546	9.0 7.65	0.546	L T	0.4–0.7	0.125–1.7	1,2 3,4
Potassium dihydrogen phosphate (KDP), KH_2PO_4	Uniaxial	1.512	1.470	0.546	7.65 17.8	0.546	L T	0.436, 0.578	0.250–1.7	5 4,6
Potassium dideuterium phosphate (KD*P or DKDP), KD_2PO_4	Uniaxial	1.508	1.468	0.546	2.98 8.00	0.546	L T	0.5–0.75	0.19–2.15	7 4
Potassium dihydrogen arsenate (KDA), KH_2AsO_4	Uniaxial	1.571	1.521	0.546	6.43	0.546	L	0.436, 0.578	0.246–>0.75	5
Ammonium dihydrogen arsenate (ADA), $NH_4H_2AsO_4$	Uniaxial	1.578	1.522	0.589	13.0	0.546	L	....	0.260–>0.75	8
Rubidium dihydrogen arsenate, RbH_2AsO_4	Uniaxial	1.559	1.520	0.546	5.54	0.546	L	0.436, 0.578	....	5
Rubidium dihydrogen phosphate, RbH_2PO_4	Uniaxial	......	......	.....	7.3	0.546	L	....	....	8
Barium titanate, $BaTiO_3$	Uniaxial	2.437	2.365	0.546	0.38	0.546	T	....	....	9
Cubic perovskite:										
Barium titanate, $BaTiO_3$	Isotropic	2.43	......	Visible	0.66[e] $T > 120°C$	0.546	T	0.4–1.0 ($g_{11}-g_{12}$)	~0.4–7.5	10,11
Strontium titanate, $SrTiO_3$	Isotropic	2.38	......	0.633	~13[e]	......	T	....	0.39–6.8	12
Potassium tantalate, $KTaO_3$	Isotropic	2.24	......	0.633	0.95 at 4.2 K[e]	0.633	T	....	....	10,13
Potassium tantalate niobate (KTN), $KTa_{0.65}Nb_{0.35}O_3$	Isotropic	2.29	......	0.633	0.58[e]	0.633	T	0.4–2.0 ($g_{11}-g_{12}$)	0.39–6	14,15
Lead magnesium niobate, $Pb_3MgNb_2O_9$	Isotropic	2.56	......	0.633	......	......	T	....	0.5–5	16
Rhombohedral perovskite:										
Lithium niobate (LN), $LiNbO_3$	Uniaxial	2.297	2.208	0.60	2.940 2.5	0.633	T L	....	0.4–5	17 18
Lithium tantalate (LT), $LiTaO_3$	Uniaxial	2.183	2.188	0.60	2.800	0.633	T	....	0.45–5 est	6,19
Monoclinic:										
Calcium pyroniobate, $CaNb_2O_7$	Biaxial	1.97, 2.16, 2.17		.....	4.55	0.633	T	....	....	20
Orthorhombic tungsten bronze:										
Barium sodium niobate, $Ba_2NaNb_5O_{15}$	Uniaxial	2.326, 2.324, 2.221		0.633	1.570	0.633	T	....	0.4–5	21
Tetragonal tungsten bronze:										
Potassium lithium niobate, $(K_2O)_{0.3}(Li_2O)_{0.7-x}(Nb_2O_5)_x$ $0.515 \leq x \leq 0.55$	Uniaxial	2.28 $x = 0.52$	2.13	0.633	1.350 $0.525 \leq x \leq 0.55$	0.633	T	....	....	22

Potassium lithium niobate, $K_{0.6}Li_{0.4}NbO_3$ or $K_3Li_2Nb_5O_{15}$	Uniaxial	2.277	2.163	0.633	0.930	0.633	T	....	0.4-5	23
Potassium sodium barium niobate $K_xNa_{1-x}Ba_2Nb_5O_{15}$, $0.70 \leq x \leq 0.90$	Uniaxial	2.315	2.230	0.633	1.410 $x = 0.8$	0.633	T	....	....	24,25
Potassium strontium niobate, $KSr_2Nb_5O_{15}$	Uniaxial	~2.25	~2.25	0.633	0.43	0.633	T	....	....	26
Strontium barium niobate, $Sr_xBa_{1-x}Nb_2O_6$, $0.25 \leq x \leq 0.75$	Uniaxial	2.312	2.299	0.633	0.037 $x = 0.75$	0.633	T	....	....	27,28
Cubic zinc blende:										
Cadmium telluride, CdTe	Isotropic	2.6	......	1-30	4.9	1.0	T	3.39,10.6, 23.35,27.95 ($n_o{}^3r_{41}$)	1-30	29-31
Cuprous chloride, CuCl	Isotropic	1.996	......	0.535	6.2 6.2	0.546	L T	....	0.4-20	32,33 33,34
Cuprous bromide, CuBr	Isotropic	2.16	......	0.535	.....	....	T	0.525,0.675 ($n_o{}^3r_{41}$)	....	35
Gallium arsenide, GaAs	Isotropic	3.42	......	1.0	4.5	1.0-1.7	T	10.6 (r_{41})	1.0-15	10,36-39
Gallium phosphide, GaP	Isotropic	3.350	......	0.60	13	0.546	T	....	0.6-4.5	10,40
Zinc sulfide, ZnS	Isotropic	2.364	......	0.60	10.2	0.546	T	0.404-0.644 (r_{41})	....	41
Zinc selenide, ZnSe	Isotropic	2.654	......	0.546	7.1	0.5475	T	....	....	10,42,43
Zinc telluride, ZnTe	Isotropic	3.06	......	0.589	2.3	0.590	T	0.59-0.69 10.6(r_{41})	0.57-52	39,44
Cubic eulytine:										
Bismuth germanate, $Bi_4(GeO_4)_3$	Isotropic	2.07	......	.....	14.8	0.543	T	0.453,0.616	0.35-6	45
Body-centered:										
Bismuth germanium oxide, $Bi_{12}GeO_{20}$	Isotropic	2.55	......	0.51	12.0	0.666	T	....	....	6,46
Tetragonal:										
Cadmium mercury thiocyanate, $Cd[Hg(SCN)_4]$	Uniaxial	......	......	.....	15.6	0.633	T	0.476	....	47
Hexagonal wurtzite:										
Cadmium sulfide, CdS	Uniaxial	2.493	2.511	0.60	8.6	0.589	L	....	0.5-16	48
Hexagonal:										
Crystalline quartz, SiO_2	Uniaxial	1.544	1.553	0.589	325	0.546	T	....	0.12-4.5	8
Cubic:										
Hexamethylenetetramine (HMT or hexamine), $N_4(CH_2)_6$	Isotropic	1.591	......	0.589	8.3	0.5475	T	....	0.35-2.25	49,50

[a] When values of $V_{\lambda/2}$ are given for more than one direction in the crystal, the minimum value is reported here.
[b] L = longitudinal; T = transverse.
[c] Other wavelengths at which half-wave voltage has been measured.
[d] 1 = Carpenter (1950); 2 = Trevelyan (1969); 3 = Dore (1967); 4 = Hookabe and Matsuo (1970); 5 = Ott and Sliker (1964); 6 = Spencer et al. (1967); 7 = Sliker and Burlage (1963); 8 = West and Carpenter (1963); 9 = Johnston and Weingart (1965); 10 = Johnston (1965); 11 = Johnston (1971); 12 = Sonin and Perfilova (1968); 13 = Geusic et al. (1963); 14 = Geusic et al. (1964); 15 = Chen et al. (1966); 16 = Bonner et al. (1967); 17 = Lenzo et al. (1966*b*); 18 = Hulme et al. (1969); 19 = Lenzo et al. (1966*c*); 20 = Holmes et al. (1965); 21 = Geusic et al. (1967, 1968); 22 = Smith et al. (1971*b*); 23 = Van Uitert et al. (1967, 1968); 24 = Burns et al. (1969); 25 = Smith et al. (1971*a*); 26 = Giess et al. (1967); 27 = Lenzo et al. (1967); 28 = Venturini et al. (1968); 29 = Stafsudd et al. (1967); 30 = Kiefer and Yariv (1969); 31 = Johnson (1968); 32 = West (1953); 33 = Sterzer et al. (1964); 34 = Rivera et al. (1967); 35 = Belyaev et al. (1965); 36 = Ho and Buhrer (1963); 37 = Turner and Kaminow (1963); 38 = Walsh (1966); 39 = Kaminow (1968); 40 = Nelson and Turner (1968); 41 = Namba (1961); 42 = McQuaid (1962); 43 = Kojima et al. (1969); 44 = Sliker and Jost (1966); 45 = Nitsche (1965); 46 = Lenzo et al. (1966*a*); 47 = Hundelshausen (1971); 48 = Gainon (1964); 49 = McQuaid (1963); 50 = Heilmeier (1964).
[e] Quadratic effect; $V_{\lambda/2}$ for first π rad retardation in a 1-cm cube.

optic effect (Dore, 1967; Trevelyan, 1969; Hookabe and Matsuo, 1970), and modulators of this type can be obtained commercially (*Laser Focus,* 1974). Unfortunately, both the longitudinal and transverse ADP-type modulators have angular fields of view that are restricted for thick samples, as mentioned above. Billings (1952) discusses ways of increasing the angular field in the absence of an applied voltage by minimizing or effectively canceling the natural retardation of ADP-type uniaxial crystals.

ADP, KDP, and KD*P are water-soluble and fragile, but they can be cut and polished without difficulty. Excellent, large (approximately 5 cm), strain-free crystals of all three materials are available. KD*P is frequently supplied in a form that is 90 percent deuterated, which changes its half-wave voltage only slightly. The 90 or 100 percent deuterated KD*P materials are much superior to ADP in that they require only about one-third the voltage necessary for ADP to produce a half-wave retardation (in the longitudinal mode). However, ADP has been successfully used also in a variety of applications (see, for example, Takasaki, 1961, 1962, 1963, 1964, 1966; Takasaki et al., 1964*a,b*).

When using a KD*P crystal as a variable retardation plate on an ellipsometer in this laboratory, we found that the retardation changed as a function of time if a fixed dc bias was applied to the crystal. This problem could be eliminated by using square-wave ac modulation and phase-sensitive detection. Kaminow (1964) has shown that heating effects can be deleterious when using KDP crystals with high-intensity laser beams (power levels greater than 1 W). The crystal is heated throughout its volume but cooled only on its surface, so that the resulting thermal gradient produces strains which severely degrade the electrooptic performance. These effects are independent of the crystal dimensions. In the event that one has crystals with very low absorption, Hook and Hilberg (1971) have shown how to greatly reduce the effect of reflection losses from the crystal faces in a high-power *Q*-switched laser by making the faces parallel within a few seconds of arc and aligning the crystals very accurately with the lasing axis.

The perovskites barium titanate, strontium titanate, potassium tantalate, and lead magnesium niobate and the mixed crystal potassium tantalate niobate (KTN), listed in Table 17, belong to the group of oxygen-octahedra ferroelectrics (Wemple and DiDomenico, 1969) and possess Curie temperatures at which their crystal structure and properties change. The Curie temperature of barium titanate is near 120°C, so that at room temperature the material exists in a tetragonal form, is ferroelectric, and exhibits a linear electrooptic effect. Since crystals of good optical and electrical quality are available, but since $BaTiO_3$ is uniaxial and has a large residual birefringence, modulators or retarders must have very parallel end faces.

Strontium titanate and potassium tantalate have Curie temperatures close to absolute zero, while the Curie temperature of KTN is about 10°C. Thus, these materials are cubic at room temperature and have a quadratic electrooptic effect. (The half-wave voltages in Table 17 are for the first π rad retardation.) Significantly smaller voltages can be obtained and linearity can be approached if the crystals are given a fixed dc bias to make them operate at higher retardations. A drawback to this method of operation is that the materials are extremely temperature-sensitive, so that the temperature must be very carefully controlled. Furthermore, the dc-bias field leads to space-charge effects that eventually reduce the internal biasing field (Kaminow and Turner, 1966*a,b*). The mixed crystal KTN looks very promising for efficient operation at room temperature because of its extremely low half-wave voltage. However, so far it has not been possible to grow single crystals of the material with sufficient uniformity, optical quality, and electrical resistivity for device applications. Lead magnesium niobate also has a large quadratic electrooptic effect, and unlike KTN it does not have electronic-conduction and compositional-inhomogeneity problems.

Lithium niobate and lithium tantalate are also perovskites belonging to the group of oxygen-octahedra ferroelectrics. Their Curie temperatures are about 1470 and 890 K, respectively (Kaminow and Turner, 1966*a,b*), so that at room temperature they exist in the tetragonal form, are ferroelectric and piezooptic, and exhibit a transverse electrooptic effect comparable in magnitude to the longitudinal effect in KD*P. Lenzo et al. (1966*b*) have measured half-wave voltages for various directions of the applied field and light propagation in $LiNbO_3$; the smallest is given in Table 17. The longitudinal electrooptic effect has also been investigated (Hulme et al., 1969), and the $V_{\lambda/2}$ value is comparable to the smallest transverse value. Techniques have been developed to grow large single crystals (approximately 1 by 1 by 3 cm) of good optical and electrical quality and to make them into single-domain material. Because the Curie temperature (near which point the electrical properties are susceptible to change by rough handling) is very high, the crystals can be cut, polished, and handled without creating additional domains, an advantage over $BaTiO_3$. In the crystallographic orientation for which the half-wave voltage is the smallest, the natural birefringence of $LiNbO_3$ is very temperature-sensitive, so that the temperature of the material must be very carefully controlled. There is also a problem when subjecting the material to high-intensity laser beams. Refractive-index inhomogeneities can be introduced, which cause the light beam passing through the crystal to diverge (Chen, 1969; Ashkin et al., 1966). Annealing the crystals near the Curie temperature to reduce the sensitivity to radiation damage does not seem to be effective.

Lithium tantalate is similar to lithium niobate in its crystal structure and electrical and optical properties, and it is also sensitive to damage by intense laser radiation. Some success has been achieved by annealing samples near the Curie temperature with an electric field applied along the optic axis (Chen, 1969). Calcium pyroniobate is closely related chemically and structurally to $LiNbO_3$ and $LiTaO_3$. It crystallizes in the monoclinic point group 2 (single twofold axis of symmetry) and is optically biaxial negative.

A theory based on the assumption that the linear electrooptic effect is fundamentally a biased quadratic effect has been proposed to explain the electrooptic properties of oxygen-octahedra ferroelectrics (Wemple and DiDomenico, 1969). These include materials with perovskite and also tungsten-bronze structures (some of the latter are also listed in Table 17). Particular interest has been shown in the tungsten-bronze group because these materials exhibit large linear electrooptic and second-harmonic-generation coefficients, are easy to grow, and do not suffer from laser damage. Burns (see Burns and Smith, 1968; Burns, 1969) has published two review articles on the properties and improvements of tungsten-bronze ferroelectrics. The material $Sr_xBa_{1-x}Nb_2O_6$ should be especially noted since for $x = 0.75$ the half-wave voltage is only 37 V. If an l/d ratio of 6 is used, the half-wave voltage can be reduced to about 8 V.

The cubic crystals having zinc blende structure are all isotropic in the absence of an applied field. They can exhibit either a longitudinal or a transverse electrooptic effect. Since they are isotropic, they are suitable for applications requiring large acceptance angles, e.g., in the infrared, and several have been used at 10.6 μm. However, small amounts of birefringence due to strain or nonuniformity can cause appreciable depolarization of a light beam. The theory of electrooptic modulation with cubic crystals has been given by Buhrer et al. (1963), and wide-angle intensity-modulation characteristics for cubic crystals are discussed by Lee (1969).

The materials in the cubic zinc blende group are all semiconductors; preparation, band structure, and electrical properties of many have been extensively studied, but there is little information on the electrooptic properties. Of all the crystals in this group, only gallium arsenide and zinc telluride are readily available with dimensions of the order of 1 cm. Zinc sulfide and cuprous chloride occur naturally but are generally not of suitable quality. Synthetic cuprous chloride crystals usually have strains when grown from the melt and deteriorate on exposure to normal atmospheric humidity. Synthetic cubic zinc sulfide generally contains stacking faults. Zinc selenide is impractical to use as an optical modulator because of its small breakdown voltage and large half-wave voltage. It also has a large change in refractive index with applied voltage.

The crystals in the last group have miscellaneous crystal structures. Bismuth germanate is weakly electrooptic, and single crystals of good quality and dimensions of a few centimeters can be grown from a melt (Nitsche, 1965). Its good optical transmission from 0.4 to 6 μm makes modulation of light sources over a wide range of wavelengths possible. Since there are many isomorphs of various chemical compositions, it may be possible to find some with lower half-wave voltages. Bismuth germanium oxide is the first tetrahedrally coordinated cubic crystal to have been synthesized in a reasonable size and with good optical quality (Spencer et al., 1967). It is photoconductive, highly optically active, weakly electrooptic, and strongly elastooptic and has a moderately strong piezoelectric effect and unusually good ultrasonic properties. It is not ferroelectric (Spencer et al., 1967). Cadmium sulfide, with a hexagonal wurtzite structure, is optically uniaxial and is readily available in single crystals with dimensions of the order of 1 cm.

Various other materials should be mentioned. The interest in crystalline quartz comes from its availability and excellent optical properties rather than from a large electrooptic effect. Crystals of quartz are optically active, and this complicates use of the material as an electrooptic modulator. Hexamethylenetetramine (HMT) is of scientific interest (because it is a molecular crystal) and of technological interest because of the magnitude of its electrooptic effect. The material is inexpensive, and crystals can be grown with ease. However, it appears that large crystals, in particular those grown from solution, do not exhibit a large electrooptic effect, presumably because of dislocations (Heilmeier, 1964). Miller et al. (1971) suggest that electrooptic effects may be enhanced at wavelengths near electronic resonances. They show that enhancement is indeed observed in the compound $Na_3Li(CrO_4)_2 \cdot 6H_2O$.

Ferroelectric electrooptic ceramics have been added to the rapidly growing list of electrooptic materials (Land and Thacher, 1969; Thacher and Land, 1969). Polycrystalline rhombohedral lead zirconate-titanate ceramics possess one of two types of electrooptic properties, depending on the nominal grain diameter. In coarse-grained ceramics the light-scattering properties depend on the orientation of the ceramic polar axis and are essentially independent of the ferroelectric remanence state. Fine-grained ceramics are birefringent, and their light-transmission characteristics are similar to those of optically uniaxial crystals. Both coarse- and fine-grained ceramics have the property that localized areas as small as 25 by 25 μm can be switched independently without affecting the light-transmission characteristics of the surrounding areas.

Ferroelectric glass-ceramic systems containing various niobates or barium titanate as the crystalline phase are another new type of electrooptic material (Borrelli, 1967; Borrelli and Layton, 1969). A transverse quadratic electrooptic effect has been observed in a wide range of niobate transparent glass-ceramic systems. When a dc field is applied, the electrooptic effect becomes linear and has been observed in both longitudinal and transverse modes.

Further information on electrooptic modulators can be obtained from the excellent review article by Kaminow and Turner (1966*a,b*), the book by Kaminow (1974), and the comprehensive articles by Billard (1966) and Spencer et al. (1967). Ellis and Walton (1971) have compiled a bibliography including electrooptic, acoustooptic, magnetooptic, mechanical, and miscellaneous types of modulators. Many uses of electrooptic crystals have been outlined by Ploss (1969). The theory of the electrooptic effect in uniaxial crystals has been given by Billings (1949*b*), and a description of the properties of crystals can be found in Nye (1957). References to specific materials are given in Table 17. This field is a rapidly growing one and may be expected to change markedly in the next few years.

89. Piezooptic Modulators The piezooptic effect is analogous to the electrooptic effect but pressure rather than an electric field is used to produce a change in the birefringence of a material. When some isotropic materials are stressed at right angles to the direction of the light beam, they become uniaxial and produce a phase retardation that is proportional to the stress. Variable retardation plates of stressed fused quartz (Ennos, 1963*a,b*) and lithium fluoride (Metcalf and Baird, 1966) have been used for the ultraviolet, beyond the transmission region of mica. An advantage of making plates which have a fixed or slowly varying retardation from piezooptic rather than electrooptic materials is that the former are highly elastic, so that the retardation remains constant under a fixed load. The retardation produced by some electrooptic modulators will change with time if a fixed dc voltage is applied (Par. 88). Second, piezooptic devices have a very large angular aperture in comparison with electrooptic devices. In a stressed isotropic material, the change in retardation is proportional to the optical-path length in the material and varies as sec θ, where θ is the angle of incidence on the material. In choosing a plate to be stressed, it is important to select one that has no residual strain, so that the entire field will be uniform in the absence of applied stress (Ennos, 1963*a*).

In addition to using piezooptic materials as slowly varying retarders, they can be used for modulators by setting up acoustic standing waves in a block of the material (Jasperson and Schnatterly, 1969; Kemp, 1969). For this application, only a comparatively small amount of power is needed to produce a given retardation because of the Q of the material. If there were no acoustic losses, an oscillation would persist indefinitely and a freely suspended block would generate modulated birefringence of more or less arbitrary amplitude by itself. The piezooptic modulators of Jasperson and Kemp were made from blocks of fused quartz coupled to piezoelectric crystalline-quartz transducers; oscillations produced in the transducers were coupled directly into the fused quartz, and both parts vibrated at their resonant frequencies. With a fused-quartz block 10 cm thick oscillating at 50 kHz, a maximum phase retardation of π at 4000 Å could be produced with only a 20-V peak-to-peak input signal (Jasperson and Schnatterly, 1969).

In addition to the piezooptic modulators described above, two other groups have used similar types of modulators (Billardon and Badoz, 1966*a,b*; Mollenauer et al., 1969). However, both these groups used composite resonators in which only a fraction of the total mass of vibrating material was in the optical element. Thus, the resonant frequency and the effective Q of each system were determined by the nonoptical elements. Hauser et al. (1963) used a glass plate under strain as a moderate-speed optical shutter.

Other materials besides glass, fused quartz, and lithium fluoride can be used in piezooptic applications. Cubic materials are preferable since they are isotropic and have no residual birefringence (except for strains or other defects). Calcium fluoride has been suggested for ultraviolet-transmitting $\lambda/4$ plates (Kemp, 1969). Giardini (1957) has measured the stress-optical coefficients of strontium titanate, a cubic perovskite, which becomes uniaxial (optically negative) when stressed along a crystallographic axis. Other cubic materials whose piezooptic properties have been measured are listed by Giardini (1957). A very sensitive method for measuring stress-optical coefficients has been proposed by Michael (1968). Although there is no general theory which satisfactorily explains observed stress-optical relationships, Mueller (1935) has proposed a theory that explains the observed stress-optical relationships for cubic crystals.

Acoustooptic pulse modulators are now being used to modulate laser beams at megahertz frequencies (Maydan, 1970*a*), and several review articles have been written on the subject (Maydan, 1970*b*; Minkoff et al., 1968; Gordon, 1966*a,b*). A bibliography of acoustooptic modulators has been compiled (Ellis and Walton, 1971), and photoelastic properties of selected materials have also been published (Dixon, 1967). These devices work on a different principle from the piezooptic modulators discussed above in that light is scattered out of the main beam by acoustic waves set up in the material by an appropriate transducer. Acoustooptic pulse modulators are rapidly gaining importance. However, since several excellent current review articles are available, the subject will not be further discussed here.

HALF-SHADE DEVICES

Half Shades for Plane-polarized Light

90. It is sometimes necessary to measure accurately the azimuth of a beam of plane-polarized light, i.e., the angle the plane of vibration makes with a reference coordinate system. This can be done most easily by using one of the types of polarizers described in Pars. 7 to 55 as an analyzer and rotating it to the position where the field appears the darkest. The analyzer azimuth is then exactly 90° from the azimuth of the plane-polarized beam. A more sensitive method is to use a photoelectric detector and offset on either side of the extinction position at angles where the intensities are equal. The average of these two angles is generally more accurate than the value measured directly, but care must be taken to keep the angles small so that asymmetries will not become important.

Before the advent of sensitive photoelectric detectors, the most accurate method of setting on a minimum was to use a half-shade device as the analyzer or in conjunction with the analyzer. The device generally consisted of two polarizers having their axes inclined at an angle α to each other (angle fixed in some types and variable in others) as shown schematically in Fig. 74*a*. As the device was rotated, one part of the field became darker while the other part became lighter. At the match position, both parts of the field appeared equally bright. (Sometimes the match position was determined by offsetting the half-shade analyzer on either side of the match position to produce a given intensity difference in both halves of the field.) Since the intensity at the match position is generally very dim, the terms half shade and penumbra have been associated with the devices (Heller, 1960*b*).

If possible, the half-shade angle α is chosen for maximum sensitivity with a given-intensity light source. For a very intense source, a small half-shade angle can be used to give an extremely high setting precision, the limit being about 4 seconds of arc for a half-shade angle of 0.5° (Heller, 1960*b*). With less intense

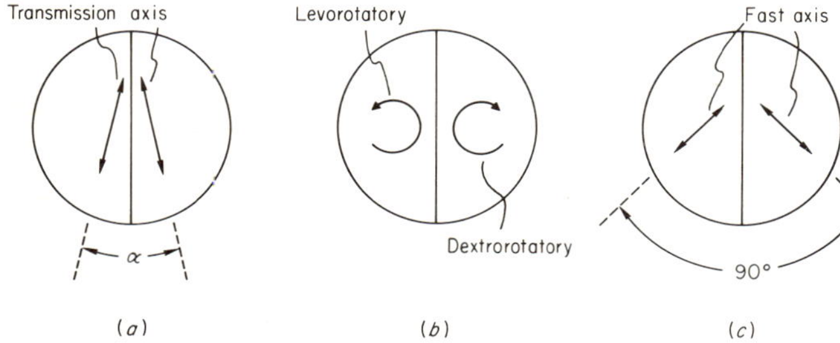

Fig. 74 Half-shade devices: (*a*) half-shade analyzer and (*b*) Nakamura biplate for determining the azimuth of a plane-polarized beam; (*c*) Bravais biplate for detecting small amounts of ellipticity.

sources, larger half-shade angles must be used, and the setting precision is not as good. Inoué and Koester (1959) discuss the optimum half-shade angle to use with a given illumination level and extinction ratio in a polarizing microscope.

Photoelectric settings can also be made with half-shade devices either by the offset method described above or by directly determining the azimuth at which the intensities in both halves of the field are equal (Mathews, 1965). There appears to be no advantage, however, in using a half-shade device with a photoelectric detector.

Although half-shade devices are usually used in monochromatic light, some types are intended to be used in white light. These are placed in front of the analyzer and derive their setting precision from a so-called sensitive tint. One of these is the Soleil biquartz (Dale, 1923*a*), which has a residual color that is extremely sensitive, turning either red or blue with a very slight rotation of the analyzer; similar devices are described by Johannsen (1968*c*). However, these devices are unsatisfactory since the optical rotation of even colorless materials is a function of wavelength, so that there is no position of the analyzer for which the tints in both halves of the field are exactly the same.

91. Jellett-Cornu Prism The simplest type of half-shade device used in monochromatic light is the Cornu modification of the Jellett prism, made by sawing a Glan-Thompson prism down the center, grinding one of the halves at an angle, and cementing the two parts back together (Heller, 1960*b*). This prism is used in place of the analyzer and passes beams of light polarized at a slight angle to each other. (Cornu used the split prism in place of the polarizer, but the principle is the same.) A similar device could also be made from a piece of high-extinction Polaroid (or other dichroic sheet polarizer). The main problem is to get the dividing line between the two halves of the field as sharp as possible so that the eye can distinguish gray areas very close to each other.

92. Lippich and Laurent Half Shades In contrast to the Jellett-Cornu prism, in which the half-shade angle is fixed, the Lippich prism (Heller, 1960*b*) has a variable half-shade angle. A small Glan-Thompson prism covering half the field is placed in front of the Glan-Thompson analyzer with its axis set at a slight (variable) angle to the axis of the main analyzer. In the Laurent variable half-shade device, a quartz half-wave plate covering part of the field is placed in front of the analyzer. Its principle of operation is explained in detail by Heller (1960*b*). The major disadvantages of variable half-shade devices such as these are that the half shade is asymmetric with changes in azimuth angle and the brightness is different in the two halves of the field because of reflections and absorption in the auxiliary plate which covers half the field.

93. Nakamura Biplate The Nakamura biplate† (Heller, 1960*b*), which works on a different principle, is a much more convenient half-shade device. It consists of pieces of right- and left-handed crystalline quartz cut perpendicular to the optic axis, each covering half the field and cemented together along an edge (Fig. 74*b*). When crystalline quartz is cut perpendicular to its optic axis, all light travels along the optic axis at the same speed, but a plane-polarized beam is rotated in either a clockwise or counterclockwise direction (when looking *against* the direction of propagation of the beam) depending on whether the quartz is right-handed (dextrorotatory) or left-handed (levorotatory). Since the Nakamura biplate is composed of equal thicknesses of the two kinds of quartz, a plane-polarized beam passing through the biplate will be changed into two beams whose planes of vibration are at an angle α to each other. The angle depends on the thickness of the plate (between 0.04 and 0.2 mm) and the wavelength of light. Table 18 gives the specific rotation in degrees per millimeter thickness of quartz as a function of wavelength. As an example, a 0.1-mm-thick Nakamura biplate used at a wavelength of 5461 Å will split a plane-polarized beam into two beams whose planes of vibration are at an angle of $2 \times 2.55° = 5.10°$ to each other. Since a Nakamura biplate is not a polarizer, it can be placed in any azimuth in front of the analyzer, and when the analyzer is set for extinction, both halves will appear equally bright. Jerrard (1954*a*) claims that a setting precision of $10^{-5}\ 2\pi$ can be obtained.

TABLE 18 Specific Rotation of Plane-polarized Light in Quartz (Jenkins and White, 1957*b*)

Wavelength, Å	Rotation, deg/mm	Wavelength, Å	Rotation, deg/mm
2265.03	201.9	5085.82	29.728
2503.29	153.9	5460.72	25.535
3034.12	95.02	5892.90	21.724
3403.65	72.45	6438.47	18.023
4046.56	48.945	6707.86	16.535
4358.34	41.548	7281.35	13.924
4678.15	35.601	7947.63	11.589
4861.33	32.761		

A Nakamura biplate has a fixed half-shade angle at a given wavelength, but the half-shade angle will be larger at the shorter wavelengths because of the increased specific rotation of quartz. This is an advantage because light sources are generally less intense and the eye is less sensitive toward the blue end of the spectrum. If devices with different half-shade angles are desired, one can have a series of Nakamura biplates of different thicknesses; the thinner ones

† Jerrard's comprehensive article (1954*a*) on the use of the Nakamura biplate should be read with care since there are several typographical errors.

will have the smaller half-shade angles (in proportion to their thicknesses). If the thicknesses of the two parts of the biplate are not identical, the analyzer azimuth for which both halves of the field appear equally bright will be displaced from the true extinction position. However, no error will be introduced if the biplate is used for all the settings, since the displacement will always be in the same sense of either increasing or decreasing azimuth from the true extinction position.

94. Savart Plate Although not strictly a half-shade device, the Savart plate (Johannsen, 1968*c*; Hammerschlag, 1969; Peek, 1971) is an extremely delicate test for small amounts of polarized light. It consists of two plates of quartz or calcite cut at 45° to the optic axis, superimposed with their principal sections at right angles to each other, and cemented with canada balsam. When such a plate is mounted in front of an analyzer, nothing is seen if the light is unpolarized. However, if the beam is only slightly polarized, parallel bands (*Savart's bands*) appear, bisecting the angle between the principal sections of the plate. These increase in intensity as the plane of vibration approaches the direction of the bands themselves. The plate is extremely sensitive and is capable of detecting the polarization of skylight or light from the sun's corona (Beckers and Wagner, 1970). For this latter application, a quarter-wave plate is placed in front of the Savart plate with its optic axis at 45° to the retardation axis of the Savart plate. With this arrangement, the fringe contrast depends only on the degree of polarization of the light and the fringe position only on the direction of the plane of vibration (Beckers and Wagner, 1970). Savart plates have also been used as achromatic wave plates in combination with diffraction gratings (Peek, 1971).

Half Shades for Elliptically Polarized Light

95. If one has elliptically polarized light and wishes to determine its ellipticity and the orientation of the major axis of the ellipse, either a quarter-wave retardation plate (Pars. 56 to 82) or a variable retardation plate such as a Babinet or Soleil compensator (Pars. 83 to 89) is generally used to change the elliptically polarized light back into plane-polarized light. It can then be extinguished with an analyzer, as described in Par. 90. Archard et al. (1952) describe three photoelectric methods which can be used to analyze elliptically polarized light. One of these requires some form of compensator; the other two do not.

Ellipticity half-shade devices are useful for detecting very small amounts of ellipticity in a nominally plane-polarized beam and hence can indicate when a compensator has completely converted elliptically polarized light into plane-polarized light. Two of these devices are the Bravais biplate and the Brace half-shade plate.

96. Bravais Biplate The Bravais biplate consists of two quarter-wave (or thinner) plates with their edges joined and their axes at 45° to the edge, as shown in Fig. 74*c*. The fast axis of one plate is parallel to the slow axis of the other plate. When placed between crossed polarizers with no sample in the field, the brightness in both halves of the biplate is the same, independent of the azimuth of the half shade relative to the azimuths of the polarizers. (The total brightness of the field, however, varies with the azimuth of the half shade.)

The operation of the Bravais biplate can be understood as follows. When a plane-polarized beam of light enters the biplate, it will emerge as two contiguous beams of elliptically polarized light having the same ellipticity and orientation but rotating in opposite senses. If the light is observed with an analyzer oriented so as to extinguish vibrations along the major axis of the ellipse, the two portions of the field will appear equally dark. When the analyzer is rotated to a different azimuth, the entire field will appear lighter but both halves will still have the same brightness. If a sample is now inserted in the beam so that elliptically polarized light is incident on the biplate, both halves of the field will no longer appear equally dark. Differences can be observed even when the ratio of the axes of the ellipse is as small as 10^{-4} (Perucca, 1935).

Because it is difficult to orient the fast axes of the two plates at exactly 90° to each other and to make a sharp dividing line between the two halves of the plate, the traditional Bravais biplate was not widely used. Tronstad (1934) suggested a modification which is much easier to construct and optically equivalent to the Bravais biplate. It consists of three separate pieces of mica. Two pieces are cut from the same sheet and superimposed with their axes parallel to cover half the field of view. An independently rotatable holder contains a third piece of mica also cut from the same sheet, which covers the whole field of view. If the fast axis of the large sheet is aligned parallel to the slow axis in the double partial sheet, the net retardation in the latter will effectively be that of a sheet of single thickness whose fast axis is at 90° to the fast axis of the large sheet. The advantage of constructing the biplate in this manner is that the parts can be oriented by observing intensity variations in the two halves of the plate when it is mounted between cross polarizers. The dividing line in the center of the field is now the edge of the double sheet rather than two edges joined together. Another modification of a Bravais biplate was suggested by Perucca (1935), who used a thin mica sheet to cover half the field and a glass plate subjected to a small uniaxial stress in place of Tronstad's mica plate covering the entire field. The usefulness of the Bravais biplate in determining azimuth, ellipticity, and sense of an elliptically polarized beam (in conjunction with a quarter-wave plate) has been discussed by Richartz (1965*a*).

97. Brace Half Shade The Brace half shade consists of an extremely thin (approximately 0.00017 mm) sheet of mica embedded in canada balsam and covering half the field. It has an almost vanishing dividing line, and although it can be used at various brightness levels, it is generally set at the position of maximum effect. When using this half-shade device, the light is not completely restored to a rectilinear vibration but to a definite small ellipticity instead (Skinner, 1925). Brace claimed that his compensator is 200 times as sensitive as the Bravais biplate and is capable of detecting retardations of $6 \times 10^{-5}\lambda$ (Johannsen, 1968*c*).

Other half-shade systems have been suggested which generally consist of several of the devices already described mounted together to form a composite system. These are described in review articles by Ramachandran and Ramaseshan (1961*b*), Jerrard (1948), and Richartz and Hsu (1949). Individual half-shade devices are described by Kent (1936), Ramaseshan (1955), Kinosita and Nomura (1958), Koester (1959*b*), and Richartz (1965*b*).

MISCELLANEOUS TOPICS

98. Polarization of Light Sources Nearly all light sources are partially polarized. Classic examples are polished metal-ribbon filaments. Worthing (1926) measured the intensity and polarization of light emitted at various angles from ribbon filaments of tungsten, molybdenum, and tantalum. He found that as the emission angle increases (measuring from the normal to the ribbon surface), the degree of polarization varies from zero at normal incidence to over 90 percent at grazing incidence. Keussler and Manogg (1960) calculated the degree of polarization to be expected from the Fresnel formulas, and their results are in excellent agreement with Worthing's measurements. Öhman (1961) studied the polarization of light emitted by thin wire filaments and found that a polarization as high as 28 percent is obtained in the red for an unpolished filament 4.8 μm thick. The plane of vibration of the electric vector is perpendicular to the direction of the filament, and the polarization appears to be nearly the same in the central part of the wire as at the limb. Öhman also showed that the degree of polarization increases with increasing wavelength and decreases with increasing filament diameter. A general review of emission polarization has been given by Sandus (1965), and polarization patterns have been measured for a tungsten ribbon filament lamp (Grum and Costa, 1974) and for a quartz bromide photolamp (Spooner, 1972). In practice it is not safe to assume that any filament lamp emits unpolarized light. Also, the degree of partial polarization usually changes with time as the lamp ages, probably because of thermal etching at the filament surface. Some arcs and even some sections of new flat-metal-filament lamps are very nearly unpolarized, but they should always be checked before making that assumption.

The polarization properties of gas lasers are well known, and polarization ratios of 99 percent or higher are not uncommon. Gas lasers may also be unpolarized, depending on their resonator geometry and whether they have Brewster angle windows. Less well understood is the polarization present in solid lasers, e.g., ruby or glass. Brunton (1964) studied the polarization of light emitted by a series of ruby rods and found the following results: (1) the polarization decreased with the angle between the C axis and the rod axis; (2) light from a 0° ruby rod was polarized; at energies well above threshold there were found to be two planes of polarization mutually at right angles (in different parts of the crystal); (3) in each of the ruby rods except the 0° one, a mean plane of polarization existed with some variation for light emitted from different regions of the crystal; (4) light which was completely polarized was emitted by a relatively small area of the transmission face of the crystal; light from remaining areas was unpolarized; (5) light from the 90° crystal was completely plane-polarized at the start of a run, but at the end of the run a high percentage of unpolarized light was present; (6) in all crystals examined the degree of polarization decreased with increasing pumping energy.

Tamai and Achiwa (1964) also investigated the polarization characteristics of a laser beam emitted by a ruby rod having the optic axis at 89° to the rod axis. They found that most of the radiation was polarized with the E vector perpendicular to the C axis. However, polarized light with E parallel to the C axis, having an intensity of about 5.5 percent of the total radiation intensity, was also observed.

A theoretical study of the polarization of light emitted by solid materials containing luminescing impurity centers has been undertaken by Ratner (1965). No comparisons were made with experimental data.

Two groups have studied the polarization of light emitted by neodymium glass lasers. Lu and Rabson (1965) measured the percent polarization and Stokes parameters of successive output spikes at 1.06 μm are found that the polarization changes from spike to spike. The percentage of polarized light appears to increase as one goes to lower pumping energies. Kruzhilin (1967) made measurements as a function of resonator geometry and found that artificial anisotropy of the resonator has a stabilizing effect on the orientation of the plane of polarization.

The helium-neon gas lasers have varying degrees of polarization. Some are completely unpolarized, and others, containing Brewster angle windows, are almost completely plane-polarized. The degree of polarization of a given gas laser depends on its resonator geometry.

99. Polarization Effects in Detectors For some applications the sensitivity of a detector, specifically a photomultiplier tube, to the plane of vibration of plane-polarized light is important. Generally a polarization-insensitive detector is desirable, but a photomultiplier detector has also been used successfully to measure small amounts of polarization of weak astronomical sources directly (Hoenig and Cutler, 1966). Side-on photomultipliers such as the RCA 931-A, 1P21, 1P22, or 1P28 have been shown to be polarization-sensitive for light nearly normally incident on the glass envelope. (The cathode surface lies at an angle of about 45° with respect to the incident beam.) Clancy (1952) measured the response of a 931-A photomultiplier vs. the angle θ the electric vector makes with the longitudinal axis of the tube. He found that this response is approximately gaussian, varying from 84 to 100 in arbitrary units, with the maximum occurring for $\theta = 90°$.

End-on photomultiplier tubes are often polarization-insensitive for light normally incident on the photocathode (Hoenig and Cutler, 1966; King and Talim, 1971; Birth and DeWitt, 1971). However, DeWitt found that an optimum azimuth of vibration can be determined giving a maximum response in a Dumont 6911 tube; this effect is also wavelength-dependent. Changes in signal output are frequently observed when end-on tubes are rotated about their symmetry axes. However, these changes cannot necessarily be attributed to polarization sensitivity since unless precautions are taken, it is difficult to separate this effect from changes occurring when a beam of nonuniform flux density is rotated on a photocathode of nonuniform response.

As the angle of incidence of the light beam on an end-on photomultiplier increases, the sensitivity to the plane of vibration also increases. The ratio of the responses for p and s polarized light (light polarized parallel and perpendicular to the plane of incidence, respectively) as a function of angle of incidence has been measured for numerous German photomultipliers, most of which had Cs_3Sb cathodes (Hora, 1960). This response ratio ρ_p/ρ_s was also a function of wavelength in the 3800- to 7000-Å wavelength region. Hoenig measured the response ratio of an RCA 6903 tube as a function of angle of incidence in the visible (light from a tungsten bulb) and found that ρ_p/ρ_s increases with increasing angle of incidence, reaching a maximum of 1.7 at 70° incidence (the limit of the measurements). Pauty et al. (1962) found a similar effect in the French Lallemand-type image converters. At an angle of incidence of 55°, ρ_p/ρ_s was about 2.2 for wavelengths longer than 2450 Å, and no variation with wavelength was found.

100. Depolarizers A large portion of this section has been devoted to ways of producing plane-polarized light. Once obtained, it is not easy to get rid of. There are only a very limited number of ways of depolarizing a light beam. True depolarization involves a complete randomization of the vibration azimuths and phases of parts of the light beam. This can be accomplished by sending the beam through an inhomogeneous nonabsorbing material in which multiple scattering occurs. Examples are powder-liquid mixtures (Boutaric and Berthier, 1947), opal glass (Hardy and Perrin, 1932*b*), hot-pressed polycrystalline Al_2O_3 or ThO_2, cold-pressed MgO, and cloudy fused quartz (Renau et al., 1967). Birefringence of the individual particles is apparently not an important depolarization mechanism in pressed polycrystalline materials (Renau et al., 1967), which implies that refraction or internal reflection within the individual crystallites is not the most important depolarization mechanism. However, for powder-liquid mixtures, the birefringence of the powder *is* important, and complete depolarization can be produced by the thinnest layers, independent of the grain size or liquid index (Boutaric and Berthier, 1947).

One might think that scattering from rough dielectric or metal surfaces would be a good way to depolarize a light beam. However, Gorodinskii (1964) showed that plane-polarized light reflected from ground-glass samples at various angles of incidence retains its polarization properties surprisingly well. The maximum degree of polarization occurs when the sum of the angles of incidence and reflection equals twice the Brewster angle. Rvachev and Polyanskii (1965) showed that a transparent plate with one ground surface is not a good depolarizer on transmission. Renau et al. (1967) measured the ratio of depolarized to total reflected light at normal incidence for various rough metal surfaces, metal surfaces coated with dielectrics, and ground-glass and rough fused-quartz surfaces. They found that the depolarization is much smaller for all these samples than for light transmitted through pressed polycrystalline materials.

In all scattering-type depolarizers there is a large loss of intensity because of light scattered out of the direct beam. Furthermore, the image quality is seriously impaired, so that scattering depolarizers are generally used in front of detectors or in other applications where the image quality is not important.

One depolarizer based on transmission through a roughened plate does retain its image-forming qualities (Peters, 1964). The plate material was KDP, and the roughness, produced by cutting the material with a diamond saw, was approximately 8 μm. The KDP plate was placed in a cell filled with mineral oil having a refractive index almost equal to the average index of KDP, so that the light-beam collimation was not noticeably impaired. The mechanism involved in this case is actually pseudo depolarization, since the device acts like a series of birefringent plates having thicknesses ranging from an integral number of wavelengths retardation to a half-integral number. Thus, even monochromatic light will appear to be completely depolarized.

Other pseudo depolarizers which introduce many different polarization states into the light beam include thick birefringent plates, such as crystalline-quartz (Holzwarth, 1965), or an oriented sheet of Mylar (Shurcliff, 1962*c*), which act as depolarizers only for light of many frequencies, and thin crystalline-quartz wedges with angles of about 1° (Jaffe et al., 1967) or 15 minutes of arc (Hughes, 1960). Portigal (1969) used 60- and 90-cm-long 5-mm-diameter, commercial-grade fused-quartz rods as depolarizers and succeeded in decreasing the polarization of a light beam by at least two orders of magnitude at a wavelength of 4358 Å with minimal intensity loss. In contrast, opal glass reduced the output by three to four orders of magnitude. Schmidt and Vedam (1971) obtained excellent depolarization of both monochromatic and polychromatic light using a 45° depolarizing prism made of α-quartz or calcite.

Several elaborate monochromatic pseudo depolarizers have been described by Billings (1951). One consists of two variable wave plates whose axes are placed at a 45° angle to each other. The wave plates vary in retardation linearly with time. Electrooptic materials such as ADP have been suggested for this application, in which case sawtooth voltages would be applied to the two plates at a frequency ratio of 2:1. Although sinusoidal excitation would be much easier experimentally, Billings has shown that perfect depolarization cannot be achieved in this way. Billings also describes measurements made on a Lyot-type depolarizer consisting of two plates of crystalline quartz in the thickness ratio 2:1, having an angle of 45° between their two fast axes. In white light this device was shown to be an excellent depolarizer although it would be ineffective in monochromatic light.

REFERENCES†

Abelès, F. (1949): *C.R.*, vol. 228, p. 553.
——— (1950*a*): *C.R.*, vol. 230, p. 1942.
——— (1950*b*): *J. Phys. Radium,* vol. 11, p. 310.
——— (1950*c*): *J. Phys. Radium,* vol. 11, p. 403.
——— (1963): *Progr. Opt.*, vol. 2, p. 251.
——— (1965): *Jap. J. Appl. Phys.*, vol. 4, suppl. 1, p. 517 (*Proc. Conf. Photogr. Spectrosc. Opt., 1964*).
Aben, Kh. K. (1962): *Opt. Spectrosc.* (*USSR*), vol. 13, p. 421.
——— (1963): *Opt. Spectrosc.* (*USSR*), vol. 14, p. 124.
Adams, L. H., and R. M. Waxler (1965): *J. Res. Natl. Bur. Stand.* (*U.S.*), vol. 69C, p. 103.
Ahrens, C. D. (1886): *J. R. Microsc. Soc.*, vol. 6, p. 397.
Alfano, R. R., and W. H. Woodruff (1966): *Appl. Opt.,* vol. 5, p. 352.
Ames, J., and A. M. D Sampson (1949): *J. Sci. Instrum.,* vol. 26 p. 132.
Ammann, E. O., and G. A. Massey (1968): *J. Opt. Soc. Am.*, vol. 58, p. 1427.
Anderson, S. (1949): *J. Opt. Soc. Am.*, vol. 39, p. 49.
Archard, J. F. (1949): *J Sci. Instrum.*, vol. 26, p. 188.
——— (1950): *J. Sci. Instrum.*, vol. 27, p. 238.
———, P. L. Clegg, and A. M. Taylor (1952): *Proc. Phys. Soc.* (*Lond.*), vol. 65B, p. 758.
——— and A. M. Taylor (1948): *J. Sci. Instrum.*, vol. 25, p. 407.
Archer, R. J. (1964): p. 255 in E. Passaglia, R. R. Stromberg, and J. Kruger (eds.), Ellipsometry in the Measurement of Surfaces and Thin Films, *Natl. Bur. Stand. Misc. Publ.* 256.
——— and G. W. Gobeli (1965): *J. Phys. Chem. Solids,* vol. 26, p. 343.
——— and C. V. Shank (1967): *J. Opt. Soc. Am.,* vol. 57, p. 191.
Ashkin, A., G. D. Boyd, J. M. Dziedzic, R. G. Smith, A. A. Ballman, J. J. Levinstein, and K. Nassau (1966): *Appl. Phys. Lett.*, vol. 9, p. 72.
Aspnes, D. E. (1971): *J Opt. Soc. Am.*, vol. 61, p. 1077.
Auton, J. P. (1967): *Appl. Opt.*, vol. 6, p. 1023.
Azzam, R. M. A., and N. M. Bashara (1971): *J. Opt. Soc. Am.,* vol. 61, pp. 600, 773, 1380.
——— and ——— (1972): *Phys. Rev.*, ser. B, vol. 5, p. 4721.
Banning, M. (1947): *J. Opt. Soc. Am.*, vol. 37, p. 792.
Barchewitz, P., and L. Henry (1954): *J. Phys. Radium,* vol. 15, p. 639.
Baumeister, P. (1961): *Opt. Acta,* vol. 8, p. 105.
———, Institute of Optics, Univ. Rochester, Rochester, N.Y. (1971): private communication.
Baumel, R. T., and S. E. Schnatterly (1971): *J. Opt. Soc. Am.*, vol. 61, p. 832.
Baxter, L., A. S. Makas, and W. A. Shurcliff (1956): *J. Opt. Soc. Am.*, vol. 46, p. 229.
Beckers, J. M. (1971): *Appl. Opt.*, vol. 10, p. 973.
——— and W. J. Wagner (1970): *Appl. Opt.*, vol. 9, p. 1933.
Belyaev, L. M., G. F. Dobrzhanskii, Yu. V. Pisarevskii, K. S. Chernyshev, and Yu. V. Shaldin (1965): *Sov. Phys. Solid State,* vol. 6, p. 2988.

† In all references to the Russian literature, volume and pages cited are for the English translation.

Benedicks, C. (1931): *Phys. Z.*, vol. 32, p. 561.
Bennett, H. E., and J. M. Bennett (1967*a*): pp. 69–78 in G. Hass and R. E. Thun (eds.), "Physics of Thin Films," vol. 4, Academic, New York.
——— and ——— (1967*b*): ibid., pp. 79–84.
———, ———, and M. R. Nagel (1961): *J. Opt. Soc. Am.*, vol. 51, p. 237.
———, D. K. Burge, R. L. Peck, and J. M. Bennett (1969): *J. Opt. Soc. Am.*, vol. 59, p. 675.
Bennett, J. M. (1964): *J. Opt. Soc. Am.*, vol. 54, p. 612.
——— (1970): *Appl. Opt.*, vol. 9, p. 2123.
———, D. L. Decker, and E. J. Ashley (1970): *J. Opt. Soc. Am.*, vol. 60, p. 1577A.
Bergstein, L. (1970): *Laser J.*, vol. 2, p. 32, January-February.
——— (1971): *J. Opt. Soc. Am.*, vol. 61, p. 665A.
——— and J. Di Sabato (1971): *J. Opt. Soc. Am.*, vol. 61, p. 666A.
——— and C. Shulman (1966): *Appl. Opt.*, vol. 5, p. 9. [See also L. Bergstein, *Appl. Opt.*, vol. 6, p. 2002 (1967) and L. V. Ioganson, ibid., p. 2001.]
Berning, J. A., and P. H. Berning (1960): *J. Opt. Soc. Am.*, vol. 50, p. 813.
Berning, P. H. (1963): pp. 78–81 in G. Hass (ed.), "Physics of Thin Films," vol. 1, Academic, New York.
Bertrand, E. (1884): *C.R.*, vol. 49, p. 538.
Bieniewski, T. M., and S. J. Czyzak (1963): *J. Opt. Soc. Am.*, vol. 53, p. 496.
Billard, J. (1966): *Rev. Phys. Appl. (Fr.)*, vol. 1, p. 311.
Billardon, M., and J. Badoz (1966*a*): *C.R.*, vol. 262B, p. 1672.
——— and ——— (1966*b*): *C.R.*, vol. 263B, p. 139.
Billings, B. H. (1949*a*): *J. Opt. Soc. Am.*, vol. 39, p. 634A.
——— (1949*b*): *J. Opt. Soc. Am.*, vol. 39, p. 797.
——— (1949*c*): *J. Opt. Soc. Am.*, vol. 39, p. 802.
——— (1951): *J. Opt. Soc. Am.*, vol. 41, p. 966.
——— (1952): *J. Opt. Soc. Am.*, vol. 42, p. 12.
——— (1963): pp. 6-39, 6-97, and 6-98 in D. E. Gray (ed.), "American Institute of Physics Handbook," 2d ed., McGraw-Hill, New York.
——— and M. A. Pittman (1949): *J. Opt. Soc. Am.*, vol. 39, p. 978.
Bird, G. R., and M. Parrish, Jr. (1960): *J. Opt. Soc. Am.*, vol. 50, p. 886.
——— and W. A. Shurcliff (1959): *J. Opt. Soc. Am.*, vol. 49, p. 235.
Birth, G. S., and D. P. DeWitt (1971): *Appl. Opt.*, vol. 10, p. 687.
Blake, R. P., A. S. Makas, and C. D. West (1949): *J. Opt. Soc. Am.*, vol. 39, p. 1054A.
Bond, W. L. (1965): *J. Appl. Phys.*, vol. 36, p. 1674.
Bonner, W. A., E. F. Dearborn, J. E. Geusic, H. M. Marcos, and L. G. Van Uitert (1967): *Appl. Phys. Lett.*, vol. 10, p. 163.
Bor, J., and L. A. Brooks (1966): *J. Sci. Instrum.*, vol. 43, p. 944.
Born, M., and E. Wolf (1965*a*): "Principles of Optics," 3d ed., pp. 30–32, 544–555, Pergamon, New York.
——— and ——— (1965*b*): ibid., p. 680.
Borrelli, N. F. (1967): *J. Appl. Phys.*, vol. 38, p. 4243.
——— and M. M. Layton (1969): *IEEE Trans. Electron. Devices*, vol. ED-16, p. 511.
Bouhet, C., and R. LaFont (1949): *Rev. Opt.*, vol. 28, p. 490.
Bouriau, Y., and J. Lenoble (1957): *Rev. Opt.*, vol. 36, p. 531.
Boutaric, A., and P. Berthier (1947): *Rev. Sci. (Paris)*, vol. 85, p. 619.
Boyd, G. D., W. L. Bond, and H. L. Carter (1967): *J. Appl. Phys.*, vol. 38, p. 1941.
Bradbury, E. M., and A. Elliott (1962): *J. Sci. Instrum.*, vol. 39, p. 390.
Bradford, A. P., G. Hass, and J. B. Heaney (1967): *Prog. Astronaut. Aeronaut.*, vol. 20, pp. 3–16.
Braybon, J. E. H., and H. G. Jerrard (1951): *J. Sci. Instrum.*, vol. 28, p. 269.
Bridges, T. J., and J. W. Kluver (1965): *Appl. Opt.*, vol. 4, p. 1121.
Bruhat, G., and L. Weil (1936): *Rev. Opt.*, vol. 15, p. 41.
Brunton, J. H. (1964): *Appl. Opt.*, vol. 3, p. 1241.
Buchman, W. W., S. J. Holmes, and F. J. Woodberry (1971): *J. Opt. Soc. Am.*, vol. 61, p. 1604.
Buckbee Mears Company, Saint Paul, Minn., M. Hass and M. O'Hara (1965): *Appl. Opt.*, vol. 4, p. 1027.
Budde, W. (1962): *Appl. Opt.*, vol. 1, p. 201.
Buhrer, C. F., L. R. Bloom, and D. H. Baird (1963): *Appl. Opt.*, vol. 2, p. 839.
Buijs, K. (1960): *Appl. Spectrosc.*, vol. 14, p. 81.
Burge, D. K., and H. E. Bennett (1964): *J. Opt. Soc. Am.*, vol. 54, p. 1428.
Burns, G. (1969): *IEEE Trans. Electron. Devices*, vol. ED-16, p. 506.

———, E. A. Giess, D. F. O'Kane, B. A. Scott, and A. W. Smith (1969): *J. Appl. Phys.*, vol. 40, p. 901.
——— and A. W. Smith (1968): *IEEE J. Quantum Electron.*, vol. QE-4, p. 584.
Carpenter, R. O'B. (1950): *J. Opt. Soc. Am.*, vol. 40, p. 225.
Cernosek, J. (1971): *J. Opt. Soc. Am.*, vol. 61, p. 324.
Chandrasekharan, V., and H. Damany (1966): *C.R.*, vol. 262B, p. 1268.
——— and ——— (1968): *Appl. Opt.*, vol. 7, p. 939.
——— and ——— (1969): *Appl. Opt.*, vol. 8, p. 671.
——— and ——— (1971): *Appl. Opt.*, vol. 10, p. 681.
Charney, E. (1955): *J. Opt. Soc. Am.*, vol. 45, p. 980.
Chelle, F. de, and H. Merdy (1967): *C.R.*, vol. 265B, p. 968.
Chen, F. S. (1969): *J. Appl. Phys.*, vol. 40, p. 3389.
Cheo, P. K., and C. D. Bass (1971): *Appl. Phys. Lett.*, vol. 18, p. 565.
———, J. E. Geusic, S. K. Kurtz, J. G. Skinner, and S. H. Wemple (1966): *J. Appl. Phys.*, vol. 37, p. 388.
Childs, W. H. J. (1956): *J. Sci. Instrum.*, vol. 33, p. 298.
Clancy, E. P. (1952): *J. Opt. Soc. Am.*, vol. 42, p. 357.
Clapham, P. B. (1971): *Opt. Acta*, vol. 18, p. 563.
———, M. J. Downs, and R. J. King (1969): *Appl. Opt.*, vol. 8, p. 1965 [See also P. B. Clapham, *Thin Solids Films*, vol. 4, p. 291 (1969).]
Clarke, D. (1967): *Opt. Acta*, vol. 14, p. 343.
Cole, T. T., and F. Oppenheimer (1962): *Appl. Opt.*, vol. 1, p. 709.
Collins, J. G., and W. H. Steel (1962): *J. Opt. Soc. Am.*, vol. 52, p. 339.
Conn, G. K. T., and G. K. Eaton (1954): *J. Opt. Soc. Am.*, vol. 44, p. 553.
Costich, V. R. (1970): *Appl. Opt.*, vol. 9, p. 866.
Cotton, A. (1931): *C.R.*, vol. 193, p. 268.
——— and P. Manigault (1944): *C.R.*, vol. 219, p. 597.
Craig, J. P., R. F. Gribble, and A. A. Dougal (1964): *Rev. Sci. Instrum.*, vol. 35, p. 1501.
Dale, A. B. (1923*a*): pp. 475–476 in R. Glazebrook (ed.), "A Dictionary of Applied Physics," vol. 4, Macmillan, London.
——— (1923*b*): ibid., pp. 496–497.
Damany, H. (1965): *Opt. Acta*, vol. 12, p. 95.
———, Laboratoire des Hautes Pressions, Bellevue, France (1970): private communication.
Daniels, J. M. (1967): *Rev. Sci. Instrum.*, vol. 38, p. 284.
Davis, N. M., A. R. Clawson, and H. H. Wieder (1969): *Appl. Phys. Lett.*, vol. 15, p. 213.
de Chelle, F., and H. Merdy (1967): *C.R.*, vol. 265B, p. 968.
Decker, D. L., J. L. Stanford, and H. E. Bennett (1970): *J. Opt. Soc. Am.*, vol. 60, p. 1557A.
Deer, W. A., R. A. Howie, and J. Zussman (1962*a*): "Rock-forming Minerals," vol. 3, "Sheet Silicates," pp. 1–30, Wiley, New York.
———, ———, and ——— (1962*b*): ibid., pp. 258–262.
DeShazer, L. G., Dept. of Electrical Engineering, Univ. Southern California, Los Angeles (1971): private communication.
Destriau, G., and J. Prouteau (1949): *J. Phys. Radium*, ser. 8, vol. 10, p. 53.
Dévé, C. (1954): "Optical Workshop Principles," trans. from the French by T. L. Tippell, p. 295, Hilger & Watts, London.
Distler, G. I., E. I. Kortukova, A. V. Kotov, V. N. Lebedeva, and V. S. Chudakov (1967): *Opt. Spectrosc. (USSR)*, vol. 23, p. 71.
———, A. V. Kotov, E. I. Kortukova, and V. N. Lebedeva (1966): *Opt. Spectrosc. (USSR)*, vol. 20, p. 306.
Ditchburn, R. W. (1963*a*): "Light," 2d ed., pp. 595–616, Interscience, New York.
——— (1963*b*): ibid., pp. 483–485.
Dixon, R. W. (1967): *J. Appl. Phys.*, vol. 38, p. 5149.
Dobrowolski, J. A. (1959): *J. Opt. Soc. Am.*, vol. 49, p. 794.
Dore, M. (1967): *IEEE J. Quantum Electron.*, vol. QE-3, p. 555.
Drechsel, L. (1957): *Naturwiss.*, vol. 44, p. 533.
——— (1958): "Jenaer Jahrbuch 1958," pt. 1, pp. 55–62, Carl Zeiss, Jena.
——— (1962): *Z. Wiss. Photogr., Photophys. Photochem.*, vol. 56, p. 215.
Dreyer, J. F. (1947): *J. Opt. Soc. Am.*, vol. 37, p. 983A.
Drummond, D. G. (1936): *Proc. R. Soc. (Lond.)*, vol. 153, p. 328.
Dumont, F. J. (1970): *J. Opt. Soc. Am.*, vol. 60, p. 719A.
——— and R. N. Smartt (1969): *J. Opt. Soc. Am.*, vol. 59, p. 1541A.
Duverney, R. (1959): *J. Phys. Radium*, vol. 20, suppl. 7, p. 66A.
Edwards, D. F., and M. J. Bruemmer (1959): *J. Opt. Soc. Am.*, vol. 49, p. 860.

Einsporn, E. (1936): *Phys. Z.*, vol. 37, p. 83.
Ejiri, A. (1967): *J. Phys. Soc. Jap.*, vol. 23, p. 901.
Ellerbroek, J., and J. T. Groosmuller (1926): *Phys. Z.*, vol. 27, p. 468.
Elliott, A., and E. J. Ambrose (1947): *Nature*, vol. 159, p. 641.
———, ———, and R. Temple (1948): *J. Opt. Soc. Am.*, vol. 38, p. 212.
Ellis, B., and A. K. Walton (1971): *Infrared Phys.*, vol. 11, p. 85.
Ellis, J. W., and J. Bath (1938): *J. Chem. Phys.*, vol. 6, p. 221.
Emberson, R. M. (1936): *J. Opt. Soc. Am.*, vol. 26, p. 63.
Ennos, A. E. (1963*a*): *Opt. Acta*, vol. 10, p. 105.
——— (1963*b*): *J. Sci. Instrum.*, vol. 40, p. 316.
——— and K. W. Opperman (1966): *Appl. Opt.*, vol. 5, p. 170.
Feussner, K. (1884): *Z. Instrumentenkd.*, vol. 4, p. 41.
Fisher, W. A. P. (1948): *Nature*, vol. 161, p. 848.
Foster, L. V. (1938): *J. Opt. Soc. Am.*, vol. 28, pp. 124, 127.
Foucault, L. (1857): *C.R.*, vol. 45, p. 238.
Françon, M., S. Mallick, and J. Vulmière (1965): *J. Opt. Soc. Am.*, vol. 55, p. 1553.
Gainon, D. J. A. (1964): *J. Opt. Soc. Am.*, vol. 54, p. 270.
Gates, J. W., and R. J. King (1962): *Opt. Acta*, vol. 9, p. 13.
Gaudefroy, C. (1929): *C.R.*, vol. 189, p. 1289.
Gee, A. E., and H. D. Polster, (1949): *J. Opt. Soc. Am.*, vol. 39, p. 1044.
Geusic, J. E., S. K. Kurtz, T. J. Nelson, and S. H. Wemple (1963): *Appl. Phys. Lett.*, vol. 2, p. 185.
———, ———, L. G. Van Uitert, and S. H. Wemple (1964): *Appl. Phys. Lett.*, vol. 4, p. 141.
———, H. J. Levinstein, J. J. Rubin, S. Singh, and L. G. Van Uitert (1967): *Appl. Phys. Lett.*, vol. 11, p. 269.
———, ———, ———, ———, and ——— (1968): *Appl. Phys. Left.*, vol. 12, p. 244.
Giardini, A. A. (1957): *J. Opt. Soc. Am.*, vol. 47, p. 726.
Giess, E. A., G. Burns, D. F. O'Kane, and A. W. Smith (1967): *Appl. Phys. Lett.*, vol. 11, p. 233.
Gieszelmann, E. L., S. F. Jacobs, and H. E. Morrow (1969): *J. Opt. Soc. Am.*, vol. 59, p. 1381 [erratum in ibid., vol. 60, p. 705 (1970)].
Gillham, E. J., and R. J. King (1961): *J. Sci. Instrum.*, vol. 38, p. 21.
Glan, P. (1880): *Carl's Repert.*, vol. 16, p. 570.
Glazebrook, R. T. (1883): *Phil. Mag.*, ser. 5, vol. 15, p. 352.
Gobrecht, H., and A. Bartschat (1959): *Z. Phys.*, vol. 156, p. 131.
Goranson, R. W., and L. H. Adams (1933): *J, Franklin Inst.*, vol. 216, p. 475.
Gordon, E. I. (1966*a*): *Appl. Opt.*, vol. 5, p. 1629.
——— (1966*b*): *Proc. IEEE*, vol. 54, p. 1391.
Gorodinskii, G. M. (1964): *Opt. Spectrosc. (USSR)* vol. 16, p. 59.
Grechushnikov, B. N. (1962): *Opt. Spectrosc. (USSR)*, vol. 12, p. 69.
——— and I. P. Petrov (1963): *Opt. Spectrosc. (USSR)*, vol. 14, p. 160.
———, G. D. Shnyrev, and I. P. Petrov (1964): *Sov. Phys. Cryst.*, vol. 9, p. 345.
Greenler, R. G., K. W. Adolph, and G. H. Emmons (1966): *Appl. Opt.*, vol. 5, p. 1468.
Groosmuller, J. T. (1926): *Z. Instrumentenkd.*, vol. 46, p. 563.
Grosse, W. (1890): *Z. Instrumentenkd.*, vol. 10, p. 445.
Grum, F., and L. F. Costa (1974): *Appl. Opt.*, vol. 13, p. 2228.
Grunstra, B. R., and H. B. Perkins (1966): *Appl. Opt.*, vol. 5, p. 585.
Haase, M. (1961): *Zeiss-Mitt.*, vol. 2, p. 173.
Hadni, A., E. Decamps, and P. Delorme (1958): *J. Phys. Radium*, vol. 19, p. 793.
———, ———, D. Grandjean, and C. Janot (1960): *C.R.*, vol. 250, p. 2007.
Hall, A. C. (1963): *Appl. Opt.*, vol. 2, p. 864.
Halle, B. (1908*a*): *Dtsch. Mech. Z.*, no. 1, pp. 6–7, Jan. 1.
——— (1908*b*): *Dtsch. Mech. Z.*, no. 2, pp. 16–19, Jan. 15.
Hamm, R. N., R. A. MacRae, and E. T. Arakawa (1965): *J. Opt. Soc. Am.*, vol. 55, p. 1460.
Hammerschlag, R. H. (1969): *Opt. Acta*, vol. 16, p. 491.
"Handbook of Chemistry and Physics" (1955): 37th ed., p. 2063, Chemical Rubber, Cleveland, Ohio.
Hardy, A. C. (1935): *J. Opt. Soc. Am.*, vol. 25, p. 305.
——— and F. H. Perrin (1932*a*): "The Principles of Optics," p. 611, McGraw-Hill, New York.
——— and ——— (1932*b*): ibid., p. 605.
Harihan, P., and D. Sen (1960): *J. Sci. Instrum.*, vol. 37, p. 278.
Harle, G. A. (1950): *Nature*, vol. 166, p. 149.

Harrick, N. J. (1959): *J. Opt. Soc. Am.*, vol. 49, pp. 376, 379.
——— (1964): *J. Opt. Soc. Am.*, vol. 54, p. 1281.
——— (1967): "Internal Reflection Spectroscopy," Interscience, New York.
Harris, S. E., E. O. Ammann, and I. C. Chang (1964): *J. Opt. Soc. Am.*, vol. 54, p. 1267. [See also E. O. Ammann, ibid., vol. 56, pp. 943, 952, and 1081 (1966) and E. O. Ammann and J. M. Yarborough, ibid., vol. 57, p. 349 (1967).]
Hartman, R. E. (1951): *J. Opt. Soc. Am.*, vol. 41, p. 244.
Hartnack and Prazmowski (1866): *Ann. Chim. Phys.*, ser. 4, vol. 7, p. 181.
Hass, M., and O'Hara, M. (1965): *Appl. Opt.*, vol. 4, p. 1027.
Hauser, S. M., L. S. Smith, D. G. Marlowe, and P. R. Yoder, Jr. (1963): *Appl. Opt.*, vol. 2, p. 1175.
Heath, D. F. (1968): *Appl. Opt.*, vol. 7, p. 455.
Heavens, O. S. (1955): "Optical Properties of Thin Solids Films," p. 92, Butterworths, London.
——— (1964): pp. 210–213 in G. Hass and R. E. Thun (eds.), "Physics of Thin Films," vol. 2, Academic, New York.
Heilmeier, G. H. (1964): *Appl. Opt.*, vol. 3, p. 1281.
Heller, W. (1960*a*): p. 2147 in A. Weissburger (ed.), "Physical Methods of Organic Chemistry," vol. 1, pt. 3, Interscience, New York.
——— (1960*b*): ibid., pp. 2215–2226.
Hellerstein, D. (1963): *Appl. Opt.*, vol. 2, p. 801.
Hertz, J. H. (1959): *Exp. Tech. Phys.*, vol. 7, p. 277.
Hidalgo, A., J. Pastor, and J. M. Serratosa (1962): *J. Opt. Soc. Am.*, vol. 52, p. 1081.
Hilton, A. R., and C. E. Jones (1966): *J. Electrochem. Soc.*, vol. 113, p. 472.
Hinson, D. C. (1966): *J. Opt. Soc. Am.*, vol. 56, p. 408.
Ho, L., and C. F. Buhrer (1963): *Appl. Opt.*, vol. 2, p. 647.
Hoenig, S. A., and A. Cutler (1966): *Appl. Opt.*, vol. 5, p. 1091.
Holl, H. B. (1963): The Reflection of Electromagnetic Radiation, vols. 1 and 2, *U.S. Army Missile Command, Redstone Arsenal, Ala., Rep.* RF-TR-63-4.
Holmes, C. H., E. G. Spencer, A. A. Ballman, and P. V. Lenzo (1965): *Appl. Opt.*, vol. 4, p. 551.
Holmes, D. A. (1964*a*): *J. Opt. Soc. Am.*, vol. 54, p. 1115.
——— (1964*b*): *J. Opt. Soc. Am.*, vol. 54, p. 1340.
——— (1965): *J. Opt. Soc. Am.*, vol. 55, p. 209.
——— and D. L. Feucht (1967): *J. Opt. Soc. Am.*, vol. 57, p. 466.
Holzwarth, G. (1965): *Rev. Sci. Instrum.*, vol. 36, p. 59.
Hook, W. R., and R. P. Hilberg (1971): *Appl. Opt.*, vol. 10, p. 1179.
Hookabe, K., and Y. Matsuo (1970): *Electron. Lett.*, vol. 6, p. 550.
Hora, H. (1960): pp. 514–610 in P. Görlich (ed.), "Jenaer Jahrbuch 1960," pt. 2, Gustav Fischer, Jena.
Horton, J. A., and W. N. Hansen (1967): *Anal. Chem.*, vol. 39, p. 1097.
Horton, V. G., E. T. Arakawa, R. N. Hamm, and M. W. Williams (1969): *Appl. Opt.*, vol. 8, p. 667.
Hughes, R. H. (1960): *Rev. Sci. Instrum.*, vol. 31, p. 1156.
Huldt, L., and T. Staflin (1959): *Opt. Acta*, vol. 6, p. 27.
Hulme, K. F., P. H. Davies, and V. M. Cound (1969): *Electron. Lett.*, vol. 5, p. 171.
Hundelshausen, U. V. (1971): *Phys. Lett.*, vol. 34A, p. 405.
Hunter, W. R. (1965*a*): *Jap. J. Appl. Phys.*, vol. 4, suppl. 1, p. 520 (*Proc. Conf. Photogr. Spectrosc. Opt., 1964*).
——— (1965*b*): *J. Opt. Soc. Am.*, vol. 55, p. 1197.
Huot de Longchamp, M. (1947): *Rev. Opt.*, vol. 26, p. 94.
Hyde, W. L. (1948): *J. Opt. Soc. Am.*, vol. 38, p. 663A.
———, New York Univ., Bronx N.Y. (1970): private communication.
Inoué, S., and W. L. Hyde (1957): *J. Biophys. Biochem. Cytol.*, vol. 3, p. 831.
——— and C. J. Koester (1959): *J. Opt. Soc. Am.*, vol. 49, p. 556.
Ioffe, S. B., and T. A. Smirnova (1964): *Opt. Spectrosc. (USSR)*, vol. 16, p. 484.
Ioshpa, B. A., and V. N. Obridko (1963): *Opt. Spectrosc. (USSR)*, vol. 15, p. 60.
Ives, H. E., and H. B. Briggs (1936): *J. Opt. Soc. Am.*, vol. 26, p. 238.
Jacquinot, P. (1954): *J. Opt. Soc. Am.*, vol. 44, p. 761.
Jaffe, J. H., H. Jaffe, and K. Rosenheck (1967): *Rev. Sci. Instrum.*, vol. 38, p. 935.
Janot, C., and A. Hadni (1963): *J. Phys. Radium*, vol. 24, p. 1073.
Jasperson, S. N., and S. E. Schnatterly (1969): *Rev. Sci. Instrum.*, vol. 40, p. 761.
———, D. K. Burge, and R. C. O'Handley (1973): *Surf. Sci.*, vol. 37, p. 548.
Jenkins, F. A., and H. E. White (1957*a*): "Fundamentals of Optics," 3d ed., p. 229, McGraw-Hill, New York.

——— and ——— (1957*b*): ibid., p. 574.
——— and ——— (1957*c*): ibid., pp. 517–518.
——— and ——— (1957*d*): ibid., pp. 545, 556–557.
Jeppesen, M. A. (1958): *J. Opt. Soc. Am.,* vol. 48, p. 629.
Jerrard, H. G. (1948): *J. Opt. Soc. Am.,* vol. 38, p. 35.
——— (1949*a*): *J. Opt. Soc. Am.,* vol. 39, p. 1031.
——— (1949*b*): *J. Sci. Instrum.,* vol. 26, p. 353.
——— (1950*a*): *J. Sci. Instrum.,* vol. 27, p. 62.
——— (1950*b*): *J. Sci. Instrum.,* vol. 27, p. 164.
——— (1951): *J. Sci. Instrum.,* vol. 28, p. 10.
——— (1952): *J. Opt. Soc. Am.,* vol. 42, p. 159.
——— (1953): *J. Sci. Instrum.,* vol. 30, p. 65.
——— (1954*a*): *J. Opt. Soc. Am.,* vol. 44, p. 289.
——— (1954*b*): *J. Opt. Soc. Am.,* vol. 44, p. 634.
Jessop, H. T., and F. C. Harris (1950): "Photoelasticity: Principles and Methods," pp. 178–180, Dover, New York.
Johannsen, A. (1968*a*): "Manual of Petrographic Methods," pp. 169, 283–285, 2d ed. (originally published in 1918), Hafner, New York.
——— (1968*b*): ibid., pp. 369–385.
——— (1968*c*): ibid., pp. 386–388.
Johnson, C. J. (1968): *Proc. IEEE,* vol. 56, p. 1719.
Johnson, W. C., Jr. (1964): *Rev. Sci. Instrum.,* vol. 35, p. 1375.
Johnston, A. R. (1965): Electro-Optic Crystals and Their Use for Light Modulation, in "Opto-Electronic Components and Devices," *AGARD Conf. Proc.* 5, September, pp. 129–164.
——— (1971): *J. Appl. Phys.,* vol. 42, p. 3501.
——— and J. M. Weingart (1965): *J. Opt. Soc. Am.,* vol. 55, p. 828.
Jones, R. C. (1956): *J. Opt. Soc. Am.,* vol. 46, p. 528.
——— (1962): *J. Opt. Soc. Am.,* vol. 52, p. 747.
———, Polaroid Corporation, Cambridge, Mass. (1970): private communication.
Kalhor, H. A., and A. R. Neureuther (1971): *J. Opt. Soc. Am.,* vol. 61, p. 43.
Kaminow, I. P. (1964): *Appl. Opt.,* vol. 3, p. 511.
——— (1968): *IEEE J. Quantum Electron.,* vol. QE-4, p. 23.
——— (1974): "An Introduction to Electrooptic Devices," Academic, New York.
——— and E. H. Turner (1966*a*): *Appl. Opt.,* vol. 5, p. 1612.
——— and ——— (1966*b*): *Proc. IEEE,* vol. 54, p. 1374.
Karl Lambrecht Corp. (1973): *Bull.* P-73, Chicago.
Käseman, E. (1962): *Optik,* vol. 19, p. 161.
Kemp, J. C. (1969): *J. Opt. Soc. Am.,* vol. 59, p. 950.
Kent, C. V. (1936): *J. Opt. Soc. Am.,* vol. 26, p. 450.
Kerr, D. W., and C. H. Palmer (1971): *J. Opt. Soc. Am.,* vol. 61, p. 450.
Keussler, V. von, and P. Manogg (1960): *Optik,* vol. 17, p. 602.
Kiefer, J. E., and A. Yariv (1969): *Appl. Phys. Lett.,* vol. 15, p. 26.
King, R. J. (1966): *J. Sci. Instrum.,* vol. 43, p. 617.
——— and S. P. Talim (1971): *J. Phys. (GB)*, ser. E, vol. 4, p. 93.
Kinosita, K., and S. Nomura (1958): *J. Phys. Soc. Jap.,* vol. 13, p. 1485.
Kizel, V. A., Yu. I. Krasilov, and V. N. Shamraev (1964): *Opt. Spectrosc. (USSR),* vol. 17, p. 248.
Koester, C. J. (1959*a*): *J. Opt. Soc. Am.,* vol. 49, p. 405.
——— (1959*b*): *J. Opt. Soc. Am.,* vol. 49, p. 560.
Kojima, C., T. Shikama, S. Kuninobu, A. Kawabata, and T. Tanaka (1969): *Jap. J. Appl. Phys.,* vol. 8, p. 1361.
Krasilov, Yu. I. (1967): *Opt. Spectrosc. (USSR)*, vol. 22, p. 267.
Křížek, M. (1963): *Czech. J. Phys.,* vol. 13B, pp. 599, 683.
Kruzhilin, Yu. I. (1967): *Opt. Spectrosc. (USSR)*, vol. 22, p. 60.
Kubo, K. (1953): *J. Sci. Res. Instrum. (Tokyo Inst. Phys. Chem. Res.)*, vol. 47, p. 1.
Kubota, H., T. Ara, and H. Saito (1951): *J. Opt. Soc. Am.,* vol. 41, p. 537.
Kudo, K., T. Arai, and T. Ogawa (1970): *J. Opt. Soc. Am.,* vol. 60, p. 1046.
Kyser, D. S., Michelson Laboratory, China Lake, Calif. (1970): private communication.
Lagemann, R. T., and T. G. Miller (1951): *J. Opt. Soc. Am.,* vol. 41, p. 1063.
Lambrecht, A., Karl Lambrecht Corp., Chicago (1969): private communication.
Lambrecht, K., Karl Lambrecht Corp., Chicago (1969): private communication.
Land, C. E., and P. D. Thacher (1969): *Proc. IEEE,* vol. 57, p. 751.
Land, E. H. (1951): *J. Opt. Soc. Am.,* vol. 41, p. 957.

——— and C. D. West (1946): pp. 160–190 in J. Alexander (ed.), "Colloid Chemistry," vol. 6, Reinhold, New York.
Landais, E. (1968): *Bull. Soc. Fr. Mineral. Cristallogr.*, vol. 91, p. 350.
Landolt, H. (1879, 1898): "Das optische Drehungsvermögen organischer Substanzen," 1879; 2d ed., Braunschweig, 1898, p. 293.
Larsen, T. (1962): *Inst. Radio Eng. Trans.*, vol. MTT-10, p. 191.
"Laser Focus 1974 Buyers' Guide," pp. 140–142, Advanced Technology Publications, Newton, Mass.
Lee, K. O., and Y. P. Neo (1971): *J. Opt. Soc. Am.*, vol. 61, p. 273.
Lee, R. W. (1969): *Appl. Opt.*, vol. 8, p. 1385.
Lenzo, P. V., E. G. Spencer, and A. A. Ballman (1966*a*): *Appl. Opt.*, vol. 5, p. 1688.
———, ———, and ——— (1967): *Appl. Phys. Lett.*, vol. 11, p. 23.
———, ———, and K. Nassau (1966*b*): *J. Opt. Soc. Am.*, vol. 56, p. 633.
———, E. H. Turner, E. G. Spencer, and A. A. Ballman (1966*c*): *Appl. Phys. Lett.*, vol. 8, p. 81.
Leurgans, P., and A. F. Turner (1947): *J. Opt. Soc. Am.*, vol. 37, p. 983A.
Lippich, F. (1882): *Wien Akad. Sitzungsber.*, ser. III, vol. 85, p. 268.
——— (1885): *Wien Akad. Sitzungsber.*, ser. III, vol. 91, p. 1059.
Loewenstein, E. V. (1961): *J. Opt. Soc. Am.*, vol. 51, p. 108.
Loferski, J. J. (1952): *Phys. Rev.*, vol. 87, p. 905.
Lostis, P. (1957): *J. Phys. Radium*, ser. 8, vol. 18, p. 51S.
Lu, S., and T. A. Rabson (1965): *Appl. Phys. Lett.*, vol. 7, p. 219.
MacNeille, S. M. (1946): U.S. Pat. 2,403,731 (July 9).
Maillard, J.-P. (1971): *Opt. Commun.*, vol. 4, p. 175.
Makas, A. S. (1962): *J. Opt. Soc. Am.*, vol. 52, p. 43.
——— and W. A. Shurcliff (1955): *J. Opt. Soc. Am.*, vol. 45, p. 998.
Marathay, A. S. (1965): *J. Opt. Soc. Am.*, vol. 55, p. 969.
——— (1966): *J. Opt. Soc. Am.*, vol. 56, p. 619.
Marcuvitz, N. (ed.) (1951): "Waveguide Handbook," pp. 218–229, 285–289, McGraw-Hill, New York.
Martin, L. C. (1948): "Technical Optics," vol. 1, pp. 196–198, Pitman, London.
Mathews, D. (1965): *Appl. Opt.*, vol. 4, p. 761.
Matsui, A., and W. C. Walker (1970): *J. Opt. Soc. Am.*, vol. 60, p. 64.
Maunders, E. A., and L. G. DeShazer (1971): *J. Opt. Soc. Am.*, vol. 61, p. 684A.
Maydan, D. (1970*a*): *J. Appl. Phys.*, vol. 41, p. 1552.
——— (1970*b*): *IEEE J. Quantum Electron.*, vol. QE-6, p. 15.
McCarthy, K. A., S. S. Ballard, and W. L. Wolfe (1963): p. 6-52 in D. E. Gray (ed.), "American Institute of Physics Handbook," 2d ed., McGraw-Hill, New York.
McCrackin, F. L. (1970): *J. Opt. Soc. Am.*, vol. 60, p. 57.
———, E. Passaglia, R. R. Stromberg, and H. L. Steinberg (1963): *J. Res. Natl. Bur. Stand (U.S.)*, vol. 67A, p. 363.
McDermott, M. N., and R. Novick (1961): *J. Opt. Soc. Am.*, vol. 51, p. 1008.
McIlrath, T. J. (1968): *J. Opt. Soc. Am.*, vol. 58, p. 506.
McIntyre, C. M., and S. E. Harris (1968): *J. Opt. Soc. Am.*, vol. 58, p. 1575.
McQuaid, R. W. (1962): *Proc. IRE*, vol. 50, p. 2484.
——— (1963): *Appl. Opt.*, vol. 2, p. 320.
Meier, R., and H. H. Günthard (1959): *J. Opt. Soc. Am.*, vol. 49, p. 1122.
Mertz, L. (1965): "Transformations in Optics," pp. 15–16, Wiley, New York.
Metcalf, H., and J. C. Baird (1966): *Appl. Opt.*, vol. 5, p. 1407.
Michael, A. J. (1968): *J. Opt. Soc. Am.*, vol. 58, p. 889.
Miller, A., A. G. Karipides, and T. M. Peltz (1971): *Appl. Opt.*, vol. 10, p. 1925.
Minkoff, J. B., W. R. Bennett, L. B. Lambert, and M. Arm (1968): *J. Appl. Phys.*, vol. 39, p. 2024
Mitchell, S. (1966): *Nature*, vol. 212, p. 65.
Mitsuishi, A., Y. Yamada, S. Fujita, and H. Yoshinaga (1960): *J. Opt. Soc. Am.*, vol. 50, p. 433.
Moeller, C. E., and D. R. Grieser (1969): *Appl. Opt.*, vol. 8, p. 206.
Mollenauer, L. F., D. Downie, H. Engstrom, and W. B. Grant (1969): *Appl. Opt.*, vol. 8, p. 661.
Mooney, F. (1952): *J. Opt. Soc. Am.*, vol. 42, p. 181.
Morris, G. C., and A. S. Abramson (1969): *Appl. Opt.*, vol. 8, p. 1249.
Mueller, A., and M. Wilhelm (1965): *J. Phys. Chem. Solids*, vol. 26, p. 2029.
Mueller, H. (1935): *Phys. Rev.*, vol. 47, p. 947.
Müller, R. (1963): *Optik*, vol. 20, p. 510.
Muller, R. H. (1969): *Proc. Symp. Recent Dev. Ellipsometry, Surf. Sci.*, vol. 16, p. 14.

Munier, J.-M., J. Claudel, E. Décamps, and A. Hadni (1962): *Rev. Opt.,* vol. 41, p. 245.

Namba, S. (1961): *J. Opt. Soc. Am.,* vol. 51, p. 76.

Nelson, D. F., and E. H. Turner (1968): *J. Appl. Phys.,* vol. 39, p. 3337.

Newman, R., and R. S. Halford (1948): *Rev. Sci. Instrum.,* vol. 19, p. 270.

Nickl, H. J., and H. K. Henisch (1969): *J. Electrochem. Soc.,* vol. 116, p. 1258.

Nicol, W. (1828–1829): *Edinb. New Phil. J.,* vol. 6, p. 83, as quoted in Johannsen (1968), p. 158.

Nitsche, R. (1965): *J. Appl. Phys.,* vol. 36, p. 2358.

Nye, J. F. (1957): "Physical Properties of Crystals," Clarendon, Oxford.

Öhman, Y. (1961): *Nature,* vol. 192, p. 254.

Oldham, W. G. (1967): *J. Opt. Soc. Am.,* vol. 57, p. 617.

O'Neill, E. L. (1963): "Introduction to Statistical Optics," pp. 133–156, Addison-Wesley, Reading, Mass.

Osterberg, H., American Optical Research Laboratory, Southbridge, Mass. (1970): private communication.

——— and A. Carlan (1957*a*): Three Element, Achromatic Quarter-Wave Plates; S. Pancharatnam's Combination, *Am. Opt. Res. Lab., Southbridge, Mass., Res. Rep.* 4500-3-1522, May 29.

——— and ——— (1957*b*): Two Element, Achromatic Quarter-Wave Plates, *Am. Opt. Res. Lab., Southbridge, Mass., Res. Rep.* 4500-3-1522, May 31.

Ott, J. H., and T. R. Sliker (1964): *J. Opt. Soc. Am.,* vol. 54, p. 1442.

Otto, A. (1969): *Optik,* vol. 29, p. 246.

Oxley, A. E. (1911): *Phil. Mag.,* ser. 6, vol. 21, p. 517.

Palik, E. D. (1963): *Appl. Opt.,* vol. 2, p. 527.

——— (1965): *Appl. Opt.,* vol. 4, p. 1017.

——— (1968): *Appl. Opt.,* vol. 7, p. 978.

——— and B. W. Henvis (1967): *Appl. Opt.,* vol. 6, p. 2198.

Pancharatnam, S. (1955*a*): *Proc. Indian Acad. Sci.,* vol. A41, p. 130.

——— (1955*b*): *Proc. Indian Acad. Sci.,* vol. A41, p. 137.

Partington, J. R. (1953): "An Advanced Treatise on Physical Chemistry," vol. 4, pp. 173–177, Wiley, New York.

Paul, B., H. Weiss, and M. Wilhelm (1964): *Solid State Electron.,* vol. 7, p. 835.

Pauty, M., P. Hartmann, and P. Vernier (1962): *C.R.,* vol. 254, p. 453.

Peek, T. H. (1971): *Appl. Opt.,* vol. 10, p. 1092.

Perkin-Elmer Corp., Instrument Division, Norwalk, Conn. (1966): Wire Grid Polarizer Accessory, Sheet no. D-454.

Perucca, E. (1935): *J. Sci. Instrum.,* vol. 12, p. 8.

Peters, C. J. (1964): *Appl. Opt.,* vol. 3, p. 1502.

Peters, C. W., T. F. Zipf, and P. V. Deibel (1953): *J. Opt. Soc. Am.,* vol. 43, p. 816A.

Pfund, A. H. (1947): *J. Opt. Soc. Am.,* vol. 37, p. 558.

Ploss, R. S. (1969): *Opt. Spectra,* vol. 3, no. 1, p. 63, January-February.

Polacoat, Incorporated, Cincinnati, Ohio (1967): *Polacoat Bull.* P-108 and P-112.

Polaroid Corporation, Cambridge, Mass. (1965): *Tech. Bull.* F 3780.

Portigal, D. L. (1969): *Appl. Opt.,* vol. 8, p. 838.

Potter, R. F. (1964): *J. Opt. Soc. Am.,* vol. 54, p. 904.

——— (1965): *Appl. Opt.,* vol. 4, p. 53.

Pridatko, G., and T. Krylova (1958): *Opt.-Mekh. Prom.,* vol. 3, p. 23.

Priebe, J. R. (1969): *J. Opt. Soc. Am.,* vol. 59, p. 176.

Provostaye, M. F. de la, and P. Desains (1850): *Ann. Chim. Phys.,* ser. 3, vol. 30, p. 158.

Rabinovitch, K., L. R. Canfield, and R. P. Madden (1965): *Appl. Opt.,* vol. 4, p. 1005.

Ramachandran, G. N., and S. Ramaseshan (1961*a*): Crystal Optics, pp. 1–54 in S. Flügge (ed.), "Handbuch der Physik," vol. 25/1, Springer, Berlin.

——— and ——— (1961*b*): ibid., pp. 39–46.

——— and ——— (1961*c*): ibid., pp. 156–158.

Ramaseshan, S. (1955): *J. Indian Inst. Sci.,* vol. 37A, p. 195.

Randall, D. D. (1954): *J. Opt. Soc. Am.,* vol. 44, p. 600.

Ratner, A. M. (1965): *Opt. Spectrosc. (USSR),* vol. 18, p. 143.

Rawlins, T. G. R. (1964): *J. Opt. Soc. Am.,* vol. 54, p. 423.

Record, F. (1948): *Nature,* vol. 162, p. 454.

Renau, J., P. K. Cheo, and H. G. Cooper (1967): *J. Opt. Soc. Am.,* vol. 57, p. 459.

Renk, K. F., and L. Genzel (1962): *Appl. Opt.,* vol. 1, p. 643.

Richards, P. L., and G. E. Smith (1964): *Rev. Sci. Instrum.,* vol. 35, p. 1535.

Richartz, M. (1965*a*): *J. Opt. Soc. Am.,* vol. 55, p. 453.

——— (1965*b*): *Optik,* vol. 22, p. 559.
——— and H.-Y. Hsu (1949): *J. Opt. Soc. Am.,* vol. 39, p. 136.
Rivera, J., L. A. Murray, and P. A. Hoss (1967): *J. Cryst. Growth,* vol. 1, p. 171.
Roberts, S., and D. D. Coon (1962): *J. Opt. Soc. Am.,* vol. 52, p. 1023.
Robin, M. B., N. A. Kuebler, and Y.-H. Pao (1966): *Rev. Sci. Instrum.,* vol. 37, p. 922.
Rohrbaugh, J. H., C. Pine, W. G. Zoellner, and R. D. Hatcher (1958): *J. Opt. Soc. Am.,* vol. 48, p. 710 [see also J. H. Rohrbaugh and R. D. Hatcher, *J. Opt. Soc. Am.,* vol. 46, p. 104 (1956) and vol. 48, p. 704 (1958)].
Rosenbaum, G., B. Feuerbacher, R. P. Godwin, and M. Skibowski (1968): *Appl. Opt.,* vol. 7, p. 1917.
Ruiz-Urbieta, M., and E. M. Sparrow (1972): *J. Opt. Soc. Am.,* vol. 62, p. 1188.
Rupprecht, G., D. M. Ginsberg, and J. D. Leslie (1962): *J. Opt. Soc. Am.,* vol. 52, p. 665.
Rvachev, V. P., and V. K. Polyanskii (1965): *Opt. Spectrosc.* (*USSR*), vol. 18, p. 594.
Salwén, A., and L. Stensland (1970): *Opt. Commun.,* vol. 2, p. 9.
Sandus, O. (1965): *Appl. Opt.,* vol. 4, p. 1634.
Sasaki, T., and H. Fukutani (1964): *Jap. J. Appl. Phys.,* vol. 3, p. 125.
Schellman, J., V. Chandrasekharan, and H. Damany (1964): *C.R.,* vol. 259, p. 4560.
———, ———, ———, and J. Romand (1965): *C.R.,* vol. 260, p. 117.
Schledermann, M., and M. Skibowski (1971): *Appl. Opt.,* vol. 10, p. 321.
Schmidt, E., and K. Vedam (1971): *Opt. Acta,* vol. 18, p. 713.
Schmieder, R. W. (1969): *J. Opt. Soc. Am.,* vol. 59, p. 297.
Schopper, H. (1952): *Optik,* vol. 9, p. 498.
——— (1953): *Optik,* vol. 10, p. 426.
Schröder, H. (1948): *Optik,* vol. 3, p. 499.
——— (1956): *Optik,* vol. 13, pp. 158, 169.
——— and R. Schläfer (1949): *Z. Naturforsch.,* vol. 4a, p. 576.
Schulz, L. G., and F. R. Tangherlini (1954): *J. Opt. Soc. Am.,* vol. 44, p. 362.
Schuster, A. (1920): "Theory of Optics," 2d ed., pp. 168–187, Arnold, London.
Shields, J. H., and J. W. Ellis (1956): *J. Opt. Soc. Am.,* vol. 46, p. 263.
Shumate, M. S. (1966): *Appl. Opt.,* vol. 5, p. 327.
Shurcliff, W. A. (1962*a*): "Polarized Light," pp. 15–29, 95–98, 109–123, Harvard University Press, Cambridge, Mass.
——— (1962*b*): ibid., pp. 43–64.
——— (1962*c*): ibid., p. 108.
——— and B. H. Billings (1963): in D. E. Gray (ed.), "American Institute of Physics Handbook," 2d ed., pp. 6-132 to 6-133, McGraw-Hill, New York.
Shustov, A. V. (1967): *Sov. J. Opt. Technol.,* vol. 34, p. 177.
Šimon, I. (1951): *J. Opt. Soc. Am.,* vol. 41, p. 336.
Skinner, C. A. (1925): *J. Opt. Soc. Am.,* vol. 10, p. 491.
Sliker, T. R., and S. R. Burlage (1963): *J. Appl. Phys.,* vol. 34, p. 1837.
——— and J. M. Jost (1966): *J. Opt. Soc. Am.,* vol. 56, p. 130.
Smartt, R. N. (1961): *J. Sci. Instrum.,* vol. 38, p. 165.
Smith, A. W., G. Burns, and D. F. O'Kane (1971*a*): *J. Appl. Phys.,* vol. 42, p. 250.
———, ———, B. A. Scott, and H. D. Edmonds (1971*b*): *J. Appl. Phys.,* vol. 42, p. 684.
Smith, F. H. (1949): *Nature,* vol. 163, p. 442.
Smith, P. H. (1969): *Proc. Symp. Recent Dev. Ellipsometry, Surf. Sci.,* vol. 16, p. 34.
Smith, S. D., T. S. Moss, and K. W. Taylor (1959): *J. Phys. Chem. Solids,* vol. 11, p. 131.
Sokolova, R. S., and T. N. Krylova (1963): *Opt. Spectrosc.* (*USSR*), vol. 14, p. 213.
Sonin, A. S., and V. E. Perfilova (1968): *Sov. Phys. Solid State,* vol. 10, p. 242.
Spencer, E. G., P. V. Lenzo, and A. A. Ballman (1967): *Proc. IEEE,* vol. 55, p. 2074.
Spooner, D. L. (1972): *Appl. Opt.,* vol. 11, p. 2984.
Stafsudd, O. M., F. A. Haak, and K. Radisavljevic (1967): *Appl. Opt.,* vol. 6, p. 1276.
Stanford, J. L., Michelson Laboratory, China Lake, Calif. (1970): unpublished data.
Steinmetz, D. L., W. G. Phillips, M. Wirick, and F. F. Forbes (1967): *Appl. Opt.,* vol. 6, p. 1001.
Stephan, G., J.-C. Lemonnier, Y. LeCalvez, and S. Robin (1966): *C.R.,* vol. 262B, p. 1272.

Sterzer, F., D. Blattner, and S. Miniter (1964): *J. Opt. Soc. Am.*, vol. 54, p. 62.
Stöber, P. (1924): *Z. Krist*, vol. 61, p. 299.
——— (1928): *Neues Jahrb. Mineral.*, vol. A57, p. 139.
——— (1930): *Chem. Erde*, vol. 6, pp. 357, 453.
Stookey, S. D., and R. J. Araujo (1968): *Appl. Opt.*, vol. 7, p. 777.
Stroke, G. W. (1963): *Phys. Lett. (Neth.)*, vol. 5, p. 45.
——— (1964): *J. Opt. Soc. Am.*, vol. 54, p. 846.
Strong, J. (1938): "Procedures in Experimental Physics," pp. 388–389, Prentice-Hall, Englewood Cliffs, N.J.
Sutton, A. M., and C. F. Panati (1969): lecture notes, Modern Optics RCA Institutes, Clark, N.J.
Swartz, J., D. K. Wilson, and R. J. Kapash (1971): High Efficiency Laser Polarizers, *Electro-Opt. 1971 West Conf., Anaheim, Calif., May.*
Takahashi, S. (1961): *J. Opt. Soc. Am.*, vol. 51, p. 441.
Takasaki, H. (1961): *J. Opt. Soc. Am.*, vol. 51, pp. 462, 463, and 1146.
——— (1962): *J. Opt. Soc. Am.*, vol. 52, p. 718.
——— (1963): *J. Opt. Soc. Am.*, vol. 53, p. 640.
——— (1964): *J. Opt. Soc. Am.*, vol. 54, p. 551.
——— (1966): *Appl. Opt.*, vol. 5, p. 759.
———, M. Isobe, T. Masaki, A. Konda, T. Agatsuma, and Y. Watanabe (1964*a*): *Appl. Opt.*, vol. 3, p. 345.
———, N. Okazaki, and K. Kida (1964*b*): *Appl. Opt.*, vol. 3, p. 833.
Tamai, T., and M. Achiwa (1964): *Jap. J. Appl. Phys.*, vol. 3, p. 60.
Tewarson, S. P. (1966): *Indian J. Phys.*, vol. 40, pp. 281, 562.
Thacher, P. D., and C. E. Land (1969): *IEEE Trans. Electron. Devices*, vol. ED-16, p. 515.
Thompson, S. P. (1877): *Proc. Phys. Soc. (Lond.)*, vol. 2, p. 185.
——— (1881): *Phil. Mag.*, ser. 5, vol. 12, p. 349.
——— (1886): *Phil. Mag.*, ser. 5, vol. 21, p. 476.
——— (1905): *Proc. Opt. Conv., 1905*, pp. 216–235.
Trevelyan, B. (1969): *J. Sci. Instrum.* [*J. Phys. (GB)*, ser. E], vol. 2, p. 425.
Tronstad, L. (1934): *J. Sci. Instrum.*, vol. 11, p. 144.
Turner, A. F. (1950): *J. Phys. Radium*, vol. 11, p. 444.
———, Bausch & Lomb Optical Company, Rochester, N.Y. (1970): private communication.
——— and P. W. Baumeister (1966): *Appl. Opt.*, vol. 5, p. 69.
Turner, E. H., and I. P. Kaminow (1963): *J. Opt. Soc. Am.*, vol. 53, p. 523A.
Twyman, F. (1952): "Prism and Lens Making," 2d ed., pp. 244, 599, Hilger & Watts, London.
Tzekhnovitzer, E. (1934): *J. Phys. Chem. (USSR)*, vol. 5, p. 1452.
Ukhanov, Yu. I. (1969): *Opt. Spectrosc. (USSR)*, vol. 27, p. 170.
Uzan, E., H. Damany, and V. Chandrasekharan (1969): *Opt. Commun.*, vol. 1, p. 221.
———, ———, and ——— (1970): *Opt. Commun.*, vol. 2, p. 273.
———, ———, and J. Romand (1965): *C.R.*, vol. 260, p. 5735.
Van Uitert, L. G., S. Singh, H. J. Levinstein, J. E. Geusic, and W. A. Bonner (1967): *Appl. Phys. Lett.*, vol. 11, p. 161.
———, ———, ———, ——— and ——— (1968): *Appl. Phys. Lett.*, vol. 12, p. 224.
Venturini, E. L., E. G. Spencer, P. V. Lenzo, and A. A. Ballman (1968): *J. Appl. Phys.*, vol. 39, p. 343.
Violino, P. (1965): *Rev. Opt.*, vol. 44, p. 109.
Wahlstrom, E. E. (1969): "Optical Crystallography," 4th ed., pp. 236–267, Wiley, New York.
Walker, M. J. (1954): *Am. J. Phys.*, vol. 22, p. 170.
Walker, W. C. (1964): *Appl. Opt.*, vol. 3, p. 1457.
Walsh, T. E. (1966): *RCA Rev.*, vol. 27, p. 323.
Walton, A. K., and T. S. Moss (1961): *Proc. Phys. Soc. (Lond.)*, vol. 78, p. 1393.
———, ———, and B. Ellis (1964): *J. Sci. Instrum.*, vol. 41, p. 687.
Weinberg, J. L. (1964): *Appl. Opt.*, vol. 3, p. 1057.
Weinberger, H., and J. Harris (1964): *J. Opt. Soc. Am.*, vol. 54, p. 552.
Weiser, G. (1964): *Proc. IEEE*, vol. 52, p. 966.
Weiss, H., and M. Wilhelm (1963): *Z. Phys.*, vol. 176, p. 399.
Wemple, S. H., and M. DiDomenico, Jr. (1969): *J. Appl. Phys.*, vol. 40, p. 735.
Wendlandt, W. W. (ed.) (1968): "Modern Aspects of Reflectance Spectroscopy," Plenum, New York.
——— and H. G. Hecht (1966): "Reflectance Spectroscopy," Interscience, New York.

West, C. D. (1945): *J. Opt. Soc. Am.*, vol. 35, p. 26.
——— (1953): *J. Opt. Soc. Am.*, vol. 43, p. 335A.
——— and R. O'B. Carpenter (1963): p. 6-188 in D. E. Gray (ed.), "American Institute of Physics Handbook," 2d ed., McGraw-Hill, New York.
——— and R. C. Jones (1951): *J. Opt. Soc. Am.*, vol. 41, p. 976.
——— and A. S. Makas (1949): *J. Opt. Soc. Am.*, vol. 39, p. 791.
Williams, W. E. (1950): "Applications of Interferometry," 4th ed., pp. 77–78, Wiley, New York.
Wolfe, W. L., S. S. Ballard, and K. A. McCarthy (1963*a*): p. 6-18 in D. E. Gray (ed.), "American Institute of Physics Handbook," 2d ed., McGraw-Hill, New York.
———, ———, and ——— (1963*b*): ibid., pp. 6-18, 6-24, and 6-28.
———, ———, and ——— (1963*c*): ibid., p. 6-84.
Wood, R. W. (1934*a*): "Physical Optics," 3d ed., p. 345, Macmillan, New York.
——— (1934*b*): ibid., p. 360.
——— (1934*c*): ibid., pp. 352–353.
——— (1934*d*): ibid., pp. 356–361.
Worthing, A. G. (1926): *J. Opt. Soc. Am.*, vol. 13, p. 635.
Wright, F. E. (1939): *Am. J. Sci.*, vol. 237, p. 736.
Wright, N. (1948): *J. Opt. Soc. Am.*, vol. 38, p. 69.
Wulff, L. (1896): *Sitz. Preuss. Akad. Wiss.*, vol. 135, p. 879.
Yamaguti, T. (1955*a*): *J. Phys. Soc. Jap.*, vol. 10, p. 219.
——— (1955*b*): *J. Opt. Soc. Am.*, vol. 45, p. 891.
———, I. Makino, S. Shinoda, and I. Kuroha (1959): *J. Phys. Soc. Jap.*, vol. 14, p. 199.
Yolken, H. T., R. M. Waxler, and J. Kruger (1967): *J. Opt. Soc. Am.*, vol. 57, p. 283.
Young, J. B., H. A. Graham, and E. W. Peterson (1965): *Appl. Opt.*, vol. 4, p. 1023.
Zernicke, F., Jr. (1964): *J. Opt. Soc. Am.*, vol. 54, p. 1215 [erratum in ibid., vol. 55, p. 210 (1965)].

Section 11

Spectroscopy

D. J. LOVELL

Optical Consultant, Barton Road, Stow, Mass.

(Numbers refer to paragraphs)

1. Introduction† Spectral information is of value in many fields of optics. The usefulness of the information varies from defining a standard of length to identifying substances involved in a chemical reaction. The purpose of this section is to list some of the more important wavelengths users of this handbook may encounter.

The electromagnetic spectrum is divided into several regions for convenience. These divisions are generally determined by the techniques used to detect and measure the radiation. Thus, radiations longer than about 1 mm are usually considered radio waves. Further divisions in this region are not of interest here. The infrared spectral region extends from wavelengths longer than those visible to the eye (about 750 nm) through those detected by thermal means. The latter limit is about 1 mm. Divisions within this region are often adopted (near, intermediate, and far), but these are arbitrary and will not be described here.

Visible radiation, or light, extends from about 400 to 750 nm. Shorter wavelengths are called ultraviolet. The lower wavelength limit is also a matter of definition. Ultraviolet radiation was originally discovered through its chemical effects, and in the traditional sense this extends to the region of a few nanometers. At these

† This section was prepared while the author was affiliated with the Massachusetts College of Optometry, Boston.

shorter wavelengths, considerable overlap is encountered with x-rays, which are distinguished by their ionizing characteristics. We generally consider x-ray radiation to extend to about 10 pm. Radiation at wavelengths shorter than this is usually attributable to inner-nuclear effects and is not of concern here.

2. Tables The regions of the electromagnetic radiation of concern to optics extend roughly over the ultraviolet, visible, and infrared radiations. We tabulate here some specific emissions of general interest in optics: Fraunhofer lines (Table 1), wavelengths useful for optical-design purposes (Table 2), wavelengths of common lasers (Table 3), and the lines and bands most characteristic of selected atoms and molecules (Tables 4 and 5).

Table 4 is arranged in alphabetical order for the selected emitters, wavelengths are given in nanometers, and the relative intensity is also shown. A comparison of relative intensities is fraught with possibilities for error. Consideration must be given to temperature, pressure, concentration, and of course the detector. The intensity is listed as a first-order value only. Table 5 is arranged in order of wavelength. In the compilation of these tables, consideration has been given to the spectral region from approximately 1 nm to 50 μm. Accordingly, only the more intense lines can be listed. These tables are intended to complement information found in other sections of the handbook. Anyone interested in a detailed list of spectral lines for use in spectroscopic determination should consult one of the references.

The primary standard of length presently adopted is the meter, which is defined as 1,650,763.73 wavelengths of a particular emission of krypton 86, in vacuum, with the discharge tube held at —20°C. Numerous secondary standards are available which are more convenient for practical use.

TABLE 1 Fraunhofer Lines

Line	Source	Wavelength, Å	Intensity	Line	Source	Wavelength, Å	Intensity
y	Atm O_2	8987.65	10	G	Fe, Ti^+	4307.912	6
x_4	Mg	8806.775	14		Ca	4307.747	3
x_3	Ca^+	8662.170	23		Ca	4226.740	20
x_2	Ca^+	8542.144	25	h	H_δ	4101.748	40
x_1	Ca^+	8498.062	20	H	Ca^+	3968.492	700
Z	Atm O_2	8226.962	20	K	Ca^+	3933.682	1,000
A	Atm O_2	7593.695	10	L	Fe	3820.436	25
a	Atm O_2	7184.526	8	M	Fe	3727.634	4
B	Atm O_2	6867.187	4	N	Fe	3581.209	30
C	H_α	6562.808	40	O	Fe	3441.019	15
a	Atm O_2	6276.607	2	P	Ti^+	3361.193	8
D_1	Na	5895.940	20	Q	Fe	3286.772	7
D_2	Na	5889.973	30	R	Ca^+	3181.276	3
D_3	He	5875.650			Ca^+	3179.342	5
	He	5875.618		r	Fe	3143.996	2
E	Fe	5270.388	4		Ti^+	3143.764	4
	Ca	5270.268	3	S_1	Ni	3101.895	3
	Fe	5269.550	8		Ni	3101.574	4
b_1	Mg	5183.619	30	S_2	Fe	3100.682	3
b_2	Mg	5172.698	20		Fe	3100.325	4
b_3	Fe^+	5169.050	4		Fe	3099.987	3
	Fe	5168.908	3		Fe	3099.896	3
b_4	Fe	5167.508	5	s	Fe	3047.614	35
	Mg	5167.328	15	T	Fe	3021.077	30
F	H_β	4861.342	30		Fe	3020.656	40
g	H_γ	4340.475	20		Fe	3020.490	20
				t	Fe, Ni	2994.436	40

TABLE 2 Spectral Lines for Optical Design

Letter	Name	Wavelength, nm
i	Hg	365
h	Hg	405
g	Hg	436
F′	Cd	480
e	Hg	546
C′	Hg	644
t	Hg	1014

TABLE 3 Common Laser Wavelengths

Wavelength, nm	Type	Wavelength, μm	Type
337.1	Pulsed N_2	1.06	Nd^+
488.0	Ar^+	2.10	YAG-Ho
514.5	Ar^+	3.39	He-Ne
540.1	Pulsed Ne	3.507	He-Xe
568.2	Kr^+	9.0	He-Xe
632.8	Ne	10.6	CO_2
647.0	Kr		
694.3	Pulsed ruby		
844.6	GaAs		

TABLE 4 Prominent Lines and Band Characteristics of Selected Atoms and Molecules

Aluminum, Al Z = 13

Wavelength, nm	Relative intensity	Wavelength, nm	Relative intensity	Wavelength, nm	Relative intensity
7.6543(2)	1,000	9.0200	1,000	185.4720	1,000
7.8327(2)	1,000	10.4047	1,000	309.2713	1,000
7.8351	1,000	10.7945	1,000	394.4032	2,000
8.5515	1,000	10.9514	1,000	396.1527	3,000
8.8170	1,000	13.0413(4)	1,000		

Aluminum oxide, AlO

The spectrum consists of a marked sequence of single-headed bands, degraded to the red, extending from 437.4 to 522.5 nm with the stronger bands at 464. 8 and 484.21 nm.

Ammonia, NH3

The strongest feature of the ammonia spectrum is an intense band extending from about 8.5 to 14 μm, with strongest bands at 10.35 and 10.75 μm. Another strong band, centered at 6.1 μm extends from 5.5 to 7 μm. Weaker bands are formed at 2.25 and 2.9 μm. There is another strong band at 791.9 nm degraded to 787.4 nm. A strong continuum occurs with maximum at 563.5 and 567.0 nm. Rather diffuse bands occur below 230 nm; the strongest of these are 208.64, 204.84, and 201.09 nm.

Argon, Ar Z = 18

Wavelength, nm	Relative intensity	Wavelength, nm	Relative intensity	Wavelength, nm	Relative intensity
46.2007	1,000	394.898	2,000	565.070	1,500
63.7282	1,000	404.442	1,200	810.369	2,000
70.0398	1,000	415.859	1,200	811.5311	5,000
74.02695	1,000	419.103	1,200	826.452	1,000
85.0602	1,000	419.832	1,200	840.821	2,000
91.97815	1,000	420.068	1,200	842.465	2,000
93.20528	1,000	425.936	1,200	852.144	2,000
104.8218	1,000	426.629	1,200	922.450	1,000
360.652	1,000	430.010	1,200	965.77	1,500
394.750	1,000	470.232	1,200	978.450	1,000

Barium, Ba Z = 56

Wavelength, nm	Relative intensity	Wavelength, nm	Relative intensity	Wavelength, nm	Relative intensity
350.112	1,000	614.172	2,000	728.027	1,000
455.4042	1,000	659.532	1,000		
553.5551	1,000	705.996	2,000		

Calcium, Ca Z = 20

Wavelength, nm	Relative intensity	Wavelength, nm	Relative intensity	Wavelength, nm	Relative intensity
393.367	600	854.209	1,000	1034.385	500
422.6728	500	866.214	1,000		

TABLE 4 Prominent Lines and Band Characteristics of Selected Atoms and Molecules (Continued)

Carbon, C $Z = 6$

Wavelength, nm	Relative intensity	Wavelength, nm	Relative intensity	Wavelength, nm	Relative intensity
3.3736	1,000	51.15225	1,000	117.5711	1,000
24.4907	1,000	53.83130	1,000	131.1363	1,000
28.9230	1,000	57.42809	1,000	133.57077	1,000
31.2422	1,000	68.73453	1,000	154.8195	1,000
37.1694(2)	1,000	90	1,000	156.14382	1,000
38.4178	1,000	97.7020	1,000	165.70078	1,000
41.9714	1,000	101.0374	1,000	193.09054	1,000
45.9633	1,000	103.70182	1,000	909.489	50

Carbon, C_2

Three strong band systems are observed in sources containing carbon or hydrocarbons. The swan bands, among the most frequently observed molecular bands, extend from 436.5 to 668.0 nm, with strong features at 436.5, 473.71, and 516.52 nm. These are degraded to the violet. The second system occurs at relatively high pressure with prominent bands degraded to the violet at 468.02 and 589.93 nm. The third system is also degraded to shorter wavelengths, with prominent bands at 360.73, 385.22, and 410.23 nm.

Carbon dioxide, CO_2

Two extensive bands are known in the ultraviolet and visible regions. One occurs in the negative glow of discharges and extends from 280 to nearly 500 nm. It consists of narrow bands degraded to the red. The second occurs in the flame of burning carbon monoxide and consists of a great number of narrow bands, not clearly degraded in either direction, on a continuous background. The bands are strongest between 350 and 500 nm, with a maximum at about 420 nm. The infrared absorbtion spectrum of the atmosphere indicates absorption bands of CO_2 centered at 2.7, 4.3, and 15 μm, the last being the strongest.

Carbon monoxide, CO

Carbon monoxide is characterized in the visible and ultraviolet regions by numerous sharp lines. In the infrared, a single band extending from 4.6 to 4.75 μm is the dominant feature.

Cesium, Cs $Z = 55$

Wavelength, nm	Relative intensity	Wavelength, nm	Relative intensity	Wavelength, nm	Relative intensity
455.5355	2,000	852.110	5,000	1002.439	1,000
459.3177	1,000	894.350	2,000	1012.3	1,200
807.902	1,000	917.224	1,000		

CH

These bands are readily excited during the combustion of hydrocarbons and in many astrophysical sources. Three systems are known at 314.3, 390.0, and 430.0 nm. The last is the strongest and is characterized by an intense band at 431.25 nm. It is characteristic of many hydrocarbon reactions such as found in comet heads and the sun's atmosphere.

TABLE 4 Prominent Lines and Band Characteristics of Selected Atoms and Molecules (Continued)

Copper, Cu $Z = 39$

Wavelength, nm	Relative intensity	Wavelength, nm	Relative intensity	Wavelength, nm	Relative intensity
159.3758	1,000	259.263	1,000	327.3962	3,000
164.2208	2,000	282.437	1,000	578.213	1,000
172.2679	1,000	324.7540	5,000		

Cyanogen, CN_2

The most distinguishing feature of this molecule is a broad band extending from about 13 to 14.2 μm with heads at 13.55 and 13.85 μm. There is also a sharp band at 4.7 μm.

Germanium, Ge $Z = 32$

Wavelength, nm	Relative intensity	Wavelength, nm	Relative intensity	Wavelength, nm	Relative intensity
303.9064	1,000				

Helium, H $Z = 2$					
58.43340	1,000	388.8646	1,000	587.5618	1,000

Hydrogen, H_2 $Z = 1$					
121.5339	1,000	656.2725	1,000	656.2849	2,000
121.5668	670				

Hydrogen, H_2

The emission spectrum has few characteristic features with no heads or groups of lines resembling the usual band structure. The system is strongest in the orange but extends throughout the whole visible spectrum.

Iodine, I $Z = 53$

Wavelength, nm	Relative intensity	Wavelength, nm	Relative intensity	Wavelength, nm	Relative intensity
20.6238	900	54.9692	900	69.5878	1,000
54.6461	900	60.8246	1,000		

Iron, Fe $Z = 26$					
247.2909	1,000	298.3572	1,000	373.7133	1,000
253.5604	1,000	299.4429	1,000	374.9487	1,000
259.8369	1,000	302.0640	1,000	385.9913	1,000
259.9396	1,000	358.1195	1,000	440.4752	1,000
259.9570	1,000	371.9935	1,000		
296.6900	1,000	373.4867	1,000		

TABLE 4 Prominent Lines and Band Characteristics of Selected Atoms and Molecules (Continued)

Iron oxide, FeO

Strong bands are observed in the orange and near infrared in the flame of an iron arc. The general appearance of the orange band is complex with particularly strong bands at 609.73, 610.969, 618.065, and 621.869 nm. A strong infrared band occurs at 811.2, 813.7, 823, and 830.2 nm.

Krypton, Kr $Z = 36$

Wavelength, nm	Relative intensity	Wavelength, nm	Relative intensity	Wavelength, nm	Relative intensity
435.548	3,000	760.15465	5,000	829.81091	5,000
465.887	2,000	810.43660	5,000	850.88736	3,000
473.900	3,000	811.29023	5,000	877.67498	5,000
557.02895	2,000	819.00570	3,000	892.86934	2,000
587.09158	3,000	826.32412	2,000	1022.146	1,000

Lead, Pb $Z = 82$

Wavelength, nm	Relative intensity	Wavelength, nm	Relative intensity	Wavelength, nm	Relative intensity
216.9994	1,000	239.379	2,500	405.7820	2,000
220.3505	5,000				

Lithium, Li $Z = 3$

Wavelength, nm	Relative intensity	Wavelength, nm	Relative intensity	Wavelength, nm	Relative intensity
13.4998	200	323.261	1,000	670.784	3,000
19.9282	300	610.364	2,000	812.652	1,000

Magnesium, Mg $Z = 12$

Wavelength, nm	Relative intensity	Wavelength, nm	Relative intensity	Wavelength, nm	Relative intensity
18.0617	1,000	31.2311	1,000	285.2129	300
23.1730	1,000	32.0999	1,000	383.826	300
27.6581	1,000	35.3094	1,000	518.362	500

Mercury, Hg $Z = 80$

Wavelength, nm	Relative intensity	Wavelength, nm	Relative intensity	Wavelength, nm	Relative intensity
253.6519	2,000	546.0740	2,000	579.065	1,000
435.835	3,000				

Methane, CH_4

A band is found in the vicinity of 3.31 μm, characterized by several sharp lines. Methane also shows extensive bands from 7.2 to 8.2 μm, centered at 7.7 μm.

Neon, Ne $Z = 10$

Wavelength, nm	Relative intensity	Wavelength, nm	Relative intensity	Wavelength, nm	Relative intensity
35.8721	1,000	48.9501	1,000	470.440	1,500
36.5594	1,000	73.5895	1,000	470.885	1,200
46.0725	1,000	77.0409	1,000	471.534	1,500
46.9820	1,000	352.047	1,000	488.492(2)	1,000

TABLE 4 Prominent Lines and Band Characteristics of Selected Atoms and Molecules (Continued)

Nickel, Ni $Z = 28$

Wavelength, nm	Relative intensity	Wavelength, nm	Relative intensity	Wavelength, nm	Relative intensity
169.2514	1,000	313.411	1,000	356.637	2,000
176.9643	1,000	341.4765	1,000	357.187	1,000
260.113	2,000	344.626	1,000	359.771	1,000
263.289	2,000	349.2956	1,000	361.046	1,000
300.249	1,000	351.5054	1,000	361.939	2,000
305.082	1,000	352.4541	1,000	712.224	1,000
310.155	1,000				

Nitric oxide, NO

The spectrum of this molecule is characterized by bands at 5.25 and 5.4 μm.

Nitrogen, N $Z = 7$

Wavelength, nm	Relative intensity	Wavelength, nm	Relative intensity	Wavelength, nm	Relative intensity
28.3579	1,000	92.3220	1,000	123.8821	1,000
33.5050	1,000	95.5335	1,000	171.8551	1,000
37.4441	1,000	108.5701	1,000	410.998	1,000
77.5965	1,000	119.95490	1,000	415.146	1,000
91.6701	1,000	120.02238	950	600.848	8,000

Nitrogen, N_2

There are a large number of band systems for this molecule. The first and second positive systems are the most readily developed. The former is degraded to the violet and extends from 503 to 1042 nm, with strong bands at 580.43, 646, 654.48, 662.36, 891.16, and 1042 nm. The latter is also degraded to shorter wavelengths and extends from 281.4 to 497.6 nm, with strongest bands at 337.13, 357.69, 375.54, 380.49, and 399.84 nm.

Nitrous oxide, N_2O

The most prominent feature of the spectrum of this molecule is a band centered at 7.77 μm, which extends from 7.5 to 8.1 μm, and with additional features at 8.5 and 8.7 μm. Another strong band extends from 16 to 17.5 μm, centered at 17 μm. Lesser bands are observed at 2.85, 3.9, and 4.5 μm.

Oxygen, O $Z = 8$

Wavelength, nm	Relative intensity	Wavelength, nm	Relative intensity	Wavelength, nm	Relative intensity
1.8969	1,000	164.366	1,000	777.1928(3)	1,000
2.1602	1,000	176.778	1,000	794.757	1,000
17.3082	1,000	436.830	1,000	844.638	2,000
62.9732	1,000	615.820	1,000		

Oxygen, O_2

The Schumann-Runge system of this molecule shows several strong bands below 190 nm, with complete absorption below 175.9 nm. The strongest bands occur at 179.2, 180.31, 181.56, 183.01, 184.58, and 186.30 nm. Several bands have been observed with indications that intensities are strongly pressure-dependent. The more intense bands occur at 384.11, 409.59, and 417.32 nm.

TABLE 4 Prominent Lines and Band Characteristics of Selected Atoms and Molecules (Continued)

Ozone, O_3

The absorption characteristics of ozone are of particular importance to terrestrial observers. There is a fairly sharp cutoff in the ultraviolet between 300 and 360 nm. There are also two strong diffuse bands at about 573 and 602 nm. In the infrared, important bands occur at 4.7, 9.6, 13.8, and 14.4 μm.

Potassium, K $Z = 19$

Wavelength, nm	Relative intensity	Wavelength, nm	Relative intensity	Wavelength, nm	Relative intensity
766.4907	9,000	769.8979	5,000		

Silicon, Si $Z = 14$

Wavelength, nm	Relative intensity	Wavelength, nm	Relative intensity	Wavelength, nm	Relative intensity
11.7860	1,000	139.3755	1,000	288.1578	500
11.8968	1,000	153.34320	1,000		
126.47374	1,000	198.8993	1,000		

Silver, Ag $Z = 47$

Wavelength, nm	Relative intensity	Wavelength, nm	Relative intensity	Wavelength, nm	Relative intensity
328.0683	2,000	520.9067	1,500	546.5487	1,000
338.2891	1,000				

Sodium, Na $Z = 11$

Wavelength, nm	Relative intensity	Wavelength, nm	Relative intensity	Wavelength, nm	Relative intensity
16.8084	1,000	37.8143	1,000	330.232	600
19.0440	1,000	40.0722	1,000	588.9953	9,000
30.8264	1,000	41.0371	1,000	589.5923	5,000
31.9638	1,000	46.3263	1,000	819.481	1,000

Strontium, Sr $Z = 38$

Wavelength, nm	Relative intensity	Wavelength, nm	Relative intensity	Wavelength, nm	Relative intensity
460.7331	1,000	707.010	1,000	1032.729	1,000

Sulfur dioxide, SO_2

The strongest features of the spectrum of this molecule are bands extending from 8 to 9.5 and from 7.1 to 7.7 μm. Sharp bands are located at 4 and 19.1 μm. A simple group of close double-headed bands, degraded to the red, is found in the region 217 to 234.3 μm.

Thorium, Th $Z = 90$

Wavelength, nm	Relative intensity	Wavelength, nm	Relative intensity	Wavelength, nm	Relative intensity
195.902	200				

Tin, Sn $Z = 50$

Wavelength, nm	Relative intensity	Wavelength, nm	Relative intensity	Wavelength, nm	Relative intensity
121.0515	2,000	283.9989	300	452.4741	500
125.1384	2,000	286.3327	300	855.260	
175.7904	5,000	317.5019	500		
189.9890	5,000	326.2328	400		

TABLE 4 Prominent Lines and Band Characteristics of Selected Atoms and Molecules (Continued)

Tungsten, W   Z = 74

Wavelength, nm	Relative intensity	Wavelength, nm	Relative intensity	Wavelength, nm	Relative intensity
430.2108	60	465.987	200		

Uranium, U   Z = 92

Wavelength, nm	Relative intensity	Wavelength, nm	Relative intensity	Wavelength, nm	Relative intensity
424.1669	40	639.545	100	644.912	100
591.540	125				

Water vapor, H_2O

Because of its importance to the transmission characteristics of the atmosphere, the spectrum of water vapor has been extensively studied. Several bands are noted, at 0.94, 1.1, 1.38, 1.87, 2.7, 3.2, 6.3, 24, 25, 26.6, and 30 μm, and beyond.

Xenon, Xe   Z = 54

Wavelength, nm	Relative intensity	Wavelength, nm	Relative intensity	Wavelength, nm	Relative intensity
462.4276	1,000	834.682	2,000	992.320	2,000
467.1226	2,000	840.919	2,000	1083.834	1,000
823.163	5,000	881.941	5,000		
828.012	5,000	979.970	2,000		

Zinc, Zn   Z = 30

Wavelength, nm	Relative intensity	Wavelength, nm	Relative intensity	Wavelength, nm	Relative intensity
213.856	800	334.5020	800	636.2347	1,000
330.2588	800				

TABLE 5 Prominent Lines and Bands in Order of Wavelength

Wavelength, nm	Relative intensity	Source	Wavelength, nm	Relative intensity	Source
1.8969	1,000	O	41.9714	1,000	C
2.1602	1,000	O	44.9065	900	Ar
3.3736	1,000	C	45.2226	900	N
4.393	900	Ar	45.9462	900	N
4.426	900	Ar	45.9521	950	N
4.449	900	Ar	45.9633	1,000	C
6.3295	900	Mg	46.0725	1,000	Ne
7.6543	1,000	Al	46.2007	1,000	Ar
7.6572	1,000	Al	46.2388	930	Ne
7.8327	1,000	Al	46.3263	1,000	Na
7.8351	1,000	Al	46.9820	1,000	Ne
8.3391	900	Na	46.9866	900	Ne
8.5515	1,000	Al	48.9501	1,000	Ne
8.8170	1,000	Al	49.1050	900	Ne
9.0200	1,000	Al	49.9530	900	N
10.4047	1,000	Al	50.8182	900	O
10.607	900	Ne	51.15225	1,000	C
10.7945	1,000	Al	52.5795	900	C
10.9514	1,000	Al	53.80801	900	C
11.7860	1,000	Si	53.81487	950	C
11.8968	1,000	Si	53.83120	1,000	C
13.0413	1,000	Al	55.4514	900	O
13.0848	1,000	Al	57.42809	1,000	C
13.1003	1,000	Al	58.43340	1,000	He
13.1441	1,000	Al	59.9598	900	O
16.8084	1,000	Na	62.9732	1,000	O
17.2935	950	O	63.7282	1,000	Ar
17.3082	1,000	O	68.73453	1,000	C
18.0617	1,000	Mg	70.0398	1,000	Ar
19.0440	1,000	Na	70.3850	900	O
19.2799	1,000	O	73.5895	1,000	Ne
19.2906	950	O	74.02695	1,000	Ar
23.1730	1,000	Mg	77.040	1,000	Ne
24.4907	1,000	C	77.5965	1,000	N
24.7205	900	N	85.0602	1,000	Ar
27.6581	1,000	Mg	85.85590	900	C
28.3420	900	N	90.41416	1,000	C
28.3470	950	N	91.6701	1,000	N
28.3579	1,000	N	91.97815	1,000	Ar
28.9230	1,000	C	92.3220	1,000	N
30.8264	1,000	Na	93.20528	1,000	Ar
31.2311	1,000	Mg	95.5335	1,000	N
31.2422	1,000	C	97.7020	1,000	C
31.2453	950	C	99.1579	900	N
31.4850	900	N	101.0374	1,000	C
31.9638	1,000	Na	103.70182	1,000	C
32.0999	1,000	Mg	104.8218	1,000	Ar
32.3310	900	Mg	108.5701	1,000	N
33.5050	1,000	N	117.5711	1,000	C
35.3094	1,000	Mg	119.95490	1,000	N
35.8721	1,000	Ne	120.02238	950	N
36.5594	1,000	Ne	121.0515	2,000	Sn
37.1694	1,000	C	121.5339	1,000	H
37.1747	1,000	C	123.8821	1,000	N
37.4204	900	N	125.1384	2,000	Sn
37.4441	1,000	N	126.47374	1,000	Si
37.8143	1,000	Na	131.1363	1,000	C
38.4032	950	C	133.57077	1,000	C
38.4178	1,000	C	139.3755	1,000	Si
40.0722	1,000	Na	153.34320	1,000	Si
41.0371	1,000	Na	154.8195	1,000	C
41.9525	950	C	155.0768	950	C

TABLE 5 Prominent Lines and Bands in Order of Wavelength (Continued)

Wavelength, nm	Relative intensity	Source
156.14382	1,000	C
159.3758	1,000	Cu
161.185	900	Al
164.2208	2,000	Cu
164.366	1,000	O
165.70078	1,000	C
169.2514	1,000	Ni
171.8551	1,000	N
172.2679	1,000	Cu
175.7904	5,000	Sn
175.9	.....	O_2
176.778	1,000	O
176.9643	1,000	Ni
177.167	900	O
179.2	.....	O_2
180.31	.....	O_2
181.56	.....	O_2
183.01	.....	O_2
184.58	.....	O_2
185.4720	.....	O_2
186.30	.....	O_2
189.9890	5,000	Sn
192.627	900	Na
198.8993	1,000	Si
201.09	.....	AlO
204.84	.....	AlO
206.238	900	I
208.64	.....	AlO
210	.....	$(CN)_2$
213.856	800	Zn
216.9994	1,000	Pb
220.3505	5,000	Pb
239.3794	4,000	Pb
247.2909	1,000	Fe
253.5604	1,000	Fe
253.6519	2,000	Hg
259.2627	1,000	Cu
259.8369	1,000	Fe
259.9396	1,000	Fe
259.9570	1,000	Fe
260.1126	2,000	Ni
263.2891	2,000	Ni
264	2,000	SO_2
282.4369	1,000	Cu
296.6900	1,000	Fe
298.3572	1,000	Fe
299.4429	1,000	Fe
300.2491	1,000	Ni
302.0640	1,000	Fe
303.9064	1,000	Ge
305.0819	1,000	Ni
309.2713	1,000	Al
310.1554	1,000	Ni
313.4108	1,000	Ni
314.3	.....	CH
323.261	1,000	Li
324.7540	5,000	Cu
327.3962	3,000	Ag
330	.....	$(CN)_2$
337.13	.....	N_2 laser
338.2891	1,000	Ag
341.4765	1,000	Ni
344.6263	1,000	Ni
349.2956	1,000	Ni
350.1116	1,000	Ba
351.5054	1,000	Ni
352.047	1,000	Ne
352.4541	1,000	Ni
356.6372	2,000	Ni
357.1879	1,000	Ni
357.69	.....	N_2
358.1195	1,000	Fe
359.7705	1,000	Ni
360.652	1,000	Ar
360.73	.....	C_2
361.0462	1,000	Ni
361.9392	2,000	Ni
371.9935	1,000	Fe
373.4867	1,000	Fe
373.7133	1,000	Fe
374.9487	1,000	Fe
375.54	.....	N_2
384.11	.....	O_2
385.22	.....	C_2
385.9913	1,000	Fe
388.8646	1,000	He
390	.....	CH
394.4032	2,000	Al
394.750	1,000	Ar
394.898	2,000	Ar
396.1527	3,000	Al
399.84	.....	N_2
404.442	1,200	Ar
405.7820	2,000	Pb
409.59	.....	O_2
410.22	.....	C_2
410.998	1,000	N
415.146	1,000	N
415.859	1,200	Ar
417.32	.....	O_2
419.103	1,200	Ar
419.832	1,200	Ar
420	.....	CO_2
420.068	1,200	Ar
425.936	1,200	Ar
426.629	1,200	Ar
427.217	1,200	Ar
430	.....	CH
430.010	1,200	Ar
435.548	3,000	Kr
435.835	3,000	Kr
436.5	3,000	C_2
436.830	1,000	O
440.4752	1,000	Fe
455.4042	1,000	Ba
455.5355	2,000	Cs
459.3177	1,000	Cs
460.7331	1,000	Sr
462.4276	1,000	X
464.82	.....	AlO
465.887	2,000	Kr
467.1226	2,000	X
468.62	.....	C_2
470.232	1,200	Ar

TABLE 5 Prominent Lines and Bands in Order of Wavelength (Continued)

Wavelength, nm	Relative intensity	Source
470.440	1,500	Ne
470.885	1,200	Ne
471.534	1,500	Ne
473.71	.....	C_2
473.900	3,000	Kr
484.21	.....	AlO
488	.....	Ar laser
488.492(2)	1,000	Ne
514.5	.....	Ar laser
516.52	.....	C_2
520.9067	1,500	Ag
540.0562	2,000	Ne laser
546.0740	2,000	Hg
546.461	900	I
546.5487	1,000	Ag
549.692	900	I
553.5551	1,000	Ba
557.02895	2,000	Kr
563.5	.....	NH_3
565.070	1,500	Ar
567.0	.....	NH_3
568.2	.....	Kr laser
573	.....	O_3
578.213	1,000	Cu
579.065	1,000	Hg
580.43	.....	N_2
585.2488	2,000	Ne
587.09158	3,000	Kr
587.5618	1,000	He
588.9953	9,000	Na
589.5923	5,000	Na
589.93	.....	C_2
600.848	800	N
602	.....	O_3
608.246	1,000	N
609.73	.....	FeO
610.364	2,000	Li
610.99	.....	FeO
614.172	2,000	Ba
615.820	1,000	O
618.05	.....	FeO
621.89	.....	FeO
632.8	.....	Ne laser
636.2347	1,000	Zn
640.2246	2.000	N
642.0	.....	Kr laser
646.85	.....	N_2
654.48	.....	N_2
656.2725	1,000	H
656.2849	2,000	H
659.532	1,000	Ba
662.36	.....	N_2
670.7844	3,000	Li
694.3	.....	Ruby laser
695.878	1,000	I
705.996	2,000	Ba
707.010	1,000	Sr
712.224	1,000	Ni
728.027	1,000	Ba
760.15465	5,000	Kr
766.4907	9,000	K
769.8979	5,000	K
787.4	.....	NH_3
791.9	.....	NH_3
794.757	1,000	O
807.902	1,000	Cs
810.369	2,000	Ar
810.43660	5,000	Kr
811.2	.....	FeO
811.29023	5,000	Kr
811.5311	5,000	Ar
812.652	1,000	Li
813.7	.....	FeO
819.00570	3,000	Kr
819.481	1,000	Na
823	.....	FeO
823.163	5,000	Xe
826.32412	1,000	Kr
826.452	1,000	Ar
828.012	5,000	Xe
829.81091	5,000	Kr
830.2	.....	FeO
834.682	2,000	Xe
840.821	2,000	Ar
840.919	2,000	X
842.465	2,000	
844.0	.....	GaAs laser
844.638	2,000	O
850.88736	3,000	Kr
852.110	5,000	Cs
852.144	2,000	Ar
854.209	1,000	Ca
866.214	1,000	Ca
877.67498	5,000	Kr
881.941	5,000	Xe
891.16	.....	N_2
892.86934	2,000	Kr
894.350	2,000	Cs
909.500	.....	C
917.224	1,000	Cs
922.450	1,000	Ar
940	.....	H_2O
965.778	1,500	Ar
978.450	1,000	Ar
979.970	2,000	Xe
992.320	2,000	Xe
1002.439	1,000	Cs
1012.3	.....	Cs
1022.146	1,000	Kr
1032.729	1,000	Sr
1034.385	500	Ca
1042	.....	N_2
1060	.....	Nd^+ laser
1083.834	1,000	Xe
1100	.....	H_2O
1380	.....	H_2O
1870	.....	H_2O

TABLE 5 Prominent Lines and Bands in Order of Wavelength (Continued)

Wavelength, μm	Source	Wavelength, μm	Source	Wavelength, μm	Source
2.1	YAG-Ho laser	4.7	O_3, $(CN)_2$	10.75	NH_3
		4.75	CO	13.55	$(CN)_2$
2.25	NH_3	5.25	NO	13.8	O_3
2.7	CO_2, H_2O	5.4	NO	13.85	$(CN)_2$
2.85	N_2O	6.1	NH_3	14.4	O_3
2.9	NH_3	7.5	SO_2	15	CO_2
3.2	H_2O	7.7	CH_4	17	N_2O
3.31	CH_4	7.77	N_2O	19.1	SO_2
3.9	N_2O	8.5	N_2O	24	H_2O
4.0	SO_2	8.7	SO_2, N_2O	25	H_2O
4.3	CO_2	9.6	O_3	26.6	H_2O
4.5	N_2O	10.35	NH_3	30	H_2O

REFERENCES

Billings, Bruce H. (ed.) (1972): Wavelengths for Spectrographic Calibration, in D. E. Grey (ed.), "American Institute of Physics Handbook," 3d ed., McGraw-Hill, New York. Lines of value for calibrating spectrometers, covering a wavelength region of 2000 Å to 25 μm.

Harrison, George R. (1939): "Wavelength Tables," Technology Press, Cambridge, Mass. These tables, probably the most universally adopted, include more than 100,000 spectrum lines from 2000 to 10,000 Å. The sensitive lines of the elements are listed separately.

Kelly, Raymond L. (1968): Atomic Emission Lines below 2000 Angstroms, *Nav. Res. Lab. Rep.* 6648, Washington, D.C., February. Lists wavelength, intensity, multiplet, configuration, term, J-J value, and reference for the several spectra of the elements from hydrogen through argon, covering a wavelength region of about 2 to 2000 Å.

Lang, D. (1961): "Absorption Spectra in the Ultraviolet and Visible Region," 11 vols., Academic, New York. Absorption spectra of many compounds are presented; the data are mainly of qualitative value.

Moore, Charlotte E. (1949): Atomic Energy Levels, 3 vols., *Natl. Bur. Stand. Circ.* 467, June 15. Provides spectra of many elements, emphasizing the configuration, designation, and ionization level.

Pearse, R. W. B., and A. C. Gaydon (1950): "The Identification of Molecular Spectra," Chapman & Hall, London. Like the MIT tables, this is a commonly adopted reference, covering approximately 2000 to 10,000 Å. Most common molecules are listed; excellent plates.

Pierson, Raymond H., Aaron N. Fletcher, and E. St. Clair Gantz (1956): Catalog of Infrared Spectra for Qualitative Analysis of Gases, *Anal. Chem.*, vol. 28, pp. 1218–1239, August. Spectra of 66 compounds in the spectral region 2 to 15 μm; the data are qualitative but provide a ready description.

Rosen, B. (ed.) (1951, 1952): "Tables des constantes et données numériques," vol. 4, "Données spectroscopiques concernant les molécules diatomiques," vol. 5, "Atlas des longuers d'onde caractéristiques des bandes d'émission et d'absorption des molécules diatomiques," Hermann, Paris. These tables extend from about 670 to 15,000 Å and provide detailed data on band structure for a large number of molecules.

Striganow, A. R., and N. S. Sventitskii (1968): "Tables of Spectral Lines of Neutral and Ionized Atoms," Plenum, New York. Wavelength, ionization level, ionization potential, transition, and J value for 22 elements from the infrared to the far-ultraviolet region.

Szymanski, Herman A. (1964): "Infrared Band Handbook," suppl. 1 and 2, Plenum, New York. Lines from 3600 to 200 cm^{-1}, with wave number, intensity, physical state, and structural group.

Zaidel, A. N., V. K. Prokof'ev, and S. M. Raiskii (1968): "Tables of Spectrum Lines," Pergamon, London. These tables embrace the region from 2000 to 8000 Å, containing 41,468 lines of 60 elements in order of wavelength. A second part, often of great convenience, contains 23,392 lines of 93 elements arranged by element.

Section **12**

The Eyes and Vision

MATHEW ALPERN

Department of Ophthalmology, University of Michigan Medical Center, Ann Arbor, Mich.

(Numbers refer to paragraphs)

1. Introduction Studies of the optical character of allegedly "normal" eyes were made by nineteenth-century physiologists, Helmholtz among them, and culminated in the award in 1911 of the Nobel prize in physiology and medicine to Gullstrand. Fortunately, a summary of this work is available in the translation of the third edition of Helmholtz's monumental "Optik," still a very important reference for the subject of this section (Helmholtz, 1924). A more recent summary by Fry (1959) may also be of interest.

2. A Schematic Eye For many purposes it is of use to have an optical system which imitates fairly accurately characteristics of the living eye defined by the first-order theory of geometrical optics. Such systems, called schematic eyes, vary greatly in complexity in the vain attempt to replicate in detail actual normal living eyes. For almost all needs, simplicity is a much greater virtue than what

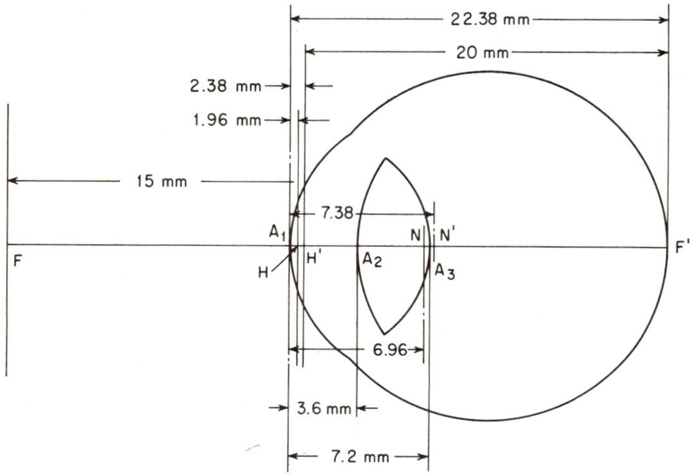

Fig. 1 Optical constants of Helmholtz's schematic eye 1 as modified by Laurance; for details see Table 1.

TABLE 1 Constants of Schematic Eye

Surface	Symbol	Radius, mm	Distance from corneal pole, mm	Refracting power, diopters
Cornea	A_1	+8	......	+41.6
Anterior lens	A_2	+10	3.6	+12.3
Posterior lens	A_3	−6	7.2	+20.5
Lens as a unit:				+30.5
Anterior principal plane	H_L	...	5.7	
Posterior principal plane	H_L'	...	6.06	
Eye as a unit:				+66.6
Anterior focal plane	F	...	−13.04	
Posterior focal plane	F'	...	22.38	
Anterior principal plane	H	...	1.96	
Posterior principal plane	H'	...	2.38	
Anterior nodal plane	N	...	6.96	
Posterior nodal plane	N'	...	7.38	

Refraction index:
 Anterior chamber n_2 .... $\frac{4}{3}$
 Lens n_3 .... 1.45
 Vitreous n_4 .... $\frac{4}{3}$
Entrance pupil position 3.04 mm from the cornea, size 1.15 times real pupil size
Exit pupil position 3.72 mm from the cornea, size 1.05 times real pupil size

can only be unattainable fidelity. The eye given here was introduced by Helmholtz and is authentic enough for most purposes. More involved examples can be found in Helmholtz's text.

The eye is shown in Fig. 1. It has only three refracting surfaces. The cornea is assumed infinitely thin, and the lens has a uniform index of refraction and only two refracting surfaces, front and back. The values for index of refraction are modified slightly from the original to give focal lengths with round numbers (Laurance, 1926). The result is a slightly more powerful lens than Helmholtz planned. The constants for Fig. 1 and Table 1 are for the unaccommodated eye. In accommodation the radius of curvature of the anterior lens surface is assumed to change to +6.0 mm while everything else remains fixed. All values are for sodium light.

The positions and magnifications of the entrance and exit pupils are computed assuming that the iris rests on the anterior lens surface and is infinitely thin.

BIOLOGICAL EYES

3. Reference Points To specify eye position, among other reasons, it is sometimes necessary to designate certain points, lines, and angles. When working with schematic eyes concepts such as nodal points and optic axis are useful, but with biological eyes they are not capable of easy and invariable operational definition. Therefore the following concepts, which are easier to define, are used.

Object of Regard. The point to which the observer directs his gaze. It is usually assumed that he fixates foveally when told to do so, but he may not and there are several straightforward and independent ways of verifying this (Alpern, 1962*a*) if it becomes necessary.

Pupillary Axis. The line connecting the center of the entrance pupil and the center of curvature of the anterior surface of the cornea.

Line of Sight. The line connecting the center of the entrance pupil and the object of regard.

Angle Lambda. Angle of intersection of the pupillary axis and the line of sight.

Center of Rotation. A point on the line of sight 13.5 mm behind the cornea. In theory a number of other operations may be used to define this point which have greater meaning for ocular kinematics, but they all give somewhat different answers. There is no fixed point in the eye or in the orbit around which the eye rotates to all the positions it can assume. The problem is discussed in detail in Alpern (1962*a*). For almost all practical problems, however, the assumption of a point defined as above is accurate enough.

Spectacle Plane. A plane tangent to the posterior surfaces of the spectacle lenses at their respective vertexes. (Properly positioned spectacles are usually tilted forward about 15° at the top with respect to parallelism with the face plane and are about 12 to 15 mm in front of the anterior vertex of the cornea. With the two eyes looking at infinity the pupillary axis of each eye should intersect the posterior vertex of the spectacle lens.)

4. Aberrations Optical characteristics of real eyes differ in important ways from those of the schematic eye. One important difference is in the third- and higher-order aberrations they demonstrate.

The longitudinal chromatic aberration has been studied by Ivanoff (1953), Wald and Griffin (1947), and Bedford and Wyszecki (1957). The latter data, which are in good agreement with the others, are given in Fig. 2.

Chromatic difference in magnification can rarely be made visible in the human eye even when particular pains are taken to try to demonstrate it (Le Grand, 1968). No good measurements of it exist. The spectral sensitivity of the retina can be relied upon in many cases to minimize the importance of both kinds of chromatic aberrations under practical viewing conditions.

Spherical aberration has been extensively measured by Ames and Proctor (1921), Ivanoff (1953), and Koomen et al. (1949), among others. The results of the various authors agree that the eye focused for infinity exhibits positive spherical aberration and the eye focused for very near distances exhibits negative spherical aberration.

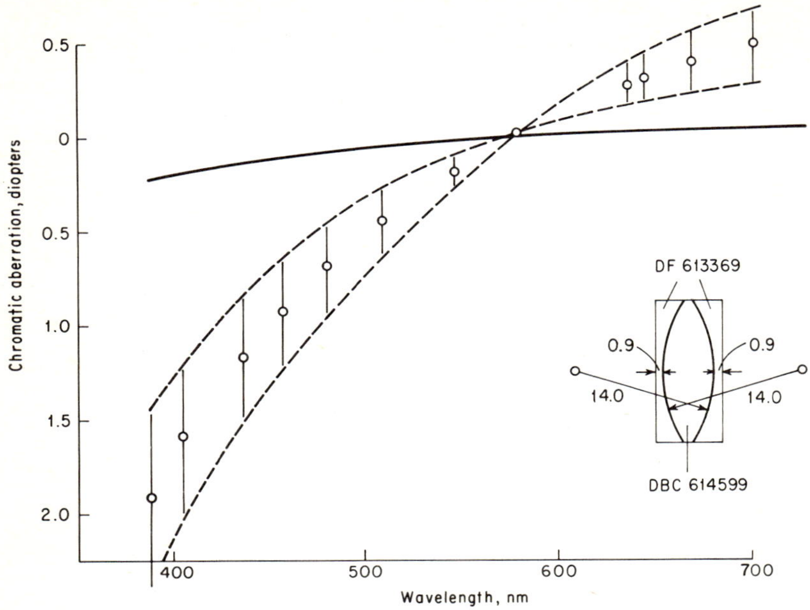

Fig. 2 Chromatic aberration of the eye. The dotted lines enclose the range of measurements on 12 observers, arbitrarily set for zero aberration at 578 nm. The circles are the weighted mean results. The solid line shows the results when the "average observer" views through the lens with characteristics in the lower right of the figure. This lens is designed to correct the chromatic aberration of the normal eye. [*Bedford and Wyszecki* (1957).]

At some intermediate distances (around 50 cm) the spherical aberration is essentially zero. The results illustrated in Fig. 3 show this effect quite clearly. Such individual differences as occur are related to the magnitude of the aberration at a given

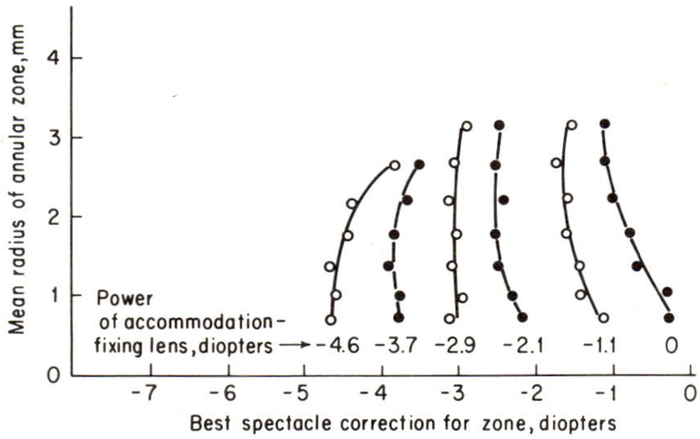

Fig. 3 Spherical aberration of the eye as measured by optimum spectacle lens used for viewing through different annular zones of the pupil at different accommodation levels. [*Koomen et al.* (1949).]

distance, as well as the precise accommodation setting of the eye at which the aberration is null. Presbyopic (see Par. 10) and aphakic observers, since they cannot accommodate, obviously will fail to show any but the positive aberration at far distances. An elegant summary of some of the problems of spherical aberration in the human eye is available (Le Grand, 1968).

Comatic Aberration. Spherical aberration of oblique pencils can easily be made visible in the human eye, e.g., in the aberrascope of Tscherning (1904). It is the most probable explanation for the loss of visual acuity which occurs when an artificial pupil is decentered off the principal line of sight (Campbell, 1958). Green (1967) measured the modulation transfer function of the eye, for light entering the eye eccentrically through the pupil. The results are moderately well fit by the transfer function of an optical system with a small amount of coma.

5. Optical Quality of the Retinal Image The third- and higher-order aberrations just discussed are the most frequently considered contributors to the deterioration in the quality of the retinal image. Marginal astigmatism due to misalignment of the axis of the various surfaces is another, but in practice one cannot easily differentiate between it and astigmatism arising from toroidal refracting surfaces (that of the anterior corneal surface being the notable exception). In either case, spherocylindrical spectacle lenses adequately correct the defect.

The combination of the various aberrations with diffraction and defocus as they relate to the human eye has been treated theoretically by Fry (1956). Since theoretical treatments of this kind are almost certain to overlook or neglect (with simplifying assumptions) some sources of blur in actual biological eyes, it is fortunate that Flamant (1955) has developed a method of measuring the line-spread function of the human eye directly. A brightly lit thin-line filament is imaged on the retina. A small amount of the light which forms this image is reflected by the fundus, emerges from the eye, and is focused by the eye's optics in the plane of the original filament. This image, however, is degraded by the double traverse of the eye's optics. If a pellicle mirror is introduced between the filament and the eye, the image of the retinal image can be analyzed by a photodetector. Flamant (1955) used a photographic plate; Westheimer and Campbell (1962), Krauskopf (1962), and Campbell and Gubisch (1966) used photomultiplier tubes.

Measurements obtained in this way give the quality of the image after degradation by the eye's optics, reflection at the fundus, and a second degradation by the eye's optics. Assuming that the fundus is a perfect diffuser and that the traverse of the eye's optics is perfectly symmetrical, Flamant developed a method for inferring from the measured line-spread function the line spread of the retinal image after a single traverse of the eye. This involves taking the Fourier transform of the data, extracting the square root, and computing the inverse transform. Line-spread measurements for different pupil sizes treated in this way are shown in Fig. 4. Each line-spread profile portrays the normalized illuminance distribution on the fundus from an infinitesimally thin bright line in optimum focus. (Small oscillations of the data points arise from unavoidable end effects, which occur during computation of the inverse transforms and do not represent real illuminance variation.) The diffraction image of a line formed by an ideal achromatic lens with light of the same spectral composition† is shown as a dotted line for each line-spread profile. The different profiles show the influence of different pupil sizes as indicated. The line spread is a minimum for 2.4-mm pupil, and for smaller-diameter pupils the spread approaches that of the diffraction-limited ideal; at 2.4 mm the curve consists of a gaussian core and an exponential "skirt," the core fitting the diffraction image and the skirt representing scatter and defects of focus. As the pupil size increases, the edge aberrations become more prominent and the exponential skirt dominates the line-spread profile more and more.

Another, and quite independent, method of evaluating the quality of the optics of the eye was developed by Arnulf and Dupuy (1960). This consists of producing an image on the fundus formed without reliance of the eye's optics and comparing it psychophysically with a similar image formed in the usual way by refraction through the optical components. The modulation transfer function of the former describes the characteristics of the retina and brain and of the latter includes

† The apparatus has characteristics such that each part of the spectrum of reflected light produces about the same relative voltage output of the photomultiplier as its luminosity coefficient; see below.

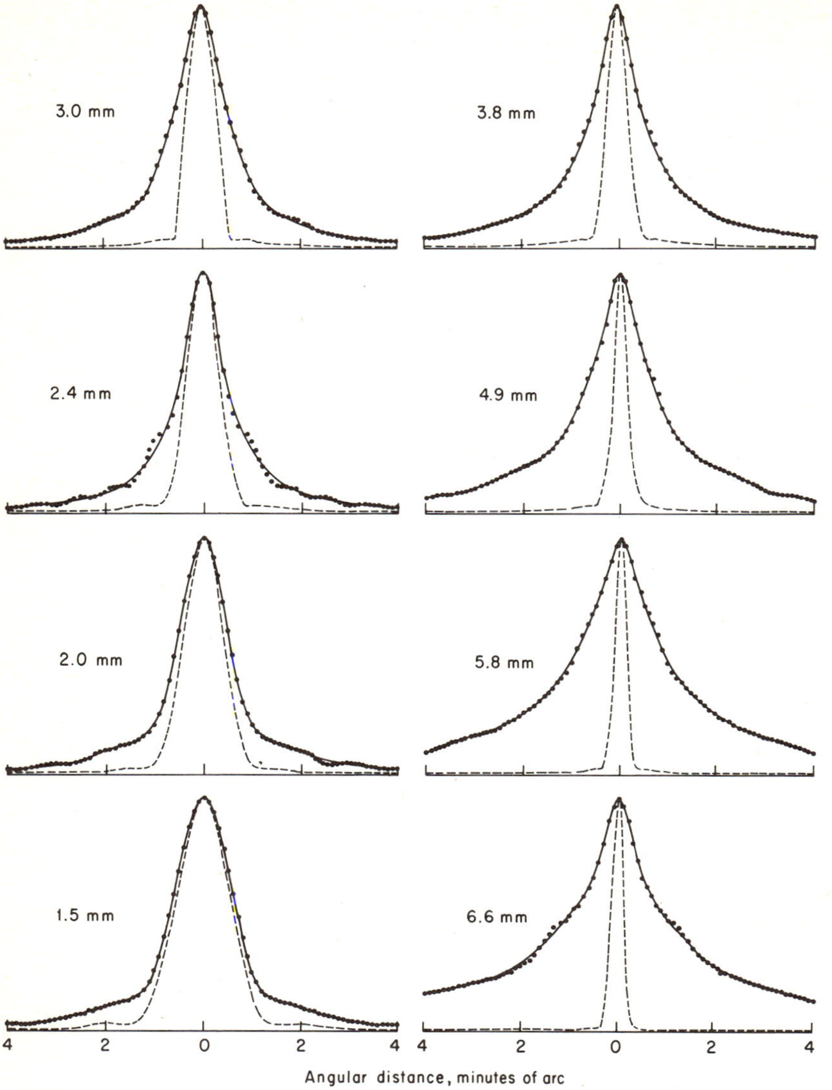

Fig. 4 Optical line-spread function of the human eye. Each curve represents the normalized distribution of illuminance occurring on the fundus for a thin-line source of light. The parameter is pupil size. The narrower curve indicates the diffraction image of a line at the given pupil diameter. [*Campbell and Gubisch* (1966).]

the entire visual system, i.e., optical components, retina, and brain. The technique for circumventing the eye's optics was developed by Le Grand (1937*a*) and later used by Byram (1944) and Westheimer (1960). It consists of forming two images from a single coherent source in the plane of the pupil. These then act as the double slit in the classical experiment of Thomas Young, producing interference fringes on the retina. Campbell and Green (1965*b*) used a laser beam to gain a good deal more intensity. A noncoherent source, viewed through a beam splitter and orthogonally polarized with respect to the laser beam, reduces the contrast of the interference fringes. By rotating a polarizer common to both beams, the contrast can be varied between 0 and 1.0 while keeping the mean luminance fixed. The measurements of modulation obtained in this way are shown as the smooth curve in Fig. 5. The open circles in the same figure show the contrast at different spatial frequencies (cycles per degree) required for threshold when the overall visual system is used (for 2-mm pupil). This latter technique was developed by Selwyn (1948), Schade (1956), and De Palma and Lowry (1962). The ordinate in this figure shows the logarithm of the reciprocal of the minimum contrast,

$$\text{Contrast} = \frac{I_{\max} - I_{\min}}{I_{\max} + I_{\min}} \tag{1}$$

required for threshold visibility. Since the open-circle measurements include the contributions made by the optics of the eye while the solid-line measurements exclude them, the differences between the two curves can be used to infer those

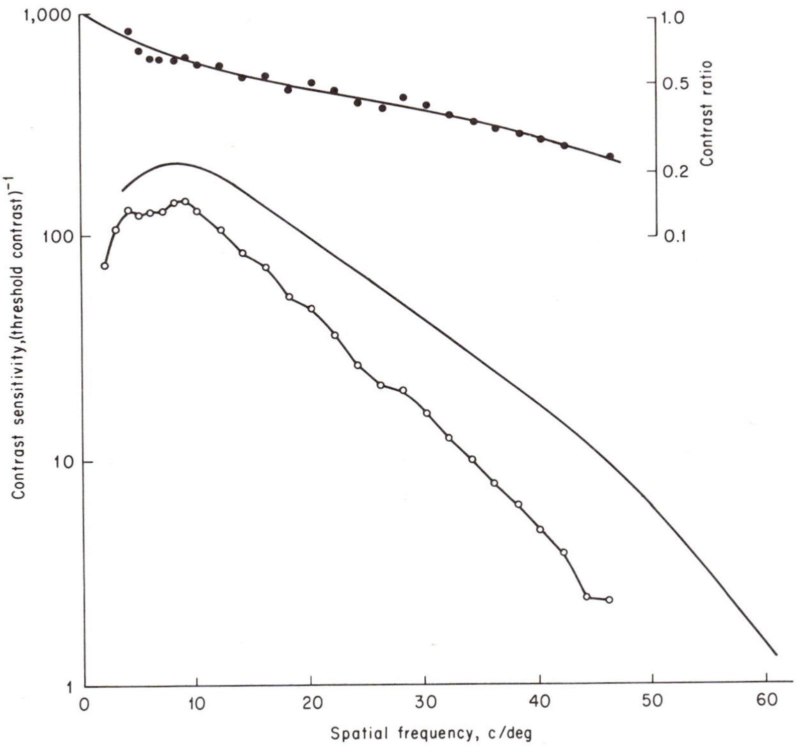

Fig. 5 Solid dots (ordinate to the right) show the reduction in contrast of the image falling on the retina due to optical defects in the eye's refractive system, according to estimates. The open circles show the contrast sensitivity of the overall system (2-mm pupil) and the solid line the measurements obtained with interference fringes, in this way obviating distortions produced by the eye's optics. The black dots are the ratios of the overall to the retinal sensitivities. [*Campbell and Green* (1965*b*).]

contributions, i.e., the log ratio of contrast sensitivity of the overall system to that of the retina and brain. This calculation is given by the filled circles in Fig. 5, which has the ordinate scale on the right. The results refer, of course, to a 2-mm pupil. Wider pupils (on theoretical grounds, as well as from Fig.

4) can be expected to have less faithful images. Figure 6 shows that this is so for the Campbell and Green calculations. The results shown here are of a similar form to the filled circles in Fig. 5 for other pupil sizes. The ordinate in Fig. 6 is linear rather than logarithmic and can be regarded as the ratio of contrast of the object to that of the image formed by the eye's optics.

The results in Fig. 6 suggest a somewhat higher quality to the optics of the eye than anticipated from the objective measurements in Fig. 4, particularly at

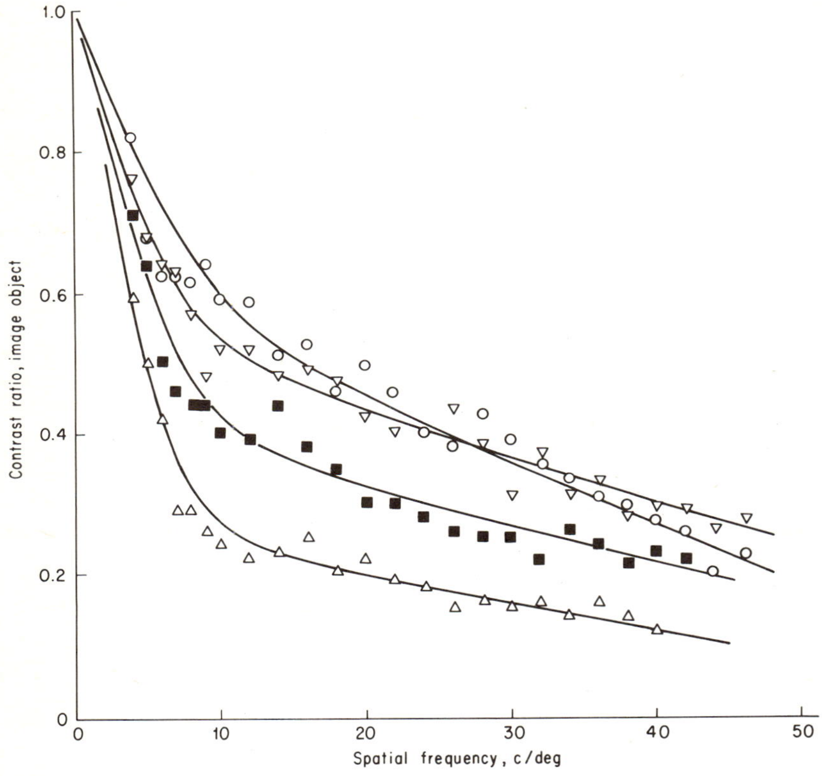

Fig. 6 Reduction in contrast of the image falling on the retina due to optical defects in the eye. The results are similar to the black dots of Fig. 5 except that the ordinate scale is linear and the parameter is pupil diameter (2, 2.8, 3.8, and 5.8 mm, from top to bottom at low spatial frequencies). [*Campbell and Green* (1956*b*).]

the high spatial frequencies. These differences are probably due to optical scatter (such as fundus halation) which is not psychophysically relevant, in part at least because of the Stiles-Crawford effect of the first kind (see Par. 9). Weale (1966) and Campbell and Gubisch (1966) found that light reflected from the fundus consists of a depolarized and a nondepolarized component. Rohler et al. (1969) regard only the latter as relevant for vision and find that the objective line-spread function of this component is considerably better than that obtained when (as in Fig. 4) both components are included in the measurement.

6. Entoptic Scatter The refracting surfaces of the eyes, the ocular media, and the fundus (Vos, 1963) scatter a certain amount of light, and this can lead to the reduction in visibility of objects. Holladay (1926*a*,*b*, 1927), Stiles (1929), Le Grand (1937*b*), among others, employ the concept of the equivalent veiling glare, i.e., the amount of luminous veil which produces the same foveal threshold elevation as a given source of peripheral glare. That what is measured under these circumstances is indeed entoptic scatter was proved directly by Rushton and Gubisch (1966), who measured the equivalence of the veil and the peripheral-glare source for foveal-pigment bleaching with the retinal densitometer. They showed that equivalences established by densitometry are equally valid for thresholds. The peripheral-glare source produces direct adaptation in the region of its focal image with time characteristics quite different from those it produces (by scatter) at the fovea, although those of the latter are identical to those produced by its equivalent veil. The fact that a bright peripheral-glare source reduces, rather than enhances, the brightness of an object viewed foveally means that other effects of glare occur in addition to physical scatter (Schouten and Ornstein, 1939). Fry and Alpern (1953) showed that the equivalent-veil concept is just as valid for this brightness *reduction* as for threshold elevation.

A good way of estimating the amount of scattered light in the eye by a given glare source at an angle θ from the line of sight (in degrees) is by the Stiles-Holladay (empirical) equation

$$B_e = \frac{aE}{\theta^n} \tag{2}$$

where B_e = equivalent veiling luminance, cd/unit area
E = illuminance produced by glare source in pupil on a plane normal to line of sight, lm/unit area
a = constant of order 10
n = constant of order 2

7. Errors of Refraction The far point of the eye is optically conjugated to the retina with accommodation fully relaxed. Errors of refraction hyperopia (or hypermetropia) and myopia are defined by far points respectively beyond and inside optical infinity. Astigmatism may exist alone or in conjunction with either hyperopia or myopia and consist of principal meridians of the eye (90° apart) with far points located at different distances.

The operational definitions of the far point are (1) objectively with a retinoscope, a modification of the Foucault knife test employing light reflected from the fundus, and (2) subjectively by the maximum amount of positive spherical lenses, i.e., convex (or least negative) required for optimum visual acuity. This end point leaves the patient on the largest-convex-lens limit of the eye's depth of field. Additional subjective tests (clock dial, cross cylinder) are used for measuring the amount of astigmatism. Details of these various tests are to be found in standard texts (Borish, 1949; Emsley, 1952; Lancaster, 1952). Other and more exotic objective and subjective optometers (Fry, 1959) have not found widespread popularity, probably because the usual methods are already quite foolproof.

The correcting lens for the error of refraction is that lens placed in its proper position before the eye the posterior focal point of which coincides with the eye's far point. Most optical instruments can be adjusted by shifting the eyepiece position so that the posterior focal point of the instrument falls at the observer's far point. If astigmatism can be neglected, the observer then need not wear spectacle lenses when viewing through the instrument. Instruments adjusted in this way for normal daytime viewing are usually quite unsuitable for viewing at low levels of illumination, since normal eyes are about 2.0 diopters more myopic when viewing under such circumstances. Individual differences are large, however, and a number of factors are thought to combine to account for the measured results: (1) the shift to scotopic from photopic spectral sensitivity to $\lambda_{max} = 510$ nm (from the value of 555 nm) so that about $0.37D$ more negative lens is required to bring the maximally visible wavelength into sharp focus; (2) the spherical aberration of the eye brought about by the wider pupil (and absence of Stiles-Crawford effect) in low light levels (Koomen et al., 1951); (3) the "crystalline" lens sets the eye's optics in a more myopic state in the absence of high-contrast resolvable details in the visual field.

So-called *night myopia* persists when a fine-detail resolvable test target controls the accommodation position of the eyes (Koomen et al.) or alternatively if a small artificial pupil holds the spherical aberration constant but the accommodation is permitted to settle at the position determined by prevailing levels of illuminance (Alpern and David, 1958) but not when both a detail target *and* an artificial pupil are employed. Apparently both spherical aberration and a change in the far point of the eye with reduction in light level are responsible for the effect in ordinary use of the eyes. Green and Campbell (1965) showed that the spherical-aberration component results from a combination of a wide pupil and the reduction of the resolving power of the retina at low luminance. They could mimic the effect at high light levels, dilating the pupil with mydriatic drops and reducing the resolving power of the retina by a decreased contrast of the test target.

A related effect is an increase in refracting power in bright empty visual fields (Westheimer, 1957). This cannot be due to any of the aberrations associated with wide pupils since the pupils are presumably narrow under such circumstances. Here it is the accommodation alone which sets the eye in a more myopic condition. These measurements also suggest why discrete sampling of night myopia among different observers shows such wide variability. Even on the same observer viewing under such conditions, the accommodation position is widely different from one moment to the next (Alpern, 1958).

8. Pupil Diameter The size of the pupil is controlled by antagonistic parasympathetically and orthosympathetically innervated smooth muscle of the iris, the sphincter and dilator, respectively (Lowenstein and Loewenfeld, 1962). A consequence of innervation by the autonomic nervous system is the enormous number of physical, physiological, and psychological variables which influence the size of the pupil. Many of these stimuli, e.g., the flux density of quanta striking the retina or the precise position in space of the object of regard, are exactly or almost exactly specifiable; but others, e.g., the emotional content of an advertising display (Hess, 1965) or the mental effort required for recall of an item stored in memory (Kahneman et al., 1967), are extremely difficult to quantify. This has a number of obvious consequences, among them the wide individual and day-to-day variability of laboratory experiments designed to measure the influence of a single variable on pupil size (Le Grand, 1957).

Small pupils at high light levels and wide pupils at low levels fulfill a number of obvious optical optimizations: they reduce aberrations when good cone vision permits high resolution of fine detail and the light levels aready exceed that required for maximum acuity; but they allow the maximum amount of available light to reach the retina at night levels when rod vision is incapable in any event of resolving the fine detail which the eye's aberrations normally degrade. Campbell and Gregory (1960) have argued from this that the pupil size assumed at any given light level is always that at which the acuity is maximum, a reasonable generalization which, however, does not account for the fact that additional quanta absorption by rods alone makes the pupil smaller at levels where the added light fails to improve the acuity. It seems unlikely that evolution of the pupil light reflex has proceeded with such a grand design.

In amphibia and teleosts the light quanta absorbed by rhodopsin within the iris itself causes the sphincter to contract (and thus the pupil to narrow) even in the absence of a retina, though it is still unclear where in the iris the rhodopsin is situated (Barr and Alpern, 1963). Whether light stimulating only the retina (and not the iris) in such animals will also cause the pupil to contract is not known, but in man and other mammals the iris photosensitivity is lost and the light reflex of the pupil is a result of absorption of light quanta in the retina alone.

It is of some practical interest to know what the pupil size is for different light levels. The experimental answer is complicated, not only by the variables already alluded to but by the fact that *position* (as well as the intensity) of bright objects in the field may influence how wide the pupil is. Hence, the most useful measurements are probably those obtained against a completely uniform background covering the whole visual field. The best data are those of Flamant (1948), who photographed the pupils of observers viewing the sky from the roof of the Institute d'Optique as the sun set. Her results are shown in Fig. 7. The discontinuity in the curves at about -4 log cd/cm^2 is probably related to the

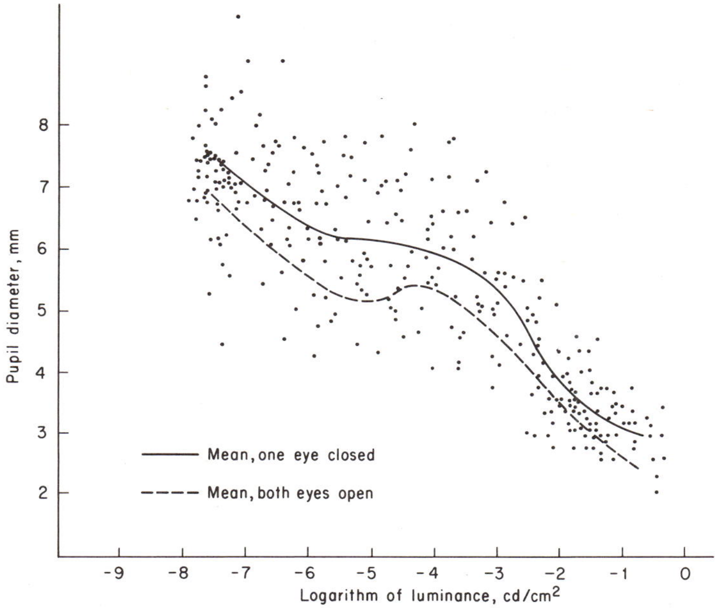

Fig. 7 Mean diameter of the pupil on 37 subjects as a function of field luminance. [*Flamant* (1948).]

beginning of a contribution of cone function to the response (ten Doesschate and Alpern, 1967*a*), but the basic physiological processes are still only poorly understood (Alpern and Campbell, 1962). For field application it is convenient to have some approximate estimate of the pupil diameter. For many of these, the simple equation of Crawford (1936), developed for a uniformly illuminated field 55° in diameter, as modified by Le Grand (1957) can be used:

$$D = 5 - 3 \tanh (0.4 \log B) \tag{3}$$

where D is the pupil diameter in millimeters and B is field luminance in candelas per square meter. Other comparable equations are discussed by Wyszecki and Stiles (1967). They all are complicated in their application by the tendency of the pupil size to decrease with age.

For normal observers the pupils of the two eyes are almost exactly the same size. The equal illumination of both retinas invariably leads to smaller pupils than those obtained if one eye is occluded. If, however, the two retinas are unequally illuminated, there is an additional complexity, particularly when the differences in the retinal illuminances are very large, such that the pupils may then be somewhat wider than if one eye were in the dark (ten Doesschate and Alpern, 1967*b*). At low and modest levels of illuminance in the steady state, the pupil response is due mainly to the excitation of rods, but at and beyond about 100 Td the response appears to be the weighted geometric mean of contributions of both rods and cones (Alpern and Campbell, 1962).

9. Directional Sensitivity of the Retina Equal amounts of light passing through different parts of the pupil but falling on the same patch of retina will not necessarily be equally effective in the intensity of visual sensation (or, for that matter, the amount of pupil contraction) produced. For cone vision, the light which passes through the center of the pupil will be about 8 times as effective as the light which passes through the edge of an 8-mm pupil. This is the Stiles-Crawford effect of the first kind, discovered in 1933 by Stiles and Crawford. The phenomenon has the optical advantage of reducing the visual importances of those edge rays which undergo aberrations in the eye's optics. Of course, in field applications the pupil light reflex is normally operative, so that for light levels high enough to require good cone vision the pupil is rarely wide enough for the effect to play an important practical role. For rod vision, the pupil may easily get wide enough, but the visual acuity for rods is poor, so that aberrated rays cannot degrade the visual sensation much and the rods are about equally sensitive to the light irrespective of which part of the pupil it passes through (Flamant and Stiles, 1948; Donner and Rushton, 1959). Obviously, it is a property of the receptors themselves rather than the eye media which gives this peculiar directional sensitivity to our sense of sight. Though of no practical importance, it is of theoretical interest for the understanding of the fundamental optical properties of cone photoreceptors and for elucidating what may account for its absence among the rods.

Typical measurements of directional sensitivity on the author's own fovea are shown in Fig. 8. There are a number of operational ways of defining the effectivity

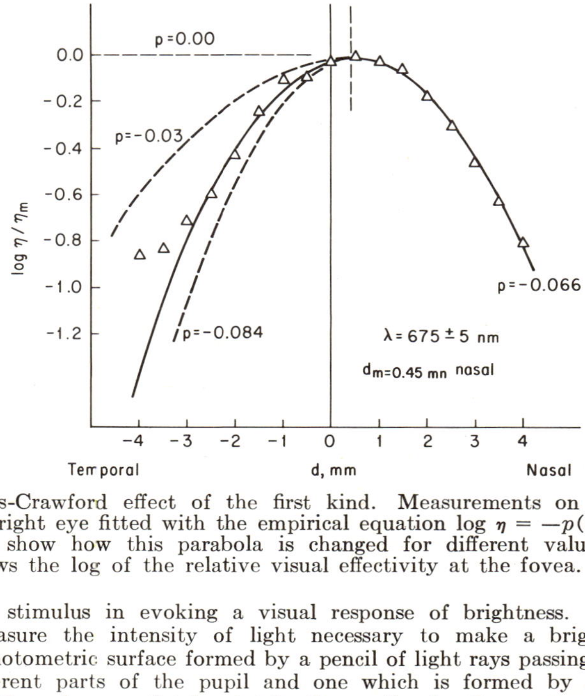

Fig. 8 Stiles-Crawford effect of the first kind. Measurements on the fovea of the author's right eye fitted with the empirical equation $\log \eta = -p(d - d_m)^2$. The dotted lines show how this parabola is changed for different values of p. The ordinate shows the log of the relative visual effectivity at the fovea.

of the light stimulus in evoking a visual response of brightness. For example, one can measure the intensity of light necessary to make a brightness match between a photometric surface formed by a pencil of light rays passing in succession through different parts of the pupil and one which is formed by a pencil that always goes through some fixed point in the pupil. Another method is to adjust the intensity of the light of a single pencil necessary for some threshold such as to eliminate, say, 30-Hz flicker or threshold visibility. The shape of the curve which describes the results of such experiments seems remarkably independent of which of these operations we use to measure it—one convincing way of verifying the generality of the phenomenon.

Results like those in Fig. 8 can be fitted by a variety of empirical equations, none of which offers much advantage over the others (Moon and Spencer, 1944; Le Grand, 1957). The equation proposed originally by Stiles is

$$\log \frac{\eta}{\eta_m} = -p(d - d_m)^2 \tag{4}$$

where subscript m refers to the position of peak effectivity and d represents the point of pupil entry in millimeters from some fixed point, e.g., the pupillary axis. This equation continues to be popular because it is simple and provides a reasonably satisfactory fit (Enoch, 1963).

In Fig. 8 the value for p of 0.066 ± 0.001 provides an adequate description of the results except (and this is typical) at the extreme temporal margin of the pupil, where Eq. (4) predicts a somewhat greater loss of sensitivity than what is found. Toraldo di Francia (1949) has provided the rationale for not trying to fit such secondary "lobes" with the same parabola that describes the rest of the curve. Evidently, values of p can be fit with a precision a good deal better than ± 0.01. The extent to which the constants p and d differ among a population of normal observers still needs empirical elucidation.

Whatever other factors influence the value for p, Stiles showed that it is remarkably sensitive to variation in the wavelength of the stimulus. In general, the fovea is less directionally sensitive, i.e., the value for p is smaller, for light in the middle of the spectrum than for light at the ends of the spectrum. The theoretical implications of this fact are clouded by the necessity of making measurements through the pupil which include transmission of the light through the lens, whose optical density is both wavelength- and position-dependent, and by the wide individual differences in these lens dependencies that exist.

The basic wavelength dependency, as first shown by Stiles (1939), is a retinal, not a medial, matter, however. The color of monochromatic light changes as the angle of incidence on the fovea changes. Brindley (1953) showed that a metameric mixture of monochromatic green and red matched to a yellow failed to do so when the light was incident on the fovea obliquely rather than normally. This has to do not merely with the wavelength variability of the brightness effect but is a genuine second directional-sensitivity property of the retina, the Stiles-Crawford effect of the second kind. The most exhaustive analysis of this latter effect was made by Enoch and Stiles (1961), who established throughout the spectrum the extent to which metameric matches are upset by passing the light through the edge of the pupil. The results on the left eye of their principal subject are shown in Fig. 9. There is a hue change toward the red and a slight desaturation for wavelengths longer than about 530 nm. For shorter wavelengths from about 530 to 505 nm, there is a reversal to supersaturation and a hue shift toward shorter wavelengths. At still shorter wavelengths the hue shift is once more toward the longer wavelength, but the color is once more supersaturated by oblique incidence. The hue shifts are readily seen from the upper right-hand diagram of Fig. 9. The abscissa in this graph shows the wave number in cm^{-1}, while the ordinate shows the change in hue in nanometers brought about by oblique retinal incidence. The convention is such that shifts in hue by oblique incidence toward longer wavelengths are plotted above the abscissa and those toward shorter wavelengths below the abscissa.

The exact theoretical interpretation of the two Stiles-Crawford effects remains obscure. The results in Fig. 9 are most easily understood by the idea that each of the three cone mechanisms—red, green, and blue—has marked directional properties which differ not only from the rods (which are almost directionally insensitive) but also from each other. What brings about these differences in directional properties of the different cone mechanisms remains one of the most fundamental unsolved problems of retinal optics. Enoch (1963) has analyzed some of the possibilities in a review article.

10. Accommodation The change of fixation from far objects to near ones requires an increase in refracting power of the eye in order to maintain optimal focus. This change in refraction is a change in accommodation. To achieve sharp focus, a considerable assist is also obtained from *miosis,* i.e., synkinetic narrowing of the pupil (Alpern et al., 1961). The related increased depth of field for a 2-mm pupil provides about $\pm 0.43D$ depth of focus (Campbell, 1957). This miosis would allow an eye focused for infinity to see objects as close as 2.27 m without change in its refraction. For sharp focus at closer distances, it becomes essential to be able to increase the refracting power of the eye as a whole. This is achieved, as was clearly demonstrated by Thomas Young (1801), by changes in the form of the lens. The details of this process are well worked out and are described elsewhere (Fincham, 1937; Alpern, 1962*b*.)

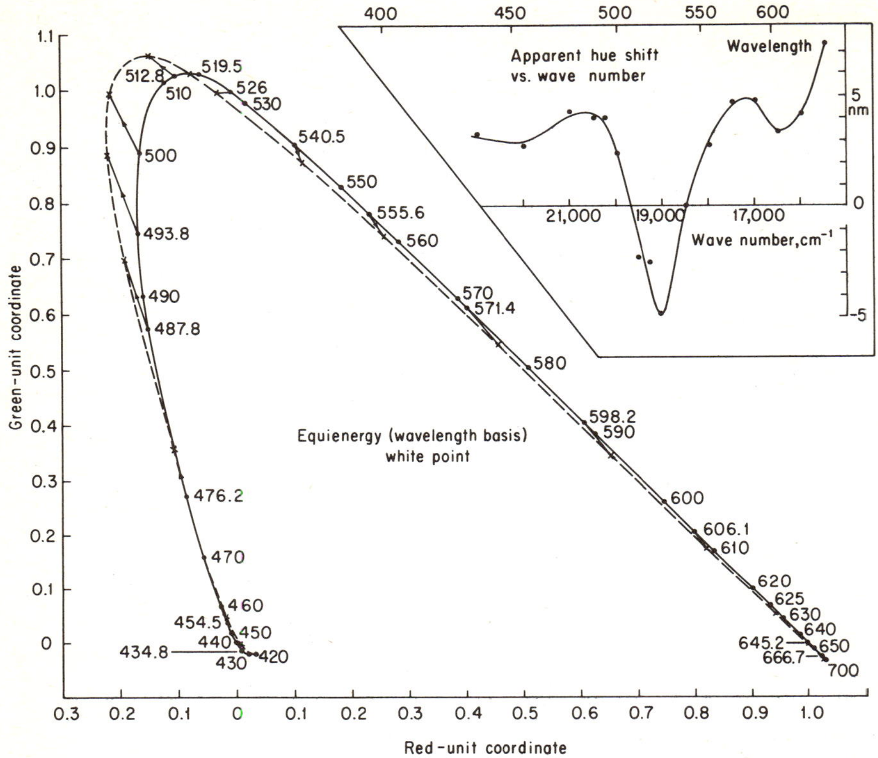

Fig. 9 Stiles-Crawford effect of the second kind. The solid line shows the spectral locus on this chromaticity diagram for central pupil entry; the dashed line shows the locus for light beam entering the eye 3.5 mm temporally from the center of the pupil on Enoch's left eye. The inset shows a derived shift in apparent hue by passing through the pupil edge. [*Enoch and Stiles* (1961).]

The contraction of the smooth intraocular ciliary muscle following parasympathetic innervation, mediated by motor end-plate release of acetylcholine, causes a relaxation in the fibers of the suspensory ligament of the lens, allowing the highly elastic lens capsule to mold the plastic lens substance into a more convex form. The local changes in lens form are due to changes in thickness of the lens capsule. The highly curved posterior lens surface changes little, if at all, in the process; the capsule is thinnest there. The capsule is thinner on the anterior surface in the center than toward the edge, and the accommodated lens bulges forward anteriorly in the central region and assumes a hyperbolic form toward the edge. The bulging central anterior surface adds some 8 or 10 diopters to the refraction of the young eye.

When viewing far objects, the relaxed ciliary muscle takes up the slack of the suspensory fibers holding the lens capsule, thus relieving the lens substance from the tension exerted by the capsule. The substance assumes the shape of the unaccommodated lens, a thinner anterior-posterior diameter and a less convex anterior surface. Innervation of orthosympathetic nerve fibers causes a further decrease in refraction (of much smaller amount, say 0.1 to 1.0D) presumably by contraction of sympathetic innervated ciliary muscle fibers antagonistic to the parasympathetic innervated ones, but perhaps also by the vasoconstriction of the ciliary-body blood vessels and the related decrease in blood volume (and its further increase in tension applied to the fibers of the suspensory ligament). The efferent—and, to a lesser extent, the afferent—nerve pathways for these effects are well worked out, but a good deal of uncertainty regarding the nerve connections within the brainstem and higher brain centers exists (see Jampel, 1959).

Nor can a dogmatic definition of the stimulus to accommodation be given (Campbell, 1959). The normal observer can detect a defocus blur and begin to respond in the correct direction with a latency of about 0.36 ± 0.09 (far to near) or 0.38 ± 0.08 s (near to far) (Campbell and Westheimer, 1960). A number of stimulus characteristics are readily grasped by the observer in his efforts to make the required response (Campbell, 1959), including retinal directional sensitivity coupled with the fine oscillatory movements of steady fixation and spherical aberration (Fincham, 1951), chromatic aberration (Fincham, 1953), and probably others, including (perhaps) the microfluctuations of accommodation (Alpern, 1958; Campbell, 1954; Guyton, 1956).

However the signals to accommodate are initiated, it is evident that the system monitors the task continuously (rather than discretely). Each change in the position of a test target causes (after a latent period) a change in the accommodative response. The accommodative response may be halted during its progress and modified according to new stimulus conditions (Campbell and Westheimer, 1960).

The ability to increase refraction of the eye by changing the form of the lens decreases with increasing age. After the age of about forty-five the difficulty has reached the point where vision of objects at normal reading distances for any length of time becomes impossible. The condition, known as *presbyopia,* is treated with additional convex lenses in the form of bifocals or reading glasses. The changes in the ability to accommodate with age above forty are illustrated in Fig. 10. The maximum changes in the refracting power of the eye are shown

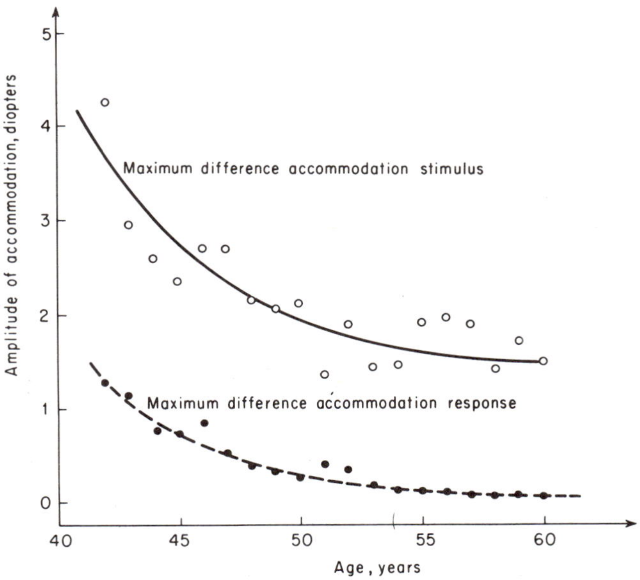

Fig. 10 Mean value of the maximum amount the eye can increase its refractive power by measurement of accommodation response (filled circles) and accommodation stimulus, according to Hamasaki et al. (1956). [*Alpern* (1962*b*).]

in the filled circles. The maximum changes in the ability to see (as measured by subjective momentary blur) are shown by the open circles. The discrepancy presumably is only partly due to depth of focus (about $0.43D$). The remainder is perhaps the momentary mustering of an increased refracting ability which cannot

Table 2 Normal Values of the Accommodative Power at All Ages from Eight to Sixty-eight (Duane, 1912)
Accommodation is given in diopters and tenths, near point being measured from the anterior focus of the eye, i.e., from a point 13 mm in front of the cornea

Age	Lower limit in normal cases	Mean value	Usual upper limit	Extreme upper limit
8	11.7	13.8	15.4	16.4
9	11.6	13.6	15.2	16.2
10	11.4	13.4	15.0	16.0
11	11.2	13.3	14.9	15.8
12	11.1	13.1	14.7	15.6
13	10.9	12.9	14.5	15.4
14	10.8	12.7	14.3	15.3
15	10.7	12.6	14.1	15.2
16	10.5	12.4	13.9	14.9
17	10.3	12.2	13.7	14.6
18	10.1	11.9	13.5	14.4
19	9.9	11.7	13.2	14.2
20	9.7	11.5	13.0	14.0
21	9.4	11.2	12.8	13.7
22	9.2	10.9	12.6	13.5
23	8.9	10.6	12.3	13.2
24	8.7	10.4	12.1	13.0
25	8.4	10.2	11.8	12.7
26	8.2	9.9	11.6	12.4
27	7.9	9.6	11.3	12.1
28	7.6	9.4	11.1	11.8
29	7.3	9.2	10.7	11.5
30	7.1	8.9	10.4	11.2
31	6.7	8.6	10.2	10.8
32	6.4	8.3	9.9	10.5
33	6.1	8.0	9.6	10.2
34	5.9	7.7	9.2	9.9
35	5.6	7.3	8.9	9.6
36	5.3	7.1	8.6	9.4
37	4.9	6.8	8.2	8.9
38	4.6	6.5	7.9	8.6
39	4.3	6.2	7.6	8.2
40	4.0	5.9	7.2	7.8
41	3.6	5.4	6.8	7.5
42	3.2	5.0	6.4	7.0
43	2.8	4.6	5.9	6.5
44	2.5	4.2	5.5	6.1
45	2.2	3.7	5.2	5.6
46	1.9	3.3	4.8	5.1
47	1.7	2.8	4.3	4.5
48	1.5	2.5	3.9	4.0
49	1.3	2.2	3.4	3.4
50	1.2	2.0	3.0	3.0
51	1.1	1.8	2.6	2.6
52	1.0	1.6	2.3	2.3

TABLE 2 Normal Values of the Accommodative Power at All Ages from Eight to Sixty-eight (Duane, 1912) (Continued) Accommodation is given in diopters and tenths, near point being measured from the anterior focus of the eye, i.e., from a point 13 mm in front of the cornea

Age	Lower limit in normal cases	Mean value	Usual upper limit	Extreme upper limit
53	0.9	1.5	2.1	2.1
54	0.9	1.4	2.0	2.0
55	0.8	1.3	1.9	1.9
56	0.8	1.2	1.8	1.8
57	0.8	1.2	1.7	1.7
58	0.7	1.2	1.7	1.7
59	0.7	1.1	1.6	1.6
60–68	0.7	1.1	1.5	1.5

be measured in the prolonged testing needed for the filled circles. Age changes of this sort for populations younger than forty are available only by methods which measure with subjective blur as a criterion (and therefore include increased depth of focus). Measurements of this kind have been made by Donders (1864) and by Duane (1912). Table 2 gives the ranges of values measured in this way by Duane.

11. The Near Response The change in focus from far to near requires simultaneous increase in refraction of the two eyes and convergence of the lines of sight for single binocular vision to be maintained. A detailed discussion of experiments designed to elucidate the relation of these quantities is systematically treated elsewhere (Alpern, 1962*a*,*b*). The increased convergence with increased accommodation occurs even in people lacking normal single binocular vision and in normal binocular observers if one eye is covered. In the latter case, the amount of so-called accommodative convergence is rarely, if ever, exactly correct for precise alignment of the two eyes. Wide population differences exist among normal observers in the amount of accommodative vergence associated with a unit change in refraction of the eye (the so-called AC/A ratio). The population mean is around $1.4 \pm 0.8°$ per diopter.

The synkinesis of increased accommodative vergence elicited by an increased accommodation is also accompanied by narrowing of the pupils of both eyes. The relation between innervation to increased accommodation and increase in accommodative vergence is linear over a wide range, as is the relation between increased innervation to accommodation and the decrease in pupil size.

We have already seen that the spectacle lens is prescribed in such a way that the eye is at the myopic limit of depth of field for distance viewing. In focusing for near, however, observers rarely increase refraction of the eye more than the minimum required for clear vision (the "lazy lag" of accommodation). Thus at near they operate at the hyperopic limit of the depth of focus. Consequently, in the process of changing fixation the eye uses one extreme for far vision, the other extreme for near, and gains the maximum advantage from its depth of field.

How the synkinesis of these three components of the near response is effected is still quite unclear. The pattern in any given individual, once established, is difficult to modify (except temporarily) even by drugs or by surgery, much less by practice effects or training. Neurological control of the processes in primates is only recently coming under investigation (Jampel, 1959). Whatever the eventual explanations for its physiology, it is evident that optically the near response provides for optimum increase in refraction, realignment of the lines of sight, and enhancement of the depth of field in the process of changing fixation from far to near objects.

12. Eye Movements The two eyes almost invariably move together even if one is occluded. Binocular synchronization is of two types according as the angle the two lines of sight form at their point of intersection does or does not change during the movement. The former is defined as a *vergence* (or disjunctive) movement, the latter as a *version* (or conjugate) movement. Both versions and vergences can occur during the same change of fixation (Alpern and Wolter, 1956; Yarbus, 1967; Westheimer and Mitchell, 1956; Rashbass and Westheimer, 1961*a*) and appear to be neurologically quite independent.

According to a stimulating (if untestable) hypothesis of Walls (1962), eye movements evolved in order to hold the retinal image of the world fixed as animals moved about in their environment. Such movements are always conjugated, never disjunctive. These ancient reflex movements—evoked by stimulation of stretch receptors in the muscles of the head and neck and of the hair cells of the vestibular apparatus, the otoliths, and the semicircular canals—still persist in modern man. But they are overlaid by a variety of other kinds of eye movement which came about as the use of the eyes became a more complex part of behavior. The evolution of a central region in the retina where visual acuity is very high (the fovea centralis) in particular has become associated with the most important varieties of eye movement found in human observers. Fixation of objects one after the other by pointing the fovea at the retinal images of these objects in turn is how we put together our sharp picture of the world around us.

Of the variety of eye movements found in man, the vergences are the newest and the reflex the oldest in evolutionary history.

Even when trying to fixate as steadily as possible, the eyes are constantly moving. The eye movements of steady fixation include (1) high-frequency tremor of 30 to 80 Hz and amplitude of from a few seconds to 2 minutes of arc; (2) slow drifts of about 5 minutes of arc of amplitude with a velocity of 1 minute/s and a duration less than 0.2 s; (3) flicks of about 5.6 minutes of arc with a velocity of about 10°/s, which may occur as often as 0.2 s or as infrequently as 3 to 4 s (for details see Alpern, 1962*a*).

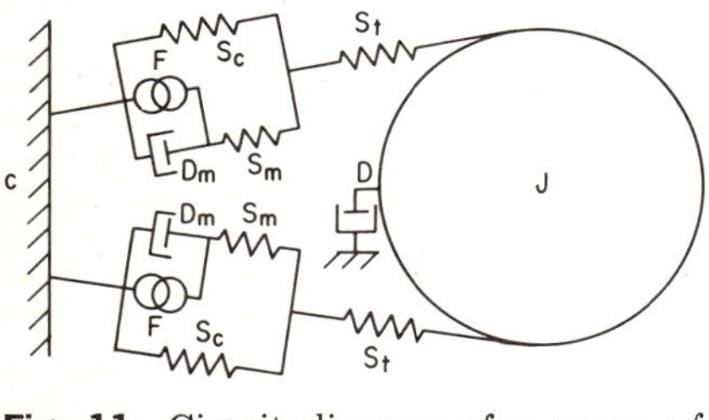

Fig. 11 Circuit diagram of a group of viscoelastic elements simulating the transfer characteristics of the eyeball-muscle system in horizontal eye rotation, showing two antagonistic muscle groups. The spring constants S_m, S_t, and S_c are 6.0, 6.0, and 3.1 kdyn·cm/deg, respectively. The constants of the damping elements D and D_m are 19 and 70 dyn·cm·s/deg. The inertial element J has the value 8.3 g·cm^2 corresponding to the moment of inertia of the eye and attachments necessary for Thomas' (1969) experiments, from which this model was derived.

It seems quite possible that at least the last two of these movements have their counterparts in the large eye movements. Before discussing the details of the latter, it is perhaps advisable to look at the mechanical properties of the eye and the orbit. Models of the eye-movement "plant" began with Westheimer's (1954) analysis of the saccadic eye movements (see below). This stimulated an extensive series of studies by Young and Stark (1963), Robinson (1964), Childress and Jones (1967), and Thomas (1969), to mention only a few. There is still a good deal of disagreement among these authors about the details of the mechanical properties of the eyeball and the orbit, so that the illustration in Fig. 11 must be taken as tentative. It is due to Thomas (1969) and is at least the most recent of the proposed models. The active force of the muscles F is shunted by the damping element D_m. (Only two muscles are shown for this horizontal section though each eye has six.) The active force is in series with the series elastic component of the muscle S_m and of the tendon S_t. The elasticity of the connective tissue in the muscle, acting in parallel with the muscle, is represented by the spring S_c. The inertia of the globe and its attachments are represented by the inertia element J, and the viscous damping forces which act on the globe from nonmuscular sources (orbital fat, Tenon's capsule, and optic nerve) are represented by the damping element D. The value of J assigned by Thomas includes a considerable factor due to a contact lens attached to the globe, the moment of inertia of which is about equal to that of the unencumbered eye moving in a rapid saccade (Robinson, 1964).

The transfer function T_m for the force applied by the muscles to eye position in this model has been computed by Robinson (1968) as

$$T_m(s) = \frac{0.291°/\text{g}}{(0.0368s + 1)(0.69 \times 10^{-5}s^2 + 0.00268s + 1)} \tag{5}$$

in which s is the Laplace operator. The model is the best available at the moment for frequencies above about 10 Hz. For frequencies below 10 Hz there is evidence that the gain rises as frequency decreases and that the transfer function should be about 0.8°/g at zero frequency. At these lower frequencies the plant appears to be heavily overdamped. For a detailed discussion of these problems see Robinson (1968), Thomas (1969), and Childress and Jones (1967).

Version Eye Movements. There are two types of version eye movements, the saccades and smooth pursuit movements. Saccadic eye movements are the abrupt changes in position of the globe on changing fixation from one object to another. Smooth pursuit movements are rotations of the globe which occur when the eyes attempt to follow a moving target. While these two movements are both conjugate, they differ in a number of other ways. Pursuit movements have been studied by Westheimer (1954*b*), Rashbass and Westheimer (1961*a*), and Robinson (1965), among others, saccadic movements by Westheimer (1954*a*, 1958), Dallos and Jones (1963), Rashbass and Westheimer (1961*a*), Yarbus (1967), Robinson (1964), and Thomas (1969). Electromyographically, there is evidence that saccadic eye movements are a consequence of intense activity of the agonist and complete inhibition of the firing of the antagonist (Alpern, 1962*a*). Smooth pursuit movements, on the other hand, are a consequence of contraction of both agonist and antagonist, the very slight differences in tension between the two being the effective force which rotates the globe. Other differences in the characteristics of smooth pursuit

TABLE 3 Comparison of Some Characteristics of the Two Types of Human Version Eye Movements (Modified from Robinson, 1965)

Characteristic	Saccade	Smooth pursuit
Latency, ms	200	125
Range of peak velocity, deg/s	10–1,000, depending upon movement amplitude	Near 0 to 25, independent of amplitude
Stimulus	Error in target position	Error in target velocity
Cerebral control	Frontal eye fields (tentative)	Occipital eye fields
Operating mode	Switching	Linear negative feedback
Type of control	Sampled data system at about 5 times/s	Continuously sampled
Need for target stimulus	No	Yes
Barbiturate toxicity (also myotonic dystrophy)	Insensitive	Sensitive
Frequency of oscillations under negative feedback, Hz	2–2.5	3.3
Amount of negative visual feedback required to assure oscillations	−5	−8

compared with those of saccadic eye movements are summarized in Table 3. The details of most of these features are given at length in Robinson (1964, 1965, 1968).

Vergence Eye Movements. These are classified according to their direction and their stimulus. There are three possible directions, horizontal, vertical, and torsional. All these kinds of vergences are evoked by a disparity in egocentric localization of the visual world seen by each eye. Typically, if the disparity is large, it results in double vision (diplopia), but small disparities excite a vergence movement which reduces the disparity to zero and realigns the lines of sight of the two eyes exactly on the object of fixation (or very nearly so). Such disjunctive movements are known as fusional movements. The important paper is by Rashbass and Westheimer (1961*a*).

Stimuli other than disparity in egocentric localization can evoke horizontal (if not vertical and torsional) vergence. The principal one is changes in innervation to accommodation. We have already discussed the link of accommodation and accommodative vergences in the near reflex. Binocular vision at any distance requires precise alignment of the lines of sight on the object of regard. In shifting from far to near, the amount of innervation to accommodation change evokes a change in accommodative convergence which differs in each individual but which usually brings the point of intersection of the lines of sight of the two eyes in the neighborhood of the near object of regard. The discrepancy from exact alignment must be made up by fusional movements. If the amplitude of fusional movements is insufficient for the discrepancy, normal single binocular vision is not possible without auxiliary prisms or bifocals and one form of strabismus can result.

Vergence movements are very slow. Their latent period (about 200 ms) is about the same as that of other eye movements, but the movement follows a slow exponential course with a constant of 300 ms or so, depending upon the amplitude (Alpern and Ellen, 1956). There is no evidence for a sampled-data system for vergences. Rashbass and Westheimer (1961*b*) found that the open-loop transfer function could be represented by a delay and an integrator with a scale factor of about 10°/s per degree of vergence error. Robinson (1968) attributes the slowness of the movement to the fact that the small gain associated with the integrator makes the closed-loop performance little better than the performance of the plant alone.

THE EFFECT OF LIGHT ON THE RETINA

13. The basic mechanism assumed for vision is the absorption of light quanta in at least one photolabile visual pigment situated in the outer segments of the retinal rods and cones. There are known to be four human photolabile visual pigments, rhodopsin in the rods and chlorolabe, erythrolabe, and cyanolabe in the cones. Evidence from microspectrophotometry (Marks, 1965) suggests that each photoreceptor (rod or cone) cell contains one and only one photolabile pigment. A very important simplifying idea is the principle of univariance; i.e., given any receptor cell, any two lights—differing or not in wavelength—will produce identical outputs from this sense cell whenever the quanta absorptions by its pigment are identical. Thus, independent of the wavelength of the light, the receptor cell can only signal how many quanta are absorbed by its pigment.

14. The Duplicity Theory The rods and cones carry out quite different physiological roles in the process of seeing. Concentrated in the peripheral retina, 20 times more numerous but with a much higher degree of neuronal convergence and with a sensitivity approaching the theoretical limit, rods are especially adapted for night vision. Cones, though they are found everywhere in the retina, achieve their highest density in the center of the fovea centralis, where there are no rods, and are especially adapted for daylight vision with good visual acuity and color vision, albeit reduced sensitivity to small amounts of light.

All human rods contain the same visual pigment, rhodopsin. It is a chromophore, the 11-*cis* isomere of vitamin A aldehyde, linked (with at least one Schiff base) to a protein (or perhaps the lipoprotein, *N*-phosphatidyl ethanolamine) *opsin*. The chemistry of rhodopsin has received extensive attention (see reviews by Brindley, 1960; Wald, 1968; Dartnall, 1962; and Abrahamson and Ostroy, 1967). Since all human rods contain only one pigment, it follows from the principle of univariance that we do not have color vision with rods alone. This is, in fact, the case. Nor do rods participate with cones in normal trichromatic color vision.

The chemistry of the visual pigments in the cones is not nearly as well understood as that of rhodopsin. Indirect evidence, the most convincing of which is due to Rushton (1968), strongly suggests that the chromophore, 11-*cis*-vitamin A aldehyde, for the rods and the three different kinds of cones is identical and that in man the three different kinds of cone pigments and the rod pigment, with their quite different absorption spectra, differ only in the character of their respective opsins.

In all known photolabile visual pigments the absorption of a light quantum results in photoisomerization of the 11-*cis*-vitamin A aldehyde into the all-trans form (Wald, 1968). Subsequent thermal reactions have been studied in detail in the extensive literature, but a discussion of them is beyond the scope of this section (see Abrahamson and Ostroy, 1967).

The normal human retina has about 120 million rods and 6 million cones. The relative proportion of red, green, and blue cones is not known, though some indirect evidence in man and direct evidence in fish (Tomita, 1968; Marks, 1965) suggests that the blue cones are distinctly less numerous (Wald, 1967), particularly in the center of the fovea, which is blue-blind. An alternative explanation (Brindley, 1954) is that blue cones have a much higher convergence than red and green cones. In addition to their roles in night and day vision, respectively, rods and cones can be distinguished functionally in the living eye by three different methods, namely, by their spectral sensitivity, their directional sensitivity, and the kinetics of their visual pigments. The second of these (the presence of the Stiles-Crawford effects in the cones and their absence in the rods) has already been discussed. The third will be discussed presently. For the moment, let us turn to the first.

15. Spectral Sensitivity The action spectrum of a receptor cell, i.e., its spectral sensitivity, depends upon the absorption spectrum of its photolabile pigment. For human rods, the former can be measured by measuring the number of quanta required for seeing a threshold flash, wavelength by wavelength through the spectrum. The latter can be inferred from difference spectra obtained by microspectrophotometry (Liebman, 1962) on retinal rods from enucleated eyes, by spectrophotometry of a solution of retinal rods, or more exactly on solution of micelles of digotonin extracts of rhodopsin, the rod visual pigment. All these methods, when appropriate corrections are made, give closely comparable results. One of these comparisons is illustrated in Fig. 12. The points are the measurements of

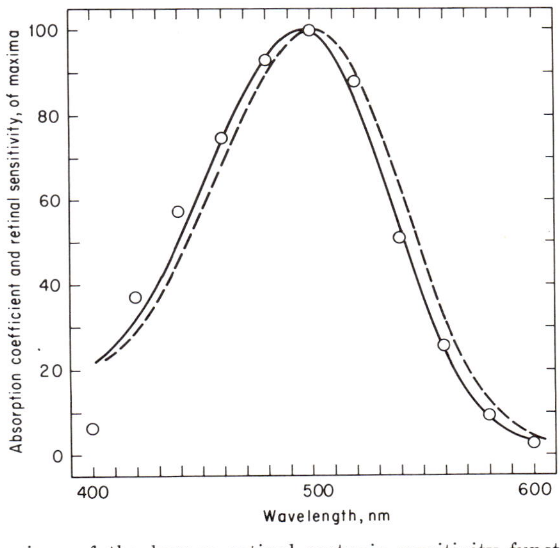

Fig. 12 Comparison of the human retinal scotopic sensitivity function (Crawford, 1949) with the absorption spectrum of human (*solid line*) and frog (*dotted line*) rhodopsin. [*From Crescitelli and Dartnall* (1953).]

absolute threshold, and the solid line is the difference spectrum of human rhodopsin digotonin micelle.

Though the measurements on rhodopsin are more objective, the psychophysical measurements on man are much more precise. The best of these, in addition to those of Crawford, are due to Flamant and Stiles (1948), who measured at each wavelength of background the amount of radiation needed to raise the threshold for a green light of wavelength 490 nm (near the rhodopsin peak) to 10 times the absolute threshold. The CIE adopted a standard rod spectral-sensitivity function in 1957 based on the most reliable data. It is shown as solid dots in Fig. 13.

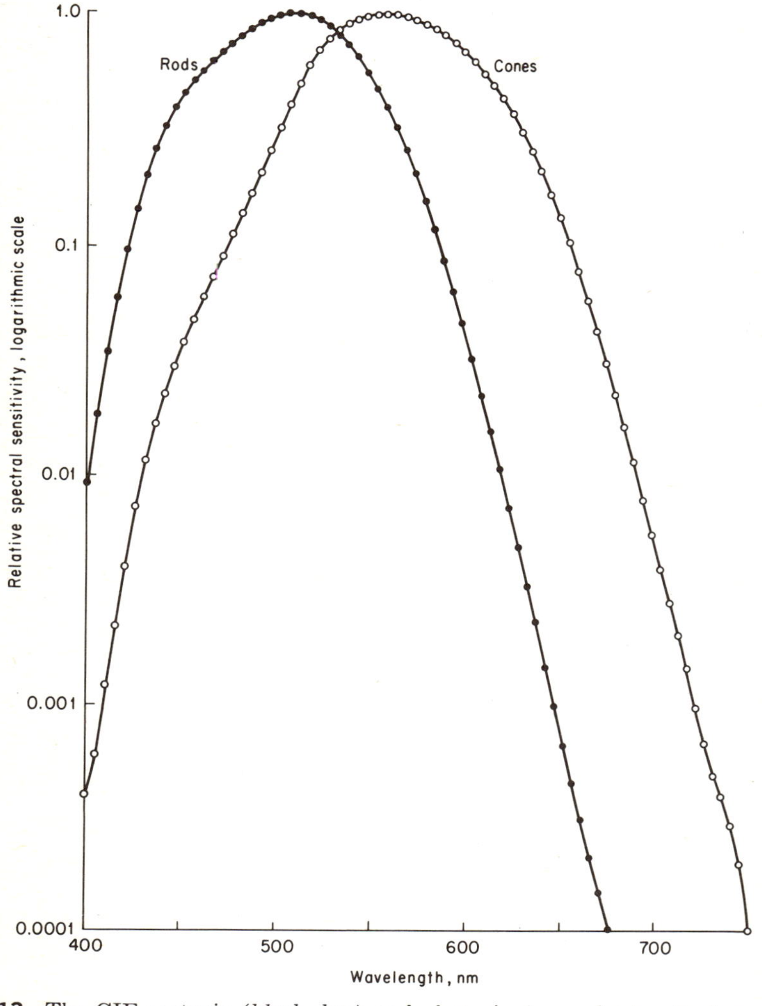

Fig. 13 The CIE scotopic (*black dots*) and photopic (*open dots*) retinal spectral-sensitivity functions. The black dots are essentially the same as the points in Fig. 12. The ordinate scale is logarithmic (not linear as Fig. 12), and each curve has been arbitrarily shifted vertically so that its maximum is set equal to 1.0. (*After M. Alpern, M. Lawrence, and D. Wolsk, "Sensory Processes." Copyright © 1967 by Wadsworth Publishing Company, Inc. Used by permission of the publisher, Brooks/Cole Publishing Company, Monterey, California.*)

The CIE photopic spectral-sensitivity (or, more properly, the spectral-luminosity) curve is also shown (as open circles) in Fig. 13. It was adapted by the CIE in 1924, pooling the results from the most careful and reliable studies. They represent measurements on the foveas and hence involve only cones. Two different criteria were used in those studies: (1) the method of equality of brightness and (2) the method of flicker photometry. Excellent reviews of the work from which the 1931 CIE curves were derived and of more recent criticism can be found in Gibson (1940) and in Le Grand (1957).

While the scotopic-spectral-sensitivity curve in Fig. 13 leads to straightforward interpretation from the principle of univariance, this is by no means the case for the photopic curve. Rods contain only a single visual pigment, rhodopsin, and its absorption spectrum must determine the scotopic-spectral-sensitivity function. But the photopic-spectral-sensitivity curve does not represent the absorption spectrum of any single human photolabile visual pigment. We know that it must be synthesized by some combination of the action spectra of three underlying cone visual pigments, but we do not yet understand how. An excellent summary of the problem is given by Wyszecki and Stiles (1967).

The nature of the absorption spectra of the underlying cone visual pigments is slowly coming to light, however. The analysis owes much to the measurement of the red and green cone pigments in normal living human eyes by Rushton (1963*a,b,c,* 1965*a,b,c*), using the method of reflection densitometry. Fortunately the fovea of certain color-blinds (protanopes and deuteranopes) reveal by this method only a single cone pigment: chlorolabe (in the protanope) and erythrolabe (in the deuteranope), but the normal fovea contains both these pigments.

The method does not lend itself to exact estimates of pigment absorption spectra, but the observation that the normal red pigment is also found in the normal (red) cones of the deuteranope and the normal green pigment in the normal (green) cones of the protanope is important. It allows careful measurement of spectral sensitivity in the red-green spectral range by psychophysical measurements on these two kinds of dichromats to be used to infer the nature of the action spectrum of the normal red and green cones. The inference is still somewhat inexact because the contribution of the blue cone pigment remains uncertain.

Technical difficulties are such that Rushton's measurements cannot reveal the blue pigment, but a variety of converging lines of evidence support the idea that it is scarce, has a relatively narrow band, and makes little or no contribution to spectral luminosity in the red-green range. Certain cone monochromats (Blackwell and Blackwell, 1961; Alpern et al., 1965) have only rods and blue cones, and their spectral sensitivity is easily measured in such cases. Evidently these blue cones have both the spectral and intensity sensitivity of normal blue cones, as measured, say, by the increment-threshold method of Stiles (1939) on the normal fovea. Using these data on such subjects for the blue cones, the foveal-luminosity curve of the protanope (Alpern and Torri, 1968*a*) for the green cones and the luminosity curve of the deuteranope (Alpern and Torii, 1968*b*) for the red cones, we have a reasonable estimate of the spectra of the three normal kinds of cones. They are shown in Fig. 14. Evidence that these ideas are along the right line is to be found from direct measurements (although still quite crude) on single cones in retinas extracted from human and primate eyes (Marks et al., 1964; Brown and Wald, 1964; Liebman and Entine, 1964).

These microspectrophotometric experiments not only provide concrete evidence for the blue pigment which cannot be demonstrated by Rushton's densitometer (see, however, Weale, 1968) but strongly suggest that there are three different kinds of cones, each with only one visual pigment. The overlying absorption spectra of the three different cone pigments, each in its own variety of cone, provide the molecular basis for trichromacy. This property of normal human foveal vision may be defined by the statement that human color vision can be completely quantified by three degress of freedom. Thus, as Brindley (1960) puts it:

> given any four lights, whether spectroscopically pure or not, it is always possible to place two of them in one half of a photometric field and two in the other half, or else three in one half and one in the other, and by adjusting the intensities of

three of the four lights to make the two halves of the field appear indistinguishable to the eye.

How the eye matches different wavelengths of the spectrum in this way has been extensively studied. The details of such measurements and the laws of color mixture are described in Sec. 9. For the present purpose it will suffice to emphasize that the cellular basis of these color-matching properties resides in the absorption spectra of three different cone pigments probably not very different from those illustrated in Fig. 14. This inference seems fairly strong on the basis of Stiles'

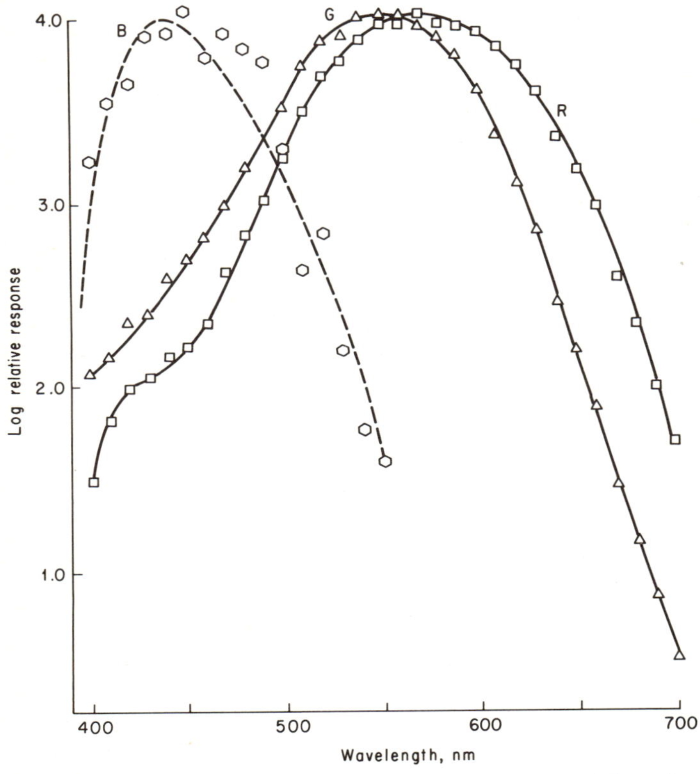

Fig. 14 The action spectra of the human cone pigments inferred from luminosity-curve measurements on color-blind subjects. Squares imply the red pigment R from luminosity of deuteranopes [*Alpern and Torii* (1968*c*)]. Triangles imply the green pigment G from luminosity of protanopes [*Alpern and Torii* (1968*b*)]. Hexagons imply the blue pigment B from three π_1 cone monochromats (unpublished data). The dotted line is Stiles' blue (π_1) mechanism. The R and G curves are uncertain for $\lambda \leq 500$ nm because of uncertainty of the contribution of blue cones to luminosity.

(1939) and Rushton's results taken in conjunction with the less exact but more direct microspectrophotometric measurements. Given this, and the principle of univariance, normal trichromacy and the linearity of the laws of color mixture are a necessary consequence.

Color-blindness may be one of several varieties, some acquired and some inherited. The former are not well understood, and the only systematic classification is in terms of the inherited defects. The latter have been extensively studied, and the details are to be found in standard texts (Wright, 1946; Le Grand, 1957; Wyszecki and Stiles, 1967). The main features are summarized in Table 4. There

TABLE 4 Inherited Color Vision Defects

	Degrees of freedom needed to quantify cone vision	Normal systems	Mystery systems	Peak wavelength of photopic luminosity curve, nm	Frequency of occurrence, %		Inherited as	Rayleigh match
					Male	Female		
Trichromacy:								$R + G = Y$
Anomalous:								
Protanomalous	3	Blue cones, green cones, rods	Red cones	540	1.0	0.02	X-linked recessive trait	$\frac{R}{R+G} > \frac{1}{2}$
Deuteranomalous	3	Blue cones, red cones, rods	Green cones	560	4.9	0.38	X-linked recessive trait	$\frac{R}{R+G} < \frac{1}{2}$

	Degrees of freedom needed to quantify cone vision	Normal systems	Absent	Peak wavelength of photopic luminosity curve, nm	Frequency of occurrence, %		Inherited as	Neutral point, nm
					Male	Female		
Dichromacy:								
Protanope	2	Blue and green cones, rods	Red cones	540	1.0	0.02	X-linked recessive trait	494
Deuteranope	2	Blue and red cones, rods	Green cones	560	1.1	0.01	X-linked recessive trait	499
Tritanope	2	Red and green cones, rods	Blue cones	555	0.002	0.001	Autosomal dominant trait	570, 400
Monochromacy:								
Blue cone	1	Blue cones, rods	Red and green cones	445, 507	?		X-linked recessive	All wavelengths
Typical (rod)	1	Rods and one variety of controversial cones	Red, green, and blue cones	507	0.003	0.002	Autosomal recessive?	All wavelengths

are three major categories (trichromats, dichromats, and monochromats), depending on whether the quantification of color vision requires three, two, or only one degree of freedom, respectively. The anomalous trichromats are distinguished from normal trichromats by the anomaloscope, a device presenting a monochromatic yellow (589 nm) on one side of a photometric field and a mixture of green (536 nm) and red (656 nm) on the other. The proportion of red $R/(R+G)$ in the mixture required to match the yellow can be varied by one control and the intensity of the yellow varied by the other. In this part of the spectrum an exact match can usually be made with only two degrees of freedom since the contribution of blue cones here is relatively trivial for most observers. Equation of the intensities of the red and green lights can be made in such a way that the normal trichromats match the yellow at values of $R/(R+G) = 0.5 \pm 0.1$.

Anomalous trichromats are divided into two groups, protanomalous and deuteranomalous, according as their settings of $R/(R+G)$ are respectively greater than or less than 0.5 ± 0.1. Since it seems quite clear that anomalous have no abnormal prereceptor losses in their ocular media (Alpern and Torii, 1968*a*), the principle of univariance leads directly to the conclusion that the action spectrum of at least one of the three cone pigments in anomalous must differ from the corresponding normal pigment. A variety of evidence (Alpern and Torii, 1968*b,c*) suggests that the anomalous pigments are in the red cones of the protanomalous and in the green cones of the deuteranomalous, the remaining cones in both cases having completely normal cone pigments. But the nature of the spectra of each of these anomalous pigments remains very much a mystery (Alpern and Torii, 1968*b,c*).

Dichromats are distinguished by matching all parts of the spectrum with two degrees of freedom. There are three known well-documented types. (Other types have been suggested without sufficient confirmation.) Each of the known types can be described by the statement that two of the normal cone pigments are present and functioning while the third normal cone pigment is absent. Thus, protanopes have normal green and blue cones but lack red cones completely. Deuteranopes lack the green cones but have normal red and blue cones. The notion that a second type of deuteranopia exists, in which both red and green cones are present but the nerve fibers are fused, now seems quite unlikely (Alpern et al., 1968). All the dichromats and anomalous trichromats discussed above inherit their defect as X-linked recessive traits, so that the incidence is much higher in males than in females. Tritanopia is a variety of dichromatism in which the normal blue pigment is missing. It is inherited not as an X-linked recessive but as an autosomal dominant trait.

The typical (rod) monochromat has very poor visual acuity and symptoms which are most easily summarized by the generalization that the retina lacks functioning cones, although there is some physiological and histological evidence for some photoreceptors other than rods (Alpern et al., 1960; Falls et al., 1965). There is a good deal of conflicting evidence regarding their nature, however. An extremely rare group of total color-blinds is the cone monochromats, of which at least two different types have been described. The blue-cone monochromats have only rods and blue cones. Visual acuity [like that of the blue cones alone (Green, 1968)] is poor. The other variety of cone monochromat, studied most extensively by Weale (1953), has normal visual acuity. Whether, as Weale claims, all four visual pigments are functioning normally in them or not is still uncertain.

16. Adaptation It is a familiar fact that our eyes are sensitive to very small differences in the intensity of lights over an enormous range of intensities. The range from the lowest level where any vision at all is possible to intensities so bright that the retina will burn is larger than 10^{13}. This is illustrated in Fig. 15. The lower level, the absolute threshold for rod vision, as we shall see, approaches the theoretical limit of any radiation-detection device. The threshold for cone (daylight) vision is 10^3 times as bright, and rod signals (which are increasing linearly with the light at these lower intensities) begin to saturate at a level $10^{4.5}$ higher still.

In the same figure are shown visual-pigment density as bleaching becomes appreciable, beginning at about 10^5 scotopic troland-seconds. The dashed line shows the fraction of remaining cone pigments, chlorolabe and erythrolabe (Rushton, 1963*c*

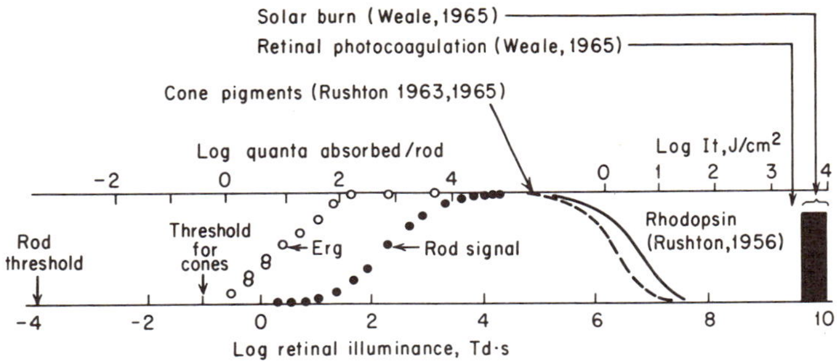

Fig. 15 The response of the retina to light flashes over the range of intensities to which it is sensitive. [*Modified from Weale* (1965).] The open circles are normalized responses of the b wave of the human electroretinogram (ERG) to full-field even illumination (100-ms flashes). The rod signals are measurements of Alpern et al. (1970.)

and 1965*b*), and the solid line the fraction of remaining rhodopsin (Rushton, 1965*a*) as light energies are increased. This figure is a dramatic way of emphasizing that the changes in visual sensitivity over the range of useful vision, i.e., the range between rod threshold and the saturation of rod signals, occur without any appreciable change in amount of visual pigments, contrary to the photochemical theories of adaptation of Hecht (1936) and Wald (1941).

Figure 16 shows the changes in visual sensitivity for rods. Plotted along the abscissa is the intensity of the background in scotopic trolands. Plotted along the ordinate is the intensity that a 200-ms test flash superimposed on this background requires at threshold for rod vision. Over most of this range the increment threshold is proportional to the background (if ΔI is the threshold intensity of 0.2-s flash and I the background intensity, both in scotopic trolands, then over most of the range $\Delta I = 0.2I$), but because the rod signals saturate at about 2000 Td·s. (Fig. 15), the rod threshold goes toward infinity (Fig. 16). Rod monochromats who have almost no functioning cones lose all contrast vision at these background intensities (Fuortes et al., 1961).

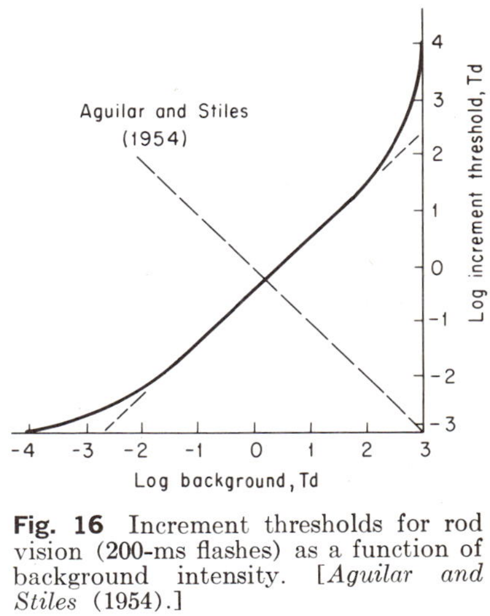

Fig. 16 Increment thresholds for rod vision (200-ms flashes) as a function of background intensity. [*Aguilar and Stiles* (1954).]

Under optimal conditions, that is 20° into the peripheral retina, 10 minutes of arc test target, of blue-green ($\lambda = 510$ nm) light near the peak of the rod-spectral-sensitivity curve and 0.001 s duration, normal observers in full dark adaptation require from 54 to 148 $h\nu$ for threshold visibility (Hecht et al., 1942). Allowing for light losses at the cornea (4 percent) and in the ocular media (50 percent) and for the fact that only about 20 percent of the quanta incident upon the rods will be absorbed by their rhodopsin, it is estimated that the absorption of from 5 to 14 quanta in a retinal area containing about 500 rods is the minimal necessary for vision. Since the chances are very small (Brindley, 1960, p. 185) that under these threshold conditions more than one quantum is absorbed by any one rod, it is evident that the absorption of one quantum of light suffices to excite a retinal rod, though several (5 to 14) rods must be almost simultaneously excited for vision to occur. Other authors (Bouman and van der Velden, 1947) have estimated that the number of nearly simultaneous rod excitations required at threshold may be as small as 2, but the evidence for this view is not compelling and it has the difficulty of accounting for further improvements which can occur in threshold visibility with various physiological conditions, e.g., hyperventilation.

Dark Adaptation. When an appreciable fraction of the visual pigments is bleached away, say by a light more intense than 7.0 log Td·s, there is a great loss of sensitivity which recovers only slowly in the dark. The curves in Fig. 17 show

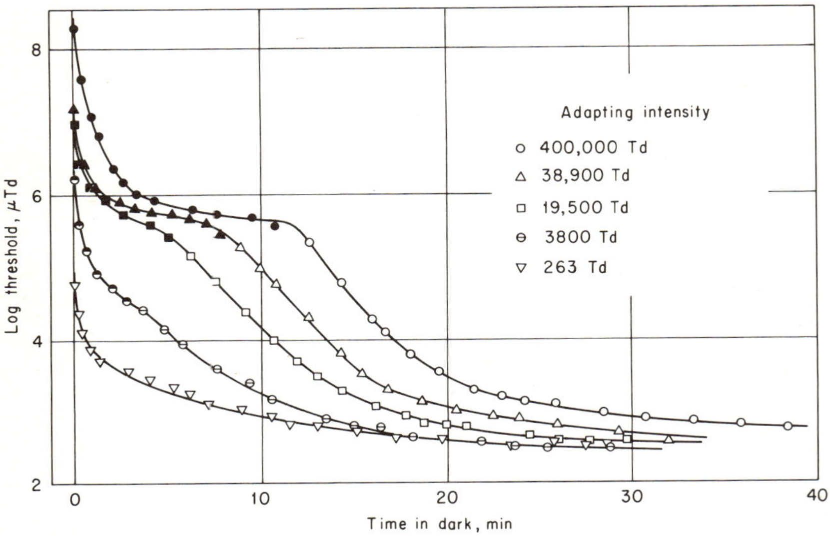

Fig. 17 The course of dark adaptation as measured with violet light following 2 min of adaptation to different intensities of retinal illuminance. The test is 5° in diameter; the region of the retina tested is 30° nasal from the fovea. The filled-in symbols indicate that a violet color was apparent at threshold; the empty symbols indicate that the flash was colorless at threshold. [*Data of Hecht et al.* (1937).]

on the ordinate the log intensity of retinal illuminance of a 5° violet test flash required for threshold visibility at different times in the dark after the bleach. The parameter in this figure is the intensity of light adaptation (2 min duration). Each dark adaptation curve is divided into two parts, the first (filled-in symbols indicating the flash was seen colored at threshold), recovering in about 6 to 10 min, due to cones (and to cones alone); the second (empty symbols indicating the flash was seen as colorless at threshold), recovering in about 30 min, due to rods (and rods alone).

The recovery of both the cone and rod log sensitivity after strong-light adaptation follows an exponential course, with time constants of about 130 s and 6 min, respectively. Rushton has shown that for both rods and cones there is a linear relation between logarithm of the threshold intensity and the amount of their respective visual pigments still in the bleached state.

This does not describe the entire process, however, because Rushton and Westheimer (1962) have shown that the threshold is elevated in adjacent rods unaffected directly by the bleaching as much as it is in their bleached neighbors. It is the state of excitation of the pool rather than that of the individual rods that determines threshold. The bleached rods signal the amount of unregenerated pigment to the pool, and the weighting of all such signals contributes to the state of adaptation of the pool, which in turn determines the threshold of all rods in the pool.

How this comes about is still only poorly understood, but one of its consequences is an equivalence between the "dark light" of bleaching and real backgrounds, either of which is capable of setting the sensitivity of the pool to a given level. Under most circumstances the "dark" light (of bleaching) and background, i.e., real, lights add linearly together, but there is evidence that they influence the sensitivity in different ways.

How do the dark light of bleaching (dark adaptation) and real background lights (light adaptation) influence sensitivity? Rushton (1965*d*) introduces the concept of the excitation pool which controls retinal sensitivity. The control can be visualized as being exercised by an automatic-gain-control device by means of which the retina manages to maintain its sensitivity to very small increments in light intensity over a wide range of real and/or equivalent (bleaching) backgrounds. The analog is a chain of resistances and capacitors in parallel connected by isolated amplifying units (Fuortes and Hodgkin, 1964). The input to the box is proportional to the light, and the output goes to the brain to signal the light, but it is also fed back into the box to increase the conductance of each stage in the chain. The conductance change increases linearly with the output. In this way, the sensitivity is regulated to the prevailing background. Rushton (1965*d*) has extended this model to human retinal adaptation and has shown that the signals from the equivalent background of bleaching may influence the sensitivity by entering the gain-control device through the feedback while real lights enter through the input. A variety of data on human vision are explained by a model of this kind containing from 4 to about 15 stages in the chain. Many, if not all, of the facts of human dark and light adaptation are best understood in this way.

17. Time Characteristics One consequence of models like the one just described is that they imply an exchange of sensitivity for speed of response. In the dark the sensitivity is high but the response slow; in the light the speed of response is fast but the sensitivity poor. Figure 18 shows for cone vision

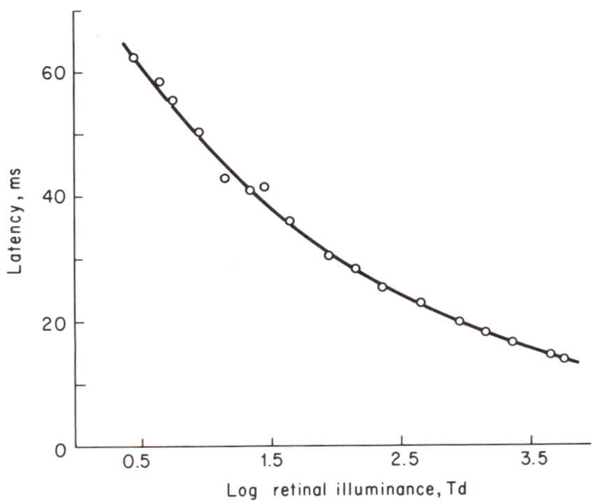

Fig. 18 The relation between latency of vision and the luminance of the background against which a dark target is viewed. The points are calculated from data on the Pulfrich pendulum (Lit, 1949) and the curve is theoretical, according to the Rushton-Hodgkin-Fuortes model of adaptation. *From M. Alpern, A Note on Visual Latency, Psychol. Rev., vol.* 75. *Copyright* 1968 *by the American Psychological Association. Reprinted by permission.*)

the relation of visual latency to the intensity of the prevailing background of illumination. These values of latency were inferred from measurements (Lit, 1949) of the differences in latency in the two eyes with differences in the background intensity in the two eyes using Pulfrich's pendulum. The curve is theoretical, based on the Rushton-Hodgkin-Fuortes model containing (in this case) four stages (Alpern, 1968). The decreased latency with increased background intensity is clear.

There are other concrete examples of the improvement of speed of response with background intensity, most notably the critical flicker frequency. This is the maximum rate of interruption of a light which is just seen to flicker. Measurements of this phenomenon are old, and an extensive literature exists (Landis, 1954). A new dimension to such studies was added by de Lange (1958*a,b*), who showed that in order to find the critical flicker frequency for any shape of interruption or modulation of the stimulus one needs to know only what he calls the ripple ratio r (the amplitude of the fundamental harmonic of the waveform divided by the mean luminance). Fourier analysis of the waveform yields r. The critical flicker frequency can then be read (provided only that $r \geq 2$ percent) directly from a curve showing what de Lange called attenuation characteristics of the eye but what has since become known as temporal-modulation transfer-function curves. Figure 19 shows curves of this kind obtained by de Lange using sinusoidally

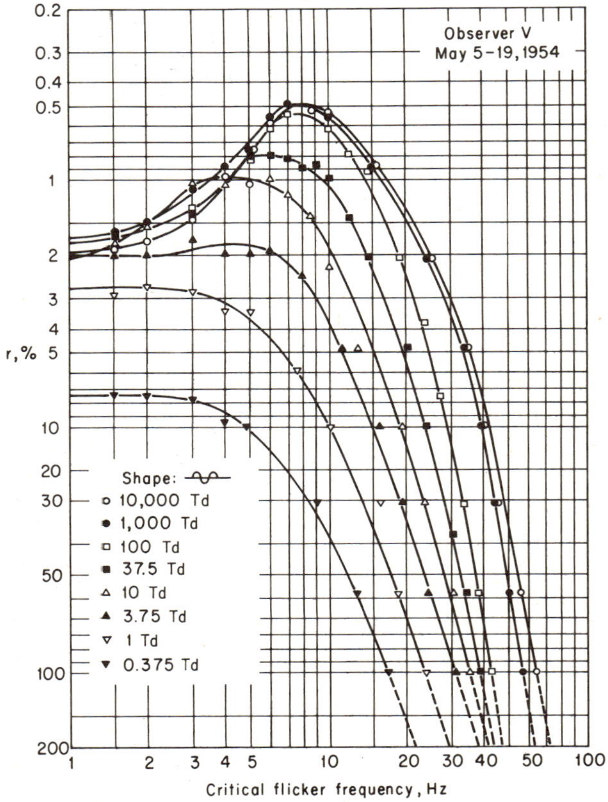

Fig. 19 This curve shows $r = (B_{\max} - B_{\min})/(B_{\max} + B_{\min})$ for detectable sine-wave flicker at a variety of frequencies with test-patch (2° central field with 60° surround) luminance varied parametrically. [*Results of de Lange* (1958*a*).]

modulated waveforms. The 2° test patch and 60° surround were matched in color (white light) and in luminance (above the critical flicker frequency). The ordinate in this figure gives the ratio

$$r = \frac{B_{\max} - B_{\min}}{B_{\max} + B_{\min}}$$

for threshold flicker detection at a given mean luminance and for a variety of frequencies. The high-frequency cutoff is readily explained by a model like that already described, in which about 10 stages of exponential delay occur between absorption of the light quanta and the changes that set off nerve impulses. It is assumed that dark adaptation involves a lengthening of the time constant of most, if not all, of the intermediate stages, by decrease of the leak conductances. What is responsible for the low-frequency cut in Fig. 19 is less clear. It is most pronounced in green light and virtually disappears in red light. The curves in Fig. 19 are largely or completely due to red and green cones; the blue cones have much lower high-frequency cutoff (about 18 Hz, according to Brindley et al., 1966).

De Lange also measured the temporal-modulation transfer function of the red-green color-sensitive system with brightness held fixed. The transition from red to green was superimposed on a metameric steady orange light, and the color-ripple ratio for detectable chromatic flicker

$$r_c = \frac{\text{red} + \text{green}}{\text{red} + \text{green} + \text{orange}}$$

was measured at a variety of frequencies and at two luminance levels. The results for the higher luminance level are shown as the x's in Fig. 20. The open circles in this same figure are the ordinary temporal-modulation transfer-function curves for the orange light similar to the curves in Fig. 19. The curve drawn through the black dots (difference between x's and o's) represents the filter action of a

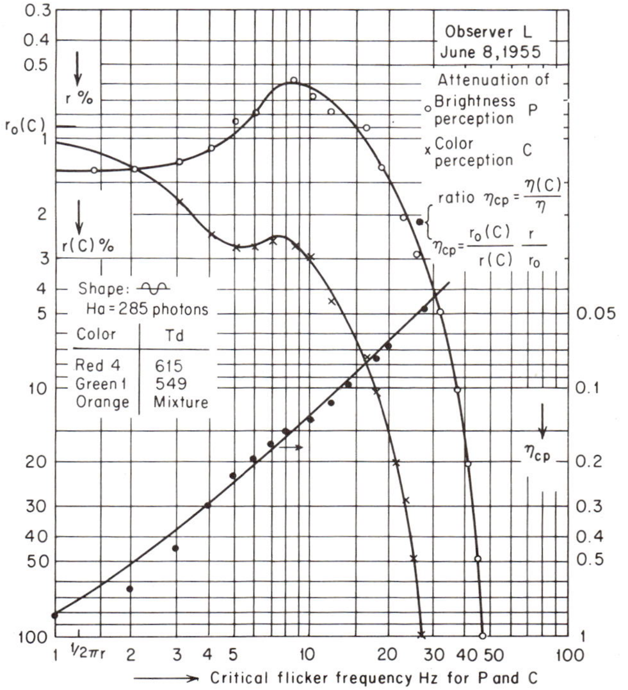

Fig. 20 Color (x's) and brightness (open circles) temporal-modulation transfer-function curves for a red and green light which are metameric to an orange. The brightness curve is for the orange metameter, the x's are the color-ripple ratio. The black points indicate the extra attenuation which color perception undergoes with respect to brightness perception. They were found by subtracting the brightness curve from the color curve. The curve through the black points is theoretical. [*Results of de Lange* (1958*b*).]

single integration process with a time constant of 123 ms. For the lower luminance a similar result was found, but three integration processes in cascade were required to fit the results. The combined time constant is roughly the same, however, as the single stage used at high levels.

The importance of de Lange's results is that they greatly simplify a vast literature describing the influence of waveform parameters such as light-dark ratio and wave shape on critical flicker frequency. Given the curves in Fig. 19, one can determine the critical flicker frequency of any periodic wave whatsoever, knowing only the ripple ratio and the mean luminance.

18. Visual Acuity The most important role of the eye does not reside in its abilities—however remarkable—to adapt to variable light intensities or to perceive subtle differences in temporal fluctuation in the light or to discriminate among shades of color. For the eye is *par excellence* a spatial sense organ, and what it does best is to localize objects in space. As with any optical instrument, it is appropriate to speak of its resolving power, i.e., the minimum angle of resolution, or of its reciprocal, *visual acuity*. Conventionally, resolving power of the eye is expressed in minutes of arc.

There are a variety of operational ways of measuring the resolving power of the eye. On any given observer they frequently give quite different answers. Two-point discrimination is referred to as the minimum separable; the smallest resolvable angular subtense of a black bar on a white field is called the minimum visible. There is also the minimum legible, i.e., the smallest angular subtense of letters on the eye chart that the patient can read. Other tasks, notably detecting the details of a checkerboard or of a black and white bar grating, are also frequently employed in the assessment of resolution.

In constructing the letters of the eye chart the Dutch ophthalmologist Snellen reasoned from the best estimates available to him for the minimum separable of the normal eye viewing under optimal conditions as 1 minute of arc; its reciprocal 1.0 is now referred to as "normal" visual acuity. He constructed standard test letters, each built in a block divided into a 5×5 square matrix. Each cell in the matrix was either black or white, the ensemble constituting a black letter on a white background. For Snellen, the normal eye would just resolve a letter when the cells of the matrix of that letter subtended 1 minute of arc at the standard test distance (20 ft or 6 m). The letters on the chart are all graded in size according to the distance that the cells of their matrix subtend 1 minute of arc. The visual acuity is then expressed as a fraction (now called the Snellen fraction), the numerator of which represents the testing distance and the denominator the distance at which the cells in the matrix of the smallest legible letter subtend 1 minute of arc.

Although the various operational tests of acuity lead to different answers, for any given observer they are all highly correlated and the various physical, physiological, and psychological variables which influence performance generally (if not invariably) do so in a similar way.

The experimental analysis of how these variables influence visual acuity is the subject of a vast literature. An excellent introduction is the review by Westheimer (1965). The contribution to the degradation in visual acuity made by the eye's optics, including changes in pupil size, have already been discussed (Pars. 4 to 6). Further losses introduced by uncorrected errors of refraction can be easily estimated (Fry, 1956; Campbell and Green, 1965*b*). Physiological variables include the intensity of retinal illuminance (Shlaer, 1937), the regions of the retina tested (Green, 1970), the photoreceptor mechanisms involved (Stiles, 1949; Green, 1968), adaptation state, contrast (Hendley, 1948), age, surround, drugs, motor factors (Westheimer, 1963) including eye movements (Miller and Ludvigh, 1962; Volkmann, 1962), orientation of the test objects (Campbell et al., 1966; Mitchell et al., 1967), binocular viewing (Campbell and Green, 1965*a*), and exposure duration (Kahneman, 1964; van Nes, 1968), to mention only the more obvious ones. Some of the factors which influence visual acuity and are less easy to understand (Flom et al., 1963), including some due to very complicated psychological variables, are covered in Westheimer (1965).

The introduction of Fourier methods in optics naturally has had an influence on the laboratory approach to the problem of visual resolution, beginning with pioneering work of Schade (1956), to which we have already referred. Sinusoidal grating targets at a fixed average luminance in which the amplitude of the sine wave is varied for threshold provides a convenient way to study visual resolution. The image of such a target is always a sine wave, and the variation of the frequency of the grating permits analysis of contrast threshold throughout the spectrum of spatial frequencies to which the eye is capable of responding. A plot of the percent modulation for threshold as a function of spatial frequency is called a modulation-transfer-function curve. Results of this kind made over a range of mean levels of retinal illuminance by van Nes (1968) are shown in Fig. 21. They are exactly analogous in the space domain to de Lange's (1958*a*)

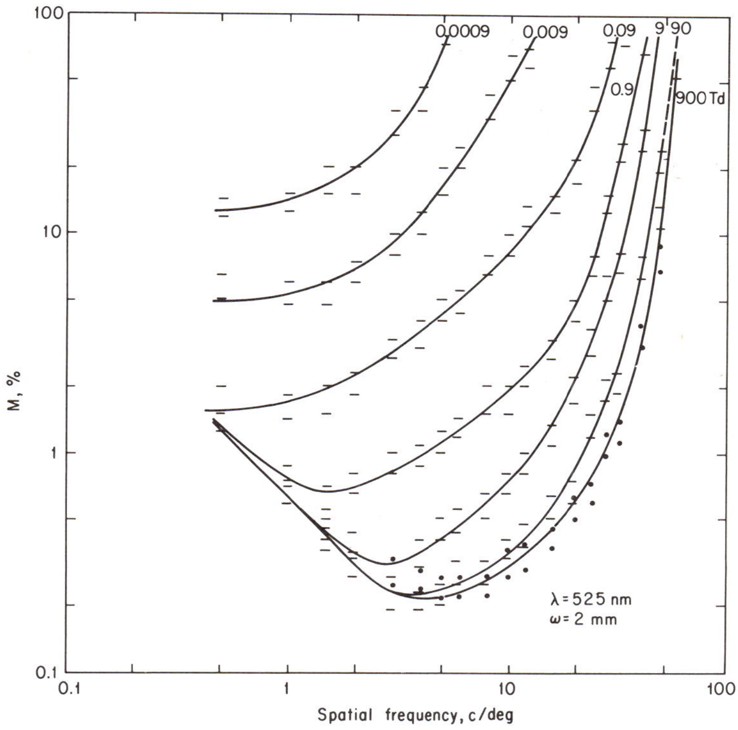

Fig. 21 Threshold-modulation transfer functions for green light (λ = 525 nm) at seven mean levels of retinal illuminance (2.0-mm diameter artificial pupil). The short horizontal lines denote the modulations at which the frequencies in question were perceived or not perceived. (These are the average results of three measurements.) For 900 Td they are replaced by dots. No difference was found between the modulation transfer function for 900 Td and that for 5,900 Td. [*Results of van Nes* (1968).]

measurements in time shown in Fig. 19 though the curves are inverted because van Nes plotted his results with increasing contrast going up and de Lange, down. The extent to which such measurements can be utilized to predict the response of the visual system to other kinds of visual-acuity targets has been tested only imperfectly. Campbell and Robson (1968) examined sine wave, sawtooth, rectangular and square-wave gratings and found fair agreement between measurement at high spatial frequencies and that predicted if the visibility depended only on the amplitude of the fundamental Fourier component of the waveform. Simple spatial models of the human visual system based on receptive fields with excitatory centers and inhibitory surrounds such as suggested say by the neutral unit of Bèkèsy (1967) and in fact by the organization of mammalian retinal ganglion cells (Enroth-Cugell and Robson, 1966; Hubel and Wiesel, 1960) allows one to take the long jump from the threshold for sine-wave gratings to the perception of visual "illusions" of the type known as Mach bands (Ratliff, 1965). The transition is fraught with implicit assumptions of physiological linearity which appear to be unrealistic, as Westheimer (1965) has emphasized, though experiments of Alpern et al. (1970) suggest that the spatial organization of the human eye may be more linear than had been supposed. An alternative view (Campbell and Robson, 1968) is that the eye analyzes the retinal image in the spatial-frequency domain in much the same way that the ear analyzes the auditory stimulus in the time domain. In this view the modulation transfer function is the envelope of large numbers of very narrow-band curves, each tuned to different spatial frequencies. How successful pushing Fourier theory to this extreme will be in finding satisfactory explanations for the large variety of phenomena of spatial vision remains to be shown.

REFERENCES

Abrahamson, E. W., and S. E. Ostroy (1967): The Photochemical and Macromolecular Aspects of Vision, *Prog. Biophys. Molec. Biol.*, vol. 17, pp. 179–315.

Aguilar, M., and W. S. Stiles (1954): Saturation of the Rod Mechanism of the Retina at High Levels of Stimulation, *Opt. Acta*, vol. 1, pp. 59–65.

Alpern, M. (1958): Variability of Accommodation during Steady Fixation at Various Levels of Illuminance, *J. Opt. Soc. Am.*, vol. 48, pp. 193–197.

——— (1962*a*): pp. 1–187 in H. Davson (ed.), "The Eye," vol. 3, Academic, New York.

——— (1962*b*): ibid., pp. 189–229.

——— (1968): A Note on Visual Latency, *Psychol. Rev.*, vol. 75, pp. 260–264.

——— and F. W. Campbell (1962): The Spectral Sensitivity of the Consensual Light Reflex, *J. Physiol.*, vol. 164, pp. 478–507.

——— and H. David (1958): Effects of Illuminance Quantity on Accommodation of the Eyes, *Ind. Med. Surg.*, vol. 27, pp. 551–555.

——— and P. Ellen (1956): A Quentitative Analysis of the Horizontal Movements of the Eye in the Experiments of Johannes Mueller, *Am. J. Ophthalmol.*, vol. 42, no. 2, PC-2, pp. 289–303.

———, H. F. Falls, and G. B. Lee (1960): The Enigma of Typical Total Monochromacy, *Am. J. Ophthalmol.*, vol. 50, pp. 996–1012.

———, G. B. Lee, and B. E. Spivey (1965): π_1 Cone Monochromatism, *Arch Ophthalmol.*, vol. 74, pp. 334–337.

———, G. L. Mason, and R. E. Jardinico (1961): Vergence and Accommodation, V: Pupil Size Changes Associated with Changes in Accommodative Vergence, *Am. J. Ophthalmol.*, vol. 52, pp. 762–767.

———, J. Mindel, and S. Torii (1968): Are There Two Types of Deuteranopes?, *J. Physiol.*, vol. 199, pp. 443–456.

———, W. A. Rushton, and S. Torii (1970): The Size of Rod Signals, *J. Physiol.*, vol. 206, pp. 193–208.

——— and S. Torii (1968*a*): Prereceptor Colour Vision Distortions in Protanomalous Trichromacy, *J. Physiol.*, vol. 198, pp. 549–560.

——— and ——— (1968*c*): The Luminosity Curve of the Protanomalous Fovea, *J. Gen. Physiol.*, vol. 52, 717–737.

——— and ——— (1968*c*): The Luminosity Curve of the Deuteranomalous Fovea, *J. Gen. Physiol.*, vol. 52, pp. 738–749.

——— and J. R. Wolter (1956): The Relation of Horizontal Saccadic and Vergence Movements, *Arch. Ophthalmol.*, vol. 56, pp. 685–690.

Ames, A., Jr., and C. A. Proctor (1921): Dioptrics of the Eye, *J. Opt. Soc. Am.*, vol. 5, pp. 22–84.

Arnulf, A., and O. Dupuy (1960): La Transmission des contrastes par le système optique de l'oeil et les seuils de contrastes rétiniens, *C.R.*, vol. 250, pp. 2757–2759.

Baker, H. D., and W. A. Rushton (1965): The Red-sensitive Pigment in Normal Cones, *J. Physiol.*, vol. 176, pp. 56–72.

Barr, L., and M. Alpern (1963): Photosensitivity of the Frog Iris, *J. Gen. Physiol.*, vol. 46, pp. 1249–1265.

Bedford, R. E., and G. Wyszecki (1957): Axial Chromatic Aberration of the Human Eye, *J. Opt. Soc. Am.*, vol. 47, pp. 564–565.

Bèkèsy, G. von (1967): "Sensory Inhibition," Princeton University Press, Princeton, N.J.
Blackwell, H. R., and O. M. Blackwell (1961): Rod and Cone Receptor Mechanisms in Typical and Atypical Congenital Achromatopsia, *Vision Res.*, vol. 1, pp. 62–107.
Borish, I. M. (1949): "Clinical Refraction," Professional Press, Chicago.
Bouman, M. A., and H. A. van der Velden (1947): The Two-Quanta Explanation of the Dependence of the Threshold Values and Visual Acuity on the Visual Angle and the Time of Observation, *J. Opt. Soc. Am.*, vol. 37, pp. 908–919.
Bridges, C. D. B. (1967): Biochemistry of Visual Processes, pp. 31–78 in M. Florkin and E. H. Stotz (eds.), "Comprehensive Biochemistry," vol. 27, Elsevier, Amsterdam.
Brindley, G. S. (1953): The Effects on Colour Vision of Adaptation to Very Bright Lights, *J. Physiol.*, vol. 122, pp. 332–350.
——— (1954): The Summation Areas of Human Colour-receptive Mechanisms at Increment Threshold, *J. Physiol.*, vol. 124, pp. 400–408.
——— (1960): "Physiology of the Retina and Visual Pathway," Arnold, London.
———, J. J. du Croz, and W. A. H. Rushton (1966): The Flicker Fusion Frequency of the Blue-sensitive Mechanism of Colour Vision, *J. Physiol.*, vol. 183, pp. 497–500.
Brown, P. K., and G. Wald (1964): Visual Pigments in Single Rods and Cones of the Human Retina: Direct Measurements Reveal Mechanisms of Human Night and Color Vision, *Science*, vol. 144, pp. 45–52.
Byram, G. M. (1944): The Physical and Photochemical Basis of Visual Resolving Power, II: Visual Acuity and the Photochemistry of the Retina, *J. Opt. Soc. Am.*, vol. 34, pp. 718–738.
Campbell, F. W. (1954): Accommodation Reflex, *Br. J. Orthop.*, vol. 11, pp. 13–17.
——— (1957): The Depth of Field of the Human Eye, *Opt. Acta,* vol. 4, pp. 157–164.
——— (1958): A Retinal Acuity Direction Effect, *J. Physiol.*, vol. 144, pp. 25P–26P.
——— (1959): The Accommodation Response of the Human Eye, *Br. J. Physiol. Opt.*, vol. 16, pp. 188–203.
——— and D. G. Green (1965*a*): Monocular versus Binocular Visual Acuity, *Nature,* vol. 208, pp. 91–192.
——— and ——— (1965*b*): Optical and Retinal Factors Affecting Visual Resolution, *J. Physiol.*, vol. 181, pp. 576–593.
——— and A. H. Gregory (1960): Effect of Size of Pupil on Visual Acuity, *Nature,* vol. 187, pp. 1121–1123.
——— and R. W. Gubisch (1966): Optical Quality of the Human Eye, *J. Physiol.*, vol. 186, pp. 558–578.
———, J. J. Kulikowski, and J. Levinson (1966): The Effect of Orientation on the Visual Resolution of Gratings, *J. Physiol.*, vol. 187, pp. 427–436.
——— and J. G. Robson (1963): Application of Fourier Analysis to the Visibility of Gratings, *J. Physiol.*, vol. 197, pp. 551–556.
——— and G. Westheimer (1960): Dynamics of the Accommodation Responses of the Human Eye, *J. Physiol.*, vol. 151, pp. 285–295.
Childress, D. S., and R. W. Jones (1967): Mechanics of Horizontal Movement of the Human Eye, *J. Physiol.*, vol., 188, pp. 273–284.
Crawford, B. H. (1936): The Dependence of Pupil Size upon External Light Stimulus under Static and Variable Conditions, *Proc. R. Soc. Lond.*, ser. B, vol. 121, pp. 376–395.
——— (1949): The Scotopic Visibility Function, *Proc. Phys. Soc. Lond.*, sec. B, vol. 62, pp. 321–334.
Crescitelli, F., and H. J. A. Dartnall (1953): Human Visual Purple, *Nature,* vol. 172, pp. 195–197.
Dallos, P. J., and R. W. Jones (1963): Learning Behavior of the Eye Fixation Control System, *IEEE Trans. Autom. Control.*, vol. AC-8, pp. 212–227.
Dartnall, H. J. A. (1962): The Photobiology of Visual Processes, pp. 321–533 in H. Davson (ed.), "The Eye," vol. 2, Academic, New York.
De Palma, J. J., and E. M. Lowry (1962): Sine-Wave Response of the Visual System, II: Sine-Wave and Square-Wave Contrast Sensitivity, *J. Opt. Soc. Am.*, vol. 52, pp. 328–335.
Donders, F. C. (1864): "On the Anomalies of Accommodation and Refraction of the Eye with a Preliminary Essay on Physiological Dioptrics," New Sydenham Society, London.
Donner, K. O., and W. A. H. Rushton (1959): Rod-Cone Interaction in the Frog's Retina Analysed by the Stiles-Crawford Effect and by Dark Adaptation, *J. Physiol.*, vol. 149, pp. 303–317.

Duane, A. (1912): Normal Values of the Accommodation at All Ages, *Trans. Sec. Ophthalmol. AMA*, vol. 53, pp. 383–391.
Emsley, H. H. (1952): "Visual Optics," 5th ed., Hatton Press, London.
Enoch, J. M. (1963): Optical Properties of the Retinal Receptors, *J. Opt. Soc. Am.*, vol. 53, pp. 71–85.
——— and W. S. Stiles (1961): The Colour Change of Monochromatic Light with Retinal Angle of Incidence, *Opt. Acta*, vol. 8, pp. 329–358.
Enroth-Cugell, C., and J. G. Robson (1966): The Contrast Sensitivity of Retinal Ganglion Cells, *J. Physiol.*, vol. 187, pp. 517–522.
Falls, H. F., J. R. Wolter, and M. Alpern (1965): Typical Total Monochromacy: A Histological and Psychophysical Study, *Arch. Ophthalmol.*, vol. 74, pp. 610–616.
Fincham, E. F. (1937): The Mechanism of Accommodation, *Br. J. Ophthalmol. Suppl.*, vol. 8, pp. 1–80
——— (1951): The Accommodation Reflex and Its Stimulus, *Br. J. Ophthalmol.*, vol. 35, pp. 381–393.
——— (1953): Defects of the Colour-Sense Mechanism as Indicated by the Accommodation Reflex, *J. Physiol.*, vol. 121, pp. 570–580.
Flamant, F. (1948): Variation du diamètre de la pupille de l'oeil en fonction de la brillance, *Rev. Opt.*, vol. 27, pp. 751–758.
——— (1955): Etude de la répartition de lumière dans l'image rétinienne d'une fente, *Rev. Opt.*, vol. 34, pp. 433–459.
——— and W. S. Stiles (1948): The Directional and Spectral Sensitivities of the Retinal Rods to Adapting Fields of Different Wave-Lengths, *J. Physiol.*, vol. 107, pp. 187–202.
Flom, M. C., G. G. Heath, and E. Takahashi (1963): Contour Interaction and Visual Resolution: Contralateral Effects, *Science*, vol. 142, pp. 979–980.
Fry, G. A. (1956): "Blur of the Retinal Image," Ohio State University Press, Columbus.
——— (1959): The Image-forming Mechanism of the Eye, pp. 647–670 in H. Magoun (ed.), "Handbook of Physiology," sec. I, "Neurophysiology: A Critical, Comprehensive Presentation of Physiological Knowledge and Concepts," vol. I, American Physiological Society, Washington, D.C.
——— and M. Alpern (1953): The Effect of a Peripheral Glare Source upon the Apparent Brightness of an Object, *J. Opt. Soc. Am.*, vol. 43, pp. 189–195.
Fuortes, M. G. F., R. D. Gunkel, and W. A. H. Rushton (1961): Increment Thresholds in a Subject Deficient in Cone Vision, *J. Physiol.*, vol. 156, pp. 179–192.
——— and A. L. Hodgkin (1964): Changes in Time Scale and Sensitivity in the Ommatidia of Limulus, *J. Physiol.*, vol. 172, pp. 239–263.
Gibson, K. S. (1940): Spectral Luminosity Factors, *J. Opt. Soc. Am.*, vol. 30, pp. 51–61.
Green, D. G. (1967): Visual Resolution when Light Enters the Eye through Different Parts of the Pupil, *J. Physiol.*, vol. 190, pp. 583–593.
——— (1968): The Contrast Sensitivity of the Colour Mechanisms of the Human Eye, *J. Physiol.*, vol. 196, pp. 415–429.
——— (1970): Regional Variations in the Visual Acuity for Interference Fringes on the Retina, *J. Physiol.*, vol. 207, pp. 351–356.
——— and F. W. Campbell (1965): Effect of Focus on the Visual Response to a Sinusoidally Modulated Spatial Stimulus, *J. Opt. Soc. Am.*, vol. 55, pp. 1154–1157.
Guyton, A. C. (1956): "Textbook of Medical Physiology," p. 686, Saunders, Philadelphia.
Hamasaki, D., J. Ong, and E. Marg (1956): The Amplitude of Accommodation in Presbyopia, *Am. J. Optom.*, vol. 33, pp. 3–14.
Hecht, S. (1936): Intensity Discrimination and Its Relation to the Adaptation of the Eye. *J. Physiol.*, vol. 86, pp. 15–21.
———, C. Haig, and A. M. Chase (1937): The Influence of Light Adaptation on Subsequent Dark Adaptation of the Eye, *J. Gen. Physiol.*, vol. 20, pp. 831–850.
———, S. Shlaer, and M. H. Pirenne (1942): Energy, Quanta, and Vision, *J. Gen. Physiol.*, vol. 25, pp. 819–840.
Helmholtz, H. V. (1924): "Helmholtz's Treatise on Physiological Optics," vol. 1, trans. from 3d German ed., J. P. C. Southall (ed.), Optical Society of America, Rochester, N.Y.
Hendley, C. D. (1948): The Relation between Visual Acuity and Brightness Discrimination, *J. Gen. Physiol.*, vol. 31, pp. 433–457.
Hess, E. H. (1965): Attitude and Pupil Size, *Sci. Am.*, vol. 212, pp. 41–54, April.
Holladay, L. L. (1926a): The Fundamentals of Glare and Visibility, *J. Opt. Soc. Am.*, vol. 12, pp. 271–319.

——— (1926*b*): The Fundamentals of Glare and Visibility, *J. Opt. Soc. Am.*, vol. 12, p. 492 (abstract).

——— (1927): Action of a Light-Source in the Field of View in Lowering Visibility, *J. Opt. Soc. Am.*, vol. 14, pp. 1–15.

Hubel, D. H., and T. N. Wiesel (1960): Receptive Fields of Optic Nerve Fibres in the Spider Monkey, *J. Physiol.*, vol. 154, pp. 572–588.

Ivanoff, A. (1953): "Les Aberrations de l'oeil: leur rôle dans l'accommodation," Editions de la Revue d'Optique Théorique et Instrumentale, Paris.

Jampel, R. S. (1959): Representation of the Near-Response on the Cerebral Cortex of the Macaque, *Am. J. Ophthalmol.*, vol. 48, no. 5, pt. 2, pp. 573–582.

Kahneman, D. (1964): Temporal Summation in an Acuity Task at Different Energy Levels: A Study of the Determinants of Summation, *Vision Res.*, vol. 4, pp. 557–566.

———, J. Beatty, and I. Pollack (1967): Perceptual Deficit during a Mental Task, *Science*, vol. 157, pp. 218–219.

Kelly, D. H. (1961*a*): Visual Responses to Time-dependent Stimuli, I: Amplitude Sensitivity Measurements, *J. Opt. Soc. Am.*, vol. 51, pp. 422–429.

——— (1961*b*): Visual Responses to Time-dependent Stimuli, II: Single-Channel Model of the Photopic Visual System, *J. Opt. Soc. Am.*, vol. 51, pp. 747–754.

——— (1962): Visual Responses to Time-dependent Stimuli, III: Individual Variations, *J. Opt. Soc. Am.*, vol. 52, pp. 89–95.

Koomen, M., R. Scolnik, and R. Tousey (1951): A Study of Night Myopia, *J. Opt. Soc. Am.*, vol. 41, pp. 80–90.

———, R. Tousey, and R. Scolnik (1949): The Spherical Aberration of the Eye, *J. Opt. Soc. Am.*, vol. 39, pp. 370–376.

Krauskopf, J. (1962): Light Distribution in Human Retinal Images, *J. Opt. Soc. Am.*, vol. 52, pp. 1046–1050.

Lancaster, W. B. (1952): "Refraction Correlated with Optics and Motility Limited to Heterophoria," Charles C Thomas, Springfield, Ill.

Landis, C. (1954): Determinants of the Critical Flicker-Fusion Threshold, *Physiol. Rev.*, vol. 34, pp. 259–286.

Lange, H. de (1958*a*): Research into the Dynamic Nature of the Human Fovea: Cortex Systems with Intermittent and Modulated Light, I: Attenuation Characteristics with White and Colored Light, *J. Opt. Soc. Am.*, vol. 48, pp. 777–789.

——— (1958*b*): Research into the Dynamic Nature of the Human Fovea: Cortex Systems with Intermittent and Modulated Light, II: Phase Shift in Brightness and Delay in Color Perception, *J. Opt. Soc. Am.*, vol. 48, pp. 784–789.

Laurance, L. (1926): "Visual Optics and Sight Testing," 3d ed., p. 452, School of Optics, London.

Le Grand, Y. (1937*a*): La Formation des images rétiniennes: sur un mode de vision éliminant les defauts optiques de l'oeil, *2e Reun. Inst. Opt. Paris.*

——— (1937*b*): Recherches sur la diffusion de la lumière dans l'oeil humain, *Rev. Opt.*, vol. 16, pp. 201–214, 241–266.

——— (1957): "Light, Colour, and Vision," trans. by R. W. G. Hunt, J. W. T. Walsh, and F. R. W. Hunt, Wiley, New York.

——— (1968): "Form and Space Vision," rev. ed., trans. by M. Millodot and G. G. Heath, Indiana University Press, Bloomington.

Liebman, P. A. (1962): In Situ Microspectrophotometric Studies on the Pigments of Single Retinal Rods, *Biophys. J.*, vol. 2, pp. 161–178.

——— and G. Entine (1964): Sensitive Low-Light Level Microspectrophotometer: Detection of Photosensitive Pigments of Retinal Cones, *J. Opt. Soc. Am.*, vol. 54, pp. 1451–1459.

Lit, A. (1949): The Magnitude of the Pulfrich Stereophenomenon as a Function of Binocular Difference of Intensity at Various Levels of Illumination, *Am. J. Psychol.*, vol. 62, pp. 159–181.

Lowenstein, O., and I. E. Loewenfeld (1962): The Pupil, pp. 231–267 in H. Davson (ed.), "The Eye," vol. 3, Academic, New York.

Marks, W. B. (1965): Visual Pigments of Single Goldfish Cones, *J. Physiol.*, vol. 178, pp. 14–32.

———, W. H. Dobelle, and E. F. MacNichol, Jr. (1964): Visual Pigments of Single Primate Cones, *Science*, vol. 143, pp. 1181–1183.

Miller, J. W., and E. Ludvigh (1962): The Effect of Relative Motion on Visual Acuity, *Surv. Ophthalmol.*, vol. 7, pp. 83–116.

Mitchell, D. E., R. D. Freeman, and G. Westheimer (1967): Effect of Orientation on the Modulation Sensitivity for Interference Fringes on the Retina, *J. Opt. Soc. Am.*, vol. 57, pp. 246–249.

Moon, P., and D. E. Spencer (1944): On the Stiles-Crawford Effect, *J. Opt. Soc. Am.*, vol. 34, pp. 319–329.

Rashbass, C., and G. Westheimer (1961*a*): Independence of Conjugate and Disjunctive Eye Movements, *J. Physiol.*, vol. 159, pp. 361–364.

——— and ——— (1961*b*): Disjunctive Eye Movements, *J. Physiol.*, vol. 159, pp. 339–360.

Ratliff, F. (1965): "Mach Bands: Quantitative Studies on Neural Networks in the Retina," Holden-Day, San Francisco.

Robinson, D. A. (1964): The Mechanics of Human Saccadic Eye Movement, *J. Physiol.*, vol. 174, pp. 245–264.

——— (1965): The Mechanics of Human Smooth Pursuit Eye Movement, *J. Physiol.*, vol. 180, pp. 569–591.

——— (1968): The Oculomotor Control System: A Review, *Proc. IEEE*, vol. 56, pp. 1032–1049.

Rohler, R., U. Miller, and M. Aberl (1969): Zur Messung der Modulationsübertragungsfunktion des lebenden menschlichen Auges im reflecktierten Licht, *Vision Res.*, vol. 9, pp. 407–428.

Rushton, W. A. H. (1956): The Difference Spectrum and the Photosensitivity of Rhodopsin in the Living Human Eye, *J. Physiol.*, vol. 134, pp. 11–29.

——— (1963*a*): A Cone Pigment in the Protanope, *J. Physiol.*, vol 168, pp. 345–359.

——— (1963*b*): The Density of Chlorolabe in the Foveal Cones of the Protanope, *J. Physiol.*, vol. 168, pp. 360–373.

——— (1963*c*): Cone Pigment Kinetics on the Protanope, *J. Physiol.*, vol. 168, pp. 374–388.

——— (1965*a*): A Foveal Pigment in the Deuteranope, *J. Physiol.*, vol. 176, pp. 24–37.

——— (1965*b*): Cone Pigment Kinetics in the Deuteranope, *J. Physiol.*, vol. 176, pp. 38–45.

——— (1965*c*): Stray Light and the Measurement of Mixed Pigments in the Retina, *J. Physiol.*, vol. 176, pp. 46–55.

——— (1965*d*): Visual Adaptation, *Proc. R. Soc. Lond.*, ser. B, vol. 162, pp. 20–46.

——— (1968): Rod/Cone Rivalry in Pigment Regeneration, *J. Physiol.*, vol. 198, pp. 219–236.

——— and R. W. Gubisch (1966): Glare: Its Measurement by Cone Thresholds and by the Bleaching of Cone Pigments, *J. Opt. Soc. Am.*, vol. 56, pp. 104–110.

——— and G. Westheimer (1962): The Effect upon the Rod Threshold of Bleaching Neighboring Rods, *J. Physiol.*, vol. 164, pp. 318–329.

Said, F. S., and R. A. Weale (1959): The Variation with Age of the Spectral Transmissivity of the Living Human Crystalline Lens, *Gerontologia*, vol. 3, pp. 213–231.

Schade, O. H., Sr. (1956): Optical and Photoelectric Analog of the Eye, *J. Opt. Soc. Am.*, vol. 46, pp. 721–739.

Schouten, J. F., and L. S. Ornstein (1939): Measurements on Direct and Indirect Adaptation by Means of a Binocular Method, *J. Opt. Soc. Am.*, vol. 29, pp. 168–182.

Selwyn, E. W. H. (1948): The Photographic and Visual Resolving Power of Lenses, I: Visual Resolving Power, *Photogr. J.*, vol. 88B, pp. 6–12.

Shlaer, S. (1937): The Relation between Acuity and Illumination, *J. Gen. Physiol.*, vol. 21, pp. 165–188.

Stiles, W. S. (1929): Effect of Glare on the Brightness Difference Threshold, *Proc. R. Soc. Lond.*, ser. B, vol. 104, pp. 322–351.

——— (1937): The Luminous Efficiency of Monochromatic Rays Entering the Eye Pupil at Different Points and a New Color Effect, *Proc. R. Soc. Lond.*, ser. B, vol. 123, pp. 90–118.

——— (1939): The Directional Sensitivity of the Retina and the Spectral Sensitivities of the Rods and Cones, *Proc. R. Soc. Lond.*, ser. B, vol. 127, pp. 64–105.

——— (1949): Increment Thresholds and the Mechanisms of Colour Vision, *Doc Ophthalmol.*, vol. 3, pp. 138–163.

——— and B. H. Crawford (1933): The Luminous Efficiency of Rays Entering the Eye Pupil at Different Points, *Proc. R. Soc. Lond.*, ser. B, vol. 112, pp. 428–450.

ten Doesschate, J., and M. Alpern (1967*a*): Effect of Photo-excitation of the Two Retinas on Pupil Size, *J. Neurophysiol.*, vol. 30, pp. 562–576.

——— and ——— (1967*b*): Influence of Asymmetrical Photo-excitation of the Two Retinas on Pupil Size, *J. Neurophysiol.*, vol. 30, pp. 577–585.

Thomas, J. G. (1969): The Dynamics of Small Saccadic Eye Movements, *J. Physiol.*, vol. 200, pp. 109–127.

Tomita, T. (1968): Electrical Response of Single Photoreceptors, *Proc. IEEE,* vol. 56, pp. 1015–1023.
Toraldo di Francia, G. (1949): The Radiation Pattern of Retinal Receptors, *Proc. Phys. Soc. Lond.,* sec. B, vol. 62, 461–462.
Tscherning, M. (1904): "Physiologic Optics: Dioptrics of the Eye, Functions of the Retina, Ocular Movements and Binocular Vision," 2d ed., trans. by C. Weiland, Keystone, Philadelphia.
van Nes, F. L. (1968): Experimental Studies in Spatiotemporal Contrast Transfer by the Human Eye, Thesis, Utrecht University, The Netherlands.
Volkmann, F. C. (1962): Vision during Voluntary Saccadic Eye Movements, *J. Opt. Soc. Am.,* vol. 52, pp. 571–578.
Vos, J. J. (1963): On Mechanisms of Glare, Thesis, Utrecht University, The Netherlands.
Wald, G. (1941): Vision: Photochemistry, pp. 1658–1667 in O. Glasser (ed.), "Medical Physics," Year Book, Chicago.
——— (1967): Blue-Blindness in the Normal Fovea, *J. Opt. Soc. Am.,* vol. 57, pp. 1289–1301.
——— (1968): Molecular Basis of Visual Excitation, *Science,* vol. 162, pp. 230–239.
——— and D. R. Griffin (1947): The Change in Refractive Power of the Human Eye in Dim and Bright Light, *J. Opt. Soc. Am.,* vol. 37, pp. 321–336.
Walls, G. L. (1962): The Evolutionary History of Eye Movements, *Vision Res.,* vol. 2, pp. 69–80.
Weale, R. A. (1953): Cone-Monochromatism, *J. Physiol.,* vol. 121, pp. 548–569.
——— (1965): High Light Intensities and Photo-chemical Reactions of Human Visual Pigments in Situ, *Cold Springs Harb. Symp. Quant. Biol.,* vol. 30, pp. 335–343.
——— (1966): Polarized Light and the Human Fundus Oculi, *J. Physiol.,* vol. 186, pp. 175–186.
——— (1968): Photochemistry of the Human Central Fovea, *Nature,* vol. 218, pp. 238–240.
Westheimer, G. (1954a): Mechanisms of Saccadic Eye Movements, *Arch. Ophthalmol.,* vol. 52, pp. 710–724.
——— (1954*b*): Eye Movement Responses to a Horizontally Moving Visual Stimulus, *Arch. Ophthalmol.,* vol. 52, pp. 932–943.
——— (1957): Accommodation Measurements in Empty Visual Fields, *J. Opt. Soc. Am.,* vol. 47, pp. 714–718.
——— (1960): Modulation Thresholds for Sinusoidal Light Distributions on the Retina, *J. Physiol.,* vol. 152, pp. 67–74.
——— (1963): Optical and Motor Factors in the Formation of the Retinal Image, *J. Opt. Soc. Am.,* vol. 53, pp. 86–93.
——— (1965): Visual Acuity, *Annu. Rev. Psychol.,* vol. 16, pp. 359–389.
——— and F. W. Campbell (1962): Light Distribution in the Image Formed by the Living Human Eye, *J. Opt. Soc. Am.,* vol. 52, pp. 1040–1045.
——— and A. M. Mitchell (1956): Eye Movement Responses to Convergence Stimuli, *Arch. Ophthalmol.,* vol. 55, pp. 848–856.
Wright, W. D. (1946): "Researches on Normal and Defective Colour Vision," Henry Kimpton, London.
Wyszecki, G., and W. S. Stiles (1967): "Color Science: Concepts and Methods, Quantitative Data and Formulas," pp. 213–214, Wiley, New York.
Yarbus, A. L. (1967): "Eye Movements and Vision," Plenum, New York.
Young, L. R., and L. Stark (1963): Variable Feedback Experiments Testing Sampled Data Model for Eye Tracking Movements, *IEEE Trans. Human Factors Electron.,* vol. HFE-4, pp. 38–51.
Young, T. (1801): On the Mechanism of the Eye, *Phil. Trans. R. Soc. Lond.,* pp. 23–88.

Section 13

Fiber Optics

WALTER P. SIEGMUND

Technical Manager, Fiber-Optics Division, American Optical Corporation, Southbridge, Mass.

(Numbers refer to paragraphs)

INTRODUCTION

1. Fiber optics is the branch of optics dealing with the transmission of light through small filamentary optical elements or fibers and generally includes the technology of their manufacture and application. The transmission of light through fibers is based on the phenomenon of total internal reflection, which occurs when light is obliquely incident on an interface between two media of different refractive index at an angle greater than the critical angle. The latter is a function of the refractive indexes, becoming smaller as the ratio of refractive indexes increases. In the absence of disturbing influences, such as scattering defects along the interface, total internal reflection is a highly efficient phenomenon, and thousands of such reflections can occur within a fiber with little loss of light. Indeed only because this is true does fiber optics exist as a practical technology.

2. Background The use of total internal reflection as a means of conveying light was well known to the designers of illuminated fountains in the nineteenth century and was the subject of an occasional scientific demonstration (Tyndall, 1854). The application of this phenomenon to small fibers was suggested in a few scientific papers in the early 1900s. The possibility of conducting images by means of flexible bundles of fibers was first proposed in two nearly simultaneous patents in the late 1920s (Baird, 1926; Hansell, 1927). A practical form of fiber optics did not emerge, however, until the concept of a transparent optical coating was proposed in 1953 (Heel, 1953). The use of such a coating provided a protected interface for total reflection and prevented leakage of light between adjacent fibers. In the absence of such a coating most fibers are so inefficient that they are of little practical use.

Fiber-optic products may be divided into two basic categories, those which are primarily light carriers and are nonimaging and those which are also imaging. In both cases, at least ideally, light is carried by individual fibers independently of neighboring fibers. Thus only the arrangement of the fiber ends distinguishes these two categories.

3. Nonimaging Products In nonimaging products the fiber ends are generally arranged in an overall area geometry (circular, rectangular, linear) to conform to a source or receiver configuration, and individual fibers within these areas are more or less randomly distributed. In some cases, e.g., where a fiber bundle is divided in two or more branches, homogeneous mixing of the fibers may be required, and this can be achieved by special techniques. Since no images are formed, image-resolving power has no significance and the fiber size is chosen largely on mechanical considerations, principally the degree of flexibility required. Typical fiber diameters range from 25 to 75 μm.

Often considered along with fiber optics, although not strictly a form thereof, are clad-rod light pipes. They can be made with the same techniques used to draw flexible optical fibers but are merely of larger diameter and thus more or less rigid depending on their diameter. They work on the same principle of total internal reflection from a protected interface and thus are very efficient light carriers. Moreover, they can often be bent by heating to conform to a predetermined path so long as the curvature is not excessive, thereby causing significant light loss at the bends.

4. Imaging Products The distinguishing characteristic of image-transmitting fiber-optic products is that the fibers are assembled so that their ends form identical arrays. Fiber bundles of this type are sometimes called *coherent bundles* to indicate that they transmit coherent, recognizable images. The name does not refer to the coherency of the transmitted radiation in the parlance of modern optics.

Imaging fiber-optic products can be divided into two groups, flexible and rigid. In flexible products, the fibers are bonded together at each end while along the major portion of their length they are loose and free to flex and twist. Only the ends need be coordinated to provide the imaging property, and the loose fibers in between can be jumbled at random (so long as they remain unbroken) without affecting the image. This is one of the unique and functionally important characteristics of fiber optics.

In rigid imaging products, unlike the flexible variety, the fibers are generally aligned at all cross sections of the fiber bundle. In the most common form of rigid bundle the fibers are fused together by heating them to the point where their claddings adhere, a process made possible by the development of glass-coated fibers. Typical forms of such fused bundles include image conduits, fiber faceplates, fiber cones, and image rotators.

5. Materials and Processes Optical fibers are made from two basic kinds of materials, glass and plastic. Glass fibers are used in most imaging products and all fused products as well as many nonimaging light guides and special assemblies. They exhibit higher light transmission, particularly in the near-infrared spectrum, provide a wider selection of refractive indexes, and have better thermal stability than plastic fibers. On the other hand, plastic fibers are less brittle and often less expensive to use, in part because they can be more readily end-finished.

All optical fibers consist of at least two components, a core and a coating, or cladding, as it is often called. The refractive index of the core must be higher than that of the coating for total internal reflection to take place at the core-cladding interface.

6. Glass Fibers In glass fibers it is general practice to use glass for both the core and the coating. The main criteria for selecting the core glass are its refractive index, transmittance, and expansion coefficient. For long fibers transmittance is of prime importance, and a medium refractive index is usually sufficient. In this case the expansion coefficient is important only in relation to that of the coating, which should have the lower coefficient to produce a strong fiber. For short-fiber applications, such as faceplates, transmission efficiency is governed more by the relative refractive indexes of the core and coating than by the glass transmittance. This calls for a core glass with a high refractive index but with less emphasis on the specific transmittance of the glass itself. In this case the choice of expansion coefficient will be governed by the application, e.g., sealing the faceplate to a cathode-ray-tube envelope with a given expansion coefficient.

All processes for making glass fibers involve heating the glass to soften it and then drawing the fiber away from the hot glass under tension. As the fiber is withdrawn, it takes on a very smooth fire-polished surface, which is essential to its optical performance. The two main methods for drawing glass-coated glass fibers are the rod-and-tube method and the double-crucible method. In the rod-and-tube method a polished rod of the core glass is inserted into a concentric tube of the cladding glass and the combination is slowly fed down into a cylindrical electric furnace which softens the glasses. The softened end of the glass setup is baited, and a fiber is withdrawn. Generally this is then wound up on a spinning drum. The fiber size is controlled by controlling the rate of feed into the furnace and the speed of the take-up drum. A size variation of less than 1 percent is readily maintained. The ratio of coating thickness to the fiber diameter is determined by the thickness of tubing used relative to the combined cross-sectional diameter of the core rod and tubing.

In the double-crucible method a special glass-melting furnace containing two concentric crucibles (usually made of platinum) with concentric orifices at the bottom are used to melt the core and cladding glasses. (The inner crucible contains the core glass.) As the glass melts in the crucibles, it flows from the orifices and can be drawn out into a fiber and again wound on a spinning drum. In this case the fiber size is determined by the orifice diameter, the glass viscosity (temperature), and the drum speed. The coating thickness is determined by the orifice design and the relative glass viscosities.

Both methods are well adapted to drawing single round fibers of a wide range of size (up to several millimeters in diameter, i.e., clad rods). Moreover, the rod-and-tube method can be used to draw fibers or rods of other than round configuration by using a core rod of the desired shape, e.g., square or hexagonal. The relative viscosities of the core and coating glasses must be appropriately matched in this instance; i.e., that of the core glass must be somewhat higher than that of the coating at the drawing temperature.

In another common form of fiber drawing, multiple fiber drawing, a number of clad rods (or large fibers) are clustered together in a selected array (round, square, hexagon, etc.) and redrawn through the same type of furnace used in rod-and-tube drawing. The resulting fiber consists of a number of distinct optical channels each with its own cladding for optical insulation but fused to its neighbors in the desired geometrical array. These are known as multiple fibers or multifibers. In this manner larger numbers of very small fibers can be precisely aligned and assembled into large-cross-sectional arrays. They are bonded together by fusing the multiple fibers under heat and pressure so that the glass coatings weld (fuse) together to form a rigid and often gastight block. Such blocks may contain millions of small accurately aligned fibers capable of transmitting images from one face to the other, as discussed in Par. 25.

7. Plastic Fibers In plastic fibers, two combinations of plastics are commonly used: (1) polystyrene for the core and polymethyl methacrylate for the coating and (2) polymethyl methacrylate for the core and a low-index fluorocarbon compound for the coating. A bundle of such fibers is often encased in a plastic sheath extruded directly over the fibers, thus making an integral flexible light guide. This material is sold in long continuous lengths like electrical cable.

Plastic fibers are made either by extrusion or by a modification of the rod-and-tube method. In extrusion it is possible to use a multiple-extrusion die and obtain several fibers simultaneously. For polymethyl methacrylate (Lucite) fibers the low-refractive-index coating is applied as a separate dipping operation.

In the modified rod-and-tube process, polystyrene is carefully prepared to remove moisture and cast into a tube of polymethyl methacrylate; the combination is drawn through a furnace into a coated fiber essentially as glass fibers are. "Fused" multiple fibers of plastic are not normally found in the trade.

8. Glossary

Acceptance Angle. The maximum incident angle measured with respect to the fiber axis at which the fiber will transport light by internal reflection.

Blemish. Localized areas in a fiber bundle which have reduced light transmission, e.g., due to broken or defective fibers.

Boule. A fused assembly of multifibers from which faceplates are cut.

Chicken Wire. A form of blemish which appears as a grid of dark lines along multifiber boundaries.

Cladding. The coating of glass or plastic surrounding the fiber core and having a lower index of refraction.

Coherent Bundles. Fiber assemblies in which the fibers are arranged in the same relationship at both ends of the assembly.

Collimated Transmittance. The transmission of a fiber-optic component with collimated input light.

Core. The inner portion of the fiber, having the higher index of refraction.

Critical Angle. The smallest angle of incidence in the medium of greater refractive index for which light is totally reflected.

Cross Talk. Light leakage from one fiber to the next by frustrated total reflection, i.e., inadequate cladding thickness.

Decollimation. Spreading of light by a fiber due to various internal and end effects.

Devitrification. The change from the vitreous (glassy) state to the crystalline state which can occur in heated glasses.

Dynamic Scanning. The technique of rapidly moving the fiber bundle during image formation to suppress the fiber pattern.

Edge Response. The measure of the ability of a fiber bundle to form an image of a knife-edge target.

Endoscope. Optical instrument often containing a fiber bundle used to view internal body cavities.

Extramural Absorption (EMA). The use of a light-absorbing material either as a secondary coating or as interstitial fibers to suppress stray light in coherent fiber bundles.

Fiber Faceplate. A fused coherent array of multifibers for use as a window in a vacuum tube.

Fiber Optics. The branch of optics dealing with the transmission of light or image information through transparent fibers.

Fiber-Optic Bundles. Assemblies of optical fibers.

Fiberscope. A flexible viewing instrument containing a coherent fiber bundle.

Fresnel Reflection. The reflection of light from the entrance or exit end faces of fibers.

Image Conduit. Rigid multifiber rod capable of image transmission.

Image Dissector. Fiber assembly having a close-packed array at one end and an open or modified array at the other.

Image Intensifier. Electrooptical tube often containing fiber-optic faceplates capable of increasing the luminance.

Image Inverter. Fiber bundle having a 180° twist to rotate the image in an optical system.

Incoherency. Image distortion due to misalignment of adjacent fibers, i.e., shear distortion.

Insertion Loss. Light lost at the entrance to a fiber bundle due to Fresnel reflection, packing fraction, and limited numerical aperture.

Lambertian Transmittance. The transmission of a fiber-optic component with diffuse (lambertian) input light.

Light Guide. Noncoherent flexible bundle of optical fibers used for light transmission.

Modulation-Transfer Function (MTF). A measure of the ability of a coherent fiber bundle (or other optical device) to transmit images with various spatial frequency components.

Monofiber. A single optical fiber as distinguished from a multifiber.

Multifiber. A cluster of optical fibers fused together into a single strand.

Noncoherent Bundles. Assemblies in which the fibers have a random relationship at each end.

Numerical Aperture (NA). A measure of the light-gathering power of a fiber-optic component (or other optical device), usually equal to the sine of the acceptance angle.

Optical Fibers. Transparent fibers, usually glass, used in fiber optics.

Packing Fraction. The ratio of the useful fiber core area to total frontal area of a fiber-optic bundle.

Resolving Power. The ability of a coherent fiber bundle to transmit detail in an image, usually given in line pairs per millimeter.

Skew Rays. Rays traveling through an optical fiber which do not intersect the fiber axis.

Thermal-Expansion Coefficient. Measure of the rate of expansion with temperature, usually given with reference to degrees Celsius; of importance in sealing fiber-optic faceplates to image tubes.

Total Internal Reflection. Reflection of light at angles of incidence greater than the critical angle between two optical media of different refractive index.

OPTICAL PROPERTIES OF FIBERS†

9. Optical fibers are basically waveguides for electromagnetic radiation at optical frequencies. This aspect of optical fibers has been investigated both theoretically and experimentally (Snitzer, 1961; Snitzer and Osterberg, 1961). Although, in principle, the transmitted radiation always consists of a combination of discrete waveguide modes, distinct mode patterns can usually be observed only in very small fibers or under special illumination conditions, or both.

10. Total Internal Reflection When light is incident at the interface between two dielectric media of different refractive index, it will be totally reflected from the higher-index side for all angles of incidence greater than the critical angle ϕ_c given by

$$\phi_c = \sin^{-1} \frac{n_2}{n_1}$$

where n_1 and n_2 are the refractive indexes of the two media. In the absence of absorption in either medium and any interface defects, this reflection is theoretically perfect. Transmission measurements on actual optical fibers indicate that the reflection coefficient exceeds 0.9999, compared with only 0.95 for a polished-metal surface.

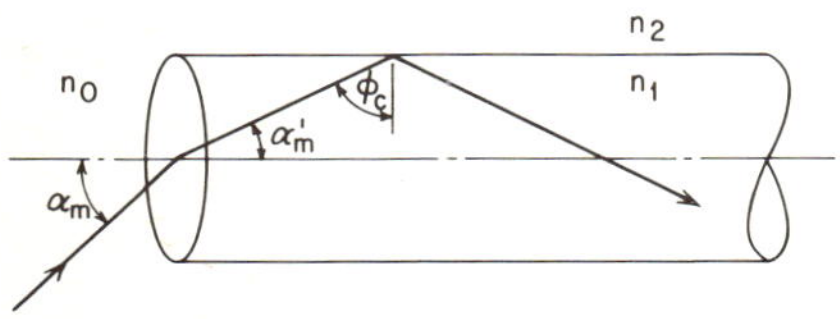

Fig. 1 Meridional ray entering the end of a straight fiber.

11. Numerical Aperture of a Fiber Take the case of a straight cylindrical fiber with a refractive index n_1 embedded in (or clad with) a medium of refractive index n_2, where $n_1 > n_2$. Let a ray of light which passes through the fiber axis (a meridional ray) be incident from a medium of refractive index n_0 onto the perpendicular end face of the fiber at an angle α. This ray will be refracted at the end face and will be totally reflected at the fiber wall if its angle of incidence is greater than the critical angle ϕ_c.

The maximum value of α for which the ray will be totally reflected can be derived from Snell's law and simple trigonometry. With reference to Fig. 1,

$$n_0 \sin \alpha_m = n_1 \sin \alpha_m' = n_1 \cos \phi_c = n_1 (1 - \sin^2 \phi_c)^{1/2}$$

Now $\sin \phi_c = n_2/n_1$. Hence,

$$\text{NA} = n_0 \sin \alpha_m = n_1 \left[1 - \left(\frac{n_2}{n_1}\right)^2\right]^{1/2} = (n_1^2 - n_2^2)^{1/2} \qquad (1)$$

By analogy with lens optics the term $n_0 \sin \alpha_m$ is called the numerical aperture (NA) of the fiber and, as in lens optics, is a measure of the light-gathering power. Strictly speaking, this expression applies only to meridional rays in an ideal fiber, and hence it is more appropriate to call this the nominal NA. Skew rays incident at angles larger than α can be conducted by fibers having a true circular cross section and, to some extent, by those with a perfect polygonal or elliptical cross section. The angle is generally called the acceptance angle of the fiber.

In practice, one finds that the limiting angle α_m is not as sharply defined in actual fibers as Eq. (1) implies. Diffraction, striae, and surface irregularities on the fiber walls all tend to decollimate the transmitted light and thereby reduce the effective NA. Inadequate insulation (cladding) between fibers can also reduce the NA, and only an actual measurement of the acceptance angle will provide a true indication of the numerical aperture. Moreover, the calculated value of the nominal NA is usually based on the refractive indexes of the starting materials from which the fiber is drawn and does not take into account any changes in

† Paragraphs 9 to 16 have been adapted from Walter P. Siegmund, pp. 3–14, chap. 1, of R. Kingslake (ed.), "Applied Optics and Optical Engineering," vol. 4, copyright 1967, Academic Press, Inc., New York, by permission of the publisher.

refractive index which may occur. Despite the uncertainties associated with this quantity, however, it is still common practice to specify the nominal NA of a fiber as one of its basic parameters. For convenience, this will be referred to simply as the NA.

The range of possible values of NA for fibers is limited only by the materials from which they can be drawn. For glass-clad glass-core fibers this range is considerable. By selecting various pairs of glasses it is possible to obtain almost any value of NA up to about 1.4. An NA greater than 1.0 can be realized, of course, only with an ambient medium of sufficiently high index n_0. This would be analogous to an immersion microscope objective. In principle, the NA can also be made arbitrarily small, but as a practical matter a low NA can be difficult to achieve owing to uncontrolled changes in refractive index during the drawing process.

Not all pairs of glasses can be arbitrarily selected, unfortunately, since one must consider other factors, including thermal and chemical compatibility, stability, and optical absorption. The choice of material may also be influenced by its availability in an appropriate form for fiber drawing. For the cladding it is usually desirable for the glass to be in the form of a tube, and ordinary commercial-grade glass tubing is commonly used. The somewhat higher optical absorption in such tubing glasses is generally of no consequence when used for the cladding. For the core glass the preferred form is a long fire-polished rod, and a few selected glasses are commercially available in this form. For other core glasses one usually has to settle for cast blocks which must be cut, ground, and polished into rods.

In selecting the type of core glass to be used, one is often confronted with a compromise between NA and transmittance. In general, the higher the core-glass refractive index the higher the NA but the lower the transmittance in the short-wavelength end of the spectrum. For applications requiring only short fibers, such as field flatteners or vacuum-tube faceplates, a higher NA combination can be used since light absorption is not very serious (except in the extreme violet and ultraviolet). For long-fiber applications, where a reasonably faithful rendition of the visible spectrum is desired, one must settle for a somewhat lower NA.

12. Behavior of Rays in a Straight Fiber The behavior of a meridional ray was considered in Par. 11. In principle, a meridional ray transmitted by a fiber will remain in the same azimuthal plane throughout its passage, emerging at the same angle α (or $-\alpha$, depending on the number of reflections) at which it entered. As a practical matter, a bundle of rays of small but finite diameter entering at angle α will gradually disperse in azimuth (because of the small skew components and the large number of reflections) so as to fill an annulus of a cone of half angle α upon leaving the fiber (Fig. 2).

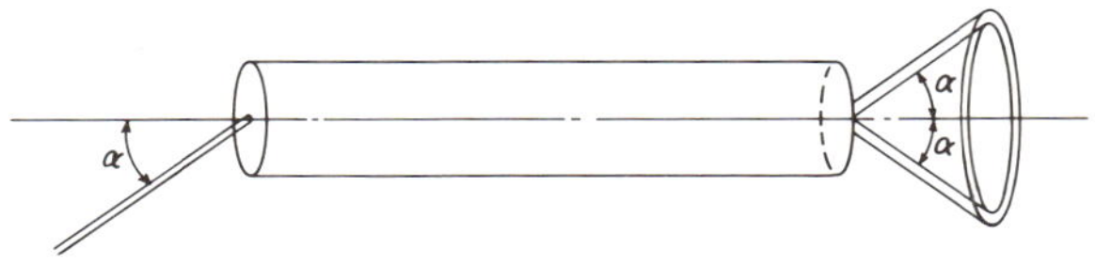

Fig. 2 Behavior of a parallel pencil of light obliquely incident on a straight fiber.

If instead of a parallel pencil of light a convergent cone is obliquely incident on the fiber, the emergent light will again be in the form of a hollow cone (Fig. 3) with a half width essentially equal to the angle of convergence of the incident cone.

If the fiber ends are not perpendicular to the axis, they act as prisms and will deflect the incident or emergent cone (Fig. 4). For a bias angle β at the end of the fiber the cone will be deflected an amount δ given by

$$\delta = \sin^{-1}\frac{n_1 \sin\beta}{n_0} - \beta \tag{2}$$

or

$$\delta \approx \frac{(n_1 - n_0)\beta}{n_0} \tag{3}$$

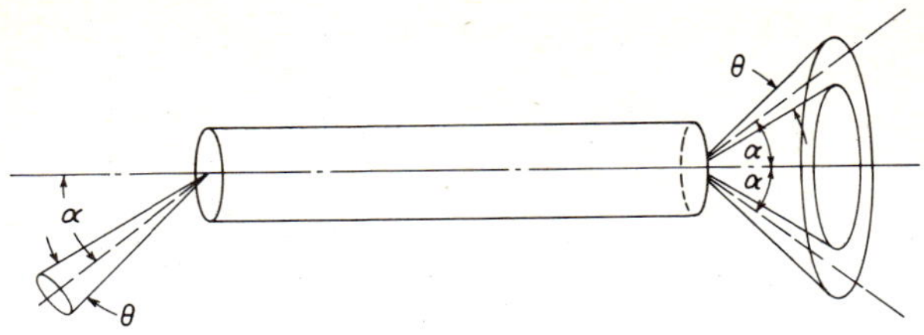

Fig. 3 Behavior of a convergent cone of light obliquely incident on a straight fiber.

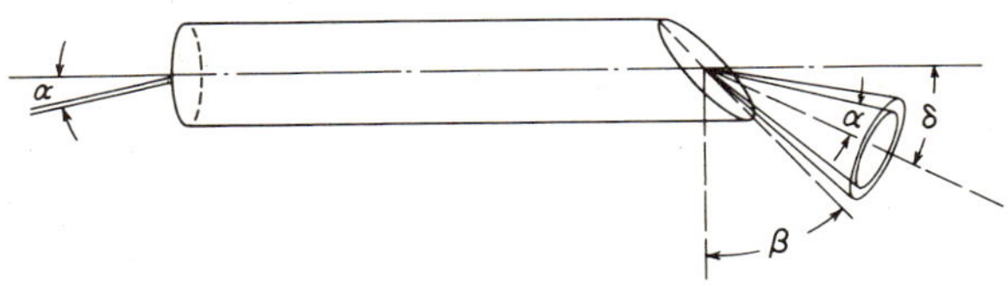

Fig. 4 Behavior of light emerging from a fiber with a beveled face.

for small angles. If the end of the fiber is in air, then $n_0 = 1$ and Eq. (3) becomes

$$\delta = (n_1 - 1)\beta \tag{4}$$

13. Behavior of Rays in a Bent Fiber It was previously stated that the effective NA of a fiber is influenced by diffraction, striae, and surface irregularities because they cause decollimation (spreading) of the light passing through the fiber. An additional cause of this decollimation and reduction of effective NA is bending of the fiber.

Regarding this effect, Potter (1960) has derived the expression

$$\Delta \cos \alpha_{\text{out}} = \frac{2dR \cos \alpha_{\text{in}}}{R^2 - (d/2)^2} \tag{5}$$

where d = fiber diameter
R = bend radius
α_{in} = half angle of incident cone
α_{out} = half angle of emergent cone

For $d \ll R$ this reduces to

$$\Delta \cos \alpha_{\text{out}} = \frac{2d}{R} \cos \alpha_{\text{in}} \tag{6}$$

Thus, for a parallel beam of light ($\cos \alpha_{\text{in}} = 1$) incident on a fiber of 0.01 cm diameter bent to a radius of 2 cm

$$\Delta \cos \alpha_{\text{out}} = 0.01$$
$$\Delta\alpha_{\text{out}} \approx 8°$$

Thus, the decollimation effects of even such a high bending ratio as 200:1 can be quite significant.

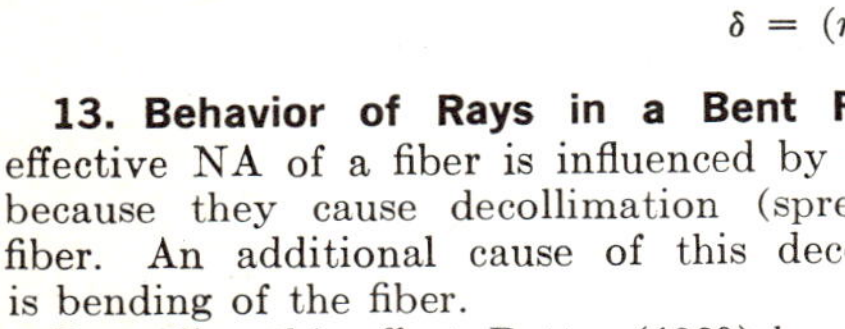

Fig. 5 Meridional-ray path in a tapered fiber.

14. Tapered Fibers The behavior of rays in fibers of varying diameter can be understood from considering the frustrum of a cone (Fig. 5). A meridional ray incident on the large end of the cone is reflected from the conical wall at progressively lower angles of incidence (at each reflection the angle of incidence is less by twice the cone angle θ) until it either escapes through the small end of the cone or falls below the critical angle, whereupon it escapes through the wall.

Conversely, for a ray entering from the small end of the cone, the angle of incidence will increase by 2θ at each reflection until the slope angle is less than θ and no further reflections take place. By the reversibility principle, light can also enter through the side of the cone when traveling from the small end and be internally reflected toward the large end. In some applications this effect can be used to advantage to inject light into a fiber array. In other instances it will only serve to dilute the signal with stray light and must be eliminated by blackening the outside wall.

The law governing the change that occurs in meridional-ray angles in passing through the cone (for small cone angles) is

$$d_1 \sin \alpha_1 = d_2 \sin \alpha_2 \tag{7}$$

where d_1 = diameter of large end of cone
d_2 = diameter of small end of cone
α_1 = meridional angle at large end
α_2 = meridional angle at small end

Since the critical angle at any point along the fiber is still fixed by the refractive indexes n_1 and n_2, the effective NA of a tapered fiber depends on the ratio of the smallest diameter along the fiber to the entrance diameter.

In a bundle of tapered fibers, the same change in acceptance angle of the light occurs. This is accompanied by a corresponding magnification (or demagnification) of the image transmitted by the bundle, which thus constitutes a fiber-optic magnifier.

15. Transmission of a Fiber The transmission of light in an optical fiber is governed by a number of factors, including the transmittance properties of the core and coating materials, the quality of the reflecting interface, residual-light leakage through the coating, and various end losses. To calculate the complete theoretical transmission of a fiber even with simplifying assumptions for fiber shape and the ideal behavior of the rays requires an elaborate integration process (Potter, 1961). A useful simplification is to treat the losses in transmission in two parts, line losses and end losses.

Line Losses. The attenuation of light along a fiber is a complex process involving absorption, scattering, and imperfect reflection. The net effect, however, can be expressed as

$$I = I_0 e^{-w(\alpha,\lambda)L} \tag{8}$$

where I_0 = incident intensity
I = transmitted intensity
L = length of fiber

$w(\alpha)$, defined as the attenuation coefficient, is a function of the angle of incidence α, the wavelength λ, and the type of fiber. As α increases, both the path length of the light in the fiber and the number of reflections increase and so does the attenuation. Thus, the effective numerical aperture should decrease with the length of the fiber. In practice, scattering of light in the fiber largely masks this effect. A typical value for w in light-guide fibers is about 0.0025 cm^{-1} for the center of the visible spectrum.

The transmission of fibers as a function of wavelength depends primarily on the transmission characteristics of the core glass and secondarily on those of the coating. Except in very small fibers the light transmitted by the fiber spends most of its time in the core glass and hence is most influenced by it. Thus, for long fibers, this glass must be carefully selected for maximum transmission. Fortunately, there are a number of water-white optical glasses available with a high enough refractive index to provide useful values of NA. Even with these glasses, however, the transmission is confined largely to the visible and near-infrared spectrum from about 400 nm to 1.7 μm, depending upon fiber length. Special glasses with a somewhat lower refractive index provide transmission out to about 350 nm, while others, such as arsenic trisulfide, extend into the infrared out to as far as 12 μm.

End Losses. The factors which make up the end losses in fibers include Fresnel reflection from the entrance and exit faces of the fiber, the packing fraction, and the effective NA. The last of these, together with the distribution characteristics of the incident light, determines the amount of light which can be accepted. The Fresnel reflection coefficient for normal incidence (and practically for angles of incidence up to 45°) can be computed from the well-known formula

$$R = \left(\frac{n_1 - n_0}{n_1 + n_0}\right)^2 \tag{9}$$

where n_1 = refractive index of core glass
n_0 = refractive index of adjacent medium
R = reflection coefficient for each face

The combined Fresnel reflection for both ends of the fiber bundle is given by

$$R_{\text{total}} = 1 - (1 - R)^2 \tag{10}$$

The values of R_{total} range from about 9 to 17 percent for core glasses with refractive indexes from 1.54 to 1.86. These losses can be reduced in some cases by low-reflection coatings.

The packing fraction is the ratio of the cross-sectional area occupied by the fiber cores to the total area. The loss due to this factor may range from only a few percent in large fibers to as much as 50 percent for small fibers. Since both the fiber-coating area and any space between fibers contribute to this loss, fused fiber bundles are more efficient in this regard than cemented bundles because the fusing process eliminates open spaces between the fibers. In principle, the lower limit of this loss factor is a function of the fiber size since the coating thickness should be essentially constant. In practice, this is usually true for small fibers when the coating thickness is held to a minimum in the interests of optical efficiency. In large-diameter fibers the coating thickness will often be greater than optically necessary as a convenience in manufacturing.

16. Image Transfer and Resolution in Fiber Bundles The image-transfer characteristics of fiber bundles involving both dynamic and static scanning have been investigated by Kapany et al. (1957) and Drougard (1964). Kapany and his coworkers point out that for static sampling by a fiber-optic bundle, i.e., where the fiber bundle is stationary with respect to the object, there is no definite cutoff spatial frequency because the spatial frequency depends upon object orientation with respect to the fiber pattern, the type of fiber pattern, i.e., square vs. hexagonal array, and constructional imperfections in the fiber bundle. Instead, they place the maximum spatial frequency between the limits $1/2d$ and $1/1.74d$, where d is the fiber diameter. In terms of high-contrast-target resolving power, these limits correspond to resolving powers of $500/d$ and $575/d$ line pairs per millimeter, respectively, where d is in micrometers. In practice, one finds significant point-to-point fluctuations due to the factors mentioned above. Moreover, since the general image quality of a static image transferred by a fiber bundle is a function of the number of defective and misaligned fibers, it can vary considerably with the individual device.

Drougard (1964) has also investigated static scanning with fiber bundles and has found that under certain orientations of the fiber pattern relative to a sine-wave bar target, frequencies greater than the nominal cutoff of $1/3d$ are passed by the bundle. On the other hand, she also concluded that one should not expect good overall image quality for spatial frequencies above $1/5d$. This again points up the uncertainty associated with the resolving power of a static fiber bundle. As a rule of thumb, a cutoff frequency of $1/2d$ or a high-contrast-target resolving power of $500/d$ line pairs per millimeter (d in micrometers) appears to work out fairly well in practice.

For dynamic scanning, i.e., where the fiber bundle is moving with respect to the object, the cutoff spatial frequency derived by Kapany et al. (1957) is given by $1.22/d$; that is, the frequency response of the fiber bundle is essentially that of a uniform disk equal to the fiber diameter. This corresponds to a high-contrast target resolving power of $1{,}220/d$ line pairs per millimeter (d in micrometers). These authors conclude that for a square fiber array, the ratio of maximum spatial frequencies for dynamic vs. static scanning is 2.44:1 while for a hexagonal fiber array the ratio is 2.12:1. Experimentally, one usually finds the ratio is only about 2:1.

In addition to this increase in resolving power, dynamic scanning greatly improves the appearance of the image by blurring out the overall structure and minor defects of the fiber bundle. The scanning amplitude required to achieve these effects was found by Kapany to be only a few fiber diameters.

17. Waveguide Properties of Fibers An optical fiber may be considered as a dielectric waveguide in which the light is propagated by total internal reflection. If the transverse dimensions of the fiber are comparable to the wavelength of the light, only certain distributions of the electromagnetic field will satisfy Maxwell's equations. This gives rise to discrete waveguide modes. Such modes are readily observed in very small fibers, but in fibers of the size used in most applications they are so numerous and overlapping that they cannot be discerned. In the application of fibers as optical communication lines, however, the waveguide properties become important, and single-mode fibers are often required.

A theoretical treatment of this subject by Snitzer (1961) and Kapany and Burke (1961) will not be repeated here. Some of the significant conclusions of the analysis follow.

1. In a dielectric waveguide all but the cylindrically symmetric modes TE_{0m} and TM_{0m} are hybrid modes; i.e., they have both electric and magnetic axial components.

2. The number of modes increases with the square of the fiber diameter.

3. All modes except the HE_{11} mode have cutoffs which are a function of wavelength, fiber diameter, and fiber NA.

4. The cutoff values for various modes are given by a characteristic equation

$$U_{mm} = \frac{\pi d}{\lambda} (n_1^2 - n_2^2)^{1/2} \tag{11}$$

where U_{mm} = mth root of cutoff condition involving nth-order Bessel function
d = diameter of fiber
λ = wavelength in free space
n_1 = refractive index of fiber core
n_2 = refractive index of cladding

This may also be written

$$u_{nm} = \frac{\pi d}{\lambda} \text{NA} \tag{12}$$

where NA is the numerical aperture of the fiber.

Values of u_{nm} (at cutoff) have been calculated for a variety of modes, and some of these are given in Table 1. Note that for the HE_{nm} modes with $n > 12$ these values apply only in the limiting case of small NA.

TABLE 1 Values of U_{nm} at Cutoff for Some Low-Order Modes

m	HE_{1m}	TE_{0m}, TM_{0m}, HE_{2m}	EH_{1m}, HE_{3m}	EH_{2m}, HE_{4m}
1	0.000	2.405	3.832	5.135
2	3.832	5.520	7.016	8.417
3	7.016	8.654	10.173	11.620
4	10.173	11.792	13.323	14.796
5	13.323	14.931	16.470	17.960

An important case is that for which only a single mode can be supported by the fiber, i.e., the HE_{11} mode. This occurs for $u_{nm} = 2.405$. Thus for

$$\frac{\pi d}{\lambda} \text{NA} < 2.405 \tag{13}$$

the fiber will support only the HE_{11} mode. For example, for a wavelength of 0.63 μm and an NA of 0.2, the fiber (core) diameter must be less than 2.4 μm.

Since a fiber of this diameter would be exceedingly difficult to handle, it is common practice to put on a very thick cladding to build the overall diameter up to 25 μm or more. For optical transmission lines, moreover, the glasses must be formulated and prepared with the utmost care to minimize light absorption.

FIBER-OPTIC PRODUCTS

18. Optical Monofiber and Clad Rod The basic building block of all fiber-optic products is clad optical monofiber and its larger-diameter version, clad rod. In principle, even this simple form can have many different specifications including core and cladding materials used, refractive indexes and NA, expansion coefficients, chemical resistance, thermal resistance, tensile strength, transmittance (including spectral dependence), diameter, cross-sectional shape, and available lengths.

In practice, fiber-optic manufacturers usually make only a few standard types of fibers, but some will make fibers of special types to order. In fact, glass monofibers are not commonly offered in the trade because of the difficulty in handling spooled fiber. Instead, fibers are usually packaged into various semifinished or finished products such as light guides. On the other hand, spooled plastic monofibers, which can be handled more easily, are available in a variety of diameters.

For general visual use, the commonly available light guides provide a convenient source of glass monofibers for experimental purposes. In these, the fibers range in diameter from about 40 to 80 μm and in length up to 6 ft. The NA typically ranges from 0.55 to 0.63, and transmittance extends from about 0.37 to 2.0 μm.

For special purposes, some manufacturers offer to make monofibers to various specifications including different glass types for improved ultraviolet or infrared transmittance and different diameters and lengths. A user having such requirements should consult the manufacturers regarding realistic specifications and costs.

Clad rod is a simple extension of the monofiber and has become available because it can be conveniently produced with machinery used in the manufacture of certain fiber-optic products. At least one manufacturer offers clad rod as a standard product in sizes of $\frac{1}{16}$-, $\frac{1}{8}$-, and $\frac{1}{4}$-in diameters and 6 or 12 in long. This material has the same basic optical characteristics as the smaller monofiber. The NA is typically 0.56, and the transmittance extends from about 0.37 to 2.2 μm. Other diameters and lengths (up to about 6 ft) can also be made available.

19. Light Guides Fiber-optic light guides are bundles of monofibers in a flexible protective sheathing used for piping light over curved paths. Since the fiber ends generally are not correlated, they cannot be used to transmit images. They are available in two basic types, (1) modular lengths and diameters with finished ends and end fittings and (2) continuous lengths of up to several thousand feet. They are made with either glass or plastic fibers.

Typical specifications for glass-fiber light guides of the first type are given in Table 2 for three of the principal manufacturers. Similar specifications for continu-

TABLE 2 Modular Light Guides

Specification	American Optical Corporation	Bausch & Lomb, Inc.	Corning Glass Works
Fiber diameter, in	0.003	0.0015	0.0025
NA	0.56	0.50	0.63
Acceptance angle, deg	68	60	78
Nominal fiber-bundle diameter, in	0.020, $\frac{1}{16}$, $\frac{1}{8}$, $\frac{1}{4}$	$\frac{1}{16}$, $\frac{3}{32}$, $\frac{1}{8}$, $\frac{3}{16}$, $\frac{1}{4}$	$\frac{1}{8}$, $\frac{3}{16}$, $\frac{1}{4}$
Length, in	12–72	3–72	12–72
Sheathing	Polyvinyl chloride	Polyvinyl chloride or strip-wound aluminum	Polyvinyl chloride over monocoil wire
Temperature rating, °F	Not specified	−40 to +200	Up to 225

ous-length light guides are given in Table 3, and those for plastic-fiber light guides are given in Table 4.

TABLE 3 Continuous Light Guides (Glass)

Specification	Owens-Corning Fiberglas	Galileo Electro-Optics Corp.	Corning Glass Works
Fiber diameter, in.	0.002	0.0025	0.018, 0.025
NA	0.54	0.66	0.63
Acceptance angle, deg.	65	82	78
Fiber-bundle diameter, in	0.031–0.096	$\frac{1}{32}$–$\frac{1}{2}$	0.045–0.093
Jacket diameter, in	0.087–0.130	0.087–0.620	0.087–0.139
Jacket material	Polyvinyl chloride	Polyvinyl chloride	Polyvinyl chloride (color-coded)
Temperature rating	To 105°C	−20 to +105°C	To 225°F

The transmission characteristics for typical glass-fiber light guides as a function of length in the visible spectrum are given in Fig. 6. The transmission extends into the near infrared out to about 1.7 μm, as shown in Fig. 7. (Note that in this figure the transmittance does not include the end losses due to Fresnel reflection and the fiber-packing fraction.)

The spectral transmission for typical plastic-fiber light guides is shown in Fig. 8. There are several strong absorption regions in the near infrared which tend to limit the application of plastic fibers in this spectral region.

Modular light guides have prefinished ends, which are ground and polished for maximum optical efficiency. The end tips are usually metallic and with a variety of configurations. There are no common standards for end-tip dimensions. Commonly used sheathings include plain polyvinyl chloride, over monocoil wire, and continuous-wound flexible metal hose; the choice depends on the degree of protection required.

Terminations for continuous light guides are made by the user, but the manufacturer can supply suitable termination hardware and recommendations for processing the ends. For glass fibers, grinding and polishing provides more than twice the efficiency of blade-cut fibers. For plastic fibers which do not fracture on cutting, the loss in blade-cut fibers is only about 15 to 25 percent over polished ends.

20. Special Light-Guide Assemblies In addition to standard modular and continuous light guides, several manufacturers offer special light-guide assemblies for particular applications. One of the unusual capabilities of a fiber bundle is that either end (or both) can readily be subdivided into two or more terminations to illuminate or pick up light from a multiplicity of locations.

The simplest form of such a bundle has two terminations at one end and one at the other and is known as a *bifurcated light guide*. It is possible for the fibers at the common end to be carefully mixed or scrambled so that light from either of the bifurcated ends is uniformly distributed over the common end. If this property is required, it should be specified to the manufacturer. Such a light guide may be used in a specialized photometer for sensing the reflectance characteristics of a surface or material placed near the common end. A light source is used to illuminate one of the separated ends, and a detector is placed at the other. The intensity (or with suitable filters, the color) of the reflected light can be determined. Such a device has been used to measure the color and thereby the oxygen content of blood inside the heart and major blood vessels.

Other forms of divided light guide are used to provide illumination for optically sensing holes in data-processing punched cards or tape. In this case, the light guide is divided into as many terminations as there are sensing positions, and light from a single lamp is uniformly distributed to each station. Photodetectors are located opposite these terminations, and the punched card or tape passes between them.

TABLE 4 Plastic Light Guides

Source	Acceptance angle, deg	Fiber size, in	No. of fibers in guide	Size, in	Guide OD, in	Length	Jacket	Temperature rating continuous, °F
Poly-Optics, Inc.†	70	0.014, 0.016	7	0.016	0.087	Continuous	Polyethylene	200
			19	0.014	0.110			
			19	0.016	0.120			
Du Pont (Crofon)	Not specified	0.010	16	. . . . .	0.087	Continuous	Polyethylene	175
			32	. . . . .	0.110			
			48	. . . . .	0.119			
			64	. . . . .	0.130			

† In addition to the standard light guides shown, this manufacturer also offers bulk clear plastic fibers in diameters from 0.003 to 0.030 in in spooled lengths from 250,000 to 4,500 ft and color-coded (jacketed fibers) in diameters from 0.020 to 0.062 in in spooled lengths from 11,000 to 1,200 ft.

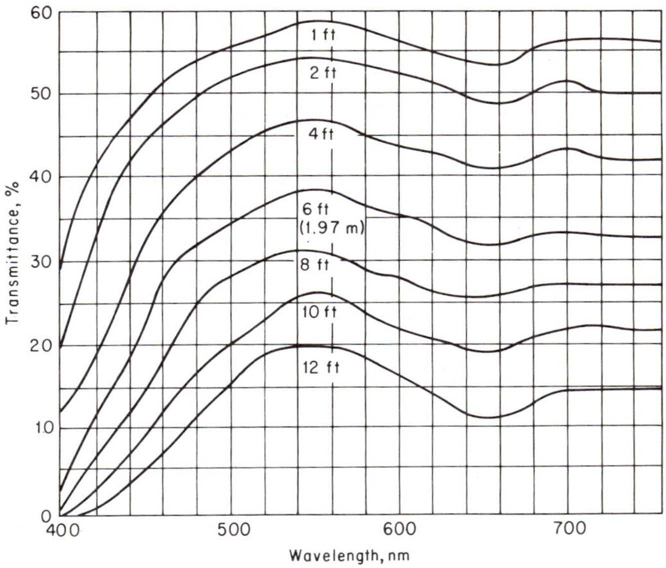

Fig. 6 Spectral transmittance for typical glass-fiber light guides.

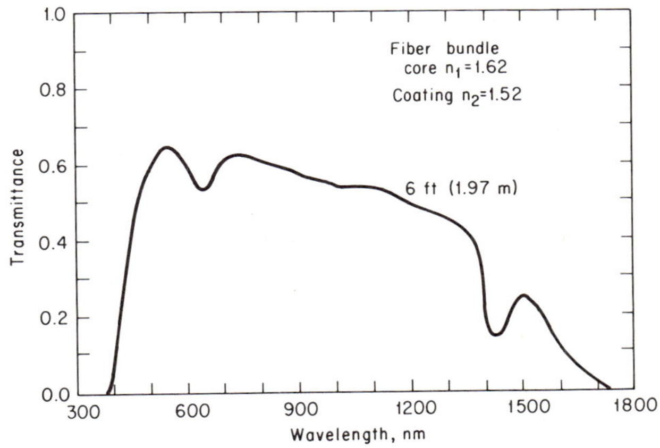

Fig. 7 Line transmittance for a 6-ft-long fiber bundle.

Fiberscopes

21. Flexible bundles of fibers in which the fiber ends are coordinated, i.e., the fibers are in an identical arrangement at both ends of the bundle, are capable of transmitting images. When such a bundle is fitted with an objective lens at one end and an eyepiece at the other, it becomes a flexible form of periscope or borescope. The word fiberscope is commonly used to designate this type of instrument; however, a number of other trade names have been coined by various manufacturers such as Imagescope, Flexiscope, Fibericon, and Fiberope to designate specialized devices.

At least three distinct types of fiberscopes available in the trade can be distinguished by their method of manufacture, monofiberscopes, multifiberscopes, and leached fiberscopes.

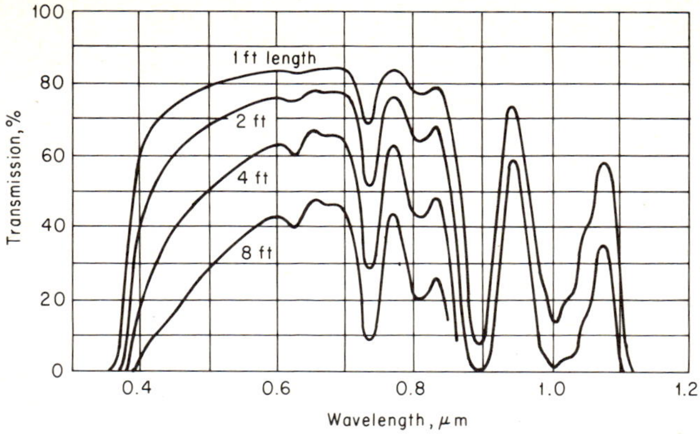

Fig. 8 Spectral transmission for plastic-fiber light guides.

22. Monofiberscopes Monofiberscopes are produced by winding a coated monofiber on a precision machine in either of two ways: (1) by continuously winding a multiplicity of layers in a manner analogous to coil winding or (2) by winding one layer at a time and assembling the desired number of layers in a separate laminating operation. In either case, the final fiberscope bundle consists of individual small monofibers bound only at the ends and separate in between. Monofibers as small as 9 μm in diameter have been used in such fiberscopes.

The principal application for these fiberscopes is in medical endoscopes for examining internal body cavities, e.g., the esophagus, stomach, lungs, and large bowel. In such instruments, the basic optical system of objective lens, fiberscope bundle, and eyepiece is supplemented by (1) a fiber light guide to supply illuminator, (2) mechanical means for remotely focusing the objective lens and for articulating (bending and twisting) the fiberscope tip, and (3) an open channel for passing air, fluids, or a biopsy forceps through the instrument. Similar instruments offered for industrial-inspection applications lack the specialized medical features.

Some of the older instruments were equipped with a small lamp at the objective end to illuminate the object; however, most instruments now use an integral fiber light guide for illumination. Light is provided either by a tungsten lamp in the handle of the instrument or by a remote illuminator connected by a fiber light guide. For endoscopic photography, special arc lamps are often used to supply intense illumination to the field of view.

Endoscopic fiberscopes are available in a wide range of sizes to fit various body cavities. Fiber bronchoscopes are about 4 mm in diameter and about 60 cm long in the flexible portion. Fiber gastroscopes and esophagoscopes are about 12 mm in diameter and vary from about 70 to 100 cm long. Fiber colonoscopes, for examining the large bowel, are also about 12 mm in diameter but run up to 180 cm long in the flexible portion.

The fiber size in monofiberscopes varies with the instrument diameter from 9 μm in the fiber bronchoscope to about 20 μm in the gastroscope. A typical fiber bundle in an endoscopic instrument contains about 25,000 to 30,000 fibers and provides adequate image quality for many diagnostic purposes.

Principal suppliers of endoscopic fiberscopes in the United States are American Cystoscope Makers, Inc., American Optical Corporation (Machida Endoscopes), and Olympus Corp. of America. Each of these suppliers also offers fiberscopes for industrial-inspection applications. Bausch & Lomb, Inc., also supply an industrial fiberscope under the trade name Flexiscope.

23. Multifiberscopes Another approach to the manufacture of fiberscopes involves the use of multifibers, which consist of several optically independent fiber elements fused together to form a single flexible strand. The multifiber, normally square and typically containing 36 elements in a 6 × 6 array, is wound on a drum to form hoop-shaped ribbons of a width corresponding to the width of fiber bundle desired. A number of such ribbons is assembled to produce the height of the fiber bundle. The bundle is then cemented at the region of assembly and cut to form a coordinated (image-carrying) fiberscope bundle. In this manner, a very large number of fiber elements can be assembled in a precise array to give high overall image-carrying capacity.

The principal application for such fiberscope bundles has been in industrial and military systems, where image-quality requirements exceed the capability of monofiberscope bundles. Bundles up to 25 mm diameter and 9 ft (275 cm) long containing approximately 5 million fiber elements have been made by this technique. The use of multifibers makes possible fiber sizes smaller than can be practicably handled as monofibers, and fiber element sizes down to 5 μm can be fabricated.

The principal sources for multifiberscope bundles and instruments are American Optical Corporation and Galileo Electro-Optics Corp. Both these manufacturers produce a basic series of 8 by 10 mm format fiberscopes in lengths from 2 to 15 ft with a protective sheathing and standard 16-mm ciné lens adapters at each end. The fiber element size is 10 μm, and nominal image-resolving power is 50 line pairs per millimeter for a high-contrast-resolving-power target. These devices can be conveniently coupled to 16-mm ciné or vidicon cameras for special photooptical applications.

24. Leached Fiberscopes The leached form of fiberscope is produced by a completely different process. A special fused multifiber rod is prepared of a large number of fiber elements each having a double coating. The inner coating serves as the optical insulation and is made of an acid-resistant glass of relatively low refractive index. The outer coating is made of a special glass which is acid-soluble. The multifiber rod is made in two or more drawing steps alternated with assembly steps to provide several thousand fiber elements in the array. The elongation resulting from the drawing process serves to align the fibers with a high degree of accuracy.

After the final drawing step, the multifiber rod is cut to the desired length, the ends are polished, and a short region at each end is coated with acid-resistant material. The rod is then soaked in an acid bath to leach out the soluble coating. This separates the fibers and produces a flexible coordinated fiberscope bundle.

The size of the bundle that can be produced in this way is about 5 mm in diameter, and the fiber size ranges from about 15 to 50 μm. The cost of manufacture of these bundles is considerably lower than for conventionally wound bundles and thus makes a number of low-cost fiberscope applications possible. Leached fiberscope bundles are currently offered either as basic image guides without lenses or as fiberscopes with objective and eyepiece lenses. The fiber size in these bundles is 20 to 40 μm, and the length runs from 18 to 24 in.

25. Fiber Faceplates Fiber-optic faceplates consist of a large number of glass fibers fused together into a vacuum-tight disk or plate of relatively large cross section compared with the thickness. Like a fiberscope, the faceplate has a coordinated fiber array capable of transmitting images; however, it is not flexible but rigid and mechanically resembles a solid plate of glass.

Fiber-optic faceplates are made by fusing together under heat and pressure a carefully aligned assembly of fibers or multifibers. They are made of glass-coated–glass-core fibers in which the coating serves both to insulate the fibers optically and to provide the fusible binder which holds the fibers together. The fused assembly is often called a *boule*.

The most common applications for fiber-optic faceplates are as windows for electrooptical vacuum tubes, such as cathode-ray tubes, television-camera tubes, and image-intensifier tubes. In these applications, the fiber faceplate serves as a means for transporting the image into or out of the tube without the use of lenses or for correcting the electron–optical-field curvature and distortion of the vacuum tube. To be useful in these applications, the faceplate must be vacuum-tight. This can be achieved by a well-controlled fusing process.

The optical and physical requirements for fiber-optic faceplates vary with the application. In image-intensifier or television-camera tubes, the image-fidelity requirements are a maximum. High optical-transmission efficiency is obtained by using fibers with a high NA (1.0 or greater). High image-resolving power (about 100 line pairs per millimeter) is obtained by using a fiber size of 5 to 6 μm. High-contrast transfer characteristics are obtained by minimizing stray light through the use of extramural absorption (EMA). Two forms of EMA are commonly employed: (1) a secondary absorbing glass coating over the clear optical coating surrounding each fiber or (2) a set of absorbing glass fibers placed interstitially among the transmitting fibers. In either case, stray light which may enter the plate through the cladding area is gradually absorbed by multiple passes through the absorbing glass before reaching the exit face of the plate.

In image-intensifier and camera-tube applications, where a photocathode is to be deposited on the faceplate, there must be no deleterious chemical interaction between the photocathode and the faceplate. The glasses used in the fibers must therefore be chosen to avoid such interaction. In particular, lead, cadmium, and antimony must be avoided.

In these applications, the faceplate is usually sealed to a metal flange, which in turn is sealed to the tube envelope. Thus, the thermal-expansion coefficient of the faceplate must be matched to that of the flange; a suitable sealing-glass frit for use in actual tube manufacture and the recommended sealing materials and procedure are normally specified by the faceplate manufacturer.

Fiber faceplates are used in cathode-ray tubes to permit direct printing of the image onto photosensitive materials without the use of lenses, to obtain a parallax-free image on the face of the cathode-ray tube, and to improve the contrast of the phosphor image, particularly in the presence of strong ambient light.

In many of these applications, the optical requirements can be somewhat less severe, particularly in large faceplates. If the cathode-ray tube is used for direct printing of photographs or digital information, a fiber size of 20 to 25 μm is usually adequate. For visual observation, a fiber size of 50 to 100 μm may suffice. In most cases, EMA is used to suppress stray light. In some direct printing applications, where a high-contrast signal is available, even the EMA may be dispensed with.

The NA of the fibers in cathode-ray-tube faceplates is lower in most cases than in image tubes. For direct printing, where there may be imperfect contact between the photosensitive material and the faceplate, an NA of 0.66 is commonly used. This provides a small amount of depth of focus. For ambient-light suppression in direct-viewing applications a still lower NA of about 0.40 may be used.

In cathode-ray tubes where a phosphor is in contact with the faceplate, the chemical-compatibility requirement is less severe, and a wide choice of glasses, including lead-containing flint glasses, can be used for the fibers. The most common method of sealing the faceplate to the tube is by frit sealing directly to the tube envelope (or auxiliary glass faceplate), usually made of KG-12 or Corning 0120 glass, using a devitrifying glass frit.

A variety of standard faceplate materials is available in the trade, mainly from two manufacturers. The specifications for these materials are given in Table 5.

Fiber-optic faceplates are available in a variety of shapes and sizes. Since the basic material behaves like solid glass, it can be cut, ground, and polished into almost any configuration. Round, rectangular, or strip plates are standard configurations, and the surfaces can be finished flat or spherical to specified radii. Special edge configurations are often provided to facilitate sealing to flanges. Faceplates are available up to about 9 in in diameter or up to 12 in long in rectangular form, but this varies with the type of plate and the manufacturer. Any prospective user should contact the manufacturer for any unusual faceplate size or configuration.

26. Image Conduit Fiber-optic image conduit is a form of fused multifiber in which the fiber elements are precisely parallel and therefore capable of transporting images. It is made by simultaneous fusing and drawing together a number of glass-coated glass fibers to form a rigid rod. If the number of fibers is very large (say over 1,000), a second drawing step to produce a multi-multifiber may be used.

Although basically rigid, image conduit can be bent by heating to conform to any desired pathway. Since the fiber size will generally be very small compared with the bending radius, practically no light is lost at the bends. Thus image

TABLE 5 Standard Fiber-Optic Faceplate Materials

Manufacturer and type	NA	Continuous stray light	Fiber size, μm	Transmission,%† Collimated	Transmission,%† Lambertian	Resolution, line pairs/mm‡	Expansion per °C × 10^7	Sealing method	Chemical compatibility	Maximum temperature 1 h, °C
American Optical:										
50C	0.88	Clear	5	92	90	114	78	Frit to Corning 8400 glass	Photocathode or phosphor	
70C	0.66	Clear	12	91	89	45	90	Frit to Corning 0120 glass	Phosphor	
70A	0.66	EMA	8	85	30	72	90	Frit to Corning 0120 glass	Phosphor	
47C	1.04	Clear	5.5	88	83	103	68	Frit to Niromet 44 metal alloy	Photocathode or phosphor	
47A	1.04	EMA	5.5	67	58	103	68	Frit to Niromet 44 metal alloy	Photocathode or phosphor	
Galileo Electro-Optics Corp.:										
A-7	0.35	Clear	6–50	...	...	10–100	70	Frit to Niromet 44 or gold	Photocathode or phosphor	500
K-2	0.66	Clear EMA	10 6	70§	28§	50 100	91–93	Frit to K-12 glass	Phosphor	460
D-11	0.93	Clear or EMA	10			50	96–98	Frit to K-12 glass	Photocathode or phosphor	525
D-14	1.0+	EMA	6	70§	55§	100	85–87	Frit to Carpenter 49 metal alloy	Photocathode or phosphor	550

† For white-light input and S-4 detector.
‡ Visually measured using high-contrast Air Force resolving-power test target.
§ Middle of visible region, typical values.

conduit should be used in place of clad rod for light piping if a sharp bend is required. Bending of small image conduit, i.e., up to about $\frac{1}{4}$ in diameter, can often be done by the user with a simple gas torch. Larger diameters usually require special equipment, and bending is best left to the manufacturer. Image conduit can also be twisted along the fiber axis to provide image rotation. Combinations of bending and twisting can be used to provide a wide range of configurations. The properties of available standard image-conduit materials are given in Table 6. In addition to the standard materials listed, image conduit with special glasses, sizes, and configurations can be obtained from either of the listed manufacturers.

TABLE 6 Fiber-Optic Image Conduits

Manufacturer and type	NA	Fiber size, μm	Conduit size, in, and shape	Conduit length,† in
American Optical:				
1C-1	0.52	12	$\frac{1}{8}$ hex	6–12
1C-2	0.52	25	$\frac{1}{4}$ hex	6–12
1C-100	0.52	12	$\frac{1}{8}$ round	6–12
Galileo Electro-Optics Corp.:				
A-7	0.35	20	0.020–2.0 square, hex, round	Up to 72
A-2	0.55	10, 20, 100	0.020–2.0 square, hex, round	Up to 72
K-1	0.66	10, 20	0.020–2.0 square, hex, round	Up to 72
D-8	1.0	3, 15	0.020–2.0 square, hex, round	6

† Supplied with ends ground and polished flat.

27. Tapered-Fiber Bundles A fused-fiber bundle can be tapered by heating and stretching one portion relative to the other. In this manner, each fiber within the bundle becomes tapered as well. An image projected on one end of such a bundle will be enlarged or reduced depending on the direction of taper. Such bundles are sometimes referred to as fiber-optic cones. The change in image size is in direct relation to the change in bundle diameter. Since the fibers change in size in the same proportion, the resolving power in line pairs per millimeter is different for the two ends of the bundle. Moreover, as pointed out in Par. 14, the NA changes inversely with the bundle diameter.

Tapered-fiber bundles are available on a special-order basis from at least two manufacturers, American Optical Corporation and Optics Technology, Inc. The specifications for ordering such a bundle should include the diameter of large end, overall length, taper ratio, fiber size at the small end, NA required at large or small end, distortional tolerances, and EMA requirements (in the absence of EMA considerable light entering the large end of the taper may appear as stray light at the small end). At the present time, an overall diameter of about 60 mm is the largest size available. A fiber size of about 5 μm at the small end represents a practical lower limit. Taper ratios of up to 20:1 are attainable, and for small ratios of up to 3:1 a length-to-diameter ratio as low as 1:1 can be provided.

TESTING FIBER OPTICS

28. The two general forms of tests applied to fiber-optic products are basic performance tests, which help to characterize the various types of products, and specialized tests designed in accordance with the individual application for which the product is intended. Only the basic test methods will be described here, e.g., those used to test fiber-optic components rather than complete systems.

29. Numerical Aperture As stated in Par. 11, the NA of a fiber-optic component is defined as

$$\mathrm{NA} = n_0 \sin \alpha_m$$

where n_0 is the index of refraction of the external medium and α_m is the angle of the input or the output cone of light (the acceptance angle).

Strictly speaking, the NA is defined by the angle at which the cone of light cuts off altogether. Since this is sometimes difficult to determine precisely, the working NA is often defined in terms of the angle at which the light input or output falls to 50 percent of its maximum (axial) value.

The NA is commonly measured by means of a goniometer, i.e., an instrument which measures relative intensity as a function of angle. This is most conveniently done on the output cone of light when the input light is diffused. Such an apparatus is shown in Fig. 9. The fiber-optic sample is illuminated with diffuse light in

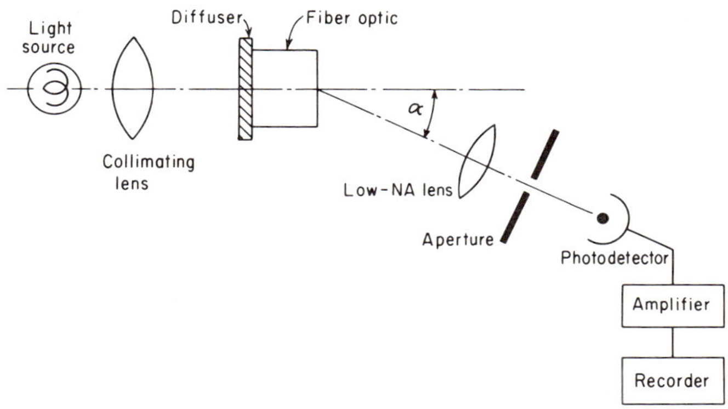

Fig. 9 Numerical-aperature test apparatus.

order to fill the fiber-optic acceptance cone completely. A piece of opal glass serves as the diffuser. The output face of the sample is placed at the center of rotation of the detector arm. The detector consists of a low-NA objective lens coupled to a photodetector through an aperture. The aperture diameter is defined by the area which is to be sampled. Aperture diameters of 1 to 3 mm are typical. The source diffuser is first scanned (as a function of angle α) without the fiber-optic component to establish a reference. The fiber-optic component is then inserted and scanned. The ratios of readings made with and without the fiber optic are plotted with respect to the angular settings, and the NA is determined using either the cutoff values or the 50 percent (or other arbitrary) criterion.

A difficulty arises when the NA of the fiber-optic component approaches or exceeds unity. The apparatus can be modified to handle this by placing a semicircular cylindrical lens with its axis at the center of rotation of the detector arm. The radius of curvature of the cylindrical lens may be about 1 in or so. To ensure optical contact of this lens and the fiber-optic component a layer of immersion oil must be used between them. In this measurement it is also important to ensure that the input-light cone is equal to or greater than the fiber-optic component NA. This can be done by using a layer of immersion oil between the source diffuser and the fiber-optic component.

For fiber-optic components which have parallel (not tapered) fibers, the NA generally is the same at both ends. For tapered or twisted components the NA may vary with direction and/or position. In these cases it is appropriate to test them in the direction and location of use.

If the input-light diffuser is removed, the input light will be collimated and the same apparatus can then be used to measure the decollimation or diffusing properties of the fiber-optic component. This will often be a function of how well the fibers are aligned within the component as well as of inhomogeneities and irregularities within individual fibers.

30. Transmission The transmission of a fiber-optic component should be measured and specified with reference to its intended use. Thus for example, one should specify both the input and output light in terms of wavelength and NA. For a close control of wavelength a spectrophotometer is often adaptable to measuring the transmittance of fiber-optic components. Where this is inconvenient or unavailable, a series of filters in conjunction with a white-light source offers a suitable substitute.

The input NA is often specified only to the extent of being either collimated or diffused. A small source at the focus of a collimating lens will provide the former, and an opal glass will provide the latter. Often one can conveniently convert collimated input light to diffuse light simply by adding a piece of opal glass at the input plane of the test object. If a specific NA is required, a lens of the corresponding NA can be used to illuminate the test object.

The means of collecting the output light by the pickup should be similarly specified. Pickup of collimated output light can be achieved by using a low-NA lens having a small aperture at its focus followed by the photodetector. Pickup of diffuse light can be achieved by using an opal glass diffusing disk in front of the photodetector. As an alternative, an integrating sphere may be used as a diffuse-light collector.

The input and output can be independently chosen to reflect the conditions of use; i.e., the input and output may be either collimated or diffuse in any combination. Two examples of experimental setups for measuring transmittance are shown in Figs. 10 and 11. In each case the measurement consists of taking

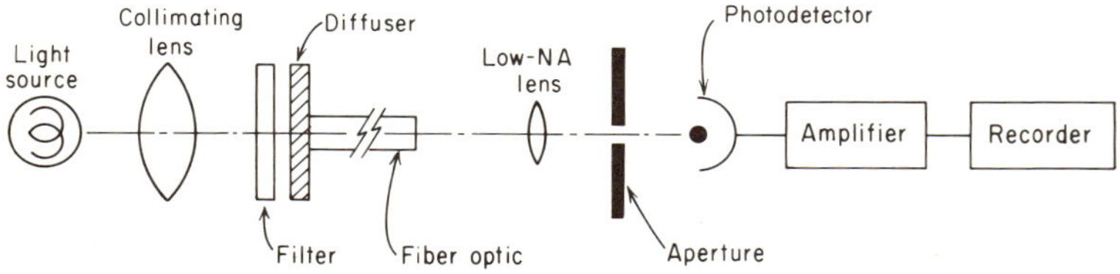

Fig. 10 Transmission-test setup. Diffuse input, collimated pickup.

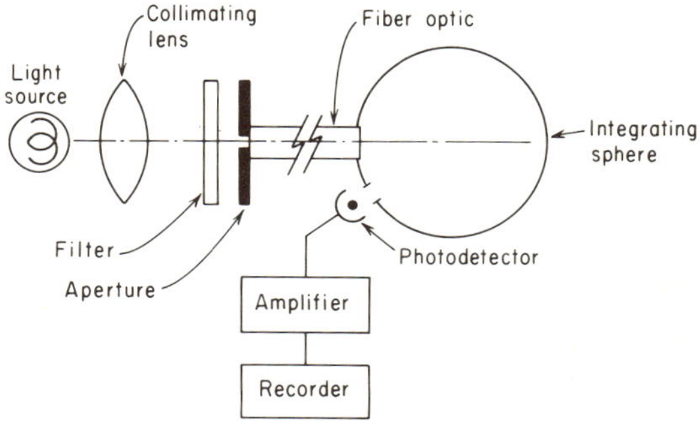

Fig. 11 Transmission-test setup. Collimated input, diffuse pickup.

readings both with and without the fiber-optic component in place and comparing the values for each spectral range. If a double-beam spectrophotometer is used, this comparison is usually automatic and the transmittance is plotted directly by the chart recorder.

The transmission values for fiber-optic components are a function primarily of the design of the fibers, the materials used, and the length. They are limited by insertion losses and internal losses. Insertion loss includes Fresnel reflection at both fiber ends, the packing density of the fiber-core area relative to the total area, and the NA. The internal losses consist of the absorption of the core glass, scattering losses, and imperfect reflection from the core-clad interface. Some estimate of the expected transmittance can often be made by taking these factors into account. Any large discrepancy between the calculated and measured values may indicate abnormal fiber breakage and/or internal fiber defects such as devitrification.

31. Image-resolving Power In addition to the previous tests, which are appropriate to both nonimaging and imaging components, imaging components require suitable tests for image fidelity. One of the simplest common tests is that of measuring the image-resolving power. This is generally done by placing a test target (such as the 1951 Air Force resolving-power test target) into direct contact with one face of the fiber-optic component and observing the other end of the component with a magnifier or lower-power microscope. The smallest target which can be resolved into discernible elements gives a measure of the limiting resolving power.

If direct contact of the target is not feasible, it can be projected onto the fiber optic by means of an auxiliary lens of sufficiently high resolving power, e.g., a microscope objective. If the magnification of the projected target image is other than unity, it is necessary to compensate the resolving-power measurement for the magnification ratio.

Resolving-power measurements on fiber-optic bundles are primarily a function of fiber size and the perfection of fiber alignment; however, they also depend on the specific location and orientation of the target relative to the fiber pattern since the latter is a discrete set of apertures rather than a continuum. For these reasons it is good practice to average a number of independent measurements to determine the resolving power. A conservative estimate of the resolving power (RP) as a function of fiber size (assuming good fiber alignment) is given by

$$\text{RP (line pairs/mm)} = \frac{500}{d}$$

where d is the fiber diameter in micrometers.

32. Modulation Transfer Function The modulation transfer function (MTF) of an optical element or system is a measure of the contrast of the transmitted image relative to that of the object and is usually given as a function of the spatial frequency of the object. The MTF can also be considered in terms of the Fourier transform of the line-spread function for the system. Since fiber-optic bundles are not spatially invariant, however, Fourier analysis cannot be strictly applied. Nevertheless, MTF measurements are often made on fiber-optic components and devices since they give more information than simple resolving-power test.

There are a number of instruments on the market for the measurement of MTF of lenses and optical systems. Generally they can be used to test fiber-optic components by using well-corrected lenses to project the modulated test object and the scanning aperture or slit onto opposite ends of the fiber array. Two basic methods are in use.

1. The image of a narrow slit (small compared to the fiber size) is projected onto the entrance face of the fiber-optic sample, and the transmitted image is projected onto a variable-spatial-frequency sine-wave test target which scans the slit image. The modulated light is picked up by a photodetector, and the signal is electronically processed and recorded to provide a direct plot of the MTF. In this case a broadband amplifier is used to cover the variable spatial (and hence temporal) frequencies.

2. A scanning target of constant temporal frequency is projected at variable optical magnification onto the entrance face of the fiber-optic sample while the transmitted image is projected onto a fixed slit followed by a photodetector and

recorder. In this case a narrow-band electronic amplifier, passing only the fundamental target frequency, converts the square-wave target into an effective sine-wave target.

In each case only a specific location and orientation of the fiber-optic component is tested at a time. This represents the static MTF case. To get a fair sampling of the overall fiber-optic component, several positions and orientations should be measured and averaged.

In principle, it is possible to obtain the dynamic MTF of a fiber-optic component by rapidly moving the component relative to the rest of the system. Nutation is an example of a suitable motion. The amplitude required is generally only a few fiber diameters, but the frequency must be high enough for a number of cycles of the nutating motion to occur during each frequency increment of the variable spatial frequency.

33. Square-Wave Response Test An alternative to MTF which is sometimes used for evaluating fiber-optic components is the square-wave response test. It is basically similar to the MTF test but the variable-spatial-frequency targets are of square-wave (black and white bars of equal spacing) rather than sinusoidal density contour. The test can often be set up using available optical-test apparatus.

In one version of the test (Fig. 12) a microscope objective is used to project the square-wave target onto the entrance face of the fiber-optic component at

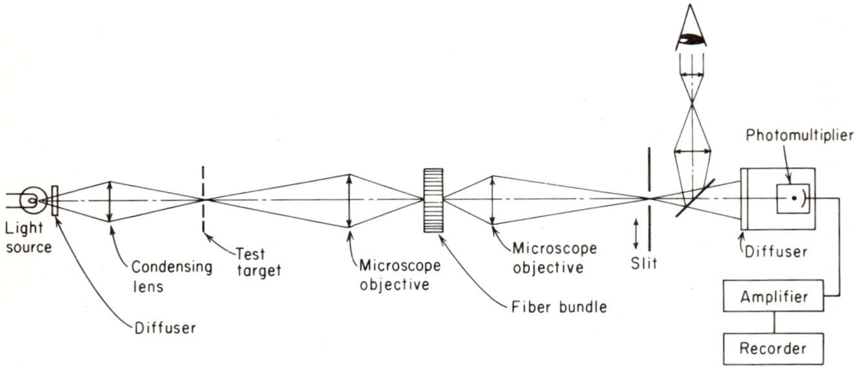

Fig. 12 Square-wave response test.

a known magnification. Another microscope objective is used to project the transmitted image of the target onto a slit aligned parallel to the bar pattern. This is followed by a photomultiplier, amplifier, and chart recorder. Either the target or the slit can then be made to scan transversely. If the target is moved, only a single line of fibers is sampled. If the fiber-optic sample is oriented so that a row of fibers falls directly under the sampling area, a maximum response is obtained. This is one version of the test commonly employed.

If the slit is traversed across the target image (and thus across many fibers), an averaging of several target positions relative to the fibers is achieved and a lower but probably more realistic response is obtained.

To carry out the test the response of the optical system is first obtained without the fiber-optic component in position using a set of square-wave targets of a suitable range of spatial frequencies. The intensity modulation of the transmitted light is recorded by the chart recorder for each frequency, and the contrast (or response) is calculated at each frequency from

$$C = \frac{I_{\max} - I_{\min}}{I_{\max} + I_{\min}}$$

This is repeated at each frequency with the fiber-optic component in place. The ratios of these values of C with and without the fiber optic are then plotted vs. frequency to get the response function. Focusing and alignment of the slit and target are critical to obtaining accurate response values, particularly at higher frequencies.

34. Edge-Response Test The edge-response test is used to evaluate the effectiveness of the stray-light absorption by the extramural absorption (EMA) in fiber-optic faceplates. In this test a microdensitometer is used to measure the falloff of intensity in the shadow of a knife-edge image transmitted by the faceplate.

A typical apparatus is shown in Fig. 13. Diffuse illumination is presented to the knife-edge target, which is placed in direct contact with the faceplate under

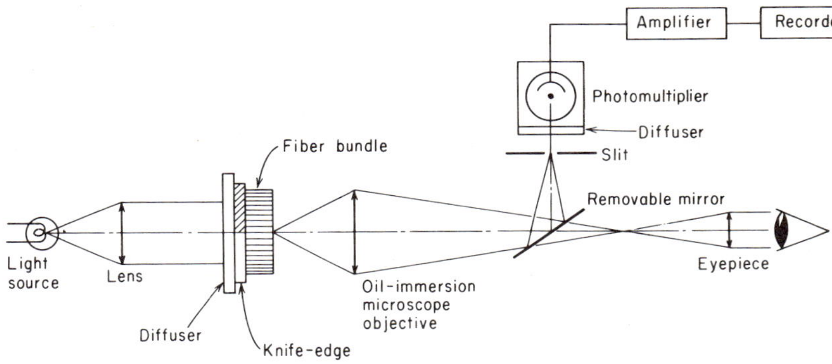

Fig. 13 Edge-response test.

test. The transmitted image is scanned with a microdensitometer having an oil-immersion microscope objective, and the light output is logarithmically amplified and recorded. Accurate alignment of the scanning slit and the knife-edge is essential.

A rapid falloff in intensity in the knife-edge shadow is a measure of effective EMA. Edge-response criteria have been set up for high-performance faceplates. A typical response criterion is shown in Table 7.

TABLE 7 Response vs. Distance from Knife-Edge Image

Distance from edge, μm	Response, %
12	4.00
25	1.00
50	0.50
125	0.25
375	0.15

35. Distortion Two types of distortion are important in imaging fiber optics, *gross distortion,* which causes the image of a straight line to curve, and *shear distortion,* which causes discrete lateral displacements, or shearing, of lines in the transmitted image. Gross distortion is caused by a long-range deformation or flow of the fibers during fabrication, while shear distortion is due to localized misalignment errors in the assembly of fibers or multifibers. Shear distortion in fiber bundles is sometimes also referred to as *incoherency*.

Both types of distortion can be measured by placing a bar target such as a Ronchi ruling in contact with one face of the fiber-optic component and measuring any displacement or curvature of the lines in the transmitted image. A contour projector is a convenient apparatus for making such measurements. The bar target should be placed in at least two (perpendicular) orientations to detect all distortions present.

36. Blemishes Blemishes in fiber-optic bundles are defined as any localized areas which do not transmit light as well as the average fiber. The unavoidable area occupied by the cladding or the normal areas between well-packed fibers are not considered blemishes.

In fiberscopes, blemishes are mainly due to three causes, imperfect packing (voids), broken fibers, and internally defective (gray) fibers. These are all of generally random character. In specifying the tolerance for such blemishes, the size, number, and distribution are generally given. Testing is often done by critically examining an enlarged photograph of the fiber bundle and counting the blemishes present.

In fused-fiber faceplates and image conduits, blemishes are due primarily to internally defective fibers. These are caused mainly by small entrapped gas bubbles, which deform the fiber walls and either scatter away the light or cut it off completely. Most of these blemishes are randomly distributed throughout the fiber material. Another form of blemish found in fused fiber-optic products is a grid of dark lines along the boundaries of some or all of the multifibers used in the assembly. This usually affects only one row of fiber elements in each multifiber along the affected boundaries and is commonly referred to as chicken wire. The tolerance criteria for blemishes in faceplates usually allow for the presence of some degree of chicken-wire blemish. The criteria are usually based on the size, number, and distribution of the random blemishes within zones of decreasing quality defined from the central area to the periphery of the plate.

37. Environmental Tests Environmental testing methods and equipment developed for other components can be applied to fiber-optic products. The usual tests are shock, vibration, temperature limits, thermal shock, and humidity. Faceplates and other fused fiber-optic components can be treated like other solid-glass components of similar dimensions and properties. Thus, temperature limits and tolerance of thermal shock are of chief interest. In plastic fibers, only the temperature limits are of particular concern. Glass fibers have somewhat greater limits on temperature but are likely to be limited by organic compounds used as adhesives or in the sheath. Vibration testing is of concern for glass-fiber devices. Humidity testing is important for glass fibers since there may be enough attack to reduce their strength. The humidity test is usually followed by flex tests to determine the effect on strength. Humidity may also cause loss of transmission in plastic fibers. Special tests for resistance to oils or other chemicals may be required if such conditions will occur.

REFERENCES

Baird, J. L. (1926): Br. Pat. 285,738.
Drougard, R. (1964): *J. Opt. Soc. Am.*, vol. 54, pp. 907–914.
Hansell, C. W. (1927): U.S. Pat. 1,751,584.
Heel, A. C. S. van (1953): *Ingenieur (Utrecht)*, vol. 65, p. 25.
Kapany, N. S., J. A. Eyer, and R. E. Keim (1957): *J. Opt. Soc. Am.*, vol. 47, pp. 423–427.
——— and J. J. Burke (1961): *J. Opt. Soc. Am.*, vol. 51, pp. 1067–1078.
Potter, R. J. (1960): Ph.D. thesis, University of Rochester, Rochester, N.Y.
——— (1961): *J. Opt. Soc. Am.*, vol. 51, pp. 1079–1089.
Snitzer, E. (1961): *J. Opt. Soc. Am.*, vol. 51, pp. 491–498.
——— and H. Osterberg (1961): *J. Opt. Soc. Am.*, vol. 51, pp. 499–505.
Tyndall, J. (1854): *Proc. R. Inst.*, vol. 1, pp. 446–448.

Section 14

Optical Properties of the Atmosphere

ROBERT A. McCLATCHEY

Infrared Physics Branch, Optical Physics Laboratory, Air Force Cambridge Research Laboratories, Bedford, Mass.

ROBERT W. FENN

Atmospheric Optics Branch, Optical Physics Laboratory, Air Force Cambridge Research Laboratories, Bedford, Mass.

JOHN E. A. SELBY

Infrared Physics Branch, Optical Physics Laboratory, Air Force Cambridge Research Laboratories, Bedford, Mass.

FREDERIC E. VOLZ

Atmospheric Optics Branch, Optical Physics Laboratory, Air Force Cambridge Research Laboratories, Bedford, Mass.

JOHN S. GARING

Director, Optical Physics Laboratory, Air Force Cambridge Research Laboratories, Bedford, Mass.

(*Numbers refer to paragraphs*)

1. Introduction† This section provides data from which attenuation or emission of radiation along a path in a given model atmosphere can be determined. The spectral region covered is from 0.25 to 25 μm. By neglecting the atmosphere above 100 km we overlook less than one-millionth of the total atmospheric mass.

We have defined five model atmospheres each for temperature, pressure, and absorbing-gas concentrations. In addition, two aerosol models have been defined. Thus, one effectively has 10 models from which to choose in order to use the data as presented without interpolation. However, interpolation schemes are described in order to allow the user to obtain optical data for intermediate conditions.

As it is not possible to cover all conceivable atmospheric models, neither is it possible to provide data for any spectral resolution. Thus, we are forced to

† The authors wish to acknowledge the many helpful suggestions made by Dr. Darrell E. Burch as a result of careful reading of the manuscript; the comments gratefully received from Dr. Cameron Cumming and Mr. Louis Elterman; and the computer programming and other analytical support given by Mr. Karekin Agazarian, Mr. Eric Shettle, and Mr. James H. Chetwynd. They also wish to thank the editors of the Handbook for permission to preprint this material as *Air Force Camb. Res. Lab. Environ. Res. Pap.* 411 AFCRL-72-0497, Aug. 24, 1972.

limit the high-resolution calculations to a few specific wavelengths. A list of laser-emission wavelengths was chosen for this purpose. We attempted to pick those lasers most commonly used in practical applications and as research tools. In addition to the laser-beam transmittance table, we have provided in graphical form a calculation scheme for determining atmospheric absorption and scattering at about a 20-wave-number resolution across the entire spectral region. This technique has been converted for a computer program completely described by Selby and McClatchey (1972). These curves, although appropriate for low-resolution transmittance calculations, would produce serious error if used for estimates of laser propagation in the atmosphere. For high-resolution curves appropriate to the laser-propagation problem see McClatchey (1971) and McClatchey and Selby (1972*a*,*b*). Much of this work is based on the compilation described in McClatchey et al. (1973).

The models defined in Pars. 2 to 4, together with the low-resolution-absorption calculation techniques presented in Pars. 9 to 20, were used to compute the radiation emitted to space and the radiation as observed looking up at the sky from the surface (sky background) for wavelengths longer than 5 μm. The radiation to space and sky backgrounds at shorter wavelengths is a complicated function of the solar zenith angle and the angle of observation, so that only a few specific examples have been given.

The effect of the refractive index on the geometrical path length and the effect of cloud and fog on the attenuation of radiation are discussed in Pars. 23 and 24.

MODEL ATMOSPHERES USED IN OPTICAL CALCULATIONS

2. Standard Model Atmospheres A series of five model atmospheres is presented in Table 1. The pressure, temperature, density, and water-vapor concentrations are those of the U.S. Standard Atmosphere Supplements, 1966, as compiled in the "Handbook of Geophysics and Space Environment" (Valley, 1965). The U.S. Standard Atmosphere, 1962, does not deviate appreciably from the midlatitude winter model shown in Table 1 for the purposes of most atmospheric-radiation calculations. The water-vapor densities above 11 km have been taken from Sissenwine et al. (1968) and ozone distributions from Valley (1965). A discussion of units and conversion can be found in Pars. 25 to 27.

3. Concentrations of Uniformly Mixed Gases Table 2 provides the data for determining the concentrations of all uniformly mixed atmospheric gases in any horizontal or vertical path. The number of $(\text{cm}\cdot\text{atm})_{\text{STP}}$ in a horizontal path at any other altitude is given by

$$(\text{cm}\cdot\text{atm})_{\text{STP}} = L\,\frac{\text{ppm}}{10}\,\frac{PT_0}{P_0T} \tag{1}$$

where L is the path length in kilometers and P_0 and T_0 refer to STP ($P_0 = 1$ atm and $T_0 = 273$ K).

The sixth column of Table 2 provides values from which the amount of gas between any two pressure levels in the atmosphere can be determined by multiplying the tabulated values by the pressure increment in millibars (Δp) times the secant of the zenith angle (see Par. 26).

4. Aerosol Distributions The two aerosol models (Table 3) describe a clear and a hazy atmosphere corresponding to a visibility of 23 and 5 km, respectively, at ground level. The aerosol size-distribution function for both models is the same at all altitudes and is similar to the one suggested by Deirmendjian (1963) for continental haze (Fig. 1), whereby the relationship between the number density $n(r)$ and the particle radius is given by

$$\frac{n(r)}{\Delta r} = \begin{cases} C_1 r^{-4} & \text{for } 0.1\ \mu\text{m} < r < 100\ \mu\text{m} \\ C_1 \times 10^4 & \text{for } 0.02\ \mu\text{m} < r < 0.1\ \mu\text{m} \\ 0 & \text{for } r < 0.02 \text{ and } r > 100.0 \end{cases} \tag{2}$$

TABLE 1 Model Atmospheres Used as a Basis of the Computation of Atmospheric Optical Properties

Ht., km	Pressure, mbar	Temp, K	Density, g/m^3	Water vapor, g/m^3	Ozone, g/m^3
			Tropical		
0	1,013	300.0	1,167	19	5.6×10^{-5}
1	904.0	294.0	1,064	13	5.6×10^{-5}
2	805.0	288.0	968.9	9.3	5.4×10^{-5}
3	715.0	284.0	875.6	4.7	5.1×10^{-5}
4	633.0	277.0	795.1	2.2	4.7×10^{-5}
5	559.0	270.0	719.9	1.5	4.5×10^{-5}
6	492.0	264.0	650.1	8.5×10^{-1}	4.3×10^{-5}
7	432.0	257.0	585.5	4.7×10^{-1}	4.1×10^{-5}
8	378.0	250.0	525.8	2.5×10^{-1}	3.9×10^{-5}
9	329.0	244.0	470.8	1.2×10^{-1}	3.9×10^{-5}
10	286.0	237.0	420.2	5.0×10^{-2}	3.9×10^{-5}
11	247.0	230.0	374.0	1.7×10^{-2}	4.1×10^{-5}
12	213.0	224.0	331.6	6.0×10^{-3}	4.3×10^{-5}
13	182.0	217.0	292.9	1.8×10^{-3}	4.5×10^{-5}
14	156.0	210.0	257.8	1.0×10^{-3}	4.5×10^{-5}
15	132.0	204.0	226.0	7.6×10^{-4}	4.7×10^{-5}
16	111.0	197.0	197.2	6.4×10^{-4}	4.7×10^{-5}
17	93.70	195.0	167.6	5.6×10^{-4}	6.9×10^{-5}
18	78.90	199.0	138.2	5.0×10^{-4}	9.0×10^{-5}
19	66.60	203.0	114.5	4.9×10^{-4}	1.4×10^{-4}
20	56.50	207.0	95.15	4.5×10^{-4}	1.9×10^{-4}
21	48.00	211.0	79.38	5.1×10^{-4}	2.4×10^{-4}
22	40.90	215.0	66.45	5.1×10^{-4}	2.8×10^{-4}
23	35.00	217.0	56.18	5.4×10^{-4}	3.2×10^{-4}
24	30.00	219.0	47.63	6.0×10^{-4}	3.4×10^{-4}
25	25.70	221.0	40.45	6.7×10^{-4}	3.4×10^{-4}
30	12.20	232.0	18.31	3.6×10^{-4}	2.4×10^{-4}
35	6.000	243.0	8.600	1.1×10^{-4}	9.2×10^{-5}
40	3.050	254.0	4.181	4.3×10^{-5}	4.1×10^{-5}
45	1.590	265.0	2.097	1.9×10^{-5}	1.3×10^{-5}
50	8.540×10^{-1}	270.0	1.101	6.3×10^{-6}	4.3×10^{-6}
70	5.790×10^{-2}	219.0	9.210×10^{-2}	1.4×10^{-7}	8.6×10^{-8}
100	3.000×10^{-4}	210.0	5.000×10^{-4}	1.0×10^{-9}	4.3×10^{-11}
			Midlatitude Summer		
0	1,013	294.0	1,191	14	6.0×10^{-5}
1	902.0	290.0	1,080	9.3	6.0×10^{-5}
2	802.0	285.0	975.7	5.9	6.0×10^{-5}
3	710.0	279.0	884.6	3.3	6.2×10^{-5}
4	628.0	273.0	799.8	1.9	6.4×10^{-5}
5	554.0	267.0	721.1	1.0	6.6×10^{-5}
6	487.0	261.0	648.7	6.1×10^{-1}	6.9×10^{-5}
7	426.0	255.0	583.0	3.7×10^{-1}	7.5×10^{-5}
8	372.0	248.0	522.5	2.1×10^{-1}	7.9×10^{-5}
9	324.0	242.0	466.9	1.2×10^{-1}	8.6×10^{-5}
10	281.0	235.0	415.9	6.4×10^{-2}	9.0×10^{-5}
11	243.0	229.0	369.3	2.2×10^{-2}	1.1×10^{-4}
12	209.0	222.0	326.9	6.0×10^{-3}	1.2×10^{-4}
13	179.0	216.0	288.2	1.8×10^{-3}	1.5×10^{-4}
14	153.0	216.0	246.4	1.0×10^{-3}	1.8×10^{-4}
15	130.0	216.0	210.4	7.6×10^{-4}	1.9×10^{-4}
16	111.0	216.0	179.7	6.4×10^{-4}	2.1×10^{-4}

TABLE 1 Model Atmospheres Used as a Basis of the Computation of Atmospheric Optical Properties (Continued)

Ht., km	Pressure, mbar	Temp, K	Density, g/m^3	Water vapor, g/m^3	Ozone, g/m^3
			Midlatitude Summer		
17	95.00	216.0	153.5	5.6×10^{-4}	2.4×10^{-4}
18	81.20	216.0	130.5	5.0×10^{-4}	2.8×10^{-4}
19	69.50	217.0	111.0	4.9×10^{-4}	3.2×10^{-4}
20	59.50	218.0	94.53	4.5×10^{-4}	3.4×10^{-4}
21	51.00	219.0	80.56	5.1×10^{-4}	3.6×10^{-4}
22	43.70	220.0	68.72	5.1×10^{-4}	3.6×10^{-4}
23	37.60	222.0	58.67	5.4×10^{-4}	3.4×10^{-4}
24	32.20	223.0	50.14	6.0×10^{-4}	3.2×10^{-4}
25	27.70	224.0	42.88	6.7×10^{-4}	3.0×10^{-4}
30	13.20	234.0	13.22	3.6×10^{-4}	2.0×10^{-4}
35	6.520	245.0	6.519	1.1×10^{-4}	9.2×10^{-5}
40	3.330	258.0	3.330	4.3×10^{-5}	4.1×10^{-5}
45	1.760	270.0	1.757	1.9×10^{-5}	1.3×10^{-5}
50	9.510×10^{-1}	276.0	9.512×10^{-1}	6.3×10^{-6}	4.3×10^{-6}
70	6.710×10^{-2}	218.0	6.706×10^{-2}	1.4×10^{-7}	8.6×10^{-8}
100	3.000×10^{-4}	210.0	5.000×10^{-4}	1.0×10^{-9}	4.3×10^{-11}
			Midlatitude Winter		
0	1,018	272.2	1,301	3.5	6.0×10^{-5}
1	897.3	268.7	1,162	2.5	5.4×10^{-5}
2	789.7	265.2	1,037	1.8	4.9×10^{-5}
3	693.8	261.7	923.0	1.2	4.9×10^{-5}
4	608.1	255.7	828.2	6.6×10^{-1}	4.9×10^{-5}
5	531.3	249.7	741.1	3.8×10^{-1}	5.8×10^{-5}
6	462.7	243.7	661.4	2.1×10^{-1}	6.4×10^{-5}
7	401.6	237.7	588.6	8.5×10^{-2}	7.7×10^{-5}
8	347.3	231.7	522.2	3.5×10^{-2}	9.0×10^{-5}
9	299.2	225.7	461.9	1.6×10^{-2}	1.2×10^{-4}
10	256.8	219.7	407.2	7.5×10^{-3}	1.6×10^{-4}
11	219.9	219.2	349.6	6.9×10^{-3}	2.1×10^{-4}
12	188.2	218.7	299.9	6.0×10^{-3}	2.6×10^{-4}
13	161.0	218.2	257.2	1.8×10^{-3}	3.0×10^{-4}
14	137.8	217.7	220.6	1.0×10^{-3}	3.2×10^{-4}
15	117.8	217.2	189.0	7.6×10^{-4}	3.4×10^{-4}
16	100.7	216.7	162.0	6.4×10^{-4}	3.6×10^{-4}
17	86.10	216.2	138.8	5.6×10^{-4}	3.9×10^{-4}
18	73.50	215.7	118.8	5.0×10^{-4}	4.1×10^{-4}
19	62.80	215.2	101.7	4.9×10^{-4}	4.3×10^{-4}
20	53.70	215.2	86.90	4.5×10^{-4}	4.5×10^{-4}
21	45.80	215.2	74.21	5.1×10^{-4}	4.3×10^{-4}
22	39.10	215.2	63.38	5.1×10^{-4}	4.3×10^{-4}
23	33.40	215.2	54.15	5.4×10^{-4}	3.9×10^{-4}
24	28.60	215.2	46.24	6.0×10^{-4}	3.6×10^{-4}
25	24.30	215.2	39.50	6.7×10^{-4}	3.4×10^{-4}
30	11.10	217.4	17.83	3.6×10^{-4}	1.9×10^{-4}
35	5.180	227.8	7.924	1.1×10^{-4}	9.2×10^{-5}
40	2.530	243.2	3.625	4.3×10^{-5}	4.1×10^{-5}
45	1.290	258.5	1.741	1.9×10^{-5}	1.3×10^{-5}
50	6.820×10^{-1}	265.7	8.954×10^{-1}	6.3×10^{-6}	4.3×10^{-6}
70	4.670×10^{-2}	230.7	7.051×10^{-2}	1.4×10^{-7}	8.6×10^{-8}
100	3.000×10^{-4}	210.2	5.000×10^{-4}	1.0×10^{-9}	4.3×10^{-11}

TABLE 1 Model Atmospheres Used as a Basis of the Computation of Atmospheric Optical Properties (Continued)

Ht., km	Pressure, mbar	Temp, K	Density, g/m^3	Water vapor, g/m^3	Ozone, g/m^3
			Subarctic Summer		
0	1,010	287.0	1,220	9.1	4.9×10^{-5}
1	896.0	282.0	1,110	6.0	5.4×10^{-5}
2	792.9	276.0	997.1	4.2	5.6×10^{-5}
3	700.0	271.0	898.5	2.7	5.8×10^{-5}
4	616.0	266.0	807.7	1.7	6.0×10^{-5}
5	541.0	260.0	724.4	1.0	6.4×10^{-5}
6	473.0	253.0	651.9	5.4×10^{-1}	7.1×10^{-5}
7	413.0	246.0	584.9	2.9×10^{-1}	7.5×10^{-5}
8	359.0	239.0	523.1	1.3×10^{-2}	7.9×10^{-5}
9	310.7	232.0	466.3	4.2×10^{-2}	1.1×10^{-4}
10	267.7	225.0	414.2	1.5×10^{-2}	1.3×10^{-4}
11	230.0	225.0	355.9	9.4×10^{-3}	1.8×10^{-4}
12	197.7	225.0	305.9	6.0×10^{-3}	2.1×10^{-4}
13	170.0	225.0	263.0	1.8×10^{-3}	2.6×10^{-4}
14	146.0	225.0	226.0	1.0×10^{-3}	2.8×10^{-4}
15	125.0	225.0	194.3	7.6×10^{-4}	3.2×10^{-4}
16	108.0	225.0	167.1	6.4×10^{-4}	3.4×10^{-4}
17	92.80	225.0	143.6	5.6×10^{-4}	3.9×10^{-4}
18	79.80	225.0	123.5	5.0×10^{-4}	4.1×10^{-4}
19	68.60	225.0	106.2	4.9×10^{-4}	4.1×10^{-4}
20	58.90	225.0	91.28	4.5×10^{-4}	3.9×10^{-4}
21	50.70	225.0	78.49	5.1×10^{-4}	3.6×10^{-4}
22	43.60	225.0	67.50	5.1×10^{-4}	3.2×10^{-4}
23	37.50	225.0	58.05	5.4×10^{-4}	3.0×10^{-4}
24	32.27	226.0	49.63	6.0×10^{-4}	2.8×10^{-4}
25	27.80	228.0	42.47	6.7×10^{-4}	2.6×10^{-4}
30	13.40	235.0	13.38	3.6×10^{-4}	1.4×10^{-4}
35	6.610	247.0	6.614	1.1×10^{-4}	9.2×10^{-5}
40	3.400	262.0	3.404	4.3×10^{-5}	4.1×10^{-5}
45	1.810	274.0	1.817	1.9×10^{-5}	1.3×10^{-5}
50	9.870×10^{-1}	277.0	9.868×10^{-1}	6.3×10^{-6}	4.3×10^{-6}
70	7.070×10^{-2}	216.0	7.071×10^{-2}	1.4×10^{-7}	8.6×10^{-8}
100	3.000×10^{-4}	210.0	5.000×10^{-4}	1.0×10^{-9}	4.3×10^{-11}
			Subarctic Winter		
0	1,013	257.1	1,372	1.2	4.1×10^{-5}
1	887.8	259.1	1,193	1.2	4.1×10^{-5}
2	777.5	255.9	1,058	9.4×10^{-1}	4.1×10^{-5}
3	679.8	252.7	936.6	6.8×10^{-1}	4.3×10^{-5}
4	593.2	247.7	833.9	4.1×10^{-1}	4.5×10^{-5}
5	515.8	240.9	745.7	2.0×10^{-1}	4.7×10^{-5}
6	446.7	234.1	664.6	9.8×10^{-2}	4.9×10^{-5}
7	385.3	227.3	590.4	5.4×10^{-2}	7.1×10^{-5}
8	330.8	220.6	522.6	1.1×10^{-2}	9.0×10^{-5}
9	282.9	217.2	453.8	8.4×10^{-3}	1.6×10^{-4}
10	241.8	217.2	387.9	5.5×10^{-3}	2.4×10^{-4}
11	206.7	217.2	331.5	3.8×10^{-3}	3.2×10^{-4}
12	176.6	217.2	283.4	2.6×10^{-3}	4.3×10^{-4}
13	151.0	217.2	242.2	1.8×10^{-3}	4.7×10^{-4}
14	129.1	217.2	207.1	1.0×10^{-3}	4.9×10^{-4}
15	110.3	217.2	177.0	7.6×10^{-4}	5.6×10^{-4}
16	94.31	216.6	151.7	6.4×10^{-4}	6.2×10^{-4}

TABLE 1 Model Atmospheres Used as a Basis of the Computation of Atmospheric Optical Properties (Continued)

Ht., km	Pressure, mbar	Temp, K	Density, g/m^3	Water vapor, g/m^3	Ozone, g/m^3
			Subarctic Winter		
17	80.58	216.0	130.0	5.6×10^{-4}	6.2×10^{-4}
18	68.82	215.4	111.3	5.0×10^{-4}	6.2×10^{-4}
19	58.75	214.8	95.29	4.9×10^{-4}	6.0×10^{-4}
20	50.14	214.1	81.55	4.5×10^{-4}	5.6×10^{-4}
21	42.77	213.6	69.76	5.1×10^{-4}	5.1×10^{-4}
22	36.47	213.0	59.66	5.1×10^{-4}	4.7×10^{-4}
23	31.09	212.4	51.00	5.4×10^{-4}	4.3×10^{-4}
24	26.49	211.8	43.58	6.0×10^{-4}	3.6×10^{-4}
25	22.56	211.2	37.22	6.7×10^{-4}	3.2×10^{-4}
30	10.20	216.0	16.45	3.6×10^{-4}	1.5×10^{-4}
35	4.701	222.2	7.368	1.1×10^{-4}	9.2×10^{-5}
40	2.243	234.7	3.330	4.3×10^{-5}	4.1×10^{-5}
45	1.113	247.0	1.569	1.9×10^{-5}	1.3×10^{-5}
50	5.719×10^{-1}	259.3	7.682×10^{-1}	6.3×10^{-6}	4.3×10^{-6}
70	4.016×10^{-2}	245.7	5.695×10^{-2}	1.4×10^{-7}	8.6×10^{-8}
100	3.000×10^{-4}	210.0	5.000×10^{-4}	1.0×10^{-9}	4.3×10^{-11}

It differs from Deirmendjian's model C in that the large-particle cutoff has been extended from 5 to 100 μm. This distribution function is normalized with $\Delta r = 1$ μm and $C_1 = 0.883 \times 10^{-3}$ (Fig. 1); i.e., the integral over the whole size range becomes 1. We need only multiply the normalized distribution function by the values of N in Table 3 to obtain the actual size distribution at any height in the atmosphere.

The refractive index for the aerosol particles (Table 4) is based on experimental data published by Volz (1972). The attenuation coefficients are then determined

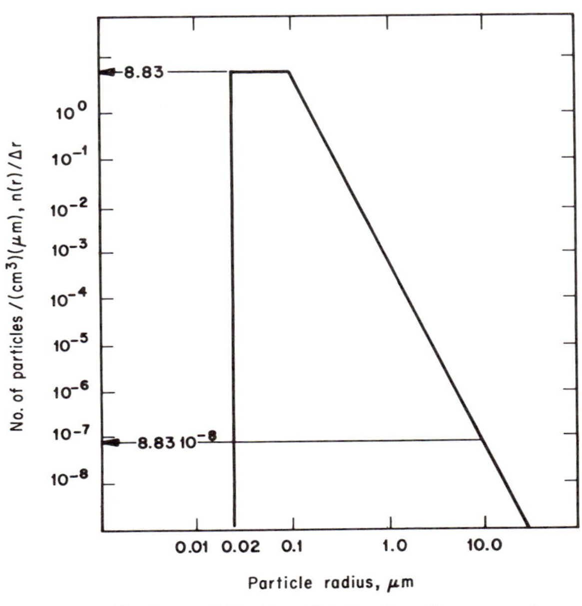

Fig. 1 Normalized particle-size distribution for aerosol models.

TABLE 2 Concentrations of Uniformly Mixed Gases

Constituent	Mol. wt.	ppm by vol.	Concentration, $(cm \cdot atm)_{STP}$ In vertical path from sea level	Concentration, $(cm \cdot atm)_{STP}$ In horizontal path at sea level	$g/(cm^2)(mbar)$	Reference
Air	28.97	10^6	8×10^5	10^5	1.02	Valley (1965)
CO_2	44	330	264	33	5.11×10^{-4}	Fink et al. (1964)
N_2O	44	0.28	0.22	0.028	4.34×10^{-7}	Birkland and Shaw (1959)
CO	28	0.075	0.06	0.0075	7.39×10^{-8}	Shaw (1968)
CH_4	16	1.6	1.28	0.16	9.01×10^{-7}	Shaw (1959)
O_2	32	2.095×10^5	1.68×10^5	2.095×10^4	0.236	Valley (1965)

TABLE 3 Aerosol Models: Vertical Distributions for a Clear and Hazy Atmosphere (Moderately Volcanic Conditions)

Altitude, km	Particle density N, particles/cm^3	
	23-km visibility clear	5-km visibility hazy
0	2,828	13,780
1	1,244	5,030
2	537.1	1,844
3	225.6	673.1
4	119.2	245.3
5	89.87	
6	63.37	
7	58.90	
8	60.69	
9	58.18	
10	56.75	
11	53.17	
12	55.85	
13	51.56	
14	50.48	
15	47.44	
16	45.11	
17	44.58	
18	43.14	
19	36.34	
20	26.67	
21	19.33	
22	14.55	
23	11.13	
24	8.82	
25	7.429	
30	2.238	
35	0.5890	
40	0.1550	
45	0.04082	
50	0.01078	
70	0.0000555	
100	0.00000001969	

as composites of 70 percent water-solubles and 30 percent dustlike substances, which can be assumed representative of continental aerosol.

The total numbers (Table 3) of aerosol particles N per unit volume for the clear atmosphere have been adjusted so that the total extinction coefficient at wavelength $\lambda = 0.55\ \mu m$ becomes identical to the values used in the atmospheric-attenuation model by Elterman (1968, 1970) at each altitude. This model corresponds to an atmospheric condition of approximately 23 km ground visibility. The hazy model is identical to the clear model for altitudes $H \geq 5$ km. For $H < 5$ km, the total number of aerosol particles increases exponentially to reach a value at ground level corresponding to a ground visibility of approximately 5 km.

FUNDAMENTALS OF ATMOSPHERIC TRANSMITTANCE

5. Transmittance of radiation in the atmosphere is complex owing to the dependence of scattering and absorption coefficients on a number of different physical properties of the atmosphere (Goody, 1964). In general, for the monochromatic transmittance of radiation along a path in the atmosphere

$$\tau = \exp(-\gamma\, \Delta L) \tag{3}$$

where γ is an attenuation coefficient and ΔL is the length of the path traversed by the radiation. The attenuation coefficient γ is given by

$$\gamma = \sigma + k \tag{4}$$

TABLE 4 Aerosol Complex Index of Refraction $n - n'i$, where n = Real Part and n' = Imaginary (Absorption) Part

Wave-length, μm	Refractive index	
	Water-soluble	Dustlike
0.20000	1.530 − 0.070*i*	1.530 − 0.070*i*
0.25000	1.530 − 0.030*i*	1.530 − 0.030*i*
0.30000	1.530 − 0.008*i*	1.530 − 0.008*i*
0.33710	1.530 − 0.005*i*	1.530 − 0.008*i*
0.48800	1.530 − 0.005*i*	1.530 − 0.008*i*
0.51450	1.530 − 0.005*i*	1.530 − 0.008*i*
0.63280	1.530 − 0.006*i*	1.530 − 0.008*i*
0.69430	1.530 − 0.007*i*	1.530 − 0.008*i*
0.86000	1.520 − 0.012*i*	1.520 − 0.008*i*
1.06000	1.520 − 0.017*i*	1.520 − 0.008*i*
1.53600	1.510 − 0.023*i*	1.400 − 0.008*i*
2.00000	1.420 − 0.008*i*	1.260 − 0.008*i*
2.50000	1.420 − 0.012*i*	1.180 − 0.009*i*
2.70000	1.400 − 0.055*i*	1.180 − 0.013*i*
3.00000	1.420 − 0.022*i*	1.160 − 0.012*i*
3.20000	1.430 − 0.008*i*	1.220 − 0.010*i*
3.39230	1.430 − 0.007*i*	1.260 − 0.013*i*
3.50000	1.450 − 0.005*i*	1.280 − 0.011*i*
3.75000	1.452 − 0.004*i*	1.270 − 0.011*i*
4.00000	1.455 − 0.005*i*	1.260 − 0.012*i*
4.50000	1.460 − 0.013*i*	1.260 − 0.014*i*
5.50000	1.440 − 0.018*i*	1.220 − 0.021*i*
6.00000	1.410 − 0.023*i*	1.150 − 0.037*i*
6.50000	1.460 − 0.033*i*	1.130 − 0.042*i*
7.20000	1.400 − 0.070*i*	1.400 − 0.055*i*
7.90000	1.200 − 0.065*i*	1.150 − 0.040*i*
8.20000	1.010 − 0.100*i*	1.130 − 0.074*i*
8.50000	1.300 − 0.215*i*	1.300 − 0.090*i*
8.70000	2.400 − 0.290*i*	1.400 − 0.100*i*
9.00000	2.560 − 0.370*i*	1.700 − 0.140*i*
9.20000	2.200 − 0.420*i*	1.720 − 0.150*i*
9.50000	1.950 − 0.160*i*	1.730 − 0.162*i*
10.00000	1.820 − 0.090*i*	1.750 − 0.162*i*
10.59100	1.760 − 0.070*i*	1.620 − 0.120*i*
11.00000	1.720 − 0.050*i*	1.620 − 0.105*i*
13.00000	1.620 − 0.055*i*	1.470 − 0.100*i*
14.80000	1.440 − 0.100*i*	1.570 − 0.100*i*
15.00000	1.420 − 0.200*i*	1.570 − 0.100*i*
17.20000	2.080 − 0.240*i*	1.630 − 0.100*i*
18.50000	1.850 − 0.170*i*	1.640 − 0.120*i*
20.00000	2.120 − 0.220*i*	1.680 − 0.220*i*
25.00000	1.880 − 0.280*i*	1.970 − 0.240*i*
27.90000	1.840 − 0.290*i*	1.890 − 0.320*i*
30.00000	1.820 − 0.003*i*	1.800 − 0.420*i*
35.00000	1.920 − 0.400*i*	1.900 − 0.500*i*
40.00000	1.860 − 0.500*i*	2.100 − 0.600*i*

where σ is the scattering coefficient and k is the absorption coefficient. Both these coefficients are defined in km^{-1}. Equation (3) is strictly valid only when applied to monochromatic radiation.

The scattering and absorption coefficients defined in Eq. (4) depend on both molecules and aerosols in the atmospheric path. If electromagnetic radiation is incident on a molecule or an aerosol, a portion of the radiation is absorbed and the rest is scattered in all directions. Thus, we must make the following definitions:

$$\sigma = \sigma_m + \sigma_a \tag{4a}$$

$$k = k_m + k_a \tag{4b}$$

where the subscripts m and a indicate molecule and aerosol, respectively. We shall endeavor in this section to provide quantitative information on all four quantities defined in Eqs. (4a) and (4b).

The molecular-scattering coefficient depends only on the number density of molecules in the radiation path, whereas the molecular-absorption coefficient is a function not only of the amount of absorbing gas but also of the local temperature and pressure of the gas. The wavelength dependence of molecular (Rayleigh) scattering is very nearly $\sigma_m \approx \lambda^{-4}$. The variation of the molecular-absorption coefficient with wavelength is much more complicated, being a highly oscillatory function of wavelength due to the presence of numerous molecular–absorption-band complexes. These band complexes result for the most part from minor atmospheric constituents. The responsible molecules are (in order of importance) H_2O, CO_2, O_3, N_2O, CO, O_2, CH_4, and N_2. All these molecules except H_2O and O_3 are assumed to be uniformly mixed by volume.

The quantities of σ_a and k_a depend on the number density and size distribution of aerosols as well as on their complex index of refraction. The amount of absorbed radiation and the distribution of scattered light can be derived theoretically for spherical particles (and some other simple forms). According to a theory developed by Mie (1908), the distribution of radiation with scattering angle ϕ can be expressed as

$$i(\phi) = I_0 \frac{\lambda^2}{4\pi^2} \frac{i_1(\phi) + i_2(\phi)}{2} \qquad \frac{\text{W}}{\text{cm}^2} \frac{\text{cm}^2}{\text{sr}} \tag{5}$$

where i_1 and i_2 represent the two polarized components of the scattered radiation (van de Hulst, 1957). They are functions of the ratio of particle size to wavelength, the particle refractive index, and the scattering angle ϕ. I_0 is the intensity of incident radiation. If there are N particles per unit scattering volume with a size distribution $n(r)$, replacing i_1 and i_2 by $\int_r i_{1,2}(r)n(r)\,dr$ will yield the integrated angular scattering intensity $I(\phi)$ from the unit volume of the particle cloud into 1 sr.

The integral $\int_0^{4\pi} I(\phi)\,d\Omega$ gives the total amount of scattered radiation, which is also equal to $I_0\sigma_a$. For a size distribution one obtains

$$I(\phi) = I_0\sigma f(\phi) \tag{6}$$

where $f(\phi)$ is the normalized scattering function, and the integral $\int_0^{4\pi} f(\phi)\,d\Omega$ becomes 1.

Figure 26 shows $f(\phi)$ for molecular scattering (wavelength-independent) and aerosol scattering for two wavelengths for the aerosol model described in Par. 3.

An absorption coefficient k_a for aerosol particles can be defined in the same manner as the scattering coefficient σ_a. The absorption and scattering coefficients due to aerosol are usually smooth functions of wavelength.

The molecular-absorption coefficient for a single spectral line can be given to good accuracy in the lower 50 km of the atmosphere by

$$k_m(\nu) = \frac{S\alpha}{\pi[(\nu - \nu_0)^2 + \alpha^2]} \tag{7}$$

where S = line intensity
α = line half width
ν_0 = frequency of line center
ν = frequency at which absorption coefficient is required

See Par. 25 for the relationship between ν and λ. The half width depends on the pressure and temperature approximately as follows:

$$\alpha = \alpha_0 \frac{P}{P_0} \sqrt{\frac{T_0}{T}} \tag{8}$$

The line intensity also depends on the temperature through the Boltzmann factor and the partition function (Goody, 1964; Penner, 1959). Above 50 km, doppler broadening should be considered.

Let us consider a plane-parallel atmosphere and a wavelength at which several lines belonging to one or more molecular constituents contribute to the absorption coefficient. The molecular-absorption coefficient at frequency ν in any one of the j layers will be given by

$$k_{mj} = \sum_i k_{mji} \tag{9}$$

where the i summation is over all molecular-absorption lines due to all absorbing species which are close enough to the frequency ν to contribute significantly to the total molecular-absorption coefficient defined in Eq. (9). Introducing the molecular scattering σ_m and the absorption k_a and scattering σ_a due to aerosols as defined in Eqs. (4*a*) and (4*b*), we have for the total attenuation coefficient in the jth layer

$$\gamma_j = k_{mj} + \sigma_{mj} + k_{aj} + \sigma_{aj} \tag{10}$$

Thus, according to Eq. (3), the transmittance of an atmosphere having j layers is given by

$$\tau(\nu) = \exp\left(-\sum \gamma_j \,\Delta L_j\right) \tag{11}$$

It should be noted from Eq. (11) that ΔL does not necessarily denote vertical distance. The problem of slant-path transmittance is easily solved by writing $\Delta L_j = \Delta z_j \sec\theta$, where Δz_j represents the increment of vertical distance. In this case, Eq. (11) becomes

$$\tau_M(\nu) = [\tau(\nu)]^M \tag{12}$$

where $M = \sec\theta$ is the secant of the angle between the vertical and the direction of observation. Thus it becomes very easy to convert vertical-transmittance values to slant-path transmittance. As θ approaches 90°, the result becomes meaningless due to the assumption of a plane-parallel atmosphere (see Par. 23).

The above discussion is valid only at a single wavelength (or very narrow wavelength interval which is known to be small compared with the width of any absorption lines in the vicinity). A sufficiently narrow spectral interval for most purposes is 0.001 wave number (cm^{-1}). The process which so limits the validity of the above relations is molecular absorption. As noted above, $k_m(\nu)$ is a highly oscillatory function of frequency, whereas $k_a(\nu)$, $\sigma_m(\nu)$, and $\sigma_a(\nu)$ are much more slowly varying functions.

To calculate the atmospheric transmittance over a large spectral interval, it is necessary to average Eq. (11) or (12) over the desired spectral interval [see Eq. (14)]. When many absorption lines are present, this can be a particularly tedious job (even on a computer). Various techniques developed to treat this problem are discussed in Pars. 9 to 20.

ATMOSPHERIC TRANSMITTANCE AT SELECTED FREQUENCIES

6. List of Laser Frequencies It is necessary to limit the number of specific frequencies at which we quote atmospheric absorption and scattering coefficients. We have therefore chosen the list of specific laser wavelengths (Table 5) at which values of these coefficients are given in Table 6.

7. Scattering and Absorption Coefficients Each atmospheric model defined in Pars. 2 to 4 has been used as a basis for calculating the absorption and scattering coefficients given in Table 6† for each laser wavelength. Numerical values for the coefficients were computed for the 0-, 1-, 2-, . . . km levels and then logarithmically interpolated for each atmospheric layer. The aerosol-attenuation coefficients given in Table 6, which correspond to the clear and hazy atmospheres defined in Par. 4, are identical for all geographical and seasonal models. Table 6 lists

† In the construction of the portions of Table 6 for $\lambda = 0.4880$, 0.5145, 0.6328, and 0.6943, the molecular absorption due to ozone was inadvertently omitted. It can be estimated from Figs. 23 and 25 (see Par. 20).

the four extinction coefficients as a function of altitude for each model: k_m and σ_m are the molecular absorption and scattering coefficients, and k_a and σ_a are the aerosol absorption and scattering coefficients. If the total attenuation coefficient is required, the four coefficients can be summed:

$$\gamma = k_m + \sigma_m + k_a + \sigma_a \tag{13}$$

8. Interpolation Procedures In general, it is permissible to interpolate linearly with respect to wavelength between values in Table 6 except for the molecular-absorption coefficient k_m. This is true because of the smooth variation of scattering coefficients and of aerosol-absorption coefficients with wavelength and the rapid spectral variations in molecular-absorption coefficients. All scattering and absorption coefficients can be interpolated linearly with respect to height. The tables have been constructed so that linear interpolation with respect to height is expected to give correct values to ±5 percent. If a given atmospheric condition differs from one of the 10 models of Pars. 2 to 4, larger errors may result.

TABLE 5 Laser Wavelengths

Wavelength, μm	Source
0.3371	Nitrogen
0.4880	Argon
0.5145	Argon
0.6328	Helium-neon
0.6943	Ruby
0.86	Gallium arsenide
1.06	Neodymium in glass
1.536	Erbium in glass
3.39225	Helium-neon
10.591	Carbon dioxide
27.9	Water vapor
337	Hydrogen cyanide

Again it should be remembered that one should never attempt to interpolate with respect to wavelength between molecular-absorption coefficient values k_m given in Table 6.

ATMOSPHERIC TRANSMITTANCE AT LOW RESOLUTION FROM 0.25 TO 25 μm

9. The calculations presented in Pars. 6 to 8 were made at infinite resolution (monochromatic calculations). In practice it is impossible to measure radiation transmitted at a single frequency. Instead one measures the transmittance $\bar{\tau}_{\Delta\nu}(\nu)$ averaged over the spectral interval $\Delta\nu$ accepted by the receiver,

$$\bar{\tau}_{\Delta\nu}(\nu) = \frac{1}{\Delta\nu} \int \tau(\nu)\, d\nu \tag{14}$$

where ν is the central frequency in the interval $\Delta\nu$. Consequently for many applications one is interested in knowing the transmittance of the atmosphere averaged over a relatively wide spectral interval, i.e., for low resolution.

Thus the term transmittance is somewhat ambiguous unless it is qualified by some indication of the spectral resolution $\Delta\nu$ over which it is averaged. This is particularly true of molecular absorption, since the absorption coefficient k_m is a rapidly varying function of frequency. Because of the rapid variation of k_m with frequency the averaged transmittance $\bar{\tau}$ does not, in general, obey the simple exponential law given in Eq. (3). That is,

$$\bar{\tau}_{\Delta\nu}(\nu) = \frac{1}{\Delta\nu} \int \exp\,[-k_m(\nu)\,\Delta L]\, d\nu \neq \exp\,[-\bar{K}(\nu)\,\Delta L] \tag{15}$$

TABLE 6 Values of Attenuation Coefficient per Kilometer as a Function of Altitude for Each Laser Wavelength and Each Atmospheric Model

Height, km	Tropical		Midlatitude				Subarctic				Clear		Hazy	
			Summer		Winter		Summer		Winter					
	Molecular absorption k_m, km^{-1}	Molecular scattering σ_m, km^{-1}	Molecular absorption k_m, km^{-1}	Molecular scattering σ_m, km^{-1}	Molecular absorption k_m, km^{-1}	Molecular scattering σ_m, km^{-1}	Molecular absorption k_m, km^{-1}	Molecular scattering σ_m, km^{-1}	Molecular absorption k_m, km^{-1}	Molecular scattering σ_m, km^{-1}	Aerosol absorption k_a, km^{-1}	Aerosol scattering σ_a, km^{-1}	Aerosol absorption k_a, km^{-1}	Aerosol scattering σ_a, km^{-1}
							$\lambda = 0.337\ \mu m$							
0	2.34×10^{-4}	8.46×10^{-2}	2.52×10^{-4}	8.64×10^{-2}	2.52×10^{-4}	9.38×10^{-2}	2.07×10^{-4}	8.83×10^{-2}	1.71×10^{-4}	9.89×10^{-2}	1.45×10^{-2}	2.30×10^{-1}	7.06×10^{-2}	1.12
0–1	2.32×10^{-4}	8.07×10^{-2}	2.52×10^{-4}	8.25×10^{-2}	2.38×10^{-4}	8.73×10^{-2}	2.16×10^{-4}	8.40×10^{-2}	1.71×10^{-4}	4.94×10^{-2}	9.88×10^{-3}	1.67×10^{-1}	4.45×10^{-2}	7.07×10^{-1}
1–2	2.27×10^{-4}	7.34×10^{-2}	2.52×10^{-4}	7.47×10^{-2}	2.16×10^{-4}	7.94×10^{-2}	2.30×10^{-4}	7.59×10^{-2}	1.71×10^{-4}	9.24×10^{-2}	4.31×10^{-3}	6.85×10^{-2}	1.68×10^{-2}	2.58×10^{-1}
2–3	2.18×10^{-4}	6.68×10^{-2}	2.57×10^{-4}	6.76×10^{-2}	2.07×10^{-4}	7.23×10^{-2}	2.39×10^{-4}	6.85×10^{-2}	1.76×10^{-4}	8.11×10^{-2}	1.84×10^{-3}	2.92×10^{-2}	5.95×10^{-3}	9.46×10^{-2}
3–4	1.93×10^{-4}	6.05×10^{-2}	2.65×10^{-4}	6.12×10^{-2}	2.07×10^{-4}	6.32×10^{-2}	2.48×10^{-4}	6.15×10^{-2}	1.85×10^{-4}	7.19×10^{-2}	8.54×10^{-4}	1.36×10^{-2}	2.17×10^{-3}	3.45×10^{-2}
4–5	1.84×10^{-4}	5.49×10^{-2}	2.74×10^{-4}	5.53×10^{-2}	2.25×10^{-4}	5.66×10^{-2}	2.61×10^{-4}	5.52×10^{-2}	1.94×10^{-4}	6.38×10^{-2}	5.32×10^{-4}	8.45×10^{-3}	7.93×10^{-4}	1.26×10^{-3}
5–6	1.75×10^{-4}	4.96×10^{-2}	2.83×10^{-4}	4.98×10^{-2}	2.56×10^{-4}	5.06×10^{-2}	2.84×10^{-4}	4.96×10^{-2}	2.03×10^{-4}	5.08×10^{-2}	3.89×10^{-4}	6.17×10^{-3}	3.89×10^{-4}	6.17×10^{-3}
6–7	1.66×10^{-4}	4.47×10^{-2}	2.99×10^{-4}	4.47×10^{-2}	2.97×10^{-4}	4.51×10^{-2}	3.06×10^{-4}	4.45×10^{-2}	2.52×10^{-4}	4.52×10^{-2}	3.13×10^{-4}	4.98×10^{-3}	3.13×10^{-4}	4.98×10^{-3}
7–8	1.62×10^{-4}	4.02×10^{-2}	3.22×10^{-4}	4.01×10^{-2}	3.51×10^{-4}	4.01×10^{-2}	3.24×10^{-4}	3.99×10^{-2}	3.38×10^{-4}	4.01×10^{-2}	3.06×10^{-4}	4.87×10^{-3}	3.06×10^{-4}	4.87×10^{-3}
8–9	1.62×10^{-4}	3.61×10^{-2}	3.44×10^{-4}	3.59×10^{-2}	4.50×10^{-4}	3.55×10^{-2}	3.87×10^{-4}	3.57×10^{-2}	5.27×10^{-4}	3.52×10^{-2}	3.04×10^{-4}	4.84×10^{-3}	3.04×10^{-4}	4.84×10^{-3}
9–10	1.62×10^{-4}	3.23×10^{-2}	3.67×10^{-4}	3.21×10^{-2}	5.99×10^{-4}	3.14×10^{-2}	4.95×10^{-4}	3.17×10^{-2}	8.19×10^{-4}	3.03×10^{-2}	2.94×10^{-4}	4.68×10^{-3}	2.94×10^{-4}	4.68×10^{-3}
10–11	1.66×10^{-4}	2.88×10^{-2}	4.14×10^{-4}	2.85×10^{-2}	7.83×10^{-4}	2.73×10^{-2}	6.44×10^{-4}	2.78×10^{-2}	1.17×10^{-3}	2.59×10^{-2}	2.82×10^{-4}	4.47×10^{-3}	2.82×10^{-4}	4.47×10^{-3}
11–12	1.75×10^{-4}	2.56×10^{-2}	4.86×10^{-4}	2.53×10^{-2}	9.96×10^{-4}	2.34×10^{-2}	8.33×10^{-4}	2.39×10^{-2}	1.57×10^{-3}	2.21×10^{-2}	2.79×10^{-4}	4.44×10^{-3}	2.79×10^{-4}	4.44×10^{-3}
12–13	1.82×10^{-4}	2.26×10^{-2}	5.76×10^{-4}	2.24×10^{-2}	1.16×10^{-3}	2.01×10^{-2}	9.89×10^{-4}	2.05×10^{-2}	1.85×10^{-3}	1.89×10^{-2}	2.75×10^{-4}	4.37×10^{-3}	2.75×10^{-4}	4.37×10^{-3}
13–14	1.87×10^{-4}	2.00×10^{-2}	6.84×10^{-4}	1.94×10^{-2}	1.27×10^{-3}	1.72×10^{-2}	1.12×10^{-3}	1.76×10^{-2}	2.03×10^{-3}	1.62×10^{-2}	2.61×10^{-4}	4.15×10^{-3}	2.61×10^{-4}	4.15×10^{-3}
14–15	1.91×10^{-4}	1.76×10^{-2}	7.74×10^{-4}	1.66×10^{-2}	1.37×10^{-3}	1.48×10^{-2}	1.25×10^{-3}	1.51×10^{-2}	2.22×10^{-3}	1.38×10^{-2}	2.51×10^{-4}	3.99×10^{-3}	2.51×10^{-4}	3.99×10^{-3}
15–16	1.96×10^{-4}	1.54×10^{-2}	8.46×10^{-4}	1.42×10^{-2}	1.55×10^{-3}	1.26×10^{-2}	1.37×10^{-3}	1.20×10^{-2}	2.45×10^{-3}	1.18×10^{-2}	2.37×10^{-4}	3.77×10^{-3}	2.37×10^{-4}	3.77×10^{-3}
16–17	2.43×10^{-4}	1.33×10^{-2}	9.56×10^{-4}	1.22×10^{-2}	1.64×10^{-3}	1.08×10^{-2}	1.51×10^{-3}	1.12×10^{-2}	2.58×10^{-3}	1.01×10^{-2}	2.30×10^{-4}	3.65×10^{-3}	2.30×10^{-4}	3.65×10^{-3}
17–18	3.33×10^{-4}	1.11×10^{-2}	1.10×10^{-3}	1.04×10^{-2}	1.73×10^{-3}	9.29×10^{-3}	1.66×10^{-3}	9.62×10^{-3}	2.60×10^{-3}	8.69×10^{-3}	2.25×10^{-4}	3.57×10^{-3}	2.25×10^{-4}	3.57×10^{-3}
18–19	4.86×10^{-4}	9.12×10^{-3}	1.25×10^{-3}	8.79×10^{-3}	1.84×10^{-3}	7.97×10^{-3}	1.71×10^{-3}	8.27×10^{-3}	2.55×10^{-3}	7.44×10^{-3}	2.03×10^{-4}	3.23×10^{-3}	2.03×10^{-4}	3.23×10^{-3}
19–20	6.98×10^{-4}	7.57×10^{-3}	1.40×10^{-3}	7.46×10^{-3}	1.84×10^{-3}	6.82×10^{-3}	1.66×10^{-3}	7.11×10^{-3}	2.42×10^{-3}	6.37×10^{-3}	1.60×10^{-4}	2.54×10^{-3}	1.60×10^{-4}	2.54×10^{-3}
20–21	8.89×10^{-4}	6.32×10^{-3}	1.48×10^{-3}	6.36×10^{-3}	1.86×10^{-3}	5.81×10^{-3}	1.58×10^{-3}	6.11×10^{-3}	2.27×10^{-3}	5.45×10^{-3}	1.17×10^{-4}	1.86×10^{-3}	1.17×10^{-4}	1.86×10^{-3}
21–22	1.06×10^{-3}	5.28×10^{-3}	1.50×10^{-3}	5.42×10^{-3}	1.80×10^{-3}	4.96×10^{-3}	1.45×10^{-3}	5.26×10^{-3}	2.08×10^{-3}	4.66×10^{-3}	8.62×10^{-5}	1.37×10^{-3}	8.62×10^{-5}	1.37×10^{-3}
22–23	1.23×10^{-3}	4.44×10^{-3}	1.47×10^{-3}	4.63×10^{-3}	1.71×10^{-3}	4.24×10^{-3}	1.34×10^{-3}	4.52×10^{-3}	1.87×10^{-3}	3.98×10^{-3}	6.54×10^{-5}	1.04×10^{-3}	6.54×10^{-5}	1.04×10^{-3}
23–24	1.40×10^{-3}	3.76×10^{-3}	1.40×10^{-3}	3.95×10^{-3}	1.61×10^{-3}	3.62×10^{-3}	1.23×10^{-3}	3.88×10^{-3}	1.66×10^{-3}	3.41×10^{-3}	5.09×10^{-5}	8.09×10^{-4}	5.09×10^{-5}	8.09×10^{-4}
24–25	1.48×10^{-3}	3.19×10^{-3}	1.33×10^{-3}	3.38×10^{-3}	1.48×10^{-3}	3.09×10^{-3}	1.12×10^{-3}	3.32×10^{-3}	1.44×10^{-3}	2.91×10^{-3}	4.15×10^{-5}	6.60×10^{-4}	4.15×10^{-5}	6.60×10^{-4}
25–30	1.24×10^{-3}	2.12×10^{-3}	1.06×10^{-3}	2.28×10^{-3}	1.10×10^{-3}	2.18×10^{-3}	8.24×10^{-4}	2.25×10^{-3}	9.75×10^{-4}	1.93×10^{-3}	2.22×10^{-5}	3.52×10^{-4}	2.22×10^{-5}	3.52×10^{-4}
30–35	6.98×10^{-4}	9.72×10^{-4}	6.12×10^{-4}	1.05×10^{-3}	5.94×10^{-4}	1.04×10^{-3}	4.86×10^{-4}	1.65×10^{-3}	5.09×10^{-4}	8.58×10^{-4}	6.33×10^{-6}	1.01×10^{-4}	6.33×10^{-6}	1.01×10^{-4}
35–40	2.79×10^{-4}	4.63×10^{-4}	2.79×10^{-4}	5.00×10^{-4}	2.79×10^{-4}	4.16×10^{-4}	2.79×10^{-4}	4.99×10^{-4}	2.79×10^{-4}	3.85×10^{-4}	1.67×10^{-6}	2.65×10^{-5}	1.67×10^{-6}	2.65×10^{-5}
40–45	1.13×10^{-4}	2.28×10^{-4}	1.13×10^{-4}	2.46×10^{-4}	1.13×10^{-4}	1.94×10^{-4}	1.13×10^{-4}	2.46×10^{-4}	1.13×10^{-4}	1.76×10^{-4}	4.38×10^{-7}	6.97×10^{-6}	4.38×10^{-7}	6.97×10^{-6}
45–50	3.60×10^{-5}	1.16×10^{-4}	3.60×10^{-5}	1.26×10^{-4}	3.60×10^{-5}	9.53×10^{-5}	3.60×10^{-5}	1.28×10^{-4}	3.60×10^{-5}	3.42×10^{-5}	1.16×10^{-7}	1.84×10^{-6}	1.16×10^{-7}	1.84×10^{-6}
50–70	9.00×10^{-6}	4.31×10^{-5}	9.00×10^{-6}	4.75×10^{-5}	9.00×10^{-6}	3.48×10^{-5}	9.00×10^{-6}	4.88×10^{-5}	9.00×10^{-6}	2.97×10^{-5}	1.04×10^{-8}	1.66×10^{-7}	1.04×10^{-8}	1.66×10^{-7}
70–100	$<10^{-6}$	3.35×10^{-6}	$<10^{-6}$	3.90×10^{-6}	$<10^{-6}$	2.56×10^{-6}	$<10^{-6}$	4.12×10^{-6}	1.80×10^{-6}	2.07×10^{-6}	3.58×10^{-11}	5.69×10^{-10}	3.58×10^{-11}	5.69×10^{-10}
							$\lambda = 0.4880\ \mu m$ (see footnote in Par. 7)							
0	$<10^{-3}$	1.82×10^{-2}	$<10^{-3}$	1.86×10^{-2}	$<10^{-3}$	2.02×10^{-2}	$<10^{-3}$	1.90×10^{-2}	$<10^{-3}$	2.13×10^{-2}	1.05×10^{-2}	1.74×10^{-1}	5.13×10^{-2}	8.48×10^{-1}
0–1		1.74×10^{-2}		1.77×10^{-2}		1.91×10^{-2}		1.80×10^{-2}		1.99×10^{-2}	7.19×10^{-3}	1.19×10^{-1}	3.23×10^{-2}	5.34×10^{-1}
1–2		1.58×10^{-2}		1.60×10^{-2}		1.70×10^{-2}		1.63×10^{-2}		1.74×10^{-2}	3.14×10^{-3}	5.18×10^{-2}	1.18×10^{-2}	1.95×10^{-1}
2–3		1.43×10^{-2}		1.45×10^{-2}		1.52×10^{-2}		1.47×10^{-2}		1.55×10^{-2}	1.34×10^{-3}	2.21×10^{-2}	4.33×10^{-3}	7.15×10^{-2}
3–4		1.30×10^{-2}		1.31×10^{-2}		1.36×10^{-2}		1.32×10^{-2}		1.37×10^{-2}	6.21×10^{-4}	1.03×10^{-2}	1.58×10^{-3}	2.61×10^{-2}
4–5		1.18×10^{-2}		1.18×10^{-2}		1.22×10^{-2}		1.19×10^{-2}		1.22×10^{-2}	3.87×10^{-4}	6.39×10^{-3}	5.77×10^{-4}	9.52×10^{-3}
5–6		1.06×10^{-2}		1.06×10^{-2}		1.09×10^{-2}		1.07×10^{-2}		1.09×10^{-2}	2.83×10^{-4}	4.67×10^{-3}	2.83×10^{-4}	4.67×10^{-3}
6–7		9.56×10^{-3}		9.54×10^{-3}		9.68×10^{-3}		9.57×10^{-3}		9.72×10^{-3}	2.28×10^{-4}	3.76×10^{-3}	2.28×10^{-4}	3.76×10^{-3}
7–8		8.61×10^{-3}		8.55×10^{-3}		8.60×10^{-3}		8.58×10^{-3}		8.62×10^{-3}	2.23×10^{-4}	3.68×10^{-3}	2.23×10^{-4}	3.68×10^{-3}
8–9		7.72×10^{-3}		7.65×10^{-3}		7.62×10^{-3}		7.66×10^{-3}		7.56×10^{-3}	2.21×10^{-4}	3.66×10^{-3}	2.21×10^{-4}	3.66×10^{-3}

9–10		6.89×10^{-3}		6.83×10^{-3}		6.73×10^{-3}		6.82×10^{-3}		6.52×10^{-3}	2.14×10^{-4}	3.54×10^{-3}	2.14×10^{-4}	3.54×10^{-3}
10–11		6.16×10^{-3}		6.09×10^{-3}		5.86×10^{-3}		5.97×10^{-3}		5.57×10^{-3}	2.05×10^{-4}	3.38×10^{-3}	2.05×10^{-4}	3.38×10^{-3}
11–12		5.46×10^{-3}		5.40×10^{-3}		5.03×10^{-3}		5.13×10^{-3}		4.76×10^{-3}	2.03×10^{-4}	3.35×10^{-3}	2.03×10^{-4}	3.35×10^{-3}
12–13		4.83×10^{-3}		4.77×10^{-3}		4.31×10^{-3}		4.41×10^{-3}		4.07×10^{-3}	2.00×10^{-4}	3.30×10^{-3}	2.00×10^{-4}	3.30×10^{-3}
13–14		4.27×10^{-3}		4.15×10^{-3}		3.70×10^{-3}		3.79×10^{-3}		3.48×10^{-3}	1.90×10^{-4}	3.14×10^{-3}	1.90×10^{-4}	3.14×10^{-3}
14–15		3.75×10^{-3}		3.53×10^{-3}		3.17×10^{-3}		3.25×10^{-3}		2.97×10^{-3}	1.82×10^{-4}	3.01×10^{-3}	1.82×10^{-4}	3.01×10^{-3}
15–16		3.27×10^{-3}		3.01×10^{-3}		2.72×10^{-3}		2.79×10^{-3}		2.54×10^{-3}	1.72×10^{-4}	2.85×10^{-3}	1.72×10^{-4}	2.85×10^{-3}
16–17		2.82×10^{-3}		2.57×10^{-3}		2.33×10^{-3}		2.41×10^{-3}		2.18×10^{-3}	1.67×10^{-4}	2.76×10^{-3}	1.67×10^{-4}	2.76×10^{-3}
17–18		2.37×10^{-3}		2.20×10^{-3}		1.99×10^{-3}		2.07×10^{-3}		1.87×10^{-3}	1.63×10^{-4}	2.70×10^{-3}	1.63×10^{-4}	2.70×10^{-3}
18–19		1.95×10^{-3}		1.88×10^{-3}		1.71×10^{-3}		1.78×10^{-3}		1.60×10^{-3}	1.48×10^{-4}	2.44×10^{-3}	1.48×10^{-4}	2.44×10^{-3}
19–20		1.62×10^{-3}		1.60×10^{-3}		1.46×10^{-3}		1.53×10^{-3}		1.37×10^{-3}	1.16×10^{-4}	1.92×10^{-3}	1.16×10^{-4}	1.92×10^{-3}
20–21		1.35×10^{-3}		1.36×10^{-3}		1.25×10^{-3}		1.31×10^{-3}		1.17×10^{-3}	8.49×10^{-5}	1.40×10^{-3}	8.49×10^{-5}	1.40×10^{-3}
21–22		1.13×10^{-3}		1.16×10^{-3}		1.06×10^{-3}		1.13×10^{-3}		1.00×10^{-3}	6.27×10^{-5}	1.04×10^{-3}	6.27×10^{-5}	1.04×10^{-3}
22–23		9.48×10^{-4}		9.93×10^{-4}		9.09×10^{-4}		9.72×10^{-4}		8.57×10^{-4}	4.75×10^{-5}	7.85×10^{-4}	4.75×10^{-5}	7.85×10^{-4}
23–24		8.04×10^{-4}		8.46×10^{-4}		7.77×10^{-4}		8.35×10^{-4}		7.32×10^{-4}	3.70×10^{-5}	6.11×10^{-4}	3.70×10^{-5}	6.11×10^{-4}
24–25		6.83×10^{-4}		7.23×10^{-4}		6.63×10^{-4}		7.14×10^{-4}		6.26×10^{-4}	3.02×10^{-5}	4.99×10^{-4}	3.02×10^{-5}	4.99×10^{-4}
25–30		4.56×10^{-4}		4.86×10^{-4}		4.42×10^{-4}		4.83×10^{-4}		4.16×10^{-4}	1.61×10^{-5}	2.66×10^{-4}	1.61×10^{-5}	2.66×10^{-4}
30–35		2.08×10^{-4}		2.24×10^{-4}		1.99×10^{-4}		2.26×10^{-4}		1.84×10^{-4}	4.60×10^{-6}	7.60×10^{-5}	4.60×10^{-6}	7.60×10^{-5}
35–40		9.90×10^{-5}		1.07×10^{-4}		8.94×10^{-5}		1.07×10^{-4}		8.28×10^{-5}	1.21×10^{-6}	2.00×10^{-5}	1.21×10^{-6}	2.00×10^{-5}
40–45		4.86×10^{-5}		5.24×10^{-5}		4.15×10^{-5}		5.28×10^{-5}		3.79×10^{-5}	3.19×10^{-7}	5.27×10^{-6}	3.19×10^{-7}	5.27×10^{-6}
45–50		2.47×10^{-5}		2.69×10^{-5}		2.04×10^{-5}		2.74×10^{-5}		1.81×10^{-5}	8.40×10^{-8}	1.39×10^{-6}	8.40×10^{-8}	1.39×10^{-6}
50–70		9.24×10^{-6}		1.01×10^{-5}		7.47×10^{-6}		1.05×10^{-5}		6.39×10^{-6}	7.58×10^{-9}	1.25×10^{-7}	7.58×10^{-9}	1.25×10^{-7}
70–100		$<10^{-6}$		$<10^{-6}$		$<10^{-6}$		$<10^{-6}$		$<10^{-6}$	2.60×10^{-11}	4.30×10^{-10}	2.60×10^{-11}	4.30×10^{-10}

$\lambda = 0.5145\ \mu m$ (see footnote in Par. 7)

0	$<10^{-3}$	1.47×10^{-2}	$<10^{-3}$	1.50×10^{-2}	$<10^{-3}$	1.63×10^{-2}	$<10^{-3}$	1.53×10^{-2}	$<10^{-3}$	1.71×10^{-2}	1.00×10^{-2}	1.66×10^{-1}	4.89×10^{-2}	8.10×10^{-1}
0–1		1.40×10^{-2}		1.43×10^{-2}		1.54×10^{-2}		1.46×10^{-2}		1.60×10^{-2}	6.84×10^{-3}	1.13×10^{-1}	3.08×10^{-2}	5.11×10^{-1}
1–2		1.28×10^{-2}		1.29×10^{-2}		1.37×10^{-2}		1.32×10^{-2}		1.41×10^{-2}	2.99×10^{-3}	4.95×10^{-2}	1.13×10^{-2}	1.87×10^{-1}
2–3		1.15×10^{-2}		1.16×10^{-2}		1.22×10^{-2}		1.19×10^{-2}		1.25×10^{-2}	1.27×10^{-3}	2.11×10^{-2}	4.12×10^{-3}	6.83×10^{-2}
3–4		1.04×10^{-2}		1.05×10^{-2}		1.09×10^{-2}		1.06×10^{-2}		1.10×10^{-2}	5.92×10^{-4}	9.81×10^{-3}	1.50×10^{-3}	2.49×10^{-2}
4–5		9.46×10^{-3}		9.51×10^{-2}		9.79×10^{-3}		9.55×10^{-3}		9.86×10^{-3}	3.68×10^{-4}	6.11×10^{-3}	5.49×10^{-4}	9.10×10^{-3}
5–6		8.55×10^{-3}		8.56×10^{-2}		8.75×10^{-3}		8.58×10^{-3}		8.80×10^{-3}	2.69×10^{-4}	4.46×10^{-3}	2.69×10^{-4}	4.46×10^{-3}
6–7		7.70×10^{-3}		7.68×10^{-2}		7.80×10^{-3}		7.71×10^{-3}		7.83×10^{-3}	2.17×10^{-4}	3.59×10^{-3}	2.17×10^{-4}	3.59×10^{-3}
7–8		6.94×10^{-3}		6.89×10^{-2}		6.93×10^{-3}		6.91×10^{-3}		6.94×10^{-3}	2.12×10^{-4}	3.52×10^{-3}	2.12×10^{-4}	3.52×10^{-3}
8–9		6.22×10^{-3}		6.16×10^{-3}		6.14×10^{-3}		6.17×10^{-3}		6.09×10^{-3}	2.11×10^{-4}	3.49×10^{-3}	2.11×10^{-4}	3.49×10^{-3}
9–10		5.55×10^{-3}		5.50×10^{-3}		5.42×10^{-3}		5.50×10^{-3}		5.25×10^{-3}	2.04×10^{-4}	3.38×10^{-3}	2.04×10^{-4}	3.38×10^{-3}
10–11		4.96×10^{-3}		4.90×10^{-3}		4.72×10^{-3}		4.81×10^{-3}		4.49×10^{-3}	1.95×10^{-4}	3.23×10^{-3}	1.95×10^{-4}	3.23×10^{-3}
11–12		4.40×10^{-3}		4.35×10^{-3}		4.05×10^{-3}		4.13×10^{-3}		3.84×10^{-3}	1.93×10^{-4}	3.21×10^{-3}	1.93×10^{-4}	3.21×10^{-3}
12–13		3.89×10^{-3}		3.85×10^{-3}		3.47×10^{-3}		3.55×10^{-3}		3.28×10^{-3}	1.91×10^{-4}	3.16×10^{-3}	1.91×10^{-4}	3.16×10^{-3}
13–14		3.44×10^{-3}		3.34×10^{-3}		2.98×10^{-3}		3.05×10^{-3}		2.80×10^{-3}	1.81×10^{-4}	3.00×10^{-3}	1.81×10^{-4}	3.00×10^{-3}
14–15		3.02×10^{-3}		2.85×10^{-3}		2.55×10^{-3}		2.62×10^{-3}		2.40×10^{-3}	1.74×10^{-4}	2.88×10^{-3}	1.74×10^{-4}	2.88×10^{-3}
15–16		2.63×10^{-3}		2.42×10^{-3}		2.19×10^{-3}		2.25×10^{-3}		2.05×10^{-3}	1.64×10^{-4}	2.72×10^{-3}	1.64×10^{-4}	2.72×10^{-3}
16–17		2.27×10^{-3}		2.07×10^{-3}		1.88×10^{-3}		1.94×10^{-3}		1.76×10^{-3}	1.59×10^{-4}	2.64×10^{-3}	1.59×10^{-4}	2.64×10^{-3}
17–18		1.91×10^{-3}		1.77×10^{-3}		1.61×10^{-3}		1.67×10^{-3}		1.51×10^{-3}	1.56×10^{-4}	2.58×10^{-3}	1.56×10^{-4}	2.58×10^{-3}
18–19		1.57×10^{-3}		1.51×10^{-3}		1.38×10^{-3}		1.43×10^{-3}		1.29×10^{-3}	1.41×10^{-4}	2.33×10^{-3}	1.41×10^{-4}	2.33×10^{-3}
19–20		1.31×10^{-3}		1.29×10^{-3}		1.18×10^{-3}		1.23×10^{-3}		1.10×10^{-3}	1.11×10^{-4}	1.84×10^{-3}	1.11×10^{-4}	1.84×10^{-3}
20–21		1.09×10^{-3}		1.10×10^{-3}		1.01×10^{-3}		1.06×10^{-3}		9.44×10^{-4}	8.09×10^{-5}	1.34×10^{-3}	8.09×10^{-5}	1.34×10^{-3}
21–22		9.08×10^{-4}		9.38×10^{-4}		8.57×10^{-4}		9.11×10^{-4}		8.07×10^{-4}	5.97×10^{-5}	9.89×10^{-4}	5.97×10^{-5}	9.89×10^{-4}
22–23		7.64×10^{-4}		8.00×10^{-4}		7.32×10^{-4}		7.83×10^{-4}		6.90×10^{-4}	4.53×10^{-5}	7.50×10^{-4}	4.53×10^{-5}	7.50×10^{-4}
23–24		6.48×10^{-4}		6.82×10^{-4}		6.26×10^{-4}		6.72×10^{-4}		5.90×10^{-4}	3.52×10^{-5}	5.84×10^{-4}	3.52×10^{-5}	5.84×10^{-4}
24–25		5.50×10^{-4}		5.83×10^{-4}		5.34×10^{-4}		5.75×10^{-4}		5.04×10^{-4}	2.88×10^{-5}	4.77×10^{-4}	2.88×10^{-5}	4.77×10^{-4}
25–30		3.67×10^{-4}		3.91×10^{-4}		3.56×10^{-4}		3.89×10^{-4}		3.35×10^{-4}	1.53×10^{-5}	2.54×10^{-4}	1.53×10^{-5}	2.54×10^{-4}
30–35		1.68×10^{-4}		1.80×10^{-4}		1.60×10^{-4}		1.82×10^{-4}		1.49×10^{-4}	4.38×10^{-6}	7.26×10^{-5}	4.38×10^{-6}	7.26×10^{-5}
35–40		7.97×10^{-5}		8.59×10^{-5}		7.20×10^{-5}		8.64×10^{-5}		6.67×10^{-5}	1.15×10^{-6}	1.91×10^{-5}	1.15×10^{-6}	1.91×10^{-5}
40–45		3.91×10^{-5}		4.22×10^{-5}		3.35×10^{-5}		4.26×10^{-5}		3.06×10^{-5}	3.04×10^{-7}	5.03×10^{-6}	3.04×10^{-7}	5.03×10^{-6}
45–50		1.99×10^{-5}		2.17×10^{-5}		1.64×10^{-5}		2.21×10^{-5}		1.46×10^{-5}	8.00×10^{-8}	1.33×10^{-6}	8.00×10^{-8}	1.33×10^{-6}
50–70		7.45×10^{-6}		8.16×10^{-6}		6.01×10^{-6}		8.45×10^{-6}		5.15×10^{-6}	7.22×10^{-19}	1.20×10^{-7}	7.22×10^{-9}	1.20×10^{-7}
70–100		$<10^{-6}$		$<10^{-6}$		$<10^{-6}$		$<10^{-6}$		$<10^{-6}$	2.48×10^{-11}	4.11×10^{-10}	2.48×10^{-11}	4.11×10^{-10}

TABLE 6 Values of Attenuation Coefficient per Kilometer as a Function of Altitude for Each Laser Wavelength and Each Atmospheric Model (Continued)

Height, km	Tropical		Midlatitude				Subarctic				Clear		Hazy	
			Summer		Winter		Summer		Winter					
	Molecular absorption k_m, km^{-1}	Molecular scattering σ_m, km^{-1}	Molecular absorption k_m, km^{-1}	Molecular scattering σ_m, km^{-1}	Molecular absorption k_m, km^{-1}	Molecular scattering σ_m, km^{-1}	Molecular absorption k_m, km^{-1}	Molecular scattering σ_m, km^{-1}	Molecular absorption k_m, km^{-1}	Molecular scattering σ_m, km^{-1}	Aerosol absorption k_a, km^{-1}	Aerosol scattering σ_a, km^{-1}	Aerosol absorption k_a, km^{-1}	Aerosol scattering σ_a, km^{-1}
	$\lambda = 0.6328\ \mu m$ (see footnote in Par. 7)													
0	$<10^{-3}$	6.31×10^{-3}	$<10^{-3}$	6.44×10^{-3}	$<10^{-3}$	6.79×10^{-3}	$<10^{-3}$	6.58×10^{-3}	$<10^{-3}$	7.37×10^{-3}	9.14×10^{-3}	1.37×10^{-1}	4.45×10^{-2}	6.69×10^{-1}
0-1		6.03×10^{-3}		6.13×10^{-3}		6.62×10^{-3}		6.26×10^{-3}		6.88×10^{-3}	6.23×10^{-3}	9.36×10^{-2}	2.81×10^{-2}	4.21×10^{-1}
1-2		5.49×10^{-3}		5.54×10^{-3}		5.91×10^{-3}		5.66×10^{-3}		6.04×10^{-3}	2.72×10^{-3}	4.08×10^{-2}	1.03×10^{-2}	1.54×10^{-1}
2-3		4.97×10^{-3}		5.01×10^{-3}		5.26×10^{-3}		5.10×10^{-3}		5.36×10^{-3}	1.16×10^{-3}	1.74×10^{-2}	3.76×10^{-3}	5.64×10^{-2}
3-4		4.49×10^{-3}		4.53×10^{-3}		4.70×10^{-3}		4.58×10^{-3}		4.75×10^{-3}	5.39×10^{-4}	8.09×10^{-3}	1.37×10^{-3}	2.06×10^{-2}
4-5		4.07×10^{-3}		4.09×10^{-3}		4.21×10^{-3}		4.11×10^{-3}		4.24×10^{-3}	3.36×10^{-4}	5.04×10^{-3}	5.00×10^{-4}	7.51×10^{-3}
5-6		3.68×10^{-3}		3.68×10^{-3}		3.76×10^{-3}		3.69×10^{-3}		3.79×10^{-3}	2.45×10^{-4}	3.68×10^{-3}	2.45×10^{-4}	3.68×10^{-3}
6-7		3.31×10^{-3}		3.31×10^{-3}		3.35×10^{-3}		3.32×10^{-3}		3.37×10^{-3}	1.98×10^{-4}	2.97×10^{-3}	1.98×10^{-4}	2.97×10^{-3}
7-8		2.99×10^{-3}		2.96×10^{-3}		2.98×10^{-3}		2.97×10^{-3}		2.99×10^{-3}	1.93×10^{-4}	2.90×10^{-3}	1.93×10^{-4}	2.90×10^{-3}
8-9		2.67×10^{-3}		2.66×10^{-3}		2.64×10^{-3}		2.66×10^{-3}		2.62×10^{-3}	1.92×10^{-4}	2.88×10^{-3}	1.92×10^{-4}	2.88×10^{-3}
9-10		2.39×10^{-3}		2.37×10^{-3}		2.33×10^{-3}		2.36×10^{-3}		2.26×10^{-3}	1.86×10^{-4}	2.79×10^{-3}	1.86×10^{-4}	2.79×10^{-3}
10-11		2.13×10^{-3}		2.11×10^{-3}		2.03×10^{-3}		2.07×10^{-3}		1.93×10^{-3}	1.78×10^{-4}	2.67×10^{-3}	1.78×10^{-4}	2.67×10^{-3}
11-12		1.89×10^{-3}		1.87×10^{-3}		1.74×10^{-3}		1.78×10^{-3}		1.65×10^{-3}	1.76×10^{-4}	2.65×10^{-3}	1.76×10^{-4}	2.65×10^{-3}
12-13		1.67×10^{-3}		1.66×10^{-3}		1.49×10^{-3}		1.53×10^{-3}		1.41×10^{-3}	1.74×10^{-4}	2.61×10^{-3}	1.74×10^{-4}	2.61×10^{-3}
13-14		1.48×10^{-3}		1.44×10^{-3}		1.28×10^{-3}		1.31×10^{-3}		1.21×10^{-3}	1.65×10^{-4}	2.48×10^{-3}	1.65×10^{-4}	2.48×10^{-3}
14-15		1.30×10^{-3}		1.23×10^{-3}		1.10×10^{-3}		1.13×10^{-3}		1.03×10^{-3}	1.58×10^{-4}	2.38×10^{-3}	1.58×10^{-4}	2.38×10^{-3}
15-16		1.13×10^{-3}		1.04×10^{-3}		9.41×10^{-4}		9.68×10^{-4}		8.82×10^{-4}	1.50×10^{-4}	2.25×10^{-3}	1.50×10^{-4}	2.25×10^{-3}
16-17		9.76×10^{-4}		8.92×10^{-4}		8.07×10^{-4}		8.34×10^{-4}		7.56×10^{-4}	1.45×10^{-4}	2.18×10^{-3}	1.45×10^{-4}	2.18×10^{-3}
17-18		8.20×10^{-4}		7.63×10^{-4}		6.91×10^{-4}		7.17×10^{-4}		6.47×10^{-4}	1.42×10^{-4}	2.13×10^{-3}	1.42×10^{-4}	2.13×10^{-3}
18-19		6.77×10^{-4}		6.51×10^{-4}		5.91×10^{-4}		6.17×10^{-4}		5.54×10^{-4}	1.28×10^{-4}	1.92×10^{-3}	1.28×10^{-4}	1.92×10^{-3}
19-20		5.62×10^{-4}		5.55×10^{-4}		5.06×10^{-4}		5.30×10^{-4}		4.75×10^{-4}	1.01×10^{-4}	1.52×10^{-3}	1.01×10^{-4}	1.52×10^{-3}
20-21		4.68×10^{-4}		4.73×10^{-4}		4.32×10^{-4}		4.55×10^{-4}		4.00×10^{-4}	7.37×10^{-5}	1.11×10^{-3}	7.37×10^{-5}	1.11×10^{-3}
21-22		3.91×10^{-4}		4.03×10^{-4}		3.69×10^{-4}		3.92×10^{-4}		3.47×10^{-4}	5.44×10^{-5}	8.17×10^{-4}	5.44×10^{-5}	8.17×10^{-4}
22-23		3.29×10^{-4}		3.44×10^{-4}		3.15×10^{-4}		3.37×10^{-4}		2.97×10^{-4}	4.13×10^{-5}	6.19×10^{-4}	4.13×10^{-5}	6.19×10^{-4}
23-24		2.79×10^{-4}		2.93×10^{-4}		2.69×10^{-4}		2.89×10^{-4}		2.54×10^{-4}	3.21×10^{-5}	4.82×10^{-4}	3.21×10^{-5}	4.82×10^{-4}
24-25		2.37×10^{-4}		2.51×10^{-4}		2.30×10^{-4}		2.48×10^{-4}		2.17×10^{-4}	2.62×10^{-5}	3.93×10^{-4}	2.62×10^{-5}	3.93×10^{-4}
25-30		1.58×10^{-4}		1.68×10^{-4}		1.53×10^{-4}		1.67×10^{-4}		1.44×10^{-4}	1.40×10^{-5}	2.10×10^{-4}	1.40×10^{-5}	2.10×10^{-4}
30-35		7.22×10^{-5}		7.76×10^{-5}		6.90×10^{-5}		7.83×10^{-5}		6.39×10^{-5}	3.99×10^{-6}	5.99×10^{-5}	3.99×10^{-6}	5.99×10^{-5}
35-40		3.43×10^{-5}		3.69×10^{-5}		3.10×10^{-5}		3.72×10^{-5}		2.87×10^{-5}	1.05×10^{-6}	1.58×10^{-5}	1.05×10^{-6}	1.58×10^{-5}
40-45		1.68×10^{-5}		1.82×10^{-5}		1.44×10^{-5}		1.83×10^{-5}		1.32×10^{-5}	2.77×10^{-7}	4.15×10^{-6}	2.77×10^{-7}	4.15×10^{-6}
45-50		3.57×10^{-6}		9.31×10^{-6}		7.06×10^{-6}		9.51×10^{-6}		6.27×10^{-6}	7.29×10^{-8}	1.09×10^{-6}	7.29×10^{-8}	1.09×10^{-6}
50-70		3.20×10^{-6}		3.51×10^{-6}		2.59×10^{-6}		3.64×10^{-6}		2.22×10^{-6}	6.58×10^{-9}	9.88×10^{-8}	6.58×10^{-9}	9.88×10^{-8}
70-100		$<10^{-6}$		$<10^{-6}$		$<10^{-6}$		$<10^{-6}$		$<10^{-6}$	2.26×10^{-11}	3.39×10^{-10}	2.26×10^{-11}	3.39×10^{-10}
	$\lambda = 0.6943\ \mu m$ (see footnote in Par. 7)													
0	8.62×10^{-2}	4.29×10^{-3}	6.44×10^{-2}	4.37×10^{-3}	1.55×10^{-2}	4.75×10^{-3}	4.07×10^{-2}	4.47×10^{-3}	4.48×10^{-3}	5.00×10^{-3}	9.14×10^{-3}	1.25×10^{-1}	4.45×10^{-2}	6.10×10^{-1}
0-1	6.84×10^{-2}	4.10×10^{-3}	5.07×10^{-2}	4.16×10^{-3}	1.25×10^{-2}	4.50×10^{-3}	3.18×10^{-2}	4.25×10^{-3}	4.19×10^{-3}	4.68×10^{-3}	6.23×10^{-3}	8.53×10^{-2}	2.81×10^{-2}	3.84×10^{-1}
1-2	4.22×10^{-2}	3.73×10^{-3}	3.03×10^{-2}	3.76×10^{-3}	8.13×10^{-3}	4.01×10^{-3}	1.88×10^{-2}	3.84×10^{-3}	3.66×10^{-3}	4.11×10^{-3}	2.72×10^{-3}	3.72×10^{-2}	1.03×10^{-2}	1.40×10^{-1}
2-3	2.73×10^{-2}	3.38×10^{-3}	1.72×10^{-2}	3.40×10^{-3}	5.15×10^{-3}	3.57×10^{-3}	1.18×10^{-2}	3.47×10^{-3}	2.56×10^{-3}	3.64×10^{-3}	1.16×10^{-3}	1.59×10^{-2}	3.75×10^{-3}	5.14×10^{-2}
3-4	1.23×10^{-2}	3.05×10^{-3}	8.67×10^{-3}	3.08×10^{-3}	3.06×10^{-3}	3.19×10^{-3}	6.78×10^{-3}	3.11×10^{-3}	1.65×10^{-3}	3.23×10^{-3}	5.39×10^{-4}	7.38×10^{-3}	1.37×10^{-3}	1.87×10^{-2}
4-5	5.24×10^{-3}	2.77×10^{-3}	4.43×10^{-3}	2.78×10^{-3}	1.48×10^{-3}	2.86×10^{-3}	3.76×10^{-3}	2.79×10^{-3}	8.60×10^{-4}	2.88×10^{-3}	3.36×10^{-4}	4.59×10^{-3}	5.00×10^{-4}	6.85×10^{-3}
5-6	3.09×10^{-3}	2.50×10^{-3}	2.06×10^{-3}	2.50×10^{-3}	7.41×10^{-4}	2.56×10^{-3}	1.95×10^{-3}	2.51×10^{-3}	3.66×10^{-4}	2.57×10^{-3}	2.45×10^{-4}	3.36×10^{-3}	2.45×10^{-4}	3.36×10^{-3}
6-7	1.56×10^{-3}	2.25×10^{-3}	1.11×10^{-3}	2.25×10^{-3}	3.60×10^{-4}	2.28×10^{-3}	9.26×10^{-4}	2.25×10^{-3}	1.54×10^{-4}	2.29×10^{-3}	1.98×10^{-4}	2.70×10^{-3}	1.98×10^{-4}	2.70×10^{-3}
7-8	7.61×10^{-4}	2.03×10^{-3}	5.87×10^{-4}	2.01×10^{-3}	1.26×10^{-4}	2.02×10^{-3}	4.32×10^{-4}	2.05×10^{-3}	7.26×10^{-5}	2.03×10^{-3}	1.93×10^{-4}	2.65×10^{-3}	1.93×10^{-4}	2.65×10^{-3}
8-9	3.54×10^{-4}	1.82×10^{-3}	2.92×10^{-4}	1.80×10^{-3}	4.42×10^{-5}	1.79×10^{-3}	1.68×10^{-4}	1.83×10^{-3}	1.27×10^{-5}	1.78×10^{-3}	1.92×10^{-4}	2.63×10^{-3}	1.92×10^{-4}	2.63×10^{-3}

9–10	1.47×10^{-4}	1.62×10^{-3}	1.44×10^{-4}	1.61×10^{-3}	1.75×10^{-5}	1.58×10^{-3}	4.64×10^{-5}	1.61×10^{-3}	8.23×10^{-6}	1.53×10^{-3}	1.86×10^{-4}	2.54×10^{-3}	1.86×10^{-4}	2.54×10^{-3}
10–11	5.28×10^{-5}	1.45×10^{-3}	6.66×10^{-5}	1.44×10^{-3}	6.87×10^{-6}	1.38×10^{-3}	1.43×10^{-5}	1.41×10^{-3}	4.63×10^{-6}	1.31×10^{-3}	1.78×10^{-4}	2.43×10^{-3}	1.78×10^{-4}	2.43×10^{-3}
11–12	1.55×10^{-5}	1.29×10^{-3}	1.95×10^{-5}	1.27×10^{-3}	5.51×10^{-6}	1.18×10^{-3}	7.70×10^{-6}	1.21×10^{-3}	2.76×10^{-6}	1.12×10^{-3}	1.76×10^{-4}	2.41×10^{-3}	1.76×10^{-4}	2.41×10^{-3}
12–13	4.63×10^{-6}	1.14×10^{-3}	4.54×10^{-6}	1.12×10^{-3}	4.07×10^{-6}	1.01×10^{-3}	4.23×10^{-6}	1.04×10^{-3}	1.59×10^{-6}	9.56×10^{-4}	1.74×10^{-4}	2.38×10^{-3}	1.74×10^{-4}	2.38×10^{-3}
13–14	1.19×10^{-6}	1.00×10^{-3}	1.18×10^{-6}	9.75×10^{-4}	1.06×10^{-6}	8.70×10^{-4}	1.10×10^{-6}	8.92×10^{-4}	$<10^{-3}$	8.18×10^{-4}	1.65×10^{-4}	2.26×10^{-3}	1.65×10^{-4}	2.26×10^{-3}
14–15	$<10^{-3}$	8.82×10^{-4}	$<10^{-3}$	8.32×10^{-4}	$<10^{-3}$	7.46×10^{-4}	$<10^{-3}$	7.65×10^{-4}		7.00×10^{-4}	1.58×10^{-4}	2.17×10^{-3}	1.58×10^{-4}	2.17×10^{-3}
15–16		7.68×10^{-4}		7.05×10^{-4}		6.39×10^{-4}		6.57×10^{-4}		6.00×10^{-4}	1.50×10^{-4}	2.05×10^{-3}	1.50×10^{-4}	2.05×10^{-3}
16–17		6.63×10^{-4}		6.02×10^{-4}		5.48×10^{-4}		5.66×10^{-4}		5.13×10^{-4}	1.45×10^{-4}	1.98×10^{-3}	1.45×10^{-4}	1.98×10^{-3}
17–18		5.57×10^{-4}		5.18×10^{-4}		4.69×10^{-4}		4.87×10^{-4}		4.39×10^{-4}	1.42×10^{-4}	1.94×10^{-3}	1.42×10^{-4}	1.94×10^{-3}
18–19		4.60×10^{-4}		4.42×10^{-4}		4.01×10^{-4}		4.19×10^{-4}		3.76×10^{-4}	1.28×10^{-4}	1.75×10^{-3}	1.28×10^{-4}	1.75×10^{-3}
19–20		3.81×10^{-4}		3.76×10^{-4}		3.44×10^{-4}		3.60×10^{-4}		3.22×10^{-4}	1.01×10^{-4}	1.38×10^{-3}	1.01×10^{-4}	1.38×10^{-3}
20–21		3.18×10^{-4}		3.21×10^{-4}		2.94×10^{-4}		3.09×10^{-4}		2.76×10^{-4}	7.37×10^{-5}	1.01×10^{-3}	7.37×10^{-5}	1.01×10^{-3}
21–22		2.65×10^{-4}		2.74×10^{-4}		2.51×10^{-4}		2.66×10^{-4}		2.36×10^{-4}	5.44×10^{-5}	7.44×10^{-4}	5.44×10^{-5}	7.44×10^{-4}
22–23		2.23×10^{-4}		2.34×10^{-4}		2.14×10^{-4}		2.29×10^{-4}		2.02×10^{-4}	4.12×10^{-5}	5.65×10^{-4}	4.12×10^{-5}	5.65×10^{-4}
23–24		1.90×10^{-4}		1.99×10^{-4}		1.83×10^{-4}		1.97×10^{-4}		1.73×10^{-4}	3.21×10^{-5}	4.39×10^{-4}	3.21×10^{-5}	4.39×10^{-4}
24–25		1.61×10^{-4}		1.70×10^{-4}		1.56×10^{-4}		1.68×10^{-4}		1.48×10^{-4}	2.62×10^{-5}	3.59×10^{-4}	2.62×10^{-5}	3.59×10^{-4}
25–30		1.07×10^{-4}		1.14×10^{-4}		1.04×10^{-4}		1.14×10^{-4}		9.80×10^{-5}	1.40×10^{-5}	1.91×10^{-4}	1.40×10^{-5}	1.91×10^{-4}
30–35		4.90×10^{-5}		5.27×10^{-5}		4.69×10^{-5}		5.32×10^{-5}		4.34×10^{-5}	3.99×10^{-6}	5.47×10^{-5}	3.99×10^{-6}	5.47×10^{-5}
35–40		2.33×10^{-5}		2.51×10^{-5}		2.11×10^{-5}		2.53×10^{-5}		1.95×10^{-5}	1.05×10^{-6}	1.44×10^{-5}	1.05×10^{-6}	1.44×10^{-5}
40–45		1.14×10^{-5}		1.23×10^{-6}		9.77×10^{-6}		1.24×10^{-5}		8.91×10^{-6}	2.77×10^{-7}	3.79×10^{-6}	2.77×10^{-7}	3.79×10^{-6}
45–50		5.81×10^{-6}		6.32×10^{-6}		4.80×10^{-6}		6.45×10^{-6}		4.26×10^{-6}	7.29×10^{-8}	9.98×10^{-7}	7.29×10^{-8}	9.98×10^{-7}
50–70		2.17×10^{-6}		2.38×10^{-6}		1.76×10^{-6}		2.47×10^{-6}		1.50×10^{-6}	6.58×10^{-9}	9.00×10^{-8}	6.58×10^{-8}	9.00×10^{-8}
70–100		$<10^{-6}$		$<10^{-6}$		$<10^{-6}$		$<10^{-6}$		$<10^{-6}$	2.26×10^{-11}	3.09×10^{-10}	2.26×10^{-11}	3.09×10^{-10}

$\lambda = 0.86\ \mu m$

0	$<10^{-6}$	1.89×10^{-3}	$<10^{-6}$	1.93×10^{-3}	$<10^{-6}$	2.09×10^{-3}	$<10^{-6}$	1.97×10^{-3}	$<10^{-6}$	2.21×10^{-3}	1.02×10^{-2}	9.77×10^{-2}	4.98×10^{-2}	4.76×10^{-1}
0–1		1.81×10^{-3}		1.84×10^{-3}		1.98×10^{-3}		1.87×10^{-3}		2.06×10^{-3}	6.98×10^{-3}	6.66×10^{-2}	3.14×10^{-2}	3.00×10^{-1}
1–2		1.64×10^{-3}		1.66×10^{-3}		1.77×10^{-3}		1.69×10^{-3}		1.81×10^{-4}	3.04×10^{-3}	2.91×10^{-2}	1.15×10^{-2}	1.10×10^{-1}
2–3		1.49×10^{-3}		1.50×10^{-3}		1.58×10^{-3}		1.53×10^{-3}		1.60×10^{-4}	1.30×10^{-3}	1.24×10^{-2}	4.20×10^{-3}	4.01×10^{-2}
3–4		1.34×10^{-3}		1.36×10^{-3}		1.41×10^{-3}		1.37×10^{-3}		1.42×10^{-4}	6.03×10^{-4}	5.76×10^{-3}	1.53×10^{-3}	1.46×10^{-2}
4–5		1.22×10^{-3}		1.22×10^{-3}		1.26×10^{-3}		1.23×10^{-3}		1.27×10^{-4}	3.76×10^{-4}	3.59×10^{-3}	5.60×10^{-4}	5.35×10^{-3}
5–6		1.10×10^{-3}		1.10×10^{-3}		1.13×10^{-3}		1.11×10^{-3}		1.13×10^{-4}	2.74×10^{-4}	2.62×10^{-3}	2.74×10^{-4}	2.62×10^{-3}
6–7		9.92×10^{-4}		9.90×10^{-4}		1.00×10^{-3}		9.93×10^{-4}		1.01×10^{-4}	2.21×10^{-4}	2.11×10^{-3}	2.21×10^{-4}	2.11×10^{-3}
7–8		8.94×10^{-4}		8.87×10^{-4}		8.92×10^{-4}		8.90×10^{-4}		8.94×10^{-4}	2.16×10^{-4}	2.07×10^{-3}	2.16×10^{-4}	2.07×10^{-3}
8–9		8.00×10^{-4}		7.96×10^{-4}		7.90×10^{-4}		7.95×10^{-4}		7.84×10^{-4}	2.15×10^{-4}	2.05×10^{-3}	2.15×10^{-4}	2.05×10^{-3}
9–10		7.15×10^{-4}		7.11×10^{-4}		6.98×10^{-4}		7.08×10^{-4}		6.76×10^{-4}	2.08×10^{-4}	1.99×10^{-3}	2.08×10^{-4}	1.99×10^{-3}
10–11		6.38×10^{-4}		6.32×10^{-4}		6.08×10^{-4}		6.19×10^{-4}		5.78×10^{-4}	1.99×10^{-4}	1.90×10^{-3}	1.99×10^{-4}	1.90×10^{-3}
11–12		5.67×10^{-4}		5.60×10^{-4}		5.22×10^{-4}		5.32×10^{-4}		4.94×10^{-4}	1.97×10^{-4}	1.88×10^{-3}	1.97×10^{-4}	1.88×10^{-3}
12–13		5.01×10^{-4}		4.95×10^{-4}		4.47×10^{-4}		4.57×10^{-4}		4.22×10^{-4}	1.94×10^{-4}	1.86×10^{-3}	1.94×10^{-4}	1.86×10^{-3}
13–14		4.43×10^{-4}		4.30×10^{-4}		3.84×10^{-4}		3.93×10^{-4}		3.61×10^{-4}	1.85×10^{-4}	1.76×10^{-3}	1.85×10^{-4}	1.76×10^{-3}
14–15		3.89×10^{-4}		3.67×10^{-4}		3.29×10^{-4}		3.37×10^{-4}		3.08×10^{-4}	1.77×10^{-4}	1.69×10^{-3}	1.77×10^{-4}	1.69×10^{-3}
15–16		3.39×10^{-4}		3.12×10^{-4}		2.82×10^{-4}		2.90×10^{-4}		2.64×10^{-4}	1.67×10^{-4}	1.60×10^{-3}	1.67×10^{-4}	1.60×10^{-3}
16–17		2.92×10^{-4}		2.67×10^{-4}		2.42×10^{-4}		2.50×10^{-4}		2.26×10^{-4}	1.62×10^{-4}	1.55×10^{-3}	1.62×10^{-4}	1.55×10^{-3}
17–18		2.45×10^{-4}		2.28×10^{-4}		2.07×10^{-4}		2.15×10^{-4}		1.94×10^{-4}	1.59×10^{-4}	1.52×10^{-3}	1.59×10^{-4}	1.52×10^{-3}
18–19		2.03×10^{-4}		1.95×10^{-4}		1.77×10^{-4}		1.85×10^{-4}		1.66×10^{-4}	1.43×10^{-4}	1.37×10^{-3}	1.43×10^{-4}	1.37×10^{-3}
19–20		1.68×10^{-4}		1.66×10^{-4}		1.52×10^{-4}		1.59×10^{-4}		1.42×10^{-4}	1.13×10^{-4}	1.08×10^{-3}	1.13×10^{-4}	1.08×10^{-3}
20–21		1.40×10^{-4}		1.42×10^{-4}		1.29×10^{-4}		1.36×10^{-4}		1.22×10^{-4}	8.25×10^{-5}	7.88×10^{-4}	8.25×10^{-5}	7.88×10^{-4}
21–22		1.17×10^{-4}		1.21×10^{-4}		1.10×10^{-4}		1.17×10^{-4}		1.04×10^{-4}	6.09×10^{-5}	5.81×10^{-4}	6.09×10^{-5}	5.81×10^{-4}
22–23		9.84×10^{-5}		1.03×10^{-4}		9.43×10^{-5}		1.01×10^{-4}		8.89×10^{-5}	4.62×10^{-5}	4.41×10^{-4}	4.62×10^{-5}	4.41×10^{-4}
23–24		8.35×10^{-5}		8.78×10^{-5}		8.06×10^{-5}		8.66×10^{-5}		7.60×10^{-5}	3.59×10^{-5}	3.43×10^{-4}	3.59×10^{-5}	3.43×10^{-4}
24–25		7.09×10^{-5}		7.50×10^{-5}		6.88×10^{-5}		7.41×10^{-5}		6.49×10^{-5}	2.93×10^{-5}	2.80×10^{-4}	2.93×10^{-5}	2.80×10^{-4}
25–30		4.73×10^{-5}		5.64×10^{-5}		4.59×10^{-5}		5.01×10^{-5}		4.31×10^{-5}	1.56×10^{-5}	1.49×10^{-4}	1.56×10^{-5}	1.49×10^{-4}
30–35		2.16×10^{-5}		2.32×10^{-5}		2.07×10^{-5}		2.35×10^{-5}		1.91×10^{-5}	4.47×10^{-6}	4.27×10^{-5}	4.47×10^{-6}	4.27×10^{-5}
35–40		1.03×10^{-5}		1.11×10^{-5}		9.27×10^{-6}		1.11×10^{-5}		8.60×10^{-6}	1.18×10^{-6}	1.12×10^{-5}	1.18×10^{-6}	1.12×10^{-5}
40–45		5.04×10^{-6}		5.44×10^{-6}		4.31×10^{-6}		5.48×10^{-6}		3.94×10^{-6}	3.09×10^{-7}	2.96×10^{-6}	3.09×10^{-7}	2.96×10^{-6}
45–50		2.56×10^{-6}		2.79×10^{-6}		2.12×10^{-6}		2.85×10^{-6}		1.88×10^{-6}	8.16×10^{-8}	7.79×10^{-7}	8.16×10^{-8}	7.79×10^{-7}
50–70		$<10^{-6}$		1.05×10^{-6}		$<10^{-6}$		1.09×10^{-6}		$<10^{-6}$	7.36×10^{-9}	7.03×10^{-8}	7.36×10^{-9}	7.03×10^{-8}
70–100				$<10^{-6}$		$<10^{-6}$		$<10^{-6}$			2.53×10^{-11}	2.41×10^{-10}	2.53×10^{-11}	2.41×10^{-10}

TABLE 6 Values of Attenuation Coefficient per Kilometer as a Function of Altitude for Each Laser Wavelength and Each Atmospheric Model (Continued)

Height, km	Tropical		Midlatitude				Subarctic				Clear		Hazy	
			Summer		Winter		Summer		Winter					
	Molecular absorption k_m, km^{-1}	Molecular scattering σ_m, km^{-1}	Molecular absorption k_m, km^{-1}	Molecular scattering σ_m, km^{-1}	Molecular absorption k_m, km^{-1}	Molecular scattering σ_m, km^{-1}	Molecular absorption k_m, km^{-1}	Molecular scattering σ_m, km^{-1}	Molecular absorption k_m, km^{-1}	Molecular scattering σ_m, km^{-1}	Aerosol absorption k_a, km^{-1}	Aerosol scattering σ_a, km^{-1}	Aerosol absorption k_a, km^{-1}	Aerosol scattering σ_a, km^{-1}
							$\lambda = 1.06\ \mu m$							
0	$<10^{-6}$	8.04×10^{-4}	$<10^{-6}$	8.20×10^{-4}	$<10^{-6}$	8.91×10^{-4}	$<10^{-6}$	8.38×10^{-4}	$<10^{-6}$	9.39×10^{-4}	1.08×10^{-2}	7.83×10^{-2}	5.28×10^{-2}	3.81×10^{-1}
0-1		7.68×10^{-4}		7.81×10^{-4}		8.43×10^{-4}		7.98×10^{-4}		8.77×10^{-4}	7.39×10^{-3}	5.34×10^{-2}	3.33×10^{-2}	2.40×10^{-1}
1-2		6.99×10^{-4}		7.06×10^{-4}		7.52×10^{-4}		7.21×10^{-4}		7.70×10^{-4}	3.22×10^{-3}	2.33×10^{-2}	1.22×10^{-2}	8.79×10^{-2}
2-3		6.33×10^{-4}		6.38×10^{-4}		6.70×10^{-4}		6.50×10^{-4}		6.82×10^{-4}	1.38×10^{-3}	9.94×10^{-3}	4.45×10^{-3}	3.22×10^{-2}
3-4		5.72×10^{-4}		5.77×10^{-4}		5.99×10^{-4}		5.84×10^{-4}		6.06×10^{-4}	6.39×10^{-4}	4.62×10^{-3}	1.62×10^{-3}	1.17×10^{-2}
4-5		5.19×10^{-4}		5.21×10^{-4}		5.37×10^{-4}		5.24×10^{-4}		5.40×10^{-4}	3.98×10^{-4}	2.87×10^{-3}	5.93×10^{-4}	4.28×10^{-3}
5-6		4.69×10^{-4}		4.69×10^{-4}		4.80×10^{-4}		4.71×10^{-4}		4.82×10^{-4}	2.90×10^{-4}	2.10×10^{-3}	2.90×10^{-4}	2.10×10^{-3}
6-7		4.22×10^{-4}		4.21×10^{-4}		4.27×10^{-4}		4.23×10^{-4}		4.29×10^{-4}	2.34×10^{-4}	1.69×10^{-3}	2.34×10^{-4}	1.69×10^{-3}
7-8		3.80×10^{-4}		3.78×10^{-4}		3.80×10^{-4}		3.79×10^{-4}		3.81×10^{-4}	2.29×10^{-4}	1.65×10^{-3}	2.29×10^{-4}	1.65×10^{-3}
8-9		3.41×10^{-4}		3.38×10^{-4}		3.36×10^{-4}		3.38×10^{-4}		3.34×10^{-4}	2.28×10^{-4}	1.64×10^{-3}	2.28×10^{-4}	1.64×10^{-3}
9-10		3.04×10^{-4}		3.02×10^{-4}		2.97×10^{-4}		3.01×10^{-4}		2.88×10^{-4}	2.20×10^{-4}	1.59×10^{-3}	2.20×10^{-4}	1.59×10^{-3}
10-11		2.72×10^{-4}		2.69×10^{-4}		2.59×10^{-4}		2.64×10^{-4}		2.46×10^{-4}	2.10×10^{-4}	1.52×10^{-3}	2.10×10^{-4}	1.52×10^{-3}
11-12		2.41×10^{-4}		2.39×10^{-4}		2.22×10^{-4}		2.26×10^{-4}		2.10×10^{-4}	2.09×10^{-4}	1.51×10^{-3}	2.09×10^{-4}	1.51×10^{-3}
12-13		2.13×10^{-4}		2.11×10^{-4}		1.90×10^{-4}		1.95×10^{-4}		1.80×10^{-4}	2.06×10^{-4}	1.49×10^{-3}	2.06×10^{-4}	1.49×10^{-3}
13-14		1.88×10^{-4}		1.83×10^{-4}		1.63×10^{-4}		1.67×10^{-4}		1.54×10^{-4}	1.95×10^{-4}	1.41×10^{-3}	1.95×10^{-4}	1.41×10^{-3}
14-15		1.66×10^{-4}		1.56×10^{-4}		1.40×10^{-4}		1.44×10^{-4}		1.31×10^{-5}	1.88×10^{-4}	1.36×10^{-3}	1.88×10^{-4}	1.36×10^{-3}
15-16		1.44×10^{-4}		1.33×10^{-4}		1.20×10^{-4}		1.23×10^{-4}		1.12×10^{-5}	1.77×10^{-4}	1.28×10^{-3}	1.77×10^{-4}	1.28×10^{-3}
16-17		1.24×10^{-4}		1.14×10^{-4}		1.03×10^{-4}		1.06×10^{-4}		9.63×10^{-5}	1.72×10^{-4}	1.24×10^{-3}	1.72×10^{-4}	1.24×10^{-3}
17-18		1.05×10^{-4}		9.72×10^{-5}		8.80×10^{-5}		9.14×10^{-5}		8.25×10^{-5}	1.68×10^{-4}	1.21×10^{-3}	1.68×10^{-4}	1.21×10^{-3}
18-19		8.63×10^{-5}		8.29×10^{-5}		7.53×10^{-5}		7.86×10^{-5}		7.06×10^{-5}	1.52×10^{-4}	1.10×10^{-3}	1.52×10^{-4}	1.10×10^{-3}
19-20		7.16×10^{-5}		7.07×10^{-5}		6.45×10^{-5}		6.75×10^{-5}		6.05×10^{-5}	1.20×10^{-4}	8.65×10^{-4}	1.20×10^{-4}	8.65×10^{-4}
20-21		5.96×10^{-5}		6.02×10^{-5}		5.51×10^{-5}		5.80×10^{-5}		5.17×10^{-5}	8.73×10^{-5}	6.31×10^{-4}	8.73×10^{-5}	6.31×10^{-4}
21-22		4.98×10^{-5}		5.14×10^{-5}		4.70×10^{-5}		4.99×10^{-5}		4.42×10^{-5}	6.44×10^{-5}	4.66×10^{-4}	6.44×10^{-5}	4.66×10^{-4}
22-23		4.19×10^{-5}		4.38×10^{-5}		4.01×10^{-5}		4.29×10^{-5}		3.78×10^{-5}	4.89×10^{-5}	3.53×10^{-4}	4.89×10^{-5}	3.53×10^{-4}
23-24		3.55×10^{-5}		3.74×10^{-5}		3.43×10^{-5}		3.69×10^{-5}		3.23×10^{-5}	3.80×10^{-5}	2.75×10^{-4}	3.80×10^{-5}	2.75×10^{-4}
24-25		3.02×10^{-5}		3.19×10^{-5}		2.93×10^{-5}		3.15×10^{-5}		2.76×10^{-5}	3.10×10^{-5}	2.24×10^{-4}	3.10×10^{-5}	2.24×10^{-4}
25-30		2.01×10^{-5}		2.15×10^{-5}		1.95×10^{-5}		2.13×10^{-5}		1.84×10^{-5}	1.66×10^{-5}	1.20×10^{-4}	1.66×10^{-5}	1.20×10^{-4}
30-35		9.20×10^{-6}		9.89×10^{-6}		8.79×10^{-6}		9.98×10^{-6}		8.14×10^{-6}	4.73×10^{-6}	3.42×10^{-5}	4.73×10^{-6}	3.42×10^{-5}
35-40		4.37×10^{-6}		4.71×10^{-6}		3.95×10^{-6}		4.73×10^{-6}		3.66×10^{-6}	1.25×10^{-6}	9.00×10^{-6}	1.25×10^{-6}	9.00×10^{-6}
40-45		2.15×10^{-6}		2.31×10^{-6}		1.83×10^{-6}		2.33×10^{-6}		1.68×10^{-6}	3.28×10^{-7}	2.37×10^{-6}	3.28×10^{-7}	2.37×10^{-6}
45-50		1.09×10^{-6}		1.19×10^{-6}		$<10^{-6}$		1.21×10^{-6}		$<10^{-6}$	8.64×10^{-8}	6.24×10^{-7}	8.64×10^{-8}	6.24×10^{-7}
50-70		$<10^{-6}$		$<10^{-6}$				$<10^{-6}$			7.79×10^{-9}	5.63×10^{-8}	7.79×10^{-9}	5.63×10^{-8}
70-100											2.67×10^{-11}	1.93×10^{-10}	2.67×10^{-11}	1.93×10^{-10}
							$\lambda = 1.536\ \mu m$							
0	1.47×10^{-4}	1.77×10^{-4}	1.52×10^{-4}	1.81×10^{-4}	1.81×10^{-4}	1.96×10^{-4}	1.74×10^{-4}	1.85×10^{-4}	2.22×10^{-4}	2.07×10^{-4}	9.24×10^{-3}	4.87×10^{-2}	4.50×10^{-2}	2.37×10^{-1}
0-1	1.31×10^{-4}	1.69×10^{-4}	1.36×10^{-4}	1.72×10^{-4}	1.62×10^{-4}	1.86×10^{-4}	1.55×10^{-4}	1.76×10^{-4}	1.98×10^{-4}	1.93×10^{-4}	6.30×10^{-3}	3.32×10^{-2}	2.84×10^{-2}	1.49×10^{-1}
1-2	1.09×10^{-4}	1.54×10^{-4}	1.11×10^{-4}	1.56×10^{-4}	1.30×10^{-4}	1.66×10^{-4}	1.27×10^{-4}	1.59×10^{-4}	1.53×10^{-4}	1.70×10^{-4}	2.75×10^{-3}	1.45×10^{-2}	1.04×10^{-2}	5.46×10^{-2}
2-3	8.99×10^{-5}	1.40×10^{-4}	9.24×10^{-5}	1.41×10^{-4}	1.04×10^{-4}	1.48×10^{-4}	1.04×10^{-4}	1.43×10^{-4}	1.20×10^{-4}	1.50×10^{-4}	1.17×10^{-3}	6.18×10^{-3}	3.80×10^{-3}	2.00×10^{-2}
3-4	7.48×10^{-5}	1.26×10^{-4}	7.62×10^{-5}	1.27×10^{-4}	8.39×10^{-5}	1.32×10^{-4}	8.54×10^{-5}	1.29×10^{-4}	9.64×10^{-5}	1.34×10^{-4}	5.45×10^{-4}	2.87×10^{-3}	1.38×10^{-3}	7.29×10^{-3}
4-5	6.14×10^{-5}	1.14×10^{-4}	6.09×10^{-5}	1.15×10^{-4}	6.82×10^{-5}	1.18×10^{-4}	6.80×10^{-5}	1.15×10^{-4}	7.68×10^{-5}	1.19×10^{-4}	3.39×10^{-4}	1.79×10^{-3}	5.06×10^{-4}	2.66×10^{-3}
5-6	5.06×10^{-5}	1.03×10^{-4}	5.09×10^{-5}	1.04×10^{-4}	5.45×10^{-5}	1.06×10^{-4}	5.69×10^{-5}	1.04×10^{-4}	6.20×10^{-5}	1.06×10^{-4}	2.48×10^{-4}	1.31×10^{-3}	2.48×10^{-4}	1.31×10^{-3}
6-7	4.12×10^{-5}	9.31×10^{-5}	4.20×10^{-5}	9.29×10^{-4}	4.38×10^{-5}	9.42×10^{-5}	4.51×10^{-5}	9.32×10^{-5}	4.95×10^{-5}	9.46×10^{-5}	2.00×10^{-4}	1.05×10^{-3}	2.00×10^{-4}	1.05×10^{-3}
7-8	3.36×10^{-5}	8.38×10^{-5}	3.35×10^{-5}	8.33×10^{-4}	3.47×10^{-5}	8.37×10^{-5}	3.69×10^{-5}	8.35×10^{-5}	3.92×10^{-5}	8.39×10^{-5}	1.95×10^{-4}	1.03×10^{-3}	1.95×10^{-4}	1.03×10^{-3}
8-9	2.75×10^{-5}	7.51×10^{-5}	2.63×10^{-5}	7.46×10^{-5}	2.77×10^{-5}	7.42×10^{-5}	2.97×10^{-5}	7.46×10^{-5}	3.04×10^{-5}	7.36×10^{-5}	1.94×10^{-4}	1.02×10^{-3}	1.94×10^{-4}	1.02×10^{-3}

9–10	2.17×10^{-5}	6.71×10^{-5}	2.15×10^{-5}	6.66×10^{-5}	2.10×10^{-5}	6.55×10^{-5}	2.35×10^{-5}	6.64×10^{-5}	2.26×10^{-5}	6.34×10^{-5}	1.88×10^{-4}	9.89×10^{-4}	1.88×10^{-4}	9.89×10^{-4}
10–11	1.77×10^{-5}	5.99×10^{-5}	1.73×10^{-5}	5.93×10^{-5}	1.65×10^{-5}	5.70×10^{-5}	1.79×10^{-5}	5.81×10^{-5}	1.65×10^{-5}	5.42×10^{-5}	1.80×10^{-4}	9.46×10^{-4}	1.80×10^{-4}	9.46×10^{-4}
11–12	1.38×10^{-5}	5.32×10^{-5}	1.38×10^{-5}	5.36×10^{-5}	1.20×10^{-5}	4.89×10^{-5}	1.36×10^{-5}	4.99×10^{-5}	1.20×10^{-5}	4.63×10^{-5}	1.78×10^{-4}	9.38×10^{-4}	1.78×10^{-4}	9.38×10^{-4}
12–13	1.12×10^{-5}	4.70×10^{-5}	1.07×10^{-5}	4.65×10^{-5}	8.87×10^{-6}	4.20×10^{-5}	9.68×10^{-6}	4.29×10^{-5}	8.64×10^{-6}	3.96×10^{-5}	1.75×10^{-4}	9.24×10^{-4}	1.75×10^{-4}	9.24×10^{-4}
13–14	8.40×10^{-6}	4.15×10^{-5}	8.01×10^{-6}	4.04×10^{-5}	6.65×10^{-6}	3.60×10^{-5}	7.45×10^{-6}	3.69×10^{-5}	6.59×10^{-6}	3.39×10^{-5}	1.67×10^{-4}	8.78×10^{-4}	1.67×10^{-4}	8.78×10^{-4}
14–15	6.87×10^{-6}	3.65×10^{-5}	6.08×10^{-6}	3.44×10^{-5}	4.81×10^{-6}	3.09×10^{-5}	5.45×10^{-6}	3.16×10^{-5}	4.69×10^{-6}	2.89×10^{-5}	1.60×10^{-4}	8.43×10^{-4}	1.60×10^{-4}	8.43×10^{-4}
15–16	5.28×10^{-6}	3.18×10^{-5}	4.30×10^{-6}	2.93×10^{-5}	3.51×10^{-6}	2.64×10^{-5}	3.85×10^{-6}	2.72×10^{-5}	3.44×10^{-6}	2.48×10^{-5}	1.51×10^{-4}	7.96×10^{-4}	1.51×10^{-4}	7.96×10^{-4}
16–17	3.76×10^{-6}	2.74×10^{-5}	3.10×10^{-6}	2.50×10^{-5}	2.59×10^{-6}	2.27×10^{-5}	2.98×10^{-6}	2.34×10^{-5}	2.33×10^{-6}	2.12×10^{-5}	1.47×10^{-4}	7.72×10^{-4}	1.47×10^{-4}	7.72×10^{-4}
17–18	2.70×10^{-6}	2.30×10^{-5}	2.29×10^{-6}	2.14×10^{-5}	1.93×10^{-6}	1.94×10^{-5}	2.18×10^{-6}	2.01×10^{-5}	1.86×10^{-6}	1.82×10^{-5}	1.43×10^{-4}	7.55×10^{-4}	1.43×10^{-4}	7.55×10^{-4}
18–19	1.85×10^{-6}	1.90×10^{-5}	1.66×10^{-6}	1.83×10^{-5}	1.40×10^{-6}	1.66×10^{-5}	1.61×10^{-6}	1.73×10^{-5}	1.38×10^{-6}	1.56×10^{-5}	1.30×10^{-4}	6.82×10^{-4}	1.30×10^{-4}	6.82×10^{-4}
19–20	1.26×10^{-6}	1.58×10^{-5}	1.21×10^{-6}	1.56×10^{-5}	1.01×10^{-6}	1.42×10^{-5}	1.21×10^{-6}	1.49×10^{-5}	$<10^{-6}$	1.33×10^{-5}	1.02×10^{-4}	5.38×10^{-4}	1.02×10^{-4}	5.38×10^{-4}
20–21	$<10^{-6}$	1.31×10^{-5}	$<10^{-6}$	1.33×10^{-5}	$<10^{-6}$	1.21×10^{-5}	$<10^{-6}$	1.28×10^{-5}		1.14×10^{-5}	7.45×10^{-5}	3.92×10^{-4}	7.45×10^{-5}	3.92×10^{-4}
21–22		1.10×10^{-5}		1.13×10^{-5}		1.04×10^{-5}		1.10×10^{-5}		9.75×10^{-6}	5.50×10^{-5}	2.90×10^{-4}	5.50×10^{-5}	2.90×10^{-4}
22–23		9.23×10^{-6}		9.66×10^{-6}		8.85×10^{-6}		9.47×10^{-6}		8.34×10^{-6}	4.17×10^{-5}	2.20×10^{-4}	4.17×10^{-5}	2.20×10^{-4}
23–24		7.83×10^{-6}		8.24×10^{-6}		7.57×10^{-6}		8.13×10^{-6}		7.13×10^{-6}	3.25×10^{-5}	1.71×10^{-4}	3.25×10^{-5}	1.71×10^{-4}
24–25		6.65×10^{-6}		7.04×10^{-6}		6.46×10^{-6}		6.95×10^{-6}		6.09×10^{-6}	2.65×10^{-5}	1.40×10^{-4}	2.65×10^{-5}	1.40×10^{-4}
25–30		4.43×10^{-6}		4.73×10^{-6}		4.31×10^{-6}		4.70×10^{-6}		4.05×10^{-6}	1.41×10^{-5}	7.45×10^{-5}	1.41×10^{-5}	7.45×10^{-5}
30–35		2.03×10^{-6}		2.18×10^{-6}		1.94×10^{-6}		2.20×10^{-6}		1.80×10^{-6}	4.04×10^{-6}	2.13×10^{-5}	4.04×10^{-6}	2.13×10^{-5}
35–40		$<10^{-6}$		1.04×10^{-6}		$<10^{-6}$		1.04×10^{-6}		$<10^{-6}$	1.06×10^{-6}	5.60×10^{-6}	1.06×10^{-6}	5.60×10^{-6}
40–45				$<10^{-6}$				$<10^{-6}$			2.80×10^{-7}	1.47×10^{-6}	2.80×10^{-7}	1.47×10^{-6}
45–50											7.37×10^{-8}	3.88×10^{-7}	7.37×10^{-8}	3.88×10^{-7}
50–70											6.65×10^{-9}	3.50×10^{-8}	6.65×10^{-9}	3.50×10^{-8}
70–100											2.28×10^{-11}	1.20×10^{-10}	2.28×10^{-11}	1.20×10^{-10}

$\lambda = 3.39225\ \mu m$

0	1.77	7.79×10^{-6}	1.83	7.95×10^{-6}	1.98	8.63×10^{-6}	1.88	8.12×10^{-6}	2.03	9.09×10^{-6}	2.22×10^{-3}	1.87×10^{-2}	1.08×10^{-2}	9.12×10^{-2}
0–1	1.75	7.44×10^{-6}	1.79	7.56×10^{-6}	1.92	8.17×10^{-6}	1.84	7.72×10^{-6}	1.97	8.50×10^{-6}	9.78×10^{-4}	8.23×10^{-3}	3.96×10^{-3}	3.33×10^{-2}
1–2	1.68	6.77×10^{-6}	1.70	6.83×10^{-6}	1.79	7.29×10^{-6}	1.75	6.98×10^{-6}	1.84	7.46×10^{-6}	4.22×10^{-4}	3.55×10^{-3}	1.45×10^{-3}	1.22×10^{-2}
2–3	1.63	6.13×10^{-6}	1.65	6.18×10^{-6}	1.75	6.49×10^{-6}	1.65	6.29×10^{-6}	1.75	6.69×10^{-6}	1.77×10^{-4}	1.49×10^{-3}	5.29×10^{-4}	4.45×10^{-3}
3–4	1.59	5.54×10^{-6}	1.60	5.59×10^{-6}	1.65	5.80×10^{-6}	1.66	5.65×10^{-6}	1.67	5.86×10^{-6}	9.38×10^{-5}	7.89×10^{-4}	1.93×10^{-4}	1.62×10^{-3}
4–5	1.54	5.02×10^{-6}	1.54	5.05×10^{-6}	1.60	5.20×10^{-6}	1.54	5.07×10^{-6}	1.62	5.23×10^{-6}	7.07×10^{-5}	5.95×10^{-4}	7.07×10^{-5}	5.95×10^{-4}
5–6	1.52	4.54×10^{-6}	1.54	4.55×10^{-6}	1.54	4.64×10^{-6}	1.61	4.56×10^{-6}	1.57	4.67×10^{-6}	4.98×10^{-5}	4.19×10^{-4}	4.98×10^{-5}	4.19×10^{-4}
6–7	1.47	4.09×10^{-6}	1.46	4.08×10^{-6}	1.54	4.14×10^{-6}	1.47	4.09×10^{-6}	1.54	4.16×10^{-6}	4.63×10^{-5}	3.90×10^{-4}	4.63×10^{-5}	3.90×10^{-4}
7–8	1.45	3.69×10^{-6}	1.52	3.66×10^{-6}	1.46	3.68×10^{-6}	1.45	3.67×10^{-6}	1.50	3.68×10^{-6}	4.77×10^{-5}	4.01×10^{-4}	4.77×10^{-5}	4.01×10^{-4}
8–9	1.46	3.30×10^{-6}	1.41	3.27×10^{-6}	1.44	3.26×10^{-6}	1.51	3.28×10^{-6}	1.47	3.23×10^{-6}	4.58×10^{-5}	3.85×10^{-4}	4.58×10^{-5}	3.85×10^{-4}
9–10	1.41	2.95×10^{-6}	1.43	2.92×10^{-6}	1.49	2.88×10^{-6}	1.40	2.92×10^{-6}	1.44	2.79×10^{-6}	4.46×10^{-5}	3.75×10^{-4}	4.46×10^{-5}	3.75×10^{-4}
10–11	1.42	2.63×10^{-6}	1.40	2.60×10^{-6}	1.41	2.51×10^{-6}	1.42	2.55×10^{-6}	1.41	2.38×10^{-6}	4.18×10^{-5}	3.52×10^{-4}	4.18×10^{-5}	3.52×10^{-4}
11–12	1.37	2.34×10^{-6}	1.39	2.31×10^{-6}	1.39	2.15×10^{-6}	1.38	2.19×10^{-6}	1.38	2.04×10^{-6}	4.39×10^{-5}	3.69×10^{-4}	4.39×10^{-5}	3.69×10^{-4}
12–13	1.40	2.06×10^{-6}	1.37	2.04×10^{-6}	1.36	1.84×10^{-6}	1.35	1.89×10^{-6}	1.34	1.74×10^{-6}	4.06×10^{-5}	3.41×10^{-4}	4.06×10^{-5}	3.41×10^{-4}
13–14	1.31	1.82×10^{-6}	1.34	1.77×10^{-6}	1.30	1.58×10^{-6}	1.34	1.62×10^{-6}	1.31	1.49×10^{-6}	3.97×10^{-5}	3.34×10^{-4}	3.97×10^{-5}	3.34×10^{-4}
14–15	1.35	1.60×10^{-6}	1.36	1.51×10^{-6}	1.29	1.36×10^{-6}	1.32	1.39×10^{-6}	1.27	1.27×10^{-6}	3.73×10^{-5}	3.14×10^{-4}	3.73×10^{-5}	3.14×10^{-4}
15–16	1.32	1.40×10^{-6}	1.28	1.29×10^{-6}	1.24	1.16×10^{-6}	1.21	1.19×10^{-6}	1.21	1.09×10^{-6}	3.55×10^{-5}	2.98×10^{-4}	3.55×10^{-5}	2.98×10^{-4}
16–17	1.21	1.20×10^{-6}	1.20	1.10×10^{-6}	1.19	$<10^{-6}$	1.19	1.03×10^{-6}	1.16	$<10^{-6}$	3.51×10^{-5}	2.95×10^{-4}	3.51×10^{-5}	2.95×10^{-4}
17–18	1.17	1.01×10^{-6}	1.14	$<10^{-6}$	1.13		1.13	$<10^{-6}$	1.09		3.39×10^{-5}	2.85×10^{-4}	3.39×10^{-5}	2.85×10^{-4}
18–19	1.10	$<10^{-6}$	1.09		1.03		1.07		1.01		2.86×10^{-5}	2.40×10^{-4}	2.86×10^{-5}	2.40×10^{-4}
19–20	9.95×10^{-1}		1.00		9.47×10^{-1}		9.97×10^{-1}		9.10×10^{-1}		2.10×10^{-5}	1.76×10^{-4}	2.10×10^{-5}	1.76×10^{-4}
20–21	8.96×10^{-1}		9.09×10^{-1}		8.64×10^{-1}		8.92×10^{-1}		8.07×10^{-1}		1.52×10^{-5}	1.28×10^{-4}	1.52×10^{-5}	1.28×10^{-4}
21–22	7.82×10^{-1}		8.14×10^{-1}		7.19×10^{-1}		7.99×10^{-1}		6.99×10^{-1}		1.14×10^{-5}	9.63×10^{-5}	1.14×10^{-5}	9.63×10^{-5}
22–23	6.55×10^{-1}		6.62×10^{-1}		6.52×10^{-1}		6.99×10^{-1}		5.86×10^{-1}		8.75×10^{-6}	7.36×10^{-5}	8.75×10^{-6}	7.36×10^{-5}
23–24	5.48×10^{-1}		6.23×10^{-1}		5.08×10^{-1}		5.92×10^{-1}		4.84×10^{-1}		6.94×10^{-6}	5.84×10^{-5}	6.94×10^{-6}	5.84×10^{-5}
24–25	4.51×10^{-1}		4.76×10^{-1}		4.41×10^{-1}		4.68×10^{-1}		3.89×10^{-1}		5.84×10^{-6}	4.91×10^{-5}	5.84×10^{-6}	4.91×10^{-5}
25–30	2.33×10^{-1}		2.63×10^{-1}		2.18×10^{-1}		2.61×10^{-1}		1.93×10^{-1}		1.76×10^{-6}	1.48×10^{-5}	1.76×10^{-6}	1.48×10^{-5}
30–35	1.93×10^{-1}		2.07×10^{-1}		1.84×10^{-1}		7.02×10^{-2}		1.71×10^{-1}		4.63×10^{-7}	3.90×10^{-6}	4.63×10^{-7}	3.90×10^{-6}
35–40	9.62×10^{-2}		1.03×10^{-1}		8.70×10^{-2}		1.04×10^{-1}		8.04×10^{-2}		1.22×10^{-7}	1.03×10^{-6}	1.22×10^{-7}	1.03×10^{-6}
40–45	4.86×10^{-2}		5.30×10^{-2}		4.14×10^{-2}		5.22×10^{-2}		3.79×10^{-2}		3.21×10^{-8}	2.70×10^{-7}	3.21×10^{-8}	2.70×10^{-7}
45–50	2.46×10^{-2}		2.67×10^{-2}		2.05×10^{-2}		2.73×10^{-2}		1.83×10^{-2}		8.48×10^{-9}	7.13×10^{-8}	8.48×10^{-9}	7.13×10^{-8}
50–70	6.75×10^{-3}		7.50×10^{-3}		4.83×10^{-3}		7.77×10^{-3}		4.51×10^{-3}		4.37×10^{-11}	3.67×10^{-10}	4.37×10^{-11}	3.67×10^{-10}
70–100	3.21×10^{-4}		3.72×10^{-4}		2.61×10^{-4}		3.91×10^{-4}		2.25×10^{-4}		1.55×10^{-14}	1.30×10^{-13}	1.55×10^{-14}	1.30×10^{-13}

TABLE 6 Values of Attenuation Coefficient per Kilometer as a Function of Altitude for Each Laser Wavelength and Each Atmospheric Model (Continued)

Height, km	Tropical		Midlatitude				Subarctic				Clear		Hazy	
			Summer		Winter		Summer		Winter					
	Molecular absorption k_m, km^{-1}	Molecular scattering σ_m, km^{-1}	Molecular absorption k_m, km^{-1}	Molecular scattering σ_m, km^{-1}	Molecular absorption k_m, km^{-1}	Molecular scattering σ_m, km^{-1}	Molecular absorption k_m, km^{-1}	Molecular scattering σ_m, km^{-1}	Molecular absorption k_m, km^{-1}	Molecular scattering σ_m, km^{-1}	Aerosol absorption k_a, km^{-1}	Aerosol scattering σ_a, km^{-1}	Aerosol absorption k_a, km^{-1}	Aerosol scattering σ_a, km^{-1}
$\lambda = 10.591\ \mu m$														
0	5.788×10^{-1}	$<1.0 \times 10^{-6}$	3.582×10^{-1}	$<1.0 \times 10^{-6}$	7.937×10^{-2}	$<1.0 \times 10^{-6}$	2.006×10^{-1}	$<1.0 \times 10^{-6}$	4.118×10^{-2}	$<1.0 \times 10^{-6}$	5.49×10^{-3}	7.97×10^{-3}	2.67×10^{-2}	3.88×10^{-2}
0–1	5.172×10^{-1}		3.256×10^{-1}		7.312×10^{-2}		1.818×10^{-1}		4.147×10^{-2}		2.41×10^{-3}	3.51×10^{-3}	9.76×10^{-3}	1.42×10^{-2}
1–2	2.845×10^{-1}		1.877×10^{-1}		5.895×10^{-2}		1.137×10^{-1}		4.002×10^{-2}		1.04×10^{-3}	1.51×10^{-3}	3.58×10^{-3}	5.20×10^{-3}
2–3	1.807×10^{-1}		1.152×10^{-1}		4.911×10^{-2}		8.152×10^{-2}		3.516×10^{-2}		4.38×10^{-4}	6.36×10^{-4}	1.31×10^{-3}	1.90×10^{-3}
3–4	9.616×10^{-2}		7.582×10^{-2}		4.043×10^{-2}		6.090×10^{-2}		3.048×10^{-2}		2.31×10^{-4}	3.36×10^{-4}	4.76×10^{-4}	6.91×10^{-4}
4–5	6.290×10^{-2}		5.544×10^{-2}		3.240×10^{-2}		4.663×10^{-2}		2.453×10^{-2}		1.74×10^{-4}	2.53×10^{-4}	1.74×10^{-4}	2.53×10^{-4}
5–6	5.019×10^{-2}		4.468×10^{-2}		2.622×10^{-2}		3.737×10^{-2}		1.932×10^{-2}		1.23×10^{-4}	1.79×10^{-4}	1.23×10^{-4}	1.79×10^{-4}
6–7	3.988×10^{-2}		3.752×10^{-2}		2.147×10^{-2}		2.861×10^{-2}		1.509×10^{-2}		1.14×10^{-4}	1.66×10^{-4}	1.14×10^{-4}	1.66×10^{-4}
7–8	3.200×10^{-2}		3.018×10^{-2}		1.728×10^{-2}		2.277×10^{-2}		1.171×10^{-2}		1.18×10^{-4}	1.71×10^{-4}	1.18×10^{-4}	1.71×10^{-4}
8–9	2.634×10^{-2}		2.378×10^{-2}		1.405×10^{-2}		1.788×10^{-2}		9.593×10^{-3}		1.13×10^{-4}	1.64×10^{-4}	1.13×10^{-4}	1.64×10^{-4}
9–10	2.074×10^{-2}		1.952×10^{-2}		1.083×10^{-2}		1.375×10^{-2}		8.932×10^{-3}		1.10×10^{-4}	1.60×10^{-4}	1.10×10^{-4}	1.60×10^{-4}
10–11	1.651×10^{-2}		1.574×10^{-2}		9.813×10^{-3}		1.195×10^{-2}		8.921×10^{-3}		1.03×10^{-4}	1.50×10^{-4}	1.03×10^{-4}	1.50×10^{-4}
11–12	1.287×10^{-2}		1.241×10^{-2}		9.484×10^{-3}		1.229×10^{-2}		8.908×10^{-3}		1.08×10^{-4}	1.57×10^{-4}	1.08×10^{-4}	1.57×10^{-4}
12–13	1.035×10^{-2}		9.534×10^{-3}		9.358×10^{-3}		1.181×10^{-2}		8.733×10^{-3}		1.00×10^{-4}	1.45×10^{-4}	1.00×10^{-4}	1.45×10^{-4}
13–14	7.382×10^{-3}		8.337×10^{-3}		9.340×10^{-3}		1.229×10^{-2}		9.109×10^{-3}		9.80×10^{-5}	1.42×10^{-4}	9.80×10^{-5}	1.42×10^{-4}
14–15	5.850×10^{-3}		8.700×10^{-3}		9.001×10^{-3}		1.222×10^{-2}		8.891×10^{-3}		9.21×10^{-5}	1.34×10^{-4}	9.21×10^{-5}	1.34×10^{-4}
15–16	4.330×10^{-3}		8.491×10^{-3}		8.749×10^{-3}		1.166×10^{-2}		8.773×10^{-3}		8.76×10^{-5}	1.27×10^{-4}	8.76×10^{-5}	1.27×10^{-4}
16–17	3.316×10^{-3}		8.364×10^{-3}		8.573×10^{-3}		1.217×10^{-2}		8.559×10^{-3}		8.65×10^{-5}	1.26×10^{-4}	8.65×10^{-5}	1.26×10^{-4}
17–18	3.556×10^{-3}		8.467×10^{-3}		8.556×10^{-3}		1.203×10^{-2}		8.324×10^{-3}		8.37×10^{-5}	1.22×10^{-4}	8.37×10^{-5}	1.22×10^{-4}
18–19	4.350×10^{-3}		8.560×10^{-3}		8.249×10^{-3}		1.199×10^{-2}		8.209×10^{-3}		7.05×10^{-5}	1.02×10^{-4}	7.05×10^{-5}	1.02×10^{-4}
19–20	5.194×10^{-3}		8.929×10^{-3}		8.011×10^{-3}		1.217×10^{-2}		7.884×10^{-3}		5.18×10^{-5}	7.52×10^{-5}	5.18×10^{-5}	7.52×10^{-5}
20–21	6.273×10^{-3}		9.186×10^{-3}		8.186×10^{-3}		1.186×10^{-2}		7.784×10^{-3}		3.75×10^{-5}	5.45×10^{-5}	3.75×10^{-5}	5.45×10^{-5}
21–22	7.471×10^{-3}		9.719×10^{-3}		8.194×10^{-3}		1.208×10^{-2}		7.523×10^{-3}		2.82×10^{-5}	4.10×10^{-5}	2.82×10^{-5}	4.10×10^{-5}
22–23	8.351×10^{-3}		1.010×10^{-2}		8.161×10^{-3}		1.208×10^{-2}		7.209×10^{-3}		2.16×10^{-5}	3.14×10^{-5}	2.16×10^{-5}	3.14×10^{-5}
23–24	9.041×10^{-3}		1.114×10^{-2}		8.107×10^{-3}		1.199×10^{-2}		7.329×10^{-3}		1.71×10^{-5}	2.49×10^{-5}	1.71×10^{-5}	2.49×10^{-5}
24–25	9.909×10^{-3}		1.112×10^{-2}		8.378×10^{-3}		1.275×10^{-2}		6.837×10^{-3}		1.44×10^{-5}	2.09×10^{-5}	1.44×10^{-5}	2.09×10^{-5}
25–30	1.203×10^{-2}		1.327×10^{-2}		8.087×10^{-3}		1.453×10^{-2}		7.238×10^{-3}		4.34×10^{-6}	6.31×10^{-6}	4.34×10^{-6}	6.31×10^{-6}
30–35	1.190×10^{-2}		1.319×10^{-2}		6.848×10^{-3}		2.007×10^{-2}		5.785×10^{-3}		1.14×10^{-6}	1.66×10^{-6}	1.14×10^{-6}	1.66×10^{-6}
35–40	1.101×10^{-2}		1.269×10^{-2}		6.714×10^{-3}		1.395×10^{-2}		5.099×10^{-3}		3.01×10^{-7}	4.37×10^{-7}	3.01×10^{-7}	4.37×10^{-7}
40–45	8.865×10^{-3}		1.063×10^{-2}		6.023×10^{-3}		1.189×10^{-2}		4.100×10^{-3}		7.92×10^{-8}	1.15×10^{-7}	7.92×10^{-8}	1.15×10^{-7}
45–50	6.038×10^{-3}		7.522×10^{-3}		4.405×10^{-3}		8.186×10^{-3}		3.082×10^{-3}		2.09×10^{-8}	3.04×10^{-8}	2.09×10^{-8}	3.04×10^{-8}
50–70	9.007×10^{-4}		1.077×10^{-3}		2.744×10^{-4}		1.097×10^{-3}		7.761×10^{-4}		1.08×10^{-10}	1.56×10^{-10}	1.08×10^{-10}	1.56×10^{-10}
70–100	1.535×10^{-5}		1.743×10^{-5}		1.580×10^{-4}		1.762×10^{-5}		1.785×10^{-5}		3.82×10^{-14}	5.55×10^{-14}	3.82×10^{-14}	5.55×10^{-14}
$\lambda = 27.9\ \mu m$														
0	>30	$<10^{-6}$	>30	$<10^{-6}$	>30	$<10^{-6}$	>30	$<10^{-6}$	>30	$<10^{-6}$	5.62×10^{-3}	2.59×10^{-3}	2.74×10^{-2}	1.26×10^{-2}
0–1	>30		>30		>30		>30		>30		2.47×10^{-3}	1.14×10^{-3}	1.00×10^{-2}	4.62×10^{-3}
1–2	>30		>30		>30		>30		>30		1.07×10^{-3}	4.93×10^{-4}	3.66×10^{-3}	1.69×10^{-3}
2–3	>30		>30		>30		>30		>30		4.48×10^{-4}	2.07×10^{-4}	1.34×10^{-3}	6.18×10^{-4}
3–4	>30		>30		>30		>30		>30		2.37×10^{-4}	1.09×10^{-4}	4.87×10^{-4}	2.25×10^{-4}
4–5	>30		>30		>30		>30		>30		1.79×10^{-4}	8.25×10^{-5}	1.79×10^{-4}	8.25×10^{-5}
5–6	>30		>30		>30		>30		$1.25 \times 10'$		1.26×10^{-4}	5.81×10^{-5}	1.26×10^{-4}	5.81×10^{-5}
6–7	>30		>30		$1.35 \times 10'$		>30		4.69		1.17×10^{-4}	5.40×10^{-5}	1.17×10^{-4}	5.40×10^{-5}
7–8	>30		$2.95 \times 10'$		4.29		$1.82 \times 10'$		1.94		1.21×10^{-4}	5.57×10^{-5}	1.21×10^{-4}	5.57×10^{-5}
8–9	$1.67 \times 10'$		$1.32 \times 10'$		1.35		6.22		3.08×10^{-1}		1.16×10^{-4}	5.34×10^{-5}	1.16×10^{-4}	5.34×10^{-5}

9–10	6.18		5.77		4.69×10^{-1}		1.49		1.97×10^{-1}		1.13×10^{-4}	5.21×10^{-5}	1.13×10^{-4}	5.21×10^{-5}
10–11	1.91		2.33		1.73×10^{-1}		4.30×10^{-1}		1.14×10^{-1}		1.06×10^{-4}	4.88×10^{-5}	1.06×10^{-4}	4.88×10^{-5}
11–12	4.86×10^{-1}		5.87×10^{-1}		1.39×10^{-1}		2.36×10^{-1}		6.90×10^{-2}		1.11×10^{-4}	5.12×10^{-5}	1.11×10^{-4}	5.12×10^{-5}
12–13	1.23×10^{-1}		1.16×10^{-1}		1.04×10^{-1}		1.33×10^{-1}		4.07×10^{-2}		1.02×10^{-4}	4.73×10^{-5}	1.02×10^{-4}	4.73×10^{-5}
13–14	2.60×10^{-2}		2.77×10^{-2}		2.68×10^{-2}		3.48×10^{-2}		2.46×10^{-2}		1.00×10^{-4}	4.63×10^{-5}	1.00×10^{-4}	4.63×10^{-5}
14–15	9.71×10^{-3}		1.28×10^{-2}		1.22×10^{-2}		1.61×10^{-2}		1.14×10^{-2}		9.43×10^{-5}	4.35×10^{-5}	9.43×10^{-5}	4.35×10^{-5}
15–16	5.05×10^{-3}		8.46×10^{-3}		7.94×10^{-3}		1.07×10^{-2}		7.44×10^{-3}		8.96×10^{-5}	4.14×10^{-5}	8.96×10^{-5}	4.14×10^{-5}
16–17	3.16×10^{-3}		6.33×10^{-3}		5.85×10^{-3}		8.10×10^{-3}		5.46×10^{-3}		8.86×10^{-5}	4.09×10^{-5}	8.86×10^{-5}	4.09×10^{-5}
17–18	2.36×10^{-3}		4.63×10^{-3}		4.20×10^{-3}		6.30×10^{-3}		3.90×10^{-3}		8.57×10^{-5}	3.96×10^{-5}	8.57×10^{-5}	3.96×10^{-5}
18–19	2.04×10^{-3}		3.80×10^{-3}		3.33×10^{-3}		4.53×10^{-3}		3.09×10^{-3}		7.22×10^{-5}	3.33×10^{-5}	7.22×10^{-5}	3.33×10^{-5}
19–20	2.13×10^{-3}		3.15×10^{-3}		2.65×10^{-3}		3.90×10^{-3}		2.42×10^{-3}		5.30×10^{-5}	2.45×10^{-5}	5.30×10^{-5}	2.45×10^{-5}
20–21	1.69×10^{-3}		2.61×10^{-3}		2.12×10^{-3}		3.13×10^{-3}		1.90×10^{-3}		3.84×10^{-5}	1.77×10^{-5}	3.84×10^{-5}	1.77×10^{-5}
21–22	2.02×10^{-3}		2.63×10^{-3}		2.07×10^{-3}		3.09×10^{-3}		1.71×10^{-3}		2.89×10^{-5}	1.34×10^{-5}	2.89×10^{-5}	1.34×10^{-5}
22–23	1.90×10^{-3}		2.20×10^{-3}		1.66×10^{-3}		2.66×10^{-3}		1.52×10^{-3}		2.21×10^{-5}	1.02×10^{-5}	2.21×10^{-5}	1.02×10^{-5}
23–24	1.84×10^{-3}		2.26×10^{-3}		1.61×10^{-3}		2.47×10^{-3}		1.35×10^{-3}		1.75×10^{-5}	8.10×10^{-6}	1.75×10^{-5}	8.10×10^{-6}
24–25	1.77×10^{-3}		2.23×10^{-3}		1.53×10^{-3}		2.35×10^{-3}		1.20×10^{-3}		1.48×10^{-5}	6.82×10^{-6}	1.48×10^{-5}	6.82×10^{-6}
25–30	1.28×10^{-3}		1.48×10^{-3}		8.61×10^{-4}		1.61×10^{-3}		7.36×10^{-4}		4.45×10^{-6}	2.05×10^{-6}	4.45×10^{-6}	2.05×10^{-6}
30–35	4.50×10^{-4}		5.15×10^{-4}		2.62×10^{-4}		5.43×10^{-4}		2.14×10^{-4}		1.17×10^{-6}	5.40×10^{-7}	1.17×10^{-6}	5.40×10^{-7}
35–40	9.05×10^{-5}		1.06×10^{-4}		5.53×10^{-5}		1.16×10^{-4}		4.08×10^{-5}		3.08×10^{-7}	1.42×10^{-7}	3.08×10^{-7}	1.42×10^{-7}
40–45	2.35×10^{-5}		2.84×10^{-5}		1.58×10^{-5}		3.16×10^{-5}		1.09×10^{-5}		8.11×10^{-8}	3.75×10^{-8}	8.11×10^{-8}	3.75×10^{-8}
45–50	6.50×10^{-6}		8.01×10^{-6}		4.67×10^{-6}		8.73×10^{-6}		3.26×10^{-6}		2.14×10^{-8}	9.89×10^{-9}	2.14×10^{-8}	9.89×10^{-9}
50–70	$<10^{-6}$		$<10^{-6}$		$<10^{-6}$		$<10^{-6}$		$<10^{-6}$		1.10×10^{-10}	5.09×10^{-11}	1.10×10^{-10}	5.09×10^{-11}
70–100											3.91×10^{-14}	1.81×10^{-14}	3.91×10^{-14}	1.81×10^{-14}

$\lambda = 337\ \mu m$

0	2.67×10^{1}	$<10^{-6}$	2.03×10^{1}	$<10^{-6}$	5.39	$<10^{-6}$	1.37×10^{1}	$<10^{-6}$	1.71	$<10^{-6}$	$<10^{-6}$	$<10^{-6}$	$<10^{-6}$	$<10^{-6}$
0–1	2.12×10^{1}		1.60×10^{1}		4.35		1.07×10^{1}		1.60×10^{-1}					
1–2	1.33×10^{1}		9.66		2.84		6.39		1.39×10^{-1}					
2–3	8.72		5.58		1.82		4.09		9.79×10^{-1}					
3–4	4.00		2.87		1.10		2.39		6.33×10^{-1}					
4–5	1.75		1.50		5.46×10^{-1}		1.36		3.45×10^{-1}					
5–6	1.06		7.17×10^{-1}		2.83×10^{-1}		7.24×10^{-1}		1.52×10^{-1}					
6–7	5.52×10^{-1}		3.97×10^{-1}		1.42×10^{-1}		3.56×10^{-1}		6.68×10^{-2}					
7–8	2.78×10^{-1}		2.17×10^{-1}		5.13×10^{-2}		1.73×10^{-1}		3.29×10^{-2}					
8–9	1.34×10^{-2}		1.12×10^{-1}		1.87×10^{-2}		6.98×10^{-2}		5.96×10^{-3}					
9–10	5.78×10^{-2}		5.72×10^{-2}		7.73×10^{-3}		2.02×10^{-2}		3.91×10^{-3}					
10–11	2.16×10^{-2}		2.75×10^{-2}		3.10×10^{-3}		6.39×10^{-3}		2.20×10^{-3}					
11–12	6.64×10^{-3}		8.43×10^{-3}		2.49×10^{-3}		3.43×10^{-3}		1.30×10^{-3}					
12–13	2.07×10^{-3}		2.05×10^{-3}		1.85×10^{-3}		1.88×10^{-3}		7.57×10^{-4}					
13–14	5.59×10^{-4}		5.41×10^{-4}		4.81×10^{-4}		4.88×10^{-4}		4.53×10^{-4}					
14–15	2.66×10^{-4}		2.47×10^{-4}		2.21×10^{-4}		2.24×10^{-4}		2.08×10^{-4}					
15–16	1.79×10^{-4}		1.61×10^{-4}		1.46×10^{-4}		1.48×10^{-4}		1.36×10^{-4}					
16–17	1.35×10^{-4}		1.20×10^{-4}		1.09×10^{-4}		1.11×10^{-4}		1.02×10^{-4}					
17–18	9.63×10^{-5}		8.73×10^{-5}		7.91×10^{-5}		8.58×10^{-5}		7.41×10^{-5}					
18–19	6.98×10^{-5}		7.00×10^{-5}		6.38×10^{-5}		6.15×10^{-5}		5.98×10^{-5}					
19–20	6.14×10^{-5}		5.59×10^{-5}		5.12×10^{-5}		5.28×10^{-5}		4.80×10^{-5}					
20–21	4.13×10^{-5}		4.44×10^{-5}		4.08×10^{-5}		4.24×10^{-5}		3.84×10^{-5}					
21–22	4.22×10^{-5}		4.33×10^{-5}		3.98×10^{-5}		4.17×10^{-5}		3.52×10^{-5}					
22–23	3.54×10^{-5}		3.45×10^{-5}		3.19×10^{-5}		3.59×10^{-5}		3.21×10^{-5}					
23–24	3.18×10^{-5}		3.33×10^{-5}		3.09×10^{-5}		3.27×10^{-5}		2.92×10^{-5}					
24–25	2.85×10^{-5}		3.17×10^{-5}		2.94×10^{-5}		2.95×10^{-5}		2.64×10^{-5}					
25–30	1.64×10^{-5}		1.74×10^{-5}		1.58×10^{-5}		1.74×10^{-5}		1.50×10^{-5}					
30–35	4.02×10^{-6}		4.32×10^{-6}		3.83×10^{-6}		4.36×10^{-6}		3.56×10^{-6}					
35–40	$<10^{-6}$		$<10^{-6}$		$<10^{-6}$		$<10^{-6}$		$<10^{-6}$					
40–45														
45–50														
50–70														
70–100														

where k_m represents the net monochromatic molecular-absorption coefficient as given in Eq. (9) and $\bar{K}$ is an average absorption coefficient, which in most cases cannot be defined when $\Delta\nu$ is much greater than the half width of a spectral line [see Eq. (8)].

On the other hand, the molecular-scattering coefficient σ_m and the aerosol scattering and absorption coefficients σ_a and k_a [see Eqs. (4*a*) and (4*b*)] are slowly varying functions of frequency, and the average transmittance (which can easily be obtained by interpolation from the monochromatic values, see Figs. 21 and 22) obeys the simple exponential law [Eq. (3)] provided that only the direct transmitted beam is being observed (see Zuev et al., 1967).

There are four basic approaches to obtaining low-resolution transmittance values for a given path through the atmosphere due to molecular absorption: (1) direct measurements over the required path, (2) measurements in the laboratory under simulated conditions, (3) line-by-line (monochromatic) calculations based on detailed knowledge of spectroscopic-line parameters (see Par. 5), which are then averaged over the required spectral interval, and (4) calculations based on band-model techniques (which use available laboratory and/or field transmittance measurements or actual line data as a basis).

From the point of view of computations, method 3 involves a considerable amount of work and computer time, and consequently method 4 has been used most frequently.

We briefly discuss the problem of continuum absorption in the atmospheric-window region (Par. 10); then we summarize the application of band models to molecular-absorption calculations (Par. 11) and present a detailed description of a graphical prediction scheme for determining the transmittance of the atmosphere due to line absorption, continuum absorption, and scattering (average over 20 cm^{-1}) for a given atmospheric path. First, the reader is provided with figures for determining the relevant absorber amounts in the required atmospheric path (Par. 12), which are then used in conjunction with the transmittance charts (Par. 19) to determine (separately) the transmittance due to (1) molecular (line) absorption (H_2O, O_3, and the uniformly mixed gases), (2) molecular (continuum) absorption (H_2O, N_2), (3) molecular (Rayleigh) scattering, and (4) aerosol extinction (absorption and scattering), as a function of frequency (wavelength).

Finally the reader must multiply together the individual transmittances at each wavelength to determine the total transmittance at that wavelength, i.e.,

$$[\bar{\tau}_{\Delta\nu}(\nu)]_{\text{total}} = [\bar{\tau}_{\Delta\nu}(\nu)]_{\substack{\text{line}\\ \text{absorption}}} [\bar{\tau}_{\Delta\nu}(\nu)]_{\substack{\text{continuum}\\ \text{absorption}}} [\bar{\tau}_{\Delta\nu}(\nu)]_{\text{Rayleigh}} [\bar{\tau}_{\Delta\nu}(\nu)]_{\text{aerosol}}$$

10. Continuum Absorption in the Atmospheric-Window Regions A number of spectral regions of weak or negligible absorption by atmospheric molecules in the infrared are generally called atmospheric windows. Various attempts have been made to determine the transparency of these window regions (Yates and Taylor, 1960; Street, 1968). Yates and Taylor concluded that the opacity in the near-infrared windows is limited at sea level by aerosol scattering. This conclusion is consistent with the results derived from our aerosol chart presented in Par. 19.

As there are no laboratory data presently available from which the molecular absorption between absorption bands can be inferred (with two important exceptions), we shall here assume that the gaps in the charts presented in Figs. 16 to 18 are real gaps of zero molecular absorption. Thus, the attenuation in these regions is dependent entirely on molecular and aerosol scattering.

However, laboratory absorption data are available for the water-vapor continuum (Burch, 1970; McCoy et al., 1969) from 8 to 13 μm and also for the pressure-induced (continuous-absorption) band of nitrogen (Bosomworth, 1965) centered near 4.3 μm. The strongest part of this nitrogen band is not too important as it is centered beneath the very strong 4.3-μm CO_2 band. However, the nitrogen absorption extends out beyond the region of CO_2 absorption near 4.0 μm and becomes predominant, accounting for 20 percent absorption in a 10-km path at sea level.

Due to the continuous nature of these absorbing regions, the transmittance follows a simple exponential law which is represented by the scale in Figs. 19 and 20.

11. Outline of Band-Model Techniques Basically, a band model assumes an array of lines having chosen intensities, half widths, and spacings, which can be adjusted to represent the line structure in some part of a real band. For a particular band model the mean transmittance can be represented by a mathematical expression (transmittance function); e.g.,

$$\bar{\tau}_{\Delta\nu}(\nu) = f[C(\nu), \Delta L, P] \tag{16}$$

where P = pressure
ΔL = effective path length (or absorber concentration)
$C(\nu)$ = frequency-dependent absorption coefficient

Several band models have been developed (Goody, 1952, 1964; Elsasser, 1942; Plass, 1958; King, 1959, 1964; Wyatt et al., 1964), the Elsasser and the Goody models being best known. The Elsasser band model consists of an array of equally spaced identical lines of Lorentz shape [see Eq. (7)]. This model has been applied to absorption bands which have a regular line structure, e.g., to some bands of CO_2, N_2O, CO, CH_4, and O_2. The Goody model, on the other hand, assumes that the band is composed of spectral lines with an exponential intensity distribution and with random spacing between lines. Again the lines are assumed to have a Lorentz shape. This model has generally been applied to bands which have an irregular line structure, e.g., to some H_2O and O_3 bands.

In practice the wavelength-dependent coefficients are determined for each absorbing gas separately, using laboratory transmittance values measured under known conditions, i.e., by solving Eq. (16) for the $C(\nu)$'s at each ν for known values of $\bar{\tau}(\nu)$, ΔL, and P.

The absorption coefficients obtained in this way are then used in the band-model transmittance function [Eq. (16)] to determine $\bar{\tau}(\nu)$ as a function of frequency ν for other values of ΔL and P for each absorber. Finally, the total mean transmittance for molecular absorption is given by the product of the mean transmittances of the individual absorbers at each frequency:

$$\bar{\tau}_{\Delta\nu}(\nu)_{\substack{\text{molecular}\\\text{absorption}}} = [\bar{\tau}_{\Delta\nu}(\nu)]_{H_2O}[\bar{\tau}_{\Delta\nu}(\nu)]_{CO_2}[\bar{\tau}_{\Delta\nu}(\nu)]_{NO_2}[\bar{\tau}_{\Delta\nu}(\nu)]_{O_3} \cdots \tag{17}$$

Exact analytical expressions have been obtained for most of the band models. However, they are sometimes difficult to use, and simpler approximations have been found to apply in two limiting conditions, which are common to all band models. These simpler expressions are the well-known *weak-line* and *strong-line* approximations (see Goody, 1964; Plass, 1958), for which the transmittance is a function of the absorber amount ΔL and the product of the pressure and absorber amount $P\,\Delta L$, respectively (for a given temperature).

The weak-line approximation, which corresponds to the exponential law, is valid when the absorption is small at the line centers (generally for high pressures and low absorber amounts). Unfortunately this case is rarely applicable to conditions existing in the terrestrial atmosphere.

The strong-line approximation is applicable where the lines are completely absorbing at their centers; the effect of increasing the amount of absorber is then confined to the edges or wings of the lines.

The regions of validity of the strong- and weak-line approximations for the Elsasser and Goody models are discussed in Plass (1958).

For practical purposes most problems fall in either the strong-line-approximation region or the intermediate regime. It is perhaps interesting to compare the functional form of the strong-line approximation for some of the models.

For the Elsasser model

$$\bar{\tau}_{\Delta\nu}(\nu) = 1 - \text{erf}\,[C(\nu)\,\Delta L\,P]^{1/2} \tag{18}$$

where erf is the error function

$$\text{erf}\,(z) = \frac{2}{\sqrt{\pi}} \int_z^\infty e^{-t^2}\,dt$$

and $C(\nu)$ is an average absorption coefficient which can be written in terms of average values of the molecular line constants defined in Par. 5.

The strong-line approximation for the Goody model is given by

$$\bar{\tau}_{\Delta\nu}(\nu) = \exp [C'(\nu)\, \Delta L\, P]^{1/2} \tag{19}$$

where $C'(\nu)$ is defined similar to $C(\nu)$ above.

King (1959) attempted to generalize the above expressions by writing

$$\bar{\tau}_{\Delta\nu}(\nu) = f[C''(\nu)\, \Delta L\, P^n] \tag{20}$$

so that when n is set equal to zero or unity, the average transmittance agrees with the weak- or strong-line approximations, respectively.

We have determined from laboratory and synthetic transmittance data,† three empirical transmittance functions based on Eq. (20) for H_2O, O_3, and the combined contributions of the uniformly mixed gases. It was found that Eq. (20) gives the best fit to laboratory and theoretical data over a wide spectral interval when $n = 0.9$ for H_2O, $n = 0.4$ for O_3, and $n = 0.75$ for the uniformly mixed gases.

The corresponding empirical transmittance functions were found to give better agreement with laboratory and synthetic transmittance data than the commonly used band models over a wide range of pressures and absorber amounts.

The procedure used for determining the parameter n and function f defined in Eq. (20) will be outlined below. By taking the logarithm of the inverse of Eq. (20) it is seen that

$$n \log P + \log \Delta L = \log f^{-1}[\bar{\tau}_{\Delta\nu}(\nu)] - \log C''(\nu) \tag{21}$$

Thus for a given frequency ν and fixed values of average transmittance $\bar{\tau}_{\Delta\nu}(\nu)$, the right-hand side of Eq. (20) becomes a constant.

A mean value for n was determined from Eq. (21) for a wide range of frequencies and several values of $\bar{\tau}$. Then for each frequency, $\bar{\tau}$ was plotted against $\log \Delta L\, P^n$, the curves superimposed, and the best mean curve determined. The mean curve thus obtained constitutes the empirical transmittance function and is displayed as the transmittance scale and associated scaling factor $\log (\Delta L\, P^n)$ in the prediction charts given in Figs. 16 to 18.

So far the discussion has been restricted to constant-pressure paths. In Pars. 12 to 18 we discuss the application of band models to variable-pressure paths.

Equivalent Constant-Pressure Paths and Absorber Amounts

12. Atmospheric paths may be divided into two types, constant-pressure paths, i.e., horizontal paths of up to about 100 km, and slant, or variable-pressure, paths. The extent of the former type is limited by the curvature of the earth. Since the transmittance functions for constant-pressure paths are simpler to handle than those for variable-pressure paths, the problem is simplified by reducing the slant path to an equivalent constant-pressure path; i.e., to a constant-pressure path which, ideally, has the same transmittance at each wavelength as the actual slant path.

We discuss here the construction and application of charts for determining the equivalent sea-level-path quantities for molecular absorption (including continuum absorption in the 4- and 8- to 13-μm region), molecular scattering, and aerosol extinction based on the model atmospheres presented in Pars. 2 to 4. Equivalent sea-level-path quantities for the U.S. Standard Atmosphere, 1962, will not differ appreciably from similar quantities provided for the midlatitude winter model. If these quantities are required for the U.S. Standard Atmosphere, 1962, see appendix B of McClatchey et al. (1972).

13. Molecular Absorption It has been assumed that for most practical purposes the average transmittance is dependent on the product $\Delta L\, P^n$, where ΔL is the absorber amount and P is the pressure. We define the equivalent sea-level amounts for each of the five model atmospheres given in Pars. 2 to 4 as the value of the above product (reduced to STP) averaged over a given atmospheric path for a particular absorber, i.e., $\Delta L_0(P/P_0)^n$, where the subscript 0 refers to STP.

Since H_2O and O_3 are distributed nonuniformly in the atmosphere, we must consider these gases separately from the uniformly mixed gases.

† Synthetic spectra here refers to monochromatic (line-by-line) transmittance calculations degraded in resolution to 20 cm^{-1}.

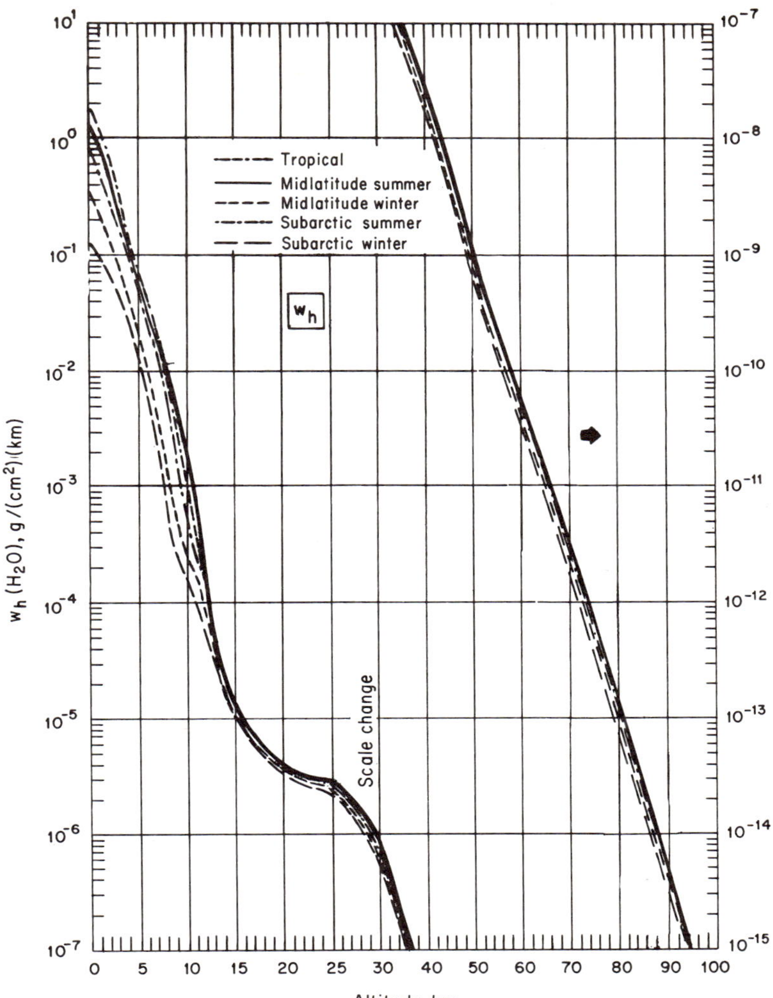

Fig. 2 Equivalent sea-level path length of water vapor as a function of altitude for horizontal atmospheric paths.

For water vapor, n is set equal to 0.9, and $\Delta L_0(P/P_0)^n = Rw(z)(P/P_0)^{0.9}$, where $w(z)$ is the water-vapor concentration at altitude z in units of grams per square centimeter per kilometer (obtained from Table 1) and R is the range in kilometers. Since $w(z)$ is given in mass units, no temperature or pressure correction is necessary to reduce this quantity to STP. The variation with altitude of

$$w(z)\left(\frac{P}{P_0}\right)^{0.9} \qquad \text{and} \qquad \int_z^\infty w(z)\left(\frac{P}{P_0}\right)^{0.9} dz$$

(denoted by w_h and w_v, respectively) is shown in Figs. 2 and 3. The subscripts h and v refer to horizontal and vertical paths, respectively.

For the uniformly mixed gases, n is set equal to 0.75, but the absorber concentration ΔL_0 (reduced to STP) for a horizontal path of range R at altitude z is given by

$$\Delta L_0 = cR\,\frac{P}{P_0}\,\frac{T_0}{T}$$

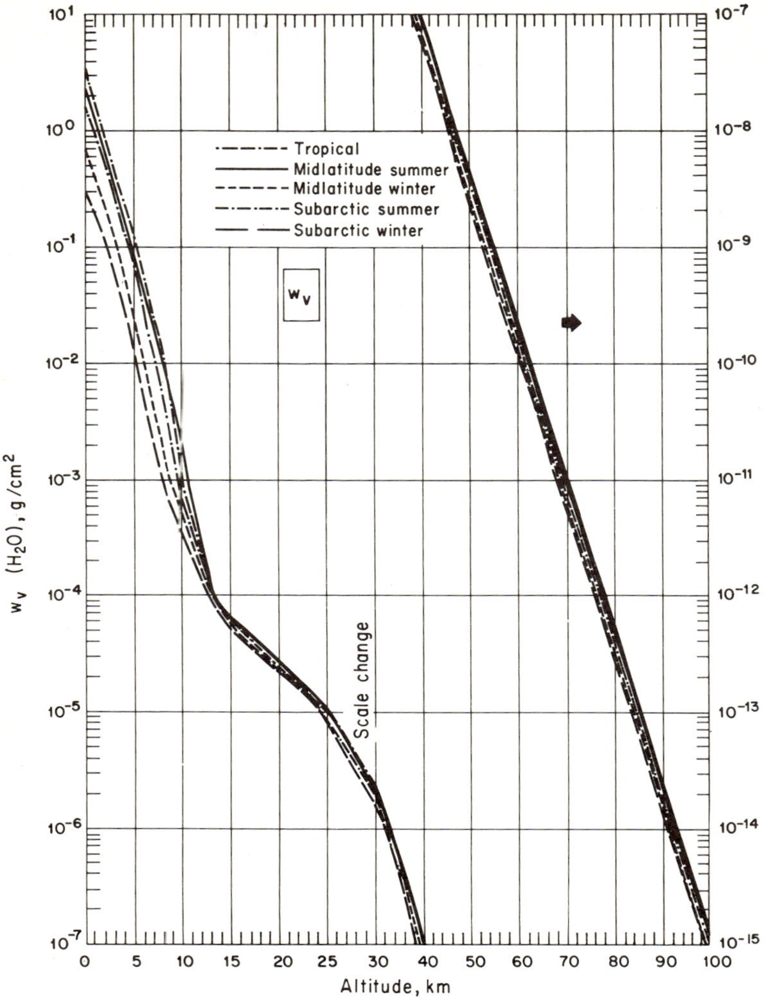

Fig. 3 Equivalent sea-level path length of water vapor as a function of altitude for vertical atmospheric paths.

where c is the fractional concentration by volume of the absorbing gas (see Table 2 and Par. 26). The terms P and T refer to the pressure and temperature at altitude z. Therefore, we have in this case

$$\Delta L_0 \left(\frac{P}{P_0}\right)^n \propto \left(\frac{P}{P_0}\right)^{1.75} \frac{T_0}{T}$$

The variation with altitude of

$$\left(\frac{P}{P_0}\right)^{1.75} \frac{T_0}{T} \qquad \text{and} \qquad \int_z^\infty \left(\frac{P}{P_0}\right)^{1.75} \frac{T_0}{T}\, dz$$

(denoted by l_h and l_v, respectively) is shown in Figs. 4 and 5.

For ozone, n is set equal to 0.4 (see Par. 11), and

$$\Delta L_0 \left(\frac{P}{P_0}\right)^n = u(z)\left(\frac{P}{P_0}\right)^{0.4}$$

where $u(z)$ is the ozone concentration in units of centimeter-atmospheres per kilometer at STP (obtained from Table 1). The variation with altitude of

$$u(z)\left(\frac{P}{P_0}\right)^{0.4} \quad \text{and} \quad \int_z^{\infty} u(z)\left(\frac{P}{P_0}\right)^{0.4} dz$$

(denoted by u_h and u_v, respectively) is shown in Figs. 6 and 7.

14. Nitrogen Continuum (4-μm) Region The absorption due to the nitrogen-collision-induced band is proportional to the square of the pressure. The effective

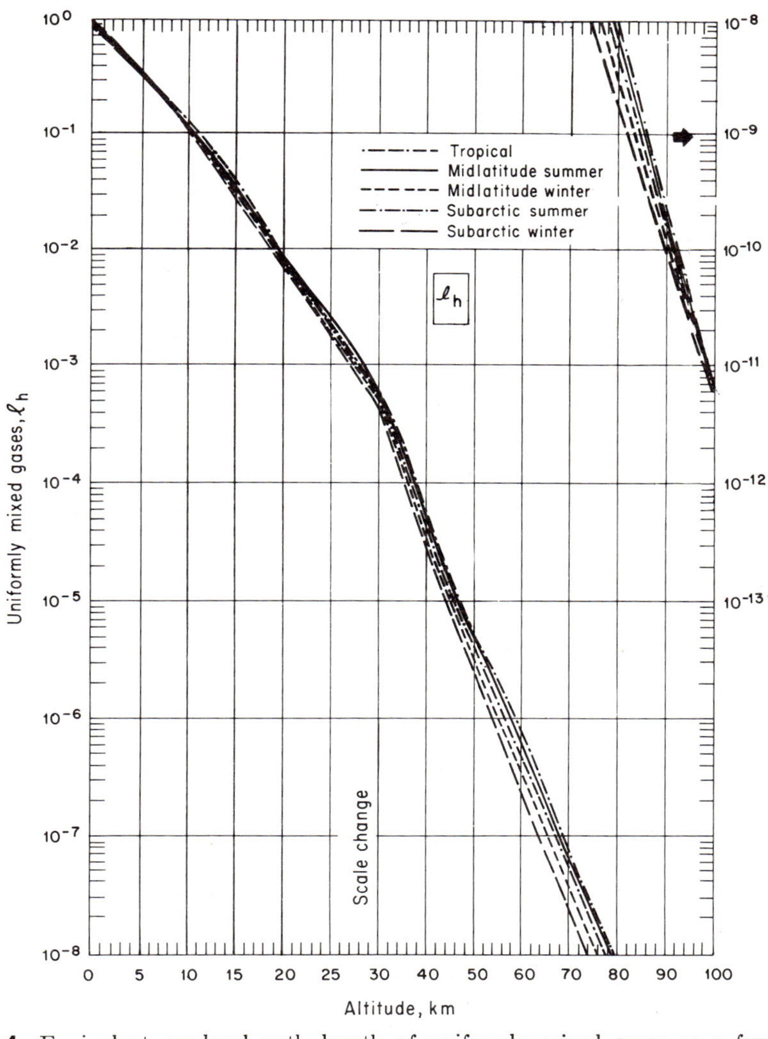

Fig. 4 Equivalent sea-level path length of uniformly mixed gases as a function of altitude for horizontal atmospheric paths.

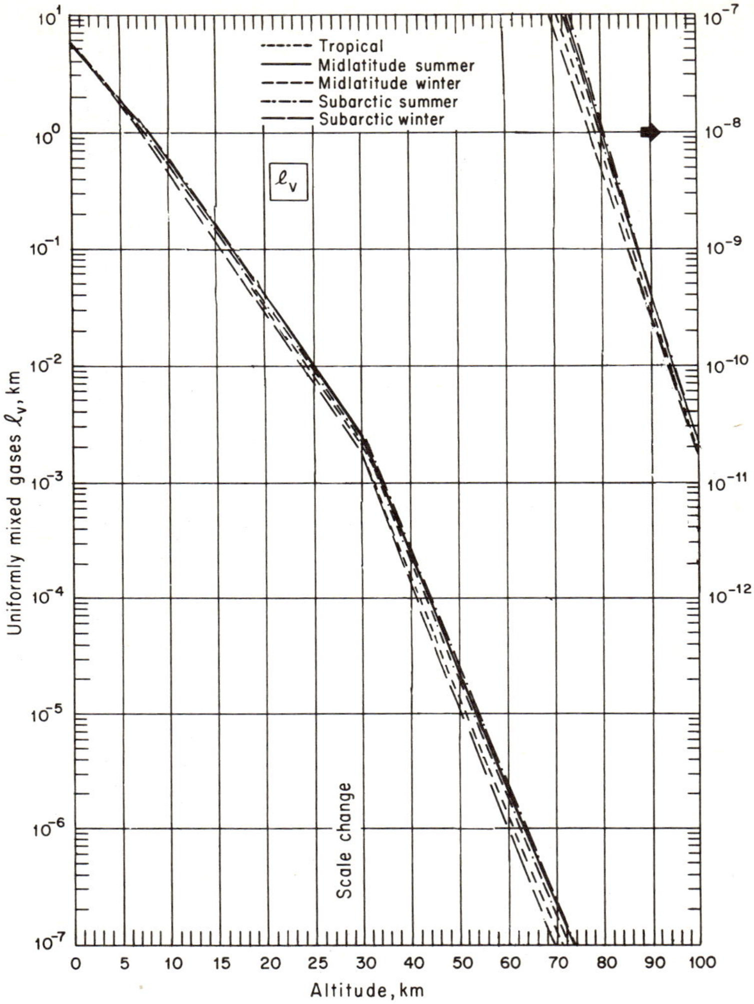

Fig. 5 Equivalent sea-level path length of uniformly mixed gases as a function of altitude for vertical atmospheric paths.

absorber amount (reduced to STP) for a horizontal path of range R at altitude z is

$$cR\left(\frac{P}{P_0}\right)^2\frac{T_0}{T}$$

The altitude variation of

$$\left(\frac{P}{P_0}\right)^2\frac{T_0}{T}$$

for horizontal paths and

$$\int_z^\infty\left(\frac{P}{P_0}\right)^2\frac{T_0}{T}\,dz$$

for vertical paths (denoted by n_h and n_v, respectively) is shown in Figs. 8 and 9.

15. Water-Vapor Continuum Laboratory measurements of the water-vapor continuum indicate that self-broadening due to water-vapor molecules is much more important than broadening due to other atmospheric molecules (Burch, 1970). The appropriate equivalent sea-level absorber amount for the water-vapor continuum is

$$b_h = \frac{1}{P_0}[p_{H_2O} + 0.005(P - p_{H_2O})]w(z)$$

for horizontal paths, where P is the total pressure and p_{H_2O} is the partial pressure of water vapor [$p_{H_2O} = 4.56 \times 10^{-5}w(z)T(z)$ atm, where the constant is given by the ratio to the universal gas constant to the molecular weight of water vapor] and $w(z)$ is the water vapor in grams per square centimeter per kilometer at altitude z obtained

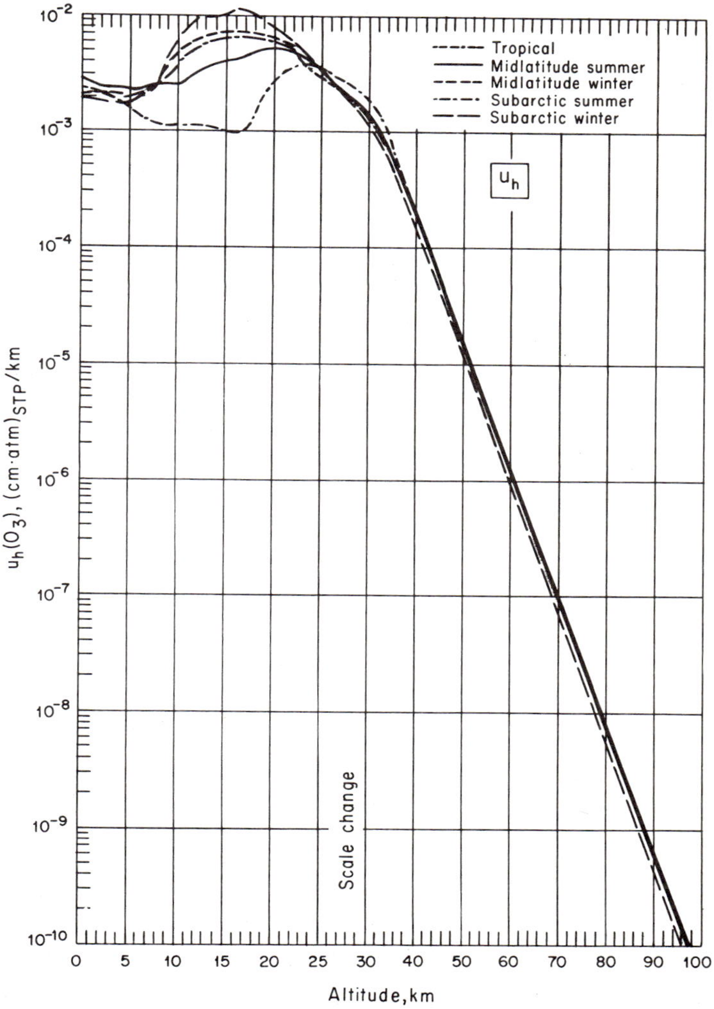

Fig. 6 Equivalent sea-level path length of ozone as a function of altitude for horizontal atmospheric paths.

from Table 1. The equivalent sea-level amount for vertical paths is given by

$$b_v(z) = \int_z^\infty \frac{1}{P_0} [p_{H_2O} + 0.005(P - p_{H_2O})]w(z)\, dz$$

The variation with altitude of these functions, denoted by b_h and b_v, respectively, is shown in Figs. 10 and 11.

16. Molecular Scattering Molecular scattering is a function of atmospheric density $\rho(z)$ (Par. 5). For each of the five model atmospheres given in Pars. 2 to 4, Figs. 12 and 13 show the variation with altitude z of $\rho(z)/\rho_0$ and $\int_z^\infty \frac{\rho(z)}{\rho_0}\, dz$ (denoted by m_h and m_v, respectively) where ρ_0 is the density of air at STP.

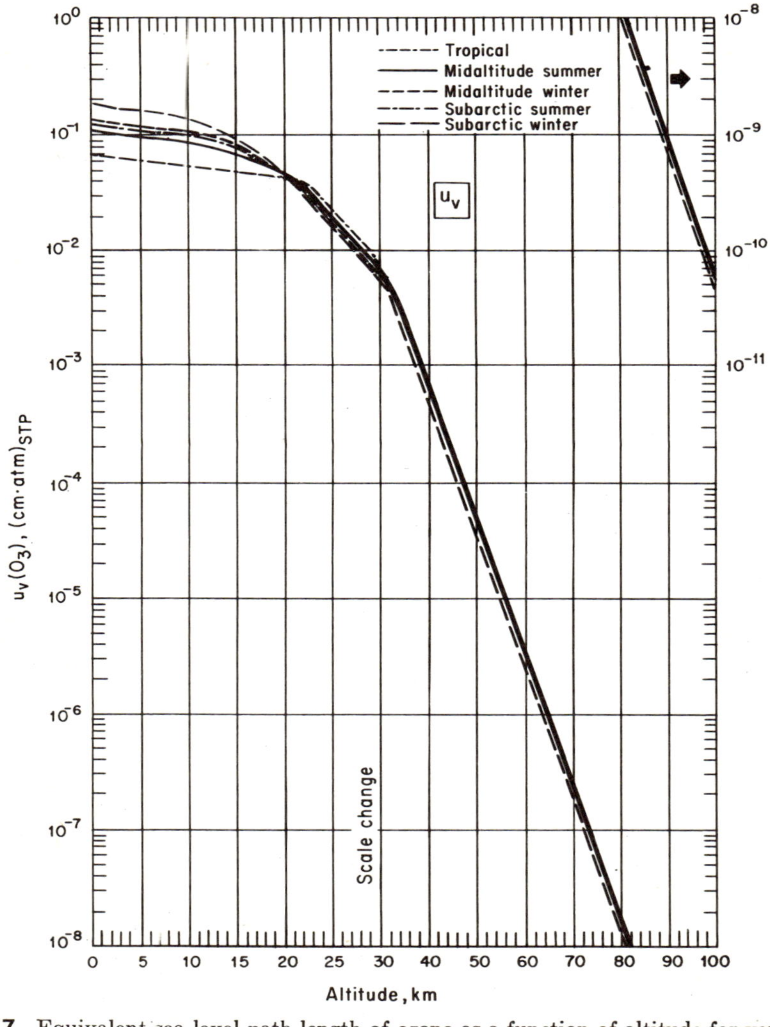

Fig. 7 Equivalent sea-level path length of ozone as a function of altitude for vertical atmospheric paths.

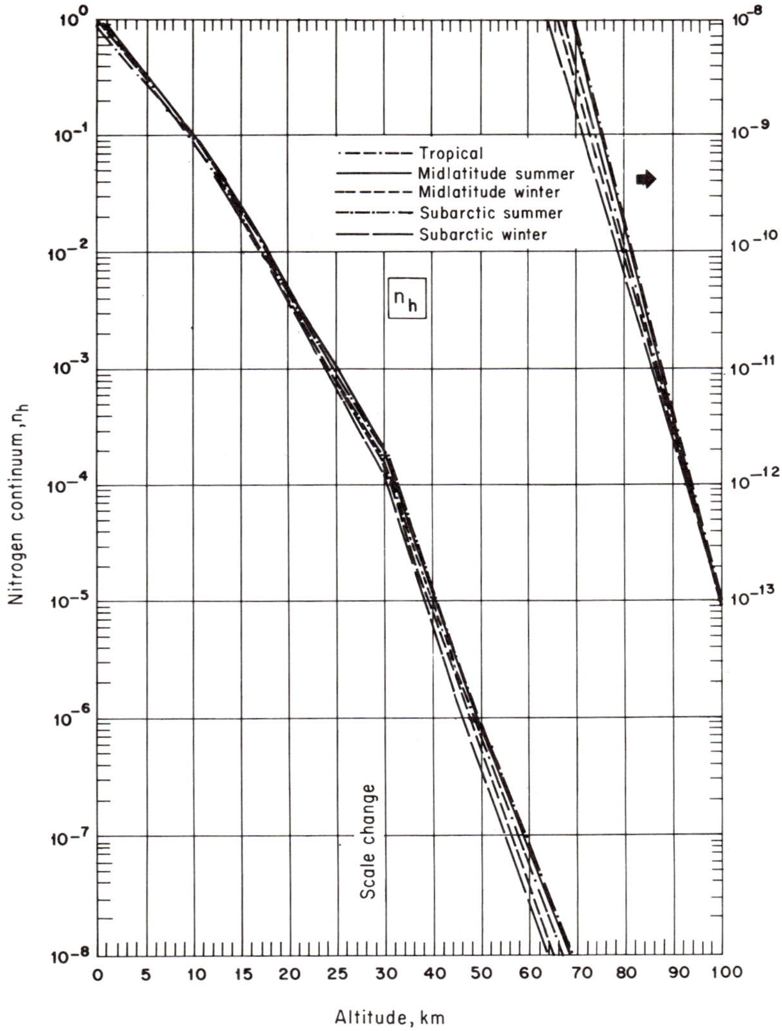

Fig. 8 Equivalent sea-level path length of nitrogen as a function of altitude for horizontal atmospheric paths.

17. Aerosol Extinction Aerosol extinction is a function of the height variation of the aerosol number density $N(z)$ for a given size-distribution function (Par. 4). Figures 14 and 15 show the variation with altitude of $N(z)/N_0$ and $\int_z^\infty \frac{N(z)\,dz}{N_0}$ (denoted by a_h and a_v, respectively) for the two hazy models given in Par. 4, where N_0 is the number density per cubic centimeter of aerosols at sea level for a visibility of 23 km.

18. Utilization of Equivalent-Path Figures We now consider how to obtain for a given atmospheric path the equivalent sea-level quantities for H_2O, O_3, the uniformly mixed gases, the nitrogen and water continua, molecular scattering, and aerosol extinction, denoted by W, U, L, N, B, M, and A respectively. At this point the reader need not be concerned with the units of these quantities, which

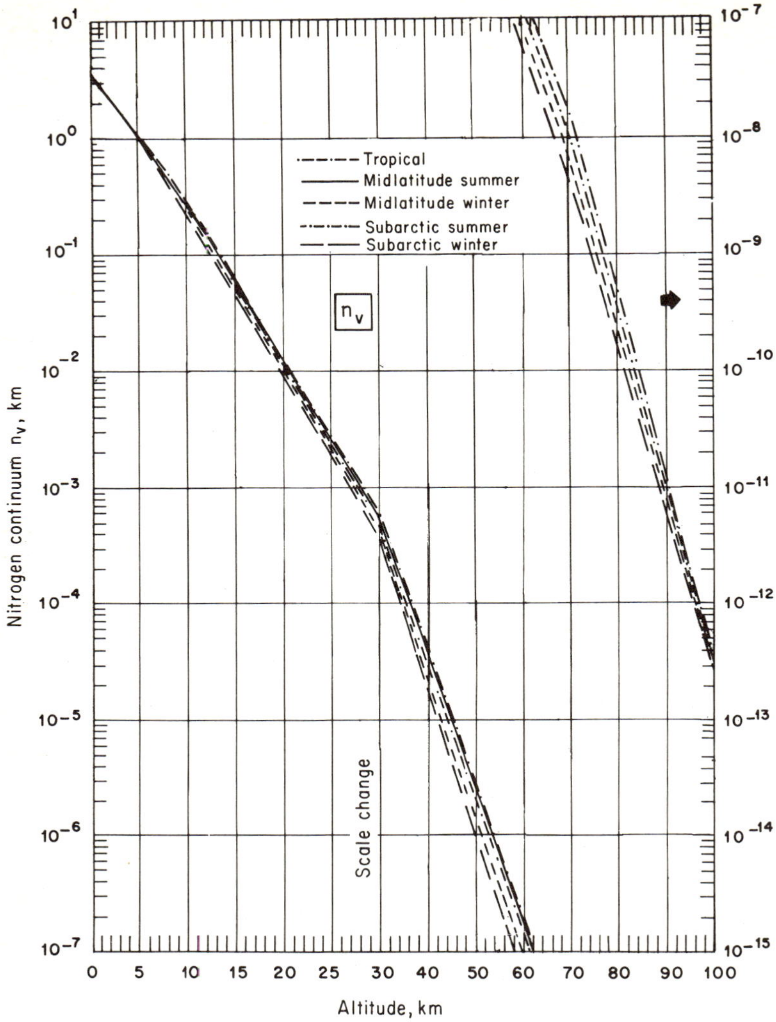

Fig. 9 Equivalent sea-level path length of nitrogen as a function of altitude for vertical atmospheric paths.

were chosen to facilitate the use of the prediction charts given in Par. 19. The subject of units will be discussed separately in Par. 27.

Figures 2 to 15 show the variation with altitude of the relevant quantities for both horizontal and vertical paths (distinguished by suffixes h and v). The use of these figures can best be illustrated by the following examples.

Horizontal Paths. To determine the equivalent sea-level quantities for a horizontal path of length R (<100 km) at an altitude z, simply multiply each of the quantities w_h, l_h, u_h, n_h, b_h, m_h, and a_h (obtained from Figs. 2, 4, 6, 8, 10, 12, 14 for the appropriate altitude z) by R (in kilometers). For example, for molecular line absorption by water vapor, W is given by

$$W = w_h(z)R \tag{22}$$

Slant Paths. The equivalent sea-level quantities (W, L, U, N, B, M, and A) for a slant path from altitude z to the top of the atmosphere at zenith angle θ are determined by multiplying each of the quantities w_v, l_v, u_v, b_v, m_v, and a_v (obtained from Figs. 3, 5, 7, 9, 11, 13, 15 for the appropriate altitude z) by sec θ (where $\theta < 80°$). For example, for molecular line absorption by water vapor

$$W = w_v(z) \sec \theta \tag{22a}$$

For $\theta > 80°$, sec θ should be replaced by $M(\theta)$ given in Fig. 32 (see Par. 23). For a slant path of range R between altitudes z_1 and z_2, the value of W for molecular line absorption by water vapor is given by

$$W = [w_v(z_1) - w_v(z_2)] \frac{R}{z_2 - z_1} \tag{23}$$

Similar expressions hold for L, U, N, B, M, and A.

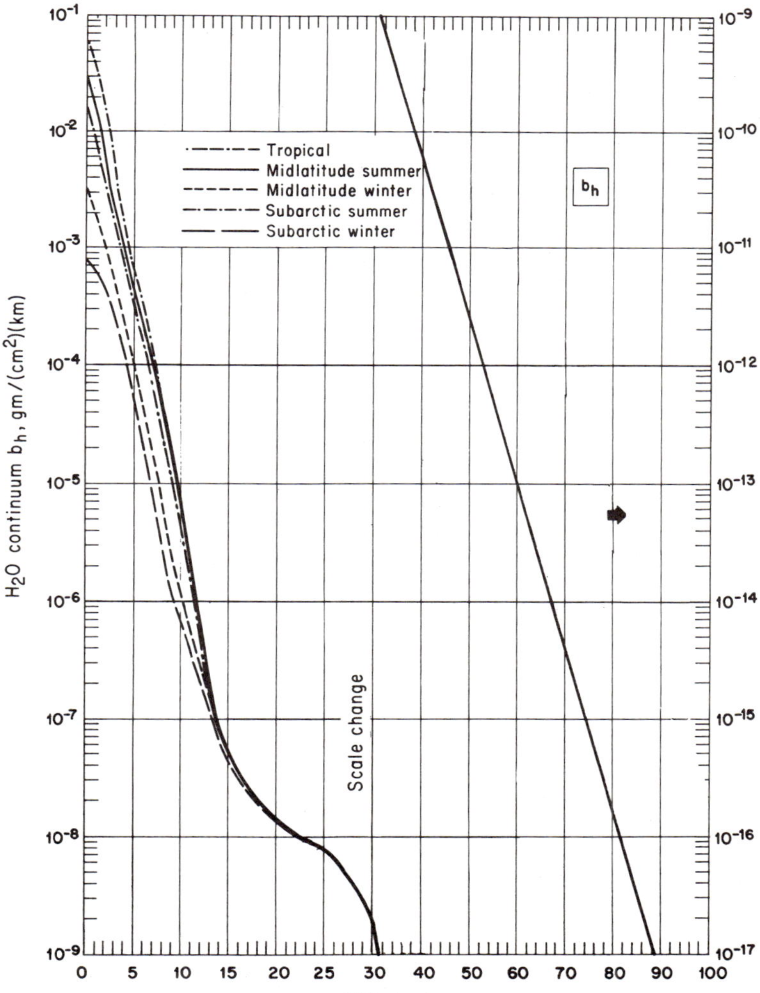

Fig. 10 Equivalent sea-level path length for water-vapor continuum as a function of altitude for horizontal atmospheric paths.

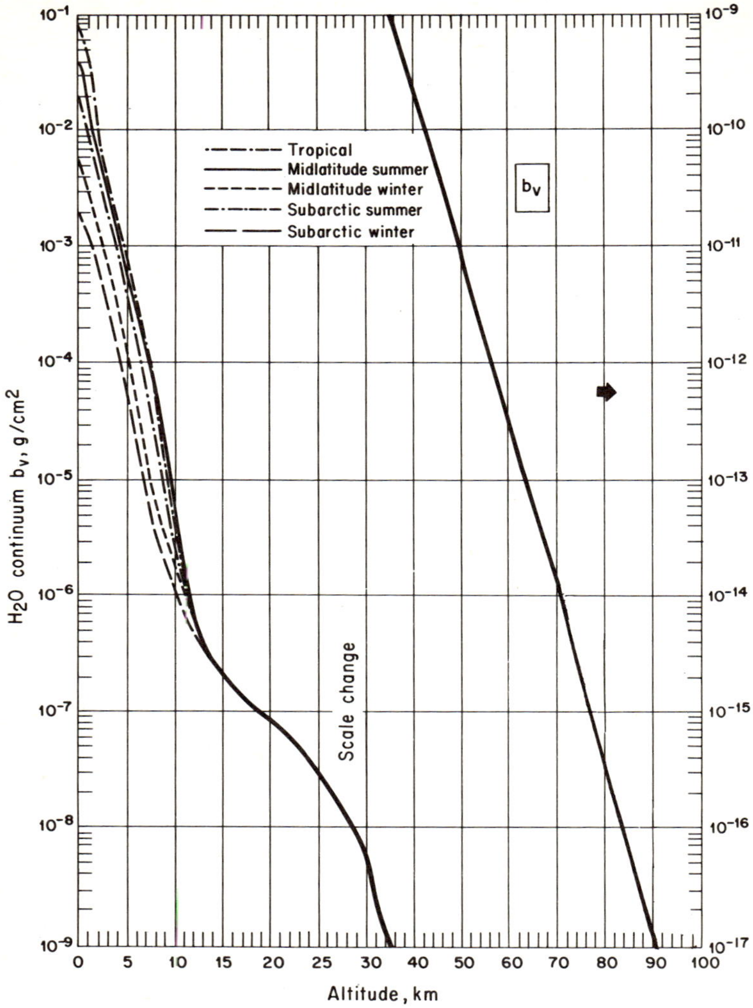

Fig. 11 Equivalent sea-level path length for water-vapor continuum as a function of altitude for vertical atmospheric paths.

Corrections which can also be applied to account for atmospheric refraction and earth curvature are summarized in Par. 23.

Having now determined W and B in grams per square centimeter, U in centimeter-atmospheres at STP, and L, N, M, and A in kilometers, one can use the prediction charts given in Figs. 16 to 22.

Description and Use of Prediction Charts

19. The purpose of this paragraph is to present a simple method of predicting atmospheric transmittance (at low resolution) which is applicable over a wide spectral interval and for a wide range of atmospheric paths. Of the methods available, the one originally devised by Altshuler (1961) is the most comprehensive that meets the above requirements, although based on fairly old experimental data. In view of this, an attempt has been made here to simplify Altshuler's method further and to update it by using more recent laboratory and field transmittance data (Burch et al., 1962, 1964, 1965*a,b,c,d,* 1967, 1968) and making some line-by-line calculations for molecular absorption (as discussed in Par. 19), and averaging over the spectral interval ($\Delta\nu = 20\ \text{cm}^{-1}$).

Charts for calculating the transmittance due to molecular line absorption by water vapor, H_2O, ozone, O_3, the uniformly mixed gases (CO_2, N_2O, CH_4, CO, O_2), the nitrogen 4-μm continuum, the water-vapor 8- to 13-μm continuum, molecular scattering, and aerosol extinction and absorption are presented in Figs. 16 to 22. It will be noted that the effects of the different uniformly mixed gases have been combined into one chart (Fig. 17) (assuming the mixing ratios given in Table 2).

Each figure shows the variation of an attenuation coefficient with respect to wavelength, relative to a horizontal datum line for a given effective sea-level path length or absorber amount. Also shown in each figure is a vertical transmittance scale associated with a set of scaling factors (for the absorber amount). To use these charts efficiently, it is necessary to trace the transmittance scale and associated scaling factors for each of the figures onto transparent paper.

If the transmittance scale is now placed on its appropriate chart so that the

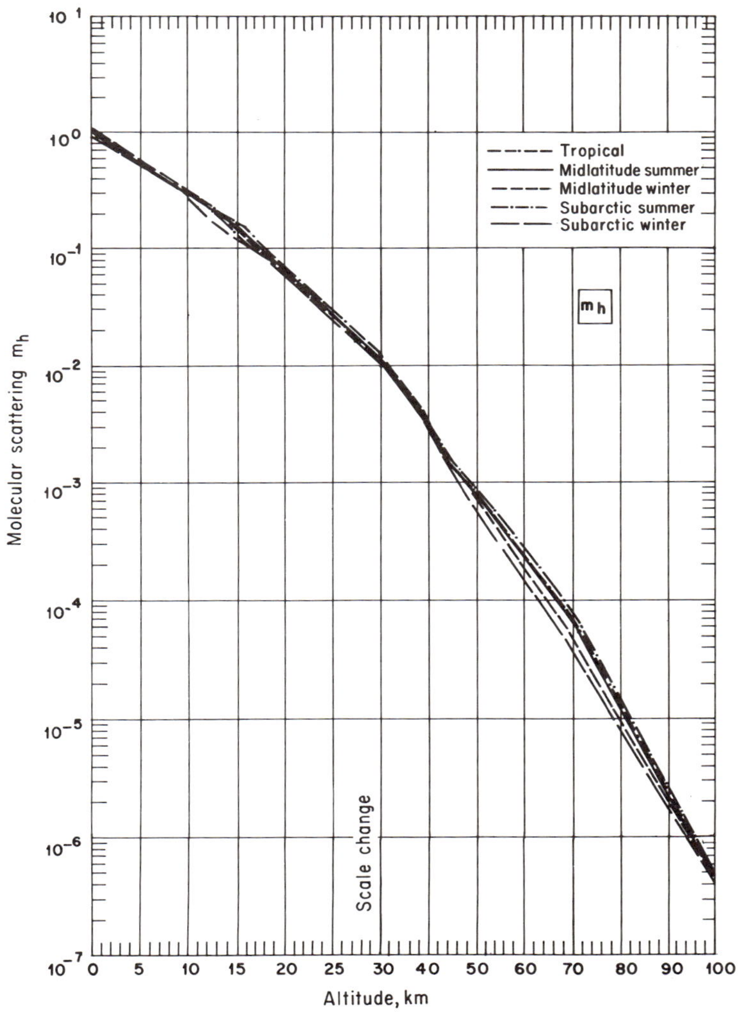

Fig. 12 Equivalent sea-level path length for molecular scattering as a function of altitude for horizontal atmospheric paths.

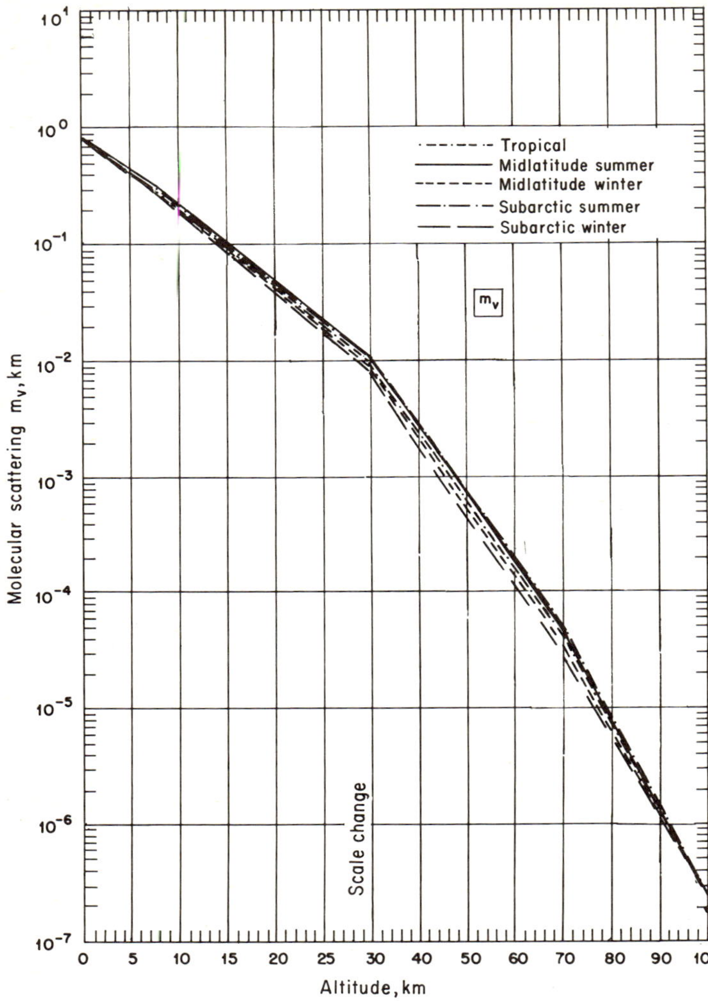

Fig. 13 Equivalent sea-level path length for molecular scattering as a function of altitude for vertical atmospheric paths.

scaling factor 1 rests on the datum line (as shown in Fig. 16*b*), one is in a position to read off the transmittance (for 1 g/cm^2 of H_2O) as a function of wavelength. Simply move the chart horizontally until the transmittance scale is vertically above the required wavelength and read off the transmittance where the scale crosses the curve. For example, in Fig. 16*b* the scale is set at 21 μm and indicates a transmittance of 0.10 for $W = 1$ g/cm^2 of H_2O.† If the transmittance is required for a different amount of absorber, for example $W = 10$ g/cm^2, just displace the transmittance scale vertically until the scaling factor 10 is coincident with the datum line. Then move the scale horizontally to the required wavelength and read off the transmittance as described above.

† This corresponds to the amount of water vapor in a 1-km path at sea level for $T = 15$°C and 78 percent relative humidity (see Par. 26).

The same procedure applies to Figs. 17 to 22, 25 and 33, where it will be noted the datum line for scaling the absorber amounts is expressed as 1 $(cm \cdot atm)_{STP}$ for O_3 (Figs. 18 and 25), 1 g/cm^2 for the water-vapor continuum (Fig. 20), and 1 km for each of the other charts.

Thus by displacing the transmittance scale horizontally according to wavelength and vertically according to the appropriate equivalent sea-level quantity (W, U, L, N, B, M, or A) for Figs. 16 to 22, one can construct a low-resolution transmittance spectrum (for $\Delta\nu = 20$ cm^{-1}).

A method for calculating the equivalent sea-level absorber amounts W, U, L, N, B, M, and A, respectively, in the appropriate units for a given atmospheric path has been described in Pars. 12 to 18 based on the 10 model atmospheres defined in Pars. 2 to 4. When the relevant meteorological parameters are known for the atmospheric path, these quantities can be calculated by an alternative method described in Par. 26.

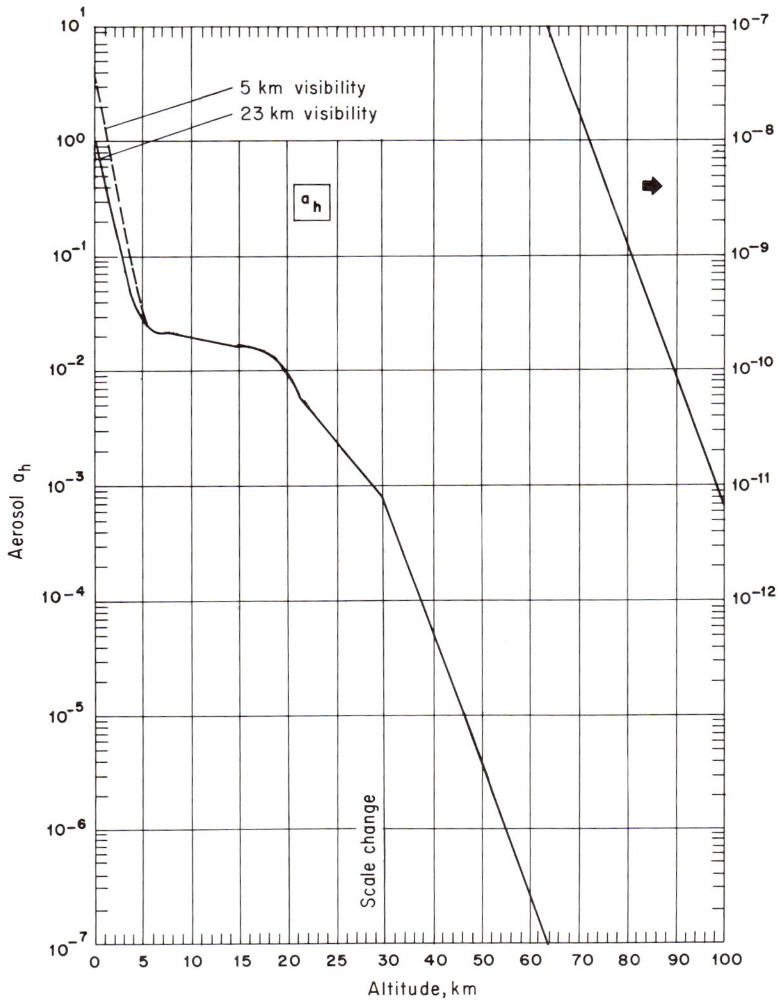

Fig. 14 Equivalent sea-level path length for aerosol extinction as a function of altitude for horizontal atmospheric paths.

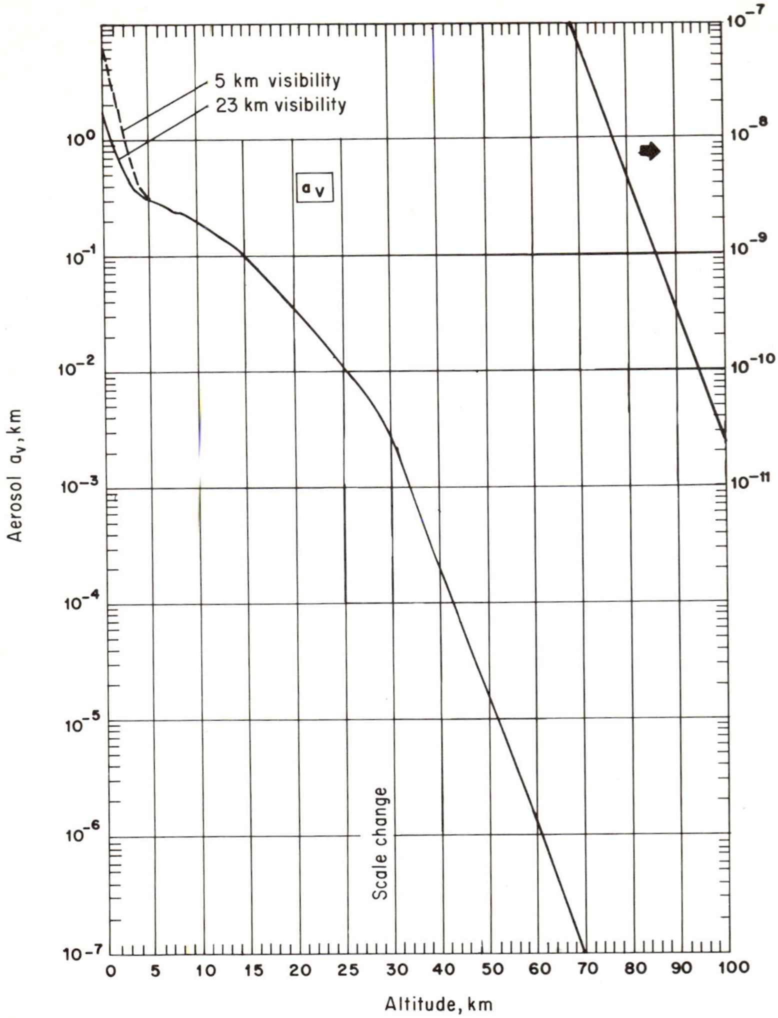

Fig. 15 Equivalent sea-level path length for aerosol extinction as a function of altitude for vertical atmospheric paths.

Absorption in the Region 0.25 to 0.75 μm

20. The only atmospheric molecule considered in this report in the spectral region from 0.25 to 0.70 μm is ozone. Weak absorption bands due to water vapor and oxygen also appear in this region but are not included here because of a lack of reliable data. Since the absorption coefficient due to ozone in this wavelength range is independent of pressure, it is sufficient to define the amount of ozone in an atmospheric path with a single parameter proportional to the total number of molecules in the path. Figure 23 gives the ozone amount as a function of altitude in $(\text{cm}\cdot\text{atm})_{\text{STP}}/\text{km}$ for each of the five atmospheric models and should be used to determine the amount of ozone in a horizontal path. Figure 24 provides total ozone amounts between the top of the atmosphere and the indicated altitudes for each model atmosphere. Ozone amounts for vertical layers within the atmosphere can be found by subtraction of values read off the figure for two different altitudes. Slant-path amounts are found by multiplication of the vertical amount by sec θ or $M(\theta)$ (see Par. 23). For a discussion of the unit $(\text{cm}\cdot\text{atm})_{\text{STP}}$, used in this section, see Par. 26.

The amount of ozone in the required atmospheric path is determined from Figs. 23 and 24. Then the atmospheric transmittance through the path is found as a function of wavelength from Fig. 25 obtained from Inn and Tanaka (1953). The application of this figure is identical to that described earlier for the prediction charts given in Par. 19.

Accuracy of Low-Resolution Transmittance Calculations. The overall accuracy in transmittance which this technique provides is better than 10 percent. The largest errors may occur in the distant wings of rather strongly absorbing bands. The reason for this error is that a unique spectral curve as given in Figs. 16 to 25 cannot be defined for all possible atmospheric paths. The curves presented in these figures were constructed for moderate atmospheric paths and will tend to overestimate the transmittance for very long paths and underestimate the transmittance for very short paths.

As the transmittance approaches 1.0, the percent error in transmittance decreases toward zero but the uncertainty in the absorptance (or emissivity) increases greatly. In fact, if the computed transmittance exceeds 0.96, the uncertainty in absorptance becomes so great that the technique should not be used for a determination of atmospheric emission or background radiation.

Error is also introduced due to the neglect of temperature dependence in the model. The spectral curves of Figs. 16 to 25 were constructed for a temperature of 273 K, and atmospheric temperatures either cooler or warmer provide some variability. This effect is not too large for normal atmospheric temperatures except in the extreme wings of bands and in bands arising from an excited lower-energy state, where the population of the energy states is quite temperature-dependent. This is usually true only for weak bands or in weak portions of stronger absorption bands, and so is not too critical for our purposes. In general, cooler temperatures will mean somewhat higher transmittance values and higher temperatures somewhat lower transmittance values.

It cannot be emphasized enough that the results obtained from these low-spectral-resolution curves should never be used in the determination of atmospheric attenuation for laser propagation. Detailed, very high spectral-resolution measurements or calculations such as are presented in Table 6 and elsewhere (McClatchey, 1971; McClatchey and Selby, 1972*a,b*) are required for this purpose.

SCATTERED SOLAR RADIATION AND INFRARED EMISSION

21. Transmitted and Reflected Sky Radiance in Clear and Cloudy Sky While the energy taken out of a beam of radiation by absorption contributes to heating of the air, the energy scattered by molecules, aerosols, or cloud droplets will be redistributed in the atmosphere. We consider only the problem of sunlight incident on a plane-parallel atmosphere. As long as only single scattering is important and a homogeneous atmosphere is assumed (aerosol–air-mixing ratio constant with height), the angular distribution of skylight at a specific wavelength can be obtained from

$$B(M,M',\phi) = I_0 \frac{M'}{M - M'} (e^{-tM'} - e^{-tM}) \frac{F(\phi)}{t_s} \tag{24}$$

where I_0 = extraterrestrial solar irradiance, W/cm^2
B = sky radiance, $\text{W/(cm}^2\text{)(sr)}$
M = secant of solar zenith angle θ
M' = secant of line of sight angle θ'
t = total optical thickness per unit air mass† $= a_v(\sigma_a + k_a) + m_v(\sigma_m + k_m)$
$t_s = a_v\sigma_a + m_v\sigma_m = t_{s,a} + t_{s,m}$
$F(\phi)$ = angular scattering function (molecular + aerosol) per unit air mass
$= f_a(\phi)\tau_{s,a} + f_m(\phi)t_{s,m}$ [see Fig. 26 for $f_a(\phi)$ and $f_m(\phi)$]

† Scattering and absorption by molecules and aerosols (see Pars. 16 and 17) for a_v and m_v, σ_a, k_a, and σ_m can be derived from Figs. 21 and 22 by taking the logarithm of the transmittance. With regard to molecular absorption k_m, Eq. (24) is valid only for monochromatic radiation (see Table 3 for k_m values).

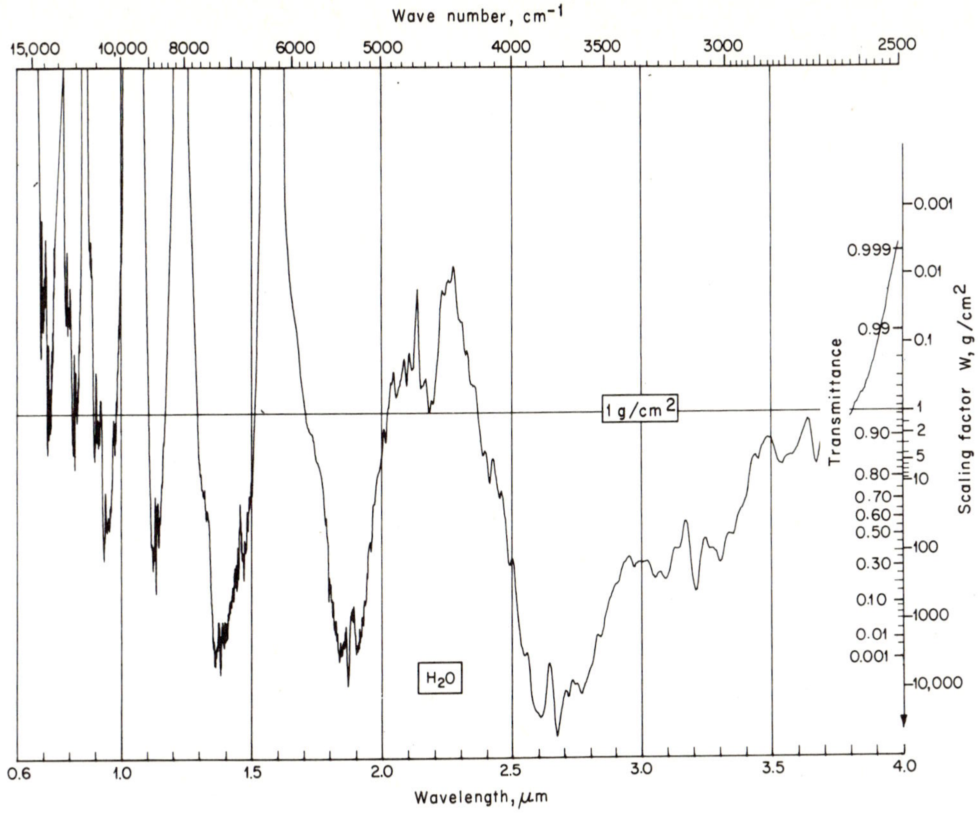

(a)

Fig. 16 Prediction chart for water-vapor transmittance: (*a*) 0.6 to 4.0 μm; (*b*) 4 to 26 μm.

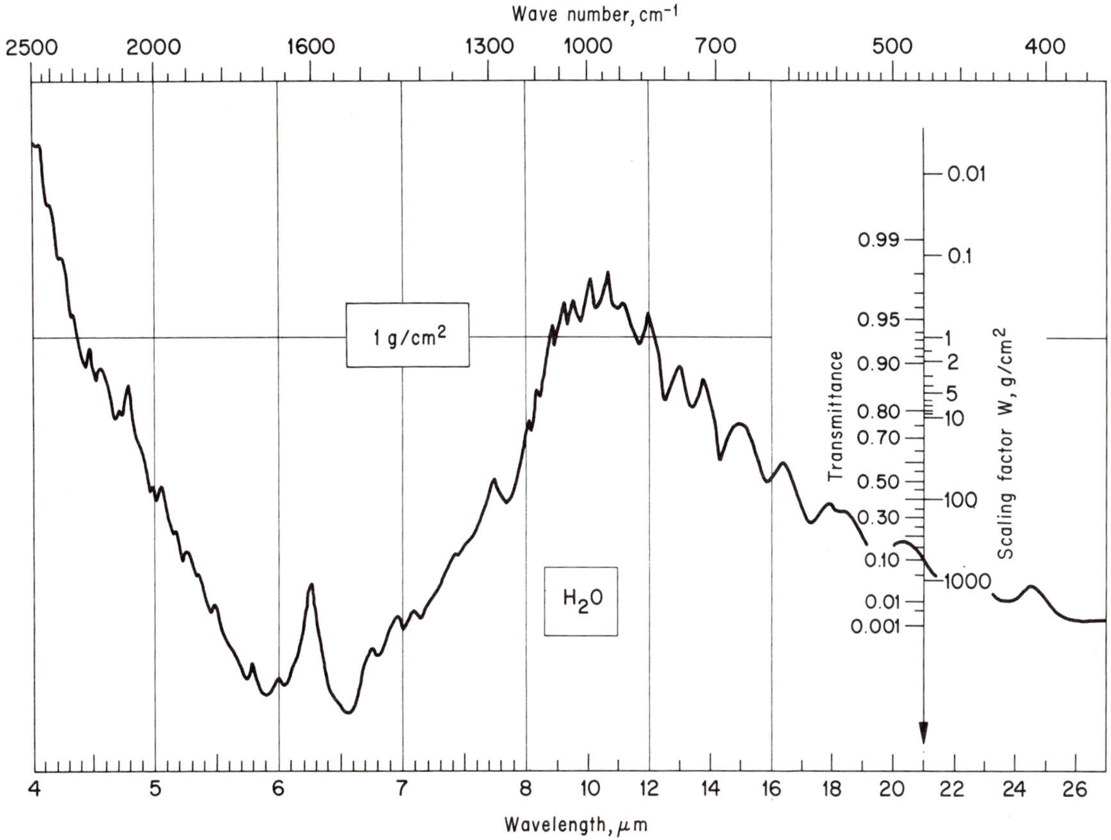

(b)

Fig. 16 *(Continued)*

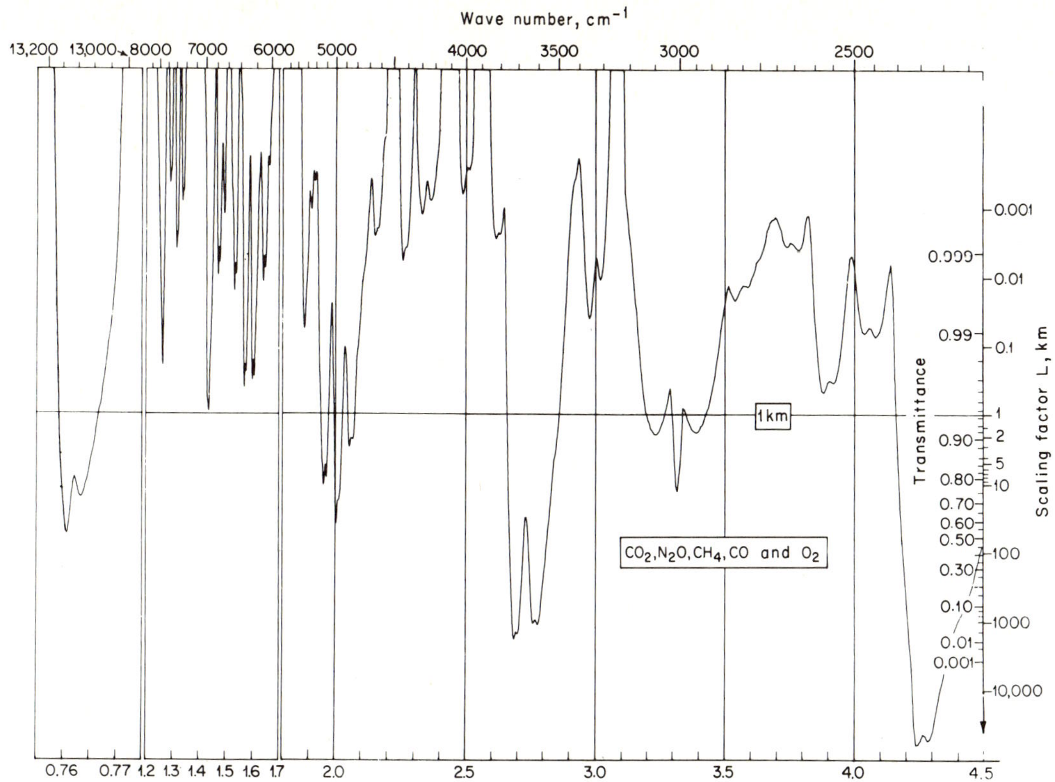

(a)

Fig. 17 Prediction chart for the transmittance of uniformly mixed gases (CO_2, N_2O, CO, CH_4, O_2): (*a*) 0.76 to 4.5 μm; (*b*) 4.5 to 19 μm.

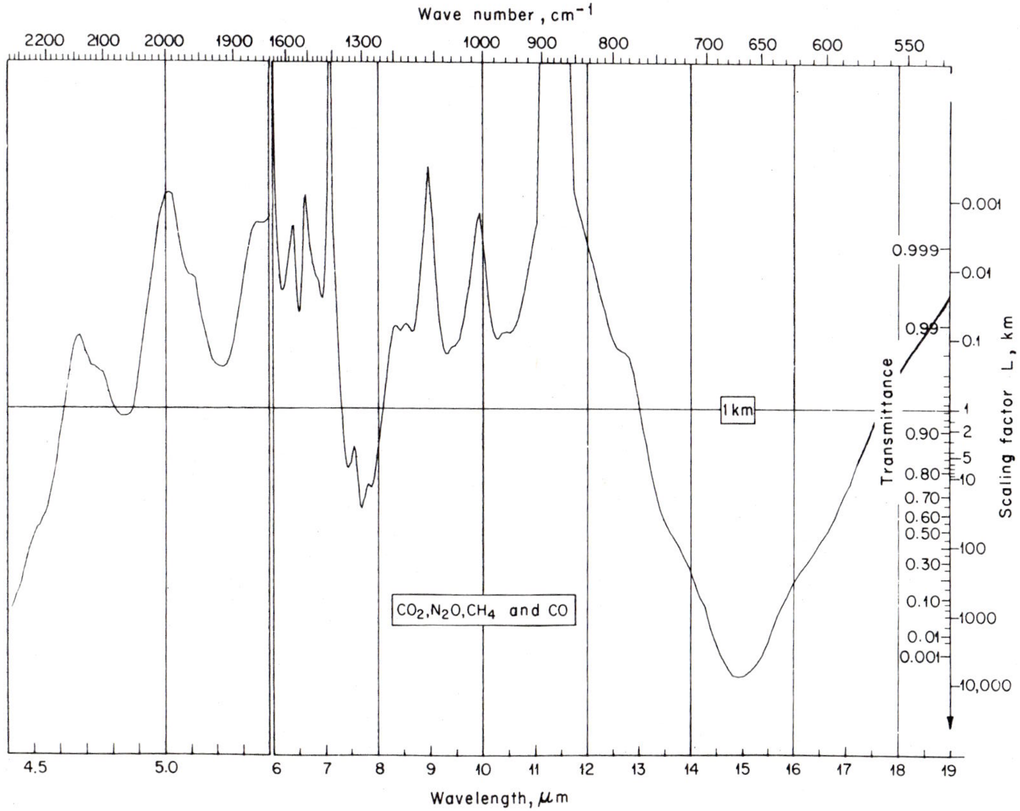

(*b*)

Fig. 17 (*Continued*)

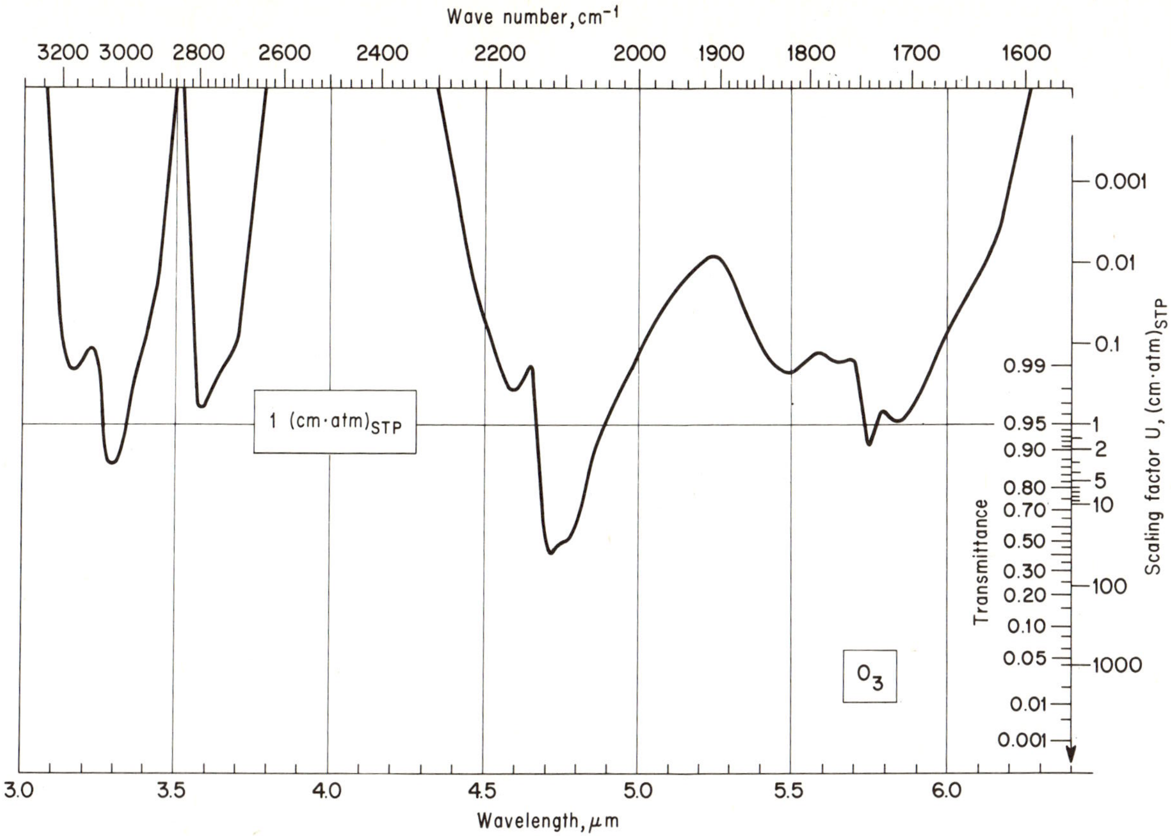

(*a*)

Fig. 18 Prediction chart for ozone transmittance: (*a*) 3 to 6.4 μm; (*b*) 8 to 18 μm.

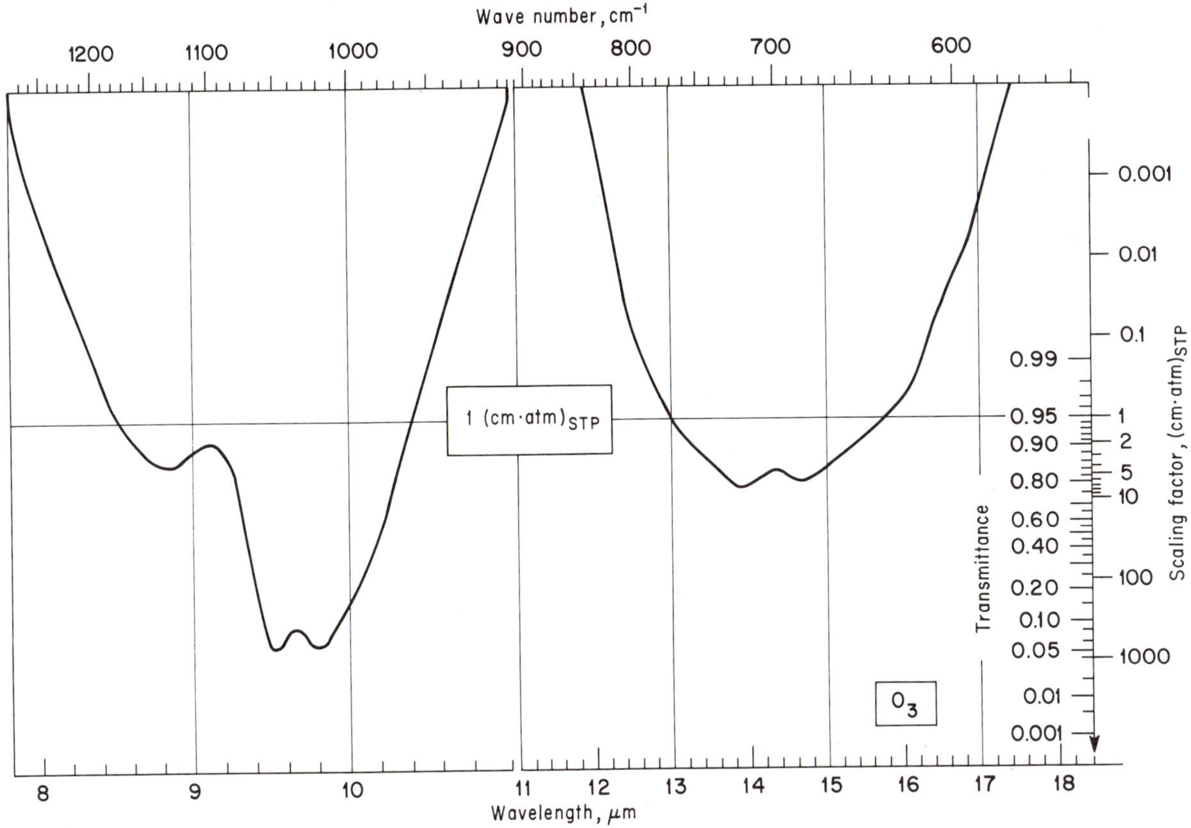

(b)

Fig. 18 *(Continued)*

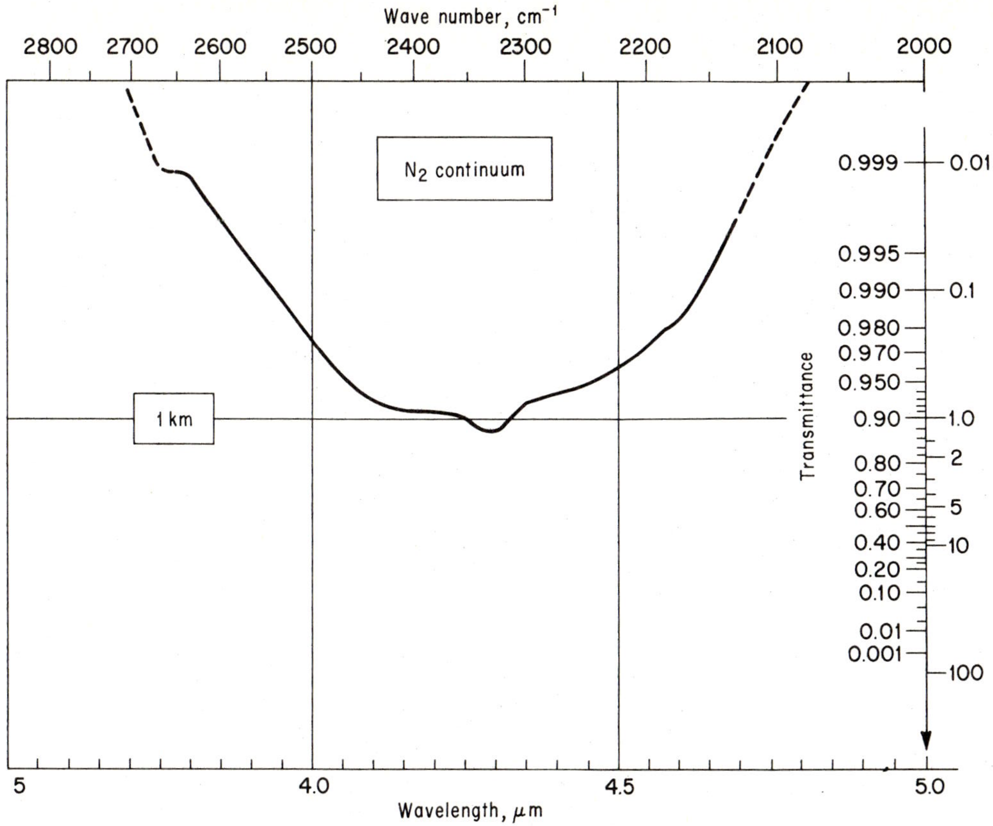

Fig. 19 Prediction chart for nitrogen transmittance.

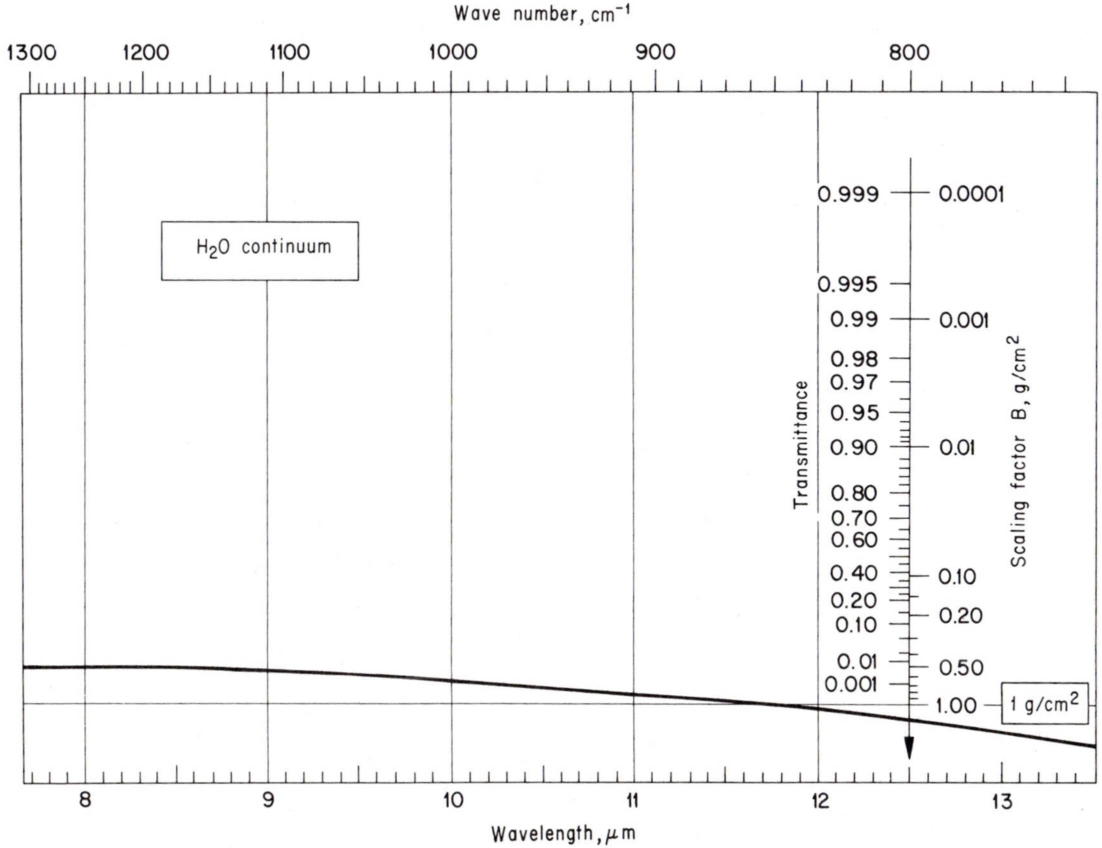

Fig. 20 Prediction chart for water-vapor continuum.

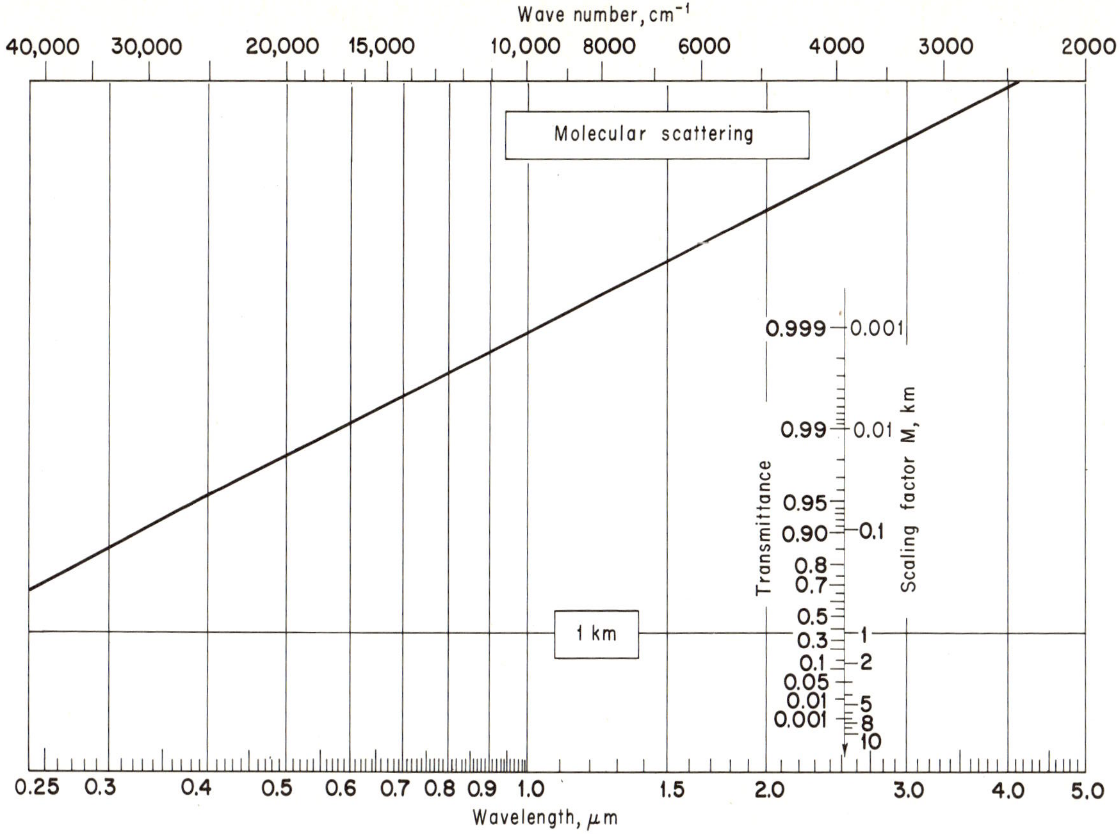

Fig. 21 Prediction chart for transmittance due to molecular scattering.

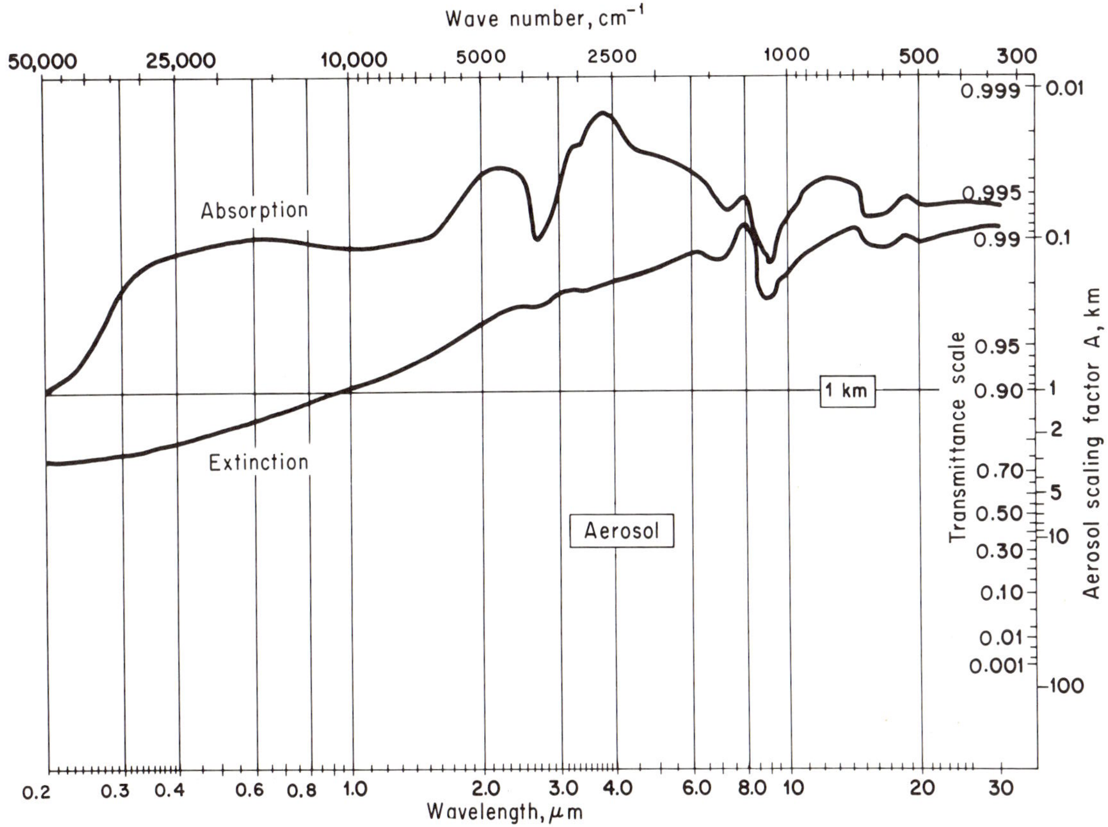

Fig. 22 Prediction chart for aerosol transmittance where $\tau_{extinction} = \tau_{scattering} \tau_{absorption}$.

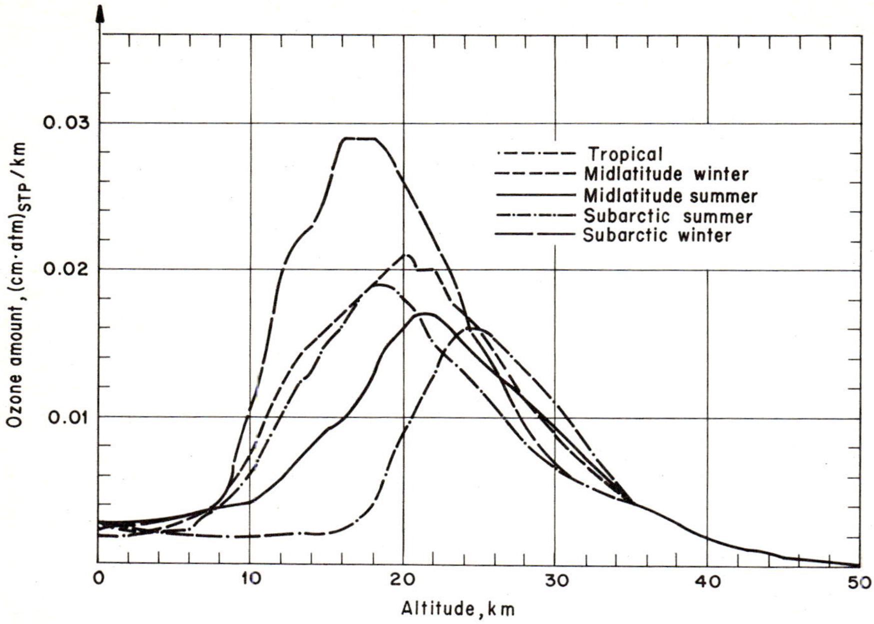

Fig. 23 Equivalent sea-level path length for ozone (0.25 to 0.75 μm) as a function of altitude for horizontal atmospheric paths.

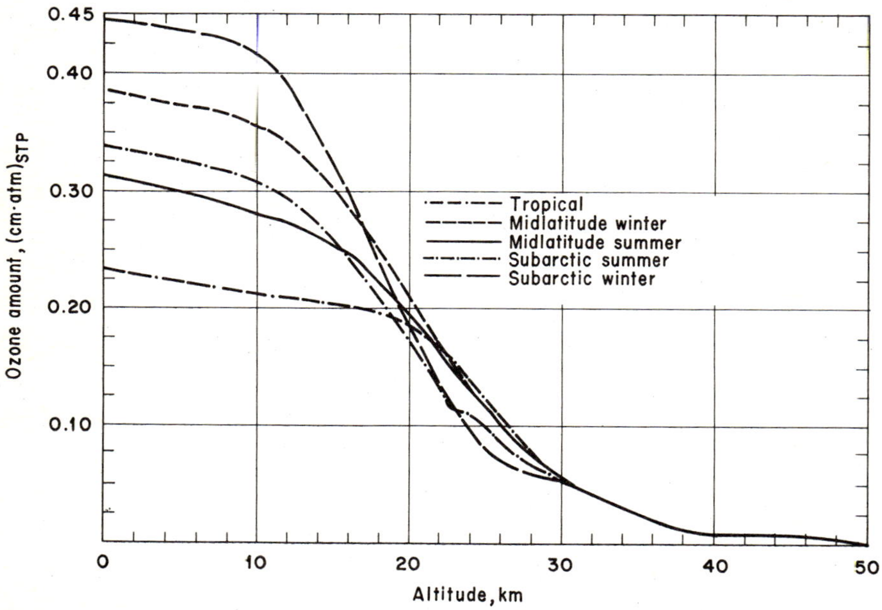

Fig. 24 Equivalent sea-level path length for ozone (0.25 to 0.75 μm) as a function of altitude for vertical atmospheric paths.

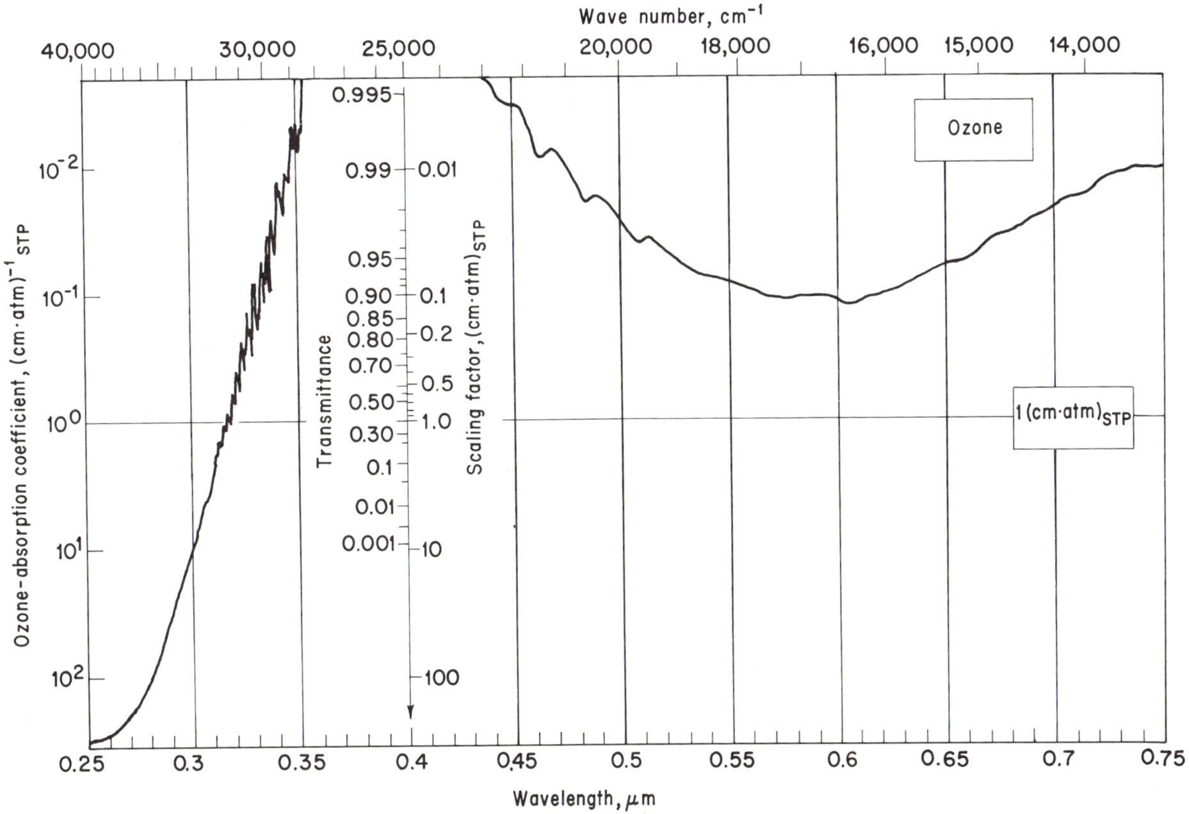

Fig. 25 Prediction chart for ozone transmittance (0.25 to 0.75 μm).

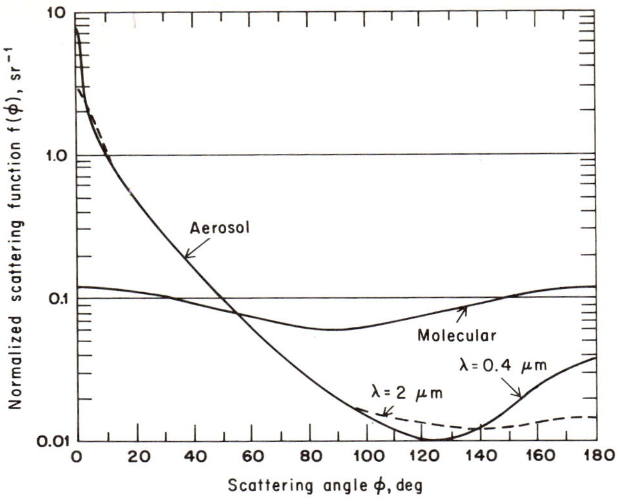

Fig. 26 Normalized scattering-phase function of aerosol and air based on the model described in Par. 4.

The scattering angle ϕ is defined by $\cos\phi = \cos\theta\cos\theta' - \sin\theta\sin\theta'\cos\Delta\Psi$, where $\Delta\Psi$ is the angular azimuth difference between the sun direction and the line of sight.

However, if tM or tM' becomes larger than 0.10, higher-order scattering can no longer be neglected; this means that illumination of the scattering volume from the sky and reflecting earth's surface becomes increasingly important. Higher-order scattering becomes dominant for tM or $tM' > 0.5$, particularly for high ground albedo. This is particularly true for the radiation leaving the top of the atmosphere (radiation to space). The general radiative-transfer equation has been solved analytically only for molecular scattering, and detailed tables of sky radiance and polarization (including the Stokes parameters) have been prepared by Coulson et al. (1960). Skylight distributions for hazy atmospheres have been computed by de Bary et al. (1965) using these tables and Eq. (24). However, computers have made possible a statistical treatment of the scattering problem in any realistic hazy or cloudy atmosphere. By this Monte Carlo method, several investigators have computed the distribution of the radiance of the hazy atmosphere, both as seen from the ground (Fig. 27*a*) and from space (Fig. 27*b*). The assumed aerosol size distribution and vertical distribution are essentially the same as in Table 3. Dotted curves of the radiance (per unit solid angle and unit incident solar flux) are for the sun in the zenith, and solid lines are for the sun at $\theta = 86.3°$.

In Fig. 27*a*, the downward, or transmitted, radiance generally increases with decreasing wavelength when the sun is at the zenith. Forward scattering causes the radiance to peak near the sun. Near the horizon, the radiance increases again, except at short wavelengths, and the albedo influence is large. At low solar elevations, downward radiance is generally much smaller, especially near the zenith at long wavelengths, and the albedo influence is significant only at short wavelengths. Since the sky-radiance values have been averaged over intervals of 0.1 in $\cos\theta$, that is, over angular intervals ranging from 5 to 25°, the radiance near the sun actually should be much higher than shown.

In Fig. 27*b*, upward or reflected radiance as seen from outside the atmosphere is strongly dependent on the albedo of the ground. Especially in the near infrared and in the nadir direction, where the atmospheric backscattered flux becomes small, the contribution of reflected ground radiance becomes dominant. This effect is less pronounced in hazier atmospheres.

The distribution of sky radiance in the ultraviolet is dominated by multiple scattering and below 0.35 μm by ozone absorption. At high solar elevations, down-

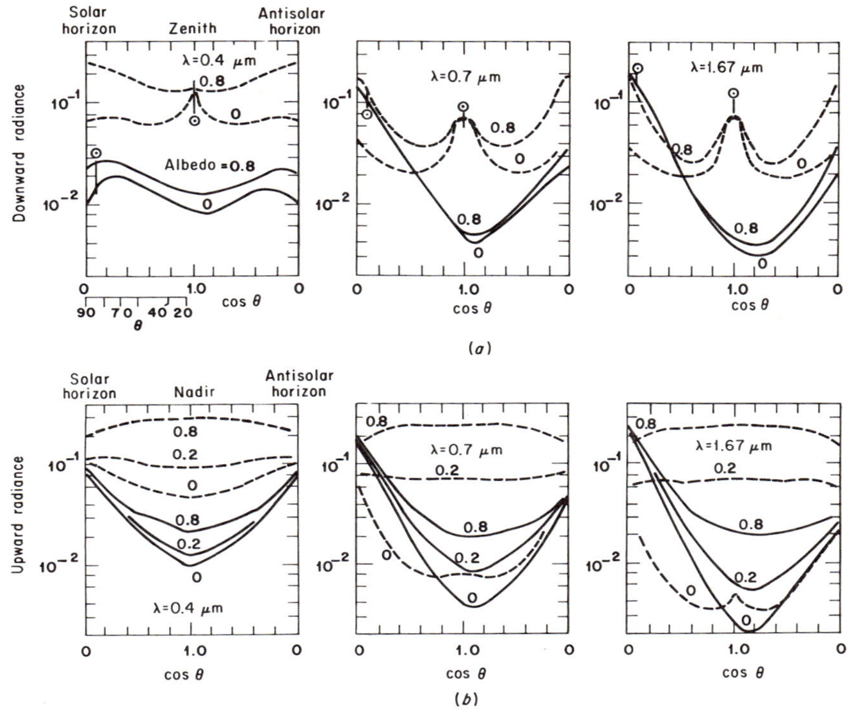

Fig. 27 (*a*) Downward and (*b*) upward radiance computed by Monte Carlo techniques. Dashed curves are for the sun at the zenith. Solid curves are for the sun at $\theta = 86.3°$. The indicated parameter is the surface albedo, and t is the total scattering optical thickness. (*After Plass,* 1958.)

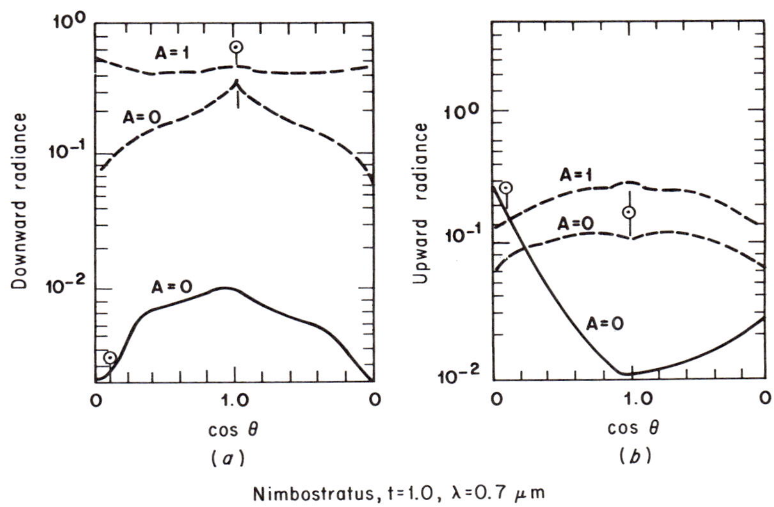

Fig. 28 (*a*) Downward and (*b*) upward radiance computed for a dense nimbostratus cloud. Dashed curves are for the sun at the zenith. Solid curves are for the sun at $\theta = 86.3°$.

ward as well as upward radiances at 0.30 μm are only about 1 percent of their respective values at 0.40 μm but independent of albedo.

For radiance from clouds, Fig. 28*a* and *b* gives values for a rather dense nimbostratus (Kattawar and Plass, 1968). Aerosol and molecular scattering have been neglected in this model because they would have little effect. The strong downward radiance (Fig. 28*a*) for high sun and low albedo is largest near the sun but with high albedo is nearly independent of the angle of observation. At low sun, the much smaller cloud radiance is largest at the zenith. The radiance of the cloud top (Fig. 28*b*) again is very large at high sun. At low sun there is little discrimination between a cloudy and a clear sky (see Fig. 27, $\lambda = 0.7$ μm).

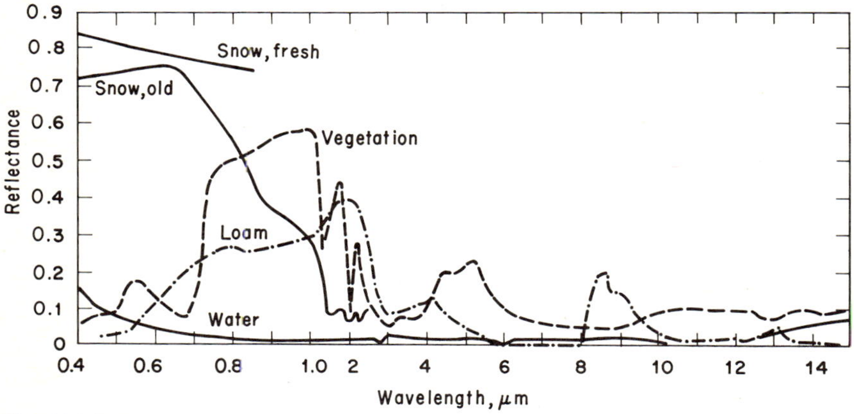

Fig. 29 Typical reflectance of water surface, snow, dry soil, and vegetation. [*Data from Krinov* (1953), *Penndorf* (1956), *and Earing and Smith* (1966).]

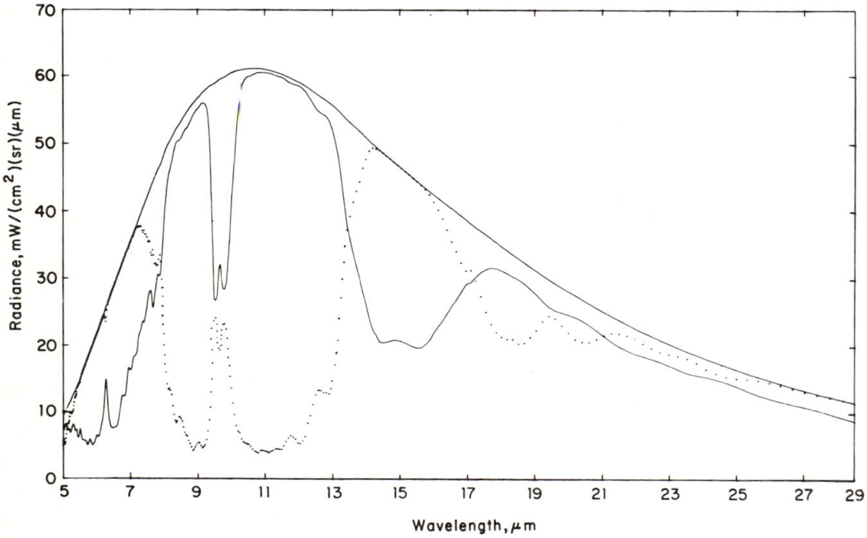

Fig. 30 Radiance as a function of wavelength computed for temperature distribution corresponding to the midlatitude winter model. Transmittance calculations are based on molecular-absorption prediction charts described in Pars. 9 to 20. Upper curve is the radiance corresponding to a blackbody at 272 K. Solid-line spectrum is the computed upward radiance. Dashed spectrum is the computed downward radiance.

An indication of the variability of the reflectance or spectral albedo of some natural objects can be obtained from Fig. 29. The data presented are typical with respect to the wavelength dependence, but absolute reflectance may vary considerably for individual objects and generally is higher for radiation near grazing incidence.

22. Infrared Emission of the Atmosphere In the infrared from 5 to 25 μm and beyond, the atmospheric emission in a collimated beam can be determined by numerical integration of the equation of radiative transfer to give the upward radiance $I\uparrow(\nu)$:

$$I\uparrow(\nu) = \int_{\tau_g}^{1.0} B(\bar{\nu},T)\, d\tau + B(\bar{\nu},T_g)\tau_g \tag{25}$$

where the integral represents the atmospheric contribution and the second term is the contribution from the ground (or lower boundary) modified by the total atmospheric transmittance τ_g between the boundary and the observer; τ is the transmittance which ranges between 1.0 at the level of observation and τ_g at the lower boundary, e.g., the earth's surface or cloud top; $B(\bar{\nu},T)$ is the Planck (blackbody) function corresponding to frequency $\bar{\nu}$ and the temperature T of an atmospheric layer, and T_g is the ground or cloud-top temperature.

To obtain the downward emission from a clear sky, the second term in Eq. (25) should be omitted, and τ_g is then the transmittance of the entire atmosphere. In a cloudy sky the subscript g refers to a cloud base.

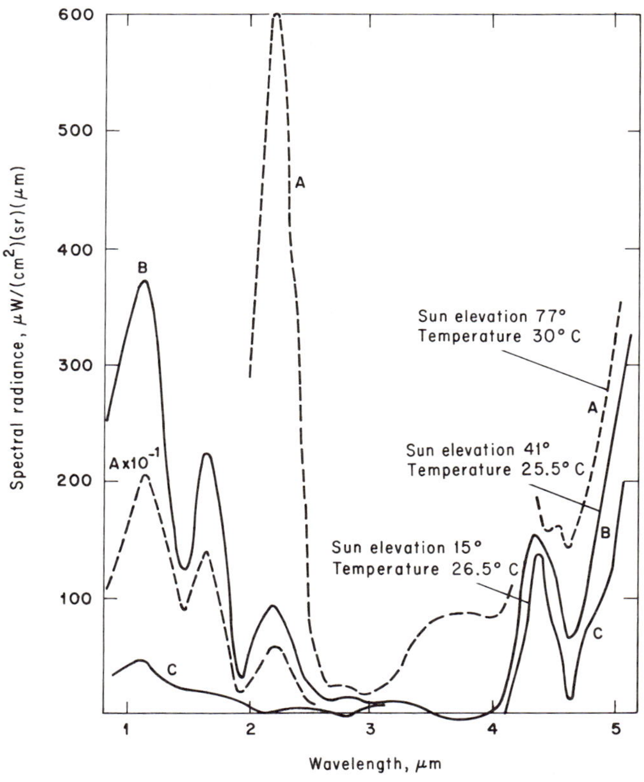

Fig. 31 Spectral radiance of a clear zenith sky showing the effect of the position of the sun on the radiation scattered below 3 μm and the effect of thermal emission at longer wavelengths. Measured at Cocoa Beach, Fla. [*Bell et al.* (1960).]

The effects of aerosols and of molecular scattering have been neglected, and the molecular-absorption charts described in Par. 11 have been applied to the water vapor, ozone, and uniformly mixed gases corresponding to the midlatitude winter model. Equation (25) was then applied using the derived transmittance values and performing the numerical integration according to a numerical analog of Eq. (25).

The results are presented in Fig. 30 for both downward and upward beam radiation. The downward-radiation case assumes the observer to be at the ground looking up at the zenith. The upward case assumes an observation point at the top of the atmosphere viewing the nadir, i.e., looking down.

It should be noted that in the upward case the major absorption features (6.3 μm H_2O, 9.6 μm O_3, and 15 μm CO_2) appear as relative minima, whereas in the downward case these same regions appear as relative maxima. The surface temperature T_g was taken to be the same as the temperature of the lowest atmospheric level in the model (272 K). The upper curve in the figure corresponds to a blackbody at this boundary temperature.

For wavelengths shorter than 5 μm, the problem of upward and downward radiance is complicated by molecular and aerosol scattering and is therefore a function of the solar zenith angle. Computation of radiation between 3 and 5 μm is particularly complicated. Some examples of measured radiation in the 1- to 5-μm region are included in Fig. 31.

REFRACTIVE EFFECTS IN THE ATMOSPHERE

23. Previously (Pars. 5 and 9 to 20) it was indicated that slant-path length could be obtained from vertical-path length through multiplication by sec θ for $\theta < 80°$. For $\theta \geq 80°$, the vertical-path length (or other concentration parameter) ΔL can be converted to slant-path length by

$$\Delta L_{\text{slant path}} = \Delta L_{\text{vertical}} M(\theta) \tag{26}$$

where $M(\theta)$ is the relative optical path, defined as the ratio of slant-path length to vertical-path length between two levels. The term $M(\theta)$ always equals sec θ for a short path near the observer and for any path if $\theta < 80°$. For a long slant path, however, $M(\theta)$ becomes smaller than sec θ because of the curvature of the atmosphere, which is partly compensated by refraction of the beam. This implies that the vertical distribution of a scattering or absorbing atmospheric gas also influences the optical path, and the slant-path length may have to be derived from Eq. (26) if high accuracy is warranted.

If curvature and refraction are considered, the general equation for the optical path between altitudes z_0 and z_1 of air of density ρ at z and scale height

$$H = \frac{z_1 - z_0}{\ln [\rho(z_0)/\rho(z_1)]}$$

may be written

$$M(\theta) = \{[\rho(z_0) - \rho(z_1)\,H]\}^{-1} \int_{z_0}^{z_1} \left\{1 - \left(\frac{r + z_0}{r + z}\right)^2 \left[\frac{n_0}{n(z)} \sin \theta\right]^2\right\}^{-1/2} \rho(z)\, dz \tag{27}$$

where n is the refractive index of air, the zero subscript refers to the lower reference level, and r is the radius of the earth.

For the molecular atmosphere and uniformly mixed atmospheric constituents, the air mass relative to the vertical air mass of the entire atmosphere measured from the ground $M(\theta)$, as computed by Kasten (1967) is presented in Fig. 32. These data are based on the ARDC model atmosphere of Minzner et al. (1959) and a refractive index of air corresponding to a wavelength of 0.7 μm. Earlier values of $M(\theta)$, for which the data by Bemporad (1907) are most often used, are in fair agreement. Kasten (1967) has shown that the average influence of latitude and season on $M(\theta)$ is less than 2 percent, but abnormal tropospheric density profiles may have larger effects. Figure 32 compares $M(\theta)$ values for

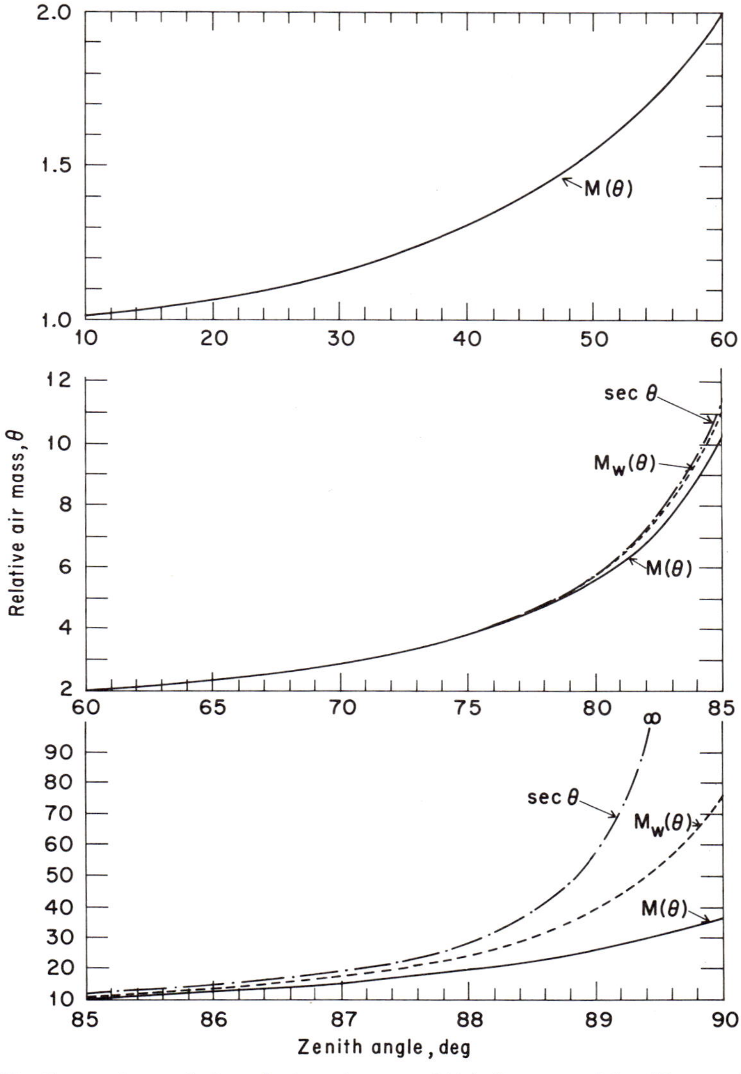

Fig. 32 Comparison of the relative air mass $M(\theta)$ [computed by Kasten (1967)] as a function of apparent zenith angle θ based on the ARDC model atmosphere (Minzner et al., 1959), with a similar quantity $M_w(\theta)$ based on a vertical water-vapor distribution according to Schnaidt (1938) and for sec θ. [Note that for $\theta < 60°$, $M(\theta) = M_w(\theta) = \sec\theta$.]

dry air with $M_w(\theta)$ values for an atmosphere containing water vapor with a distribution according to Schnaidt (1938). Values of sec θ are included for comparison.

The angular deviation between the actual and apparent solar zenith angles $\Delta\theta$ due to refractive effects is presented in Table 7.

For thin atmospheric layers (10 km or less) at high altitudes, where refractive effects can be neglected, Table 8 provides the geometric portion, i.e., the effect due to curvature only, of the slant-path correction. For example, the geometric path through a thin layer at 30 km viewed at a zenith angle of 90° is 10.35

times the geometric path in the vertical. However, the optical path through a uniformly mixed atmospheric constituent at the same viewing angle would be 36.3 (see Fig. 32). This situation may occur in the treatment of transmittance through the ozone layer or stratospheric dust layers.

TABLE 7 Mean Annual Values of $\Delta\theta = \theta_{actual} - \theta_{app}$ as a Function of Apparent Solar Zenith Angle θ

θ app, deg	$\Delta\theta$, deg	θ app, deg	$\Delta\theta$, deg
75	0.061	88	0.311
80	0.090	88.5	0.358
83	0.124	89	0.417
85	0.167	89.5	0.497
86	0.199	90	0.570
87	0.248		

EFFECT OF CLOUDS AND FOG ON ATMOSPHERIC RADIATION

24. As with aerosol extinction (Par. 17), when a beam of radiation is intercepted by a cloud or fog layer, some of the incident radiation may be absorbed and some may be scattered out of this beam, both effects thereby causing a net attenuation of the original radiation. The magnitude of the attenuation depends on the number, size distribution, and refractive index of the suspended droplets and on the thickness of the layer. Since the radiation may be scattered in all directions, including those close to that of the original beam, the resulting attenuation may also depend on the acceptance angle of the receiver.

Natural fogs and low-level clouds are composed of spherical water droplets, the refractive properties of which have been fairly well documented in the spectral region of interest here (Kislovskii, 1959; Irvine and Pollack, 1968). However, in cirrus clouds the particles consist of hexagonal or cylindrical ice crystals of various shapes, which, when mutually oriented, can give rise to anomalous scattering and polarization phenomena when radiation is propagated through them.

We will restrict ourselves here to the propagation of a collimated beam of radiation. Generally, fogs and clouds show only a small increase in attenuation with wavelength from the visible region to about 6 μm. Near 10 μm an extinction minimum occurs. However, certain infrequent types of fog characterized by a sharply peaked distribution of small droplets† show a significant increase in infrared transmittance.

These general effects can be seen from Fig. 33, which shows the wavelength variation of the extinction coefficient for a drop-size distribution representative of fair-weather cumulus cloud. The model drop-size distribution (Fig. 34), which is given by

$$n(r) = 2.373r^6e^{-1.5r} \qquad \text{cm}^{-3}/\mu\text{m} \tag{28}$$

has a maximum corresponding to a radius of 4 μm and is normalized to 100 drops per cubic centimeter (Deirmendjian, 1964).

Figure 33 was calculated and based on single-scattering theory, using the complex refractive-index values for water given by Kislovskii (1959).

It will be seen from Fig. 33 that the attenuation due to this fairly thin model cloud is rather severe; e.g., for a 100-m path length the transmittance is approximately 0.3 and 0.15 at 10 and 1 μm, respectively.

Some mention should be made, at this point, of the validity of using single-scattering theory to account for the attenuation of radiation in cloud. It has generally been assumed that calculations of attenuation based on single-scattering theory are valid for an optical thickness of less than 0.03 (van de Hulst's criterion) and that for values of optical thickness in excess of unity, secondary and multiple scattering effects become important. However, this applies to the propagation

† This phenomenon is also found near the base of stratus and cumulus clouds.

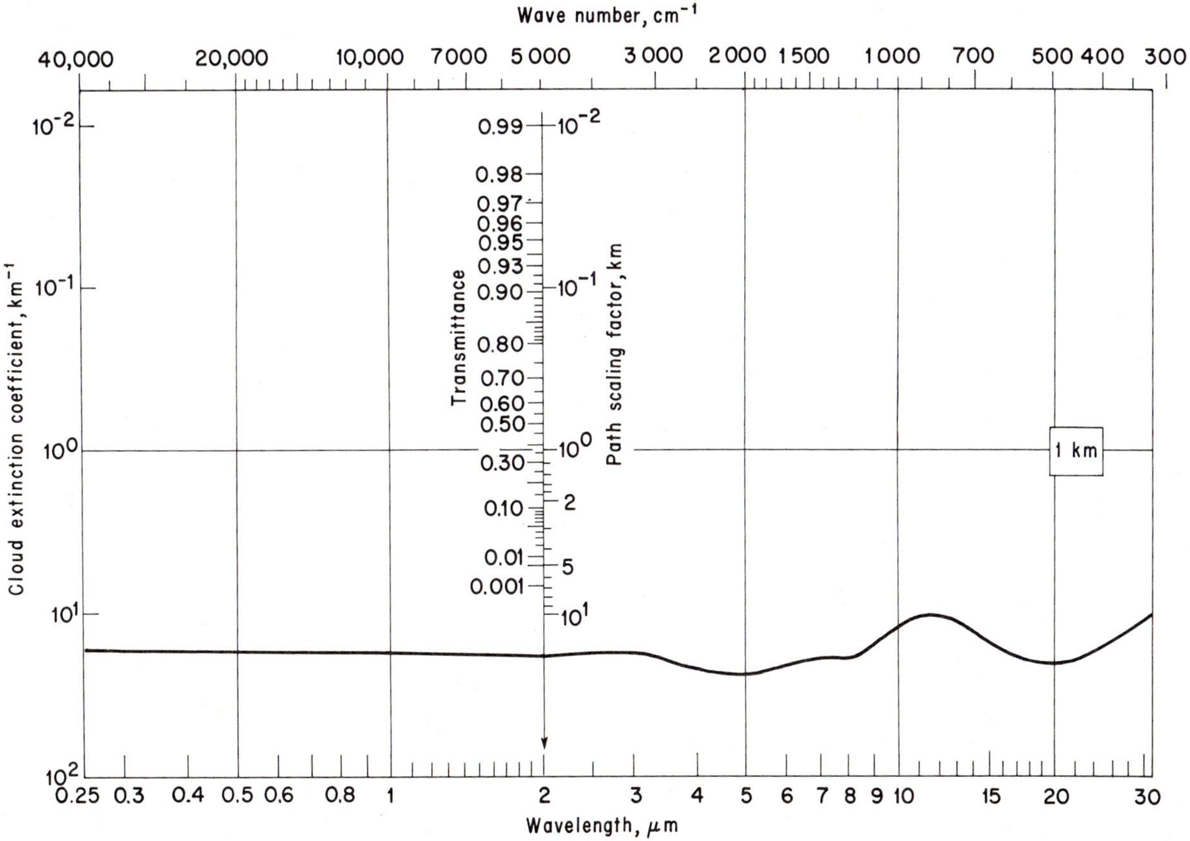

Fig. 33 The attenuation as a function of wavelength due to fair-weather cumulus cloud. [*After Deirmenjian* (1964, 1969).]

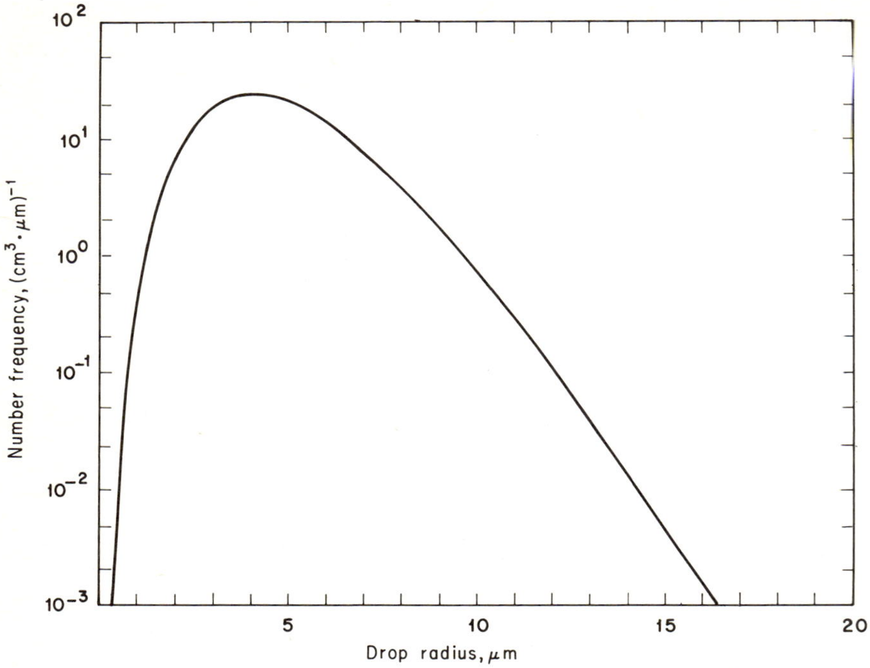

Fig. 34 Particle-size distribution for fair-weather cumulus-cloud model.

TABLE 8 Geometric Path of a Thin High-Altitude Layer as a Function of Zenith Angle θ Relative to a Vertical Path through That Layer

	Height of layer above surface, km					
θ, deg	10	20	30	40	50	60
70	2.89	2.86	2.83	2.80	2.77	2.74
75	3.78	3.71	3.63	3.57	3.50	3.44
80	5.49	5.26	5.05	4.87	4.70	4.56
85	9.66	8.51	7.70	7.08	6.60	6.20
87	13.07	10.56	9.11	8.13	7.41	6.86
88	15.17	11.58	9.74	8.57	7.74	7.12
89	17.08	12.37	10.19	8.87	7.96	7.28
89.5	17.66	12.58	10.30	8.94	8.01	7.33
90	17.88	12.66	10.35	8.97	8.03	7.34

of diffuse radiation. For a narrow beam of collimated radiation (or laser-beam propagation) it has been found (Zuev et al., 1967) that single-scattering theory can account for the observed attenuation for values of optical thickness of up to 25. In this context optical thickness is defined as the product of the extinction coefficient and path length.

UNITS AND CONVERSION FACTORS

25. Throughout this section we have used the term frequency rather loosely to signify spectral variation. In general, frequency is assumed to be synonymous with wave number, ν (cm^{-1}), which is given by $\nu = 10^4/\lambda$, where the wavelength λ is given in units of micrometers, that is, 10^{-6} m.

The subject of units used in atmospheric physics is rather confusing to the beginner, particularly where units of absorber concentration or amount are concerned. In Par. 26 we show the reader how to derive the absorber concentrations for use with the prediction charts given in Pars. 9 to 20, on the assumption that one has the relevant meteorological data available. We then discuss the units of absorber concentration and indicate how to make the appropriate conversions (Par. 27).

26. Absorber Concentration For most gases absorber concentration, denoted here by ΔL, is generally measured by the equivalent length of pure absorber in the path under the existing conditions of pressure P and temperature T. The unit centimeter-atmospheres (cm·atm) is given by $\Delta L = cRP$, where c is the fractional concentration of the absorber (see Table 2), R is the path length in centimeters, and the P is the pressure in atmospheres. More often ΔL_0 is quoted as the absorber concentration reduced to STP conditions, which is a more meaningful quantity where vertical or slant atmospheric paths are concerned.

$$\Delta L_0 = cR\frac{P(z)}{P_0}\frac{T_0}{T(z)} \qquad (\text{cm·atm})_{\text{STP}} \tag{29}$$

where $T_0 = 273$ K and $P_0 = 1$ atm. If c is given in parts per million (ppm) (see Table 2) and R is given in kilometers, the following expression can be used:

$$\Delta L_0\ (\text{cm·atm})_{\text{STP}} = 10^{-1}cR\frac{P}{P_0}\frac{T_0}{T} \tag{30}$$

For a uniformly mixed gas of fractional concentration by volume c, the number of centimeter-atmospheres at STP in a slant path from any altitude z to the limit of the atmosphere is given by

$$\Delta L_0(\theta, z \to \infty)_{\text{STP}} = cH(z)\sec\theta\frac{P(z)}{P_0}\frac{T_0}{T(z)} = cH(z)\,(\sec\theta)\frac{\rho(z)}{\rho_0} \tag{31}$$

where $H(z)$ = scale height of atmosphere above altitude z
θ = zenith angle
$\rho(z)$ = atmospheric density at altitude z
ρ_0 = atmospheric density at STP

From the hydrostatic equation, $H(z)$ can be defined as

$$P(z) = g\rho(z)H(z) \tag{32}$$

where $P(z)$ is the pressure at altitude z and g is the acceleration due to gravity (981 cm/s²). Thus, for a vertical column of air at STP, $H(0) = 7.99 \times 10^5$ cm, that is, approximately 8 km. If the scale height in centimeters is required at another altitude z, it can be determined from

$$H(z) = 1.02 \times 10^6\frac{P(z)}{\rho(z)} \tag{33}$$

where $P(z)$ is given in millibars and $\rho(z)$ is given in grams per cubic meter (see Table 1). Note from Eq. (33) that the quantity $H(z)\rho(z)$ is directly proportional to $P(z)$ and is independent of temperature. Thus, inserting $H(z)\rho(z)$ into Eq. (31), we have

$$\Delta L_0(\theta, z \to \infty)_{\text{STP}} = c(1.02 \times 10^6)P(z)\frac{\sec\theta}{\rho_0} \tag{34}$$

and for the amount of absorber between levels z_1 and z_2 (see Table 2):

$$\Delta L_0(\theta, z_1 \to z_2)_{\text{STP}} = c(1.02 \times 10^6)[P(z_1) - P(z_2)]\frac{\sec\theta}{\rho_0} \tag{35}$$

For condensable vapors, e.g., water vapor, it is more convenient to measure the total length of liquid (in centimeters) which may be precipitated out of the path (per unit area). The number of precipitable centimeters (abbreviated pr cm) of water vapor in a given path of length R cm is given by $w = \rho(T)\mathcal{H}R$, where $\rho(T)$ is the

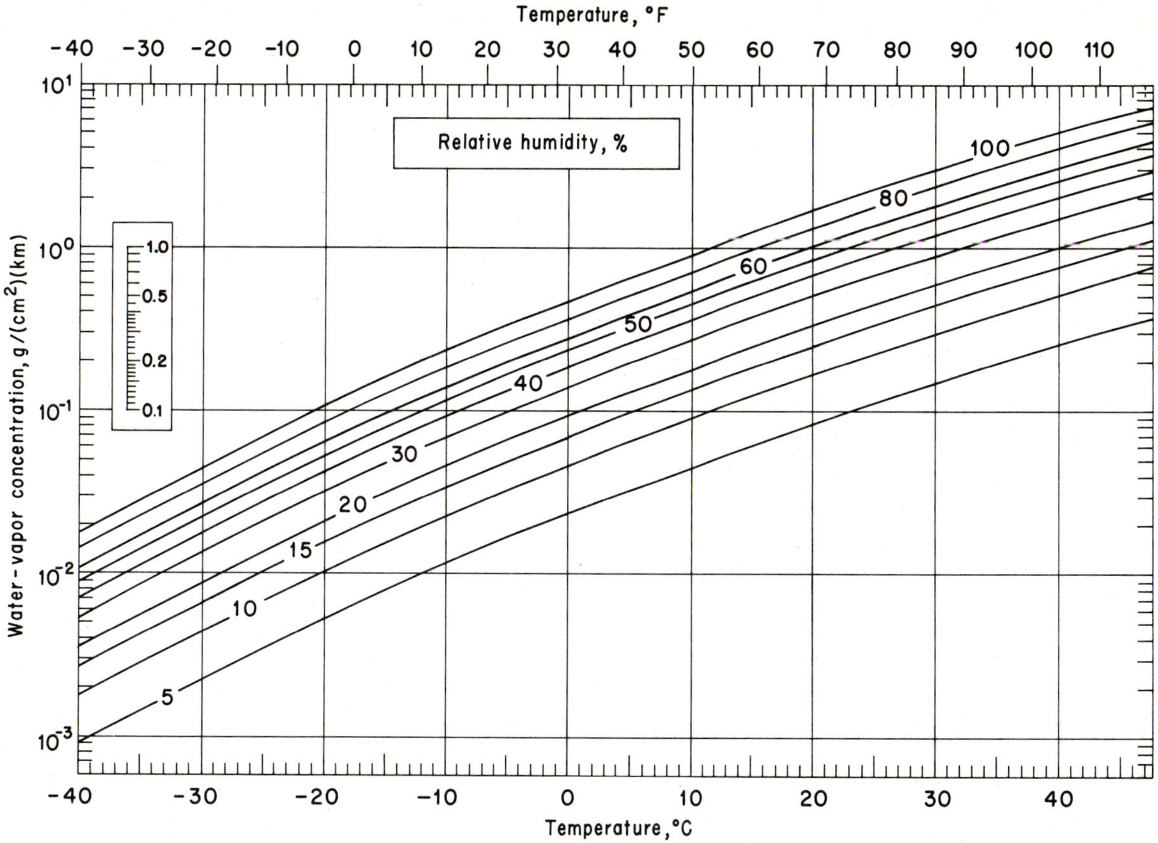

Fig. 35 Water-vapor concentration per kilometer path length as a function of temperature and relative humidity.

saturation vapor density of water at temperature T g/cm^3 and $\mathcal{H}$ is the fractional relative humidity. It will be noted that the units of precipitable centimeters and grams per square centimeter are equivalent.

Figure 35 will enable the reader to calculate the number of precipitable centimeters of water vapor per kilometer path length for a given temperature and relative humidity. For temperatures below $-40°$C the reader should use density values of water vapor measured over ice for determining water-vapor concentrations for atmospheric paths (see List, 1963).

In Table 1, water-vapor densities are given in grams per cubic meter as a function of altitude for the five model atmospheres. It may be useful to remember that for water vapor,

$$1 \text{ g/m}^3 = 0.1 \text{ g/(cm}^2\text{)(km)} = 0.1 \text{ pr cm/km}$$

27. Unit-Conversion Factors According to Avogadro's hypothesis, the molecular weight M in grams of any gas occupies 22.4 l at STP. Hence 1 cm^3 of gas at STP weighs $M/(2.24 \times 10^4)$ g. Since 1 cm·atm)$_{STP}$ is equivalent to a length of 1 cm of gas at STP per square centimeter, we can write

$$1 \text{ cm·atm of gas} = \frac{M}{2.24 \times 10^4} \text{ g/cm}^2 \tag{36}$$

where M is the molecular weight of the gas (see Table 2).

Also since 1 cm^3 of gas at STP contains $1/(2.24 \times 10^4)$ mol and 1 mol of gas contains the same number $\mathcal{R}$ of molecules, where $\mathcal{R}$ is Avogadro's number, we can write that 1 cm^3 of gas at STP contains $(6.06 \times 10^{23})/(2.24 \times 10^4) = 2.69 \times 10^{19}$ molecules or

$$1 \text{ (cm·atm)}_{STP} = 2.69 \times 10^{19} \text{ molecules/cm}^2 \tag{37}$$

Note that Eq. (37) is valid independent of the gas. Since for water vapor we can write 1 g/cm^2 in terms of (cm·atm)$_{STP}$, using Eq. (36) we also see that for water vapor

$$1 \text{ pr cm } H_2O = 3.34 \times 10^{22} \text{ molecules/cm}^2 \tag{38}$$

Thus the unit molecules per square centimeter is independent of the nature of the absorbing gas and basic to all gases. It is hoped that this unit will become more widely used and thereby help to unify absorber-concentration units for all the atmospheric constituents.

REFERENCES

Altshuler, T. L. (1961): Infrared Transmission and Background Radiation by Clear Atmospheres, *General Electric Rep.* 61 SD 199, AD-401923, December.

Bell, E. E., L. Eisner, J. Young, and R. Oentjen (1960): *J. Opt. Soc. Am.*, vol. 50, p. 1313.

Bemporad, A. (1907): *Meteorol. Z.*, vol. 24, p. 306.

Birkland, J. W., and J. H. Shaw (1959): *J. Opt. Soc. Am.*, vol. 49, p. 637.

Bosomworth, D. P., and H. P. Gush (1965): *Can. J. Phys.*, vol. 43, p. 729.

Bullrich, K. (1964): *Adv. Geophys.*, vol. 10, p. 99.

Burch, D. E. (1970): Investigation of the Absorption of Infrared Radiation by Atmospheric Gases, *Semi-annu. Tech. Rep.* U-4784, Philco Ford *Aeronutronic Rep.* Contract F19628-69-C-0263.

——— and D. A. Gryvnak (1966): Absorption by H_2O between 5045 and 14,485 cm^{-1} (0.69–1.98 μm Region), *Aeronutronic Rep.* U-3704, July.

——— and ——— (1967): Strengths, Widths and Shapes of the Oxygen Lines near 7600 Angstroms, *Aeronutronic Rep.* U-4076, May.

———, ———, and R. R. Patty (1964): Absorption by CO_2 between 4500 and 5400 cm^{-1} (2 μm Region), *Aeronutronic Rep.* U-2955, Dec. 15.

———, ———, and ——— (1965*a*): Absorption by CO_2 between 6600 and 7125 cm^{-1} (1.4 μm Region), *Aeronutronic Rep.* U-3127, June.

———, ———, and ——— (1965*b*): Absorption by CO_2 between 8000 and 10,000 cm^{-1} (1–1.25 μm Region), *Aeronutronic Rep.* U-3200, August.

———, ———, and ——— (1965*c*): Absorption by CO_2 between 5400 and 6600 cm^{-1} (1.6 μm Region), *Aeronutronic Rep.* U-3201, August.

———, ———, and ——— (1965*d*): Absorption by H_2O between 2800 and 4500 cm^{-1} (2.7 μm Region), *Aeronutronic Rep.* U-3202, September.

———, ———, and ——— (1967): Absorption by CO_2 between 7125 and 8000 cm^{-1} (1.25–1.4 μm Region), *Aeronutronic Rep.* U-3930, February.

———, ———, and ——— (1968): Absorption by CO_2 between 3100 and 4100 cm^{-1} (2.44–3.22 μm Region), *Aeronutronic Rep.* U-4132, April.

———, ———, E. B. Singleton, W. F. France, and D. Williams (1962): Infrared Absorption by Carbon Dioxide, Water Vapor and Minor Atmospheric Constituents, *Ohio State Univ. Air Force Camb. Res. Lab. Res. Contract* AF19(604)-2633.

Coulson, K. L., J. V. Dave, and Z. Sekera (1960): "Tables Related to Radiation Emerging from a Planetary Atmosphere with Rayleigh Scattering," University of California Press, Los Angeles.

de Bary, E., B. Braun, and K. Bullrich (1965): Tables Related to Light Scattering in a Turbid Atmosphere, *Air Force Camb. Res. Lab. Spec. Rep.* 33, I–III, September.

Deirmendjian, D. (1963): Scattering and Polarization Properties of Polydispersed Suspensions with Partial Absorption, in Milton Kerker (ed.), *Proc. Interdiscipl. Conf. Electromagn. Scattering, Potsdam, N.Y.*, Pergamon, New York.

——— (1964): *Appl. Opt.*, vol. 3, p. 187.

——— (1969): "Electromagnetic Scattering on Spherical Polydispersions," American Elsevier, New York.

Earing, D. G., and J. A. Smith (1966): Target Signatures Analysis Center: Data Compilation, *Air Force Avionics Lab.* AD-489-968, July.

Elsasser, W. M. (1938): *Phys. Rev.*, vol. 54, p. 126.

——— (1942): *Harv. Meteorol. Stud.* 6.

Elterman, L. (1968): UV, Visible, and IR Attenuation for Altitudes up to 50 km, *Air Force Camb. Res. Lab. Environ. Res. Pap.* 285, AFCRL-68-0153, April.

——— (1970): Vertical Attenuation Model with Eight Surface Meteorological Ranges 2 to 13 Kilometers, *Air Force Camb. Res. Lab. Environ. Res. Pap.* 318, AFCRL-70-0200.

Fink, U., D. H. Rank, and T. A. Wiggins (1964): *J. Opt. Soc. Am.*, vol. 54, p. 472.

Goody, R. M. (1952): *Q. J. R. Meteorol. Soc.*, vol. 78, p. 165.

——— (1964): "Atmospheric Radiation," vol. 1, "Theoretical Basis," Oxford University Press, London.

Gryvnak, D. A., R. R. Patty, D. E. Burch, and E. E. Miller (1966): Absorption by CO_2 between 1800 and 2850 cm^{-1} (3.5–5.8 μm Region), *Aeronutronic Rep.* U-3857, December.

Hering, W. S., and T. R. Borden (1964): Ozone Observations over North America, *OAR Res. Rep.* AFCRL-64-30, vol. 2.

Inn, E. C. Y., and Y. Tanaka (1953): *J. Opt. Soc. Am.*, vol. 43, pp. 870–873.

Irvine, W. M., and J. B. Pollack (1968): *Icarus*, vol. 8, p. 324.

Kasten, F. (1967): *Arch. Meteorol. Geophys. Bioklim.*, vol. B15, p. 62.

Kattawar, G. W., and G. N. Plass (1968): *Appl. Opt.*, vol. 7, p. 869.

King, J. I. F. (1959): *Proc. Infrared Inf. Symp.*, vol. 4, no. 1, p. 164.

——— (1964): *J. Quant. Spectr. Radiat. Transfer*, vol. 4, p. 705.

Kislovskii, D. L. (1959): *Opt. Spectrosc.*, vol. 7, p. 201.

Krinov, E. L. (1953): Spectral Reflectance Properties of Natural Formations, *Natl. Res. Counc. Can. Tech. Trans.* TT-439, Ottawa, from Russian of 1947.

List, R. J. (ed.) (1963): "Smithsonian Meteorological Tables," 6th ed., Smithsonian Institution, Washington, D.C.

McCaa, D. J., and J. H. Shaw (1967): Infrared Absorption Bands of Ozone, *Ohio State Univ. Air Force Camb. Res. Lab. Rep.* AFCRL-67-0237, February.

McClatchey, R. A. (1971): Atmospheric Attenuation of CO Laser Radiation, *Air Force Camb. Res. Lab. Environ. Res. Pap.* 357, AFCRL-71-03070.

———, W. S. Benedict, S. A. Clough, D. E. Burch, R. F. Calfee, K. Fox, L. S. Rothman, and J. S. Garing (1973): AFCRL Atmospheric Absorption Line Parameters Compilation, *Air Force Camb. Res. Lab. Environ. Res. Pap.* 434, AFCRL-TR-73-0096.

———, R. W. Fenn, J. E. A. Selby, F. E. Volz, and J. S. Garing (1970): Optical Properties of the Atmosphere, *Air Force Camb. Res. Lab. Environ. Res. Pap.* 331, AFCRL-70-0527.

———, ———, ———, ———, and ——— (1971): Optical Properties of the Atmosphere (Revised), *Air Force Camp. Res. Lab. Environ. Res. Pap.* 354, AFCRL-71-0279.

———, ———, ———, ———, and ——— (1972): Optical Properties of the Atmosphere, 3d ed., *Air Force Camb. Res. Lab. Environ. Res. Pap.* 411, AFCRL-72-0497.
——— and J. E. A. Selby (1972*a*): Atmospheric Attenuation of HF and DF Laser Radiation, *Air Force Camb. Res. Lab. Environ. Res. Pap.* 400, AFCRL-73-0312.
——— and ——— (1972*b*): Atmospheric Transmittance, 7–30 μm: Attenuation of CO_2 Laser Radiation, *Air Force Camb. Res. Lab. Environ. Res. Pap.* 419, AFCRL-72-0611.
McCoy, J. H., D. B. Rensch, and R. K. Long (1969): *Appl. Opt.*, vol. 8, p. 1471.
Mie, G. (1908): *Ann. Phys.*, vol. 24, p. 377.
Minzner, R. A., K. S. W. Champion, and H. L. Pond (1959): The ARDC Model Atmosphere, 1959, *Air Force Surv. Geophys.* 115, AFCRL-TN-59-267, August.
Penndorf, R. (1956): Luminous and Spectral Reflectance as Well as Colors of Natural Objects, *Air Force Camb. Res. Lab. Geophys. Res. Pap.* 44, AFCRL-TR-56-203, February.
Penner, S. S. (1959): "Quantitative Molecular Spectroscopy and Gas Emissivities," Addison-Wesley, Reading, Mass.
Plass, G. N. (1958): *J. Opt. Soc. Am.*, vol. 48, p. 690.
——— and G. W. Kattawar (1968): *Appl. Opt.*, vol. 7, p. 1129.
Reddy, S. R., and C. W. Cho (1965): *Can. J. Phys.*, vol. 43, p. 2331.
Schnaidt, F. (1938): *Meteorol. Z.*, vol. 55, p. 296.
Selby, J. E. A., and R. A. McClatchey (1972): Atmospheric Transmittance from 0.25–28.5 μm: Computer Code LOWTRAN 2, *Air Force Camb. Res. Lab. Environ. Res. Pap.* 491, AFCRL-72-0611.
Shaw, J. H. (1959): A Determination of the Abundance of N_2O, CO, and CH_4 in Ground Level Air at Several Locations near Columbus, Ohio, *Air Force Camb. Res. Lab. Sci. Rep.* 1, *Contract* AF19(604)-2259.
——— (1968): Monthly Report on Infrared Temperature Sounding, *Ohio State Univ. RF Proj.* 2469 *Rep.* 16, October.
Sissenwine, N., D. D. Grantham, and H. A. Salmela (1968): Humidity up to the Mesopause, AFCRL-68-0550, October.
Street, J. L. (1968): *Appl. Opt.*, vol. 7, p. 1545.
Valley, S. L. (ed.) (1965): "Handbook of Geophysics and Space Environments," Air Force Cambridge Research Laboratories, Bedford, Mass.
Van de Hulst, H. C. (1957): "Light Scattering by Small Particles," Wiley, New York.
Volz, F. E. (1972): *Appl. Opt.*, vol. 11, p. 755.
Wyatt, P. J., V. R. Stull, and G. N. Plass (1964): *Appl. Opt.*, vol. 3, p. 229.
Yates, H. W., and J. H. Taylor (1960): Infrared Transmission of the Atmosphere, *Naval Res. Lab. Rep.* 5453, June 8.
Zuev, V. E., M. V. Kabanov, and B. A. Savel'ev (1967): *Izv. Akad. Nauk SSSR Fiz. Atmos. Okeana,* vol. 3, p. 724.

Section **15**

Optical Properties of Water

JOHN E. TYLER

Visibility Laboratory, Scripps Institution of Oceanography, University of California, San Diego, Calif.

(Numbers refer to paragraphs)

1. Naturally Occurring Water Naturally occurring water, like air, is a scattering and absorbing medium. It differs from air in that the magnitudes of the optical effects are generally larger per unit length than those of air. In addition, the scattering particles found in water are predominantly irregular in shape, exhibit a broad size distribution, and are often strongly pigmented.

Numerical values of the optical properties of naturally occurring water depend upon the composition of the water and to some extent on the geometrical distribution of the light field penetrating the water. They can vary greatly, for example, from values close to those for pure water to values that would more nearly describe thin paint.

A complete general theory of radiative transfer in a scattering-absorbing medium with specific application to large bodies of naturally occurring water has been developed by Preisendorfer (1960). This theory employs the radiometric terms, units, and symbols adopted by the Optical Society of America, Committee on Colorimetry (1944). Identical basic concepts and terms were adopted by IAPSO† in 1964

† International Association for Physical Sciences of the Ocean (of IUGG).

for use in optical oceanography (Jerlov, 1968). However, the symbols differ in the two disciplines, see Table 1.

Preisendorfer's theoretical work assumes the radiant energy to be monochromatic, and the symbols and relationships introduced herein should be so interpreted. However, it is often not possible to obtain monochromatic experimental data in a real situation, and many of the data included in this section are for a specified bandwidth with specified peak wavelength. Broadband data do not conform to theory as well as monochromatic data.

Figure 1 enumerates three basic measurements of optical oceanography, $\sigma(\theta)$, $N(z,\theta,\varphi)$, and the ratio N_r/N_0, and shows diagrammatically the optical properties

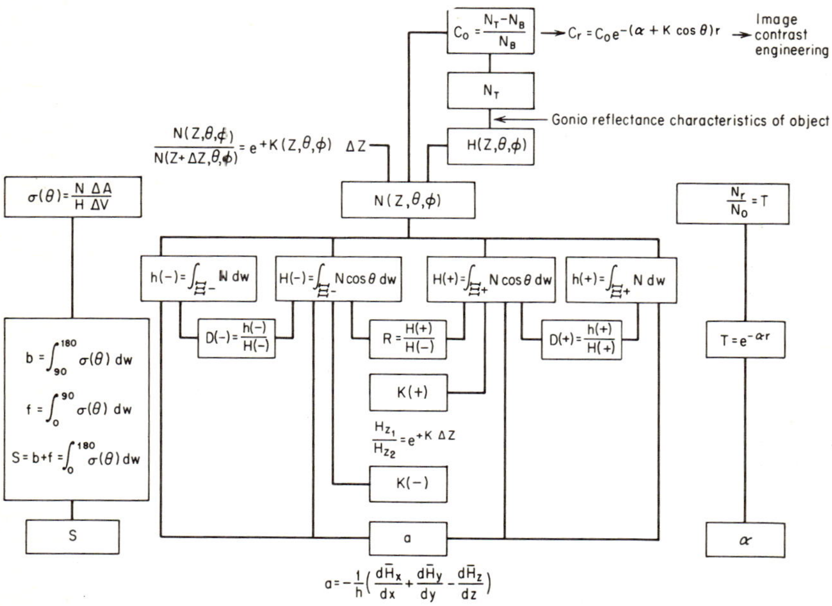

Fig. 1 Diagrammatic representation of the optical properties that can be computed from the three basic measurements. For explanation of symbols see Table 1. [*Reprinted with permission from R. J. Goulard (ed.), "Symposium on Interdisciplinary Aspects of Radiative Energy Transfer, Philadelphia, 1966." Copyright 1968, Pergamon Press, Inc.*]

that can be derived from them. The operational definition of each optical property is given in the figure.

$\int\Xi-$ = integration over the entire upper hemisphere
$\int\Xi+$ = integration over the entire lower hemisphere
$(-)$ = radiant energy moving downward through a horizontal plane
$(+)$ = radiant energy moving upward through a horizontal plane
$H(z,\theta,\varphi)$ = irradiance on a plane at any depth z and any orientation (θ,φ)
C_0 = inherent contrast of an object before attenuation through a water path
C_r = apparent contrast of an object after attenuation through a water path r
N_r = radiance of a beam of light measured after passing through path length r
N_0 = radiance of a beam of light at beginning of path = radiance in air
N_T = radiance of object seen from some distance d (more recently ${}_tN_0$ and ${}_tN_r$ have been used for inherent and apparent radiance, respectively, of an object)
N_B = radiance of infinite water background surrounding object seen from distance d (more recently ${}_bN_0$ and ${}_bN_r$ have been used for inherent and apparent radiance, respectively, of background)

z increases vertically downward. $z_2 > z_1$; hence $H_{z_1} > H_{z_2}$. $z + \Delta z > z$; hence $N(z,\theta,\varphi) > N(z + \Delta z, \theta,\varphi)$. When C_0 is calculated, N_T and N_B are inherent values of radiance, i.e., as seen when the observation distance is zero and before attenuation through the water path.

Numerical values of $\sigma(\theta)$ and N_r/N_0 and their derivatives are generally associated with a particular location and depth and specify the instantaneous optical properties of the water at that place and depth.

TABLE 1 Concepts and Symbols in Optical Oceanography†

	Unit	OSA radiometric symbol	Symbol adopted by IAPSO‡
Volume-scattering function	$(m \cdot r)^{-1}$	$\sigma(\theta)$	$\beta(\theta)$
Backward-scattering coefficient	m^{-1}	b	b_b
Forward-scattering coefficient	m^{-1}	f	b_f
Total-scattering coefficient	m^{-1}	s	b
Radiance at depth z in direction (θ,ϕ)	$W/(m^2)(\Omega)$	$N(z,\theta,\phi)$	L
Scalar irradiance, downwelling (–)	W/m^2	$h(z,-)$	
Upwelling (+)	W/m^2	$h(z,+)$	
Total scalar irradiance	W/m^2	$h(z)$	E_0
Irradiance on a horizontal plane, from			
upper hemisphere§	W/m^2	$H(z,-)$	E_d
From lower hemisphere§	W/m^2	$H(z,+)$	E_u
Distribution functions	Ratio	$D(z,\pm)$	
Reflectance function (irradiance reflectance)¶	Ratio	$R(z,-)$ or (R)	R
Diffuse-attenuation coefficient for			
irradiance, from upper hemisphere	m^{-1}	$K(z,-)$	
From lower hemisphere	m^{-1}	$K(z,+)$	
Absorption coefficient	m^{-1}	a	a
Ratio of the radiance at end of path r			
to input radiance	Ratio	N_r/N_0	L_r/L_0
Beam transmittance (for specified path length)	Ratio	T	T
Total-attenuation coefficient	m^{-1}	α	c

† Nonspecialists who undertake radiometric measurements generally apply the term *intensity* indiscriminately to all such measurements regardless of the geometry of collection or emission.

‡ International Association for Physical Sciences of the Ocean.

§ Called *downward* and *upward irradiance* by IAPSO.

¶ Called *irradiance ratio* by IAPSO.

Numerical values of $N(z,\theta,\varphi)$ describe the directional distribution (θ,φ) of radiance around a point in the underwater light field at some depth z. At great depths in a homogeneous body of water the distribution of radiance is independent of the surface lighting conditions and is a function only of α and $\sigma(\theta)$, that is, of the inherent attenuation and scattering properties of the water. At lesser depths the distribution of radiance around a point will be altered by the distribution of the light field at the boundary.

The optical properties $D(\pm)$, which are a rough one-parameter description of the shape of the radiance distribution, $R(-)$ (or R), which is the irradiance reflectance, and $K(\pm)$, which are the depth rate of decay of $H(\pm)$, respectively, all derived from $N(z,\theta,\varphi)$, will vary with the shape of the radiance distribution and hence near the surface, even in homogeneous water, will reflect to some extent the variation of radiance distribution with depth.

Numerical values of the optical properties derived from radiance distributions are also usually associated with a specific geographical location and depth and specify the instantaneous opti al properties of the water at that place and depth.

The association of hydrooptical data with a specified geographical location and depth is a matter of convenience. Numerical values of the optical properties depend upon the composition of the water body, which may be only superficially related to location or depth.

The water volumes employed for making the measurements indicated in Fig. 1 are always much larger than the size of the scattering particles. The circulation of particles through the measuring volume is therefore not ordinarily detected (thus, only statistical averages of the microscopic light field are detected). On the other hand, the volumes employed are generally small compared to many ocean features of interest in physical oceanography. These two facts determine the validity of applying radiative-transfer theory to light in the sea.

By heuristic arguments it can be shown that in a deep homogeneous body of natural water the volume-scattering function $\sigma(\theta)$ and the volume-attenuation coefficient α and all properties derived from them will be constants. Likewise the absorption coefficient a will be a constant. The radiance distribution $N(z,\theta,\varphi)$ and the integral quantities $h(\pm)$ and $H(\pm)$ will vary with depth in an exponential manner except near the boundaries of the medium. The diffuse-attenuation coefficients $K(\pm)$, reflectance function $R(-)$, and distribution functions $D(\pm)$, will also exhibit boundary effects but will asymptotically approach constant values at great depths, specifically the depth at which all radiance values experience the same rate of attenuation as a function of depth. In reality $K(\pm)$, $R(-)$, and $D(\pm)$ will be practically constant long before asymptotic depth is reached.

2. Volume-Scattering Function $\sigma(\theta)$ For monochromatic light, a single value of the volume-scattering function at angle θ can be obtained by application of the equation

$$\sigma(\theta) = \frac{N\,\Delta A}{H\,\Delta V} = \frac{J}{H\,\Delta V}$$

where H is the irradiance falling on the scattering volume ΔV and N (or J) is the measured radiance (or radiant intensity) at a specific angle θ due to scattering.

This equation can be rewritten

$$\sigma(\theta) = \frac{H_\theta r^2}{H\,\Delta V}$$

where H_θ is the irradiance on a photodetector positioned at angle θ to the forward direction and r is the distance between ΔV and the photodetector. This form of the equation is particularly useful for laboratory measurements, that is, r is small, and has been used extensively by physical chemists (McIntyre, 1964) who apply the term *Rayleigh ratio* to $\sigma(90°)$.

In situ instrumentation for measuring the volume-scattering function of natural water has been designed and constructed by Tyler (1963) and Tyler and Austin (1964) on the basis of a theoretical analysis by Pritchard and Elliott (1960). Laboratory measurements of contained water samples have generally made use of a Brice-Phoenix photometer or some modification of this instrument.

Kraut and Dandliker (1955) have measured the relative volume-scattering function for "clean" water from $\theta = 25.8°$ to $\theta = 135°$ and conclude that clean water is

TABLE 2 Absolute Spectral Values of the Rayleigh Ratio $\sigma(90°)$ for "Clean" Water and Unpolarized Incident Radiation (Kraut and Dandliker, 1955)

λ, nm	$\sigma(90°)$, cm^{-1}
365.0	6.80×10^{-6}
404.7	4.05×10^{-6}
435.8	2.89×10^{-6}
546.1	1.05×10^{-6}

almost a Rayleigh scatterer, i.e., very nearly conforms to the equation $\sigma(\theta) = 1 + \cos^2 \theta$. They also give spectral values of the Rayleigh ratio for clean water (see Table 2).

Bauer and Morel (1967) have combined their measurements with those of Jerlov (1961) and Morel (1965) (thereby mixing data for different wavelengths) to compile the relative scattering function, for the particles only, of seawater (Table 3).

TABLE 3 Relative Values of the Scattering Function $\sigma(\theta)$ and of $\sigma(\theta)$ sin θ, for the Particles Only, of Seawater (Bauer and Morel, 1967)†

θ, deg	$\sigma(\theta)/\sigma(90)$	$\frac{\sigma(\theta)\sin\theta}{\sigma(90)}$	λ	$\Delta\lambda$	Reference
1	34,400‡	585			
1.5	26,100	678			
2	19,100	668			
2.5	13,100	571			
3	8,500	442			
4	4,750	332			
5	2,990	260			
6	1,920	202	500	50	Bauer and Morel (1967)
7	1,300	154			
8	915	127			
9	670	104			
10	500§	87			
11	395	75.5			
12	314	65.3			
13	257	57.7			
14	224	54.2			
20	102	34.8	465	...	Jerlov (1961)
30	31.7	15.8			
37.5	15.7	9.55			
45	8.5	6.00			
60	3.41	2.95			
75	1.61	1.55			
90	1.00	1.00	546	...	Morel (1965)
105	0.74	0.715			
120	0.65	0.563			
135	0.65	0.460			
142.5	0.70	0.426			
150	0.87	0.435			

† Reprinted by authorization of the Centre International de la Recherche Scientifique, with permission from Georges Jobert, Directeur of *Annales de Geophysique.*

‡ Extrapolated.

§ $\sigma(10) = 402$ for the extreme of Jerlov's curve.

More recently Nyffeler (1970) has substantiated the validity of combining data for different wavelengths, as indicated above, by measuring the scattering function between 3 and 14° at three wavelengths (Fig. 2) and has also measured the scattering function between 1 and 15° with a laser beam at two wavelengths (Fig. 3).

Duntley (1963), Morel (1965), and Ivanoff and Morel (1964) have remarked on the quasi-invariable shape of the scattering function of natural water.

In Tables 4 to 7 optical bandwidth for the $\sigma(\theta)$ data was isolated by means of an EMI 9524-B multiplier phototube, a no. 61 Wratten filter, and a tungsten-iodine cycle lamp.

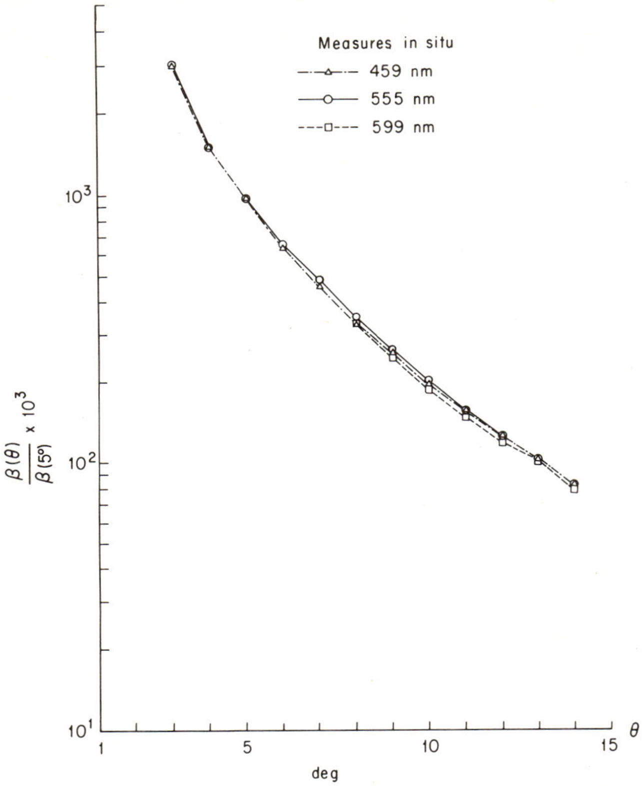

Fig. 2 The scattering function between 3 and 14° at three wavelengths. (*From F. Nyffeler, AGARD Conf. Proc.* 77, 1970, *by permission.*)

TABLE 4 Scattering-Function Data† (Reese and Tucker, 1970)
San Diego Bay, June 30, 1957; time, 2050; depth of measurement, 9 m; depth of water, 20 m

θ, deg	$\sigma(\theta)$, $(m \cdot rad)^{-1}$	θ, deg	$\sigma(\theta)$, $(m \cdot rad)^{-1}$	θ, deg	$\sigma(\theta)$, $(m \cdot rad)^{-1}$
16	1.00×10^{0}	64	1.97×10^{-2}	112	4.82×10^{-3}
19	6.79×10^{-1}	67	1.64×10^{-2}	115	4.68×10^{-3}
22	4.77×10^{-1}	70	1.35×10^{-2}	118	4.56×10^{-3}
25	3.47×10^{-1}	73	1.23×10^{-2}	121	4.49×10^{-3}
28	2.62×10^{-1}	76	1.13×10^{-2}	124	4.40×10^{-3}
31	2.01×10^{-1}	79	9.73×10^{-3}	127	4.42×10^{-3}
34	1.49×10^{-1}	82	8.85×10^{-3}	130	4.46×10^{-3}
37	1.18×10^{-1}	85	8.30×10^{-3}	133	4.56×10^{-3}
40	8.99×10^{-2}	88	7.62×10^{-3}	136	4.73×10^{-3}
43	6.93×10^{-2}	91	7.02×10^{-3}	139	4.86×10^{-3}
46	5.41×10^{-2}	94	6.47×10^{-3}	142	4.95×10^{-3}
49	4.20×10^{-2}	97	6.15×10^{-3}	145	5.14×10^{-3}
52	3.27×10^{-2}	100	5.73×10^{-3}	148	5.32×10^{-3}
55	2.90×10^{-2}	103	5.28×10^{-3}	151	5.51×10^{-3}
58	2.59×10^{-2}	106	5.00×10^{-3}	154	5.64×10^{-3}
61	2.30×10^{-2}	109	4.91×10^{-3}	157	5.55×10^{-3}

† Simultaneous measurement of beam transmittance $T \approx 0.13\ m^{-1}$ for a peak wavelength of 530 nm, with bandwidth at half-peak response equal to 54 nm.

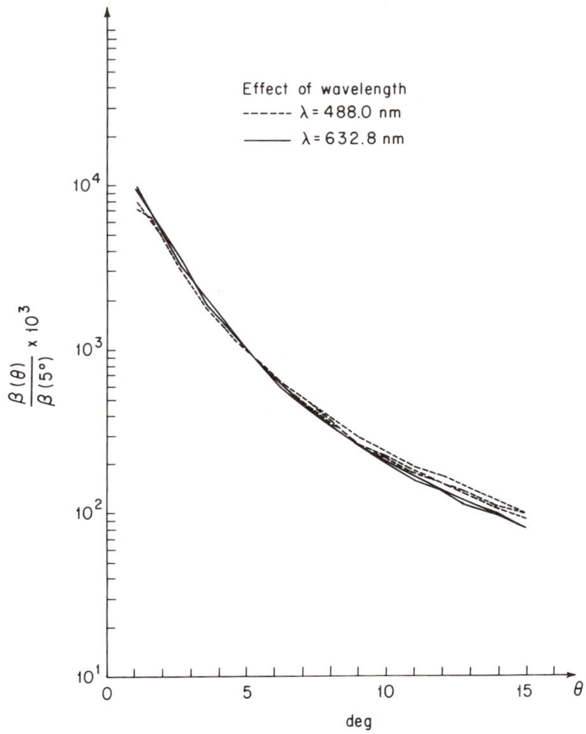

Fig. 3 The scattering function between 1 and 15° measured with a laser beam at two wavelengths. *(From F. Nyffeler, AGARD Conf. Proc.* 77, 1970, *by permission.)*

TABLE 5 Scattering-Function Data† (Reese and Tucker, 1970)
Lat. 32°16′N, lon. 117°20′W; July 21, 1966; depth of measurement, 29 m; depth of water, 1,280 m

θ, deg	σ(θ), (m · rad)$^{-1}$	θ, deg	σ(θ), (m · rad)$^{-1}$	θ, deg	σ(θ), (m · rad)$^{-1}$
16	1.23×10^{-1}	67	1.52×10^{-3}	118	4.95×10^{-4}
19	8.85×10^{-2}	70	1.36×10^{-3}	121	5.09×10^{-4}
22	5.32×10^{-2}	73	1.18×10^{-3}	124	5.28×10^{-4}
25	4.23×10^{-2}	76	1.06×10^{-3}	127	5.18×10^{-4}
28	2.95×10^{-2}	79	9.50×10^{-4}	130	5.28×10^{-4}
31	2.15×10^{-2}	82	8.21×10^{-4}	133	5.37×10^{-4}
34	1.48×10^{-2}	85	7.52×10^{-4}	136	5.41×10^{-4}
37	1.12×10^{-2}	88	6.93×10^{-4}	139	5.46×10^{-4}
40	9.17×10^{-3}	91	6.70×10^{-4}	142	5.64×10^{-4}
43	7.25×10^{-3}	94	6.38×10^{-4}	145	6.01×10^{-4}
46	5.87×10^{-3}	97	5.73×10^{-4}	148	6.24×10^{-4}
49	4.18×10^{-3}	100	5.64×10^{-4}	151	6.38×10^{-4}
52	3.40×10^{-3}	103	5.41×10^{-4}	154	6.47×10^{-4}
55	2.88×10^{-3}	106	5.14×10^{-4}	157	6.65×10^{-4}
58	2.38×10^{-3}	109	5.41×10^{-4}	160	6.70×10^{-4}
61	2.04×10^{-3}	112	5.00×10^{-4}		
64	1.76×10^{-3}	115	4.95×10^{-4}		

† Simultaneous measurement of beam transmittance $T \approx 0.68\ m^{-1}$ for a peak wavelength of 530 nm, with bandwidth at half-peak response equal to 54 nm.

TABLE 6 Scattering-Function Data† (Reese and Tucker, 1970)
Lat. 32°31′N, lon. 117°31′W; Aug. 23, 1967; depth of measurement, 38 m; depth of water, 1,200 m

θ, deg	$\sigma(\theta)$, $(m \cdot rad)^{-1}$	θ, deg	$\sigma(\theta)$, $(m \cdot rad)^{-1}$	θ, deg	$\sigma(\theta)$, $(m \cdot rad)^{-1}$
16	2.53×10^{-2}	67	4.01×10^{-4}	118	1.72×10^{-4}
19	1.63×10^{-2}	70	3.46×10^{-4}	121	1.73×10^{-4}
22	1.17×10^{-2}	73	3.34×10^{-4}	124	1.74×10^{-4}
25	8.49×10^{-3}	76	3.01×10^{-4}	127	1.73×10^{-4}
28	5.92×10^{-3}	79	2.69×10^{-4}	130	1.78×10^{-4}
31	4.27×10^{-3}	82	2.47×10^{-4}	133	1.86×10^{-4}
34	3.30×10^{-3}	85	2.40×10^{-4}	136	1.99×10^{-4}
37	2.60×10^{-3}	88	2.22×10^{-4}	139	2.06×10^{-4}
40	1.99×10^{-3}	91	2.02×10^{-4}	142	2.14×10^{-4}
43	1.55×10^{-3}	94	1.91×10^{-4}	145	2.17×10^{-4}
46	1.23×10^{-3}	97	1.81×10^{-4}	148	2.21×10^{-4}
49	9.36×10^{-4}	100	1.78×10^{-4}	151	2.28×10^{-4}
52	7.34×10^{-4}	103	1.77×10^{-4}	154	2.38×10^{-4}
55	6.51×10^{-4}	106	1.74×10^{-4}	157	2.38×10^{-4}
58	5.83×10^{-4}	109	1.71×10^{-4}	160	2.45×10^{-4}
61	5.41×10^{-4}	112	1.68×10^{-4}		
64	4.63×10^{-4}	115	1.67×10^{-4}		

† Simultaneous measurement of beam transmittance $T = 0.89\ m^{-1}$ (average) for a peak wavelength of 524 nm, with bandwidth at half-peak response equal to 52 nm.

TABLE 7 Scattering-Function Data† (Reese and Tucker, 1970)
Lat. 32°31′N, lon. 117°31′W; Aug. 23, 1967; depth of measurement, 737 m; depth of water, 1,200 m

θ, deg	$\sigma(\theta)$, $(m \cdot rad)^{-1}$	θ, deg	$\sigma(\theta)$, $(m \cdot rad)^{-1}$	θ, deg	$\sigma(\theta)$, $(m \cdot rad)^{-1}$
16	2.61×10^{-3}	67	1.45×10^{-4}	118	1.14×10^{-4}
19	1.90×10^{-3}	70	1.28×10^{-4}	121	1.17×10^{-4}
22	1.45×10^{-3}	73	1.24×10^{-4}	124	1.18×10^{-4}
25	1.03×10^{-3}	76	1.18×10^{-4}	127	1.22×10^{-4}
28	7.89×10^{-4}	79	1.12×10^{-4}	130	1.26×10^{-4}
31	6.84×10^{-4}	82	1.10×10^{-4}	133	1.32×10^{-4}
34	5.41×10^{-4}	85	1.09×10^{-4}	136	1.42×10^{-4}
37	4.40×10^{-4}	88	1.06×10^{-4}	139	1.51×10^{-4}
40	3.67×10^{-4}	91	1.05×10^{-4}	142	1.56×10^{-4}
43	3.17×10^{-4}	94	1.03×10^{-4}	145	1.58×10^{-4}
46	2.88×10^{-4}	97	1.02×10^{-4}	148	1.63×10^{-4}
49	2.43×10^{-4}	100	1.03×10^{-4}	151	1.68×10^{-4}
52	2.16×10^{-4}	103	1.03×10^{-4}	154	1.72×10^{-4}
55	2.05×10^{-4}	106	1.03×10^{-4}	157	1.76×10^{-4}
58	1.91×10^{-4}	109	1.07×10^{-4}	160	1.77×10^{-4}
61	1.77×10^{-4}	112	1.08×10^{-4}		
64	1.60×10^{-4}	115	1.10×10^{-4}		

† Simultaneous measurement of beam transmittance $T = 0.92\ m^{-1}$ (average) for a peak wavelength of 524 nm, with bandwidth at half-peak response equal to 52 nm.

3. Radiance Distribution $N(z,\theta,\varphi)$ Radiance in a direction θ, φ is given by

$$N(\theta,\varphi) = \frac{P}{\Omega A}$$

Different authors use different definitions for θ and φ and also measure z as positive toward the zenith or toward the nadir, depending on their sphere of interest (air or water) (Preisendorfer, 1964).

TABLE 8 Simultaneous Data for the Scattering Coefficient and Total-Attenuation Coefficient Obtained in Clean Ocean Water, July 1971, Tongue-of-the-Ocean, Bahama Islands†

	α, m^{-1}	S, m^{-1}	K, m^{-1}	Approx. depth of water, ft
July 9	0.192	0.095	....	42
July 13	0.151	0.029	0.064	3,500
July 14	0.165	0.035	....	3,500
Spectral bandwidth, $\Delta\lambda$, nm	95	74	Photopic	
Peak wavelength, nm	530	514	~555	

† T. Petzold, Scripps Institute of Oceanography, 1971, unpublished data.

Table 9 describes, in relative values, the radiance distribution at seven depths under clear, sunny conditions at Lake Pend Oreille, Idaho. Table 10 describes, in relative values, the radiance distribution under completely overcast conditions at the same lake. The data in Tables 9 and 10 were obtained with spatial resolution of 6.6° cone angle and optical bandwidth at half the peak transmittance of 64 nm, with peak transmittance of 480 nm.

Radiance-distribution data for specified planes or directions in ocean water have been published in graphical form by Jerlov and Fukuda (1960), Sasaki et al. (1960, 1962), and Sasaki (1964). The relative-radiance distribution within the same vertical plane, taken first in air and then at several depths below the surface of unruffled water, is shown in Fig. 4.

Curve A shows results of observations taken in air. From $\theta = 0°$ to $\theta = 74°$ the curve reports the radiance of the overcast sky; at $\theta = 74°$ the dark mountainous skyline is recorded, and this is followed by detail representing patches of dark rock or snow. From $\theta = 90°$ to $\theta = 102°$ the reflection of the mountains is recorded, followed by reflection of the sky disturbed by slight wave action. At 162° the instrument's own shadow image is recorded.

Curve B shows the radiance distribution in the same direction as that in curve A but from a depth of 2.5 cm below the water surface. In the zenith direction, B shows higher radiance than A because the space light contributed by 2.5 cm of water is greater than the absorption of this same path. (It is also possible that some change took place in the zenith overcast during the short time between curves A and B.) Optical compression of the sky tends to keep the radiance at a high level for the first 40°. The skyline itself is very much compressed and can no longer be resolved, but the edge of the Snell circle of light appears sharp and distinct at 48°. Beyond 48° the instrument measures the radiance of the short path to the surface plus a longer reflected path that looks down into the water at an angle equal to $180° - \theta$. This accounts for the changes in slope of the curve between 48 and 88°. Beyond 88° the path radiance is due entirely to space light from one direction only, which decreases steadily as the tilt angle increases to 180°.

Curve C shows the radiance distribution in the same direction as curves A and B but from a depth of 3.05 m. Some detail can still be observed. The steep slope between 40 and 48° is the remaining change at the edge of the Snell

TABLE 9 Radiance Distribution under Clear Sunny Sky (Tyler, 1960)†

Tilt angle θ, deg	Azimuth angle Φ, deg									
	0	20	40	60	80	100	120	140	160	180
	Depth, 4.24 m; Sun Altitude, 56.6°									
0	204,000	204,000	204,000	204,000	204,000	204,000	204,000	204,000	204,000	204,000
10	541,000	481,000	374,000	286,000	220,000	174,000	139,000	119,000	108,000	104,000
20	4,300,000	1,320,000	545,000	277,000	168,000	118,000	93,000	79,600	72,100	69,100
30	7,980,000	1,100,000	401,000	198,000	123,000	87,200	69,400	59,700	54,500	52,400
40	573,000	427,000	234,000	135,000	90,500	68,300	56,300	48,700	44,100	42,300
50	207,000	164,000	106,000	69,500	49,300	37,700	31,000	26,500	23,800	23,200
60	114,000	91,800	66,400	47,700	35,300	27,300	22,300	19,000	17,100	16,600
70	61,300	55,800	45,100	34,300	26,700	21,200	17,500	14,800	13,200	12,400
80	41,500	38,200	31,500	25,100	20,100	16,100	12,900	11,300	10,000	9,460
90	26,900	25,300	21,500	17,600	14,200	11,700	9,840	8,560	7,820	7,480
100	17,000	16,000	13,900	11,700	9,940	8,590	7,620	6,860	6,390	6,140
110	11,200	10,700	9,280	8,060	7,180	6,490	5,990	5,600	5,340	5,220
120	7,430	7,170	6,630	6,020	5,590	5,250	5,000	4,800	4,670	4,590
130	5,360	5,220	4,950	4,710	4,520	4,320	4,180	4,040	4,010	3,990
140	4,230	4,170	4,040	3,960	3,850	3,780	3,710	3,650	3,620	3,600
150	3,570	3,560	3,520	3,480	3,440	3,380	3,340	3,300	3,280	3,260
160	3,250	3,250	3,250	3,250	3,250	3,240	3,240	3,240	3,230	3,230
170	3,110	3,110	3,110	3,110	3,110	3,110	3,110	3,110	3,110	3,110
180	3,050	3,050	3,050	3,050	3,050	3,050	3,050	3,050	3,050	3,050
	Depth, 10.4 m; Sun Altitude, 56.6°									
0	127,000	127,000	127,000	127,000	127,000	127,000	127,000	127,000	127,000	127,000
10	274,000	258,000	215,000	166,000	129,000	103,000	86,800	78,300	73,300	71,400
20	1,970,000	610,000	284,000	158,000	101,000	72,000	57,300	49,600	46,500	45,400
30	2,540,000	472,000	208,000	110,000	68,500	49,200	39,000	33,400	30,900	30,300
40	298,000	207,000	118,000	71,000	46,900	34,400	27,200	22,900	20,300	19,200
50	118,000	104,000	69,900	45,900	31,300	23,000	17,700	14,800	13,500	13,000
60	55,900	51,400	41,000	29,400	20,400	14,700	11,600	9,670	8,700	8,360
70	32,500	30,100	24,000	17,900	13,200	10,100	8,150	6,800	6,100	5,860
80	18,100	17,200	14,700	11,400	8,740	6,900	5,660	4,800	4,320	4,150
90	10,600	10,100	8,700	7,020	5,630	4,640	3,960	3,530	3,320	3,270
100	6,500	6,160	5,540	4,800	4,000	3,380	3,000	2,730	2,590	2,540
110	4,320	4,120	3,750	3,320	2,880	2,500	2,260	2,110	2,030	2,010

120	2,940	2,830	2,640	2,440	2,220	1,970	1,840	1,750	1,700	1,690
130	2,020	1,980	1,930	1,860	1,770	1,720	1,630	1,570	1,540	1,530
140	1,630	1,620	1,610	1,570	1,540	1,500	1,450	1,420	1,390	1,390
150	1,410	1,410	1,400	1,370	1,350	1,330	1,310	1,280	1,260	1,260
160	1,240	1,240	1,240	1,240	1,240	1,240	1,240	1,240	1,230	1,230
170	1,180	1,180	1,180	1,180	1,180	1,180	1,180	1,180	1,180	1,180
180	1,180	1,180	1,180	1,180	1,180	1,180	1,180	1,180	1,180	1,180
Depth, 16.6 m; Sun Altitude, 56.6°										
0	59,100	59,100	59,100	59,100	59,100	59,100	59,100	59,100	59,100	59,100
10	121,000	111,000	93,100	74,100	59,900	48,700	41,000	36,500	34,000	33,200
20	350,000	183,000	109,000	71,500	49,400	36,900	29,100	24,400	21,900	21,200
30	385,000	169,000	86,200	53,000	36,300	26,700	20,400	16,700	14,600	13,800
40	88,200	75,900	53,500	35,700	24,500	17,900	13,600	10,300	9,380	8,780
50	45,300	41,000	31,700	21,700	14,900	11,100	8,550	6,860	5,880	5,490
60	24,400	22,500	17,400	12,500	9,030	6,950	5,500	4,590	3,960	3,670
70	12,400	11,200	9,100	7,230	5,670	4,520	3,460	3,040	2,670	2,530
80	6,750	6,360	5,490	4,510	3,640	2,990	2,500	2,130	1,870	1,750
90	4,050	3,770	3,280	2,790	2,340	1,970	1,660	1,440	1,310	1,250
100	2,360	2,280	2,070	1,800	1,540	1,330	1,180	1,070	990	964
110	1,520	1,470	1,350	1,210	1,090	982	891	832	796	778
120	1,010	995	946	884	816	758	711	684	666	656
130	740	732	717	685	658	633	606	584	572	564
140	595	587	581	566	550	536	522	512	510	510
150	502	499	497	495	486	480	472	465	462	462
160	451	451	451	450	447	443	439	434	433	433
170	422	422	422	422	422	422	422	422	422	422
180	418	418	418	418	418	418	418	418	418	418
Depth, 29.0 m; Sun Altitude, 56.6°										
0	9,630	9,630	9,630	9,630	9,630	9,630	9,630	9,630	9,630	9,630
10	14,300	13,300	12,000	10,600	9,460	8,410	7,540	6,850	6,330	6,080
20	22,100	16,300	12,100	9,530	7,730	6,440	5,490	4,750	4,220	3,980
30	20,000	13,500	9,280	6,990	5,560	4,620	3,900	3,320	2,880	2,680
40	9,970	8,090	6,080	4,740	3,820	3,110	2,590	2,180	1,820	1,660
50	5,110	4,610	3,750	3,040	2,490	2,030	1,650	1,350	1,100	993
60	2,780	2,490	2,070	1,730	1,440	1,210	1,010	842	708	651
70	1,440	1,330	1,140	986	839	719	618	530	455	425
80	799	739	657	579	505	444	389	343	303	281
90	470	447	412	365	323	287	254	225	202	191

† Originally published by the University of California Press; reprinted by permission of The Regents of the University of California.

TABLE 9 Radiance Distribution under Clear Sunny Sky (Tyler, 1970) (Continued)

Tilt angle θ, deg	Azimuth angle Φ, deg									
	0	20	40	60	80	100	120	140	160	180
	Depth, 29.0 m; Sun Altitude, 56.6° (*continued*)									
100	293	278	256	231	210	192	175	159	148	143
110	191	185	172	159	150	138	129	121	113	111
120	135	130	123	116	110	105	98.9	94.2	90.0	88.6
130	99.2	97.0	93.6	90.1	87.3	84.1	80.5	78.3	76.4	75.5
140	79.6	78.4	76.4	74.6	73.4	72.1	70.5	69.1	67.8	66.5
150	68.3	67.6	66.3	65.8	64.7	63.7	62.8	61.8	61.3	60.6
160	60.4	59.9	59.6	59.1	58.9	58.5	58.1	58.0	57.4	57.3
170	56.2	56.2	56.2	56.2	56.2	56.2	56.2	56.2	56.2	56.2
180	55.2	55.2	55.2	55.2	55.2	55.2	55.2	55.2	55.2	55.2
	Depth, 41.3 m; Sun Altitude, 56.6°									
0	1,380	1,380	1,380	1,380	1,380	1,380	1,380	1,380	1,380	1,380
10	1,650	1,600	1,510	1,410	1,320	1,250	1,180	1,120	1,070	1,050
20	1,680	1,580	1,420	1,270	1,120	1,010	910	822	750	721
30	1,260	1,180	1,049	921	811	716	631	557	496	474
40	859	800	700	614	533	465	408	361	322	306
50	510	478	426	376	333	294	259	229	202	191
60	300	285	254	222	194	170	149	132	117	111
70	165	156	143	126	113	101	90.2	80.6	73.2	70.8
80	88.9	84.9	79.1	72.5	66.6	60.8	55.7	50.9	47.3	45.2
90	51.3	50.4	47.7	44.9	41.7	38.6	36.0	33.4	31.4	30.4
100	33.5	32.5	30.7	28.9	27.2	25.6	24.1	22.8	21.7	20.9
110	21.8	21.5	20.9	19.9	19.0	18.1	17.4	16.7	16.0	15.7
120	15.8	15.8	15.4	14.9	14.3	13.8	13.3	12.9	12.5	12.3
130	12.3	12.2	12.0	11.7	11.5	11.2	11.0	10.8	10.6	10.5
140	10.2	10.1	10.0	9.90	9.78	9.65	9.57	9.45	9.33	9.02
150	8.75	8.73	8.66	8.63	8.57	8.53	8.46	8.40	8.35	8.34
160	7.88	7.88	7.88	7.88	7.87	7.87	7.87	7.87	7.86	7.86
170	7.53	7.53	7.53	7.53	7.53	7.53	7.53	7.53	7.53	7.53
180	7.43	7.43	7.43	7.43	7.43	7.43	7.43	7.43	7.43	7.43

Depth, 53.7 m; Sun Altitude, 56.6°										
0	202	202	202	202	202	202	202	202	202	202
10	219	218	212	205	194	184	173	163	157	155
20	194	192	180	168	156	143	132	123	116	113
30	139	137	130	120	110	98.8	89.8	82.7	77.6	75.5
40	88.7	87.2	82.9	77.0	70.7	63.9	58.2	53.3	49.9	48.6
50	52.7	52.3	50.1	47.1	43.7	39.6	36.2	33.4	31.2	30.0
60	32.0	31.2	29.6	27.4	25.1	23.1	21.3	19.9	18.8	18.4
70	17.8	17.6	16.9	16.0	14.9	13.8	12.8	12.0	11.4	11.2
80	10.3	10.2	9.87	9.32	8.80	8.28	7.82	7.48	7.14	7.02
90	6.27	6.17	5.98	5.70	5.50	5.27	5.06	4.87	4.67	4.58
100	4.00	3.98	3.91	3.78	3.63	3.49	3.37	3.26	3.18	3.17
110	2.67	2.65	2.63	2.60	2.54	2.46	2.41	2.34	2.31	2.30
120	1.99	1.98	1.97	1.94	1.91	1.86	1.82	1.79	1.76	1.75
130	1.54	1.53	1.53	1.52	1.51	1.48	1.47	1.46	1.45	1.44
140	1.30	1.30	1.30	1.29	1.28	1.28	1.28	1.27	1.27	1.27
150	1.16	1.16	1.16	1.15	1.15	1.15	1.15	1.15	1.15	1.15
160	1.05	1.05	1.05	1.05	1.05	1.05	1.05	1.05	1.05	1.05
170	1.00	1.00	1.00	1.00	1.00	1.00	1.00	1.00	1.00	1.00
180	0.990	0.990	0.990	0.990	0.990	0.990	0.990	0.990	0.990	0.990
Depth, 66.1 m; Sun Altitude, 56.6°										
0	28.8	28.8	28.8	28.8	28.8	28.8	28.8	28.8	28.8	28.8
10	29.8	29.6	29.2	28.4	27.7	26.9	26.1	25.3	24.5	24.1
20	26.1	25.7	24.7	23.8	22.8	21.6	20.6	19.6	18.9	18.5
30	19.7	19.3	18.5	17.5	16.5	15.6	14.6	13.8	13.1	12.8
40	12.7	12.4	11.9	11.3	10.6	10.1	9.44	8.87	8.41	8.19
50	7.64	7.54	7.20	6.83	6.46	6.09	5.70	5.40	5.14	5.01
60	4.43	4.34	4.18	3.98	3.78	3.58	3.38	3.20	3.08	2.98
70	2.46	2.44	2.37	2.29	2.20	2.11	2.01	1.92	1.86	1.81
80	1.43	1.42	1.39	1.35	1.29	1.25	1.19	1.14	1.11	1.09
90	0.858	0.855	0.839	0.817	0.788	0.757	0.727	0.700	0.677	0.661
100	0.531	0.525	0.518	0.506	0.496	0.482	0.468	0.453	0.445	0.439
110	0.357	0.355	0.349	0.342	0.334	0.325	0.317	0.311	0.305	0.301
120	0.252	0.250	0.247	0.244	0.241	0.239	0.235	0.233	0.231	0.226
130	0.186	0.184	0.184	0.181	0.180	0.179	0.177	0.175	0.173	0.172
140	0.146	0.145	0.145	0.144	0.144	0.143	0.142	0.141	0.141	0.139
150	0.124	0.124	0.124	0.122	0.122	0.121	0.121	0.120	0.120	0.120
160	0.108	0.108	0.108	0.108	0.108	0.108	0.108	0.108	0.108	0.108
170	0.100	0.100	0.100	0.100	0.100	0.100	0.100	0.100	0.100	0.100
180	0.0975	0.0975	0.0975	0.0975	0.0975	0.0975	0.0975	0.0975	0.0975	0.0975

TABLE 10 Radiance Distribution, Overcast Sky (Tyler, 1970)†

Tilt angle θ, deg	Azimuth angle Φ, deg									
	0	20	40	60	80	100	120	140	160	180
	Depth, 6.1 m; Sun Altitude, 40.0°									
0	321,000	321,000	321,000	321,000	321,000	321,000	321,000	321,000	321,000	321,000
10	343,000	342,000	338,000	332,000	325,000	319,000	311,000	306,000	302,000	300,000
20	327,000	326,000	321,000	312,000	302,000	290,000	281,000	272,000	267,000	265,000
30	277,000	275,000	268,000	258,000	243,000	229,000	216,000	207,000	202,000	200,000
40	199,000	196,000	189,000	179,000	167,000	152,000	141,000	132,000	127,000	126,000
50	96,800	95,700	92,500	87,300	81,700	74,900	69,500	65,900	63,600	62,900
60	53,500	52,800	51,100	48,800	45,600	42,800	39,700	37,500	36,100	35,400
70	32,500	32,100	31,300	30,000	28,200	26,100	24,100	22,800	21,900	21,600
80	20,000	19,700	19,100	18,300	17,100	15,800	14,700	13,900	13,500	13,400
90	12,600	12,400	12,000	11,500	10,800	9,950	9,170	8,710	8,380	8,310
100	8,060	7,980	7,780	7,490	7,080	6,640	6,190	5,860	5,670	5,620
110	5,620	5,580	5,450	5,250	4,970	4,670	4,390	4,180	4,050	4,010
120	4,130	4,080	3,990	3,850	3,700	3,500	3,330	3,200	3,120	3,090
130	3,200	3,190	3,150	3,070	2,970	2,860	2,740	2,630	2,570	2,550
140	2,650	2,640	2,590	2,540	2,480	2,390	2,300	2,240	2,190	2,180
150	2,300	2,290	2,280	2,250	2,200	2,150	2,090	2,040	2,000	1,980
160	2,070	2,070	2,060	2,030	2,000	1,960	1,910	1,890	1,890	1,880
170	1,860	1,860	1,860	1,860	1,860	1,860	1,860	1,860	1,860	1,860
180	1,830	1,830	1,830	1,830	1,830	1,830	1,830	1,830	1,830	1,830
	Depth, 18.3 m; Sun Altitude, 40.0°									
0	26,800	26,800	26,800	26,800	26,800	26,800	26,800	26,800	26,800	26,800
10	27,500	27,400	27,200	26,900	26,600	26,000	25,200	24,600	24,100	23,900
20	23,600	23,400	23,000	22,500	21,900	21,000	20,100	19,400	18,800	18,600
30	17,500	17,400	17,200	16,800	16,200	15,400	14,400	13,400	12,900	12,600
40	11,900	11,800	11,600	11,100	10,600	9,800	9,080	8,460	8,040	7,860
50	7,520	7,460	7,310	7,090	6,750	6,250	5,710	5,240	4,880	4,720
60	4,600	4,540	4,440	4,280	4,070	3,740	3,440	3,230	3,050	2,970

70	2,780	2,750	2,690	2,590	2,450	2,300	2,130	1,990	1,910	1,860
80	1,650	1,640	1,590	1,550	1,470	1,380	1,280	1,220	1,160	1,150
90	1,040	1,020	995	948	892	830	784	745	722	713
100	645	638	627	603	583	557	533	514	502	500
110	454	449	439	426	409	391	378	366	357	356
120	334	331	324	315	302	289	280	272	268	265
130	258	255	250	243	235	228	218	214	209	209
140	206	205	202	198	193	188	185	182	179	179
150	176	175	174	170	167	164	162	161	161	160
160	156	156	156	155	152	151	150	149	147	147
170	145	145	145	145	145	145	145	145	145	145
180	141	141	141	141	141	141	141	141	141	141

Depth, 30.5 m; Sun Altitude, 40.0°

0	2,430	2,430	2,430	2,430	2,430	2,430	2,430	2,430	2,430	2,430
10	2,500	2,490	2,470	2,430	2,380	2,310	2,210	2,120	2,070	2,030
20	2,190	2,170	2,130	2,080	2,010	1,910	1,790	1,680	1,600	1,570
30	1,690	1,670	1,640	1,600	1,550	1,420	1,300	1,180	1,090	1,050
40	1,150	1,130	1,090	1,050	998	927	840	757	702	676
50	688	683	670	649	629	596	544	498	466	453
60	413	409	400	389	375	353	332	302	284	275
70	254	252	247	240	231	217	203	189	178	173
80	153	151	148	144	136	130	121	114	110	108
90	93.9	93.4	91.3	88.6	85.2	81.1	75.9	72.3	69.7	68.4
100	61.2	60.7	59.7	57.9	56.0	53.4	50.6	48.3	46.4	45.3
110	42.9	42.7	41.8	40.7	39.3	37.8	35.8	34.2	33.3	32.7
120	31.1	30.9	30.5	29.8	29.1	28.1	27.2	26.3	25.8	25.4
130	23.8	23.7	23.5	23.1	22.8	22.2	21.6	21.2	21.0	20.9
140	20.4	20.3	20.0	19.8	19.5	19.0	18.4	18.1	17.6	17.5
150	18.1	18.0	17.8	17.5	17.3	16.9	16.7	16.5	16.3	16.3
160	16.2	16.1	16.1	16.1	16.0	15.8	15.5	15.3	15.2	15.2
170	14.9	14.9	14.9	14.9	14.9	14.9	14.9	14.9	14.9	14.9
180	14.8	14.8	14.8	14.8	14.8	14.8	14.8	14.8	14.8	14.8

† Originally published by the University of California Press; reprinted by permission of The Regents of the University of California.

TABLE 10 Radiance Distribution, Overcast Sky (Tyler, 1970) (Continued)

Tilt angle θ, deg	Azimuth angle Φ, deg									
	0	20	40	60	80	100	120	140	160	180
	Depth, 42.8 m; Sun Altitude, 40.0°									
0	250	250	250	250	250	250	250	250	250	250
10	254	252	250	247	242	239	233	227	223	221
20	229	226	223	216	209	201	190	182	176	174
30	175	173	167	161	155	146	138	131	125	123
40	119	117	113	109	104	99.1	92.5	87.4	83.9	82.6
50	76.0	74.9	73.4	70.4	67.0	63.6	59.8	56.9	54.7	53.8
60	45.7	45.3	44.0	42.6	40.8	39.1	38.2	36.6	35.4	34.9
70	27.6	27.2	26.6	25.9	24.9	24.0	22.9	22.0	21.2	20.9
80	16.8	16.6	16.2	15.8	15.4	14.7	14.1	13.6	13.2	13.1
90	10.5	10.4	10.1	9.87	9.55	9.21	8.87	8.58	8.38	8.20
100	6.82	6.78	6.61	6.48	6.28	6.06	5.82	5.61	5.40	5.35
110	4.66	4.63	4.55	4.46	4.36	4.24	4.08	3.95	3.86	3.82
120	3.46	3.42	3.39	3.33	3.27	3.16	3.08	3.02	2.96	2.94
130	2.67	2.66	2.65	2.61	2.58	2.52	2.45	2.39	2.34	2.32
140	2.22	2.22	2.20	2.16	2.13	2.10	2.06	2.01	1.99	1.97
150	1.93	1.93	1.93	1.92	1.91	1.89	1.88	1.86	1.83	1.82
160	1.76	1.76	1.76	1.76	1.76	1.76	1.76	1.76	1.76	1.76
170	1.67	1.67	1.67	1.67	1.67	1.67	1.67	1.67	1.67	1.67
180	1.65	1.65	1.65	1.65	1.65	1.65	1.65	1.65	1.65	1.65

Depth, 55.0 m; Sun Altitude, 40.0°										
0	29.6	29.6	29.6	29.6	29.6	29.6	29.6	29.6	29.6	29.6
10	29.9	29.7	29.3	28.8	28.2	27.6	27.0	26.5	25.9	25.7
20	24.7	24.6	24.0	23.4	22.8	22.2	21.5	20.9	20.5	20.4
30	18.8	18.5	18.1	17.7	17.3	16.6	16.0	15.4	14.9	14.6
40	13.0	12.8	12.6	12.3	11.9	11.6	11.2	10.8	10.5	10.4
50	8.11	8.00	7.86	7.72	7.54	7.32	7.07	6.86	6.69	6.61
60	4.99	4.96	4.85	4.76	4.64	4.55	4.42	4.28	4.19	4.17
70	3.00	2.98	2.93	2.87	2.81	2.75	2.69	2.63	2.59	2.56
80	1.90	1.88	1.84	1.82	1.78	1.75	1.71	1.67	1.65	1.63
90	1.19	1.18	1.16	1.15	1.12	1.09	1.07	1.05	1.04	1.03
100	0.765	0.761	0.749	0.736	0.723	0.708	0.689	0.678	0.665	0.661
110	0.512	0.509	0.503	0.495	0.488	0.481	0.469	0.460	0.452	0.449
120	0.363	0.360	0.358	0.355	0.352	0.346	0.341	0.337	0.333	0.331
130	0.273	0.272	0.271	0.268	0.266	0.264	0.259	0.256	0.254	0.253
140	0.215	0.215	0.214	0.213	0.210	0.208	0.206	0.204	0.203	0.203
150	0.181	0.181	0.181	0.181	0.181	0.181	0.181	0.181	0.181	0.181
160	0.158	0.158	0.158	0.158	0.158	0.158	0.158	0.158	0.158	0.158
170	0.146	0.146	0.146	0.146	0.146	0.146	0.146	0.146	0.146	0.146
180	0.143	0.143	0.143	0.143	0.143	0.143	0.143	0.143	0.143	0.143

circle of light. The dark band at 60° exhibited in B is nearly obscured by space light; in C, between 50 and 80°, it can just be seen together with its brighter rim.

Curve D shows the radiance distribution in the same direction as the other curves but from a depth of 6.1 m. At this depth all fine detail is lost, leaving only the gross changes in radiance between zenith and nadir settings.

4. Radiance *K* Values A radiance K value for a specified direction can be calculated from values of radiance at two or more depths, as indicated in Figs. 1 and 5. They specify the rate of loss or gain of radiance for specific directions in the underwater light field (see Tables **11** and **12**).

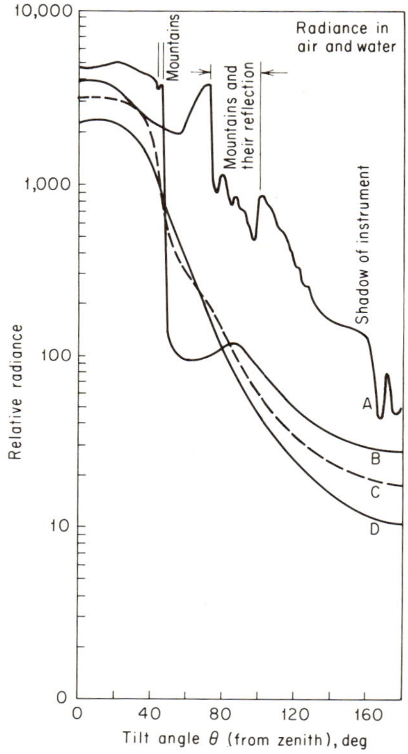

Fig. 4 Relative-radiance distribution within the same vertical plane, in air and at several depths (see text discussion). [*Tyler* (1958).]

Relative radiance in three fixed directions as a function of depth is shown in Fig. 5.

5. Transmittance of Collimated Light per Unit Length From Fig. 1, $N_r/N_0 = T = e^{-\alpha R}$, where N_r is the measured radiance after passing through a path length r and N_0 is the radiance input at the beginning of the path.

Transmittance is most accurately measured with a cylindrically restricted beam of artificial light of small cross section so arranged that the aperture stop of the projection system is equal in size to its exit pupil. The aperture stop of the detector system, located at the exit pupil of the projector, is then made slightly larger than the exit pupil of the projection system. This arrangement obviates the index-of-refraction correction that might otherwise be necessary in comparing measurements in air and in water and minimizes the amount of forward-scattered radiant energy collected by the detector. Authors generally measure T and report the total attenuation coefficient α. [Some authors use base 10 loga-

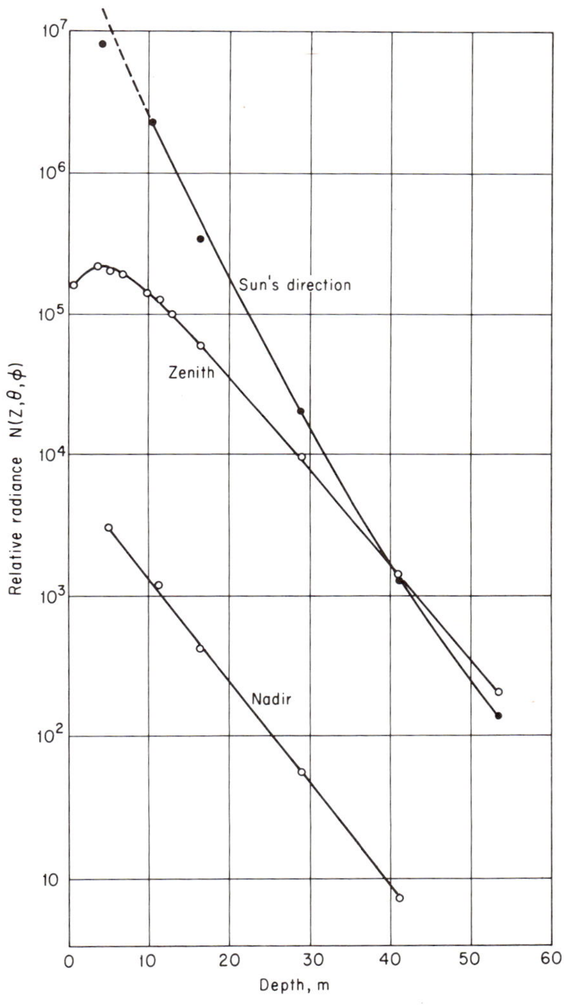

Fig. 5 The slopes of these curves specify, at any point, the radiance K value for the direction indicated. The transformation of the complex radiance distribution found near the surface to its asymptotic shape requires enhancement of the radiance in some directions and attenuation in others. [*Tyler* (1960). *Originally published by the University of California Press; reprinted by permission of The Regents of the University of California.*]

rithms, e.g., Joseph (1961), who also uses the ratio of the radiance transmitted per meter of ocean water divided by the radiance transmitted per meter by the clearest water observed as his value of T.]

The most recent laboratory determinations of the spectral transmittance of various

TABLE 11 Radiance K Values, Clear Sunny Sky (Tyler, 1970)†

Tilt angle θ, deg	Azimuth angle Φ, deg									
	0	20	40	60	80	100	120	140	160	180
	Depth, 7.33 m; Sun Altitude, 56.6°									
0	0.0764	0.0764	0.0764	0.0764	0.0764	0.0764	0.0764	0.0764	0.0764	0.0764
10	0.110	0.101	0.0896	0.0879	0.0863	0.0846	0.0846	0.0676	0.0627	0.0607
20	0.127	0.125	0.105	0.0909	0.0824	0.0804	0.0784	0.0764	0.0712	0.0679
30	0.185	0.136	0.106	0.0958	0.0942	0.0925	0.0932	0.0942	0.0919	0.0886
40	0.106	0.117	0.112	0.105	0.106	0.111	0.117	0.122	0.126	0.131
50	0.0912	0.0751	0.0682	0.0679	0.0738	0.0817	0.0912	0.0945	0.0919	0.0942
60	0.115	0.0948	0.0791	0.0794	0.0906	0.100	0.106	0.110	0.109	0.111
70	0.104	0.101	0.103	0.106	0.114	0.119	0.124	0.125	0.125	0.121
80	0.134	0.130	0.124	0.129	0.135	0.137	0.133	0.138	0.136	0.133
90	0.150	0.149	0.147	0.149	0.150	0.149	0.147	0.143	0.138	0.134
100	0.156	0.155	0.150	0.144	0.148	0.151	0.151	0.149	0.147	0.144
110	0.155	0.154	0.147	0.144	0.148	0.155	0.157	0.158	0.156	0.155
120	0.154	0.152	0.150	0.147	0.150	0.158	0.161	0.163	0.163	0.161
130	0.158	0.157	0.153	0.151	0.152	0.150	0.153	0.153	0.155	0.155
140	0.154	0.153	0.150	0.150	0.149	0.151	0.152	0.153	0.155	0.155
150	0.152	0.151	0.151	0.151	0.152	0.152	0.153	0.153	0.155	0.154
160	0.157	0.157	0.157	0.157	0.157	0.157	0.157	0.157	0.156	0.156
170	0.156	0.156	0.156	0.156	0.156	0.156	0.156	0.156	0.156	0.156
180	0.154	0.154	0.154	0.154	0.154	0.154	0.154	0.154	0.154	0.154
	Depth, 10.42 m; Sun Altitude, 56.6°									
0	0.100	0.100	0.100	0.100	0.100	0.100	0.100	0.100	0.100	0.100
10	0.121	0.119	0.113	0.110	0.106	0.103	0.0991	0.0958	0.0939	0.0925
20	0.197	0.160	0.131	0.110	0.0994	0.0945	0.0942	0.0958	0.0968	0.0961
30	0.246	0.152	0.125	0.107	0.0991	0.0965	0.0994	0.103	0.107	0.108
40	0.151	0.140	0.121	0.108	0.106	0.109	0.115	0.126	0.125	0.127
50	0.122	0.114	0.0988	0.0951	0.0971	0.101	0.105	0.110	0.114	0.177
60	0.125	0.115	0.110	0.110	0.111	0.112	0.114	0.119	0.119	0.122
70	0.131	0.131	0.131	0.127	0.126	0.125	0.131	0.129	0.130	0.129
80	0.148	0.146	0.142	0.140	0.138	0.136	0.133	0.136	0.137	0.137
90	0.154	0.155	0.153	0.150	0.146	0.144	0.144	0.144	0.146	0.145
100	0.160	0.158	0.155	0.151	0.151	0.151	0.151	0.151	0.151	0.150
110	0.162	0.160	0.157	0.155	0.154	0.153	0.155	0.155	0.154	0.154

120	0.163	0.160	0.158	0.156	0.157	0.157	0.158	0.158	0.157	0.157
130	0.161	0.159	0.157	0.157	0.157	0.156	0.157	0.157	0.158	0.158
140	0.160	0.159	0.158	0.158	0.158	0.159	0.159	0.159	0.159	0.158
150	0.159	0.159	0.159	0.158	0.159	0.158	0.159	0.159	0.159	0.159
160	0.160	0.161	0.161	0.159	0.161	0.161	0.162	0.163	0.163	0.163
170	0.162	0.162	0.162	0.162	0.162	0.162	0.162	0.162	0.162	0.162
180	0.162	0.162	0.162	0.162	0.162	0.162	0.162	0.162	0.162	0.162
Depth, 16.58 m; Sun Altitude, 56.6°										
0	0.123	0.123	0.123	0.123	0.123	0.123	0.123	0.123	0.123	0.123
10	0.147	0.145	0.139	0.133	0.127	0.123	0.118	0.115	0.115	0.115
20	0.214	0.178	0.154	0.136	0.125	0.117	0.115	0.114	0.115	0.115
30	0.243	0.178	0.152	0.136	0.125	0.119	0.116	0.117	0.119	0.120
40	0.164	0.160	0.148	0.136	0.128	0.125	0.125	0.126	0.129	0.131
50	0.150	0.145	0.136	0.127	0.121	0.118	0.119	0.121	0.125	0.128
60	0.152	0.147	0.141	0.135	0.130	0.127	0.124	0.126	0.129	0.131
70	0.153	0.152	0.150	0.144	0.140	0.137	0.136	0.135	0.136	0.137
80	0.160	0.160	0.157	0.153	0.149	0.146	0.142	0.142	0.142	0.142
90	0.164	0.164	0.160	0.157	0.153	0.150	0.148	0.147	0.148	0.149
100	0.165	0.164	0.162	0.159	0.156	0.154	0.153	0.152	0.152	0.152
110	0.165	0.164	0.162	0.152	0.156	0.156	0.156	0.155	0.156	0.156
120	0.162	0.162	0.162	0.160	0.159	0.158	0.159	0.159	0.160	0.160
130	0.161	0.161	0.161	0.160	0.160	0.160	0.160	0.160	0.160	0.160
140	0.161	0.161	0.161	0.161	0.161	0.160	0.160	0.161	0.161	0.161
150	0.161	0.161	0.161	0.161	0.161	0.161	0.161	0.161	0.161	0.161
160	0.161	0.162	0.162	0.162	0.162	0.163	0.163	0.163	0.163	0.163
170	0.163	0.163	0.163	0.163	0.163	0.163	0.163	0.163	0.163	0.163
180	0.163	0.163	0.163	0.163	0.163	0.163	0.163	0.163	0.163	0.163
Depth, 29.0 m; Sun Altitude, 56.6°										
0	0.152	0.152	0.152	0.152	0.152	0.152	0.152	0.152	0.152	0.152
10	0.174	0.172	0.167	0.160	0.154	0.148	0.143	0.141	0.140	0.140
20	0.216	0.192	0.176	0.163	0.153	0.145	0.140	0.137	0.136	0.137
30	0.232	0.201	0.178	0.164	0.154	0.146	0.141	0.137	0.137	0.136
40	0.136	0.188	0.176	0.164	0.155	0.148	0.142	0.136	0.136	0.136
50	0.182	0.180	0.175	0.164	0.154	0.147	0.140	0.138	0.136	0.136
60	0.178	0.175	0.171	0.163	0.156	0.150	0.146	0.144	0.142	0.142
70	0.175	0.173	0.168	0.164	0.159	0.154	0.148	0.147	0.146	0.145
80	0.176	0.175	0.172	0.167	0.162	0.158	0.154	0.151	0.149	0.148
90	0.177	0.175	0.171	0.167	0.163	0.159	0.155	0.153	0.151	0.151

† Originally published by the University of California Press; reprinted by permission of The Regents of the University of California.

TABLE 11 Radiance K Values, Clear Sunny Sky (Tyler, 1970) (Continued)

Tilt angle θ, deg	Azimuth angle Φ, deg									
	0	20	40	60	80	100	120	140	160	180
	Depth, 29.0 m; Sun Altitude, 56.6° (*continued*)									
100	0.172	0.172	0.171	0.167	0.163	0.160	0.157	0.156	0.155	0.155
110	0.172	0.171	0.169	0.166	0.164	0.162	0.159	0.158	0.158	0.158
120	0.168	0.168	0.167	0.165	0.164	0.162	0.161	0.161	0.161	0.161
130	0.166	0.166	0.165	0.165	0.164	0.163	0.162	0.161	0.161	0.161
140	0.165	0.164	0.164	0.164	0.163	0.163	0.162	0.161	0.162	0.163
150	0.164	0.164	0.164	0.164	0.163	0.163	0.163	0.162	0.162	0.162
160	0.164	0.164	0.164	0.164	0.163	0.163	0.163	0.162	0.162	0.162
170	0.163	0.163	0.163	0.163	0.163	0.163	0.163	0.163	0.163	0.163
180	0.163	0.163	0.163	0.163	0.163	0.163	0.163	0.163	0.163	0.163
	Depth, 41.3 m; Sun Altitude, 56.6°									
0	0.156	0.156	0.156	0.156	0.156	0.156	0.156	0.156	0.156	0.156
10	0.169	0.166	0.163	0.159	0.157	0.155	0.153	0.151	0.149	0.148
20	0.191	0.179	0.170	0.163	0.158	0.154	0.151	0.148	0.145	0.144
30	0.200	0.185	0.172	0.164	0.158	0.155	0.152	0.149	0.146	0.144
40	0.191	0.183	0.174	0.166	0.161	0.157	0.154	0.150	0.145	0.143
50	0.185	0.181	0.174	0.168	0.163	0.159	0.155	0.149	0.144	0.141
60	0.180	0.177	0.172	0.167	0.164	0.160	0.156	0.151	0.147	0.144
70	0.178	0.175	0.170	0.167	0.163	0.158	0.157	0.153	0.149	0.147
80	0.176	0.173	0.170	0.167	0.164	0.161	0.158	0.155	0.152	0.149
90	0.174	0.173	0.171	0.168	0.165	0.161	0.158	0.155	0.152	0.151
100	0.173	0.171	0.169	0.166	0.164	0.162	0.162	0.157	0.155	0.154
110	0.172	0.171	0.169	0.166	0.165	0.163	0.161	0.159	0.157	0.157
120	0.170	0.170	0.167	0.165	0.164	0.163	0.161	0.160	0.158	0.158
130	0.168	0.168	0.166	0.165	0.164	0.163	0.162	0.161	0.160	0.160
140	0.167	0.166	0.165	0.164	0.164	0.163	0.162	0.162	0.161	0.160
150	0.165	0.165	0.164	0.164	0.163	0.162	0.162	0.161	0.161	0.160
160	0.164	0.164	0.163	0.163	0.163	0.162	0.162	0.162	0.162	0.162
170	0.163	0.163	0.163	0.163	0.163	0.163	0.163	0.163	0.163	0.163
180	0.163	0.163	0.163	0.163	0.163	0.163	0.163	0.163	0.163	0.163

Depth, 53.7 m; Sun Altitude, 56.6°										
0	0.156	0.156	0.156	0.156	0.156	0.156	0.156	0.156	0.156	0.156
10	0.163	0.161	0.159	0.157	0.156	0.155	0.154	0.153	0.153	0.152
20	0.168	0.166	0.163	0.160	0.157	0.154	0.153	0.151	0.149	0.148
30	0.168	0.166	0.163	0.160	0.157	0.154	0.152	0.149	0.147	0.146
40	0.170	0.168	0.164	0.161	0.158	0.155	0.152	0.150	0.147	0.146
50	0.169	0.167	0.165	0.162	0.159	0.157	0.154	0.151	0.148	0.147
60	0.170	0.169	0.166	0.162	0.159	0.156	0.153	0.150	0.147	0.146
70	0.169	0.168	0.166	0.162	0.159	0.156	0.154	0.150	0.148	0.147
80	0.166	0.165	0.163	0.161	0.159	0.157	0.155	0.154	0.151	0.150
90	0.165	0.164	0.163	0.163	0.161	0.158	0.157	0.156	0.155	0.155
100	0.167	0.166	0.165	0.163	0.161	0.160	0.159	0.158	0.157	0.156
110	0.166	0.165	0.165	0.164	0.163	0.162	0.162	0.161	0.160	0.159
120	0.167	0.167	0.167	0.166	0.165	0.164	0.163	0.162	0.161	0.161
130	0.169	0.168	0.169	0.168	0.168	0.168	0.167	0.167	0.166	0.166
140	0.172	0.172	0.171	0.171	0.171	0.171	0.171	0.170	0.170	0.169
150	0.172	0.172	0.172	0.172	0.172	0.172	0.172	0.172	0.172	0.171
160	0.174	0.174	0.174	0.174	0.174	0.174	0.174	0.174	0.174	0.174
170	0.175	0.175	0.175	0.175	0.175	0.175	0.175	0.175	0.175	0.175
180	0.176	0.176	0.176	0.176	0.176	0.176	0.176	0.176	0.176	0.176

Depth, 59.9 m; Sun Altitude, 56.6°										
0	0.157	0.157	0.157	0.157	0.157	0.157	0.157	0.157	0.157	0.157
10	0.161	0.161	0.160	0.160	0.157	0.156	0.153	0.151	0.150	0.150
20	0.162	0.162	0.160	0.158	0.155	0.153	0.150	0.149	0.147	0.147
30	0.158	0.159	0.158	0.156	0.153	0.149	0.147	0.145	0.144	0.143
40	0.157	0.157	0.157	0.156	0.150	0.149	0.147	0.145	0.144	0.144
50	0.156	0.157	0.157	0.156	0.155	0.151	0.149	0.147	0.146	0.145
60	0.159	0.159	0.158	0.155	0.153	0.151	0.149	0.148	0.146	0.147
70	0.159	0.159	0.159	0.157	0.155	0.153	0.150	0.148	0.147	0.147
80	0.159	0.159	0.158	0.156	0.155	0.153	0.152	0.152	0.150	0.150
90	0.160	0.159	0.158	0.157	0.157	0.157	0.157	0.157	0.156	0.157
100	0.163	0.163	0.163	0.162	0.161	0.160	0.159	0.159	0.159	0.160
110	0.162	0.163	0.163	0.164	0.164	0.164	0.164	0.163	0.164	0.164
120	0.167	0.167	0.168	0.168	0.167	0.166	0.165	0.165	0.164	0.166
130	0.171	0.171	0.172	0.172	0.172	0.171	0.172	0.172	0.172	0.172
140	0.177	0.178	0.178	0.178	0.178	0.178	0.178	0.178	0.178	0.179
150	0.181	0.181	0.181	0.181	0.181	0.182	0.182	0.183	0.183	0.182
160	0.184	0.184	0.184	0.184	0.184	0.184	0.184	0.184	0.184	0.184
170	0.187	0.187	0.187	0.187	0.187	0.187	0.187	0.187	0.187	0.187
180	0.188	0.188	0.188	0.188	0.188	0.188	0.188	0.188	0.188	0.188

TABLE 12 Radiance K Values, Overcast Sky (Tyler, 1970)†

Tilt angle (θ)	Azimuth angle (Φ)									
	0	20	40	60	80	100	120	140	160	180
	Depth, 12.2 m; Sun Altitude, 40.0°									
0	0.204	0.204	0.204	0.204	0.204	0.204	0.204	0.204	0.204	0.204
10	0.207	0.207	0.206	0.206	0.205	0.206	0.206	0.207	0.207	0.207
20	0.215	0.216	0.216	0.215	0.214	0.215	0.216	0.217	0.218	0.218
30	0.226	0.226	0.225	0.224	0.222	0.222	0.223	0.224	0.224	0.226
40	0.230	0.230	0.228	0.228	0.226	0.225	0.225	0.225	0.226	0.227
50	0.209	0.209	0.208	0.206	0.204	0.204	0.205	0.208	0.210	0.212
60	0.201	0.201	0.200	0.199	0.198	0.200	0.200	0.201	0.203	0.203
70	0.201	0.201	0.201	0.201	0.200	0.199	0.199	0.200	0.200	0.201
80	0.204	0.203	0.204	0.202	0.201	0.200	0.200	0.200	0.201	0.201
90	0.204	0.204	0.204	0.204	0.204	0.203	0.201	0.201	0.201	0.201
100	0.207	0.207	0.206	0.206	0.205	0.203	0.201	0.199	0.199	0.198
110	0.206	0.206	0.206	0.206	0.205	0.203	0.201	0.200	0.199	0.199
120	0.206	0.206	0.205	0.205	0.205	0.204	0.203	0.202	0.201	0.201
130	0.206	0.207	0.207	0.208	0.208	0.207	0.207	0.205	0.205	0.205
140	0.209	0.209	0.208	0.209	0.209	0.208	0.206	0.206	0.205	0.205
150	0.210	0.210	0.210	0.211	0.211	0.210	0.209	0.208	0.206	0.206
160	0.211	0.211	0.211	0.210	0.210	0.209	0.208	0.208	0.209	0.208
170	0.209	0.209	0.209	0.209	0.209	0.209	0.209	0.209	0.209	0.209
180	0.209	0.209	0.209	0.209	0.209	0.209	0.209	0.209	0.209	0.209
	Depth, 18.3 m; Sun Altitude, 40.0°									
0	0.200	0.200	0.200	0.200	0.200	0.200	0.200	0.200	0.200	0.200
10	0.201	0.202	0.201	0.201	0.201	0.202	0.203	0.204	0.204	0.205
20	0.205	0.205	0.205	0.205	0.205	0.206	0.207	0.208	0.209	0.210
30	0.208	0.208	0.208	0.208	0.207	0.208	0.209	0.212	0.214	0.215
40	0.211	0.211	0.211	0.210	0.210	0.208	0.210	0.211	0.212	0.214
50	0.202	0.202	0.201	0.201	0.199	0.198	0.199	0.200	0.201	0.202
60	0.199	0.199	0.199	0.198	0.197	0.196	0.196	0.197	0.198	0.199
70	0.198	0.198	0.198	0.198	0.197	0.196	0.196	0.196	0.197	0.197
80	0.199	0.199	0.199	0.198	0.198	0.197	0.196	0.197	0.196	0.197

90	0.200	0.200	0.200	0.199	0.198	0.197	0.196	0.196	0.196	0.196
100	0.199	0.199	0.199	0.199	0.198	0.197	0.197	0.196	0.197	0.197
110	0.199	0.199	0.199	0.199	0.198	0.197	0.197	0.197	0.196	0.197
120	0.200	0.200	0.199	0.199	0.198	0.197	0.196	0.196	0.196	0.196
130	0.200	0.200	0.200	0.200	0.199	0.198	0.198	0.197	0.196	0.196
140	0.199	0.199	0.198	0.198	0.198	0.197	0.197	0.197	0.197	0.197
150	0.198	0.198	0.198	0.198	0.198	0.198	0.197	0.197	0.196	0.196
160	0.198	0.198	0.198	0.197	0.197	0.197	0.196	0.196	0.197	0.197
170	0.197	0.197	0.197	0.197	0.197	0.197	0.197	0.197	0.197	0.197
180	0.196	0.196	0.196	0.196	0.196	0.196	0.196	0.196	0.196	0.196
Depth, 30.5 m; Sun Altitude, 40.0°										
0	0.191	0.191	0.191	0.191	0.191	0.191	0.191	0.191	0.191	0.191
10	0.192	0.192	0.192	0.192	0.192	0.192	0.191	0.192	0.191	0.191
20	0.189	0.190	0.190	0.190	0.190	0.190	0.191	0.191	0.191	0.191
30	0.188	0.189	0.189	0.190	0.190	0.190	0.189	0.189	0.190	0.189
40	0.188	0.189	0.189	0.189	0.189	0.188	0.187	0.187	0.186	0.186
50	0.188	0.188	0.188	0.188	0.188	0.187	0.186	0.185	0.183	0.183
60	0.188	0.188	0.188	0.188	0.188	0.186	0.184	0.183	0.182	0.182
70	0.188	0.189	0.188	0.188	0.187	0.186	0.185	0.184	0.184	0.183
80	0.187	0.188	0.187	0.187	0.186	0.185	0.184	0.183	0.183	0.183
90	0.188	0.187	0.187	0.186	0.185	0.184	0.183	0.182	0.182	0.182
100	0.186	0.185	0.186	0.185	0.185	0.184	0.184	0.184	0.185	0.185
110	0.187	0.187	0.186	0.186	0.185	0.185	0.185	0.185	0.185	0.185
120	0.186	0.187	0.186	0.186	0.183	0.184	0.184	0.184	0.183	0.184
130	0.186	0.186	0.185	0.185	0.184	0.184	0.183	0.183	0.183	0.184
140	0.185	0.185	0.185	0.185	0.184	0.184	0.183	0.184	0.184	0.184
150	0.184	0.184	0.184	0.183	0.182	0.182	0.182	0.182	0.183	0.183
160	0.183	0.183	0.183	0.183	0.182	0.182	0.182	0.181	0.181	0.181
170	0.182	0.182	0.182	0.182	0.182	0.182	0.182	0.182	0.182	0.182
180	0.182	0.182	0.182	0.182	0.182	0.182	0.182	0.182	0.182	0.182

† Originally published by the University of California Press; reprinted by permission of The Regents of the University of California.

TABLE 12 Radiance K Values, Overcast Sky (Tyler, 1970) (Continued)

Tilt angle (θ)	Azimuth angle (Φ)									
	0	20	40	60	80	100	120	140	160	180
	Depth, 42.8 m; Sun Altitude, 40.0°									
0	0.180	0.180	0.180	0.180	0.180	0.180	0.180	0.180	0.180	0.180
10	0.181	0.181	0.181	0.181	0.181	0.181	0.180	0.179	0.179	0.179
20	0.183	0.183	0.183	0.183	0.183	0.182	0.181	0.179	0.178	0.177
30	0.184	0.184	0.184	0.184	0.184	0.182	0.180	0.177	0.175	0.175
40	0.183	0.183	0.182	0.182	0.181	0.179	0.176	0.174	0.172	0.171
50	0.181	0.182	0.182	0.181	0.181	0.180	0.177	0.175	0.173	0.173
60	0.180	0.180	0.180	0.180	0.179	0.178	0.176	0.174	0.172	0.171
70	0.181	0.181	0.181	0.181	0.180	0.178	0.177	0.175	0.173	0.172
80	0.179	0.179	0.179	0.178	0.177	0.176	0.174	0.172	0.172	0.171
90	0.178	0.178	0.178	0.178	0.177	0.176	0.174	0.173	0.172	0.171
100	0.179	0.179	0.179	0.178	0.178	0.177	0.175	0.174	0.173	0.172
110	0.181	0.181	0.181	0.180	0.179	0.178	0.177	0.177	0.176	0.175
120	0.182	0.182	0.182	0.181	0.180	0.180	0.179	0.178	0.178	0.177
130	0.183	0.183	0.182	0.182	0.182	0.181	0.181	0.181	0.180	0.180
140	0.186	0.186	0.186	0.185	0.185	0.185	0.184	0.183	0.183	0.182
150	0.188	0.188	0.188	0.187	0.187	0.186	0.185	0.185	0.184	0.184
160	0.190	0.189	0.189	0.189	0.189	0.188	0.188	0.187	0.187	0.187
170	0.190	0.190	0.190	0.190	0.190	0.190	0.190	0.190	0.190	0.190
180	0.190	0.190	0.190	0.190	0.190	0.190	0.190	0.190	0.190	0.190

	Depth, 48.9 m; Sun Altitude, 40.0°									
0	0.175	0.175	0.175	0.175	0.175	0.175	0.175	0.175	0.175	0.175
10	0.175	0.175	0.175	0.176	0.176	0.177	0.176	0.176	0.176	0.176
20	0.182	0.181	0.182	0.182	0.181	0.180	0.178	0.177	0.175	0.175
30	0.182	0.183	0.182	0.181	0.179	0.178	0.176	0.175	0.174	0.174
40	0.180	0.181	0.179	0.178	0.177	0.175	0.173	0.171	0.170	0.170
50	0.183	0.183	0.183	0.181	0.179	0.177	0.175	0.173	0.172	0.172
60	0.181	0.181	0.180	0.179	0.178	0.176	0.176	0.175	0.175	0.174
70	0.182	0.181	0.181	0.180	0.179	0.177	0.175	0.174	0.172	0.172
80	0.178	0.178	0.178	0.177	0.176	0.174	0.172	0.172	0.170	0.171
90	0.178	0.178	0.177	0.176	0.176	0.174	0.173	0.172	0.171	0.170
100	0.179	0.179	0.178	0.178	0.177	0.176	0.175	0.173	0.171	0.171
110	0.181	0.181	0.180	0.180	0.179	0.178	0.177	0.176	0.175	0.175
120	0.185	0.184	0.184	0.183	0.183	0.182	0.180	0.180	0.179	0.179
130	0.187	0.187	0.187	0.187	0.186	0.185	0.184	0.183	0.182	0.181
140	0.192	0.192	0.191	0.190	0.190	0.189	0.189	0.189	0.187	0.186
150	0.194	0.194	0.194	0.193	0.193	0.192	0.192	0.190	0.190	0.189
160	0.198	0.198	0.198	0.198	0.198	0.198	0.198	0.198	0.198	0.198
170	0.200	0.200	0.200	0.200	0.200	0.200	0.200	0.200	0.200	0.200
180	0.201	0.201	0.201	0.201	0.201	0.201	0.201	0.201	0.201	0.201

types of water for visible radiation are those of Sullivan (1963) and Hulburt (1945).

Because of the very considerable error that can result from the inclusion in the measurement of spurious scattered light, published values of the total attenuation coefficient (Hulburt, 1945) are considered to have their greatest reliability at long wavelengths, where absorption predominates over scattering. At shorter wavelengths scattering assumes an increasingly important role, and the inclusion of spurious scattered light in the measurement would result in progressively greater

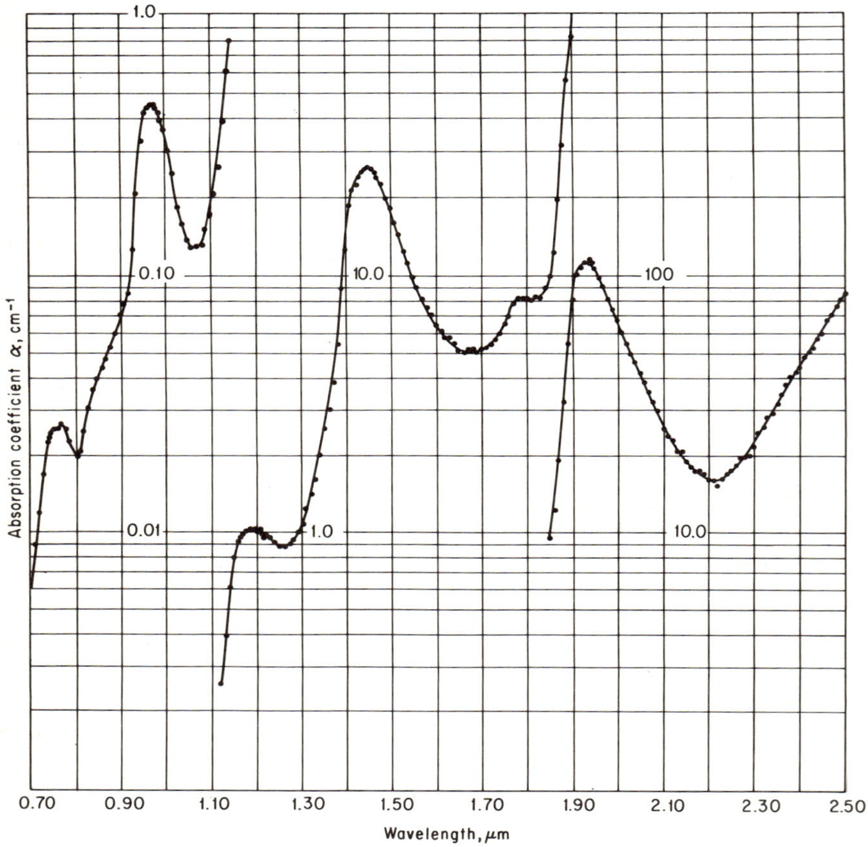

Fig. 6 Near-infrared absorption for distilled water obtained by beam-transmittance measurements. [*Curcio and Petty* (1951).]

errors, with reported values of the attenuation coefficient sinking further and further below the real value.

Similarly, published values for the absorption coefficient (Sullivan, 1963) obtained from beam-transmittance measurements would also be most reliable at the long wavelengths. In this case the inclusion of spurious scattered light would also cause increasingly great errors toward the wavelengths of maximum transmittance, with the reported absorption coefficient rising above the real value.

In theory $\alpha = a + s$, and from this point of view the absorption coefficient a cannot be greater than the total attenuation coefficient α (see Tables 13 and 14).

TABLE 13 Total-Attenuation Coefficient α, m^{-1} (Hulburt, 1945)

λ, nm	Distilled	Coastal†	Bay	λ, nm	Distilled	Coastal†	Bay
400	......	0.394		550	0.047	0.203	0.323
410	......	0.360		560	0.053	0.210	0.323
420	......	0.331	0.805	570	0.066	0.223	0.323
430	......	0.298	0.705	580	0.074	0.248	0.331
440	......	0.270	0.628	590	0.120	0.264	0.365
450	0.040	0.255	0.556				
460	0.036	0.234	0.512	600	0.197	0.291	0.429
470	0.036	0.220	0.475	610	0.243	0.351	0.465
480	0.0365	0.213	0.447	620	0.265	0.387	0.476
490	0.037	0.206	0.419	630	0.280	0.399	0.488
				640	0.292	0.414	0.500
500	0.038	0.201	0.388	650	0.308	0.425	0.518
510	0.039	0.200	0.368	660	0.335	0.448	0.542
520	0.040	0.199	0.351	670	0.375	0.479	0.563
530	0.042	0.199	0.337	680	0.406	0.510	0.589
540	0.044	0.200	0.331	690	0.467	0.565	0.630
				700	0.576	0.625	

† Numerical values read from fig. 2 of Hulburt (1945).

TABLE 14 Absorption Coefficient, m^{-1}, for Distilled Water and Artificial Seawater Obtained from Beam-Transmittance Measurements (Sullivan, 1963)

λ, nm	Distilled water	Artificial Seawater	λ, nm	Distilled water	Artificial Seawater
400	0.058		665	0.419	0.419
410	0.047		670	0.425	0.422
420	0.038		675	0.438	0.432
430	0.040		680	0.447	0.450
440	0.032		685	0.467	0.474
450	0.033		690	0.511	0.501
580	0.109	0.133	695	0.569	0.570
585	0.119	0.146	700	0.648	0.647
590	0.172	0.167	705	0.750	0.745
595	0.214	0.208	710	0.904	0.908
600	0.272	0.265	715	1.114	1.107
605	0.293	0.289	720	1.375	1.372
610	0.293	0.299	725	1.756	1.759
615	0.300	0.301	730	2.309	2.308
620	0.309	0.317	735	2.608	2.600
625	0.305	0.317	740	2.698	2.729
630	0.320	0.312	745	2.712	2.753
635	0.325	0.320	750	2.683	2.698
640	0.334	0.326	755	2.696	2.714
645	0.339	0.336	760	2.685	2.727
650	0.351	0.357	770	2.519	2.540
655	0.387	0.389	780	2.346	2.376
660	0.407	0.405	790	2.052	2.097

The beam transmittance (percent per meter) for California coastal water as a function of depth to 1,000 m has been obtained with a very sophisticated instrument (Hughes and Austin, 1965; Gilbert 1967) (see Fig. 7). The low transmittance within the top 100 m is typical of ocean waters (see Gilbert and Rue, 1968). The optical bandwidth used for the measurements in Fig. 7 closely approximates photopic.

6. Horizontal Irradiance and the Diffuse-Attenuation Coefficient The underwater irradiance falling on a horizontal plane facing the zenith [downwelling irradiance $H(-)$] and on a horizontal plane facing the nadir [upwelling irradiance $H(+)$]

TABLE 15 In Situ Data for the Total-Attenuation Coefficient for Selected Depths and Spectral Bandwidths, July 1971, Tongue-of-the-Ocean, Bahama Islands†

Wratten filter no.	98	45	61	21	25
Peak wavelength, nm . . .	450	495	530	580	611
Bandwidth at half peak transmittance, nm . . .	51	52	95	47	33
Depth	Total-attenuation coefficient α, m^{-1}				
5	0.105	0.111	0.145	0.235	0.355
10	0.105	0.113	0.148	0.235	0.355
20	0.111	0.116	0.150	0.236	0.355
50	0.116	0.111	0.140	0.234	0.350
100	0.083	0.078	0.112	0.202	0.320
200	0.061	0.052	0.089	0.180	0.301
240	0.059	0.051	0.089	0.180	0.301

† T. Petzold, Scripps Institute of Oceanography, unpublished data.

TABLE 16 In Situ Data for the Total-Attenuation Coefficient α, m^{-1}, Obtained at Eight Locations, July 1971, Tongue-of-the-Ocean, Bahama Islands†
Optical bandwidth, 95 nm; peak wavelength, 529 nm

Depth, m	Location							
	1	2	3	4	5	6	7	8
1	0.148	0.151	0.140	0.140	0.163	0.211	0.143	0.149
5	0.148	0.149	0.140	C.140	0.160	0.198	0.151	0.163
10	0.148	0.149	0.142	0.151	0.160	0.198	0.151	0.165
20	0.151	0.151	0.143	0.151	0.154	.....	0.154	0.150
40	0.147	0.143	0.140	0.140	0.137	.....	0.151	0.147
60	0.143	0.137	0.138	0.139	0.142	.....	0.138	0.159
100	0.116	0.122	0.117	0.116	0.105	.....	0.124	0.110
250	0.089	0.083	0.084	0.087	0.084	.....	0.083	0.084
Simultaneous $K(-)$ data, m^{-1}			0.058	0.056	.....	.....	0.064	

† T. Petzold, Scripps Institute of Oceanography, unpublished data.

have attracted the interest of scientists since 1922, and the literature of oceanography and limnology abounds with tables of data on this property and on the diffuse-attenuation coefficient, which is commonly calculated from such data by means of the equation

$$H_{z_2} = H_{z_1} e^{-K\,\Delta z}$$

Many of the early instruments did not measure true irradiance, nor were they capable of truly monochromatic spectral isolation. The usefulness of many of these data is therefore limited, and the data must be used with caution. Recent advances in the design of underwater optical integrators by Smith (1969) provide greater accuracy for this measurement.

Spectral values of the diffuse-attenuation coefficients [$K(+)$ and $K(-)$] have been obtained in various types of natural water by Tyler and Smith (1970). These can be used together with spectral values of irradiance (Arveson et al., 1969) to compute approximate values of spectral irradiance at specified depths underwater.

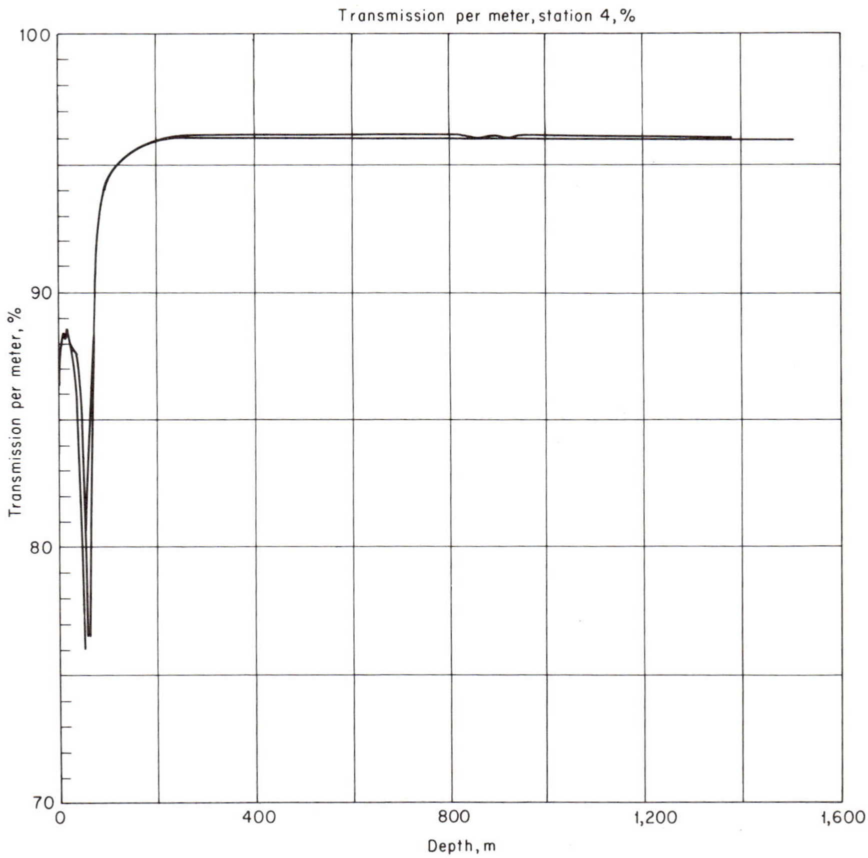

Fig. 7 Beam transmittance for California coastal water as a function of depth. [*Gilbert and Rue* (1967).]

Because of the dependence of the value of $K(-)$ on the distribution of radiance in the light field and the fact that this distribution varies with depth, the value of $K(-)$ for any wavelength will vary with optical depth K_z even in homogeneous water. The effect is greatest near the surface, and for accurate irradiance prediction an appropriate value of $K(-)$ must be used.

7. Irradiance Reflectance Spectral data on irradiance reflectance for several locations have been published by Tyler and Smith (1970). Due to boundary effects, irradiance reflectance is generally larger near the surface of a homogeneous water body and decreases with depth (Tyler et al., 1971).

TABLE 17 Spectral Values for $K(-)$ and $K(+)$ for Three Types of Water[a]
Spectral resolution 5.6 nm[b]

Wavelength λ, nm	Crater Lake[c] $K(-)$, m^{-1}	Buzzards Bay[d] $K(+)$, m^{-1}	Nantucket Shoal[e] $K(-)$, m^{-1}	Wavelength λ, nm	Crater Lake[c] $K(-)$, m^{-1}	Buzzards Bay[d] $K(+)$, m^{-1}	Nantucket Shoal[e] $K(-)$, m^{-1}
357				542	0.0674	0.1867	0.1517
362	0.0504	0.611[f]	0.543	547	0.0725	0.1863	0.1474
367	0.0479	0.624	0.519	552	0.0748	0.1833	0.1418
371	0.0481	0.618	0.495	557	0.0785	0.1730	0.1367
376	0.0410	0.607	0.451	561	0.0816	0.1692	0.1383
381	0.0425	0.601	0.425	566	0.0850	0.1646	0.1315
386	0.0391	0.566	0.470	571	0.0912	0.1615	0.1366
391	0.0394	0.545	0.371	576	0.1016	0.1807	0.1558
396	0.0392	0.530	0.351	581	0.1124	0.1664[f]	0.1823
401	0.0385	0.504	0.336	586	0.1256	0.207	0.1965
406	0.0387	0.500	0.325	591	0.1360	0.237	0.226
411	0.0386	0.491	0.321	596	0.1699	0.280	0.251
416	0.0377	0.472	0.305	601	0.1943	0.344	0.317
420	0.0375	0.457	0.288	605	0.204	0.379	0.337
425	0.0371	0.444	0.290	610	0.206	0.382	0.342
430	0.0366	0.422	0.289	615	0.208	0.393	0.350
435	0.0379	0.407	0.288	620	0.221	0.399	0.359
440	0.0363	0.391	0.273	624	0.230	.....	0.367
445	0.0365	0.372	0.252	629	0.306	0.422	0.370
450	0.0356	0.350	0.261	634	0.321	0.435	0.379
455	0.0358	0.337	0.246	639	0.350	0.432	0.403
459	0.0354	0.325	0.236	644	0.383	0.449	0.415
464	0.0353	0.307	0.231	649	0.395	0.454	0.443
469	0.0352	0.294	0.207	654	0.453	0.456[f]	0.445
474	0.0356	0.280	0.210	658	0.447	0.496	0.493
479	0.0362	0.270	0.1958	663	0.463	0.494	0.514
484	0.0368	0.258	0.1893	668	0.471	0.436	0.540
488	0.0379	0.249	0.1807	673	0.496	0.395	0.550
493	0.0393	0.245	0.1815	678	0.500	0.383	0.549
498	0.0420	0.244	0.1679	683	0.544	0.410	0.555
503	0.0446	0.220	0.1624	687	0.580	0.465	0.582
508	0.0512	0.217	0.1701	692	0.676	0.492	0.615
513	0.0555	0.216	0.1626	697	0.700	0.646[f]	0.651
518	0.0570	0.214	0.1597	702	0.832	0.548[f]	0.704
523	0.0580	0.203	0.1599	707	0.884	0.754[f]	0.780
527	0.0602	0.1993	0.1570	712	0.997[f]	1.050[f]	0.815
532	0.0629	0.1896	0.1465	716	1.507[f]	.....	1.023[f]
537	0.0655	0.1872	0.1527	721	.....	.....	0.877[f]

[a] Unpublished data from R. C. Smith, Scripps Institute of Oceanography, 1969.
[b] See Tyler and Smith (1970), pp. 10–13.
[c] Crater Lake, 43°56′N, 122°05′W. Data taken July 26, 1969. These data are average values of nine determinations obtained over a depth range from 1.5 to 98.6 m.
[d] Buzzards Bay, 41°30′N, 70°50′W. Data taken Aug. 13, 1969. These values represent the water mass between depths 0.2 and 4.7 m.
[e] Nantucket Shoal, 41°23′N, 69°31′W. Data taken Aug. 21, 1969. These values represent the water mass between 1.5 and 8.5 m.
[f] Questioned value.

TABLE 18 Spectral Values of Irradiance Reflectance for Two Types of Water[a]

Spectral resolution 5.6 nm[b]

Wave-length	Gulf of California[c]	Crater Lake[d]	Wave-length	Gulf of California[c]	Crater Lake[d]
362	......	0.0798	518	0.0144	0.0153
367	0.0110	0.0812	523	0.0142	0.0139
371	0.0110	0.0814	527	0.0140	0.0132
376	0.0113	0.0810	532	0.0146	0.0119
381	0.0112	0.0830	537	0.0132	0.0111
386	0.0117	0.0818	542	0.01250	0.00970
391	0.0117	0.0842	547	0.01120	0.00924
396	0.0118	0.0860[e]	552	0.00984	0.00854
401	0.0128	0.0750[e]	557	0.01040	0.00808
406	0.0121	0.0785[e]	561	0.00958	0.00735
411	0.0122	0.0725	566	0.01000	0.00661
416	0.0124	0.0723	571	0.00889	0.00553
420	0.0125	0.0710	576	0.00843	0.00552
425	0.0122	......	581	0.00584	0.00393
430	0.0129	0.0676	586	0.00575	0.00315
435	0.0137	0.0630	591	0.00449	0.00338
440	0.0128	0.0598	596	0.00389	0.00261
445	0.0142	0.0568	601	0.00295	0.00202
450	0.0145	0.0550	605	0.00273	0.00196
455	0.0158	0.0542	610	0.00218	0.00173
459	0.0154	0.0542	615	0.00199	0.00161
464	0.0162	0.0523	620	0.00206	0.00147
469	0.0171	0.0489	624	0.00225	0.00141
474	0.0179	0.0460	629	0.00195	0.00117
479	0.0171	0.0429	634	0.00177	0.00103
484	0.0198	0.0406	639	0.00146	0.00099
488	0.0196	0.0364	644	0.00085[e]	0.00106
493	0.0193	0.0325	649	0.00056[e]	0.00067[e]
498	0.0174	0.0267	654	0.00015[e]	0.00059[e]
503	0.0166	0.0244	658	....	0.00045[e]
508	0.0157	0.0192	663	....	0.00035[e]
513	0.0143	0.0162	668	....	0.00016[e]

[a] Unpublished data from R. C. Smith, Scripps Institute of Oceanography.

[b] See Tyler and Smith (1970), pp. 10–13.

[c] Gulf of California, 28°41′N, 112°16′W. Tables give the fraction of daylight irradiance reflected at a depth of 5 m. Water analysis:

Depth, m	Chlorophyll concentration, mg/m^3	Total pigments, mg/m^3	Seston, mg/m^3
2	0.451	0.524	850
8	0.611	0.655	610

[d] Crater Lake, 43°56′N, 122°05′W. Data taken July 26, 1969. Tables give the fraction of irradiance reflected at a depth of 6 m (daylight irradiance only). Chlorophyll concentration below detection levels.

[e] Questioned value.

8. Distribution Factors Distribution factors have been calculated from radiance-distibution data by Tyler et al. (1959), and by Tyler and Preisendorfer (1962) (Table 19). These are probably insufficient data to establish the range of values of $D(+)$ and $D(-)$ in various types of water.

At the present writing a complete experimental documentation of the optical properties of ocean or lake water does not exist for any single type of water. However, Tyler et al. (1971) have combined experimental data for clear water from several sources and have computed tables of theoretically consistent values for the remaining optical properties of a clear water type (Table 20).

TABLE 19 Distribution Factor, Lake Pend Oreille, Idaho (Tyler and Preisendorfer, 1962)
Optical bandwidth for all data = 64 nm with peak transmittance of 480 nm

Depth, m	R_∞	$D(-)$	$D(+)$	$K(-)$, m^{-1}
Overcast Sky				
6.1	0.0221	1.29	2.77	0.216
18.3	0.0250	1.33	2.82	0.206
30.5	0.0266	1.33	2.77	0.189
42.8	0.0279	1.34	2.79	0.180
55.0	0.0258	1.34	2.94	0.178
Clear Sunny Sky				
4.24	0.0215	1.247	2.704	0.129
10.4	0.0184	1.288	2.727	0.153
16.6	0.0204	1.291	2.778	0.174
29.0	0.0227	1.313	2.781	0.169
41.3	0.0235	1.315	2.757	0.165
53.7	0.0234	1.307	2.763	0.158
66.1	0.0190	1.308	2.947	0.154

In this table the values given for $K(-)$ and R and the value of S at 550 nm are experimental. Spectral values of S have been derived on the assumption that a portion of S varies with λ in accordance with the λ^{-4} law, whereas the remaining portion of S is considered to be constant with wavelength. The proportions at 550 nm in Table 20 are based on Morel (1968) but have been varied systematically in Table 20 as indicated in the crossheadings. The spectral values for α a, $D(-)$, and $D(+)$ have been calculated on the assumption that the radiance distribution $N(\theta,\varphi)$ can be adequately represented by the expression

$$N(\theta,\varphi) = \frac{N(90^\circ)}{\left[1 + \frac{K(-)\cos\theta}{\alpha}\right]^{\rho}}$$

where ρ is a function of wavelength.

TABLE 20 Data and Computed Values for Clear Water (Tyler et al., 1971)

λ, nm	R	$K(-)$, m^{-1}	α, m^{-1}	a, m^{-1}	s, m^{-1}	$D(-)$	$D(+)$
Molecular Scattering = 0.0021 m^{-1}, Particle Scattering = 0.0329 m^{-1}, at 550 nm							
360	0.5300×10^{-1}	0.0687	0.08891	0.04434	0.04456	1.33323	2.52431
370	0.5646×10^{-1}	0.0650	0.08451	0.04133	0.04318	1.34179	2.51411
380	0.5796×10^{-1}	0.0625	0.08161	0.03947	0.04214	1.34597	2.51112
390	0.5903×10^{-1}	0.0600	0.07889	0.03768	0.04121	1.34985	2.51104
400	0.5946×10^{-1}	0.0590	0.07744	0.03700	0.04044	1.35031	2.50865
410	0.5802×10^{-1}	0.0580	0.07631	0.03657	0.03974	1.34792	2.51489
420	0.5601×10^{-1}	0.0570	0.07531	0.03620	0.03911	1.34475	2.52436
430	0.5350×10^{-1}	0.0565	0.07477	0.03624	0.03852	1.33976	2.53469
440	0.5051×10^{-1}	0.0560	0.07441	0.03634	0.03806	1.33413	2.54893
450	0.4649×10^{-1}	0.0557	0.07432	0.03674	0.03757	1.32590	2.56877
460	0.4250×10^{-1}	0.0558	0.07463	0.03744	0.03719	1.31687	2.58941
470	0.3851×10^{-1}	0.0565	0.07545	0.03861	0.03684	1.30637	2.60973
480	0.3448×10^{-1}	0.0570	0.07622	0.03968	0.03653	1.29583	2.63464
490	0.2900×10^{-1}	0.0582	0.07790	0.04162	0.03627	1.28002	2.67352
500	0.2300×10^{-1}	0.0620	0.08188	0.04589	0.03599	1.25728	2.71534
510	0.1698×10^{-1}	0.0695	0.08936	0.05358	0.03577	1.22803	2.75945
520	0.1400×10^{-1}	0.0735	0.09349	0.05791	0.03557	1.21230	2.79128
530	0.1200×10^{-1}	0.0780	0.09784	0.06250	0.03533	1.19918	2.80999
540	0.9999×10^{-2}	0.0820	0.10196	0.06682	0.03513	1.18637	2.83941
550	0.8492×10^{-2}	0.0888	0.10853	0.07353	0.03500	1.17312	2.85264
560	0.7293×10^{-2}	0.0960	0.11546	0.08061	0.03485	1.16121	2.86234
570	0.5896×10^{-2}	0.1100	0.12892	0.09418	0.03474	1.14412	2.86550
580	0.4499×10^{-2}	0.1330	0.15109	0.11651	0.03457	1.12346	2.86101
590	0.3249×10^{-2}	0.1730	0.18992	0.15542	0.03450	1.10020	2.84329
600	0.2100×10^{-2}	0.2430	0.25863	0.22424	0.03438	1.07541	2.82385
610	0.1701×10^{-2}	0.2750	0.29039	0.25611	0.03428	1.06709	2.82694
620	0.1449×10^{-2}	0.3020	0.31722	0.28303	0.03419	1.06135	2.82892
630	0.1159×10^{-2}	0.3310	0.34636	0.31225	0.03411	1.05552	2.84351
640	0.8598×10^{-3}	0.3600	0.37587	0.34181	0.03405	1.04983	2.87363
650	0.6204×10^{-3}	0.3950	0.41135	0.37736	0.03399	1.04429	2.90458
660	0.4002×10^{-3}	0.4300	0.44720	0.41328	0.03392	1.03885	2.95613
670	0.2201×10^{-3}	0.4700	0.48835	0.45449	0.03386	1.03322	3.03189
680	0.1029×10^{-3}	0.5050	0.52474	0.49094	0.03379	1.02819	3.13872
690	0.4499×10^{-4}	0.5500	0.57092	0.53714	0.03377	1.02373	3.24978
700	0.1700×10^{-4}	0.6300	0.65181	0.61812	0.03368	1.01913	3.36059
λ^{-4} Scattering = 0.00421 m^{-1}, Particle Scattering = 0.0308 m^{-1}, at 550 nm							
360	0.5299×10^{-1}	0.0687	0.09743	0.04366	0.05376	1.35357	2.57219
370	0.5654×10^{-1}	0.0650	0.09208	0.04073	0.05135	1.36091	2.55711
380	0.5794×10^{-1}	0.0625	0.08827	0.03896	0.04931	1.36323	2.55023
390	0.5898×10^{-1}	0.0600	0.08474	0.03725	0.04748	1.36535	2.54625
400	0.5949×10^{-1}	0.0590	0.08247	0.03661	0.04586	1.36431	2.53957
410	0.5804×10^{-1}	0.0580	0.08071	0.03622	0.04448	1.36040	2.54283
420	0.5599×10^{-1}	0.0570	0.07912	0.03591	0.04321	1.35563	2.54951
430	0.5346×10^{-1}	0.0565	0.07803	0.03599	0.04204	1.34918	2.55716
440	0.5050×10^{-1}	0.0560	0.07726	0.03611	0.04114	1.34245	2.56914
450	0.4648×10^{-1}	0.0557	0.07676	0.03654	0.04022	1.33304	2.58682
460	0.4253×10^{-1}	0.0558	0.07667	0.03726	0.03941	1.32290	2.60488
470	0.3850×10^{-1}	0.0565	0.07717	0.03846	0.03871	1.31136	2.62359
480	0.3453×10^{-1}	0.0570	0.07762	0.03955	0.03807	1.30001	2.64610
490	0.2898×10^{-1}	0.0582	0.07903	0.04152	0.03750	1.28313	2.68376
500	0.2298×10^{-1}	0.0620	0.08279	0.04581	0.03698	1.25971	2.72419
510	0.1699×10^{-1}	0.0695	0.09000	0.05350	0.03649	1.22977	2.76590
520	0.1399×10^{-1}	0.0735	0.09394	0.05785	0.03608	1.21348	2.79626
530	0.1198×10^{-1}	0.0780	0.09815	0.06246	0.03569	1.19993	2.81383

TABLE 20 Data and Computed Values for Clear Water (Tyler et al., 1971) (Continued)

λ, nm	R	$K(-)$, m^{-1}	α, m^{-1}	a, m^{-1}	s, m^{-1}	$D(-)$	$D(+)$
λ^{-4} Scattering = 0.00421 m^{-1}, Particle Scattering = 0.0308 m^{-1}, at 550 nm (*continued*)							
540	0.1000×10^{-1}	0.0820	0.10210	0.06680	0.03529	1.18675	2.84087
550	0.8492×10^{-2}	0.0888	0.10853	0.07353	0.03500	1.17312	2.85264
560	0.7302×10^{-2}	0.0960	0.11533	0.08063	0.03469	1.16094	2.86038
570	0.5900×10^{-2}	0.1100	0.12870	0.09422	0.03447	1.14361	2.86245
580	0.4499×10^{-2}	0.1330	0.15077	0.11659	0.03418	1.12275	2.85689
590	0.3247×10^{-2}	0.1730	0.18952	0.15555	0.03396	1.09933	2.83809
600	0.2098×10^{-2}	0.2430	0.25817	0.22442	0.03374	1.07456	2.81817
610	0.1701×10^{-2}	0.2750	0.28990	0.25632	0.03357	1.06624	2.82053
620	0.1451×10^{-2}	0.3020	0.31668	0.28326	0.03341	1.06049	2.82183
630	0.1160×10^{-2}	0.3310	0.34577	0.31250	0.03327	1.05465	2.83585
640	0.8599×10^{-3}	0.3600	0.37519	0.34211	0.03307	1.04891	2.86481
650	0.6197×10^{-3}	0.3950	0.41062	0.37768	0.03294	1.04340	2.89527
660	0.3999×10^{-3}	0.4300	0.44643	0.41360	0.03282	1.03803	2.94593
670	0.2200×10^{-3}	0.4700	0.48753	0.45481	0.03271	1.03248	3.02068
680	0.1029×10^{-3}	0.5050	0.52383	0.49127	0.03256	1.02751	3.12534
690	0.4496×10^{-4}	0.5500	0.56997	0.53746	0.03250	1.02313	3.23491
700	0.1699×10^{-4}	0.6300	0.65081	0.61843	0.03238	1.01862	3.34402
λ^{-4} Scattering = 0.00842 m^{-1}, Particle Scattering = 0.0266 m^{-1}, at 550 nm							
360	0.5295×10^{-1}	0.0687	0.11528	0.04276	0.07252	1.38130	2.64562
370	0.5650×10^{-1}	0.0650	0.10767	0.03994	0.06773	1.38702	2.62368
380	0.5799×10^{-1}	0.0625	0.10178	0.03825	0.06352	1.38762	2.61046
390	0.5900×10^{-1}	0.0600	0.09652	0.03662	0.05989	1.38788	2.60135
400	0.5945×10^{-1}	0.0590	0.09272	0.03605	0.05666	1.38517	2.59024
410	0.5801×10^{-1}	0.0580	0.08963	0.03571	0.05392	1.37957	2.58943
420	0.5599×10^{-1}	0.0570	0.08678	0.03544	0.05133	1.37288	2.59173
430	0.5353×10^{-1}	0.0565	0.08465	0.03555	0.04909	1.36491	2.59545
440	0.5048×10^{-1}	0.0560	0.08293	0.03574	0.04718	1.35615	2.60408
450	0.4646×10^{-1}	0.0557	0.08157	0.03621	0.04535	1.34492	2.61837
460	0.4248×10^{-1}	0.0558	0.08084	0.03696	0.04387	1.33335	2.63406
470	0.3849×10^{-1}	0.0565	0.08057	0.03820	0.04237	1.32015	2.64870
480	0.3449×10^{-1}	0.0570	0.08043	0.03932	0.04111	1.30728	2.66828
490	0.2898×10^{-1}	0.0582	0.08129	0.04132	0.03997	1.28905	2.70274
500	0.2300×10^{-1}	0.0620	0.08460	0.04562	0.03897	1.26449	2.74054
510	0.1700×10^{-1}	0.0695	0.09136	0.05335	0.03801	1.23330	2.77948
520	0.1400×10^{-1}	0.0735	0.09489	0.05773	0.03716	1.21594	2.80630
530	0.1199×10^{-1}	0.0780	0.09879	0.06238	0.03641	1.20154	2.82090
540	0.9993×10^{-2}	0.0820	0.10237	0.06677	0.03560	1.18736	2.84436
550	0.8492×10^{-2}	0.0888	0.10853	0.07353	0.03500	1.17312	2.85264
560	0.7294×10^{-2}	0.0960	0.11508	0.08068	0.03439	1.16028	2.85748
570	0.5899×10^{-2}	0.1100	0.12824	0.09431	0.03393	1.14252	2.85665
580	0.4498×10^{-2}	0.1330	0.15009	0.11675	0.03333	1.12118	2.84794
590	0.3251×10^{-2}	0.1730	0.18870	0.15579	0.03291	1.09763	2.82704
600	0.2099×10^{-2}	0.2430	0.25727	0.22478	0.03248	1.07287	2.80618
610	0.1700×10^{-2}	0.2750	0.28885	0.25677	0.03208	1.06440	2.80689
620	0.1449×10^{-2}	0.3020	0.31557	0.28377	0.03180	1.05862	2.80739
630	0.1159×10^{-2}	0.3310	0.34455	0.31307	0.03148	1.05276	2.81969
640	0.8607×10^{-3}	0.3600	0.37387	0.34271	0.03116	1.04708	2.84710
650	0.6198×10^{-3}	0.3950	0.40924	0.37831	0.03092	1.04166	2.87656
660	0.4002×10^{-3}	0.4300	0.44489	0.41428	0.03060	1.03633	2.92455
670	0.2200×10^{-3}	0.4700	0.48590	0.45549	0.03041	1.03096	2.99723
680	0.1030×10^{-3}	0.5050	0.52209	0.49192	0.03016	1.02615	3.09852
690	0.4501×10^{-4}	0.5500	0.56806	0.53811	0.02995	1.02189	3.20375
700	0.1699×10^{-4}	0.6300	0.64882	0.61908	0.02974	1.01756	3.30963

TABLE 20 Data and Computed Values for Clear Water (Tyler et al., 1971) (Continued)

λ, nm	R	$K(-)$, m^{-1}	α, m^{-1}	a, m^{-1}	s, m^{-1}	$D(-)$	$D(+)$
λ^{-4} Scattering = 0.01684 m^{-1}, Particle Scattering = 0.01816 m^{-1}, at 550 nm							
360	0.5295×10^{-1}	0.0687	0.15190	0.04178	0.11012	1.41209	2.73894
370	0.5651×10^{-1}	0.0650	0.13957	0.03905	0.10052	1.41695	2.71040
380	0.5794×10^{-1}	0.0625	0.12960	0.03744	0.09216	1.41655	2.69262
390	0.5899×10^{-1}	0.0600	0.12072	0.03587	0.08484	1.41565	2.67827
400	0.5949×10^{-1}	0.0590	0.11365	0.03533	0.07832	1.41203	2.66233
410	0.5798×10^{-1}	0.0580	0.10785	0.03504	0.07281	1.40496	2.65812
420	0.5600×10^{-1}	0.0570	0.10250	0.03481	0.06769	1.39687	2.65624
430	0.5352×10^{-1}	0.0565	0.09824	0.03496	0.06328	1.38732	2.65641
440	0.5049×10^{-1}	0.0560	0.09453	0.03518	0.05934	1.37669	2.66054
450	0.4651×10^{-1}	0.0557	0.09154	0.03569	0.05585	1.36379	2.67119
460	0.4254×10^{-1}	0.0558	0.08909	0.03649	0.05260	1.34991	2.68161
470	0.3846×10^{-1}	0.0565	0.08746	0.03777	0.04968	1.33461	2.69257
480	0.3452×10^{-1}	0.0570	0.08673	0.03890	0.04783	1.32108	2.71109
490	0.2901×10^{-1}	0.0582	0.08592	0.04097	0.04494	1.29959	2.73754
500	0.2301×10^{-1}	0.0620	0.08823	0.04531	0.04291	1.27293	2.77102
510	0.1700×10^{-1}	0.0695	0.09403	0.05306	0.04096	1.23965	2.80477
520	0.1399×10^{-1}	0.0735	0.09679	0.05751	0.03928	1.22043	2.82604
530	0.1199×10^{-1}	0.0780	0.10001	0.06222	0.03779	1.20454	2.83416
540	0.1000×10^{-1}	0.0820	0.10287	0.06669	0.03617	1.18863	2.84996
550	0.8492×10^{-2}	0.0888	0.10853	0.07353	0.03500	1.17312	2.85264
560	0.7302×10^{-2}	0.0960	0.11456	0.08077	0.03379	1.15903	2.85063
570	0.5899×10^{-2}	0.1100	0.12734	0.09450	0.03283	1.14030	2.84468
580	0.4498×10^{-2}	0.1330	0.14877	0.11709	0.03168	1.11803	2.83002
590	0.3251×10^{-2}	0.1730	0.18716	0.15629	0.03087	1.09417	2.80577
600	0.2101×10^{-2}	0.2430	0.25552	0.22551	0.03001	1.06942	2.78200
610	0.1700×10^{-2}	0.2750	0.28686	0.25768	0.02917	1.06065	2.77909
620	0.1451×10^{-2}	0.3020	0.31335	0.28484	0.02850	1.05466	2.77602
630	0.1160×10^{-2}	0.3310	0.34222	0.31424	0.02797	1.04886	2.78659
640	0.8603×10^{-3}	0.3600	0.37136	0.34397	0.02738	1.04325	2.81096
650	0.6198×10^{-3}	0.3950	0.40652	0.37968	0.02684	1.03794	2.83691
660	0.4000×10^{-3}	0.4300	0.44192	0.41572	0.02619	1.03277	2.88014
670	0.2200×10^{-3}	0.4700	0.48275	0.45691	0.02584	1.02777	2.94836
680	0.1030×10^{-3}	0.5050	0.51864	0.49332	0.02531	1.02324	3.04159
690	0.4498×10^{-4}	0.5500	0.56444	0.53946	0.02497	1.01933	3.13991
700	0.1700×10^{-4}	0.6300	0.64488	0.62045	0.02442	1.01531	3.23574

REFERENCES

Arveson, J. C., R. N. Griffin, Jr., and B. D. Pearson, Jr. (1969): *Appl. Opt.*, vol. 8, pp. 2215–2232.

Bauer, D., and A. Morel (1967): *Ann. Geophys.*, vol. 23, fasc. 1, pp. 109–123.

Curcio, J. A., and C. C. Petty (1951): *J. Opt. Soc. Am.*, vol. 41, pp. 302–304.

Dawson, L. H., and E. O. Hulburt (1934): *J. Opt. Soc. Am.*, vol. 24, pp. 175–177.

Duntley, S. Q. (1963): *J. Opt. Soc. Am.*, vol. 53, pp. 214–233.

Gilbert, G., and R. O. Rue (1967): *U.S. Nav. Ordnance Test Stn. Tech. Publ.* 4343, China Lake, Calif.

——— and ——— (1968): *Nav. Weapons Cent. Tech. Publ.* 4542, China Lake, Calif.

Hughes, R. S., and R. W. Austin (1965): *U.S. Nav. Ordnance Test Stn. Tech. Publ.* 3748, China Lake, Calif.

Hulburt, E. O. (1945): *J. Opt. Soc. Am.*, vol. 35, pp. 698–705.

Ivanoff, A., and A. Morel (1964): *C. R.*, vol. 258, pp. 2873–2876.

Jerlov, N. G. (1961): *Medd. Oceanogr. Inst. Goteborg*, vol. 8, no. 11, pp. 4–39.

——— (1968): "Optical Oceanography," chap. 1, Elsevier, London.

——— and M. Fukuda (1960): *Tellus*, vol. 12, pp. 348–355.

Joseph, J. (1961): *Dtsch. Hydrogr. Z. Ergaenzungsh.*, Reihe B. (4°), Nr. 5.

Kraut, J., and W. B. Dandliker (1955): *J. Chem. Phys.*, vol. 23, p. 1544.

McIntyre, D. (1964): *J. Res. Natl. Bur. Stand.*, vol. 68A, no. 1, pp. 87–95.

Morel, A. (1965): *Ann. Geophys.*, vol. 21, fasc. 2, pp. 281–284.
——— (1968): *Cah. Oceanogr.*, vol. 20, p. 291.
Nyffeler, F. (1970): *AGARD Conf. Proc. 77 Electromagn. Sea,* AGARD-CP-77-70, Hartford House, London.
Optical Society of America, Committee on Colorimetry (1944): *J. Opt. Soc. Am.*, vol. 34, pp. 245–266.
——— (1953): "The Science of Color," p. 229, Crowell, New York.
——— (1962): Editor's Page, *J. Opt. Soc. Am.,* vol. 52, p. 490.
Preisendorfer, R. W. (1960): *Symp. Radiant Energy Sea, Int. Union Geod. Geophys. Monogr.* 10.
——— (1964): pp. 51–59, in "Physical Aspects of Light in the Sea: A Symposium," University of Hawaii Press, Honolulu.
Pritchard, B. S., and W. G. Elliott (1960): *J. Opt. Soc. Am.,* vol. 50, p. 191.
Reese, J. W., and S. P. Tucker (1970): *Nav. Undersea Res. Dev. Cent. Tech. Pub.* NUC TP 203, October.
Sasaki, T. (1964): pp. 19–24 in "Physical Aspects of Light in the Sea: A Symposium," University of Hawaii Press, Honolulu.
———, S. Watanabe, G. Oshiba, and N. Okami (1960): *Rec. Oceanogr. Works Jap. Spec. No.* 4, pp. 197–205, March.
———, ———, ———, ———, and M. Kajihara (1962): *Bull. Jap. Soc. Sci. Fish.*, vol. 28, pp. 489–496.
Smith, R. C. (1969): *J. Mar. Res.*, vol. 27, pp. 241–351.
Sullivan, S. A. (1963): *J. Opt. Soc. Am.*, vol. 53, pp. 962–968.
Tyler, J. E. (1958): *J. Mar. Res.*, vol. 16, pp. 96–99.
——— (1960): *Univ. Calif. Scripps Inst. Oceanogr. Bull.,* vol. 7, pp. 363–412.
——— (1963): *J. Appl. Opt.*, vol. 2, p. 245.
——— and R. W. Austin (1964): *Appl. Opt.*, vol. 3, p. 613.
——— and R. W. Preisendorfer (1962): in "The Sea," vol. 1, Wiley-Interscience, New York.
———, W. H. Richardson, and R. W. Holmes (1959): *J. Geophys. Res.,* vol. 64, pp. 667–673.
——— and R. C. Smith (1970): "Measurements of Spectral Irradiance Underwater," Gordon and Breach, New York.
———, ———, and W. H. Wilson, Jr. (1972): *J. Opt. Soc. Am.,* vol. 62, pp. 83–91.

Section 16

Optical Instruments for Metrology

HERBERT E. TORBERG
Electro-Optical Division, Kollmorgen Corporation
Northampton, Mass.

WILLIAM J. ROWAN
Electro-Optical Division, Kollmorgen Corporation
Northampton, Mass.

J. RICHARD VYCE
Optical Systems Division,
Itek Corporation, Lexington, Mass.

(Numbers refer to paragraphs)

INTRODUCTION†

1. The optical-metrology industry is generally conceded to have originated in England just before World War II. Optical tooling was born in response to a requirement of the aircraft industry for a nonsagging reference line to replace taut wires in large wing jigs and fixtures. Its advantages triggered its widespread

† Paragraphs 1 to 41 and 64 to 90 were written by Herbert E. Torberg and William J. Rowan.

use and a proliferation of types of instruments. This section discusses only the basic instruments.

Specific references to representative instruments do not represent a comprehensive survey of available instruments; a listing does not imply that the instrument named is superior in any way to a similar instrument not named.

2. Application and Use Optical-metrology instruments are useful over a range of applications in the laboratory, fabrication shop, shipyard, and rocket-launch site, in addition to their long-standing application to astronomy, navigation, construction, and land surveying. Any of the following requirements may be satisfied by the instruments described in this section: establishment of a direction or a line in space, establishment of three or more points in line with each other, or any number of points parallel to, or a measured distance from, another line; any number of points in, or a measured deviation from, a given direction; reference to true horizontal or true vertical; measured rotation about a line of sight; accurate measure of distance or relative distance; location or direction of a line or plane in precise geometric relation to another line or plane. Most of these depend on availability of a clear line of sight. Typical applications are shown in Table 1, and in the discussion of individual instruments.

3. Scope This section presents a survey of optical-metrology instrument types, primarily those readily available and commonly used, along with well-established concepts of other instruments often tailored to particular applications. Metrology in this instance includes noncontact measure of linear displacement from a reference line or angular deviation from a direction, as commonly encountered at higher levels of manufacture. It does not, in general, include those techniques and their instruments applicable to manufacture or inspection of either optical or mechanical detail parts, e.g., test plates, profile projectors, rotary tables, toolmakers' microscopes, optical comparators in general, or those uniquely associated with astronomy, navigation, or land surveying.

The basic alignment instruments using image-forming optical systems generally exist or are feasible in both manual and automatic versions. Other instruments and techniques based on the optical phenomena of interference and diffraction are of increasing importance and potential for finer measurements.

4. Definitions General optical terms are defined in Sec. 2; metrology instruments and techniques are defined in the text of this section. Other terms of interest to optical metrology are defined as follows.

Actinic Radiation. That portion of total radiation within the passband of the optical system and at wavelengths at which the (particular) sensor is excited.

Alignment. The process or state of causing or finding points, lines, or planes to be particularly located or oriented with respect to a reference line, or a measure of displacement or deviation therefrom.

Alignment Periscope. A device for displacing a line of sight, usually by twice deviating it 90°.

Altitude. The vertical angle from the earth-tangent plane to a celestial body. Similar angles to terrestrial objects are usually called elevations.

Automatic Alignment. A state of alignment (see *Alignment*) automatically monitored to produce an electric signal sensible and proportional to misalignment, actuating corrective action or recording.

Autoreflection. An angular alignment technique involving observation of the virtual image of a telescope in a distant reflector.

Azimuth. An angle in the earth-tangent plane. In astronomy, navigation, and land surveying, referred either to true or to magnetic north.

Cursor. A visible device or pattern adjacent to, or apparently in, an image plane by means of which image displacement can be estimated or measured; usually movable and often connected to an externally indicating mechanism.

Deviation. Change in direction of a line of sight.

Displacement. Translation of a line of sight without deviation.

Elevation. In vehicle-referenced systems, an angle in the vehicle-vertical plane from base, e.g., deck, to object (see *Altitude*).

Horizontal. An attitude parallel to the earth-tangent plane as defined by gravity sensors.

TABLE 1 Applications of Instruments
Numbers refer to paragraphs of this section

To define, measure, or measure with respect to:	Applicable instruments, accessories, and other techniques
Reference line	Alignment telescope, 15–21
	Laser alignment device, 64–66
	Zone plates, 80
	Axicon, 81
	Alignment interferometer, 71–74
	Any instrument incorporating line-of-sight telescope
Level line	Sight level, 27–32
	Laser alignment device plus gravity reference, 64–66
	Any line-of-sight telescope plus gravity reference
Reference direction	Autocollimator, 8–14
	Pointing interferometer, 75–78
	Alignment telescope in autoreflection, 15–21
	Any instrument with autocollimating capability
Plane by intersecting lines	Cross-axis-telescope alignment transit, 38–41
	Rotating zone plate, 80
	Any reference-line device with fixed pentaprism
Plane by generation	Theodolite, 22–26
	Alignment transit, 38–41
	Any reference-line device with rotating pentaprism
Plane normal to line	Any reference-line device with rotating pentaprism or alignment transit
Large vertical angle	Theodolite, 22–26
Large horizontal angle	Alignment transit, 38–41
Body rotation about an axis normal to a reference direction	Autocollimator or substitute, 8–14
	Theodolite, 22–26
	Alignment transit, 38–41
	Pointing interferometer, 75–78
Body rotation about a line in a reference direction	Twist autocollimator, 33–37, 52–53
	Automatic polarimeter, 54–55
	Three-axis instruments, 56–59
Distance	Automatic optical rangefinder, 60–61
	Automatic optical probe, 62–63
	Laser-interferometer distance-measuring instrument, 67–70

Line of Sight. A straight line from instrument to object as intentionally modified by deviation and displacement; extension of an instrument's optical axis. As defined by a line-of-sight telescope, the locus of all points to which the center of the reticle is projected as system focus is varied from minimum object distance to infinity.

Mechanical Axis. The centerline of an optical instrument as defined by its mounting surfaces, i.e., where the optical axis of an alignment instrument ought to be.

Own Reticle. The instrument reticle as viewed directly, as distinct from its reflected image in autocollimating.

Pentareflector (Pentaprism, Pentamirror). An optical constant deviator for 90° deviation in the plane of the mirror normals; two mirrors inclined to each other at 45° or a prism with equivalent faces.

Plunge (or Transit) and Reverse. Rotating the telescope of a transit or theodolite 180° about its vertical axis then 180° about its horizontal axis so as to point in the original direction.

Porro Reflector (Porro Prism, Porro Mirror). An optical constant deviator for 180° deviation in the plane of the mirror normals; two mirrors at 90° to each other or an equivalent prism.

Retroreflector. A prism or arrangement of mirrors which will deviate a ray 180°, making the reflected ray parallel to the incident ray. A Porro reflector is a retroreflector in one plane. A cube-corner reflector (three faces mutually perpendicular) is a retroreflector in space.

Striding Level. A spirit level in a precision mount designed to sit astride a cylinder.

Traverse. A direction angle in the elevated plane of the line of sight, usually a small angle. A very small traverse angle can be determined by linear measure in the image plane, along the apparent horizontal.

Twist. Rotation about the line of sight.

Vertical. The direction of the gravity vector (as opposed to that established by horizon-scanning or other approach). Almost all optical alignment uses gravity as an assumed constant.

Vignetting. Loss of illumination; darkening at the edge of an optical image (field) due to loss of edge bundles. Severe vignetting results in reduction of field.

MANUAL ALIGNMENT INSTRUMENTS

5. Nature of Alignment Alignment to some degree is required in the construction of almost anything. In many cases the requirements can be met by the

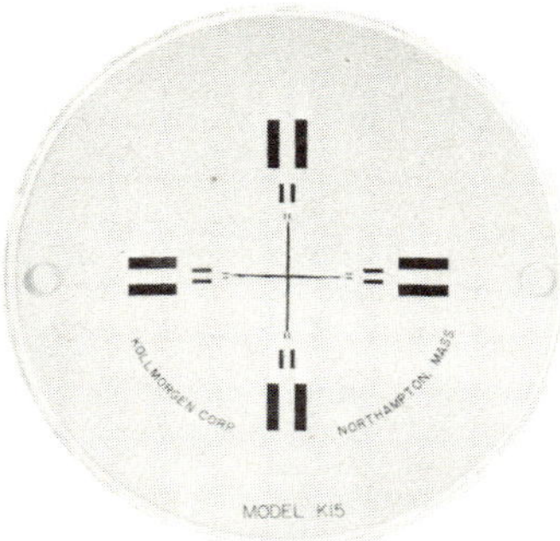

Fig. 1 Alignment target. (*Kollmorgen Corp.*)

mechanic's eye, with the aid of simple straightedges, squares, and scales. Larger structures require sighting by eye and stretched strings or wires; while plumb lines and spirit levels provide for alignment relative to gravity, therefore to the local horizontal plane. Optical alignment equipment and concepts extend the alignment and measuring capabilities to new dimensions, both to precise alignment of larger structures and to minute increments of measure of smaller structures.

6. Scales and Targets Autocollimators and pointing interferometers operate with high-quality mirrors, laser systems with live detectors, and automatic alignment equipment with retroreflectors. Manually operated line-of-sight telescope devices, such as alignment telescopes, require targets to identify line center and scales to make offset measurements beyond the range of the instrument's optical micrometer. Both normally carry patterns adapted to the setting of filar-bifilar reticles at varying target distances. Typical alignment targets and scales are shown in Figs. 1 and 2.

An optical tooling scale comprises a series of paired-line (or bounded white-space) targets, repeated at precise intervals which are less than the off-center measuring range of the telescope's optical micrometer. The scales shown are divided into tenths of an inch. Alignment telescopes use integral or detachable autoreflection targets to establish normality to a mirror. A typical autoreflection target is shown in Fig. 3.

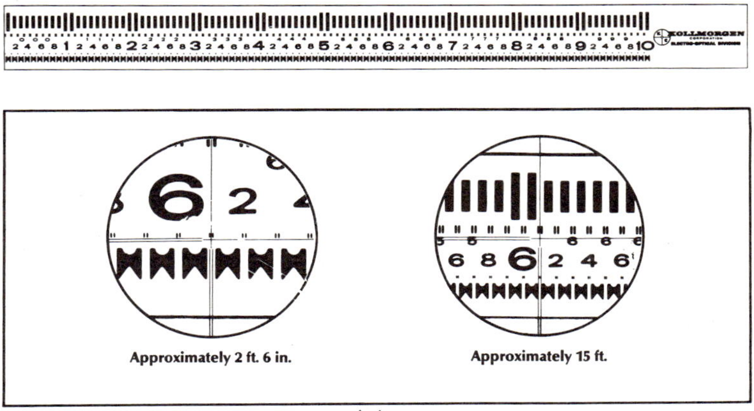

(a)

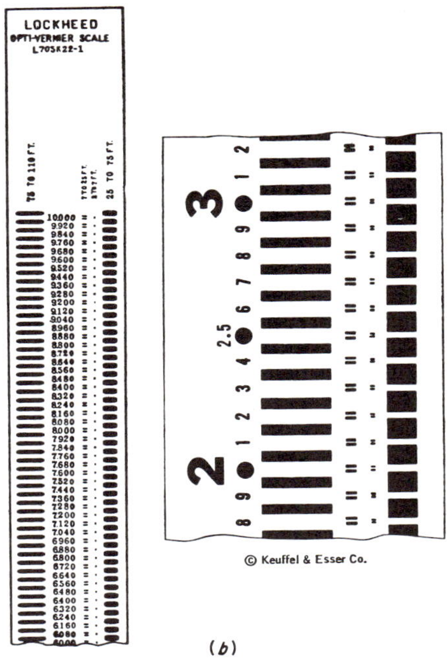

(b)

Fig. 2 Optical tooling scales. (*a*) As seen in a nominal 30-power alignment telescope. (*Kollmorgen Corp.*) (*b*) (*Lockheed Aircraft Corp., Keuffel & Esser Co.*)

7. Visual Acuity in Alignment The cursor and/or optical-micrometer setting criteria of almost all manual alignment instruments include the positioning of a single line between two lines or in a black-bounded white space (or vice versa), which experiment and experience show can be done with greater accuracy and repeatability than any other form of mark matching. The eye's *vernier acuity,* or ability to establish a relationship between such marks, is independent of line length above a minimum subtense and, within wide limits, of illumination. While the settings made by an experienced observer are highly repeatable, individual

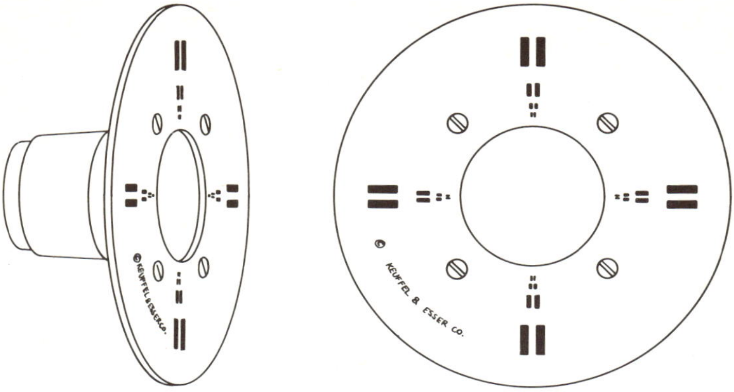

Fig. 3 Autoreflection target. (*Copyright by Keuffel & Esser Co.*)

physiological differences produce effects whose net result, called *observer bias,* is demonstrable and measurable.

An observer's visual acuity, attention, and motivation are directly and adversely affected by physical discomfort or fatigue. A well-designed alignment operation or experiment involving alignment instruments must consider this often ignored factor, particularly if many readings are to be taken in the same position.

Autocollimator

8. Description An autocollimator is an optical device for fine measurement of small angles. Used with a plane mirror, a plane autocollimator is sensitive to rotation of either the mirror or itself about any axis perpendicular to its own optical axis. It is not sensitive either to rotation about its own optical axis or to translation within the dimensions of the mirror. By reference to a crossline reticle, rotation about a sensitive axis not parallel to either line is resolved into components about two axes parallel to the reticle lines. The means of measuring is normally provided for one or both of these axes.

The autocollimator projects a beam of collimated light from an illuminated reticle. Light reflected from a mirror set normal to the collimated bundle is focused by the objective lens and forms a second image of the reticle directly superimposed on the first. If the mirror is rotated slightly about any axis perpendicular to the autocollimator's optical axis, the return image will be displaced on the image plane by a linear distance proportional to the angle of rotation. While practical considerations limit the effective range of an autocollimator, this displacement is theoretically independent of the distance between the autocollimator and the mirror and as such is the salient characteristic of an autocollimator as a metrology instrument.

Autocollimators are designed so that image displacement can be read directly as the angle of mirror tilt. An autocollimator may be a single-axis instrument, i.e., capable of reading mirror tilt about one axis only, or dual-axis.

Two-axis measurements can also be made by rotating a barrel-mounted single-axis instrument, but in that case reliance is placed on the alignment of the optical axis to the mechanical axis of the tube and on the accuracy of 90° rotation. Many commercially available autocollimators, like other barrel-mounted alignment instruments, are packaged for most of their length in a barrel compatible with National Aerospace Standard† (NAS) optical-tooling-mount bore diameters (nominally 2.2500 in). A typical mounting arrangement is shown in Fig. 4.

† Aerospace Industries Association of America, Inc., Standard NAS-900.

In essence, an autocollimator is a telescope having an illuminated reticle in the infinity-focus image plane of the objective lens, parfocal with a measurement image plane in which the appropriate cursors are mounted. In a bright-field instrument, a beam splitter is normally employed to introduce field illumination silhouetting the reticle, while in a bright-line configuration the reticle will usually be edge-lit. Micrometer-screw-driven cursors (one for a single-axis instrument and two for a dual axis) customarily having both linear and lead-screw-shaft dial graduations, are mounted as close as possible to the primary image plane or in

Fig. 4 Autocollimator and mount (NAS barrel). (*Kollmorgen Corp.*)

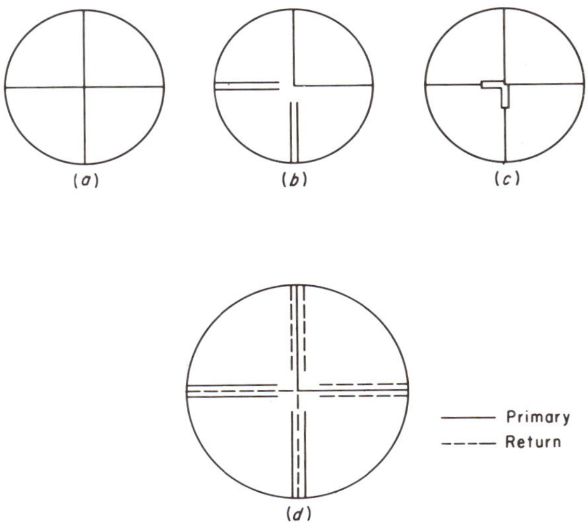

Fig. 5 Reticle patterns: (*a*) crosslines, (*b*) filar-bifilar, (*c*) combination, (*d*) filar-bifilar in coincidence autocollimation.

an auxiliary image plane. Although other configurations have been marketed, a conventional instrument has linear graduations of 1 minute of arc, or some common division of a minute, in the field of view, and seconds and divisions of seconds totaling one or more linear scale increments on a dial which is part of the cursor lead screw or directly coupled to it. The eyepiece can usually be focused to accommodate the observer.

The reticle pattern may be simple crosslines, a filar-bifilar, or a combination of both, as shown in Fig. 5. For all readings except coincidence autocollimation, plain crosslines measured by a bifilar cursor are acceptable. For accurate coincidence autocollimation from a plane mirror, however, a filar-bifilar is desirable. A combination reticle, also shown, provides for both applications. The optical system of a filar-bifilar instrument is customarily so arranged that autocollimation from a plane reflector will cause the return image of the reticle to appear inverted and reverted.

Operator control of the reticle illumination level is desirable and usually provided. Representative optical schematics are shown in Fig. 6.

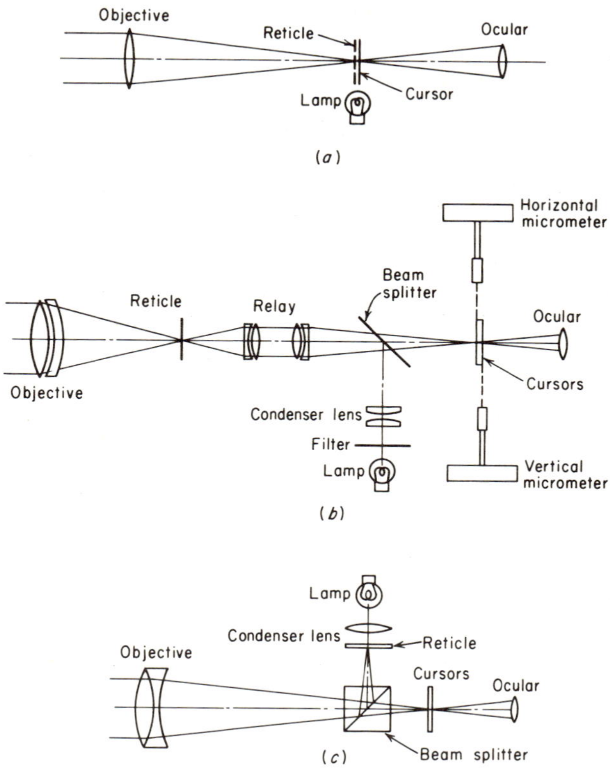

Fig. 6 Autocollimators: (*a*) edge-lit reticle, (*b*) cursors in auxiliary image plane, reticle illuminated by beam splitter, (*c*) offside reticle, introduced with illumination via beam splitter.

9. Basic Principles A plane autocollimator measures the tilt of a plane reflector about one or both of its axes of sensitivity, with respect to its own axis, by scaling the linear displacement between an original point in the image plane and its reflected image. This linear distance is directly proportional to the angle between the collimated bundle emanating from the axial point on the reticle and its reflection from the mirror or to the angle between the instrument's optical axis and the mirror normal.

A simple conceptual autocollimator comprising a point source of light in the center of the image plane at the infinity focus of a perfect objective lens would project a single bundle of perfectly collimated light the diameter of the objective lens clear aperture. Reflected by a perfect mirror normal to the bundle, the return image is formed as an Airy disk concentric with the source. Under these conditions and in stable and homogeneous air the effective range (distance) of the instrument is limited only by atmospheric attenuation. A practical manual autocollimator requires an extended reticle, with its resultant field of view, both to define the optical axis and to provide angular operating range. Such an instrument is subject to limitations of operating distance and angular range at distance due primarily to loss of field, as described below.

Each of an infinity of points on the extended reticle illuminates the objective lens and thereby projects a bundle of collimated light (Fig. 7) concentric with its own projection through the center of the objective (thin lens). One, and only one, of this infinite number of points by definition lies on the optical axis

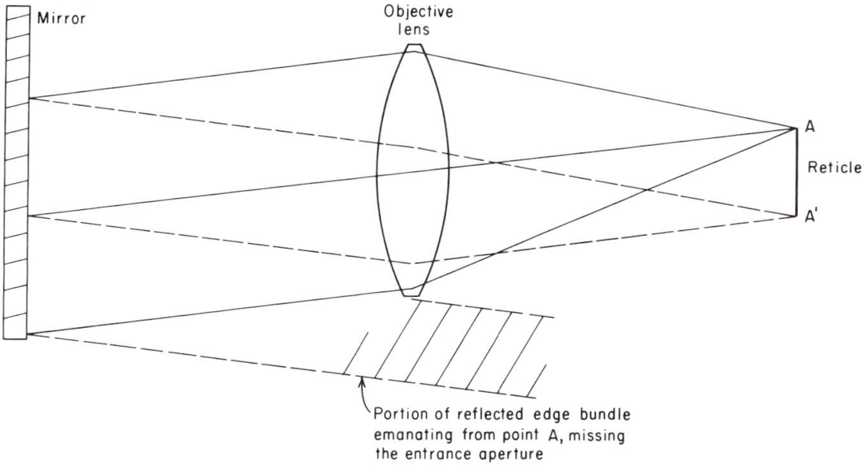

Fig. 7 Vignetting.

and therefore projects a bundle of collimated light concentric with the optical axis. This bundle is analogous to that projected by the point-source reticle in the hypothetical example above and exhibits the same characteristics. The autocollimator's output is therefore a cone of solid angle equal to the field angle of the reticle, composed of an infinity of collimated bundles, each in turn emanating from a different point on the reticle. The output cone is truncated at the objective's aperture stop.

Only the axial bundle as reflected will be totally returned to the objective lens from a mirror set normal to it. As the distance between the autocollimator and such a mirror is increased, the reflected edge bundles, which are not normal to the mirror, begin to miss the objective lens. This reduced illumination results in vignetting of the edge of the image and is followed by total extinction of edge points as the distance is increased further. The image of any point on the reticle is lost when its return bundle completely misses the objective aperture.

The limit of the angular field returned by a mirror normal to the optical axis vs. the distance to the mirror is given by

$$\phi = \tan^{-1}\frac{h}{l} \tag{1}$$

where ϕ = field limit (half angle)
$h = d/2$
d = diameter of entrance aperture (usually objective lens)
l = distance from entrance aperture to mirror measured along optical axis

Equation (1) is in terms of the true field of the instrument. Linear scale and cursor graduations are customarily labeled in terms of mirror tilt, or one-half the above value.

The angular range of the instrument similarly diminishes as the distance between the autocollimator and the mirror is increased. Allowing for some field illumination beyond the return image of the axial point or (in one plane only) reference line, the practical working limit lies somewhat inside the theoretical limit. The theoretical limit is reached at a given distance when the image of the axial point is extinguished, i.e., when the reflected central or axial bundle of collimated light completely misses the objective aperture. At that angle of mirror tilt, the mark to which the cursor must be set has disappeared. There will still be a returned field; however, without definition of the axis it is generally of little value. For values within the designed field, the limiting angle of mirror tilt (θ) that can be measured at a distance is given by

$$\theta = \tfrac{1}{2} \tan^{-1} \frac{d}{l} \tag{2}$$

where l is the distance from the entrance aperture to the nearest part of the mirror intercepted by the axial bundle and d is the diameter of the entrance aperture (normally the objective lens). With adequate reticle illumination, an

TABLE 2 Representative Autocollimators

Manufacturer	Typical model: Single axis	Typical model: Dual axis
Davidson Optronics, Inc.	D-638	D-656
Hilger & Watts, Ltd.	TA-1	TA-51
Kollmorgen Corp.	K-221	K-222
Perkin-Elmer Corp.	CS3-24	
Nippon Kogaku, K. K. (Nikon)		Model 6
E. Leitz, Wetzlar		Autocollimator

autocollimator of 1.8-in-diameter objective aperture (NAS barrel) has a practical working range of at least 150 ft, at which distance its effective angular range is approximately ±1.25 minutes of arc. Table 2 lists typical instruments.

10. Substitute Instruments In addition to autocollimators as such, a number of other alignment instruments, notably some alignment telescopes, can be converted into autocollimators of somewhat lesser capability by adjusting to infinity focus or adding autocollimating attachments. Autocollimation, as an operating technique, may be employed with other instruments having illuminated reticles and capable of infinity focus, e.g., some theodolites, making due allowance in each case for the nature of the instrument's readouts.

11. Applications Autocollimators are routinely used for aligning airframe and missile jigs and fixtures, for monitoring rocket-launcher attitude, for aligning rocket-tracking instrumentation, for transferring alignment between inertial navigators, and similar applications, as well as for the more prosaic applications of calibrating rotary angular indexing and measuring devices against optical cubes and polygons.

With an appropriately mounted mirror, an autocollimator is a primary instrument for checking the flatness or parallelism of machine-tool ways, for measuring the flatness of surface plates, and for various other machine-shop and inspection purposes. The applications of autocollimators to experimental mechanical measure are limited only by the ingenuity of the user. Mounted or integral mirrors at significant locations on a test article permit measurement of minute angular deflections to complement the linear measuring capability of strain gages.

12. Techniques Definition of the optical axis, i.e., the primary or own reticle, may or may not be visible in the observer's field of view, depending on the location of the reticle in the optical system. When it is visible, direct and verifiable measurement can be made by means of the cursors between own reticle and its reflected image, for the absolute measure of a mirror normal's deviation from the optical axis. When it is not visible, such absolute measurement can be inferred from the cursor scale by assuming that the origin is at numerical zero, or it can be measured as the average of the original reading and the deviation indicated when the autocollimator is rolled 180° about its mechanical axis. In either case (visible or invisible primary reticle) the relative difference between two mirror normals or between successive discrete orientations of the normal to one mirror can be measured directly and positively. For certain other applications, however, capabilities and applicable techniques differ because of this same characteristic. Coincidence autocollimation, i.e., pointing the instrument so that its optical axis is normal to a mirror or other reflective surface, is possible only with an instrument permitting the user to see his own reticle along with its reflected image.

Similarly, the normal to an imperfectly aligned mirror can be defined within an autocollimator as the point of intersection of two cursors each displaced from the primary reticle line toward its return image one-half the distance between them. This capability is of value, for example, in comparing the orientation of a mirror normal to the projected axis of another collimator, since the indicated displacement of a projected reticle is one-half the angle between the optical axes of the projecting and receiving instruments.

Autocollimators in which the own reticle is not visible can be used for the same purposes only by relying on calibration for own-axis location or by rolling the instrument, with attendant possible error.

13. Accessories An autocollimator will function with any plane reflector of suitable optical quality. For most purposes a front surface mirror flat to one-eighth wave of sodium light, or better, is recommended. Various configurations of mounted mirrors are available for specific purposes, e.g., surface-plate testing. A pool of oil or mercury, properly designed, can be used as a horizontal mirror. A number of pendulous-weight gravity reference mirrors are available.

Porro reflectors can be used to eliminate sensitivity and alignment requirements on one axis when only the other is of interest. Optical cubes or polygons will define angles of rotation. In general, the line of sight can be deviated or displaced by such devices as pentareflectors, retroreflecting or displacement periscopes, or optical wedges. Their use will not interfere with measurement of rotation of the final reflector, provided that they are arranged to have a predictable and stable effect on the direction of the line of sight and neither introduce dioptric power nor intolerably diminish the reflected light. Attention must be paid to the stability and rigidity of the mounting system, since the autocollimator cannot distinguish between its own rotation and that of the mirror.

14. Sensitivity and Accuracy An autocollimator can distinguish between successive positions of a return image in the field of view if the difference between them is large enough to result in repeated readably different settings of a micrometer-driven bifilar cursor. This value, empirically determined, is generally accepted to be less than 2.5 μm (2.5×10^{-6} m). Return-image linear displacement for a given mirror tilt depends on objective focal length, as

$$\Delta r = 2f\,\Delta\theta \tag{3}$$

where Δr = change in radial dimension or linear displacement
f = focal length
$\Delta\theta$ = change in mirror orientation or mirror tilt, rad

On that basis, an autocollimator of 25 cm focal length will discern mirror tilt of 1 second of arc.

Factors affecting sensitivity and accuracy include the following:

1. Effective portion of the objective aperture when only part is forming the return image. A well-designed instrument will be relatively insensitive to partial-aperture deviation.

2. Uniformity of reticle illumination.
3. Focus of objective.
4. Parfocality of multiple-image planes.
5. Thermal compensation of design.
6. Backlash or lost motion in cursor-drive mechanism relative to micrometer-dial readout.
7. Orthogonality of reticle lines.
8. Parallelism of cursor pattern to reticle lines.
9. Stability of reticle position:
   a. Long-term stability for a reticle out of the observer's field of view.
   b. Short-term stability in any case.
10. Observer contribution:
   a. Visual acuity (including effects of uncorrected astigmatism).
   b. Manual dexterity.
   c. Technique and discipline.
   d. Individual directional bias in cursor setting.
11. Procedures, e.g., verification of reticle stability, design of experiment or measurement, application of statistical methods to data reduction.

The magnitudes of these effects vary. An experienced observer, using a 110-cm equivalent-focal-length autocollimator with proper technique under ideal conditions of observer comfort and in stable ambients should be able to repeat readings to an average value of 0.1 second of arc with a maximum deviation of 0.2 to 0.3 second of arc.

The inherent accuracy of an autocollimator is most nearly realizable when the full entrance aperture is employed, in conjunction with a nearby, near-normal mirror. The ideal reflector is always large enough to return to the autocollimator's entrance aperture all the light that the geometry of the setup permits. In many cases, however, reflector size may be limited by other requirements, making the reflector, in effect, an entrance aperture stop. When readings of such a reflector may use different portions of the autocollimator's entrance aperture, the effect of this condition on indicated alignment must be considered.

For 10 percent of aperture, deviation may range from 2 seconds of arc across the entrance aperture to a much larger number, depending on the instrument. An instrument's susceptibility may be qualitatively assessed by moving a card with an approximately ½-in-square hole between a fixed nearby mirror and the autocollimator and observing the effects of scanning across the objective aperture.

Alignment Telescope

15. Description An alignment telescope is an optical instrument for observing and measuring vertical and horizontal linear displacements from a line of sight. It is a sighting, or line-of-sight, telescope, having, in addition to a basic telescope's objective lens and eyepiece, a reticle (to define the optical axis with respect to the field of view) and internal focusing. Alignment telescopes can usually be focused from very near object distances to infinity. Some can be focused down to the front surfaces of their own objective lenses. Most instruments have telescopic magnification between 30× and 50× at infinity focus. Magnification diminishes and the angular field correspondingly increases as the telescope is focused on nearer objects.

Some instruments, commonly called *line scopes,* are complete as described above. They display vertical and horizontal linear displacements from the line of sight but lack measuring means other than target scales.

Generally an alignment telescope is understood to have micrometer measuring capability along one or two axes in the image plane. Most alignment telescopes are packaged for much of their length in 2¼-in-diameter NAS barrels, to fit standard optical-tooling mounts (Fig. 8).

16. Basic Principles An alignment telescope must be capable of being focused on targets at ranges from very close to the objective lens to at least the maximum practical working distance. Focusing is usually accomplished internally, i.e., by axial motion of an internal lens element or assembly, between the objective lens and the eyepiece, which remain fixed. The focusing capability is required so that each of a line of targets can be observed independently, with respect to the telescope's reticle.

As focus is changed, the projection of the reticle cross hairs on the field of view must track a true, straight line. This is the line to which targets are adjusted or, in the case of a measuring instrument, from which their lateral displacement is measured. The precision and trueness of the axial-guidance mechanism is crucial to system accuracy, since it relates the telescope axis as defined by the reticle at one object distance to that defined at another.

Measuring instruments are equipped with micrometers on one or two axes. Image-plane cursors cannot be used as they are in the autocollimator; the field angle subtended by a given linear displacement will vary with target distance despite optical compensation which tends to reduce that effect. The micrometer dials are usually graduated in thousandths of an inch and in some cases are readable to 0.005 in. Each micrometer drives an optical displacement mechanism that affects one axis only. In some cases the vertical- and horizontal-displacement mechanisms are separate optical elements; in others a single element is moved in two directions. A displacement mechanism must be located in the optical system so that the image translation resulting from its operation is truly related to the

Fig. 8 Alignment telescope. (*Kollmorgen Corp.*)

scale of the target at any range. Thus the observed effect is large when the distance to the target is short. It diminishes exponentially as distance increases, becoming zero at infinity.

Displacement of a target center from the line of sight is measured by translating the target center to the telescope cross hairs and reading the micrometer indication. Measuring ranges of typical instruments vary from ± 0.050 to ± 0.100 in.

17. Optical Micrometer The measuring capability of the alignment telescope, other than by relatively crude direct observation of a scale in the target plane, is supplied by the optical micrometer. Its function is to displace the optical axis, or line of sight, laterally (either horizontally or vertically) without deviating it, i.e., without changing its direction. This is usually accomplished by one of the means discussed below. In the planoparallel block (tilt-plate type) (Fig. 9) the flatness of the faces causes all parallel incident rays to be refracted to the same extent, regardless of their distances from the optical axis. Parallelism of the faces ensures that each emergent ray will resume the direction it had at incidence and that each ray incident at a given angle will travel the same distance in glass and thus be displaced the same amount. This type of optical micrometer is usually located in front of the telescope's objective lens, between the target and the objective.

Rotation through a small angle about one axis of a block of sufficient thickness will displace or translate the line of sight proportionately. Displacement of the line of sight is not linear with respect to rotation of the glass plate. In addition, for a given line-of-sight displacement, the required angle of rotation is a function of plate thickness. Therefore, depending on the parameters selected, mechanical compensation to the rate of plate rotation vs. micrometer travel may be required to maintain a reasonable correspondence between line-of-sight displacement and micrometer indication.

Tilt-plate optical micrometers are available as separate assemblies to fit various alignment instruments built around line-of-sight telescopes.

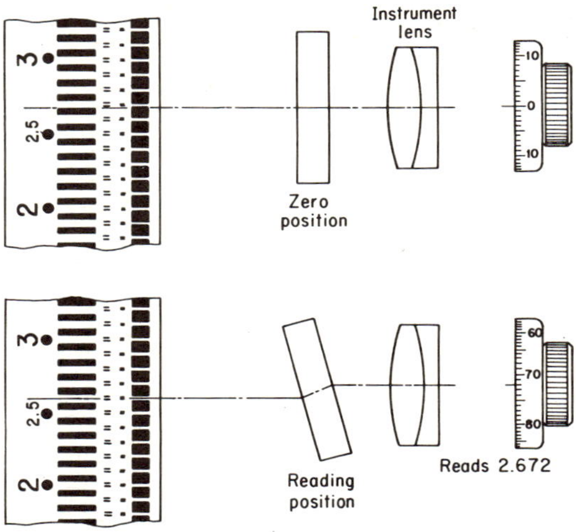

Fig. 9 Tilt-plate optical micrometer. (*Copyright Keuffel & Esser Co.*)

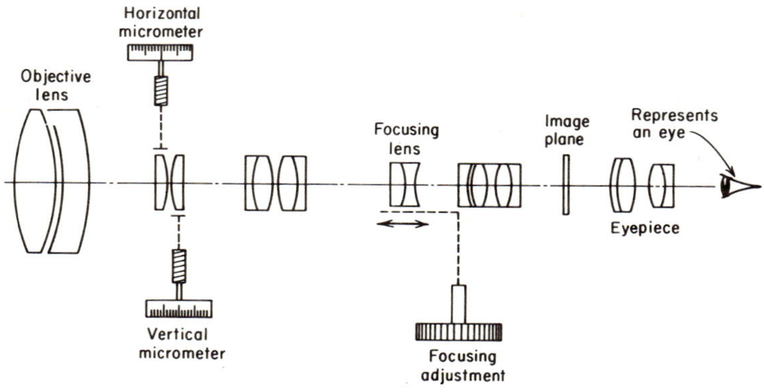

Fig. 10 Alignment telescope. (*Kollmorgen Corp.*)

Other types include the thick meniscus lens employed by Keuffel & Esser in their alignment telescope. It is located in collimated light (object at infinity). Translated along two axes separately, this device produces on each axis the same result as the tilt-plate optical micrometer. In this case displacement of the line of sight is linear with respect to micrometer rotation. The internal optical micrometer of the Kollmorgen alignment telescope (Fig. 10) comprises two auxiliary lenses in the image-forming optical system. These are translated, one vertically and one horizontally, by appropriate micrometer mechanisms. This translation, linearly related to micrometer travel, results in proportional displacement of the line of sight.

18. Applications The primary function of an alignment telescope is to establish a reference line in space, to display criteria for alignment to that line, and to measure displacement from it. This capability is directly applicable to alignment of reference points or to the measure of their misalignment.

Intersecting straight lines define a plane. Intersecting alignment-telescope lines of sight may define a reference plane. A plane may also be swept by rotating a pentaprism or other right-angle reflector around the line of sight.

Alignment telescopes are primary optical tooling instruments which find application across the board in precision manufacture of large structures, such as aircraft and rockets, and in some particular situations in shipbuilding and repair, e.g., shaft-bearing alignment, in plant machinery installation and repair, and verification of large machine-tool alignment and flatness.

19. Techniques Some techniques used with alignment telescopes are briefly described. Detailed instructions may be obtained from the various equipment manufacturers, who often also provide consultation on particular applications. A number of books on optical tooling, listed in the references, further explain the following techniques.

TABLE 3 Representative Telescopes

Manufacturer	Model
Line-of-sight telescope:	
Brunson Instrument Co.	81
Alignment telescope:	
Brunson Instrument Co.	83
Davidson Optronics, Inc.	D-275
Keuffel & Esser Co.	71 2020
Kollmorgen Corporation	K-122
E. Leitz	Alignment telescope
Rank Taylor Hobson	112/636/3970
Tokyo Optical Co., Ltd. (Topcon)	AT-1

Leveling. Referring a series of target positions to a horizontal line of sight, regarding vertical displacement only. The level of the alignment telescope can be established by adjusting its mounting to the indications of a striding level on the barrel, autocollimating or autoreflecting from a pendulous vertical mirror, autocollimating or autoreflecting from a liquid (horizontal) mirror by pentaprism, or collimating to a previously leveled instrument.

Bucking In. Aligning a telescope to a line defined by the centers of two existing targets by repositioning the telescope until the two targets read the same displacement. Also, pointing and training in autoreflection or autocollimation and translating as required to establish a line of sight normal to a fixed mirror.

Aligning. Adjusting a line of transparent targets or the work or fixture supporting the targets to a line of sight as defined by the center of the reticle or to a fixed displacement from the center of a fixed alignment telescope by focusing on each in turn.

Measuring Alignment. Determining displacement of each of a series of targets from the reference line of sight by using the optical micrometer.

Collimation. The alignment of one infinity-focused, reticle-equipped optical instrument to another by angular adjustment until their optical axes are parallel, although not necessarily coincident, as indicated by coincidence of their reticle centers. The projection instrument's reticle must be sufficiently illuminated to be visible in the receiving instrument's field, along with the receiver's own reticle.

Angular Alignment. The alignment of a reflector to be normal to a line of sight. A plane reflector may be made normal in one or both orthogonal planes containing the line of sight, a Porro reflector in the plane of its roof axis only. Alignment may be made by autocollimation, in which the telescope is focused to infinity and the reflected image of its own illuminated reticle is brought into coincidence with its primary reticle by angular adjustment of the reflector. With some instruments an auxiliary lens enables the optical micrometers to measure angular misalignment like an autocollimator.

Alignment may also be by autoreflection, in which the virtual image of an autoreflection target on the front of the alignment telescope is observed in the mirror and the mirror is adjusted until the target and the telescope cross hairs coincide. This method of angular alignment is a less accurate substitute for autocollimation. It depends for accuracy on sufficient distance to the mirror and on the built-in alignment of the autoreflection target center to the telescope's optical axis. The autoreflection target may be a pattern on the foremost glass surface of the instrument, which is common, or a detachable target concentric with the barrel (Fig. 3). Angular alignment of a succession of mirror positions at different distances requires refocusing, introducing another possible source of error.

Generation of a Plane. A plane normal to the optical axis can be generated by rotating a pentaprism about the optical axis.

20. Accessories Alignment telescopes require suitable targets. Due to the variety available, targets are included in the list of accessories. The following accessories are commonly available and useful with alignment telescopes as with other optical tooling instruments:

- Instrument stands and mounts
- Targets and target stands
- Linear scales and stands
- Autoreflection target (built-in in some instruments)
- Autocollimating eyepiece (to provide reticle illumination and/or a more suitable reticle)
- Autocollimating auxiliary lens (to make system focus infinity while retaining internal focus such as to make the optical micrometer read seconds of arc)
- Mounted mirrors of various configurations
- Porro prisms
- Pentaprisms
- Striding levels
- Vertical gravity mirrors
- Horizontal (liquid) mirrors
- Optical micrometer (if not included in the instrument)

21. Sensitivity and Accuracy The alignment telescope's sensitivity to target displacement from the line of sight is a function of target distance. As the distance to the target is increased, the apparent size of the target in the field of view diminishes, although less than it would if magnification were not increasing. With apparent size the visible effect of the optical micrometer also diminishes, since it represents linear measure in the true scale of the target plane. As a rule of thumb, the sensitivity of the instrument is accepted to be 1 part in 200,000 vs. object distance. Differential linear dimensions in the target plane should be detectable at that level; i.e., if there were no other errors in the system, displacement from the optical reference line could be measured to 1/200,000 of the object distance. The overall accuracy of lateral measure from the reference line of sight will usually be slightly less, due to variations in the direction of the line of sight with change of focus. A quality instrument should preserve the direction of the line of sight within 1 second of arc over the focusing range.

Alignment-telescope working distance is limited only by the accuracy required and the size and accuracy of the target. Good illumination is always essential to effective target discrimination.

The eyepiece must be focused first on the reticle while the field of view is obscured (as by a white card). System focus is then adjusted, either to focus on the target or to infinity (for autocollimation) to obtain coincidence of the reticle and objective focal planes. Departure from coincidence will introduce parallax, another source of error. By the same token, the lack of parallax may be accepted as proof of coincident focus. Parallax may be seen as relative motion between the image and the reticle pattern when the eye is translated up and down and right and left through the limits of the exit pupil.

Theodolite

22. Description A theodolite is a geodetic instrument for the measurement of fine angles in at least the horizontal plane and usually in the vertical plane as well. Geodetic theodolites may include a number of additional features such as an inclined telescope for celestial observations and an image-plane, or traverse, micrometer. For metrology purposes the basic horizontal and vertical angular-measuring capabilities are most useful. The addition of an illuminated reticle suitable for autocollimation is often desirable and sometimes essential. The configuration of a theodolite is basically that of the surveyor's transit, i.e., a tripod or stand-mounted instrument carrying a focusing telescope in an altitude-azimuth mount on a leveling base with full-circle horizontal and vertical scales. The lower portion of the instrument carries the horizontal scale and the vertical-axis spindle as well as the supports for the vertical scale and the elevation axis, which in turn carries the telescope; this lower portion must be capable of being leveled and of retaining level to a high degree of accuracy. A theodolite differs from a surveyor's transit principally in the accuracy of its construction and the degree of fineness with which it can be set and the scales read.

Among modern refinements are optical scale-reading systems and compact folded main optical systems. Coincidence optical scale-reading systems counter the effects of scale eccentricity. By simultaneous presentation they average indications from diametrically opposite points on the vertical and horizontal circles.

The (usually) internally focusing telescope is of necessity a line-of-sight telescope.

As in the alignment telescope, the optical axis of the instrument, as defined by the reticle in the field of view, must remain constant as the instrument changes focus to permit angular measurement between objects at different distances.

Within the constraints imposed by the necessary axes, circles, telescope, and the required ability to transit, plunge, and rotate, theodolites take many forms. Compact instruments are generally the most adaptable to metrology.

23. Basic Principles Aside from the refinements to the theodolite which are primarily of value in geodesy and survey work, a theodolite measures angles in two planes as the difference between the readings of the horizontal and vertical scales when the instrument is pointed at one object and the readings of the same scales when the instrument is pointed at a second object. Pointing the instrument may consist of setting the reticle cross hairs in a particular manner on the object in the true field of view or, often in metrology, with the telescope set at infinity focus and the reticle sufficiently illuminated, establishing coincidence autocollimation (see Par. 9) from a particular mirror.

The inherent accuracy ordinarily characteristic of a theodolite is high, e.g., the trueness of running of both the vertical and horizontal spindles, the orthogonality or mutual squareness of the vertical, horizontal, and telescope axes, and the angular accuracy and concentricity to the appropriate axes of the horizontal and vertical scales. In addition the designed capability of plunging and rotating the telescope (rotating both principal axes 180°) makes averaging observations possible, thus canceling systematic errors due to residual nonorthogonality.

24. Representative Instruments Theodolites are normally classified by their precision of readout and presumed accuracy. Most manufacturers make one or more instruments in each class. Typical examples are given in Table 4.

25. Applications A theodolite can be used for any application involving accurate angular measure in the vertical and horizontal planes, either by pointing to targets or autocollimating from mirrors. It can also be used to measure deviation from a near-vertical plane established by two fixed reference mirrors at near 90° separation. Deviation may be read directly with an image-plane micrometer or indirectly by rotation about the (normally vertical) azimuth axis. In the latter case, deviation from the plane is approximately equal to the azimuth dial reading multiplied by the secant of the altitude angle. (When the instrument is not level, azimuth and altitude are properly called *train* and *elevation,* respectively.)

26. Sensitivity and Accuracy The inherent sensitivity and accuracy of a theodolite depend on design and construction of the instrument. Realization of the inherent accuracy depends on ambient conditions, and the application of appropriate techniques.

Design factors include fineness of graduation of scales, magnification and illumination of scales, rigidity and stability, thermal compensation, provision for coincidence reading, line-of-sight telescope and reticle design, optical-system parameters, and slow-motion drives. Construction factors include perpendicularity and concentricity of mounted scales to appropriate axes, scale linearity, orthogonality of axes, and general quality control. Use factors include stabilization in the environment, qualification of the operator, application of error-canceling techniques, and reduction of data.

The accuracy, even the readability, of theodolites ranges from the order of 1 minute of arc to a small fraction, for example, $\frac{1}{20}$ second of arc. Theodolites are often categorized as first-, second-, third-, fourth- (or other) order, based on their capabilities. The terms belong to geodetic-survey triangulation. The allowable average error of closure in the triangles of the several systems is defined by the (U.S.) Federal Board of Surveys and Maps as:

TABLE 4 Representative Theodolites

Manufacturer	Model
Kern & Co., Ltd.	DKM-3
Wild Heerbrugg	T-3
Hilger & Watts, Ltd. (Microptic)	No. 2
Kern & Co., Ltd.	DKM-2
Wild Heerbrugg	T-2
F. W. Breithaupt & Sohn (Teins)	No. 1
Hilger & Watts, Ltd. (Microptic)	No. 1
Kern & Co., Ltd.	DKM-1
Nippon Kogaku, K. K. (Nikon)	NT-2
Wild Heerbrugg	T-1

First order, 1 second of arc
Second order, 3 seconds of arc
Third order, 6 seconds of arc
Fourth order, 6 seconds of arc but yielding computed side accuracy equal to third order

Since the allowable error is a total error, required instrument accuracies are presumed to be higher but are not rigidly defined. First-order theodolites usually read directly to $\frac{1}{10}$ second of arc and allow estimation of smaller angles. Optical tooling and other metrology applications are usually best served by first- and second-order theodolites.

Figure 11 shows a compact first-order theodolite.

Sight Level

27. Description A sight level or tilting level (Fig. 12) is an instrument for establishing a horizontal line of sight and for measuring small vertical displacements from it. It consists of an accurate line-of-sight telescope on a vertical spindle, gravity references (usually a circular spirit level and a sensitive vial-type spirit level), means of making the spindle approximately vertical, and means of slightly elevating and depressing the line of sight above and below a line nomal to the spindle axis. Optical-tooling sight levels are commonly provided with fine adjustments and in some cases readouts in both horizontal- and vertical-plane rotations. Horizontal-plane (azimuth or vertical axis) rotation is normally unrestricted; vertical-plane rotation (tilting) about an axis orthogonal to the vertical axis is usually restricted to a very small angle. Tilting is usually controlled by

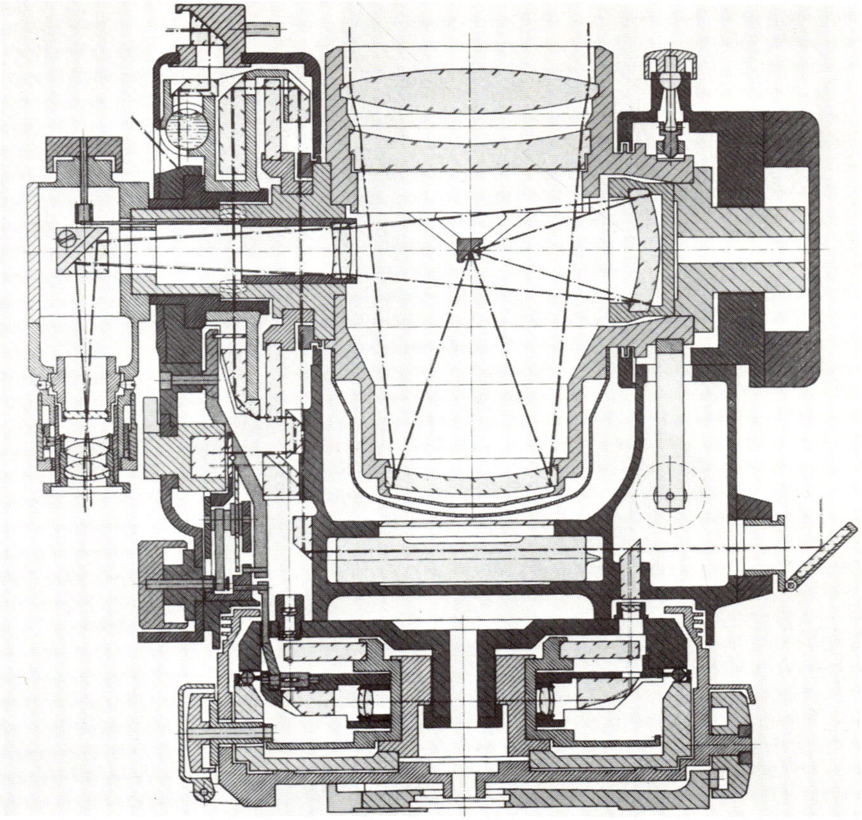

Fig. 11 Theodolite. (*Kern & Co., Ltd.*)

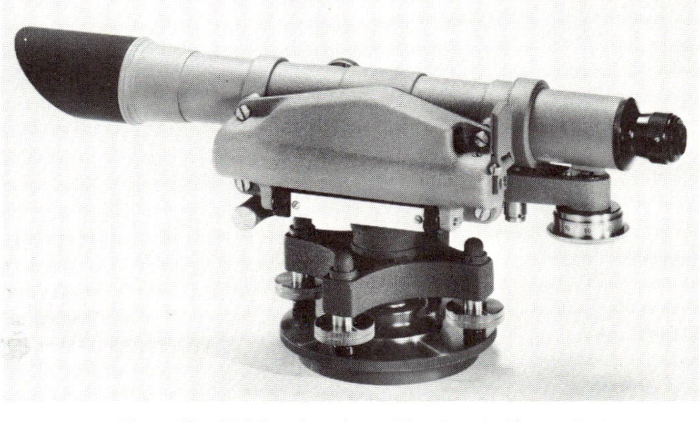

Fig. 12 Tilting level. (*Keuffel & Esser Co.*)

a fine lead screw or micrometer type of adjustment mechanism at a distance from the horizontal axis. A combination clamp-lock and lead-screw adjustment is commonly provided on the vertical axis to permit fine adjustment but not fine readout in azimuth. The sight level may be equipped with such optical-tooling aids as an optical micrometer and an autoreflection target. (The alignment telescope is discussed in Pars. 15 to 21.)

The telescope and the sensitive vial level (usually with an optical coincidence-reading system, see Fig. 13) are mounted on the horizontal axis. The horizontal axis is carried on a deck which has limited azimuth adjustment from the main rotation member about the vertical axis. The telescope is driven about the horizontal axis with respect to this deck, which also carries a spherical spirit level for coarse leveling. The entire assembly is mounted on a tripod or instrument-stand leveling base.

The engineer's Y and dumpy levels, ancestors of the shop sight level, have been largely superseded in the field by automatic levels, a number of which are

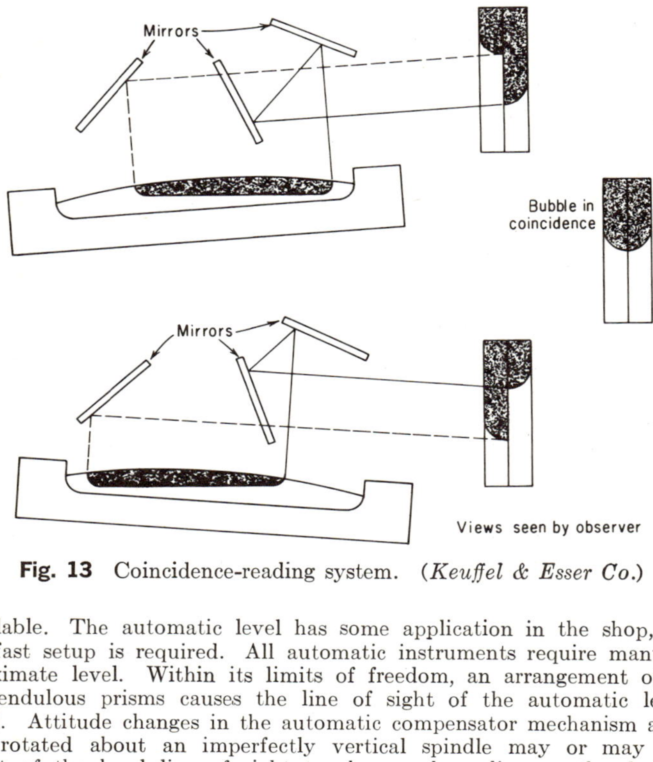

Fig. 13 Coincidence-reading system. (*Keuffel & Esser Co.*)

now available. The automatic level has some application in the shop, especially where a fast setup is required. All automatic instruments require manual setting to approximate level. Within its limits of freedom, an arrangement of self-positioning pendulous prisms causes the line of sight of the automatic level to be horizontal. Attitude changes in the automatic compensator mechanism as the telescope is rotated about an imperfectly vertical spindle may or may not cause the height of the level line of sight to change, depending on the design of the instrument. When this condition may exist, the point about which the line of sight is tilted in the vertical plane does not lie on the vertical axis in all circumstances of adjustment and focus. Thus the height of a Y- or dumpy-level line may vary from one horizontal direction to another.

28. Basic Principles A level line is determined by adjusting the telescope in a vertical plane to the indication of the sensitive vial level at each azimuth orientation from a plane normal to the spindle axis, provided that the spindle axis is first adjusted to closely approximate the vertical. Approximate vertical is established by reference to the coarse (usually spherical) level at several pairs of diametrically opposite azimuth settings of the rotating support.

The definition of a level line of sight relies on the existing relationship between the optical axis of the telescope and the indication of the sensitive vial level. Procedures for verifying and adjusting this relationship are given in the references.

The spindle axis is not required to be perfectly vertical since the line of sight can be leveled independently at each azimuth orientation. Because the horizontal axis intersects the vertical axis, this releveling does not affect the height of the line of sight of the tilting level.

Table 5 lists representative instruments.

TABLE 5 Representative Sight Levels

Manufacturer	Typical model
Sight level:	
Brunson Instrument Co.	545
Keuffel & Esser Co.	71 3010
Nippon Kogaku, K. K. (Nikon)	E-5
Automatic level:	
F. W. Breithaupt & Sohn	4301
Kern & Co., Ltd.	GK-1-A
Nippon Kogaku, K. K. (Nikon)	AP
Carl Zeiss	Ni 2

29. Substitute Instruments The leveling function can be performed by instruments other than sight levels, with varying degrees of convenience and accuracy. Jig transits, theodolites, or alignment telescopes may all be employed, with appropriate separate gravity references where required.

30. Applications The basic functions of a level are to define a horizontal line of sight at a constant height for all azimuth orientations and to measure vertical displacements from that line at varying distances. Ancillary functions, available in some instruments, include measurement of lateral (horizontal) displacement from line or plane, angular alignment by autoreflection, and measurement of small angular displacements from the horizontal.

31. Accessories Since the sight level can be described as a special-purpose alignment telescope, most of the accessories listed in Par. 20 are applicable to the sight level as well.

32. Sensitivity and Accuracy A quality sight level is considered capable of establishing a level line of sight to an accuracy of 1 second of arc. As in the alignment telescope, the accuracy of projection of the level line of sight in depth depends on the accuracy of the focusing slide mechanism in the telescope. Operating accuracy, sensitivity, and applicable precautions may be taken as equal to those of the alignment telescope (Par. 21).

The effect of curvature of the surface of the earth is normally disregarded in shop-range optical tooling. However, when the range is sufficient and the working accuracy high enough, it should be borne in mind that local gravity vectors are inclined away from each other 1 second of arc for approximately each 101.3 ft of horizontal distance.

Twist Autocollimator

33. Description The twist autocollimator is an adaptation of the autocollimator to the measurement of rotation about the autocollimator's optical axis. A twist autocollimator could be identical in concept and structure to a standard tilt-angle-reading autocollimator except that the reticle and cursor must be configured to be sensitive to image rotation rather than to translation in the image plane.

34. Basic Principles A roof-edge reflector (a Porro prism or Porro mirrors) is necessary to make object rotation visible as rotation of the returned image of the reticle. The returned image rotates about its own center through twice the relative rotation between the autocollimator and the Porro reflector. At the same time the center of the returned image rotates about the reticle center through an angle equal to the relative rotation between the autocollimator and the Porro reflector at a radius proportional to the nonorthogonality of the roof edge to the optical axis. The above effects can be observed with any autocollimator; a twist autocollimator must be designed to measure them.

35. Representative Instruments The twist autocollimator exists as a design concept which has been applied in particular instances, rather than as an off-the-shelf component.

36. Applications The more sensitive (2×) rotation of field will normally be the object of measurement; however, the (1×) rotation about the reticle center may be of value in resolving ambiguities when the extent of rotation to be measured exceeds ±90°. When an eccentric return is permitted or desired, provision must be made for translating the rotary cursor to maintain concentricity with the returned image.

The operating distance of such an instrument is circumscribed by a limitation inherent in all autocollimators, i.e., the reduction of field with increasing distance to the mirror, even for an on-axis return.

37. Sensitivity and Accuracy A manual twist autocollimator is relatively insensitive but can accept measured-body rotation up to 360° without ambiguity. Sensitivity is a constant to the designed maximum operating distance; beyond that it is a function of distance to the reflector, since the field diameter of the returned image governs the circumferential displacement of an edge of the pattern for a given rotation.

By the criteria given in Par. 14 for the standard autocollimator, circumferential motion equal to 2.5 μm of linear motion at the edge of the field, due to return-image rotation, should be detectable. This can be translated to angular sensitivity as shown below. The following relationships hold: detectability of relative rotation between roof edge and autocollimator is a function of the linear radius of the returned image, assuming a matching cursor; the operating range (maximum distance to reflector) depends on the field angle subtended by the reticle.

For example, a twist autocollimator having a rotary cursor to match a returned image at approximately $\frac{1}{10}$ in (2.5 mm) radius and an objective of appropriate focal length would not exceed the detectivity

$$\theta = \tfrac{1}{2} \tan^{-1} \frac{d}{r} \tag{4}$$

where θ = minimum detectable rotation of Porro
d = minimum detectable linear increment = 2.5 μm
r = radius of cursor = 2.5 mm, corresponding to $f \tan \alpha$
f = objective focal length
α = half-field angle of reticle pattern

$$\theta = \tfrac{1}{2} \tan^{-1} \frac{2.5 \times 10^{-6}\,\text{m}}{2.5 \times 10^{-3}\,\text{m}} = 1.75 \text{ minutes of arc}$$

Alignment Transits

38. Description This broad category embraces a number of instruments derived from the surveyor's transit. Alignment transits include jig transits (Fig. 14), transit squares, and the unique cross-axis-telescope transit square. Their basic function is to establish a plane, either vertical or normal to a line of sight. Some can measure angles in that plane. The common features of alignment transits include a leveling base, a vertical axis of rotation, a line-of-sight telescope mounted on horizontal-axis trunnions, which in turn are carried on the vertical axis, and, most important, a means of determining the orientation of the horizontal axis. The latter may take any form including (1) a fixed mirror in factory-set relation (normal) to the elevation axis; (2) an axle-mounted mirror, near normal, rotating with the telescope in elevation; this type permits verification of the normality of the mirror to the elevation axis or a measure of its misalignment; (3) a hollow axis with a partially silvered reflector at one end; this type permits establishing more than one

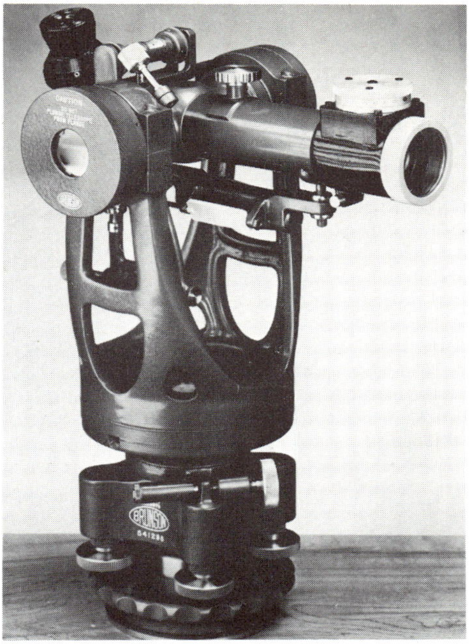

Fig. 14 Jig transit. (*Brunson Instrument Co.*)

transit normal to a line of sight; or (4) a line-of-sight telescope in the horizontal axis.

The transit telescope and the cross-axis telescope may possess all the features of the line-of-sight telescope in other applications, e.g., internal focusing and autoreflection target. Either or both may also carry a displacement optical micrometer or provision for autocollimating.

39. Basic Principles An alignment transit is the most nearly universal of all optical instruments for shop metrology. The principles underlying its operation are to some degree those of the theodolite, the sight level, the alignment telescope, and the autocollimator, depending on the chosen mode of application and the accessories employed. Its primary function remains the generation of a vertical plane or a plane normal to a line of sight.

TABLE 6 Representative Transits

Manufacturer	Model
Brunson Instrument Co.	71
Keuffel & Esser Co.	71 1010
Tokyo Optical Co. (Topcon)	JT-2
Optical transit square:	
Brunson Instrument Co.	76

40. Application The transit may be used as a sight level, alignment telescope, or pointing autocollimator by the methods described for the basic instruments. In its plane-generating application, it is leveled and adjusted in azimuth to autocollimate the autocollimator or to autoreflect to the alignment telescope, sight level, or another transit defining either the reference direction or the line of sight. The theodolite's plunge and rotate (or transit) technique is used to define each line in the swept plane, to eliminate potential errors due to nonorthogonality of the transit's intersecting axes. Horizontal rotation is constrained to 180° by indexing to the axle mirror, to autocollimate or autoreflect either to previously aligned instruments or to the same instrument twice, once directly from an outside surface and, after horizontal rotation, through a hollow elevation axis from the inside of the same surface.

Displacement from the swept plane can be measured using an auxiliary optical micrometer with alignment-telescope techniques.

41. Sensitivity and Accuracy Deviations from the generated plane can be measured to the order of 0.001 in at 17 ft, or about 1 second of arc. Factors involved in the accuracy of definition of a normal plane include line-of-sight telescope factors, horizontal-axis wobble, vertical-axis orientation, autocollimating or autoreflecting errors in setting azimuth, and setting error in plane-defining mirrors or points. By the accepted accuracy of first-quality metrology instruments, each of these may contribute the equivalent of 1 second of arc. However, the concurrent existence of all maximum errors having the same sign of effect is unlikely. Where high accuracy is required, a statistical approach to determination of probable error (see Par. 90) should be taken.

AUTOMATIC AND PHOTOELECTRIC ALIGNMENT INSTRUMENTS†

Fundamentals of Automatic Alignment

42. Introduction Both the nature of the measurement and the property of light explicitly utilized in the measurement are important in characterizing the automatic metrological instruments to be discussed. The following instruments will be considered: automatic autocollimator, automatic alignment telescope, automatic twist autocollimator, automatic polarimeter, automatic optical range finder, and ranging optical probe. This list specifically excludes those instruments whose operation is based on interference and therefore depend on the coherence property of light. These instruments are discussed in Pars. 67 to 78.

The autocollimator, alignment telescope, and twist autocollimator are imaging instruments in which the error in precise superposition of an automatic reticle and its autoreflected image is measured radiometrically and can be correlated with misalignment of an appropriate autoreflecting element on the body being observed. These instruments depend on the straight-line propagation of light to ensure a known relationship between autoreflecting-element movement and the corresponding image movement and on the short wavelength of light to obtain good image quality from a reasonable aperture size.

The polarimeter measures radiometrically the departure of a polarized beam of light from the orientation of minimum transmission through a polarizing analyzer. This departure can be correlated with a rotation about the line of sight of the source of polarized light on the body being observed. The polarizing element is generally the final element in the quasi-collimated transmitted beam, and the analyzer is the first element in the receiver. Polarimeter measurement is essentially independent of image quality and straight-line propagation and depends only on the polarization property of light.

The optical range finder measures distance to an appropriate reflector as the round-trip transit time of a light beam and depends principally on the known velocity of light. In some cases the actual transit time of optical pulses is measured, while in the geodimeter, distance is determined by the number of wavelengths of radio-frequency-modulated light between the instrument and the reflector.

The various optical probes utilize several techniques to measure very small changes in distance to a nearby body, all based on the convergence of light at a relatively steep angle to a focal point at the body. Their operation depends on the straight-line propagation of light and its short wavelength.

The most important distinction between manual and automatic alignment instruments is the ability of the latter to provide an electric signal indicative of error

† Paragraphs 42 to 63 were written by J. Richard Vyce.

in the alignment relationship of interest. Implicit in this automatic measurement ability are two major factors.

1. An automatic optical alignment instrument must possess a means of sensing its alignment condition relative to a remote body of interest. This suggests the transmission of appropriately coded light between the body and the instrument, an element on the body to modify the transmitted light to an extent indicative of the amplitude and direction of misalignment, a means of analyzing the light received to develop an optical error signal, and finally a detector to convert the optical error signal into an electric signal for convenient amplification and processing.

2. The instrument must be able to detect an alignment error of some minimum magnitude. Thus, the signal corresponding to the desired minimum detectable error must be detectable in the presence of all sources of noise in the instrument and in the environment.

To provide a realistic framework for discussion of the operating basis of automatic alignment instruments, the automatic autocollimator will be investigated in some detail. The subsequent discussion of other instruments will generally be patterned after that of the automatic autocollimator.

43. Error-Signal Generation *Light Coding, Transmission, and Analysis.* The most important feature of any automatic alignment instrument is how its trans-

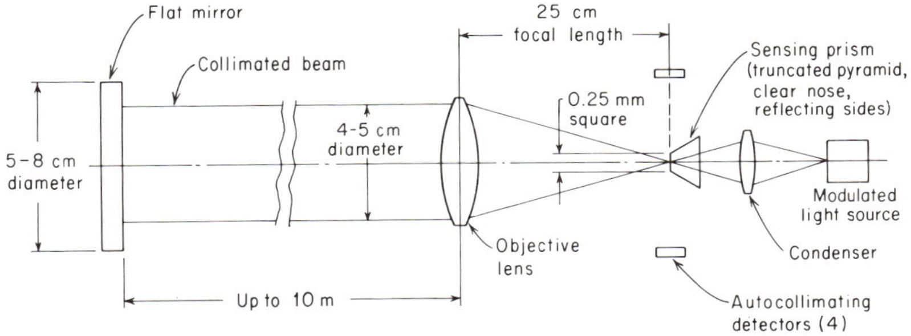

Fig. 15 Operating arrangement of two-axis autocollimator and typical dimensions.

mitted light is coded to allow measurement of the alignment of interest. (Coding here designates the spatial or temporal information imposed on the transmitted light beam that is modified by interaction with the observed alignment elements.) The model to be used for all alignment situations in this discussion is that of an alignment instrument comprising both transmitter and receiver elements with a common optical aperture and an optical element on the body being observed that serves to modify the information on the incident light beam in relation to the degree of misalignment and to reflect the light back to the alignment instrument.

The automatic autocollimator measures body orientation about axes normal to the line of sight. Its operation (Figs. 15 and 16) is based on (1) collimation by the objective lens of the light from an illuminated reticle (sensing prism) and transmission of the collimated beam to a flat mirror on the body; (2) reflection of the beam back to the autocollimator aperture by the flat mirror; (3) autocollimation, or formation of the reflected image of the reticle in the plane of the reticle; and (4) determination of the lateral displacement between the reticle and its autocollimated image by measuring the amount of light reflected from each face of the sensing prism. The relation between image displacement Δr, mirror tilt $\Delta\theta$, and objective focal length f is $\Delta r = 2f\,\Delta\theta$, as shown for the manual autocollimator in Eq. (3). Thus, detecting a 5-μrad (1 second of arc) mirror tilt using an automatic autocollimator with a 25-cm focal-length objective means that 2.5 μm of lateral image displacement must be detected. This degree of detectivity is quite readily attainable. One evident advantage of the automatic autocollimator is that image displacement corresponding to a given tilt angle is independent of operating distance.

The problem is how most effectively to configure the light source, reticle, and detector (referred to here as the automatic reticle) for best automatic autocollimator performance. A number of solutions are embodied in the various commercial and military autocollimators, and the configuration illustrated is representative.

Because of its ability to discern detail in an extended field, the eye can readily detect displacement in reticles of simple line pattern when used as a displacement detector in a visual automatic autocollimator; the nonimaging detectors most desirable in automatic autocollimators are simple, although highly sensitive, radiant-power detectors. They therefore can detect image displacement only if it causes a change in the level of light they receive. Hence the automatic reticle must be configured to produce such an energy change in the presence of autocollimated image displacement. As shown in Figs. 15 and 16, lateral displacement of the autocollimated image causes an imbalance of the light falling on opposing detectors. The difference in their signals indicates the magnitude of displacement, and the polarity indicates its direction.

This configuration performs all the functions required of an automatic reticle in an automatic autocollimator. Coding of the transmitted beam is the spatial

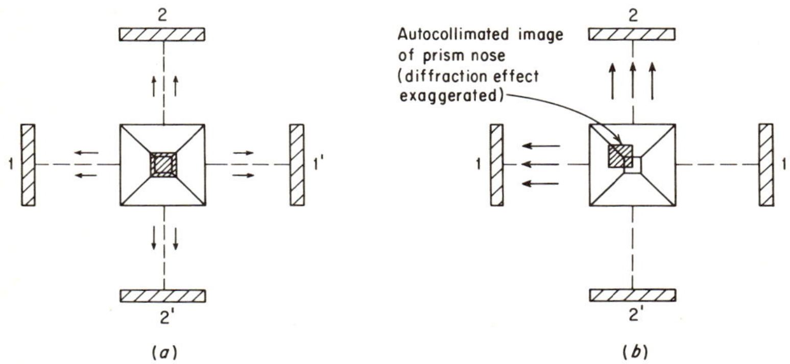

Fig. 16 Signal generation for two-axis automatic autocollimator: (*a*) null condition: small, balanced amount of light to all detectors due to diffraction and aberration; (*b*) two-axis error condition: light energy reflected to detectors 1 and 2 triggers electric signals.

distribution of light from the precisely defined prism nose; light transmission is via the collimated beam; interaction of the incident beam with the mirror on the monitored body results in a deflection of the reflected beam; this in turn causes a displacement of the autocollimated nose image; the displacement is analyzed by the sensing prism, which isolates the light in the image falling outside the nose, and is detected when the isolated light is reflected to the detectors. Finally the direction of displacement is indicated by the polarity of the detector output.

Signal-generation processes in other automatic alignment instruments are fundamentally similar to that of the automatic autocollimator.

Reticle Illumination. Since the radiation utilized in automatic optical alignment instruments is light, their radiation sources are simply light sources. In addition to emitting optical radiation, alignment-instrument light sources may have to meet a number of other requirements, including long life, good shock and vibration resistance, low cost, and modulation capability.

The tungsten-filament lamp is the most common light source in automatic alignment instruments, although mercury-vapor, glow-discharge, and recombination-radiation diodes are also employed. The important light-source radiometric characteristics are size and uniformity of the illuminated area, directional and temporal radiance uniformity, and high radiance with good efficiency.

Most alignment instruments require for operation a small brightly illuminated reticle pattern or spot. In such cases, the light source is usually imaged directly onto the reticle by a condenser lens large enough to ensure that the light from the reticle will fill the instrument objective lens.

For error sensing all instruments whose performance depends on good image quality utilize edges within or at the periphery of the illuminated pattern, which should be uniformly illuminated if erratic performance is to be avoided. It is also important that the objective-lens aperture be uniformly illuminated to avoid degradation of the quality of the projected image.

The factor of interest in radiometric analysis is lamp radiance N in watts per square centimeter per steradian, which gives the total amount of radiant power per unit area emitted in a solid angle of 1 sr. Because lamp radiance and detector sensitivity vary with wavelength, both must be known over the spectral region of interest to determine the actinic sensitivity of the source-detector combination.

For most applications, radiance N varies from about 30 W/(cm^2)(sr) for a derated tungsten filament to 3,000 W/(cm^2)(sr) for a small high-pressure xenon lamp. Actinic radiance, which is effective within the spectral bandwidth of typical detectors, covers the range from about 1 to 1,000 W/(cm^2)(sr) for the same sources. Light-emitting diodes have much lower radiance than more conventional sources.

Transmitted-Light Modulation. The automatic autocollimator has been discussed thus far as a dc instrument in which the change in light level at the detectors caused by mirror rotation is converted into an electric signal and amplified as required to produce an output signal of sufficient amplitude. Although this mode of operation is feasible and is employed in some older instruments, there are two problems with dc operation: (1) In most operating environments there is ambient dc light which can enter the autocollimator and, if nothing else, cause a background signal that degrades error resolution. More frequently, the light level is unsteady or fluctuating, and these level changes produce an erroneous output signal. (2) Solid-state detectors exhibit internal noise that increases at low frequencies and has a maximum at direct current.

Both the ambient-light and detector dc-noise problems are solved by modulating the transmitted light at an electrically convenient carrier frequency that should be well separated from normal power frequencies and their harmonics. In this case the detected signal is typically ac amplified with moderate filtering and then synchronously demodulated to dc, where very narrow band filtering can be done. Because this final narrow-band filtering effectively occurs around the ac carrier frequency, all ac and dc noise outside this band is effectively eliminated. The resulting automatic-autocollimator error resolution is greatly improved, and the instrument is substantially immune to ambient light.

Instrument light sources can be modulated in numerous ways, including direct source modulation, electrooptical modulation, mechanical chopping, etc. All have advantages and disadvantages, but the trend is to directly modulated light-emitting diodes and to nonmoving electrooptical modulators.

All other automatic alignment instruments also benefit from transmitted-light modulation for the same reasons, but in most other instruments the modulation is an essential part of the instrument concept.

In the automatic optical probe, modulation is required for the identification of overlapping images formed by multiple lens apertures; in the automatic alignment polarimeter it is required for scanning a plane of polarization through an angle greater than the range of minimum polarizer transmission; and in the automatic optical range finder it is necessary to divide the transmitted light into the temporal segments required for distance measurement. Thus, optical-signal modulation is an essential function in all high-performance automatic alignment instruments.

Detectors. Photoelectric detectors are used to convert optical signals into electric signals. We have already shown how the automatic autocollimator is configured to develop optical error signals in the presence of mirror misalignment. Five detector characteristics are of importance in converting these error signals from optical to electric: spectral response, quantum efficiency, time constant, intrinsic noise level, and gain.

Spectral response indicates the range of optical wavelengths to which the detector responds. Quantum efficiency is a measure of the number of initial photoelectrons (always ≤ 1) released by each incident photon in the optical signal. Since detector spectral response does not normally have sharply defined limits and quantum efficiency is usually variable with wavelength, these two characteristics are normally presented together in graphical form as spectral-sensitivity curves.

Detector time constant is only occasionally important since most detectors have sufficiently fast response to permit 1,000-Hz bandwidth and the bandwidth normally required in most alignment instruments is less than this except for the automatic optical range finder.

Intrinsic detector noise, usually expressed as equivalent watts of incident radiation for a given measuring bandwith, sets a lower limit on the optical signal that can be detected under given conditions. Finally, detector gain is the specification of the detector output signal that results from a given optical input signal.

At a wavelength of 550 nm, 1 W of light generates a current of about 0.5 A in a detector of unity quantum efficiency and no internal amplification.

Of all detectors with sensitivity in the wavelength range from 0.5 to 1 μm commonly used in alignment instruments, the two most frequently employed are the photomultiplier and the silicon photodiode. The photomultiplier has much higher gain produced by high internal amplification, lower noise, and wider frequency response than the silicon photodiode but is large, requires a stable supply of high voltage, and exhibits variable gain and fatigue in operation. The silicon photodiode is extremely small and convenient to use, has adequate gain and spectral response and low noise and frequency response for most applications, and requires no power. It tends to be temperature-sensitive, but this tendency can be greatly reduced by appropriate output-circuit design. Its low output-signal level occasionally necessitates a sophisticated low-noise preamplifier to avoid degradation of signal-to-noise ratio.

Because the various alignment instruments typically have optical signal levels of no less than 10^{-11} W/μrad, the silicon photodiode with its noise level of 10^{-12} W or less is a satisfactory detector in all cases with the possible exception of the automatic optical range finder in which extremely fast response is necessary. In practice, a large number of laboratory alignment instruments are still made with photomultipliers, partly because of their historical precedence, partly because they can be less expensive and permit convenient gain variation by high-voltage adjustment, and partly because they readily tolerate the low optical gain inherent in many instrument designs, particularly those using beam splitters or resulting from unusually high attenuation in the external optical path. Nevertheless, military alignment instruments have employed solid-state detectors for many years, and the trend is strongly in that direction for all instruments.

Electronics. Although most automatic alignment instruments use electronics for light modulation and detector-signal processing, the electronics (beyond the detectors) usually need not critically affect instrument performance. This is so because of the relatively high signal-power level that should exist at the detector output in a properly designed instrument; i.e., the intrinsic output noise level of both photomultipliers and most good solid-state detectors is well above the Johnson noise level at the input of a well-matched low-noise preamplifier. So long as preamplifier noise is lower than intrinsic detector noise, the electronics need not degrade instrument performance (Fig. 17).

The principal electronic signal-processing functions are amplification, impedance transformation, filtering, and heterodyning (see above). Amplification and impedance transformation are required to bring the detector output signal to a level and impedance compatible with normal display and recording devices or servoamplifiers.

Heterodyning and filtering are required for extraneous-noise rejection, as discussed under *Transmitted-Light Modulation.* Alignment instruments transmit light preferably modulated at a frequency well apart from power frequencies and their harmonics to minimize susceptibility to external dc and ac light and high enough to avoid most low-frequency detector noise. Narrow-band filtering is desirable to reduce photon and electronic noise and atmospherically induced signal fluctuations. Heterodyning is required to convert from the modulation frequency to a standard frequency or direct current. The need for narrow-band filtering usually dictates that the conversion be to direct current. Because alignment-error directional information is usually given by the ac error-signal phase, preservation of this information as dc polarity requires synchronous demodulation of the ac signal.

These electronic processes are straightforward and may be implemented in numerous ways. Because there is generally no technical reason for electronic noise to limit alignment-instrument performance, its limitation by this source is a result either of a conscious economic decision or poor design.

44. Accuracy The evaluation of any alignment instrument's accuracy is most naturally separated into short- and long-term behavior. Short-term behavior is what can be called error resolution or detectivity. Long-term behavior is usually called stability. The boundary between the two is wide and strongly influenced by the dynamic nature of each alignment situation. A useful guide to their dividing line is a period equal to 1,000 time constants: performance deviations within that period are attributed to the instrument's detectivity; deviations over longer periods are attributed to stability.

An instrument's detectivity and stability ultimately must be proved by experiment. We consider here the analytic factors involved in the synthesis of an instru-

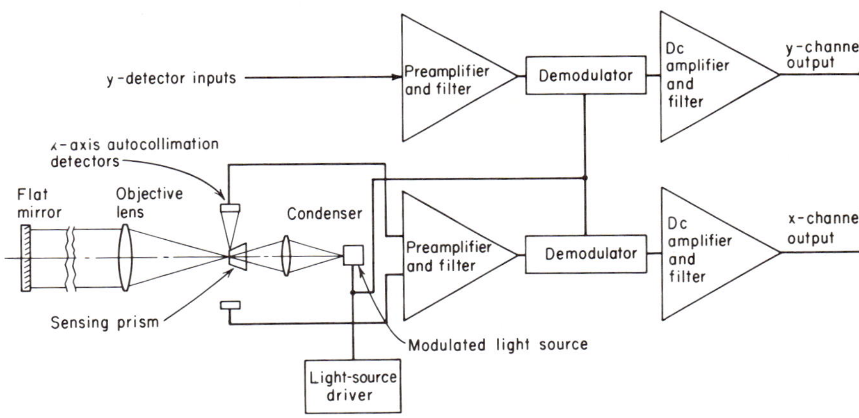

Fig. 17 Functional diagram for two-axis automatic autocollimator; y axis normal to the page; y detectors not shown.

ment to meet a given accuracy specification. Design synthesis to obtain given detectivity invariably involves radiometric analysis to determine instrument sensitivity, comparison with analytic or estimated noise, and adjustment of design parameters to ensure an adequate signal-to-noise ratio at the required detectivity level. Design synthesis to ensure instrument stability consists of electrooptical analysis to determine stray light and electrooptical balance characteristics, optomechanical analyses to determine the dimensional and optical variation caused by expected thermal disturbances, and adjustment of design parameters to maintain long-term null shift within an acceptable level.

Radiometric Analysis. When the automatic-autocollimator configuration, its basis of operation, and the magnitude of optical effects produced by a given alignment error are known, it is possible to obtain a quantitative understanding of automatic-autocollimator behavior by the calculation of optical signal levels and detectivity-limiting noise levels, commonly known as radiometric analysis.

In radiometric analysis it should be understood that most alignment instruments are limited in their detectivity not by fundamental noise factors, which can be accurately calculated, but by extraneous and frequently incalculable noise sources. These include internal stray light, background light, relevant variations in the intervening medium (usually the atmosphere), and detector and electronic noise levels which are above fundamental limits. Thus radiometric analysis need not (and in many cases cannot) be especially precise: order of magnitude is often sufficient, and only rarely is accuracy of better than a factor of 2 called for.

A typical automatic autocollimator has the operating geometry shown in Fig. 15. Its operation is based on sensing the lateral displacement of the autocollimated illuminated reticle pattern caused by body-mirror tilt. The process of radiometric analysis, as indicated in Fig. 18, involves the following steps:

1. Selection of a light source and determination of its radiance in watts per square centimeter per steradian at the illuminated reticle.
2. Calculation of the geometric transmission of the optical system, a gain factor $\ll 1$ involving the solid angle subtended by the objective lens and/or collimating mirror.
3. Determination of optical-system transmission, a gain factor <1 covering the losses at reflective and refractive surfaces in glass, from filters and beam splitters and in the optical path.
4. Calculation of the illuminated image area in square centimeters swept out by 1 μrad of mirror tilt.
5. Multiplication of steps 1 to 4 to obtain total automatic-autocollimator radiometric sensitivity in watts per microradian.
6. Determination of the source actinic efficiency, a gain factor <1 covering the integrated product of lamp spectral radiance and detector spectral sensitivity indicative of the portion of incident radiation converted to photoelectrons in the detector.

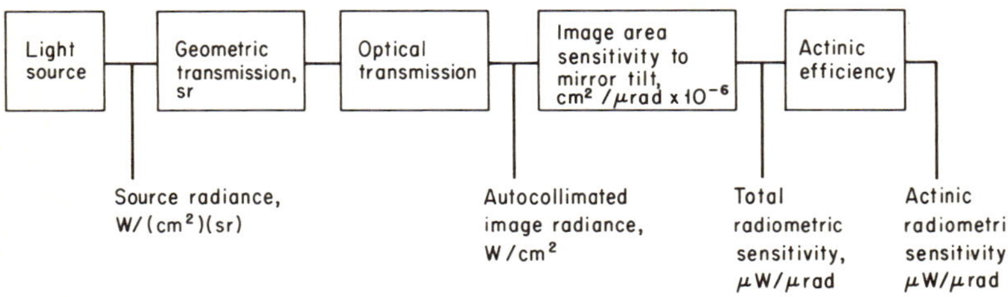

Fig. 18 Autocollimator radiometric-analysis procedure.

Intrinsic automatic-autocollimator behavior is established by steps 2 to 4, but automatic-autocollimator radiometric sensitivity can be varied by the appropriate choice of light source and detector.

When this procedure is applied to a typical automatic autocollimator with a 5-cm-diameter aperture and a derated tungsten-filament lamp designed for working at a distance of 500 cm or less, we find that its total radiometric sensitivity is about 10^{-6} W/μrad. Its actinic radiometric sensitivity is about 10^{-7} W/μrad with a silicon photodiode and about 10^{-9} W/μrad with a photomultiplier. Actinic radiometric sensitivity can be measured directly at the automatic-autocollimator detector output if the autocollimating mirror can be offset through a known angle. For silicon photodiodes, which have a radiometric sensitivity of about 0.5 A/W, the measured sensitivity of the automatic autocollimator above would be about 50 nA/μrad.

Noise. An instrument's radiometric sensitivity alone tells us little about its performance, for with adequate electronic gain, a given level of noise-free radiant signal can be converted to any desired magnitude of electric output signal. The ability to do this usefully is restricted, however, by the presence of noise, which invariably accompanies the signal and prevents the detection of misalignment below a noise threshold.

There are two basic classifications of noise sources in using alignment instruments: those associated with the instrument and those associated with the environment. Instrumental noise sources consist principally of photon noise in the radiometric signal, photon noise in internal stray light, photoelectron noise in the currents resulting from these two radiometric fluxes, modulation noise caused by internal reflection of scanned or amplitude-modulated transmitted light, detector noise, electronic amplifier noise, etc. These noise sources are characterized by fixed absolute magnitudes for a given set of operating conditions; i.e., it is possible in the presence only of instrumental noise sources to say that an instrument has a definite detectivity, or ability to detect a given small misalignment. Detectivity is frequently defined as the magnitude of misalignment whose signal is equal to the noise level in the integrating time chosen for the instrument.

Environmental optical noise sources consist of background light and any signal-varying effects in the medium through which the alignment takes place. Background light, if constant, introduces only photon noise, but it can be the source of greater noise if the level observed by the instrument fluctuates. The one regularly observed signal-varying effect of an environmental medium is atmospheric shimmer. This factor causes both reduction of image quality and image movement, thereby degrading the detectivity of alignment instruments whose performance depends on image quality and position.

Polarizing instruments are essentially immune to this environmental noise source but suffer instead from a somewhat variable polarizing-rotation interaction between the atmosphere and the earth's magnetic field. The important point about the signal-varying effect of the alignment medium that distinguishes it from all other noise sources is that the magnitude of its effect is independent of the instrument. Regardless of the radiometric sensitivity or intrinsic detectivity of various automatic autocollimators, when the atmosphere causes, say, 10 μrad of image motion, all automatic autocollimators are degraded to that level of detectivity. The actual magnitude of atmospheric disturbance covers a range of 1,000 to 1 or more, allowing 10^{-2} μrad detectivity in special short-range laboratory automatic autocollimators and only 10 μrad detectivity (or worse) in poor outdoor situations.

Let us look again at the automatic autocollimator which exhibits an actinic sensitivity of 10^{-7} W/μrad with silicon photodiodes. The most fundamental automatic autocollimator noise source is photon fluctuations in the diffracted portion of the autocollimated image that must fall on the detectors at null. Its magnitude is about 10^{-13} W. An automatic autocollimator having this as the largest noise source would have a detectivity of 10^{-13} W per 10^{-7} W/μrad $= 10^{-6}$ μrad. Internal stray light or external background light could cause more noise than diffraction, but (unlike diffraction) these noise sources usually are controllable.

The intrinsic noise level of silicon photodiodes is below 10^{-12} W and limits automatic-autocollimator detectivity to 10^{-5} μrad.

This exceedingly good detector-limited detectivity is well beyond the smallest angle known to have been measured and is certainly far better than needed in normal applications. Moreover, although electronic noise need not degrade it, internal stray light, modulation light, and atmospheric shimmer practically make this level of detectivity unattainable and probably even unapproachable.

The best reported automatic-autocollimator detectivity, about 10^{-2} μrad, was obtained with an unusual automatic autocollimator which observed the autocollimating mirror from about 2 cm. Despite the short working distance and an enclosed air path, atmospheric shimmer was the limiting noise source.

What this says is that the automatic autocollimator is unlikely to be limited in performance by noise from diffraction, detectors, or electronics but most probably by noise from the atmosphere. In a poorly designed instrument, stray light or electronic noise could be larger than that of the atmosphere. An extremely bright background can also cause more noise than atmospheric shimmer.

As a guide to automatic-autocollimator design, these considerations indicate that silicon photodiodes and photomultipliers, if properly applied, will not limit automatic-autocollimator performance and are therefore more than adequate for any application. There is some justification for using the brightest possible light source, however, in any application where the atmosphere is not already the performance limitation. This follows from the fact that the automatic-autocollimator signal increases directly with lamp radiance whereas the noise from internal light and diffraction increases only by the square root of the radiance and background light noise presumably is fixed. Automatic-autocollimator signal-to-noise ratio and detectivity therefore improve with the square root of lamp radiance if internal light noise dominates and improve directly if background noise dominates. In general, the practical performance limit is achieved in any alignment situation when the optical signal is large enough to reduce the detectivity limits of all other noise sources below that of atmospheric perturbation.

Stability. Alignment-instrument stability is functionally defined as the long-term change in zero indication, or null shift, in the absence of change in alignment between the instrument housing and the monitored optical element. Stability is subject to two principal sources of perturbation, electrooptical and optomechanical. Electrooptical instability arises from a change in the balance of dual elements, e.g., two light sources or two detectors, which combine to generate an error signal, or from a change in the balance of stray or background light on two detectors. The specific effect of balance change is to produce a nonzero signal from previously balanced background light or diffracted light. Cancellation of this signal to regain a null output requires a true error signal of opposite sign. The misalignment of the monitored optical element needed to generate the canceling error signal is the magnitude of electrooptical null shift.

Some alignment instruments exhibit a null at a condition of alignment that departs some amount from the ideal, usually because of internal stray light. This so-called null offset is observed as an indicated error signal with the alignment path obstructed. Even if no further balance change occurs, the presence of null offset makes the instrument susceptible to null shift as operating conditions change. The null-offset signal caused by internal stray light, for example, has a fixed magnitude independent of external conditions, but the magnitude of misalignment required to cancel the offset depends on the external-alignment-path optical gain, which in many cases can vary.

Conversely, two unbalanced detectors in the absence of stray or background light can produce a null output under varying conditions of alignment, depending on the optical quality of the monitored optical elements, defocusing, or use of different parts of the instrument aperture. Here the root of the null shift is the degradation in image quality caused by poor-quality monitored elements or defocus or by increased diffraction if less than the full instrument aperture is involved in image formation. The poorer image spreads more light onto the detectors, causing a larger unbalance signal, which requires a correspondingly larger misalignment or null shift for its cancellation.

There are a number of ways of avoiding significant sensitivity to electrooptical null shift by appropriate instrument design. Use of a single light source and detector in an automatic autocollimator, for example, almost completely eliminates the likelihood of electrooptical null shift. However, this approach complicates automatic-autocollimator design by requiring a chopper or electrooptical modulator to obtain directional error information. Fortunately, even if dual elements are employed, a properly designed instrument can achieve an extremely high degree of electrooptical null stability.

Optomechanical instability is normally caused by internal or external temperature changes, which influence the dimensions of mechanical and optical parts and the optical properties of refractive elements. Null shift in this case results from movement of parts and optical deviation relative to the instrument mounting surface and from defocus.

The usual approach to maximizing optomechanical stability involves minimizing internal thermal sources and using high-stability materials of low and matching expansivity, optical designs with loose alignment tolerances, kinematic element mounting, design symmetry, etc. Generally, instruments using refractive lenses cannot be made as stable as those using reflective lenses, because the glass in refractors is of high and dissimilar expansivity and cannot be matched to normal cell metals. Reflectors and their critical structure, on the other hand, can be made completely of optical materials having expansivity less than 10×10^{-9} per Celsius degree. But reflectors have disadvantages, including a central aperture obstruction in most cases, that make them less desirable for many applications. They are rare among commercial instruments and have been used only in special high-value situations. Thus excellent optomechanical stability is available but at high cost.

It is difficult to make credible stability measurements to better than 1 μrad, and alignment-instrument stability, when specified at all, usually falls in the range of 1 to 10 μrad. Therefore the important consideration in applying alignment instruments is to recognize when stability is necessary at or below this level and to take steps to ensure its presence to the requisite degree.

45. Measuring Mechanisms Most automatic alignment instruments have inherent radiometric linearity of worse than 1 percent of full scale and require careful balance and calibration to achieve even that. If more accurate measurement of alignment change is desired, a more complex automatic reticle must be incorporated; more commonly a measuring mechanism, manual or automatic, is added. Typically, the measuring mechanism maintains the alignment instrument nulled as the reference element deviates by moving either the automatic reticle or an optical element that varies the optical relation of the reticle to the objective in an accurately specified way. Alignment change is indicated on a dial or meter; in fully automatic instruments it is also indicated electrically by an analog or digital output signal.

46. Two-Axis Operation In the automatic autocollimator (and the automatic alignment telescope in a comparable way) it is frequently necessary to measure mirror deviation about two axes in real time. For this purpose many automatic autocollimators are made as two-axis units with a two-axis automatic reticle and two measuring mechanisms. One version of automatic reticle for a two-axis automatic autocollimator is illustrated in Fig. 16.

The two orthogonal pairs of detectors provide sensitivity to mirror deviation in any direction, and measuring mechanisms aligned with the detector pairs can measure the deviation in terms of its components about these two axes.

Alignment instruments other than the automatic autocollimator and automatic alignment telescope inherently measure only a single degree of body misalignment, and two-axis operation is inapplicable to them.

47. Performance Evaluation When considering the acquisition of an alignment instrument it is usually possible to arrange a demonstration or even a loan of candidate instruments. The results of performance tests during such a trial will significantly influence the purchase decision, and the relation of the tested characteristics to real needs will determine eventual user satisfaction with the instrument.

Fortunately, several easily performed tests provide a measure of the most critical aspects of instrument behavior. The most important instrument characteristic in most cases is error resolution. If we take the automatic autocollimator as an example, error resolution is measured by the following procedure.

1. Adjust the autocollimating mirror to produce null output from the automatic autocollimator.

2. Adjust readout gain to obtain a readable peak-to-peak null noise output equal to, say, plus or minus the smallest graduation on the readout meter (this can be done precisely on meters with calibrated gain settings by adjusting the gain to obtain a large reading from noise and then scaling down by the appropriate factor).

3. Adjust the mirror to produce a full-scale output and determine the mirror angle in microradians.

4. The error resolution is given by

$$\text{Error resolution} = 2^{-3/2}\,\frac{\text{mirror movement}}{\text{meter graduations}}\ \mu\text{rad rms}$$

The second characteristic to be measured is internal noise. To do this

1. Adjust the mirror to produce a null.

2. Adjust readout gain to obtain a moderate reading (say ±10 percent full scale) from the null noise.

3. Block the light path to the mirror or remove the mirror and record the meter output. Normal behavior in this test is a centered reduction in noise level because atmospheric shimmer is the largest noise source in a well-designed automatic autocollimator. Undesirable behavior is a constant or increased output, which indi-

cates large internal noise, and a change in the center position of the output, which indicates large unbalanced internal noise that represents a source of variable null shift.

4. Extinguish the autocollimator light source and record the meter output. This is a measure of detector and electronic noise, which should be far lower than null noise. It also indicates whether internal noise and unbalance are caused by stray internal light or by the detectors and electronics.

Instruments whose internal noise and/or null shift are equal to or greater than, say, 10 percent of null noise under normal operating conditions with a full-sized mirror are of questionable value, for there will be many situations in which external gain is greatly reduced by an unavoidable small mirror or optical losses when internal noise will prevent accurate measurement.

The next test measures the degree of collimation and/or instrument optical quality.

1. Adjust a mirror of known high quality to null the autocollimator and adjust the meter to obtain a centered full-scale peak-to-peak reading from null noise.

2. At the mirror aperture center a dark opaque mask with an approximately one-fourth-diameter opening and record the meter reading.

3. Calibrate autocollimator sensitivity with the mask centered on the mirror by recording the meter output after introducing a known mirror movement.

4. With the autocollimator nulled, move the mask to other portions of the mirror aperture and record the meter readings.

Desirable behavior is the complete absence in steps 2 and 4 of significant null shift as indicated by decentering of the meter reading. The presence of null shift indicates that the automatic autocollimator is defocused, the objective lens is of poor quality, or the automatic autocollimator has poor internal optical geometry which causes variation in the signal produced by light coming from different parts of the aperture.

Another critical test is that of null stability.

1. Place the automatic autocollimator and autocollimating mirror on a stable and massive surface (usually a granite surface plate) in a stable environment and adjust to a null.

2. Record the meter reading over a long period of time. Any significant change in the null indication is undesirable.

Susceptibility to a variety of external influences can be tested more or less readily in the setup described above by varying, say, external light level and frequency, power-line voltage and frequency, temperature, etc.

Finally, autocollimator measuring range and gain linearity can be tested readily as follows.

1. Null the autocollimator with an adjustable mirror.

2. Adjust the mirror to appropriate angles on both sides of null over the full autocollimator range of sensitivity and record the corresponding meter readings. The linear range should be large enough to cover the anticipated measuring-range and/or acquisition-range requirements, and the linearity should be sufficient to indicate the magnitude of a large angle off null with the desired accuracy.

If manual or electromechanical error-nulling mechanisms are incorporated, their accuracy can be tested as above.

The frequency response of static instruments can be tested as follows.

1. Adjust the autocollimating mirror to appropriate angles off the autocollimator null.

2. At each angle, using a semicircular chopper driven by an adjustable-speed geared motor (capable of chopping over the frequency range of interest), locate the chopper in the autocollimating path, run the chopper at appropriate speeds, and record the corresponding autocollimator output on a high-speed recorder.

To test servo-nulling instruments the autocollimating mirror must be driven at amplitudes and frequencies of interest by a dynamic deviation mechanism. Amplitude response vs. frequency can be determined directly from the speed and recorder data.

These tests measure all important performance characteristics of the automatic autocollimator. They also relate directly to analogous tests for all other automatic alignment instruments. The corresponding tests for other instruments are readily defined and the necessary test procedures easily evolved.

Automatic Autocollimator

48. Description and Operating Principles The automatic autocollimator measures the orientation of a remote body about axes normal to the line of sight, as discussed in Par. 13. A plane mirror is the only autocollimating element that can be used with a two-axis autocollimator. In single-axis alignment situations the degree of freedom about the unused axis is frequently exploited by utilizing instead a roof reflector aligned with its roofline normal to the axis of interest. This provides great freedom in autocollimator location in the plane containing the alignment axis, but roofline normality to the alignment axis must be increasingly accurate as the autocollimator departs further from normality to the alignment axis to avoid cross-coupling error (equal to the product of the two misalignment angles). The roof reflector can be constructed as a prism or made of two mirrors. Highest reflectivity is obtained from a total-internal-reflection prism, but its angular field about the roofline is small because the nominal 45° incidence at the roof surfaces is close to the critical angles of typical glasses.

49. Operation Potential applications for the automatic autocollimator encompass a wide range of performance requirements. The performance that has been or can be achieved under normal conditions will be reviewed first, and a discussion of the variations required to cope with unusual conditions follows.

Normal automatic autocollimator applications may be defined as those involving:

1. Operation at distances up to 10 m
2. Operation in an enclosed air-filled space
3. An automatic autocollimator aperture of 5 cm diameter
4. An autocollimating mirror of high reflectivity at least as large as the automatic-autocollimator aperture
5. No intervening material in the optical path
6. An absence of significant mirror translation
7. Relatively small ambient-temperature variation
8. Insignificant external light
9. Limiting error resolution from 0.1 to 1 μrad
10. Operating periods of less than 24 h
11. A signal bandwidth of less than 10 Hz
12. An acquisition range of 1 to 10 mrad
13. A required measuring accuracy of 1 to 0.1 percent of full scale
14. No significant shock or vibration

The normal performance limitation under these conditions is air shimmer. Depending on temperature variations in the air, air movement, the optical-path length, whether it is enclosed, and the signal bandwidth, limiting error resolution can be as little as 10^{-2} μrad or as large as 10 μrad. Generally it is not too difficult to get from 1 to 10^{-1} μrad error resolution if the automatic autocollimator is capable of it.

The commonest special automatic autocollimator operating conditions or requirements are:

1. Long operating distance
2. Large acquisition range
3. Large lateral mirror translation
4. Operation in vacuum
5. High stability
6. Fast response
7. Multimirror monitoring

Requirements 1 to 3 are typically met by utilizing an automatic reticle of small angular extent and increasing the autocollimator aperture appropriately. For situations where the resulting automatic-autocollimator aperture would be excessively large, alternative solutions are available that produce the optical effect of laterally translating the autocollimator as required. Operation in a vacuum of $>10^{-4}$ Torr is feasible if the automatic autocollimator has no moving parts to suffer lubrication failure or electrical parts to suffer corona arcing. At higher vacuum, where convective heat transfer diminishes, suitable conduction paths are necessary to remove heat from the automatic autocollimator. Automatic autocollimators with refractive objective lenses must be refocused for use in a vacuum.

High stability over a long operating period and in the presence of temperature variation is usually available only from special automatic autocollimators. The major sources of instability are absolute expansivity of the automatic autocollimator, relative expansion with resultant decentering of the objective, light-source output change, and detector gain variation. High-stability automatic autocollimators should therefore utilize a reflective objective made of low-expansivity optical materials, preferably mounted in a body of the same material, a regulated long-life lamp, and temperature-compensated solid-state detectors.

Fast automatic autocollimator response (>1,000 Hz) requires principally that the modulation frequency be much higher than the desired response. For this purpose an automatic autocollimator with a dc light source and a high-frequency modulator or one with a light source directly modulated at high frequency is satisfactory.

Multimirror monitoring requires that the automatic autocollimator be capable of accommodating the multimirrors optically and distinguishing their signals. One way of providing mirror identification is by time-sharing, using an appropriate chopper at the mirrors. Another possibility is spectral separation utilizing spectral bandpass filters on the mirrors and matching dichroic beam splitters in the automatic autocollimator, with separate detectors for each channel.

Automatic Alignment Telescope

50. Description and Operating Principles The automatic alignment telescope, the translational counterpart of the automatic autocollimator, measures remote-body translation normal to the line of sight. Among the ways in which this measurement can be made is that illustrated in Fig. 19. Here, an instrument essentially identical

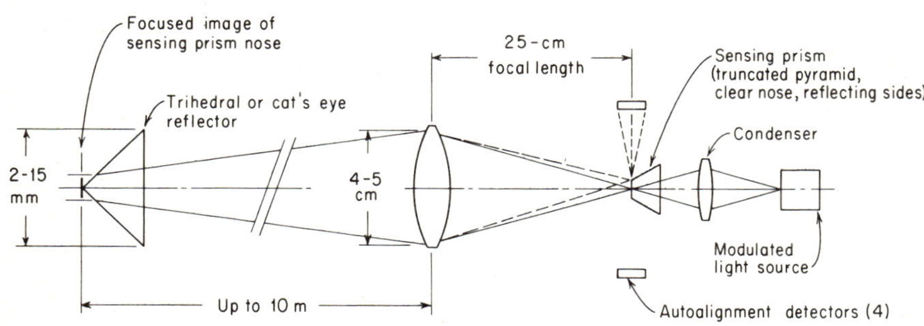

Fig. 19 Operating arrangement of two-axis automatic alignment telescope.

to the autocollimator projects light to a trihedral retroreflector on the monitored body, and the light is reflected back to the alignment telescope. Unlike that of the autocollimator, however, the alignment-telescope beam is not collimated. Instead the reticle prism nose is focused on the trihedral. This produces the two effects essential to automatic alignment telescope operation:

1. An autoreflected image of the nose is formed in the plane of the nose and is superimposed on the nose when the trihedral is centered on the alignment-telescope axis.

2. The nose image displaces laterally when the trihedral moves off the telescope axis.

The relation between image displacement Δr and trihedral displacement $\Delta \rho$ at a working distance R with objective focal length f is

$$\Delta r = \frac{2f\,\Delta \rho}{R} \tag{5}$$

which is quite analogous to the corresponding relation for the autocollimator [Eq. (3)]. In fact, if an angular displacement $\Delta\gamma = \Delta\rho/R$ is defined, it is seen that

$$\Delta r = 2f\,\Delta\gamma \tag{6}$$

which has exactly the form and magnitude of Eq. (3). Consequently, the automatic alignment telescope can be configured identically to the automatic autocollimator; its displacement-detecting process is the same as that of the autocollimator, and its sensitivity to angular displacement is the same as that of the autocollimator to mirror tilt. Its signal-processing electronics are the same as those for the autocollimator (Fig. 17).

The one element unique to the automatic alignment telescope is the trihedral retroreflector. A retroreflector is a device that reflects all radiation parallel to its direction of incidence. The best-known retroreflectors are the cat's-eye and the cube corner, or trihedral. Their retroreflection property is combined with a folding property which causes eccentric rays (the cat's-eye nodal point and the trihedral apex are their respective centers) to emerge diametrically opposed to their point of incidence. It is this property that causes decentration of the autoreflected nose image when the nose image is eccentric to the retroreflector and is the source of the factor 2 in Eqs. (5) and (6).

Although the cat's-eye and the trihedral exhibit similar retroreflective properties, the trihedral is more common and is more frequently used as an alignment-telescope retroreflector. The trihedral can be made both as a prism and as an air-path mirror set, the latter being rare. Both retroreflectors have large, geometrically limited angular fields if the prism trihedral has reflective mirror coatings. If a smaller angular field is permissible, the trihedral can be used as a total-internal-reflection element with significantly higher net reflectivity so long as the reflecting surfaces remain clean.

51. Operation Potential applications for the automatic alignment telescope are more limited than for the automatic autocollimator for several reasons. One is the generally greater significance of body orientation than its lateral position with respect to a line. In addition, autocollimators are more flexible instruments in their ability to autocollimate mirrors over a wide range of distances without adjustment, whereas the alignment telescope must be refocused for each operating distance with the attendant possibility of introducing error. Finally, most automatic-alignment-telescope tasks can be performed with a varying degree of accuracy by laser alignment systems.

Where then does the automatic alignment telescope fit into the spectrum of alignment instruments? Relative to laser alignment devices the automatic alignment telescope has certain advantages. It is single-ended; i.e., all active optical and electronic functions are performed in the instrument, and only a passive optical element is required on the monitored body. The automatic alignment telescope typically is constructed of long-life components that permit longer unattended life than attainable by lasers. Lasers also exhibit beam wander, which sets a lower limit on their alignment accuracy. Most important, however, is the ready availability of automatic-alignment-telescope capability from focused versions of many automatic autocollimators. Thus any automatic autocollimator of appropriate configuration can be used as an automatic alignment telescope simply by focusing it on a trihedral attached to the body to be monitored. When a suitable automatic autocollimator is already owned, this temporary conversion to an automatic alignment telescope is far less expensive and more convenient than the purchase of a laser alignment system.

Normal automatic-alignment-telescope applications involve the same general constraints as the automatic autocollimator (Par. 49), and the expected performance falls into the same specific range of 0.1 to 1 μrad (10^{-7} to 10^{-6} m resolution at a distance of 1 m). Because of their almost identical bases of operation, air shimmer is the common performance limiting factor for autocollimators and alignment telescopes and under adverse conditions can decrease automatic-alignment-telescope error resolution to 10 μrad or worse. Under ideal atmospheric conditions and in a vacuum, alignment-telescope error resolution can improve well beyond 0.1 μrad.

All comments in Par. 45 on the use of measuring mechanisms (manual or automatic) to obtain good specific accuracy apply fully to the automatic alignment telescope, as does the discussion of two-axis operation in Par. 46.

Because automatic alignment telescopes have been much rarer than automatic autocollimators, there is a much smaller body of experience from which to segregate normal and unusual operating conditions. Some unusual conditions are similar to those of the automatic autocollimator (long operating distance, operation in vacuum, high stability, fast response) and are met by similar design adaptations. Conditions requiring adaptations unique to the automatic alignment telescope are large acquisition field, large axial translation, and multiple trihedral monitoring.

Over a very wide angular range of incidence the trihedral will retroreflect incident light to the alignment telescope. Acquisition over a large field of trihedral lateral position therefore requires simply that the alignment telescope be capable of scanning the field systematically to ensure eventual illumination of the trihedral.

Large axial translations of the trihedral call for automatic-alignment-telescope refocus (preferably automatic) without shifting the axis (as in the visual-alignment telescope). Automatic focusing is not known to have been implemented in alignment telescopes; however, techniques such as those employed in the optical probe make it feasible.

Multitrihedral monitoring involves monitoring a number of trihedrals dispersed along the line of sight of the telescope. In view of the operating principle of the automatic alignment telescope, it is clear that the trihedrals cannot be monitored simultaneously and must be observed in time sequence with the telescope successively focused on each. Moreover, to observe one trihedral it may be necessary to remove all trihedrals between the alignment telescope and the one being monitored. However, because the nearer trihedrals can be smaller than the more distant ones (because of the corresponding variation in projected prism nose size) and the automatic alignment telescope can be made with an aperture much larger than the trihedrals, it is feasible in some situations to observe any one of several aligned (and appropriately sized) trihedrals despite the presence of the others simply by focusing on the one of interest. Spectral coding and other techniques can also be employed to separate the returns from several trihedrals, as in Par. 49.

The automatic alignment telescope is a relatively uncommon automatic instrument that serves an increasingly necessary alignment function. Alignment-telescope capability can be obtained in many cases by refocusing an automatic autocollimator and observing a trihedral.

Automatic Twist Autocollimator

52. Description and Operating Principles The twist autocollimator measures body rotation about the line of sight. Its operation (Fig. 20) is based on sensing

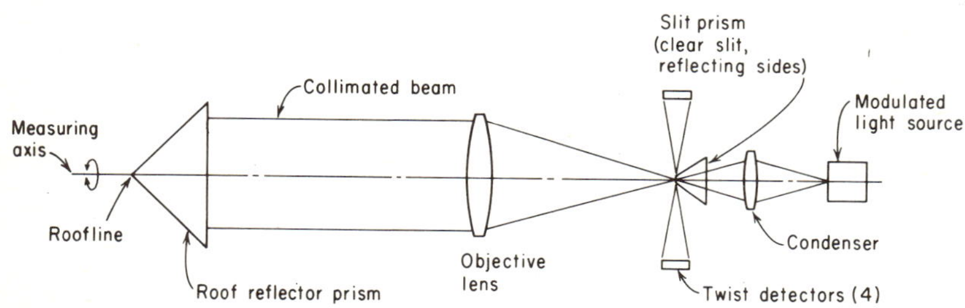

Fig. 20 Operating arrangement of twist autocollimator.

the rotation of an autocollimated image caused by rotation about the line of sight of an autocollimating roof reflector whose roof is normal to the line of sight. Here the relation between roof-prism rotation $\Delta\phi$ and image rotation $\Delta\beta$ is

$$\Delta\beta = 2\Delta\phi \tag{7}$$

The maximum displacement of the autocollimated image Δr at image angular height α with objective focal length f is

$$\Delta r = \alpha f \, \Delta\beta = 2\alpha f \, \Delta\phi \tag{8}$$

or smaller than that in the automatic autocollimator [Eq. (3)] by the factor α, which normally has a magnitude of 10^{-3} to 10^{-2} rad. Moreover, the basic image motion is a rotation about the optical axis, and the coding and analyzing functions involve more complex mechanization than in the automatic autocollimator or automatic alignment telescope.

The simplest automatic-twist-autocollimator derivation (see Fig. 20) from the automatic autocollimator consists of replacing the sensing prism with a slit prism whose slit is parallel to the roofline of the roof prism and rearranging the four automatic autocollimator detectors into two sets of diagonally paired detectors connected to produce positive and negative signals when illuminated (Fig. 21). As the observed body rotates about the line of sight, one-speed roof-prism rotation causes two-speed rotation of the autocollimated slit image in the same direction.

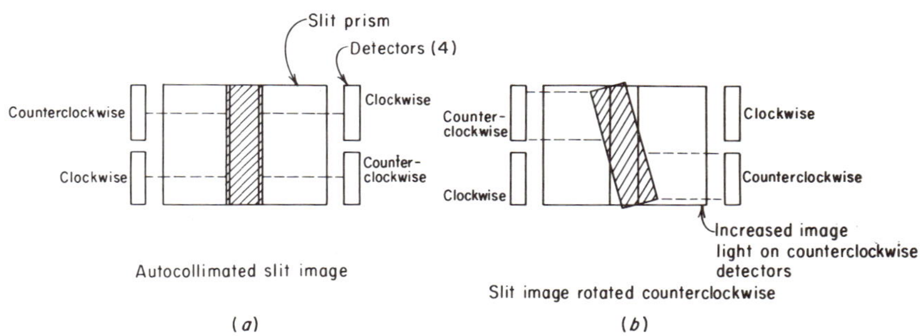

Fig. 21 Signal generation in twist autocollimator: (*a*) nulled (unrotated); (*b*) counterclockwise misalignment.

A counterclockwise rotation of the roof prism causes the slit-image rotation shown in the figure (observed from behind the prism), which illuminates the coupled counterclockwise detectors. Clockwise rotation has the opposite effect.

Thus, measurement of body rotation about the line of sight with the automatic twist autocollimator involves sensing autocollimated image rotation rather than translation, which requires a somewhat more complex automatic reticle than the automatic autocollimator. In addition, because the image displacement for a given body rotation is reduced by the factor $10^{-3} < \alpha < 10^{-2}$ rad from that in the automatic autocollimator, the sensitivity of the automatic twist autocollimator is far lower (2.5×10^{-3} to 2.5×10^{-2} μm displacement for a 5-μrad body rotation) and therefore requires greater care in electrooptical noise reduction. Furthermore, diffraction and atmospherically induced image motion have the same magnitude as in the autocollimator, but their equivalent angular magnitude is much larger in the twist autocollimator because of its low sensitivity. Nevertheless, the automatic twist autocollimator is a direct descendant of the automatic autocollimator and its functioning is based on the same processes and concepts, differing only in the source and nature of image motion and its magnitude for a given body rotation.

Functionally, the automatic twist autocollimator therefore involves generally the same optical and electronic components as the automatic autocollimator (Fig. 17). The modulated autocollimated slit image when rotated is incident on one of the two diagonal detector pairs which are connected with opposing polarity. The net detector output signal is null when the slit image is precisely parallel to the slit (Fig. 21*a*) as a result of cancellation of the signals from light spread onto the detectors by diffraction and aberrations. In the presence of roof-prism rotation (Fig. 21*b*) the more strongly illuminated pair of detectors produces a net ac signal whose magnitude is proportional to roof rotation and whose phase indicates the direction of rotation. This signal is ac-amplified and filtered, synchronously demodulated, and dc-amplified and filtered to provide a proportional, polarity-reversible dc output signal of appropriate gain and impedance.

The factors unique to the automatic twist autocollimator are the monitored roof prism, which produces rotation of the autocollimated slit image, the detector geometry, which permits sensing this rotation while remaining insensitive to nonsignal light, and a large nonsignal light level. For good performance certain constraints must be placed on the relation between the automatic twist autocollimator and roof prism. Null alignment exists when the roof apex is parallel to the twist-autocollimator slit. In this condition of alignment the autocollimated slit image is precisely superimposed on the slit as desired. Small translation of the prism and rotation about the roofline have no effect on the image. Rotation about an axis normal to the roofline and the line of sight causes the slit image to move along the slit. By making the illuminated portion of the slit longer than the detector span, a moderate amount of such movement is tolerable without error. It is only necessary therefore to maintain compatibility between the twist-autocollimator misalignment tolerances (which are controllable in design) and the degree of extraneous misalignment to avoid error and ensure that the twist autocollimator senses only the twist misalignment of interest.

The roof reflector used with the automatic twist autocollimator can be constructed as a prism or made of two mirrors, as in the automatic autocollimator.

The electronics must deal with a signal containing, besides the desired information, a component at twice the modulation frequency, due to optical diffraction, which is larger than a 1-μrad error signal by a factor of approximately 10^4, and a broad atmospheric-shimmer component of similar magnitude. Extracting a usable error signal from so much background requires filtering the ac signal, synchronous rectification to polarity-reversible direct current, dc filtering, and occasionally more sophisticated techniques.

The signal-to-noise ratio can be improved by using long integrating times in the dc filter, but there is usually a practical filtering limit set by the need to detect misalignment in a relatively short time.

53. Operation The automatic twist autocollimator can be used to measure body rotation about the line of sight under the general conditions of Par. 49, where the roof reflector is maintained reasonably well aligned about the other axes. Performance under these conditions, unlike that of the automatic autocollimator, is fundamentally limited by photoelectron noise to between 0.1 and 1 μrad error resolution. This is a result of the much lower error sensitivity of the automatic twist autocollimator combined with a diffracted-light level equal to that in the automatic autocollimator. Moreover, because there is so little image movement for a given amount of image twist, the image movement produced by air shimmer causes a much higher relative noise level than in the automatic autocollimator.

In an evacuated environment or in space, the automatic twist autocollimator can provide error resolution of 1 μrad or better. In the atmosphere, however, air shimmer will normally degrade error resolution to much worse than 10 μrad unless the optical path is well shielded.

In some applications twist alignment may be necessary along a devious optical path involving bends and possibly windows. Because its performance depends on image quality, the twist autocollimator can tolerate these extraneous elements without error so long as they are of good optical quality and introduce no image rotation of their own.

The automatic twist autocollimator is an uncommon alignment instrument capable of making a difficult measurement. Extraordinary care is required in its construction and use to achieve the level of accuracy in twist measurement so easily obtained from an automatic autocollimator in tilt measurement.

Automatic Alignment Polarimeter

54. Description and Operating Principles The automatic alignment polarimeter measures body rotation about the line of sight by sensing change in the polarization characteristics of autocollimated light produced by a polarizing element on the body. (Because operation of an autocollimating automatic alignment polarimeter is somewhat complicated to explain, the simpler transmitter-receiver automatic alignment polarimeter will serve as the model for the present explanation.) A polarizing transmitter on the observed body, consisting of a light source, a collimating lens, a coding polarizer, and a polarizing modulator (Fig. 22), sends a polarized beam of light to a receiver. Hence, neglecting the modulator, an analyzing polarizer serves to reject essentially all the incident light when its polarizing direction is perfectly orthogonal to the coding polarizer but transmits small amounts of light as the polarizers depart from orthogonality. The relationship between the incident light I_0, transmitted light ΔI, and twist misalignment $\Delta\theta$ is

$$\Delta I = I_0 \sin^2 \Delta\theta \approx I_0 \Delta\theta^2 \tag{9}$$

The transmitted light is usually collected by a condenser lens and focused on a detector in the receiver.

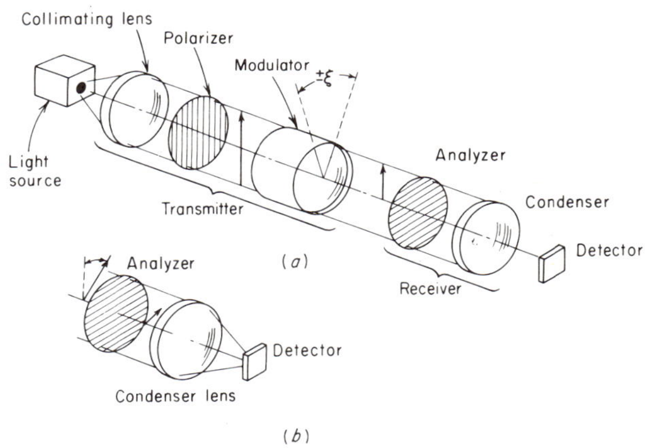

Fig. 22 Operating arrangement of alignment polarimeter: (*a*) aligned: incident energy is completely absorbed when polarized precisely normal to the analyzer; (*b*) twist error present: nonnormal component of incident energy is transmitted by the analyzer.

This simple automatic alignment polarimeter therefore possesses quantitatively the ability to make the desired measurement of body rotation about the line of sight. It exhibits an unusual characteristic, however, in the nonlinearity of its input-output relation [see Eq. (9)]. Not only is this generally undesirable in a measuring instrument, but, as will be shown later, in the alignment polarimeter it prevents the resolution of adequately small error. Both problems are solved by the introduction of a polarizing modulator, preferably located at the transmitter, which serves to oscillate the transmitter polarization direction through an angle ξ, much larger than the maximum error angle $\Delta\theta$. Its effect is to linearize the input-output relation such that

$$\Delta I = 2I_0\xi \, \Delta\theta \tag{10}$$

and to permit the resolution of extremely small angular error.

The automatic alignment polarimeter thus departs greatly from the previous instruments in its operating principle and the embodiment of its automatic reticle but nevertheless involves the same basic functions in its operation, applied to polarization rather than image formation. Because operation is based on polarization, performance is essentially independent of image quality. Consequently, there is no geometric automatic reticle, as in the previous instruments, where diffraction of an autocollimated image at null spreads light onto detectors and thereby introduces a fundamental performance limitation.

There is a counterpart effect in the polarimeter, however. All polarizers depart from perfect extinction of transmitted light when crossed, probably as a result of molecular structural irregularities. The best available ratio of minimum to maximum transmission, the so-called *extinction coefficient,* is about 10^{-6} and obtains in crystalline polarizers. From Eq. (9) the $\Delta\theta^2$ relation between transmitted light and polarizer rotation indicates that an extinction coefficient of 10^{-6} corresponds to a rotation uncertainty of about ± 1 mrad, or about 3 minutes of arc (Fig. 23). Indeed, the most sophisticated visual or static (unmodulated) photoelectric polarimeters are barely able to achieve 1 minute of arc setting accuracy. It is this irreducible level of transmitted light which in an automatic alignment polarimeter causes the same limitation in twist-angle resolution as diffraction does in an automatic twist autocollimator.

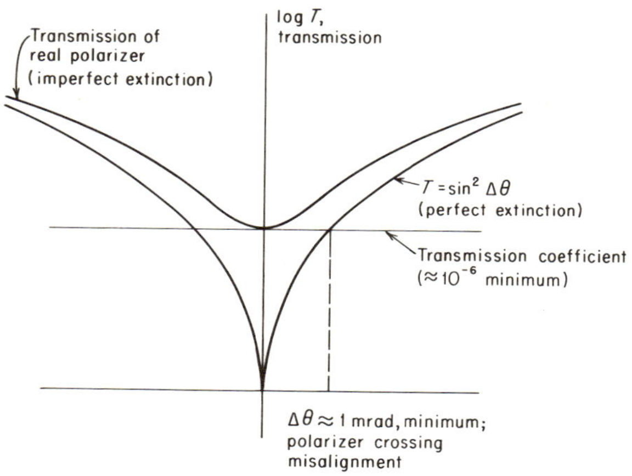

Fig. 23 Null characteristics of polarizer.

Nevertheless, polarimeter error resolution is not noise-limited but functionally limited at the ± 1-mrad level by imperfect extinction, and greatly improved accuracy is attainable by twist-angle modulation. In essence, by oscillating the polarization direction through an angle much larger than the extinction range, the resulting transmission signal permits locating the transmission minimum within a small fraction of the extinction range. Oscillation of the polarization direction can be provided by direct oscillation of the polarizers or by any of a number of polarization modulators. Proper application of this modulation technique converts imperfect extinction from a first-order source of error to a higher-order source of noise and allows error-resolution improvement of the order of 10^{-3} or more.

Because of its distinctive optical character, dependent on sensing a deviation from orthogonality between the direction of linear photon vibrations transmitted by the transmitter and receiver polarizers, the automatic alignment polarimeter differs significantly in its configuration and the nature of its critical elements from instruments whose operation is based on sensing the position or orientation of an image. The imaging portions of the automatic alignment polarimeter are required only to transmit light efficiently from the transmitter to the receiver and need not be of high quality. More critical are the polarizers and the polarizing modulator.

Polarizer requirements call for linear polarization with very low extinction coefficient, high transmission in the spectral region of interest, and good stability. Best results are obtained from natural-crystal polarizing materials such as calcite. Although fragile and expensive, this and similar materials are essential for accurate twist-error resolution because of their unequaled extinction and stability.

Polarizing modulators are available based on electrooptical and magnetooptical effects. Application of an electric or magnetic field along or across a volume of certain solids and liquids induces double refraction in the material, which converts entering linearly polarized light into elliptically polarized light. An axial magnetic field induces optical activity in appropriate materials, causing so-called Faraday rotation of the direction of entering polarization. The latter is preferable for alignment-modulation purposes because the light remains linearly polarized, its direction oscillating through a small angle.

Using Faraday angular oscillation of linearly polarized light as an example (Figs. 22 and 23), as the polarization direction of incident light oscillates about the crossed direction at the analyzer, the small uncrossed ac component is transmitted. If the oscillation is symmetrical about the orthogonal direction, a symmetrical optical signal reaches the detector, causing a pure-harmonic electric signal. But as the oscillation becomes asymmetric due to twist error, the transmitted light on each half cycle is unbalanced, causing a fundamental error signal [Eq. (10)].

Typically the harmonic background signal is a factor of 10^4 or more greater than the fundamental signal from an error of 1 μrad. This is similar to the

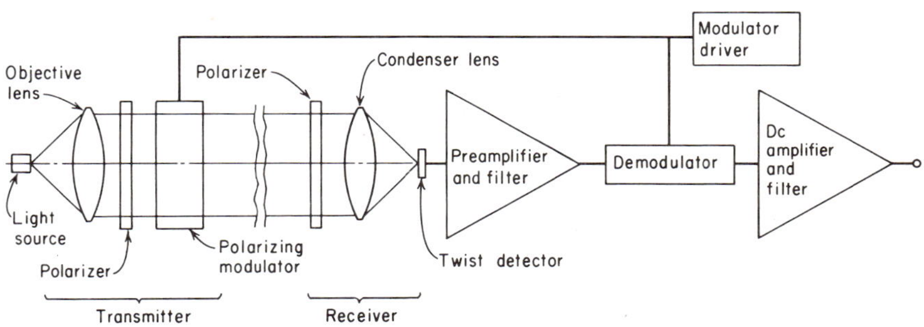

Fig. 24 Functional diagram of alignment polarizer.

diffracted-light background signal in the twist autocollimator, and the solution to extracting the signal similarly consists of ac and dc filtering and possibly more sophisticated techniques, in all cases restricted by the need to detect dynamic errors.

Automatic-alignment-polarimeter electronics (Fig. 24) therefore perform generally the same functions and thus resemble those of the automatic twist autocollimator. The detector signal is ac-amplified, ac-filtered, synchronously rectified to polarity-reversible direct current, dc-amplified, dc-filtered, and further amplified and impedance-matched to appropriate output devices.

55. Operation The automatic alignment polarimeter can be used to measure body rotation about the line of sight under the general conditions described in Par. 49, where the receiver polarizer is reasonably well aligned about the other axes. Performance under these conditions, as with the automatic twist autocollimator, is fundamentally limited by photoelectron noise from 10^{-1} to 1 μrad error resolution. This follows from the low misalignment sensitivity of the alignment polarimeter combined with the large modulation light level, comparable to twist-autocollimator diffraction. For a geometrically and radiometrically similar twist autocollimator and alignment polarimeter the two instruments have equivalent noise-limited resolution in the absence of shimmer and polarizing disturbances, a fact that is surprising in view of the great difference in their sensing mechanisms.

Unlike the twist autocollimator, however, in which the background light consists of roughly equal modulated diffracted light and wide-band shimmer light, the alignment-polarimeter background is almost completely polarization modulation background, which is more readily attenuated than wide-band background. Under given operating conditions in the atmosphere, the alignment polarimeter therefore will have the better performance, and by a wide margin, unless twist-autocollimator air shimmer can be reduced to a negligible level by careful enclosure of the optical path and insulation against temperature changes.

Nevertheless, the alignment polarimeter is not unequivocally superior to the twist autocollimator, even in the convenient single-ended form normally employed. One negative characteristic is extreme sensitivity to refractive and reflective optics in the optical path. These generally depolarize the transmitted light, thereby greatly increasing the effective extinction coefficient and noise, and rotate the plane of rotation, both by a variable amount that is highly sensitive to the specific optical arrangement.

Moreover, air is sufficiently affected by the earth's and other magnetic fields to rotate the plane of polarization by several microradians over normal operating distances. The effect varies with alignment direction. This reduces the usefulness of the alignment polarimeter as an absolute alignment device where the effect is constant but unknown and reduces measurement accuracy where it is variable.

Finally, the natural-crystal polarizers required for best performance are fragile and less stable optomechanically than the materials employed in an automatic twist autocollimator.

Thus, an unexpected complementary symmetry between the automatic twist autocollimator and automatic alignment polarimeter allows one or the other to perform best under given conditions. Where the alignment path is straight and free of extraneous optics or must be kept unshielded, and where the environment is reasonably benign, and no significant variable polarization rotation effects exist, the automatic alignment polarimeter is probably the best choice. But if path bending or folding is required, if there are extraneous optics, or stability in a harsh environment is necessary and air shimmer can be held to a sufficiently low level, the automatic twist autocollimator would be preferable.

One serious practical disadvantage of the alignment polarimeter is the inability to check the absolute condition of alignment in the laboratory visually in a direct way. Whereas the roof prism for a twist autocollimator can easily be provided with an autocollimating reference surface precisely normal to the roofline, making and checking a corresponding surface normal to the polarizing direction of a polarizing prism requires an automatic polarimeter and a complex procedure.

Three-Axis Alignment Instruments

56. Background By combining compatible tilt- and twist-measuring techniques it is possible with a single instrument to measure the total angular deviation of a body in terms of its components about the line of sight and two orthogonal axes normal to the line of sight. Such three-axis alignment instruments can be configured in four ways; automatic instrument designs may be based on all combinations of geometric and polarizing tilt and twist measurement. Of these four, the ones most readily implemented are autocollimator plus twist autocollimator (Fig. 25), polarizing autocollimator plus alignment polarimeter, and autocollimator plus alignment polarimeter (Fig. 26). Polarizing autocollimator plus twist autocollimator is not known to have been attempted and appears to have little merit. Because the polarizing autocollimator is rare compared with the geometric autocollimator and the latter is easily combined with both the twist autocollimator and alignment polarimeter but the former is not, only geometric-autocollimator-based three-axis instruments will be considered in detail.

The use of a three-axis alignment instrument is motivated primarily by operational not economic considerations. In some alignment situations complete body angular deviation must be measured from one instrument location, and constraints on space, weight, or other factors applied to the body reference element, the alignment path, or the alignment instrument force the use of one instrument and one line of sight for all measurements. Economically, although a single three-axis alignment instrument made in production quantities might be less expensive than separate tilt and twist instruments, its present custom character makes it more expensive, but the growing need for triaxial measurement is bound to stimulate the development and increased use of three-axis alignment instruments.

Functionally, the choice of measuring techniques to utilize in a three-axis alignment instrument for a given application must be based on the relative performance of geometric and polarizing twist-alignment techniques in the specific alignment environment.

The general instrumental characteristics of the automatic-autocollimator-based three-axis alignment-instrument concepts are evident from Figs. 25 and 26 and from the preceding paragraphs on specific alignment techniques. The unusual body-reference elements and automatic reticles required by three-axis alignment instruments have not previously been considered, however, and therefore will bear examination.

57. Three-Axis Autocollimator The autocollimator–twist-autocollimator three-axis alignment instrument (Fig. 25) combines automatic-autocollimator and auto-

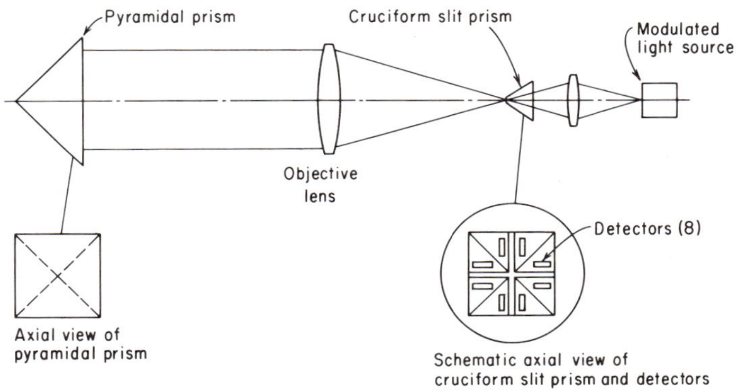

Fig. 25 Operating arrangement of three-axis autocollimator.

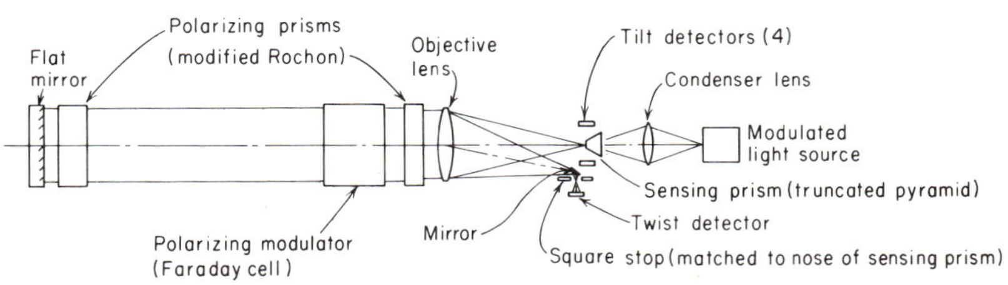

Fig. 26 Operating arrangement of three-axis autocollimator-polarimeter.

matic-twist-autocollimator techniques. The autocollimator usually utilizes a flat reference mirror and senses lateral translation of the autocollimated image caused by mirror tilt. The twist autocollimator usually utilizes a roof reference reflector and senses autocollimated image rotation. A compatible combination of the two that does not depend on separation of the transmitted light into dual channels (by color or polarization) comprises a cruciform slit prism with eight detectors in the three-axis alignment instrument and a concave pyramidal reflector on the observed body. The cruciform slit prism is simply a two-axis version of the automatic-twist-autocollimator slit prism, and the pyramidal reflector similarly is a two-axis version of the roof reflector.

Two independent autocollimated cruciform images of the slit prism are formed by the two effectively independent and orthogonal roof reflectors in the pyramid. Both images rotate through twice the twist misalignment angle, whereas only one image shifts horizontally and one vertically by twice the amount corresponding to the associated tilt misalignment. Image motion therefore contains information about the direction and magnitude of misalignment in all three axes.

Tilt information in two axes is extracted as the difference between the detector signals on opposite sides of the two slits. Twist information is extracted as the difference between the detector signals on the clockwise and counterclockwise sides of the four slit arms. The operation of the autocollimator–twist-autocollimator three-axis alignment instrument is thus readily decomposed into automatic-autocollimator and automatic-twist-autocollimator elements, with behavior equivalent to that of those instruments. Therefore all instrumental and operational factors pertaining to the automatic autocollimator and automatic twist autocollimator also apply to the autocollimator–twist-autocollimator three-axis alignment instrument.

58. Autocollimator plus Alignment Polarimeter The autocollimator–alignment-polarimeter three-axis alignment instrument (Fig. 26) combines automatic-autocollimator and automatic-alignment-polarimeter techniques. Here, polarization and polarizing modulation are directly superimposed on a two-axis automatic autocollimator without affecting its autocollimating operation but adding twist information. Most of the two-axis automatic autocollimator functional elements, consisting of a small illuminated source, a collimating objective, and a retroreflector (the flat mirror when autocollimated) are required in a single-ended automatic alignment polarimeter in any case. Because the polarizers and polarizing modulator required for the automatic alignment polarimeter are of good optical quality, they have negligible effect on automatic-autocollimator operation. The automatic autocollimator and automatic alignment polarimeter are therefore quite compatible and permit efficient integration in a common instrument.

With the exception of the polarization modulation in the collimated beam, which has no significant effect on the automatic autocollimator, the autocollimator's operation in the autocollimator–alignment-polarimeter three-axis alignment instrument is exactly as described in Par. 48.

The single-ended automatic alignment polarimeter utilizes polarizing prisms which transmit both directions of polarization, one on axis and the other directed several degrees off axis (Fig. 26). Its operation involves initial polarization of the axially collimated beam, oscillation of the direction of polarization by the polarizing modulator (in synchronization with the automatic autocollimator amplitude modulation), and transmission through the nominally parallel polarizer at the body, as distinct from the crossed polarizer in double-ended automatic-alignment-polarimeter operation described in Par. 54. The light on its first passage through the parallel body polarizer is essentially unaffected for small angles of oscillation and misalignment.

After reflection back through the body polarizer the light is axially directed, linearly polarized in the direction of the body polarizer, and practically unmodulated. (The small modulation angle has no effect when the polarizers are nearly parallel.) Passage through the polarizing modulator again oscillates the plane of polarization. Most of the light passes axially through the parallel alignment-polarimeter polarizer, again with little effect from the polarization modulation, and this light of almost undiminished intensity contains tilt information from the autocollimating mirror that is detected by the autocollimator detectors at the sensing prism.

The small orthogonal polarized component introduced by the polarizing modulator and twist misalignment and containing the desired twist information is diverted to an off-axis image point in the focal plane by the alignment-polarimeter polarizing prism, where it is detected by the alignment-polarimeter detector. Thus, autocollimator and alignment-polarimeter functions are combined without compromise in this version of a three-axis alignment instrument, and its operational characteristics are as described for the automatic autocollimator and automatic alignment polarimeter in Pars. 49 and 55.

59. Polarizing Autocollimator plus Alignment Polarimeter The polarizing-autocollimator–alignment-polarimeter three-axis alignment instrument combines polarizing tilt measurement, a technique of proved feasibility but rare application, and polarizing twist measurement. Although it provides tilt measurement comparable to an automatic autocollimator, the constraints and advantages of operational use of the automatic alignment polarimeter for twist measurement (Par. 55) here apply to tilt measurement also. Thus, with all measurements based on polarization, the automatic polarizing-autocollimator–alignment polarimeter three-axis alignment instrument is relatively intolerant to folds or extraneous optics in the line of sight on the one hand but is essentially immune to air shimmer on the other. It appears to be the best three-axis alignment instrument for use in situations involving severely degraded seeing conditions in the optical path.

The polarizing autocollimator plus alignment polarimeter tends to be more complex and expensive than other three-axis alignment instruments because the automatic polarizing autocollimator is more complex than the automatic autocollimator. For this reason, which undoubtedly accounts for its rarity, and because the principle of automatic-polarizing-autocollimator operation is not easily described, this version of the three-axis alignment instrument will not be considered further.

Automatic Optical Range Finder

60. Description and Operating Principles The automatic optical range finder measures the distance to a body from the round-trip transit time of light. Another method of measuring range involves triangulation, as in visual military and camera range finders. Although it is not known to have been done at long ranges, the triangulation method of range measurement presumably could be automated under cooperative conditions. An example of this for short-range use is the optical probe (Pars. 62 and 63).

If we restrict the discussion to optical range finders based on transit time, the functions of coding and analysis now involve simply the detection of light pulses separated by very short time intervals. In unambiguous unipulse range measurement the automatic optical range finder sends to the observed body a light pulse of short rise time and relatively short duration which turns on a high-speed timing counter as it departs. The reflected pulse turns off the counter, and the number of counts accumulated during the transit interval is proportional to the body range.

It is difficult to achieve the desired short optical-pulse rise time and length and an adequately high reflected light level for clean unipulse measurement. This difficulty can be overcome by using continuous-wave high-frequency modulation of the outgoing light, with which the reflected light is continuously compared in phase. Any one modulation wavelength, however, is susceptible to ranging ambiguity equal to any integral multiple of that wavelength. This ambiguity is readily resolved by use of three or more judiciously chosen modulation wavelengths.

Thus, the automatic optical range finder is a fundamentally simple time or temporal-phase measuring device, more similar to radar (it is sometimes called optical radar) than to the previous measuring instruments, which require more complex spatial or polarization coding and analysis elements.

The basic automatic optical-range-finder relation gives the time resolution Δt corresponding to given range resolution ΔR

$$\Delta t = \frac{2\Delta R}{c} \tag{11}$$

where c is the velocity of light. By relatively straightforward techniques it is possible to achieve specific measuring accuracy of less than 10^{-6} in routine work and considerably better with the most sophisticated instrumentation. Because the velocity of light is known to an accuracy of only about 10^{-6}, instruments capable of this specific accuracy provide maximum measuring accuracy at present. Operating distances of 100 km and more are feasible, and with lasers and large-aperture telescopes the distance to the moon has been measured from a 0.1-m^2 retroreflector.

A typical automatic optical-range-finder configuration (Fig. 27) comprises a light source, a high-frequency or pulse modulator, a relatively large aperture objective (if a nonlaser source is employed), a detector, and a phase comparator.

Some form of retroreflector (trihedral, cat's-eye, etc.) is employed as the target at the body in most cooperative situations, but satisfactory measurements can be made from diffuse surface reflection by powerful instruments. In measuring a distance to a given accuracy or sensing a small change in distance, the automatic optical range finder must detect a phase shift between the transmitted and returned light corresponding to the desired distance resolution. Assuming a modulation frequency of 30 MHz and a resulting wavelength of 10 m, resolving to, say,

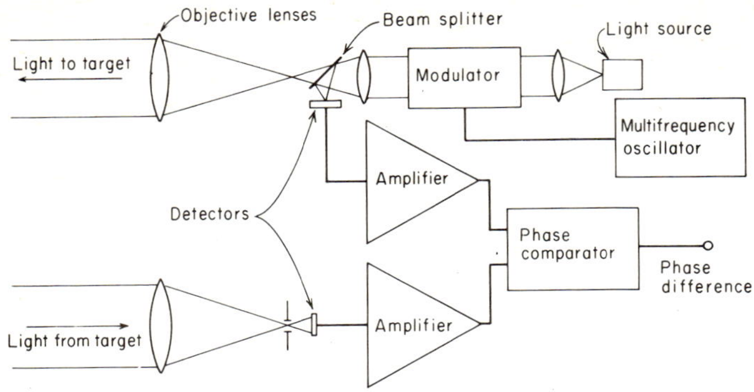

Fig. 27 Schematic diagram of optical range finder.

10^{-6} at 10 km, or to ± 1 cm, requires detecting electric phase shift of 6 mrad or about 20 minutes of arc, which is readily accomplished.

61. Operation The automatic optical range finder is inherently a long-range instrument because of limitations in the minimum electric phase shift or time delay that can be detected. With a minimum error resolution of a few millimeters typical of the best commercial instruments, their specific accuracy of 10^{-6} can be maintained down to only a few kilometers. For high-accuracy measurement of distances up to about 1 m a normal spectral line-emission fringe-count interferometer can be utilized, and a laser fringe-count interferometer can measure distances up to hundreds of meters.

Automatic optical-range-finder performance can be limited either by extraneous light during the day or by component or photon noise at night. By using a laser source and a narrow-band spectral filter, ambient light can be virtually eliminated as a source of noise. Similarly, a powerful laser source combined with appropriately large optics can decrease photon noise and the effects of component noise by several orders of magnitude, shifting the accuracy limitation from noise to detector response time or to uncertainty in the velocity of light.

In addition to instrumental and light-velocity uncertainties, air has an index of refraction that differs from that in a vacuum by about 3×10^{-4}, and under standard conditions this difference is subject to significant change with temperature, pressure, and humidity. Thus, although automatic optical-range-finder specific accuracy can be improved beyond 10^{-6}, as discussed, the uncertainty in atmospheric index of refraction frequently limits range-measurement accuracy to that level or worse in terrestrial applications.

The automatic optical range finder and interferometers are generally considered to be distance-measuring instruments, capable of accurately measuring either the total distance to a body (optical range finder) or large departures of a body from an initial position (interferometers). In many cases, however, only null sensing is desired, i.e., an indication of the smallest departure of the body from a given position relative to the sensor. In addition to the above instruments, which naturally have this capability, the optical probe (Pars. 62 and 63) provides null sensing up to distances of several meters with specific resolution of 10^{-6} or less.

Automatic Optical Probe

62. Description and Operating Principles The automatic optical probe or, more precisely, the optical noncontacting surface sensor, uses triangulation to measure at short distances extremely small changes in the axial position of a noncooperative body. From a number of such existing or proposed devices, one of the true nulling versions with high accuracy will be discussed for illustrative purposes.

Automatic-optical-probe operation is based on sensing the departure from coincidence at the observed body surface of two images of an illuminated spot formed by converging beams symmetrically disposed about the surface normal. The spot images are coincident when the surface is at the external automatic-optical-probe focus, but they separate laterally and in opposite directions as the surface approaches or departs from the automatic optical probe. Two detectors are arranged to detect the imbalance in reflected light resulting from image separation, the difference in their output providing the distance error signal.

An appropriate automatic reticle (Fig. 28) consists of a light source that illuminates the nose of a sensing prism through a condenser lens with aperture stops such that the light emanating from the nose fills only diametrically opposed edge apertures of the objective lens. These form the two beams which converge to coincident images of the nose at the external focus. The converging beams are identified by a chopper at the condenser aperture stop, which sends light alternately

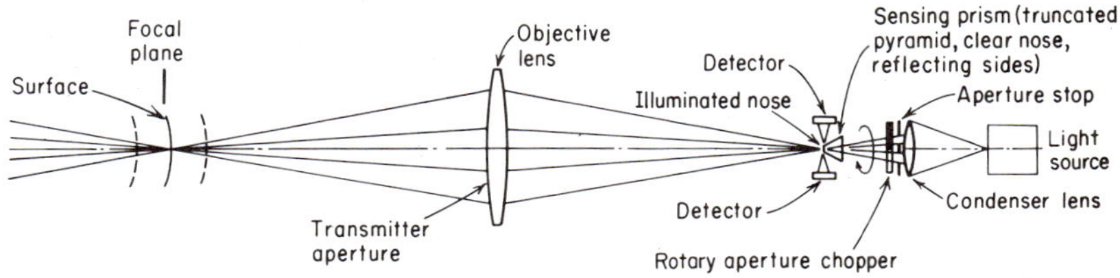

Fig. 28 Operating arrangement of optical probe.

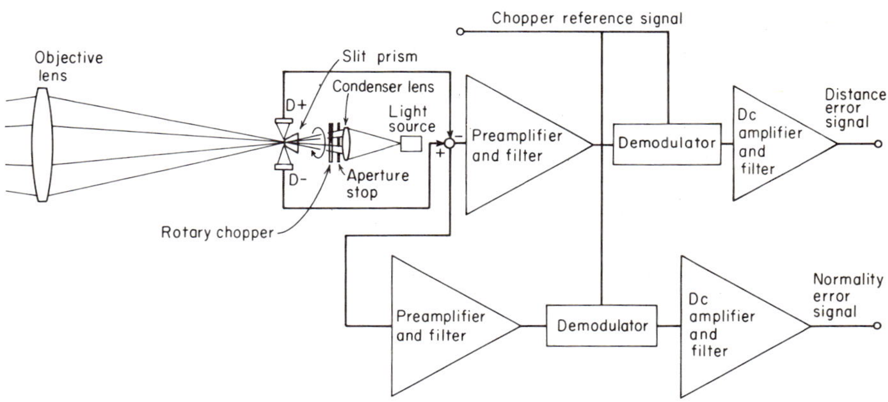

Fig. 29 Functional diagram of optical probe.

to each beam. Nose-image separation is sensed by the two detectors, which receive light through the full objective aperture.

For small surface axial displacement from the focal plane, there is a small lateral departure of the nose images from perfect coincidence, and the autoreflected images at the sensing prism correspondingly overlap opposite sides of the prism nose. Light from one beam therefore falls on one detector, and vice versa, the relations reversing as the surface moves closer to or farther from the automatic optical probe. In the presence of a distance error, then, each detector receives light whose magnitude is proportional to the error magnitude and whose timing identifies the illuminated beam and hence the error direction. The usual electronic processes of ac amplification and filtering, synchronous demodulation, and dc amplification and filtering are utilized to obtain low-noise dc output signals whose magnitude and polarity indicate the magnitude and direction of error (Fig. 29).

Because automatic-optical-probe working distances normally are short ($\ll$1 m), the beam-convergence angles can be relatively large with moderate-sized optics.

For a given range error the spot separation $\Delta\sigma$ that must be sensed to resolve a range error ΔR with a beam convergence angle ψ is

$$\Delta\sigma = 2\psi\,\Delta R \tag{12}$$

If $\psi = 0.1$ rad, the lateral image separation that must be sensed to resolve 1-μm range error is 0.2 μm, which is roughly one-tenth the magnitude of image displacement in the automatic autocollimator caused by 5-μrad mirror tilt. Various versions of automatic optical probe of several configurations have demonstrated distance-error resolution from 0.1 μm down to 1 nm on diffuse and specular surfaces.

The automatic optical probe thus utilizes basic automatic-autocollimator techniques to perform a measuring function very different from that of the autocollimator. It is more complicated, however, because of its need for two converging beams whose optical signals must be identifiable by simple detectors, calling for coding that is most simply accomplished by temporal beam chopping, and by the lower signal levels resulting from the smaller useful image motion and reduced reflectivity of typical nonmirror surfaces. It is simpler than the fringe-count interferometer and has the significant advantage over most noncontacting sensors of being able to utilize an extremely small reflecting area (0.1 mm diameter or less) on the observed surface from a distance of several centimeters.

63. Operation There are several ways of applying the optical probe in measurement. Since the probe has a relatively linear response to distance error up to the point at which spot separation is a large fraction of the original spot size, it can be used as a static measuring device with a specific accuracy of about 1 percent. With somewhat more complex probe designs, better specific accuracy and digital output may be possible.

In cases for which a wider measuring range is necessary, the probe must be employed with a measuring machine having appropriate characteristics. If the measurements required are unidimensional and the surface departure from a regular form (not necessarily flat) does not exceed the optical probe's normality tolerance under the circumstances, the probe can be operated in a fixed orientation and will provide an error signal that may be used to maintain its fixed separation from the surface by manual or automatic adjustment.

Where the surface-slope variation exceeds the probe's normality tolerance, the orientation of the automatic optical probe with respect to the measuring machine must be capable of an appropriate range of adjustment under the control of a normality-error signal that is conveniently obtained from the sum of the detector signals (Fig. 29).

The automatic optical probe is a useful sensor wherever the distance to a very small surface area must be sensed remotely with high accuracy by a small, simple instrument.

Laser Alignment Device†

64. Description The laser alignment device is an alternative to the alignment telescope. A typical unit consists of a continuous-discharge laser mounted in, concentric to, and coaxial with a heavy-walled steel cylinder finished to the NAS barrel diameter (Par. 8). The operating system includes a means of detecting the centroid of the laser beam at any point along its path. The means (usually photosensors in quadrature array) are located at the points of alignment (as alignment targets are with an alignment telescope) or in the laser device itself, in which case the equipment must also include one or more plane mirrors (for angular alignment) or retroreflectors (for displacement detection). Auxiliary equipment includes the laser power supply and displacement readout system.

The laser beam, visible on whatever intercepts it, takes the place of the line of sight of the alignment telescope. The accuracy of setting or measurement

† Paragraphs 64 to 70 are based on information provided by Perkin-Elmer Corp.

is a function of beam symmetry and of the detection electronics system and thus does not depend on the skill or visual acuity of the observer.

65. Basic Principles Lasers are relatively new sources of light energy, light which has uniquely high intensity, monochromaticity, and coherence. Lasers for alignment are normally of the helium-neon gas type. They produce a continuous beam of light which is straight and bright over long distances, yet they operate at low power levels which may be considered safe.

A typical source comprises a helium-neon plasma tube and dielectrically coated quartz mirrors mounted in a suitable housing. The ends of the plasma tube are Brewster angle windows or internal reflectors. The laser light, which is the red line of ^{20}Ne, passes through a partially reflective spherical mirror at the projector end of the laser and is expanded and collimated to some useful beam diameter (generally 10 mm for metrology purposes and 20 mm for civil-engineering uses). Power output is of the order of 1.3 mW at 6328 Å.

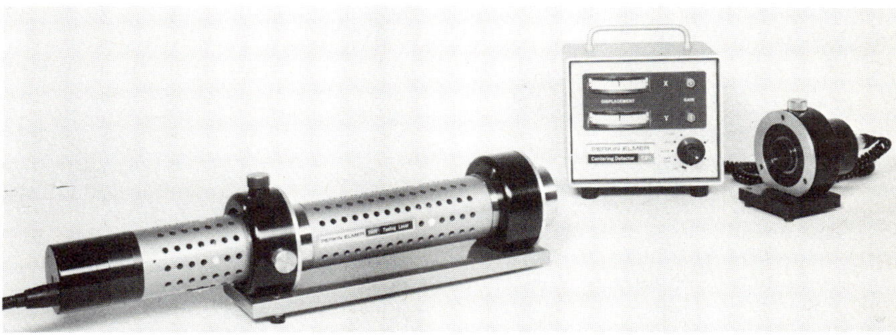

Fig. 30 Components of laser alignment device. *(Perkin-Elmer.)*

The collimated light energy has a gaussian distribution of intensity. This permits precise detection of the beam centroid and accurate measurement of displacement from the center by means of photoelectric sensors, a capability necessary for alignment work in the mechanical industries. The laser source is built into a cylindrical housing adaptable to standard optical-tooling mounts. It is constructed so that the beam diameter is concentric with the locating diameters on the housing exterior and the beam is projected parallel to the axis of the locating diameters.

At least one manufacturer impresses high-frequency modulation on the continuous-wave laser and makes the detector responsive only to the modulation frequency, to ensure discrimination of the signal in any ambient. Figure 30 shows an alignment-laser source mounted in a stand and the laser power supply. Also shown is a photodetector target with its power supply and readout unit. The photodetector unit contains four photocells in quadrature array. Opposing photodetectors are wired in a bridge circuit, so that any offset between the centroid of laser-beam energy and the center of the detectors produces a differential signal directly related to the amount of offset. The quadrature array permits simultaneous indication of linear displacement in two orthogonal directions perpendicular to the axis of the laser beam.

Another arrangement of laser source and photodetector, called an autoreflecting laser, is shown in Fig. 31. Here the photocells are built into the projector end of the laser housing to sense the deviation or offset of the laser beam after reflection back along its path from a target mirror. Used with a plane mirror as shown, this arrangement will sense angular misalignment; with a corner retroreflector it senses displacement along two axes.

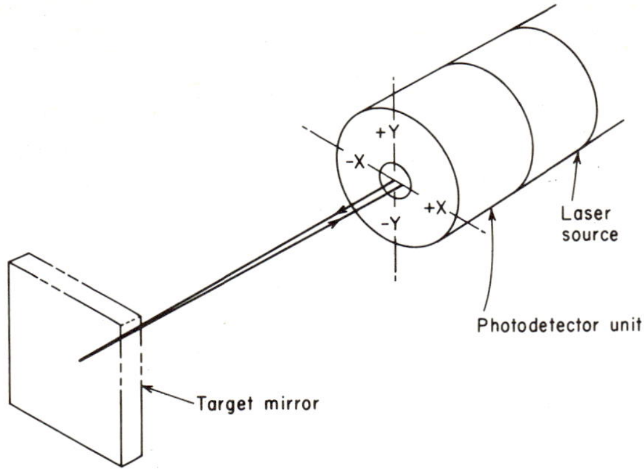

Fig. 31 Autoreflecting laser alignment system. (*Perkin-Elmer.*)

66. Applications and Accuracy Applications of laser alignment devices to metrology include:

1. Alignment of large aircraft jigs
2. Leveling machine-tool beds
3. Evaluating the geometry of machine-tool motions
4. Alignment of large workpieces for machining
5. Flatness measurement of machine tables and ways
6. Measurement of nutation in rotary devices
7. Alignment of propulsion equipment, beams, hatches, torpedo tubes, etc., in ships
8. Monitoring ship-hull movements in sea trials

Over a distance of 200 ft, a laser alignment device can theoretically measure to 0.0025 in.

TABLE 7 Representative Lasers for Alignment

Manufacturer	Model
Perkin-Elmer Corp.	5600, 5650, 5700
Keuffel & Esser Co.	71-2600

The laser beam is visible on any object, making rapid alignment of the beam to a surface or path possible or alignment of objects to the beam. A target grid or photoelectric meter can be read without eye fatigue and by more than one person at a time. Deviations can be determined and adjustments made at the target location by the same operator. The high intensity of laser light permits broad use of the pentaprism. Figures 32 and 33 show how a pentaprism can be employed to deflect the beam 90° for the determination of squareness. Given rotational capability, the pentaprism can also be used to establish a reference plane to which large surfaces can be compared or in which radial alignments can be established. Some applications require planes to be established parallel or perpendicular to the direction of gravity. For these purposes, an autoreflecting laser is positioned so that its beam is normal to a vertical or a horizontal leveling mirror, depending upon the desired orientation of the plane, and a rotating, beam-splitting pentaprism is used to generate the plane. Adequate light energy is available for simultaneous monitoring of the laser beam with respect to the leveling mirror and alignment or measurement work in the perpendicular plane. The laser beam can also be translated by a tilt-block optical micrometer (see Par. 17). By

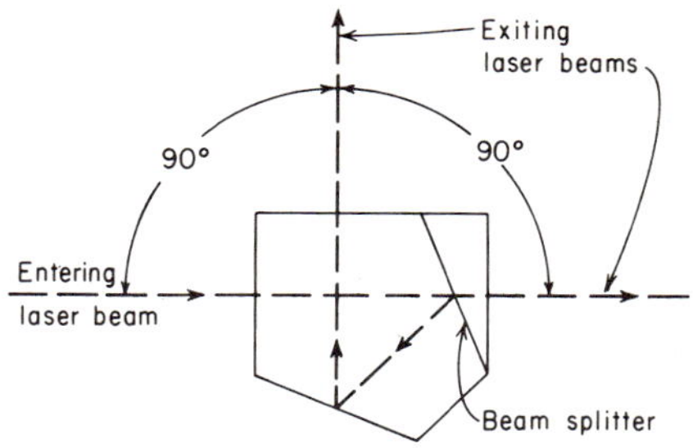

Fig. 32 Passage of a laser beam through a pentaprism.

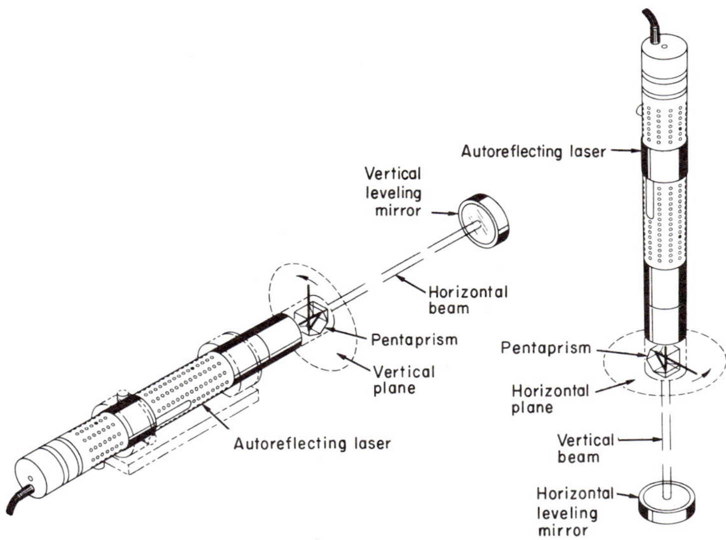

Fig. 33 Generation of planes. (*Perkin-Elmer.*)

means of a micrometer installed on a null-indicating autoreflecting laser, lateral displacement of a retroreflector can be measured from the laser location with (geometrically) twice the sensitivity of a remote detector.

INTERFEROMETRIC METROLOGY INSTRUMENTS

Laser-Interferometer Distance-Measuring Equipment

67. Description A laser interferometric distance-measuring system, commonly called a laser interferometer, is a system for measuring distance, at shop and machine-tool ranges, in increments of a fraction of a wavelength of light. Due to ambient atmospheric effects on the wavelength and temperature effects on materials and structures, various compensation factors are required. These may be applied manually or automatically. No two systems are identical. To represent the genre, a Perkin-Elmer laser interferometer is described, comprising the following:

1. A continuous-wave helium-neon gas laser source providing a stabilized, single-frequency beam of light

2. An interferometer, attached to the laser source, which converts into electric outputs the phase shift between a portion of the laser beam projected along

the path of measurement and a portion used as a fixed reference within the interferometer

3. An adjustable retroreflector prism capable of movement along the measurement path to return the projected laser beam to the interferometer

4. Transducers for monitoring atmospheric temperature and pressure and material temperature

5. An integral power supply and computer for conversion of inputs into outputs for visual and other presentation of length measurements in standard English or metric units

68. Basic Principles The representative system functions as follows. As shown in Fig. 34, the laser interferometer projects a beam of coherent monochromatic light from the laser source, through the distance to be measured, to the retroreflector prism, which is attached to the object whose distance is to be measured. Reflected

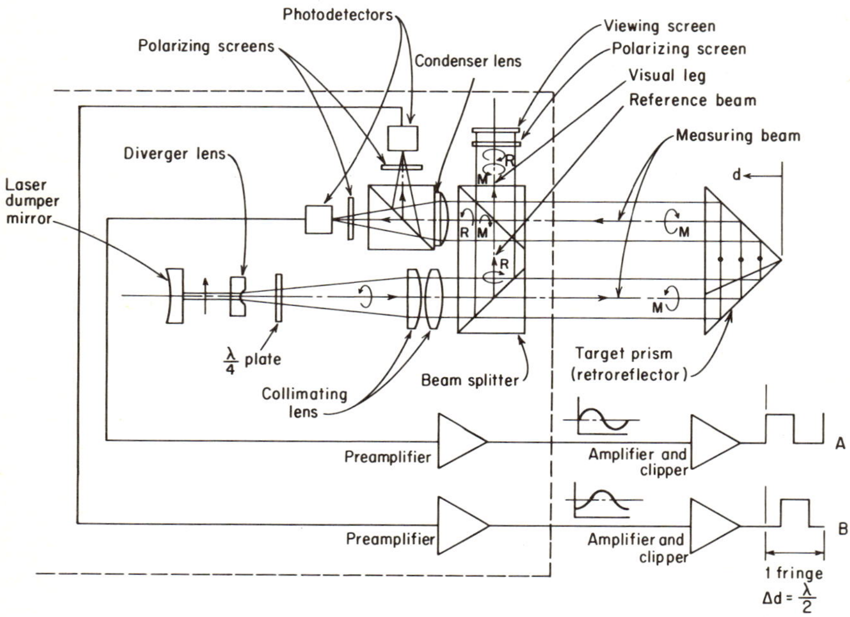

Fig. 34 Laser interferometers. (*Perkin-Elmer.*)

by the prism back to the interferometer on a path offset from the outgoing path, the measuring beam is combined with the reference beam so that its intensity is augmented, unchanged, or attenuated depending on its phase relationship to the reference beam, producing a pattern of fringes. Displacement of the target reflector along the beam a distance less than half a wavelength of the source will alter the phase relationship. A shift of one fringe with respect to a reference point within the interferometer occurs when the path length of the measuring beam changes by one wavelength. This represents a movement of the target equal to one-half the wavelength of the source light, or approximately 12.45 μin.

The combined reference and reflected laser-beam images and fringes are presented on a viewing screen at the interferometer and are also sensed by two photodetectors. A sine-wave electric signal is obtained from each detector, one cycle per fringe. The arrangement of the photodetection system is such that the two signals are 90° out of phase; the direction of target motion determines which signal is leading.

The 90° out-of-phase, or A quad B, output signals from the interferometer go to a computer-display module, which converts them into square waves. Because of the 90° phase difference, a bidirectional counter in the computer counts quarter fringes, equivalent to target displacement of one-eighth wavelength (about 3 μin) and also senses the direction of target motion. The total number of quarter fringes counted is multiplied by an appropriate factor in the computer to yield the actual distance the target mirror moved in inches or centimeters.

The computer accepts compensation factors for air pressure, temperature, and relative humidity, which affect the wavelength of the measuring light beam, as well as a compensation factor for the temperature of the material to which the target mirror is affixed. The indicated measurement is thus reduced to standard metrology temperature (68°F or 20°C). Information from pressure and temperature transducers is continuously supplied to the computer, which appropriately modifies the fringe-count distance indications. Compensation for relative humidity, which has small effect on wavelength, is manually inserted in the computer. Machine or part temperature compensation can be introduced automatically or manually or omitted, depending on circumstances and with a commensurate effect on accuracy.

The laser source is a dc-excited helium-neon laser (output wavelength 6328 Å); its construction achieves the requisite frequency and directional stability in the following manner. The end reflectors are rigidly mounted at the ends of an Invar resonator housing. The housing (cavity) is so shaped and suspended as to minimize

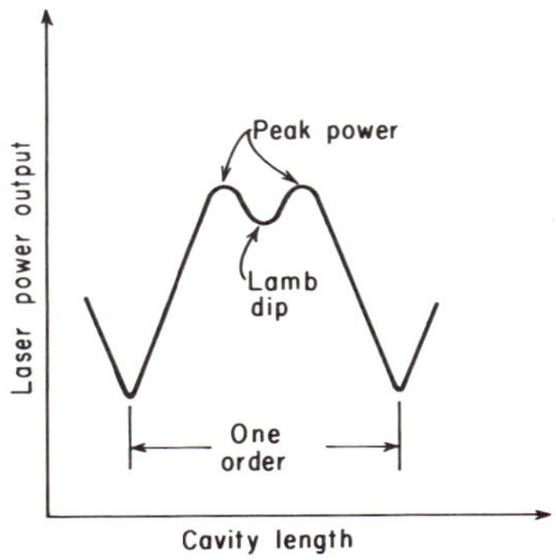

Fig. 35 Lamb dip waveform.

the effects of mechanical vibrations along the laser-beam path. The laser tube is thermally isolated from the cavity by a heat shield, which uses the outer housing as its heat sink. The entire cavity is contained in a temperature-controlled oven thermally isolated from ground.

The laser-output mirror is mounted to the resonator housing through a piezoelectric crystal. A small amount of energy is allowed to pass through the rear reflector to a photodiode serving as the sensing element in a laser-frequency stabilization loop. The photodiode aids in monitoring the output power level of the laser and therefore in setting the laser to the correct frequency for both open- and closed-loop operation. The processed photodiode output is utilized as the error signal which maintains the cavity length by varying the length of the piezoelectric element.

The offset beam design of the interferometer allows freedom from feedback to the source, so that the laser can be operated in a closed-loop mode. Closed-loop operation is achieved by locking the resonator frequency to the Lamb dip in power output which appears at the center of the helium-neon emission line. (A Lamb dip, see Fig. 35, is detected as a small decrease in power output between two neighboring peaks while a between-orders minimum appears as a sharp break in power output.) In an open-loop mode, a tuning control allows tuning of the cavity by applying a signal to the piezoelectric crystal. The crystal expands or contracts, thus varying the cavity length and the output frequency.

The laser-output mirror is corrected to provide a 1-mm collimated output beam. This beam passes through a half-wave plate within a 10-power beam expander, making the final output a polarized beam 10 mm in diameter. The beam is expanded to this dimension to minimize diffraction spreading, allowing interferometer operation at long distance. The larger-diameter beam also enables the system to tolerate greater lateral misalignment of the retroreflector.

The number of reflections in the interferometer is such that orthogonally polarized light interferes at the second beam splitter, producing linearly polarized light. This plane of polarization rotates 180° per half-wavelength movement of the target retroreflector.

Fringes are detectable only when the interference is observed through a plane Polaroid. It is possible to obtain proper phasing beween the A and B electrical outputs by physically rotating the Polaroid material.

The basic interferometer module, with its associated electronics, is located in a housing attached to the laser and is in an environment that is essentially temperature-controlled. Preamplifier electronics are in the temperature-controlled environment of the interferometer head. Integrated-circuit linear preamplifiers having a 500-kHz bandwidth permit retroreflector movement of up to 6 in/s. Visual perception of the measuring and reference beams, for purposes of alignment, is possible at a viewing screen on the interferometer.

TABLE 8 Representative Interferometers

Manufacturer	Model
Perkin-Elmer Corp.	5900
Hewlett-Packard Co.	5525A
Optomechanisms, Inc.	770

69. Applications Principal applications of laser interferometers are for evaluating positioning accuracy of machine tools and measuring machines, calibrating positioning systems and secondary length standards, and as linear-displacement transducers in machine tools and measuring machines.

70. Accuracy Basic accuracy of the system described herein is ±1 ppm +1 count over distances up to 150 ft in industrial-shop environments. Plus 1 count is the degree of resolution of the instrument, equal to one-eighth wavelength. Thus, in English units, the basic accuracy is $\pm(1\ \mu\text{in/in} + 3\ \mu\text{in})$. For a displacement of 50 in, for example, the basic accuracy would be 0.000053 in; for 500 in, 0.000503 in, etc. Its accuracy, as well as the great measuring range afforded by the laser, provides capability for many industrial and scientific requirements previously best served by end standards and incremental scales. Accuracies greater than 1 part in 10^6, to nearly 1 part in 10^7, are possible with laser interferometry because of the stability of single-frequency laser sources. However, the measuring environment for such accuracy must be stable, and material or machine temperature must be held close to the standard 68°F (20°C).

Alignment Interferometer

71. Description Precise alignment of points at a distance of several miles is the unique capability of an interferometric arrangement called an alignment interferometer. As described by Saunders (1963*a*,*b*, 1964), the system comprises point light sources at the points to be aligned and a receiver which can detect displacement from its optical axis of any of these points when illuminated one at a time.

As shown in Fig. 36, the receiver consists of a pair of Koester's double-image prisms (see Par. 76) and a beam splitter, polarizers or quarter-wave plates, and an objective, either refracting (as shown) or reflecting. Eyepieces are not required but may be used to magnify the fringe pattern. The two double-image prisms and the beam splitter are so arranged that in effect both prisms are centralized on the optical axis. Their internal interfaces (dividing planes) are in line with the axis and rotationally oriented about it at right angles to each other.

72. Basic Principles (See also page 52 in Sec. 2.) The extension into space of the effective line of intersection of the dividing planes of the two prisms defines the reference line with respect to which displacement of the point sources can be detected. A number of light sources can be aligned to the interferometer axis, or vice versa, to an accuracy which is a function of aperture and distance to the source. The instrument is sensitive to the difference in optical-path lengths from the source to opposite extremities of the entrance aperture, at right angles to the parting planes of the two Koester's prisms. The prisms are so situated with respect to the focus of the objective lens that the light from either extremity of the aperture is supplied to only one side of the appropriate prism. The beams are recombined in the prism, producing an interference pattern caused by the above difference, indicating in-plane or other alignment. When both axes indicate in-plane, the point is aligned to the axis. The double-image prisms, in which the internal interface equally transmits and reflects incident light, serve as the beam combiners in planes 90° apart. The set of two prisms and the beam splitter constitutes an alignment interferometer if either the source or the prism output is polarized. Differential polarization in the prism, with an unpolarized light source, will produce

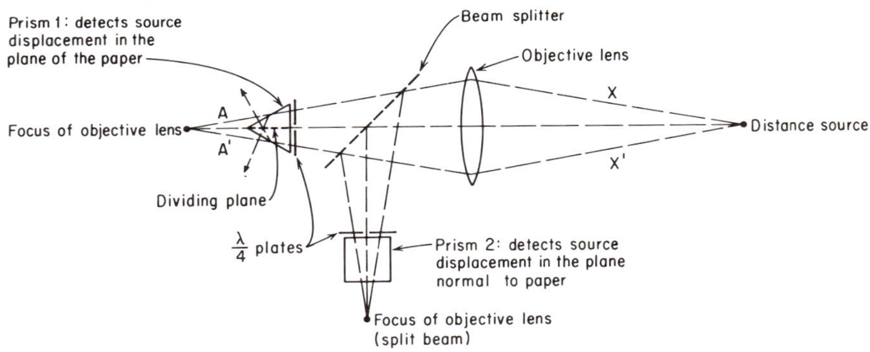

Fig. 36 Alignment interferometer. The dividing plane of the double-image prism is a semireflecting optical film, which serves as the beam combiner in this application. The interference patterns may be viewed at either A or A' with respect to each prism. When the source is moved off the axis, the change in relative lengths of X and X' produces a change in the interference patterns. (*J. B. Saunders.*)

two sets of fringes approximately so related that the bright areas of one set coincide with the dark areas of the other set, obscuring both. Polarization will permit the use of either set, depending on rotational orientation of the polarizer, at the expense of half the light. Quarter-wave plates cut at 45° to their optical axes and oriented with their axes at right angles effect a relative phase shift of 180° between light in the principal planes of polarization, thereby causing reinforcement rather than cancellation of fringe patterns. A slight wedge angle is built into the double-image prisms to produce acceptable fringe spacing for visual use.

The optical sensitivity of the alignment interferometer is proportional to the aperture. Since the Koester's prisms are difficult to manufacture in large sizes, a more practical large-aperture instrument can be obtained by using a large objective with relatively small prisms. Without an objective lens the prism requires either an auxiliary collector or a spherical face.

No commercial version of this instrument is known to be available.

73. Applications The primary application of the alignment interferometer appears to be in bringing into line three or more well-separated points. If the order of interference observed can be determined and the distance is known, both angular misalignment and linear displacement from the reference line can be established. Saunders (1963*b*) suggests applying the instrument to the guidance of a satellite-mounted astronomical telescope, when it would become functionally a pointing interferometer, essentially a two-axis receptor-only version of the instrument described in Par. 75. Automated versions of the basic instrument are feasible.

74. Accuracy Under conditions in which the geometry of the setup is the determining factor, i.e., in a completely stable and homogeneous medium or in a vacuum, an alignment interferometer of 1 in aperture can detect lateral displacement from the reference line of less than 1.5 mm at 1 mi distance. A 10-in-aperture instrument can make the same determination to less than 0.15 mm, or approximately 0.006 in.

Alignment accuracy is closely approximated by

$$d = \frac{XN\lambda}{A} \tag{13}$$

where d = lateral displacement of source
X = axial distance to source
N = observed order of interference
λ = wavelength of source
A = aperture of interferometer

Pointing Interferometer

75. Description The pointing interferometer, an instrument for measuring angles as small as 0.02 second of arc, comprises a Michelson type two-beam interferometer, using a Koester's double-image prism having equal internal optical paths. It senses minute rotation of a mirror about an axis normal to the plane containing both projected beams by the shift of the interference pattern presented to the observer. The pattern shifts because of the change in the relative lengths of the two optical paths as the mirror is rotated about a sensitive axis. Like an autocollimator, the pointing interferometer is insensitive to rotation of the mirror about its own normal or to lateral or axial translation of the mirror. A representative instrument, the Kollmorgen pointing interferometer, is designed to mount on the barrel of any alignment telescope with an NAS $2\frac{1}{4}$-in standard barrel. It uses both the magnification and the (necessarily internal) optical micrometer of the alignment telescope to divide the scale unit (one fringe, representing mirror rotation of 1 second of arc) into increments as small as 0.02 second of arc. Contact or precise alignment with the alignment telescope is not required.

76. Basic Principles (See also page 52 in Sec. 2.) The instrument, shown schematically in Fig. 37, employs both a white-light source (incandescent lamp) and a monochromatic source (mercury lamp) introduced coaxially and parfocally via a filter-reflector. Light from the two sources is collimated by an objective lens and fed to the Koester's prism, from which it is projected as twin parallel beams. The double-image prism comprises two 30-60-90 plane prisms cemented along the long side, forming a truncated equilateral plane prism with a beam-splitting bisecting perpendicular. Its critical attributes are that the two internal light paths are equal for light incident on either of the constituent prism hypotenuses, whether transmitted or reflected by the beam splitter, and the exit beams are parallel. Reflected by a mirror normal to their common axis, both beams retrace their paths. A portion of the returned light exits from the constituent-prism hypotenuse not used for entry. This output, containing both beams, exhibits interference patterns caused by their mutual interaction. These are reflected into the telescope and thus to the observer. Optical-path lengths are constant within the instrument, changing relative to each other only as mirror rotation causes one external air path to lengthen and the other to shorten.

The change in relative path length, a function of beam separation and mirror rotation, causes the fringe pattern to shift in a direction normal to the pattern lines. Beam separation is chosen so that mirror rotation of 1 second of arc will create a path-length difference of one-half wavelength of mercury light, which in turn will cause the pattern to shift one fringe. The pattern contracts as the mirror normal is deviated in either direction, in the plane of the twin beams, from parallelism with the interferometer beams. However, the sensitivity gradient (one fringe per second of arc) remains constant. Measurement of a larger angle within the instrument's 2,000-second range is therefore a matter of counting fringes as they pass the reference mark. Photoelectric counting can be used.

The inherent stability of the Koester's prism makes a field-operating interferometer practical. The basic measuring fringes are provided by the monochromatic (green) output of the mercury lamp. The addition of a small amount of white light (rheostat-controlled by the operator) will identify the achromatic (zero-order) fringe. The angle between the mirror and the plane of the beam splitter (common side of constituent prisms) is uniquely defined by the zero-order fringe, providing a reference point for measuring mirror rotation.

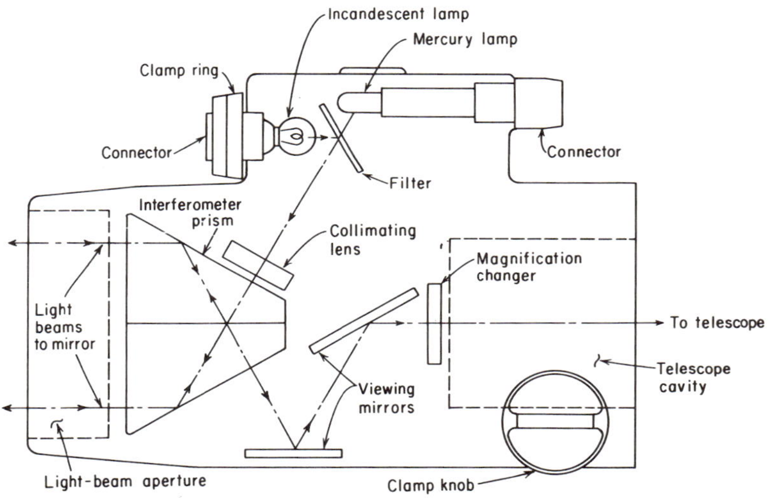

Fig. 37 Pointing interferometer. (*Kollmorgen Corp.*)

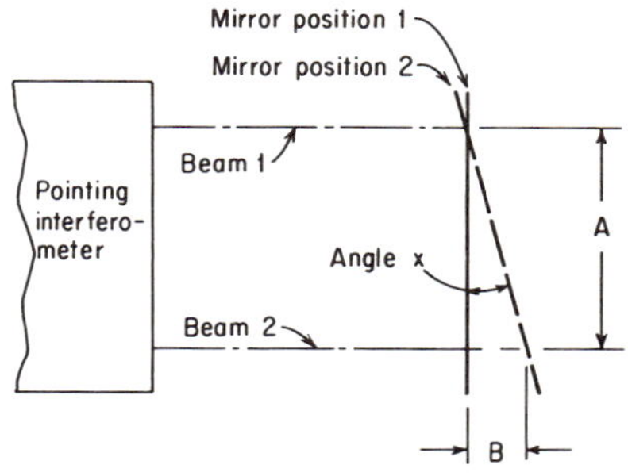

Fig. 38 Sensitivity of a pointing interferometer. (*Kollmorgen Corp.*)

$$\tan x = \frac{B}{A}$$

where B equals one-half wavelength for the passage of one fringe; therefore for one fringe passage

$$\tan x = \frac{\lambda/2}{A} \qquad \text{or} \qquad x = \tan^{-1}\frac{\lambda/2}{A}$$

77. Applications The pointing interferometer is a rotation-sensing device which exceeds the sensitivity and accuracy of an autocollimator and may thus be used in most autocollimator applications. However, a two-beam interferometer is sensitive to changes in the length of the light path of one beam relative to the other for any reason. Therefore it can be used to measure differences of refractive index, absolute refractive index when combined with physical measurement, or thickness of thin films; with auxiliary fixtures and components it can be used to measure absolute pressure of a gas, sphericity of ball-bearing balls and like components, and other mechanical properties.

In all applications the air path should be kept as short as possible. Since homogeneity and stability of the air are critical to operation, some form of enclosure for the air path may be required.

78. Sensitivity and Accuracy The inherent accuracy of the pointing interferometer stems from the stability of the wavelength of mercury light and from the rigidity and relative dimensional stability of the structure. The interferometer measures the tangent of tilt rather than the angle directly; however, over the measuring range the difference between the tangent and the angle is insignificant. At 2,400 seconds from the zero-order fringe, the angle error would be 0.005 percent. Angular sensitivity is shown in Fig. 38.

OTHER DEVICES FOR METROLOGY

79. The basic optical metrology functions of alignment and angular measurement have been accomplished by more exotic means than the equipment and techniques described above and with equal or greater precision. Brief descriptions of linear zone plates, axicons, and moire-fringe devices are given as representative of this category. Details of these and other schemes can be found in the references.

80. Zone Plate Linear zone plates, and/or circular diffraction gratings, illuminated through a pinhole, will indicate alignment or misalignment of a distant reticle center to the line of centers of the pinhole and zone plate. A transmissive linear zone plate may be a metal-coated glass plate with a number of concentric rings (in effect, circular slits) inscribed at equal increments of radius through the metal coating. The eye approximately in line, with or without an eyepiece, sees a pattern of concentric interference fringes due to diffraction at the slits. A reticle may be aligned to the fringe pattern, or since the pinhole and reticle define a line, the zone plate may be aligned to it, or its displacement measured, at the reticle. The eyepiece serves merely to magnify the reticle and the fringe pattern. Circular diffraction gratings may replace the zone plates, providing better illumination, among other improvements.

Intersecting straight lines can be defined by using two light sources and reticles. A single zone plate rotating about its central axis normal to the plane defined by the measured lines at something higher than an eye-detectable flicker frequency will function as the zone plate in both systems simultaneously, thus assuring intersection of the lines.

81. Axicon An axicon, as the name implies, is a refracting element of such spherical aberration as to image a point source as an axial line. It may take the form of a ring, a cylinder, a cone, a spherical lens, or some combination of those. McLeod (1954) defines the axicon as any figure of revolution that by reflection or refraction or both, will bend light from a point source on the axis of the figure of revolution so as to cross the axis not at a point, as would be the case with a lens, but along a continuous line of points along a substantial extent of the axis.

A common form of axicon is a plano-convex shallow cone. McLeod (1954) describes the positioning of a satisfactorily bright image 100 ft from a 6-in-diameter cone, illuminated by an on-axis 0.003-in-diameter white source 100 ft behind. A cross hair could be located to within 0.001 in.

Axicons, for their special properties, may be used as objectives in common optical instruments, such as telescopes, projectors, and autocollimators. An axicon autocollimator, illuminated by a pinhole source, projects both parallel rays and an axial line into space. It refocuses reflected parallel rays via a beam splitter, for the observer, much as an autocollimator with a plane-image-forming objective does. It can also, and at the same time, detect linear misalignment of illuminated points, thereby simultaneously serving as autocollimator and alignment telescope.

82. Moire-Fringe Devices Moire fringes are the effect of superimposing a repetitive design, such as a grid or diffraction grating, on the same or a different design to produce a pattern distinct from its components. Sensitivity to relative motion between the two base patterns can be high. The availability of electronic fringe counters and low-cost diffraction gratings, both linear and radial, has estabilshed moire-fringe counting as a practical technique for digital measurement and for automatic control of machine tools.

Moire-fringe techniques can also be used directly for topographical contour mapping of large and small objects, ranging from irregular shapes in three dimensions to nearly perfect flats. Another direct application is in registration of an overlay on a substrate, where highly accurate positioning is required.

PRECISION OF MEASUREMENT AND ERROR ANALYSIS

Justification for Random-Error Analysis

83. Errors in the indications of optical-metrology instruments under ideal conditions of operation may be classified as systematic or random. Systematic, or assignable-cause, constant errors are inherent in the system comprising the instrument and the operator. Examples are errors due to nonorthogonality of supposedly orthogonal axes (or similarly, nonparallelism), to eccentricity of supposedly concentric parts, to backlash in driven-indicator mechanisms, to nonlinearity of scales, and to other mechanical deficiencies, as well as operator bias.

Systematic errors can be compensated for by calibration and the corresponding correction of data, as well as by the use of error-canceling techniques. The latter include:

1. Plunge and reverse with transit-type instruments
2. Rolling telescopes about their mechanical axes
3. Sliding scale, i.e., use of different parts of scales to counter nonlinearity
4. Approaching the final setting for each reading always in the same direction of mechanism rotation, to justify application of operator-bias correction and/or dead-band compensation

The remaining errors, under ideal conditions, are random. The population of all random errors is assumed to follow a normal, or gaussian, distribution, which permits the application of statistical methods to data already corrected for systematic error.

84. Precision and Accuracy Precision refers to the magnitude of random errors in a set of data; accuracy refers to the magnitude of systematic errors. High precision denotes small random errors; high accuracy denotes small systematic errors.

The distinction between precision and accuracy can be illustrated by the rifle-target analogy. A tight cluster of bullet holes several cluster diameters away from the bull's-eye represents high precision, i.e., repeatability or small deviation from the mean, but with low accuracy, i.e., subject to a constant or systematic error such as misaligned sights. On the other hand, an open pattern randomly distributed but centered on the bull's-eye represents high accuracy with low precision, i.e., free of systematic or other constant errors but having a large standard deviation from the mean. Finally, a tight cluster centered on the bull's-eye represents high precision with high accuracy.

Observations and Errors

85. Under less than ideal conditions, e.g., necessary extreme environments and/or unavoidable operator discomfort or fatigue, error-canceling techniques should still be valid. However, compensation based on calibrations made under ideal conditions may well be invalid, and the magnitude of random errors will probably increase. The relationships and effects of static instrument deformities will probably change, due to thermal effects, to the extent of requiring calibration in the environment, while vibration may affect the precision of setting by a combination of mechanical, physiological, and psychological effects. Combined with additional effects on the operator, those of the above which are not basically or conditionally systematic will increase the range of the data, therefore also the standard deviation of the sample and the number of readings required to minimize their effect on accuracy.

The extent to which any of these calibrations and operating or data-reduction techniques is applicable in any instance will depend on circumstances and on the accuracy required. A rule of thumb suggests that the overall measuring capability should be finer by a factor of 10 than the greatest allowable departure from the nominal of the quantity measured.

An approach to statistical data reduction is given in Baumeister and Marks (1967); a general discussion of the accuracy of measurement appears in Juran (1962).

86. Best Estimate and the Mean Where systematic errors have been compensated and only random errors are presumed to remain, the arithmetic mean is the best value obtainable from n equally careful, independent measurements of a single unknown made under the same conditions; n measurements yield n simultaneous equations in one unknown, as

$$x = x_1 \qquad x = x_2 \qquad \cdots \qquad x = x_n$$

where x is an unknown quantity, $x_1, x_2, \ldots, x_n$ are the first, second, . . . , nth readings and the arithmetic mean

$$x_0 = \frac{x_1 + x_2 + \cdots + x_n}{n}$$

The difference between the arithmetic mean and any observed value is called the deviation from the mean or the residual of the observed value with respect to x_0, expressed as

$$v_1 = x_0 - x_1 \qquad v_2 = x_0 - x_2 \qquad v_n = x_0 - x_n$$

For a given set of observations, the arithmetic mean satisfies the criterion of the method of least squares for identifying the best estimate of the true value of x, which requires that the sum of the squares of the deviations from the estimated value of x of the individual observed values be less than the sum of the squares of the deviations of the individual observed values from any other possible value of x.

87. Standard Deviation The standard deviation s of a sample (a given set of observed values) is given by

$$s = \sqrt{\frac{\Sigma v^2}{n}} \tag{14}$$

i.e., the square root of the average value of the squares of the residuals, or deviations from the mean, of the observed values. With regard to random errors only, the greater the number of observations n, the more closely the standard deviation s of the sample approaches the standard deviation σ of the population. The following proportions of the total number of observations are included within limits bounding the mean set by various factors of σ:

$\pm 0.674\sigma$	50%
$\pm\sigma$	68.27%
$\pm 2\sigma$	95.45%
$\pm 3\sigma$	99.73%

The odds that a single observation will fall within $\pm 3\sigma$ limits are 370:1. As a practical matter, using the standard deviation of a sizable sample as an approximation of the standard deviation of the population, it is wise to examine for accidental or nonsystematic assignable-cause errors individual observations which fall outside the interval of the arithmetic mean $\pm 3\sigma$.

Paragraph 88 gives methods of evaluating probable error. Bessel's formulas are exact; Peters' are approximate but more convenient. Probable error is that magnitude of deviation whose probability of being exceeded is one-half.

88. Probable Error Probable error of a single observation of a series:

$$r = \frac{0.6745}{(n-1)^{1/2}} (v_1^2 + v_2^2 + \cdots + v_n^2)^{1/2} \qquad \text{Bessel}$$

or approximately

$$r = \frac{0.8453}{[n(n-1)]^{1/2}} (|v_1| + |v_2| + \cdots + |v_n|) \qquad \text{Peters}$$

The notation $|v_1|$ denotes the absolute value of v_1, that is, its numerical value without regard to algebraic sign. The statement that the probable value of $x = x_0 \pm r$ means that from the information available the true value of x is just as likely to be within $x_0 \pm r$ as outside, provided that r is small.

Probable error of the mean:

$$r_0 = \begin{cases} \dfrac{0.6745}{[n(n-1)]^{1/2}} (v_1^2 + v_2^2 + \cdots + v_n^2)^{1/2} & \text{Bessel} \\ \dfrac{0.8453}{n(n-1)^{1/2}} (|v_1| + |v_2| + \cdots + |v_n|) & \text{Peters} \end{cases}$$

As with the single observation, $x = x_0 \pm r_0$ means that the true value of x is just as likely to lie within $x_0 \pm r_0$ as outside it, provided r_0 is small. Note that $r_0 = r/n^{1/2}$; r and r_0 indicate relative precision (not accuracy). For convenience, the factors $0.6745/(n-1)^{1/2}$, $0.8453/[n(n-1)]^{1/2}$, $0.6745/[n(n-1)]^{1/2}$, and $0.8453/n(n-1)^{1/2}$ are given for values of n in Table 9.

As an example, from the table, the probable error of the mean of ten readings is $0.0711(\Sigma v^2)^{1/2}$, or 0.0711 times the square root of the sum of the squares of the deviations from the mean of the individual readings if Bessel's formula is used. It is $0.0282(\Sigma v)$ by Peters' approximation.

89. Acceptability of Data Values derived from the above relationships can be tested for reasonableness, based on the distribution of residuals, as follows.†

Arrange the residuals (deviations from the mean) in order of magnitude without regard to sign and count the number y of residuals which are numerically less than some arbitrarily assigned value a; divide y by n, the total number of observations, and divide a by r, the probable error of a single observation. Do this for various values of a and compare the results with Table 10, which gives the standard distribution of residuals, as found from experience from a large number of different series of observations. In particular, the number of residuals numerically less than r should be about equal to the number numerically greater than r (if n is large). If any large discrepancy appears, the series of observations should be regarded as unsatisfactory.

Errors in Systems

90. Metrological setups are often in fact systems; i.e., more than one instrument contributes both systematic and random error to any determination. When individual-instrument systematic errors have been canceled or otherwise removed from the data, presumed random residual errors for which rms values have been derived may be combined to yield system error by the root-sum-square (rss) formula, to the same confidence level as the rms values. That is,

$$\text{rss} = \sqrt{\sigma_1^2 + \sigma_2^2 + \cdots + \sigma_n^2}$$

† Adapted from Baumeister and Marks (1967) with permission.

TABLE 9 Factors for Computing Probable Error†

n	Bessel		Peters	
	$\frac{0.6745}{(n-1)^{1/2}}$	$\frac{0.6745}{[n(n-1)]^{1/2}}$	$\frac{0.8453}{[n(n-1)]^{1/2}}$	$\frac{0.8453}{n(n-1)^{1/2}}$
2	0.6745	0.4769	0.5978	0.4227
3	0.4769	0.2754	0.3451	0.1993
4	0.3894	0.1947	0.2440	0.1220
5	0.3372	0.1508	0.1890	0.0845
6	0.3016	0.1231	0.1543	0.0630
7	0.2754	0.1041	0.1304	0.0493
8	0.2549	0.0901	0.1130	0.0399
9	0.2385	0.0795	0.0996	0.0332
10	0.2248	0.0711	0.0891	0.0282
11	0.2133	0.0643	0.0806	0.0243
12	0.2034	0.0587	0.0736	0.0212
13	0.1947	0.0540	0.0677	0.0188
14	0.1871	0.0500	0.0627	0.0167
15	0.1803	0.0465	0.0583	0.0151
16	0.1742	0.0435	0.0546	0.0136
17	0.1686	0.0409	0.0513	0.0124
18	0.1636	0.0386	0.0483	0.0114
19	0.1590	0.0365	0.0457	0.0105
20	0.1547	0.0346	0.0434	0.0097
21	0.1508	0.0329	0.0412	0.0090
22	0.1472	0.0314	0.0393	0.0084
23	0.1438	0.0300	0.0376	0.0078
24	0.1406	0.0287	0.0360	0.0073
25	0.1377	0.0275	0.0345	0.0069
26	0.1349	0.0265	0.0332	0.0065
27	0.1323	0.0255	0.0319	0.0061
28	0.1298	0.0245	0.0307	0.0058
29	0.1275	0.0237	0.0297	0.0055
30	0.1252	0.0229	0.0287	0.0052
31	0.1231	0.0221	0.0277	0.0050
32	0.1211	0.0214	0.0268	0.0047
33	0.1192	0.0208	0.0260	0.0045
34	0.1174	0.0201	0.0252	0.0043
35	0.1157	0.0196	0.0245	0.0041
36	0.1140	0.0190	0.0238	0.0040
37	0.1124	0.0185	0.0232	0.0038
38	0.1109	0.0180	0.0225	0.0037
39	0.1094	0.0175	0.0220	0.0035
40	0.1080	0.0171	0.0214	0.0034
45	0.1017	0.0152	0.0190	0.0028
50	0.0964	0.0136	0.0171	0.0024
55	0.0918	0.0124	0.0155	0.0021
60	0.0878	0.0113	0.0142	0.0018
65	0.0843	0.0105	0.0131	0.0016
70	0.0812	0.0097	0.0122	0.0015
75	0.0784	0.0091	0.0113	0.0013
80	0.0759	0.0085	0.0106	0.0012
85	0.0736	0.0080	0.0100	0.0011
90	0.0715	0.0075	0.0094	0.0010
95	0.0696	0.0071	0.0089	0.0009
100	0.0678	0.0068	0.0085	0.0008

† Reprinted from T. Baumeister and L. S. Marks, "Standard Handbook for Mechanical Engineers," 7th ed., 1967, by permission of McGraw-Hill Book Company.

TABLE 10 Standard Distribution of Residuals†

a = any positive quantity
y = number of residuals which are numerically $< a$
r = probable error of a single observation
n = number of observations

$\frac{a}{r}$	$\frac{y}{n}$	Diff	$\frac{a}{r}$	$\frac{y}{n}$	Diff
0.0	0.000	54	2.5	0.908	13
0.1	0.054	53	2.6	0.921	10
0.2	0.107	53	2.7	0.931	10
0.3	0.160	53	2.8	0.941	9
0.4	0.213	51	2.9	0.950	7
0.5	0.264	50	3.0	0.957	6
0.6	0.314	49	3.1	0.963	6
0.7	0.363	48	3.2	0.969	5
0.8	0.411	45	3.3	0.974	4
0.9	0.456	44	3.4	0.978	4
1.0	0.500	42	3.5	0.982	3
1.1	0.542	40	3.6	0.985	2
1.2	0.582	37	3.7	0.987	3
1.3	0.619	36	3.8	0.990	1
1.4	0.655	33	3.9	0.991	2
1.5	0.688	31	4.0	0.993	6
1.6	0.719	29			
1.7	0.748	27	5.0	0.999	
1.8	0.775	25			
1.9	0.800	23			
2.0	0.823	20			
2.1	0.843	19			
2.2	0.862	17			
2.3	0.879	16			
2.4	0.895	13			

† Reprinted from T. Baumeister and L. S. Marks, "Standard Handbook for Mechanical Engineers," 7th ed., 1967, by permission of McGraw-Hill Book Company.

may be applied with the same accuracy as that with which the data-derived standard deviation of the sample s represents the standard deviation of the population σ. Assuming that $s = \sigma$, rss will be determined to the same confidence level as that to which the values of σ are expressed, for example, σ, 2σ, or 3σ.

REFERENCES

Baird, D. C. (1962): "Experimentation," Prentice-Hall, Englewood Cliffs, N.J.

Barry, B. A. (1964): "Engineering Measurements," Wiley, New York.

Baumeister, T., and L. S. Marks (1967): "Standard Handbook for Mechanical Engineers," 7th ed., McGraw-Hill, New York.

Beers, Y. (1962): "Introduction to the Theory of Error," Addison-Wesley, Reading, Mass.

Davis, R. E., and F. S. Foote (1953): "Surveying Theory and Practice," 4th ed., McGraw-Hill, New York.

Dyson, J. (1960): Circular Diffraction Gratings as Alignment Devices, in P. Mollet (ed.), "Optics in Metrology," Pergamon, New York.

Ehrhardt, O. W. (1966): New Face of Metrology: Its Impact, or Measuring to Millionths, *Tool Manuf. Eng.*, August.

Juran, J. M. (ed.) (1962): "Quality Control Handbook," 2d ed., McGraw-Hill, New York.

Kingslake, R. (1965): "Applied Optics and Optical Engineering," Academic, New York.

Kissam, P. (1962): "Optical Tooling," McGraw-Hill, New York.

Lovins, G. H. (1964): High Precision Pointing Interferometer, *Appl. Opt.*, vol. 3, no. 7, p. 883, July.

McLeod, J. H. (1954): The Axicon: A New Type of Optical Element, *J. Opt. Soc. Am.*, vol. 44, p. 592.

——— (1960): Axicons and Their Uses, *J. Opt. Soc. Am.*, vol. 50, p. 166.

Meadows, D. M., W. O. Johnson, and J. B. Allen (1970): Generation of Surface Contours by Moire Patterns, *Appl. Opt.*, p. 942, April.

Minkowitz, S. (1968): Industrial Laser Interferometry, *Opt. Spectra*, May-June.

Optical Alignment (Basic), *NASA Publ.* PB 183592, 1969.

Oster, G. (1969): "Science of Moire Patterns," Edmund Scientific Co., Barrington, N.J.

Saunders, J. B. (1963*a*): An Alignment Interferometer, *Appl. Opt.*, vol. 2, no. 5, May.

——— (1963*b*): An Alignment Interferometer, *J. Res. Natl. Bur. Stand.*, vol. 67C, no. 4, October-December.

——— (1964): An Alignment Interferometer, *J. Res. Natl. Bur. Stand.*, vol. 3, no. 7, p. 883, July.

Schede, R. W. (1967): Interferometers for Use as Integral Parts of Machine Tools, *IEEE Trans. Ind. Gen. Appl.*, vol. IGA-3, no. 4, p. 328, July-August.

Steel, H. W. (1960): Axicons with Spherical Surfaces, in P. Mollet (ed.), "Optics in Metrology," Pergamon, New York.

Takasaki, H. (1970): Moire Topography, *Appl. Opt.*, p. 1457, June.

Tsuruta, T., Y. Itoh, and S. Anzai (1970): Moire Topography for the Measurement of Film Flatness, *Appl. Opt.*, p. 2802, December.

Van Heel, A. C. S. (1950): High Precision Measurements with Simple Equipment, *J. Opt. Soc. Am.*, vol. 40, p. 809.

——— (1961): Modern Alignment Devices, *Progr. Opt.*, vol. 1.

Section 17

Optical Modulators

EDWARD HARTFIELD
Itek Corporation, Lexington, Mass.

BRIAN J. THOMPSON
The Institute of Optics, University of Rochester, Rochester, N.Y.

(Numbers refer to paragraphs)

1. Introduction† For more than a century, there has been a need for methods of controlling the intensity or amplitude of a light beam. For example, the early measurements of the velocity of light required the light to be modulated; the modulation was accomplished by using a rotating toothed wheel.

Today the requirement is for modulating devices that can operate at 100 MHz or higher. Proposed applications for communications between satellites require operation in the gigahertz range. The laser has provided us with a high-intensity monochromatic light source capable of many applications in communications, display, and data reading and recording when a useful modulator is also available.

In today's terminology the toothed wheel and the devices that merely interrupt the beam are referred to as choppers and are hence an elementary form of modulator (Par. 24). The Kerr cell (Par. 14) has replaced the toothed wheel in velocity-of-light experiments; it can provide modulation frequencies two orders of magnitude greater and has become an important general-purpose modulator.

Electrooptic modulators first became of interest in the late 1940s following theoretical and experimental studies with ammonium dihydrogen phosphate and potassium dihydrogen phosphate. These modulators depend upon the Pockels effect, which had been known for many years but was too small for practical utilization in older materials. Material studies continue, and other materials are now also of interest, including lithium niobate, lithium tantalate, and barium sodium niobate (Par. 6).

A parallel effort has also taken place in acoustooptic modulators. Liquid cells containing alcohol or water have been used, and solid modulators of glass, fused silica, and sapphire have been developed. Newer materials again are starting to play a role, particularly α-iodic acid and lead molybdate (Par. 15).

Finally, magnetooptical effects have been considered based primarily upon the Faraday and Kerr effects, but only limited use has been made of these results to date.

Table 1 summarizes the important parameters of the more practical modulating devices: Further information on these materials can be found in the appropriate paragraphs.

ELECTROOPTIC MODULATORS

Definition

2. An electrooptic modulator is a device utilizing the phenomenon of induced electrical birefringence in certain materials such that an incident plane-polarized wave will become elliptically polarized upon passage through the material, in response to an applied voltage across the material.

Types

3. Electrooptic modulators utilizing the Pockels effect may be classified as either longitudinal or transverse. In longitudinal modulators, the electric field is parallel to the direction of light propagation; hence semitransparent or ring-electrode structures are required. Transverse modulators operate with the electric field orthogonal to the direction of light propagation, thus avoiding placing the electrodes in the

† Section 17 was written while both authors were with Technical Operations, Inc., Mountainview, Calif.

TABLE 1 A Summary Table of the More Generally Useful Materials for Light Modulators

Electrooptic material	ϵ	$(El)_{\lambda/2}$ for $\lambda = 0.633\ \mu m$, V	Temp., °C	tan δ	Optical transmission, μm	Light-propagation direction	Electric field direction
ADP	56	7,670	Room	0.006 at 9.2 GHz	0.25–1.1	45°c	$\parallel b$
KDP	21	19,000	Room	0.0074 at 9.2 GHz	0.25–1.1	$\parallel a$	$\parallel c$
KD*P	50	6,940	Room	0.11 at 9.2 GHz	0.25–1.6	45°a	$\parallel c$
$LiNbO_3$	84	4,000	25	†	0.45–4.5	$\parallel c$	$\parallel a$
$BaNaNb_5O_{15}$	51	1,720	30	...	0.45–4.5	$\parallel b$	$\parallel c$
$LiTaO_3$	43	2,700	Room	0.002 at 0.1 GHz	0.45–4.5	$\parallel a$	$\parallel c$

Acousto-optic material	Polarization direction‡		Figure-of-merit ratio	Useful optical wavelength range, μm	Approximate ultrasonic attenuation, dB/cm at 500 MHz
	Acoustic wave	Optical wave			
Fused silica	Long	$\perp$	1.0	0.2–2.5	3
$LiNbO_3$	Long, [11$\bar{2}$0]	...	4.64	0.5–4.5	<0.05
α-HIO_3	Shear, a–b	a–c	11.3	0.5–4.5	2.5
α-HIO_3	Long, b	a–c	27.6	0.5–4.5	2.5
GaP	Long, [110]	$\parallel$	29.5	0.6–10	<1
α-HIO_3	Long, a	a–c	31.8	0.5–4.5	2.5
α-HIO_3	...	a–b	55.2	0.5–4.5	2.5
H_2O	Long	$\perp$	106	0.2–0.85	—

† See Fig. 3.
‡ Notation in accordance with present IRE conventions.

path of light. (For review articles see Kaminow and Turner, 1966;† Spencer et al., 1967.) Kerr cells are an excellent example of this transverse class.

4. Longitudinal Modulators In longitudinal modulators, the electric field is applied parallel to the direction of the optic (z) axis (see Fig. 1), and the incoming plane-polarized wave is aligned parallel to either the x or y crystal axes. With no voltage applied, the indexes of refraction for the ordinary and extraordinary beams are identical along the optic axis. With an applied voltage, the crystal becomes birefringent, and a pair of axes are induced at 45° to the crystallographic axes. The plane-polarized wave is resolved into two orthogonally polarized components, along the direction of the induced axes. Here the extraordinary component undergoes a retardation with respect to the ordinary component, resulting in an elliptically polarized output from the modulator assembly.

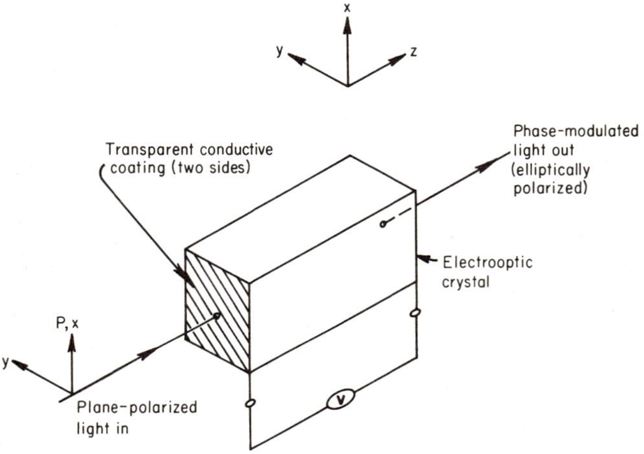

Fig. 1 Longitudinal KDP modulator.

The retardation ϕ, for the longitudinal modulator of Fig. 1, is given by

$$\phi = \frac{n_0{}^3 r_{63} V}{\lambda} \tag{1}$$

where ϕ = number of wavelengths of retardation
n_0 = ordinary index of refraction of electrooptic material
r_{63} = electrooptic coefficient
V = applied voltage across material
λ = wavelength of incident radiation

For the value $\phi = \frac{1}{2}$, V is defined as the half-wave voltage $V_{\lambda/2}$. At a given wavelength, the half-wave voltage for a longitudinal modulator is independent of crystal dimensions but increases with increasing wavelengths. Table 2 lists half-wave voltages for the important electrooptic materials.

The allowable operating voltages and the frequency range are both functions of the modulator construction. Generally, ring-type electrode structures permit the greatest bandwidth, ranging from direct current to beyond 40 MHz. However, higher frequencies generally require operation at reduced voltages. Typical modulator capacitances range between 5 and 40 pF.

For intensity-modulated light, extinction is defined as the ratio of I_{max}/I_{min} measured at the output of the analyzer. In general, I_{max} will be less than the input intensity because of absorption, reflection, and scattering losses. The minimum

† This review is an excellent in-depth treatment of the theory and practical implementation of electrooptic effects. Since it contains a very full bibliography (125 references), we refer the reader to this list of references instead of repeating them here.

TABLE 2 Characteristics of Longitudinal Modulators (after Goldstein, 1968)

Material	Electrooptic constant r_{63}, pm/V	Typical half-wave voltage at 546 nm, kV	Approximate n_0
Ammonium dihydrogen phosphate (ADP) . . .	8.5	9.2	1.526
Potassium dihydrogen phosphate (KDP)	10.5	7.5	1.51
Potassium dideuterium phosphate (KD*P) . . .	26.4	2.6–3.4	1.52
Potassium dihydrogen arsenate (KDA)	10.9	6.4	1.57
Rubidium dihydrogen phosphate (RDP)	11.0	7.3	
Ammonium dihydrogen arsenate (ADA)	5.5	13	1.58

transmittance can be limited by one or more of the following factors: beam-divergence angle, residual crystal birefringence, crystal thickness and inhomogeneity, electric field uniformity, background scattered light, and polarizer-analyzer alignment. In practice, extinctions ranging from 100:1 to > 1,000:1 are attainable with monochromatic low-divergence laser beams.

5. Transverse Modulators Transverse modulators operate with the electric field perpendicular to the direction of light propagation. Figure 2 depicts a transverse modulator that has been built with materials such as lithium tantalate, $LiTaO_3$ (Denton et al., 1967). The electric field E is applied parallel to the z axis, and

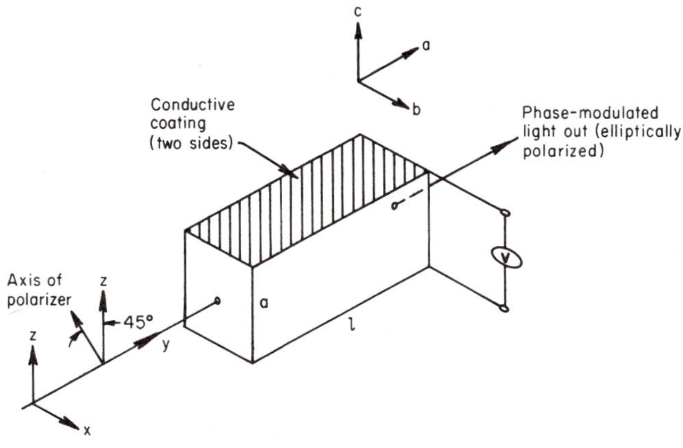

Fig. 2 Transverse $LiTaO_3$ modulator. [*After Denton et al.* (1967).]

the light is propagated along the y axis a distance l with the input polarization plane at 45° to the z axis. The half-wave voltage for this modulator is given by

$$V_{\lambda/2} = \frac{(El)_{\lambda/2}}{l/a} \tag{2}$$

The quantity $(El)_{\lambda/2}$ is the half-wave field-distance product and is defined for a cube of a given material. It is a function of the crystal orientation, incident wavelength, and operating temperature. For the $LiTaO_3$ modulator of Fig. 2, $(El)_{\lambda/2} \approx$ 2,700 V at a wavelength $\lambda = 633$ nm. In transverse modulators, the half-wave voltage is a function of the crystal dimensions and, specifically, the aspect ratio l/a, the ratio of crystal length to the electrode separation a. Generally, l/a is made large to minimize drive-voltage requirements. In some commercial modulators, l/a may be as high as 100. Generally, the transverse-modulator capacitance is directly proportional to modulator length, since most crystals have a square cross section, the most efficient shape for a circular beam. The maximum frequency places an upper limit on device capacitance, and hence upon the length, except for cases where the modulator is constructed in a traveling-wave configuration, such that the optical and microwave indexes are closely matched.

Traveling-wave modulators have been built with response extending to 1 GHz (Peters, 1965). Capacitive types are capable of a similar bandwidth. Generally, either transit-time or dielectric-heating considerations place an upper limit on modulator bandwidth. In most cases, the system bandwidth is limited by the driving amplifier rather than the modulator itself.

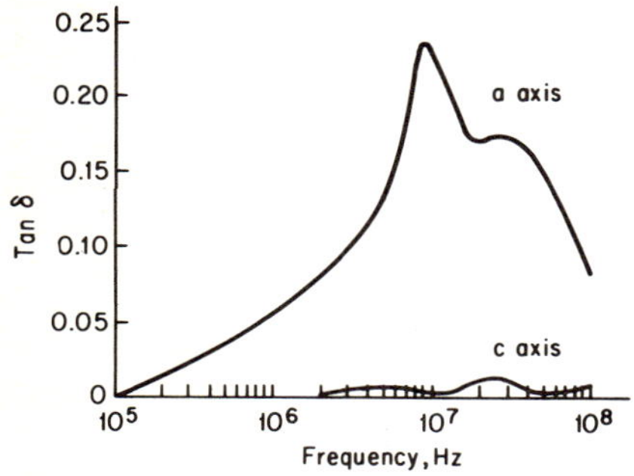

Fig. 3 Loss-tangent variation of $LiNbO_3$ with frequency at 25°C. [*After Nassau et al.* (1966).]

Electrooptic Materials

6. Hundreds of materials, both solid and liquid, exhibit the electrooptic effect, but only a relatively small handful have proved useful for practical applications. From time to time, new materials are discovered that appear promising. At this stage, the problem of growing better and larger crystals becomes paramount. Two principal problems associated with optical quality are variations of refractive index due to one or more of the following: residual strains, included impurities, changes in stoichiometry, growth faults, and damage to the crystal by the laser-beam intensity. Tables 3 to 6 include both the older, reliable materials and a few of the latest materials that have some commercial availability. The variation of loss tangent with frequency is given in Fig. 3 for $LiNbO_3$. At this writing the materials of prime interest are ADP, KDP, KD*P, KTN, $LiNbO_3$, $LiTaO_2$, $Ba_2NaNb_5O_{15}$, $Sr_xBa_{1-x}Nb_2O_6$, CuCl, ZnSe, hexamethylenetetramine, and selenium.

TABLE 3 Characteristics of KDP, KD*P, and ADP from a Variety of Sources

	KH_2PO_4 (KDP)	KD_2PO_4 (KD*P)	$NH_4H_2PO_4$ (ADP)
Electrooptic constant at 546 nm, pm/V:			
$r_{41}{}^T$	8.77	8.8	24.5
$r_{41}{}^S$	. . . .	. . . .	. . . .
$r_{63}{}^T$	10.5	26.4	8.5
$r_{63}{}^S$	9.7	. . . .	5.5
Dielectric constant:			
$\epsilon_{33}{}^T/\epsilon_0$	21	50	15.4
$\epsilon_{33}{}^S/\epsilon_0$	21	48	13.8
$\epsilon_{11}{}^T/\epsilon_0$	42	. . . .	56
$\epsilon_{11}{}^S/\epsilon_0$	42	. . . .	56
Loss tangent, constant strain at 9.2 GHz:			
$\tan \delta_3$	0.0075	0.11	0.006
$\tan \delta_1$	0.0045	0.025	0.007
Curie temperature, K	122	221	148
Piezoelectric constant, pm/V:			
d_{14}	+1.3	. . . .	−1.5
d_{36}	+21	58	+48
Elastooptic constant:			
p_{66}	−0.068	. . . .	−0.075
p_{44}	. . . .	. . . .	−0.051
Maximum safe operating temperature, °C	100	100	80

TABLE 4 Characteristic of $LiTaO_3$ and $LiNbO_3$ from a Variety of Sources

Characteristic	$LiTaO_3$	$LiNbO_3$
Electrooptic constant, pm/V:		
$r_{13}{}^S$	7	8
$r_{33}{}^S$	30.3	28
$r_{51}{}^S$	20	28
$r_{51}{}^T$	......	32
$r_{22}{}^S$	1	3.4
$r_{22}{}^T$	......	3.3
$r_c{}^S$	28	21
$r_c{}^T$	22	19
Dielectric constant:		
$\epsilon_{33}{}^T/\epsilon_0$	49	26
$\epsilon_{33}{}^S/\epsilon_0$	43	23
$\epsilon_{11}{}^T/\epsilon_0$	45	80
$\epsilon_{11}{}^S/\epsilon_0$	41	42
Loss tangent	0.002 at 100 MHz	See Fig. 3
Curie temperature	883 K	1210°C
Elastooptic constant, pm²/N:		
p_{11}	......	0.032
p_{12}	......	0.063
p_{13}	......	0.069
p_{33}	......	0.061
p_{31}	......	0.153
p_{41}	......	0.136

	pm/V	Stress, C/m²	Strain, pC/N
Piezoelectric constant:			
d_{15}	26	...	68
d_{22}	6.4	...	21
d_{31}	...	...	1.0
d_{33}	19.8	...	6.0
e_{15}	...	3.7	
e_{22}	...	2.5	
e_{31}	...	0.2	
e_{33}	...	1.3	

TABLE 5 Characteristics of GaAs and CuCl from a Variety of Sources

	GaAs	CuCl
Electrooptic constant, pm/V	$r_{41} = 1.0$ at $\lambda = 1\ \mu m$	$r_{41}{}^S = 4.3$ $r_{41}{}^T = 6.1$
Dielectric constant	11–12	7.5
Loss tangent	<0.0005	0.001
Half-wave voltage (unity aspect ratio)	13.3 kV at $\lambda = 1\ \mu m$	

7. Optical Transmittance The optical transmittance of electrooptic materials covers the visible spectrum and the near infrared. Figures 4 to 6 give transmittance curves of seven crystals that have transparent regions beyond 0.2 μm. All these transmittance curves contain losses due to Fresnel reflection as well as absorption and scattering. In most cases, the use of antireflection coatings can reduce the Fresnel reflection losses to 1 percent or less, depending upon the optical bandwidth, significantly improving transmittance, especially for the high-index materials. The

TABLE 6 Characteristics of $Ba_2NaNb_5O_{15}$ from a Variety of Sources

Electrooptic constant, pm/V, at λ = 633 nm	$\lvert n_3{}^3 r_{33}\rvert = 620$	$\lvert n_1{}^3 r_{13}\rvert = 230$	$\lvert n_2{}^3 r_{23}\rvert = 170$
Dielectric constant at 30°C	$\epsilon_x = 246$	$\epsilon_y = 242$	$\epsilon_z = 51$
	532 nm	633 nm	1060 nm
Refractive index (transparent from 0.45–5 μm):			
n_x	2.373	2.326	2.263
n_y	2.370	2.324	2.261
n_z	2.256	2.221	2.175
Curie temperature, °C		560	
Half-wave voltage (unity aspect ratio, λ = 633 nm)	Electric field ∥ z	light ∥ x	1.57 kV
	Electric field ∥ z	light ∥ y	1.72 kV

magnitude of the Fresnel reflection loss through a medium having two planes of index discontinuities can be determined from

$$T = \frac{2n}{n^2 + 1} \tag{3}$$

where T is the transmittance and n is the index of refraction.

Historically, ammonium dihydrogen phosphate (ADP) and potassium dihydrogen phosphate (KDP) were the earliest materials grown with good to excellent optical quality. Lithium niobate is a recent material, first offered commercially with excellent optical properties in 1968. Potassium tantalate niobate (KTN) is not

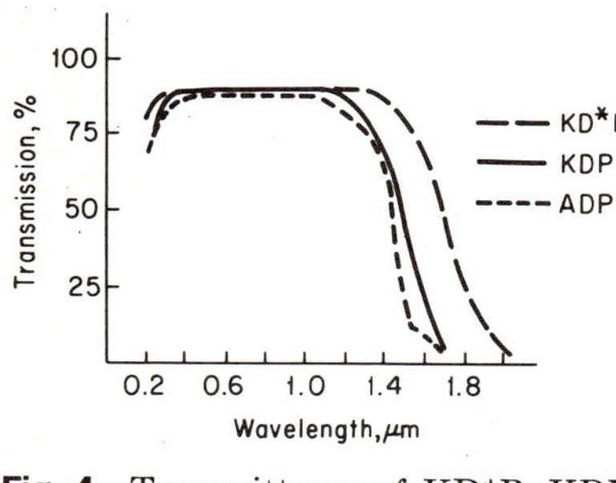

Fig. 4 Transmittance of KD*P, KDP, and ADP. [*After Goldstein* (1966).]

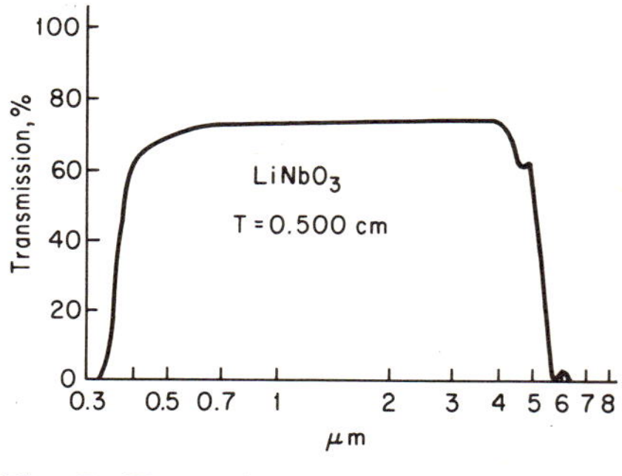

Fig. 5 Transmittance of $LiNbO_3$. [*After Nassau et al.* (1966).]

yet available in large specimens having good optical quality. Barium sodium niobate is a new composition, first offered commercially free of striae in early 1969. It is transparent between 0.45 and 5 μm.

8. Optical Damage When exposed to sufficiently intense light or laser beams, some materials develop refractive-index inhomogeneities. This phenomenon, called optical damage, is a function of both the power density and the radiation wavelength, as well as time of exposure. For example, $LiNbO_3$ has been used successfully in laser Q-switching applications at 1.06 μm with no optical damage. In the visible region, however, a few milliwatts per square millimeter of power are sufficient to cause damage at room temperature. Heating $LiNbO_3$ to a temperature of

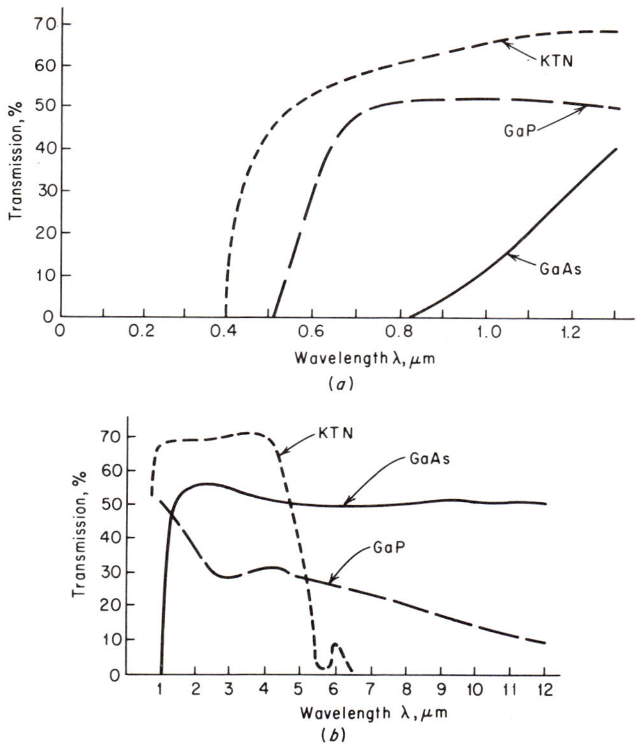

Fig. 6 Transmittance of gallium arsenide, GaAs; gallium phosphide, GaP; and potassium tantalate niobate (KTN). [*After Hirschmann* (1967).]

160°C removes the optical-damage effects and prevents further damage. Other compositions [ADP, KDP, potassium dideuterium phosphate (KD*P), gallium arsenide, GaAs] do not exhibit this damage phenomenon.

9. Refractive Index The refractive indexes of some electrooptic materials are listed in Tables 7 to 9.

General Characteristics

10. Transfer Characteristic The transmitted laser-beam intensity through the modulator of Fig. 2 is given by

$$\frac{I}{I_{\max}} = \sin^2 \frac{\phi_v + \phi_t + \phi_c}{2} \tag{4}$$

where I = transmitted intensity
$I_{\max}$ = maximum transmitted intensity
ϕ_v = voltage-induced phase retardation
ϕ_t = temperature-induced phase retardation
ϕ_c = natural-birefringence phase retardation

For an isothermal system or for modulators having no temperature-dependent birefringence, ϕ_t is zero. For modulators exhibiting natural birefringence, the transmitted intensity becomes

$$\frac{I}{I_{\max}} = \sin^2 \frac{\phi_v + \phi_c}{2} \tag{5}$$

TABLE 7 Refractive Indexes in Air of KDP and ADP (after Zernike, 1964)

Wave-length, μm	Ordinary ray	Extra-ordinary ray	Ordinary ray	Extra-ordinary ray
0.200	1.621	1.563	1.648	1.587
0.300	1.545	1.498	1.563	1.512
0.400	1.524	1.480	1.540	1.492
0.500	1.514	1.472	1.530	1.483
0.600	1.509	1.468	1.524	1.478
0.700	1.505	1.465	1.519	1.475
0.800	1.502	1.463	1.515	1.473
0.900	1.499	1.462	1.512	1.471
1.000	1.496	1.461	1.509	1.469
1.100	1.493	1.459	1.505	1.467
1.200	1.490	1.458	1.502	1.466
1.300	1.487	1.457	1.498	1.464
1.400	1.483	1.456	1.495	1.463
1.500	1.480	1.455	1.491	1.461
1.600	1.476	1.454	1.486	1.459
1.700	1.472	1.453	1.482	1.457
1.800	1.468	1.452	1.478	1.456
1.900	1.464	1.451	1.473	1.454
2.000	1.460	1.450	1.468	1.452

TABLE 8 Refractive Indexes of KD*P, Potassium Dihydrogen Arsenate (KDA), and Ammonium Dihydrogen Arsenate (ADA) from a Variety of Sources

Wave-length, Å	KD*P		KDA		ADA	
	n_o	n_e	n_o	n_e	n_o	n_e
6563	.....	.....	1.563	1.514	1.572	1.518
5893	.....	.....	1.567	1.518	1.576	1.522
5461	1.508	1.468				
4861	.....	.....	1.576	1.525	1.586	1.530
4078	1.518	1.477				

TABLE 9 Refractive Indexes of $LiNbO_3$ and $LiTaO_3$ (after Boyd et al., 1964; Bond, 1965)

Wave-length, μm	$LiNbO_3$		$LiTaO_3$	
	n_o	n_e	n_o	n_e
4.00	2.1155	2.0553	2.0335	2.0377
2.00	2.1974	2.1250	2.1066	2.1115
1.00	2.2370	2.1567	2.1391	2.1432
0.80	2.2571	2.1745	2.1538	2.1578
0.70	2.2716	2.1874	2.1652	2.1696
0.60	2.2967	2.2082	2.1834	2.1878
0.50	2.3410	2.2457	2.2160	2.2205
0.45	2.3780	2.2772	2.2420	2.2468

Some modulators utilize pairs of crystals to cancel the natural birefringence. If the cancellation is complete, the phase retardation becomes a function of voltage alone, and for the linear electrooptic effect is expressed by

$$\phi_v = \pi \frac{V}{V_{\lambda/2}} \tag{6}$$

or

$$\frac{I}{I_{\max}} = \sin^2\left(\frac{\pi}{2}\frac{V}{V_{\lambda/2}}\right) \tag{7}$$

If the modulator is operated between parallel polarizers, the transmittance becomes

$$\frac{I}{I_{\max}} = \cos^2\left(\frac{\pi}{2}\frac{V}{V_{\lambda/2}}\right) \tag{8}$$

The transmitted intensity through the modulator in the absence of an applied ac signal voltage can be adjusted so that the modulator will operate from a bias point. In this case, Eq. (5) becomes $I/I_{\max} = \sin^2[(\phi_v + \phi_b)/2]$, where ϕ_b = bias retardation. Many applications involve setting the bias at the minimum-transmission point or the 50 percent transmission point. However, biasing at any point on the sine-squared transmission curve is possible. The bias control is either a dc electric potential or an optical phase-shifting element. The dc-potential method is generally the simplest; however, it does impose a bandwidth limitation. An optical phase-shifting element allows the electric bandwidth to extend to direct current. However, it generally introduces additional optical tolerances which limit the extinction performance of the modulator if not closely controlled. Optical phase shifters depend upon birefringence, and the optical phase shift ϕ is given by

$$\phi = \frac{2\pi}{\lambda}(n_o - n_e)l \tag{9}$$

where n_o and n_e are the ordinary and extraordinary indexes of refraction of the birefringent material.

A useful arrangement for optical bias is a quarter-wave plate ($\phi = \pi/2$) in a Sénarmont configuration (see, for example, Johnston and Weingart, 1965). In the Sénarmont method, the varying ellipticity of the light produced by the modulator is converted into a plane-polarized wave of varying azimuth.

Bias can be used to compensate or control natural or temperature-dependent birefringence. In controlling the temperature-dependent birefringence, sampling of the intensity output is required to generate feedback signals for drift correction. Because the drift depends upon relatively long thermal time constants, the sampling frequency can be quite low.

11. Piezoelectric Effect All crystals that exhibit the linear electrooptic (Pockels) effect are also piezoelectric. The application of an electric field will generally produce strains in these crystals, which in turn change the principal indexes of refraction. Thus, an additional phase shift is superimposed on the "true" electrooptic effect. The relation is generally of the form†

$$r_{\alpha k}{}^T = r_{\alpha k}{}^S + P_{\alpha\beta} d_{k\beta} \tag{10}$$

where α, β, etc., take values 1 through 6 while i, j, k take values 1 through 3. The term $p_{\alpha\beta}d_{k\beta}$ represents the strain contribution (secondary electrooptic effect) to the constant-stress electrooptic coefficient $r_{\alpha k}{}^T$. Here $p_{\alpha\beta}$ represents the elastooptical coefficient, and $d_{k\beta}$ is the piezoelectric coefficient. Also, α, β, and k are subscripts referring to the optical indicatrix. At high frequencies, the electrooptic coefficient is lowered to the constant-strain value $r_{\alpha k}{}^S$. At odd multiples of the fundamental resonance additional resonances can occur.

The induced birefringence ΔB is

$$\Delta B_\alpha = r_{\alpha k}{}^T E_k \tag{11}$$

† Strictly true under dc field conditions only.

where E_k is the applied electric field. From Eqs. (10) and (11) it can be seen that the strain-induced birefringence will be minimized if $p_{\alpha\beta}d_{k\beta}/r_{\alpha k}^T \ll 1$. Literature values for the pertinent constants are used to compute the strain birefringence. The computed ratio $p_{44}d_{14}/r_{41}^T$ for ADP amounts to 0.0031. For the ADP r_{63} electrooptic coefficient, the $p_{66}d_{36}/r_{63}^T$ ratio is 0.43, or some two orders of magnitude higher. The $p_{66}d_{36}/r_{63}^T$ ratio for KD*P amounts to 0.16. From these values, it can be seen that ADP has a significantly lower secondary electrooptic effect for r_{41}^T. Thus, the response of an r_{41} ADP modulator can be flat to within 0.05 dB.

12. Figure of Merit The electric power required to operate a modulator is proportional to CV^2. For applications where loss-tangent effects are negligible, this may be used to derive a figure of merit, defined as $\epsilon(El)_{\lambda/2}^2$. Where the loss-tangent heating produces appreciable temperature gradients in a crystal, a more useful figure of merit is given by $\epsilon(El)_{\lambda/2}^2 \tan\delta$. For example, Table 10 shows the figure of merit

TABLE 10 Figure-of-Merit Comparison for Two Electrooptic Materials

	$\epsilon(El)_{\lambda/2}^2$	$\epsilon(El)_{\lambda/2}^2 \tan\delta$
ADP	$3,250 \times 10^6$	19.5×10^6
KD*P	$2,420 \times 10^6$	266×10^6

based upon both definitions for the two common materials ADP and KD*P. Although KD*P requires less electric drive power, it dissipates within the crystal 13 times as much electric power as ADP. Hence, where loss-tangent heating effects predominate, ADP is the preferable of these two materials.

13. Currently Available Modulators Table 11 lists the pertinent parameters of currently available electrooptic modulators. These data are compiled from information obtained from a variety of companies.† Many available modulators have a four-crystal configuration, illustrated in Fig. 7. Double refraction is compensated

† For a listing of specific companies the reader is referred to the current issue of the Optical Industry and Systems Directory.

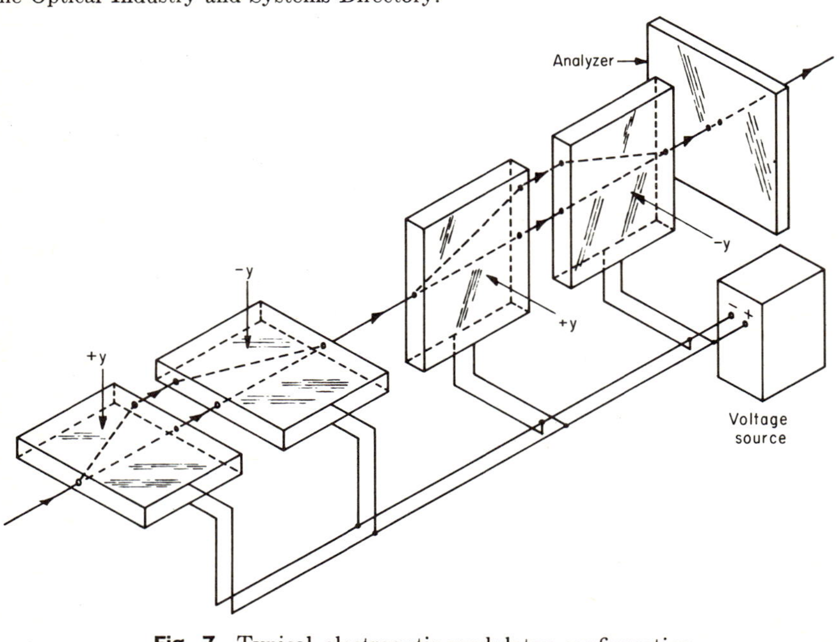

Fig. 7 Typical electrooptic-modulator configuration.

TABLE 11 Parameters of Currently Available Modulators

Type	Capacitance, pF	Half-wave voltage	Bandwidth	Insertion loss, %
Pockels, longitudinal,				
KDP or KD*P.	35-50	3.3-7.5 kV	0-18 kHz	20
	15-35	9-11.2 kV	0-3 MHz	15
KDP, KD*P, or ADP.	25-35	4.4-14.5 kV	0-2 MHz	40
Transverse, ADP	75-175	85-390 V at 633 nm	0-50 MHz	5
$LiNbO_3$	5-15	70 V	0-20 MHz	5

by the use of pairs, and two pairs with axes rotated 90° reduce temperature-birefringence effects.

Kerr Cells

14. The Kerr cell depends upon the Kerr electrooptic effect, a quadratic effect induced in many solids and liquids by an electric field. With the light propagation at right angles to the electric field, the optical phase difference resulting from the induced birefringence is

$$\phi = \frac{2\pi}{\lambda}(n_e - n_o)l \tag{12}$$

where l is the geometrical path length along the direction of propagation. The relation between optical phase difference and field strength is

$$\phi = 2\pi K E^2 l \tag{13}$$

where K is the Kerr constant (a function of wavelength and temperature). It may be either positive or negative. Table 12 lists the Kerr constant and other

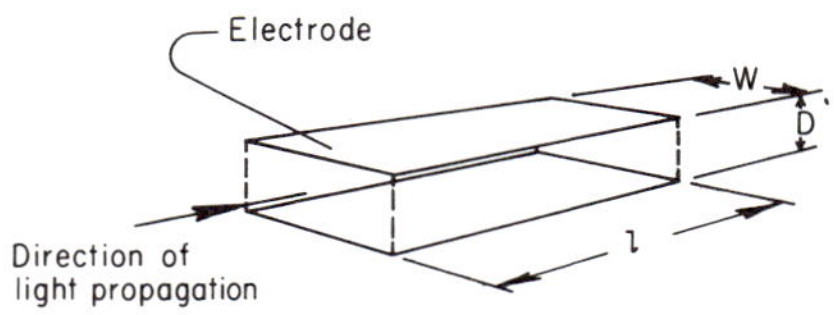

Fig. 8 Parallel-plate Kerr cell geometry.

data for 12 liquids. The value of the Kerr constant at other wavelengths can be computed from

$$\frac{K(\lambda_2)}{K(\lambda_1)} = \frac{\lambda_1}{\lambda_2} \tag{14}$$

For $\phi = \pi$ rad, the half-wave cell voltage is defined by

$$V_{\lambda/2} = \frac{d}{\sqrt{2Kl}} \tag{15}$$

where E of Eq. (13) and V are related by $E = V/d$ (see Fig. 8). Table 13 gives electrical requirements for Kerr cells of two specific sizes and several fluids.

The transmission of the cell between crossed polarizers is given by

$$\frac{I}{I_0} = \sin^2\left[\frac{\pi}{2}\left(\frac{V}{V_{\lambda/2}}\right)^2\right] \tag{16}$$

TABLE 12 Measured Characteristics of Twelve Large-Kerr-Constant Liquids (after Schenck, 1966)

Liquid	Formula	Kerr constant K, esu × 10^{-6} at 538 nm	Static dielectric constant ϵ	Molecular relaxation time, ps	Dc resistivity, Ω·cm	Short-wavelength cutoff, nm	Constant-geometry ratio		
							$\frac{V_2}{V_1}$	$\frac{C_2}{C_1}$	$\frac{E_2}{E_1}$
Ethyl cyanoacetate	$NCCH_2COOC_2H_5$	4.4	28.2	54.0	1.40×10^6	310	2.74	0.78	5.8
o-Dichlorobenzene	$C_6H_4Cl_2$	4.6	10.4	21.5	2.6×10^9	290	2.68	0.29	2.1
Nitrobenzene	$C_6H_5NO_2$	33.0	36.0	42.2	5.5×10^7	420	1.00	1.00	1.00
Benzaldehyde	C_6H_5CHO	7.5	18.9	25.8	4.1×10^6	380	2.10	0.53	2.3
m-Nitrotoluene	$NO_2C_6H_4CH_3$	17.8	28.0	58.1	2.2×10^7	440	1.37	0.78	1.5
Acetophenone	$CH_3COC_6H_5$	6.1	18.0	45.4	3.2×10^6	380	2.32	0.50	2.7
Acetonitrile	CH_3CN	4.7	38.11	12.1	1.4×10^6	215	2.65	1.08	7.6
Acrylonitrile	CH_2CHCN	7.8	32.5	12.0	4.9×10^5	300	2.05	0.90	3.8
Propionitrile	CH_3CH_2CN	3.5	29.4	25.4(?)	3.5×10^5(?)	250	3.07	0.82	7.7
α,α,α-Trifluorotoluene	$C_6H_5CF_3$	2.3	9.77	14.5	1.1×10^9	290	3.80	0.27	3.9
Benzonitrile	C_6H_5CN	12.1	26.3	36.4	3.0×10^6	310	1.64	0.73	2.0
Phenyl acetonitrile	$C_6H_5CH_2CN$	2.6	19.5	36.4	3.0×10^6	310	3.58	0.54	6.9

TABLE 13 Electrical Requirements for Specific Types of Kerr Cells (after Schenck, 1966)

Cell fluid	Dimen-sion†	Quarter- and Half-wave voltage, kV					
		5500 Å		7000 Å		10,600 Å	
		$V\lambda/2$	$V\lambda/4$	$V\lambda/2$	$V\lambda/4$	$V\lambda/2$	$V\lambda/4$
Nitrobenzene	*a*	48.9	37.3	55	42	67.6	51.7
	b	6.2	4.5	7	5	8.6	6.2
o-Dichlorobenzene .	*a*	120.2	98.8	135	111	166.1	136.5
	b	16.5	12.0	18.5	13.3	22.8	16.3
Acrylonitrile.	*a*	97.9	74.8	110	84	135.3	103.3
	b	12.5	8.9	14	10	17.2	12.3
Benzonitrile	*a*	79.2	60.5	89	68	109.5	83.6
	b	10.3	7.1	11.5	8	14.2	9.8
Acetonitrile	*a*	128.0	99.7	136	112	167.3	137.8
	b	16.0	12.0	18.5	13.3	22.8	16.3

† *a* = 1.25 by 2.00 in; *b* = 0.12 by 0.25 in.

where the plane of polarization is at 45° to the electric field. Since a Kerr material has no natural birefringence, no additional terms are required in Eq. (16). When it is desired to modulate rather than shutter light, the Kerr cell is generally biased to its 50 percent transmission point with a dc electric field, thus utilizing the most linear portion of the transmission characteristic. When a full half-wave voltage is applied, the intensity of the third harmonic is 12 percent of the fundamental. To keep the third harmonic to less than 1 percent of the fundamental, the applied voltage must not exceed 30 percent of the half-wave voltage. Modulation up to frequencies of 2 GHz has been accomplished utilizing a carbon disulfide, CS_2, Kerr cell in a traveling-wave configuration.

ACOUSTOOPTIC MODULATORS

15. The principles of acoustooptical devices for the modulation of light (as well as for the related technologies of laser-beam scanning and optical delay lines) are based on the scattering of light by sound waves. Specifically, refraction and diffraction effects are observed when light passes through a transparent material at right angles to a high-frequency sound field propagating in the same medium. The interaction of the light beam with the acoustic beam can result in optical-beam deflection and modulation of the polarization, phase, frequency, or amplitude of the optical energy.

The frequency range of ultrasonic waves lies from the upper limit of the ability of the human ear to hear, at about 10^5 s^{-1}, to frequencies on the order of 10^9 s^{-1}. The actual wavelength of course depends upon the velocity of the waves in the medium in which they travel.

The scattering of light by sound is called Brillouin scattering (Brillouin, 1922). The diffraction into a series of orders of light incident normally on an acoustic diffraction grating in a liquid medium is termed the Debye-Sears (1932) effect (after their verification of acoustic diffraction in a variety of materials including benzene, carbon tetrachloride, glycerin, and toluene). The equivalent effect in a solid medium is called the Raman-Nath effect, following the extensive work of these two authors in 1935 and 1936. These effects disappear at high sound frequencies unless the cell is very short. However, when the modulator is designed so that the optical beam has a significantly long interaction path and the angle of incidence of the optical beam is displaced from the normal to the acoustic beam, all the input light is diffracted; this is Bragg angle diffraction. Finally, ultrasonic refraction occurs when the optical beam is small compared with the acoustic wavelength and the entire optical beam is thus deflected over a range of angles determined by the acoustic amplitude.

The impetus in recent years in this field has resulted from the discovery of the laser. Ultrasonic techniques were first used to modulate the output of a ruby laser in 1962 (DeMaria and Gagosz). Other work quickly followed (DeMaria et al., 1963; Bayer et al., 1963; Tsuya, 1964), each group using different experimental arrangements.

Acoustooptic devices have been used for light modulation in a variety of experimental systems, and the basic principles of acoustooptical deflection and modulation devices have been described in a number of review articles (Gordon, 1966; Flinchbaugh, 1967; Crocetti et al., 1968). The use of acoustooptic devices requires the design and fabrication of broadband efficient acoustical transducers, which can present an additional problem (see, for example, Foster and Rozgony, 1966).

Debye-Sears and Raman-Nath Modulators

16. Consider a plane wavefront of light of wavelength λ and angular frequency ω incident normally on a medium confined to a rectangular shape and of length d and index n_0. A plane compressional wave of wavelength Λ and angular frequency

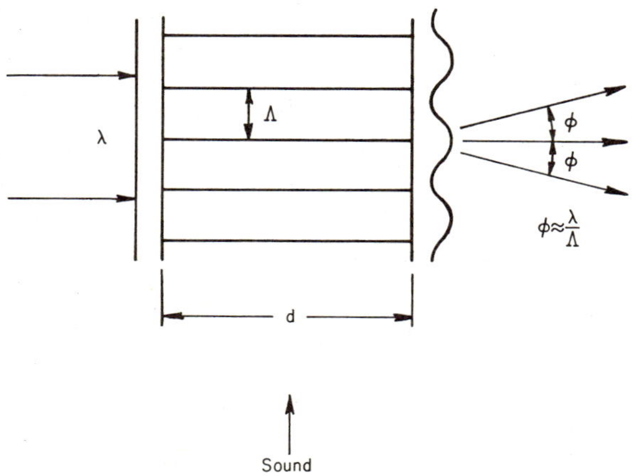

Fig. 9 Diffraction of light by ultrasound.

ω_s propagates at right angles to the incident plane wave. This compressional wave creates a periodic index variation inside the medium, the distance between two successive planes of maximum density being Λ (see Fig. 9). The compressional wave is moving with a velocity v, which is very much smaller than the velocity of light, and hence to a first approximation the index variations are stationary. The resultant effect is that the plane wave of light is incident normally upon a stationary phase grating. Hence ideally we may write the varying index in the medium

$$n_0 + \Delta n \sin \frac{2\pi}{\Lambda} x \tag{17}$$

where x is the position coordinate. In fact the time variations should not be ignored, and hence we have

$$n_0 + \Delta n \cos \left(\omega_e t - \frac{2\pi}{\Lambda} x \right) \tag{18}$$

The grating will thus give rise to diffraction orders consisting of a zero order and first order at angles ϕ given by

$$\sin \phi = \pm \frac{\lambda}{\Lambda} \tag{19}$$

The real situation differs from the ideal in that other diffraction orders are present, given by

$$\sin \phi = \frac{N\lambda}{\Lambda} \qquad N = 0, \pm 1, \pm 2, \pm 3, \ldots \tag{20}$$

Hence for a given λ the angular separation of the orders decreases with increasing Λ, and thus if Λ is sufficiently large, the orders will lie so close together that they are not resolved. This explains why the diffraction effect is not observed when ordinary sound waves are illuminated with visible light. Willard (1949) has pointed out that Debye-Sears and Raman-Nath diffraction occurs only if

$$D^2\Lambda^2(\Delta n)^2 \leq \frac{n}{16} \tag{21}$$

where D is the width of the light beam.

An approximate relationship for the relative intensities in the various orders has been developed by Raman and Nath (1935; 1936). Specifically, the intensity I_N in the Nth order is given by

$$I_N = J_N{}^2[\psi_m(x)] \tag{22}$$

where J_N is the Nth-order Bessel function and $\psi_m(x)$ is the peak light deviation, given by

$$\psi_m(x) = \frac{2\pi}{\lambda} d\bar{n}(x) \tag{23}$$

where $\bar{n}(x)$ is the maximum incremental change in optical index produced by the progressing plane monochromatic ultrasonic wave and d is the thickness of the medium. There is also a doppler shift of the frequency of the light in the various orders, which was first detected by Debye et al. (1934). These frequency shifts can conveniently be demonstrated by optical heterodyning techniques (see, for example, Cummins et al., 1962).

In liquids, ultrasonic propagation can usefully be obtained only by compressional waves since an ideal liquid cannot support shear. However, in solid media the process is more complicated, and shear and compressional waves are possible. Furthermore the polarization must be considered because of stress birefringence. Hence in solid light modulators it is important to determine the shape and orientation of the index ellipse as a function of the spatial strain distribution produced by the acoustic wave.

17. Figures of Merit A number of figures of merit have been suggested in the literature, being determined by the optical index of refraction n, the appropriate component of the photoelastic tensor p, the mass density ρ, and the acoustic phase velocity ν.

The first figure of merit used by several authors (Smith and Korpel, 1965; Spencer et al., 1966) is defined as

$$F_1 = \frac{n^6p^2}{\rho\nu^3} \tag{24}$$

The parameter characterizes the fraction of incident optical power scattered by an acoustic beam of given power and cross-sectional dimension. However, as Gordon (1966) points out, it takes no account of the acoustic frequency range or bandwidth over which scattering is achieved. It should be pointed out here that p, ν, and ρ exhibit no extreme variations from one material to the next. Hence one looks for materials with good optical and acoustic qualities and large refractive index. The optical and acoustic qualities, however, are not included as variables in any of the figures of merit.

With these thoughts in mind Gordon (1966) has suggested a new figure of merit

$$F_2 = \frac{n^7p^2}{\rho\nu} \tag{25}$$

However, F_1 is still pertinent when bandwidth is not of concern.

With this new figure of merit lithium niobate is 11 times better than fused quartz for longitudinal waves. There is also a strong dependence on the free-space wavelength of the light λ_0, since the n and p tend to increase with λ_0 and for ZnO and CdS very drastic changes occur in p near the band gap (Tell et al., 1965). Table 14 contains some of these figures of merit for a variety of materials.

Bragg Angle Modulation

18. This type of modulator is designed so that there is a long interaction path between the sound and the incident light whose incidence is displaced from the normal. Figure 10 illustrates the situation. The necessary condition for the angle of incidence is given by

$$\sin \theta = \frac{1}{2} \frac{\lambda}{n\Lambda} \tag{26}$$

where θ is the Bragg angle. In the arrangement shown in Fig. 9 the diffracted beam is the minus first order.

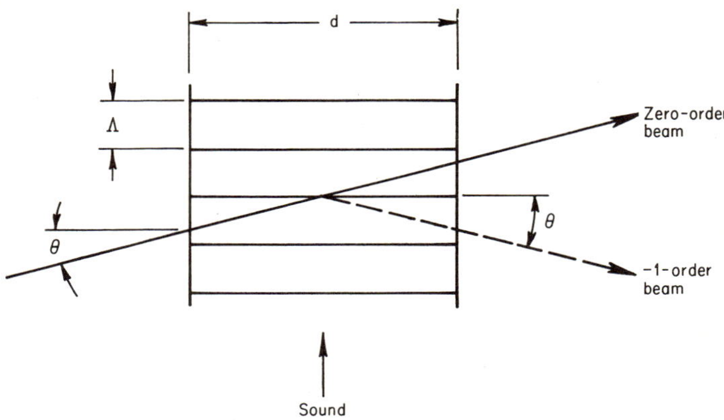

Fig. 10 Bragg angle diffraction by ultrasound.

Following Willard (1949) and Hance and Parks (1965), we use the wavefront-intercept parameter N defined as the number of ultrasonic wavelengths crossed by a point on the wavefront. Hence

$$N = \frac{d \tan \theta}{\Lambda} \tag{27}$$

Since $\lambda/\Delta \ll 1.0 \tan \theta$, $\sin \theta = \theta$ and thus

$$N \approx \frac{1}{2} \frac{\lambda d}{n\Lambda^2} \tag{28}$$

In the limit $N \rightarrow \infty$ the intensities in the zero and minus-first-order beam have a simple form

$$I_0 = \cos^2 \frac{\psi(x)}{2} \qquad I_{-1} = \sin^2 \frac{\psi(x)}{2} \tag{29}$$

where $\psi(x)$ is given by Eq. (7). The effect of finite values of N is to reduce the maximum value of I_0 and I_{-1} from their idealized value of unity. Hence, a condition can be stated so that the intensity is greater than, say, 0.95. Hence, the requirement in the parameters can be stated as

$$\frac{\lambda f_0^2 d}{nV^2} \geq 2 \tag{30}$$

TABLE 14 Figures of Merit for Acoustooptic Devices (after Dixon, 1967; Pinnow and Dixon, 1968)

Material	λ, μm	n	Polarization and direction†		Figure of merit	
			Acoustic wave	Optical wave	$\frac{n^6p^2}{\rho v^3} \times 10^{-18}$	$\frac{n^7p^2}{\rho v} \times 10^{-7}$
Fused quartz	0.63	1.46	L	⊥	1.51	7.89
			T	∥ or ⊥	0.467	0.963
GaP	0.63	3.31	L, [110]	∥	44.6	590
			T, [110]	∥ or ⊥, [010]	24.1	137
GaAs	1.15	3.37	L, [110]	∥	104	925
			T, [100]	∥ or ⊥, [010]	46.3	155
TiO_2	0.63	2.58	L, [11$\bar{2}$0]	⊥, [001]	3.93	62.5
$LiNbO_3$	0.63	2.20	L, [11$\bar{2}$0]	....	6.99	66.5
YAG	0.63	1.83	L, [100]	∥	0.012	0.16
			L, [110]	⊥	0.073	0.98
YIG	1.15	2.22	L, [100]	⊥	0.33	3.94
$LiTaO_3$	0.63	2.18	L, [001]	∥	1.37	11.4
As_2S_3	0.63	2.61	L	⊥	433	762
	1.15	2.46	L	∥	347	619
SF-4	0.63	1.616	L	⊥	4.51	1.83
β-ZnS	0.63	2.35	L, [110]	∥, [001]	3.41	24.3
			T, [110]	∥ or ⊥, [001]	0.57	10.6
α-Al_2O_3	0.63	1.76	L, [001]	∥, [11$\bar{2}$0]	0.34	7.32
CdS	0.63	2.44	L, [11$\bar{2}$0]	∥	12.1	51.8
ADP	0.63	1.58	L, [100]	∥, [010]	2.78	16.0
			T, [100]	∥ or ⊥, [001]	6.43	3.34
KDP	0.63	1.51	L, [100]	∥, [010]	1.91	8.72
			T, [100]	∥ or ⊥, [001]	3.83	1.57
H_2O	0.63	1.33	L	....	160	43.6
Te	10.6	4.8	L, [11$\bar{2}$0]	∥, [0001]	4,400	10,200
α-HIO_3	0.63	.....	L–*a*	*a–c*	48.2	
				b–c	20.8	
				c–a	46.0	
			L–*b*	*a–c*	41.6	
				b–c	58.9	
				c–a	32.8	
			L–*c*	*a–b*	83.5	
				b–a	77.5	
				c–a	63.0	
			Shear *a–b*	*a–c*	17.1	

† L = longitudinal; T = transverse.

where $V/f_0 = \Lambda$. For $\lambda = 6.328 \times 10^{-5}$ cm and water as the ultrasonic medium ($n = 1.33$ and $V = 1.5 \times 10^5$ cm/s)

$$df_0^2 \geq 950$$

where f_0 is in megahertz. To minimize f_0 then, d should be as large as possible.

The advantages of a Bragg angle modulator are that it requires less sound power and all the diffracted light appears in a single beam.

Practical Acoustooptic Modulators

19. Many of the materials listed in Table 14 have been used to make laboratory modulators for research purposes. Dixon (1967) has also listed the photoelastic constants of many of these materials. Among the most interesting newer solid materials are α-iodic acid (Kurtz et al., 1968; Pinnow and Dixon, 1968) and lead molybdate. The only commercially available acoustooptic modulators at the present are made with schlieren-grade fused silica.

MAGNETOOPTIC EFFECTS

20. Faraday Effect In the presence of a magnetic field, certain substances become optically active. Faraday discovered that the plane of polarization of plane-

TABLE 15 Verdet Constants for Some Common Materials†

Substance	T, °C	V for 5893 Å, minutes of arc/(Oe)(cm)
Water	20	0.0131
Glass, phosphate crown	18	0.0161
Light flint	18	0.0317
Carbon disulfide, CS_2	20	0.0423
Phosphorus, P	33	0.1326
Quartz (perpendicular to axis)	20	0.0166

† From F. A. Jenkins and H. E. White, "Fundamentals of Optics." Copyright 1957. Used with permission of McGraw-Hill Book Company.

polarized light is rotated through an angle θ when transmitted through glass in a direction parallel to the magnetic field. This effect has since been observed in solids, liquids, and gases.

The amount of rotation is given by

$$\theta = VHl \tag{31}$$

where V = Verdet constant
H = field strength
l = thickness

The Verdet constant is defined as the rotation per unit path per unit field strength. In general, the sign of the Verdet constant is positive for diamagnetic materials and negative for paramagnetic and ferromagnetic materials.

The direction of rotation of the polarization plane and the direction of current in the surrounding coil are the same. Therefore, the direction of rotation is the same for light passing in the opposite direction. Hence, the rotation can be multiplied by reflecting the light back and forth in the substance. In naturally optically active crystals, a reversal of the light path cancels the rotation.

Some values of the Verdet constant are given in Table 15.

The Faraday effect in glass has been utilized in laser systems for optical isolators, through which light can pass in only one direction. In an isolator, a rotation of 45° is required. This usually requires several centimeters of glass in fields of 5,000 G. Table 16 gives the Verdet constant for a number of glasses for varying wavelengths. All Verdet constants increase at shorter wavelengths.

The Faraday effect is also being investigated for possible applications in computer mass memories. Thin films (200 to 3000 Å) of manganese bismuth, MnBi, are utilized as the Faraday element. Thin films of gadolinum iron garnet have also been used successfully. Under fields of 4 to 5 kOe, the rotation is 7×10^5 deg/cm. Packing densities of at least 10^6 bits/cm^2 and a readout rate of 10^8 bits/s are predicted.

21. Kerr Effect The Kerr magnetooptic effect is characterized by a change in the state of polarization of plane-polarized light when reflected from the polished pole of an electromagnet. The light becomes elliptically polarized. It is most readily observed when the electric vector of the incident light is either parallel or perpendicular to the plane of incidence. This eliminates the ordinary effect of elliptical polarization that occurs by reflection of plane-polarized light from metals.

TABLE 16 Verdet Constant V for Faraday Rotation Glasses (after Snitzer, 1966)

Glass	λ, μm	T, K	V, minutes of arc/(Oe)(cm)	Useful λ, μm
Schott SFS-6 (Pb-Si)	0.700	300	0.071	1.45–2.0
	1.000	300	0.032	
Schott SF-6 (Pb-Si)	0.632	300	0.053	0.45–2.0
AO As-S	1.100	300	0.065	0.9–8
TeO_2, 20% PbO, 80%	0.7	300	0.128	
Ce^{3+}-P	0.5	300	−0.326	0.4–2.0
		24	−2.57	
	0.7	300	−0.132	
Pr^{3+}-P	0.7	300	−0.123	0.65–0.85
Tb^{3+}-P	0.7	300	−0.150	0.5–1.4
Pr^{3+}-B	0.67	300	−0.243	0.65–0.85
Pr^{3+}-Al-Si	0.7	300	−0.198	0.65–0.85
		196	−0.655	
Tb^{3+}-Al-Si	0.7	300	−0.216	0.5–1.4
		196	−0.800	
Dy^{3+}-Al-Si	0.6	300	−0.272	0.5–0.67
		196	−1.03	

22. Miscellaneous Effects The propagation of a wave through a substance in a direction perpendicular to an applied magnetic field produces an induced birefringence that is similar to that observed in uniaxial crystals for propagation at right angles to the optic axis. The wave is resolved into two plane-polarized orthogonal components (parallel and perpendicular to the applied field lines) having indexes n_1 and n_2. The optical phase difference for a geometrical length l of a substance is

$$\phi = \frac{2\pi}{\lambda}(n_1 - n_2)l \tag{32}$$

The magnitude of ϕ is proportional to the magnetic field strength squared. The law is

$$\phi = 2\pi CB^2 l \tag{33}$$

where C is a factor dependent upon both wavelength and temperature. For transparent liquids at 20°C and 589 nm, C is typically on the order of 10^{-12} $(G^2\cdot cm)^{-1}$ and may be either positive or negative. The effect just described is known as the Cotton-Mouton effect. A special case observed in gases near resonant frequencies is called the Voigt effect.

The Zeeman effect is the influence of a magnetic field on a source of radiation. This was first described as a broadening of spectral lines in the presence of a magnetic field. It was shown that a line is split into three plane-polarized components when the line of observation is perpendicular to the magnetic field. With the observation along the field, two circularly polarized components result. For additional information reference is made to a text (e.g., Baldwin, 1969).

At present the Faraday effect is most likely to find widespread use in practical applications.

MODULATION OF SEMICONDUCTOR-DIODE LASERS

23. Semiconductor-diode lasers are pumped by passing current through them; hence modulation of the output is obtained by direct modulation of the current. This is a very attractive feature of diode lasers. Furthermore the optical lifetimes are low, resulting in high modulation frequencies. For example, efficient modulation has been reported out to 20 MHz for operation below threshold. The optical delay line is much smaller above threshold, and 25 percent modulation has been reported at 9 GHz (Konnerth and Lawza, 1964). For a recent review of semiconductor-diode lasers and their characteristics see Minden (1968). Problems arise in the use of modulated diode lasers in distortion of the signal. The light output in the near threshold is a nonlinear function of the current, and the current is a nonlinear function of the voltage (Vallese et al., 1963). Special low-impedance power supplies and transmission lines are also required because the diode presents a resistive load of less than 1 Ω when operated in the region of its linear current-voltage characteristics.

The performance is also affected by junction heating caused by a current passing through the diode. Hence with a modulated input a frequency shift of the laser spectrum is noted, usually by the appearance of different modes as the temperature changes.

MECHANICAL MODULATORS

24. Mechanical modulators, or choppers, as they are often called, can provide an approximately rectangular, triangular, or other shaped waveforms. However, they cannot be used in the modulation of light to produce rapidly varying modula-

TABLE 17 Guide to References for Electrooptic Materials

Ammonium dihydrogen phosphate (ADP):	Lithium tantalate, $LiTaO_3$: (continued)
Phillips (1966)	Lenzo et al. (1966*b*)
Davis and Vedam (1966)	Chen (1969)
Adhav (1969)	Levinstein et al. (1967)
Goldstein (1968)	Riesz and Biazzo (1969)
Zernike (1964)	Denton et al. (1967)
Potassium dihydrogen phosphate (KDP):	Bond (1965)
Duguay and Hansen (1968)	Barium sodium niobate, $Ba_2NaNb_5O_{15}$:
Phillips (1966)	Warner et al. (1969)
Davis and Vedam (1966)	Strontium barium niobate, $Sr_xBa_{1-x}Nb_2O_6$:
Adhav (1969)	Lenzo et al. (1967)
Goldstein (1968)	Potassium tantalate niobate (KTN):
Zernike (1964)	van Raalte (1967)
Potassium dideuterium phosphate (KD*P):	Cuprous chloride, CuCl:
Phillips (1966)	Feldman and Horowitz (1969)
Adhav (1969)	Zinc telluride, ZnTe:
Goldstein (1968)	Sliker and Jost (1966)
Lithium niobate, $LiNbO_3$:	Hexamethylenetetramine:
Zook et al. (1967)	Lee (1969*a*)
Chen (1969)	Selenium:
Levinstein et al. (1967)	Schneider and Vedam (1970)
Lenzo et al. (1966*a*)	General references:
Fay (1969)	Varghese (1967)
Nassau et al. (1966)	Ammann and Yarborough (1966)
Electro-Opt. Syst. Des. (1970)	Ammann (1966*a,b*)
Boyd et al. (1964)	Kemp (1969)
Lithium tantalate, $LiTaO_3$:	Adhav (1969)
Sugibuchi et al. (1968)	Lee (1969*b*)
Smith (1967)	Marks (1969)

tion for reading, recording, or display. The area of application is mainly in error-signal generation in tracking and pointing. A chopper consisting of rotating sectors and reticles can produce modulation in phase, frequency, and amplitude. An excellent detailed discussion of the use of reticles can be found in the literature (Biberman, 1966). Many techniques can be used to produce the required chopping action, including toothed and slotted wheels, tuning forks, and rotating polarizers. The upper frequency is limited to about 25 kHz for tuning forks. Higher frequencies can, of course, be obtained with high-speed rotating mirrors.

REFERENCES

25. The material properties given in the tables and figures have been derived from a variety of sources. Additionally, there is a wealth of information available in the current and past literature that cannot be included here. To avoid constant repetition of the source and for the convenience of the reader, references are given in Table 17 for certain electrooptic materials. They are limited to pertinent articles that have appeared since the Kaminow and Turner review of 1966. Articles too general for inclusion under a specific material are listed under general references.

Adhav, R. S. (1969): *J. Opt. Soc. Am.*, vol. 59, p. 4.

Ammann, E. D. (1966*a*): *J. Opt. Soc. Am.*, vol. 56, p. 7.

——— (1966*b*): *J. Opt. Soc. Am.*, vol. 56, p. 8.

——— and J. M. Yarborough (1966): *J. Opt. Soc. Am.*, vol. 56, p. 12.

Baldwin, G. C. (1969): "An Introduction to Nonlinear Optics," Plenum, New York.

Bayer, E., K. M. Hellwege, and G. Schaack (1963): *Phys. Lett.*, vol. 5, p. 33.

Biberman, L. M. (1966): "Reticles in Electro-optical Devices," Pergamon, New York.

Bond, W. L. (1965): *J. Appl. Phys.*, vol. 36, p. 1674.

Boyd, G. D., R. C. Miller, K. Nassau, W. L. Bond, and A. Savage (1964): *Appl. Phys. Lett.*, vol. 5, p. 234.

Brillouin, L. (1922): *Ann. Phys.*, vol. 17, p. 88.

Chen, F. S. (1969): *J. Appl. Phys.*, vol. 40, p. 8.

Crocetti, C. P., E. J. Calucci, A. D. Rugari, and C. M. Blank (1968): p. 319 in H. R. Luxenburg and R. L. Kuehn (eds.), "Display Systems Engineering," McGraw-Hill, New York.

Cummins, H., N. Knable, L. Ganpel, and Y. Yeb (1962): *Appl. Phys. Lett.*, vol. 2, p. 62.

Davis, J. A., and K. Vedam (1966): *J. Opt. Soc. Am.*, vol. 56, p. 11.

Debye, P., H. Sack, and F. Coulon (1934): *C. R.*, vol. 198, p. 922.

——— and F. W. Sears (1932): *Proc. Natl. Acad. Sci. Wash.*, vol. 18, p. 409.

DeMaria, A. J., and R. Gagosz (1962): *Proc. IRE*, vol. 50, p. 1522.

———, ———, and G. Barnard (1963): *J. Appl. Phys.*, vol. 34, p. 453.

Denton, R. T., F. S. Chen, and A. A. Ballman (1967): *J. Appl. Phys.*, vol. 38, p. 1611.

Dixon, R. W. (1967): *J. Appl. Phys.*, vol. 38, p. 5149.

Duguay, M. A., and J. W. Hansen (1968): *Appl. Phys. Lett.*, vol. 13, p. 178. *Electro-Opt. Syst. Des.*, July 1970, p. 9.

Fay, Homer (1969): *J. Opt. Soc. Am.*, vol. 59, p. 11.

Feldman, A., and D. Horowitz (1969): *J. Opt. Soc. Am.*, vol. 59, p. 11.

Flinchbaugh, D. E. (1965): *J. Acoust. Soc. Am.*, vol. 37, p. 975.

——— (1967): *Laser Focus*, September, p. 25.

Foster, N. F., and G. A. Rozgony (1966): *Appl. Phys. Lett.*, vol. 8, p. 221.

Goldstein, R. (1968): *Laser Focus*, February.

Gordon, E. I. (1966): *Appl. Opt.*, vol. 5, p. 1629, and *IEEE J. Quantum Electron.*, vol. 2, p. 104.

Hance, H. B., and J. K. Parks (1965): *J. Acoust. Soc. Am.*, vol. 38, p. 14.

Hirschmann, E. (1967): *NASA Tech. Note* D-3775.

Jenkins, F. A., and H. E. White (1957): "Fundamentals of Optics," McGraw-Hill, New York.

Johnston, A. R., and J. M. Weingart (1965): *J. Opt. Soc. Am.*, vol. 55, p. 7.

Kaminow, I. P., and E. H. Turner (1966): *Appl. Opt.*, vol. 5, p. 10.

Kemp, J. C. (1969): *J. Opt. Soc. Am.*, vol. 59, p. 8.

Konnerth, K., and C. Lawza (1964): *Appl. Phys. Lett.*, vol. 4, p. 143.

Kurtz, S. K., J. T. Perry, and J. G. Bergman (1968): *Appl. Phys. Lett.*, vol. 12, p. 186.

Lee, R. W. (1969*a*): *J. Opt. Soc. Am.*, vol. 59, p. 12.
——— (1969*b*): *Appl. Opt.*, vol. 8, p. 1385.
Lenzo, P. V., E. G. Spencer, and A. A. Ballman (1967): *Appl. Phys. Lett.*, vol. 11, p. 23.
———, ———, and K. Nassau (1966*a*): *J. Opt. Soc. Am.*, vol. 56, p. 5.
———, E. H. Turner, E. G. Spencer, and A. A. Ballman (1966*b*): *Appl. Phys. Lett.*, vol. 8, p. 81.
Levinstein, H. J., A. A. Ballman, R. T. Denton, A. Ashkin, and J. M. Dziedzic (1967): *J. Appl. Phys. Lett.*, vol. 38, p. 8.
Marks, A. M. (1969): *Appl. Opt.*, vol. 8, p. 1397.
Minden, M. T. (1968): chap. 5 in S. L. Marshall (ed.), "Laser Technology and Applications," McGraw-Hill, New York.
Nassau, K., H. J. Levinstein, and G. M. Loiacono (1966): *J. Phys. Chem. Solids*, vol. 27, p. 989.
Optical Industry and Systems Directory, 1974.
Peters, C. J. (1965): *Proc. IEEE*, vol. 53, p. 455.
Phillips, R. A. (1966): *J. Opt. Soc. Am.*, vol. 56, p. 5.
Pinnow, D. A., and R. W. Dixon (1968): *Appl. Phys. Lett.*, vol. 13, p. 156.
Raman, C. V., and N. S. Nath (1935): *Proc. Indian Acad. Sci.*, vol. A2, p. 405.
——— and ——— (1936): *Proc. Indian Acad. Sci.*, vol. A3, p. 75.
Riesz, R. P., and M. R. Biazzo (1969): *Appl. Opt.*, vol. 8, p. 1393.
Schenck, W. J. (1966): *ISA Trans.*, vol. 5, p. 14.
Schneider, W. C., and K. Vedam (1970): *J. Opt. Soc. Am.*, vol. 60, p. 6.
Sliker, T. R., and J. N. Jost (1966): *J. Opt. Soc. Am.*, vol. 56, p. 1.
Smith, R. T. (1967): *Appl. Phys. Lett.*, vol. 11, p. 146.
Smith, T. M., and A. Korpel (1965): *IEEE J. Quantum Electron.*, vol. 1, p. 283.
Snitzer, E. (1966): *Appl. Opt.*, vol. 5, p. 10.
Spencer, E. G., P. V. Lenzo, and A. A. Ballman (1967): *Proc. IEEE*, vol. 55, p. 12.
———, ———, and K. Nassau (1966): *IEEE J. Quantum Electron.*, vol. 2, p. 69.
Sugibuchi, K., H. Tsuya, and Y. Fujino (1968): *Appl. Phys. Lett.*, vol. 13, p. 107.
Tell, B., J. M. Worlock, and R. J. Martin (1965): *Appl. Phys. Lett.*, vol. 6, p. 123.
Tsuya, N. (1964): *Bull. Am. Phys. Soc.*, vol. 9, p. 281.
Vallese, L. M., M. King, and C. A. Perry (1963): *Proc. IEEE*, vol. 51, p. 1252.
van Raalte, J. A. (1967): *J. Opt. Soc. Am.*, vol. 57, p. 5.
Varghese, V. C. (1967): *J. Opt. Soc. Am.*, vol. 57, p. 11.
Warner, A. W., M. Onoe, and G. A. Coquin (1967): *J. Acoust. Soc. Am.*, vol. 42, p. 1223.
———, G. A. Coquin, A. H. Metzler, and J. L. Fink (1969): *Appl. Phys. Lett.*, vol. 14, p. 34.
Willard, G. W. (1949): *J. Acoust. Soc. Am.*, vol. 21, p. 101.
Zernike, F., Jr. (1964): *J. Opt. Soc. Am.*, vol. 54, p. 1215.
Zook, J. D., D. Chen, and G. N. Otto (1967): *Appl. Phys. Lett.*, vol. 11, p. 159.

Appendixes†

† The editors wish to acknowledge the assistance of Robert M. Burley, Baird-Atomic, Inc., and William W. Striker, Honeywell Radiation Center, in gathering the material for these appendixes.

‡ For conversion of illuminance units and luminance conversion factors, see Tables 8 and 10 of Sec. 1.

Appendix A

Physical Constants and Units

1. Fundamental Constants The values for the fundamental constants given in Table A1 were compiled by E. R. Cohen and B. N. Taylor under the auspices of the Task Group on Fundamental Constants of the Committee on Data for Science and Technology (CODATA), International Council of Scientific Unions, and have been officially adopted by CODATA.

The complete table of values appears in *Dimensions/NBS,* vol. 58, no. 1, January 1974; *J. Phys. Chem. Ref. Data,* vol. 2, no. 4, p. 741, 1973; and *CODATA Bull.* 11, December 1973.

2. The International System of Units The International System of Units (SI), established in 1960 by the General Conference of Weights and Measures under the Treaty of the Meter, is based on the meter (m) for length, defined as 1,650,763.73 wavelengths in vacuum corresponding to the transition $2p_{10} - 5d_5$ of krypton 86; the kilogram (kg) for mass, defined as the mass of the prototype kilogram at Sèvres, France; the second (s) for time, defined as the duration of 9,192,631,770 periods of the radiation corresponding to the transition between the two hyperfine levels of the ground state of cesium 133; the kelvin (K) for temperature, defined as 1/273.16 of the thermodynamic temperature of the triple point of water; the ampere (A) for electric current, defined as the current that, if flowing in two infinitely long parallel wires in vacuum separated by one meter, would produce between the wires a force of 2×10^{-7} newton per meter of length; and the candela (cd) for luminous intensity, defined as the luminous intensity of 1/600,000 square meter of a blackbody at the temperature of freezing platinum.

TABLE A1 Fundamental Constants[a]

Constant	Formula[b]	Symbol	Value[c]	Uncertainty, ppm	Unit: SI	Unit: cgs[d]
Speed of light in vacuum	.....	c	299792458(1.2)	0.004	m/s	10^2 cm/s
Permeability of vacuum	.....	μ_0	4π = 12.5663706144	.....	10^{-7} H/m	
Permittivity of vacuum	$\frac{1}{\mu_0 c^2}$	ϵ_0	8.854187818(71)	0.008	10^{-12} F/m	
Fine-structure constant	$\left[\frac{\mu_0 c^2}{4\pi}\right]\left(\frac{e^2}{\hbar c}\right)$	α	7.2973506(60)	0.82	10^{-3}	10^{-3}
		α^{-1}	137.03604(11)	0.82		
Elementary charge	.....	e	1.6021892(46)	2.9	10^{-19} C	10^{-20} emu
			4.803242(14)	2.9	.....	10^{-10} esu
Planck constant	.....	h	6.626176(36)	5.4	10^{-34} J · s	10^{-27} erg · s
	$\frac{h}{2\pi}$	$\hbar$	1.0545887(57)	5.4	10^{-34} J · s	10^{-27} erg · s
Avogadro constant	.....	N_A	6.022045(31)	5.1	10^{23} mol^{-1}	10^{23} mol^{-1}
Electron rest mass	.....	m_e	9.109534(47)	5.1	10^{-31} kg	10^{-28} g
			5.4858026(21)	0.38	10^{-4} u	10^{-4} u
Proton rest mass	.....	m_p	1.6726485(86)	5.1	10^{-27} kg	10^{-24} g
			1.007276470(11)	0.011	u	u
Ratio of proton mass to electron mass	.....	$\frac{m_p}{m_e}$	1,836.15152(70)	0.38		
Neutron rest mass	.....	m_n	1.6749543(86)	5.1	10^{-27} kg	10^{-24} g
			1.008665012(37)	0.037	u	u
Electron charge-to-mass ratio	.....	$\frac{e}{m_e}$	1.7588047(49)	2.8	10^{11} C/kg	10^7 emu/g
			5.272764(15)	2.8	.....	10^{17} esu/g
Magnetic flux quantum	$[c]^{-1}\left(\frac{hc}{2e}\right)$	Φ_0	2.0678506(54)	2.6	10^{-15} Wb	10^{-7} G/cm
		$\frac{h}{e}$	4.135701(11)	2.6	10^{-15} J · s/C	10^{-7} erg · s/emu
			1.3795215(36)	2.6	.....	10^{-17} erg · s/esu

TABLE A1 Fundamental Constants[a] (Continued)

Constant	Formula[b]	Symbol	Value[c]	Uncertainty, ppm	Unit	
					SI	cgs[d]
Josephson frequency-voltage ratio	...	$\frac{2e}{h}$	4.835939(13)	2.6	10^{14} Hz/V	
Quantum of circulation	...	$\frac{h}{2m_e}$	3.6369455(60)	1.6	10^{-4} J · s/kg	erg · s/g
		$\frac{h}{m_e}$	7.273891(12)	1.6	10^{-4} J · s/kg	erg · s/g
Faraday constant	N_Ae	F	9.648456(27) 2.8925342(82)	2.8 2.8	10^4 C/mol ...	10^3 emu/mol 10^{14} esu/mol
Rydberg constant	$\left[\frac{\mu_0 c^2}{4\pi}\right]^2 \left(\frac{m_e e^4}{4\pi^3 \hbar c}\right)$	R_∞	1.097373177(83)	0.075	10^7 m^{-1}	10^5 cm^{-1}
Bohr radius	$\left[\frac{\mu_0 c^2}{4\pi}\right]^{-1} \left(\frac{\hbar^2}{m_e e^2}\right) = \frac{\alpha}{4\pi R_\infty}$	a_0	5.2917706(44)	0.82	10^{-11} m	10^{-9} cm
Classical electron radius	$\left[\frac{\mu_0 c^2}{4\pi}\right] \left(\frac{e^2}{m_e c^2}\right) = \frac{\alpha^3}{4\pi R_\infty}$	$r_e = \alpha \lambda\!\!\!^{-}_C$	2.8179380(70)	2.5	10^{-15} m	10^{-13} cm
Bohr magneton	$[c]\left(\frac{e\hbar}{2m_e c}\right)$	μ_B	9.274078(36)	3.9	10^{-24} J/T	10^{-21} erg/G
Compton wavelength of electron	$\frac{h}{m_e c} = \frac{\alpha^2}{2R_\infty}$	λ_C	2.4263089(40)	1.6	10^{-12} m	10^{-10} cm
		$\lambda\!\!\!^{-}_C = \frac{\lambda_C}{2\pi} = \alpha a_0$	3.8615905(64)	1.6	10^{-13} m	10^{-11} cm
Compton wavelength of proton	$\frac{h}{m_p c}$	$\lambda_{C,p}$	1.3214099(22)	1.7	10^{-15} m	10^{-13} cm
		$\lambda\!\!\!^{-}_{C,p} = \frac{\lambda_{C,p}}{2\pi}$	2.1030892(36)	1.7	10^{-16} m	10^{-14} cm

Quantity	Formula	Symbol	Value	Uncertainty (ppm)	SI	cgs
Compton wave-length of neutron	$\frac{h}{m_n c}$	$\lambda_{C,n}$	1.3195909(22)	1.7	10^{-15} m	10^{-13} cm
		$\bar{\lambda}_{C,n} = \frac{\lambda_{C,n}}{2\pi}$	2.1001941(35)	1.7	10^{-16} m	10^{-14} cm
Boltzmann constant	$\frac{R}{N_A}$	k	1.380662(44)	32	10^{-23} J/K	10^{-16} erg/K
Stefan-Boltzmann constant	$\frac{\pi^2 k^4}{60\hbar^3 c^2}$	σ	5.67032(71)	125	10^{-8} W/(m^2)(K^4)	10^{-5} erg/(s)(cm^2)(K^4)
First radiation constant	$2\pi hc^2$	c_1	3.741832(20)	5.4	10^{-16} W · m^2	10^{-5} erg · cm^2/s
Second radiation constant	$\frac{hc}{k}$	c_2	1.438786(45)	31	10^{-2} m · K	cm · K
Gravitational constant	....	G	6.6720(41)	615	10^{-11} m^3/(s^2)(kg)	10^{-8} cm^3/(s^2)(g)
Ratio kx unit[e] to angstrom	$\Lambda = \frac{\lambda(\text{Å})}{\lambda(\text{kxu})}$	Λ	1.0020772(54)	5.3		
Ratio Å*[f] to angstrom	$\Lambda^* = \frac{\lambda(\text{Å})}{\lambda(\text{Å}^*)}$	Λ^*	1.0000205(56)	5.6		

[a] Abstracted from New Values for the Fundamental Constants, *Dimensions/NBS*, vol. 58, no. 1, p. 4, January 1974, by permission.

[b] Where formulas for constants are given, the relations are written as the product of two factors. The second factor, in parentheses, is the expression to be used when all quantities are expressed in cgs units with the electron charge in electrostatic units. The first factor, in brackets, is to be included only if all quantities are expressed in SI units. With the exception of the auxiliary constants which are taken to be exact, the uncertainties of these constants are correlated, and therefore the general law of error propagation must be used in calculating additional quantities requiring two or more of these constants.

[c] Numbers in parentheses are the one standard-deviation uncertainties in the last digits of the quoted value computed on the basis of internal consistency. The unified atomic mass scale $^{12}C \equiv 12$ has been used throughout.

[d] To avoid separate columns for electromagnetic and electrostatic units, both are given under the cgs heading. When using these units, the elementary charge e in the symbol column should be understood to be replaced by e_m or e_e, respectively.

[e] $\lambda(\text{Cu}K\alpha_1) \equiv 1.537400$ kxu.

[f] $\lambda(\text{W}K\alpha_1) \equiv 0.2090100$ Å*.

TABLE A2 Defined Values and Conversion Factors†

Atomic mass unit (u)	1/12 the mass of an atom of the ^{12}C nuclide
Standard acceleration of free fall	9.806 65 m/s^2; 980.665 cm/s^2
Standard atmosphere	101,325 N/m^2; 1,013,250 dyn/cm^2
Thermochemical calorie	4.184 J; 4.184 × 10^7 ergs
International Steam Table calorie	4.1868 J; 4.1868 × 10^7 ergs
Liter	0.001 m^3
Mole (mol)	Amount of substance comprising as many elementary units as there are atoms in 0.012 kg of ^{12}C
Inch	0.0254 m; 2.54 cm
Pound, avdp	0.45359237 kg; 453.59237 g

† *Natl. Bur. Stand. Misc. Publ.* 253, rev. ed., May 1967.

TABLE A3 Unit Prefixes

Adopted by International Committee on Weights and Measures

Multiple or sub-multiple	Prefix	Abbre-viation	Multiple or sub-multiple	Prefix	Abbre-viation
10^{12}	tera	T	10^{-2}	centi	c
10^{9}	giga	G	10^{-3}	milli	m
10^{6}	mega	M	10^{-6}	micro	μ
10^{3}	kilo	k	10^{-9}	nano	n
10^{2}	hecto	h	10^{-12}	pico	p
10	deka	da	10^{-15}	femto	f
10^{-1}	deci	d	10^{-18}	atto	a

Note: Compound prefixes should not be used, e.g., not $\mu\mu$ but p, not mμ but n.

Appendix B

Conversion Factors

TABLE B1 Common Fractions of Inches to Decimals and Millimeters†

Common fractions	Decimal	mm	Common fractions	Decimal	mm
1/64	0.0156	0.3969	33/64	0.5156	13.0969
1/32	0.0313	0.7938	17/32	0.5313	13.4938
3/64	0.0469	1.1906	35/64	0.5469	13.8906
1/16	0.0625	1.5875	9/16	0.5625	14.2875
5/64	0.0781	1.9844	37/64	0.5781	14.6844
3/32	0.0938	2.3813	19/32	0.5938	15.0813
7/64	0.1094	2.7781	39/64	0.6094	15.4781
1/8	0.125	3.1750	5/8	0.625	15.8750
9/64	0.1406	3.5719	41/64	0.6406	16.2719
5/32	0.1563	3.9688	21/32	0.6563	16.6688
11/64	0.1719	4.3656	43/64	0.6719	17.0656
3/16	0.1875	4.7625	11/16	0.6875	17.4625
13/64	0.2031	5.1594	45/64	0.7031	17.8594
7/32	0.2188	5.5563	23/32	0.7188	18.2563
15/64	0.2344	5.9531	47/64	0.7344	18.6531
1/4	0.250	6.3500	3/4	0.750	19.0500
17/64	0.2656	6.7469	49/64	0.7656	19.4469
9/32	0.2813	7.1438	25/32	0.7813	19.8438
19/64	0.2969	7.5406	51/64	0.7969	20.2406
5/16	0.3125	7.9375	13/16	0.8125	20.6375
21/64	0.3281	8.3344	53/64	0.8281	21.0344
11/32	0.3438	8.7313	27/32	0.8438	21.4313
23/64	0.3594	9.1281	55/64	0.8594	21.8281
3/8	0.375	9.5250	7/8	0.875	22.2250
25/64	0.3906	9.9219	57/64	0.8906	22.6219
13/32	0.4063	10.3188	29/32	0.9063	23.0188
27/64	0.4219	10.7156	59/64	0.9219	23.4156
7/16	0.4375	11.1125	15/16	0.9375	23.8125
29/64	0.4531	11.5094	61/64	0.9531	24.2094
15/32	0.4688	11.9063	31/32	0.9688	24.6063
31/64	0.4844	12.3031	63/64	0.9844	25.0031
1/2	0.500	12.7000	1	1.000	25.4000

† From William L. Wolfe (ed.), "Handbook of Military Infrared Technology," GPO, Washington, 1965.

TABLE B2 Length†

	XU	Å	nm	μm	mil	mm	cm	in	ft	yd	m	fath	km	mi	nmi	AU	lt y	parsec
1 x-ray unit (XU)	1	10^{-3}	10^{-4}	10^{-7}	3.937×10^{-9}	10^{-10}	10^{-11}	3.937×10^{-12}	3.281×10^{-13}	1.094×10^{-13}	10^{-13}							
1 angstrom (Å)	10^{3}	1	0.1	10^{-4}	3.937×10^{-6}	10^{-7}	10^{-8}	3.937×10^{-9}	3.281×10^{-10}	1.094×10^{-10}	10^{-10}							
1 nanometer (nm)	10^{4}	10	1	10^{-3}	3.937×10^{-5}	10^{-6}	10^{-7}	3.937×10^{-8}	3.281×10^{-9}	1.094×10^{-9}	10^{-9}							
1 micrometer (μm)	10^{7}	10^{4}	10^{3}	1	3.937×10^{-2}	10^{-3}	10^{-4}	3.937×10^{-5}	3.281×10^{-6}	1.094×10^{-6}	10^{-6}							
1 mil (0.001 in)†	2.54×10^{8}	2.54×10^{5}	2.54×10^{4}	25.4	1	0.0254	2.54×10^{-3}	10^{-3}	8.333×10^{-5}	2.778×10^{-5}	2.54×10^{-5}							
1 millimeter (mm)	10^{10}	10^{7}	10^{6}	10^{3}	39.37	1	0.1	0.0394	3.281×10^{-3}	1.094×10^{-3}	10^{-3}							
1 centimeter (cm)	10^{11}	10^{8}	10^{7}	10^{4}	3.937×10^{2}	10	1	0.3937	0.0328	0.0109	0.01							
1 inch (in)	2.54×10^{11}	2.54×10^{8}	2.54×10^{7}	2.54×10^{4}	10^{3}	25.4	2.54	1	0.0833	0.0278	0.0254							
1 foot (ft)	3.048×10^{12}	3.048×10^{9}	3.048×10^{8}	3.048×10^{5}	1.20×10^{4}	304.8	30.48	12	1	0.3333	0.3048	0.1667	3.048×10^{-4}	1.894×10^{-4}	1.646×10^{-4}			
1 yard (yd)	9.144×10^{12}	9.144×10^{9}	9.144×10^{8}	9.144×10^{5}	3.60×10^{4}	914.4	91.44	36	3	1	0.9144	0.50	9.144×10^{-4}	5.682×10^{-4}	4.937×10^{-4}			
1 meter (m)	10^{13}	10^{10}	10^{9}	10^{6}	3.937×10^{4}	10^{3}	100	39.37	3.281	1.094	1	0.5468	10^{-3}	6.214×10^{-4}	5.400×10^{-4}			
1 fathom (fath)	......	......	......	......	......	......	......	......	6	2	1.829	1	1.829×10^{-3}	1.136×10^{-3}	9.875×10^{-4}			
1 kilometer (km)	......	......	......	......	......	......	......	......	3,280.8	1,093.6	10^{3}	546.8	1	0.6214	0.5400	6.689×10^{-9}	1.057×10^{-13}	3.24×10^{-14}
1 statute mile (mi)	......	......	......	......	......	......	......	......	5,280	1,760	1.609×10^{3}	880	1.609	1	0.8690	1.077×10^{-8}	1.690×10^{-13}	5.221×10^{-14}
1 nautical mile (nmi) (international)	......	......	......	......	......	......	......	......	6,076.1	2,025.4	1 852	1,012.7	1.852	1.151	1	1.239×10^{-8}	1.957×10^{-13}	6.008×10^{-14}
1 astronomical unit (AU)	......	......	......	......	......	......	......	......	......	......	......	......	1.495×10^{8}	9.289×10^{7}	8.072×10^{7}	1	1.580×10^{-5}	$4.849 \times 10^{-}$
1 light-year	......	......	......	......	......	......	......	......	......	......	......	......	9.460×10^{12}	5.880×10^{12}	5.11×10^{12}	6.294×10^{4}	1	0.3065
1 parsec	......	......	......	......	......	......	......	......	......	......	......	......	31×10^{12}	19.2×10^{12}	16.64×10^{12}	2.062×10^{5}	3.263	1

† The mil (0.001 in) is not to be confused with the angular mil (see Table B12).

TABLE B3 Inches and Millimeters†

Decimals of an Inch to Millimeters

	0	1	2	3	4	5	6	7	8	9
0.0		0.254	0.508	0.762	1.016	1.270	1.524	1.778	2.032	2.286
0.1	2.540	2.794	3.048	3.302	3.556	3.810	4.064	4.318	4.572	4.826
0.2	5.080	5.334	5.588	5.842	6.096	6.350	6.604	6.858	7.112	7.366
0.3	7.620	7.874	8.128	8.382	8.636	8.890	9.144	9.398	9.652	9.906
0.4	10.160	10.414	10.668	10.922	11.176	11.430	11.684	11.938	12.192	12.446
0.5	12.700	12.954	13.208	13.462	13.716	13.970	14.224	14.478	14.732	14.986
0.6	15.240	15.494	15.748	16.002	16.256	16.510	16.764	17.018	17.272	17.526
0.7	17.780	18.034	18.288	18.542	18.796	19.050	19.304	19.558	19.812	20.066
0.8	20.320	20.574	20.828	21.082	21.336	21.590	21.844	22.098	22.352	22.606
0.9	22.860	23.114	23.368	23.622	23.876	24.130	24.384	24.638	24.892	25.146

Millimeters to Decimals of an Inch

	0.	1.	2.	3.	4.	5.	6.	7.	8.	9.
0		0.0394	0.0787	0.1181	0.1575	0.1968	0.2362	0.2756	0.3150	0.3543
1	0.3937	0.4331	0.4724	0.5118	0.5512	0.5906	0.6299	0.6693	0.7087	0.7480
2	0.7874	0.8268	0.8661	0.9055	0.9449	0.9842	1.0236	1.0630	1.1024	1.1417
3	1.1811	1.2205	1.2598	1.2992	1.3386	1.3780	1.4173	1.4567	1.4961	1.5354
4	1.5748	1.6142	1.6535	1.6929	1.7323	1.7716	1.8110	1.8504	1.8898	1.9291
5	1.9685	2.0079	2.0472	2.0866	2.1260	2.1654	2.2047	2.2441	2.2835	2.3228
6	2.3622	2.4016	2.4409	2.4803	2.5197	2.5590	2.5984	2.6378	2.6772	2.7165
7	2.7559	2.7953	2.8346	2.8740	2.9134	2.9528	2.9921	3.0315	3.0709	3.1102
8	3.1496	3.1890	3.2283	3.2677	3.3071	3.3464	3.3858	3.4252	3.4646	3.5039
9	3.5433	3.5827	3.6220	3.6614	3.7008	3.7402	3.7795	3.8189	3.8583	3.8976

† From T. Baumeister and L. S. Marks, "Standard Handbook for Mechanical Engineers," 7th ed., McGraw-Hill, New York, 1967, by permission.

TABLE B4 Area

	cm^2	in^2	ft^2	m^2	yd^2	Acre	km^2	mi^2	nmi^2
1 square centimeter (cm^2)	1	0.1550	1.076×10^{-3}	10^{-4}	1.196×10^{-4}				
1 square inch (in^2)	6.452	1	6.944×10^{-3}	6.452×10^{-4}	7.716×10^{-4}				
1 square foot (ft^2)	929	144	1	0.0929	0.1111	2.296×10^{-6}	9.290×10^{-8}	3.587×10^{-8}	2.709×10^{-8}
1 square meter (m^2)	10^4	1,550	10.76	1	1.196	2.471×10^{-6}	10^{-6}	3.860×10^{-7}	2.914×10^{-7}
1 square yard (yd^2)	8,361	1,296	9	0.8361	1	2.066×10^{-6}	8.361×10^{-7}	3.228×10^{-7}	2.438×10^{-7}
1 acre	.....	......	4.356×10^4	4,046	4,840	1	4.046×10^{-3}	1.562×10^{-3}	1.179×10^{-3}
1 square kilometer (km^2)	.....	......	1.076×10^7	10^6	1.196×10^6	247.1	1	0.3861	0.2916
1 square mile (mi^2)	.....	......	2.788×10^7	2.590×10^6	3.098×10^6	640	2.590	1	0.7551
1 square nautical mile (nmi^2)	.....	......	3.692×10^7	3.430×10^6	4.102×10^6	847.5	3.430	1.324	1

TABLE B5 Volume

	cm^3	in^3	fl oz	l	qt	Imp qt	pt	Imp pt	gal	Imp gal	ft^3
1 cubic centimeter (cm^3)	1	0.061023	0.03381	9.9997×10^{-4}	1.0567×10^{-3}	8.7985×10^{-4}	2.1134×10^{-3}	1.7601×10^{-3}	2.6417×10^{-4}	2.1992×10^{-4}	3.531×10^{-4}
1 cubic inch (in^3)	16.387162	1	0.5541	0.0163868	0.017316	0.01442	0.034632	0.0288	4.329×10^{-3}	3.6039×10^{-3}	5.787×10^{-4}
1 U.S. fluid ounce (fl oz)	29.5737	1.80469	1	0.0295729	0.03125	0.02602	0.0625	0.0521	7.8125×10^{-3}	6.5039×10^{-3}	1.044×10^{-3}
1 liter (l)	1,000.027	61.025	33.8147	1	1.0567	0.87985	2.1134	1.7601	0.26418	0.2199	0.3531
1 U.S. quart (qt)	946.4	57.749	32	0.94633	1	0.8326	2	1.6656	0.25	0.2081	0.0334
1 Imperial quart (Imp qt)	1,136.5846	69.354	38.432	1.1365	1.201	1	2.402	2.0004	0.3003	0.250	0.0401
1 U.S. liquid pint (pt)	473.2	28.875	16	0.473167	0.5	0.41632	1	0.8328	0.125	0.1041	0.0167
1 Imperial pint (Imp pt)	568.299	34.678	19.2155	0.5683	0.6005	0.4999	1.2008	1	0.1501	0.1250	0.0201
1 U.S. gallon (gal)	3,785.4	231	128	3.78533	4	3.3306	8	6.6622	1	0.8326	0.1336
1 Imperial gallon (Imp gal)	4,546.1	277.42	153.722	4.546	4.8038	3.9998	9.6076	8.001	1.201	1	0.1605
1 cubic foot (ft^3)	2.8317×10^4	1,728	957.5	28.317	29.92	24.9143	59.84	49.8334	7.481	6.229	1

TABLE B6 Pressure†

	dyn/cm²	g/cm²	kg/m²	mm Hg	atm	lb/in²	lb/ft²	ton/in²	ton/ft²
1 dyne per square centimeter (dyn/cm²)	1	0.0010197	0.010197	7.5006×10^{-4}	9.8692×10^{-7}	1.4504×10^{-5}	0.0020886	0.7252×10^{-8}	1.0443×10^{-6}
1 gram per square centimeter (g/cm²)	980.665	1	10	0.73556	9.6784×10^{-4}	0.014223	2.0482	7.112×10^{-6}	1.0241×10^{-3}
1 kilogram per square meter (kg/m²)	98.0665	0.1	1	0.073556	9.6784×10^{-5}	0.0014223	0.20482	7.112×10^{-7}	1.0241×10^{-4}
1 millimeter of mercury (mm Hg)	1,333.22	1.3595	13.595	1	0.001315	0.019337	2.7845	8.668×10^{-6}	1.3923×10^{-3}
1 atmosphere (atm)	1.01325×10^{6}	1033.2	1.0332×10^{4}	760	1	14.696	21,162	0.0073480	1.0581
1 pound per square inch (lb/in²)	68,947	70.307	703.07	51.715	0.068046	1	144	5×10^{-4}	0.0720
1 pound per square foot (lb/ft²)	478.80	0.48824	4.8824	0.35713	4.7254	0.0064444	1	3.4723×10^{-6}	5×10^{-4}
1 ton per square inch (ton/in²)	1.3789×10^{8}	140,620	1.406200	103,430	136.092	2,000	288,000	1	144
1 ton per square foot (ton/ft²)	9.5760×10^{5}	976.48	9,764.8	71,826	0.94509	13.889	2,000	0.0064444	1

† From William L. Wolfe (ed.), "Handbook of Military Infrared Technology," GPO, Washington, 1965.

TABLE B7 Force †, ‡

	dyn	N	g	kg	pdl	lb	ton
1 dyne (dyn)	1	10^5	0.0010197	1.0197×10^{-6}	7.2330×10^{-5}	2.2481×10^{-6}	1.1241×10^{-9}
1 newton (N)	10^5	1	101.97	0.10197	7.2330	22.481	1.1241×10^{-4}
1 gram (g)	980.665	0.0980665	1	0.001	0.070932	2.2046×10^{-3}	0.11023×10^{-5}
1 kilogram (kg)	980665	9.80665	1000	1	70.9032	2.2046	0.11023×10^{-2}
1 poundal (pdl)	1.3825×10^4	0.13825	14.098	0.014098	1	0.031081	1.5492×10^{-5}
1 pound (lb)	4.4482×10^5	4.4482	453.59	0.45359	32.174	1	0.0005
1 ton (2,000 lb)	8.8964×10^8	8896.4	907180	907.180	64548	2000	1

† Conversion between gravitational and absolute units of force is based on the values of g of 980.665 cm/s and 32.174 ft/s.
‡ From William L. Wolfe (ed.), "Handboook of Military Infrared Technology," GPO, Washington, 1965.

TABLE B8 Torque†

	dyn · cm	kg · m	lb · ft	lb · in	pdl · ft
1 dyne centimeter (dyn · cm)	1	1.0197×10^{-8}	7.3757×10^{-8}	8.8511×10^{-7}	2.3731×10^{-6}
1 kilogram meter (kg · m)	9.8066×10^7	1	7.23305	86.799	232.72
1 pound foot (lb · ft)	1.3558×10^7	0.1383	1	12	32.174
1 pound inch (lb · in.)	1.1298×10^6	0.01152	0.08333	1	2.6811
1 poundal foot (pdl · ft)	4.2140×10^5	4.2970×10^{-3}	0.03108	0.37299	1

† From William L. Wolfe (ed.), "Handbook of Military Infrared Technology," GPO, Washington, 1965.

TABLE B9 Mass†

	mg	g	kg	ton (metric)	oz	lb	ton (avdp)
1 milligram (mg)	1	10^{-3}	10^{-6}	10^{-9}	35.2739×10^{-6}	2.20462×10^{-6}	1.10231×10^{-9}
1 gram (g)	10^3	1	10^{-3}	10^{-6}	35.2739×10^{-3}	2.20462×10^{-3}	1.10231×10^{-6}
1 kilogram (kg)	10^6	10^3	1	10^{-3}	35.2739	2.20462	1.10231×10^{-3}
1 ton (metric)	10^9	10^6	10^3	1	35273.9	2204.62	1.10231
1 ounce (oz)	28349.5	28.3495	0.0283495	28.3495×10^{-6}	1	0.0625	3.125×10^{-3}
1 pound (lb)	453592	453.592	0.453592	0.453592×10^{-6}	16	1	0.0005
1 ton (avdp)	0.907185×10^9	907,185	907.185	0.907185	32,000	2000	1

† From William L. Wolfe (ed.), "Handbook of Military Infrared Technology," GPO, Washington, 1965.

TABLE B10 Velocity†

	cm/s	m/s	km/s	in/s	ft/s	mi/s	nmi/s	km/h	in/h	ft/h	mi/h	nmi/h
1 centimeter per second (cm/s)	1	10^{-2}	10^{-5}	3.9370×10^{-1}	3.2808×10^{-2}	6.2137×10^{-6}	5.3996×10^{-6}	3.6000×10^{-2}	1.41732×10^{3}	1.1811×10^{2}	2.2369×10^{-2}	1.9438×10^{-2}
1 meter per second (m/s)	10^{2}	1	10^{-3}	3.9370×10^{1}	3.2808	6.2137×10^{-4}	5.3996×10^{-4}	3.6000	1.41732×10^{5}	1.1811×10^{4}	2.2369	1.9438
1 kilometer per second (km/s)	10^{5}	10^{3}	1	3.9310×10^{4}	3.2808×10^{3}	6.213×10^{-1}	5.3996×10^{-1}	3.6000×10^{3}	1.41732×10^{8}	1.1811×10^{7}	2.2369×10^{3}	1.9438×10^{3}
1 inch per second (in/s)	2.54	2.54×10^{-2}	2.54×10^{-5}	1	8.3332×10^{-2}	1.5783×10^{-5}	1.3715×10^{-5}	9.1440×10^{-2}	3.6000×10^{3}	3.000×10^{2}	5.6817×10^{-2}	4.93725×10^{-2}
1 foot per second (ft/s)	3.0480×10^{1}	3.0480×10^{-1}	3.0480×10^{-4}	12	1	1.8940×10^{-4}	1.6458×10^{-4}	1.09728	4.3200×10^{4}	3.6000×10^{3}	6.81804×10^{-1}	5.9247×10^{-1}
1 (statute) mile per second (mi/s)	1.6093×10^{5}	1.6093×10^{3}	1.6093	6.3360×10^{4}	5280	1	8.6898×10^{-1}	5.7936×10^{3}	2.28096×10^{8}	1.9008×10^{7}	3.6000×10^{3}	3.12824×10^{3}
1 nautical mile per second (nmi/s)	1.85195×10^{5}	1.85195×10^{3}	1.85195	7.29134×10^{4}	6,076.1033	1.15078	1	6.67187×10^{3}	2.62488×10^{8}	2.1874×10^{7}	4.1428×10^{3}	3.6000×10^{3}
1 kilometer per hour (km/h)	2.7758×10^{1}	2.7758×10^{-1}	2.7758×10^{-4}	1.09287×10^{1}	9.1072×10^{-1}	1.72486×10^{-4}	1.49886×10^{-4}	1	3.9370×10^{4}	3.2786×10^{3}	6.20948×10^{-1}	5.3959×10^{-1}
1 inch per hour (in/h)	7.05053×10^{-4}	7.05053×10^{-6}	7.05053×10^{-9}	2.7759×10^{-4}	2.3132×10^{-5}	4.38114×10^{-9}	3.8071×10^{-9}	2.54000×10^{-5}	1	8.3333×10^{-2}	1.57720×10^{-5}	1.37056×10^{-5}
1 foot per hour (ft/h)	8.46064×10^{-3}	8.46064×10^{-5}	8.46064×10^{-8}	3.3311×10^{-5}	2.77584×10^{-4}	5.25737×10^{-8}	4.56852×10^{-8}	3.0480×10^{-4}	12	1	1.89394×10^{-4}	1.64467×10^{-4}
1 (statute) mile per hour (mi/h)	4.4672×10^{1}	4.4672×10^{-1}	4.4672×10^{-4}	1.7588×10^{-1}	1.4667	2.7759×10^{-4}	2.4122×10^{-4}	1.6093	6.3360×10^{4}	5280	1	8.6889×10^{-1}
1 nautical mile per hour (nmi/h)	5.1407×10^{1}	5.1407×10^{-1}	5.1407×10^{-4}	2.02397×10^{-1}	1.6878	3.1944×10^{-4}	2.7759×10^{-4}	1.8519	7.2913×10^{4}	6.07606×10^{3}	1.15078	1

† From William L. Wolfe (ed.), "Handbook of Military Infrared Technology," GPO, Washington, 1965.

TABLE B11 Acceleration†

	m/s²	m/min²	m/h²	m/(h)(s)	in/s²	in/min²	in/h²	ft/s²	ft/min²	ft/(min)(s)	mi/s²	mi/min²	mi/h²	mi/(h)(s)
1 meter per second per second (m/s²)	1	3.6000×10^{3}	1.29600×10^{7}	3.6000×10^{3}	3.9370×10^{1}	1.41732×10^{5}	5.10235×10^{8}	3.28083	1.18110×10^{4}	1.9685×10^{2}	6.21373×10^{-4}	2.23694	8.05299×10^{3}	2.23694
1 meter per minute per minute (m/min²)	2.77778×10^{-4}	1	3.6000×10^{3}	1	1.09361×10^{-2}	3.93700×10^{1}	1.41732×10^{5}	9.11343×10^{-4}	3.28083	5.46806×10^{-2}	1.72604×10^{-7}	6.21373×10^{-4}	2.23694	6.21373×10^{-4}
1 meter per hour per hour (m/h²)	7.71605×10^{-8}	2.77778×10^{-4}	1	2.77778×10^{-4}	3.03781×10^{-6}	1.09361×10^{-2}	3.93700×10^{1}	2.53151×10^{1}	9.11343×10^{-4}	1.51890×10^{-5}	4.79454×10^{-11}	1.72604×10^{-7}	6.21373×10^{-4}	1.72604×10^{-7}
1 meter per hour per second [m/(h)(s)]	2.77778×10^{-4}	1	3.60000×10^{3}	1	1.09361×10^{-2}	3.9370×10^{1}	1.41732×10^{5}	9.11343×10^{-4}	3.28083	5.46806×10^{-2}	1.726035×10^{-7}	6.21373×10^{-4}	2.23694	6.21373×10^{-4}
1 inch per second per second (in/s²)	2.54000×10^{-2}	9.14400×10^{1}	3.29184×10^{5}	9.14400×10^{1}	1	3.6000×10^{3}	1.29600×10^{7}	8.33333×10^{-2}	3.00000×10^{2}	5.00000	1.57829×10^{-5}	5.68183×10^{-2}	2.04546×10^{2}	5.60966×10^{-2}
1 inch per minute per minute (in/min²)	7.055556×10^{-6}	2.54000×10^{-2}	9.14400×10^{1}	2.5400×10^{-2}	2.77778×10^{-4}	1	3.60000×10^{3}	2.31481×10^{-5}	8.33333×10^{-2}	1.38889×10^{-3}	4.38413×10^{-9}	1.57829×10^{-5}	5.68183×10^{-2}	1.57829×10^{-5}
1 inch per hour per hour (in/h²)	1.95988×10^{-9}	7.05556×10^{-6}	2.54000×10^{-2}	7.05556×10^{-6}	7.71605×10^{-8}	2.77778×10^{-4}	1	6.43004×10^{-9}	2.31481×10^{-5}	3.85803×10^{-7}	1.21781×10^{-12}	4.38413×10^{-9}	1.57829×10^{-5}	4.38413×10^{-9}
1 foot per second per second (ft/s²)	3.04800×10^{-1}	1.09728×10^{3}	3.950208×10^{6}	1.09728×10^{3}	12	4.32000×10^{4}	1.55520×10^{8}	1	3.60000×10^{3}	6.00000×10^{1}	1.89394×10^{-4}	6.81820×10^{-1}	2.45455×10^{3}	6.81820×10^{-1}
1 foot per minute per minute (ft/min²)	8.46667×10^{-5}	3.04800×10^{-1}	1.09728×10^{3}	3.04800×10^{-1}	3.33333×10^{-3}	12	4.32000×10^{4}	2.77778×10^{-4}	1	1.66667×10^{-2}	5.26096×10^{-8}	1.89394×10^{-4}	6.81820×10^{-1}	1.89394×10^{-1}
1 foot per minute per second [ft/(min)(s)]	5.08000×10^{-3}	1.82880×10^{1}	6.58368×10^{4}	1.82880×10^{1}	2.0000×10^{-1}	7.20000×10^{2}	2.5920×10^{6}	1.66667×10^{-2}	60	1	3.15657×10^{-6}	1.13637×10^{-2}	4.09092×10^{1}	1.13637×10^{-2}
1 mile per second per second (mi/s²)	1.60934×10^{3}	5.79364×10^{6}	2.71931×10^{10}	5.79364×10^{6}	6.33600×10^{4}	2.28096×10^{8}	8.211456×10^{11}	5.28000×10^{3}	1.90080×10^{7}	3.16800×10^{5}	1	3.6000×10^{3}	1.29600×10^{7}	3.6000×10^{3}
1 mile per minute per minute (mi/min²)	4.47040×10^{-1}	1.60934×10^{3}	7.55364×10^{6}	1.60934×10^{3}	1.76000×10^{1}	6.33600×10^{4}	2.28096×10^{8}	1.46667	5.28000×10^{3}	8.8000×10^{1}	2.77778×10^{-4}	1	3.6000×10^{3}	1
1 mile per hour per hour (mi/h²)	1.241778×10^{-4}	4.47040×10^{-1}	2.09823×10^{3}	4.47040×10^{-1}	4.88889×10^{-3}	1.76000×10^{1}	6.33600×10^{4}	4.07407×10^{-4}	1.46667	2.44444×10^{-2}	7.71605×10^{-8}	2.77778×10^{-4}	1	2.77778×10^{-4}
1 mile per hour per second [mi/(h)(s)]	4.47040×10^{-1}	1.60934×10^{3}	7.55364×10^{6}	1.60934×10^{3}	1.76000×10^{1}	6.33600×10^{4}	2.28096×10^{8}	1.46667	5.28000×10^{3}	8.80000×10^{1}	2.77778×10^{-4}	1	3.60000×10^{3}	1

† From William L. Wolfe (ed.), "Handbook of Military Infrared Technology," GPO, Washington, 1965.

TABLE B12 Angles

	Second	Minute	Artillery mil†	Milliradian	Degree	Radian	Quadrant	Circumference
1 second of arc (″)........	1	1.6667×10^{-2}	4.9383×10^{-3}	4.8481×10^{-3}	2.7778×10^{-4}	4.8481×10^{-6}	3.0864×10^{-6}	7.7161×10^{-7}
1 minute of arc (′)........	60	1	2.963×10^{-1}	2.9089×10^{-1}	1.6667×10^{-2}	2.9089×10^{-4}	1.8518×10^{-4}	4.6296×10^{-5}
1 artillery mil†...........	202.5	3.375	1	.98175	5.625×10^{-2}	9.8175×10^{-4}	6.250×10^{-4}	1.5625×10^{-4}
1 milliradian (mrad).......	206.27	3.4378	1.0186	1	5.7296×10^{-2}	10^{-3}	6.3662×10^{-4}	1.5916×10^{-4}
1 degree (°)..............	3.60×10^{3}	60	17.778	17.453	1	1.7453×10^{-2}	1.1111×10^{-2}	2.7778×10^{-3}
1 radian (rad)............	2.0627×10^{5}	3.4378×10^{3}	1.0186×10^{3}	10	57.296	1	6.3662×10^{-1}	1.5916×10^{-1}
1 quadrant................	3.240×10^{5}	5.40×10^{3}	1.60×10^{3}	1.5708×10^{3}	90	1.5708	1	2.50×10^{-1}
1 circumference...........	1.296×10^{6}	2.160×10^{4}	6.4001×10^{3}	6.2832×10^{3}	360	6.2832	4	1

† The artillery mil is used on the scales and reticles of U.S. Army and Air Force instruments, except in the coast artillery, where degrees and hundredths of a degree are used. The infantry mil is used on all U.S. Navy instruments. It is defined as that angle whose tangent is exactly 0.001.

TABLE B13 Degrees, Minutes, and Seconds to Radians†

Degrees						Minutes		Seconds	
0°	0.00000 00	60°	1.04719 76	120°	2.09439 51	0′	0.00000 00	0″	0.00000 00
1	0.01745 33	61	1.06465 08	121	2.11184 84	1	0.00029 09	1	0.00000 48
2	0.03490 66	62	1.08210 41	122	2.12930 17	2	0.00058 18	2	0.00000 97
3	0.05235 99	63	1.09955 74	123	2.14675 50	3	0.00087 27	3	0.00001 45
4	0.06981 32	64	1.11701 07	124	2.16420 83	4	0.00116 36	4	0.00001 94
5	0.08726 65	65	1.13446 40	125	2.18166 16	5	0.00145 44	5	0.00002 42
6	0.10471 98	66	1.15191 73	126	2.19911 49	6	0.00174 53	6	0.00002 91
7	0.12217 30	67	1.16937 06	127	2.21656 82	7	0.00203 62	7	0.00003 39
8	0.13962 63	68	1.18682 39	128	2.23402 14	8	0.00232 71	8	0.00003 88
9	0.15707 96	69	1.20427 72	129	2.25147 47	9	0.00261 80	9	0.00004 36
10	0.17453 29	70	1.22173 05	130	2.26892 80	10	0.00290 89	10	0.00004 85
11	0.19198 62	71	1.23918 38	131	2.28638 13	11	0.00319 98	11	0.00005 33
12	0.20943 95	72	1.25663 71	132	2.30383 46	12	0.00349 07	12	0.00005 82
13	0.22689 28	73	1.27409 04	133	2.32128 79	13	0.00378 15	13	0.00006 30
14	0.24434 61	74	1.29154 36	134	2.33874 12	14	0.00407 24	14	0.00006 79
15	0.26179 94	75	1.30899 69	135	2.35619 45	15	0.00436 33	15	0.00007 27
16	0.27925 27	76	1.32645 02	136	2.37364 78	16	0.00465 42	16	0.00007 76
17	0.29670 60	77	1.34390 35	137	2.39110 11	17	0.00494 51	17	0.00008 24
18	0.31415 93	78	1.36135 68	138	2.40855 44	18	0.00523 60	18	0.00008 73
19	0.33161 26	79	1.37881 01	139	2.42600 77	19	0.00552 69	19	0.00009 21
20	0.34906 59	80	1.39626 34	140	2.44346 10	20	0.00581 78	20	0.00009 70
21	0.36651 91	81	1.41371 67	141	2.46091 42	21	0.00610 87	21	0.00010 18
22	0.38397 24	82	1.43117 00	142	2.47836 75	22	0.00639 95	22	0.00010 67
23	0.40142 57	83	1.44862 33	143	2.49582 08	23	0.00669 04	23	0.00011 15
24	0.41887 90	84	1.46607 66	144	2.51327 41	24	0.00698 13	24	0.00011 64
25	0.43633 23	85	1.48352 99	145	2.53072 74	25	0.00727 22	25	0.00012 12
26	0.45378 56	86	1.50098 32	146	2.54818 07	26	0.00756 31	26	0.00012 61
27	0.47123 89	87	1.51843 64	147	2.56563 40	27	0.00785 40	27	0.00013 09
28	0.48869 22	88	1.53588 97	148	2.58308 73	28	0.00814 49	28	0.00013 57
29	0.50614 55	89	1.55334 30	149	2.60054 06	29	0.00843 58	29	0.00014 06
30	0.52359 88	90	1.57079 63	150	2.61799 39	30	0.00872 66	30	0.00014 54
31	0.54105 21	91	1.58824 96	151	2.63544 72	31	0.00901 75	31	0.00015 03
32	0.55850 54	92	1.60570 29	152	2.65290 05	32	0.00930 84	32	0.00015 51
33	0.57595 87	93	1.62315 62	153	2.67035 38	33	0.00959 93	33	0.00016 00
34	0.59341 19	94	1.64060 95	154	2.68780 70	34	0.00989 02	34	0.00016 48
35	0.61086 52	95	1.65806 28	155	2.70526 03	35	0.01018 11	35	0.00016 97
36	0.62831 85	96	1.67551 61	156	2.72271 36	36	0.01047 20	36	0.00017 45
37	0.64577 18	97	1.69296 94	157	2.74016 69	37	0.01076 29	37	0.00017 94
38	0.66322 51	98	1.71042 27	158	2.75762 02	38	0.01105 38	38	0.00018 42
39	0.68067 84	99	1.72787 60	159	2.77507 35	39	0.01134 46	39	0.00018 91
40	0.69813 17	100	1.74532 93	160	2.79252 68	40	0.01163 55	40	0.00019 39
41	0.71558 50	101	1.76278 25	161	2.80998 01	41	0.01192 64	41	0.00019 88
42	0.73303 83	102	1.78023 58	162	2.82743 34	42	0.01221 73	42	0.00020 36
43	0.75049 16	103	1.79768 91	163	2.84488 67	43	0.01250 82	43	0.00020 85
44	0.76794 49	104	1.81514 24	164	2.86234 00	44	0.01279 91	44	0.00021 33
45	0.78539 82	105	1.83259 57	165	2.87979 33	45	0.01309 00	45	0.00021 82
46	0.80285 15	106	1.85004 90	166	2.89724 66	46	0.01338 09	46	0.00022 30
47	0.82030 47	107	1.86750 23	167	2.91469 99	47	0.01367 17	47	0.00022 79
48	0.83775 80	108	1.88495 56	168	2.93215 31	48	0.01396 26	48	0.00023 27
49	0.85521 13	109	1.90240 89	169	2.94960 64	49	0.01425 35	49	0.00023 76
50	0.87266 46	110	1.91986 22	170	2.96705 97	50	0.01454 44	50	0.00024 24
51	0.89011 79	111	1.93731 55	171	2.98451 30	51	0.01483 53	51	0.00024 73
52	0.90757 12	112	1.95476 88	172	3.00196 63	52	0.01512 62	52	0.00025 21
53	0.92502 45	113	1.97222 21	173	3.01941 96	53	0.01541 71	53	0.00025 70
54	0.94247 78	114	1.98967 53	174	3.03687 29	54	0.01570 80	54	0.00026 18
55	0.95993 11	115	2.00712 86	175	3.05432 62	55	0.01599 89	55	0.00026 66
56	0.97738 44	116	2.02458 19	176	3.07177 95	56	0.01628 97	56	0.00027 15
57	0.99483 77	117	2.04203 52	177	3.08923 28	57	0.01658 06	57	0.00027 63
58	1.01229 10	118	2.05948 85	178	3.10668 61	58	0.01687 15	58	0.00028 12
59	1.02974 43	119	2.07694 18	179	3.12413 94	59	0.01716 24	59	0.00028 60
60	1.04719 76	120	2.09439 51	180	3.14159 27	60	0.01745 33	60	0.00029 09
Degrees						Minutes		Seconds	

† From E. S. Allen, "Six-Place Tables," 7th ed., McGraw-Hill, New York, 1947, by permission.

TABLE B14 Radians to Degrees, Minutes, and Seconds†

Radians	Degrees	Degrees, min., sec.	Radians	Degrees	Degrees, min., sec.
.00001	.000573	0° 0′ 2.1″	.01	.572958	34′ 22.6″
2	1146	4.1″	2	1.145916	1° 8′ 45.3″
3	1719	6.2″	3	1.718873	1° 43′ 7.9″
4	2292	8.3″	4	2.291831	2° 17′ 30.6″
.00005	.002865	0° 0′ 10.3″	.05	2.864789	2° 51′ 53.2″
6	3438	12.4″	6	3.437747	3° 26′ 15.9″
7	4011	14.4″	7	4.010705	4° 0′ 38.5″
8	4584	16.5″	8	4.583662	4° 35′ 1.2″
9	5157	18.6″	9	5.156620	5° 9′ 23.8″
.00010	.005730	0° 0′ 20.6″	.10	5.729578	5° 43′ 46.5″
2	11459	41.3″	.2	11.459156	11° 27′ 33.0″
3	17189	1′ 1.9″	.3	17.188734	17° 11′ 19.4″
4	22918	1′ 22.5″	.4	22.918312	22° 55′ 5.9″
.0005	.028648	0° 1′ 43.1″	.5	28.647890	28° 38′ 52.4″
6	34377	2′ 3.8″	.6	34.377468	34° 22′ 38.9″
7	40107	2′ 24.4″	.7	40.107046	40° 6′ 25.4″
8	45837	2′ 45.0″	.8	45.836624	45° 50′ 11.8″
9	51566	3′ 5.6″	.9	51.566202	51° 33′ 58.3″
.0010	.057296	0° 3′ 26.3″	1.0	57.295780	57° 17′ 44.8″
2	.114592	6′ 52.5″	2	114.591559	114° 35 29.6″
3	.171887	10′ 18.8″	3	171.887339	171° 53′ 14.4″
4	.229183	13′ 45.1″	4	229.183118	229° 10′ 59.2″
.005	.286479	0° 17′ 11.3″	5	286.478898	286° 28′ 44.0″
6	.343775	20′ 37.6″	6	343.774677	343° 46′ 28.8″
7	.401070	24′ 3.9″	7	401.070457	401° 4′ 13.6″
8	.458366	27′ 30.1″	8	458.366236	458° 21′ 58.4″
9	.515662	30′ 56.4″	9	515.662016	515° 39′ 43.3″

† From E. S. Allen, "Six-Place Tables," 7th ed., McGraw-Hill, New York, 1947, by permission.

TABLE B15 Decimal Equivalents of Minutes, Seconds, and Common Fractions†

From minutes and seconds into decimal parts of a degree

0'	0°.0000	0''	0°.0000
1	.0167	1	.0003
2	.0333	2	.0006
3	.05	3	.0008
4	.0667	4	.0011
5'	.0833	5''	.0014
6	.10	6	.0017
7	.1167	7	.0019
8	.1333	8	.0022
9	.15	9	.0025
10'	0°.1667	10''	0°.0028
1	.1833	1	.0031
2	.20	2	.0033
3	.2167	3	.0036
4	.2333	4	.0039
15'	.25	15''	.0042
6	.2667	6	.0044
7	.2833	7	.0047
8	.30	8	.005
9	.3167	9	.0053
20'	0°.3333	20''	0°.0056
1	.35	1	.0058
2	.3667	2	.0061
3	.3833	3	.0064
4	.40	4	.0067
25'	.4167	25''	.0069
6	.4333	6	.0072
7	.45	7	.0075
8	.4667	8	.0078
9	.4833	9	.0081
30'	0°.50	30''	0°.0083
1	.5167	1	.0086
2	.5333	2	.0089
3	.55	3	.0092
4	.5667	4	.0094
35'	.5833	35''	.0097
6	.60	6	.01
7	.6167	7	.0103
8	.6333	8	.0106
9	.65	9	.0108
40'	0°.6667	40''	0°.0111
1	.6833	1	.0114
2	.70	2	.0117
3	.7167	3	.0119
4	.7333	4	.0122
45'	.75	45''	.0125
6	.7667	6	.0128
7	.7833	7	.0131
8	.80	8	.0133
9	.8167	9	.0136
50'	0°.8333	50''	0°.0139
1	.85	1	.0142
2	.8667	2	.0144
3	.8833	3	.0147
4	.90	4	.015
55'	.9167	55''	.0153
6	.9333	6	.0156
7	.95	7	.0158
8	.9667	8	.0161
9	.9833	9	.0164
60'	1.00	60''	0°.0167

From decimal parts of a degree into minutes and seconds (exact values)

0°.00	0'	0°.50	30'
1	0' 36''	1	30' 36''
2	1' 12''	2	31' 12''
3	1' 48''	3	31' 48''
4	2' 24''	4	32' 24''
0°.05	3'	0°.55	33'
6	3' 36''	6	33' 36''
7	4' 12''	7	34' 12''
8	4' 48''	8	34' 48''
9	5' 24''	9	35' 24''
0°.10	6'	0°.60	36'
1	6' 36''	1	36' 36''
2	7' 12''	2	37' 12''
3	7' 48''	3	37' 48''
4	8' 24''	4	38' 24''
0°.15	9'	0°.65	39'
6	9' 36''	6	39' 36''
7	10' 12''	7	40' 12''
8	10' 48''	8	40' 48''
9	11' 24''	9	41' 24''
0°.20	12'	0°.70	42'
1	12' 36''	1	42' 36''
2	13' 12''	2	43' 12''
3	13' 48''	3	43' 48''
4	14' 24''	4	44' 24''
0°.25	15'	0°.75	45'
6	15' 36''	6	45' 36''
7	16' 12''	7	46' 12''
8	16' 48''	8	46' 48''
9	17' 24''	9	47' 24''
0°.30	18'	0°.80	48'
1	18' 36''	1	48' 36''
2	19' 12''	2	49' 12''
3	19' 48''	3	49' 48''
4	20' 24''	4	50' 24''
0°.35	21'	0°.85	51'
6	21' 36''	6	51' 36''
7	22' 12''	7	52' 12''
8	22' 48''	8	52' 48''
9	23' 24''	9	53' 24''
0°.40	24'	0°.90	54'
1	24' 36''	1	54' 36''
2	25' 12''	2	55' 12''
3	25' 48''	3	55' 48''
4	26' 24''	4	56' 24''
0°.45	27'	0°.95	57'
6	27' 36''	6	57' 36''
7	28' 12''	7	58' 12''
8	28' 48''	8	58' 48''
9	29' 24''	9	59' 24''
0°.50	30'	1°.00	60'

0°.000	0''.0
1	3''.6
2	7''.2
3	10''.8
4	14''.4
0°.005	18''
6	21''.6
7	25''.2
8	28''.8
9	32''.4
0°.010	36''

Common fractions

8 ths	16 ths	32 nds	64 ths	Exact decimal values
			1	.01 5625
		1	2	.03 125
			3	.04 6875
	1	2	4	.06 25
			5	.07 8125
		3	6	.09 375
			7	.10 9375
1	2	4	8	.12 5
			9	.14 0625
		5	10	.15 625
			11	.17 1875
	3	6	12	.18 75
			13	.20 3125
		7	14	.21 875
			15	.23 4375
2	4	8	16	.25
			17	.26 5625
		9	18	.28 125
			19	.29 6875
	5	10	20	.31 25
			21	.32 8125
		11	22	.34 375
			23	.35 9375
3	6	12	24	.37 5
			25	.39 0625
		13	26	.40 625
			27	.42 1875
	7	14	28	.43 75
			29	.45 3125
		15	30	.46 875
			31	.48 4375
4	8	16	32	.50
			33	.51 5625
		17	34	.53 125
			35	.54 6875
	9	18	36	.56 25
			37	.57 8125
		19	38	.59 375
			39	.60 9375
5	10	20	40	.62 5
			41	.64 0625
		21	42	.65 625
			43	.67 1875
	11	22	44	.68 75
			45	.70 3125
		23	46	.71 875
			47	.73 4375
6	12	24	48	.75
			49	.76 5625
		25	50	.78 125
			51	.79 6875
	13	26	52	.81 25
			53	.82 8125
		27	54	.84 375
			55	.85 9375
7	14	28	56	.87 5
			57	.89 0625
		29	58	.90 625
			59	.92 1875
	15	30	60	.93 75
			61	.95 3125
		31	62	.96 875
			63	.98 4375

† From T. Baumeister and L. S. Marks, "Standard Handbook for Mechanical Engineers," 7th ed., McGraw-Hill, New York, 1967, by permission.

TABLE B16 Mean Solar Time

	ns	μs	ms	s	min	h	d	y
1 nanosecond (ns)	1	10^{-3}	10^{-6}	10^{-9}	1.6667×10^{-11}	2.7778×10^{-13}	1.1574×10^{-14}	3.1689×10^{-17}
1 microsecond (μs)	10^{3}	1	10^{-3}	10^{-6}	1.6667×10^{-8}	2.7778×10^{-10}	1.1574×10^{-11}	3.1689×10^{-14}
1 millisecond (ms)	10^{6}	10^{3}	1	10^{-3}	1.6667×10^{-5}	2.7778×10^{-7}	1.1574×10^{-8}	3.1689×10^{-11}
1 second (s or ″)	10^{9}	10^{6}	1000	1	1.6667×10^{-2}	2.7778×10^{-4}	1.1574×10^{-5}	3.1689×10^{-8}
1 minute (min or ′)	6.000×10^{10}	6.000×10^{7}	6.000×10^{4}	60	1	1.6667×10^{-2}	6.9444×10^{-4}	1.9013×10^{-6}
1 hour (h)	3.600×10^{12}	3.600×10^{9}	3.600×10^{6}	3.600	60	1	4.1667×10^{-2}	1.1408×10^{-4}
1 day (d)	8.644×10^{13}	8.644×10^{10}	8.644×10^{7}	8.644×10^{4}	1440	24	1	2.7379×10^{-3}
1 year (y)	3.1557×10^{16}	3.1557×10^{13}	3.1557×10^{10}	3.1557×10^{7}	5.2596×10^{5}	8.7659×10^{3}	3.6524×10^{2}	1

TABLE B17 Temperature†

The general formulas are

$$°F = (°C \times \tfrac{9}{5}) + 32 \qquad \text{and} \qquad °C = (°F - 32) \times \tfrac{5}{9}$$

The absolute temperature is defined by the statement that it is directly proportional to the pressure-volume product of an ideal gas. On the Celsius scale the absolute temperature is equal to $273.16 + t$°C. A temperature of 50°C is then equal to $273.16 + 50 = 323.16$ degrees Celsius absolute (°C abs) or kelvins (K). On the Fahrenheit scale the absolute temperature is equal to $460 + t$°F and is termed degrees Rankine (°R).

The numbers in boldface type refer to the temperature (in either Celsius or Fahrenheit degrees) which it is desired to convert into the other scale. If converting from Fahrenheit degrees to Celsius degrees, the equivalent temperature is in the left column, while if converting from degrees Celsius to degrees Fahrenheit, the equivalent temperature is in the column on the right. Interpolation factors are printed in the last two columns.

Given: 200°C.
Required: Equivalent temperatures in degrees Fahrenheit and Rankine and kelvins.
Solution: 200°C = 392°F
200°C = 200 + 273.16 = 473.16 K
200°C = 392 + 460 = 852°R

C		F	C		F	C		F	C		F
−262	**−440**		−15.0	**5**	41.0	12.8	**55**	131.0	66	**150**	302
−257	**−430**		−14.4	**6**	42.8	13.3	**56**	132.8	71	**160**	320
−251	**−420**		−13.9	**7**	44.6	13.9	**57**	134.6	77	**170**	338
−246	**−410**		−13.3	**8**	46.4	14.4	**58**	136.4	82	**180**	356
−240	**−400**		−12.8	**9**	48.2	15.0	**59**	138.2	88	**190**	374
−234	**−390**		−12.2	**10**	50.0	15.6	**60**	140.0	93	**200**	392
−229	**−380**		−11.7	**11**	51.8	16.1	**61**	141.8	99	**210**	410
−223	**−370**		−11.1	**12**	53.6	16.7	**62**	143.6	104	**220**	428
−218	**−360**		−10.6	**13**	55.4	17.2	**63**	145.4	110	**230**	446
−212	**−350**		−10.0	**14**	57.2	17.8	**64**	147.2	116	**240**	464
−207	**−340**		−9.44	**15**	59.0	18.3	**65**	149.0	121	**250**	482
−201	**−330**		−8.89	**16**	60.8	18.9	**66**	150.8	127	**260**	500
−196	**−320**		−8.33	**17**	62.6	19.4	**67**	152.6	132	**270**	518
−190	**−310**		−7.78	**18**	64.4	20.0	**68**	154.4	138	**280**	536
−184	**−300**		−7.22	**19**	66.2	20.6	**69**	156.2	143	**290**	554
−179	**−290**		−6.67	**20**	68.0	21.1	**70**	158.0	149	**300**	572
−173	**−280**		−6.11	**21**	69.8	21.7	**71**	159.8	154	**310**	590
−169	**−273**	−459.4	−5.56	**22**	71.6	22.2	**72**	161.6	160	**320**	608
−168	**−270**	−454	−5.00	**23**	73.4	22.8	**73**	163.4	166	**330**	626
−162	**−260**	−436	−4.44	**24**	75.2	23.3	**74**	165.2	171	**340**	644
−157	**−250**	−418	−3.89	**25**	77.0	23.9	**75**	167.0	177	**350**	662
−151	**−240**	−400	−3.33	**26**	78.8	24.4	**76**	168.8	182	**360**	680
−146	**−230**	−382	−2.78	**27**	80.6	25.0	**77**	170.6	188	**370**	698
−140	**−220**	−364	−2.22	**28**	82.4	25.6	**78**	172.4	193	**380**	716
−134	**−210**	−346	−1.67	**29**	84.2	26.1	**79**	174.2	199	**390**	734
−129	**−200**	−328	−1.11	**30**	86.0	26.7	**80**	176.0	204	**400**	752
−123	**−190**	−310	−0.56	**31**	87.8	27.2	**81**	177.8	210	**410**	770
−118	**−180**	−292	0	**32**	89.6	27.8	**82**	179.6	216	**420**	788
−112	**−170**	−274	0.56	**33**	91.4	28.3	**83**	181.4	221	**430**	806
−107	**−160**	−256	1.11	**34**	93.2	28.9	**84**	183.2	227	**440**	824
−101	**−150**	−238	1.67	**35**	95.0	29.4	**85**	185.0	232	**450**	842
−95.6	**−140**	−220	2.22	**36**	96.8	30.0	**86**	186.8	238	**460**	860
−90.0	**−130**	−202	2.78	**37**	98.6	30.6	**87**	188.6	243	**470**	878
−84.4	**−120**	−184	3.33	**38**	100.4	31.1	**88**	190.4	249	**480**	896
−78.9	**−110**	−166	3.89	**39**	102.2	31.7	**89**	192.2	254	**490**	914
−73.3	**−100**	−148	4.44	**40**	104.0	32.2	**90**	194.0	260	**500**	932
67.8	**−90**	−130	5.00	**41**	105.8	32.8	**91**	195.8	266	**510**	950
−62.2	**−80**	−112	5.56	**42**	107.6	33.3	**92**	197.6	271	**520**	968
−56.7	**−70**	−94	6.11	**43**	109.4	33.9	**93**	199.4	277	**530**	986
−51.1	**−60**	−76	6.67	**44**	111.2	34.4	**94**	201.2	282	**540**	1004
−45.6	**−50**	−58	7.22	**45**	113.0	35.0	**95**	203.0	288	**550**	1022
−40.0	**−40**	−40	7.78	**46**	114.8	35.6	**96**	204.8	293	**560**	1040
−34.4	**−30**	−22	8.33	**47**	116.6	36.1	**97**	206.6	299	**570**	1058
−28.9	**−20**	−4	8.89	**48**	118.4	36.7	**98**	208.4	304	**580**	1076
−23.3	**−10**	14	9.44	**49**	120.2	37.2	**99**	210.2	310	**590**	1094
−17.8	**0**	32	10.0	**50**	122.0	38	**100**	212	316	**600**	1112
−17.2	**1**	33.8	10.6	**51**	123.8	43	**110**	230	321	**610**	1130
−16.7	**2**	35.6	11.1	**52**	125.6	49	**120**	248	327	**620**	1148
−16.1	**3**	37.4	11.7	**53**	127.4	54	**130**	266	332	**630**	1166
−15.6	**4**	39.2	12.2	**54**	129.2	60	**140**	284	338	**640**	1184

† From R. H. Perry, 2d ed., "Engineering Manual," McGraw-Hill, New York, 1967, by permission.

TABLE B17 Temperature (Continued)

C		F	C		F	C		F	C		F
343	**650**	1202	704	**1300**	2372	1066	**1950**	3542	1371	**2500**	4532
349	**660**	1220	710	**1310**	2390	1071	**1960**	3560	1377	**2510**	4550
354	**670**	1238	716	**1320**	2408	1077	**1970**	3578	1382	**2520**	4568
360	**680**	1256	721	**1330**	2426	1082	**1980**	3596	1388	**2530**	4586
366	**690**	1274	727	**1340**	2444	1088	**1990**	3614	1393	**2540**	4604
371	**700**	1292	732	**1350**	2462	1093	**2000**	3632	1399	**2550**	4622
377	**710**	1310	738	**1360**	2480	1099	**2010**	3650	1404	**2560**	4640
382	**720**	1328	743	**1370**	2498	1104	**2020**	3668	1410	**2570**	4658
388	**730**	1346	749	**1380**	2516	1110	**2030**	3686	1416	**2580**	4676
393	**740**	1364	754	**1390**	2534	1116	**2040**	3704	1421	**2590**	4694
399	**750**	1382	760	**1400**	2552	1121	**2050**	3722	1427	**2600**	4712
404	**760**	1400	766	**1410**	2570	1127	**2060**	3740	1432	**2610**	4730
410	**770**	1418	771	**1420**	2588	1132	**2070**	3758	1438	**2620**	4748
416	**780**	1436	777	**1430**	2606	1138	**2080**	3776	1443	**2630**	4766
421	**790**	1454	782	**1440**	2624	1143	**2090**	3794	1449	**2640**	4784
427	**800**	1472	788	**1450**	2642	1149	**2100**	3812	1454	**2650**	4802
432	**810**	1490	793	**1460**	2660	1154	**2110**	3830	1460	**2660**	4820
438	**820**	1508	799	**1470**	2678	1160	**2120**	3848	1466	**2670**	4838
443	**830**	1526	804	**1480**	2696	1166	**2130**	3866	1471	**2680**	4856
449	**840**	1544	810	**1490**	2714	1171	**2140**	3884	1477	**2690**	4874
454	**850**	1562	816	**1500**	2732	1177	**2150**	3902	1482	**2700**	4892
460	**860**	1580	821	**1510**	2750	1182	**2160**	3920	1488	**2710**	4910
466	**870**	1598	827	**1520**	2768	1188	**2170**	3938	1493	**2720**	4928
471	**880**	1616	832	**1530**	2786	1193	**2180**	3956	1499	**2730**	4946
477	**890**	1634	838	**1540**	2804	1199	**2190**	3974	1504	**2740**	4964
482	**900**	1652	843	**1550**	2822	1204	**2200**	3992	1510	**2750**	4982
488	**910**	1670	849	**1560**	2840	1210	**2210**	4010	1516	**2760**	5000
493	**920**	1688	854	**1570**	2858	1216	**2220**	4028	1521	**2770**	5018
499	**930**	1706	860	**1580**	2876	1221	**2230**	4046	1527	**2780**	5036
504	**940**	1724	866	**1590**	2894	1227	**2240**	4064	1532	**2790**	5054
510	**950**	1742	871	**1600**	2912	1232	**2250**	4082	1538	**2800**	5072
516	**960**	1760	877	**1610**	2930	1238	**2260**	4100	1543	**2810**	5090
521	**970**	1778	882	**1620**	2948	1243	**2270**	4118	1549	**2820**	5108
527	**980**	1796	888	**1630**	2966	1249	**2280**	4136	1554	**2830**	5126
532	**990**	1814	893	**1640**	2984	1254	**2290**	4154	1560	**2840**	5144
538	**1000**	1832	899	**1650**	3002	1260	**2300**	4172	1566	**2850**	5162
543	**1010**	1850	904	**1660**	3020	1266	**2310**	4190	1571	**2860**	5180
549	**1020**	1868	910	**1670**	3038	1271	**2320**	4208	1577	**2870**	5198
554	**1030**	1886	916	**1680**	3056	1277	**2330**	4226	1582	**2880**	5216
560	**1040**	1904	921	**1690**	3074	1282	**2340**	4244	1588	**2890**	5234
566	**1050**	1922	927	**1700**	3092	1288	**2350**	4262	1593	**2900**	5252
571	**1060**	1940	932	**1710**	3110	1293	**2360**	4280	1599	**2910**	5270
577	**1070**	1958	938	**1720**	3128	1299	**2370**	4298	1604	**2920**	5288
582	**1080**	1976	943	**1730**	3146	1304	**2380**	4316	1610	**2930**	5306
588	**1090**	1994	949	**1740**	3164	1310	**2390**	4334	1616	**2940**	5324
593	**1100**	2012	954	**1750**	3182	1316	**2400**	4352	1621	**2950**	5342
599	**1110**	2030	960	**1760**	3200	1321	**2410**	4370	1627	**2960**	5360
604	**1120**	2048	966	**1770**	3218	1327	**2420**	4388	1632	**2970**	5378
610	**1130**	2066	971	**1780**	3236	1332	**2430**	4406	1638	**2980**	5396
616	**1140**	2084	977	**1790**	3254	1338	**2440**	4424	1643	**2990**	5414
621	**1150**	2102	982	**1800**	3272	1343	**2450**	4442	1649	**3000**	5432
627	**1160**	2120	988	**1810**	3290	1349	**2460**	4460			
632	**1170**	2138	993	**1820**	3308	1354	**2470**	4478			
638	**1180**	2156	999	**1830**	3326	1360	**2480**	4496			
643	**1190**	2174	1004	**1840**	3344	1366	**2490**	4514			
649	**1200**	2192	1010	**1850**	3362						
654	**1210**	2210	1016	**1860**	3380						
660	**1220**	2228	1021	**1870**	3398						
666	**1230**	2246	1027	**1880**	3416						
671	**1240**	2264	1032	**1890**	3434						
677	**1250**	2282	1038	**1900**	3452						
682	**1260**	2300	1043	**1910**	3470						
688	**1270**	2318	1049	**1920**	3488						
693	**1280**	2336	1054	**1930**	3506						
699	**1290**	2354	1060	**1940**	3524						

Interpolation Factors

0.56	**1**	1.8	3.33	**6**	10.8
1.11	**2**	3.6	3.89	**7**	12.6
1.67	**3**	5.4	4.44	**8**	14.4
2.22	**4**	7.2	5.00	**9**	16.2
2.78	**5**	9.0	5.56	**10**	18.0

TABLE B18 Density-Transmittance†

Locate the density reading in the table. The whole number at the extreme left plus the decimal number at the top of the column in which the reading is located is the percent transmission of the density. For example, the percent transmission of a density of 1.29 is 5.1 percent. For a density that appears more than once in the table, use the middle density number. For example, the percent transmission of the density 0.44, which appears 9 times, is 36.3; the percent transmission of the density 0.32 is either 47.8 or 47.9.

Trans-mit-tance, %	Density									
	0	0.1	0.2	0.3	0.4	0.5	0.6	0.7	0.8	0.9
0		3.00	2.70	2.52	2.40	2.30	2.22	2.15	2.10	2.05
1	2.00	1.96	1.92	1.89	1.85	1.82	1.80	1.77	1.74	1.72
2	1.70	1.68	1.66	1.64	1.62	1.60	1.59	1.57	1.55	1.54
3	1.52	1.51	1.50	1.48	1.47	1.46	1.44	1.43	1.42	1.41
4	1.40	1.39	1.38	1.37	1.36	1.35	1.34	1.33	1.32	1.31
5	1.30	1.29	1.28	1.28	1.27	1.26	1.25	1.24	1.24	1.23
6	1.22	1.21	1.21	1.20	1.19	1.19	1.18	1.17	1.17	1.16
7	1.15	1.15	1.14	1.14	1.13	1.13	1.12	1.11	1.11	1.10
8	1.10	1.09	1.09	1.08	1.08	1.07	1.07	1.06	1.06	1.05
9	1.05	1.04	1.04	1.03	1.03	1.02	1.02	1.01	1.01	1.00
10	1.00	1.00	0.99	0.99	0.98	0.98	0.97	0.97	0.97	0.96
11	0.96	0.95	0.95	0.95	0.94	0.94	0.93	0.93	0.93	0.92
12	0.92	0.92	0.91	0.91	0.91	0.90	0.90	0.90	0.89	0.89
13	0.89	0.88	0.88	0.88	0.87	0.87	0.87	0.86	0.86	0.86
14	0.85	0.85	0.85	0.84	0.84	0.84	0.84	0.83	0.83	0.83
15	0.82	0.82	0.82	0.82	0.81	0.81	0.81	0.80	0.80	0.80
16	0.80	0.79	0.79	0.79	0.78	0.78	0.78	0.78	0.77	0.77
17	0.77	0.77	0.76	0.76	0.76	0.76	0.75	0.75	0.75	0.75
18	0.74	0.74	0.74	0.74	0.73	0.73	0.73	0.73	0.73	0.72
19	0.72	0.72	0.72	0.71	0.71	0.71	0.71	0.71	0.70	0.70
20	0.70	0.70	0.69	0.69	0.69	0.69	0.69	0.68	0.68	0.68
21	0.68	0.68	0.67	0.67	0.67	0.67	0.67	0.66	0.66	0.66
22	0.66	0.66	0.65	0.65	0.65	0.65	0.65	0.64	0.64	0.64
23	0.64	0.64	0.63	0.63	0.63	0.63	0.63	0.63	0.62	0.62
24	0.62	0.62	0.62	0.61	0.61	0.61	0.61	0.61	0.60	0.60
25	0.60	0.60	0.60	0.60	0.59	0.59	0.59	0.59	0.59	0.59
26	0.58	0.58	0.58	0.58	0.58	0.58	0.57	0.57	0.57	0.57
27	0.57	0.57	0.57	0.57	0.56	0.56	0.56	0.56	0.56	0.56
28	0.55	0.55	0.55	0.55	0.55	0.54	0.54	0.54	0.54	0.54
29	0.54	0.54	0.53	0.53	0.53	0.53	0.53	0.53	0.53	0.52
30	0.52	0.52	0.52	0.52	0.52	0.52	0.51	0.51	0.51	0.51
31	0.51	0.51	0.51	0.50	0.50	0.50	0.50	0.50	0.50	0.50
32	0.49	0.49	0.49	0.49	0.49	0.49	0.49	0.49	0.48	0.48
33	0.48	0.48	0.48	0.48	0.48	0.47	0.47	0.47	0.47	0.47
34	0.47	0.47	0.47	0.46	0.46	0.46	0.46	0.46	0.46	0.46
35	0.46	0.45	0.45	0.45	0.45	0.45	0.45	0.45	0.45	0.44
36	0.44	0.44	0.44	0.44	0.44	0.44	0.44	0.44	0.43	0.43
37	0.43	0.43	0.43	0.43	0.43	0.43	0.42	0.42	0.42	0.42
38	0.42	0.42	0.42	0.42	0.42	0.41	0.41	0.41	0.41	0.41
39	0.41	0.41	0.41	0.41	0.40	0.40	0.40	0.40	0.40	0.40
40	0.40	0.40	0.40	0.40	0.39	0.39	0.39	0.39	0.39	0.39
41	0.39	0.39	0.39	0.38	0.38	0.38	0.38	0.38	0.38	0.38
42	0.38	0.38	0.38	0.37	0.37	0.37	0.37	0.37	0.37	0.37
43	0.37	0.37	0.37	0.36	0.36	0.36	0.36	0.36	0.36	0.36
44	0.36	0.36	0.35	0.35	0.35	0.35	0.35	0.35	0.35	0.35
45	0.35	0.35	0.34	0.34	0.34	0.34	0.34	0.34	0.34	0.34
46	0.34	0.34	0.33	0.33	0.33	0.33	0.33	0.33	0.33	0.33
47	0.33	0.33	0.33	0.32	0.32	0.32	0.32	0.32	0.32	0.32
48	0.32	0.32	0.32	0.32	0.32	0.31	0.31	0.31	0.31	0.31

† Eastman Kodak Co., "Wratten Filters," 21st ed., Rochester, N.Y., 1962. Copyright; reproduced by permission.

TABLE B18 Density-Transmittance (Continued)

Trans- mit- tance, %	Density									
	0	0.1	0.2	0.3	0.4	0.5	0.6	0.7	0.8	0.9
49	0.31	0.31	0.31	0.31	0.31	0.31	0.30	0.30	0.30	0.30
50	0.30	0.30	0.30	0.30	0.30	0.30	0.29	0.29	0.29	0.29
51	0.29	0.29	0.29	0.29	0.29	0.29	0.29	0.29	0.29	0.28
52	0.28	0.28	0.28	0.28	0.28	0.28	0.28	0.28	0.28	0.28
53	0.28	0.28	0.27	0.27	0.27	0.27	0.27	0.27	0.27	0.27
54	0.27	0.27	0.27	0.27	0.26	0.26	0.26	0.26	0.26	0.26
55	0.26	0.26	0.26	0.26	0.26	0.25	0.25	0.25	0.25	0.25
56	0.25	0.25	0.25	0.25	0.25	0.25	0.25	0.25	0.25	0.24
57	0.24	0.24	0.24	0.24	0.24	0.24	0.24	0.24	0.24	0.24
58	0.24	0.24	0.24	0.23	0.23	0.23	0.23	0.23	0.23	0.23
59	0.23	0.23	0.23	0.23	0.23	0.23	0.22	0.22	0.22	0.22
60	0.22	0.22	0.22	0.22	0.22	0.22	0.22	0.22	0.22	0.22
61	0.21	0.21	0.21	0.21	0.21	0.21	0.21	0.21	0.21	0.21
62	0.21	0.21	0.21	0.21	0.20	0.20	0.20	0.20	0.20	0.20
63	0.20	0.20	0.20	0.20	0.20	0.20	0.20	0.20	0.19	0.19
64	0.19	0.19	0.19	0.19	0.19	0.19	0.19	0.19	0.19	0.19
65	0.19	0.19	0.19	0.19	0.18	0.18	0.18	0.18	0.18	0.18
66	0.18	0.18	0.18	0.18	0.18	0.18	0.18	0.18	0.17	0.17
67	0.17	0.17	0.17	0.17	0.17	0.17	0.17	0.17	0.17	0.17
68	0.17	0.17	0.17	0.17	0.16	0.16	0.16	0.16	0.16	0.16
69	0.16	0.16	0.16	0.16	0.16	0.16	0.16	0.16	0.16	0.16
70	0.15	0.15	0.15	0.15	0.15	0.15	0.15	0.15	0.15	0.15
71	0.15	0.15	0.15	0.15	0.15	0.15	0.14	0.14	0.14	0.14
72	0.14	0.14	0.14	0.14	0.14	0.14	0.14	0.14	0.14	0.14
73	0.14	0.14	0.14	0.13	0.13	0.13	0.13	0.13	0.13	0.13
74	0.13	0.13	0.13	0.13	0.13	0.13	0.13	0.13	0.13	0.13
75	0.12	0.12	0.12	0.12	0.12	0.12	0.12	0.12	0.12	0.12
76	0.12	0.12	0.12	0.12	0.12	0.12	0.12	0.11	0.11	0.11
77	0.11	0.11	0.11	0.11	0.11	0.11	0.11	0.11	0.11	0.11
78	0.11	0.11	0.11	0.11	0.11	0.10	0.10	0.10	0.10	0.10
79	0.10	0.10	0.10	0.10	0.10	0.10	0.10	0.10	0.10	0.10
80	0.10	0.10	0.10	0.10	0.09	0.09	0.09	0.09	0.09	0.09
81	0.09	0.09	0.09	0.09	0.09	0.09	0.09	0.09	0.09	0.09
82	0.09	0.09	0.08	0.08	0.08	0.08	0.08	0.08	0.08	0.08
83	0.08	0.08	0.08	0.08	0.08	0.08	0.08	0.08	0.08	0.08
84	0.08	0.07	0.07	0.07	0.07	0.07	0.07	0.07	0.07	0.07
85	0.07	0.07	0.07	0.07	0.07	0.07	0.07	0.07	0.07	0.07
86	0.07	0.06	0.06	0.06	0.06	0.06	0.06	0.06	0.06	0.06
87	0.06	0.06	0.06	0.06	0.06	0.06	0.06	0.06	0.06	0.06
88	0.06	0.05	0.05	0.05	0.05	0.05	0.05	0.05	0.05	0.05
89	0.05	0.05	0.05	0.05	0.05	0.05	0.05	0.05	0.05	0.05
90	0.05	0.04	0.04	0.04	0.04	0.04	0.04	0.04	0.04	0.04
91	0.04	0.04	0.04	0.04	0.04	0.04	0.04	0.04	0.04	0.04
92	0.04	0.04	0.04	0.03	0.03	0.03	0.03	0.03	0.03	0.03
93	0.03	0.03	0.03	0.03	0.03	0.03	0.03	0.03	0.03	0.03
94	0.03	0.03	0.03	0.03	0.02	0.02	0.02	0.02	0.02	0.02
95	0.02	0.02	0.02	0.02	0.02	0.02	0.02	0.02	0.02	0.02
96	0.02	0.02	0.02	0.02	0.02	0.02	0.01	0.01	0.01	0.01
97	0.01	0.01	0.01	0.01	0.01	0.01	0.01	0.01	0.01	0.01
98	0.01	0.01	0.01	0.01	0.01	0.01	0.01	0.01	0.01	0.00
99	0.00	0.00	0.00	0.00	0.00	0.00	0.00	0.00	0.00	0.00
100	0.00	0.00	0.00	0.00	0.00	0.00	0.00	0.00	0.00	0.00

TABLE B19 Electron Energy

	J	Erg	eV	Hz	Wave number
1 joule (J)	1	10^7	6.2419×10^{18}	1.509×10^{33}	5.035×10^{22}
1 erg	10^{-7}	1	6.2419×10^{11}	1.509×10^{26}	5.035×10^{15}
1 electron volt (eV)	1.60207×10^{5}	1.60207×10^{12}	1	2.418×10^{14}	8.066×10^{3}
1 hertz (Hz)	6.625×10^{-34}	6.625×10^{-27}	4.135×10^{-15}	1	3.336×10^{-11}
1 wave number	1.986×10^{-23}	1.986×10^{-16}	1.240×10^{-4}	2.998×10^{10}†	1

† Velocity of light.

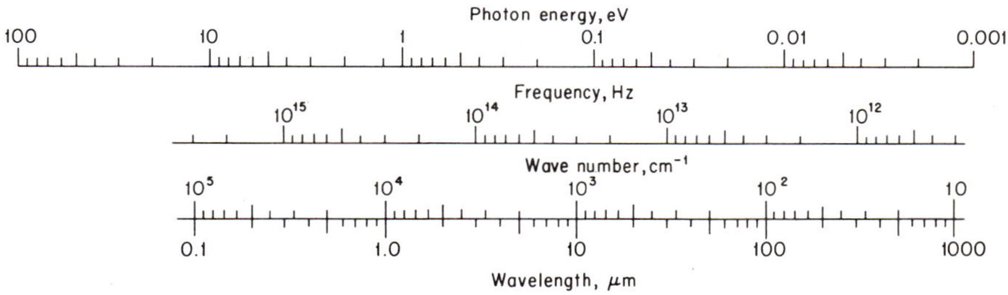

Fig. B1 Electron energy conversions.

See Tables 8 and 10 of Sec. 1 for luminance conversion factors and conversion of illuminance units, respectively.

Appendix C

Lens Data

The axial position of a lens is generally specified in terms of the location of the vertex of one of the lens surfaces. To locate a lens properly in a mechanical mount requires knowledge of the axial separation between the vertex and the

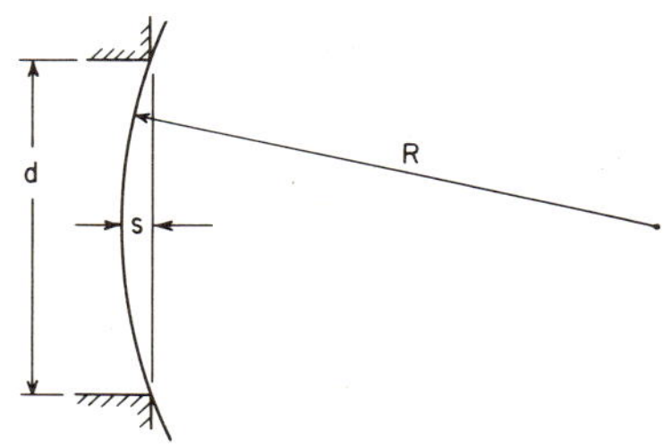

Fig. C1 Sagitta; s = sagitta, R = radius of curvature of spherical surface, and d = diameter of ring of contact with mounting shoulder.

ring of contact between the curved surface of the lens and a locating shoulder. On a spherical surface this is called the *sagitta,* or "sag."

$$s = R - \sqrt{R^2 - \left(\frac{d}{2}\right)^2}$$

Table C1 gives values for s/R for a range of values of d/R. To obtain the sagitta for given values of R and d:

1. Determine the ratio d/R.
2. Obtain the ratio s/R from the table.
3. Multiply the given value of R by the ratio s/R thus obtained.

TABLE C1 *s/R*†

	0	1	2	3	4	5	6	7	8	9
0.00	0.000000	0.000000	0.000000	0.000001	0.000002	0.000003	0.000005	0.000006	0.000008	0.000010
	0.000013	0.000015	0.000018	0.000021	0.000025	0.000028	0.000032	0.000036	0.000041	0.000045
	0.000050	0.000055	0.000061	0.000066	0.000072	0.000078	0.000085	0.000091	0.000098	0.000105
	0.000113	0.000120	0.000128	0.000136	0.000145	0.000153	0.000162	0.000171	0.000181	0.000190
	0.000200	0.000210	0.000221	0.000231	0.000242	0.000253	0.000265	0.000276	0.000288	0.000300
0.05	0.000313	0.000325	0.000338	0.000351	0.000365	0.000378	0.000392	0.000406	0.000421	0.000435
	0.000450	0.000465	0.000481	0.000496	0.000512	0.000528	0.000545	0.000561	0.000578	0.000595
	0.000613	0.000630	0.000648	0.000666	0.000685	0.000703	0.000722	0.000741	0.000761	0.000780
	0.000800	0.000820	0.000841	0.000861	0.000882	0.000904	0.000925	0.000947	0.000968	0.000991
	0.001013	0.001036	0.001059	0.001082	0.001105	0.001129	0.001153	0.001177	0.001201	0.001226
0.10	0.001251	0.001276	0.001301	0.001327	0.001353	0.001379	0.001405	0.001432	0.001459	0.001486
	0.001514	0.001541	0.001569	0.001597	0.001626	0.001654	0.001683	0.001713	0.001742	0.001772
	0.001802	0.001832	0.001862	0.001893	0.001924	0.001955	0.001986	0.002018	0.002050	0.002082
	0.002115	0.002147	0.002180	0.002214	0.002247	0.002281	0.002315	0.002349	0.002383	0.002418
	0.002453	0.002488	0.002524	0.002559	0.002595	0.002632	0.002668	0.002705	0.002742	0.002779
0.15	0.002816	0.002854	0.002892	0.002930	0.002969	0.003008	0.003047	0.003086	0.003125	0.003165
	0.003205	0.003245	0.003286	0.003327	0.003368	0.003409	0.003450	0.003492	0.003534	0.003577
	0.003619	0.003662	0.003705	0.003748	0.003792	0.003835	0.003880	0.003924	0.003968	0.004013
	0.004058	0.004104	0.004149	0.004195	0.004241	0.004287	0.004334	0.004381	0.004428	0.004475
	0.004523	0.004571	0.004619	0.004667	0.004716	0.004764	0.004814	0.004863	0.004913	0.004962
0.20	0.005013	0.005063	0.005114	0.005164	0.005216	0.005267	0.005319	0.005371	0.005423	0.005475
	0.005528	0.005581	0.005634	0.005687	0.005741	0.005795	0.005849	0.005904	0.005958	0.006013
	0.006068	0.006124	0.006180	0.006236	0.006292	0.006348	0.006405	0.006462	0.006519	0.006577
	0.006635	0.006693	0.006751	0.006809	0.006868	0.006927	0.006986	0.007046	0.007106	0.007166
	0.007226	0.007287	0.007347	0.007409	0.007470	0.007531	0.007593	0.007655	0.007718	0.007780
0.25	0.007843	0.007906	0.007970	0.008033	0.008097	0.008161	0.008226	0.008290	0.008355	0.008421
	0.008486	0.008552	0.008618	0.008684	0.008750	0.008817	0.008884	0.008951	0.009019	0.009086
	0.009154	0.009223	0.009291	0.009360	0.009429	0.009498	0.009568	0.009638	0.009708	0.009778
	0.009848	0.009919	0.009990	0.010062	0.010133	0.010205	0.010277	0.010350	0.010422	0.010495
	0.010568	0.010642	0.010715	0.010789	0.010864	0.010938	0.011013	0.011088	0.011163	0.011238
0.30	0.011314	0.011390	0.011466	0.011543	0.011620	0.011697	0.011774	0.011851	0.011929	0.012007
	0.012086	0.012164	0.012243	0.012322	0.012401	0.012481	0.012561	0.012641	0.012721	0.012802
	0.012883	0.012964	0.013046	0.013127	0.013209	0.013291	0.013374	0.013457	0.013540	0.013623
	0.013706	0.013790	0.013874	0.013959	0.014043	0.014128	0.014213	0.014298	0.014384	0.014470
	0.014556	0.014642	0.014729	0.014816	0.014903	0.014990	0.015078	0.015166	0.015254	0.015343
0.35	0.015432	0.015521	0.015610	0.015699	0.015789	0.015879	0.015970	0.016060	0.016151	0.016242
	0.016333	0.016425	0.016517	0.016609	0.016701	0.016794	0.016887	0.016980	0.017074	0.017167
	0.017261	0.017356	0.017450	0.017545	0.017640	0.017735	0.017831	0.017927	0.018023	0.018119
	0.018216	0.018313	0.018410	0.018507	0.018605	0.018703	0.018801	0.018900	0.018998	0.019097
	0.019197	0.019296	0.019396	0.019496	0.019597	0.019697	0.019798	0.019899	0.020001	0.020102
0.40	0.020204	0.020306	0.020409	0.020511	0.020614	0.020718	0.020821	0.020925	0.021029	0.021133
	0.021238	0.021343	0.021448	0.021553	0.021659	0.021765	0.021871	0.021978	0.022084	0.022191
	0.022299	0.022406	0.022514	0.022622	0.022730	0.022839	0.022948	0.023057	0.023166	0.023276
	0.023388	0.023496	0.023607	0.023717	0.023828	0.023940	0.024051	0.024163	0.024275	0.024388
	0.024500	0.024613	0.024726	0.024840	0.024953	0.025067	0.025182	0.025296	0.025411	0.025526
0.45	0.025641	0.025757	0.025873	0.025989	0.026105	0.026222	0.026339	0.026456	0.026574	0.026691
	0.026809	0.026928	0.027046	0.027165	0.027284	0.027404	0.027523	0.027643	0.027763	0.027884
	0.028005	0.028126	0.028247	0.028369	0.028490	0.028612	0.028735	0.028858	0.028980	0.029104
	0.029227	0.029351	0.029475	0.029599	0.029724	0.029849	0.029974	0.030099	0.030225	0.030351
	0.030477	0.030603	0.030730	0.030857	0.030985	0.031112	0.031240	0.031368	0.031497	0.031625
0.50	0.031754	0.031883	0.032013	0.032143	0.032273	0.032403	0.032534	0.032665	0.032796	0.032927
	0.033059	0.033191	0.033323	0.033456	0.033589	0.033722	0.033855	0.033989	0.034123	0.034257
	0.034391	0.034526	0.034661	0.034797	0.034932	0.035068	0.035204	0.035341	0.035477	0.035614
	0.035752	0.035889	0.036027	0.036165	0.036303	0.036442	0.036581	0.036720	0.036860	0.037000
	0.037140	0.037280	0.037421	0.037562	0.037703	0.037844	0.037986	0.038128	0.038270	0.038413
0.55	0.038556	0.038699	0.038842	0.038986	0.039130	0.039274	0.039419	0.039564	0.039709	0.039854
	0.040000	0.040146	0.040292	0.040439	0.040586	0.040733	0.040880	0.041028	0.041176	0.041324
	0.041472	0.041621	0.041770	0.041920	0.042069	0.042219	0.042370	0.042520	0.042671	0.042822
	0.042973	0.043125	0.043277	0.043429	0.043582	0.043734	0.043888	0.044041	0.044195	0.044349
	0.044503	0.044657	0.044812	0.044967	0.045123	0.045278	0.045434	0.045590	0.045747	0.045904
0.60	0.046061	0.046218	0.046376	0.046534	0.046692	0.046851	0.047009	0.047169	0.047328	0.047488
	0.047648	0.047808	0.047968	0.048129	0.048290	0.048452	0.048614	0.048776	0.048938	0.049101
	0.049263	0.049427	0.049590	0.049754	0.049918	0.050082	0.050247	0.050412	0.050577	0.050743
	0.050908	0.051074	0.051241	0.051407	0.051574	0.051742	0.051909	0.052077	0.052245	0.052414
	0.052582	0.052751	0.052921	0.053090	0.053260	0.053431	0.053601	0.053772	0.053943	0.054114
0.65	0.054286	0.054458	0.054630	0.054803	0.054976	0.055149	0.055322	0.055496	0.055670	0.055844
	0.056019	0.056194	0.056369	0.056545	0.056721	0.056897	0.057073	0.057250	0.057427	0.057604
	0.057782	0.057960	0.058138	0.058317	0.058495	0.058674	0.058854	0.059034	0.059214	0.059394
	0.059575	0.059755	0.059937	0.060118	0.060300	0.060482	0.060665	0.060847	0.061030	0.061214
	0.061397	0.061581	0.061765	0.061950	0.062135	0.062320	0.062505	0.062691	0.062877	0.063064
0.70	0.063250	0.063437	0.063625	0.063812	0.064000	0.064188	0.064377	0.064565	0.064755	0.064944
	0.065134	0.065324	0.065514	0.065705	0.065896	0.066087	0.066278	0.066470	0.066662	0.066855
	0.067048	0.067241	0.067434	0.067628	0.067822	0.068016	0.068211	0.068406	0.068601	0.068797
	0.068992	0.069189	0.069385	0.069582	0.069779	0.069976	0.070174	0.070372	0.070571	0.070769
	0.070968	0.071168	0.071367	0.071567	0.071767	0.071968	0.072169	0.072370	0.072571	0.072773

† Prepared by Itek Scientific Computation Center, Lexington, Mass.

TABLE C1 s/R **(Continued)**

	0	1	2	3	4	5	6	7	8	9
0.75	0.072975	0.073178	0.073380	0.073583	0.073787	0.073990	0.074194	0.074399	0.074603	0.074808
	0.075014	0.075219	0.075425	0.075631	0.075838	0.076045	0.076252	0.076459	0.076667	0.076875
	0.077083	0.077292	0.077501	0.077711	0.077920	0.078130	0.078341	0.078551	0.078762	0.078974
	0.079185	0.079397	0.079609	0.079822	0.080035	0.080248	0.080462	0.080675	0.080890	0.081104
	0.081319	0.081534	0.081749	0.081965	0.082181	0.082398	0.082615	0.082832	0.083049	0.083267
0.80	0.083485	0.083703	0.083922	0.084141	0.084360	0.084580	0.084800	0.085020	0.085241	0.085462
	0.085683	0.085905	0.086127	0.086349	0.086572	0.086795	0.087018	0.087242	0.087466	0.087690
	0.087914	0.088139	0.088365	0.088590	0.088816	0.089042	0.089269	0.089496	0.089723	0.089951
	0.090179	0.090407	0.090635	0.090864	0.091094	0.091323	0.091553	0.091783	0.092014	0.092245
	0.092476	0.092707	0.092939	0.093172	0.093404	0.093637	0.093870	0.094104	0.094338	0.094572
0.85	0.094807	0.095042	0.095277	0.095512	0.095748	0.095985	0.096221	0.096458	0.096696	0.096933
	0.097171	0.097409	0.097648	0.097887	0.098126	0.098366	0.098606	0.098846	0.099087	0.099328
	0.099570	0.099811	0.100053	0.100296	0.100538	0.100782	0.101025	0.101269	0.101513	0.101757
	0.102002	0.102247	0.102493	0.102739	0.102985	0.103231	0.103478	0.103726	0.103973	0.104221
	0.104469	0.104718	0.104967	0.105216	0.105466	0.105716	0.105966	0.106217	0.106468	0.106720
0.90	0.106971	0.107224	0.107476	0.107729	0.107982	0.108236	0.108489	0.108744	0.108998	0.109253
	0.109509	0.109764	0.110020	0.110277	0.110533	0.110790	0.111048	0.111306	0.111564	0.111822
	0.112081	0.112340	0.112600	0.112860	0.113120	0.113381	0.113642	0.113903	0.114165	0.114427
	0.114689	0.114952	0.115215	0.115479	0.115743	0.116007	0.116272	0.116537	0.116802	0.117068
	0.117334	0.117600	0.117867	0.118134	0.118401	0.118669	0.118938	0.119206	0.119475	0.119744
0.95	0.120014	0.120284	0.120555	0.120826	0.121097	0.121368	0.121640	0.121912	0.122185	0.122458
	0.122732	0.123005	0.123279	0.123554	0.123829	0.124104	0.124380	0.124656	0.124932	0.125209
	0.125486	0.125763	0.126041	0.126319	0.126598	0.126877	0.127156	0.127436	0.127716	0.127997
	0.128278	0.128559	0.128840	0.129122	0.129405	0.129688	0.129971	0.130254	0.130538	0.130822
	0.131107	0.131392	0.131677	0.131963	0.132249	0.132536	0.132823	0.133110	0.133398	0.133686

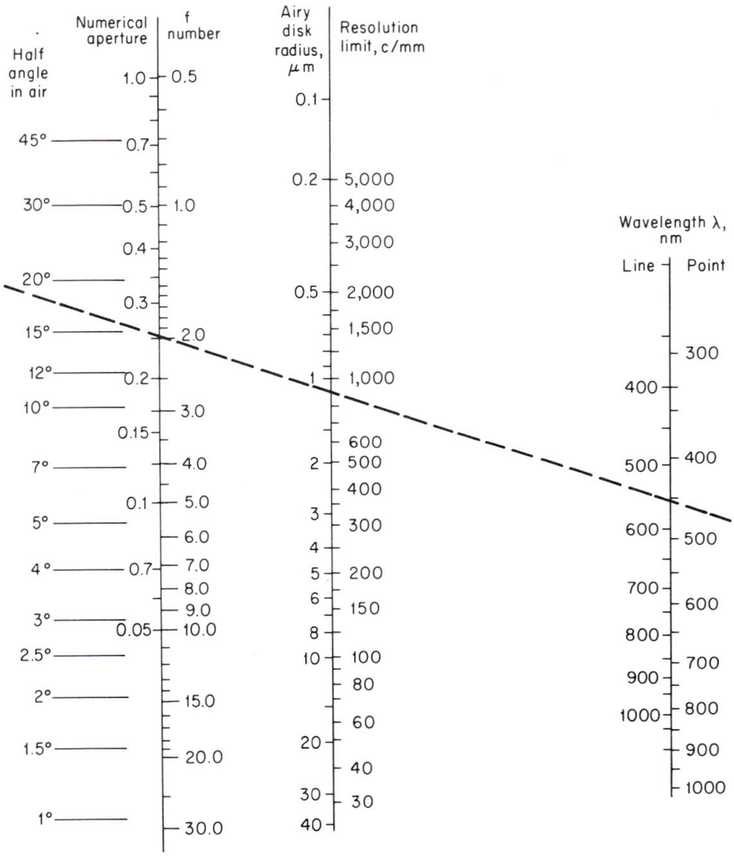

$$\text{Numerical aperture} = \frac{1}{2(\text{f number})} = \sin(\text{half angle})$$

$$\text{Airy disk radius} = \frac{1.22\,\lambda(\text{f number})}{10^3}$$

$$\text{Resolution limit, c/mm} = \frac{10^6}{\lambda(\text{f number})}$$

Fig. C2 Theoretical limit of resolution and radius of Airy disk for a perfect lens. (*Modulation Transfer Data for Kodak Films, Kodak Sales Service Pamphlet P-49. Copyright; reproduced by permission.*)

Appendix D

Standard Screw Threads

The following tables of screw-thread data have been extracted from ANSI Standard B1.1-1960, Unified Screw Threads, and ANSI Standard B1.10-1958, Unified Miniature Screw Threads.

The Unified Standard for screw threads (B1.1-1960) is the basic American standard for fastening types of screw threads. It is in agreement with formal recommendations of the International Organization for Standardization, is agreed to by representatives of standardization bodies of Canada, the United Kingdom, and the United States, and provides for screw-thread interchangeability between these three nations.

The standard for Unified Miniature Screw Threads (B1.10-1958) is intended for general-purpose fastening screws and similar uses in watches, instruments, and miniature mechanisms. This series covers a diameter range from 0.39 to 1.40 mm and has the endorsement of the American-British-Canadian Conference of April 1955 as the basis for a unified standard among the inch-using countries.

For more detailed description and data, reference should be made to the above standards which may be purchased from:

American National Standards Institute, Inc.
1430 Broadway
New York 10018

TABLE D1 Basic Dimensions of Coarse-Thread Series, UNC and NC†

Size	Basic major diameter, in	Threads per inch	Basic pitch diameter,‡ in	Minor diameter external threads, in	Minor diameter internal threads, in
1 (0.073)§	0.0730	64	0.0629	0.0538	0.0561
2 (0.086)	0.0860	56	0.0744	0.0641	0.0667
3 (0.099)§	0.0990	48	0.0855	0.0734	0.0764
4 (0.112)	0.1120	40	0.0958	0.0813	0.0849
5 (0.125)	0.1250	40	0.1088	0.0943	0.0979
6 (0.138)	0.1380	32	0.1177	0.0997	0.1042
8 (0.164)	0.1640	32	0.1437	0.1257	0.1302
10 (0.190)	0.1900	24	0.1629	0.1389	0.1449
12 (0.216)§	0.2160	24	0.1889	0.1649	0.1709
$\frac{1}{4}$	0.2500	20	0.2175	0.1887	0.1959
$\frac{5}{16}$	0.3125	18	0.2764	0.2443	0.2524
$\frac{3}{8}$	0.3750	16	0.3344	0.2983	0.3073
$\frac{7}{16}$	0.4375	14	0.3911	0.3499	0.3602
$\frac{1}{2}$	0.5000	13	0.4500	0.4056	0.4167
$\frac{9}{16}$	0.5625	12	0.5084	0.4603	0.4723
$\frac{5}{8}$	0.6250	11	0.5660	0.5135	0.5266
$\frac{3}{4}$	0.7500	10	0.6850	0.6273	0.6417
$\frac{7}{8}$	0.8750	9	0.8028	0.7387	0.7547
1	1.0000	8	0.9188	0.8466	0.8647
$1\frac{1}{8}$	1.1250	7	1.0322	0.9497	0.9704
$1\frac{1}{4}$	1.2500	7	1.1572	1.0747	1.0954
$1\frac{3}{8}$	1.3750	6	1.2667	1.1705	1.1946
$1\frac{1}{2}$	1.5000	6	1.3917	1.2955	1.3196
$1\frac{3}{4}$	1.7500	5	1.6201	1.5046	1.5335
2	2.0000	$4\frac{1}{2}$	1.8557	1.7274	1.7594
$2\frac{1}{4}$	2.2500	$4\frac{1}{2}$	2.1057	1.9774	2.0094
$2\frac{1}{2}$	2.5000	4	2.3376	2.1933	2.2294
$2\frac{3}{4}$	2.7500	4	2.5876	2.4433	2.4794
3	3.0000	4	2.8376	2.6933	2.7294
$3\frac{1}{4}$	3.2500	4	3.0876	2.9433	2.9794
$3\frac{1}{2}$	3.5000	4	3.3376	3.1933	3.2294
$3\frac{3}{4}$	3.7500	4	3.5876	3.4433	3.4794
4	4.0000	4	3.8376	3.6933	3.7294

† Extracted from American Standard Unified Screw Threads (ANSI B1.1-1960) with the permission of the publisher, the American Society of Mechanical Engineers.

‡ British: effective diameter.

§ Secondary sizes.

TABLE D2 Basic Dimensions of Fine-Thread Series UNF and NF†

Size	Basic major diameter, in	Threads per inch	Basic pitch diameter,‡ in	Minor diameter external threads, in	Minor diameter internal threads, in
0 (0.060)	0.0600	80	0.0519	0.0447	0.0465
1 (0.073)§	0.0730	72	0.0640	0.0560	0.0580
2 (0.086)	0.0860	64	0.0759	0.0668	0.0691
3 (0.099)§	0.0990	56	0.0874	0.0771	0.0797
4 (0.112)	0.1120	48	0.0985	0.0864	0.0894
5 (0.125)	0.1250	44	0.1102	0.0971	0.1004
6 (0.138)	0.1380	40	0.1218	0.1073	0.1109
8 (0.164)	0.1640	36	0.1460	0.1299	0.1339
10 (0.190)	0.1900	32	0.1697	0.1517	0.1562
12 (0.216)§	0.2160	28	0.1928	0.1722	0.1773
$\frac{1}{4}$	0.2500	28	0.2268	0.2062	0.2113
$\frac{5}{16}$	0.3125	24	0.2854	0.2614	0.2674
$\frac{3}{8}$	0.3750	24	0.3479	0.3239	0.3299
$\frac{7}{16}$	0.4375	20	0.4050	0.3762	0.3834
$\frac{1}{2}$	0.5000	20	0.4675	0.4387	0.4459
$\frac{9}{16}$	0.5625	18	0.5264	0.4943	0.5024
$\frac{5}{8}$	0.6250	18	0.5889	0.5568	0.5649
$\frac{3}{4}$	0.7500	16	0.7094	0.6733	0.6823
$\frac{7}{8}$	0.8750	14	0.8286	0.7874	0.7977
1	1.0000	12	0.9459	0.8978	0.9098
$1\frac{1}{8}$	1.1250	12	1.0709	1.0228	1.0348
$1\frac{1}{4}$	1.2500	12	1.1959	1.1478	1.1598
$1\frac{3}{8}$	1.3750	12	1.3209	1.2728	1.2848
$1\frac{1}{2}$	1.5000	12	1.4459	1.3978	1.4098

† Extracted from American Standard Unified Screw Threads (ANSI B1.1-1960) with the permission of the publisher, the American Society of Mechanical Engineers.

‡ British: effective diameter.

§ Secondary sizes.

TABLE D3 Basic Dimensions of Extrafine Thread Series UNEF and NEF†

Size, in		Basic major diameter, in	Threads per inch	Basic pitch diameter,‡ in	Minor diameter, in	
Primary	Secondary				External threads	Internal threads
	12 (.216)	0.2160	32	0.1957	0.1777	0.1822
$\frac{1}{4}$	0	0.2500	32	0.2297	0.2117	0.2162
$\frac{5}{16}$		0.3125	32	0.2922	0.2742	0.2787
$\frac{3}{8}$		0.3750	32	0.3547	0.3367	0.3412
$\frac{7}{16}$		0.4375	28	0.4143	0.3937	0.3988
$\frac{1}{2}$		0.5000	28	0.4768	0.4562	0.4613
$\frac{9}{16}$		0.5625	24	0.5354	0.5114	0.5174
$\frac{5}{8}$		0.6250	24	0.5979	0.5739	0.5799
	$\frac{11}{16}$	0.6875	24	0.6604	0.6364	0.6424
$\frac{3}{4}$		0.7500	20	0.7175	0.6887	0.6959
	$\frac{13}{16}$	0.8125	20	0.7800	0.7512	0.7584
$\frac{7}{8}$		0.8750	20	0.8425	0.8137	0.8209
	$\frac{15}{16}$	0.9375	20	0.9050	0.8762	0.8834
1		1.0000	20	0.9675	0.9387	0.9459
	$1\frac{1}{16}$	1.0625	18	1.0264	0.9943	1.0024
$1\frac{1}{8}$		1.1250	18	1.0889	1.0568	1.0649
	$1\frac{3}{16}$	1.1875	18	1.1514	1.1193	1.1274
$1\frac{1}{4}$		1.2500	18	1.2139	1.1818	1.1899
	$1\frac{5}{16}$	1.3125	18	1.2764	1.2443	1.2524
$1\frac{3}{8}$		1.3750	18	1.3389	1.3068	1.3149
	$1\frac{7}{16}$	1.4375	18	1.4014	1.3693	1.3774
$1\frac{1}{2}$		1.5000	18	1.4639	1.4318	1.4399
	$1\frac{9}{16}$	1.5625	18	1.5264	1.4943	1.5024
$1\frac{5}{8}$		1.6250	18	1.5889	1.5568	1.5649
	$1\frac{11}{16}$	1.6875	18	1.6514	1.6193	1.6274

† Extracted from American Standard Unified Screw Threads (ANSI B1.1-1960) with the permission of the publisher, the American Society of Mechanical Engineers.
‡ British: effective diameter.

TABLE D4 Basic Dimensions of Unified Miniature Screw Threads†
Boldface sizes are preferred

Size designation	Pitch, mm	Threads per inch	Basic major diameter		Basic pitch diameter		Basic minor diameter	
			mm	in	mm	in	mm	in
0.30 UNM	**0.080**	**318**	**0.300**	**0.0118**	**0.248**	**0.0098**	**0.217**	**0.0085**
0.35 UNM	0.090	282	0.350	0.0138	0.292	0.0115	0.256	0.0101
0.40 UNM	**0.100**	**254**	**0.400**	**0.0157**	**0.335**	**0.0132**	**0.296**	**0.0117**
0.45 UNM	0.100	254	0.450	0.0177	0.385	0.0152	0.346	0.0136
0.50 UNM	**0.125**	**203**	**0.500**	**0.0197**	**0.419**	**0.0165**	**0.370**	**0.0146**
0.55 UNM	0.125	203	0.550	0.0217	0.469	0.0185	0.420	0.0165
0.60 UNM	**0.150**	**169**	**0.600**	**0.0236**	**0.503**	**0.0198**	**0.444**	**0.0175**
0.70 UNM	0.175	145	0.700	0.0276	0.586	0.0231	0.518	0.0204
0.80 UNM	**0.200**	**127**	**0.800**	**0.0315**	**0.670**	**0.0264**	**0.592**	**0.0233**
0.90 UNM	0.225	113	0.900	0.0354	0.754	0.0297	0.666	0.0262
1.00 UNM	**0.250**	**102**	**1.000**	**0.0394**	**0.838**	**0.0330**	**0.740**	**0.0291**
1.10 UNM	0.250	102	1.100	0.0433	0.938	0.0369	0.840	0.0331
1.20 UNM	**0.250**	**102**	**1.200**	**0.0472**	**1.038**	**0.0409**	**0.940**	**0.0370**
1.40 UNM	0.300	85	1.400	0.0551	1.205	0.0474	1.088	0.0428

† Extracted from American Standard Unified Miniature Screw Threads (ANSI B1.10-1958) with the permission of the publisher, the American Society of Mechanical Engineers.

INDEX